Jane's

AVIONICS

Edited by Edward Downs

Twentieth Edition

2001-2002

Total number of entries 3,347 New and updated entries 818
Total number of images 1,570 New images 139

ISBN 0 7106 2311 9
"Jane's" is a registered trade mark

In the USA and its dependencies
Jane's Information Group Inc, 1340 Braddock Place, Suite 300, Alexandria, Virginia 22314-1651, USA

When Every Move Is Precise
Singapore Technologies Aerospace
Creating Value for You - Worldwide
We spare no effort to equip your aircraft with the latest and the best technology. From customised avionics upgrades to enhanced structural integrity, our leading-edge design engineering and development expertise assures your fleet performance precision.
• aircraft maintenance & modification
• component/engine repair & overhaul
• engineering & materials services
Singapore Technologies Aerospace
A company of Singapore Technologies Engineering
www.stengg.com

Contents

ADMINISTRATION

Publishing Director: Alan Condron,
e-mail: Alan.Condron @janes.co.uk

Managing Editor: Simon Michell, e-mail: Simon.Michell@janes.co.uk

Content Production Manager: Anita Slade,
e-mail: Anita.Slade@janes.co.uk

Application Support Manager: Ruth Simmance,
e-mail: Ruth.Simmance@janes.co.uk

Content Editing Manager: Jo Fenwick, e-mail: Jo.Fenwick@janes.co.uk

Pre-Press Manager: Christopher Morris, e-mail: Christopher.Morris@janes.co.uk

Team Leaders: Sharon Marshall, e-mail: Sharon.Marshall@janes.co.uk
Neil Grace, e-mail: Neil.Grace@janes.co.uk

Database Editor: Peter van Dooren, e-mail: Peter.vanDooren@janes.co.uk

Content Update: Jacqui Beard, Information Collection Assistant
Tel: (+44 20) 87 00 38 08 Fax: (+44 20) 87 00 39 59
e-mail: yearbook@janes.co.uk

Jane's Information Group Limited, Sentinel House, 163 Brighton Road,
Coulsdon, Surrey CR5 2YH, UK
Tel: (+44 20) 87 00 37 00 Fax: (+44 20) 87 00 37 88
e-mail: jav@janes.co.uk

SALES OFFICE

Send enquiries to: Julie Reeder - Group Sales Manager
Jane's Information Group Limited, Sentinel House, 163 Brighton Road,
Coulsdon, Surrey CR5 2YH, UK
Tel: (+44 20) 87 00 37 00 Fax: (+44 20) 87 63 10 06
e-mail: info@janes.co.uk

Send USA enquiries to: Robert Loughman - Vice-President Product Sales
Jane's Information Group Inc, 1340 Braddock Place, Suite 300,
Alexandria, Virginia 22314-1651, USA
Tel: (+1 703) 683 37 00 Fax: (+1 703) 836 00 29 Telex: 6819193
Tel: (+1 800) 824 07 68 Fax: (+1 800) 836 02 97
e-mail: order@janes.com

Send Asia enquiries to: David Fisher - Group Sales Manager
Jane's Information Group Asia, 60 Albert Street, 15-01 Albert Complex,
Singapore 189969
Tel: (+65) 331 62 80 Fax: (+65) 336 99 21
e-mail: info@janes.com.sg

Send Australia/New Zealand enquiries to: David Moden - Business Manager
Jane's Information Group, PO Box 3502, Rozelle Delivery Centre,
New South Wales 2039, Australia
Tel: (+61 2) 85 87 79 00 Fax: (+61 2) 85 87 79 01
e-mail: info@janes.thomson.com.au

ADVERTISEMENT SALES OFFICES

Australia: Richard West (UK Head Office)

Austria: Dr Uwe Wehrstedt (See Germany)

Benelux: Steven Soffe (UK Head Office)

Channel Islands: Steven Soffe (UK Head Office)

Eastern Europe: Dr Uwe Wehrstedt (See Germany)

Egypt: Steven Soffe (UK Head Office)

France: Patrice Février
BP 418, 35 avenue MacMahon,
F-75824 Paris Cedex 17, France
Tel: (+33 1) 45 72 33 11 Fax: (+33 1) 45 72 17 95
e-mail: patrice.fevrier@wanadoo.fr

Germany: MCW Dr Uwe Wehrstedt
Hagenbrite 9, D-06463 Ermsleben, Germany
Tel: (+49 34) 74 36 20 91 Fax: (+49 34) 74 36 20 90

Greece: Steven Soffe (UK Head Office)

India: Joni Beeden (UK Head Office)

Ireland: Joni Beeden (UK Head Office)

Israel: Oreet Ben-Yaacov
Oreet International Media, 15 Kinneret Street,
IL-51201 Bene Berak, Israel
Tel: (+972 3) 570 65 27 Fax: (+972 3) 570 65 26
e-mail: oreet@oreet-marcom.com

Italy and Switzerland: Ediconsult Internazionale Srl
Piazza Fontane Marose 3, I-16123 Genoa, Italy
Tel: (+39 010) 58 36 84 Fax: (+39 010) 56 65 78
e-mail: ediconsult@iol.it

Korea, South: Young Seoh Chinn
JES Media International, 2nd Floor ANA Building,
2571 Myungil-Dong, Kangdong-Gu, Seoul 134 070, South Korea
Tel: (+82 2) 481 34 11 Fax: (+82 2) 481 34 14
e-mail: jesmedia@unitel.co.kr

Middle East: Steven Soffe (UK Head Office)

Pakistan: Joni Beeden (UK Head Office)

Rest of the World: Joni Beeden (UK Head Office)

Russian Federation and Associated States (CIS): Simon Kay
33 St John's Street, Crowthorne, Berkshire RG45 7NQ, UK
Tel: (+44 1344) 77 71 23 Fax: (+44 1344) 77 58 85
e-mail: crowkay@msn.com

Saudi Arabia: Steven Soffe (UK Head Office)

Scandinavia: Gillian Thompson
The Falsten Partnership, PO Box 21175,
London N16 6ZG, UK
Tel: (+44 20) 88 06 23 01 Fax: (+ 44 20) 88 06 81 37
e-mail: falsten@dial.pipex.com

Singapore: Richard West (UK Head Office)

South Africa: Stephen Judge (UK Head Office)

Spain: Julio de Andres
Via Exclusivas SL, Viriato 69SC, E-28010 Madrid, Spain
Tel: (+34 91) 448 76 22 Fax: (+34 91) 446 01 98
e-mail: j.a.deandres@viaexclusivas.com

Thailand: Joni Beeden (UK Head Office)

Turkey: Richard West (UK Head Office)

United Arab Emirates: Steven Soffe (UK Head Office)

UK - London: Joni Beeden (UK Head Office)

UK - South, South East, South West, North West, Wales and West Scotland:
Steven Soffe (UK Head Office)

UK - North East, East, East Scotland: Joni Beeden (UK Head Office)

Senior Key Accounts Manager: Richard West
Jane's Information Group Limited, Sentinel House, 163 Brighton Road,
Coulsdon, Surrey CR5 2YH, UK
Tel: (+44 1892) 72 55 80 Fax: (+44 1892) 72 55 81
e-mail: richard.west@janes.co.uk

Advertising Sales Admin Manager: Christine Watkins
Jane's Information Group Limited, Sentinel House, 163 Brighton Road,
Coulsdon, Surrey CR5 2YH, UK
Tel: (+44 20) 87 00 38 50 Fax: (+44 20) 87 00 38 59
e-mail: christine.watkins@janes.co.uk

Senior Advertisement Sales Executive: Stephen Judge
Jane's Information Group Limited, Sentinel House, 163 Brighton Road,
Coulsdon, Surrey CR5 2YH, UK
Tel: (+44 20) 87 00 38 53 Fax: (+44 20) 87 00 37 44
e-mail: stephen.judge@janes.co.uk

Advertising Sales Executive: Joni Beeden
Jane's Information Group Limited, Sentinel House, 163 Brighton Road,
Coulsdon, Surrey CR5 2YH, UK
Tel: (+44 20) 87 00 39 63 Fax: (+44 20) 87 00 37 44
e-mail: joni.beeden@janes.co.uk

Advertising Sales Executive: Steven Soffe
Jane's Information Group Limited, Sentinel House, 163 Brighton Road,
Coulsdon, Surrey CR5 2YH, UK
Tel: (+44 20) 87 00 39 43 Fax: (+44 20) 87 00 37 44
e-mail: steven.soffe@janes.co.uk

USA and Canada: Ronald R Lichtinger III, Advertising Sales Director
Jane's Information Group Inc, 1340 Braddock Place, Suite 300,
Alexandria, Virginia 22314-1651, USA
Tel: (+1 703) 683 37 00 Fax: (+1 703) 836 55 37
e-mail: lichtinger@janes.com

North Eastern USA and East Canada: Harry Carter
Jane's Information Group Inc, 1340 Braddock Place, Suite 300,
Alexandria, Virginia 22314-1651, USA
Tel: (+1 703) 683 37 00 Fax: (+1 703) 836 55 37
e-mail: carter@janes.com

South Eastern USA: Kristin D Schulze
PO Box 270190,
Tampa, Florida 33688-0190, USA
Tel: (+1 727) 524 77 41 Fax: (+1 727) 524 75 62
e-mail: kristin@intnet.net

Western USA and West Canada: Richard L Ayer
127 Avenida del Mar, Suite 2A,
San Clemente, California 92672, USA
Tel: (+1 949) 366 84 55 Fax: (+1 949) 366 92 89
e-mail: ayercomm@earthlink.net

Administration:
USA and Canada: Shanee Johnson
Jane's Information Group Inc, 1340 Braddock Place, Suite 300,
Alexandria, Virginia 22314-1651, USA
Tel: (+1 703) 683 37 00 Fax: (+1 703) 836 55 37
e-mail: shanee.johnson@janes.com

UK and Rest of World: Sue Tucker
Jane's Information Group Limited, Sentinel House, 163 Brighton Road,
Coulsdon, Surrey CR5 2YH, UK
Tel: (+44 20) 87 00 37 42 Fax: (+44 20) 87 00 38 59
e-mail: sue.tucker@janes.co.uk

How to use *Jane's Avionics*

At the beginning of this edition of *Jane's Avionics* we have added some extra pages to help the reader. The first is this 'How to use *Jane's Avionics*' which gives an overview of what is contained in each section and an explanation of the three categories into which the various entries have been divided.

The reader will also find technical overviews, which are intended to provide useful background information regarding some of the core technologies employed in civil and military avionics systems.

These pages, together with the Foreword and Contents, are printed on coloured paper to make it easier for readers to find the pages they need.

Jane's Avionics contains information on airborne electronic equipment for both military and civil aircraft. The criteria for inclusion is that the equipment should have a significant electronic content, be in development, production or service and be operated in manned aircraft. Avionics equipment designed specifically for unmanned aerial vehicles (UAVs) is covered in *Jane's Unmanned Aerial Vehicles and Targets*, although this distinction is becoming increasingly difficult to make in light of the development of Unmanned Combat Air Vehicles (UCAVs), which may be regarded as combat aircraft operated by a remote, ground-based pilot. Items which are applicable to both manned and unmanned aircraft will continue to be found in *Jane's Avionics*.

This edition of *Jane's Avionics* is divided into two main sections - one for civil avionics and the other for military avionics. Each of these main sections has been divided into two subsections for ease of presentation. In the civil section, all items except data recording are integrated in one sub-section, while data recording equipment has been allocated a small subsection of its own. Military avionics has also been divided into two subsections; the first is essentially concerned with aircraft control and navigation, the second with displays and targeting equipment.

The reader will find that the subsection *'Military Displays and Targeting Systems'* is in colour; this change to the format of *Jane's Avionics* reflects the widespread use of full colour displays in civil and military cockpits.

Within each subsection entries are arranged alphabetically by country of manufacture. For each country, entries are attributed to a contractor, and contractors are listed alphabetically. The entries for each contractor are also listed alphabetically. Entries are structured as follows:

Title.

Narrative: contains information on function, development history, and technical description.

Specification: lists the main technical parameters and dimensions.

Operational status: contains information on development/production/service status.

Contractor: details of the prime contractor.

In addition to the main text, *Jane's Avionics* contains three indexes: a contractors' index; an alphabetical index; and an index that correlates entries to contractors.

The contractors' index contains details of the name, address and, where available, the telephone/fax number/e-mail/website of each contractor represented in the book.

The Alphabetical index lists every entry in alphabetical order.

The Manufacturers' index lists entries alphabetically by contractor.

To help users of this title evaluate the published data, Jane's Information Group has divided entries into three categories. A full list of all entries indicating their current status is provided in the index.

● ***VERIFIED*** The editor has made a detailed examination of the entry's content and checked it's relevancy and accuracy for publication in the new edition to the best of his ability

● ***UPDATED*** During the verification process, significant changes to content have been made to reflect the latest position known to Jane's at the time of publication.

● ***NEW ENTRY*** Information on new equipment and/or appearing for the first time in the title.

All new pictures are dated with the year of publication. New pictures this year are dated 2001. Some are followed by a seven digit number for ease of identification by our image library.

Total number of entries	3,347	New and updated entries	818
Total number of images	1,570	New images	139

DPA DIRECTORY & DATABASE PUBLISHERS ASSOCIATION MEMBER

British Library Cataloguing-in-Publication Data.
A catalogue record for this book is available from the British Library.

Printed and bound in Great Britain by Biddles Limited, Guildford and King's Lynn

Alphabetical list of advertisers

Glossary of acronyms and abbreviations

A	Ampère
AAA	Anti-aircraft artillery
AAAA	Advanced architectures for airborne arrays
AACMI	Autonomous air combat manoeuvring installation
AACS	Airborne advanced communications system
AADS	Airborne active dipping sonar
AAED	Advanced airborne Expendable Decoy
AAICP	Air-to-air interrogator control panel
AAIP	Analogue autoland improvement programme
AAU	Audio amplifier units
AAW	Anti-air warfare
AAWWS	Airborne adverse weather weapon system
ABCCC	Airborne battlefield command and control centre
ABICS	Ada-based integrated control system
ABIS	All-bus instrumentation system
ABL	AirBorne laser
AC	Alternating current
ACAS	Airborne collision avoidance system
ACARS	Automatic communications and reporting system
ACC	1. Avionics computer control 2. Axis-controlled carrier
ACCS	Airborne computing and communications system
ACE	1. Actuator control electronics 2. Advanced communication engine 3. Autonomous combat (manoeuvres) evaluation
ACEM	Aerial camera electro-optical magazine
ACI	Armament control indicator
ACIDS	Automated communications and intercom distribution system
ACIS	1. Armament control indicator set 2. Advanced cabin interphone system
ACLS	Automatic carrier landing system
ACM(I)	Air combat manoeuvring installation
ACMS	1. Aircraft condition monitoring system 2. Armament control and monitoring system 3. Avionics control and management system
ACNIP	Auxiliary (or advanced) communication navigation and identification panel
ACP	1. Advanced controller processor 2. Armament control panel 3. Audio converter processor
ACP(C)	Automatic communications processor (control)
ACS	Armament control system
ACT	Airborne crew trainer
ACTIVE	Advanced control technology for integrated vehicles
ACU	1. Adaptive control unit 2. Airborne computer unit 3. Annotation control unit 4. Antenna control unit 5. Audio control unit 6. Auxiliary control unit
ADAAPS	Aircraft data acquisition, analysis and presentation system
ADACS	Airborne digital automatic collection system
ADAD	Air defence alerting device
ADAS	1. Airborne data acquisition system 2. Auxiliary data annotation set
ADAU	Auxiliary data acquisition unit
ADC	1. Advanced data controller 2. Air data computer
ADDS	Advanced digital dispensing system
ADECS	Advanced digital engine control system
ADELE	Alerte détection et localisation des emetteurs
ADELT	Automatically deployed emergency locator transmitter
ADF	Automatic direction-finder
ADI	1. Attitude director indicator 2. Azimuth display indicator
ADID	Aircraft data interface device
ADIRS	Air data and inertial reference system
ADIRU	Air data inertial reference unit
ADLGP	Advanced dataLink for guided platforms
ADM	1. Advanced development model 2. Air data module
Ad-Me	Advanced metal evaporated
ADOCS	Advanced digital optical control system
ADR	Accident data recorder
ADS	1. Audio distribution system 2. Automatic dependent surveillance 3. Airlifter defence systems
ADS-B	Automatic dependent surveillance broadcast
ADS-C	Automatic dependant surveillance contract
ADSU	Air data sensor unit
ADTU	Auxiliary data transfer unit
ADU	1. Air data unit 2. Annotation display unit
ADVCAP	Advanced capability
ADVICE	Acoustic data vessel identification, classification and explanation
AEC	Automatic exposure control
AEGIS	Airborne early warning/ground environment integration segment
AELS	Airborne electronic library system
AERA	Automated en-route air traffic control
Aero-H	Aeronautical high-gain antenna
Aero-I	Aeronautical intermediate-gain antenna
AES	Aeronautical earth station
AESOP	Airborne electro-optical special operations payload
AETMS	Airborne electronic terrain-mapping system
AEU	Airborne electronics unit
AEW	Airborne early warning
AF	Audio frequency
AFA	Audio frequency amplifier
AFC	Automatic frequency control
AFCAS	Automatic flight control augmentation system
AFCS	Automatic flight control system
AFDAS	Aircraft fatigue data analysis system
AFDS	Automatic flight director system
AFH	Advanced fibre heater
AFIS	Airborne flight information system
AFMCS	Advanced flight management computer system
AFMS	Automatic flight management system
AFSAT	Air force satellite
AFSATCOM	Air force satellite communications
AFTI	Advanced fighter technology integration
AFV	Armoured fighting vehicle
AGC	Automatic gain control
AGES	Air-to-ground engagement simulator
AGINT	Advanced GPS inertial navigation technology
AGL	Above ground level
AGPPE	Advanced general purpose processor element
AGRA	Automatic gain ranging amplifier
AGREE	Advisory group on the reliability of electronic equipment (US)
AGTV	Active gated television
AGU	AIRLINK gateway unit
AHEAD	Attitude, heading and rate of turn indicating system
AHIP	Army helicopter improvement programme
AHRS	Attitude and heading reference system
AHS	Attitude heading (reference) system
AI	Air interception
AIBU	Advanced interference blanker Unit
AICS	Airborne integrated communications system
AIDC	Aero industry development centre
AIDS	Aircraft integrated data suite (or system)
AIG	ACAS implementation group
AIM	1. Air intercept missile 2. Air traffic control beacon/IFF/Mk XII system
AIME	Autonomous integrity monitored extrapolation
AIMES	Avionics integrated maintenance expert system
AIMS	1. Advanced integrated MAD system 2. Air traffic control radar beacon, IFF, Mk 12 transponder system 3. Aircraft information management system 4. Aircraft integrated monitoring system
AINS	1. Aided inertial navigation system 2. Airborne inertial navigation system
AIP	Australian industrial participation
AIPT	Advanced image processing terminal
AIRSTAR	Airborne surveillance and target acquisition radar
AIT	Advanced intelligence tape
AIU	1. Armament interface unit 2. Automatic ignition unit
AJB	Audio junction box
AKU	Avionic keyboard unit
ALARMS	Airborne laser radar mine sensor
ALAT	French army light aviation corps
ALB	Airborne LIDAR bathymeter
ALE	Automatic link establishment
ALFS	Airborne low-frequency sonar
ALLTV	All light level television
ALRAD	Airborne laser range-finder and designator
AM	Amplitude modulation
AMAC	Airborne multi-application computer
AMC	Advanced micro-electronics converter
AMCS	Airborne missile control system
AME	1. Amplitude modulation equivalent 2. Angle measuring equipment
AMHMS	Advanced magnetic helmet-mounted sight
AMICS	Adaptive multidimensional integrated control system
AMIDS	Advanced missile detection system
AMIMU	Advanced multisensor inertial measurement unit
AMLCD	Active matrix liquid crystal display
AMMCS	Airborne multiservice/multimedia communications system
AMP	Advanced modular processor
AMPA	Advanced mission planning aid
AMRAAM	Advanced medium-range air-to-air missile
AMRS	Advanced maintenance recorder system
AMS	1. Airborne maintenance subsystem 2. Avionics management system
AMSAR	1. Airborne multirole steerable array radar 2. Airborne multirole multifunction Solid-state active array radar
AMSL	Above mean sea level
AMSS	Aeronautical mobile satellite service
AMU	1. Audio management unit 2. Auxiliary memory unit 3. Avionics management unit
ANC	Active noise cancellation
ANDVT	Advanced narrowband digital voice terminal
ANMI	Air navigation multiple indicator
ANSI	American national standards institute
ANVIS	Aviator's night vision imaging system
AOA	1. Airborne optical adjunct 2. Angle of attack
AOC	Assumption of control (message)
AOCM	Advanced optical countermeasures
AOPT	Advanced optical position transducer
APALS	Autonomous precision approach and landing system
APAR	Advanced phased-array radar
A/PDMC	Aircraft products data management computer
APHIDS	Advanced panoramic helmet interface demonstrator system
APIRS	Aircraft piloting inertial reference system
APR	Auto power reserve
APS	1. Adaptive processor system 2. Aircraft position sensor 3. Altitude position sensor
APSP	Advanced programmable signal processor
APU	Auxiliary power unit
AQL	Advanced quick look
ARBS	Angle rate bombing set
ARCADS	Armament control and delivery system
ARDS	Airborne radar demonstrator system
ARI	Azimuth range indicator
ARIA	Advanced range instrumentation aircraft
ARIC	Airborne radio and iIntercom control
ARIES	Airborne recorder for IRLS and EO sensors
ARINC	Aeronautical radio incorporated
ARJS	Airborne radar jamming system
ARM	Anti-radiation missile
ARPA	Advanced Research Projects Agency (US)
ARPS	Advanced radar processing system
ARSA	Advisable radar service area
ARTCC	Air route traffic control centre
ARTI	Advanced rotorcraft technology integration
ARWS	Advanced radar warning system
AS	Anti-spoofing
ASAC	Airborne surveillance airborne control
ASAP	Airborne shared aperture programme
ASARS	1. Airborne search and rescue system 2. Advanced synthetic aperture radar system
ASAS	Airborne separation assurance system
ASC	Airborne strain counter
ASCB	1. Aircraft standard communications bus 2. Avionics standard communications bus
ASCII	American Standard Code for Information Interchange
ASCOT	Aerial survey control tool
ASCTU	Air supply controller/test unit
ASDC	Armament signal data converter
ASE	AutoStabilisation equipment
ASETS	Airborne seeker evaluation test set
ASGC	Airborne surveillance ground control
ASI	ARINC standards interface
ASIC	Application specific integrated circuit
ASIT	Adaptive surface interface terminals
ASMD	Anti-ship missile defence
ASMU	Avionics system management unit

ASNI	Ambient sea noise indication
ASP	1. Advanced signal processor 2. Airborne signal processor 3. Aircraft systems processor
ASPIS	Advanced self-protection integrated suite
ASPJ	Airborne self-protection jammer
ASPRC	Associative processor
ASR	Advanced special receiver
ASSG	Acoustic sensor signal generator
AST	Airborne surveillance testbed
ASTAR	Airborne search target attack radar
ASTE	Advanced strategic and tactical expendables
ASTOR	Airborne standoff radar
ASU	Acoustic simulation unit
ASUW	Anti-surface warfare
ASV	Anti-surface vessel
ASVW	Anti-surface vessel warfare
ASW	Anti-submarine warfare
ASWAC	Airborne surveillance warning and control
AT^3	Advanced tactical targeting technology
ATA	1. Actual time of arrival 2. Advanced tactical aircraft
ATAC	1. Applied technology advanced computer 2. Air transportable acoustic communication
ATAL	Appareillage de télévision sur aéronef léger
ATARS	Advanced tactical air reconnaissance system
ATC	Air traffic control
ATCRBS	Air traffic control radar beacon system
ATCS	Advanced tactical communications system
ATDS	Airborne tactical data system
ATE	Automatic test equipment
ATF	Advanced tactical fighter
ATH	Autonomous terminal homing
ATHS	Automatic target handover system
ATIRCM	Advanced threat infra-red countermeasures
ATIS	Automatic terminal information service
ATLANTIC	Airborne targeting low-altitude navigation thermal Imaging and cueing
ATLIS	Automatic tracking laser illumination system
ATM	Air traffic management
ATN	Aeronautical telecommunications network
ATOPS	Advanced transport operations system
ATR	1. Air transport tracking 2. Automatic target recognition
ATRJ	Advanced threat radar jammer
AVAD	Automatic voice alert device
AVDAS	Airborne video data acquisition system
AVE	Airborne vehicle equipment
AVICS	Air vehicle interface and control system
AVTR	Airborne videotape recorder
AWACS	Airborne warning and control system
AWADS	Adverse weather aerial delivery system
AWARE	Advanced warning of active radar emissions
BBU	Battery back-up unit
BCSV	Bearing compartment scavenge value
BDHI	Bearing, distance and heading indicator
BDI	Bearing distance indicator
BER	BIT error rate
BFDAS	Basic flight data acquisition system
BFO	Beat frequency oscillation
BIT	Built-in test
BITE	Built-in test equipment
BPI	Bits per inch
BPSK	Bi-phase shift keyed
B-RNAV	Basic area NAVigation
BSIN	Bus system interface unit
BSP	Barra side processor
BT	Bathythermograph
BTH	Beyond the horizon
BTU	Basic terminal unit
BVCU	Bleed valve control unit
BVR	Beyond visual range
C^3CM	Command, control and communications countermeasures
C^3I	Command, control, communications and intelligence
C^4I	Command, control, communications, computers and intelligence
C/A	Coarse acquisition
CAA	Civil Aviation Authority (UK)
CAB	Common avionics baseline
CACTCS	Cabin air conditioning and temperature control system
CAD	Computer-aided design
CADC	Central air data computer
CAINS	Carrier aircraft inertial navigation system
CAMBS	Command activated multiBeam sonobuoy
CAMEL	Cartridge active miniature electromagnetic
CARA	Combined altitude radar altimeter
CARABAS	Coherent all radio band sensing
CARS/DGS	Contingency airborne reconnaissance system/deployable ground station
CASS	1. Command active sonobuoy system 2. Consolidated automatic support system
CAT	CATegory
CBL	Control by light
CC	Countermeasures computer
CCD	Charge coupled device
CCG	Communication control group
CCIL	Continuously computed impact line
CCIP	Continuously computed impact point
CCIR	Comité Consultatif International des Radiocommunications
CCRP	Continuously computed release point
CCS	Conformal countermeasures system
CCTWT	Coupled cavity travelling wave tube
CCU	1. Cockpit control unit 2. Common control unit 3. Communication control unit
CDC	Cabin display computer
CDI	Course deviation indicator
CDIRRS	Cockpit display of infra-red reconnaissance system
CDM	Collaborative decision making
CDNU	Control and display navigation unit
CDR	Critical design review
CDTI	Cockpit display of traffic information
CDU	1. Cockpit display unit 2. Control and display unit
CEAT	Centre d'Essais Aeronautique de Toulouse
CEDAM	Combined electronic display and map
CEP	Circular error of probability
CEV	Centre d'Essais de Vol
CFAR	Constant false alarm rate
CFD	Chaff and flare dispensing
CFDCU	Chaff and flare dispenser control unit
CFDIU	Centralised fault display interface unit
CFDS	Centralised fault display system
CFIT	Controlled flight into terrain
CG	Centre of gravity
CGCC	Centre of gravity control computer
CHAALS	Communications high-accuracy airborne location system
CHT	Cylinder head temperature
CI	Control Indicator
CIBS-M	Common integrated broadcast service-modules
CI-F	Control indicator - front
CIP	Common integrated processor
CIRCE	Cossor interrogation and reply cryptographic equipment
CIRTEVS	Compact infra-red television system
CI-S	Control indicator - side
CIS	1. Commonwealth of Independent States 2. Control indicator set
CIT	Combined interrogator/transponder
CITS	Centrally integrated test system
CIU	1. Cockpit interface unit 2. Control interface unit
CLASS	Coherent laser radar airborne shear sensor
CLDP	Convertible laser designation pod
CLDS	Cockpit laser designation system
CM	COMSEC module
CMD	1. Colour multipurpose display 2. Countermeasures dispenser
CMDR	Card maintenance data recorder
CMDS	Countermeasures dispensing system
CMF	Central maintenance function
CMLSA	Commercial microwave landing system avionics
CMOS	Complementary metal oxide silicon
CMP	Central maintenance panel
CMRA	Cruise missile radar altimeter
CMRS	Crash/maintenance recorder system
CMS	Computer module system
CMT	Cadmium mercury telluride
CMUP	Conventional mission upgrade programme
CMWS	Common missile warning system
C-Nite	Cobra nite
CN2H	Conduit nuit, second-generation, helicopters
CNC	Communication/navigation equipment controls
CNI	Comunications, navigation and identification
CNI-MS	CNI management system
CNIU	Communications/navigation interface unit
CNS	Communications navigation surveillance
CNS/ATM	Communications navigation surveillance/ air traffic management
CNS/ATN	Communications navigation surveillance/ aeronautical telecommunications network
CODAR	Correlation detection and ranging
COIL	Chemical oxygen iodine laser
COMAC	Cockpit management computer
COMED	Combined map and electronic display
COMINT	Communications intelligence
COMPASS	Compact multipurpose advanced stabilised system
COMSEC	Communications security
COTIM	Compact thermal imaging module
COTS	Commercial-off-the-shelf
CP	Computer processor
CPA	Cabin public address
CPC	Cabin pressure controller
CPCS	Cabin pressure control system
CPDLC	Controller-pilot dataLink communications
CPM/P	Command post modem/processor
CPR	Covert penetration radar
CPS	Covert penetration system
CPU	Central processing unit
CPU-F	Control panel unit - front
CPU-S	Control panel unit - side
CPVR	Crash protected video recorder
CR	Countermeasures receiver
CRE	Communications radar exciter
CRPA	Controlled reception pattern antenna
CRT	Cathode ray tube
CS	Communications subsystem
CSA	Control stick assembly
CSAS	Command and stability augmentation system
CSC	1. Communication system controller 2. Compass system controller
CSCG	Communications system control group
CSD	Common strategic doppler
CSMU	Crash survivable memory unit
CSS	1. Computer support system 2. Complimentary satellite system
CSU	1. Communications switching unit 2. Control status unit 3. Crew station unit
CSVR	Crash survivable voice recorder
CT	1. Control transmitter 2. Crew member terminal
CTAS	Chrysler Technologies airborne systems
CTC	Cabin temperature controller
CTS	Central tactical system
CTT/H-R	The commanders' tactical terminal/ hybrid-receive only
CTU	Control terminal unit
CU	Control unit
CUGR	Cargo utility GPS receiver
CV	Aircraft carrier
CVFDR	Cockpit voice and flight data recorder
CVR	Cockpit voice recorder
CVR/DFDR	Cockpit voice recorder/digital flight data recorder
CW	Continuous wave
CWI	Continuous wave illuminator
CWS	Control wheel steering
DADC	Digital air data computer
DAFCS	Digital automatic flight control system
DAFD	Digital autopilot/flight director
DAFICS	Digital automatic flight inlet control system
DAI	DCMS audio interface
DAIS	Digital avionics information system
DAIRS	Distributed-architecture infra-red sensor
DAMA	Demand assigned multiple access
DAMS	Drum auxiliary memory sub-unit
DAP	Downlink aircraft parameters
DAPU	Data acquisition and processing unit
DAR	1. Direct access recorder 2. Drone anti-radar
DARP	Digital audio record and playback
DARPA	Defense Advanced Research Projects Agency (US) (now ARPA)
DARTS	Digital airborne radar threat simulator
DARU	Data acquisition and recording unit
DASH	Display and sight helmet
DASS	Defensive aids subsystem
DAT	Digital audio tape
DAU	1. Data acquisition unit 2. Digital amplifier unit
dB	decibel(s)
DBC	DCMS bus coupler
DBI	DCMS bus interface
dBm	decibels × 10^{-3}
DC	Direct current
DCC	Digital computer complex
DCI	DCMS crew member interface
D/CM	Diagnostic and condition monitoring
DCMS	Digital communication management system
DCPS	Data collection and processing system
DCSU	Dual crew station unit
DCU	1. Data collection unit 2. Digital computer unit
DDI	1. Data display indicator 2. Digital display indicator
DDIC	Digital display indicator control
DDPS	Digital display processing system
DDS	Direct digital synthesis
DDU	Disk drive unit
DDVR	Displayed data video recorder
DEC	Digital electronic control
DECS	Digital engine control system
DECU	Digital engine control unit
DED	Data entry display

DEEC	Digital electronic engine control
DEFCS	Digital electronic flight control system
DEMON	Demodulation of noise
DES	Data encryption standard
DEU	Display electronics unit
DF	Direction-finding
DFAD	Digital feature analysis data
DFDAU	Digital flight data acquisition unit
DFDR	Digital flight data recorder
DFGC	Digital flight guidance computer
DFIR	Deployable flight incident recorder
DFLCC	Digital flight control computer
DFU	Deployable flotation unit
DG	Directional gyro
DGA	Displacement gyro assembly
DGNS	Differential global navigation system
DGPS	Differential GPS
DIANE	Détection identification analyse des nouveaux emetteurs
DICASS	Directional command activated sonobuoy system
DICU	Display interface control unit
DID	Data insertion device
DIFAR	Directional acoustic frequency analysis and recording
DIFM	Digital instantaneous frequency measurement
DII	DCMS interphone interface
DII COE	Defense information infrastructure common operating environment
DIL	Digital integrated logic
DIM	Dispense iterface microprocessor
DIRCM	Directional infra-red countermeasures
DITACS	Digital tactical system
DITS	Digital information transfer system
DITU	De-icer timer unit
DIU	Data interface unit
DLMS	Digital land mass system
DLPP	DataLink pre-processor
DLS	Data loader system
DLT	Digital linear tape
DMA	Direct memory access
DMDG	Digital map display generator
DME	Distance measuring equipment
DME-P	Distance measuring equipment - precision
DMG	Digital map generator
DMM	Data management module
DMS	Data multiplexer sub-unit
DMU	1. Data management unit 2. Digital master unit
DOA	Direction of arrival
DoD	Department Of Defense (US)
DOLE	Detection of laser emissions
DOLRAM	Detection of laser, radar and millimetric threats
DOP	Digital onboard processor
DOS	Disk operating system
DP	Display processor
DPCM	Digital pulse code modulation
DPG	Data processor group
DP/MC	Display processor/mission computer
DPRAM	Dual-port random access memory
DPS	Data processing system
DPSK	Digital phase shift keying
DPU	Digital processing unit
DRAM	Dynamic random access memory
DRC	Data recording cartridge
DRD	Digital radar display
DRFM	Digital radio frequency memory
DRU	Data retrieval unit
DSCS	Defense satellite communication system
DSDC	Digital signal data converter
DSP	1. Day surveillance payload 2. Digital signal processing 3. Digital signal processor
DSS	Data storage set
DSU	1. Data storage unit 2. Digital switch unit 3. Dynamic sensor unit
DSUR	Data storage unit receptacle
DTC	Data transfer cartridge
DTD	Data transfer device
DTE	Data transfer equipment
DTED	Digital terrain elevation data
DTF	Digital tape format
DTIU	Data transfer interface unit
DTM	Data transfer module
DTM/D	Digital terrain management and display
DTN	Data transfer network
DTS	1. Data terminal set 2. Data transfer system 3. Digital terrain system
DTU	1. Data transfer unit 2. Display terminal unit
DU	Display unit
DUCK	DCMS universal configuration key
DV/A	Doppler velocimeter/altimeter
DVI	Direct voice input
DVOF	Digital vertical obstruction file
DVRS	Display video recording system
DVS	Doppler velocity sensor
EADI	Electronic attitude director indicator
EAMED	Eventide airborne multipurpose electronic display
EAP	1. Emergency audio panel 2. Experimental aircraft programme
EAR	Electronically agile radar
EAROM	Electrically alterable read-only memory
EARS	ECI airborne relay system
EASU	Engine analyser and synchrophase Unit
EATCHIP	European Air Traffic Control Harmonisation and Integration Programme
EATMS	European Air Traffic Management System
EAU	Engine analyser unit
ECA	Electronic control amplifier
ECAC	European Civil Aviation Conference
ECAM	Electronic centralised aircraft monitor
ECB	Electronic control box
ECCM	Electronic counter-countermeasures
ECDU	Enhanced control and display unit
ECIPS	Electronic combat integrated pylon system
ECL	Emitter coupled logic
ECM	Electronic countermeasures
ECNI	Enhanced communications, navigation and identification
ECOP	Electronic co-pilot
ECP	Engineering change proposal
ECR	Electronic combat and reconnaissance
ECS	Environmental control system
ECU	1. Electronics control unit 2. Environmental control unit 3. Exercise control unit
EDAU	Extended data acquisition unit
EDC	Error detection and correction
EDIU	Engine data interface unit
EDM	Engine data multiplexer
EDTS	Expanded data transfer system
EDU	1. Electronic display unit 2. Engine diagnostic unit
EEC	Electronic engine controls
EEMS	Electrostatic engine monitoring system
EEPROM	Electrically erasable programmable read-only memory
EEZ	Economic exclusion zone
EFCS	Electronic flight control system
EFCU	Electrical flight control unit
EFDARS	Expansible flight data acquisition and recording system
EFIS	Electronic flight instrumentation system
EFMCS	Enhanced flight management computer system
EGAC	Enhanced general avionics computer
EGI	Embedded GPS-inertial
EGNOS	European Geostationary Navigation Overlay System
EGT	Exhaust gas temperature
EHF	Extra high frequency
EHSI	Electronic horizontal situation indicator
EIA	Electronic Industries Association (US)
EICAS	Engine indication and crew alerting system
EID	Emitter identification
EIS	Electronic instrument system
EIU	1. Electronic interface unit 2. Engine interface unit
EL	Electroluminescent
ELAC	Elevator and aileron computer
ELB	Emergency locator beacon
ELF	Electronic location finder
ELINT	Electronic intelligence
ELIOS	ELINT identification and operating system
ELIPS	Electronic integrated protection shield
ELMS	Electrical load management system
ELS	1. Electronic library system 2. Emitter location system
ELT	Emergency locator transmitter
E-MAGR	Enhanced-miniaturised airborne GPS receiver
EM	Electromagnetic
EMC	Electromagnetic compatibility
EMD	Engineering model derivative
EMDU	Enhanced main display unit
EMI	Electromagnetic interference
EMMU	Engine monitor multiplexer unit
EMP	Electromagnetic pulse
EMS	Entry monitor system
EMSC	Engine monitoring system computer
EMTI	Enhanced moving target indicator
EMU	Engine monitoring unit
EMux	Electronic multiplexing
EO	Electro-optic
EOB	Electronic order of battle
EOCM	Electro-optical countermeasures
EOIVS	Electro-optical/infra-red viewing system
EOSS	Electro-optical sensor system
EOVS	Electro-optical viewing system
EPAD	Electrically powered actuator design
EPI	Engine performance indicator
EPLD	Electronic programmable logic device
EPMS	Electrical power management system
EPR	Engine pressure ratio
EPROM	Erasable programmable read-only memory
EPRT	Engine pressure ratio transmitter
EQAR	Expanded quick access recorder
ERAPS	Expandable reliable acoustic path sonobuoy
EROM	Erasable read-only memory
ERP	Effective radiated power
ERS	Electronic resource system
ERWE	Enhanced radar warning equipment
E/SA	Embedded and special application
ESC	Engine supervisory control
ESG	Electrostatically suspended gyro
ESM	Electronic support measures
ESP	1. Expandable system programmer 2. Expendable signal processor
ESS	Exercise support system
ESU	Electronic storage unit
ETA	Estimated time of arrival
ETE	Estimated time en route
ETIPS	Electrothermal ice protection system
ETMP	Enhanced terrain masked penetration
ETPU	Engine transient pressure unit
EU	Electronics unit
EUROCAE	European Organisation for Civil Aviation Electronics
EVM	Engine vibration monitor
EVS	Enhanced vision Sensor
EW	Electronic warfare
EWAAS	End-state WAAS (wide area augmentation system)
EWMS	Electronic warfare management system
EWMU	EW management unit
EWPI	Electronic warfare prime indicator
EXCAP	Expanded capability
FAA	Federal aviation administration (US)
FAADC²I	Forward area air defence command, control and intelligence
FAC	1. Flight augmentation computer 2. Forward air controller
FACS	Fully automatic compensation system
FACTS	FLIR augmented cobra TOW sight
FADEC	Full authority digital engine control
FAF	Final approach fix
FAFC	Full authority fuel control
FAM	Final approach mode
FAMIS	Full aircraft management/inertial system
FANS	Future air navigation system
FAP	Final approach
FBL	Fly-by-light
FBS	Fly-by-speech
FBW	Fly-by-wire
FCC	Flight control computer
FCDC	Flight control data concentrator
FCMC	Flight control and monitoring computer
FC/NP	Fire control/navigation panel
FCPC	Flight control primary computer
FCS	Flight control system
FCSC	Flight control secondary computer
FCU	Flight control unit
FDAMS	Flight data acquisition management system
FDAU	Flight data acquisition unit
FDE	Fault detection and exclusion
FDEP	Flight data entry panel
FDIU	Flight data interface unit
FDM	Frequency division multiplex
FDMU	Flight data management unit
FDP	Flight data panel
FDR	Flight data recorder
FDR/FA	Flight data recorder/fault analyser
FDS	Flight director system
FDT	Flight-deck terminals
FDU	Flight data unit
FET	Field effect transistor
FEWSG	Fleet electronic warfare support group
FFSP	Full function signal processor
FFT	Fast fourier transform
FGCP	Flight guidance control panel
FINAS	Ferranti inertial nav/attack system
FIRAMS	Flight incident recorder and aircraft monitoring system
FIRMU	Flight incident recorder memory unit
FIS	Flight information service
FL	Foot-lambert
FLAG	Four-mode laser gyro
FLAGSHIP	Four-mode laser gyro software hardware implemented partitioning
FLASH	Folding light acoustic systems for helicopters
FLIR	Forward-looking infra-red
Flops	Floating point operations per second
FM	Frequency modulated
FMA	Flight mode annunciator

FMC	1. Flight management computer 2. Forward motion compensation
FMCS	1. Fatigue monitoring and computing system 2. Flight management computer system
FMCW	Frequency modulated continuous wave
FMGC	Flight management and guidance computer
FMGS	Flight management and guidance system
FMICW	Frequency modulated interrupted continuous wave
FMS	Flight management system
FMU	Flight management unit
FOAEW	Future organic airborne early warning
FOC	Full operational capability
FOV	Field of view
FPD	Flat-panel display
FPMU	Fuel pump monitoring unit
FQIS	Fuel quantity indication system
FRPA	Fixed reception pattern antenna
FRP	Federal (US) radionavigation plan
FRS	Fighter/reconnaissance/strike
FSA/CAS	Fuel savings advisory and cockpit avionics system
FSAS	Fuel saving advisory system
FSC	Fuel savings computer
FSD	Full-Scale deflection
FSK	Frequency shift keying
FSRS	1. Flight safety recording system 2. Frequency selective receiver system
FSS	Flight service station
ft	Feet
FT	Fault tolerant
FT-ADIRS	Fault tolerant air data inertial reference system
FT-ADIRU	Fault tolerant air data inertial reference unit
FTC	Fast time constant
FTI	Fixed target indication
FTIT	Fan turbine inlet temperature
FTRG	Fleet tactical readiness group
G	Giga = 1,000,000,000
GaAs	Gallium arsenide
GANS	Global access navigation safety
GAS	Global positioning adaptive antenna system
GATM	Global air traffic management
GATR	Ground air transmit receive
GATS	GPS-aided targeting system
GBIB	Ground-based integrity broadcast
GCA	Ground collision avoidance
GCAS	Ground collision avoidance system
GCR	Ground clutter reduction
GDE	Graphics differential engine
GDP	Graphics drawing processor
GEM	GPS embedded module
GEN-X	Generic expendable
GES	Ground earth station
GGP	GPS guidance package
GHz	Giga hertz
GIC	GPS integrity channel
GIC	GPS/WAAS integrity channels (GPS global positioning system) (WAAS wide area augmentation system)
GIG	GPS integration guidelines
GIS	Geographic information system
GIT	General interface terminal
GLINT	Gated laser illuminator for night television
GLONASS	Global orbital navigation satellite system
GMR	Ground mapping radar
GMT	Greenwich mean time
GMTI	Ground moving target indication
GNC	General navigation computer
GNLS	GPS navigation and landing system
GNSS	Global navigation satellite system
GNSSU	Global navigation satellite sensor unit
GP&C	Global positioning and communication
GPCDU	General purpose control display unit
GPIN	Global positioning laser inertial navigation
GPIRS	Global positioning/inertial reference system
GPIRU	Global positioning inertial reference unit
GPS	Global positioning system
GPSSU	Global positioning system sensor unit
GPVI	Graphics processor video interface
GPWS	Ground proximity warning system
GRCS	Guardrail common sensor
GRE	Ground readout equipment
GRIS	Guardrail common sensor interoperability system
GS	Groundspeed
GSDI	Groundspeed and drift indicator
GSE	Ground support equipment
GSM	GPS sensor module
GTAR	GEC Thomson airborne radar
GTRE	Gas-Turbine Research Establishment (UK)
GUI	Graphics user interface
h	hour(s)
HAC	Anti-tank variant of Tiger helicopter
HACLCS	Harpoon airborne command, launch and control system
HAD	Hybrid analogue digital
HADAS	Helmet airborne display and sight
HADS	Helicopter air data system
HAINS	High-accuracy inertial navigation system
HAP	Escort variant of Tiger helicopter
HAPS	Helicopter acoustic processing system
HARM	High-speed anti-radiation missile
HBR	High bit rate
HDD	Head-down display
HDDR	Head-down display recorder
HDG	Heading
HDS	Hard disk subsystem
HDTV	High-definition TV
HER	Harsh environment recorder
HERALD	Helicopter equipment for radar and laser detection
HEU	HUD electronics unit
HF	High frequency
HFAC	Helicopter flight advisory computer
HFDU	High-frequency data unit
HgCdTe	Mercury cadmium telluride
HGS	Holographic guidance system
HHTI	Hand-held thermal imager
HHUD	Holographic head-up display
HIADC	High-integration air data computer
HIBIRD	Helicopter identification by infra-red detection
HIBRAD	Helicopter identification by radar detection
HICU	HIPSS interface control unit
HIDAS	Helicopter integrated defensive aids system
HIDEC	Highly integrated digital engine control
HIDSS	Helmet integrated display sight system
HIPAS	High-performance active sonar
HIPSS	Helicopter integrated power and switching system
HIRES	High resolution
HIRF	High-intensity radiated field
HIRNS	Helicopter infra-red navigation system
HIRS	Helicopter infra-red system
HISAR	Hughes integrated surveillance and reconnaissance system
HIT	Hughes improved terminal
HITMORE	Helicopter installed television monitor recorder
HLD	Head level display
HLWE	Helicopter laser warning equipment
HMCU	Hydraulic monitoring computer unit
HMD	Helmet-mounted display
HMDD	Helmet-mounted display device
HMFU	Hydromechanical fuel unit
HMU	Hydromechanical unit
HNVS	Helicopter night vision system
HOCAS	Hands-on collective and stick
HOFIN	Hostile fire indicator
HOPS	Helmet optical position sensor
HOS	Helitow observation system
HOTAS	Hands-on throttle and stick
HOWLS	Hostile weapons location system
HP	High pressure
HPA	High-power amplifier
HPAG	High-power amplifier group
HRP	Headset receptacle panel
HSDB	High-speed databus
HSI	Horizontal situation indicator
HSI	Hyperspectral imagery
HSIC	High-speed integrated circuit
HSSL	Helicopter self-screening launcher
HTC	Hover trim control
HTNS	Helicopter tactical navigation system
HTTB	High-technology testbed
HUD	Head-up display
HUDC	Head-up display computer
HUDWAC	Head-up display and weapon aiming computer
HUMC	Health and usage monitoring computer
HUMS	1. Health and usage monitoring system 2. Health and usage monitoring and sensing
HVPSU	High-voltage power supply unit
Hz	Hertz
IAC	Integrated avionics computer
IAD	Integrated antenna detector
IAHFR	Improved airborne high-frequency radio
IAHFR/NOE	Improved airborne high-frequency radio nap of the Earth
IAM	Initial approach mode
IAMS	Integrated armament management system
IAPS	Integrated avionics processing system
IAS	Indicated air speed
IB	Interconnecting box
IBS	Integrated broadcast service
I/C	Interface and control
ICAAS	Integrated controls and avionics for air superiority
ICAO	International civil air traffic organisation
ICAP	Increased capability
ICCP	Integrated communications control panel
ICDU	1. Integrated control and display unit 2. Intelligent control display unit
ICE	Improved combat efficiency
ICMS	Integrated countermeasures system
ICNIA	Integrated communications navigation identification avionics
ICNIS	Integrated communications navigation identification set
ICP	Integrated control panel
ICS	1. Intercommunications system (or set) 2. Internal countermeasures set
ICSM	Integrated conventional stores management
ICSM/GPS	Integrated conventional stores management/global positioning system
ICU	1. Interface computer unit 2. Interface converter unit 3. Interstation control unit
ICW	Interrupted continuous wave
ID	Identification
IDACS	Integrated digital audio control system
IDAP	Integrated defense avionics program
IDAS	Integrated design automation system
IDECM	Integrated defensive electronic countermeasures
IDF	Instantaneous direction-finding
IDL	Interoperable datalink
IDM	Inductive debris monitor
IDP	Imagery display processor
IDS	1. Infra-red detection set 2. Interdiction Strike
IEEE	Institute of Electrical and Electronic Engineers (US)
IEU	Interface electronics unit
IEW	Integrated electronic warfare
IEWCS	Intelligence and electronic warfare common sensor
IF	Intermediate frequency
IFF	Identification friend or foe
IFFCP	Identification friend or foe control panel
IFM	Instantaneous frequency measurement
IFM/SHR	IFM superheterodyne receiver
IFMU	Integrated flight management unit
IFOG	Interferometric fibre optic gyro
IFR	Instrument flight rules
IGAC	Israeli general avionics computer
IHADSS	Integrated helmet and display sighting system
IHAS	Integrated hazard avoidance system
IHDTV	Intensified high-definition television
IHEWS	Integrated helicopter electronic warfare suite
IHU	Integrated helmet unit
IHUMS	Integrated health and usage monitoring system
IIDS	Integrated instrumentation display system
IIS	Infra-red imaging system
IJMS	Interim JTIDS message system
ILS	Instrument landing system
IMA	Integrated modular avionics
IMC	1. Instrument meteorological conditions 2. Image motion compensation
IMSS	Integrated multisensor system
IMU	Inertial measurement unit
in	inch(es)
INACP	Integrated navigation aids control panel
INEWS	Integrated electronic warfare system
INS	Inertial navigation system
INU	Inertial navigation unit
I/O	Input/output
IOC	1. Initial operational capability 2. Input output computer
I/P	Identification position
IP	Intermediate pressure
IPADS	Improved processing and display system
IPCS	Intelligent power control system
IPEC	Inflight passenger entertainment and communications
IPT	Intelligent power terminal
IPU	Interface processor unit
IR	Infra-red
IRCCD	Infra-red charge coupled device
IRCM	Infra-red countermeasures
IRFIS	Inertial referenced flight inspection system
IRIS	1. Infra-red imaging subsystem 2. Integrated radar imaging system
IRLS	Infra-red linescanner
IRMS	Integrated radio management system
IRP	Interphone receptacle panel
IRS	Inertial reference system
IRST(S)	Infra-red search and track (system)
IRU	Inertial reference unit

IR/UV	Infra-red/ultraviolet
IRV	Infra-red vision
IRVAT	Infra-red video automatic tracking
ISA	Instruction set architecture
ISAHRS	Improved standard attitude and heading reference system
ISAR	Inverse synthetic aperture radar
ISB	Independent sideband
ISBA	Inertial sensor-based avionics
ISC	Intercommunications set control
ISDS	IRCM self-defence system
ISIS	Integrated strike and interception system
ISLS	Interrogation side-lobe suppression
ISO	International standardisation organisation
ISP	Intelligence, surveillance and reconnaissance
ISS	Integrated sensor system
ISU	Intercommunications set control unit
ITAR	Integrated terrain access and retrieval system
I-TOW	Improved tube-launched optically tracked wire-guided (missile)
IUMS	Integrated utilities management system
IVMMS	Integrated vehicle mission management system
IVSC	Integrated vehicle subsystem controls
IVSI	Instantaneous vertical speed indicator
IWAAS	Initial WAAS (wide area augmentation system)
JAR	Joint airworthiness requirement
JASS	Joint airborne SIGINT system
JAST	Joint advanced strike technology
JDAM	Joint direct attack munition
JIAWG	Joint Integrated Avionics Working Group (US)
Joint STARS	Joint surveillance and target attack radar system (also JSTARS)
JPALS	Joint precision approach landing system
JPATS	Joint primary aircraft training system
JPO	Joint Program Office (US)
JSAF LBSS	Joint SIGINT avionics family low- band subSystem
JTIDS	Joint tactical information distribution system
JTRS	Joint tactical radio system
JTW	Joint targeting workstation
k	1,000
K	Kelvin
KAPS	Kollsman auto-schedule pressurisation System
kbit	Kilobit
KBU	KeyBoard unit
kbyte	Kilobyte
kg	Kilogram
KFD	Key fill device
kHz	Kilohertz
kips	Thousand instructions per second
kops	Thousand operations per second
kt	Nautical miles per hour (knot)
KTD	Key transfer device
L1	GPS carrier frequency (1227.6 MHz)
L2	GPS carrier frequency (1575.42 MHz)
LAAP	Low-altitude autopilot
LAAS	Local area augmentation system
LAASH	LITEF analytical air data system for helicopters
LAAT	Laser augmented airborne TOW
LADGNSS	Local area differential global navigation satellite system
LADGPS	Local area differential global positioning system
LAEO	Low-altitude electro-optical
LAHRS	Kearfott low-cost altitude heading reference system
LAIRS	1. Light aircraft reconnaissance system 2. Loral advanced imaging radar system
LAMPS	Light airborne multipurpose system
LAN	Local area network
LANA	Low-altitude night attack
LANE	Low-altitude navigation equipment
LANTIRN	Low-altitude navigation and targeting infra-red for night
LASS	Low-altitude surveillance system
LASTE	Low-altitude safety and target enhancement
LCC	Leadless ceramic chip-carrier
LCD	Liquid crystal display
LCF	Low cycle fatigue
LCINS	Low-cost inertial navigation system
LCLU	Landing control logic unit
LCM	Lance-cartouches modulaire
LCOS(S)	Lead computing optical sight (system)
LCTAR	Le Centre Thomson d'Applications Radar
LCWDS	Low-cost weapon delivery system
LDT/SCAM	Laser detector and tracker/strike camera
LDU	Launcher decoder unit
LEA	Leurre electromagnetique actif
LECOS	Light electronic control system
LED	Light emitting diode
LED-RHA	Light emitting diode-recording head assembly
LESM	Lightweight electronic support measures
LF	Low frequency
LGCIU	Landing gear computer and interface unit
LH	Light helicopter
LHN	LITEF helicopter navigation
LHR	LITEF helicopter reference
LICAS	Low-intensity conflict aircraft system
LIDAR	Light detection and ranging
LIMAR	Laser imaging and ranging
LINS	Laser inertial navigation system
LIP	Limited installation programme
LISCA	Leurre infrarouge à signature et cinématique adaptée
LITDL	Link 16 interoperable tactical data link
LLLTV	Low-light level television
LLTV	Low-light television
LNA	Low-noise amplifier
LO	Local oscillator
LOCUS	Laser obstacle cable unmasking system
LOFAR	Low frequency omnidirectional acoustic frequency Analysis and recording
LORAN	Long-range aid to navigation
LORES	Low resolution
LOROP	Long-range oblique photography
LOS	Line of sight
LOSI	Line of sight indicator
LP	Low pressure
LPC	Linear predictive coding
LPCBA	Low-pressure compressor bleed actuator
LPD	Low probability of detection
LPHUD	Low-profile head-up display
LPI	Low probability of interception
LQA	Link quality analysis
LRCU	Landing roll-out control unit
LRM	Line-replaceable module
LRMTS	Laser ranger and marked target seeker
LRU	Line-replaceable unit
LSB	Lower sideband
LSI	Large scale integration
LST	Laser spot tracker
LTD/R	Laser target designator/ranger
LTR	Loop transfer recovery
LVDT	Linear variable differential transformer
LWA	Laser warning analyser
LWF	Lightweight fighter
LWR	Laser warning receiver
m	Metre
M	1,000,000 or mega
M-ADS	Modified automatic dependent surveillance
MACC	Multi-application control computer
MACS	Multiple application control system
MAD	Magnetic anomaly detector
MADAR	1. Maintenance analysis, detection and recording 2. Malfunction detection, analysis and recording
MADC	1. Micro air data computer 2. Miniature air data computer
MADGE	Microwave aircraft digital guidance equipment
MAESTRO	Modular avionics enhancement system targeted for retrofit operations
MAG	Micromachined accelerometer gyro
MAGR	Miniature airborne GPS receiver
MAHRU	Microflex attitude and heading reference unit
MAP	Missed approach point
MAP	Modular airborne processor
MARA	Modular architecture for real-time applications
MARE	Miniature analogue recording electronics
MARS	1. Modular airborne recording system 2. Multiapplication recorder/ reproducer system
MASINT	Measurements and signatures intelligence
MASTER	Military aircraft satcoms terminal
MASPA	Minimum aviation system performance standards
MATSS	Mobile aerostat tracking and surveillance system
MAW	1. Missile approach warner 2. Mission adaptive wing
MAWS	Missile approach warning system
mb	Millibar
MBAT	MultiBeam array transmitter
Mbit	Megabit
Mbyte	Megabyte
MC	Mission computer
MCCP	Main communications control panel
MCDU	Multifunction control and display unit
MCE	Modular control equipment
Mcops	Million complex operations per second
MCT	1. Mercury cadmium telluride 2. Metal oxide semiconductor controlled thyristor
MCU	1. Management control unit 2. Master control unit 3. Missile control unit 4. Modular concept unit 5. Multifunction control unit
MCW	Modulated continuous wave
MDF	Mission data file
MDG	Map display generator
MD(G)T	Mission data (ground) terminal
MDI	Multifunction display indicator
MDL	Mission data loader
MDP	1. Maintenance data panel 2. Modular display processor
MDRI	Multipurpose display repeater indicator
MDT	Mission data terminal
MDTC/P	Mega data transfer cartridge with processor
MDTS	Mission data transfer system
MEA	Minimum en route altitude
MEC	Modular electronics concept
MEECN	Minimum essential emergency communications network
MERLIN	Modular ejection-rated low-profile imaging for night
MESA	Minimum emergency safe altitude
METR	Multiple emitter targeting receiver
MFCD	Multifunction colour display
MFD	Multifunction display
MFD(S)	Multifunction display (system)
MFHD	Multifunction head-down display
MFMS	Military flight management system
MGR	Miniature GPS receiver
MHDD	Multifunction head-down display
MHz	Megahertz
MIAMI	Microwave ice accretion measurement instrument
MICNS	Modular integrated communications and navigation system
Micro-AIDS	Micro-aircraft integrated data system
MIDA	Message interchange distributed application
MIDS	Multifunction information distribution system
MIGITS	Miniature integrated GPS/INS tactical system
MIL-SPEC	Military specification
MILSTAR	Military strategic and tactical relay
MIL-STD	Military standard
MIMU	Multisensor inertial measurement unit
min	Minute(s)
MIPS	Million instructions per second
MIRLS	Miniature infra-red linescan system
MIRTS	Modular infra-red transmitting system
MLI	Mid-life improvement
MLPRF	Modular low-power radio frequency
MLS	Microwave landing system
MLU	1. Mid-life update 2. Monitor and logic unit
MLV	Memory loader and verifier
mm	millimetre(s)
MM	Mission manager
MMC	Mission management computer
MMIC	Monolithic microwave integrated circuit
MMLSA	Military microwave landing system avionics
MMMS	Multimission management system
Mmo	Maximum operating mach number
MMP	Maintenance monitoring panel
MMR	MultiMode receiver
MMRS	MultiMission radar system
MMS	Mast-mounted sight
MMSS/R	Mobile mass storage system model R
MMW(R)	Millimetric wave (radar)
MNOS	Metal nitride oxide silicon
MNPS	Minimum navigation performance standards
MNS	Modular navigation system
MOA	Minimum operating altitude
MODAR	Modular aviation radar
MODAS	Modular data acquisition system
MODIR	Modulated infra-red jammer
MONOHUD	Monocular head-up display
Mops	Million operations per second
MOPS	Minimum operating performance standard
MOS	Metal oxide silicon
MOSFET	Metal oxide silicon field effect transistor
MOSP	Multimission optronic stabilised payload
MP	Mission planner
MPA	Maritime patrol aircraft
MPCD	Multipurpose colour display
MPD	Multipurpose display
MPHD	Multipurpose head-down
MPM	Multipurpose modem
MPPU	Multipurpose processor unit
MPS	Mission planning subsystem
MQLF	Mobile quick-look facility
MR	Maritime reconnaissance

MRAALS	Marine remote area approach landing system
MRAAM	Medium-range air-to-air missile
mrad	milliradian
MRLG	Monolithic ring laser gyro
MRT	1. Miniature receive terminal 2. Multirole turret
MRTI	Multirole thermal imager
MRTU	Multiple remote terminal unit
MRU	1. Maintenance recorder unit 2. Mobile reporting unit
ms	millisecond
m/s	metres per second
MSA	Minimum safe altitude
MSAR	Miniature synthetic aperture radar
MSCADC	Miniature standard central air data computer
MSD	1. Map storage display 2. Mass storage device
MSF	Mission support facility
MSI	Multispectral imagery
MSIP	1. Multi(national) staged improvement programme 2. MultiStaged improvement programme
MSIS	Multisensor stabilised integrated system
MSP	Mission system processor
MSPS	Modular self-protection system
MSR	1. Marconi secure radio 2. Modular strain recorder 3. Modular survivable radar
MSS	Maritime surveillance system
MSTAR	Moving and stationary target acquisition and recognition
MSU	1. Maintenance station unit 2. Mass storage unit 3. Mode selector unit
MSWS	MultiSensor warning system
MTAS	Millimetric target acquisition system
MTBF	Mean Time between failures
MTBR	Mean Time between repairs
MTBUR	Mean Time between unscheduled repairs
MTI	Moving target indicator
MTIS	Modular thermal imaging sight
MTL	Magnetic tape loader
MTTR	Mean time to repair
MTU	Magnetic tape unit
MUST	Multimission UHF satcom transceiver
Mux	Multiplexer
mV	Millivolt
MVP	Military VME processor
mW	Milliwatt
MWR	Microwave radiometer
NAS	Navigation attack system
NASA	National Aeronautics and Space Administration (US)
NASH	Navigation and attack system for helicopters
NAT	National air traffic
NATO	North Atlantic Treaty Organisation
nav/ccm	navigation and communication
NavHARS	Navigation, heading and attitude reference system
NavWASS	Navigation and weapon aiming subsystem
NBC	Nuclear, biological and chemical
NCS	Network control station
NCU	Navigation computer unit
ND	Navigation display
NDB	Non-directional beacon
Nd:YAG	Neodymium/yttrium aluminium garnet
NEACP	National emergency airborne command post
Ng	Gas generator RPM
Ni/Cd	Nickel/cadmium
NIDJAM	Navigation/identification deception jammer
NIIRS	National imagery interpretability rating scale (US)
NILE	NATO improved link 11
NIS	NATO identification system
NITE-OP	Night imaging through electro-optics
NIU	Navigation interface unit
n mile(s)	Nautical mile(s)
NMOS	Negative metal oxide semiconductor
NMS	Navigation management system
NMU	Navigation management unit
NOCUS	North continental US
NOE	Nap of the earth
NOR	Logic circuit usable as either AND/OR
NOTAR	No tail rotor
NOVRAM	Non-volatile random access memory
NPR	No power recovery
NPRM	Notice of proposed rule making
NPU	Navigation processor unit
NQIS	Navigation quality inertial sensor
NRT	Near real-time
ns	Nanoseconds
NSA	National Security Agency (US)
NSIU	Navigation switching interface unit
NSP	Night surveillance payload
NSSL	National Severe Storms Laboratory (US)
NTDS	Naval tactical data system
NTS	Night targeting system
NV	1. Night vision 2. Non-volatile
NVG	Night vision goggles
NVG/HUD	Night vision goggles/head-up display
NV/HUD	Night vision/head-up display
NVIS	Night vision imaging systems
NVM	Non-volatile memory
NWDS	Navigation and weapon delivery system
OADS	Omnidirectional air data system
OAS	Offensive avionics system
OASYS	Obstacle avoidance system
OAT	Outside air temperature
OBEWS	OnBoard electronic warfare simulator
OBI	Omni-bearing indicator
OBTEX	OnBoard targeting experiments
OCU	Optical control unit
ODIN	Operational data interface
ODS	1. Operational debrief station 2. Optical disk system
ODU	Optical display unit
OEM	Original equipment manufacturer
OFP	Operational flight programme
OHU	Optical head unit
OMMS	Oxygen mask-mounted sight
OMS	1. Onboard maintenance system 2. Operational management system
OMT	Onboard maintenance terminal
OOOI	Out/off/on/in
OQAR	Optical quick access recorder
OQPSK	Offset quadrature phase shift keyed
ORS	Offensive radar system
OSA	Operational support aircraft
OSC	Optical sensor converter
OSF	Optronique secteur frontal
OSU	Omega/VLF sensor unit
OTAR	Over the air re-keying
OTCIXS	Officer in tactical command information exchange xubsystem
OTH(T)	Over the horizon (Target)
OTIS	Optronic tracking and identification system
OTPI	On-top position indicator
OWL	Obstacle warning ladar
OWL/D	Optical warning, location and detection
OWS	Obstacle warning system
P	Precision
P³I	Preplanned product improvement
PA	1. Pilot's associate 2. Public address
PAC	1. Precision attitude control 2. Public address set control
PACA/IAGSS	Precision attitude control augmentation/improved air-to-ground sight system
PACIS	Pilot aid and close-in surveillance
PAL	Phase alternation line
PAM	Pulse amplitude modulation
PAMIR	Passive airborne modular infra-red
PAR	Power analyser and recorder
PAS	Performance advisory system
PAT	Pilot access terminal
PAWS	Passive airborne warning system
PBDI	Position, bearing and distance indicator
PBIL	Projected bomb impact line
PC	1. Personal computer 2. Printed circuit 3. Pulse compression
PCB	Printed circuit board
PCM	1. Power converter module 2. Pulse code modulation
PCSB	Pulse coded scanning beam
PCU	Pilot's control unit
PDC	Programme development card
PDC/PMM	Programme development card/ performance monitor module
PDES	Pulse Doppler elevation scan
PDNES	Pulse Doppler non-elevation scan
PDR	Programmable digital radio
PDS	Portable data store
PDU	Pilot's display unit
PE	Processing element
PENETRATE	Pass no enhanced navigation with terrain referenced avionics
PEO	Program executive officer
PEP	Peak envelope power
PERP	Peak effective radiated power
PFCS	Primary flight control (or computer) system
PFD	Primary flight display
PFD/ND	Primary dlight display/navigation display
PFIS	Portable flight inspection system
PI	Process (or programme) instruction
PICC	Processor interface controller and communication
PIDS	Pylon integrated dispenser station
PILOT	Pod integrated localisation, observation, transmission
PIN	Positive-intrinsic-negative (semiconductor)
PIRATE	Passive infra-red airborne track equipment
PISA	Pilot's infra-red sighting ability
PITS	Passive identification and targeting system
PIU	1. Processor interface unit 2. Pylon interface unit
PLAID	Precision location and identification
PLB	Personal locator beacon
PLGR	Precision lightweight GPS receiver
PLL	Phase-locked loop
PLRS	Precision location reporting system
PLS	Personnel location system
PLSS	Precision location strike system
PLU	Programme load unit
PMA	Projected map sssembly
PMAWS	Passive missile approach warning system
PMF	Processeur militaire Français
PMM	Performance monitor module
PNVS	Pilot night vision sensor
PODS	Portable data store
POET	Primed oscillator expendable transponder
PPI	Plan position indicator
PPM	1. Preprocessor module 2. Programmable processing modules
pps	Pulse per second
PPS	1. Photovoltaic power system 2. Precise positioning service (GPS)
PRF	Pulse recurrence frequency
PRI	Pulse repetition interval
PRIDE	Pulse recognition interval de-interleaving
PRNav	Precision area navigation
PROM	Programmable read-only memory
PRU	Performance reference unit
PSC	Performance seeking control
PSCS	Photographic sensor control system
PSI	Pounds per square inch
PSK	Phase shift keyed
PSP	Programmable signal processor
PSU	1. Power supply unit 2. Power switching unit
PTC	Pack temperature controller
PTM	Pressure transducer module
PTMU	Pressure and temperature measurement unit
PTP	Programmable touch panel
PTT	Press to transmit
PULSE	Precision up-shot laser steerable equipment
PVD	Paravisual director
PVI	Pilot-vehicle interface
PVS	Pilot's vision system
P/Y	GPS precision code
QAR	Quick access recorder
QDM	Magnetic heading to runway
QEC	Quadrantal error correction
QPSK	Quadrature phase shift keyed
QR	Quadrant receiver
RA	1. Relay assembly 2. Resolution advisory
RADC	Rome Air Development Center (US)
RAE	Royal Aerospace Establishment (UK)
RAF	Royal Air Force (UK)
RAID	Redundant array of independent disks
RAIM	Receiver autonomous integrity monitoring
RAM	Random access memory
RAMS	1. Racal avionics management system 2. Removable auxiliary memory set
RAMU	Removable auxiliary memory unit
RAN/RAWS	Royal australian navy role adaptable weapons system
RAPPORT	Rapid alert programmed power management of radar targets
RA/TA	Resolution advisory/traffic advisory
RA/VSI	Resolution advisory/vertical speed indicator
RC	Resistance capacitance
RCS	Radar cross-section
RCU	1. Remote-control unit 2. Rudder control unit
RDAS	Reconnaissance data annotation set
RDC	Remote data concentrator
RDI	Radar Doppler à impulsions
RDM	Radar Doppler multifunction
RDP	Range Doppler profile
RDU	Remote Display unit
REACT	Rain echo attenuation compensation technology
REU	Remote electronics unit
REWTS	Responsive electronic warfare training system
RF	Radio frequency
RFA	Royal fleet auxiliary
RFI	Radio frequency interference
RFP	Request for proposals
R/FPU	Recorder/film processor unit

RFTDL Range-finder target designator laser
RFU Radio frequency unit
RGB Red, green, blue
RGS Recovery guidance system
RHA Recording head assembly
RHWR Radar homing and warning receiver
RIMS Replacement inertial measurement system
RIS Reconnaissance interface system
RISC Reduced instruction set computer
RLG Ring laser gyro
RMI 1. Radio magnetic indicator 2. Remote magnetic indicator
RMPA Replacement maritime patrol aircraft
RMR Remote map reader
RMS 1. Reconnaissance management system 2. Root mean square
RNav Area navigation
RNP/ANP Required navigation performance/ actual navigation performance
RNS Radar navigation system
RO Range only
RODS Ruggedised optical data system
ROM Read-only memory
ROMAG Remote map generator
RPA Rotocraft pilot's associate
RPG Receiver processor group
rpm revolutions per minute
RPU Receiver processor unit
RPV Remotely piloted vehicle
RRU Remote readout unit
RSC Remote switching control
RSIP Radar system improvement programme
R/T Receiver/transmitter
RTA Receiver-transmitter-antenna
RTCA Radio Technical Commission For Aeronautics (US)
RTCA FFSC Radio Technical Commission For Astronautics' Free Flight Select Committee (US)
RTD Real-time display
RTIC Real-time information into cockpit
RTMM Removable transport media module
RTT Radio telemetry theodolite
RTU Radio tuning unit
RVDT Rotating variable differential transformer
RVSM Reduced vertical separation minima
RWR Radar warning receiver
RWS Range-while-search

s second(s)
SA 1. Selective availability 2. Situation awareness
SAAHS Stability augmentation and attitude hold system
SAARU Secondary attitude and air data reference unit
SA/AS Selective availability/anti-spoofing
SAC Strategic air command (US)
SACT Signal acquisition conditioning terminal
SADANG Système acoustique d'Atlantique nouvelle génération
SAFCS Standard automatic flight control system
SAFFIRE Synthetic aperture fully focused imaging radar equipment
SAFIS Semi-automatic flight inspection system
SAHRS Standard attitude and heading reference system (Kearfott)
SAIRS Standardised advanced infra-red sensor
SAM 1. Situation awareness mode 2. Surface-to-air missile
SAMIR Système d'alérte missile infra rouge
SAMSON Special avionics mission strap-on (pod)
SAR 1. Search and rescue 2. Signal acquisition remote 3. Synthetic aperture radar
SARPs Standard and recommended practices
SARSAT Search and rescue satellite aided tracking
SASS Small aerostat surveillance system
SAT 1. Situational awareness technology 2. Static air temperature
SATCOM Satellite communications
SATURN Second generation of anti-jam tactical UHF radios for NATO
SAU 1. Safety and arming unit 2. Signal acquisition unit
SAW Surface acoustic wave
SAWS Silent attack warning system
SBC Single board computer
SC Single card
SCADC Standard central air data computer
SCAR Sistemi de control de armamento
SCAT Speed command of attitude and thrust
SCAT-1 Special category 1
SCD Signal command decoder
SC-DDS Sensor control-data display set
SCDL Surveillance and control datalink
SCI Serial communication interface
SCNS Self-contained navigation system
SCP Single card processing
SCS Sidewinder control system
SCT Single-channel transponder
SCU 1. Station control unit 2. Stores control unit 3. Supplemental control unit
SDC 1. Signal data computer 2. Signal data converter 3. Situation display console
SDCS Satellite data communications system
SDCU Smoke detection control unit
SDP Signal data processor
SDS Satellite data system
SDU 1. Satellite data unit 2. Smart display unit
SEAD Suppression of enemy air defences
SEAFAC Systems engineering avionics facility
SEC Spoiler and elevator computer
SECU Spoilers electronic control unit
SELCAL Selective calling
SEM Standard electronic module
SEMA 1. Smart electromechanical actuator 2. Special electronic mission aircraft
SEMC Standard electronic memory cartridge
SENAP Signal emulation of aero-navigation and landing
SEOS Stabilised electro-optical system
SEP Spherical error probability
SETS Severe environment tape platform
SEU Sight electronics unit
SFCC Slat flap control computer
SFDP Smart flat-panel display
SFDR Standard flight data recorder
SGU Signal generator unit
SHARP 1. Standard hardware aquisition and reliability programme 2. Strapdown heading and attitude reference platform
SHF Super high frequency
SHIP Software/hardware implemented partitioning
SHR Superheterodyne receiver
SHUD Smart head-up display
SICAS Secondary surveillance radar (SSR) improvements and collision avoidance systems
SID 1. Sensor image display 2. Standard instrument departure
SIF Selective identification facility
SIFF Successor identification friend or foe
SIGINT Signals intelligence
SIMOP Simultaneous operation (of collocated RF sets)
SIMS Signal identification mobile system
SINCGARS Single-channel ground/air radio system
SIRFC Suite of integrated RF countermeasures
SIS Superhet IFM subsystem
SIT Silicon intensified target
SITREP Situation report
SIU 1. Sensor interface unit 2. Sidewinder interface unit
SKE Station-keeping equipment
SLAM Standoff land attack missile
SLAMMR Side-looking airborne modulated multimission radar
SLAR Side-looking airborne radar
SLIR Side-looking infra-red
SLOS Stabilised long-range observation system
SMA Surface movement advisor
SMDU Strapdown magnetic detector unit
SMEU Switchable main electronic unit
SMP Stores management processor
SMS Stores management system
SMT Surface mount technology
SMU Systems management unit
SMWP Standby master warning panel
S/N Signal/noise ratio stress against number of alternating load cycles to failure
SOCUS South continental US
SOF Special operations force
SOJ Standoff jamming
SOLL Special operations low level
SOTAS Standoff target acquisition system
SPASYN Space synchro
SPEES Système pour l'elévation de l'endommagement structural
SPEW small platform electronic warfare (system)
SPEWS Self-protection electronic warfare system
SPHERIC System for protection of helicopters by radar and infra-red countermeasures
SPI Short pulse insertion
SPILS Spin prevention and incidence limiting system
SPN/GEANS Standard precision navigator/gimballed electrostatic aircraft navigation system
SPO System program office (US)
SPRITE Signal processing in the element
SPS Standard positioning service (GPS)
SPT Signal processing tools
SPU 1. Signal processing unit 2. Stores power unit
SRA 1. Shop repair assembly 2. Shop replaceable assembly
SRAM Static random access memory
SRC Surveillance radar computer
SRFCS Self-repairing flight control system
SRL Systems research laboratories
SRPC Strain range pair counter
SRS 1. Sonobuoy reference system 2. Superhet receiver subsystem 3. Survival radio set
SRU 1. Scanner receiver unit 2. Shop replaceable unit
SS System status
SSB Single sideband
SSCVFDR Solid-state combined voice and flight data recorder
SSCVR Solid-state cockpit voice recorder
SSFDR Solid-State flight data recorder
SSICA Stick sensor and interface control assembly
SSID Solid-state ice detector
SSQAR Solid-state quick access recorder
SSR Secondary surveillance radar
SSTI Stabilised steerable thermal imager
SSU Sensor surveying unit
STANAG Standard NATO agreement
STAR 1. Signal threat analysis and recognition 2. Standard terminal arrival route
STARS 1. Small tethered aerostat relocatable system 2. Stand-off target attack radar system
START Solid-state angular rate transducer
STC 1. Sensitivity time control 2. Supplemental type certificate (US) 3. Swept time constant
STCA Short-term conflict alert
STEVI Sperry turbine engine vibration indicator
STIRS Strapdown inertial reference system
STIS Stabilised thermal imaging system
STORMS Stores management system
STP Status test panel
STR Sonar transmitter/receiver
STRAP Straight through repeater antenna performance
STS Support and test station
STTE Special-to-type test equipment
STU Satellite terminal Unit
SUAWACS Soviet union airborne warning and control system
SUM Structural usage monitor
SWIP Systems weapons improvement programme

T^2A Total terrain avionics
TA Traffic advisory
TACAMO Take charge and move out
TACAN Tactical air navigation
TACCO Tactical co-ordinator
TACDS Threat adaptive countermeasures dispensing system
TACNAVMOD Tactical navigation modification
TADIXS Tactical data information exchange subsystem
TADIXS-B Tactical data information exchange system broadcast
TADS Target acquisition designation sight
TADS/PNVS Target acquisition and designator set/ pilot night vision sensor
TADS/PNVS Targeting air data system/pilot's night vision system
TAFIM Technical architecture for information management
TAIMS Three-axis inertial measurement system
TANS Tactical air navigation system
TAOM Tactical air operations module
TARPS Tactical aircraft reconnaissance pod system
TAS True airspeed
TAT Total air temperature
TAWS Terrain awareness and warning system
TBCP Telebrief control panel
TBMS Tactical battlefield management subsystem
TCA Terminal control area
TCAS Traffic alert and collision avoidance system
TCCP Take command control panel
TCS Television camera set
TDM 1. Tactical data modem 2. Time division multiplex
TDMA Time division multiple access
TDMS Tactical data management system
TDS Tactical data system
TDU Test display unit
TED Transferred electron device
TEMS Turbine engine monitoring system
TEORS Tactical electro-optical reconnaissance system
TERCOM Terrain contour matching
TEREC Tactical electronic reconnaissance

TERPROM	Terrain profile matching
TESS	Threat emitter simulator system
TEVI	Turbine engine vibration indicator
TEWS	Tactical electronic warfare system
TF	Terrain-following
TF/TA²	Terrain-following, terrain-avoidance, threat-avoidance
TFEL	Thin film electroluminescent
TFPRT	Thin film platinum-resistant thermometer
TFR	Terrain-following radar
TFT	Thin film transistor
TFTS	Terrestrial flight telecommunication system
TIALD	Thermal imaging and laser designator
TIBS	Tactical information broadcast service
TICM	Thermal imaging common modules
TIFS	Total in-flight simulator
TILS	Tactical instrument landing system
TINS	Thermal imaging navigation set
TIP	Technical improvement programme
TIRRS	Tornado infra-red reconnaissance system
TIS	Traffic information service
TISEO	Target identification system electro-optical
TIU	Time insertion unit
TJS	Tactical jamming system
TLR	Target locating radar
TM	Transverse magnetic
TNR	Tornado nose radar
TODS	Tactical optical disk system
TOF	Trigger-on-failure
TOPMS	Take-off performance monitoring system
TOW	Tube-launched optically tracked wire-guided (missile)
TP	Tactics planner
TPU	Transmitter processing unit
T/R	Transmitter/receiver
TRAAMS	Time reference angle of arrival measurement system
TRAC-A	Total radiation aperture control-antenna
TRANSEC	Transmission security
TRAP	Tactical related applications
TRF	Tuned radio frequency
TRIXS	Tactical reconnaissance intelligence exchange service
TRN	Terrain reference navigation
TRSB	Time reference scanning beam
TRU	Transmitter/receiver unit
TSD	Tactical situation display
TSFC	Thrust specific fuel consumption
TSO	Technical service order (US)
TSPJ	Tornado self-protection jammer
TSSAM	Tri-service standoff attack missile
TTC	Tape transport cartridge
TTFF	Time to first fix
TTL	Transistor/transistor logic
TTS	Time to station
TTTS	Tanker transport training system
TTU	Triplex transducer unit
TWMS	Tactical weapons management system
TWS	Track-while-scan
TWT	Travelling wave tube
TV	Television
UAC	Universal avionics computer
UAV	Unmanned aerial vehicle
UCS	Utilities control system
UDL	The universal datalink
UFCD	Up-front control display
UFCP	Up-front control panel
UFD	Up-front display
UFDR	Universal flight data recorder
UHF	Ultra high frequency
UHU	Multipurpose escort/anti-armour variant of Tiger helicopter
UKIRCM	United Kingdom infra-red countermeasures
UK MoD	United Kingdom Ministry of Defence
ULAIDS	Universal locator airborne integrated data system
URR	Ultra reliable radar
USAF	United States Air Force
USB	Upper sideband
USN	United States Navy
USTS	UHF satellite tracking (or terminal) system
UTM	Universal transverse mercator
UV	Ultraviolet
V	Volt
Vl	Relative velocity of flow on the underside of an aerofoil
VADR	Voice And data recorder
VAMP	VHSIC avionics modular processor
VATS	Video augmented tracking system
VAWS	Voice alarm warning system
VBM	Volatile bulk memory
VCR	Video cassette recorder
VCS	Video camera system
VDA	Versatile drone autopilot
VDL	VHF (very high frequency) datalink
VDM	Visual display module
VDU	Visual display unit
VEMD	Vehicle and engine management display
VFR	Visual flight rules
VG	Vertical gyro
VGG	Visual graphics generator
V/H	Velocity/height (ratio)
VHF	Very high frequency
VHSIC	Very high-speed integrated circuit
VID	Virtual image display
VIEWS	Vibration indicator engine warning system
VIGIL	Vinten integrated infra-red linescan
VISTA	Variable stability in-flight simulator test aircraft
VITS	Video image tracking system
VLA	Vertical line-array
VLAD	Vertical line-array DIFAR
VLC	Very low clearance
VLF	Very low frequency
VLLR	Very light laser range-finder
VLSI	Very large scale integration
VMO	Maximum permitted operating speed
VMS	Vehicle management system
VMU	1. Velocity measuring unit 2. Voice message unit
Vnav	Vertical navigation
VNE	Never to be exceeded speed
Vocodor	Voice encoder/decoder
VOGAD	Voice-operated gain adjustment device
VOR	VHF omnidirectional range
VOR/Loc	VHF omnidirectional range/locator
VORTac	VHF omnidirectional range/tacan
VOS	Voice-operated switch
Vox	Voice keying or activation
VPU	Voice processor unit
VPU(D)	Voice processor unit (with data mode)
Vr	Radial velocity
VRD	Virtual retinal sisplay
Vref	Typical approach speed
VRU	Vertical reference unit
Vs	Stalling speed
VSD	Video symbology display
VSI	Vertical speed indicator
VSRA	Vertical/short take-off and landing research aircraft
V/STOL	Vertical and short take-off and landing
VSVA	Variable stator vane actuator
VSW	Verification software
VSWR	Voltage standing-wave ratio
VTA	Voice terrain advisory
VTAS	Visual target acquisition system
VTM	Voltage tunable magnetron
VTO	Volumetric top off
VTOL	Vertical take-off and landing
VTR	Videotape recorder
V/UHF	Very/ultra high frequency
W	Watt
WAAS	Wide area augmentation system
WAC	Weapon aiming computer
WACCS	Warning and caution computer system
WAGS	Windshear alert and guidance system
WAN	Wide area network
WASP	Weasel attack signal processor
WBC	Weight and balance computer
WBS	Weight and balance system
WCCS	Wireless communication and control system
WCMS	Weapons control and management system
WCP	Weapon control panel
WDA	Weather display adaptor
WDC	Weapon delivery computer
WDIP	Weapon data input panel
WDNS	Weapon delivery and navigation system
WEAC	Weapons computer
WFOV	Wide field of view
WIU	Weapon interface unit
WNC	Weapons and navigation computer
WPU	Weapon processing unit
WRA	Weapon-replaceable assembly
WSIP	Weapons system improvement programme
WSW	Windshear warning
WSW/RGS	WindShear warning/recovery guidance system
WTC	Windshield temperature controller
YIG	Yttrium indium garnet
ZTC	Zone temperature controller

The Electromagnetic Spectrum

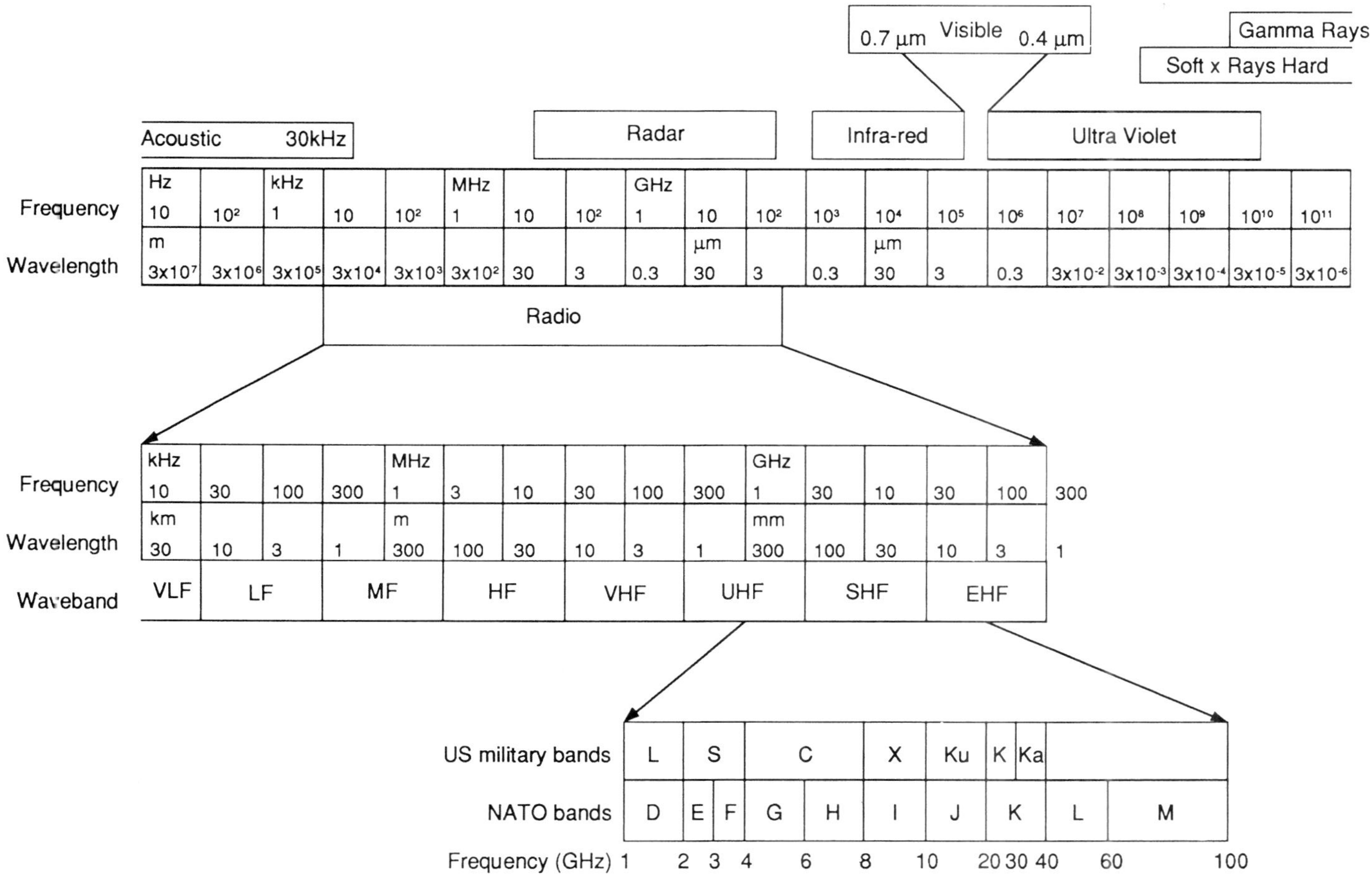

The Joint Electronics Type Designation System (JETDS)

The Joint Electronic Type Designation System (JETDS) is an unclassified US military system designed to identify, by a series of numbers and letters, the use of an equipment and its major components. The system was formerly known as the AN system. A typical example of the system would be:

AN/FPS-5A

In this example the prefix AN indicates that the type number has been assigned in the JETDS system. It does not necessarily mean that the army, navy or air force uses the equipment.

The next letter (F) gives the type of installation where the equipment is used such as fixed, mobile, shipborne and so on. See Table A for a complete listing of installations. The following letter (P) indicates the type of equipment (radar, teletype, radio and so on). See Table B for a complete listing of equipment types. The letter before the dash (S) gives the purpose of the equipment such as transmitting, receiving, detecting and so on. See Table C for a complete listing of equipment purposes. The number after the dash (5) is the model number. The final letter (A) gives notification when a model has been modified but can be interchanged with, or substituted for, the equipment in its original state. If the modified system cannot be interchanged with, or substituted for the original, a new type designation is assigned.

Table A Installation

A	Piloted aircraft
B	Underwater mobile, submarine
C	Air transportable (inactive)
D	Pilotless carrier
F	Fixed, ground
G	Ground, general
K	Amphibious
M	Ground, mobile
P	Pack or portable
S	Water surface craft
T	Ground, transportable
U	General utility
V	Ground, vehicular
W	Water, surface and underwater
Z	Piloted and pilotless vehicle combination

Table B Type of Equipment

A	Infrared
B	Pidgeon (inactive)
C	Carrier (wire)
D	Radiac
E	Nupac (inactive)
F	Photographic
G	Telegraph or teletype
I	Interphone and public address
J	Electromechanical
K	Telemetering
L	Countermeasures
M	Meteorological
N	Sound in air
P	Radar
Q	Sonar and underwater sound
R	Radio
S	Special types or combinations of types
T	Telephone (wire)
V	Visual and visible light
W	Armament
X	Facsimile or television
Y	Data processing

Table C Purpose of Equipment

A	Auxiliary assemblies (inactive)
B	Bombing
C	Communications
D	Direction-finder or reconnaissance/ surveillance
E	Ejection or release
G	Fire control or searchlight directing
H	Recording or reproducing
K	Computing
L	Searchlight control (inactive)
M	Maintenance and test assemblies
N	Navigation aids
P	Reproducing (inactive)
Q	Special, or combination of purposes
R	Receiving, passive detecting
S	Detecting, or range and bearing search
T	Transmitting
W	Control (automatic flight control or remote control)
X	Identification and recognition
Y	Surveillance (search, detect and tracking) and control (fire control, air control)

FREE ENTRY/CONTENT IN THIS PUBLICATION

Having your products and services represented in our titles means that they are being seen by the professionals who matter – both by those involved in procurement and those working for the companies that are likely to affect your business. We therefore feel that it is very much in the interests of your organisation, as well as Jane's, to ensure your data is current and accurate.

- **Don't forget –** You may be missing out on business if your entry in a Jane's book, CD-ROM or Online product is incorrect because you have not supplied the latest information to us.
- **Ask yourself –** Can you afford not to be represented in Jane's printed and electronic products? And if you are listed, can you afford for your information to be out of date?
- **And most importantly –** The best part of all is that your entries in Jane's products are TOTALLY FREE OF CHARGE.

Please provide (using a photocopy of this form) the information on the following categories where appropriate:

1. Organisation name: ______
2. Division name: ______
3. Location address: ______
4. Mailing address if different: ______
5. Telephone (please include switchboard and main department contact numbers, e.g. Public Relations, Sales, etc.): ______
6. Facsimile: ______
7. E-mail: ______
8. Web sites: ______
9. Contact name and job title: ______
10. A brief description of your organisation's activities, products and services: ______
11. Jane's publications in which you would like to be included: ______

Please send this information to:
Jacqui Beard, Information Collection, Jane's Information Group,
Sentinel House, 163 Brighton Road, Coulsdon, Surrey, CR5 2YH, UK
Tel: (+44 20) 87 00 38 08
Fax: (+44 20) 87 00 39 59
E-mail: yearbook@janes.co.uk

Copyright enquiries:
Contact: Keith Faulkner
Tel/Fax: (+44 1342) 30 50 32
E-mail: keith.faulkner@janes.co.uk

Please tick this box if you do not wish your organisation's staff to be included in Jane's mailing lists ❑

JAV

Night vision goggles and image intensifiers

INTRODUCTION

This overview of Night Vision Goggles (NVG) and Image Intensifiers (IIs), the first in a series of technical papers to be integrated into *Jane's Avionics*, is intended to serve two purposes for the reader:

- To provide a technical overview to potential users of Night Vision Devices (NVD).
- To provide useful background information to engineers and technicians working in the field of night vision as to the operational employment of NVD.

MILITARY AIR OPERATIONS BY NIGHT

Nowadays, it is taken as read that military air operations, be they ground attack or air-air, both for fixed and rotary wing aircraft, will be carried out continuously, i.e. 24hr operations. However, this 'holy grail' of warfare, enabling pressure to be kept up on an enemy and allowing no time for re-organisation or resupply, is a comparatively modern concept for ground attack operations, where the need to 'see' the ground as well as the target is of paramount importance. During the late 1970s, the advent of aircraft such as the Panavia Tornado GR1 and General Dynamics F-111, equipped with Terrain Following Radar (TFR), allowing blind flight at extremely low level, gave commanders the ability to prosecute operations on a 24-hour basis. However, while aircraft equipped with TFR were truly all-weather capable, weapon delivery at night was based purely on the ability of the Ground Mapping Radar (GMR) of the aforementioned aircraft to discriminate the target, which could prove a severe limitation against small, mobile targets. Further, TFR systems, with their inherent need for an integrated autopilot system, are a costly solution for many air forces and are comparatively noisy in electro-magnetic terms when faced with sophisticated air defence systems.

The use of Electro-Optic (EO) sensors offers a cheaper, totally passive solution to night operations. Although only able to offer pilots the capability to operate visually (as opposed to TFR operations, which are all-weather), Night Vision Goggles (NVG), operating in the 0.6 to 0.9 µm waveband, complemented by Forward Looking InfraRed (FLIR), operating in the 8 to 12 µm or 3 to 5 µm wavebands bands, affords pilots the ability to visually acquire and deliver weapons against all types of target, including mobile targets such as armour. The complimentary nature of NVGs and FLIR enables pilots to continue operating when one of the sensors is degraded due to atmospheric attenuation (more of which later).

However, operations at night bring their own problems - humans, by definition, are day creatures and are thus not able to readily adapt to night operations. While some claim that EO sensors 'turn night into day', this is not strictly true, to the extent that night flying techniques while utilising EO sensors are sufficiently different to daytime operations that unique training methods are required to maintain proficiency and, perhaps more importantly, safety. Loss of peripheral vision, degraded depth perception (the ability to determine relative position of objects in the scene) and distance estimation while operating with NVGs conspire to deny pilots much of the raw data they process routinely while manually flying with reference to ground features, thus degrading situational awareness and spatial orientation.

While many air forces equipped with day/night capable multi-role aircraft claim 24-hour capability, the reality is that day and night operations are difficult to mix effectively for a typical fighter squadron, since such squadrons are invariably organised in a 'traditional' manner, with insufficient manpower and support to maintain continuous operations. So, during peacetime, squadrons generally conduct either day flying or night flying during any given period. However, recent operations in the Gulf and Bosnia/Kosovo have highlighted these deficiencies, resulting in increases in manpower (pilots and groundcrew) and support (airframes and spares) for night attack squadrons sent into theatre.

In addition, dedicated night attack training is made difficult due to the fact that (at least in Europe) it is inevitably carried out close to urban areas, and at times which are generally severely restricted by the need to minimise disturbance to the slumbers of the civilian population.

While many countries see the acquisition of EO sensors as a relatively 'cheap' route to 24 hour operations for their air forces, it should be noted that the sensors are really only part of the capability: for instance, effective operational employment of NVGs is greatly dependent on effective internal and external lighting, both NVIS (Night Vision Imaging System) compatible (i.e. filtered, to block the majority of IR radiation which causes the NVG image to 'gain down' and thus degrade the scene viewed by the pilot) and tactical (purely IR lighting visible only with IIs - used to facilitate identification of elements within a formation, or to friendly ground forces equipped with suitable NVGs). If an aircraft is to be modified for EO operations, the modification of the lighting suite should be considered as important as the integration of the sensors themselves (although this has certainly not been the case in many cases). In addition, development of specific night tactics and operational procedures, supervised pilot conversion and regular and effective training are also pivotal to the acquisition of a true night capability.

In order to fully appreciate the strengths and limitations of IIs and NVGs, we should first examine some of the characteristics of electromagnetic radiation and its propagation through the atmosphere:

CHARACTERISTICS AND MEASUREMENT OF LIGHT

The energy that all EO devices receive, and the resulting visible image seen by the operator, is all electromagnetic energy. Light, or optical radiation, may be characterised in two ways:

- As particles, called photons
- As waves, which are propagated through a medium, which, for our purposes, is air.

The particle theory of light describes the emission of light from a source, such as the moon. The intensity of this energy, which is useful in dealing with the amount of reflected light available for IIs, can be measured as the amount of light that strikes an object or surface at some distance from a source. When considering units for the measurement of incident and reflected light, we usually employ photometric units, as opposed to the perhaps more recognised radiometric units. Reflected light, or luminance, is normally expressed in terms of foot-lamberts (fl), while the available light, or illuminance, is expressed in terms of lux, which is lumens per square metre (lm/m2). Other commonly used SI units and their definitions are shown at Figure 1.

Figure 1 - Photometric Units (SI)

Definition	**Photometric Units** Name	Unit (SI)
Energy	Luminous energy	Lumen-sec
Energy per unit time (power)	Luminous flux	Lumen
Power input per unit area	Illuminance	Lumen/m^2 (lux)
Power emitted per unit area	Luminous exitance	Lumen/m^2
Power emitted per unit solid angle	Luminous intensity	candela
Power emitted per unit solid angle per unit projected area	Luminance	Candela/m^2

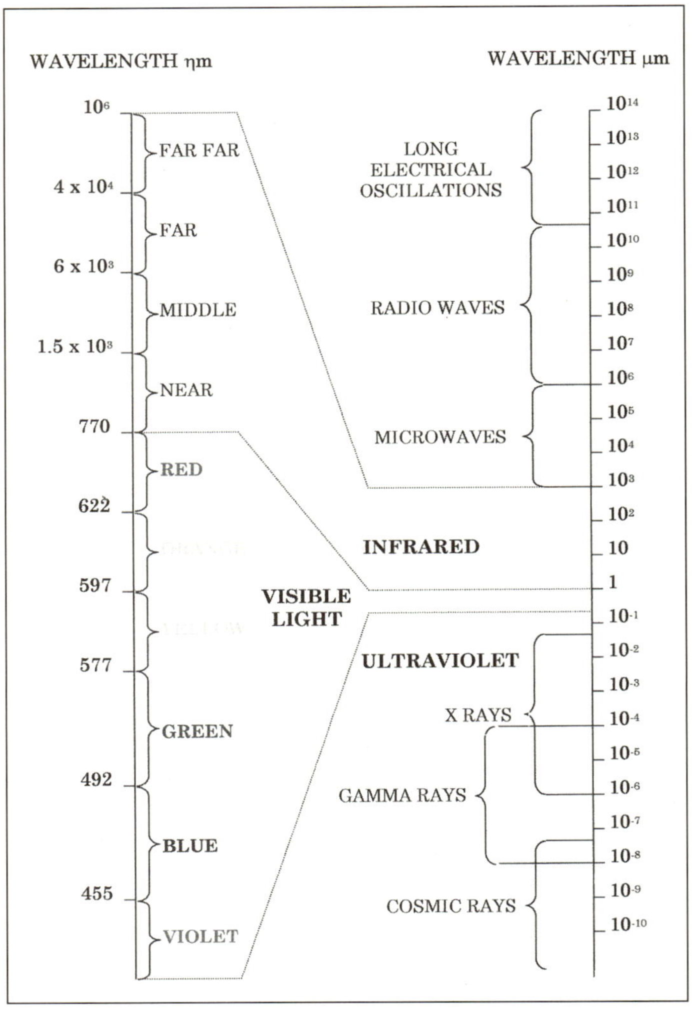

Figure 2 – The Electromagnetic Spectrum **2001**/0110032

The wave theory of light characterises light in terms of wavelength, frequency and velocity, enabling us to fully describe the propagation of light through the air, or an optical system such as the human eye, conforming to the well-known equation l = e/¦, where e is the speed of light (for our purposes, a constant), l is the wavelength of the radiation and ¦ is the frequency. Thus wavelength is inversely proportional to frequency.

For a pilot to exploit the night by utilising the available illuminance, then he needs an imaging system which can interpret the visual scene across a broad range of wavelengths and present it to the pilot within the visual spectrum. This is exactly what the Night Vision Goggle is designed to do for night aviators. As we shall see in more detail later, these devices are particularly sensitive to reflected energy in the near- IR band, which is where the major contributors (moonlight and starlight) to the natural illuminance in the night sky are at their most intense.

The Electromagnetic Spectrum

Many different types of radiation are present in the atmosphere, such as light, heat and radio waves. Just as a radio must be tuned to receive specific frequencies, II devices, such as NVGs, and the human eye, are sensitive to different wavelengths (or bands) of the electromagnetic spectrum. These bands are inherently similar in nature, and as such, can best be related by their position in the entire electromagnetic spectrum, as shown in Figure 2. The optical band, covered by visible light, is a relatively small portion of the entire spectrum. All these radiations obey the same laws of reflection, refraction, diffraction and polarisation, and, as previously mentioned, the velocity of propagation can be considered to be constant. Thus they differ from one another only in their frequency. Visible light is bounded on the short wavelength side by UV radiation and, on the long wavelength side, by IR radiation.

The human visual system, operating in the visible range, and NVG, operating in the visible and near IR range, are dependent on reflected energy, which requires scene illumination. When a scene is irradiated by an external source such as the sun, it is observed primarily by reflected energy. This holds true for near ultra-violet, visible and near IR wavelengths.

SOURCES OF NIGHT ILLUMINATION

There are several natural sources of illumination available at night:

- Residual Sunlight
- Moonlight
- Starlight
- Airglow

The primary source of daytime illumination is the sun. The illumination produced by the sun on a horizontal surface on the earth depends on its angle above the horizon (elevation angle). However, due to refraction effects, the sun also provides light when it is below the horizon, therefore residual sunlight is also of interest to the night aviator. Significant times are civil twilight, when many tasks in the cockpit can be achieved with unaided vision and there is generally too much residual sunlight for NVG operations, nautical twilight, when the sun's effects steadily decrease to the extent that normal NVG operations are possible and finally, astronomical twilight, when all residual sunlight has disappeared from the night sky and normal NVG operations are possible given other forms of natural or artificial illumination.

After the sun has fully set, the most significant source of light is the moon. The illuminance provided by the moon depends on its elevation and phase, and knowledge of these is important in planning NVG operations.

To the casual observer, the most obvious moon characteristic is that it changes phase regularly. The phase of the moon is described either by its age, or by the percentage of the illuminated disc. For the purposes of night operations, it is useful to consider the four lunar phases and their characteristics:

New Moon. The new moon phase is characterised by very low light levels that limit useful NVG flying conditions.

First Quarter. Relatively good light levels are available for NVG operations with the best period in the early part of the evening.

Full Moon (Second Quarter). This phase is characterised by very high light levels once the moon rises above the horizon.

Third Quarter. Relatively good light levels are available for NVG operations, although optimum lighting periods progress towards the early morning.

After rising, the moon continues to increase its elevation, which is dependant on its declination and the observer's latitude. The former depends on the position of the moon in its orbit, and its plane of motion. This characteristic is important when considering those moon phases where the elevation is critical for providing sufficient light.

However, the moon is not the only source of natural light in the night sky. Starlight also provides illuminance, which is equivalent to about 25%- 30% of the total light available from a moonless night sky. The majority of stars emit a significant amount of radiation between 0.8 and 1.0 μm, which means that a great deal of their output is invisible to the human eye, but falls within the sensitivity band of the latest generation of NVGs, termed Generation III (Gen III) NVGs (0.6 to 0.9 μm). This sensitivity facilitates night operations in starlight only conditions.

Perhaps surprisingly to anyone who has never worn NVGs, the greatest portion (approx. 40%) of natural light in a moonless night sky comes from an effect called airglow, which originates in the upper atmosphere and is caused by solar radiation, which produces emissions from atmospheric particles.

The remainder of night illumination comes from other sources such as aurora, luminous patterns of light which sometimes appear in the night sky at high latitudes.

Having described the main sources of natural night illumination, we should now attempt to illustrate how the modern NVG's response is tuned to respond in the wavelengths that characterise these phenomena. Figure 3 shows approximate curves for the three main contributors to illumination in the night sky, namely starlight, moonlight and airglow, as a function of wavelength. Note that starlight and airglow are mainly present in the near IR band, which means they are largely invisible to the unaided eye, but fall well within the sensitivity of NVGs, as shown by the overlaid response curves for the unaided eye and NVGs.

Having discussed the sources of natural illumination in the night sky, defined the position of their radiations in the overall electromagnetic spectrum and described the tuning of NVGs

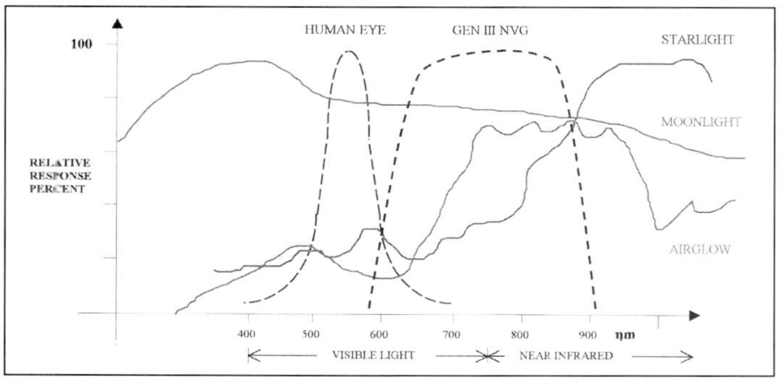

Figure 3 – Response of the Human Eye vs. NVG/Sources of Night Illumination **2001**/0110033

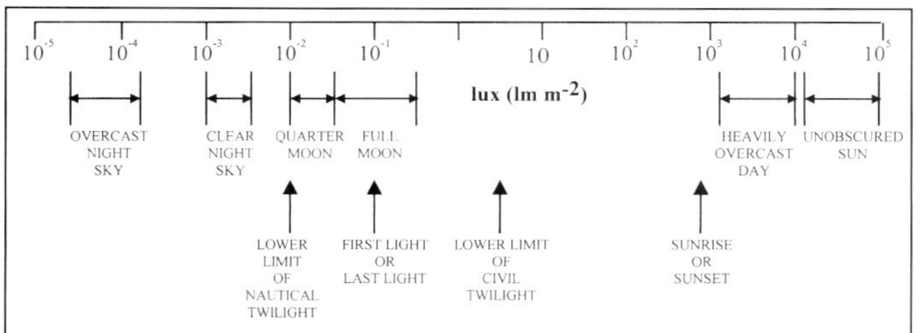

Direct Sunlight	10^5 lux
Full Daylight (Not Direct Sunlight)	10^4
Overcast Day	10^3
Twilight	10
Deep Twilight	1
Full Moon	10^{-1}
Quarter Moon	10^{-2}
Moonless, Clear Night Sky	10^{-3}
Moonless, Overcast Night Sky	10^{-4}

Figure 4 – Luminance Levels **2001**/0110034

response to their characteristic wavelengths, it would be useful to consider the practical application of all of this theory!

By considering the positions of the sun and moon, and prevailing weather conditions (cloud cover), a theoretical luminance level can be calculated. Figure 4 shows illuminance levels for various moon phases and cloud conditions.

It should be noted that, while the luminance levels shown in Figure 4 serve as a useful guide to mission planning for NVG operations, they should only be considered as a guide, since the contribution of other factors, particularly the contribution of cultural lighting (described later) to the scene, are not considered. Accordingly, actual light levels encountered may be considerably greater than predicted in urban areas, or drastically less if cloud predictions were in error.

While it is obvious that accurate information concerning luminance is critical to the safe conduct of night attack sorties, calculation of light levels can seem an extremely complicated proposition for the night aviator; however, mission planning is facilitated by computer programmes, such as the UK Met Office Illumination Program and the US WinANVIS light level prediction program, which calculate the orientation of the sun and moon, factor for prevailing weather (cloud cover) and predict the illumination levels for any point on the earth's surface at any time.

Cultural Lighting

As witnessed during recent conflicts, such as allied operations in Kosovo, there are very many sources of man-made light in the overall night scene. In the absence of natural lighting, e.g. when operating under a thick cloud cover, cultural light from urban areas, reflected from the cloud base, can provide significant illumination for NVGs. As a result, low-level operations will often be possible even when there is no natural illumination available, although care should be taken to make use of such lighting without subjecting pilots' NVGs to adverse conditions. If NVGs are pointed directly at brightly lit areas, this may saturate NVGs, causing them to 'gain down' and the wearer will effectively be blinded. If the area of bright light is small point (such as a single street light), then this will be shown as a so-called 'halo' around the light, leading to a loss of contrast in the surrounding image.

Recent conflicts such as the Gulf War and Bosnia/Kosovo have shown that 'blackout' conditions, as practised during WWII, are not guaranteed during modern air warfare and the presence of cultural lighting in the target area should still be considered, particularly for Military Operations in Urban Terrain (MOUT).

ATMOSPHERIC TRANSMISSION

Having considered available sources of illumination for IIs, we should now consider the attenuation mechanisms within the atmosphere. For all of the wavelengths of interest, there are components within the atmosphere that will refract, scatter and absorb radiation. The transmission of electromagnetic energy through the atmosphere is dependent upon the concentration and distribution of those attenuating atmospheric constituents, which is, in turn dependent on meteorological conditions. Therefore, the atmosphere itself is a limiting factor for transmission.

Scattering occurs due to the presence of molecules and particles of matter in the atmosphere (aerosols). Scattering causes attenuation of an incident beam of radiation because in the scattering process the energy is redistributed in all directions of propagation.

Another form of attenuation of radiation is absorption. Water is the primary absorbing molecule for IR and visible radiation. The severity of scattering and absorption is a function of radiation wavelength and particle size, and it is often difficult to determine which is the primary attenuation mechanism. However, for particular wavebands, 'windows' exist where the transmittance of the atmosphere is consistently high, therefore modern EO systems are designed to operate at wavelengths within these windows. Figure 5 shows the three windows within the visible/IR range of wavelengths where atmospheric transmittance is high, together with examples of associated EO systems found within *Jane's Avionics*.

The design of all EO sensors utilise one, or more, of the windows in the near, mid and far IR regions of the spectrum. Modern NVGs utilise the window that extends from the deep red area of the visible range into the near IR region (0.6-3.0 mm). The other two windows are both in the IR portion of the spectrum. One is the 3-5 mm band, which is located in the mid-IR region; this band is associated mostly with missile seeker heads, although more recently with modern targeting FLIR systems, such as the AN/AAQ-27. The other window is in the 8-12 mm band, which lies in the far IR; this band is normally used for navigation and surveillance purposes in systems such as the AN/AAR-50 NAVFLIR and TIALD.

WEATHER AND VISIBILITY

Due to the generally similar operating wavelengths of the unaided eye (0.4 - 0.7 μm) and modern NVGs (0.6 - 0.9 μm), the attenuating effects of weather phenomena on night illuminance may be considered to be largely as would be seen by the naked eye during the day. The following summarises some of the major weather effects and highlights the subtle differences associated with the higher wavelength of operation of NVGs:

Cloud

Clouds are highly variable in their attenuating effects. Water, the main contributory constituent in low-level clouds, can be found in all states, thus making judgement of overall degradation of sensors difficult. Clouds reduce the natural night sky illumination to an extent dependent on the amount of cloud coverage and the overall density and/or thickness of the clouds. It follows that a thick, overcast

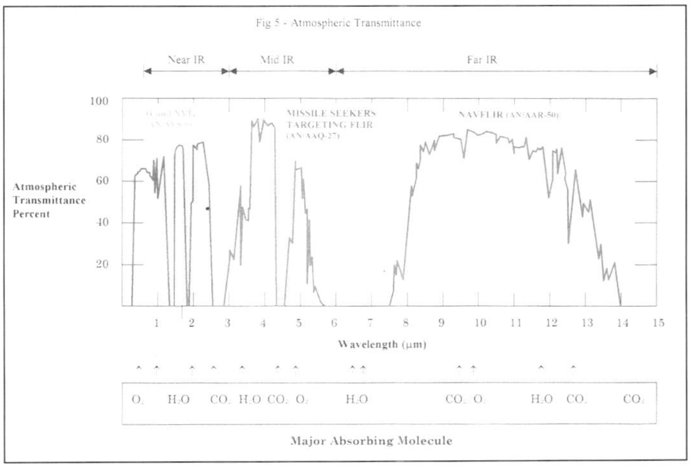

Figure 5 – Atmospheric Transmittance **2001**/0110035

layer of cloud will reduce the ambient light much more than a thin, broken layer.

Fog
The effect of fog on NVGs is similar to that of clouds. In meteorological terms, only their distances from the ground differentiate fog and cloud. When a fog layer is very thin (10m or so) and it is made up of very small droplets, NVGs operating in the near IR wavelength may be able to see through it. However, it should be noted that encountering fog and poor visibility during flight at low level is a particularly hazardous aspect of NVG operations. When viewing the NVG scene, a decrease in the intensity of ground lights, possibly coupled with increased intensity of halos around ground lights is an indication that moisture in the air is high and that fog may be forming.

Rain
The effect of rain on NVG scene is difficult to predict, since it is usually associated with cloud, which, as discussed, is a major attenuator of the available natural illumination. In isolation, individual droplet size and the overall density of droplets will determine the level of attenuation. Similarly to rain viewed with the unaided eye, heavy rain (large droplet size and high density) obstructs light transmission and the low light levels associated with it can be very noticeable when flying at low level. However, light and medium rain (lower density and much smaller droplet size) may be almost invisible to NVGs.

Snow
Snowflakes are generally large in comparison to the wavelength of visible and near-IR energy; therefore snow, sleet and hail will all significantly attenuate illumination levels. As for the unaided eye, under these conditions the NVG scene will be unusable.

Battlefield Obscuration
The battlefield will yield many obscurants to NVG performance, such as exhaust smoke, fires and debris thrown into the air by explosions. While these effects are inevitably most severely felt at low level, some obscurants, such as oil fires during the Gulf War, can affect medium level operations. While operational planners will tend to route aircraft missions away from areas of intense ground warfare, sometimes aircraft will be tasked into such areas as part of Close Air Support (CAS) or Battlefield Air Interdiction (BAI) missions. In such cases, pilots should be aware that degradation of NVGs (indeed, most aircraft sensors) may be sudden and severe, with similar actions required to those on encountering a volcanic dust cloud.

INTERPRETATION OF THE NVG SCENE

In this context, contrast is a measure of the luminance difference between two or more surfaces. Assuming a constant level of illumination, contrast as perceived by the NVGs is dependent upon the differing reflectivities, or albedos, of each object and on the amount of texture, or roughness, in the scene. Although NVGs rely on reflected light from the scene, just as the naked eye, it should be noted that the II tubes are sensitive **mainly** in the near IR wavelengths, therefore we should consider the reflectivities, or albedos, of objects in the night scene at those higher wavelengths, to gain a true understanding of the NVG scene.

While it would be impossible to quantify all significant objects which make up the night environment in terms of how they are perceived on Night Vision Goggles, the following is intended to provide a broad perspective of the NVG scene:

Roads
Generally, roads show up well on NVGs, due to their contrast (large albedo difference) with flanking terrain. Roads that cut through heavily forested areas are also easily recognised if they are visible through the tree canopy. Light coloured concrete roads/motorways are easily identified under most medium to high light conditions, whereas asphalt roads can be difficult to identify because the dark surface absorbs available light.

Water
There is very little contrast between the land and a body of water in low light conditions. As the light level increases, land-water contrast increases and reflected moonlight is easily detected. It is usually reflected moonlight that confirms that a dark patch on the ground, as viewed through NVGs, is actually water!

Fields and Forests
Ploughed fields look dark on NVGs due to the roughness of the surface, which absorbs most of the incident light. Despite their low albedos, cultivated fields provide for a 'patchwork' effect which provides for excellent contrast with surrounding roads and other terrain features.

Trees have different albedos depending on their type (deciduous or coniferous) and the time of the year. Large areas of unbroken forest can mask underlying terrain and generally provide poor contrast, although small areas of forest, divided by open fields, cultural objects and roads contribute to the generally excellent contrast levels found in European countries.

Desert
Open desert, without significant vegetation or cultural objects, produces a very bland image on NVGs due to lack of texture in the terrain resulting in poor contrast in the NVG scene. During bright light conditions, mountain ranges can be easily identified due to intense shadowing, although these shadows can obscure other less significant features which can represent a serious hazard to pilots flying at low level in such terrain.

Snowfields
Snow has an extremely large albedo, reflecting approximately 90% of incident light. As previously discussed, under high levels of illuminance, this can result in too much light entering the NVG, causing a decrease in gain and loss of resolution. However, during typical winter conditions in temperate climates such as Europe, snowfall is usually associated with significant cloud cover, therefore snow cover can be extremely useful increasing illuminance.

However, significant amounts of snow lying on the ground can increase overall light levels to the point where the NVG reduces gain considerably, thus decreasing resolution. This, coupled with drastically reduced contrast associated with complete snow cover can make for subtly hazardous flying conditions at low level. Conversely, and similarly to forest areas, as a snowfield becomes more broken, with roads, fields and trees becoming discernible, contrast levels increase and provide excellent conditions for low level flight on NVGs.

NIGHT VISION GOGGLES (NVGs)

NVGs have been in widespread service for years, but only recently have pilots of fast jet tactical aircraft used NVGs for operational purposes. All NVGs operate on the same basic principles. Image intensifier tubes produce a bright monochromatic (green) EO image of the outside world in light conditions too low for normal vision. Unlike FLIR, however, which uses the far IR area of the electromagnetic spectrum to create an image based on object temperature and emissivity relative to its surroundings, NVGs use the red and near IR area of the spectrum to produce their own unique visible image of the world. This image is based on the relationship between the amount of light present, referred to as illuminance, and the amount of light which is reflected from objects in the scene, referred to as luminance or brightness. Under defined luminance conditions, the NVGs are able to provide navigation and terrain avoidance references for pilots.

Night Vision Devices - Background
Generally, image intensifier technology is expressed in terms of first, second, third and, latterly, fourth generation systems:

First generation (Gen I) systems were used in starlight scopes used by snipers in Vietnam. These devices featured gains in the range of 40,000 to 60,000 by means of a three-stage cascade configuration of simple intensifier tubes contained in the scope. First generation tubes were extremely durable and offered extended life, but were not suitable for aviation applications due to a number of factors, which included blooming (a tendency for the tube to washout if a bright light source appears anywhere in the device FoV), a high voltage requirement and, most significantly, their size and weight.

Second generation (Gen II) image intensifiers introduced the Micro Channel Plate (MCP). The invention of the MCP was the breakthrough that facilitated miniaturisation of intensifier tubes to the extent necessary for helmet-mounted systems for use in fast jet

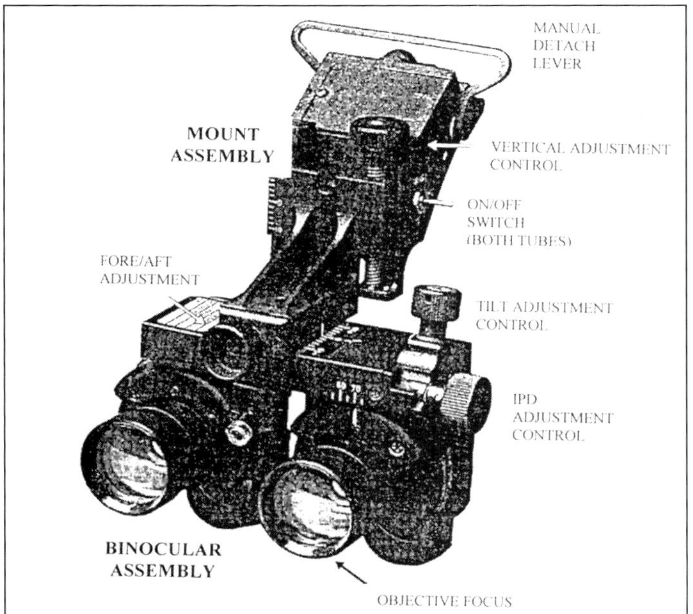

Fig 11 – NVG Adjustment Controls – CATSEYE Type 2

Figure 11 – NVG Adjustment Controls - CATSEYE Type 2 **2001**/0110039

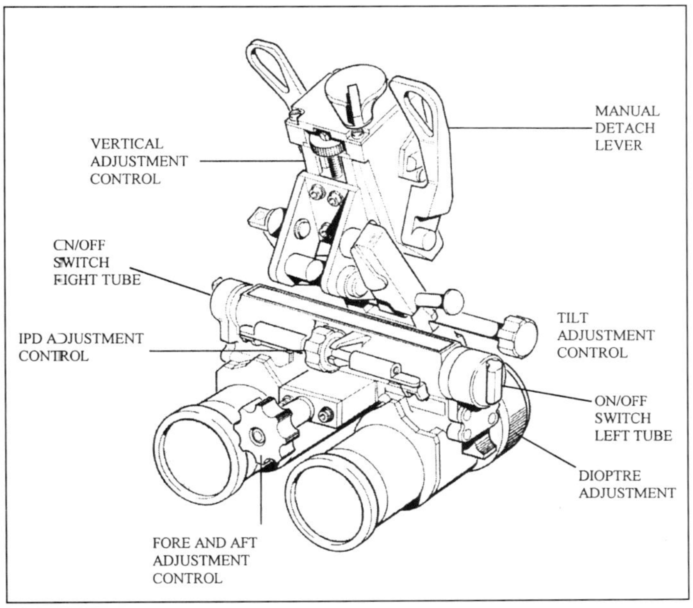

Figure 12 – NVG Adjustment Controls - NIGHTBIRD Type 1 **2001**/0110040

- Fore/Aft Adjustment (dioptre) - Dioptre adjustment is achieved by moving the eyepiece lens assembly axially, relative to the monocular housing. Normally, dioptre adjustments will be required for each operator prior to use. Adjustment to a known personal setting is also possible by means of a graduated scale.

Ejection Safety - Autoseparation

Fast jet aircrew helmets are designed to exacting specifications in terms of mass and balance, to ensure optimum protection during the ejection sequence. However, until the advent of integrated helmets, none were designed from the outset to include the mass of a set of NVGs attached to the front, which obviously upsets the balance and centre of gravity of the helmet assembly, resulting in possible complications for the pilot should the goggles be in place during ejection. As a result, pilots were obliged to takeoff and land with the goggles stowed, since it was judged that the pilot would have insufficient time to manually doff the goggles prior to ejection during these flight phases. This presented some quite serious operational limitations, since airfields would have to maintain full lighting for aircraft movements, thus presenting an easy target for enemy air attack, and pilots would have to allow sufficient time to recover full night vision prior to landing during poor weather conditions.

In order to address the problem of unpremeditated ejection when wearing NVGs, the UK and US both developed autoseparation systems, which sensed movement of the ejection seat at the

Figure 13 – NVG PS/BIT unit as fitted to a Tornado GR4 **2001**/0110045

beginning of the sequence and fired a small charge to detach the goggles from the pilot's helmet. While the UK developed and has fielded autoseparation for all of its fast jet aircraft, the US developed a system (the AN/AVS-8), but never fielded the system.

Considering the UK system as fitted to the martin Baker Mk10 ejection seat as fitted to RAF Tornado aircraft and aircrew using the GEC-Ferranti NIGHTBIRD Gen III NVG, operation of the system is as follows: During unpremeditated ejection, the NVG assembly will be released from the Helmet Mounting Plate (HMP) automatically. A gas motor assembly, located within the NVG mounting assembly, will actuate the release mechanism for the NVG, when power is supplied from the Power Supply/BIT Unit (PS/BIT) (Figure 13). The wiring to carry the power connects from the PS/BIT Unit to the Pilot's Equipment Connector (PEC), then up through a modified connector to the helmet cableform; the connections are passed through on spring loaded contacts at the helmet mounting plate and then to the gas motor.

NVG DEVELOPMENTS

While II and NVG development has been rather slow in the UK in recent years, with greater emphasis placed on research into uncooled FLIR technologies, the two major II manufacturers in the US, ITT and Litton, have made some significant improvements in II performance as the definition of the new 'fourth generation' (Gen IV) specification becomes defined.

NVG Limitations

Before outlining the upcoming developments in NVG technology, we should first examine the major limitations, as perceived by pilots, of NVGs. By far the most important limitation of operating with NVGs is the nominal 40° FoV, which seriously degrades situational awareness and forces an exaggerated eye scan to build up a mental picture of immediate NVG scene. Close behind in terms of importance, resolution of the current range of NVGs also presents serious limitations to their tactical employment, with degraded depth perception and distance estimation posing problems for pilots, particularly in the low level arena. While there are other issues to be addressed regarding future operational applications of NVGs, including Laser Eye Protection (LEP) and ejection safety, it was the issues of goggle FoV and resolution which were considered of highest importance in developing future goggles.

Resolution and Field-of-View

In common with most imaging systems, FoV and resolution for NVGs are directly linked; for any goggle/tube pairing, if resolution is increased, then FoV will proportionately decrease and vice versa. Therefore, in order to increase resolution for the system, the most significant limiting factors for the tube must be addressed. The principle limiting factors governing tube resolution are the distance between the individual MCP channels (termed pitch - see Figure 14), the distance between the photocathode and the MCP and the distance between the MCP and phosphor screen. Of these, MCP pitch is the most powerful factor in terms of improving resolution. Channel spacing has been reduced from 12 μm in Gen II tubes to 6 μm in Gen

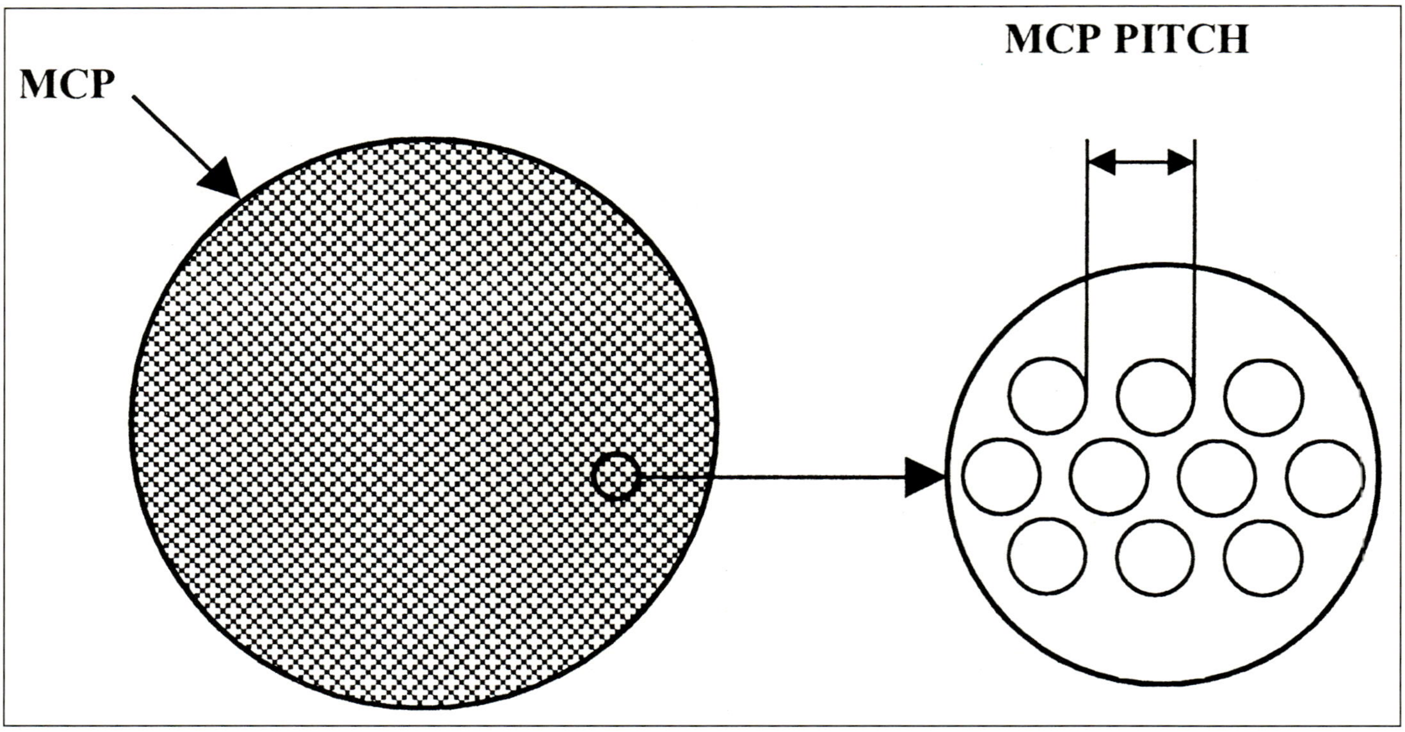

Figure 14 – MCP Pitch **2001**/0110041

III tubes, corresponding to over 6 million channels in an 18 mm OMNIBUS 4 specification tube. This has resulted in improvements in sensitivity up to the current 64 lp/mm of the latest goggles. As manufacturing techniques improve, the MCP pitch will undoubtedly decrease further (perhaps as low as 4 μm), allowing for greater resolutions (up to 125 lp/mm), and/or a reduction in tube diameter (and thus weight of the goggle).

Having described the ongoing work into improving resolution, it is apparent that trading resolution for increased FoV is not an option. Therefore, a means of increasing the total FoV without affecting system resolution was sought by manufacturers. As part of a US Small Business Innovative Research (SBIR) programme, the Night Vision Corporation was awarded a contract to investigate Wide FoV (WFoV) NVG technologies and fabricate a proof of concept demonstrator. This resulted in the Panoramic Night Vision Goggle (PNVG) Phase 1 SBIR, which consisted of an arrangement of four II tubes mounted on an aircrew helmet in a Type 2 indirect viewing goggle producing a 100° horizontal by 40° vertical FoV.

The work of Phase 1 SBIR led to Phase II SBIR, which ended in July 1999 with 12 working models of two types of PNVG delivered. PNVG I is a fast jet Type 2 NVG, with four II tubes mounted along the brow of a standard HGU-55P aircrew helmet. The outer tubes are canted outward and the tubes are connected to biaxial eyepieces. The result is a panoramic NVG with a 100 × 40° FoV, with resolution as for the donor AN/AVS-9 tubes. Later prototypes were intended to incorporate an electro luminescent (EL) display for symbology or imagery. The relatively flush fit of the Type 2 arrangement of the system means that the NVG can be retained during ejection.

The PNVG is intended to interface with the helmet and the aircraft in the same way as the Joint Helmet Mounted Cueing System (JHMCS). Installed on any JHCMS-compatible aircraft, the PNVG will be able to display flight data and sensor imagery. The aircraft's avionics and sensors will give the pilot cueing information to find targets, and the pilot will be able to designate targets with head movements. Infrared imagery can be displayed on an inset panel using a 640 × 480 EL source.

Another version of the PNVG, designated PNVG II, was designed for pilots of helicopters and transport aircraft. This differed from PNVG I in being a Type 1 system, consisting of the familiar AN/AVS-9 with two further tubes grafted onto the outside of the goggle to provide for the wide horizontal FoV.

The PNVG I underwent operational evaluation at Nellis AFB, using F-15 and F-16 aircrew, in the USA last year, with greatly increased situational awareness reported. This positive reaction led to the development of the Integrated Panoramic Night Vision Goggle (IPNVG), with the design aims of resolution equal to or better than the AN/AVS-9, compatibility with standard aircrew spectacles and based around a 16 mm II tube. Other key objectives were:

- Wide FoV
- Laser Eye Protection/Laser Hardening
- Ejection/Crash/Ground Egress Safety
- Fit/Comfort
- Reduced Halo
- Increased Image Quality
- Integrated Symbology/Imagery Display
- Field Supportability
- Reliability/Maintainability/Affordability

The IPNVG features an 86° × 36° FoV, a LEP visor and II tube hardening; Type 1 and Type 2 variants for army and air force use have been developed. However, as of this issue, it is unclear as to which of these prototypes (if any) will be selected by the US armed forces for further development.

High and Low Light Performance

While increased II tube performance at low light levels has been steadily increased with improvements to signal to noise ratio (SNR) resulting from increased photocathode sensitivity, problems with high light level performance have recently come to the fore in recent conflicts involving MOUT. As previously discussed, Gen III tubes utilise a metal oxide film in order to protect the photocathode and extend the life of the goggle. The deletion of this barrier film, made possible by the integration of an Autogated power supply, which fulfils the duties of the barrier film, is claimed by the manufacturers

TYPE	PHOTOCATHODE	MCP PITCH	PHOTO RESPONSE μA/lumen	RESOLUTION lp/mm
Gen II	S-25 Multi Alkali	12 micron	100 - 400	22 - 36
Gen III	GaAs	9 micron (OMNI III)	600 - 1800	45
		6 micron (OMNI IV)		64
"Gen IV'	GaAs	4 - 6 micron	Up to 2200	> 64

to improve both high and low light performance by directly improving the SNR.

Halo Size

Halo size is a function of the spacing between the photocathode and the MCP. This gap determines the perceived size of all halos within the FoV, while the intensity of the halo is a function of the overall system gain; hence halos are more intense under overall low light conditions when tube gain is high. While halo size increased in Gen III tubes due to increased bias voltage requirements increasing the distance between the MCP and photocathode, the introduction of new technologies such as Autogated power supplies has enabled this gap to decreased with resultant reduction in halo size.

The Future

The development of NVG technologies for tactical aircraft has become entwined with that of integrated helmets, such as the JHMCS and the Striker display helmet for Eurofighter Typhoon. The Charge Coupled Device (CCD) has matured rapidly, to the extent that it is now seen as a viable alternative to current II technologies in terms of night vision. Compact and lightweight compared to traditional II tubes, CCDs may represent the next stage in the evolution of night vision for fast jet, since they are more easily integrated into a single day/night capable helmet than systems such as the IPNVG, which is designed to interchange with a separate day display module.

Aselsan M930 aviators' Night Vision Goggles, note offset mount for Cobra attack helicopter application (Aselsan) ***2001***/0103875

Foreword

Welcome to the twentieth edition of *Jane's Avionics* and my first as editor. The 2001-2002 edition presents an evolution of the format changes introduced in the previous edition by my predecessor, Chris Johnson. Changes in format are therefore minor; most notable is the deletion of entries concerning generic weapons control. The advent of integrated cockpits and the MIL-STD-1553B databus has resulted in most aircraft exercising weapons control via Hands on Throttle and Stick (HOTAS) and MultiFunction Display (MFD), thus obviating the need for dedicated weapon control panels.

Most noteworthy in this edition is the introduction of colour into the main body of the book. *Military display and targeting systems* is presented in colour in recognition of the fact that most modern military aircraft now utilise colour Active Matrix Liquid Crystal Displays (AMLCD), with much effort devoted to the use of colour to enhance readability and interpretation. In describing such systems, the illustration of colour displays in black and white would deny the reader much intrinsic information, particularly concerning such issues as choice of colour for various categories of information, therefore colour illustration has been introduced where it will be most effective. I hope to extend the use of colour to *civil displays* in the next edition.

As the new editor of *Jane's Avionics*, it is my aim not only to continue to detail up-to-date product information in the world of civil and military avionics, but also to incorporate technical overviews of major core technologies, which I hope will provide useful background to a broad range of readers. Drawing on personal experience in the field of electro-optics development for military fast jet operations, the first of these features, *Night Vision Goggles and Image Intensifiers* provides information on the development, integration and utilisation of Night Vision Devices (NVD) in military aircraft.

Jane's Avionics 2001-2002 can also be found online. This has two advantages for the reader who requires the most up-to-date information. First, *Jane's Avionics* is available via the internet on subscription. Second, the update period is reduced from annually to monthly, making the data more timely - a significant benefit to the reader and to the contractors whose products are described.

Company takeovers and mergers

The past year has continued the consolidation of the avionics industry. Most notable was the merger of Aerospatiale Matra SA, Contrucciones Aeronauticas SA (CASA) and DaimlerChrysler Aerospace AG to form the European Aeronautic Defence and Space Company (EADS), listing at the Frankfurt, Madrid and Paris stock exchanges on the 10th July 2000. By revenue, EADS is the third largest aerospace company in the world. Interestingly, the new company also finds itself in the curious position of holding significant interests in two major competing combat aircraft programmes, namely Eurofighter Typhoon and Rafale - only time will tell how this potential conflict of interest will be resolved!

Earlier in the year Thomson-CSF completed its takeover of Racal Defence and on the 6th December 2000 became Thales. The company has changed the trading names of all of its subsidiary companies, including the former Racal Defence, Pilkington Optronics and Shorts Missile Systems, to Thales.

While I have taken every care to try to allocate items of equipment to their new owners and to use the appropriate new designations, in some cases changes have come to me too late to be fully incorporated. To that end, while some equipment may be listed under former company names, I have endeavoured to ensure that contractors' contact addresses correctly reflect current status.

Developments in civil avionics

Continued growth in civil air travel, exemplified by the launch of a new class of 'Super Jumbo', the Airbus A380, a 550 seat,

Neither of the competing Joint Strike Fighter prototypes feature a traditional Head Up Display (HUD). Helmet Mounted Display (HMD) systems will provide for vastly enhanced situational awareness, encompassing all navigation and attack functionality and allowing the pilot to fully exploit the off-boresight capabilities of modern air-to-air and air-to-ground weapons (Lockheed Martin) **2001**/0109232

twin-deck design capable of flying over 8,000 n miles, is set to place even greater pressure on the already congested airspace in Europe and the US. In response, work is continuing on the problems of implementing the Communications Navigation Surveillance/Air Traffic Management (CNS/ATM) concept, which promises to make the traditional 'flight plan' obsolete, with routing between airports virtually direct, thus reducing flight time, delays and operating costs. While the enabling technology seems to be well defined, there is still some debate as to how airspace will be managed, how sectors will be controlled and where the responsibility for aircraft separation will reside within an ATM system. We must wait and see if these problems can be resolved, so that the concept of 'free flight' can be turned into a reality.

In the meantime, the immediate problem of airspace congestion, particularly in the North Atlantic, has prompted improvements in Traffic alert and Collision Avoidance Systems (TCAS). The recent implementation of TCAS Change 7 (sometimes called Airborne Collision Avoidance System, ACAS II), a software update, features upgraded collision estimation calculation, which eliminates nuisance Traffic Alerts (TAs) and the more serious Resolution Advisories (RAs) for aircraft operating under Reduced Vertical Separation Minima (RVSM). In addition, the system is fully compatible with Automated Dependent Surveillance Broadcast (ADS-B) requirements,

The recently announced Airbus A380, a 550 seat twin-deck 'super jumbo', is capable of flying over 8,000 n miles (Airbus Industrie) **2001**/0109233

including 100 n mile surveillance and 40 n mile interrogation range capability, which is a vital component of the CNS/ATM concept.

In parallel with the introduction of enabling technologies for 'free flight', significant enhancements to air traffic safety are being made with the introduction to service of systems such as the Enhanced Ground Proximity Warning System (EGPWS), which will predict terrain and obstacles ahead of the aircraft to provide more accurate and timely warning to crews, and research into enhanced vision systems, which, when combined with a Head Up Display (HUD), will enable aircraft without Category 3 landing capability to land more safely in adverse weather conditions.

The incorporation of military-derived systems, such as enhanced vision systems and HUDs, into civil aircraft may increase as military aircraft such as the Airbus A400M transport are introduced into service. With such large degrees of commonality (between the A400M and the civil Airbus family), previous cost barriers to the integration of systems such as FLIR would be removed, since most of the integration would effectively have already been accomplished. This would enable airlines with a requirement for such systems to specify them at an affordable cost. While this may seem like wasted effort in these days of Category 3B autoland capability, it should be noted that a great many airports will never qualify for such operations due to terrain constraints; in such cases, a 'pseudo' Category 2/3 capability would be most beneficial.

Developments in military avionics

While the previous discussion regarding the problems in implementing 'free flight' would seem to have little bearing on military operations, in reality similar concerns have been voiced over recent NATO operations in Bosnia and Kosovo. Increasingly, military air operations are conducted in close proximity to active civilian airspace, which has served to highlight the inherent incompatibility of military aircraft systems with civilian navigation and identification requirements. While these problems are being addressed by the US Air Force Global Access Navigation Safety/Global Air Traffic Management (GANS/GATM) programme, the stakes will shortly be raised as technology matures enabling the first Unmanned Combat Air Vehicles (UCAV) to be introduced into service. Essentially aircraft without a pilot, stealthy UCAV are currently being developed which will be capable of not only reconnaissance and target identification/designation, but also long-range interdiction and precision delivery of smart weapons.

Assuming true Artificial Intelligence is some way off, control of these UCAVs will be man-in-the-loop via datalink, either directly by Line of Sight (LOS), or via third party (communications platform or satellite). Considering the operational employment of such systems, problems with the integration of armed UCAV into civilian airspace, during transit or recovery to base, will potentially be significant, to say the least. The integrity of control commands, bandwidth problems and data latency are all very real issues which must be addressed before UCAV can realise their potential as a true alternative to manned aircraft. Recent experience with the Northrop Grumman RQ-4A Global Hawk during the Kosovo conflict and more recent milestone range and altitude tests has provided much data and this programme will continue to be an invaluable proving ground for enabling technologies. However, flight at 65,000 ft above civilian airspace by an unarmed reconnaissance drone is an entirely different matter to the direct penetration of the airways structure by a UCAV armed with a couple of 2,000 lb-class weapons!

The pending selection of a targeting pod by the USAF for its Block 50 F-16s is intended to provide a true multirole capability for the aircraft, which will include destructive Suppression of Enemy Air Defences (SEAD) using precision guided weapons. Viewed as a real 'force multiplier', the stated requirement is for a pod capable of aiming weapons such as the Joint Direct Attack Munition (JDAM), the AGM-154 Joint Stand-off Weapon (JSOW) and inventory Laser Guided Bombs (LGBs), utilising a third-generation FLIR and a laser capable of operating at 40,000 ft. Possible contenders are Terminator/ATFLIR, LANTIRN 2000 and Litening II.

In the field of electro-optics, while many had forewarned of the demise of II technology in favour of uncooled Forward Looking Infra-Red (FLIR) systems which offered considerably lighter weight over traditional cooled systems, the development of Night Vision Goggles for aircrew has continued in the US. Both ITT and Litton are claiming 'fourth-generation' performance from their latest 'unfilmed' Image Intensifier (II) tubes, featuring autogated power supplies and reduced Halo. For an overview of NVG technology, see the accompanying *Night Vision Goggles and Image Intensifiers*. Recently, the US Navy committed to Gen IV NVGs for their fixed wing aircraft.

As of April 2001, the contenders in the Joint Strike Fighter (JSF) competition are still locked in battle, with both Lockheed Martin and Boeing having flown their respective demonstrators. Interestingly, both aircraft feature Helmet-Mounted Displays (HMDs), without any form of fixed HUD. Flight, navigation and weapon aiming information will be projected directly onto the pilot's visor, thereby significantly increasing situational awareness by allowing totally 'heads-out' (of the cockpit) operation and realising the full off-boresight capability of both

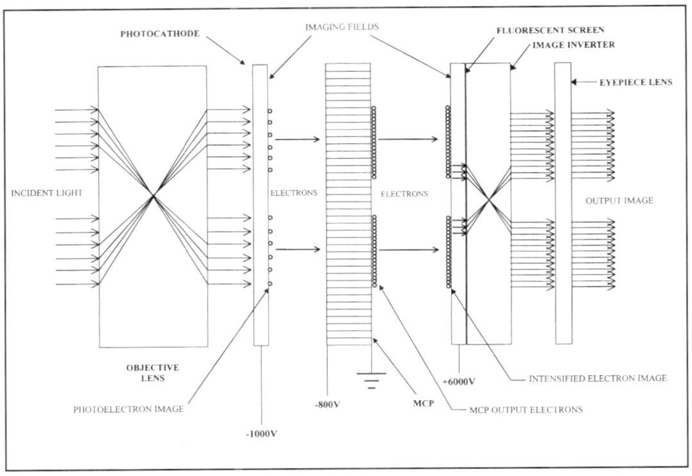

Figure 10 – Basic Tube Anatomy **2001**/0110037

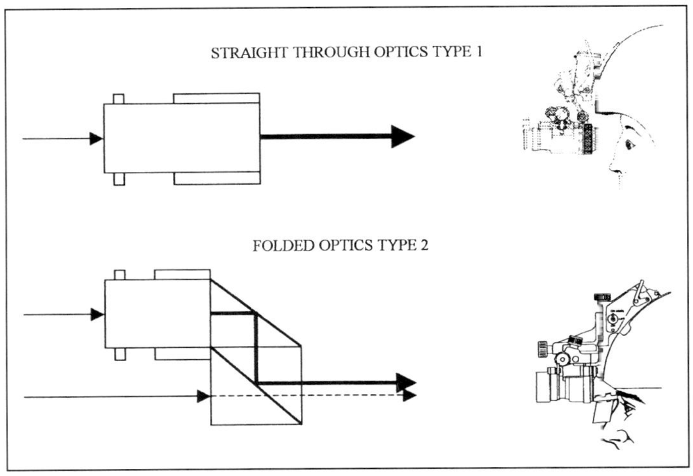

Figure 10 – NIGHTBIRD and CATSEYE NVGs **2001**/0110038

(OMNIBUS 4/5) provide for 25 mm of eye relief (the distance between the eye and the eyepiece lens) at 40° Field of View (FoV), enabling spectacles to be worn and facilitating view beneath the goggle (at cockpit instrumentation).

Types of NVG

NVGs fall into two main categories - Type 1, or straight-through optics, are the most common type of NVG in use today, characterised by ANVIS (USAF) and NIGHTBIRD (UK RAF) and Type 2, or folded optics, characterised by CATSEYE (Used in the UK until superceded by the NIGHTBIRD and in the US by the USMC, now superceded by the AN/AVS-9). Figure 9 shows NIGHTBIRD and CATSEYE NVGs, while Figure 10 illustrates the different optical paths of Type 1 and Type 2 NVGs.

The reason for the development of two entirely different designs of NVG can be traced back to the approaches of engineers in the US and

Figure 9 - Nightbird NVG **2001**/0110042

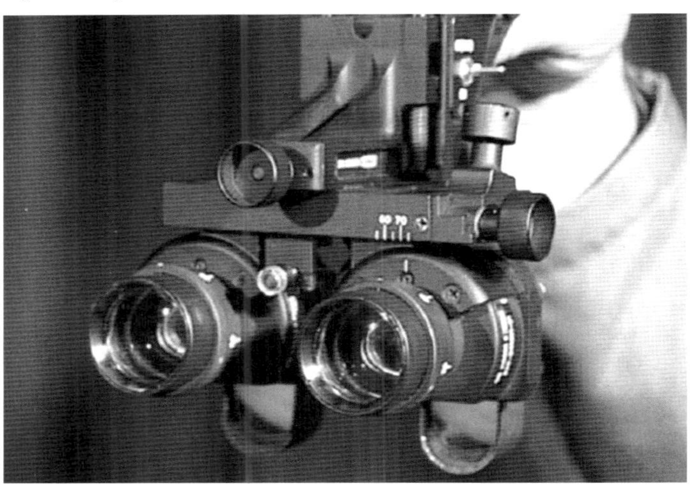

Figure 9 - Catseye NVG **2001**/0110043

UK. In the US, the first NVG designed for aviation use was the binocular AN/AVS-6, which was developed from the binocular AN/PVS-5, an NVG designed for vehicle drivers. The AN/PVS-5 is an enclosed system (i.e. fully covers the face) utilising a single II tube, with the optical path split into two, providing an image for each eye. The AN/AVS-6 is a logical development of that system, utilising two II tubes without any faceplate to allow for viewing around and below the goggle at flight instrumention.

In the UK however, engineers designed an entirely new type of goggle aimed purely at the fast jet cockpit. The resulting Ferranti (now BAE Systems) CATSEYE NVG featured two separate optical paths, one an intensified image from the II tube and the other a direct view of the outside world. The reason for this approach was to minimise interference from the II image when the pilot viewed raster FLIR imagery in the Head Up Display (HUD). A sensor on the goggle (the sensor can be seen situated between the objectives of the CATSEYE goggle in Figure 9) removed the II image when the goggle was aligned with the HUD, reinstating the image when the goggle moved out of line. A three-position power switch enabled pilots to select/deselect this Auto Scene Reject (ASR) feature. This system allowed the pilot to view FLIR imagery and/or the outside world without the II image and provided for a much better view underneath the goggle at the aircraft Head-Down instruments and displays. The major drawback to the CATSEYE system was the relatively poor 30° FoV, compared to 40° for the AN/AVS-6, although the system found favour with the USMC, who used the CATSEYE goggle (US designation MXU-810/U) in all their fixed wing fast jet aircraft until only recently replacing it with the AN/AVS-9 Type 1 NVG.

Offering a wider FoV and a generally brighter image, the Type 1 NVG is now the dominant version in service. In the US, the original AN/AVS-6 has been replaced by the AN/AVS-9, which, although largely similar in appearance, offers increased image resolution and brightness. In the UK, the CATSEYE NVG was used briefly in the Harrier aircraft until it was replaced by the Type 1 NIGHTBIRD NVG, which is similar to the AN/AVS-6 in terms of performance.

NVG Adjustment

- The adjustments and controls for Type 1 (NIGHTBIRD) and Type 2 (CATSEYE) NVGs are shown are shown in Figures 11 and 12. Note the similarity in positioning of the major controls.
- Vertical Height Adjustment - This raises and lowers the monocular optical axis with respect to the helmet datum.
- Interpupillary Distance (IPD) - This moves the individual monocular assemblies equal distances from a fixed central datum position. A scale allows pre-setting of the IPD to a known personal setting.
- Tilt Adjustment - The monocular assemblies, which are linked together mechanically, will elevate or depress from the nominal. By convention, forward rotation of the knob tilts the NVG downwards.
- Fore and Aft Adjustment (Eye Relief) - Fore and aft adjustment is carried out by rotating the adjustment knob located at the front between the monocular assemblies.
- Objective Focus - For the CATSEYE NVG, this control focuses the goggle for distance. For the NIGHTBIRD NVG, the objective lens assembly is fixed focus at between 20 m and infinity.

aircraft. However, the gain associated with second generation tubes is somewhat lower than first generation tubes at 20,000 to 30,000. Second generation tubes are still somewhat susceptible to blooming when exposed to a bright light source but this tendency is minimised by the function of the MCP. The MCP allows the saturation to be confined to individual channels instead of the entire FoV as was the case in first generation tubes. This localised saturation appears on the intensified image as a "halo" effect around any bright light source. Image contrast is drastically reduced inside the halo, and the overlapping of halos from multiple light sources removes scene detail between the lights. Although now superceded by third (and soon fourth) generation tubes, it should be noted that second generation tubes remain in widespread use amongst many ground forces.

Most fast jet aircraft today employ third generation (Gen III) tubes in their NVGs. There are two major differences between second and third generation tubes: the S-20 multi-alkali photocathode of second generation tubes has been replaced by a Gallium Arsenide photocathode, and a metal oxide film has been applied to the MCP. The Gallium Arsenide photocathode surpasses the photosensitivity of the multi-alkali photocathode beyond 550 ?m. This means that third generation tubes are far more sensitive in the region where near IR radiation from the night sky is plentiful, and the 800 to 900 ?m peak sensitivity range demonstrated by the tube has a five to seven times greater photon rate than in the visible region. This translates directly to increased gain for NVGs utilising these tubes.

The addition of a metal oxide film to the MCP in third generation tubes greatly extends their service life over that of second generation tubes. This is due to the fact that the life of an intensifier tube is largely a function of the lifetime of the photocathode. End of life for a photocathode is primarily caused by ion bombardment which is given off by the MCP as it is struck by electrons from the photocathode. This bombardment results in a gradual loss of photocathode luminous efficiency. The higher the light input, the more ions are generated, and the shorter the life expectancy of the tube. For this reason the NVGs should never be exposed to any bright light. The addition of the metal oxide film to the MCP in third generation tubes helps to offset this problem. The film is transparent to electrons, so they pass from the photocathode to the MCP, but not to the ions. The ions are "trapped" in the metal oxide film and thus prevented from contaminating the photocathode. One disadvantage encountered with the introduction of the metal oxide film is an increase in the bias voltage requirement between the photocathode and the MCP. This, in turn, requires increased spacing between the two components to prevent arcing. Ultimately this spacing results in an increased halo size, as compared to second generation performance, when viewing bright light sources. Depending on light levels encountered during the life of the intensifier, another limiting factor for the lifetime of the tube is a gradual drop in secondary electron emissions through the MCP.

The most recent evolution of the image intensifier is often termed 'fourth generation' (Gen IV) tube (although this definition has not been formalised yet), which utilises an autogated power supply, which matches the applied power to the gain of the tube, thus avoiding saturation and possible damage which the tube is exposed to high light levels. This has enabled the manufacture of a tube without the metal oxide film ion barrier, since autogating provides protection for the MCP, and with a decreased gap between the MCP and photocathode, since bias voltage requirement is less. These modifications, together with advances in MCP manufacturing techniques enabling smaller pitches to be achieved, has resulted in a tube which offers a significant performance advantage over third generation tubes in both high and low light conditions.

Due to export restrictions imposed on potential fourth generation tube technologies, a hybrid tube, termed Gen III+, combining a standard third generation tube with an autogating power supply, has been developed. Claimed to offer near 'Gen IV' performance, NVGs based on this technology are available to a much wider range of users.

Night Vision Goggles - Theory of Operation

IIs consist of three main components - a photocathode, a Micro Channel Plate (MCP) and a phosphor screen (Figure 8):

Photocathode. Photocathode developments represent the principal difference between Gen II and Gen III intensifier tubes. Generation II equipment utilises a multi alkali compound whereas a Generation III tube has a Gallium Arsenide coated photocathode. As already discussed, the photocathode converts photons into electrons under the influence of an electrical field and these electrons are accelerated into the MCP channels. The photocathode is fitted with a brightness output control, called Bright Source Protection (BSP) circuit, which limits the number of electrons leaving the photocathode by reducing the voltage between the output of the photocathode and the input to the MCP. This function is performed automatically.

Micro Channel Plate. Electrons collide with the sides of the channels, Figure 7, and result in an output of, typically, 1000 electrons for each electron that enters. Consequently, the current leaving the MCP is 1000 times higher than that leaving the photocathode. The MCP is extremely thin (approximately one millimetre), therefore IIs using them have become known as 'wafer type' intensifiers. The MCP is encased in an 18 mm tube, although recent advances in MCP technology have facilitated the development of 16 mm tubes without loss of resolution. Within this diameter, there are up to 6 million glass tubes, or channels (in the latest US OMNIBUS IV tubes), uniformly tilted at approximately 5-6° (see Figure 7) to ensure that input electrons strike the sides of the tubes. The MCP also has a means of protection from high light levels, the Automatic Brightness Control (ABC). As the name suggests, the circuit automatically adjusts MCP voltage to maintain image brightness at a constant level for a wide range of ambient illumination levels.

Phosphor Screen. The phosphor screen converts the magnified number of electrons produced by the MCP back into photons, thus producing an intensified, visible image. The choice of phosphor is governed by a requirement for resolution, eye readability and persistence (of the resultant image). The 'green' image seen through most NVDs is due to the phosphor used; in US Gen III OMNIBUS 2/3 goggles, the P-20 phosphor produces a saturated green image, whereas Gen III OMNIBUS 4/5 goggles utilise a P-43 phosphor which produces a green image with a yellow tint.

The final components in the optical train are the image inverter and the eyepiece lens. The fibre-optic image inverter, sometimes referred to as a twister, consists of a bundle of millions of fibre-optic strands, which rotates through 180° to re-orient the intensified image. The image inverter also collimates the image for correct positioning in relation to the wearer's eye.

The eyepiece lens is the last optical component in a NVG. It focuses the visible image onto the retina of the viewer and incorporates a limited dioptre adjustment to allow for some correction for individuals' vision. However, the eyepiece does not correct for astigmatism, therefore the wearer should still wear any prescribed corrective spectacles. Modern Gen III goggles

Figure 6 – NIGHTBIRD Gen III NVG **2001**/0110044

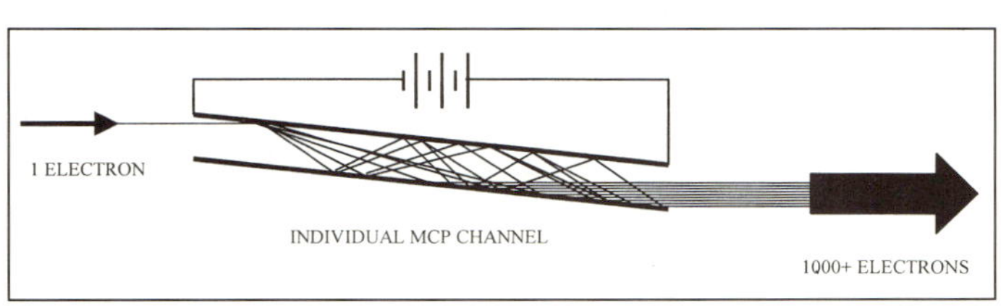

Figure 7 – Individual MCP Channel **2001**/0110036

air-to-air and air-to-ground weaponry. Considering Field of View and symbology requirements across the board, the development of these systems will undoubtedly pose some interesting problems in terms of human factors engineering.

Of currently deployed HMD systems, the Joint Helmet Mounted Cueing System (JHMCS), integrated into US F-22, F-15, F-16 and F-18 aircraft, employs a day-only module (the day module is designed to interchange with the Panoramic Night Vision Goggle module for night operations). The next generation of systems, such as the JSF HMD and the Eurofighter Typhoon Integrated Helmet, will offer day and night capability in one unit. While the CCD cameras integrated into the Eurofighter helmet do not yet offer the same level of performance as the latest II tubes, CCD technology is advancing at such a rate that the gap may well be bridged by the time the helmet is introduced into service, and all in a much lighter package.

Hopefully, while this HMD technological 'feeding frenzy' is going on, someone will keep the interests of the poor pilot in mind! Fit, comfort and, above all, (ejection) safety should not be relegated in the face of extracting the last ounce of performance from a helmet-mounted display system. In order to aim weapons effectively, the position of the pilot's eye relative to weapon sightline must be known to a high level of accuracy, therefore a consistent fit, resulting in a stable platform for the display, is critical to the performance of an HMD. Since the average fighter pilot's head varies in size, the achievement of a consistent and comfortable fit can present significant problems.

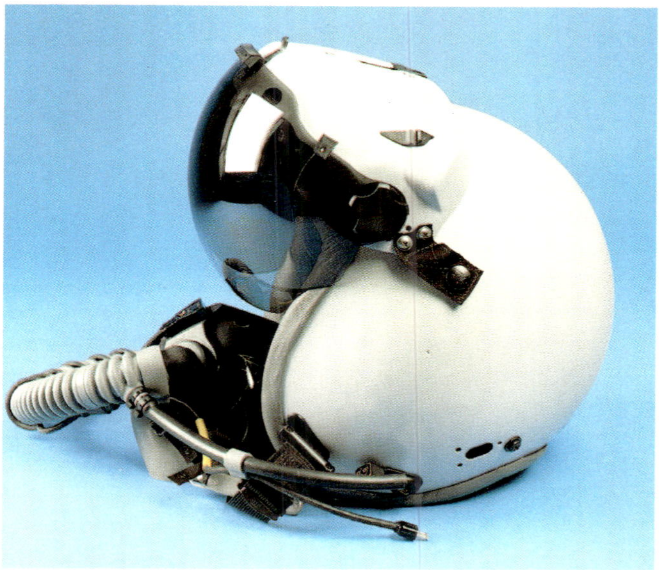

The Joint Helmet Mounted Cueing System (JHMCS) display module shown fitted to a USAF HGU-55P aircrew helmet. The display unit is day-only, and is designed to be interchanged with the Panoramic Night Vision Goggle (PNVG) module for night operations **2001**/0109234

Acknowledgements

Firstly, I would like to convey my sincere thanks to Chris Johnson, the previous editor of *Jane's Avionics,* for his help in preparing me for the job and his efforts in handing over a thoroughly researched and well-presented work. In 'learning' the job, life has been made much easier by the fact that updates to existing entries have, in many cases, been minimal or, indeed, nugatory!

I am most grateful to the many contractors who replied to my request for help in checking the data presented on their products in the previous edition of *Jane's Avionics*; comments, new information, and pictures have been most useful in the preparation of this edition. I am also most grateful to the many staff of Jane's Information Group who have guided me through the publishing process, particularly, Belinda Cunningham, who dealt with all the mailing activities and Gary Ransom who compiled the pictures provided; also the Content Editing department for typesetting and proofreading, and the compositors.

Edward Downs
Editor, *Jane's Avionics*
Coulsdon
April 2001

Jane's

Users' Charter

This publication is brought to you by Jane's Information Group, a global company with more than 100 years of innovation and an unrivalled reputation for impartiality, accuracy and authority.

Our collection and output of information and images is not dictated by any political or commercial affiliation. Our reportage is undertaken without fear of, or favour from, any government, alliance, state or corporation.

We publish information that is collected overtly from unclassified sources, although much could be regarded as extremely sensitive or not publicly accessible.

Our validation and analysis aims to eradicate misinformation or disinformation as well as factual errors; our objective is always to produce the most accurate and authoritative data.

In the event of any significant inaccuracies, we undertake to draw these to the readers' attention to preserve the highly valued relationship of trust and credibility with our customers worldwide.

If you believe that these policies have been breached by this title, you are invited to contact the editor.

A copy of Jane's Information Group's Code of Conduct for its editorial teams is available from the publisher.

INVESTOR IN PEOPLE

CIVIL AVIONICS (INCLUDING COTS, PRODUCTS AND DUAL-USE CIVIL/MILITARY SYSTEMS)

Civil/COTS, CNS, FMS and displays
Civil/COTS data management

CANADA

Allstar and Superstar Wide Area Augmentation System - Differential GPS (WAAS-DGPS) receivers and Smart Antenna

The Allstar and Superstar receivers are a new line of low-cost Global Positioning System (GPS) receiving units designed to track the US Federal Aviation Administration (FAA) Wide Area Augmentation System (WAAS) signal to provide levels of accuracy similar to those of the Differential GPS (DGPS) without the need for extra beacons or receivers.

At the same time, BAE Systems Canada Inc is introducing a new line of Smart Antennas. There is an RS-232 seven-pin version and an RS-422 12-pin connector version to optimise compatibility with other host systems. The RS-422 version is available with a 12-pin connector mounted either on the chassis of the Smart Antenna or at the end of a 1 ft cable extending from a central 1 in wide aperture.

The Smart Antenna contains the high-performance 12-channel all-in-view Superstar GPS receiver. It measures 115 (Ø) × 90 (H) mm, and consumes only 1.8 W.

Operational status

The Allstar receiver was launched in September 1999, the Smart Antenna in October 1999, and the Superstar receiver at the end of 1999.

Contractor

BAE Systems, Canada Inc.

VERIFIED

CMA-900 GPS navigation/flight management system

The CMA-900 provides full-performance GPS navigation and extensive flight management features, including company routes and worldwide ARINC 424 navigation databases, SID/STAR navigation, GPS instrument approaches, offset tracks, and autopilot-coupled holding patterns and procedure turns. As a multisensor navigator, the CMA-900 integrates other approved navigation sensors, and offers several navigation modes including GPS, DME/DME, DME/VOR, Omega/VLF and optional INS/IRS. This enables the system to provide seamless navigation for all phases of flight with maximum accuracy, integrity and availability. Other features include RNP/ANP, RTA and fuel management functions, and auto-tuning of navigation and communications radios. Pre-planned product improvement to provide Vertical Navigation (VNav) and other Flight Management System (FMS) features was a design criterion.

The system is authorised to TSO-C129, Class A1 and fully compliant with both the required and desired performance requirements of the FAA Notice N8110.57. Designed for operation in the CNS/ATM environment, the CMA-900 incorporates growth capacity for future datalink, and ATN compatibilities, as well as local and wide-area augmentation differential GPS (LAAS/WAAS), and GIC for precision approaches and landings. This makes the CMA-900 compatible with any of Eurocontrol's possible P-RNAV requirements.

The CMA-900 incorporates a 12-channel GPS Sensor Module (GSM), which is directly derived from the GPS sensor unit CMA-3012. This system has been selected for the Boeing 777, 737 and Airbus 330/340 series and has provisions to meet all RTCA requirements for SCAT-I landing equipment. The DGNS capability includes pseudo-ILS approaches and hardware provisions for ARINC 429 or RS-232 interfaces with the local and wide-area datalink receivers and a CMA-2014 multipurpose display. This is a 14-line 24 character fully ARINC 739-compatible colour AMLCD display.

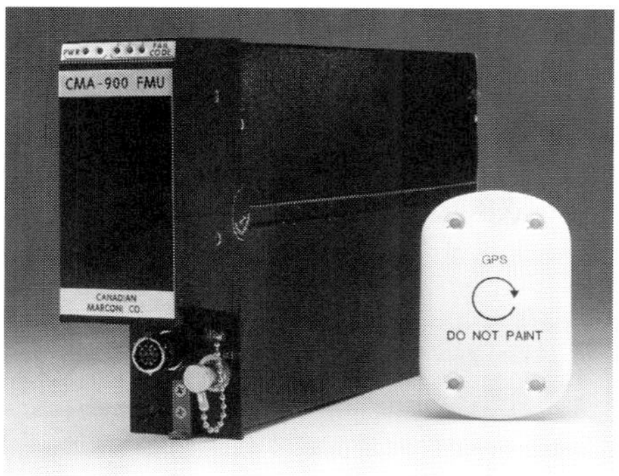

The CMA-900 FMU and GPS antenna ***1999***/0051579

The technology used in the GSM includes integrity-related features such as proprietary algorithm for high-performance, real-time, satellite Fault-Detection and Exclusion (FDE) including Receiver Autonomous Integrity Monitoring (RAIM), and software written in Ada to critical category standards. Performance is provided by high-dynamic, all-in-view reception of up to 12 channels (including Inmarsat integrity overlay) and carrier phase tracking. The all-in-view tracking and very fast acquisition capabilities of the CMA-900 will be particularly important, in the future, to offset any satellite blanking by the aircraft during terminal-area and precision approach manoeuvres. For aircraft currently equipped with older DME receivers, the DME/DME mode can be implemented using either an ARINC 709 scanning DME, or a digital DME sensor dedicated to the CMA-900.

The CMA-900 operating procedures conform to airline practices, including keyboard and page layouts, full scratchpad, line-select and menu functions. This approach results in fleet-wide commonality in crew interface and a very flexible system with the ability to meet the FANS CNS/ATM environment without keyboard redesign. It also ensures excellent transfer of training as pilots progress through a mixed fleet of aircraft.

The CMA-900 is equipped with the optimum suite of analogue and digital interfaces for retrofit applications, including outputs to conventional and Electronic Flight Instruments (EFIS), suitably modified weather radars, autopilot and flight director systems, and inputs from other navigation sensors, and both analogue and digital air data and heading systems. Data loader interfaces are in accordance with ARINC 615.

Other capabilities include full ARINC 739 compatibility and ARINC 429 file transfer protocols, which allow the FMU to interface with a number of aircraft systems, including ACARS and other ARINC 739-compatible CDUs. The CMA-900 also supports ARINC 429 fuel-computer inputs for fuel management

The CMA-900 Multifunction control display unit ***1999***/0051580

functions in conjunction with operator inputs of fuel on board, and so on.

Specifications

Dimensions:
(FMU) 200 × 57 × 324 mm
(MCDU) 171 × 146 × 214 mm
(GA/AEU) 19 × 74 × 119 mm
Weight:
(FMU) 3.63 kg
(MCDU) 4 kg
(GA/AEU) 0.5 kg
Power supply: 28 V DC, 73 W (max)
Accuracy:
(horizontal) 33 ft
(vertical) 130 ft
(velocity) 1 kt
(time) 2 μs
Reliability:
(FMU) 10,000 h MTBF
(CDU) 8,000 h MTBF

Operational status

Installed and operating on: B-707, B-727, B-737-200, B-747-200/300, Dash 8, DC9-30, DC9-50, DC-10, EMB-110, Gulfstream G2, MD-81, MD-82, MD-83 and MD-87.

Contractor

BAE Systems, Canada Inc.

VERIFIED

CMA-2055 instrumentation displays

The CMA-2055 family of self-contained, multifunction cockpit displays will accept digital, analogue and discrete inputs from helicopter sensors, transducers, switches and databusses, convert them into digital format, and display the resultant data in a number of different formats. The type and number of inputs can be optimised to meet the needs of a specific airframe, and include features such as engine data; caution/warning annunciations; engine condition trend monitoring; power up and continuous BIT; maintenance functions; rotor track/balance; and vibration monitoring.

Specifications

Dimensions: 314.6 (W) × 152 (H) × 211.2 (D) mm
Weight: 6.3 kg, including all options
Display size: two display screens, each 114 × 114 mm (4.5 × 4.5 in)
Interfaces: Analogue sensors; I/O discrete; RS-422 databusses; RS-232 for ground support equipment.
Environment: RTCA DO-178A.

Operational status

In production and service on MD-900 series Explorer, and Kazan Ansett helicopter.

Contractor

BAE Systems, Canada Inc.

VERIFIED

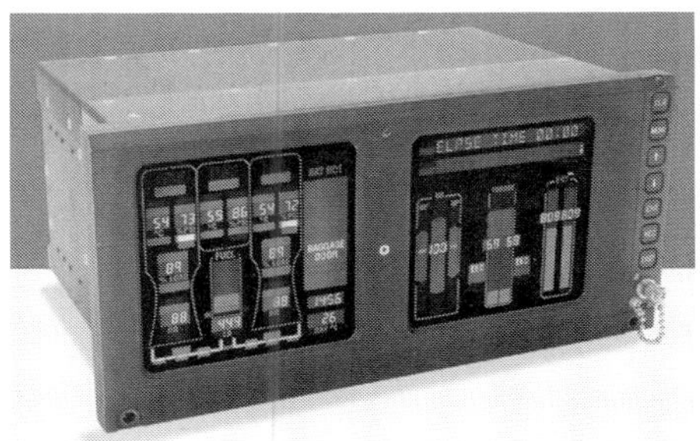

CMA-2055 instrumentation display ***1999***/0051654

CMA-2068 - MultiFunction Display (MFD)

The CMA-2068 MultiFunction Display (MFD) is a high-resolution colour Active Matrix Liquid Crystal Display (AMLCD) system for military and commercial applications. The CMA-2068 MFD employs an open architecture for intelligent processing and display of graphic, alphanumeric and video information.

An integrated bezel incorporates 24 programmable soft (line) keys and two rocker keys for operator interface and control.

Specifications

Display: full colour active matrix LCD, VGA 640 × 480 pixels
6 × 8 in (152 × 203 mm); other display sizes available
Interfaces:
MIL-STD-1553B BC/RT;
ARINC 429/629;
Discrete input/output;
DC/AC analogue;
Thermocouple, resistance bulb and so on;
RS-422 or RS-232;
RS-170A colour video (optional);
Processor: Intel family of CPUs;
Graphics/symbol generator;
Applications memory – 4 Mbyte Flash, 512 kbyte static RAM (expandable to 2 Mbyte);
64-256 kbyte RAM;

Operational status

In development.

Contractor

BAE Systems, Canada Inc.

VERIFIED

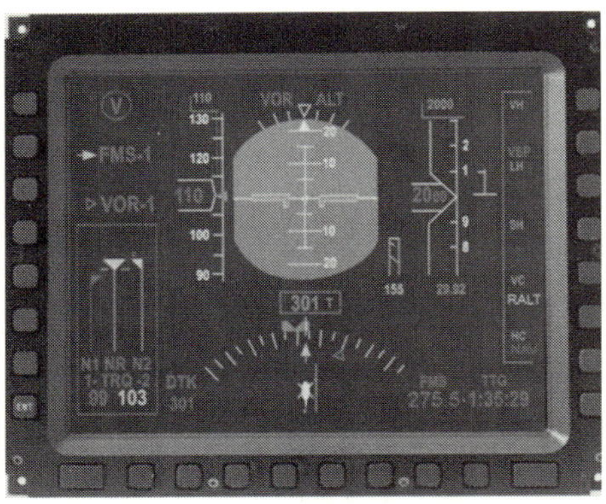

CMA-2068 multifunction display **1999**/0051655

CMA-2082 Avionics Management Systems (AMS)

CMA-2082 Avionics Management Systems (AMS) are a family of microprocessor-controlled Cockpit Display Units (CDUs) that can interface with a wide range of aircraft avionics equipment via either MIL-STD-1553B or ARINC 429 digital databusses. Each AMS has spare card slots which permit interfacing to avionic systems requiring discrete, analogue, synchro or other interfaces. The CMA-2082 AMS family of CDUs can act as either bus controllers or remote terminals on the MIL-STD-1553B databus. Each configuration of the CMA-2082 CDUs incorporates a powerful central processing unit based on the Intel 80486 CPU family with extensive onboard memory.

The CMA-2082D avionics management system

The CMA-2082A/2082A-5 AMS configurations utilise a Thin Film ElectroLuminescent (TFEL) flat panel display with contrast enhancement/bandpass filters to achieve both sunlight-readability and night vision goggle compatibility. The display is made up of active picture elements arranged in an X-Y matrix. The CMA-2082A has a 3 × 5 in (76 × 127 mm) active display area; and the CMA-2082A-5 has a 3 × 5 in (76 × 127 mm) active display area. The TFEL display permits viewing angles of up to 150° in any axis (greater than 45° on edges by the bezel) for optimum cross cockpit visibility.

A comprehensive keyboard with 29 alphanumeric data entry keys, two rocker keys for 'bright/dim', 'previous/next' functions and 8 to 10 soft (line) keys forms an integral part of the front panel. The keyboard functions are assigned by the system software and are displayed on the screen.

The CMA-2082D AMS utilises an Active Matrix Liquid Crystal Display (AMLCD), that is sunlight readable, with filters to ensure night vision goggle compatibility. The CMA-2082D has an active display area of 4 × 4 in (102 × 102 mm) with a pixel array 512 × 512 RGB × 64 grey shades. Viewing angle up to ± 45°.

The CMA-2082D has a comprehensive keyboard comprising 53 data keys, 12 soft (line) keys and two rocker keys.

Typical applications of the various CMA-2082 CDU configurations are control of Communication/Navigation/Identification (CNI) systems; FLIR/radar operating modes; Aircraft Survivability Equipment (ASE); weapons systems and digital maps.

Specifications

Dimensions:
(CMA-2082A) 5.75 (W) × 7.12 (H) × 6.7 (D) in; 146 × 181 × 170 mm
(CMA-2082A-5) 5.75 (W) × 9.4 (H) × 6.7 (D) in; 146 × 239 × 170 mm
(CMA-2082D) 5.75 (W) × 9.38 (H) × 8.24 (D) in; 146 × 230 × 209 mm

Weight*:
(CMA-2082A) 10 lb; 4.54 kg
(CMA-2082A-5) 12 lb; 5.44 kg
(CMA-2082D) 11 lb; 5.0 kg
*Dependent on whether spare card slots are populated

Display Size:
(CMA-2082A) 3 × 3 in; 76 × 76 mm; 192 × 192 pixels
(CMA-2082A-5) 3 × 5 in; 76 × 127 mm; 192 × 320 pixels
(CMA-2082D) 4 × 4 in; 102 × 102 mm; 512 × 512 RGB × 64 grey shades

Display capacity:
(CMA-2082A) 12 lines; 21 characters per line
(CMA-2082A-5) 20 lines; 21 characters per line

Visibility: readable in 10,000 ft candles of incident light. ANVIS compatible.

Memory:
CMA-2082A-5: 512 kbyte UV-PROM (max); 2 Mbyte Flash EPROM (max); 256 kbyte RAM
CMA-2082D 2 Mbyte Flash; 512 kbyte Static RAM
(programmable via MIL-STD-1553B or RS-422 interfaces)

Environment: MIL-E-5400T, Class 1

Operational status

The CMA-2082 AMS family of CDUs is in production and service.

Various configurations of the CMA-2082A AMS are in service with the US Army UH-60Q Medevac helicopter; US Air Force Special Operations Forces MH-60G, MH-53J, MC-130E, and AC-130H; US Navy ES-3A, E-2C; Belgian Mirage V; and Canadian Forces UTTH.

The CMA-2082D has been selected for a German military programme.

Planned product improvements to the CMA-2082 AMS family of CDUs includes integrating a GPS receiver and speech recogniser for direct voice control.

Contractor

BAE Systems, Canada Inc.

VERIFIED

CMA-2102 high-gain satcom antenna system

The CMA-2102 high-gain satcom antenna system is designed to support the Inmarsat® Aero-H and Aero-H+ satellite communications services, which provide aircraft with simultaneous two-way, digital voice and real-time data communications capability for flight crew, cabin crew and passengers. It provides a suitable baseline for the forthcoming Communications Navigation Surveillance/Air Traffic Management (CNS/ATM) system.

The CMA-2102 is a single, phased-array, electronically steered, top-mounted, high-gain antenna, with a single Beam Steering Unit (BSU) and Diplexer/Low Noise Amplifier (D/LNA) which provides hemispherical coverage in an extremely reliable installation. Since the CMA-2102 is mounted on the top of the aircraft, its coverage pattern does not suffer from

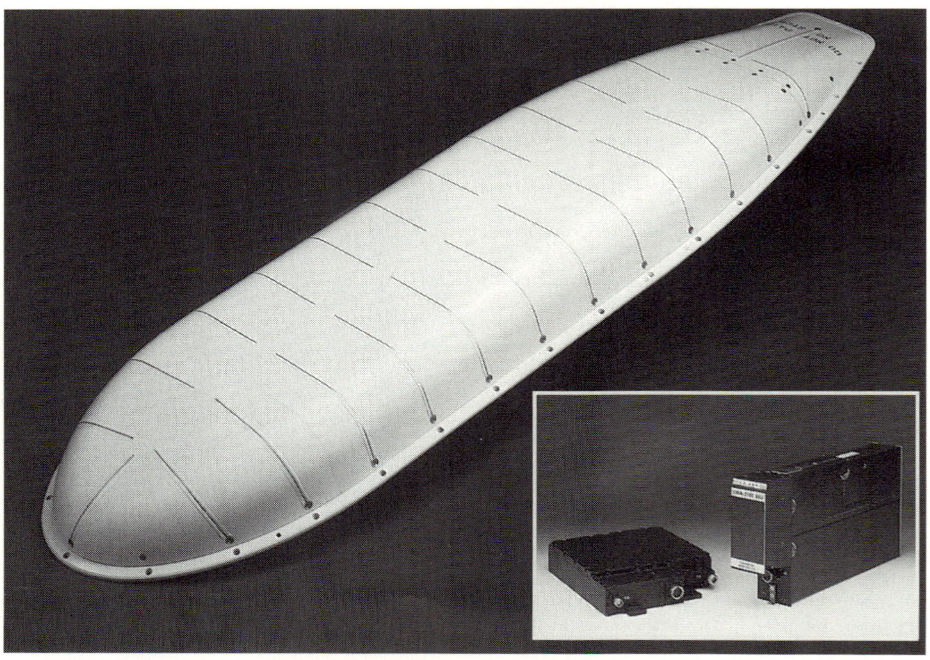

The CMA-2102 high-gain satcom antenna system with (inset) the low-noise amplifier (left) and beam-steering unit (right) 0018149

keyholes/blindspots, and installation is greatly simplified. The system conforms to ARINC Characteristic 741.

The CMA-2102 covers 360° in azimuth and –5 to +90° in elevation and operates over the frequency range 1,525 to 1,660.5 MHz.

Specifications

Gain: between 12 dBiC and 17 dBiC over 90% of Inmarsat hemisphere; minimum of 9 dBiC over 100% of Inmarsat hemisphere
Axial ratio: less than 6 dB for all steering angles and all frequencies of operation within coverage region
Dimensions:
(BSU) 2 MCU
(antenna) 1,702 × 470 × 126 mm
(D/LNA) 281 × 197 × 50 mm
Weight:
(BSU) 2.7 kg
(antenna) 27.9 kg
(D/LNA) 3 kg
Power consumption:
(D/LNA) 6 W
(BSU/antenna) 46 W

Operational status

CMA-2102 systems have been selected by many major airlines and VIP/military operators, including American Airlines, Air France, Air Canada, Cathay Pacific, China Airlines, EVA Airways Corporation, Japan Airlines, KLM, Lufthansa, Quantas, Singapore Airlines, Saudi Arabian Airlines, Swissair, United Airlines and several others. Commissioned on the following aircraft types (TCs and/or STCs): Boeing (707/727/737/747/757/767/777); DC-10; MD-11 and MD-90; and Airbus (A300/319/320/321/330/340). BAE Systems, Canada claims that the CMA-2102 has claimed over 70 per cent of the wide-body aircraft market share.

In the military role, the CMA-2102 has been selected for a variety of tactical helicopters for attack, combat search and rescue and anti-submarine warfare roles. Installations include the AH-1P, Bell CF-UTTH, Denel Rooivalk, Eurocopter Tiger, E-4B, F-27/28, GKN Westland and Agusta Cormorant and P-3C.

Contractor

BAE Systems, Canada Inc.

VERIFIED

CMA-2200 intermediate-gain satcom antenna system

The CMA-2200 intermediate-gain satcom antenna system design supports the requirements of the new-generation Inmarsat Aero-I satellite communication service. The Inmarsat-3 satellite provides aircraft with telephony, fax and real-time data communications and has been developed to meet the communication requirements of short-/medium-haul regional and corporate jet operators.

The CMA-2200 Intermediate Gain Antenna (IGA) is a derivative of the proven technology and architecture of the CMA-2102 High-Gain Antenna (HGA). The CMA-2200 is a top-mounted, linear array, electronically steered antenna, which conforms to ARINC Characteristic 761.

The CMA-2200 system consists of the antenna and Diplexer/Low Noise Amplifier (D/LNA). The traditional Beam Steering Unit (BSU) has been eliminated, the steering of the IGA being performed directly by the terminal equipment provided by leading manufacturers.

Specifications

Coverage: >95% of the Inmarsat hemisphere
Frequency:
(receive) 1,525-1,559 MHz
(transmit) 1,626.5-1,660.5 MHz
Gain 8.5 dBiC typical; 6.0 dBiC min
Dimensions:
(D/LNA) 280 × 197 × 50 mm
(IGA) 759 × 97 × 109 mm
Weight:
(D/LNA) 3.0 kg
(IGA) 2.7 kg
Power consumption:
(D/LNA) 6 W
(IGA) 7 W

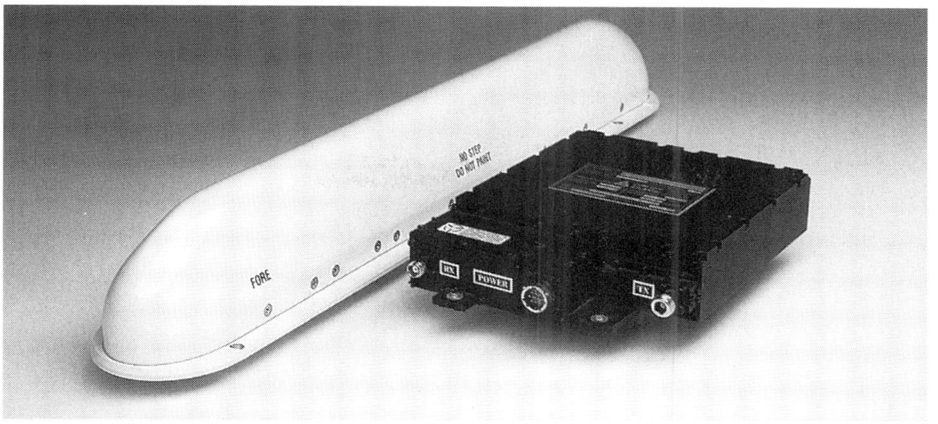

The CMA-2200 intermediate-gain satcom antenna system with the antenna unit (left) and diplexer/low noise amplifier (right) 0018148

Operational status

Inmarsat's rigorous AERO-I IGA testing was completed in August 1997. CMA-2200 black-label production shipsets started deliveries in January 1998. The CMA-2200 system has been selected for the Honeywell/Racal MCS-3000/6000 and MCS-7000 satcom systems.

Contractor

BAE Systems, Canada Inc.

VERIFIED

CMA-3000 Flight Management System (FMS)

The CMA-3000 FMS is an evolutionary step in the company's navigation and GPS products. The CMA-3000 is a single component hybrid made up of the development of CMA-900 GPS/FMS and CMA-2014 Mk III MultiControl Display Unit. The system is integrated into a single control display unit with the CMA-900's multisensor navigation and optional external GPS sensor functions. The system is designed to provide helicopters, regional and commuter aircraft with a versatile GPS-based flight management system housed in one line replaceable unit, featuring a multipurpose control and display function; a full-function embedded navigator and a radio management system.

The basic CMA-3000 features eight ARINC 429 inputs and three ARINC 429 outputs, eight discrete inputs, four discrete outputs and one RS-422 I/O port which makes it suitable for most helicopter applications.

For retrofit into aircraft with analogue interfaces, the CMA-3000 can be complemented with an external analogue adaptor unit. This configuration provides a cost-effective solution while still preserving a common pilot interface.

The CMA-3000 is designed for multisensor RNAV operation during the worldwide transition to GPS primary means navigation. Since certification approvals have already been obtained for the CMA-900, the CMA-3000 GPS/FMS will support installation approval for GPS primary means oceanic/remote operations. In addition, the CMA-3000 will be authorised for GPS-based instrument approaches under TSO-C129 Class A1. This full-function navigation system can also be used with a wide variety of navigation sensors.

The CMA-3000 uses state-of-the-art colour Active-Matrix Liquid Crystal Display (AMLCD) technology. It features a 5 in diagonal display screen with 14 lines of 24 characters each. The system includes a full alphanumeric keyboard; 12 line-select keys; 15 function keys; and nine dedicated annunciators.

The CMA-3000 display and back-lighting are sunlight readable, and compliance with MIL-L-85762A for Class B Night Vision Goggle (NVG) operation is available as an option.

The system's operating procedures conform to current airline practices of 'glass cockpit' aircraft, including keyboard and page layouts and full scratchpad, line-select and menu functions. The CMA-3000 conformity allows fleet-wide commonality in the pilot interface. All navigation and radio tuning functions utilise the same scratchpad/line-select crew interface philosophy.

Waypoint, navigation and guidance information is generated in both geographic and track-related reference frames. This information is clearly visible to the pilot on the unit's display. In addition, the system will output the information to flight instruments and autopilot/flight director systems. Its capabilities include:

(1) Flight planning and route creation, selection and modification;
(2) Complete oceanic, en-route, terminal and non-precision approach navigation and guidance;
(3) GPS instrument approaches;
(4) Continuous and manually initiated predictive FDE;
(5) Outputs to Electronic Flight Instrument Systems (EFISs), and digital autopilot and flight director systems;
(6) Direct-to and leg/course intercept navigation, holding patterns, DME arcs, procedure turns, and offset tracks;
(7) Automatic leg change with fly-by and fly-over leg transitions;
(8) Time and fuel management, including Required Time of Arrival (RTA) computation and display;
(9) Required and Actual Navigation Performance (RNP/ANP) computation and display;
(10) Search pattern navigation;
(11) ARINC 615-3 data loading capability for software maintenance and database update;
(12) Sensor status information display;
(13) Navigation and communication radio tuning;
(14) Digital map display interface to support route exchange and positioning;
(15) Kalman Filter Integration of GPS/AHRS (INS) (option);
(16) Compliance with all relevant RTCA and TSO requirements;
(17) Operation in severe (100 V/m) high-intensity radiated field (HIRF) environments.

Specifications

Dimensions: 172 (H) × 146 (W) × 160 (D) mm
Mass: 3.0 kg
Power requirements: 28 V DC, 70 W (max)

Operational status

In service.

Contractor

BAE Systems, Canada Inc.

VERIFIED

CMA-3012 Global Navigation Satellite Sensor Unit (GNSSU)

The CMA-3012 GNSSU meets the requirements of primary means oceanic/remote area operation as specified by FAA order 8110.60. The Sureflight software package is available to complement all CMC GPS receivers for primary means navigation flight planning and dispatch. Key characteristics include:

(1) Twelve simultaneous channels, all of which can be used for continuous satellite tracking, and any two of which are assignable as GPS/WAAS Integrity Channels (GIC)
(2) Comprehensive end-to-end receiver Built-In Test (BIT)
(3) Carrier phase tracking
(4) Differential GPS (SCAT 1) functionality

CMA-3012 GPS sensor unit 0015339

(5) Full compliance with TSO-129A B1/C1 and RTCA DO-208 sensor requirements
(6) Growth provisions for WAAS.

Specifications

Receiver: 12 parallel channels
Frequency: L1, 1,575.42 MHz, C/A code
Time to first fix: 95% confidence 75 s max
Time to reacquisition: 5 s max
Accuracy:
(horizontal position) 22.5 m, 95%, S/A off
(differential) <2.4 m, 95% (optional)
(altitude) 30 m, 95% S/A off
(velocity) 0.1 kt, 95%, S/A off
Position update: 5 s (optional)
Dimensions: 66 × 216 × 241 mm
Weight: 2.55 kg
Input power: 18 to 36 V DC, 20 W max
MTBF: 45,000 h
Inputs: 8 ARINC 429, 1 RS-232
Outputs: 3 ARINC 429, 1 RS-232; 1 28 V valid discrete; three 1 Hz time marks

Operational status

FAA approved as primary means of oceanic/remote area navigation. Installed with Racal RNav 2 system on helicopters operating over the North Sea.

Since 1993, over 2,500 units have been installed in numerous air transport and general aviation aircraft types.

Contractor

BAE Systems, Canada Inc.

VERIFIED

ADT-200A aeronautical data terminal

The ADT-200A aeronautical data terminal has been designed principally as an aeronautical transportation fleet management tool, utilising satellite communications. The communications system is based on Inmarsat Standard C communications format and is capable of two-way communications consisting of position reporting, general messaging and the sending of coded messages. The fleet communications control is performed through a network control centre which allows communications with the individual ADT-200A terminals. Position information is derived from a GPS receiver and encoded along with a terminal identification code in the transmit signal.

The ADT-200A is compliant with all applicable mechanical and electrical airborne regulations. The system consists of a compact keyboard/display unit, modular transceiver and antennas.

The keyboard/display unit is easily mounted in the cockpit or elsewhere in the aircraft to alert the operator to incoming messages. It enables the operator to read messages under any lighting condition and to respond via a keyboard.

The transceiver houses the transmitter, GPS position locator and power supply unit. The transceiver provides a link between the operator's keyboard/display unit and the antenna, sends and receives messages and transmits the position of the aircraft as determined by the GPS.

The antennas are rugged, lightweight and weather resistant with a durable aerodynamic cover. They are used for both L-band satellite data communications and for receiving position signals from the GPS system. Once installed, the antennas are automatically aligned with the satellite.

Specifications

Dimensions:
(transceiver) 190.5 × 210.8 × 324.1 mm
(MDS antenna) 25.4 × 127 mm diameter
(GPS antenna) 19.1 × 76.2 mm diameter
(keyboard/display) 267 × 140 × 51 mm
Weight:
(transceiver) 7.5 kg
(MDS antenna) 0.6 kg
(GPS antenna) 0.1 kg
(keyboard/display) 0.9 kg
Power supply: 22-32 V DC
Frequency:
(transmit) 1,626.5-1,660.5 MHz
(receive) 1,530-1,559 MHz
Channel spacing: 5 kHz

Contractor

CAL Corporation.

VERIFIED

AMT-100 aeronautical mobile terminal

The AMT-100 aeronautical mobile terminal provides air-to-ground and ground-to-air voice communications to and from any location on the globe via the Inmarsat satellite system. The unit communicates through aeronautical standard ground stations operated by Inmarsat signatories.

The AMT-100 is an all-in-one system providing a single-voice channel with direct dial capability for automatic global operation. The system, consisting of a transceiver, high-power amplifier and non-obtrusive high-gain antenna, has been designed for corporate aviation aircraft.

The transceiver assembly handles all protocols, performs voice and data modulation and handles frequency conversion and Doppler correction. It also contains the antenna control processor. The high-power amplifier is a reliable solid-state unit designed for a wide range of aircraft installations. The antenna assembly is a high-gain tailfin-mounted steerable antenna which provides full azimuth and elevation coverage without gaps. The antenna may be pointed using aircraft navigation system inputs.

Specifications

Dimensions:
(transceiver) 381 × 190.5 × 317.5 mm
(amplifier) 127 × 190.5 × 317.5 mm

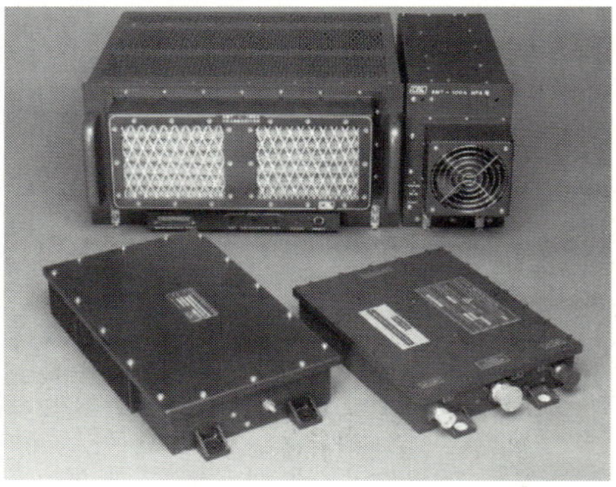

The AMT-100 provides voice communications with the public telephone network via the Inmarsat satellite system

(diplexer/LNA) 431.8 × 50.8 × 198.1 mm
(antenna drive assembly) 355.6 × 63.5 × 190.5 mm
Weight:
(transceiver) 14.5 kg
(amplifier) 7.7 kg
(diplexer/LNA) 2.72 kg
(antenna drive assembly) 3.17 kg
Power supply: 28 V DC

Operational status

The first AMT-100-equipped Bombardier Challenger was in service by the end of 1992. Full access approval was gained from Inmarsat in August 1993.

Contractor

CAL Corporation.

VERIFIED

JS-100A aeronautical satcom system

CAL has developed the JS-100A satcom system specifically for use on smaller aircraft. The unit has been engineered to minimise impact on airframe and performance.

The system consists of a transceiver assembly, high-power amplifier and antenna assembly. The transceiver assembly handles all protocols, performs voice and data modulation and handles frequency conversion and Doppler correction. It also contains the antenna control processor. The high-power amplifier is a reliable solid-state unit designed for a wide range of aircraft installations. The antenna assembly is a high-gain steerable antenna which provides full azimuth and elevation coverage without gaps. The antenna may be pointed using aircraft navigation system inputs or a signal strength measurement from the transceiver.

Specifications

Dimensions:
(transceiver) 317.5 × 317.5 × 190.5 mm
(high-power amplifier) 127 × 317.5 × 195.6 mm
(antenna) 1,371.6 × 203.2 × 200.7 mm
Weight:
(transceiver) 12.25 kg
(high-power amplifier) 5.44 kg
(antenna) 15.42 kg
Power supply: 115 V AC, 400 Hz or 28 V DC

Contractor

CAL Corporation.

VERIFIED

Contaminant and Fluid Integrity Measuring System (C/FIMS™)

The Contaminant and Fluid Integrity Measuring System (C/FIMS™) comprises a Control and Display Unit (CDU), a Central Control Unit (CCU) and up to eight flush-mounted sensors, depending on aircraft configuration. The electrical properties of the substance detected by each sensor are evaluated by the CCU and a corresponding output advisory signal is displayed on the CDU.

C/FIMS™ is designed to operate independently as a stand-alone system. The system is activated through the aircraft weight on the wheels. Built-in Test (BIT) status signals are provided to inform the flight crew if a critical fault is detected.

C/FIMS™ is an enhanced software version of the previously developed Clean Wing Detection System

(CWDS™). The sensing technology detects, discriminates and measures the thickness of the contaminant and layers of contaminants on the surface and specifically frost, ice, snow and de/anti-icing fluids; measures and indicates the integrity of the de/anti-icing fluid; predicts the onset of fluid failure and the imminence of ice adhesion and accurately measures aircraft skin temperature.

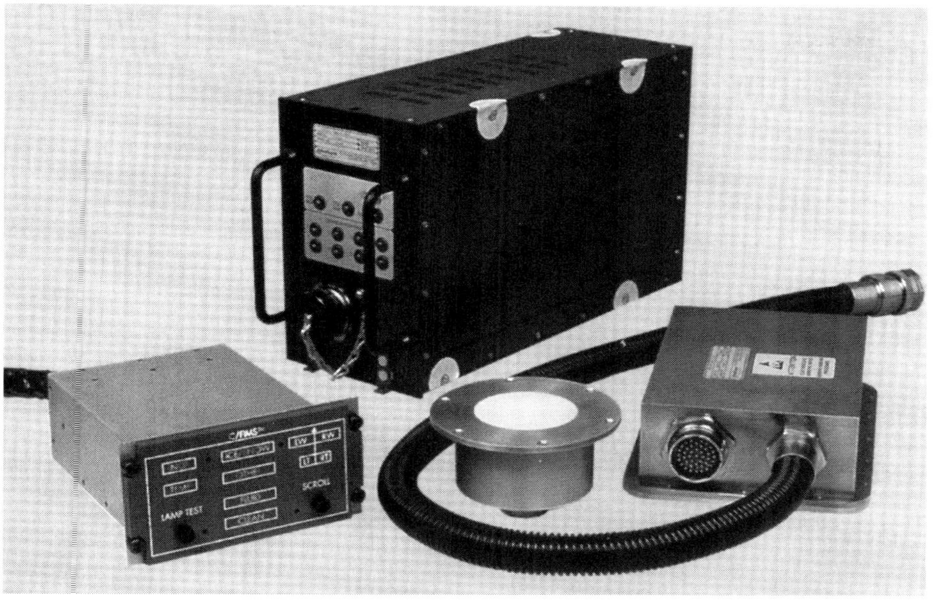

Contaminant and Fluid Integrity Measuring System (C/FIMS™) 0018229

Operational status

Operationally evaluated.

Contractor

Honeywell Aerospace Canada.

VERIFIED

ETIPS™ Electro-Thermal Ice Protection Systems

ETIPS™ is a new technology family of ice protection systems engineered to keep critical surfaces free of ice, frost and snow.

Typically, an ETIPS™ installation comprises Advanced Fibre Heater (AFH) assemblies, electronic controllers, and control/display panels as may be required by the application. AFH assemblies are typically fitted to upper wing surfaces, engine intakes and helicopter rotor blades.

ETIPS™ is automatic in operation for on-ground operation only, without pilot intervention.

Operational status

Installation certified for MD-80 series aircraft under FAA Supplementary Type Certification (STC) SA6061MM.

Contractor

Honeywell Aerospace Canada.

VERIFIED

APS-504(V) series radar

The APS-504(V) series of airborne search radars has been designed primarily for maritime surveillance applications. They can be installed in either a fixed- or rotary-wing aircraft. In addition to coastal and offshore surveillance missions, these radars can be used for weather avoidance, low-resolution land mapping and navigation.

The APS-504(V)2 is the commercial version of the AN/APS-504 which was developed specifically for Canadian Forces' Tracker aircraft. It uses the same 100 kW peak power I/J-band magnetron and transmitter pulsewidths of 0.5 and 2.4 μs. The system consists of a two-axis antenna unit with parabolic antenna, transmitter/receiver unit, analogue PPI display unit and radar control unit.

The APS-504(V)3 was developed to include an improved transmitter/receiver and a digital signal processor and scan converter that produce a ground-stabilised PPI display in a high-resolution 875-line video format. Navigation and cursor data are overlayed on the non-fading radar video, which can also be recorded and played back. Several sizes of flatplate antennas are available, mounted on a two-axis pedestal.

The APS-504(V)5 is the most advanced of the APS-504 family. It employs a TWT-based transmitter with wideband frequency agility, high-ratio pulse compression, scan-to-scan integration and digital signal processing to enhance the detection of sea surface targets, including targets with radar cross-sections as small as 1 m² in Sea State 3. The APS-504(V)5 can be configured to meet various installation and performance requirements.

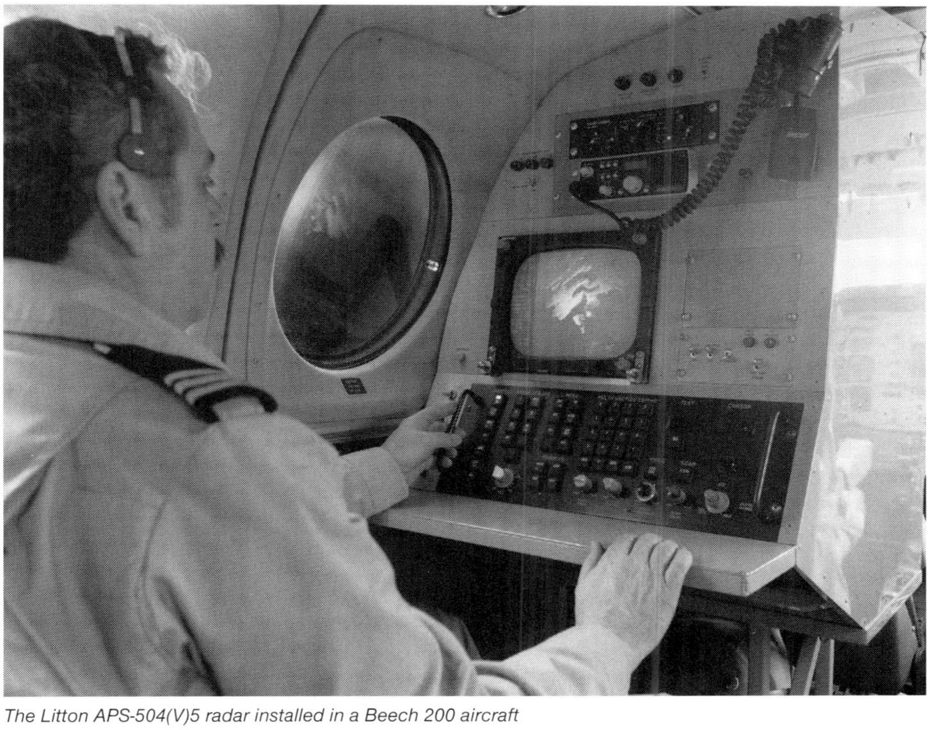

The Litton APS-504(V)5 radar installed in a Beech 200 aircraft

Specifications

(APS-504(V)5)
Weight: (with 3-axis pedestal) 180 kg
Power supply: 115 V AC, 400 Hz, 3 phase, 1.5 kVA
28 V DC, 15 A
Frequency: 8.9-9.4 GHz (16 frequencies)
Peak power: 6.6 kW
Antenna: flatplate and parabolic (various sizes available)
Scan rate: 7.5-120 rpm automatically selected
Display: RS-343 875-line video
Range scales: 5.5-370 km

Operational status

The APS-504(V)2 is no longer in production. There were 53 of this version produced.

The APS-504(V)3 is no longer in production. There were 25 of this version produced.

The APS-504(V)5 is still in production. It has been installed in aircraft ranging from twin-engined turboprops, such as the Beech 200, to larger aircraft such as the Boeing 737. More than 65 of this version have been produced.

Contractor

Litton Systems Canada.

VERIFIED

IRIS synthetic aperture radar

Available for both civil and military applications, the Integrated Radar Imaging System (IRIS) is an airborne synthetic aperture radar reconnaissance system.

A major advantage afforded by the IRIS is real-time tactical operation. The full-resolution, full-swath, onboard processing, unique to the IRIS, enables the simultaneous viewing of reconnaissance images in the air and, via downlink, on the ground. In a battlefield or other active military situation, this capability allows the integration of the airborne reconnaissance system with a ground-based, tactical command and communications system to co-ordinate and control ground forces and air strike support. Intelligence derived from the IRIS reconnaissance images on the position of opposing forces can be used by field commanders to direct ground operations and airborne support within seconds of acquisition of the reconnaissance data.

Two operational modes are provided, with the capability of switching in flight from a wide swath to a high-resolution mode in seconds. Images are produced from distances of up to 100 km and altitudes of up to 49,000 ft, providing standoff range capability.

The IRIS consists of three segments: an airborne segment, a transportable ground segment and a precision analysis facility. The standard equipment for these segments is augmented by options for image display, storage and interpretation. Drawing on these options, the IRIS can be configured for a variety of user requirements.

The airborne segment consists of an imaging synthetic aperture radar with onboard digital processing, downlink transmission, high-density magnetic tape data recording and image production capabilities. Images are processed in real time and displayed on a video monitor and hard-copy paper strip. The airborne segment has been packaged in compact, rugged modules for application in high-performance executive-size turboprop and jet aircraft. In addition, independent processors are included for simultaneous fixed and moving target imaging. Moving and fixed targets can be displayed independently or superimposed in colour in a single image.

The transportable ground segment consists of a downlink receiver, digital data processor and tactical workstation. The tactical workstation is rugged, transportable and inexpensive – ideal for rapid deployment in critical reconnaissance areas. It features high-density magnetic tape data recording, continuous paper strip printing, continuous image and frame image video display capabilities.

The precision analysis facility provides radar interpreters with the ability to analyse reconnaissance data received via downlink or recorded on board the aircraft and the transportable ground segment. The facility can archive and retrieve multiple data sets from the airborne and transportable segments, as well as maps and other interpretation aids entered at the facility. The precision analysis facility allows for in-depth interpretation of imagery, change detection, production of map mosaics and preparation of precision hard-copy film products.

Specifications

Weight: 400 kg
Frequency: 9,375 MHz
Average transmitter power: 160 W
Platform speed: 550 kt (max)
Antenna length: 1.0 m standard, 1.4 m high gain

Operational status

In production.

Contractor

MacDonald Dettwiler & Associates Ltd.

VERIFIED

InterVOX™ intercom systems

AMS4X/AA9X multi-user audio controller **1999**/0044774

AA80 InterVOX™ intercom system

The AA80 InterVOX™ is an advanced intercom system for general aviation flying, using Northern Airborne Technology's exclusive InterVOX™ and ANF™ (Active Noise Filtering) technology to produce high audibility under difficult and noisy conditions. Split bus operation and 'automatic emergency switching' during power failure are standard features. Short circuit protection is provided to ensure continued operation in the event of a headset failure. Music and ICS muting are also provided to eliminate interference with ATC reception. InterVOX™ provides full boom microphone operation for pilot and co-pilot, plus ICS operation for up to two passengers.

Specifications

Dimensions: 33 × 66 × 142 mm
Weight: 0.31 kg
Power: 11-32 V DC, 180 mA

AA85 InterVOX II™ intercom system

The AA85 InterVOX II™ is a high quality voice-activated intercom using Northern Airborne Technology's InterVOX™ and ANF™ technology, designed as a direct replacement for the earlier AA80 series intercoms. A single unit provides services for two, four or six place installations and it provides pilot/crew/all modes to include or isolate other crew and passengers from the services available. The provision of independent microphone circuits reduces noise and other interference.

Specifications

Dimensions: 33 × 66 × 142 mm
Weight: 0.31 kg
Power: 12-30 V DC, 600 mA
Displays: 2 colour LED
Approvals: TSO-C50c pending

Contractor

Northern Airborne Technology Ltd.

VERIFIED

AA85 InterVOX II ™ intercom system **1999**/0044773

Multi-user audio controllers

Northern Airborne Technology Ltd produces an extensive range of multi-user audio controllers in the AMS4X/AA9X series, including:

AMS42/42F

The AMS42 is designed for VFR/IFR/OAS/forestry/corporate/fleet use. The AMS42F is specifically intended for forestry application. The AMS42/42F systems are dedicated dual-channel audio controllers with VOX ICS operation and additional internal switch programming; these controls will work in both LH or RH pilot configurations; support is provided for the pilot, co-pilot and five passengers; five transceiver positions are provided. Functions include: COM 1, COM 2, FM 1, FM 2, AUX, PA transceivers, RX and ICS lever controls.

AMS43/43H/43P

The AMS43 series is designed for VFR/IFR/OAS/forestry/corporate/fleet use. All models represent a dedicated configuration of the AA95 series audio controller with LIVE/KEYED/VOX ICS operation; support is provided for the pilot, co-pilot and four passengers; five transceivers and a PA position are provided, as well as six additional switched receivers and two tape inputs; emergency operation and three-level alerting are also featured in these controllers. The AMS43H has special panel legends for Canadian and Australian operators with HF. The AMS43HP has an internal switch to select between PA Key out or Alert 3 in. Functions include: COM 1, COM 2, AUX, FM 1, FM 2, PA transceivers, NAV 1/2, ADF 1/2, DME/MKR receivers.

AMS44

The AMS44 is designed for police/forestry/charter/corporate/fleet use. It is a dedicated version dual-channel audio controller with VOX ICS operation and front panel Nav Aid selections. It will work in either LH or RH pilot configurations, with support for pilot, co-pilot and five passengers. It includes five transceiver positions, as well as five additional audio sources, together with ICS Tie line and Direct inputs. Functions include: COM 1, COM 2, FM 1, FM 2, AUX, PA transceivers, NAV 1, NAV 2, ADF, DME MKR RX switches.

AA12

The AA12 is designed as a VFR audio controller. It provides audio to six crew positions with VOX, LIVE and KEYED ICS modes. Functions include: COM 1, COM 2, COM 3, PA transceivers, NAV, ADF and DME/MKR Nav Aids. Further models in the series include: AA12-001, AA12-002, AA12T-002.

AA92H

The AA92H series are custom dual-channel controllers based on the AMS42 design. Models available include: AA92H-407, AA92H-410, AA92H-411.

AA94

The AA94 series are custom dual-channel controllers based on the AMS44 design. Models available include: AA94-400 (special), AA94-426, AA94-427 (law enforcement).

AA95

The AA95 series are functional equivalents of the AA90 series incorporate a 500 mW headset driver to meet USFS/OAS requirements. Models available include: AA95-512 (police), AA95-824 (police), and a variety of AA-95 emergency medical crew models.

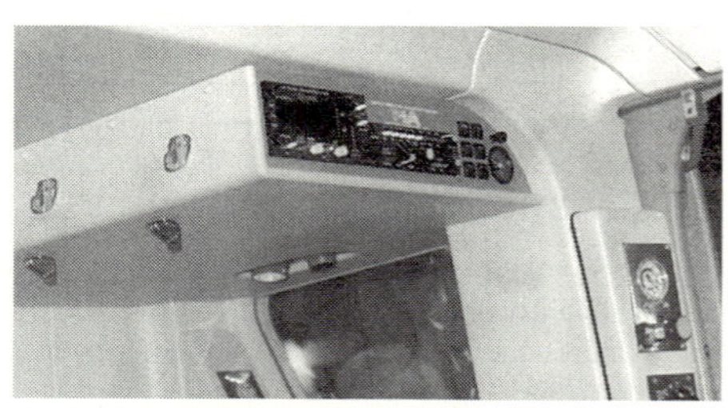

Typical installation of AA9X multi-user audio controller **1999**/0044775

AA96

The AA96 series is again a variant of the AA90 design, customised for tactical (AA96-001) and USFS (AA96-400) operations.

AA97

The AA97 is also a variant of the AA90 design, optimised for law enforcement operations, with individual control of each of six transceivers. Functions include VHF 1/2, TAC 1/2/3/4 and PA transceivers, NAV 1, NAV 2, ADF 1, DME switched Nav Aids.

N335

The N335 is Northern Airborne Technology's 'next generation' dual-user audio panel. It combines new technology with the modular design concept of the N301A family, plus the flexibility and control of the AA97. The N335 supports two operators with either high or low impedance systems and contains a selectable, fully redundant back-up power supply, audio amplifiers, ICS bus, and secondary volume potentiometers. Functions include: NAV 1, NAV 2, DME 1, DME 2, ADF, MKR, DF, VHF 1, VHF 2, UHF, FM1, FM 2, FM3, GSM.

Contractor

Northern Airborne Technology Ltd.

VERIFIED

Single-user audio controllers

Northern Airborne Technology Ltd produces a range of single-user audio controllers including the AA24, AA25 and N301A.

AA24 and AA25

The AA24 and AA25 provide high performance audio to a single-user in a network with VOX, LIVE and KEYED Inter-Communication System (ICS) capability. The small size and extensive functions make these models ideal for single-seat aircraft. Inputs are fully floating, including a dedicated alerting input, and units can work with either high impedance headsets (models AA24-001 and AA25-001) or low impedance headsets (models AA24-801 and AA25-801).

Both models provide the following services: COM1, COM 2 COM 3, COM 4 + PA, NAV, ADF, DME, MKR and AUX, ICS, RX and VOX controls.

The AA25 also provides independent radio volume adjustment for up to six sources.

Specifications

Dimensions:
AA24: 38.1 × 146.05 × 149.35 mm
AA25: 66.8 × 146.05 × 149.35 mm
Weight:
AA24: 0.5 kg
AA25: 0.68 kg
Power 22-32 V DC, 0.4 A

N301A

The N301A single-user audio controller updates the earlier A301-6 and N301 systems. It is designed for use in new installations where single- or multiple-user networks of 10+ stations are required. User controlled features include split RX and ICS volume controls, a live or 'hot microphone' ICS function switch, an additional

AA24 single-user audio controller **1999**/0044776

N301A single-user audio controller **1999**/0044778

RX input for the AUX TX position, and an extra Nav Aid position. Additional internal inputs have been added to improve threat alerting capability. TSO approval pending.

Specifications

Dimensions: 66.68 × 146.05 × 93.98 mm
Weight: 0.75 kg
Power: 22-32 V DC, 0.17 A

Contractor

Northern Airborne Technology Ltd.

VERIFIED

Tac/Com™ tactical FM communications systems

Northern Airborne Technology Ltd has designed a range of Tac/Com tactical FM communications systems in consultation with law enforcement, emergency medical services and forestry agencies. The design aim was to minimise problems associated with complex multiradio installations and multimission aircraft such as panel space, simultaneous operation, inexperienced users, complex operational modes and future upgrades.

The Tac/Com™ family is a series of configurable multiple radio control heads, agile transceivers and supplemental equipment. The modular control heads are capable of operating either Northern Airborne Technology or vendor radios. The Northern Airborne Technology radios are synthesised agile transceivers covering VHF and UHF bands and providing multiple scanning modes, agile synthetic guard, priority scanning, CTCSS and \ DPL squelch control, and they are DCS encryption ready. Supplemental equipment includes: encoders/decoders, DTMF keypads, relay-simulcast controllers and antennas.

All Tac/Com NT series transceivers are harness and tray compatible and all Tac/Com 250/350 series control heads can be retrofitted into existing C722, C962 or C1000 (3 in dzus) openings.

Tac/Com II™ control heads

Tac/Com II control heads provide operators with multi-radio operational capabilities, control of up to four radios, with each radio supporting up to 128 channels. Models available include: TH250 (two-channel), TH350 (three-channel) and TH450 (four-channel). Variants of these options are available to meet customer requirements.

Tac/Com™ NT series transceivers

The Tac/Com NT series of transceivers offer varying frequency bands from 29-869.975 MHz to support most VHF and UHF FM applications. The transceivers support multiple scanning modes including priority scanning (guard), high and low power Tx modes and CTCSS and DPL tone capability. VHF transceivers support 5, 6.25 and 12.5 kHz channel spacing and a multimode (AM and FM) version is available for use in the UK and Australia.

NTX series transceivers

The NTX generation of FM transceivers are the latest evolutionary step within Northern Airborne Technology's series of Tac/Com remote-control airborne radios. Using microprocessor controlled digital technology, the NTX radio is capable of high speed scanning, high volume channel storage, and it is compatible with the existing Tac/Com II control head. Depending on the model, optional features can include a two-channel dedicated real-time guard (USFS compatible), selectable bandwidth filters, and DF outputs.

A major advantage of the NTX transceiver is very compact size: 2.0 in wide (1/4 ATR), 5.1 in high and 10.5 in deep, with a total weight of under 2.5 kg, including undertray.

Specifications

Model NTX138
Frequency coverage: 138.000-173.995 MHz
Channels: 126, plus 2 guard, channel memory
Channel increments: 2.5 kHz
RF power: 1 or 10 W selectable
Weight: 2.3 kg
Power: 28 V DC
Compliance: USFS

Model NTX066
Frequency coverage: 66.000-87.995 MHz
Channels: 128 channel memory
Channel increments: 2.5 kHz
Weight: 2.3 kg
Power: 28 V DC

NPX panel-mount series FM transceivers

The NPX panel-mount series of FM transceivers are designed as very small size (2 in high), light weight (1.36 kg) stand-alone radios for single-mission users. The NPX series is a synthesised, frequency-agile, transceiver that provides easy access to all pre-programmed channels and the ability to modify or add channels in the air. Each of the 100 memory channels can store a receive frequency with CTCSS tone, a transmit frequency with CTCSS tone, scan function and alpha-numeric identifier on a 32-character two-line LED display. Transmit power of 1 or 10 W can be selected . DTMF encoding and direct keyboard programming can be added using Northern Airborne Technology's DTE12 datapad. Panel lighting can be 28, 14 or 5 V DC controlled, with dimmer. NVG compatible lighting is optional.

Specifications

Model NPX138/138N
Frequency coverage: 138.000-173.995 MHz
Channels: 100 channel memory, optional 2-channel guard
Channel increments: 5/6.25 kHz
Channel selectable: 25 or 12.5 kHz (Model NPX138N)
RF power: 1 or 10 W selectable
Weight: 1.36 kg
Power: 28 V DC
Compliance: USFS

Model NPX066
Frequency coverage: 66.000-87.995 MHz
Channels: 100 channel memory
Channel increments: 5/6.25 kHz
Weight: 1.36 kg
Power: 28 V DC

Contractor

Northern Airborne Technology Ltd.

VERIFIED

NPX panel-mount series FM transceivers **1999**/0044779

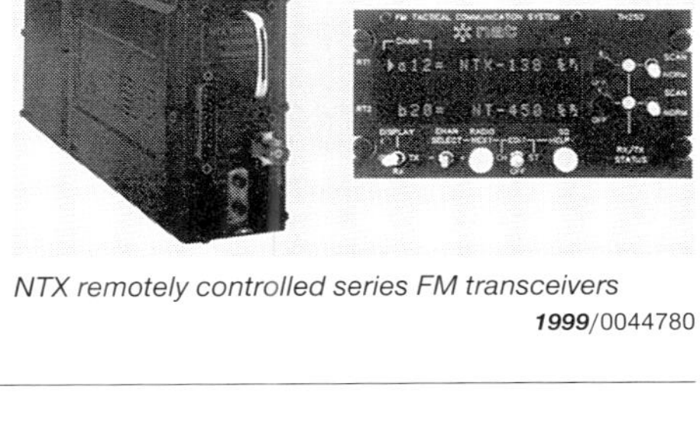

NTX remotely controlled series FM transceivers **1999**/0044780

Access/A A710/A711 airborne audio systems

The Technisonic Access/A system is based on an integrated family of modular, configurable, panel-mounted controls, coupled with external special function units, which produce a complete airborne communications suite. This audio system architecture supports extensive alerting, warning and signalling functions, voice message storage, partitioned crew services, and interface capabilities.

Access/A provides interconnection and interface facilities between airborne crews and their radio systems, as well as providing internal intercom, internal and external paging, airframe alerting, music and entertainment functions. Access/A is designed for applications like emergency medevac, forestry aircraft, search and rescue, customs and emergency services operations.

Access/A A710 and A711 controls are designed to be continuously expanded, relabelled and upgraded. Up to ten crew stations, and up to six crew members per crew station can be supported within a single aircraft. Many special modes and functions such as voice alerting, voice storage, NVG compatibility can be implemented.

Access/A A710/A711 stations support both speaker and headset installations, and provide for internal paging as well as interfaces to commercially available external paging systems.

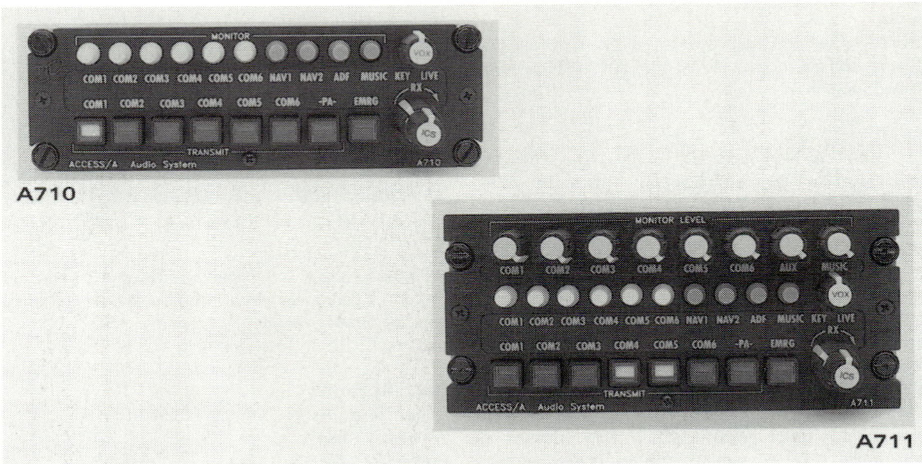

The Technisonic Industries Limited Access/A A710 and A711 airborne audio systems **2000**/0062785

Specifications

Power: 28 VDC, 350 mA
Dimensions:
(A710 dzus panel) 146 (W) × 47.6 (H) × 154.2 (D) mm
(A711 dzus panel) 146 (W) × 66.7 (H) × 154.2 (D) mm
Weight:
(A710) 1.09 kg
(A711) 1.36 kg
Features:
1- 6 users per crew station
direct push-button Tx function
optional voice alerting, with priority
optional voice message storage/replay
six transceiver operation, plus PA
headset and speaker outputs
VOX, PTT or live ICS modes
individual Tx enunciation
split ICS/Rx volume controls
emergency Rx/Tx capability, including boom microphone
changeable front panel legends
configurable ICS loops
simulcast operation
Qualification: RCTA C50c

Operational status

In wide use.

Contractor

Technisonic Industries Limited.

VERIFIED

The Technisonic Industries Limited TFM-138B airborne VHF FM transceiver **2000**/0062786

Specifications

	TFM-138	TFM-138B
Frequency:	138.000 to 174.000 MHz	138.000 to 174.000 MHz
Tuning increments:	2.5 kHz	2.5 kHz
Operating mode:	F3E simplex or semi-duplex	F3E simplex or semi-duplex
Channel spacing:	12.5, 25 or 30 kHz	12.5, 25 or 30 kHz
Memory channels:	100	120
Output power:	1 W or 10 W	1 W or 10 W
Dimensions:	203.2 × 76.2 × 146.05 mm	203.2 × 76.2 × 146.05 mm
Weight:	1.4 kg	1.5 kg
Power:	28 V DC, 2 A	28 V DC, 2 A
Guard receiver:	two channel synthesised	two channel synthesised
Certifications:	FCC and DOC type approved; DO-160c	FCC and DOC type approved; DO-160c

TFM-138 series airborne VHF FM transceivers

The Technisonic TFM-138 series of airborne VHF FM transceivers utilise frequency synthesis techniques to provide FM communications on every currently available channel within the VHF FM high band. These radios cover the entire band from 138 to 174 MHz in 2.5 kHz increments. Data entry and function control are via a front panel 12-button keypad. Operating frequency and other related data are presented on a 48-character two line LED matrix display, available in green or red.

Technisonic FM transceivers can be operated in 'direct entry' or 'simplex' mode by keying in the operating frequency, or can function without restriction on any split frequency pair within the band. The TFM-138 features 100 preset memory positions, each of which is capable of storing a receive frequency, a transmit frequency, a separate CTCSS tone for each receive and transmit frequency and an alpha numeric identifier for each channel. The TFM-138A and TFM-138B feature similar storage capability, but provide 120 channels of preset memory and offer the additional capability of allowing for DPL or DCS coded squelch operation. The TFM-138 and TFM-138B allow either 25 kHz wide band or 12.5 kHz narrow band operation on any or all of the preset channels. Data can be entered into any of the preset non-volatile memory positions for both main and guard channels via the front panel keyboard. Data on stored channel settings is instantly available.

Technisonic FM transceivers feature a synthesised two channel guard receiver, a DTMF encoder for signalling during transmit, and a scan function which will scan any or all of the frequencies stored in the preset memory.

The TFM-138 series consists of four models: the TFM-138, TFM-138A, TFM-138B and TFM-138C.

TFM-138A specifications are identical to TFM-138B, with the exception that the TFM-138A does not offer 12.5 kHz operation. TFM-138C specifications as per TFM-138B, but with the addition of voice encryption.

Operational status

In wide use.

Contractor

Technisonic Industries Limited.

VERIFIED

TFM-403 airborne UHF FM transceiver

The Technisonic Industries TFM-403 UHF FM transceiver utilises frequency synthesis techniques to provide FM communications on currently available channels within the public safety, forestry, government agency, and general service UHF FM band. The TFM-403 covers from 403.000 to 512.000 MHz in 2.5 kHz steps. Data entry and function control are via a front panel, 12-button keypad. Operating frequency and other related data are presented on a 48-character, two line LED matrix display, available in green or red.

The TFM-403 transceiver can be operated in the 'direct entry' or 'simplex' mode by keying in the desired operating frequency, or without restriction on any split frequency pair within the band. There are 120 memory positions, each of which can store a Rx and Tx frequency, a separate CTCSS and/or a DPL/DCS (Digitally Coded Squelch) tone for each receive and transmit frequency, and a nine-character alpha numeric identifier for each channel. Data is entered into the preset memory positions for main and guard channels via the front panel keyboard, or downloaded from a PC. All stored data is available for instant recall.

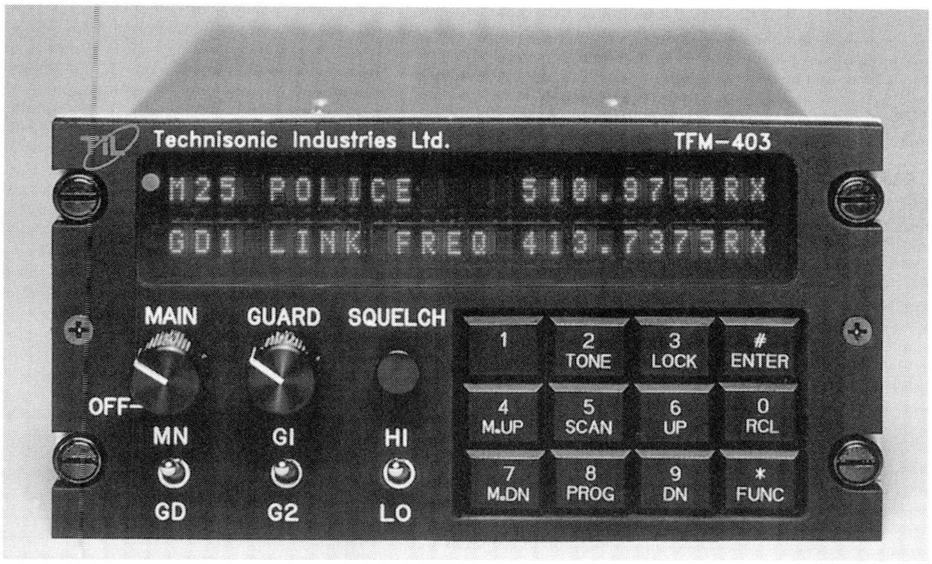

The Technisonic Industries Limited TFM-403 airborne UHF FM transceiver **2000**/0062787

The TFM-403 transceiver features a synthesised, programmable, two-channel guard receiver, a DTMF encoder for signalling during transmit, display of CTCSS tone frequency as well as EIA identifier and/or display of a DPL/DCS code, and a scan/priority scan function that can scan any or all of the channels stored in preset memory. Additional operator selectable capabilities include a 90 second transmitter time out, a keyboard lockout feature and a direct/repeat function for simplex operation. The TFM-403 is dzus mounted.

Specifications

Frequency: 403.000 to 512.000 MHz
Tuning increments: 2.5 kHz
Operating mode: F3E simplex or semi-duplex
Channel spacing: 20 or 25 kHz
Memory channels: 120
Output power: 1 W or 10 W
Dimensions: 203.2 × 76.2 × 146.05 mm
Weight: 1.4 kg
Power: 28 V DC, 2A
Guard receiver: two channel synthesised
Certifications: FCC and DOC type approved

Operational status

In wide use.

Contractor

Technisonic Industries Limited.

VERIFIED

TFM-500 airborne VHF/UHF FM transceiver

The Technisonic TFM-500 airborne VHF/UHF FM transceiver utilises frequency synthesis techniques to provide FM communications on currently available channels within the general radio service VHF FM high band and general service UHF FM band. The VHF module covers from 138.000 to 174.000 MHz in 2.5 kHz increments, while the UHF module covers from 403.000 to 512.000 MHz, also in 2.5 kHz increments. Data entry and function control are via a front panel, 12-button keypad. Operating frequencies and other related data are presented on a 96 character, four line LED matrix display, available in either green or red.

The TFM-500 can be operated in the 'direct entry' or 'simplex' mode by keying in the desired operating frequency. It can also function without restriction on any split frequency pair within either band, and offers 'cross band' repeat capability. The unit features 400 preset memory positions (200 VHF and 200 UHF) each capable of storing a receive frequency, a transmit frequency, a separate CTCSS tone for each receive and transmit frequency, an alpha numeric identifier for each channel, and the additional ability to provide DPL or DCS coded squelch operation on each channel. The TFM-500 can operate with either 25 or 12.5 kHz bandwidth on all memory channels. Operating data can be entered via the front panel or from a PC. Data can also be downloaded to a PC.

The TFM-500 includes a synthesised two channel VHF or UHF guard receiver, a DTMF encoder for signalling during transmit, and a scan function which will scan any or all of the frequencies stored in the memory. A remote control head is offered (RC-500) which provides for slaved operation of the main transceiver from a remote location, allowing a second position in the aircraft to exercise frequency control. Both UHF and VHF operating frequency as well as guard frequency can be controlled from the remote position. Active frequency is displayed on both local and remote displays. The TFM-500 is dzus mounted.

Operational status

In wide use.

Contractor

Technisonic Industries Limited.

VERIFIED

Specifications

	VHF module	UHF module
Frequency:	138.000 to 174.000 MHz	403.000 to 512.000 MHz
Tuning increments:	2.5 kHz	2.5 kHz
Operating mode:	F3E simplex or semi-duplex	F3E simplex or semi-duplex
Channel spacing:	12.5, 25 or 30 kHz	12.5, 25 or 30 kHz
Memory channels:	200	200
Output power:	1 W or 10 W	1 W or 10 W
Dimensions:	overall TFM-500 dual band transceiver: 203.2 × 95.3 × 146.05 mm	
Weight:	1.4 kg	1.4 kg
Power:	28 V DC, 2 A	28 V DC, 2 A
Guard receiver:	two channel synthesised	two channel synthesised
Certifications:	FCC and DOC type approved; DO-160c	FCC and DOC type approved; DO-160c

The Technisonic Industries Limited TFM-500 airborne VHF/UHF FM transceiver **2000**/0062788

Gyrostabilised video, infra-red, laser and film surveillance systems

WESCAM™ Inc produces a range of gimbal-mounted thermal imaging and film camera surveillance systems, together with associated datalink systems, controls and installation kits. They are suitable for installation in a variety of air and ground platforms, including fixed-wing aircraft, helicopters, UAVs and land/sea vehicles. Their primary roles include civil/military surveillance, airborne law enforcement, environmental studies and sport/entertainment filming. A number of models are available, designated by their gimbal diameter (inches) and number of sensors (e.g. model 12 DS having a diameter of 12 in and being fitted with a dual sensor):

WESCAM™ 12DS

The model 12DS is WESCAM's smallest gyrostabilised camera, designed specifically to meet airborne law enforcement requirements. Weighing less than 23 kg with the optional Smartlink Interface Unit (SIU), the 12 in diameter dual sensor camera features a high-resolution two field of view Indium Antimonide (InSb) staring array Thermal Imager (TI) and a colour CCD Daylight TV camera with × 14 zoom lens. The active gyrostabilisation and vibration isolation enables the Model 12DS to operate with less than 35 micro-radians Line-of-Sight (LOS) jitter. The Model 12DS can also be integrated with aircraft radar, navigation and map displays, GPS and other sensor systems.

WESCAM™ 12DS200

The model 12DS200 is a longer range variant of the 12DS, designed specifically to meet the needs of Unmanned Aerial Vehicles (UAVs), with application on rotary and fixed wing aircraft. With the inclusion of a third Field-of-View (FoV) long range optic, the system provides for long range detection, recognition,

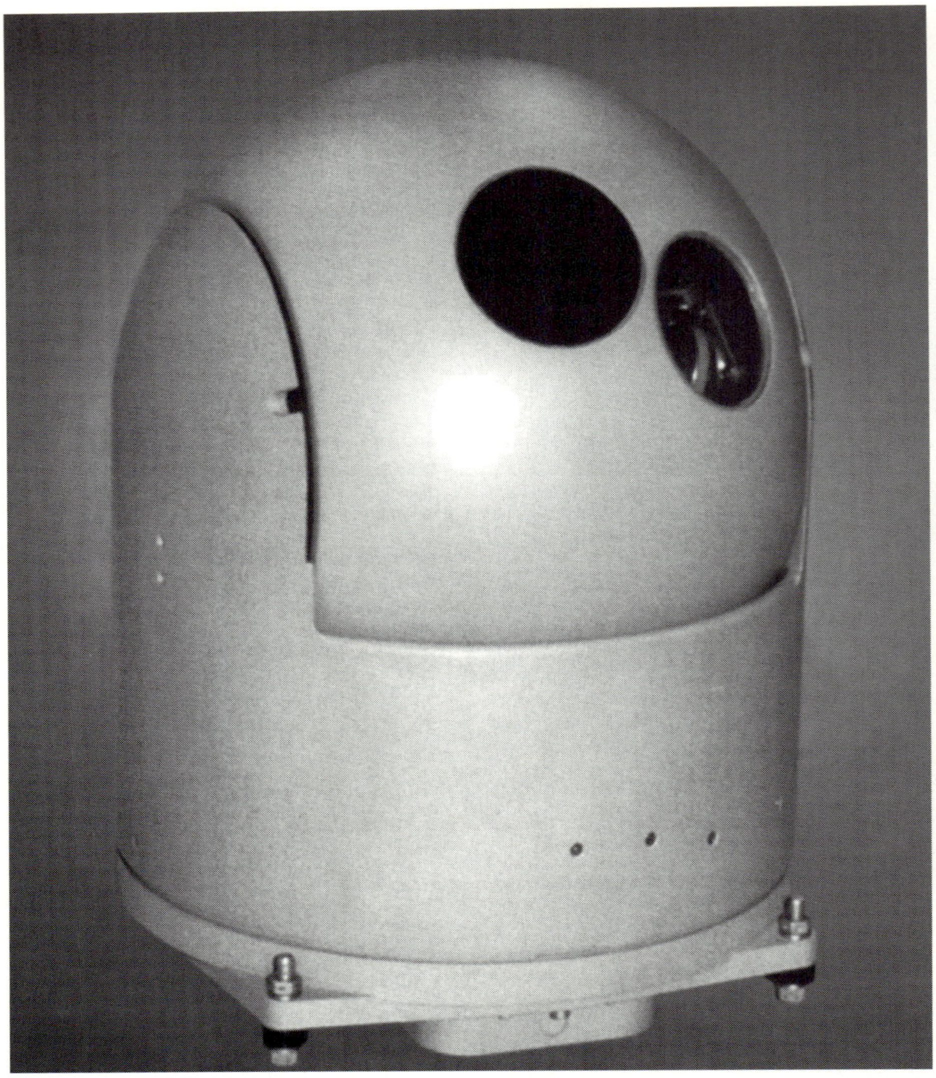

Model 12DS200 **2001**/0099733

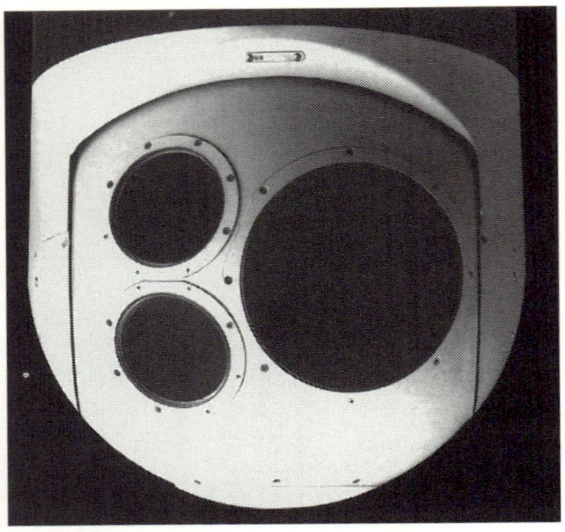

Model 14TS 0002205

AN/AAQ-501 0002206

identification and tracking of vehicles or personnel by day or night and in poor weather conditions.

Specifications

Gimbal

Two axis inner (pitch/yaw) and two axis outer (azimuth/elevation) active gyro stabilisation.

Weight: 18.2 kg

Dimensions: 305 mm (diameter) × 370 mm (height)

Power Supplies: 28 V DC

Azimuth/Elevation Slew Rate: Max >90°/s

Azimuth Range: Continuous 360°

Elevation Range: +90 to −120°

Line of Sight Jitter: <35 microradians RMS

Thermal Imager (TI)

Mid Wavelength Infra Red (MWIR) InSb staring Focal Plane Array (FPA)

Spectral Range: 3-5 μm

Resolution: 256 × 256

Cooling: Stirling Cycle cooler

FoV: 25° (H) × 25° (V) @ 17 mm

7.3° (H) × 7.3° (V) @ 60 mm

2.2° (H) × 2.2° (V) @ 200 mm (12DS200)

Colour Daylight CCD Camera

1-CCD Sony XC-999 in conjunction with 14 × (10 × 12DS200) continuous zoom lens

Format: NTSC or PAL

Resolution: 470 TV Lines/460 TV Lines (PAL)

Focal Lengths: 16 mm to 160 mm

FoV: 23° (H) × 17° (V) @ 16 mm

2.3° (H) × 1.7° (V) @ 160 mm

WESCAM™ 14TS/QS

Extensively deployed on UAVs, the Model 14TS can accommodate three sensors and the Model 14QS four sensors. The range of sensor payloads available for the Model 14 include: a 3-5μm 6 FoV TI, which is available in Platinum Silicide (PtSi) or InSb detector formats; a colour daylight CCD TV camera (DLTV) with 955 mm long range spotter lens; a colour daylight CCD TV camera with 10 × zoom lens, and an 'eye-safe' laser range-finder (LRF).

Specifications

Gimbal

Two axis inner (pitch/yaw) and two axis outer (azimuth/elevation) active gyro stabilisation.

Weight: 34 kg

Dimensions: 360 mm (diameter) × 420 mm (height)

Power Supplies: 28 V DC

Azimuth/Elevation Slew Rate: Max >90°/s

Line-of-Sight (LOS) Range: Continuous 360°

LOS Tilt Range: +90 to −120°

LOS Tilt: TI/DLTV zoom: +10 to −120°

DLTV Spotter/LRF: +30 to −120°

Line of Sight Jitter: <35 microradians RMS

Thermal Imager (TI) - InSb

Mid Wavelength Infra Red (MWIR) InSb staring Focal Plane Array (FPA)

Spectral Range: 3-5 μm

Resolution: 256 × 256

FoV: 1 × , 2 × (optical)

40.9° × 40.9° @ 11 mm

20.1° × 20.1° @ 22 mm

6.3° × 6.3° @ 70 mm

3.1° × 3.1° @ 140 mm

1.6° × 1.6° @ 280 mm

0.8° × 0.8° @ 560 mm

Thermal Imager (TI) - PtSi

Mid Wavelength Infra Red (MWIR) PtSi staring Focal Plane Array (FPA)

Spectral Range: 3-5 μm

Resolution: 512 × 512

FoV: 1 × 2 × (optical)

40.9° × 31.3° @ 19 mm

20.2° × 15.5° @ 38 mm

10.9° × 8.4° @ 70 mm

5.4° × 4.2° @ 140 mm

2.7° × 2.1° @ 280 mm

1.4° × 1.0° @ 560 mm

Colour Daylight CCD Camera with Spotter

1-CCD Sony XC-999

Format: NTSC or PAL

Resolution: 470 TV Lines/460 TV Lines (PAL)

Focal Length: 955 mm

FoV: 0.38° (H) × 0.29° (V)

Colour Daylight CCD Camera with Zoom Lens

1-CCD Sony XC-999 in conjunction with 10 × continuous zoom lens

Focal Lengths: 16 mm to 160 mm

FoV: 23° (H) × 17° (V) @ 16 mm

2.3° (H) × 1.7° (V) @ 160 mm

Eye Safe LRF (14QS only)

Laser: Erbium glass, 1.54 μm wavelength

Pulse Rate: 1 Hz

Range/Accuracy: 49,995 m (claimed)/± 5 m

WESCAM™ 16DS-A/16DS-W/16DS-M

Models 16DS-A (AGEMA) and 16DS-W (Westinghouse - now Northrop Grumman) dual sensor systems provide high-resolution daylight television images together with 8-12 μm thermal imaging. The Model 16DS-A is optimised for environmental inspection, while the Model 16DS-W is configured to meet military specifications, with features such as higher temperature sensitivity for target detection. Model 16DS-M differs in that it incorporates a 3-5 μm PtSi staring array TI. All of the Model 16 series include automatic gain adjustment, remote switching of field of view and electronic zoom (optional on 16DS-W).

Specifications

Gimbal

Two axis inner (pitch/yaw) and two axis outer (azimuth/elevation) active gyro stabilisation.

Weight: 34 kg (16DS-A), 39 kg (16DS-W), 44.1 kg (16DS-M)

Dimensions: 400 mm (diameter) × 510 mm (height)

Power Supplies: 28 V DC

Azimuth/Elevation Slew Rate: 60°/s (typical)

Azimuth Range: Continuous 360°

Elevation Range: +90° to -120° (-180° 16DS-M)

Line of Sight Jitter: <35 microradians RMS

Thermal Imager (TI) - 16DS-A

AGEMA THV 1000 TI, SPRITE 5-bar focal plane

Spectral Range: 8-12 μm

Cooling: Stirling Cycle cooler

FoV: 20° (H) × 13° (V) (Wide)

5.0° (H) × 3.3° (V) (Narrow)

Thermal Imager (TI) - 16DS-W
Northrop Grumman Micro-FLIR TI, SPRITE 8-bar focal plane
Spectral Range: 8-12 μm
Cooling: Stirling Cycle cooler
FoV: 10.0° (H) × 7.5° (V) (Wide)
3.0° (H) × 2.3° (V) (Narrow)

Thermal Imager (TI) - 16DS-M
Mid Wavelength Infra Red (MWIR) PtSi 640 × 480 pixel Focal Plane Array (FPA) TI
Spectral Range: 3-5 μm
Cooling: Stirling Cycle cooler
FoV: 36.9° (H) × 28.6° (V) (Wide)
11.15° (H) × 8.5° (V) (Intermediate)
2.85° (H) × 2.2° (V) (Narrow)
Features: Automatic gain control, polarity switching, electronic zoom (2 × /4 ×), image freeze

Colour Daylight CCD Camera - 16DS-A, 16DS-W
1-CCD Sony XC-999
Format: NTSC or PAL
Resolution: 470 TV Lines/460 TV Lines (PAL)

Colour Daylight CCD Camera - 16DS-M
3-CCD Sony XC-003 in conjunction with 14 × 10.5 Fujinon zoom lens
Format: NTSC or PAL
Resolution: 570 TV Lines (NTSC/PAL)
Lens FoVs: 3.0° (H) × 9.8° (V) @ 21 mm
0.93° (H) × 0.7° (V) @ 294 mm

WESCAM 16SS-A (AN/AAQ-501)
Model 16SS-A is a precision-stabilised infra-red single sensor which detects, identifies and tracks distant objects in total darkness or poor weather. The sensor used is the AGEMA THV1000, 5-bar SPRITE focal plane, 8-12 μm system, providing Fields-of-View (FoVs) as for the Model 16DS-A. Originally 'Commercial-Off-The-Shelf' (COTS), the hardware was designed to meet both military and commercial aviation requirements and has been given the military nomenclature AN/AAQ-501. Major operational roles are surveillance and search and rescue.

WESCAM™ 16SS320/16SS750/16SS725
The Model 16SS320 single-sensor daylight broadcast TV system delivers high-resolution TV broadcast-quality images for electronic newsgathering applications, surveillance and observation, suitable for nosemounting on small fixed and rotary wing aircraft. The sensor is a daylight broadcast 8-320 mm colour camera with a 20 × zoom lens with imaging to 4,000 ft (typically). The Model 16SS750 is also a single-sensor daylight broadcast TV system, suited to long stand-off range applications. The sensor is a three-CCD colour video camera coupled to a 10.5-750 mm lens. This model is available in two configurations, the 16SS725 or 16SS750, with × 33 or × 36 magnification.

WESCAM™ 20TS/QS
The Model 20TS/QS is a triple or quad-sensor (IR/two daylight cameras/optional LRF) surveillance and observation system designed for high performance fixed and rotary wing aircraft, such as the Lockheed P-3 Orion and Eurocopter AS355. TI options include an 8-12m IR imager with 3 FoV optics, or a 3-5m InSb FPA with 4 FoV optics. The system also includes 2 daylight colour cameras, one with a low magnification zoom lens and the other a high magnification 4 FoV spotter scope.

Model 16SS750 0002207

Model 24SS 0002208

Model 16DS-M ***2001***/0099732

WESCAM™ 24SS
The Model 24SS is a single-sensor daylight broadcast TV system designed for broadcast and live news applications, also suitable for long stand-off range maritime patrol and surveillance operations from mid-sized fixed and rotary wing aircraft, such as Bell 206. The sensor is a daylight broadcast 33-1600 mm colour camera with a 24 × zoom lens.

Operational status
All models in production and widely used for broadcast news, maritime surveillance and reconnaissance applications from a variety of fixed and rotary wing aircraft, including UAVs.

Contractor
Wescam Inc.

UPDATED

CHINA, PEOPLE'S REPUBLIC

BGZ-1 combined precision altimeter

The BGZ-1 combined precision altimeter combines the functions of precision altitude measuring, preselected altitude, altitude warning and parallel altitude reporting. It senses the total pressure and static pressure signals independently to implement precision altitude measuring and altitude indicating. It has the functions of preselected altitude and over-altitude warning.

Specifications

Performance data:
Operating range: −1,600 to 89,900 ft
Accuracy: <4%
Preselected altitude:
Operating range: 0-89,900 ft
Resolution: 100 ft
Power supply:
Operating power supply: 27 V DC ±10%, 10 W
Illuminating power supply: 27 V DC, 4 W
Dimensions: 80 × 80 × 220 mm
Weight: 2.3 kg

Contractor

Chengdu Aero-Instrument Corporation.

VERIFIED

BGZ-1 combined precision altimeter
0002356

Colour weather radar

The China Leihua Electronic Technology Research Institute's colour weather radar was designed, with assistance from Rockwell Collins, for civil aircraft. The radar system incorporates a variety of new technologies such as the narrowed waveguide planar slotted array, high-reliability transmitter, miniaturised logic-controlled receiver, and digital colour indicator. The transmitter employs a pulse-modulated magnetron. It also has ground-mapping features.

Specifications

Weather avoidance range: >500 km (for strong thunderstorm areas)
Scan range:
(azimuth) ±60°
(elevation) ±15°
Antenna stabilisation range:
(roll) ±30°
(pitch) ±10°
Weight: 20 kg (no cables)
Antenna beamwidth: <6°

Contractor

China Leihua Electronic Technology Research Institute.

VERIFIED

JDF-2HF SSB transceiver

JDF-2HF is a solid-state automatic tuning SSB airborne radio set designed for fighter/interceptors, bombers, transport aircraft and helicopters. Operating in the 2 to 11.999 MHz range, the transceiver's modes are A3J USB and LSB, A2J USB CW and LSB CW, and A3 AM. The full set consists of a transceiver, an antenna tuner and a preset control box. Power output ranges from 15 to 50 W.

Contractor

China National Electronics Import & Export Corporation.

VERIFIED

Model 265 radar altimeter

The Model 265 is a nanosecond pulse radar altimeter. It can be used on various types of civil or military high-speed jet aircraft, helicopters and unmanned aerial vehicles. It outputs high-precision and continuous true height signals to navigation, autopilot and fire-control systems.

The Model 265 features a large altitude range, high accuracy, high reliability and stability, small volume and the use of the front edge of the tracking pulse to measure the shortest range between the aerial and the ground, without interference. The system incorporates a self-test function and provides warning signals at preset heights.

Specifications

Weight: 3 kg
Altitude: 0-4,920 ft
Accuracy: ±3 ft or 3% of true height
Resolution: ±0.5 ft or 0.1% of height

Operational status

Model 265 radar altimeters have been used on the J7-IIM and J8-II fighters, Y-12 transport aircraft, the J-8 trainer and the CKIE low-altitude pilotless target.

Contractor

China National Electronics Import & Export Corporation.

VERIFIED

CZECH REPUBLIC

SENAP – signal emulator of aeronavigation and landing

The SENAP system was developed as a substitution system to replace the RSBN navigation system. Replacement of RSBN by SENAP releases the 700 to 1,000 MHz band for other civil purposes. The system utilises a GPS receiver, and can integrate VOR/DME, Loran, ILS and other navigation data sources.

SENAP retains the RSBN pilot control, and the original RSBN operational capabilities and accuracy: range 0 to 500 km; azimuth 0 to 360°. SENAP complies with ICAO standards. Operation in either SENAP or RSBN modes can be selected by the pilot.

Specifications

Dimensions/Weight:
Basic unit 319 × 194 × 157 mm/5.8 kg
Adaptor 319 × 194 × 90.5 mm/5.8 kg
Power supply: 27 V DC.

Operational status

A version is available for fitment to Su-22 Mk4 aircraft.

SENAP
0001309

Contractor

Elektrotechnika-Tesla Kolin a.s.

VERIFIED

LUN 3520 LPR 2000 VHF/UHF airborne transceiver system

The LUN 3520 LPR 2000 VHF/UHF airborne transceiver system covers the frequency bands 118.000 to 155.975 MHz and 222.000 to 399.975 MHz, with VHF channel spacing at 8.33 and 25 kHz and UHF channel spacing at 25 kHz UHF.

The system comprises a range of different units, including:

LUN 3520.11-8 transmitter/receiver, including guard receiver capabilities at 121.5 and 243 MHz.

LUN 3520.24-8 control box, designed for remote control from a single position, in conjunction with the audio switch box LUN 3520.44, with the ability to select from 20 pre-set channels.

LUN 3520.40-8 control box, designed for remote control from a single position, with the ability to tune the frequency in use, and to select from 20 pre-set frequencies.

LUN 3520.41-8 control box, designed for remote control from a single position, with the ability to tune the frequency in use.

LUN 3520.42-8 control box, designed for remote control from two positions, in conjunction with audio switch box LUN 3520.52 or LUN 3520.53; it provides for operation of the aircraft intercom system, as well as control of the radio and its tuned frequency.

LUN 3520.43-8 control box, designed for remote control from two positions, it provides control of the aircraft intercom and other radio signals aboard the aircraft, such as the radio compass, but not control of the radio transmitter.

LUN 3520.44-8 control box, designed for remote control from one location; in conjunction with the control box LUN 3520.24 it is possible to tune the frequency in use.

LUN 3520.45-8 control box, designed for remote control from a single position, with the ability to tune the frequency in use.

LUN 3520.47-8 control box, designed for remote control from two positions, it provides for operation of the aircraft intercom and control of other radio signals aboard the aircraft, but not control of the transmitter.

LUN 3520.48-8 control box, designed for remote control from two positions, it provides for operation of the aircraft intercom and control of other radio signals aboard the aircraft, including control of the transmitter.

LUN 3520.49-8 control box, designed for remote control from two positions, it provides for operation of the aircraft intercom and control of other radio signals aboard the aircraft, but not control of the transmitter.

LUN 3520.50-8 audio switch box, designed for audio control of two transmitters and the aircraft identification system.

LUN 3520.50-8 audio switch box, designed for audio control of the three transmitters and four radio navigation systems, including ADF and DME.

LUN 3520.53-8 audio switch box, designed for audio control of two transmitters, the aircraft identification systems and other navigation devices, including TACAN, VOR and ADF.

LUN 3520.60-8 audio frequency unit to interconnect third and fourth users to the aircraft intercom system, and to generate a warning tone at 500 Hz.

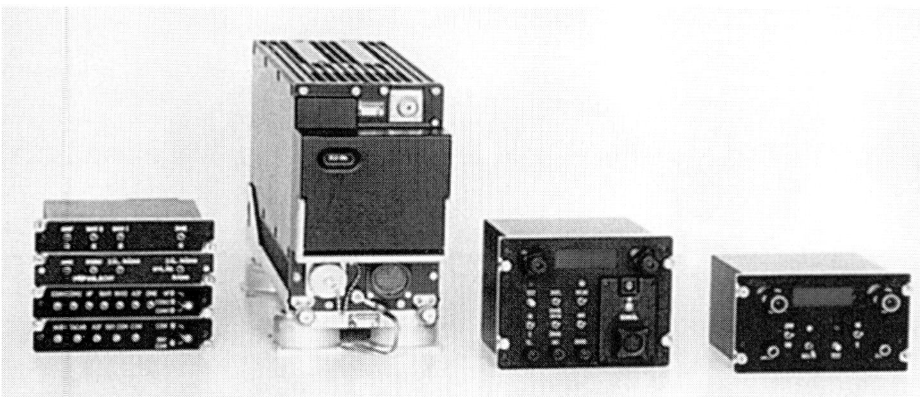

Representative units of the LUN 3520 LPR 2000 system showing, from the left, a stack of audio control boxes, the transmitter/receiver (LUN 3520.11-8), control box (LUN 3520.43-8), and control box (LUN 3520.45-8)
***2000**/0062177*

Specifications

Frequency ranges:
(VHF) 118.000 to 155.975 MHz
(UHF) 220.000 to 399.975 MHz
Channel spacing:
(VHF) 8.33 and 25 kHz
(UHF) 25 kHz
RF power output:
(full) 16 W
(reduced) 6 ±2 W
Power: 27 V DC, 25 W on receive, 220 W on transmit
Temperature range: −60 to + 60°C

Operational status

Certified to International Civil Air traffic Organisation (ICAO) standards in 1998.

Contractor

MESIT pristroje spol, s r. o.

VERIFIED

LUN 3526 VHF airborne transceiver

The LUN 3526 VHF airborne transceiver is designed to meet new European airspace requirements for 8.33 kHz channel spacing; some of the operational parameters can be reprogrammed by the user through a PC.

Specifications

Frequency range: 117.980 to 137.990 MHz , (optional 108.000 to 156.000 MHz)
Channel spacing: 8.33 and 25 kHz
Transmitter power: 10 W
Modulation: A3E
Memory channels: 20 channels
Dimensions: 82.5 × 82.5 × 281.6 mm
Weight: 1.4 kg
Power: 28 V DC

Contractor

MESIT pristroje spol, s r. o.

VERIFIED

Airspeed indicators

Mikrotechna Praha a.s. manufactures a range of 3 in airspeed indicators, including the models: LUN 1106, LUN 1106.XX-8, LUN 1107.XX-8, LUN 1113.XX-8, LUN 1114; LUN 1115, LUN 1116, LUN 1117, and UL 20 series. All are metal pressure-capsule instruments. All series utilise pitot-static pressure. In general, options are available within the following specifications:

Specifications

Range: 0-350 kt, or 0-400 mph, or 0-600 km/h
Lighting: none; 5 V DC; 14 V DC; 27 V DC
Case: ARINC 408 3 ATI; round MS-33638 (AS); round MS-33638 (AS) short

LUN 1114 airspeed indicator 0018214

Dial layout: single- or dual-scale and range marked to customer requirements
Approvals: TSO C2d

Leading details of particular models:
LUN 1114
LUN 1114 is calibrated 0-300 kt, and indicates maximum allowable airspeed; it is TSO-C46a, and RTCA/DO 160C approved.

UL 20 series
The UL 20 series are designed specifically for use in ultralight aircraft. Both 3 and 2 in versions are available.

Contractor

Mikrotechna Praha a.s.

VERIFIED

Altimeters

Mikrotechna Praha a.s. manufactures a range of five altimeters, including the models: LUN 1125, 1127, 1128, 1129 and UL 10 series. Altimeters can be manually adjusted to variances in barometric pressure; temperature is automatically compensated by a bi-metallic element.

Specifications

Range: −1,000 to +20,000 ft; −1,000 to +35,000 ft; −1,000 to +50,000 ft and −1,000 to +80,000 ft (LUN 1125 only)
Lighting: none; 5 V DC, 14 V DC; 27 V DC
Case: ARINC 408 3 ATI; round MS 33549 (AS); ARINC 408 2 ATI (LUN 1129)

UL 10-10 altimeter 0018213

Dial layout: mbar/in Hg; mbar; in Hg
Approvals: TSO C10b

UL 10 series
The UL 10 series is designed for sports and ultralight aircraft. Both two and three in versions are available, with scales calibrated in inches Hg and millibars.

LUN 1125.xxxxxx encoding altimeter
The LUN 1125.xxxxxx barometric counter pointer altimeter has three moving drums, which show tens of thousands, thousands and hundreds of feet. The pointer indicates 1,000 ft per revolution on a scale calibrated at intervals of 20 ft. The counter pointer display is actuated by a high performance mechanism

with a high stability capsule. A built-in vibrator minimises friction and optimises accuracy. All instruments can be customised to meet specific requirements.

Specifications

Operating range:
–1,000 to +35,000 ft, or –1,000 to 50,000 ft, or –1,000 to +80,000 ft
Barometric scale: 950 to 1,050 mbar (28.1 to 31.0 in Hg) and 920 to 1,050 mbar (27.2 to 31.0 in Hg)
Dimensions: ARINC 408 3ATI, round MS33549 (AS)
Dial layout: mbar/in Hg, mbar, in Hg
Output: encoding output according – ICAO code or baro output
Lighting: 5 VDC, 14 VDC or 27 VDC in colour white, NVIS 28 VDC
Qualification: TSO C10b, C88a, RTCA/D0-160C

Contractor

Mikrotechna Praha a.s.

UPDATED

The LUN 1125.xxxxxx encoding altimeter
***1999**/0051656*

Barometric altimeter LUN 1124

The LUN 1124 barometric altimeter indicates the absolute or the relative altitude of the aircraft using barometric pressure. It is equipped with integral lighting.

Specifications

Weight: 1.1 kg
Power supply: 5 V DC
Range: 0-10 km
Scale adjusting range: 790-1,050 mbar

Contractor

Mikrotechna Praha a.s.

VERIFIED

Barometric altimeter LUN 1127.XXXX

The LUN 1127.XXXX barometric altimeter is a three-pointer metal-capsule instrument.

Barometric altimeter LUN 1127.XXXX ***0002357***

Specifications

Weight: 0.5 kg
Operating range: –1,000 to +35,000 ft, or –1,000 to +20,000 ft
Min increment: 20 ft
Barometric scale: 946-1,050 mbar
Qualifications: TSO-C10b, RTCA/DO-160C

Contractor

Mikrotechna Praha a.s.

VERIFIED

Combined airspeed indicator with machmeter LUN 1170.XX-8

The LUN 1170.XX-8 can be used for simultaneous measurement of the indicated and true airspeed, including Mach number.

It can indicate one mach number and up to three IAS values, and can be equipped with white or NVIS integrated lighting.

Specifications

Ranges:
Indicated airspeed: 100-1,200 km/h
True airspeed: 300-1,200 km/h
Mach number: 0.5-0.9
Ceiling: 15 km
Dimensions: ARINC 408 3ATI, square MS, round MS
Weight: 0.9 kg
Power supply: 5 V or 28 V DC

Contractor

Mikrotechna Praha a.s.

VERIFIED

Combined airspeed indicator with machmeter LUN 1170.XX-8 ***1999**/0051657*

Gyroscopic horizon LUN 1241

The LUN 1241 gyroscopic horizon is a direct-indicating instrument, which indicates pitch and roll angle. The artificial horizon is equipped with an arresting device which can be set by a tie rod within the range +10 to –7°. The main display indicates angles between +20 and –20°. Greater values of pitch, up to ±75° are shown in an indication window. Failure of the gyro supply or the arresting functions is indicated by a warning flag. Internal lighting is provided, including NVIS.

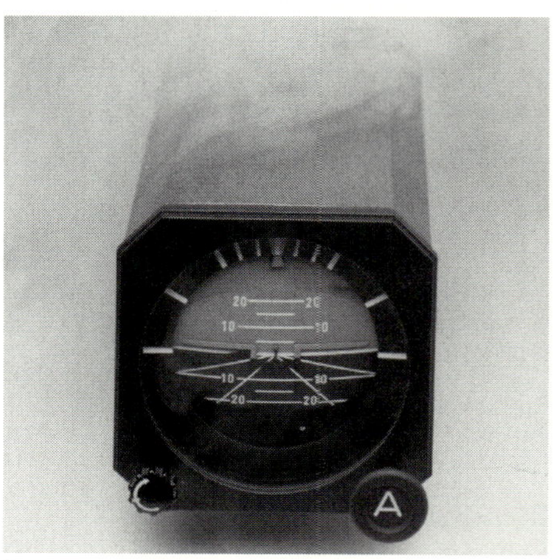

Gyroscopic horizon LUN 1241 ***0001352***

Specifications

Dimensions: ARINC 3 ATI round
Weight: 1.6 kg
Power supply: gyro: 27 V DC, or 14 V DC; lighting: 5 V DC/AC, 14 V DC/AC, or 28 V DC/AC
Range: ±70° pitch; 360° roll
Accuracy: ±0.5°, or ±1.0°
Qualifications: TSO C4c and RTCA/DO-160C

Contractor

Mikrotechna Praha a.s. ***VERIFIED***

Maximum allowable airspeed indicators LUN 1114.XXXX

The LUN 1114.XXXX airspeed indicator is a metal pressure-capsule instrument for indication of actual and maximum allowable airspeed.

Specifications

Range: 0-300 kt
Dimensions: ARINC 3 ATI
Weight: 0.5 kg
Power supply: 5 V or 28 V DC
Qualification: TSO-C46a, RTCA/DO-160C

Contractor

Mikrotechna Praha a.s. ***VERIFIED***

Vertical speed indicators

Mikrotechna Praha a.s. manufactures a number of vertical speed indicators including the LUN 1144, 1149, 1184 and UL 30 series. They are metal-pressure capsule instruments.

Specifications

Ranges: 0-2,000; 0-3,000; 0-4,000; 0-6,000 ft/min
Dimensions: ARINC 3 ATI, or MS 33549 (AS) round; LUN 1184 ARINC 2ATI
Weight: 0.6 kg
Power: 5 or 27 V DC
Approvals: TSO C8b or C8d, RTCA/DO 160C

UL 30 series

The UL 30 series of vertical speed indicators is specifically designed for sport aircraft; both 3 and 2 in variants are available, with scales calibrated in ft/min or m/s.

Contractor

Mikrotechna Praha a.s. ***VERIFIED***

Vertical speed indicator LUN 1184 ***1999**/0051658*

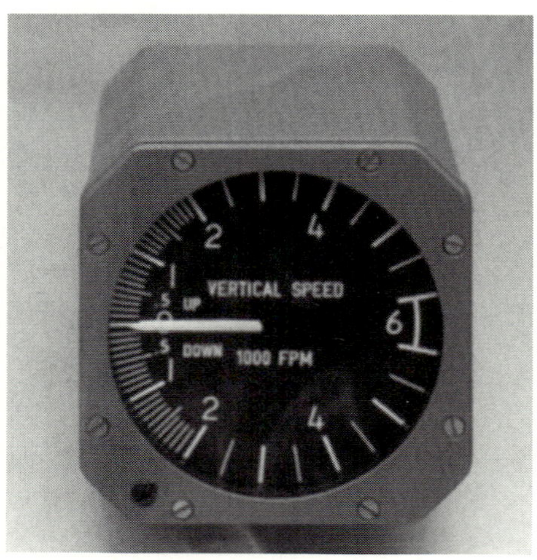

Vertical speed indicator LUN 1149 0018215

DENMARK

SIT/ISIT low-light level television system

JAI's intensified low-light television system is essentially a man-portable low-light level television system designed for use as electronic newsgathering equipment by broadcasting organisations. It has, however, been applied to a number of airborne military, paramilitary and civil roles, particularly aboard helicopters. These include reconnaissance, fisheries patrol and inspection, police night surveillance activities, mountain rescue and night electronic newsgathering. In such applications, the camera may be shoulder-mounted for use from a helicopter cabin footstep, or used with a Ronford 15S mounting head, combined with a helicopter shockmount to counteract the effects of rotor-induced vibration.

The system operates over a wide range of lighting conditions, down to moonlight and starlight illumination levels. For helicopter work, the system is supplied with a 4.5 in (114 mm) viewfinder and a double pan bar for use with the 15S head. A full range of accessories is available and special equipment for outside broadcast applications includes: a video transmitter/receiver operating in the 2 GHz band, a real-time video image processing system and a man-portable videotape recording unit. The company also supplies a range of motorised zoom lens assemblies which meets the special optical requirements of low-light television systems. The manufacturer claims that, with these lenses, low-light television systems can operate over a range of high or low ambient lighting conditions.

Specifications

Weight: variable, according to lens system fitted, but typically approximately 6 kg (including camera, lens system and Ni/Cd belt-mounted battery power supply).

Operational status

In production and service.

Contractor

JAI A/S.

VERIFIED

Airborne pollution surveillance system

The TERMA surveillance system provides integration of a wide variety of different surveillance sensors, navigation equipment, and video systems to monitor oil and other polluting agents at sea; the system can also be used to monitor fishing violations, to map ice and for search and rescue missions.

The system can include any, or all, of the following elements, according to customer requirement:

Side-Looking Airborne Radar (SLAR)

The I-band SLAR has become the primary long-range sensor for oil pollution surveillance, typically covering a 20 n mile swath from preferred search altitudes. An oil slick is detected by the variation in the reflected radar signals between oil-covered water and normal sea-water. In applications like ice mapping and ship surveillance, the SLAR system covers a 40 n mile swath.

Infra-Red/Ultra-Violet (IR/UV) scanner

The IR/UV scanner is provided for close-range imagery and allows a general assessment of the oil slick to be made, as the aircraft passes overhead of the slick first detected by the SLAR.

The infra-red system can be operated both day and night. It provides information on the spreading of oil and indicates the variations of thickness within the slick. The ultra-violet sensor is only used during day-light. It maps the complete area covered with oil, irrespective of thickness.

The TERMA Elektronik airborne pollution surveillance system operator console and observer console (inset)
2000/0062179

MicroWave Radiometer (MWR)

The scanning MWR system is used for oil thickness measurements and quantification, enabling clean-up operations to be optimised. The MWR measures microwave responses from the sea surface at two different wavelengths: in J- and Ka-band.

Video and photo cameras

Video cameras are used to secure evidence of oil pollution, fishery violation and other illegal actions. The scenes taken with the camera can be recorded on a videotape or stored as still photographs in the computer system. Real-time navigation data is integrated into the picture. The video camera can be normal colour, a Low Light Level TeleVision (LLLTV), and/or an infra-red sensor.

For photographic documentation, a hand-held camera with a real-time annotation can be integrated into the system.

Data downlink

Data downlink equipment is used for transmission of real-time or stored data to a ground station or ship. Mission control aboard a ship is often more efficient when images of the actual situation are made by an aircraft flying at high altitudes.

Data storage

The information from microwave and optical sensors can be recorded either on a standard videotape recorder or on a high-resolution digital tape recorder. The digitally stored sensor information can be further analysed on the ground in an image processing system.

Display console

The TERMA concept is to utilise only one common control. Typically, it will comprise a 14 in colour Sensor Image Display (SID), 10 in colour Map Display (MD) and a 10 in colour Control Panel Display (CPD).

The SID is used to display data from the sensor currently in use; stored data can also be displayed. Real-time navigation data can be integrated into the bottom of the SID format. By means of a trackerball, the operator can annotate real-time data.

The MD provides the operator with an outline map of the area under surveillance. The MD is integrated with the video system, so that the operator can overlay map and video data. The map can be zoomed and the operator can annotate data with symbols.

The CPD facilitates the operation of all surveillance sensors, stored data and video systems, using logical menus.

Specifications

SLAR

Ranges: 20 and 40 n miles
Interfaces: 625 PAL Y/C
Navigation interfaces: ARINC, RS 422, RS 232
Power: 28 V DC, 60 W
Antenna: 2 single or 1 double
(horizontal beamwidth) $< 0.6°$
(vertical beamwidth) $19 \pm 3°$
(gain) > 33 dB

(polarisation) HH or VV
(dimensions) 3,658 mm wide
Transceiver:
(frequency) 9,375 ± 30 MHz
(PRF) 0 to 2,000 Hz
(pulse-width) 0.25 or 0.6 microseconds
(peak power) 20 kW ± 1 dB

MWR
Dimensions: 500 (L) × 450 (W) × 740 (H) mm
Weight: 40 kg
Antenna:
(frequencies) 11.2 and 37.5 GHz
(polarisation) horizontal
(aperture) 350 mm
(beamwidth) 5 and 1.5°
(normal search altitude) 800 m
(foot print) 70 × 70 m and 20 × 20 m
(swath width) 1,600 m
(high-resolution altitude) 400 m
(foot print) 35 × 35 m and 10 × 10 m
(swath width) 800 m
(scan method) sinusoidal perpendicular linescan
(aircraft speed) maximum 100 m/s
Receiver:
(bandwidths) 250 MHz and 1 GHz
(tuning range) ±1 GHz at 37.5 GHz
Consoles:
(operator console dimensions) 1,150 (L) × 575 (W) × 1,400 (H) mm
(operator console weight) 145 kg
(observer console dimensions) 1,150 (L) × 775 (W) × 1,190 (H) mm
(observer console weight) 150 kg

Operational status
In widespread use.

Contractor
TERMA Elektronik AS.

VERIFIED

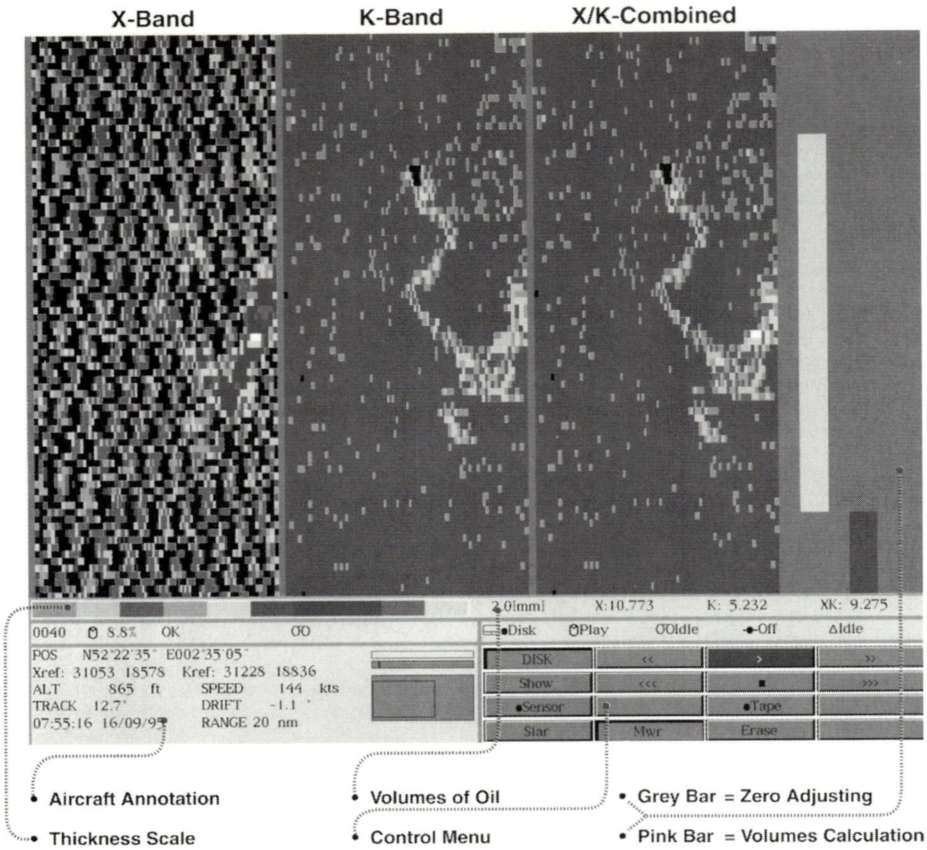

The TERMA Elektronik microwave radiometer image display ***2000***/0062182

MicroWave Radiometer (MWR)

The MWR is designed to measure the thickness and quantity of oil in oil spills in all weather conditions, both during day and night. By using two frequencies, the MWR can quantify oil thickness from approximately 0.1 to 7 mm, thus covering most situations of interest.

The MWR system comprises a processor, which converts the measured brightness temperatures directly into values of thickness, that are displayed as a rolling image on a colour display. By integrating the data over the entire spill, the total volume of oil spill is calculated. All data is stored on a digital tape recorder. The MWR can be integrated with other sensors to provide a complete oil pollution surveillance system (see separate entry).

The system is designed for the widest possible swath width, in practice twice the flight altitude. Search altitude is normally 800 m, but 400 m altitude can be used for high-resolution requirements.

The MWR operates on two frequencies (37.5 and 10.7 GHz). The 37.5 GHz channel ensures fine resolution of thin oil layers. The 10.7 GHz channel measures greater oil thicknesses without ambiguity, thus ensuring correct calculation of oil spill volumes. Since radiometers are sensitive broadband microwave receivers, they are susceptible to jamming and interference from external sources. In consequence, the 37.5 GHz channel is made tunable by 1 GHz, which enables the input channel of the MWR to eliminate sources of interference; the use of two widely different frequencies (Ka- and J-bands) further reduces the possibility of interference or failure.

The signals from the MWR receiver are digitised and converted to brightness temperatures. Together with the position data from the scanning antenna, brightness temperature maps corresponding to the two frequencies are generated and stored. These maps are displayed in real time on the monitor as a rolling colour coded image. The oil thickness is then calculated , and displayed as an oil thickness map. The total volume is similarly calculated and displayed.

Specifications
Frequencies: 10.7 and 37.5 GHz
Polarisation: horizontal
Aperture: 350 mm
Beamwidths: 5 and 1.5°
Normal search altitude: 800 m
(foot print) 70 × 70 m and 20 × 20 m
(swath width) 1,600 m
High-resolution altitude: 400 m
(foot print) 35 × 35 m and 10 × 10 m
(swath width) 800 m
Scan method: sinusoidal perpendicular linescan
Aircraft speed: maximum 100 m/s

The TERMA Elektronik microwave radiometer hardware unit ***2000***/0062181

Operational status
In service.

Contractor
TERMA Elektronik AS.

VERIFIED

Aeronautical CAPSAT

Aeronautical CAPSAT uses British Telecom's Standard Positioning Service (SPS), a worldwide fax, telex and data satellite service. It provides automated Global Positioning System (GPS) position reporting and two-way messaging capability between aircraft and ground controllers located anywhere in the world within Inmarsat coverage.

Aeronautical CAPSAT enables ground controllers to monitor the position of helicopters from a central location. Since helicopters typically operate at low altitudes, the use of satellites provides a more reliable two-way communications link than other methods of transmission.

The system can be programmed to send regular, automatic GPS position reports to a PC connected to the worldwide telephone and data network. The system's data format is compatible with those of the international Future Air Navigation System (FANS) and Automatic Dependence Surveillance (ADS).

The ground controller also has a satellite connection, making the system ideal for operations in areas where the communications infrastructure is unreliable or non-existent. The accuracy of the satellite GPS is integrated into Aeronautical CAPSAT and the system is geared towards safety and is very user-friendly.

Able to report the aircraft's position automatically every 2 minutes, or on direct request by ground operators, no physical aircrew input is needed. In the event of a forced landing, ground operators will have rapid access to accurate position data that is not limited by geographical position.

Fully two-way, the system can operate certain modes in near-realtime and lets aircrew alter flight plans on the basis of the latest meteorological information. They can also make immediate changes to forward maintenance and refuelling plans and report aircraft position, regardless of where it is in the world.

The system comprises a small aerodynamic fin antenna; an internal transceiver and amplifier subsystem; a compact message terminal with a full QWERTY keyboard and integrated LCD (Liquid Crystal Display); and a mini printer unit for receiving hard copies of messages.

Ground operators connect to the telephone or data network via a modem or when in remote regions, to British Telecom's Goonhilly ground earth station using a portable satcoms unit. The CAPSAT Manager PC software is a full featured fleet and solo aircraft tracking program that lets ground operators monitor flight paths on a moving map display and communicate directly with aircrew, over the satellite link, via their onboard system.

Operational status

Supplied to the Royal Air Force, the UK Customs and Excise, and to the UK Meteorolgical Research Flight.

Contractor

Thrane & Thrane A/S.

VERIFIED

TT-3000M Aero-M single-channel aeronautical satcom

The TT-3000M Aero-M satcom is the airborne version of the Inmarsat Mini-M system, of which Thrane & Thrane have supplied over 15,000 worldwide.

TT-3000M Aero-M is a single-channel system, which provides voice, secure voice, fax and data transmission capabilities and can accommodate two handsets, for use from the cockpit or passenger cabin.

The TT-3000M system comprises the TT-3068A Inmarsat Aero-M satcom receiver and the TT-5006A antenna (see separate entries for the TT-3068A and the TT-5000.

Contractor

Thrane & Thrane A/S.

VERIFIED

TT-3024A Inmarsat-C aeronautical capsat

The compact TT-3024A aeronautical Inmarsat-C/GPS transceiver is designed for automatic data reporting and message transfer of position reports, performance data and operational mesages on a global basis, from sea level to 55,000 ft and from 70° north to 70° south.

The TT-3024A operates through the established network of Inmarsat and GPS satellites with interconnection to the international telex, fax and packet switched data networks, offering fast and reliable transfer of information, 24 hours per day.

The integrated GPS receiver calculates position, altitude, speed and heading every second, used for automatic Doppler compensation and transfer of position status reports to any predefined air traffic control or other receiver at specified intervals.

Data collecting equipment may be connected to the TT-3024A via an RS-422/423 port, enabling all selected data to be transferred as data reporting packages every 2 minutes or whenever a special event occurs.

Operational messages, weather and flight plan information as well as passenger messages may be transferred to/from any telex or data subscriber via the international networks.

Specifications

General specifications: the TT-3024A receiver meets or exceeds all Inmarsat specifications for the Inmarsat-C aero system and all relevant GPS specifications
Antenna: integrated Inmarsat-C/GPS omnidirectional antenna, RHC polarised
Figure-of-merit (G/T): –23 dB/K at 5° elevation
EIRP: 12 dBW min at 5° elevation
Transmit frequency: 1,626.5-1,646.5 MHz
Receive frequency: Inm-C 1,530.0-1,545.0 MHz, GPS 1,575.42 MHz
Channel spacing: 5 kHz
Modulation: 1,200 symbols/s BPSK
Data rate: 600 bits/s
TX message channel: TDMA and FDMA, interleaved code symbol
Position reporting: built-in 5-channel GPS receiver with Lat/Long/alt/speed/track calculation, update rate 1 s
Position accuracy: C/A code with 96 m spherical error probability
Initial stabilisation: 15 min at max Doppler shift correction and GPS almanac update
Solid-state storage: 256 kbyte RAM memory
Onboard message/data interface: RS-422/423, 110-9,600 bps and Centronics parallel
Navigational interface: RS-422/423 V.10 Special for interface to onboard navigational systems
Roll and pitch: min ±25° from level flight
Altitude: MSL-55,000 ft
Airspeed: full Doppler compensation to min 620 kt
Power: floating 10.5-32 V DC, 9.5 W Rx, 80 W Tx
Dimensions:
(antenna) 114 × 274 × 98 mm
(LNA/HPA) 161 × 213.9 × 49.5 mm
(electronics assembly) ¼ ATR short
Weights:
(antenna) 0.75 kg
(LNA/HPA) 2.3 kg
(electronics assembly) 2.5 kg

Operational status

Full US FAA and Inmarsat approval for aeronautical use.

Contractor

Thrane & Thrane A/S.

VERIFIED

TT-3068A Inmarsat Aero-M system

The TT-3068A Aero-M, together with the TT-5006A antenna (see TT-5000 series Inmarsat Aero-I system entry), are designated the TT-3000M Aero-M system. This system provides true independent and worldwide global interconnection telephone call, fax and data access, with Inmarsat spot-beam operation.

The Aero-M is a single-channel satcom that is operated from the handset or from the two-wire DTMF phone interface.

Integration of the Navigational Reference System (NRS) makes the Aero-M totally independent of aircraft systems.

Specifications

EIRP: 14 dBW
Coverage volume: >85%
Operating frequencies:
receive: 1525.0 to 1559.0 MHz
transmit: 1626.5 to 1660.5 MHz
Channel spacing: 1.25 kHz
Voice: 4.8 kbps AMBE (3.6 kbps voice, 1.2 kbps FEC)
Asynchronous data rate: max 2.4 kbps
Phone 1 interface: RS-485, 4-wire handset
Phone 2 interface: 2-wire 600 Ω CCITT Rec. G. 473, standard DTMF telephone
Fax interface: 2-wire 600 Ω CCITT Rec. G. 473, T.30 Group-III Fax, 2.4 kbps, RJ-11 jack
Data interface: Serial EIA standard RS-423, Hayes compatible, max 19.2 kbps
Printer interface: Serial EIA standard RS-423 max 19.2 kbps
Power: 10.5-32 V DC
Roll and pitch: minimum ±25° from level flight
Altitude: MSL-55,000 ft
Airspeed: Full Doppler compensation to minimum 620 kt groundspeed
Airspeed acceleration: minimum ±1 *g*
TT-5006A Aero-I satcom antenna:
(dimensions): 560 × 149 × 129 mm
(weight): 0.68 kg

Contractor

Thrane & Thrane A/S.

VERIFIED

TT-3608F Aero-C capsat printer unit

The TT-3608F printer unit for the Aero-C capsat transceiver TT-3024A, is designed as a compact and ruggedised printer for use in aeronautical installations.

The TT-3608F is connected to the capsat transceiver through a single two-wire RS-485 cable which may be extended up to 300 m. The TT-3608F may be used both as the general message printer for the capsat system as well as a local printer for the terminal/PC connected to the transceiver.

The TT-3608F includes a receive mail indication and LOGIN status, enabling 'non-visible' installation of the capsat transceiver electronics.

An RS-232/422 front panel connector allows direct interface to any PC compatible or dedicated terminal for message preparation and status/system set-up.

The TT-3608F may be located at the rear of the centre console in the aircraft cockpit, enabling easy access and in-flight message reception.

Up to eight TT-3608F printers may be connected to the Aero-C transceiver for distributed printing.

Specifications

Printer: 112 mm 9 × 7 dot thermal printer with 2 motors, 40/80 char per line, 1 line/s, selectable printing orientation
Indication: LED indicators for receive mail, system LOGIN and low paper
Capsat interface: two-wire RS-485 interface
Terminal/PC interface: RS-232/RS-422 serial interface from front panel connector
Power: 10.5-32 V floating DC, 2.5 W standby, 7 W printing
Altitude: MSL-55,000 ft
Dimensions: 117.5 × 213.9 × 57.5 mm
Weight: 1.3 kg

Contractor

Thrane & Thrane A/S.

VERIFIED

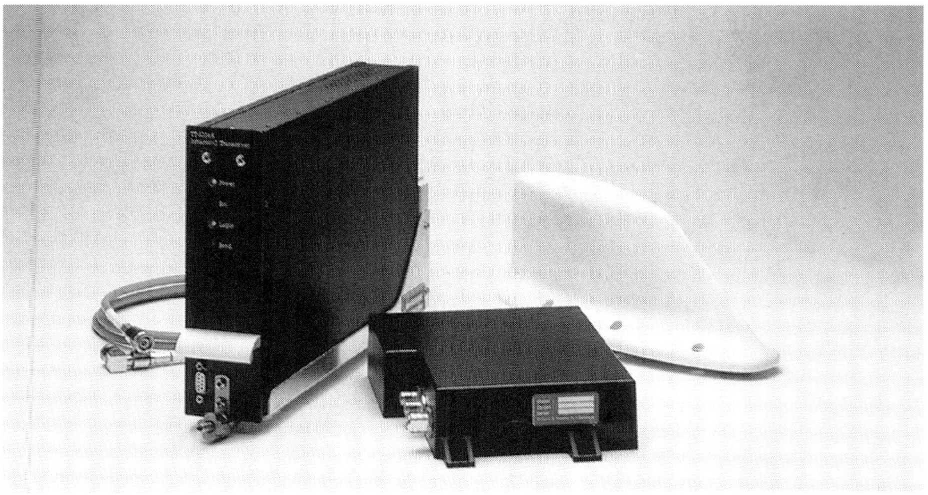

TT-3024A Inmarsat-C aeronautical capsat 0022245

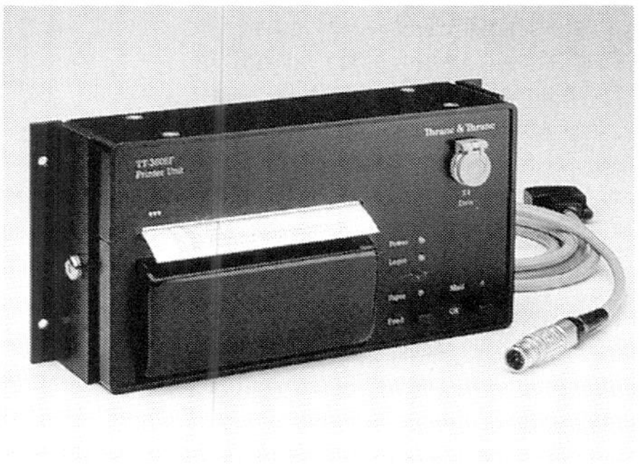

TT-3608F Aero-C capsat printer unit 0022246

TT-5000 series Inmarsat Aero-I system

The Thrane & Thrane compact, lightweight, low-power TT-5000 series Aero-I system functions as a high-quality multiple-channel communication centre, for telephone calls, fax prints, data transfers or e-mail messages to any destination.

The Aero-I system is designed to be an integrated part of the CNS/ATM system, offering three simultaneous channels (two voice/fax or data channels and one packet data channel). The built-in Cabin Telephone Unit (CTU) provides up to four handsets and two fax/phone/modem ports.

Integration of an optional Navigational Reference System (NRS) utilising 12-channel GPS data and 3-D flux gate technology, makes the TT-5000 series totally independent of aircraft systems. However, aircraft navigational systems can also be connected to create an integrated aircraft system.

Specifications

General specifications: the TT-5000 system meets or exceeds current and proposed Inmarsat specifications for the Inmarsat Aero-I system
Figure-of-merit (G/T): –19 dB/K min
EIRP: >16.5 dBW total
Coverage volume: >85%
Operating frequencies:
(receive) 1,530.0-1,559.0 MHz
(transmit) 1,626.5-1,660.5 MHz
(GPS) 1,575.42 MHz
Channel spacing: 2.5 kHz
Antenna system: Jet Sat-97 mechanically steered antenna, or CAL ANT-30 electronically steered antenna
Navigational interfaces: stand-alone with NRS (option) or ARINC 429 IRS bus
Altitude: MSL to 55,000 ft
Airspeed: full antenna tracking and Doppler compensation to min Mach 1.0

TT-5033A SDU
Channels: 3 (2 voice/fax/modem data, 1 packet data). Voice 4.8 kbytes/s, fax and modem data 2.4 kbytes/s, packet data 0.6/1.2 kbytes/s
Dimensions: 2 MCU ARINC 600
Weight: 3.6 kg
Power: +28 V DC, 20 W

TT-5010A HPA
Dimensions: 260 × 243 × 74 mm
Weight: 4 kg
Power: +28 V DC, 25-100 W
Output power: 18 W linear

TT-5012A DLNA
Dimensions: 254 × 193 × 50 mm
Weight: 2.2 kg

TT-5620A handset
Display: 2 × 12 character LCD
Key pad: 21 keys

TT-5000 series Inmarsat Aero-I system 0018940

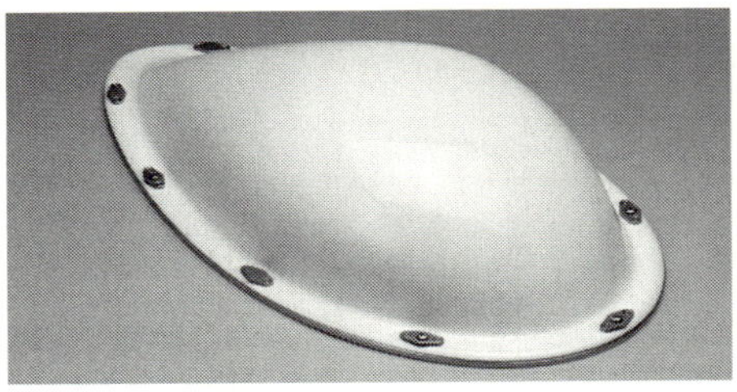

TT-5002A/B Aero-I satcom antenna **1999**/0044781

Interface: 4 wire and RS-485
Power: +12 V DC, 0.2 A from SDU

TT-5622A handset
WH-10 AlliedSignal

Antenna options:

TT-5002A/B Aero-I satcom antenna
Dimensions: 600 × 336 × 127 mm (TT-5002A)
222 diameter × 119.5 mm (TT-5002B)
Weight: 4.8 kg (TT-5002A)
3.8 kg (TT-5002B)
Steering: mechanically steered, 2-axis

TT-5004A Aero-I satcom antenna
Dimensions: 886 × 117 × 86 mm
Weight: 2.6 kg
Steering: electronically steered phased array

TT-5006A Aero-I satcom antenna
Dimensions: 560 × 149 × 124 mm
Weight: 1.9 kg
Steering: mechanically steered, 2-axis

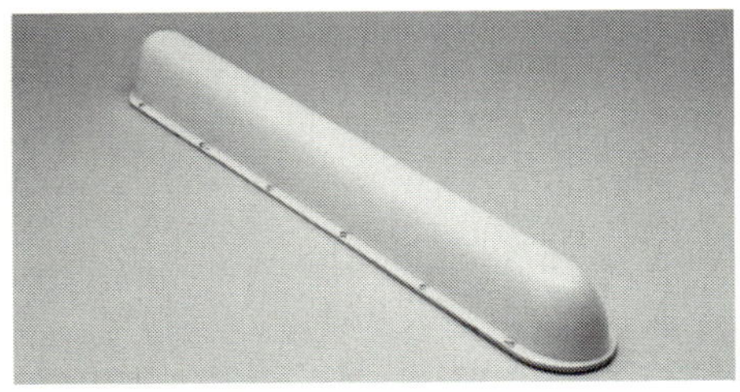

TT-5004A Aero-I satcom antenna **1999**/0044782

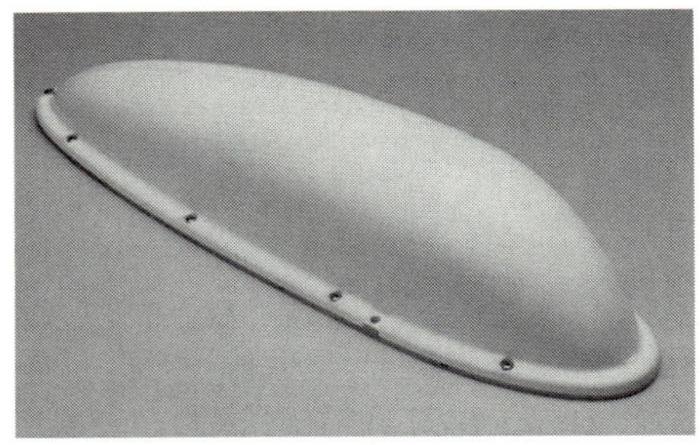

TT-5006A Aero-I satcom antenna **1999**/0044783

TT-5006 OmniPless antenna
TT-5008A NRS antenna
Dimensions: 38 × 243 × 129 mm
Weight: 0.68 kg

Contractor

Thrane & Thrane A/S.

VERIFIED

FRANCE

COSPAS-SARSAT emergency locator transmitters - A06 range

The A06 range of COSPAS-SARSAT emergency locator transmitters includes the following types:

(1) For ED-62 requirements; A06-A06V1 models;
(2) For TSO requirements; A06T model;
(3) For TSO and ED-62 requirements; A06V2-606 models.

The A06 range is packaged in an emergency distress orange container. The transmitter includes a three-postion switch; red high-intensity light and buzzer; antenna connector; three-frequency antenna; remote control (whip, or aircraft skin antenna). Activation is automatic by G-switch.

Transmissions are on 406 MHz COSPAS-SARSAT frequency at 5 W power, and on the 121.5 and 243 MHz international distress frequencies, at 100 mW power.

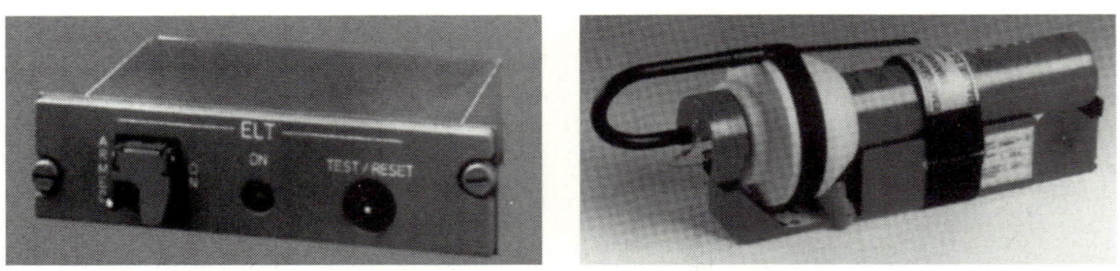

COSPAS-SARSAT emergency locator transmitters - A06 range 0015341

A remote-control unit option is available for Airbus cockpit fit and as a custom option for other aircraft types. Transmitter weight is 1.3 kg.

An updated variant has been developed for Lufthansa that provides direct identification of the aircraft; aircraft identification data is entered via a new programming module developed for the purpose.

Contractor

CEIS TM - LCD Division.

VERIFIED

51RV-4/5DF VOR/ILS receivers

The 51RV-4DF VOR/ILS receiver forms one part of a flight inspection system, providing test signals and measurement data for flight inspection of VOR localiser and glide slope installations. It is a direct retrofit for ARINC 547 VOR/ILS receivers.

All parameters that need to be accessed are available on the front panel and no wiring harness modification is required. The 51RV-4DF features digital bearing output filtering of the basic bearing information, remote control by two out of five binary control, remote control by serial information in ARINC 429 binary broadcast form, dual conversion of 200 VOR/Loc channels and 40 glide slope channels, dual-instrumentation circuits with internal comparison and level monitoring and BITE.

The 51RV-5DF is identical to the 51RV-4DF except that it meets ICAO FM immunity requirement.

Operational status

In production for numerous airport and military calibration authorities worldwide.

Contractor

Rockwell-Collins France, Blagnac.

VERIFIED

The Rockwell Collins France 51RV-4/5 DF VOR/ILS receiver 0015342

DF-301E direction-finder

The DF-301E (military designation OA-8697 or OA-8397A) is a solid-state direction-finder which utilises digital electronic circuitry to achieve improvements in bearing accuracy, acquisition speed and stability. Operating in the UHF and/or VHF frequency range of 100 to 400 MHz, automatic direction-finding capability is provided within one unit, giving cost, space and weight savings.

Used in conjunction with associated receiver and bearing indicator, the electronically commutated antenna provides relative bearing information to the UHF/VHF signal source transmitter. In the airborne environment, the DF-301E is used for course navigation, or to determine the relative direction of another transmitting aircraft. The unit meets MIL-E-5400 Class 2 requirements for environmental conditions. The unit weighs 3.4 kg and measures 88 × 134 mm.

The Rockwell-Collins France DF-301 EF direction finder 0015343

Operational status

In production. Over 8,000 units are in operation.

Contractor

Rockwell-Collins France, Blagnac.

VERIFIED

ETC-40X0F centralised control system

The ETC-40X0F provides frequency and mode control of any radio communication and radio navigation equipment in the aircraft. It is a fully modular architecture system allowing easy adaptation to the avionics configuration of the aircraft. Main features are tandem seat operation, compatibility with any type of radio com/nav equipment, flexibility, modularity, colour display, easy installation and BITE.

ETC-40X0F basic configuration is composed of two EDU-40X0F control and display units and one MPU-40X0F bus concentrator.

ETC-40X0F can control and display ARINC 429 as well as any other type of equipment. ARINC 429-equipped systems are directly linked to the EDU-40X0F, and non-ARINC 429 radios are linked through dedicated interfaced boards, within MPU-40X0F.

The ETC-40X0F allows the operator a global display of all com/nav equipment in the aircraft. For each piece of equipment, essential information is displayed throughout three levels of pages. Top level pages display active and preset frequencies. Technical parameters, such as mode, channel and test, are available on two additional pages. The system provides simultaneous access to all avionics equipment through each of the two EDU-40X0F units.

The MPU-40X0F unit houses the ARINC 429 bus concentrator and interface boards for non-ARINC 429 equipment. ETC-40X0F is also offered with colour LCD technology and embedded GPS.

Specifications

Dimensions:
(EDU-40X0F) 146 × 242 × 105 mm
(MPU-40X0F) 322 × 124 × 193 mm
Weight:
(EDU-40X0F) 3.6 kg
(MPU-40X0F) 4.5 kg
Power supply: 28 V DC, 4 A max
Radio interfaces: 6 ARINC 429, 12 non- ARINC 429
Programmable channels: 16 per radio
Altitude: up to 55,000 ft
NVG compatible

Operational status

In production for Eurocopter Cougar Mk 1, Mk 2, Fennec, Raytheon Hawker 800 XP and Lockheed Martin C-130 aircraft.

Contractor

Rockwell-Collins France.

VERIFIED

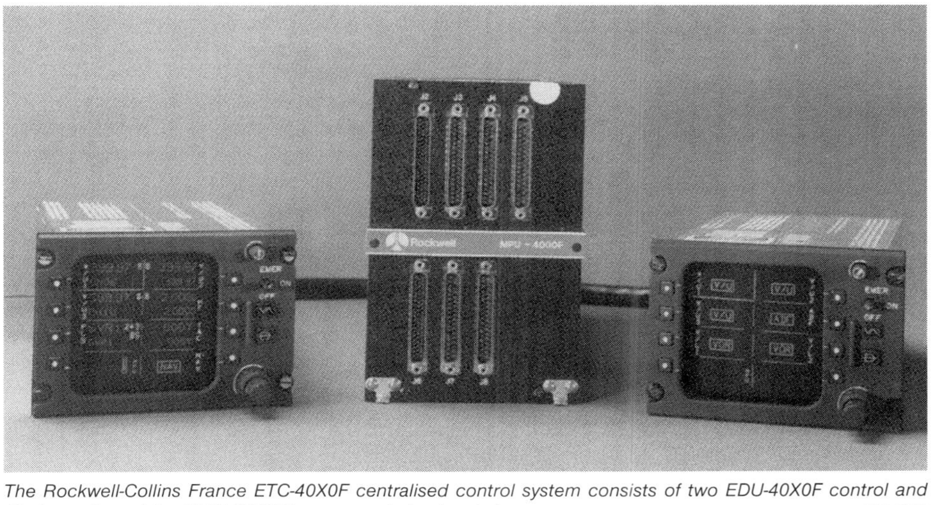

The Rockwell-Collins France ETC-40X0F centralised control system consists of two EDU-40X0F control and display units and the MPU-40X0F bus concentrator (centre) 0011849

VIP communications suite

The VIP communications suite provides voice, computer data, telex (TTY) and fax air-to-ground communications over different media such as satellite, HF or UHF radio.

VIP message handling facilities allow users to edit, store, retrieve and modify messages, and encrypt or decrypt them off line. The system provides an automatic store-and-forward capability including logging, routeing and relaying. It supports a variety of message types and formats. Multiple compression algorithms speed up message transmission, a useful capability when operating in poor propagation conditions. Routeing is on the basis of criteria such as urgency, medium availability, cost and priority.

The VIP communications suite integrates the Rockwell MDM-2501, multiwaveform adaptive high-speed modem. This offers a dozen different standard waveforms, including the MIL-STD-188-110A single-tone waveform for speeds up to 2,400 bits/s with forward error correction on 300 to 3,000 Hz HF channels. This waveform uses automatic multipath compensation.

The system uses the FED-STD-1052 multimedia link protocol. It can operate over half- or full-duplex channels. The protocol supports automatic adaptivity to optimise transmission performance in terms of throughput and bit error rate.

The VIP communications suite also features MIL-STD-188-141A and FED-STD-1045A ALE.

The system has different levels of automation. In a fully automatic configuration the VIP communications suite will direct dial, choosing the appropriate media and performing SELCALL and ALE functions.

The VIP communications suite is integrated into a high-impact magnesium alloy case and is based on the 486 Intel processor with up to 500 Mbytes of hard disk. A TEMPEST version is available.

Contractor

Rockwell-Collins France.

VERIFIED

ELT 90 series Emergency Locator Transmitters

Satori makes three ELTs:

ELT 90-ELT 92 a bi-frequency (121.5/243 MHz) automatic fixed/portable equipment conforming with EUROCAE ED-62/RTCA DO-182

ELT 96-406 a tri-frequency (121.5/243/406 MHz) automatic fixed/portable equipment conforming with EUROCAE ED-62/RTCA DO-182

ELT 96S-406 a tri-frequency (121.5/243/406 MHz) automatic fixed/portable equipment conforming with EUROCAE ED-62/RTCA DO-204.

Specifications

Dimensions: 83 × 103 × 215 mm
Weight: 1.52 kg
Output power: ELT 90-ELT 92: 100 mW PERP 48 h
ELT 96 (both models): 5 W PERP 48 h.

Contractor

Satori.

VERIFIED

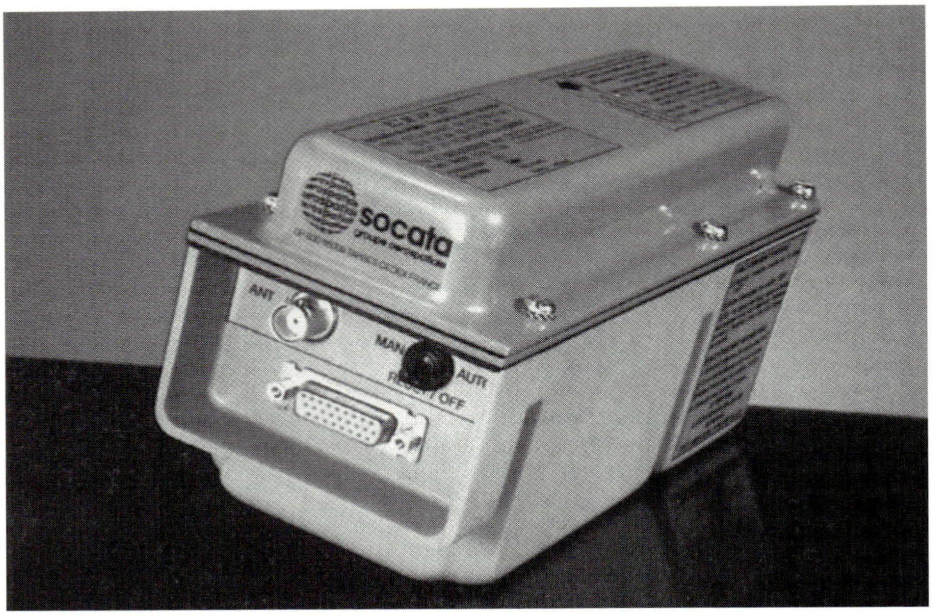

ELT 96-406 emergency locater transmitter ***0001194***

KANNAD 406/121 series automatic Emergency Locator Transmitters (ELTs)

SERPE-IESM (Société d/Etudes et de Réalisation de Protection Electronique-Informatique Electronique Sécurité Maritime) manufactures a range of ELTs for land, sea and air application. The KANNAD 406 series all provide services covering the three international distress frequencies: 406.025/121.5/243 MHz. SERPEIESM also produces the KANNAD 121 AF a two-frequency (121.5/243 MHz) unit for light aircraft applications.

General aviation ELTs

KANNAD 406 AF

The KANNAD 406 AF automatic fixed Cospas-Sarsat ELT is designed to be installed near the tail of the aircraft, and to be connected to an outside antenna; it is a three-frequency (406.025/121.5/243 MHz) unit; overall dimensions: 181 × 107 × 93 mm; weight: 1.18 kg; six-year battery replacement life. A remote-control panel (on option) located in the cockpit allows manual activation and self-test of operational parameters. A buzzer mounted in the ELT warns the pilot should an activation occur. Programming can be done automatically via the ELT front panel, to define nationality and registration marking, aircraft designator and ELT serial number up to 4096, aircraft ICAO 24 bit address, aircraft serial number. Activation can be initiated by: automatic G-switch, manual operation, or remotely (from the cockpit). The unit is certified to EUROCAE ED62 and FAA TSO-C91a/TSO-C126.

KANNAD 406 AF-H

The KANNAD 406 AF-H automatic fixed Cospas-Sarsat ELT is designed for flat installation onboard helicopters (rather than the normal 45° installation). It is a three-frequency (406.025/121.5/243 MHz) unit; overall dimensions: 181 × 107 × 93 mm; weight: 1.19 kg; six-year battery replacement life. A remote-control panel (on option) located in the cockpit allows manual activation and self-test of operational parameters. A buzzer mounted in the ELT warns the pilot should an activation occur. Programming can be done automatically via the ELT front panel, to define nationality and registration marking, aircraft designation and ELT serial number up to 4096, aircraft ICAO 24 bit address, aircraft serial number. Activation can be initiated by: automatic G-switch, manual operation, or remotely (from the cockpit). The unit is certified to EUROCAE ED62 and FAA TSO-C91a/TSO-C126.

KANNAD 121 AF

The KANNAD 121 AF automatic fixed ELT is a two-frequency (121.5/243 MHz) unit designed for light aircraft, for installation near the tail, connected to an external antenna; it has a Morse capability; overall dimensions: 181 × 107 × 93 mm; weight: 1.18 kg; six year battery replacement life. A remote-control panel (on option) located in the cockpit allows manual activation and self-test of operational parameters. A buzzer mounted in the ELT warns the pilot should an activation occur. Programming can be done automatically via the ELT front panel, to define aircraft tail number in Morse code. Activation can be initiated by: automatic G-switch, manual operation, or remotely (from the cockpit). The unit is certified to EUROCAE ED62 and FAA TSO-91a/TSO-C126 capability.

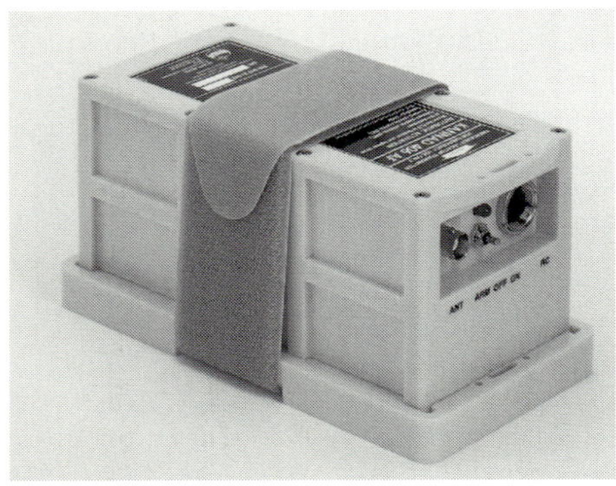

KANNAD 406 AF ELT ***1999***/0051582

Air transport aviation ELTs

KANNAD 406 AP

KANNAD 406 AP automatic portable Cospas-Sarsat ELT is designed to be installed near the tail of the aircraft and to be connected to an external antenna, when removed and connected to the auxiliary antenna, the KANNAD 406 AP becomes a survival beacon. It is a three-frequency (406.025/121.5/243 MHz) unit; overall dimensions: 285 × 107 × 93 mm; weight: 1.290 kg; six-year battery replacement life. A GPS or ARINC 429 interface can be added to load the position of the aircraft in the ELT. In case of activation, the position is transmitted to the Cospas-Sarsat LEO and GEO satellites, together with aircraft identification. A remote-control panel (on option) located in the cockpit allows manual activation and self-test of operational parameters. A buzzer mounted in the ELT warns the pilot should an activation occur. Programming can be done automatically via the ELT front panel, to define nationality and registration marking, aircraft designator and ELT serial number up to 4096, aircraft ICAO 24 bit address, aircraft serial number, activation can be initiated by: automatic G-switch, manual operation, or remotely (from the cockpit). Certified to EUROCAE ED62 and FAA TSO-C91a/TSO-C126 capability; it also complies with JAR-OPS 1.820 regulation.

KANNAD 406 ATP

The KANNAD 406 ATP automatic Cospas-Sarsat ELT is designed to be installed near the tail of the aircraft and

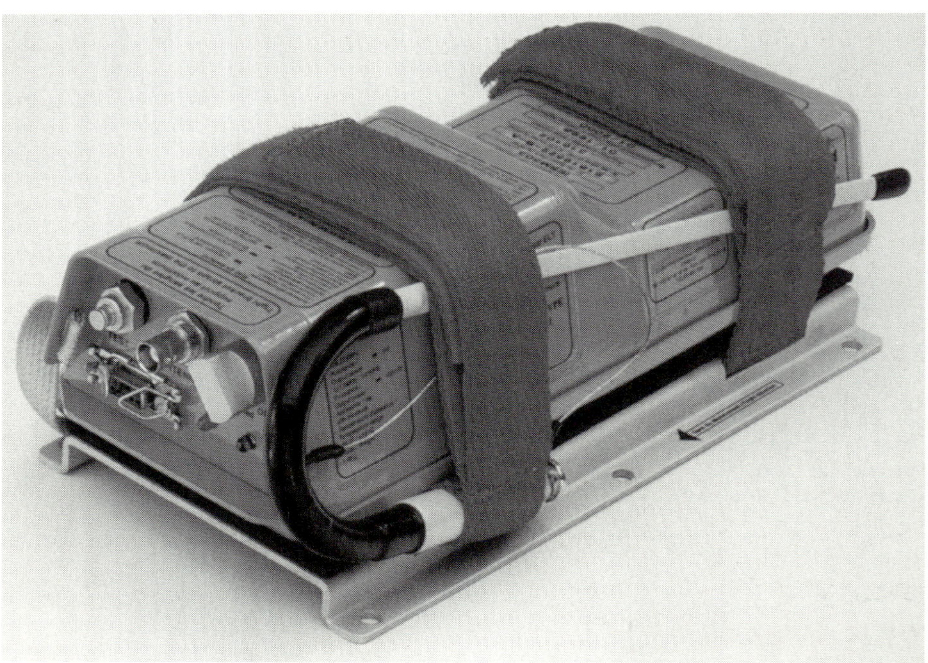

KANNAD 406 ATP ELT ***1999***/0051583

to be connected to an external antenna, when removed and connected to the auxiliary antenna, the KANNAD 406 AP becomes a survival beacon. It is a three-frequency (406.025/121.5/243 MHz) unit; overall dimensions: 300 × 150 × 95 mm; weight: 2.155 kg; six-year battery replacement life. In case of activation, the position is transmitted to the Cospas-Sarsat LEO and GEO satellites, together with aircraft identification. A remote-control panel (on option) located in the cockpit allows manual activation and self-test of operational parameters. A buzzer mounted in the ELT warns the pilot should an activation occur. Programming can be done automatically via the ELT front panel, to define nationality and registration marking, aircraft designator and ELT serial number up to 4096, aircraft ICAO 24 bit address, aircraft serial number. Activation can be initiated by: automatic G-switch, manual operation, or remotely (from the cockpit). The unit is certified to EUROCAE ED62 and FAA TSO-C91a/TSO-C126 capability.

Remote-control units

SERPE-IESM offers two remote-control units for cockpit control of the KANNAD 406 ELT units: RC 200 for fitment in light aircraft and general aviation types; RC 300 for use in air transport aircraft.

Antennas

Rod and blade antennas manufactured by Rayan (Chelton Group) and a blade antenna manufactured by Sensor Systems are offered with the KANNAD 406 ELTs; rod antennas are used for low speed operation, blade antennas for high speed operation.

Contractor

SERPE-IESM.

VERIFIED

AFDS 95-1 and -2 Automatic Flight Director System

The AFDS 95-1 and -2, designed for VFR and IFR operation with light and medium helicopters, are modular systems consisting of a basic two- to four-axis autopilot computer, flight director computer and associated equipment. In the event of a flight director computer failure, the system is reconfigured in its basic mode, which allows the pilot to complete the mission.

By using inputs from sensors, the autopilot computer provides references for automatic control relative to attitude and heading. The basic autopilot functions are cyclic and yaw damping, long-term pitch and roll attitude hold and long-term heading hold in cruise and hover. The fourth axis facility is intended for operators needing transition and hover capability. The autopilot computer drives the electromechanical actuators and, the trim and artificial feel actuators. The digital, panel-mounted flight director interfaces with the navigation equipment and permits the selection of operating modes, allowing en route and approach applications, search and rescue, or anti-submarine warfare operations. It also provides co-ordination of the collective, yaw and cyclic coupling and longitudinal and lateral speed control functions when using Doppler signal data.

The AFDS 95-1 and -2 flight control systems are transparent, therefore the pilot's control inputs are detected and the attitude hold terms removed to keep the autopilot from resisting pilot control in manoeuvring flight The AFDS 95-1 system is designed for analogue operation, while the AFDS 95-2 is for the digital environment. The AFDS 95-1 system is FAA certified in VFR and single-pilot IFR configurations.

Operational status

The AFDS 95-1 system is in service on the B 222, AB 212 and A 109 K2 helicopter and in production. Bell Helicopters has selected the AFDS 95-2 system for its Model 427 light twin-turbine helicopter operating in the single pilot IFR configuration. The AFDS 95-2 has also been selected for the Agusta 109 M.

Contractor

Sextant.

VERIFIED

The Sextant AFDS 95-2 automatic flight display system ***2000***/0079260

Airbus new glass cockpit

Airbus Industrie, through Aerospatiale, has selected Sextant to supply the new glass cockpit for Airbus Industrie's A318/A319/A320/A321 and A330/A340 families.

This new system is based on six large new-generation active matrix LCD panels. Offered as a forward fit or retrofit option on all A318/A319/A320/A321 and A330/A340 aircraft, the new system will be standard equipment on the new A340-500/600. It is common to these two aircraft families, ensuring a high degree of operational flexibility for all airlines.

The new system is slated for certification in mid-2000, with the first delivery following shortly afterwards for a US Airways A320.

It also features an open design, which means that airlines can adapt it to changes in the air traffic environment (CNS/ATM), as well as video requirements (applicable to A340-600 only).

Contractor

Sextant.

VERIFIED

ATC-TCAS II control panel

The ATC-TCAS II control panel combines the independent control of two mode 'S' transponders with the LCD display of ATC code, and the selection of the TCAS II modes. It equips the A319/320/321 and A330/340 aircraft families. A version, which is certified on A300-B4/300-600/310 including TCAS test activation, is available for general cases of TCAS II retrofit installations.

Operational status

In production.

Contractor

Sextant.

VERIFIED

The Sextant ATC-TCAS control panel 0001355

Attitude director indicators

Sextant produces a wide range of attitude director indicators for both civil and military use.

The current range provides a wide choice of instruments and features flight director command bars, failure flag indicators and warning lights for decision height alerting. Some instruments include a basic yaw indicator for additional assistance in asymmetric flight.

Operational status

Current military programmes include Transall C-160, Dassault Aviation Atlantique 2, Aerospatiale Nord 262 and Eurocopter SA 341 and SA 342 helicopters. Civil-standard attitude director indicators are fitted in Airbus Industrie's A300/A310.

Contractor

Sextant.

VERIFIED

Autoflight system for the A330 and A340

The autoflight system for the Airbus A330 and A340 is composed of two flight management guidance and envelope computers, a flight control unit and three multipurpose control and display units.

The flight control unit is used for short-term control of the aircraft autopilot and to select the display modes.

The multipurpose control and display units are installed on the centre pedestal in the cockpit. They are used for long-term control of the aircraft, initialisation of the fuel management system and to provide the interface between the aircrew and the maintenance system, ACARS, IRS and GPS.

The autoflight system can operate under autocontrol, using references computed by the flight management guidance and envelope computers on the basis of data selected by the aircrew through the multipurpose control and display units, or under manual control.

Operational status

In production for the Airbus A330 and A340.

Contractor

Sextant.

VERIFIED

Automatic Dependent Surveillance (ADS) unit

The Sextant Automatic Dependent Surveillance (ADS) unit was introduced in 1993. The reduction of separation distance, and thus the creation of more airspace, depends on establishing the exact position of an aircraft in space and time. To do this, the flight data used by the pilot must be processed and transmitted to a receiving station. The ADS system features the automatic transmission of this information and immediate reporting of data to the airline.

ADS will extend surveillance capability to areas not covered by radar. By adding a Secondary Surveillance Radar (SSR), the service can be extended to terminal areas and high-density continental airspace.

ADS is expected to provide a number of benefits to both pilots and airlines, including significantly enhanced flight safety, extension of ATC services to oceanic regions and areas not covered by radar, reduced separation minima, and highly accurate surveillance protected against interference in high-density airspace.

The Sextant approach is two-fold: refining the ADS concept through a stand-alone computer designed for current aircraft; and developing an embedded ADS function for the avionics suites on future aircraft.

Contractor
Sextant.

VERIFIED

Centralised Fault Display Interface Unit (CFDIU) for the A320

The concept of a Centralised Fault Display Interface Unit (CFDIU), pioneered for the A320, suits the need for effective maintenance tools that enable troubleshooting down to the faulty LRU. The CFDIU collects and processes all built-in test equipment messages from the main aircraft avionics subsystems. Maintenance information can be displayed for diagnosis either by the flight crew or by ground personnel.

The CFDIU receives all failure message output and stores it for subsequent use. It also performs a correlation check between messages, separated by a short time interval, that are assumed to have been generated by the same fault. It is possible for the CFDIU to transmit messages to the ground during flight by a datalink, or printout all the in-flight messages after the aircraft has landed.

The CFDIU is packaged in a 4 MCU ARINC 600 box and consumes 25 W at 28 V DC.

Operational status
In service in the Airbus A320.

Contractor
Sextant.

VERIFIED

Digital Automatic Flight Control System (AFCS) for the A300, A310 and A300-600

Digital processors replace the largely analogue elements of the original automatic flight control system in the Airbus Industrie A300 wide-body airliner, bringing the standard of this AFCS up to that of an almost wholly digital system. This is available in all A300 production with the forward-facing crew compartment. The system was first flown on the A300 in December 1980 and entered service with Garuda Indonesian Airways in January 1982. It is certified for Cat IIIB operation. The newer A310, the first of which entered service in April 1983, has the new digital automatic control system as standard.

The digital automatic control system provides the flight augmentation functions of pitch trim in all modes of flight, yaw damping, including automatic engine failure compensation when the autopilot is engaged, and flight envelope protection. It has a comprehensive complement of autopilot and flight director modes that permit automatic operations from take off to landing and roll out, a thrust control system which operates throughout the flight envelope, and a derate capability. It contains protection features against excessive angle of attack and has a fault isolation and detection system for line maintenance.

The AFCS also integrates wind-shear detection as a combination of vertical and horizontal shear components, with annunciation and flight guidance functions.

Design has been in accordance with ARINC 600 and 700 characteristics and has led to the adoption of ARINC 429 databusses between the automatic control system processors and sensors. Four to six processors are used, comprising two flight augmentation computers, one or two flight control computers (the second unit being necessary only if Cat III automatic landing capability is required), together with a thrust control computer. A second thrust control option is available, and the system also includes a flight control unit providing pilot interface with the autopilot/flight director and autothrottle functions, and a thrust rating panel which allows crew access to thrust limit computations.

A further new item of equipment is an engagement unit, with pitch trim and yaw damper engage levers, autothrottle arm and engine trim controls, which is mounted in the flight deck roof panel. Two pitch and roll dynamometric rods are also used as control wheel steering sensors. There are two trim actuators, an autothrottle actuator, a coupling unit on each engine and two further dynamometric rods connected to the throttle control linkage.

Flight deck controller equipment has been revised and a new autopilot/flight director and autothrottle mode selector is installed in the centre glareshield. Variable data can be entered by rotating selector knobs and shown by liquid crystal display readouts. The various modes are engaged by push-buttons. Modes available are altitude capture and hold, heading select, profile to capture and maintain vertical profiles and thrust commands from the flight management system, localiser, landing and speed reference. Autothrottle modes include delayed flap approach, speed/Mach number select, and engine N1 or engine pressure ratio selection.

The thrust rating panel is mounted above the centre pedestal. This has comprehensive controls permitting thrust levels to be selected, depending on operating mode and providing for selection of such facilities as derated thrust take off.

The fault isolation and detection system has a dedicated maintenance/test panel which, on a two-line by 16-character display, provides written alert messages based on fault information from automatic testing activities. This is conducted in all LRUs and includes fault isolation, tests to check for correct operation after maintenance action and on the ground automatic landing availability checks. Up to 30 faults from six flights can be stored and retrieved.

The system is jointly produced by Sextant (as prime contractor), Smiths Industries Aerospace (UK) and Bodenseewerk Gerätatechnik GmbH (Germany).

Computer units
1 or 2 flight control computer(s) (10 MCU size)
1 or 2 thrust control computer(s) (8 MCU size)
2 flight augmentation computers (8 MCU size)
Control units
flight control unit (glareshield)
thrust rating panel (centre panel)
FAC/ATS engagement unit (roof panel)
Maintenance/test panel

Other units
2 pitch dynamometric rods
2 roll dynamometric rods
2 trim actuators
autothrottle actuator
2 engine coupling units
2 engine dynamometric rods

Operational status
In service in the Airbus A300-600ST and A310.

Contractor
Sextant.

VERIFIED

Digital VOR/DME indicator

The Sextant VOR/DME instrument displays navigation data derived from suitable ground beacons that has been processed by a flight management system using the Rockwell Collins Pro Line, or an ARINC 429 digital databus. It requires 115 V AC/400 Hz or 28 V DC input and weighs less than 2.1 kg.

Operational status
In production.

Contractor
Sextant.

VERIFIED

Dracar digital map generator

Designed for modern aircraft, the Dracar digital map generator provides information on terrain for both head-up and head-down displays. In automatic mode, images are computed to take account of aircraft position and show the terrain as it would appear through the front of the aircraft. In manual mode the point of view can be slewed manually to fore and aft, left and right and up and down. Images may be superimposed on the HUD, either as a direct profile comparison with the outside terrain or in conjunction with FLIR imagery. The map generator can also display a plan view of terrain and false colour can be added to provide a ground collision avoidance mode.

The database contains Digital LandMass System (DLMS) data or raster paper charts. The synthesised three-dimensional image can be either standard map imaging with better visual quality or use an expert system for integration of mapping data to optimise the trajectory.

Specifications
Resolution: 512 × 512 pixels
Coverage:
(1:100,000 scale) 500,000 km^2
(1:250,000 scale) 3 million km^2

Operational status
In development.

Contractor
Sextant.

VERIFIED

Airbus A310 uses the digital automatic flight control system

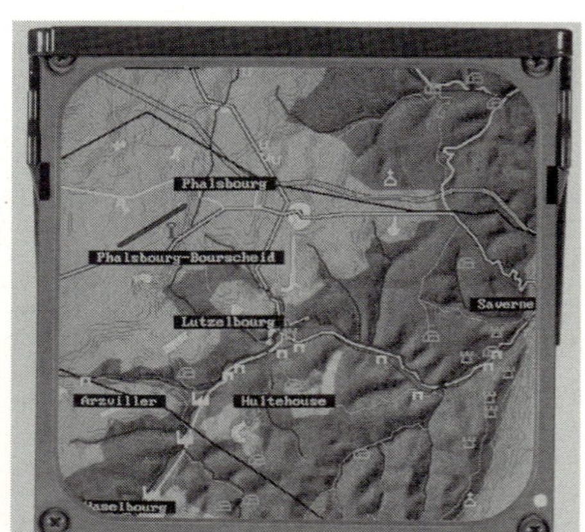

Dracar digital map display

Electronic Flight Instrument System for the A310/A300-600 (EFIS)

The electronic flight instrument display designed for the A310 uses six identical 6.25 × 6.25 in (159 × 159 mm) shadow-mask colour CRTs. Each pilot has on the flight panel in front of him a pair of EFIS instruments: a primary flight display CRT which replaces the conventional electromechanical attitude director indicator and a navigation display CRT which supplants the earlier horizontal situation indicator and weather radar display. Each pair of CRTs is driven by a single symbol generator and a third system acts as a hot standby. Also associated with each EFIS pair is a control/display unit. Two more displays on the centre panel represent the Electronic Centralised Aircraft Monitor (ECAM) display, presenting information on aircraft systems in any phase of flight and schematic diagrams of the hydraulic or electrical systems, for example, to supervise their operation. One of them is normally reserved for warnings, the other for systems. The ECAM system operates independently of the pilot's EFIS instruments, being driven by two symbol generators (one operational, the other a hot spare) and a single control unit mounted on the throttle box. All six CRTs are interchangeable, reducing the number of spares required. Frame repetition frequency is 70 Hz.

Sextant has continued to refine the system. Colour stability and visibility of the displays in high-ambient lighting conditions have been demonstrated, the latter at 100,000 lux instead of the 85,000 lux specified. Nine colours are used on the primary flight display, seven are employed for symbology to avoid confusion and blue and brown represent the sky and earth. A uniformly smooth sky unmarred by the raster scan was achieved by slightly out of focus imaging and has proved particularly acceptable to the eye. The three-dimensional effect familiar to pilots who have used electromechanical primary flight displays has been achieved by masking certain symbols when they would normally be occulted by others mounted further forward in the instruments. Sextant and Airbus Industrie have, thereby, devised a presentation which is instantly recognisable as that of a typical primary flight display, but one that can also utilise the vast quantity of newly available digital data circulating within an aircraft on the databus. Airbus Industrie and Sextant have added the following information to the periphery of the instrument: a moving speed scale along the left-hand edge with its standard symbols generated by computer; a selected pressure altitude readout along the right-hand edge; an autopilot and autothrottle mode annunciator along the top edge; and radio altimeter heights along the lower edge of the instrument.

For the navigation display, classic compass card symbology has been retained, again in conjunction with the flexibility that digital sensing and processing affords. Its use is innovative in that it creates new symbology to suit each phase of flight, in the form of an electronic map display with five distance scales marking the course to be followed with radio waypoints and their identifying codes. On selection by the pilot, a three-colour weather map can be superimposed on the navigation map, resulting in the saving of panel space that would be required by a separate weather CRT and the more comprehensible integration of weather and navigation information. The warning and system status CRTs on the centre panel show information previously unavailable to flight crews. In addition to special alarms, failure warning is given by presenting the crew with synoptic displays of failed systems, indications of vital actions to be followed to meet the difficulty and the effects on other systems.

Airbus A310 flight deck

Different software is used for each flight director to eliminate common-mode faults. Two separate software teams developed the two sets of software in two different locations. Software was implemented in machine code.

Operational status

In production for the Airbus A310 and A300-600.

Contractor

Sextant.

VERIFIED

Electronic Flight Instrument System for the A320 (EFIS)

In August 1984, Sextant was selected to develop and build the electronic flight instrument system for the Airbus Industrie A320 airliner. The system for this aircraft is an integration of the EFIS and ECAM CRT suites installed on the earlier A310, although with a number of important differences.

As with the A310 there are six CRT displays, but their disposition on the instrument panels is different. While on the earlier aircraft the flight director and navigation director are stacked vertically with the two ECAM instruments side-by-side on the centre panel, on the A320 the pilots' CRTs are ranged side by side, with the ECAM displays situated one above the other. The CRTs themselves are larger, 7.25 × 7.25 in (184 × 184 mm), compared with 6.25 × 6.25 in (159 × 159 mm), so that more symbology can be displayed without congestion. Each CRT can show at least seven colours. The upper ECAM, for the first time, shows primary engine parameters, replacing the 10 conventional electromechanical dial instruments on the A310. The lower ECAM will normally display systems information, such as electrical or hydraulic circuits, to show the location of faults. Confidence in the electronic display generated by A310 experience has resulted in the decision to delete most of the traditional electromechanical dial and pointer instruments, retaining just a few as a back-up in the most important functions.

As with the A310 system, all six CRTs are identical and interchangeable. However, whereas the A310 employs five symbol generators to drive them, the A320 uses only three and, with more complex functions, they are called display management computers (DMCs). All three are identical. The CRTs and computers are grouped within very advanced architecture, permitting extensive redundancy. In the event of a CRT, sensor or computer failure the system automatically reconfigures itself, with top priority given to flight commands and engine warning. At the same time there has been a significant weight decrease from 153 to 115 kg. As with the A310 system, elimination of software errors in the pilot's flight director instruments is accomplished by the use of two independent development teams. However, unlike the earlier equipment, software is implemented in high-order language.

Operational status

In production for the Airbus A320.

Contractor

Sextant.

VERIFIED

Electronic instrument system for the A330/A340

In 1989, Sextant was selected to develop the electronic instrument system for the Airbus A330 and A340. The system is an integration of the EFIS and ECAM CRT suites. The EFIS system integrates the flight director and the navigation director with a weather radar display. The upper ECAM shows the primary engine parameters while the lower ECAM displays 15 system pages such as electrical, hydraulic, fault location,

The electronic instrument system for the A330 and A340 includes six shadow-mask colour CRTs

doors, bleed, brake circuits, auxiliary power unit, secondary engines, and so on.

The system uses six identical 7.25 × 7.25 in (184 × 184 mm) shadow-mask colour CRTs and three display management computers. In the event of a CRT, sensor or computer failure, the system is automatically reconfigured with top priority given to flight commands and engine warnings.

Specifications

Weight: 10 kg
Display dimensions: 184.1 × 184.1 × 355.6 mm

Operational status

In production.

Contractor

Sextant.

UPDATED

EVR 716/750 enhanced VHF Data Radio (VDR)

The Sextant EVR 716/750 (enhanced VHF Radio) is a new VHF AM transceiver designed to comply with 25 and 8.33 kHz channel spacing in the VHF band; it is immune to FM radio signals in accordance with ICAO annex 10.

The EVR 716 offers 25 and 8.33 kHz channel spacing for voice communications and 25 kHz channel spacing data transmission capability with an external modem (MSK/CSMA 2.4 kbytes/s).

The EVR 750 VHF Data Radio (VDR) offers, in addition to the EVR 716, ARINC 750 data modes compatible with the ATN protocols (31.5 kbytes/s) with internal modems.

The EVR 716 complies with ARINC 716 and the EVR 750 with ARINC 750. The two systems share a common hardware configuration and the EVR 716 can be upgraded to the EVR 750 capability by software only.

Specifications

Dimensions: 3 MCU
Weight: 4.8 kg
Power supply: 28 V DC
Temperature: –15 to +70°C

ARINC specifications

ARINC 716
ARINC 750

Operational status

Sextant's EVR 716-01 VHF Data Radio (VDR) is certified on the Airbus A319/A320/A321 family and on the A340. It is also field-approval certified on B-737/-747-400/-757/-767/-777 aircraft and supplemental type certified on B-737-400 and NG aircraft. The EVR 716 is part of a line of new-generation digital VDR systems developed by Sextant.

The EVR 716-01 enables airlines to comply with Europe's 8.33 kHz channel spacing requirements, as well as future data communications regulations within the scope of FANS ATN VDL mode A and 2 (Aeronautical Telecommunications Network). Sextant's VDR system can easily be upgraded to meet new FANS digital transmission mode requirements.

Operating modes

Mode ident	Main features	EVR-716	EVR-750
0A	Analogue voice, 25 kHz channel spacing	Compliant	Compliant
0B	Analogue voice, 8.33 kHz channel spacing	Compliant	Compliant
1A	Data MSK-AM 2.4 kbyte/s CSMA non-persistent modem in the (C)MU Standard ACARS mode	Compliant	Compliant
1B	Data MSK-AM 2.4 kbyte/s CSMA P-persistent modem in the EVR ICAO VDL mode 1	Require 750 software upgrade	Compliant
1C	Data MSK-AM 2.4 kbyte/s CSMA non-persistent modem in the EVR ACARS mode for CMU/ATSU interface	Require 750 software upgrade	Compliant
2	Data D8PSK 31.5 kbyte/s CSMA P-persistent modem in the EVR ICAO VDL mode 2	Require 750 software upgrade	Compliant
3*	*Digital voice and data* D8PSK 31.5 kbyte/s TDMA *mode defined in* RTCA DO-224	*Hardware provisions*	*Hardware provisions*
4*	*Data* D8PSK 31.5 kb/s *or* GMSK 19.2 kbyte/s TDMA *mode defined in* STDMA *standard*	*Hardware provisions*	*Hardware provisions*

* This mode is not yet validated and adopted by ICAO

Sextant EVR 716/750 enhanced VHF radio 0001172

The VDR has been selected as standard fit on the Dash 8 Series 400 regional aircraft. Sextant has now booked orders for more than 3,000 VDRs.

An enhanced version of the VDR will be able to operate in future VHF datalink (VDL) Mode 2.

Contractor

Sextant.

UPDATED

FDS-90 Flight Director System

The FDS-90 system comprises an attitude director indicator and a navigation coupler/computer unit.

The H140 attitude director indicator can operate autonomously using self-contained gyros and power inverters. It uses a ball-type real-world display and has a three-cue command capability. There are annunciators for go-around, decision height and flight director mode monitor. The gyro can be caged.

The B152 coupler/computer is a smal,l panel-mounted unit, which interfaces the attitude director indicator to the navigation equipment. It may have up to 11 push-buttons that permit selection of different flight director operating modes:

HDG: captures and tracks the heading selected on the horizontal situation indicator
NAV or V/L: captures and tracks VOR and ILS localiser beam for short-range navigation or FMS data for long-range navigation. It has the ability to interface with Tacan, Loran or Doppler sensors, depending upon the mission
BC: tracks the back course localiser
BALT: maintains the baro-altitude existing at the time of selection
GS: captures and tracks an ILS glide slope beam
VS: maintains the vertical speed that exists at the time of engagement
IAS: maintains the airspeed that exists at the time of engagement

In addition to the various functions incorporated in the FDS-90, that are more particularly concerned with en route and approach applications at cruise speeds, the search and rescue and anti-submarine warfare coupler functions are specific to hover mode and to the various transition phases. These include:

APP 1: from the initial conditions of height above 200 ft and IAS higher than 60 kt it provides the simultaneous commands for a descent to 150 ft min and deceleration to 60 kt
APP 2: from the final APP 1 conditions it provides the commands required to control the helicopter's descent down to the height selected by the pilot and deceleration to hover
HOV: holds the zero lateral and longitudinal groundspeeds provided by the Doppler radar or the zero sonar cable angles, depending on the submode selected by the pilot
CLB: from the initial conditions of height above 200 ft and IAS higher than 60 kt it provides the commands for a climb at 300 ft/min and acceleration up to 60 kt.

The SAR and ASW coupler is connected to the ADI for manual use in the FD mode or to the basic stabilisation system (Ministab, AFDS 95-1 or any other AFCS) for automatic operation.

Specifications

Dimensions:
(attitude director indicator) 4 ATI
(nav coupler/computer) 3 ATI standard case
Weight:
(attitude director indicator) 2.5 kg
(nav coupler/computer) 1.3 kg
Power supply:
28 V DC, <20 W (attitude director indicator)
28 V DC, <15 W (nav coupler/computer)
Environmental: DOO 160 C, TSO'd

Operational status

In production and in service, on S-61 (HH3-F) and B206.

Contractor

Sextant.

VERIFIED

Flight Management and Guidance System (FMGS) for the A320

The autopilot, flight director and flight management functions are integrated in a single system in the A320, called the Flight Management and Guidance System (FMGS). It reduces the number of LRUs by 60 per cent and the volume and weight by 50 per cent in comparison with earlier systems such as that in the A310.

Two flight augmentation computers in 8 MCU boxes provide dual-channel command signals for yaw damping, rudder trim, rudder travel limiting and flight envelope protection. The entire system is monitored by fault detection and isolation software, with warning indications and appropriate vital actions and responses being automatically displayed on the Electronic Centralised Aircraft Monitor (ECAM) CRT displays.

The computer has about 2.5 Mbytes of non-volatile memory and takes in data from the Air Data and Inertial Reference System (ADIRS), radio navigation aids, radio altimeter and the Full Authority Digital Engine Control (FADEC). Through interfaces with the crew via the flight control unit, Multifunction Control and Display Unit (MCDU), the FMGS provides autoland, autothrust, autopilot cruise, four-dimensional navigation control, flight management and performance management in the vertical profile.

Operational status
In production for the Airbus A320.

Contractor
Sextant.

UPDATED

The A320 flight-deck showing flight management and guidance system controls on the glare shield (British Airways)

Fly-by-wire systems for the A330 and A340

As on the A320, the A330 and A340 flight controls are hydraulically actuated and electrically or mechanically controlled. Pilot controls in the cockpit consist of two sidesticks, conventional rudder pedals, mechanical pitch trim and electrical rudder trim.

Electrical flight control is achieved by seven computers of three different types: three Flight Control Primary Computers (FCPC), which are in charge of generating control laws and controlling surfaces; two Flight Control Secondary Computers (FCSC), which are also in charge of controlling surfaces; and two Flight Control Data Concentrators (FCDC), which interface the flight control system with other aircraft systems to provide an isolation function.

Each of the two FCSCs can control the power elements used to activate the aircraft control surfaces. In normal operation, the FCSC achieves spoiler control, rudder trim control and rudder travel limiting. In back-up mode, as in the case of a failure of the FCPC, the FCSC achieves aileron control, elevator control and yaw damping.

The FCDC performs data concentration, warning and maintenance functions. In data concentration, the FCDC transmits information, such as control surface positions, to the display management computers and to the flight data interface unit. In warning, the FCDC indicates flight control failure status to the flight warning computers and to the display management computer. For maintenance the FCDC isolates and memorises the flight control system failures and interfaces with the centralised maintenance computer.

Operational status
In production for the Airbus A330 and A340.

Contractor
Sextant.

VERIFIED

Fuel Control and Monitoring Computer (FCMC) for the A330 and A340

The A330 and A340 have two fuel tanks in each wing, one in the tailplane and one in the fuselage.

The A330 and A340 fuel control and monitoring system is composed of about 96 fuel height probes and other sensors for temperature, densitometers and level detectors installed in the fuel tanks, one to three refuelling panels and two identical Fuel Control and Monitoring Computers (FCMC).

The FCMC in the A330 and A340 concentrates functions previously scattered in distinct computers in the A310 and A320 such as fuel quantity measurements, Hi-Lo measurement and centre of gravity control.

The role of the FCMC is to measure the fuel quantity that is available on the aircraft and to indicate it to the aircrew. It also controls automatic refuelling of tanks to preselected levels and monitors associated pumps and valves, together with monitoring and controlling the fuel temperature in the tanks and fuel tank utilisation sequence. The FCMC also detects too low or too high levels in tanks and manages the associated warnings, controls fuel distribution for wing load alleviation and controls the centre of gravity of the aircraft by transferring fuel from one tank to another.

Operational status
In production for the Airbus A330 and A340.

Contractor
Sextant.

VERIFIED

Gyro horizons

Sextant gyro horizons have been installed on virtually all commercial transport aircraft. Today, approximately 150 airlines and 30 air forces use Sextant gyro horizons as stand-by attitude reference instruments.

The Series H3XX gyro horizons are 3 ATI-sized and weigh about 1.5 kg. They make extensive use of modern alloys, have built-in static inverters and gyro speed monitors and Sextant-patented simplified erection and anti-spin devices. The basic versions include H301 (drum display), H321 (sphere display) and H341 (sphere display with ILS capability). 2 ATI and 4 ATI versions are also in production.

Operational status
In production and in service.

Contractor
Sextant.

VERIFIED

Icare map display and Mercator Remote Map Reader for the Mirage 2000N (RMR)

Sextant, under a French Ministry of Defence contract, developed, for the Dassault Mirage 2000N aircraft, an integrated electronic map system. It uses two full-colour display units that collect data from the electronic Mercator Remote Map Reader (RMR) radar and the aircraft symbol generation systems.

Operational status
In production. The first Icare systems entered operational use in 1986 with the French Air Force.

Contractor
Sextant.

VERIFIED

IMS Integrated Modular avionics Systems

Sextant has developed a new IMS family of Integrated and Modular avionics Systems to offer regional and business aircraft operators an integrated system for new-build and upgrade aircraft.

The IMS family capabilities are based on innovative avionics technologies – open design and a wide range of functions and performance – adapted to each market segment's specific needs. Sextant is developing different versions to provide avionics systems that enhance the intrinsic qualities of regional and business aircraft and ensure long-term customer and product support to regional jets, business aircraft and new generations of airliners.

The first application is the IMS-100 on the Bombardier de Havilland Dash 8-400 regional turboprop, certified in 1999.

Typically, IMS configurations comprise: a new radio-communications and navigation unit; smart LCD units;

The cockpit of the Bombardier de Havilland Dash 8-400 incorporating the Sextant IMS **2000**/0079257

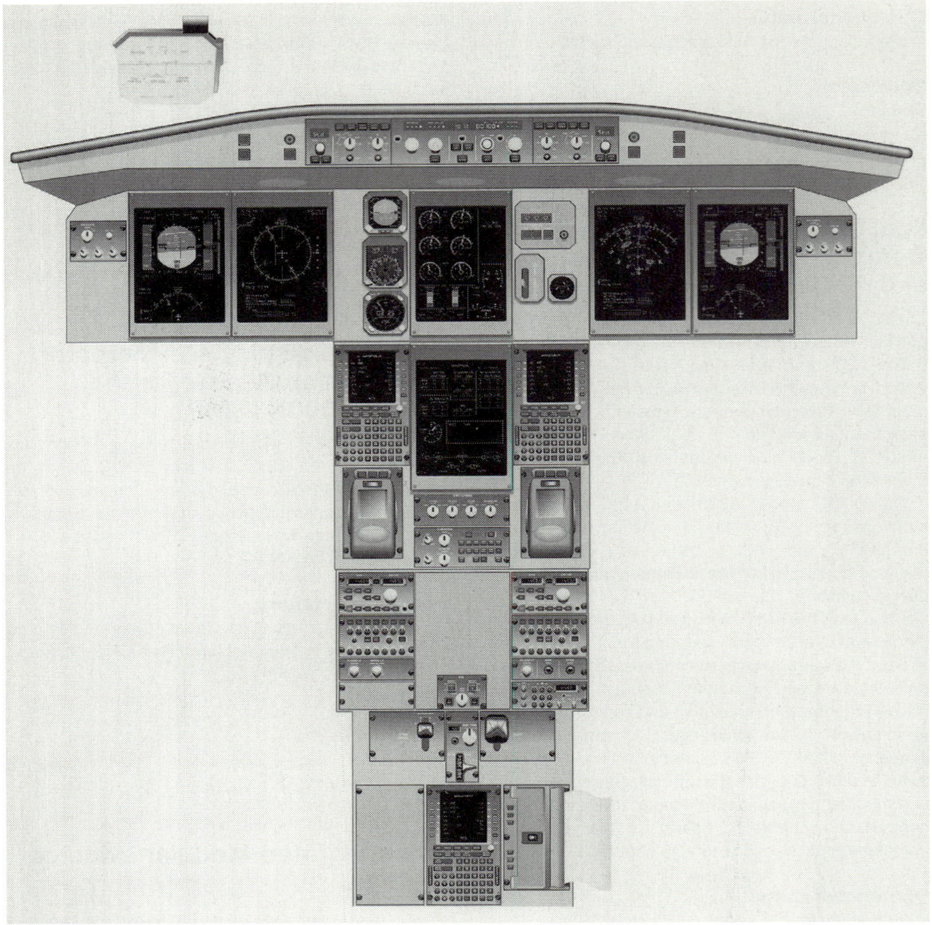

IMS 300 for air transport-aircraft **1999**/0051585

fibre optic gyro AHRS; FMS; and a centralised diagnostics and maintenance interfacing system.

Contractor

Sextant.

VERIFIED

Integrated Electronic Standby Instrument (IESI)

The Sextant solid-state Integrated Electronic Standby Instrument (IESI) is designed to replace conventional pneumatic altimeter, airspeed and gyroscopic horizon indicators with a single LRU standard 3 ATI box. It is intended for use in civil aircraft (commercial transport, commuter and business type helicopters).

The IESI incorporates design growth capability to display: SSEC, ILS (localiser and glide path), back course, slip-heading, dual-baroset scales and Mach number to meet customer requirements.

Two configurations are available: a single LRU configuration to display data from its three internal functions, together with two remote and interchangeable Sextant air data modules; a single LRU configuration dedicated to attitude, which acts as a standby horizon indicator.

Specifications

Dimensions: 3 ATI case according to ARINC 408A standard

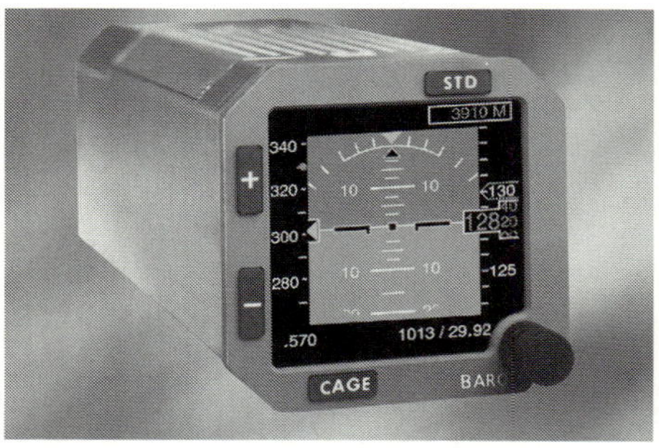

Integrated Electronic Standby Instrument (IESI) **1999**/0051659

Weight: 1.9 kg
Altitude: according to TSO C10c (AS 8009A)
Power supply: 28 V DC (emergency bus)
Certification: TSO-C10c (altitude), -C2d (airspeed), -C4c (attitude), -C95 (machmeter), and -C113 (airborne multipurpose electronic displays)

Operational status

In production since late 1997. Selected for: Canadair Regional Jet Series 700, Continental Business Jet, Embraer ERJ-135/-145 and CRJ 700. It is available as an option on the Q-400 aircraft and is also available on Airbus A318/A319/A320/A321 aircraft, where it is known as the Integrated Standby Instrument System (ISIS).

Contractor

Sextant.

VERIFIED

LCD engine indicator

The LCD engine indicator uses a 4 ATI size liquid crystal display screen with the same technology as Sextant's 3 ATI VSI/TCAS instrument. The basic engine instrument comprises three screens which display data from the FADEC system, secondary sensors, fuel management computer and APU.

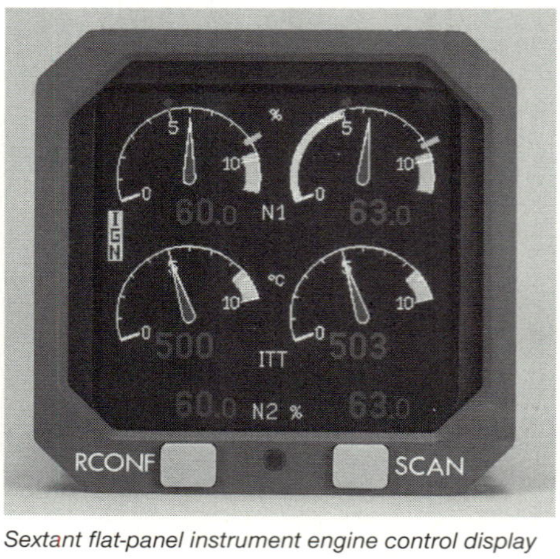

Sextant flat-panel instrument engine control display

Operational status

In production. Selected by Dassault for the Falcon 2000 and Falcon 50 EX.

Contractor

Sextant.

VERIFIED

MEGHAS® new-generation avionics suite for helicopters

Sextant and SFIM Industries have developed a new generation of avionics systems for helicopters – MEGHAS®. Based on an integrated family of equipment, this new avionics system is designed to satisfy the specific requirements of new helicopters up to the 6-tn class. MEGHAS® has been certified for the Eurocopter EC 120, EC 135, EC 155 and Ecureuil B3 helicopters.

MEGHAS® handles all main helicopter functions: autopilot, guidance and navigation, engine and vehicle control, onboard maintenance, radio communications and radio navigation.

By reducing the crew's workload, MEGHAS® allows helicopter crews to accomplish their missions (EMS, SAR, surveillance and offshore transport) quickly and efficiently. The system's modular design facilitates adding new mission functionality, as well as providing weight savings, lower cost of ownership and higher reliability. Fewer parts and proven technologies provide a significant increase in MTBF compared with conventional systems.

The flexibility of the MEGHAS® architecture also means that it supports all flight configurations (VFR, IFR, single-pilot and dual-pilot), in compliance with international regulations (FAR, JAR 27 or 29).

The major components in the MEGHAS® avionics system include:

- Central Panel Display System (CPDS). The CPDS comprises the Vehicle and Engine Management Display (VEMD) for single-engine helicopters, used in conjunction with the Caution and Advisory Display (CAD) for twin-engine helicopters.

  The VEMD comprises two active matrix LCDs. It replaces up to a dozen conventional indicators and allows immediate verification of both vehicle and engine parameters. The design complies with all High-Intensity Radiated Field (HIRF) and lightning requirements (FAR 29 category A).

  The CAD uses a single active matrix LCD. In conjunction with the VEMD, it manages all fuel functions and warnings. The CAD also operates as a back-up for main engine parameters in case of VEMD failure.
- Flight Control Display System (FCDS). Sextant has developed two 'smart' displays for the FCDS, featuring very high resolution, multifunction colour LCDs; the SMD45H (4 × 5 in LCD) and the SMD68 (6 × 8 in LCD), with integrated NVG cooling compatibility and video system. These smart displays show both primary flight and navigation information. For specific mission functions, including FLIR, map generator and weather radar, the video functions in these displays allow display of this information using control units on the SMD45H and SMD68CVN.
- Automatic Flight Control System (AFCS). The AFCS is designed and built by SFIM Industries. It includes a SFIM Industries attitude and heading reference system (APIRS) and a Sextant ADS 3000 air data system. Use of a removable memory module means that aircraft installation is easy and only a single calibration flight is required.

  The AFCS can be upgraded, changing it from a simple stability augmentation system to a complete four-axis autopilot system coupled to a GPS receiver for auto-approach.
- Health and Usage Monitoring System (HUMS). The HUMS system is designed by SFIM Industries, and uses the Miscellaneous Flight Data Acquisition Unit (MFDAU).
- Navigation functions, including a GPS receiver and SAR (Search And Rescue) capability.
- Centralised Radio Control (CRC), comprising radio

Meghas new-generation avionics suite for helicopters ***1999***/0051586

MEGHAS® helicopter avionics suite on the Eurocopter EC 155 ***2000***/0079256

communications, radio-navigation and identification control systems.

Operational status

Applications of MEGHAS® can be found in the following programmes:

- The EC 120 uses the VEMD. It was certified in mid-1997, with initial deliveries starting at the end of that year, at the rate of about 100 units per annum.
- The Ecureuil B3 modernisation programme uses the VEMD to replace all instruments on the central engine and vehicle control console. The installation was certified in late-1997 and deliveries started early in 1998, at a rate of about 60 VEMDs per year.
- In August 1998, the EC 135 with the Central Panel Display System (CPDS), comprising the VEMD, CAD and FCDS was certified for IFR Category A dual-pilot operation. Deliveries at the rate of 40 systems per year started in October 1998. IFR Category A single-pilot certification with the AFCS was planned for the end of 1999. The basic FCDS should be upgraded to offer FLIR, map generator and weather radar functions, handled through the SMD68CVN display.
- The EC 155, with CPDS, FCDS and AFCS was certified for IFR dual-pilot operation in March 1999. Deliveries of production equipment started at the end of 1998. This programme represents deliveries of 15 systems per year. Single-pilot certification with a four-axis dual AFCS was expected for the end of 1999. The basic FCDS will be upgraded to include mission functions (FLIR, mapping and weather radar), handled through the SMD68CVN display.

Contractors

Sextant.
SFIM Industries (SAGEM Group).

VERIFIED

MultiMode Receiver (MMR) TLS 755

The Sextant MultiMode Receiver (MMR) is a new landing and precision approach sensor that provides navigation, en-route GPS, ILS, MLS and GLS functions in one LRU. In addition, the MMR provides position, time and velocity in a permanent GPS navigation mode.

The receiver has been designed according to ARINC 755 specifications for dual or triplex installations providing Cat III operations. The MMR is immune to FM radio signals in accordance with ICAO annex 10.

The MMR offers a high degree of integrity and reliability due to the digital technology used in the design.

The TLS 755 receivers are ARINC 755 compliant and integrate an ARINC 743 GPS sensor that includes WAAS (Wide Area Augmentation System), LAAS (Local Area Augmentation System) and GLONASS capabilities.

Specifications

Dimensions: 3 MCU
Frequency bands:
108-112 MHz for ILS localiser
329-335 MHz for glide slope
108-118 MHz for datalink DGNSS
5 GHz (option) for MLS or datalink
Channel spacing: 50 kHz for datalink (compliant 25 kHz)
Modulation: AM
DPSK 10 kbps/15kbps
D8PSK 31.5 kbps
Weight: 5.4 kg
Power supply: 115 V AC, 400 Hz single phase
Temperature range: –40 to +70°C
Compliances:
ILS ARINC 710-9
MLS ARINC 727
GPS ARINC 743A
MMR ARINC 755

Operational status

In production. TSO qualified for all Airbus and Boeing aircraft; certified on Airbus A319/A320/A319/A330/A340 and Boeing B-737NG/-747-400/-777. In production. Selected by over 30 airlines, with more than 2,800 units ordered by late 1999.

Contractor

Sextant.

UPDATED

Nadir 1000 integrated navigation and mission management system

The Nadir 1000 is an integrated navigation and mission management computer developed to provide mission assistance for military and civil light helicopters over both land and sea. The Nadir 1000 is designed to provide multisensor navigation from Doppler, radio navigation and GPS, for flight management, navigation and mission management, and system links for SAR, hover and ASM roles.

The Nadir 1000 weighs 3 kg in the basic version and consists of two modules: a front panel module and a processing/power supply module. Optionally, analogue/digital and input/output modules can be added.

Operational status

In production. Selected for several Eurocopter export programmes. More than 100 systems delivered for Fennec and Super Puma.

Contractor

Sextant.

VERIFIED

The Nadir 1000 navigation and mission management system for helicopters 0015351

Nadir Mk 2 navigation/mission management system

The Nadir Mk 2 is a multipurpose processing and display system that can store details of up to 100 waypoints and the characteristics of up to 100 VOR/DME stations. It is not limited to Doppler but can operate with many other navigation sensors. Nadir Mk 2 comprises a four MCU central processing unit and a general purpose control/display unit. It can provide the following:

1. Navigation management functions based on Doppler, VOR/DME, Tacan, Omega/VLF, GPS, heading sensor, inertial sensor and air data inputs.
2. Flight management functions, with guidance for

optimum cruise conditions, fuel and weight management and engine monitoring; air data computations involving speed, altitude and outside air temperature.
3. Interface functions with autopilot, radar and navigation indicators.

Nadir Mk 2 may be used in a dual-system configuration, in which one computer is responsible for navigation and flight management, while the second deals with weapons and aircraft management.

Operational status
In production. Nadir Mk 2 has been selected for 30 Eurocopter helicopter programmes and for fixed-wing aircraft operations.

Contractor
Sextant.

VERIFIED

Pitot probe and angle-of-attack sensors

Sextant manufactures high-performance Pitot probes and angle-of-attack sensors for all Airbus jetliners. Long a supplier of Pitot probes for the A300, A310, A319, A320, A321, A330 and A340, Sextant's new-generation Pitot probe was certified for the A319/320/321 in March 1997 and the A330/340 in April 1998. This new product offers higher resistance to water ingestion and improved anti-icing performance compared to previous-generation probes and gives operators product commonality for these two jetliner families.

Sextant also manufactures angle-of-attack sensors for Airbus jetliners. To meet airlines' requirements for enhanced maintainability, Sextant has provided vane replacement on these sensors. In the past, replacing a vane meant that the new vane had to be wind-tunnel re-calibrated, but starting at the end of September 1998, Sextant's product marking concept means that the vane can be replaced, without requiring a wind-tunnel re-calibration.

Contractor
Sextant.

VERIFIED

SMD 45 H liquid crystal smart multifunction display

The SMD 45 H was specially designed for helicopter applications and is currently proposed for multiple light helicopter upgrade programmes.

The SMD 45 H is a 'Smart' full-colour multifunction display using a 4 × 5 in Active Matrix Liquid Crystal Display (AMLCD) providing excellent viewability under any aircraft cockpit ambient light conditions.

The SMD 45 H integrates in one panel-mounted LRU all necessary functions required for stand-alone operation such as systems bus interface, data processor and graphics generator.

The SMD 45 H high-resolution display also provides imagery from various aircraft sensors with synthetic flight symbology overlay.

Specifications
Active screen size: 4 × 5 in
Resolution: 640 × 480 pixels
Brightness and contrast: >150 fL in white

SMD 45 H liquid crystal smart multifunction display ***1999***/0051661

Aircraft interface: ARINC 429 bus (6 in, 4 out) or Mil-STD-1553B Bus (optional)
Power: 28 V DC, 50 W
Overall dimensions:
(front panel) 185 × 134 × 23 mm (without push-buttons)
(case size) 185 × 201 × 166 mm
Weight: 2.8 kg
Video interface: 1 input STANAG 3 350 C
Environmental: DO 160 C
NVG-Compatible

Operational status
Fitted on BK 117C2, EC 120, EC 135, EC 155 and Ecureuil B3 helicopters.

Contractor
Sextant.

UPDATED

SMD 66 integrated multifunction display

The SMD 66 is a multipurpose electronic display developed for use on helicopter and fixed-wing aircraft flight decks. The full-colour high-resolution shadow-mask fully integrated display is capable of providing stroke and raster image for primary flight, navigation and tactical displays and systems, and engine monitoring. The image measures 6 × 6 in (152 × 152 mm). The display features brightness automatic setting up to 8,000 ft candles, to improve legibility, and is compatible with night vision goggles.

Operational status
In production for the Super Puma Mk 2; also being supplied for the Indian Air Force Jaguar upgrade programme.

Contractor
Sextant.

VERIFIED

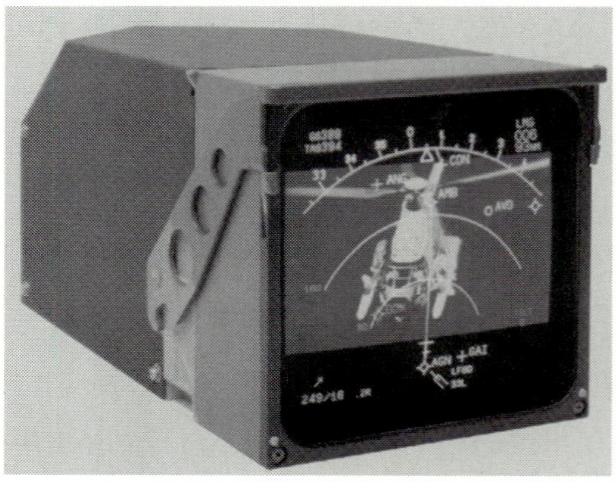

Sextant SMD 66 integrated multifunction display

SMD 68 CVN liquid crystal multifunction display

The SMD 68 H is a 'smart' full-colour multifunction display designed for helicopter applications. It uses a 6 × 8 in Active Matrix Liquid Crystal Display (AMLCD) providing excellent viewability under all aircraft cockpit ambient light conditions.

The SMD 68 H integrates all the necessary functions required for stand-alone operation (including databus interface, data processor and graphics generator) in one panel-mounted LRU.

As an option, the SMD 68 H can also display imagery from aircraft sensors such as FLIR, map, weather radar together with synthetic flight symbology overlay.

Specifications
Active screen size: 6 × 8 in
Resolution: 1,344 × 1,008 pixels
Brightness and contrast: 0.4 to 315 cd/m^2
Aircraft interface: ARINC 429
Power: 28 V DC, 130 W
Dimensions: 192 (W) × 267 (H) × 202 (D) mm
Weight: 6.5 kg
Cooling: internal fan
Video interface: (STANAG 3 350 C or B) 1 input
Environment: DO 160 C
Reliability: > 5,000 flight hours
Options: filtered NVG compatibility

Operational status
Selected for BK 117C2, C-130, Casa C 295, Dash 8-400, EC 135 and EC 155.

Contractor
Sextant.

UPDATED

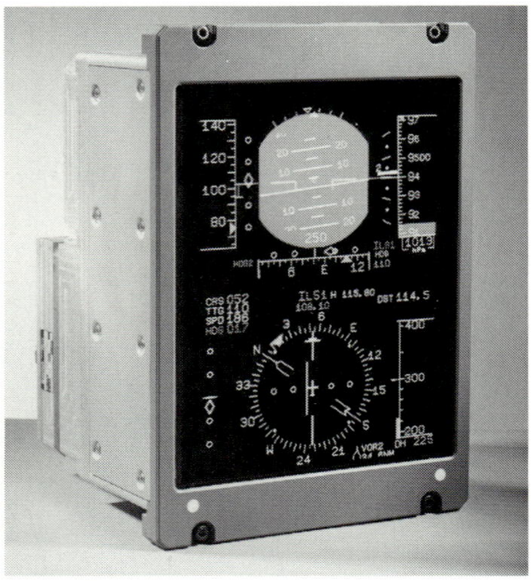

Sextant SMD 68 H liquid crystal smart multifunction display ***1999***/0051662

TCAS Resolution Advisory/Traffic Advisory (RA/TA) VSI

Sextant's Resolution Advisory/Traffic Advisory (RA/TA) VSI combines the vertical speed and Traffic alert Collision Avoidance System (TCAS) information on a 3 ATI instrument using an active matrix type full-colour. The instrument is capable of acquiring digital, analogue and pneumatic signals, enabling Sextant to propose a unique part number for a given airline. The RA/TA VSI is Sextants' first application of its flat-panel instrument programme. This flat panel instrument programme aims to replace all the 3 ATI and 4 ATI electromechanical instruments.

Operational status
In production. Over 50 airlines and aircraft manufacturers of the air transport, regional and business aircraft segments have chosen the 3 ATI Vertical Speed Indicator/Traffic alert Collision Avoidance System (VSI/TCAS) liquid crystal flat-panel instrument.

Contractor
Sextant.

UPDATED

The Sextant TCAS RA/TA VSI is shown at the top left of this group

TMV 544 forward view repeater display

Sextant has developed what it terms a 'forward view repeater', which reproduces the pilot's field of view for the benefit of rear-seat occupants. The TMV 544 was developed in response to conclusions that some shortfalls in the instruction of pilots in advanced training aircraft resulted from the instructor's inability to scan the view directly ahead because it was obscured by the pupil's ejection seat headrest. The system is based on a video camera that films the head-up display and outside world from the front compartment and reproduces it on a television monitor in the rear compartment.

Operational status

In production.

Contractor

Sextant.

VERIFIED

Topscreen avionics for basic training aircraft

Sextant is developing Topscreen, a comprehensive display system for basic trainers such as the Pilatus PC-7 and PC-9, and the Embraer Tucano.

Sextant will supply 37 instruments for each of the EMB 312F Tucano trainers deployed by the French Air Force. Basic indicators are 3 ATI size ADI and HSI units using liquid crystal displays (EADI and EHSI).

Sextant is also supplying seven instruments for each Pilatus PC 7 Mk II Astra trainer deployed by the South African Air Force. Basic indicators in this case are 4 ATI size EADI and EHSI units.

Contractor

Sextant.

VERIFIED

VEMD Vehicle and Engine Management Display

The VEMD is designed for both single- and twin-engined helicopters. It is interfaced to engine and vehicle sensors and linkable to data concentrators as well as FADEC.

It displays the information on two 5 in (127 mm) diagonal displays. This allows several display modes: normal operation mode (first limitation indication, vehicle information), reversionary mode performance, health monitoring flight report (overlimit detection) and maintenance modes. The dual architecture both in terms of processing and displays, the high level of failure detection and the presence of reversionary modes provide a high level of availability.

The VEMD displays the information on a twin-active matrix colour LCD display unit with a wide viewing angle and provides high readability in any ambient lighting conditions (compatible with filtered NVG). In normal operation the upper matrix displays engine information and the lower one, vehicle information. If one channel fails, the main information is displayed on the remaining display.

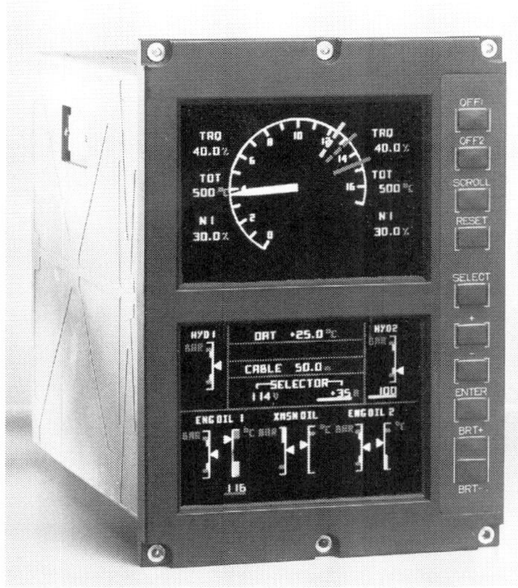

Sextant vehicle and engine management display
***2000**/0079261*

The VEMD can also display health monitoring information such as engine cycle count, engine power check, BIT results. This information is stored in a non-volatile memory and is readable on the display in maintenance mode.

Each channel of the dual architecture of the VEMD (display, processing, power supply) performs the acquisition of each parameter and cross-checks its consistency with the other channel. This monitoring activity, added to the self-test of each channel, provides a high level of failure detection and low level of erroneous data display.

The VEMD has dual processing and display systems interfaced to engine and vehicle sensors: thermocouples (TOT); frequencies (N1, N2, NR); voltages (torque, fuel quantity, oil pressure, pressure, gearbox oil pressure); resistors (oil temperature, OAT, gearbox oil temperature); voltages (voltmeter, ammeter) pressure sensor as well as numeric interfaces for FADEC (RS-422, ARINC 429, EIA 485). It generates the power supply required for the sensors as well as discrete outputs.

Specifications

Useful screen size: (3 × 4 in) × 2
Lighting in day conditions: 0.4 to 100 Cd/mL
Dimensions: 156 × 212 × 205 mm
Max weight: 3 kg
Power consumption: 44 W
Power supply: 28 V DC
MTBF: 5,000 h
No cooling required

Operational status

Fitted on BK 117C2, EC 120, Ecureuil B3, EC 135, EC 155.

Contractor

Sextant.

VERIFIED

Warning and maintenance system for the A330 and A340

The A330 and A340 warning and maintenance system is composed of two major subsystems: the flight warning system (FWS) and the central maintenance system (CMS).

The FWS presents the aircrew with visual and warning messages related to failures. It also indicates the seriousness of each failure and the corrective actions to be taken.

The system consists of two system data acquisition concentrators (SDACs), which acquire data then generate signals which go to three display management computers (DMCs), and/or two flight warning computers (FWCs). The DMCs use the SDAC signals to generate displays of systems pages and engine parameters on two Electronic Centralised Aircraft Monitor (ECAM) display units. The FWCs generate alert messages, memos, aural alerts and synthetic voice messages for ECAM display, together with radio altitude callouts, decision height callouts and landing speed increments. Routine control of ECAM displays is exercised via an ECAM control panel.

The onboard maintenance system generates and displays maintenance data on the multipurpose control and display units (MCDUs) for use by aircrew and ground maintenance personnel.

The system consists of two centralised maintenance computers (CMCs), three MCDUs and one onboard printer.

The CMS operates in two main modes: In flight, the CMS operates in Normal (Reporting) mode, and records and displays failure messages transmitted by each system built-in test equipment (BITE). On the ground, the CMS operates in Interactive (Menu) mode, and allows the connection of any BITE directly with the MCDU in order to display maintenance data or to initiate a test.

Development and production is conducted jointly by Sextant and Aerospatiale.

Operational status

In service on the Airbus A330 and A340.

Contractors

Sextant.
Aerospatiale.

UPDATED

AFCS 85 autopilot/flight director system

The basic AFCS 85 system is a full-time two-axis series autopilot, which is the first step in a building block concept to provide automatic stabilisation for light helicopters.

The basic version provides long-term hands-off VFR capability including heading select mode, baro-altitude and airspeed hold.

Adding the FDC 85 flight director coupler and pitch and roll autopilot monitor gives single pilot IFR capability by providing facilities such as automatic lateral and vertical guidance and by driving command bars on the attitude director indicator.

The basic system weighs 6.8 kg.

Operational status

In production. The system has been chosen for the Alouette II, Bell 407, single- and twin-engine Ecureuil, and Gazelle helicopters.

Contractor

SFIM Industries (SAGEM Group).

VERIFIED

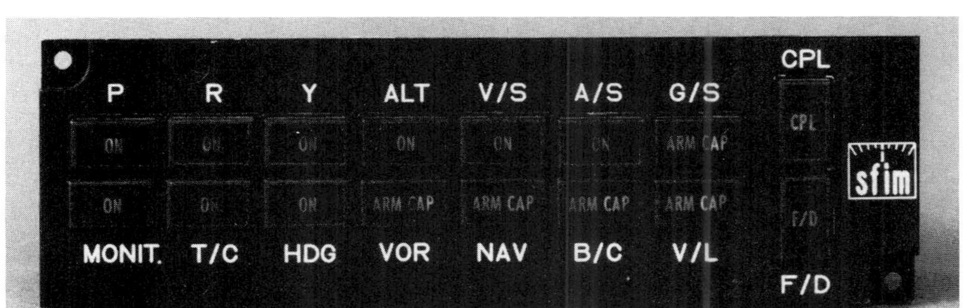

Controller for the SFIM Industries 85T31 automatic flight control system

AFCS 155 autopilot/flight director system

The AFCS 155 is a series duplex, fail-passive autopilot. Optimised for both single and dual pilot IFR operations, covering the whole flight envelope from hover to VNE, it is IFR-certified on the Eurocopter Super Puma and Dauphin helicopters and can be integrated with SFIM couplers. These include the FDC 85 three- or four-axis and FDC 155 four-axis digital coupler. In the duplex configuration each channel has its own power supplies, sensors and interconnections.

Basic functions include long-term attitude and heading hold, turbulence compensation, collective link mode and autotrim. All autopilot configurations satisfy single pilot IFR requirements and there are three upper modes: heading select, altitude hold and airspeed hold.

Fly-through or transparent handling characteristics allow the pilot to make quick attitude and heading changes while benefiting from dynamic damping. On releasing the controls the autopilot returns to long-term stabilised flight at the previously set attitudes. If new attitude settings are required the pilot can either use the stick release button on the cyclic pitch grip or can change the settings at a slow rate by using the stick-top four-way 'beep' trim button. The 'beep' trim button is also used to alter the reference airspeed slowly when the automatic airspeed hold mode has been selected. All versions of the system include automatic trim, which keeps the series actuators centred so that the autopilot has full authority.

Additional flight director coupler facilities provide IFR automatic navigation, radio navigation and approach capabilities, including steep approach MLS beam capture and track. There are also additional modes for anti-submarine warfare and offshore or search and rescue operations. The latter includes automatic pattern following, automatic up and down transitions to and from selected radio altimeter height, Doppler/radio height or sonar cable hover and low Doppler speed automatic hold. Compatible couplers include FDC 85 three- and four-axis navigation and approach couplers (the latter is Cat II certified), and the FDC 155 all mission digital coupler.

Specifications

Weight:
(autopilot computer) 8 kg
(servo amplifier) 4.2 kg
(autopilot control box) 1.6 kg
(4 actuators) 1 kg each
(3 trim servos) 1.3 kg each
(barosensor) 1.5 kg
(FDC 85) 2 kg
(FDC 155) 8 kg
(flight director coupler box) 2 kg
(collective pitch motor) 2 kg

Operational status

In production and installed on Eurocopter AS 332 Super Puma and AS 365N Dauphin helicopters.

Contractor

SFIM Industries (SAGEM Group).

VERIFIED

AFCS 165/166 flight director systems

The AFCS 165/166 and variants are four-axis flight control and flight director systems intended for medium and heavy-capacity helicopters. The power components are either electromechanical or electro-hydraulic, the sensors being digital or analogue. A dual/dual microprocessor system provides fail-passive and fail-operational capabilities without performance loss.

The basic AFCS 165/166 comprises Cat. II navigation and approach modes and allows single pilot IFR operations. A modular system approach is also possible. Modes relating to specific missions such as ASW, SAR or anti-tank are offered as an option, software integrated.

In addition to the monitoring and safety functions, the AFCS is fitted with preflight test, a computer self-test and built-in first and second line maintenance test facilities.

Operational status

Both flight directors are in production. The AFCS 165 has been selected by Eurocopter for the Super Puma Mk II, and the AFCS 166 has been chosen for the Advanced Light Helicopter developed by Hindustan Aeronautics.

Contractor

SFIM Industries (SAGEM Group).

VERIFIED

AFCS 166 control unit as fitted in the naval version of the Hindustan Advanced Light Helicopter (ALH)

APIRS Aircraft Piloting Inertial Reference strapdown Sensor

APIRS is SFIM Industries latest AHRS. It utilises advanced fibre optic gyros and silicon accelerometers to achieve a rugged design making it suitable for both commercial and military fixed- and rotary-wing aircraft.

APIRS provides attitude, heading, angular rates and linear acceleration data to the automatic flight control system and cockpit displays via ARINC 429 digital databuses. Optional interfaces include: the Honeywell ASCB bus, analogue, and three-wire synchro outputs.

APIRS is well suited to be a replacement for mechanical AHRS and vertical and directional gyros in older avionics systems.

Specifications

Dimensions: 289 × 105 × 114 mm
Weight: 3.2 kg
Power supply: 28 V DC, <30 W

Operational status

In production; selected for the following programmes:
- Eurocopter EC 135 and EC 155
- NH 90 NATO helicopter
- Hindustan ALH helicopter
- DASH 8 – 400
- DASH 8 100/200 and 300 upgrade
- Dassault Super Etendard upgrade

Contractor

SFIM Industries (SAGEM Group).

VERIFIED

APM 2000 autopilot module

The APM 2000 is the core of the Automatic Flight Control Subsystem for the new generation of helicopters, that permit single-pilot IFR operations.

The functions of the APM 2000 are: stability and control augmentation; long-term attitude hold; and AP modes of flight. Depending on the configuration, the system can operate either in simplex APM operation or in duplex APMs. The APM 2000 is a fail-passive digital computer that uses highly integrated technology in the modular concept known as 'Meghas®'. The APM 2000 is the basis of a computer family that can meet requirements from two to four-axis autopilot system implementation. Two identical APM 2000 modules can be integrated to build a fail-operational configuration, to improve the mission reliability.

The APM 2000 can interface either with the Electro-Hydraulic Actuators (EHA) through a SAS 2000 computer or directly with Smart Electro-Mechanical Actuators (SEMA).

A typical APM 2000 installation could include one APM 2000, two trim actuators, one to four SEMA actuators, one APMS, two AHRS (digital), and two ADU. According to version, the APM 2000 could be connected to flight display systems, radio navigation systems, a Doppler system, a MFDAU/HUMS, or one APM 2000 (duplex configuration).

Specifications

Dimensions: 337 × 183 × 38 mm
Power supply: 16 to 32.2 V, 25 W
Weight: 2 kg
Environmental: DO 160 C, TSO

Operational status

In production. Eurocopter Deutschland has selected the APM 2000 system for the EC 135, and the EC 145; it has also been selected by Eurocopter for the EC 155.

Contractor

SFIM Industries (SAGEM Group).

VERIFIED

The SFIM Industries APIRS AHRS (Attitude Heading Reference System) 0001322

SFIM Industries APM 2000 autopilot module 0001398

FCS 60B automatic Flight Control System

The FCS 60B is designed for commuter aircraft to provide ease of operation for flight profiles ranging from initial climb out to Cat II ILS or MLS approach.

In the BAE Systems Advanced Turboprop, the system consists of a pair of three-axis autopilot/flight director fully digital computers, which drive dual primary and trim servos. The FCS 60B also acts as a back-up fly-by-wire system for the three axes. This mechanisation allows independent and redundant en route operation of the pilot and co-pilot flight directors, as well as cross-monitoring for Cat II approaches.

Operational status

Certified on the BAE Systems Advanced Turboprop aircraft.

Contractor

SFIM Industries (SAGEM Group).

VERIFIED

Units of the SFIM Industries FCS 60B autopilot and automatic flight control system for commuter aircraft

SAS 2000 Stability Augmentation System

The SAS 2000 computer (Stability Augmentation System) is used to control the stability of helicopters. It is configured as two separate optional equipment packages in order to offer the maximum customer flexibility and choice: yaw SAS and/or pitch and roll SAS.

The yaw SAS consists of an integrated yaw rate gyro and control law computer, which provides command signals to an electromechanical series actuator driving the input lever of the tail rotor hydraulic boost.

The equipment consists of: a Fibre Optic Gyro (FOG) and Smart Electro-Mechanical Actuator (SEMA). The FOG measures the yaw angular rate and is used to provide damping about the vertical axis. In addition to the rate signal, the unit is also used to implement the control laws and output signals to feed the SEMA.

The SEMA is a high performance electromechanical actuator using modern brushless motors with rear earth permanent magnets. It incorporates a complete position servo feedback loop using Hall effect sensors within the same housing.

SFIM Industries pitch and roll SAS computer 0001396

The pitch and roll SAS computer is designed to be interfaced with a conventional attitude sensor, the SAS analogue pitch and roll computer controls the slaving loop of an electro hydraulic actuator. The pitch and roll SAS computer interconnects a full digital auto pilot module to the EHAs. In the VFR configuration the pitch and roll SAS computer is connected to helicopter generation, EHAs, trim actuator units (pitch and roll), cyclic stick grip and instrument panel switches, a remote vertical gyroscope or panel-mounted gyro horizon and pitch/roll axes solenoids.

Specifications

FOG
Mass: 0.75 kg
Software: RTCA DO 178B
Power supply: 16 V DC to 32.2 V DC
Environmental: DO 160 C

SEMA
Input: ARINC 429
Adjustable working stroke (± 2.5 and 8 mm)
Working speed up to 18 mm/s
Stall load: > 50 N
Limit load: > 4000 N
Mass: < 0.8 kg

Pitch and roll SAS computer
Dimensions: 200 × 122 × 112 mm
Weight: 1.3 kg
Power supply: 28 V, 20 W
Environmental: DO 160 C

Operational status

In production, and installed on Eurocopter EC 135.

Contractor

SFIM Industries (SAGEM Group).

VERIFIED

Audio-Radio Management System (ARMS)

The TEAM Audio-Radio Management System (ARMS) is designed for aircraft and helicopter applications. It is used to manage radio-communication, radio-navigation, the aircraft identification system, and the aircraft intercom system, together with several other functions. The basic ARMS comprises: two Audio-Radio Control and Display Units (ARCDU); and one Audio Management Unit (AMU). Optionally, the audio capability can be specified separately, in which case the system would include only two radio control and display units (RCDU), or the system could have more than two ARCDU and audio control panels, located either in the cockpit or elsewhere in the aircraft when fitted in special mission aircraft.

ARMS replaces the standard radio and audio control panels, reducing acquisition, installation, maintenance and logistic costs. ARMS provides immediate access to the main functional capabilities controlled, and the ARCDU is fitted with a high-quality colour LCD active matrix.

ARMS control functions are as follows: radio communication; radio navigation; the intercom system and aircraft identification. Optional functions include: datalink; public address; radio auto-tuning; aural warning; SELCAL; GPS or DGPS management. Additionally, ARMS can be integrated with other major avionic systems including: the navigation system; anti-collision system; centralised diagnostic system; flight management system; flight data acquisition unit.

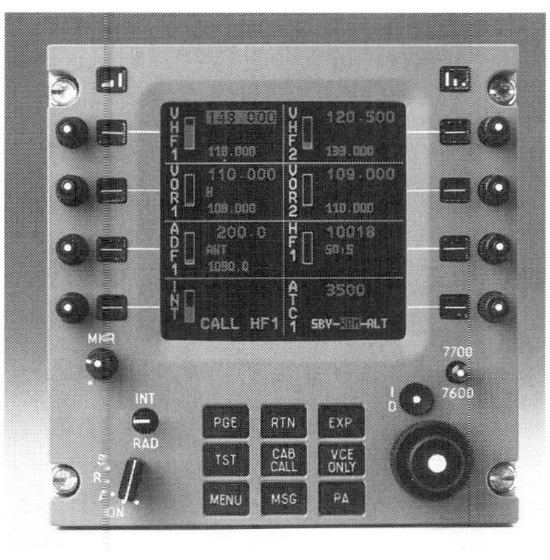

ARCDU front panel ***1999***/0044784

Specifications

ARCDU dimensions: 146 (W) × 150 (D) mm × H (Height (H) can be variable, greater than 105 mm, to meet installation requirements)
ARCDU weight: <3 kg
Interfaces: MIL-STD-1553; ARINC 429; other interfaces to meet customer requirements

Contractor

TEAM.

VERIFIED

CP3938 audio management system

The TEAM CP3938 audio management system is a single line replaceable unit, which enables two pilots to operate the following radio services; up to four communications radios; up to four radio navigation systems; and the aircraft emergency warning systems. In addition, it provides inter-communication between: the two pilots; the pilots and up to three passengers.

The TEAM CP3938 audio management system ***1999***/0044785

The CP3938 was designed for the EC120 helicopter, but can be fitted on many other helicopter types. It replaces the TEAM SIB27 (TB27) without any modification to the helicopter wiring.

The front panel is customised, using removable labels, to match the radio communication, radio navigation and warning sets fitted to the aircraft.

The transmission/reception channels and warnings adaptation is customised to the equipment fitted on the aircraft at installation.

Specifications

Power: 22-32 V DC, 12 W
MTBF: 16,000 h
Bandwidth: 300-5,000 Hz
Dimensions (H × W × D): 57 × 136 × 170 mm
Weight: 0.955 kg

Contractor

TEAM.

VERIFIED

SAVIB69 audio management system

SAVIB69 audio management system provides control of the aircraft's external and internal communications systems. It includes the ADAV aural warning function with 265 synthetic voice messages. SAVIB69 comprises: one audio management unit CTA 3488 (including voice alert); one audio control panel (BCA 3487) for each cockpit, one fuselage ground connection; and one telebriefing connection.

Each control panel provides an automatic transmit/receive function on the interphone channel; the ADAV aural warning system memory capacity of 265 messages corresponds to a global alert time of 5 minutes. The system permits the selection of: up to four communication channels with volume control; up to six navigation channels with volume control; four listening channels without volume control; one telebriefing connection; one double track audio recording system; 24 discrete signals corresponding respectively to one alert.

SAVIB69 audio management system 0011848

Each audio control panel transmits the status of its selector switches to the audio management unit via ARINC 429 busses. Each audio management unit incorporates up to two independent audio signal processors. The ADAV mode provides the crew with a prioritised listening monitior on a fixed frequency.

SAVIB69 is connected to the aircraft MIL-STD-1553B databus.

Specifications

BCA

Bandwidth: 300-5,000Hz ±3 dB
Dimensions: 65 (H) × 146 (W) × 110 (D) mm
Weight: 0.7 kg
Power: 17-32 V DC, 10 W
MTBF: 10,000 flight hours

CTA

Bandwidth: 300-5,000Hz ± 3 dB
Dimensions: 194 (H) × 60 (W) × 320 (D) mm
Weight: 3 kg
Power: 17-32 V DC, 20 W
MTBF: 4,500 flight hours

Operational status

Designed for Rafale, both single-seat and two-seat versions.

Contractor

TEAM.

VERIFIED

SELCAL airborne selective calling system

The SELCAL system is a selective calling device which enables a ground operator to call an aircraft without the pilot having to constantly monitor a particular frequency. The SELCAL system comprises: a signal processing unit; a code selection unit; visual and audio annunciators.

The SELCAL system can monitor any HF or VHF radio receiver installed on an aircraft. On receipt of a signal it lights a channel indicator and sounds an alarm. The system comprises five radio communication channels operating simultaneously and continually. The code selection panel is recommended as part of both the SIB54 (A320) and the SIB73 (A330340) systems.

Options: code selection panel BC2065; SELCAL call signal indicator BC2066.

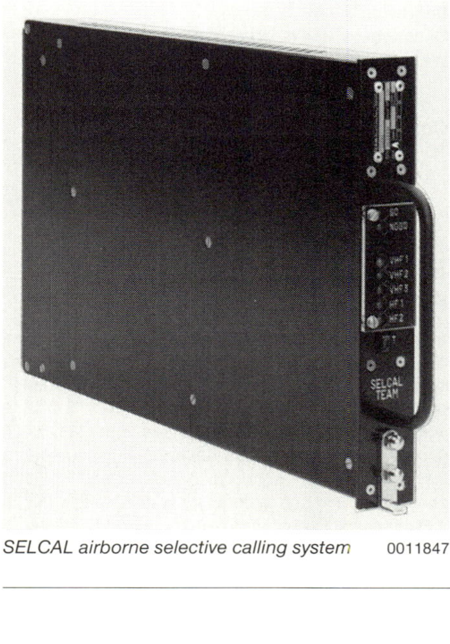

SELCAL airborne selective calling system 0011847

Specifications

Standards: ARINC 714 and ARINC 531
Approvals: TSO C59 according to DO 160C and DO 178A

Operational status

SELCAL model 2253 is currently installed on the following aircraft: Aerospatiale/Aeritalia ATR42; Airbus Industrie A300, A310; Boeing 737-500, 747-400, 757, 767, 777, MD-80; BAE Systems 146; Fokker F-100. The SC2539 model is currently installed on Boeing 737 first-generation aircraft.

Contractor

TEAM.

VERIFIED

SIB31/43/45/54/66/73/85 audio management systems

The SIB family of audio management systems provides crew members with an interphone link (in multicrew aircraft) and enables each one to operate the following facilities:

SIB31: radio communications; radio navigation; emergency warning system
SIB43: radio communications; radio navigation; warning system (option: public address)
SIB45: radio communications; radio navigation; emergency warning system
SIB54: radio communications; radio navigation; interphone; public address
SIB66: radio communications; radio navigation; interphone
SIB73: radio communications; radio navigation; interphone; public address; emergency warning system
SIB85 SATCOM: radio communications (including satcom); radio navigation; interphone; public address; emergency warning system.

The systems comprise the following units:

SIB31: junction box BJ1977; three to five main audio control panels CP1976
SIB43: one Remote Control Audio Unit (RCAU); up to three Audio Control Panels (ACP2531); hand microphone, headphones, speakers, jack panel
SIB45: junction box BJ2620A; three to five main audio control panels CP2618A
SIB54: up to five audio control panels ACP2788; 1 × audio management unit AMU2790; 1 × SELCAL code selection panel BC2065
SIB66: up to four audio control panels BC3438; 1 × ground crew panel PC2748; 1 × SELCAL code selection panel BC2065
SIB73: up to five audio control panels ACP2788AC01; 1 × audio management unit AMU3514; 1 × SELCAL code selection panel BC2065; 2 × speakers HP3520
SIB85 SATCOM: up to five audio control panels; ACP2788AF01; 1 × audio management unit AMU3514; 1 × SELCAL code selection panel BC2065; 2 × speakers HP3520.

Facilities provided include:

SIB31: 4 × Tx/Rx selective tunable; 4 × navigation channels tunable; 3 × listening channels
SIB43: up to six communications channels with volume control; 10 navigation channels; voice recorder; Ground Proximity Warning System (GPWS); Flight Warning Computer (FWC)

SIB31 audio management system 0011846

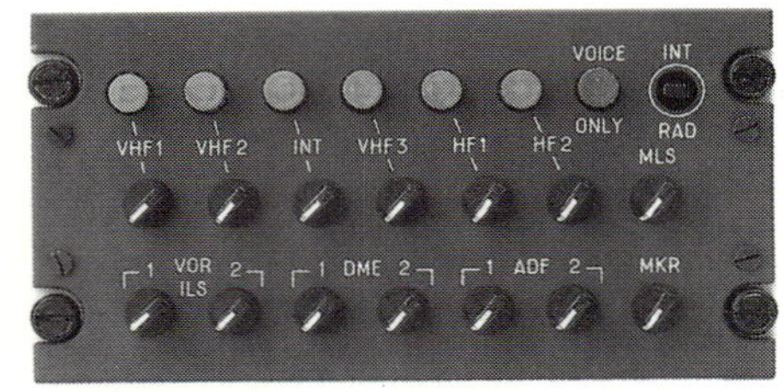

SIB43 audio management system **1999**/0044786

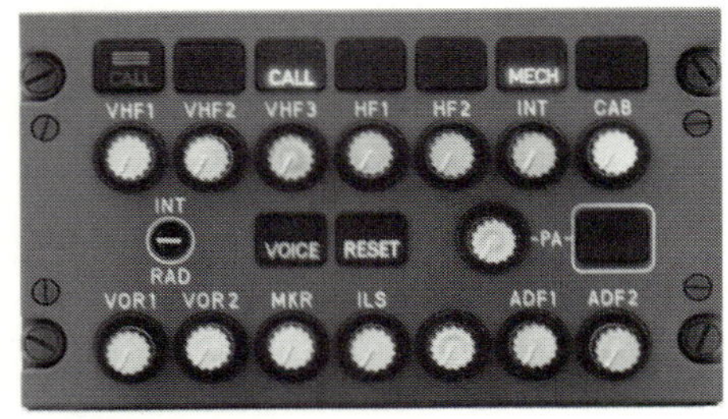

SIB73 audio management system 0011845

SIB85 SATCOM audio management system 0011844

SIB45: 6 × Tx/Rx selective tunable; 8 × navigation channels tunable; 3 × listening channels
SIB54: up to five audio channels; up to 12 navigation channels; one each public address/intercom/cabin channel
SIB66: six audio channels; six navigation channels; 2 × interphone channels; 1 × general call channel
SIB73: up to six audio channels; up to 11 navigation channels; one each public address/intercom/cabin channel
SIB85 SATCOM: up to seven audio channels (including 2 × satcom channels); up to six navigation channels; one each public address/intercom/cabin channels.

Operational status

Systems have been installed on the following aircraft:

SIB31: Dauphin SA365C; Ecureuil AS355F; Puma SA330J; Super Puma SA332L1 helicopters
SIB43: ATR42, ATR72
SIB45: Super Puma SA332 helicopters
SIB54: Airbus A320 aircraft
SIB66: Transall aircraft of the French Air Force Transport Air Command (COTAM)
SIB73: Airbus A330/340 aircraft
SIB85 SATCOM: Airbus A330/340 aircraft equipped with VOICE SATCOM from the cockpit.

Contractor

TEAM.

VERIFIED

AHV-16 radio altimeter

The AHV-16 microprocessor-based radio altimeter has been built with a reprogrammable memory and is designed for commuter and military aircraft. It has both digital and analogue outputs to ARINC 552 and ARINC 429 standards. It can interface with existing digital and analogue avionic equipment and with electromechanical instruments as well as electronic flight instrument systems.

Specifications

Dimensions:
(indicator) 3 ATI
(transmitter/receiver unit) 230 × 90 × 90 mm
Weight:
(indicator) 1.2 kg
(transmitter/receiver unit) 2.0 kg
Power supply: 28 V DC
(R/T unit) 15 W,
(indicator) 8 W
Accuracy 1 ft ±2% of altitude
Reliability: >5,000 h MTBF

Operational status

In production and service in transport aircraft and helicopters.

Contractor

Thales (formerly Thomson-CSF Communications).

VERIFIED

TLS-2040 MultiMode Receiver (MMR)

The TLS-2040 MultiMode Receiver is a new VOR/ILS receiver designed to replace the existing ARINC 547 VOR/ILS receivers. It fully complies with ICAO Annex 10 FM immunity requirements and includes a VHF receiver for the DGPS datalink. It is form-and-fit exchangeable with existing ARINC 547 receivers.

The TLS-2040 receiver can optionally be configured with Marker, MLS, DGPS and GPS functions by simply adding modules.

The TLS-2040 utilises the latest technology developed for commercial aviation and military programmes. The TLS-2040 belongs to the TLS-2000 product line.

Specifications

Basic functions: VOR/ILS/VHF receiver for DGPS
Options: Marker, MLS, DGPS and GPS
Dimensions: ½ ATR short
Weight: <4 kg
Power: 28 V DC
Interfaces: As defined by ARINC 547, MIL-STD-1553B, ARINC 429

Contractor

Thales (formerly Thomson-CSF Communications).

VERIFIED

Ground Collision Avoidance System (GCAS)

Designed and manufactured by Thomson-CSF DETEXIS, the GCAS is jointly marketed and supported worldwide by Teledyne Controls and Sextant Avionique. It is being marketed to meet airline requirements for Ground Proximity Warning Systems (GPWS) and Enhanced Ground Proximity Warning Systems (EGPWS), as well as Terrain Awareness Warning System (TAWS) requirements.

GCAS incorporates a worldwide digital terrain database (like EGPWS) and an alert algorithm that includes aircraft performance potential by incorporating attitude and aircraft type-specific and configuration data (for example, true aircraft flap settings, actual aircraft weight and actual engine performance capabilities). Audio and visual caution and warning messages are similar to those found in GPWS. GCAS is designed as a near form/fit replacement for existing GPWS / EGPWS systems and to be linked to aircraft navigation or GPS data. GCAS reverts to basic GPWS mode if a reliable navigation signal is lost. The digital GCAS can be operated without a display, a feature intended to encourage retrofits on older aircraft, although it is intended to provide display data on appropriate weather radar or EFIS screens, using ARINC 453 protocols. An upgraded 3-D display, for future aircraft, such as Airbus A3XX, is planned.

GCAS calculates aircraft time to clear obstacles, rather than time to impact. In making its calculations, GCAS demands no more than 0.5 *g* manoeuvre, 75 per cent airframe load limit and 90 per cent take-off power.

Modes of operation include: Collision Prediction and Alerting (CPA) - true look ahead capability based on predicted aircraft flight path and knowledge of terrain environment - advanced and reliable ground hazard warning and caution - alert computation based on aircraft climb capability.

Basic modes of operation include: (ICAO Annex 6) modes 1 & 2 (back-up mode); modes 3, 4 and 5; callouts and bank angle modes.

Specifications

Dimensions: 2 MCU
Weight: 2.8 kg
Power: 28 V DC, 25 W or 115 V AC

Operational status

The system has been under evaluation at the French Flight Test Centre in an Airbus A300 simulator since 1995 and on board a Dassault Falcon 2000 since 1997. Initial certification in the Falcon 2000 was in 1998 under European Joint Airworthiness Authority (JAA) rules, followed by US FAA certification. Availability for commercial air transport users was anticipated for early 2000. GCAS will meet the future FAA NPRM and ICAO Amendment of Annex 6 for TAWS and be compliant with the new TSO C-151.

Contractor

Thales (formerly Thomson-CSF DETEXIS).

VERIFIED

JETSAT Aero-I satcom

Designed and manufactured by Thomson-CSF DETEXIS, the satcom Aero-I system, JETSAT, is marketed and supported worldwide by Sextant.

The JETSAT is designed to operate under the Inmarsat 3 satellite spot beams, offering cockpit and cabin voice, fax and data services. This fully integrated system provides five channels (four voice/fax/PC data and one packet data), and meets FANS operational requirements. It comprises one IGA (Intermediate Gain Antenna) and two LRUs (Line Replaceable Units): one HLD (High power amplifier/Low noise amplifier/Diplexer) and one SDU (Satellite Data Unit). The JETSAT is a compact, lightweight system offering maximum installation flexibility, with no operational constraint on LRU location (other than the antenna).

Specifications

Dimensions:
2 LRUs each of 4 MCU
(antenna) 650 × 336 × 140 mm
Weight: 18 kg

Operational status

JETSAT has been selected for a number of Falcon and Bombardier Global Express and Challenger aircraft.

Contractor

Thales (formerly Thomson-CSF DETEXIS).

VERIFIED

GERMANY

AD-FIS Flight Inspection System

The AD-FIS is a highly integrated, digital flight inspection system which utilises advanced methods of real-time data processing and satellite navigation technologies. AD-FIS is compliant with ICAO recommendations and performs calibration of all ground-based navigation aids including, ILS, MLS, TACAN, NDB, DME, PAR, SSR, VHF/UHF Comms, VDF and UDF.

The system consists of the following subsystems: flight inspection receivers; position reference sensors; data processing equipment; operator console and aircraft interface.

Data from the system's sensors, which include airborne receivers and ground-based position reference equipment, is interfaced to the computer via an ARINC 429 databus. Data from ground-based sensors is transmitted to the aircraft via a UHF datalink. Sensor information received by the computer is stored in raw, unprocessed format on a magneto optical disk, allowing future access to all sensor data, and is simultaneously processed by the computer. Processed data is displayed, in real time, to the operator in graphical format via the operator's LCD screens, and finally printed on the system's colour printer.

The system hardware is housed in a compact rack design. The rack houses the AD-VC6 computer, flight inspection receivers, colour printer, operator station and additional items such as a spectrum analyser. Ground-based reference equipment includes a highly accurate laser tracking system and DGPS station.

Specifications

Dimensions: vary according to customer specifications
Weight: (approximate) 150 kg. Varies according to system design

Operational status

In production and in service.

Contractor

Aerodata Flugmesstechnik GmbH.

VERIFIED

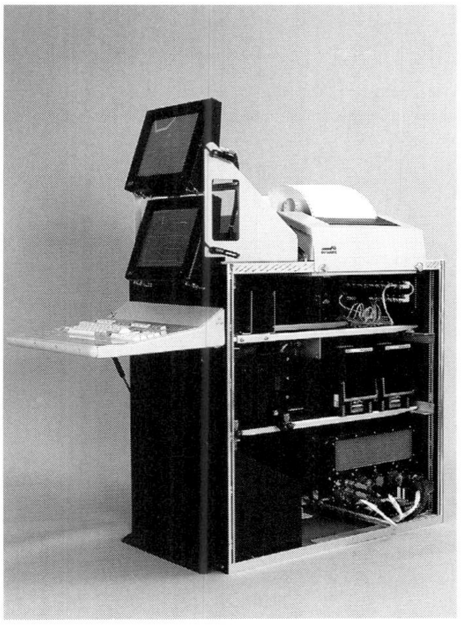

AD-FIS Flight-Inspection System
0005428

AeroNav® Integrated Navigation System (INS)

AeroNav® is a TSO C-129a certified GSP navigation system. It provides a 12-channel GPS receiver, integrated with an Inertial Measurement Unit (IMU) for improved navigation performance and Aircraft Autonomous Integrity Monitoring (AAIM). Flight management software with Jeppesen database supports the crew during mission planning and in flight. AeroNav®'s open interface architecture provides links to moving maps like the Dornier DKG-3 and special mission equipment, for example FLIR, or communications equipment (Satcom or GSM).

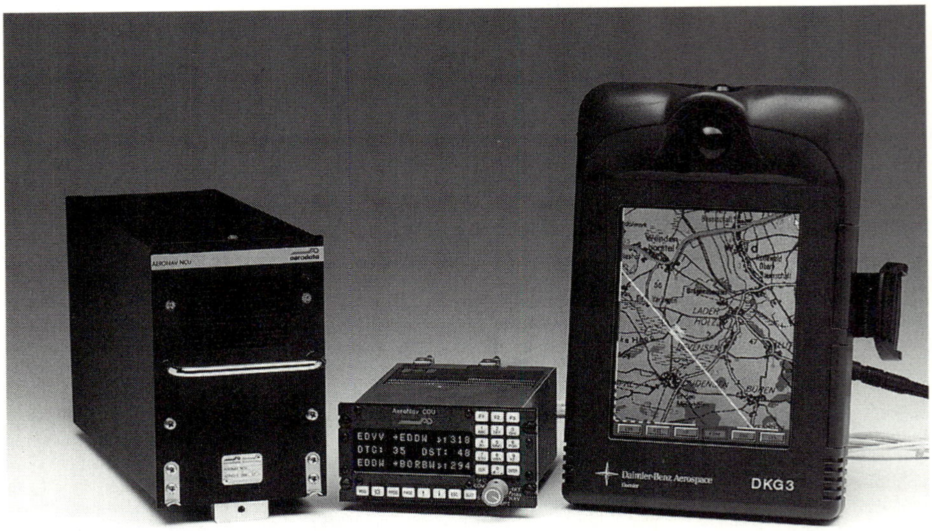

AeroNav® is certified for B-RNAV and IFR non-precision approaches.

Specifications

Dimensions:
NCU (Navigation and Communication Unit): 214 (H) × 126 (W) × 360 (D) mm (½ ATR short)
CDU (Control and Display Unit): 66 (H) × 146 (W) × 185 (D) mm
antenna: diameter 88.9 mm, height 14.5 mm
Weight:
NCU 5.9 kg
CDU 1.4 kg
antenna: 0.15 kg

Operational status

In production and in service.

Contractor

Aerodata Flugmesstechnik GmbH.

VERIFIED

AeroNav® Integrated Navigation System (INS)
0005427

3300 series VHF navigation systems

The series 3300 navigation systems have been designed for use in fixed- or rotary-wing aircraft, and have been certified for high-altitude operation. The series includes a variety of VOR/ILS receivers, indicators, special application converters and a glide slope receiver, which can be combined to tailor systems to meet the exact requirements of specific installations. Interfaces are provided for most types of electromechanical or electronic HSI, RMI and CDI displays.

Complete VOR/LOC receiver, VOR/LOC converter and glide slope receiver systems are housed in a single compact unit, which complies with ARINC standards. They do not require the individual control boxes and remote receiver units with interconnecting cables, or external forced-air cooling, which are typical of older navigation receiver designs in this performance class.

Clear LCD displays, which are easily read in the brightest sunlight show both active and standby frequencies. Remote tuning via databus can be provided, and tandem tuning heads are available for dual-cockpit installations. Built-in test functions for the displays and for the proper operation of the receivers and converters are standard features.

The NR 3300 navigation receivers are certified to the stringent requirements of all applicable FAA TSO, LBA, ICAO and RTCA specifications.

Specifications

Dimensions: 146 × 47.5 × 225 mm (excluding clearance for connectors and cables)
Weight:
(NR 3300-(1)/(3)) 1 kg
(NR 330-(2)/(4)) 1.2 kg

NR3300 and Indicator

Operational status

In service and production.

Contractor

Becker Avionic Systems.

VERIFIED

ADF 3500 system

In addition to the standard frequency range of 190 to 1,799.5 kHz, the ADF 3500 also receives the international maritime distress frequency of 2,182 ±5 kHz, making it suitable for search and rescue, and other offshore operations. It is certified for high-altitude operation in turbine-powered aircraft.

The complete system is housed in a single compact unit, which complies with ARINC standards. It does not require remote boxes, interconnecting cables, or external forced-air cooling.

Clear LCD displays, which are easily read in the brightest sunlight show both active and standby frequencies. Preselection of a standby frequency enables instant switch over to a second station for position fixing by cross bearings. A low-profile combined sense and loop antenna is suitable for mounting on high-speed aircraft. Becker RMI converters are available to enable the system to drive most types of indicators.

The ADF 3500 systems are certified to the requirements of applicable FAA TSO, JTSO, RTCA, EUROCEA and FTZ specifications.

RA 3502 ADF receiver system 0001323

Specifications

System model numbers	System components
ADF 3502-(1)	Standard version AD 3502 receiver AN 3500 antenna ID 3502 indicator
ADF 3503-(1)	For RMI with standard synchro input AD 3502 receiver AN 3500 antenna AC 3503-(1) converter
ADF 3504-(1)	For RMI with standard 2.5 V DC Sin/Cos input AD 3502 receiver AN 3500 antenna AC 3504-(1) converter
ADF 3504-(2)	For RMI with standard 5-10 V DC Sin/Cos input AD 3502 receiver AN 3500 antenna AC 3504-(2) converter
CU 3502-(01)	Tandem Control Unit can be added to the system to enable dual control operation

Dimensions
AD 3502 receiver: 146 × 47.5 × 245 mm
Weight: 1.0 kg
ID 3502 indicator: 82.55 × 82.55 × 135 mm
Weight: 0.5 kg
AN 3500 antenna: 190 × 54 × 330 mm
Weight: 1.7 kg

Supply voltage:	25 - 30 V DC (20 V DC in emergency) 26 V AC, 400 Hz, for AC 3503-(1) converter
Power consumption:	At 27.5 V DC, without panel lights
ADF 3502	650 mA
ADF 3503	1.15 A
ADF 3504	0.6 A
Frequency Ranges:	190-1,799.5 kHz and 2,182 kHz ± 5 kHz
Channels Spacing:	500 Hz
Bearing accuracy:	<3 ° at 70 μV 190-850 kHz <8 ° at 70 μV >850 kHz

Contractor

Becker Avionic Systems.

VERIFIED

AirScout navigation and moving map system

The Becker AirScout moving map display is designed for use on aircraft and helicopters. It uses GPS data to indicate position, velocity and current track information on a full-colour electronic moving map display. Some AirScout systems are integrated with GPS, others are not.

The core of the system is the Jeppesen database which includes data on all airports, navaids and communication frequencies.

Scanned images can be read into the system from any available source, including maps, satellite imagery, aerial photography and FLIR.

The AirScout consists of one panel-mounted full-colour LCD with flexible graphic capabilities, a dzus-mounted Control Unit (CU) and ¼ ATR short LRU, which houses the GPS sensor, processor and mass memory. The display can be mounted in the panel, or specially configured for use in helicopters on a side panel where it can be folded away when not in use. AirScout is also available as a self-contained, single-block, unit.

The display is swivel mounted to provide a wider viewing angle and to reduce glare. The display is readable in bright sunlight.

The AirScout family is based on the following systems:
AirScout Map (without GPS)
AirScout Voyager (with GPS)
AirScout Executive
AirScout Professional (pentium)

Specifications

Dimensions:
Display: 165 × 200 × 72 mm
Control unit: 47.5 × 145 × 127 mm
Main LRU: ¼ ATR
Weights:
Display: 1.45 kg
Control unit: 0.48 kg
Main LRU: 1.95 kg
Power supply: 28 V DC,1 A
GPS sensor: 12 channels; WAAS and P-code GPS sensor optional; DGPS receiver optional

Operational status

In service with military, police and medical services in Europe and the USA.

Contractor

Becker Avionic Systems.

VERIFIED

AR 3202 VHF transceiver

The single block unit AR 3202 VHF transmitter/receiver is a member of Becker's 3000 Series Prime Line avionic systems which feature microprocessor control. It provides 760 channels in the VHF band which extends from 118 to 136.975 MHz for civil aircraft. This range can be extended to 144 or 152 MHz for military aircraft. The system features a non-volatile memory and solid-state switches which eliminate mechanical contacts, giving increased ruggedness and reliability. Opto-electronic switching is employed for frequency selection. Frequency generation and display are microprocessor-controlled. The display comprises two liquid crystal presentations, one of which indicates the active channel while the other shows a preselected frequency.

The complete system is housed in a single compact unit, which complies with ARINC standards. It does not require the remote boxes, interconnecting cables, or external cooling of older designs.

Additionally the set can be operated using the CU 3202 remote controller. This is particularly useful in a tandem-seat trainer where any frequency selection made in one cockpit is displayed in the other.

The transmitter output power is 20 W and the AR 3202 is suitable for both fixed-wing aircraft and helicopters.

Specifications

Dimensions: 47.5 × 146 × 225 mm
Weight: 1.3 kg

The Becker Avionic Systems AirScout fuel management system application ***2000***/0081494

Frequency ranges:
Standard: 118.0-136.975 MHz
Option 1: 118.0-144.0 MHz
Option 2: 118.0-152.0 MHz
Channel spacing: 25 kHz
Power supply: 25-30 V DC; 3.5 A (transmit), 0.24 A (standby). Certified to FAA TSO C37c, TSO C38c and ICAO Annex 10 specifications.

Operational status

In production and in service.

Contractor

Becker Avionic Systems.

VERIFIED

AR 3209 VHF transceiver

There are two versions of the single block unit AR 3209 transceiver, the AR 3209-(09) 5 W system and the AR 3209-(11) 10 W system.

General features include: storage memory for up to 20 of 760 channels; front panel adjustment of squelch, side tone and intercom; two sunlight-readable LCDs.

The AR 3209 is an ideal retrofit for the Becker COM 2000 and is cable, plug and form/fit compatible.

Specifications

Dimensions: 146 × 47.5 × 229 mm
Weight: 1.2 kg
Radio frequency: 118.0-136.975 MHz
Channel spacing: 25 kHz
Memory channels: 20
Power supply: AR 3209-(09): 13.75 V DC
AR 3209-(11): 27.5 V DC
Certifications: TSO C37d, TSO C38d, and ICAO Annex 10

Operational status

In production.

Contractor

Becker Avionic Systems.

VERIFIED

AR 4201 VHF-AM transceiver

The small lightweight AR 4201 offers 760 channels and is certified for use in VFR and IFR equipped aircraft. It is ideal for installation in gliders and home-built and small single-engined aircraft due to its limited power requirement and 57 mm round format.

The equipment has a transmit power of 5 to 7 W. A standby frequency and 99-channel memory are available and can be easily programmed and recalled. The AR 4201 features intercom, panel lighting, voltage indicator, an RF input, automatic test routines, a serial interface and an optional temperature indication.

Two dynamic and two standard microphones can be connected without any alterations. The display shows the active frequency and either the standby frequency, the memory channel used or the supply voltage and also the external temperature (with the optional temperature sensor).

Becker Avionic Systems AR 3209 0001174

Becker Avionic Systems AR4201 VHF-AM transceiver
2001/0098258

Using the RS-232 interface, all functions can be remotely operated. With this standard feature, the AR 4201 can be integrated into future Becker Systems.

Specifications

Dimensions: 192 × 60.6 × 60.6 mm
Weight: 0.67 kg
Power supply: 12.4-15.1 V DC;
(transmit) <2.5 A
(standby) <0.07 A
Frequency range: 118.0-136.975 MHz.
Channel spacing: 25 kHz
Memories: 99
Certifications: TSO C37d, TSO C38d, and ICA Annex 10

Contractor

Becker Avionic Systems.

UPDATED

AS 3100 audio selector and intercommunication system

A member of the Becker 3000 Series avionic systems, the AS 3100 controls four transmitters and up to six receivers. By addition of an auxiliary unit, a further six receiver units can be added to the audio chain. In different versions the AS 3100 is capable of either voice-operated switch or push-to-talk operation with all other stations, voice filter for ADF and navigation systems, connection of various types of microphone and emergency operation and providing redundancy for the transmitter/receiver operation. The intercom amplifier has a common bus connecting up to six cabin and three cockpit stations. A cockpit voice recorder output is incorporated.

The system is of modular construction and may be tailored to precise customer requirements. It is suitable for fixed-wing aircraft and helicopters. To achieve maximum adaptability, a full range of sub-units has been developed to extend the function of the main AS 3100 controller to a full cabin communication and passenger entertainment system. An intercom amplifier permits communication between passengers and crew in noisy aircraft such as helicopters, and a service station allows communication between the crew and flight attendants as well as a public address facility. A tape player, used in conjunction with the public address amplifier, is the basis of passenger entertainment through headsets or loudspeakers. The public address amplifier includes one mono or stereo amplifier and a double-tone gong which operates when activated by the fasten seat belts or no smoking signs.

An external jack box, which is normally installed in the wheel well or any other readily accessible location, permits communication between cockpits or flight deck and ground crew during starting and departure checks.

All units operate from a 28 V DC supply.

Digital Voice Communication System DVCS 5100

The DVCS 5100 is composed of the Audio Selector Unit REU 5100 ( ) and the ACU 5100 ( ).

The system can drive up to 8 radios transceivers or 7 transceiver plus one public address amplifier(PA) is capable of monitoring up to 8 transceivers and 8 navigation receivers where the volume is individually adjustable and monitoring up to 6 fixed inputs.

The 10 internally generated aural warnings are activated by discrete control lines.

Aircraft Intercommunication (IC) in HOT MIKE, VOX or PTT activated mode, the IC volume can be adjusted separately.

Optical and / or acoustical IC call and quit functions. Connecting / disconnecting of different intercom circuits between cockpit and operator or passenger is possible.

BITE to test all controls , indications and the microcontrollers. Optical and acoustical test result output.

Switchover to SLAVE mode for training applications or in case of partial defect.

Automatic switchover to EMERgency operation in case of failure of power supply. The system is fitted with a serial interface for programming the unit to customer or aircraft requirements, by manufacturer or approved maintenance workshop.

Specifications

Width: 146.1 mm
Height: 76.2 mm
Depth: approx. 110 mm

Standard ARINC 8 HE:
Mounting: D-ZUS
Weight: = 600 g

Remote Electronics Unit REU 5100 - ( ):
Length: 320.5 mm
Width: 57.2 mm

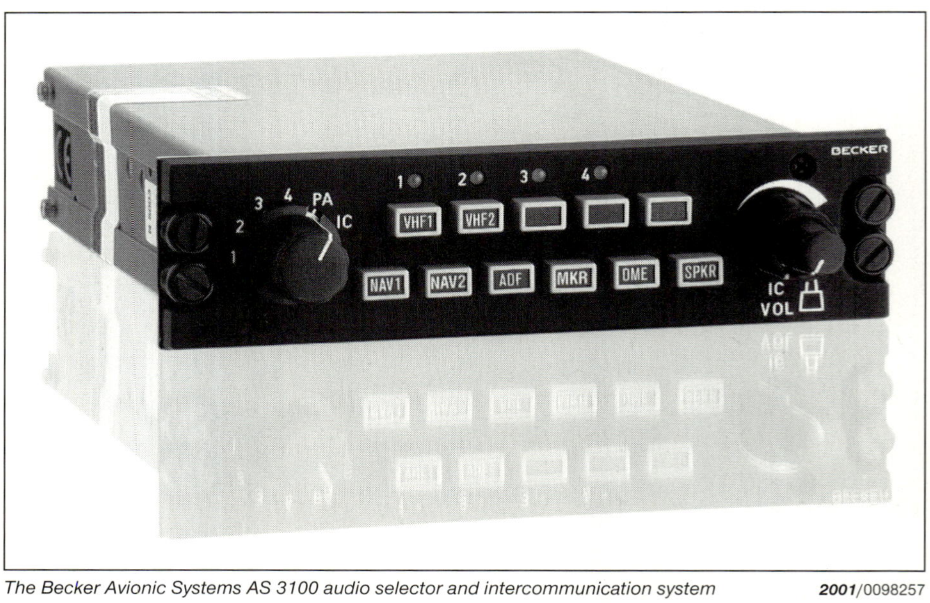

The Becker Avionic Systems AS 3100 audio selector and intercommunication system ***2001***/0098257

Height: 193.5 mm
Standard: 1/4 ATR short
Mounting: ATR Fixture
Weight: = 1,8 kg

Environmental conditions:
Operating temperature:
(ACU) -20 to +55° C
(REU) -40 to +55° C
(short time) +70° C
Storage temperature: -55 to +85° C
Altitude: = 50 000 ft (15 200 m)
Vibration: S+U
Acceleration: 12 g
Shock: (operational) 6 g / 11 ms half-sine wave
(crash safety) 15 g / 11 ms half-sine wave
Other environmental conditions acc. to DO-160D

Applicable documents
RTCA DO-214 audio systems characteristics
EUROCAE / RTCA ED 12B / DO-178B; Level C Software
EUROCAE / RTCA ED 14D / DO-160D environmental conditions
JTSO C50c audio performance
Mil-Spec L 85762A NVG compatibility (Option)

Dimensions:
(main control unit) 38 × 146 × 35 mm
(auxiliary unit) 29 × 146 × 26 mm
(cassette player) 57 × 146 × 170 mm
(service station) 210 × 66 × 115 mm
(public address amplifier) 129 × 45 × 245 mm
(external jack box) 117 × 80 × 80 mm
Weight:
(main control unit) 0.8 kg
(auxiliary unit) 0.2 kg
(cassette player) 1 kg
(service station) 1 kg
(public address amplifier) 0.8 kg
(external jack box) 0.6 kg

Operational status

In production and in service.

Contractor

Becker Avionic Systems.

UPDATED

ATC 3401 Mode A/C transponder

The single-block unit ATC 3401 is certified in accordance with FAA TSO C47c, Class 1A, for the highest level of unrestricted service and can report altitudes up to 62,700 ft.

The ATC 3401 transponder displays 4,096 identification codes in its left window and can display its operating mode or flight level reporting altitude in its right window. A stored VFR code, such as 1200, can be quickly recalled with a VFR push-button. Provisions are included to enable IDENT to be controlled from a button on the control stick and for power to be turned on and off from an external control, if required.

The complete system is housed in a single compact unit, which complies with ARINC standards. It does not require remote boxes, interconnecting cables, or external forced-air cooling.

To facilitate dual-transponder installations, provisions are included for automatic transfer of one transponder to the standby mode, when the other transponder is selected for normal mode operation. This prevents both transponders transmitting at the same time.

The ATC 3401 Mode A/C transponder is a member of the Becker PrimeLine family of avionics equipments.

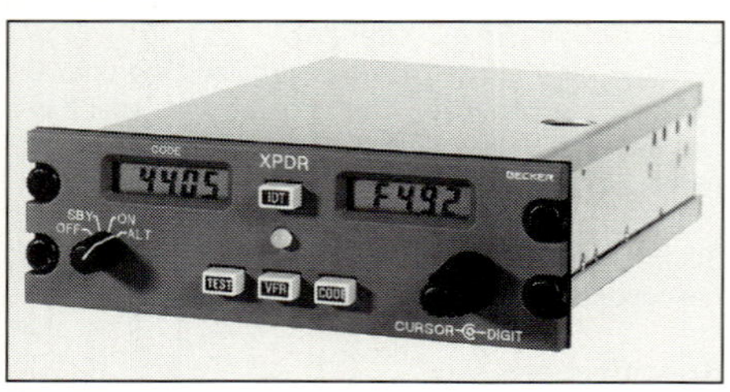

The Becker Avionic Systems ATC 3401 Mode A/C transponder ***2000***/0081497

Specifications
Dimensions: 146 × 47.5 × 217 mm (not including clearance for connectors and cables)
Weight: 1.2 kg
Power supply: 10-32 V DC; at 28 V DC, 0.8 A normal, 0.3 A standby; at 14 V DC, 1.5 A normal, 0.5 A standby
Transmitter frequency: 1,090 MHz ±0.3 MHz
Receiver frequency: 1,030 MHz ±0.2 MHz
Modes: A, A + C

Operational status
In service.

Contractor
Becker Avionic Systems.

VERIFIED

COM 5200 series VHF communication systems

The COM 5200 systems flexible architecture allows the VHF communication radios installation to be customised to provide the required combination of output power and cost. Changes in output power can be made later without extensive additional installation costs. The system is intended for installations where minimum panel space is to be used. The remote controlled system consists of the CU 5209 (control unit) and the remote-controlled RT 3209 (-11) VHF transceiver. The lightweight CU 5209 control unit fits into a standard 2.25 in (57 mm) round instrument panel cut-out, and is only 2.5 in (63.5 mm) deep. Lightweight remote transceiver units (RT 3209 (-11)), with output powers from 5 to 20 W are mated with this control head, to complete the systems and can be installed at any convenient place in the aircraft.

The CU 5209 control unit uses a high-contrast, double line LCD display, which is readable in bright sunlight.

Both active and standby frequencies are displayed, and can be transferred by a single stroke of the 'Flip-Flop' button. Up to 99 preset frequencies can be entered from the front panel and stored in non-volatile memory. A BITE automatically checks and monitors the system, to ensure proper operation and to facilitate maintenance. All systems will be JTSO certified for either VFR or IFR use in all types of fixed-wing and rotary-wing aircraft, to most ICAO requirements for VHF radio. The COM 5200 systems can be combined with other Becker Compact-Line avionics systems, such as VOR/ILS navigation receivers and transponders, which have similar control units. The remote transceiver units can be controlled by other types of CDU or FMS devices and can be integrated into flight management systems.

Specifications
CU 5209 for RT 3209 Control unit
CU 5202 for RT 3202
RT 3209 (-11) ≥10 W (CW) transceiver
RT 3202 =16 W (CW) transceiver
Frequency range: 118.000-136.975 MHz
Channel number: 760
Channel spacing: 25 kHz
Dimensions excluding connector: CU 5209
CU 5209 and CU 5202: 61.3 × 61.3 × 62 mm
RT 3202: 134 × 50 × 265 mm
RT 3209: 134 × 50 × 243 mm
Power supply: 28 V DC;
Transmit:
(RT 3202) ≤4 A
RT3209 (-11) ≤2.5 A
Standby: ≤0.25 A

Contractor
Becker Avionic Systems.

UPDATED

Becker remote-controlled transceiver COM 5200 series VHF communications systems **1999**/0044787

Digital Voice Communication System DVCS 5100

The DVCS 5100 is a member of the Becker PrimeLine communication and navigation family; it comprises the Audio Selector Unit (ASU) Remote Electronics Unit REU 5100 ( ) and the Audio Control Unit ACU 5100 ( ). The system can control up to eight radio transceivers, or seven transceivers plus one Public Address (PA) amplifier; it is also capable of monitoring up to eight transceivers and eight navigation receivers, where the volume is individually adjustable, and of monitoring up to six fixed input signals.

The 10 internally generated aural warnings are activated by discrete control lines. While controlling the aircraft intercommunication facilities in 'hot microphone' VOX or PTT modes, the volume can be adjusted independently.

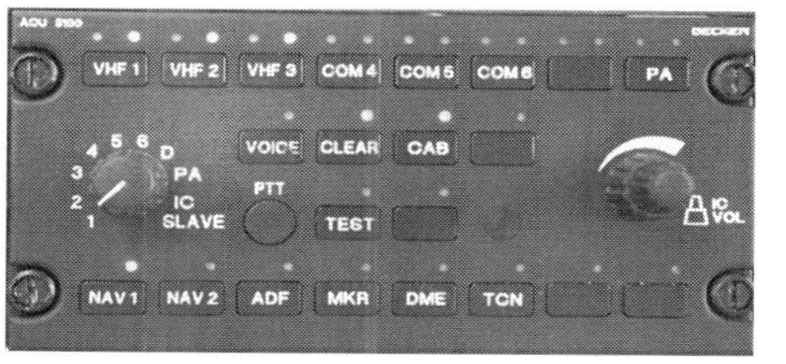

The Becker Avionic Systems digital voice communication system DVCS 5100 **2000**/0062183

Optical and/or acoustic call and quit functions and separate intercom circuits between the cockpit and passenger cabin are provided.

Automatic switch over to emergency operation is provided in case of failure of the power supply. The system is fitted with a serial interface for programming customer or aircraft requirements; such customising actions need to be performed at the vendor's facility or an approved workshop.

Switch over to SLAVE mode is available for training applications, or in case of partial defect. Full BITE is provided.

Specifications
Audio Control Unit ACU 5100 ( )
Dimensions: 146.1 (W) × 76.2 (H) × 110 (D) mm
Weight: 0.6 kg
Mounting: dzus

Remote Electronics Unit REU 5100 ( )
Dimensions: 320.5 (L) × 57.2 (W) × 193.5 (H) mm, ¼ ATR short
Weight: 1.8 kg

Contractor
Becker Avionic Systems.

VERIFIED

IN 3300 series VOR/LOC/GS indicators

The IN 3300 series indicators are precision engineered instruments which display VOR, localiser and glide slope deviation information for en route navigation and approaches.

Either 5, 14 or 28 V lighting is provided and non-reflective glass is used to ensure reliable readability under all operating conditions.

The model IN 3300-(10) contains a course selector and display, rectilinear VOR/LOC and glide slope cross-pointers and warning flags, and a TO-FROM indicator. The IN-3300-(3) also contains a built-in marker beacon receiver and automatically photocell-dimmed indicator lamps for airway marker, outer marker and inner marker beacon indications.

IN 3300 series indicators can be combined with Becker's NR 3320/30 navigation receivers to form a dependable, lightweight, easily installed VHF navigation system.

The IN 3300 series indicators are approved for operation to 50,000 ft, and are certified to the rigorous requirements of all applicable FAA and RTCA specifications.

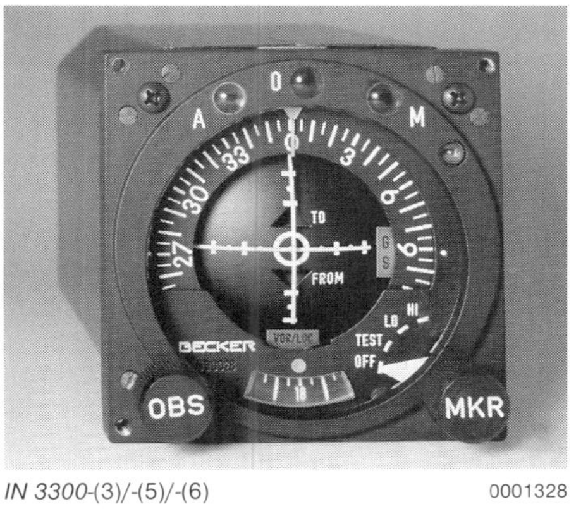

IN 3300-(3)/-(5)/-(6) 0001328

Specifications

Dimensions: 82.55 × 82.55 × 130 mm (excluding clearance for connectors and cables)
Weight:
(IN 3300-(3)) 850 g
(IN 330-(10)) 820 g

Operational status

In production and in service.

Contractor

Becker Avionic Systems.

VERIFIED

IN 3360-(2)-B compact VOR/LOC/GS indicator

The IN 3360-(2)-B indicator is a precision engineered instrument which displays VOR, localiser and glide slope deviation information for en route navigation and approaches.

The indicator contains a course selector and display, VOR/LOC and glide slope cross-pointers, warning flags, and a TO-FROM indicator. It provides all steering information needed for IFR flying, and is ideal for use in aircraft with limited panel space, or for use as part of an 'emergency bus' avionics package for large aircraft.

The IN 3360-(2)-B indicators are approved for operation to 40,000 ft, and are certified to the requirements of TSO C52a.

Specifications

Dimensions: 60 × 60 × 110 mm
Weight: 0.40 kg

Operational status

In production and in service.

Contractor

Becker Avionic Systems.

VERIFIED

IN 3360-(2)-B VOR/LOC/GS indicator 0001329

NR 3300 series VHF navigation receivers

The NR 3320-(02)-(01) and NR 3330-(02)-(01) single-block units are new navigation receivers designed as retrofit systems for the Becker NAV 2000 series receivers with which they are pin compatible without mechanical modification.

The NR 3320-(02)-XXX provides a composite navigation signal with glide slope receiver capability, while the NR 3330-(02)-XXX provides its composite navigation output without the glide slope.

Active and standby (preset) frequency read out is given on two sunlight-readable LCDs.

Specifications

Dimensions: 126 × 46 × 186 mm
Weight: 1.2 kg
Power supply: 13.5 V AC/27.5 V DC
NAV Receiver:
Frequency range: 108-117.950 MHz
Channels: 200
Channel spacing: 50 kHz
Memory channels: 20
Glide slope Receiver:
Frequency range: 329.150-335 MHz
Channels: 40
Channel spacing: 150 kHz

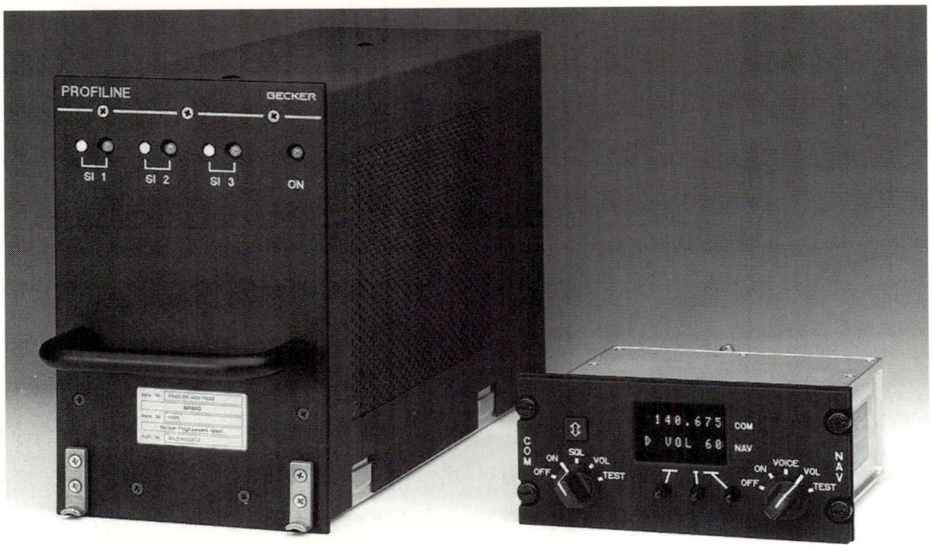

Becker ProfiLine NR900 receiver and CU900-(1) control unit 0005424

Contractor

Becker Avionic Systems.

VERIFIED

PrimeLine communications and navigation system for business aviation

The Becker PrimeLine communications and navigation system fulfils the need of the business aviation market for a complete dzus rail-mounted family of Cat I avionic products. The basic PrimeLine products are completely housed in single units, requiring no remote boxes. Products in the range include:

AR 3202 20 W airborne VHF transceiver, AR 3209 10 W airborne VHF transceiver;
NR 3320/30 VOR/LOC/GS navigation receivers;
IN 3300 series VOR/ILS navigation indicators, including a 3 in VOR/LOC/GS CDI, a 3 in VOR/LOC CDI, and a 2.25 in VOR/LOC/GS CDI;
ADF 3502/3/4 automatic direction-finder systems;
RMI 3337 radio magnetic indicator;
HSI 421 (4 in) and HSI 8131 (3 in) horizontal situation indicators;
Audio selector and indicator systems;
ATC 3401 transponder;
AirScout Moving map system.

Operational status

In service.

Contractor

Becker Avionic Systems.

VERIFIED

PrimeLine II communication and navigation system

The PrimeLine II is a remotely controlled system consisting of COM, NAV, ADF and ATC systems. It is designed for installations where there is minimum available panel space. It utilises small, lightweight, control units, which fit into standard 2¼ in (57 mm) round instrument panel cut-outs. The units are only 2½ in deep.

The connection between the control units and the transceivers is achieved by use of two prefabricated cable harnesses. The interfaces to all COM and NAV modules are standard and universal. This minimises installation time and eases the addition and replacement of individual modules as desired, without the need of additional cables or major changes.

Operational status

Certification JTSO has been achieved.

A completely digital control panel will replace the existing conventional control units to ensure upgrade potential for the future.

Contractor

Becker Avionic Systems.

VERIFIED

ProfiLine communication and navigation system

The ProfiLine communications and navigation system comprises:

1. The CU900-(1) dzus-width control unit with ARINC 429 and ARINC 410 interface for COM transceivers and NAV receivers (optionally NVG-compatible).
2. The NR900 VOR/ILS receiver is designed as a retrofit replacement for the Rockwell Collins 51RV-1C, within a ½ ATR short (ARINC 4-4A) housing. It is based on a modern design with integrated BITE, which also tests the HF section. It complies with the new ICAO annex 10 requirements for FM immunity. The NR900 has an ARINC 410, an ARINC 429 control interface, and a built-in marker receiver. A DGPS receiver and a MIL-STD-1553 interface will be available optionally in the future.

Contractor

Becker Avionic Systems.

VERIFIED

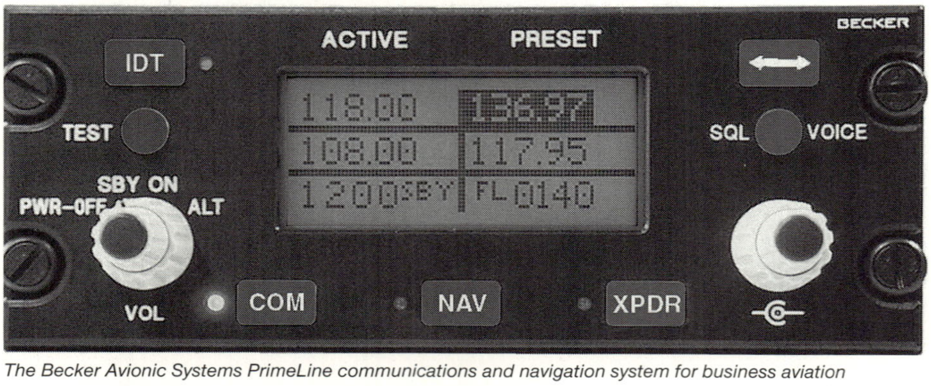

The Becker Avionic Systems PrimeLine communications and navigation system for business aviation
2000/0081496

Remote-Control ATC 2000-(3)-R system

The remote-control ATC 2000-(3)-R system consists of the CU 5401 (control unit) and the remotely controlled ATC 3401-1-R transponder.

This small, lightweight air traffic control transponder is designed for installation where instrument panel space is limited. Its CU 5401 control unit is designed to be mounted into a standard 57 mm (2¼ in) round instrument panel cut-out, and is only 63.5 mm (2½ in) deep. The 300 W mode A/C transmitter and receiver are housed in a separate unit, which can be installed remotely in the aircraft.

The unit uses a clear, high-contrast, double line LCD, which is readable under all lighting conditions including bright sunlight. When an altitude encoder is connected, the reported flight level is displayed below the transponder code, to verify correct operation of the entire system.

The standard 4096 identification codes are selected by the rotary selector switch.

Two preset identification codes, for VFR flight or other purposes, can be entered from the front panel and stored in non-volatile memory for instant recall by a single key stroke.

The ATC 2000-(3)-R transponder is suitable for all types of fixed- and rotary-wing aircraft. It operates from both 14 and 28 V input power and is certified for both VFR and IFR operations. It complies with the requirements of JTSO C74c, class 1A and can report altitudes up to 62,700 ft. Low power consumption and small size and weight make it suitable for use as a standby transponder.

The ATC 2000-(3)-R transponder can be combined with other Becker CompactLine avionics systems, such as the COM 5200 or AR 4201 VHF transceivers or the NAV 5300 VOR/ILS navigation systems, which have similar control units. The transponder remote receiver/transmitter unit can be controlled by other types of CDU or FMS.

A special version of the ATC 2000-(3)-R, designated the ATC 2000-(3)-R62, is available for high-altitude operation up to 60,000 ft, fitted with a special pressure box.

Specifications

Dimensions: CU 5401 61.3 × 61.3 × 62 mm
ATC 3401-(1)-R 134 × 50 × 253 mm
Power supply: 10-32.2 V DC; 1.1 A at 14 V; 0.55 A at 28 V
Transmitter frequency: 1,090 MHz
Receiver frequency: 1,030 MHz
Modes: A and C
Control interface: RS-422

NAV 5300 VOR/ILS navigation system 0001325

Contractor

Becker Avionic Systems.

VERIFIED

Remotely-controlled NAV 5300 VOR/ILS navigation systems

The NAV 5300 systems are part of the Becker Compact Line of equipments that can be customised at installation to customer requirements. They consist of the CU 5301 (control unit) and the remotely-controlled RN 33XX navigation receiver.

The system is ideally suited to installations where minimum panel space is to be used. They utilise a small, lightweight CU 5301 control unit, which fits into a standard 57 mm round instrument panel cut-out, and is only 63.5 mm deep. Lightweight remote receivers, with VOR/LOC or VOR/ILS beacon capabilities are mated with this control unit, to complete the systems.

The receivers can be installed at any convenient place in the aircraft. The CU 5301 control unit uses a clear, high-contrast, double line LCD display, which is readable under all lighting conditions, even bright sunlight.

Both active and standby frequencies are displayed, and can be transferred by a single stroke of the 'flip-flop' button. Up to 99 preset frequencies can be entered from the front panel, and stored in non-volatile memory. Parallel outputs are provided for automatic DME channelling. The steering signals and flag drive outputs are compatible with most commonly used CDI, HSI, flight director, and autopilot systems.

All systems will be JTSO certified for either VFR or IFR use in all types of fixed-wing and rotary-wing aircraft, and comply with ICAO requirements for VHF radios.

The NAV 5300 systems can be combined with other Becker Compact Line avionics systems, such as VHF transceivers and ATC transponders which have similar control units. The remote receiver units can be controlled by other types of CDU or FMS.

Specifications

RN 3330-(1): VOR/LOC receiver and converter
RN 3320-(1): VOR/LOC receiver and converter and GS receiver
CU 5301: control unit
RM 3300-( ): converters to drive RMI
Dimensions:
CU 5301 61.3 × 61.3 × 62 mm
RN 3320/30 134 × 50 × 243 mm
RM 3300-( ) 134 × 50 × 214 mm
Operation voltage:
10 to 32 V DC 26 V AC, 400 Hz, 10 mA (RMI converter)
Power consumption:
At 28 V DC.
without panel lights
RN 3330-(1) 250 mA
RN 3329-(1) 330 mA
Frequency ranges:
VOR/LOC 108.00-117.95 MHz
200 channels
Glide slope 329.15-335.00 MHz
40 channels
Channel spacing
VOR/LOC 50 kHz
Glide slope 150 kHz

Contractor

Becker Avionic Systems.

VERIFIED

Becker Avionic Systems ATC 2000-(3)-R system 0001197

RM 3300 series RMI converters

The RM 3300 series converters enable Sine/Cosine or composite video signals from navigation receivers to be used to drive RMI indicators and other display devices which require XYZ synchro inputs.

By varying the circuitry of the internal modules in the RM 3300 family, the converters can be used to provide inputs for XYZ synchros per ARINC 407, Sin/Cos DC or Sin/Cos AC indicators. There is even a model which simply functions as a synchro amplifier, to enable low-power synchro signals to drive corresponding high-power synchros.

These converters are approved for installation in unpressurised areas of aircraft operating up to 50,000 ft. There are no altitude restrictions for installations in pressurised areas.

Specifications

Dimensions: 54 × 139 × 214 mm (excluding mating connector)
Weight: 0.75 kg

Operational status

In production and in service.

Contractor

Becker Avionic Systems.

VERIFIED

RMU 5000 radio remote-control unit

The RMU 5000 is a member of the Becker PrimeLine communication and navigation family; it provides control of up to three separate radios from a single compact control head; two variants are available, allowing the customer to select which of the COM/NAV/XPDR [RMU 5000 (1)] or COM/NAV/ADF [RMU 5000 (2)] grouping he wishes to control.

The Becker Avionic Systems RMU 5000 radio remote-control unit ***2000***/0062184

The RMU 5000 provides for three redundant power supply inputs, ensuring that if one bus or power source is lost, all radios remain constantly available and operating. Full redundancy can be obtained through an external panel-mounted switch for IFR operation with dual RMU installation.

Contractor

Becker Avionic Systems.

VERIFIED

SAR-ADF 517 Search And Rescue - Aircraft Direction-Finder

Becker Avionic Systems' new SAR-ADF 517 is able to locate beacons transmitting on the 406.025 MHz COSPAS/SARSAT emergency frequency, as well as the VHF (121.5 MHz) and UHF (243 MHz) SAR frequencies.

The 406.025 MHz frequency transmits a 450 ms digital pulse every 50 s. Traditional ADF systems rely on a continuous swept tone from the beacon to ensure reliable homing, but the digital pulse from the 406.025 MHz beacon is too brief for the ADG to secure lock. Accordingly, Becker changed its rescue navigation aids to the new SAR-ADF 517 version, which is equipped with the following frequencies: 121.5 MHz, plus an adjustable training frequency near 121.5 MHz; Channel 16 for sea rescue; 243 MHz, plus an adjustable training frequency near 243 MHz; 406.025 MHz, plus a training frequency.

In the 406.025 MHz mode, either 121.5 or 243 MHz can be selected to cover automatically the 49.5 s lapse time between two digital pulses.

The hardware for the system comprises an 80 mm diameter control and display unit connected to a remote antenna mounted on the underside of the aircraft.

Specifications

Method of bearing: Doppler principle (frequency of rotation 3 kHz, cw/ccw)
Accuracy: ±5° rms

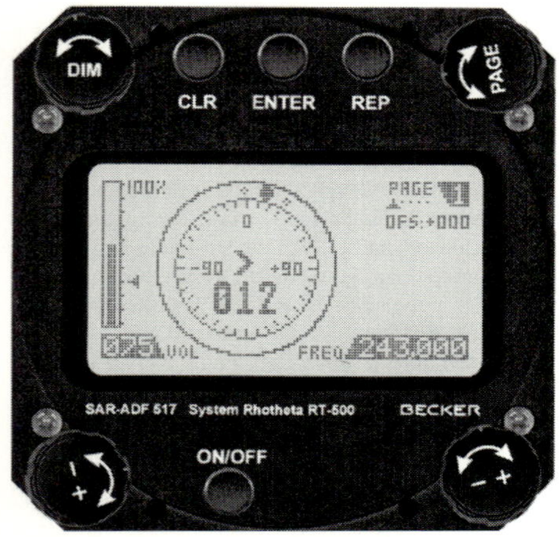

The Becker Avionic Systems SAR-ADF 517 ***2000***/0062185

Received frequency:
(VHF) 121.500 MHz, 123.100 MHz
(channel 16 marine band) 156.8 MHz
(UHF) 243.000 MHz, 243.500 MHz
(COSPAS/SARSAT) 406.025 MHz, 406.028 MHz
Polarisation: vertical
Polarisation error: <5° at 60° vectorial field rotation
Cone of confusion: approximately 30° measured to the vertical
ELT identification: by direction of audio sweep tone, frequency range 300 to 1,600 Hz and repetition rate 250 to 500 ms
Weights:
(display unit) 0.25 kg
(DF antenna) 2.0 kg
Dimensions:
(display unit) 82 (W) × 82 (H) × 35 (D) mm
(DF antenna) 270 (diameter) × 185 (L) mm
Power: 12-32 V DC, 400 mA

Contractor

Becker Avionic Systems.

VERIFIED

Flight Control Unit (FCU) for the Airbus A319/A320/A321

The FCU for the Airbus A319, A320 and A321 represents a smart, multipurpose control and display unit, interfacing between the pilot and the autoflight and electronic flight instrumentation systems. Installation in the glareshield and separate controls for pilot and co-pilot for communicating with the primary instrumentation system enable the crew to work head-up. The FCU allows the pilot to engage autopilot, flight director and autothrust systems, and to set flight altitude, speed and course.

The FCU consists of two independent computers with automatic switchover to ensure redundant signal processing. All data exchange between FCU and external systems is done via a discrete interface and ARINC 429 serial datalink. Contrast and illumination of the displays and panels are adaptable to extreme environmental conditions, from direct exposure to sunlight at high altitude to the special requirements of night approaches.

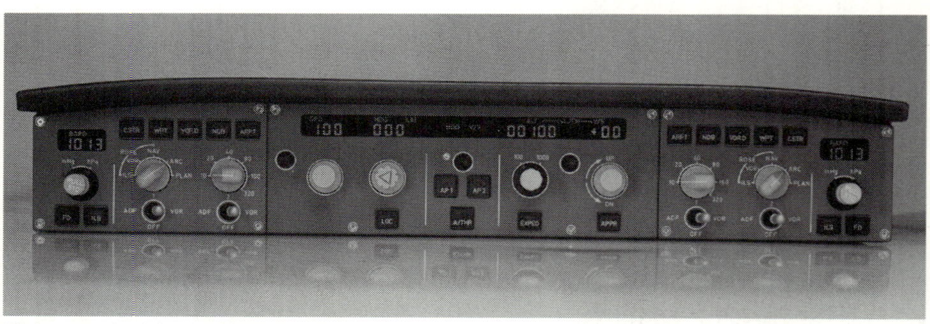

Flight control unit for the Airbus A319, A320 and A321

Application of specifically designed LCDs to display set values allows wide viewing angles with high contrast and sunlight readability, even with the pilots wearing polarised sunglasses. Set values are introduced into the FCU using optical encoders. These are incremental opto-electronic devices with a notched input and push-pull capability.

LED keys for the selection of operating modes are based on push-buttons with redundant lighting. Each push-button shows a green confirmation bar and a white illuminated legend. Misreading of non-illuminated bars in full sunlight is prevented by integrated layers of optical filter coating. The push-buttons have double poles to ensure precise operation and repeatable tactile properties.

Operational status

In production for the Airbus A319, A320 and A321.

Contractor

Bodenseewerk Gerätetechnik GmbH/BGT.

VERIFIED

DKG 3 and DKG 4 digital map display system for helicopters

The Defence and Civil Systems business unit of DaimlerChrysler Aerospace AG (Dasa/Munich), at Dornier GmbH in Friedrichshafen, has developed the cockpit map display systems, DKG 3 and DKG 4, destined for use in aircraft and helicopters respectively. These compact electronic devices replace conventional printed maps.

DKG 3

The digital map display system DKG 3, in the form of kneeboard equipment, supports the navigation and communication of helicopter crews by displaying colour maps in various scales, the helicopter's own position, flight route and other flight and mission-related information. In the future, the DKG 3 will make it possible to transmit and receive text files and situation data as graphical overlays, via datalink.

In navigation mode, the current position of the air vehicle on the map is real-time controlled by an on-board navigation system or a GPS, as an option. The map is automatically oriented in flight direction. This specific feature of the DKG 3 reduces the operator's response time especially when flying at low altitude. This avoids misunderstanding between the pilot and the navigator about left and right.

Map and flight planning data are stored in PCMCIA format on two memory cards which are inserted in lateral slots in the DKG 3. By pressing a button, the operator can switch easily between the stored maps and, additionally, with a zoom function, magnify or scale down the selected map detail.

With the DKG 3, flight planning can be autonomously performed or changed either on the ground or in the helicopter.

Another primary task of the ground station is providing the logistics for the digital map data. The Geogrid Map Preparation Software (MAPS) generates all the necessary cartographic data adaptations. This cartographic data, which can be stored on a CD-ROM, is loaded, interactively assembled and written on the DKG 3 memory cards. The complete area of the Federal Republic of Germany, with scales of 1:200,000 (survey map of Germany) and 1:500,000 (ICAO map of Germany) can be stored on such a data carrier (mini hard disk with 170 Mbyte, PCMCIA). Currently, mini hard disks are available with a capacity of up to 340 Mbyte.

The DKG 3 has a coloured, 10.4 in diagonal liquid crystal display (LCD). Its removable cover provides antiglare against direct sunlight and reflections. In combination with a filter glass, it is fully compatible for operation with night-vision goggles.

The new DKG 3 is used under VFR conditions. As the device does not interfere with flight guidance and control, it is categorised as a 'non-flight-safety-critical' system. The map display system is already in service with the German Federal Border Guard, police helicopter squadrons, the air rescue service of Germany's automobile association (ADAC) and industrial companies. A first lot is in operation with the medium-lift CH-53 helicopters of the German armed forces.

DKG 4

The DKG 4 was specifically designed for retrofitting in existing helicopters. It equips the new EC 135 helicopters of the police helicopter squadrons from Bavaria and Mecklenburg-Vorpommern. This equipment is designed for fixed installation in the helicopter cockpit and has a large separate map display (typically 6 × 8 in). DKG 4 is the tactical centre for operating the mission suite of the helicopter.

System facilities and modes of operation include:

- Planning and navigation.
- Real-time heading-up presentation of true moving map.
- Various map scales and continuous zoom-in/zoom-out functions.
- Memory cards for map and mission data storage and transfer.
- Tactical symbology.
- Graphical data communication.

The DKG 4 has interfaces with the following aircraft systems and sensors:

- The navigation system.
- Tactical radio.
- External sensors, such as camera, FLIR, Radio Direction Finder (RDF).
- The HELLAS Obstacle Warning System (OWS).
- A remote-control system and helicopter displays.

Specifications

Dimensions: 180 × 146 × 76 mm
Power: 28 V DC, 20 W
Data storage:
(mission data): typically 8 MB flash
(map data): up to 340 MB flash, and up to 1 GB hard drive.

Operational status

In service with various German police services and the German Federal Border Guard. The model DKG 4 equips the Federal Border Guard fleet of EC135 and EC155 helicopters.

Contractor

DaimlerChrysler Aerospace AG, Defense and Civil Systems.

UPDATED

HiVision radar

Impairment of pilots' vision in fog, rain and similar poor weather conditions, is a significant reason for aircraft collisions on taxiways and for aircraft crashes near airports. DaimlerChrysler Aerospace AG and the Deutsches Zentrum für Luft- und Raumfahrt e.V. (DLR) have developed a system concept, called Enhanced Vision System (EVS), to minimise such effects.

EVS comprises both optical detection devices and a millimetre (35 GHz) radar. The role of the radar imaging component is to provide the required capability in darkness, rain and fog, when the performance of the optical component is seriously impaired.

DaimlerChrysler Aerospace AG is responsible for the radar element; both conventional and electronically steerable antennas are being trialled. DLR is responsible for flight testing and display integration. The intention is for the EVS to meet the requirements of both civil and military aircraft and helicopter systems.

Operational status

In development testing, with comparative evaluation of the merits of both components of the EVS being conducted between 1999 and 2002.

Contractors

DaimlerChrysler Aerospace AG, Defense and Civil Systems.
Deutsches Zentrum für Luft- und Raumfahrt e.V. (DLR).

VERIFIED

Cockpit warning system

This unit is designed to indicate both warning and caution messages in the aircraft cockpit. Some 16 warning captions provide sunlight readable indications. When illuminated they appear as red letters on a black background and extinguished they remain black. For night operation the brightness can be dimmed by input voltage variation.

When the unit receives a warning signal, the corresponding warning caption is illuminated and also the unit activates the external master warning indication. The external master warning indication turns off when no warning caption is active or can be reset by an external switch.

If required, a timer makes the master warning lamp flash. The master caution inputs can be activated by this unit. They only extinguish when the reset button has been pressed. A lamp test switches on all warning captions in the panel as well as the external master warning lamps.

Every warning caption can be changed individually from the front of the unit without removing the whole system from the instrument panel. Every caption is backlit by two incandescent lamps.

Specifications

Power: 28 V DC nominal, 900 mA
Weight: 0.81 kg

Contractor

Diehl Luftfahrt Electronik GmbH.

VERIFIED

Diehl cockpit warning system 0018204

EuroNav III task management system

The EuroNav III task management system is part of the overall EuroNav AH Guidance System. EuroNav AH is a mission control management system for up to 500 helicopters, consisting of the fixed EuroNav z Central Control Station and up to 500 helicopter EuroNav III task management systems. The system utilises the GPS navigation system to determine helicopter positions, and all participants are linked via appropriate communications facilities.

The EuroNav z Central Control Station collects, evaluates, and stores data on all aspects of the operation, maintenance, management, and financing of the helicopter fleet concerned.

Each helicopter is fitted with the EuroNav III task management system, which comprises an RN4-2.9 navigation computer, with its database and navigation program; a DYNT51-1.7 display 5.1 in; a CLN14-1.8 control unit; an SBN-2.8 switch box, and a GPS antenna.

The EuroNav III task management system is operated by the air observer, who can use the system to

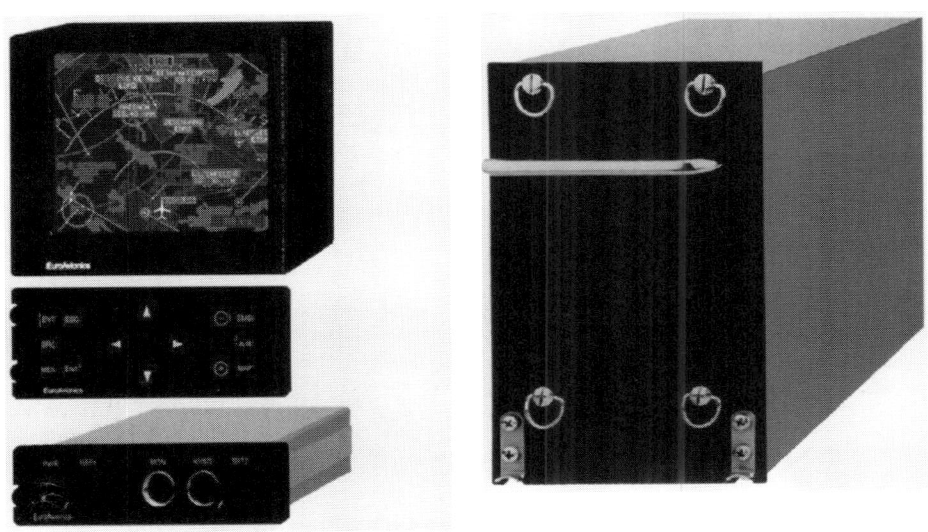

EuroNav III task management system, showing (top to bottom) the monitor, control unit, switch box and navigation computer 0015357

perform all operational actions required for the following types of task: flight planning; flight execution; flight data recording; emergency situation recovery.

The navigation computer is based on the high end industrial standard and includes a 12-channel GPS receiver with capability to accept DGPS correction data. It is housed in a ½ ATR box, and includes 2 × HDD, up to 1.3 Gbit each.

The standard monitor provides a full-colour 5.1 in display, with 320 × 240 pixel resolution. As options, 6.4 and 10.4 in displays with 640 × 480 pixels resolution and colour depth from 4 to 256 are available. The control unit and switch box are used to control and interconnect all system components.

Operational status

Widely used by police forces and search-and-rescue organisations in many countries of Western Europe and Asia.

Contractor

EuroAvionics Navigationssysteme GmbH & Co.

VERIFIED

LCR-92 μAHRS Attitude and Heading Reference System

The LCR-92 is an extremely small and light strapdown reference system using LITEF fibre optic gyros. It provides pitch, roll, magnetic heading information and angular rates around the aircraft body axes, replacing conventional vertical/directional gyro installations in one single box.

The system features full compatibility with modern digital cockpit instruments with an ARINC 429 databus. Analogue outputs for use with conventional mechanical indicators are available as an option.

Specifications

Dimensions: 278 × 102 × 128 mm
Weight: 2.1 kg
Power supply: 28 V DC, <25 W
Accuracy (2σ): heading 1°.
Reliability: >6,000 h MTBF

Operational status

In production for business jets, turboprops and helicopters.

Contractor

LITEF GmbH.

VERIFIED

LCR-92 μAHRS Attitude and Heading Reference System 0001330

LCR-93 μAHRS Attitude and Heading Reference System

The LCR-93 is a new member of the μAHRS family based on the LCR-92. It has the same housing and dimensions and offers the same interface configurations. The mechanical layout and connectors ensure mechanical plug-in interchangeability with the LCR-92.

In addition to the LCR-92 capabilities, the LCR-93 standard version provides body axis referenced accelerations as well as inertial altitude and inertial vertical speed. The LCR-93 performance is increased by using air data augmentation.

The advanced version LCR-93V combines GPS and air data inputs to provide the necessary outputs to Head-Up Display (HUD) with a speed vector indication.

Specifications

Dimensions: 278 × 102 × 128 mm
Weight: 2.2 kg (2.5 kg with synchro board)
Accuracies (95%):

	Basic	Normal	
Attitude	0.3°	0.3°	static
	1.0°	0.5°	dynamic
Heading	1.0'	1.0'	static
	2.0'	2.0'	dynamic
DG mode	<5°/h	<5°/h	
Angular rates	0.1°/s	0.1°/s	1% max
Acceleration	5 mg	5 mg	1% max

Operational status

Selected by Cessna as standard AHRS for the Excel and replaces the LCR-88 in the Pilatus PC-9.

Contractor

LITEF GmbH.

VERIFIED

LCR-98, VG/DG replacement system

The LCR-98 is a strapdown Attitude and Heading Reference System (AHRS), specifically designed to replace Vertical and Directional Gyros (VG/DG) in aircraft like the Dassault Falcon 20, Canadair Challenger 601 and Gulfstream GII and GIII.

Featuring Fibre Optic Gyro (FOG) technology, the LCR-98 offers improved aircraft attitude and heading performance, together with higher reliability and lower cost of ownership.

The LCR-98 is designed as form, fit and function replacement of the VG-311, additionally incorporating the directional gyro function as a substitute for C9-C11 and MHRS-type gyros, using existing aircraft wiring and re-routeing the DG harness.

Operating the LCR-98 is virtually transparent to the pilot but with a tenfold increase in Mean Time Between Failure (MTBF) over mechanical VG/DGs.

Specifications

Dimensions: 219 (L) × 216 (W) × 155 (H) mm
Weight (max): 6 kg
Installation: fits into VG-311 cradle, using existing AC wiring and connectors
Power: 115 V AC, 400 Hz, 70 W with full external load
MTBF: >12,000 h
Outputs:
Synchro pitch, roll, 2 × headings
Analogue 50 and 200 mV AC pitch and roll
Discretes attitude and heading warning and interlocks
Accuracy:
Attitude 0.5° static, 1.0° dynamic, (95%)
Heading
- Slaved 0.75° (rms)
- DG 5.0°/h (95%)
Qualification:
RTCA/DO-160C including HIRF and lightning,
RTCA/DO-178 A, Level 1 for flight critical software
Certification: TSO C4c, C5e and C6d

Operational status

Certified in 1998.

Contractor

LITEF GmbH.

VERIFIED

MT-Ultra moving map system

The MT-Ultra moving map display is a single-box unit, designed for installation in the cockpit in a standard 6¼ in radio component space. It includes a highly integrated navigation computer and display system, which presents navigational data to the pilot on a full VGA resolution, sunlight readable, screen.

The MT-Ultra moving map display includes a comprehensive worldwide navigational database, supplied by MOVING-TERRAIN, with a 12-channel GPS receiver system. Control is by single alphanumeric keys located around the edge of the display, to minimise workload in high stress conditions. Features that can be selected by the pilot in flight include: zoom in/zoom out, display orientation, instant display of current position, flight planning and replanning facilities.

Data loading and updating is by CD-ROM, and the system can be integrated with other onboard GPS receiver systems if required. A quick release capability is provided to optimise the installation, operation and flight planning options.

Specifications

Dimensions: 158 (W) × 150 (H) × 48 (D) mm
Weight: 1.15 kg
Power: 12-28 V DC, 16 W
Screen: TFT colour display 6.5 in diagonal
Resolution: 640 × 480 pixels
Viewing angle: > 60°

Operational status

In production.

Contractor

MOVING-TERRAIN® Air Navigation Systems GmbH.

VERIFIED

VP 7 flight computer and variometer

The VP 7 flight computer and TE-compensated pressure transducer variometer is designed for use in gliders. It includes the following features:

1. A large graphic display and a liquid crystal display which present analogue variometer indication, speed-to-fly, digital averaging and trend indication data.
2. Precise altimeter presentation for 0-12,000 m with 1 m resolution.
3. Speed indication with resolution of 1 km/h.
4. A GPS-receiver.
5. 250 user programmable turn points with 32 tasks.
6. An airports database, including gliderports.
7. A comprehensive range of navigation and flight aids, specific to glider operation.
8. Optional features including automatic setting of radio frequencies, wind calculation through the compass with direction and speed indication, connection to a GNSS flight recorder.

Contractor

Peschges Variometer GmbH.

VERIFIED

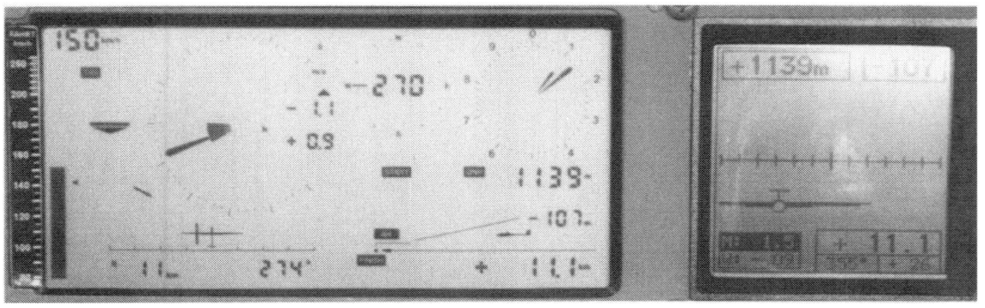

The Peschges Variometer GmbH VP 7 flight computer and variometer
2000/0085271

FSG 70/FSG 71 M VHF/AM transceivers

The FSG 70/FSG 71M VHF COMM transceivers are designed, as updates of the Walter Dittel FSG 50 and FSG 60M transceivers, for direct panel mounting into 2.25 in (57 mm) diameter instrument panels, without shock mounts.

Frequency coverage is 118.000 to 136.975 MHz, in 760 channels with 25 kHz increments; 1,600 channel capacity spanning 118.000 to 157.975 MHz is available as an option for government use.

Standard features include: liquid crystal frequency display, transmit light, separate microphone inputs for low level dynamic and high level standard microphones, adjustable microphone sensitivity, intercom, auxiliary audio input, transmitter sidetone and audio filter. The FSG 71M includes panel preset 10 channel electronic memory capability.

Specifications

Frequency range: 118.000-136.975 MHz (118.000-157.975 MHz option)
Channel spacing: 25 kHz
Transmitter power: 6 W into 50 Ω
Audio power: 8 W
Power supply: 13.8 V DC +10/−20%
Dimensions (W × H × D): 63 × 61 × 191 mm
Weight: FSG 70: 0.74 kg, FSG 71M: 0.80 kg

Operational status

Due to ultra-low standby power consumption (27 mA) and wide supply voltage range these transceivers are suitable for sailplane, ultralight and other low power/weight applications, as well as conventional aircraft and helicopters.

Contractor

Walter Dittel GmbH Luftfahrtgerätewerk.

VERIFIED

Walter Dittel's FSG 90 VHF/AM transceiver display
1999/0044789

Walter Dittel's FSG 71M VHF/AM transceiver display
1999/0044788

FSG 90/FSG 90F VHF/AM transceivers

The FSG 90/FSG 90F VHF COMM transceivers are designed as dual-mode 25/8.33 kHz channel spacing transceivers, for direct panel mounting into 2.25 in (57 mm) diameter instrument panels.

Frequency coverage is 118.000 to 136.975 MHz, in 760 channels with 25 kHz-only channel spacing, and 2,278 channels in 25/8.33 kHz dual mode operation. For governmental applications an extended frequency range of 118.000 to 149.975 MHz with 3,838 channels in dual mode is available. 99 memory channels are available in 25/8.33 kHz dual mode and another 99 memory channels in 25 kHz-only mode.

Three display modes are provided: frequency only, frequency/memory channel, active/standby frequency change.

Specifications

Frequency range: 118.000-136.975 MHz (118.000-149.975 MHz option)
Channel spacing: 25/8.33 kHz dual mode and 25 kHz only
Transmitter power: 6 W into 50 Ω
Audio power: 8 W
Power supply: 11.0-16.5 V DC
Dimensions (W × H × D): FSG 90: 63 × 58 × 200 mm
FSG 90F: 63 × 58 × 230 mm
Weight: FSG 90: 0.80 kg
FSG 90F: 1.1 kg

Operational status

Designed for all aircraft and helicopter initial-fit and upgrade applications, as main or standby communications systems. A higher power version, the FSG90H1, producing 10 W RF power at 50 Ω entered production in the 4th quarter of 1999.

Contractor

Walter Dittel GmbH Luftfahrtgerätewerk.

VERIFIED

INDIA

Global Positioning System (GPS) receiver models 9405/9405A

Aerospace Systems Private Ltd (ASL), a company promoted by the TATA Group in India and Satellite Tracking Systems Inc in the USA is engaged in the design and development of GPS receivers for air, land and sea use; its aircraft models are the 9405 for military fighters and the 9405A for commercial aircraft and benign environment military application such as maritime patrol aircraft.

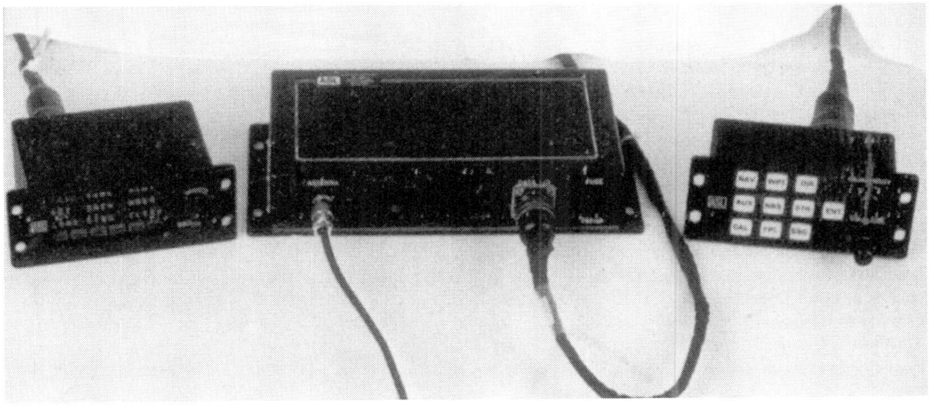

The Aerospace Systems Private Ltd (ASL) model 9405 GPS receiver
2000/0064377

The ASL receivers are multichannel units that use standard navigation databases, as well as user-defined data tables. ASL claims that perhaps the unit's most important function is its ability to save information for later use. Flexible navigation features such as multiple datum, route plans, time plans, range and bearing, course deviation indication, emergency navigation, navigation calculations, waypoints and route library data are all in-built features.

The GPS receivers typically comprise the receiver, processor board, power supply, display, keyboard, display/keyboard controller and antenna (the antenna being externally sourced).

Specifications

Receiver: 6-channel; L1 frequency; C/A code
Accuracy:
(position) 30 m without SA; 100 m with SA2; 5 m with differential corrections (optional)
(velocity) 0.5m/s
Position update: 1/s
Acquisition time: 60 s with warm start; about 5 min from cold
Display: 3-line × 16-character LED
Keyboard: 10 backlit keys; 2 rotary knobs
Power: 19 - 36 V DC
Interface: one RS-232 port
Antenna: microstrip patch antenna with preamp
Dimensions:
(GPS receiver) 52 (H) × 235 (L) × 130 (W) mm
(display) 63 (H) × 145 (W) × 85 (D) mm
(keyboard) 68 (H) × 145 (W) × 55 (D) mm
Weight: 3 kg (total)
Operating temperature: –40 to +85°C
Environmental: MIL-STD-810E
EMI/EMC: MIL-STD-461C

Operational status

Development contract awarded by the GPS task force, constituted by the Department Of Electronics (DOE) and Defence Research and Development Organisation (DRDO) under the aegis of the Future Air Navigation System (FANS) programme of the government of India.

The development programme took 4.5 years. An initial production batch of 100 receivers was delivered to the Indian Air Force (IAF) for installation in fighter aircraft.

Contractor

Aerospace Systems Private Ltd (ASL)

UPDATED

ARC-610A automatic direction-finder

The all-solid-state modular construction ARC-610A automatic direction-finder covers the frequency range 190 to 1,700 kHz in steps of 0.5 kHz to indicate on an LED display the direction of a ground beacon. It has 10 preset channels and includes self-test. The system consists of a receiver and a controller.

Specifications

Dimensions:
(receiver) 288 × 60.5 × 200 mm
(controller) 146 × 66 × 146 mm
Weight:
(receiver) 5 kg
(controller) 1 kg
Power supply: 27.5 V DC, 2 A
Bearing accuracy: ±2° excluding quadrantal error
Temperature range: –55 to +55°C
Altitude: up to 70,000 ft

Contractor

Hindustan Aeronautics Ltd.

VERIFIED

Hindustan's ARC-610A automatic direction finder ***1999***/0051590

ARC 1610A automatic direction-finder

The ARC 1610A automatic direction-finder provides bearing information of the known ground beacons operating in medium frequency along with code reception to identify the selected ground beacons.

The ADF system ARC 1610A is a compact lightweight system which provides accurate, stable bearing information with a high degree of reliability and maintainability using hybrid technology. The system is compatible with ARINC 570.

Specifications

Frequency range: 190-1,700 kHz
Frequency indicator: LED display
Modes of operation: ADF and ANT
Bearing accuracy: ±2°
Hunting: ±2°
Bearing resolution: 6 s
Power input: 27.5 V DC, 2 A

Hindustan's ARC 1610A automatic direction-finder 0051591

Preset channels: 10
Audio output: 100 mW across 600 Ω
Dimensions:
(receiver) 250 × 195 × 90 mm
(controller) 163 × 14.6 × 66 mm
Weight:
(receiver) 4 kg
(controller) 1 kg

Contractor

Hindustan Aeronautics Ltd.

VERIFIED

GNS-642 Global Navigation System

The GNS-642 system tracks all satellites in view to provide complete positional and navigation data. It is able to handle up to 100 waypoints and 10 routes.

Specifications

Receiver: 5 channel, L1 frequency
Sensitivity: –160 dBW, SNR 10 dB
Horizontal position accuracy: 25 m (RMS)
Vertical position accuracy: 50 m (RMS)
Velocity accuracy: 0.15 kt (RMS)

Hindustan's GNS-642 Global Navigation System 0002441

Output data: position, time, bearing, course and distance to waypoint, ground speed, ETA
Routes: 10 routes, 10 waypoints per route
Output display: 3 lines, 16 characters LED
Dimensions and weight:
(receiver) 124 × 120 × 120 mm; weight <1.5 kg
(display unit) 146 × 56.5 × 90 mm; weight <0.5 kg
Power supply: 27.5 V DC, 20 W
Interfaces: RS-232 with NMEA-0183

Contractor

Hindustan Aeronautics Ltd.

VERIFIED

UHS 190A UHF homing system

The UHS 190A UHF homing system for fixed-wing aircraft and helicopters is used in conjunction with the normal communications transceiver for locating ground transmitters and personal rescue beacons. It operates in the frequency range 225 to 399.975 MHz with two UHF antennas. The system provides an accuracy of ±5° for homing. The system has the capability to drive two homing indicators.

The UHS 190A consists of a UHF homing adaptor for receiving input signals from the two UHF antennas, homing controller for processing the ADF audio signal from the V/UHF communication set and a homing indicator for indicating the relative bearing of the ground station.

Specifications

Dimensions:
(adaptor) 146 × 98 × 32 mm
(controller) 146 × 51 × 105 mm
(indicator) 10 × 100 × 50 mm
Weight:
(adaptor) 0.3 kg
(controller) 0.55 kg
(indicator) 0.45 kg
Power supply: 22-31 V DC

Operational status

The UHS 190A is fitted on Indian Navy Sea King, Chetak, Il-38 and Dornier aircraft.

Contractor

Hindustan Aeronautics Ltd.

VERIFIED

INTERNATIONAL

Altimeter type 140500

The type 140500 altimeter measures and presents data on relative height. It provides visual warning (by black and white flag) when the aircraft approaches the ground. A manual pressure setting facility is provided, and both red and white dial lighting variants are available.

Altitude is displayed by a counter for tens of kilometres, and by two pointers for tens and hundreds of metres. A built-in vibrator is used to minimise friction in the mechanism.

By customer request, the setting and indication of atmospheric pressure can be defined in units of hPa.

Specifications

Altitude measuring range: –500 to +15,000 m
Measurement error:
(at 0 m height) ±10 m
(at 12,000 m height) ±100 m
Dimensions: 65 × 65 × 160 mm
Weight: 0.92 kg
Power: +27 V DC, 1.5 W; 5.5 V AC, 3 VA

Contractors

ELAN Elektronische und Anzeiger GmbH, Germany.
CINAVE Companhia de Instrumentos de Navegaçao Aeronáutica Lda, Portugal.
France Aerospace SARL, France.
International Aerospace Inc, USA.

VERIFIED

Aria - EFIS-95 - airborne integrated avionics system

Aria - EFIS-95 is a joint venture development between NIIAO Institute of Aircraft Equipment, Moscow and Honeywell Aerospace of the USA to develop an airborne integrated avionics system for the Be-200 medium-capacity amphibian aircraft.

The modular nature of the system is such that it could also be applied to a large number of other advanced civil aircraft, including: Il-96-300, Il-114, Tu-204/-214/-324/-330/-334, and Yak-142M; it could also be applied as an upgrade to Il-86, Tu-154/-154M and Yak-42.

NIIAO is primarily responsible for software development and aircraft integration management, while Honeywell (and its subsidiaries) is the supplier of most of the system hardware (except the air data system), including: FMS, EFIS, GPWS, FDE, AHRS (LITEF LCR-88), radar (Bendix King RDR-4B). Provision is being made for the system datalinks to accommodate CNS/ATM requirements.

Specifications

System weight:
(Be-200 class aircraft) 400 kg
Power:
(Be-200 class aircraft) 6 kW
MTBF: 20,000 h

Contractors

NIIAO Institute of Aircraft Equipment, Moscow.
Honeywell Aerospace, USA.

VERIFIED

Automatic Flight Control System (AFCS) for Concorde

BAE Systems jointly developed with Sextant the Concorde Automatic Flight Control System (AFCS), which comprises eight separate subsystems with no fewer than 38 individual units. The system provides fully automatic control of the aircraft from take-off, through the entire flight envelope and down to approach and landing phases. The system is duplex-monitored and fail-operational to meet the operational requirements of ICAO Cat IIIA and is cleared for automatic landing with 15 ft decision height and 200 m runway visual range. It was the first AFCS to use a push-button controller fitted into the flight deck glareshield.

Operational status

In service in British Airways and Air France Concorde aircraft.

Contractors

BAE Systems.
Sextant.

VERIFIED

Automatic flight control system for Concorde 0001403

AWIN Aviation Weather INformation system

Historically, pilots have had access to current strategic weather information only during flight planning while still on the ground. Once airborne, weather updates are obtained by means of weather radar, radio or ACARS (Aircraft Communications Addressing and Reporting System) datalink. Now, under a new NASA aviation safety initiative, a team has been assembled to develop a system that will bring cost-effective strategic and tactical weather displays on to the flight decks of the world's airline fleets.

Led by Honeywell Inc, Commercial Aviation Systems, the team has been awarded a NASA contract to create a worldwide aviation weather distribution and display system called AWIN – Aviation Weather INformation system. The system will be based on an open architecture, tailored to a wide range of aircraft and capable of continued growth in the 21st century. Information will be gathered from a variety of sources – satellite images, ground weather services, other aircraft en route, processed, distributed and displayed in the cockpit in real time. The result is intended to be a commercially viable worldwide infrastructure for obtaining and disseminating critical weather information that will increase the overall safety of the aviation industry.

Pilots typically plan en route course deviations using 'text-based' weather information and graphic weather radar showing precipitation 15 to 25 minutes ahead. Integrating this kind of short-term weather picture with longer-term weather reports can be awkward and distracting. The result is often a route change that is inefficient, increasing pilot workload as the weather deteriorates and fuel reserves dwindle.

Under the NASA programme, Honeywell will be able to bring together the core competencies needed from multiple companies and integrate them to develop an optimal weather information package.

The result is intended to be an advanced system that allows each aircraft to be both a source and a user of weather information. In the cockpit, there will be easy-to-read, real-time displays that can show weather many miles ahead. Team responsibilities are as follows:

- Honeywell leads the team, serving as the system integrator and furnishing the primary development facilities. A worldwide data centre will be created and Honeywell will supply the airborne server and strategic displays.
- ARINC Incorporated provides weather data integration, communications and distribution to all ground stakeholders such as Airline Operations Centres (AOC). ARINC will also distribute information originating from the Honeywell Data Centre.
- Weather Services International (WSI) will provide both raw and value-added domestic US weather data.
- National Centre for Atmospheric Research (NCAR) will provide advanced weather data products to inform pilots about turbulence, icing, cloud ceiling and visibility, convection and winds.
- AlliedSignal Corporation will provide a tactical data display solution via the Enhanced Ground Proximity Warning System (EGPWS).
- United Airlines will provide end-use application, input and simulator support.
- COMSAT will provide satellite communications services. COMSAT is a leading provider of satellite services to the aeronautical industry and was a major contributor in the original Cockpit Weather Information Network (CWIN) project from which the current programme derives. COMSAT designed the broadcast satellite capability that will make the AWIN system economically possible.
- Kavouras Inc will provide real-time, high-resolution Nexrad radar and text information for cockpit displays.
- Swissair will provide an end-user application in the form of an MD-11 tri-jet as well as feedback and operational expertise from the perspective of a European airline.
- Société Internationale de Télécommunications Aéronautiques (SITA) will provide worldwide international communications service using both VHF radio-based ACARS and satellite communications.
- National Weather Service Aviation Weather Center (NWS/AWC) will provide raw and value-added products for both US national and international aviation-specific weather.

Operational status

The effort from the AWIN team will refine and build on the work already done in the earlier Cockpit Weather INformation (CWIN) programme, particularly in the area of system integration and multiple, redundant worldwide data coverage. Special emphasis will also be given to understanding the critical issues of the 'computer/human interface'. The development phase is planned to be five years.

Contractor

Honeywell Inc, Commercial Aviation Systems (consortium leader).

VERIFIED

Bearing, distance, heading indicator type 14100-2C

The type 14100-2C bearing, distance, heading indicator is a digital instrument developed to provide anti-collision data using the TACAN air-to-air mode. The indicator uses data from multiple, digital and analogue sources to provide bearing and distance information.

The design includes a compass repeater, VOR/ADF bearing pointer, VOR or TACAN bearing pointer, TACAN and DME range, TACAN and alarm flags.

Operational status

The type 14100-2C indicator was developed for the German Tornado retrofit programme, and FAA qualified on B-727 and C-130 aircraft.

Contractors

ELAN Elektronische und Anzeiger GmbH, Germany.
CINAVE Companhia de Instrumentos de Navegaçao Aeronáutica Lda, Portugal.
France Aerospace SARL, France.
International Aerospace Inc, USA.

VERIFIED

Type 14100-2C bearing, distance, heading indicator
2000/0062186

CMA-3012/3212 GPS sensor units

The CMA-3012 and CMA-3212 are designed for certification under anticipated new rules which will allow the use of GPS as a sole means of navigation. Key characteristics for this application include 12 receive channels, all of which can be used for continuous satellite tracking and any two of which are assignable as GPS integrity channels; comprehensive end-to-end receiver BITE; carrier phase tracking; differential GPS input and growth provisions for GPS/GLONASS.

The CMA-3012 and CMA-3212 conform to ARINC 743 and 429-12, DO-160C, 178B and 208, TSO C129 and MIL-STD-810. Inputs consist of eight ARINC 429 and an RS-232. Outputs consist of three ARINC 429, three 1 kHz time marks and RS-232 and 28 V valid discrete.

Specifications

Dimensions:
(CMA-3012) 66 × 220 × 240 mm
(CMA-3212) 199.6 × 57.1 × 388.6 mm
Weight:
(CMA-3012) 3.2 kg
(CMA-3212) 3.6 kg
Power supply: 18-36 V DC, 20 W (max)
Frequency: 1,575.42 MHz, C/A code
Accuracy (95%):
24 m (horizontal)
30 m (vertical)
0.3 kt (velocity)
Reliability: 40,000 h MTBF

Operational status

TSO-C129 B1/C1 was received in mid-1995, and SCAT I implementation was completed in mid-1996. The system has been certified for Primary Means operation to FAA N8110.57. Ground preflight software (SureFlight) is also available.

Contractors

BAE Systems Canada Inc.
Honeywell Inc, Commercial Aviation Systems.

VERIFIED

DME 83200 Distance Measuring Equipment

The DME 83200 indicator processes digital signals derived from a DME transmitter/receiver to display distance from a selected DME ground station.

The DME 83200 is housed in a standard non-pressurised case. A DIM control knob provides self-test.

The readout is a four-digit assembly featuring a seven-segment filament display or optional fibre optic display; a decimal point is provided to give nautical miles reading in 0.1 n mile steps up to 399.9 miles. An invalidity flag is shown as four dashes in the absence of a valid signal from R/T, or in the event of a malfunction within the indicator data processing circuitry.

Specifications

Dimensions: ARINC ½ 3ATI; 41.9 (H) × 85.6 (W) × 146 (D) mm
Weight: 0.43 kg
Altitude: 35,000 ft maximum
Display range: 0 to 399.9 n miles in 0.1 n mile steps
Power: 27.5 V DC, 0.35 A

Contractors

ELAN Elektronische und Anzeiger GmbH, Germany.
CINAVE Companhia de Instrumentos de Navegaçao Aeronáutica Lda, Portugal.
France Aerospace SARL, France.
International Aerospace Inc, USA.

VERIFIED

Flight Management System (FMS) for Airbus

Airbus Industrie and Aerospatiale have jointly selected Sextant and Smiths Industries Aerospace to supply the new Flight Management System (FMS) for the Airbus A318/319/320/321 and A330/340 families of aircraft. Airlines will be able to select this FMS to meet their forward fit and retrofit needs for Communication/Navigation/Surveillance with Air Traffic Management (CNS/ATM) capability.

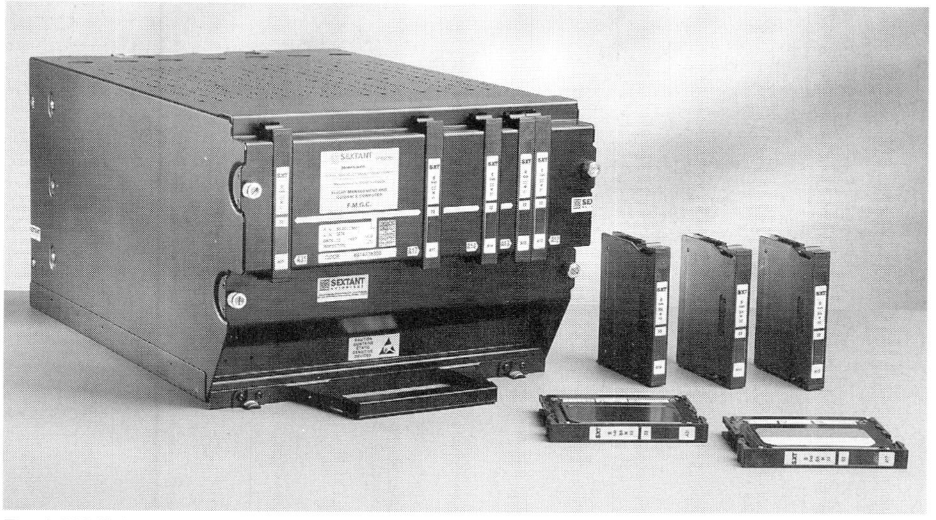

The A 320 flight management and guidance computer **2000**/0079259

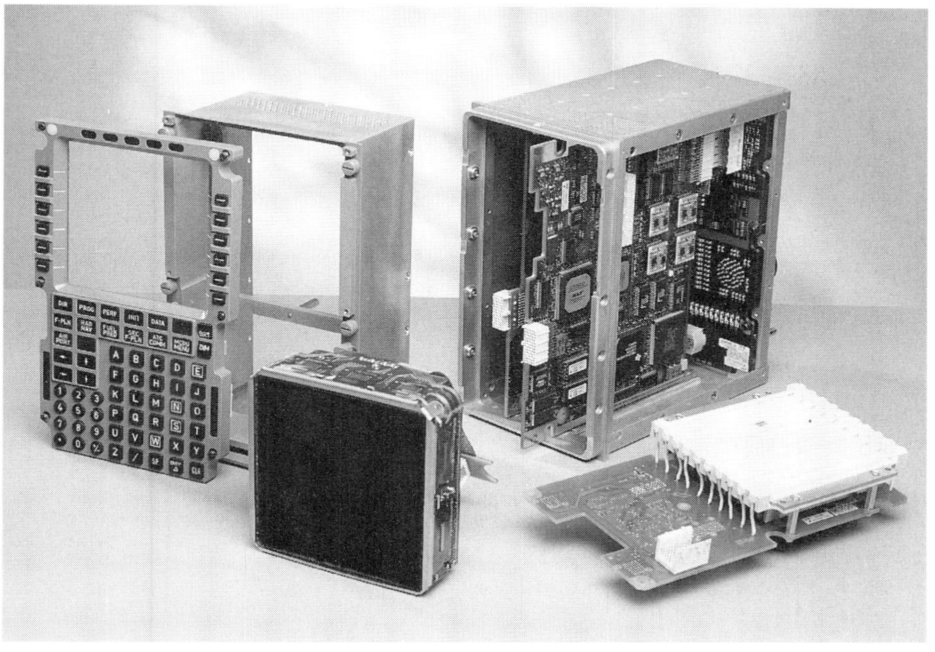

The A 320 flight management and guidance control system Multipurpose Control and Display Unit **2000**/0079258

The new FMS will be common to the A318/319/320/321 and A330/340 aircraft families, thereby enhancing commonality between these types. It will incorporate new-generation hardware and software, and ensure a single interface (Sextant) for the Flight Management and Guidance Computer (FMGC) for the A318/319/320/321 and the Flight Management, Guidance and Envelope Computer (FMGEC) for the A330/340. Both the FMGC and FMGEC are also included in the Sextant auto flight system and display system.

The Sextant/Smiths Industries Aerospace FMS is based on two FM boards (integrated in the FMGC/FMGEC) each of which features its own high-speed processor:

(1) The Flight Management Processor board (FMP) which processes the 'core FM' functions (navigation, flight plan management, trajectory predictions, lateral and vertical guidance, performances advisory and so on);

(2) The Interface and Display Processor board (IDP), which groups and gathers all the interface functions (MCDU display and control, EFIS display, processing of the inputs/outputs, datalink processing, interfaces with the FG(E) – Auto Flight Control and so on);

The FMS includes a new colour Multipurpose Control and Display Unit (MCDU), common to the A318/319/320/321 and A330/340 aircraft families and featuring an active matrix LCD flat-panel display. It provides the following operational features:

(1) A temporary flight plan that allows systematic assessment of flight plan revisions. The temporary flight plan will provide full predictions for all lateral and vertical revisions and will allow multiple flight plan revisions.

(2) A Required Time of Arrival (RTA) function that benefits from the high accuracy of the fully integrated lateral/vertical trajectory and that is optimised on all flight phases (tolerance control will be selectable from 0 to 30 seconds).

(3) Increased flexibility and commonality across the fleet for datalink applications, using tables that can be loaded with ACARS MU and ATSU (protocol messages, prompts and triggers are airline-customised).

(4) Increased flexibility provided by loadable and cross-loadable data bases, including the navigation data base.

Contractors

Sextant.
Smiths Industries Aerospace.

UPDATED

Flight monitoring system for Ilyushin Il-96M/T aircraft

Rockwell Collins has signed contracts with the Russian State Research Institute for Aviation Systems (GosNIIAS) and the Elara Joint Stock Company in Cheboksari for co-production of the Il-96M/T's Aircraft Systems Central System (ASCS). The system will perform flight engineering functions and will provide a two-member flight crew capability for the long-range, four-engine airliner produced by the Ilyushin Aircraft Association.

Operational status

Development.

Contractors

Rockwell Collins.
Elara Joint Stock Company, Cheboksari.
GosNIIAS Russian State Research Institute for Aviation Systems.

VERIFIED

Global Navigation Satellite Sensor Unit (GNSSU)

Honeywell and Canadian Marconi Company are collaborating in the production of the Global Navigation Satellite Sensor Unit (GNSSU). The 12-channel GNSSU tracks all GPS satellites in view to provide better than 25 m position accuracies. It offers sensor computational errors of no more than 1.5 m, receiver autonomous integrity monitor and ARINC 743 design. The GNSSU provides accurate worldwide oceanic, en route and approach navigation, simplified pilot interface, time and position for automatic dependent surveillance and inertial reference system integration.

Increased capabilities result from combining the GNSS data in an 18-state Kalman filter inside the laser IRU. With this, the inertial error of 2 n miles/h is bounded by the accurate GNSS satellite measurements. There is continued system accuracy during periods of less than four satellites and continued integrity with less than five satellites. The high-frequency inertial sensor measurements integrated with the high-accuracy GNSS measurements provide the optimum navigation solution. The GNSS integration in the IRS eliminates the possibility of an added 25 m track error that could occur with blending of GNSS and IRS in the FMS. The GNS/IRS is designed to enable calibration of the inertial sensors after sole source GNSS certification.

Specifications

Dimensions: 63.5 × 215.9 × 241.3 mm
Weight: 3.18 kg
Accuracy:
(position) 25 m
(velocity) 1 kt
(time) 2 ms
Reliability: 55,000 h MTBF predicted

Operational status

The GNSSU has been selected as the standard option on the Boeing 777. It was TSO'd by the FAA in January 1994. Southwest Airlines has ordered the GNSSU for its Boeing 737-700 aircraft.

Contractors

Honeywell Inc, Commercial Aviation Systems.
Canadian Marconi Company.

VERIFIED

Horizontal Situation Indicator (HSI) type 29800

The type 29800 HSI is designed to display data from standard analogue, digital and ARINC sources including VOR, ADF, TAC, DME, ILS and MLS systems. It meets civil and military standards.

The standard size is 4 × 4 in (101 × 101 mm), but other sizes are available ranging from 4 × 5 in to 3 × 3 in, by using different display formats.

Specifications

Altitude: –1,000 to +50,000 ft
Dimensions: 101 (W) × 101 (H) × 229 (D) mm
Weight: 3.3 kg
Power: 28 V DC, 250 mA; 26 V AC, 400 Hz, 250 mA

Contractors

ELAN Elektronische und Anzeiger GmbH, Germany.
CINAVE Companhia de Instrumentos de Navegaçao Aeronáutica Lda, Portugal.
France Aerospace SARL, France.
International Aerospace Inc, USA.

VERIFIED

Type 29800 horizontal situation indicator
2000/0062187

ILS-85 Instrument Landing System

The ILS-85 is designed to integrate with the EFIS-85, FCS-85 and other systems in the -85 series utilised by the Il-96-300 and Tu-204 aircraft.

Specifications

Frequency range:
(localiser) 108.00-111.95 MHz
(glideslope) 329.15-335.00 MHz
Channels:
(ILS mode) 40
(LS-50 mode) 20
Precision:
(ILS mode) ICAO Cat III
(LS-50 mode) Cat II
Receiver dimensions: 3 MCU
Power: 115 V AC, 400 Hz, 30 VA
Weight:
(receiver) 5.5 kg
(antenna) 0.53 kg
MTBF: 5,000 h

Contractors

NIIAO Institute of Aircraft Equipment, Moscow.
Ukraine Research Institute of Radio Equipment, Kiev.

VERIFIED

MCS 3000/6000 aeronautical satellite communications system

Honeywell and Racal Avionics are teamed for the development, manufacture and marketing of a multichannel satellite communications system for commercial aircraft, fully compatible with ARINC 741 and ARINC 761 for Aero-I.

The Honeywell/Racal MCS 3000/6000 systems provide a three- or six-channel full-duplex voice and data communications capability supporting such functions as Airline Communications And Reporting System (ACARS), Automatic Dependent Surveillance (ADS) and flight deck and passenger telephone and fax communications between an aircraft and the ground.

The MCS 6000 airborne terminal comprises a Satellite Data Unit (SDU), Radio Frequency Unit (RFU) and High-Power Amplifier (HPA), and may be interfaced to a variety of high-gain phased-array antenna subsystems and voice/data communications devices.

The SDU performs the functions of system controller, data modulation and demodulation, data synchronisation and decoding and voice coding/ decoding.

The RFU performs the functions of down converting the received L-band (NATO D-band) signals to a lower frequency for input to the digital processing circuits and up converting the modem output signals to the L-band (NATO D-band) transmit frequency for each operational channel. The RFU operates in full-duplex mode, simultaneously supporting both receive and transmit functions at all times.

The HPA is a linear power amplifier which provides the gain to generate the required output power. The output power of the HPA is under the control of the SDU which receives data from the ground station commanding an increase or decrease in output power to maintain the satellite signal at a satisfactory level. The HPA is available in either an ARINC 741 or ARINC 761 configuration.

The system operates at L-band (NATO D-band) frequencies. Signals are relayed via the Inmarsat space segment satellites, linking in to the ground telecommunications network through a series of dedicated Ground Earth Stations (GES). The satellite and GES networks combine to provide a worldwide communications service.

MCS 3000/6000 systems support both air-to-ground and ground-to-air communications. The systems are capable of supporting 9.6 kbytes/s voice, 4.8 kbytes/s fax, 2.4 kbytes/s PC modem and 10.5 kbytes/s packet data services for ARINC 741 applications; the systems will support 4.8 kbytes/s voice, 2.4 kbytes/s fax, 2.4 kbytes/s PC/modem and 1.2 kbytes/s packet data services for ARINC 761 (Aero-I) applications. Typical applications for these services break down into the areas of passenger services, airline operational and administrative services and air traffic control.

Passenger services include telephone, fascimile, PC and value added data services such as catalogue sales, hire car reservations and duty free sales.

Airline operational and administrative services include the ACARS datalink supporting engineering, operational and cabin management functions.

Air traffic control uses include Automatic Dependent Surveillance aircraft position reporting.

MCS 3000/6000 operates in full accordance with Inmarsat specifications and type approval has been received on all major wide-body aircraft types.

Aero-I is an upgraded capability of the MCS 3000/ 6000 to allow narrow-body aircraft with intermediate-gain antennas to utilise the new Inmarsat third-generation Aero-I spot-beam services. As an extension to the Aero-I upgrade, existing high-gain MCS equipment can be upgraded, via a 'Service Bulletin'; this takes advantage of 'evolved Aero-H' services, which provide a similar range of passenger global communications for telephone, fax and pc-data, as Aero-I. Using 'evolved Aero-H' in spot-beam coverage allows the operator to benefit from reduced service charges and 4.8 kbytes/s voice CODECs (digital transmission).

Honeywell/Racal MCS 3000/6000 aeronautical satellite communications system showing both the six-channel MCS-6000 system (radio frequency unit (left), satellite data unit (centre) and high-power amplifier (right)); and the three-channel MCS-3000 system (satellite data unit (left) high-power amplifier (right)) 0002115

Specifications

Dimensions:
(SDU) 6 MCU; weight 10.73 kg
(RFU) 4 MCU; weight 7.68 kg
ARINC 741 (HPA) 8 MCU; weight 12.86 kg (Aero-H)
ARINC 761 (HPA) 4 MCU: weight 7.05 kg (Aero-I)
Power supply: 115 V AC, 400 Hz; or 28 V DC
Output power: 60 W typical
Frequency:
(transmit) 1,626.5-1,660.6 MHz
(receive) 1,530-1,553 MHz

Operational status

Over 1,600 installations have been completed on all major wide-bodied aircraft types and on many top-of-the-range executive jets.

Integration has been completed with all major antenna, ACARS and passenger telephone equipment vendors.

In March 1996, the Racal/Honeywell team was named as the US government's provider of choice for multichannel SATCOM using the MCS-3000/6000 systems, featuring STU-III secure voice, access to Microsoft-Mail, 9,600 bits/s fax and other capabilities, in conjunction with the Tecom T-4000 High-Gain Antenna System and the Honeywell CM-250 Communications Management Units. The MCS-6000 is also fitted to the Advanced Range Instrumentation Aircraft (ARIA) RC-135s.

System upgrades that became available in May 1998 include new MCS-3000i/+ and MCS-6000i/+ variants which, in addition to supporting Aero-H, will be able to support Aero-I and Aero-H+.

The Aero-H+ system (Aero I and H) will use the same high-gain antenna as Aero-H but, with the capability of using spot-beam satellites, it offers a potentially lower service cost. The Racal/Honeywell Aero-I systems operate in the spot-beams of the new-generation Inmarsat-3 satellites. All services offered with Aero-H are available on Aero-H+ and Aero-I.

The Racal/Honeywell team has certified the Canadian Marconi CMA-2200 antenna as its exclusive

Aero-I antenna. All FAA certification work for the MCS 3000/6000 Aero-I Boeing 737-800 installation has been completed, and launch customers for the MCS 3000/ 6000 on the Boeing 737-800 are Hainan Airlines and Royal Air Maroc.

Aero-I has also gained its first European certification onboard the UK Ministry of Agriculture's Fisheries Patrol aircraft – a turboprop Cessna 406.

By mid-1999, Aero H+ had been certified on Airbus A330, Boeing 747-400 and Global Express aircraft.

Contractors

Honeywell Inc, Commercial Aviation Systems.
Racal Avionics Ltd.

VERIFIED

MCS-7000 aeronautical satellite communications system

Honeywell and Racal Avionics have launched the MCS-7000 as their latest-generation enhanced satellite communications system for commercial airliners and business jet aircraft, based on their existing MCS-3000 and MCS-6000 systems. The new MCS-7000 provides up to seven channels of voice/data communications and the capability of Aero-H, Aero-H+ or Aero-I (spot-beam) services, depending on the High-Power Amplifier (HPA) and antenna configuration. The MCS-7000 satcom systems consist of only two units, a 6 MCU-size Satellite Data Unit (SDU) and a 4 MCU-size High-Power Amplifier (HPA) with embedded Beam Steering Unit functionality.

The Aero-H+ system will use the same high-gain antenna as Aero-H but, with the capability of using spot-beam satellites, it offers a potentially lower cost. The Aero-I system operates in spot beams of the new-generation Inmarsat-3 satellites. Spot beams have lower power requirements and therefore smaller, lower-power HPAs and smaller Intermediate-Gain Antennas can be used. The selected antenna is the Canadian Marconi CMA-2200.

All services currently offered with Aero-H systems are available on Aero-H+ and Aero-I including cockpit voice (allowing instantaneous communication with operations, maintenance and air traffic control); passenger telephony, passenger fax, news and weather broadcasts; interactive passenger services and a PC data capability.

MCS-7000 is the standard production system, available from June 1999, providing several additional features, including an optional internal PBX phone system with digital handsets. For aircraft which are not equipped with an Inertial Reference System (IRS), a new Signal Conditioning Unit (SCU) supports satcom operation and a proprietary interface supports the imminent Complementary Satellite Systems, including Low Earth Orbit (LEO) and Medium Earth Orbits (MEO). Maximum flexibility is provided for operators to choose Inmarsat services for the cockpit (safety/ATC services) and an alternative service for passenger requirements (voice/fax).

Company data also uses the terminology MCS-7000+ and MCS-7000i when describing this system. It is also claimed that, in addition to supporting existing Inmarsat Aero-H, Aero-H+ and Aero-I, the MCS-7000 system has growth to support the Complementary Satellite Systems (CSS).

Operational status

MCS-7000 is the standard production system and was available from June 1999. Launch customer is Hainan Airlines on its Boeing 767 aircraft.

MCS-7000i is also being supplied by Honeywell, via its Defense Avionics Systems division, to the US Air Force, together with the CM-950 and CM-950 VIA communications management units, to meet Global Air Traffic Management (GATM II) upgrade programme requirements.

Contractors

Honeywell Inc, Commercial Aviation Systems.
Racal Avionics Ltd.

UPDATED

Multi Link 2000 M-ADS helicopter satellite communications system

The Multi Link 2000 Modified Automatic Dependent Surveillance (M-ADS) system has been developed by Kongsberg Defence & Aerospace for installation in helicopters operated by Helikopter Service and Norsk Helikopter.

The M-ADS system comprises the following system elements:

(1) Airborne equipment consisting of an M-ADS unit; Satellite TRansceiver (STR); low-noise amplifier and antenna, linked to Inmarsat satellite
(2) Inmarsat satellite and ground station, linked to ATC centre
(3) Air Traffic Control (ATC) centre to set up ADS-Contracts (ADS-C) with the aircraft, so that current position and intent can be displayed.

System functionality includes:

(1) ADS - Contract (ADS-C) and ADS - Broadcast (ADS-B).
(2) Dual air/ground datalink for safety critical messages: ADS, controller-pilot datalink communication and flight information service.
(3) VHF air/ground and air/air communication 9VDL Mode 4/STDMA).
(4) Inmarsat aeronautical satellite air/ground communication.
(5) ICAO compliant end-to-end data integrity and transfer reliability from built-in Aeronautical Telecommunication Network (ATN) protocols.

The system transmits position, waypoint, groundspeed and other aircraft information from air to ground via the Inmarsat satellite system, providing a means of accurate position reporting. This information, which is automatically transmitted at regular intervals, will provide pinpoint accuracy and keep air traffic controllers up to date with the aircraft's position and intentions.

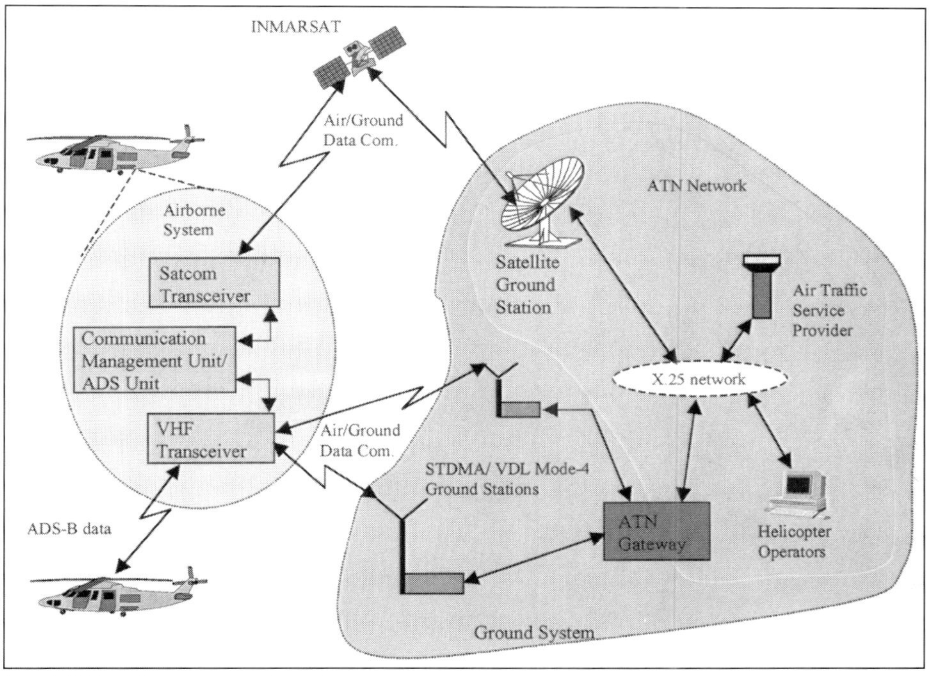

The Multi Link 2000 system equipment **2001**/0089531

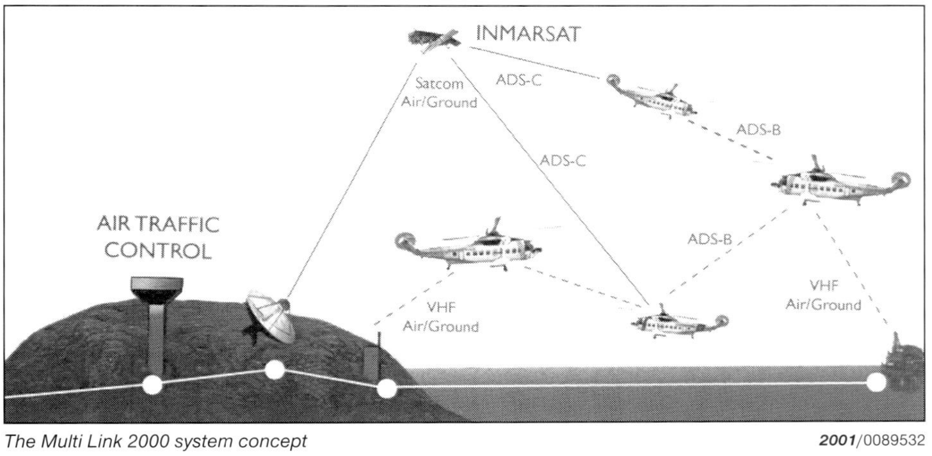

The Multi Link 2000 system concept **2001**/0089532

Specifications

SDU: 2 MCU weighing <4.5 kg
HPA: 2 MCU weighing <4.5 kg
ADSU: 2 MCU weighing <3 kg

Operational status

The M-ADS system is installed in Eurocopter Super Puma AS332L/LI/L2 and AS365N-2 Dauphin, Bell 214ST and Sikorsky S-61N and S-67 helicopters. Racal Avionics delivers the Satellite TRansceivers (STRs).

Contractors

Kongsberg Defence & Aerospace AS.
Racal Avionics Ltd.

UPDATED

Rate of climb indicator type 130500

The type 130500 rate of climb indicator measures vertical speed, to a limit of 60 m/s; lighting can be red or white.

Specifications

Vertical speed: 0 to ±60 m/s
Dimensions: 65 (W) × 65 (H) × 160 (D) mm
Weight: 0.6 kg
Power: +27 V DC, 1.5 W; 5.5 V AC, 3 VA

Contractors

ELAN Elektronische und Anzeiger GmbH, Germany.
CINAVE Companhia de Instrumentos de Navegaçao Aeronáutica Lda, Portugal.
France Aerospace SARL, France.
International Aerospace Inc, USA.

VERIFIED

Satcom conformal antenna subsystem

This Satcom antenna is designed for aeronautical communications in the L band, to meet the specific needs of the airline and general aviation industry. It is compliant with ARINC 741 and meets DO-160C.

The equipment is composed of two side-mounted conformal antennas, two Beam-Steering Units (BSUs) located inside the aircraft and the associated Diplexer/ Low-Noise Amplifier (D/LNA) assemblies.

In order to provide a Satcom system with no operational limitations, an antenna with optimised RF characteristics has been designed. A large number of radiating elements arranged in an optimum pattern offers a better combination of high-gain and low-sidelobe levels. The thin profile of the High-Gain Antenna (HGA) results in a negligible drag penalty of less than 0.02 per cent of total aircraft drag and the antenna subsystem is adaptable to all high-gain Satcom avionics subsystems available or under development.

Specifications

Dimensions:
(High-Gain Antenna × 2) 566.4 × 495.3 × 7.6 mm
(Beam-Steering Unit × 2) 342.9 × 261.6 × 88.9 mm
Weight:
(High-Gain Antenna × 2) 7.5 kg
(Beam-Steering Unit × 2) 8.5 kg
(Diplexer/LNA × 2) 3 kg

Operational status

The antenna has received Inmarsat multichannel access approval with no restrictions.

The antenna system is certified on Airbus A300, A310, A330 and A340; Boeing 707, 737-300, 747-400, 767, MD-11, L-1011 and MD-80 aircraft, and the Falcon 900. More than 400 Satcom antenna systems have been ordered by over 20 major airlines.

Contractors

Honeywell Aerospace, Electronic & Avionics Lighting.
Thomson-CSF DETEXIS.

VERIFIED

SCS-1000 Mini-M aeronautical satcom

The Racal/Honeywell satcom consortium have signed an agreement with OmniPless (Pty) Ltd of South Africa to be responsible for the exclusive marketing, distribution and product support of the SCS-1000 Mini-M aeronautical satcom worldwide. The OmniPless SCS-1000 Mini-M is a complementary entry system to the MCS 3000/6000/7000 product-line of aeronautical satcom systems manufactured by Racal/Honeywell.

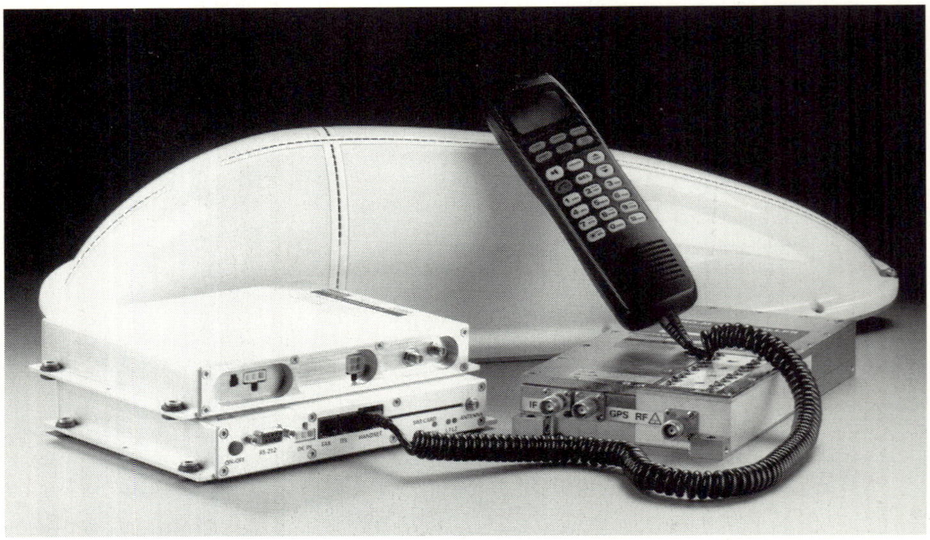

The SCS-1000 Mini-M is based on the Inmarsat land-mobile Mini-M and is intended for small to medium-sized business jet and turboprop operators. Small size offers low cost and flexible installation. The product is type-approved for use with the spot beam capabilities of the latest generation Inmarsat-3 satellites.

The SCS-1000 is a single-channel system, which supports standard (or secure) voice, fax or personal computer data transmission. All antenna steering control sensors, GPS and attitude sensors are integral to the antenna, making them independent of other aircraft systems. Remaining system components comprise an Antenna Control Unit (ACU), Power Supply Unit (PSU), telephone unit and handset.

Contractors

OmniPless (Pty) Ltd.
Racal Avionics Limited.
Honeywell Inc, Commercial Aviation Systems.

UPDATED

The SCS-1000 Mini-M satcom system
2001/0103885

SLS 2000 Satellite Landing System

In January 1995, Honeywell and Pelorus Navigation Systems Inc of Calgary, Canada, teamed to develop and manufacture a Differential GPS (DGPS) ground reference station called the Satellite Landing System (SLS). The SLS allows for improved all-weather operations that reduce delays and operating costs while maintaining high integrity and safety. It can provide Special Cat I (SCAT I) capability to all runway ends within a 30 n mile radius, making it more cost effective than traditional landing aids that are limited to one runway end. It ensures local airport control and enhances satellite coverage and it can provide the flexibility to design approaches that minimise flight time and meet noise abatement objectives.

SLS will be Cat I and II capable with growth to Cat III, and SLS will enable variable geometry precision approaches and departures.

SLS-1000 and SLS-2000

The SLS ground station is available in two configurations, SLS-1000 and SLS-2000. Both systems comprise three major subsystems: ground reference station, Remote Satellite Measurement Units (RSMUs) and VHF (Very High Frequency) datalink transmitter.

The SLS-1000 unit is a fail-safe system that is designed continuously to perform self tests to determine its 'health'. If it detects a problem, it will notify the operator and any aircraft in the area that it is not capable of sending accurate data.

The SLS-2000 unit is a fail-operational system that is not affected by single component failures. The system operator is notified of a component failure and can call for service while the unit continues to operate.

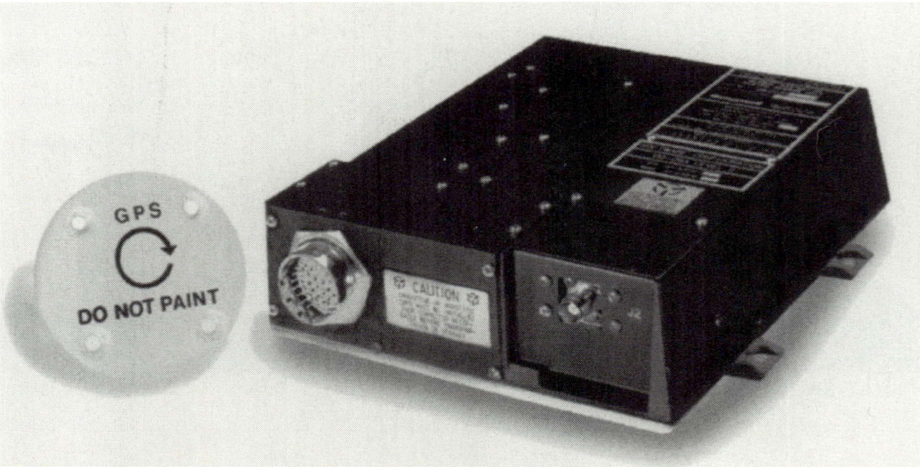

Antenna Differential Global Navigation Satellite Sensor Unit (DGNSSU) 0001405

The SLS-1000 and SLS-2000 are self-calibrating. The systems are designed with performance monitors, which eliminate the need for periodic flight checks. Both systems come with a fault-tolerant power supply and battery back-up to ensure continuity of service in the event of a power outage.

Airborne equipment complement

Although Honeywell is involved with design of the ground-based element of SLS-2000, Honeywell alone is designing and manufacturing the airborne element.

For the forward fit market, Honeywell and Pelorus are working with industry and aircraft manufacturers to design the optimum solution for future aircraft.

In the retrofit market, most aircraft will need some combination of the following systems: Flight Management System (FMS) or SLS controller; VHF datalink receiver; Differential Global Navigation Satellite Sensor Unit (DGNSSU); Analogue Interface Unit (AIU).

Maintaining the high integrity of the system is critical to certification and safety. To do this, the Honeywell DGNSSU, designed to ARINC 743 standards, calculates and directs the flight controls. The DO 178B level B, critical-level software, ensures that the system will perform the calculations correctly.

Annunciations and specific interface requirements are resolved on an aircraft-by-aircraft basis.

The FMS or SLS controller tunes the VDL-500 to the SLS ground station datalink frequency at a given airport. The range error corrections and path points received by the VDL-500 are sent to the DGNSSU.

The DGNSSU makes the necessary corrections and calculates the approach path.

The approach path, based on SLS position information, is transmitted to the flight controls as an ILS lookalike signal. The approach is flown by the flight controls using this input.

VDL-500 Flight Management System (FMS) 0001404

Operational status

In September 1998, a Continental Airlines MD-80 flew the inaugural flight of the SLS 2000 system using it for precision landings at commercial airports.

Contractors

Honeywell Inc Commercial Aviation Systems.
Pelorus Navigation Systems Inc.

VERIFIED

Timearc 6 GPS navigation management system

The Timearc 6 GPS navigation management system provides accurate position information from an advanced six-channel GPS receiver designed with the latest technology for military and commercial applications. The Jeppesen® NavData card contains information on all airports, VORs, NDBs and intersections in Europe or North America. The three-line 16-character LED alphanumeric display, which is extra bright and adjustable in intensity, is easy to read and is equipped with a left/right cursor indexer.

The Timearc 6 uses GPS satellites for position calculation and navigation and provides worldwide continuous and precise navigation data, offering operation under dynamic conditions up to 10 *g*, time to first fix of less than 1 minute and optional precise position data with differential GPS.

The associated mission planning system provides the ability to augment the Jeppesen® NavData card with user-defined waypoints and individual mission data. In addition, besides Universal Transversal Mercator (UTM), positions can be displayed in various local grids, including British National Grid and Swiss Grid to compare directly with local maps. Optionally, the Control and Display Unit (CDU) is equipped with an ARINC 429 interface, and the CDU features SID, STAR, and approaches capability.

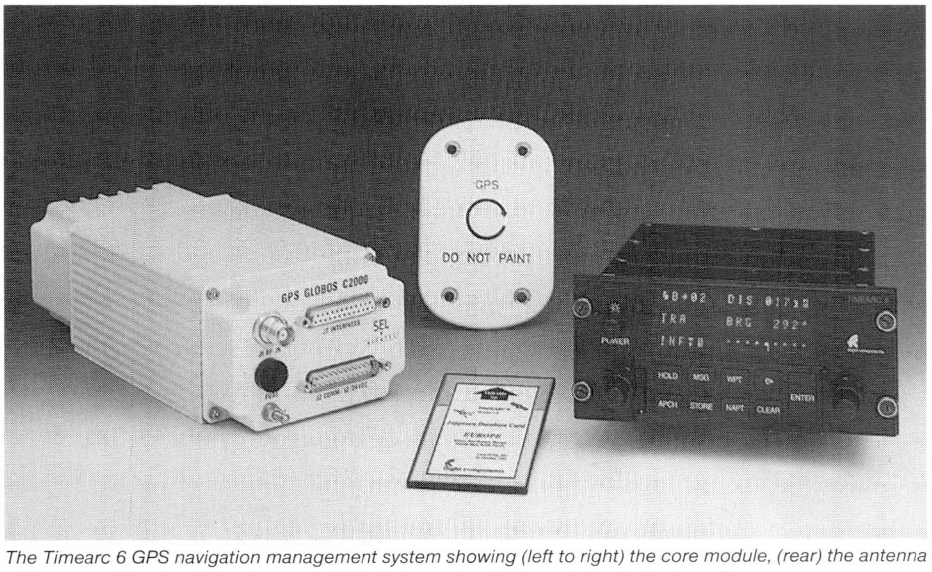

The Timearc 6 GPS navigation management system showing (left to right) the core module, (rear) the antenna with preamplifier, (front) the Jeppesen® NavData card and the control and display unit

Specifications

Dimensions: 66 × 146 × 177 mm
Weight: 1.37 kg
Power supply: 10-40 V DC, 0.26 A at 28 V
Temperature range: −20 to +55°C
Altitude: up to 35,000 ft
Accuracy - 3-D position (95%):
(SPS not degraded) 51 m
(SPS with selective availability) 174 m
(DGPS without selective availability) 10 m
(DGPS with selective availability) 20 m

Contractors

Aerodata GmbH.
Flight Components AG.
Sextant.

VERIFIED

Timearc Visualizer multifunction digital display system

The Timearc Visualizer multifunction digital display system is a versatile display system for enhanced situational awareness and mission management; it displays moving map, video inputs, electronic library and waypoint-find options.

A large, flat panel, full-colour, high-resolution active matrix (TFT) LCD driven by the Digital Map Display Generator (LRU) is used to combine and use mapping data from multiple raster data sources.

A special Smart Point Track Stick, integrated into the flat panel LCD, allows the user to scroll over the entire activated map and to pinpoint any desired position. A special window shows the digital readout of such a position in latitude/longitude, or any other selected grid system. A pinpointed position can also be used for an instant transfer into the GPS Navigation Management System or for manually-initiated direct navigation steering.

The Timearc Visualizer does not include a GPS engine, but most stand-alone GPS navigation systems or GPS sensors can be interfaced with the Visualizer, provided they possess a free serial interface either RS-232/422/485 or ARINC 429.

Special purpose software available includes an integrated Electronic Library System (ELS) and 'Waypoint' or 'Street Find' options. The Visualizer version 3.0 displays moving maps and video in real time. The active matrix LCD is available in two models: the standard LCD, and a high-performance LCD for optimal sunlight readability.

Specifications

Colour Display (AMLCD)

Dimensions: 280 × 199 × 52 mm

Resolution: 640 × 480

Colours: 256

Weight: approx 1.9 kg

Electrical:

interfaces: keyboard

Display:

screen size: 214 mm

Data Generator (DMD/LGM)

Dimensions: 96 × 163 × 318 mm

Weight: approx 3.0 kg

Electrical:

input voltage: 10-40 V DC

power consumption: 25 W (max)

interfaces: RS-232/422, ARINC 429 (optional)

Operational status

Widely used by police forces, search-and-rescue organisations, forestry/oil/gas industries in helicopter and fixed-wing aircraft installations.

Contractors

Flight Components AG.

Dallas Avionics.

UPDATED

Timearc Visualizer
0018196

Vertical reference indicator type 41500

The type 41500 vertical reference indicator is a self-contained aircraft indicator that provides pitch and roll information.

Specifications

Dimensions: 65 (W) × 65 (H) × 170 (D) mm

Weight: 1.1 kg

Power: 115 V AC, 400 Hz, 200 mA

Contractors

ELAN Elektronische und Anzeiger GmbH, Germany.

CINAVE Companhia de Instrumentos de Navegaçao Aeronáutica Lda, Portugal.

France Aerospace SARL, France.

International Aerospace Inc, USA.

VERIFIED

Visual Guidance System (VGS)

BAE Systems has utilised its C-17 and combat aircraft head-up display technology to develop the Visual Guidance System (VGS), previously termed the HUD 2020 and HUD 2022. The HUD 2020 system is for corporate jet operators and the HUD 2022 for air transport aircraft. In these developments BAE Systems has teamed with Honeywell Commercial Aviation Systems.

The VGS displays essential, stroke and raster, flight information to the pilot in his forward field of view, thus providing:

(1) Improved situational awareness
(2) Navigation data and precise flight path guidance
(3) Warning alerts, including TCAS, TAWS, weather radar and windshear warnings
(4) Improved assurance during ground operations on the runway and taxiways in adverse conditions.

The VGS comprises 4 electronic units and a mounting tray:

(1) OverHeadUnit (OHU). The OHU contains the CRT, drive circuitry and optical system to project the image on to the combiner assembly. It has a 30 × 25° field of view
(2) Combiner Assembly (CA): The CA consists of a lightwieght glass combiner, an ambient light sensor and an integral control panel. The combiner overlays the critical flight symbology projected by the OHU onto the real world image
(3) Display Guidance Computer (DGC). The DGC receives data from the various aircraft systems and uses this data to generate symbology and command guidance, which is then supplied to the OHU
(4) HUD Annunciator Panel (HAP). The HAP is located at the first officer's station and enables the 'pilot not flying' to monitor the status of the VGS
(5) The mounting tray carries the OHU and CA in a customised tray above the pilot's position.

Operational status

The VGS was granted US FAA full Supplemental Type Certification (STC) for use on the Boeing 737-800 in October 1999. UK Civil Aviation Authority (CAA) approval for use of the VGS on all UK registered Boeing 737-600/-700/-800 aircraft was granted in January 2000. Approvals cover VGS use in Cat IIIa landings with the auto-throttle on or off. VGS is also approved for single engine Cat III approach. The VGS can be used to fly the aircraft manually to a decision height of 50 ft when Runway Visual Range (RVR) is as low as 600 ft

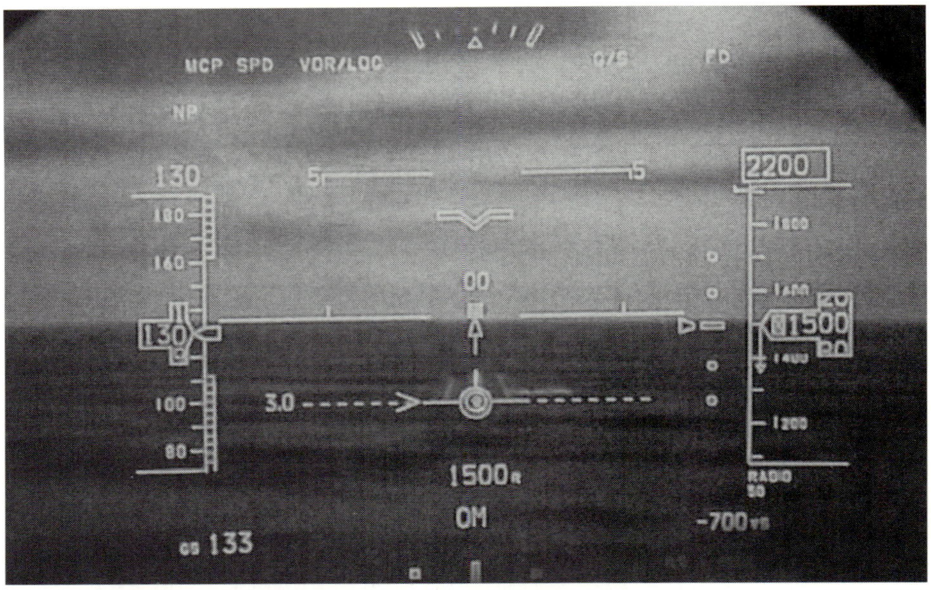

The visual guidance system display **2000**/0081733

The visual guidance system cockpit installation **2000**/0080277

and for take-off guidance at RVR of 300 ft. The US FAA STC work was accomplished with American Airlines, which has selected the system for its fleet of 737-800 aircraft.

Development work on an Enhanced Visual Guidance System (EVGS) is underway, and BAE Systems has successfully conducted proof of concept trials to interface VGS with both millimetric wave radar and infra-red sensors.

Contractors

BAE Systems.

Honeywell Inc, Commercial Aviation Systems.

VERIFIED

XK 516D HF airborne voice/data radio

The XK 516D airborne radio is designed for use in commercial aircraft. The system provides conventional voice and high-speed data air-to-ground, ground-to-air and air-to-air communications over long distances. The data communication is suitable for aircraft operational and administrative communications, as well as air traffic communications. The XK 516D is a joint development by Rohde and Schwarz and Honeywell Aerospace.

The XK 516D consists of the XK 516D1 transceiver and the FK 516/517 antenna coupler. The data modules which provide the high-speed data function are fully integrated within the transceiver. The voice/ data therefore fits within the space of a conventional radio and additional space for the data capability is not needed.

The functioning of the equipment is controlled by the integrated test system in which a number of functions are continuously monitored. After the test routine has been triggered, any faulty module will be located and indicated. BITE results are reported to the onboard CFDS/CMC system via two ARINC 429 busses. Interfaces to the central maintenance systems of Airbus and Boeing aircraft are implemented in the radio, featuring one part number for nearly all aircraft types.

The XK 516D is designed to meet the requirements of ARINC 719 for the voice function and ARINC 753/ 635 for the data function. The integrated data communication capability meets the specifications of ARINC 753 and 635 and high-speed data communication up to 1,800 bits/s user rate is provided. The data capability also means that operators can obtain the benefits of ACARS and the Aeronautical Telecommunications Network (ATN) beyond limited VHF coverage, using the XK 516 D1 transceiver.

To provide full compatibility between existing and new equipment and aircraft wiring, multiwire serial interface to ARINC 753, conventional ARINC 719 control lines and a single wire coaxial interface between transceiver and the antenna coupler are available. This provides interchangeability between existing voice transceivers and the coupler.

The antenna coupler is a digitally tuned coupler with tuning times, typically of less than 3 seconds. The learn mode can provide even shorter tuning times of several hundred milliseconds.

The FK 516 antenna coupler is designed to tune suppressed shunt/notch antennas that are common on the fin structure of most modern aircraft, while the FK 517 antenna coupler is designed to tune the tailcone antennas fitted to older aircraft.

Specifications

RF power: 400 W PEP
RF range: 2 - 30 MHz
Tuning increments: 100 Hz
Data rate: 150, 300, 600, 1,200, 1,800 bits/s
Dimensions: 6 MCU
Weight: < 12 kg
Power: < 1 kW

Contractors

Honeywell Aerospace, Electronic and Avionics Lighting.

Rohde & Schwarz GmbH & Co KG.

VERIFIED

ISRAEL

D-Map digital moving map system

Designed for a two-seater cockpit, Elbit's digital moving map system (D-Map) single- or twin-map generators are capable of producing numerous types of images presented on a MultiFunction Colour Display (MFCD). Various map modes and features are controlled by keys integrated as part of the display. The system also features a Mass Storage Device (MSD) for storing databases such as a 300 × 300 km map in three different scales, obstacle overlay and library. The powerful graphic unit can draw raster and vector maps simultaneously.

Using a pointing device switch, the pilot can prepare and display his emergency flight plan in seconds. Working at a ground station, it is possible to plan an entire mission in advance, incorporating tactical mission data, flight plan, obstacles and communication data. The accumulated data is subsequently downloaded to the MSD or directly to the Map Display Generator (MDG) via serial communication channels. The end result is a significant increase in flight safety and reduction in pilot workload.

The system is capable of operating in two major display modes: as a digitised map display or aerial photographs, in which a digitised paper map in a range of linear scales from 1:50,000 to 1:2 million is displayed; or as a digital map display in which the display is a pseudo two-dimensional terrain model consisting of coloured altitude planes and vector type ground features calculated from a digital terrain elevation database.

The image presented on the MFCD is clear, crisp and easily interpreted. The system also offers the user tools such as zoom, heading up, scale and freeze for map manipulation. The online computing power of the MDG can calculate the line of sight between any given positions or point altitudes for threat coverage or safety corridors for routes and approaches.

Specifications

Dimensions:
(MDG) 137 × 198 × 472 mm
(MFCD) 216 × 216 × 400 mm
(MSD) 225 × 140 × 303 mm
Weight:
(MDG) 10 kg
(MFCD) 11 kg
(MSD) 6 kg
Transmits: flight plans prepared on maps
Video:
(RS-170) R, G, B
(RS-422) serial channel
(RS-232) obstacles database update
Options: MIL-STD-1553B mux bus, NVG-HUD

Operational status

D-Map has been selected by Rega, a prominent air ambulance service, for the A 109 EMS helicopter and has already been selected by the Israeli Air Force for the CH-53 helicopter modernisation programme.

D-Map has also been selected for the V-22 Osprey, where EFW Inc, an Elbit subsidiary in the USA, will act as prime contractor in the contract.

In July 1999, D-Map was selected by the Swiss Air Force for its Super Puma Mk 1 helicopter fleet of 12 aircraft.

Contractor

Elbit Systems Ltd.

VERIFIED

ITALY

Lisa-4000 strapdown inertial reference unit

The Lisa-4000 performs the basic functions of flight reference and navigation. It provides heading and attitude for display and autopilot and high-speed low-latency anti-aliasing filtered body angular rate and linear acceleration data for autopilot inner loop stability augmentation.

Navigation is performed by coupling the Lisa-4000 with GPS, Doppler radar and air data systems. The Lisa-4000 also outputs high-speed inertial velocity data required for accurate weapon release. It therefore provides all the inertial parameters required for autopilot, navigation, display and weapon delivery.

The Lisa-4000H version is a full-aided inertial navigator with the option of a MIL-STD-1553B or ARINC 429 interface. It can also be supplied with additional synchro outputs to drive back-up cockpit instruments.

Specifications

Dimensions: ½ ATR short
Weight: 6.2 kg
Power supply: 28 V DC, 90 W unregulated
Cooling: 0 to 71°C
Accuracy (RMS):
(heading) 0.2° + compass
(pitch and roll) 0.2°
(body rates-PQR) 0.15°/s, 200 Hz, 2.5 ms latency
(body rates-XYZ) 5 *m*g, 200 Hz, 2.5 ms latency
(horizontal velocity) 2 m/s
(vertical velocity) 1 m/s
(position) 1% of distance travelled
(altitude) 100 ft + 1%
Inputs: GPS, Doppler velocities, barometric altitude, true airspeed, compass, up to 100 waypoints
Outputs: bearing to destination, distance to go, time to go, wind speed and direction, groundspeed

Operational status

The Lisa-4000B is in production for the A 129 attack helicopter. The Lisa-4000D is operating on the UK Royal Navy EH 101 helicopter. The civil qualified Lisa-4000E is for the EH 101 civil variant. The GPS coupled Lisa-4000G is in production for an RPV. The Lisa-4000H is in production for Italian Navy AB-212, AB-412, SH-30 helicopters, and the EB version on Romanian MiG-21. A fibre optic gyro, designated Lisa-FG has been selected for the MB-339A Mid-Life Upgrade (MLU).

Contractor

Litton Italia SpA.

VERIFIED

JAPAN

Flat-panel display

The flat-panel display uses a high-resolution liquid crystal display now under development for use in advanced aircraft cockpits. It features a full-colour display with adjustable high-brightness back-light, to give exceptional contrast. The system is lighter and thinner, with less power consumption, than conventional CRTs.

Contractor

Japan Aviation Electronics Industry Ltd.

VERIFIED

Pulse radar altimeters

JAE pulse radar altimeters are used in military and civil fixed-wing aircraft and helicopters. The systems consist of a transmitter/receiver, an antenna and an indicator. The unit is provided with a low warning light or audio signal to provide a warning when the aircraft reaches a preset altitude.

Operational status

The APN-171 is fitted to the Shin Meiwa PS-1, Sikorsky HSS-2, Kawasaki/Vertol KV-107, Kawasaki P-2J and Kawasaki C-1. The APN-194 is fitted to the Lockheed P-3C and the JARN-P2 is fitted to the Boeing-Vertol CH-47, Bell AH-1S and Mitsubishi LR-1.

Contractor

Japan Aviation Electronics Industry Ltd.

VERIFIED

Central warning display

The central warning display has been developed for commercial aircraft. It is designed to work with two master warning lights which can effectively alert pilots and identify potential problems or hazards.

The indication panel is a liquid crystal display which can indicate a maximum of 10 warning items in red, amber or green. When more than 10 items occur at the same time, up to 50 additional warnings are listed by a scrolling function. The LCD has a high back-light unit consisting of halogen lamps, allowing the manually or automatically dimmable display to be read under sunlight conditions. The red and amber warning legends can be tailored to customer requirements by a software change. Dual redundancy is incorporated in the power supply, input interface circuit, digital computer and LCD display.

Specifications

Dimensions: 134 × 125 × 250 mm
Weight: 2.5 kg

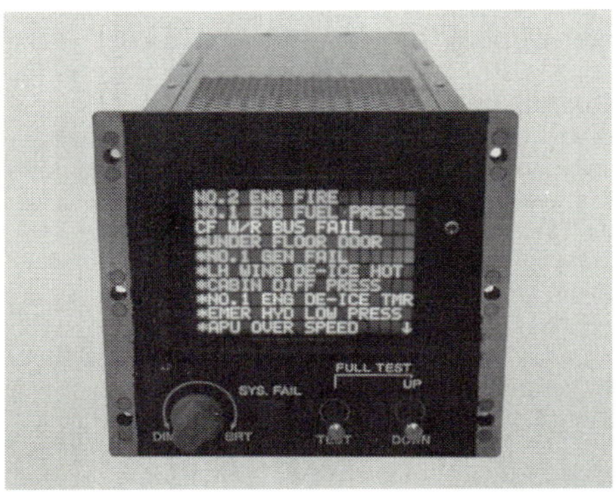

Koito central warning display for commercial aircraft

Display size: 55 × 75 mm
Power supply: 16-30 V DC, 1.3 A at 28 V DC
Temperature range: –40 to +70°C
Altitude: –100 to 25,000 ft
Reliability: >5,000 h MTBF

Contractor

Koito Manufacturing Company Ltd.

VERIFIED

Colour liquid crystal display

The colour liquid crystal display is a bright display which is composed of a high-quality colour filter with excellent colour purity and brilliant back-up illumination with uniform brightness. The display is an eight-colour type using the three primary colour filters of red, green and blue and can easily provide stationary coloured pictures at low cost. In addition, it can increase display information by changing a monochrome liquid crystal display into a multicoloured display.

Contractor

Koito Manufacturing Company Ltd.

VERIFIED

Electroluminescent display

Koito has developed two types of ElectroLuminescent (EL) displays: the AC powder EL display and the thin-film EL display. Both types are thin lightweight surface light emitting devices and are considered ideal for displays due to features such as the uniform brightness on the light emitting surface, high reliability and lack of heat generation.

The AC powder EL display is a flexible type of surface light emitting device which can be bent. It is formed by applying fluorescent and transparent conducting layers on aluminium foil and then laminating a transparent plastic sheet. It is very resistant to excessive vibration and load and is built into passenger aircraft cabin aisles, in aircraft such as the Boeing 767, as floor marker lamps.

In the AC powder EL display the ultra-slim flat light source provides extremely uniform and soft illumination. Composed of organic materials with emphasis on plastic film, it is easily bendable and will provide illumination from a curved surface. Lightweight, tough and durable to vibration and shock, it does not suffer filament breakages as in conventional bulbs. As an EL element, it provides cool light without generating heat.

The thin-film EL display consists of an insulating layer, luminous layer and electrodes applied on the glass substrate by vacuum evaporation. It is far brighter than the AC powder EL display.

It can be built easily into portable apparatus and the high precision of electrode stripes and thin-film combine to provide an easily visible display with high resolution. Since it is an all-solid-state self-luminous element, it has a higher stability and longer life. The response speed of the display is about the same as a CRT display and so it can display letters, numbers, graphs and so on, in real time. The thin-film EL display is available as a dot-matrix or segment display.

Contractor

Koito Manufacturing Company Ltd.

VERIFIED

Direction-finding and receiving system

The direction-finding and receiving system is designed to detect SHF frequencies between 6.4 and 7.1 GHz used for video transmissions from ground stations and employs a directional antenna to track the transmission automatically. The system operates up to a maximum aircraft speed of 155 kt and a maximum altitude of 20,000 ft as an emergency communications system.

The system consists of separate direction-finding and receiving antennas, a DF signal processor, DF control panel and monitor on which is displayed the angle of ground transmissions relative to the aircraft to an accuracy of ±5°. One or more optional antenna directing systems may be added, to enable the aircraft to relay transmissions.

Contractor

Tokimec Inc.

VERIFIED

The Tokimec direction-finding and receiving system with the receiving and DF antennas mounted underneath the ventral surface of an S-76 helicopter just aft of the nosewheel. An antenna directing system is also mounted on each side of the cabin

3 ATI flat-panel display

The 3 ATI flat-panel display features an active matrix liquid crystal display and an integrated symbol generator. It is an electronic display that provides graphics and/or characters via an ARINC 429 interface. It has highly integrated modules, incorporating customised ASICs for high-speed graphic processing and SMDs for reduced size.

Specifications

Dimensions: 3 ATI
Weight: 1.5 kg
Power supply: 27.5 V DC, 17 W (42 W with heater)
Display size: 57.7 × 57.7 mm, 480 × 480 RGBW Quad dots
Viewing angle: 35° vertical × 90° horizontal
Contrast ratio: >70
Brightness: >100 fL

Operational status

Under development.

Contractor

Tokyo Aircraft Instrument Co Ltd.

VERIFIED

3 ATI flat-panel display

FMPD-10 EFIS package

FMPD-10 is an Electronic Flight Instrument System (EFIS) package both for commercial and military aircraft. The standard package consists of four smart type displays, two display interface processors and two display control panels. The displays show EHSI, EAI, EHSI/EAI composite graphics, WXR graphics, waypoint map, and fuel information on 3.7 × 3.7 in AMLCD glasses, and also provides hover mode data for helicopter application.

Operational status

In production and in service.

Contractor

Tokyo Aircraft Instrument Co Ltd.

VERIFIED

LK-35 series standby altimeter

The standby altimeter is an exceptionally accurate pressure-actuated instrument with a –1,000 to 50,000 ft range and P/N LK-35-1 and P/N LK-35-9 provide two resolved outputs of barometric setting. A three-drum counter supported by ball bearings provides digital indication in tens of thousands, thousands and hundreds of feet, supplemented by a pointer which indicates altitude in hundreds of feet.

A knob on front of the altimeter provides the means for setting barometric pressure which is shown in inches of mercury and mb on two sets of four-digit counters.

LK-35 series standby altimeter

Specifications

Altitude range: –1,000 to +50,000 ft
(LK-35-8) –2,000 to +50,000 ft
Barometric range: 22-31.02 inHg
(LK-35-7/LK-35-8) 22-31.99 inHg
(LK-35-9) 22-30.99 inHg
Accuracy certifications: FAA/TSO-10b
Lighting: white (unfiltered)
Electrical input:
(vibrator) 28 V DC
(lighting) 5 V AC or DC
(resolved) 26 V, 400 Hz
Weight:
(LK-35-1, -9) 1.45 kg
(LK-35-2, -4, -5, -7) 1.3 kg
(LK-35-8) 1.5 kg

Operational status

In production for B737-400, B757, B767, DC-9.

Contractor

Tokyo Aircraft Instrument Co Ltd.

VERIFIED

SFPD-20 Smart Flat-Panel Display (SFPD)

The SFPD-20 Smart Flat-Panel Display (SFPD) is a self-contained 4 × 5 in (102 × 127 mm) display unit that can replace conventional electromechanical instruments or be used in new aircraft with digital databusses like the ARINC 429. The SFPD, having an active matrix liquid crystal display, provides HSI, ADI, engine data and map information.

Specifications

Dimensions: 127 × 101.6 × 165.1 mm
Weight: 2.5 kg
Power supply: 27.5 V DC, 25 W (53 W with heater)
Display area: 92.7 × 74.2 mm
Viewing angle: 50° vertical by 90° horizontal
Contrast ratio: >100
Brightness: >100 ft-lamberts

Operational status

In production for commercial aircraft.

Contractor

Tokyo Aircraft Instrument Co Ltd.

VERIFIED

FMPD-10 EFIS package 0001358

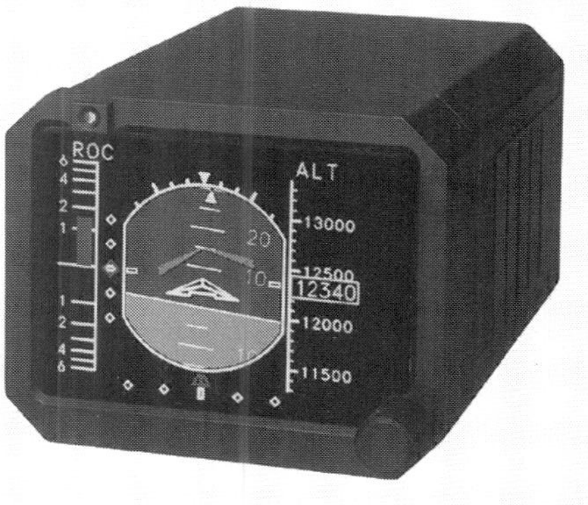

SFPD-20 display

Satellite communication system

Toshiba has developed a satellite communication system which enables air-to-ground international telephone calls from commercial airliners via the Inmarsat satellite service. This was the first type-approved multichannel operation aeronautical satellite communications system in the world. The onboard system consists of radio transceiver equipment for communication with the Inmarsat satellites via a separate antenna and terminal equipment that includes cordless telephones and a control display unit. For passenger use, up to three telephones can be installed in an aircraft and one of the three telephone channels can be used in the cockpit. A 600 bits/s data channel is available for ACARS data communication for automatic dependent surveillance.

Toshiba has developed LSIs to achieve the component miniaturisation necessary for light compact equipment suitable for installation in aircraft.

The radio transceiver consists of a high-power amplifier, radio frequency unit and satellite data unit. The equipment meets ARINC 741 international standards for electronic equipment used in commercial airlines. It can receive and send digital voice signals at 9,600 bits/s, allowing up to three digital voice lines and one 600 bits/s digital data line to be connected to the system.

Specifications

Dimensions:
(high-power amplifier) 8 MCU
(radio frequency unit) 4 MCU
(satellite data unit) 6 MCU

Operational status

The equipment has received Supplemental Type Certification approval for the Boeing 747-400 from the US FAA; certification for use in other aircraft is pending. The equipment is used by All Nippon Airways.

Contractor

Toshiba Corporation.

VERIFIED

NORWAY

PolyCom secure communication system

The Kongsberg Defence & Aerospace PolyCom communications system is designed to provide essential secure communication services for helicopter operations in the North Sea Norwegian Shelf, including: Search & Rescue (SAR): Emergency Medical Services (EMS); policing, security and special forces operations; and fire fighting.

The PolyCom system comprises a VHF/UHF transceiver helicopter/base unit that can be fitted into a helicopter to provide airborne command and co-ordination functions, or located at a base unit, and a VHF/UHF helicopter/man portable transceiver for local use.

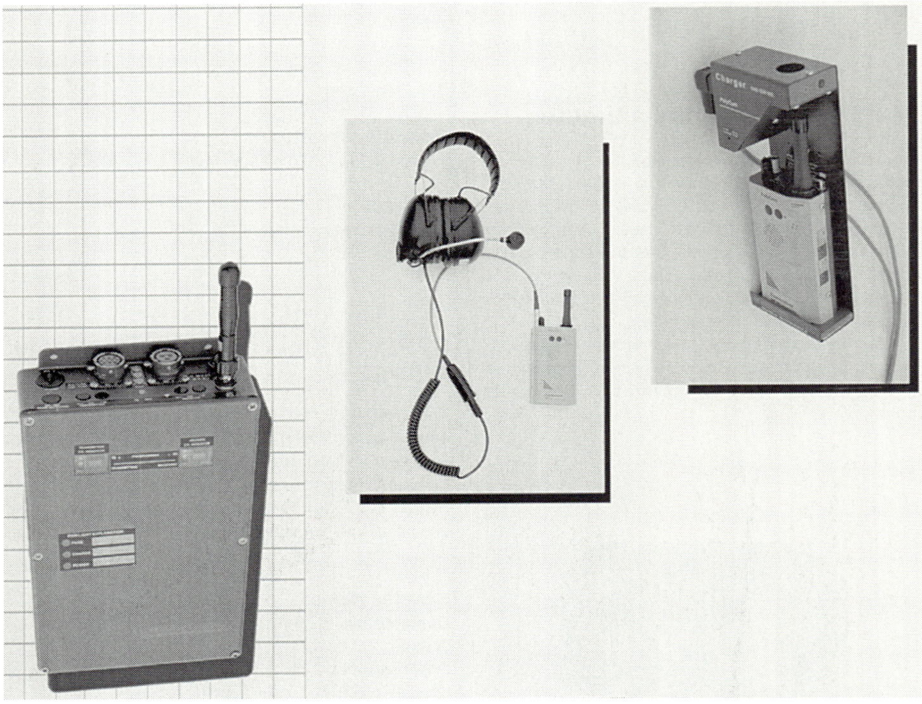

The Kongsberg Defence & Aerospace PolyCom VHF/UHF secure communications system **2000**/0062789

The helicopter/base unit is a complete radio transceiver providing a full duplex helicopter intercom network, installed permanently, or as a 'ready to go' strap-in system. It can also be used as a stand alone mobile repeater for a local communication network.

The helicopter/man portable transceiver features a special Press To Talk (PTT) switch for handsfree operation in SAR and other outside-helicopter uses.

Specifications

Helicopter/base VHF/UHF transceiver
Frequency coverage: 136-174/403-470 MHz
Channel spacing: 2.5, 20.0 and 25.0 kHz
Channels: 16
Dimensions: 244 × 147 × 55 mm
Weight: 2 kg
Qualifications: RTCA DO-160C, ETS 300 086

Helicopter/man portable VHF/UHF transceiver
Frequency coverage: 136-174/403-470 MHz
Channel spacing: 12.5, 20.0 and 25.0 kHz
Channels: 16
Operating temperature: –25 to +55°C
Weight: 0.5 kg
Qualifications: 1P67

Operational status

In service with the Norwegian Helikopter Service and other Norwegian Shelf helicopter operators.

Contractor

Kongsberg Defence & Aerospace.

VERIFIED

Navaids flight inspection system

The Navia Aviation navaids flight inspection system is a digitally computerised system for the checking of radio navigation aids. The basic system is configured with a powerful 32-bit computer and very high-capacity data storage for flight data. The computer performs all the routine operations such as sensor tuning, recorder parameter set-up, pilot guidance, final report printout and sensor parameter calibration. In addition, the computer performs real-time processing, storage and colour presentation, which greatly improves the speed of inspection procedures.

The system consists of three components: the digitally computerised airborne console, the ground-based position reference system and the maintenance and ground support system.

Optimised integration of data presentation for the airborne controller leads to a compact panel design. Due to the console design, operators are able to face forward even for installations in the smallest aircraft.

The basic system may be used for checking VOR, DME, NDB, VDF, PAR, ILS, SSR, markers and VHF comms. Optionally, UDF, Tacan, MLS, SLS and UHF communications may be checked.

Specifications

Dimensions: 500 × 610 × height (mm) dependent on configuration
Weight:
(console) 95 kg
(rack) 75-95 kg
Power supply: 28 V DC, 15-25 A
26 V AC, 400 Hz
115 V AC, 400 Hz, 40-350 VA
Crash certification: FAR 25

Contractor

Navia Aviation AS, a subsidiary of Northrop Grumman Electronic Sensors and Systems Sector.

VERIFIED

The operator console for the navaids flight inspection system **1999**/0051592

POLAND

AP-2 digital autopilot

The AP-2 digital autopilot was designed and produced according to FAR23 regulations and the AS402A document. It is a two-axis flight control and flight director system for light aircraft.

The autopilot provides pitch and roll attitude stabilisation and control, altitude stabilisation, heading stabilisation, interception and tracking of VOR radials and localiser and glide slope beam interception and tracking and localiser back course mode for ILS approaches. Two basic modes of operation are available: automatic flight control for basic autopilot functions and flight director mode.

The AP-2 includes the main processing unit, selector panel, servo amplifiers, optional diagnostic console and optional optical flight director indicator.

The device may be equipped with two different operators' consoles. The basic console, the autopilot selector panel, enables the pilot to engage or disengage the autopilot, select any mode of operation and receive signals about correct or incorrect autopilot operation. The optional console provides additional diagnostic functions.

The AP-2 autopilot generates analogue signals for the flight director indicator and digital signals for the external mode annunciator and other onboard instruments such as the flight recorder.

Specifications

Dimensions:
(flight computer) 160 × 75 × 280 mm
(selector panel) 160 × 75 × 40 mm
Weight: 9.7 kg total typical
Power supply: 27.7 V DC
(flight director) 0.8 A
(flight control) 3.5 A

Operational status

The present version of the AP-2 is adjusted for Piper Seneca II aircraft and Edo-Aire flight instruments.

Contractor

ATM Inc.

VERIFIED

RUSSIAN FEDERATION AND ASSOCIATED STATES (CIS)

SPPZ ground proximity and warning systems

The SPPZ ground proximity warning systems can be used in all types of passenger and transport aircraft equipped with flight navigation systems that have digital information exchange. There are two models: the SPPZ-85 and SPPZ-2.

The SPPZ systems compute data obtained from the following systems to produce their warnings; the radio altimeter; air data computer system; ILS or MLS receiver; onboard inertial navigation system; landing gear and flap sensors. The SPPZ-2 system also utilises data from the flight management system and flight control system.

Warning data provided by the systems is as follows: excessive sink rate; excessive terrain closure rate; negative climb rate after take-off or missed approach; insufficient terrain clearance at landing with wrong configuration; inadvertent descent below glide slope; excessive difference in absolute altitude and pressure height; inadvertent flight into dangerous windshear (SPPZ-2 only).

SPPZ ground proximity warning systems 0011883

Specifications

Outputs: 2 analogue; 40 voice; ARINC 429
Power: 115 V AC, 400 Hz, 20 VA (SPPZ-85), 25 VA (SPPZ-2)
Dimensions: 2 MCU
Weight: 3.5 kg

Operational status

Fitted to: An-70, Il-96, Il-114, Tu-204 and Tu-334 aircraft.

Contractor

AeroPribor-Voskhod Joint Stock Company.

VERIFIED

VBM mechanical barometric altimeters

AeroPribor-Voskhod VBM mechanical barometric altimeters are designed on a unified constructional base, and they have a built-in vibrator for reducing friction in the mechanism. Altitude is displayed by a counter calibrated in kilometres. The altimeters are produced in three separate colour options: integral red; red/white; and white dial lighting.

Operational status

Widely used.

Contractor

AeroPribor-Voskhod Joint Stock Company.

VERIFIED

VBZ-SVS electronic barometric altimeter

The VBZ-SVS electronic altimeter is a new-generation instrument that combines the functions of air data computer, altitude indicator and flight level deviation alert in one unit.

The Liquid Crystal Display (LCD) indicates: the current value of relative height, local barometric pressure at ground level, assigned flight level, warning of deviation from flight level by 60 to 150 m (flashing frame of display), warning of deviation from flight level by more than 150 m (illuminated frame on the display), warning of flight below 1,000 m.

AeroProbor-Voskhod Joint Stock Company VBM mechanical barometric altimeters 0018208

Specifications

	VBM-1	VBM-1A	VBM-2/ VBM-2Ø	VBM-3	VBM-R
Altitude range	–500 to 10,000 m	–500 to 5,000 m	–500 to 15,000 m	0 to 30,000 m	–500 to 30,000 m
Measurement error, at altitude (m)	± 10	± 10	± 10	± 10	± 10
0	± 15	± 15	± 20	± 35	± 100
5,000	± 30		± 40	± 50	
9,000			± 55	± 60	
12,000				± 300	
19,000					
Atmospheric pressure range	700-1,080 kPa	700-1,080 kPa	700-1,080 kPa	700-1,080 kPa	700-1,080 kPa
Dimensions	85 × 85 × 190 mm	85 × 85 × 190 mm	85 × 85 × 240 mm	95 × 85 × 190 mm	65 × 65 × 160 mm
Weight	1.2 kg	1.2 kg	1.4 kg	1.6 kg	1.0 kg
Power (all systems):	27 V DC, 1.5 W; 5.5 V AC, 2.2 VA				

Specifications

Inputs:		
Static pressure, kPa	115.5 to 1,074	
Total pressure, kPa	115.1 to 1,150	
Stagnation temperature, °C	–60 to 99	
Outputs:	**Range:**	**Permitted error:**
True altitude (QNH), m	–503 to 15,240	±4.6 (at –503 m), ±24.4 (at 15,240 m)
Relative altitude (QFE), m	0 to 15,240	±6.1 (at 0 m), ±24.4 (at 15,240 m)
Rate of climb, m/s	± 102	0.15 or 5%
Indicated airspeed, m/s	55.5 to 832	±9.3 (at 55.5 m/s), 1.85 (at 832 m/s)
True airspeed, m/s	185 to 1,108	±7.4
Mach number	0.1 to 1.0	–
Flight level	0 to 15,000	
Dimensions, mm	85 × 85 × 235	–
Mass, kg	1.6	–

Vertical speed data and assigned flight level are exchanged with other onboard systems by 32-bit serial data code as per ARINC 429. Flight level deviations and built-in test data are also reported routinely.

Contractor

AeroPribor-Voskhod Joint Stock Company.

VERIFIED

VBZ-SVS electronic barometric altimeter ***2000***/0062742

ARK series radio compasses

The ARK series of radio compasses comprises the ARK-19, ARK-25 and ARK-UD.

Model ARK-19 covers the frequency band 150 to 1299.5 kHz, while model ARK-25 covers the band 150 to 1750 kHz; frequency coverage for the ARK-UD is not defined but it is claimed to track both FM and pulse signals.

All models are designed for both fixed- and rotary-wing aircraft installation.

Contractor

Airon Complect.

VERFIED 21/03/01

DME/P-85 navigation system

The DME/P-85 navigation system provides slant range, in kilometres and nautical miles, from DME/N, TACAN and DME/P ground stations.

Specifications

Weight:
(interrogator) 9.5 kg
(protuding antenna) 0.33 kg
(suppressed antenna) 0.55 kg
Measurement range: 300 nm; 555 km
Frequency range: 960-1,213 MHz
Channels: 352
Dimensions: 4 MCU
Power: 115 V AC, 400 Hz, 70 VA

Operational status

Development contract award in 1985. Standard fit in Russian airliners.

Contractors

All-Russia Research Institute of Radio Equipment.
NIIAO Institute of Aircraft Equipment.

VERIFIED

VOR-85 VHF Omni Range system

VOR-85 provides identification of ground VOR and marker beacons, and magnetic azimuth relative to VOR ground beacons.

Specifications:

Dimensions: 3 MCU
Weight:
(receiver) 5 kg
(navigation antenna) 1.7 kg
(marker antenna) 0.9 kg
Frequency range:
(VOR) 108.00-117.95 MHz
(marker) 75 MHz
Channels (VOR): 160
Accuracy (95%): 0.5°
Power: 115 V AC, 400 Hz, 30 VA

Operational status

Development contract award in 1985. Standard fit in Russian airliners.

Contractors

All-Russia Research Institute of Radio Equipment.
NIIAO Institute of Aircraft Equipment.

VERIFIED

AGB-96, AGB-98 and AGR-100 horizon gyros

The AGB-96, AGB-98 and AGR-100 horizon gyros are designed for aircraft pitch/roll indication.

Contractor

AviaPribor Corporation.

VERIFIED

AGR-29 and AGR-81 standby gyros

The AGR-29 and AGR-81 standby horizon gyros are designed for aircraft pitch/roll indication. The AGR-29M can be used as a remote indicator of pitch and roll from another vertical sensor.

AviaPribor AGR-81 (left) and AGR-29 (right) standby horizon gyros 0018206

Specifications

Compliance: ENLGS P 8.1.2

	AGB-96	**AGB-98**	**AGR-100**
Readiness time	2.0 min	2.0 min	2.0 min
Angular range			
(roll)	±360°	±360°	±360°
(pitch)	±85°	±85°	±85°
Power			
(gyro) all systems:	27 V DC		
(sensors) all systems:	36 V AC, 400 Hz, 0.8 A		
Temperature range all systems:	–20 to +70°C		
Dimensions	105 × 105 × 250 mm	85 × 85 × 250 mm	61 × 61 × 220 mm
Weight	2.2 to 2.5 kg	1.8 to 2.0 kg	1.5 kg

AviaPribor AGR-100, AGB-98, AGB-96 horizon gyros (left to right) 0018207

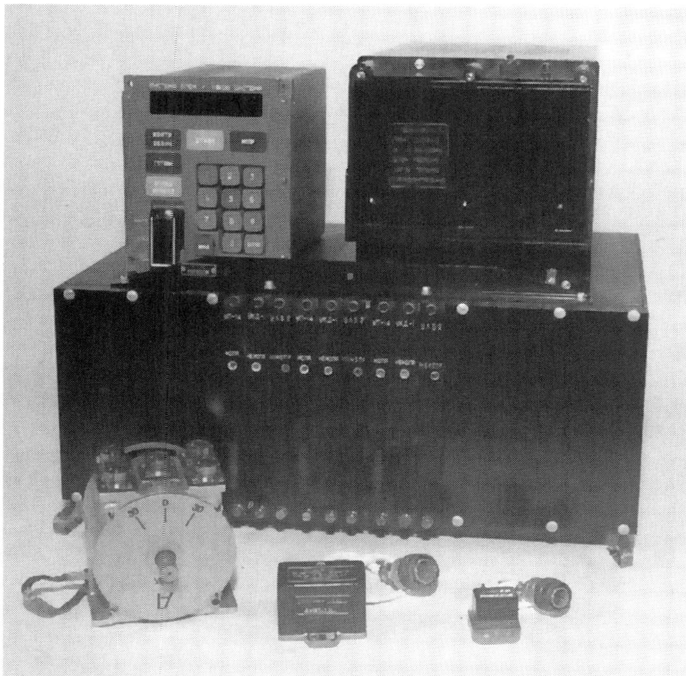

ASShU-334 fly-by-wire flight control system
0018242

ASUU-96 automatic control and stability augmentation system 0018241

Specifications

Compliance: ENLGS P 8.1.2

	AGR-29	AGR-81
Readiness time	2.0 min	2.0 min
Angular range		
(roll)	±360°	±360°
(pitch)	±360°	±360°
Power		
(gyro)	27 V DC	
(sensors)	36 V AC, 400 Hz, 1.0 A	
(back lighting)	6 V DC	
Temperature range	–30 to +70°	
Dimensions	105 × 105 × 250 mm	85 × 85 × 250 mm
Weight	up to 3.1 kg	2.2 to 2.5 kg

Contractor

AviaPribor Corporation.

VERIFIED

Airspeed and altitude indicators

AviaPribor produces a range of airspeed and altitude indicators that conform with ARINC 429 data interface requirements. The photograph shows some representative examples; the specifications provide representative data.

Contractor

AviaPribor Corporation.

VERIFIED

Specifications

	VMC altimeters	USC airspeed indicators
Measurement range	0-10,000 m	80-800 km/h
Error	±10 to ±30 m	5.5 to 10 km/h
Dimensions	86 × 86 × 250 mm	86 × 86 × 230 mm
Weight	2.0 kg	1.5 kg

AviaPribor USC airspeed and VMC altitude indicators 0018205

ASShU-334 fly-by-wire flight control system

The ASShU-334 fly-by-wire flight control system is a digital and analogue system designed to provide stability and control for regional aircraft during manual, automatic and override manual flight conditions. System components are as follows:

- stability and control computers: 4 units
- analogue units: 2 units
- pre-flight maintenance panel: 1 unit
- linear acceleration: 6 units
- angular rate sensors: 18 units
- position sensors: 7 units
- transformer units: 3 units

Specifications

No of redundant channels: 6
Weight: 90 kg
Power: 500 W

Operational status

Fitted to the An-334 twin turbofan medium-range airlifter aircraft.

Contractor

AviaPribor Corporation.

VERIFIED

ASUU-96 automatic control and stability augmentation system

The ASUU-96 control and stability augmentation system is a digital-analogue system, designed for the Il-96 aircraft; it comprises the following units:

- control and stability computer unit: 5 units
- control monitor unit: 4 units
- actuator unit: 1 unit
- pre-flight maintenance panel: 1 unit
- linear accelerator transducer: 4 units
- triplex position pickoff: 8 units
- angular rate sensor: 12 units

Specifications

No of redundant channels: 4
Weight: 200 kg
Power: 115 V AC, 400 Hz, 2,000 VA

Operational status

Fitted to Il-96 wide-bodied airliner.

Contractor

AviaPribor Corporation.

VERIFIED

I-21 inertial navigation system

The I-21 inertial navigation system is designed to autonomously determine flight/navigation data, and provide it to the aircraft system for generation of steering commands between defined waypoints. It conforms to ARINC 61.

Specifications

Navigational errors:
(geographical co-ordinates (in 10 h)) 37 km (max)
(ground and speed components) 12.6 km/h
(true heading) 0.2 + 0.025°/°C
(angles of roll/pitch, gyro heading) 0.1°

Readiness time: <15 min at 20°C
MTBF: 1,200 h
Power: 500 VA (1,500 VA (max)
Weight: <50 kg

Operational status

Fitted to An-124, An-224, Il-62, Il-76, Tu-154, Tu-160 aircraft

Contractors

Aviapribor Corporation.
Ramenskoye Design Company AO RPKB.

VERIFIED

I-21 inertial navigation system 0015367

ИМ-7, ИМ-8 LCD multifunction displays

Aviapribor has developed two new, full colour LCD multifunction displays designed to interface with modern navigation/FMS systems and display: flight/navigation, radar, EFIS/EICAS/EIS, caution, warning and aircraft/system status data, and aural warnings.

Contractor

Aviapribor Corporation.

NEW ENTRY

Specifications

	ИМ-7	**ИМ-8**
Display	Colour LCD	Colour LCD
Useful screen size	101 × 101 mm	159 × 159 mm
Interface	ARINC-429, ARINC-708, RS-422	ARINC-429, ARINC-708
Compliance	DO-160	DO-160
MTBF	10,000 h	10,000 h
Dimensions	130 × 130 × 250 mm	203 × 230 × 280 mm
Weight	5 kg	8 kg
Electrical	27 V DC	115 V AC
Power consumption	50 W 100 W with heating	100 W 230 W with heating

ИМ-68 multifunction display unit

Aviapribor's ИМ-68 multifunction display unit is based on a full colour LCD and is designed to receive, process and present flight/navigation data, alert messages and TV-sensor data for aircraft and helicopters.

Specifications

Display: 211 × 158 mm full colour LCD
Dimensions: 200 × 270 × 290 mm
Weight: 9 kg
Electrical: 115 V AC, 400 Hz
Power consumption: 110 W (230 W with heating)
Interface: ARINC-429, RS-232, MIL-STD-1553B
Compliance: DO-160
MTBF: ≥ 5,000 h

Contractor

Aviapribor Corporation.

NEW ENTRY

Aviapribor's ИМ-68 multifunction display unit
2001/0103867

BINS-85 inertial navigation system

Aviapribor's BINS-85 inertial navigation system was designed for the An-70 transport aircraft.

Specifications

Error performance:
Position: ≤3.7 km/flight hour
Groundspeed: 14.4 km/h
True heading: ≤0.4°
Roll and pitch: ≤0.1° (navigation mode)
Status ready time:
(at 0 to +55°C) ≤10 min
(at 0 to –20°C) ≤18 min
Temperature range: –20 to +55°C
Digital outputs: 32 parameters
Power: 115 V AC, 400 Hz, 200 VA (max) 27 V DC, 180 W
Mean time between failures: 5,000 h
Dimensions: 319 × 322 × 194 mm
Weight: ≤20 kg

Operational status

In service on the An-70 transport aircraft.

Contractor

Aviapribor Corporation.

UPDATED

BINS-90 Inertial Navigation System

The BINS-90 INS is designed for passenger and freight aircraft, with the ability to compute up to 40 navigation parameters. The system has been designed around three LG-1 laser gyros and three AK-6 quartz accelerometers. The BINS-90 can be coupled to a GPS/GLONASS receiver, which dramatically improves the overall accuracy of the system navigation solution.

BINS-90 inertial navigation system ***2001***/0103861

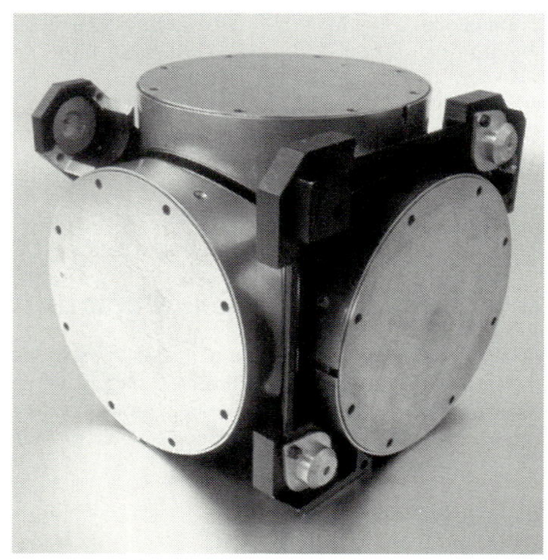

LG-1 Laser Gyro ***2001***/0103862

Specifications

Accuracies:
Position: <3.7 km/flight hour (200 m with GPS/GLONASS support)
Speed: <4 m/s (0.2 m/s with GPS/GLONASS support)
Status ready time:
<10 min (autonomous mode)
<3 min (with correction)
Temperature range: –60 to +60°C
Altitude: up to 12,500 m (41,000 ft)
Angular speeds (all axes): up to 200°/s
Digital outputs: 40 parameters
Power: 115 V AC, 400 Hz, 200 VA (max) 27 V DC, <80 W
Mean time between failures: 5,000 h
Dimensions: 194 × 194 × 470 mm
Weight: 16 kg (max)

Contractor

Aviapribor Corporation.

NEW ENTRY

GINS-3 Gravimetric Inertial Navigation System

GINS-3 is designed for two applications: stand-alone navigation of aircraft; and aerogravimetry. System components are the gravimetric inertial unit, radio altimeter, air data system, satellite navigation system, electronics unit, and computer.

GINS-3 gravimetric inertial navigation unit 0015366

Specifications

Navigation errors:
(position co-ordinates, independent of flight time and distance) 0.2 to 10 km
(ground and vertical speed components) 0.1 to 1 m/s
(altitude above sea level) 1 to 5 cm
Aerogravimetry error of gravitational anomaly in free air: up to 0.7 mGal
Scales of maps produced: 1/200,000 and more
Power: 27 V DC, 1,000 VA (max)
Weight: up to 120 kg

Contractor

Aviapribor Corporation.

UPDATED

IM-3, IM-5, IM-6, IGM multifunction displays

Aviapribor produces a number of multifunction displays to provide: flight/navigation information; radar data; EFIS/EICAS/EIS advisory, caution, warning and status data, and aural warnings.

Operational status

The displays illustrated are fitted to the Tu-204 and Il-96 twin and four turbofan airliners. They are also suitable for other aircraft and helicopters.

Contractor

Aviapribor Corporation.

UPDATED

Specifications

	IM-3	IM-5	IM-6	IGM
Display technology	CRT stroke	CRT raster	CRT raster	AMLCD matrix
Useful screen size	159 × 159 mm	159 × 159 mm	83 × 107 mm	101 × 101 mm
Pixels pitch	0.15/0.3 mm	0.3/0.3 mm	0.2/0.3 mm	0.2 mm
Refresh rate	50/80 Hz	40/80 Hz	40/80 Hz	80 Hz
Colours displayed	8	8 (16)	16	8 (16)
Allowable ambient illumination	70,000 lx	60,000 lx	75,000 lx	75,000 lx
Viewing angle	±53°	±53°	±53°	±40°
Dimensions	203 × 230 × 356 mm	203 × 222 × 356 mm	130 × 145 × 390 mm	130 × 130 × 230 mm
Weight	17.5 kg	12.5 kg	6 kg	4 kg
Power	200 VA	100 VA	60 VA	90 VA

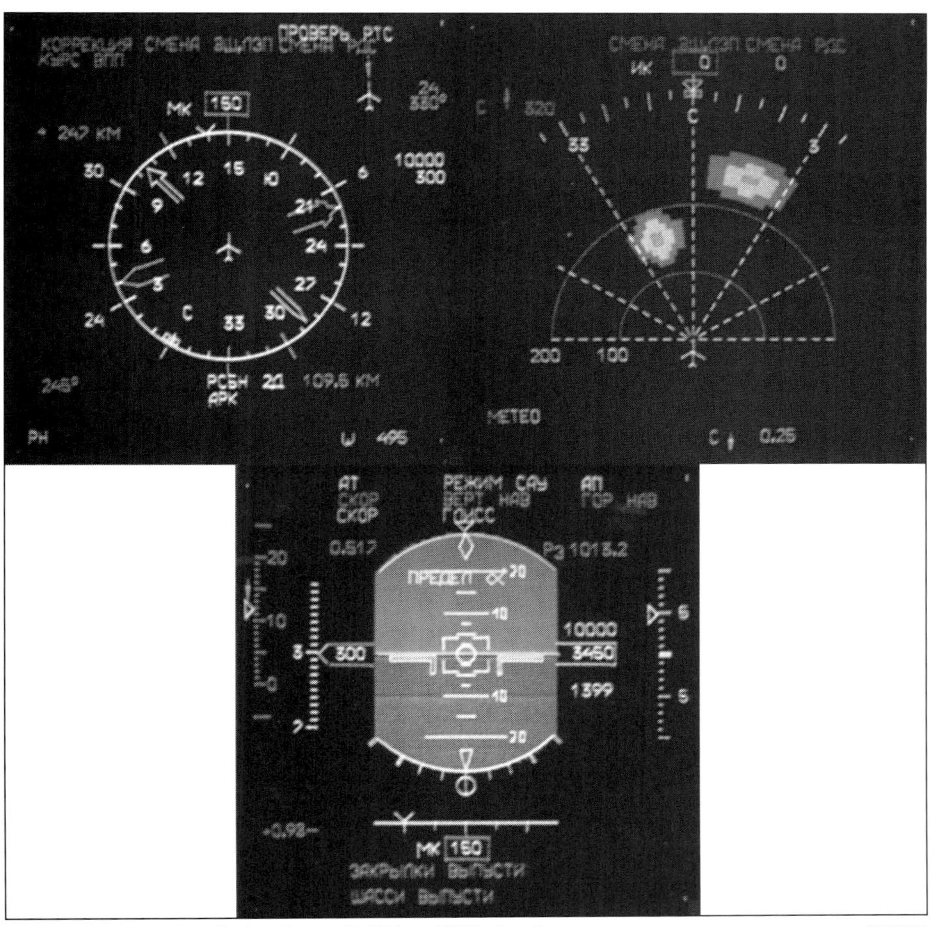

Aviapribor multifunction displays for the Tu-204 and Il-96 aircraft 0018203

IRM-1 radio compass indicator

The IRM-1 radio compass indicator is designed to display data derived from the MKS-1 compass and the ADF system.

Specifications

Compliance: ENLGS, p.82.1 g
Interfaces: MKS-1 compass; KR87 digital ADF receiver (AlliedSignal)
Heading, selected course and ADF bearings: 360°
Error: ±1.5° heading; ±2.0° ADF bearing
Power: 27 V DC
Temperature range: –20 to +55°C
Dimensions: 85 × 85 × 150 mm
Weight: 1.1 kg

MKS-1 compact compass system
The MKS-1 compact compass system has two operating modes; gyro and manual slaving; automatic magnetic slaving. The system comprises: GK heading gyro: KU-1 slaving unit; ID-6-1 magnetic field sensor.

Specifications

Compliance: AP-23; ENLG-S, p.82.1.d
Warm-up time: 2 min
Gyromagnetic heading measurement error: 2°
Gyro drift: 0.4°/min
Power: 27 V DC
Operating temperature range:
(GK-1) –55 to +55°C
(KU-1) –20 to +55°C
(ID-6-1) –60 to +150°C
Dimensions:
(GK-1) 115 × 200 × 145 mm
(KU-1) 55.4 × 55.4 × 94 mm
(ID-6-1) 102 × 60 mm
Weight:
(GK-1) 2.2 kg
(KU-1) 0.35 kg
(ID-6-1) 0.6 kg

Contractor

Aviapribor Corporation.

UPDATED

IRM-1 radio compass indicator 0018202

Multifunction Control and Display Unit (MCDU)

Aviapribor's Multifunction Control and Display Unit (MCDU) provides all the required functionality for pilot interaction with a modern Flight Management System (FMS), including route, waypoint and radio aid management and airport standard approach and departure procedures. The unit features a full colour Active Matrix Liquid Crystal Display (AMLCD) and facilities for ARINC-429 and RS-232 interfaces.

Specifications

Display: 100 × 100 mm full colour AMLCD
512 × 512 colour triads
1,024 × 768 pixels
Viewing angles ± 45° horizontal, + 35° to –10° vertical (CR ≥10:1)
Dimensions: 146 × 228 × 266 mm
Weight: 7 kg
Power consumption: 60 W
Interface: ARINC-429, RS-232
MTBF: ≥ 10,000 hrs

Aviapribor's Multifunction Control and Display Unit (MCDU) **2001**/0103863

Contractor

Aviapribor Corporation.

NEW ENTRY

NV-1 navigation computer

The NV-1 navigation computer is designed to provide a flight management interface for the pilot, including flight/navigation systems, radio aids and satellite navigation systems (GPS and/or GLONASS).

Specifications

Processor: 80486DX4-100
Interface:
ARINC-429
RS-232 (2 channels)
GPS (24 channels)
Electrical: 27 V DC
Power: 60 W (max)
Dimensions: 318 × 194 × 90.5 mm
MTBF: 10,000 h

Contractor

Aviapribor Corporation.

NEW ENTRY

NV-1 navigation computer **2001**/0103869

SNS-2 satellite navigation receiver

The SNS-2 is a 24-channel (12 GLONASS, 12 GPS) satellite navigation receiver which computes and outputs the following object movement and positional data:

- latitude and longitude
- altitude
- groundspeed and vertical speed
- track
- UTC time
- time marker

SNS-2 is an integrated system, capable of receiving and processing signals from all visible GPS and/or GLONASS constellations. The system is capable of automatic measurement of raw data, almanac and ephemeris download, without preliminary initialisation. These features enable a reduction in selective availability for GPS and satisfactory precision, integrity and availability characteristics for the demands of typical Required Navigation Performance (RNP).

SNS-2 has built-in Receiver Autonomous Integrity Monitoring (RAIM) functionality, utilising data from all tracked satellites and differential capability when RTCM-104 data is available; the system fulfils the technical requirements for onboard GNSS equipment as a primary navigation aid, adopted by the Russian Federal Aviation Service in August 1999. SNS-2 can be installed on all types of aircraft.

Specifications

Output data
Position (2Σ):
44 m (GLONASS + GPS/GLONASS)
100 m (GPS)
Altitude (2Σ):
66 m (GLONASS + GPS/GLONASS)
15 m (GPS)
Groundspeed: 5 m/s
UTC time: 0.01 s
Track: 20 arc min (W > 50 km/h)
Marker: 50 ns

Interface
Input: GLONASS and GPS L1, 1 × ARINC 429 for differential corrections, 1 × ARINC 429 for control commands
Output: 2 × ARINC 429, time marker in accordance with ARINC 743A

Physical
Antenna frequency: 1,575.42 to1,609.0 ±18.0 MHz
Antenna unit weight: 0.8 kg
Amplifier unit weight: 0.4 kg
Satellite receiver weight: 2.65 kg
Power: 27 V DC; 10 W
Readiness time: 2.5 min

Contractor

Aviapribor Corporation.

NEW ENTRY

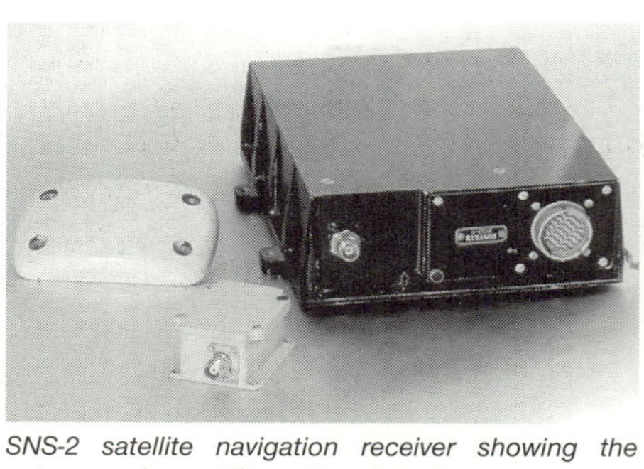

SNS-2 satellite navigation receiver showing the antenna unit, amplifier unit and receiver **2001**/0103871

SNS-3 satellite navigation receiver

The SNS-3 is a 24-channel (12 GLONASS, 12 GPS) satellite navigation receiver which computes and outputs the following object movement and positional data:

- latitude and longitude
- altitude
- groundspeed and vertical speed
- track
- UTC time
- time marker

SNS-3 also offers full flight planning capability, enabling the system to act as a stand-alone navigation system. In common with the SNS-2, SNS-3 is an integrated system, capable of receiving and processing signals from all visible GPS and/or GLONASS constellations. The system is capable of automatic measurement of raw data, almanac and ephemeris download, without preliminary initialisation. These features enable a reduction in selective availability for GPS and satisfactory precision, integrity and availability characteristics for the demands of typical Required Navigation Performance (RNP).

SNS-3 has built-in Receiver Autonomous Integrity Monitoring (RAIM) functionality, utilising data from all tracked satellites and differential capability when RTCM-104 data is available; the system fulfils the technical requirements for onboard GNSS equipment as a primary navigation aid, adopted by the Russian Federal Aviation Service in August 1999.

SNS-3 also able to accept navigation data from other aircraft systems, such as AHRS, ADS or other guidance systems; in the case of satellite information loss, the system is able to maintain a navigational solution for up to 15 minutes without reacquisition. The system features an alphanumeric display and controls to allow for the display and control of all system modes. SNS-3 can be installed on all types of aircraft.

Specifications

Output data
Position (2Σ):
44 m (GLONASS + GPS/GLONASS)
100 m (GPS)
Altitude (2Σ):
66 m (GLONASS + GPS/GLONASS)
15 m (GPS)
Groundspeed: 5 m/s
UTC time: 0.01 s
Track: 20 arc min (W > 50 km/h)
DTG: 100 m
Track deviation: 100 m
Marker: 50 ns

Interface
Input: GLONASS and GPS L1, 1 × ARINC 429 for differential corrections, 1 × ARINC 429 for control commands, 2 × ARINC 429 for AHRS/ADS/INS data
Output: 2 × ARINC 429, dual analogue channel for track deviation, time marker in accordance with ARINC 743A

Physical
Antenna frequency: 1,575.42 to 1,609.0 ±18.0 MHz
Antenna unit weight: 0.8 kg
Amplifier unit weight: 0.4 kg
Satellite receiver weight: 3.5 kg
Power: 27 V DC; 10 W
Readiness time: 2.5 min

Contractor

Aviapribor Corporation.

NEW ENTRY

SNS-3 satellite navigation receiver **2001**/0103872

VSUP-85 flight control computer system

The VSUP-85 flight control computer system is designed to generate control data for all phases of flight of modern transport aircraft types; it comprises two types of line-replaceable units: flight control computer (three units) and cockpit control unit (one unit).

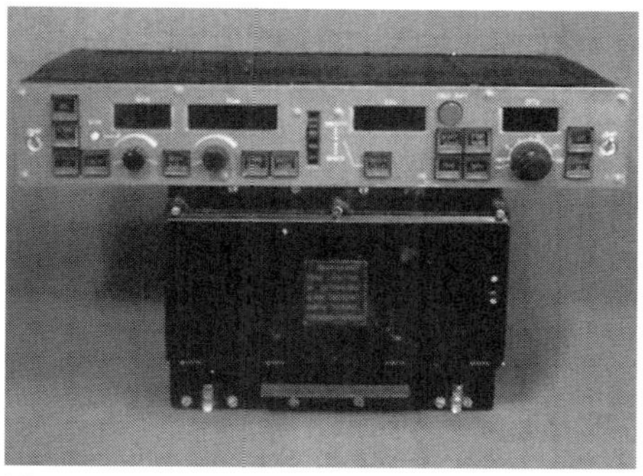

VSUP-85 flight control computer system 0018239

Specifications

Compliance: NLGS, DO-160
Weight:
(Tu-204) 51.5 kg
(Il-96) 36 kg
Power: 115 V AC, 400 Hz, 525 VA
27 V DC, 30 W
36 V AC, 400 Hz, 4.5 VA
6 V AC, 400 Hz, 25 VA
MTBF:
(computer) 5,000 h
(control unit) 6,000 h

Operational status

Fitted to the Tu-204 twin turbofan medium-range airliner, and the Il-96 wide-bodied airliner. The -85 element of the designator indicates that the original development contract was awarded in 1985.

Contractor

Aviapribor Corporation.

UPDATED

VSUPT-334 flight and thrust control computer system for Tu-334

The VSUPT-334 flight and thrust control computer system performs both flight control and thrust control functions for the Tu-334 medium-range airlifter and its twin turbofan ZMKB Progress D-436T1 engines; it comprises the following line-replaceable units: flight/thrust computer (three units), thrust unit, cockpit control unit, thrust actuator.

Specifications

Compliance: NLGS, DO-160
Weight: 67 kg
Power: 115 V AC, 400 Hz, 525 VA
27 V DC, 400 W
36 V AC, 400 Hz 4.5 VA
6 V AC, 25 VA
MTBF:
(computer) 5,000 h
(cockpit control unit) 6,000 h
(thrust actuator) 5,000 h

Operational status

Fitted to the Tu-334 twin turbofan medium-range airlifter.

Contractor

Aviapribor Corporation.

UPDATED

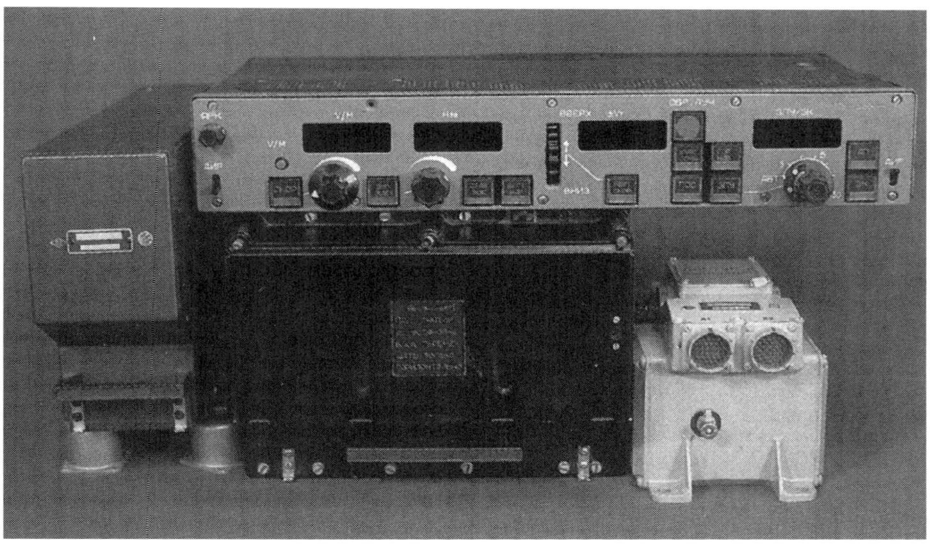

VSUPT-334 flight and thrust control computer system for Tu-334 0018238

VSUT-85 thrust control computer system

The VSUT-85 thrust control computer system is designed to provide automatic control/hold of indicated airspeed or Mach number by controlling power plant thrust on Tu-204 and Il-96 aircraft; it comprises the following line-replaceable units: thrust control computer (two units), thrust control, thrust actuator.

Specifications

Compliance: NLGS, DO-160
Weight:
(Tu-204) 34 kg
(Il-96) 38 kg
Power: 115 V AC, 400 Hz, 325 VA
(Tu-204) 27 V DC, 432 W
(Il-96) 486 W
MTBF:
(computer BVUT) 5,000 h
(thrust actuator PRT-204) 5,000 h
(thrust control panel PUT-3) 8,000 h

Operational status

Fitted to the Tu-204 twin turbofan medium-range airlifter, and Il-96 four turbofan, wide-bodied, airliner. The -85 element of the designator indicates that initial development contract award was in 1985.

Contractor

Aviapribor Corporation.

UPDATED

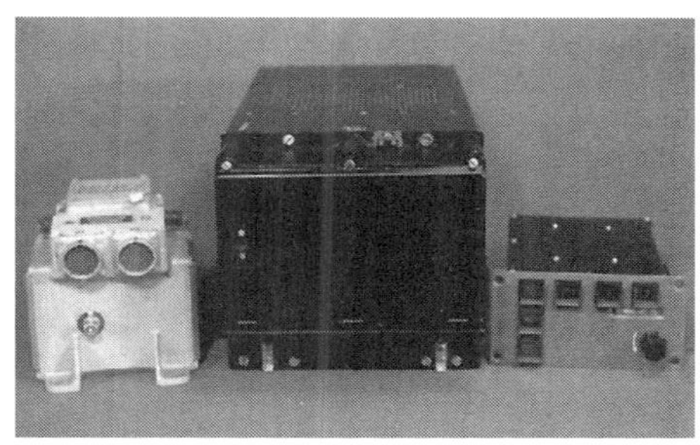

VSUT-85 thrust control computer system 0018237

PV-95 multifunctional control and display unit

The PV-95 display computer unit is designed for display of real-time piloting and navigation data; it comprises a CISC architecture processor module (MV58); channel multiplexing module (MO 52) to GOST 26765.52-87 and MIL-STD-1553B standards; channel module (MD 51) to GOST 18977-79 and ARINC 429 bus standards; graphics control module; secondary voltage module (MN 95). The PV-95 system can be readily configured to customer requirements.

Specifications

Screen: monochrome EL flat panel
Active display area: 96 × 77 mm
Resolution: 320 × 256 pixels
Viewing angle: not less than 160°
Display:
(alphanumeric) 14 lines of 24 characters
(graphics) vectors and arcs
Maximum ambient lighting: not less than 75,000 lx
Keys: 75 keys with night backlight
Processor: CISC, 32 bit
Instruction system: Electronica MPK-1839
Clock frequency: 10 Hz
Operands: 8, 16, 32, 64
Internal transfer rate: 5 Mips
Memory:
(RAM) 128 Kbyte
(ROM) up to 4,096 Mbyte
Programmable timers: up to 4
Input/output:
(ARINC 429) 18/6
(discrete) 16/8
(MIL-STD-1553B) 2
BIT: yes
Dimensions:
(overall) 123 × 224 × 255 mm
(front panel) 146 × 228.6 × 23.4 mm
Power: +27 V DC, 50 W

Contractor

Electroavtomatika OKB.

VERIFIED

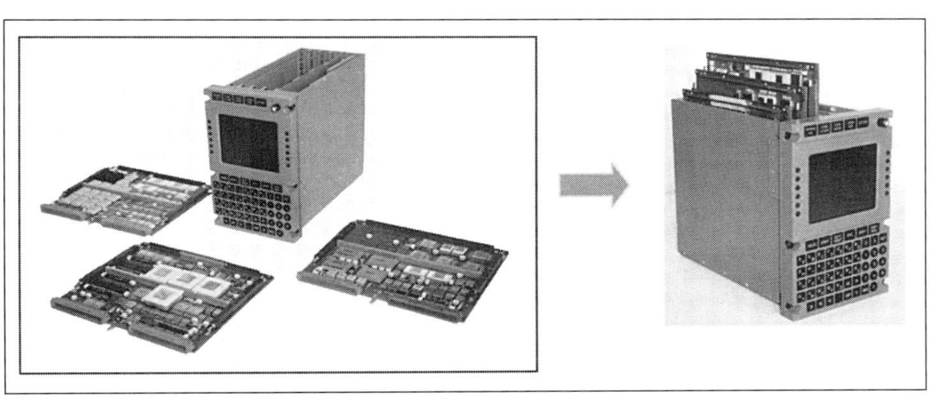

The Electroavtomatika PV-95 multifunctional control and display unit
2000/0062188

SEI-85 and KISS-1-1M multifunction displays

The ElectroPribor Kazan Plant manufactures more than 700 types of instruments, systems and displays for installation in all types of military and civilian fixed- and rotary-wing aircraft built throughout the CIS. The photograph below shows the flight deck of the Tu-204 aircraft cockpit which incorporates the SEI-85 and KISS-1-1M electronic multifunction displays to present navigation, weather radar and aircraft/engine performance data. Similar displays are used in the Il-96-300 aircraft.

Operational status

In service and in production.

Contractor

ElectroPribor Kazan Plant.

VERIFIED

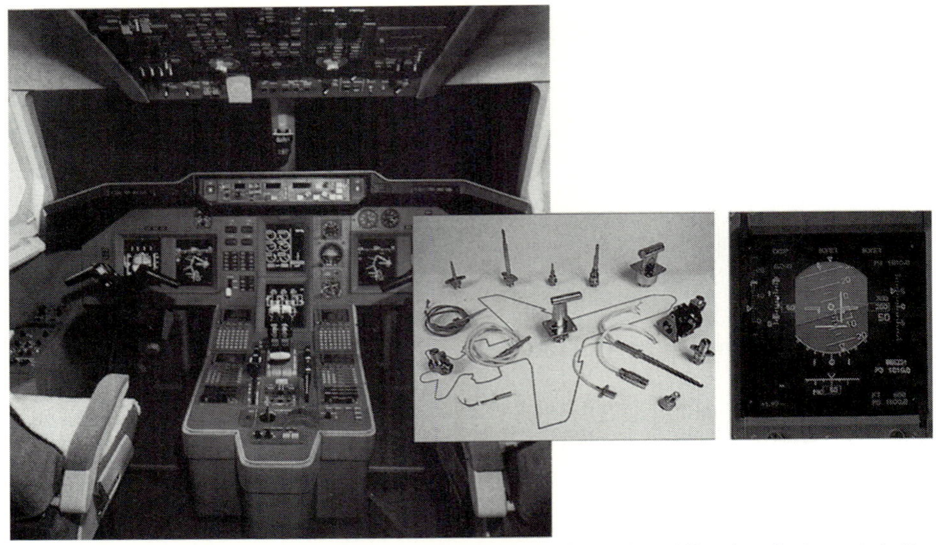

The Tu-204 aircraft cockpit incorporates SEI-85 and KISS-1-1M electronic multifunction display and alerting systems, manufactured by ElectroPribor Kazan Plant 0018201

Tachometer Indicators

ElectroPribor Kazan Plant manufactures 14 types of tachometer for measuring engine shaft speed, calibrations can be either in revolutions per minute (rpm) or per cent.

Specifications

ITE-2TB
Operating temperature range: –60 to 60°C
Reading range: 10-110%
Error: +0.5%
Weight: 1.30 kg

ITE-2
Operating temperature range: –60 to +60°C
Reading range: 10-110%
Error: +0.5%
Weight: 0.95 kg

ElectroPribor Kazan Plant tachometers 0018200

2TE15-1M
Operating temperature range: –60 to +60°C
Reading range: 1,000-15,000 rpm
Error: +75 rpm
Weight: 0.90 kg

Operational status

In service and in production for a wide variety of military and civilian fixed- and rotary-wing aircraft throughout the RFAS.

Contractor

ElectroPribor Kazan Plant.

VERIFIED

Temperature gauges

ElectroPribor Kazan Plant manufactures 10 types of temperature gauges and thermocouples for measurement of gas flow temperatures in turbojet and turboprop engines, and in the cylinder-head of piston engines.

Specifications

ITG-1
Measuring temperature range: +200 to +1,100°C
Error: +12°C
Weight: 0.76 kg

TBG-1
Measuring temperature range: +300 to +900°C
Error: +7°C
Weight: 0.76 kg

ElectroPribor Kazan Plant temperature gauges 0018199

TUT-9
Measuring temperature range: –40 to +300°C
Error: +14°C
Weight: 0.25 kg

Operational status

In service and in production for a wide variety of military and civilian fixed- and rotary-wing aircraft throughout the RFAS.

Contractor

ElectroPribor Kazan Plant.

VERIFIED

Aircraft Systems Control System (ASCS) - Electronic Flight Engineer

GosNIIAS has developed the Electronic Flight Engineer: Aircraft Systems Control System (ASCS), as an outgrowth of its work with Rockwell Collins and Smiths Industries on the following elements of the Il-96 M/T avionics development programme: the Integrated Display System (IDS), Flight Management Computer System (FMCS), Automatic Flight Control System (AFCS), Thrust Management Computer (TMC) and Central Maintenance Computer System (CMCS); and on experience gained in production of the Collins Traffic alert and Collision Avoidance System (TCAS).

The Electronic Flight Engineer is intended for aircraft systems automation and serviceability/maintenance enhancement. The automation element permits reduction of the flight crew to two.

The Electronic Flight Engineer automates control of over 20 aircraft systems, including: hydraulics, electrical power, air conditioning, pressure regulation and other systems. The system comprises three computer units (ASCC) and eight MFUs.

Operational status

Certified by the US FAA in early 1999.

Contractor

GosNIIAS State Research Institute of Aviation Systems.

VERIFIED

MU-19 multichannel differential Monitoring Unit

The MU-19 multichannel differential monitoring unit provides continuous tracking of all GLONASS and GPS satellites in view, keeping all stored data in memory. It can also be used to monitor the quality of navigation data transmitted via satellites, and to provide differential corrections to satellite data. The MU-19 also monitors the discrepancy between GLONASS and GPS timing data, and determines the discrepancy between the PZ-90 and WGS-84 geodetic co-ordinates at the point of monitoring.

The MU-19 system is controlled via keyboard and display.

Specifications

Measurement accuracy:
(using GLONASS satellites)
pseudo ranges: 1-3 m
pseudo range rates: 1-2 cm/s
(using GPS satellites (S/A on)
pseudo ranges: 30 cm
pseudo range rates: 30 cm/s
Position accuracy in non-differential mode:
(using GLONASS satellites)
latitude and longitude: better than 5-7 m
altitude: better than: 8-10 m
(using GPS satellites)
latitude and longitude: better than 25-35 m
altitude: better than 45-55 m
Interfaces: 2 × RS-232C ports
Power: 220 V AC; 25 W
Dimensions:
(MU-19 receiver) 345 × 260 × 200 mm
(antenna) 90 × 90 × 40 mm
Mass:
(MU-19 receiver) 6 kg
(antenna) 0.3 kg

Contractor

GosNIIAS State Research Institute of Aviation Systems.

VERIFIED

A-744 radio-navigation receiver

The A-744 is an integrated GLONASS/GPS navigation receiver that determines position, time, velocity (horizontal and vertical components) in real time. It is designed to be used either independently, or as part of an integrated aircraft system. It can be commanded to GPS only mode, or to GLONASS-GLONASS/GPS mode.

Specifications

Dimensions: 281 × 191 × 64 mm
Weight: 2.7 kg
Power supply: 27 V DC, 18 W
Antenna: active, non-protruding
Interfaces: GOST 18977-79 PTM 1495-75 (ARINC 743A, 429) multiplex channel GOST.26765.52-87 (MIL-STD-1553B)
Number of channels: 6 (universal)
Operating frequency:
GLONASS, EA code: 1598-1616 MHz (F1)
GPS, C/A code: 1575.42 MHz (L1)
Output data: position, velocity, time, quasi-range and Doppler shift measurement
Position error (2e) m:
GLONASS: 45 horizontal/70 vertical
GPS (SA including GLONASS/GPS): 100 horizontal/160 vertical
DGPS/Diff.GLONASS: 6 horizontal/10 vertical
Time to first fix: less than 150 s
Dynamic characteristics: (not more than)
(velocity) 500 m/s
(acceleration) 40 m/s^2
(acceleration increment) 30 m/s^2/s
Temperature limits:
(operation) –50 to +50°C
(storage) –55 to +85°C
Antenna dimensions/weight:
active antenna: 110 mm diameter, 41 mm height, 0.31 kg
passive antenna: 80 × 80 mm, 33 mm height, 0.27 kg

Contractor

Leninetz Holding Company.

VERIFIED

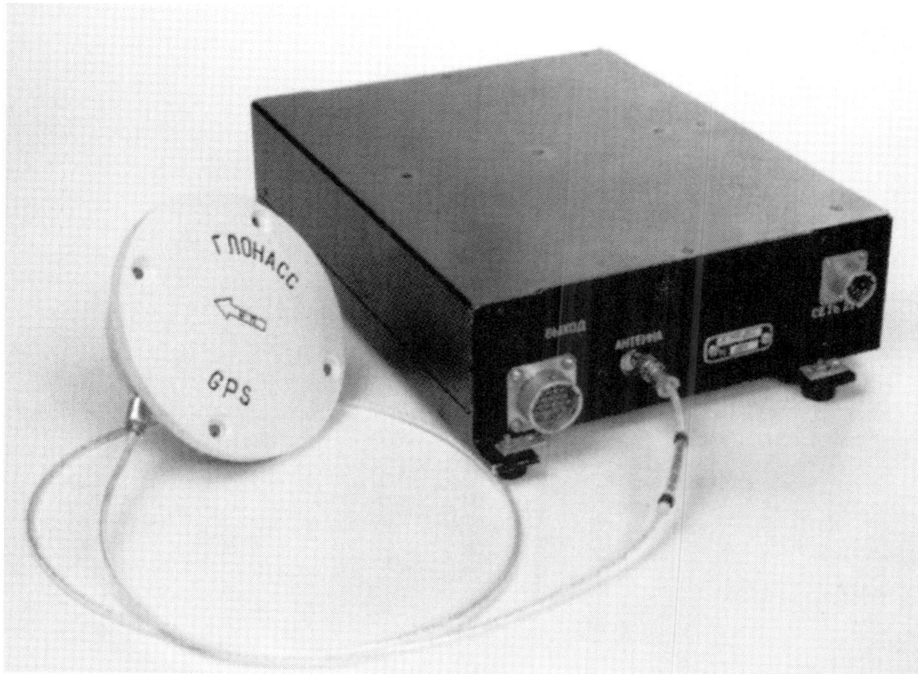

A-744 radio-navigation receiver 0005425

Duet dual-band weather/ navigation radar

Duet is a dual-band (centimetric/millimetric) weather/ navigation radar intended for civil aircraft. Basic operating modes include:

in centimetre-band operation: weather, turbulence, windshear, map

in millimetre-band operation: map, landing/takeoff, taxi

Specifications

Range scales:
(cm-band) 640, 320, 160, 80, 40, 20, 10 km
(mm-band) 20, 10, 4, 1 km
Range resolution:
(cm-band) 150 m
(mm-band) 7.5 m
Scan sector: ±90, ±45, ±25°
Scan rate: 45°/s
Azimuth resolution:
(cm-band) 3°
(mm-band) 0.7°
Antenna diameter:
(two-band) 760 mm
(one-band) 610, 560, 508, 457, 380 mm
Form factor of transceivers: 8 MCU ARINC-600
Display: Duet display; aircraft multifunctional display; head-up display
Weight: two-band radar 50 kg

Contractor

Leninetz Holding Company.

VERIFIED

High-resolution airborne radar

Leninetz has developed an airborne, high-resolution, forward-looking, pulse radar designed to fulfil the following roles: detection, resolution and tracking of land and sea targets; low-altitude obstacle avoidance; navigation and landing assistance in poor visibility conditions. It is also described as being suitable for search and rescue functions, and is said to be suitable for aircraft or helicopter installation.

Specifications

Detection ranges:
buildings/bridges/small ships: 15-20 km
vehicles: 6-15 km
small boats/rafts: 4-5 km
power lines: 5-6 km (support masts); 3 km power cables
men: 1.5 km
Range resolution: 8 m
Azimuth resolution: 30 minutes of angle
Scan angles: ± 45° azimuth; +10 to 20° elevation
Dimensions:
antenna and transmitter/receiver: 500 × 350 × 700 mm
data handling unit: 300 × 200 × 400 mm
Weight: 40 kg
Power consumption: 0.8 kW

Operational status

Specifications confirmed during flight test.

Contractor

Leninetz Holding Company.

VERIFIED

Neva multifunction civil weather/ navigation radar

The Neva radar is designed for heavy- and medium-sized civil/commercial aircraft. The system includes three operating modes:

1. Weather/turbulence indication, providing horizontal and vertical sections, four-colour display of rain, isolation of windshear and turbulence areas and automatic warnings.

2. Obstacle avoidance, providing high/medium/low altitude modes, three-colour display and automatic warnings.

Neva multifunction radar for civil aviation 0002446

3. Ground mapping, providing panoramic and detailed ground images, beam sharpening up to four times and navigation interfaces.

Specifications

Wavelength: 3.2 cm
Range scales: 600, 300, 150, 75, 40, 20, 10 km
Azimuth coverage: ±90, ±30°
Elevation coverage: ±15°
Weight: 50 kg
Power consumption: 115 V, 400 Hz, 300 V A
Antenna diameter: 762 or 559 mm, with polarisation selectivity and monopulse
Display: colour TV, 16 cm

Contractor

Leninetz Holding Company.

VERIFIED

VID-95 antenna 0002352

VID-95 approach/landing radar

VID-95 is a short-range, high-resolution, approach and landing radar, intended for operation in ICAO Cat II and IIIA conditions, at poorly equipped airfields. The radar operates at 8 mm wavelength, using a high-speed electronically scanning beam, and coherent received signal processing to control landings from 200 m altitude, provided the aircraft is positioned on the glide path using ILS, satellite or other navigation means.

Specifications

Runway detection range:
(VFR conditions) >5 km
(ICAO Cat II & IIIA) >3 km
Frequency: 35.8 GHz
Pulse repetition frequency: 10,000 Hz
Scan sector: ±45°
Drift angle compensation: 34°
Beam scan rate: 440°/s
Antenna beamwidth:
(horizontal) 0.5°
(vertical) 8°
Weight: 50 kg

Contractor

Leninetz Holding Company.

VERIFIED

CH-3301 GLONASS/GPS airborne receiver

NAVIS receivers are able to process signals from both GLONASS and GPS satellites, thereby increasing accuracy and reliability by comparison with receivers that process only one source of data.

Specifications

Memory: 500 waypoints, 50 routes
Display: alphanumeric LCD display; 20 symbols on each of 2 lines
I/O: 2 input/output ports RS-232C; 1 output channel in ARINC 429 format; 1 analogue barometric input
Antenna: to ARINC 743A
Dimensions: 384 × 159 × 51 mm
Weight: 2.4 kg
Power: 27 V DC; 115 V AC, 400 Hz
Temperature:
(receiver) –20 to +50°C
(antenna) –55 to +55°C

Contractor

NAVIS.

VERIFIED

EFIS-85 electronic flight instrument system

EFIS-85 is an integrated navigation, flight information and cathode ray tube display system. It comprises four displays, three symbol generators and two control panels. The panels are interchangeable and may function as Pilot Flight Displays (PFDs) or Navigation Displays (NDs).

PFDs display attitude, altitude and speed information, flight director and heading information; it also displays glide slope and localiser data, and information from automated systems.

NDs display navigation data, radio navigation data, weather radar and TCAS data.

Each symbol generator can support the operation of both ND and PFD for one pilot or both pilots' displays can be fed from the same symbol generator.

All units have built-in-test facilities, together with redundancy of image generation and display options.

Specifications

Display
Dimensions: 203 × 230 × 356 mm
Weight: 17.3 kg
Power: 250 W

Symbol generator
Dimensions: 194 × 127 × 324 mm
Weight: 7.0 kg
Power: 120 W

Control panel
Dimensions: 146 × 176 × 200 mm
Weight: 2.5 kg
Power: 25 W

Operational status

Development contract award in 1985. Fitted to Il-96-300 and Tu-204 airliners.

Contractors

NIIAO Institute of Aircraft Equipment.
Ulyanovsk Instrument Design Office.

VERIFIED

FCS-85 flight control system

FCS-85 is part of a standard digital avionics system designed for aircraft automatic control during flight and for director control during take-off and approach.

In conjunction with other avionics control systems FCS-85 provides control functions including:

(1) in conjunction with TCS-85 thrust control system, automatic control in altitude and heading, automatic hold of barometric altitude;
(2) with the control column, control of pitch, heading and roll;
(3) automatic approach and automatic landing in compliance with ICAO Cat IIIA requirements using ILS and MLS radio beacons;
(4) automatic approach to decision height in compliance with ICAO Cat II requirements using ILS, MLS, Landing System-50 (LS-50) radio beacons;
(5) automatic director approach to decision height in compliance with ICAO Cat I requirements using ILS, MLS, LS-50, SRNS radio beacons;
(6) automatic monitoring of FCS operation;
(7) prevention of exceedance conditions.

FCS-85 comprises the following LRUs: flight control computers (FCC-1) 3 linked units, each 8 MCU, ARINC 429 compatible; control panel (CP-56) 1 unit.

Specifications

FCC-1
Power: 115 V AC, 400 Hz, 150 VA; 36 V AC, 400 Hz, 1.5 VA; 27 V DC, 30 W
Weight: 12 kg
MTBF: 5,000 h

CP-56
Power: 115 V AC, 400 Hz, 75 VA; 6 V AC, 400 Hz, 12 VA; 27 V DC, 1 W
Weight: 8 kg
MTBF: 6,000 h

Operational status

Development contract award in 1985. Fitted to Il-96-300 and Tu-204 airliners.

Contractors

NIIAO Institute of Aircraft Equipment.
Moscow Institute of Electromechanics and Automatics.

VERIFIED

FILS Fault Isolation and Localisation System

FILS is the top level, of three levels, of onboard BITE and fault isolation system. It integrates the BITE activities and data management and storage of all major flight management systems, such as EFIS, FCS, TCS.

FILS is designed to: process fault diagnosis and isolation; output fault data on the databus and displays and ground test facilities; store data on faults for up to 10 flights and 30 faults per flight in non-volatile memory; input data on faults into aircraft utility systems.

FILS is a single LRU, with alphanumeric display, mounted in the flight engineer position, that complies with ARINC 604 and 734.

Specifications

ROM capacity: 40 kwords
Non-volatile memory: 16 kwords
Inputs; up to 25 channels (ARINC 429)
Outputs: up to 6 channels (ARINC 429); up to 12 discrete instructions
Weight: 10 kg
Power: 115 V DC, 400 Hz, 60 VA
MTBF: 5,000 h

Operational status

Development contract award in 1985. Fitted to Il-96-300 and Tu-204 airliners.

Contractor

NIIAO Institute of Aircraft Equipment.

VERIFIED

FMS-85 flight management system

As part of the standard avionics system for the Il-96-300 and Tu-204 aircraft, FMS-85 provides the following function: generation of information and control signals for 4-D navigation and control of en route flight, SID/STAR procedures, data on navigation aids, optimisation of fuel use.

The FMS-85 comprises the following LRUs: TsVM 80-40001 flight management computer; POOI-85M multifunction control and display unit (MCDU). The MCDU is used to input navigation and flight control data to the central computer and their output to the EFIS; the control of the FMS-85 modes of operation, and control of the I42-1S long-range navigation and satellite navigation systems.

Specifications

RAM capacity: 19 kwords
Non-volatile memory capacity: 236 kwords
ROM capacity: 128 kwords
Inputs; 72 channels; 32 discretes
Outputs: 10 channels; 8 discretes
Power: 115 V AC, 400 Hz, 190 VA; 27 V DC, 2.5 W
Weight: 15 kg

Operational status

Development contract award in 1985. Fitted to Il-96-300 and Tu-204 airliners.

Contractor

NIIAO Institute of Aircraft Equipment.

VERIFIED

MIKBO series of compact integrated avionic systems

The MIKBO series of compact integrated avionic systems is designed and produced by the Aircraft Instrument-making Establishment MIKBOTRON, which is part of NIIAO, the Institute of Aircraft Equipment.

The MIKBO series is specially designed for use in superlight aircraft. Included in the series are the following systems:

MIKBO-2

MIKBO-2 is in development for use in superlight single-engined aircraft operating in VFR conditions in the daytime, under visual meteorological conditions in the local area and during route and regional transit flights. MIKBO-2 comprises: magnetic compass MKB-90; VHF radio; microcomputer; AMLCD; ASI; engine speed indicator; cylinder head temperature sensor; integrated instrument panel. Total weight <3 kg; dimensions 350 × 150 × 150 mm; power 12 V DC <5 W.

MIKBO-21

MIKBO-21 is designed for superlight single-engined aircraft operating in VFR conditions in daytime under visual meteorological conditions in the local area. MIKBO-21 comprises: ASI YC-150; barometric altimeter CD-10; variometer BP-5; magnetic compass KI-13; cylinder head temperature sensor; and engine shaft speed indicator. Dimensions 300 × 180 × 200 mm; power 12 V DC, 1.5 W.

MIKBO-22

MIKBO-22 is in preproduction for use in superlight single-engined aircraft operating in VFR conditions in the daytime, under visual meteorological conditions in the local area and during route and regional transit flights. MIKBO-22 comprises: barometric altimeter BD-10; variometer BP-5; magnetic compass MKB-90; ASI; cylinder head temperature sensor/indicator; engine shaft speed indicator; speed indicator, VHF radio. Weight < 5 kg; dimensions 350 × 230 × 200 mm; power 12 or 27 V DC, 13 W.

MIKBO-23

MIKBO-23 is in preproduction for use in superlight twin-engined aircraft operating in VFR conditions in the daytime, under visual meteorological conditions in the local area and during route and regional transit flights. MIKBO-23 comprises: barometric altimeter BD-10; variometer BP-5; magnetic compass MKB-90; ASI; cylinder head temperature sensor/indicator; engine shaft speed indicator; speed indicator, slip indicator; main rotor speed indicator (MIKBO-23AJ only); VHF radio 'ptakha'. Weight 6 kg; power 12 or 27 V DC, <13 W.

MIKBO-32

MIKBO-32 is in design for use in superlight and light primary training aircraft certified to AP-23 or FAR-23 operating in VFR conditions from unequipped airfields. MIKBO-32 includes: a multifunction instrument panel with two flat panel displays and databus integrating: VHF radio, ADF, radio altimeter, heading computer, engine and air data sensors, magnetic compass, and CVR/FDR. Weight <30 kg; power <300 W.

MIKBO-43

MIKBO-43 is in design for use in light multipurpose aircraft, including amphibians, operating in IFR conditions from unequipped airfields in all geographical areas, as well as off water in wave heights up to 0.5 m. MIKBO-43 includes the same equipment as MIKBO - 32, plus: multifunction weather radar and emergency COSPAS-SARSAT radio beacon. Weight <75 kg; power <700 W.

Contractor

NIIAO Institute of Aircraft Equipment.

VERIFIED

Short-range airborne radio navigation systems

RadioPribor manufactures airborne short-range navigation systems designed to provide navigation data derived from ground-based beacons and ILS systems.

New generations of the equipment, designated A-317 and A-324, comprise a built-in computer (A-313), which interfaces with long-range terrestrial and satellite navigation systems to provide a comprehensive navigation capability.

Operational status

The short-range radio navigation system is in service, fitted to aircraft delivered to Eastern Europe, Asian and African states.

Contractor

Production Association RadioPribor.

VERIFIED

ARINC 429 series of electromechanical indicators

The Ramenskoye Design Company has designed a series of flight and navigation instruments as per ARINC 429. They include a Flight Command Indicator (FCI), Horizontal Situation Indicator (HSI) and two Radio Magnetic Indicators (RMI-3 and RMI-5).

Specifications

Number of input lines:
(FCI and HSI): 6
(RMI-3): 7
(RMI-5): 5
Number of output lines:
(FCI and HSI): 1
Number of indicated parameters:
(FCI and HSI): 9
(RMI-3): 5
(RMI-5): 3
Dimensions:
(FCI and HSI): 120 × 120 × 240 mm
(RMI-3): 85 × 120 × 240 mm
(RMI-5): 85 × 95 × 200 mm
Weight:
(FCI and HSI): 5 kg
(RMI-3): 3.5 kg
(RMI-5): 2 kg
Power: 27 V DC

Contractor

Ramenskoye Design Company AO RPKB.

VERIFIED

BINS-TVG strapdown Inertial Navigation System (INS)

The Ramenskoye Design Company BINS-TVG strapdown INS is manufactured as a monoblock, which comprises three accelerometers and three hemispherical resonance gyroscopes rigidly coupled with the monoblock casing. The system electronics utilise large-scale integrated circuits. The monoblock assembly can be integrated with a dual-mode GPS/GLONASS navigation receiver, or supplied separately.

The BINS-TVG system is intended for civil and military application on aircraft and helicopters.

Contractor

Ramenskoye Design Company AO RPKB.

VERIFIED

BINS-TWG strapdown inertial navigation system

The BINS-TWG strapdown inertial navigation system is designed to provide flight and navigation data for civil

BINS-TWG strapdown inertial navigation system
***1999**/0051594*

Specifications

	INS alone	with GPS/GLONASS
Accuracies (2 sigma):		
(co-ordinates)	1.85 km/h	20 m
(velocity)	0.6 m/s	0.1 m/s (3 sigma)
(roll/pitch)	0.05°	0.05°
(heading)	0.07°/h	0.07°/h
Warm-up time	3 min	3 min
Data output format	ARINC 429	ARINC 429
MTBF	5,000 h	5,000 h
Dimensions	7 MCU	8 MCU
Weight	14 kg	15 kg

and military aircraft and helicopters. It is constructed as a monoblock comprising an inertial sensor unit consisting of three accelerometers and three hemispherical resonance gyros. The unit includes integral GPS/GLONASS capability.

Contractors
Ramenskoye Design Company AO RPKB.
Ramensky Instrument Engineering Plant JSC.

VERIFIED

HG 1150BE01 strapdown Inertial Reference Unit (IRU)

The HG 1150BE01 contains three ring laser gyros and three pendulous accelerometers of Honeywell design in a strapdown configuration to provide primary attitude, heading, body angular rates, body linear accelerations, velocity and position information. Outputs conform to ARINC 704 standards.

Contractors
Ramenskoye Design Company AO RPKB.
Ramensky Instrument Engineering Plant JSC.

VERIFIED

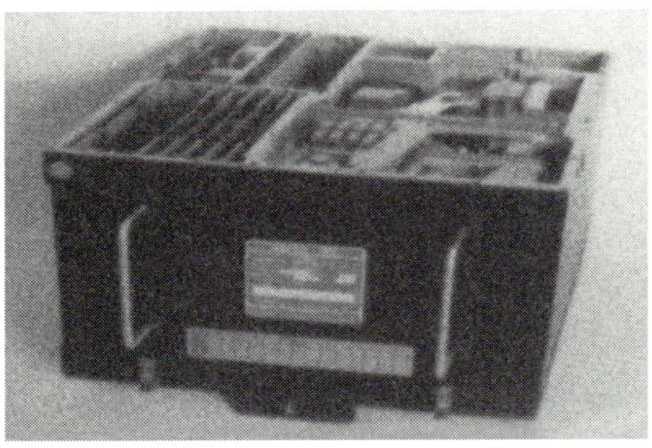

HG 1150BE01 strapdown Inertial Reference Unit (IRU) **1999**/0051595

Inertial Measurement Unit (IMU)

The IMU is a functionally independent unit providing measurement of angular rates in pitch, roll and yaw, together with linear accelerations along longitudinal, lateral and vertical axes.

The IMU contains two three-axis dynamically tuned gyros and pendulous accelerometers.

Contractors
Ramenskoye Design Company AO RPKB.
Ramensky Instrument Engineering Plant JSC.

VERIFIED

Inertial Measurement Unit (IMU) **1999**/0051593

INS-80 and INS-97 Inertial Navigation Systems

The INS-80 system is made as a single unit, comprising gyrostabilised platform, platform electronic units, a computer and I/O devices.

Two free dynamically tuned gyroscopes and three accelerometers are mounted on the gyrostabilised platform. The electronic units are mounted on separate PCBs.

INS-80 inertial navigation system 0015368

The INS-97 system also includes an integral Satellite Navigation System (SNS), plus antenna.

Optionally, the INS-80 and INS-97 systems can include magnetic heading measurement and display.

Specifications
Position error: 3.7 km/h
Velocity error: 2.0 m/s
Roll, pitch and heading error: 0.1°
Readiness time: 1 to 10 min
Volume: 18 dm³
Weight: 19 kg

Operational status
Fitted to the Su-30 aircraft.

Contractors
Ramenskoye Design Company AO RPKB.
Ramensky Instrument Engineering Plant JSC.

VERIFIED

Integrated navigation and control system

This integrated navigation and control system provides the following outputs:

(1) Vehicle attitude, angular velocities and linear accelerations;
(2) Flight parameters such as altitude, vertical speed, angles of attack and slip angles;
(3) Speeds and co-ordinates using GPS/GLONASS;
(4) Automatic and flight director control of the aircraft;
(5) Fatigue monitoring, warning of critical flight conditions, voice and aural warnings;
(6) Flat screen display (2) of flight, navigation, radar, weather, map and aircraft data.

A multiprocessor computer integrates data from all relevant sensor systems.

Specifications
Position error: 80 m
Velocity error: 0.3 m/s
Attitude error:
roll, pitch: 0.5°
heading: 1.0°
Readiness time: 10 min
Temperature range: +60 to −60°C
Linear acceleration: 5*g*
Weight: 28 kg
Interface: ARINC 429

Contractor
Ramenskoye Design Company AO RPKB.

VERIFIED

MFI multifunction active matrix liquid crystal displays

The Ramenskoye Design Company produces a range of MFI AMLCDs designed for presentation of raster enhanced graphic TV and other image data. The display specified below forms the display part of the Ramenskoye Design Company's SINUS integrated navigation, flight management and display system. Other MFI displays, designated MFI-2, MFI-3, MFI-9 and MFI-10 are also specified below.

Specifications
SINUS MFI display
Display size: 130 × 130 mm
Pixels: 864 × 864
Resolution: 0.15 × 0.15 mm
Viewing angles:
(horizontal) ±45°
(vertical) +35 to −10°
Power: 115 V AC, 400 Hz
Dimensions: 205 × 205 × 205 mm
Weight: <7 kg

MFI-2
Display size: 130 × 100 mm
Pixels: 640 × 480
Resolution: 0.15 × 0.15 mm
Viewing angles:
(horizontal) ±60°
(vertical) +55 to −35°
Brightness: 300 cd/m²
Colours: ≥4,096
Power: 115 V AC/400 Hz or +27 V DC
Dimensions: 220 × 280 × 280 mm
Weight: ≤10 kg

MFI-3
Display size: 211 × 159 mm
Pixels: 640 × 480
Resolution: 0.15 × 0.15 mm
Viewing angles:
(horizontal) ±60°
(vertical) +55 to −35°

MFI active matrix liquid crystal display 0018198

Integrated navigation and control system **1999**/0051596

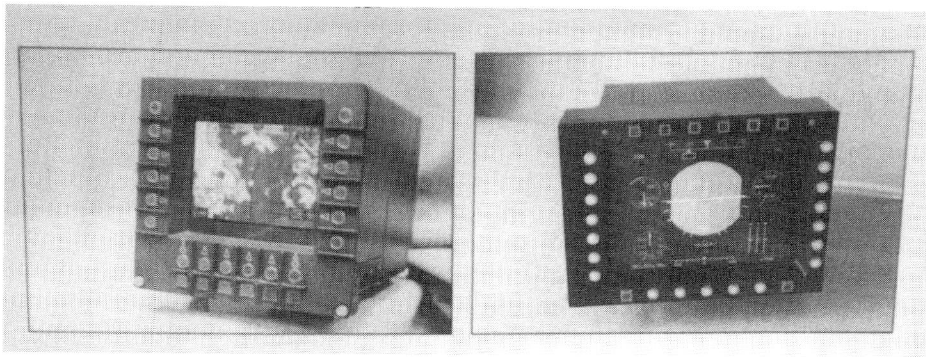

Ramenskoye Design Company MFI-9 and MFI-10 AMLCDs ***1999***/0054302

Brightness: 300 cd/m^2
Colours: >4,096
Power: 115 V AC/400 Hz or +27 V DC
Dimensions: 220 × 280 × 280 mm
Weight: ≤10 kg

MFI-9
Display size: 170 × 130 mm
Pixels: 640 × 480
Resolution: 0.27 × 0.27 mm
Viewing angles:
(horizontal) ±50°
(vertical) +30 to −50°
Brightness: 200 cd/m^2
Colours: >4,096
Power: 115 V AC/400 Hz or +27 V DC
Dimensions: 210 × 230 × 180 mm
Weight: ≤7 kg

MFI-10
Display size: 211 × 159 mm
Pixels: 640 × 480
Resolution: 0.33 × 0.33 mm
Viewing angles:
(horizontal) ±60°
(vertical) +55 to −35°
Brightness: 500 cd/m^2
Colours: 4,096
Power: 115 V AC/400 Hz or +27 V DC
Dimensions: 220 × 280 × 180 mm
Weight: ≤8 kg

Contractor
Ramenskoye Design Company AO RPKB.

UPDATED

MFI-2 and MFI-3 MultiFunction Active-Matrix Liquid-Crystal Display (AMLCD) Indicators

The Ramenskoye Design Company MFI-2 and MFI-3 AMLCDs are designed for real-time presentation of overlaid alphanumeric and graphical data from multiple sources using embedded processors.

Specifications
MFI-2
Display size: 130 × 100 mm
Pixels: 640 × 480
Resolution: 0.15 × 0.15 mm
Viewing angles:
(horizontal) ±60°
(vertical) +55 to −35°
Brightness: 300 cd/m^2
Number of colour shades: 4,096
Power: 115 V AC, 400 Hz; or +27 V DC
Dimensions: 220 × 280 × 280 mm
Weight: 10 kg

MFI-3
Display size: 211 × 159 mm
Pixels: 640 × 480
Resolution: 0.15 × 0.15 mm
Viewing angles:
(horizontal) ±60°
(vertical) +55 to −35°
Brightness: 300 cd/m^2
Number of colour shades: 4,096
Power: 115 V AC, 400 Hz; or +27 V DC
Dimensions: 220 × 280 × 280 mm
Weight: 10 kg

Contractor
Ramenskoye Design Company AO RPKB.

VERIFIED

MultiFunction Control Displays (MFCDs)

Ramenskoye Design Company produces a range of MFCDs, designated PS-2, PS-3 and PS-5.

Specifications
PS-2 MFCD
Prompt panel: 4 lines with 21 symbols
Data panel: 10 lines with 21 symbols
External lighting: up to 61,000 lx
Screen colour: green
Brightness adjustment: yes
Number of keys: 25
On-screen touch keys: 15
Power: 115 V AC, 400 Hz, 100 VA
Dimensions:
(front panel) 170 × 200 × 85 mm
(electronic panel) 220 × 25 × 130 mm
Weight: 7 kg

PS-3 MFCD
Prompt panel: 2 lines with 21 symbols
Data panel: 8 lines with 21 symbols
External lighting: up to 61,000 lx
Screen colour: green
Built-in processor: yes
Brightness adjustment: yes
Number of keys: 36 (8 multifunctional)
Power: 115 V AC, 400 Hz, 50 VA (2-phase)
Dimensions: 147 × 102 × 278 mm
Weight: 4.5 kg

PS-5 MFCD
Prompt panel: 2 lines with 21 symbols
Data panel: 10 lines with 21 symbols
External lighting: up to 61,000 lx
Screen colour: green
Number of keys: 33 (10 multifunctional)
Displayed frame storage: up to 256
Power: 115 V AC, 400 Hz, 50 VA
Dimensions:
(front panel) 170 × 200 × 85 mm
(electronic panel) 220 × 25 × 130 mm
Weight: 8.5 kg

Contractor
Ramenskoye Design Company AO RPKB.

VERIFIED

NS BKV-95 integrated Navigation System

The NS BKV-95 integrated navigation system performs three functions: strapdown attitude and heading reference system, GPS/GLONASS receiver, navigation computer. The system is provided with control displays to customer specification.

Specifications
Dimensions: 388 × 124 × 194 mm (4 MCU)
Weight: 8 kg
Power: 27 V DC
Outputs: GOST 18977-79 Sup 3; (ARINC 429)
Accuracy (2 Sigma):
(position) 100 m
(groundspeed) 2 km/h
(roll/pitch angles) 0.25°
(magnetic heading) 1°

NS BKV-95 integrated navigation system 0015369

Ramenskoye Design Company Multifunction Control Displays ***1999***/0054303

Operational status
Suitable for all types of civil aircraft.

Contractors
Ramenskoye Design Company AO RPKB.
Ramensky Instrument Engineering Plant JSC.

VERIFIED

SBKV-2V strapdown Attitude and Heading Reference System (AHRS)

The SBKV-2V strapdown Attitude and Heading Reference System (AHRS) is designed for application on civil and military fixed- and rotary-wing aircraft, as part of their flight and navigation systems, or as a standalone unit. Pre-planned product improvements include provision of a second processor for special purposes, including GPS, SAR and other improvements to navigation and traffic management capability.

Specifications
Accuracy (2σ):
(roll and pitch angles): 0.25°
(magnetic heading): 1.0°
(gyroscopic heading): 0.5°/h
Dimensions: 413 × 130 × 200 mm
Weight: 9 kg
Power: 27 V DC, 80 W; 115 V AC, 60 VA

Contractor
Ramenskoye Design Company AO RPKB.

VERIFIED

SINUS integrated navigation, flight management and display system

SINUS is an integrated navigation and flight management system, designed for the aircraft and helicopter retrofit market. It comprises systems that perform the normal functions of an integrated navigation and flight management system, including inertial navigation, satellite navigation (using both GLONASS and NAVSTAR), air data and altimeter integration. The inertial flight computer integrates these functions with radar and FLIR data, via an airborne graphics system to provide navigation, radar and flight management data to the pilot by any combination of the following methods: multifunction AMLCD, head-up display, helmet-mounted display, and aural warnings. The SINUS system also provides appropriate outputs for automatic control of aircraft flight controls and engine systems.

Specifications
Navigation errors:
(inertial mode) 1 n mile/h
(satellite mode) 50 m
(FLIR mode) 20-100 m
Velocity error: 0.3 m/s
Angle errors:
(roll/pitch) 0.5°
(yaw) 1°
Weight: 28 kg

Operational status
Designed for full internal, or partial podded, fit for upgrade of military aircraft and helicopters such as MiG-21, Su-24, Mi-8, Mi-24.

Contractors
Ramenskoye Design Company AO RPKB.
Ramensky Instrument Engineering Plant JSC.

VERIFIED

SINUS integrated navigation, flight management and display system **0015370**

I-42-1L strapdown integrated inertial navigation system

The I-42-1L integrated inertial navigation system comprises an inertial unit based on triple GL-1 laser gyros and triple A-L2 accelerometers, together with a proprietary combined GPS-GLONASS/NAVSTAR unit as per ARINC 743A.

Specifications
Accuracy in GPS mode:
(position): 50-200 m
(speed): 5 km/h
(pitch and roll): 0.04°
(true heading): 0.4°
Power: 27 V DC and 115 V AC, 400 Hz

Operational status
Fitted to civil aircraft including Il-76 and Tu-204.

Contractor
Ramensky Instrument Engineering Plant.

VERIFIED

I-42-1S strapdown inertial navigation system

The I-42-1S strapdown inertial navigation system is designed to provide flight and navigation parameters as per ARINC 429, as part of an integrated flight and navigation system. The I-42-1S comprises triple KM-11-1A laser gyros and triple A-L2 accelerometers.

Specifications
Accuracy:
(present position): 3.7 km/h
(ground speed components): 22.5 km/h
(true heading): 0.4°
(roll and pitch): 0.1°
Weight: 43 kg
Power: 115 V AC, 400 Hz

Contractor
Ramensky Instrument Engineering Plant.

VERIFIED

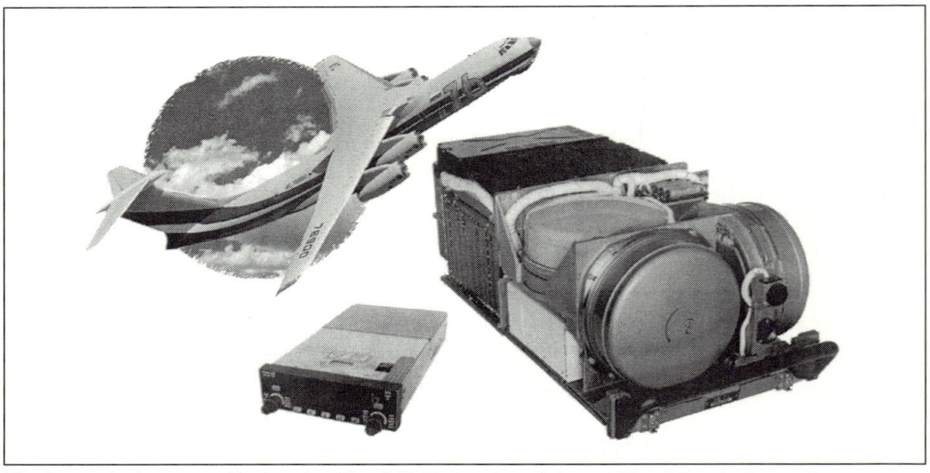

The Ramensky Instrument Engineering Plant I-42-1L strapdown integrated inertial navigation system **2000**/0079228

MK compass system

The MK compass system is designed to provide gyromagnetic, magnetic and true heading. It consists of a gyromagnetic course unit (BGMK-6A series 1), a magnetic variation selector (ZMS-3 series 2) and a flux gate detector (ID-6 series 1). Flight accuracy of gyromagnetic heading data is ±1.0°. Weight is 6 kg.

Operational status
Being offered for MiG-21 modernisation and for civil application.

Contractor
Ramensky Instrument Engineering Plant.

VERIFIED

PKP-72 and PKP-77 flight directors and PNP-72 compass

The PKP-72 and PKP-77 flight director indicators and the PNP-72 compass are analogue instruments that have found wide application in Russian aircraft.

The Ramensky Instrument Engineering Plant MK compass system **2000**/0080654

PKP-72 and PKP-77 flight directors and PNP-72 compass 0018197

Specifications

	PKP-72	PKP-77	PNP-72
Roll error	0.7°	0.7°	
Pitch error	0.7°	0.7°	
Yaw error			1.0°
Azimuth			0.5 to 1.5 km

Contractor

Ramensky Instrument Engineering Plant JSC.

VERIFIED

RMI-3 radio magnetic indicator

The RMI-3 radio magnetic indicator displays magnetic heading, relative bearing from two radio beacons, distance from two radio beacons. The display is software programmable. Communication interfaces are digital, as per ARINC 429.

The Ramensky Instrument Engineering Plant RMI-3 radio magnetic indicator **2000**/0079227

Specifications

Accuracy:
(bearing): ±1.5°
(distance): ±0.1 km
Dimensions: 85 × 120 × 240 mm
Weight: 3.0 kg
Power: 27 V DC

Operational status

Fitted to Antonov An-125 Ruslan.

Contractor

Ramensky Instrument Engineering Plant

VERIFIED

Standby Horizontal Situation Indicator (HSI) INP-RD

The Ramensky Instrument Engineering Plant standby horizontal situation indicator displays aircraft position, heading and pitch with respect to the radio beacon and four cardinal points. It comprises a sealed electromechanical unit and the control and monitor unit (BUK-14).

The Ramensky Instrument Engineering Plant standby horizontal situation indicator **2000**/0080655

Specifications

Accuracy:
(bearing): 1.0°
(distance 0-25 km): ±1 km
(distance 25-999 km): ±3 km
Slaving rate of pointers:
(current heading): 30°/s
(radio beacon bearing): 60°/s
(selected heading): 40°/s
(distance) 15 km/s
Readiness time: 1 minute (maximum)
Weight: 2.3 kg

Operational status

Fitted to Su-30 military aircraft, as well as civil aircraft.

Contractor

Ramensky Instrument Engineering Plant.

VERIFIED

STRIZH gyrostabilised magnetic compass

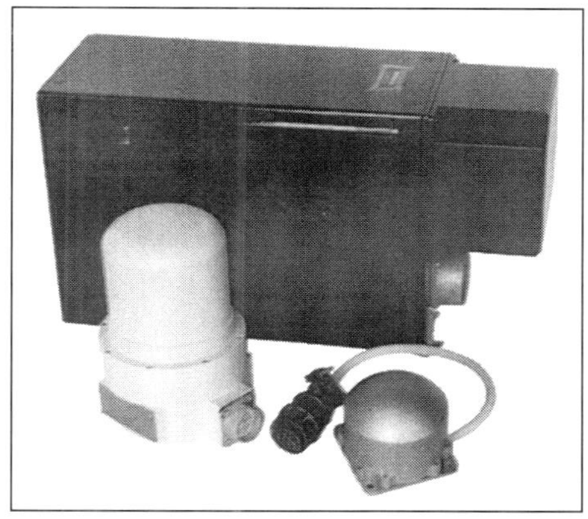

The Ramensky Instrument Engineering Plant STRIZH gyrostabilised magnetic compass **2000**/0079232

The STRIZH gyrostabilised magnetic compass is designed for small civil aircraft. It updates and improves on the capability of earlier products available from the Ramensky Instrument Engineering Plant.

Specifications

Gyromagnetic heading accuracy (2σ): 1.0°
Readiness time: 3 min
Weight: 3.5 kg
Operating conditions:
(flight speed): up to 1,000 km/h
(flight altitude): up to 15,000 m
(axial angular velocity): up to 200°/s
(linear acceleration): up to 4 *g*
(heading angles): 0-360°
(roll angles): 0-180°
(pitch angles): 0-90°
(latitude): 0-75°

Contractor

Ramensky Instrument Engineering Plant.

VERIFIED

KSU-821 flight control system

The KSU-821 flight control system is a four-channel redundant digital system fitted to the MiG-AT aircraft.

Contractor

Russkaye Avionica Joint Design Bureau.

VERIFIED

MFI-68 multifunction display

The MFI-68 is designed for equipping newly developed and upgraded aircraft. It presents data from onboard systems and sensors in the form of colour symbol – graphic information, as well as TV data. It incorporates 22 multifunctional control keys.

Specifications

Screen type: colour liquid-crystal panel CS8362
Screen size: 211 × 158 mm
Pixels: 640 × 480
Pixel type: vertical band RGB
Number of bits per colour subpixel: 5
Greyscale: 64
Image contrast: 80:1

Viewing angles:
(horizontal) ±60°
(vertical) –5 to +35°
Interfaces:
ARINC: 3 input channels, 1 output channel
GOST 26675.52-87 (MIL-STD-1553B): 1
GOST 7845-79 (STANAG 3350B): 1
Power supply: +27 V, 115 V AC/400 Hz
Power: <60 W (<180 W with heaters)
Operation temperature: –40 to +85°C
Dimensions: 260 × 200 × 260 mm
Weight: <6.5 kg

Contractor
Russkaya Avionica Joint Design Bureau.

VERIFIED

MultiFunctional Control Panel (MFCP)

The MFCP is designed for control of navigation and communications facilities via ARINC-429 and MIL-STD-1553B datalinks. Control is by a number of multifunctional keys, placed to the left and to the right of the electroluminescent screen.

Specifications
Screen type: electroluminescent
Screen size: 96 × 77 mm
Interfaces:
GOST 18977-79 (ARINC-429): 3 input channels, 1 output channel
GOST 26675.52-87 (MIL-STD-1553B): 1
Power: +27 V, <30 W
Dimensions: 127 × 215 × 145 mm
Weight: <2.2 kg

Contractor
Russkaya Avionica Joint Design Bureau.

VERIFIED

STR7-4 fuel quantity and flow metering system

The STR7-4 fuel quantity and flow metering system fulfils the following purposes: computing aircraft total fuel mass during refuelling and in flight, computing and indication of fuel mass transferred in flight, indication of fuel mass in the aircraft tanks, and low fuel-remaining warnings, indication and fuel mass flow, control of fuel usage, fuel system BITE.

System line-replaceable units include: nine capacitance-type fuel quantity sensors; three fuel flow transmitters; 13 level switches; six temperature sensors; two electromechanical indicators; two control and monitoring panels; four electronic converter units.

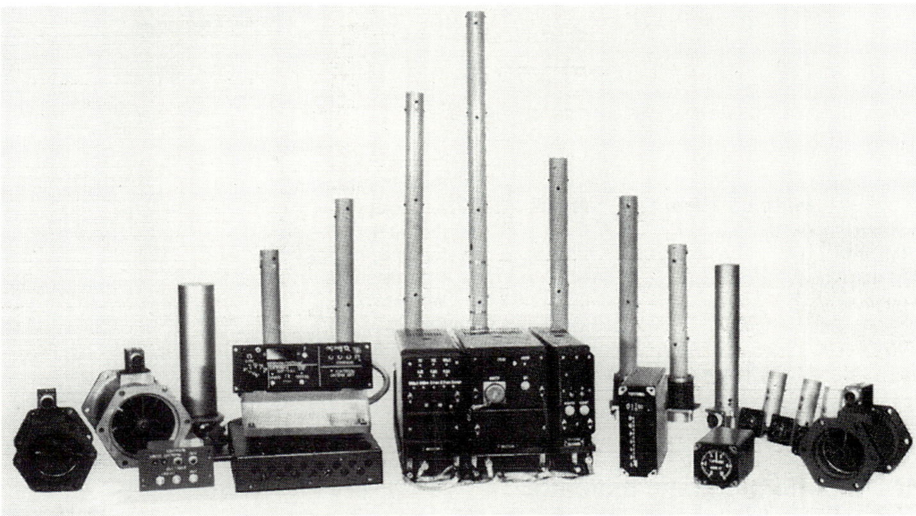

STR7-4 fuel quantity and flow metering system 0018235

Specifications
Fuel mass error: ±2%
Fuel level error: ±5 mm
Data exchange: ARINC 429
Unit design: ARINC 404
Mass:
(excl transmitters) 25 kg
(incl transmitters) 38 kg
Power: 27 V DC; 115 V AC, 400 Hz

Contractor
TekhPribor State Enterprise.

VERIFIED

SUIT8-10 fuel management and indicating system

The SUIT8-10 fuel management and indicating system provides the following: measurement and indication of fuel mass in aircraft tanks, and total aircraft fuel mass, control of fuelling and fuel use, warning of critical fuel reserves and unbalance.

Composition of the system for the Il-114 aircraft is as follows: six capacitance-type fuel quantity transmitters; six fuel quantity compensator transmitters; six level switches; three electronic converter units; two indicators; one control panel.

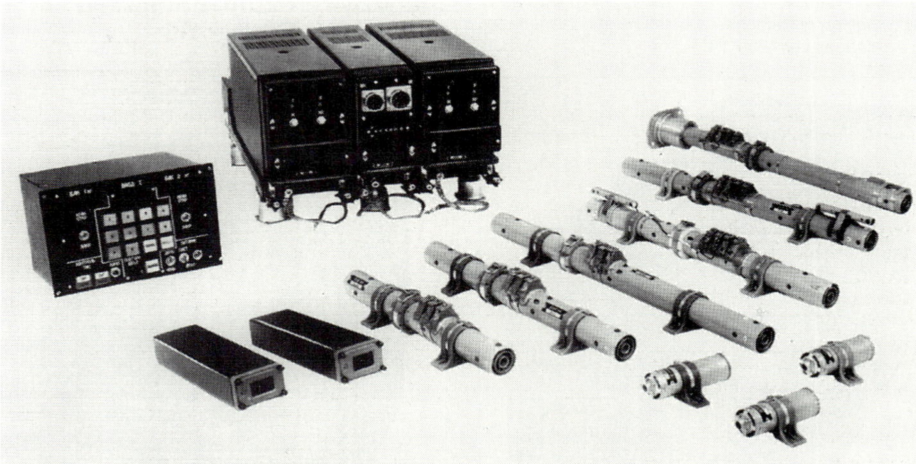

SUIT8-10 fuel management and indicating system 0018234

Specifications
Fuel mass error: ±1.5%
Fuel level error: ±5 mm
Data exchange: ARINC 429
Unit design: ARINC 404
Mass:
(excl transmitters) 20 kg
(incl transmitters) 26 kg
Power: 27 V DC; 115 V AC, 400 Hz

Contractor
TekhPribor State Enterprise.

VERIFIED

Advanced Moving Map System (AMMS)

The AMMS combines Flight Management System (FMS) and navigation display functions to support complex Search And Rescue (SAR) and special mission roles.

The AMMS includes an independent source of navigation information (GPS or GPS + GLONASS), together with its own processor and data storage system, a full colour high-resolution display, and both digital and analogue input/output ports.

Route planning is performed directly onto the AMMS, which simplifies this operation and reduces preparation time. All types of maps are displayed, including: aeronautical, topographic and marine; they can be used separately, or in combination. Planning takes into account wind and current where SAR operations are planned. Automatic planning of five SAR routes can be accommodated, taking into account aircraft performance data.

Terrain alerts and ground proximity time is determined on the basis of vector charts, mathematical models of the terrestrial surface and precise co-ordinates from the satellite navigation system.

The AMMS is designed to accommodate the Automatic Dependent Surveillance - Broadcast (ADS-B) concept. It is connected to the SoTDMA/VDL-4 transponder to provide ADS-V functions. Symbols meet TCAS and RTCA/DO-243 requirements. Information on air traffic is displayed in combination with electronic maps, route and GPWS information. Datalinks can be established for the exchange of data with ATC services and other aircraft. The AMMS can be used as a terminal for INMARSAT satcom.

Functions of the AMMS include:

(1) determination of position, track, speed and time via GPS or GPS + GLONASS
(2) horizontal and vertical navigation worldwide
(3) integration with the navigation and FMS
(4) provision of warnings if the track departs from the flight plan or FMS demands
(5) provision of Terrain Awareness and Warning System (TAWS) data
(6) provision of general aircraft warnings

(7) cartographic support for all navigation functions, AMMS supports Jeppesen NavData® databases, and claims that its own database The Transas Electronic Chart Database is the second largest in the world, comprising over 5,500 marine and topographic vector digitised charts
(8) substitution of electromechanical instruments
(9) cross telling of data to the second pilot's flight director
(10) automatic saving of data for debriefing purposes

Contractor

Transas Marine Limited, Transas Aviation Limited.

VERIFIED

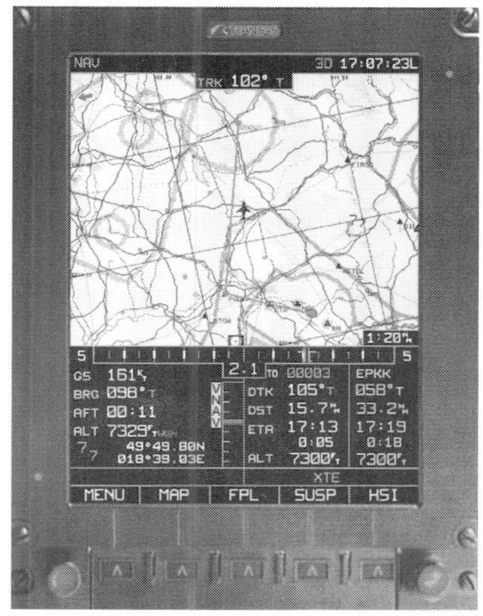

The Transas Aviation Limited advanced moving map system navigation display format
2000/0062790

SOUTH AFRICA

Display systems

The Avitronics family of Active Matrix Liquid Crystal Displays (AMLCDs) includes MultiFunction Colour Displays (MFCDs) and MultiFunction Keyboards (MFKs), that can be configured to customer requirements.

The MFCD is built around a 5 in diagonal, full colour, active matrix display. The display bezel hosts the custom function keys as well as the brightness and control Man/Machine Interface (MMI) requirements. The display modes include alphanumeric characters, symbols and graphics. The MFCD includes programming and built-in test capabilities.

Advanced Thin Film Transistor (TFT) AMLCD technology is used to provide high-quality brightness and contrast, together with low volume and power consumption, and Night Vision Goggle (NVG) operation.

Software design permits the user to define font-set and symbol characters, including special bitmap symbols that can be downloaded via the external connector.

Specifications

Power requirements: 28 V DC (accepts 18-32 V)
Data interfaces: MIL-STD-1553B, RS-422, RS-485, ARINC 429
Environmental:
MIL-STD-810E
MIL-STD-461C
DO 160C
Software design: RTCA DO178B

Contractor

Avitronics (Pty) Ltd.

UPDATED

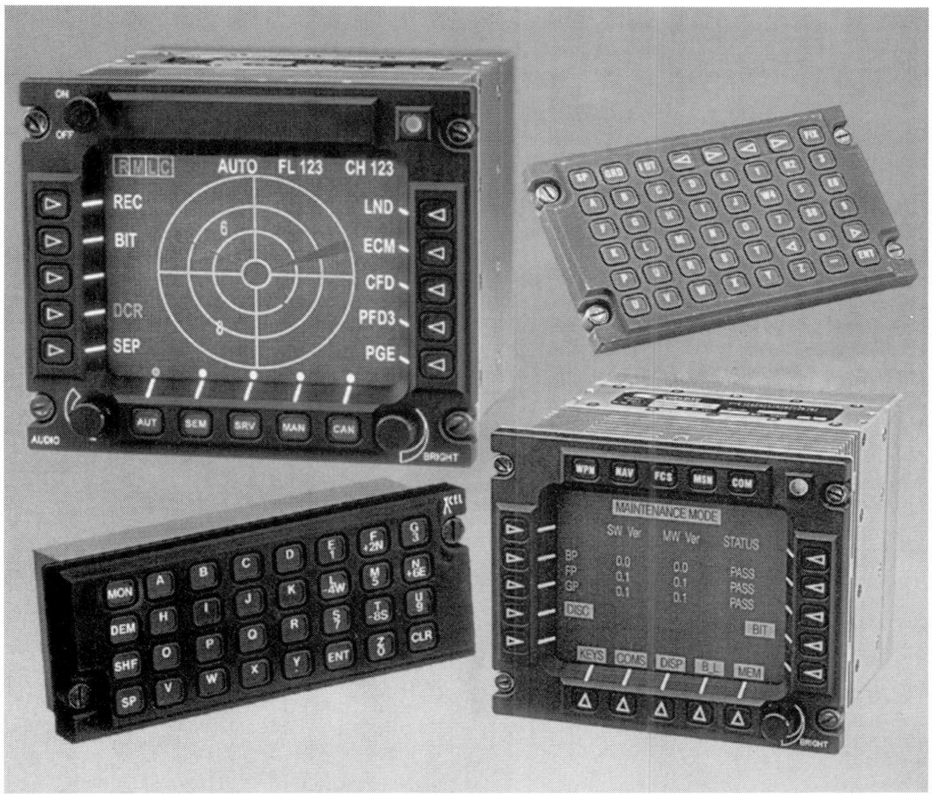

Avitronics multifunction display systems showing top left a Threat Display and Control Unit (TDCU), bottom right the display showing maintenance data and multifunction keyboard options ***2001***/0089528

SWEDEN

MFID 68 MultiFunction Integrated Display

Originally developed for the Saab JAS 39 Gripen, the MFID 68 is a fully integrated, multi-mode liquid crystal colour display which may be applied to a wide range of aircraft cockpits.

It features a 6.2 × 8.3 in (158 × 211 mm) display surface with a high-resolution NVIS-compliant screen with 20 push-buttons and manual/automatic brightness/contrast controls.

The display comprises computer, interfaces and graphics generator with a digital map. Standard interfaces are used to connect the system to the host aircraft avionics suite, with facilities to implement user-defined functions through software.

The MFID 68 is designed to provide a high-performance interface to the pilot. It is included as part of the EP-17 display system for batch 3 and export variants of the Saab Gripen and ordered for inclusion in the planned mid-life upgrade of the Swedish Air Force JA 37 Viggen.

MFID 68 MultiFunction Integrated Display
1999/0051668

Specifications

Display size: 158 × 211 mm (6.2 × 8.3 in)
Resolution: 600 × 800 pixels
Display computer #1: Ericsson real-time computer, 1 Mb program memory, 640 kb data memory, Pascal D80
Display computer #2: Power PC with internal VME bus 16 Mb memory, Ada or Pascal D80

Operational status

Under development for the JAS 39 Gripen and in production for the JA 37 Viggen.

Contractor

Ericsson Saab Avionics AB.

UPDATED

R3 GP&C (Global Positioning & Communication) transponder

The R3 GP&C transponder is designed for air-to-air and air-to-ground navigation. The transponder consists of a GPS receiver for navigation and a VHF radio which communicates with other transponders. The VHF radio link is a self-organising robust Time Division Multiple Access (TDMA) datalink synchronised by the very accurate time pulses derived from the GPS satellite system.

The datalink is capable of transmitting more than 2,000 position reports per minute. The limited radio range for VHF dictates a cellular system. This means that time slots in the TDMA datalink can be used freely in each cell. The total capacity of the system is therefore very large.

The GP&C system operates in two modes: autonomous and polled. In autonomous mode, the slot allocation process is controlled in each transponder by utilising a sophisticated algorithm to ensure that interference between slots is avoided. In polled mode, the slot allocation process is controlled by a base transponder and the mobile transponders are commanded to use a certain slot in the TDMA frame and reporting rate. Messages consist of position messages and text messages. The position messages include identity, latitude and longitude, altitude, velocity, course, navigation mode and GPS time. Text messages are 40 characters long and may be broadcast, without acknowledgement, or addressed with acknowledgement. The transponder is connected to display systems on board aircraft and to a network on the ground for ATC services.

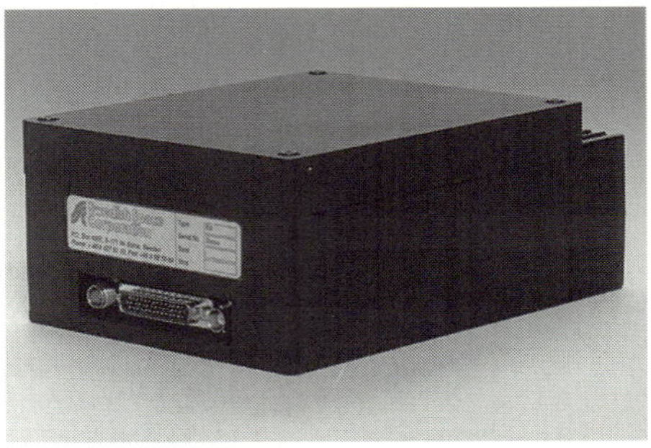

R3 GP&C global positioning & communication transponder demonstration model 0015371

Differential corrections (DGPS) are also transmitted on this network.

Specifications

Dimensions: 152 × 88 × 210 mm
Weight: 2.8 kg
Power supply: 14 or 28 V DC
Frequency: 136-174 MHz
Channel spacing: 25 kHz

Operational status

Tested in several European countries for ADS-B and fleet management applications.

Contractor

GP&C Sweden AB.

VERIFIED

TopEye survey system

Saab claims that its TopEye survey system utilises a unique laser range-finder technique, integrated with the inertial navigation system of the host helicopter and digital GPS position data to produce an effective near-realtime topographical survey system.

The laser range-finder can measure up to four different survey points in one and the same laser pulse, enabling TopEye to exactly define the structure of the area surveyed, and to produce a highly accurate digital terrain model, without recourse to photogrammetry. Accurate to within one centimetre, the laser range-finder scans across the track of the helicopter by measuring the distance to the ground with up to 6,000 laser pulses per second. Four different distances can be identified and recorded by a single laser pulse, while CCD cameras simultaneously take pictures from the helicopter, both vertically and at a 45° oblique angle. The helicopter INS data and digital GPS data is used to support post-mission processing that produces the digital terrain product data.

Operational status

The airborne system can be installed on a Squirrel AS 350 helicopter within two hours, with no structural modifications. An underslung pod contains the complete data acquisition system (camera plus laser range-finder).

Contractor

Saab Survey Systems.

VERIFIED

AMR 345 VHF/UHF transceiver

The SaabTech (formerly CelsiusTech) AMR 345 is a small VHF/UHF AM/FM panel-mounted radio for speech and data communications, covering the frequency range 104 to 162 and 223 to 408 MHz, with 25 kHz channel spacing. It has an ECCM capability and is designed to operate in a secure voice system. Transmitter output power is 10 W in the AM mode and 15 W in FM.

The transceiver memory has a flexible storage capacity for up to 1,000 preset channels. These channels can be used for immediate access to all frequencies at all airbases. The keyboard is used to select the preset frequencies, as well as for manual setting. A BIT facility performs a go/no go test in the aircraft.

An autonomous system is formed by the transceiver, a 28 VDC power supply, a headset and an antenna. It is also operable with other systems in the aircraft by means of a serial data control link. The microprocessor control allows easy adaptation of the transceiver (including the front panel function) to any system configuration. Preset channels and data back-up are stored in an electrically reprogrammable non-volatile memory.

The AMR 345 panel-mounted VHF/UHF transceiver - part of the communication system for JAS 39 Gripen Batch 2 aircraft

The transceiver is hardened to ensure full operation after EMP exposure. ECCM capability is provided by use of an external ECCM control unit which can be designed in accordance with customer requirements.

The transceiver is designed to meet applicable MIL standards.

Specifications

Dimensions: (transceiver) 230 × 146 × 76 mm
Weight: 3.5 kg
Frequency accuracy: 5 ppm (–40 to +70°C)

Operational status

Production complete. The AMR 345 is part of the communication system in JAS 39 Gripen Batch 2 aircraft for the Royal Swedish Air Force. Batch 3 and export aircraft do not include this transceiver. The AMR 345 was also selected to re-equip the Saab 105 (SK 60), Saab J32 Lansen, Saab SK 35 Draken, C-130 Hercules, Eurocopter Super Puma helicopter and the CASA C-212 transport aircraft. All of these deliveries are now complete. As of September 2000, the system was no longer actively marketed by SaabTech.

Contractor

SaabTech Electronics AB.

UPDATED

Global Positioning and Communication (GP&C) system

The Global Positioning and Communication system (GP&C) is a system designed for radio transmission of position data among a large community of users. Each user broadcasts positioning data on the radio network and thus every user is able to keep track of all other users within radio range. The basic GP&C system comprises the GP&C transponder, antennas and a presentation unit.

The core of the system is the GP&C transponder which within itself comprises a satellite receiver, a radio transceiver and a communications computer. The satellite receiver is a commercially available GPS receiver, both standard and military receivers can be integrated in the transponder. The radio transceiver and the communications computer with special software are unique units specially manufactured for integration in the transponder. The special software, executed in the communications computer, implements the actual Self-Organising Time Division Multiple Access (SOTDMA) transmissions on the radio network. To achieve global synchronisation in the SOTDMA network, the time signal from the GPS receiver is used by each user to determine when data shall be transmitted on the radio network.

In the basic GP&C system position data from its own GPS receiver is both passed to the local presentation unit as well as broadcasted on to the radio network where it is received by other users. In a similar fashion, GPS position from other users is received by the radio processed by the communications computer and passed to the local presentation unit.

For efficient use of the radio channel, a specially designed version of Time Division Multiple Access (TDMA) is used, SOTDMA. During operation every unit participating in the radio network is assigned or selects a free slot which is a fixed duration of time in the SOTDMA cycle. One cycle equals 1 minute and is currently divided into 2,250 slots. A unique feature in the SOTDMA concept is the ability to handle units entering and leaving the radio network without two or more units transmitting in the same slot. An enhanced version of the software is currently under test where capacity is extended to 4,500 slots per 1 minute cycle.

Primarily designed for exchange of position information, the system is capable of transmitting all types of information embedded in GP&C messages. The system can handle 256 different types of messages of which the majority is available for the user to define in his specific application. Information that does not fit in to one slot are composed in a special long message by chaining several slots together.

In applications where enhanced position accuracy is required, differential corrections can be transmitted over the radio network, thus providing all users with a position accuracy well under 5 m.

The GP&C system gives an overview of the air traffic situation. The GP&C display is shown here in a Swedish CAA aircraft

Operational status

At present, all systems in operation are operating on the VHF frequency band, but radio units for UHF and SHF are currently being evaluated.

Contractor

SaabTech Electronics AB (formerly CelsiusTech Electronics AB).

VERIFIED

LINCS T3L/M airborne transponder

LINCS is a datalink system designed for radio transmission of position data, text and other data among a large community of users. Users transmit their position and receive position data from other users within the radio horizon. By using text messages between aircraft and ground control, the user can get pre-departure clearance, weather information and so on, to enhance flight safety and assist collision avoidance.

The LINCS system derives position mainly from the Global Positioning System (GPS), using GP and C (Global Positioning and Communication) technology and the STDMA (Selforganising Time Division Multiple Access) datalink. Based on the 1 m accuracy of differential GPS positioning, LINCS supports both ADS-B (Automatic Dependent Surveillance - Broadcast) and ADS-C (Automatic Dependent Surveillance - Contract).

LINCS equipment comprises the LINCS T3L/F ground station and LINCS T3L/M airborne transponder. The LINCS T3L/M airborne transponder comprises: a 16-channel/25 kHz VHF radio transponder; 12-channel GPS receiver (with RAIM); VHF and satellite antennas, communication computer; and control/display unit.

Specifications

GPS receiver: Ashtech G 12 all-in-view 12-channel, 10 Hz position update

VHF transceiver: complies with I-ETS 300 113 packet radio

VHF 112-140 MHz (or 136-174 MHz) (trial frequency 136.95 MHz)

FM/GMSK modulation

9,600 bps data rate

1-25 W power output

16-channels/25 kHz

dynamically controlled refresh rate

Operational status

A full-scale test has been conducted at Landvetter airport, Gothenburg, Sweden. The system was used as an ADS and TCAS (Traffic alert and Collision Avoidance System). Both civil and military applications are being evaluated. LINCS is a joint-venture product from Transponder Tech AB, owned by Saab Dynamics AB and SaabTech (formerly CelsiusTech) Electronics AB.

Contractor

Transponder Tech AB.

UPDATED

LINCS T3L/M airborne transceiver (left) and T3L/F ground station (right)
1999/0044804

SWITZERLAND

Altimeters, types 3A and 3H, 2 in

Thommen 2 in altimeters are barometric instruments. The counter-pointer display is actuated by a high-performance mechanism with a high-stability beryllium capsule. A built-in vibrator minimises friction and optimises accuracy in use. Temperature is compensated by a bimetallic element.

All instruments are offered customer-specific (see type designation) and are mounted for easy servicing.

The counter has two moving drums, which show tens of thousands and thousands of feet. Three fixed zeros are printed on the display. The pointer indicates 1,000 feet per revolution on a scale calibrated at intervals of 20 ft. Warning flags are displayed on the counter between 10,000 and 0 ft respectively below 0 ft.

The barometric counter indicates the pressure in mbar or in Hg, or in both units of measurement. The display uses a single four-digit scale when indicating barometric pressure in mbar or in Hg, and a dual scale when indicating pressure in mbar/in Hg. QNH/QFE adjustment is provided by a knob on the front face of the instrument. It has built-in end stops.

Altimeters, type 3A and 3H, 2 in 0018189

An electrical indication of pressure (3H43, 3H42 models only) can be provided by potentiometer geared to the baro-scale mechanism.

The altimeters are installed in lightweight aluminium ARINC, square and round cases.

Lighting is offered in different voltages and different colours. All instruments are available with NVG-compatible lighting.

Operational status

The 2 in altimeters are installed as primary or standby instruments in civil and military aircraft.

Contractor

Revue Thommen AG.

VERIFIED

Encoding altimeters, types 3A and 3H, 3 in

Thommen 3 in encoding altimeters are totally self-contained, requiring no electrical power to drive the altitude mechanism. The altimeters incorporate an optical encoder providing height reporting output. The counter-pointer display is actuated by a high-performance mechanism with a high stability beryllium capsule. A built-in vibrator minimises friction and optimises accuracy in use. Temperature is compensated by a bimetallic element.

The counter has three moving drums, which show tens of thousands, thousands and hundreds of feet. The pointer indicates 1,000 ft per revolution, on a scale calibrated at intervals of 20 ft. Warning flags are displayed on the counter between 10,000 and 0 ft respectively, below 0 ft.

The barometric-counter indicates the pressure as four digits as single baro in mbar or in Hg or as dual baro in mbar/in Hg. Adjustment for QNH/QFE is provided by a knob on the front, built-in stops are included.

There is an optical electronic system supplying altitude data in digital form, ICAO coded (Gillham) for SSR pressure altitude transmission in 100 ft increments, to standard accuracy TSO-C88.

Altimeters type 3A and 3H, 3 in 0018188

Encoding Altimeters are provided with a code ON/OFF flag alarm, which indicates any encoding power malfunction.

An electrical indication of pressure (3H67/3H66 only) can be provided by potentiometer geared to the baro-scale mechanism.

The altimeters are installed in lightweight aluminium ARINC, square and round cases.

Lighting is offered in different voltages and different colours. All instruments are available with NVG-compatible lighting.

Operational status

The 3 in encoding altimeters are installed as primary or standby instruments in civil and military aircraft.

Contractor

Revue Thommen AG.

VERIFIED

MACH/Airspeed Indicators (MAI), type 5, 3 in

The 3 in MACH airspeed indicators are pneumatically operated instruments deriving indicated airspeed and MACH information from the pilot-static sources. The combined displacements of the airspeed pointer and the MACH-disk determine the MACH number, which is indicated simultaneously with airspeed by the airspeed pointer. The control relays for the outside warning system (Vmo and Mmo) are actuated by optical detection devices. All instruments are offered customer-specific.

The MACH airspeed indicators are installed in lightweight aluminium ARINC and square cases.

Lighting is offered in different voltages and different colours. All instruments are available in NVG-compatible lighting.

Operational status

The 3 in MAI are installed as primary or standby instruments in civil and military aircraft.

MACH/airspeed indicators, type 5, 3 in 0018187

Contractor

Revue Thommen AG.

VERIFIED

Vertical Speed Indicator (VSI), type 4A16, 3 in

A pointer indicates vertical speed on a fixed dial. Zero adjustment used to compensate internal stresses in the mechanism, is operated by turning the whole mechanism frame in relation to the case/dial assembly.

The Vertical Speed Indicators are installed in lightweight aluminium ARINC, square or round cases.

Lighting is offered in different voltages and different colours. All instruments are available in NVG-compatible lighting.

Operational status

The 3 in Vertical Speed Indicators are installed as primary instruments in civil and military aircraft.

Contractor

Revue Thommen AG.

VERIFIED

Vertical Speed Indicator (VSI) Type 4A16, 3 in 0018186

Engine Interface and Vibration Monitoring Units (EIVMU)

Vibro-Meter's EIVMUs are based on use of Surface Mounted Device (SMD) technology and Commercial-Off-The-Shelf (COTS) components to produce specialised modules that are integrated into engine interface and condition monitoring systems. Products include the BSCD-120 engine interface and monitoring unit, and the EVM-260 engine vibration monitoring unit.

The BSCD-120 acquires and monitors all vital engine parameters, outputs processed information on ARINC 429 digital buses for the cockpit display unit and Flight Data Recorder (FDR); it also generates alarm signals for the cockpit display and FDR, and performs self test routines.

The EVM-260 receives signals from engine-mounted accelerometers

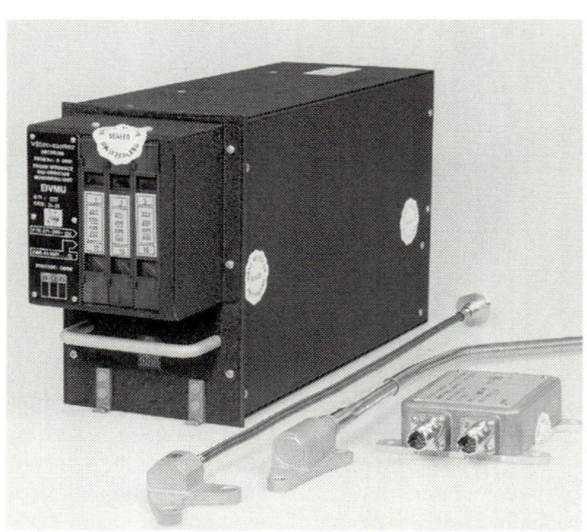

The Vibro-Meter EIVMU, as supplied for Airbus A330/A340 aircraft, showing remote charge converter unit and associated engine accelerometers **1999**/0054212

The Vibro-Meter BSCD-120 and EVM-260, as supplied for the Tu-204-120 aircraft **1999**/0054213

Operational status

Vibro-Meter's EIVMU is installed in all Airbus Industrie A330 and A340 aircraft. Most recently, it has been confirmed for the A340-500 and -600 aircraft. Vibro-Meter has also completed the certification test programme for their latest series of generic engine vibration monitoring systems on the Boeing Next Generation 737-600/-700/-800 family of aircraft. Units have also been approved for use on the Tupolev Tu-204-120 aircraft with Rolls-Royce RB211-535B4 engines.

Contractor

Vibro-Meter SA.

VERIFIED

TURKEY

CDU-900 flight management system - Control Display Unit

The ASELSAN CDU-900 is a cockpit control and display unit for fixed-wing and rotary-wing applications. The CDU-900 provides processing and interface control for all flight management functions, including navigation guidance, flight management, communication and navigation systems management, status monitoring and Built-In-Test (BIT) capabilities.

The CDU-900 serves as the primary avionics computer for communications control, navigation/guidance, equipment status monitoring and MIL-STD-1553 data bus control. The system provides the main aircrew interface for flight management and INS or GPS/INS navigation/management in a reliable, low risk/cost design.

CDU-900 operations are executed using a full alphanumeric keypad, arrow keys, function keys and eight line select keys on the front panel of the unit. The CDU-900 performs guidance computations, data management and subsystem control functions in the FMS-800 system.

Specifications

Display: 220 × 170 pixels, NVIS compatible
Processor: Intel 80486 DX-2, 50 MHz
Dimensions: 181 × 146 × 168 mm (7.125 × 5.75 × 6.6 in)
Keyboard: full alphanumeric with 7 generic key functions
Cooling: convection
Interface: Dual MIL-STD-1553B databus, RS-232, ARINC 429, discretes
Weight: 4.5 kg (10 lb)
Power: 28 V DC, 34 W (max)
Compliance: DO-160C, MIL-STD-810D, MIL-STD-461D
MTBF: 5,800 h

Contractor

Aselsan Inc, Microelectronics, Guidance and Electro-Optics Division.

NEW ENTRY

UNITED KINGDOM

FV-0100 and FV-0300 FlightVu external airborne video cameras

FlightVu FV-0100 and FV-0300 cameras are designed to be fitted to the exterior of commercial aircraft to provide flight-deck crew with an exterior view of the aircraft and its surroundings, both in flight and on the ground. Its role is to enhance safety and security, providing the crew with the opportunity to assess security incidents, to assist them in ground handling or to check for exterior damage, fire or other occurrences. Used with the AD Aerospace FlightVu Video Data Recorder (FVDR), it can also provide instant playback of any hazardous incident to aid evaluation of actions.

The video picture can also be made available to the passengers, as part of the in-flight entertainment, throughout all phases of flight. FlightVu camera data can also be integrated with moving map displays and other in-flight entertainment or advertising.

The cameras are hermetically sealed and available for simple fitting virtually anywhere around the exterior of the aircraft. The low-profile, aerodynamically shaped optical window ensures viewing through all weather conditions. Automatic de-icing is provided, together with nitrogen purging for mist-free operation.

Monochrome or colour options are available.

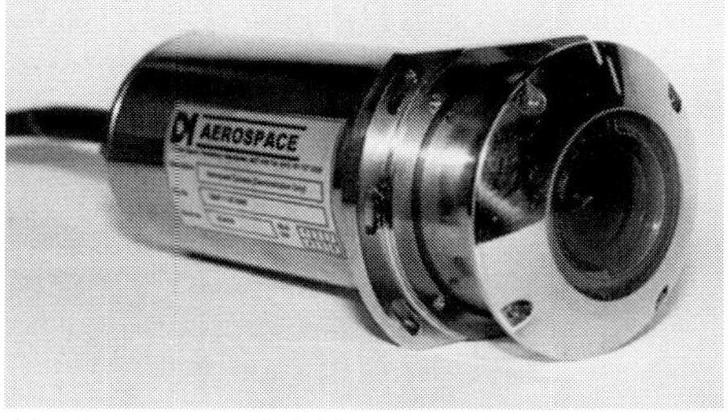

AD Aerospace FlightVu external airborne video camera ***1999***/0051008

Specifications

Camera:
Field-of-view: 85 × 69° maximum; 14 × 11° minimum
Sensor: ½ in CCD technology
Dimensions: 129.8 mm long, 57.9 mm diameter
Weight: 0.8 kg
Power: 28 V DC, 3.5 W
Altitude: to 50,000 ft
Temperature: –55 to +40° C

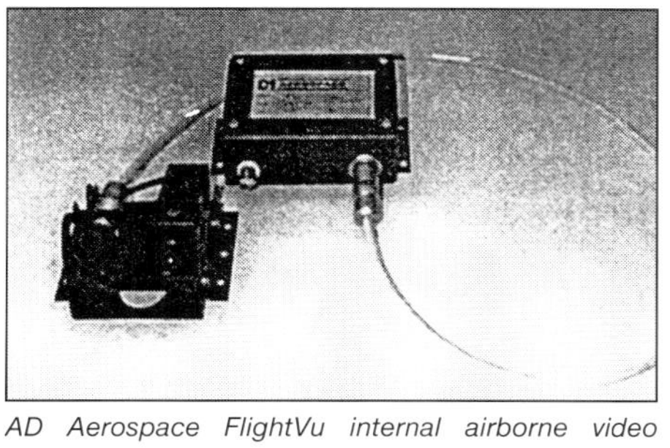

AD Aerospace FlightVu internal airborne video camera ***1999***/0051009

FlightVu video data recorder:
Dimensions: standard ARINC 600 4MCU
Power: 115 V AC, 400 Hz or 28 V DC to RTCA Do160 specification

Contractor

AD Aerospace Ltd.

VERIFIED

FV-0210 FlightVu internal airborne video camera

The FV-0210 internal airborne video camera is designed for use by general aviation to fulfil a variety of possible tasks such as instrument monitoring, crew observation, crew training, security and entertainment. It can be installed where required in the cockpit. Both monochrome and colour versions are available.

Specifications

Field of view: 73 × 56°
Sensor: ¼ in CCD technology
Dimensions:
camera: 80 × 80 × 25 mm
power supply unit: 120 × 80 × 35 mm
Weight:
camera: 0.15 kg
power supply unit: 0.25 kg
Power: 28 or 12 V DC, 2 W
Altitude: to 15,000 ft
Temperature: –15 to +55° C

Contractor

AD Aerospace Ltd.

VERIFIED

Autonomous Landing Guidance (ALG)

The ALG system is an onboard imaging radar giving civil or military aircraft the ability to land and taxi in low-visibility conditions without the need for other Category II or III automatic landing systems.

Contractor

Alenia Marconi Systems.

VERIFIED

Active noise control

BAE Systems has developed an active noise control system which lowers the noise within the aircraft cabin to provide a reduction in the fatigue level for the cabin staff and an improvement in the comfort level for the passengers. Primarily targeted at turboprop aircraft, it can be fitted to any aircraft where tonal noise is considered to be intrusive.

All major tonal noise sources present within a particular spectrum are identified and monitored via an array of microphones discretely placed within the aircraft trim and connected to a central controller. Antiphase signals are then calculated using powerful Digital Signal Processors (DSPs), which then control the tonal noise present in the cabin via a set of loudspeakers also discretely placed in the aircraft trim.

Flight trials have taken place on a number of aircraft including the Jetstream 41, for which the system has now been selected. The system can either be specified as an initial fit or retrofitted.

Trials have also been conducted on the Lockheed Martin C-130 military transport where significant reductions in noise level were demonstrated in both the cockpit and cargo bay areas.

Operational status

System developed to production standard and selected for Jetstream 41.

Contractor

BAE Systems.

VERIFIED

BAE Systems active noise control 0001409

AD120 VHF/AM radio

Originally designed for civil aviation applications by the King Radio Corporation in the USA, BAE Systems has requalified this VHF/AM system for military roles and, manufacturing under licence, markets the system under the designation AD120. It is installed in most UK military fixed-wing aircraft and helicopters and has also been supplied to a number of overseas customers.

Covering the frequency band 108 to 137 MHz, the AD120 provides channel spacing of 25 kHz. Services provided are double sideband AM voice communication. Power output can be varied between 10 and 20 W. The system is an all-solid-state design of modular construction and is designed for easy installation and maintenance. It is also exceptionally simple to operate. The system's standard remote controller possesses only five controls: an on-off switch, volume control, a test button and two rotary switches for frequency selection. Tuning is instantaneous. Automatic squelch and gain control eliminate the need for manual adjustment.

The self-test facility may be used during operation as a confidence check.

Specifications

Dimensions:
(transmitter/receiver) 60 × 127 × 315 mm
(controller) 146 × 47 × 90 mm

Weight:
(transmitter/receiver) 2.3 kg
(controller) 0.6 kg

Operational status

In production. More than 1,200 systems are currently in service.

Contractor

BAE Systems.

VERIFIED

AD660 Doppler velocity sensor

The AD660 Doppler sensor is packed into a single unit containing the antenna, transmitter/receiver, tracker, and digital and analogue inputs and outputs, all under microprocessor control. The transmitter is a Gunn diode generating about 200 mW of power at 13.325 GHz. The signal is frequency modulated, the actual frequency being selected by the microprocessors to give the best return signal in any set of conditions, eliminating for example height hole effects. The system employs four beams, any three of which provide satisfactory performance, the fourth being used to enhance operation at extremes of altitude and to provide a self-check function. The standard output for groundspeed and drift angle is an ARINC 429 serial digital data highway, but ARINC 582 and 561 digital and ARINC 407 synchro outputs can also be provided.

Specifications

Dimensions: 379 × 237 × 132 mm

Operational status

No longer in production. The AD660 was an option on Boeing 727 and 737 air transport aircraft, adopted by British Airways and Lufthansa among other operators. It was selected by CASA for its C-101DD Aviojet advanced trainer/ground attack aircraft.

Contractor

BAE Systems.

UPDATED

CONTRAN® - VHF radio anti-blocking system

CONTRAN® is a dual-function avionics system, the primary purpose of which is to prevent simultaneous or conflicting VHF radio transmissions between aircraft and Air Traffic Control.

CONTRAN® operates by monitoring any transmissions which are in progress on a specific frequency as a pilot tries to initiate a transmission. It will then, if necessary, inhibit the attempted transmission and make the pilot aware that it has done so. The pilot can then retransmit after the preceding transmission has been completed.

The unit's secondary purpose is to guard against unintentional transmissions by alerting the flight crew to an onboard 'stuck microphone' which is then automatically terminated if appropriate.

The BAE Systems AD120 VHF radio

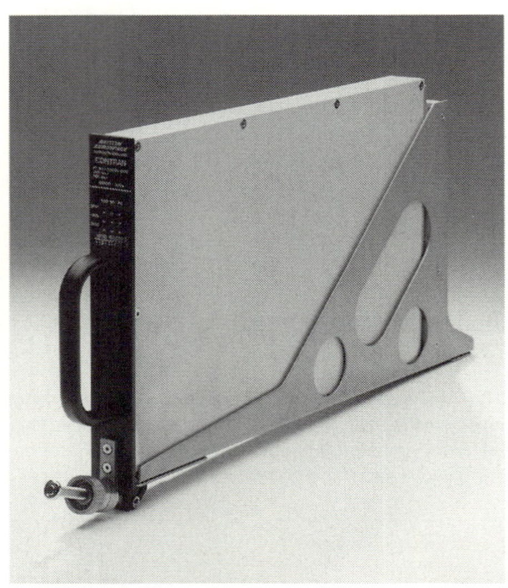

CONTRAN-VHF radio system 0001179

Specifications

Standards: FAA TSO-C122, TSO-C128; RTCA DO207, DO209; Eurocae ED67, ED68
Size: 1 MCU Standard ARINC 600 Mounting
Weight: <1.9 kg

Operational status

CONTRAN® was first ordered by Britannia Airways Ltd for its entire fleet of Boeing 757/767 aircraft. Deliveries commenced in 1996.

Contractor

BAE Systems.

VERIFIED

Heli-Tele television system for helicopters

The Heli-Tele is a broadcast-standard television surveillance system designed for mounting on helicopters. The system provides long-range real-time airborne surveillance to meet the requirements of police, military, paramilitary, civil and other security forces.

Heli-Tele consists of a colour camera, with a high-magnification zoom lens, mounted on a gyrostabilised, steerable, platform. The operator can steer the camera and adjust the magnification to display any selected area of the ground scene. Video information is transmitted to any number of ground stations via an air-to-ground microwave link with a range of over 90 km. For night or poor visibility surveillance, a standard, modular, thermal imaging sensor package may be installed in place of the TV camera payload, with a turnaround time of only 30 minutes.

Operational status

In service. The Heli-Tele has been certified by the UK's Civil Aviation Authority and is fitted to many types of helicopter.

Contractor

BAE Systems.

VERIFIED

Primary flight computer for Boeing 777 0001415

Primary Flight Computers (PFCs) for the Boeing 777

The three Primary Flight Computers (PFCs) form the core element of the Fly-By-Wire (FBW) system for the Boeing 777. The complete Primary Flight Control System (PFCS) provides control in all three axes together with protection and compensation facilities.

The FBW system provides full time control in pitch, roll and yaw axes with pilot commands being input to the PFCs via a triplex ARINC 629 bus and Actuator Control Electronics (ACE) units. Pilot commands are provided from conventional control column and pedal inputs via multiple position transducers to the ACEs which convert the signals into digital form for transmission to the PFCs.

The PFCs compute the pilot commands for control of a single tabbed rudder surface, two elevators, a single all moving tailplane, an outboard aileron, a flaperon and seven spoiler surfaces on each wing. These computed commands are then routed via the ACEs to the appropriate electrohydraulic actuator. The computation also implements the C* control laws, together with the protection and compensation facilities for ease of crew workload.

All three primary flight computers are identical and contain three dissimilar computing lanes. The secondary redundancy management of these lanes allows fault tolerant operation in the event of failures, Within each primary flight computer, each lane uses a different 32-bit microprocessor. Three different compilers are used to convert the Ada high order software design into the microprocessor's object code. This level of dissimilarity provides protection against residual compiler and processor errors.

Extensive use has been made of ASICs ensuring a low component count and high reliability. This high reliability, coupled with the fault tolerant architecture, provides benefits to the operator through reduced life cycle costs, improved maintainability and increased system availability.

Operational status

In service on the Boeing 777 aircraft.

Contractor

BAE Systems.

VERIFIED

ADELT CPT-600/609 Automatically Deployed Emergency Locator Transmitter

The basic ADELT CPT-600 system comprises the beacon, carrier and ejection mechanism, deployment battery and control panel. The beacon is contained within the carrier and mounted externally at a suitable location. The control panel provides for testing of system integrity, system arming, crew activation, and deployment confirmation. Various remote activation devices may be incorporated in the installation design, such as frangible, float, inertia, saline, or hydrostatic switches. At least one sensor should couple with the deployment battery to avoid dependency on the aircraft electrical supply. When operated, whether by crew or remote sensor activation, a small cartridge triggers the release of two coiled springs, which eject the beacon safely away from the aircraft. Once upright and in the water, the beacon will automatically transmit homing signals on 121.5/243 MHz, while the transponder reacts to aircraft or ship's radar, pinpointing its position on the search vessel's radar screen.

A derivative of the ADELT CPT-600, the CPT-609 is comprised of a homing transmitter on 121.5/243 MHz, radar transponder on 9 GHz compatible with aircraft or ship's radar, and satellite transmitter operating on 406 MHz.

The beacon is programmed with the country code and the aircraft registration marking or radio call sign. When the beacon is activated by deployment into water, the satellite transmitter transmits its programmed information as a burst of coded signals to the orbiting COSPAS/SARSAT satellites receiving on 406 MHz. The message is stored by the satellite and downloaded to the nearest local user terminal ground station. Here the signal is processed to obtain latitude and longitude of the aircraft in distress and its identity. This in turn is passed to a mission control centre which routes the information to the rescue co-ordination centre nearest the incident, from where search and rescue forces will be sent. Accuracy of the 406 MHz signal is to within 1 to 2 km, although testing has proved accuracies of 0.2 km can be achieved.

Once in the general vicinity of the beacon, SAR teams can locate the scene with the aid of the 121.5/243 MHz homing transmitter and the 9 GHz radar transponder which will guide SAR forces to within 10 m of the beacon.

The ADELT CPT-600

The ADELT CPT-609 can be field programmable, greatly reducing the cost and impact of changing aircraft identity. The CPT-609 can be reprogrammed at remote locations without the need for costly and inconvenient disassembly for replacement of EPROMS.

Contractor

Caledonian Airborne Systems Ltd.

VERIFIED

CPT 110 Course Deviation Indicator (CDI) system

Caledonian Airborne Systems Ltd supply the CPT 110 CDI as a flat panel, fitted in the location of the aircraft registration, as part of its B-RNAV installation, together with a customised installation for the KLN 900 radar, using ARINC 429 interface.

Contractor

Caledonian Airborne Systems Ltd.

VERIFIED

The CPT 110 CDI alongside the registration
1999/0051669

Scorpio 2000 series datalink systems

Scorpio 2000 is one of a series of expandable datalink systems which accept host aircraft and designated target global position data from the radar/navigation system databus and translates, encodes and up-converts them for transmission via radio link to ground station.

Typically, the radar used with the system is the AlliedSignal RDR-1500B, and the message data can be secured to meet user requirements. Options include: GPS internal (or external DGPS) interfaces, flat panel display and keyboard control.

Specifications

Radio link: VHF, power to suit requirements
Data interface/speed: ARINC 419, 9,600 baud
Weight: 2.4 kg
Power: 28 V DC

Scorpio 2000 series datalink, with AlliedSignal RDR-1500B radar display 0011835

Contractor

Caledonian Airborne Systems Ltd.

VERIFIED

Track While Scan (TWS) system and airborne radar indicator unit

Track-While-Scan (TWS) System

The Caledonian Airborne Systems Ltd TWS system interfaces with the AlliedSignal RDR-1500B radar and navigation systems to enable on-screen target track and marker placement. TWS, radar and Scorpio 2000 datalink system control is by point-and-click using the joystick control. The system is capable of tracking up to 20 independent targets, providing information on present position (PP), ground speed and heading. Aircraft, target and marker data are overlaid on the radar indicator unit; operator prompts and messages are presented on the lower screen. In addition, the TWS system has the capability to insert 50 user-defined geostationary markers and store 100 screen images.

Airborne Radar Indicator Unit

The Caledonian Airborne Systems Ltd airborne radar indicator unit is a direct solid-state replacement for existing CRT-based indicators. It handles a variety of radar, composite and VGA video standards and provides a high-resolution flicker-free display. The 10.4 in multifunction display depicted serves as a direct replacement for the AlliedSignal IN-1502A radar indicator, and as the display for the Caledonian Airborne Systems Ltd Track-While-Scan (TWS) system. The design can, however, be tailored to suit displays from 4 to 16 in sizes and a number of video standards. The radar display can also be used as a multifunction display for other sensor data, in which application it is able to provide map overlay and to operate in night vision goggles conditions.

Specifications

Display resolution:
800 × 600 pixels (SVGA) in a 264 mm diagonal
upgrade path to: 1024 × 768 pixels (XVGA)
Video standards:
RGB computer graphics (up to SVGA) with separate H & V syncs
composite and S-video (Y/C)
CCIR/PAL & RS-170/NTSC scaled or native formats
upgrade path to : computer graphics (up to XVGA) and RDR-1500B radar video
Colours: 262,144
Luminance:
1000 Cd/m^2
upgrade path to: 1500 Cd/m^2
Keypad: 8 keys including 5 programmable
Interface: RS422
Power: 28 V DC at 105 W; 5V DC at 1 W
Weight: 5 kg
Dimensions: 268 × 221 × 131 mm

Contractor

Caledonian Airborne Systems Ltd.

UPDATED

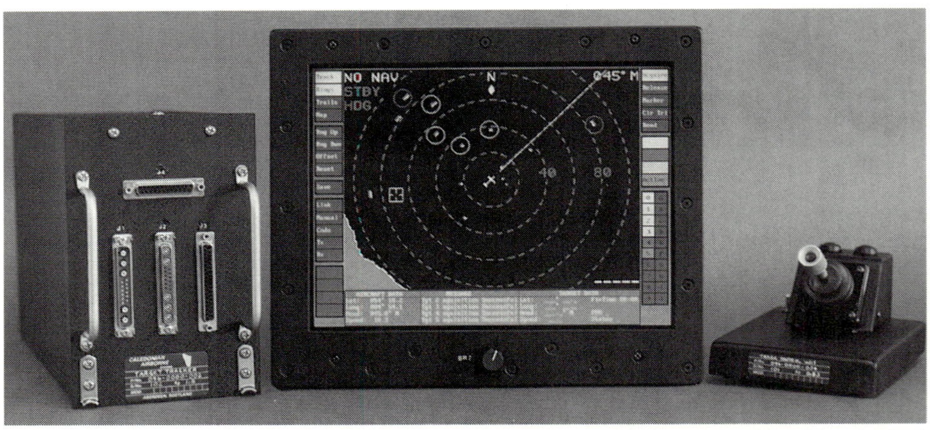

Track-while-scan system processor (left), digital solid-state radar indicator unit (centre) and joystick control unit (right)
0018195

805 series UHF transceivers

The 805 series is a fully synthesised UHF transceiver covering the frequency range 225 to 399.975 MHz. The state-of-the-art design provides both AM and FM communications, together with a dedicated guard receiver to monitor the 243 MHz distress frequency. When used with the Chelton 715-7 control unit and a suitable UHF antenna, such as the Chelton type 16-1 or any other Chelton multiband antenna, the 805-1 offers a complete AM/FM communications system suitable for connection to the aircraft audio system. The 805-2 variant provides a datalink capability.

The use of software variables in the design permits the implementation of user-specific requirements through simple reprogramming. This ability, coupled with the inbuilt growth facilities, enables the design to be readily configured for a wide variety of applications addressing interfaces such as microphone sensitivity and audio output levels. These variations are indicated by a suffix to the transceiver part number. A typical system comprises the following units: 805-1 UHF transceiver; 715-7 control unit; 715-7S slave control unit; 27601 mounting tray; optional remote fill gun.

The 715-7 control unit provides control of the transceiver's operating mode and frequency. Selection of the operating frequency may be either by direct entry of the desired frequency or by recalling a stored channel. Up to 100 frequencies can be stored in the transceiver. Each stored channel can be assigned a separate transmit and receive frequency for half-duplex operation and compatibility with satellite communication systems. Alternatively the transceiver can be controlled by a simple command set supplied via an RS-422 databus.

For users who wish to change the stored frequencies on a regular basis a remote fill gun is available. A computer program running on a PC-compatible computer permits the desired frequencies to be assigned to any channel number. The database is then downloaded to the 715-7 controller in a few seconds via an infra-red link using the remote fill gun. This enables a complete fleet of aircraft to be updated in a matter of minutes without the need to remove any equipment from the aircraft.

For installations requiring dual-control units, a slave control unit, the 715-7S, is offered. When the main control unit is in control, the slave control units acts as a remote readout unit. Control of the system can be passed from the main control unit to the slave control unit and the main control unit will then act as the remote readout unit. Control units can be provided with lighting

Chelton 805 series UHF transceivers

options, including NVG compatibility, to match cockpit specifications.

The datalink communications transceiver variant, 805-2, is rated to provide a 100 per cent transmit duty cycle.

805 series transceivers are compatible with Chelton 930 Series direction-finders.

Specifications

Transmitter
Power output: 10 W nominal
Frequency response:
(voice) 300 Hz to 3 kHz at +3 dB
(data) 100 Hz to 10 kHz at +3 dB
Sidetone level: 5.0 Vrms into 150 ohms

Main receiver
Sensitivity: <3.0 µV for 10 dB (S+N)/N at 30% AM
<1.5 µV for 10 dB SINAD at +3 kHz FM
Audio response:
(AM voice) 300 Hz to 3kHz at +3 dB
(FM voice) 300 Hz to 3 kHz at +3 dB
(adf) 100 Hz to 6 kHz at +3 dB
(AM data) 100 Hz to 10 kHz at +3 dB
(FM data) 100 Hz to 10 kHz at +3 dB

Guard receiver
Sensitivity: <3.0 V for 10 dB (S+N)/N at 30% AM
<1.5 V for 10 dB SINAD at +3 kHz FM

Contractor

Chelton (Electrostatics) Ltd.

VERIFIED

905 series VHF transceivers with integral guard receiver

The 905-2 is a fully synthesised VHF air and marine band transceiver covering the frequency range 118.000 to 173.975 MHz.

The design provides both AM and FM communications, together with a dedicated guard receiver to monitor the 121.5 MHz distress frequency. When used with the Chelton 715-25 control unit and a suitable VHF antenna such as the Chelton type 21-38-3 or any other Chelton multiband antenna, the 905-2 offers a complete AM/FM communications system suitable for connection to the aircraft audio system. The equipment has been designed to operate with DSC controllers.

Automatic Test Equipment (ATE) employed during the test of the equipment enables calibration data to be stored inside the transceiver ensuring performance can be maintained throughout its life.

The use of software variables in the design permits the implementation of user specific requirements through simple reprogramming. This ability, coupled with the inbuilt growth facilities, enables the design to be readily configured for a wide variety of applications addressing interfaces such as microphone sensitivity and audio output levels. These variations are indicated by a suffix to the transceiver part number. A typical system comprises the following units:

905-2 VHF transceiver
715-25 control unit
21-38-3 antenna
27601 mounting tray

The control unit provides control of the transceiver's operating mode and frequency. Selection of the operating frequency may be either by direct entry of the desired frequency or by recalling a stored channel. Each stored channel can be assigned a separate transmit and receive frequency for half-duplex operation. Alternatively, the transceiver can be controlled by a command set supplied via an RS-422 databus.

Chelton's 905 series VHF transceivers ***1999***/0044805

For installations requiring dual control units, a slave control unit, the 715-25S type, is available. When the main control unit is in control the slave control unit acts as a remote readout unit. Control of the system can be passed from the main control unit to the slave control unit and the main control unit will then act as the remote readout unit. Control units can be provided with lighting options, including NVG compatibility, to match cockpit specifications.

905 series transceivers are compatible with Chelton 930 series direction-finders.

Specifications

Transmitter
Power output: 16 W AM/25 W FM nominal
Transmit duty cycle: 25%
Frequency response:
Voice: 300 Hz to 2.5 kHz at –3 dB
Data: 100 Hz to 10 kHz at –3 dB

Main receiver
Sensitivity:
<3.0 µV for 10 dB (S + N)/N at 30% AM
<1.5 µV for 10 dB SINAD at ±3 kHz FM
Selectivity:
narrowband > ±8.0 kHz at –6 dB
>60 dB at ±25 kHz
wideband > ±15.0 kHz at –6 dB
>60 dB at ±50 kHz
Audio Response:
AM voice 350 Hz to 2.5 kHz at ≤ –6 dB
FM voice 350 Hz to 2.5 kHz at ≤ –6 dB
ADF 100 Hz to 6 kHz at ≤ –3 dB
AM data i/p 100 Hz to 10 kHz at ≤ –3 dB
FM data i/p 100 Hz to 10 kHz at ≤ –3 dB

Guard receiver
Sensitivity:
<3.0 µV for 10 dB (S + N)/N at 30% AM
<1.5 µV for 10 dB SINAD at ±3 kHz FM
Selectivity:
> ±15.0 kHz at –6 dB
>60 dB at ±50 kHz
Audio Response: 350 Hz to 2.5 kHz at ≤ –6 dB

Contractor

Chelton (Electrostatics) Ltd.

VERIFIED

915 series VTAC transceivers

The 915-1 is a fully synthesised VTAC transceiver covering the frequency range 30.000 to 87.975 MHz.

The design provides FM communications, together with a dedicated AM detector to provide a demodulated output for an associated direction-finding or homing system.

When used with a Chelton 715-22 control unit and a suitable VTAC antenna, such as the Chelton 12-227 pin diode antenna and 7-915 pin 227 logic unit, the 915-1 offers a complete FM communications system suitable for connection to the aircraft audio system.

Automatic Test Equipment (ATE) employed during the test of the equipment enables calibration data to be stored inside the transceiver ensuring performance can be maintained throughout its life.

The use of software variables in the design permits the implementation of user specific requirements through simple reprogramming. This ability, coupled with the inbuilt growth facilities enables the design to be readily configured for a wide variety of applications addressing interfaces such as microphone sensitivity and audio output levels. These variations are indicated by a suffix to the transceiver part number. A typical system comprises the following units:

915-1 VTAC transceiver
715-22 control unit
12-227 antenna
7-915 pin 227 logic unit
27601 mounting tray
Optional remote fill gun.

Chelton's 915 series VTAC transceivers 30-88 MHz ***1999***/0044806

The 715-22 control unit provides control of the transceiver's operating mode and frequency. Selection of the operating frequency may be either by direct entry of the desired frequency or by recalling a stored channel. Up to 100 frequencies can be stored in the transceiver. Each stored channel can be assigned a separate transmit and receive frequency for half-duplex operation. Alternatively, the transceiver can be controlled by a command set supplied via an RS-422 databus.

For users who wish to change the stored frequencies on a regular basis, a remote fill gun is available. A computer program running on a PC-compatible computer permits the desired frequencies to be assigned to any channel number. The database is then downloaded to the 715-22 controller in a few seconds via an infra-red link using the remote fill gun. This enables a complete fleet of aircraft to be updated in a matter of minutes without the need to remove any equipment from the aircraft.

For installations requiring dual control units, a slave control unit, the 715-25S type, is available. When the main control unit is in control the slave control units act as a remote readout unit. Control of the system can be passed from the main control unit to the slave control unit and the main control unit will then act as the remote readout unit. Control units can be provided with lighting options, including NVG compatibility, to match cockpit specifications.

915 series transceivers are compatible with Chelton 930 series direction-finders.

Specifications

Transmitter

Power output: 15 W nominal
Transmit duty cycle: 25%
Frequency response:
voice: 300 Hz to 2.7 kHz at –3 dB
data: 100 Hz to 10 kHz at –3 dB

Main receiver

Sensitivity:
<1.0 μV for 10 dB SINAD at ±3 kHz FM
<3.0 μV for 10 dB (S + N)/N at 30% AM
Selectivity:
narrowband > ±7.5 kHz at –6 dB
>60 dB at ±25 kHz
wideband > ±15.0 kHz at –6 dB
>60 dB at ±50 kHz
Audio Response:
FM voice 300 Hz to 2.7 kHz at –3 dB
ADF 100 Hz to 6 kHz at –3 dB
AM data 100Hz to 10 kHz at –3 dB

Contractor

Chelton (Electrostatics) Ltd.

VERIFIED

930 series direction-finding system

The 930 series V/UHF direction-finding antennas enable a standard AM communications receiver to be converted into a 360° direction-finding system. These antennas are suitable for installation on fixed- or rotary-wing aircraft as well as maritime, land-mobile and ground installations. With the adaptor plate, P/N 23468, the 931-1 antenna is essentially a drop-in replacement for the DF301E direction-finder. The 930-1 is a replacement for the AN18 direction-finder.

A typical system interconnection utilises an existing transmitter/receiver. The RF changeover relay enables the transmitter/receiver to be switched between the direction-finding antenna and the normal communications antenna. The action of the direction-finding antenna is to modulate the received RF signal, the sense and depth of this modulation being related to the direction of arrival of the received signal. The receiver demodulates the modulation and passes the resulting audio back to the direction-finding antenna for processing. After processing in the antenna unit, the resulting bearing is available on an ARINC 407 output to drive a synchro indicator. An optional ARINC 429 or RS-422 databus output capable of driving an electronic display can also be specified.

The System 930 has been specifically designed to be compatible with direction-finding on pulsed tone personal locator beams (PLBs), including SARSAT beacons, when used with a SARSAT-compatible receiver such as Chelton 7-28-406.

Operational status

930 series of direction-finding antennas has already been installed on the following platforms: Agusta AB412; BAE Systems (Operations) Limited Nimrod; Bell 212, 214ST, 412SP and CH146; Cessna 416 and Caravan; DHC Dash 8; ECD UH1D; ECF Puma and Super Puma; Learjet; Britten Norman Islander; PZL Mielec AN28; Shrike Commander; and Westland Lynx.

Contractor

Chelton (Electrostatics) Ltd.

VERIFIED

Series 7-202 HF airborne receiving system

The Series 7-202 HF receiving system comprises a Type 7-202-1 HF receiver and Type 19-197 active antenna. The system is designed to provide worldwide reception of transmissions.

The Type 7-202-1 is a fully synthesised HF receiver capable of receiving transmissions anywhere in the long, medium or short wavebands. The Type 19-197 is a high-efficiency low-noise active antenna providing continuous coverage of the long, medium and short wave and VHF/FM broadcast bands. The antenna is protected against low-level lightning strikes.

The receiving system is fully automatic in operation and, once the desired service has been selected, it will automatically search out and tune to the best available transmission anywhere in the world. A database programmed into the receiver contains details of the coverage areas of all the available transmitters together with their frequencies and times of transmissions. The receiver, by interrogating the aircraft's navigational computer, obtains latitude and longitude, together with time of day. Using this information, the receiver tunes through available transmissions and, using its inbuilt signal quality measuring circuits and algorithm, determines which transmission offers the best quality and tunes the receiver to the appropriate frequency. The resulting audio is then output for distribution over the aircraft audio system. The inbuilt microprocessor continuously monitors the quality of the received signal and, should the quality deteriorate for any reason, the radio will automatically retune to any better quality transmission which may be available.

The receiver may be programmed with the worldwide coverage information for two different services which are selected by a single control line. Programming of the receiver's database to define the transmissions available, together with their broadcast times and frequencies, is accomplished with a user-friendly database program incorporating a graphical user interface. This program runs on an IBM PC and, once completed, may be downloaded directly into the receiver, enabling any operator to adapt the receiver to meet the specific needs of passengers and routes. A trace program built into the receiver records the selected frequencies and received signal strengths, thus enabling fine tuning of the database to ensure optimum performance.

Specifications

Weight:
(antenna) 2.5 kg
(receiver) 1.8 kg
Power supply: 28 V DC, 500 mA max
Frequency: 150 kHz to 108 MHz
Channel spacing: 5 kHz and 9 kHz

Contractor

Chelton (Electrostatics) Ltd.

VERIFIED

VLT 15-10-6MK3/VLR 15-16-1 helicopter video downlink

The video downlink has been designed for use as a helicopter-to-ground system, either for surveillance or outside broadcast use. The VLT 15-10-6MK3 transmitter is a multichannel 10 W unit. Any 16 frequencies within a 180 MHz bandwidth may be pre-programmed within the band 2.2 to 2.6 GHz. The VLR 15-16-1 is its matching 16 channel receiver.

A 3.5 GHz version is also available and the 2.5 GHz unit may be upgraded to 3.5 GHz standard.

Operational status

ECS helicopter video downlink systems are approved by the UK Home Office and Civil Aviation Authority; they are in use by some 20 UK police forces.

Contractor

ECS Enterprise Control Systems Limited.

VERIFIED

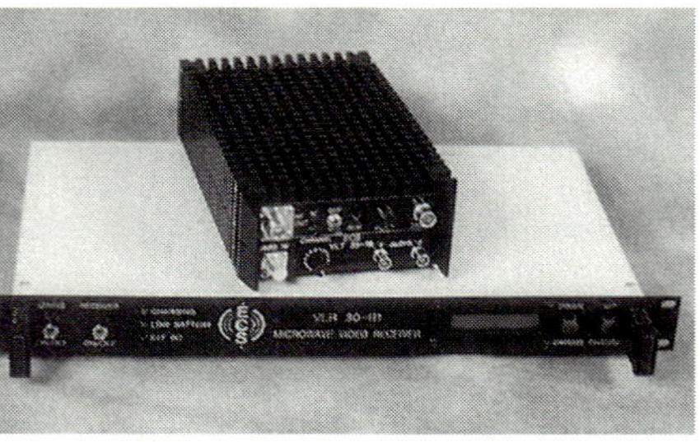

VLT 15-10-6 MK3/VLR 15-16-1 helicopter video downlink 0011834

Angle of attack system

The angle of attack system is designed to measure and display angular airflow direction relative to a fixed datum, such as the horizontal fuselage datum.

The system consists of the FTRP100 Airstream Direction Detector (ADD) and an Angle of Attack Indicator (AAI).

The FTRP100 ADD utilises non-contact inductive resolvers to sense angular position of the detector probe relative to the fixed datum.

The detector probe is equipped with internal heaters for anti-icing and moisture control plus drainage at the base to prevent water ingress. The aerodynamic design and performance creates positive pressure at the input slots, thus preventing sand and dust entering the detector probe tube.

The rear housing contains either one or two non-contact inductive resolvers which sense the position of the detector probe and can be specified to output information in either digital or analogue format.

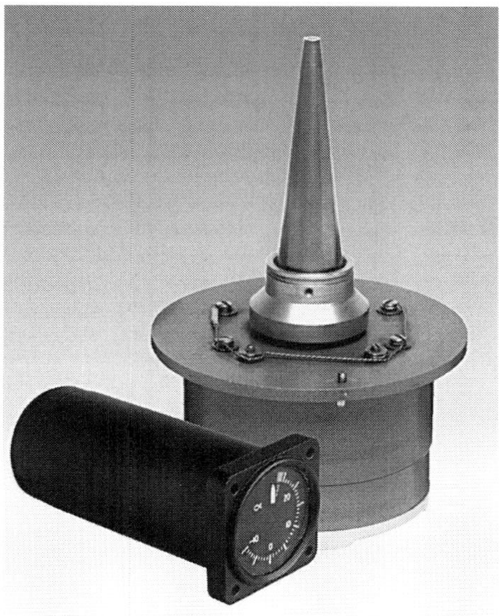

Ferranti Technologies' angle of attack system
0018223

The AAI is an easy to read indicator displaying relative angular airflow in increments of 1° from –5 to +25°C. Provision can be made for markers on the display to highlight optimum angle of attack and stall conditions.

Contractor

Ferranti Technologies Limited.

VERIFIED

Attitude indicators - FH series

Attitude indicators of the FH series are electrically driven gyroscopic instruments which display aircraft attitude in two axes by a spherical type presentation.

The product range encompasses instruments with 2, 3, and 4 in displays, all with a variety of colour and lighting options and a choice of AC or DC electrical input. An integral slip indicator may also be specified.

All instruments are hermetically sealed and feature alternative panel angle options, manual caging and automatic gyro control during accelerated flight.

Operational status

In production for fixed- and rotary-wing aircraft for both military and commercial applications. FH series instruments are specified as primary or standby instrumentation in many aircraft including Jaguar, Tornado, Nimrod, Hawk, Chinook, Super Lynx, Bell 412 and Eurocopter AS 350.

Contractor

Ferranti Technologies Limited.

VERIFIED

The Ferranti Technologies' FTS20 series turn and slip indicators **2000**/0062850

Turn and slip indicators - FTS20 series

The Ferranti Technologies' FTS20 series turn and slip indicators are single degree of freedom gyroscopes, that operate from a 28 V DC nominal supply.

The indicators are housed in a 56 mm diameter case with a 60 mm square mounting flange. The instrument is available with optional integral lighting, which can be night vision goggles compatible.

The rate of turn pointer is driven by a rate gyroscope, which is powered by an alternating current supply. Aircraft slip is indicated by a ball-in-tube inclinometer. The alternating current for the rate gyroscope is derived from an integral inverter circuit. The direct current supply is monitored by a warning flag, which operates when voltage drops below 14 V. Versions are available which provide voltage outputs for differing rates of turn.

Specifications

Scale reading: zero; rate 1 (180°/min); rate 2 (360°/min)
Scale colours: white on black
Altitude rating: 10,700 m (35,000 ft)
Dimensions:
Case diameter: 56 mm
Flange size: 60 mm square
Case length (excluding connector); 148 - 172 mm
Mass: 0.7 - 0.75 kg

Contractor

Ferranti Technologies Limited.

VERIFIED

Ferranti Technologies' attitude indicators FH22, FH30, FH32 and FH40 (left to right) 0018193

Engine displays

Engine Display Units (EDU)

Two redundant EDUs display all necessary information for the safe operation of turbine engines, including fuel quantity and usage calculations. Engine parameters are normally displayed in both analogue and digital format using colour to highlight exceedance conditions. Displayed parameters include: Torque (TQ); Inter Turbine Temperature (ITT); power turbine rpm (NP); gas generator rpm per cent of maximum (NG); VACuum suction (VAC); oil temperature/pressure; fuel quantity; Fuel Flow (FF); Fuel At Destination (FAD); Time To Destination (TTD); and Outside Air Temperature (OAT). Each EDU also has a selectable reversionary mode where all the data from both screens is compressed and displayed on one screen.

Meggitt Avionics Light Off Detectors **1999**/0051673

Data Acquisition Unit (DAU)

Meggitt Avionics produces a range of DAUs to acquire engine and fuel data, which is transmitted to cockpit displays via ARINC 429 databusses.

Sterling indicators

The Sterling range of general purpose 2 in indicators use interchangeable personality modules to interface with a wide range of sensors. High reliability, accuracy and low maintenance is guaranteed by a single moving

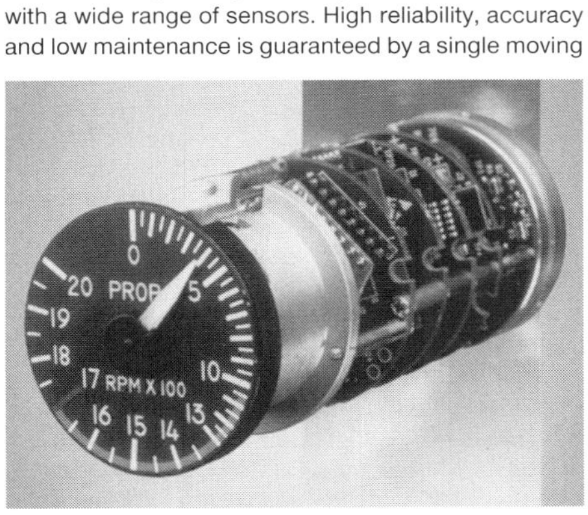

Meggitt Avionics Sterling indicator **1999**/0051672

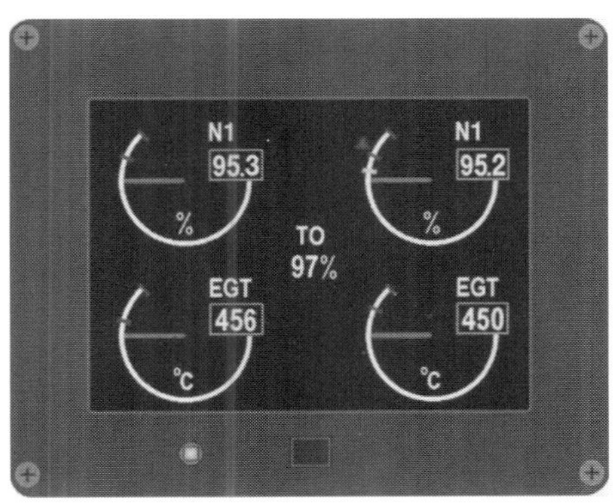

Meggitt Avionics Engine Display Units
1999/(top) 0051670/(above) 0051671

part design. The pointer is mounted directly to a precision stepper motor shaft. Use of a microprocessor allows customised dial faces and non-volatile storage of up to 45 hours of flight data. A solid-state digital display can be added to the dial face.

Light Off Detectors (LOD)

For monitoring afterburner ignition on military aircraft, the LOD uses an ultra-violet sensitive Geiger-Muller tube that views the afterburner flame through a port in its liner. Critical attributes of this sensor are immunity to sunlight and rapid detection times despite exposure to afterburner temperatures and pressures.

Contractor

Meggitt Avionics.

VERIFIED

Flight displays

Primary Flight Display (PFD)

The PFD displays all the information that is needed to fly the aircraft. It receives data from various external sensors in order to display pitch and roll attitude, rate of climb, airspeed, heading, ILS (loc/GS), side slip and barometric setting. If the PFD fails, all this data will be presented on the navigation display.

Navigation Display (ND)

The ND displays all of the required navigation modes; compass rose (HSI), MAP and ARC. Data is received

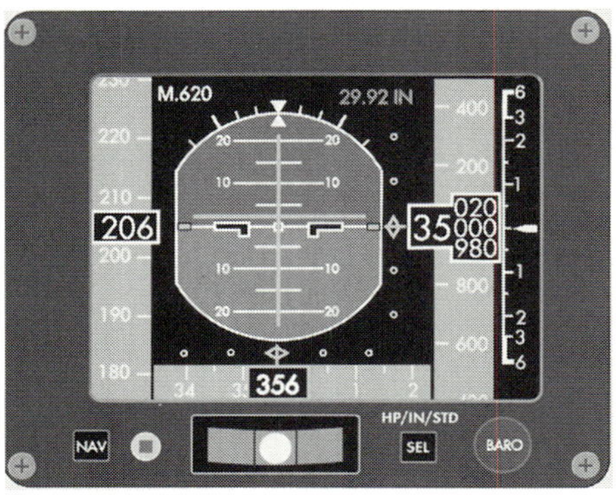

Meggitt Avionics Primary Flight Display ***1999***/0051674

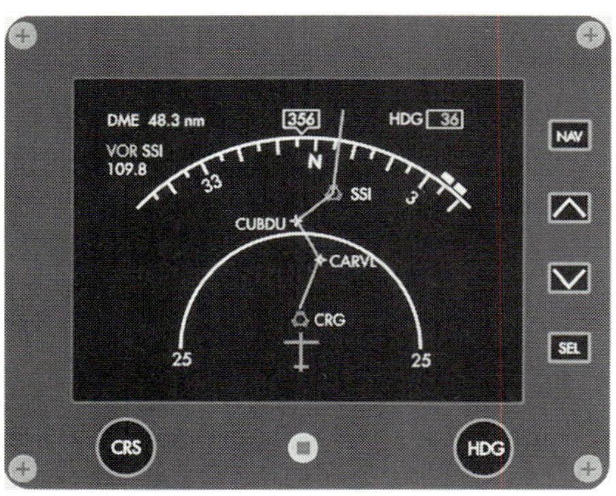

Meggitt Avionics Navigation Display ***1999***/0051675

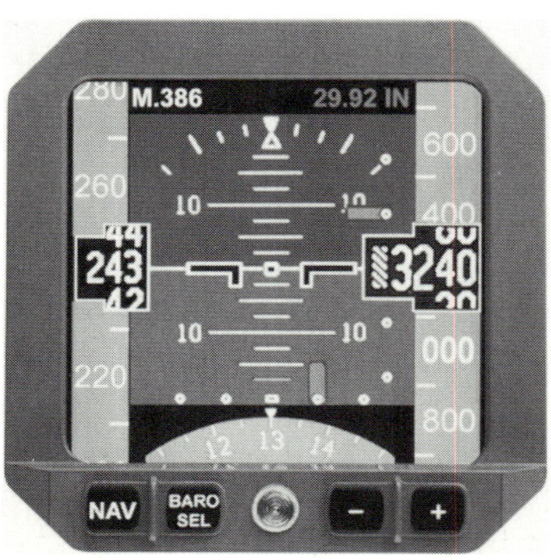

Meggitt Avionics Secondary Flight Display System™ ***1999***/0051676

Meggitt Avionics Secondary Navigation Display™ ***1999***/0051677

from a number of systems including GPS, radio receivers (VOR and ADF), and the heading system. Outputs for the autopilot are provided. Mode, heading and course selection controls are on the bezel of the navigation display.

Secondary Flight Display System™ (SFDS)

The SFDS meets the requirement for standby flight information – attitude, altitude and airspeed – replacing two or three conventional electromechanical cockpit standby instruments with a single 3ATI cockpit instrument. The system comprises the Secondary Flight Display, which provides on a colour Active Matrix Liquid Crystal Display (AMLCD), flight information in a similar and compatible format to modern CRT or AMLCD primary flight displays. Within the SFDS are solid-state inertial sensors and microprocessor systems to measure aircraft pitch and bank attitudes. A second Line Replaceable Unit (LRU), the Air Data Unit (ADU) is connected to the appropriate pitot and static pressure ports and includes solid-state silicon pressure sensors, and a microprocessor to provide digital air data (altitude, indicated air speed and Mach No) to the display.

The ADU can be installed conveniently close to the pitot/static ports to minimise the problems of long pneumatic pipes such as dynamic lags and leaks and eliminates the need to route pipes to the instrument panel.

The air data computations include corrections for known errors in the static pressure source and thus provide air data measurements as accurate as those on the primary flight displays.

The Mark II SFDS is being introduced in early 2000. It provides the following enhancements: video quality graphics; 2.5 million colours to match primary display characteristics (with customer selectable colour options); higher update rates; higher resolution; a heading option; and full anti-aliasing.

Secondary Navigation Display™ (SND)

The solid-state 3ATI SND cockpit instrument interfaces with ADF, VOR and DME radios together with the magnetic flux gate to provide navigation and heading data on a very high quality display. The SND is a complete navigation display.

Series 35000 electronic clock

Utilising a high quality Liquid Crystal Display (LCD), this 3ATI clock is entirely solid-state. In GPS mode, the clock receives its time signal via GPS to give the highest accuracy. For aircraft without GPS, or if GPS is not available, the clock operates in Manual mode using its integral high accuracy internal time standard to display time and date. Universal Time Co-ordinate (UTC) is received from the GPS on ARINC 429. Available outputs for other aircraft systems include time and date in either GPS or Manual mode.

Operational status

The Primary Flight Display (PFD) and Navigation Display (ND) form part of the New Piper Malibu Meridian primary flight display system.

The Secondary Flight Display System™ (SFDS) has been selected by British Airways for retrofit in its B737-300 and B737-400 aircraft and the Mark II version is being supplied from March 2000. The Meggitt SFDS has also been selected for: the Lockheed Martin JSF

Meggitt Avionics series 35000 electronic clock ***1999***/0051678

demonstrator aircraft; the US Air Force C-130/C141B and U-2 upgrade programmes and for the US Air Force C-5 fleet; the South African Air Force C-130 upgrade; for the Royal Australian Air Force Hawk Lead-In Fighter (LIF) programme and for export upgrades to the Chinook CH-47SD helicopter (initial customers being the Royal Air Force in the UK and the Republic of Singapore Air Force). Meggitt is building the 1,000th SFDS in January 2000.

The Secondary Navigation Display™ (SND) is in production for Gulfstream IVSP and Gulfstream V aircraft.

Contractor

Meggitt Avionics.

VERIFIED

Primary flight display system for New Piper Malibu Meridian

The New Piper Aircraft company has selected Meggitt Avionics to provide the primary flight display system for its new Malibu Meridian aircraft. The integrated system is designated MAGIC™. Principle components for the dual-seat installation are: the Primary Flight Displays (PFDs); Navigation Displays (NDs); Engine Instrument Display System (EIDS); Data Acquisition Units (DAUs); and Air Data Attitude Heading Reference System (ADAHRS).

Operational status

The first MAGIC™ system was delivered to The New Piper Aircraft company in June 1999. Equipment certification was planned for December 1999, with production deliveries beginning in February 2000.

Contractor

Meggitt Avionics.

VERIFIED

Solid-state air data instruments

Altimeters

The range of solid-state 3ATI altimeters is based on silicon pressure transducers and custom Liquid Crystal Displays (LCDs) with unique features to meet requirements for primary flight displays. Configurations are available for civil and military applications and as primary, primary with reversion to standby, standby or repeater variants.

Meggitt Avionics 3 ATI altitude alerter ***1999***/0051680

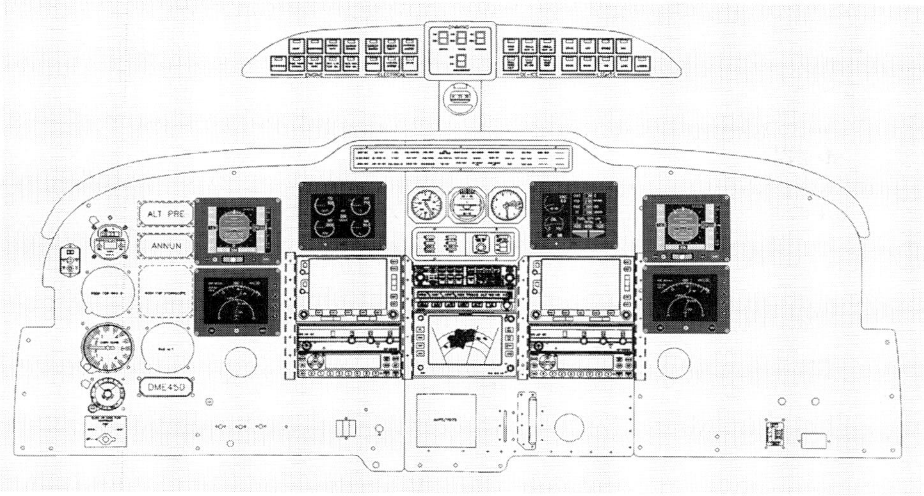

Flight deck of the Malibu Meridian, showing Meggitt Avionics flight display units ***1999***/0051683

Meggitt Avionics 3 ATI altimeter ***1999***/0051679

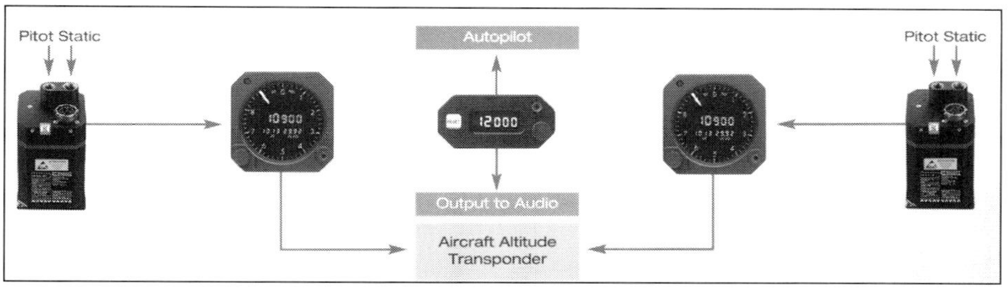

Meggitt Avionics reduced vertical separation minimum system ***1999***/0051682

Meggitt Avionics 3 ATI airspeed and Mach/airspeed indicator ***1999***/0051681

Altitude alerter
The Meggitt Avionics altitude alert unit is a solid-state design, utilising a high quality Liquid Crystal Display (LCD) housed in a half 3ATI case.

The pre-selected altitude is displayed on the five digit, seven segment, LCD by adjusting the rotary knob on the front of the unit.

The visual amber altitude alert warning is provided both on this unit and also provides an output to illuminate an amber warning on the Meggitt Avionics primary altimeter via ARINC 429 databus interface. The altitude alert unit monitors the displayed altitude on the primary altimeter (via ARINC 429 databus) and compares that with the altitude set on the alert unit.

Airspeed and Mach/airspeed indicators
With integral pitot and static silicon pressure transducers, microprocessor technology and Liquid Crystal Displays (LCDs), these units provide high quality display of airspeed or Mach/airspeed complete with all markers and bugs. Repeater variants are also available.

Air data and attitude sensors
Meggitt Avionics produces a range of solid-state air data and attitude sensors to provide inputs, via ARINC 429 databuses, to their display systems including: an Air Data Unit (ADU) to provide accurate altitude and airspeed data; and Attitude Reference Unit (ARU) to provide attitude data; and ADAHRS an Air Data Attitude Heading Reference System, which outputs pitch and roll attitude, pitch, roll and yaw rates, altitude, rate of change of altitude (IVSI), airspeed and heading.

Reduced Vertical Separation Minimum (RVSM) system
The RVSM system utilises the above sensor and display units (as depicted in the diagram below) to meet the requirements of full RVSM compliance with a system that is easily installed, and offers low procurement cost. It reduces errors by utilising the latest sensing and processing technology, combined with improvements resulting from advantages achieved by locating the Air Data Units adjacent to the pitot and static probes.

Operational status
The RVSM system was selected by the Royal Air Force in the UK for its VC-10 aircraft to meet NATS RVSM requirements.

Standby air data and inertial sensing devices are being supplied to Rockwell Collins for incorporation into the US Marine Corps UH-1Y and AH-1Z helicopter upgrade programmes.

Contractor
Meggitt Avionics.

VERIFIED

Integrated Mission Equipment System (IMES)

The first IMES to be installed on a non-military aircraft was delivered in August 1998 for installation into a UK Police BK117 aircraft.

The IMES integrates various role equipment together to enhance mission capability and ease observer workload in highly demanding situations. In this particular installation, the IMES has also been fitted with a Voice Recognition System (VRS) to enable voice control of selected mission equipment.

IMES is designed to provide simple integration of various items of role equipment manufactured by different suppliers. The system provides a loop-through facility that allows the equipment to be operated in the normal manner as individual components, or by a central control system using one piece of existing role equipment as the 'Master Control.'

On the BK117 the following equipment is integrated through the IMES system:
EuroNav III Task Management and Moving Map System ('Master Control');
FLIR Systems 400 Dual Sensor Camera System;
Video Recorder;
OCTEC Autotracker with Scene Lock and upgraded vehicle tracking mode;
NATS radio systems;
CRT Monitor;
Skyquest 12.1 in LCD monitor;
Searchlight.

This IMES system is fitted with a voice recognition unit that translates voice input commands from the observer to digital control commands for each item of equipment. The main purpose of direct voice control is to allow the observer to concentrate on a task and undertake equipment control without having to look at or touch the role equipment in question.

During all control phases, the observer is presented with messages on his FLIR video screen telling him what the IMES has understood and is doing. The messages stay on the video monitor for five seconds, and individual control of any unit can still be implemented manually at any time.

The Direct Voice Control (DVC) system is similar in technology to that being used in the new Eurofighter Typhoon programme. Voice command recordings have been supervised and implemented at DERA, Farnborough by its speech research unit. At present, the voice recognition system is specific to individual voices, but a new voice recognition card from Octec will allow the system to recognise commands after a brief pre-flight introduction, making the system much more flexible.

IMES is capable of sending observers voice and/or video screen messages that are generated from the EuroNav moving map or other role equipment. For example the EuroNav can be progammed with microwave link receiving sites, and by using its ground terrain mapping and interface to the radalt, can calculate the required height above ground that the aircraft needs to be to enable successful transmission of microwave signals.

When the microwave link is switched ON, IMES tells the EuroNav to switch on its microwave height analysis feature, and if the aircraft is too low, IMES can send a voice and/or video text message to advise the observer of the potential problem.

The functionality of voice feedback can be used for a number of features including powerline proximity warnings, and video tape ending message.

IMES is totally integrated with both the FLIR and map screens and can send data to both screens. Currently under trial is a system that orientates digital mapping information to the FLIR picture and sends digital road name information to the FLIR screen as an overlay onto the FLIR video image. This enables the observer to easily identify the road he is viewing on the FLIR screen without having to look at the map screen individually.

A new generation 14.5 in flat panel, high brightness video screen that allows suitable size picture-in-picture viewing is used. Typically a video image taking up the main screen with an inset moving map picture. At the touch of a button the two images reverse.

The EuroNav moving map system has a tracker feature whereby the observer simply types in a radial being given by the Tracker device in the aircraft. This enables the Tracker information to be displayed graphically on the map screen. This function is being automated.

Contractors

Octec Limited.

Skyquest Aviation.

VERIFIED

Central warning unit

The central warning unit has been designed and manufactured for use on modern fighter aircraft. The unit houses all the control circuitry and displays to provide 70 illuminated captions, along with audio tone and synthetic voice generation.

Each unit incorporates 28 primary red channels and 42 secondary amber channels. Additional features include day/night mode dimming of caption illumination, test facilities for active and non-active sensor channels, ground activation of certain channels with override facility for test purposes and generation of audio horn tones and synthetic voice.

Various combinations of caption are available in the range, while all variants incorporate a self-illuminating Betalight panel over the switches. Electrical connections are made via a pair of multipin connectors. The unit is capable of total NVG-compatibility.

Specifications

Dimensions: 134 × 165 × 165 mm
Power supply: 22.5-30.5 V DC normal (16-30.5 emergency)
Temperature range: −40 to +70°C
Altitude: up to 40,000 ft

Contractor

Page Aerospace Ltd.

VERIFIED

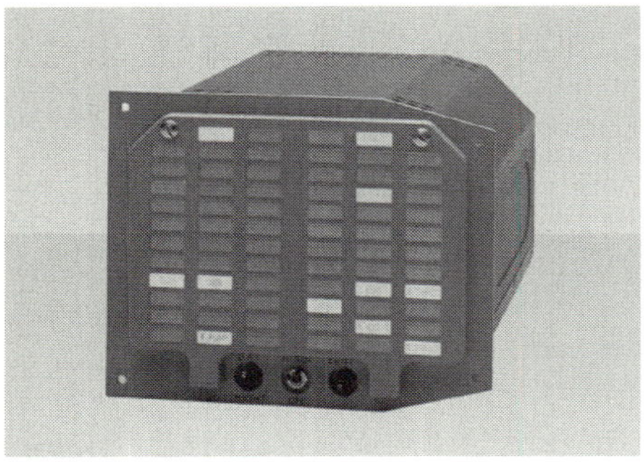

Page Aerospace central warning unit

Standby Master Warning Panel (SMWP)

The Standby Master Warning Panel (SMWP) provides the pilot with essential red primary warning indications in the event of any failure of the standard warning system. It is designed for use in glass cockpits, so the one-piece display matches the appearance of glass instruments.

Standby warning panel

Page Aerospace two- and three-terminal light modules

The display panel has 12 red warning captions, lit by LEDs, which flash until acknowledged. Different captions can be included, to satisfy customer requirements, up to 16 characters in length. Brightness control, to switch between day sunlight-readable and night illumination levels, is achieved by a Dim push-switch mounted directly under the display.

Situated adjacent the Dim switch is a rotary mode switch to enable selection of either automatic or manual operation and test. The test position checks more than 97 per cent of the circuit.

In view of its essential role the SMWP is made fault tolerant, so that no single fault will cause the loss of more than one warning. Even in the event of power supply failure, the unit will still operate in a degraded mode.

Specifications

Dimensions: 87.8 × 100.3 × 152 mm
Weight: 0.73 kg
Power supply: 28 V DC, <22 W (max)
Temperature range: −15 to +55°C
Altitude: up to 45,000 ft
Reliability: >100,000 h MTBF

Contractor

Page Aerospace Ltd.

VERIFIED

Two- and three-terminal light modules

Designed for use in single or stacked multichannel configuration, the filament lamp indicator/annunciator modules are a development of a module used extensively in aircraft central warning systems. The modules, available in two- and three-terminal arrangements, provide a low-cost, lightweight and compact alternative to other devices.

Both types of modules incorporate a clip-on caption frame within which a wide range of sunlight-readable blank or engraved caption screens may be contained. The three-terminal module may be used with two screens, one for each lamp, for systems where dual lamping is not essential. In these instances a lamp-holder separator screen may be fitted in a groove provided. Various alternative mounting arrangements facilitate single or multiple panel or stacked installation.

Specifications

Dimensions: 28 × 10.9 × 29 mm
Weight: 0.018 kg
Temperature range: −40 to +55°C
Altitude: up to 60,000 ft

Contractor

Page Aerospace Ltd.

VERIFIED

Voice, tone and display warning systems

Page Aerospace designs and manufactures a wide range of alerting systems for civil and military aircraft. These include centralised and decentralised warning systems, synthesised programmable tone and voice systems, flight mode annunciators, attention alerting devices and indicator modules.

In recent years the company has developed a screen finishing process which results in high levels of sunlight legibility.

For night flying operations the company has developed a night vision goggle-compatible range of screens and bezels for displays and instruments. These screens effectively filter out the infra-red content from illuminated cockpit devices and minimise flooding of the goggles.

Other areas of research and development cover high-quality synthesis of audio tones and voice. Units are currently being evaluated by the Defence Evaluation Research Agency, Farnborough and aircraft manufacturers in order to assess the optimum man/machine interface.

Operational status

Systems are now used in the central warning panel installations of the Saab 340, BAE Systems 146 and ATP and Raytheon Hawker 800 aircraft.

Contractor

Page Aerospace Ltd.

VERIFIED

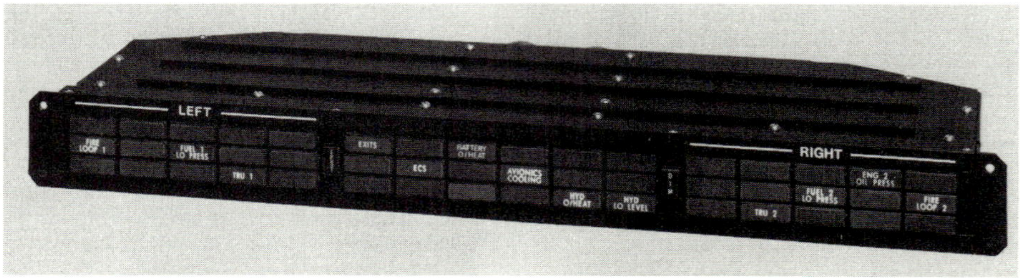

Page central warning panel for the BAe ATP

EICAS/EIDS Engine Instrument Crew Alerting System/Engine Instrument Display System

The EICAS/EIDS instruments use the latest technology to record engine conditions and provide indications on AMLCDs (Active Matrix Liquid Crystal Displays) that immediately alert flight crews to abnormal values.

Specifications

Dimensions: 134 × 135 × 242 mm
Weight: 3.4 kg
Power: 18-32 V DC, 50 W (nominal), 70 W (maximum)
Display area: 112 × 84 mm
Grey levels: 64

Contractor

Penny & Giles Aerospace Ltd.

VERIFIED

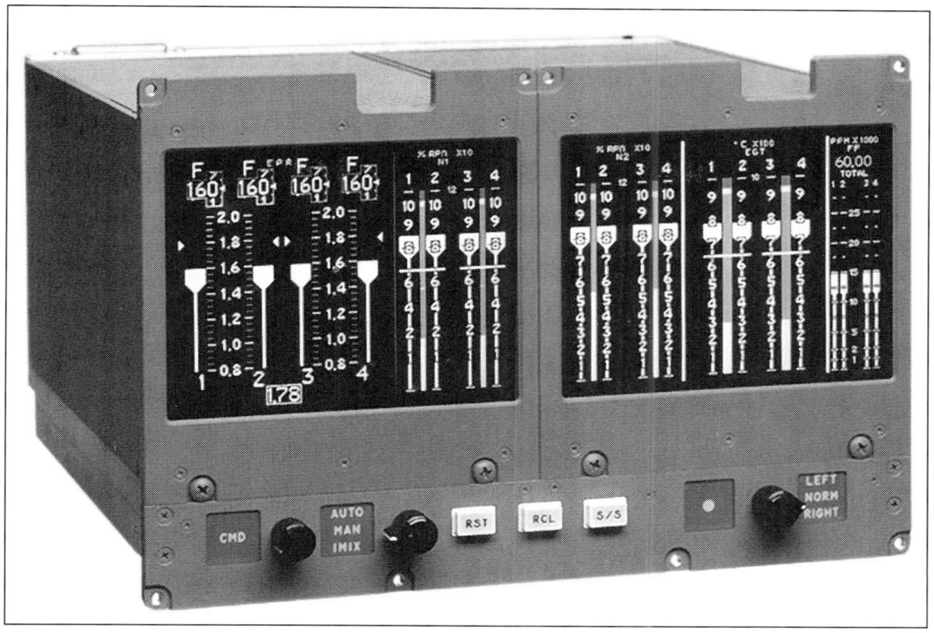

Engine Instrument Crew Alerting System (EICAS)

Engine Instrument Display System (EIDS)

In-Step digital technology step-motor aircraft engine instruments 0002457

Ice and snow detection system

The ice and snow detection unit monitors ice conditions in low-speed aircraft. By having a fast response time and being totally independent of airspeed, this system is suited to helicopter applications.

The unit uses bleed-air from the compressor to cause air to flow through the detector even at zero airspeed. Information is displayed as liquid water content per cubic metre.

Specifications

Weight:
(detector) 0.9 kg
(meter) 0.34 kg
Airspeed: 0-250 kt
Altitude: sea level to 15,000 ft
Reliability: 10,850 h MTBF

Penny & Giles ice detection system

Operational status

A version incorporating snow detection and digital outputs has been selected for the EH 101 helicopter where automatic control of the anti-de-ice systems is provided.

Contractor

Penny & Giles Aerospace Ltd.

VERIFIED

In-Step digital technology step-motor aircraft engine instruments

B&D Instruments and Avionics Inc, a subsidiary of Penny & Giles Aerospace Ltd, has designed microprocessor-controlled micro-stepping motors to reduce the maintenance cost of aircraft engine instruments.

Both 2 in MS style instruments and 2 in ATI style instruments are available, to display the following engine values: torque, propeller speed (Np), gas generator speed (Ng), propeller oil pressure/temperature, engine oil pressure/temperature and exhaust gas temperature.

In-Step exhaust gas temperature (EGT) instrument 0002456

Contractor

Penny & Giles Aerospace Ltd.

VERIFIED

Automatic Voice Alert Device (AVAD)

The Automatic Voice Alert Device (AVAD) stores prerecorded human speech in digitised form and operates under microprocessor control to assemble messages from words or phrases held in a vocabulary store. Since the voice is not synthesised, it can be either male or female in any language to provide the appropriate degree of stress and urgency for any situation. Racal says that high-quality reproduction ensures that the voice remains recognisable and intelligible under all conditions.

Message priority order, repetition rate and volume are programmable to individual requirements to provide the maximum information to aircrew without disrupting their primary tasks. Racal claims that existing discrete audio warning systems can be replaced by AVAD with a minimum of installation cost and complexity.

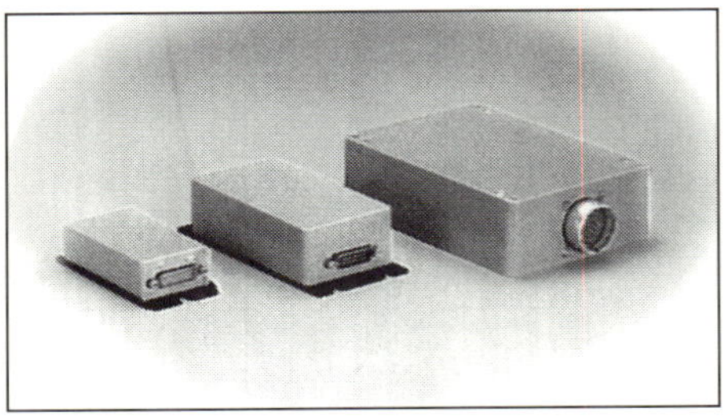

Racal Acoustics AVAD units V694, V695 and V697 (right to left) ***1999***/0044841

The units in the current AVAD range are the V694, V695 and V697.

V694

The AVAD V694 is designed to integrate and control audio alerts keyed by signals from system sensors and push-buttons. The message format is completely flexible and the system conforms to ARINC 577 audible warning systems and ARINC 726 flight warning computer systems for civil aviation applications.

Up to 4 minutes of prerecorded speech can be encoded and stored as a vocabulary of messages, phrases, words and tones. Messages can be constructed using words and phrases drawn from the unit memory under software control. This allows the total duration of all messages to exceed the vocabulary by a considerable amount. Up to 15 message channels are based on positive or zero volt keying. Four channels can respond to variable DC voltage inputs, offering a sensitivity of up to 256 logic switching steps across the input range. A stabilised 5 V DC output is available for use as a reference for an external potentiometer. Typical variable readouts might be aircraft altitude, electrical system voltages or engine pressure ratios.

Unit operation and vocabulary store may be defined by the user and program software allows control of each message cycle. Each message has normal, regrade and test priority values assigned to it. The output sequence is then controlled by the regrade and test inputs.

There are two audio outputs, one for a telephone or headset and the other for driving a loudspeaker direct. A variant can be provided where the loudspeaker drive output is removed and replaced by an auxiliary audio input. This auxiliary input facilitates the summing of existing aircraft audio warnings with the single AVAD output for routeing into the aircraft audio integration/intercommunication control system. This AVAD configuration is particularly suitable for helicopter use. The nominal level of output can be adjusted by individual potentiometers offering a 20 dB range, with the relative volume of each message or message format under software control. The complete message format, including repetition rate, pauses, message inhibit/de-inhibit and tones, may be similarly controlled. A key input is provided which inhibits any message in progress without affecting other keyed messages. A message will remain inhibited until its sensor input is removed and reapplied. Alternatively, the inhibit may be removed after a specified time if the warning message keyline remains activated during this period.

Full test/reset facilities provide fast checking of unit operation. Correct operation of the microprocessor and amplifier circuitry and control program status are continuously self-monitored. In the event of a fault condition, both audio outputs are automatically inhibited and a fail output is provided to activate an indicator to warn of unit failure.

V695

The V695 is a smaller seven-channel unit with microprocessor control. It offers all the programming flexibility of the V694 but has a storage capacity of 13.5 seconds. An auxiliary audio input is provided.

V697

The V697 provides a single output for applications in which only one warning such as 'fire' or 'pull up' is required. A second keyline may be used to provide a system enable/disable input. The V697 can be enabled by either positive or zero volt keying and the resultant warning output can be either a continuous or one message cycle per key event.

All the above units provide a bandwidth of 100 Hz to 4 kHz. Audio distortion is less than 4 per cent. Volume is preset and adjustable by 20 dB. The V694 is also programmable to different levels for each message and variation during messages.

Specifications

Dimensions:
(V694) 165 × 115 × 51 mm
(V695) 134 × 61 × 32 mm
(V697) 110 × 50 × 28 mm
Weight:
(V694) 0.95 kg
(V695) 0.2 kg
(V697) 0.14 kg
Power supply: 28 V DC
Temperature range: –40 to +70°C

Operational status

In production and in worldwide service with a variety of fixed- and rotary-wing aircraft including, UK Royal Air Force Jaguar fixed-wing aircraft and Puma and all Sea King helicopters.

Contractor

Racal Acoustics Ltd.

VERIFIED

RA690 analogue Communication Control Systems (CCS)

The RA690 series of CCS equipment provides an all-analogue solution that utilises either distributed facilities or a central amplifier with remote station boxes for crew interface. The RA690 is extensively used in military and civil aircraft worldwide.

The conventional configuration of distributed amplifiers, filters and volume controls is designed using the latest technologies to ensure the most cost effective performance possible. When configured to utilise a central amplifier and remote passive station boxes, the single point audio is routed within the central amplifier, with selection and control implemented via analogue control lines between these major units. Both architectures are built using functional blocks to produce a customised configuration for the application required. The functional blocks include Voice Operated Switches (VOS), attenuators, power supplies, muting and intercom override and interface matching networks.

Facia panel engraving and lighting details (including NVG type), switching and electrical characteristics are configured on each station box type to suit the particular requirements of the user.

Additional flexible RA690 system capability is provided by S690 Audio Selector Panels (ASPs) and D690 switch boxes. The S690 ASP allows an increase in the number of receiver audio inputs which can be accommodated. D690 switch boxes enhance the facilities of the microphone switching, intercom networks and other CCS functions.

Varying combinations of station boxes, ASPs and switch boxes can be installed at each crew position. Where required, the ASP and switch box functions can be integrated within the station box to form a single unit.

The RA690 Series includes a range of general purpose amplifiers and other units. These include specially designed amplifiers to ensure systems conform to Civil Aviation Authority (CAA) specifications 11 and 15 for Cockpit Voice Recorder (CVR) and passenger address systems. Additionally, existing CCS installation performance can be enhanced with use of A6916 units providing noise tracking level control and voice operated switching and a power supply for active noise reduction headsets.

The RA690 range is further complemented by headset connection jack boxes and specially configured audio balance units to ensure correct interface between the CCS and various Tx/Rx equipments.

Operational status

The RA690 CCS is widely used in military and civil aircraft worldwide. Some 960 noise tracking level control and voice-operated switches have been ordered by the UK Royal Air Force for Sea King helicopters.

Contractor

Racal Acoustics Ltd.

VERIFIED

RA800 digital Communication Control Systems (CCS)

The RA800 CCS is designed to meet present and foreseeable requirements for all aircraft types. The RA800 provides centralised audio control, interfacing to other aircraft systems and sophisticated intercom networks for the flight crew. It is built on a foundation of intelligent modules, which enables systems to be configured for the particular requirements of the aircraft platform.

The core of the RA800 is the Communication Audio Management Unit (CAMU). All audio routeing is contained within the CAMU: this increases immunity to electromagnetic interference by considerably reducing the volume and complexity of aircraft cabling. Other advantages include ease of installation, weight saving and improved reliability gained through significant cable reduction.

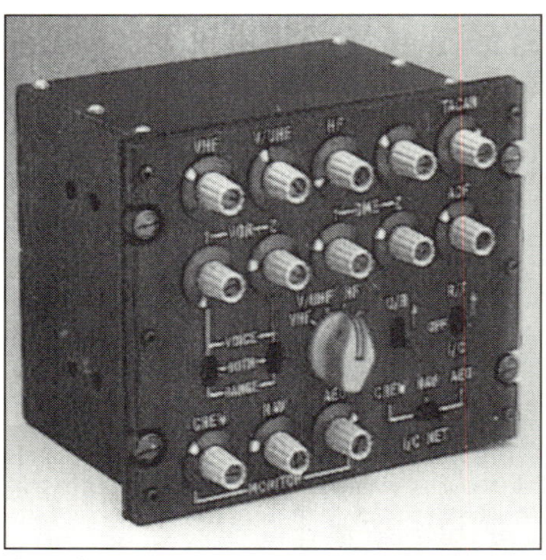

Units of the Racal Acoustics' RA690 analogue CCS

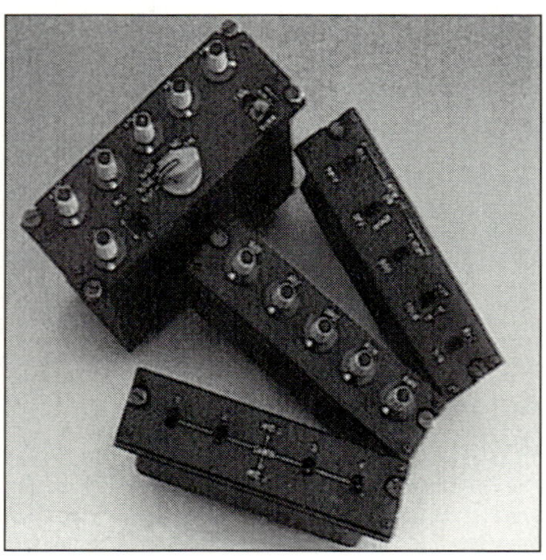

1999/0044844

Racal Acoustics' digital RA800 CCS units ***1999***/0044839

The crew interface to the system is via Audio Control Panels (ACPs) for the control functions and, in some variants, Headset Electronics Units (HEUs) for microphone and telephone audio. Connections form the ACPs and HEUs to the CAMU are via digital links.

A Tempest variant, RA800 'Light', employs Digital Signal Processing of all audio and control functions, together with fibre optic interconnection between major system elements. The fibre optic connection between the CAMU and the ACP, and the CAMU and the HEU, is a serial datalink. The use of fibre optic cables further improves the immunity to electromagnetic interference hazards and ensures secure TEMPEST operations. A schematic diagram of a typical RA800 'Light' is shown.

For RA800 equipment configured to use traditional copper wire interfaces, the connection between the ACP and the CAMU is an ARINC 429 bus.

RA800 is microprocessor controlled with centralised or distributed intelligence dependent upon the platform. Where an aircraft is required to undertake a variety of roles, the RA800 system can readily be software reconfigured on board.

The system interfaces to navigation, radio and warning equipment either directly or by means of the aircraft avionics bus (MIL-STD-1553).

Sophisticated BIT is carried out within all units that make up the RA800 CCS system. This data is collated and can be sent to the aircraft maintenance system.

Voice and/or Tone warnings are produced either from an embedded module within the RA800 CAMU or from a stand-alone unit.

Operational status

Designed for civil and military aircraft applications. In service on BAE Systems Hawk aircraft. Ordered for UK Royal Air Force EH 101 and Chinook helicopters, SAAF C-130B aircraft, Royal Australian Air Force Lead-in-Fighter aircraft, and Canadian SAR helicopters.

Contractor

Racal Acoustics Ltd.

UPDATED

Airborne real-time datalink

Racal Avionics has integrated its satcom and navigation/mission management products to provide a complete real-time datalink package, which includes both the ground segment and the airborne Line Replaceable Units (LRUs). The airborne LRUs utilised are: the multifunction Control Display Navigation Unit (CDNU) as the operator interface and the Satellite TRansceiver (STR) as the communications system. The primary function of the CDNU is to provide aircraft navigation and subsystem control. It is also the airborne datalink end system, providing Automatic Dependent Surveillance (ADS) and Controller to Pilot DataLink Communication (CPDLC), which also allows free text messages and pre-formatted messages to be sent to a variety of ground destinations. It displays messages received from the ground and allows the user to reply to those messages. The CDNU also has an Emergency ADS Mode.

ADS reports are initiated from the ground, via an ADS contract, and include: present position, altitude, velocity, and aircraft intent. All ADS contracts are transparent to the pilot.

The STR equipment which supports packet data services - Data 2 (ACARS) and Data 3 (ATN FANS X.25) communications over the worldwide Inmarsat satellite system, comprises two LRUs: the combined Satellite Data Unit (SDU)/Radio Frequency Unit (RFU) and a High Power Amplifier (HPA).

The Inmarsat system utilises a digital format for real-time full-duplex data communications at up to 1,200 bps. Being a low-gain system, the STR operates via a 0 dBic omnidirectional antenna and therefore requires no other navigation input for beam-steering purposes.

The ground terminal allows the operator to initiate ADS contracts with numerous aircraft and displays the requested information on a map and status window. The terminal also allows text messages to be sent to selected aircraft and displays messages received from aircraft. Any aircraft that initiates an Emergency ADS message is highlighted on the map with an audio alert.

Operational status

In September 1999, the UK Royal Air Force initiated a 3-month real-time, helicopter tracking trial of the system, using Search And Rescue (SAR) Sea King Mk 3A helicopters. An automatic call is made by the helicopter system to the Rescue Co-ordination Centre (RCC). Call interval is determined by the RCC, but 60 seconds is being used for trial purposes. The call declares helicopter present position, altitude, velocity and operational intent.

The system being trialled is an evolutionary development of the Norwegian Modified-Automatic Dependent Surveillance (M-ADS) programme.

Contractor

Racal Avionics Ltd.

VERIFIED

UK Royal Air Force Sea King Mk 3A search and rescue helicopters participating in the airborne real-time datalink trial ***2000***/0080260

Groundspeed/drift meter Type 9308

The Type 9308 standard aircraft indicator size unit shows groundspeed within 3.5 kt at 100 kt or 5 kt at 300 kt. Drift is registered within 0.5° of true value. If the signal is lost, the last measured groundspeed and drift are displayed. The Type 9308 is compatible with Doppler 71 and 72 sensors.

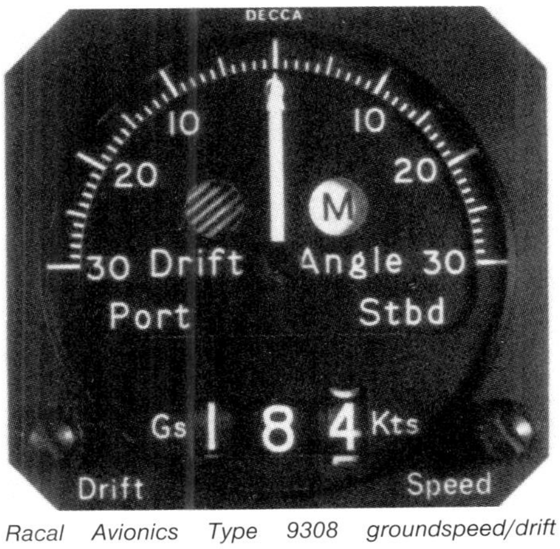

Racal Avionics Type 9308 groundspeed/drift presentation

Operational status
In production.

Contractor
Racal Avionics Ltd.

VERIFIED

Hovermeter Type 80564B

The Hovermeter Type 80564B is similar to the Type 9306 in capability and appearance, but is compatible with Doppler 80 and 90 series sensors.

Contractor
Racal Avionics Ltd.

VERIFIED

The position bearing and distance indicator for the Racal Avionics Doppler 80

Hovermeter Type 9306

The Type 9306 standard aircraft indicator size unit displays along heading velocity over the range –10 to 20 kt with an error not greater than 1 kt and vertical velocity range ±500 ft/min with a maximum error of 40 ft/min. The Type 9306 is compatible with Doppler 71 and 72 sensors.

Operational status
In production.

Contractor
Racal Avionics Ltd.

VERIFIED

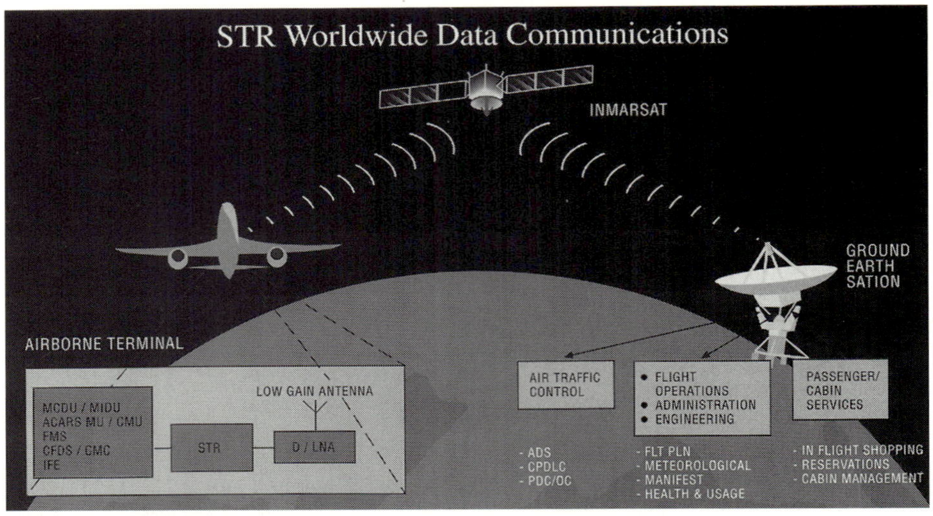

STR worldwide data communications 0011830

STR Satellite TRansceiver

The STR Satellite TRansceiver is a single channel satcom system which supports packet data services – Data-2 (ACARS, AFIS, A622 FANS) and Data-3 (ATN FANS, X-25 cabin) communications – over worldwide Inmarsat satellite system.

The STR is an Inmarsat Aero 'L' Class 1 system, which meets the integrity and real-time communication requirements necessary to support Air Traffic Services, both today and in the future, while simultaneously supporting operational, administrative and passenger datalink services.

Applications typically supported by the STR are:

(a) Operational/fleet management services: flight planning data, departure times, ETAs, position reports, manifest and engineering data
(b) Communications/Navigation Surveillance (CNS) services: FANS applications including: Automatic Dependent Surveillance (ADS), Controller to Pilot DataLink Communications (CPDLC) and Pre-Departure and Oceanic Clearance (PDC/OC) via A622 FANS, and via ATN FANS as a growth option
(c) Passenger and cabin management data services - X-25 based.

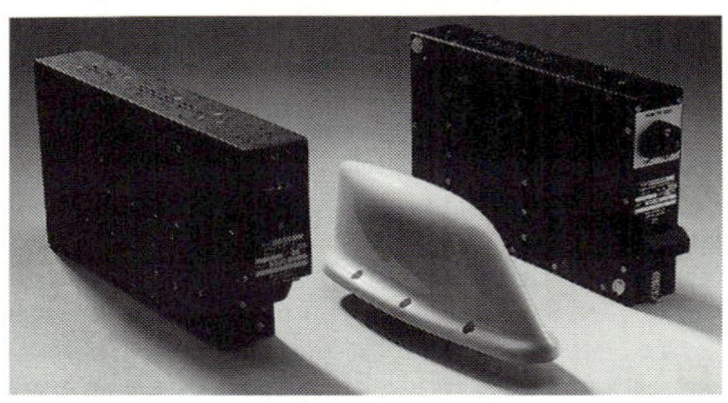

Racal Avionics STR Satellite TRansceiver and omnidirectional antenna 0011829

The STR utilises the proven technology of the Racal/Honeywell MCS 3000/6000 multichannel systems, reconfigured into two compact LRUs: a Satellite Data Unit/Radio Frequency Unit (SDU/RFU) and High Power Amplifier (HPA). The system interfaces to an omni-directional low-gain antenna and to a variety of cockpit and cabin communication management devices including: ACARS Mus/CMUs, AFIS, FMS, and In-Flight Entertainment (IFE).

Specifications
Data rate: 600, 1,200 bps
Dimensions: each unit; 2 MCU
Weight: each unit; <4.5 kg
Power: 115 V AC, 400 Hz, <275 V A

Operational status
In production. Selected by Kongsberg Defence & Aerospace for the Modified-Automatic Dependent Surveillance (M-ADS) North Sea System. The STR also forms part of the Racal Avionics airborne real-time datalink fitted to UK Royal Air Force Sea King Mk 3A Search And Rescue (SAR) helicopters.

Contractor
Racal Avionics Ltd.

VERIFIED

LandStar GPS

The Racal Survey LandStar satellite-based differential GPS system is available in Europe, Middle East, Southern Africa, North and South America, Australia, New Zealand and Indonesia. LandStar's precise positioning capability may now be used in a wide range of applications in these regions including surveying, mineral exploration, forestry and agriculture and also for transport and emergency services.

The European service follows the launch of Italsat F2 communications satellite which transmits differential corrections to users throughout an area from the Western Atlantic to East of Moscow and from the Straits of Gibraltar to Scandinavia.

Users of the LandStar America system receive differential GPS corrections via the AMSC (American Mobile Satellite Corporation) L-band (NATO D-band) spot beam geostationary satellite. Differential corrections are obtained from a network of 11 precisely co-ordinated reference stations which ensure effective accuracy and coverage throughout the region. The data is gathered at Racal's LandStar hub in Houston where it is processed and monitored for quality control. It is then uplinked to the satellite via the AMSC facility in Reston, Virginia.

The Australian LandStar DGPS service uses the Optus B1 geostationary satellite to broadcast the differential corrections generated at nine reference stations over the Australian continent. By increasing the transmitted signal strength Racal has extended the coverage to include the additional countries of Indonesia, New Zealand and Papua New Guinea.

The LandStar differential GPS service was developed by Racal from experience gained on SkyFix, which is the world's most extensive DGPS network. The power of the LandStar satellite spot beam transmissions have made it possible to incorporate the LandStar receiver and antenna within a small, lightweight package suitable for backpack, vehicle and light aircraft applications. The unit processes and

LandStar differential GPS supporting geomagnetic surveying

receives the differential messages, including wide area multireference signals, together with proprietary quality control and other messaging. Data is provided in industry standard RTCM Sc 104 format so that LandStar can be used with virtually every differential capable GPS receiver. Accuracies achieved are typically better than 1 m.

LandStar receivers are available in an OEM or a stand-alone form which may have an integral GPS unit fitted as an option.

Operational status
Available.

Contractor
Racal Survey Ltd.

VERIFIED

Ruggedised and long scale meters

The SIFAM range of ruggedised and long scale meters is available to existing designs or can be modified to user specifications. Movements can also be incorporated into custom enclosures or be supplied for assembly. Movements are available in three sizes: 24, 27 and 33 mm diameter, with fully magnetically shielded movements, sapphire jewels and tungsten carbide pivots. Instrument error is typically ± 1.5 per cent.

Operational status
In production for the Westland EH 101 helicopter.

Contractor
SIFAM Instruments Limited.

UPDATED

SIFAM ruggedised and long scale meters
2000/0081488

SARFIND cockpit display unit

The SARFIND cockpit display unit presents GPS data to the aircrew. It employs an aviation standard sunlight-readable vacuum fluorescent display screen which is able to display up to 40 characters across two lines. The unit is a standard aircraft instrument panel fit. An ARINC version is under development.

The display consists of two screens that scroll at 10 second intervals. Screen 1 gives latitude, longitude, quality of fix and age of the data. Screen 2 gives time of the fix and user identification. The display has a wide viewing angle with filtered display characters. In addition, there is an audio/visual alarm which is activated on receipt of position data.

Contractor
Signature Industries Ltd.

VERIFIED

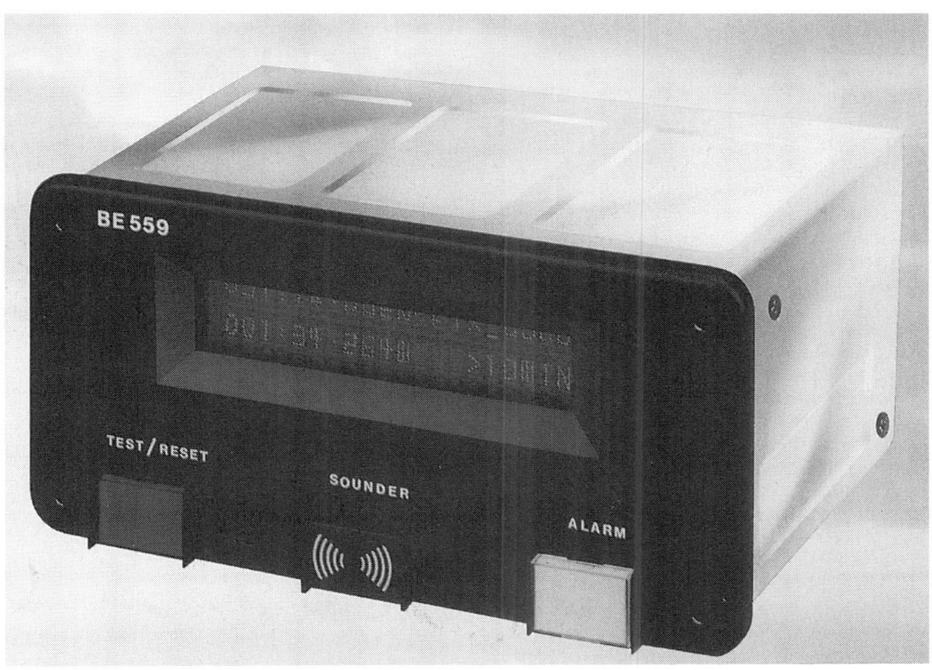

SARFIND cockpit display unit

Colour Skymap and Colour Tracker

Colour Skymap provides a GPS-based colour moving map. Colour Tracker is the repeater equivalent of Colour Skymap and has been specifically engineered to interface with existing GPS and LORAN receivers.

Both products can be installed into a standard 6.25 in radio stack and will automatically display pilot position within 30 seconds of switch on, relative to geographic features, controlled airspace, airports, danger/restricted areas and other aeronautical data points. The heart of Colour Skymap and Colour Tracker is the high-resolution, ultra-high contrast, active matrix TFT colour LCD, which offers clarity over a wide viewing angle and varying light conditions. This feature, coupled with the ability to rotate the display through 180°, provides complete mounting flexibility within the cockpit, enabling installations to be optimised for left- or right-handed operation.

Colour Skymap and Colour Tracker both come with a worldwide navigational capability; the regions supported are the Americas, Atlantic International and Pacific International. Each front-loading PCMCIA data card contains the regional Jeppesen® data, the Skyforce geographic database and the entire operating software. It is this approach that allows Skyforce to provide periodic operating system enhancements during the normal Jeppesen® updating process.

In addition to their extensive user-defined waypoint storage and flight planning capability, Colour Skymap and Colour Tracker offer many other features, including simplified joystick data entry and aeronautical feature interrogation, user-defined airfields, split-screen navigation modes including HSI, an on-screen CDI, auto zoom, extended track, emergency search, emergency airport minimum criteria, map customisation, map configuration (North Up or Track Up), choice of map datum and turning point arrival, airspace and marker annunciation.

Specifications

Screen: 127 mm diagonal active matrix TFT LCD
Dimensions: 158 × 101 × 175 mm
Weight: 1.5 kg

GPS Receiver (Colour Skymap only)
Receiver: 8-channel parallel, simultaneous tracking
Acquisition:
12 s (almanac, position, time and ephemeris known)
43 s (almanac, position and time known)
Reacquire: 1.5 s
Accuracy:
15 m (without S/A)
1 to 5 m with differential option

Operational status

In service in both the general aviation and military aviation market. IFR approved.

Contractor

Skyforce Avionics Ltd.

VERIFIED

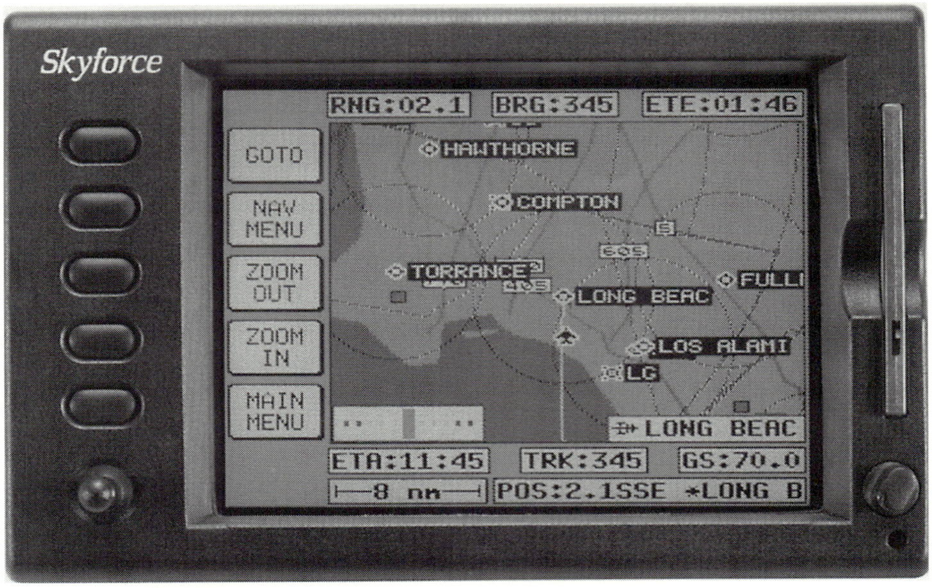

Colour Skymap and Colour Tracker 0001362

Specifications

	Display Head	Control Panel	Processor Unit
Standard configuration	5 in active matrix TFT LCD 320 × 240 pixels with 5 soft keys and joystick	combined PCMCIA card on/off control and display/ keyboard connections	133 MHz Pentium 3.2 Gbytes GPS input video output
Dimensions	160 × 112 × 60 mm	146 × 65.5 × 150 mm	75 × 170 × 230 mm
Weight (approx)	0.65 kg	0.6 kg	1.5 kg
Options	10 in active matrix TFT LCD, 640 × 480 pixels, instead of, or additional to 5 in display	keyboard; split PCMCIA card reader	video input; ARINC 429; MIL-STD-1553B

Contractor

Skyforce Avionics Ltd.

VERIFIED

Skymap II™ and Tracker II™

The Skymap II™/Tracker II™ system is a monochrome variant of the Skymap/Tracker product line, which can be knee, yoke, swivel, gimbal, panel or rack mounted. The monochrome reflective supertwist liquid crystal display providing sunlight readability and wide-angle viewing.

Skymap II™ provides GPS moving map precision navigation worldwide. Tracker II™ is the repeater equivalent of Skymap II™, and has been engineered to interface with existing GPS and Loran receivers.

Skymap II™ and Tracker II™ include interchangeable data modules that contain regional Jeppesen data, Skyforce geographic data, and the entire operating software. This architecture permits Skyforce to offer periodic operating system enhancements within the normal Jeppesen updating process. A worldwide navigational capability is available in three cassettes to cover the Americas, Atlantic International and Pacific International. Operator facilities are similar to those listed for Colour Skymap/Tracker.

Specifications

Screen: 127 mm diagonal high contrast reflective supertwist back-lit LCD (128 × 240 pixels)
Dimensions: 158 × 115 × 35 mm
Weight: 0.65 kg

GPS Receiver (Skymap II™ only)
Receiver: 8 channel parallel, simultaneous tracking
Acquisition: 12 s (almanac, position, time and ephemeris known); 43 s (almanac, position and time known)
Reacquire: 1.5 s
Accuracy: 15 m (without S/A); 1-5 m with differential option

Skymap II™ Version 2.00+
The following additional features were added in summer 1997: Aeronautical data - Jeppesen Plus+; cartographic data version 2.00+; operating system

Observer™ moving map task management system

The Skyforce Observer™ is a fully integrated moving map task-management system that has been designed to create a paperless cockpit for an airborne observer in military, paramilitary, emergency or utility services operations. The system is capable of displaying and navigating over a number of highly detailed maps down to street/house level, whether they be in a raster-scanned or digital vector format. The maps are layered, allowing the customer to switch instantly from layer to layer. It is configured for GPS input.

Observer™ interfaces with any display, which offers a VGA or composite video input (NTSC/PAL).

Using a special-purpose user-defined facility, customers can introduce databases into the system via a PCMCIA card. These databases can be configured to hold any type of information whether it be text, pictorial or both, thereby allowing the user to store operation orders, daily tasking, or to highlight special waypoints. With a flight logging capability that can be used for post-flight analysis and a search-and-rescue function, Observer™ can be used for both operational and fleet management tasks.

Skyforce Observer™ moving map task management system 0018192

version 2.00 Plus+ to offer a choice of three languages: German, French and English selectable from the set-up menu.

Operational status

Amongst many military and civil installations, Skymap II™ is fitted to the Hawk aircraft of the Royal Air Force Red Arrows Display Team. At the beginning of 1997, new features added included: HSI, E6B calculator and flight logging capabilities, and map detail was improved. In summer 1997, Skymap II™ was enhanced to the Skymap II™ Version 2.00+ standard.

Contractor

Skyforce Avionics Ltd.

VERIFIED

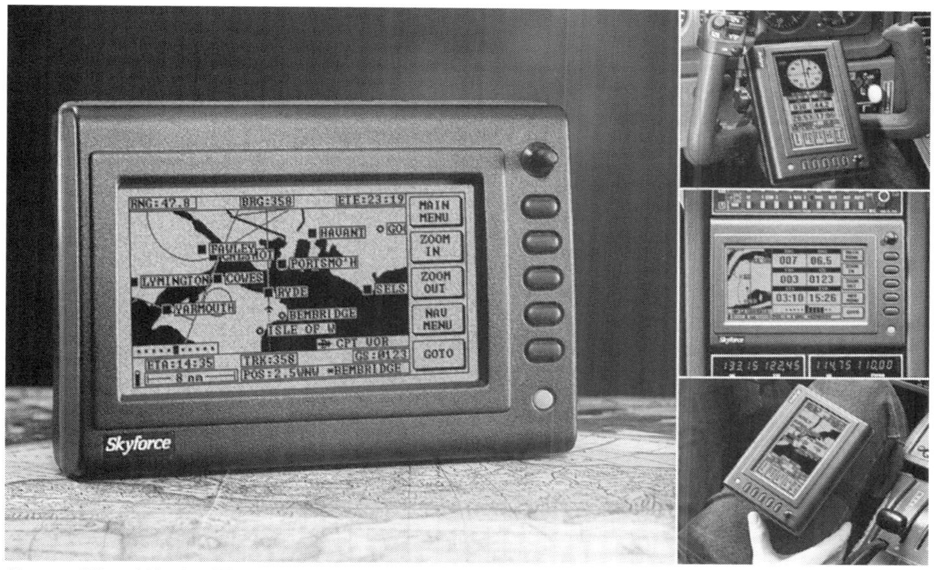

Skymap II™ and Tracker II™ display and optional mounting configurations 0018190

Skymap IIIC™ and Tracker IIIC™

Skymap IIIC™ and its Tracker IIIC™ companion are full-screen colour updates of Skymap IIC™ and Tracker IIC™, which again can be knee, yoke, swivel, gimbal, panel or rack mounted.

Skymap IIIC™ utilises an integral eight-channel parallel GPS receiver, while Tracker IIIC™ requires an input from an external GPS or Loran receiver.

Standard language options for the text now include English, French, German and Spanish.

Specifications

Screen: 127 mm (5 in) diagonal
Dimensions: 158 × 115 × 65 mm
Weight: 0.9 kg
GPS receiver (Skymap IIIC™ only):
channels: 8 parallel
acquisition: 12 s, almanac, position, time and ephemeris known
43 s, almanac, position and time known
Reacquire: 1.5 s
Accuracy: 15 m (without S/A); 1-5 m with differential option
Internal database: geographical; Jeppesen®; updates (1/3/6 or 12 months); English, French, German and Spanish
Waypoints: 500 user-defined waypoints and 25 user-configurable airfields for non-ICAO listed airfields
Routes: 99 reversible routes, with up to 99 turning points

Contractor

Skyforce Avionics Ltd.

VERIFIED

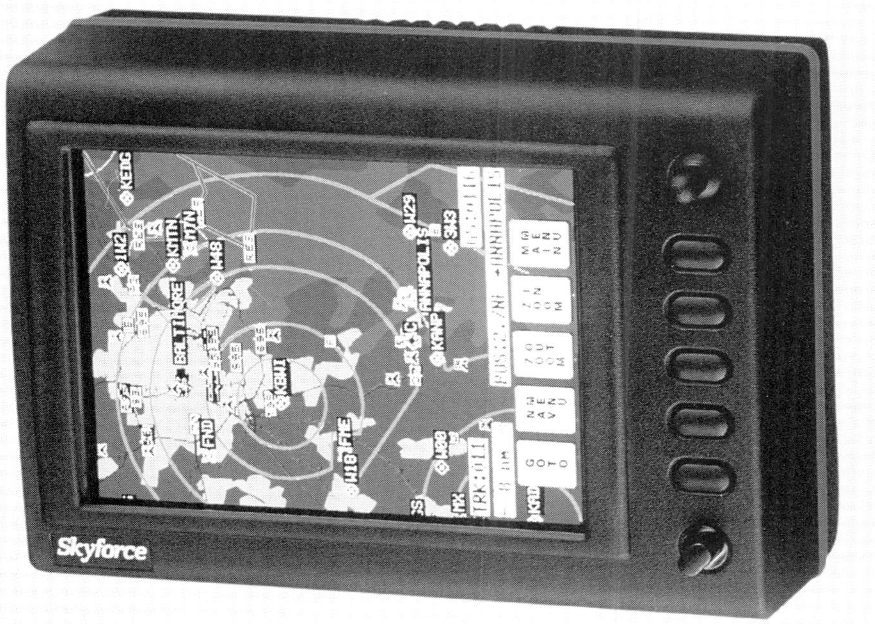

Skymap IIIC™ moving map colour display ***1999***/0051684

Ruggedised flat panel LCD aircraft monitors

Skyquest Aviation makes a range of LCD aircraft monitors including the following sizes: 5.0, 5.8, 10.4, 12.1, 14.5 and 17.7 in. Data provided here relates to the 12.1 in product.

The 12.1 in monitor features full PAL or NTSC capability. It is a high-brightness, sunlight-readable display, with switch selectable NVG option. It is suitable for both panel mounting and free standing installation.

Specifications

Viewing area: 12.1 in diagonal
Dimensions: 313 × 243 × 48.4 mm
Contrast ratio: >120:1
Brightness: 870 cd/m^2 (sunlight-readable)
Resolution: full PAL
Pixels: 800 × 600
TV lines: 730
Grey scales: 64 to 256
Colour capability: 262,144 colours
Operating temperature range:
(standard unit) –5 to +60°C
military unit (with heater) –40 to +60°C
military unit (with heater and cooler) –40 to >+60°C
Power: built in 28 V DC

Contractor

Skyquest Aviation.

UPDATED

Skyquest 12.1 in high-brightness display used by UK police in an AS355 helicopter ***1999***/0051685

503 series Emergency Location Transmitters (ELTs)

The 503 ELT system provides full-frequency coverage, including 121.5, 243 and 406.025 MHz with optional inclusion of transmission of last known GPS co-ordinates. It also offers 'full remote extended range operation', incorporating a bi-directional G switch, together with both fixed or portable ELT function, in one unit.

Specifications

Frequency: 121.5, 243.0 and 406.025 MHz
Peak Effective Radiated Power (PERP): 0.1 W at 121.5/243.0 MHz; 5.0 W at 406.025 MHz
Transmission duration: 24 h (min) at 5 W PERP; 48 h (min) at 0.1 W PERP
Repetition rate: 520 ms every 50 s
Activation: manual via the cockpit remote switching panel, or automatically by bidirectional G switch
Dimensions: 255 × 105 × 45 mm
Weight: 1.3 kg
Battery life: 6 years

503-15 series ELT interface unit

The 503-15 series ELT interface unit provides last known GPS co-ordinates on ELT, ADELT or CPI. The ELT interface unit is a stand-alone unit, which utilises an RS-232 interface to continually update the ELT, ADELT or CPI. Other interfaces are available optionally. The freedom of a stand-alone unit allows full integration with the aircraft's flight management computer, while retaining the ability to program the ELT/ADELT or CPI via a preset 24 bit aircraft address without requiring specialist maintenance operations at equipment exchange.

Contractor

H R Smith.

VERIFIED

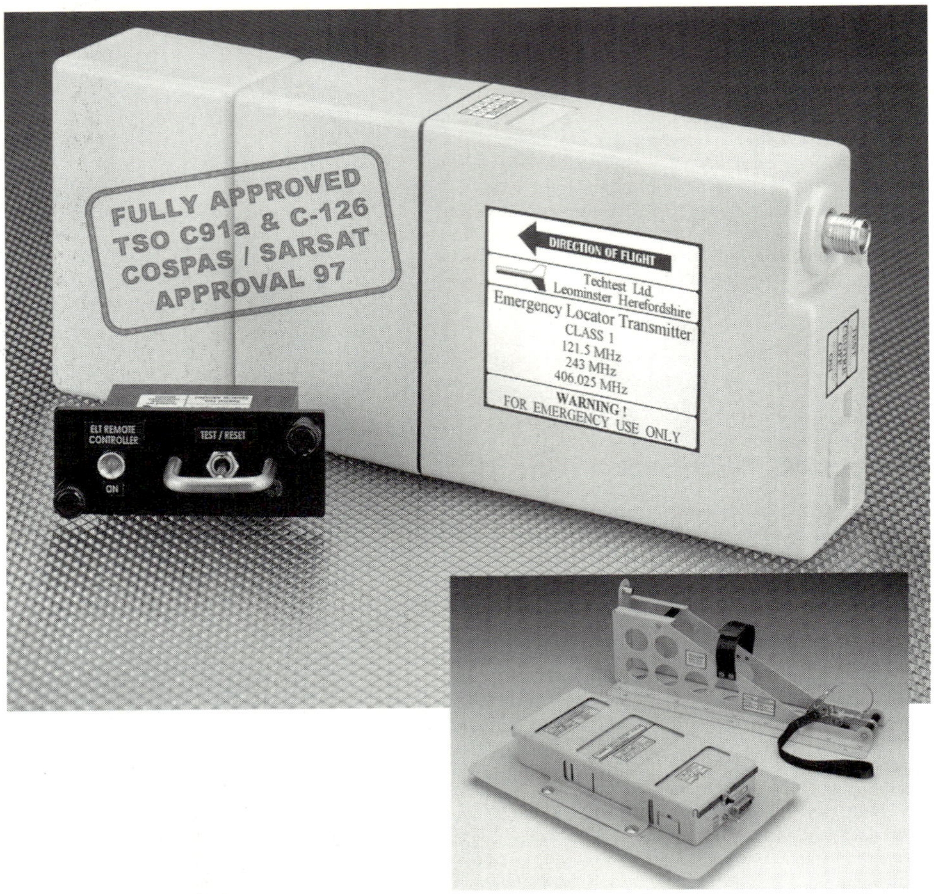

The 503 series emergency location transmitter ***2000***/0062847

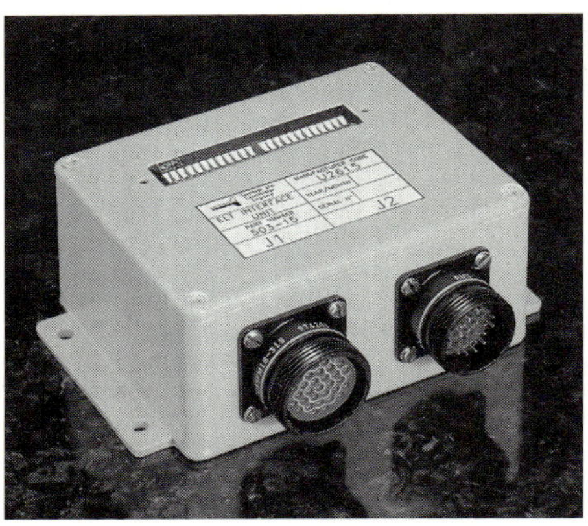

The 503-15 series emergency location transmitter interface unit ***1999***/0051603

RAF Rescue helicopter with Series 406 Homer and Series 500 personal locator beacon 0001180

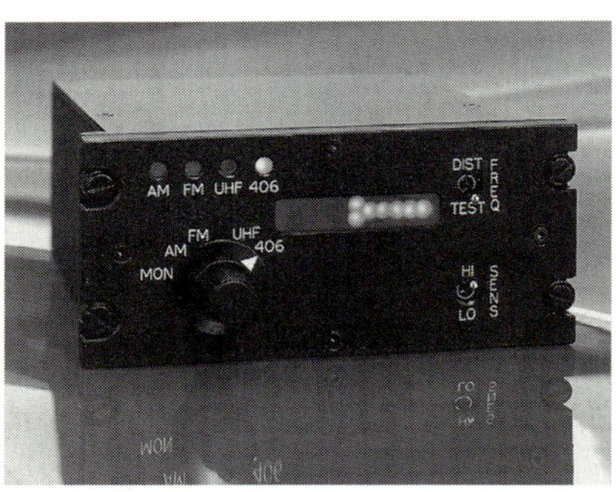

Series 406-053 NVIS-compatible homing system 0001181

SAR homing systems

Series 406

Fully compatible with the latest 406.025 MHz COSPAS/SARSAT emergency locator beacons, the 406 system is a complete self-contained unit interfacing with a single pair of antennas to provide 'left/right' steering information against a transmission source.

Designed to monitor four distress frequencies: 121.5, 156.8, 243.0 and 406.025 MHz, the unit processes the information and displays it on an analogue indicator. Adjacent-frequency test modes are provided.

Specifications

Radio frequencies: 121.5, 156.8, 243.0, 406.025 MHz
Dimensions: 146 × 66.6 × 155 mm
Weight: 0.9 kg

Series 406 derivatives

406-1 Homer: The initial system covers 121.5, 156.8, 243.0 MHz. Fitted to both rotary- and fixed-wing aircraft.
406-2 Homer: Identical in function to the 406-1 Homer, with the addition of 406.025 MHz capability. Both standard and night vision compatible; additional remote indicators optional.
406-3 Homer with extended frequency range capability: With a remote controller can be extended over the whole 100-400 MHz band, while retaining the 406-2

Homer features. Can be interfaced to ARINC 429 or 1553 databus.

406-053 Full NVIS-compatible homing system: Four frequency capability: 121.5, 156.8, 243.0, 406.025 MHz, plus additional frequencies at ±1 MHz of each distress frequency. 5 V or 28 V Green NVIS; choice of white or red lighting for non-NVIS units.

Series 407

The series 407 remote indicators are for use where operators need a lightweight indicator either as a secondary instrument for navigators' use or where panel space precludes use of the 406 series. The 407 provides a visual indication of the 'left/right' steering information to track a transmission source, the directional data is indicated only when a valid 'homing' signal is received from the 406 homing unit.

Specifications

Dimensions: 86 × 48 × 123 mm
Weight: 0.3 kg

Contractor

H R Smith.

VERIFIED

5ATI electronic display system for B-747 classic aircraft

Designed to replace the electromechanical ADI/HSI and engine instruments, the Smiths Industries display system for the Boeing 747-100, -200 and -300 aircraft is based on a 5ATI display unit with an Active Matrix Liquid Crystal Display (AMLCD). The system is fully compatible with present and future aircraft systems and provides significantly improved reliability and operational flexibility. The fully integrated display suite comprises seven identical display units, two EHSI control panels and an engine displays control panel all mounted on the aircraft's flight instrument panel. Data is exchanged in ARINC 429 format. Weather radar aand TAWS modes are also being provided as additional display modes.It is available as a fully integrated flight display system, as an EHSI/EADI only, or as an engines-only installation.

Operational status

The complete system has been integrated into the B747-200/300 series aircraft.

Contractor

Smiths Industries Aerospace.

VERIFIED

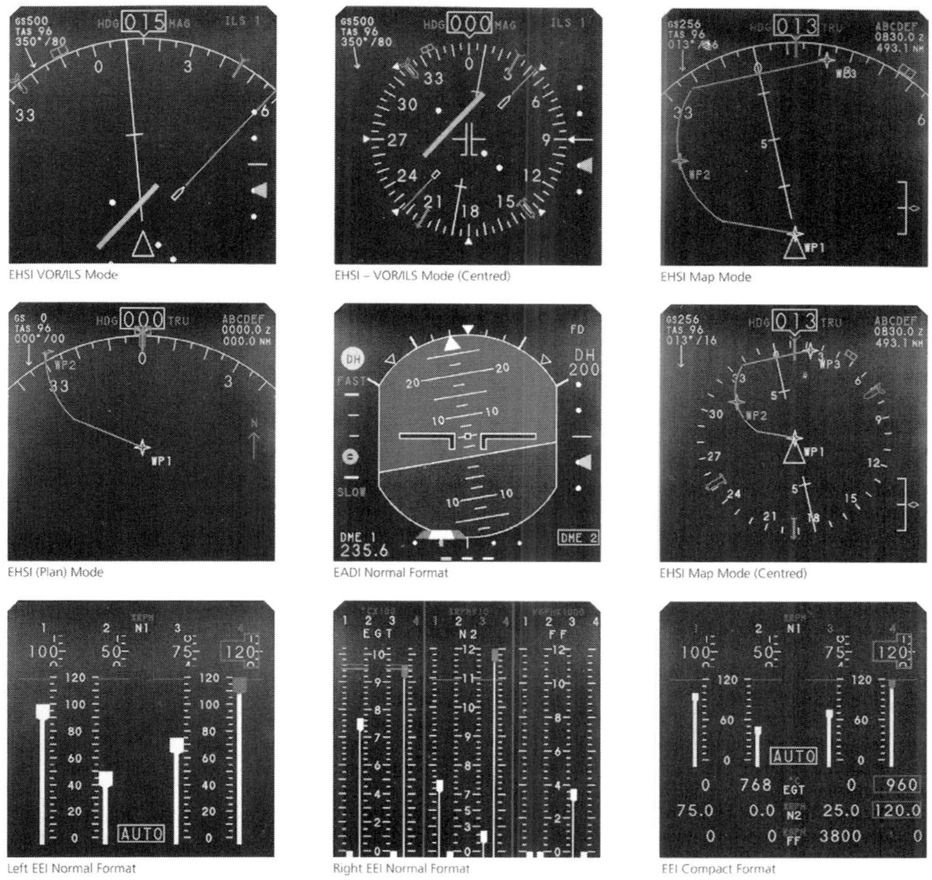

The Smiths Industries 5ATI display upgrade for B747-100/-200/-300 aircraft **1999**/0051686

300 RNA series Horizontal Situation Indicators (HSI)

The 300 RNA range of Horizontal Situation Indicators (HSIs) is intended for civil and military fixed-wing aircraft and helicopters. Each instrument consists of a mainframe, synchro frame and electronics. Large-scale integrated circuits are used for signal processing and synchros are used to drive the various displays.

The instruments can be used in Nav, Tac, App or ADF modes. The range display is a four-digit electronic module at the top of the instrument face reading up to 999 n miles. Another numeric display is used to indicate the setting of the command track pointer. A knob is provided for selecting the relevant runway QDM on the command track pointer. These instruments provide complete ILS information as well as Tacan and ADF displays.

Operational status

304/305 RNA HSIs are used in the BAE Systems Nimrod and 748 aircraft as integral parts of the SFS6 flight systems. The 309 RNA HSI equips Royal Navy Westland Sea King helicopters and Royal Air Force BAE Systems Hawk trainers. The 306/307 RNA is used in the SEPECAT Jaguar and BAE Systems Strikemaster. The 330 series HSI is used in the Royal Air Force's Panavia Tornado aircraft and Chinook helicopters.

Contractor

Smiths Industries Aerospace.

UPDATED

Active Matrix Liquid Crystal Displays (AMLCDs)

Smiths Industries' flat panel Active Matrix Liquid Crystal Displays (AMLCDs) are compact, lightweight, high-resolution colour displays, designed to meet military and commercial requirements, for both new-build and retrofit aircraft and helicopter installations. Smiths Industries has production contracts for AMLCDs in all its display facilities, both in the UK and USA.

3 ATI flat panel instruments

Selected for the JPATS aircraft, 3 ATI flat panel instruments provide a cost-effective, state-of-the-art means of providing graphics in a high-visibility display. Features include construction based on five modules (one dispay card and four flexible I/O cards); up to 16 colours at 130 ft Lamberts illumination, reconfiguration by pilot-operated switch, ARINC 429 interface, NVIS option. Applications include air data instruments, engine instruments, system status displays and standby displays.

These displays have been selected by Raytheon for the JPATS aircraft; they are also used for the V-22 standby flight display and for the F/A-18 E/F engine fuel display. Civil applications of the 3 ATI display include the Boeing 717.

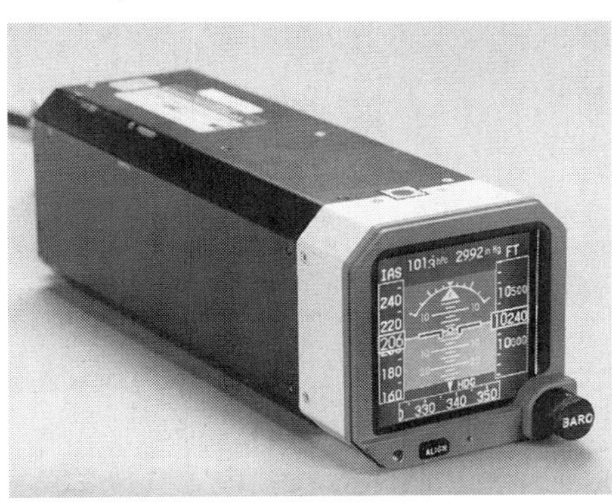

Smiths Industries' integrated standby instrument system **1999**/0051687

Integrated Standby Instrument System (ISIS)

Selected for the MD-90, B717-200, CR5-200, Learjet 31/45/60, ISIS combines the functions of three electromechanical standby instruments into a single-box, drop-in solution. This self-sensing flat panel instrument, with integral air data and/or altitude sensors, provides increased reliability and reduces the space required in the cockpit panel.

5 ATI flat panel instruments

Smiths Industries' 5 ATI design provides a fully self-contained solid-state instrument capable of directly interfacing to either analogue or digital sensors and presenting primary flight and/or engine information using graphics symbology. The 5 ATI flat panel design uses standard mountings, with no additional requirements for remote symbol generators or cooling air - they are, therefore, suited to upgrade and retrofit EFIS requirements, as well as new-build installations. Features include self-contained interface and graphics processing; analogue and/or digital ARINC 429 or MIL-STD-1553B interfaces; EADI and EHSI formats; ADI/HSI and engine instrument formats; video option (for example HUD), GPS/CNS/ATM compatibility; passive cooling.

Applications include retrofit and OEM installations, standby displays, message displays, video monitor displays.

These displays have been selected for the Royal Australian Air Force Hawk aircraft. Each twin-cockpit aircraft will have up to six identical display units to show flight, navigation, weapon and system symbology, plus digital map and sensor displays. Fitted to some Boeing 747 aircraft.

Smiths Industries' 3 ATI airspeed indicator display (left), engine instrument display (centre), and airspeed indicator (right) 0001367/0001368/0001369

Smiths Industries' 5 ATI attitude director (left) and map/weather display (right) 0001365/0001366

Multifunction Control Display Unit (MCDU)
Smiths Industries' flat panel MCDU employs a large full-colour AMLCD which complements modern military and civil cockpits. Features include: flexible reconfiguration of front panel keys; passive cooling; alternative ARINC 429, MIL-STD-1553, RS-422 databusses. Applications include GPS; CNS/ATM; ACARS; weather radar; FMS; Satcom/radio control; video, cockpit and maintenance displays

Selected by the UK MoD for the Nimrod MRA.4 aircraft upgrade.

5 × 6 in Electronic Display Unit (EDU)
Smiths Industries' 5 × 6 in Electronic Display Unit (EDU) presents engine and utility system parameters on a multicolour, flat panel AMLCD. Selected for the BAE Systems' Hawk and the Eurofighter Typhoon.

MultiFunction Glareshield Display (MFGD)
Smiths Industries' MultiFunction Glareshield Display (MFGD) is designed for glareshield installation to provide the pilot with continuous peripheral awareness weather in head-up or head-down attitude. The unit utilises full-colour AMLCD and is suited to the display of tactical messages in the CNS/ATM environment. Boeing has selected this display as a Para Visual Director (PVD) for the B777.

Smiths Industries' multifunction control display unit 0001364

Contractor
Smiths Industries Aerospace.

VERIFIED

Datalink Control Display Unit (DCDU)

Airbus Industrie will certify FANS systems for A330s and A340s incorporating Smiths Industries Datalink Control Display Units (DCDUs). The DCDU will also be available for retrofit to Airbus A319s, A320s and A321s.

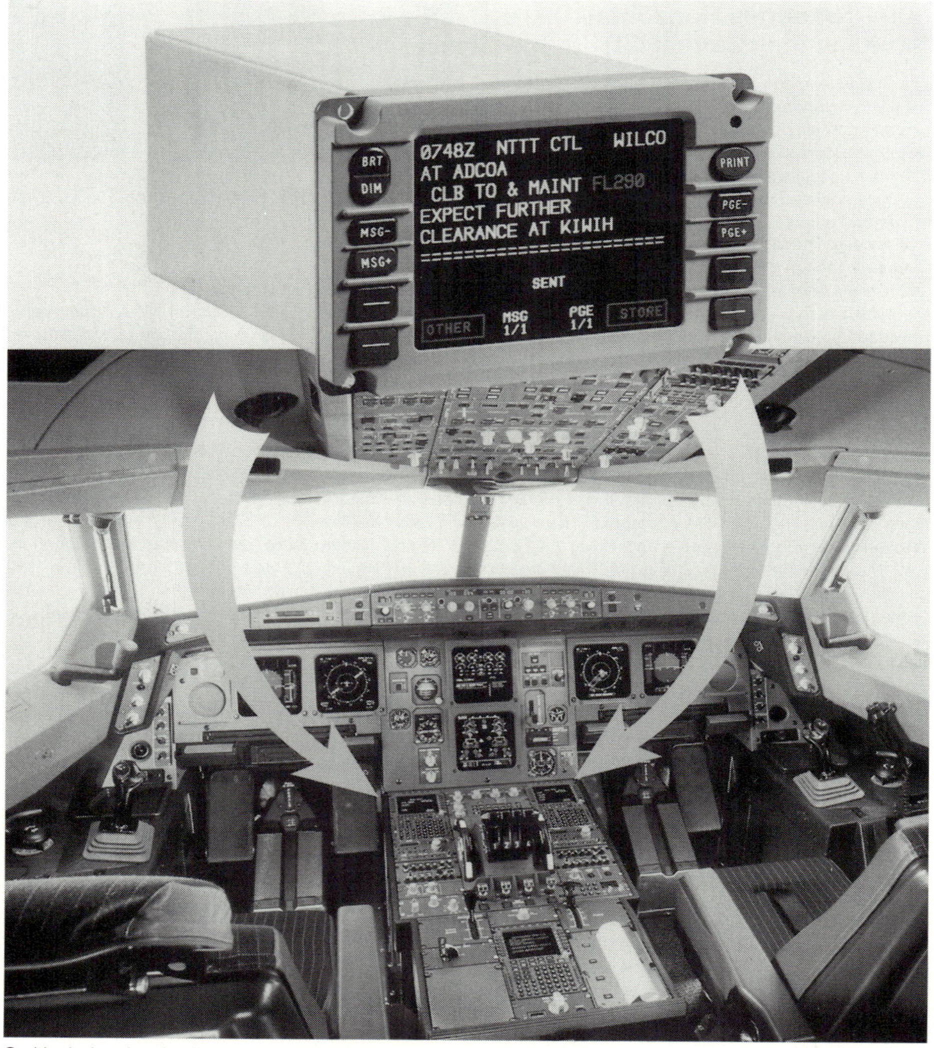

Smiths Industries datalink control display unit Airbus installation ***1999***/0051688

The DCDU is a development of the company's existing AMLCD technology. The DCDU is a 4 × 3 in, full colour, landscape display. Two DCDUs are fitted to each aircraft – one for each pilot. The DCDU is a cost-effective way of providing datalink messages to the pilot. There are plans to incorporate this unit into many aircraft and as retrofits to avoid costly modifications to the current display systems.

Operational status

A330/340 certification April 2003.

Contractor

Smiths Industries Aerospace.

VERIFIED

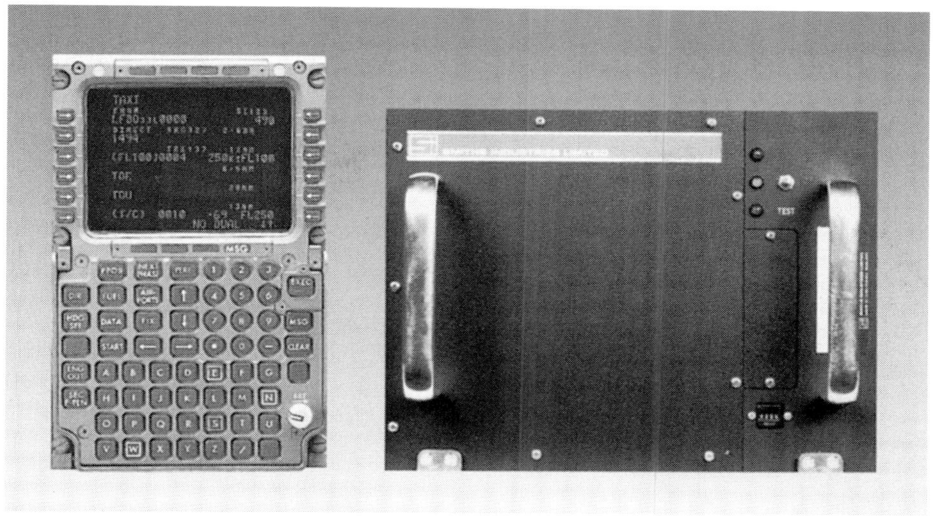

Enhanced flight management computer system for Airbus A300-600 and A310. ***1999***/0051635

Electrical Load Management System (ELMS) for B-777 aircraft

ELMS provides a comprehensive range of functions including distribution and protection of primary and secondary electrical power, and control of aircraft utilities sub-systems. The system also interfaces with the cockpit controls and displays via an ARINC 629 databus for control switching and system status reporting.

The Boeing 777 ELMS has three primary panels and four secondary panels: left, right and auxiliary power and ground handling/service distribution, left power, right power, and standby power.

The design of the ELMS features compact primary panels with line replaceable contactors. Three of the secondary power management panels incorporate dual ARINC 629 bus interfacing, dual processing elements and line replaceable electronic and switching relay modules. The design makes the maximum use of multiple redundancy techniques and immunisation against high intensity radio frequencies. Contributing to the integrity of the system is the incorporation of technology already proved on the Eurofighter Typhoon and Longbow Apache load management systems.

Operational status

In production for, and in service on Boeing 777. A development of the B-777 ELMS has been selected for use in the B767-400ER.

Contractor

Smiths Industries Aerospace.

VERIFIED

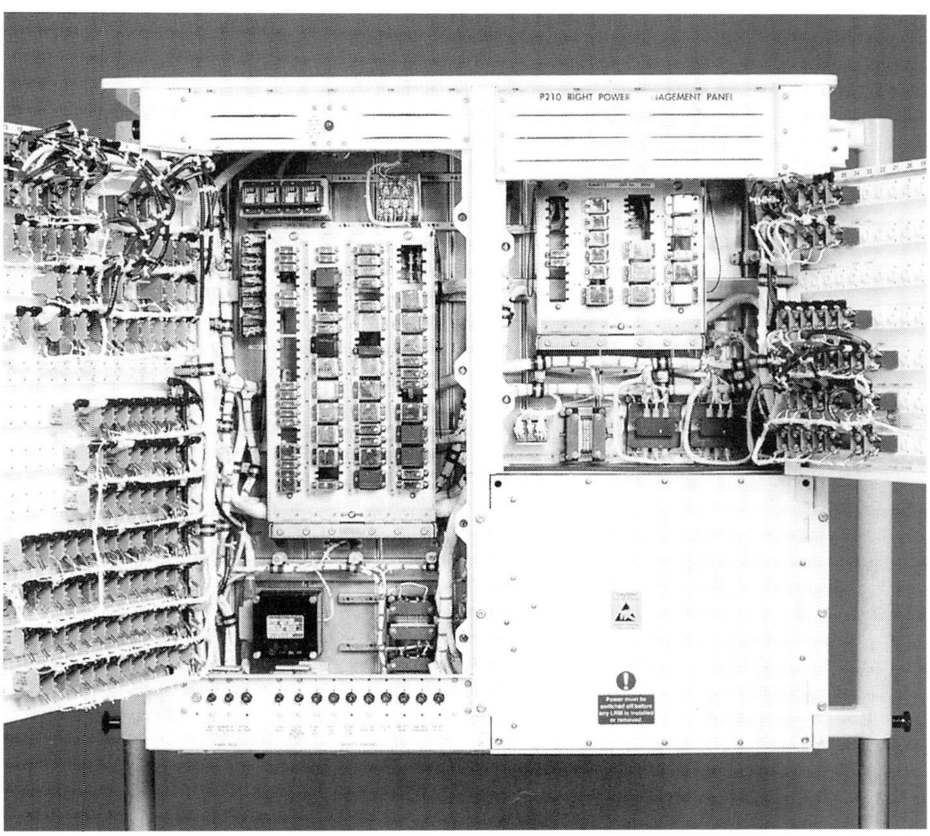

Smiths Industries ELMS for the Boeing 777 0003120

Flight Management Computer System (FMCS)

Conforming to the full ARINC 702 specification, and a standard option on the Airbus A310 and A300-600 aircraft, this Flight Management Computer System (FMCS) is the prime interface between crew and aircraft and enables optimum performance to be achieved from take off to final approach. Main functions include flight planning, navigation, performance optimisation, flight guidance (with coupling to autopilot and autothrottle) and display processing. The operational procedures create a working routine which is easy to implement and is similar for all phases of flight, optimising the factors affecting flight profile to give greater economy of fuel consumption, flight time and aircrew workload.

The system design is based on a parallel multiprocessing arrangement of microprocessors within the flight management computer unit. This technique permits high-processing capability and gives the flexibility to accommodate future expansion of functions and procedures. Two sets of dual 16-bit microprocessors - one dedicated to navigation, the other to performance functions - provide overall throughput of over 1 Mops. Additional microprocessors are dedicated to input/output and database control functions. A bubble memory provides 256 kwords of memory for navigation and performance database storage. There is provision for up to 56 discrete inputs and 16 discrete outputs, plus 32 input and 12 output ARINC 429 channels. The system contains its own built-in test routines which constantly monitor system operations and fault detection.

The crew interface is with the control/display unit which has a 14 lines by 24 character CRT format. The bottom line can be used for scratchpad entries. A full alphanumeric keyboard is provided, together with function keys and 12-line select keys adjacent the CRT. Self-contained built-in test provides a cued step-by-step test of all push-buttons, annunciators and the CRT display. For routine operations, most of the information is defaulted from the navigation database, requiring a minimum of manually entered data.

The Enhanced Flight Management Computer System (EFMCS) supersedes the FMCS. This provides one million words of EEPROM memory for navigation and performance database storage, replacing the 256 kwords of bubble memory in the FMCS. Further reliability and functional improvement are provided, including the facility for interfacing to an ARINC 615 high-speed data loader. The EFMCS is in the preproduction stage.

Singapore Airlines was the launch customer for the second-generation enhanced management computer for six new build Airbus A310 and retrofit of the existing fleet of 15 A310s. The enhanced system offers four times the current database capacity; an essential feature for extended route operations and improved performance and reliability.

Specifications

Dimensions:
(computer unit) 8 MCU
(control/display unit) 267 × 229 × 146 mm
Weight:
(computer unit) 12.7 kg
(control/display unit) 6.3 kg
Power:
(computer unit) 200 W
(control/display unit) 87 W

Operational status

In production. Airline customers include Kuwait Airways, Saudia, Air France, Sabena, Nigeria Airways, Air India, Cyprus Airways, Air Nuigini, Singapore Airlines, Air Algerie and Monarch Airlines. The Smiths Industries FMCS has also been chosen for the Boeing E-6A of the US Navy.

Contractor

Smiths Industries Aerospace.

VERIFIED

Fuel quantity gauging and indication systems

Smiths Industries designs and manufactures electronic equipment and systems for the measurement, management and indication of fuel in civil and military aircraft. Current fuel system applications include large civil transport aircraft such as the Boeing 777 and Airbus family; commuter aircraft and business jets such as the Raytheon Hawker 800/1000, BAE Systems 146 and Jetstream 41 and 61 aircraft; and military aircraft such as the BAE Systems Hawk, AMX and Eurofighter Typhoon. Smiths Industries is also a partner in collaborative programmes, including for the Airbus A300, A310, A319 and A320.

Analogue and digital displays

Fuel contents indicators range from simple moving coil analogue types to servo-pointer and digital multitank displays driven from an ARINC 429 databus. Internal illumination is optional and dial presentation and colour are displayed to specification. Solid-state LED or LCD indicators can be used to provide numeric, analogue or graphical presentations of fuel quantity.

Analogue and digital signal processors

The analogue output from capacitance probes can be processed by entirely analogue means and used to drive either analogue or digital indicators. Where higher accuracy or additional facilities are required, probe signals can be digitised and processed digitally to provide outputs of fuel mass in ARINC 429 or MIL-STD-1553 formats. All analogue and digital processors incorporate BITE, which performs levels of self-test varying from basic confidence checks to comprehensive system testing and calibration.

Digital Fuel Quantity Indication System (FQIS)

The FQIS developed for the Airbus A300, A310, A319 and A320, incorporates advanced digital computing technology which brings improved accuracy and reliability to the system compared with analogue equipment installed on earlier aircraft. BITE, failure recording and a recall facility are included.

The system computer is a two-channel (dual-redundant) digital data processing unit which calculates the fuel mass and performs control, monitoring and display functions. The architecture ensures that no single fault can cause loss of the gauging function; if one channel fails, the second channel assumes the primary function with no loss of performance. Operational software for both channels is embedded into an On-Board Removable Memory Module (OBRMM) fitted on the front of the computer. Software upgrades can be introduced without dismantling the computer or removing it from the aircraft.

Refuelling (and de-fuelling) is controlled from a Preselector Unit on the refuel panel. Once the operator has selected the fuel mass required, the system fills the tanks simultaneously to save time and automatically distributes the fuel to maintain the physical balance of the aircraft. Pre-selected and actual fuel mass are displayed continuously throughout the refuelling process.

Fuel level sensing systems

Fuel level sensing systems provide an accurate and safe means to detect fuel levels. They feature a low-cost, solid-state fluid level sensor connected to a separate switch unit which can either stand alone or be incorporated into another unit within the fuel system. The lightweight sensor unit is small enough to be mounted on a tank wall or fuel gauging probe and is immune to temperature effects over a wide operating range. Applications include high- and low-level warning indication or control, automatic shut-off switching for refuelling, automatic control of liquid transfer and sequential draining and filling of tanks. Fuel level sensors may also be applied to other fluids such as oil or hydraulic fluid.

Ultra-sonic and capacitance tank probes

Fuel height within tanks can be measured using either capacitance or ultra-sonic probes. Capacitance probes can be used in analogue or digital fuel systems. Systems using ultra-sonic probes are digital throughout. It is usual to install several probes in each tank so that fuel levels can be gauged accurately over a wide range of aircraft attitudes and fuel contents. Non-linear height and volume characteristics of fuel tanks can be accommodated either by mechanically profiling the linear electrodes of the probes or, in digital systems, by incorporating appropriate software in the processor.

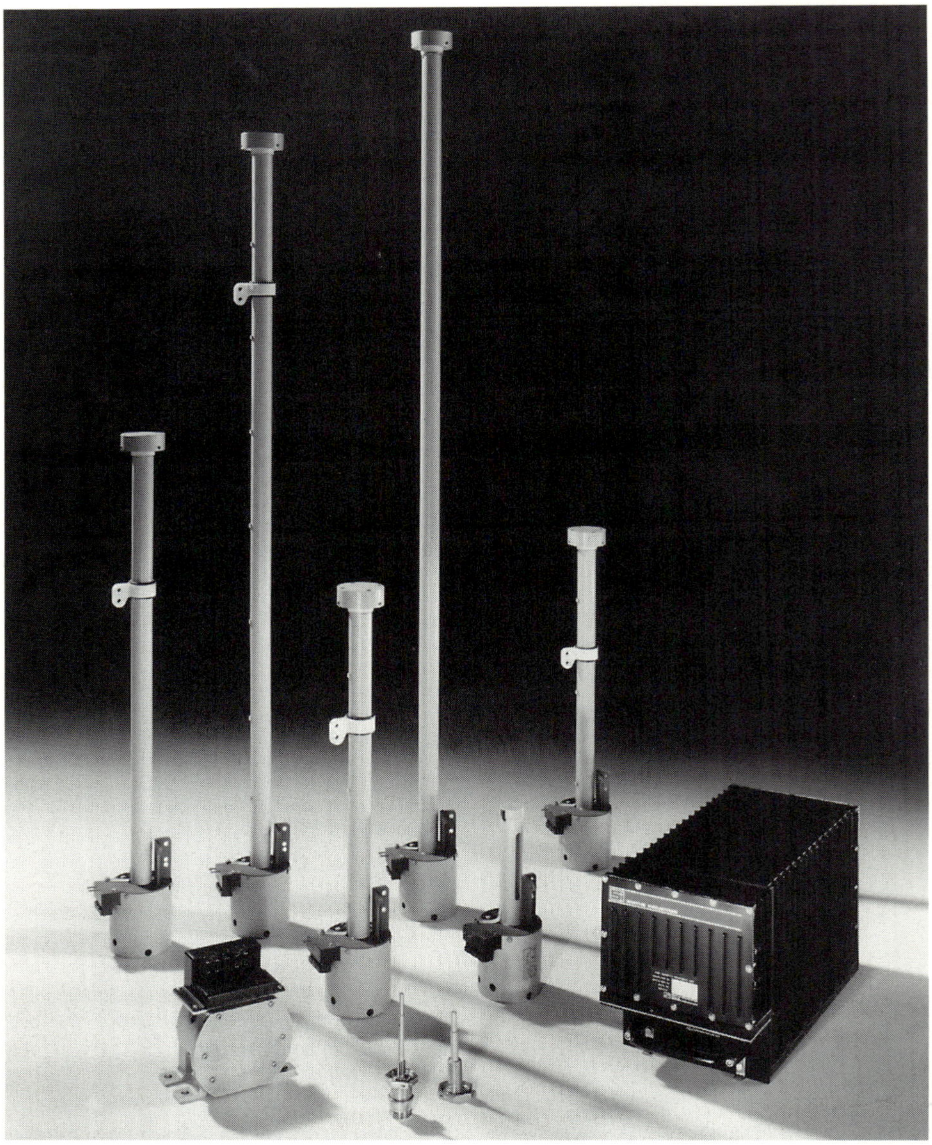

Ultra-sonic fuel quantity gauging system 0001423

'Smart' probes

Smiths Industries Aerospace 'smart' probes incorporate signal processing electronics. The processed signal from each probe is sent to the central fuel computer using Remote Interface Units (RIU) connected to the aircraft's databus. The advantage of this probe is that it permits use of a more distributed avionics architecture, and reduces wiring requirements by use of the databus. The system is in production for and in service in EH101; it has been selected for Boeing VCAV CDA, Eurofighter Typhoon and BAE Systems Nimrod MRA4.

Ultra-sonic Fuel Quantity Indication System (FQIS)

This digital system uses ultra-sonic sound to provide highly accurate measurement of fuel quantity in the B-777. Fifty-four sensors are distributed among the Boeing 777's wing and fuselage tanks. A central processor addresses each sensor individually and computes fuel volume and mass in each tank. Each fuel tank has a densitometer and a water detector, and one of the wing tanks is fitted with a fuel temperature sensor. Fuel management data is provided to the crew via an ARINC 629 databus linked to the cockpit displays. The data includes: quantity in each tank; total quantity; fuel imbalance and low-level warnings. There is a complete built-in test facility and the total system is designed to have a high level of tolerance to faults. In addition to the cockpit display of information, there is a refuel control and display panel linked to the central processor via an ARINC 429 databus.

Hawker Horizon Fuel Management System

Smiths Industries Aerospace is responsible for designing and delivering the complete fuel system for the Raytheon Hawker Horizon aircraft.

Operational status

In production and in service in a large number of aircraft types.

Contractor

Smiths Industries Aerospace.

VERIFIED

Ground roll director system

The Smiths Industries ground roll director system provides pilots with head-free guidance information. It meets the requirements for ground roll guidance in Cat IIIB weather conditions. It has been designed so that pilots can easily revert to para-visual guidance when forward visual references disappear in conditions of deteriorating visibility such as drifting fog, RVR reporting failure, or differing fog densities.

The system is based on the Para-Visual Director (PVD) concept developed during the early 1960s. A display unit is positioned in the glareshield directly in front of each pilot. During head-free operation, while concentrating on external visual cues, the pilot immediately registers any movement of the black and white bands in his peripheral vision and makes instinctive corrections to the azimuth steering controls without looking directly at the display unit. For the Boeing 777, an Active Matrix Liquid Crystal Display (AMLCD) replaces the original electromechanical 'barber's pole' indicator, providing a 100 × 25 mm usable area that offers the additional use as a multifunction display for messages such as those associated with an ATC datalink or FMC.

Specifications

Dimensions:
(display) 204 × 63.6 × 33.8 mm
(computer) 321 × 193.5 × 61.5 mm
Weight:
(display) 0.55 kg
(computer) 2 kg

Operational status
In production for the Lockheed Martin TriStar L-1011 and Boeing 747-400, 757, 767, 777 and MD-11.

Contractor
Smiths Industries Aerospace.

VERIFIED

HRA series radar altimeter

The HRA series radar altimeter unit meets all requirements for a highly accurate measurement and flight-deck indicator on civil and military aircraft. The outputs are suitable for use with flight control systems including those used during automatic landing. Other applications include terrain-following and avoidance, reconnaissance and anti-submarine warfare.

Specifications
Dimensions: 203 × 150 × 96 mm
Weight: 2.9 kg
Power supply: 115 V AC, 400 Hz, 28 V DC
Frequency: 4.3 GHz
Pulse repetition frequency: 10 kHz
Radiated power: 100 W peak (nominal)
Altitude: 0-5,000 ft, 0-2,500 ft, 0-1,000 ft
Output signals: analogue
Accuracy: ±3 ft or 3%
Manoeuvrability:
(typical) pitch and roll ±35° within stated accuracy
Temperature range:
(operating) −40 to +71°C
(non-operating) −54 to +95°C

Operational status
In service.

Contractor
Smiths Industries Aerospace.

VERIFIED

KTX series radar altimeter

The KTX series is a lightweight pulse radar altimeter developed for applications requiring low-volume installation and is particularly suitable for executive aircraft, low-cost military aircraft and missiles. It incorporates a high proportion of integrated and solid-state circuitry, as well as the operational features of other Smiths Industries radar altimeters.

Specifications
Dimensions: 152 × 102 × 86 mm
Weight: 1.36 kg
Power supply: 28 V DC
Frequency: 4.3 GHz
Pulse repetition frequency: 8 kHz ± 2 kHz
Radiated power: 50 W peak
Altitude: 0-5,000 ft, 0-2,500 ft, 0-1,000 ft
Output signals: analogue
Accuracy: ±3 ft or 3%
Manoeuvrability:
(typical) pitch and roll ±30° within stated accuracy
Temperature range:
(operating) −55 to +70°C
(non-operating) −55 to +90°C

Operational status
In service.

Contractor
Smiths Industries Aerospace.

VERIFIED

LED dot matrix displays

In parallel with the development of CRT-based electronic flight instrument system displays, Smiths Industries produced a family of engine performance indicators using a hybrid DC-servo pointer/LED readout. The servo-driven pointer moves over a conventional circular scale, while an emissive electro-optic drum counter constructed from a dot matrix of high-brightness LED offers a display which is fully compatible with the CRT and light emissive displays now being introduced in civil and military aircraft.

The readout is of the rolling drum type, with values appearing to roll smoothly upwards or downwards. Compared with LED bar instrumentation, this device is much less affected by single LED failures and very accurate interpolation of data is possible due to the apparent motion in the display. A 7 × 5 matrix 4.5 mm high is used, with three blank rows of LEDs between each figure, but a 9 × 5 matrix giving 5.2 mm high characters and with reduced character spacing is an alternative. In place of a mechanical failure flag the LED matrix displays N_1 and N_2 and the word 'Fail' if an indication is outside limits or invalid.

To minimise the number of moving parts a microprocessor is used to convert the input signal to digital form to drive the counter and then convert the digital signal back to analogue in order to drive a conventional pointer DC servomotor. The instrument is contained in a 2 in (51 mm) round or 2 ATI case, with integral analogue display lighting and an ambient light sensor which automatically adjusts LED brightness.

Operational status
In production. LED instruments are used on Boeing 737-300/400/500 aircraft to display N_1 and N_2 engine speeds, Exhaust Gas Temperature (EGT) and fuel flow/fuel used for the CFM56 power plant. They are also supplied for the Airbus A310 and A300-600 aircraft and the Saab JAS 39.

Contractor
Smiths Industries Aerospace.

VERIFIED

LED engine and system displays

Smiths Industries' solid-state instrument systems include primary engine displays, aircraft system displays and annunciator panels, and are direct replacements for electromechanical instruments. They are ideal for both new-build and existing aircraft, where they can be retrofitted by easy change of panels and without the need for mechanical or electrical modifications.

Each display system presents information in formats familiar to the aircrew, who thus require minimal training in their use. These displays provide significant benefits in terms of weight, power consumption, reliability, ease of maintenance and cost of ownership.

Primary engine display
The primary engine and aircraft displays are each contained in a single unit which, for ease of maintenance, comprises three modules: a display and associated dimmers, a printed circuit board assembly and a power supply module. Dial markings and legends associated with the displayed parameters are printed on a glass lens which is held in place over the LED displays. Both units are housed in conveniently sealed, lightweight cases with connectors mounted on the rear. Interconnections between the modules and rear connectors are by flexible tape wiring.

For optimum integrity the primary engine display incorporates a separate processor for each engine parameter. Four power supplies are provided to ensure system integrity. The power supplies are configured as two independent pairs so that a failure will result only in partial loss of the parameters of one engine.

Secondary or system display
In the secondary or system display the required reliability is achieved by using multiple processors and power supplies with the parameters to be processed suitably arranged between them.

Brightness is controlled automatically to achieve comfortable viewing over the whole range of ambient lighting conditions on the flight deck, from total darkness to direct sunlight.

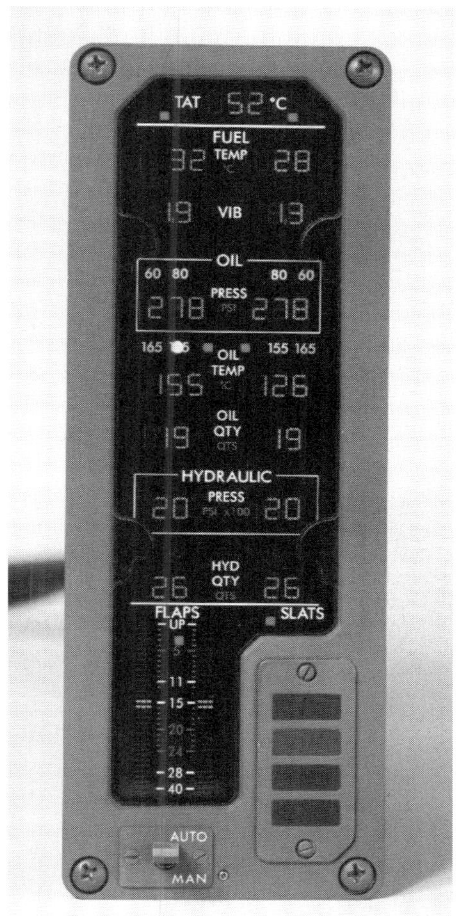

Smiths Industries' LED engine and system display

Smiths Industries engine indicators for the display of engine speed (left), exhaust gas temperature (centre) and fuel flow (right)

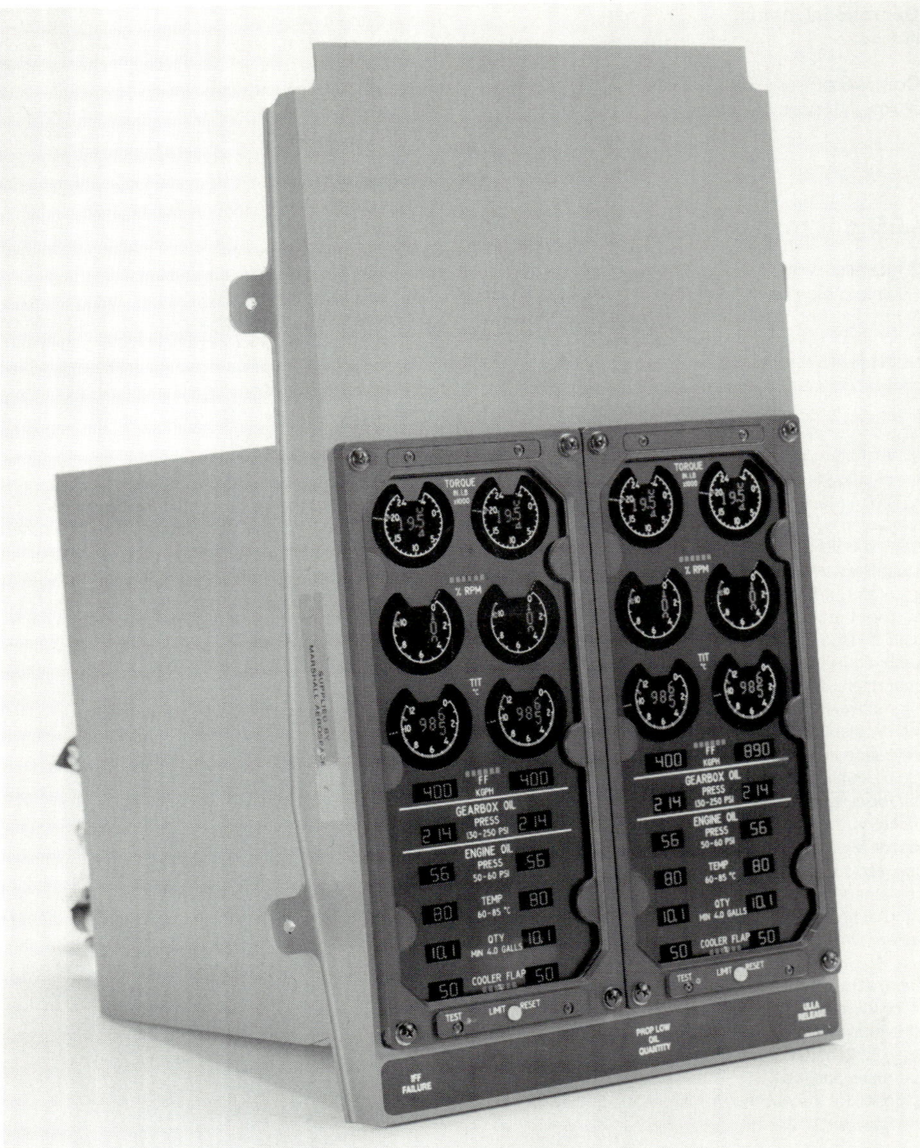

Smiths Industries' Engine Instrumentation Display System (EIDS) for C-130 and P-3 aircraft ***1999***/0051689

Comprehensive BITE facilities are a feature of all Smiths Industries LED display systems. On power-up, a test sequence is initiated automatically. Test routines can be initiated by either the pilot or the maintenance crew.

Annunciator panel
The overhead annunciator panel is a solid-state replacement for existing discrete annunciator panels. Capable of displaying up to 120 cautionary messages, six warnings and 15 advisory indications, it can be installed in any MD-80 series aircraft without the need for electrical or mechanical modifications. On new-build aircraft in particular it makes possible considerable savings in wiring, connectors and weight. Caution messages are displayed on two dot matrix LED panels. Each panel can display six simultaneous cautions of up to 20 characters in length. Both message displays can be scrolled up or down so that a total of 120 different cautions can be viewed.

Eight push-button switches with integral filament lamps operate in conjunction with the two LED message panels. Each switch is associated with a particular aircraft system. Whenever a caution input is received the corresponding lamp is lit so that a system caution is annunciated even if both LED message panels are full.

Normally, cautions received from all aircraft systems are displayed on the message panels in either chronological or priority order as required. If the pilot prefers, he can use the push-button to select one particular aircraft system so that only cautions associated with that system will be displayed.

The two panels are arranged to display a block of 12 consecutive messages. If a fault occurs on one panel then all messages are automatically made available on the remaining panel while the faulty panel remains blank.

A new development of the annunciator panel is the master warning and caution system selected for the MD-90. This system, comprising a separate display and control unit, identifies and presents more than 200 messages.

Operational status
Fitted in Boeing 737, MD-80 and MD-90, and BAE Systems 146 and Jetstream 41 aircraft. Also available as a developed system for C-130 and P-3 aircraft retrofit.

Engine Instrumentation Display System (EIDS)
Smiths Industries has developed, in conjunction with the UK Royal Air Force, a new Engine Instrumentation Display System (EIDS) for the C-130K using LED technology. It has also been selected for use in the US Coast Guard P-3 Orion aircraft.

The EIDS is a direct replacement for the existing 32 electromechanical instruments of existing C-130s and P-3s. The new technology provides significant benefits in terms of weight, power consumption, reliability and cost of ownership. The retrofitting of the EIDS also releases valuable panel space for potential display upgrade programmes.

The EIDS presents primary parameters in a pointer/counter format with secondary parameters in a numeric format. These provide one easily read display requiring minimum crew re-training. The system provides reliability at least 20 times better than existing indicators. It is a COTS design based on more than 2,000 systems with over 10 million hours in service on commercial aircraft.

Operational status
Under development as part of a UK MoD contract.

Contractor
Smiths Industries Aerospace.

VERIFIED

LED standby engine display panel

Smiths Industries Aerospace supply a standby engine indicator display panel for the Boeing 757 and 767.

The display of engine parameters is by a light emitting diode in a seven-bar format and incorporates four numeric readouts for each engine - N_1 and N_2, EGT and EPR. Eight engine displays are housed in a 3 ATI case.

Operational status
In production for the Boeing 757 and 767.

Contractor
Smiths Industries Aerospace. ***VERIFIED***

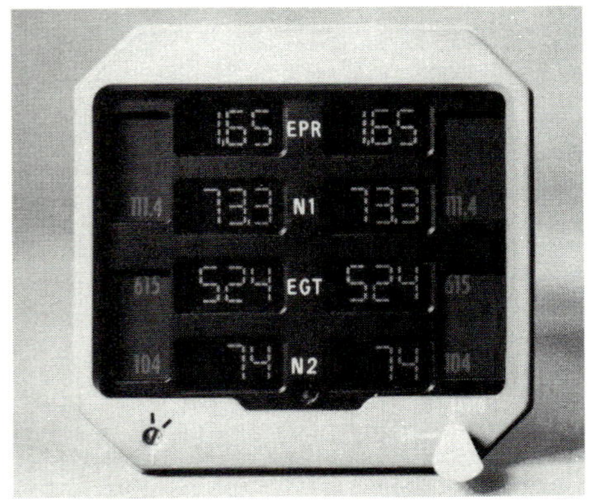

Standby engine display panel for the Boeing 757 and 767

Project ICHTHUS

A consortium of UK companies, led by Smiths Industries Aerospace, has launched a three year research and technology demonstrator project in the developing field of integrated health and usage monitoring systems for aircraft.

Supported by the UK Department of Trade and Industry (DTI) Aerospace and Defence Industries Directorate, Project ICHTHUS (Integrated and CoHerent Technology for aircraft Health and Usage Support systems) will examine a number of emerging technologies for monitoring engines, and fixed- and rotary-wing aircraft systems and structures. In particular it will address the integration issues surrounding such technologies.

The project's partners are BAE Systems, Airbus and GKN Westland Helicopters as aircraft constructors; Rolls-Royce as engine manufacturer; and Smiths Industries Aerospace, BAE Systems, Avionics, and Ultra Electronics as equipment manufacturers. The consortium will consult major operators in assessing the economic implications of the technologies investigated.

The project will be supported by the Department of Trade and Industry (DTI), under its Civil Aircraft Research and Demonstration Programme (CARAD).

Contractor
Smiths Industries Aerospace (leading a consortium of UK companies).

VERIFIED

SEP10 Automatic Flight Control System (AFCS)

The SEP10 automatic flight control system is fitted to the British Aerospace 146. It provides three-axis control or stabilisation and incorporates a pitch and roll two-axis autopilot, elevator trim, flight director and yaw damping facilities. It uses simple, well-proved control laws and the minimum of sensors. There is also a synchronise control facility which allows the pilot to disengage the autopilot clutches and sensor chasers temporarily and to manoeuvre the aircraft manually, so adjusting the data of the basic and manometric autopilot modes.

The autopilot is based on rate control laws. Pitch and roll rate signals are derived from ARINC three-wire attitude references, thus eliminating the need for rate gyros. Other ARINC standard interfaces accept a wide range of sensor inputs, including those from barometric and radio navaid sensors, and allow systems to be tailored to suit operators' needs. The autopilot computer uses digital computing techniques to provide outer loop control and to organise the mode logic, and has capacity to accommodate optional facilities. Analogue computing is used for the inner loop stabilisation computing, servo-drive amplifiers and safety monitors.

The system can be supplied with either a parallel acting yaw damper, which uses a rotary servomotor to drive the rudder and rudder pedals, or a duplicated series yaw damper which drives linear actuators in series with the rudder control run. In each case, the yaw damping system is self-contained and consists of an analogue yaw computer, sensor and the relevant actuator or servo motor.

Flight director computations are performed within the digital section of the autopilot computer, which can supply commands to V-bar or split-axis flight directors. The flight director and autopilot share common mode selection and outer loop guidance but, if desired, they can be operated independently.

Emphasis has been placed on maintainability and ease of testing, both for the installed system and for individual units in the workshop. Routine testing is designed to confirm correct functioning of safety devices, the tests being performed by operating a test-button in conjunction with buttons on the mode selector. Modular construction has been used extensively to ensure that faulty equipment can be corrected and recertified easily and quickly.

The following descriptions of individual LRUs outline the operation of a full SEP10 system.

Autopilot controller The autopilot controller, in addition to providing autopilot and yaw damper engage or disengage controls, also includes pitch rate and roll angle selectors, and the elevator and rudder trim indicators. Engagement of the autopilot and yaw damper is confirmed by the illumination of a legend within each selector. Pitch control uses a spring-centred lever which has a non-linear feel so that minor adjustments can be made instinctively. Roll control is accomplished by rotation of the control knob, which remains offset by a displacement proportional to the roll angle demanded in the basic mode, but returns automatically to the central position on selection of an alternative mode.

Mode selector There are 11 push-button switches, each illuminating as mode indicators for the selection of both autopilot and flight director functions; control mode engagement is confirmed by the illumination of a white triangle on the appropriate button.

A turbulence facility is included to soften flight disturbances in turbulent air. This reverts the autopilot to the basic stabilisation mode and at the same time reduces the overall gain of the system. Autopilot and engagement lights are provided so that the engagement state of the system can be seen on the mode selector. There is also provision for remote mode indication.

Autopilot computer The autopilot computer receives both analogue and logic information from sensors, controllers and selectors and processes it to formulate the pitch and roll axis demands and the flight director commands. The majority of autopilot computing is performed digitally, although analogue techniques are used to provide pitch and roll stabilisation and authority limitations. Correct functioning of the computer safety circuits is verified by a test facility at a convenient remote station.

Yaw computer This unit takes short-term damping information from a yaw rate gyro, a lateral accelerometer and the roll VRU; it drives the series rudder actuator to provide yaw damping and turn co-ordination. For aircraft types requiring a parallel damper, such a system is available.

Altitude selector The altitude selector provides facilities for altitude preselect mode as well as the normal altitude alerting functions. Altitude information is obtained from either a servo altimeter or an air data source. Selected altitude is presented on a counter display. A warning flag obscures this display in the event of a power failure or absence of altitude valid signal, and a test facility allows checking of the associated audio-visual signals and altitude preselect function.

Air data unit Where there is a requirement for a Mach hold facility to secure better fuel economy, the basic airspeed sensor can be replaced with an air data unit providing the necessary extra outputs.

Monitor computer For operation to Cat II weather minima, this unit computes the performance monitor functions necessary to provide a fail-safe pitch channel. It independently monitors autopilot pitch rate, localiser and glide slope deviation and provides outputs that can be used to disconnect the autopilot and provide warnings to the pilot. The computer is completely independent of the autopilot, and a self-test facility allows a check to be made on the correct operation of all the monitoring functions.

Operational status
Selected for the BAE Systems 146.

Contractor
Smiths Industries Aerospace.

VERIFIED

SEP20 Automatic Flight Control System (AFCS)

Both the autostabiliser and the autopilot in the SEP20 AFCS are fully digital and achieve levels of reliability and repeatability higher than was possible with earlier analogue systems. In addition, digital technology allows the mode of operation to be modified according to varying flight conditions or aircraft configurations. If required, the pilot can uncouple the autopilot and fly the aircraft using flight director commands provided by the system.

A comprehensive built-in test facility is incorporated, providing output data on the status of the system. Maintenance and release test functions are included, together with preflight safety checks, in-flight monitoring and the ability to identify a faulty LRU quickly.

The full AFCS comprises two identical digital Flight Control Computers (FCC), a Pilot's Control Unit (PCU), a Dynamic Sensor Unit (DSU) and a Hover Trim Control (HTC) unit for helicopter SAR applications. Extensive use is made of ARINC 429 both for external communications and for communications within the system. The equipment is designed to suit widely differing primary roles, such as anti-submarine operations or civil passenger transport, necessitating exacting safety standards.

The two identical FCCs are packaged in a 6 MCU configuration. Each comprises nine printed circuit cards incorporating four microprocessors of two widely used types - the Intel 80286 and Motorola 68000. Each microprocessor has been programmed independently to minimise the possibility of common mode faults.

The PCU enables the pilot to select the required AFCS control mode and displays the state of mode engagement. The unit is divided into two segregated sections in order to maintain integrity and fault survivability.

To achieve the level of sensor signal redundancy for failure survival in the yaw axis, the DSU incorporates a yaw rate gyro, lateral accelerometer and normal accelerometer. Output signals are provided to ARINC 429 digital format.

For helicopter applications the HTC is integral with the winch controller and provides the winchman with limited authority control of the aircraft through the AFCS hover trim mode.

Operational status
Selected for the Westland/Agusta EH 101 helicopter.

Contractor
Smiths Industries Aerospace.

VERIFIED

'Smart' throttle actuators

The 'Smart' throttle actuators for the Boeing 777 represent a new generation design using 'Application Specific Integrated Circuit' (ASIC) technology. ASIC reduces volume, weight and the number of wiring connectors by integrating the avionics within the electromechanical actuator of each throttle lever.

The smart actuator of each throttle lever receives digital commands from the 777s Airplane Information Management System (AIMS). The actuators drive the throttles to the commanded position as well as feeding back throttle positional data to the AIMS.

Operational status
In production for, and in service on, Boeing 777 and 737.

Contractor
Smiths Industries Aerospace.

VERIFIED

STS 10 full flight regime autothrottle

Designed for the Boeing 737-300 and now installed as standard equipment, the Smiths Industries autothrottle has been developed from the highly successful system supplied to Boeing for the 727-200 and 737-200 aircraft. In 1990, the STS 10 was fitted to the 737-500 and is capable of operating with intermixed engine situations.

A single unit will fit either 737-300, 737-400 or 737-500. It interfaces with flight management systems, digital air data systems, inertial reference systems and digital autopilots and uses advanced digital techniques for higher reliability, easier maintenance and lower cost of ownership.

The system comprises a digital computer with independent electromechanical drive to each throttle lever. The computer, which is housed in a single ½ ATR long box, accepts analogue and digital information from sensors and systems on board the aircraft. After processing this data the computer generates outputs to drive servo-actuators which adjust the position of each throttle lever independently, so achieving optimum engine performance. A further output from the same computer drives the fast/slow indicators on the ADIs or EFIS displays.

The autothrottle includes a number of unique features designed to enhance performance and promote flight safety. A particular feature of the system is the ability to override the actuator drive and adjust the throttle levers manually, without the pilot applying more force than he would normally use in manual operation.

To achieve precise control throughout the full flight regime, the autothrottle computer continuously monitors all the necessary engine and aircraft parameters and adjusts the thrust in accordance with the prevailing flight conditions. Protection is included to prevent exceeding predetermined N1 engine limits and maximum aircraft incidence.

The system includes damping controls which are designed to minimise throttle activity during normal flight conditions.

If a large change in vertical windspeed occurs during the approach a command is inserted which enables the system to achieve the required level of thrust more quickly.

With the launch of the Boeing 737-600/700/800 the autothrottle computer is being repackaged into a smaller lighter case. By the use of ASIC technology the autothrottle computer is being re-engineered into a ⅜ ATR Short case. The software embodied in the new unit will remain largely unchanged but is adapted for the differences in airframe performance and changed engine characteristics. The full flight regime features of the current B737-300/400/500 autothrottle computer will be embodied in the new autothrottle computer for the 737-600/700/800.

The single largest change is that the 737-600/700/800 aircraft will use FADEC controlled engines in the same way as the 777. This similarity has lead to a change in the servo drive for the autothrottle system. In the 737-300/400/500 it was necessary to drive the cables which routed the throttle level commands to the engine. In FADEC controlled aircraft it is only necessary to drive the pilots' throttle levers and therefore the technology used in Autothrottle Servo Motor (ASM) developed for the 777 has been adapted for use in the 737-600/700/800. The main change for Smiths Industries was the need to adapt the 777 unit into a smaller envelope for the new 737 variants.

Operational status
In production. The unit is standard fit on 737-300, 737-400, 737-500 and 737-700 for which a new ⅜ ATR unit will be produced.

Contractor
Smiths Industries Aerospace.

VERIFIED

NavSymm DR5-96S UHF differential datalink system

The NavSymm DR5-96S differential datalink system is designed to work with the NavSymm Sharpe XR6 GPS receiver to provide a complete solution to the differential GPS requirement.

The NavSymm DR5-96S operates at 9,600 bps, enabling the receiver to accept RTCM SC-104 (version 2.1) messages and return positional information down the same datalink. It can be synchronised to GPS time so that mobile location data can be retrieved and returned at scheduled times.

The NavSymm DR5-96S can also be used to transmit raw measurement data in real-time positioning systems in order to achieve accuracies down to a few centimetres. The radio is tolerant of noisy RF environments, and is available with two software-selectable, predefined, factory-set frequencies in the FCC licence-free band of 450 to 470 MHz.

Specifications

RF range: 450-470 MHz; 2 programmable synthesised channels
Channel spacing: 12.5 or 25 kHz
Transmit power: 2 W
Transmit data rate: 9,600 or 4,800 bps
Operating mode: half duplex
Dimensions: 175 × 80 × 57 mm
Weight: 1 kg
Power: 10-32 V DC; 1.2 W receiver; 13.5 W transmit

Operational status

Available.

Contractor

Symmetricom Ltd.

VERIFIED

NavSymm DR5-96S UHF differential datalink system
0011832

NavSymm DR5-RDS VHF differential datalink system

By receiving GPS differential corrections from Differential Corrections Services provided over the Radio Data System (RDS) the NavSymm DR5-RDS is able to provide DGPS accuracy over a wide area without establishing an RTCM correction transmitter.

The RDS is a system of transmitting data over an inaudible FM sub-carrier with normal communications.

The unused capacity of some RDS channels is used to transmit DGPS correction data. The NavSymm DR5-RDS decodes these signals and outputs them in standard RTCM SC 104 form to the GPS receiver.

Specifications

Frequency range: 87.5-107.9 MHz, 100 kHz steps, agile PLL
Scanning performance: 10 to 40 seconds to detect all the RDS stations of the FM band
Dimensions: 175 × 78 × 74 mm
Weight: 0.77 kg
Power: 11-32 V DC; 0.6 W

Operational status

Available

Contractor

Symmetricom Ltd.

VERIFIED

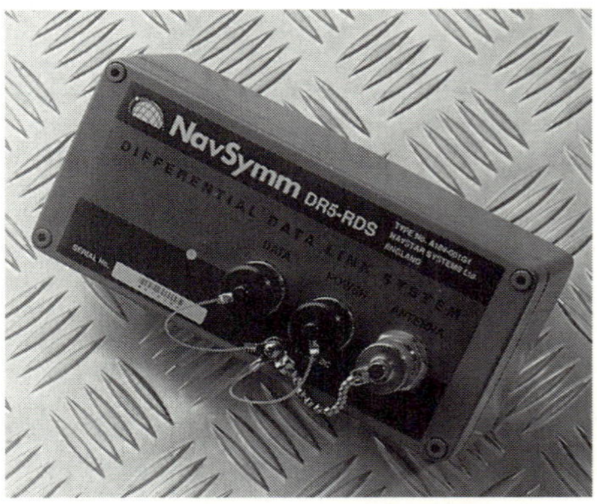

NavSymm DR5-RDS VHF differential datalink system
0011831

NavSymm Sharpe XR6 12-channel GPS receiver

The NavSymm Sharpe XR6, 12-channel receiver, is a totally redesigned unit based on the earlier NavSymm XR5 series; it offers much improved levels of accuracy and performance. The NavSymm Sharpe XR6 receiver provides two-way communication on all three ports for transfer of information at up to 20 Hz. Satellite tracking reacquisition is achieved in under 1 second, at acceleration rates of up to 4 *g*. It also incorporates an event marker allowing other equipment to demand position information. The event marker is activated in one of two ways, either by the arrival of a pulse or an ASCII string. In the latter case, the position information is incorporated with the ASCII string and either output on a data port or stored in the internal memory. Uplink of Ephemeris data is also possible via a datalink using RTCM messages thus ensuring optimum performance in highly-dynamic situations.

NavSymm Sharpe XR6 12-channel GPS receiver
0015373

The NavSymm Sharpe XR6 has been designed with open system architecture, offering access to many types of raw data for system integration purposes. It can also be used as a base station for the purpose of RTCM messages.

Sharpe XR6 RPS

The Sharpe XR6 Relative Position System (RPS) is an advanced-capability variant of the XR6 receiver that employs a special mode of differential GPS (or RTK GPS) processing. The RTK output provides relative position accurate to 2 cm, once per second with a latency of 1 – 2 seconds; real-time output provides sub -metre accuracy at 10 Hz with a latency of 80 ms.

Specifications

Receiver: 12-channel C/A code, L1 frequency
Update rate: 10 Hz
Max speed: 1,000 kt
Acceleration: 4 *g*
Time to first fix: 20 s (with current Ephemeris)
Accuracy RMS (PDOP<3):
(position) <15 m stand alone
(with DGPS) <2 m beacon
Velocity: 0.03 m/s (differential mode)
Time output: 1 pps +/-100 ns
Latency, navigation mode: 80 ms
Dimensions: 175 × 80 × 57 mm
Weight: <1 kg
Power: 11-32 V DC; 8 W

Operational status

Available. The Sharpe XR6 receiver is also available as a single card for use by OEMs.

Contractor

Symmetricom Ltd.

VERIFIED

N-250 Flight Management System (FMS)

The Flight Management System (FMS) for the IPTN N-250 aircraft provides all the control and monitoring functions for the hydraulically driven flight control system for the N-250 aircraft. It provides full authority fly-by-wire system control for the elevators, ailerons, rudder, spoilers and flaps. Mechanical back-up, by way of a standby actuation system, is also provided to control the critical elevator and aileron surfaces.

A control/monitor function is used in the N-250 FMS. Each lane uses a different processor board: an Intel 80C196 processor and a Motorola 6802 processor, and associated dissimilar software. This encourages improved system integrity and graceful degradation, without reducing system performance, and reduces common mode failures.

Nine identical Electronic Control Units (ECUs) control the five different types of aircraft surfaces. A master pin coding arrangement allows the ECU automatically to lock into the appropriate software program for the system to which it is connected. Each ECU transmits signals such as surface position, system status and fault indication to the EICAS system via an ARINC 429 databus. These compact modules are totally interchangeable between the different actuator subsystems, encouraging a low ECU spares level, low maintenance and improved aircraft despatch rates.

The N-250 FMS is a lightweight, easy to install, expandable system. It can be used on aircraft with medium to large technology content.

Specifications

Dimensions: ARINC 600 1 MCU
Weight: 1.8 kg (max)

Operational status

In development.

Contractor

TRW Aeronautical Systems, Lucas Aerospace.

VERIFIED

Versatile electronic engine controller

Developed for the Pratt & Whitney PW305 turbofan engine, this is a versatile digital engine control unit which can be integrated with minimal redesign on a wide range of aero engines and airframes. Reduced unit cost from volume production, together with minimal non-recurring costs and benefits from reliability growth in related applications, combine to make this an attractive and cost-effective unit for both military and civil applications.

Operational status

In service in the Raytheon Hawker 1000, Learjet 60 and Astra Galaxy aircraft.

Contractor

TRW Aeronautical Systems, Lucas Aerospace.

VERIFIED

Flight deck warning system

The flight deck warning system is a microprocessor-based system suitable for a wide range of aircraft types from small business and commuter aircraft to large airliners. It comprises an electronic monitoring system Central Warning Panel (CWP) and audio warning system.

The electronic monitoring system monitors aircraft systems and sensors providing appropriate alerts to the flight crew via the CWP.

Use of a microprocessor results in a system of compact size, low weight and high flexibility. It enables alert signals to be prioritised and reduces spurious alerts through the use of time delays. Changes to the system can be made simply by altering the software. An important feature of the central warning electronics is the self-test facility which can be activated during the preflight phase and is in continual operation during flight. When a failure is detected in the unit, a warning is given to the pilot on the CWP and BITE indicators isolate the fault to board level. A failure of the microprocessor does not render the system inoperable as warnings will still be indicated on the flight deck.

The CWP is a dedicated display incorporating hidden legend annunciators. It is constructed from identical display modules which can be assembled to suit the application.

Entirely separate from the central warning system, the audible warning system monitors critical aircraft functions and can provide up to 10 warning tones to the flight crew headphones. This high-integrity system has duplicated outputs and dual power supplies to minimise loss of warning under fault conditions.

Operational status
In service on the Avro RJ.

Contractor
Ultra Electronics Ltd, Controls Division.

VERIFIED

Landing Gear Control and Interface Unit (LGCIU) for Airbus

The LGCIU is in service on the Airbus A319, A320, A321, A330 and A340. The LGCIU controls the extension and retraction of landing gear and doors, provides an instantaneous on-wheels indication and monitors the position of cargo door locks and flap position. It interfaces with the fault warning and centralised aircraft maintenance system, providing status signals, built-in test and data to the maintenance recording systems.

Landing gear system status is monitored using a two-wire proximity sensor system. Interface with other aircraft systems is through an ARINC 429 databus.

Operational status
In service on the Airbus A319, A320, A321, A330, A340. In development for the A340-500/600.

Contractor
Ultra Electronics Ltd, Controls Division.

VERIFIED

Propeller Electronic Control (PEC)

The PEC is being developed for the de Havilland Dash 8-400. The PEC carries out safety-critical propeller control functions, including pitch control, synchrophasing and speed control. Additionally, the PEC controls routine feather and auto-feather operations, and the Automatic Take-off Thrust Control System (ATTCS). In the event of an engine failure at take-off, ATTCS minimises crew workload by auto-feathering the propeller on the failed power plant and up-trimming the propeller on the healthy engine, swapping roles at each power-up dual-channel active/standby control lanes ensure redundancy. Control lanes are microcontroller based.

Fault monitoring and detection indicated to the FADEC through an RS-422 link. The PEC also signals the FADEC to prevent dispatch in the event of certain fault conditions.

Operational status
In service on Dash 8-400.

Contractor
Ultra Electronics Ltd, Controls Division.

VERIFIED

UltraQuiet Active Noise and Vibration Control (ANVC) systems

The UltraQuiet Active Noise and Vibration Control systems for passenger aircraft cabins reduce noise by duplicating the primary sound field with an additional secondary sound field in anti-phase. The UltraQuiet system can operate over the complete flight envelope including take-off and landing, reduce the harmonics of the blade passing frequency, track the variation in the blade passing frequency which occurs during air turbulence and when the aircraft banks, and reduce beat caused by poor propeller synchrophasing.

A typical system comprises 24 loudspeakers, 48 microphones and an electronic controller. The loudspeakers are lightweight devices specially designed for optimum performance at low frequencies. They are positioned in the trim beside the seats and in the roof. The exact positions are determined by measurement and practicalities of the trim design. The loudspeakers each have a low-power amplifier which is mounted on, or adjacent to, the loudspeaker. This gives maximum flexibility when deciding the number of loudspeakers required. The microphones are positioned in the rear of the trim. The electronic controller consists of printed circuit boards inside a ½ ATR short ARINC 600 case. The unit has a comprehensive internal built-in test system.

The performance of the system is optimised for cabin wide noise reduction, with no areas where noise is unacceptable. The system compensates automatically for the changes in sound field caused by staff and passengers moving around the cabin. The system typically achieves attenuations of 8 to 12 dBA throughout the flight envelope.

Operational status
UltraQuiet Active Tuned Vibration Attenuators (ATVAs) are the latest addition to Ultra Electronics' portfolio of active noise control systems. The ATVAs are used in place of loudspeakers, and produce the anti-phase sound-field by vibrating the fuselage of the aircraft. This reduces both cabin noise and vibration. The launch customer for the ATVA system is de Havilland. The system is standard on the Beech 1900D, Dash-8Q Series 100, 200, 300 and 400; Saab 340B+; Saab 2000 and King Air 350. It is optional on the Bombardier Challenger and other King Air aircraft.

Contractor
Ultra Electronics Ltd, Noise & Vibration Systems.

VERIFIED

UltraQuiet Active Noise and Vibration Control System

Aircraft moving coil indicators

Weston Aerospace manufactures a range of moving coil indicators for the aircraft industry, in varying case sizes from 1 to 2 in, of modular construction, with integral lighting. Indicators are available to display parameters such as: electrical parameters, temperature, pressure, position, torque and speed. Options for indication of one, two or three parameters on each instrument are available.

Specifications

Accuracy:
direct reading (mA or V)
±1.5% to ±4% full scale deflection
indirect reading (pressure, temperature) ±4.5% full-scale deflection
Display: black dial/white markings as standard; colour as required

Contractor

Weston Aerospace.

VERIFIED

Weston Aerospace aircraft moving coil indicators
0018185

UNITED STATES OF AMERICA

Liquid Crystal Displays (LCDs)

Aerospace Display Systems, Inc produces a range of customised LCDs for civil and military (NVG compatible) use, including the Autopilot Readout Device (ARD) used on A310, A320 and A321 aircraft; the Tension Skew Indicator (TSI) used in the MH-53E Super Stallion helicopter during minesweeping operations; the Vertical Torque Indicator (VTI) used in commercial helicopters.

Contractor

Aerospace Display Systems, Inc.

VERIFIED

Aerospace Display Systems tension skew indicator
1999/0051690

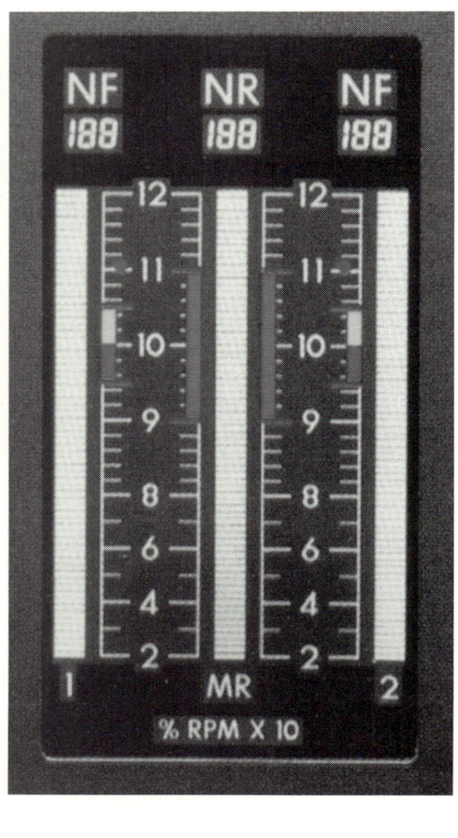

Aerospace Display Systems vertical torque indicator
1999/0051691

Aerosonic 2 and 3 in instruments

Aerosonic produces a range of 2 in (51 mm) and 3 in (76 mm) instruments including 2 in ASIs, 2 in altimeters, 2 in vertical velocity indicators, 3 in general aviation ASIs, altimeters and vertical speed indicators.

All indicators come in ARINC, military, round or clamp mounting and comply with TSO C-10b and TSO C-88a or military type 32/A requirements. Options include NVIS capability.

Contractor
Aerosonic Corporation.

VERIFIED

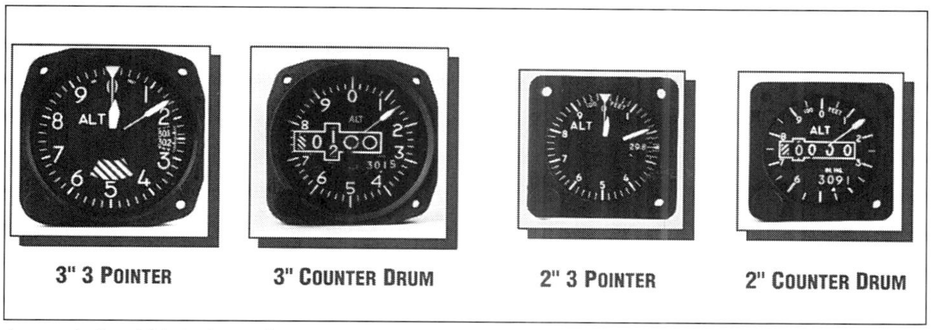

Aerosonic 2 and 3 in instruments 0001371

Airborne vibration monitoring system

The airborne vibration monitoring system is designed for engine rotor imbalance monitoring and onboard two-plane engine trim balancing. The system consists of two major components: a remote charge converter and a signal conditioner.

Vibration data is transmitted to the cockpit for display via an ARINC 429 or 629 databus. Advanced digital signal processing techniques provide accurate vibration magnitude and phase management. Utilising in-flight data provides superior trim balance information, allowing accurate one shot engine balancing and eliminating requirements for engine ground runs for balance verification. The system simplifies engine balance operations by providing specific maintenance instructions. Extensive BIT capabilities isolate faults between accelerometers, the two LRUs and the interconnection cabling.

For system retrofit applications, an integrated signal conditioner, incorporating the functions of a remote charge converter, is available in the ARINC 429 configuration.

Specifications
Dimensions:
(signal conditioner) ARINC 600 3 MCU
(remote charge converter) 56 × 61 × 114 mm
Weight:
(signal conditioner) 2.8 kg
(remote charge converter) 0.45 kg

Operational status
Selected for the Boeing 777.

Contractor
Ametek Aerospace Products.

VERIFIED

Airborne vibration monitoring system for the Boeing 777

LM Series compact indicators

The LM Series indicators are a low-cost response to the FAA FAR 121.343 requirements for in-flight recording of engine parameters. Standard LM Series features include back-lighted dial and pointer, 360° dial arc, BIT, smooth pointer movements, modular construction for easy servicing, compact 3.5 in (88.9 mm) long design and 40,000 h MTBF.

The simple mechanical design incorporates only one moving part - the pointer - so there are no brushes or wipers to wear out. Surface-mounted PCB design helps to ensure long, trouble-free service life for the electronic components. The indicators are designed for display of primary and secondary engine parameters in retrofit as well as new production aircraft installations.

Optional features available include flight recorder inputs, high-accuracy redundant digital display, RS-232 interface, output warning signals and exceedance and trend monitoring.

Operational status
LM Series indicators are currently installed in Fokker F27 and Mitsubishi YS-11 aircraft.

Contractor
Ametek Aerospace Products.

VERIFIED

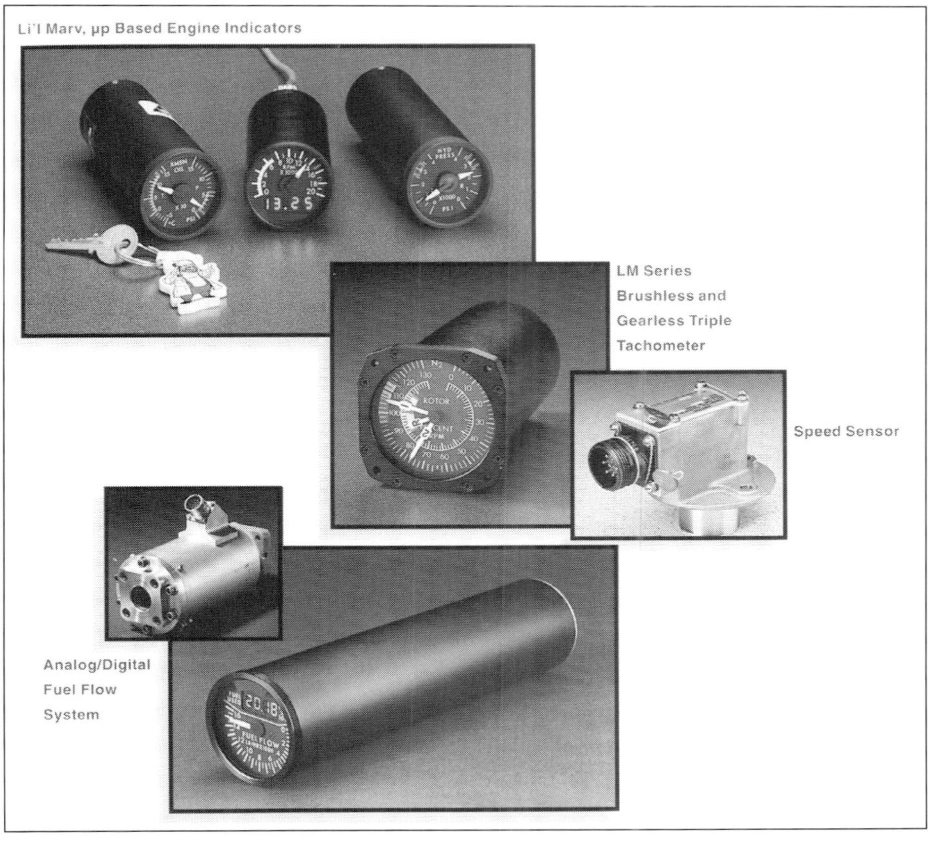

Ametek's LM Series indicators
1999/0051692

Sentinel® instrument system

The Sentinel® instrument system is designed to meet the need for fully integrated aircraft systems providing a complete solution from signal interface to cockpit display. The system consists of multiple cockpit Electronic Display Units (EDUs) and a dual-redundant Data Acquisition Unit (DAU). All signals are acquired and transmitted to the displays in digital format by both halves of the DAU which can be mounted in any convenient location in the aircraft.

Typical system applications include primary and secondary engine instruments and caution/advisory panels. Auxiliary functions such as fuel and electrical power management can be incorporated on additional selectable screens or can be displayed on a separate display.

The 117.8 × 83.8 mm (4.4 × 3.3 in) active matrix liquid crystal display provides crisp full-colour graphics in a presentation which can be custom-designed for any application. Graphics and fonts can be created and changed without impact on hardware. Individual symbols and fonts can be selected from a large variety of existing styles or can be tailored to meet specific needs.

Bezel-mounted soft keys provide the pilot with access to lower-level functions and allow input of information such as barometric pressure or target set points. Menu options direct the pilot to the desired information. Solid-state design provides high reliability to meet demanding requirements. The Sentinel® can display up to 256 colours simultaneously, selectable from a palette of 4,096 colours. Colours can be easily selected and matched to a standard chromatic co-ordinate. Easily replaceable fluorescent lamps are used to provide a uniformly lit 125 ft-lambert average white brightness, suitable for viewing in direct sunlight.

Specifications

Weight: 2.49 kg
Power supply: 28 V DC or 115 V AC, 25 W without heater

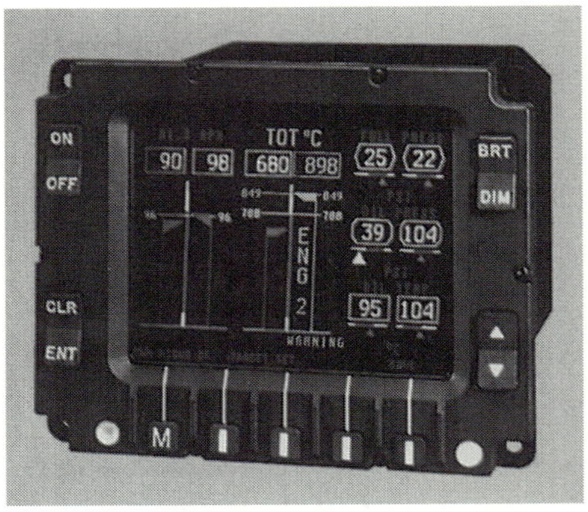

Sentinel® instrument system is installed in the Agusta A 109 helicopter

Temperature range: –40 to +70°C
Display: 111.8 × 83.8 mm, 960 × 234 pixels

Operational status

Currently installed on the Agusta A 109 helicopter.

Contractor

Ametek Aerospace Products.

VERIFIED

Standby engine indicator

The standby indicator provides a display of four engine parameters for each engine. The digital displays are part of an LCD, having white characters on a black background. These are easily readable in direct sunlight, as well as at dusk or night, when a separate lighting circuit provides a high-brightness, high-contrast presentation.

A special electronics design eliminates digit toggling during static conditions and provides fast response and sequential counting during dynamic conditions. The display update rate is variable, causing the display to simulate a mechanical counter and giving the flight crew a sense of the rate change of the parameter. When a parameter reaches a programmable limit, the respective display will flash to communicate the warning. If more than one parameter is over limit, they will flash synchronously to avoid confusion.

Loss of signal for any of the inputs is detectable and results in a display of three dashes. Loss of power results in a blank display. BITE permits mechanics to determine if a fault resides outside the indicator. Thus a high mean time between unscheduled removals is achieved.

Specifications

Dimensions: 82.8 × 103.1 × 188.7 mm
Weight: 1.59 kg
Power supply: 10-32 V DC, 2.8 W per channel

Contractor

Ametek Aerospace Products.

VERIFIED

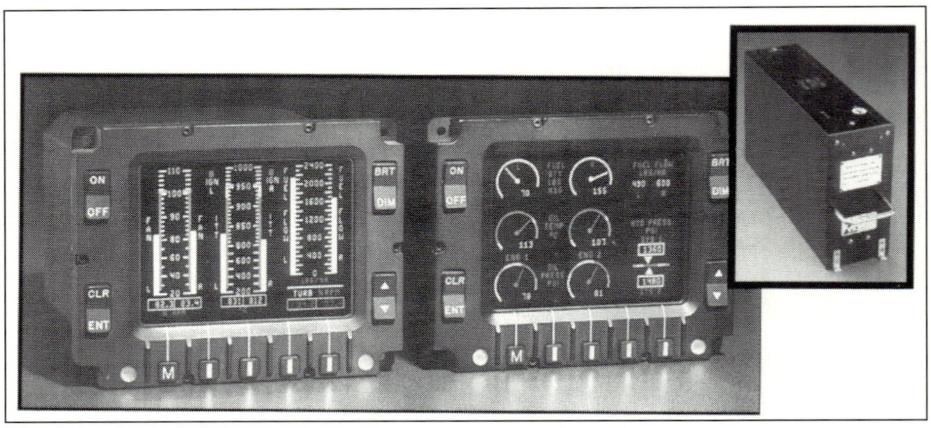

The Sentinel® electronic display units and a data acquisition unit
2000/0081843

DC-1590 series magnetic Digital Compass

The DC-1590 series magnetic digital compass provides encoded heading data. The display is in red, sunlight-readable, 0.5 in (13 mm) high, three-digit numerals and features leading zero suppression. Brightness is adjustable and, to prevent display jitter, display rate is also fully adjustable. A test switch is provided to test all digit segments. Display readings are in 1° increments. No warm-up time is required.

The compass sensor unit is a high-quality liquid-filled instrument, fully illuminated for night viewing. All-solid-state to ensure reliable performance, it can be installed up to 38 m from the display. A set of East/West offset switches, mounted internally, provides unlimited compensation for magnetic variation. This remains in operation permanently until readjusted, allowing compensated operation to give true heading values. At the same time, a set of East/West offset push-button switches, mounted on the display, allows input compensation for variation changes.

The DC-1590 may be used to drive remote displays, autopilots, plotters and other electronic navigation equipment through a binary encoded decimal output connector.

Specifications

Weight: 4.5-6.8 kg depending on model

Contractor

Arc Industries Inc.

VERIFIED

DC-2200TM series magnetic Digital Compass

The DC-2200TM series magnetic digital compass is a twin display system. One display shows true heading, while the other shows magnetic heading. The operator can at any time add or remove any amount of variation for course correction.

Contractor

Arc Industries Inc.

VERIFIED

FMS 5000 Flight Management System

The FMS 5000 consists of a multichain master independent Loran capable of tracking two Loran chains and up to 12 ground stations simultaneously. All world Loran chains are available, including the US mid-continent NOCUS/SOCUS chains. The FMS 5000 automatically selects the strongest master and secondary Loran stations, providing hands-off operation.

Three remote GPS receiver options are available for the FMS 5000. A five-, six- or 12-channel tracking receiver can be supplied with the FMS 5000, or added in the future. Together, the Loran and GPS receivers continuously scan up to 12 Loran ground stations and all satellites in view. The FMS 5000 displays the navigation solution with instant exchange of sensor positioning, providing hands-off Loran/GPS operation.

The FMS 5000 interfaces with analogue instruments, fuel computers, air data computers, moving maps, autopilots, CDI/HSI and annunciators. An optional ARINC 429 interface allows coupling to EFIS systems.

A two-line 40-character sunlight-readable LED display shows bearing, distance, groundspeed, CDI, waypoint identity, altitude and track. Waypoints can be located by identity, city, local proximity or adjacent waypoints.

The FMS 5000 uses a Jeppesen NavData card containing 40,000 waypoints. The North America coverage extends from Alaska through Central America. The International card contains all areas outside North America. Worldwide navigation, combining North America and International data, is available on a single World card. Airports, VORs, NDBs, terminal and en route intersections are stored on the crew updatable NavData card. Special use airspace includes floors and ceilings, TCAs, ARSAs, ATAs, MOAs and restricted, prohibited and alert penetration warnings.

Coupled to an optional Mode C encoder interface, pressure altitude with a pilot input barometric correction is displayed. VNav becomes automatic with known present altitude and known waypoint elevations. An altitude hold advisory will notify the pilot of altitude deviation.

Contractor

Arnav Systems Inc.

VERIFIED

FMS 7000 Flight Management System

The FMS 7000 is a small dzus-mounted Flight Management System (FMS) for business and commuter aircraft. It is designed to provide the pilot with a comprehensive primary or separate secondary navigation tool and database facility. It can be configured with a Loran sensor and a high-precision GPS sensor. The FMS 7000 uses sensor assessments of signal quality to provide the best position information.

A standard Jeppesen North American or International database with over 60,000 aviation facilities is contained on a high-capacity NavData card that can be easily updated. Each card provides all worldwide port-of-entry airports with runway lengths in excess of 6,000 ft, plus all hard-surfaced airports, facilities, frequencies and navigation information for the North American or International geographic areas. An optional Jeppesen World NavData card provides comprehensive worldwide navigation on a single card.

The FMS 7000 automatically builds SID and STAR route waypoints, transition routes, crossing altitudes, arrival routes and arrival frequencies. Jeppesen terminal navigation is also included on the World NavData card.

The Arnav advanced communication network, Arnet, and the ARINC Standards Interface (ASI) have been developed specifically for the FMS 7000. Arnet is a high-speed serial communication device that facilitates integration of other avionics such as autopilot, air data and fuel computers. ASI allows the FMS 7000 to communicate with virtually all cockpit systems, including analogue and digital format DME, EFIS, HSI and flight directors. ARINC 429, ARINC 419, six-wire ARINC 568, synchro, MIL-STD-1553 and several DME controls or data formats are supported.

Cross-talk capability and system redundancy is optional on the FMS 7000. Tandem control/display units allow the co-pilot to view all flight, fuel and air data information and perform waypoint search routines.

The Arnav GridNav mission management system program is a software option for Loran, GPS and combination FMS. The GridNav option allows the pilot to plot and fly a grid pattern of user-specified dimensions using a minimum number of waypoints and with simplified navigation programming. An event trigger provides a precise timed pulse to a camera shutter or target drop release.

Sensor options for the FMS 7000 include Loran only, GPS only or both Loran and GPS. The Loran-based system automatically selects the proper Loran chains and stations based on aircraft location. The FMS tracks up to 12 Loran stations and automatically selects the signal geometry that provides the best fix. Loran is approved for IFR flight and permits the pilot to file IFR direct when within the boundaries of Loran coverage.

The GPS-based receiver is differential ready and conforms to TSO C-129 certification. The FMS 7000 may be populated with either a five-channel or 12-channel GPS receiver. Both work in any weather and are not subject to precipitation static, low-frequency thunderstorm emissions or any other weather interference. Through a powerful microprocessor, the FMS 7000 tracks all satellites in view and then selects the satellite geometry to acquire the most precise fix. Signal acquisition and tracking are continuous throughout all dynamics of flight. Initial time to fix a reliable position is less than 1 minute and position is updated every second.

Both Loran and GPS sensor options can be installed in the FMS 7000 to provide multisensor blended mode navigation. The combined sensor output can be used for IFR direct navigation when operating within Loran coverage areas and when TSO is achieved. The FMS has TSO C-60b for IFR en route navigation and terminal navigation, TSO C-115a multisensor certifications and Loran approach certification.

Specifications

Dimensions:
(Loran/GPS LRU) ¼ ATR short
(CDU) 57.1 × 139.7 × 127 mm
Weight:
(Loran/GPS LRU) 1.82 kg
(CDU) 0.68 kg
(Loran/GPS antennas) 0.64 kg
Power supply: 11-35 V DC, 15 W (max)
Temperature range:
(LRU) −55 to +70°
(CDU) −20 to +70°C
Altitude: up to 55,000 ft
Certification: TSOs C-60b, C-115a, C-44a, C-106
Environmental: Do-160c Cats A1/D2, MIL-STD-167

Operational status

The FMS 7000 is installed in a wide range of civil and military aircraft and helicopters.

Contractor

Arnav Systems Inc.

VERIFIED

GPS-506 Global Positioning System receiver

The GPS-506 is a six-channel GPS continuous tracking remote sensor which interfaces with the R-50i Loran C receiver (see later item). The R-50i has a European database. The R-50i Loran/GPS is approved by the US FAA for IFR operations.

Operational status

In service.

Contractor

Arnav Systems Inc.

VERIFIED

MFD 5000 cockpit management system

The MFD 5000 cockpit management system is an interactive graphics map, terrain and obstruction proximity system, air data system and EICAS, all presented on a single sunlight-readable multifunction display. It displays the aircraft position on a comprehensive 200,000 waypoint plan view chart. Jeppesen NavData, including airports, runways, frequencies, TCA/ARSA boundaries, VORs, NDBs, airways, intersections, SIDs/STARs and approaches combines with digital elevation mapping, geography, hydrography, manmade obstructions, highways and more, presenting VFR and IFR charts in scale levels ranging from the airport to hemisphere. Custom databases for EMS and vehicle tracking are also available.

The MFD 5000 monitors up to 35 engine and airframe conditions. By interfacing with a host of aircraft systems, the EICAS constantly scans for out-of-range conditions, decreasing pilot workload while increasing safety. All caution advisories are cross-referenced, providing audio and visual alerts to the MFD 5000 and external annunciators. The EICAS and map are fully interactive, to deliver crew advisories on position, environment, navigation, fuel management, air data and engine and airframe conditions.

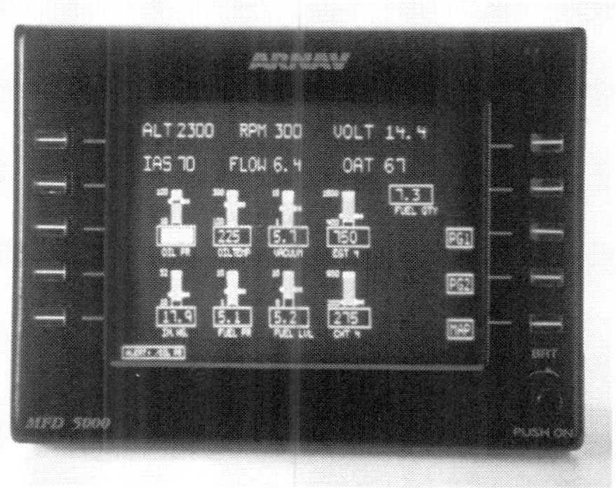

Arnav MFD 5000 displays both navigation and aircraft systems data

The EICAS couples to many existing aircraft systems including analogue instruments, rpm, voltage, oil temperature, encoding altimeters, fuel computers, Loran/GPS receivers, other engine monitoring systems or a host of Arnav transducers. The EICAS contains several pages of information with three assignable priority levels. For mixture purposes, the hottest EGT and CHT are automatically presented on the main display.

The remote LRU 5020 EICAS computer measures all channels 10 times/s to ensure constant accuracy. Bright columns display operating ranges with a digital value written below. Pilot programmable alarms advise on out-of-range conditions. When an alert condition occurs, the display switches to the failed system, followed by the associated checklist.

Through altitude encoders and Loran/GPS receivers, the MFD 5000 compares the aircraft altitude with the digital terrain-mapping database. When below minimum safe altitude, the terrain/obstruction proximity system advises the pilot of ground proximity, taking account of both terrain and manmade obstructions.

The Arnav FMS 5000/7000 and STAR 5000 navigation management systems are interactive with the MFD 5000. All flight planning and search routines can be programmed through the FMS 5000/7000 for display on the MFD 5000. The remote Arnav GPS-505, GPS-506 and GPS-512 receivers also interface with the MFD 5000, eliminating the need for a panel mount navigator. Other Loran and GPS receivers with RS-232 protocol are compatible.

Contractor

Arnav Systems Inc.

VERIFIED

MFD 5200 colour AMLCD MultiFunction Display

Designed for the general aviation market, the MFD 5200 colour AMLCD display is a lightweight, low-cost version of those installed in jet transport aircraft.

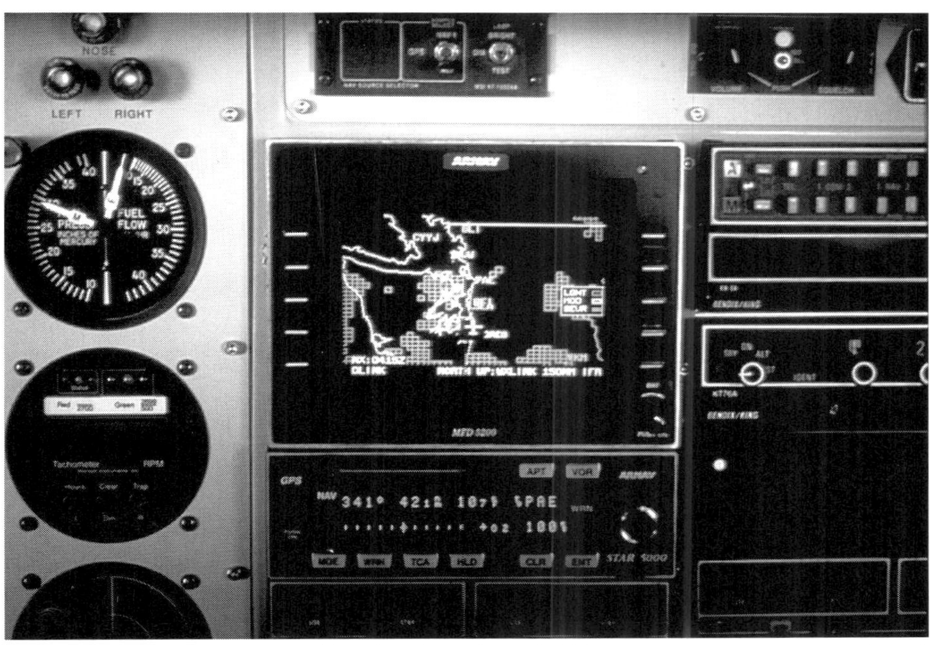

MFD 5200 showing moving map and Nexrad weather radar via WxLink 0018179

The system consists of a low-profile 1.15 kg remote-mounted symbol generator and 1.3 kg colour AMLCD control display unit. The sunlight-readable 5 in diagonal display provides situational awareness from onboard systems including navigation, air data, engine and airframe, as well as environmental elements such as the weather. The basic function of the MFD 5200 is moving map navigation.

Position data input from any GPS or Loran shows as an airplane or helicopter icon over a digitised aviation chart. The flight plan automatically uploads from the Arnav STAR 5000 or FMS 5000 GPS navigators or AlliedSignal KLN 89-89B and KLN 90-90B GPS navigators, reducing pilot workload. It will interface to the next-generation BFGoodrich WX-500 Stormscope®, to display thunderstorms, lightning strikes, and building storms in relation to the aircraft position and flight plan. It is FAA approved for depiction of broadcast Nexrad precipitation radar and METAR weather reports. An optional VIPER (Video InPut EncodeR) board converts the MFD 5200 from its digital moving map role, to a composite NTSC host for video or infra-red camera display.

Operational status
Certified under a multiple STC, the MFD 5200 is now operating in Sikorsky, Bell, Mooney, Cessna, Piper, and Raytheon aircraft.

Contractor
Arnav Systems Inc.

VERIFIED

Navision 50 moving map

The Navision 50 is a general aviation version of the Arnav airborne moving map. It interfaces with the R-50 and R-50i Loran C and is compatible with most general aviation Lorans. With an adjustable zoom feature ranging from 1 to 1,000 n miles, the pilot can view airport diagrams including runway layout and runway numbers, or look up to 1,000 n miles ahead on the flight path. The display can be viewed north up or it can be aligned with the aircraft ground track.

About the size of a 2 in thick approach plate, the Navision 50 is portable and can be used anywhere in the cockpit. Also included is the Jeppesen North American NavData card and flight planning mode for developing flight plan legs shown in columnar form or graphically. The user waypoint feature allows pilot entry of mountain peaks or other high terrain and actual elevations can be entered to be shown on the map.

The 76 × 127 mm (3 × 5 in) electroluminescent display is easily read in bright sunlight and may be dimmed for night flight. The Navision 50 includes many of the features found on the Navision 1000 which was designed for larger aircraft.

Contractor
Arnav Systems Inc.

VERIFIED

R-50 Loran C receiver

The R-50 IFR Loran C is a development of the successful AVA-1000 and the performance has been improved to permit operation in areas of poor Loran coverage, such as the North Slope of Alaska, Bermuda, the Caribbean and parts of the mid-West. The R-50

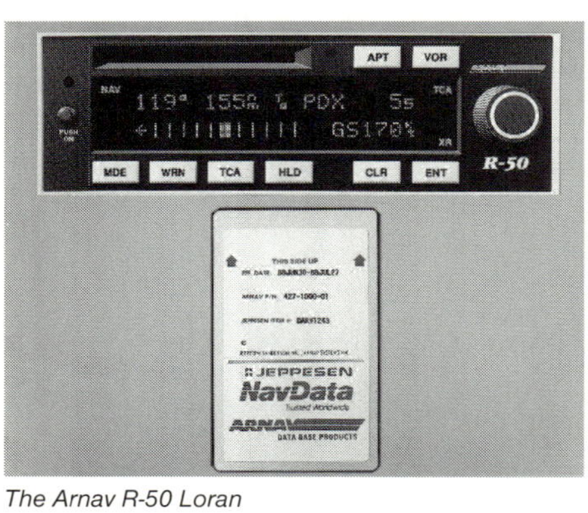

The Arnav R-50 Loran

includes the Jeppesen NavData slip-in database card which provides information on airports, VORs, NDB intersections and special use airspace alerts. The panel-mounted system offers extended range, automatic chain nomination, flight planning and 100 flight plan waypoints and can accommodate up to 1,500 waypoints. Present position is shown on a two-line 40-character dot matrix display and outputs are provided to an autopilot, a CDI and to equipment using RS-232 signal format.

Operational status
In service.

Contractor
Arnav Systems Inc.

VERIFIED

R-50i Loran C receiver

The IFR and TSO C60b approved multichain R-50i Loran C simultaneously tracks up to 12 Loran stations and can be used as a primary means of en route and terminal IFR navigation. Up to 120 waypoints can be stored and flown as flight plans. A sunlight-readable two-line 40-character LED display provides bearing, distance, waypoint identifier, course deviation indication, track guidance, groundspeed and suggested vertical navigation.

The Jeppesen NavData card database contains North American airports, VORs, NDBs and intersections and can store 150 custom waypoints. The database can be accessed by identification or partial city name or identifier.

Navigation management features include vertical navigation, extended range, true airspeed/winds aloft calculations, fuel range and minimum safe altitudes.

The R-50i can instantly display the 15 closest airports, VORs or intersections at the touch of a button. Incorporated in the system is a software feature called Pilot Prompt which guides the pilot through every operation with simple messages. Included with each R-50i is a home power supply and simulation mode for flight planning and training.

Specifications
Dimensions: 50.8 × 158.75 × 271 mm
Weight: 1.13 kg
Power supply: 10-35 V DC, 1.5 A

Contractor
Arnav Systems Inc.

VERIFIED

STAR 5000 panel mount GPS

The STAR 5000 blends GPS technology with the Arnav R-50i IFR-certified Loran. The panel mount GPS receiver delivers five-channel continuous and parallel tracking of all satellites in view. Its onboard network interfaces with an array of avionics including analogue instruments, fuel computers, air data computers, moving maps, autopilots, CDI/HSI and annunciators.

A two-line 40-character sunlight-readable LED display shows bearing, distance, groundspeed, course deviation, waypoint identification, altitude and track. Waypoints can be located by identification, city, local proximity or proximity to other waypoints. The system uses a 40,000 waypoint Jeppesen NavData card, known as the Gold Card because of the wealth of information. Jeppesen NavData is offered for three geographical areas. The North America card coverage extends from Alaska to Central America. The International card covers all areas outside North America. Worldwide navigation combining the data on the other two cards is available on a single World card.

Airports, VORs, NDBs, terminal and en route intersections are stored on the pilot updatable NavData card. Special use airspace includes floors and ceilings and the card provides TCA, ARSA, ATA, MOA, restricted, prohibited and alert area penetration warnings. Coupled to an optional Mode C encoder interface, pressure altitude with a pilot input barometric correction is displayed. VNav becomes automatic with known present altitude and known waypoint elevations. An altitude hold advisory will notify the pilot of an altitude deviation.

Standard features of the STAR 5000 include nearest airport, VOR, intersection search, true airspeed, wind and fuel calculations, MSA/MESA, 150 waypoint flight planning and 300 user waypoints and the pilot's choice of the 40,000 waypoint North America or International NavData cards. The STAR 5000 has UTM and military grid reference co-ordinates as a standard feature. Options include the altitude encoder interface, GridNav mission management software, ARINC 429 interface and an air data computer.

Contractor
Arnav Systems Inc.

VERIFIED

System 6 avionics system

System 6 is a series of products designed for integration into a complete avionics system. Products include flight management system, multi-engine display, datalink, dispatch terminal, engine monitor and data recorder.

The FMS5000 and FMS7000 flight management systems will complement any avionics suite and operate in conjunction with GPS and Loran C to provide position to within 3 m anywhere in the world. They feature Loran receiver only, GPS receiver only or both Loran and GPS receivers, differential GPS receivers, fuel/air data computer, ARINC 429 interface, single or tandem CDU operations, sunlight-readable LED display, anti-reflective optic filter, NVG optic filter, function keys and turn/push knob control, Jeppesen worldwide NavData card and SAR GridNav mission management.

The multifunction display shows position on a moving map and ranges from high-contrast monochrome to full-colour panel mount displays.

The datalink provides the capability to send and receive position reports, encrypted or open messages, text or graphic databases and monitored engine health data. It can track up to 400 ground vehicles and aircraft and integrates VHF, UHF, HF, microwave transmissions and satellite communications. A built-in differential GPS sensor provides high accuracy for position reporting.

The dispatch terminal is a central control facility that can be located on the aircraft or on the ground. Standard IBM PC equipment and System 6 software perform vehicle tracking, data communication, archiving and dissemination of data.

The engine monitor measures up to 70 piston or turbine engine components and environmental conditions on the aircraft. Actual operations and set tolerance exceedances are measured 10 times per second. This information can be shown on the multifunction display, transmitted via the datalink to a dispatch terminal and recorded on the data recorder.

The data recorder is a small, ruggedised aircraft-quality computer capable of recording mission plans, position history, engine health and environmental conditions on a PCMCIA storage card for playback on a standard IBM computer.

Specifications
Dimensions:
(FMS5000) 51 × 159 × 255 mm
(FMS7000) 57.2 × 139.7 × 127 mm
(MFD 5000 monochrome display) 120.6 × 159 × 226.1 mm

Units in the System 6 avionics system

(MFD 5100 colour display) 120.6 × 159 × 226 mm
(MFD 5200 colour display) (see previous entry)
LRU 5010 computer/recorder) 500.8 × 159 × 235 mm
(radio transceiver/tracking unit) 184.2 × 69.9 × 190.5 mm

Operational status
The FMS7000 and datalink have been flight tested in an Erlanger Bell 412 medical transport helicopter.

Contractor
Arnav Systems Inc.

VERIFIED

ELT200HM Emergency Locator Transmitter

The ELT200HM is designed for helicopter application, it can be mounted horizontally, and weighs only 0.87 kg. Dimensions are only 64 × 69.9 × 158.8 mm.

The ELT200HM incorporates a 6-axis 'G' switch and a Lithium battery pack that gives a 4-year replacement life.

The FAA granted TSO C91A approval for the ELT200HM in July 1999, with shipments beginning in September 1999.

Contractor
Artex Aircraft Inc.

UPDATED

ELT Emergency Locator Transmitters

Artex Inc manufactures a series of FAA approved emergency locator transmitters that meet TSO C91a and TSO C126. The ELTs are designed for installation in fixed- or rotary-wing aircraft. The ELTs for fixed-wing aircraft activate via a single-axis 'G' switch and the ELTs for helicopters activate via a six-axes 'G' switch module.

Artex also manufactures an ELT-to-NAV interface for use with the ELT 110-406 systems. This interface allows the ELT to couple via ARINC 429 to GPS, Loran or FMC thus broadcasting latitude/longitude as part of the 406 MHz digital message. The overall message includes: type of aircraft, owner, emergency contact, country code, serial number of ELT, ELT manufacturer, latitude/longitude position. Models available include the following:

Specifications
All models provide 50 mW output at 121.5/243 MHz for 50 hours. In addition the ELT 110-406 systems operate at 5 W on 406.025 MHz for 24 hours.

Operational status
Widely fitted by fixed- and rotary-wing aircraft manufacturers.

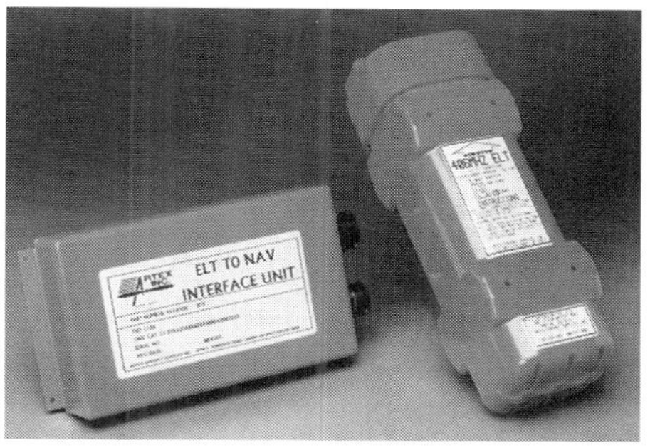

ELT to NAV interface and ELT 406 **1999**/0051606

Contractor
Artex Aircraft Supplies Inc.

VERIFIED

Model number specifications	ELT-200 /200HM	ELT 110-4	ELT 110-6	ELT-100HM	ELT-200HM	ELT 110-406NAV	ELT 110-406NAV/HM
Application	General aviation, fixed-wing, helicopters	General aviation, Commercial aviation	General, Military & Commercial aviation	Helicopters, general, commercial & military	Helicopters, general, commercial & military	General, corporate helicopters military & commercial aviation	General corporate, helicopters military & commercial aviation
Frequency	121.5/243.0 MHz	121.5/243.0 MHz	121.5/243.0 MHz	121.5/243.0 MHz	121.5/243.0 MHz	121.5/243/ 406.025 MHz	121.5/243/ 406.025 MHz
Approvals	TSO C91a	TSO C91a	TSO C91a	TSO C91a	TSO C91a	TSO C126	TSO C126
Interface capability	N/A	N/A	N/A	GPS/Loran	GPS/Loran	GPS, interface ARINC 429, most RS 232	GPS, interface ARINC 429, most RS 232

Area navigation control/display for the L-1011 TriStar

This system, installed on forward sections of the L-1011 throttle console so that it can be viewed and operated by both pilots, comprises a centrally mounted electronic automatic chart with identical Control/Display Units (CDUs) situated alongside a control unit and an electronic unit for the chart display. The CDUs present pages of alphanumeric information as required, while the electronic chart shows pictorially the route being flown, with positions of waypoints, destination airfields and navigation beacons marked.

The control unit has a scale selector that can shrink or expand the navigation situation shown on the chart, a slew control that can move the display bodily about the screen or rotate it (the alphanumeric symbols, however, remaining upright), mode selector permitting north up, track up or look ahead display and an RNav switch that can select data from one or other of the navigation computers on the aircraft. The chart can store 1,000 symbols and display 500 of them at any one time, it can also act as a back-up to the CDUs, presenting commanded alphanumerics as required.

The CDUs can display 12 lines of information with 17 characters per line and can access and display data stored in an aircraft computer, for example, route details, flight plans, en route navaid frequencies and terminal area procedures such as standard departures and arrivals.

The system incorporates continuous failure monitoring and self-test and the computing section furnishes a self-test pattern for complete end-to-end checks.

Specifications
Dimensions:
(control unit) 147 × 69 × 160 mm
(chart) 216 × 231 × 384 mm
(electronic unit) 124 × 196 × 368 mm
(CDU) 147 × 229 × 305 mm

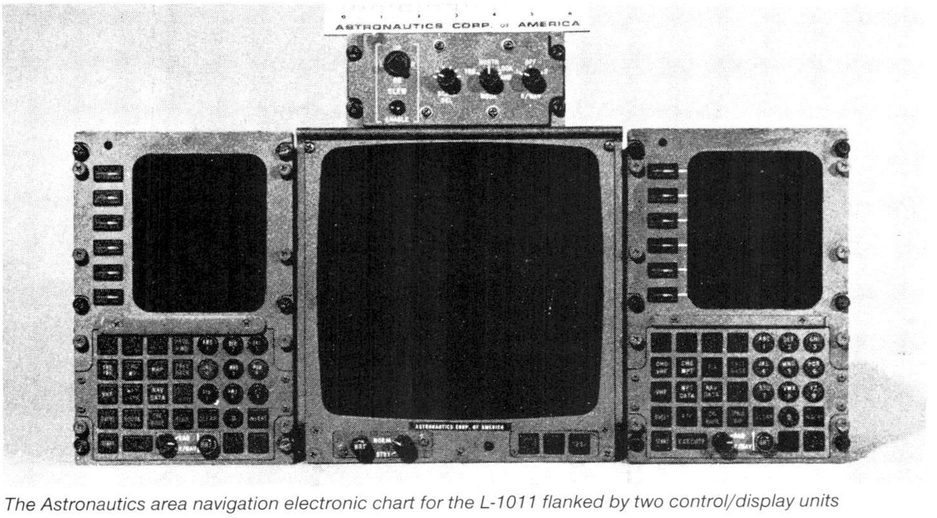

The Astronautics area navigation electronic chart for the L-1011 flanked by two control/display units

Weight:
(control unit) 0.68 kg
(chart) 12.25 kg
(electronic unit) 6.35 kg
(CDU) 8.17 kg
Power:
(chart) 200 VA
(electronic unit) 90 VA
(CDU) 90 VA

Operational status
In service in the L-1011 TriStar.

Contractor
Astronautics Corporation of America.

VERIFIED

Colour AMLCD MultiFunction Displays (MFDs)

The Astronautics Corporation of America produces a range of colour AMLCD MFDs for the military and civil aircraft markets, including: 3 × 4; 3.5 × 4.5; 4 × 4; 5 × 5; 8 × 6; 6.25 × 6.25; and 10 × 8 in sizes. All are designed with modular construction, and all modules can be replaced without adjustment. All models are NVG compatible, and provide high-resolution, high-brightness performance.

A specific application is the Astronautics 5 ATI 4 × 4 in Electronic Flight Instrument (EFI), fitted to the S-61. It is pin-programmable as either an ADI or HSI. It also includes software that allows integration of other display formats such as map, TCAS, and colour weather radar. This instrument is TSO'd by the FAA and can be used in military or civil applications.

Operational status
The 5 × 5 in, 4 × 4 in, and 5 ATI displays are utilised in A-4, B-747, C-130, F-5, Kfir, OV-10, S-61, Tornado ECR and Tornado IDS aircraft, and on the A 129 Mangusta and Boeing 530MG helicopters. The 8 × 6 in

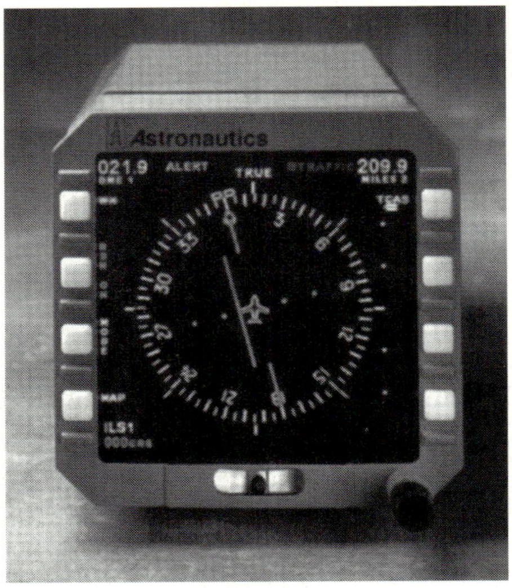

The Astronautics electronic flight instrument in HSI mode 0018178

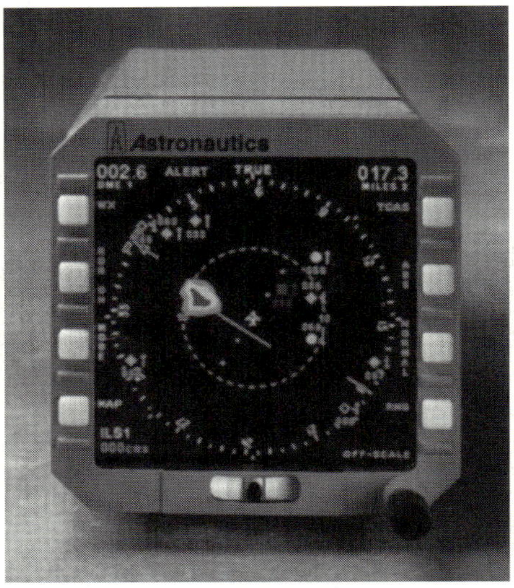

The Astronautics electronic flight instrument in TCAS mode 0018177

touchscreen display is flying in a DC-10 in-service evaluation programme; and the EFI is flying in the revenue service on a United Airlines DC-10 aircraft.

Contractor

Astronautics Corporation of America.

VERIFIED

Engine Instrument and Caution Advisory System (EICAS) display

Astronautics' EICAS is a VGA-compatible, low-power, sunlight readable, and NVG-compatible Active Matrix Liquid Crystal Display (AMLCD). Modular, solid-state construction and programmable symbols make the display readily adaptable to a wide range of input/output applications. Large non-volatile memory storage capacity with easy direct downloading simplifies the transfer of engine data to maintenance records. A separate serial test port can be used to amend parameters to meet user requirements. When two displays are installed, a cross-talk channel provides complete redundancy.

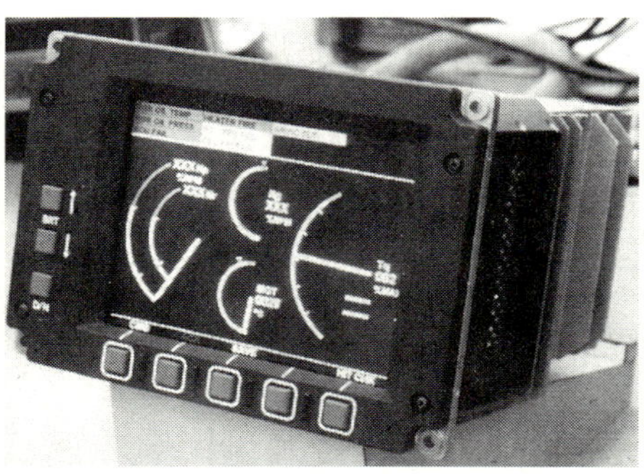

Astronautics engine instrument and caution advisory system display **2000**/0064370

Specifications

Processor: Pentium
Viewing area: 129.5 × 96.5 mm
Resolution: 640 × 480 pixels
Viewing angle:
(horizontal) ±45°
(vertical) +30 to –10°
Brightness: 150 fL
Multifunction keys: 5 programmable soft keys, plus 2 for brightness and 1 for day/night
MTBF: 10,000 h
Dimensions: 203.2 × 139.7 × 203.2 mm (190.5 × 152.4 × 203.2 mm with PCMIA)
Weight: 3.64 kg
Certifications: DO-16-C, DO-178B Level A
TSO: C113, C43b, 47a

Operational status

In service, applications include A-119, Ayers Loadmaster and UH-1H.

Contractor

Astronautics Corporation of America.

VERIFIED

Flight director autopilot systems

The three-cue flight director system for helicopters combines a versatile flight director computer with a three-command bar ADI, dual-bearing pointer HSI and a multimode computer controller. The system provides ILS, VOR and ADF approach capability.

The three-cue system adds a collective command steering bar to the pitch and roll command steering bars used in two-cue systems. Continuous altitude, airspeed, vertical speed, VOR/ILS and optional Doppler and altitude alert inputs permit the flight director computer to respond to pilot-selected flight modes. A pilot can execute VOR/ILS intercepts, glide slope intercepts, vertical and airspeed holds, deceleration rates and altitude hold. The computer provides automatic intercept and tracking of VOR, glide slope and localiser, and initiation of deceleration for ILS approaches. Pilot and co-pilot displays and controls are effected by optional slaved horizontal situation indicators and transfer controls.

The indicators are 127 mm (5 in) units, hermetically sealed with a dry nitrogen/helium atmosphere. Direct current servos used in these units are claimed to result in considerably less heat dissipation, less power drain, higher torque and greater reliability than typical AC servoed units.

The mode controller has four switches for mode, nav select, vertical speed and airspeed selections. An optional remote dual-course selector can be added to supplement the basic controller.

The flight director computer accepts inputs from the mode controller and an array of sensors. It computes the pitch, lateral and collective commands required to adhere to selected and/or scheduled flight parameters and displays the commands on the attitude director indicator.

Specifications

Dimensions:
(ADI) 127 × 133 × 194 mm
(HSI) 127 × 108 × 174 mm
(flight director computer) 59 × 194 × 319 mm
(mode controller) 146 × 105 × 127 mm
(remote course selector) 146 × 32 × 165 mm
Weight:
(ADI) 3.2 kg
(HSI) 2.7 kg
(flight director computer) 2.9 kg
(mode controller) 1.6 kg
(remote course selector) 0.7 kg
Power supply:
115 V AC, 400 Hz, 45 VA
28 V DC, 0.5 A
5 V AC, 8 VA (for lighting)

Operational status

In production.

Contractor

Astronautics Corporation of America.

VERIFIED

Three-axis autopilot for the 500/530 helicopter

Astronautics has developed an autopilot for light helicopters which has been FAA certified and is available for Boeing 500D, 500E and 530 helicopters. The autopilot provides full three-axis control to reduce pilot fatigue. There are seven basic operating modes, plus hands-off stabilisation. In basic attitude retention mode, the helicopter can be flown hands-off not only in straight and level flight but also during climbs, descents and turns. The desired attitude will be held even in autorotation and all turns are automatically co-ordinated. In altitude hold mode the desired altitude is maintained to within ±20 ft. The altitude can be captured with vertical velocities as high as 1,000 ft/min. Following high-vertical velocity climb or descent the helicopter will smoothly change pitch attitude and capture the selected altitude. In heading hold mode the pilot may select the desired heading either before or after the mode is engaged. All turns to the new heading are automatically co-ordinated at a bank angle of 20°. Hands-off hover capability is provided either in or out of ground effect. Heading hold may be engaged at any time during hover. Any new desired heading is entered by moving the 'bug' on the heading gyro, and yaw damping is provided throughout the autopilot flight envelope.

Operational status

In production and in service in Boeing 500/530 helicopters.

Contractor

Astronautics Corporation of America.

VERIFIED

The Astronautics three-axis helicopter autopilot system

FlightMax Flight Situation Displays

Avidyne Corporation has developed a new range of fully FAA-certified Flight Situation Displays (FSD), designed to interface with a wide range of Flight Management (FM), sensor and display systems. The FlightMax series of displays offer the following common features:

- Vector-graphic moving map with colour-contoured terrain base (when interfaced with a suitable GPS/FMS)
- BF Goodrich Stormscope® WX 500 Lightning interface
- IFR & VFR charts (Jeppesen NavData for North America with optional worldwide NavData), including Terminal Area Charts (TACs)
- Traffic collision avoidance information (Ryan TCAD™ (Traffic and Collision Alert Device), BF Goodrich Skywatch™ Traffic Advisory System (TAS), or conventional Traffic Alert and Collision Avoidance Systems (TCAS I). Traffic data is displayed on a full-size, full-color display with standard traffic symbology, providing for increased situational awareness for the pilot.

The open-architecture design of the system will also accommodate enhancements including datalink graphical weather via Echo Flight or AirCell, and long-range traffic awareness via ADS-B (Automatic Dependent Surveillance Broadcast) and TIS (Traffic Information System).

Avidyne has integrated three new radar-enhancing features into the FlightMax 850, 750 and 650. *BeamView*, *TiltView* and *AutoTilt* provide additional information on the radar display, which help pilots more easily track and interpret weather radar data. Existing FlightMax 740 systems can also be upgraded to these new features:

- *BeamView* displays a sweeping "flashlight beam" which corresponds to the beam width of the radar antenna. By displaying the radar beam width as a function of range, pilots can more easily discern targets that are separated by a distance of less than the beam width and which appear on the screen as a single target.
- *TiltView* computes the relative altitude of the radar beam at each range, and displays this information adjacent to each range arc on the display scale in thousands of feet. At a glance, *TiltView* gives pilots the relative altitude of any displayed target, and eliminates the need for cumbersome radar tilt calculations. *TiltView* allows pilots to easily determine aircraft altitude relative to terrain or cloud cell tops.
- *AutoTilt* automatically adjusts the antenna tilt angle to compensate for changes in aircraft altitude and selected range, which eliminates much of the workload associated with normal tilt management.

Avidyne FlightMax 850 showing current weather data obtained via AirCell datalink interface overlaid on the moving map
2001/0099737

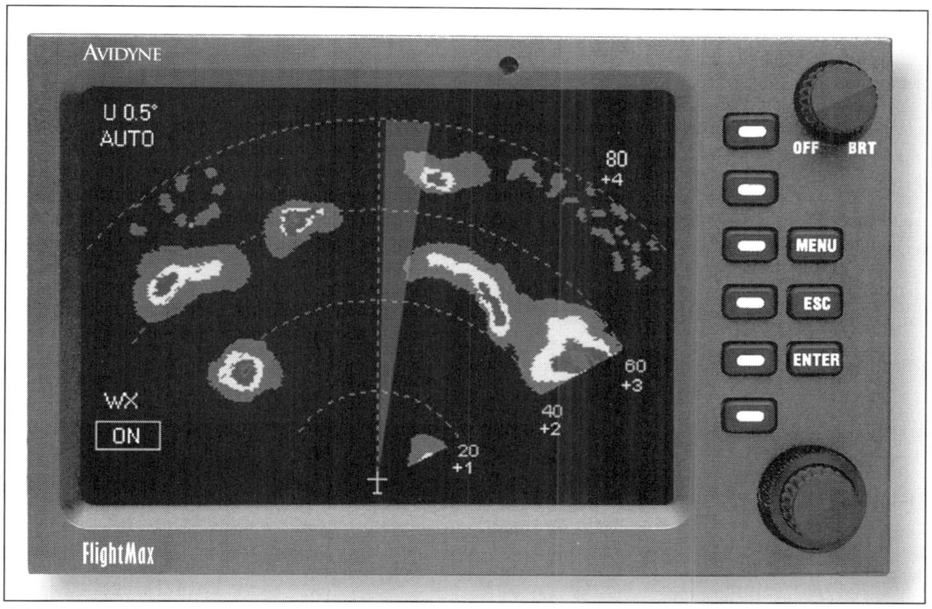

Avidyne FlightMax 850 showing BeamView, TiltView and AutoTilt radar enhancements
2001/0099738

FlightMax 450

The FlightMax 450 is designed to provide a low cost situational awareness capability for non-radar aircraft.

FlightMax 650

Tailored for aircraft with RDR 130/150/160 radar, the FlightMax 650 replaces existing radar indicators and provides all the functionality of the existing radar system, with the added benefits of the FlightMax family.

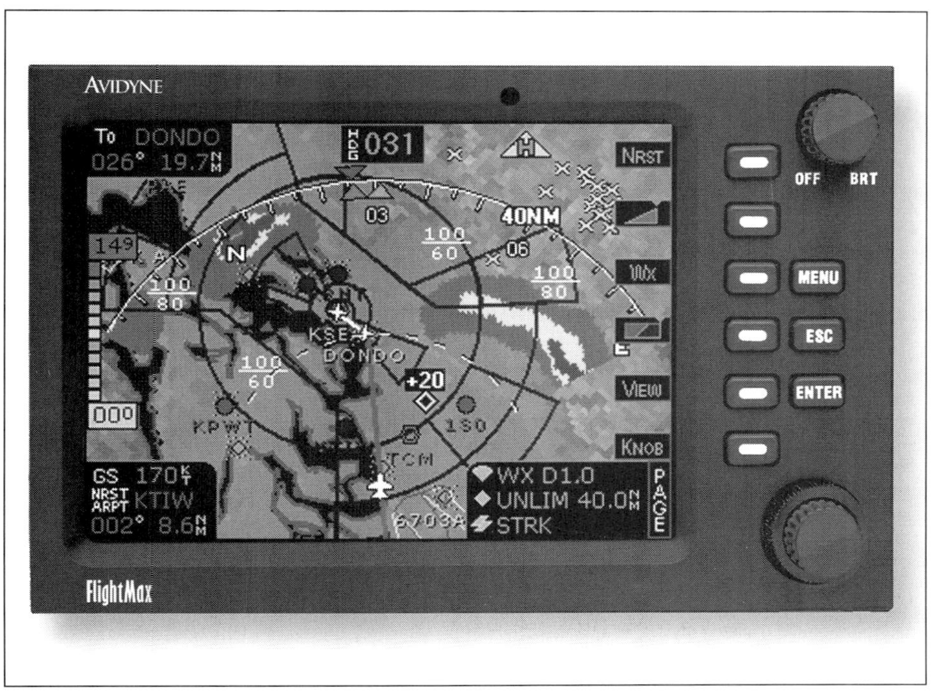

FlightMax 750

Primarily targeted at turbo props, cabin class twins and radar-equipped high performance singles, the FlightMax 750 includes interfaces for Bendix/King RDR 2000 and RDS 81/82NP radar systems. The system can replace most new and used Bendix/King radar indicators including support for Vertical Profile™.

FlightMax 850

The FlightMax 850 is designed to replace Collins WXR 250/270/270A/300 and Bendix RDR 1100/1200/1300 radar indicators which are found primarily in corporate aircraft, such as the Bombardier Aerospace Learjet, Cessna Citation and Beech King Air, as well as many regional airliners. The system also provides an optional interface for Honeywell's entire family of Enhanced Ground Proximity Warning Systems (EGPWS), providing a versatile Terrain Awareness and Warning System (TAWS) display solution. As an added feature, the FlightMax 850 provides traffic and terrain alerts as text messages with one-button pop-up access to the respective traffic or terrain warning display, allowing the pilot complete avoidance protection, even when viewing other displays.

Avidyne FlightMax 850 showing customisable data blocks to top and bottom left, with weather, terrain and traffic collision avoidance data overlaid on the moving map
2001/0099736

Specifications

Display:
5 inch diagonal, colour Active-Matrix LCD
Sunlight readable (150fL)
500:1 dimming range,
320 × 234 pixels, 65,536 colours
Size:
FSD: (W × D × H) 15.88 × 30.48 × 10.16 cm
Optional CD-ROM Data Loader: (W × D × H) 15.88 × 25.40 × 3.18
Weight:
FSD: > 3.64 kg
CD-ROM Data Loader: 0.68 kg
Power: 11 - 35 V DC, 5.0 Amp @ 14 V
GPS Interface: RS-232 or GAMA 429 high and low speed (optional on FlightMax 450)
Cooling: Forced Air Required
Operating altitude: Up to 25,000 ft. (cabin pressure altitude)
Operating temperature: –20 to +55°C
TSO compliance: TSO C113, TSO C110a, TSO C63c, TSO C118, TSO C147

Operational status

Avidyne received FAA Technical Standard Order (TSO) approval of its FlightMax 850, FlightMax 750, FlightMax 650, and FlightMax 450 Flight Situation Displays (FSD) during the third quarter of 2000. This is the fifth generation of Avidyne's planned series of certified products, which address situational awareness and safety for business and commercial aircraft operators. This certification includes seven new radar interfaces, with full overlay capability, including terrain and water base maps. Other interfaces include the enhanced versions of the Honeywell Mark V, Mark VI, Mark VII, Mark VIII, and KGP 560 GA-EGPWS, and utilise Honeywell' s KCPB picture bus for optimum display clarity and resolution.

Simultaneously, Rockwell Collins announced an agreement to offer the FlightMax 850 FSD as an additional solution for operators of turbine-powered and entry-level jet aircraft. FlightMax will provide an integrated display solution for Collins Traffic Alert and Collision Avoidance Systems (TCAS II) and Collins weather radar systems, including legacy analog systems.

FlightMax FSDs are in production and in service on a wide range of corporate aircraft and regional airliners.

Contractor

Avidyne Corporation.

NEW ENTRY

Angle Of Attack (AOA) system

The Angle Of Attack (AOA) system is designed to assist pilots to obtain optimum aircraft performance on descent, approach and landing. It also provides useful data during instrument flying, in-flight turbulence, navigation and low-speed flight. The system consists of the AOA transmitter, AOA control panel indicator and AOA approach indexer.

The AOA transmitter is the foundation of the system. It senses the airflow direction at the side of the aircraft's fuselage. Specially equipped with heaters for de-icing and moisture control, the transmitter has a unique drainage feature to prevent water intake in the air or on the ground.

The AOA indicator is an easy-to-read indicator which displays aircraft lift information on a graded scale from 0 to 1.0. 0 represents zero lift and 1.0 represents 100 per cent lift, or stall. Combined with the system's flap position information, the AOA control panel indicator's display is valid for all flap configurations. The indicator also operates the indexer and the fast-slow pointer on flight director systems.

Mounted on the aircraft's glareshield, the AOA approach indexer is a three-light/three-colour unit that instructs the pilot on the best speed of approach, based on the angle of attack.

Operational status

The system is fitted on US Air Force fighter aircraft, US Navy and US Marine Corps carrier-based aircraft and aircraft such as the Astra, Avanti, Beechjet, Citation, Falcon, Gulfstream, Jetstar, Sabreliner, Starship and Westwind.

Contractor

Avionics Specialties.

VERIFIED

PAR Power Analyser and Recorder

The PAR Power Analyser and Recorder is an intelligent turbine health monitor designed to analyse and record aircraft operation and display to the pilot the right information at the right time.

PAR automatically displays the DC bus voltage which remains displayed until the engine is started. Once the engine has been started the outside air temperature and computed density altitude are displayed to provide the pilot with performance guidelines. The PAR measures temperatures, N_1 and N_2 turbine speeds, torque, rotor speed or propeller rpm, outside air temperature, pressure altitude, airspeed and fuel flow.

Should an exceedance of any parameter occur, the display will show 'Exceedance in progress' and the memory will record the exceedance and the value of all other parameters at the time the exceeding parameter is at its peak. Time, date and engine S/N will be included for identification. Pressure altitude and outside air temperature will also be recorded for every exceedance.

The PAR system comprises a computer and pilot's display panel in a quarter-dwarf ATR-style case with an RS-232 communication port for downloading recorded data to a hand-held printer or personal computer. Information may be transmitted via telephone modem. The system also includes a torque pressure transducer and independent sensors for outside air temperature and pressure altitude. The pilot's display will indicate when memory is 80 per cent full or greater. If the signal is ignored, recording will continue over the oldest information.

Specifications

Dimensions: 57.1 × 152.4 × 203.2 mm
Weight: 2.27 kg
Power supply: 12-32 V DC
Temperature range: –40 to +75°C
Memory capacity: 175 events

Contractor

Avionics Specialties.

VERIFIED

TCAS/IVSI collision avoidance display

The collision avoidance/vertical speed indicator (TCAS/IVSI) unit gives collision avoidance messages that are easily understood. The dial of the indicator is fashioned like a traditional IVSI unit for easy recognition. The indicator's pointer and failure flag are prominent on the dial so they can be seen at a glance. Resolution Advisories (RAs) are indicated in red and green segment lamps around the perimeter of the dial. Preventive RAs are indicated in red and corrective RAs in red and green.

Housed in a 3 ATI case, the TCAS/IVSI will replace current vertical speed indicator units. The TCAS/IVSI provides inertial-led features plus integral lighting which conforms to most airline lighting requirements. The instrument requires connection to the aircraft's static pressure and electrical systems.

Using the TCAS/IVSI display as part of a collision avoidance system allows combination with equipment from other manufacturers to complete the TCAS II system.

Contractor

Avionics Specialties.

VERIFIED

5163-1 SELCAL decoder

The purpose of the SELective CALling (SELCAL) system is to permit exclusive calling of individual aircraft over normal radio channels that link the ground station to that aircraft. The system operates with HF and VHF ground-to-air transmitters and receivers and does not interfere with the normal operation of communications except when the SELCAL is performing its calling function.

Each SELCAL-equipped aircraft is assigned a four letter identifier which is used when the ground station wishes to contact it. The SELCAL decoder is designed to respond only to the identifier for which it is set. When the identifier is received by the aircraft, the decoder actuates a signal indicator in the form of a lamp, bell, chime or any combination of these. Typically, the signal will be annunciated on an audio control panel microphone select switch.

The 5163-1 SELCAL decoder is a rack-mounted, 16 tone decoder designed for the ICAO and ARINC standard system. It may be installed in non-pressurised and non-temperature controlled locations on aircraft up to altitudes of 55,000 ft. The unit is housed in a 1 MCU ARINC 600 package.

Operational status

In service.

Contractor

AVTECH Corporation.

VERIFIED

7522 series VHF comm radio tuning panel

The AVTECH 7522 series VHF comm radio tuning panel updates ARINC 500 (analogue) and ARINC 700 (digital) series radios to provide the new 8.33 kHz frequency spacing standard, required in order to meet certain European ATC requirements after 1 January 1999. The new panel has two connectors: a 55-pin connector used exclusively for ARINC 500 series radios, and a 24-pin connector for ARINC 700 series radios. The 500 series radios are tuned with a modified 2 × 5 coding. The 700 series radios are tuned via an ARINC 429 databus. This approach allows airlines with mixed fleets (500 and 700 series radios) to stock only one type of tuning panel. The current range of 760 channels at 25 kHz spacing is increased to 2,280 channels at 8.33 kHz spacing.

Operational status

TSO approved September 1997. Selected by a number of airlines to meet new European standards.

FAA STC approval was granted in January 1998 for Rockwell Collins VHF-700B, VHF-900B (digital) and 618M-5 (analogue) radios with AVTECH's 7522-1-2 control panels.

Contractor

AVTECH Corporation.

VERIFIED

Audio selector panels

The AVTECH audio selector panel provides complete selection and volume control of audio communications for cockpit crews of transport aircraft. These panels are custom-designed and manufactured specifically for each aircraft application, whether supplied as standard equipment by the airframe manufacturer or specified as buyer-furnished equipment by the purchasing airline.

The panels are designed to minimise cross-talk and spurious noise, while maximising fidelity and reliability.

Operational status

In production and in service with Boeing 727, 737, 747, 757, 767, DC-10, MD-80 and MD-90 aircraft and a number of other aircraft types, including the Beech T34C, Bell 204A, Canadair CL-41, Cessna Citation and Citation Jet, de Havilland DHC-8, Embraer EMB-312 and Learjet 20, 30 and 50 Series.

Contractor

AVTECH Corporation.

VERIFIED

Cockpit Voice Recorder (CVR) audio mixer

The Cockpit Voice Recorder (CVR) audio mixer is designed to comply with FAA regulations mandating cockpit voice recorders for all multi-engined turbine-powered aircraft that require two crew members and seat six or more passengers. The CVR audio mixer sums and routes audio signals from microphones, headphones and speakers to the cockpit voice recorder. It also provides hot microphone biasing, adjustable channels for balancing audio levels and paired pins to make installation easy and economical. Both single- and dual-station units are available. The systems are qualified to TSO C50c requirements.

Operational status

In service in various business aircraft.

Contractor

AVTECH Corporation.

VERIFIED

Digitally Controlled Audio System (DCAS)

At the heart of each DCAS is the Digital Signal Processor (DSP). DSP circuitry gives flight crews crisp drift-free audio communications with few electronic parts and dramatically increased reliability over analogue systems, extending system flexibility through software add-on features.

AVTECH DSP systems consist of a Remote Electronics Unit (REU), plus an Audio Control Panel (ACP) for each user. In operation, control switch selections are made at the ACP, then multiplexed and sent to the REU. All analogue audio signals are also sent to the REU where they are filtered and converted to digital signals. The resulting digital audio is selected, amplified, filtered and summed in the DSP circuit according to control selections. After processing, the digital audio signal is converted back to analogue for distribution to the user. All processing is done at microprocessor speeds resulting in real-time audio communication with crisp digitally processed quality. Advanced features include a digital control bus between each flight deck ACP, BITE to monitor and report digitally on system operational integrity and advanced data reporting via an ARINC 629 interface to other aircraft systems. Jack panels, headsets, microphones and speakers provide audio signal input/output, completing the system.

Operational status

In service, widely used.

Contractor

AVTECH Corporation.

VERIFIED

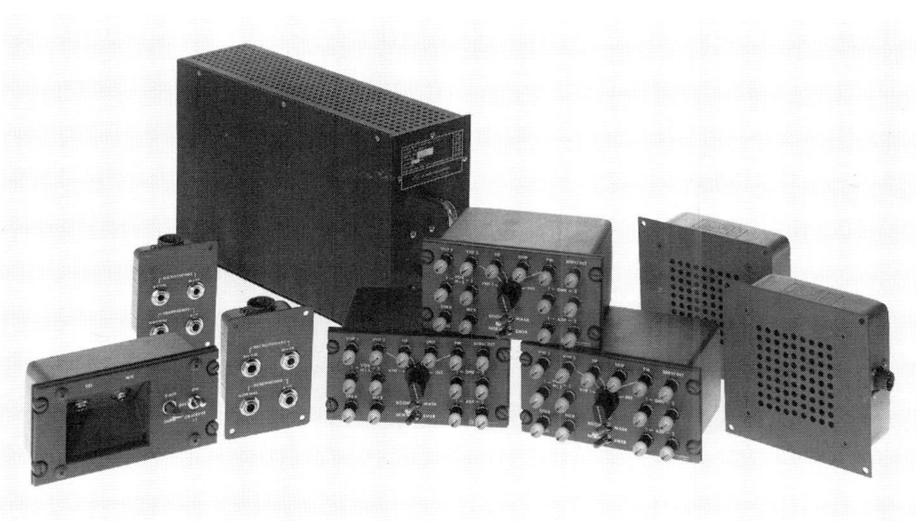

AVTECH 5060-1 audio selector panels for the de Havilland DHC-8

Display Switching Unit (DSU) for Terrain Awareness and Warning Systems (TAWS)

AVTECH's DSU is designed to be a low-cost alternative to replacing the entire display system in aircraft implementing TAWS capability for the first time. AVTECH's DSU is designed to interface with Honeywell's Enhanced Ground Proximity Warning System (EGPWS) terrain presentation on Honeywell UDI Weather Radar Displays (Primus models 200, 300, 300SL, 400 440, 450, 500, 650, 660, 700, 800, 870, 880, 90 and AVQ30).

The DSU switches up to three sources of display information and facilitates overlays:

1. EGPWS data is in an ARINC 708A/453 bus format; the DSU reformats the data into UDI format;
2. Traffic alert Collision Avoidance System (TCAS); the DSU passes UDI data;
3. Data/navigation/lightning data; the DSU passes through UDI data.

The hardware is qualified to DO-160D and the software to DO-178B. The MTBF is predicted to be 60,000 hours and the DSU is described as 'fail-safe' to the extent that it passes through UDI data, and does not interface with audible warnings and therefore does not deactivate when a failure occurs.

Operational status

AVTECH received FAA TSO C105 approval for the DSU during the second quarter of 2000.

Contractor

AVTECH Corporation.

UPDATED

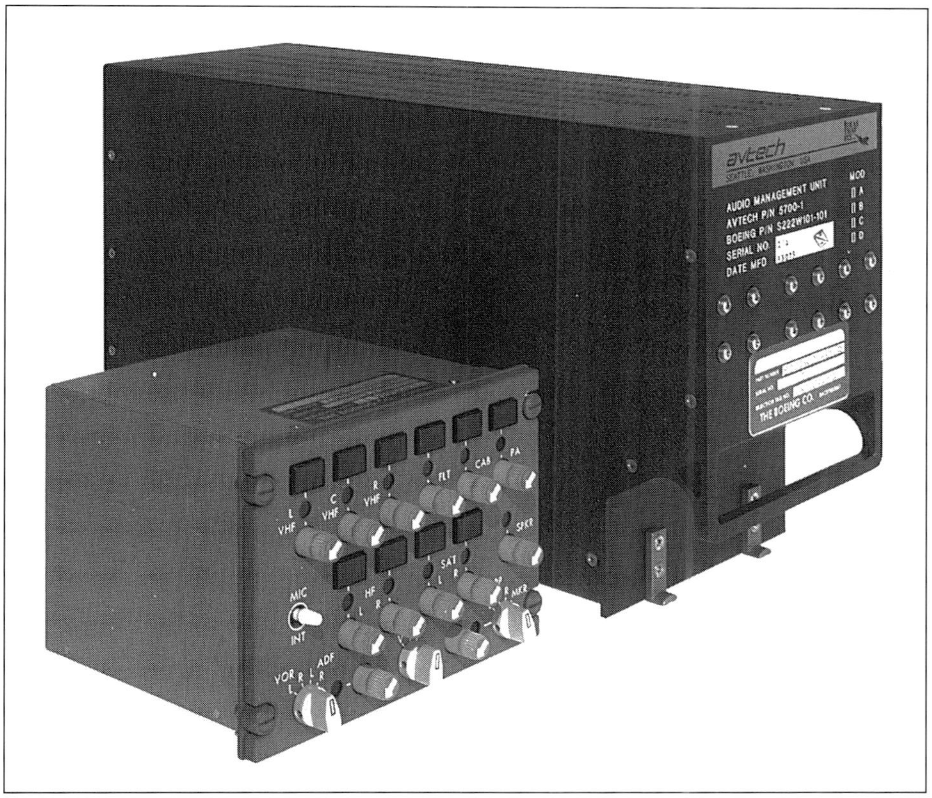

The digitally controlled audio system for the Boeing 777

GRD-2116 overwater Doppler navigation system

The GRD-2116 Doppler velocity sensor/navigation system is designed to provide optimum performance during both overwater and overland operations.

Improvement in received signal-to-noise is accomplished by increasing transmitter power, using a low-noise receiver amplifier and transmitting and receiving RF energy via separate electrical apertures to improve isolation. Low sidelobes and power management are used to achieve low probability of intercept.

Mechanically, the unit is the same size and weight as the AN/ASN-157, has the same mounting configuration and the same ARINC 575 and MIL-STD-1553 interfaces. A digital frequency tracker provides rapid signal acquisition and accurate hover.

Contractor

BAE SYSTEMS Aerospace Inc, CNI Division.

UPDATED

M3 GPS Approach

The Northstar M3 GPS Approach uses 12-channel, parallel tracking, with user-replaceable FliteCards to provide an IFR (Instrument Flight Rules) navigation system. Used in conjunction with the Northstar SmartComm it changes the M3 Approach into a GPS/Comm. The SmartComm automatically organises the 40 closest frequencies relative to a calculated position. With frequencies displayed on the M3 Approach the SmartComm is designed for aircraft with limited panel space or for those in need of an extra communications system.

The M3 has received TSO C-129, Class A1 approval for en route, terminal and non-precision approach operations.

The M3 is designed so that the pilot only has to select the approach and follow the indicator. The M3 Approach's auto-sequencing feature provides seamless transition through legs of the entire approach, complete with each leg of procedure turns and holds programmed into the database. If radar vectors are received to the final approach course, the pilot is able to select 'Vectors to Final' and automatically pass intermediate waypoints, while retaining situational awareness relative to final approach course and final approach fix.

Specifications

Channels: 12 continuous, parallel-tracking
Navigation accuracy:
15 m RMS (30 m 2DRMS)
100 m 2DRMS with S/A activated
Navigation update rate: 1 s
Time to fix: 1 min (typical)
Operating modes:
2D Nav, 4 satellites visible
3D, 5 or more satellites
Automatic cold start: neither time nor position input required
Voltage: 10-35 V DC negative ground
Power: 35 W nominal
Size: 159 × 299 × 51 mm
Weight:
(with mounting tray) 1.9 kg

The Northstar M3 GPS Approach **1999**/0051611

Operational status

Listed as recommended equipment in the American Champion Aircraft Corporation list of avionics options.

Contractor

BAE Systems North America.

VERIFIED

SmartComm intelligent frequency management

SmartComm is an intelligent position-based communications management system. SmartComm organises all database frequencies relative to present position. A standby stack keeps track of the last five frequencies used. The local category list of Centre, Approach, Atis and so on, allows the selection of frequencies by type with up to eight frequencies in each list.

All new Northstar navigators feature SmartComm and include fully operational SmartComm software. Optional 760-channel remote-mounted transceivers instantly tune the frequency selection directly from the Northstar navigator, tuning it into a GPS/comm unit.

Specifications

Nominal voltage: 13.75 V DC
Current (transmit): <2.5 A
Frequency range: 118 to 136.975 MHz
Channels: 760
Weight: 0.55 kg
Dimensions: 70 × 61 × 210 mm

Operational status

In production.

Contractor

BAE Systems North America.

VERIFIED

VFR GPS-60

The 12-channel, parallel-tracking GPS-60 provides all-weather, worldwide navigation. The VFR-only GPS-60 offers the same features and accuracy found in the M3 IFR GPS (except GPS approaches), at a VFR price. The GPS-60 provides immediate access to flight information, such as distance and bearing to destination; ground speed; NDBs, VORs, Victor and Jet airways, and intersections; winds aloft, and the Class B and Class C airspaces.

The Northstar GPS-60 will interface directly with modern CDIs, HSIs, autopilots and moving maps.

Every Northstar GPS-60 comes complete with a user-updateable FliteCard containing a comprehensive

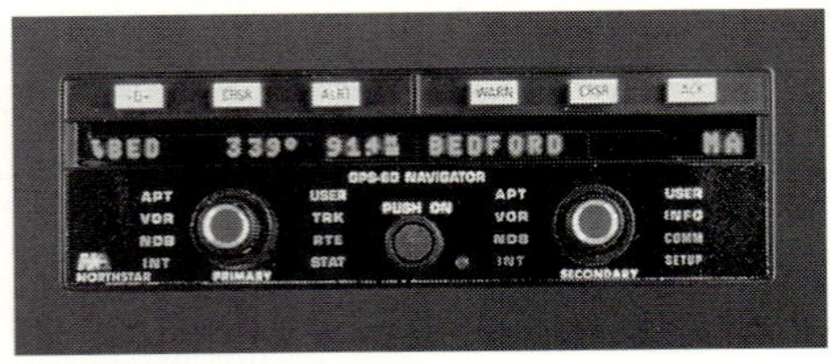

The Northstar VFR GPS-60 **1999**/0051612

Jeppesen database of over 8,000 US public airports. With Northstar's SmartComm intelligent frequency software, the GPS-60 becomes a GPS/Comm. The SmartComm enables the GPS-60 to organise, tune and display the 40 closest frequencies relative to calculated position. Upgrading to the SmartComm is a designed capability for aircraft with limited panel space or in the need of an extra communication system. Alternative databases available for the GPS-60 include North America, International, and helicopter, all of which include private airports in the United States in addition to public-use airports.

Specifications

Type: L1 frequency, C/A code (SPS)
Multichannel: continuous tracking
Navigation accuracy:
15 m RMS (30 m 2DRMS)
100 m 2DRMS with S/A activated
Navigation update rate: 1 s
Time to first fix: 1 min (typical)
Operating modes:
2D Nav, 3 satellites visible
3D Nav, 4 or more satellites
Automatic cold start: neither time nor position input required
Annunciator output: warn, parallel offset, waypoint alert, VFR
Serial ports: RS-422/RS-485
Power: 10-36 V DC at 14 W
Size: 159 × 298 × 51 mm
Weight: 2 kg

GPS antenna

Type: low-profile patch with integral L1 preamplifier
Length: 87 mm
Width: 56 mm
Weight: 0.14 kg
Mounting: surface-mounts to top of aircraft; requires 12.5 mm cut out in aircraft skin

Contractor

BAE Systems North America.

VERIFIED

Airborne cameras

Ball's specialised airborne cameras give pilots a variety of exterior views, enhancing safety and security.

Ball designed the lightweight, charge-coupled device camera, both colour and monochrome, specifically for the flight environment. An electronic sensitivity control ensures clear images in both bright sunlight and low-light flight conditions. The environmentally sealed, self-contained camera is engineered to withstand extreme temperatures, shock and vibration.

The camera operates with electronic shutter control, improved low-light sensitivity and existing aircraft lighting to give flight crews optimal vision, day or night.

Camera features include an environmentally sealed aluminium housing, adaptable for a variety of applications; lens sizes from 2.8 to 25 mm; electronic shutter dynamic range greater than 100,000 to 1; sapphire with ITO window coating for reliable de-icing/defogging; electronic shutter control for automatic low-to high-light level operation.

Operational status

Selected by Boeing Commercial Airplane Group for a 10-year contract to outfit its entire fleet of 777-300s with airborne cameras, as a ground manoeuvring system.

Contractor

Ball Aerospace and Technologies Corp.

VERIFIED

AIRLINK antenna system for satcoms

AIRLINK low- and high-gain antenna systems are designed for use with INMARSAT satellite communications.

The high-gain antenna system uses two conformal, electronically steered, phased-arrays in a side-mounted architecture. This configuration yields superior coverage with minimal aerodynamic drag penalties. The high-gain antenna system is fully approved for multichannel data, voice and data applications.

The low-gain antenna system consists of a single-blade antenna and is used for low-speed data applications. It is ideally suited as a back-up for the high-gain system.

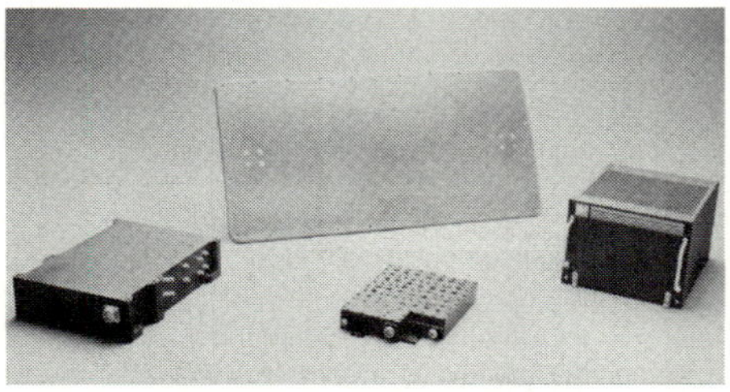

The Ball AIRLINK antenna system showing the conformal array (above) and beam-steering unit (left), diplexer/low-noise amplifier (centre) and high-power amplifier (right)

Ball has recently introduced the AIRLINK Gateway Unit (AGU) which provides the digital signal processing necessary for operating with INMARSAT's circuit mode data channel. This channel provides users with a host of applications at the 9.6 kbits/s rate and can also provide secure satellite communications.

Specifications

Dimensions:
(antenna array) 407 × 813 × 9.5 mm
(beam-steering unit) 89 × 264 × 343 mm
(diplexer/low-noise amplifier) 51 × 198 × 282 mm
(high-power amplifier) 193 × 257 × 925 mm
(AIRLINK Gateway Unit) 7 MCU
Weight:
(antenna array) 7.1 kg
(beam-steering unit) 8.4 kg
(diplexer/low-noise amplifier) 3 kg
(high-power amplifier) 20 kg
(AIRLINK) Gateway Unit) 13.5 kg
Power supply: 115 V AC, 400 Hz, single phase
Frequency: 1,530-1,559 MHz, 1,626.5-1,660.5 MHz

Operational status

Selected by United Airlines and British Airways for Boeing 777 aircraft and by Scandinavian Airline System for Boeing 767-300s. Also selected by the United States government for VIP/SAM fleet.

Contractor

Ball Aerospace and Technologies Corp. ***VERIFIED***

Fire Detection Suppression system (FiDS)

The BFGoodrich Aerospace FiDS is an integrated system which includes smoke detectors, flight deck panel, maintenance bay panel, wiring, plumbing, and halon bottles. It is designed to satisfy the requirements of FAR Pt 28.855.

The system provides 60 second detection and 60 minute suppression. Should smoke be detected the existing master caution warning light illuminates both on the glare shield and on the flight deck panel, and an audible warning is provided to the flight deck crew. Despatch reliability is assured using redundant dual-loop smoke detectors operating independently in each zone and redundant electronic circuitry.

FiDS is adaptable to several models of aircraft including: Boeing 727, 737 and DC-9, and is supplied as a turnkey installation kit.

Contractor

BFGoodrich Aerospace Aircraft Integrated Systems.

VERIFIED

Fuel Quantity Indicating Systems (FQIS)

BFGoodrich Aerospace provides fuel quantity indicating systems for both commercial and military aircraft, fixed- and rotary-wing including: Boeing, Airbus, Saab, Embraer, de Havilland, Lockheed Martin, Sikorsky, Bell, and other foreign OEMs. These fuel quantity indicating systems provide high reliability, ease of maintenance and guaranteed 1 per cent accuracy.

Retrofit FQIS systems similar to the production configurations have been chosen by numerous airlines for their Boeing 747-200, -300, and 757/767 cargo and passenger aircraft.

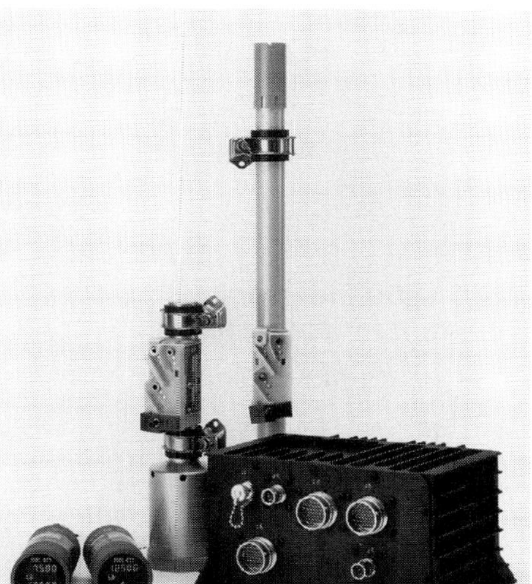

Boeing 737-700 FQIS tank unit hardware 0018231

Retrofit fuel quantity indicating system for Lockheed Martin C-130 aircraft 0018232

The in-tank hardware for the fuel quantity indicating systems on these aircraft include all new harnesses, fuel quantity sensors, compensators for providing signal variation in dielectric constant of the fuel, plus densitometers which provide direct fuel density measurement information to the processor. Flight deck and refuelling panel indicators provide visual verification of total fuel and individual tank quantities. The processor unit controls the entire fuel system by calculating fuel quantity, controlling the refuelling sequence and performing extensive health monitoring on both wiring and electronics. The processor provides messages on operational health and history.

Other programmes for which BFGoodrich produces specific retrofit fuel quantity indicating systems are: Boeing 727, Lockheed Martin C-130 and P-3.

BFGoodrich provides fuel management and measurement for the Boeing B-1B, Northrop Grumman B-2 and the Airbus A330/340. In addition to measuring the fuel quantity, these systems control pumps and valves to transfer the fuel between tanks thus maintaining centre of gravity for optimum performance.

Contractor

BFGoodrich Aerospace Aircraft Integrated Systems.

VERIFIED

ADI-330 self-contained Attitude Director Indicator

The ADI-330, with a full electrical erection system, combines the reliability and safety of a case-contained gyro with the convenience and operational features of a remote gyro. Ideally suited for large corporate and transport aircraft, the ADI-330 provides accurate, reliable pitch and roll information under all normal conditions. When used in conjunction with an emergency power supply, it also serves as an efficient, long-running standby attitude reference.

Electrical erection automatically provides 20°/min fast erect during initial power-up, eliminating the need for manual caging or uncaging during preflight checks. The unit also includes a bezel-mounted push-button for fast erect on demand.

The face presentation of the ADI-330 has been designed to be visually integrated into cockpits which use EFIS electronic displays. The style, colours and integral incandescent lighting of the instrument have been chosen to provide an attractive, consistent and familiar attitude information presentation.

Designed for 28 V DC, the unit provides useful attitude information down to 18 V and features power failure monitoring. Estimated MTBF for the ADI-330 is 4,500 operating hours and it is TSO C4c/C4d qualified.

Contractor

BFGoodrich Aerospace Avionics Systems.

VERIFIED

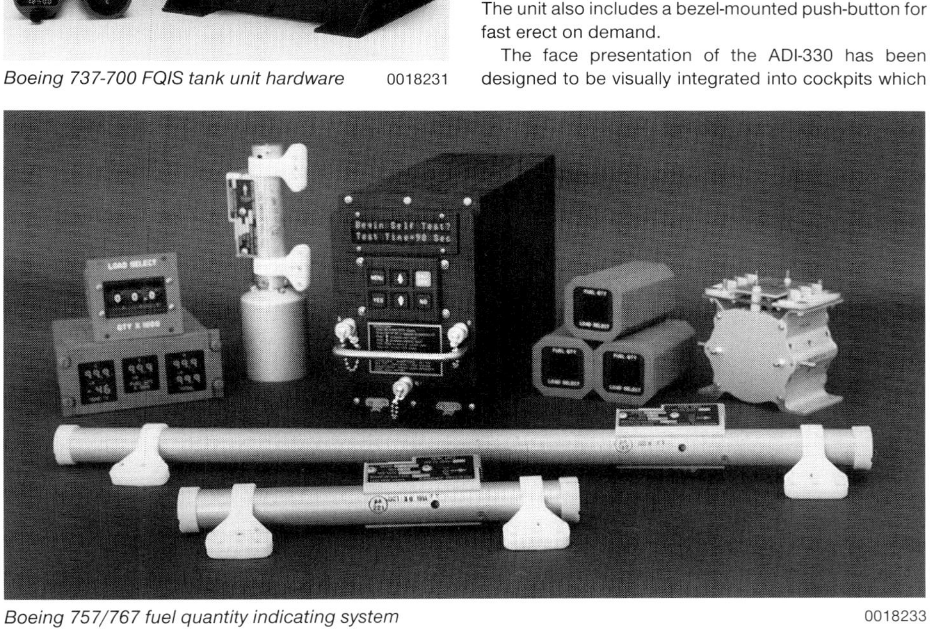

Boeing 757/767 fuel quantity indicating system 0018233

ADI-330/331 self-contained Attitude Director Indicator

The ADI-330/331 is a 3 ATI compact, completely self-contained instrument which features glide slope and localiser cross-pointers, integral inclinometer for slip and skid indication and mechanical erection with manual caging. The ADI-330 accepts analogue glide slope/localiser signals; the ADI-331 accepts ARINC 429 digital glide slope/localiser signals. Direct mechanical linkage eliminates electrical servo response lag.

The ADI-330/331 features 9 minutes of usable attitude information after complete power loss, 18 to 30 V DC power supply, blue/white internal lighting, Power Off warning and GS/Loc signal validity flags. The design is compatible with EFIS displays and the instrument is qualified to TSO C4c.

Contractor

BFGoodrich Aerospace Avionics Systems.

VERIFIED

ADI-332/333 self-contained Attitude Director Indicator

The ADI-332/333 is a 3 in (76.2 mm) compact, completely self-contained attitude director indicator featuring glide slope and localiser cross-pointers with back course, integral inclinometer for slip and skid indications and mechanical erection with manual caging. The ADI-332 accepts an analogue glide slope/localiser signal; the ADI-333 has an ARINC 429 digital interface. Direct mechanical linkage eliminates electrical servo response lag.

The ADI-332/333 features 9 minutes of usable attitude information after complete power loss, a selectable switch for BC/ILS/OFF modes, 18 to 30 V DC power supply, blue/white internal lighting, Power Off warning and glide slope/localiser signal validity flags and self-contained lateral and fore and aft acceleration compensation. The design is compatible with EFIS displays and the instrument is qualified to TSO C4c.

Contractor

BFGoodrich Aerospace Avionics Systems.

UPDATED

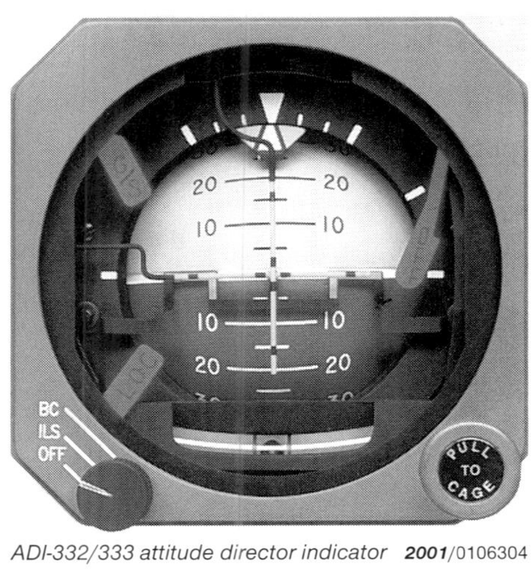

ADI-332/333 attitude director indicator **2001**/0106304

VOR (left) and ILS (right) displays on the ADI-335 standby attitude and navigation indicator ***2001***/0106400

ADI-335 standby attitude and navigation indicator

The ADI-335 standby attitude and navigation indicator provides roll and pitch attitude information by electromechanical means. In addition, the instrument contains both en route VOR/DME and ILS navigation displays. Both the navigation modes contain appropriate validity flags and the indicator will interface with an ARINC 429 bus.

The ADI-335 is qualified to TSO C4c and C52a in accordance with DO-160C.

Contractor

BFGoodrich Aerospace Avionics Systems.

UPDATED

ADI-350 Attitude Director Indicator

The ADI-350 is a self-contained gyro indicator with flight director needles, rate of turn and slip indication. Synchro pick-offs are provided for remote indicators, radar stabilisation and flight control functions. Inputs are needed to operate the flight director needles and the rate of turn indicator.

Specifications

Dimensions: 82.04 × 82.04 × 228.60 mm
Weight: 2.72 kg
Power supply: 115/208 V AC, 400 Hz, 3 phase or 28 V DC

Operational status

In production. Applications include the Northrop Grumman A-6E, KA-6D and EA-6B, Boeing F/A-18, Canadair CP-140, Rockwell B-1B and Bell OH-58D. The ADI-350V has also been selected for use in the upgrade programme for the F-14 Super Tomcat.

Contractor

BFGoodrich Aerospace Avionics Systems.

VERIFIED

AI-803/804 2 in standby gyro horizon

The AI-803/804 is a 2 in standby gyro horizon for high-performance military and general aviation aircraft. The self-contained package eliminates the need for additional electronic components. After complete loss of external power, attitude information remains available for 9 minutes.

Specifications

Dimensions: 2 ATI
Weight: 1.1 kg
Power supply: 115 V AC, 400 Hz or 28 V DC

Operational status

In production and widely deployed in civil and military aircraft.

Contractor

BFGoodrich Aerospace Avionics Systems.

VERIFIED

AIM 205 Series 3 in directional gyros

Low cost and lightweight characterise the AIM 205 Series of 3 in directional gyros. They are precise flight instruments with a low drift rate, designed to provide the pilot with a constant azimuth reference free from the instability inherent in magnetic compasses. A high-speed electric rotor, mounted in a universal gimbal system, maintains angular momentum to overcome normal bearing friction and establishes a gyroscopically stable datum reference relative to space.

Specifications

Dimensions: 200.7 mm long
Weight:
(205-1) 1.13 kg
(205-2) 1.32 kg
Power supply: 14 V DC, 28 V DC or 115 V AC, 400 Hz, single phase
Temperature range: –30 to +50°C
Environmental: DO-160A

Contractor

BFGoodrich Aerospace Avionics Systems.

VERIFIED

AIM 520 Series 2 in attitude gyros

Compact design and lightweight characterise the AIM 520 Series electric attitude gyros. The 2 in case makes the series suitable for use as a secondary attitude gyro and for those aircraft where panel space is at a premium.

Specifications

Dimensions: 157.5 mm long
Weight: 0.77 kg nominal
Power supply: 28 V DC
(starting) 0.67 A
(running) 0.45 A
Temperature range: –30 to +50°C
Certification: TSO C4c

Contractor

BFGoodrich Aerospace Avionics Systems.

VERIFIED

AIM 1100 3 in self-contained attitude indicator

The AIM 1100 is designed as a form, fit and function replacement for the AIM Model 305. It is characterised by its lightweight and rugged design. Engineering enhancements include a proprietary bearing design improving rotor life and an improved pointer bar resulting in better performance and reliability under vibration.

It is designed for helicopters and general aviation aircraft operating in high duty cycle environments. It is available in 14 or 28 V DC form with an optional slip/skid indicator and fixed or trimmable pitch airplane symbol.

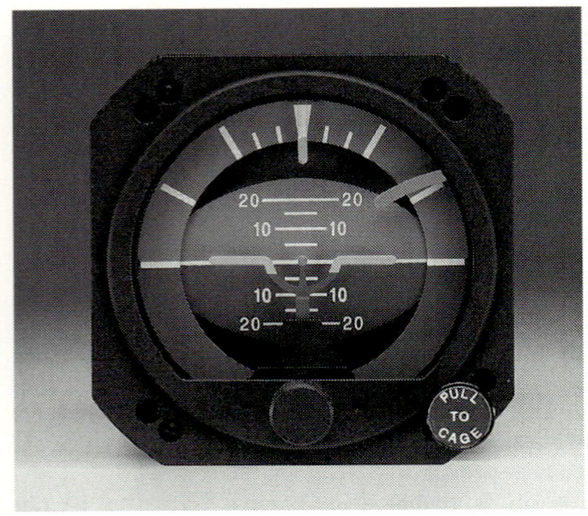

AIM 1100 3 in self-contained attitude indicator 0018174

The AIM 1100 is certified to FAA TSO C4c, RTCA DO-160C, Section 8.0, Vibration Curves S and P.

Contractor

BFGoodrich Aerospace Avionics Systems.

VERIFIED

DG-710 Directional Gyro system

The DG-710 provides three operational modes: slaved DG, free DG and compass. The microprocessor-controlled gyro permits such features as extensive built-in test and instantaneous slaving.

Standard features include two isolated synchro outputs and autopilot interlock.

Specifications

Dimensions: 155 × 139 × 167 mm
Weight: 3.36 kg
Power supply: 18 to 32.2 V DC

Operational status

In production. The DG-710 is in service in the UK Royal Air Force Tucano trainer.

Contractor

BFGoodrich Aerospace Avionics Systems.

NEW ENTRY

EHSI-4000 Electronic Horizontal Situation Indicator (EHSI)

Derived from the HSI-3000, the EHSI-4000 (Electronic Horizontal Situation Indicator) was designed to replace older electromechanical HSIs and developed specifically for the requirements of business aircraft environments. The EHSI-4000 allows for pilot-commanded heading and course inputs on a single flat panel 3 ATI Active Matrix Liquid Crystal Display. The unit features basic functions for selection among four modes: ILS, NAV, ILS/NAV, VOR, DME, and FMS.

Contractor

BFGoodrich Aerospace Avionics Systems.

NEW ENTRY

EHSI-4000 electronic horizontal situation indicator ***2001***/0103880

GH-3000 Electronic Standby Instrument System (ESIS)

The GH-3000 electronic standby instrument system replaces all three traditional standby instruments with a single, fully digital, flat panel AMLCD. The 3 ATI size, self-contained inertial measurement cluster eliminates the need for a mechanical gyro. Designed to DO-160C, the GH-3000 requires 28 V DC and weighs 1.59 kg. External systems interfaces provide ILS/VOR/DME/FMS/magnetic heading and a compact remote air data computer at less than 0.62 kg supplies airspeed and altitude.

Specifications
Dimensions: 3 ATI
Weight: 1.59 kg
Power: 28 V DC
Certification: DO-160C

Operational status
TSO certification for conformance to the following standards granted in March 1997: TSO-C2d/-C4c/-C10b/-C34e/-C36e/-C113. Available for production delivery. Certified in Challenger 604, Falcon 50 Gulfstream IV and V, Raytheon Hawker 800XP and Eurocopter AS 365N2.

Selected by the US Army Special Forces for its MH-47E and MH-60K helicopters, and by Bell Helicopter for the 609 Tiltrotor programme.

Contractor
BFGoodrich Aerospace Avionics Systems.

UPDATED

GH-3000 electronic standby instrument system
***2001**/0103879*

HSI-3000 Electronic Horizontal Situation Indicator (EHSI)

The design technology for the HSI-3000 is based on the GH-3000 Electronic Standby Instrument System (ESIS), which was certified in 1997 for display of attitude, air data, and navigation data.

With the inclusion of a graphics processor and anti-aliasing graphics, the HSI-3000 provides an easily read display and high reliability. It incorporates an Active Matrix Liquid Crystal Display (AMLCD) that is visible in direct sunlight, as well as a dark cockpit, and is NVE compatible. A built-in photocell on the front bezel, when used in conjunction with the setting of the aircraft databus, automatically adapts to ambient lighting conditions.

Designed to replace older electromechanical horizontal situation indicators, the HSI-3000 provides distance, bearing, and course information from on-board INS and TACAN in a 3ATI display.

Operational status
The HSI-3000 was selected by Lockheed Martin Tactical Aircraft Systems (Fort Worth) for the US Air Force F-16 Block 40/50 Common Configuration Implementation Program (CCIP). The F-16 Block 40/50 CCIP is upgrading approximately 750 aircraft. In this application, the HSI-3000 will interface, via the MIL-STD-1553 databus, to the new upgraded TACAN chosen for the CCIP, as well as other existing analogue equipment.

Contractor
BFGoodrich Aerospace Avionics Systems.

UPDATED

LandMark TAWS 8000 Terrain Awareness and Warning System

Providing a continuous colour overview of the surrounding terrain, the LandMark™ TAWS Class B system offers predictive warning functions using present position from a GPS receiver, altitude information and aircraft configuration. The system then compares that information to a terrain and obstacle database. Both aural and visual warnings are issued whenever potential Controlled-Flight-Into-Terrain (CFIT) situations arise.

The LandMark™ system can be displayed on any TAWS-compatible ARINC 453 EFIS system or radar indicator, or, when coupled with the BFGoodrich RGC250 Radar Graphics Computer, terrain can also be viewed on an aircraft's existing compatible weather radar indicator. The RGC250 also offers several other advantages such as improved terrain contouring, runway and obstacle depictions, traffic data and lightning display information, all shown through the radar indicator display.

As a forward looking terrain avoidance system, the LandMark™ TAWS Class B system issues warnings on premature descents, reduced terrain clearance and imminent terrain impact. In addition, the system will warn of imminent contact with the ground (GPWS modes) for excessive rates of descent, negative climb rates or altitude loss after takeoff, with a voice callout of 'five hundred' when the aircraft descends to 500 ft above the nearest runway.

Certified to TSO-C151a Class B and DO-178B Level C, the LandMark™ System operates at altitudes up to 55,000 ft.

Specifications
Weight: 2.01 kg
Power supply: 28 V DC
Temperature range: –55 to +70°C

Contractor
BFGoodrich Aerospace Avionics Systems.

NEW ENTRY

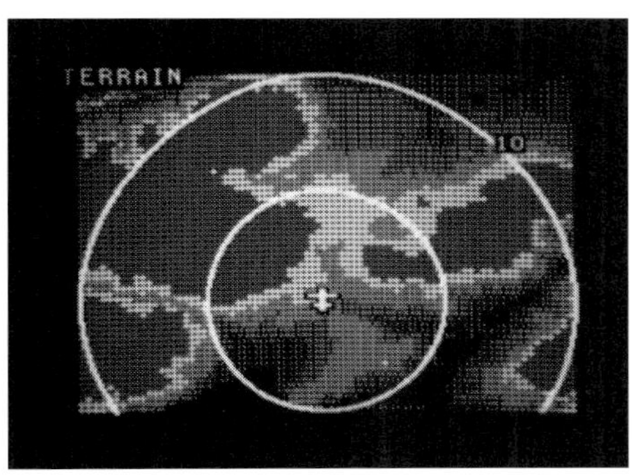

LandMark™ TAWS 8000 display ***2001**/0098573*

SKYWATCH® and SKYWATCH® HP traffic advisory system

SKYWATCH® is an active surveillance system that operates as an air-to-air or ground-to-air interrogation device. It is derived from the TCAS 791 system as a more affordable alternative to full TCAS systems for helicopter operators and general aviation aircraft.

When replies to SKYWATCH® Mode C-type transponder interrogations are received, the responding aircraft's range, bearing, relative altitude and closure rate are computed to a fixed position and traffic conflicts are predicted. Visual targets are displayed using TCAS-like symbology, with aural traffic alerts.

The multifunction capability of the SKYWATCH® system makes it possible to share a 3 ATI cathode ray tube display with late-model WX-1000 Stormscope weather mapping systems. SKYWATCH® and Stormscope display functions are selected via a remote panel-mounted switching device. When operating in the Stormscope mode, the control display unit will temporarily switch to the SKYWATCH® view if an intruder aircraft is detected which poses an immediate collision threat.

SKYWATCH® HP is a development of SKYWATCH®, with a 20 n mile surveillance range and a 15 n mile or more display range (depending on the selected display interface). SKYWATCH® HP also has enhanced display options such as an ARINC 429 EFIS output and the ability to display traffic information on a wide variety of MultiFunction Displays (MFDs) and weather radar indicators using the RGC250 radar graphics computer (see separate entry). SKYWATCH® HP is ADS-B equipped in anticipation of the Future Air Navigation System (FANS) requirements.

Specifications
Tracking capability: up to 30 targets
Display range: 2 and 6 n miles
Range accuracy: ±0.05 n miles (typical)
Bearing accuracy: 5° RMS (typical)
Altitude resolution: ±200 ft
TRC receiver/transmitter:

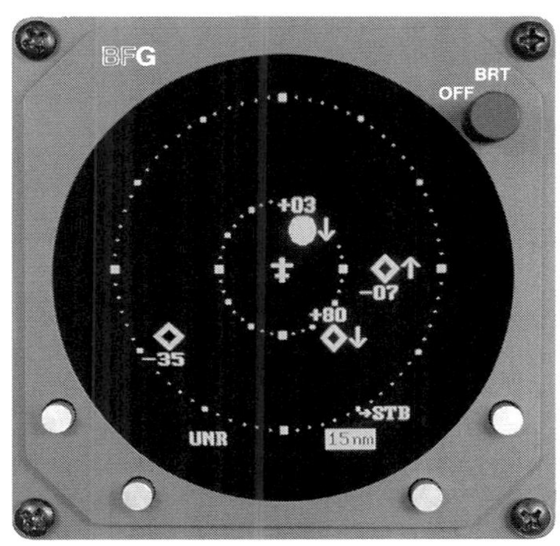

SKYWATCH® HP display unit ***2001**/0103878*

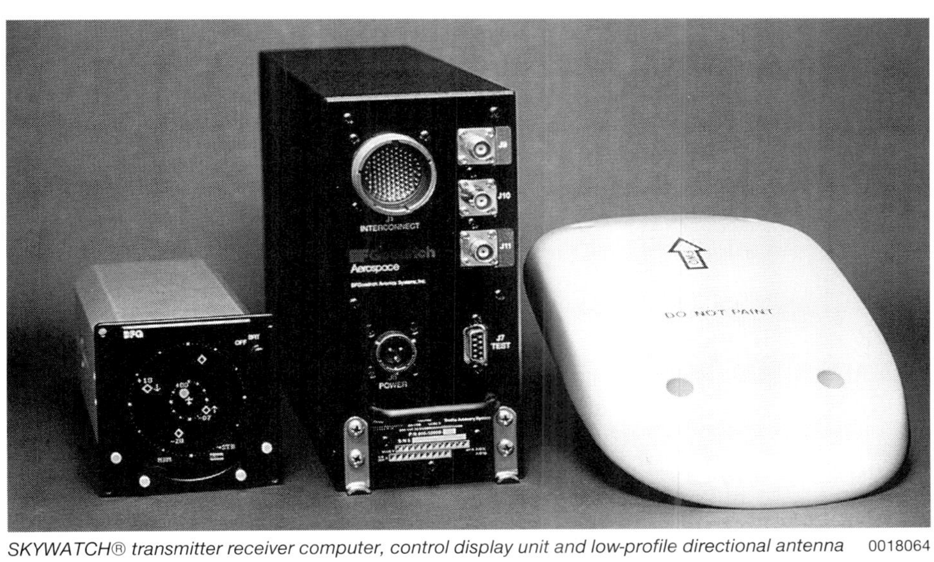

SKYWATCH® transmitter receiver computer, control display unit and low-profile directional antenna 0018064

Dimensions: ARINC standard 404A ⅜ ATR short
Weight: 4.06 kg
Power required: 11-34 V DC
Multifunction control display unit:
(display) raster scan CRT
(resolution) 256 × 256 pixels
(dimensions) 3 ATI × 209.3 mm
(weight) 1.03 kg
NY164 L-band (NATO D-band) directional antenna:
(dimensions) 279.4 × 158.8 × 35.6 mm
(weight) 1.04 kg

Operational status
In production.

Contractor
BFGoodrich Aerospace Avionics Systems.

UPDATED

Smartdeck™ integrated flight displays and control system

BFGoodrich Avionics Systems' SmartDeck™ integrated flight displays and control system is an integrated avionics suite, designed to enhance flight safety through the application of leading edge technology, human factors engineering, and 'smart' systems integration. SmartDeck™ economically provides general aviation pilots with an electronic situational display with primary flight cues, together with moving map, weather, traffic and terrain information on 10 in diagonal flat panel displays. SmartDeck™ technology fuses data from all available aircraft sources to enhance pilots' situational awareness with 'Highway-In-The-Sky' and moving map presentations. SmartDeck™ will also monitor engine and aircraft systems for early detection of potentially hazardous situations.

Operational Status
Under development.

Contractor
BFGoodrich Aerospace Avionics Systems.

NEW ENTRY

Stormscope® WX-500 weather mapping system

The Stormscope® WX-500 weather mapping system has been designed to interface with the new generation of multifunction displays currently manufactured by Advanced Creations, ARNAV, Avidyne, Archangel,

Stormscope® WX-900 weather mapping system display ***2001***/0098571

Eventide, and Skyforce Avionics. All sensor functions are controlled through the multifunction display.

The WX-500 with its advanced digital ranging algorithms combines the most popular features associated with the Series I and Series II systems to provide precision mapping of electrical discharges which are associated with thunderstorm activity.

There are two modes of operation – cell mode and strike mode. It operates in 25, 50, 100 and 200 n mile ranges, and displays a 360° view of electrical activity and a forward 120° view.

Heading stabilisation, a strike rate indicator, and continuous (and operator initiated) self-tests are other features of the unit.

Specifications
Dimensions:
antenna 25.4 × 87.6 × 174 mm
WX-500 processor w/o tray 130.8 × 44.6 × 228.6 mm

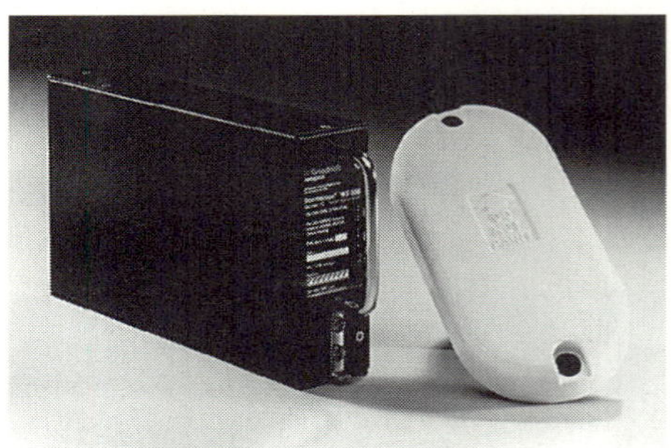

Stormscope® WX-500 weather mapping system 0018063

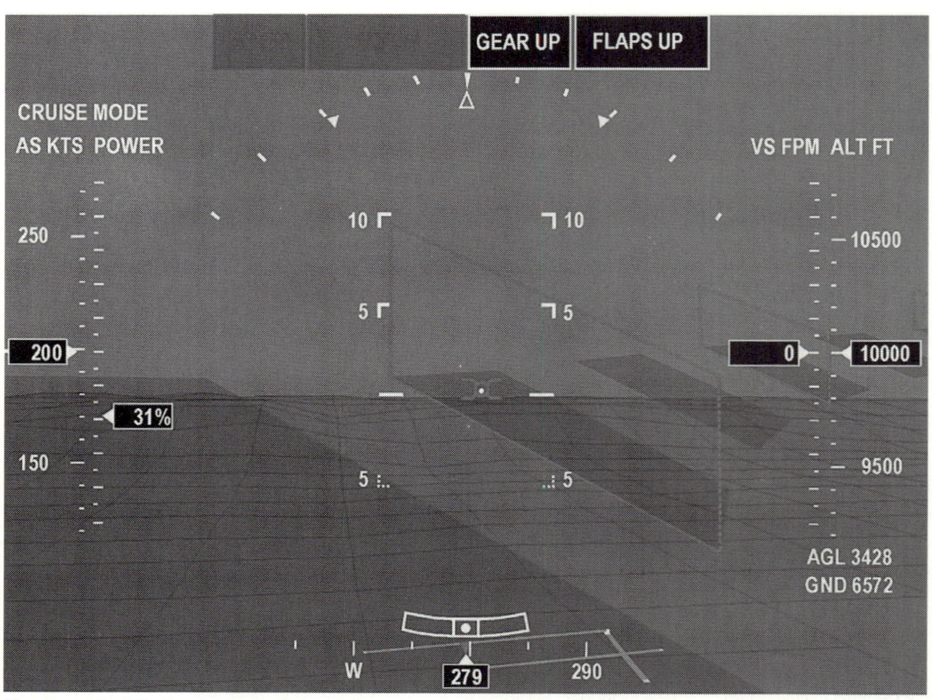

The Smartdeck™ pilots' primary flight display, showing flight parameters, aircraft configuration, flight guidance and terrain information ***2001***/0103881

Weight:
antenna 0.42 kg
WX-500 processor: 1.12 kg
Power supply: 11-32 V DC
Temperature range: –55 to +70°C
Altitude: 55,000 ft MSL

Contractor
BFGoodrich Aerospace Avionics Systems.

VERIFIED

Stormscope® WX-900 weather mapping system

The Stormscope WX-900 weather mapping system maps electrical discharges and thunderstorm activity, clearly alerting crews to storms containing lightning.

Unlike radar, the WX-900 displays the electrical discharges associated with cumulus and mature and dissipating thunderstorms. It provides a full 360° view of weather or 180° during monitor modes, with pilot-selectable ranges of 25, 50 and 100 n miles (46, 92 and 185 km). The Supertwist LCD is a self-contained panel-mount unit with electroluminescent back-lighting.

The WX-900 performs a self-test after turn-on, then performs tests continuously during system operation. The system features push-button selection of view, pilot initiated self-test programme, brightness adjustment, time mode which provides elapsed flight time and approach timer functions, battery monitor mode which monitors the aircraft electrical system and a noise analyser mode. An integral service menu includes strike test, spectrum analyser and board test modes.

Specifications
Dimensions:
(antenna) 254 × 87.6 × 167 mm
(display/processor) 86 × 86 × 192 mm
Weight:
(antenna) 0.42 kg
(display/processor) 0.71 kg
Power supply: 10.5-32 V DC, 8 W
Temperature range:
(display) 0 to +55°C
(antenna) –55 to +70°C
Altitude: up to 20,000 ft

Contractor
BFGoodrich Aerospace Avionics Systems.

UPDATED

Stormscope® WX-950 weather mapping system

The Stormscope® WX-950 weather mapping system is the newest addition to the Stormscope® family. The WX-950 provides two modes of operation – cell mode and strike mode. It uses a high-resolution CRT within its 3 in ATI display.

The WX-950 operates in 25, 50, 100 and 200 n mile ranges, displays a 360° view of electrical activity in all directions and a forward 120° view that doubles the resolution for analysing activity ahead.

When configured with a compatible heading system, the WX-950 provides heading stabilisation. A discharge rate indicator, integrity indicator and built-in self tests are other features of the unit.

Specifications
Dimensions:
(antenna) 25.4 × 87.6 × 174 mm
(display/processor) 3 ATI
Weight:
(antenna) 0.38 kg
(display/processor) 1.3 kg
Power supply: 11-32 V DC
Temperature range: –20 to +55°C
Altitude: 55,000 ft

Contractor
BFGoodrich Aerospace Avionics Systems.

VERIFIED

Stormscope® WX-1000/ WX-1000+/WX-1000E weather mapping systems

The Stormscope WX-1000 and WX-1000+ series are weather mapping systems that map electrical discharges. Time and date information, stopwatch and timing information and checklists can be selected from the main menu. All units feature a continuous self-test with error messages that assure accurate weather detection.

The WX-1000 has pilot-selectable ranges of 25, 50, 100 and 200 n miles (46, 92, 185 and 370 km), and provides either 360° or 120° viewing options. The high-resolution CRT display provides a sunlight-readable storm activity picture that can be used in tandem with conventional radar systems to determine the severity and stages of storms. Electrical discharge information is acquired and stored on all ranges simultaneously. Other features provide six programmable checklists, each containing a maximum of 30 lines with up to 20 characters per line. The WX-1000 can be upgraded to the WX-1000+.

The WX-1000+ is a heading-stabilised WX-1000 which displays digital heading information in degrees when operating in the weather-only modes and a flag advisory in the event of heading source malfunction. The WX-1000+ can be upgraded to the 'E' version with navaid option, to display navigational information from a Loran or GPS.

The WX-1000E with Navaid AN/AMS-2 simultaneously displays thunderstorm information overlaid with a course line to 10 Loran or GPS-generated waypoints. Other features include a Loran/GPS-generated Course Deviation Indicator (CDI) and the display of six of 14 user-selectable Loran or GPS-generated flight parameters. Full flight plan within the selected range can be displayed. In military service, the WX-1000E with Navaid is catalogued as the AN/AMS-2

The WX-1000E EFIS Interface allows thunderstorm activity to be displayed on a standard EFIS display. Thunderstorm data is transmitted via an ARINC 429 standard databus.

Stormscope Series II The WX-1000E with Navaid and WX-1000E EFIS Interface are also incorporated into the Stormscope Series II.

Specifications

Dimensions:
(processor) 86 × 124 × 322 mm
(display) 86 × 86 × 210 mm
(antenna) 29 × 114 × 256 mm
Weight:
(processor) 3.02 kg
(display) 1.03 kg
(antenna) 0.91 kg
Power supply: 10.5-32 V DC, 28 W
Temperature range:
(display) –20 to +70°C
(processor/antenna) –55 to +70°C

Contractor

BFGoodrich Aerospace Avionics Systems.

VERIFIED

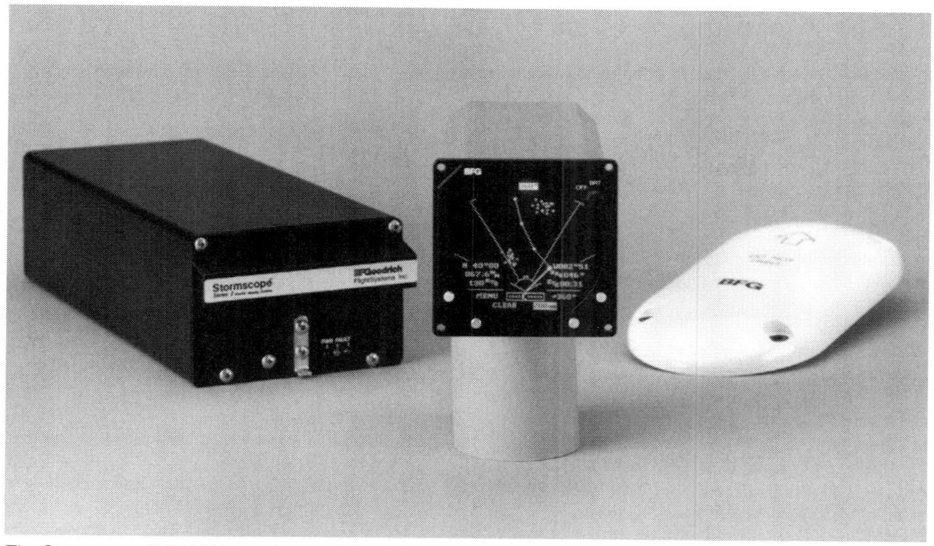

The Stormscope® WX-1000 series consists of (left to right) the processor, display and antenna

TCAS 791 traffic alert and collision warning system

The TCAS 791 traffic alert and collision warning system consists of the TRC 791 receiver/transmitter, CD605 control display unit, NY156 L-band (NATO D-band) directional antenna and NY152 L-band (NATO D-band) omnidirectional antenna. It has been designed to operate as a TCAS 1 for regional airliners and other aircraft covered by the FAA mandate.

Working as an active air-to-air interrogation device, the TCAS 791 interrogates other airborne transponders in the surrounding airspace. It computes bearing, range, altitude and closure rates to plot traffic location and predict collision threats.

Utilising a combiner directional antenna, TCAS 791 determines bearing by a time difference across four poles. Distance and relative altitude is determined by comparing transponder Mode C to the intruder's transponder Mode C.

Using the dedicated CD605 3 in ATI enhanced visibility control/display unit (optional), traffic information is displayed utilising recognisable standard TCAS symbology. Traffic is displayed out to the horizontal range selected – 5, 10 or 20 n miles – and within ± 2,700 ft altitude relative to aircraft (normal mode). While the TCAS 791 tracks up to 35 intruder aircraft at a time, it automatically displays the eight highest-priority targets representing the greatest threats.

Should another aircraft present a collision threat, both visual and aural messages alert the pilot, allowing sufficient time to make visual contact with the threat aircraft and take appropriate action.

In addition to being able to select horizontal range displayed, TCAS 791 offers four relative altitude operational modes for 'Look Up-Look Down' capability.

Ground Mode: Displays aircraft out to selectable ranges of 5 or 10 n miles and at altitudes up to 9,000 ft. This mode is particularly useful for checking potential traffic at uncontrolled airports.

Normal (NRM) Mode: Targets are displayed within ±2,700 ft altitude relative to the aircraft. This is the mode normally used in cruise.

Above (ABV) Mode: Targets are displayed up to 9,000 ft above and 2,700 ft below aircraft altitude. This mode would typically be set just before take-off to look for traffic during departure and climb-out.

Below (BLW) Mode: Targets are displayed up to 2,700 ft above and 9,000 ft below aircraft altitude. This mode is typically set before initiating a rapid descent from cruise altitude.

Specifications

Dimensions:
(TRC791 Receiver/transmitter)
193 × 157.5 × 345.4 mm
(CD605 control/display unit) 3 ATI short × 223.8 mm
(NY152 omnidirectional antenna)
57.1 × 47.2 × 68.1 mm
(NY156 directional antenna) 273.4 × 158.7 × 35.6 mm
Weight:
(TRC791 receiver/transmitter) 8.51 kg
(CD605 control/display unit) 1.36 kg
(NT152 omnidirectional antenna) 0.14 kg
(NT156 directional antenna) 1.04 kg
Power supply: 28 V DC, 95 W
Tracking: up to 35 targets
Range: 50 km nominal
Accuracy:
(range) ±0.1 km typical
(bearing) 5° RMS typical
(altitude) ±100 ft

Operational status

BFGoodrich claims that TCAS 791 was the first TCAS I to be TSO'd and the first to receive full unrestricted STC. STC approvals have been received for a large number of regional airline and corporate aircraft types.

Contractor

BFGoodrich Aerospace Avionics Systems.

VERIFIED

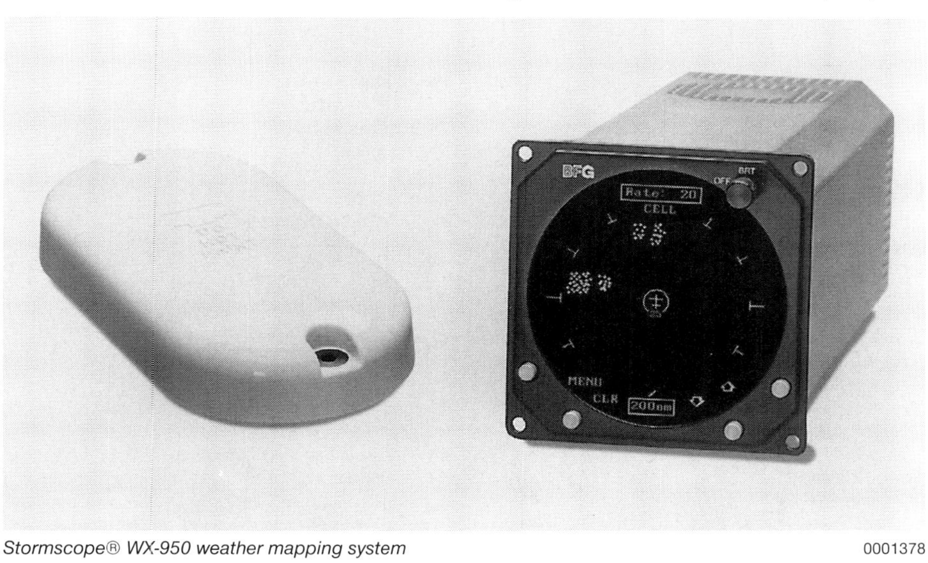

Stormscope® WX-950 weather mapping system

0001378

The TCAS 791 is designed for regional airliners

Boeing phased-array antenna

The Boeing Company has developed an electronically steered high-performance, low-profile phased-array communication antenna that will help revolutionise mobile satellite communication by increasing the data flow by thousands of times over current capabilities. This will make possible high data-rate retrieval, in-flight entertainment and many other high-bandwidth applications that have been unavailable on mobile platforms hitherto.

For commercial carriers or military aircraft, the phased-array communication antenna offers the ability to provide operators with more information options.

In use, an antenna beam is directed to acquire a BSS satellite emitting signals from a stationary orbit above the Earth's equator. The antenna beam location is controlled electronically to acquire and then track the satellite of interest. Once the antenna beam locks on to the satellite, broadcast signals are brought aboard the aircraft where they are decoded by the receiver and distributed to operators (or passengers).

The 1,500+ element antenna measures approximately 2 ft by 3 ft and is 1 in thick. Unlike conventional, mechanically steered antennas that are bulky and slow to switch between satellites, the Boeing phased-array antenna steers beams electronically, permitting instantaneous connections between satellites and mobile platforms.

In June 1996, Boeing flight-tested a prototype phased-array antenna. The tests demonstrated the antenna system's ability to automatically acquire and track broadcast service satellites and display video data on board the aircraft while in flight.

The Boeing phased-array antenna 0011826

Also in June 1996, the antenna was installed on a Boeing C-135 US Air Force avionics testbed aircraft to support a series of Joint Warrior Interoperability Demonstration (JWID '96 and '97) exercises.

In November 1996, the antenna was installed on a private 757 business jet. With the antenna on board, the aircraft is able to receive live television as well as business data.

Operational status

Boeing installed its first production phased array antenna on the E-3 707 AWACS test system-3 aircraft in June 1998, in preparation for its participation in the US Air Forces' Expeditionary Force Experiment (EFX '98).

To support the same exercise the antenna was also fitted on a C-135 US Air Force testbed aircraft and a KC-135 aircraft.

Contractor

The Boeing Company.

VERIFIED

Windshear Alert and Guidance System (WAGS)

WAGS detects changes between air data and inertial data. When this rate of change becomes excessive the pilot will be alerted aurally and visually.

WAGS is active below 1,500 ft AGL. It is a stand-alone box that interfaces with the flight guidance system to provide flight director and thrust setting guidance. The system will couple with the autopilot and autothrottle systems to provide optimal flight path guidance through a hazardous windshear.

Operational status

In service in the MD-88 and MD-11. WAGS can be retrofitted to all MD-80 aircraft.

Contractors

The Boeing Company.
Honeywell Inc, Commercial Aviation Systems.

VERIFIED

Altitude Preselect/Alerter 1D960

The 1D960 Altitude Preselect/Alerter provides precise altitude preselecting and/or alerting capabilities, in addition to ATC (100 ft resolution) or RS-232 (1 ft resolution) altitude encoding. The display is a sunlight-readable dot matrix format.

During climb or descent, the desired capture altitude is entered on the 1D960 with the rotary switch. The 1D960 is armed by pressing the 'AP Arm' switch on the face of the unit. The Altitude Preselect/Alerter compares the selected altitude with the information provided by the encoding altimeter. As the altitude approaches to within 1,000 ft of selected altitude the display will flash and give two audible tones. Upon closure of 200 ft from selected altitude the display stops flashing and gives one audible tone. At selected altitude the autopilot will smoothly transition to altitude mode and the 1D960 will change from Arm Mode to display current encoder altitude. If the aircraft deviates more than 200 ft from selected altitude the display will flash and give one audible tone.

The 1D960 Altitude/Preselect/Alerter can be installed with the two-axis Century 2000 or Century 41 Flight Control Systems, or used as a stand-alone altitude alerter.

Specifications

Dimensions: 37 × 81 × 229 mm
Panel cutout: 3 in ATI
Weight: 0.61 kg
Power: 14 or 28 V DC

1D960 altitude preselect/alerter 0001429

Contractor

Century Flight Systems Inc.

VERIFIED

Century I autopilot

The Century I autopilot is an all-electric, rate-based lightweight single-axis wings level/heading system. An electric actuator in the aileron circuit provides the control power for attitude stabilisation and pilot commanded, knob controlled turn rates of up to 200°/min. A tilted rate gyro inside a standard 3 in (76 mm) case senses roll rate and rate of turn, for both instruments and servo, and the system can be slaved to VOR/Loc or panel-mounted GPS. Century I can also be used as a back-up to the Century IIB, III or IV vacuum/electric systems (see later items), sharing the same roll servo.

Specifications

Dimensions: 89 × 89 × 194 mm
Weight: 3.17 kg
Power supply: 14 or 28 V DC, 1.25 A

Operational status

In production.

Contractor

Century Flight Systems Inc.

VERIFIED

Century IIB autopilot

A single-axis development of the Century I, the Century IIB has two panel-mounted instruments, a directional gyro with a heading bug and an attitude gyro as sensors, and additional navigation options. The system includes a solid-state computer, control panel and roll servo providing lateral stabilisation, roll command and heading select. An optional radio coupler permits heading select, VOR capture and tracking, increased sensitivity for VOR approaches and localiser capture, as well as GPS capture and tracking. An HSI may be substituted for the directional gyro.

Specifications

Dimensions:
(directional gyro) 85.6 × 82.6 × 161.8 mm
(control unit) 92.3 × 50.8 × 108 mm
Weight: 4 kg
Power supply: 14 V DC, 2 A or 28 V DC, 1.5 A

Operational status

In production.

Contractor

Century Flight Systems Inc.

VERIFIED

Century III autopilot

The Century III two-axis autopilot comprises directional and attitude gyro panel-mounted instruments, control unit, computer/amplifier and electric servos. It is claimed to be approved for more makes and models of aircraft than all of its competitors combined, and to be the preferred altitude hold system for dealer-installed retrofit installations in a wide range of singles and twins. The standard features include separate roll and pitch engagement, altitude hold, pitch and roll command and automatic or manual electric pitch trim. The optional radio navigation coupler is identical to that of the Century IIB.

Specifications

Dimensions:
(control unit) 127 × 57.2 × 63.5 mm
(panel instruments) as for Century IIB
Weight: 8.9-10.9 kg depending on options
Power supply: 14 or 28 V DC, 4.5 A

Operational status

In production.

Contractor

Century Flight Systems Inc.

VERIFIED

Century 41 autopilot/flight director

The first Century autopilot to feature digital processing, the Century 41 is a two-axis system with built-in VOR/Loc/GS couplers and outputs to drive a single cue V-bar flight director. It employs both position and rate signals to command the flight control servos. Other features include synchronised pitch attitude and altitude hold modes, a pitch modifier, automatic and manual electric trim and automatic preflight check schedule. VOR/Loc/GS capture intercepts are tailored to groundspeed, intercept angle, wind direction and distance from the ground station. The VOR tracking circuitry incorporates gain reduction to ensure 'soft' passage around the station.

When using the go-around mode, the pilot merely has to press a single button, clean up the aircraft and add power; the autopilot flies to a calibrated pitch-up attitude appropriate to the single engine safety speed for that particular type of aircraft, and turns on to a new heading preset by the pilot. The NSD-360A slaved or unslaved HSI may be substituted for the standard directional gyro. Both vacuum and electric variants are available.

The Digital Altitude Preselector/alerter (DAP) model ID960 is available as an option for all Century 41 installations.

Specifications

Dimensions: 3 ATI panel displays
Weight (installed): 16.2-17.5 kg

Operational status

In production.

Contractor

Century Flight Systems Inc.

VERIFIED

Century 2000 autopilot/flight director

The Century 2000 is the company's newest model. The panel-mounted one-, two- or three-axis modular design autopilot provides economical upgrading from any of several levels of capability.

Any Century 2000 can be expanded to include fully coupled roll and pitch plus yaw damper three-axis autopilot with flight director and altitude preselect by installing expansion kits. The basic system is roll axis with heading and built-in VOR/Loc coupler. Automatic proportional pitch trim and glide slope coupler are standard with the pitch axis.

The Century 2000 has a number of automatic performance and safety features such as soft track and internal circuit checks. The system will interface with the NSD-360A HSI and any other ARINC HSI. The flight director steering horizon is the single-cue delta/vee presentation. Non-flight director and heading only directional gyros are also compatible.

The Digital Altitude Preselector/alerter (DAP) model ID960 is available as an option for all Century 2000 installations.

Century 2000 autopilot/flight director 0001432

Specifications

Dimensions:
(panel space) 158.3 × 57 mm
Weight: 8.3 kg
Power supply: 14 or 28 V DC, 4.5 A

Operational status

In production.

Contractor

Century Flight Systems Inc.

VERIFIED

NSD Series Horizontal Situation Indicators (HSI)

Century offers four NSD Series HSI models:

NSD360A slaved HSI has all the features available in an HSI, matched with the simplicity of air gyro operation.

NSD360A slaved HSI with RMI bootstrap provides accurate heading information to a variety of other flight instrument displays.

NSD360A non-slaved HSI provides effective HSI performance.

NSD1000 HSI does not require a remote gyro, but combines the power of a remote gyro with the light weight, reliability and simple installation of a self-contained one-box instrument. Slaving is a standard feature. RMI bootstrap is optional.

The NSD HSIs feature 360° heading presentation, rectilinear course deviation indicator, full-view glide slope indicator, masking glide slope warning flag, 45° tick marks, referencing heading bug, failed gyro warning flag, free gyro mode, gyro caging knob, lost power warning flag, discrete nav warning flag and RNav and Loran compatibility. They also feature autopilot outputs for heading and course, continuously caged heading and course selection knobs, reference aircraft and heading lubber line, diffused incandescent perimeter lighting and course arrow with reciprocal indicators. The slaved models have built-in slaving indicator and automatic magnetic gyro slaving. The NSD1000 includes a built-in electric gyro.

Specifications

Dimensions: 85.6 × 85.6 × 220.7 mm
Weight: 2.09 kg
Power supply: 14 or 28 V DC

Contractor

Century Flight Systems Inc.

VERIFIED

The Century NSD1000 HSI (left) and NSD360A HSI (right)

Flexcomm C-1000/C-1000S control head

The C-1000/C-1000S control head is designed for use with Wulfsberg's Flexcomm communication system. It is capable of controlling the RT-30, RT-118, RT-138, RT-138F, RT-406F and RT-450 model radios. Flexcomm provices FM communications in low-band VHF, high-band VHF, UHF and VHF AM.

The C-1000/C-1000S is a fully frequency-agile control unit which provides thumbwheel control of all available Flexcomm channels and provides for the storage of up to 30 preset channels for simplex or semi-duplex operation. Any of the channels may be changed by the operator. However, this capability may be disabled if the operator so desires. It also provides control of CTCSS tones in both receive or transmit. The C-1000 has edge lighting, using either 5 or 28 V. When used with a complete Flexcomm system it automatically selects the appropriate radiotelephone unit depending on the desired frequency.

The system also contains full discrete switches for use with external devices such as antennas and DTMF coders.

The 'S' model has been modified to enable/disable the Motorola DVP encryption system.

Specifications

Dimensions: 146 × 76 × 191 mm
Weight: 1.2 kg
Preset channel memory: 30 channels

Operational status

In production and in service.

Contractor

Chelton Avionics Inc, Wulfsberg Electronics Division.

VERIFIED

Flexcomm communication systems

Wulfberg's Flexcomm systems are synthesised AM/FM communications systems designed specifically for use in fixed- and rotary-wing aircraft. The communications system comprises a control head (C-5000 or C-1000), single-band or multiband transceivers (RT-30, RT-138F, RT-406F and RT-5000) and antennas. All components operate on 28 V DC and have 5 V DC and or 28 V DC lighting.

Flexcomm I transceivers are 10 W FM single-band radios, each covers a particular band: the VHF LO-band of 29.7 to 49.9975 MHz with the RT-30, the VHF HI-band of 138 to 174 MHz with the RT-138F, and the UHF band 406 to 512 MHz with the RT-406F. Options include: a 'guard receiver'; a 'guard receiver' with CTCSS receive tone function; and for special applications, the RT-138F and RT-406F can be modified for additional sensitivity – this option precludes the use of a 'guard frequency'. Encryption is available using Motorola DVP with a C-1000 or C-5000 control head.

The Flexcomm II AM/FM transceiver, the RT-5000, covers all frequencies between 29.7 and 960 MHz.

C-1000 and C-5000 control heads are used to control multiple receivers, thus providing the system with the power to act as a command and control system performing relays, simulcast, repeater and full-duplex operations.

Operational status

In production and in service.

Contractor

Chelton Avionics Inc, Wulfsberg Electronics Division.

VERIFIED

Flexcomm II C-5000 control head

The C-5000 is the newest generation control head from Wulfsberg Electronics. This unit can control and independently monitor any combination up to three of the following transceivers: RT-30, RT-118, RT-138F, RT-406F, RT-450, RT-5000, RT-7200 and RT-9600F.

The C-5000 includes channel identification via alphanumeric or frequency display, simultaneous control of up to three RT systems, built-in dual-tone multifrequency, built-in dual-independent microphone inputs which allow independent operation of two RTs, multiple control head installation capability, simulcast, relay and repeater modes, programme scan and a 350-channel memory. Each channel has individual transmit and receive frequencies and selection of CTCSS and DCS tones is possible with the RT-138F, RT-406F and RT-9600F models.

The operator can monitor main and Guard receiver audio from all transceivers simultaneously and the C-5000 has independent volume control of the active or selected RT system and secondary RT systems permitting the systems to perform relays, simulcasts, repeater and full-duplex operations. The communications management controller is easy to operate, with vacuum fluorescent display, which is easily readable in bright sunlight, and optional NVG-compatibility.

Operational status

Available.

Contractor

Chelton Avionics Inc, Wulfsberg Electronics Division.

VERIFIED

RT-30 VHF/FM transceiver

The RT-30 transceiver can be operated with either the C-1000 or C-5000 control head.

The RT-30 FM transceiver uses a digital frequency synthesiser to provide FM communications over the frequency range 29.7 to 49.99 MHz. There are no band-spread limitations; the receiver can operate at one frequency extreme with the transmitter at the other.

Fully solid state, it also provides 32 sub-audible CTCSS tones. An available system includes a single channel Guard receiver operating anywhere in the band. Separate audio inputs and outputs are provided for use with external CTCSS tones, tone bursts and DTMF encoders.

Specifications

Dimensions: 111 × 266 × 127 mm
Weight: 3.4 kg
Power output: 10 W continuous
Frequency: 29.7-49.99 MHz
Channels: 20 kHz under Part 90, capable of 10 kHz incremental tuning
Pre-set channels: 350
Temperature range: –40 to +60°C

Operational status

In production and in service.

Contractor

Chelton Avionics Inc, Wulfsberg Electronics Division.

VERIFIED

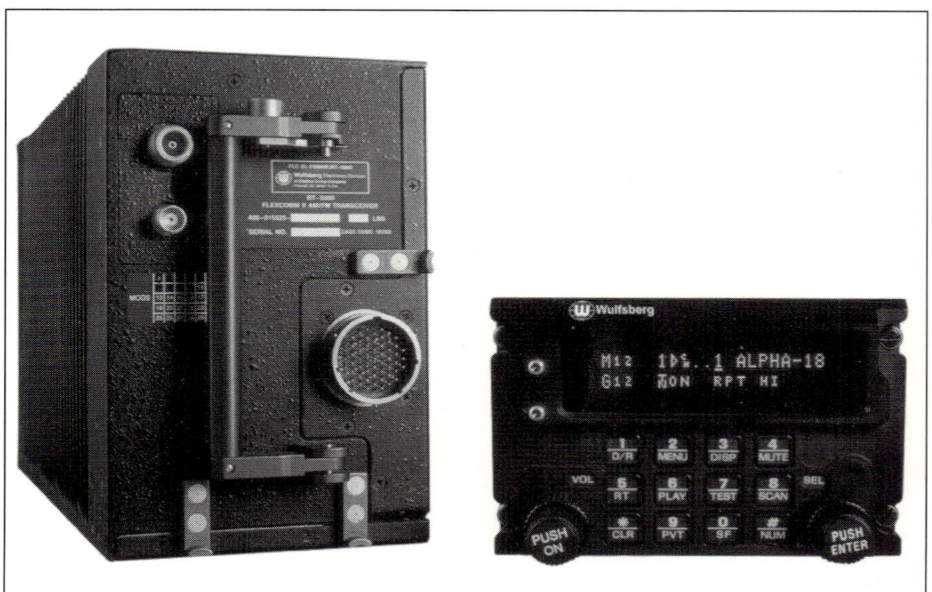

Wulfsberg Electronic's Flexcomm II RT-5000 transceiver (left) and C-5000 control head (right) ***1999***/0044816

RT-138F VHF/FM transceiver

The RT-138F transceiver can be operated with either the C-1000 or C-5000 control head.

The RT-138F uses a digital frequency synthesiser to provide FM communications over the frequency range 138 to 173.975 MHz. There are no band-spread limitations; the receiver can operate at one frequency extreme with the transmitter at the other. Of fully solid-state construction, it also provides 32 subaudible CTCSS tones. A single-channel Guard receiver is available which can operate anywhere in the band.

Separate audio inputs and outputs are provided for use with external CTCSS tones, tone bursts, DTMF encoders, voice scramblers, data and so on.

Specifications

Dimensions: 111 × 266 × 127 mm
Weight: 3.4 kg
Power output: 10 W continuous
Frequency: 138-173.975 MHz
Channels: 25 or 30 kHz under Part 90, capable of 2.5 kHz incremental tuning
Pre-set channels: 350
Temperature range: –40 to +50°C

Operational status

In production and in service.

Contractor

Chelton Avionics Inc, Wulfsberg Electronics Division.

VERIFIED

RT-406F UHF/FM transceiver

The RT-406F transceiver can be operated with either the C-1000 or C-5000 control head.

The RT-406F uses a digital frequency synthesiser to provide FM communications over the frequency band 406 to 512 MHz. There are no band-spread limitations.

Fully solid state, the system also provides 32 sub-audible CTCSS tones and a single-channel Guard receiver is available operating anywhere in the band.

Separate audio inputs and outputs are provided for use with the external CTCSS tones, tone bursts, DTMF encoders, voice scramblers, data and so on.

Specifications

Dimensions: 111 × 266 × 127 mm
Weight: 3.4 kg
Power output: 10 W continuous
Frequency: 406-512 MHz
Channels: 25 kHz under Part 90, capable of 12.5 kHz incremental tuning
Pre-set channels: 350
Temperature range: –40 to +60°C

Operational status

In production and in service.

Contractor

Chelton Avionics Inc, Wulfsberg Electronics Division.

VERIFIED

RT-5000 AM/FM transceiver

The RT-5000 AM/FM transceiver covers the frequency band 29.7 to 960 MHz. When controlled by the C-5000 control head, the RT-5000 can be used in conjunction with other Wulfsberg transceivers to perform cross-band relays between any bands from 29.7 to 960 MHz. The transceiver is tuned using a robust proprietary serial bus designed specifically for the air environment. DF audio output is standard, together with encryption outputs.

Options include an AM/FM guard receiver covering 29.7 to 960 MHz; single channel FM guard receivers; CTCSS and DCS tone squelches are standard on both the main and guard receivers.

Specifications

Dimensions: 355.6 × 124.0 × 194.1 mm
Weight: 6.81 kg
Power output: 10 W FM nominal; 15 W AM nominal

Frequency: 29.7-960 MHz
AM/FM: 29.7-399.9987 MHz
FM only: 400-960 MHz
Channels: 12.5/20/25/30/50 kHz
Tuning: 1.25 kHz steps
IF bandwidths: 12.5 kHz narrow band; 25 kHz standard band; 35 kHz wide band; 70 kHz extra-wide band.
Temperature range: –40 to +60 °C

Operational status
In production and in service.

Contractor
Chelton Avionics Inc, Wulfsberg Electronics Division.

VERIFIED

RT-7200 VHF/FM transceiver

The RT-7200 transmitter/receiver is an air-to-ground communications system suitable for a wide range of fixed-wing aircraft or helicopters. It provides FM operation over a choice of 7,200 channels within the band 138 to 174 MHz at either 25 or 50 kHz increments but can be tuned to the much lower incremental value of 5 kHz. Like other equipment in the range, the RT-7200 provides a semi-duplex facility with automatic push-to-talk tuning.

In simplex operation, frequency selection may be made on an individual dial-up basis using conventional thumbwheel controls; a digital readout on the C-722 remote controller confirms the selection. Alternatively, up to 15 channels are available on a programmable preselection basis. Power output is operator selectable at either 1 or 10 W.

The system has an optional two-channel built-in Guard receiver which conforms to the same general specification as the main equipment's receiver section. There is no frequency separation restriction if the optional Guard receiver is fitted.

Automatic signal-to-noise squelch, with manual override, is provided and a separate input is included for encoder, voice scrambler, data or other systems. Construction is all-solid-state and digital synthesis is employed for frequency generation.

Specifications
Dimensions: 127 × 320 × 127 mm
Weight: 4.21 kg
Power output: 1 or 10 W selectable
Frequency: 138-174 MHz
Channels: 7,200
Channel spacing: 25 or 50 kHz (5 kHz tunable)

Operational status
No longer in production, but still in service.

Contractor
Chelton Avionics Inc, Wulfsberg Electronics Division.

VERIFIED

RT-9600/RT-9600F VHF/FM transceiver

The RT-9600/RT-9600F is virtually identical to the RT-7200 system in its general technical specification. The major difference is in its frequency coverage and channel capacity. It covers the frequency range 150 to 174 MHz but provides a total of 9,600 channels in this band. Normal frequency spacing is 25 or 50 kHz but the RT-9600F (also known as the AN/ARC-513) is capable of tuning to a finer 2.5 kHz incremental spacing.

Specifications
Dimensions: 127 × 320 × 132 mm
Weight: 4.21 kg
Frequency: 150-174 MHz
Channels: 9,600
Channel spacing: 25 or 50 kHz (2.5 kHz tunable)

Operational status
No longer in production, but still in service.

Contractor
Chelton Avionics Inc, Wulfsberg Electronics Division.

VERIFIED

The RT-7200 VHF radio with the C-722 controller

DC-COM Model 500 voice-activated panel-mount intercom

The DC-COM Model 500 voice-activated panel-mount intercom was designed to provide simultaneous dual radio monitor/transmit capability. It provides three modes of operation: COM 1 in which pilot and co-pilot can transmit and receive on COM 1 radios; COM 2 in which the pilot can transmit and receive on COM 1, and the co-pilot can transmit and receive on COM 2; COM 3 in which the pilot and co-pilot can transmit and receive on COM 2 radio.

Facilities provided include: the ability to use up to six aviation headsets; individual VOX circuits for each headset; voice-operated intercom; pilot priority transmission; true stereo capability; fail-safe operation; optional ATC/AUX switches; horizontal or vertical installation.

Specifications
Dimensions: 43.2 × 71.1 × 152.4 mm
Weight: 0.43 kg
Power: 11-30 V DC @ 150 mA maximum
Temperature range: –40 to + 70° C
Total headsets: 6
Total output power: 300 mW with 6 headsets

Contractor
David Clark Company Incorporated.

UPDATED

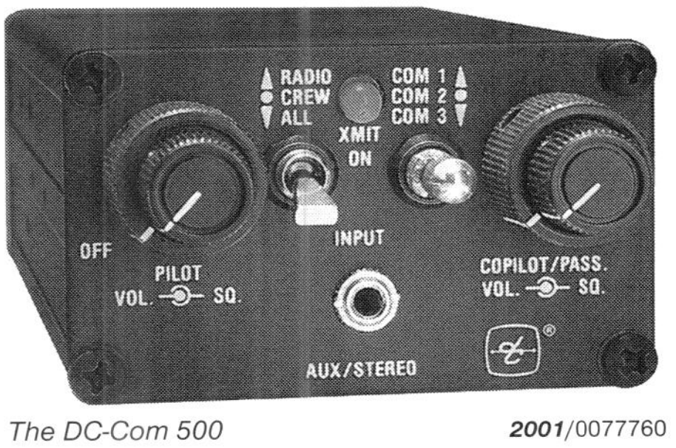

The DC-Com 500 **2001**/0077760

Performance management system

The Delco performance management system was initially developed for use in the Boeing 747 as a retrofit system and as an option on the 747-200 and 747-300.

It is a two-box system with the addition of an engine interface unit and a switching unit. The first device processes engine discrete signals, such as fuel flow, into a format usable by the computer. The switching unit handles all switching of autopilot, autothrottle and instrument signals from standard aircraft configuration to performance management control.

The performance management system achieves closed-loop control of the aircraft by providing commands to the autopilot for vertical steering, and either directly or via the inertial navigation system to the autopilot for lateral steering and the autothrottle for engine thrust control.

Delco has also adapted the performance management system for all models of the DC-9, although the engine interface unit is not required and, in the case of the MD-80, the switching unit is not needed. The system has also been adopted for the DC-10 and Boeing 727 airliners.

Specifications
Dimensions:
(computer unit) ½ ATR long
(control/display unit) 146 × 114 mm high
(engine interface unit) ¼ ATR long
(switching unit) ½ ATR long
Weight:
(computer unit) 13.2 kg
(control/display unit) 2.7 kg
(engine interface unit) 10 kg
(switching unit) 10 kg
Power:
(computer unit) 200 W
(control/display unit) 38 W
(engine interface unit) 4.2 W
(switching unit) 110 W

Operational status
In service on Boeing 727, 747, and DC-9, DC-10 and MD-80 aircraft.

Contractor
Delco Electronics.

VERIFIED

61000 series temperature exceedance monitoring

The DIAMOND J alumel/chromel 61000 series Thermocouple Engine Instruments with exceedances provide a back-lit analogue pointer with a seven-segment digital display. The seven-segment digital display can be obtained in up to four digits. The signal is processed digitally throughout the instrument to a high-resolution stepper motor. There are 2, 2.5 and 3 in round, or ATI, cases available. The pointer imperceptibly steps in increments of 0.45° in a smooth sweeping motion (2 in instrument); 0.225° per step for the 2.5 and 3 in variants.

An example of DIAMOND J's 61000 series Instruments

The DIAMOND J product line includes instrumentation for both fixed-wing and helicopter applications. The full complement of instruments include single-, dual- and triple-function instruments from one basic design. In many cases, only the dial face and software differentiate one instrument from another.

The DIAMOND J line of microprocessor stepper motor instruments does not require periodic calibration or adjustment during the life of the instrument. The high-impedance input of the instrument eliminates the need for in-line balancing resistors. The instrument's only moving parts are contained in the stepper motor/pointer shaft assembly and are calibrated in the electronic circuitry design. The stepper motor has a step increment of 0.45°; this is imperceptible to the human eye at normal distances between the pilot and the instrument panel. Dial expansion is offered to enhance resolution in areas of critical temperature. Accuracy is ±5°C.

Exceedance functions include the following:

(1) Start-up exceedance: the temperature and time over preset parameters are monitored and stored;
(2) Flight operations exceedance: engine conditions exceeding preset parameters are monitored and stored;
(3) Exceedance display: at any time that an exceedance is detected the display flashes twice per second. The display is normalised when engine conditions are within tolerance, but the exceedance reading can only be eliminated by resetting the instrument out of the instrument panel;
(4) Pilot's pre-warning: the display can be used to flash once per second if an exceedance is approached (above a preset value) to enable the pilot to avoid an exceedance;
(5) Data retrieval and reset: on power-up the instrument plays back all data in non-volatile memory; a reset unit must be used to purge data from memory.

Specifications

Power supply: 28 V DC, will operate from 8-36 V DC continuously, as well as accepting 100 V+ spikes
Power consumption: 6 W
Temperature range: –40 to +70°C
Certification: TSO-C43b

Contractor

DIAMOND J Inc.

VERIFIED

Microtrac II engine vibration monitor

Microtrac II is a microprocessor-based engine vibration signal conditioner. Vibration is detected by piezoelectric accelerometers mounted at sensitive regions in the engines. The electrical signals proportional to vibration are transmitted by the accelerometers to the signal conditioner in the avionics bay.

The Microtrac II employs digital signal processing and high-performance digital finite impulse response filters. The unit provides data which is processed by both programmable digital broadband and digital narrowband filters. The narrowband tracking filter is controlled by the output from a tachometer on the engine to isolate the vibration frequency. The narrow bandwidth means that the signal-to-noise ratio of the final vibration indication is high, and the digital filter's transient response is fast so that it can track the fundamental vibration frequencies of each of the engine's rotors during rapid accelerations.

Digital data is input and output via an ARINC 429-compatible serial digital databus. Analogue data is output as voltage or current for flight deck meter display and flight recording systems. A variety of engine vibration data can be recorded in flight and stored in non-volatile memory for later use in determining engine health. This flight history is output over the ARINC 429 bus.

All Microtrac II models provide in-flight phase angle data which can be used by maintenance personnel for engine trim balancing. The phase and vibration amplitude data can be stored in the unit or in separate data acquisition systems. The in-flight balancing measurement can be accomplished more accurately than on the ground and repeated ground run-ups are avoided, saving time and fuel.

The system uses a generic approach to produce a standard hardware unit which will satisfy the specifications for most engine/aircraft configurations. Software is then developed to satisfy the specific needs of each individual engine/aircraft configuration. This software is incorporated into a personality module which is fitted to the front of the Microtrac II.

The design utilises ASICs and digital processing circuitry to reduce the number of components required and simplify the mechanical packaging. The easily accessible front panel connector provides ARINC 429, conditioned wideband velocity and tachometer outputs, and analogue discrete inputs to control operation of the unit. The front panel LED display is used to display generated BITE faults and stored flight history information.

Specifications

Dimensions: 93.2 × 193.5 × 320 mm
Weight: 2.8 kg
Reliability: 30,000 h MTBF

Operational status

Standard equipment on Boeing aircraft, including the MD-11. Microtrac II has also been selected as the EVM signal conditioner on the MD-90 and as the vibration monitor for the Tay-engined Boeing 727.

Contractor

Endevco.

VERIFIED

Argus moving map displays

Argus moving map displays present all vital navigation and position information on an instantly readable display in front of the pilot. By minimising head-down time and giving the pilot more time to manage the aircraft, the Argus makes flying safer. The display can interface with, and receive its navigation data from most GPS, FMS and Loran C navigation systems. Approvals have also been obtained for the display of Ryan TCAD Model 9900B+ TCAS data and BFGoodrich Stormscope® WX-500 weather mapping data. Eventide also makes available hardware adaptors which allow Argus units to display information from older Stormscope models, including the WX-10, 10A, 11 and 1000E (with 429 EFIS option).

The Argus 3000 is designed for VFR use in light single- and twin-engined aircraft. Like the Argus 5000, it interfaces with most popular GPS, FMS and Loran C receivers. Its database contains over 11,000 landing facilities, 6,500 navaids and every special use airspace, including TCAs and ARSAs.

The Argus 5000's comprehensive, versatile moving map display shows all TCAs, ARSAs, navaids and landing facilities. At the touch of a button, the information submode presents detailed information about any on-screen facility from its own field-replaceable database. The Argus 5000 also provides a convenient digital readout of bearing or radial and distance to any selected facility.

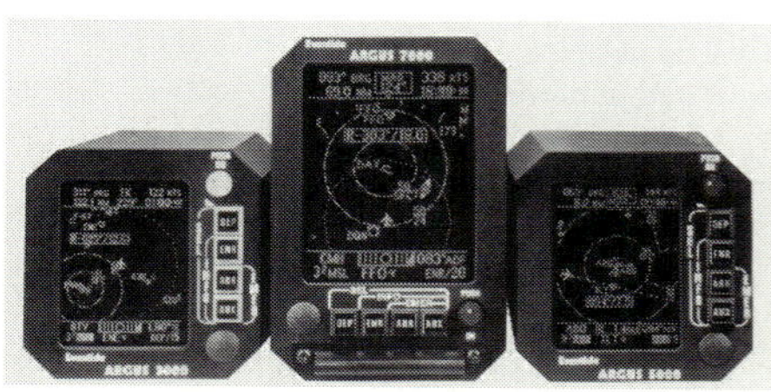

Argus moving map displays showing (left to right) the 3000, 7000 and 5000 models

The larger Argus 7000 provides all the information contained in the Argus 5000 on a bigger display. The Argus 7000 fits in a similar tray to the Argus 3000 and 5000, but has 2.3 times the screen area of the standard size Argus models. It can be updated without the need to remove it from the instrument panel.

A North American database, containing navigation information for airports, navaids and special use airspace within Canada, United States, Mexico, Central America and the Caribbean is available for use with the Argus 7000/CE, 7000, 5000/CE and 5000. An international database that gives the same worldwide coverage is available for all Argus maps.

The CE and Enhanced monochrome Argus maps are internet compatible for database updates and software upgrades at significant savings.

The Argus family of moving maps offers Enhanced software at the production level, or as an upgrade to existing maps. Enhanced software provides a variety of features: low cost internet database/software upgradeability; full flight planning capabilities; internal demo mode; terminal mode; name/location database search capabilities; Victor and Jet airways; weather display capabilities; collision avoidance display capabilities.

With the RMI adaptor, the Argus moving map display can indicate up to two ADF or VOR pointers, as with a traditional RMI display, and also it provides a digital bearing. This mode is approved for ADF or VOR approaches.

The RMI adaptor can be remotely mounted or plugged directly into the back of the Argus. An RMI/ARINC adaptor incorporates a data converter that allows Argus to accept ARINC 419 and ARINC 429 data from flight management systems, inertial management systems, and VLF Omega systems.

The Argus 5000/CE and 7000/CE units display colour screen graphics and add several new hardware enhancements including flight recording, an internal barometric pressure sensor (pressurised cockpits require an adaptor to enable this feature), a rotary encoder for easy data entry and convenient database updates.

Advanced display technology allows the Argus 5000/CE and 7000/CE to present graphics in red, green and yellow, while retaining the sharp, bright, sunlight-readable qualities of the monochrome Argus models (which continue to be available). With these new colour Argus models, the pilot can select a colour scheme to display a variety of graphic data – including the ability to 'colour code' Class B and Class C airspace.

The built-in barometric pressure sensor gives these Argus models new capabilities which provide the pilot with greater situational awareness; the aircraft's current altitude is taken into account when displaying restricted areas.

Flight recording is another new feature. The Argus will record current latitude, longitude, time, date, altitude, groundspeed and track and heading in its non-volatile memory for up to 10 hours of flight, which can be 'played back' in real or compressed time on the Argus screen.

With the Argus 5000/CE and 7000/CE, updating the database is easier and less expensive. Updates are available over the Internet, on floppy disk and on PC (PCMCIA) Cards.

Argus can interface with, and receive its navigation data from, most GPS, FMS and Loran C navigation systems. The units are available with a US, North American, and international database. All databases contain navigation information for airports, navaids and special use airspace. The Argus 7000, 7000/CE, 5000, and 5000/CE are TSOd and can be IFR approved. The Argus 3000 is TSOd as well, when equipped with the international database.

Specifications

Dimensions:
(Argus 3000/5000/5000/CE) 81.3 × 81.3 × 269.4 mm
(Argus 7000/7000/CE) 81.3 × 121.9 × 273 mm
Weight:
(Argus 3000/5000) 1.6 kg
(Argus 5000/CE) 1.5 kg
(Argus 7000) 2 kg
(Argus 7000/CE) 1.8 kg
Power supply: 11-33 V DC, 15 W

(RMI Adaptor)
Dimensions: 79.4 × 79.4 × 69.34 mm
Weight:
(RMI) 0.37 kg
(RMI/ARINC) 0.41 kg
Power supply: 11-33 V DC, 2-3 W
Environmental: RTCA-DO-160B

Contractor

Eventide Avionics.

UPDATED

Eventide Airborne Multipurpose Electronic Display (EAMED)

The Eventide Airborne Multipurpose Electronic Display (EAMED) is a powerful compact digital avionics development platform. It is a self-contained 16-bit avionics computer with high memory capacity which includes a CPU board, power supply, video board and CRT and a comprehensive set of input/output peripheral connections, including RS-232C and RS-422. EAMED displays both text and graphics on a high-resolution sunlight-readable screen. The unit fits in a standard panel cutout and hardware meets the DO-160B specification.

EAMED has a wide range of applications, including onboard communications, data acquisition, reconfigurable indicators and interactive situation displays.

Specifications

Dimensions: 76 × 76 × 266.7 mm
Weight: 1.6 kg
Power supply: 11-33 V DC, 15 W

Contractor

Eventide Avionics.

VERIFIED

HGS-2000 Head-Up Guidance System

Flight Dynamics has pioneered the application of the head-up display for commercial, military and corporate transport aircraft with the development of a Cat IIIa HGS. The system comprises five main components: the overhead unit integrating a CRT and lens assembly; a combiner containing the holographic element; a high-integrity computer; a drive electronics unit; a pilot's control panel.

The HGS is aimed at improving safety and performance by offering a system capable of operating to lower weather minima and as an economic alternative to Cat III automatic landing systems. In July 1984, Flight Dynamics received FAA approval for its HGS for manually flown Cat IIIa approaches down to a runway minimum range of 700 ft and decision height of 50 ft in the Boeing 727. This was the first manual system certified by the FAA for the demanding low-visibility environment. In August 1987, the HGS was also certified for windshear detection and recovery guidance and, in 1990, was approved by the FAA for low-visibility take-offs down to 300 ft RVR.

The system projects flight guidance symbology, focused at infinity, on the holographic combiner. Along with basic flight information such as airspeed, altitude, course and heading, the HGS displays inertial flight path and acceleration, providing the sensitivity and accuracy required for Cat III operations. Safety is also improved, as industry studies have demonstrated that projected flight path and precise energy management greatly improves the pilot's situational awareness and aircraft control, particularly in difficult or unexpected conditions.

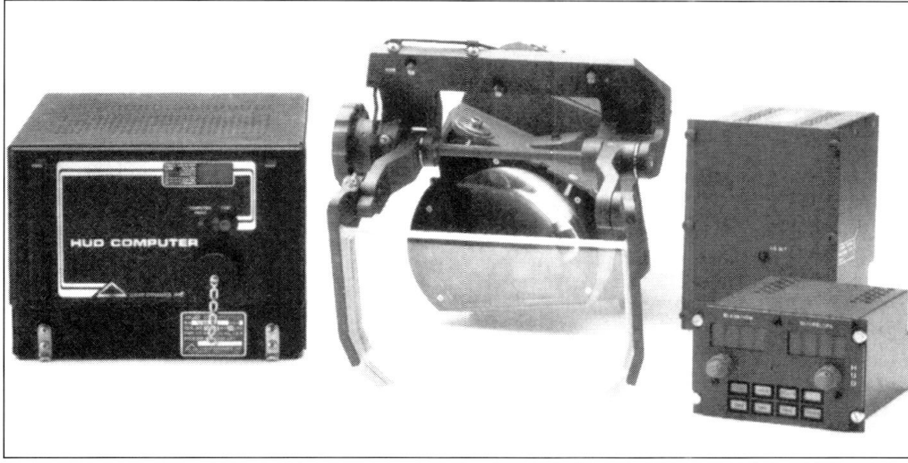

Components of the holographic guidance system 0106406

The combiner provides the pilot with a full 30 × 24° field of view, a feature especially useful in high-cross-wind conditions. By comparison, military fighter HUDs typically have 20 × 15° fields of view. The holographic technology improves both the reflectivity of the projected symbology and the transmissivity of real-world details as seen by the pilot.

Flight Dynamics' holographic head-up guidance system 0106405

Flight Dynamics is currently evaluating the use of the HGS to obtain even lower take-off and landing minima. The commercial transport industry has also expressed interest in combining a fail-passive autoland system with the fail passive HGS to achieve Cat IIIb capability. This hybrid landing system should be certifiable to 300 ft runway visual range and would combine the benefits of an automatic landing system with the projected flight path and head-up advantages of the HGS.

Alaska Airlines, the first carrier to equip its 23-aircraft Boeing 727 fleet with the HGS, received FAA operational approval to operate in revenue service down to 50 ft decision height in late 1988 and the system has now been cleared for 300 ft RVR (Runway Visual Range) take-offs. In October 1989, Alaska Airlines conducted the world's first manually flown Cat IIIa landings with passengers on board.

Most recently, Flight Dynamics is proposing integration of infra-red sensor imagery to improve pilot situational awareness during climb-out and descent phases of flight.

Operational status

Flight Dynamics claims that its line of Head-Up Guidance Systems has received about 60 per cent of the Boeing 737 New Generation orders. More widely, it has sold to a large number of airlines, corporate users, and military transport operators, flight training companies and NASA. The HGS displays flight information and guidance on a holographic combiner in the pilot's normal head-up field of view. Aircraft equipped with the Flight Dynamics' HGS include the Falcon 900 EX, Falcon 2000, Saab 2000, Boeing 727 and 737, de Havilland Dash 8, Bombardier Regional Jet, Dornier 328 and the Lockheed Martin C-130J (two units per aircraft - one for each pilot).

The designation HGS-2350 has been used for the installation in the Easyjet B737-300, and HGS-2850 for the Falcon 2000 installation. It is reported that about 50 per cent of all Falcon 2000 buyers are selecting the HGS, together with about 80 per cent of Falcon 900EX customers. Latest in the series of Flight Dynamics' head-up guidance systems is the HGS-4000, for which US FAA supplemental type certification is expected early in 2000, before delivery to Delta Airlines as launch customers for its new-generation Boeing 737-600/700/800 aircraft.

The HGS-4000 will incorporate enhanced features aimed at improving pilot situational awareness, including rapid recovery from unusual attitude or in-flight situations. The enhanced features will be available for retrofit to earlier HGS models. On 1 April 1999, Rockwell Collins acquired 100 per cent ownership of Flight Dynamics, and now incorporates the HGS into its standard cockpit product lines.

Contractor

Flight Dynamics.

VERIFIED

FV-2000 civilian Head-Up Display (HUD) systems

FV-2000 HUD

The basic element of the FV-2000 HUD system is the FV-2000 Mission Display Processor (MDP). It is a highly capable, low cost, airborne processing display system. As a HUD system's computational core, it is designed to interface with virtually any combination of analogue or digital equipment. The 40 MHz 68030 microprocessor generates both flight and military displays.

The FV-2000 civilian HUD comprises the FV-2000 MDP, together with an FV-2000 overhead optical unit that provides a 30° field of view. The FV-2000 HUD for civilian aircraft displays the following types of information: the velocity vector, the conformal pitch scale, and an accelerate/speed cue. These three represent the energy-state of the aircraft.

The FV-2000 HUD incorporates both expanded (conformal) and compressed (non-conformal) pitch scales and has the ability to display a velocity vector or flight path marker. The velocity vector feature shows the aircraft's actual ground track and projects the touchdown point. It also shows the immediate effects of windshear. The accelerate/speed cue and all of the navigational and flight information is also depicted on an easy-to-read display. The system has been certified as a primary flight display and can be retrofitted in aircraft with either electromechanical instruments or EFIS, creating fleet commonality for flight operations.

The system comprises the optical unit, combiner glass, control panel and HUD computer and weighs 9.98 to 14.25 kg. Each Flight Visions FV-2000 HUD is specifically designed for seamless integration into each aircraft type for which it is certified. The control panel is mounted in the cockpit pedestal or panel. Other features include TCAS, GPWS, runway overlay, raster capability and an RS-170 video port. The processor is certified for use outside the pressure hull. The FV-2000 contains self-diagnostics through the control panel and maintenance pages are accessible through an RS-232 port.

All Flight Visions HUDs employ holographic combiners to offer maximum transmissivity and contrast ratio.

The FV-2000 HUD is certified as primary attitude display, providing redundant information in the event of a head-down ADI failure. The FV-2000 system is currently certified and flying on various types of business aircraft, from twin engined turboprops to international jet aircraft. It is available for all Beech King Airs, the Bell 230, Cessna Citation 500 and 550, Dassault Falcon 50, Gulfstream III and IV and the Lear 55.

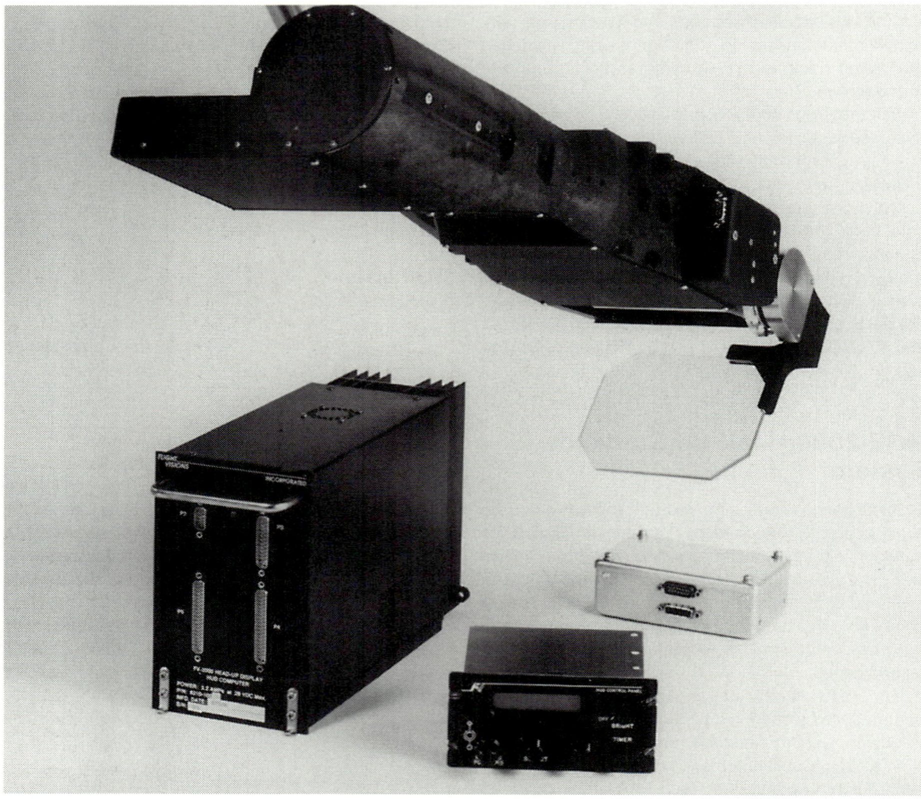

Flight Visions FV-2000 HUD components (top to bottom, clockwise order): straight-mounted overhead optical unit, grid amp, control panel, HUD computer
0001464

FV-2000E HUD

The FV-2000E HUD is designed to meet Enhanced Vision System (EVS) display requirements. It comprises the FV-2000 optical unit and the FV-2000E Head-Up Display Symbol Generator (HSG).

The current FV-2000 system already incorporates design elements to accept Enhanced Vision Sensor (EVS) technologies including FLIR and MilliMetre Wave Radar (MMWR). After new EVS sensors become available, existing systems can be updated by replacing the FV-2000 remote-mounted computer with an FV-2000 PSDU. The FV-2000's overhead optical unit already has raster capability to display all EVS available. The FV-2000, when integrated with EVS FLIR/MWR systems will be designated FV-2000E.

The FV-2000 Power Supply Deflection Unit (PSDU) combines enhanced display generation modules from the FV-3000 with the less expensive computing engine of the FV-2000. It is designed for users that do not need the processing power and flexibility of the FV-3000 but still need a HUD symbol generator.

The FV-2000E gives the pilot the ability to complete an approach with Category II or III weather conditions to a Category I runway. When these EVS sensors become available, existing FV-2000 systems can be quickly updated by exchanging the FV-2000 MDP with the FV-2000E computer. The FV-2000s overhead optical unit already has raster capability necessary for displaying all types of enhanced vision images.

The FV-2000 HUD installed in a Falcon 50

Operational status

The FV-2000 is US FAA certified for Falcon 50, Gulfstream III and IV, Learjet 55, Citation 550, all Beech King Air models and the Bell 230 helicopter. Certifications for about 20 more aircraft types are expected. The FV-2000 will be available for most turbojet-powered business aircraft.

The FV-2000 PSDU, display processor together with the Sparrow Hawk™ HUD, has been selected by Israel Aircraft Industries for upgrade of seven Kfir aircraft, with options for a further 12 systems.

Contractor

Flight Visions Inc.

VERIFIED

Airborne microwave transmission systems

FLIR Systems Inc airborne microwave transmission systems are designed to meet the requirements of government forces (police, customs, SAR) and TV companies to transmit high data rate TV and FLIR pictures from helicopters to ground stations; they are compatible with most surveillance systems, including all of those manufactured by Broadcast and Surveillance Systems Ltd (BSS) and FLIR Systems Inc.

Two models are available: the medium-range system, capable of transmitting good quality live pictures from air tc ground up to 56 km; the long-range system, which can achieve 96 km.

The system comprises: a 1 W RF transmitter, 20 or 30 W ERP power amplifier, 4 dBi circularly polarised omnidirectional antenna actuator, CAA-certified to lower below the helicopter during flight to provide unobstructed 360° coverage. Pilot controls allow full control of the antenna system.

The system is normally supplied for operation around a nominal frequency of 2.4 GHz, but all bands from 0.6 to 7.5 GHz are possible.

Contractor

FLIR Systems Inc.

VERIFIED

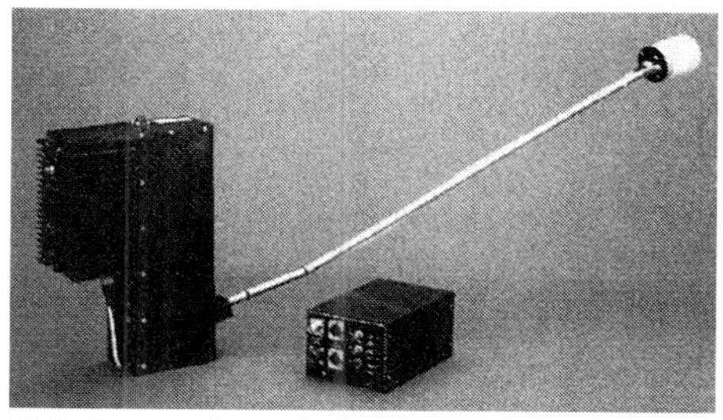

FLIR Systems Inc airborne microwave transmission system, showing controllable antenna 0001465

Cockpit control panels

G6990 series TCAS panels

Gables TCAS panels are available for all Boeing commercial aircraft types in both keyboard and rotary layouts. The G6990-51 is certified for Boeing New Generation aircraft. The G6990-40 has been selected by British Airways for installation on its A320 aircraft.

G6993-08 and G6992-22 have been selected by United Parcels Service, Airborne Express and Federal Express for use in the Cargo Airline Association ADS-B programme, installed on DC-9 and B-727 aircraft.

Gables has teamed with Honeywell Inc, Commercial Aviation Systems, to develop an ATC/TCAS (Mode-S/ATCRBS) control panel that is compatible with the new Honeywell TCAS 1500 system.

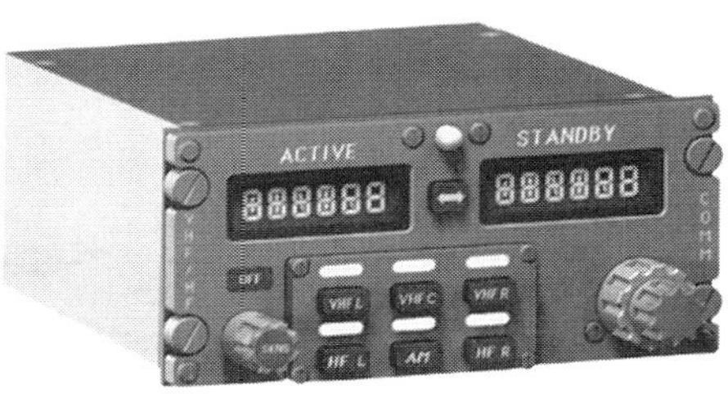

Gables' G7404 radio tuning panel 0022250

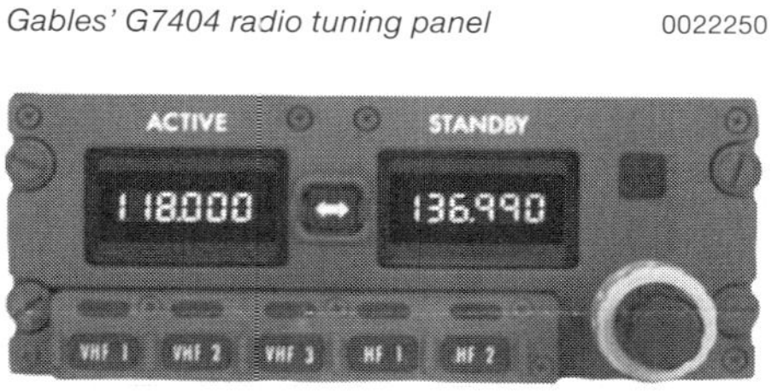

Gables' G7409 radio control panel 0022249

G7130-03 ATC/TCAS control panel

The G7130-03 control panel has been certified for the Falcon-900 aircraft, compatible with the Rockwell Collins 822-0078-005 radio.

G7185 series cargo smoke detection and G7183/4 fire suppression control panel

Gables' series of smoke detection and suppression control panels are designed to alert crew members to a fire event in ample time to avoid a catastrophic situation.

A specific design criterion of these units is the avoidance of false alarms, which has been a problem with smoke detection systems. The control panels include LED annunciators.

The G7185, integrated cargo smoke detection/suppression control panel, automatically switches from dual- to single-loop operation. This switch is made without affecting the dual-loop mode of additional detectors that have not suffered a fault. Depending upon where the smoke is detected, the control panel installed in a two-bay aircraft will automatically arm the extinguishing system in either the forward or aft bay of the aircraft. The panels in aircraft with three cargo bays will automatically arm the system in the forward, mid or aft bays of the aircraft.

AirTran Airways has selected the G7183 smoke detection control panel and the G7184 cargo fire-extinguishing control panel for use on its fleet of DC9 aircraft. Continental Airlines has chosen the G7185 smoke detection control panel for use on its MD80 aircraft.

Gables' G7130-03 ATC/TCAS general aviation control panel 0022248

G7400 series VHF control panels

Continuing VHF control panel programmes include six variants of the G7400 series for Boeing 737, 757 and 767 aircraft. Included in this series is the G7406, which is capable of tuning both ARINC 566A and 716/750 series radios. The G7404 radio tuning panel is now available in three variants for use on Boeing 747-400 aircraft. Gable's radio tuning panels have also been certified on Boeing 737 New Generation aircraft. The G7409 control radio panel has been selected for the Boeing MD-10 aircraft.

G7406-25 has been selected to provide VHF comm control panels with 8.33 kHz channel spacing for the US Air Force KC10 fleet; installation began in April 1999.

G7400 series control panels are also certified on the Russian Aviation Register.

G7500-01 VHF navigation control panel

The G7500-01 panel is standard on Boeing 737 New Generation aircraft.

Contractor

Gables Engineering Inc.

VERIFIED

GI 102A and GI 106A Course Deviation Indicators (CDI)

The GARMIN GI 102A and GI 106A are panel-mounted Course Deviation Indicators (CDIs) with rectilinear meter movements and contain integral GPS, NAV, and VLOC mode annunciators. The GI 102A has a VOR/LOC/GPS needle, TO/FROM indicator, and NAV warning flag. The GI 106A adds a GS needle and flag. Both units contain an integral resolver for OBS feedback. The GI 102A and GI 106A units also feature anti-reflective lenses and are backlit.

Garmin GI 102A course deviation indicator (Garmin International) ***2001***/0099757

Specifications

Dimensions: 82.6 (W) × 82.6 (H) × 120.7 (D) mm
Weight: 0.64 kg
Display lighting: 4.3 W at maximum intensity, 14 V DC @ 0.31 A or 28 V DC @ 0.155 A (typical)
Altitude: 55,000 ft (operating)
Temperature: –55 to +70°C
Cooling: convection

Operational status

In production.

Garmin GI 106A course deviation indicator (Garmin International) ***2001***/0099758

Contractor

Garmin International.

NEW ENTRY

GMA 340 audio panel

The GMA 340 audio panel features LED-illuminated push-button controls that allow audio selection of both NAV and Comm audio. Large buttons activate the Comm microphone and audio for up to three communication transceivers. Another feature of the GMA 340 is Split Comm, which allows the pilot to transmit/receive on Comm 1 and the co-pilot to transmit/receive on Comm 2. Speaker selection is available to monitor the aircraft radios or to make PA announcements. There are two unswitched inputs for telephone ringers, altitude alert warnings, or other warning tones. In addition, the GMA 340 includes a six-place VOX intercom with three selectable modes of isolation, dual stereo music inputs, and independent pilot, co-pilot, and passenger volume control. Each microphone input has a dedicated VOX circuit to ensure that only the active microphone is heard when squelch is broken. Special cabin noise de-emphasis circuitry enhances cockpit communications. Also standard is the three-light Marker Beacon Receiver/Indicator with high/low sensitivity selection and SmartMute™ marker audio muting. Operation is available in 14 or 28 volts without voltage converters or dropping resistors. Installation is simplified with the built-in intercom and marker beacon, and the GMA 340

Garmin GMA 340 audio panel (Garmin International) **2001**/0099746

also features photocell dimming of annunciators, pilot fail-safe mode connection to Comm 1, and FAA TSO approval.

Specifications

Audio panel: 3 × transceiver inputs; 5 × receiver inputs; 2 × unswitched inputs
Intercom: 6 × positions: pilot, co-pilot, 4 × passengers
Headphones: 3 × stereo: pilot, co-pilot, passengers
Speaker outputs: 1, pilot selectable
Marker beacon receiver: RF crystal-controlled at 75 MHz
Dimensions: 172.7 × 159.77 × 33 mm
Power supply: 11-33 V DC
TSO compliance: TSO-C50c; TSO-C35d

Contractor

Garmin International.

UPDATED

GNC 250 and GNC 300 TSO GPS/Comm

GNC 250: VFR GPS/Comm

The panel-mounted GNC 250 comprises a GPS sensor and a fully-TSO'd, 760-channel VHF communications radio. Supplemented with a transponder, a GNC 250 serves as a complete avionics package for VFR operation. The GNC 250 includes Garmin's MultiTrac8™ receiver, which provides clear, accurate navigation information linked to a choice of three navigation databases (Americas, International or Worldwide) and Jeppesen® information.

The GNC 250 incorporates a TSO'd, full-function digital transceiver. Standard features include flip-flop frequency selection - with frequency prompts from take-off to touchdown - automatic squelch and a 'cursor select' function, which enables selection of a standby frequency from the GPS database.

GNC 300: IFR-certified GPS/Comm

The GNC 300 adds IFR certification to the nav capabilities of the GNC 250. Complying with the FAA's TSO C129 A1, the GNC 300 offers non-precision approach, en route and terminal GPS capability under Instrument Flight Rules (IFR). Like the GNC 250, the GNC 300 incorporates a VHF communications radio. It presents its information on the same 4-line, 80-character vacuum-fluorescent display. Also, like the GNC 250, the GNC 300 is ideal for either new aircraft installations or as an update to the current panel.

Incorporating the same MultiTrac8™ GPS receiver, the GNC 300 streamlines 'direct-to' navigation. Its comprehensive Jeppesen® database offers airport, runway, VOR, NDB, intersection and communication frequency information for either the western hemisphere (Americas option) or the rest of the world (International option). Ample room for trip planning, waypoint storage and a 'nearest airports' list provides additional capability. Sectorised, altitude-sensitive SUA warnings are also included.

The GNC 300 also features a Nav/Com page that automatically provides frequencies from preloaded flight plans.

Fully-TSO'd, the GNC 300 also incorporates a powerful 760-channel digital transceiver for audio communications. As with the GNC 250, the GNC 300's Comm section includes such standard features as flip-flop frequency selection, automatic squelch and 'cursor select' function.

Both the GNC 250 and 300 interface with most flight control, EFIS, HSI and other cockpit systems.

Operational status

Both the GNC 250 and GNC 300 have been superceded, although both remain in service and widely used in general aviation aircraft.

Contractor

Garmin International.

UPDATED

GNC 250XL GPS/Comm system

The GNC 250XL combines the (superceded) GNC 250 with very high-resolution moving map graphics and communication capability. The system combines the GPS 150XL12-channel global positioning system (GPS) with a 760 channel VHF transceiver. The GNC 250XL also features front-loading data cartridges for easy Jeppesen map updating.

Specifications

Dimensions: 159 × 147 × 51 mm
Weight: 1.16 kg
Power supply: 10-15 V dc or 115 V ac (with optional AC adapter)
Display: 80 × 240 DSTN display
GPS receiver: PhaseTrac12™ 12-parallel channel
Database: Jeppesen Americas, International or Worldwide
VHF transceiver: 760-channel (aviation band 118-136.975 MHz), 5 W
Certification: TSO C37d Class 4 (VHF transmitter), TSO C38d Class C (VHF receiver), AC 20-138 (VFR) (GPS receiver)

Operational status

The GNC 250XL has been available since May 1997.

Contractor

Garmin International.

UPDATED

GNC 300XL TSO GPS/Comm system

In May 1998, Garmin International announced the introduction of two new panel-mounted navigators – the GPS 155XL TSO and the GNC 300XL TSO. Both products are IFR approach certified under TSO C129a A1 and include high-definition moving maps using DSTN LCD (Double Super Twist Nematic Liquid Crystal Display) technology and approach-certified IFR GPS receivers. The yellow-on-black display, which is reversible to black-on-yellow, is the same as that used on the earlier GPS 150XL and GNC 250XL systems. The GNC 300XL TSO employs Garmin's 12-parallel channel GPS receiver, which allows up to1,000 user waypoints and 20 reversible flight plans with up to 31 waypoints each, and includes a NavData® navigation card.

Specifications

Dimensions: 159 × 147 × 51 mm
Weight: 1.16 kg
Power supply: 10-15 V DC or 115-230 V AC (with optional AC adaptor)
Display: 80 × 240 DSTN display
GPS receiver: PhaseTrac12™ 12-parallel channel
Database: Jeppesen Americas, International or Worldwide
VHF transceiver: 760-channel (aviation band 118-136.975 MHz), 5 W
Interface: ARINC 429, RS 232
Certification: TSO C37d Class 4 (VHF transmitter), TSO C38d Class C (VHF receiver), TSO C129a Class A1 (en route, terminal and approach)

Contractor

Garmin International.

UPDATED

GNC 420 Comm/Nav/GPS system

The GNC 420 is a 'one-box' Comm/Nav/GPS with full-colour LCD map graphics. It is capable of WAAS/LAAS (Wide Area Augmentation System/Local Area Augmentation System) operation. The map display offers a map page, default Nav page, and a NAVCOM page. The Nav capability includes conventional VOR, glideslope and localiser facilities. The COMM element features a 10 W, VHF transceiver, with both 760-channel (25 kHz channel spacing) and 2,280-channel (8.33 kHz channel spacing) capability.

The system utilises the Jeppesen Americas, International or Worldwide NavData card, and offers 1,000 user-defined waypoints, with 20 reversible flight paths, of up to 31 waypoints each. The GNC 420 is TSO C129a certified for a non-precision approach.

Garmin GNC 420 Comm/Nav/GPS (Garmin International) **2001**/0099753

Garmin GNS 430 Comm/Nav/GPS (Garmin International) **2001**/0099749

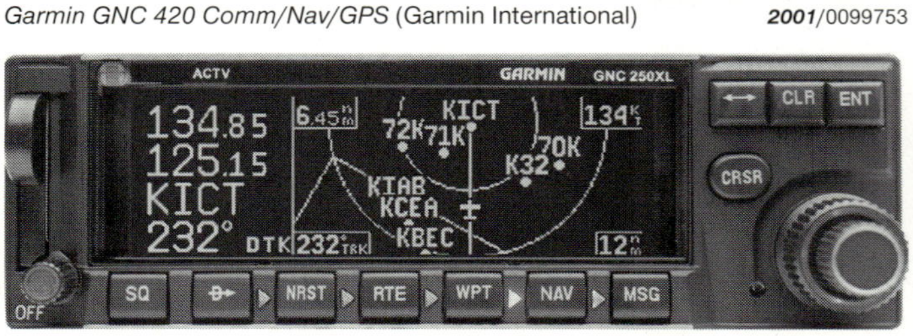

Garmin GNC 250XL GPS/Comm system (Garmin International) **2001**/0099747

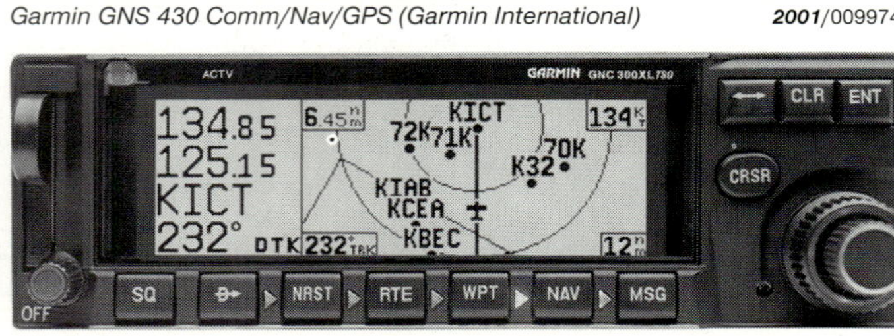

Garmin GNC 300XL TSO GPS/Comm system (Garmin International) **2001**/0099748

Specifications
Dimensions: 159 (W) × 67.3 (H) × 279.4 (D) mm
Weight: 2.64 kg
Power supply: 27.5 V DC
Display: Colour LCD
GPS receiver: PhaseTrac12™ 12-parallel channel
Accuracy: 15 m (position), 0.1 kt (velocity) RMS (steady state)
Database: Jeppesen Americas, International or Worldwide
VHF transceiver: 760-channel (25 kHz spacing) or 2,280-channel (8.33 kHz spacing), 10 W
Interface: ARINC 429, RS 232
Certification: TSO C129a Class A1 (en route, terminal and approach), TSO C37d Class 4 and 6 (transmit), TSO C38d Class C and E (receive)

Operational status
In production and in service.

Contractor
Garmin International.

NEW ENTRY

GNS 430 Comm/Nav/GPS

The GNS 430 is a 'one-box' Comm/Nav/GPS with full-colour LCD map graphics. It is capable of WAAS/LAAS (Wide Area Augmentation System/Local Area Augmentation System) operation. The map display offers a map page, default Nav page, and a NAVCOM page. The Nav capability includes conventional VOR, glideslope and localiser facilities. The COMM element features a 10 W, VHF transceiver, with both 760-channel (25 kHz channel spacing) and 2,280-channel (8.33 kHz channel spacing) capability.

The GNS 430 utilises the Jeppesen Americas or International NavData card, and offers 1,000 user-defined waypoints, with 20 reversible flight paths, of up to 31 waypoints each.

When combined with Garmin's GMA 340 audio panel and GTX 320 transponder, the GNS 430 produces a stack of avionics reduced to a footprint only 5.5 in high.

Specifications
Dimensions: 165.1 (W) × 67.3 (H) × 279.4 (D) mm
Weight: 3 kg
Power supply: 27.5 V DC
Display: Colour LCD
GPS receiver: PhaseTrac12™ 12-parallel channel
Database: Jeppesen Americas, International or Worldwide
VHF transceiver: 760-channel (aviation band 118-136.975 MHz), 10 W
Interface: ARINC 429, RS 232
Certification: TSO C37d Class 4 (VHF transmitter), TSO C38d Class C (VHF receiver), TSO C129a Class A1 (en route, terminal and approach), TSO C40c (VOR), TSO C36e (LOC), TSO C34e (glideslope)

Operational status
Announced in May 1998.

Contractor
Garmin International.

UPDATED

GNS 530 Comm/Nav/GPS system

The GARMIN GNS 530 is a 'one-box' WAAS (Wide Area Augmentation System) upgradeable IFR GPS, Comm, VOR, LOC and glide-slope system, with colour moving map. The TSO qualified VHF Comm facility offers a choice of 25 kHz or 8.33 kHz spacing, for a 760 or 2,280 channel configuration, respectively. A Jeppesen database (which can be updated with front-loading data cards) contains all airports, VORs, NDBs, airway intersections, FSS, approach, SIDs/STARs and SUA information.

The GNS 530 features a 5 in colour display which separates land data, terminal areas, route, and approach information for easy scanning and reduced pilot workload, a feature more commonly found in commercial FMS systems.

Garmin GNS 530 Comm/Nav/GPS (Garmin International) ***2001***/0099754

Specifications
Dimensions: 159 (W) × 117 (H) × 279.4 (D) mm
Weight: 3.9 kg
Power supply: 27.5 V DC
Display: Colour LCD
GPS receiver: PhaseTrac12™ 12-parallel channel
Accuracy: 15 m (position), 0.1 kt (velocity) RMS (steady state)
Database: Jeppesen Americas, International or Worldwide
VHF transceiver: 760-channel (25 kHz spacing) or 2,280-channel (8.33 kHz spacing), 10 W
Interface: ARINC 429, RS 232
Certification: TSO C129a Class A1 (en route, terminal and approach), TSO C37d Class 4 and 6 (transmit), TSO C38d Class C and E (receive), TSO C40c (VOR), TSO C36e (LOC), TSO C34e (GS)

Operational status
In production.

Contractor
Garmin International.

NEW ENTRY

GPS 150/GPS 150XL Global Positioning System

The Garmin GPS 150 Global Positioning System includes an extensive Jeppesen database and features a fully dimmable display which provides three full 20 character lines of information in a 50.8 mm (2 in) pane space. The vacuum fluorescent dot matrix display and an advanced optical filter provide readability from almost any angle, even in daylight.

The GPS 150 features a front-loading Jeppesen data card for airport (including runway lengths, approach and localiser frequencies), VOR, NDB, airway intersection, airspace, MSA and FSS information. The system features single button-press access to all information, including continuously updated lists of the nearest waypoints, including up to 1,000 user-defined waypoints. Some 20 reversible flight plans, with up to 31 waypoints each, can be stored.

A built-in rechargeable battery will automatically provide up to two hours of power in the event of aircraft power loss. Checklist and message functions are included to reduce pilot workload.

Additional features include a 'direct to' function to set an instantaneous course to any waypoint, an AutoStore function to build routes easily while en route, Offset Navigation for parallel tracking and AutoSearch for quick search and rescue grid activation.

The GPS 150 continuously tracks and uses up to eight satellites with 1-second updates and horizon-to-horizon coverage. The unit has a built-in back-up

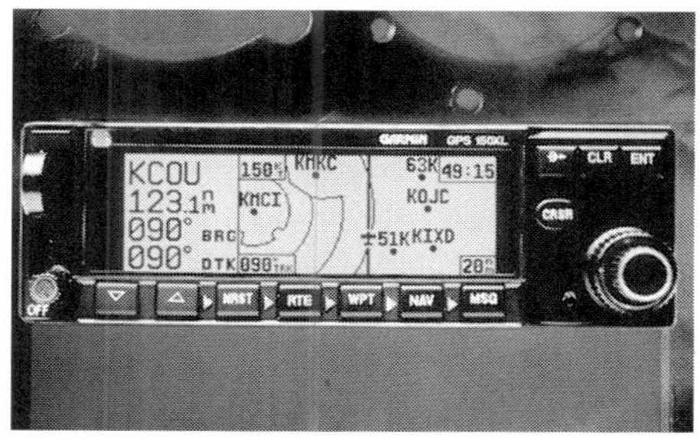

Garmin GPS 150XL global positioning system 0015377

battery that can provide up to 4 hours of operation, even if aircraft electrical power is lost.

The GPS 150XL updates the GPS 150 by adding very high-resolution moving map graphics, a 12-parallel channel GPS receiver and a DSTN (Double Super Twist Nematic) display, which doubles the number of pixels available to greatly improve the quality of the display.

Specifications
Dimensions: 159 × 100 × 51 mm
Weight: 0.95 kg (unit with battery pack)
Power supply: 10-15 V DC or 115-230 V AC
Display: Vacuum fluorescent dot matrix (GPS 150) or 80 × 240 DSTN (GPS 150XL) display
GPS receiver: MultiTrac8™
Interface: RS 232
Database: Jeppesen Americas or International

Operational status
The GPS 150 and GPS 150XL models have been superceded by the GPS 155XL TSO, although both remain in service in a wide range of general aviation aircraft.

Contractor
Garmin International.

UPDATED

GPS 155 TSO Global Positioning System

The GARMIN GPS155/165 TSO are the world's first GPS receivers to be fully certified to FAA TSO-C129, Class A1 standards for non-precision approach use, allowing flight under IFR from takeoff to touchdown with GPS as the primary navigation reference.

The GPS 155 TSO stores navigation fixes for over 4,500 approaches, from the initial fix to the missed approach holding point, while Receiver Autonomous Integrity Monitoring (RAIM), a critical requirement for IFR approach certification, confirms adequate reception of satellite signals. The unit features a three-line vacuum-florescent dot matrix display that provides exceptional readability from almost any angle and in direct sunlight. A Jeppesen database provides access to navigation, communication, airport, airspace, approach, departure and arrival route information from a front-loading NavData® card. The GPS 155 TSO also stores up to 1,000 user waypoints and 20 reversible flight plans of up to 31 waypoints each (with capacity for user comments on up to 250 data items).

The GPS 155 TSO is a standard 6.25 in, Mark-width box, designed for easy rack-mounting in most light aircraft. The system interfaces with flight control, EFIS, HSI, moving map, altitude encoder, fuel management, or other systems, while a back-up battery prevents system loss in the event of an electrical failure.

In February 1995, the GPS 155 TSO received FAA A1 certification, clearing it for en route, terminal and non-precision approach IFR operations, and for BRNAV operations in 1998.

Specifications

Dimensions: 159 × 51 × 140 mm
Weight: 1.17 kg
Power supply: 10-33 V DC, 115-230 V AC with battery charger
Battery life: up to 2 h
Temperature range: –30 to +70°C
GPS receiver: MultiTrac8™
Accuracy: 15 m (position), 0.1 kt (velocity) RMS (steady state)
Interface: RS 232, RS 422, ARINC 429
Certification: TSO C129 Class A1 (en route, terminal and approach)

Operational status

No longer in production. The GPS 155 TSO has been superseded by the GPS 155XL TSO, although a large number of units remain in service in general aviation aircraft.

Contractor

Garmin International.

UPDATED

GPS 155XL TSO Global Positioning System

The GPS 155XL TSO is a panel-mounted receiver which replaces the superseded GPS 150 and GPS 155 TSO (see previous items). The unit is IFR approach certified under TSO C129a A1 and includes a very high-definition moving map. The moving map on the GPS 155XL TSO is displayed via the familiar Garmin DSTN (Double Super Twist Nematic) LCD yellow-on-black display found in the VFR-only GPS 150XL and GNC 250XL. A photocell in the display automatically controls the intensity of the backlight and will reverse the display from black-on-yellow to yellow-on-black for maximum contrast in daylight or night time viewing. Users may also make these adjustments manually.

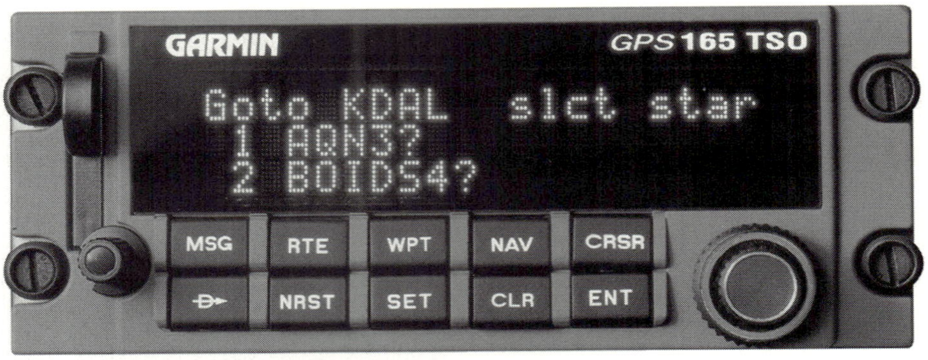

Garmin GPS 165 TSO GPS (Garmin International) ***2001***/0099752

The database in the GPS 155XL TSO includes non-precision approaches for US airports, all published SIDs and STARs, 1,000 user-defined waypoints and 20 reversible routes with up to 31 waypoints in each. It contains an extensive, updatable, Jeppesen database providing detailed information on airports, VORs, NDBs, intersections, comms frequencies, runways, FSS and MSAs. In Emergency Search mode, the system lists the nine nearest airports, VORs, NDBs, or user-defined waypoints, together with the two nearest FSS with frequencies and the two nearest ARTCC frequencies. The unit also includes a built-in Ni/Cd battery back-up, capable of providing emergency power in the event of aircraft electrical power failure.

The GPS 155XL TSO can fully interface with flight controls, EFIS, HSI, moving map, altitude encoder, fuel management and other aircraft systems.

Specifications

Dimensions: 159 × 147 × 51 mm
Weight: 0.77 kg
Power supply: 10 to 33 V DC or 115 to 230 V AC (with optional AC adapter)
Display: 80 × 240 DSTN display
GPS receiver: PhaseTrac12™ 12-parallel channel
Performance:
  Accuracy: 15 m (position) RMS, 0.1 kt (velocity) RMS (steady state)
  Update rate: 1 Hz, continuous
  Acquisition time: 15 s (warm), 45 s (cold)
Database: Jeppesen Americas or International
Interface: ARINC 429, RS 232
Certification: TSO C129a Class A1 (en route, terminal and approach)

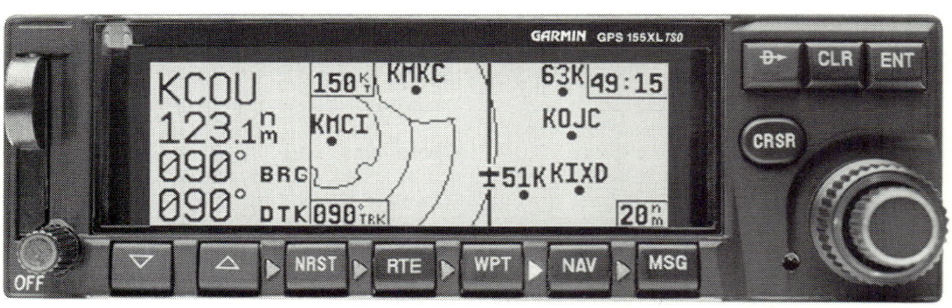

Garmin GPS 155XL TSO Global Positioning System (Garmin International) ***2001***/0105002

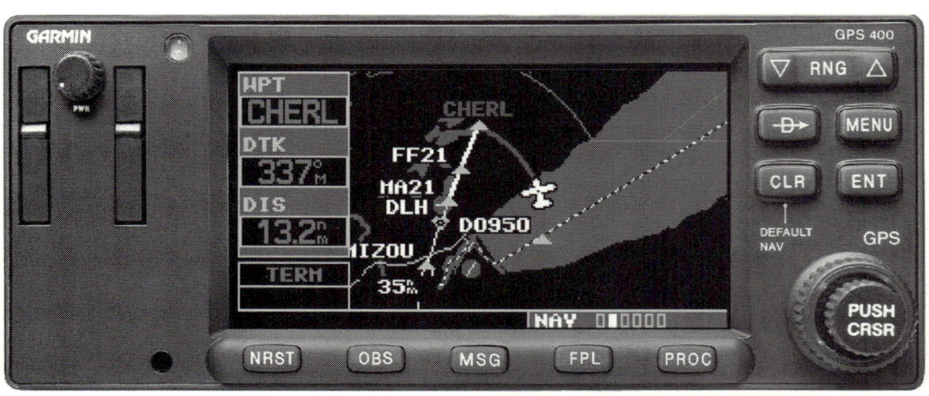

Garmin GPS 400 global positioning system (Garmin International) ***2001***/0099756

Contractor

Garmin International.

NEW ENTRY

GPS 165 TSO Global Positioning System

The GPS 165 TSO is the dzus-rail configuration of the GPS 155 TSO (see earlier item). The GPS 165 TSO tracks and uses up to eight satellites continuously, providing 1 second updates with horizon-to-horizon coverage.

System features include 1,000 user-defined waypoints, 20 reversible routes with up to 31 waypoints each, single key 'direct to' operation, Jeppesen NavData stored on an updatable removable data card, Auto Search search and rescue grid activation function, optional remote battery pack back-up providing 2 hours of operation in the event of aircraft power failure and interfaces for ARINC 429, RS-232 and RS-422.

In April 1995, the GPS 165 TSO achieved FAA C129 A1 certification for IFR approach operations, and for BRNAV operations in 1998.

Specifications

Dimensions: 146 × 57 × 144 mm
Weight: 0.97 kg
Display: High intensity (600 fL) dot matrix fluorescent
Power supply: 10-33 V DC, 115-230 V AC with battery charger
Battery life: up to 2 h
Temperature range: –30 to +70°C
GPS receiver: MultiTrac8™
Accuracy: 1-5 m (position), 0.1 kt (velocity) RMS (steady state)

Operational status

The GPS 165 TSO is in production and in service in a wide range of general aviation aircraft.

Contractor

Garmin International.

UPDATED

GPS 400 Global Positioning System

The GARMIN GPS 400 combines a global positioning system with a full-colour moving map for enhanced situational awareness. The map features a built-in all-land database that shows cities, roads/highways, railways, rivers, lakes and coastlines, in addition to a Jeppesen database. Thanks to a high-contrast colour display, the information can be easily read from wide viewing angles even in direct sunlight. Like the GNC 420, the GPS 400 is TSO C129a certified for a non-precision approach.

Specifications

Dimensions: 159 (W) × 67.3 (H) × 279.4 (D) mm
Weight: 3 kg
Power supply: 11-33 V DC
Display: Colour LCD
GPS receiver: PhaseTrac12™ 12-parallel channel
Accuracy: 15 m (1.5 m with differential correction) (position), 0.1 kt (velocity) RMS (steady state)
Database: Jeppesen Americas, International or Worldwide
Interface: ARINC 429, RS 232
Certification: TSO C129a Class A1 (en route, terminal and approach)

Garmin GTX 320 IFF transponder (Garmin International) **2001**/0099745

Garmin GTX 327IFF transponder (Garmin International) **2001**/0099755

Operational status
In production.

Contractor
Garmin International.

NEW ENTRY

GTX 320 IFF transponder

The Garmin GTX 320 is a compact, panel-mounted, solid-state, 200 W, Class 1A transponder. It fits into existing installations, as a replacement upgrade for earlier-generation cavity tube transponders, to provide improved reliability and to eliminate warm-up time.

Specifications
Transmitter power: 200 W
Transmitter frequency: 1090 MHz
Receiver frequency: 1030 MHz
Weight: 0.95 kg
Dimensions: 208 (L) × 159 (W) × 42 (H) mm
Power: 11-33 V DC, 12 W
TSO compliance: C74c Class 1A
Mode A: 4,096 codes
Mode C: 100 ft increments –1,000 to +63,000 ft

Contractor
Garmin International.

UPDATED

GTX 327 digital transponder

The GARMIN GTX 327 is a solid-state Mode C digital transponder, which features a 200 W transmitter, a DSTN Liquid Crystal Display, a numeric keypad and a dedicated VFR button, which facilitates rapid selection of 1200/VFR squawking. The GTX 327 also offers timing and display functions such as flight time and count-up and count-down timers, as well as current pressure altitude.

Specifications
Dimensions: 159 (W) × 42 (H) × 208 (D) mm
Weight: 0.95 kg
Power supply: 11-33 V DC, 15 W
Display: DSTN LCD
Transmitter frequency: 1090 MHz
Transmitter power: 200 W nominal
Receiver frequency: 1030 MHz
Mode A capability: 4096 ident codes
Mode C capability: 100 ft increments from -1,000 to 63,000 ft
Certification: TSO C74 Class 1A, certified to 50,000 ft

Operational status
In production.

Contractor
Garmin International.

NEW ENTRY

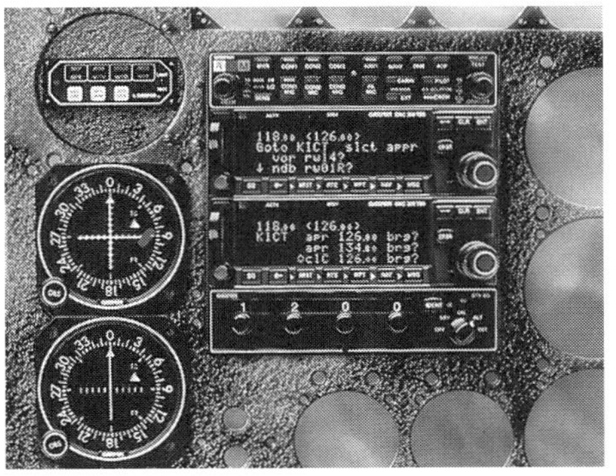

Garmin IFR stack 0015379

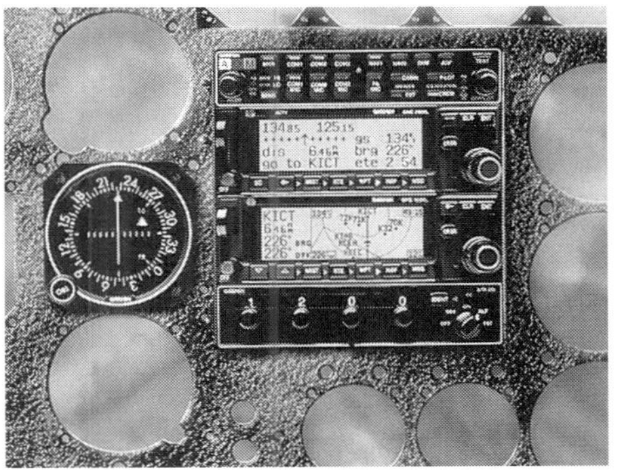

Garmin VFR stack 0015380

IFR/VFR avionics stacks

Garmin has integrated its products into IFR and VFR avionics stacks comprising the following units:

IFR avionics stack
GMA 340 TSO'd audio panel
Dual GNC 300 TSO IFR-certified GPS/Comm
GTX 320 Class 1A transponder
MD 41 annunciator
GI 102A course deviation indicator
GI 106A course deviation indicator with glideslope

VFR avionics stack
GMA 340 TSO'd audio panel
GNC 250XL GPS/Comm with moving map
GPS 150XL GPS/Comm with moving map
GTX 320 Class 1A transponder
GI 102A course deviation indicator

Contractor
Garmin International.

UPDATED

Alpha 12 VHF radio

The Genave Alpha 12 is a panel-mounted VHF/AM transmitter/receiver for light aircraft. It has a low power consumption, particularly suiting it to aircraft with a limited electrical generation capacity such as gliders, certain agricultural aircraft or home-builts. It provides 12 channels in the band 118 to 135.975 MHz at a channel spacing of 25 kHz. Transmitter power output is a nominal 4 W carrier, with 3.3 W minimum.

Features include a MOSFET, track-tuned front end and crystal intermediate frequency filtering. A light-emitting diode is incorporated to act as a transmit indicator.

Specifications
Dimensions: 63 × 165 × 254 mm
Weight: 1.81 kg

Operational status
In production and in service.

Contractor
Genave Inc.

VERIFIED

Alpha 100 VHF radio

The Genave Alpha 100 is a panel-mounted VHF/AM transmitter/receiver providing 100 channels in the band 118 to 127.9 MHz at a channel spacing of 100 kHz. Transmitter power output is 8 W peak power, 2 to 3 W carrier. Construction is fully solid state and the receiver section is of the double conversion, superheterodyne type and is crystal controlled. Facilities include a manually adjustable squelch disable and automatic gain control. Frequencies are selected by means of a dual-knob selector with digital readout and a light-emitting diode is employed as a transmit indicator.

The system has a low power requirement, in common with other Genave equipment, making it suitable for aircraft with little or no electrical generation capacity.

Specifications
Dimensions: 165 × 63 × 228 mm
Weight: 1.82 kg

Operational status
In production and in service.

Contractor
Genave Inc.

VERIFIED

Alpha 720 VHF radio

The Genave Alpha 720 is a panel-mounted VHF/AM transmitter/receiver providing 720 channels in the 118 to 135.975 MHz band at a channel separation of 25 kHz. Transmitter output power is 4 W nominal. It is designed for the general aviation and light aircraft market.

The system is a single crystal unit using digital phase-locked synthesis techniques for frequency generation. Construction is fully solid state, with extensive employment of integrated circuitry. Features include a transformerless series modulator in the transmitter section, a single conversion receiver and field effect transistor front end and mixer circuitry. Facilities include automatic squelch disable and active impulse noise limitation to reduce external interference effects. Channel selection is performed by use of a dual-control frequency selector knob on the equipment's front casing; the selection is confirmed by a dimmable incandescent readout display.

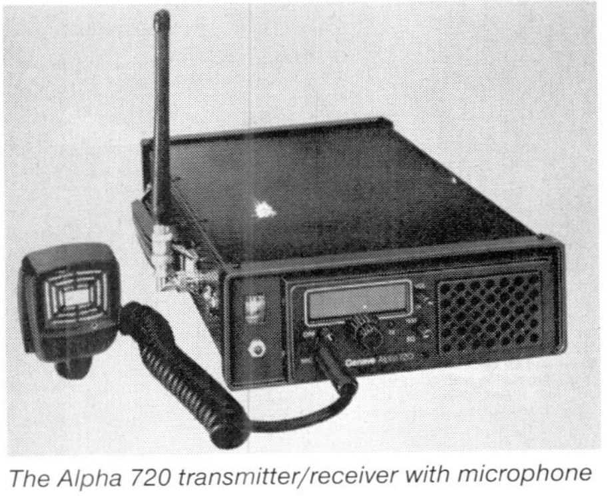

The Alpha 720 transmitter/receiver with microphone

Like many Genave products, the Alpha 720 is a low power consumption system suitable for aircraft with limited electrical power. A variant, the Man-Pack, designed for portable use, is produced for gliders, home-builts and agricultural aircraft without electrical systems.

Specifications
Dimensions: 63 × 165 × 254 mm
Weight: 1.81 kg

Operational status
In production and in service.

Contractor
Genave Inc.

VERIFIED

GA/1000 VHF nav/com system

The Genave GA/1000 is a VHF/AM nav/com system for light aircraft. It can be panel mounted as a single unit or, alternatively, installed so that the VOR/Loc indicator section can be retained as a panel instrument with the control head mounted elsewhere in the cabin.

The system's communications section covers 720 channels in the VHF band 118 to 135.975 MHz at 25 kHz spacing. The independent navigation receiver covers the band 108 to 117.95 MHz with 50 kHz separation and covers 200 navigation channels (160 VOR and 40 localiser). The use of separate receivers for communications and navigation functions permits radio operation without disrupting reception of navigation signals. Both sections employ hot filament digital readout displays for confirmation of the frequency selection. This is carried out in each case through use of dual-frequency selector knobs on the front case of the control unit.

Transmitter output power is a nominal 4 W carrier signal and a radio frequency actuated LED is incorporated as a transmit indicator. Communications receiver gain is automatically controlled and automatic squelch with manual disable is also provided. Audio output may be either through a 4 ohm speaker for which a 3 W output is available, or into 600 ohm earphones from a 100 mW output. The same values are applicable to the audio outputs from the navigation receiver section. In VOR mode, transmissions from the system's transmitter section cause no visible deflection of the course deviation indicator needle. Both VOR and localiser have ARINC-standard autopilot outputs.

The GA/1000 uses solid-state integrated circuitry and a single crystal digital synthesiser is employed for frequency generation. Receiver demodulation circuitry is of the single conversion type. A range of antenna systems and associated antenna coupler units is available to match differing aircraft installation requirements and a speaker muting relay is also obtainable.

Operational status
In production and in service.

Contractor
Genave Inc.

VERIFIED

ADF-2070 Automatic Direction-Finder (ADF)

The ADF-2070 is a panel-mounted unit designed for the general aviation sector. It is claimed to be able to receive signals from exceptionally long ranges and has two sensitivity settings: extended range reception and conventional ADF which is used primarily on approach. This extended-range facility is provided by a coherent detection feature which results in good reception characteristics with high immunity from thunderstorms and other static interference. Continuous digital tuning ensures lock on to the desired frequency.

The system also features a blade antenna which serves both the communications radio and the ADF systems. The ADF sensor is installed in the base of the blade and feeds signals to the receiver via a small amplifier which is mounted adjacent to the antenna blade but within the aircraft skin. It is claimed that this configuration provides nearly twice the gain of other combination antenna units, results in a reduction in cable length and has no impedance matching requirement.

Provision is made for correction of quadrantal error either on the ground or while airborne. The receiver output can drive either a standard ADF indicator with rotatable azimuth card or an HSD-800 horizontal situation indicator.

Specifications
Dimensions: 45 × 159 × 234 mm
Weight: 2.63 kg

Operational status
In service.

Contractor
Honeywell Aerospace, Electronic & Avionics Lighting.

VERIFIED

Airborne Flight Information System (AFIS)

Honeywell Aerospace has expanded the capabilities of its GNS-1000 flight management system and GNS-500A Series 4 and Series 5 navigation systems into a full airborne flight information system, with comprehensive facilities for flight plan creation before flight and amendment in flight. The Airborne Flight Information System (AFIS) is also available for the GNS-X flight management system.

Before flight the pilot can access the Global Data Center via a computer, to obtain flight planning, wind and en route weather information. This information is recorded on a mini-computer disk, which is later loaded into the AFIS system via an onboard data transfer unit. During flight, the comparison of planned and actual flight plan information can be viewed at any time on a control and display unit.

During flight, the air-to-ground and ground-to-air datalink can be used to obtain additional wind and weather information, send and receive flight-related messages and update flight plans.

Onboard elements of the AFIS comprise:
(1) A Data Management Unit (DMU), which includes receiver/transmitter;
(2) A Data Transfer Unit (DTU);
(3) A single-unit High-Power Amplifier/Lower-Noise Amplifier (HPA/LNA);
(4) An antenna;
(5) An SDCS satellite communication unit.

Operational status
Fully operational.

Contractor
Honeywell Aerospace, Electronic & Avionics Lighting.

VERIFIED

AIRSAT™ satcom systems

Honeywell Aerospace, Electronic & Avionics Lighting is developing a group of AIRSAT satcom system products to provide worldwide telephone communication services to aircraft flight deck crew, cabin staff and passengers, via the 66-satellite Iridium network.

AIRSAT 1 satcom system
AIRSAT 1 provides single-channel worldwide telephone services; it comprises a single-channel 3 MCU-sized Iridium Transceiver Unit (ITU), a back-lit digital handset and a low-gain, top-mounted ARINC 761-type blade antenna.

AIRSAT 5 and 8 multichannel satcom system
AIRSAT 5 and AIRSAT 8 systems provide five-channel and eight-channel worldwide passenger/cockpit voice, fax and data communications; each comprises two 4 MCU-sized processor units, a diplexer and a blade antenna, all ARINC 761 compliant.

Specifications
Radio frequency: 1,616.0-1,626.5 MHz
Modulation: QPSK
Operation: full duplex
Dimensions/weight:
AIRSAT 1: ITU: 3 MCU less than 6.82 kg
AIRSAT 5 and 8: STU-105 (5-channel): 4 MCU/7.28 kg
STU-108 (8-channel): 4 MCU/8.18 kg
HPA-105/108: 4 MCU/9.09 kg
DNLA-105/108: 279.4 × 203.2 × 50.8 mm/2.27 kg
ANT-100: 108 × 279.4 × 120.7 mm/1.36 kg

Operational status
AIRSAT 1 was available second quarter 1999. AIRSAT 5 and 8 were available fourth quarter 1999.

In July 1999, Lockheed Martin Tactical Aircraft Systems and Honeywell Aerospace, Electronic

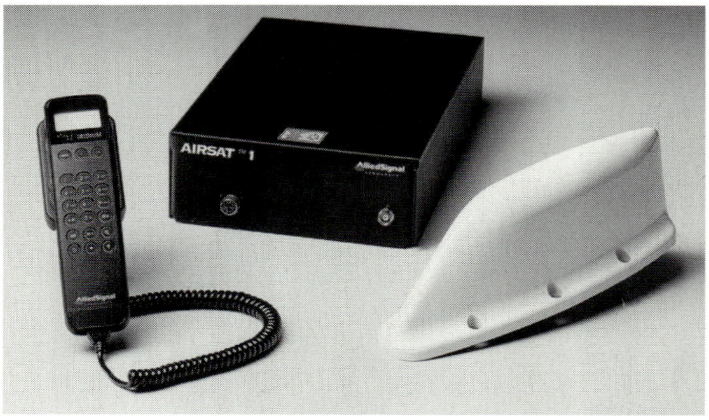

The Honeywell Aerospace AIRSAT™ 1 satcom system ***1999***/0044812

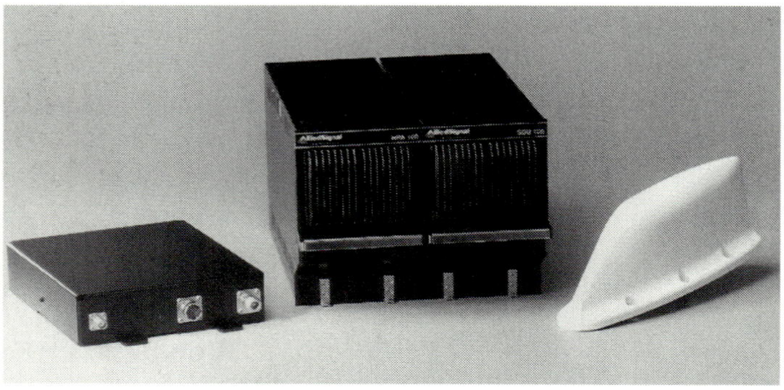

The Honeywell Aerospace AIRSAT™ 5 and 8 multichannel satcom system ***1999***/0044813

& Avionics Lighting entered into a co-operative agreement to demonstrate two-way satellite communication on a tactical fighter using the Iridium satellite and the AIRSAT 1 system. This activity is part of the overall US Air Force effort to provide Real-Time Information into the Cockpit (RTIC).

Contractor
Honeywell Aerospace, Electronic & Avionics Lighting.

VERIFIED

ALA-52A radio altimeter

The ALA-52A radio altimeter is a lightweight solid-state digital low-range unit which utilises a simplified microprocessor-based design.

The ALA-52's capabilities are achieved by an advanced microprocessor which handles all data computations, including the application of correction factors for aircraft installation delay; the control of tracking filter gain bandwidth characteristics; collection and processing of the beat frequency count representing altitude information; the output of the altitude data for display via the ARINC 429 interface. Flag logic and monitor levels are also controlled by the microprocessor to reference criteria defined in the firmware.

As an added confidence factor, the ALA-52A utilises a second microprocessor of differing design architecture, to compute and verify altitude information by independent comparison.

One of the ALA-52A's other major advantages is its ability to perform continuous automatic self-calibration. By utilising a continuous feedback loop comprising the transmitter, quartz bulk-wave-delay device, a crystal reference and the modulator, the unit not only monitors the slope of the transmission but also maintains proper calibration. It complies with ICAO Annex 10.

Specifications
Weight: 4.54 kg
Altitude: up to 50,000 ft

Operational status
In production.

Contractor
Honeywell Aerospace, Electronic & Avionics Lighting.

VERIFIED

AN/ARC-199 HF radio

The AN/ARC-199 is a solid-state HF communications system providing 280,000 channels in the frequency band 2 to 30 MHz. It incorporates power management and low probability of intercept with selectable power output of 4, 40 or 150 W PEP. Twenty channels can be preselected for instantaneous recall and use. Automatic recognition of incoming messages by a selective addressing system is used.

The radio is microprocessor controlled and contains a dual MIL-STD-1553B databus interface. Remote control can be provided through this interface, as well as through a dedicated keyboard/CRT display controller. The equipment can operate with both voice and data formats and can be used 'in clear' or with encryption devices. The CRT display is compatible with NVG. The antenna coupler is able to tune a variety of antennas throughout the frequency range, including long open wires, grounded wires, whips and shunt antennas.

Specifications
Weight: 14.5 kg

Operational status
In service on US Army helicopters.

Contractor
Honeywell Aerospace, Electronic & Avionics Lighting.

VERIFIED

ARINC 700 series digital integrated avionics

The ARINC 700 series Communications, Navigation and Identification (CNI) equipment began as part of the US industry-wide momentum in the 1970s to develop new technology systems for the transport aircraft to replace the Boeing 707 and DC-8. Honeywell Aerospace was successful in winning the competition to provide the CNI suite for the Boeing 767 launched by United Airlines in July 1978 and the slightly later Boeing 757. The current ARINC 700 series consists of:
ALA-52A radio altimeter
DFA-75A automatic direction-finder
RIA-35A ILS receiver
RTA-44A VHF communication transceiver
RVA-36A VHF navigation receiver
SMA-37A distance measuring equipment
TRA-67 Mode S transponder.

Operational status
In production and in service in Boeing 757 and 767 aircraft.

Contractor
Honeywell Aerospace, Electronic & Avionics Lighting.

VERIFIED

DFA-75A ADF receiver

The DFA-75A utilises advanced LSI and microprocessor designs to achieve a greater level of dependability, accuracy and performance. While it uses advanced techniques to deliver performance advantages over its predecessors, its design is based on proven and efficient components which increase reliability and reduce weight and complexity.

The DFA-75A meets or exceeds all ARINC 712 characteristics for form, fit and function. One of the ways it meets the requirement for improved installations is by using a combined loop/sense antenna and digital interface with the receiver. Another major improvement is its ability to deliver more stable bearings in the presence of thunderstorms and during severe ionospheric conditions. This is achieved through a scheme of signal modulation, coherent demodulation and adaptive digital filtering from the antenna. The DFA-75A also provides automatic bearing adjustments to compensate for circular and quadrantal errors. Adjusted bearing data is smoothed by a second order digital filter.

Microprocessor monitoring permits a higher level of fault isolation diagnostics to be performed within the unit during BIT. The flight fault memory provides a non-volatile memory with the capacity to store 10 faults in each of 64 individual flight segments. It complies with ICAO Annex 10.

Specifications
Weight: 4.175 kg
Frequency: 190-1,750 kHz
Channel spacing: 0.5 kHz

The DFA-75A ADF receiver

Operational status
In production.

Contractor
Honeywell Aerospace, Electronic & Avionics Lighting.

VERIFIED

DFS-43 direction-finder system

The DFS-43 is an all-digital automatic direction-finding system. It is a lightweight system consisting of the DF-431 receiver, CD-432 control display unit and AT-434 combined loop and sense antenna.

The DF-431 receiver provides accurate reception of en route NDBs, locator outer markers and commercial AM broadcast stations. It is available with front- or rear-mounted connectors.

The CD-432 control display unit provides for a dual-frequency readout, with one active and one standby. Frequencies are alternated via the frequency transfer button. The unit provides frequency control from 190 to 1,860 kHz, with half or whole kHz incremental spacing. It also features a non-volatile frequency memory which retains the last frequency used, eliminating the possibility of frequency loss due to power interruptions.

The DFS-43 automatic direction-finder employs full time monitoring and self-testing of key functions such as the power supply, synthesiser lock, receiver lock and signal processing.

The AT-434 combined loop/sense antenna system is designed specifically for use with the DFS-43 system.

Specifications
Dimensions:
(antenna) 146 × 152.4 × 332.7 mm
(control display unit) 63.5 × 79.4 × 63.5 mm
(front mount receiver) 101.6 × 101.6 × 279.4 mm
(rear mount receiver) 101.6 × 101.6 × 320.5 mm
Weight:
(antenna) 1.72 kg
(control display unit) 0.27 kg
(front mount receiver) 2.36 kg
(rear mount receiver) 2.36 kg
Power supply: 18-33 V DC, 0.6 A at 28 V

Contractor
Honeywell Aerospace, Electronic & Avionics Lighting.

VERIFIED

DMA-37A DME interrogator

The DMA-37A is a fast-scan DME interrogator which meets all ARINC 709 characteristics. The design utilises advanced all-digital technologies to provide pilots with accurate and reliable DME and Tacan slant range information which can be transmitted via the ARINC 429 bus for use by both the visual display instruments and AFCS.

The DMA-37A features a centralised microprocessor system which offers superior signal processing capabilities and handles all unit control monitoring functions. Channel and control information is received

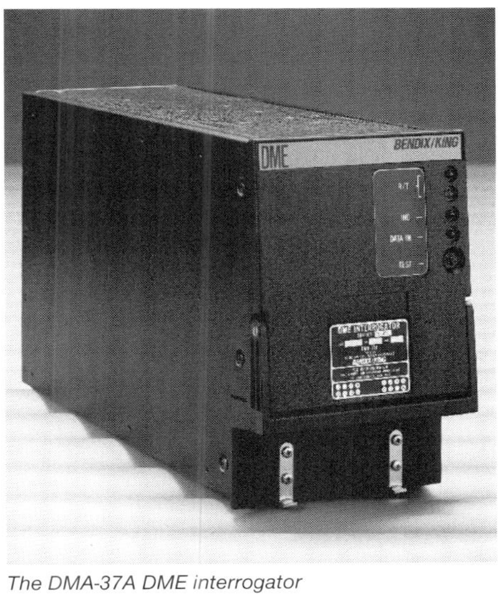

The DMA-37A DME interrogator

through one of two selectable ARINC 429 frequency/function data input ports.

The primary processor automatically decodes the Morse code signal received from the channel selected for Ident both for single channel timing and when in the multiscan mode. A dedicated second microprocessor translates the logic signal representing the dots and dashes into alphanumeric characters before being formatted into ARINC 429 for transmission on the output ports by the main CPU.

The unit's microprocessor also helps to simplify maintenance through a rigorous self-monitoring and self-diagnostic routine. To further reduce down time, the fault memory and BITE are interfaced with the central fault display system and access to strategic points within the unit is provided via the automatic test equipment connector. It complies with ICAO Annex 10.

Specifications

Weight: 5.85 kg
Frequency:
(transmitter) 1,025-1,150 MHz
(receiver) 962-1,213 MHz
Channel spacing: 1 MHz

Operational status

In production.

Contractor

Honeywell Aerospace, Electronic & Avionics Lighting.

VERIFIED

The Series 3 and EFS 10 electronic flight instrument systems on the Citation III

DMS-44 Distance Measuring System

The DMS-44 is a digital, solid-state, dual-transmitter, distance-scanning measuring system which can receive three stations simultaneously. It is a lightweight all-digital system consisting of the DM-441 transmitter/receiver and the SD-442 sector display.

The all-solid-state transmitter provides the capability simultaneously to scan stations for Nav 1 and Nav 2 and a third station which is transparent to the pilot for computation. It utilises two separate microprocessors and a video processor. The master processor provides signal processing, control computations and analogue range information. The slave processor performs the digital input/output generation, including the input of frequency-tuning interfaces. The range processor can lock on to data in less than 200 ms and provides an LSB accuracy of better than 0.01 n miles. The system is available with either front or rear connector mounts.

The SD-442A panel-mounted selector/display unit provides full-time display of distance and groundspeed to the selected VORTac. The active Nav is annunciated below the distance readout.

The DMS-44 employs full-time self-monitoring of key circuits such as the synthesiser, receiver, transmitter, power supply and master processor.

Specifications

Dimensions:
(selector/display) 82.6 × 39.4 × 63.5 mm
(front mount transmitter/receiver) 127 × 101.6 × 279.4 mm
(rear mount transmitter/receiver) 127 × 101.6 × 320.6 mm
Weight:
(selector/display) 0.204 kg
(front mount transmitter/receiver) 2.77 kg
(rear mount transmitter/receiver) 3.22 kg
Power supply: 18-33 V DC, 1.2 A

Contractor

Honeywell Aerospace, Electronic & Avionics Lighting.

VERIFIED

EFS 10 Electronic Flight instrument System

The EFS 10 electronic flight instrument system is designed to interface directly with the Gold Crown III multisensor KNS 600 flight management system and KFC 400 flight control system. It uses a dual-stroke and raster display writing technology to create symbols and backgrounds, allowing each stroke/raster module to concentrate fully on a single display to make it brighter and easier to read.

A typical five-tube EFS 10 installation consists of identical and interchangeable symbol generators for each pilot's EADI and EHSI, plus one for the multifunction display. This provides multiple reversionary options for all displays in the event of a failure. Display units are available in 127 × 152.4 mm (5 × 6 in) horizontal format or 127 × 127 or 152.4 × 152.4 mm (5 × 5 or 6 × 6 in) square formats.

The EFS 10 includes full-time self-diagnostics and automatic internal testing of key circuits and sensor inputs, which may also be initiated by the pilot. A more comprehensive test routine allows checks down to circuit board level.

Contractor

Honeywell Aerospace, Electronic & Avionics Lighting.

VERIFIED

EFS 40/EFS 50 Electronic Flight instrument Systems

The EFS 40 is a 101.6 mm (4 in) electronic flight instrument system designed for operators of light jet and turboprop aircraft and turbine helicopters. The EFS 50 is a 127 mm (5 in) system. They are modular systems comprising an Electronic Attitude Director Indicator (EADI) and Electronic Horizontal Situation Indicator (EHSI). The package can be installed in two-, three-, four- or five-tube configurations.

The system's basic building block is the EHI. This unit incorporates a navigation mapping capability and modular design with interfaces for analogue and digital inputs. Added functions include a weather map display and joystick controller, plus a wide range of sensor input interfaces. The EHI is available with either a combination control/display unit or display with separate control panel.

The interchangeable ED 461/462 in the EFS 40 measures 106 × 106 × 238.8 mm and weighs 2.275 kg. The ED 551A in the EFS 50 measures 117.3 × 133.4 × 238.8 mm and weighs 3.76 kg. The remote SG 465 EADI/EHSI symbol generator unit fits into a ⅜ ATR short box. Control panels are offered as a built-in mode controller in the ED 461 or as the remote CP 468 mode controller for the ED 462 display unit. Standard interfaces are provided for most of the avionics, flying control and long-range navigation systems in current use in business aviation aircraft.

Contractor

Honeywell Aerospace, Electronic & Avionics Lighting.

VERIFIED

EHI 40 Electronic Horizontal situation Indicator

The EHI 40 is a 101.6 mm (4 in) electronic horizontal situation indicator which has been designed to offer reductions in size, weight and cost. It includes a built-in or separate mode controller, navigation mapping capability and modular design with interfaces for analogue and digital inputs. Added functions include a weather map display and joystick interface, and a wide range of sensor input interfaces. The system consists of a control display unit, symbol generator and associated navigation sensors. The EHI 40 is available with either a combination control display unit or traditional display with separate control panel.

During 1990, a companion attitude director indicator was introduced. The complete 4 in EFIS is designated EFS 40. The system is also available in a 5 in format, designated the EFS 50.

Specifications

Dimensions:
(display unit) 106 × 106 × 238.8 mm
(symbol generator) 193.5 × 57.2 × 320.5 mm
Weight:
(display unit) 2.275 kg
(symbol generator) 3.64 kg
Power supply: 28 V DC

Contractor

Honeywell Aerospace, Electronic & Avionics Lighting.

VERIFIED

Engine performance indicators

Honeywell Aerospace makes a series of engine performance indicators for business, commuter and military aircraft. The indicators provide a continuous analogue display of critical engine variables, such as fan and core rotation speeds, turbine temperature and fuel flow. Configurations with circular or square cross-section are available.

Solid-state circuitry is employed throughout. The indicators are unaffected by the large fluctuations in electrical supply during engine starts.

Specifications

Power supply:
12-34 V DC, 120 mA nominal, 450 mA (max)
5 V DC, 250 mA for lighting
Input signals:
(rotation) monopole pulse
(turbine temperature) chromel-alumel thermocouple per NBS Monograph 125
(fuel flow) second harmonic selsyn mass flow transmitter, 115 V AC 400 Hz reference

Accuracy:
(rpm) (0 to +55° C ambient) ±0.25% at 100% rpm
(–30 to +70° C ambient) ±0.5% at 100% rpm
(temperature) (0 to +55° C ambient) ±5° C at 900° C
(–30 to +70° C ambient) ±10° C at 900° C
(fuel flow) (0 to +55° C ambient) ±1% of full-scale
(–30 to +70° C ambient) ±2% of full-scale
Rpm range: 0-110%
Turbine temperature range: +100 to +1,000° C
Fuel flow range: 0-2,300 lb/h
Response: 3 s full-scale slew
Temperature range:
(operating) –30 to +55° C
(short term) –30 to –70° C
(ambient extreme) –65 to +70° C
Shock: 6 *g* for 0.011 s
Vibration: 5-3,000 Hz, 0.02 in amplitude, limited to ±1.5 *g*
Humidity: 95%, +70°C for 6 h, cool to +38°C for 18 h

Operational status

In production. Typical applications are the Raytheon Hawker 125-700, Gates Learjet 35/36 and 55, Lockheed Martin JetStar 2, CASA CN-235 and Taiwanese XAT-3.

Contractor

Honeywell Aerospace, Electronic & Avionics Lighting.

VERIFIED

Enhanced Airborne Collision Avoidance System (ACAS II/TCAS II)

Enhanced ACAS II is a development of the Honeywell Aerospace CAS 81 TCAS II, designed for commercial air transport aircraft. CAS 81 can be readily upgraded to the Enhanced ACAS II standard.

Enhanced ACAS II meets the full range of ACAS II and Mode S regulatory requirements and, in addition, incorporates increased surveillance range options, better bearing accuracy, improved reliability and an advanced communication datalink.

ACAS II is represented by the TCAS II design requirements, upgraded to the Change 7 software standard.

The Enhanced ACAS II modular architecture assures easy upgrades to future industry-defined systems such as the Airborne Separation Assurance System (ASAS) and Automatic Dependent Surveillance-Broadcast (ADS-B).

Enhanced ACAS II and Mode S is a vital component of the Communications, Navigation, Surveillance/Air Traffic Management (CNS/ATM) concept, embodying the Future Aeronautical Navigation System (FANS) concept and the military GANS/GATM requirements. Honeywell Aerospace also produces a derivative military variant of the Enhanced ACAS II concept, designated Enhanced-TCAS (E-TCAS).

Additional growth provisions include integration with the Enhanced Ground Proximity Warning System (EGPWS) technologies to provide enhanced situational awareness.

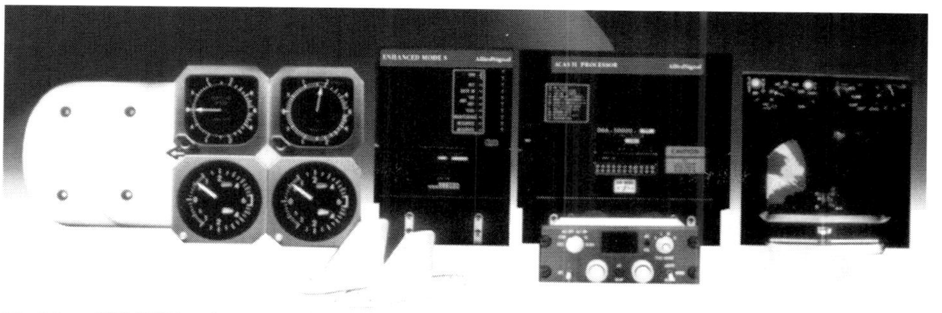

The future CNS/ATM environment will feature the use of extended ADS-B range and advanced datalink systems using GPS, ATN and Mode S technologies that the Enhanced ACAS II is designed to fulfil. **2000**/0081846

Significant capabilities of the Enhanced ACAS II and Mode S transponder system include:
(1) ADS-B squitter capabilities, including flight ID, velocity and GPS position.
(2) 100 n mile ADS-B extended surveillance range.
(3) 40 n mile active interrogation range.
(4) Hybrid surveillance.
(5) Internal event recording.
(6) Cockpit Display of Traffic Information (CDTI).
(7) Display range exceeding 100 n miles.
(8) Enhanced datalinking functions.
(9) Growth capability to meet future requirements.
(10)Compliance with relevant ARINC and ICAO requirements.

System components can be selected to meet specific requirements, but include the following, derived from the CAS 67A and CAS 81A ranges:
(a) IVA-81A Traffic Advisory/Vertical Speed Indicator (TA/VSI).
(b) IVA-81B Resolution Advisory/Vertical Speed Indicator (RA/VSI)
(c) IVA-81C and IVA-81D TA/VSIs.
(d) PPI-4B colour indicator.
(e) CTA 81 ATC/TCAS control panels.
(f) Enhanced TRA 67 A Mode S transponder.
(g) MST 67A transponder.
(h) Enhanced TPA 81A ACAS/TCAS processor unit.
iKFS 578A controller
(j) ANT 81A antenna.

Operational status

Honeywell Aerospace has produced civil variants configured with both ARINC and non-ARINC form factors. It also produces the military variant designated E-TCAS. Items of equipment from these three separate development lines can be mixed and matched to meet specific customer requirements.

One product line being marketed specifically is designated Enhanced CAS 67A ACAS II for the corporate and regional aviation markets. CAS 81A can be upgraded for the air transport market, and E-TCAS has been contracted by the US Air Force for designated transport and tanker aircraft types.

Contractor

Honeywell Aerospace, Electronic & Avionics Lighting.

VERIFIED

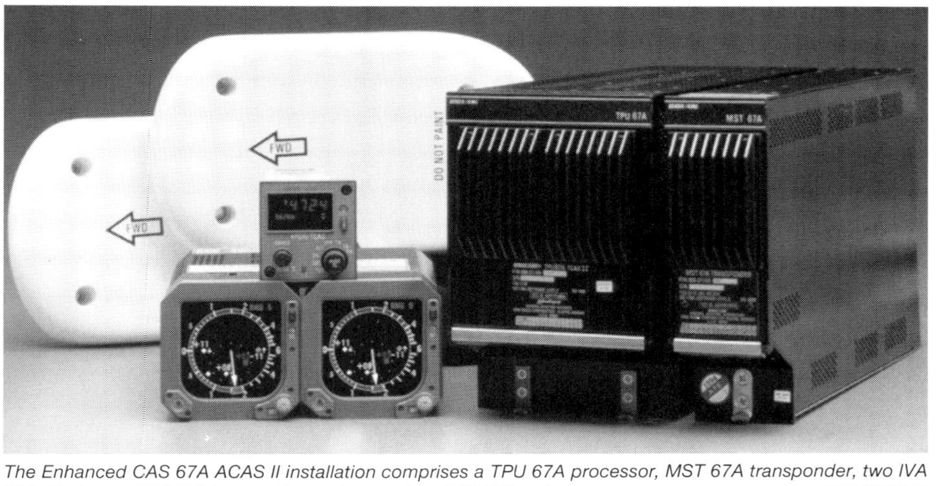

The Enhanced CAS 67A ACAS II installation comprises a TPU 67A processor, MST 67A transponder, two IVA 81D displays, a KFS 578A controller and one or two ANT 67A antennas **2000**/0081847

Enhanced Ground Proximity Warning System (EGPWS)

The EGPWS includes all traditional GPWS functions, but also has a proprietary worldwide terrain database. Referencing aircraft location from a navigation system, the EGPWS can display nearby terrain and provides aural warnings approximately 60 seconds in advance of a terrain encounter, compared with 10 seconds for a traditional GPWS.

EGPWS provides the following advancements on the GPWS:
(1) Look-ahead alerting algorithms.
(2) Multiple radio altimeter inputs.
(3) Significant reduction in unwanted warnings.
(4) Landing-short alerting algorithms.
(5) Optionally, Honeywell Aerospace also plans to include an embedded 12-channel GPS receiver to provide an upgrade capability to aircraft that have no existing GPS or FMS (Flight Management System) output available.

The following variants of the EGPWS are (or will shortly be) available:
(a) The Mk V air transport version for aircraft with digital data interfaces.
(b) The Mk VI regional aircraft version, which is similar to the Mk V but smaller and lighter; it has a regional terrain database (rather than global) and lacks windshear capability.
(c) The Mk VII air transport version is similar to the Mk V, but for aircraft that have analogue data interfaces.
(d) The Mk VIII is expected to include an integrated 75 × 75 mm Terrain Awareness and Display System (TADS), that can be retrofitted into the space occupied by an existing altimeter.

TADS will also be available for the Mk V, VI and VII variants from 1999 (it can be fitted as a replacement for an ARINC 453 altimeter). TADS displays threatening terrain on a navigation display (or on Wx radar display for non-EFIS aircraft) utilising the existing ARINC 453 databus. TADS gives the pilot visual detection of flight paths that could lead to conflict with precipitous terrain; it provides timely alerts with a visual assessment of terrain situation; it provides synchronised aural alert messages; it utilises present and projected aircraft position with built-in terrain databases.

The following modes are available with the EGPWS:

Mode 1: excessive descent rate alert/warning. Mode 1 provides pilots with alert/warning for high-descent rates into terrain. Typical warning time exceeds 20 seconds.

Mode 1 increases alert/warning for sink rate near the runway threshold. The pilot receives a timely alert for rapidly building sink rates exceeding 1,000 ft/min near the runway (earlier-generation GPWS computers provided little, if any, warning).

Mode 1 reduces nuisance alerts when visually repositioning down on the glideslope. The pilot has an additional margin to safely reposition the aircraft.

Mode 2: excessive closure rate to terrain warning. The same technology breakthroughs, that allow for the display of threatening terrain, permit advanced, virtual 'look ahead' warnings. 'Caution-Terrain' and 'Terrain Ahead' alerts based on position data and terrain database precede the GPWS warning.

Warning for excessive closure rate to terrain has been dramatically increased by using the speed of the aircraft to expand the warning envelope.

When compared with systems which do not have speed enhancement, warning times are typically

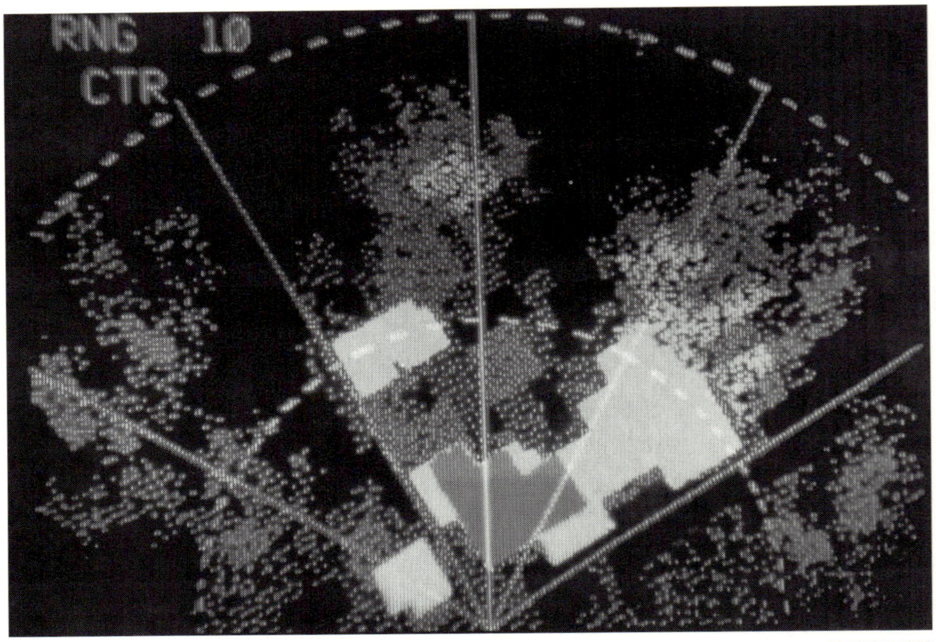

EGPWS terrain display ***1999***/0044978

doubled for inadvertent flight into mountainous terrain during initial approach or descent.

Terrain alert precedes the *pull-up* and will continue after a *pull-up* alert until 300 ft of barometric altitude has been gained to help encourage the pilot to climb and avoid any following terrain.

This warning mode has not only improved warning time, but has minimised the chance of unwanted warnings.

On final approach, the decreased speed and landing configuration of the aircraft desensitise the warning envelope to allow the aircraft to land at airports situated on terrain without nuisance warnings.

Mode 3: alert to descent after take-off. Mode 3 alerts the pilots to an inadvertent descent into terrain after take-off or missed approach. The alert is given after significant barometric altitude loss has occurred and allows considerable margin for third segment acceleration and flap retraction even under engine-out conditions.

After the aircraft has climbed to a safe altitude, based on climb performance and time, this mode is automatically switched out and replaced by an alert/warning floor below the aircraft based on speed and aircraft configuration.

During initial climb, additional warning protection is provided against a shallow, accelerating climb into rising terrain.

Possible nuisances from premature mode switching caused by terrain undulations or faulty radio altimeter unlock are eliminated.

Mode 4: insufficient terrain clearance. Mode 4 will warn the pilots of insufficient terrain clearance during climbout, cruise, initial descent or approach. This warning mode is especially valuable when the aircraft's flight path, relative to terrain, is insufficient to develop excessive closure rate or descent rate warnings.

Warning times have been essentially doubled over earlier generations of GPWS computers by automatically expanding the warning envelope as the speed of the aircraft increases. Similar to Mode 2, the warning envelope on approach gradually collapses as speed decreases, the landing gear is lowered and the flaps selected. The wording of the warning is also changed to correlate with the phase of flight and the actual cause of alarm. Warning time as compared to the earlier generations of GPWS is greatly improved for situations such as initial descent or approach where landing gear is extended for additional drag. Speed will automatically expand the warning envelope, providing additional warning time.

Terrain Clearance Floor uses navigation position data coupled with an airport/runway database to provide valuable protection in the landing configuration.

Mode 5: alert to inadvertent descent below glideslope. Mode 5 is automatically armed when the pilot selects an ILS frequency and selects gear down. The warning envelope contains two boundaries: a 'soft' alerting region and a 'hard' alert region. Both boundaries are a function of glideslope deviation.

When the aircraft penetrates the 'soft' alerting region, the audio level of glideslope voice is 6 dB below other GPWS computer voices. The voice repetition rate increases as deviation below glideslope increases. If the aircraft subsequently enters the 'hard' alerting region, the *glideslope* voice audio level increases to equal that of other voices. Below 150 ft of radio altitude, the amount of glideslope deviation required to produce an alert is increased to reduce nuisance alerts which could be caused by close proximity to the glideslope transmitter.

Mode 5 can be inhibited by pressing either of the cockpit *below G/S* lights to permit deliberate descent below the glideslope in order to utilise the full runway under certain conditions. Any other warning always has priority over a *glideslope* alert. Possible nuisances from false back course glideslope signals are automatically eliminated.

Mode 6: altitude callouts and excessive bank angle alert. Altitude callouts, which have become a popular feature with many airlines on new aircraft, are available in the system. They are pin-selectable at the rear connector and there are 32 menus available for the purpose of increasing altitude awareness on final approach.

Tones are also available. Automatic audio level increase is available when windshield rain removal is in use. Honeywell Aerospace recommends the use of a few automatic callouts near the runway and a 'smart' callout, which would rarely be heard, for most ILS landings.

Bank angle can be used to alert crews of excessive roll angles. The bank angle limit tightens from 40° at 150 ft AGL to 10° at 30 ft AGL to help alert the crew on landing of excessive roll corrections which might result in wingtip or engine damage. *Bank angle* is also useful to help alert the pilot of severe overbanking which might occur from momentary disorientation during initial climb-out.

Mode 7: Windshear detection and annunciation. Visual and aural windshear warnings are given for windshears that significantly degrade the performance of the aircraft. Optional visual and aural caution alerts can be given for increasing performance windshears that may be a signature of microbursts.

The detection level is automatically varied by outside air temperature and lapse rate, change in flight path, relationship to glideslope, height above ground, bank angle and approach stall margin. This advances the alert/warning time and improves the margin against unwanted alerts or warnings.

The final mode capability is 'alert envelope modulation'. This permits aircraft position data to be used to reduce nuisance warnings at particular problem airports, by modifying algorithms for specific local terrain peculiarities.

EGPWS is a form/fit replacement for the existing Mk V and MK VII GPWS, but some additional cable work is required.

Specifications

Dimensions: 2 MCU
Weight: 3.2 kg
MTBF: 15,000 h

Operational status

In production. Selected by a large number of air transport operators for fitment to Boeing and Airbus aircraft.

In December 1997, major US air carriers announced a voluntary programme to install EGPWS, in a six year programme extending until the year 2003, affecting about 4,500 aircraft operated by US major and national airlines. US regional airlines are not part of this voluntary programme, nor are charter carriers.

This initiative pre-empts a planned US FAA notice of proposed rule making for the fitment of EGPWS to US long-haul and most regional aircraft.

The Mk V, Mk VI and Mk VII EGPWS are in production. The Mk VIII availability was planned for late 1999.

Military versions of the Mk V and Mk VII are also available.

The EGPWS has been integrated with the RDR-4B forward-looking windshear radar; the combined system has been successfully flight tested for certification on the Boeing 777.

In June 1999, Honeywell Aerospace announced that it had joined with Sikorsky Aircraft to engineer a new

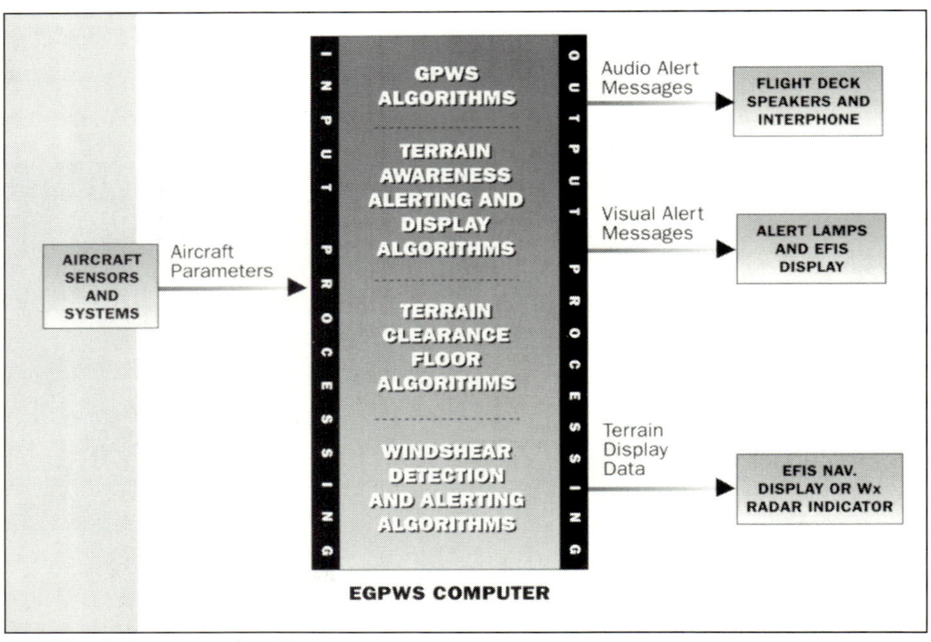

EGPWS typical system configuration ***1999***/0044979

version of the EGPWS for helicopters. The system will have several features specifically designed for helicopters including a digital moving map, which will show the helicopter's height above terrain with colour or shading contrasts, and wires and other obstructions will be overlayed on the map and conspicuously marked. The EGPWS will also provide visible and audible warnings in advance of threatening terrain or obstacles. The helicopter EGPWS will also feature high-resolution terrain data at locations away from airports to cater for the requirement to operate close to the ground at distant locations. Flight tests will be conducted using an S-76 helicopter from Sikorsky. Certification is expected in the third quarter of 2000.

By June 1999, more than 6,000 EGPWS had been ordered. The system is standard on Bombardier Global Express, Dassault Falcon 900EX and Gulfstream V, and an option on other business aircraft such as the Cessna Citation X and Excel, Hawker 800XP and Learjet 45.

Contractor

Honeywell Aerospace, Electronic & Avionics Lighting.

VERIFIED

Honeywell's ETCAS has been selected for the USAF transport fleet (Lockheed Martin) ***2001***/0103882

Enhanced TRA 67A transponder

The Enhanced TRA 67A is a new enhanced Mode S transponder designed to meet new European requirements and other anticipated regulations worldwide. It meets Change 7 and Downlinked Aircraft Parameters (DAPs) requirements adopted for European and other nations by the International Civil Aviation Authority (ICAO). It also performs Automated Dependent Surveillance-Broadcast (ADS-B) functions in anticipation of future needs.

In compliance with the DAPs requirement, the Enhanced TRA 67A will broadcast such data as roll angle, groundspeed, true airspeed, magnetic heading and indicated airspeed. Change 7 is an enhancement to the software of Terrain alert and Collision Avoidance Systems (TCAS) and related systems, which has been mandated in Europe. A Mode S transponder is an essential part of a TCAS.

The Enhanced TRA 67A will help facilitate ADS-B by broadcasting digital information such as aircraft location (from the airplane's global positioning system), aircraft altitude, velocity and acceleration. This information, when received by other aircraft that are also equipped with ADS-B, can be used to extend the range and enhance the accuracy of traffic information provided by the TCAS.

Operational status

The US Federal Aviation Authority (FAA) granted Technical Standard Order (TSO) approval for the Enhanced TRA 67A in June 1999; specific aircraft certifications will follow.

Contractor

Honeywell Aerospace, Electronic & Avionics Lighting.

VERIFIED

ETCAS - Enhanced Traffic alert and Collision Avoidance System

Honeywell has designed and developed an Enhanced Traffic Alert & Collision Avoidance System (ETCAS), to provide military aircraft operators with an extended surveillance range and the capability to coordinate formation flying in addition to standard TCAS operations. ETCAS is the only system providing this capability currently certified by the FAA and international aviation authorities. The FAA, US Air Force and international authorities have granted ETCAS the frequency approvals necessary to ensure smooth integration into military operations and prevent frequency overlap with commercial aviation operations.

ETCAS provides two modes of operation. The basic mode is ACAS II which is the same as TCAS II with Change 7.0 software and is RVSM compatible. In addition to the standard TCAS functions of situational awareness, traffic alert and resolution advisories, the ETCAS provides a formation mode. This mode allows aircraft operators to locate, identify, rendezvous with and formate with aircraft equipped with a variety of identification systems, including Identification Friend and Foe (IFF), Modes 1, 2, 3 and 4, Mode A, Mode C and Modes S transponder equipped private, commercial and military aircraft. In formation mode, the ETCAS performs interrogation surveillance to a range of 40 nautical miles with azimuth coverage of 360° and up to 12,700 feet above and below the aircraft. The system can mark and identify formation members and rendezvous aircraft by displaying the aircraft's Mode A code next to its symbol, clearly distinguishing it from other traffic within ETCAS range. In addition to the Mode A codes, the system provides unique symbology for formation members. The symbols remain until deselected manually or until the ETCAS is commanded to switch to normal TCAS operation. ETCAS information can be displayed on a variety of cockpit instruments, including radar displays, navigation displays, electronic flight displays and traffic alert/vertical speed indicators (depending on aircraft type and cockpit configuration). While functioning in formation mode, ETCAS continues to monitor the airspace and maintain standard TCAS II collision avoidance capability.

Utilising IFF/Mode-S transponders, the ETCAS provides the operator with continuous TCAS surveillance and protection while maintaining concurrent IFF reply capability. Honeywell will introduce ETCAS with increased surveillance range capability in 2001. Utilising hybrid surveillance technology, ETCAS will track like-equipped aircraft to 100 n miles. The company claims this increased range surveillance will be achieved without increasing transmission power. This feature will be available as an upgrade to existing ETCAS systems.

The US Air Force and Lockheed Martin have let a contract to replace the current station-keeping equipment for providing tanker rescue/rendezvous mission and refuelling capability with a modified version of ETCAS.

Operational status

ETCAS has been designated as standard equipment on US Air Force transport aircraft.

Contractor

Honeywell Aerospace, Electronic & Avionics Lighting.

UPDATED

FCS-60 series 3 digital Flight Control System

The designation FCS-60 embraces a family of four flight control systems with performance to suit different categories of regional commuter and business aircraft.

The FCS-60 is basically a three-axis system under microprocessor control. The simplest system available has one channel per axis, weighs 14 kg and is driven by a single air data sensor. Navigation and sensor units are controlled by a single fail-passive flight controller, the outputs from which operate pilot and co-pilot flight director instruments. The system can be expanded to dual-simplex or duplex configuration, the second of which may be certified for Cat II operations. Both have dual air data sensors and the second version has dual-drive motors and three microprocessors. The system provides pitch stabilisation with automatic trim, roll stabilisation, heading hold, yaw damping and turn co-ordination. The yaw damper can be used independently of the other channels. Additional modes provide lift compensation during turns, compensation for turbulence and pilot-commanded inputs.

Operational status

In production.

Contractor

Honeywell Aerospace, Electronic & Avionics Lighting.

VERIFIED

FCS-870 automatic Flight Control System

Honeywell Aerospace has combined integrated circuit technology with several new automatic flight control features to produce a system which offers optimum performance over a wide range of general aviation aircraft. The FCS-870 is an autopilot with flight director and independent yaw damper options which form the basis of many options. It is designed for installation in a broad range of aircraft types, from heavy singles to most turboprop-powered types. The system meets or exceeds the TSOs for these classes of aircraft.

The complete FCS-870 consists of the cockpit instruments (including flight controller, attitude director indicator, horizontal situation indicator and mode annunciator) and a remote-mounted computer amplifier and servos. The yaw damper option adds a side-slip sensor, a panel-mounted turn and slip indicator and a remote yaw servo.

Flight controller This is a small panel-mounted unit used to select the desired operational modes of the system. All nomenclature on the panel is back-lit for easy night viewing.

Mode annunciator This can be mounted in any convenient head-up panel location so that the pilot can monitor the autopilot or flight director functions in use. The unit also alerts the pilot to some fault and armed system conditions.

Computer amplifier This is the main flight control system unit and is an all-solid-state device which houses all the lateral and longitudinal computational circuitry, power supplies and altitude transducer. It also contains the calibration circuits which ensure compatibility with specific aircraft sensor and output requirements, plus an additional circuit board providing input signals for flight director operations. Lateral and longitudinal data circuits are segregated on opposite sides of the unit, so reducing the amount of inter-wiring and augmenting reliability and serviceability. Relays have been eliminated and all heat-generating components are near the outside of the unit to improve cooling.

Servos Advanced design servos provide the greatest torque required for the highest-performance aircraft likely to use the system. Three similar units are used for pitch, roll and trim control.

Yaw damper The independent yaw damper provides positive turn co-ordination and rudder control in all flight conditions. In twin-engined types the yaw damper is claimed to provide substantial assistance in maintaining directional control during an engine-out sequence.

Instrumentation The company recommends integration with the following flight instruments: DH-886A 4 in director horizon indicator or DH-841V 3 in director horizon indicator, plus HSD-880 4 in or HSD-830 3 in horizontal situation indicators.

Specific features of the flight control system are command turn (half- or full-rate co-ordinated turns can be initiated by rotating a knob); full pitch integration (provides smooth capture of desired altitude and eliminates standoff errors); automatic altitude preselect; pitch synchronisation (keeps elevator surfaces aligned with trim to eliminate disengagement disturbances); control wheel steering (manoeuvre to desired attitude with button depressed and then release to leave flight control system maintaining pilot's demand); coupled go-around, automatic crosswind correction; all angle intercepts and pitch rate command and manual or automatic glideslope capture.

Specifications

Weight:
(basic autopilot) 8.94 kg
(with 3 in FD/HSI) 14.11 kg
(with 4 in FD/HSI) 17.08 kg
(optional yaw damper) 3.33 kg
TSO compliance: C9C, C52A, DO160.

Operational status

In service.

Contractor

Honeywell Aerospace, Electronic & Lighting Systems.

VERIFIED

FPD 500 Flat Panel Display system

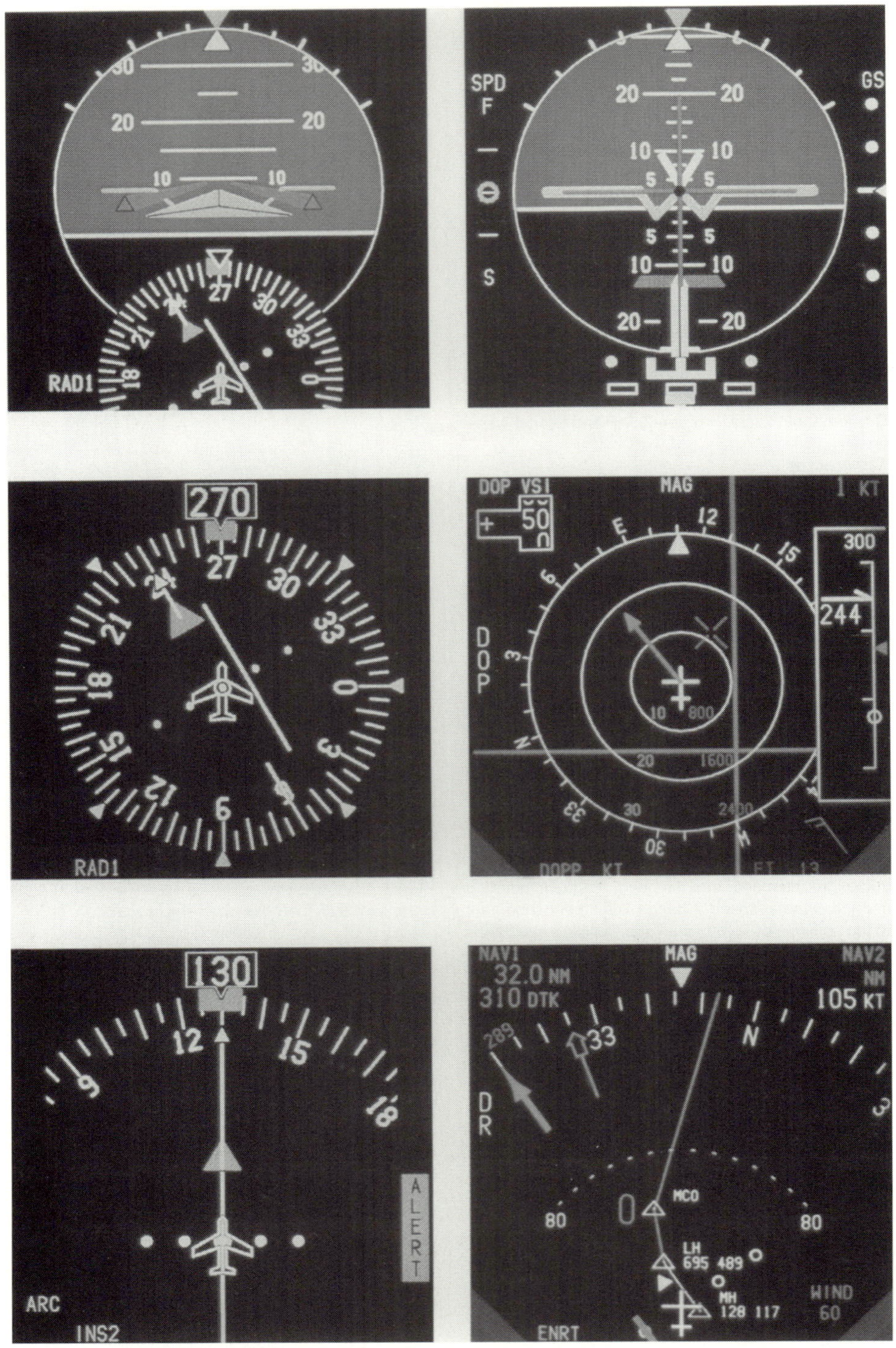

FPD 500 flat panel display system 0001377

The FPD 500 is an Active Matrix Liquid Crystal Display (AMLCD) that can be used directly to replace the four electromechanical primary flight displays. It has an adaptor designed to plug-in readily to existing electromechanical instrument wiring harnesses. Its form factor matches that of most 5 in instruments.

The FPD 500 can provide the functions for primary and secondary flight displays, navigation, engine displays and other multifunctional use. It can be used for a number of applications including Attitude Director Indicator (ADI); Horizontal Situation Indicator (HSI); moving-map, radar/windshear display; or TCAS indicator.

Specifications

Dimensions: ARINC 408 5ATI
Weight: 2.61 kg
Power: 28 V DC, or 115 V AC/400 Hz; 42 W typical, 60 W max
Resolution: 192 × 256 DPI (min)
NVG-compatible TSO: DO-160C

Operational status

In production. US Federal Aviation Administration (FAA) approved for B 727-200 aircraft.

Contractor

Honeywell Aerospace, Electronic & Avionics Lighting.

VERIFIED

Global Star 2100 Flight Management System (FMS)

Global Star 2100 provides the capability to meet the requirements of the Future Air Navigation System (FANS) and Aeronautical Telecommunications Network (ATN) systems. Through the ATN network, the onboard ATN router will connect the aircraft with the ground via VHF, satcom, HF and Mode S datalinks with the Automatic Dependent Surveillance (ADS) position reporting and Air Traffic Control (ATC) conflict resolution system. Global Star 2100 is also compatible, through ARINC 739 connections, with conventional AFIS and ACARS systems.

The Global Star 2100 system combines a navigation computer, multichannel IFR-approach-certified GPS receiver, navigation database, flight management system functionality and a Multifunction Control and Display Unit (MCDU), with 5.5 in display into a single compact unit. A two-box installation can also be provided in which all functions except the MCDU are integrated into a separate LRU that can be located in the avionics bay.

The Global Star 2100A is a derivative system, designed for installations where cockpit and avionics bay space is limited; it is available in both single-box and two-box variants, but with a 4 in display.

The internal GPS receiver provides real-time and predictive Receiver Autonomous Integrity Monitoring (RAIM) and Fault Detection and Exclusion (FDE). In dual configurations, Global Star 2100 satisfies the requirements for primary means of navigation using GPS alone.

Global Star 2100 provides exceptional flight planning capabilities, with complete SID/STAR procedures, airways, en-route manoeuvres, and 'direct-to' capabilities. Operators are able to store up to 200 flight plans with as many as 100 waypoints, stored in non-volatile flash memory.

Specifications

Display: full-colour LCD flat panel
Display diagonal: 5.5 in
Lines: 12 lines × 24 characters
Dimensions: 180.8 (H) × 146.05 (W) × 199.4 (L) mm
Weight: 3.64 kg
Inputs:
(analogue): fuel flow, air data, heading, VOR/DME
(digital): air data, heading, VOR/DME, weather radar (joystick), radio frequencies
Outputs:
(analogue): HSI course and bearing, crosstrack and vertical deviation, to/from, autopilot steering, annunciators
(digital): EFIS/flight director, autopilot, radio tuning

Operational status

The Global Star 2100 system is reported to be standard fit in the Cessna Citation Excel, de Havilland DHC-8 Series 400, and Avro RJ, and to be an option on the Bombardier Learjet 45.

Contractor

Honeywell Aerospace, Electronic & Avionics Lighting.

VERIFIED

GNS-1000 Flight Management System (FMS)

The GNS-1000 flight management system offers a choice of flight management capabilities including GPS, VLF/Omega, DME/DME, VOR/DME and IRS. It houses both a five-channel GPS receiver that continuously tracks up to five satellites simultaneously and a VLF/Omega sensor. A single module formats the system to existing interface requirements.

Specifications

Dimensions: 384.6 × 195.1 × 124.5 mm
Weight: 9.58 kg

Contractor

Honeywell Aerospace, Electronic & Avionics Lighting.

VERIFIED

GNS-X Flight Management System (FMS)

The GNS-X flight management system combines a Navigation Management Unit (NMU) and a CDU to provide comprehensive navigational capability. The internal database has 4 Mbits of memory and the GNS-X has interfaces with aircraft navigation systems (GPS, Omega, INS and so on) and aircraft systems (autopilot, air data system, AFIS and so on). Internal DME/DME and DME/VOR processing is standard. Also featured is an internal Loran C, frequency management and automatic fuel flow.

Specifications

Dimensions:
(NMU) ¼ ATR short
Weight:
(NMU) 3.2 kg
(CDU) 2.9 kg

Operational status

In production from early 1988. The GNS-X is certified on the King Air 200 and Falcon 900.

Global Star 2100 flight management system multifunction control display unit **2000**/0081855

Contractor

Honeywell Aerospace, Electronic & Avionics Lighting.

VERIFIED

GNS-XL/GNS-XLS Flight Management Systems (FMS)

The GNS-XL and GNS-XLS systems are derived from the earlier GNS-X series of Global Wulfsberg systems. Both are compact, single-box, flight management systems that incorporate eight-channel GPS, and feature flat-panel liquid crystal displays. Both provide control of aircraft navigation sensors, communication, radio and fuel management. With full analogue/digital interfaces, they are well suited for both new programmes, and upgrade/retrofit requirements.

The receiver incorporates RAIM for enhanced reliability, and to meet FAA TSO C129 Class A1/B1/C1 requirements, enabling the operator to make GPS-derived IFR approaches as well as en route and terminal navigation. Both systems include Fault Detection and Exclusion (FDE), enabling them to be used as primary means of navigation during transoceanic and remote area operations. Position and velocity data is accepted from internal and external sensors, using a special navigation filter to generate a composite system position. As well as the built-in GPS receiver, the system includes a VORTAC Position Unit (VPU) processor, which automatically selects the best available DME/DME and VOR/DME measurements from VOR/DME, VORTAC, TACAN and ILS DME units. External interfaces include those for VLF/Omega, Inertial Reference Systems (IRS), or Inertial Navigation System (INS).

Data communication interfaces include onboard aircraft systems such as the air data computer, compass system, altitude pre-selector, EFIS, weather radar, autopilot, communication and navigation radios and the fuel flow system, together with a full AFIS (Airborne Flight Information System), and optional Satellite Data Communications (SDC) connection.

The GNS-XLS features a full flight planning capability for both area navigation (RNAV) and vertical navigation (VNAV), and it simplifies frequency management.

The GNS-XL and GNS-XLS systems have growth potential for Wide Area Augmentation System (WAAS) and Differential GPS.

Specifications

GNS-XL
Display: full-colour 5½ in diagonal display
Dimensions: 181 × 146 × 200 mm
Weight: 3.64 kg
Inputs:
(analogue) fuel flow, air data, heading, VOR/DME
(digital) air data, heading, VOR/DME, weather radar, EFIS, radio frequencies
Outputs:
(analogue) HSI course and bearing, XTK and vertical deviation, to/from, autopilot steering, annunciators
(digital) EFIS/flight director, autopilot, radio tuning

GNS-XLS
Display: full-colour 4 in diagonal
Lines: 10 lines × 22 characters
Dimensions: 114 (H) × 146 (W) × 165 (D) mm
Weight: 3.18 kg
Inputs: analogue and digital: fuel flow, air data, heading, weather radar, EFIS, VOR/DME/DME

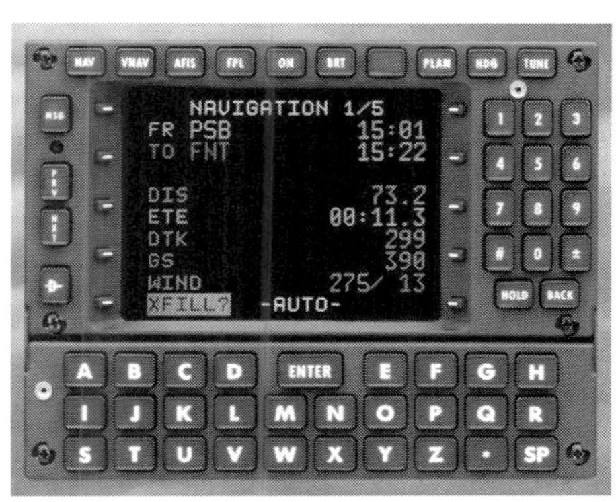

GNS-XLS flight management system multifunction control display unit **2000**/0081854

Outputs: analogue and digital: EFIS/flight director, autopilot, radio tuning

Operational status

GNS-XL is certified by the US FAA to TSO C-129 Class A1; it can be upgraded for compatibility with FANS. It has been standard equipment on Citation Ultra business jets from January 1997.

GNS-XLS was certified for Fault Detection and Exclusion (FDE) by the US FAA in November 1996. Selected by BAE Systems Asset Management as a navigation upgrade for the BAe 146 fleet.

Contractor

Honeywell Aerospace, Electronic & Avionics Lighting.

VERIFIED

Gold Crown nav/com avionics family

In the late 1960s, the Gold Crown range of remote-mounted (that is, panel-mounted flight deck instruments driven by rack-mounted processing amplifiers) equipment was introduced for the larger piston and turbine twins then being developed and to complement the Silver Crown range. In 1972, a more advanced family of digital equipment appeared and began to take a large share of this market. The most recent range is Gold Crown III, unveiled in 1981. Designed for top-of-the-line corporate and commuter twins, turboprops and turbine helicopters, this completely redesigned family incorporates custom LSI and microprocessor technology. The remote-mounted units are some 40 per cent smaller and 30 per cent lighter than their predecessors. Being of solid-state design the systems are substantially more reliable.

The Gold Crown III avionics system has been modified to be compatible with night vision goggles. Existing gas discharge displays have been replaced with dichroic liquid displays. This line of equipment is often specified by overseas customers for use in military transport and training aircraft.

The Gold Crown III family comprises the following equipment:

KAA 955 audio amplifier/control
KDF 806 digital automatic direction-finder
KDM 706 distance measuring equipment
KFC 200 flight control system
KFC 250 autopilot/flight director system
KFC 325 autopilot/flight director system
KFC 400 autopilot/flight director system
KHF 950 HF single-sideband transceiver
KHF 990 HF SSB transceiver
KNI 582 indicator
KNR 634 digital navigation receiver
KNS 81 integrated nav/RNav system
KQI 553A pictorial navigation indicator
KTR 908 VHF communications transceiver
KTR 909 UHF communications transceiver
KXP 756 transponder
MST 67A Mode S transponder.

More detailed descriptions of some of these systems can be found under the appropriate section headings in this book.

Operational status

In production and service.

Contractor

Honeywell Aerospace, Electronic & Avionics Lighting.

VERIFIED

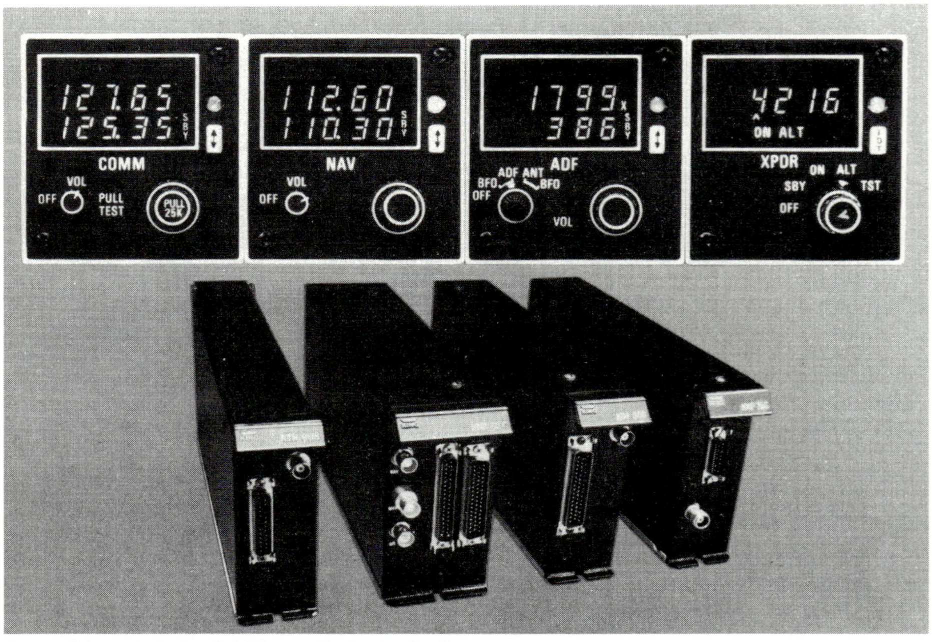

Some of the units in the Gold Crown III avionics family

GPWS Mk II Ground-Proximity Warning System

The Mk II Ground-Proximity Warning System (GPWS) computer is designed for aircraft wired to ARINC 594 standard and is suitable for service in a wide cross-section of commercial, military or business aircraft. The Mk II model was claimed to be the first GPWS to use a Mach/airspeed input and therefore to have a much faster response time than previous GPWS computers. It was the first such system to offer voice alerts which specifically identified each warning mode, and the first to offer a warning mode for minimum approach conditions.

Warning modes for the Mk II unit are generally the same as for the Mk V digital GPWS although there are differences between the warning times and the warning envelopes themselves.

Specifications

Dimensions: ¼ ATR short
Weight: 3.63 kg (max)

Operational status

In service.

Contractor

Honeywell Aerospace, Electronic & Avionics Lighting.

VERIFIED

GPWS Mk V Digital Ground-Proximity Warning System

The Mk V digital ground-proximity warning system computer is designed for service with aircraft equipped with ARINC 700 avionics. It provides the flight crew with back-up warning for seven potentially dangerous situations including windshear conditions.

Alerts and warnings are provided by steady or flashing visual indications and by audible warnings.

The Mk V ground proximity warning syste
1999/0044980

Each audio warning is also annunciated to identify the particular situation such as excessive descent rate, excessive closure rate to terrain, significant altitude loss after take-off, insufficient terrain clearance, excessive descent below glideslope, altitude call-outs and windshear detection.

Windshear detection and annunciation are provided by the Mk V computer. When the computer detects an impending windshear situation, an optional amber light is turned on in the cockpit. If the aircraft experiences further windshear severity, a red warning light is displayed along with a voice message 'windshear' repeated three times. The windshear alert function takes priority over other GPWS alerts.

Other alerts are repeated twice. If the aircraft's performance continues to degrade, the message is repeated. A particular advantage of the variety of voice alerts is that its operationally orientated warnings permit confirmation by cross-checking of the panel instruments. Diagnosis of flight warnings can thus be quickly carried out and corrected. The speed of the ground-proximity warning system envelopes has been increased, providing longer warning times.

The system contains a number of features to assist in test maintenance and repair procedures. These include a non-volatile memory which stores both steady-state or intermittent faults occurring over the last 10 flight sectors and which can be erased only when the unit is removed from the aircraft for bench work. The accepted test procedure is programmed within the computer and a simple test fixture is all that is required to re-address computer output data back into the computer itself. An alphanumeric display on the front of the unit can be used to isolate faults and indicate specific LRUs which require replacement. Faults can be isolated on the bench to board level.

The GPWS complies with ARINC 600 standards and its subcomponents are grouped by circuit function on plug-in/fold-out removable printed circuit boards with easily removed captive hardware. Latitude and longitude are used to modify warning boundaries at certain locations to reduce nuisance probability, or increase available warning time.

Specifications

Dimensions: 2 MCU
Weight: 3.2 kg
MTBF: 30,000 h

Operational status

In production and in service on a wide range of digital avionic air transport and corporate aircraft.

Contractor

Honeywell Aerospace, Electronic & Avionics Lighting.

VERIFIED

GPWS Mk VI Ground-Proximity Warning System

The Mk VI ground-proximity warning system is designed for regional, commuter, corporate and business turbojet and turboprop aircraft. It operates in six modes: excessive descent rate alert and warning; excessive closure rate to terrain; alert to descent after take-off; alert to insufficient terrain clearance; alert to inadvertent descent below glideslope; and altitude call-outs and bank angle alert. The Mk VI system is said to cost 40 per cent of the earlier Mk II GPWS.

The system has been refined to delete unwanted warnings by reducing glideslope and terrain clearance floor limits to trigger warnings at altitudes down to 750 ft above ground level on approach or to 925 ft above ground level with ILS acquired.

Manual functions have been added to minimise the chance of false warnings due to flapless landings or other operational modes, when landing at airports with unique terrain features or in the event of incompatible terrain clearance during approach and departure procedures.

Specifications

Dimensions: 152.4 × 76.2 × 260.4 mm
Weight: 1.9 kg

Operational status

In production and in service.

Contractor

Honeywell Aerospace, Electronic & Avionics Lighting.

VERIFIED

The Mk VI ground-proximity warning system
1999/0044981

GPWS Mk VII Ground-Proximity Warning System

The Mk VII warning computer is designed for all existing and new aircraft with ARINC 500 analogue avionics as a replacement for the Mk I and Mk II ARINC 594 ground-proximity warning computers. The improved Ground-Proximity Warning System (GPWS) dynamics provide the advantages of increased warning times, prioritisation of aural warnings and reduction in nuisance warnings in the cockpit, while implementing a cost-effective windshear warning system. The computer has a common part number used across a wide range of aircraft types, minimising the investment in spares.

The Mk VII warning computer meets the requirements of FAA AC 25-12 for windshear detection and alerting. It uses the existing ARINC GPWS interface, with additional signals provided through a second connector, for windshear detection, optional recovery guidance and custom altitude call-outs.

The GPWS features ground-proximity warning and glideslope alerting; altitude call-out menus; bank angle alerting; reduced audio cockpit clutter; improved take-off monitoring for noise abatement procedures; automatic adjustment of warning modes for ILS approaches; windshear detection and annunciation; optional wincshear recovery guidance; verbal annunciation of system faults and front replaceable software modules. It fits into existing GPWS rack space.

Specifications

Dimensions: ¼ ATR short
Weight: 2.72 kg (max)
Power supply: 115 V AC, 400 Hz, single phase, 15.7 W nominal
Environmental: DO-160B
Reliability: 15,000 h MTBF

Operational status

In production and in service.

Contractor

Honeywell Aerospace, Electronic & Avionics Lighting.

VERIFIED

HF datalink

The HF datalink service is an extension of the VHF Aircraft Communications Addressing and Reporting System (ACARS). It provides an aeronautical data communication link beyond the line of sight limitations of VHF, with the potential for worldwide coverage.

The avionics system consists of an airborne HF Data Unit (HFDU) incorporating an HF modem and a datalink processor which implements the air-ground link and network access protocols. The system makes use of the aircraft's existing HF radio and antenna. Data transmission rates are comparable to those of low data rate Satcom.

The HFDU provides interface between the ACARS MU and HF and VHF radios. In VHF mode, it is transparent to the MU and VHF radios. In the HF mode, there is no operator involvement. The HFDU performs automatic search and selection of HF frequency, controls tuning and keying of the HF radio and employs the HF modem to send or receive ACARS messages. The 4 MCU enclosure provides HF modem and airborne datalink processor, power supply plus backplane and two spare slots for enhancements such as ATN.

Operational status

The HF datalink has been certified on an American Airlines Boeing 767-300 aircraft.

Contractor

Honeywell Aerospace, Electronic & Avionics Lighting.

VERIFIED

IHAS 5000/IHAS 8000 Integrated Hazard Avoidance System

The IHAS systems are designed to improve situational awareness and safety.

IHAS 5000 is designed for general aviation aircraft, particularly for piston-powered aircraft not fitted with weather radar. Integrating the four major airborne safety systems - position awareness, weather avoidance, traffic advisories and terrain awareness/warning - the IHAS 5000 presents safety and situational awareness information on a large colour display.

The IHAS 5000 system comprises the following items of equipment:

(1) The KDR 510 datalink radio to deliver textual information and colour weather graphics over a high-speed, real-time, VHF DataLink (VDL) mode 2 datalink.
(2) The KT 73 datalink Mode S transponder. This system receives traffic information uplinked from Air Traffic Control (ATC) ground stations, and presents a picture of nearby traffic on the KMD 550 MultiFunction Display. At present this service is only available in the USA.
(3) The KMD 550 MFD is a full-colour MFD, which displays data on a 5 in diagonal screen. The KMD 550 MFD includes intuitive controls and an extensive aeronautical and cartographic database, to present detailed obstacle and terrain elevation shading data.

An optional addition to the IHAS 5000 system is offered by the KMH 880 unit. KMH 880 is a fully integrated terrain avoidance and traffic system that combines an active traffic sensor capability and Enhanced Ground-Proximity Warning System (EGPWS) to deliver traffic advisories of the same type as those provided to airlines by the TCAS I system.

IHAS 8000 is designed for business aviation users, and it incorporates the KMD 850 MFD, instead of the KMD 550 MFD. The KMD 850 MFD includes both analogue and digital weather radar interfaces.

Specifications

KMD 550
Dimensions: 158 (W) × 101 (H) × 241 (D) mm
Weight: 2.3 kg
Screen size: 5 in diagonal
Power: 10-33 V DC, 25 W

KMH 880
Dimensions: 115 (W) × 153 (H) × 357 (D) mm
Weight:
(display): 4.1 kg
(antenna): 0.5 kg
Power: 28 V DC

Operational status

Delivery of the first IHAS components was due to start in 2000, with complete system availability from 2001.

Contractor

Honeywell Aerospace, Electronic & Avionics Lighting.

VERIFIED

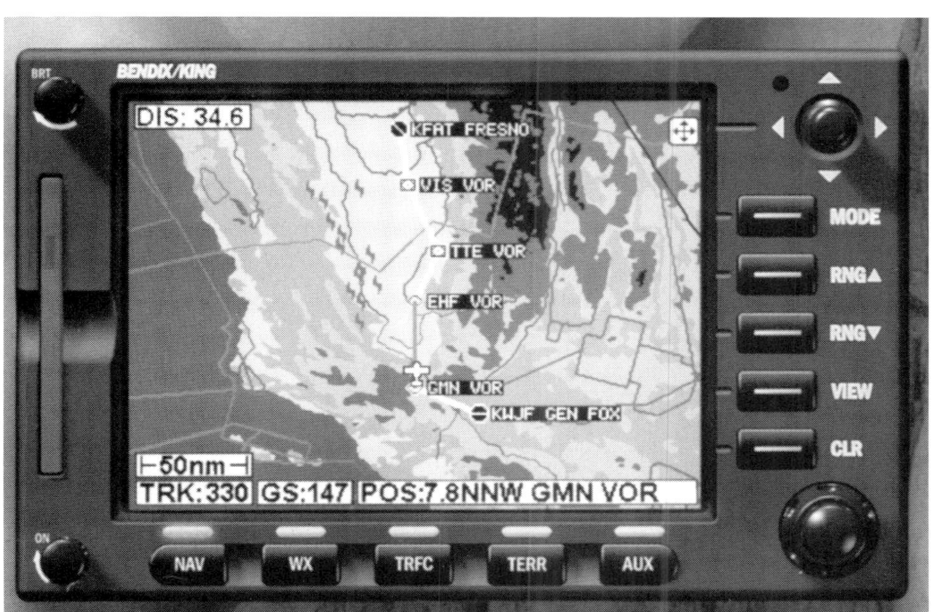

IHAS 5000 integrated hazard warning system
2000/0081845

KAP 100 Silver Crown autopilot

The KAP 100 is a panel-mounted single-axis, wings level, digital flight control system which includes a KG258 horizontal reference indicator and KG107 directional compass. An optional slaved compass system can be substituted for the latter item. Options include manual electric trim, control wheel steering and a yaw damper. Lateral modes include heading select, navigation tracking, approach and localiser back course modes. The system was announced in June 1982 and is certified on several single-engined aircraft types.

Specifications

KAP 100 (KG 107 directional gyro, no yaw damper)
Weight: 4.9 kg
Power supply: 14 V DC, 3.1 A or 28 V DC, 1.6 A

Operational status

In service.

Contractor

Honeywell Aerospace, Electronic & Avionics Lighting.

VERIFIED

KAP 150 Silver Crown autopilot

The KAP 150 is a panel-mounted two-axis digital autopilot providing pitch and lateral control facilities. It is integrated with standard flight instrument packages which are available from the company's range of products. Slaved compass and remote mode annunciator options are provided. Autopilot modes include pitch hold, heading select, altitude hold, navigation tracking, approach, glideslope and localiser back course, vertical trim and control wheel steering. A yaw damper is available as an optional extra. The system is certified on several single-engined aircraft types.

Specifications

KAP 150 (KG 107 directional gyro, no yaw damper)
Weight: 8.2 kg
Power supply: 14 V DC, 5.1 A
28 V DC, 2.5 A

Operational status

In production and in service.

Contractor

Honeywell Aerospace, Electronic & Avionics Lighting.

VERIFIED

KAP 150H helicopter digital flight control system

The KAP 150H two- or three-axis autopilot is designed to meet the needs of single-engine turbine-powered helicopter operators in emergency medical transport, law enforcement, pipeline patrol or a variety of other uses. A compact panel-mounted system, the KAP 150H is engineered for easy console installation.

A derivative of the KAP 150, the KAP 150H autopilot operates in heading select, altitude hold, vertical trim and control wheel steering modes, providing workload reduction for a single pilot during VFR cruise operation. The KAP 150H also reduces pilot fatigue, enhancing safety and delivering a smooth ride for passengers.

System options include a yaw axis and either a standard directional gyro or the AlliedSignal KCS 55A compass system, an electrically slaved 76 mm (3 in) horizontal situation indicator.

Specifications

Weight: 9.07 kg
Power supply: 6.5 V AC, 3.5 V DC

Operational status

Certified for the Bell 206B JetRanger and LongRanger series of helicopters.

Contractor

Honeywell Aerospace, Electronic & Avionics Lighting.

VERIFIED

KFC 150 Silver Crown autopilot/ flight director system

The KFC 150 is a KAP 150 autopilot with a (76 mm) air-driven gyro, flight instrument package. Single-cue V-bar presentation is used on the flight director. Also included in the package is a KCS 55A slaved compass system with KI 525A pictorial navigation indicator. The KFC 150 provides pitch attitude hold, altitude hold, flight director, heading select, navigation, approach, glideslope, back course, vertical trim and control wheel steering.

Specifications

KFC 150 (without yaw damper)
Weight: 11.5 kg
Power supply: 14 V DC, 8.7 A or 28 V DC, 4.4 A

Operational status

In production and in service.

Contractor

Honeywell Aerospace, Electronic & Avionics Lighting.

VERIFIED

KFC 200 Silver Crown autopilot/ flight director system

The KFC 200 automatic flight control system is suitable for a wide range of single- and twin-engined light aircraft. It comprises a two-axis autopilot with flight director instruments and can be configured as the full KFC 200 or the lower-cost KAP 200 system. Both variants have a two-axis autopilot, the low-cost option providing wings-level and pitch attitude hold with altitude, navigation, approach, localiser, back course and heading select modes. The larger variant additionally includes go-around and control wheel steering modes.

Flight instrumentation options include the KG 258 flight command indicator and KI 525A horizontal situation indicator for the low-cost KAP 200, or the same horizontal situation indicator with a KI 256 flight command indicator in the more comprehensive KFC 200 system. Manual electric pitch trim facilities are included and options include slaved gyro and yaw damper installations.

Specifications

KFC 200 (without yaw damper)
Weight: 12.8 kg
Power supply: 14 V DC, 15.5 A or 28 V DC, 9.5 A

Operational status

In production.

Contractor

Honeywell Aerospace, Electronic & Avionics Lighting.

VERIFIED

KFC 250 Gold Crown autopilot/ flight director

The KFC 250 system is effectively the computation and control function of the Silver Crown KFC 200 in conjunction with the (108 mm) KFC 300 flight director. A solid-state computer generates flight director commands in parallel with three-axis autopilot control signals. The system comprises a KAP 315 mode annunciator, KCL 310 flight director, KPI 552 pictorial navigation indicator (essentially a horizontal situation indicator), KAS 297 altitude selector, KC 290 mode selector and KC 291 yaw mode controller.

Specifications

Weight: 20 kg
Power supply: 115 V AC, 400 Hz, 80 VA
28 V DC, 11 A
26 V AC, 400 Hz, 42 VA

Operational status

In production for high-performance piston twins and medium-size business turboprops.

Contractor

Honeywell Aerospace, Electronic & Avionics Lighting.

VERIFIED

KFC 325 digital flight system showing (top left) the EFS 40 EADI, (bottom left) the EFS 40 EHSI and (right) the KMC 321 digital mode selector

KFC 275 flight control system

The KFC 275 digital flight control system is designed primarily for piston twins. This system uses the same KCP 220 autopilot computer with four microprocessors and the same KDC 222 air data sensor as the KFC 325. However, the KFC 275 uses a different flight instrument system and (76 mm) electromechanical instruments, including the KI 256 V-bar flight command indicator and the KCS 55A slaved pictorial navigation indicator system.

The KMC 221 mode controller provides mode selection and annunciation for most modes, along with the KAP 185A mode annunciator. Annunciations are provided for both armed and coupled modes when appropriate.

Like the KFC 325, the KFC 275 can be configured with optional altitude/vertical speed preselect. There is, however, a choice of either the KEA 130A three pointer encoding altimeter or the KEA 346 counter-drum pointer servoed altimeter to go with two different versions of the KAS 297 altitude/vertical speed preselector.

Operational status

In production.

Contractor

Honeywell Aerospace, Electronic & Avionics Lighting.

VERIFIED

KFC 325 digital flight control system

The KFC 325 is a three-axis digital flight control system designed to suit high-performance turboprop aircraft. An extensive preflight test of full-time monitors ensures system integrity while airspeed-compensated control maintains proper response to changing aircraft configurations. Manual electric trim speed is also adjusted to aircraft speed.

The remote-mounted flight computer contains four microprocessors, with one dedicated to each of the following functions: roll, pitch, yaw damp and logic. In addition to providing these computations, the microprocessors provide extensive preflight test of pitch, roll and accelerometer monitors which ensures system integrity during flight operations.

The KFC 325 is configured with a 102 mm (4 in) electromechanical ADI and HSI with growth provisions for interface with the EFS-10, EFS-40 and EFS-50 electronic flight instrument system EADI and EHSI displays. The electromechanical instruments include the KCI 310A rotating sphere flight command indicator and the KPI 553A pictorial navigation indicator with radiate of DME distance, groundspeed and TTS plus radar altitude and distance from 1,000 ft AGL to touchdown.

In addition to such standard modes as altitude hold (ALT), heading select (HDG), nav (VOR/RNav), approach (APR), glideslope (GS), reverse localiser (BC), control wheel steering (CWS), indicated airspeed (IAS), hold and yaw damp (YD), the KFC 325 also has the standard comfort modes of soft ride and half bank. Altitude/vertical speed preselect is optional. The optional KAS 297C altitude and vertical speed selector is a panel-mounted unit which can interface with a KEA 346 counter-drum pointer servoed altimeter.

The KFC 325 also includes as standard equipment the KDC 222 air data sensor which provides altitude and airspeed as well as normal and sideslip inputs to the flight control system.

The servos use capstan assemblies which may be left in the aircraft should servo repair be required, thus allowing the aircraft to remain rigged.

Operational status

In production and in service.

Contractor

Honeywell Aerospace, Electronic & Avionics Lighting.

VERIFIED

KFC 400 digital flight control system

Intended for turboprop and turbine-powered general aviation aircraft, the KFC 400 is a dual-channel system designed to fail-passive. The system is microprocessor driven and when used in conjunction with the KNS 660 navigation and frequency management system can provide full three-dimensional flight guidance with automatically scheduled climb and descent profiles.

The system includes a digital air data computer providing altitude, airspeed, vertical speed and Mach number via an ARINC 429 digital databus, and it can also drive a range of electromechanical flight instruments. Continuous fault monitoring is incorporated.

Operational status

In production. It has been certified as part of the avionics suite on the Beechjet 400.

Contractor

Honeywell Aerospace, Electronic & Avionics Lighting.

VERIFIED

KFC 400 digital flight control system

KHF 950 HF radio

The KHF 950 has been designed specifically for fixed-wing aircraft. The system offers a choice of control head to suit the user's requirement: the KCU 1051 with Automatic Link Establishment (ALE), the KCU 951 all-digital remote controller, or the smaller-sized KFS 951. The system uses the remote KAC 952 power amplifier/antenna coupler, and the remote KTR 953 receiver exciter. Both the KAC 952 and KTR 953 are designed to operate up to 55,000 ft in an unpressurised environment when using a grounded antenna.

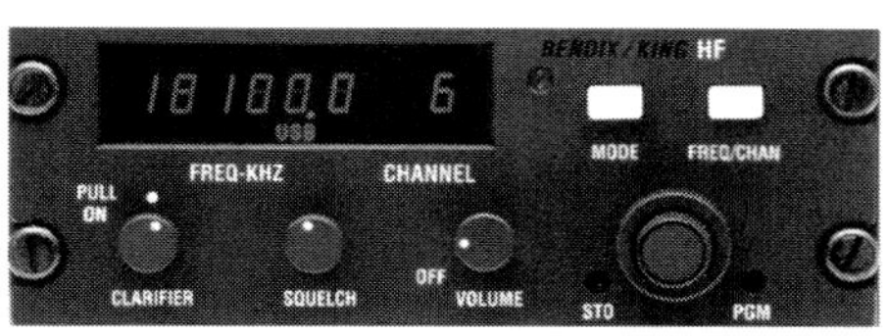

The dzus-mounted KCU 951control/display for use with the KHF 950 HF radio **2000**/0081851

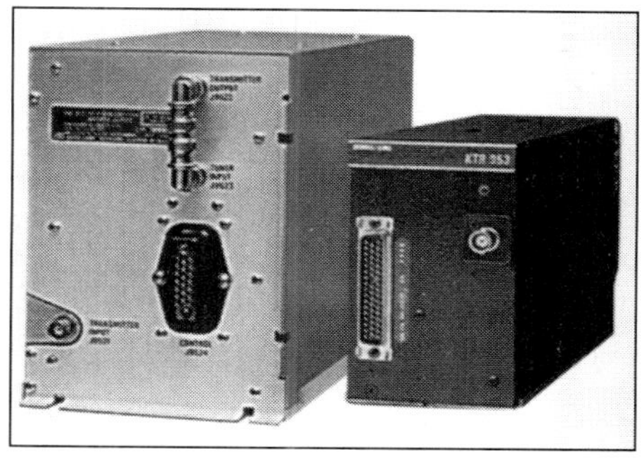

The KAC 952 power amplifier and antenna coupler, and the remote KTR 953 receiver/exciter used in the KHF 950 HF radio **2000**/0081852

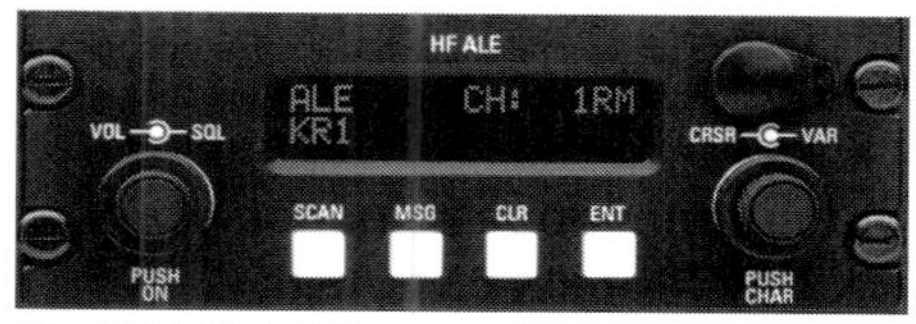

The KCU 1051 ALE control display unit used in the KHF 950 ALE and KHF 990 ALE HF radios **2000**/0081853

Frequency coverage is from 2.000 to 29.999 MHz and offers 280,000 frequencies at 100 Hz spacing. The system operates in USB, LSB and AM modes. Transmitter output in each SSB mode is 150 W PEP and 35 W average over the full frequency range.

The KHF 950 employs synthesised frequency generation techniques and uses microprocessor control for easy in-flight operation. The KCU 1051 searches for and finds the best HF frequency to use, while providing the following system features: 100 preset channel memory, fast and simple manual frequency tuning, memory that keeps track of the antenna tuning, data transmission and compliance with US federal ALE standards.

Operation may be either simplex, for normal air traffic or similar communication, or semi-duplex which permits patch-through into public utility telephone circuits. Provision is also made for a SELCAL facility and the dedicated circuits enable continuous SELCAL monitoring to be maintained without having to select the AM mode.

A feature of the KHF 950 is its automatic antenna tuning capability, an operation carried out by simply keying the microphone. The system will operate satisfactorily on antennas only 10 ft long. It will also tune to fixed-rod aerials and towel-rail antennas, and can operate from shunt and notch antenna systems.

An optional controller is the KFS 954 which is also panel-mounted and measures only 14.5 cm^2. This unit contains storage for all 176 ITU maritime radiotelephone channels plus additional preselected simplex air traffic or conventional airborne communication channels, but offers only 19 preset channels. By preprogramming the ITU channels, it is possible for the operator to call any radiotelephone station without having to select the separate transmit and receive channels manually. The operator merely selects the radiotelephone mode and the required channel. The KHF 950 also interfaces with teletype and facsimile systems and a dual-installation equipment allows dual-frequency reception from a single antenna.

Specifications

Weight: 9.16 kg

Operational status

In production and in service. The system has been widely adopted and has been installed in aircraft such as Bombardier Challenger, Gulfstream G III, Lear 55, Citation III and Falcon 50. The KHF-950 has also been selected by the US Army for helicopters.

Contractor

Honeywell Aerospace, Electronic & Avionics Lighting.

VERIFIED

KHF 970 HF radio

The KHF 970 is a derivative of the KHF 950 HF/SSB airborne radio which has been adapted specifically for military use. Selection of the system for nap of the earth helicopter communications, an application for which HF is now recognised as being superior to VHF, was announced by the US Army in 1982. A surface vehicle version for military use has been designated KVR 980.

The KHF 970 covers the HF band from 2 to 29.999 MHz, providing 280,000 channels at 100 Hz increments. Transmitted output power is 150 W. Preset channel selection from a non-volatile memory is also provided but, unlike the KHF 950, all channels and frequencies are selected on a keyboard and displayed on a CRT.

Other features include scanning of preset channels and provision for multiple selective addressing, frequency link analysis and for automated communications operating instructions.

Operational status
In production.

Contractor
Honeywell Aerospace, Electronic & Avionics Lighting.

VERIFIED

KHF 990 HF radio

The KHF 990 HF radio is a helicopter system which draws on technology used in the company's airborne KHF 950 and marine KMC 95 systems. It provides 280,000 channels in the 2 to 30 MHz HF band at frequency increments of 100 Hz. Modulation is in SSB mode and transmitted output power is 150 W PEP.

The KHF 990 has been optimised for helicopter operation. It uses the miniature KFS 594 controller, a KAC 992 combined antenna coupler/probe antenna and a remotely located KTR 993 receiver/exciter/power amplifier. This combination provides a fully capable yet lightweight system. Like the KHF 950, this system has also been updated to offer Automatic Link Establishment (ALE) using the KCU 1051 control display unit.

The KFS 594 controller provides access to 176 permanently programmed ITU marine radiotelephone channels and to 19 programmable channels which may be selected or retuned by the pilot. The KAC 992 is an automatic, digital antenna coupler which is self-contained in the end of a probe antenna system. It may be mounted externally or internally with only the probe portion of the antenna protruding from the aircraft.

The KTR 993 receiver/exciter/power amplifier can be mounted in any convenient location within the helicopter with no restrictions on proximity to the other two units. It meets the TSO requirements for explosion-proof, drip-proof and salt-spray categories.

The KSU 1051 provides automatic selection of the best available frequency, 100 preset channel memory, fast and simple manual frequency tuning, memory that keeps track of antenna tuning, data transmission and compliance with US federal ALE standards.

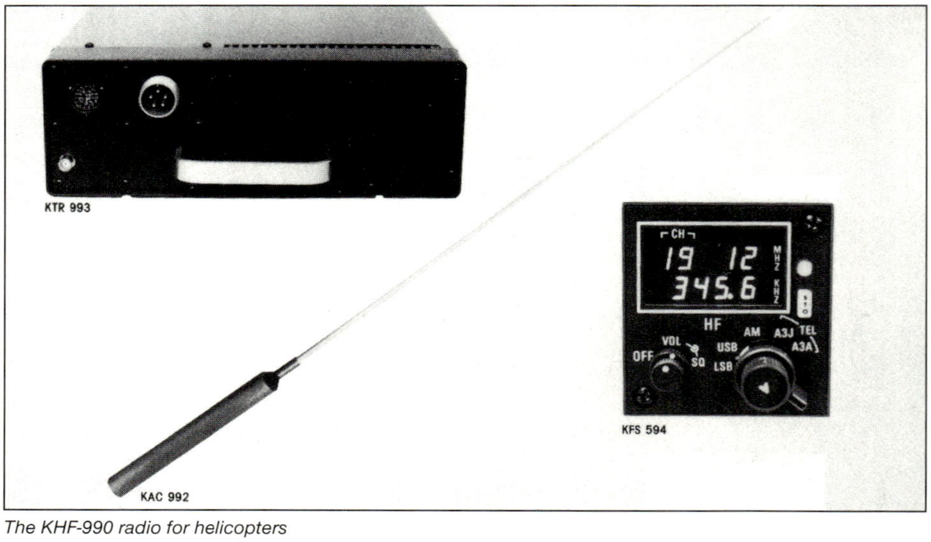

The KHF-990 radio for helicopters

Specifications
Weight: 9.9 kg

Operational status
In production.

Contractor
Honeywell Aerospace, Electronic & Avionics Lighting.

VERIFIED

KLN 35A GPS/KLX 135A GPS/COMM systems

The panel-mounted KLN 35A GPS incorporates moving map graphics, a high-visibility display and a choice of customised Jeppesen NavData databases - including coverage for America, Atlantic and Pacific areas.

The moving map, useful for providing situational awareness, displays Special-Use Airspace (SUA) boundaries and provides SUA alerting. All three databases contain appropriate Flight Service Station (FSS) and Air Route Traffic Control Center (ARTCC) frequencies, and airport runway data.

The KLN 35A presents this information via an advanced double super-twist nematic LCD, offering improved viewing in direct sunlight and extended side-to-side visibility.

The KLX 135A offers greater capability. Incorporating all the same GPS performance and features, the KLX 135A GPS/COMM also integrates a TSO'd, 760-channel, Very High-Frequency (VHF) communications radio with its navigation functions.

A new capability developed specifically for the KLX 135A, QuickTune, allows the pilot to enter the standby COMM frequency directly from the GPS database, saving effort and reducing the chances of making an entry error.

Specifications
KLN 35A GPS
Dimensions: 158.7 × 50.8 × 289.1 mm
Weight: 0.94 kg
Power requirements: 11-33 V DC

KLX 135A GPS/COMM
Dimensions: 158.7 × 50.8 × 289.1 mm
Weight: 2 kg
Power requirements: 14 V DC (28 V DC with available KA 39 voltage converter)
VHF communications transceiver transmitter power: 5 W (min) (7 W (nominal))

Contractor
Honeywell Aerospace, Electronic & Avionics Lighting.

VERIFIED

KLN 88 Loran navigation system

The KLN 88 is a multichain Loran navigation system certified to TSO C60b standards. It tracks up to eight Loran stations in up to four different chains simultaneously.

The database, covering the United States, Canada, Mexico, Central America and the Caribbean, contains over 40,000 elements and includes all public use and military airports with runways of 1,000 ft or longer, VORs, NDBs, published intersections and outer markers, plus up to 250 user-defined waypoints. In addition, ARTCC and special use airspace boundaries are outlined. Virtually every airport communications, flight service station, ATIS and navaid frequency in North America is included. The information, provided by Jeppesen, is contained in a cartridge which plugs directly into the back of the KLN 88. Information is updated by direct replacement of the cartridge every 28 days. The 83.8 mm (3.3 in) split-screen CRT display allows two whole pages of data to be viewed simultaneously and includes a built-in moving map graphics facility.

Specifications
Dimensions: 160.3 × 50.8 × 334 mm
Weight: 2.82 kg
Power supply: 11-33 V DC, 2.5 A (max)

Contractor
Honeywell Aerospace, Electronic & Avionics Lighting.

VERIFIED

KLN 89/KLN 89B GPS navigation systems

The KLN 89 is a panel-mounted, eight-channel, GPS-based navigation system with a database that can be updated by the pilot. The KLN 89B adds IFR-certified en route, terminal, and approach capability. A basic system comprises panel-mounted unit, altitude input, and KA 92 antenna. Among additional components that may be added to increase capabilities are an external Course Deviation Indicator (CDI) or HSI; RMI, some Shadin or ARNAV fuel management systems; several external moving maps, and certain Shadin air data systems.

Specifications
Dimensions: 160.3 × 50.8 × 272.3 mm
Weight: 1.16 kg
Power: 11-33 V DC at 2.5 A
TSO (KLN 89B only): C129 Class A1

Contractor
Honeywell Aerospace, Electronic & Avionics Lighting.

VERIFIED

KLN 89/89B GPS navigation systems 0001335

KLN 90B approach-certified GPS navigation system

Together with the capability to perform non-precision GPS approaches, the KLN 90B navigation system offers an easy-to-read CRT map display and a comprehensive Jeppesen database. Designed to meet the FAA's C129 A1 specifications, the KLN 90B features an improved eight-channel parallel GPS receiver for even more reliable satellite tracking. Other enhancements include a more pilot-friendly interface and an expanded database, complete with SID and STAR waypoints and approaches.

Providing all the benefits of satellite-derived input - worldwide coverage, a high degree of accuracy and immunity to atmospheric disturbance - the KLN 90B harnesses the power of GPS to make non-precision approaches easier for the pilot.

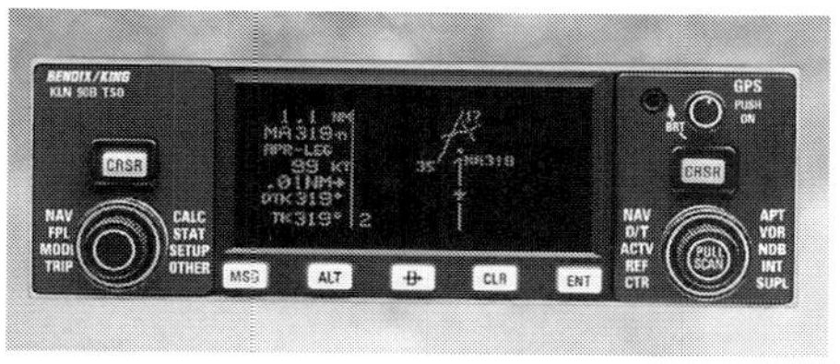

KLN 90B approach-certified GPS navigation system
0001336

Navigation pages were designed specifically for use during approaches, to provide the greatest amount of information with the least amount of effort. In fact, most necessary and en route information is presented concisely and logically on one large, easy-to-read screen. This page also incorporates a moving map, to show progress throughout the approach. Additional approach-specific pages are readily accessed, significantly reducing pilot workload during this critical phase of flight.

KLN 90B database information includes airport data (identifier, name, runway length/surface/lighting, instrument approach availability, customs, types of fuel, oxygen, landing fees, and other data comparable to that found in an airport services guide, and a runway diagram for airports providing published runway threshold co-ordinates); communications frequencies (ATIS, clearance, ground control, tower, CTAF, advisory and various VFR frequencies); VOR information; NDB information; intersection information (data about low altitude, high altitude, approach and SID/STAR intersections, with outer markers and compass locators; ARTCC and FIR boundaries and frequencies); flight service station frequencies and locations; minimum safe altitudes; special-use airspace.

Specifications

Dimensions: 160.3 × 50.8 × 334 mm
Weight: 2.66 kg
Temperature range: –40 to +70°C
Altitude range: up to 50,000 ft
Power inputs: 11-33 V DC at 2.5 A (max)
TSO: C129 A1

Contractor

Honeywell Aerospace, Electronic & Avionics Lighting.

VERIFIED

KLN 94 colour GPS receiver

The KLN 94 colour GPS navigator/moving-map receiver and display is a new product that is compatible with other products in the Silver Crown and Silver Crown Plus range. Designed to save installation cost and space while upgrading to a colour display and replacing both GPS and Loran receivers, it is a direct plug-in replacement for the KLN 89B. It includes a comprehensive aeronautical database including airports, VORs, NDBs, intersections and special-use airspace. It provides IFR GPS capability (TSO C129a A1) for en-route, terminal and non-precision approaches; vector to final approach; range/map capability; special procedures capability; and a quick-tune interface to the KX 155A NAV/COMM.

Specifications

Dimensions: 160.3 (W) × 50.8 (H) × 272.3 (D) mm

Contractor

Honeywell Aerospace, Electronic & Avionics Lighting.

VERIFIED

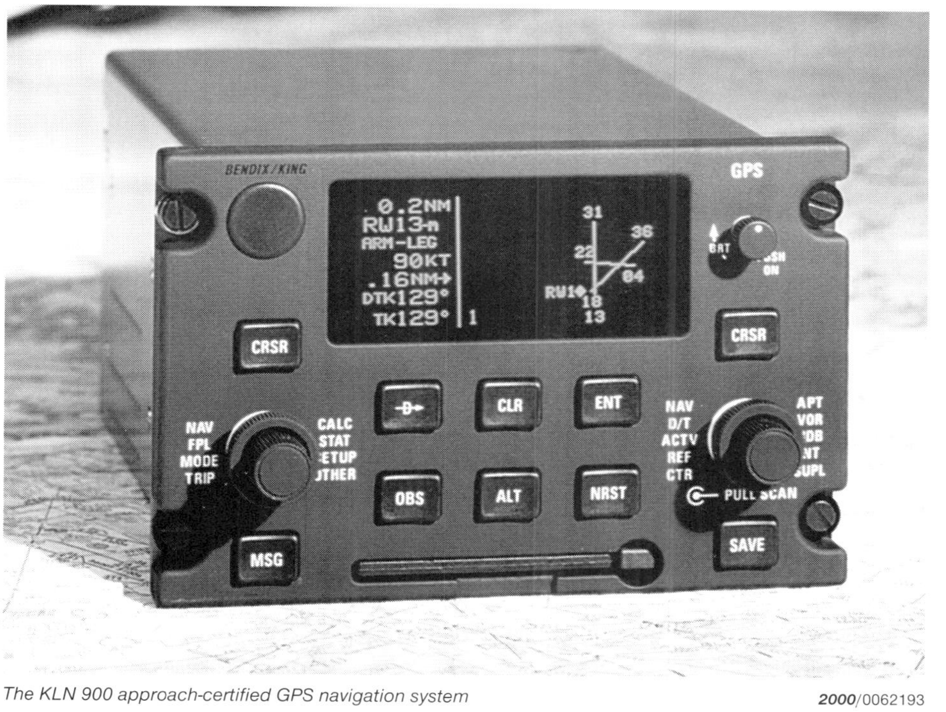

The KLN 900 approach-certified GPS navigation system ***2000***/0062193

KLN 900 approach-certified GPS navigation system

The KLN 900 approach-certified GPS navigation system meets Basic aRea NAVigation (B-RNAV) requirements for European Civil Aviation Conference (ECAC) airspace when used with AlliedSignal's PreFlight™ software version 2.0 or later. Versions of the KLN 900 will also meet primary means oceanic/remote operation requirements.

The KLN 900 is designed for customers who wish to upgrade the navigation capability of their aircraft, replace outdated systems such as LORAN and RNAV equipment, install an affordable back-up or lower-cost alternative to a Flight Management System (FMS).

The KLN 900 is a dzus-mount, eight-channel unit. It can be IFR certified for en-route, terminal and approach GPS-based navigation. The pilot is able to load his database and flight plan data with a front panel diskette. The unit can be upgraded to provide a precision approach capability. It features a monochrome Cathode Ray Tube (CRT) display, and it can handle a wide variety of analogue and digital inputs and outputs.

A basic installation comprises a dzus-mounted KLN 900 unit, database cartridge, and KA 92 antenna. Among the additional components that can be connected directly to increase the KLN 900's capability are an external Course Deviation Indicator (CDI); a Horizontal Situation Indicator (HSI); Radio Magnetic Indicator (RMI); fuel management system; moving-map display; and air data system.

Specifications

Dimensions: 146 (W) × 95 (H) × 240 (L) mm
Weight: 2.0 kg
Power: 11-33 V DC, 3.3 A
TSO: C129 A1

Contractor

Honeywell Aerospace, Electronic & Avionics Lighting.

VERIFIED

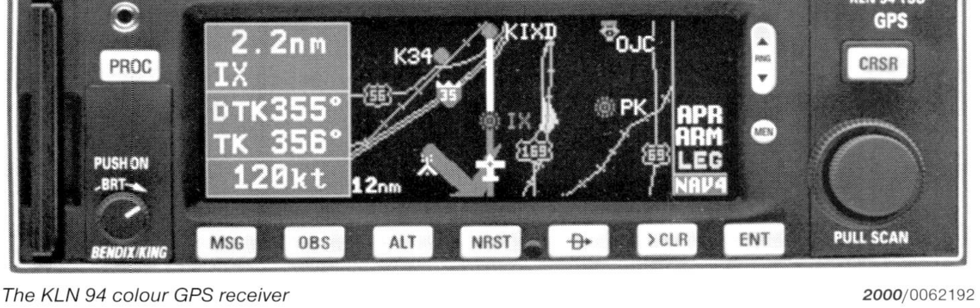

The KLN 94 colour GPS receiver ***2000***/0062192

KMA 20 audio control unit

Part of the Silver Crown panel mount avionics range, the Bendix King KMA 20 is a compact, flip-switch audio control unit providing the functions of marker beacon receiver, isolation amplifier and audio control panel. Designed principally for the general aviation sector, the solid state KMA 20 is fully TSO qualified. The standard model can accommodate up to two transceivers and four external receivers, while the International model can accommodate an extra transceiver. Other system features include:

- isolation amplifier facilitating combination of all receiver audio inputs into a single cockpit speaker with 40 dB of isolation between each radio
- different features and combinations of marker light presentation, AUTO switch, COM/NAV/ADF/HF/DME selections
- available with white or red lighting
- 8 W output power

Specifications

Dimensions: 160 × 41 × 157 mm
Weight: 1.05 kg
Electrical: 27.5 V DC
Compliance: TSO C50b

Operational status

In production and in service.

Contractor

Honeywell Aerospace, Electronic & Avionics Lighting.

NEW ENTRY

KMA 24 audio control system

Part of the Silver Crown panel mount avionics range, the Bendix King KMA 24 is a compact, lightweight system for the integrated control of a number of radio communication and navigation systems. Designed principally for the general aviation sector, the solid state KMA 24 is fully TSO qualified.

The KMA 24 can control up to three transmitter/receivers and six receivers, including an internal marker beacon receiver for which it contains an automatically dimmed three-light presentation. Other system features include:

- pushbutton audio selector panel, speaker and headphone isolation amplifiers and a marker beacon receiver
- can be integrated into any 500 ohm output audio system
- different features and combinations of marker light presentation, AUTO, 2 × COM, 2 × NAV, DME, ADF, HF, TEL selections available

- two unswitched inputs for use as altimeter warning and telephone ringer
- outputs for ramp hailer and passenger address or intercom
- isolation amplifiers are provided for headphones and speaker to provide isolation even when the same source is selected
- three isolated 16 ohm resistors provided for transceiver speaker output loads
- 14V or 28 V DC operation

Specifications
Dimensions: 173 × 33 × 159 mm
Weight: 0.77 kg
Temperature: -20° to +55° C
Electrical: 14 or 28 V DC
Compliance: TSO C35d, C50b

Operational status
In production and in service.

Contractor
Honeywell Aerospace, Electronic & Avionics Lighting.

VERIFIED

KMA 24H audio control system

Part of the Silver Crown panel mount avionics range, the KMA 24H is a variant of the KMA 24 system (see previous entry) where the internal marker beacon facility is replaced by a five-station hot microphone intercom and associated volume control. There are two variants of the KMA 24H, the -50/54 and -70/71. The following summary outlines common system features and differences:

Common Features
- pushbutton audio selector panel with separate speaker and headphone isolation amplifiers
- can be integrated into any 500 Ω output audio system
- interphone communication including five microphone audio inputs for up to five intercom stations
- two unswitched inputs for use as altimeter warning and telephone ringer
- speaker outputs for ramp hailer and passenger address or intercom
- PA mute provided to mute passenger background music systems when a microphone is keyed
- 14 or 28 V DC operation
- TSO qualified

KMA 24H -50/54
- similar to KMA-24 but does not include marker beacon presentation and includes intercom functions/controls
- night vision goggle compatible version available
- designed for use in both single- and dual-audio panel installation
- control up to three transceivers and six receivers

KMA 24H -70/71
- similar to KMA-24H -50/54 but includes voice operated intercom, separate alternate action capability and keyed activation of up to five stations
- capability for pilot and co-pilot to isolate out of the intercom system through separate independent amplifiers
- up to five transceivers and five receivers can be controlled in -71 version
- up to four transceivers and six receivers can be controlled in -70 version
- user can select VOX, keyed or hot mic intercom through adjustments of the VOX control on the front panel
- incorporates summing amplifier to combine received audio with the pilots microphone for installations where a voice recorder is necessary

Specifications
Dimensions: 173 × 33 × 159 mm (-50/54), 165 × 33 × 160 mm (-70/71)
Weight: 0.77 kg
Temperature: –20° to +55°C (+70°C -70/71)
Electrical: 14 or 28 V DC
Compliance: TSO C50b (-50/54), C50c (-70/71)

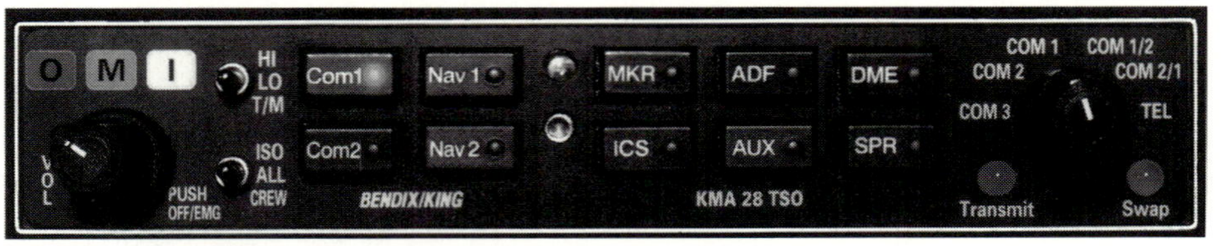

The Bendix King KMA 28 audio control system **2001**/0103877

Operational status
In production and in service.

Contractor
Honeywell Aerospace, Electronic & Avionics Lighting.

NEW ENTRY

KMA 26 audio control system

Part of the Silver Crown panel mount avionics range, the Bendix King KMA 26 is an audio control unit with inputs for up to three transceivers, nine switched audio inputs and four unswitched inputs. Further features of the system are:
- audio amplifier
- six-station intercom with three modes - pilot, crew, and all
- self-contained marker beacon receiver with 3-lamp display
- high and low sensitivity/light test switch for marker lamps
- marker mute button to temporarily suppress marker audio while passing over a beacon
- individual volume controls for crew and passengers
- four separate microphone squelch circuits to reduce background noise
- self-contained mode switching
- Emergency (EMG) position on MIC selection switch connects microphone and headphones directly to COMM 1 in the event of a failure
- accepts inputs from two music entertainment sources; entertainment audio is automatically muted when communication audio is detected.

Specifications
Dimensions: 191 × 33 × 159 mm
Weight: 0.77 kg
Altitude: up to 50,000 ft
Compliance: TSO C35d, C50c

Operational status
In production and in service.

Contractor
Honeywell Aerospace, Electronic & Avionics Lighting.

NEW ENTRY

KMA 28 audio control system

Part of the Silver Crown panel mount avionics range, the Bendix King KMA 28 is an audio control unit with five receiver (2 NAV, ADF, DME and AUX) and four transceiver inputs (3 COM and one approved cell phone), together with a six-place stereo intercom. Further features of the system are:
- audio selector panel with 6-place intercom
- 'IntelliVox' feature samples ambient noise and adjusts threshold for clip-free voice communications
- dual high-fidelity stereo inputs
- two split modes (COM 1/2 and 2/1)
- 3-light marker beacon
- stereo entertainment inputs

Specifications
Dimensions: 173 × 33 × 159 mm
Weight: 0.5 kg
Altitude: up to 50,000 ft
Temperature: –22° to +55°C
Electrical: 11 to 33 V DC, 1.5 A
Compliance: TSO C35d, C50d, RTCA DO-160C, -178B level D, -143, -214

Operational status
In production and in service.

Contractor
Honeywell Aerospace, Electronic & Avionics Lighting.

NEW ENTRY

KMD 150 colour multifunction display/GPS

The KMD 150 colour multifunction display incorporates a built-in GPS receiver. It forms part of the Bendix/King Silver Crown Plus™ system. The KMD 150 includes an aeronautical database, together with a comprehensive roads and rivers cartographic package, and an interface for Stormscope 500 radar data. It includes a joystick control for data entry and more information request.

A variant that does not include the GPS is available for customers wishing to integrate with an existing GPS capability.

The KMD 150 colour multifunction display/GPS **2000**/0062195

Specifications

Screen size: 5 in (127 mm) diagonal
Dimensions: 158 (W) × 101 (H) × 175 (D) mm
Weight: 1.5 kg
Power: 10-33 V DC, 10 W
Environmental standard: DO 160C

Contractor

Honeywell Aerospace, Electronic & Avionics Lighting.

VERIFIED

KMD 550/KMD 850 MultiFunction Displays (MFDs)

The KMD 550 is a 4 in, full-colour, MFD designed for piston- and light turbine-engined aircraft that are not equipped with radar. The KMD 850 also includes a digital and analogue weather radar interface, for use in aircraft so fitted.

Both units provide a moving-map function that displays airports, special-use airspace, DMEs, VORs, and other features such as terrain elevation.

It is also possible to interface the display to datalinked information, and thus to upgrade the onboard capability to the Integrated Hazard Avoidance System (IHAS) standard, due to be introduced by Honeywell Aerospace in 2000.

IHAS capability allows datalink to the display of weather information, terrain information deriving from EGPWS and air traffic information deriving from Mode S sources, to provide comprehensive situational awareness.

The KMD 550 MFD is utilised in the IHAS 5000 offered for general aviation users, while the KMD 850 is fitted to IHAS 8000 available for business aviation aircraft.

Contractor

Honeywell Aerospace, Electronic & Avionics Lighting.

VERIFIED

KN 62A/63/64 digital DMEs

The compact KN 62A digital DME is a fully self-contained 200-channel system in the Silver Crown range. It features an all-solid-state transmitter and four LSI chips and requires only 33 mm of panel height.

The KN 62A can be channelled remotely through almost any navigation equipment receiver, or tuned directly with its own frequency selection knobs. Dual channelling makes two DME frequencies available at all times. The self-dimming digital display provides information on simultaneous DME distance, groundspeed and time-to-station, or DME distance and internally selected frequency.

The KN 62A meets the US FAA's TSO performance and environmental standards; the KN 64 offers similar performance but is not TSO'd.

The KN 63 is also TSO'd, and can be integrated with the KNS 81 integrated Nav/RNav to increase sophistication.

Specifications

Weight: 1.18 kg
Power supply: 11-33 V DC, 15 W

Contractor

Honeywell Aerospace, Electronic & Avionics Lighting.

VERIFIED

KNR 665 Gold Crown integrated area navigation system

A member of the Gold Crown family of avionics for the business, corporate turboprop and jet upper end of the general aviation spectrum, the KNR 665 can operate in conjunction with the KFC 300 flight director and auto-pilot system.

The KNR 665 can memorise and display up to 10 waypoints and/or VORTac stations. Waypoint information in Jeppesen format includes VORTac frequency, outbound and reciprocal courses and waypoint bearing and distance. All waypoint information is displayed simultaneously for ease of reference. Individual parameters for any waypoint can be changed without resetting other data; the information is inserted by push-button and checked on a scratchpad before being entered into the system. An autocourse facility provides air traffic control clearance to a particular waypoint by pressing the 'autocourse' button, so that the route to that particular waypoint is instantly computed and displayed. The new course is automatically displayed on the horizontal situation indicator. The system can be checked by a push-button.

Operational status

In service.

Contractor

Honeywell Aerospace, Electronic & Avionics Lighting.

VERIFIED

KNS 80 Silver Crown integrated navigation system

The KNS 80 comprises a 200-channel VOR/Loc receiver, 200-channel digital DME, 40-channel glideslope receiver and digital RNav computer that can store up to four VOR/Loc frequencies and waypoints. Information is displayed on a full-width light-emitting diode display, and the system can operate in conjunction with an ARINC horizontal situation indicator or course deviation indicator. Extensive use is made of Large Scale Integration (LSI) technology, the single LSI chip accomplishing the work that would have required 40 conventional integrated circuits. As with other members of the Silver Crown family, the KNS 80 is aimed at the lower end of the general aviation sector.

Specifications

Dimensions: 160 × 76 × 305 mm
Weight: 2.7 kg
Power supply: 11-33 V DC, 25 W
Distance to next waypoint: 199.9 n miles in 0.1 n mile increments, 0.1° angle increments selectable on display

Contractor

Honeywell Aerospace, Electronic & Avionics Lighting.

VERIFIED

KNS 81 Silver Crown integrated navigation system

A development of the KNS 80, the KNS 81 integrated navigation system can accommodate up to nine waypoints and embodies a number of new features. A remotely mounted DME enables a KDI 572 indicator to be positioned directly in front of the pilot, providing simultaneous digital readouts of distance, groundspeed and time to either a VORTac station or an RNav waypoint. There is simultaneous display of bearing, distance and frequency waypoint details on the KNS 81 panel for easy programming and updating of navigation information. A radial push-button permits a rapid bearing check to a chosen VORTac or RNav waypoint, displayed on the DME panel indicator in place of groundspeed and time to next waypoint. A check push-button permits a rapid cross-check of bearing and distance from the VORTac without disturbing other navigation instrument settings. A radio magnetic indicator output to a KI 229 or KNI 582 indicator gives an accurate bearing to a selected RNav waypoint or VORTac station.

Specifications

Dimensions: 160.3 × 5.1 × 291.2 mm
Weight: 2 kg
Power supply: 11-33 V DC, 15 W
Number of waypoints: 9
Distance to next waypoint: 199.9 n miles in 0.1 n mile increments; 0.1° angle increments

Contractor

Honeywell Aerospace, Electronic & Avionics Lighting.

VERIFIED

KNS 660 flight management system

Aimed at the turbine-powered corporate and commuter sector of general aviation, KNS 660 is the designation for a family of great-circle navigation management systems to work in conjunction with Gold Crown III avionics and other compatible units.

The KNS 660 has its own database, ARINC 429 input/output formats, analogue processing, air data function, vertical navigation computation, frequency management and optional Omega/VLF sensor. It has a choice of two control/display units, the 152 mm (6 in) KCU 568 or the 114 mm (4.5 in) KCU 567 for installations with limited space. The KNS 660 will handle signals from VOR/DME, Omega/VLF, Loran C, INS, GPS, AHRS (Attitude Heading and Reference System) and compass. The KTU 709 Tacan system can replace the DME if required.

An optional receiver/processor providing a choice between Omega/VLF and Loran C is now available. A potential growth area is MLS.

Operational status

Production began in July 1984. The KLN 670 GPS Navstar sensor was developed and certified as part of the KNS 660 system in 1987.

Contractor

Honeywell Aerospace, Electronic & Avionics Lighting.

VERIFIED

KR 86 digital ADF

The all-solid-state KR 86 ADF offers positive and precise crystal digital tuning between 200 and 1,750 kHz and includes BFO for LF stations which broadcast an unmodulated signal. Constant audio output over wide variations of signal strength is assured by automatic gain control. The KR 86 is compact and self-contained and is easily fitted in a standard panel cutout. Self-test is provided.

Contractor

Honeywell Aerospace, Electronic & Avionics Lighting.

VERIFIED

KR 87 ADF system

The basic KR 87 system includes the KR 87 receiver, KI 227 indicator, KA 44B combined loop and sense antenna, and mounting racks and connectors. The all-solid-state receiver operates on DC voltages between 11 and 33 V and draws only 12 W of power. Electronic timers in the KR 87 provide aids to flight management as a flight or elapsed timer.

The KR 87 features a crystal filter for better long-range reception, coherent detection circuitry to lock on to weak stations, electronic tuning using a microprocessor and an LSI single-crystal digital frequency synthesiser circuit, EAROM non-volatile storage of frequencies during shutdown or power interruptions and a fold-out modular construction for easy access to all circuits and components. The display is self-dimming.

Contractor

Honeywell Aerospace, Electronic & Avionics Lighting.

VERIFIED

KRA 10A radar altimeter

The KRA 10A radar altimeter is a low-cost system suitable for independent use or in combination with King Silver Crown avionics equipment and tailored to general aviation requirements. There is a standard facility for presetting decision height which produces a visual and aural warning on reaching the set altitude. Antennas suitable for flat and sloping skin installations are available. The KRA 10A is an all-solid-state system with short warm-up time and can be fitted with an auxiliary output to interface with flight director and autopilot installations.

Specifications
Dimensions:
(indicator) 100 × 83 × 83 mm
(transmitter/receiver unit) 79 × 89 × 203 mm
(antenna) approx 100 × 100 mm aperture
Weight:
(indicator) 0.4 kg
(transmitter/receiver unit) 0.9 kg
(antenna) 0.4 kg
Power supply: 28 V DC, 6 VA
Altitude: up to 2,500 ft
Accuracy:
(0-100 ft) 5 ft
(100-500 ft) 5%
(>500 ft) 7%

Operational status
In production.

Contractor
Honeywell Aerospace, Electronic & Avionics Lighting.

VERIFIED

KRA 405 radar altimeter

Part of the King Gold Crown avionics range, the KRA 405 is an all-solid-state radar altimeter suitable for twin-engine general aviation and regional airliner types. The KRA 405 interfaces with King KPI 553A HSI and KFC 300 autopilot to give smooth tracking of the glideslope beam. It can provide indications from 2,000 ft above ground level and a usable output is available from 2,500 ft above ground level for ground-proximity warning system operation. Separate transmit and receive horn antennas are used.

Specifications KRA 405
Dimensions:
(indicator) 83 × 83 × 170 mm
(transmitter/receiver unit) 83 × 133 × 296 mm
(antennas) each 178 mm diameter
Weight:
(indicator) 0.8 kg
(transmitter/receiver unit) 2.9 kg
(antennas) 1.2 kg total
Power supply: 28 V DC, 24 VA
Frequency: 4,300 MHz
Altitude: up to 2,500 ft
Accuracy:
(0-500 ft) 5%
(>500 ft) 7%

The KRA 405B updates the KRA 405 by reducing the number of primary circuit boards from seven to two, reducing overall weight of the receiver/transmitter by 50 per cent, and adding updated software. System elements are KRA 405B receiver/transmitter; KNI 415 (fixed-wing) or KNI 416 (rotary-wing) indicator; two KA 54A antennas; optional CM2000 configuration module.

The KTR 909B transceiver and KFS 599B control head **2000**/0081849

Specifications KRA 405B
Dimensions:
KRA 405B: 279 × 76 × 90 mm
KNI 415/416: 170 × 83 × 83 mm
KA 54A: 93 × 89 × 19 mm
Weight:
KRA 405B: 1.36 kg
KNI 415/416: 0.77 kg
KA 54A: 0.09 kg (each)
Altitude: up to 2,500 ft
Accuracy:
(0-500 ft) ±5 %
(>500 ft) ±7%

Operational status
The KRA 405B is in production.

Contractor
Honeywell Aerospace, Electronic & Avionics Lighting.

VERIFIED

KTR 908 VHF transceiver

The KTR 908 is a remote-mounted airborne VHF communications transceiver with a standard frequency range of 118 to 135.975 MHz and channel spacing of 25 kHz. An optional extension to 151.975 MHz is available. Power output is 20 W from a 28 V DC power supply and storage of active and standby frequencies is provided. It is operated from a cockpit-mounted KFS598 controller which requires 57 mm panel space.

Specifications
Dimensions:
(transceiver) 147 × 45 × 299 mm
Weight:
(transceiver) 1.6 kg

Operational status
In production and in operational service with the US Army and National Guard.

Contractor
Honeywell Aerospace, Electronic & Avionics Lighting.

VERIFIED

KTR 909/909B UHF transceivers

The all-solid-state KTR 909 transceiver operates in the 225 to 399.975 MHz range in 25 kHz increments and is capable of 10 W of transmitter power. Weighing only 2.09 kg, the compact system includes the KTR 909 remote-mounted transceiver and the Gold Crown III KFS 599A control head. The KFS 599A is offered in two versions: with standard gas discharge display or with ANVIS NVG-compatible display.

The KFS 599A can be tuned by dialling in the desired operating frequency or by selecting any of the 20 user-programmable channels.

The KTR 909 is capable of operating with tandem KFS 599A control heads designated as master and slave, which is ideally suited for training applications. Additionally, one version of the KTR 909 can be tuned via compatible radio management systems, such as the RMS 555 or FMS systems with frequency management.

Dual-monitoring capability allows the KTR 909 to monitor either the main receiver, the Guard receiver or both simultaneously.

The KTR 909 offers 1,000 Hz tone modulation used in DF operations. In addition, it provides an ADF mode which allows the unit to perform the tuning function for remote ADF systems. This function, when interfaced with peripheral DF equipment, permits an ADF indicator to be used as a bearing indicator for DF operations.

The KTR 909B differs from the KTR 909 in that it incorporates a dedicated guard frequency, monitoring 243 MHz via a dedicated second receiver, while the KTR 909 scans other frequencies when the unit is not in use. The KTR 909B incorporates an ARINC 429 bus interface, and is controlled by the KFS 599B control head.

Specifications
RF power: NTL 8 W (10 W nominal)
Frequency range: 225.000 - 399.975 MHz
Channel spacing: 25 kHz

Operational status
In production and in service.

Contractor
Honeywell Aerospace, Electronic & Avionics Lighting.

VERIFIED

KTU 709 Tacan

Particularly suited to corporate aircraft, where weight, cost and power requirements are notably important, the KTU 709 Tacan transmitter/receiver is based on Honeywell Aerospace's extensive DME experience and the extensive use of Large Scale Integration (LSI) technology; the system specifically is based on the new-generation KDM 706 distance measuring equipment. This Tacan provides bearing, slant range, range rate and time to station or waypoint information to a KDI 572 control/indicator unit. Transistors provide a 250 W peak-to-peak output for a typical range of 250 n miles and the system covers 252 channels; all tuning is done electronically, using a digital frequency synthesiser designed around a Honeywell Aerospace LSI chip.

Specifications
Dimensions: 76.2 × 127.0 × 260.4 mm
Weight: 2.6 kg
Frequency:
(transmit) 1,025-1,150 MHz
(receive) 962-1,213 MHz
Number of channels: 250
Reliability: 2,000 h MTBF design

Contractor
Honeywell Aerospace, Electronic & Avionics Lighting.

VERIFIED

KX 125 nav/com system

The self-contained and fully TSO'd Bendix King KX 125 is compact nav/com system which utilises a back-lit LCD for its display. It operates on all 760 comm frequencies, from 118 to 136.975 MHz in 25 kHz steps. These are displayed on the left display for both active and standby frequencies. The nav receiver offers 200 VOR/Loc frequencies in 50 kHz steps. These are displayed on the right display for both active and standby frequencies. The KX 125's middle display shows the course deviation when the CDI mode is selected. In the bearing mode, the middle window displays three-digit bearing-to-station information. Selecting the radial mode allows radial information to be shown.

Whenever a transmitter has been activated continuously for more than 35 seconds, the unit reverts to receive mode and the comm frequency displays flash to alert the pilot to a stuck microphone.

A built-in audio amplifier is standard. Other features include autopilot interfacing and channeling of separate DME and glideslope receivers. The KX 125 can also drive an external CDI.

Specifications
Dimensions: 160 × 520 × 264 mm
(6.352 × 2.052 × 10.380 in)
Weight: 1.76 kg (3.88 lb)
Electrical: 14 V DC (can be operated on 28 V DC with KA-39 voltage converter)
Power output: ≥5 W
Temperature: −20° to +55°C
Altitude: up to 50,000 ft
Compliance: TSO C36e, C40c, C38c, C37c, RTCA DO-160B, -178A, -186 -195, -196

Contractor
Honeywell Aerospace, Electronic & Avionics Lighting.

UPDATED

KX 155 nav/com system

The Bendix King KX 155 is self-contained nav/com transceiver which utilises a solid state, gas discharge digital display and a push button frequency flip-flop with display of active and standby nav and com frequencies. The main features of the system are:
- 720 or 760 communications frequencies
- 200-channel nav transceiver
- 40-channel glideslope receiver (option)
- built-in audio amplifier (option)
- 14 or 28 V DC power supply option
- 10 W power output
- 25 or 50 kHz com receiver selectivity
- fully TSO qualified

The KX 155 can be used with the KI 203/208 VOR/LOC or the KI 204/209 VOR/LOC/GS indicators.

Specifications
Dimensions: 159 × 520 × 258 mm
(6.25 × 2.05 × 10.16 in)
Weight: 2.16 to 2.5 kg (4.75 to 5.5 lb) depending on options fitted
Electrical: 14 or 28 V DC
Power output: ≥10 W
Temperature: −20° to +55° C
Altitude: up to 50,000 ft
Compliance: TSO C34c (GS option), C36c, C37b C38b, C38c, C40a, RTCA DO-131, -132 (GS option) -153, -156, -157

Contractor
Honeywell Aerospace, Electronic & Avionics Lighting.

NEW ENTRY

KX 155A nav/com system

The Bendix King KX 155A nav/com transceiver is a development of the KX 155 (see previous entry), offering the following enhanced or standard fit (where optional on the KX 155) features:
- 760 communications frequencies available with 32 programmable channels
- automatic microphone shut down if activated for more than 33 seconds (stuck microphone)
- bearing-to-station mode and radial-from-station mode
- elapsed time and approach timer
- built in 4 Ω audio amplifier
- internal CDI
- 28 V DC power only
- full backlighting of bezel nomenclature and control

The KX 155A can be used with the KI 203/208 VOR/LOC or the KI 204/209 VOR/LOC/GS indicators, or as part of the KCS 55A compass system with the KI 525A HSI (KN 72 VOR/LOC converter required).

Specifications
Dimensions: 159 × 508 × 258 mm
(6.25 × 2.00 × 10.16 in)
Electrical: 28 V DC
Power output: ≥10 W
Temperature: −20° to +55°C
Altitude: up to 50,000 ft
Compliance: TSO C34e, C36e, C37d, C38d, C40a, C40c, RTCA DO-131, -153

Contractor
Honeywell Aerospace, Electronic & Avionics Lighting

UPDATED

KX 165 nav/com system

The Bendix King KX 165 is self-contained nav/com transceiver which utilises a solid-state, gas discharge digital display and a push button frequency flip-flop with display of active and standby nav and com frequencies. The KX 165 is similar in specification to the KX 155 (see previous entry) but includes a built-in VOR converter and radial display. The main features of the system are:
- 720 or 760 communications frequencies
- 200-channel nav transceiver
- 40-channel glideslope receiver (option)
- built-in VOR converter
- digital display of radial from VOR or VORTAC in lieu of standby NAV frequency
- 14 or 28 V DC power supply option
- 10 W power output
- 25 or 50 kHz com receiver selectivity
- fully TSO qualified

The KX 165 can be used with the KI 202 VOR/LOC or the KI 206 VOR/LOC/GS indicators.

Specifications
Dimensions: 159 × 520 × 258 mm
(6.25 × 2.05 × 10.16 in)
Weight: 2.16 to 2.5 kg (4.75 to 5.5 lb) depending on options fitted
Electrical: 14 or 28 V DC
Power output: ≥10 W
Temperature: −20° to +55°C
Altitude: up to 50,000 ft
Compliance: TSO C34c (GS option), C36c, C37b, C38b, C40a, RTCA DO-131, -132 (GS option), -153, -156, -157, -160

Contractor
Honeywell Aerospace, Electronic & Avionics Lighting.

NEW ENTRY

KX 165A TSO nav/com system

The Bendix King KX 165A TSO nav/com transceiver is a development of the KX 165 (see previous entry). The unit is fully TSO qualified, with European/JTSO qualification underway. System features include:
- 25 or 8.33 kHz channel spacing
- 32-channel com memory
- internal glideslope
- stuck microphone reset
- radial and bearing display
- elapsed time and approach timer
- composite nav output
- HSI output (left/right/GS/flags)
- internal audio amplifier
- electronic CDI
- 10 W transmitter
- 28 V DC power
- full backlighting of bezel nomenclature and control

Specifications
Dimensions: 159 × 508 × 258 mm
(6.25 × 2.00 × 10.16 in)

Contractor
Honeywell Aerospace, Electronic & Avionics Lighting.

NEW ENTRY

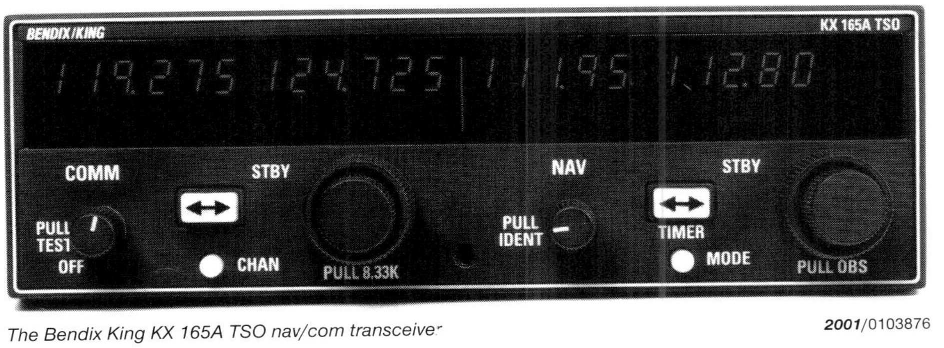

The Bendix King KX 165A TSO nav/com transceiver **2001**/0103876

KXP 756 Gold Crown III solid-state transponder

The KXP 756 is a third-generation system incorporating modern avionics techniques such as large-scale integrated circuitry, microprocessor data programming and all-solid-state transmitter design to provide reliability and simplicity of operation. It operates on Modes A, B and C up to 70,000 ft and can reply on any one of 4,096 preselected codes. Information is provided on one or two 2¼ in square gas discharge digital displays which are automatically adjusted in brightness by a photocell, for maximum visibility under all light conditions. The system is controlled by two concentric knobs – one for mode selection and the other for code selection – and incorporates identification, VFR code and self-test functions.

Specifications

Dimensions:
(control unit) 146.7 × 21.9 × 23.5 mm
(remote unit) 298.45 × 50.8 × 134 mm
Weight:
(control unit) 0.31 kg
(remote unit) 1.727 kg
Power supply: 11-13 V DC
Altitude: up to 60,000 ft
Frequency:
(receive) 1,030 MHz
(transmit) 1,090 MHz

Operational status

In production.

Contractor

Honeywell Aerospace, Electronic & Avionics Lighting.

VERIFIED

KY 96A and KY 97A VHF communications transceivers

The KY 96A and KY 97A VHF communications transceivers are identical in all respects except that the KY 96A operates at 28 V, while the KY 97A operates at 14 V. Frequency coverage is from 118 to 136.975 MHz at 25 kHz spacing.

Both the active and a standby frequency are displayed on the illuminated liquid crystal display and the set is switched between the two by a single button push. Up to nine channels can be programmed into the non-volatile memory. The transceivers feature audio levelling, so that weak signals are automatically amplified and strong signals are muted, and are equipped with an audio amplifier to drive a speaker for those installations not equipped with an audio panel.

Specifications

Dimensions: 33 × 158.8 × 266.7 mm
Weight: 1.32 kg
Power output: 5 W min
Frequency: 118-136.975 MHz
Temperature range: –20 to +55°C

Operational status

In production and in service.

Contractor

Honeywell Aerospace, Electronic & Avionics Lighting.

VERIFIED

The KY 96A VHF communications transceiver

KY 196A/KY 196B and KY 197A VHF transceivers

During 1987, the KY 196 and 197 transceivers were updated, with the addition of some new features, and redesignated KY 196A and KY 197A. The KY 196A has since been further upgraded to the KY 196B standard by the addition of class 4 and class 5 operation to the class 3 standard already available in the KY 196A. The KY 196B is also able to meet the ICAO EUR Region's 8.33 kHz channel spacing requirement providing a total of 2,280 channels in the range 118.000 to 136.990 MHz.

The KY 196A is a compact lightweight panel-mounted transmitter/receiver particularly suitable for light aircraft. It covers the VHF band from 118 to 136.975 MHz in which range 760 channels are provided; channels are selectable at increments of 25 or 50 kHz. One of the features of the update is the expansion of the frequency coverage by 1 MHz at the top end of the range.

A principal feature of the system is that a second frequency, in addition to the one in use, may be stored for immediate selection. Both the operating and standby frequencies are presented on a self-dimming gas discharge display in a window on the front panel. When the standby channel is selected the former operating channel is entered into the standby store. Non-volatile storage, provided by an electrically alterable read-only memory chip, ensures both frequencies remain stored when the power supply is off or disconnected. No separate memory power supply is required.

Solid-state construction is employed throughout. The system is microprocessor-controlled and digital synthesis techniques are used for frequency generation. A MOSFET RF amplifier and mixer stage is used to provide clear signal reception.

Other improvements in the upgrade include a bigger selection knob, lighted push-buttons, pilot programmable lighting and dimming levels and a facility which detects a stuck microphone and stops transmission after 2 minutes.

Specifications

Dimensions: 160.3 × 34.3 × 258.5 mm
Weight: 1.27 kg
Power supply:
(KY 196A/KY 196B) 28 V DC
(KY 197A) 14 V DC
Power output:
(KY 196A/KY 196B) 16 W
(KY 197A) 10 W
Temperature range: –20 to +55°C

Operational status

In production and in service.

Contractor

Honeywell Aerospace, Electronic & Avionics Lighting.

VERIFIED

KY 196B panel-mounted VHF transceiver

In June 1999, the KY 196B panel-mounted VHF transceiver was announced as an update of its earlier KY 196A model to comply with new European requirements for 8.33 kHz spacing between channels. Transfer from 25 kHz to 8.33 kHz operation is by pull-out of the frequency selection knob, increasing channels available from 760 to 2,280.

The KY 196B is primarily designed for light business jets and turboprop aircraft that fly at high altitude but often have limited space for avionics; it fits a standard 6.3 in 'mark width' opening for panel-mounted avionics. It can also be used as a plug-in replacement for the KY 196A with no modifications to wiring or mounting rack.

AlliedSignal also produces three remote-mounted transceivers with 8.33 kHz channel spacing for larger transport and regional aircraft: the RTA-83A and RTA-83B and the RTA-44D digital voice and data radio.

Contractor

Honeywell Aerospace, Electronic & Avionics Lighting.

VERIFIED

Mark II Communications Management Unit (CMU)

The Mark II CMU is designed to meet the needs of current Aircraft Communications Addressing and Reporting System (ACARS) management requirements and Future Air Navigation System (FANS) Air Traffic Management (ATM) regime. The unit can meet the requirements of both older analogue aircraft and newer digitally managed aircraft to meet the complete needs of airlines with a single part number item.

The Mark II CMU is designed around an Intel 486DX processor to provide a flexible format that can be configured by the user to meet specific needs for screen formats, downlink/uplink format and control in accordance with ARINC 429, encryption/decryption and other functions.

Specifications

Dimensions: 4MCU in accordance with ARINC 600
Weight: 5.44 kg
Power: 30 W, 115 V AC or 28 V DC
MTBF: > 20,000 h

Contractor

Honeywell Aerospace, Electronic & Avionics Lighting.

VERIFIED

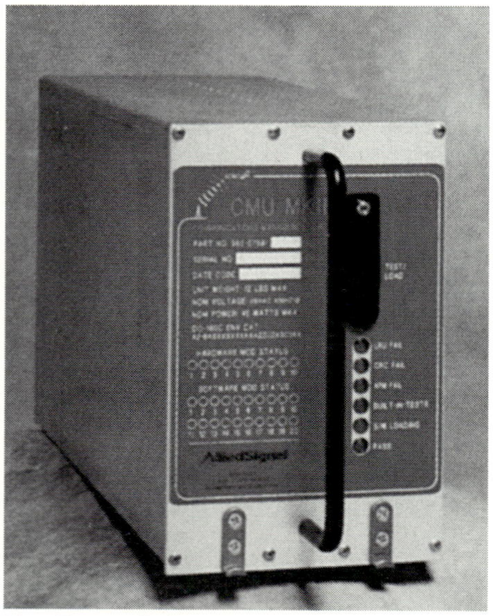

The Honeywell Aerospace, Electronic & Avionics Lighting Mark II communications management unit
2000/0062198

MST 67A Mode S transponder

A third-generation Mode S transponder, the compact MST 67A offers all the capabilities of heavier airline-type units in a much smaller package. It incorporates a number of patent pending features, including such advances as a 16-bit microprocessor, programmable gate array digital signal processing and SAW technology. Fully TSO'd, the remote-mounted MST 67A is equipped with standard ARINC 400 series connectors.

A choice of control heads allows the MST 67A to fit virtually any corporate or regional airliner class cockpit. Featuring a photocell-equipped gas discharge display, the KFS 578A control unit supplies ARINC 429 data to all versions of the system. The KFS 578A can also serve as the aircraft's TCAS controller. For aircraft already equipped with a dzus-mount transponder control panel, the CTA 81A is available as a drop-in replacement. Fully compatible with ARINC 718 and ARINC 735 and providing many of the same interfacing and control functions as the KFS 578A, the CTA 81A features a high-contrast liquid crystal display.

In its non-diversity version, the transponder uses a bottom antenna only, for operators who do not anticipate installing TCAS in the aircraft but wish to ensure compliance with ATC reporting standards. A non-diversity MST 67A is fully compliant with air-to-ground/ground-to-air datalink applications.

The diversity version of the MST 67A uses inputs from two antennas, mounted top and bottom of the aircraft. Required for TCAS/ACAS operations, the diversity option provides the aircraft with air-to-air datalink communications capability.

Enhanced BIT features constant monitor transponder status. A bidirectional interface between the transponder and the control unit also enhances diagnostic capabilities. With the test mode selected on the control panel, internally diagnosed problems can be viewed in real time and information stored for as many as the nine previous flights, in non-volatile memory, can be reviewed.

Specifications

Dimensions:
(MST 67A transponder) 57.2 × 381 × 193.8 mm
(KFS 578A control unit) 53.1 × 57.2 × 187.5 mm
(CTA 81A control panel) 146.1 × 57.2 × 119 mm
Weight:
(MST 67A transponder) 3.86 kg
(KFS 578A control unit) 0.45 kg
(CTA 81A control panel) 0.82 kg

Operational status

The MST 67A transponder forms part of Honeywell Aerospace's latest lines of TCAS/ACAS systems.

Contractor

Honeywell Aerospace, Electronic & Avionics Lighting.

VERIFIED

Multi-input Interactive Display Unit (MIDU)

The MIDU is designed in accordance with ARINC 739 Multipurpose Control Display Unit (MCDU) screen formats and keyboard operation. It enables flight crew to manage the input, output and display of up to 11 separate systems. Examples of MIDU subsystem control include Aircraft Communications Addressing and Reporting System (ACARS) management unit; Aircraft Condition Monitoring Systems (ACMS) data management unit; satellite communications system; central maintenance computer.

The MIDU achieves a full data transfer capability using a credit card-sized removable memory (1 - 120 Mbyte capacity units available) for mass storage and data transfer in association with appropriate Airborne Data Loader (ADL) and Quick-Access Recorder (QAR) systems. MIDU supports the following standard aircraft interfaces: ARINC 429, ARINC 573 (QAR) and RS-422.

The VHF Radio Management Panel (RMP) application enables the MIDU to act as a VHF radio controller in accordance with ARINC 716, in which mode it supports conventional frequency (and new European) frequency spacing, displays active and standby frequencies, and provides certifiable voice-mode switching between VHF voice and ACARS data.

The MIDU display is a colour Active Matrix Liquid Crystal Display (AMLCD) with 14-line × 24-character format in accordance with ARINC 739.

Specifications

Dimensions: 114.3 (H) × 146.1 (W) × 165.1 (D) mm
Display area: 61.0 (H) × 73.7 (W) mm
Weight: 2.95 kg
Power: 40 W, 115 V AC or 28 V DC

Contractor

Honeywell Aerospace, Electronic & Avionics Lighting.

VERIFIED

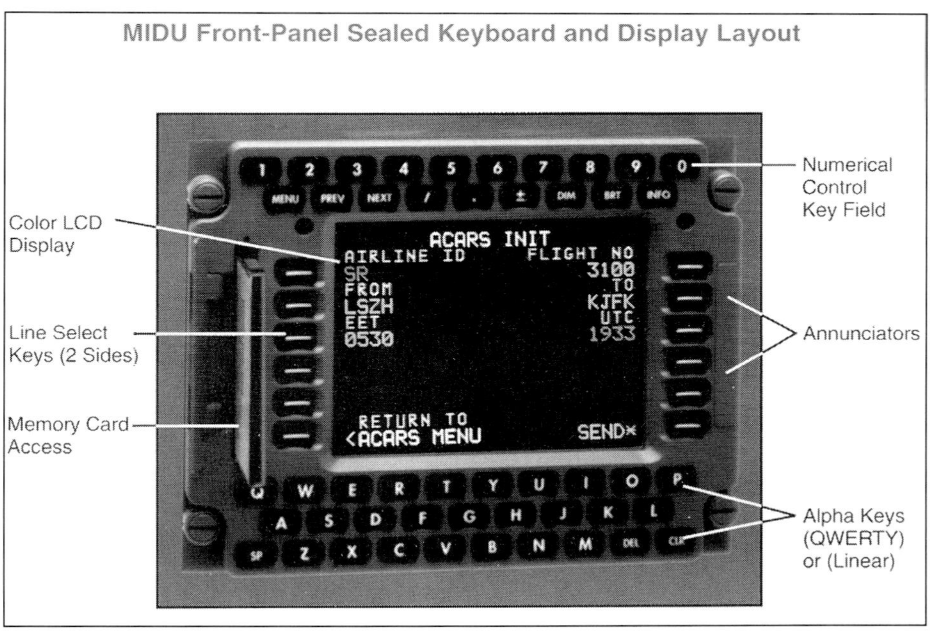

The Honeywell Aerospace, Electronic & Avionics Lighting multi-input interactive display unit showing typical display formats **2000**/0062200

Quantum™ line communication and navigation systems

The Quantum™ line of communication and navigation equipment comprises:
ALA-52B radio altimeter
DFA-75B automatic direction-finder
DMA-37B distance measuring equipment
MMR multimode receiver
RIA-35B instrument landing system
RTA-44D VHF data radio
RVA-36B VOR

Quantum™ line features extend reliability over previous designs, with a 30,000 hour MTBF and a guarantee on Mean Time Between Unscheduled Removals. It also offers commonality of numerous parts, assemblies and modules.

Quantum™ line equipment utilises more than 80 per cent of components and software that are common across the line. Quantum™ line features include 486 main processor; common monitor processor in the ILS and radio altimeter; liquid crystal display; packaging to ARINC 650; HIRF/lightning protection; ICAO FM immunity; ETOPS. A common front LCD panel permits easier maintenance with easily understood fault message displays. A flash card is provided for onboard data loading, rapid evaluation of system performance and data recording. A PC-compatible maintenance access port allows for diagnostics and fault isolation on the aircraft. Upgraded mechanical packaging improves protection for electromagnetic interference, reduces heat rise and ensures structural integrity.

Performance has been improved by providing 8.33 kHz channel spacing on the VHF data radio and raising the radio altimeter range to 5,000 ft.

The Quantum™ line is composed of the RIA-35B ILS receiver, DMA-37B DME interrogator, DFA-75B ADF receiver, ALA-52B radio altimeter, RTA-44D VHF Data Radio (VDR), and MMR 0018183

Operational status

Widely employed on civil aircraft.

Contractor

Honeywell Aerospace, Electronic & Avionics Lighting.

VERIFIED

RDR-4A/RDR-4B Forward Looking WindShear radar (FLWS)

Chosen by Boeing as standard equipment for the 767 and 757 transports, the RDR-4A was designed to meet ARINC 708 requirements. The I/J-band system features a solid-state transmitter and line of sight antenna with split-axis performance and is compatible with the EFIS flight decks of the Boeing 767, 757, MD-80, DC-10, Airbus A310, and Lockheed Martin L-1011 transports and other designs. The range is 592 km.

The RDR-4B incorporates forward-looking windshear detection and avoidance capabilities. The windshear detection capability is easily incorporated into existing RDR-4A radars, without form or fit changes to the installation.

The RDR-4B is a Doppler weather radar that measures actual horizontal windspeed, using reflections from the moisture that is always present in the atmosphere, and penetrates weather systems and detects microbursts embedded in rain. It provides specific windshear locations on a radar PPI presentation, giving 30 to 60 seconds of advance warning, on a display free from interference such as ground clutter. The RDR-4B can display turbulence up to 46 miles ahead and 90° left or right. It can display windshear up to 5¾ miles ahead up to 40° left or right.

The system operates automatically any time the aircraft is below 2,300 ft AGL, although the mode is selectable at any time.

The RDR-4B radar display is a map-like presentation that shows areas of weather and their locations relative to the aircraft. Light rain is shown in green, moderate rain in yellow, and heavy rain in red. Turbulence is magenta and windshear is a symbol called a windshear icon - a pattern of red and white arc-shaped stripes.

The RDR-4B can generate audible windshear warnings and alerts whenever the aircraft is less than 1,500 ft AGL. During approach, windshear ahead within 1.7 miles causes a 'go around' warning; in the landing phase, this warning is reduced to 0.6 miles. During take-off, windshear within 3.4 miles generates a warning.

The latest variant of the RDR-4B is designated the RDR-4B Predictive WindShear Radar (PWSR) and uses a computer processor to identify patterns of air movement associated with windshear, and to provide a full minute's warning to the crew.

From late 1999, a further development of the RDR-4B radar became available. Automatic tilt setting of the antenna angle (also known as auto-tilt) is provided by interfacing the RDR-4B with the Enhanced Ground Proximity Warning System (EGPWS) to predict the optimum setting for the antenna tilt. The ability to integrate the RDR-4B with the EGPWS is being

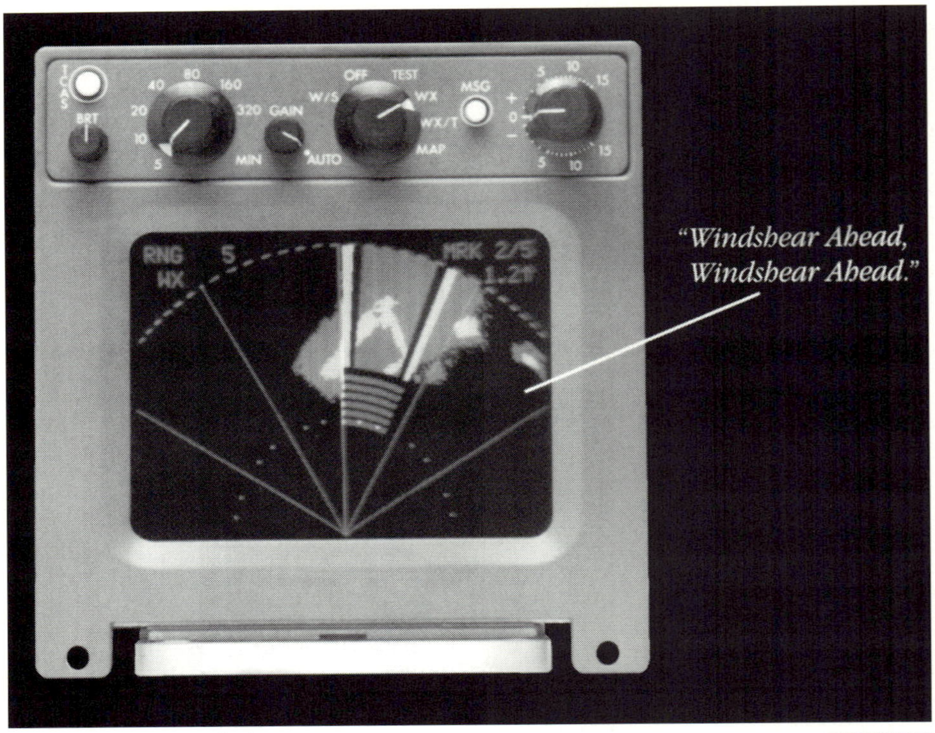

The RDR-4B windshear display ***2000***/0081848

designated Terrain Data Correlation (TDC). The advantage to the pilot is that the combined capability supplements EGPWS data where database coverage is poor, lowers minima at airports with precipitous terrain, alerts pilots to obstacles not in the EGPWS database, and automatically updates the database.

From 2000, an enhanced high-altitude turbulence detection capability will be added, based on using advanced windshear processing technology with a higher pulse transmission rate to measure turbulence ahead of the aircraft.

Specifications

RDR-4B receiver/transmitter
Transmitter frequency: 9,345 MHz
Transmitter power: 125 W nominal
Maximum detection ranges:
(weather and map modes): 320 n miles
(turbulence mode): 40 n miles
(windshear): 5 n miles
Pulse-width:
(radar modes): 6 and 18 μs interlaced
(windshear mode): 1.5 μs
Pulse repetition rate:
(weather and map modes): 380 Hz
(turbulence mode): 1,600 Hz
(windshear mode): 6,000 Hz
Dimensions: 8 MCU
Weight: 13.2 kg

PPI-4B colour indicator
Display size: 3.3 × 4.3 in
Display ranges and marks: 5/2, 10/5, 20/5, 40/10, 80/20, 160/40, 320/80 n miles
Display colours
(radar mode):
green: level 2 rainfall
yellow: level 3 rainfall
red: level 4 rainfall and above
magenta: turbulence
(windshear mode):
red and black icon overlays standard weather display
yellow flashlight beams mark end of icon
(TCAS overlay):
red: resolution advisory
yellow: traffic advisory
white: proximate and non-traffic threat
(EGPWS terrain data):
green: no terrain threat
yellow: caution terrain
red: warning terrain
Weight: 6.7 kg

Antenna and drive
Antenna:
(REA-4B/DAA-4A): 30 in flat plate, gain 35 dB, beamwidth 2.9°, weight 13.7 kg
(REA-4A/DAA-4B): 24 in flat plate, gain 33 dB, beamwidth 3.6°, weight 5.5 kg

Operational status

More than 6,000 RDR-4A radars are in service with numerous operators.

The RDR-4B was first certified in September 1994, and first use of the RDR-4B in commercial service was on a Continental Airlines Boeing 737-300 in November 1994. Over 1,500 RDR-4B radars are now in service.

The RDR-4B Predictive WindShear Radar (PWSR) was granted a supplemental type certification by the US FAA for the Boeing 737-300 and 737-500 aircraft in September 1999. United Airlines has ordered 495 for its existing fleet of late-generation airplanes including Airbus A319 and A320, Boeing 737-300, 737-500, 747-400, 757, 767, and 777 models. United Airlines has specified that all its new aircraft are to be delivered with factory-installed RDR-4B radars.

Contractor

Honeywell Aerospace, Electronic & Avionics Lighting.

VERIFIED

RDR-1400 weather/multifunction radar

The RDR-1400 radar is designed for commercial helicopters, particularly those associated with the large international offshore oil and gas industry. It differs from almost all other Honeywell Aerospace weather radars by having a beacon interrogator that exploits the increasing use of portable radar beacons in these industries. The system is suitable for search and rescue, surveillance, aerial survey work and law enforcement, as well as rig servicing.

The original monochromatic RDR-1400 has now been joined by a colour radar version with greater performance. The following operational modes are available:

Beacon navigation. The growing popularity of portable beacons is supported by several special RDR-1400 capabilities. A beacon signature is denoted by a short line or oblique on the display, the actual location of the device being determined by the middle of the line; the pilot can overlay beacon returns on the weather map. The beacon's discrete code can be displayed for positive identification, an important factor when the pilot is trying to locate a specific rig in a drilling farm where numerous rigs may be transponder-equipped. Beacon detection range is up to 296 km depending on altitude.

Beacon Trac. This mode, peculiar to Honeywell, generates and displays on the weather radar screen an inbound course to the discrete beacon. This course line can be rotated 360° about the beacon by rotating the horizontal situation indicator course selector, thus allowing the pilot to choose a convenient course to the beacon.

Obs Trac. This mode provides another course-following option. When in a weather or search mode, a track line or course-bearing cursor can be generated from the aircraft position and controlled by the horizontal situation indicator course selector to provide a course line to the chosen target. The Obs Trac heading is displayed digitally in the lower right-hand corner of the indicator. Left/right deviations can be determined by comparing heading information to this number and by observing the movement of the track line in relation to background targets.

Search. Three search modes are available. Search 1 has special sea clutter rejection circuitry to detect objects such as small boats or buoys down to the minimum range. Search 2 is for precision ground-mapping in situations where high target resolution is important. Search 3 includes maximum return clutter and can detect and track oil slicks.

Specifications

Dimensions:
(transmitter/receiver) 127 × 159 × 352 mm
(control/display unit) 159 × 159 × 276 mm
Weight:
(305 mm antenna) 15.47 kg
Frequency: I/J-band
Power output: 10 kW
Antenna size: 305 or 457 mm flat plate
Antenna scan angle: 120° or 60°
Display size: 110 × 85 mm
Qualification: TSO 63b

Operational status

In service in civil and military helicopters.

Contractor

Honeywell Aerospace, Electronic & Avionics Lighting.

VERIFIED

RDR-1400C colour weather and search and rescue radar

The RDR-1400C is a weather detection system. It has a 445 km maximum display range, giving the user time to plan weather avoidance manoeuvres. For clear, detailed close-ups, two modes permit selection of full-scale ranges of either 1 n mile or 0.5 n mile, enhancing safety and precision of movement. The Target Alert feature flashes a warning whenever third-level (red) weather areas are detected up to 46 km beyond the selected range.

Different surveillance missions require different capabilities, so the RDR-1400C provides three specialised search modes. Search 1 incorporates special sea clutter rejection circuitry to help detect small boats or buoys down to a minimum range of 270 m.

Search 2 is designed for precision ground-mapping, where high target resolution is important. Search 3 mode, which includes normal ground-mapping, can also be used to detect and track sea-surface phenomena such as oil slicks.

The RDR-1400C complies with TSO C-102, enabling land or sea approaches in 200 ft ceiling, ½ mile visibility minima. Its beacon tracking mode permits operation with either current beacon codes or the newer DO-172/16 format, changing with the push of a button.

In BeaconTrac® mode and equipped with a graphics interface unit, the RDR-1400C can automatically show target latitude and longitude along with NAV sensor data on a moving map display. A white course line, rotatable 360° around the centre of a target beacon, offers a digital flight path to the beacon from any direction. Left/right flight deviation cues are provided by a deviation control bar located on the indicator.

Delivering 10 kW of power, the system excels in the detection of small targets. For example, the RDR-1400C can locate a 9 m target in Sea State 2 from up to 18 km away, while operating at a typical helicopter altitude of 152 m.

Specifications

Dimensions:
(receiver/transmitter) 127 × 158.8 × 352.4 mm
(indicator) 158.8 × 158.8 × 276.2 mm
Weight:
(receiver/transmitter) 6.58 kg
(indicator) 5.22 kg
Power requirements:
4.2 A, 27 V AC
3 A, 115 V AC, 400 Hz
TSO compliance: C-102
Frequency: I/J-band
RF power output: 10 kW
Scan angle: 120 or 60°
Scan rate: 28°/s
Display range/marks: 0.5/0.125; 1/0.25; 2/0.5; 5/1.25; 10/2.5; 20/5; 40/10; 80/20; 160/40; 240/60 n miles
Min tracking range: 270 m
Beacon range: line of sight to 300 km

Operational status

In June 1999, the RDR-1400C radar was selected by CASA as the standard-fit radar for its new-production C-295 and CN-235 aircraft, as part of an extensive avionics package awarded to Honeywell Aerospace, Electronic & Avionics Lighting. The RDR-1400C will also be offered by CASA for CN-235 aircraft retrofit installations.

Contractor

Honeywell Aerospace, Electronic & Avionics Lighting.

VERIFIED

RDR 2000 Vertical Profile® weather radar

The RDR 2000 is a Vertical Profile® weather radar. In addition to normal weather radar features, the RDR 2000 adds a vertical display of weather, enabling the pilot to monitor storm development, by observing the rate of growth of the tops of storm cells. In addition, it is possible to check the angle of a storm cell's leading edge to see which way it is heading. Four levels of colour are used to display information.

Typically, an RDR 2000 radar would be paired with the GC 360A radar graphics unit and a moving-map GPS unit, such as the KLN 90B or KLN 900 or CDU-Xls Flight Management System (FMS) to provide a comprehensive display capability.

Specifications

Dimensions:
254 or 309 mm antenna
Power output: 3.5 kW nominal (rated 4.0 kW)
Horizontal scan: 100° at 25°/s
Vertical scan: 60°
Weather Avoidance: 400 km typical for 309 mm antenna system

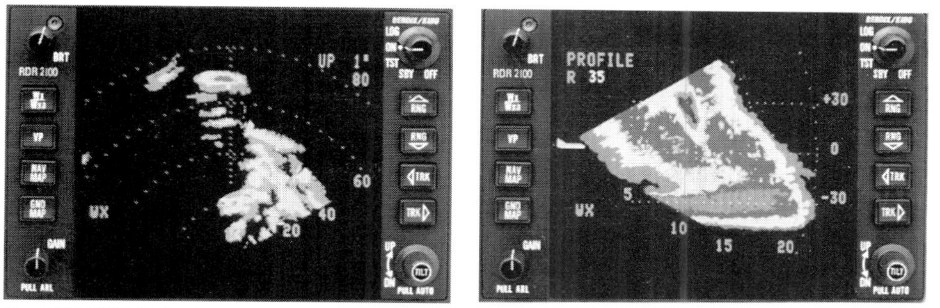

The RDR 2100VP displays a storm cell in traditional form in the left hand display and in vertical profile mode on the right hand display, as would be seen by the pilot on a split-screen display **2000**/0062202

Range scales: 10,20,40,80,160 and 240 n miles
Weight: 12.02 kg

Operational status

In production.

Contractor

Honeywell Aerospace, Electronic & Avionics Lighting.

VERIFIED

RDR 2100 Vertical Profile® weather radar

The RDR 2100 Vertical Profile® four-colour, digital, weather radar includes sophisticated features such as Automatic Range Limiting (ARL), which alerts pilots to the possibility of a foreground storm cell blocking the picture of a more distant cell.

For aircraft equipped with a MultiFunction Display (MFD), the RDR 2100 can provide split-screen view to show both vertical and horizontal views simultaneously, sector scan to provide more frequent updates, and a logarithmic scale to show nearby storms in large detail while still allowing the pilot to see distant storms.

The Vertical Profile system scans vertically at the azimuth selected by the pilot using a track line. This enables the pilot to examine the angle of a storm cell's leading edge to determine the direction of movement, to check the radar tops and to distinguish between ground and weather returns. The radar is fully stabilised to ±30° combined pitch and roll. Four levels of colour are available with switchable range scales of 5, 10, 20, 40, 80, 160, 240 and 320 n miles.

The system is fully EFIS compatible using ARINC 429 and ARINC 453 databus structures.

Specifications

Transmit power: 6 kW
Weight: 12.02 kg
Antenna 305mm (standard); 254 mm (optional)
Horizontal scan: 120°
Vertical scan: 60°

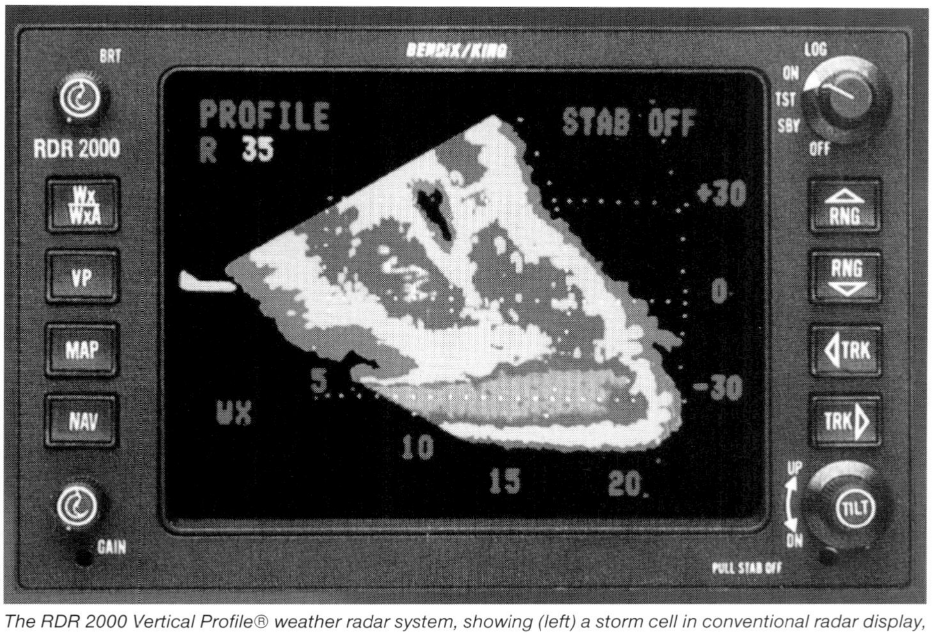

The RDR 2000 Vertical Profile® weather radar system, showing (left) a storm cell in conventional radar display, and (right) in the vertical profile mode **2000**/0081850

Contractor

Honeywell Aerospace, Electronic & Avionics Lighting.

VERIFIED

RDR 2100VP Vertical Profile weather radar system

The RDR 2100VP is the flagship radar of Honeywell Aerospace's new series of vertical profile radars. It functions equally well as a stand-alone unit, or in conjunction with an electronic display system, either flat panel or EFIS.

In addition to offering the standard features of a four-colour, solid-state, digital radar, it includes a vertical display of weather information, selected by a push-button selection (Vertical Profile). The pilot can thus monitor the rate at which individual storm cells are developing, and by checking the angle of the cell's leading edge, determine the way the weather is moving. If the aircraft is fitted with multifunction displays, both the horizontal and vertical modes can be displayed simultaneously, by choosing a split-screen display.

The radar's full-width scan covers 120°, but a faster update 60° scan is available, and also an auto-step feature, in which horizontal scans are made at 4° intervals across a ±10° vertical span. A logarithmic weather display permits the pilot to watch close-in weather activity, while monitoring more distant weather patterns.

Configuration options for the Horizontal Situation Indicator (HSI) include standard weather returns in a 120° sector scan overlaid with navigation courses, navaids, waypoints, airports and other key data.

Honeywell Aerospace claims a three-fold improvement in magnetron life by comparison with current systems in service, and suggests that the RDR 2100VP radar is ideal for regional airline use.

Specifications

Antenna/receiver/transmitter
Performance index:
(10-in antenna) 213.9 ±2.5 dB
(12-in antenna) 216.7 ±2.5 dB
Displayed weather ranges: 5, 10, 20, 40, 80, 160, 240, 320 n miles
Weather colours: 5, including black
Vertical profile scan angle: ±30°
Ground map variable gain: 0 to –20dB (configurable at installation)
Peak output power: 6.0 kW, nominal
Pulse-width: 4 μs
Pulse repetition frequency: 106.5 ±5 Hz
Antenna scan angle: 100 or 120° (configurable at installation)
Antenna scan rate: 25°/s
Tilt angle, manual: –15 to +15°
Stabilisation: ±30° combined pitch, roll and tilt
Stabilisation adjustments (non-volatile): stored in configuration module (part of aircraft installation)
Power: 28 V DC, 3.0 A
Weight: 4.5 kg
Altitude: 55,000 ft
Temperature range: –55 to +70°C
TSO: C63c Class 7

Indicator IN-862A
Display ranges:
(weather) 5, 10, 20, 40, 80, 160, 240, 320 n miles
(navigation only) up to 1,000 n miles
Power: 28 V DC, 2.0 A
Weight: 4.43 kg

Altitude: 35,000 ft unpressurised
Temperature range: –20 to +55°C
Dimensions: 114.3 (H) × 162.8 (W) × 344.7 (L) mm
TSO: C63c

Contractor
Honeywell Aerospace, Electronic & Avionics Lighting.

VERIFIED

RDS-81 weather radar

The RDS-81 is a four-colour radar. The fourth colour, magenta, depicts the very heaviest areas of precipitation, enabling the system to provide a more clearly defined picture of where potentially severe weather is located. It also helps depict the steep rainfall gradients that are often associated with heavy turbulence or windshear. A selectable weather alert mode causes the magenta returns to flash on and off continuously as a further warning of intense storm cell activity. The 254 or 305 mm flat plate antenna is fully stabilised to keep it parallel to the horizon or at the selected tilt angle and automatic compensation is provided up to 25° of combined pitch and roll.

The digital control/display unit features a 127 mm display coated with a polarised optical filter to provide good colour, contrast and definition even in extreme lighting conditions. Range scales of 10, 20, 40, 80, 160 and 240 n miles (18.5, 37, 74, 148, 296 and 444 km) are available and a different combination of pulse-width and PRF for each scale optimises radar transmission characteristics to give good long-range performance and high resolution at short ranges.

The RDS-81 can be combined with the GC 318A radar graphics unit and the KNS 81 integrated Nav/RNav system to provide a multifunction display with graphics, checklists and EFIS.

Specifications
Dimensions:
(control/display unit) 104 × 159 × 267 mm
Weight:
(control/display unit) 4.3 kg
(antenna/receiver/transmitter) 4.3 kg
Power supply:
(control/display unit) 28 V DC
Power output: 1 kW peak nominal
Frequency: 9,345 MHz
Pulsewidth: 0.6 to 11.5 μs depending on range
PRF: 128 to 1,026 Hz depending on range

Operational status
No longer in production.

Contractor
Honeywell Aerospace, Electronic & Avionics Lighting.

VERIFIED

RDS-82 radar

The RDS-82 is a digital, I/J-band radar. It has a four-colour display. Magenta indicates rainfall in excess of 51 mm/h and attenuation circuits automatically adjust the radar's sensitivity. The system is based on ARINC 429 digital data handling, so that all information and commands between the radar head and the control/display unit are carried on a twisted cable-pair. The antenna can be a 254 or 305 mm flat plate, scans 90° at the rate of 30°/s and radiates 1 kW. The system operates up to 55,000 ft without pressurisation.

Operational status
Launched in 1983. No longer in production.

Contractor
Honeywell Aerospace, Electronic & Avionics Lighting.

VERIFIED

RDS-84 Series 3 radar

One of two digital colour radars announced in April 1984, the RDS-84 is an addition to the company's range of Series 3 third-generation digital avionics for general aviation. It is intended for heavy piston twins and light jet types. Providing a 120° antenna scan and weighing 9 kg, the RDS-84 features a combined transmitter/receiver and antenna like the RDS-82 introduced in 1983.

Operational status
No longer in production.

Contractor
Honeywell Aerospace, Electronic & Avionics Lighting.

VERIFIED

RDS-84VP weather radar

The RDS-84VP (Vertical Profile) digital quadra colour radar takes advantage of the benefits of digital technology, to provide more precise readings of rainfall rates and a continuous bright picture of the weather and the navigation situation. Microprocessor technology allows the RDS-84VP to process more information than previous systems. All radar controls are multiplexed on an ARINC 429 control bus. Piezoelectric crystal oscillators provide a precise time reference to enable the RDS-84VP to track distant storms with great accuracy.

Another key advantage of the digital avionics design of the RDS-84VP is the system's built-in diagnostic capability, which allows it to serve as its own test set. The microprocessor-driven monitoring also alerts the pilot to any faults in the system with messages on the radar indicator.

The RDS-84VP employs autopulse, a technique that provides accurate displays of storm cells at all ranges. Autopulse enables the radar to optimise pulsewidth and PRF for each different radar range. This provides optimum long-range performance and high resolution at short range where it is most needed. Autopulse enables the RDS-84VP to operate with highly efficient peak output power.

Three features help the RDS-84VP to keep storms in perspective. The RDS-84VP uses Sensitivity Time Control (STC) logic to increase the amplifier gain for signals returning from greater distances. This means the storms that are further away will still have properly delineated areas of green, yellow, red and magenta. STC prevents storms from seeming to grow in intensity and size as they get closer. STC works only up to a specified range based on the point where amplifier gain has been increased to the maximum. The RDS-84VP also has an extended STC feature to compensate for weak signals returning from beyond the STC range. These weak signals are assigned display colours on a different schedule to ensure that areas of intense rainfall will be properly shown as red or yellow. The RDS-84VP also compensates for radar signal attenuation. Radar energy sent out by the transmitter antenna may be partially absorbed by one storm before it reaches a second one. To help optimise the display of a second storm, special weather attenuation compensation circuitry increases receiver gain as needed to pick up storms hidden in the shadows of closer storms.

Specifications
Dimensions:
(IN 842A control/display unit) 114 × 163 × 290 mm
(GC 360A radar graphics unit) 163 × 34 × 342 mm
Weight:
(antenna/receiver/transmitter) 5.8 kg
(IN 842A control/display unit) 4.44 kg
(GC 360A radar graphics unit) 1.77 kg
Power supply: 28 V DC, 2.7 A
Power output: 1.3 kW peak
Frequency: 9,345 MHz
Pulsewidth: 0.8 to 15.3 μs depending on range
PRF: 128 to 1,086 Hz depending on range
Antenna size: 305 or 457 mm flat plate
Antenna scan: 120° at 30°/s
Antenna tilt: ±15°

Operational status
The RDS-84VP was designed for business jets and turboprops. No longer in production.

Contractor
Honeywell Aerospace, Electronic & Avionics Lighting.

VERIFIED

RDS-86 Series 3 quadra radar

The RDS-86 addition of Series 3 digital avionics radar for business aircraft was announced in April 1984. It was intended for heavy corporate jets such as the Dassault Falcon 900, Gulfstream IV and Canadair CL-600/601 Challenger. Like the RDS-84, the RDS-86 Series 3 radar has a combined transmitter/receiver and antenna and displays four colours (as indicated by the description quadra). The system has automatic range limitation, whereby areas giving signal returns of an unreliably low level are painted blue as a warning; antenna stabilisation, to maintain a steady picture during climb and descent; and a long-range navigation mode in which data up to 1,850 km ahead can be displayed, in addition to eight selectable distance scales ranging from 9 to 592 km.

The system weighs 10 kg.

Operational status
No longer in production.

Contractor
Honeywell Aerospace, Electronic & Avionics Lighting.

VERIFIED

RIA-35A ILS receiver

The RIA-35A is an ARINC 710-compatible ILS receiver, part of the CNI700 digital nav/com family. It uses dual microprocessors and custom large-scale integration design, with continuous monitoring of all unit subassemblies. It complies with ICAO Annex 10.

Specifications
Dimensions: 102 × 124 × 330 mm
Weight: 2.9 kg

Operational status
In service. In September 1985, AlliedSignal (now Honeywell) announced that the RIA-35A had been selected as standard on the Gulfstream G-IV aircraft and in June 1986 the RIA-35A was named as standard equipment on the Airbus A320. It is in service with numerous airlines worldwide. No longer in production.

Contractor
Honeywell Aerospace, Electronic & Avionics Lighting.

VERIFIED

RMA-55B Multi-Mode Receiver (MMR)

The RMA-55B Multi-Mode Receiver (MMR) is a member of the Quantum Line of communications and navigational equipment. Building on the modular architecture of the product line, the MMR ensures full compatibility with current and future landing systems.

The RMA-55B meets industry-defined sensor requirements for Category III Instrument Landing Systems (ILS), including requirements for ICAO Annex 10 FM immunity and initially en route/non-precision Global Positioning System (GPS) approaches. In a single 3 MCU unit, the MMR replaces the functions previously performed by an ARINC 710 receiver and an ARINC 743 GPS receiver. In addition, it provides for incremental growth to Microwave Landing Systems (MLS) and Differential GPS (DGPS) as requirements evolve.

Specifications
Power: 115 V AC
Weight: 5.44 kg (ILS/GPS/MLS)
Dimensions: 324.0 (L) × 90.9 (W) × 194.0 (H) mm
Form factor: 3 MCU per ARINC specification 600
Cooling: forced air per ARINC specification 600
Temperature operation: –15 to +70°C
Warm-up period: stable operating within 1 min after application of power
Frequency selection: serial digital in accordance with ARINC specification 429
Certification: DO-160C

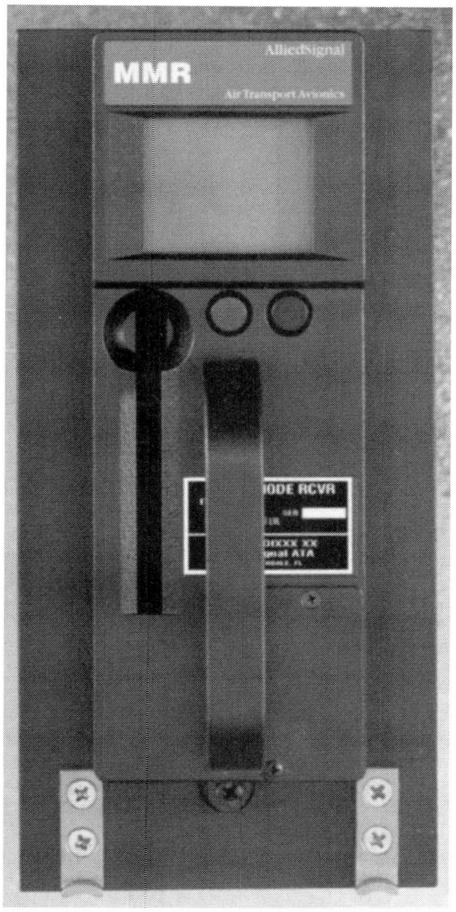

The RMA-55B multi-mode receiver 2000 ***2000***/0062203

Contractor
Honeywell Aerospace, Electronic & Avionics Lighting.

VERIFIED

RMS 555 radio management system

The RMS 555 offers high-technology features and performance in a compact RMU 556 control/display unit, driven by a ¼ ATR dwarf KDA 557 data adaptor. Utilising a 3.6 in diagonal CRT display, the RMU 556 provides a multitude of different pages, including the normal active frequency, memory and diagnostic pages.

Line item push-buttons provide quick access to each of the frequencies to be selected. Positive detent concentric knobs are used to provide secure input of frequency, channel or codes.

In a dual-RMS installation, the pilot and co-pilot can each tune all radios in the system, including the cross-side radios. Each RMS is capable of handling 17 different pieces of equipment via an ARINC 429 databus, including three VHF comms, dual navs, dual Tacans, dual ADFs, dual DMEs, dual ATCRBs or Mode S transponders, dual MLS and TCAS.

Standby frequencies for comm, nav and ADF allow for flip-flop tuning. The memory capability of the RMS 555 system also allows the pilot to store up to 20 pilot-programmable frequencies for each VHF comm, 10 for each ADF and 10 for each Tacan.

Specifications
Dimensions:
(data adaptor) 99 × 85.9 × 231 mm
(control/display unit) 152.4 × 62.6 × 231 mm
(configuration module) 53.3 × 40.5 × 38.6 mm
Weight:
(data adaptor) 1.87 kg
(control/display unit) 1.46 kg
(configuration module) 0.046 kg
Power supply: 28 V DC, 1.25 A

Contractor
Honeywell Aerospace, Electronic & Avionics Lighting.

VERIFIED

RTA-44A VHF transceiver

The RTA-44A is designed to be a highly reliable VHF transceiver for sharp clear voice communications. Many of the RTA-44A's performance capabilities come from a single microprocessor which is utilised in the computer architecture of the transceiver for high firmware reliability, a minimum parts count and reduced power usage. The microprocessor also provides the desired flexibility for offset operation and power-on channel tuning.

The transmitter consists of linear amplifier stages connected by microstrip impedance matching sections. These eliminate conventional interstage coupling capacitors and inductors along with their inherent failures and misadjustments.

The receiver features a programmable RF attenuator controlled by an off-channel signal sensor which automatically varies the attenuation to maintain a maximum desired/undesired signal input to the preselector. This makes the rejection of signals possible, even when an onboard transmitter, isolated by 30 dB, is being operated.

The microprocessor reduces maintenance time by monitoring pertinent modes throughout the transceiver, allowing a thorough fault analysis and fault isolation routine to be conducted down to subassembly/module level. Failed subassemblies are identified by an alphanumeric code displayed on the HEX LED display.

Three dedicated buffers and an 8-bit analogue-to-digital converter are utilised in the computer architecture to monitor all events external to the microcomputer for BITE signal processing.

Specifications
Weight: 4.1 kg
Frequency: 118-137.975 MHz
Number of channels: 800 (25 kHz spacing)

Operational status
No longer in production.

Contractor
Honeywell Aerospace, Electronic & Avionics Lighting.

VERIFIED

RTA-44D VHF Data Radio (VDR)

The RTA-44D VHF Data Radio (VDR) is part of the new Quantum Line of communications and navigation equipment.

The VDR, an airborne VHF digital communications transceiver, provides clear voice and data communication among onboard aircraft systems, with other aircraft and with ground-based operations.

The RTA-44D VDR provides standard 8.33 kHz channel capability to fulfil the new European airspace requirement. The unit is fully interchangeable with ARINC 716 communications transceivers, providing retrofit compatibility and satisfying spares pooling requirements.

In an Aviation VHF Packet Communications (AVPAC) system environment, the RTA-44D VDR operates as a simple transceiver with an analogue interface to the ACARS Management Unit (MU) or as a Minimum Shift Keying (MSK) modem. The RTA-44D VDR interfaces with an ARINC 758 Communications Management Unit (CMU Mark 11).

In an Aircraft Communications Addressing and Reporting System (ACARS) environment, the RTA-44D VDR operates as a simple transceiver with an analogue interface to the ACARS Management Unit (MU) or as a Minimum Shift Keying (MSK) modem. The RTA-44D VDR interfaces with an ARINC 724/724B ACARS MU in this environment.

The RTA-44D VDR requires an antenna for RF inputs and outputs, a control head or radio management panel, an audio input source and output sink for analogue voice functions. It may also be connected to a Central Maintenance Computer (CMC) to transfer maintenance data.

In addition to existing ARINC 750 CSMA kbps datalink capability (Mode 2), it features interfaces that allow for Mode-3 TDMA digitised voice capability, soon to be required for operation within US airspace. Processing power and reserve interfaces are built in to provide for the evolving Mode 4 STDMA technology, which may be required to complement Mode-S technology for ADS-B applications.

To ensure the RTA-44D VDR has the functional capability to meet CNS/ATM datalink requirements, the unit contains an Intel 80486SX control processor with 512 kbytes of RAM and 512 kbytes of ROM and two TMS320C31 Digital Signal Processing (DSPs), each with 128 kbytes of RAM.

Current implementation of the ARINC 716 VHF COMM radio and maintenance functions uses only the Main Processor Module (containing the 80486SX processor and a single TMS320C31).

The addition of the second DSP module, combined with the 80486SX and the first DSP provides sufficient power for all foreseeable datalink requirements.

The RTA-44D VDR is designed to the following standards and specifications:

(1) ARINC 716 airborne VHF communications transceiver
(2) ARINC 750 airborne VHF data radio
(3) RTCA DO-186 'Minimum Operational Performance Standards (MOPS) for airborne radio communications equipment operating within the radio frequency range 117.975-137.000 MHz'
(4) RTCA DO-207 'MOPS for devices that prevent blocked channels used in two-way radio communications due to unintentional transmissions'
(5) EUROCAE ED-23B 'minimum performance specification for airborne VHF communications equipment operating in the frequency range 117.975-137.000 MHz'

The RTA-44D VDR operational modes are ARINC 716 voice, 716 data, and 750 data. The 716 voice and 716 data modes emulate the two operation modes of an ARINC 716 VHF COMM. In the 750 data mode, the RTA-44D VDR bridges ACARS traffic between the CMUNDR 429 bus and the internal 2.4 kbps MSK modem, or bridges AVPAC traffic between the CMU/VDR 429 bus and the 2.4 kbps MSK or the 31.5 kbps D8PSK modems.

Specifications
Power: 27.5 V DC
Weight: 3.76 kg
Dimensions: 319.02 (L) × 90.9 (W) × 194.06 (H) mm
Form factor: 3 MCU per ARINC specification 600
Cooling: forced air per ARINC specification 600
Temperature, operating: –55 to +70°C
Warm-up period: stable operating within 1 minute after application of power
Frequency selection: serial digital in accordance with ARINC specification 429
Certification: O C37d Class 3 & C38d Class C; DO-160C; DO-186; DO-178B; DO-207; ICAO Annex 10 FM Immunity

Contractor
Honeywell Aerospace, Electronic & Avionics Lighting.

VERIFIED

RTA-83A VHF tranceiver

The RTA-83A VHF transceiver is part of the new line of communication equipment developed to meet the European airspace requirements for 8.33 kHz channel spacing and ICAO Annex 10 FM interference immunity. It meets all ARINC 566A specifications and is fully interchangeable with other ARINC 546 and 566A series transceivers.

The RTA-83A provides VHF voice and data communication between onboard aircraft systems, to other aircraft and to ground-based systems. It operates as a standard double sideband AM analogue voice transceiver. In an ACARS environment, it interfaces to either the ACARS Management Unit (MU) or Communications Management Unit (CMU), providing analogue Minimum Shift Keying (MSK) data capability.

Specifications
Power: 27.5 V DC; receive 1.0 A, transmit: 8.0 A
Weight: 4.5 kg
Form factor: ½ ATR short per ARINC specification 404
Cooling: forced air per ARINC specification 404
Temperature, operating: –55 to +70°C
Warm-up period: stable operation within 1 minute after application of power
Frequency range: 118.000 to 136.992 MHz

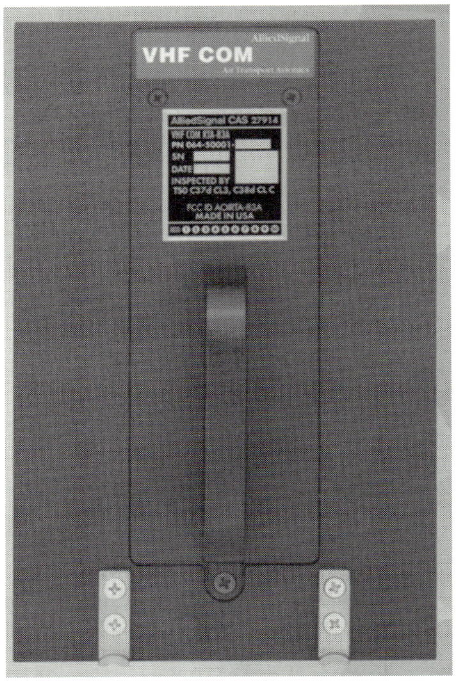

The RTA-83A VHF transceiver 2000 **2000**/0062204

Channel spacing: 8.33 kHz or 25 kHz
Frequency selection: 2 out of 5 per ARINC specification 410 or serial digital per ARINC specification 429
Certification: TSO C37d Class 3/5 and C38d Class C/E; O-186; DO-160C

Contractor

Honeywell Aerospace, Electronic & Avionics Lighting.

VERIFIED

RTA-83B VHF transceiver

The RTA-83B VHF transceiver is part of the new line of communication equipment developed to meet the European airspace requirements for 8.33 kHz channel spacing and ICAO Annex 10 FM interference immunity.

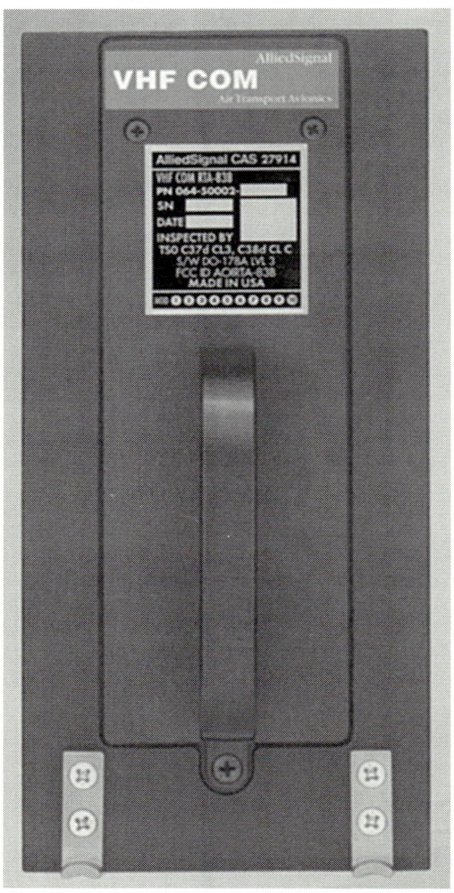

The RTA-83B VHF transceiver **2000**/0062205

It meets all ARINC 716 specifications and is fully interchangeable with the RTA-44A and other ARINC 716 series transceivers.

The RTA-83B provides VHF voice and data communication between onboard aircraft systems, to other aircraft and to ground-based systems. It operates as a standard double sideband AM analogue voice transceiver. In an ACARS environment, it interfaces to either the ACARS Management Unit (MU) or Communications Management Unit (CMU), providing analogue Minimum Shift Keying (MSK) data capability.

Specifications

Power: 27.5 V DC; receive: 1.0 A, transmit: 8.0 A
Weight: 4.0 kg
Form factor: 3 MCU per ARINC specification 600
Cooling: forced air per ARINC specification 600
Temperature, operating: –55 to +70°C
Warm-up period: stable operation within 1 minute after application of power
Frequency range: 118.000 to 136.992 MHz
Channel spacing: 8.33 kHz or 25 kHz
Frequency selection: serial digital per ARINC specification 429
Certification: TSO C37d Class 3/5 and C38d Class C/E; DO186; DO160C

Contractor

Honeywell Aerospace, Electronic & Avionics Lighting.

VERIFIED

RVA-36A VOR/marker receiver

The RVA-36A airborne VOR/marker receiver is designed in accordance with ARINC 711. The design provides a highly accurate digital omni-bearing output in ARINC 429 format.

With the advent of pure digital interface and reliable microcomputers, the RVA-36A was designed to make optimum use of a microprocessor, or microcomputer, which permits features and operational flexibility that could not have been achieved in earlier generation equipment. Signal processing, control, monitoring and detailed fault analysis are all under the control of a programmable microprocessor.

Some of the features offered in the RVA-36A which are attributable to the microprocessor are improved accuracy, comprehensive monitoring, detailed fault isolation and precise predictable and more selective receiver design. In addition, BITE monitors the performance of all assemblies and, if a fault occurs, it is stored in EEPROM and can be sent to the central fault display system. It complies with ICAO Annex 10.

Specifications

Dimensions: 324.1 × 90.93 × 194.06 mm
Weight: 4.6 kg
Frequency: 108-117.95 MHz
Channel spacing: 50 kHz

Operational status

In service.

Contractor

Honeywell Aerospace, Electronic & Avionics Lighting.

VERIFIED

Satellite Data Communications System (SDCS)

The Satellite Data Communications System (SDCS) uses the INMARSAT network to send and receive information worldwide. Interfaced to Airborne Flight Information System (AFIS), it provides worldwide communication capabilities.

It provides two-way unrestricted message forwarding to another SDCS-equipped aircraft, the Global Data Center, a fax machine, an auto answer terminal and other service providers such as BASEOPS International, Air Routeing International, Jeppesen Dataplan, Universal Weather and Aviation and MEDLINK. It also enables provision of worldwide weather information and flight planning/flight plan filing.

Specifications

Weight:
(HPA/LNA) 2.04 kg
(antenna) 0.4536 kg
(SCU) 2.72 kg
Power supply: 27.5 V DC
(transmit) 4.5 A
(receive) 0.5 A

Contractor

Honeywell Aerospace, Electronic & Avionics Lighting.

VERIFIED

Series III Integrated Digital Avionics System

The Series III integrated digital avionic equipment was announced in 1982. Based on ARINC 429 databusses, the system was also designed for compatibility with a new line of EFIS displays being produced.

At the centre of the Series III system is the EFIS display, which is produced in three sizes – 158 mm (6.25 in) square, 127 × 152 mm (5 × 6 in) and 120.7 × 127 mm (4.75 × 5 in) – and with what Honeywell claims to be a 'unique stroke writing method'. Typical configuration, at least for the higher-performance turbine types, would be two CRTs each for pilot and co-pilot and two shared EICAS displays for systems and navigation.

The current Series III family comprises:
DFS-43 automatic direction-finder
DMS-44 distance measuring equipment
TRS-42 transponder.
VCS-40 VHF communication transceiver
VNS-40 VHF navigation receiver
More detailed descriptions of some of these systems can be found under the appropriate section headings in this book.

Operational status

In service. The DFS-43 and DMS-44 are still listed as in production. An initial installation on board a Cessna Citation III was completed in December 1984, having been certified by the FAA earlier in that year. The five-system suite was chosen by Fokker in May 1985 as the standard communication, navigation and identification package for its Fokker 50 airliner.

Contractor

Honeywell Aerospace, Electronic & Avionics Lighting.

VERIFIED

Silver Crown nav/com avionics family

The Silver Crown designation applies to a family of Communications, Navigation and Identification (CNI) and autopilot equipment for single- and light twin-engine aircraft up to turboprop size and for light helicopters. The various units are self-contained, with processing and presentation or indication mechanisms accommodated together in single panel-mounted boxes without the need for remote units. Initial members of the range appeared in the 1960s, and subsequent additions have made Silver Crown into a very comprehensive suite. In 1980 the series was upgraded by the introduction of some units incorporating digital technology.

The Silver Crown family consists of a large number of alternative units:
KA 134 audio control console/KR 22 marker beacon receiver
KAP 100 single axis autopilot
KAP 150 two-axis autopilot
KAP 200 two-axis autopilot
KCN 90 GPS navigation system
KFC 150 two- or three- axis digital flight control system
KFC 200 two- or three- axis flight control system
KGS 55A compass system
KMA 24/24H audio control system with three-light marker beacon receiver or five-station intercom
KN 53 navigation receiver
KN 62A digital DME
KN 63 digital DME

The cockpit of the Beech Bonanza with all Honeywell Aerospace Silver Crown avionics

KNS 80 integrated navigation system (VOR/Loc receiver, digital DME, digital RNav computer, glideslope receiver)
KNS 81 integrated navigation system (VOR/Loc/glideslope receiver, RNav computer)
KR 86 automatic direction-finder
KR 87 digital automatic direction-finder
KRA 10A radar altimeter
KT 70 Mode S transponder
KT 76A transponder
KT 96 radiotelephone
KX 125 nav/com transceiver
KX 155/165 nav/com transceivers
KY 92 VHF communication transceiver
KY 196/KY 197 communication transceiver.

More detailed descriptions of some of these units can be found under the appropriate section headings in this book.

Operational status

In production for, and in service with, a wide range of light single- and twin-engined aircraft.

Contractor

Honeywell Aerospace, Electronic & Avionics Lighting.

VERIFIED

TCAS Systems: CAS 66A/CAS 67A/CAS 81

CAS 66A TCAS I

The CAS 66A TCAS I features a computerised processor which interrogates the ATC transponders and then displays the positions of up to 30 surrounding aircraft. Two antennas are used, mounted on the top and bottom of the aircraft, to minimise blind spots.

Equipped with the power of a TCAS II, the CAS 66A display can be optimised for maximum situational awareness. It offers crew-selectable ranges of 3, 5, 10, 15, 20 and 40 n miles. A selectable altitude display window gives three perspectives on traffic: normal view; upward view (2,700-8,700 ft above); downward view (2,700-8,700 ft below). CAS 66A was designed to meet TCAS II standards, and features bearing accuracy of ±3°, rather than the ±30° TCAS I requirement.

If the transponders of surrounding aircraft are not able to report altitude, the CAS 66A provides position only. If they are able to provide altitude, CAS 66A reports this – either relative or absolute – and trend data.

CAS 66A provides coverage even in non-radar environments, and eliminates the need for Loran or GPS sensors. Eliminating only the capacity to show Resolution Advisories (RAs), the CAS 66A's display options include the choice of several weather radar indicators, any compatible Electronic Flight Instrument System (EFIS), or a dedicated Traffic Advisory (TA) display. CAS 66A interfaces through the ARINC 429 standard databus. CAS 66A can be upgraded to TCAS II standard.

CAS 67A TCAS II

CAS 67A system components include a processor which calibrates the antennas, choice of one or two directional antennas, a Mode S transponder, to provide datalink communications and a cockpit display. Display options include an EFIS or weather radar interface, a combination Resolution Advisory/Vertical Speed Indicator (RA/VSI) which provides RAs in TCAS mode, but otherwise functions as a conventional VSI, and a combination flat-panel electronic Traffic Advisory/Vertical Speed Indicator (TA/VSI) which integrates TAs, RAs and VSI on one instrument. The CAS 67A system interrogates the transponders of surrounding aircraft, determining whether they are Mode A-, Mode C- or Mode S-equipped. If the intruder has Mode A, CAS 67A will provide position data. If the intruder has Mode C or Mode S the CAS 67A can show altitude and altitude trend data. If the intruder has Mode S the two transponders automatically establish communication and co-ordinate resolution manoeuvres for both aircraft.

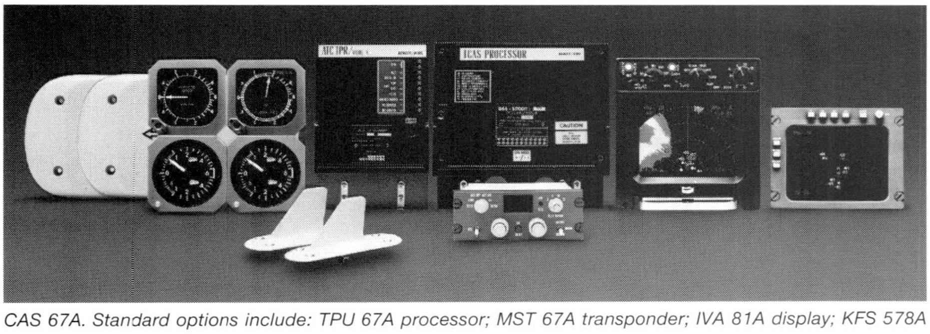

CAS 67A. Standard options include: TPU 67A processor; MST 67A transponder; IVA 81A display; KFS 578A controller; ANT 67A antennas; IVA 81B electromechanical display

CAS 81 TCAS II

Developed to meet the needs of airline operation, the CAS 81 TCAS II offers corporate aviation technology identical to the system in use today by most of the major airline carriers. The CAS 81 TCAS II system comprises a TCAS processor and a receiver/transmitter that supports all necessary surveillance and CAS logic functions. In addition, the CAS 81's processor serves as the central point of co-ordination for TCAS signal input and output, including aircraft systems interfaces, display drivers and audio outputs. Its modular design allows for flexibility in reconfiguring the processor I/O section to meet the needs of differing aircraft installations.

Patented interferometry processing provides for true omnidirectional reception of bearing data on the first reply from an intruder's transponder. It also provides bearing accuracy of ±3°, versus the ±15° FAA requirement. While especially important in crowded airspace, this increased accuracy helps minimise bearing errors due to an intruder's relative elevation, and reduces the effect of reflections from aircraft structures. Another feature of the CAS 81's processor is its ability to compensate for signal mismatches between antenna elements. This automatic antenna calibration also helps to ensure increased bearing accuracy.

The central processor also co-ordinates beam steering, eliminating the need for a separate antenna steering unit. A typical CAS 81 installation features two four-element directional antennas, which provide extremely accurate signal reception from any direction and altitude within surveillance range.

Other key components include a Mode S transponder to co-ordinate air-to-air communication between approaching aircraft. This co-ordination ensures that the Resolution Advisories (RAs) issued by CAS 81 result in the proper complementary manoeuvres for the two aircraft. The Mode S transponder functions in the same manner as a Mode C ATCRBS transponder in providing altitude and identity information to ATC ground stations.

Operational status

CAS 66A, CAS 67A and CAS 81A TCAS systems are in service with a large number of airlines worldwide. CAS 66A and CAS 67A are designed to meet the requirements of business aviation. CAS 81A is designed for commercial transport types.

These systems have recently been replaced in production by their Airborne Collision Avoidance System II (ACAS II) counterparts. ACAS II is a design standard that fulfils TCAS II requirements, and incorporates Change 7 software, that is a worldwide requirement to reduce the possibility of airborne collisions inherent in the earlier TCAS design standard.

Honeywell Aerospace has already progressed beyond the formal requirements of ACAS II, and is now producing equipment to a standard which it designates as Enhanced ACAS II/TCAS II (see separate entry).

Contractor

Honeywell Aerospace, Electronic & Avionics Lighting.

VERIFIED

TPR 2060 transponder

The TPR 2060 is a lightweight, compact air traffic control transponder designed for light aircraft and general aviation. It responds automatically to Mode A and Mode C interrogations and, with a suitable encoding altimeter input, will transmit aircraft altitude information with the normal reply pulses. A Mode B capability is optionally available for use in areas employing Mode B interrogation.

The TPR 2060 features special DME suppression circuitry to prevent interference between the transponder and DME installations when the antennas for the two systems are sited in close proximity. The system also permits transmission of a special identification pulse for a 20 second period by an ident button on the front panel. A reply lamp remains lit during this time to reassure the user that the transponder is identing.

Self-test facilities are incorporated. During self-test operation, the unit's coding and decoding circuits are exercised in the same manner as they would be during actual radar interrogation. The unit, which may be panel, console or roof mounted, is in a single case and is of large-scale integrated circuit-type construction.

Specifications

Dimensions: 45 × 160 × 215 mm
Weight: 1.18 kg

Operational status

In service.

Contractor

Honeywell Aerospace, Electronic & Avionics Lighting.

VERIFIED

TRS-42 ATC transponder system

The TRS-42 ATC transponder system is a digital 325 W solid-state transponder for positive identification in the ATC environment. It consists of the TR-421 transmitter/receiver and the CD-422 control display unit.

The TR-421 transmitter/receiver solid-state design gives 4,096 codes of operation, plus Modes A, B and Mode C altitude reporting when connected to an encoding altimeter. The unit utilises a single chip microprocessor which ensures code data validity and display. To increase system reliability the TR-421 utilises a dual-transmitter design. Under normal operating conditions the dual transmitters work together to provide a full 325 W of power. If one of the transmitters fails, the unit would continue to function, although at a reduced power capability. This feature is especially important in single transponder installations, where the loss of the transmitter would leave no identification capability.

The CD-422 control display unit has the capability to control a dual-transponder installation via a single control head. The selection is made by simply pressing the selector button on the front panel. In a single transponder installation, this button is not provided. The CD-422 also provides an annunciation of the letters ID whenever the transponder replies to an interrogation. When the mode selector is in the VFR position, the active transponder is channelled to the VFR 1200 code. This code may be preprogrammed according to other international VFR codes.

The TRS-422 provides full-time self-testing along with a pilot-selectable TEST mode. The self-testing monitors all key circuits such as the transmitter, receiver, encoder, decoder, video processor and central processor.

Specifications

Dimensions:
(control display unit) 63.5 × 79.38 × 63.5 mm
(front connector transmitter/receiver) 10.16 × 10.16 × 27.94 mm
(rear connector transmitter/receiver) 10.16 × 10.16 × 32.05 mm
Weight:
(control display unit) 0.27 kg
(front connector transmitter/receiver) 2.31 kg
(rear connector transmitter/receiver) 2.73 kg
Power supply: 18-33 V DC, 0.9 A nominal

Contractor

Honeywell Aerospace, Electronic & Avionics Lighting.

VERIFIED

VCS 40 VHF communication system

The VCS 40 VHF communication system is an all-digital transceiver employing microprocessor-controlled circuitry. It incorporates a full complement of Honeywell Aerospace III performance features including 20 W solid-state transmitter, continuous transmit capability at reduced power, white-on-black dichroic liquid crystal frequency display, 760-channel operation at 25 kHz spacing over 118 to 136.975 MHz, optional 1,360 channel extension up to 151.975 MHz, built-in SELCAL and ACARS capability and automatic self-test and diagnostics.

The VCS 40 consists of the VC 401B digital transceiver and the panel-mounted CD 402B control/display unit.

Specifications

Dimensions:
(transceiver) 104 × 101.6 × 333.6 mm
(control/display unit) 63.5 × 80 × 63.5 mm
Weight:
(transceiver) 2.78 kg
(control/display unit) 0.27 kg
Power supply: 18-33 V DC
(transmit) 4.5 A
(receive) 0.5 A
Altitude: up to 55,000 ft

Contractor

Honeywell Aerospace, Electronic & Avionics Lighting.

VERIFIED

VNS-41 VHF navigation system

The VNS-41 navigation system is a digital VOR/ILS receiver and processor system providing VOR, localiser, glideslope and marker beacon reception. It is a lightweight system consisting of the VN-411 receiver and the CD-412 panel-mounted control display unit.

The VN-411 receiver contains the VOR/Loc, glideslope and marker beacon receiver and processors. It employs full-time self-test monitoring of the key internal circuits such as the power supply, synthesisers and automatic gain control and all receivers and converter circuits. Advanced signal processing provides VOR accuracy within 1°, as well as steady navigation signals.

The CD-412 control display unit provides for a dual-frequency readout, one active and one standby. The frequencies are alternated via the frequency transfer button. The CD-412 displays 200 channels with 50 kHz spacing from 108 to 117.95 MHz. It also features a non-volatile frequency memory which retains the last frequency used, eliminating the possibility of frequency loss due to power interruptions. The CD-412 can display digitally either the bearing or radial to the selected VOR in the standby window.

Specifications

Dimensions:
(control display unit) 63.5 × 79.4 × 63.5 mm
(front connector receiver) 101.6 × 101.6 × 279.4 mm
(rear mount receiver) 101.6 × 101.6 × 320.5 mm
Weight:
(control display unit) 0.27 kg
(front connector receiver) 2.04 kg
(rear mount receiver) 2.81 kg
Power supply: 18-33 V DC, 0.8 A
Frequency:
(navigation receiver) 108-117.95 MHz
(glideslope receiver) 329.15-334 MHz
Channel spacing:
(navigation receiver) 50 kHz
(glideslope receiver) 150 kHz
Number of channels:
(navigation receiver) (VOR) 160, (Loc) 40
(glideslope receiver) 40

Operational status

In production.

Contractor

Honeywell Aerospace, Electronic & Avionics Lighting.

VERIFIED

Warning computer Mk VII

The warning computer Mk VII is a VLSI technology single ¼ ATR computer that can provide windshear detection and alerting, recovery guidance and advanced ground proximity in a single integrated economical flight safety system. The Mk VII is designed for all existing and new aircraft with ARINC 500 analogue avionics. It offers many benefits to owners of analogue aircraft.

Specifications

Dimensions: ¼ ATR short
Weight: 2.63 kg
Power supply: 115 V AC, 400 Hz, single phase, 11.5 W (max)
Reliability: 30,000 h MTBF

Operational status

In production and in service on Boeing and Airbus aircraft. In service on US Navy P-3C, E-2, C-2 and S-3 aircraft.

Contractor

Honeywell Aerospace, Electronic & Avionics Lighting.

VERIFIED

AA-300 radio altimeter

The AA-300 radio altimeter consists of RA-315 and RA-335 indicators, RT-300 transmitter/receiver and AT-220, -221 or -222 antenna.

The RT-300 transmitter/receiver is a solid-state unit offered in three optional configurations for different outputs.

The RA-315 indicator has a servo-controlled pointer display of radio altitude up to 2,500 ft. Below 500 ft the scale is expanded to enhance readability. There is an adjustable decision height bug and an amber decision height warning lamp. The RA-335 is similar to the RA-315 but is configured for helicopters, having a range of 0 to 1,500 ft. Below 200 ft the scale is expanded to improve readability.

Specifications

Dimensions:
(RA-315 and -335) 3 ATI × 1,143 mm
(RT-300) 104 × 116 × 281 mm
(AT-220) 63 × 159 × 142 mm
Weight:
(RA-315 and -335) 0.7 kg
(RT-300) 2 kg
(AT-220) 0.3 kg
Power supply: 21-32 V DC, 0.5 A
Accuracy:
(RT-300) 0-100 ft ± 3 ft, 100-500 ft ± 3%, 500-2,500 ft ± 4%
(RA-315) 0-100 ft ± 5 ft, 100-500 ft ± 5%, 500-2,500 ft ± 7%
(RA-335) 0-100 ft ± 5 ft, 100-500 ft ± 5%, 500-1,500 ft ± 7%

Operational status

In production.

Contractor

Honeywell Inc, Commercial Aviation Systems.

VERIFIED

Advanced Air Data System (AADS)

Honeywell's AADS introduces a new range of air data systems, designed to meet Reduced Vertical Separation Minimum (RVSM) requirements. Latest products in the range include the AZ-252 Air Data Computer (ADC); AZ-960 ADC; AM-250 barometric altimeter; and integrated multifunction probe.

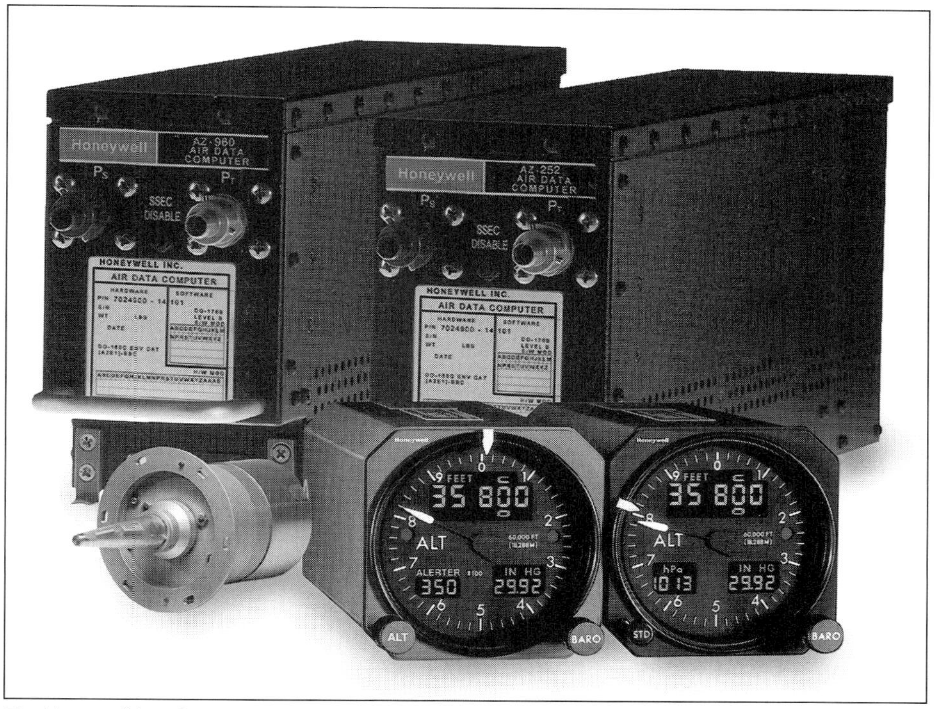

The Honeywell Inc, Commercial Aviation Systems' advanced air data systems, comprising AZ-252 ADC (top right), AZ-960 ADC (top left), AM-250 barometric altimeter (bottom right) and integrated multifunction probe (bottom left)
2000/0081857

Operational status

In August 1999, an installation comprising dual AZ-960 ADCs, dual BA-250 barometric altimeters and an AL-801 altitude alerter was certified for RVSM operation in Gulfstream II and IIB aircraft.

Contractor

Honeywell Inc, Commercial Aviation Systems.

VERIFIED

Advanced Common Flightdeck (ACF) for DC-10

Honeywell Commercial Aviation Systems worked with Boeing and Federal Express to install a new 'Advanced Common Flightdeck' (ACF) on the Federal Express company's fleet of DC-10 trijets.

The cockpit upgrade is part of a long-term, two-phase project that could eventually convert many DC-10 transports to advanced technology freighter aircraft.

The 'glass cockpit' programme will give the upgraded aircraft the latest in cockpit and control systems technology, including MD-11-derived Category IIIB autoland capability and navigation systems/sensors, as well as commonality with new aircraft coming off assembly lines.

The ACF is based on Honeywell's integrated cockpit design for the two-crew MD-11, with six across 8 × 8 in displays showing all flight and systems information.

The electronic flight instruments, based on Honeywell's liquid crystal 'flat panel' displays, will actually take the MD-10 cockpit a generation beyond the MD-11.

The ACF derives from the aircraft system controllers originally designed for the MD-11. This system manages the functions of all major aircraft systems including the hydraulic and fuel systems, eliminating the need for a flight engineer as a crew member.

The ACF also uses a new generation of high-speed, high-capacity computers built around Honeywell's Versatile Integrated Avionics (VIA) architecture.

The VIA architecture is based on shared computing resources and robust partitioning developed by Honeywell for its Integrated Modular Avionics (IMA) concept.

The VIA system design for the DC-10 upgrade will integrate multiple functions in the VIA computers, including displays; flight management; central aural warning.

Three of the VIA computers, coupled with a pair of new aircraft interface units linking them to the aircraft systems, replace 22 separate computers in the existing DC-10 design. With other changes in the avionics bay, an overall saving of approximately 1,000 lb in removed equipment is predicted as a result of the ACF modification.

The VIA-based ACF concept is also being designed into the new MD-95 100-passenger twin jet.

Operational status

A successful first flight, complete with Category IIIB automatic landing was made in April 1999. FedEx placed 79 firm orders, with options for 40 more. Certification was completed during 2000.

Contractor

Honeywell Inc, Commercial Aviation Systems.

UPDATED

Advanced GNS/IRS integrated navigation system

Honeywell has integrated laser inertial functions with the Global Navigation Satellite System (GNSS). The advanced GNS/IRS offers high reliability, long life, low power consumption, small size, light weight, fast alignment, full performance for alignment up to 78° latitude and improved BITE.

The advanced 4 MCU IRS is 60 per cent smaller, 40 per cent lighter and uses 50 per cent of the power of Honeywell's 10 MCU systems, while meeting the same performance specifications. The heart of the system is the Ring Laser Gyro (RLG). The advanced IRS is fully provisioned for GNSS integration. The blending of these two systems into one navigation solution offers precise navigation accuracies.

The GNSS unit will continuously track all satellites in view to give accuracies of 25 m or better with selective availability switched off. The GNSS unit offers growth potential for Wide Area Augmentation System (WAAS) and Local Area Augmentation System (LAAS) as well as the capability for differential GPS. Further enhancements under development include capabilities to support precision approaches to unimproved runways, automatic dependent surveillance and sole-means navigation.

Specifications

Dimensions: 124.5 × 317.5 × 193 mm
Weight: 12.25 kg
Accuracy:
(navigation) (IRS) 2 n miles/h,
(GNS/IRS hybrid) 25 m
(velocity) (IRS) 12 kt
(GNS/IRS hybrid) 0.3 kt
(attitude) 0.1°
(heading) 0.4°

Contractor

Honeywell Inc, Commercial Aviation Systems.

VERIFIED

Aero-I SATCOM satellite communications system

The Honeywell/Racal Aero-I SATCOM satellite communications system provides operators with flexible access to the Inmarsat Aero-I spotbeam service for multichannel voice and data, as well as single-channel data in the continuous global coverage regions.

The three-channel systems consist of a Satellite Data Unit (SDU) and a High-Power Amplifier (HPA). MCS-3000 Aero-I system capability can be expanded to the six-channel MCS-6000 Aero-I version simply by adding a 4-MCU Radio Frequency Unit (RFU). The addition of the Intermediate Gain Antenna and a Diplexer/Low Noise Amplifier completes the system. External beam steering, required by Aero-H, is not necessary with Aero-I, since the new 20W HPAs contain this function.

The Honeywell Inc, Commercial Aviation Systems' advanced common flightdeck for the M-10 freighter
2000/0081862

Operational status

The Aero-I system complements the established Inmarsat Aero-H service, which has become a standard installation on long haul, wide-body aircraft since its introduction in 1990. All services currently offered with Aero-H systems are available on Aero-I, including:

- Cockpit voice, allowing instantaneous communication with operations, maintenance and air traffic control
- Passenger telephony
- Passenger fax
- News and weather broadcasts
- Interactive passenger services
- PC data capability

The Aero-I SATCOM system gained its Supplemental Type Certificate (STC) from the US Federal Aviation Administration on the Next-Generation Boeing 737-800 in November 1998. The aircraft are operated by launch customers Royal Air Maroc and Hainan Airlines; the system will utilise the ground stations of the Satellite Aircom Consortium, which have been upgraded to provide worldwide coverage for the Aero-I service.

These certifications directly correspond with the Aero-I capabilities of Inmarsat ground earth stations. Inmarsat Aero-I service operates in the spot beams of the new generation Inmarsat-3 satellites. Since spot beams have lower power requirements, smaller, lower power HPAs can be used and smaller Intermediate Gain Antennas can be installed for operation.

In addition to supporting passenger communications, it is intended that Aero-I should become the first compact SATCOM system to comply with the International Civil Aviation Organization's (ICAO) Standards and Recommended Practices (SARPS). These form a comprehensive set of specifications for aeronautical satellite communications when used for cockpit and safety of flight applications, such as Air Traffic Control and Future Air Navigation System (FANS) technologies.

Contractors

Honeywell Inc, Commercial Aviation Systems.
Racal Avionics.

NEW ENTRY

AHZ-800 Attitude Heading Reference System (AHRS)

The AHZ-800 is the next generation Attitude Heading Reference System (AHRS) designed for high performance and high reliability while attaining lower power dissipation and reduced size and weight. This is accomplished through the use of advanced manufacturing techniques, such as very large-scale integration and application specific integrated circuits, and fibre optic rate-sensing advanced sensor technology.

Honeywell has developed a practical interferometric fibre optic gyro sensor that replaces the heavier less reliable spinning iron rate-sensors used in the conventional AHRS. The advent of the interferometric fibre optic gyro sensor makes the AHZ-800 a truly solid-state device. The AHZ-800 is a 4 MCU package that outputs attitude, heading and rate data on ARINC 429 and ASCB digital buses. The attitude and heading source approaches the performance of an inertial reference system, but at a significantly lower cost.

Operational status

The AHZ-800 is in service with the Dornier 328 regional airliner.

Contractor

Honeywell Inc, Commercial Aviation Systems.

VERIFIED

AIMS Airplane Information Management System

Honeywell's AIMS architecture, for the B-777 is an advanced avionics system designed to meet very high requirements for functionality, maintainability and dispatch reliability.

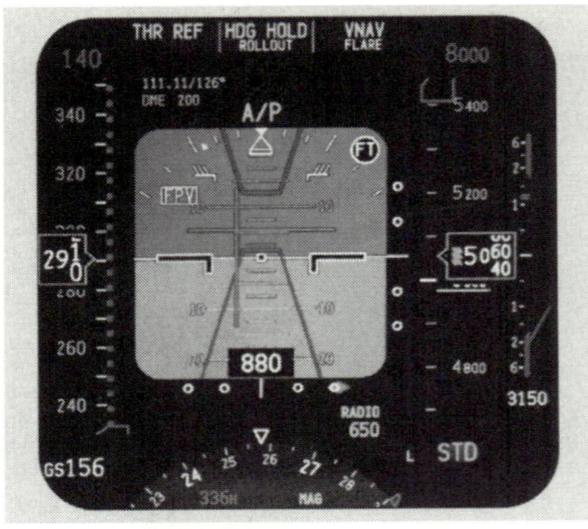

AIMS Primary flight display 0001436

The AIMS consists of dual integrated cabinets that contain all of the central processing and input/output hardware to perform the following functions: flight management; displays; navigation; central maintenance; airplane condition monitoring; flight deck communications; thrust management; digital flight data; engine data interface; data conversion gateway.

By eliminating the need for separate Line Replaceable Units (LRUs) for each subsystem - each with its own power supply, processor, chassis, operating system, utility software, input/output ports and built-in test - the AIMS concept saves significant weight, space and power consumption on board the airplane while improving overall system reliability and maintainability.

Displays

The 777's flight deck features six 'D' size (8 × 8 in) Active Matrix Liquid Crystal Displays (AMLCDs) in a horizontal layout similar to the 747-400. These advanced technology screens display primary flight, navigation and engine information with automatic reversion capability.

The flight management function for the Boeing 777 takes advantage of Honeywell's mature systems developed for the 757, 767 and 747-400 aircraft, while at the same time providing greatly improved performance, functional capability and growth potential for the Future Air Navigation System (FANS).

The system includes three passively cooled Multifunction Control Display Units (MCDUs) featuring full-colour flat panel LCDs. The MCDU controls the flight management function and is capable of controlling other ARINC subsystems.

The Honeywell CMF consolidates the maintenance activity of virtually all systems on the B-777. A primary objective is to reduce maintenance by improving fault isolation and detection capability, thus reducing the number of spares needed as well as minimising dispatch delays and cancellations.

The B-777's ACMF enables the flight crew to make informed decisions about the state of the aircraft. The system continuously monitors engines and aircraft systems and generates reports that can be customised to individual airline needs.

The DCMF/FDCF is the intermediate system of the airborne datalink infrastructure. It serves as the router between various airborne applications, the printer, the onboard Local Area Network devices and ground-based datalink applications via various air-ground subnetworks (such as VHF radio and SATCOM).

It also functions as the crew interface to the datalink system by means of cockpit displays, the CDU and cursor control device. It supports airline programmable ACARS applications.

The Boeing 777 AIMS provides capability for the airlines to customise various aircraft systems such as the Airplane Condition Monitoring Function (ACMF) as well as for flight management, flight deck and data communications, central maintenance and the electronic checklists. This capability is enabled by the AIMS philosophy of fault containment and software partitioning.

Operational status

In service on Boeing 777 aircraft.

Contractor

Honeywell Inc, Commercial Aviation Systems.

UPDATED

Air Data Inertial Reference System (ADIRS)

The ARINC 738 Air Data Inertial Reference System (ADIRS) is an air data computer combined with an Inertial Reference System (IRS). Together with the Pegasus flight management system and TCAS 2000 collision avoidance system, the ADIRS forms the WorldNav™ system.

The ADIRS provides complete ARINC inertial reference system outputs including primary attitude and heading, body rates, acceleration, groundspeed, velocity and aircraft position and ARINC 706 air data outputs, which include altitude, true airspeed, Mach number, air temperature and angle of attack.

Each ADIRS is equipped with three air data inertial reference units, one control display unit and eight air data modules mounted remotely, adjacent to the pitot and static pressure sensors.

The air data reference electronics and the laser gyro inertial reference units are packaged in a 10 MCU box or 4 MCU box and require a nominal power of 109 W. The unit meets the functional requirements of ARINC 738 and the environmental requirements of DO-160b. On the A320, aircraft maintenance is simplified by extensive reporting of ADIRS LRU Operational Status to the centralised fault data system.

The air data module requires a nominal power of 1.8 W. The unit meets the environmental requirements of DO-160b and features a solid-state pressure transducer.

The control display unit is packaged in accordance with ARINC 738 and requires a nominal power of 5 W exclusive of warning lights. The unit meets the functional requirements of ARINC 738 and the environmental requirements of DO-160b. It features a liquid crystal display.

Specifications

Dimensions:
(ADIRU) 322.6 × 322.6 × 193 mm
(ADM) 50.8 × 76.2 × 152.4 mm
(CDU) 170.2 × 146 × 152.4 mm

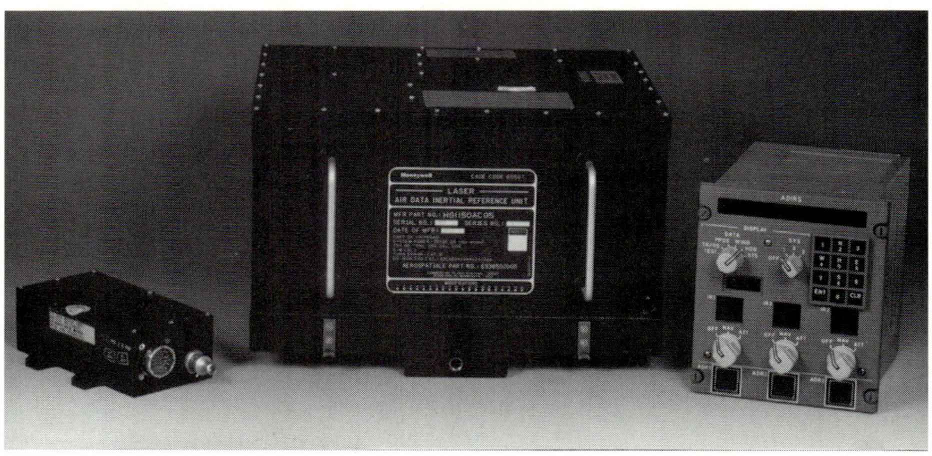

The Honeywell ADIRS air data inertial reference system showing (left) the air data module, (centre) the integrated air data/inertial reference unit and (right) the integrated control display unit

Weight:
(ADIRU) 19.5 kg (10 MCU box)
or 10 kg (4 MCU box)
(ADM) 0.63 kg
(CDU) 2.27 kg

Operational status
ADIRS equips Airbus A319, A320, A330 and A340 aircraft.

Fault Tolerant Air Data Inertial Reference System (FT-ADIRS)
The Fault Tolerant Air Data Inertial Reference System (FT-ADIRS) consists of a Fault Tolerant Air Data Inertial Reference Unit (FT-ADIRU), a Secondary Attitude and Air data Reference Unit (SAARU) and six Air Data Modules (ADMs). The FT-ADIRU provides attitude and heading data for inertial navigation as well as air data computations. The SAARU provides a back-up source of attitude and air data computations. The ADMs provide both the FT-ADIRU and SAARU with three redundant sources of static and pitot pressure data.

Operational status
The FT-ADIRS is standard on the Boeing 777 aircraft.

Contractor
Honeywell Inc, Commercial Aviation Systems.

VERIFIED

AM-250 barometric altimeter

The AM-250 integrates Honeywell's silicon pressure sensor with an indicator produced by Ametek Aerospace Products Inc. It meets Reduced Vertical Separation Minimum (RVSM) requirements.

In addition to full analogue and digital displays, the AM-250 has two displays to indicate barometric offset in hectoPascals or inches/millimetres of mercury. A rotary encoder and push-button switch allow the entry of barometric offset and auto barometric set. The altimeter is internally lit.

The AM-250 is available in three versions: one provides basic ARINC 429 output labels; another supports True AirSpeed (TAS) for FMS/GPS systems; the third provides an altitude preselect function. The unit can be configured to display height in feet or metres. It includes annunciators for altitude deviation and failure warning.

Specifications
Altitude range: −1,999 to +60,000 ft
Altitude display – accuracy: ±2 ft
Altitude display – pointer: ±10 ft
Barometric setting range: 16.00 to 32.00 InHg
Barometric setting accuracy: ±1 ft
Dimensions: 3 ATI, length 170.2 mm
Weight: 1.36 kg
Power:
(meter) 28 V DC, 12 W
(lighting) 5 V DC, 2.75 W

Operational status
In production for CitationJet among other aircraft.

Contractors
Honeywell Inc, Commercial Aviation Systems.
Ametek Aerospace Products Inc.

VERIFIED

Automatic Flight Control System (AFCS) for the A 109

The Agusta A 109 flight control system has been designed to reduce pilot workload and improve reliability and safety at a realistic cost. A typical installation includes two autopilot (helipilot) systems for redundancy, which may be used with or without a flight director system.

This duplex system consists of one directional and two vertical gyros, controller panel, two computer units and five series actuators, two each for pitch and roll and one for yaw control. A trim computer and two trim actuators for pitch and roll adjustments are also part of the system.

The Honeywell Inc, Commercial Aviation Systems CD-820 **2000**/0081859

A flight director facility can be incorporated but requires the addition of a mode controller and navigation receiver inputs, plus ADI and HSI. Flight director functions include heading, vertical speed, barometric altitude and airspeed select, go-around procedure selection, and VOR, ILS, MLS, Omega/VLF, RNav system and Loran or Tacan coupling.

Operational status
In service in the Agusta A 109 helicopter.

Contractor
Honeywell Inc, Commercial Aviation Systems.

VERIFIED

CD-820 Control Display unit

The CD-820 Flight Management System (FMS) CD-820 control display unit is a replacement for the CD-800, CD-810 and CD-815 units.

CD-820 has a large, full colour, Active Matrix Liquid Crystal Display (AMLCD) with an actual writing area of 13.8 in (320 × 240 pixels). CD-820 also interfaces with Honeywell cabin entertainment systems, such as OneView and Airshow.

When used with Honeywell's FMS version 6.0 software, the CD-820 offers additional features such as weather, when interfaced with Teledyne's Telelink, as well as with Honeywell's EGPWS. Single-button selection of video, graphics and Air Traffic Control (ATC) offers the pilot easy access to camera video, weather and CNS/ATM ATC data.

An additional feature of the CD-820 is incorporation of a new 'fn' key, offering operators an easy way to correct entries into the flight management system, altering flight plans or making immediate updates to ATC requirements.

Operational status
In production.

Contractor
Honeywell Inc, Commercial Aviation Systems.

UPDATED

Classic Navigator

Honeywell's Classic Navigator is designed as a cost-effective solution to the performance and reliability problems of mechanical spinning gyro ARINC 561 Inertial Navigation Systems (INS).

Honeywell's Classic Navigator combines the Honeywell/Trimble HT-9100 navigation management system with an all-digital Attitude Heading Reference System (AHRS). The Classic Navigator system combines the benefits of Global Navigation Satellite Sensor (GNSS) technology with established airline flight management operational procedures, to meet CNS/ATM functionality requirements.

The AM-250 barometric altimeter **2000**/0062206

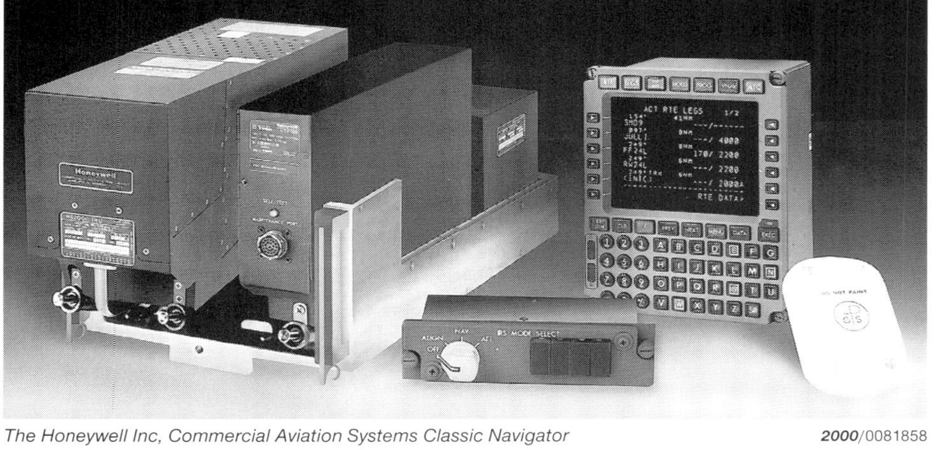

The Honeywell Inc, Commercial Aviation Systems Classic Navigator **2000**/0081858

Operational status

Available for such classic aircraft as the A300-B4, B-747-100/-200/-300, DC-10 and L-1011.

Contractor

Honeywell Inc, Commercial Aviation Systems.

VERIFIED

CM-950 Communications Management Unit (CMU)

In July 1998, Honeywell added the ARINC 758-compliant CM-950 CMU to its WorldNav CNS/ATM product line, designed to support future 'free flight' airborne datalink protocols and applications, as a successor to ACARS management units.

The CM-950 CMU is fully partitioned, allowing airlines to modify user-defined functions without the need for costly and time consuming recertification.

Data loading is available by means of a standard ARINC 615 data loader. For high speed loading, the CMU is equipped with a PCMCIA interface and is provisioned for future Ethernet data loading capability.

The CM-950 CMU provides an easy upgrade path to future airline requirements, and combined with Honeywell's ground-based Airline Maintenance and Operations Support System (AMOSS), it provides a complete end-to-end product for any operator interested in a turnkey solution.

Operational status

The US Air Force has awarded Honeywell an Indefinite Delivery/Indefinite Quantity (ID/IQ) contract to supply a Communications Management Unit (CMU) and satellite communications system (satcom) to support its Global Air Traffic Management (GATM II) upgrade programme, up to a total quantity of 1,681 MCS-7000I satcom units and 3,554 CM-950 units by 2007.

Contractor

Honeywell Inc, Commercial Aviation Systems.

VERIFIED

Honeywell's CM-950 communications management unit ***1999***/0044821

Data Nav V navigation/checklist display system

Honeywell's Data Nav V turns any Honeywell colour weather radar screen into a versatile en route navigation map or aircraft checklist. It allows pilots to select navigation waypoints and instantly create new waypoints by positioning an electronic designator 'bug' anywhere on the radar screen.

The Data Nav systems are designed to add navigation map and/or checklist display capability to all Honeywell colour weather radar indicators. The latest version, Data Nav V, features an ARINC 429 interface with a variety of Flight Management System (FMS) and long-range navigation systems, enhanced on-screen information and simplified pilot operation. Operators can compile their own aircraft-specific checklists on a personal computer and load them quickly and easily into the Data Nav computer, using the new optional Honeywell Standard Checklist software program.

The ARINC 429 interface links the Data Nav V system, not only with the Honeywell FMZ-2000 FMS system, but those of other makes that provide ARINC 429 outputs, as well as with certain other long-range navigation systems. The FMS or long-range navigation system computer will lay out a flight plan, convert it to ARINC 429 digital data format and transmit it to the Data Nav V computer, which generates navigation map, course, heading, latitude-longitude position and waypoint symbology for display on the Honeywell weather indicator. In addition, Data Nav V, using FMS or nav system inputs, will display VORTAC locations, track lines, projected course, distance to waypoints and ground speed.

En route, the navigation display creates a moving picture of the aircraft's position as preselected courses are flown. Combined with the radar weather display, it gives a complete visual presentation of selected course and weather conditions ahead.

Data Nav V presents a full menu - up to 200 custom preprogrammed pages of normal and emergency preflight procedures and cockpit checklists, together with performance data, operating notes and other selected data.

The Data Nav V Electronic Checklist simplifies the routine checklist procedure and permits quick retrieval of emergency checklists, reducing the chance of errors and oversight by giving positive indication of a checked procedure. Because the system remembers checklist position, a pilot may instantly switch back and forth between weather/map display and the previously selected checklist.

Data Nav V provides nav map and checklist capabilities for both EFIS and non-EFIS equipped aircraft.

It can be used with a two or four-tube EFIS system, where the Honeywell radar indicator serves as an additional display to enhance overall system capability.

The Data Nav V system is available in three configurations, with panel-mounted controllers for selecting nav map with weather only, weather and checklists only, or either map/weather or checklist displays. Data Nav V is a direct replacement for the Honeywell Data Nav III system, when using the most common FMS interfaces.

Operational status

Data Nav I, II and III no longer in production. Replaced with Data Nav V, in production.

Contractor

Honeywell Inc, Commercial Aviation Systems.

UPDATED

DU-870 CRT Display Unit

Honeywell produces an 8 × 7 in (203 × 177.8 mm) CRT display that provides all the information found on larger displays, yet fits in smaller cockpits. The DU-870 streamlined display is designed for the cockpits of twin-turbine aircraft and is featured in the Primus 1000 and 2000 advanced avionics system.

The DU-870 is the result of aircraft panel size and fit surveys conducted by Honeywell to determine the optimum size, shape and screen area required to meet market needs. The Primus 1000 and 2000 display system architecture provides the ability to view any display format on the DU-870. A micro symbol generator is integrated into the DU-870 to reduce aircraft wiring.

As with the other components of the Primus 1000 and 2000 system, hardware techniques such as surface-mount technology, very large-scale integration and application specific integrated circuits are incorporated to minimise weight and volume. The DU-870 also features self-contained cooling for improved reliability and quiet operation.

Contractor

Honeywell Inc, Commercial Aviation Systems.

VERIFIED

EDZ-605/805 electronic flight instrument systems

The EDZ-605 and EDZ-805 electronic flight instruments are intended for corporate aircraft and regional airliners. The EDZ-605 is a 5 in (127 mm) system, while the EDZ-805 is a 6 in (152 mm) system. The products are the result of a new symbol generator and software changes. The generator has increased stroke writing and memory capabilities, while software updating has provided a variety of cosmetic improvements to the display.

The Electronic Attitude Director Indicator (EADI) features an enlarged sphere presentation with linear pitch tape, improved single-cue aircraft symbol and stroke filled single-cue command bar. A digital T-bar airspeed and/or angle of attack presentation has also been added. Other improvements include a stroke filled roll pointer, larger roll indices, shorter horizon indices and colour reversal on glide slope, angle of attack, rate of turn and expanded localiser scales. In addition, the 5 in (127 mm) EADI is now available in a truncated sphere presentation.

The Electronic Horizontal Situation Indicator (EHSI) features a stroke filled lubber line, colour reversal on the glide slope scale and weather radar mode annunciation.

The ED-605/ED-805 are compatible with most analogue systems and all digital systems.

Operational status

In production. Chosen for many business aircraft.

Contractor

Honeywell Inc, Commercial Aviation Systems.

VERIFIED

The Honeywell EHSI for corporate aircraft and regional airliners

EDZ-705 Electronic Flight Instrument System (EFIS)

The EDZ-705 Electronic Flight Instrument System (EFIS) is tailored to meet the needs of specific helicopter applications from executive use to search and rescue missions. The system provides the ability to display standard search patterns as provided by various navigation sources, collective cue symbology which allows the pilot to follow flight director collective cue demands, hover display symbology which allows the pilot to track and maintain target location in relation to the helicopter position in a search and rescue environment and four-axis helicopter flight director mode annunciation.

Since panel space is at a minimum, the EDZ-705 offers 5 × 5 in (127 × 127 mm) displays.

Contractor

Honeywell Inc, Commercial Aviation Systems.

VERIFIED

Electronic Flight Instrument System (EFIS) for the MD-80 aircraft

The Electronic Flight Instrument System (EFIS) gives a pictorial presentation of all primary flight instrument data. In addition to displaying conventional ADI and HSI

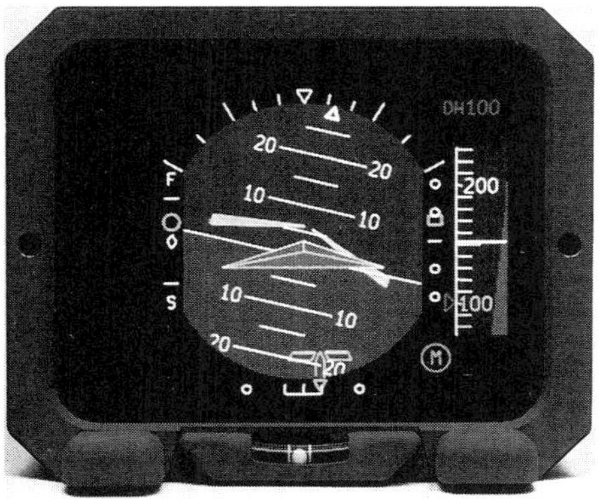

The Honeywell MD-80 EFIS

information, these easily installed indicators also display weather information, advisory maps and radio altitude data. Maximum flexibility is achieved with numerous pilot-selectable display formats. The compact, easy-to-read system is programmed with standard commercial display formats and can also be tailored to specific customer requirements.

Specifications

Dimensions: 129 × 154 × 267 mm
Weight: 4.8 kg
Power supply: 115 V AC, 400 Hz, 65 W

Operational status

In production and in service in MD-80 Series aircraft.

Contractor

Honeywell Inc, Commercial Aviation Systems.

VERIFIED

FMZ-2000 Flight Management System (FMS)

The FMZ-2000 is Honeywell's current generation Flight Management System, designed for applications in corporate and regional airliners. The system is available in two configurations: a stand-alone navigation computer, the NZ-2000, or a circuit card in the Honeywell IC-800 integrated avionics computer. Both configurations are compatible with the Avionics Standard Communication Bus (ASCB) and ARINC 429 system architecture. They include similar operational benefits and feature a full colour Control and Display Unit (CDU) and associated DL-900 data loader (see separate entry).

Both FMZ-2000 configurations provide a premium user interface including an easy-to-read keypad with select keys and a colour display. In addition, the system provides a worldwide navigation database including Standard Instrument Departures (SIDs), Standard Terminal Arrival Routes (STARs) and approach procedures. Multiple sensor inputs, including GPS, IRS, and DME/DME positioning are accommodated. The system provides vertical navigation in climb and descent, multiple holding patterns, and a non-precision approach capability with high precision waypoints.

The system also includes, Honeywell's patented algorithm, SmartPerf™. SmartPerf™#153; effectively 'learns' aircraft-specific performance. The SmartPerf™ function enables the navigation computer to learn the performance of a specific aircraft, thus affording operators access to performance calculations previously available only to aircraft equipped with a dedicated performance management computer. Using the aircraft's performance database, and both entered and sensed atmospheric data, SmartPerf™ provides performance calculations of time, fuel and predicted altitude at all waypoints; time, fuel, and distance to top of climb and top of descent; maximum endurance targets; optimum cruise altitude; time and distance to bottom of step climb; figure of merit indicating accuracy of fuel calculations and predictions for stored flight plans. Pilots have the option of manually entering altitude and calibrated airspeed or Mach speed constraints.

The FMZ-2000 also performs coupled vertical guidance to multiple, three-dimensional waypoints. Using multiple waypoint vertical navigation allows operators to enter altitudes for each waypoint, both climbing and descending. From this data, the navigation computer displays precise altitude crossings of all predefined waypoints and a complete vertical profile, including top of climb and top of descent on compatible Electronic Flight Instrument Systems (EFIS).

When equipped with Global Positioning System (GPS) sensors and receiving valid GPS data, the FMZ-2000 will navigate entirely by GPS and still maintain input from all available sensors. This affords operators the accuracy of GPS with the integrity and safety provided by a multiple sensor system.

The latest upgrade, known as FMZ-2000 Version 5.X, is available for both Primus 2000 and NZ-2000 based systems. Version 5.X has gained TSO approval for the Citation X, ATR 42 and Gulfstream V. The Version 5.X upgrade features a Pentium processor together with a larger 16 Mbyte database memory, providing additional capacity for future expansion to meet CNS/ATM requirements. Software improvements to be offered include: parallel database loading, terminal area speed targets, SLS/LAAS compatibility, multiple flight plans on one data loader disk and support of 8.33 kHz communications tuning for European airspace. Take-Off and Landing Data (TOLD) software enhancements, which reduce pilot workload by automating the computation of take-off and landing data, can also be selected by the customer for specific aircraft types.

To make training even easier, Honeywell has introduced optional Personal Computer (PC)-based training for the FMZ-2000. This product will allow pilots and maintenance personnel to accomplish self-paced, interactive training using a CD-ROM-based instruction set. Training is designed in modules and can now be accomplished anywhere the customer has access to a suitably equipped personal computer.

Operational status

TSO'd in a number of aircraft versions, the stand-alone system is currently standard equipment on the Bombardier Global Express, Gulfstream IV, Dassault Falcon 900B, and Raytheon Hawker 800XP. It is optional on the Dassault Falcon 2000 and the Falcon 50EX, Cessna Citation VII, and Embraer 145. it is also STC'd in a number of retrofit applications including the Challenger 600/601, Hawker 800/1000, Boeing 727, Cessna Citation V and Beechjet 400. The integrated system is standard equipment on the Gulfstream V, Cessna Citation X, Dassault Falcon 900EX and Dornier 328.

Contractor

Honeywell Inc, Commercial Aviation Systems.

UPDATED

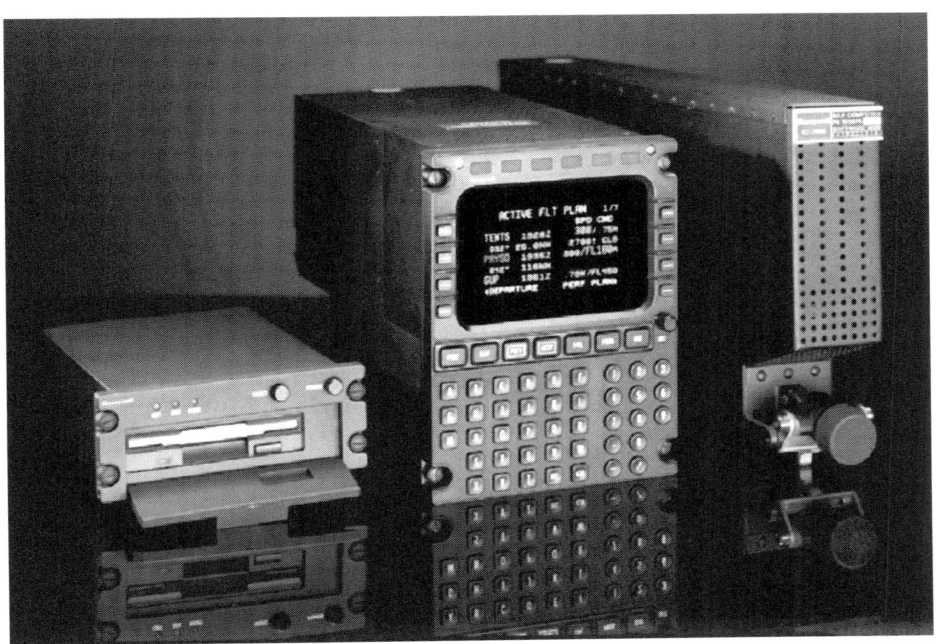

Honeywell's FMZ-2000 flight management system with DL-900 data loader, Control Display Unit and NZ-2000 Navigation Computer **2001**/0103883

HT1000 GNSS navigation management system

The HT1000 system consists of three Line-Replaceable Units (LRUs): Navigation Processor Unit (NPU); Multifunction Control and Display Unit (MCDU); Antenna Coupler Unit (ACU);

The NPU is the primary processing and control unit of the HT1000. It contains a 12-channel GPS receiver, multiple microprocessors, a navigation database and external system interfaces.

The GPS receiver receives and processes GNSS signal data from the antenna coupler unit to compute aircraft position and velocity. Using its 8 Mbyte navigation database and GNSS position, the NPU performs all the functions necessary to provide aircraft guidance through the waypoints of a selected flight plan.

Built for adaptability in all applications, the NPU features analogue, digital and discrete interfaces to support all aircraft types. The NPU functionality is packaged in a small, lightweight 2-MCU size unit.

Data entry and display to the pilot are accomplished through the use of a colour ARINC 739 compatible MCDU. The MCDU controls the user input and display function of the HT1000, as well as other ARINC 739 subsystems. The MCDU offers high reliability and a flexible presentation. The keyboard includes a 66-key full alphanumeric keyboard, dedicated function keys and 12 line-select keys. The display is a 5.5 in diagonal, multicolour active-matrix liquid crystal flat panel.

The ACU receives, amplifies, conditions and sends GPS signals to the GPS receiver in the NPU. The ACU contains an omnidirectional flat microstrip antenna with integral preamplifier and is referred to as 'active' since it performs the first stage of signal amplification. The ACU permits ACU to NPU cable runs in excess of 100 ft, which enhances the ease and flexibility of installation design. Moreover, the ACU is designed to optimise the GPS receiver performance and has improved filtering to prevent SATCOM interference.

The HT1000 is designed to achieve Federal Aviation Administration TSO C129, Class 1A certification, permitting GPS navigational use for Instrument Flight Rules (IFR) en route, terminal and approach operations. The HT1000 includes Receiver Autonomous Integrity Monitoring (RAIM).

The HT1000 can accept inputs from auto-tuned Distance Measuring Equipment (DME) and Inertial Navigation System (INS) to provide multisensor navigation per TSO C115a.

The HT1000 will implement Required Navigation Performance/Actual Navigation Performance (RNP/ANP) similar to the Honeywell FANS 1 software. RNP is the required performance accuracy of a particular segment of airspace. ANP is a measure of the uncertainty in the position estimate of the system.

On aircraft with enhanced flight guidance equipment, dual HT1000s will allow the airlines to apply to operate under the 'G' flight plan category.

The HT1000 is a full-flight regime lateral and vertical navigation system with interfaces to the aircraft flight instruments, flight director/autopilot and ARINC 739 compatible systems through the MCDU. Based on the

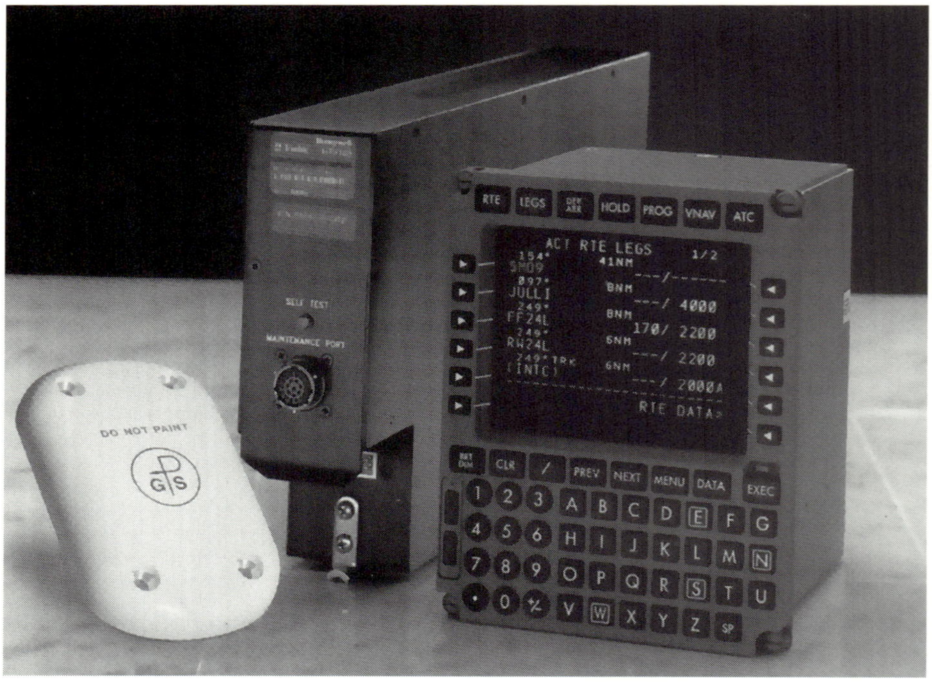

HT1000 GNSS navigation management system showing antenna coupler unit (left) navigation processor unit (centre) and multifunction control and display unit (right) 0015382

programmed flight plan, the NPU provides coupled lateral guidance. Using the flight plan and fuel flow inputs, the NPU provides advisory climb, en route and descent vertical guidance.

The system is designed to take advantage of a pilot's knowledge of current air transport FMS operation, thereby providing 12 line-select keys, standard ARINC 739 keys, dedicated function keys and a scratchpad field for data input and messages.

A flight planning function allows pilots to create new flight plans via traditional air transport methods or through company route entry.

The HT1000 system's 12-channel GPS receiver provides continuous all-in-view satellite tracking. The digital receiver computes position updates five times per second, measures position to 15 m and measures velocity to 0.1 kt. The GPS receiver is capable of receiving and utilising digital differential corrections for applications requiring increased accuracy and integrity. Slant range information received from DME-DME is used for position calculation. DME-DME position is automatically utilised as the primary source in the event of a GPS failure.

A configuration module, installed on the mounting tray, reduces maintenance actions by eliminating connector strapping. The module is programmable via the MCDU (or separate connector port) enabling individual aircraft type definitions to be easily accomplished.

An installation checkout and maintenance selection uses the MCDU display to show system installation status as well as inputs. Maintenance and historical BITE information is also available for downloading.

A single NPU has the capability to interface with two MCDUs and provide each crew member with input/output capability. Dual HT1000 system hardware consists of two each of the LRUs listed in the system description. With dual units, each system can be interfaced with its onside instrumentation and by request, data entry from offside units can be transferred between units. In the event of sensor failure, the offside data can be used in both systems.

Operational status

The HT1000 is in production. Growth options include ARINC 724B ACARS datalink. ARINC 610A SimSoft, Wide Area Augmentation System (WAAS) en route, WAAS Cat I approach, Automatic Dependent Surveillance (ADS), Local Area Differential GPS (LADGPS) (special) Cat I, II approach, and Global Orbiting Navigation Satellite System (GLONASS) capability.

Contractors

Honeywell Inc, Commercial Aviation Systems.
Trimble Aerospace Products.

VERIFIED

HT 9000 GPS navigation system

The Honeywell/Trimble HT 9000 remote-mounted GPS navigation system is designed to meet the demanding certification, operational and maintenance requirements of today's commercial, air transport and corporate operators. The HT 9000 is FAA TSO C-129 (A1) approved for IFR en route, terminal and non-precision approach procedures as well as primary oceanic and remote means of navigation. The nine-channel all-in-view Navigation Processor Unit (NPU) interfaces to most digital and analogue aircraft systems. It also provides the user with the ability to fly all 24 ARINC leg types with the accuracy and precision of GPS. The HT 9000 provides the option of a compact, ARINC, or an FMS-like multifunction Control Display Unit (CDU).

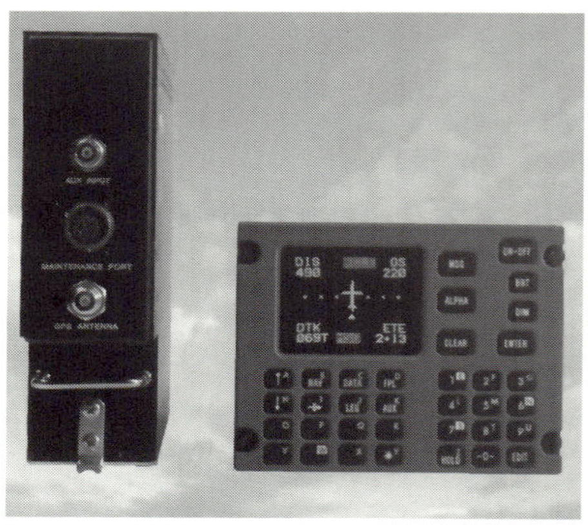

HT 9000 GPS navigation system showing the navigation processor unit (left) and control/display unit (right) 0015381

Contractors

Honeywell Inc, Commercial Aviation Systems.
Trimble Aerospace Products.

VERIFIED

HT 9100 GNSS navigation management system

The HT 9100 system is a full flight regime lateral navigation system, designed to provide state-of-the-art navigation performance. The system combines the benefits of Global Navigation Satellite Sensor (GNSS) technologies and innovations with established airline Flight Management System (FMS) operational procedures.

The HT 9100 is capable and certified for RNP/ANP (Required Navigation Performance/Actual Navigation Performance), similar to the Honeywell 747-400 FANS 1 software. (RNP is the required performance accuracy of a particular segment of airspace. ANP is a measure of the uncertainty in the position estimate of the system).

The HT 9100 is designed to permit GPS navigation under Instrument Flight Rules (IFR) conditions for en route, terminal area and approach operations. In addition, the system complies with FAA Notice N8110.60, permitting GPS as the sole navigation system for oceanic and remote operations.

The system shipset consists of a Navigation Processor Unit (NPU), Multifunction Control and Display Unit (MCDU) and an Antenna Coupler Unit (ACU). The NPU is the primary processing and control unit of the HT 9100. It contains a 12-channel Global Positioning System (GPS) receiver, Central Processing Units (CPUs) and navigation database for flight management functions. The MCDU controls the user input and display function. Data entry and display to the pilots are accomplished through the use of a 66-key alphanumeric keyboard with 12 line-select keys and 5.5 in diagonal multicolour active-matrix liquid crystal flat panel dislay. The ACU receives, amplifies, conditions and sends GPS signals to the receiver in the NPU.

Operational status

HT 9100 GNSS navigation management system has received Technical Standard Order (TSO C129 A1) approval from the US Federal Aviation Administration. This approval encompasses Standard Instrument Departures (SIDs) Standard Terminal Arrival Routes (STARs), GPS overlay and GPS approaches as well as Company routes, J-routes and V-route capability.

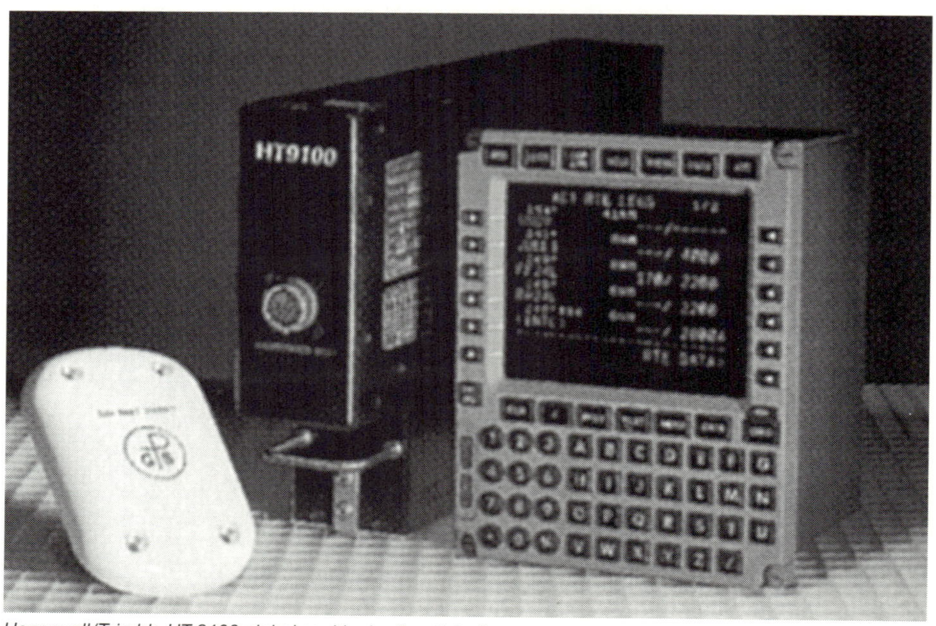

Honeywell/Trimble HT 9100 global positioning/inertial reference system 0003108

In addition, the HT 9100 has also received Supplemental Type Certificate (STC) approval from the FAA on 727, DC-10, L-1011 and MD-80 aircraft. The certification gives operators the ability to fly IFR (Instrument Flight Rules), supplemental en route, terminal, non-precision approaches and primary oceanic/remote navigation with the system.

The Honeywell/Trimble team forged a strategic alliance in June 1996 and announced in December a launch order for more than 500 HT 9100s from American Airlines. Many other orders have followed.

In 1998, the HT 9100 global navigation management system obtained FAA certification on the classic B 737-200 for B-RNAV (Basic aRea NAVigation) to meet Eurocontrol requirements in 1999. Royal Air Maroc was the launch 737 customer.

Contractors

Honeywell Inc, Commercial Aviation Systems.
Trimble Aerospace Products.

UPDATED

HUD 2020 Head-Up Display system

Honeywell's HUD 2020 Head-Up Display (HUD), developed in partnership with BAE Systems (formerly GEC-Marconi) is designed to aid the situational awareness of civilian pilots during all flight phases by enabling focus to be maintained outside and ahead of the aircraft. The HUD 2020 electro-optical overhead unit generates an image onto a lightweight combiner, mounted in front of the pilot's eyeline, which provides real-time flight and aircraft performance information. The large 30° × 25° Field-of-View (FoV) facilitates enhanced pilot awareness during low-visibility approaches, providing full information on aircraft energy state, flightpath, runway orientation and touchdown point, thus facilitating Category II certification and providing a path towards Category IIIa operations.

The symbology set for the HUD 2020 was designed in partnership with Gulfstream. The system is fully compatible with the entire suite of Honeywell-standard avionics aboard Gulfstream GIV, GIV-SP and GV aircraft, offering common control methodology, mode annunciation and operation. The system consists of the CB-200 Combiner, EO-200 Overhead Unit, HG-200 Display Guidance Computer (DGC) and the DC-884 Display Controller.

The HUD 2020 is also capable of supporting enhanced vision sensor (for example FLIR) inputs which would further enhance the capability of the system in Category III weather conditions.

Operational status

In production.

The HUD 2020 system, comprising the CB-200 combiner mounted onto the EO-200 overhead unit, with the DC-884 display controller mounted below on the front panel (Honeywell) **2001**/0105279

Contractor

Honeywell Inc, Commercial Aviation Systems.

NEW ENTRY

Integrated Global Positioning/ Inertial Reference System (GPIRS)

The integrated Global Positioning/Inertial Reference System (GPIRS) combines the best of global positioning and inertial reference systems to provide very accurate worldwide navigation.

The inertial reference system is upgraded to an integrated GPIRS by adding GPS processing software in the inertial reference unit and coupling it to an ARINC 743 GPS sensor unit.

The GNSSU (Global Navigation Satellite Sensor Unit) is a remote-mounted unit that provides all the functions necessary for either integrated or stand-alone configurations. All existing Honeywell IRUs can be modified into GPIRUs which are one-way interchangeable with existing IRUs.

The GPS sensor unit receives satellite data using a 12-channel design. Data are received from all satellites in view, with updates once per second. The all-digital multiple correlator design allows for satellite tracking during periods of low signal-to-noise. This enables the receiver to track satellites to a 0° elevation angle.

Specifications

Dimensions:
(IRS) 317.5 × 320 × 198.1 mm
(GPSSU) 190.5 × 215.9 × 55.9 mm
Weight:
(IRS) 19 kg
(GPSSU) 2.27 kg
Time to first solution:
(IRS) 10 min
(GPSSU) 4 min typical
Accuracy:
(position) (IRS) 2 n miles/h 95%
(GPSSU) 25 m
(velocity) (IRS) 8 kt
(GPSSU) 1.8 kt
(time) (GPSSU) 350 ns
Reliability:
(IRS) 5,000 h MTBF
(GPSSU) 20,000 h MTBF

Operational status

Selected as standard equipment by BAE Systems for its smaller air transport aircraft and in a dual fit for the Boeing MD 90-30.

Contractor

Honeywell Inc, Commercial Aviation Systems.

VERIFIED

Laseref inertial reference system

A derivative of the Lasernav inertial navigation system, Laseref is intended as an attitude and heading reference or primary sensor, for flight and navigation management equipment. Laseref shares many of the Lasernav modules and assemblies.

The solid-state, strapdown sensor generates present position, groundspeed, heading and windspeed and direction for flight management systems and navigation equipment, attitude and heading for flight instruments, weather radar stabilisation and autopilots. It replaces vertical and directional gyros, compass systems, fluxgate sensors and other independent navigation equipment with self-contained sensors and computing circuits to provide digital outputs, together with the ARINC 407 synchro outputs still needed by current generation avionics in ARINC 429 format. As with Lasernav II, alignment time is greatly reduced in comparison with gimballed inertial systems; typically 2½ to 10 minutes depending on latitude.

GPIRU configurations are available which contain additional electronics that blend inertial data with satellite information obtained from the Honeywell GPS sensor unit.

The Laseref system comprises two components: an inertial reference unit containing the sensing and computing elements, and a mode select unit that provides power to the former and governs its operation. Data insertion is through the control/display unit of the flight management system used in conjunction with the Honeywell system. The company offers an optional device, the inertial sensor display unit, as an alternative means of initialisation. This unit has a small display for reading out inertially computed present position, groundspeed, wind data and heading. Comprehensive built-in test equipment is provided; the system performs a rapid preflight self-test and monitors its operation throughout flight. Like Lasernav II, Laseref can be installed in an unpressurised bay.

Specifications

Dimensions:
(inertial reference unit) 322 × 324 × 193 mm
(mode select unit) 146 × 38 × 63.5 mm
(inertial sensor display unit) 146 × 114 × 167 mm
Weight:
(inertial reference unit) 21.1 kg
(mode select unit) 0.45 kg
(inertial sensor display unit) 2.3 kg
Power supply:
(total) 115 V AC, 400 Hz, 137 W
or 28 V DC

Operational status

In service on the Gulfstream II, Raytheon Hawker 800 and other business aircraft, also the Boeing 727-200 and DC-10.

Contractor

Honeywell Inc, Commercial Aviation Systems.

VERIFIED

Laseref II inertial reference system

Laseref II performs the same functions as Laseref and uses the same inertial sensor assembly, but in addition is designed to interface with new-generation digital avionics, such as flight management systems, using the Aircraft Standard Communications Bus (ASCB). The Laseref II may also interface with aircraft having the ARINC 429 digital databus. Laseref II has been selected as the standard factory IRS installation on the Dassault Falcon 900, Gulfstream IV and Canadair CL-601-3A.

GPIRU configurations are available which contain additional electronics that blend inertial data with satellite information obtained from the Honeywell GPS sensor unit.

Specifications

Dimensions:
(inertial reference unit) 322 × 324 × 193 mm
(mode select unit) 146 × 38 × 62.5 mm
Weight:
(inertial reference unit) 21.1 kg
(mode select unit) 0.45 kg
(inertial sensor display unit) 2.3 kg
Power supply:
(total) 115 V AC, 400 Hz, 137 W
or 28 V DC

Operational status

In production. Laseref II is standard in a dual configuration on the Gulfstream IV, Dassault Falcon 900 and Canadair Challenger 601. Other applications include the Raytheon Hawker 800, Cessna Citation III and de Havilland Dash 8.

Contractor

Honeywell Inc, Commercial Aviation Systems.

VERIFIED

Laseref III inertial reference system

The Laseref III all-digital laser system is 60 per cent smaller, 45 per cent lighter and uses 50 per cent less power than its predecessor, the Laseref II. The heart of the Laseref III IRU is a smaller ring laser gyro sensor. In addition to the new sensor, Honeywell is utilising surface mount technology, very large-scale integration, application specific integrated circuits, more powerful and faster processing and enhanced software in the new system. The Laseref III IRU includes integrated GPS processing which further enhances position and velocity data with a hybrid blending of raw inertial and satellite data.

The Laseref III IRU is pin-for-pin compatible with the Laseref II IRU and can be installed in the latter's 10 MCU tray using a mechanical adapter. It operates with the same mode-select unit and optional Lasertrak navigation display unit as existing systems.

Specifications

Dimensions: 124.5 × 320 × 198.1 mm
Weight: 11.79 kg
Align time: 2.5-10 min
Accuracy:
(navigation) 2 n miles/h
(velocity) 12 kt
(attitude) 0.1°
(heading) 0.4°

Operational status

In production since September 1991. The Laseref III has been selected for the Dornier 328 regional airliner and the upgrade of the Dassault Falcon 2000. Certified on the Raytheon Hawker 1000 in a dual configuration.

Contractor

Honeywell Inc, Commercial Aviation Systems.

VERIFIED

Laseref SM inertial reference system

The Laseref SM inertial reference system comprises the Mode Select Unit (MSU) and the Inertial Reference Unit (IRU). The MSU selects the IRU mode of operation and displays operational status messages on the annunciator panel. The IRU contains the laser inertial components, a processor and associated electronics and BIT. The INU is designated to add a GPS navigation processor card to incorporate an integrated GPS receiver.

By configuring the Laseref SM with an appropriate flight management system, a comprehensive special mission system is created. The combination of these units results in the ability to perform missions requiring special patterns, automatic camera control and computed air release point, in addition to standard flight management functions.

Laseref SM interfaces provide required data to a variety of special mission equipment. Standard ARINC interfaces output essential flight data to digital and analogue flight instruments, autopilots, radars, sensors and other special devices. Additional outputs are provided for special interface requirements.

The Laseref SM IRU is designed to add, as a growth option, an integrated GPS receiver. By adding one card to the IRU, a GPS PreProcessor Module (PPM) and an antenna, the standard Laseref SM becomes a fully integrated inertial/GPS system. The PPM receives satellite data using a two-channel fast sequencing design. Data is received from all satellites in view, up to a maximum of eight. The low signal-to-noise/fast sequencing design allows useful satellite tracking to 0° elevation angle with rapid acquisition of satellite data. Pseudo range and pseudo range rate data is transmitted from the PPM to the IRU where the added GPS navigation processor card processes the pure GPS solution. This card also contains the Kalman filter that blends the GPS and inertial data to provide the GPS hybrid solution.

Specifications

Dimensions:
(inertial) 317.5 × 320 × 198.1 mm
(GPS) 152.4 × 177.8 × 50.8 mm
Weight:
(inertial) 21.32 kg
(GPS) 1.36 kg
Reaction time: 2.5-10 min
Accuracy:
(position) (inertial) 0.8 n miles/h CEP
(GPS) 25 m SEP
(velocity) (inertial) 10 ft/s
(GPS) 0.1 m/s
(time) (GPS) 350 ns
Reliability:
(inertial) 5,000 h MTBF
(GPS) 20,000 h MTBF

Contractor

Honeywell Inc, Commercial Aviation Systems.

VERIFIED

Laser Inertial Reference System (IRS)

The world's first production ring laser gyro Inertial Reference System (IRS) was chosen by Boeing as part of the avionics package common to both the 767 and 757. It was also selected for the Boeing 737-300/400/500, MD-80 and MD-11, and the Airbus A320, A330 and A340, Fokker 100, and BAE Systems 146-300.

The strapdown configuration is so called because the gyrostabilised platform of current conventional inertial navigation and attitude reference systems is replaced by three ring laser gyro units mounted rigidly to the aircraft and at right angles to one another. The laser gyro detects and measures angular rates of motion by measuring the frequency difference between two contrarotating laser beams made to circulate (hence the term ring) in a triangular cavity by mirrors. When the units are at rest the distances travelled by each beam are the same, as are the frequencies. When the unit rotates, one path lengthens while the other shortens and so a frequency difference is established proportional to the rate of rotation of the unit. The difference is measured and processed digitally in ARINC 704 format as aircraft attitude in pitch, roll and yaw.

Since the accelerometers are mounted rigidly in the box, their signals are related to aircraft axes and have to be processed to convert them to the external inertial reference frame necessary to provide navigation and flight control information and guidance.

A strapdown system has no moving parts to wear, fail or become misaligned; no gimbals, torque motors, spin-motors, slip-rings, or resolvers, and no scheduled maintenance, realignment or recalibration requirements are anticipated. A typical installation comprises three inertial reference units (containing the sensing and computing elements) and a display unit. The Honeywell laser device is contained within a low expansion, triangular glass block, with a 34 cm path length. It has demonstrated a MTBF of 20,000 hours during more than 50 million flight hours.

Specifications

Dimensions: 4 MCU or 10 MCU
Weight:
(4 MCU) 12.24 kg
(10 MCU) 19.5 kg
Power:
(4 MCU) 44 W
(10 MCU) 86 W
Outputs: primary attitude information to displays and Automatic Flight Control Systems (AFCS), linear accelerations, velocity vector and angular rates to AFCS, wind shear detection and energy management, magnetic heading for displays and AFCS and long-range navigation data

ARINC 704
Accuracy (10 h flight - 95% probability):
(position) 2 n miles/h
(velocity) 12 kt
Self-test: BIT (initiated and continuous) detects 95% of failures with 95% confidence level
Reliability: >5,000 h MTBF predicted

Operational status

In production for Boeing 767, 757 and 737-300/400/500, MD-11 and MD-80 Series aircraft; and Airbus A300-600, A310, A320, A330 and A340.

Contractor

Honeywell Inc, Commercial Aviation Systems.

VERIFIED

Lasernav II navigation management system

Introduced in 1984, Lasernav II is an inertial system for airlines and general aviation which combines self-contained, strapdown laser inertial position and aircraft motion sensors with externally sensed radio signals to provide a single efficient integrated guidance package. The notable advantages of previous Honeywell laser inertial systems were maintained in Lasernav II: 2½ to 10 minutes alignment time dependent upon latitude, three to four times the reliability of earlier systems and reduced size, weight and power consumption. The system can also be mounted in an unpressurised environment.

Position data obtained from the Honeywell GPS sensor unit and VOR/DME and Omega/VLF stations is blended with inertial position information using high-speed digital computing techniques. DME/DME updating is obtained by way of an auto-tuning function that requires no pilot inputs. VOR/DME updating is obtained by manually tuning the radios. Triple inertial navigation system mixing combines inertial data from two other Lasernav II systems to calculate a composite inertial position.

In addition to normal navigation functions, Lasernav II can replace all conventional attitude and heading sensors including vertical and directional gyros, flux valves and compass controllers. In a typical dual installation, the total box count can be reduced by up to 14 separate boxes, resulting in weight savings of up to 60 kg and a greatly simplified installation.

Lasernav II comprises a navigation management unit and a control/display unit. An internal non-volatile memory can store 20 flight plans of up to 20 waypoints each. A Global Wulfsberg NDB-2 worldwide database (required for DME auto-tuning) facilitates the automatic flight planning feature that requires the pilot to input only the departure and destination points. A great circle route is computed, intermediate waypoints are selected and range and bearing to nearest VOR/DME is displayed, all automatically.

Specifications

Dimensions:
(navigation management unit) 322 × 324 × 193 mm
(control/display unit) ARINC 562
Weight:
(navigation management unit) 22.1 kg
(control/display unit) 3.2 kg
Power supply:
(total) 115 V AC, 400 Hz, 160 W
28 V DC, 19 W
Accuracy (95% probability):
(position) 2 n miles/h
(velocity) 8 kt
(heading) 0.4°
(pitch and roll) 0.1°

Operational status

In service. Installations approved include Cessna Citation III, Gulfstream II and III, Dassault Falcon 50 and Canadair Challenger CL-600 and CL-601 business jets, and Boeing 737-200 airliners.

Contractor

Honeywell Inc, Commercial Aviation Systems.

VERIFIED

Lasernav laser inertial navigation system

Introduced during early 1983, Lasernav® exploits the strapdown inertial reference system developed for the Boeing 767 and 757, but also has facilities that make it suitable for long-range business and corporate jet aircraft.

In addition to normal navigation functions, Lasernav can replace all customary attitude and heading sensors, including compass system components such as the flux detector, resulting in a reduction of up to 14 separate boxes. This is said to result in a weight saving of up to 60.3 kg, and a volume reduction of up to 50 per cent by comparison with equivalent dual installations in other systems.

Lasernav comprises two units; an inertial navigation unit and a control/display unit. The memory can store the co-ordinates of up to 255 waypoints in 20 routes and up to 20 waypoints per flight plan, for immediate recall. In dual installations each system can store a different flight plan, but can share it with the other system if required. For international routes or long overwater sectors, the system can use a Global Wulfsberg NDB-2 database and only the departure point and destination co-ordinates need to be inserted.

Lasernav then computes a great circle route, selects and identifies air traffic reporting points and lists bearing and distance to the nearest VOR/DME on the eight-line by 14-character control/display unit.

A notable advantage is the sharp reduction in alignment time; the interval from switch-on to ready is as little as 2½ minutes at the equator and 10 minutes at 60° latitude.

Specifications

Dimensions:
(inertial navigation unit) 322 × 324 × 193 mm
(control/display unit) ARINC 561
Weight:
(inertial navigation unit) 21.1 kg
(control/display unit) 3.2 kg
Power supply:
(total) 115 V AC, 400 Hz, 146 W
28 V DC, 28 W

Operational status

In service.

Contractor

Honeywell Inc, Commercial Aviation Systems.

VERIFIED

Lasertrak navigation display unit

A companion device to the Honeywell Laseref, Laseref II and Laseref III inertial reference systems, the Lasertrak navigation display unit provides back-up waypoint navigation using positions from up to three IRS. In the event of a flight management system failure the Lasertrak allows continued navigation along the planned route. Accepting up to 10 waypoints, Lasertrak computes and displays desired track and cross-track error for the intended course.

The Lasertrak can be used in flight management and inertial reference systems architecture when flight dispatch with a failed FMS is desired.

Specifications

Dimensions: 146 × 114 × 152 mm
Weight: 2.3 kg
Power supply: 27 V DC, 10 W

Operational status

Production deliveries started in March 1987. Installation certified in the Dassault Falcon 50 and 900, Gulfstream IV and Canadair Challenger CL-601-3A.

Contractor

Honeywell Inc, Commercial Aviation Systems.

VERIFIED

LSZ-860 lightning sensor system

Honeywell's newest lightning sensor, the LSZ-860, is an upgraded version of the LSZ-850. The LSZ-860 senses both visible and high-energy invisible electrostatic and electromagnetic disturbances caused by electrical discharge activity within a 200 n mile radius around the aircraft. When lightning occurs, the system carefully analyses the discharge and creates the proper symbol for display on most Primus colour radar indicators or on most Honeywell EFIS/MFD displays. The system's computer rapidly and accurately determines the rate of vertical lightning in a fixed geographical area and then displays the centre of that area with the lightning rate symbol. Wide bandwidth, extensive signal processing and lightning stroke recognition algorithms ensure a more accurate display. Extraneous signal filtering minimises noise which would otherwise clutter and confuse interpretation.

The LSZ-860 system computes the location and lightning rate for up to 50 thunderstorm areas. The computer tracks the location of each of these areas. To ensure accurate tracking of lightning areas, all displays are both heading and velocity stabilised to keep the symbol over the same ground position regardless of aircraft manoeuvring.

The system gathers information in a full 360° pattern around the aircraft, even in the standby mode. Thus, the lightning sensor system is always ready to present the weather picture in either the full 360° display mode without a radar overlay or in the sector display mode that can include a radar and navigation data overlay.

Three distinct levels of lightning rate are computed for display, each depicted by a unique lightning rate symbol. The symbols represent the vertical lightning rate-of-occurrence. The symbol location is the average position of the lightning that occurred in the previous two minutes, and is displayed inside the radius of the range selected. Each lightning symbol represents the centre of a circular area with a radius dependent on the range selected (8 n mile radius at a range of 25 n mile, 18 n mile radius at a range of 100 n miles and 30 n mile radius at a range of 200 n miles). Whenever any lightning activity is detected at any range, the lightning sensor computer will place a magenta lightning alert symbol at the proper bearing and at the end of the selected range for five seconds.

Specifications

Dimensions: (LP-860) 375 × 194 × 61.5 mm; (AT-850) 294 × 31.8 × 154 mm
Weight: (LP-860) 3.06 kg; (AT-850) 1.14 kg
Power supply: 28 V DC, 28 W

Operational status

LSZ-860 was available from January 1998.

Contractor

Honeywell Inc, Commercial Aviation Systems.

VERIFIED

MLZ-850 Microwave Landing System (MLS) receiver

The stand-alone MLZ-850 consists of an ML-850 digital MLS receiver, the CM-850 controller and one or two AT-851/852 antennas. The small AT-851/852 antenna is normally located on the nose section of the aircraft centreline. A second antenna is mounted on the aft lower surface of the fuselage and is required on fixed-wing aircraft to be able to receive the MLS on all headings. All connections between the antennas and the receiver are with conventional coaxial cable in common use. An optional antenna-mounted preamplifier is used for aircraft requiring long cable runs.

The pilot selects azimuth and elevation angles on the CM-850 cockpit controller which is also used to tune one of the 200 MLS channels and auto-tune the DME.

The MLZ-850 is compatible with conventional cockpit displays, providing analogue deviation outputs to ADIs, HSIs and compatible flight control systems. An ARINC 429 formatted digital output provides interfaces with other digital technology systems such as EFIS. BITE provides a preflight test of the digital receiver system.

Specifications

Dimensions:
(receiver) 354 × 84 × 99 mm
(controller) 139.7 × 60.4 × 66.7 mm
(blade antenna) 63.5 × 38.1 × 19 mm
(flush mount antenna) 6.6 × 76.3 mm diameter
Weight:
(receiver) 2.09 kg
(controller) 0.454 kg
(blade antenna) 0.1 kg
(flush mount antenna) 0.18 kg
Power supply: 28 V DC, 15 W
Frequency: 5,031-5,090.7 MHz
Channels: 200

Operational status

In production.

Contractor

Honeywell Inc, Commercial Aviation Systems.

VERIFIED

Pegasus Flight Management System (FMS)

Honeywell's new-generation Pegasus FMS had its first flight on a Boeing MD-90 in May 1997 and was certified on the Boeing 757/767 and MD-90 in March 1998.

Pegasus is designed to bring the benefits of the emerging Communications, Navigation, Surveillance/ Air Traffic Management (CNS/ATM) environment to the operators of air transport aircraft.

Pegasus is designed to be used across multiple aircraft hardware platforms, including: Airbus A320, A330, A340; Boeing 717, 757, 767, MD-11 and MD-90.

Honeywell claim that Pegasus offers more than 25 times the throughput capacity and 16 times more total memory than the previous-generation FMS, making it ideal for powering the functionality that will be required in the coming Free Flight era.

Pegasus is being marketed by Honeywell as one of the major components of its WorldNav™ package to meet the CNS/ATM environment. The other main elements of the WorldNav™ package are: the Honeywell Traffic alert and Collision Avoidance System TCAS 2000 and the Honeywell Air Data/Inertial Reference System (ADIRS), or the ANSIR 2000 Air Data/Inertial Reference Unit.

Operational status

First flight May 1997; first certified March 1998. Selected by a large number of airlines for both Airbus and Boeing fleets. US FAA certification for the A330/ A340 is scheduled for February 2000, and for the A320 family in March 2001.

Contractor

Honeywell Inc, Commercial Aviation Systems.

VERIFIED

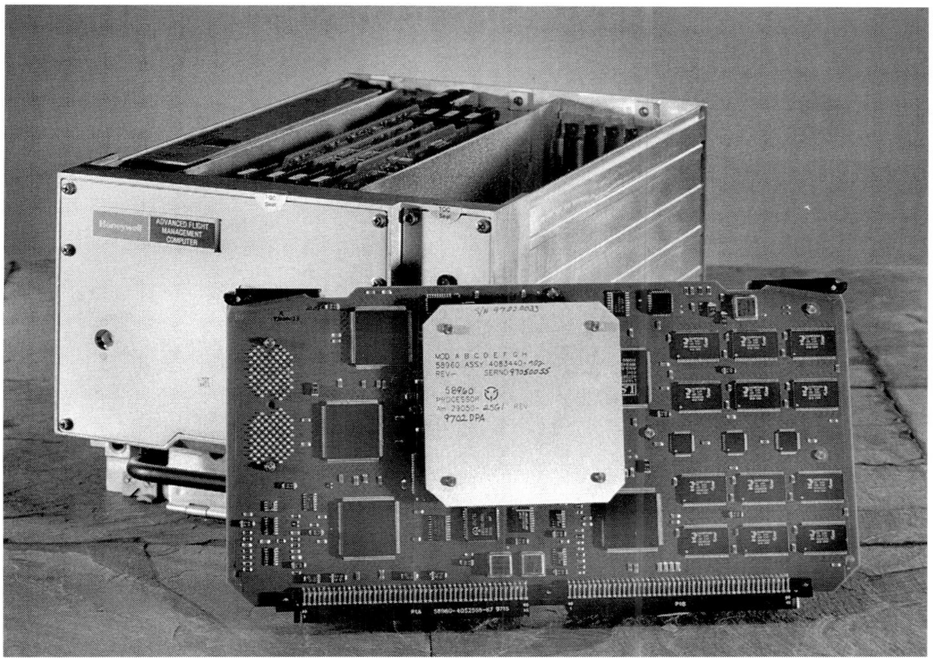

Honeywell's new-generation Pegasus flight management system ***1999***/0051636

Primus II radios

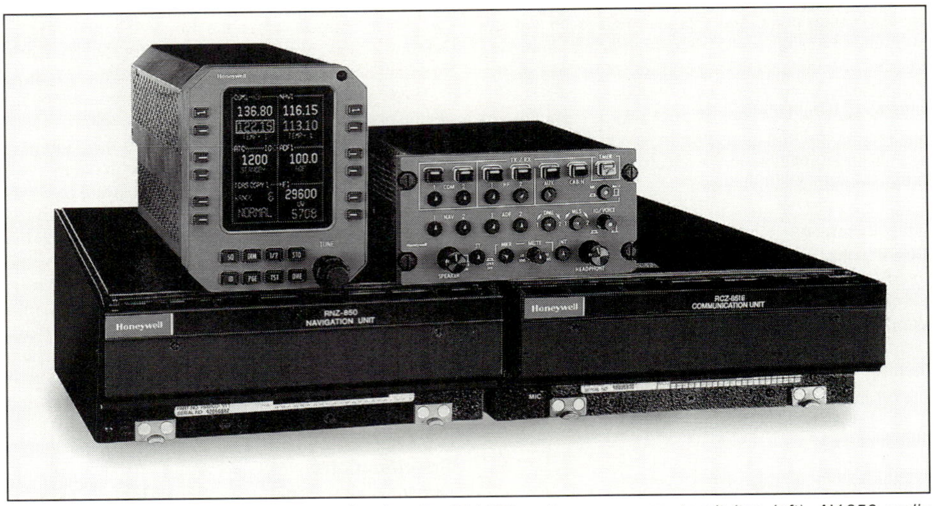

The Honeywell Primus II radio system showing the RM-850 radio management unit (top left), AV-850 audio control panel (top right), RNZ-850 navigation unit (bottom left) and RCZ-851E communication unit (bottom right)
0001185

Honeywell introduced the Primus II series of radios in April 1987; they are aimed at the business aircraft and regional airliner segment of the market. The system incorporates VLSI technology and digital bus tuning and control, together with centralised radio management and a digital audio system.

Primus II radios are controlled and display their information via a radio system bus which is the Sperry-developed Avionics Standard Communications Bus (ASCB) formatted for radios. The system comprises the full-colour RM-850 series radio management unit, RNZ-850 series integrated navigation unit, RCZ-851E series integrated communications unit and AV-850 audio control unit.

The radio management unit provides control over operating modes, frequencies and codes for all units in the system. Five dedicated windows support the com, nav, transponder, ADF and MLS functions. The unit also provides BITE control and readout.

The integrated navigation unit contains a VHF navigation receiver, DME transceiver modules and ADF receiver module. The VHF navigation receiver houses the functions of VOR/localiser receiver, glide slope receiver and marker beacon receiver to provide an ILS that meets Cat II low-approach requirements. The DME transceiver is a six-channel scanning DME that simultaneously tracks selected DME channels and two preselect navigation frequencies. It meets the initial approach mode accuracy requirements of the P-DME specification with an accuracy of better than 100 ft and can also operate on W, X, Y or Z DME channels. The extended range ADF module can receive low-frequency NDBs below 200 kHz as well as the marine emergency band of 2.181 to 2.183 MHz.

The integrated communications unit incorporates separate VHF communication transceiver and transponder modules. Optional transponders include Mode A/C, Mode S and Mode S with diversity. The communication transceiver operates across the entire 118 to 152 MHz frequency range, but for civil use the upper limit can be reduced to 136 MHz.

The MLS receiver operates as an extension of the integrated navigation unit and can interface with standard digital and analogue outputs.

The digital audio system receives digitised audio from the other units via one high-speed digital bus from each side, providing immunity to noise and virtual elimination of cross-talk. It can control 16 or more audio signals and multiple audio panels may be installed in the aircraft. It is available in three- or four-row versions, with a variety of layouts.

From May 1998, modifications became available to provide the 8.33 kHz communication bandwidth required for European flights from 1 January 1999 as well as upgrade kits for all existing Primus II radios.

Operational status
In production and in service.

Contractor
Honeywell Inc, Commercial Aviation Systems.

VERIFIED

Primus 700/701 Series surface mapping, beacon and colour weather radar

A three-box system including receiver/transmitter, indicator and antenna, the Primus 700/701 is compatible with certain Honeywell EFIS. Its powerful 10 kW magnetron transmitter has six pulse-widths, seven bandwidths and four PRFs for maximum performance on all ranges in all modes. Ten selectable range scales from 0.5 to 300 n miles (1 to 556 km) provide optimal range scales in every condition.

Five antenna sizes from 10 to 24 in make the Primus 700/701 system suitable for any airframe.

The Primus 700/701 dual-EFIS interface capability and antenna sweep time-sharing in effect make two radars available to the crew throughout the flight. Each crew member can select their own range, mode, gain and tilt display on their EHSI.

With the radar indicator, pilots will have the Honeywell features of a variable range mark and azimuth cursor with digital distance and bearing readouts. A new menu function allows the indicator switches to do double duty, controlling infrequently used features such as heading display on/off. Without a radar indicator installed, the system includes one or more WC-700 radar controllers.

Primus 700/701 weather radar features include a four-colour display of rainfall intensity. On ranges of 50 n miles (93 km) or less, turbulence detection shows areas where there is moderate or stronger levels of turbulence.

The Honeywell Rain Echo Attenuation Compensation Technique (REACT) safety feature performs three distinct functions. First, it maintains target calibration by compensating for attenuation caused by intervening rainfall. Returns remain properly calibrated for the storm behind the storm. Second, REACT advises pilots of areas where target calibration cannot be maintained even with maximum compensation. For those areas, REACT changes the screen background to blue, warning that calibration is no longer possible and attenuation may be hiding areas of severe weather. Third, any target displayed in the blue field will appear in magenta to alert the pilot of its probable severity.

The system also includes Honeywell weather radar features of ground clutter reduction and target alert.

The radar indicator interfaces with Honeywell's LSZ-860 lightning sensor system and Data Nav for complete severe weather avoidance, navigation and checklist capability.

High-resolution and high-sensitivity mapping modes are available, designated GMAP 1, which is used to locate small targets at sea, and GMAP 2, which provides ground mapping over land or coastline. Display modes include three display colours, which differ from those used for the weather display and pilot-selectable sea clutter reduction. Range scales of 0.5, 1, 2.5, 5, 10, 25, 50, 100, 200 and 300 n miles are available and the shortest range scale provides a resolution of 55 ft.

The Primus 701 includes a beacon capability which makes low-visibility approaches possible where standard navaids may not be available. It also allows air-to-air rendezvous. The Primus 701 operates in radar only, beacon only or both beacon and radar modes. Beacon targets are shown in contrasting colours in weather- and ground-mapping modes.

Primus 880 weather radar
0001215

Primus 700/701 BITE includes comprehensive and continuous internal fault monitoring and a menu function which allows access to monitor pages. A non-volatile memory records internal fault data for later retrieval by maintenance technicians.

Operational status

In production. Primus 700 has been selected by the US Army as a retrofit for 55 MH-60 and CH-47 helicopters, and by Boeing as the standard factory-installed radar in the Chinook CH-47SD.

Primus 701, which includes beacon capabilities, has been selected by Cougar Helicopters for its Eurocopter AS 332 Super Puma helicopters.

Contractor

Honeywell Inc, Commercial Aviation Systems.

UPDATED

Primus 880, 660 and 440 weather radars

The Primus 660 and 880 weather radars are high-power (10 kW) successors to the Primus 650 and 870 systems, respectively. The Primus 440 is designed as a powerful, reliable weather radar for light-class business aircraft.

All three systems have a stabilised antenna, up to 24 in for the Primus 880, and are packaged in a Transmitter/Receiver/Antenna (TRA) architecture that weighs 6.36 kg. Each system is compatible with Honeywell's LSZ-860 Lightning Sensor System and may be displayed on either the Electronic Flight Instrument System (EFIS) or on a dedicated weather radar indicator.

The Primus 880 features Doppler turbulence detection pulse pair processing that detects spectrum spreading caused by turbulence within any storm cell, regardless of rainfall rates. Once detected, turbulent areas are displayed in white on all ranges up to 50 n miles, allowing pilots to safely manoeuvre around potentially hazardous weather.

For the first time on a Primus radar system, Primus 880 also features Built-In Test Equipment (BITE) on two of the most important components of the system, the transmitter and receiver, providing a complete RF loop-back which continuously tests the transmitter power and receiver sensitivity and reports any faults to the pilot.

Other features of the Primus 880 include Honeywell's exclusive Rain Echo Attenuation Compensation Technique (REACT) which alerts pilots to storms hidden behind other storms; Target Alert (TA) which notifies pilots of potentially hazardous weather directly in front of the aircraft; Altitude Compensated Tilt (ACT), which allows detection of weather that may affect the aircraft en route and reduces the amount of tilt management performed by the pilot; Ground Mapping (GM), which serves as a navigation aid by depicting terrain features not available in this clarity and detail with lower-power radars.

Honeywell's optional LSZ-860 Lightning Sensor System overlays lightning information on to the precipitation/turbulence display to provide a very powerful severe weather detection capability. The LSS accurately displays the position and lightning rate of up to 50 storm cells at the same time.

In July 1999, an interface became available to integrate the Primus 440, 660 and 880 radars to the Honeywell Aerospace, Electronic & Avionics Lighting Mk VII Enhanced Ground Proximity Warning System (EGPWS).

Operational status

In production.

Contractor

Honeywell Inc, Commercial Aviation Systems.

VERIFIED

Primus 1000 integrated avionics system

The Primus 1000 integrated avionics system is designed for mid-sized business jets and regional turboprops. Advanced processing technologies and the integration of key functions result in increased capability, reliability and flexibility while achieving significant reductions in size, weight, power requirements and installation costs.

The Primus 1000 is based on the IC-600 integrated avionics computer which combines a display processor, flight director, Cat II fail-passive autopilot and EICAS processor in a single ½ ATR box. The Primus 1000 utilises an ARINC 429 architecture and is composed of an EFIS electronic display system incorporating from two to five 8 × 7 in (203.2 × 177.8 mm) large format displays, single or dual flight director and an optional single or dual fail-passive autopilot. The other components of the system

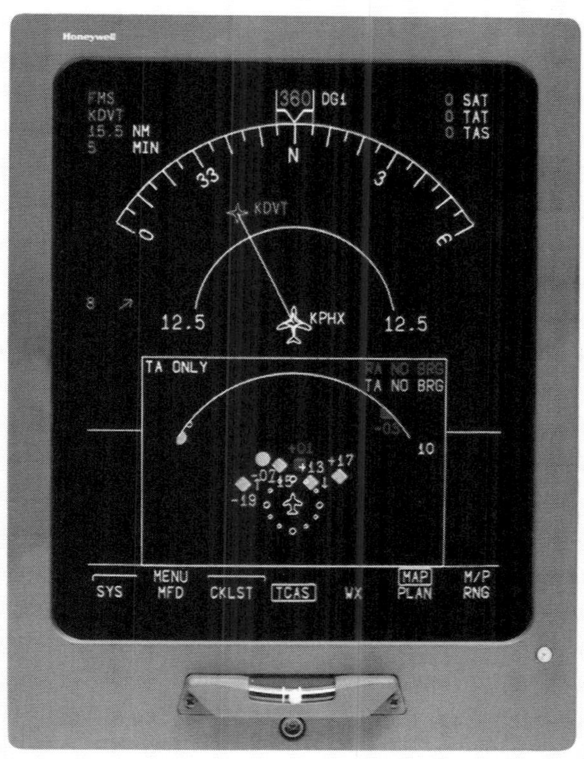

Primus 1000/2000 EFIS 8 × 7 in Navigation Display (ND), showing MAP and TCAS information
2001/0099082

The Primus 1000 integrated avionics system installed in the Embraer-145 aircraft, showing a five display EFIS/ Engine Instrument and Crew Advisory System (EICAS) using DU-870 8 × 7 in colour displays 0001387

include AZ-840 micro air data computer, separate vertical and directional gyros or the AH-800 fibre optic AHRS, Primus II radio system and the Primus 650 weather radar system. Standard options are Honeywell's TCAS, MLS, lightning sensor system, Primus 870 turbulence detection weather radar, Primus 700 or Primus 450 weather radar and the Laseref III inertial reference system.

Operational status

The Primus 1000 was selected for the Lear 45 in September 1992 and later for the Citation Bravo, Citation V Ultra and Embraer ERJ-135 and ERJ-145. A Primus 1000 system selected for the Sino Swearingen SJ30-2 in April 1996 specified dual IC-600 computers, AZ-850 all-digital Micro Air Data Computers (MADC), Primus II digital integrated radios and Primus 650 weather radar.

Contractor

Honeywell Inc, Commercial Aviation Systems.

UPDATED

Primus 2000 advanced avionics system

The Primus 2000 advanced avionics system is designed for twin-turbine aircraft in the business and regional airliner markets. Its small size and weight are the result of using the most advanced technology in components, packaging techniques and systems design. The system incorporates surface-mount technology, very large-scale integrated circuits, application specific integrated circuits and high-density multilayer circuit boards. The Primus 2000 also offers flexibility and maximum growth potential through the use of an advanced architecture built around Honeywell's Avionics Standard Communications Bus (ASCB).

Interconnection between all major systems is accomplished by the ASCB, which provides both the total data handling capacity and requires a lower wire count than the one-way ARINC 429 standard. The bidirectional databus has critical level capability that eliminates the need for numerous dedicated lines between avionics systems. The ASCB architecture enhances system level availability, allowing the ASCB to be the sole means of interconnect for most avionics subsystems.

The Primus 2000 is composed of an electronic display system incorporating from two to six large screen 8 × 7 in (203 × 177.8 mm) display units and associated control panels, an integrated avionics computer containing electronic display processors, a fault warning computer, a fail-operational/fail-passive automatic flight control system and an optional flight management system, micro air data computers, attitude and heading reference system or inertial reference system, data acquisition units, Primus II radio and Primus 650 weather radar systems and the avionics standard communications bus. Honeywell's flight management system, Traffic alert and Collision Avoidance System (TCAS), Global Positioning System (GPS), Microwave Landing System (MLS), lightning sensor system, Primus 870 turbulence detection radar and laser inertial reference system are offered as standard options.

Operational status

PRIMUS 2000 was first certified on the Fairchild Dornier 328 in 1998, with initial deliveries in 1999. Also now certified on the Cessna Citation X and on the Dassault Falcon 900EX, in a five, 8 × 7 in, full colour EFIS configuration, including two MFDs, two primary flight displays and an EICAS display. Certification also planned for the Falcon 900C, with first deliveries expected during 2000.

Contractor

Honeywell Inc, Commercial Aviation Systems.

UPDATED

The Honeywell Primus 2000 integrated avionics system on the Citation III aircraft

Primus 2000XP integrated avionics cockpit

Honeywell's Integrated Avionics Computer IC-800 is standard on the Global Express in a triple configuration. It combines the function of five major subsystems into a compact, lightweight package that is 75 per cent smaller than the combined volume of the units it has replaced.

The Primus® 2000XP integrated cockpit was designed specifically for the Global Express. Its high-accuracy air data system precisely measures speed, altitude, temperature and other critical flight information, while six 8 × 7 in displays keep the pilot fully informed about all aspects of the flight.

Primus 2000XP standard features include:

(1) six-tube Electronic Flight Instrument System (EFIS)/Engine Instrument and Crew Advisory System (EICAS) using DU-870 8 × 7 in colour displays;
(2) dual fail-operational autopilot;
(3) dual integrated NZ-2000 Flight Management Systems (FMS);
(4) triple long-range Nav source (two FMS and LaserTrak);
(5) Global Navigation Satellite Sensor Unit (GNSSU) Global Positioning System (GPS);
(6) dual full flight regime autothrottle triple Integrated Avionics Computers (IAC);
(7) triple AZ840 Micro Air Data Computers (MADC);
(8) triple LASEREF III Inertial Reference Systems (IRS);
(9) quad Ametek DA-810 Data Acquisition Units (DAU);
(10) Primus 880 colour weather radar;
(11) Central Aircraft Information Maintenance System (CAIMS);
(12) dual Primus II Integrated Nav/Com/Ident radios;
(13) TCAS II Traffic Alert and Collision Avoidance System.

Optional avionics for the Global Express include: a third integrated FMS; second integrated GPS; third VHF com; lightning sensor system; three or six channel SATCOM, featuring worldwide phone, fax and data satellite communication.

The IC-800 features include: two symbol generators; fault warning computer; tone and voice generator; Radio System Bus (RSB) interface; Avionics Standard Communication Bus (ASCB) controller and interface; flight guidance computer; flight management computer; autothrottle control and power supply.

The exceptional system integration within the IAC is made possible by the lastest technologies, including: Very Large Scale Integration (VLSI); Surface Mount Technology (SMT) and advanced manufacturing techniques to reduce circuit card size. Single cards within two of the IACs contain entire FMS computers. The IACs also house the electronics for automatic flight control, autothrottle and systems monitoring.

Operational status

Understood to have been certified on the Bombardier Global Express in 1998.

Contractor

Honeywell Inc, Commercial Aviation Systems.

VERIFIED

The Primus Epic® integrated avionics display system for the Fairchild 728 JET aircraft. **2000**/0081860

Primus Epic™ avionics system

The heart of the Primus Epic™ avionics system is the Virtual Backplane Network™ developed exclusively for this system. This architectural concept blends the cabinet-based modular capabilities of the Honeywell 777 AIMS system with the aircraft-wide network capabilities of the Primus 2000 system. The architecture allows a very high degree of system integration and scalability by allowing all data generated by any function to be globally available within the system.

A key component of the Virtual Backplane Network is the open architecture afforded with the bidirectional Avionics Standard Communication Bus (ASCB). The ASCB continues the evolution of the ASCB used in many business and regional aircraft. The Primus Epic™ version of the ASCB provides the throughput equivalent of 100 high-speed ARINC 429 databusses. The Primus Epic™ bus is claimed to significantly reduce wire weight, power use and installation cost, while greatly increasing the capacity to support aircraft utility systems control and other future aviation requirements.

The Primus Epic™ system hardware is built on the Modular Avionics Unit (MAU) derived from Honeywell technology, developed for the Boeing 777. Computing modules within the MAUs and flat panel displays utilise Honeywell's Digital Engine Operating System (DEOS™), which allows the different aircraft and functions to run simultaneously and independently. The full size 8 × 10 in colour flat panel liquid crystal displays, in a two- to six-display configuration, feature new functionality within a point-and-click Graphical User Interface (GUI) environment, which supports moving maps, ground-based weather, real-time video and electronic pilot manuals.

Pilots may choose traditional interfaces or new cursor control devices, including touchpad, joystick, light pen or tracker ball to interact with on-screen 'soft key' controls. In future, Primus Epic™ will offer voice command as a control option for some functions.

The initial utilities integrated into the Primus Epic system via the MAU were the landing gear control and anti-skid braking systems, but the intention is to integrate a wide range of utility systems including: air conditioning and environmental control; electrical power distribution; fire protection; fuel; hydraulics, ice protection; lights; oxygen; APU; engine vibration and others.

A variant of the system, designated Primus Epic® CDS Retrofit, became available in June 1999. This system includes an upgraded Control Display System

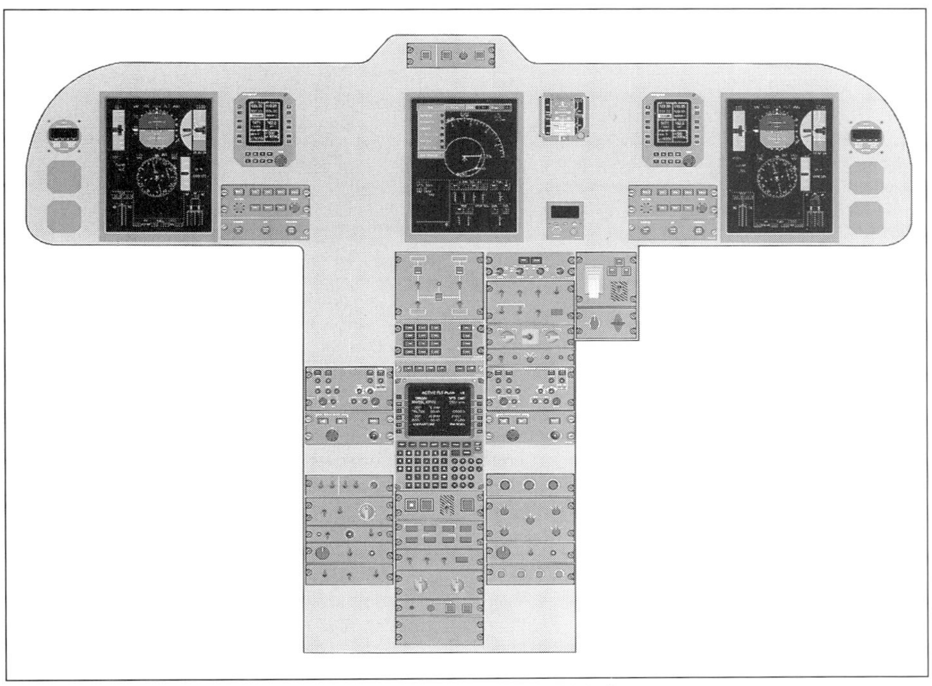

The Primus Epic® integrated avionics display system for the AB 139 helicopter. **2000**/0081861

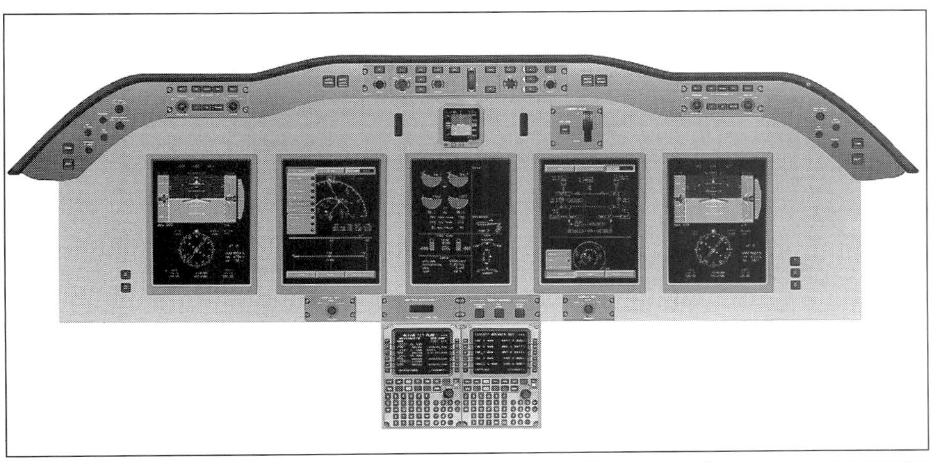

The Primus Epic® integrated avionics display system for the Hawker Horizon aircraft. **2000**/0081856

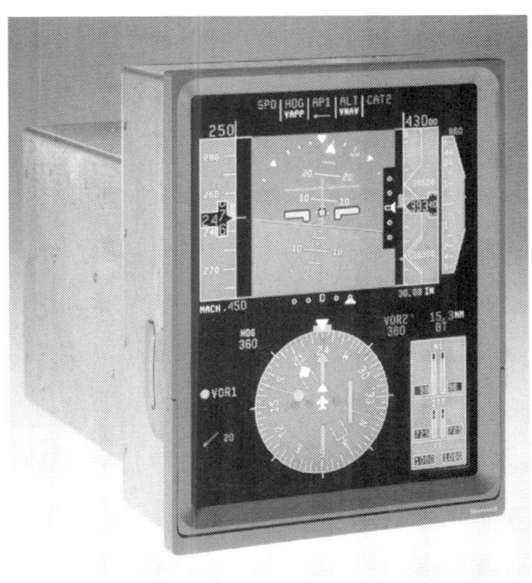

The DU-1080 flat-panel display selected for the Hawker Horizon 5-display Primus Epic™ installation **1999**/0051695

(CDS), using 8 × 10 inch LCD Primary Flight Displays (PFDs) and MultiFunction Displays (MFDs). It is designed around the Primus 1000 and Primus 2000 systems and utilises the IC-1080 integrated computer, with optional internal FMZ-2000 Flight Management System (FMS), DU-1080 LCD display units and display controllers. It includes built-in growth capabilities to support future CNS/ATM requirements and to further enhance system features with products such as Flight Management Systems (FMS), Global Positioning Systems (GPS), Satcom. GPS Landing Systems (GLS) and Traffic alert and Collision Avoidance Systems (TCAS).

Operational status

Launch customer will be the Hawker Horizon from Raytheon Aircraft Company. The installation will comprise a five-display integrated system with two flight situational displays, two multifunction displays, two cursor control devices and glareshield controllers; the display unit chosen for the system is the DU-1080 flat panel colour liquid crystal display unit (8 × 10 in). The Hawker Horizon is expected to make its first flight in late 1999, with certification in 2001.

Primus Epic™ has also been selected by Fairchild Aerospace for the 728/928 JET family, by Embraer for the ERJ170/190 aircraft, and by Agusta-Bell for the AB139 helicopter. The Primus Epic configuration for the 728/928 JET aircraft family comprises three MAUs and five 8 × 10 in colour flat-panel displays. In the case of the AB 139 helicopter, four configurations are available: a basic VFR system; a 3-axis AFCS IFR installation; a 4-axis AFCS IFR installation; and a Search And Rescue (SAR) version.

Contractor

Honeywell Inc, Commercial Aviation Systems.

VERIFIED

RCZ-852 diversity Mode-S transponder

The Honeywell RCZ-852 diversity Mode S transponder offers a small, light package that is optimised for corporate aircraft and regional airline applications. The RCZ-852 implements all currently defined Mode-S functions with provision for future growth. Current Mode-S transponders are used in conjunction with TCAS and ATCRBS to identify and track aircraft position, including altitude. This system transmits and receives digital messages between aircraft and air traffic control. The datalink provides positive and confirmed communications more efficiently than current voice systems.

The Honeywell design meets future needs by including growth capability to support the functions defined by CNS/ATM (Communications, Navigation, Surveillance/Air Traffic Management).

The Traffic Alert and Collision System (TCAS) is fully supported, including 'diversity' (top and bottom) antenna ports. Diversity provides reliable RF communication links between both ground-based and airborne interrogators. The transponder incorporates a TCAS II interface and is designed to be compatible with all TCAS II systems conforming to ARINC 718/735 characteristics.

The RCZ-852 transponder is an ICAO Level 3 system with growth to Level 4. Level 3 means that it will transmit and receive standard length (112 bit) datalink messages for 'COMM A' and 'COMM B' and receive 16-segment extended length datalink messages for 'COMM C'.

Honeywell has included full Built-in Test Equipment (BITE) and self-test capabilities to provide maximum reliability whilst minimising maintenance costs. The BITE system separates aircraft installation and aircraft system failures external to the transponder minimising the time required to return removed units to service.

Specifications

Dimensions: 107 × 84 × 318 mm
Weight: 2.3 kg
Power: 28 V DC

Contractor

Honeywell Inc, Commercial Aviation Systems.

VERIFIED

RD-350J Horizontal Situation Indicator (HSI)

The RD-350J Horizontal Situation Indicator (HSI) provides heading, two DME distances, radio navigation information via a displacement bar, indications of selected course and heading and to/from indications for VOR operations. Flags give indication of failures.

A nav mode annunciator takes the form of a rotary display in the centre of the compass card, controlled by VOR/Loc valid, Loc tuned, to/from and back course selected signals. The annunciator displays a symbol to indicate selected modes and can also indicate when radio data is invalid.

Specifications

Dimensions: 127 × 127 × 226 mm
Weight: 4.2 kg
Power supply: 26 V AC, 400 Hz, 6.4 W

Operational status

In production.

Contractor

Honeywell Inc, Commercial Aviation Systems.

VERIFIED

RD-700 series Horizontal Situation Indicators (HSI)

The RD-700 series has been developed for applications in new aircraft or retrofit installations. High-torque, low-power flag and shutter movements eliminate sticking displays and low-power devices coupled with open card construction result in low heat dissipation and power demands. All indications are conventional in presentation and numerical indicators feature standard readouts. The following list summarises the presentation and displays of each instrument:

RD-700 horizontal situation indicator

Displays all standard navigation radio inputs, compass system and ARINC 561 INS data including digital readout of drift angle and groundspeed.

Specifications

Dimensions: 5 ATI
Weight: 3.9 kg
Power supply: 115 V AC, 400 Hz
or 26 V AC, 400 Hz

RD-700A horizontal situation indicator

All standard navigation radio and compass data are presented together with dual digital DME readout. The RD-700A HSI features new flag, shutter and annunciator mechanisms and improved packaging to enhance reliability. Honeywell claims a 44 per cent reduction in power consumption compared with earlier designs.

Specifications

Dimensions: 5 ATI
Weight: 3.9 kg
Power supply: 115 V AC, 400 Hz
or 26 V AC, 400 Hz

RD-700C horizontal situation indicator

In addition to radio and compass navigation data, this instrument includes ARINC 561 INS data and digital readout of time and distance to waypoint and groundspeed.

Specifications

Dimensions: 5 ATI
Weight: 3.9 kg
Power supply: 115 V AC, 400 Hz
or 26 V AC, 400 Hz

RD-700D horizontal situation indicator

Although very similar to the RD-700C, this presentation does not include the time to waypoint counter.

Specifications

Dimensions: 5 ATI
Weight: 3.9 kg
Power supply: 115 V AC, 400 Hz
or 26 V AC, 400 Hz

RD-700F horizontal situation indicator

Displays all standard data including to/from, drift angle and digital readout of groundspeed and distance to waypoint.

Specifications

Dimensions: 127 × 127 × 216 mm
Weight: 3.9 kg
Power supply: 115 V AC, 400 Hz
or 26 V AC, 400 Hz

RD-700G horizontal situation indicator

In addition to all standard navigation, radio, compass and ARINC 561 INS inputs, this instrument presents digital readout of drift angle, distance to waypoint and groundspeed, although the drift and groundspeed presentation is at the lower area of the indicator rather than the more usual upper region of the instrument.

Specifications

Dimensions: 5 ATI
Weight: 3.9 kg
Power supply: 115 V AC, 400 Hz
or 26 V AC, 400 Hz

RD-700M horizontal situation indicator

In addition to standard information displays, the RD-700M uses 11-position low-power magnetic wheels for dual DME displays. A new thermal design includes a more efficient heat-sink mounting for high-power components and heat sensitive capacitors. Electronic and mechanical sections are segregated for maintenance access, while open-board packaging in the electronic section facilitates troubleshooting.

Specifications

Dimensions: 127 × 127 × 218 mm
Power supply: 115 V AC, 400 Hz
or 26 V AC, 400 Hz

Contractor

Honeywell Inc, Commercial Aviation Systems.

VERIFIED

RD-800 series Horizontal Situation Indicators (HSI)

The RD-800 Horizontal Situation Indicator (HSI) features three digitally driven servoed displays in conjunction with two four-digit gas tube displays showing time and distance to waypoints. Microprocessor control gives improved versatility in navigational data processing.

Specifications

Dimensions: 5 ATI
Weight: 4 kg
Power supply: 115 V AC, 400 Hz
or 26 V AC, 400 Hz

RD-800J horizontal situation indicator

In the RD-800J HSI the readout of true airspeed is provided by conventional counter displays and for ease of interpretation the command bars are colour identified.

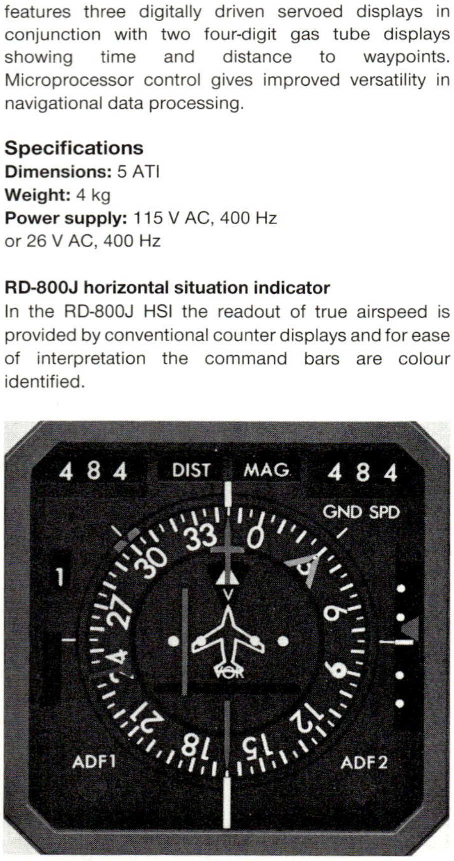

The Honeywell RD-850 horizontal situation indicator

Specifications

Dimensions: 5 ATI
Weight: 4 kg
Power supply: 115 V AC, 400 Hz
or 26 V AC,400 Hz

RD-850 horizontal situation indicator

The RD-850 HSI features the most up-to-date applications of instrument technology including microprocessor control. Coloured display elements are included together with distance to go and groundspeed counters. Automatic direction-finder annunciators are fitted in the lower instrument area.

Specifications

Dimensions: 5 ATI
Weight: 4.7 kg
Power supply: 115 V AC, 400 Hz
or 26 V AC, 400 Hz

Contractor

Honeywell Inc, Commercial Aviation Systems.

VERIFIED

Secondary Attitude and Air data Reference Unit (SAARU)

The SAARU, a fail-safe and highly reliable device, operates as a secondary system to the fault-tolerant air data inertial reference unit (see earlier item).

SAARU measures the aircraft's linear and rotational motions and computes air data measurements to provide fail-safe secondary attitude and air data reference information. The 10 MCU device also provides digital attitude and air data reference information to the cockpit LCD standby displays.

Operational status

Selected for the Boeing 777 aircraft.

Contractor

Honeywell Inc, Commercial Aviation Systems.

VERIFIED

SFS-980 digital flight guidance system for the MD-80

Used exclusively on the MD-80, the SFS-980 is an automatic flight guidance system based on relatively few LRUs, especially in respect of the range of functions performed. In addition to a comprehensive selection of conventional autopilot/flight director operating modes, the system is FAA cleared for Category IIIA automatic landings of 50 ft decision height and 700 ft runway visual range and has a full-time autothrottle. The large-scale use of digital computing has led to the installation of a comprehensive built-in self-monitoring and maintenance diagnosis capability. Major LRUs and functions are:

Digital Flight Guidance Computer (DFGC) There are two identical 1 ATR long (13.15 kg) DFGC units individually capable of handling all system functions including fail-passive automatic landing. Within each unit analogue/digital conversions take place and the complete system occupies 31 boards against a maximum capacity of 51. Each digital processor has 30 kwords of read-only memory and 4 kwords of RAM. In addition to all autopilot/flight director processing, each unit also generates thrust rating indication signals, plus maintenance data storage and status test panel data. The aircraft can have a head-up display for take-off and go-around data presentation and guidance signals for this unit are also generated in each DFGC.

Flight Guidance Control Panel (FGCP) This unit fits in the centre glareshield and contains mode selection and control functions for full-time autothrottle, both flight directors, the autopilot and altitude alerting. Pilots may select which DFGC controls all functions. Autothrottle speed/Mach number, selected heading, vertical speed and selected altitude readouts are shown on seven-segment incandescent lamps which may be dimmed by a control knob on the bottom of the panel. The speed/Mach number knob is a three-position control, while the heading selection is one of two concentric knobs. The outer knob provides selection of maximum bank angle for all autopilot/flight director lateral modes except Loc. The inner knob is a four-position device providing heading and autopilot/flight director heading mode selection. The three-position Alt knob provides altitude selection and autopilot/flight director altitude preselect mode arming.

VHF/nav control panels Two panels on either side of the FGCP allow selection of VORTac station frequencies and courses. Displays are incandescent lamp readouts.

Flight Mode Annunciator (FMA) One FMA on each pilot's panel provides instrument failure warning for ILS, attitude, heading, automatic landing, auto-trim or instrument monitor functions. These indicate which autopilot or flight director system is engaged and warn of autothrottle or autopilot disconnects. The units also annunciate which autothrottle, flight director and/or autopilot modes are armed and in which mode the system is currently controlling.

Autopilot duplex servo drive Three duplex servos drive the ailerons, elevators and rudder. The duplex servo functions only during ILS, land or go-around mode. A linear actuator provides normal yaw damping functions in other flight regimes.

Each servo has two separate DC electric motors whose outputs are summed in a differential gear train with a single output. Each servo sends position and rate feedback signals to the DFGC, where servo models monitor actual servo operation. The duplex servo design provides fail-passive protection against hard over manoeuvres. A fault in one servo channel is cancelled by the mechanical velocity summing conducted in the dual-servo differential.

Autothrottle/speed control system Conducted in the DFGC, this function provides fast/slow attitude director indicator commands throughout the entire flight regime. During take-off and go-around it provides pitch guidance for the flight director, autopilot and optional head-up display. The speed control system provides a speed margin above stall (alpha speed) for all autothrottle modes, in addition to the autopilot stall protection feature.

The autothrottle system automatically prevents excursions beyond maximum operating airspeed and Mach values, slat and flap placard speeds and engine EPR limits. It also keeps the aircraft at or above alpha speed for prevailing flap/slat position and angle of attack, using limit data stored in the DFGC solid-state memory.

Automatic reserve thrust enhances safety, operational economy and noise reduction. In the event of an engine failure, as indicated by a difference of more than 30 per cent between engine N_1 (fan) speeds simultaneously with slats extended indication, the good engine thrust is automatically boosted by about 4 per cent. Automatic reserve thrust also allows use of less than maximum certified thrust for take-offs and go-arounds without corresponding reduction in gross take-off weight, with benefit to fuel and maintenance costs.

Windshear computer Initial production units were delivered for use on Delta Air Lines MD-88 aircraft in 1988.

Operational status

In service in MD-80 series aircraft.

Contractor

Honeywell Inc, Commercial Aviation Systems.

VERIFIED

SkyLink Total Aircraft Information System (TAIS)

Honeywell Aviation Services and Swissair have agreed to undertake a joint project, known as SkyLink, involving development and flight testing of a system for airline and operational data communications. SkyLink will link the aircrew to the airline flight planning facilities and other operational data sources, and it will provide data communications services to the cabin crew and the passengers.

SkyLink will include both on-ground and in-air communications elements (fitted to a Swissair A321 aircraft); the project will initially be implemented at Zurich Airport.

The on-ground portion will comprise a high data rate RF-based ground datalink communication path while the aircraft is in the gate area.

The in-air portion will comprise a satcom-based airborne datalink between the aircraft and the ground, and the aircraft TAIS equipment.

TAIS will include: the recently developed Honeywell Aircraft Wireless LAN (Local Area Network) Unit (AWLU) and antenna pair; a Honeywell Airborne Server Router (ASR); an interface to the existing Honeywell/Racal aeronautical satellite communications system and other aircraft avionics systems; cockpit and cabin displays; and aircraft wiring.

Operational status

An agreement between Honeywell and Swissair called for an in-service evaluation and subsequent purchase of systems for Swissair's MD-11 and Airbus fleet.

Contractor

Honeywell Inc, Commercial Aviation Systems.

UPDATED

SP-150 automatic flight control system for the 727

The SP-150 automatic flight control system embodies an automatic landing function and is used extensively in Boeing 727 airliners. It superseded the earlier SP-50 automatic landing system and employs more solid-state components for rate sensing and integration. The system has a dual-channel configuration which meets FAA Category IIIA requirements of 50 ft decision height and 700 ft runway visual range.

A wide range of stabilisation, attitude hold and external sensor steering modes is available in the basic autopilot. Additionally there is provision for area navigation steering and for a radio altimeter input, so control law gains can be varied during the approach. Fail-operational performance is assured by the dual-computer configuration and a dual-channel yaw damper is also part of the overall flight control system.

Computational equipment used in the system includes two ½ ATR pitch computers, two ⅜ ATR yaw damper couplers and a single ¾ ATR roll control computer.

Operational status

In service in the Boeing 727.

Contractor

Honeywell Inc, Commercial Aviation Systems.

VERIFIED

SP-177 automatic flight control system for the 737

The SP-177 is an integrated digital/analogue automatic flight control system for the Boeing 737 twinjet airliner providing Cat. IIIA automatic landing capability. Main elements of each installation are two flight control computers and a glareshield-mounted controller. These are associated with an AD-300C ADI and an RD-800J HSI for each pilot. The system can be integrated with a performance management system and autothrottle.

Each computer performs pitch and roll computation for autopilot and flight director functions. The system has been configured so that pilot involvement is minimised and to ensure the flexibility of the system is available with simple and logical crew control operations. The pitch axis uses pitch attitude and rate, altitude rate, vertical acceleration and longitudinal acceleration for stabilisation. Vertical acceleration is blended with the altitude rate signal from the air data system to provide filtered altitude rate for altitude acquisition, altitude hold, vertical speed and glide slope control. Radio altitude and radio altitude rate are used to command automatic landing flare. Additionally, there is a flight director take-off mode which, based on flap setting and an angle of attack submode, keeps speed above the stall.

The design aims to provide high maintainability and reliability, has built-in self-test features and incorporates the minimum number of LRUs. Operation of the built-in test facilities can be selected from the flight deck and provides a readout on a performance data computer system display.

Specifications

Dimensions:
(flight control computer × 2) 1 ATR long
(mode control panel) 440 × 74 × 348 mm
System weight: 38 kg
Digital interfaces: ARINC 429
Power: 160 W

Operational status

In service. The SP-177 system is installed in Boeing 737-200s.

Contractor

Honeywell Inc, Commercial Aviation Systems.

VERIFIED

SP-300 digital automatic flight control system for the 737-300

The SP-300 is essentially an all-digital version of the hybrid analogue/digital SP-177 automatic flight control system designed for earlier versions of the Boeing 737-200. The previous system employed analogue circuitry to compute the safety critical Cat IIIA automatic landing functions and digital circuits for en route flight control. The configuration of the SP-300 meets the requirements for fail-passive Cat IIIA automatic landing and independent computation for Cat II flight director approach. To meet the safety criteria for approach and landing, the system has a dual/dual configuration; each of the two control channels having two different processors with different software. In this way the possibility of common mode failures and software errors is reduced to a very low level.

The mode control panel on the glareshield provides centralised control for all autopilot, flight director and autothrottle functions.

Operational status

In service in the Boeing 737-300.

Contractor

Honeywell Inc, Commercial Aviation Systems.

VERIFIED

SPZ-1 autopilot/flight director for the 747-100/200/300

The SPZ-1 system, with minor changes, is fitted to 100, 200 and 300 versions of the Boeing 747 airliner. It comprises the pitch and roll functions and associated electromechanical ADI and HSI flight directors. Boeing is the design authority, Honeywell making the equipment to Boeing's specifications and drawings. The current SPZ-1 is a three-channel system, pitch and roll computations being accommodated in three separate boxes. The two sets of flight director instruments are normally driven by different computers but can be switched as necessary in the event of a failure. Each of the three channels has its own set of attitude, air data and other sensors and the entire system is designated as fail-operational (defined as the situation whereby no single failure will cause the performance of the system to fall below the limits required by the autoland manoeuvre).

Three SPZ-1 configurations have been developed. As originally planned, the system incorporated two channels in pitch and roll and was designated as a fail-passive (no single failure will cause a hardover control demand) Cat II system. The two channels are compared one with another by means of a single box Monitor and Logic Unit (MLU). This system also incorporates a one-box automatic stabiliser trim. From this was developed the basic three-channel fail-operational system, and with it the 747 became the first US transport aircraft to be certified for Cat IIIA autoland. The upgrading was accomplished partly by adding the third autopilot, partly by changes in the mode select panel and by the addition of three Landing Control Logic Units (LCLUs) which replace the MLU. In 1976, the autoland capability was extended to include automatic roll-out control along the runway, representing Cat IIIB. This capability was gained by replacing the LCLU with a Landing Roll-out Control Unit (LRCU) and by adding a microprocessor-controlled built-in test schedule.

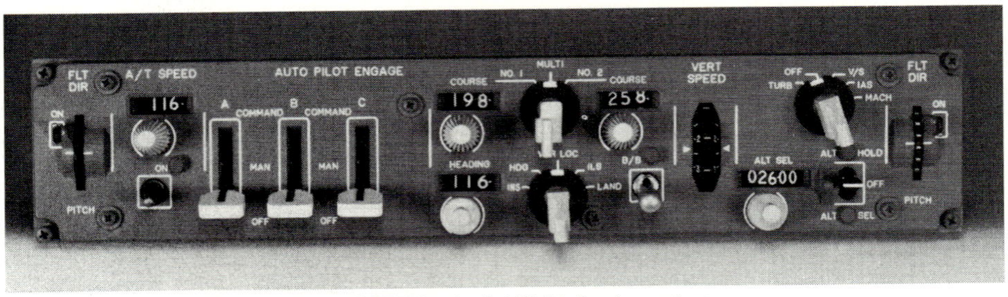

Glareshield controller for the Honeywell SPZ-1 autopilot/flight director system

The performance of the SPZ-1 has since been enhanced by the Analogue Autoland Improvement Programme (AAIP). This was launched in 1977 with the initial aim of minimising the number of disconnects being experienced at ILS capture. It was expanded in 1980 to include optimisation of the ILS tracking capability and touchdown footprint, go-around and landing flare performance.

Specifications

Dimensions:
(roll computer × 3) ⅜ ATR each
(pitch computer × 3) ⅜ ATR each
(autostabiliser) ⅜ ATR
(LRCU × 2) ¾ ATR each
(LCLU × 2) ¾ ATR each
Weight:
(roll computer × 3) 7.7 kg each
(pitch computer × 3) 8.1 kg each
(autostabiliser) 4.5 kg
(LRCU × 2) 14.1 kg each
(LCLU × 2) 10.4 kg each

Operational status

All three configurations in service in Boeing 747-100, -200 and -300 aircraft.

Contractor

Honeywell Inc, Commercial Aviation Systems.

VERIFIED

SPZ-200A autopilot/flight director system

The SPZ-200A autopilot/flight director system is available for corporate business aircraft such as the Beech King Air and BAE Systems (Operations) Limited Jetstream 31. It is available with either 4 or 5 in (102 or 127 mm) instruments and includes an air data system. Options include vertical navigation control.

Operational status

In service in Jetstream 31. No longer in production.

Contractor

Honeywell Inc, Commercial Aviation Systems.

VERIFIED

SPZ-500 automatic flight control system

The SPZ-500 is an integrated autopilot/flight director system suitable for corporate aircraft that provides all the necessary pilot interface controls, air data computation and control servos to fly a selected flight profile automatically. It integrates with companion flight director and air data systems.

The autopilot is a full-time system which provides continuous control through all phases of climb, cruise and descent, and with a full complement of lateral and vertical modes. These may be flown automatically by engaging the SPZ-500 or manually by following computed steering commands presented on the flight director instruments. The latter also enables the pilot to monitor autopilot performance.

The SPZ-500 contains an air data system to provide information over a wide range of flight profiles.

The SPZ-500 is a full three-axis autopilot with a yaw damper. It has acceleration and rate limiting circuits to provide smooth autopilot performance without compromising positive control action. Turn entry and exit is smooth and by programming the roll rate limit as a function of selected mode, rates are matched to the required manoeuvres.

Control of the autopilot for basic stabilisation and attitude command is provided through the autopilot controller. Engaging the system with no flight director mode selected causes the aircraft to maintain the existing pitch attitude, roll to wings level and then hold the existing heading. With a navigation or vertical path mode selected, engaging the autopilot automatically couples the selected mode. When the autopilot is engaged the yaw damper is automatically engaged to provide yaw stabilisation through control of the rudder. When the autopilot is not engaged, the yaw damper may be used separately to assist the pilot during manual flight.

The autopilot controller includes the turn knob and pitch wheel, allowing the pilot to insert pitch and roll commands manually. The amount of bank or pitch change is proportional to the command selected. The soft ride engage button reduces autopilot gains for smoother operation in turbulence, and the automatic elevator trim annunciators show any out of trim condition. The autopilot may be preflight checked with the test button.

Touch control steering allows the pilot to take control of the aircraft momentarily without disengaging the system. The pilot can push the button on the control wheel and manually change the aircraft flight path. While the touch control steering button is pressed, the autopilot synchronises on the existing aircraft attitude. On releasing the button the system holds the new attitude and resumes the coupled flight mode.

Several 4 or 5 in (102 or 127 mm) flight director instrumentation sets can be integrated with the flight control system. If the Honeywell ADZ-241/242 air data system is installed this includes Honeywell air data instrumentation.

Specifications

Autopilot
Dimensions:
(controller) 67 × 146 × 114 mm
(computer) 194 × 71 × 321 mm
(3 servos and mounts) 100 × 129 × 224 mm
(normal accelerometer) 51 × 25 × 61 mm
(trim servo) 56 × 84 × 175 mm
Weight: 13.39 kg total

Air data system
Dimensions:
(computer) 193 × 124 × 361 mm
(altimeter) 83 × 83 × 159 mm
(VNav computer/controller) 38 × 83 × 272 mm
(vertical speed indicator) 83 × 83 × 140 mm
(Mach/airspeed indicator) 83 × 83 × 186 mm
Weight: 9.44 kg total

Flight director system
Dimensions:
(ADI (AD-650B unit)) 129 × 129 × 223 mm
(HSI (RD-650B unit)) 103 × 129 × 208 mm
(remote controller) 38 × 146 × 66 mm
(computer) 194 × 71 × 321 mm
(mode selector) 48 × 146 × 114 mm
(rate gyro) 46 × 52 × 95 mm
Weight: 10.16 kg total

Operational status

In service. SPZ-500 variants are available for the Dassault Falcon 10 and 20, Citation I, II, III, V and VI,

Cheyenne II, Cessna Conquest, Mitsubishi Marquis and Solitaire business jets.

Contractor
Honeywell Inc, Commercial Aviation Systems.

VERIFIED

SPZ-600 automatic flight control system

The SPZ-600 automatic flight control system is designed specifically for long-range high-performance business jets and is a complete dual-channel system from the sensors to the servos. All roll, pitch and (optional) yaw channels are fully operational at all times and the performance of each is continuously monitored and compared. If a failure that resulted in a hardover manoeuvre occurs in any channel, it is immediately shut down, resulting in single channel operation in that axis only. Performance status is continuously displayed on the autopilot status panel, as well as on the autopilot master warning annunciator. Honeywell says that the unique monitoring system and duplex servo design have been thoroughly tested in transport aircraft. Each of the duplex servos incorporates two independent servo motors that operate from signals applied by their own autopilot channels. The common tie to a single control surface is accomplished through a mechanical differential gear mechanism.

In the event of a system fault, the master warning annunciator flashes amber and may be cancelled by pressing the annunciator. The status panel then indicates the system has automatically disconnected the malfunctioning channel and is a single channel in that axis only. The status panel also enables the manual selection of single channel operation in any axis and has provisions for testing the dual-autopilot channels and monitoring circuits before flight.

Control of the autopilot for basic stabilisation and attitude command is provided through the autopilot controller. Engaging the system with no flight director mode selected causes the aircraft to maintain the existing pitch attitude, roll to wings level and hold the existing heading. With a navigation or vertical path mode selected, engaging the autopilot automatically couples the selected mode. When the autopilot is engaged the yaw damper is automatically in use, when not engaged the yaw damper may be selected separately to assist the pilot during manual flight. The autopilot controller also has a turn knob and pitch wheel, which allow the pilot to insert pitch and roll commands manually. The amount of bank or pitch change is proportional to the command selected. A soft ride engage button reduces autopilot gains for smoother operation in turbulence and a couple button selects which flight director is driving the autopilot.

Touch control steering allows the pilot to take control of the aircraft momentarily without disengaging the system. He pushes the touch control steering button on the control wheel and manually changes the flight path as desired. While the button is pressed, the autopilot synchronises with the existing aircraft attitude and on releasing the button the system holds the new attitude or resumes the coupled flight mode.

The flight director system uses standard 5 in (127 mm) instruments and there is a choice of displays, with either split cue or V-bar directors and vertical scale differences. The ADZ-E 242 air data system includes a full set of appropriate instruments, including altitude alert controller and true airspeed/temperature indicator.

Specifications
Autopilot
Dimensions:
(controller) 69 × 146 × 114 mm
(status/switching panel) 48 × 146 × 114 mm
(2 duplex servos and brackets) 106 × 205 × 298 mm
(computer) 194 × 71 × 321 mm
(yaw actuator) 54 diameter × 232 mm
(normal accelerometer) 31 × 25 × 61 mm
Weight: 14.17 kg total

Flight director system
Dimensions:
(ADI (AD-650B unit)) 129 × 129 × 224 mm
(HSI (RD-650B unit)) 103 × 129 × 203 mm
(remote controller) 38 × 146 × 66 mm
(computer) 194 × 71 × 321 mm
(mode select) 47 × 146 × 114 mm
(rate gyro) 46 × 52 × 65 mm
Weight: 10.16 kg total

Air data system
Dimensions:
(computer) 193 × 125 × 361 mm
(altimeter) 83 × 83 × 159 mm
(vertical speed indicator) 83 × 83 × 146 mm
(Mach/airspeed indicator) 83 × 83 × 186 mm
(VNav computer/controller) 38 × 87 × 500 mm
Weight: 9.44 kg total

Gyro references
Dimensions:
(vertical gyro) 157 × 165 × 238 mm
(directional gyro) 191 × 154 × 229 mm
(flux valve and compensator) 121 diameter × 73 mm
Weight: 6.08 kg total

Operational status
In service. SPZ-600 variants are available for the Fokker F27, Raytheon Hawker 125-700, Bombardier Challenger, Gulfstream III and Cessna Citation aircraft.

Contractor
Honeywell Inc, Commercial Aviation Systems.

VERIFIED

HSI and ADI flight director displays, computers and controllers are part of the Honeywell SPZ-600 automatic flight control system

SPZ-700 autopilot/flight director for the Dash 7

Chosen as the standard autopilot/flight director combination for the de Havilland Dash 7 transport, the SPZ-700 is a full-time three-axis system for Cat II approaches. It provides all usual vertical and lateral flight director modes, together with RNav and MLS guidance. These vertical modes are altitude hold, altitude select, vertical speed hold and IAS hold, while the MLS mode provides for the steep approach path appropriate to STOL aircraft. The lateral modes are standard heading, navigation, ILS, back course, VOR approach and RNav.

The system is controlled via a single control panel that annunciates functions engaged. Manual demands can be fed in through the pitch trim wheel and a proportional roll knob. For manual flight a yaw damper can be chosen independently of the autopilot.

Operational status
In service in the de Havilland Dash 7 aircraft.

Contractor
Honeywell Inc, Commercial Aviation Systems.

VERIFIED

SPZ-4000/4500 automatic flight control system

Specifically designed for turboprop aircraft, the SPZ-4000/4500 system provides the accuracy, smoothness and reduced weight and volume advantages of a digital control system. It also offers more control authority than an analogue system and has comprehensive self-test capability. Autopilot, flight director and air data functions are integrated into a single system.

The functions of the three-axis autopilot and flight director are combined in a single flight control computer, and lightweight flight control servos are used. Engaging the autopilot without selection of flight director mode causes the aircraft to maintain the existing pitch attitude and to roll wings level on the existing heading. With a navigation or vertical path mode selected, engaging the autopilot automatically couples the selected mode. The yaw damper is automatically engaged in autopilot operations and can be manually engaged to assist in hands-on flying.

Control functions include pitch wheel and turn knob inputs for pitch and roll commands. There is a preflight autopilot check facility and an annunciator to draw attention to out of trim conditions.

Operational status
The SPZ-4500 is in service on the BAE Systems (Operations) Limited Jetstream 41, C-160 Transall, Convair 5800 and Beech King Air 200.

Contractor
Honeywell Inc, Commercial Aviation Systems.

VERIFIED

SPZ-5000 integrated avionics system

The SPZ-5000 integrated avionics system is designed for light business jets and regional airline or business turboprops. Advanced processing technologies and the integration of key functions result in increased capability and flexibility while achieving significant reductions in size, weight, power requirements and installation costs.

The SPZ-5000 is based on the IC-500 display/guidance computer which integrates the EFIS symbol generator, the flight director and the autopilot into a ½ ATR size box. The SPZ-5000 utilises an ARINC 429-based architecture and is composed of an EFIS electronic display offering a choice of either 5 × 5 in (127 × 127 mm) or 5 × 6 in (127 × 152.4 mm) size displays (the latter will incorporate integrated air data displays as an option), single or dual flight director and an optional single or dual fail-passive autopilot. The SPZ-5000 utilises the full range of Honeywell's advanced sensors including the AZ-840 micro air data computer or the AZ-429 air data sensor, separate vertical and directional gyros or the AH-800 fibre optic AHRS, Primus II radio system and the Primus 650 weather radar system. Standard options are Honeywell's TCAS, MLS, lightning sensor, Primus 870 turbulence detection weather radar, Primus 700 or Primus 450 weather radar and the Laseref III inertial reference system.

Operational status
In production. First certified on the Cessna Citation Jet 525 in October 1992.

Contractor
Honeywell Inc, Commercial Aviation Systems.

UPDATED

SPZ-7000 series digital flight control system

The SPZ-7000 is Honeywell's fourth-generation helicopter autopilot, and is claimed to be the first digital pitch, roll, yaw and collective four-axis system to receive civil certification. The system is microprocessor-based and has two flight computers providing full autopilot and stability augmentation, dual-flight director mode selectors, automatic preflight and en route testing features and comprehensive diagnostic circuits.

A version of the system designated SPZ-7300 was chosen by Agusta for search and rescue versions of the A 109, AB 212 (a licence-built version of the Bell 212) and AB 412. The SPZ-7300 couples a helipilot system with a digital flight path computer added to provide important new functions for the search and rescue mission. In addition to the standard flight director modes, it generates a two-stage decelerating approach to the hover, hover augmentation and the ability to hover while coupled to a Doppler navigation reference for better hover performance.

Operational status

The system was certified aboard the Sikorsky S-76 Mk II in November 1983 (now also certified on the S-76B model) and in December 1984 was specified by the airframe company as the factory standard option for this upgraded version of the corporate helicopter; in the S-76 Mk II the dual-channel system permits Instrument Meteorological Conditions (IMC) operation with one pilot. The system has also been certified on the Eurocopter AS 365N helicopter and for the Bell 222UT utility transport helicopter. All versions in production.

Contractor

Honeywell Inc, Commercial Aviation Systems.

VERIFIED

The Honeywell SPZ-7000 digital automatic flight control system is installed in the Eurocopter AS-365N helicopter

SPZ-8000 all-digital autopilot for fixed-wing aircraft

SPZ-7600 integrated SAR avionics

The SPZ-7600 is a system designed to fill the civil search and rescue role for operations in all weather conditions worldwide. It includes dual FZ-706 flight control computers, helicopter optimised EDZ-705 EFIS displays with special SAR symbology, the Primus 700 Series surface mapping/weather/beacon radar and solid-state AA-300 radio altimeter systems with circuitry and displays specifically for rotary-wing applications.

The SPZ-7600 is built around the single pilot IFR technology of the SPZ-7000 (see item above), proven in service on Sikorsky S-76 helicopters in global conditions.

During critical operations, the SPZ-7600 can be programmed to execute an automatic hands-off approach to hover and auto-hover with velocity hold at an electronically pinpointed datum.

Operational status

In production. Certified for Bell 412 and Sikorsky S-76 helicopters.

Contractor

Honeywell Inc, Commercial Aviation Systems.

VERIFIED

Honeywell SPZ-8500 integrated avionics system on Gulfstream V flight deck 0003303

SPZ-8000 digital flight control system

The SPZ-8000 digital flight control system was designed with emphasis on corporate aircraft such as the Raytheon Hawker 800. The system is unusual among its kind in having a bidirectional databus - the Honeywell ASCB avionics standard communications bus - and includes a flight management system accommodating both lateral and vertical guidance. Honeywell believes that its ASCB system provides both the higher refresh rates needed for flight control applications and a greater flexibility in use than the one way ARINC 429 standard. The system is built around two FZ-800 flight computers and is fail-operational.

Operational status

In production. The first aircraft to be certified was the de Havilland Canada Dash 8. This was followed by the Raytheon Hawker 800 and 1000, Aerospatiale/Alenia ATR 42, Cessna Citation III and the Dassault Falcon 900. The Gulfstream G-IVSP and Canadair Challenger CL-601-3A followed in 1987. Latest standard installations include the ATR-72 and Citation VII and Gulfstream IV.

In the Gulfstream IV installation, the SPZ-8000 system integrates: the Electronic Flight Instrument System (EFIS); the Engine Instrument and Crew Alerting System (EICAS); Digital Flight Control System (DFCS); Flight Management System (FMS); Laseref II Inertial Reference System (IRS); and Primus 870 Weather Radio System.

Contractor

Honeywell Inc, Commercial Aviation Systems.

VERIFIED

SPZ-8500 integrated avionics system

Honeywell's SPZ-8500 integrated avionics system for the Gulfstream V aircraft includes six 8 x 8 in EFIS/EICAS CRT displays with system synoptic pages; dual

IC-800 integrated avionics computers providing dual, fail operational autopilot/flight directors with coupled go-around and Flight Path Angle (FPA) modes; FMZ-2000 Flight Management System (FMS) with worldwide navigation database, including airfield departure and approach procedures and coupled vertical guidance. Other system features include triple micro air data computers, triple Laseref III inertial reference systems and Primus 870 Doppler turbulence detecting weather radar.

Operational status

Certified on Gulfstream V in April 1997.

Contractor

Honeywell Inc, Commercial Aviation Systems.

UPDATED

TCAS II (TCAS 2000) Traffic Alert and Collision Avoidance System

TCAS 2000 is Honeywell's latest TCAS II system. Compared with the earlier TCAS II, it is smaller and lighter, offers double the range and increases the computer capacity by 350 per cent. TCAS 2000 generates advisory information on targets up to 160 km away and can provide this information to other TCAS II-equipped aircraft to co-ordinate manoeuvres.

TCAS 2000 provides for standard TCAS II surveillance up to 32 km for ATCRBS-(Mode A/C) (Air Traffic Control Radar Beacon System) equipped aircraft and up to 64 km for Mode S-equipped aircraft. As an option, TCAS 2000 can provide for extended range suveillance of up to 160 km for Mode S-equipped aircraft. TCAS 2000 is designed to handle closure rates of up to 1,200 kt and vertical rates of 10,000 ft/min. TCAS 2000 computes range, relative altitude, and bearing of nearby transponder-equipped aircraft and visually and aurally alerts pilots of potential collisions, recommending the least disruptive vertical manoeuvre for safe separation. Warning of potential collisions occur at least 20 to 30 seconds before predicted convergence, with more warning at higher altitudes.

The Honeywell TCAS 2000 consists of a computer unit, Mode S transponder, control panel, resolution and traffic advisory displays, and antennas. It incorporates

Honeywell TCAS 2000 displays: typical EFIS (above) and VSI/TRA (below) 0018069

Honeywell SPZ-8500 EFIS 8 × 8 in Primary Flight Display (PFD) ***2001***/0099081

Honeywell RT-950 TCAS computer unit (6 MCU) (left) and RT-951 TCAS computer unit (4 MCU) (right) 0018068

software for Change 7 requirements, and forms an integral part of the Honeywell 'WorldNav' concept.

The computer unit performs airspace suveillance, intruder tracking, traffic display, threat assessment, collision threat resolution and TCAS co-ordination. It uses data from airframe and other systems to change performance parameters for varying altitudes and aircraft configurations. Collision avoidance algorithms supplied by the FAA are used to determine whether a track aircraft is a threat and, if so, the best avoidance manoeuvre.

The Mode S transponder is specially designed for the air traffic control systems of the 1990s. It performs the functions of existing Mode A and Mode C transponders and provides data exchange between TCAS-equipped aircraft. It also communicates with ground-based Mode S sensors which set TCAS sensitivity levels based on traffic density. The transponder can transmit and receive on either the top or the bottom aerial to optimise signal strength and reduce interference.

The control panel selects and controls all TCAS elements including the computer, Mode S transponder, displays and conventional ATCRBS or second Mode S transponder. It includes a transponder failure lamp and four-character LED display for transponder codes which are set with concentric rotary switches. A variety of displays may be used for TCAS information. The Traffic Advisory (TA) and Resolution Advisory (RA) may be displayed on a colour flat panel display which integrates vertical speed indication (VSI/TRA). The display is packaged in a 3 ATI-sized indicator. The TCAS also interfaces with EFIS systems to display traffic and resolution advisory information in an integrated display format.

The colours used in TCAS are amber for alert, red for resolution advisory, and blue for non-hazardous traffic. The Honeywell AT-910 directional antenna features electronic sidelobe suppression and amplitude ratio tracking. The low-profile four-element antennas are mounted on the top and bottom of the fuselage and are capable of transmitting in four selectable directions and receiving omnidirectionally. The antenna transmits at 1,030 MHz and receives at 1,090 MHz.

TCAS 2000 is available in two sizes to meet most upgrade needs, as well as to forward fit a wide variety of aircraft. It is available in both 6 MCU (RT-950) and 4 MCU (RT-951) packages. Both versions offer 28 V DC power connections, while the 6 MCU version also offers a 115 V AC connection. The 6 MCU version is fully compatible in form, fit and function with current Honeywell TCAS II installations, while the 4 MCU version only requires the smaller tray for compatibility.

TCAS 2000 is being marketed for military use. In this application it is being offered with a new version of the XS-950 Mode S transponder which includes IFF (Identification Friend or Foe) capability, developed by Honeywell with TRW, known as XS-950 S/I.

Specifications

Max range: 80 n miles to meet future Communications, Navigation Surveillance/Air Traffic Management (CNS/ATM) requirements
Display ranges: 5, 10, 20, 40 and 80 n miles
Tracks: 50 aircraft tracks (24 within 5 n miles)
Closing speed: 1,200 kt max
Vertical rate: 10,000 ft/min max
Normal escape manoeuvres: climb or descend rates; vertical speed limits
Enhanced escape manoeuvres: increased climb or descend; reversed direction of climb or descend

Operational status

Certified for a wide range of air transport, regional and corporate aircraft. In production and in service with many carriers.

TCAS 2000/Mode S has been selected by Uzbekistan Airways for fitment to Ilyushin aircraft in 1999, with further options for 2000.

TCAS 2000 has been selected by the US Naval Air Systems Command for fitment to its C-2A Greyhound, VP-3C Orion, KC-130 F/R/T Hercules aircraft between 1999 and 2003.

TCAS 2000 is also listed as a preferred item under the US Air Force Global Air Traffic Management (GATM) equipment list and forms an integral part of the Honeywell 'WorldNav' concept.

Contractor

Honeywell Inc, Commercial Aviation Systems.

VERIFIED

TCAS 1500 Traffic alert and Collision Avoidance System

TCAS 1500 is a new low cost TCAS variant for turbo prop and business jet aircraft and helicopter operators who need TCAS I capability for traffic awareness and safety of flight.

TCAS 1500 is derived from TCAS 2000. It uses the same transmitter/receiver unit, modified for TCAS I operation. The TCAS 1500 system includes TCAS computer, transmitter/receiver, one directional antenna, a system controller, three ATI display units and a Diversity Mode S transponder capable of meeting the Eurocontrol Automatic Dependent Surveillance - Broadcast (ADS-B) and Downlink Aircraft Parameters (DAP) functionality requirements.

The system generates advisories up to 15.7 n miles away and can give these advisories on any transponder-equipped aircraft transmitting an altitude code. For aircraft equipped with ATCRBS (Air Traffic Control Radar Beacon Systems), the system can display targets up to 20 n miles and with Mode S equipped aircraft, the range is up to 40 n miles.

Specifications

Display ranges 6, 12, 20, 40 n miles, can be upgraded to 120 n miles with TCAS 2000
Tracks up to 50 (24 within 5 n miles)
Closing speed 1,200 kt max
Vertical rate 10,000 ft/min

Contractor

Honeywell Inc, Commercial Aviation Systems.

VERIFIED

Versatile Integrated Avionics (VIA)

VIA is a mature, flexible, general purpose processor developed for the commercial airline market. VIA has direct application for military aircraft and offers a true Commercial-Off-The-Shelf (COTS) solution.

The VIA design lends itself to both new production (for example the Boeing 717) and retrofit (Federal Express MD-10) applications. Since the beginning of 1999, Honeywell has been selected by several customers, including the US military, to put the VIA system on several military aircraft, including the C-5, KC-10 and E-6.

The baseline VIA system consists of a common chassis and power supply hosted in an ARINC standard 8 MCU Line Replaceable Unit (LRU). Internal cards provide generic computing and I/O communication. The innovative ARINC 659 backplane and construction of each of the cards provides high integrity, high-availability processing and communications. For the ARINC 659 backplane, I/O data is transmitted up to four times, once on each of the Ax, Ay, Bx and By busses. The ARINC 659 data rate is approximately 60 Mbits/s.

The baseline VIA contains one Core Processing Module (CPM) with dual AMD29050 RISC microprocessors executing identical code on a cycle-for-cycle basis; two I/O controller Modules (IOM) providing redundant I/O interfaces; and one high-integrity commercial power supply.

For military applications, the VIA contains a data transfer module for MIL-STD-1553B bussing with provisions for VIA operations either as a bus controller, back-up bus controller or remote terminal.

VIA equipment configuration ***1999***/0051696

The baseline VIA has provisions for two additional core processing modules, graphics generators to drive AMLCDs, and three additional I/O modules should increased computational and/or I/O capability be necessary for specific applications.

Solutions involving two or more VIAs use a Cross Channel Link (CCL) providing a communication tie between VIA processors.

The VIA design uses an open system architecture to support customer or third party participation in software development and follow-on support. ARINC standards and FAA-approved tools allow VIA customers to develop their own software to provide unique customer solutions.

The overall VIA design is based on ARINC 651 design guidance for IMA. The VIA external interfaces are commercial and military standards such as AEINC 429 and MIL-STD-1553B. The VIA backplane is designed to ARINC 659.

A single VIA, operating in a partitioned mode, provides solutions to multiple avionics functions previously provided by separate LRUs. Robust partitioning is designed into the VIA hardware and software to ensure the VIA will operate as a virtual LRU for a selected time frame. Virtual LRU operations are possible for flight management, integrated displays, COMM/NAV/Surveillance (CNS) management, flight controls, central caution and warning, central maintenance diagnostics, built-in test and other functions. The partitioning scheme guarantees virtual LRU operation by ensuring that a functional partition is not allowed to affect another partition's code, I/O or data storage area; a partition is not allowed to consume shared processor or I/O resources to the exclusion of another partition; and a single failure of common hardware is not allowed to prevent safe flight and landing.

The VIA is designed to easily accommodate future Global Air Traffic Management (GATM) and FADEC military upgrades by using the partitioning concept to add the desired performance in a separate partition or to an existing partition.

Specifications

Dimensions: 203.2 (H) × 304.8 (W) × 406.4 (D) mm
Weight: 20.45 kg

Operational status

VIA is a second-generation Honeywell product based on an Integrated Modular Avionics (IMA) design for the Boeing 777 aircraft. The IMA was first certified by the FAA in April 1996. The Boeing 777 IMA design was repackaged into a discrete LRU called VIA. VIA certification by the FAA was achieved in October 1997 on the Boeing 737-700 series aircraft.

The latest application of the system is designated VIA 2000. In the Boeing 717, VIA 2000 supplies the following systems: the Category IIIa Auto Flight System, including windshear detection and stall warning (the system can be upgraded to Cat IIIb capability); dual Air Data/Inertial Reference Systems (ADIRS); and the central fault display system.

VIA is in production. Applications include both new production aircraft (Boeing 717, 737-700 and MD-90) and retrofit applications (Federal Express MD-10). VIA has been selected for the C-5 Avionics Modernisation Programme (AMP) to meet Global Air Traffic Management (GATM) requirements for operating worldwide missions in the 21st century; other applications include the KC-10 and E-6.

Contractors

Honeywell Inc, Commercial Aviation Systems.
Honeywell Defense Avionics Systems.

UPDATED

VHF DataLink System (DLS)

At Airshow China '98 Honeywell and Aviation Industries of China (AVIC) signed a Memorandum of Understanding to support Communication, Navigation Surveillance/Air Traffic Management (CNS/ATM) in China. Honeywell will develop a VHF DLS for Chinese government aircraft. The system includes a new integrated Communications Management Unit and VHF Data Radio (CMU/VDR).

The complete Honeywell VHF DLS comprises the Honeywell integrated CMU/VDR, Honeywell/Trimble HT9100 GPS navigator and an ARINC 740/744-compliant ACARS printer.

Operational status

The programme is in three stages: demonstration of the system on three aircraft; installation on a large number of Chinese government transport aircraft; repackaging for smaller aircraft for the remainder of the government aircraft fleet. This final stage is to be completed by 2005.

Contractor

Honeywell Inc, Commercial Aviation Systems.

VERIFIED

Windshear systems

Honeywell windshear systems provide detection, alert and guidance in a single unit with two levels of detection - 'caution' and 'warning'. During take-off and approach, the most critical phases of flight, the systems offer an angle of attack reference on the flight director. The ADI also gives an immediate pitch cue to help in exiting a shear. With extensive filtering and an automatic compensation for aircraft manoeuvres and configuration changes, the windshear systems integrate safety with reliability.

Key windshear features enhance the value of the system. The low installation cost is complemented by nearly universal compatibility. Modification of existing aircraft systems is not required and self-contained sensors reduce the proliferation of system configurations. Pin-programmable for multi-aircraft application, the windshear system meets FAA reliability requirements with a 99.9 per cent availability rating and an undetected failure ratio of .00001.

Honeywell windshear systems are available in two configurations: as a stand-alone unit installed in a ⅜ ATR short box or as a system integrated with the advanced flight management computer system or flight control computer for new airliners.

Specifications

Dimensions: ⅜ ATR short
Weight: 6.8 kg
Power: 22 W
Reliability: 20,000 h MTBF

Operational status

In production and standard fit on Avro 146/RJ, Boeing 727, 737 and 747, Fokker 100 and F28 aircraft, Lockheed Martin L-1011, Boeing DC-8, DC-9, MD-11, MD-80 and MD-90 aircraft.

Contractor

Honeywell Inc, Commercial Aviation Systems.

VERIFIED

WorldNav™ CNS/ATM avionics

Honeywell has developed its WorldNav™ CNS/ATM avionics product line to meet the requirements of the emerging Communications, Navigation, Surveillance/Air Traffic Management (CNS/ATM) operating environment. Central to Honeywell's WorldNav concept is the Pegasus flight management system; other items of equipment included in the WorldNav concept were initially the TCAS 2000 Collision Avoidance System, the ADIRS Air Data Inertial Reference System.

Most recent addition to the WorldNav system is the new CM-950 Communications Management Unit (CMU) to provide new technology needed to implement the Aeronautical Telecommunications Network (ATN).

Operational status

The three initial elements of the WorldNav CNS/ATM system are in production, and this suite of WorldNav CNS/ATM avionics has been selected by a consortium of Latin American airlines to equip their fleet of A319 and A320 aircraft for future CNS/ATM operations.

More recently in mid-1999, Honeywell signed EVA Air as the launch customer for the CMU. The initial installation is to be for new Boeing 747-400 freighter aircraft. Follow-on installations are to be for EVA Air's complete fleet of existing Boeing 747-400,767 and MD-11 aircraft.

Contractor

Honeywell Inc, Commercial Aviation Systems.

VERIFIED

XS-950 Mode S ATDL (Air Transport DataLink) transponder

The Honeywell XS-950 transponder was designed for the air transport market and meets all ARINC 718 requirements. The XS-950 implements all currently defined Mode S functions with provision for future growth. Current Mode S transponders are used in conjunction with TCAS and ATCRBS to identify and track aircraft position, including altitude. This system transmits and receives digital messages between aircraft and air traffic control. The datalink provides positive and confirmed communications more efficiently than current voice systems.

The Honeywell design meets future needs by including growth capability to support the functions defined by CNS/ATM (Communications, Navigation, Surveillance/Air Traffic Management), and is a constituent element of the Honeywell 'WorldNav' concept, together with the Traffic alert and Collision Avoidance System (TCAS-2000), Pegasus Flight Management System (FMS), and Satcom Aero-H, Aero-H+ and Aero-I products.

The TCAS is fully supported, including 'diversity' (top and bottom) antenna ports. Diversity provides reliable RF communication links between both ground-based and airborne interrogators. The transponder incorporates a TCAS II interface and is designed to be compatible with all TCAS II systems conforming to ARINC 718/735 characteristics.

The Mode-S ATDL transponder is an ICAO Level 4 system with growth to Level 5. Level 4 means that it will transmit and receive standard length (112 bit) datalink messages for 'COMM A' and 'COMM B' and transmit and receive 16-segment extended length datalink messages for 'COMM C' and 'COMM D'.

Honeywell has included full Built-in Test Equipment (BITE) and self-test capabilities to provide maximum reliability while minimising maintenance costs. The XS-950 interfaces to all air transport OEM onboard maintenance systems.

A new version of the XS-950 Mode S datalink transponder, incorporating IFF (Identification Friend-or-Foe), has been developed by Honeywell with TRW, for use with TCAS 2000 by the military. The divisions concerned are Honeywell Defense Avionics Systems, Albuquerque and TRW Avionics Systems Division, San Diego.

The new version is designated XS-950 Mode S/IFF transponder (XS-950S/I). It enables military aircraft to operate within civil airspace and to meet the requirements for reduced separation. It also provides IFF reporting required in military operational airspace. All current military IFF functionality is provided in the XS-950 S/I. Future growth in the XS-950 S/I will parallel changes in the commercial sector, thus affording the military future cost savings. The XS-950 S/I can be controlled with a multifunction control display over an ARINC 429 or MIL-STD-1553 bus, or by a special purpose control panel that provides control of the Mode S functions, the IFF function and the TCAS.

The XS-950 S/I can perform all the functions of the existing Air Traffic Control Radar Beacon System (ATCRBS), including Selective Identification Features (SIF), Modes 3/A and C operation. The XS-950 S/I transponder also meets military IFF Mode 1, 2 and 4 requirements, and it can transmit and receive extended-length Mode S digital messages. Growth capability to support CNS/ATM (Communications Navigation Surveillance/Air Traffic Management) Mode-S Level 5 is provided. Other features of the XS-950 S/I include US FAA-specified antenna diversity for simultaneous operation with both top and bottom antennas; interfaces to IFF Mode 4 crypto computer (KIT-C); ICAO Level 4 datalink transponder capability providing extended uplink/downlink message throughput; ICAO Level 5 upgrade capability by software change alone; transmit power 795 W maximum peak pulse, 316 W minimum, 400 W nominal.

Growth features of the XS-950S/I, planned for the third quarter of 2000, include GPS squitter (ADS-B) positional data for more accurate TCAS operation, and extended reception range of TCAS proximity alerts and ADS-B squitter data (lat/long/altitude/flight ID/Mode S address).

Specifications

Dimensions:
(XS-950): 124.5 × 194 × 325 mm
(XS-950S/I): 124.5 × 193 × 386 mm
Weight:
(XS-950): 5.7 kg
(XS-950S/I): 5.2 kg
Power: 115 V AC or 28 V DC

Operational status

The XS-950 is in production. The XS-950S/I received US FAA TSO approval in September 1999. Honeywell Defense Avionics Systems received a contract to supply the US Navy and US Marine Corps with systems for aircraft including the C-2, C/KC-130 and VP-3; first deliveries were in September 1999.

Contractor

Honeywell Inc, Commercial Aviation Systems.
Honeywell Inc, Defense Avionics Systems.
TRW Avionics Systems Division.

UPDATED

HG7500/HG8500 Series radar altimeters

HG7500 configurations available include analogue and/or digital altitude, altitude trips and ARINC 552A. The JG107X height indicator interfaces with the HG7500 and HG8500.

The HG8500 is a form, fit and function replacement for the HG7500. It has a solid-state transmitter and gallium arsenide receiver, and has been qualified to very stringent environmental and EMI requirements.

Options available on the HG8500 include transmitter power management, low-altitude performance in poor antenna installations and unique outputs to meet existing field applications.

Specifications

Dimensions: 137 × 83 × 83 mm
Weight: 1.3 kg
Power supply: 28 V DC, 16 W

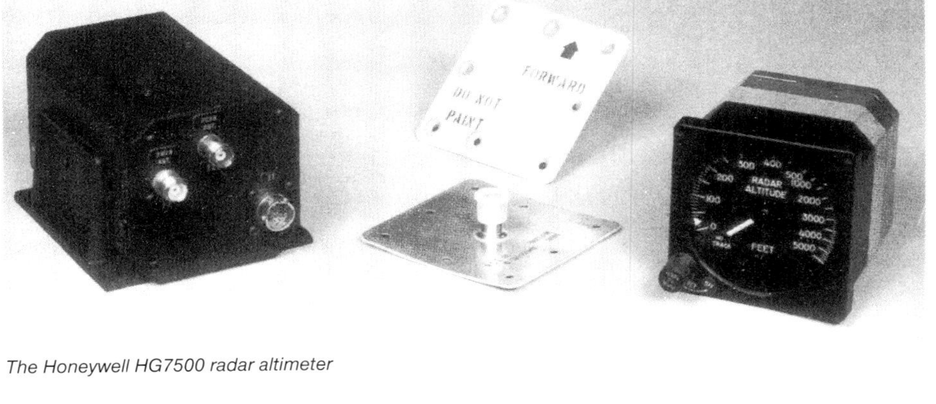

The Honeywell HG7500 radar altimeter

Altitude:
(HG7502/HG8502) 0-2,500 ft
(HG7505/HG8505) 0-5,000 ft
(HG7508/HG8508) 0-8,000 ft
Accuracy: ±3 ft ±3% analogue altitude
±3 ft ±1% digital altitude

Operational status

In service on various helicopters and commercial airliners.

Contractor

Honeywell Inc, Sensor and Guidance Products.

VERIFIED

SRS 1000 Attitude and Heading Reference System (AHRS)

The SRS 1000 strapdown Attitude and Heading Reference System (AHRS) uses advanced technology to meet modern AHRS requirements with very low life-cycle cost. A 15-second reaction time, independent of ambient temperature or vibration, is claimed.

There are two units: the attitude heading reference unit and compass controller unit. The attitude heading reference unit accepts magnetic compass, air data and Doppler inputs to provide attitude, heading, body rates and accelerations, groundspeed and drift angle information to other aircraft systems. The compass controller provides systems information and control functions to the crew.

An inertial measurement unit within the reference unit contains two flexure suspended gyros and two toroidal accelerometers for the X and Y axes, plus a standard force-feedback accelerometer for the Z axis.

Operational status

No longer in production. The system is standard equipment on the Airbus Industries A300 and A310 wide-body airliners.

Contractor

Honeywell Inc, Sensor and Guidance Products.

UPDATED

MIAMI ice detection system

The Microwave Ice Accretion Measurement Instrument (MIAMI) permits the accurate measurement of ice accumulating in critical areas of the aircraft.

As an ice warning system, the MIAMI alerts the pilot to the earliest initiation of ice growth. As little as 0.076 mm of ice can be detected, illuminating a warning light on the annunciator panel. Then the preprogrammed microprocessor takes over and both computes icing rate in in/min and indicates in the cockpit digital display the ice thickness.

If required, the system can be used to activate or deactivate the de-icer equipment on aircraft and missile systems and works equally well with all types of de-icing or anti-icing equipment.

The system transducer element consists of a resonant surface waveguide. The resonant frequency of this varies according to the amount of ice accreted and so a relationship between ice thickness and frequency shift can be established. It is said that this type of transducer has the advantage of not requiring an external and frangible probe and the transducer may be profiled to conform to the contour of the mounting surface.

The transducer can be mounted anywhere on the aircraft, including rotors or wings, and the system can be protected against sand or rain erosion. The system is microprocessor-controlled and a single microprocessor unit can control any number of transducers.

Operational status

In production. Installed on the Cessna T303 Crusader.

Contractor

Ideal Research & Development Corporation.

VERIFIED

Avionic display modules

ImageQuest Active Matrix Liquid Crystal Displays (AMLCDs) are designed for avionic applications, and feature amorphous Silicon Thin Film Transistors (TFTs), thermal control of AMLCD and back-light, and VGA format.

Contractor

ImageQuest Technologies Inc.

VERIFIED

Specifications

Parameters	4 × 4 in avionic display	6.24 × 8.31 in avionic display	10.4 in avionic touchscreen display
Active display area	101.76 × 101.6 mm	211.2 × 158.4 mm	211.2 × 158.4 mm
Module dimensions	127 × 127 × 76.2 mm	284 × 184.4 × 73.5 mm	300 × 246 × 55 mm
Weight	1.42 kg	3.5 kg	3.41 kg
Pixel configuration	RGB Delta Triad	RGB stripe	RGB stripe
Pixel array	480 × 480	640 × 480	640 × 480
Luminance	200 fL	>200 fL	>50 fL
Grey shades	256	64	16
Power supply	12 to 28 V DC, 29 W	12 to 28 V DC, 29 W	12 V DC, 5 W

Reduced Vertical Separation Minimum (RVSM) air data system

The IS&S RVSM system is designed to be fitted to the widest possible number of aircraft types as a retrofit or new build system. In addition to satisfying RVSM requirements, retrofitted aircraft receive an upgraded air data system.

The system is a triplex installation, integrating three high-precision altitude sensing devices which continuously monitor and compare three sources of computed pressure altitude. Compared to conventional RVSM solutions using dual Central Air Data Computers (CADC), the triplex system ensures that an aircraft will not be excluded from RVSM airspace in the event of a single air data component malfunction.

The IS&S system provides full triplex operation and, when selected with the IS&S Digital Air Data Computer (DADC), it provides the following functions: visual wind-shear warning, true airspeed display, angle of attack display, stall warning, and take-off monitor. It also warns of excessive deviation, monitors CADC performance, provides an altitude alerting function with minimum descent altitude mode for preselect and voice warnings, and exceeds North Atlantic corridor altitude reference requirements.

IS&S claims that the advantage of its system is that the altimeters and airspeed indicators, in their self-sensing mode, act in combination as independent air data computers. In contrast to conventional systems, the pilot using an IS&S RVSM system knows where a failure or deviation has occurred and still has two remaining systems to satisfy RVSM requirements.

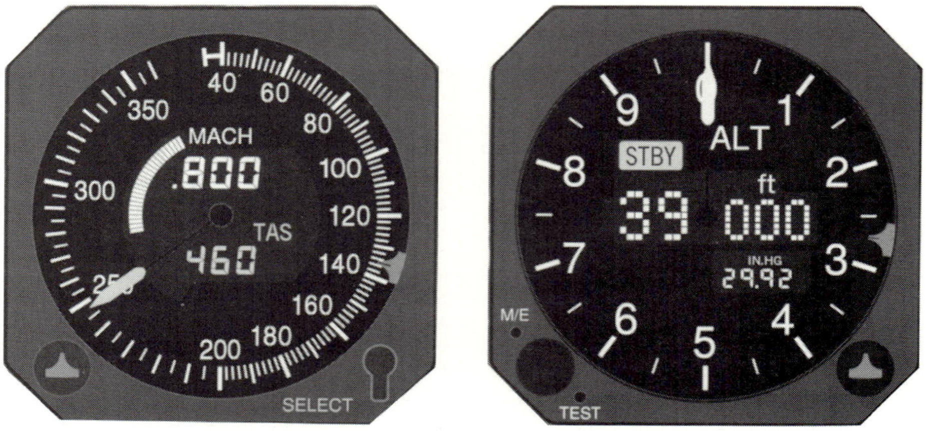

The IS&S reduced vertical separation minimum air data system airspeed and altimeter displays **2000**/0064371

In normal operation, the altimeters and airspeed indicators function as repeaters displaying information from the DADC's air data sensors. They are, however, also integrated with the aircraft air data sensors and are equipped with their own internal transducers. They constantly compare the DADC signal with their self-sensed information and, if a difference is detected, the altimeter and airspeed indicator automatically switch into their standby mode and display self-sensed information instead of that from the DADC.

System components comprise:
one DADC, two dual-mode solid-state altitude indicators, two dual-mode solid-state airspeed indicators, one multifunction preselect altitude alerter.

The system is fully compliant with North Atlantic RVSM, Free Flight, and Europe 2002 requirements.

Operational status

Over 500 systems installed on 13 different programmes by summer 1999. Customers include regional and business aircraft, Cargo Airline Association operators and the US Navy, for its C-9 aircraft.

Contractor

Innovative Solutions & Support Inc (IS&S).

VERIFIED

Very large format AMLCD

The new Innovative Solutions & Support Inc (ISS) very large format (14.6 × 11.1 in viewing area), flat screen, Active Matrix Liquid Crystal Display (AMLCD) replicates the traditional 'T-shaped' airspeed, attitude, altitude and HSI data presentation, but without the cluttered vertical tapes of typical EFIS systems. Angle Of Attack (AOA) and Vertical Speed Indication (VSI) are also presented, and notable features are the size, brightness, resolution (1,280 × 1,024 pixels), contrast (200:1), off-boresight (±80°) viewing, and anti-aliasing qualities.

These capabilities have been obtained by bringing together 'enabling technologies' from the commercial electronics market, and utilising them in the context of aviation without the constraints imposed by use of conventional single-purpose black-box displays. As well as use of the latest commercial glass capabilities, modern very high throughput (1.2 billion instructions per second) commercial processors are utilised in dual redundant, dual channel, configuration to provide the computing power required to support the display and symbology sets envisaged. The display is linked to a remote 3MCU electronics unit.

Additional data such as VOR, TCAS, engine data, flight path data, and even threat warning data could be added in the space available, to meet customer requirements.

Operational status

Launch customer for the ISS AMLCD was the Pilatus PC-12. However, ISS believes the real opportunity for this display is the military training-aircraft market, because the display can be redefined to display any combination of conventional formats, simply by software change, and thus to represent any operational aircraft required, resulting in much reduced operational conversion flying requirements. Certification for supplementary and primary flight use for the PC-12 was expected to be completed by the end of 1999.

Contractor

ISS Innovative Solutions & Support Inc.

UPDATED

The ISS 14.6 × 11.1 in viewing area AMLCD

***1999**/0051697*

EDM-700 family of Engine Data Management systems

The EDM-700 family is a complete engine data recording system, it records not only Exhaust Gas Temperature (EGT) and Cylinder Head Temperature (CHT) as some systems do, but all 24 engine temperatures and engine pressures, plus RPM and per cent horsepower.

Single- and twin-engined installations are available with Record, 'Snapshot' and Alarm modes of operation.

Contractor

JP Instruments.

VERIFIED

EDM-700 Engine Data Management systems

0001441

31400 pneumatic altimeter

The Kollsman 31400 pneumatic altimeter is a 2 in ATI indicator designed to meet the requirements of TSO C10b. The altimeter is used as a standby indicator for aircraft equipped with EFIS. The indicator measures and displays accurate pressure altitude. A single pointer affixed to a handstaff makes one revolution per 1,000 ft and is read against a dial graduated in 20 ft increments. The handstaff simultaneously drives a counter-drum in synchronism with the pointer.

The instrument provides a digital display with a baro correction range of 28.1 to 31 in mercury or 950 to 1,050 millibars. Operating range of the instrument is −1,000 to +50,000 ft. The indicator receives inputs of static pressure. It is integrally lit with 5 V white or blue white lighting.

Operational status

The indicator is in production and is used in a number of aircraft including the de Havilland Dash 8.

Contractor

Kollsman Inc.

VERIFIED

44929 digital pressure altimeter

The Kollsman 44929 is a 3 in ATI altimeter meeting the requirements of TSO C10b and TSO C88a. This digital altimeter senses atmospheric pressure changes, displays altitude with high accuracy, and generates an altitude reporting signal. The altimeter is designed specifically for aircraft that do not require static pressure source correction.

The instrument is driven solely by atmospheric pressure acting on dual aneroid diaphragms. No electrical power is required to operate the pneumatic mechanism. The encoding feature of the altimeter is provided by means of an optical encoder which uses light emitting diode light sources and phototransistor detectors. A code disc is driven by the main shaft of the altimeter mechanism and rotates between the light sources and photo detector array to provide the encoded information to the transponder.

Operating range of the instrument is −1,000 to +50,000 ft.

Operational status

In production and used on many types of subsonic aircraft.

Contractor

Kollsman Inc.

VERIFIED

46650 fuel flow/fuel used indicator

The 46650 fuel flow/fuel used indicator family was designed to replace older electromechanical indicators directly. It features a single microcontroller, four printed circuit boards, a DC torque converter which drives the pointer and an LED digital display. With a quoted MTBF of over 20,000 hours, the unit is highly reliable. In addition, the circuit boards and major components are easy to assemble, permitting quick repairs.

A pointer, which swings in a 220° arc below the LED display, indicates the total fuel flow. The LED shows the

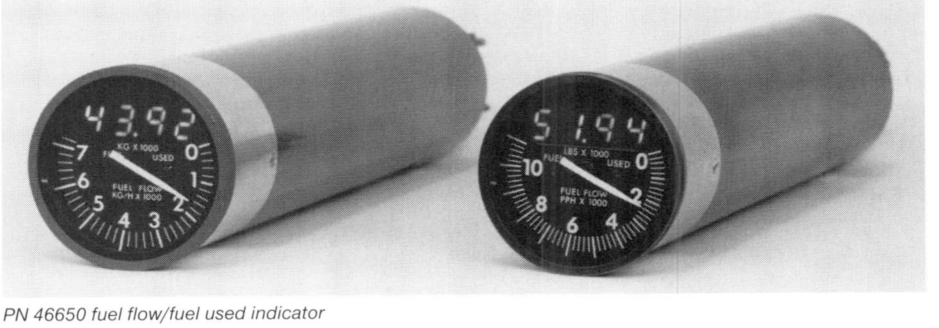

PN 46650 fuel flow/fuel used indicator

fuel used. The indicator can be used either with old fuel flow transducers or the new motorless transducers.

The 5 V aircraft instrument lighting system controls the bezel lighting. Based on bezel lighting values, the microprocessor controls the intensity of the digital display. Input power is 115 V AC, 400 Hz. In the event of power failure, a non-volatile memory saves the fuel used readings for a minimum of 30 days. A self-test feature verifies indicator integrity.

Operational status

The fuel flow/fuel used indicator is used on Boeing 727, 737, DC-8 and DC-9.

Contractor

Kollsman Inc.

VERIFIED

47174-( ) Resolution Advisory/ Vertical Speed Indicator (RA/VSI)

The 47174-( ) Resolution Advisory/Vertical Speed Indicator (RA/VSI) was designed to meet wide applications for Traffic alert and Collision Avoidance Systems (TCAS) installations. It is fully compliant with the TCAS guidelines set out in ARINC 735 and will interface with all TCAS equipment.

A single resolution advisory arc is composed of 52 bi-colour surface-mounted LEDs. These LEDs are individually addressed by an embedded microcontroller and provide a high-resolution and flexible RA display.

The modular design of the instrument satisfies all VSI interfaces. These VSI inputs include an internal pneumatic sensor, ARINC 429, ARINC 565, ARINC 575 and Manchester code. Options are available to accommodate specific VSI interfaces and 28 V DC aircraft power. This flexibility allows an operator to utilise a single indicator that will be universal throughout the whole fleet.

Specifications

Power supply: 115 V AC, 400 Hz or 28 V DC

Operational status

The Kollsman RA/VSI is installed in AMR Eagle Saab 340 aircraft.

Contractor

Kollsman Inc.

UPDATED

The Kollsman resolution advisory/vertical speed indicator

48660-( ) EGT indicator

The 48660-( ) solid-state Exhaust Gas Temperature (EGT) indicator has been designed as a direct replacement for older electromechanical indicators. The installation requires no panel or harness modifications. The EGT indicator interfaces with the existing aircraft chromel/alumel Type K thermocouple harness. A single microcontroller processes the thermocouple signal and executes all commands within the indicator. Output for an Airborne Integrated Data System (AIDS) is a standard feature with each indicator.

In the normal operating mode, the EGT indicator displays temperature on both the dial and the three-digit LED display. When sensed temperature drives the pointer into the yellow or caution zone, an amber warning light illuminates to warn of an impending over temperature condition. If the maximum allowable temperature is exceeded, the indicator stores the greatest over temperature value. The blue recall push-button is then illuminated. The duration of the over temperature event is calculated and stored.

The stored information is retrieved by depressing the recall push-button, and the over temperature value and duration are shown on the LED display. When the aircraft over-limit reset switch is activated the blue recall push-button is extinguished. Non-volatile memory, provided through an EEPROM, does not require an internal or external battery when aircraft power is off.

Operational status

Selected for the Federal Express DC-10-10 and DC-10-30 fleet.

Contractor

Kollsman Inc.

VERIFIED

49200 series digital altimeter systems

49200-101 RVSM Altitude Measuring System (AMS)

The Kollsman 49200-101 RVSM (Reduced Vertical Separation Minima) AMS offers an RVSM solution in a single integrated package, designed to provide a retrofit capability to meet the increased accuracy requirement for flying in RVSM airspace. This unit uses separate highly accurate pitot and static transducers contained in a single 3ATI package, a Total Air Temperature (TAT) probe, and Static Source Error Correction (SSEC) processing to meet RVSM accuracy. An altitude pre-select/alerter function with autopilot output is also built into the unit to meet the RVSM altitude alerter accuracy requirement. The unit features an advanced Liquid Crystal Display (LCD) and the familiar needle pointer positioned by a high-reliability, dual-channel, brushless DC repeater; both the analogue and digital scales are calibrated in both feet and metres.

49200-301 Advanced Servo Altimeter Alerter System (ASAAS)

The Kollsman 49200-301 Advanced Servo Altimeter Alerter System is targeted for retrofit on older commercial airliners. It is designed to provide a retrofit capability to meet the increased accuracy requirement for flying in RVSM airspace. Northwest Airlines has selected the system for its Boeing B747-100/200 aircraft. This product is a very accurate altimeter, with accuracy better than ±20 ft from −1,000 to 50,000 ft. In addition, it has a built-in altitude alerter function and interfaces with existing air data computers and autopilot flight director systems. Should there be an electrical malfunction or failure, the unit automatically reverts to a standby mode, using its own self-contained, highly accurate solid-state pressure transducers. Like the 49200-101 RVSM AMS, this unit also features an advanced Liquid Crystal Display (LCD) and the familiar needle pointer positioned by a high-reliability, dual-channel, brushless DC repeater; both the analogue and digital scales are calibrated in both feet and metres. In this unit both feet and metres can be simultaneously displayed on a separate four-digit LCD.

The Kollsman 49200-101 altitude measuring system
1999/0051698

The Kollsman 49200-301 advanced servo altimeter alerter system ***1999***/0051699

The Kollsman 49200-401 barometric altimeter
1999/0051700

49200-401 barometric altimeter

The Kollsman 49200-401 barometric altimeter displays precision pressure altitude developed via a highly accurate internal solid-state transducer, in accordance with TSO C10b, from −1,000 to +53,000 ft. Barometric altitude is provided to the encoder/transponder through an 11-bit, parallel interface which represents the current display altitude as defined by TSO C88a. All functions are integrated into a single 3ATI unit that will retrofit into existing aircraft.

Contractor

Kollsman Inc.

VERIFIED

Cockpit Multifunction Displays and Display Electronics Units (MFDs/DEUs) for V-22 Osprey aircraft

The MFDs/DEUs are ruggedised, full colour, 6 × 6 in, beam index CRT cockpit displays and display electronics units for the V-22 Osprey aircraft. The suite comprises four Multi Function Displays (MFDs) and two Display Electronics Units (DEUs) per aircraft. The displays are NVG-compatible and provide graphics/ symbology overlays on digital map, FLIR and radar sensor video inputs. There are three independent channels of high-performance display processing.

Operational status

Boeing EMD contract awarded 1994; LRIP contract awarded 1996.

Contractor

L-3 Display Systems.

VERIFIED

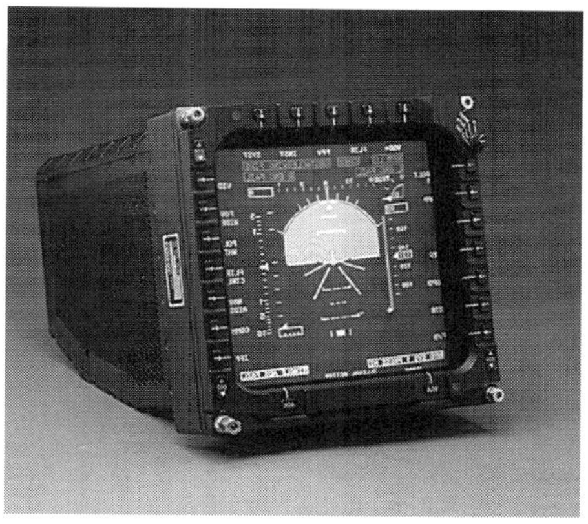

L-3 Display Systems multifunction display for the V-22 Osprey aircraft 0018167

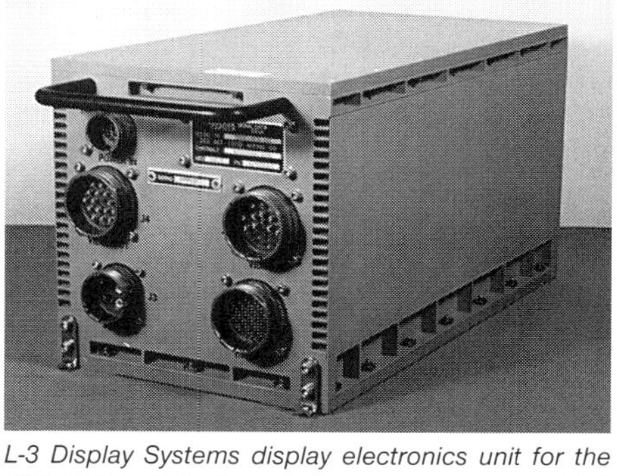

L-3 Display Systems display electronics unit for the V-22 Osprey aircraft 0018166

CCU-800 Cockpit Control Unit

The CCU-800 Cockpit Control Unit is designed for use in flight test, trials and certification work. It converts Pulse Code Modulation (PCM) data generated by trials equipment into a display form that allows the pilot to instantaneously verify flight parameters required for safety of flight or to confirm adequate completion of the planned mission.

The CCU-800 is a rugged, sunlight readable, full colour (4,096 colours), high resolution (640 × 480 pixels) 6.4 in liquid crystal display. A microprocessor and PCM demodulator allows third party processor controlled equipment (recorders and control systems) to be controlled during flight, and to present the pilot with real-time data on the status of planned tasks.

The CCU-800 comprises three main items: the CSP-800 Cockpit Switch Panel, the CDP-800 Cockpit Display Panel and the CEU-800 Cockpit Electronics Unit. The CCU-800 is suitable for civil and military use, and is currently in use on a major military flight test programme. Dimensions of the CCU-800 are: 269.2 (H) × 257.3 (W) × 196.9 (D) mm. It meets MIL-STD-810E and MIL-STD-416D.

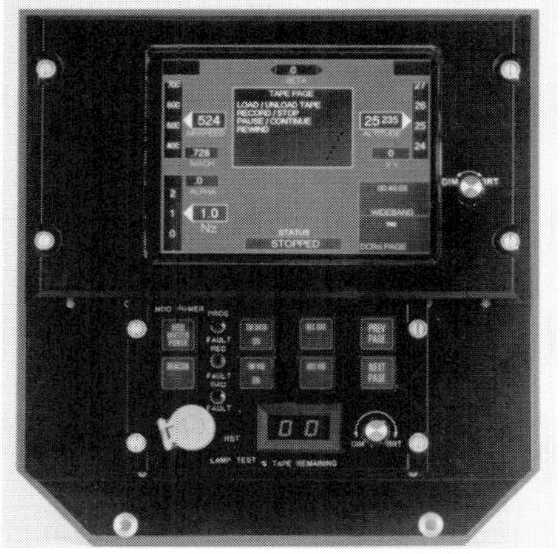

The L-3 Telemetry-East CCU-800 Cockpit Control Unit ***1999****/0051693*

Contractor

L-3 Telemetry-East.

VERIFIED

IEC 9001 GPS Navigation and Landing System (GNLS)

The IEC 9001 GPS Navigation and Landing System (GNLS) is a stand-alone, coarse acquisition (C/A) code, differential GPS-based system intended for installation in Flight Management System (FMS) – equipped commercial transport aircraft.

For en route navigation, the 9001 GNLS provides GPS position in accordance with ARINC 743A format to the flight management computer. For the approach mode, it calculates and provides glideslope and localiser deviations to the Electronic Flight Instrumentation System (EFIS) and Flight Control Computer (FCC) based on the self-contained approach navigation database and preselected FMS approach path.

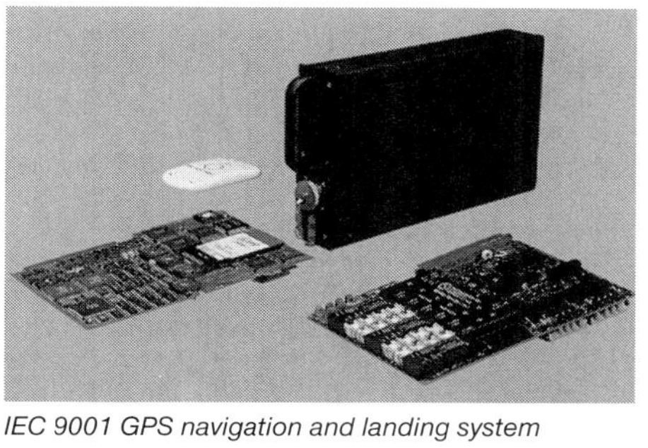

IEC 9001 GPS navigation and landing system ***1999****/0051615*

The system inputs/outputs conform to ARINC 429 low-speed and high-speed data as defined in ARINC 710-9 for the ILS receiver. The 9001 supports en route, terminal, and approach navigation accuracy and integrity requirements.

The 9001 GNLS can be enhanced with a kinematic upgrade to enable CAT-II/IIIb landings. It can also be upgraded to accept and process GLONASS signals.

Specifications

Navigational signal: L1 C/A code (SPS)
Receiver: 12 hardware channels
Time to first fix: <60 s
Dynamics: 900 kt max velocity
Accuracy:
100 m 2σ rms (selective availability on)
5 m utilising differential GPS corrections
Antenna: active conformal meeting ARINC 743A
Environment: fully compliant to environmental specifications of DO160c
Dimensions: 2 MCU
Weight: 2.9 kg
Power: 45 W @ 28 V DC

Contractor

L-3/Interstate Electronics Corporation.

VERIFIED

IEC 9002/9002M Flight Management Systems (FMS)

The IEC 9002/9002M FMS provides complete LNav and approach capabilities, and accurate navigation in all phases of flight using the 12-channel GPS receiver, which accepts differential GPS corrections. It features a control display unit with a 4 in diagonal, sunlight-readable, high-resolution, 16-colour LCD and can be enhanced with a kinematic upgrade to provide Cat. III landing accuracy. The unit provides rapid GPS satellite acquisition (time to first fix 2 minutes). It is equipped with Jeppesen worldwide navigation database including SIDs, STARs, GPS instrument approaches, airports with runways greater than 4,000 ft, high- and low-altitude airways, intersections, VHF navaids and NDBs, and provides for 400 flight plans (company routes or pilot-defined) of 100 waypoints each; 2,000 pilot-defined waypoints.

IEC 9002/9002M is (S)CAT-I DGPS compatible and GLONASS and WAAS upgradable. It is a full-featured GPS flight management system intended for installation in all types of aircraft. It accepts differential GPS corrections required for (S)CAT-I operations. Accuracies in the 0.5 m range, allowing Cat. III operations, are attained when the IEC 9002 is enhanced with its kinematic upgrade and used with the compatible Model 8000 DGPS ground station.

L-3/Interstate Electronics Corporation IEC 9002/9002M flight management system 0018230

Specifications

Navigational signal: GPS L1 C/A code (SPS)
Receiver type: 12 GPS hardware channels
Time to first fix: 2 min
Dynamics: 900 kt (max) velocity
Accuracy:
100 m 2σ rms (selective availability on)
5 m using differential GPS correction

Operational status

The IEC 9002/9002M has been certified by the FAA to TSO C129 Class A1, aboard the B737-200, L-100/C-130, C-12 D/J, UC-12B/F/M and T-44A. The unit is capable of Direct-Y acquisition in heavy jamming environments. Standard features include: an FAA TSO certified P (Y) code input capability; GRAM upgrade capability without TSO impact; anti-jam performance; ARINC 739 compliance; 50 per cent throughput reserve for GATM/FANS functional growth.

The 9002/9002M has been selected by several airlines, and over 100 units have been delivered.

Contractor

L-3/Interstate Electronics Corporation.

VERIFIED

RC30 aerial camera system

The current model RC30 was introduced in 1993 together with two new high-resolution lens cones and a serial connection to external computer and sensor systems. A further enhancement is the PAV30 gyrostabilised camera mount, introduced in 1995.

The RC30 camera system includes the following features:

(1) FMC; the FMC counteracts forward image motion in a range from 1 mm/s to 64 mm/s
(2) PAV30 with Angular Motion Compensation (AMC) is a camera mount with built-in gyros and sensors. It also functions as a sensor of aircraft attitude angles which are recorded in ASCOT (Aerial Survey COntrol Tool)
(3) S-type lens cones with AWAR resolutions of 107 lp/mm (30/4 NAT-S) and 123 lp/mm (15/4 UAG-s). The continuously adjustable shutter speeds go from 1/100 s to 1/1,000 s and also the aperture between f/4 and f/22 is continuously adjustable
(4) PEM (automatic real-time exposure control). The PEM sensor is optimised for aerial photography and covers the full range of film types (B/W, B/W-Infra-red, Color-negative and -positive, Falsecolor-Infra-red)
(5) EDI (external data interface), standard interface between the camera and the ASCOT as well as the link to GPS receivers, aircraft navigation systems, or mission computers
(6) A complete array of interchangeable filters for all exposure conditions, absorption of unwanted spectral ranges, optimising light distribution in the image plane, correcting colour-balance of film emulsion

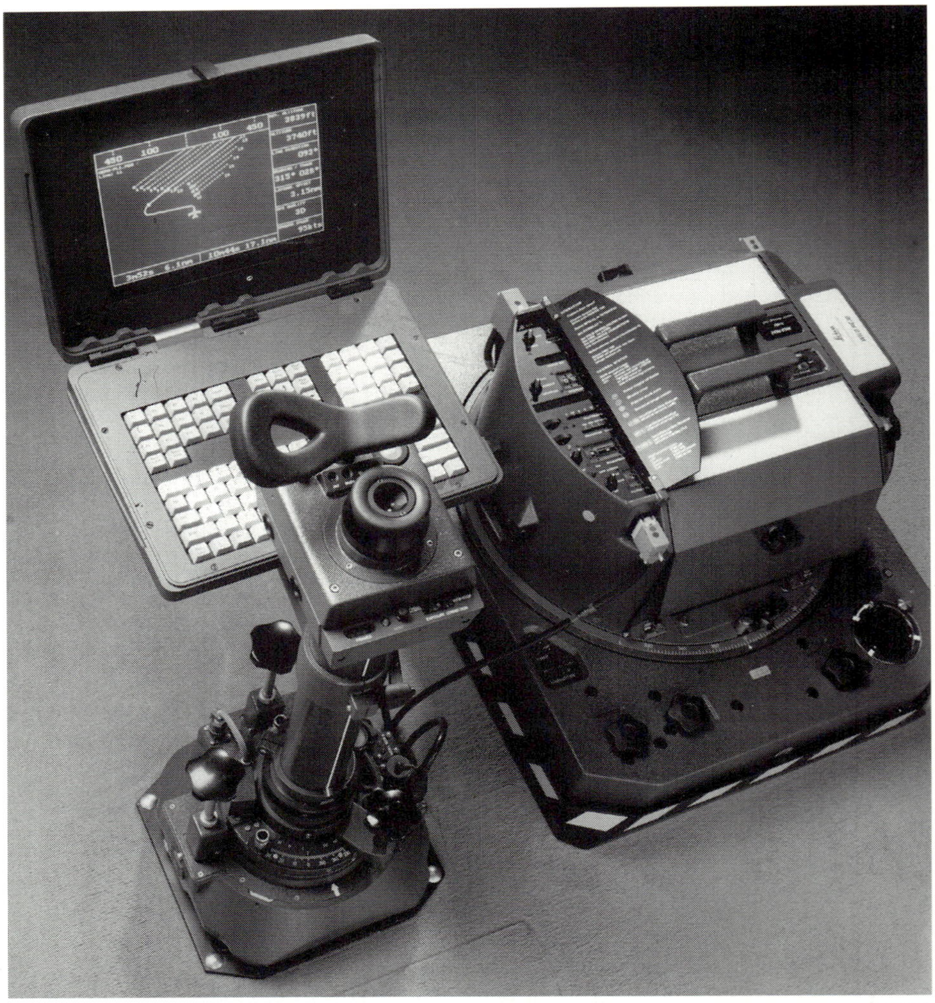

RC30 aerial camera system 0018332

ASCOT is a hardware/software system with an integrated 12-channel Leica GPS receiver to enable reliable survey flight navigation and provide the following additional functions:

(1) interactive survey flight planning in the office or in flight
(2) guidance of the pilot and the operator/navigator on the flight line and during turns, as well as autosearch for nearest route
(3) automatic exposure release at predefined locations (pinpoint photography)
(4) automatic in-flight data annotation on imagery (freely definable layout)
(5) interface to camera mount PAV30 for drift signals provided by external gyros or other sensors and recording of residual angles of gyrostabilised mount
(6) logging of GPS raw data and camera data for post-processing, mission evaluation and reporting
(7) output to kinematic GPS post-processing software FLYKIN.

SKI-AERO (kinematic GPS post-processing software) to determine camera projection centres and to transfer data to aerotriangulation packages such as ORIMA, HATS PATB/GPS-Aero, ALBANY or BLUH.

RC30: microprocessor control, external connections and communication.

Specifications

Camera mount: PAV30 gyrostabilised mount
Stabilisation range in: Pitch and roll ±5°
Drift ±30°
Typical residual angular motion: < 0.3°/s
Typical residual deviation from vertical: < 0.2°/s
Navigation computer:
ACU30 with internal GPS
Internal Silicon Disk
up to 8 RS-232 interfaces
2 monitors (pilot/navigator)
Leica 9400 GPS, L1 and Leica 9500 GPS, L1 + L2 12 channels, DGPS, RTCM
Navigational accuracy: < 100 m
DGPS accuracy: < 5 m
Lens cones:
(15/4 UAG-S wide angle) 90°, f/4
AWAR: 123 lp/mm
(30/4 NAT-S normal angle) 55°, f/4
(AWAR) 107 lp/mm
(8.8/4 SAGA-F super wide angle) 120°, f/4
Film cassettes: single lightweight daylight cassettes

Operational status

In service.

Contractor

LH Systems LLC.

UPDATED

LTN-101 FLAGSHIP® global positioning, air data, inertial reference system

Four-mode Laser Gyro (FLAG) Software/Hardware Implemented Partitioning (SHIP) is designed for a wide range of applications. It can be used in single, dual or triple installations as an Inertial Reference System (IRS), combined IRS and Global Positioning System (GPS), Air Data Inertial Reference System (ADIRS) or ADIRS and GPS. At switch on, the system automatically recognises aircraft type and configures itself for either an ARINC 704 or 738 installation. An adaptor tray allows FLAGSHIP's® 4 MCU ADIRU to fit directly into a 10 MCU rack without system or rack modification.

FLAGSHIP® integrates navigational functions and offers reductions in size, weight and power by eliminating the need for external air data computers and their interconnections. This also leads to savings in spares and maintenance and yields significant increases in reliability.

The GPS-IRS integration is an ideal combination because the two functions are highly complementary. The self-contained IRS contributes to GPS dynamic performance by facilitating satellite acquisition and tracking. The GPS in turn supplies ultra-precise position and velocity to the IRS, whose inertially derived position and velocity accuracies degrade with time. The GPS also makes possible inertial alignment during taxi or flight and furnishes long-term correction data which is used to calibrate the IRS inertial sensors.

The FLAGSHIP system consists of the LTN-101 Air Data Inertial Reference Unit (ADIRU), Sextant Air Data Module (ADM), Global Positioning System Sensor Unit (GPSSU), Mode Select Unit (MSU) and Control Display Unit (CDU). The ADIRU contains the inertial instrument package and performs all system computations with the exception of GPS sensor calculations. Critical air data and inertial reference functions are hardware partitioned to facilitate fault containment. The four-mode laser gyro requires no dithering and is free of conventional ring laser gyro lock-in and other dithering associated errors. The Litton A-4 accelerometer triad completes the sensor package. Surface mount devices and ASIC contribute to an ADIRU 60 per cent smaller than other ARINC 738 systems. Designed for a variety of environments, the ADIRU will operate up to 18 hours without cooling air.

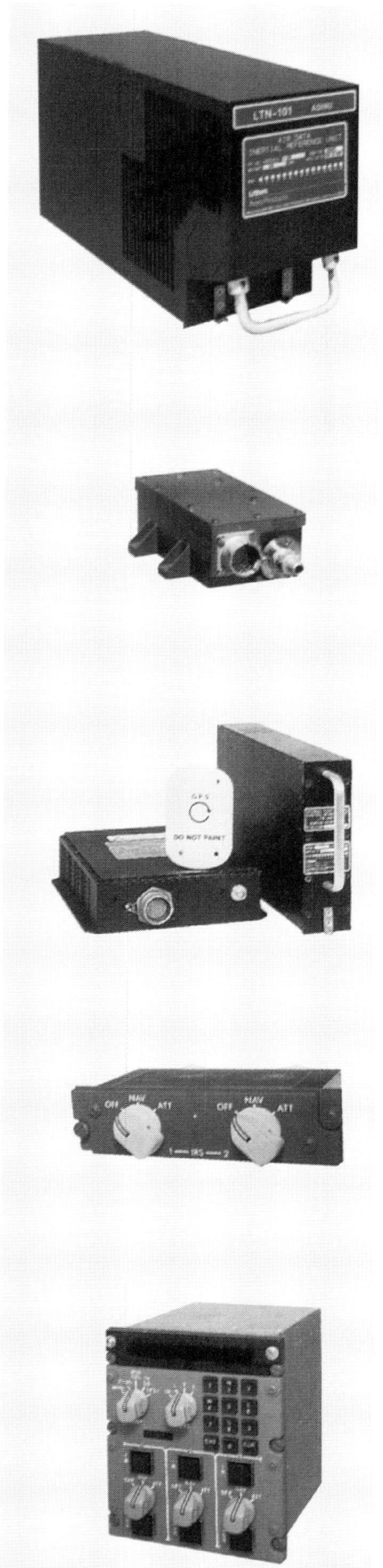

LTN-101 FLAGSHIP showing system units; Air Data Inertial Reference Unit (ADIRU) - ARINC 738 Air Data Module (ADM) - ARINC 738 Global Navigation System Sensor Unit (GNSSU) - ARINC 743 Mode Select Unit (MSU) - Control Display Unit (CDU) 0015385

The ADMs interface air data sensors with the ADIRU. Using an aneroid capsule and resonating quartz blade sensor, the ADM converts static and dynamic pressure into electrical signals. These signals are temperature corrected, converted to ARINC 429 format and transmitted to the ADIRU on a digital databus.

The stand-alone GPSSU is a third-generation Litton design. The eight-channel continuous tracking receiver features enhanced integrity monitoring and rapid time to first fix. GPS usability is maximised by early acquisition of low-elevation satellites and minimal loss of satellite reception during aircraft manoeuvres. GPS outputs of position, velocity, time and raw satellite data are supplied to the ADIRU and other avionics. Both ARINC 743 configurations are offered: a 2 MCU avionic bay-located unit using an antenna with an internal preamplifier or a remote GPS sensor designed for installation near a passive antenna.

The MSU is a switching device used to apply power, annunciate system operating modes and indicate when the system is running on battery power.

The optional CDU, offered as a flight management computer back-up, provides a keyboard for initialisation data and a data display for auxiliary readout. Rotary switches select individual system modes. Push-buttons and annunciators allow inertial and air data output databusses to be turned off by the operator and indicate faults.

Detailed module BIT history is stored on each module. This includes the identity of the aircraft, ADIRU and other modules and LRUs in the system. System BIT history is stored on the computer module, along with its own history. If this module is replaced, system history is transferred to the new module.

Specifications

Dimensions:
(ADIRU) 4 MCU
(GPSSU) (rack mount) 2 MCU
(remote) 64 × 216 × 241 mm
(ADM) 145 × 97 × 53 mm
(MSU) 89 × 146 × 76 mm
(CDU) 171 × 146 × 152 mm
Weight:
(ADIRU) 12.3 kg
(GPSSU) 3.6 kg
(ADM) 2.4 kg
(MSU) 2.7 kg
(CDU) 10 kg

Operational status

Selected by many airlines for some 130 aircraft, including Airbus A319, A320, A321, A330 and A340 and Canadair RJ-100 and RJ-700 aircraft. Also selected for the Ilyushin Il-96M, Tu-204-200, Saab 2000, CL-604 and An-28.

In June 1999, Lufthansa selected the LTN 101 FLAGSHIP Global Navigation Air Data Inertial Reference Unit (GNADIRU) for all Lufthansa Airbus aircraft delivered over the next 10 years. Each aircraft will be equipped with three FLAGSHIP systems and will include Autonomous Integrity Monitored Extrapolation (AIME®) technology, which provides a new software algorithm that solves the GPS integrity problem and offers the potential of sole means of navigation with GPS accuracy for commercial aircraft. During flight, AIME® continuously analyses satellite and inertial signals. If the data's integrity is compromised, AIME automatically uses the navigation history to maintain accuracy and integrity. Litton claims that tests have demonstrated that AIME® is superior to RAIM in terms of solving GPS integrity problems.

Contractor

Litton Aero Products.

UPDATED

LTN-2001 global positioning system

The LTN-2001 C/A code GPS provides continuous worldwide precision three-dimensional navigation data and offers a means to upgrade and enhance the performance of ARINC 561, 599, 704 and 738 navigation systems. The eight-channel continuous tracking receiver is a mature third-generation Litton design. It maximises GPS usability and features advanced integrity monitoring to deliver performance superior both to sequencing type receivers and those with fewer channels.

The system tracks low-elevation satellites as they rise above the horizon and continues to track them until they disappear, minimises loss of satellite reception during aircraft manoeuvres, exceeds the ARINC 743 external interference specification and resists multipath reception caused by terrain features and aircraft surfaces. The LTN-2001 integrates with upgradeable navigation systems and provides in-flight alignment, continuous sensor calibration and error bounding capability for inertial navigation systems, plus GPS positional accuracy and worldwide capability.

The LTN-2001 provides GPS outputs of position, velocity, altitude, time, pseudo range and delta range to

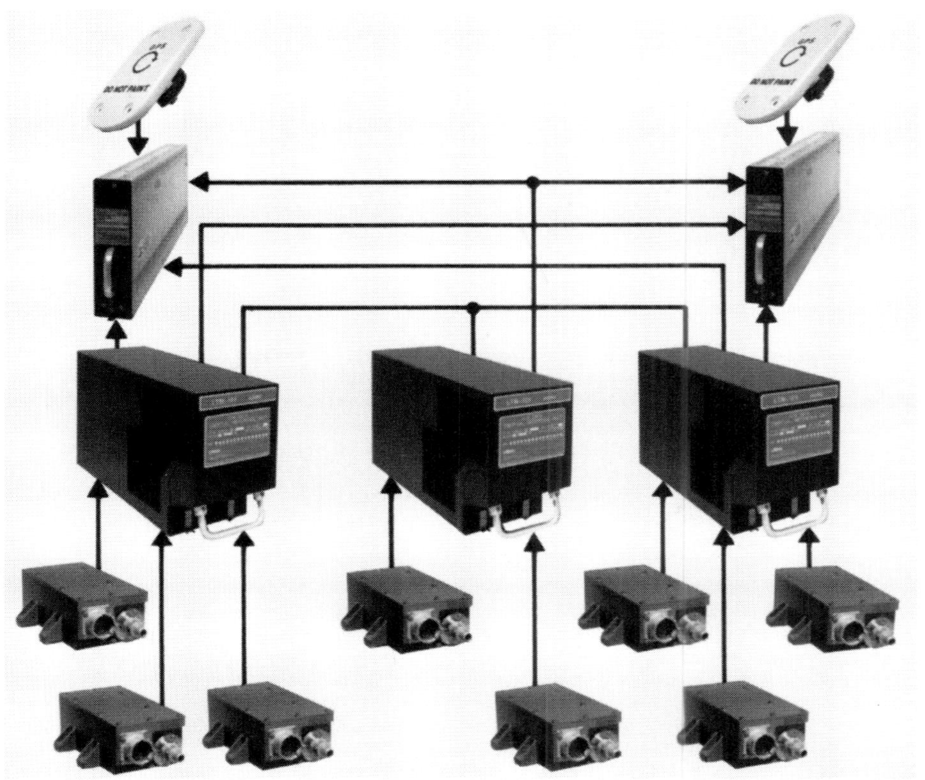

LTN-101 FLAGSHIP in a triplex configuration 0015386

inertial navigation systems, flight management systems, smart CDUs and other avionics via ARINC 429 high- and low-speed databusses. An RS-422 serial output is available as an option. Both ARINC 743 configurations are available: a 2 MCU avionic bay-located unit and antenna with internal preamplifier or a remote sensor unit designed for installation within 10 ft of a passive antenna.

Specifications

Dimensions:
(rack) 2 MCU
(remote) 63.3 × 215.9 × 241.3 mm
Weight: 3.63 kg
Power: 15 W
Accuracy (2 DRMS): 100 m
Antenna: active or passive conformal

Contractor

Litton Aero Products.

VERIFIED

LTN-72 inertial navigation system

The LTN-72, introduced in 1972, is a self-contained, all-weather, worldwide navigation system for commercial aircraft that is independent of any ground-based aids. Designed to incorporate area navigation facilities, the INS provides continuous position, navigation and guidance data.

The system comprises three units: Mode Selector Unit (MSU), Control/Display Unit (CDU) and Inertial Navigation Unit (INU). The MSU is used to energise and align the system before flight and to select navigation or attitude reference modes of operation. The CDU permits the crew to enter present position and waypoint co-ordinates, select track steering and display information generated by the system. The INU houses the gimbal structure with its gyros and accelerometers, associated electronics, power supply and data converter.

Ease of maintenance has been emphasised; for example, the principal mechanical elements - gyros and accelerometers - can be removed and replaced in 20 minutes using only screwdrivers. The gimbals are cantilevered, permitting the servo electronics to be mounted directly on the platform. This permits the use of flexible leads instead of slip-rings in some cases, improving reliability. The platform has only two slip-rings compared with four on a conventional platform.

The LTN-72 has very extensive self-test and failure detection facilities. It complies with ARINC 561 in that the probability of an undetected failure in attitude during the last 30 seconds before touchdown is less than 1 in 10^6. Again, an analogue output test feature permits tests not only of the INS but also of the flight instruments by driving them to various test readings.

The LTN-72R inertial navigation system is a development of the LTN-72, with automatic radio position update, automatic Omega position update and triple system mixing capability. It may be operated in the area navigation mode, using range and bearing information from selected VORTac stations or from a combination of Omega transmitters, providing very high accuracy independent of time.

The system uses newly developed gyros, platform and accelerometers and a new expansible C-4000 digital computer.

The LTN-72RL is an advanced, worldwide inertial navigation system able to automatically update itself by radio navigation fixes. It has a control/display unit which functions as an intelligent data terminal and incorporates a five-line by 16-character light-emitting diode display for presentation of operator-entered or computer-processed data. Waypoint data and VOR/DME locations can be prestored in the computer and the system contains an algorithm of magnetic variation that can be used to compute magnetic heading, track and desired track independently of the aircraft compass system; this algorithm is limited to latitudes between 60° north and 60° south. The prestored database contains information specified by the operator on selected VHF navaids, airports and some high-altitude waypoints. This bulk data is programmed in read-only memory.

A section of electrically alterable memory is allocated for particular waypoints or fixes not contained in the standard databases; up to 160 routes with an average of 20 waypoints per route can be stored in this way and recalled for use at any time. The total number of waypoints in all routes is limited to 3,200.

Specifications

Dimensions:
(inertial navigation unit) 267 × 219 × 507 mm
(control/display unit) 146 × 114 × 157 mm
(mode selector unit) 146 × 38 × 51 mm
Weight:
(inertial navigation unit) 26.8 kg
(control/display unit) 2.3 kg
(mode selector unit) 0.5 kg
Power supply: 115 V AC, 400 Hz
Number of waypoints: 9 to 99 plus remote entry capability
Waypoint offset capability: up to 399 n miles worldwide
Display: 7-segment incandescent numerals for all ARINC terms, together with display of INS parameters on flight director, horizontal situation or remote indicators
Inputs: self-contained inertial guidance. The avionics interface is designed to ARINC 561 and 575 and compatible with all flight directors and autopilots
Outputs: actual track, track angle error, cross-track and desired track, plus access to computer during flight for great circle distance computations

Operational status

In service. The Anglo-French Concorde is one application of the LTN-72. The LTN-72RL was initially certified in July 1981 for Saudia Airlines Boeing 747s.

Contractor

Litton Aero Products.

VERIFIED

LTN-90-100 ring laser gyro inertial reference system

The LTN-90-100 comprises an inertial reference unit, a mode selector unit and an inertial sensor display unit. At the heart of the system, and contained within the inertial reference unit, are the Ring Laser Gyros (RLGs) which measure rotation accelerations and rates about the three aircraft axes, and the three single-axis accelerometers that measure accelerations and rates along the aircraft axes. The RLG system and accelerometers are mounted at right angles to each other and are rigidly secured to the case.

Unlike RLGs which are based on a triangular light path, the Litton LG-8028 units use a square path configuration, with a 28 cm path length. The reason for this, says Litton, is that for the same scale factor a square is smaller than a triangular gyro, resulting in a more compact overall sensor assembly. The square gyro is said to produce less backscatter at each reflection because of the improved 45° angle of incidence, versus 30° for a triangle, which makes for lower random noise.

The substantially greater data processing power required by strapdown systems over conventional gyro-based inertial navigation equipment is provided, in the case of the LTN-90-100, by three Zilog 8000 microprocessors, each of which is assigned a particular task. This computing power is also employed to correct temperature variations. Traditionally, gyroscopes are susceptible to changes in temperature, causing inaccuracies, and the normal approach to dealing with this effect is to maintain a constant temperature by trickle heating. In the LTN-90-100 the effect is compensated by mathematically modelling the way in which the characteristics of the system and accelerometers vary with changing temperature and then by applying a correction to their output signals.

Easy maintenance is a key advantage claimed by Litton. It is achieved by functional partitioning, so that each circuit board has all the components needed to support a major activity, more effectively isolating faults and simplifying troubleshooting. This partitioning is achieved by reducing the number of electronic plug-in modules from between 15 and 20 to 7.

Litton describes the LTN-90-100 as a sensor rather than (in the case of navigation) as a complete system and this is the way in which it is used in the A310. With inputs only from an air data computer, the LTN-90-100 can provide accurate outputs of attitude, heading, present position, drift angle, ground track, flight path angle, groundspeed, vertical velocity and windspeed, as well as aircraft angular rates and linear accelerations. Alternative control/display and mode selector unit configurations can be supplied to optimise specific aircraft requirements.

The A310 has a triple-redundant Litton LTN-90-100 installation

Specifications

Inertial reference unit
Dimensions: 194 × 322 × 318 mm
Weight: 19.9 kg
Power: 110 W
Cooling: ARINC 600
MTBR: 2,500 h

Mode selector unit
Dimensions: 38 × 146 × 51 mm
Weight: 0.45 kg
Power: negligible
Cooling: none
MTBR: 50,000 h

Inertial sensor display unit
Dimensions: 114 × 146 × 152 mm
Weight: 2.27 kg
Power: 15 W
Cooling: none
MTBR: 15,000 h

Accuracy (95%):
(heading) 0.4°
(pitch and roll) 0.1°
(position) 2 n miles/h
(groundspeed) 8 kt
(flight path angle) 0.4°
(body rates) 0.1%
(body accelerations) 0.01 *g*
Reaction time: 10 min

Operational status

In production for Airbus A310 and A300-600 and US Navy Boeing E-6A TACAMO.

Contractor

Litton Aero Products.

VERIFIED

LTN-92/LTN-92E Inertial Navigation Systems (INS)

The LTN-92 provides proven technology in a standard ARINC 561 1 ATR box. The inertial sensors and much of the electronics are the same as in the LTN-90-100 inertial reference unit, which is currently flying on the Airbus A310, Gulfstream III, Canadair Challenger and other high-technology aircraft. The LTN-92 has been chosen for installation on the Cathay Pacific Boeing 747-300s and -200s as well as for retrofit on its Lockheed L-1011 aircraft. Other customers include Air China, Continental, Federal Express, Hawaiian and Western airlines.

The LTN-92 INS comprises three separate units: the Inertial Navigation Unit (INU), the Control/Display Unit (CDU) and the Mode Select Unit (MSU). Three 28 cm ring laser gyros and a triad of force rebalanced accelerometers comprise the instrument cluster of the INU. The instrument electronics digitise the instrument control signals and the instrument outputs for easy microprocessor interface. There are three microprocessors used to process the instrument data, perform input and output interface functions and perform navigation calculations. As heat is not required and the latest technology parts are used, power consumption is reduced to a maximum of 175 W total operating power. The CDU has a five-line light-emitting diode dot matrix display and keyboard, providing the

INU interface with the crew. The MSU controls the operational mode of the INS.

The LTN-92 is pin-for-pin compatible with existing ARINC 561 INS systems and ARINC 571 IRS systems. In addition there are three 429 high-speed digital output busses, nine 575/429 low-speed input busses, 12 programmable synchro outputs, four synchro inputs, four analogue DC or AC two-wire outputs, DME pulse pair input and 2 × 5 radio tuning line interfaces. This allows the LTN-92 to interface with existing avionics equipment installations as well as being a part of new installations.

Other capabilities are programmed into the LTN-92 to make it a versatile navigation unit. The system will accept RNav, Tacan, triple INS mixing or manual position updates for improved performance. With an external database or using the internal 16 × 16 k EEPROM data storage the crew may programme a flight plan and the INS will steer the aircraft along the entered flight plan. An interface to a weather radar system is provided for modification of the flight plan during flight. Intersystem communication is utilised to check system performance and send flight plan and initialisation data between systems. An extensive software built-in test monitor program of more than 150 performance checks is continually run to ensure the validity of the computed data.

The LTN-92 has been designed for future growth capability. Space has been provided for incorporation of an air data computer and GPS. The system meets the existing requirements of the Required Navigation Performance (RNP) - 10 rules for up to 12.5 hours flight with GPS augmentation, and offers a growth path to full CNS/ATM navigation in the first decade of the 21st century.

Specifications (LTN-92)

Dimensions:
(inertial navigation unit) 257 × 218.9 × 507.5 mm
(control/display unit) 146 × 114.3 × 157.5 mm
(mode selector unit) 146 × 38.1 × 50.8 mm
(battery unit) 129 × 193.8 × 365.5 mm
Weight:
(inertial navigation unit) 25.85 kg
(control/display unit) 2.27 kg
(mode selector unit) 0.45 kg
(battery unit) 12.24 kg
Power supply: 115 V AC, 400 Hz, single phase
ARINC 561 accuracy characteristics:
(position) 2 n miles/h (95%)
(pitch, roll and attitude) 0.05°
(heading) 0.4°
(groundspeed) ±8 kt
(vertical velocity) 30 ft/min
(body angular rates) 0.1°/s
(body accelerations) 0.01 *g*

Operational status

In production and service. Applications include Boeing 747, DC-8 and DC-10, Gulfstream II and III, Lear 35, Lockheed Martin L-1011 and the US Presidential Boeing 747-200. The French Air Force has selected the LTN-92 in a dual configuration for installation in 10 C-130 aircraft. Over 2,500 systems have been installed on 25 aircraft types.

Contractor

Litton Aero Products.

UPDATED

LTR-81-01 and LTR-81-02 Attitude and Heading Reference Systems (AHRS)

The LTR-81-01 is an advanced design LTR-81 AHRS providing the full set of ARINC 705 digital outputs plus a complete set of analogue outputs capable of interfacing with analogue aircraft systems. Greater accuracy, higher reliability and a significant weight reduction are achieved without the need for vertical, directional or rate gyros, body-mounted accelerometers and compass couplers. The LTR-81-01 and LTR-81-02 are identical, except that the LTR-81-01 has two additional analogue output cards.

The LTR-81-01/02 accepts inputs from the magnetic compass, air data computer and VOR/DME. The system outputs consist of attitude, heading, groundspeed, vertical speed, drift angle, flight path angle and linear accelerations and angular rates. The LTR-81-01/02 mechanisation ensures no degradation of accuracy during aircraft manoeuvres. Two advanced K-273 two-degree-of-freedom tuned rotor gyroscopes are incorporated in the design. Three high-accuracy accelerometers complete the inertial instruments.

The LTR-81-01/02 computer module dual-processor mechanisation which has been implemented in the system offers numerous advantages including fewer components, improved monitoring, lower power consumption and commercially available support hardware and test equipment.

The LTR-81-01/02 features modular construction, permitting ready access, and simple removal for ease of maintenance. Single layer printed circuit plug-in modules permit simpler repair and result in reduced repair time. BITE technology has been implemented, including redundant comparison checks of all gyro loops.

The system will continue to perform accurately after brief power interruptions and is not adversely affected during absence of air data or VOR/DME inputs for periods of several minutes.

The LTR-81-01 has been certified as a critical system in accordance with FAR 25.1309 and its software is certified to RTCA DO-178.

Designed with future growth as a primary consideration, the system is capable of interfacing with all analogue and digital equipment envisioned to the turn of the century, including GPS.

Specifications

Dimensions: 8 MCU
Weight:
(LTR-81-01) 13.15 kg
(LTR-81-02) 10.89 kg
Power supply: 115 V AC, 400 Hz, single phase
(LTR-81-01) 100 W
(LTR-81-02) 85 W
Alignment time: 45 s
Accuracy (95%):
(heading) 2°
(pitch and roll) 0.5°
(groundspeed) 8 kt
(flight path angle) 1°
(body rates) 0.1°/s or 1%
(body accelerations) 0.01 *g*
Reliability: >7,000 h MTBF

Operational status

Selected by over 50 airlines with over 2,000 units delivered. Applications include the Airbus A300, Boeing MD-82, MD-83, MD-87 and MD-88, Fokker 50 and 100, British Aerospace ATP, and business jets and helicopters.

Contractor

Litton Aero Products.

VERIFIED

LTR-97 fibre optic gyro system

The LTR-97 is a Vertical Gyro/Directional Gyro (VG/DG) replacement system which uses Fibre Optic Gyro (FOG) technology and Application Specific Integrated Circuits (ASIC) in a strapdown configuration to provide superior aircraft attitude and heading and full supportability.

Ideal for retrofit, the LTR-97 offers users of aircraft such as the Boeing 727, 737, DC-9 and MD-8X a cost-effective, advanced technology replacement for their mechanical VG/DG systems with up to 10 times the Mean Time Between Failure (MTBF). The LTR-97 is designed as a plug and play replacement for earlier systems, with no requirements for aircraft wiring changes, or crew training.

A high-speed processor calculates attitude and heading by integrating these rate signals from the FOG and the roll and pitch level sensors. Coupling to the magnetic flux valve is provided through the existing slaving system. Slaved magnetic heading is output on two synchros with automatic switching to free DG mode if the slaving signal is lost.

Specifications

Weight: 5 kg
Dimensions: 256.5 (W) × 154.9 (H) × 248.9 (L) mm
Power: 115 V AC, 64 W (115 VA) – full load
Reliability: MTBF >12,000 flight hours
Accuracies:
Attitude (static): 0.5°
(dynamic) 1.0°
Heading:
(slaved mode) dynamic 0.75°
(DG mode) dynamic 1.3° /h (excluding earth rate)
Outputs:
(synchro) Pitch, roll, 2 headings
(analogue) 50/200 mV AC pitch and roll
(discretes) ATT and HDG warns, 6 interlocks
6° roll discrete and interlock
Qualifications:
(RTCA D) 160 C including HIRF and lightning
DO-178A, Level 1 for flight critical software
Certifications: TSO C4c, C5e, C6d

Operational status

Currently FAA certified STCs for the B727-100/-200, B737-200, DC-9, and MD-80.

LTR-97 Heading and Attitude System (HAS)

The LTR-97 HAS uses FOG technology in a strapdown configuration to provide improved attitude and heading data.

Designed for retrofit applications, the LTR-97 HAS provides a replacement system for VG/DG mechanical systems on classic aircraft such as the A-300, B-747, DC-10 and L-1011.

Based on Litton's LTR-97 VG/DG replacement system, the LTR-97 HAS provides up to 10 times the MTBF of mechanical VG/DG systems. The LTR-97 HAS uses existing aircraft wiring and can be installed in about 20 minutes.

Specifications

Weight: 9.09 kg
Dimensions: per ARINC 561: 256.54 (W) × 269.24 (H) × 508.0 (L) mm
Power: 115 V AC, 100 W
MTBF: >10,000 h
Attitude accuracy:
(static): 0.5°
(dynamic): 1.0°
Heading accuracy: DG mode, dynamic: 1.3°/h
Outputs:
(synchro): 3 pitch, 3 roll, 1 heading
(analogue): 50/200 mV AC pitch and roll
(discretes): ATT and HDG warnings
Qualifications: RTCA DO-160C including HIRF and lightning: DO-178A, level 1 for critical software
Certifications: TSO-C4c, -C5e, -C6d

Contractor

Litton Aero Products.

VERIFIED

LN-260 FOG INS/GPS system

Litton Guidance & Control Sytems new LN-260 highly integrated GPS/Inertial system provides accurate navigation and motion compensation to all classes of fixed and rotary-wing aircraft. The LN-260 is the aircraft production system based on the Litton GPS Guidance Package (GGP) programme.

The GGP programme is DARPA-sponsored to develop a technology core of low-cost, modular, Fibre Optic Gyro (FOG) and miniature GPS-based guidance systems that can be easily configured to support a broad spectrum of US Department of Defense platforms. Phase I of the GGP programme developed two prototype demonstration units for field testing. US Army testing of the Phase I units was successfully completed in the M-981 in 1995 as was tactical fighter testing in the F/A-18 by the US Navy in 1996. In April 1997, Litton was contracted to complete Phase II and build production hardware.

The LN-260 is a complete integrated navigation system with a 12-channel P(Y) code GPS. The fully integrated, tightly coupled GPS inertial design provides better positioning performance, quicker response time

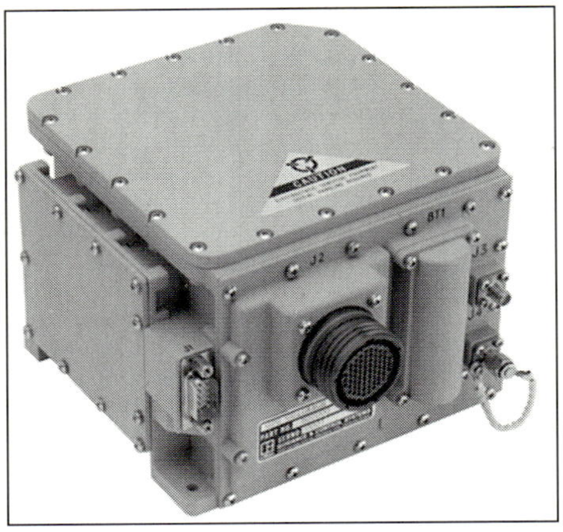

LN-260 FOG INS/GPS system ***1999***/0051616

and higher A/J performance than either older GPS receiver updates or earlier embedded INS/GPS systems. The modular open system architecture provides for easy adaptation to customer specific requirements.

Specifications

	GPS Aided	**Inertial**
Position:	<10m CEP	
4 min GC	<20m after 4 min loss of GPS	<0.8 - 5.0 nmi/hr
Stored heading		30 sec
SINS DL		5 min
Hand-set		<5 nmi/hr (>5 min)
Velocity:	<0,03 m/s rms	<2.0 - 5.0 fps rms
Attitude and heading:	<0.2° rms	<0.05 - 0.1° rms

Contractor
Litton Guidance & Control Systems.

VERIFIED

Model 1044 altimeter

The Model 1044 radar altimeter was designed and developed to be a low-cost, lightweight, very accurate, high-rate production unit. The altimeter transmits and receives a very low-RF signal that is Phase Shift-Key (PSK) modulated using a pseudo-random code. The PSK technique was selected because it provides the equipment with the capability to measure over a wide range of altitudes from 0 to 10,000 ft with low power and fewer parts than other types of radar altimeters. A commanded built-in test verifies that the unit is operational by testing 98 per cent of all the altimeter functions.

The altimeter comprises two subassemblies. The microwave subassembly contains a filter board assembly and four microwave hybrids: the transmitter, receiver, dielectric resonant oscillator and video amplifier. The microwave hybrids provide increased reliability with reduced parts count. The electronic assembly contains an analogue processor, digital processor and power supply.

Contractor
Litton Guidance & Control Systems.

VERIFIED

Model 2100 Doppler Velocimeter/ Altimeter (DV/A)

The DV/A combines both an altimeter and a Doppler velocity sensor into one compact package. The DV/A consists of two subassemblies: the antenna assembly and the Doppler processor board. The antenna assembly is mounted in the vehicle structure. The Doppler processor board is mounted in a navigation system box. The DV/A measures vehicle velocities and slant range. The output format to the navigation computer is in RS-422.

Specifications
Dimensions:
(antenna assembly) 170 × 386 × 52.1 mm
(Doppler processor board) 10 × 235 × 140 mm
Power: 27 W
Temperature range: –32 to +71°C

Contractor
Litton Guidance & Control Systems.

VERIFIED

Bearing/Distance Horizontal Situation Indicators (BDHSI)

The BDHSI-421 was designed to display heading, VOR, ILS, ADF, DME, Tacan or long-range navigation information. Several variations are available that can be configured to the operator's requirements. This indicator is a 4 ATI form factor.

The BDHSI-423 was designed as a successor to the BDHSI-421, displaying heading, VOR, ILS, ADF, DME, Tacan or long-range navigation information, but with the capability of a second bearing pointer. The unit is compatible with any medium- or high-performance fixed-wing aircraft or helicopter.

Operational status
The BDHSI-421 is no longer in production. A variant of this indicator is fitted to UK Royal Air Force Tucano aircraft.

The BDHSI-423 is currently fitted on the Agusta A 109 and Eurocopter BK-117 helicopters.

Contractor
Litton Special Devices.

VERIFIED

Digital Bearing/Distance/heading Indicators (BDI)

The BDI-300A, a dual-switched RMI with dual-digital DME display is designed to interface directly with new-generation avionics transmitting ARINC 429 data. The indicator is a 3 × 4 in (76.2 × 102.6 mm) ATI form factor.

The BDI-302 digital bearing/distance/heading indicator

The BDI-302A, a dual-switched RMI with dual-digital DME display, is designed to interface directly with new-generation avionics, transmitting ARINC 429 or Collins CSDB, RS-422 format. The indicator is a 3 × 4 in (76.2 × 102.6 mm) form factor.

The indicator is compatible with any high-performance fixed-wing aircraft or helicopter.

Specifications
Dimensions:
(BDI-300A) 106.7 × 86.4 × 190.5 mm
(BDI-302A) 106.7 × 86.4 × 177.8 mm
Weight:
(BDI-300A) 0.27 kg
(BDI-302A) 1.4 kg
Temperature range:
(BDI-300A) –15 to +55°C
(BDI-302A) –15 to +70°C
Altitude: up to 15,000 ft
TSO: C-6a, C-66b

Operational status
The BDI-300A is fitted to the Gulfstream IV and Bombardier Challenger aircraft.

The BDI-302A is fitted on the Dassault Falcon 2000 and the Raytheon Hawker 1000 aircraft.

Contractor
Litton Special Devices.

VERIFIED

Digital radio magnetic indicators

The RMI-303a is a 3 in (76.2 mm) form factor indicator which displays magnetic heading from either of two ARINC 429 sources or from a flux valve. Annunciators are provided to identify the source of the heading information. When in the normal mode, ARINC 429 heading is displayed from the primary heading source. On loss of valid primary heading, the indicator automatically reverts to the alternative ARINC 429 input. If this source also becomes invalid, the indicator automatically reverts to the standby heading mode, receiving heading information from the flux valve. A mode switch is provided to manually select the desired heading source. In addition to the heading display, two bearing pointers are provided. Each pointer is switchable between its respective ARINC 429 ADF or VOR source.

The RMI-321 is fitted in the Falcon 50, Learjet 60 and Yak-142

Although designed for the Gulfstream V, the RMI-303A is compatible with any high-performance aircraft.

The RMI-321 is a dual-switched pointer Radio Magnetic Indicator (RMI) in a 3 in (76.2 mm) form factor. The unit is designed to interface with the later generation avionics transmitting signals in ARINC 429, Collins CSDB, sin/cos or ARINC synchro formats for ADF, VOR or heading information.

Specifications
(RMI-321)
Weight: 1.41 kg (max)
Power supply: 27.5 V DC, 0.65 A (max)
Accuracy:
(card) ±1°
(pointer) ±2°
Temperature range: –15 to +70°C
Altitude: –1,000 to 15,000 ft
Environmental: DO-160C

Operational status
Fitted in the Dassault Falcon 50, Learjet 60 and Yak-142.

Contractor
Litton Special Devices.

VERIFIED

Emergency Locator Transmitter (ELT) Model 952-21

The Model 952-21 ELT is a triple frequency unit containing a 121.5/243 MHz homing transmitter and a 406.025 MHz satellite transmitter.

The 406.025 MHz transmitter contains an unique digital coded message that can be received by polar orbiting satellites that are part of the COSPAS/SARSAT system.

Operational status

Selected by the US Air Force for 580 aircraft of its C/KC-135 fleet.

Contractor

Litton Special Devices.

VERIFIED

Horizontal Situation Indicators (HSI)

The HSI-315 is a 3 in (76.2 mm) form factor indicator displaying heading from either ARINC 429 or ARINC synchro input, ADF bearing from a DC sin/cos source and a manually selected course pointer. The unit contains a built-in VOR/Loc converter.

The unit is compatible with any high-performance aircraft.

The HSI-415 is a 3 in (76.2 mm) form factor indicator capable of functioning as an HSI or switchable to perform as an RMI. In the HSI mode, the unit displays aircraft heading, bearing from either ARINC synchro or DC sine/cos source, manually selected course, VOR/Loc deviation, VOR to/from indication and glide slope deviation. In the RMI mode, the course pointer is continuously motorised to a position that will centre the VOR deviation bar and indicate to the station. With the simultaneous display of ADF bearing, the HSI-415 provides all the functions of an RMI.

The unit is compatible with any high-performance fixed-wing aircraft or helicopter.

The HSI-8131 was designed as a 3 ATI form factor instrument compatible with most ARINC analogue directional gyros. It is applicable to all rotary- and fixed-wing aircraft.

Specifications

(HSI-415)
Dimensions: 3 ATI form factor
Weight: 1.59 kg (max)
Power supply: 27.5 V DC, 1.5 A (max)
Temperature range: –30 to +70°C
Altitude: –1,000 to 55,000 ft
Environmental: DO-160B

Operational status

The HSI-315 is fitted on the Citation X, Citation V Ultra and Citation II Bravo aircraft.

The HSI-415 is fitted on the Learjet 31A, Sikorsky S-76 and Agusta A 109.

The HSI-8131 is no longer in production.

Contractor

Litton Special Devices.

VERIFIED

Liquid Crystal Display indicators

Litton solid-state Liquid Crystal Display (LCD) indicators employ advanced microprocessor circuitry with liquid crystal technology. Interface can be accomplished with analogue and digital signals from inputs of thermocouples, strain gauges, potentiometers and LVDTs. The indicators can be adapted for direct replacement of electromechanical indicators.

The LCDs function over a wide temperature range and work well under harsh environmental conditions. They can be grouped together in one display or installed as separate indicators and can be supplied in round scale formats or vertical bar graphs in various sizes. Digital augmentation is available to provide greater accuracy when required. Displays can be back-lit with electroluminescent, incandescent lighting or LEDs, and can be made NVG-compatible.

Formats available include a compact space-saving design as small as 1.5 in (38 mm), vertical scales with or without digital display for torque, tachometer, pressure or temperature display and large formats available as a single-pointer, dual-pointer or single-bar display augmented with a digital readout.

Operational status

In use in military, commercial and business fixed- and rotary-wing aircraft.

Contractor

Litton Special Devices.

VERIFIED

Radio Magnetic Indicators (RMIs)

The Model 3100 is a dual-switched Radio Magnetic Indicator (RMI) in a 3 in (76.2 mm) form factor. The VOR pointers will accept AC sine/cos bearing information and the ADF pointer will accept ARINC synchro information. The slaved heading card accepts ARINC synchro data bearing information for selected ADF or VOR stations. The unit is TSO'd and is applicable to any fixed-wing or helicopter application.

The Model 3100 RMI is fitted in the Jetstream 61 and de Havilland Dash 8

The Model 3337 is a dual-switched Radio Magnetic Indicator (RMI) in a 3 in (76.2 mm) form factor. The pointers in VOR position will accept either AC sine/cos or ARINC synchro information. In the ADF position, the pointers will accept either DC sine/cos or ARINC synchro signals. The slaved heading card will accept ARINC synchro signals.

Specifications

Weight: 1.13 kg
Power supply: 28 V DC
(Model 3100) 0.55 A (max)
(Model 3337) 0.75 A (max)
Accuracy:
(card) ±1°
(pointers) ± 2°
Temperature range:
(Model 3100) –30 to +55°C
(Model 3337) –30 to +70°C
Altitude:
(Model 3100) –1,000 to 20,000 ft
(Model 3337) –1,000 to 40,000 ft
Environmental : DO-138

Operational status

The Model 3100 is fitted in the Jetstream 61, de Havilland Dash 8 and other aircraft.

The Model 3337 is fitted on the ATR-42 and ATR-72, Saab 340, UK Royal Air Force Tucano, Jetstream 41 and corporate and commuter aircraft, as well as helicopters.

Contractor

Litton Special Devices.

VERIFIED

CNS-12™ ACARS/GPS/ADS communication, navigation and surveillance system

The Magellan CNS-12™ features a WAAS-capable 12-channel GPS receiver with Receiver Autonomous Integrity Monitoring (RAIM), an optional D8PSK receiver card for differential GPS and aircraft navigation technology to meet the ICAO VDL requirements. The GPS, together with the ARINC VHF Aircraft Communications Addressing and Reporting System (ACARS) two-way datalink, provides timely and accurate Automatic Dependent Surveillance (ADS) position reports from any aircraft via the ACARS network.

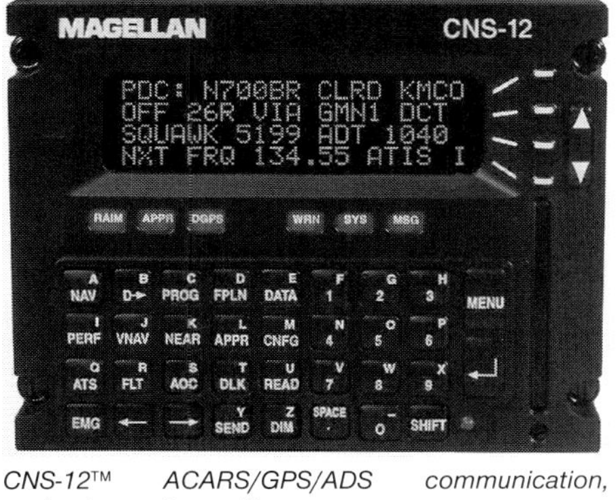

CNS-12™ ACARS/GPS/ADS communication, navigation and surveillance system 0015387

The CNS-12™ employs a PCMCIA card capable of storing information on nearly 21,000 airports, navaids and waypoints. The Jeppesen database card contains detailed information on airports, VORs, DMEs, ILS/DMEs, VORTacs, Tacans and NDBs, as well as en route and terminal intersections. GPS overlay approaches, SIDs and STARs are included on the expandable database card. Additionally, a pilot-defined waypoint database stored in RAM can accommodate up to 500 waypoints. The information is used to create up to 50 flight plans with up to 40 waypoints in each. User flight plans and waypoints can be stored on a programmed PCMCIA card. The active flight plan and stored waypoints can be selected from the internal database, user datacard or Jeppesen NavData card and displayed in a variety of ways. En route, approach and vertical navigation is displayed on the CNS-12™ display.

A second, dedicated D8PSK VHF receiver for differential GPS precision approach information is optional. The receive only module will make precision differential GPS approaches possible as the ground infrastructure becomes available.

The CNS-12™ enables two-way communication for sending and receiving clearances, text messages, system essential messages, weather reports, Out/Off/On/In (OOOI) reports, engine data and other operational messages over the GLOBALink/CNS network. It also automates predeparture clearance and delivery of flight papers. The CNS-12™ communications element relays digital Automatic Terminal Information Services (ATIS) arrival and departure messages for any airport during flight.

The two-way VHF datalink complies with ARINC 745-2 for ADS. The CNS-12™ will automatically transmit data derived from its integrated GPS to air traffic service centres, providing surveillance beyond the radar horizon.

Specifications

Dimensions: 146 × 107.9 × 203.2 mm
Weight: 3.08 kg
Power supply: 14 or 28 V DC, 7 A (max)
Temperature range: –15 to +55°C
Altitude: up to 15,000 ft
Frequency: 108-136.975 MHz
Accuracy:
(position) 15 m RMS
(with DGPS) 8 m RMS
(velocity) 0.1 kt (with DGPS)
(time) UTC to nearest μs

Operational status

In service on a variety of commercial aircraft, CNS-12™ links via ARINCs GLOBALink/CNS datalink service and can be integrated with the ADS system.

Contractors

Magellan Systems Corporation, ARINC.

VERIFIED

ADF 841 Automatic Direction-Finder

The ADF 841 combines the functionality of frequency storage, transfer and timing with simple operation. It features frequency storage, frequency transfer, crystal clear audio, frequency memory, a gas discharge display and automatic dimming. Since both the mode and timer functions are selected through the same switch, pilot workload is minimised, especially in IFR conditions.

Supplied with a combined loop sense antenna and indicator with rotatable azimuth card, the ADF 841 employs coherent detection which both reduces interference and increases the reception range.

Specifications

Dimensions:
(receiver) 158.7 × 38.1 × 279.4 mm
(indicator) 82.5 × 82.5 × 14.3 mm
(antenna) 152.4 × 292.1 × 10.2 mm
Weight:
(receiver with tray) 1.5 kg
(indicator) 0.34 kg
(antenna) 1.6 kg
Power supply: 11-32 V DC, 1 A at 13.73 V

Operational status

In production.

Contractor

Narco Avionics Inc.

VERIFIED

The Narco Avionics ADF 841 automatic direction-finder

AT 150R ATC transponder

The Narco AT 150R transponder is a self-contained panel-mounted unit which meets the FAA's C74c Class 1A TSO specification for ATC transponders. Featuring full 4,096 Mode A and A/C code capability up to 30,700 ft in 100 ft increments, the AT 150R is compatible with leading encoding altimeters and blind encoders such as the Narco AR 850 altitude reporter.

The AT 150R is both a transmitter and receiver which responds automatically to ground-based radar interrogation. The transponder reply is displayed on ATC radar displays as two short parallel lines which fill in when the system squawks ident.

Specifications

Dimensions: 159 × 45 × 286 mm
Weight: 1.68 kg
Power supply: 14 V DC, 1.6 A or 28 V DC, 1.6 A with adaptor
Power output: 250 W nominal
Frequency:
(transmit) 1,090 MHz ± 3 MHz
(receive) 1,030 MHz

Operational status

In service. The AT 150R direct replacement transponder is a direct replacement for all Narco AT 50, AT 50A and AT 150 transponders.

The Narco Avionics AT 150R transponder

***1999**/0044990*

Contractor

Narco Avionics Inc.

VERIFIED

COM 211 VHF/UHF transceiver

The COM 211 was designed for aircraft operations requiring both VHF and UHF capability in one box. It features: frequency memory channels, flip-flop frequency transfer, non-volatile memory, acoustic audio levelling, seamless transition between VHF and UHF, and active and standby frequency display.

Specifications

Frequency range:
(VHF): 118.000-136.975 MHz, in 25 kHz steps
(UHF): 225.000-399.975 MHz, in 25 kHz steps
Memory channels: 10
Output power:
(VHF): 8 W nominal
(UHF): 10 W nominal
Dimensions: 160.3 × 34.3 × 273.7 mm
Weight: 1.32 kg
Power:
(receive) 28 V DC 0.5 A
(transmit) 6.0 A

Operational status

Development complete.

Contractor

Narco Avionics Inc.

VERIFIED

The Narco Avionics COM 211 VHF/UHF transceiver

0011819

COM 810/811 series VHF radios

Narco's COM 810 and COM 811 models are solid-state microprocessor-controlled communication systems designed principally for light and general aviation aircraft. Each covers the VHF band from 118 to 136.975 MHz in which it provides 760 channels, two of which are preselectable; one is for active and the other for standby use. The 810 and 811 systems are essentially similar except that the former is designed for operation from a 13.75 V DC supply and the latter from a 27.5 V DC supply.

The active and standby frequencies are presented on LED displays which are automatically dimmed during darkness by a built-in photocell circuit. The legend XMT is illuminated when the microphone is keyed for transmission.

New frequencies may be entered in either the active or standby positions when desired; an illuminated arrow indicates which section has been selected for new frequency entry. Frequency selection is completed by use of a concentric tuning control, the outer part of which makes frequency readout changes at the rate of 1 MHz per detent and the inner part providing kHz changes at 25 kHz per detent. Clockwise rotation increases the numerical value of the frequency selection and counter-clockwise rotation decreases it. A transfer switch is used to exchange selected frequencies between active and standby modes.

An optional feature is a connection which enables the last entered frequencies to be retained in the system memory when the radio is inactive. This requires a trickle current of 0.1 mA from the aircraft's battery. If this circuit is not connected, then the radio automatically retunes to the 121.5 MHz internationally designated emergency frequency in the active mode and to the 121.9 MHz ground control frequency in standby mode the next time it is switched on. In the event of a display failure, the radio automatically reverts to these frequencies and may be retuned to the desired channel by counting the detent clicks of the tuning control.

Built-in automatic squelch control, deactivated by use of a pull/test switch, maintains audio silence until a signal is received. Automatic audio-levelling in both transmitter and receiver allows all signals to be heard at the same level regardless of modulation. A built-in 10 W amplifier, provision for multiple audio inputs and intercommunication facilities are also included.

Latest in the range is the TSO'd COM 810+R, which directly replaces all Narco COMs from the COM 11 to the COM 120 and the COM 810/811 and COM 810+/811+.

Specifications

Dimensions:
(transmitter/receiver) 159 × 38 × 279 mm
Weight:
(transmitter/receiver) 1.3 kg
(mounting tray) 0.34 kg
Frequency: 118-136.975 MHz
Channels: 760
Power output: 8 W nominal

Operational status

In production and in service.

Contractor

Narco Avionics Inc. ***VERIFIED***

CP 136 and CP 136M audio control panels

Belonging to the Centerline range, the CP 136 and the slightly larger 136M variant provide fully solid-state control of all radios, headsets and loudspeakers using push-button control and LED selection display. The units provide 10 W across 4 ohms to speakers or 50 MW to 600 ohms headphones.

Specifications

Dimensions: 28 × 159 × 213 mm
Weight:
(CP 136) 0.82 kg
(CP 136M) 0.91 kg
Power supply: 13.75 or 27.5 V DC

Operational status

CP 136 no longer in production; the CP 136M is still being produced.

Contractor

Narco Avionics Inc.

VERIFIED

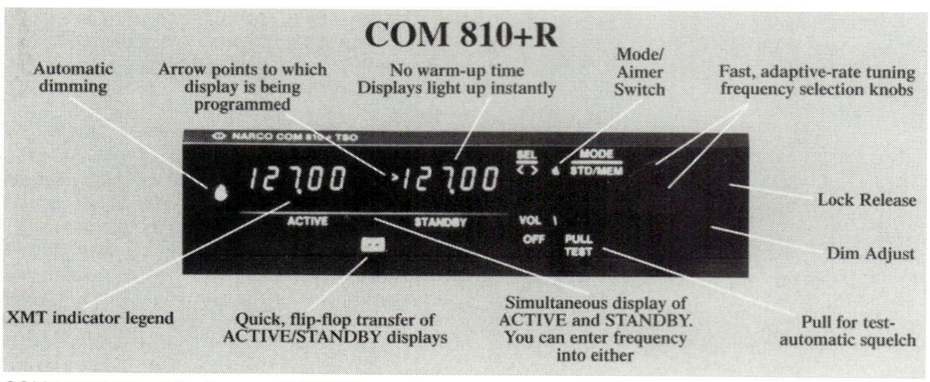

COM 810+R 'Direct Replacement' COM Transceiver

0001186

DME 890 Distance-Measuring Equipment

DME 890 distance-measuring equipment features instant lock on, remote channelling capability, high-visibility gas discharge displays, microprocessor-based electronics and automatic display dimming. It is RNav compatible and has a two-frequency storage capability. Simultaneous display of distance, groundspeed and time to station are provided.

A flexible unit, the DME 890 can be remotely channelled by the Mk 12D and Mk 12E nav/com, Nav 824/825 or other navigation receivers. The DME 890 is ideally suited for use with the Narco NS 801 RNav system.

Specifications

Dimensions: 158.1 × 38.1 × 279.4 mm
Weight: 1.7 kg
Power supply: 11-32 V DC, 15 W
Number of channels: 200
Display range: 160 n miles in 0.1 n mile increments
Accuracy:
(range) ±0.1 n mile
(groundspeed) ±3 kt

Operational status

In service.

Contractor

Narco Avionics Inc.

VERIFIED

IDME 891 DME and VOR/ILS indicator

The Narco IDME 891 is a remotely channelled DME and VOR/ILS indicator. Designed for use with the Mk 12D or E Nav/Com, Nav 825 navigation receiver, NS 801 RNav and the CP 136M audio panel/marker beacon receiver, the IDME 891 simplifies navigation and requires very little panel space.

The IDME 891 combines VOR/ILS/DME and marker beacon lights all in one package and fits the standard 3 in instrument panel hole. The IDME 891 matches the Narco DME-890 for power, accuracy and lock on capability. The readout is a high-intensity display which indicates in increments of 0.1 n mile in the range mode and 1 kt in the groundspeed mode. Course indication indicator, glide slope, operational flags and marker beacon lights are standard.

Specifications

Weight:
(IDME-891) 1.18 kg
(antenna) 0.21 kg
Power supply: 11-32 V DC
Channels: 220
Display range: 0-160 n miles
Accuracy:
(range) ±0.1 n mile
(groundspeed) ±5 kt or 5% (whichever is greater)

Contractor

Narco Avionics.

VERIFIED

KWX 56 digital colour radar

The KWX 56 system is an inexpensive pitch/roll stabilised digital three-colour radar.

The system comprises two units: a 5 in (127 mm) diagonal KI 244 or KI 248 panel-mounted high-contrast black matrix display and a KA 126 or KI 128 combined antenna/transmitter/receiver. The latter can be stabilised using a flight director or vertical gyro; the addition of roll-stabilisation eliminates screen blanking caused by ground returns during medium or steep turns. The flat plate antenna has a diameter of either 10 or 12 in (254 or 305 mm), and can be supplied either unpressurised or pressurised for altitudes of up to 20,000 and 50,000 ft (6,100 and 15,200 m) respectively.

The KGR 356 radar graphics unit combines with the KWX 56 to give a self-contained panel-mounted area navigation system. With the KGR 356 in the Nav mode, the radar can display a weather plot with the superimposed location of the active VORTac and waypoints stored in the memory.

Specifications

Dimensions:
(antenna/receiver/transmitter) 254 mm diameter × 160 mm (control/indicator)
314 × 159 × 121 mm
Weight:
(antenna/receiver/transmitter) 4.27-4.61 kg
(control/indicator) 3.9 kg
Power supply: 28 V DC ±10%, 3 A max
Power output: (peak) 7.5 kW nominal, 6 kW min
Frequency: 9,375 MHz
Pulsewidth: 3.75 μs nominal
PRF: 109 Hz nominal
Ranges: 10, 20, 40, 80, and 160 n miles (18.5, 37, 74, 148 and 296 km) by rotary switch
Display: conventional colours. In mapping mode red becomes magenta, green becomes cyan and yellow remains unchanged
Stabilisation: from KI 256 flight director and ARINC standard vertical gyros associated with most autopilot/flight director systems, the KWX 56 can be used with Century 41, Century IV, Cessna ARC 400, 800 and 1000 series autopilots

Operational status

In production and in service.

Contractor

Narco Avionics Inc.

VERIFIED

KWX 58 digital colour radar

The KWX 58 is similar to the KWX 56 but includes additional features that are particularly suited to turbine-powered aircraft including 592 km range, and a 'target alert' feature to show significant weather beyond the display range. Weather plots are shown in green, yellow, red and magenta, according to rainfall density, on a high-contrast, non-fading black matrix screen. The system uses a 254 or 305 mm diameter planar-array roll/pitch-stabilised antenna radiating 7.5 kW. Narco claims that the penetration compensation feature gives a more accurate detection of storms behind closer areas of rainfall. Display ranges are 5, 10, 20, 40, 80, 160 and 320 n miles (9, 18.5, 37, 74, 148, 296 and 592 km). The KWX 58 system weighs 8.5 kg.

Specifications

Dimensions:
(antenna/receiver/transmitter) 254 mm diameter × 160 mm
(control/indicator) 314 × 159 × 121 mm
Weight:
(antenna/receiver/transmitter) 4.27-4.49 kg
(control/indicator) 3.9 kg
Power supply: 28 V DC ±10%, 3 A max
Power output:
(peak) 7.5 kW nominal, 6 kW min
Frequency: 9,375 MHz ± 30 MHz
Pulsewidth: 3.75 μs nominal
PRF: 109 Hz nominal

Operational status

In production and in service.

Contractor

Narco Avionics Inc.

VERIFIED

Mk 12 series NAV/COM receivers

The Narco Mk 12 is a VHF communications/navigation radio system designed principally for light and general aviation aircraft. The communications section is a transmitter/receiver covering the VHF band from 118 to 136.975 MHz and encompassing 760 channels. Two channels are preselectable, one for active and the other for standby use. Transmitter power output is nominally 8 W.

The navigation section consists of a VOR/Loc receiver covering the VHF band from 108 to

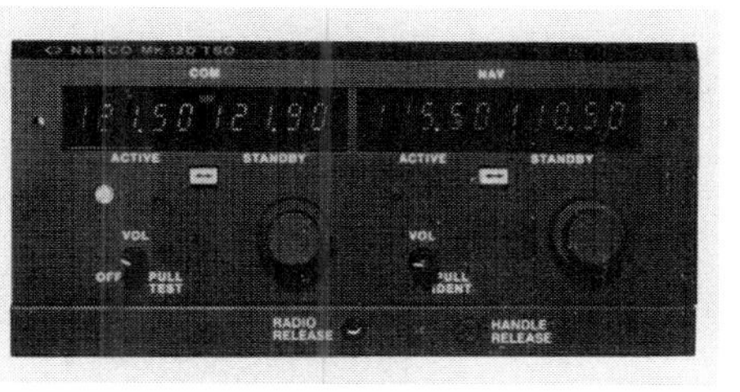

Narco Avionics Mark 12D/2 NAV/COM receiver
0001342

Narco Avionics Mark 12D+ NAV/COM receiver
0015388

117.95 MHz providing 200 channels; again, two of these are preselectable for active and standby use.

In each section, new frequencies may be entered into the standby mode at any time and transfer buttons are activated to exchange the selected frequency between active and standby positions.

Navigation section output will drive compatible horizontal situation indicators, area navigation systems, VOR radio magnetic indicators or the system's companion ID 824 VOR/Loc indicator. It will also drive the ID 825 VOR/ILS indicator and, for full ILS capability, a combined 40-channel glide slope receiver, covering the band 329.15 to 335.0 MHz, is available.

Mark 12D+

The TSO'd Mark 12D+ significantly reduces cockpit workloads with its 10 COM frequency storage, digital radial readout in NAV, and full 760 COM channels.

The Mark 12D+ is outwardly similar to the Mark 12D with the exception of a mode selector knob which is located beneath the COM display windows. The new radio retains the standby and active frequency display and flip-flop transfer features, but allows the user to program each of the 10 channels. The frequency selector also doubles as the channel selector.

Active and standby navigation frequencies are displayed in the NAV window. Except during entry of a standby NAV frequency, the right position of the NAV window always displays a digital radial readout from the VORTAC (or dashes if the signal is too weak).

Specifications

Dimensions: 159 × 64 × 279 mm
Weight:
(nav/com with glide slope receiver) 2.0 kg
(without glide slope receiver) 1.9 kg
(mounting tray) 0.34 kg

Operational status

The Mark 12D+ updates the earlier Mark 12 A to D models.

Contractor

Narco Avionics Inc.

VERIFIED

NAV 122D and NAV 122D/GPS self-contained NAV receiver indicators

The NAV 122D and NAV 122D/GPS self-contained NAV receiver indicators are being relaunched as technology updates of the earlier discontinued 3 in NAV instruments carrying the NAV 122 nomenclature. The new NAV 122D and NAV 122D/GPS instruments will be form/fit replacements for all earlier models from the NAV 122 back to the NAV 12 (except that replacement of the NAV 12 model will require an interconnect cable).

The new NAV 122D includes self-contained VOR/Localiser/Glideslope receiver and converter, with VOR/Glideslope indicator and Marker Beacon lights; DME channelling and autopilot interface. It utilises, as a self-contained item, the following elements: 200

Narco Avionics NAV 122D self-contained NAV receiver indicator 0015389

channel NAV receiver, VOR, Localiser and Glideslope Indicator, with Marker Beacon Lights.

The new NAV 122D/GPS optional version additionally includes the resolver required to interface with GPS, where VOR-style, 'Left/Right' indication is required for GPS IFR approval.

Specifications

VOR/Loc receiver: 108.00-117.95 MHz
Glideslope receiver: 329.150-335.00 MHz
External interfaces:
(DME channelling) 2 out of 5
(autopilot) left/right, to/from, up/down, glideslope flag
(Marker lamps) inputs for amber, white, blue lamps

Operational status

TSO'd units available from the first quarter of 1999.

Contractor

Narco Avionics Inc.

UPDATED

NS9000 multisensor navigation system

The NS9000 provides multisensor functionality by blending signals from its integral VOR/DME and GPS receivers, and interfaces with conventional VOR/Loc/GS indicators.

The NS9000 overcomes the inherent weakness of GPS and VOR/DME receivers (scalloping, P-static and satellite unavailability).

To derive latitude/longitude positions, the NS9000 selects the 10 closest VORs and 10 closest DMEs from its Jeppesen database. Based on geometry and distance, the NS9000 then selects the two VORs and two DMEs with the highest accuracy potential. It then automatically scans these stations and computes up to four fixes. The NS9000 then blends those resulting fixes with the position data derived from its five-channel GPS sensor (and Loran if it is RS-232 equipped). Three microcomputers reject any fix which varies significantly from others.

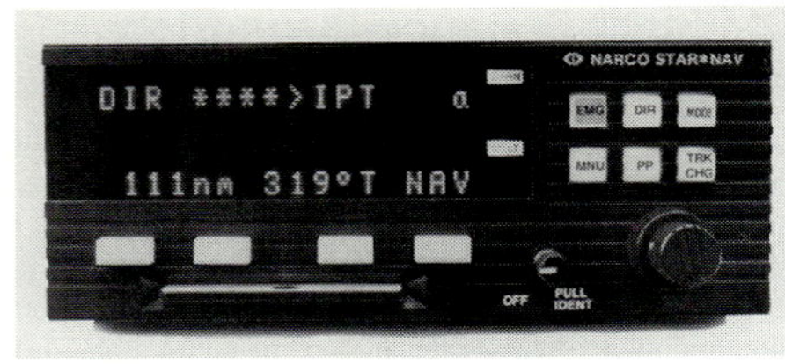

*The Narco Avionics NS9000 STAR*NAV multisensor navigation system* 0001343

It integrates information from GPS, VOR/DME, and Loran to provide the most accurate navigation. Operation is entirely automatic after selection of the waypoint. Frequency only has to be entered when switching from multisensor mode to standard mode, to enter an ILS frequency for example. This enables use of the 40-channel localiser and glide slope receiver for precision approach.

The Narco NS9000 is FAA TSO'd and STC'd for non-precision and full-precision approach. It is also compliant with ICAO FM immunity requirements and received full Basic RNAV (BRNAV) approval in November 1997.

Contractor

Narco Avionics Inc.

VERIFIED

Liquid Crystal Control Display Units (LCCDU)

The LCCDU is a microprocessor-based communications control and display unit. It provides monaural or binaural monitoring capability of up to 30 channels for each ear as required to each crew position equipped with an LCCDU, and it provides the microphone interface to the Communications Control Unit (CCU).

Through the full function keyboard, the LCCDU controls transmit and receive functions including the control and reassignment of all communication assets, crypto assignments, and Built-In Test (BIT) readouts.

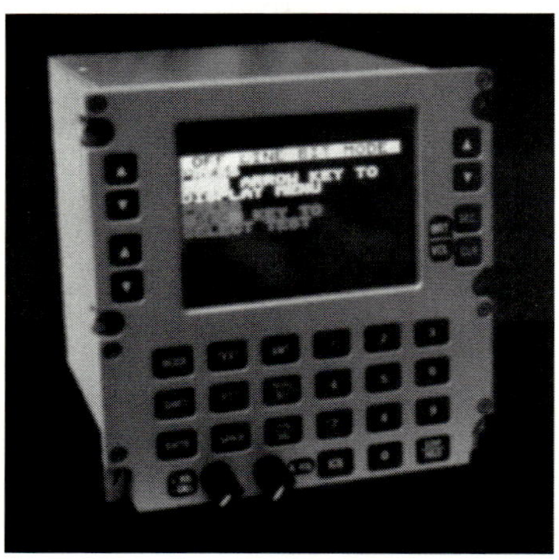

Liquid Crystal Crew Display Unit (LCCDU) 0018162

Complete status of the communications assets for each operator is available through the eight-colour liquid crystal display. Each LCCDU is identical and completely interchangeable with any other position location. The CCU may be programmed in real time if necessary to allow or disallow access of any given communications asset by any given crew position via the LCCDU.

The integral display intensity control is capable of adjusting the display brightness from off to full intensity without external controls or potentiometers. The LCCDU interfaces with the CCU via a serial RS-422 control data twisted-shielded wire pair.

Options include: tone generation crypto/modem switching; dual crew binaural audio; cross-band or in-band radio relay.

Specifications

Dimensions: 165.1 × 146 × 152.4 mm
Weight: 2.73 kg
Power: 28 V DC, 28 W

Contractor

Palomar Products Inc.

VERIFIED

Model 8000 Multifunction Display (MFD)

The Model 8000 MFD is fully compatible with AN/AAQ-16 night vision systems for either new installation of for retrofit purposes. It is also compatible with Generation II and Generation III night vision goggles.

Model 8000 Multifunction sunlight readable display 0001393

The Model 8000 is sunlight readable, with a display output greater than 200 fL (up to 400 fL) and a contrast ratio of 6:1 @ 10,000 fc ambient illumination.

The front control panel features 14 push-buttons with 14-bit encoded outputs, day/night/off switch, contrast and brightness adjustment, display size, centre and symbol brightness adjustment, Video 1 or 2 select. Front panel functions can be customised for specific requirements.

Specifications

Weight: < 6.4 kg
Largest viewable display area: 122 × 162 mm

Contractor

Palomar Products Inc.

VERIFIED

AMLCDs

Planar manufactures a range of AMLCD displays for military, industrial and commercial applications. They include ElectroLuminescent (EL) flat panel displays, miniature Active Matrix ElectroLuminescent (AMEL), 1,000 lines per in units suitable for Head-Mounted Displays, Active Matrix Liquid Crystal Displays (AMLCD), as well as full colour CAT-based avionic systems.

The Planar 11 × 8 in landscape orientation flat panel display is an amorphous silicon AMLCD full colour display for use in environmentally demanding applications that require maximum readability, rugged construction, and a wide temperature range.

Specifications

Model AM5: 5.0 × 5.0 in
Active display area: 127 × 127 mm
Pixel configuration: RGGB Quad
Display resolution:
Full colour: 480 × 480
Grey levels: 255
Viewing angles (20:1 contrast ratio):
horizontal: ±57°
vertical: +40°; –55°
NVIS compatibility (optional): NVIS-B MIL-STD-L-85762 Class B

Model AM6: 6.25 × 6.25 in
Active display area: 159 × 159 mm
Pixel configuration: RGGB Quad
Display resolution:
Full colour: 512 × 512
Grey levels: 255
Viewing angles (20:1 contrast ratio):
horizontal: ±57°
vertical: +40°; –54°
NVIS compatibility (optional): NVIS-B MIL-STD-L-85762 Class B

Model: 6.0 × 8.0-Q/6.0 × 8.0-S/8.0 × 6.0-S in
Active display area:
157 × 211/157 × 211/211 × 157 mm
Pixel configuration: RGGB Quad (Q model); RGGB Quad (S model)
Display resolution:
Full colour: 600 × 800
Monochrome: 1,200 × 1,600
Grey levels: 64/256

Viewing angles:
horizontal: >±45°
vertical: >+35°; –15°
NVIS compatibility (optional): NVIS-B MIL-STD-L-85762 Class B

Model: 11.0 × 8.0-Q
Active display area: 276 × 200 mm
Pixel configuration: RGGB Quad
Display resolution:
Full colour: 1,536 × 1,120
Monochrome: 3,072 × 2,240

Operational status

In December 1999, Planar Systems Inc announced a new 3 ATI MLCD, suitable for both civil and military flight-deck use in Attitude Direction Indicator (ADI), Horizontal Situation Indicator (HSI) and engine data applications.

Contractors

Planar Advance Inc.
Planar Systems Inc.

UPDATED

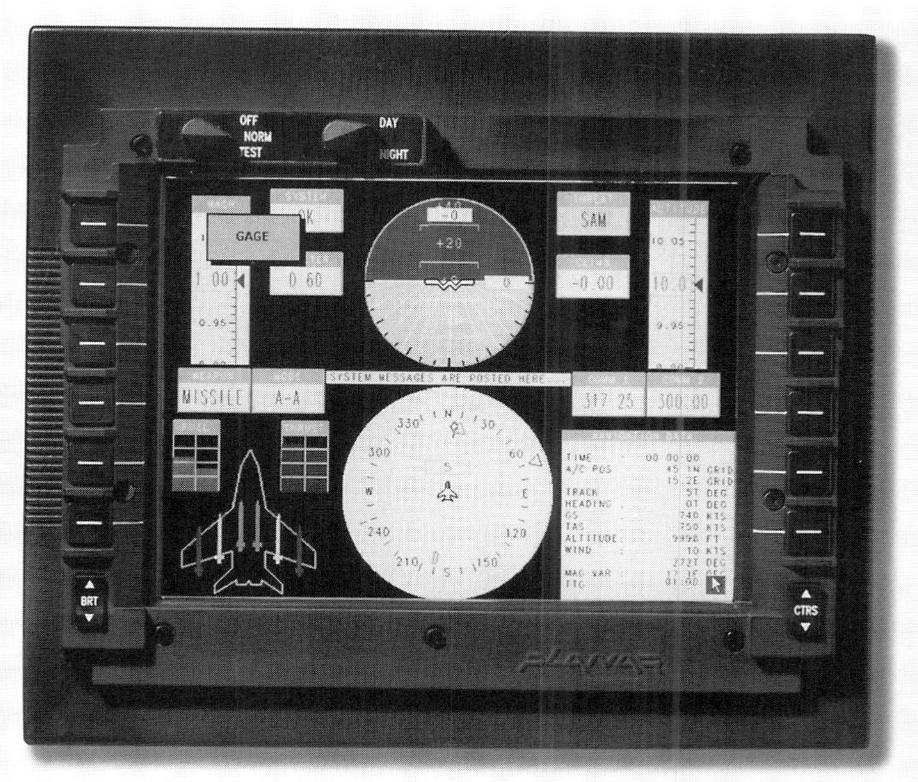

Planar AMLCD
0001392

B-2 cockpit with Planar displays
0003322

12-Channel Miniature Precision Lightweight GPS Receiver (PLGR) Engine (MPE)

Rockwell Collins' 12-channel Miniature Precision Lightweight GPS Receiver (PLGR) Engine (MPE) is a small, lightweight, second-generation Global Positioning System (GPS) receiver which provides precise positioning capabilities for military navigation, communications, timing tracking and designating systems.

The MPE is based upon the NightHawk 12-channel signal processor, Phoenix RF front end, ACE fast acquisition engine hardware and mature GPS receiver software; the system is designed to meet a wide variety of applications.

The MPE has been designed to use the same mating connectors and mounting footprint as the Rockwell Collins 5-channel MPE-I, while providing increased functionality.

The MPE transmits position, velocity and timing information via both RS-232 and CMOS serial interfaces. MPE is derived directly from PLGR, MPE-I and PLGR II, therefore it is compatible with existing PLGR, MPE-I and PLGR II integration protocols. The main features of the system are:

- PLGR-II performance in a small size for embedded applications
- 12-channel parallel, P/Y-code GPS receiver
- Selective Availability/Anti-Spoofing (SAAS)
- Direct-Y acquisition
- Military anti-jam capability
- 499 Waypoints
- Military Grid Reference System (MGRS), Universal Transverse Mercator (UTM), universal stereo polar graphic (UPS), ECEF, BNG, ITMG
- Timing data: 1 pps (pulse per second), HaveQuick (HQ)

Specifications

Frequency:
L1/L2 dual-frequency tracking
L1 – C/A, P/Y
L2 – P/Y
Dynamics:
Velocity: 1,200 m/s max
Acceleration: 9 *g* max
Time accuracy: 100 ns
Positional accuracy:
SDGPS: < 2 m CEP
WAGE: < 4 m CEP
PPS: < 12 m CEP
Velocity accuracy: 0.03 m/sec RMS (steady rate)
MTBF: > 40,000 h
Dimensions: 106 × 68 × 16 mm
Weight: 71 gm (max)
Power supply: +5 and +3.3 V DC, 1.2 W (typical)
Temperature range: –40 to +85°C

Operational status

Production scheduled for February 2001.

Contractor

Rockwell Collins.

NEW ENTRY

51Z-4 marker beacon receiver

The Rockwell Collins 51Z-4 marker beacon receiver automatically provides aural and visual indication of passage over airway and instrument landing system marker beacons. The system is approved for Category II approaches in a number of Rockwell Collins' all-weather avionic system certifications. Operating at a frequency of 75 MHz, the receiver sensitivity can be varied between two preadjusted levels through a cockpit Hi-Lo switch. The Hi position is used to gain early indication of a marker beacon, the Lo position is then subsequently used closer to the beacon for a sharper position fix. Alternatively, the receiver sensitivity is continuously variable from the cabin or flight-deck by means of a potentiometer.

The system is of all-solid-state construction and contains triple-tuned circuitry for the rejection of spurious signals generated by television and FM broadcast transmitters. It is designed for three-lamp indication but may be easily modified for single-lamp operation by removal of a resistor and wiring the three lamp outputs together. In either type of operation, outputs can operate two sets of indicator lamps in parallel.

An optional self-test facility causes the internal generation of 3,000, 1,300 and 400 Hz marker signals which are detected in sequence by the receiver. The indicators light in order and corresponding aural tones are also generated.

The unit is suited to retrofit installation since it is mechanically and electrically interchangeable with a number of other marker beacon receivers, including the 51Z-2 and 51Z-3 units. An associated marker beacon antenna, the 37X-2 system, is also available. Designed for operation with the 51Z-4 and other compatible receivers, the antenna is plastic-filled and sealed to reduce the effects of precipitation static. This unit, which weighs less than 0.45 kg, can be mounted without cutting into the airframe structure and has negligible drag.

Specifications
Dimensions: ¼ ATR short low
Weight:
(without self-test option) 1.36 kg
(with self-test option) 1.47 kg

Operational status
In service.

Contractor
Rockwell Collins.

UPDATED

95S-1A direct conversion receiver

Designed for communications and surveillance use, the Collins 95S-1A direct conversion receiver spans the 5 kHz to 2,000 MHz range in 1 Hz steps. The remotely tunable receiver uses state-of-the-art Direct Conversion Receiver DSP technology to attain low-noise, spurious-free, high-performance reception of AM, FM, SSB and ISB signals from 100 Hz to 300 KHz bandwidth.

Direct conversion architecture uses a single mixer with a single oscillator tuned to the desired frequency, converting a signal directly to baseband – the zero frequency IF signal. Rockwell Collins claims that this technology reduces spurious signal generation, and that it improves reliability due to lower part counts.

Software upgrades are installed to internal flash memory via the serial control port.

Operational status
In service.

Contractor
Rockwell Collins.

UPDATED

Rockwell Collins 95S-1A direct conversion receiver
0011824

618M-3 and 618M-3A VHF radio

The 618M-3 radio and its variant, the 618M-3A, are air transport-category VHF transceivers of 25 W nominal transmitter output. They were developed by Rockwell Collins as retrofit replacements for the earlier 618M-1 radio and for similar ARINC 546 and 566 systems as a response to the introduction of 25 kHz spacing between channels for VHF air traffic control communication.

Coverage of the VHF band by the 618M-3 is from 118 to 135.975 MHz, while the 618M-3A version coverage extends from 116 to 151.975 MHz. The former provides

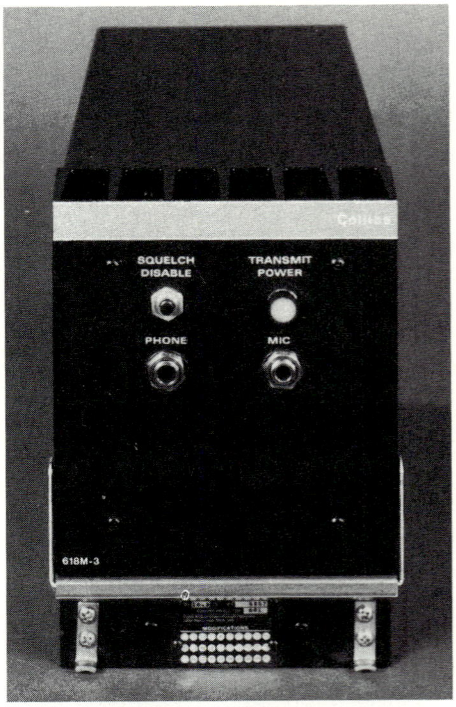

The Rockwell Collins 618M-3 HF radio

720 channels, the latter 1,440 channels, in each case in 25 kHz incremental steps. Besides basic voice communications, each system also possesses data and SELCAL facilities.

Frequency generation is supplied from a digitally controlled frequency standard. Low component density and use of solid-state circuitry is a feature and heat-sinking with cooling vanes assists in maintaining low transmitter temperatures. All of these factors result in improved reliability and the calculated MTBF is greater than 4,000 hours.

Speech compression ensures good intelligibility irrespective of the user's voice characteristics or microphone technique and overmodulation is avoided by input signal amplitude limiting. Carrier-to-noise, carrier override squelch control and automatic gain control are also incorporated.

Specifications
Dimensions: ½ ATR short
Weight: 4.53 kg
Temperature range:
(with cooling air) –54 to +55°C
(without cooling air) –54 to +30°C

Operational status
In service.

Contractor
Rockwell Collins.

UPDATED

628T-1 HF/SSB radio

The 628T-1 is an HF/SSB transceiver for long-range transport aircraft operating on overwater routes or other areas over which reliable extended range communications are required. It provides USB, AM voice and data communications on any of 24,200 channels, at 1 kHz increments in the 2.8 to 26.999 MHz band. Tuning is automatically controlled through a remote-control unit. Nominal transmission power is 200 W peak power in SSB or 100 W average in compatible AM.

This all solid-state system uses digital synthesis techniques for frequency generation. High stability is maintained through temperature compensation of the frequency standard.

Mechanical design is aimed at maximising maintainability. The transmitter/receiver is housed in a case with hinged tray and fold-out doors for easy accessibility. Plug-board circuitry is used extensively and bench-testing is simplified through the provision of a built-in test connector at the rear of the casing. Special provision is made for the replacement of the power amplifier transistors without the necessity of removing the entire power amplifier board.

Transmitter cooling is achieved through a heatsink and filtered forced air flow, while the receiver section relies on conventional convection cooling and is not dependent on a cool air supply.

An AM SELCAL facility is provided by means of a special audio output through which SELCAL signals are monitored irrespective of the selected operating mode. Options include automatic antenna tuning couplers, to permit antenna performance optimisation over the frequency spectrum covered by the 628T-1, and a 999W-1/A1 adaptor unit which permits interchangeability with the Rockwell Collins 618T-2/5 transceivers without disturbance to aircraft wiring, racks, connectors, antenna couplers or frequency selector.

Specifications
Dimensions: ¾ ATR short
Weight: 13.6 kg

Operational status
In service.

Contractor
Rockwell Collins.

UPDATED

628T-2 HF radio

The 628T-2 is based on the 628T-1 and to some extent it is also based on the earlier 618T-2/5 equipment which is still to be found on many transport aircraft. It uses much of the operational and design experience derived from these systems. The 628T-2, however, was designed to offer certain advantages over the earlier equipment, principally higher power output and extended coverage of the HF band, hence a greater number of channels and more options in terms of operating modes.

Transmitter output power is 400 W peak power and full coverage of the HF band, from 2 to 30 MHz, permits use of up to 280,000 channels at 100 Hz increments or 28,000 channels at a separation of 1 kHz.

Operational modes include USB, LSB, AM, carrier wave and data. Full 400 W peak output is available in the sideband modes with 125 W average in compatible AM and 125 W nominal in continuous wave mode.

In mechanical and electronic design, the 628T-2 is very similar to the 628T-1 although Rockwell Collins has incorporated a number of improvements with regard to sensitivity, cross-modulation elimination, intermediate frequency translation and heat dissipation in the power amplifier stage. All are aimed at extension of performance and enhancement of reliability.

Specifications
Dimensions: 6 MCU to ARINC 600
Weight: 12.72 kg

Operational status
In service. High Frequency DataLink (HFDL) upgrade kit available.

Contractor
Rockwell Collins.

UPDATED

628T-3 and 628T-3/A HF radios

Like the 628T-2, the 628T-3 and 628T-3/A are based on the 628T-1 design and represent the latest standard in this series. The principal difference between the 628T-1/2 and the 628T-3( ) is the 28 V DC power requirement of the latter. The principal difference between the 628T-3 and the 628T-3/A variant is the narrower bandwidth intermediate frequency filtration of the latter, resulting in heightened selectivity over part of the operating spectrum and slight differences in audio response.

As in the case of the 628T-2, each system provides full coverage of the 2 to 30 MHz HF band and offers 280,000 channels at 100 Hz increments. There is, however, no option for a smaller number of channels at

1 kHz separation; transmitter output is 200 W peak power.

Operating modes are voice, and voice and data in both upper and lower sidebands, with compatible AM and carrier wave in USB only. SELCAL facilities are as those of the 628T-1.

Specifications

Dimensions: ¾ ATR short
Weight: 11.36 kg

Operational status

In service.

Contractor

Rockwell Collins.

UPDATED

714E HF control units

The 714E control units used in conjunction with the 618T series transmitter/receivers provide remote selection of the available 28,000 channels. Frequencies are indicated in a direct reading digital display and can be selected in 1 kHz increments throughout the 2 to 29.9999 MHz range. Frequency selection is accomplished by rotating four knobs until the desired frequency appears in the window. A function selector and radio frequency sensitivity adjustment are included.

The 714E-2 control unit may be used either in new installations or as a replacement for the 614C-2 control unit in retrofit applications.

The 714E-3 is used with equipment operated in the continuous wave or data mode.

Specifications

Dimensions: 146 × 65 × 159 mm
Weight: 0.9 kg

Operational status

In service.

Contractor

Rockwell Collins.

VERIFIED

860F-4 digital radio altimeter

The 860F-4 employs large-scale integrated technology and microprocessors but has an analogue output for driving existing non-digital instruments. It has been designed to ARINC 552/552A and can be used as a reference for Cat IIIA automatic landings. Rockwell Collins says that the digital technology gives a 30 per cent reduction in parts and 40 per cent saving in weight compared with the earlier non-digital 860F-1 system. The equipment can be used from –20 to 2,500 ft. The outputs comply with ARINC 429 (digital) and ARINC 552/552A (DC analogue).

Specifications

Dimensions: ½ ATR short
Weight: 5.4 kg
Power supply: 115 V AC, 400 Hz, 50 VA
Frequency: 4,300 MHz

Contractor

Rockwell Collins.

VERIFIED

ADF-60A Pro Line Automatic Direction-Finder (ADF)

The ADF-60A is a lightweight and rugged ADF operating from 190 to 1,749.5 kHz in 500 Hz steps. Two combined loop and sense antennas are available: the ANT-60A for single system installations or the ANT-60B for dual system installations.

Specifications

Dimensions:
(receiver) ⅜ ATR short/dwarf
(ANT-60A) 419 × 216 × 41 mm
(ANT-60B) 604 × 269 × 25 mm
Weight:
(receiver) 1.9 kg
(ANT-60A) 1.5 kg
(ANT-60B) 2.3 kg

Operational status

In service.

Contractor

Rockwell Collins.

UPDATED

ADF-462 Pro Line II Automatic Direction-Finder (ADF)

The ADF-462 is an all-digital ADF operating from 190 to 1,799 kHz together with the 2,182 kHz distress frequency. Two antennas are available: the ANT-462A for single system installations or the ANT-462B for dual system installations. The ADF-462 is compatible only with CSDB or ARINC 429 controls. Outputs are also serial digital and are provided in both CSDB or ARINC 429 characteristics. In addition, a DC sine/cosine output is available for interfacing with conventional RMIs.

Specifications

Dimensions:
(receiver) ⅜ ATR short/dwarf
(ANT-462A) 452 × 218 × 41 mm
(ANT-462B) 604 × 269 × 25 mm
Weight:
(receiver) 1.7 kg
(ANT-462A) 1.5 kg
(ANT-462B) 2.3 kg

Operational status

In production and service.

Contractor

Rockwell Collins.

VERIFIED

ADF-700 Automatic Direction-Finder (ADF)

Designed in accordance with ARINC 712, the ADF-700 automatic direction-finder was introduced in 1980. It is based on experience gained with the earlier DF-203 and DF-206 receivers. These technical advances were pioneered by Rockwell Collins and incorporated into draft ARINC characteristic 712, which also provided for ARINC 429 interfaces and an integral loop/sense antenna. Centralised fault monitoring is now incorporated into the design consistent with ARINC 604 characteristics.

In addition to mechanical and reliability improvements introduced by the new characteristic, a significant performance improvement came with the change from analogue to digital technology. In the ADF-700 all bearing signal baseband processing is performed digitally. The reference and bearing signals are converted into digital form using a 12-bit CMOS analogue-to-digital converter and these are subsequently handled by an Intel 8086 16-bit microprocessor. The ARINC 429 input/output functions are performed by an Intel 8049 processor in conjunction with a Rockwell Collins universal asynchronous transmitter/receiver.

The advent of third-generation microprocessors permits the introduction of digital ADF signal processing to improve accuracy and reliability, a significant advantage being the elimination of mechanical adjustments. The system incorporates improved self-test capabilities as a result of using the 8086 microprocessor to control a test sequence incorporating a digital test signal synthesis. The same power supply subassembly is also used in the VOR-700 and ILS-700, leading to reductions in spares inventories and maintenance costs.

Specifications

Dimensions: 2 MCU per ARINC 600, 604
Weight: 3.4 kg
Power supply: 115 V AC, 380 to 420 Hz (1 sigma)
Frequency: 190 to 1,750 kHz
Channel spacing: 0.5 kHz
Modes: ANT-aural receiver, ADF navigation, CW/MCW
Tuning: ARINC 429 dual serial bus
Accuracy: better than 0.9° with ARINC 712 antenna in 35 µV/m field, exclusive of antenna error
Compliance: ARINC 712, 600, 429; TSO C41c; DO-142, -160A; ED14A, EUROCAE

Operational status

In production and in service in a wide range of commercial aircraft, including Boeing 747, 757/767, Airbus A310, 319/320/321 330/340 and Fokker 100.

Contractor

Rockwell Collins.

UPDATED

AHS-85 strapdown Attitude and Heading reference System

Designed for business aircraft and regional airlines, the AHS-85 attitude and heading reference system is aimed at a market largely ignored before 1983. The emphasis is on simplicity and all sensing and processing functions are grouped in one unit. The system uses a sensor package which eliminates all common high failure rate components and uses rotors that spin at one-seventh the speed of conventional gyros. The primary sensors are two piezoelectric multisensors, each with four independent pick-offs, which determine two axes of angular rate and linear acceleration data, thus providing one axis of redundant information for system monitoring. These devices are complemented by a flux detector mounted independently in a portion of the airframe without magnetic disturbances. The replacement of any sensing element does not require adjustment to the installation and compass compensation remains unchanged if the AHS-85 is replaced.

Processed data is compatible with flight instruments, autopilot systems, weather radar stabilisation and flight data recording. True airspeed can be accepted from a remote air data system. The system provides digital and analogue outputs and three-axis body rate and acceleration data and does not suffer directional gyro gimbal errors.

Specifications

Dimensions: 124.5 × 193 × 320 mm
Weight: 6.9 kg
Power supply: 28 V DC, 50 W
Accuracy:
(pitch and roll) 0.5° nominal,
1° during manoeuvres
(heading) 1° nominal,
2° during manoeuvres

Operational status

In service.

Contractor

Rockwell Collins.

UPDATED

AHS-3000 Attitude Heading System

The AHS-3000 Attitude Heading System is Rockwell Collins' first avionics application of Digital Quartz Gyro (DQG) technology, developed by Systron Donner. This DQG's small, lightweight, low-power, solid-state micromachined gyroscope is fabricated chemically from material similar to that used in quartz watches. It has proven to be reliable in demanding aerospace, automotive, railroad and maritime applications.

Because the construction has no moving mechanical parts, the AHS-3000 has an estimated mean time between failures of more than 10,000 h - up to × 5 improvement over most existing spinning-wheel and fibre optic systems.

The attitude heading computer used in the AHS-3000 demands only one-third of the weight, size and power requirements of the current AHRS computer.

Operational status

The AHS-3000 was developed for integration with the Rockwell Collins' Pro Line 21 avionics system. The AHS-3000 is in production and in service.

Contractor

Rockwell Collins.

VERIFIED

ALT-50/55 Pro Line radio altimeters

The ALT-50 and ALT-55 Pro Line radio altimeters, the first with a range of 0 to 2,000 ft, the other 0 to 2,500 ft, have been designed for business aircraft. Both types provide decision height annunciation for Cat II landings. The decision height annunciators can be set at any desired altitude and both instruments can interface with high-performance flight directors and autopilots. The DRI-55 indicator is offered with a numeric readout of radar altitude and decision height.

Dual ANT-52 antennas are included.

Specifications

Dimensions:
(transmitter/receiver) ⅜ ATR short dwarf
(indicators) 77 × 77 × 152 mm
(antenna) 25 × 25 × 20 mm
Weight:
(transmitter/receiver) 2.54 kg
(indicator) 0.7 kg
(antenna) 0.1 kg
Power supply: 28 V DC

Operational status

In production and in service.

Contractor

Rockwell Collins.

VERIFIED

AMS-850 avionics management system

The AMS-850 avionics management system provides centralised avionics management, including control of EFIS modes, weather radar, TCAS, radio tuning and flight management, in one convenient location. The system offers consistent, intuitive menu-driven operation that facilitates training, minimises entry errors and simplifies avionics management.

An integral part of the Pro Line 4 system, the system takes advantage of Pro Line 4 multifunction displays to provide head-up operation, with navigation maps and tables and other avionics information shown on the MFD. Integration of the AMS-850 as part of the Pro Line 4 system reduces weight and size and simplifies interconnections. Flight management computers are packaged as line-replaceable modules, centrally housed in the Rockwell Collins integrated avionics processing system. Pro Line 4 integration allows the system to perform diagnostic functions, identifying any malfunctioning avionics unit and displaying the information on the MFD. The system also offers complete LRU fault history and detailed analysis of system status.

The menu-driven operation of the AMS-850 simplifies flight crew route planning. Flight plans may be entered waypoint by waypoint or selected from up to 99 previously defined and stored routes. En route, terminal area and non-precision approach navigation as well as primary means navigation in oceanic and remote areas are provided by the system database, and new waypoints may be added within a flight plan or appended. When editing a flight plan, pressing the line select key next to the displayed waypoint automatically brings up the edit page on the CDU for consistent operation. Pilot-defined waypoints can be created by plotting a path on the system's present position map using the joystick, and direct-to is also available for straight point to point routeing.

VNav capability is available with the advanced Collins system, fully integrated with the autopilot, navigation and air data systems for smooth operation. The AMS-850 is available with a full alpha keyboard to accommodate a wide range of applications.

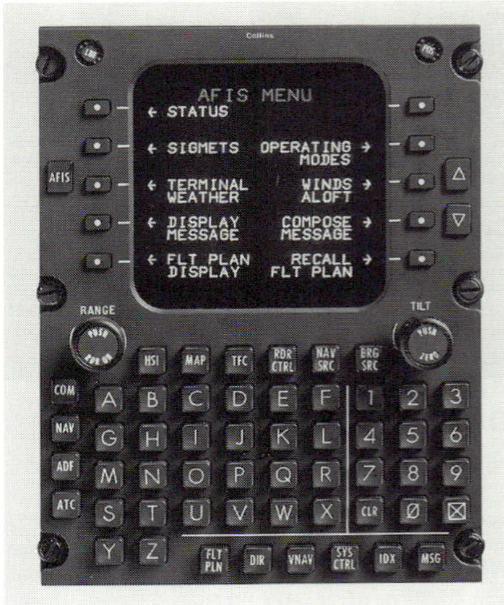

The AMS-850 avionics management system offers full AFIS capability

The AMS-850 is available with AFIS compatibility, offering flight crews access to flight planning, weather information, SIGMET advisories and messaging capabilities. The AMS-850 also provides straightforward fuel management, computing fuel used, fuel remaining and endurance at the present fuel flow. The system's worldwide database is contained on disks that are updated every 28 days, which contain global VHF navigation and airport information.

Operational status

FAA Supplemental Type Certification on Beechjet 400A awarded in December 1996 and on Starship in October 1997. FAA TSO C129 Class B1 and Technical Order 8110.60 criteria for IFR operation to make non-precision approaches with GPS as primary means of navigation.

Contractor

Rockwell Collins.

VERIFIED

APS-65 autopilot

Introduced in 1982, the APS-65 was claimed to be the first digital autopilot offered for turboprop types with dual microprocessor computation. The basic operating modes are: roll hold, pitch hold, heading hold, navigation mode, approach mode, indicated airspeed hold, vertical speed hold, climb and descent, altitude select and go-around. Additionally, when climbing, the autopilot can be programmed to fly the aircraft at optimum efficiency, thus providing fuel savings.

A new control technique for the pilot to use when introducing pitch, altitude, airspeed and vertical speed hold changes has been incorporated. This employs a rocker switch that can be operated to select precise alterations. Altitude can be adjusted in increments of 25 ft, pitch in increments of 1°, airspeed in increments of 1 kt and vertical speed in increments of 200 ft/min. The new technique has been incorporated to reduce pilot workload and permit smoother flying. Operational safety is enhanced through the dual processor configuration, which ensures that no single equipment or software fault can result in exceeding preset limits or a malfunction in more than one axis of control. Built-in monitoring ensures automatic disengagement when a fault is detected and appropriate diagnostics assist in fault detection in any element. The system is lighter and has a lower parts count than comparable analogue systems; higher reliability is attributed to these features.

The autopilot is suitable for the heavier business turboprop types such as the King Air and can be integrated with electromechanical or electronic flight director systems and navigation systems produced by Rockwell Collins.

Specifications

Weight: 10 kg total

Operational status

In service. First to be certified was the Beech King Air in April 1983. The APS-65 has now been chosen for 28 types of turboprop business aircraft. The King Air was noteworthy as the first general aviation all-digital turboprop to be certified. Its Rockwell Collins avionics suite included APS-65 autopilot, EFIS-85 electronic flight instrument system, ADS-80/85 air data system and nav/com equipment.

Contractor

Rockwell Collins.

UPDATED

APS-85 digital autopilot

The APS-85 autopilot completes the Pro Line 2 family of digital avionics for general and business aviation. APS-85 is an entirely digital, fail-passive autopilot with dual-redundant flight guidance computers. It has been certified for Cat II operation; Cat IIIA landings will be possible with some growth.

The main flight panels of the Raytheon Hawker 800 with the Rockwell Collins APS-85 autopilot mode select panels fitted on the coaming in front of the pilot and co-pilot

Apart from its digital nature, the APS-85 has a number of features not found on earlier Rockwell Collins general aviation autopilots, including the APS-80. For example it includes climb and descent modes. When selected in the climb mode, the system flies the aircraft according to an airspeed or Mach number schedule appropriate to that type of aircraft and its weight at take-off. In the descent mode the system sets up a rate of descent tailored to the aircraft manual. Three diagnostic modes - report, input and output - are incorporated. In the first, faults in the flight control computers are displayed to the crew. In the second, the system reads out information from outside entering the system, for example from the air data system or control surface position sensors. In the final mode, the system provides particular flight control computer outputs for examination. The diagnostics on the APS-85 are self-contained and do not require additional test equipment.

Specifications

Weight: 14.0 kg total

Operational status

In production and service. It has been certified as the autopilot on the Saab 340, the Raytheon Hawker 800, Falcon 50B, Falcon 20F, Learjet 55, CL-600/601, Astra and DC-8.

Contractor

Rockwell Collins.

VERIFIED

Automatic Flight Control and Augmentation System for the Fokker 100

The Rockwell Collins Automatic Flight Control and Augmentation System (AFCAS), Electronic Flight Instrument System (EFIS) and MultiFunction Display System (MFDS) provide the Fokker 100 with advanced operating capabilities in the areas of automatic landing, display integration and engine and aircraft annunciation.

The AFCAS integrates cruise autopilot, yaw stability augmentation, fail-operative automatic landing, fail-operative thrust management and envelope protection functions into a single system. Certified for Category IIIB conditions, the AFCAS will give the Fokker 100 automatic landing capabilities down to a decision height of 50 ft and a runway visual range of 150 m.

Operational status

The Fokker 100 has demonstrated Category IIIB automatic landing capabilities using AFCAS.

Contractor

Rockwell Collins.

VERIFIED

CMU-900/APM-900 datalink Communications Management Unit/Aircraft Personality Module

Rockwell Collins' CMU-900 is a datalink communication management unit specifically designed to provide a seamless transition with minimum cost of ownership as the worldwide Data Link System evolves from ACARS to ATN. The system manages datalink communications over all three air/ground subnetworks (VHF, HF and SATCOM) simultaneously. The CMU-900 supports all aircraft platforms with common hardware and core software and AOC that reduces the configuration management and administrative burden placed upon an airline operating numerous aircraft types and configurations. It is developed and certified to DO-160C and DO-178B, Level D for current ACARS and FANS-1 routing operation, and will be certified to Level C standards to allow the CMU to function as an ARINC 758 ATS end system (Level 2D).

The CMU-900 provides a powerful User Application Development capability allowing airlines the option of performing their own datalink non-essential application maintenance and modifications. It minimises software update costs by providing isolated partitions for user-modifiable and non-modifiable, as well as aircraft-specific and aircraft-independent functions. In common with Rockwell Collins' predecessor DLM-700B/C and DLM-900 series of management units, the CMU-900 minimises cost of ownership by maintaining exceptional levels of reliability.

Datalink communication unit/system configurations are dependent upon aircraft type, airline operation and ground system requirements. Customised AOC software configurations are available upon request.

The CMU-900 is also capable of supporting an ARINC 724B backplane wiring configuration, to facilitate installation in existing ACARS provisions.

The APM-900 Aircraft Personality Module complements the CMU-900 in a true ARINC 758 CMU installation. The APM-900 provides a means of permanently storing aircraft-unique parameters necessary for the initialisation and operation of the CMU, including the ICAO address, registration, aircraft type and airline code. The CMU provides power to the APM, program the configuration data upon initial installation and read the configuration data upon power-up.

When combined with the Rockwell Collins VHF, HF and SATCOM (Aero-H and Aero-I) radios, the advanced communications capabilities of the CMU-900 and APM-900 create a comprehensive data communications product line for both forward-fit and retrofit applications.

An option providing a Type III PCMCIA slot, able to host a single Type III or a combination of multiple Type I and Type II modules, has been designed and built for the CMU. PCMCIA mass memory devices may be utilised to support custom applications requiring capture and removal for subsequent off-line analysis of large amounts of data.

An Ethernet card has been designed and built for the CMU to provide a 10/100-BaseT Ethernet link from the CMU to external peripherals. This future optional interface is intended for use with integrated information systems.

A Cross-Talk Interface that communicates with a second CMU-900 is planned and will be offered within the timeframe necessary to support applications requiring dual CMU redundancy.

Specifications

Dimensions: 4 MCU per ARINC 600
Weight: 5.5 kg
Power supply: 115 V AC, 400 Hz, with 28 V DC optional
Power usage: 35 W (nominal); 2.1 W (standby)
Temperature range: –40 to +55°C
Altitude: up to 55,000 ft

Operational status

The CMU-900 and APM-900 received type certification aboard a Delta Airlines Boeing 737-800 aircraft in May 1999. Rockwell Collins claims that the CMU-900/APM-900 pairing is the first ARINC 758-compliant CMU to receive Federal Aviation Administration (FAA) certification; Rockwell Collins also claims that it was the first to receive black label approval for flight testing aboard all Boeing 737-X, 747-400, 757 and 767 production aircraft models.

CMU-900 certifications were expected later in 1999 for Boeing 747-400 and 767 aircraft. Supplemental type certifications were being processed for the 737-400, 747-100/200, 757, 767 and MD-80 aircraft in Summer 1999. Deliveries were being made in 1999 to the US Air Force for the C-17 GATM update programme.

In February 2000, the US Air Force contracted Rockwell Collins to implement and demonstrate complete end-to-end commercial Airline Operational Control datalink and Very high-frequency DataLink (VDL) Mode 2 capability on C/KC-135 aircraft as part of the GATM programme. CMU-900 software was modified and a datalink module added to the AN/ARC-210 V/UHF radio to provide the required capability. The datalink module will initially provide VDL Mode 2 functionality and ARINC 750 compliant interfaces to the dual CMU-900 units. The added module also provides the necessary hardware to support VDL Mode 3 via a future software upgrade.

Contractor

Rockwell Collins.

UPDATED

Datalink systems

Rockwell Collins datalink systems are designed for commercial and military transports and conform to ARINC 724 and 597. Their associated control panels have full alphanumeric keyboards with a two-line, 32-character light-emitting diode display. Information can also be displayed on a compatible weather radar screen.

These datalinks can control VHF radios, act as a checklist memory and provide the interface between flight and performance management systems and ground-based data transmitters. The operator can change the software and the units have comprehensive built-in test facilities and a continuous GMT clock facility.

The DLC-800 is a menu-driven control/display designed to be used with the Rockwell Collins DL-700 datalink systems. The unit is completely compatible with other datalink management units which use the ARINC 429 databus and can be retrofitted into existing installations. Airline programmable software enables an airline to create distinctive symbols, display formats and messages for use on the DLC-800. The unit has a high-resolution infra-red touch-input system and a wide viewing angle for easy reading by both pilots. The display is easily readable in direct sunlight. Key actions can be highlighted by using inverse characters in the form of dark on light background. Messages can flash on the screen to attract attention.

Management units are available in two formats. The DLM-700 management unit is in accordance with ARINC 724, 597 and 600. The 597A-1 is in accordance with ARINC 404A and 597.

Specifications

Dimensions:
(control unit) 114 × 146 × 120 mm
(DLM-700 management unit) 4 MCU
(579A-1 management unit) ? ATR short
Weight:
(control unit) 1.2 kg
(DLM-700 management unit) 4 kg
(597A-1 management unit) 3.6 kg
Temperature range: –15 to +70°C
Altitude: 55,000 ft

Operational status

In production.

Contractor

Rockwell Collins.

VERIFIED

Digital RMI/DME indicators

Designed to interface with both the current ARINC 700 sensors and the older-generation analogue avionics, this new series of combined Radio Magnetic Indicators and Distance Measuring Equipment (RMI/DME) forms a family of instruments that share many common features including electrical, servo, thermal, packaging and lighting methods. Use of a patent digital encoder/driver module, common to each instrument and each channel within it, permits packaging techniques which greatly improve the effectiveness of the heat transfer arrangements. The encoder/driver module consists of an 11-bit digital encoder, DC motor and associated drive circuits, all under microprocessor control. Each of the six channels is isolated from its neighbour for integrity and each has its own single chip microcomputer. Each channel monitors its own faults and activates its own flag or shutter and output signal.

The members of the digital RMI/DME instrument family comprise:

RMI-733A radio magnetic indicator: a compact lightweight VOR/ADF selectable three-servo instrument. A version designated RMI-733A is a three-servo instrument designed to display information from an ADF receiver.

RDMI-743 radio distance magnetic indicator: this instrument features liquid crystal DME readouts for better reliability and readability. The three-servo unit is selectable to either VOR/ADF or VOR only.

RDMI-743A radio distance magnetic indicator: with a magnetic wheel DME display, the RDMI-743A is a four-servo VOR instrument.

Operational status

All in production.

Contractor

Rockwell Collins.

VERIFIED

DLC-800 interactive touchscreen control display

The DLC-800 interactive touchscreen control display is part of the Rockwell Collins Airline Communication And Reporting System (ACARS) product line. It is designed to make data entry for datalink information simpler and error free.

The DLC-800 features a high-resolution flat-panel liquid crystal display with an infrared touch input system and reconfigurable menu formats. The menus can be input from either the DLM-700 or DLM-700B management unit. Other features include highlighted key actions, flashing messages and a wide range of graphic capabilities.

The high-contrast ratio screen is easily read in direct sunlight and at a wide viewing angle for both pilots. Inverse dark video characters on a light screen also enhance readability and message legibility. Larger text sizes can be used to highlight menu titles, while smaller sizes can list detailed information such as flight plans and weather reports.

Aircraft system communication with the DLC-800 is accomplished via ARINC 429. The DLC-800 can be readily tailored to meet individual requirements.

Future growth potential includes graphics capability for pictorial and symbol displays such as weather plots. Other growth options include ATC communications such as clearances and route maps and Satcom compatibility.

Specifications

Dimensions: 114 × 146 × 178 mm
Weight: 2.1 kg
Power supply: 115 V AC, 400 Hz, single phase, 25 W

Operational status

It was reported that Air Canada ordered the DLC-800 for their Airbus A320s.

Contractor

Rockwell Collins.

UPDATED

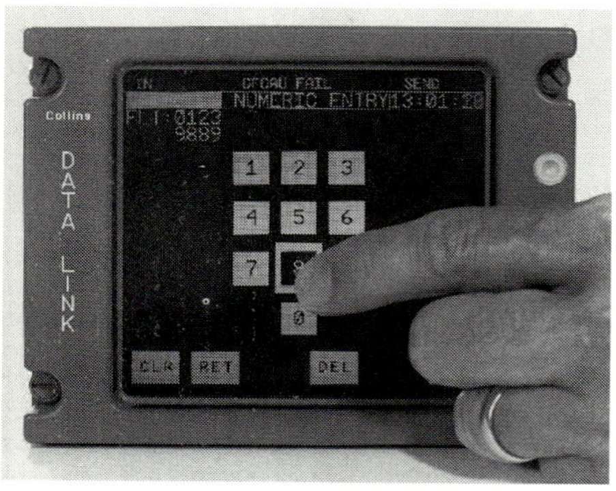

The Rockwell Collins DLC-800 control display shows highlighted keys to confirm correct data entry

DLM-900 datalink system

The Rockwell Collins DLM-900 is a next-generation datalink management unit specifically designed to provide an affordable and seamless transition as the global datalink system evolves from ACARS to ATN. The DLM–900 transmits and receives messages through existing airborne communications systems (VHF, SATCOM and HFDL) via SITA, AVICOM Japan, or ARINC ground networks. The system meets the requirements of ARINC characteristic 724B. It also meets the environmental requirements of DO–160C and software requirements of DO–178B, Level D. The unit provides easy upgrade via software data load to add Aeronautical Telecommunications Network (ATN), Future Air Navigation System (FANS) and other future capabilities as they come online. Software update costs are minimised by providing isolated partitions for essential (Level C) and non-essential (Level D) functions as well as aircraft-specific and aircraft-independent functions. A powerful user application development capability allows airlines to perform their own datalink non-essential application maintenance and modifications. This unit was designed with sufficient processor throughput, memory capacity and interface capabilities to host future functionalities and provides a spare full-depth card slot to accommodate growth.

Specifications

Dimensions: 4 MCU per ARINC 600
Weight: 5.5 kg
Power supply: 115 V AC, 400 Hz, with 28 V DC optional
Power usage: 35 W (nominal); 2.1 W (standby)
Temperature range: –40 to +55°C
Altitude: up to 55,000 ft

Operational status

In production.

Contractor

Rockwell Collins.

UPDATED

DME-42/442 Pro Line II DME systems

The DME-42/442 are all-digital DME systems which can provide complete information on up to three DME stations using a single transceiver. They were designed for business or commuter aircraft where only a single DME facility is needed, but where additional information is useful. Station information is presented on the associated IND-42 display.

The systems can provide data to show 'distance to station' up to 300 n miles, time to station up to 120 minutes and groundspeed up to 999 kt. They also decode the station identifier and provide for display.

The DME-442 is compatible only with CSDB or ARINC 429 controls and displays. The DME-42 can directly replace the earlier DME-40 system.

Specifications

Dimensions:
(R/T) ½ ATR short/dwarf
(IND-42) 42 × 86 mm
Weight:
(R/T) 2.4 kg
(IND-42) 0.41 kg
Power supply: 28 V DC, 0.7 A, plus 0.3 A for display
Altitude: up to 70,000 ft

Operational status

In production and in service.

Contractor

Rockwell Collins.

VERIFIED

DME-900 Distance Measuring Equipment

The DME-900 distance measuring equipment has been designed to provide highly accurate distance data outputs in all modes of operation. Error sources in the interrogator are identified and reduced to the lowest level. A highly efficient parallel ranging technique allows the DME-900 to perform less than 15 interrogations, half the allowable maximum average. Only two output devices are required to achieve a 700 W nominal power output. The synthesiser/driver operates at L-band transmit channel frequencies, significantly reducing the parts count and circuit complexity. In addition, optimum filtering methods reduce naturally occurring noise contained in the DME/Tacan signal.

Specifications

Dimensions: 4 MCU
Weight: 6 kg
Power supply: 115 V AC, 380-420 Hz, 35 W (max)
Frequency:
(transmit) 1,025-1,150 MHz
(receive) 962-1,213 MHz
Channels: 252
Range: –1 to 320 n miles
Accuracy: ±0.1 n mile

Operational status

In production and in service.

Contractor

Rockwell Collins.

UPDATED

EFIS-84 Electronic Flight Instrument System

The EFIS-84 is a 4 in electronic flight instrument system offering full capability and high reliability at an affordable cost. The EFIS-84 pairs an EADI and an EHSI to provide the same capabilities as the Collins EFIS-85 and EFIS-86 systems.

The system is designed for applications ranging from helicopters to regional airliners and corporate aircraft. Retrofit of existing 4 in electromechanical systems with the EFIS-84 offers significant operational, reliability and cost of ownership benefits.

The EFIS-84 offers bright clear attitude and navigation information in easy-to-interpret formats. Attitude is displayed in full sky presentation which provides an attitude area more than twice as large as that of electromechanical instruments. A race-track attitude display, with a circular attitude depiction similar to the familiar electromechanical ADI, may also be selected as an installation option. A customer choice of V-bar or cross pointer steering commands is also offered.

A full selection of modes is offered on the EHSI, including the traditional compass rose, arc and map. Significant operational advantages are offered by the extensive map capability, particularly to pilots operating in terminal areas. A key benefit of the EFIS-84 is the ability to display weather radar information integrated with the navigation map. The system is compatible with the Collins WXR-350 or the advanced TWR-850 turbulence weather radar.

System flexibility allows interface with a variety of analogue and digital aircraft sensors, as well as a broad range of long-range navigation options. The optional multifunction display offered as part of the EFIS-84 provides additional system capability, including expanded map display and checklists.

Operational status

In service. Certified on the Beech 1900D regional airline aircraft.

Contractor

Rockwell Collins.

UPDATED

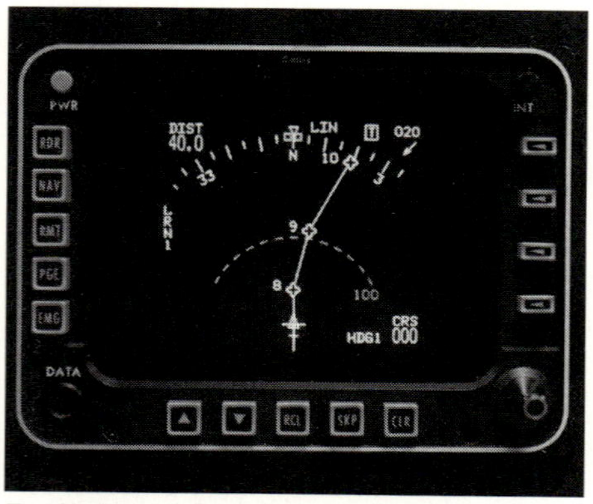

The Rockwell Collins multifunction display offered as an EFIS-84 option

EFIS-85 Electronic Flight Instrument System

A full EFIS-85 system comprises dual 5 × 5 in (127 × 127 mm) EADIs and EHSIs with one set of equipment for each pilot, a shared multifunction display on the centre panel and mode controls. The

EADI and EHSI CRTs can display all the information traditionally associated with electromechanical flight director instruments as well as weather radar patterns, navigation maps, performance data and navigation waypoints. Presentation is very flexible. The EHSI-85 can show conventional HSI information on a circular scale or it can expand just the forward sector of the display and show weather maps.

The EFIS-85 CRT incorporates a three-gun assembly, a shadow-mask, a faceplate with phosphor coating and a glass envelope to enclose the elements. The in-line electron gun assembly provides improved convergence and mechanical rigidity and the high-resolution shadow-mask gives four to six times better resolution than that of a domestic television set because the phosphor dots are so much closer together. The displays use both stroke and raster writing. The high-intensity stroke writing of symbols, in conjunction with contrast enhancement filters, enables displays to be read even in full sunlight. Primary colours are red, blue and green with easy synthesis of several derivative colours, including white.

The new advanced map displays are now available with EFIS-85 and -86 electronic flight instrument systems. These include heading up, north up, aircraft centred, north up max view and plan displays. The maps display airports and navaids from navigation systems including the UNS-1A or GNS-1000.

Operational status

In service. The aircraft used for approval trials was a Dassault Falcon 100; certification was announced in December 1982.

Rockwell Collins has certified EFIS-85 on over 50 types of aircraft.

Contractor

Rockwell Collins.

UPDATED

EFIS-86 Electronic Flight Instrument System

In 1983, Rockwell Collins introduced the EFIS-86 Advanced, an electronic flight instrument system with five 6 × 6 in (152 × 152 mm) or 5 × 6 in (127 × 152 mm) CRT displays for top of the range general aviation and commuter aircraft. It comprises dual EADIs, dual EHSIs and a single MFD. The two pilots each have an EADI and an EHSI and share the centrally mounted MFD display.

The CRTs combine raster and stroke writing. The EADI provides conventional ADI information, together

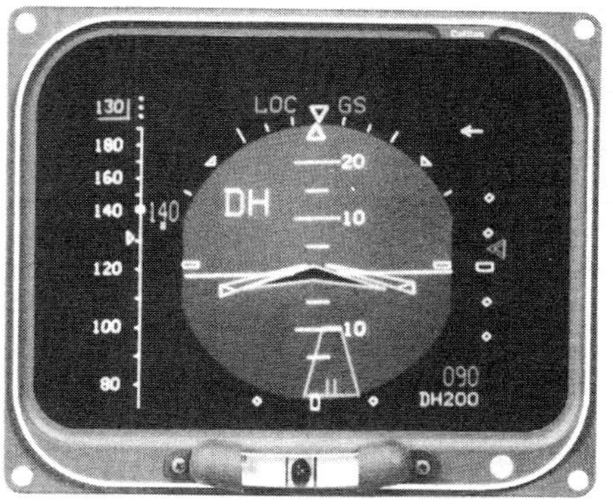

The Rockwell Collins 6 × 5 in EADI with primary airspeed display (above) and weather imagery (below)

with airspeed, airspeed trend and multisource vertical and lateral deviation information. The addition of air data information is facilitated by the increase in the display area of the EFIS-86 which ensures the display remains legible. The large EHSI makes it possible to consolidate all conventional HSI, navigation and weather information directly in front of the pilot, thereby reducing the need to scan other instruments. Three EHSI modes can be selected: full compass rose (as in a conventional HSI), expanded sector display or sector display with weather radar paints.

The system is linked with long-range navigation and flight management systems to show additional information such as an expanded display of HSI, radar and navigation data from the flight instruments either in combination or separately for more detailed examination. The display can store and show preselected lists of waypoints or up to 100 pages of preprogrammed data such as checklists and emergency procedures, written in chapter form for ease of input and retrieval. The display can also act as a standby flight instrument in the event of an EADI or EHSI CRT failure.

An EFIS-86C has been certified on the Canadair Challenger CL-601.

Operational status

In service. EFIS-86 Advanced has been certified in the Dassault Falcon 50 and 200, with airspeed indication on the ADI as the primary speed indication. The aircraft and their EFIS systems have been approved to Cat II operation, the first such approval for an EFIS-equipped business jet. New display symbology has been developed for two Dassault aircraft with speed scale and autopilot mode annunciation. Other systems are flying on the Embraer EMB-120, Gulfstream III, Saab 340 and the Astra Jet.

Contractor

Rockwell Collins.

UPDATED

EFIS-700 for the Boeing 737, 757 and 767

The EFIS-700 electronic flight instrument system for the Boeing 737, 757 and 767 comprises an EADI and an EHSI for each of the two pilots. Each pair of instruments has an associated mode control panel. The system provides all the functions associated with earlier electromechanical ADI and HSI flight director instruments, and in addition shows map and flight plan data, weather patterns, radio height, automatic flight control modes, autoland, system status, windshear and flight path information on 7 × 6 in EHSI (178 × 152 mm) and 5 × 6 in EADI (127 × 152 mm) CRTs.

These instruments utilise bright, three-gun, robust, shadow-mask CRT technology providing eight colours. The traditional red, blue and green associated with CRTs is augmented by magenta, yellow, green, cyan and white. When used in conjunction with a contrast enhancement filter, the high-resolution CRTs provide bright displays that are readable under all flight deck lighting conditions.

Each of the two pairs of EFIS instruments in an aircraft is driven by its own symbol generator, but a third standby generator is retained as a spare which can be switched in as necessary on failure of a dedicated unit. These symbol generators utilise the Rockwell Collins AAMP (Advanced Architecture MicroProcessor), implemented with the latest Very Large-Scale Integration (VLSI) and CMOS technology.

The EADI presents primary attitude information, together with pitch and roll steering commands. Secondary data is also shown such as groundspeed, autopilot and autothrottle mode. In order to keep the display uncluttered, information is switched out as soon as it is not needed, for example, instrument landing system and radio height symbols are absent during cruise, appearing only during the final approach.

The EHSI depicts the horizontal position of the aircraft in relation to selected flight data and a map of the navigation features in the vicinity of the aircraft at any given time. Aircraft track, trend vector information and desired flight plan are also displayed. This allows rapid and accurate manual or automatic flight path correction. Other information can be displayed, such as windspeed and direction, vertical deviation from a selected profile and time to the next navigation waypoint. Weather patterns can be superimposed on the navigation picture.

Operational status

In production. Installed in the Boeing 737, 757 and 767.

Contractor

Rockwell Collins.

UPDATED

EFIS-1000 for the Fokker 100

In 1984, Rockwell Collins was chosen to design a third-generation 6 × 7 in (152 × 178 mm) EFIS-1000 system for the Fokker 100. The system utilises six integrated, full-colour CRT displays; each pilot has a Primary Flight Display (PFD) mounted over a Navigation Display (ND) on the instrument panel. These are also two multifunction displays. EFIS control panels mounted on each side of the glareshield allow individual control of CRT brightness, mode of operation and weather radar display. The multifunction display system controller is mounted on the centre pedestal for use by both crew members. Remote light sensors behind the windscreen help to control the automatic brightness function so that the displays are readable under all flight deck lighting conditions.

Operational status

In service in the Fokker 100.

Contractor

Rockwell Collins.

VERIFIED

EHSI-74 Electronic Flight Instrument System (EFIS)

EHSI-74 is for the general aviation single- and light twin-turboprop market and comprises a 4 × 4 in (102 × 102 mm) EHSI designed to work in conjunction with the company's ADI-84 attitude director indicator and an information display/radar navigation centre, the Rockwell IND-270 CRT, which is used in conjunction with the company's WXR-270 weather radar. The system is completed by a DCP-270 display control panel.

The IND-270 CRT can present up to 128 pages of easily accessed pilot programmable text and navigation information from a Collins LRN-85 long-range navigation system. The equipment is programmed by a portable data reader that can load the system with performance tables such as cruise/consumption, emergency checklists and other alphanumeric information. Chapter by chapter indexing facilitates input and retrieval. In addition to the information stored within the DCP-270, many pages of data are available from the LRN-85.

The EHSI-74 no longer requires a US STC as do other EFIS in most installations.

Operational status

In production and in service. Certified on the Commander 690, King Air 200, Cessna 441, King Air F90, Bonanza, Beech 100, Mitsubishi MU-2 and dual installation for the Learjet Model 35A. Recently the system was installed in the Chinese Y-7.

Contractor

Rockwell Collins.

VERIFIED

Engine Indication and Crew Alerting System for the Boeing 757 and 767 - EICAS-700

The Rockwell Collins Engine Indication and Crew Alerting System, EICAS-700 is standard equipment on the 757 and 767 airliners. The EICAS system comprises two multicolour CRT displays, two computers and a single selector panel. The display unit is identical to the EHSIs on the pilot's display panels, though rotated

Layout of the flight panels in the Boeing 757 and 767. Each pilot has a set of two EFIS displays mounted one above the other. The two EICAS displays on the centre panel can be monitored by either pilot

through 90° in its function as an engine indicating system.

Each aircraft has two EICAS CRTs mounted one above the other on the centre instrument panel where they can be monitored by the two pilots. On the centre panel the top EICAS CRT is programmed to display primary engine information such as engine pressure ratio, fan speed and exhaust gas temperature as electronic symbols representing traditional circular scale and pointer instruments, together with cautionary information, for example, wheel-well overheat or failure of a yaw damper in the flight control system. The lower EICAS display shows lower priority information such as compressor speed, fuel flow and oil temperatures, pressures and tank contents.

In the case of the failure of one EICAS display, priority information automatically switches to the other CRT, and the dual-redundant computer installation permits both CRTs to be driven from one unit.

The multicolour CRT displays in the EICAS configuration measure 7 × 6 in (178 × 152 mm) and are driven by one of the two computers, the other acting as a hot spare.

Operational status

In service in Boeing 757 and 767 aircraft.

Contractor

Rockwell Collins.

UPDATED

FCC-105-1 automatic flight control system

Based on the FCC-105 system, the FCC-105-1 automatic flight control system is applicable to the Sikorsky S-76 and H-76 helicopters and has been in production since 1977. It offers a wide range of features up to full three-axis stability augmentation with turn co-ordination, pitch, roll and yaw attitude and altitude/airspeed hold modes. Each system amplifier has independent pitch, roll and yaw channels contained in individual modules which can be selected separately. This building block principle permits customers to select as many features as necessary to meet their particular requirements.

The system comprises a stability augmentation system amplifier, yaw switch, heading hold amplifier, airspeed switches, control panels, linear electromechanical actuators, indicator panel, rate gyros, cyclic switches and airspeed hold amplifier.

Specifications

Dimensions
(main processor) 190 × 318 × 113 mm
Weight
(main processor) 3.4 kg
Power: 35 W

Operational status

In production for the Sikorsky S-76 and H-76 helicopters.

Contractor

Rockwell Collins.

VERIFIED

FCS-110 flight control system for the Lockheed Martin L-1011

The FCS-110 analogue flight control system, a combined project involving both Rockwell Collins and BAE Systems North America, has been installed in the majority of Lockheed Martin L-1011 TriStar airliners. A comprehensive set of autopilot modes, including Cat IIIB automatic landing capability, is incorporated. It is installed in conjunction with the FD-100 flight director system. All work on the autoland system was done by Lear Astronics, which was also responsible for other elements of the system. In conjunction with a Hamilton Standard flight management system, the FCS-110 permits flight path optimisation. Two-dimensional optimisation was certified in 1972 and three-dimensional in 1977.

Operational status

In service in the Lockheed Martin L-1011.

Contractors

Rockwell Collins.
BAE Systems North America, Aircraft Controls.

UPDATED

FCS-240 flight control system for the Lockheed Martin L-1011-500

The FCS-240 advanced digital flight control system was designed to replace the FCS-110 fitted to Lockheed Martin L-1011 aircraft. It is a dual/dual integrated autopilot/flight director system and was initially tailored to the Lockheed Martin TriStar 500 wide-body transport. Compared to its analogue predecessor the number of computing units has diminished from seven to two and benefits are weight savings, improved reliability and lower cost of ownership. Cat IIIB automatic landing capability is included with a comprehensive set of conventional autopilot modes.

Operational status

In service on the Lockheed Martin L-1011-500.

Contractor

Rockwell Collins.

VERIFIED

FCS-700 flight control system for the 767/757

The fully digital FCS-700 triplex autopilot and flight director system, with fail-operational automatic landing capability, was designed for Boeing 767 and 757 transports.

Development sprung from the FCS-111X experimental flight control system which Rockwell Collins and Boeing evaluated during the late 1970s. The fully digital architecture permits integration of comprehensive self-test and failure protection monitoring.

Autopilot modes include control wheel steering, automatic cruise hold/select modes for heading, altitude, vertical speed and airspeed/Mach number, approach modes with back course capability and automatic approach and landing. The latter facility includes automatic flare control, roll-out guidance and coupled go-around. Computed flight director steering is available in non-coupled flight, including take-off. Turn co-ordination and dutch-roll damping is provided. Dual or triplex control computer configurations can be used.

Steering commands from a navigation computer, using a wide variety of navigation sensors, will permit completely automatic control of preplanned vertical and lateral flight profiles. Large-scale integrated circuit technology is used in ARINC 429 bus interface drives and digital multiplexers and in several areas of each flight control computer, which is based on the Rockwell Collins CAP-6 processor configuration. Computer operating speed is 300 kops and high-order language programming is used throughout the system.

Maintainability improvements are claimed from the MCDP-701 maintenance central display panel which uses a microprocessor to perform control, display and data management tasks. In-flight system failures are indicated by the maintenance control display panel and fault data is stored for up to 10 flights.

Operational status

In production for Boeing 757 and 767 aircraft.

Contractor

Rockwell Collins.

VERIFIED

Computers and glareshield controller for the Rockwell Collins FCS-700 automatic flight control system used in the Boeing 757 and 767

FCS-700A autopilot/flight director system for the 747-400

The FCS-700A is an enhanced derivative of the autopilot on the Boeing 757 and 767 and is designed to provide LRU interchangeability between those two aircraft types and the 747-400. It is a Cat IIIB fail-operative system tailored to the control requirements of the 747-400 and is usable within the complete flight envelope for autopilot and autoland functions including automatic landings with a decision height of zero feet and a runway visual range of 600 m.

The FCS-700A consists of three flight control computers and a mode control panel. Each computer has dual Rockwell Collins Adaptive Processor System (APS) microprocessors for increasing processor capacity to allow for future growth in autopilot capability for MLS.

The system replaces 14 dedicated LRUs on previous 747 aircraft, saving considerable space and weight.

Operational status

In service in Boeing 747-400 aircraft.

Contractor

Rockwell Collins.

VERIFIED

FD-108/109 flight director system

Essentially similar systems, the FD-108 uses 4 in (102 mm) attitude director and horizontal situation indicator instruments and the FD-109 uses 5 in (127 mm) instruments. Two categories of each are available, suffixed Y and Z, the latter incorporating air data and additional features. Either flight director system can be integrated with many types of modern autopilots and both versions are used on current production airliners.

The attitude director indicator has a V-bar display and a flat tape background which provides linear pitch information through ±90°. Outputs are provided for annunciations of all submodes and approach progress indications. A conventional compass rose horizontal situation indicator is provided with an electronic digital display of distance to go and a mechanical course readout.

Operational status

In service. No longer in production.

Contractor

Rockwell Collins.

VERIFIED

FD-110 flight director system

The FD-110 system uses 5 in (127 mm) attitude director and horizontal situation indicator instruments. Several configurations are available, making them suitable for individual airliner, company and autopilot combinations. Configurations can use cross-pointer or V-bar attitude director indications, various ARINC system interface compatibilities or area navigation and inertial navigation system integration. The equipment is used in many current airliner types.

Operational status

In service.

Contractor

Rockwell Collins.

VERIFIED

FDS-84 Pro Line flight director system

Featuring separate 4 in (102 mm) attitude director and horizontal situation indicator displays, the FDS-84 Pro Line system is compatible with Rockwell Collins FCS-80 and FCS-105 autopilot/flight control systems. It is a comprehensive system suitable for high-performance business and executive aircraft and commuter airliners.

The attitude director indicator uses a flat tape attitude display background and has V-bar steering command symbology. A radio altimeter readout is optional in this unit. The horizontal situation indicator uses a compass rose presentation with an electronic readout which can show distance, time to go or groundspeed information. The equipment is fully compatible with ARINC standard VOR, DME, INS and Omega/VLF sensors.

Operational status

In service. No longer in production.

Contractor

Rockwell Collins.

VERIFIED

FDS-85 Pro Line flight director system

Based on the 5 in (127 mm) ADI-85 attitude director indicator and HSI-85 horizontal situation indicator instruments, and compatible with the Rockwell Collins APS-80 autopilot, the FDS-85 Pro Line system is suitable for high-performance business aircraft and commercial airliners. It is the company's most comprehensive mechanical flight director system. Flat tape attitude indication, V-bar steering commands and electronic readout of distance, time to go, speed or elapsed time are standard. A separate HCP-86 heading/course control panel is used.

Specifications

Dimensions:
(ADI-85) 131 × 131 × 206 mm
(HSI-85) 131 × 112 × 229 mm
(HCP-86) 146 × 38 × 152 mm
Weight:
(ADI-85) 3.2 kg
(HSI-85) 3.3 kg
(HCP-86) 0.28 kg
Power supply:
26 V AC, 400 Hz, 44 VA
28 V DC, 260 mA
5 V AC/DC, 10 W for lighting
Temperature range: –20 to +70°C
Altitude: up to 35,000 ft
Cooling: convection

Operational status

In service. No longer in production.

Contractor

Rockwell Collins.

VERIFIED

FDS-255 Flight Display System

The Rockwell Collins Flight Display System (FDS-255) is a 5ATI colour flat-panel display offering an extremely wide viewing angle. The FDS-255 is a key element of the Rockwell Collins Flight2 Systems. It can replace existing instruments, combine functions of existing instruments, or provide the entire display system in a new or retrofit application. The active matrix liquid crystal design and proven Rockwell display circuitry give the user a highly reliable and easily maintainable flight instrument. FDS-255 is programmed in Ada. User specific display formats can be developed and downloaded over the ARINC 429 bus.

Featuring a 'smart head' design, the FDS-255 contains internal graphics generation, analogue and digital input/output and weather radar. In a dual installation, reversionary colour, full-motion video is provided with a video card slot.

The FDS-255 video with 64 grey scales will display FLIR, digital map and TCAS; colour, full-motion video is provided via an optional video card. The wide viewing angle permits easy cross-cockpit viewing.

Because of the highly integrated display design, Rockwell is able to provide 17.7 sq in of usable viewing area in a 5 ATI format. The FDS-255 has eight standard modes, but can easily be tailored to each customer's specific application.

Flight/navigation operating modes include: ADI, HSI, PFD, ARC, radar-map, windshear, map and hover. TSO versions are also available.

FDS-255 Flight Display System

Specifications

Dimensions: 129 × 129 × 222 mm
Weight: 3.8 kg (max)
Power supply: 28 V DC, 50 W typical
(See also Flight2 Systems entry)

Operational status

In production for the US Air Force C/KC-135 Pacer Crag and US Navy P-3. Deliveries began in 1996. Also in production for C-12, C-130, KC-10, MH-60, P-3, VC-10 and various helicopter and commercial aircraft applications.

Contractor

Rockwell Collins.

VERIFIED

FDS-2000 Flight Display System

Rockwell Collins' FDS-2000 Flight Display System provides the aircraft interface, data processing, display processing and display control to replace existing cockpit electromechanical indicators. The FDS-2000 replaces both the Attitude Director Indicator (ADI) and the Horizontal Situation Indicator (HSI), with control provided by the Display Select Panels (DSP). The FDS-2000 provides flight crews with primary flight display instrumentation incorporating attitude, heading, flight director guidance and other features available depending on the capabilities of installed avionics.

The new 5 in LCDs offer substantial cost savings over comparable new electromechanical flight instruments. Collins claims that, when compared with predecessor CRT flight instruments, the LCD requires up to 50 per cent less power, weight and space, while doubling the graphics capabilities with displays that are sharper, brighter and featuring greater readability under direct sunlight.

The main features of FDS-2000 are as follows:

- Maximum active display surface available in 5ATI format
- Fits into standard 5 in electromechanical installations

Rockwell Collins' FDS-2000 AMLCD is claimed to utilise 50 per cent less power, weight and space than conventional aircraft instrumentation **2001**/0105290

- Full complement of flight instrument software provides ADI, HSI and PFD (combined ADI/HSI formats)
- High-resolution active matrix LCD
- Designed for optimum flight deck viewing
- Capable of displaying EGPWS data
- Provides a platform for future digital ARINC 429 I/O growth.

Operational status
In production.

Contractor
Rockwell Collins.

NEW ENTRY

FIS-70 Pro Line flight instrumentation system

Compatible with the Rockwell Collins APS-80 or AP-106A autopilots, the FIS-70 Pro Line comprises the 4 in (102 mm) ADI-70 attitude director indicator and the HSI-70 horizontal situation indicator instruments and is suitable for high-performance turboprop and business jet aircraft. It is essentially a low-cost version of the FDS-84 system and excludes the digital distance/course readouts and radio altimeter display options. All other FDS-84 features are incorporated.

Specifications
Dimensions:
(ADI-70) 106 × 106 × 210 mm
(HSI-70) 106 × 106 × 229 mm
Weight:
(ADI-70) 2.1 kg
(HSI-70) 2.2 kg
Power supply:
26 V AC, 400 Hz, 1.04 A
28 V DC, 0.2 A
28 V DC, 0.34 A for lighting.
Temperature range: –15 to +70°C
Altitude:
(ADI-70) -1,000 to 50,000 ft
(HSI-70) up to 35,000 ft
TSOs:
(ADI-70) C3b, C4c, C52a
(HSI-70) C6c, C52a

Operational status
In service. No longer in production.

Contractor
Rockwell Collins.

VERIFIED

FMR-200X multimode weather radar

The Rockwell Collins FMR-200X Flight Multimode Weather Radar System is a standard Non-Developmental Item/Commercial Off-The-Shelf (NDI/COTS) ARINC 708A X-band coherent Doppler colour weather radar system. This system provides full precipitation detection, turbulence detection, forward-looking windshear detection and an additional active skin paint mode, capable of detecting tanker-size aircraft at ranges up to 15 nm.

New multimode weather radar software has been designed and certified to RTCA DO-178B level D standards. The system has been qualified to meet environmental standards described in RTCA DO-160C (with a few exceptions). MIL-STD-461D is used for radiated emissions, antenna spurious and harmonics. All operations other than the skin paint mode, sixteen-level mapping and minor display and control bus modifications are defined in accordance with ARINC 708A (airborne weather radar with forward-looking windshear detection capability). This system is based on the Collins WXR-700X Radar System that has over 6,600 installations worldwide.

The WXR-700X originally entered service in 1980 and was the first Air Transport airborne radar to incorporate a Doppler turbulence detection feature. In 1986 Rockwell Collins began a co-operative programme with NASA to expand the system's capability to detect windshear/microburst events ahead of the host aircraft.

Specifications
WRT-701X
Receiver/transmitter

Size	8 MCU per ARINC 600
Weight	14.1 kg (maximum) including R/T mount
Cooling	Mount based fan
Power	200 W nominal at 115 V ac 400 Hz

WCP-701
Mode control panel

Size	10.3 H x 22.6 W x 23.6 D mm
Weight	0.77 kg (maximum)
Cooling	Natural convection
Power	5 W (maximum) (powered from R/T)

WMA-701X
Antenna pedestal

Size	Per ARINC 708
Weight	12.7 kg (maximum) (including flatplate)
Cooling	Natural convection
Power	100 W nominal at 115 V ac 400 Hz

WFA-701X
Flatplate antenna

Size	110 x 134 mm
Weight	3.1 kg (maximum)
Beamwidth	2.5° elevation x 3.5° azimuth
Sidelobe Performance	>30 dB

Operational status
The FMR-200X is being installed on USAF KC-135 aircraft as part of the 'Lightning Bolt' procurement streamlining initiative.

Contractor
Rockwell Collins.

NEW ENTRY

Dual FMS-4100 flight management systems in the Canadair Regional Jet aircraft

FMS-4100 Flight Management System

The FMS-4100 Flight Management System (FMS) is designed for the new generation of regional airline aircraft and is available to operators of Canadair and Saab aircraft.

The FMS-4100 is a fully integrated part of the Rockwell Collins Pro Line 4 avionics system standard on the Canadair Regional Jet and Saab 2000. FMS-4100 computers are packaged as LRUs housed in the system's Integrated Avionics Processing System (IAPS). Pro Line integration of the FMS minimises weight and volume requirements and enhances diagnostics capability. The full-sized control/display unit provides a large, seven-colour, sunlight-readable CRT with line select keys and a full alphanumeric keyboard.

The FMS-4100 offers crews standardised cockpit operation, with an ARINC CDU with scratchpad data entry. Preflight and general flight planning are available: a primary flight plan is used for active guidance and a secondary flight plan can be stored as an alternative and activated if required. Up to 1,000 standard routes, each with up to 100 waypoints, can be stored in the system database. This contains navaids, waypoints, NDBs, airports, airport reference points and runway thresholds. The FMS-4100 also includes SIDs, STARs, parallel offsets, high- and low-altitude airways and holding patterns.

Navigation information is shown on Pro Line 4 MultiFunction Displays (MFDs). Present position and planning maps are displayed on the MFD, in addition to text pages for progress, nav status, position summary, VOR/DME status and Loran status. The FMS-4100 also provides complete radio tuning capability and is both AFIS- and ACARS-compatible.

Operational status
In service.

Contractor
Rockwell Collins.

UPDATED

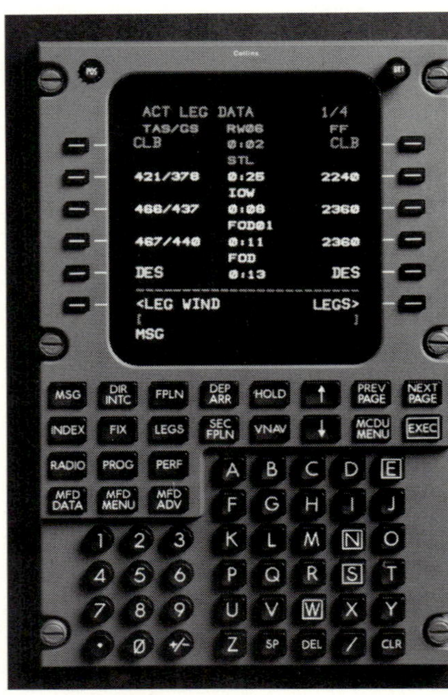

CDU for FMS 4100/4200 **2001**/0105286

FMS-4200 Flight Management System

Significant software and hardware upgrades to the Rockwell Collins FMS-4100 Flight Management System have created the new FMS-4200 Flight Management System.

Non-precision GPS approach capability, coupled or advisory VNav guidance, thrust and fuel management and ACARS/AFIS™ compatibility are among the in-flight features of the new system.

The navigational database for the new FMS-4200 has been expanded from 4 Mbytes to 12 Mbytes to facilitate international corporate operations that rely on the memory-intensive demands of the WorldWide Navigational Data Base (WWNDB).

Through integration with the GPS-4000 Global Positioning System sensor, the new FMS-4200 calculates navigational solutions based on GPS signals from all satellites in view, making GPS-based RNAV non-precision approaches possible worldwide.

New technologies developed for the FMS-4200 include the ability to provide full-profile coupled or advisory vertical navigation guidance for climb, cruise and descent. The advisory VNav system displays situational guidance - vertical speed, required vertical speed and vertical deviation - eyes-up on a multifunction display. Vertical information - top of climb, top of descent, vertical constraints and descent rates - is displayed on the control display unit and the multifunction display. During climb or descent, advisory vertical cues are presented on the primary flight display, including glidescope, vertical speed target and next vertical constraint.

Contractor

Rockwell Collins.

VERIFIED

FMS-6000 series Flight Management Systems

FMS-6000

The FMS-6000 provides integrated multisensor navigation, flight plan modification and execution, sensor control, multifunction display map support and steering/pitch commands to the flight control system. When integrated with the Rockwell Collins GPS-4000 navigation system, it forms the AVSAT 6000 system, which in turn can be integrated with the Pro Line 4 avionics system to create a complete avionics package. The FMS-6000 features a cathode ray tube control display unit.

FMS-6100

The FMS-6100 features a liquid crystal display control display unit. It is optimised to fulfil the navigation requirements of Falcon 2000, 50 EX, 50 and 20 aircraft. Integrated with the GPS-4000 Global Positioning System sensor, the FMS-6100 provides data for SIDs, STARs and multisensor RNAV and VOR approaches and crews can automatically execute non-precision GPS approaches and GPS overlay approaches. The system also provides automatic FMS-to-ILS transfers to reduce pilot workload in terminal areas.

Operational status

Certified aboard the Canadair Challenger 604, Falcon 20 and Falcon 2000.

Contractor

Rockwell Collins.

VERIFIED

FMS navigation systems

The Rockwell Collins FMS is a family of satellite-based precision navigation systems designed for all phases of flight, including take-off, en route, approach and landing. It is grouped into series, to match the mission of the aircraft and caters for long-range corporate jets, medium to light jets and turboprops and 30- to 100-seat regional airline passenger aircraft.

At the core of each FMS series is a Rockwell Collins GPS sensor. The capabilities of a flight management system are incorporated within the navigation system.

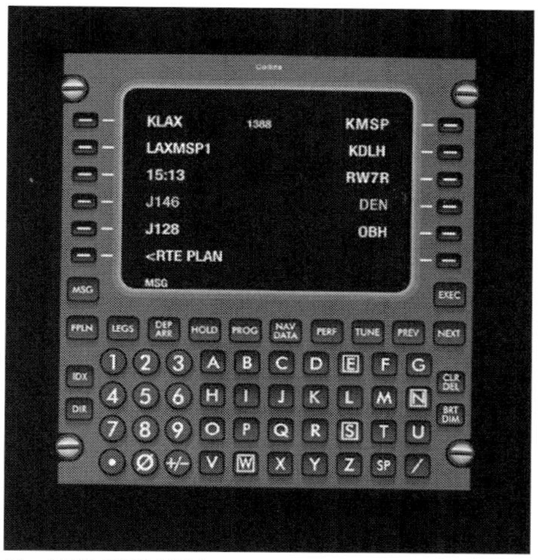

The FMS 3000 CDU

The avionics GPS engine features 12-channel satellite tracking, including those comprising the Wide Area Augmentation System, providing satellite coverage sufficient to meet the FAA-required navigation performance and precision approach criteria; a new software code to ensure certification to DO-178B Level A; Receiver Autonomous Integrity Monitoring (RAIM) and predictive RAIM, and differential correction computation to qualify the receiver for Cats I and II.

The FMS series provides the capability to calculate position and navigation information accurately anywhere in the world; provide VNav guidance and comprehensive pilot annunciations; fly complete airways, SIDs, STARs and approach legs, and predict fuel consumption. FMS also significantly reduces pilot workload by automating many calculations and functions. Information is presented on multifunction display pages, allowing easy, efficient movement through the system. Pilots can remain eyes-up during all phases of flight, resulting in enhanced situational awareness and greater control. Flight planning is simplified to the point where even the most complicated procedures can be generated with a few keystrokes. FMS is a component system of the technology required for the planned 'free flight' regime in which aircraft en route separations will be reduced.

FMS 3000

The FMS 3000 is designed for light jet and turboprop aircraft. It provides the capability for en route, terminal and non-precision approach and lateral/vertical navigation, and includes the growth potential to support precision approach computations.

The 3000 series installation consists of the CDU-3000 Control/Display Unit, FMC-3000 Flight Management Computer (FMC) and FMS GPS sensor. Updated navigation information may be uploaded in the field and 100 flight plans, with 100 waypoints per plan, can be stored in the navigation database.

FMS 3000 also provides access to ground-based messaging, weather data and flight plan transmission through the Airborne Flight Information System (AFIS). Free form messages can be sent to and from the CDU. Frequently used messages can be formatted and stored.

FMS 4000

The FMS 4000 meets the needs of regional airline aircraft, providing worldwide GPS navigation, simplifying cockpit management and enhancing flight crew and aircraft performance. In addition to GPS, long-range navigational inputs such as IRS/AHRS, VLF/Omega, VOR and DME can be used to provide accurate determination of aircraft position. Coupled with flight management technology, the system provides the integrity, flexibility and operational capabilities required by regional airline operators.

FMS 4000 provides the capability for preflight and general flight planning. Complete lateral and vertical navigation is available. Radio sensor management is also available. Two flight plans are available: the primary flight plan is used for active guidance, while a secondary flight plan can be stored as an alternative and activated if needed. Full time and fuel management performance are provided, with fuel prediction available. Up to 1,000 standard company routes, each with up to 100 waypoints, can be stored in the system's database. The database contains navaids, waypoints, non-directional beacons, airports, airport reference points and runway thresholds. Flight planning includes the capability to execute SIDs, STARs, airways and holding patterns.

Once a flight plan is activated, it is presented graphically on a MultiFunction Display (MFD). The MFD also displays navaids, intersections, airports, terminal waypoints and non-directional beacons. In addition, the MFD can present complete operational status, progress and flight plan summary text.

FMS 4000 comprises three LRUs: the CDU-4100 Control/Display Unit, FMC-4200 Flight Management Computer (FMC) and GPS-4000 GPS sensor. The FMC-4200 is packaged as a line-replaceable module, housed in the Collins Integrated Avionics Processing System (IAPS). Pro Line 4 integration allows the FMS 4000 to perform self-diagnostic functions and informs the Pro Line 4 maintenance diagnostic computer of internal faults.

The FMS 4000 has been certified for use on the Saab 2000 aircraft.

FMS 5000

Designed specifically around the needs of medium to light jet and turboprop aircraft, the FMS 5000 centralises control of EFIS, navigation, weather radar, TCAS and radio management in one conventional location to maximise cockpit space. It includes features such as the ability to calculate accurately position and navigation information anywhere in the world, provide VNav guidance and comprehensive pilot annunciations, fly complete airways, SIDs, STARs and approach legs, predict fuel consumption, and offer the right combination of performance, efficiency and ease of use required to perform in complex and demanding operating environments.

FMS 5000 combines GPS technology with flight management capability to provide satellite-based precision navigation systems for take-off, en route, terminal area, approach and landing operations. Other long-range navigational inputs include IRS/AHRS, VLF/Omega, VOR and DME. These may be used to provide the most accurate determination of aircraft position. Vertical navigation is available in each phase of flight. Included is the ability to execute a vertical direct-to, generating a vertical path from the aircraft's current position to a designated altitude.

FMS 5000 also provides access to ground-based messaging, weather data and flight plan transmission through the Airborne Flight Information System (AFIS). Free-form messages can be sent to and from the CDU. Frequently used messages can be formatted and stored.

FMS 5000 comprises three LRUs: the CDU-5000 Control/Display Unit, FMC-5000 Flight Management Computer (FMC) and AVSAT GPS sensor. The FMC is packaged as a line-replaceable module, housed in the Collins Integrated Avionics Processing System (IAPS). Pro Line 4 integration allows the FMS 5000 to perform self-diagnostic functions and informs the Pro Line 4 maintenance diagnostic computer of internal faults.

FMS 6000

The FMS 6000 is designed for long-range business aircraft. It features the ability to calculate position and navigation information accurately anywhere in the world, provide VNav guidance and comprehensive pilot annunciations, fly complete airways, SIDs, STARs and approach legs and manage fuel consumption.

FMS 6000 comprises three LRUs: the CDU-6000 or CDU-6100 Control/Display Unit, FMC-6000 Flight Management Computer and GPS-4000 sensor. It can be configured for single, dual or triple operation by installing the desired number of FMC modules and the same number of control/display units. It is designed to be integrated with the Pro Line 4 avionics system.

FMS 6000 is designed for worldwide navigation, including polar navigation. Long-range inputs in addition to GPS can be used, including IRS/AHRS, VLF/Omega, VOR and DME to provide the most accurate determination of position. It contains a global database of current information used for both flight planning and navigation.

Global Navigation Landing Unit (GNLU)

To meet the requirements of the Air Traffic Management (ATM) system, Rockwell Collins is offering the GNLU system (see separate entry on GLU/GNLU multimode receivers). The system defined by GNLU is of a single navigation unit integrating all the

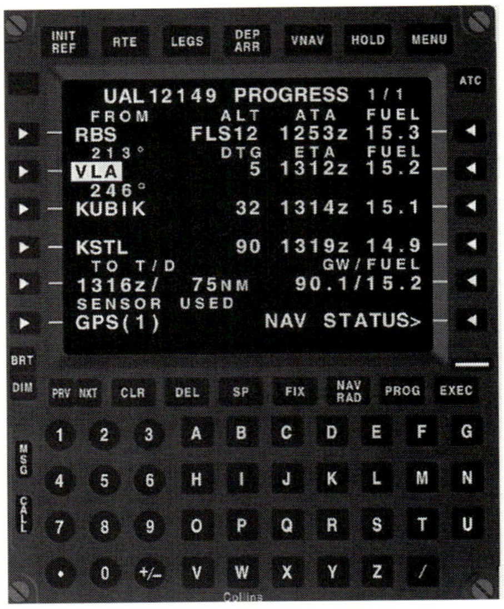

GNLU multifunction control/display unit 0015375

GNLU-900 GNSS 0015376

GNSS-based (Global Navigation Satellite Systems) en route, terminal and landing system capabilities required to operate in a global Communications Navigation Surveillance/Air Traffic Management (CNS/ATM) environment.

The GNLU contains an advanced GPS receiver (FAA order 8110.60 compliant) with the required integrity to support precision GNSS-based Landing System (GLS) approach certification to Cat I and II levels.

Because the role of the GPS sensor is so important, Rockwell Collins has designed and developed a specialised GPS engine for use in air transport system applications. Optional MLS modules, developed in conjunction with DaimlerChrysler Aerospace, and optional ILS modules can be integrated within the GNLU to provide classic aircraft with analogue interfaces. This is an efficient update path for compliance with new landing sensor requirements, such as FM immunity.

As part of the total Rockwell Collins FMS package, the GNLU adds substantially more than GPS receiver inputs to existing flight management capabilities; GNLU is the primary processing and control unit of the GPS/FMS system. The FMS supports complete lateral navigation and provides vertical guidance on approach.

The complete system comprises the cockpit-mounted Multifunction Control/Display Unit (MCDU), the remote 4-MCU sized GNLU, and a GPS antenna.

The GNLU is structured to add the interfaces and processing needed for precision GNSS approaches, using an integral uplink data receiver to simplify installation. The GNLU navigation computer has the volume and processing power to support the Wide Area Augmentation System (WAAS) and Local Area Differential GNSS (LADGNSS) receiver and modem capability; all essential WAAS hardware is provided in the baseline configuration. The system also provides ACARS and ARINC 622 datalink interfaces and is provisioned to interface with ATN compliant architectures.

Specifications (GNLU-900 GNSS)

GPS performance:
(channels) 12-channel, all-in-view tracking
(frequency) 1,575.42 MHz (L1) C/A code
Accuracy:
(horizontal) 23 m (4.5 m with DGPS); 1.5 m/s (velocity)
(altitude) 30 m (6.0 m with DGPS); 1.2 m/s (velocity)
Dimensions: 4 MCU per ARINC 600
Weight: 7.3 kg
Power: 28 V DC

Contractor

Rockwell Collins.

VERIFIED

GLU/GNLU-900 series MultiMode Receivers (MMR)

The Rockwell Collins 900 series MMR provides two or more landing system standards. Both the ILS and GNSS functions are basic to the landing receiver, while the MLS and GLS functions are categorised as options. The GLU-900 series is designed for use in digital aircraft, while the GNLU-900 series is designed for 'classic' analogue types. Retrofit with the GNLU-900 series MMR permits 'classic' aircraft to operate in accordance with European BRNAV requirements.

The MLS design provides ILS lookalike interfaces to the existing aircraft autopilot and display systems and accommodates requirements for both dual-dual and triplex auto-land architectures. It also offers the accuracy, reliability and integrity required for critical performance in Cat III landings.

The MMR supports existing all-weather landing capability (GLU (ILS) and the GNLU (ILS/VOR)) and capabilities can be expanded as the industry moves toward the future GLS.

The pilot interface to the system is mechanised through a multipurpose control display unit which uses standard ARINC operational philosophy. This allows interoperability training benefits to mixed fleet operators, and future integration of the datalink control function.

The GPS flight management processor and navigation database software can be loaded on the flight line, simplifying system growth.

Specifications

GNLU-900
Dimensions: 4 MCU per ARINC 600
Weight: 7.3 kg approx
Power: 29 V DC
TSO: C129 B1
Qualification: RTCA DO-160 C, DO-178 B

GLU-900
Dimensions: 3 MCU
Weight: 3.86 kg
Power: 115 V AC, 400 Hz
TSO: -C36e, -C34e, (A2D2 YBA (BCL) EIXXXXXZEAEZYZLXX
General: DO-192; DO-195; DO-160C; FCC part 15, EUROCAE ED-46; ED-47A

GPS performance
Channels: 12-channel, all-in-view tracking
Frequency: 1,575.42 MHz (L1) C/A code transmissions
Sensitivity:
(acquisition) −121 dBm
(tracking) −125 dBm
Accuracy (95%):
(horizontal position) 23 m (4.5 DGPS)
(altitude) 30 m (6.0 m DGPS)
(horizontal velocity) 1.5 m/s
(vertical velocity) 1.2 m/s
(time) 100 ns
Time to fix first (95%): 75 s max with valid initialisation; 10 min, max without valid initialisation

MLS performance
Channels: 200, per ICAO Annex 10
Antenna connections: 3 (2 passive, 1 active)
Datalink frequency (optional): C-band

ILS performance
Frequency range: LOC 108.10 to 11.95 MHz; GS 329.15 to 335.0 MHz
Channel spacing: LOC 50 kHz; GS 150 kHz
Navigation outputs: ARINC 429

Operational status

The GLU-900 series MMR has been certified on a wide range of Airbus and Boeing late-model aircraft.

The GNLU-900 series MMR has been certified for a wide range of retrofit applications including Boeing-737, -747, -757, -767, -777 aircraft and Airbus Industrie A330/A340 and Avro RJ jetliners.

A derivative variant, the GNLU-910 GPS/FMS MMR, has been certified by the European Joint Aviation Authority for use on A300 aircraft, together with the FPI-955 LCD flat panel display as part of a system upgrade to meet European Basic Area Navigation (BRNAV) route requirements, and to provide Global Positioning System (GPS)-based landing approaches. The GNLU-910 GPS/FMS MMR can be integrated with existing analogue sensors.

The GNLU-920 MMR, which will have VHF omnidirectional range and ILS capabilities, is being fitted, as a double GNLU-920 unit fit, into nine US Air Force RC-135 aircraft to meet Global Access Navigation Safety/Global Air Traffic Management (GANS/GATM) requirements, and to comply with ICAO Annex 10 requirements for FM immunity. Aircraft fit was due to begin in late 1999 and to continue throughout 2000.

Yet another variant, the GNLU-945, forms part of the US Air Force's C-5 and KC-135 Global Air Traffic Management (GATM) programme. Each C-5 aircraft will be equipped with two GNLU-945 MMRs, and each MMR will include the following capabilities: VOR, ILS, and MLS providing for Cat III operations. Deliveries are scheduled to begin in 2002 and continue until 2004.

Many thousands of units have been delivered and are on order.

Contractor

Rockwell Collins.

UPDATED

GPS-4000/4000A

The Collins GPS-4000 enables aircraft equipped with flight management capability to perform GPS-based en route, terminal area and non-precision approach navigation, as well as primary means of oceanic/remote operations.

The GPS-4000 is part of the continued development of the Collins AVSAT satellite-based communication and navigation system, providing operators with the advanced capabilities required to operate within the evolving CNS/ATM environment.

The 2 MCU GPS-4000 sensor processes the transmissions of up to 12 GPS satellites simultaneously, calculating navigation solutions based on information from all satellites in view. A minimum of four satellites with acceptable geometry, or three satellites plus calibrated barometric altitude, are required to calculate navigation solutions.

The heart of the GPS-4000 is the Collins GPS Engine. The self-contained 12-channel GPS receiver outputs Earth-Centred Earth-Fixed (ECEF) position, velocity and GPS time once a second. This information is used by the flight management system to calculate a flight plan-based navigation solution. In addition, Predictive Receiver Autonomous Integrity Monitoring (PRAIM) allows crews to determine whether the satellite geometry at the destination airport will support approach at the planned time of arrival.

The GPS-4000 is designed to support sole means of navigation for en route, terminal and non-precision approach in today's Air Traffic Control environment. It is also capable of being upgraded via service bulletins to provide precision approach capability as the infrastructure evolves, accommodating use of the Wide Area Augmentation System (WAAS) and Local Area Augmentation System (LAAS) in support of precision approach operations.

Operational status

The Rockwell Collins GPS-4000 Global Positioning System sensor has been certified for use on the Beechjet 400A, Canadair RJ, Learjet 60 and Saab 2000.

Contractor

Rockwell Collins.

VERIFIED

HF-121A/B/C HF radios

The HF-121/A/B/C family of HF radios have been installed in a wide variety of commercial and military aircraft in the United States and in foreign countries. Earlier versions of the HF-121 product line have received military nomenclature designation of AN/ARC-153, -157, -191(V), -207(V), AN/URC-91 and -97(V) by the United States and AN/ARC-512 for other nations.

The HF-121 family can transmit and receive both data and voice signals, providing 280,000 channels covering the full HF-band from 2 to 30 MHz and operating in USB, LSB, ISB and AME modes. Transmitted power output level is selectable at either 100, 500 or 1,000 W.

The latest member of the family is the HF-121C (JETDS designation AN/ARC-230 – see separate entry), a high-performance radio system designed for HF applications requiring voice and data operations, including 400 W PEP/Average transmit power. Compatible with the requirements of MIL-STD-188-203-IA for Link 11/TADIL A, the HF-121C has been optimised for tactical digital data communications and simultaneous operation (SIMOP) of multiple radio sets with minimum frequency and antenna separation. Embedded Automatic Link Establishment (ALE), MIL-STD-188-110A Modem and ARINC 714-6 SELCAL capabilities are also available.

The basic radio set is partitioned into a receiver-exciter with integral pre/post-selector on a mounting shelf and a power amplifier-power supply on a mounting shelf. The receiver-exciter can be used independently of the power amplifier. Serial control via RS-232 or MIL-STD-1553B is available.

Description

The HF-121C HF radio set is composed of the 671Z-3() receiver/exciter, 549E-1 Power Amplifier, 913Z-5 control and the applicable antenna coupler:

- **671Z-3() receiver/exciter**
  The receiver/exciter is composed of a DSP based receiver/exciter-modem, advanced microprocessor control and an agile digitally tuned bandpass filter.
- **1549E-1 power amplifier**
  The power amplifier is a solid-state design that provides a power output of 400 W PEP/Average.
- **913Z-5 control**
  The control provides remote control and monitor of the 671Z-3() R/E, the 549E-1 PA and associated coupler using an RS-232 serial interface. In addition to the basic control functions, this control will also provide the control and monitor functions for Automatic Link Establishment (ALE).
- **Antenna coupler**
  The HF-121C HF radio set will interface with any of the standard product line AN/ARC-190 (V) antenna couplers; CU-2275(V)/ARC-190(V) and CU-2314(V). Additional AN/ARC-190 antenna couplers are available for unique platform requirements.

System applications

The HF-121C radio set is compatible with standard Air Transport Rack (ATR) unit dimensioning. Each unit mounts in an equipment tray for ease of removal and maintenance.

The antenna coupler is a pressurised unit, which may be mounted in pressurised or unpressurised areas of the aircraft.

The transmitter is capable of continuous transmission in all modes at ambient temperatures up to +55°C at sea level. Modes such as SSB voice which yield high peak to average power ratios (8 dB or greater) are claimed to show no degradation in power output up to +55°C and 50,000 ft. Short periods of operation up to +71°C are permitted in all modes. For combinations of duty cycle, peak-to-average power ratios, altitude and temperatures that exceed the thermal capability of the transmitter, power output is reduced to a power level at which it can safely operate without equipment damage.

SIMOP operation

The HF-121C radio set is designed to provide high-performance operation in the presence of collocated communication systems. A digitally tuned, four-pole bandpass filter provides filtering in the receive and transmit path to optimise the system's Simultaneous Operation (SIMOP) performance in collocated environments.

To achieve optimum receiver performance in a collocated situation, the level of interfering signals must be reduced to an acceptable level relative to the desired signal strength. A preselector will improve performance in the areas of out-of-band intermodulation distortion, cross-modulation, image rejection, IF rejection, reciprocal mixing and high-voltage protection.

When used in the transmit path, the bandpass filter is useful in reducing the levels of the out-of-pass band broadband noise, spurious outputs and harmonics that are generated in the receiver/exciter-modem.

Link 11

Link 11 is a digital data communication network interconnecting dispersed elements of a task force that provides for the exchange of tactical data among communication systems. The exchange of data uses a 16-tonne composite signal generated by the system's Data Terminal Set (DTS).

The HF-121C radio set has been optimised to provide compatibility with the signal characteristics of Link 11 as defined in MIL-STD-188-203-1A, Interoperability and Performance Standards for Tactical Digital Information Link. The performance enhancements provided by the DSP technologies used in the 671Z-3( ) R/E enhances the ability of the system to be compatible with Link 11 data transmission and reception.

SELCAL

The HF-121C radio set provides for the demodulation of selective calling waveforms compatible with ARINC 714-6. The SELCAL address is selected from the control unit.

Multimode modem

The HF-121C radio set provides an internal modem compatible with the waveforms of MIL-STD-188-110A and STANAG 4285. The waveforms of MIL-STD-188-110A include the narrow- and wide-shift FSK and the serial (single tone) mode. An RS-232C data port is provide to interface with an external data source.

Automatic Link Establishment (ALE)

The traditional method of operating HF communication systems requires manual frequency-time planning and co-ordination in addition to skilled radio operators who are needed to perform the frequency selection, frequency monitoring and link establishment. The HF-121C radio set, operating under the control of ALE, will provide automated selective calling, preset channel scanning and real-time channel propagation evaluation to achieve automatic connectivity.

The automatic connectivity of ALE provides improved HF communications reliability with less user training, including: automatic frequency management, automatic link establishment, automatic link confirmation and automatic disconnect. ALE improves the connectivity between communication systems by monitoring multiple frequencies and selecting the best calling frequency. This selection is based on real-time analysis of the quality of the frequency channel. If desired, the ALE system can operate in a 'radio silent' mode, which allows the channel quality to be continuously updated, but which prevents transmissions from the system except by operator intervention.

The sequence of events in an ALE call may be summarised as follows:

The operator selects an individual address to call and then keys the system. The system then automatically;

(1) Selects the best preset channel
(2) Determines the availability of selected channel
(3) Transmits the address data sequence
(4) Receives the automatic handshake - link confirmation
(5) Breaks the squelch
(6) Signals the user to begin communications
(7) Monitors system key activity
(8) Returns the system to the muted scan mode when communications are completed

The basis for ALE frequency selection is the Link Quality Analysis (LQA) that is continuously performed on the channels within the ALE scan list. The LQA is determined by analysing the signal characteristics (signal-to-noise ratio and delay distortion) of the data signal used in sounding or initiating an ALE call. A database of the LQA values is generated, which ranks the quality of the ALE frequencies associated with each of the addressee's mission directories. The entries in the LQA database are continually updated, with automatic downgrading of old entries.

Operational status

In production and in service in US Navy and US Air Force aircraft.

Contractor

Rockwell Collins.

UPDATED

HF-230 HF radio

The Rockwell Collins HF-230 radio is for use in fixed-wing aircraft and helicopters. It provides 280,000 channels at 100 Hz channel spacing between 2 and 29.9999 MHz. All 176 ITU radio-telephony channels are preprogrammed, giving phone-patch capability over very long ranges wherever this facility is available. Lower sideband operation is possible for international or maritime communications.

The system comprises a TCR-230 transceiver, PWR-230 power amplifier and a range of antenna couplers. A DSA-220 adaptor permits two such systems to be operated in the same aircraft.

The radio features 40 pilot programmable channels and, when selected, channel number and frequency are displayed.

The CTL-230 display unit forms part of the HF-230 and features gas-discharge symbology.

An automatic probe antenna coupler, the PAC-230, is available for helicopter applications.

Specifications

Weight: 11.1 kg
Power output: 100 W PEP
Temperature range: –55 to +70°C
Altitude: up to 55,000 ft (with pressurised antenna coupler)

Operational status

In production and in service.

Contractor

Rockwell Collins.

VERIFIED

The Rockwell Collins HF-230 HF radio and ITU radio-telephony transceiver

HF-9000 series HF radios

The HF-9000 series is a series of lightweight HF radios developed to meet the HF communications requirements of commercial business jets and military aircraft ranging from helicopters to high-performance fighters.

The initial emphasis was on the development of a system for light fixed- and rotary-wing tactical aircraft. That system, designated the HF-9000 and the AN/ARC-217(V) in its military form, is now in production. The AN/ARC-217 employs modular design combined with fibre optics, microprocessor technology, digital synthesisers and couplers, combined with MIL-STD-1553B or ARINC 429 control.

The basic communications equipment includes an HF receiver/transmitter with a 175 W HF power amplifier/antenna coupler. A 200 W version is also available. System design is such that it can be configured with a control unit for panel mounting in the cockpit or for a MIL-STD-1553B control system. All control and status information transferred between the

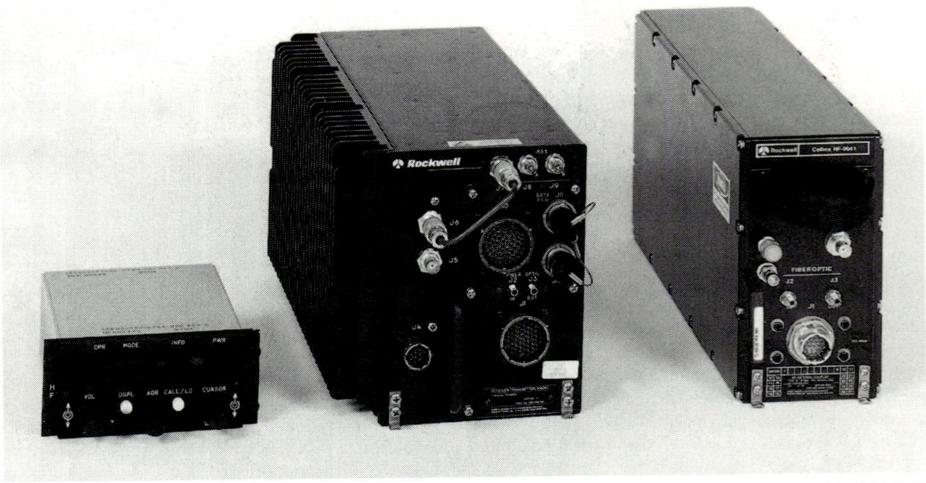

The HF-9000D digital signal processing HF radio ***2000***/0079231

transceiver and the power amplifier/coupler is transferred through a small fibre optic cable, permitting fast exchange of large amounts of data between the two units.

The system can be operated in simplex or duplex modes over the frequency range from 2 to 29.9999 MHz in 100 Hz increments in both Upper SideBands and Lower SideBands (USB, LSB) voice and data, Amplitude Modulation Equivalent (AME) and Continuous Wave (CW) modes, with growth to HF datalink operations. Up to 99 programmable pre-set channels can be stored in a non-volatile memory and each memory channel can store separate receive and transmit modes and frequencies. The transceiver uses a direct digital frequency synthesiser for rapid frequency changes with microprocessor control to improve stability. The antenna coupler is designed to permit rapid tuning of a wide range of antennas in a variety of aircraft.

HF-9000D

The latest variant to be added to the HF-9000 series is the HF-9000D, a new lightweight Digital Signal Processor (DSP)-based communications system that provides a single integrated system solution to current and future HF voice and data communications requirements. Offering significantly improved reliability over earlier systems, the HF-9000D series are designed for use on a broad range of military fixed-wing and rotary-wing airborne, transportable, shipboard and fixed site applications. The integrated multimode system provides data communication capability over HF modems, video imaging system, secure voice devices and data encryption devices, while continuing to provide voice HF communications. Embedded system functionality includes MIL-STD-188-141A Automatic Link Establishment (ALE), MIL-STD-188-110A data modem functionally, MIL-STD-188-148 ECCM capabilities, Independent SideBand (ISB) data operation and ARINC 714-6 SELCAL decoding with growth capability for future HF waveforms. The HF-9000D is compatible with the requirements of MIL-STD-203-1A for LINK 11 (TADIL A) data communications. Built-in test equipment (BITE) is utilised for diagnostic testing and monitoring.

Specifications

Dimensions (typical):
(HF-9010 control unit) 67 × 146 × 149 mm
(HF-9087D transmitter/receiver) 193 × 172 × 320 mm
(HF-9040 coupler) 192 × 96.5 × 320 mm
Weight (typical):
(HF-9010 control unit) 1.2 kg
(HF-9087D transmitter/receiver) 9.5 kg
(HF-9040 coupler) 3.6 kg

Operational status

In service with Gulfstream IV aircraft and with the Royal Australian Air Force as a replacement for the 618T HF radio on aircraft such as the C-130 Hercules. A total of over 2,000 sets have been sold.

Contractor

Rockwell Collins.

UPDATED

HF-9500 series HF radios

Rockwell Collins' new 400 W HF-9500 HF Airborne Radio system is a versatile, digital signal processor based high-frequency radio communications system. It provides a single integrated system solution to current and future HF voice and data communications requirements for military and commercial fixed-wing and rotary-wing airborne, transportable, shipboard and fixed-site applications. The integrated multimode system provides data communications capability over HF with modems, video imaging systems, secure voice devices and data encryption devices, while continuing to provide voice HF communications capability. Embedded system functionality includes MIL-STD-188-141A Automatic Link Establishment (ALE), MIL-STD-188-110A data modem functionality, Independent SideBand (ISB) data operation and ARINC 714-6 SELCAL decoding and growth capability for future HF waveforms. The HF-9500 is compatible with the requirements of MIL-STD-188-203-1A for Link 11 (TADIL A) data communications. The HF-9500 system's compact design and modular construction is ideally suited for retrofit of existing HF systems requiring functionality upgrade.

Specifications

Dimensions (typical):
(HF-9515 control unit) 146 × 163 × 114 mm
(HF-9550 transmitter/receiver) 257 × 444 × 194 mm
(HF-9545 coupler) 129 × 347 × 192 mm
Weight (typical):
(HF-9515 control unit) 1.2 kg
(HF-9550 transmitter/receiver) 20.5 kg
(HF-9545 coupler) 7.7 kg

Operational status

In production. The HF-9500 radio system is the standard HF replacement radio for the P-3 Orion retrofit upgrade programs for the US Navy and the Royal Australian Air Force, where it is designated AN/ARQ-57.

Contractor

Rockwell Collins.

NEW ENTRY

HF messenger

Rockwell Collins' HF messenger permits personal computers to exchange digital information over a standard HF radio. HF messenger profile software provides all the functions required to send and receive digital data and files between the personal computer and the radio. It also provides all the functions required to send and receive digitised signals over the HF propagation medium.

Normally, computer networks transfer data via the normal ground-based communications infrastructure: telephone lines, fibre optic cables, microwave links and satellite channels. However, for airborne communications, terrestrial-based links are not possible and satellite channels may be too costly or impractical. Under these circumstances, the aircraft HF radio may be used to provide the required connectivity.

HF messenger is also delivered with a Point to Point Protocol (PPP) type client interface. With this PPP interface, the HF node can be connected to any commercial standard TCP/IP router/remote access server to provide a transparent switched connection over HF - TCP/IP network supporting commercial standard e-mail applications.

HF messenger is based on the protocol defined in the NATO STANAG 5066 (an international public standard for HF data communications developed by NC3A NATO agency). The system can be installed in any commercial PC running Windows NT. It can be associated to any external HF modem and radio by loading the appropriate equipment control driver. The HF modem must be capable of supporting at least one of the following waveforms:

- MIL-STD-188-110A
- STANAG 4285
- STANAG 4529
- MIL-STD-188-110B Appendix C

The HF radio must also meet the general performance requirements of MIL-STD-188-141A.

Added features for HF radios equipped with ALE

HF messenger provides drivers for use with Rockwell Collins HF radios with embedded ALE functionality. Automatic Link Maintenance (ALM) controls the transmission performance to maintain optimum data rate. It automatically and independently adapts the transmission parameters to any node (data rate, interleaver, frame length) to maximise user data throughput.

HF server

HF messenger defines a profile for a HF server providing, when associated with an external HF modem and radio, a gateway for automatic and transparent HF data interconnection facilities for local and/or remote POP3 or Z-modem.

Multiservice provider

HF Messenger provides connectionless transmissions between several HF node client applications for broadcast, multicast, or point-to-point with ARQ services. Dedicated point-to-point transmissions between two specific client applications (hard links) can also be established. The hard link connected clients can then either use the entire HF connection resources or share those resources with other clients on the same node, providing maximum use of HF resources.

Operational status

In production.

Contractor

Rockwell Collins.

NEW ENTRY

HFS-700 HF radio

The Rockwell Collins HFS-700 is an HF transmitter/receiver intended for service aboard transport aircraft which have a long-range communication requirement. Designed and developed in accordance with ARINC 719, the HFS-700 draws largely upon operational experience gained with the company's earlier 628T-1 and 628T-2/5 systems.

Covering the full HF band from 2 to 30 MHz, the HFS-700 provides 28,000 channels at increments of 1 kHz and uses ARINC 429 serial channel selection. Operating modes are USB, LSB, AM equivalent, data and CW. Nominal transmitted power outputs are 400 W PEP in SSB modes, 125 W in compatible AM and 125 W in CW. A special audio output permits SELCAL monitoring to continue in all mode settings.

The system is of all solid-state construction and extensive use is made of CMOS and linear integrated circuits. The digital frequency generation synthesiser is locked to a highly accurate reference standard.

Features include a Digital Information Transfer System (DITS) tuning interface, new installation concept connectors and cooling with a blower. Particular attention has been paid to control of excessive temperature during abnormal operating conditions and a special design of heatsink has been employed to maintain output transistors within derated temperature limits. Additionally, a dissipation detector reduces drive to lower levels under high-dissipation conditions.

Mechanical design features include plug-in boards with hinged tray and fold-out doors to allow easy access to components during servicing without use of card extenders.

Specifications

Dimensions: 6 MCU per ARINC 719600
Weight: 12.7 kg
Temperature range: –55 to +70°C
Altitude: up to 40,000 ft

Operational status

In production and in service. High-Frequency DataLink (HFDL) upgrade kits available.

Contractor

Rockwell Collins.

VERIFIED

HFS-900D HF DataLink (HFDL) radio

The HFS-900D is designed to provide the air transport industry with next-generation HF voice and data communications service. It provides all traditional HF modes of operation, including USB, LSB, CW, AME and analogue data, with ARINC 635/753 standard High Frequency DataLink (HFDL) modes compatible with the ATN network. The HFS-900D provides the means to process, transmit and receive data and analogue voice and is designed to operate on frequencies spaced 100 Hz apart in the 2 to 30 MHz band. For compatibility with existing ARINC 719 installations, the HFS-900D provides Single SideBand (SSB) voice, Amplitude Modulated Equivalent (AME), Continuous Wave (CW), Selective Calling (SELCAL) and analogue data functions. The unit contains an internal data modem and controller, and voice transmission is compatible with current SSB HF transceivers. Data transmission is compatible with ground HF transmitting and receiving systems using conventional HF transceivers and ARINC 635 compliant modems and controllers. The HFS-900D is ARINC 634 system design guideline compliant and offers improved built-in test capability and software dataload capability in accordance with ARINC 615. It is compatible with a new high-speed digital antenna coupler (CPL-920D) and a new-ARINC 753 coupler interface to improve communication with the coupler. The HFS-900D offers HFDL through two different upgrade paths, a service bulletin/upgrade kit for 628T-2A, HFS-700, HFS-900 transceivers, or a new HF-900D data radio manufactured with the service bulletin.

Specifications

Dimensions: 6 MCU
Weight: 12 kg
Power supply: 115 V AC, 3 phase
Frequency: 2-29.9999 MHz
Channels: 280,000 in 100 Hz increments
Emissions:
(receive) AM, USB/LSB/AME, data, CW
(transmit) USB/LSB/AME, data, CW
Temperature range: –55 to +70°C
Altitude: up to 50,000 ft

Operational status

In production. Also available are HFDL upgrade kits for Rockwell Collins radios already in service, including 628T-2A, HFS-700 and HFS-900 radios.

Contractor

Rockwell Collins.

VERIFIED

ILS-900 Instrument Landing System receiver

The ILS-900 instrument landing system is a high-integrity ILS receiver with software verified to DO-178A Level 1. This system performs to Cat III fly-by-wire failure and monitor requirements in dual and triplex autopilot installations.

The Rockwell Collins displays and flight control system developed for the Boeing 747-400.

The receiver has been designed for Cat III dual-autopilot operations which place heavy reliance on internal monitoring within the ILS receiver. A high degree of monitoring integrity has been included in the ILS-900 and system reliability has been improved with digital technology. In addition, the system features audio mute in cruise mode and automatic Morse decoding. Absolute partitioning between BIT and the primary monitoring deviation circuitry has been achieved by using four separate processors.

Specifications

Dimensions: 3 MCU
Weight: 3.86 kg
Power supply: 115 V AC, 400 Hz, single phase
Frequency:
(localiser) 108.1-111.95 MHz
(glide slope) 329.15-335 MHz
Channel spacing:
(localiser) 50 kHz
(glide slope) 150 kHz
Temperature range: –55 to +71°C

Operational Status

In production and in service.

Contractor

Rockwell Collins.

UPDATED

Integrated display system for the Boeing 747-400

Rockwell Collins has developed an integrated electronic colour display system for the Boeing 747-400. A Rockwell Collins digital flight control and central maintenance computer (CMC) also are standard on the aircraft. The display system features six 8 in (210 mm) square colour CRT displays. Any one of these units can be selected to show either EFIS or EICAS data.

The new technology gives the 747-400 only 38 per cent of the cockpit lights, gauges and switches compared with older 747 aircraft and leads to better aircraft availability and significant reductions in crew workload.

The IDS-7000 integrated display system utilises the third-generation hardware design first certified in the EFIS-1000 system. Each display unit contains all drive and symbol generation electronics necessary to perform any of the display tasks. Each pilot has a Primary Flight Display (PFD) and Navigation Display (ND) mounted in a side-by-side configuration. All primary air data and heading information is included on the PFD displays. EICAS functions are provided on two displays mounted in an over/under configuration in the centre panel.

The display system also includes three electronics interface units (EIUs) which function as data conversion and collection for EICAS displays, message processing, snapshot information recording, exceedance data recording and CMC-7000 central maintenance computer data collection. The central maintenance computer, in conjunction with the display units, provides ground maintenance crews with maintenance page displays. The display units, EIUs and CMCs are all capable of being programmed on the aircraft without removal of the equipment.

Operational status

In production for the Boeing 747-400.

Contractor

Rockwell Collins.

VERIFIED

Integrated Information System (I²S)

The aim of the Rockwell Collins I²S is to replace paper with technology in commercial aircraft. This is achieved by automating the transfer of data between the flight deck and the airline's dispatch operations on the ground. Rockwell Collins is currently teamed with Condor Flugdienst, the charter affiliate of Lufthansa, to participate in a year's trial of the concept to link an aircraft-based intranet to airline terminal area databases.

Some of the features of the I²S include:

(1) Communication via a new high data rate transfer medium on the ground (wireless gatelink), together with HF, VHF and Satcom for air-to-ground links
(2) A high-frequency datalink between the aircraft and the airline's information system to transmit intranet data including navigation databases, flight plans, graphical weather data and maintenance data
(3) In-flight entertainment services.

The first products in the I²S product line have successfully completed DO-160D airworthiness testing. These units include, the MAU-2000 (Microwave Airborne Unit) and PMAU-2000 (Pilot's Microwave Airborne Unit), which together facilitate wireless Local Area Netwok (LAN) connectivity onboard the aircraft. They also provide for wireless terminal area gatelink from the aircraft to the airline's Information Technology (IT) system and provide a wirelss LAN environment inside the aircraft.

Contractor

Rockwell Collins.

VERIFIED

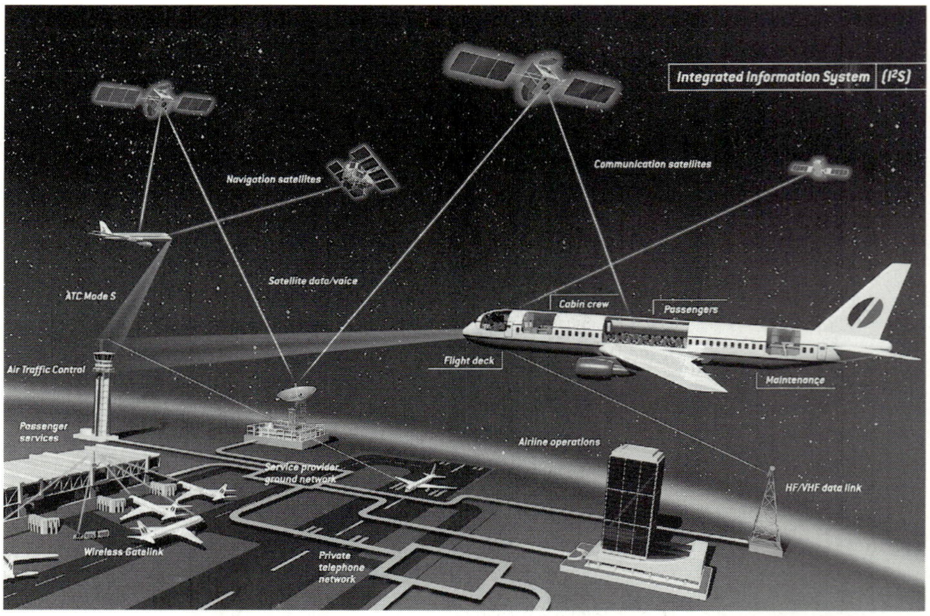

The Rockwell Collins integrated information system concept **2000**/0079230

Integrated Processing Centre (IPC)

The Rockwell Collins IPC provides cost-effective, adaptable core processing capability and is the primary integrating component of the Flight2 architecture. The architecture and modular components of the IPC are shared across all three Rockwell Collins businesses (Air Transport, Business and Regional, and Government).

The IPC is a direct derivative of the Collins Integrated Avionics Processing Systems (IAPS), currently flying in over 1,000 commercially certified aircraft, including the US Air Force's T-1A. It uses modular architectural elements that are compliant with AFR part 25 civil certification standard.

The IPC configuration for each specific aircraft platform is achieved through a building-block approach, which uses a suite of modules that are standard components across the Rockwell Collins business units. These modules perform centralised application processing as well as input/output data concentration from peripheral subsystems and devices. An avionics-qualified dual-redundant, full-duplex 100 base-Tx Ethernet Local Area Network (LAN) configured in a hub-and-spoke topology forms the next-generation data transfer system for all inter- and intra-IPC data transfers.

The IPC modular hardware is designed to allow LTM replacement at the operator level even under power-on (hot swap) conditions. This LRM concept lowers mean repair time and reduces spare requirements, resulting in significantly lower life-cycle costs. The hardware design is full dual path (dual power supplies, dual LAN, dual I/O) to provide increased levels of aircraft availability. The LRM is based on commercially available standards that easily support the integration of third party designs. In addition, the LRM is sized to accommodate standard VME form factor cards. A number of cabinet sizes are available to accommodate a wide variety of module configurations. Growth slots are provided to support embedded functions such as Terrain Avoidance Warning System (TAWS), digital map, autopilot, communication management or Head-Up Display (HUD) processing.

The IPC provides a software architecture comprised of three software protocol layers for device drivers, system services and applications. Each layer of the architecture strictly enforces open commercial Application Program Interfaces (APIs). The layered protocol design insulates software applications from changes to the underlying processing resources or network devices. The architecture's adherence to open commercial standards also allows for the introduction of third-party developed applications.

Execution of multiple software applications on the same hardware resource is also supported in the architecture. This feature allows a processing resource to host multiple applications of different critical levels and functions, thereby allowing future changes and capability upgrades to be certified on an incremental basis, without having to perform regression testing on unrelated applications.

Contractor

Rockwell Collins.

VERIFIED

LRA-900 low-range radio altimeter

The LRA-900 low-range radio altimeter includes enhancements to a digitally controlled variable bandwidth filter which provides improved noise rejection and leading-edge tracking. Signal microprocessors perform automatic calibration that continuously compares the received ground return signal frequency with the frequency produced by separate precise delay lines in each channel. This results in fine resolution, providing accuracy to support autoland flare and touchdown computations. Resolution at touchdown is quoted as better than 1.2 in.

Low-component count and low-stress level circuits give the LRA-900 high reliability and quick, easy and effective onboard fault isolation. A comprehensive self-test monitor in the system is capable of determining operational status to a 99 per cent confidence level.

The LRA-900 is designed to fulfil all new environmental requirements. The system is protected for Cat K lightning and Cat Y High-Intensity Radiated Fields (HIRF). In addition, it is fully functional over 200 ms of power interruptions. It meets ARINC 707, 429 and 600, TSO-C87, DO-160C, DO-155 and DO-178A.

Specifications

Dimensions: 3 MCU
Weight: 4.3 kg
Power supply: 115 V AC, 380-420 Hz, 24 W nominal
Frequency: 4,300 ± 25 MHz
Altitude: –20 to 5,000 ft
Accuracy: ±1 ft or 2% of indicated altitude
Temperature range: –40 to +70°C

Operational Status

In production and in service.

Contractor

Rockwell Collins.

UPDATED

Micro Line nav/com family

In the early 1970s, Rockwell Collins introduced the first members of a family of panel-mounted navigation and communication units, called Micro Line, for light single- and twin-engined aircraft. The rationale for this step was the lack of any agreed standard or specification for avionics systems appropriate to this type of aircraft, which gave rise to considerable compatibility or interchangeability problems between equipment from different manufacturers. By contrast, the airlines already had the very comprehensive ARINC characteristics to ensure compatibility. Intended for low-budget operation and for aircraft with limited internal space and weight allowance, the various Micro Line units each contained the relevant processing and indication circuits and mechanisms within single boxes, so that they can be mounted as self-contained items on the pilot's panel without the need for any associated remote rack-mounted equipment.

The Micro Line family comprises:
ADF-650A automatic direction-finder
AMR-350 audio control unit/marker beacon receiver
ANS-351 area navigation system
AUD-250 audio control unit
DME-450/451 DME
GLS-350 ILS glide slope receiver
IND-350/351 course deviation indicators
MKR-350 marker beacon receiver
PWC-150 28/14 V power converter
TDR-950 transponder
VHF-251/253 communication transceiver
VIR-351 navigation receiver.

Operational status

In service. No longer in production.

Contractor

Rockwell Collins.

VERIFIED

Pro Line 4 integrated avionics system

The Pro Line 4 system includes large side-by-side electronic flight displays that present primary flight, navigation and engine information. Autopilot, flight direction and centralised avionics maintenance functions are provided by the system's integrated avionics processing system. The heart of the Pro Line 4 avionics line is control central, the integrated avionics processing system.

Functions performed by control central include processing of inputs from sensors located throughout the aircraft, and computation of commands and displays for autopilot functions, flight management system and electronic displays. Instead of a number of units performing flight control functions, eight modules in the control central work together to serve as the aircraft central processing system. The small size of the modules allows control central to provide fail/safe integrity, with quadruple redundancy of critical

Rockwell Collins Pro Line 4 EICAS display
2001/0105284

Pro Line 4 as installed in the Challenger 604 - a six-tube configuration, showing primary flight (× 2), navigation, weather radar and engine and systems information **2001**/0105285

functions. Internal communications between critical functions are protected, eliminating exposed external wires and busses. If one module fails, the system automatically reconfigures to carry on operation without interruption.

Advances in component and manufacturing technology allow circuit boards in control central to perform the functions of up to the eight boards normally used. Surface-mounted devices, automatic insertion of components and manufacture control have allowed a considerable reduction in circuit board size.

As well as control central, Pro Line 4 includes EFIS and EICAS displays, radio tuning units and a TWR-850 Doppler weather radar.

Operational status

In production. Pro Line 4 is currently installed in the Beechjet 400A, Canadair Regional Jet and Challenger 604, Dassault Falcon 20, 50 and 2000, IAI Astra SPX, Galaxy Business Jet, Learjet 60 and the Saab 2000 regional airliner.

Contractor

Rockwell Collins.

UPDATED

Pro Line 21 CNS

Pro Line 21 CNS is a new product line planned by Rockwell Collins to meet the demands of the new Communications Navigation Surveillance/Air Traffic Management (CNS/ATM) era. Products planned for the Pro Line 21 CNS include:

VHF-4000 VHF voice/data communication transceiver

The VHF-4000 will provide both voice and data, at 31.5 kbps, to deliver uplinked automated digital messages, flight plan changes and graphical weather depictions. Initially, the VHF-4000 will include Mode 2 capability, but the unit has been designed to also accommodate Modes 3 and 4 without major modification.

NAV-4000 navigation receiver

The NAV-4000 navigation receiver combines the VOR, localiser, glide slope, marker beacon and ADF functions.

RIU-4000 CNS manager

The RIU-4000 CNS manager interfaces all the radios with the rest of the avionics system and also provides high-quality audio for the flight deck as well as digital communication management.

DME-4000 distance measuring unit

The DME-4000 will be coupled to the NAV-4000 to provide a comprehensive terrestrially based navigation capability.

SAT-4000 and SAT-6000 satellite communication systems

The SAT-4000 and SAT-6000 systems will provide global AERO-I and AERO-H/I services.

HF-900D HF communication system

The HF-900D will be available for back-up voice and data communication worldwide, if needed.

SNS-4000 satellite navigation system

Design parameters of the SNC-4000 will be finalised when final specifications have been produced by the regulatory agencies, but the plan is that it will support appropriate en-route navigation and approach requirements.

TDR-4000 Mode S transponder

The TDR-4000 will be the successor to the TDR-94 to add Levels 4 and 5 capabilities, when extended datalink capabilities are defined and implemented.

Operational status

Pro Line 21 CNS is in development. The first set of new radios, to be available in 2002, are the VHD-4000 VHF voice/data communication receiver, the NAV-4000 navigation receiver, the DME-4000 distance measuring equipment and the RIU CNS manager. The remaining functions will be available later, as the requirements for them are defined.

Contractor

Rockwell Collins.

UPDATED

Pro Line 21 integrated avionics system

Advances in man/machine interface engineering and incorporation of large-format active matrix liquid crystal displays (AMLCDs) are the key elements of the Pro Line 21 avionics system, introduced in 1995 for corporate and regional aircraft.

Pro Line 21 flight decks are custom-configured with two to five adaptive flight displays that utilise a mix of AMLCD formats, including a 6 × 8 in and 8 × 11 in active area display. In addition to primary flight and navigation information, this LCD technology allows clear presentation of approach plates, terrain maps, real-time video and other highly detailed - and even three-dimensional - graphics that deliver flight operations information to pilots in innovative formats developed and refined by pilots.

Pro Line 21's key subsystems include the Rockwell Collins FMS 3000 satellite-based precision navigation and communication system with GPS-4000 Global Positioning System sensor, an advanced-technology AHRS and an advanced flight control system with fail-passive autopilot. Also standard are solid-state weather radar and Pro Line radio sensors, including transponder, TCAS and DME. An optional maintenance system displays current LRU status, fault history and diagnostic data on the multifunction display.

Operational status

Pro Line 21 has been selected for the Raytheon Premier I, which also includes the new Rockwell Collins AHS-3000 Attitude Heading Reference System (AHRS); the IAPS-3000 Integrated Avionics Processing System; the AFC-3000 Automatic Flight Control System; and a complete radio and radar package. The system has also been selected for the Bell Agusta 609 civil tilt-rotor, which will feature three 6 × 8 in AMLCDs including two Primary Flight Displays (PFDs) and one MultiFunction Display (MFD), the Bombardier Continental, Citation CJ1 and CJ2 business jets.

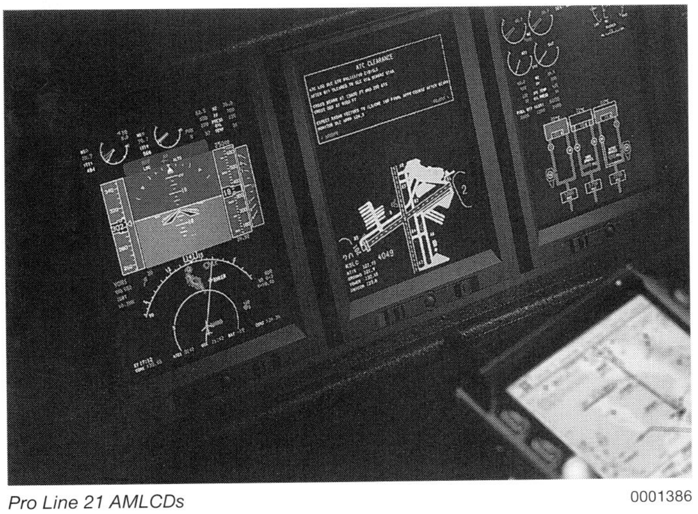

Pro Line 21 AMLCDs 0001386

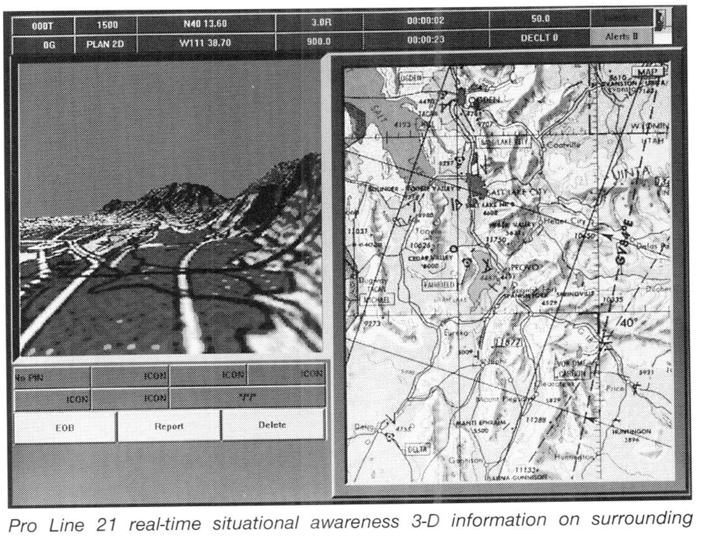

Pro Line 21 real-time situational awareness 3-D information on surrounding airspace and terrain

Rockwell Collins Pro Line II installed in a Falcon 50 **2001**/0105283

Contractor

Rockwell Collins.

UPDATED

Pro Line II digital nav/com family

Acknowledging the advances in technology, notably in signal processing, since the appearance of Pro Line in 1970, Rockwell Collins decided in the late 1970s to develop a replacement. The result was Pro Line II, the first members of which (VHF communications transceiver, VHF navigation transceiver and DME) appeared in January 1983. Other units were added and there now exists a complete range of Pro Line II equipment.

Pro Line II boxes contain analogue/digital and digital/analogue circuits so that individual units of the earlier family can be exchanged on a one-for-one basis without change to the aircraft wiring or racks. Microprocessors within the new units are programmed to accept either analogue or digital frequency tuning arrangements.

The Pro Line II family consists of:

ADS-82 air data system
AHS-85 attitude/heading reference system
APS-65 autopilot
APS-85/95 autopilot
CTL-22 communication control unit
CTL-23 comm/nav control unit
CTL-32 navigation control unit
CTL-62 automatic direction-finder control unit
CTL-92 transponder control unit
DME-42 DME receiver
EFIS-85/86 electronic flight instrument systems
EHSI-74 electronic flight instrument system
IND-42 DME control unit
MCS-65 compass system
TWR-850 turbulence weather radar system
VHF-21/22 VHF communication transceiver
VIR-32/33 navigation receiver.

Operational status

In production and service on a wide range of regional and business aircraft.

Contractor

Rockwell Collins.

UPDATED

Pro Line nav/com family

Introduced in 1970, the Pro Line series avionics was intended for medium and large general aviation piston and turbine-engined twins. The family was originally designated Low-Profile to emphasise the compact size and form factor of individual units but, in 1975, was renamed Pro Line in recognition of its acceptance by professional pilots in regional airlines, and by the defence forces. Unlike the self-contained members of the Micro Line family, Pro Line systems comprise panel-mounted indicators and controls driven by or controlling separate rack-mounted processing and computing boxes. The size and form factor was laid down by Rockwell Collins, there being no industry-wide agreement on packaging for general aviation electronics, in contrast to the highly defined ATR standards governing the characteristics of equipment for airlines. In the military field, Pro Line is found on a wide range of aircraft, not only those equivalent in size and performance to general aviation types, but also on attack and surveillance aircraft such as the A-4, F-5E, C-130 and E-3 AWACS.

The Pro Line family comprises:

346B audio control/isolation and speaker amplifiers
ADF-60 automatic direction-finder
ALT-50/55 radio altimeters
AP-105 autopilot
AP-106A autopilot
APS-80 autopilot
BDI-36 bearing/distance indicator
CTL series control heads
DME-40 distance measuring equipment
FDS-84 flight director system (comprising FD-108 flight director and FIS-70 flight instrument system)
FDS-85 flight director system (comprising FD-109 flight director and associated horizontal situation indicator)
FPA-80 flight profile advisory system
HF-230 HF transceiver
PN-101 pictorial navigation system
TDR-90 transponder
VHF-20A/B communication transceivers
VIR-30A/M and VIR-31A/H navigation receivers
WXR-220/270/300/350 colour weather radars.

Operational status

Production is being phased out in favour of Pro Line II avionics, but equipment remains in wide service. Well over 225,000 Pro Line boxes and controls have been manufactured and sold.

Contractor

Rockwell Collins.

VERIFIED

Rockwell Collins HF radios

The following table summarises the features of the major products in the Rockwell Collins' family of HF radio products on opposite page:

Contractor

Rockwell Collins.

NEW ENTRY

RTU-4200 series Radio Tuning Units

The RTU-4200 Radio Tuning Unit (RTU) provides centralised control of VHF comms, VOR, ILS, DME, ADF, transponder and TCAS. Integration with a Flight Management System (FMS) also allows tuning of radios via either the FMS control/display unit or the RTU, with the tuned frequency always appearing on the RTU display. Each RTU can control all radio sensors. It also has the capability to store 20 preset comm and 20 preset nav frequencies.

Brightness of the active matrix liquid crystal display can be controlled by the aircraft's master dimming control bus or by crew member adjustment on the RTU. Parallax compensation is provided and control settings and radio diagnostics are stored in non-volatile memory for availability after power shutdown.

Operational status

The RTU-4200 is standard equipment on Gulfstream IV SP business jet aircraft. The RTU-4210 has been selected by Cessna for the Citation X business jet. The RTU-4280 has been selected by Gulfstream for the Gulfstream V business jet. The RTU-4200 has been certified aboard the Sikorsky S-76B helicopter.

Contractor

Rockwell Collins.

UPDATED

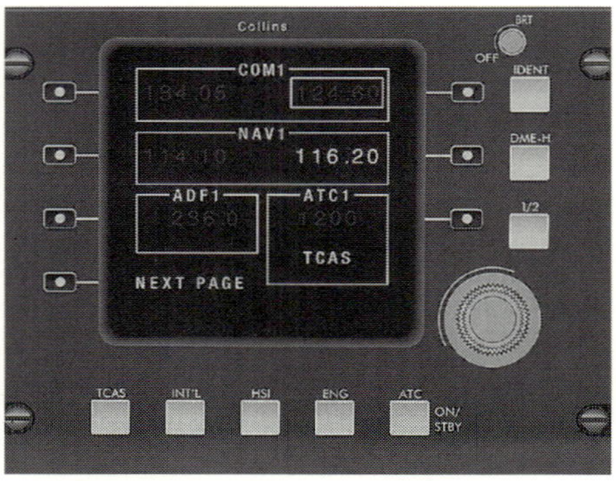

The RTU-4280 has been selected for the Gulfstream V

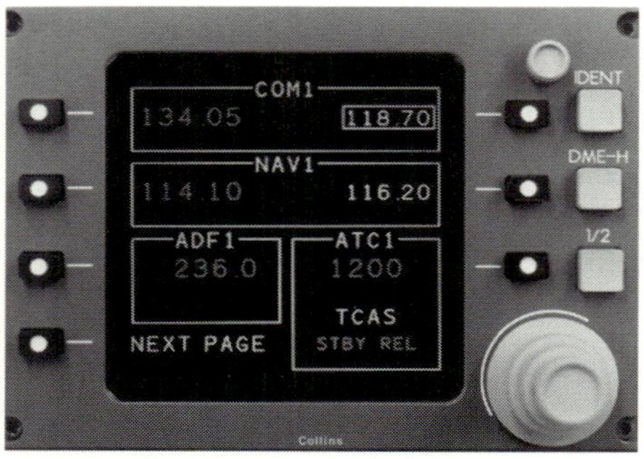

The RTU-4200 radio tuning unit **2001**/0105282

SAT-900/901 satellite communications systems

The SAT-900/901 satellite communications systems operate at L-band through the INMARSAT organisation space segment. Both systems support a low-speed digital link for the transmission of ACARS/AIRCOM datalink messages when out of range of terrestrial VHF stations. Rockwell Collins satcom systems are currently operational with the DLM-700B and management units from other commercial datalink equipment suppliers.

In addition to low-speed data, the SAT-901 can support a digital voice channel for use by either the cockpit or cabin. The selection of the single channel for voice or data is made via the MCDU or MIDU in the cockpit. Cabin communications require the use of the optional CTI-901 cabin telecommunications interface unit. The CTI-901 provides the industry standard CEPT E1 interface between the SDU-901 and the cabin telephone system.

The SAT-900/901 systems are compatible with all ARINC 741 low- and high-gain antenna subsystems. A low-gain antenna can be used with either system to support low-speed data communications, but a high-

Parameter	**AN/ARC-190(V)**	**HF-121C (ARC-230)**	**HF-9000**	**HF-9000D**	**HF-9500**	**AN/ARC-220 and VRC-100**
Typical Platform	fixed wing	fixed wing	fixed and rotor wing	fixed and rotor wing	fixed wing	ARC-220: rotor wing VRC-100: vehicular / 2 person lift
DSP based processing	no	yes	no	yes	yes	yes
ALE: MIL-STD-188-141A and FED-STD-1045	yes, via external CP-2024()	yes, embedded DSP	yes, specific version	yes, embedded DSP	yes, embedded DSP	yes, embedded DSP
ALE: MIL-STD-188-141B	growth	growth	no	growth	growth	yes
ALE link protection	yes	yes	no	yes	yes	yes
Quick ALE	growth	growth	no	growth	growth	yes
ECCM: MIL-STD-188-148A	yes, via external CP-2024()	growth	yes, specific ECCM version	growth	growth	yes, embedded DSP
Data modem external MIL-STD-188-110A	yes, data mode	yes, data mode	yes, data mode	yes, data mode	yes, data mode	yes, data mode
Data Modem Internal MIL-STD-188-110A Narrow shift FSK, Wide shift FSK, Serial single tone STANAG 4285	no	yes, embedded DSP	no	yes, embedded DSP	yes, embedded DSP	yes, embedded DSP
SELCAL, ARINC 719 audio via external LRU:	yes, dedicated AM audio output	yes, dedicated AM audio output	yes, dedicated AM audio output	yes, dedicated AM audio output	no	no
SELCAL, ARINC 714-6 decoder	yes, embedded in CP-2024()	yes, embedded DSP	no	yes, embedded DSP	yes, embedded DSP	growth, embedded DSP
HFDL: ARINC 635/753	yes, embedded in CP-2024()	yes, external HFDU, embedded as growth	yes, external HFDU	yes, external HFDU	yes, external HFDU	yes, external HFDU
FCC Type accepted	yes, specific version	no	yes, specific version	no	no	no
FAA TSO certified	yes, specific version	no	yes, specific version	no	no	no
RF output power: Multitone:	400 W Pk	400 W Pk	175/200 W Pk	200 W Pk	400 W Pk	175 W Pk
RF output power: Single tone (RTTY):	400 W Pk and Avg	400 W Pk and Avg	100 W Pk and Avg	100 W Pk and Avg	200 W Pk and Avg	100 W Pk and Avg
Input power required:	115 VAC, 3 Ph, 400 Hz	115 VAC, 3 Ph, 400 Hz	+28 VDC	+ 28 VDC	115 VAC, 3 Ph, 400 Hz	+28 VDC
Control interface:	discrete CNTL (RS-422), RS-422, and MIL-STD-1553B	discrete CNTL (RS-232), RS-232, and MIL-STD-1553B	discrete CNTL (Fiber Optic), ARINC 429, MIL-STD-1553B	discrete CNTL (Fiber Optic), MIL-STD-1553B	discrete CNTL (RS-422), RS-422, and MIL-STD-1553B	discrete CNTL (RS-422), RS-422, and MIL-STD-1553B
Secure device interface:	KY-100, ANDVT	KY-100, ANDVT	KY-100, ANDVT	KY-100, ANDVT	KY-100, ANDVT	KY-100, ANDVT
Environment:	MIL-E-5400T	MIL-E-5400T	MIL-E-5400T	MIL-E-5400T	MIL-E-5400T	MIL-E-5400T
Altitude	Class 1	Class 1	Class 1	Class 1	Class 1	< 25,000 Ft
Temperature	Class 1	Class 1B	Class 1, 1B (version specific)	Class 1B	Class 1B	Class 1B
Cooling:	external, forced air	external, forced air	convection cooled	convection cooled	external, forced air or self cooled with duty cycle	convection cooled
SIMOP Transmit HF to HF:	optional, F-1535, F-1602 LRUs	4 pole BPF embedded	optional, 4 pole HF-9060 LRU	growth, optional 4 pole LRU	2 pole BPF embedded	no
SIMOP Transmit HF to VHF/UHF:	embedded LP filter in CPLR	embedded LP filter in CPLR	no	no	embedded LP filter in CPLR	embedded LP filter in CPLR
SIMOP receive:	optional F-1535, F-1602 LRUs	4 pole BPF embedded	optional 4 pole HF-9060 LRU	growth, optional 4 pole LRU	2 pole BPF embedded	no
Link 11 external modem MIL-STD-188-230-1A:	no	compatible, link 11 mode	no	compatible, data mode	compatible, data mode	no
RF emission modes	USB/LSB voice or data, AME, or CW	USB/LSB/ISB voice, data, link 11, or RTTY AME, or CW	USB/LSB voice USB/LSB, data AME, or CW	USB/LSB voice USB/LSB/IS, data AME, or CW	USB/LS voice USB/LSB/ISB, data, AME or CW	USB/LSB voice USB/LSB, data AME, or CW
Duty cycle, 100%, all service conditions:	continuous	continuous	continuous, voice	continuous, voice	continuous	2:1, receive: transmit
# of channels:	280,000	280,000	280,000	280,000	280,000	280,000
Frequency Range:	2.0000 to 29.9999 MHz	2.0000 to 29.9999 MHz	2.0000 to 29.9999 MHz	2.0000 to 29.9999 MHz	2.0000 to 29.9999 MHz	2.0000 to 29.9999 MHz
Syllabic squelch:	yes	yes	yes	yes	yes	yes
Rapid tune antenna coupler:	yes	yes	yes	yes	yes	yes
BIT to LRU/SRU level:	yes	yes	yes	yes	yes	yes
Spare card slot for growth:	no	yes	yes	yes	yes	yes
Data fillable:	yes	yes	yes	yes	yes	yes
Field software reprogrammable:	yes, with HFDL only	no	no	yes	yes	yes

gain antenna subsystem is required with the SAT-901 for voice communications.

Each system is made up of a Satellite Data Unit (SDU), Radio Frequency Unit (RFU), High-Power Amplifier (HPA) and ARINC 741 antenna subsystem. The SAT-900 consists of an SDU-900, RFU-900 and HPA-900. The SAT-901 uses the same RFU and HPA as the SAT-900, along with the voice capable SDU-901.

The SAT-900 is compatible with INMARSAT Data-1 protocols; the SAT-901 is compatible with INMARSAT Data-2 and Voice-2 protocols.

Specifications

Dimensions:
(SDU-901) 190 × 193 × 325 mm
(RFU-900) 125 × 193 × 325 mm
(HPA-900) 125 × 193 × 325 mm
Weight:
(SDU-901) 8.84 kg
(RFU-900) 8.3 kg
(HPA-900) 6.8 kg
Power supply: 115 V AC, 400 Hz
Power output: 40 W continuous
Frequency:
(receive) 1,530-1,559 MHz
(transmit) 1,625.5-1,660.5 MHz
Synthesiser step size: 2.5 kHz
Doppler correction: within 20 Hz over ± 2 kHz
Modulation type: A-BPSK

Operational status

No longer in production. Both the US FAA and the Australian CAA have certified the SAT 900 system. Northwest Airlines ordered the system for its Boeing 747s. United Airlines and Japan Airlines also ordered it for the Boeing 747-400.

Contractor

Rockwell Collins.

VERIFIED

SAT-906 Aero-H and –L satellite communications system

Rockwell Collins' SAT-906 Aero-H and -L satellite communications system provides multichannel voice, facsimile and data capability. The voice channels can be used for cockpit or cabin communications while the data channel will support low- and high-speed data using INMARSAT Data 2 or Data 3 protocols. The SAT-906 complies with ARINC 741 and 746 characteristics and can be configured for up to six channels through the addition of plug-in modules. The system provides voice channels and CEPT-E1 interface to the passenger telephone system without the need for any additional LRUs. The SAT-906 also provides DTMF capability on two analogue audio lines.

The SAT-906 is capable of low- and high-speed data, cockpit voice and passenger PC, fax and cabin telephone services when mated with a high-gain antenna subsystem.

The SAT-906 units provided by Collins are the SDU-906, RFU-900, HP-900 and the HPA-901A. The SAT-906 is compatible with antennas provided by several manufacturers.

The SDU-906 provides the interface to all other aircraft systems and includes the modems, codecs and protocol support for communication with ground earth stations.

The RFU-900 consists of a wideband L-Band to IF down-converter for receive operation and a wideband IF to L-Band up-converter for transmit operation. The RFU operates in full duplex mode, receiving L-Band signals in the range of 1,530 MHz to 1,559 MHz.

The HPA-900 is a Class C amplifier that can provide single-channel low-speed data for a one-channel system or back-up for a multichannel system.

The HP-901A is a Class A amplifier that is required for multichannel operation through the INMARSAT space segment.

An optional SIU-900 provides interface between legacy ARINC 561 INS systems and the SDU-906. The SIU-900 converts data from synchro and ARINC 419 interfaces to ARINC 429 in a format required for SATCOM antenna pointing.

An optional CIU-906 provides an on-board PBX capability supporting multiple conventional two-wire telephones. The CIU-906 may also be provided with special codecs to allow operation at STU-III secure telephones through the INMARSAT system.

The SAT-906 is compatible with all service suppliers that support INMARSAT Data-2, Data-3 and Voice-2 protocols. The system has been installed on VC-25A, B747SP, C-9C, VC-137B/C, C-22, C-141, C-135, C-32A, Indian Airforce B737 and Japanese Defence Agency B747 aircraft.

Rockwell Collins series 700 avionics. From left to right the units are the HFS-700, VHF-700, PAU-700, VOR-700, ILS-700, ADF-700, DME-700, LRA-700 and TPR-700

Specifications

Dimensions:
(SDU-906) 6 MCU
(RFU-900) 4 MCU
(HPA-900) 4 MCU
(HPA-901) 8 MCU
(CIU-906) 4 MCU
Weight:
(SDU-906) 14.5 kg
(RFU-900) 5.4 kg
(HPA-900) 6.8 kg
(HPA-901) 13.6 kg
(CIU-906) 5.5 kg
Power supply:
(SDU-906) 115 V AC, 400 Hz, 180 W
(RFU-900) 60 W
(HPA-900) 200 W (maximum)
(HPA-901) <300 W
(CIU-906) 100 W

Operational status

In production and in service on a wide variety of civilian and military aircraft.

Contractor

Rockwell Collins.

UPDATED

SATCOM 5000

Rockwell Collins SATCOM 5000 is a small, light satellite communications system that provides the full range of telecommunications services conforming to Inmarsat's Aero-I specifications (Aero-I delivers telephone, fax and real-time data services for both cockpit and passenger communications).

SATCOM 5000 is designed for operation within spot beam coverage areas of the Inmarsat-3 satellites, especially the northern hemisphere. It offers up to five voice-data channels over 75 per cent of the earth's surface and a single channel in global beam coverage. It is designed to accommodate in-flight exchange of operational and air traffic management data envisioned for Communication Navigation Surveillance/Air Traffic Management (CNS/ATM) and Global Air Navigation Safety/Global Air Traffic Management (GANS/GATM) initiatives.

Operational status

SATCOM 5000 has been integrated and trialled on a Challenger 604 widebody business jet.

Contractor

Rockwell Collins.

VERIFIED

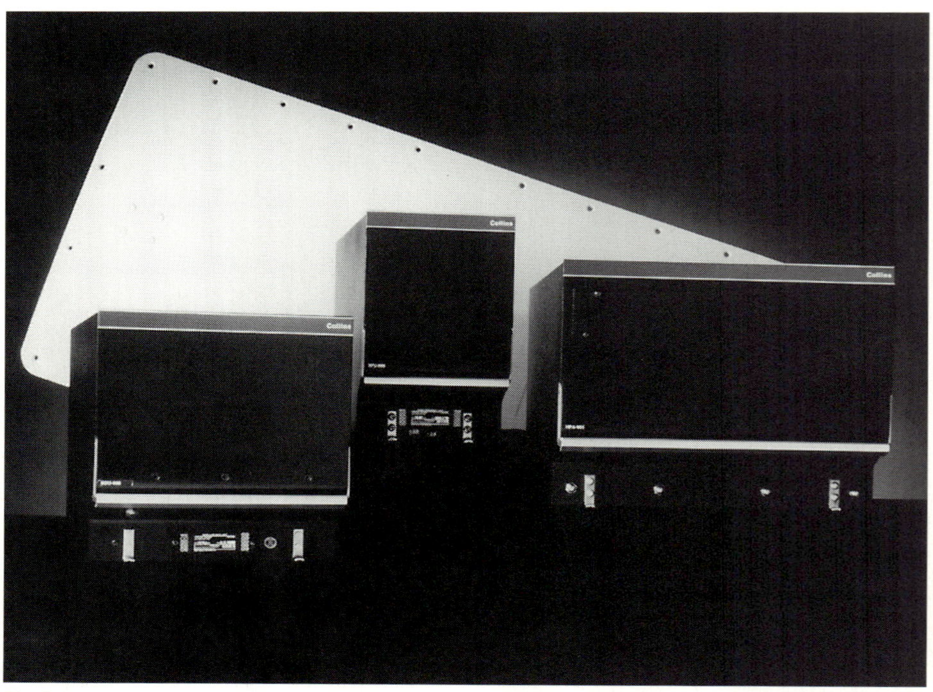

The Rockwell Collins Satcom 906 communications system showing (left to right) the SDU-906 satellite data unit, the RFU-900 Radio Frequency Unit and the HPA 901 High-Power Amplifier 0001184

Series 500 avionics family

Developed for commercial aircraft, the Series 500 avionics family is based on ARINC 500 characteristics and is the final step in the company's range of analogue equipment. Though still analogue in nature, more advanced solid-state circuits together with a small number of components per function (a lower parts count) make for a substantial weight saving and greater reliability over previous equipments. Series 500 boxes are available on a one-for-one replacement basis for earlier systems.

The Series 500 family comprises:
51RV-4 VOR/ILS receiver
51Y-7 (DF-206) automatic direction-finder
54W-1 comparator warning monitor
346D-2/2B passenger address amplifier
490S-1 HF antenna coupler
618M-3 VHF transceiver
621A-6A air traffic control transponder
618T-1/2/3 HF transceiver
860E-4/860E-5 DME systems
860F-4 digital radio altimeter
Datalink system (comprising DLC-700 control unit and 597A-1 management unit)
FD-110 flight director system (comprising FMC-28 flight mode controller, 562A-5F5 flight computer, 329B-8J attitude director indicator, and 331-8K horizontal situation indicator)
ILS-70 instrument landing system receiver.

Operational status

In production and in service.

Contractor

Rockwell Collins.

VERIFIED

Series 700 digital avionics

Design of the Series 700 digital avionics for commercial aircraft dates back to the early 1970s when US industry in general began planning for the new generation of transports then in prospect. The new aircraft emerged as the Airbus A300 series and the Boeing 767 and 757. The digital systems for these and later projects were defined by the ARINC characteristics of the 700 series. This series was introduced in 1978, launch year of the first of the new transport aircraft, the Boeing 767, and is now offered on the 747-400 and A320. The Series 700 avionics provides centralised fault monitoring capability as specified in ARINC 604.

Together with the usual range of navaids and other electronic devices, the Series 700 introduced an electronic flight instrument system based on television-style CRTs and a similar display suite for engine indication and warning.

The Series 700 family comprises:
ADF-700 automatic direction-finder
DME-700 DME
HF-700 HF radio transceiver
ILS-700 instrument landing receiver
LRA-700 low-range radio altimeter
PAU-700 passenger address amplifier
TPR-720 transponder
VHF-700 VHF radio transceiver
VOR-700 VHF navigation receiver

Operational status
In production and service.

Contractor
Rockwell Collins.

VERIFIED

Series 900 avionics system

Series 900 avionics include VHF, ADF, VOR, ILS, DME, altimeter, transponder, weather radar, TCAS and satellite communications systems. They meet industry and FAA requirements for equipment certified to Criticality Level 2. This requires tolerance to aircraft environments including 200 ms extended power interrupts, immunity to high-intensity radiated fields and conformity to the stringent environmental requirements of DO-160C. Conformity to the strict software documentation standards of DO-178C and the higher VHF FM interference levels required by ICAO Annex 10 are also included.

Series 900 is packaged in ARINC standard LRUs. Only the high-power systems require forced-air cooling, due to improvements in heat dissipation techniques. All other systems can be passively cooled.

All software is in Ada. External software loading capability has been enhanced and expanded to allow easy shop or on-aircraft modification.

Series 900 units are interchangeable with Series 700 products. This allows fleet commonality and use of the same top-level test equipment.

Operational status
In production and in service for Airbus A320/330/340 and Boeing 747-400, 757, 767 aircraft.

Contractor
Rockwell Collins.

VERIFIED

Standby Instruments for the Boeing 777

Rockwell Collins is supplying flat-panel colour LCD standby indicators for the Boeing 777. These 3 × 3 in instruments include the attitude indicator, airspeed indicator and altimeter. They are passively cooled.

Specifications
Dimensions: 76.2 × 76.2 × 215.9 mm
Reliability: >15,000 h MTBF

Operational status
Certified on Boeing 777. In production.

Contractor
Rockwell Collins.

VERIFIED

Synthetic Vision Information System (SVIS)

Rockwell Collins has been selected as leader of a team formed by the US National Aeronautics and Space Administration (NASA) for the development of synthetic vision technology. The SVIS will provide cockpit displays to give flight crews clear views of their surroundings in bad weather and darkness. The contract award will fund advancement of technology for air transport and business and regional aircraft applications.

Rockwell Collins's SVIS will offer enhanced situational awareness, real-time guidance and predictive alerting. It will also provide improved flight planning capabilities, intended to prevent crews from landing short, flying too close to the terrain, or inadvertently landing on an occupied runway. SVIS will provide intuitive guidance cues to reduce pilot workload and to allow pilots to compensate for errors made at the strategic level.

The SVIS is intended to raise the crew's situational so potential dangers are avoided before Terrain Awareness Warning Systems (TAWS) warnings become necessary. However, the SVIS is not a warning system. It will rely on traditional subsystems including Traffic alert and Collision Avoidance Systems (TCAS), Ground Proximity Warning Systems (GPWS), Flight Management Systems (FMS) and displays to provide warning functions.

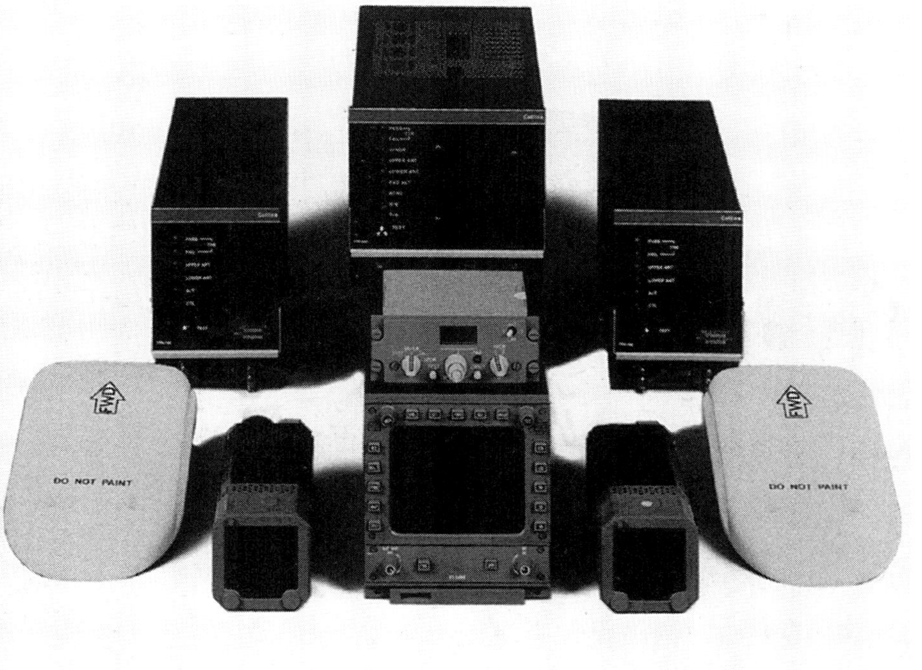

Units of the Rockwell Collins TCAS systems 0018065

The SVIS will be based on the Rockwell Collins Pro Line 21 architecture and it will present data on high-resolution Liquid Crystal Displays (LCDs). Software and/or hardware upgrades will make it possible to retrofit the concepts to Electronic Flight Instrument System (EFIS) and ElectroMechanical (EM) display systems on existing aircraft. In addition to the LCDs, Head-Up Display (HUD) formats are also being developed.

Operational status
Rockwell Collins serves as the team leader for this NASA aviation safety programme. Other team members include: American Airlines, Boeing, Delft University of Technology, Embry-Riddle Aeronautical University, the US Federal Aviation Administration (FAA), Flight Dynamics, Jeppesen and NASA.

SVIS is expected to be available for application in air transport and business jet markets within five years, from mid-1999.

Contractor
Rockwell Collins.

VERIFIED

TCAS II Traffic alert and Collision Avoidance Systems

Rockwell Collins Business and Regional Systems currently produces the TCAS-94 and TCAS 4000, TCAS II systems. Both systems are suitable for installation in business jets and turboprops, as well as for airliners.

TCAS-94
TCAS-94 tracks up to 150 targets simultaneously and displays as many as 30 at once, with the ability to switch between a short-range display in high-traffic areas and an extended long-range display. An integral part of the system is the advanced TDR-94D Mode S transponder, specifically designed to be lighter in weight and smaller in size than ARINC-compliant transponders used on larger transport aircraft.

Designed to complement specific cockpit requirements and different operator preferences, the Rockwell Collins TCAS-94 presents a variety of display configurations, whether the aircraft is equipped with electromechanical instrumentation, an Electronic Flight Instrument System (EFIS), or advanced Pro Line 4 or Pro Line 21 avionics.

TCAS-94 comprises the following major items or equipment:

R/T processor: TTR-920 or TTR-921
Top directional antenna: TRE-920
TCAS bottom antenna:
(omni): 237 Z-1
(directional, optional): TRE-920
Mode S R/T (diversity): TDR-94D
Mode S antennas (2): ANT-42
Control:
(TCAS): CTL-92T
(Mode S): CTL-92
(combined): TTC-920G
Displays
Traffic Advisory (TA): alternatives
(EFIS (MFD)): MFD-85C
EFD-4076
EFD-4077
Resolution Advisory (RA):
(EFIS (PFD)): EFD-4076
EFD-4077
Resolution Advisory - alternative Kollsman VSI/RAL:
(pneumatic, ARINC 575, VSI, 28 V DC): 47174-003
(pneumatic, Manchester, VSI, 28 V DC): 47174-004
(ARINC 429, VSI, 28 V DC): 47174-005
(Combined VSI/RA/Traffic)
(optional): TVI-920D
Power: 28 V DC; 115 V AC, 400 Hz

TCAS 4000
New hardware and software upgrades developed for TCAS 4000 support the addition of Automatic Dependent Surveillance-Broadcast (ADS-B) capability to the existing TCAS/Mode S collision avoidance function. These software upgrades and performance enhancements - defined under the FAA's latest Change 7.0 software standards - set the standard for next-generation, GPS-referenced air traffic management

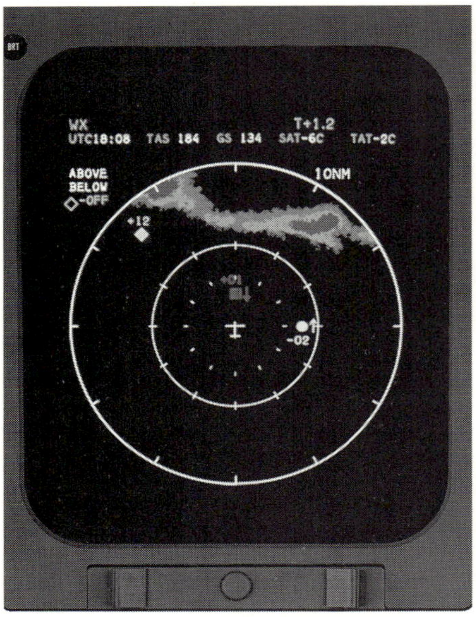

Pro Line 4 MFD format with weather and traffic
1999/0044991

Pro Line 4 PFD format with resolution advisory
1999/0044992

concepts, and ultimately the direct routing efficiencies of 'free-flight'.

With ADS-B, each aircraft uses its transponder to periodically broadcast its identity, altitude and location as defined by GPS co-ordinates. On request from air traffic control, or other TCAS-equipped aircraft, the ADS-B system will transmit additional data such as heading, next FMS waypoint and vertical speed.

Key features of the TCAS 4000 system include:

(1) DO-185A, Change 7.0 software, providing new verbal commands and display symbology, plus enhanced target surveillance in high-density areas above Fl 180;
(2) aural and visual warnings 15 to 35 seconds before convergence;
(3) increased display range greater than 100 n/miles;
(4) tracking of up to 60 aircraft targets;
(5) operation at up to 1,200 kt closing speed;
(6) enhanced escape/co-ordination manoeuvres;
(7) Comm D/Level 4 datalink for Mode S transponder;
(8) universal fit with analogue or digital interfaces;
(9) compact 4-MCU R/T design (optional 6 MCU standard unit available); AC/DC power in single unit.

The heart of the TCAS 4000 system is the 4 MCU TTR-4000 TCAS II Receiver/Transmitter, which incorporates all radar surveillance and computer processing functions. In addition, the standard TCAS 4000 system consists of top- and bottom-mounted TCAS antennas, Mode S transponder, L-band antennas, a TCAS/Mode S control panel and cockpit displays. Rockwell Collins offers a variety of TCAS cockpit display options, from integrated EFIS displays,

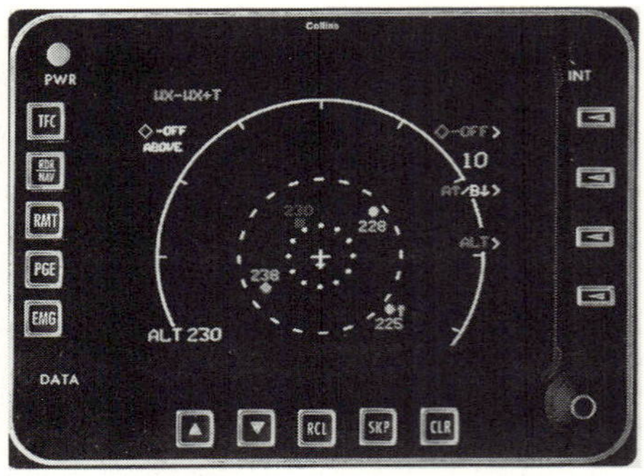

MFD-85C with traffic ***1999***/0044993

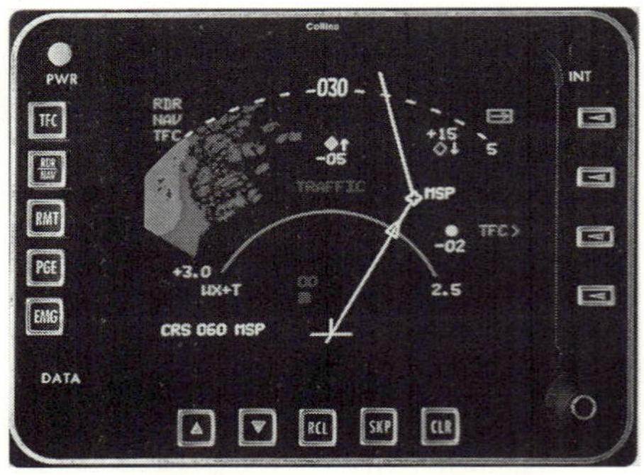

MFD-85C with navigation, weather and traffic
1999/0044994

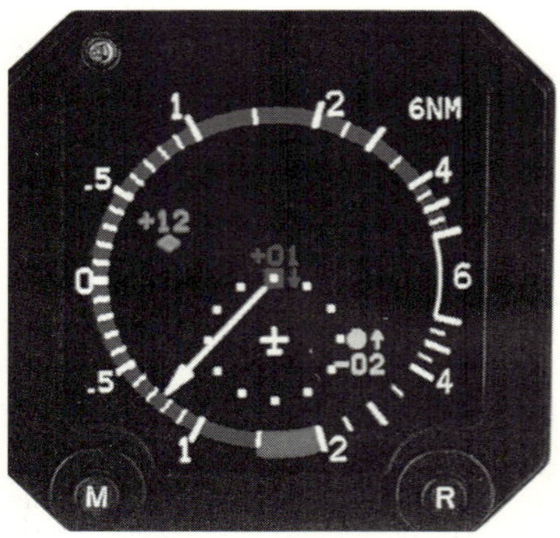

TVI-920D TCAS VSI/RA/TA indicator ***1999***/0044996

including Pro Line 4 and Pro Line 21 avionics, to individual TCAS instruments.

Major TCAS II Line Replaceable Units (LRUs)
TDR-94/94D Mode S transponders
The TDR-94 provides non-diversity operation, datalink capability.

The TDR-94D provides full-diversity operation: two receivers for top- and bottom-mounted antennas, receiver selection based on better signal, required in TCAS II/IV to ensure data exchange between two aircraft for manoeuvre co-ordination; TCAS interface compatibility; datalink capability.

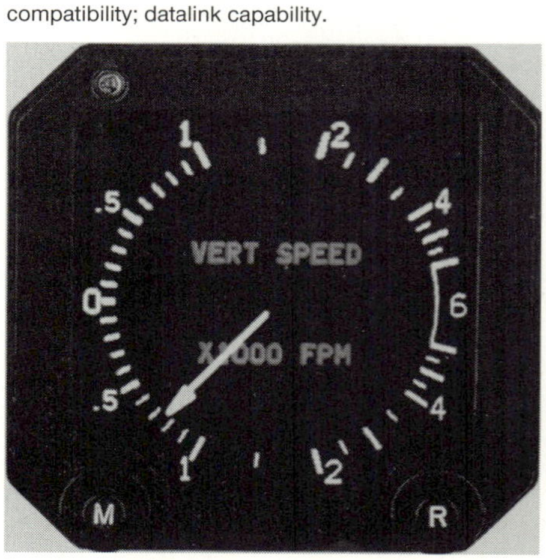

TVI-920D vertical speed indicator ***1999***/0044995

Certified to Class 2A and Class 3A air-ground datalink. The system supports DAAPS, ADS-B extended squitters and ACAS requirements.

TTR-920 Receiver/Transmitter (R/T) processor
Solid-state construction. No forced air cooling required. Automatic antenna calibration allowing non-critical installation. Whisper/Shout attenuator uses unique selective modulation scheme that reduces transmitter power and number of interrogations; each TCAS transmission is increased in power in 1 dB steps, up to 26 dB to create a wave of interrogations, with sidelobe suppression to limit the number of responses. Four-channel antenna beam-steering is included, that provides 5° RMS accuracy, eliminating the need for an external LRU.

TVI-920D VSI/RA/TA display
The TVI-920D integrates TCAS II RA and TA displays with a conventional vertical speed indicator display using colour LCD flat-panel technology. There are three display modes: vertical speed, TAs and proximate traffic; vertical speed, RAs, TAs and proximate traffic; vertical speed, pop-up mode - traffic information displayed only when a TA/RA exists.

TVI-920D is a direct replacement for existing 3 ATI VSIs, and they can integrate with all common VSI sources - pneumatic, analogue and digital.

Operational status
Rockwell Collins Business and Regional Systems' TCAS II systems have been selected by nearly 100 major airlines and regional carriers.

Contractor
Rockwell Collins.

VERIFIED

TDR-90 transponder

The TDR-90 is an air traffic control Mode A and C transponder with 4,096 codes and an altitude reporting capability of up to 126,000 ft when used with an encoding altimeter. It is a remotely controlled system designed primarily for general aviation.

The system has a transmitter output power of 325 W nominal (250 W minimum) on a frequency of 1,090 MHz. Positive sidelobe suppression is incorporated in order to provide a cleaner paint on the interrogator's trace. Two-way mutual suppression avoids interference with DME. Another feature is a strip-line duplexer to control receiver front-end noise while retaining high sensitivity and frequency stability regardless of antenna matching.

A built-in test facility for both the transmitter and receiver functions is included. Test signals are injected at just above the minimum sensitivity level to ensure that receiver, decoder, encoder and transmitter are functioning correctly.

The system's CTL-92 control unit has two-knob code selection, ident, self-test, standby and altitude reporting on/off controls. An optional system selection switch can also be incorporated for use in dual installations. The display is of the gas-discharge type. The TDR-90 electronic unit can, however, interface with most conventional transponder controllers as well as the CTL-92 unit.

Specifications
Dimensions: ¼ ATR short
Weight: 1.59 kg

Operational status
In production by Rockwell Collins Business and Regional Systems, and in service.

Contractor
Rockwell Collins.

VERIFIED

TPR-900 ATCRBS/Mode S transponder

Operating with the Air Traffic Control Radar Beacon System (ATCRBS) Mode A and C interrogators, as well as Mode S, the TPR-900 transponder is also compatible with ARINC 735 TCAS systems. It has interfaces for dual Gilham, synchro, ARINC 429 and ARINC 575 input interface ports and is compatible with several types of barometric altimeters.

The solid-state TPR-900 transmitter meets FAA requirements for Class 4 (Comm D) datalink. Up to four segments of downlinked extended length messages can be sent by the transmitter at nominally 450 W power with expansion to 16 segments for Level 5 operation. Reported altitude, discrete address, maximum airspeed, sensitivity control, TCAS control data and Mode S ground station identification are provided as outputs from the system.

The TPR-900 operates with a diversity selection function using two receivers to improve air-to-air surveillance for TCAS. When a signal is received at the two antennas located on top and bottom of the aircraft, diversity selection determines which provides the stronger interrogation signal. The proper reply, depending on the type of interrogation, is then transmitted through the most efficient transmission antenna. Transponder operation is unaffected by aircraft position in relation to other aircraft or ground stations.

The system conforms to the requirements of DO-181A Change 1 and is designed to fulfil all new environmental requirements. It has protection for Cat K lightning and Cat U High-Intensity Radiated Fields (HIRF). In addition, the TPR-900 is fully functional for up to 200 ms of power interruptions.

Specifications

Dimensions: 125 × 193 × 325 mm
Weight: 5.6 kg
Power supply: 115 V AC, 400 Hz, 45 VA
Frequency: 1,090 MHz
Temperature range: –40 to +70°C

Operational status

Updated by the TPR-901, but still in production.

Contractor

Rockwell Collins.

VERIFIED

TPR-901 ATCRBS/Mode S transponder

The TPR-901 is an ARINC 718, 4 MCU transponder, which operates with older air traffic control radar beacon interrogators in Mode A and Mode C, and with newer sensors that operate with discretely addressed interrogations in Mode S. The transponder is also compliant with ICAO standards and recommended practices for airborne collision avoidance systems (ACAS II, also known as TCAS Change 7.0). Interfaces are provided for dual Gillham encoding altimeters, as well as synchro, ARINC 429 and ARINC 575 air data systems. Diversity antenna ports for reliable air-to-air TCAS surveillance are standard.

The all-solid-state TPR-901 meets all US FAA requirements for Comm A, B and C digital datalink and ICAO Level 3 operations. Growth is provided to support Comm D, Levels 4 and 5 in future. Provisions are also included for direct connection to GPS and FMS facilities for extended squitter operation in Mode S and identification and altitude reporting to support Automatic Dependent Surveillance - Broadcast (ADS-B).

The TPR-901 has a data-loading interface for on-aircraft software updates. Provision is made for transmission of flight identification or radio call sign to meet future European regulatory requirements, and Downlink of Aircraft Parameter Sets (DAPS), a requirement for future CNS/ATM (Communications Navigation Surveillance/Air Traffic Management) operations. DAPS includes datalink of magnetic heading, airspeed, roll angle, track angle rate of change, vertical rate, track angle and ground speed for ATC computations of flight trajectory and air traffic management.

The TPR-901 is the latest Rockwell Collins transponder, updating the TPR-720 and TPR-900, which both remain in production.

The TPR-901 is qualified to the latest environmental and software verification specifications, including high-intensity radiated fields, lightning and power interruptions to 200 ms, without upset or damage.

Specifications

Dimensions: 125 × 193 × 325 mm
Weight: 5.6 kg
Power: 115 V AC, 400 Hz, 45 VA

Contractor

Rockwell Collins.

VERIFIED

TWR-850 Turbulence detecting Weather Radar

The TWR-850 uses solid-state electronics technology, has an integrated transmitter/receiver antenna and weighs less than 9 kg. The radar picture is displayed on a CRT which is part of the EFIS, using one or two WXP-850 control panels to select the various operating modes and ranges; two control panels enable pilot and co-pilot to select independent displays. Auto-tilt and ground clutter suppression are additional features which ease the important task of tilt management to discriminate weather from ground returns. A feature, offered to the general aviation and regional airline sectors for the first time, is picture rotation as the aircraft turns. Maximum weather detection range is 550 km, turbulence can be detected out to 85 km.

Specifications

Frequency: I-band, 9,345 MHz
Range: 550 km

Operational status

In production by Rockwell Collins Business and Regional Systems. In service.

Contractor

Rockwell Collins.

VERIFIED

VHF-21/22/422 Pro Line II VHF Radios

Designed primarily for general aviation aircraft of all types, the Rockwell Collins VHF-21/22/422 transmitter receivers are remotely controlled, rack-mounted sets with 20 W transmitter output. They are available in four versions. The A equipment covering the VHF band from 118 to 136.975 MHz and the B variant from 118 to 151.975 MHz. These A and B units utilise 25 kHz channel spacing, and versions with broader receiver bandwidths are available. The C equipment covers the VHF band from 118 to 136.975 MHz and the D variant the 118 to 151.975 MHz band. The C and D variants provide both 8.33 and 25 kHz channel spacing.

The radios use digital synthesis frequency generation techniques and are of all solid-state construction. They provide automatic carrier and phase noise squelch and automatic gain control and are designed to drive cabin audio systems of all types. Principal attractions are low weight, compactness and the low power consumption of 6.5 A during transmission. Consequently they require no forced-air supply and electronic section cooling is carried out by a combination of heatsink and convective air flow.

The VHF-21 can directly replace the earlier VHF-20 series radios. The VHF-422 is compatible only with CSDB or ARINC 429 controls.

Either hard or soft mounting may be used and all connections are made through a single connector on the rear of the casing.

Specifications

Dimensions: 84 × 86 × 355 mm
Weight: 2.1 kg

Operational status

In production and in service.

Contractor

Rockwell Collins.

VERIFIED

VHF-700A/B transceivers

The Rockwell Collins VHF-700A/700B is a VHF AM transmitter/receiver designed principally for air transport use which covers the band 118 to 135.975 MHz with channel spacing of 25 kHz. Typical transmitted power output is 25 W.

The system, which has been designed and developed to ARINC 716 specifications, has a number of advanced features including a new receiver design which eliminates mutual interference between equipment even when up to three transmitter/receivers are used in the same aircraft. A complete end-to-end self-test facility, which checks 99 per cent of all critical components, is built-in. The ARINC 429 serial databus is also tested to ensure that the system is receiving valid information from the remote tuning selector and centralised ARINC 604 fault monitoring capability.

The VHF-700 is microprocessor-controlled, of all solid-state construction and employs a temperature compensated crystal oscillator as the reference standard for its digital frequency generation synthesiser.

VHF-700B

The VHF-700B adds 8.33 kHz channel spacing capability to the VHF-700 range.

Specifications

Dimensions: 3 MCU short
Weight: 4.1 kg
Temperature range: –55 to +71°C
Altitude: up to 50,000 ft

Operational status

The VHF-700 is widely used on Boeing and Airbus airliners. It was first offered in 1997.

Rockwell Collins has received US FAA supplemental type certification for the VHF-700B radio with the AVTECH 7522-1-2 control panel.

Contractor

Rockwell Collins.

UPDATED

VHF-900/VHF-900B transceivers

The Rockwell Collins VHF-900 hardware was developed for Boeing CAT III GPS autoland procedures. The VHF-900B transceiver was designed to meet the January 1999 European requirement for 8.33 kHz channel spacing. It was selected by Quantas Airways to retrofit its fleet of Boeing 747-400 aircraft.

VHF-900 hardware is also being evaluated by Rockwell Collins as the basis of a system to provide graphical weather data to aircraft flight decks, using a VHF radio datalink operating at 31.5 kbps, at which speed a weather data bitmap file can be uplinked in 3 to 5 seconds (the same file, uncompressed, would require 15 minutes on an ACARS link).

Operational status

Rockwell Collins has received US FAA supplemental type certification for the VHF-900B radio with the AVTECH 7522-1-2 control panel.

Contractors

Rockwell Collins.

VERIFIED

VIR-32/33/432/433 Pro Line II navigation receivers

The VIR-32/33/432/433 are digital VOR/ILS navigation receivers designed for business or commuter aircraft. The VIR-32 can directly replace the VIR-30A or be installed in an all-digital aircraft. Several versions of the VIR-32 are available, but all are identical in format, the variations being selected by making the appropriate wiring connections. The system can receive 200 VOR/localisers and the associated 40 glide slope channels.

The VIR-432/433 is compatible only with CSDB or ARINC 429 controls.

Specifications

Dimensions: ATR short/dwarf
Weight: 2 kg
Power supply: 28 V DC, 1.4 A
Altitude: up to 70,000 ft

Operational status

In service.

Contractor

Rockwell Collins.

UPDATED

VOR-700A VOR/marker beacon receiver

The VOR-700A VHF omnidirectional range/marker beacon receiver incorporates digital technology combined with the background derived from previous industry standards such as the Collins 51RV-2, 51RV-4 and 51Z-4 VORs. The system was designed in accordance with ARINC 711.

All bearing signal baseband processing in the VOR-700A is accomplished digitally. The 30 Hz reference and variable signals are converted into digital form using a 12-bit CMOS analogue-to-digital converter and are thereafter handled by an Intel 8086 16-bit microprocessor. The ARINC 429 input/output functions are performed by an Intel 8048 microprocessor in conjunction with a Collins universal asynchronous transmitter/receiver large-scale integration-based circuit. Additional functions, such as self-test, auto-calibration and monitoring, are also conducted digitally.

Digital processing improves the accuracy of measuring bearings by reducing the effects of temperature variation and ageing. Implementation of 30 Hz bandpass filters in firmware, compared with previous analogue methods, permits improved tracking of ground station modulation frequency variations, increased navigation sensitivity and better rejection of undesired components in the modulation of received signals.

The VOR-700A parts count has been reduced by 40 per cent compared with the most recent analogue technology VOR systems such as the 51Z-4 marker beacon receiver and the VOR portion of the 51RV-4. The system is fully compliant with ARINC 711 and extends many parameters of previous generation equipment. In particular, bearing measurement accuracy is five times better and there are 40 per cent fewer adjustments. Centralised fault monitoring capability is offered.

Specifications

Dimensions: 3 MCU per ARINC 600
Weight: 4.5 kg
Power supply: 115 V AC, 400 Hz, 30 VA
Frequency:
(VOR) 108-117.95 MHz
(marker beacon) 75 MHz
Channel spacing:
(VOR) 50 kHz

Operational status

In production and in service.

Contractor

Rockwell Collins.

UPDATED

VOR-900 VOR/marker beacon receiver

The VOR-900 maximises bearing accuracy by minimising the effects of temperature variation and ageing by using digital techniques. Implementation of 30 Hz bandpass filters in firmware, instead of analogue techniques, provides improved tracking of ground station modulation frequency variations. It also contributes to increased navigation sensitivity and better rejection of undesired components of received signal modulation.

Low parts count, quality components and straightforward design all contribute to reliable operation of the VOR-900. The system accommodates ground station and environmental anomalies. Functional test is provided by employing microprocessors to control a test sequence using digital test signal synthesis. LRU and card testing are simplified by an organised modular design which results in easily tested functional modes.

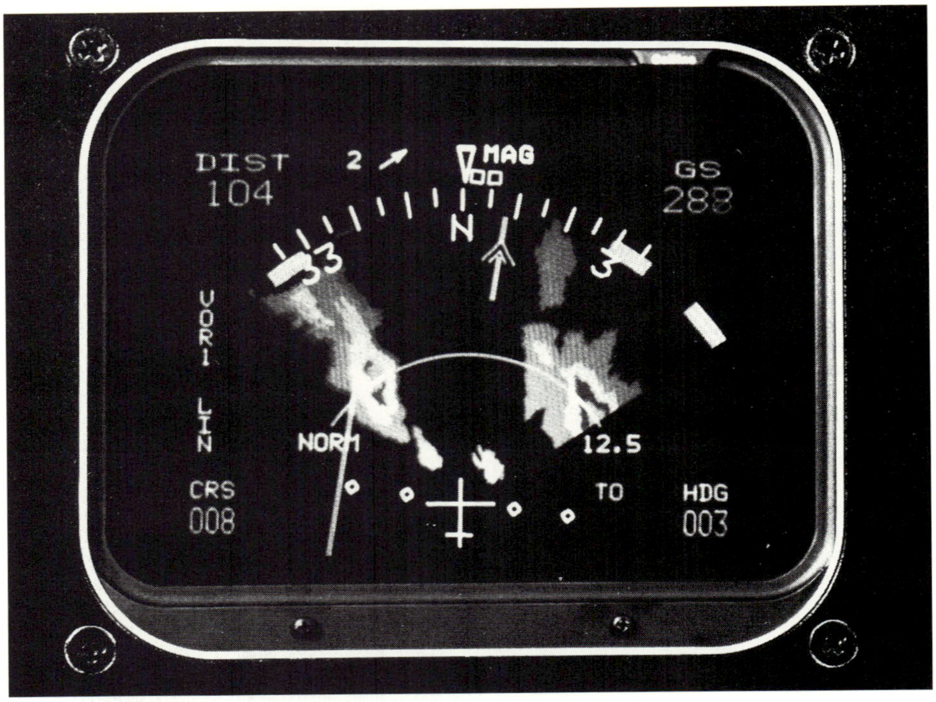

Rockwell Collins WXR-350 weather radar display

Specifications

Dimensions: 3 MCU per ARINC 600
Weight: 4.08 kg
Power supply: 115 V AC, 400 Hz
Frequency:
(marker beacon) 75 MHz
(VOR) 108-117.95 MHz
Channel spacing:
(VOR) 50 kHz
Temperature range: –55 to +71°C

Contractor

Rockwell Collins.

UPDATED

VP-110 voice encryption device

The VP-110 is a voice encryption device designed for use with airborne radio communications systems. The unit is packaged in a ½ ATR short unit and requires 28 V DC. A companion unit, the VP-100, performs the same function for fixed-station radios. Although aimed primarily at HF radios, the system will work equally well on both VHF and UHF narrowband equipment and ordinary telephone lines with the addition of a TA-110 adaptor.

The equipment is intended for a range of uses such as law enforcement, business, diplomatic, government agency and selected military voice transmission applications. It has been designed for the encryption of sensitive transmissions.

The system eliminates all syllabic content in the encrypted mode while retaining clear voice quality and recognition. For transmission, the voice is converted into analogue signals and divided into low- and high-band frequency ranges. It is then encoded and transmitted in a random mode with regard to time and frequency.

Public keying is provided, enabling private conversations between two stations without prior manual exchange of a recognition code. In this method of communication, the operator selects the mode and the two units exchange a set of numbers, using a complex mathematical algorithm, which in effect establishes a signature for connecting private conversations. Eight codes, or key variables, can be entered in the unit microprocessors, providing 10×7^{19} code possibilities.

Rockwell Collins produces two versions of the VP-110: one for the US market and one for export. The former uses a Data Encryption Standard (DES) algorithm while the latter is provided with a Rockwell-developed algorithm. An Over-The-Air Rekeying (OTAR) option is available.

Operational status

In production and in service.

Contractor

Rockwell Collins.

VERIFIED

VP-116 voice encryption device

Rockwell Collins' VP-116 voice encryption device is fully compatible with narrowband communications channels including UHF, VHF, HF-SSB and telephone. The VP-116 is an upgrade of the earlier VP-110 system. The main features of the system are:

- Time and frequency division voice encryption
- 10^{19} encryption key variables
- Excellent voice quality and voice recognition
- Reliable in-band FSK synchronisation
- Data encryption option
- Storage of eight key variables
- Extensive self-test capability
- Clear/secure and local/remote operation
- Over-the-air Re-keying (OTAR) of key variables
- Dual algorithm capability
- Night Vision Goggle (NVG) compatibility
- Exportable version available

High voice security is provided by a patented combination of time division and frequency scrambling. The scrambling process is implemented entirely with digital techniques and is controlled by a proprietary non-linear algorithm. This algorithm uses a 64-bit key variable, which provides 10^{19} different combinations. Up to eight key variables may be stored in the VP-116 at one time, allowing flexibility of key management. The multiple key variables may be used for different time periods of for different nets.

The VP-116 has analogue audio interfaces for easy integration into a wide variety of communications systems. Transmit audio, receive audio and push-to-talk key are the only three signals from the communications system that need to be routed through the VP-116. In the clear mode, both transmit and receive audio signals bypass the voice privacy circuits to provide clear voice transmission and reception. In the privacy mode, if a private signal is not being processed, any clear signals received will automatically bypass the voice privacy circuits to provide clear voice reception.

Operational status
In production.

Contractor
Rockwell Collins.

NEW ENTRY

WXR-350 weather radar

Rockwell Collins introduced the WXR-350 weather radar in 1984. It is intended to work specifically in conjunction with electronic flight instrument systems and detects four levels of precipitation, the fourth being displayed in magenta. There is also a path attenuation correction alert function which informs pilots of possible weather cells which are hidden behind heavy rainfall areas. Antennas, of 12 and 18 in, are available to suit light turboprops to heavy corporate jets. The WXR-350 employs a proven magnetron with nominal peak power of 5 kW to give avoidance ranges of up to 300 n miles.

Operational status
In service on a wide variety of aircraft. No longer in production.

Contractor
Rockwell Collins.

UPDATED

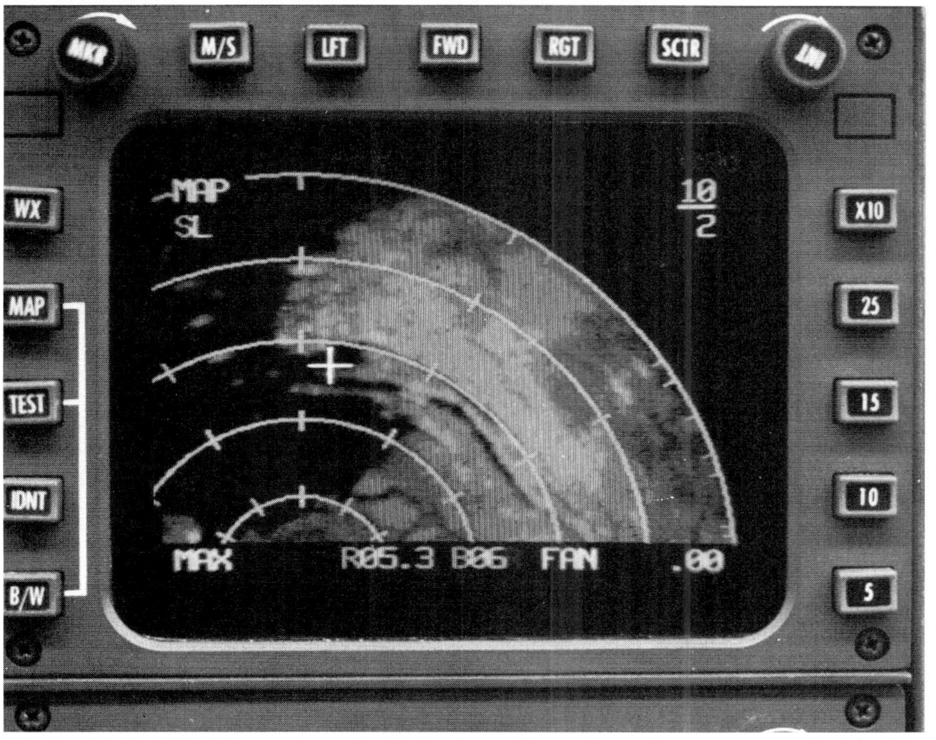

A Rockwell Collins WXR-700 ground-mapping display shown offset to give maximum coverage to the right of the aircraft's track

WXR-700( ) Forward-Looking Windshear radar (FLW)

Rockwell Collins' WXR-700( ) colour weather radar system is an all solid–state weather radar system suitable for new or retrofit applications with dedicated radar indicators or with electronic flight instruments. More recently, Rockwell Collins has added a Forward-Looking Windshear (FLW) detection capability to the WXR-700( ) radar system. This capability, per ARINC 708A characteristic, provides forward looking windshear detection out to 5 nm and turbulence detection to 40 nm. In addition to commercial airline applications, Rockwell Collins was selected to provide the US Air Force with the first Commercial-Off-The-Shelf (COTS) FLW radar with skin paint function in conjunction with the PACER CRAG programme.

The WXR-700( ) comprises four units: a slotted array flat-plate antenna (for good sidelobe reduction), a microprocessor-controlled transmitter/receiver, a display unit and a control unit. The microprocessor control system in the transmitter/receiver supervises all control and data transfers, programmes and controls the RF processes such as pulsewidth, bandwidth and PRF selection and directs antenna scan and stabilisation. The unit also contains circuits to reduce ground clutter suppression when operating in the weather mode. An optional feature is pulse-pair Doppler processing, whereby, with the addition of a single circuit board to the transmitter/receiver, the horizontal velocity of rainfall can be sampled. This technique is recommended by the NSSL as being particularly suitable for the analysis of storm cells.

The CRT indicator uses a high-resolution shadow-mask tube with a multicolour display scheme. The CRT provides alphanumeric identification of radar modes and incorporates annunciators and controls. The receiver has a sufficiently wide dynamic range to detect the Z-5 and Z-6 levels of rainfall that indicate a high probability of hail.

It provides both visual and aural alerts of windshear events occurring up to 90 seconds ahead of the aircraft flight path. The radar is automatically activated below 1,200 ft AGL and scans a detection path of 30° either side of the aircraft centreline.

The WXR-700( ) is available at 9.3 GHz X-band frequency as the WXR-700X.

The WRT-701X Receiver/Transmitter (R/T) features a crystal controlled multipulsewidth transmitter and a truly coherent receiver with programmable bandwidth signal processing. These features provide pulse-pair processing to allow ground clutter suppression, path attenuation compensation, Doppler turbulence detection and provide a high-resolution picture when used with an EHSI or 5ATI LCD. Microprocessor control achieves superior stabilisation performance (with both digital and analogue attitude inputs), flexible installation configuration and programmable growth for future requirements.

Quality enhancements have been implemented in the WRT–701X R/T. Improvements include a single-channel multiplier assembly, redesigned power supply and amplifier and 150 W reduction in power consumption, while maintaining 150 W peak power output.

In support of OEM decisions to include FLW radar on most production aircraft, Collins has introduced the -612, -623, and -633 versions of the WRT-701X FLW radar containing the newly required EGPWS interface, enhanced BITE and enhanced fault logging memory.

Key features of the FLW radar enhancements include:

- Added EGPWS interface for display alert prioritisation
- Revised Boeing FLW parameters:
- Elimination of 'advisory' icon displays from 3 to 5 nm
- Expansion of takeoff caution alert inhibits to 80 kt to 400 ft AGL
- Maintain takeoff warning alert inhibits from 100 kt to 50 ft AGL
- Expanded 'tilt code' fault message reporting
- Data loadable software via RS–232
- Backwards compatible with existing FLW installations.

Specifications

for C- and X-band systems (equivalent to NATO G- and I-band)

Dimensions:
(antenna) ARINC 708
(transmitter/receiver) 8 MCU per ARINC 600
(indicator) ARINC 708 Mk II
(control unit) 146 × 67 × 152 mm

Weight:
(antenna) 12.25 kg
(transmitter/receiver) 12.25 kg
(indicator) 8.16 kg
(control unit) 1.04 kg

Frequency:
(C-band) 5,440 MHz
(X-band) 9,330 MHz

Power output:
(C-band) 200 W
(X-band) 100 W

PRF:
(C-band) 180-1,440
(X-band) 180-1,440

Pulsewidths:
(C-band) 2-20 μs
(X-band) 1-20 μs

Range:
(C-band) 445 km
(X-band) 593 km

Operational status
In production and service. The WXR-700 FLW radar has received final type certification approvals for installation in Airbus A-319, -320, -321, -330 and -340. It has also been certified on Boeing aircraft including Boeing -737, -747, -757, -767 and -777.

The US Navy has selected the WXR-700 FLW and Rockwell Collins' FDS-255 flight display system for its C-9 aircraft upgrade.

A special 'missionised' version of the commercial WXR-700 radar, designated the FMR-200X (see separate entry), is being fitted to US Air National Guard (ANG) C/KC-135 PACER CRAG upgrade aircraft as a replacement for the AN/APN-59 radar. The FMR-200X radar will provide a colour weather system incorporating FLW, skin paint detection and limited ground map capability.

A derivative version of the WXR-700, designated WXR-701X, upgraded with Doppler turbulence detection has been selected by the UK Ministry of Defence for the upgrade of its C-130K aircraft.

Contractor
Rockwell Collins.

UPDATED

NeoAV IIDS Integrated Instrument Display Systems

The Rogerson Kratos NeoAV Integrated Instrument Display Systems comprise a family of cockpit displays using advanced Active Matrix Liquid Crystal Display (AMLCD) technology. IIDS can efficiently replace up to 38 conventional instruments, as well as the caution advisory system. Standard features include engine instrument indication, fuel quantity measurement indications, hydraulic systems indications, electrical system indications, caution, warning, and advisory messages, and outside air temperatures. IIDS also has FADEC or EEC compatibility, trend monitoring and weight and balance synoptic indications.

Special helicopter capabilities include mast torque indications and chip detector warning outputs.

Functions are monitored and managed with colour AMLCDs, featuring redundancy and maintenance capabilities. Multiple reversionary display modes operated by bezel-mounted buttons enable the system to be 'fail operational' in case of in-flight malfunctions. Built-In Test Equipment (BITE) is available. Metric or English displays are available, as is built-in non-volatile memory for engine history and maintenance logs. Engine performance instruments are typically presented on one display, while aircraft systems instruments are on a second display. Solid-state components, redundant architecture and derating assure high reliability. IIDS also offers significant reductions in weight, heat and power consumption.

Operational status

IIDS displays have been certified for the Bell 430, Sikorsky S76 and Canadair CL415. Certification is in process for the Casa 212, 235 and 295, Bell 412 and 427, as well as for Zeppelin Airship. Rogerson Kratos IIDS installations will be factory standard on all Bell twin-engined helicopters (Models 212, 412, 427 and 430).

Contractor

Rogerson Kratos, a Rogerson Aircraft Corporation subsidiary.

VERIFIED

The Rogerson Kratos NeoAV IIDS display system was fitted to the Bell 427 helicopters as it entered flight test
0022211

Bell 430 cockpit showing both the IIDS Integrated Instrument Display System and the Model 550 EFIS Electronic Flight Instrument System
0018163

NeoAV Model 500 EFIS Electronic Flight Instrument System

The NeoAV Model 500 is a flat-panel electronic flight instrument system that provides EFIS control and reversionary (cross side) switching and ADI/ASI composite mode; it features a fully self-contained computer-controlled symbol generator using colour active matrix liquid displays. It incorporates a high-speed 32-bit microprocessor-based system with BIT and flight recording capability. Each of the units is interchangeable and networked for maximum redundancy.

Operational status

Certified by the US FAA on the B 727-100 in 1994 and on the Bell Model 430 in 1996. The NeoAV Model 500 replaces existing HSI/ADI instruments in Federal Express Boeing 727 and DC-10 aircraft. Also selected for the Boeing 707 and 737, Gulfstream GII and Bell Helicopter Model 430 helicopter. Selected for the Bell Model 427.

Contractor

Rogerson Kratos, a Rogerson Aircraft Corporation subsidiary.

VERIFIED

NeoAV Model 550 EFIS Electronic Flight Instrument System

The NeoAV Model 550 is an Active Matrix Liquid Crystal Display (AMLCD) Electronic Flight Instrumentation System (EFIS). The system is a microprocessor-based, self-contained unit that includes the symbol generator, Colour Active Matrix Liquid Crystal Display (CAMLCD) and interface hardware/software. The system design also incorporates full-time, Built-In Test (BIT) operated by a high-speed 32-bit microprocessor. The NeoAV 550 EFIS also includes a manual, self-test feature for the pilots and maintenance personnel. Faults are

NEOAV 500 displays are shown here in a Boeing 727 cockpit

continuously recorded and stored in non-volatile memory for review after the aircraft has landed. Each NeoAV 550 component in the system is interchangeable (single part number) and is networked for maximum redundancy. Both analogue and digital interfaces are available, offering efficient, 'plug in' display interchangeability. Self-contained (bezel-mounted) EFIS control or remote controllers are available as options.

Operational status

NeoAV was certified on the Boeing 727 in 1994. The Bell Helicopter 430 was certified in 1996, follow-on certifications were for the Bell 412 and Bell 427. Rogerson Kratos EFIS installations will be factory standard on all Bell twin-engined helicopters (Models 212, 412, 427, and 430). Other certifications include the Agusta 109 as a standard factory fit in 1998, and the US Air Force C18B EFIS upgrade in 1998.

Contractor

Rogerson Kratos, a Rogerson Aircraft Corporation subsidiary.

VERIFIED

Angle of attack computer/ indicator

The angle of attack computer/indicator provides the pilot with a continuous display of aircraft lift information on a decimal scale, with 1.0 representing the stall. The display is valid regardless of bank angle, aircraft weight or wing configurations.

The computer/indicator face is scaled red below 1.1 V, amber from 1.1 V to 1.3 V and black from 1.3 V to maximum speed. It also features a settable bug and index slaved to each other, which can be set for a desired airspeed target between 1.2 V and 1.5 V. Centring and maintaining the angle of attack pointer within the index will result in the selected speed target.

The system computer drives the ADI fast/slow pointer, a 2 in round angle of attack indicator and 1Safe Flight's speed indexer lights. The panel-mounted angle of attack indicator displays aircraft lift information. An approach reference is provided and the display is valid for all flap positions.

Operational status

The system has been certified on the Raytheon Hawker 800.

Contractor

Safe Flight Instrument Corporation.

VERIFIED

AutoPower® automatic throttle system

The role of the Safe Flight Instrument Corporation AutoPower® system is to provide aircrew with precise control of Indicated Air Speed (IAS) throughout flight with minimum crew workload. The system comprises four main components: a glareshield-mounted digital Indicated Air Speed (IAS) display; an AutoPower computer; a clutch pack; and a set of yoke-mounted increase/decrease switches. An AutoPower engage switch, cockpit annunciators and circuit breakers are also required. The AutoPower computer contains all the electronics necessary for system operation. The clutch pack contains one clutch per throttle and a servo drive motor.

AutoPower can be disengaged by pressing the Go-Around switch. In addition, the pilot can override the AutoPower system at any time, by moving the throttles with normal throttle control force.

Contractor

Safe Flight Instrument Corporation.

VERIFIED

N1 computer and display

The Safe Flight Instrument Corporation N1 computer system continuously monitors altitude and air temperature from the aircraft's air data computer, together with anti-ice and environmental control system mode logic from the applicable aircraft systems. The N1 computer interpolates and displays the N1 thrust setting in real time for takeoff, climb, cruise and go-around, each based on the airplane flight manual performance charts. The panel-mounted display instantly displays the appropriate target N1 thrust setting for each flight mode selected by the crew.

Selecting takeoff/go-around will display the flight manual target N1 thrust setting for takeoff. Once airborne, select N1 thrust settings for climb, cruise, or go-around with the mode switch, and the system continuously displays the appropriate N1 thrust setting schedule for the selected mode and given conditions.

Safe Flight Instrument Corporation N1 computer displays **2000**/0081842

Contractor

Safe Flight Instrument Corporation.

VERIFIED

Performance computer system

Safe Flight's performance computer system is designed to provide increased accuracy in obtaining the best balance between fuel economy and airspeed, while reducing crew workload in the operation of business jets. The system computes and displays the speed appropriate to maximum specific range under the prevailing conditions. Alternatively long-range speed, percentage of maximum specific range being achieved, optimum cruise altitude or several other parameters may be selected as the baseline. Inputs from various aircraft sensors, together with stored performance data, are used in these computations which take into account all external factors such as the effects of fuel burn or change of wind velocity appropriate to the type of aircraft. The computer is based on an Intel 8085 microprocessor and has 28 kbytes of read-only memory and 2.25 kbytes of RAM. The computer can supply data to compatible autothrottles or thrust management systems produced by Safe Flight which are controlled by means of a control/display unit.

Specifications

Dimensions:
(computer) ⅜ ATR short
(control/display unit) 57 × 146 × 121 mm
Weight:
(computer) 3.4 kg
(control/display unit) 0.55 kg
Power supply: 115 V AC, 400 Hz

Operational status

In production and in service.

Contractor

Safe Flight Instrument Corporation.

VERIFIED

Powerline detection system

Safe Flight Instrument Corporation powerline detection system is designed to warn helicopter pilots of imminent approach to powerlines. It depends on the ability of an electromagnetic detection system to detect the presence of electromagnetic radiation at powerline frequency. It provides both visual and audio warnings to the pilot.

A threshold warning light illuminates and remains alight as long as the powerline is sensed. An unmistakable audio alert, which clicks like a Geiger counter, is heard through the audio system and increases in frequency as the helicopter approaches the powerline. The audio alert can be muted on the display unit, and a mute light also illuminates. The audio alert remains silent unless the powerline detection signal strengthens, signifying that the helicopter is getting closer to the powerline.

Safe Flight claims that its system will provide 300 to 500 ft detection on powerlines, and that the detector used is not triggered by other electromagnetic sources such as radio, radar and TV systems.

Operational status

In development. Production was forecast for late 1999.

Contractor

Safe Flight Instrument Corporation.

VERIFIED

Recovery Guidance System (RGS)

Safe Flight offers recovery guidance, an optional enhancement of the basic windshear warning system. With the combined WindShear Warning/Recovery Guidance System (WSW/RGS), as soon as the warning occurs continuously, computed pitch guidance for recovery is displayed on the flight director command bars. The system also provides pitch guidance for take-off and go-around on the same instrument, to maintain pilot familiarity with use of the system and promote confidence in it. The pilot does not have to change his flight scan or depart from accustomed procedures during the crucial emergency escape manoeuvre. The company maintains that, by following the command bars, the best possible climb profile to maximise the performance capabilities of the aircraft will be achieved.

The system is armed automatically, even if the flight director is turned off, by the windshear warning system's alert output, but only becomes operative, displaying pitch guidance for recovery, when the pilot activates the GA switch. System logic may be programmed to perform these switching and display functions automatically.

Safe Flight's RGS displays pitch attitudes up to, but not in excess of, the stick shaker target. However, the RGS can be programmed to display stick shaker target information on the ADI slow/fast indicator alongside the pitch guidance display on the command bars. With this optional function, when the pilot activates the GA switch for recovery guidance, the slow/fast indicator changes from a speed mode to shaker mode with the slow bar representing shaker target. The pilot then has a continuous visual indication of his margin to stick shaker. Internal system monitoring and a self-test function ensure system reliability.

Operational status

Recovery guidance is presently available in combination with the windshear warning system in a single ¾ ATR box, or where Safe Flight's Speed Command of Attitude and Thrust (SCAT) system is desired (or already installed), through tie-in of the windshear warning and SCAT system computers.

Contractor

Safe Flight Instrument Corporation.

VERIFIED

Windshear warning system

Safe Flight's airborne windshear warning system provides a voice alert to the crew of high-performance aircraft at the start of an encounter with hazardous low-level windshear. The system is operative during take-off and approach and is a computer-based device which, using conventional sensing elements, resolves the two orthogonal components of a wind gradient with altitude and provides a threshold alert that an aircraft is encountering a potentially hazardous situation. The vectors concerned are horizontal windshear and downdraught drift angle.

Horizontal windshear is derived by subtracting groundspeed acceleration from airspeed rate. The latter term is obtained by passing airspeed analogue data from the airspeed indicator or the air data computer through a high-pass filter. Longitudinal acceleration is sensed by a computer integral accelerometer, the output of which has been summed with a pitch attitude reference gyro to correct for the acceleration component due to pitch. A correction circuit is employed to cancel any errors due to prolonged acceleration. This circuit has a 'dead band', equivalent to 0.2° of pitch, which prevents correction for airspeed rates of less than 0.1 kt/s. Summed acceleration and pitch signals are fed through a low-pass filter, the output from which is summed with the airspeed rate signal to give horizontal windshear.

The vertical computation for downdraught drift angle is developed through the comparison of measured normal acceleration with calculated glide path manoeuvring load. Flight path angle is determined by subtracting the pitch attitude signal from an angle of attack signal sensed by the stall warning flow sensor. This is fed to a high-pass filter and from there to a multiplier to which the airspeed signal has been applied. Thus, the flight path angle rate, corrected for airspeed, provides the computed manoeuvring load term. This is compared in a summing junction with the output of a normal computer integral accelerometer and the failure of the two values to match is the indication of acceleration due to downdraught. The acceleration, when integrated, is the vertical wind velocity and is further divided by the airspeed signal to compute the downdraught angle.

The outputs of both horizontal and vertical channels are determined solely by the atmospheric conditions and ignore manoeuvres that do not increase the total energy of the aircraft. Windshear correctly compensated by increased engine thrust shows no change in airspeed in the presence of an inertial acceleration as thrust is applied. If, however, the shear goes uncorrected, an acceleration or deceleration becomes apparent. Similarly, in the case of the vertical component, a vertical displacement compensated by the crew shows a positive flight path angle rate in a downdraught with less than the computed incremental normal acceleration. Correspondingly, in a downdraught for which the drift angle is allowed to develop, a negative angle rate at a near constant 1 *g* results in the same computation and output. This is important to the crew as it eliminates the possibility that their actions in anticipating or countering windshear might well mask the condition as far as the warning system is concerned.

Both downdraught drift angle and horizontal windshear signals are combined and the resulting output fed through a low-pass filter to the system computer. This provides two output signals: a discrete alert and, through a voice generator, an audio alert. Warning output is set at a threshold of –3 kt/s for horizontal shear and –0.15 rad downdraught drift angle, or for any combination of the two components which, acting together, would provide an equivalent signal level. According to Safe Flight, any wind condition requiring additional thrust equivalent to 0.15 *g* to maintain glide path and airspeed will result in a non-stabilised approach.

A crossover network is employed to sense zero crossovers of the combined windshear warning signal and this is sampled every 25 seconds. If the warning signal does not pass through a band close to zero, the network automatically provides failure indication, alerting the crew to the fact that the unit is inoperative. A self-test function activated by the pilot is also built into the system.

The company has now entered into a licensing agreement with Boeing Commercial Airplanes for the further exploitation of windshear warning technology. This agreement provides for the licensing to Boeing of Safe Flight's existing and pending patents and proprietary data in the areas of windshear detection, alert and escape guidance.

The agreement will facilitate the incorporation of windshear warning capabilities on new models of Boeing commercial aircraft. It is anticipated that the technology will be an added feature on aircraft including the 737-300, 757, 767 and 747-400 models. A version will also be offered for retrofitting on Boeing aircraft currently in service.

Specifications

Dimensions: ¼ ATR
Weight: 2.72 kg

Operational status

In production and service. Certified for Cessna Citation III, Falcon Jet Falcon 50, and Raytheon Hawker 800.

Contractor

Safe Flight Instrument Corporation.

VERIFIED

Integrated radio control panel

The integrated radio control panel is designed to provide the pilot and/or co-pilot with the means to manage all communications equipment on board a commercial or military aircraft. Two interface units are available: MIL-STD-1553 for military applications and ARINC 429 for commercial aircraft.

The primary display medium is a CRT which displays mode and frequency information. In addition to the CRT, several dedicated controls are provided for those functions which require immediate access. The unit has a non-volatile memory for power-off data retention.

Specifications

Dimensions: 203 × 146 × 200 mm
Weight: approx 4.5 kg

Contractor

SCI Systems Inc.

VERIFIED

AMS-2000 Altitude Management and alert System

The Shadin AMS-2000 Altitude Management and alert System uses information from Mode-C altitude encoders to deliver: time-based altitude alerting, rather than fixed altitude buffers, to notify the pilot 15 seconds before an altitude target or limit is reached - regardless of rate of climb or descent; automatic calculation and display of density altitude; real-time display of instantaneous vertical speed without the inherent lag of the aircraft's static system; calculation and display of aircraft and engine performance percentage.

Specifications

Dimensions: 38 × 80 × 146 mm
Power: 10 to 28 V DC, at 300 max
Weight: 0.25 kg

Contractor

Shadin Co Inc.

VERIFIED

AMS-2000 Altitude Management and alert System
0001445

DigiData fuel/airdata system

DigiData provides fuel management, navigation and airdata functions including:

E-6B data: Pressure; altitude; density altitude; outside and true air temperature; wind aloft; wind component (speed and direction); indicated and true airspeed; ground speed; Mach; and instantaneous vertical speed.

Fuel flow: Left and right fuel flow; fuel used; fuel remaining; fuel to and reserves at destination; range; cruise efficiency (nautical mileage); and endurance.

Navigation: Heading; ground track; and magnetic variation.

DigiData is connected to aircraft systems to measure Indicated AirSpeed (IAS), pressure altitude, Outside Air Temperature (OAT), heading and fuel flow. From these raw data, TAS and temperature data are calculated by the microprocessor and combined with data from the navigation receiver to calculate automatically wind aloft, fuel needed to destination, specific range and density altitude.

Pressure altitude data is automatically used by most GPS receiver manufacturers to substitute for the fourth satellite range if it is not in view to provide 24-hour, three-dimensional position accuracy with only three satellites, without manual entry of altitude.

The engine fuel flow interface makes the fuel management data totally dynamic, basing it on real-time, winds aloft readings.

With an external Shadin ARINC 429 converter, DigiData can also drive EFIS displays and multisensor navigational management systems.

Specifications

Dimensions: 3.125 (round) × 6½ in (deep) (79 (round) × 165 mm (deep))
Power: 9-35 V DC, 1 W nominal
Weight:
(indicator) 623.7 g
(transducer) 453.6 g
Flow rate: up to 450 gal/h
Max usable fuel: 1,800 gal
Accuracy: ±1%
Operating temperature: –20 to +55°C
Input: pitot pressure, static pressure, outside air temperature, heading synchro, fuel flow
Output: IAS, TAS, ground speed, Mach, P.ALT, D.ALT, OAT, TAT, wind aloft, wind component, fuel flow, fuel used, fuel remaining, cruise efficiency, endurance, left fuel used, right fuel used, IVS, heading, track, magnetic variation
Output format: RS-422/RS-232
ARINC 429 (optional)
Compatible receivers: ARNAV (R-15, R-30, R-40, R-50, STAR 5000, FMS 5000, FMS 7000); Trimble (2000, 2000A, 2100, 3000, 3100); AlliedSignal (KLN90, KLN90A, KLN88); II Morrow (604, 612, 614, 618); Northstar; Garmin; Magellen

Contractor

Shadin Co Inc.

VERIFIED

Aircraft intercoms

Sigtronics Corporation makes a range of aircraft, panel-mount, intercom systems; -4 and -400 systems indicate 4-place versions; -6 and -600 systems indicate 6-place versions; the -800 is an 8-place variant:

SAS-440/-640 auto squelch panel-mounted intercom series
These intercoms virtually eliminate the need to constantly readjust the squelch during flight. The SAS-440 has radio priority to assure that the only voice heard by air traffic control is that of the crew. SAS-440 supports four headsets. SAS-640 supports six headsets.

SCI-4/SCI-6 dual squelch voice activated intercoms
Retrofit system for SPA-400/SPA-600 versions, with 'pilot', 'crew' and 'all' functions.

Sigtronics SAS-440 auto squelch panel-mounted intercom **1999**/0044831

Sigtronics SDB-800 intercom system **1999**/0044832

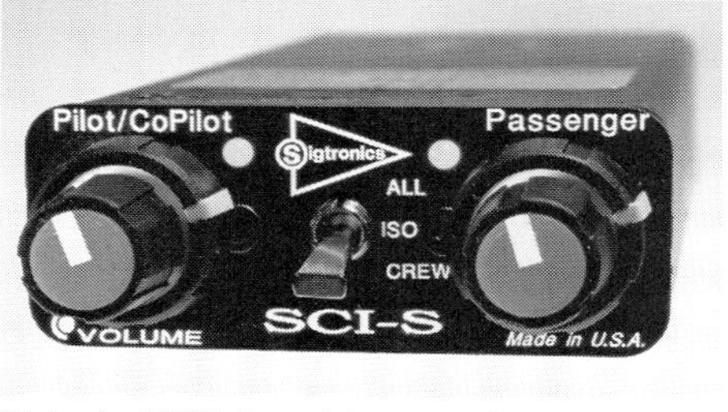

Sigtronics SCI-S stereo intercom system **2000**/0062741

SDB-800 dual audio panel intercom
Effectively, a dual SPA-400/-600 installation can support two pilots, and up to eight headsets.

SPA-400/600 intercom series
An industry standard intercom for many years, the SPA-400/600 has radio priority, and a pilot fail-safe feature, which ensures that the pilot will always hear the radios, even if the intercom is set to off.

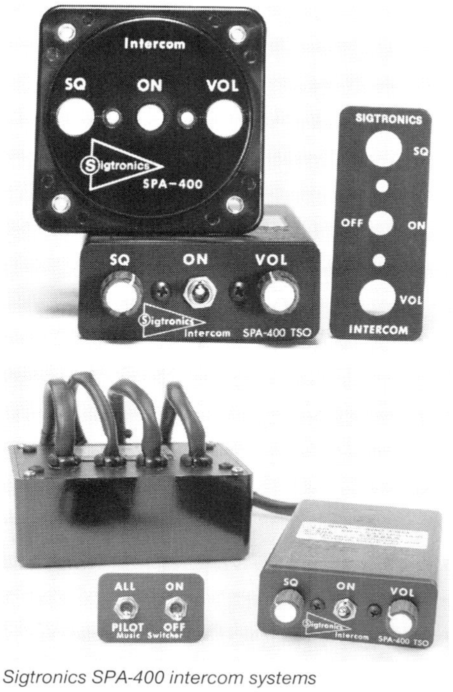

Sigtronics SPA-400 intercom systems **1999**/0044833/0044834

SPA-400N/-600N intercom series
Specially designed version of the SPA-400/-600 series for very high noise cockpits; helicopters; warbirds and ultralights.

ST-400/-600 stereo intercom series
A full stereo version of the SPA-400/-600 series.

ST-440/-640 stereo intercom series
A full stereo version of the ST-440/-640 series

SCI-S stereo dual squelch voice activated intercom
A 6-place, stereo version of the SCI-6 system.

Contractor
Sigtronics Corporation.

VERIFIED

SAMS-1000 audio management system

SAMS-1000 is an audio management system designed for use with up to six radios and five navigation receivers, in addition to siren, intercom and public address functions. Each operator can independently monitor any or all radios. SAMS-1000 includes a 'hot microphone' or push-to-talk intercom; voice-operated intercom is an option. SAMS-1000 is designed for use in airborne law enforcement, and search and rescue operations.

Contractor
Sigtronics Corporation.

VERIFIED

Sigtronics SAMS-1000 audio management system **1999**/0044830

2180 series mini-Control Display Unit (CDU)

The Smiths Industries 2180 series mini-Control Display Unit is a general application CDU which features a seven-colour raster display and has an RS-422 interface. The unit has a full alphanumeric keyboard and a unique colour display which is small in size and offers high resolution, making it ideal for use with navigation and communication systems including GNSS.

CRT units are to be replaced by colour AMLCD 4 in diagonal units (see 2880 series CCDU).

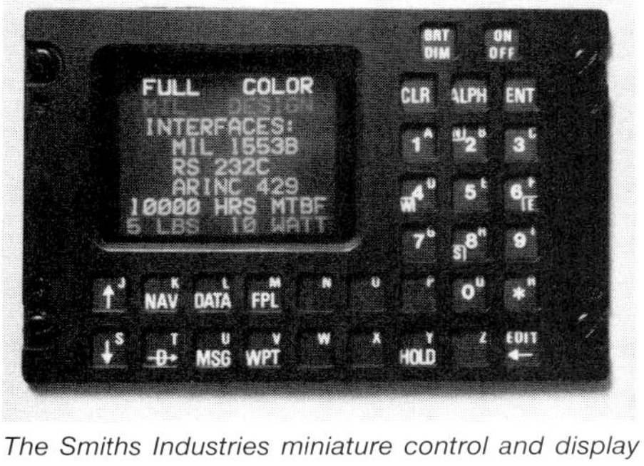

The Smiths Industries miniature control and display unit

Specifications
Dimensions: 95 × 146 × 203 mm
Weight: 2.2 kg

Operational status
In production.

Contractor
Smiths Industries Aerospace.

VERIFIED

2584 series colour flat panel Multipurpose Control Display Unit (MCDU)

Smiths Industries 2584 series flat panel MCDU is an ARINC 739A compatible control/display unit. The MCDU is designed as an interface unit to support a number of aircraft applications on varying types and is standard production on Boeing 737 next generation airplanes.

The broad utility of the MCDU permits applications as a control/display unit for the Global Positioning System (GPS), ACARS, Flight Management System (FMS), Digital Flight Recorder (DFR), or Satcom. The unit communicates using ARINC 739 protocol.

The MCDU can receive software upload data via a standard ARINC 615 airborne data loader to

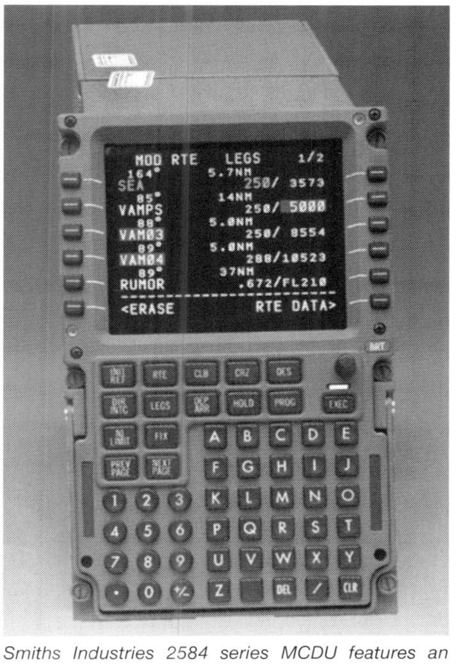

Smiths Industries 2584 series MCDU features an Active Matrix Liquid Crystal Display (AMLCD) **1999**/0051705

accommodate modifications to the operational flight programme.

Specifications

Input/output: ARINC 739 compatible
7 input ports (hi/low speed)
2 output ports (hi/low speed)
discretes: 4 input and 1 output
ARINC 429 compatible
Display: full colour AMLCD display
Brightness: 95fL standard (automatic brightness control)
Resolution: 648 × 532 colour elements
14 rows of 24 characters
Active display size: 96.8 (W) × 79.5 (H) mm
Control display unit for:
FMS
Digital Flight Recorder
NAV radio tuning
Datalink
ACARS
Satcom
Dimensions: 228.6 (H) × 146.1 (W) × 284.5 (D) mm
Weight: 4.1 kg
Power: 115 V AC, 400 Hz
Keyboard: 3 keyboard FANS options available

Contractor

Smiths Industries Aerospace.

VERIFIED

2600/2610 series electronic chronometers

The 2600/2610 series electronic chronometers utilise two dichroic LCDs to display GMT, chronograph, elapsed time and day and date. Smiths digital clocks utilise a temperature compensated crystal oscillator for extreme accuracy. Trickle current from the aircraft's hot bus maintains functionality with bus power off.

A variant of the 2600 is the 2620 which has been selected for the Boeing 777 and features ARINC 429 input and output, GPS and ASIC technology.

Operational status

In production. Smiths digital clocks are standard fit on the Airbus A320 and A340, BAE Systems Avro 85/100, Boeing 727, 737, 747, 757, 767 and 777 Series, MD-80, MD-11 and MD-90, and Fokker 50/70 and 100, IPTN N-250.

Contractor

Smiths Industries Aerospace.

VERIFIED

Smiths Industries 2620 digital clock as fitted to Boeing 777

2619 series GPS digital chronometer

The 2619 series GPS digital chronometer provides display of primary time of day and date reference for the flight crew as well as a chronograph and elapsed time function. Three transilluminated digital liquid crystal displays provide elapsed time, UTC (day/date) and chronograph time.

In primary mode the 2619 series GPS clock receives and processes an external ARINC 429 GPS time source and provides an ARINC 429 output signal of time and date for use by other aircraft systems. The GPS time continually synchronises the internal time base. Should the ARINC 429 GPS input fail the clock will automatically use its internal time base.

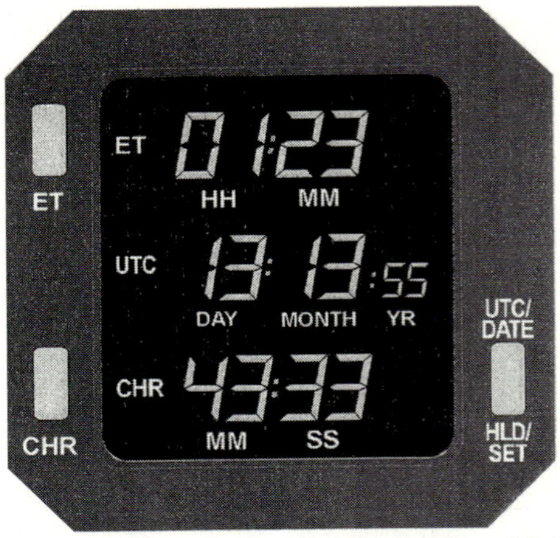

Smiths Industries 2619 series GPS digital chronometer ***1999***/0051706

The clock provides three time keeping modes, UTC mode (00:00 to 23:59), ET (00:00 to 99:59) and CHR (000 to 999) function, and manual mode.

The elapsed time is controlled by a push-button switch which activates the start of ET time, holds CHR time and resets CHR time.

An additional feature of the 2619 series GPS clock is the date display mode. In this mode the clock will display the day, month and year. The date display and output signals are automatically adjusted to achieve an update commensurate with the update of the UTC display. In addition, bissextile (leap) year correction for the date display output signal is accomplished automatically by the clock.

Operational status

In production in 2000.

Contractor

Smiths Industries Aerospace.

VERIFIED

2850 series alerting altimeter - 3ATI AMLCD

The 2850 series alerting altimeter is designed to be installed as: a Reduced Vertical Separation Minima (RVSM) update instrument; as a replacement for electromechanical servo altimeters; or as a multifunction unit.

The primary altimeter displays barometrically corrected altitude in feet on a large AMLCD display. An additional altitude readout in metres is added by pushing a bezel mounted switch. Baro correction counters display in both inches of mercury and millibars. A 'quick set' feature sets 29.92/1,013 baro settings at the push of a button. The unit accepts ARINC 429 digital signals from all standard Digital Air Data Computers (DADC).

The unit incorporates an integral altitude alert function. A bezel push-button mode switch permits selection of alert set mode or the baro set mode. Based on the mode selected, the rotary knob sets the alert altitude or baro setting. No separate altitude alerter unit is required.

Smiths Industries 2850 series alerting altimeter showing 'alert set' mode ***1999***/0051707

The 'altitude alert' setting appears on the flat panel display below the dual barometric displays. Alert annunciation is performed both on the display using a 'yellow' caution alert, and through remote aural and visual alerts, which are activated by discrete output from the altimeter. Formats for the display retain flexibility due to software control.

Additional mode/features options include a baro corrected output to other aircraft systems, NVIS B lighting and a settable MDA (Minimum Descent Altitude) bug.

Contractor

Smiths Industries Aerospace.

VERIFIED

2880 series AMLCD Compact-Control and Display Unit (CCDU)

The 2880 series Compact-Control and Display Unit (CCDU) features a 4 in diagonal colour AMLCD with a wide viewing angle. The 2880 series CCDU is graphic-capable and replaces the 2180 series Mini-CDU (CRT model).

The CCDU employs a full alphanumeric backlit keyboard with either electroluminescent or incandescent lighting. The display is sunlight readable and an automatic brightness feedback loop is provided. The base version of the Mini-CDU is housed in a deep case, 88.9 (H) × 146.1 (W) × 200.7 (D) mm.

The Mini-CDU provides a flexible interface unit for a number of applications including navigation systems, flight control systems and communication systems. NVIS-B capability is an option.

Operational status

The 2880 series CCDU complements Smiths Industries AMLCD MCDU, multifunction 3ATI electronic display, 5ATI primary flight display and datalink DCDU products.

Contractor

Smiths Industries Aerospace.

VERIFIED

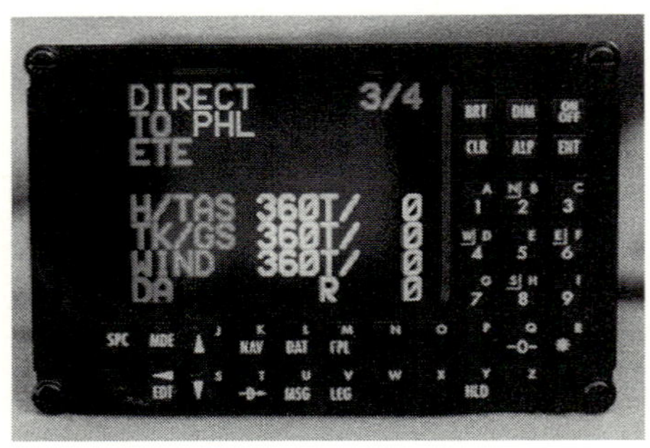

Smiths Industries 2880 series AMLCD Compact-Control and Display Unit (CCDU) ***1999***/0051708

2882 Series AMLCD Multifunction Control Display Unit (MCDU)

The 2882 Series flat-panel AMLCD Multifunction Control Display Unit (MCDU) provides a cockpit interface unit with an enhanced display technology, updated ARINC 739 format, advanced electronics and flexible system interface. The unit is designed to facilitate the integration of numerous functions such as GPS, RNav, ACARS, performance advisories and airport maps, and to help conserve space on the flight deck by the elimination of dedicated control heads.

The flat-panel MCDU provides a display of alphanumeric and graphics information, together with a keyboard for crew selection of modes and data entry. A set of dedicated function keys provides the crew with immediate access to particular pages and a set of dedicated alphanumeric keys allows the pilot to enter data. The ARINC 429 digital data enters through the rear connector and is processed by the graphics system processor module where the data is decoded and then displayed on the active matrix LCD. The unit

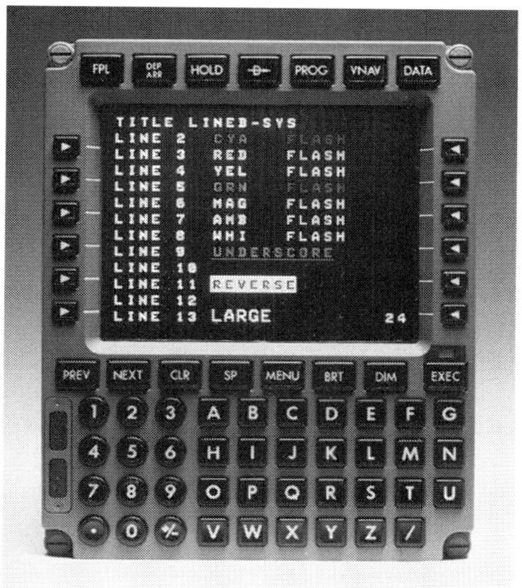

Smiths Industries 2882 MCDU

contains standard RS-422 high-speed input and output. An optional MIL-STD-1553 interface unit is available as are NVIS-B and low-temperature operation options.

Specifications

Dimensions: 146 × 181 × 101.6-177.8 mm
Weight: 2.72 kg (approx)
Display: 139.7 mm diagonal
14-lines by 28-characters

Operational status

In production and in service. Current platforms include: A300/310, ATR-42/72, BAe 146/RJ-85, B-727, B-747, C-130, C-141, DC-8, DC-10, E-6, King Air, MD-80, Nimrod MRA 4, VC-25 Air Force One and Two.

Contractor

Smiths Industries Aerospace.

VERIFIED

2888 series AMLCD Cockpit Display Unit (CDU)

The 2888 series CDU provides a cockpit control unit that interfaces via a primary RS-422 port with the Flight Management System (FMS). Further interfaces on RS-422 include graphics file transfer, graphical weather and telelink. RGB and NTSC video inputs are also available.

The system is based on open architecture PC-based microprocessor technology. Software is developed to DO-178B, and hardware to DO-160D.

Specifications

Dimensions: 146 × 181 × 101.6 -177.8 mm
Weight: 2.72 kg
Display: 140 mm diagonal

Operational status

In production.

Contractor

Smiths Industries Aerospace.

UPDATED

2889 series Mode Select Panel (MSP)

The Smiths Industries 2889 series Mode Select Panel (MSP) provides modern display technology with traditional flight control unit functions in a compact package. This solid-state unit is designed for use as a pilot interface control unit with high reliability and flexible page formats, including graphics if desired.

The launch programme for the 2889 series MSP is an intermediate sized helicopter which will be entering service in 1999.

The 2889 series MSP is a derivative of the Multifunction Control Display Unit (MCDU) product line which is in service on both military and civil programmes.

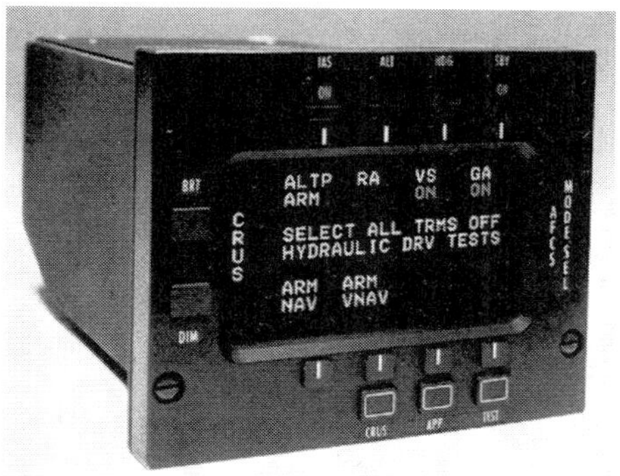

Smiths Industries 2889 series Mode Select Panel
***1999**/0051709*

The 2889 series MSP provides pilot/operator keyboard and display interfaces to: Flight Control Computer (FCC); Aircraft System Computer (ASC); Automatic Flight Control System (AFCS); and in an optional configuration can provide a back-up Engine Instrument Display (EID).

The pilot accesses functions and modes using the high-contrast lighted keyboard. The pilot receives information from the AMLCD display, which is used to display text information. The 2889 series MSP communicates with the FCC, ASC or AFCS utilising an ARINC 429 digital data bus.

The 2889 series MSP is able to receive external software upload data for updating the operational flight programme. NVG capability is optional.

Operational status

In production.

Contractor

Smiths Industries Aerospace.

VERIFIED

AMLCD 3 ATI HSI/ADI

Smiths Industries has been involved in Active Matrix Liquid Crystal Display (AMLCD) technology since its inception in the aerospace industry. Various display sizes have been integrated into a number of AMLCD products. Due to the multifunction nature of AMLCDs, a variety of applications has been developed, including digital triple torque, airspeed, altitude, vertical speed, collision avoidance, engine parameters and aerodynamic surface or control position indication presented via MIL-STD-1553B, ARINC 429 digital bus or integrated interfaces.

Specifications

Qualifications: DO-160D and DO-178B
Interfaces: synchro interface added in 2000 to facilitate retrofit to older civil and military aircraft.

Operational status

In production for Cessna Citation Jet 1 and Citation Jet 2 aircraft.

Contractor

Smiths Industries Aerospace.

VERIFIED

Digital fuel gauging systems

Smiths Industries has had its 2300 Series digital fuel quantity indicators in service for some years on an increasing variety of civil and military aircraft. The system is standard fit on the Boeing 737, 747-200/300 and 777.

Smiths Industries has developed an integrated fuel quantity system in a single LRU. This device performs automatic refuelling, valve control, preselection of fuel quantity and digital data output for EFIS displays.

Operational status

In addition to supplying Boeing, for the 727, 737 and older 747s, Smiths Industries has certified the system for retrofit to a variety of aircraft including Boeing DC-8, DC-9 and DC-10 and the Fokker F28.

Smiths Industries digital fuel gauging repeater and master indicators for the Boeing 747

Military retrofits have been completed on the A-4 Skyhawk, C-130 Hercules, C-141 and F-4 Phantom.

Contractor

Smiths Industries Aerospace.

VERIFIED

Flight Management System (FMS)

The Flight Management System (FMS) is used for commercial air transport applications. Currently it is available in single or dual configuration for the Boeing 737 and in a triple configuration for the Ilyushin Il-96. The system is designed to meet the new airspace requirements specified by ICAO in their instrumentation plans for communication navigation and surveillance for Air Traffic Management. This includes communications with an adaptable datalink to allow each airline to customise its airline communications; navigation with the latest in required navigational performance/actual navigation performance methodology certified to use GPS and surveillance with the building blocks to implement functions such as ATC clearance entry and ADS reporting.

The flight management computer system has a single high-power 32-bit microprocessor and contains options for 4, 8 or 16 Mbytes of on-aircraft loadable EEPROM for storage of the operational flight program and navigation database. Up to two million bytes of memory is dedicated to the operator's navigation database and can be structured according to the customer's specification. A wide range of options is available from a simple listing of navaids and airports to a detailed coverage of the airline's operating routes, SIDs, STARs and gate assignments. Smiths Industries offers airlines a choice of using Jeppesen, Racal or Swissair for the navigation database update service.

The crew interface is via a CRT control and display panel on which 14 by 24 character lines of information can be displayed. Aircraft lateral and vertical profile data is presented and critical information is highlighted by reverse video presentation. The bottom line of the CRT is used as a scratchpad for crew entries.

The system can be coupled to the autopilot and autothrottle for automatic profile tracking and energy management. Fuel savings in the range 4 to 14 per cent are predicted in normal operations.

The system is available to all Boeing 737 operators either in a dual-unit or single-unit installation. The dual unit manages all performance, navigation and approach functions, as well as improving dispatch availability. The dual-unit format can be installed in place of the current single unit. The computer is a passively cooled 4 MCU form factor that is half the weight, uses one fifth the power and is over five times faster than competitive flight management systems. The system has spare memory and processor throughput to accommodate the functions of the CNS/ATM implementation programme and the Global Air Traffic Management (GATM) requirements of the US Department of Defense.

Specifications

Dimensions:
(display) 267 × 146 × 229 mm
(computer) 320 × 257 × 193 mm (4 MCU)
Weight:
(display) 8.2 kg
(computer) 7.7 kg
Power required:
(total) 75 W
Reliability: 18,000 h MTBF (predicted)

The Smiths Industries flight management system is available in single or (as shown here) in dual configurations

Operational status

In production for, and in service on, Boeing 737-300, -400, -500, -600, -700 and -800 aircraft, in single or dual configurations.

This FMS has also been selected for Airbus aircraft in co-operation with Sextant. Elsewhere, a triple configuration has been selected for the Ilyushin IL-96M, and in US military service the system has been selected for the US Air Force VC-25 aircraft (Air Force One and Two – 2 converted Boeing 747-200 aircraft) and on the US Navy E-6 aircraft.

Contractor

Smiths Industries Aerospace.

VERIFIED

TCAS RA Vertical Speed Indicators

Smiths Industries has developed special vertical speed indicators for use with TCAS I and TCAS II Traffic alert and Collision Avoidance Systems. The units are form, fit and function interchangeable with most existing VSIs. Designed for ARINC 735 compatibility, the units may be used with TCAS II systems now entering service.

The 2074 Series instruments are outwardly similar to the Smiths Industries 2070 Series VSIs already in service with many airlines. The 2074 Series is intended for those applications where traffic advisories will be presented through some other medium such as a weather radar. Resolution advisories are presented through red and green LED 'eyebrows'. Green

Smiths Industries Model 2074 RA VSI

segments indicate those vertical speeds which will maintain safe separation from other aircraft, while red depicts potentially dangerous vertical speeds.

The 2074 flat-panel vertical speed/TCAS I instrument provides TCAS I data in addition to acting as a standard VSI. This instrument has successfully interfaced with both the BFGoodrich and AlliedSignal TCAS I systems. An ARINC 429 channel provides good readability with a full-colour display.

Operational status

The 2074 RA VSI is certified for use with AlliedSignal, Honeywell and Rockwell Collins TCAS systems. Over 800 units are currently in operation. The 2074 provides a dedicated RA display capability with TCAS 2000 (Mark 7) compliance.

Contractor

Smiths Industries Aerospace.

VERIFIED

LCD altitude selector/alerter

The new S-TEC LCD altitude selector/alerter is a compact, lightweight unit combining the computer and programmer in a single package. It is fully TSOd to C9c standards.

This system allows the pilot to preselect altitudes and rates of climb/descent through use of the autopilot, and also provides an altitude alert mode, decision height mode, altitude readout from the encoder, and barometric calibration in inches of mercury, or in millibars.

Specifications

Power required: 14/28 V DC
Weight: 0.6 kg
Dimensions: 41 × 87 × 172 mm

Operational status

In production.

Contractor

S-TEC Corporation, a Meggitt Avionics Company.

UPDATED

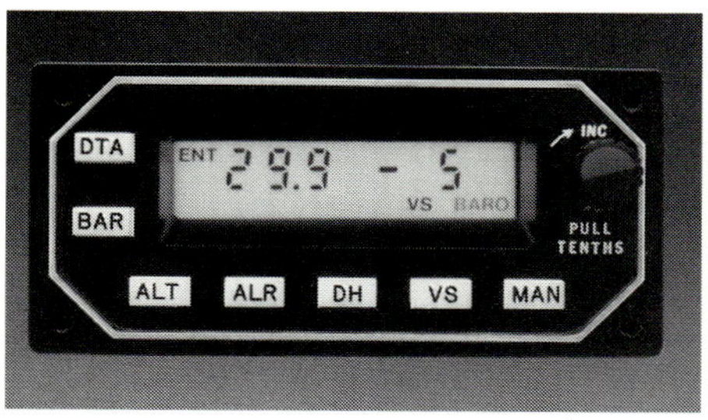

LCD altitude selecter/alerter 0001449

Single-cue flight director system

S-TEC's single-cue flight director system provides for steering horizon operation both when the autopilot is engaged, and when it is not.

When the autopilot is engaged, the flight director provides a visual display of what the computers are telling the autopilot servos to do. When used during manual flight, it provides the pilot with steering indications that integrate information from several signal sources, dramatically reducing the complexity of the instrument scan. Flight director operation is automatic when the autopilot is engaged.

S-TEC's single-cue flight director is similar in appearance to other systems in order to provide pilots with a measure of conformity, and to reduce error during operation. It contains a fixed delta-shaped symbol, bright orange in colour, which represents the airplane. Pitch and roll attitudes are displayed by a movable attitude field, coloured with a blue sky, and an earthtone ground, separated by a thin white line.

In a half-circle above the attitude field are the bank indexes - the centre or 'level' index, represented by a white inverted triangle, with bank indexes indicated by white vertical lines on either side of the centre index. They represent, 10, 20, 30 and 60° left and right bank angles. Four white lines above, and two below the white horizontal line on the attitude field represent various pitch attitudes: 5° increments in pitch-up to a maximum of 20°; and 10° increments in pitch down attitude, to a maximum of 20°.

The command bars, painted yellow, display computed bank and pitch commands. They rotate around the attitude fields to command climb, descent, left and right bank angles. In manoeuvring, the airplane

Single-cue flight director system 0001448

symbol is 'flown into' the command bars, until the two are accurately aligned, which satisfies the command.

When operating in approach mode, operation basically is the same as in other autopilot modes. When the glide slope is captured, the command bars indicate the commands to be satisfied. The airplane symbol is 'flown into' the command bars to satisfy the commands.

Specifications

Power required: 14/28 V DC
Weight: 1.4 kg
Dimensions: 89 × 89 × 176 mm

Operational status

In production.

Contractor

S-TEC Corporation, a Meggitt Avionics Company.

UPDATED

ST-180 HSI slaved compass system

The ST-180 system combines a magnetically slaved gyroscopic compass with a VOR/Localiser and glide slope display. The ST-180 was designed specifically for rotary-wing and high performance fixed-wing aircraft. This HSI slaved compass system offers a single convenient display which provides the pilot with all necessary information about the aircraft's position relative to ground-based navigational aids. Simultaneous indications are provided for selected course, course deviation and selected heading. Glide slope deviation is displayed when an active ILS frequency has been selected. The ST-180 HSI Slaved Compass System was developed and approved for helicopter operations.

The ST-180 consists of the Horizontal Situation Indicator (HSI), the remote electric gyro, the magnetic flux sensor and the model slaving panel. The gyro employs electromechanical erection thereby eliminating the performance and reliability degradation frequently experienced with air-erected gyros.

Critical circuits within the HSI are continuously monitored to minimise potential erroneous information. Should the flux sensor fail, the system may be switched to a free-gyro mode. Should a gyro failure be detected, the system may be transitioned into the 'Automatic Emergency Mode' (AEM). In AEM, the compass card is controlled by the Flux Sensor and behaves similarly to a normal wet compass. The pilot can continue to use the HSI.

Specifications

FAA TSO: C34e; C36e; C40c; C6d; C9c; C52a
Input power: 27.5 V DC (supplied by remote gyro)
Compass card accuracy: ±1°
Dimensions: ARINC specification 408 ATI-3
Weight: 1.4 kg

Operational status

In production.

Contractor

S-TEC Corporation, a Meggitt Avionics Company.

UPDATED

Stability and Control Augmentation System (SCAS) for helicopters

The S-TEC SCAS for rotary-wing aircraft provides a new and innovative approach to stabilising the helicopter in flight while enhancing controllability.

The S-TEC SCAS is a full-time series system which augments the pilot's control of the aircraft from hovering flight throughout the normal manoeuvring flight regime all the way to the landing sequence. It improves basic aircraft control harmony and reduces outside disturbances to the desired flight path thus reducing pilot workload and improving the quality of flight. During operation the system is completely transparent to the pilot.

Options to the basic SCAS will include a force trim system which will restrain the cyclic control, allowing the pilot freedom, for short periods, to accomplish other cockpit management duties such as the operation of avionics and other aircraft systems. The force trim system can be converted to parallel trim actuator operation in the upgrade to a complete autopilot system.

This primary system, in addition to being available to improve flight characteristics for the aircraft in which it is installed, becomes the first component of what will be a complete family of modular helicopter flight control systems. As full featured systems are added to the S-TEC rotary-wing product line, the SCAS will become the nucleus of the building block upgradability to a complete flight control system.

Growth to VFR flight control systems will require only the addition of an autopilot circuit card and transducers and the conversion of the force trim system into parallel trim actuators. The upgrade will incorporate full authority autopilot functions of IAS Hold, Altitude Hold, Heading Select/Hold and VOR/LOC/GPS navigation intercept and tracking.

Contractor

S-TEC Corporation, a Meggitt Avionics Company.

UPDATED

System 20/30/30ALT/40/50/60/65 autopilots

The basic System 20 rate autopilot consists of a combined programmer/computer/annunciator in the turn co-ordinator and an electrically operated roll servo. It permits both straight and turning flight, while inputs from a radio beacon provide the means for VOR tracking for en route navigation, localiser tracking for approach and reverse or back course tracking. The system can be upgraded to include a heading mode by adding the optional directional gyro.

The System 30 has all the foregoing functions, but adds an altitude hold mode and elevator trim indicators. The optional directional gyro can be added to provide heading hold. An accelerometer confers pitch stability and automatically disconnects the system in the event of a pitch axis fault. In addition, the system incorporates a preflight test feature that tests internal limiter circuitry.

The System 30 ALT is an altitude only autopilot which is rate based operating off accelerometers in a pitch computer module.

The System 40 rate autopilot consists of a combined programmer/computer/annunciator in a single box, a turn co-ordinator and an electrically operated roll servo. It permits both straight and turning flight, while inputs from a radio beacon provide the means for VOR tracking for en route navigation, localiser tracking for approach and reverse or back course tracking. The system can be upgraded to include a heading mode by adding the optional directional gyro.

The System 50 has all the foregoing functions, but adds an altitude hold mode and elevator trim indicators. The optional directional gyro can be added to provide heading hold. An accelerometer confers pitch stability and automatically disconnects the system in the event of a pitch axis fault. In addition the system incorporates a preflight test feature that tests internal limiter circuitry.

The System 60 is a rate autopilot using rate of change of aircraft motion instead of attitude to generate demands to the flying controls. The System 60 is a single- or two-axis system (60-1 and 60-2) comprising an electric turn co-ordinator, 3 in (76 mm) air-driven directional gyro, mode programmer/annunciator, pitch and/or roll guidance computers and servos, master switch and control wheel disengage switch and altitude transducer.

The System 65 autopilot has the same performance standards as the System 60 but includes automatic elevator pitch trim, liquid crystal display programmer and a remote annunciator.

A liquid crystal display altitude selector/alerter combines the computer and programmer in a single package.

System 60 PSS provides a cost-effective way of adding vertical flight control to virtually any single axis autopilot system. It provides altitude hold, glide slope coupling and vertical speed capabilities. The system does not interface with existing roll axis autopilots, but complements them. Being self-contained the System 60 PSS does not connect to the aircraft pneumatic or vacuum systems.

Operational status

In production.

Contractor

S-TEC Corporation, a Meggitt Avionics Company.

UPDATED

System 55X autopilot

The System 55X is a pure rate-based autopilot. It combines programmer, computer, annunciator and servo amplifier functions into one panel and is the first S-TEC autopilot designed specifically to be integrated into the aircraft radio panel.

The System 55X roll axis has heading select, VOR/Localiser front and back course intercept and tracking. It can be interfaced with GPS or Loran systems and all radio couplers, together with GPSS roll steering, are standard fit.

Operational status

In production.

Contractor

S-TEC Corporation, a Meggitt Avionics Company.

UPDATED

TEC LINE VHF-251A communications receiver

The TEC LINE VHF-251A 760-channel communications receiver is a direct enhanced replacement for the VHF-250/251/251S and 251E communications transceivers.

Features include: 760 channels in the 118 to 136.975 MHz band; 10 W power; non-volatile 10 frequency memory; failed display emergency mode; stuck-microphone protection.

Specifications

FAA TSO: C37c, C38c, DAO-160c
Dimensions: front panel: 79.25 × 66.29 × 43.69 mm
chassis: 80.77 × 67.41 × 316.2 mm
Weight: 1.75 kg

Contractor

S-TEC Corporation, a Meggitt Avionics Company.

UPDATED

VIR-351 VHF-nav receiver

The S-TEC VIR-351 VHF navigation receiver is a solid-state, 200-channel unit which provides VOR/Loc deviation. VOR/Loc flag output, VOR to/from information to the 350 series indicators, and is capable of tuning the GLS-350 glide slope receiver and DME-451 Distance Measuring Equipment system.

Specifications

FAA TSO: C40ac, C36c class D, DO-138
Dimensions: 79 × 66 × 316 mm
Weight: 1.4 kg

Operational status

In production.

Contractor

S-TEC Corporation, a Meggitt Avionics Company.

UPDATED

ASB-500 HF/SSB radio

The Sunair ASB-500 and the associated ACU-150D have been specifically designed for aircraft and helicopters requiring a large number of operational frequencies but where space and weight are limiting factors. It covers the HF band and provides USB, AM and optionally LSB modes of operation.

The radio comprises a transceiver, remote control and automatic antenna coupler. The controller is panel-mounted and includes a six-digit LED frequency display and illuminated status and antenna coupler tuning monitors. The coupler is solid-state and tunes extremely quickly with a 10-channel last-tuned memory. The radio provides 100 W output power and is certified to the relevant FCC and FAA TSOs.

Specifications

Dimensions:
(transceiver) 123.8 × 193.7 × 39.4 mm
(controller) 146 × 127 × 66.7 mm
(antenna coupler) 177 × 152 × 305 mm
Weight:
(transceiver) 6.6 kg
(controller) 0.8 kg
(antenna coupler) 3.9 kg
Power supply: 27.5 V DC
(receive) 2.7 A
(transmit) 13 A
Frequency: 2-17.9995 MHz
Power output: 100 W PEP
Channels: 32,000 at 500 Hz spacing
Temperature range: –46 to +55°C
Altitude: up to 30,000 ft

Operational status

The ASB-500 and ACU-150D are available for general aviation and helicopter applications.

Contractor
Sunair Electronics Inc.

VERIFIED

ASB-850A HF/SSB radio

Sunair's ASB-850A is a multipurpose synthesised military HF transceiver. It is a particularly light and compact system designed for light fixed-wing aircraft and helicopters operating in tactical roles. The system is remotely controlled from a miniature panel-mounted unit which contains an LED frequency selection display. It operates in USB, LSB and AME modes.

The ASB-850A is of all solid-state construction and is claimed to be of exceptionally robust design and manufacture.

A new high-speed automatic antenna coupler is incorporated within the system and this unit can tune the antenna to the frequency selected in 1 second or less for initial tuning and in a matter of milliseconds for tuning to the last 10-tuned channels memory.

Specifications
Dimensions:
(transceiver) 470 × 124 × 241 mm
(amplifier/antenna coupler) 448 × 152 × 235 mm
(controller) 172 × 146 × 64 mm
Weight:
(transceiver) 7.9 kg
(amplifier/antenna coupler) 7.9 kg
(controller) 0.6 kg
Power supply: 27.5 V DC
(receive) 2 A
(transmit) 17 A
Frequency: 2-29.9999 MHz
Channels: 280,000 at 100 Hz spacing
Temperature range: –46 to +71°C
Altitude: up to 30,000 ft

Operational status
In service. Production ended in 1994. A spares and support service is continuing as long as is possible and practical.

Contractor
Sunair Electronics Inc.

VERIFIED

Sunair ASB-500/ACU-150D HF radio set

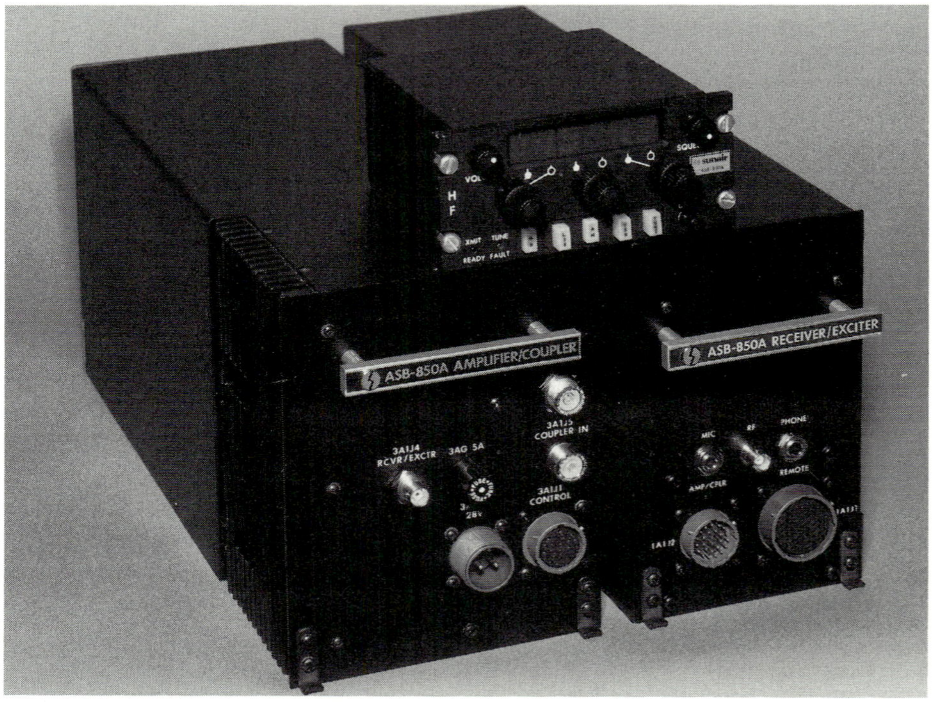

Sunair ASB-850A HF/SSB airborne transceiver

TeleLink TL-608 datalink system

The TeleLink TL-608 digital datalink system meets the needs of the corporate or regional airline pilot by providing a variety of features and versatile connectivity options in a small package. Numerous bidirectional communications media are supported, including: VHF ACARS, satellite, airborne telephone and other, to meet the communication requirements of the flight crew and the passengers.

Although derived from ACARS, the TL-608 is designed to take advantage of emerging datalink technologies and open architecture software to provide the operator with a choice of service providers and communications media.

Specifications
Dimensions: 1 MCU: 25.4 × 194 × 386.8 mm
Weight: 1.8 kg
Power: 28 V, 10 W
Inputs: ARINC 429 receivers, RS-232/422 receivers, discretes, VHF modem, telephone modem
Outputs: ARINC 429 transmitters, RS-232/422 transmitters, discretes, VHF modem, telephone modem
Interfaces: FMS, CDU, DAU, printer, dataloader, maintenance terminal, laptop PC
Communications media: VHF radio, airborne telephone, Satcom, Mode S

Operational status
Selected by Bombardier Inc as the standard option Communications Management Unit (CMU) for the Global Express aircraft, where it will interface with the Honeywell flight management system and Satcom to provide data and voice coverage worldwide.

Contractor
Teledyne Controls, Business & Commuter Avionics.

VERIFIED

TeleLink helicopter datalink

The Teledyne TeleLink helicopter datalink supports a variety of bidirectional communication media, including: ACARS, satellite and telephone. Capabilities include: flight operations data exchange; GPS-based position reporting; uplink/downlink of FDR/HUMS data; custom messaging. Interfaces include: ARINC 429, 739, CSDB and RS-232/422 I/O.

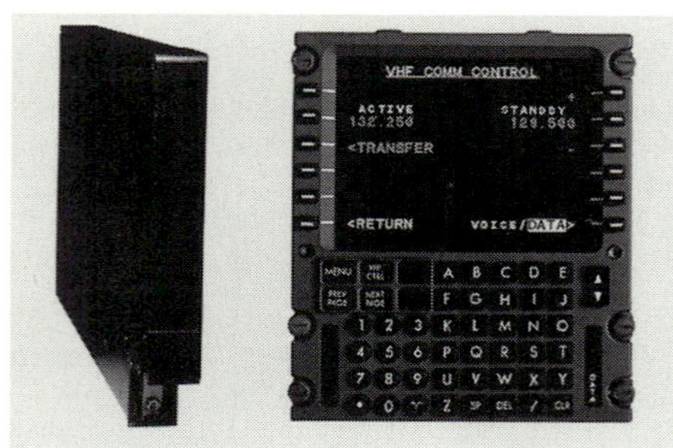

TeleLink helicopter datalink 0011820

Contractor
Teledyne Controls, Business and Commuter Avionics.

VERIFIED

ProCom 4 aircraft intercom

The Telex ProCom 4 aircraft intercom is compact and offers optional panel-mounting configurations. It provides noise-free voice-activated communications for pilot, co-pilot and up to three passengers. It has provisions for optional connection of a music/auxiliary source and tape recorder.

The ProCom 4 is equipped with a master squelch control and individual squelch circuits with trimmers for each user. With the ProCom 4 only the microphone of the person talking is hot, resulting in less noise. In addition, using separate squelch trimmers solves adjustment problems caused by such things as different ambient noise levels at different microphones throughout the cockpit, several types of microphone being used for different user voice levels. In the event of an intercom failure, the pilot can still use the radio.

Specifications
Weight: 0.28 kg
Power supply: 12-28 V DC, 125 mA at 28 V

Contractor
Telex Communications Inc.

VERIFIED

Aerial surveyor

The Aerial surveyor is a complete survey system which precisely identifies the position of the aircraft at the time an aerial photograph is taken. Measurements are made with respect to a referenced station, such as another Aerial surveyor or geodetic surveyor, and then post-processed. Knowing the precise position of the aircraft reduces the need for expensive ground control.

Measurement rates as fast as every half second minimise the effect of the aircraft dynamics and are ideal for efficient aerial photogrammetry. Interface capability is provided through dual RS-232 ports. Synchronisation of cameras is possible by utilising the 1 pps output and/or the event marker which allows camera shutter operation to be directly linked to precise GPS time and positioning.

On board the aircraft the receiver/datalogger will operate at aircraft speeds as high as 700 kt and maintain measurement integrity at accelerations up to 2 *g*. A preamplifier, which allows connection to an aircraft antenna, is provided.

Optional features, like the external frequency input or RTCM input, provide the flexibility to design a system which meets exact requirements.

Specifications
Dimensions: 300 × 350 × 130 mm
Weight: 7.2 kg
Power supply: 10.5-35 V DC, 8.5 W
Temperature range: –20 to +55°C

Contractor
Trimble Navigation Ltd, Aerospace Products.

VERIFIED

AT 3000 altitude digitiser

The AT 3000 is an all-solid-state blind encoding altimeter designed to interface with the TRT 250D transponder and most other modern Mode C transponders. The addition of the AT 3000 to the transponder provides an altitude reporting capability conforming to FAR 91-36B. The Federal Aviation Administration has granted a TSO C88 approval for the AT 3000.

Specifications
Dimensions: 42 × 65 × 159 mm
Weight: 0.23 kg
Temperature range: –20 to +55°C
Altitude: –1,000 to 30,000 ft ±50 ft

Operational status
In service.

Contractor
Trimble Navigation Ltd, Aerospace Products.

VERIFIED

Jet Call

Jet Call is an ARINC ground-to-air selective calling system utilising thumbwheel coding with four buttons in full view. Sixteen available tones provide over 10,000 possible combinations and take about 10 seconds to set. The Jet Call uses no wire jumpers or remote switches. Two or five decoder channels are available to handle up to three VHF comms and two HF transceivers. The Jet Call is TSO'd to FAA C059. It uses all solid-state circuitry and switched capacitor filters with high inherent stability and has low power consumption and easy installation, using standard Mil type D connectors with insertable and removable pins.

Specifications
Dimensions: 127 × 57 × 321 mm
Weight: 1.45 kg

Operational status
In service.

Contractor
Trimble Navigation Ltd, Aerospace Products.

VERIFIED

TDF 100D Automatic Direction-Finder (ADF)

The Trimble TDF 100D automatic direction-finding system comprises three units: the TDF 100D panel-mounted receiver/control unit, the TA 10 combination sense and loop antenna and the TDI 10 ADF indicator.

The TDF 100D receiver/control unit tunes from 200 to 1,800 kHz in 1 kHz increments and also provides capability for receiving the marine distress frequency on 2,182 kHz. The front panel controls provide on/off volume control, a three-position toggle switch for selection of ADF, ANT and BFO modes, a push-push filter switch to clarify identification signals and digital frequency selector switches. The unit is designed to mount beside the Terra transponder, comms or navigation units.

The TA 10 ADF antenna contains a loop antenna, sense antenna and amplifier circuits for each antenna. It operates between temperatures of –55 and +70°C at altitudes up to 50,000 ft. The antenna may be top- or bottom-mounted and provides for quadrantal error adjustments.

The TDI 10 indicator mounts in any 3 in instrument housing. The indicator unit contains the power supply for the system, requiring 11 to 33 V DC at a nominal 8.5 W. A heading knob positions the compass card manually, while the pointer is driven by a stepper motor. Bearing accuracy is ±3° and the pointer will move from 175° off bearing to the correct bearing in less than 5 seconds.

Specifications
Dimensions:
(indicator) 139.7 × 83.8 × 83.8 mm
(receiver/control unit) 79.4 × 40.6 × 261.6 mm
(antenna) 203.2 × 134.6 × 66 mm
Weight:
(indicator) 0.59 kg
(receiver/control unit) 0.68 kg
(antenna) 0.64 kg

Contractor
Trimble Navigation Ltd, Aerospace Products.

VERIFIED

TN 200D navigation receiver

The TN 200D is a companion to the TX 760 nav/com receiver. It provides navigation and audio signals to operate Tri-Nav, Tri-Nav-C and most other VOR/ILS indicators. An optional 40-channel glide slope receiver can be added at any time and becomes a part of the TN 200D. Provision for remote channelling of DME systems is provided when the glide slope option is installed. A digital frequency synthesiser generates all 200 navigation channels from 108.00 to 117.95 MHz with digital channel selection.

Specifications
Dimensions:
264.6 × 159 × 84 mm (with mounting tray)
Weight: 2.09 kg
Power supply:
(transmit) 13.75 V DC, up to 2.5 A
Frequency:
(nav) 108-117.95 MHz at 50 kHz spacing
(glide slope) 329.15-335 MHz (40 channels)

Operational status
In service.

Contractor
Trimble Navigation Ltd, Aerospace Products.

VERIFIED

TRA 3000 radar altimeter

The TRA 3000 is a lightweight, panel-mounted radar altimeter for general use on executive and light aircraft and RPVs. The combined transmitter/receiver/antenna unit can be fuselage- or wing-mounted and used during all phases of flight. All indicators provide a decision height warning system. The top of the line TRI 40 indicator also includes a landing gear position indication.

Specifications
Dimensions:
(indicator) 35 × 88 × 189 mm
(transmitter/receiver/antenna) 192 × 126 × 25 mm
Weight:
(indicator) 0.27 kg
(transmitter/receiver/antenna) 0.68 kg
Frequency: 4,300 MHz
Altitude: 40-2,500 ft
Accuracy: 40-100 ft ± 5 ft, 100-500 ft ± 5%, 500-2,500 ft ± 7%

Operational status
In service.

Contractor
Trimble Navigation Ltd, Aerospace Products.

VERIFIED

TRA 3500 radar altimeter

The TRA 3500 is a lightweight radar altimeter for general use on executive and light aircraft, helicopters and RPVs. It consists of a remote receiver/transmitter unit with two microstrip antennas which may be mounted on the fuselage or wings. The indicator provides visual and aural warnings.

Specifications
Dimensions:
(receiver/transmitter) 76 × 76 × 174 mm
(antennas) 11 × 106 × 174 mm
(indicator) 35 × 88 × 189 mm
Weights:
(receiver/transmitter) 1.48 kg
(indicator) 0.34 kg
Frequency: 4,300 MHz
Altitude: 0-2,500 ft
Accuracy: 0-100 ft ± 5 ft, 100-500 ft ± 5%, 500-2,500 ft ± 7%

Operational status
In service.

Contractor
Trimble Navigation Ltd, Aerospace Products.

VERIFIED

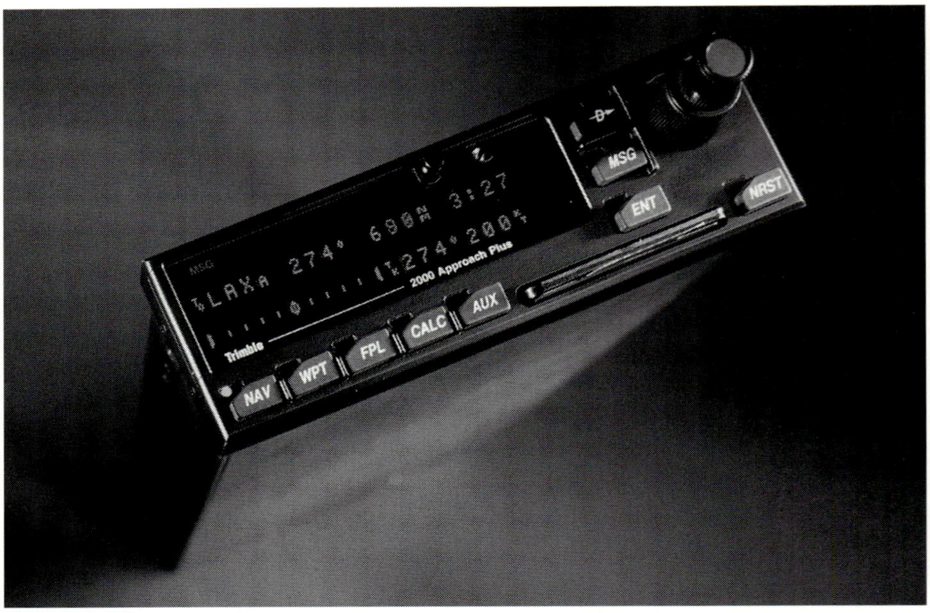

Trimble 2000 Approach Plus airborne IFR GPS navigator 0015390

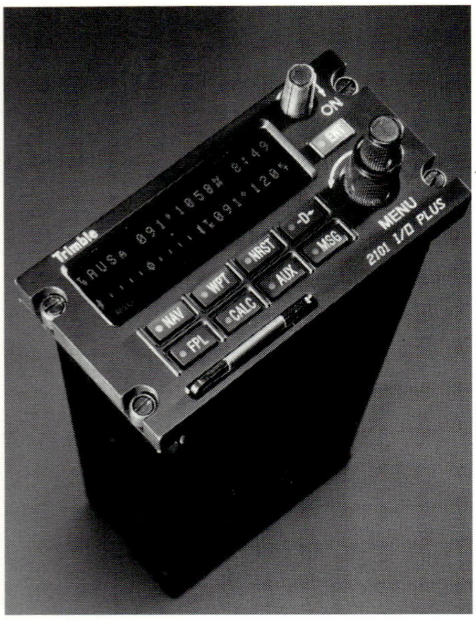

Trimble 2101 I/O Approach Plus airborne IFR GPS navigation system 0015391

Trimble 2000 Approach Plus airborne IFR GPS navigation system

The Trimble 2000 Approach Plus, certified to C-129 (A1), may be used for supplemental en route, terminal and approach IFR operations.

The Trimble 2000 Approach Plus has been designed to minimise pilot workload during execution of approach procedures. The pilot is automatically prompted to enable the approach at 30 n miles from the destination airport, an action that requires a single press of the Enter key. Procedure turns are fully automatic, no holding or suspension of procedures is required. All system functions required during the approach are automatically selected; all messages are automatically presented and removed. If a missed approach is required, the pilot selects the Direct key when located at a position anywhere from inside the Final Approach Fix (FAF) to beyond the Missed Approach Point (MAP). Flight path presentation is accurate and only database waypoints are presented.

The Trimble 2000 Approach Plus incorporates an advanced 12-channel GPS receiver. It continuously tracks all satellites in view, determining and displaying a new position five times every second and measuring speed to better than a tenth of a knot. It calculates and displays position, bearing, distance, ground speed and ground track and the estimated time en route on a high-contrast LED display. An electronic Course Deviation Indicator (CDI) and a patented track angle error display (TKE) are included. The Trimble 2000 Approach continuously checks accuracy using Receiver Autonomous Integrity Monitoring (RAIM) and predicts RAIM conditions for the approach arrival time.

Installation of the Trimble 2000 Approach Plus is simplified. Only four external annunciators are required; Message (MSG), Waypoint (WPT), Approach (APR), and Hold (HLD). Connection to an external CDI/HSI for navigation and autopilot interface is available. External switching of resolvers is not required, resulting in substantially reduced installation costs. When integrated to an optional air data computer, Trimble 2000 Approach Plus will display true air speed, density altitude, pressure altitude and calculate and display winds aloft. The system can automatically sequence through up to 40 flight plans with up to 40 waypoints each (maximum waypoint total is 400), show the nearest airport, plan vertical descents, display minimum safe altitudes and provide a variety of other functions.

The Trimble 2000 Approach Plus incorporates Jeppesen's NavData card, including airports, approaches, SIDS, STARS, VORs, intersections, NDBs, airspace boundaries and MEAs.

Specifications

Type: 12-channel receiver, L1 frequency, C/A code, continuous all-in-view tracking
Acquisition time: 1.5-3.5 min
Position update rate: 5 times/s
Dynamics: 800 kt (4 *g* tracking)
Accuracy:
(position) 15 m RMS
(velocity) 0.1 kt steady-state
(altitude) 35 m RMS (msl)
(time) UTC to nearest μs
Computation range: Great Circle: 0-999 n miles
Distance resolution: 0-9.99 n miles: in 0.01 n mile increments; 10-99.9 n miles in 0.01 n mile increments; 100 n miles + in 1.0 n mile increments
Interfaces:
EAI Standards, RS-422
CDI, flags and external annunciators, altitude input
GPS antenna: Omnidirectional flat microstrip with integral preamp. SMA connector
Display:
LED: 2 lines of 20 characters each
High-intensity, orange alphanumeric
Dimensions:
(receiver) 158.75 × 274.32 × 50.8 mm
(antenna) 95.25 × 101.6 × 19.05 mm
Weight:
(receiver) 0.92 kg
(antenna) 200 g
Power: 10-32 V DC, negative ground, 0.9 A at 14 V DC, 0.5 A at 28 V DC, 12 W
Compliance: FAA TSO C-129 (A1); FAA AC 20-138

Operational status

No longer produced.

Contractor

Trimble Navigation Ltd, Avionics Products.

VERIFIED

Trimble 2101 I/O Approach Plus airborne IFR GPS navigation system

The Trimble 2101 I/O Approach Plus is the full input/output version of Trimble's GPS Dzus rail mount navigator series. Certified to TSO C-129(A1), it may be used for supplemental en route, terminal and approach IFR operations.

The 2101 I/O Approach Plus is also approved for 8110.60 GPS as Primary Means of Navigation for Oceanic Operations and BRNAV European Airspace Operations.

The Trimble 2101 I/O Approach Plus minimises pilot workload during approaches. It automatically selects all the required navigation functions and automatically presents and removes all messages.

At the heart of the Trimble 2101 I/O Approach Plus is an advanced 12-channel GPS receiver with a defined upgrade path to WAAS capability. It continuously tracks all satellites in view, determining and displaying a new position four times per second and measuring speed to better than a tenth of a knot. It calculates and displays position, bearing and distance to waypoint, ground speed and ground track and estimated time en route. It includes an electronic Course Deviation Indicator (CDI) and a patented track angle error (TKE) display. The 2101 I/O Approach Plus presents the data on a high-contrast LED display. A Night Vision Goggles (NVG) compatible variant is available, which uses specially made glass, filters and polarisers.

The 2101 I/O Approach Plus continuously monitors its accuracy via Receiver Autonomous Integrity Monitoring (RAIM). It also predicts RAIM conditions for the approach arrival time.

The Trimble 2101 I/O Approach Plus is easy to install. Installation requires only four external annunciators: Message (MSG), Waypoint (WPT), Approach (APR) and Hold (HLD). Expensive resolvers are not required.

The 2101 I/O Approach Plus supports CDIs, HSIs, altimeters, RMIs, fuel flow gauges and many other flight instruments.

When integrated with an air data computer, the Trimble 2101 I/O Approach Plus displays true air speed, density altitude and pressure altitude and calculates and displays winds aloft and applies current wind ETE and ETA calculations. It also automatically sequences through up to 40 waypoints in a flight plan, shows the nearest airport, plans vertical descent profiles and performs a variety of other functions.

The Trimble 2101 I/O Approach Plus incorporates Jeppesen's NavData card - an avionics database that includes airports, approaches, SIDS, STARS, VORs, NDBs, intersections, airspace boundaries and more.

A Precision aRea Navigation (PRNav) version will be available in late 2000. The military version is known as the Cargo Utility GPS Receiver (CUGR).

Specifications

Type: 12-channel receiver, L1 frequency, C/A code, continuous all-in-view tracking
Acquisition time: 1.5-3.5 min
Position update rate: 1 time/s
Dynamics: 800 kt (4 *g* tracking)
Accuracy:
(position) 15 m RMS
(velocity) 0.1 kt steady-state
(altitude) 35 m RMS (msl)
Computation range: Great Circle: 0-999 n miles
Distance resolution: 0-9.99 n miles: in 0.01 n mile increments; 10-99.9 n miles in 1.0 n mile increments
GPS antenna: Omnidirectional flat microstrip with integral preamp. TNC connector
Display:
LED: 2 lines of 20 characters each
High-intensity, orange alphanumeric
Dimensions: 146.05 × 196.85 × 76.2 mm
Weight: 1.26 kg
Power: 10-32 V DC, negative ground, 1.0 A at 14 V DC, 0.5 A at 28 V DC, 14 W

Contractor

Trimble Navigation Ltd, Aerospace Products.

VERIFIED

Trimble 8100 advanced GPS navigation system

The Trimble 8100 advanced GPS navigation system is designed to meet the certification, operational, and maintenance standards required for commercial, air transport and corporate operations.

The Trimble 8100 is an advanced nine-channel GPS receiver with a defined upgrade path to 12-channels and WAAS capability. This system provides four-dimensional aircraft guidance for timed great circle and vertical navigation through selected flight plan waypoints. Trimble's 8100 provides transitions between all phases of flight navigation, including oceanic, en route, terminal, and non-precision approach.

Certified to FAA TSO-C129 and RTCA DO-208, Trimble's 8100 provides flight planning information using Company Routes, J-Routes, SIDS, STARS and approaches, and supports all 23 ARINC 424 leg types. The system is also approved as a primary means of navigation for remote and oceanic operations.

Trimble's 8100 input/output capability allows a wide range of analogue and digital interface options. It can also be used as a stand-alone ARINC 743 sensor, allowing a one-box GPS solution for an entire aircraft fleet. The Trimble 8100 is capable of interfacing to a DME transceiver to support multisensor navigation (TSO-C115a) and European P-RNAV.

Flight planning and navigation information is entered on one of two different CDUs (Compact and ARINC). Each CDU features a seven-colour display on which alphanumeric and symbolic data, including a graphical CDI presentation, are displayed.

Trimble's 8100 is Differential GPS (DGPS) capable, in anticipation of future Cat. I, II and III approach and landing certifications.

Specifications

Navigation Processor Unit:
Dimensions: ARINC 2 MCU (¼ ATR short)
368.3 × 196.6 × 572 mm
Weight: 2.73 kg
Power: 28 V DC input
Control Display Unit: (compact)
Dimensions: 203.2 × 95.3 × 146.1 mm
Weight: 2.07 kg
General: 9-channel, L1 frequency (1,575 MHz), C/A code, digital GPS receiver, tracks all satellites in view
Inputs: Synchros: TAS, HDG, ALT, pitch, roll ARINC 429:10 (high or low speed)
Fuel flow
Discretes: 32
Outputs: Synchros: 3 programmable, (synchro or SIN/COS)
Roll steering; left-right; glide slope
ARINC 561/568
ARINC 429: 5 (high or low speed), GAMMA compatible
Discretes: 16
Database: Expanded Jeppesen database (available via ARINC 615, or separate data loader)
Compliant:
TSO-C129, RTCA DO-208, TSO-C115a
Acquisition time: 1 min, nominal
Update rate: 5 Hz
Dynamics: 0-800 kt (3 *g* acceleration)
Time: Universal co-ordinated time to the nearest μs
Accuracy typical:
(position) 15 m RMS)
(altitude) 35 m RMS
(velocity) 0.1 kt RMS steady rate

Contractor

Trimble Navigation Ltd, Avionics Products.

VERIFIED

Trimble 8100 advanced GPS navigation system 0001345

Tri-Nav C VOR indicator with Loran C course deviation

A gas discharge display indicator used in conjunction with most brands of navigation radios with or without glide slope capability, the Tri-Nav C displays one VOR/Loc course or at the flip of a switch the display changes to show deviation from the selected Loran course. The Tri-Nav C presents horizontal deviation information in increments calibrated to 10° full-scale right and left.

Specifications

Dimensions: 82.6 × 82.6 × 114.3 mm
Weight: 0.57 kg

Operational status

In service.

Contractor

Trimble Navigation Ltd, Aerospace Products.

VERIFIED

Tri-Nav VOR/Loc indicator

A gas discharge display indicator used in conjunction with most brands of navigation radios, with or without glide slope capability, the Tri-Nav displays two VOR/Loc courses with glide slope deviation.

Specifications

Dimensions: 82.6 × 82.6 × 114.3 mm
Weight: 0.57 kg

Operational status

No longer produced.

Contractor

Trimble Navigation Ltd, Avionics Products.

VERIFIED

TRT 250D transponder

The TRT 250D transponder is a lightweight solid-state transponder which meets FAA Modes A and C requirements. Operated by easy-to-use backlit push-buttons, the TRT comes in a flat-pack mounting. The TRT 250 is compatible with all current production height encoding altimeters.

Specifications

Dimensions: 41 × 81 × 254 mm
Weight: 0.77 kg
Frequency:
(receive) 1,030 MHz
(transmit) 1,090 MHz

Operational status

No longer produced.

Contractor

Trimble Navigation Ltd, Avionics Products.

VERIFIED

TX 760D communications transceiver

The all solid-state TX 760D communications transceiver features a 5 W transmitter and includes voice-activated intercom capability. It has 760 channels over the frequency range 118 to 136.975 MHz (including 10 memory channels), a planar gas discharge display, dual displays for active and standby, single-knob tuning and digital frequency synthesis.

Specifications

Dimensions: 79.4 × 41.3 × 290.8 mm
Weight: 0.68 kg
Power supply: 13.75 V DC
(standby) 0.325 A
(transmit) 2 A

Contractor

Trimble Navigation Ltd, Aerospace Products.

VERIFIED

Flat Panel Integrated Display (FPID) control panels

Universal's FPID control panels provide for flexibility in the installation of the associated flat panel integrated displays.

The Reversion/Test/DH Control Panel provides reversion control, attitude and radio altimeter test, and DH set functions for both sides. One activation of the Revert switch causes a composite ADI/HSI display to appear on both upper and lower FPIDs.

The Display Control Panel is designed for use in aircraft that already have a remote heading/course selector panel. It provides source, display and range selections.

For pilots who desire a remote heading/course panel centrally located, the Course/Heading Control Panel provides independent course selection to each side, with single heading selection driving both heading bugs.

The Radar Control Panel, together with the FPID MFD, will take the place of a cockpit radar indicator and provides full control functionality.

Contractor
Universal Avionics Systems Corporation.

VERIFIED

Flat Panel Integrated Displays (FPIDs)

Universal Avionics produces a range of colour active-matrix liquid-crystal FPIDs. Current models include the Electronic Flight Instrument EFI-340 (3ATI), EFI-450 (4ATI), EFI-500 (4 × 5ATI), EFI-550, EFI-600 (5ATI), EFI-640 (5 × 6ATI), MultiFunction Display MFD-640, EFI-710 (6ATI), Primary Flight Display PFD-840 and PFD-890. Details of three representative models are given below.

Electronic Flight Instruments EFI-550 and EFI-640
The EFI-550 supports advanced display of HSI/ADI and other data. It has a full anti-aliased graphics capability and it supports both analogue and digital interfaces. It provides horizontal viewing angles of ±60° and electronically adjustable vertical viewing angle, with +45 to −10° vertical coverage, relative to the normal. Sunlight readability with greater than 10,000:1 dimming range is provided.

Specifications
(EFI-550 and EFI-640)
Bezel dimensions:
(EFI-550) 121.9 × 131.6 mm
(EFI-640) 131.5 × 156.8 mm
Display size:
(EFI-550) 85.1 × 110.7 mm
(EFI-640) 98.1 × 130.6 mm
Overall dimensions:
(EFI-550) 116.8 (H) × 126.5 (W) × 200.7 (D) mm (EFIS-85 standard)
(EFI-640) 5 × 6ATI
Weight:
(EFI-550) 3.64 kg
(EFI-640) 3.86 kg
MTBF: >7,500 h
Power: 28 V DC, 90 W

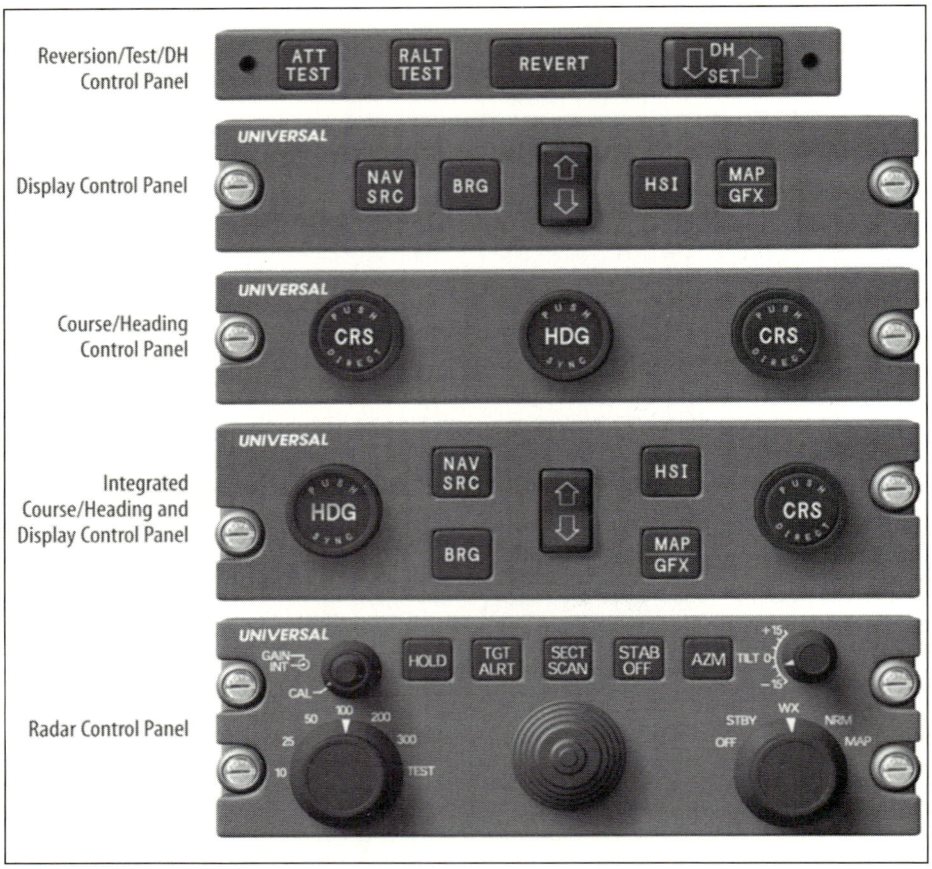

The Universal Avionics Systems' flat panel integrated display control panels **2000**/0062210

Supported functions: HSI, ADI, navigation, map, flight plan, weather radar, TAWS, EGPWS, TCAS, ADS-B, UniLink messages, traffic information service, cross cockpit and same-side reversionary, warnings and annunciators.

The Universal Avionics Systems' EFI-550 colour active-matrix liquid-crystal flat panel integrated display **2000**/0062207

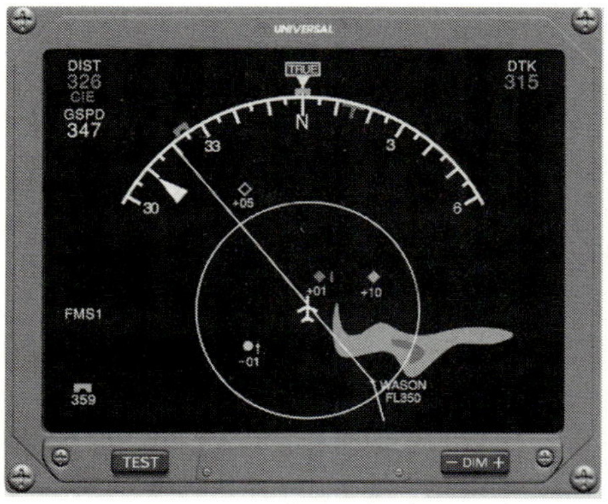

The Universal Avionic Systems' EFI-640 colour active-matrix liquid-crystal flat panel integrated display **2000**/0062208

MultiFunction Display MFD-640
The MFD-640 is designed for the display of moving maps, weather radar, terrain awareness, TCAS and other data. It is similar to the EFI-640 in construction and optical capabilities, but includes a comprehensive set of display control features that the EFI-640 does not have.

Specifications
(MFD-640)
Bezel dimensions: 131.5 × 156.8 mm
Display size: 98.1 × 130.6 mm
Overall dimensions: 5 × 6ATI
Weight: 3.86 kg
MTBF: >7,500 h
Power: 28 V DC, 90 W
Supported functions: Navigation, map, flight plan, weather radar, TAWS, EGPWS, TCAS, ADS-B, UniLink messages, traffic information service, cross cockpit and same-side reversionary, warnings and annunciators.

Contractor
Universal Avionics Systems Corporation.

VERIFIED

The Universal Avionics Systems' MFD-640 colour active-matrix liquid-crystal flat panel integrated display **2000**/0062209

GPS-950 sensor

The GPS-950 GPS sensor has been specifically designed to interface with UNS flight management and navigation management systems.

The GPS-950 utilises a five-channel receiver designed to process continuously pseudo range and range rate measurements from the four or more satellites having the best geometry and to output a navigation solution which includes three-dimensional position and velocity, time and Operational Status. When necessary, the fifth channel is used to gather ephemeris data from all the satellites in view, allowing rapid adaptation to obscuration. Otherwise, it is used to gather navigation data. In degraded constellation geometries, the GPS-950 will continue to provide two-dimensional navigation, using measurements from three satellites along with either GPS-corrected pressure or baro-corrected altitude.

Operation of the GPS-950 is controlled through and by the UNS systems with no additional control or display required. The GPS-950 provides outputs for up to three, and can accept inputs from two, UNS systems. Digital communications via low-speed ARINC 429 format are utilised between the sensor and UNS systems. These can integrate position data from the GPS-950 sensor with other long- and short-range navigation sensors to derive a highly accurate best computed position.

Integrity of the GPS position solution is monitored by both the GPS-950 sensor and the UNS systems with appropriate messages provided to the pilot.

Specifications
Dimensions:
(sensor) 194 × 25.1 × 386.8 mm
(antenna) 119.4 × 73.7 × 57.2 mm
Weight:
(sensor) 1.5 kg
Power supply: 18-32 V DC, 20 W

Operational status
Qualified in March 1994 to TSOs C-115a and C-129 B/C3.

Contractor
Universal Avionics Systems Corporation.

VERIFIED

GPS-1000 sensor

The GPS-1000 long-range navigation sensor features a 12-channel engine which tracks all satellites in view simultaneously and includes real-time and predictive Receiver Autonomous Integrity Monitoring (RAIM) capabilities. It features high acquisition and tracking sensitivity, carrier phase smoothing and position updates every second and will be able to accept pseudo-range correction signals from local area differential GPS in the future.

The GPS-1000 can be certified for en route and terminal operations, and non-precision approaches.

Specifications

Dimensions: 194.1 × 25.1 × 386.8 mm
Weight: 1.6 kg
Power supply: 18-32 V DC, 0.255 A at 28 V
Certification: TSO C-129 Class B1/C1

Operational status

In production.

Contractor

Universal Avionics Systems Corporation.

VERIFIED

GPS-1200 sensor

The GPS-1200 long-range navigation sensor utilises a more highly advanced 12-channel GPS engine than the GPS-1000. It features real-time and predictive Receiver Autonomous Integrity Monitoring (RAIM), satellite fault detection and exclusion capability and carrier phase tracking, and is Wide Area Augmentation System (WAAS) ready. It also includes local area differential GPS signal integration capability. Besides en route and terminal operations, the GPS-1200 can be certified as one of the two required long-range sensors for NAT MNPS navigation across the North Atlantic. It can also be certified for non-precision approaches and incorporates DO-178B critical level software which will be required for precision approach certification in the future.

Specifications

Dimensions: 194.1 × 25.1 × 386.6 mm
Weight: 1.6 kg
Power supply: 18-32 V DC, 0.255 A at 28 V
Certification: TSO C-129 Class B1/C1

Operational status

In production.

Contractor

Universal Avionics Systems Corporation.

VERIFIED

LCS-850 Loran C sensor

The LCS 850 Loran C sensor has been specifically designed to interface with UNS flight management and navigation management systems with no additional control or display required. Using a true multiple station solution, the LCS-850 provides accurate latitude/longitude position information to the UNS management systems. It tracks up to eight stations in a multiGRI position solution, automatically selecting the best signals from multiple chains. Position is calculated using a minimum of three stations on one GRI or two stations on each of two GRIs. Notch filters reduce interference and provide increased performance in weak Loran signal areas.

The LCS-850 meets the rigorous standards of TSO C-60b and also meets both system and accuracy requirements for RNav operations in the US National Airspace System and has NAT MNPS airspace approval.

Specifications

Dimensions: 194.1 × 25.1 × 386.6 mm
Weight: 1.6 kg
Power supply: 18-32 V DC, 0.225 A at 28 V

Operational status

In production.

Contractor

Universal Avionics Systems Corporation.

VERIFIED

MultiMissions Management System (MMMS)

The MMMS incorporates the advanced technology, system design, features and capabilities included in the UNS line of Flight Management Systems (FMS). In addition, the MMMS includes special interfaces, and the ability to fly the following six distinct types of pattern: rising ladder, expanding square, racetrack, sector search, orbit and border patrol. From the Control Display Unit (CDU), the operator is able to select the pattern and define the specific parameters appropriate to the mission. The type of pattern is graphically displayed, together with active/interrupted status and leg sequencing. Track, bearing, time and distance are numerically displayed as referenced to appropriate pattern waypoints throughout mission operations. Patterns can be activated, cancelled, or interrupted at any time. The pattern being flown can be interrupted and a new pattern selected, as required.

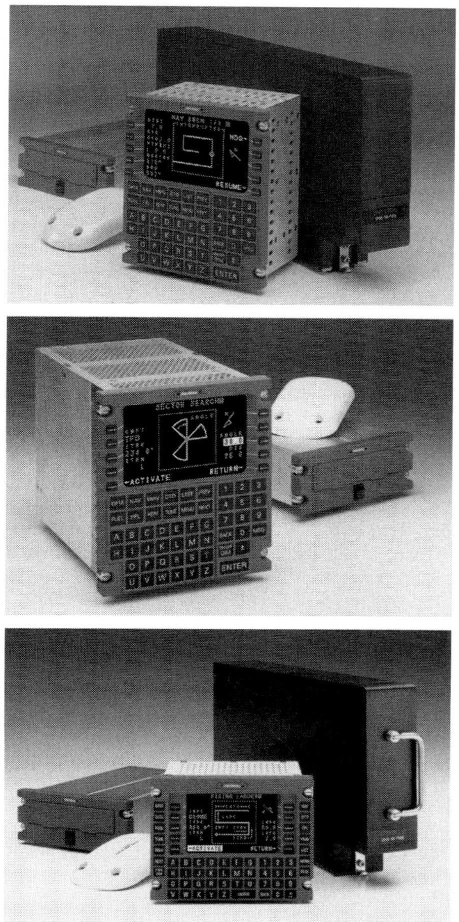

Examples of MMMS patterns displayed on the UNS-1D, UNS-1C and UNS-1K systems ***1999***/0051643

The MMMS systems interface with many of the aircraft systems, sensors and cockpit displays, including TACAN, radar, Doppler, EFIS and other systems. Additionally, the integral 12-channel GPS receiver meets the requirements for primary means of navigation remote/oceanic and provides seamless navigation throughout all phases of flight including non-precision approaches and special mission patterns worldwide. Flight management features include full SID, STAR and approach procedures, airways, advanced VNAV, holding patterns, heading mode, 3-D approach mode, fuel management and other facilities.

The MMMS specialised software is available for the UNS-1C, UNS-1D and UNS-1K FMS.

Contractor

Universal Avionics Systems Corporation.

VERIFIED

System-1 integrated avionics

System-1 is a new generation of integrated avionics; it includes Universal's UNS-1 Flight Management System (FMS); Terrain Awareness and Warning System (TAWS); Flat Panel Integrated Displays (FPIDs); Universal Cockpit Display (UCD); Data Transfer Unit (DTU); and UniVision cabin information system.

UNS-1 Flight Management System (FMS)
At the heart of the system is Universal's UNS-1 FMS (see separate entry). Worldwide, it provides a flight management capability from take-off to landing, with both lateral and vertical guidance, using position data from integral 12-channel GPS receivers that complement external navigation sensors. The UNS-1 database can include full procedures for Standard Instrument Departures (SIDs), Standard Terminal Arrival Routes (STARs) and three-dimensional approaches. Both lateral and pseudo glideslope outputs provide the pilot with ILS look-alike, IFR certified and autopilot coupled, guidance on the FMS Control Display Unit (CDU) for every approach.

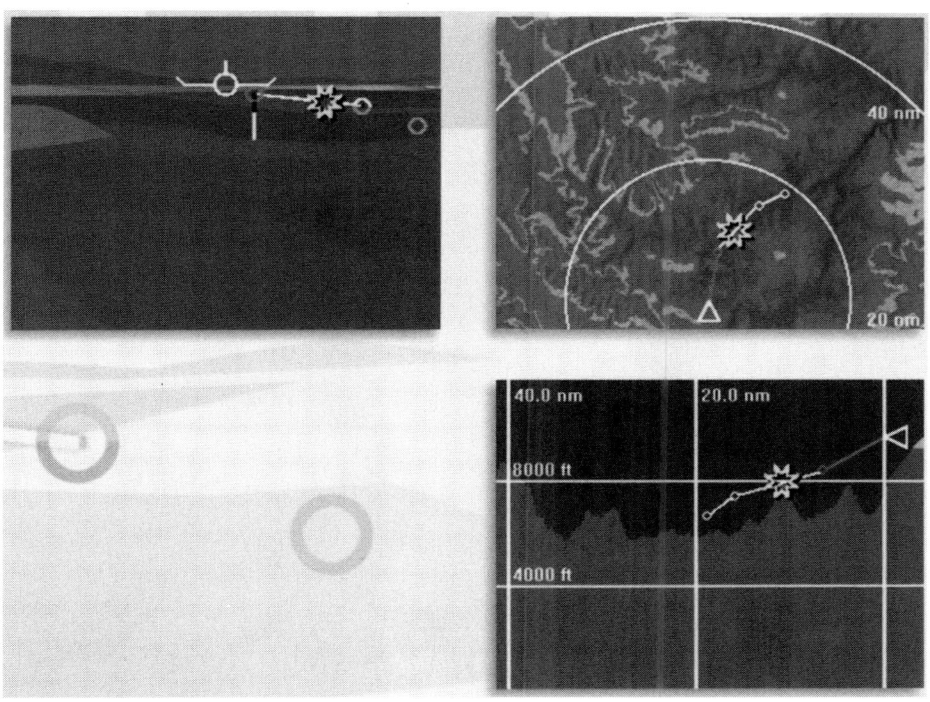

The Universal Avionics Systems' cockpit display and representative data formats top left: 3-D perspective view; top right: plan view; bottom right: profile view ***2000***/0062213

Terrain Awareness and Warning System (TAWS)
Universal's TAWS system (see separate entry) provides terrain situational awareness relative to current and predicted aircraft position, together with an advanced Ground Proximity Warning System (GPWS). The system provides alert information to the flight crew both visually and aurally. Using a worldwide terrain map database, TAWS provides outputs for display of terrain in several views including profile and perspective views on video-capable displays, such as the FPIDs, FMS CDU and UCDs that form part of System-1. ARINC 453 outputs are also available for use with weather radar displays.

Flat Panel Integrated Displays (FPIDs)
Universal offers a range of colour FPIDs to present System-1 data. They provide anti-aliased graphics capabilities and ±60° horizontal viewing angle with vertical electronic steering for best performance. Standard ARINC sizes are available for Primary Flight Data (PFD) and Navigation Data (ND); MultiFunction Data (MFD) including selectable moving maps, weather radar, weather map, TAWS, Enhanced Ground Proximity Warning System (EGPWS) data, and Terrain alert and Collision Avoidance System (TCAS) data; engine performance and other data. Current models include EFI-340 (3ATI), EFI-450 (4ATI), EFI-500 (4 × 5ATI), EFI-550, EFI-600 (5ATI), EFI-640 (5 × 6ATI), MFD-640, EFI-710 (6ATI), PFD-840 and PFD-890. See separate entries for details of Universal's FPIDs and FPID control panels.

Universal Cockpit Display (UCD)
Universal's UCD comprises a supplemental, 10 in, touchscreen, display terminal for the cockpit, together with a remote-mounted computer unit. The UCD provides the pilot with access to electronic displays, including JeppView™ terminal area charts as well as procedural checklists. It also accommodates video inputs to provide full TAWS display capabilities. See separate entry.

Data Transfer Unit (DTU-100)
The DTU-100 utilises 100 Megabyte Zip[R] disks. It is utilised for updating the FMS navigation database and the TAWS terrain database via a 10 MHz ethernet connection.

UniVision multimedia cabin information system
The UniVision system provides a communication and multimedia access and presentation capability.

Contractor
Universal Avionics Systems Corporation.

VERIFIED

Terrain Awareness and Warning System (TAWS)

Universal's TAWS provides terrain situational awareness relative to current and predicted aircraft position. This 'look ahead' capability is displayed in three views: Plan, Profile and 3-D perspective. Each view includes the display of the flight plan and flight path intent in conjunction with a detailed display of the surrounding terrain. Full system display benefits are realised using video-capable devices such as Universal's Flat Panel Integrated Displays (FPIDs), its FMS CDU and its new touchscreen cockpit terminal. ARINC 453 outputs of terrain in plan view only are available for other MFDs and weather radar displays.

Based upon information from the FMS, the Air Data Computer (ADC), radio altimeter and ILS, the TAWS system is able to determine the aircraft's state and intent, and to provide warnings and alerts well in advance of potential hazards that could result in a Controlled Flight Into Terrain (CFIT). Warnings are provided if any part of the entered flight plan were to pose a threat.

The TAWS system also provides alerts in accordance with standard Ground Proximity Warning System (GPWS) functionality modes. A selectable option is available for bank angle alerting as well.

The TAWS computer is housed in a 2 MCU-sized LRU weighing about 3.2 kg. The worldwide terrain database, stored in flash memory, contains 30 arc sec elevation data with up to 6 arc sec data at mountainous airports. TAWS data is presented as part of Universal's System-1, network of integrated flight management products.

Contractor
Universal Avionics Systems Corporation.

VERIFIED

UniLink air-to-ground two-way datalink

UniLink is designed to be interfaced with and controlled through Universal Avionics Systems Corporation's colour flat-panel Flight Management Systems, which include the UNS-1B plus, UNS-1C, UNS-1Csp, and UNS-1D. The UniLink menu software integrates with the UNS FMS and provides access for sending and receiving data and graphics. Flight plans can be uplinked through UniLink and loaded directly into the FMS. Position reports and other data from the FMS can be automatically downlinked.

UniLink has been designed to support several communications media including VHF, telephony and Satcom. Other media, such as HF, will be added as they become supported by the Aeronautical Telecommunications Network (ATN).

UniLink has been designed to support all ACARS message types, including triggered events such as OOOI (Out, Off, On, In) and planned interface with Digital Flight Data Acquisition Units (DFDAUs).

The UniLink module is available as a model UL-600 housed in a 1-MCU sized unit, supporting single-, dual- or triple-FMS installations; it will also be available as a separate PCB for the UNC-1C, UNS-1Csp, and UNS-1D FMS.

Universal Avionics Systems Corporation claims that the combined UNS-1 and UniLink suite fulfils the evolving Communications/Navigation/Surveillance (CNS) routing and communication requirements in the future Aircraft Traffic Management (ATM) system.

Specifications
(UL-600)
Dimensions: 1-MCU
Weight: 1.47 kg
Power: 28 V DC, 5 W
Memory: 2 Mbytes FLASH memory; 1 Mbyte SRAM; 32-bit controller
Interfaces: VHF modern: 1 input/1 output; telephony; 1 input/1 output; ARINC 429: 8 input/3 output; RS-422/232: 8 input/8 output; RS-232 diagnostics/load port: 1 input/1 output; discretes: 16 input/16 output; configuration module 1 input/2 output

Contractor
Universal Avionics Systems Corporation

VERIFIED

UniLink air-to-ground two-way datalink 0011857

Universal Cockpit Display (UCD)

The Universal Cockpit Display (UCD) features a slimline 10 in touchscreen cockpit display terminal which provides the pilot access to electronic terminal area charts, checklists, Terrain Awareness and Warning System (TAWS) displays and other data.

JeppView™ charts and associated NOTAMs and airport information can be loaded onto the remote-mounted UCD Computer (UCDC). Departure and arrival airport information supplied from the Flight Management System (FMS) will prompt the display of the associated charts. Manual searches are accomplished through entry of airport name or ICAO identifier. In addition to normal zoom functions, a touch zoom feature allows the pilot to view a selected area on a chart at a higher magnification.

Aircraft specific procedural checklists can be downloaded or pilot defined. Colours indicate the checked status of each item. Notes can also be incorporated. Checklists are presented in index-chapter format yielding quick access for both normal, abnormal and emergency situations.

The 780 × 1,024 pixel, high-resolution UCD Terminal (UCDT), also accommodates video inputs and supports display of two simultaneous views from the TAWS including selection of plan, profile and 3-D perspective displays.

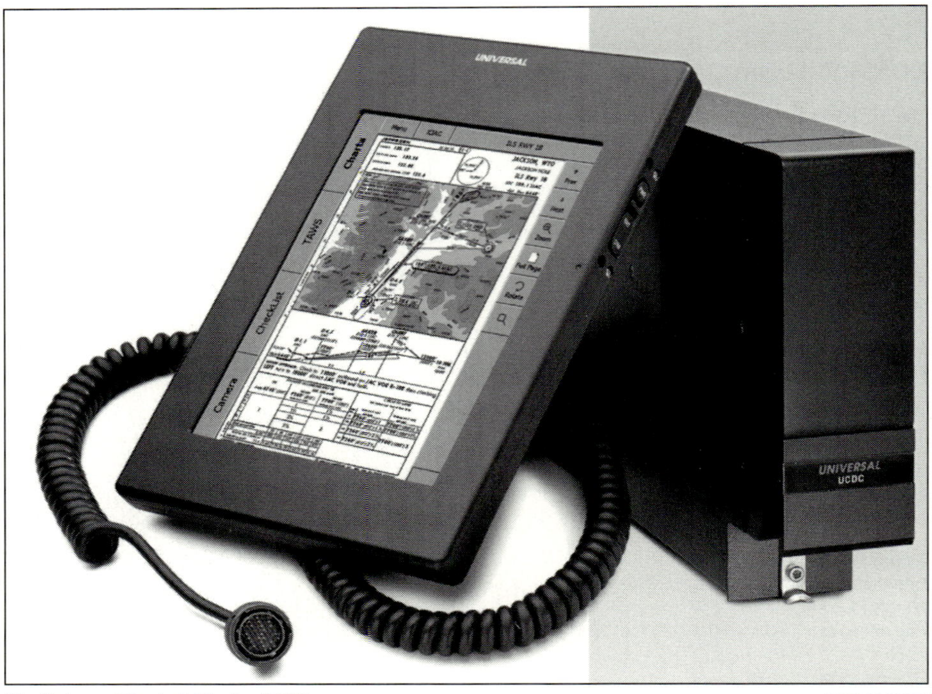

The Universal Cockpit Display (UCD) **2000**/0062215

Growth potential includes access to maintenance manuals and other reference materials, UniLink and UniVision interfaces, display for external video cameras and overlay of aircraft position on charts.

Specifications

UCDC

Dimensions: 2 MCU
Weight: 3.4 kg

UCDT

Dimensions: 213.4 × 304.8 × 22.4 mm
Weight: 1.8 kg
Display: 6.1 × 8.1 in, 10 in diagonal, 780 × 1,024 pixels

Contractor

Universal Avionics Systems Corporation.

VERIFIED

UNS-1B Flight Management System (FMS)

The UNS-1B FMS comprises a Control/Display Unit (CDU), Navigation Computer Unit (NCU) and Data Transfer Unit (DTU) for uploading and downloading navigation database information. The CDU is available with a 127 mm colour display, 10 line-select keys, 10 function keys and a full alphanumeric keyboard.

The UNS-1B uses position data from long- and short-range navigation sensors to determine the best computed position. Automatic scanning DME/DME/DME positioning with slant range error correction, as well as en route Rho/Theta, is computed. Vertical navigation, three-dimensional approach mode - including GPS approaches - and holding pattern are also included. All 20 leg types can be flown, including heading to altitude, radial intercept, DME arcs and procedure turns - all manoeuvres required to fly complete SIDs, STARs and holding pattern procedure accurately. The UNS-1B FMS accepts fuel flow data from up to four engines.

Analogue and digital outputs are provided to flight directors, autopilots, EFIS, multifunction displays and radar navigation displays. Digital communications are provided using the ARINC 429/571/561/575 formats.

The UNS-1B has a 3.1 Mbyte database capacity, equivalent to 200,000 waypoints, and can include SIDs, STARs and approaches. It also includes airports, navaids, en route and terminal waypoints and airways in the database. The Jeppesen database is formatted on 3.5 in diskettes and is updated on a 28-day cycle. Pilot-defined data can include 200 routes, with up to 98 waypoints in each for a total of 3,000 waypoints, 100 arrivals/departures, 100 approaches and 100 runways. Built-in batteries prevent memory loss when the unit is removed from the aircraft.

Additional system features include a heading mode for direct control of aircraft heading through the CDU, bank angle commands correlated to altitude and turn anticipation to eliminate overshoots. Options include frequency management for centralised control of aircraft navigation and communication radios, AFIS interface capabilities for airborne ground link information and aircraft specific performance data functions.

The latest variant is UNS-1B plus. It utilises a colour flat-panel control/display. The 2MCU-sized navigation computer unit houses the new ASCB-interface printed circuit board. This board can also be installed in the UNS-1C and UNS-1D models as retrofits for the Challenger 601-3A and Falcon 900 aircraft.

UNS-1C all-in-one flight management system

The UNS-1B plus with GPS-1200 sensors is certified for GPS operation in en route, terminal and approach phases of flight (TSO C1156 and C129 Class B1/C1), and meets the requirements for primary means of navigation in oceanic/remote airspace.

UNS-1B with GPS-1000 12-channel receiver, RAIM and Fault Detection and Exclusion (FDE) has received certification by the Civil Aviation Authority of New Zealand for GPS primary means of navigation in remote/oceanic airspace.

Specifications

Dimensions:
(CDU) 169 × 146 × 200.2 mm
(NCU) 194 × 57 × 388 mm
(DTU) 53.8 × 146 × 205 mm
Weight:
(CDU) 3.45 kg
(NCU) 3.45 kg
(DTU) 1.47 kg
Power supply: 27.5 V DC, 60 W (max)
26 V AC, 400 Hz, 1 VA

Operational status

In production.

Contractor

Universal Avionics Systems Corporation.

VERIFIED

UNS-1C Flight Management System (FMS)

The UNS-1C Flight Management System (FMS) is an all-in-one unit that includes an integral 12-channel GPS receiver with real-time and predictive Receiver Autonomous Integrity Monitor (RAIM). It is designed for integration with the advanced all-digital avionics suites of the next-generation aircraft, such as the Lear 45, Falcon 900EX, Falcon 2000, IAI Galaxy and Astra SPX.

The UNS-1C updates the UNS-1B and enhancements include: a 5 Mbyte navigation database, full-colour active matrix liquid crystal flat-panel display and the latest surface mount technology available, providing a more compact design and much higher reliability.

Specifications

Dimensions:
(UNS-1C) 162 × 146 × 228 mm
(data transfer unit) 53.8 × 146 × 205 mm
Weight:
(UNS-1C) 3.63 kg
(data transfer unit) 1.47 kg
Power supply: 27.5 V DC, 50 W (max)
26 V AC, 400 Hz, 1 A

Operational status

The UNS-1C is available for retrofits offered as a turnkey update for the Dash 8, selected by Horizon Air for its 25 new Dash 8 Series 200 aircraft. It has also been certified on the Bell 430 helicopter.

A more compact 'all-in-one unit' version, designated UNS-ICsp (special package), is available as a special option on the Citation, Ultra and Citation VII.

The UNS-1C is also available in MultiMission Management System (MMMS) versions. The MMMS includes the ability to fly six mission patterns and to accommodate EFIS, Doppler, radar, video and FLIR data.

The UNS-1C is being replaced by the UNS-1E, so-called Super FMS from February 2000. (See separate entry.)

Contractor

Universal Avionics Systems Corporation.

VERIFIED

UNS-1D Flight Management System (FMS)

The UNS-1D is an upgraded UNS-1C with new Control Display Unit (CDU) and new remote Navigation Computer Unit (NCU) with internal 12-channel GPS receiver. The UNS-1D is also available in MultiMission Management System (MMMS) versions. The MMMS includes the ability to fly six patterns and to accommodate EFIS, Doppler, radar, video and FLIR data.

Operational status

Selected for retrofit to US Navy C-9B/DC-9 transport aircraft to enable them to meet CNS/GATM, BRNAV and ACAS requirements.

The UNS-1D is being replaced by the UNS-1F so-called Super FMS from February 2000. (See separate entry.)

Contractor

Universal Avionics Systems Corporation.

VERIFIED

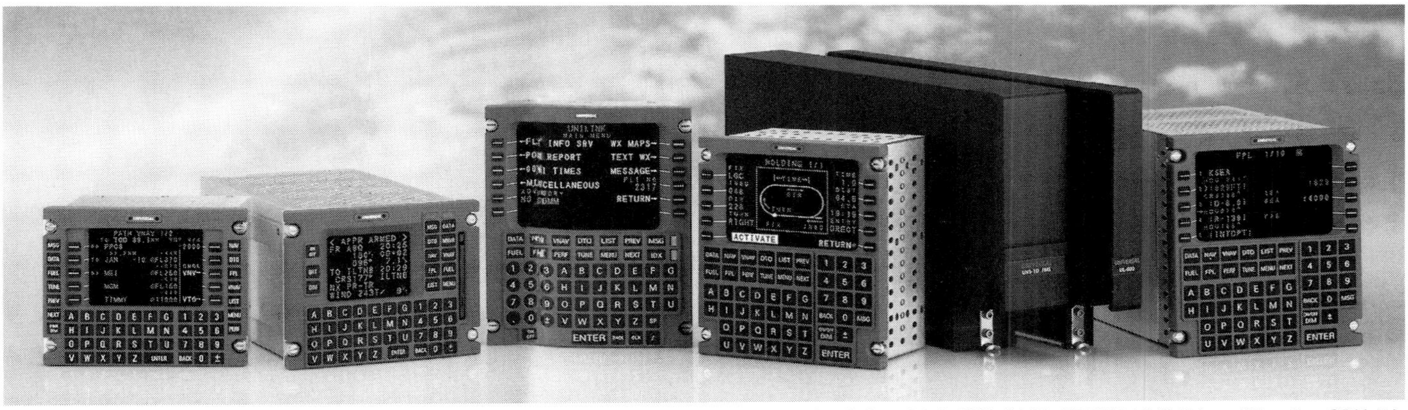

The picture shows a range of products manufactured by the Universal Avionics Systems Corporation (left to right): UNS-1K 4 in FPCDU, UNS-1M, multifunction CDU, 5 in FPCDU, UNS-1D NCU, UniLink UL-600, UNS-1C. Separate entries describe the other products **1999**/0051644

UNS-1E, UNS-1F, UNS-1L 'super' Flight Management Systems (FMS)

UNS-1E, UNS-1F and UNS-1L FMS supersede the UNS-1C, UNS-1D and UNS-1K respectively. The new systems are described as 'super' FMSs by Universal Avionics. Each system contains an integral 12-channel combined GPS/GLONASS receiver providing increased accuracy, integrity and availability for navigation en route and on non-precision approach worldwide. To allow for future software enhancements, each 'super' FMS incorporates a new 32-bit processor, which provides 25 times the computational performance of the current systems. The program memory has been increased fourfold, and the navigation database capacity has been increased to 32 Mbytes. The new FMSs also feature an Ethernet communication port for interfacing to Universal's new line of System-1 products.

The UNS-1E features an integrated 5 in display. The UNS-1F comprises a remotely mounted 2 MCU navigation computer, and separate 5 in control display unit. The UNS-1L includes a 2 MCU navigation computer unit and a separate 4 in control display unit. The 5 in displays are graphics- and video-capable and support weather and TAWS displays.

Operational status

All systems are available from February 2000.

Contractor

Universal Avionics Systems Corporation.

VERIFIED

UNS-1K Flight Management System (FMS)

The UNS-1K is positioned between the higher end UNS-1C and UNS-1D systems and the lower placed UNS-1M navigation management system produced by Universal Avionics Systems Corporation. The UNS-1K utilises the same software, SCN 602, as the UNS-1C and UNS-1D systems and features 10 line select keys and large 4 in colour flat-panel display. The control display unit is housed in a standard 4.5 in tall × 5.75 in wide Dzus-mounted unit 3.25 in deep. The navigation computer unit is 2 MCU in size, and includes an integral 12-channel GPS receiver. It also includes the Multi-Missions Management System (MMMS).

The system will meet emerging Required Navigation Performance/Actual Navigation Performance requirements around the world, including new European B-RNAV requirements.

System features include flight planning with full SID/STAR procedures, airways and approaches. A best computed position is based upon inputs from the integral GPS receiver, auto-scanning DME, and the operator's complement of external navigation sensors. The system will fly all ARINC 424 leg types. En route manoeuvre capabilities include a dedicated direct-to function, FMS heading commands, PVOR tracking, coupled VNAV with computed top of descent and vertical direct-to commands. Procedural holding patterns and approaches along with their transitions and missed approach procedures are contained in the database. The Approach Mode provides IFR-approved, pseudo-localiser, pseudo-glide slope guidance to any airport making all approaches look like an ILS. Fuel Management and optional Frequency Management functions are available and the MultiMission Management System (MMMS) is available on the UNS-1C/-1D models. A standby power-off mode retains flight plan and fuel initialisation data for up to 8 hours. A comprehensive test mode is incorporated to facilitate installation check out and return to service.

The integral 12-channel GPS receiver provides real-time and predictive Receiver Autonomous Integrity Monitoring (RAIM), automatic Fault Detection and Exclusion (FDE), step detection, and manual satellite deselection capabilities. The UNS-1K is certified for GPS operation en route, terminal and approach phases of flight and meets the GPS navigation operational approvals and Minimum Navigation Performance Specifications (MNPS) for navigation in the North Atlantic Track (NAT) airspace. The system can also be approved under FAA Notice 8110.60 for primary means of navigation in remote/oceanic airspace using GPS alone in conjunction with Universal Avionics' off-line PC-based Flight Planning and RAIM Fault Detection and Exclusion programme. The UNS-1K will also be compatible with Universal Avionics' GLS-1250 GPS landing system (currently in flight test) which will provide future growth by adding GPS precision approach capability.

The UNS-1K will also interface with Universal Avionics' new UniLink air-to-ground datalink. Adding UniLink provides the operator with such capabilities as predeparture clearances, flight plan up/downlinking, oceanic clearances, position reporting, digital ATIS, messages and text weather, through the UNS-1K control display unit. The UniLink UL-600 is housed in a 1 MCU size unit which can provide datalink information through several communications media including VHF, telephony and satcom systems. The UNS-1K with UniLink combine to provide the operator with full capability Communication/Navigation/Surveillance (CNS) avionics suite.

Operational status

Both the UNS-1K and UniLink are available. UNS-1K is certified TSO C129a Class A1/B1/C1 and C115b, with Supplemental Type Certification (STC) on Boeing 737-200. Selected for a number of regional aircraft types.

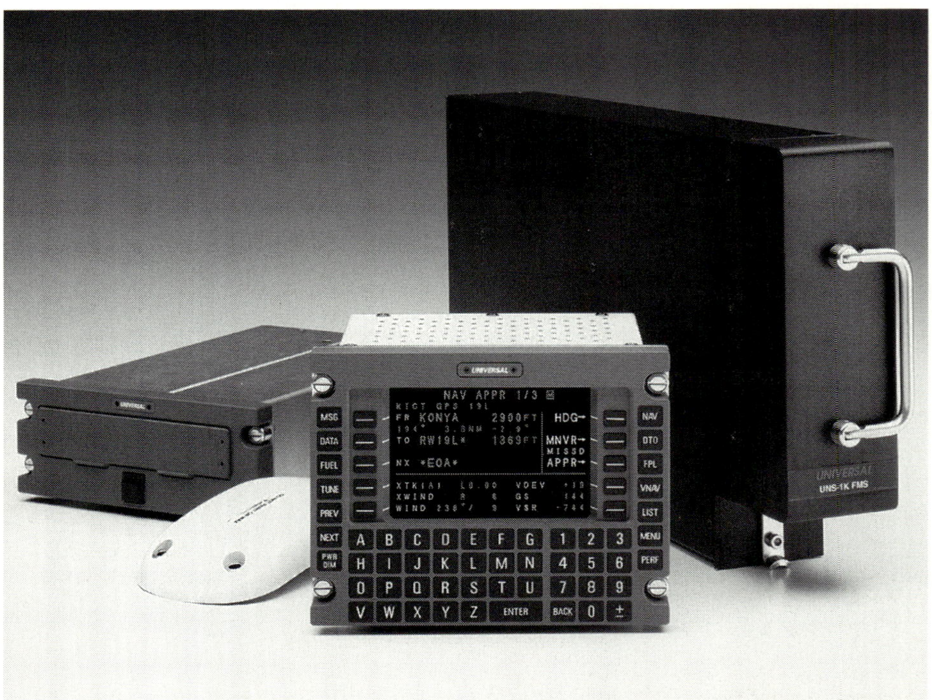

UNS-1K flight management system 0018224

The UNS-1K is being replaced by the UNS-1L so-called Super FMS from February 2000. (See separate entry.)

Contractor

Universal Avionics Systems Corporation.

VERIFIED

UNS-1M navigation management system

The all-in-one UNS-1M is a full navigation management system, self-contained in a control/display unit, which meets the standards of TSO C-129 A1/B1/C1. It features a flat-panel display, integral 12-channel GPS receiver with Receiver Autonomous Integrity Monitor (RAIM) and full alphanumeric keyboard. Smart auto-scanning DME/DME/DME provides continuous DME updating. Three external long-range sensor inputs accommodate combinations of Omega/VLF, Loran C, inertial and GPS sensors. The UNS-1M provides digital and analogue outputs for autopilot, mechanical HSI and EFIS systems. The unit's three-dimensional approach mode provides guidance on approach and is IFR certified for non-precision GPS, RNav, VOR and VOR/DME approaches. The fuel management system uses DC analogue sensors to provide real-time data on fuel weight, gross weight and landing weight. A worldwide Jeppesen navigation database is stored on, and updated via, a non-volatile flash memory card. The database includes airports, VORs, DMEs, VOR/DMEs, ILS/DMEs, VORTacs, Tacans, NDBs and en route and terminal waypoints.

Specifications

Dimensions: 114 × 146 × 241 mm
Weight: 2.81 kg
Power supply: 19-32 V DC, 35 W (max) at 27.5 V
26 V AC, 400 Hz

Operational status

In production. Selected by Executive Airlines for ATR-42 and ATR-72 aircraft.

Contractor

Universal Avionics Systems Corporation.

VERIFIED

UNS-764 Omega/VLF sensor

The UNS-764 Omega/VLF sensor was designed to operate in conjunction with the Universal's UNS-1 flight management and navigation management systems, working either alone or in combination with inertial, GPS or Loran C sensors. The system comprises an antenna and receiver, the control/display unit of the UNS-1 system providing the necessary management functions. Gate-array and surface-mount technologies combine to reduce the size and weight of the UNS-764 system over previous Omega/VLF sensors. Operation is automatic, control being exercised via the UNS-1. Five independent receive channels enable all available Omega, and up to eight VLF, stations to be monitored simultaneously.

Specifications

Dimensions: 194.1 × 56.9 × 387.6 mm
Weight: 2.9 kg
Power supply: 28 V DC, 20 W (max)

Operational status

In production.

Contractor

Universal Avionics Systems Corporation.

VERIFIED

UNS 764-2 GPS/Omega/VLF sensor

The UNS 764-2 is a combination GPS/Omega/VLF long-range navigation sensor. It incorporates a 12-channel GPS receiver that features real-time and predictive Receiver Autonomous Integrity Monitoring (RAIM), satellite fault detection and exclusion capability, carrier phase tracking. It is Wide Area Augmentation System (WAAS) ready.

The Omega/VLF receiver features five independent channels which track all available Omega and up to eight VLF stations simultaneously.

Specifications

Dimensions: 194.1 × 56.9 × 387.6 mm
Weight: 3.6 kg
Power supply: 28 V DC, 30 W nominal
Certification: TSO C-129 Class B1/C1, C-94a

Operational status

In production.

Contractor

Universal Avionics Systems Corporation.

VERIFIED

UNS-RRS radio reference sensor

The UNS-RRS radio reference sensor has been designed to provide DME, VOR and Tacan radio data for the UNS-1 flight and navigation management system. It is remotely tuned through the UNS-1 CDU via an ARINC 429 databus and each RRS can support two UNS-1 systems.

Specifications

Dimensions: 194.1 × 56.9 × 387.6 mm
Weight:
(receiver) 352 kg
Power supply: 28 V DC, 1 A (max)

Operational status

In production.

Contractor

Universal Avionics Systems Corporation.

VERIFIED

ADS-B for US Cargo Airline Association

ADS-B (Automatic Dependent Surveillance-Broadcast) is 'automatic' in that it requires no radar interrogation or pilot input. It is 'dependent' on the Global Positioning Systems (GPS), and it is a 'surveillance' system in that any other aircraft or ground station equipped with a receiver and display can monitor transmitting aircraft.

ADS-B 'broadcasts' position and other useful information to aircraft and designated ground stations. At the same time, the aircraft's own receiver can listen to all other aircraft within a 100-mile radius and see their information on a cockpit display.

ADS-B provides high-resolution visual information, including: ADS-B traffic data; three-dimensional position and navigation data; up-linked flight and weather information; and up-linked traffic information services.

By adapting satellite navigation and broadband datalink technology, ADS-B enables any aircraft to both send and receive highly accurate position data, giving every pilot an electronic picture of surrounding traffic. On the same screen it is possible to access moving map positional displays, flight information (including graphical weather), navigational information and other proprietary information. ADS-B supersedes the need for voice contact with ATC.

ADS-B is a technology driver for the 'Free Flight' concept.

In 1996, the Cargo Airline Association (CAA) (not to be confused with the various Civil Aviation Authorities) appointed II Morrow, now UPS Aviation Technologies (UPS AT), to act as project manager for its ADS-B development project.

The UPS Aviation Technologies solution comprises the following equipment: a flat-panel Cockpit Display of Traffic Information (CDTI) where traffic, weather/terrain and navigation information is integrated. Symbology, pioneered by NASA, displays traffic with heading, flight ID, and a velocity vector showing ground track; additionally, when the pilot selects a specific aircraft, information such as closure rate and ground speed give increased awareness of potential threat. The planned architecture includes a basic system for general aviation comprising Apollo navigator and display together with a broadband datalink; while for air transport aircraft a more comprehensive fit is proposed comprising CDTI, data fusion processor, Mode-S, keypad, 1090 receiver, VDLM4 and UAT.

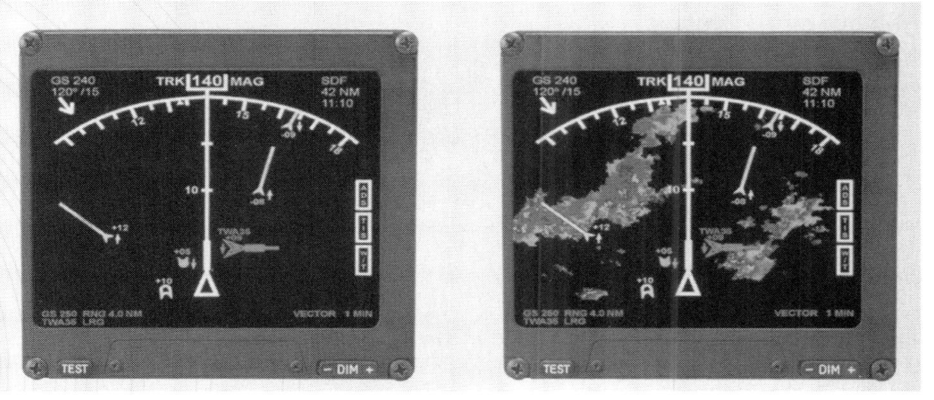

The UPS Aviation Technologies' ADS-B CDTI display, showing traffic, with (right) and without terrain data (left)
1999/0044811

Operational status

In 1996, ADS-B development for the CAA started. In June 1998, the first aircraft fit was tested. A 12-aircraft flight test and evaluation with three cargo companies (Airborne Express on four × DC-9 aircraft and Fedex and UPS on four Boeing 727 aircraft each) was conducted in late 1998. Equipment fit began in 1999, for a total fleet of 800 aircraft.

Contractor

UPS Aviation Technologies, Inc.

VERIFIED

Apollo 360 GPS moving map

The Apollo 360 GPS moving map display slides into a standard 3⅛ in round instrument hole. The back-lit LCD display can be configured to display standard numeric information or a moving map which shows position relative to flight plan and track, airspace boundaries, and nearby waypoints. Fitted as standard is an extensive Jeppesen database that includes public use airports, VORs, NDBs, intersections and all special-use airspace. Satellite tracking is provided by an eight-channel parallel sensor. Both the database and the operating software can be updated by means of a serial data port without removing the unit from the cockpit panel. Coupled with the optional Apollo eight-channel GPS receiver the 360 moving map can be transformed into a GPS navigator.

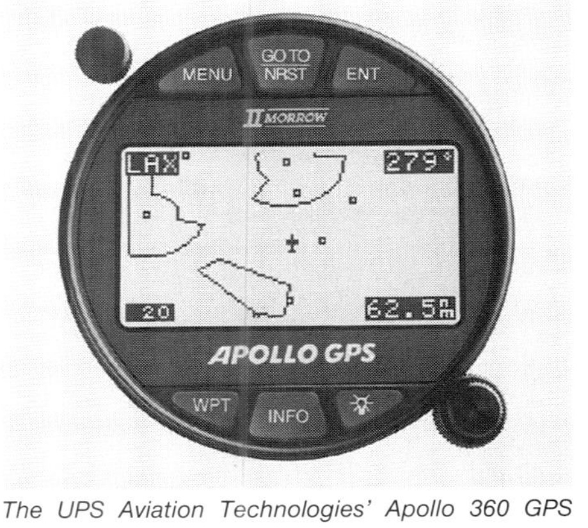

The UPS Aviation Technologies' Apollo 360 GPS moving map

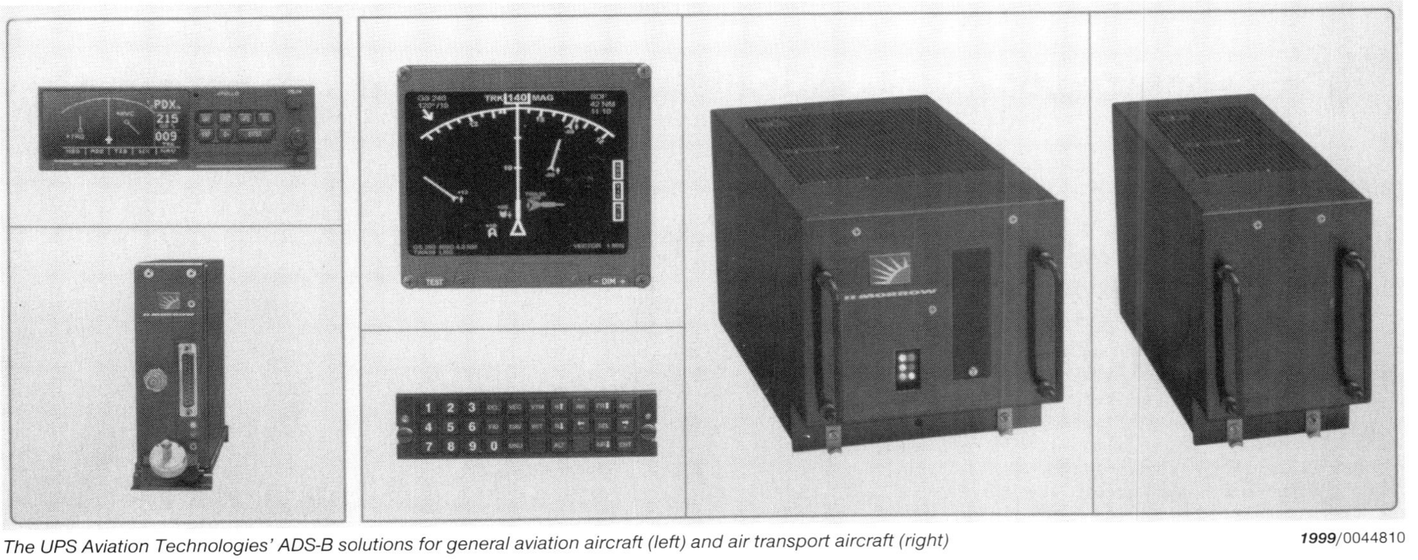
The UPS Aviation Technologies' ADS-B solutions for general aviation aircraft (left) and air transport aircraft (right)
1999/0044810

Specifications

Dimensions: 3⅛ in (79 mm) (diameter), 200.4 mm (deep)
Weight: 1.36 kg
Power supply: 10-40 V DC, 3 W nominal
Temperature range: –10 to +55°C
Altitude: up to 55,000 ft
Channels: 8 parallel
Accuracy: 15 m rms (100 m rms with S/A)

Contractor

UPS Aviation Technologies, Inc.

VERIFIED

Apollo 2101 GPS Navigation Management System (NMS)

The Apollo 2101 GPS NMS features eight-channel parallel GPS and RAIM; it is certified TSO C129 for IFR en route and non-precision approach. Designed for fixed- or rotary-wing aircraft, commercial, corporate and military, it features a full alphanumeric keyboard and high-contrast LED, with NVG option.

The integrated database provides instant data on airports, VORs, NDBs and intersections. GPS approach data is automatically displayed for destination during flight planning.

The Apollo 2101 fits any dzus rail and requires no external OBS wiring. The NAVNET interface, high-speed serial bus and ARINC-429/-561 support provide compatibility with CDI, MSI and moving map displays as well as air-data, ACARS and EFIS systems.

C115A multisensors certification allows use of additional positioning and data systems to improve safety.

Specifications

Dimensions:
(2101 NMS computer) 146 × 76 × 135 mm
(2102 keypad) 146 × 38 × 139.7 mm
Weight: 1.7 kg
Channels: 8 parallel
Accuracy:
(horizontal) 15 m rms (100 m rms with S/A)
(vertical) 156 m rms with S/A
(velocity) 0.5 m/s

Operational status

In production.

Contractor

UPS Aviation Technologies, Inc.

VERIFIED

The UPS Aviation Technologies' Apollo 2101 GPS NMS 0001334

Apollo GX50, GX55, GX60 and GX65 advanced navigation and communication solutions

Apollo GX50 advanced navigation solutions

The Apollo GX50 is an approach-certified, eight-channel, GPS receiver that provides a high-definition moving map display on a sunlight-viewable screen, that indicates the aircraft's position relative to airports, runways, VORs, NDBs, intersections and SUAs. The Apollo GX50 provides a 'smart approach' feature and operation is streamlined by the use of 'smart keys' allowing wanted/unwanted information to be added/

The UPS Aviation Technologies' Apollo GX50 advanced navigation solutions ***1999***/0051604

The UPS Aviation Technologies' Apollo GX60 advanced navigation and communications solutions ***1999***/0051605

deleted with a single button action. The GX50 is TSO-C129a Class A1 approved for IFR non-precision approach operation.

Apollo GX55 advanced navigation solutions

The Apollo GX55 is an enhanced version of the Apollo GX50 that is TSO-C129 Class A2 approved for IFR en route and terminal operation. The Apollo GX55 is also a pin-for-pin tray-compatible replacement for panel-mounted Apollo Loran and Flybuddy GPS units.

Apollo GX60 advanced navigation and communication solutions

The Apollo GX60 is UPS Aviation Technologies' top of the range advanced navigation and communication solution. It is approved to the following standards: TSO-C129a Class A1 GPS; TSO-C37d transmit; TSO-C38d receive; and TSO-C128 stuck mic, with 'smart' key control of cockpit operations. It features the same eight-channel GPS receiver outfit as the GX50/55, with 'smart key' control and 'smart approach' features, plus a 760-channel VHF 8 W communications transceiver.

Apollo GX65 advanced navigation and communication solutions

The Apollo GX65 offers essentially the same features as the GX60, but without non-precision approach capability.

Specifications

Dimensions:
(all models): 50.8 (H) × 158.8 (W) × 282.6 (D) mm
Weight:
GX50/GX55: 1.179 kg
GX60/GX65: 1.409 kg
GPS receiver: (all models): 8-channel
Accuracy: (all models):
Lat/Long position: 15 m, rms typical
25 m, SEP, without SA
100 m, 2 Drms with SA
Comm receive: (GX60/GX65): 760 channels, 118.000-136.975 MHz
Comm transmit: (GX60/GX65): 8 W

Contractor

UPS Aviation Technologies, Inc.

VERIFIED

Apollo MX20 multifunction display

The MX20 multifunction display is a 6-in diagonal, colour, Active Matrix Liquid Crystal Display (AMLCD). It has a 65,000 colour, full-VGA capability and is sunlight readable. Display features include VFR and IFR en-route charts, lightning strike and terrain awareness displays. It can be interfaced with other Apollo and GPS receivers. The open architecture design provides for future growth, including weather and traffic datalinks, TCAS and radar interfaces. It is TSO approved to TSO-C110a and TSO-C113.

Specifications

Dimensions: 127 (H) × 158.75 (W) × 203.2 (D) mm

Contractor

UPS Aviation Technologies, Inc.

VERIFIED

Apollo SL30 nav/comm

The Apollo SL30 nav/comm is fully qualified to appropriate TSOs. It is identical in dimensions to UPS Aviation Technologies' other Slimline products and includes the same 760-channel transceiver that is used in the SL40 comm and the SL60 GPS/comm systems. It also includes a 200-channel navigation receiver, which uses a fully digital solid-state design. The SL30 has built-in localiser and glideslope receivers. The navigation receiver has the ability to track both the active and standby navigation channels. It will automatically decode the VOR Morse code signal and display the station ID.

When interfaced with other Apollo GX GPS units, Apollo SL30 nav/comm will maintain a list of the ten nearest VORs and the frequencies of the destination airport.

Specifications

Weight: 1.5 kg

Contractor

UPS Aviation Technologies, Inc.

VERIFIED

Apollo SL40 Comm/SL50 GPS Navigator/SL60 Comm/GPS

The Apollo SL (SlimLine) products manufactured by UPS Aviation Technologies are identical in size and shape (1.3 × 6.25 (standard) × 10.5 in (H × W × D), but have different controls to operate the services provided.

Apollo SL40 Comm

The Apollo SL40 Comm is a TSO-approved VHF transceiver. It provides the following capabilities: 760 × 25 kHz channels (118.000 to 136.975 MHz); 8 W transmit power output; 12 W audio amplifier; two-place voice-activated intercomm; standby frequency monitor

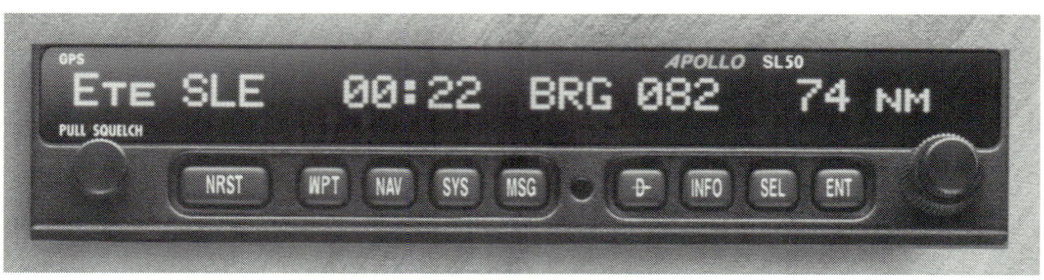

The UPS Aviation Technologies' Apollo SL50 GPS Navigator ***1999***/0044808

The UPS Aviation Technologies' Apollo SL60 Comm/GPS ***1999***/0044809

function; US National Weather Service weather channel (up to seven channels); 16-frequency memory store; interface to the UPS Aviation Technologies' 2001 navigation management system, and other UPS Aviation Technologies' products. The SL40 Comm uses a single-line 32-character Light Emitting Diode (LED) display. In the Comm version, only the left half of the display is used to display the active and standby frequencies.

Apollo SL50 GPS Navigator

The Apollo SL50 GPS Navigator includes a large waypoint database with information about airports, VORs, NDBs, intersections, other airport facilities and space for 200 pilot-entered waypoints. Other capabilities include: the ability to process 10 reversible flight plans; special-use airspace alerts; parallel track; minimum en route safe altitude and minimum safe altitude. Navigation displays include: Lat/Long (to 0.01 minute); bearing and distance to waypoint; ground speed and track angle; desired track and distance between waypoints; cross track error; 'TO' waypoint identification, ETAs. The Apollo SL50 can be directly integrated with the Apollo SL40 Comm and 360 Moving Map. The Apollo SL50 features an eight-channel parallel GPS receiver, with Jeppesen database.

The Apollo SL50 GPS Navigator can be installed for VFR or IFR operations; it is approved to FAA TSO C-129 A2 requirements for en route and terminal operations.

Apollo SL60 Comm/GPS

The Apollo SL60 Comm/GPS combines the Comm features of the Apollo SL40 and navigation/GPS features of the Apollo SL50 GPS Navigator.

Operational status

All models are available.

Contractor

UPS Aviation Technologies, Inc.

VERIFIED

Apollo SL40, SL50 and SL60 slimline nav/comm series

The first three models of the slimline series, the SL40 comm transceiver, SL50 GPS navigator and SL60 GPS/comm, share the 33 × 159 mm form factor and sunlight-readable Light Emitting Diode (LED) display technology. They utilise the advanced technology of UPS Aviation Technologies' high-end GPS receivers.

The SL40 comm transceiver, (8 W carrier transmit power) is TSO certified and features direct access to National Weather Service information and automatic standby frequency monitoring. The SL50 GPS navigator is TSO C129 approved for en route and terminal operations. The SL60 GPS/comm carries these features, plus a navigation database to automatically select communication frequencies throughout the flight plan.

Specifications

Comm radio: 118-136.975 MHz; 760 channels; 10 W; 2 × 8 RF memory.

GPS receiver: 30 reversible flight plans with 30 legs; 200 user defined waypoints; 8-channel parallel GPS; certified TSO C129 class A2 for en route and terminal approach.

Operational status

Believed to be still in production.

Contractor

UPS Aviation Technologies, Inc.

UPDATED

Apollo SL70 transponder

The SL70 is fully certified to Mode A/C transponder standards and is identical to other UPS Aviation Technologies' Slimline products. It utilises a solid-state design, with no cavity tube. The output is 300 W. It displays 'squawk' code and altitude and it also has an altitude hold feature. Pressure altitude information can be provided to the SL70 via a 'grey' code, or via serial data. The serial data output is provided to allow interface with other avionics. Like all other UPS Aviation Technologies' products, the SL70 accepts 10 to 40 V DC input.

Contractor

UPS Aviation Technologies, Inc.

VERIFIED

CIVIL/COTS DATA MANAGEMENT

AUSTRALIA

Aircraft Fatigue Data Analysis System (AFDAS)

BAE Systems, Australia's Aircraft Fatigue Data Analysis System (AFDAS), directly monitors and permanently records all relevant fatigue load spectra occurring at selected locations on aircraft structures.

AFDAS provides information for the planning of aircraft maintenance repairs and spares, fatigue predictions, assessment of fatigue damage or crack growth damage or crack status and a basis for correlation of those features with operational usage patterns.

Central to the AFDAS concept is a computer database of actual flight loads. Each aircraft is fitted with strain gauges permanently installed at critical locations in its structure. An airborne unit, the Strain Range Pair Counter (SRPC) automatically processes and stores the information from the gauges. During aircraft servicing, this data is transferred to a floppy disk by means of a portable data readout computer.

The AFDAS comprises up to 12 transducers within the aircraft, including strain gauge bridges and accelerometers, the SRPC, the ground-based portable data readout computer and data analysis software subsystem.

The strain gauge transducers are mounted in the aircraft at the critical strain points together with the SRPC. The SRPC monitors the output from each transducer and extracts range pairs which provide the basis required to compute the fraction of the fatigue life of a structure that has been used and the extent of potential crack growth. The conditioned amplified transducer signals are repetitively sampled and digitised, under the control of the Central Processor Unit (CPU). The CPU then performs additional processing to extract range pairs which are counted and stored in a non-volatile memory. Data from individual flights can be stored separately in memory before performing a data readout operation.

The ground-based portable readout computer is connected to the SRPC when a memory full indication is provided by the SRPC. This computer interrogates the SRPC and transfers fatigue data gathered since the last readout operation on to a mini-floppy disk. The SRPC memory is then cleared in readiness for continued data acquisition.

A portable terminal is also used to initialise the SRPC with information such as aircraft tail number and time and date. This terminal also allows a wide range of diagnostic routines built into the SRPC to be initiated. The SRPC processing algorithms can be tailored to suit the requirements of individual users.

Specifications

Dimensions: 139 × 275 × 154 mm
Weight: 4.5 kg
Power supply: 28 V DC, 21 W
Data storage: 64 kbytes
Interface: RS-232
Environmental: MIL-STD-810C

Operational status

AFDAS is currently in service with Royal Australian Air Force aircraft including the PC-9, F-111C and F/A-18.

Contractor

BAE Systems, Australia Ltd.

VERIFIED

CANADA

CMA-2060 Data Loader System (DLS)

The CMA-2060 DLS provides a general purpose solution to data storage and transfer problems in military avionics applications. The system consists of a removable 1 to 2 Mbyte non-volatile Data Transfer Cartridge (DTC) and a Data Transfer Unit (DTU). The DLS provides real-time data exchanges with an RS-422 databus for functions such as recording single or multiple data and/or maintenance information for subsequent downloading to ground support equipment, and interfacing with ground support equipment for the downloading of flight data and/or the uploading of flight configuration, subsystem configuration and mission profile data.

Specifications

Dimensions:
(DTU) 44 × 146 × 168 mm
(DTC) 83 × 18 × 131 mm
Weight:
(DTU) 1.36 kg
(DTC) 0.45 kg
Power supply: 28 V DC, 1 A (nominal)

Operational status

In production and service. Selected by the Canadian DND and installed on the CH-146 Griffon and CC-130.

Contractor

BAE Systems, Canada Inc.

VERIFIED

CMA-2071 Structural Usage Monitor (SUM)

The CMA-2071 Structural Usage Monitor (SUM) acquires and processes structural usage data from rotary- and fixed-wing aircraft. The system comprises a single LRU mounted on the aircraft which monitors aircraft sensors in real time and stores the acquired data digitally in a large, solid-state, non-volatile memory. The recorded data is classified according to type and flight regime and then compressed and stored in the form of time history, histograms and aircraft header information.

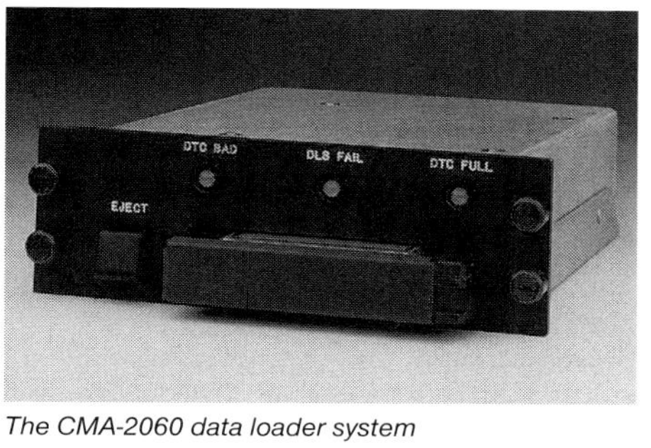

The CMA-2060 data loader system

When interfaced with a crash survivable memory unit, the CMA-2071 can provide flight data and cockpit voice recording in accordance with ED-55/56. Associated with the SUM is the CMA-2081 Ground Support Equipment (GSE), the purpose of which is to retrieve the acquired data from the onboard SUM, process it, and provide display readouts to facilitate maintenance functions such as predictive requirements and the determination of aircraft structural integrity. The GSE also allows for single or two-point SUM sensor calibration, updating of SUM system and parameter configuration data, display of SUM BIT results and the transfer of flight data from the GSE to remote post-processing facilities.

Specifications

Dimensions: ½ ATR
Weight: 2.3 kg
Power supply: 28 V DC, 40 W
Processor: CMOS MIL-STD-1750A or i80846 CPU with memory management
Interfaces: DC analogue, AC analogue, discretes, thermocouple, resistance bulb, pressure, bridge strain gauge, synchro, pulse, vibration, MIL-STD-1553B, RS-422, RS-232C
Inputs: up to 250

Operational status

In production. Selected for the Canadian DND for installation on the CC-130.

Contractor

BAE Systems, Canada Inc.

VERIFIED

CMA-2074 Data Interface Unit (DIU)

The CMA-2074 DIU is a small, rugged, highly adaptable system for interfacing discrete, digital and analogue sensors to MIL-STD-1553B and ARINC 429 avionic busses for a variety of fixed-wing aircraft and helicopters. The DIU provides Ada software-controlled acquisition, processing and multiplexing of avionics data from multiple interfaces in either stand-alone or dual-redundant applications and provides 1553B bus control, remote terminal and non-1553B avionic subsystem control. The unit can be software-configured to specific customer requirements and expanded to include functions such as maintenance recording and exceedance detection.

A typical application alllows for interfacing, filtering, signal processing and data acquisition of up to 256 inputs. These are digitised, scaled, validated and multiplexed on to one or more serial busses. This data is then available to all avionics systems connected to the same serial bus. Data received on the same MIL-STD-1553B bus or another can be reformatted as required by the DIU. Commands received via the avionics busses can be used to control DIU analogue and digital outputs. Up to 256 inputs and outputs can be accommodated, exclusive of avionic busses.

The DIU is contained in a ½ ATR short (4 MCU) enclosure; other enclosure options are available. Ada operational software allows the unit to be adapted for use in many aircraft types and configurations. The DIU provides complete closed-loop BIT and performance monitoring of all system components. Program software can be modified by data upload via either the RS-422A GSE interface or the 1553B interface. By multiplexing many signals on to a MIL-STD-1553B or ARINC 429 bus, aircraft wiring and installation costs, as well as weight, are greatly reduced.

Specifications

Dimensions: 127 × 185 × 305 mm
Weight: 4.3 kg (max)
Power supply: 16-36 V DC, 30 W (max) (MIL-STD-704A)
Interfaces:
(digital) MIL-STD-1553B, 3 RS-422A/232, 4 ARINC 429 inputs/2 outputs per card
(analogue) 2-wire differential/28 channels per card, 4 synchro and 4 LVDT channnels per card, 12 channel synchros per card, 6 strain gauge channels per card.

Frequency, pulse, ratiometric, variable reluctance, piezoelectric, capacitance and thermocouple available (discrete) single-wire input/56 channels per card, single-wire outputs
Environmental: MIL-E-5400 Class 1A, MIL-STD-461 Parts1/2 EMI/EMC
Reliability: >5,000 h MTBF

Operational status

In production and service. Variants of the CMA-2074 are currently in service and installed on the US Coast Guard HC-130/KC-130 and US Air Force MH-53J Pave Low III and KC-135 Speckled Trout aircraft.

Contractor

BAE Systems, Canada Inc.

VERIFIED

CMA-2074MC Mission Computer

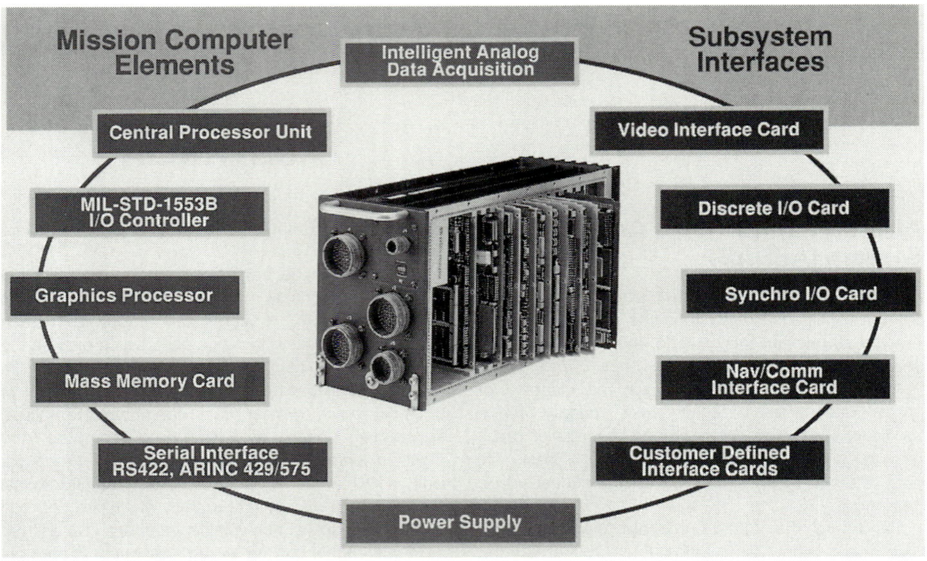

CMA-2074 MC mission computer with illustrative interfaces ***1999***/0051279

The CMA-2074 Mission Computer is a powerful and highly flexible system which can be configured for a variety of roles in integrated avionic architectures. The modular design concept enables the CMA-2074 MC to accommodate MIL-STD-1553B, ARINC-429 and RS-422/232 digital databusses as well as analogue, discrete, synchro, strain gauge and video interfaces.

The CMA-2074 MC has been fielded for the following applications:

Colour Display Generator (CDG) - EH 101 naval helicopter
The CDG accepts input video signals from eight mission sensors and two spare inputs. The CDG manages the tactical database information received over the MIL-STD-1553B databus and can generate tactical maps. Three video channels are available, each with its own independent 2-D/3-D colour graphics generator. Two analogue joystick interfaces are also available. Each of the output channels can be independently selected by the operators to include the tactical map overlaid by one or more video signals from the mission sensors to provide composite multivideo and graphic images overlaid and mixed on to three independent colour displays.

Integrated Electronic Warfare Mission Processor (IEWMP) - MH-53J helicopter
The IEWMP provides functional enhancements to the MH-53J helicopter Electronic Warfare (EW) suite. The IEWMP implements intelligent fail-safe selection and control of aircraft EW systems, and also provides dual MIL-STD-1553B bus interfacing which allows communication between the EW databus and the integrated avionics databus.

Load Monitoring System (LMS) - Canadian Forces CC-130
The LMS interfaces to strain gauges and other aircraft sensors. The loads on the aircraft structure are measured to provide aircraft structural on-conditioning monitoring. A flight data recorder function is also provided, with outputs to the MIL-STD-1553B bus, a data loader system and a Crash Survivable Memory Unit (CSMU).

Communication Navigation Interface Shipset (ISS) - USN/USMC/USCG C-130
The dual ISS provides the interface between the legacy analogue aircraft systems and the MIL-STD-1553B digital databus avionics in the cockpit. This architecture enables control of non-MIL-STD-1553B compatible navigation and communications systems from a central Control Display Unit (CDU) on the flight deck.

Engine Indicating Crew Alerting System (EICAS) Interface Unit (IU) - KC-135 Test Platform (Speckled Trout)
Dual interface unit systems interface analogue and discrete sensor data from the engine and aircraft systems. This data is processed and output via a MIL-STD-1553B digital databus to the cockpit colour EICAS. The interface unit provides the bus controller function in this application. Secondary interface unit functions are to provide the pilots with procedural checklists, and to record engine parameter exceedances for post-flight analysis and maintenance rectification.

Specifications

Dimensions: 140 (W) × 193 (H) × 365 (D) mm
Weight: 7.3 kg
Processor: 32 bit intel 80486 (25 or 50 MHz)
Memory: 512 kbyte EEPROM; 1Mbyte Flash EPROM; 256 kbyte RAM (spare slots for additional memory)

Operational status

In production and service.

Contractor

BAE Systems, Canada Inc.

VERIFIED

AN/URT-43 recorder locator system

The DRS Flight Safety & Communications AN/URT-43 recorder locator system is an automatically deployable emergency beacon for fixed-wing aircraft that enables rapid location and rescue of survivors. Optional features include 406 MHz COSPAS/SARSAT capability and a solid state flight data and voice recorder. The AN/URT-43 system comprises: a composite airfoil, which encloses the radio transmitter, transmitter battery and flight recorder (if installed); a mounting tray including the release mechanism; crash detection sensors (frangible switches); hydrostatic switches; and a cockpit control unit for the testing of the system that can also be used for manual deployment. If a flight recorder is installed, the system will include an Aircraft Monitoring Unit (AMU), which provides the interface between the aircraft sensor, voice link and the recorder.

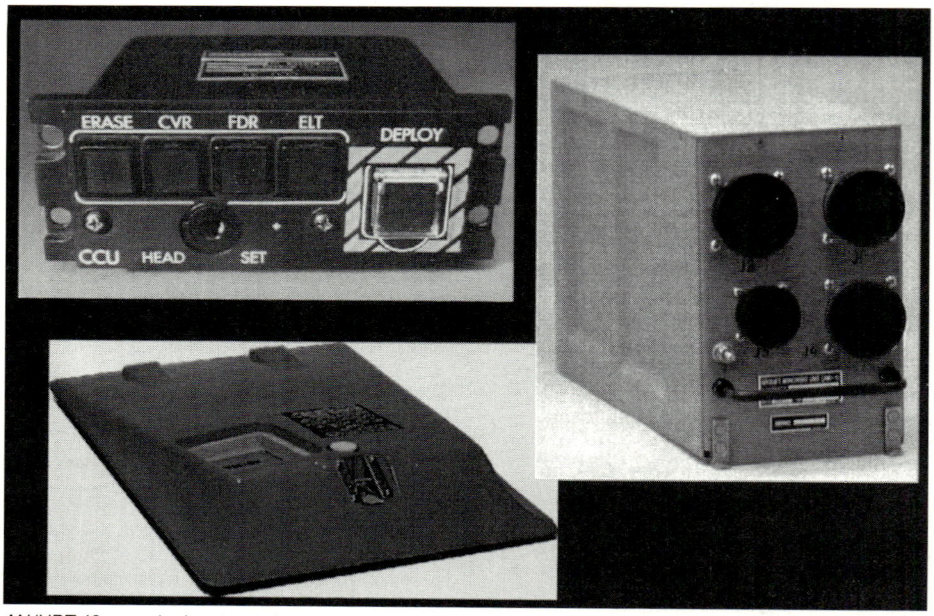

AN/URT-43 recorder locator system ***1999***/0051342

The beacon meets the international requirements for 121.5 and 243 MHz transmitters and is COSPAS/SARSAT compatible. The signal is detectable up to approximately 80 n miles (depending on receiver altitude and terrain) and provides direction-finding capability at 50 n miles. The system can be configured as a single- or-dual-frequency transmitter with a selection of 121.5, 243 and 406 MHz.

Specifications

Airworthiness certification:
TSO C-91a, C-123a, C-124a
EUROCAE ED-55, ED-56, ED-56A
RTCA DO-183, ED-62
CAA specifications 11, 16, 18

Cockpit Voice Recorder (CVR)
Audio channels conform to ED-56A functional requirements:
1 channel 150 to 6,000 Hz bandwidth;
3 channel 150 to 3,500 Hz bandwidth;
1 h storage for each channel

Flight Data Recorder (FDR)
Conforms to ED-55 requirements;
25 hours of data storage

Storage capability	**Recording time**
4-channel audio	1 hour per channel
Flight data	25 hours
ATC data link, HUMS and auxiliary data	expandable

Interface options
Flexible flight data acquisition options:
Direct from databus: 8 channel ARINC 429 or 2 dual-redundant MIL-STD-1553
analogue acquisition suite: discretes, analogue, synchro, frequency, thermocouple
Industry standard ARINC 573/717 interface

Emergency Locator Transmitter (ELT)
Operates on any combination of any two of 121.5, 243 and COSPAS/SARSAT-compatible 406 MHz

Weight
Total system: 17.5 kg
ELT only: 12.5 kg
Airfoil dimensions: 607 × 758.5 × 111 mm
Aircraft Monitoring Unit (AMU): ½ ATR
Power: 28 V DC, 30 W max

Contractor
DRS Flight Safety and Communications, a DRS Technologies Canada Company.

VERIFIED

Deployable Recorder Systems

DRS Flight Safety and Communications' (DRS FS&C) deployable flight data and cockpit voice recorder systems are designed to eliminate survivability problems associated with fixed flight incident recorders by separating the memory module and locator beacon from the aircraft at the onset of a crash.

At initiation, the aerofoil section of the system deploys and displaces from the immediate area of the crash, landing with controlled impact and able to float indefinitely on water. Simultaneously, an Emergency Locator Transmitter (ELT) is activated which, operating on both VHF and UHF Guard frequencies, assists search and rescue forces to rapidly find the downed host aircraft. A 406 MHz beacon is under development that will further enhance the system's effectiveness by allowing for immediate notification and localisation via the COSPAS/SARSAT satellite system.

The system features flexible flight data acquisition options via ARINC 429 or MIL-STD 1553 databus and incorporates an optional solid-state memory flight data and cockpit voice recorder that provides additional data, both for crash investigation and maintenance analysis and training purposes.

Specifications
Physical:
Weight: Total system, 17.5 kg; ELT only, 12.5 kg
Aerofoil dimensions: 607 x 758.5 x 111 mm
Aircraft Monitoring Unit (AMU): ½ ATR
Power: 28 V DC, max 30 W

Interface Options:
Databus: 8 channel ARINC 429 or 2 dual-redundant MIL-STD 1553
Analogue Acquisition Suite: Discretes, analogue, synchro, frequency, thermocouple
Standard ARINC 573/717 interface

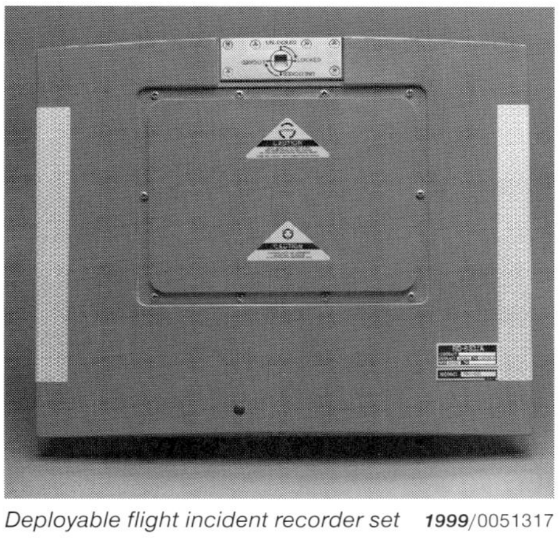

Deployable flight incident recorder set **1999**/0051317

Emergency Locator Transmitter (ELT):
Operating frequencies: 121.5 and 243 MHz (COSPAS/SARSAT compatible 406 MHz under development)
Battery pack conforms with TSO-C97

Cockpit Voice Recorder (CVR):
Audio channels conform to ED-56A functional requirements
- 1 channel 150 - 6000 Hz bandwidth
- 3 channel 150 - 3500 Hz bandwidth
- 1 hour storage each channel

Flight Data Recorder (FDR):
Conforms to ED-55 requirements
- 25 hours data storage
- 4-channel audio recording 1 hour per channel
- Expandable ATC datalink, HUMS and auxiliary data recording

Environmental:
System conforms to RTCA/DO-160C, MIL-E 5400 and MIL-STD 810E

Operational status
Installed on US Navy F/A-18C/D aircraft and selected for the US Navy F/A-18E/F variants. The system is employed on more than 550 types worldwide, including the Lockheed Martin C-130 Hercules, P-3 Orion and German Air Force and Navy Tornado aircraft.

Contractor
DRS Flight Safety and Communications, a DRS Technologies Canada Company.

UPDATED

Emergency Avionics Systems (EAS)

DRS Flight Safety and Communications' family of products integrates the functionality of a Flight Data Acquisition Unit (FDAU), Flight Data Recorder (FDR), Cockpit Voice Recorder (CVR) and Emergency Locator Transmitter (ELT), into a deployable beacon offering highly versatile survival and recovery capability. In a deployable system, the beacon increases the survivability of flight and voice data by avoiding the intense destructive forces which occur during a crash by automatically deploying from the airframe away from the accident site.

The beacon, which contains the solid-state crash-protected data recorder memory and ELT, immediately initiates transmission of search and rescue distress signals, providing immediate location of the accident site, thereby assisting early recovery of survivors, protection of valuable flight/voice data and reduced search and rescue and recovery costs. The beacon floats on water indefinitely in the event of an over-water incident. The beacon transmits at a frequency of 121.5 and 243 MHz. As an option, a 406 MHz beacon can be chosen. This allows identification and messaging information to be added to the distress signal to significantly improve localisation of the downed aircraft via the COSPAS/SARSAT system.

The EAS3000 can be configured as a Combi-Cockpit Voice/Flight Data Recorder (CVR/FDR) or optionally a dedicated FDR or CVR. In each configuration the beacon contains the Emergency Locator Transmitter (ELT). An ELT-only configuration (ELB3000) is available for customers who do not require CVR/FDR capability. The ELB3000 can be upgraded with CVR/FDR capability when required. A Cockpit Control Unit (CCU) enables preflight checks to be performed on EAS3000 integrity. The CCU also allows manual deployment of the Beacon Airfoil Unit (BAU).

By combining the recorder and locator functions, installation, operation and maintenance costs are reduced. In addition, the EAS3000 offers significant weight advantages over conventional installations of multiple, fixed onboard systems.

The EAS3000 provides full data acquisition capability enclosed in a ½ ATR short Data Acquisition Unit (DAU) accommodating various inputs, including: ARINC-429; MIL-STD-1553; analogue; discrete; thermocouple; synchro; and RVDT. The DAU contains all sensor interfaces and processor electronics for data inputs plus audio input for the CVR. Provision is made for receiving HUMS data.

Specifications
Cockpit voice recorder
Functionally compliant to ED-56A
4 audio channels, 1 h storage per channel
Bulk erase option, ARINC 757 compatible

Flight data recorder
Compliant to ED-55
Selection of data acquisition options: MIL-STD-1553 bus; ARINC-429; direct to sensor interface suite: analogue, synchro, discrete, pulse, thermocouple; 25 h storage; multiple data storage configurations; ARINC 747 compatible

Recorder system
Solid-state
ED-56A compliant record controls
Download and playback maintenance system allows: FDR data analysis; audo reply: BITE system

EAS3000 (left), cockpit control unit (centre), and data acquisition unit (right) 0015298

Emergency locator transmitter
DO-183 compliant
Operates on 121.5 MHz and 243 MHz
Option for 406 MHz
COSPAS/SARSAT compatible
Automatically deployable
BITE system

Compliance:
Transport Canada/Civil Aviation Authority (CAA) appliance approval, including TSOs: C123 (CVR); C124 (FDR); and C91a (ELT)

Operational status

Deployable emergency avionics systems are currently installed on the following fixed-wing aircraft and helicopters: C-5, C-9B, C-135, CC-115, CC-130, CP-140, E-3A AWACS, E-4A, E-6A, F/A-18, P-3 (A, B and C variants), T-43A, and Tornado aircraft; 212/412, CH-46, CH-47, Dauphin, EH 101, H-3, HH-1H, HH-1N, Lynx, Puma, S-61, Super Lynx and Super Puma helicopters. The EAS3000 is being fitted on all variants of the EH101, the Kaman SH-2G for Australia, and the UK Royal Navy Sea King helicopter.

EAS-3000F

The new EAS-3000F is a modular, deployable beacon system that incorporates advanced technology in a single crash-survivable unit. It was developed specifically as a flight safety system for fixed-wing applications, incorporating the latest technology in deployable aircraft monitoring and data acquisition systems. The system integrates a cockpit voice and flight data recorder with a crash-survivable emergency locator beacon for fast recovery of flight data and an increased success rate for search and rescue teams.

The EAS-3000F is released automatically from the aircraft's outer surface during an incident and immediately emits a locator beacon for recovery by search and rescue teams. In water, the system floats indefinitely. On land, it escapes the intense destructive forces that occur during a crash by separating from the aircraft at the time of impact. The recovered data, which provide detailed information of the events during an incident, are utilised for accident investigation, training, aircraft and avionics design and manufacture, and flight safety procedure development.

The design of the EAS-3000F system is based on more than 30 years of experience in deployable data storage technology for emergency beacons, cockpit voice recorders and flight recording systems. DRS has supplied over 4,000 similar units worldwide for combat jet, military and commercial transport and rotary-wing aircraft. The recovery rate of these systems exceeds 95 per cent of the incidents reported. The retrieval rate of data in recovered systems has been 100 per cent.

Operational status

The EAS-3000F has been selected by the Canadian Department of National Defence to upgrade the capabilities of the CP-140 long-range maritime patrol aircraft.

Contractor

DRS Flight Safety and Communications, a DRS Technologies Canada Company.

VERIFIED

AN/AYK-23(V) airborne military computer

The AN/AYK-23(V) is a 32-bit general purpose computer capable of being configured to suit many real-time applications. Its militarised construction, to MIL-E-5400 Class 1A, makes it suitable for severe airborne environments.

Using Motorola's 68030 technology and militarised VME architecture, the AN/AYK-23(V) is a functional superset, and a form, fit replacement for the AN/AYK-10(V) computer used on the S-3B. Additional acoustics system WRAs captured by the AN/AYK-23(V) include one PDP, two drums, and one drum power supply. The original CMS-2 and Ultra-32 Tactical Mission Programme was translated and recompiled to the 68030, while the original acoustics software was rewritten in Ada and compiled to 68030.

The AN/AYK-23(V) consists of a set of multilayer printed circuit cards including central processors, graphics processors, monitor drive cards, global memory cards, maintenance interface cards, input video cards, and six input/output interface card types. Up to 72 of these circuit cards can be configured in the system. The AN/AYK-23(V) provides user-friendly and efficient maintenance diagnostics and debug capabilities through a basic operator panel and remote terminal interface. All WRAs are front-removable. With its open architecture, the AN/AYK-23(V) is capable of significant future performance and input/output growth.

The AN/AYK-23(V) is designed to control time-critical real-time systems such as command and control, fire control, electronic warfare and countermeasures, communications, navigation, acoustics and radar.

Specifications

Dimensions: 906 × 1,513 × 463 mm
(module) 152 × 229 mm
Weight:
(baseline) 125 kg
(max configuration) 193 kg
Power supply: 115 V AC, 3 phase, 400 Hz
(baseline) 635 W
(max configuration) 3,000 W

Operational status

In May 2000, the US Navy placed an order for 17 sets for its S-3B aircraft, for delivery by November 2002.

Contractor

Lockheed Martin Canada.

VERIFIED

AN/UYK-507(V) high-performance computer

The AN/UYK-507(V) computer is a 16/32-bit general purpose computer capable of being configured to suit many real-time applications. Its militarised construction, to MIL-E-16400, makes it suitable for severe environments.

Using ASIC technology and high-performance architecture, the AN/UYK-507(V) is functionally compatible with the AN/UYK-502(V), AN/UYK-505(V) and AN/UYK-44 standard military computers.

While emulating these earlier processors, the AN/UYK-507(V) computer provides additional features for object code optimisation and system performance improvements. Expanded memory reach, included as part of the processor design, allows absolute addressing to four billion memory locations. Five Mips performance is achieved with 90 per cent cache hits out of the 64 kbytes of cache memory. Higher performance rates are achievable with increased cache hit rates.

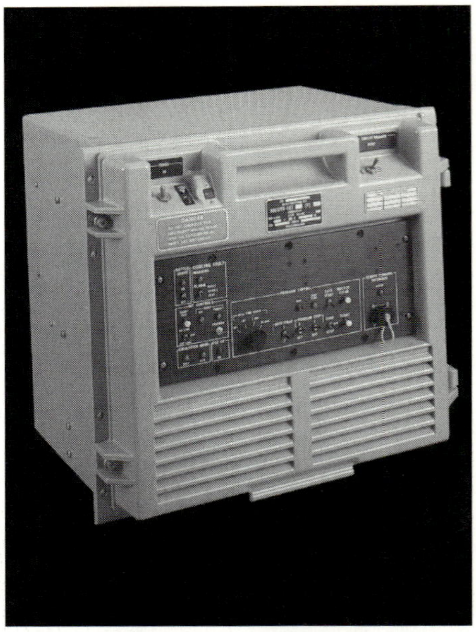

The Lockheed Martin, Canada AN/UYK-507(V) high-performance computer

The AN/UYK-507(V) consists of a set of multilayer printed circuit cards including a central processor, input/output processor, maintenance processor, semiconductor memory, 13 input/output interface types and a power supply. The AN/UYK-505(V) provides user-friendly and efficient maintenance diagnostics and debug capabilities through a basic operator panel with a remote terminal interface.

The AN/UYK-507(V) features an open architecture and can be easily incorporated into existing applications since it continues to use the AN/UYK-502(V) cabinet, power distribution system and input/output interface modules. This allows present systems to achieve a significant performance improvement by a simple substitution of the present AN/UYK-502(V) processor/memory nine-card set with the new AN/UYK-507(V) processor/memory three-card set.

Either as a stand-alone unit or embedded in the system, the AN/UYK-507(V) is designed to control time-critical real-time systems such as command and control, fire control, electronic warfare, communications, navigation, logistics and naval tactical data systems. Other uses are radar and signal processing, message handling, information management and air traffic control.

Operational status

Dimensions: 444.5 × 482.6 × 469.9 mm
(module) 171 × 228 mm
Weight: 52.6 kg (max)
Power supply: 115 V AC, single phase, 575 W

Contractor

Lockheed Martin Canada.

VERIFIED

CHINA, PEOPLE'S REPUBLIC

9416 air data computer

The 9416 air data computer is used on the K-8 Karakorum trainer. It features an Intel 8086 processor and combines with CAIC's 9414 preselector and 9415 altimeter into an air data system.

Operational status

In production and in service.

Contractor

Chengdu Aero-Instrument Corporation (CAIC).

VERIFIED

MD-90 Central Air Data Computer (CADC)

The MD-90 CADC is a product of co-development by Chengdu Aero-Instrument Corporation and Honeywell Inc of the USA and features a 32-bit Motorola MC68332 microprocessor.

Operational status

First flight test was in 1993 and FAA certification in October 1994.

Contractor

Chengdu Aero-Instrument Corporation (CAIC).

VERIFIED

SS/SC-1, -1A, -1B, -1G, -2, -4, -5, -10, -11 air data computers

This series of air data computers are all configured for use in fighter aircraft of the J-7 type and trainers such as the K-8 Karakorum. They take data from pressure, temperature and attitude sensors, process it using Intel 8086 series processors and pass the data to the navigation, weapons management and flight control systems, and to other Chengdu flight displays.

Operational status

In production and in service.

Contractor

Chengdu Aero-Instrument Corporation (CAIC).

VERIFIED

Y7-200B Air Data System (ADS)

The Y7-200B ADS is designed for the Y7-200B aircraft, and it features the following components: 8903 air data computer; 8904 altimeter; 8905 airspeed indicator; 8908 vertical speed indicator; 89008 altitude preselector/altitude alerter; total air temperature/static temperature/true airspeed indicator.

The Y7-200B ADS provides outputs to the FMS, EFIS and navigation systems, and aural warnings/ground proximity warnings to the pilot.

Operational status

In production and in service.

Contractor

Chengdu Aero-Instrument Corporation (CAIC).

VERIFIED

SS/SC-5 air data computer and K-8 Karakorum trainer 0002361

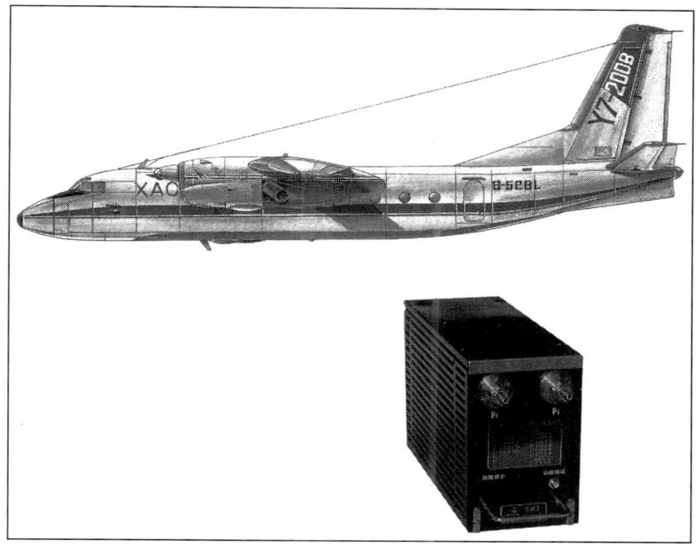

Y7-200B aircraft and Air Data System 0002345

MD-90 CADC and MD-90 aircraft 0001275

CZECH REPUBLIC

AMOS Aircraft Monitoring System

AMOS is designed to provide acquisition, processing and storage of information for the L159 light attack aircraft; it monitors airframe fatigue data; flight data; accident data; pilot response data. There is an optional facility for in-flight transmission of the acquired data.

AMOS comprises a Flight Data Acquisition Unit (FDAU); crash-protected Flight Data Recorder (FDR-159); Function Signalling Panel (FSP); Flight Data Entry Panel (FDEP); Ground Evaluation Equipment (GEE); and Ground Support Unit (GSU).

Using the GEE/GSU, it is possible to replay the mission in 3-D, as well as to analyse and evaluate aircraft system data.

Contractor

VZLU-SPEEL Ltd, Aeronautical Research and Test Institute - Special Electronics.

VERIFIED

Specifications

	FDR-159	FDAU	FSP
Dimensions	185 × 134 × 330 mm	240 × 260 × 390 mm	145 × 32 × 100 mm
Weight	<8 kg	<10 kg	<0.35 kg
Power	28 V DC, <8 W	28 V DC, <55 W	28 V DC, <10 W
Memory capacity	16 Mbyte (custom)	4 Mbyte (custom)	
MTBF	>5,000 flight h	>3,000 flight h	>20,000 flight h
Communications interfaces	RS-232; RS-422; ARINC 429; MIL-STD-1553B		
Compliances	MIL-STD-810E; EUROCAE ED-55; TSO-C124		

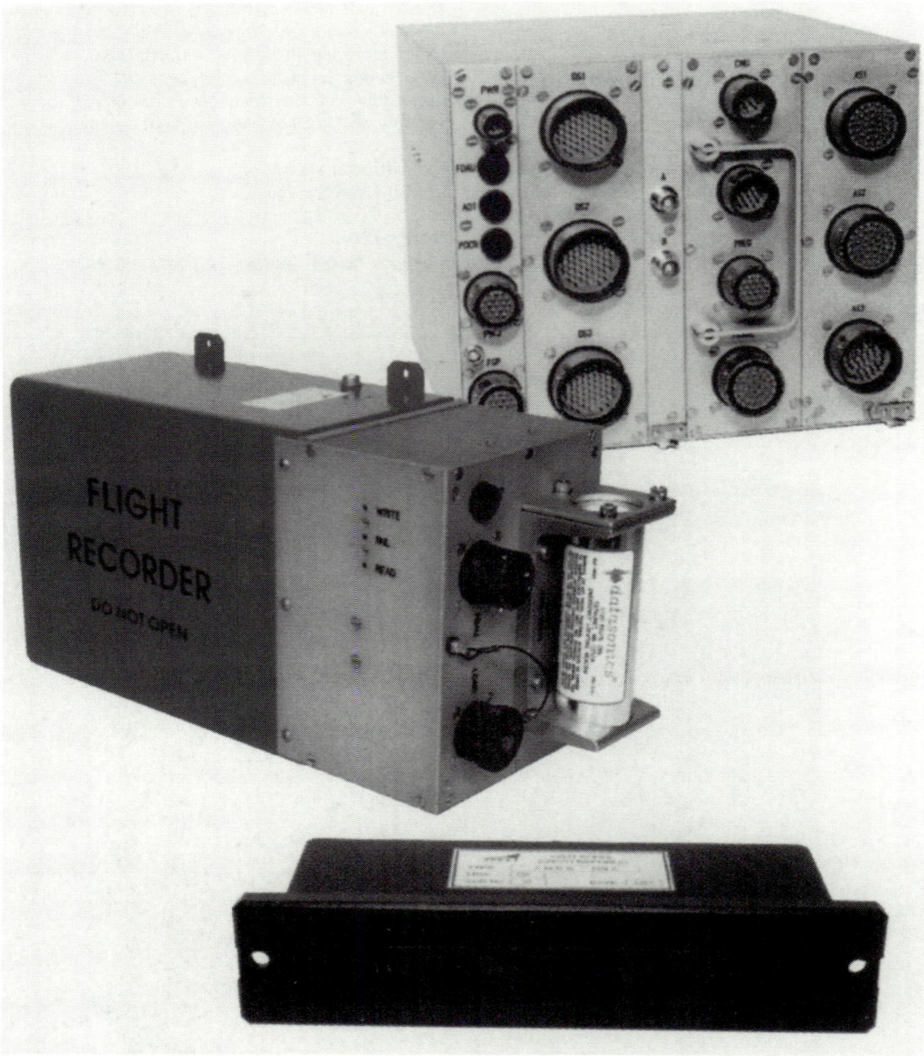

VZLU-SPEEL's AMOS aircraft monitoring system, showing FDAU, FDR-159, and FSP 0015299

CVR-M1 Cockpit Voice Recorder

The CVR-M1 cockpit voice recorder is an all solid-state unit, designed to be a direct replacement for the MARS-BM mechanical tape-based recorder equipping Sokol PZL W-3A helicopters. The aircraft wiring remains unchanged, only the mechanical fixings are replaced.

The CVR-M1 comprises two individual modules: the connection and communication module; and the solid-state crash-protected memory unit. These modules are fixed together to form a single line replaceable unit. The stored voice data can be replayed using the CVR-M1 or downloaded and replayed using a standard PC, equipped with multimedia capability and the voice replay software supplied by VZLU-SPEEL Ltd.

Specifications

Dimensions: 232 × 124 × 122 mm
Weight: 7 kg max
Power: 28 V DC, 15 W
Recording medium: solid-state flash memory
Recorded audio channels: 2 crew microphones (100-3,500 Hz); 1 area microphone (100-6,800 Hz)
Recorded time channel: 1
Recording capacity: 48 Mbyte – corresponds to the last flight hour
Interface: RS-485/10 Mbps
MTBF: 10,000 h
Compliance: MIL-STD-810E, TSO C124, ED-56A
Operating temperature: –55 to +60°C

Contractor

VZLU-SPEEL Ltd, Aeronautical Research and Test Institute – Special Electronics.

VERIFIED

Flight Data Recorders (FDRs)

VZLU-SPEEL has developed a family of solid-state FDRs that are designed to be direct replacements for the mechanical crash recorders in Russian aircraft,

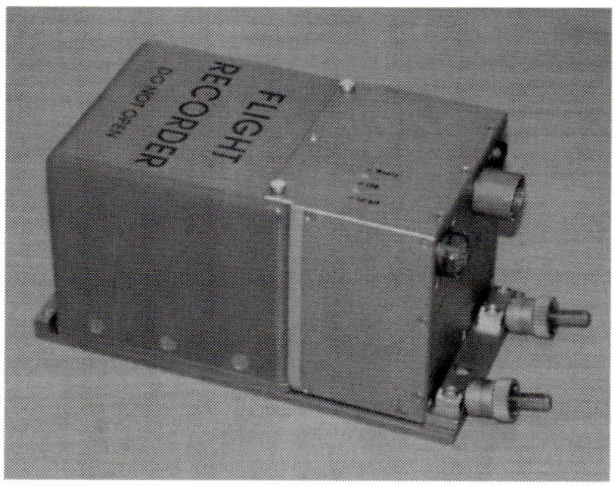

VZLU-SPEEL's CVR-M1 cockpit voice recorder
1999/0051318

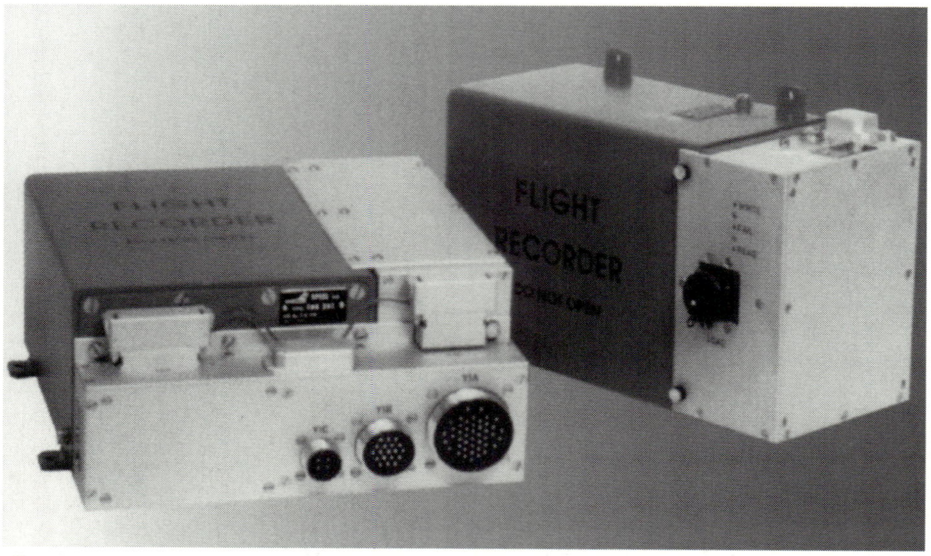

VZLU-SPEEL's FDR-39 (left) and FDR-59 (right) 0015300

where SARPP-12, TESTER U3 (2T-3M), TESTER U3L (M2T-3), BUR-1 (ZBN1) and PARES systems are used.

VZLU-SPEEL's FDR recorders include the following models:

FDR-39H for Mi-17 and Mi-24 helicopters
FDR-39T for L39, L410 and MiG-21 aircraft
FDR-59A for L59MS aircraft
FDR-59B for SOKOL helicopters and L410 UVP aircraft
FDR-59T for MiG-29, Su-22, and Su-25 aircraft

The above FDRs are supported by the PMU Portable Memory Unit for downloading and by the Ground Evaluation Equipment (GEE) for evaluating data.

FDR-39H and FDR-39T solid-state flight data recorders
The FDR-39H and FDR-39T solid-state FDRs are designed as replacements for the SARPP-12DM (IM) and SARPP-12 mechanical crash recorders respectively. They comprise three modules: power supply; measuring unit; and solid-state memory. Download can be accomplished via RS-232C interface to a portable Notebook, or via RS-422 interface to Speel's Portable Unit (PMU-F).

FDR-59A and FDR-59B solid-state flight data recorders
The FDR-59A and FDR-59B solid-state FDRs are designed as replacements for the PARES 59E and the ZBN-1 from the BUR-1 system mechanical tape crash recorders respectively. They comprise two modules, connection and communication module, and solid-state memory unit. Both have 4 Mbyte recording capacity, corresponding to approximately eight flight hours.

Specifications

	FDR-39H and FDR-39T	FDR-59A and FDR-59B
Dimensions	258 × 241 × 117 mm	185 × 125 × 264 mm
Weight	8 kg max	7 kg max
Power	28 V DC, <9 W	28 V DC, <15 W
Interfaces	RS 232C; RS-422	RS-232C; RS-422
MTBF	10,000 h	10,000 h

Contractor
VZLU-SPEEL Ltd, Aeronautical Research and Test Institute - Special Electronics.

VERIFIED

FRANCE

CP 1654 airborne life monitor

The CP 1654 calculates the residual engine service life of the main engine components. It provides definition of a short- and medium-dated schedule for engine components' removal before shipment to the second-level maintenance shop as well as the planning applicable to the fourth-level maintenance of the modules.

Operational status
In production.

Contractor
ELECMA, the Electronics Division of SNECMA.

VERIFIED

The ELECMA CP 1654 airborne life monitor for the M 53-P2 engine

DS4100 compact airborne digital cartridge recorder

The DS4100 is a ruggedised family of products based on commercially available and proven technology.

Design applications are listed as:

(1) electro-optical, infra-red and synthetic aperture radar (SAR) reconnaissance imagery data acquisition and processing
(2) acoustic and sonar data acquisition and processing
(3) system and flight test data acquisition and processing.

Storage capacity per cartridge is greater than 72 Gbyte. The sustained data rate is 240 Mbits/s (480 Mbits/s burst). Interface flexibility includes a basic 8-bit parallel + clock capability and an optional multichannel analogue/digital/databus capability, including data acquisition and formatting. Time code is simultaneously added in digital form.

Specifications
Record time:
(at 240 Mbps): 40 min
(At 80 Mbps): 120 min
Dimensions: 195 (H) × 300 (W) × 254 (D)
Weight: 11 kg, excl DDC
Power: 28 V DC, < 120 W (160 W with pre-heaters)

Operational status
The DS4100 has been trialled in a Mirage 2000 and Enertec states that there have been French sales for an undisclosed purpose. The UK Defence Evaluation Research Agency evaluated the system in 2000.

Contractor
Enertec.

UPDATED

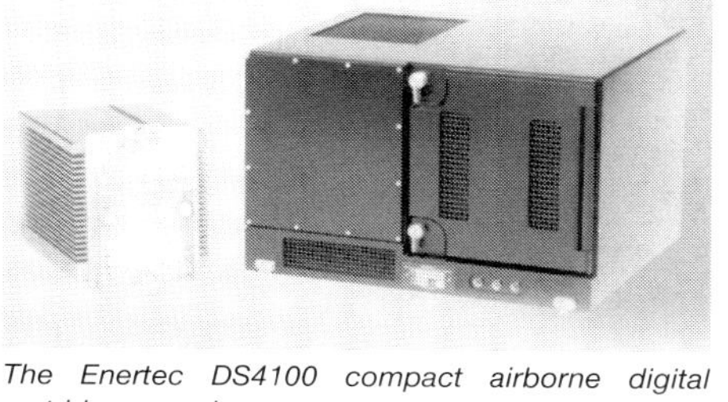

The Enertec DS4100 compact airborne digital cartridge recorder **2000**/0080289

DV 6410 series airborne digital recorders

The DV 6410 series of airborne recorders includes the DV 6410 and the DV 6210. Data rates are 10 to 240 Mbits/s for the DV 6410 and 5 to 120 Mbits/s for the DV 6210. Recording time varies between 25 hours at 5 Mbits/s and 24 minutes at 240 Mbits/s.

The DV 6410 series has a storage capacity of 350 Gbits on a 19 mm D1-M cassette in a format compatible with ANSI IDI. It is a compact, fighter- and space-proof unit. The DV 6410 also provides for two auxiliary channels and a full remote-control interface.

Specifications
Dimensions:
(record-reproduce unit) 314 × 240 × 520 mm
(power supply unit) 314 × 140 × 302 mm
Weight:
(record-reproduce unit) 35 kg
(power supply unit) 12 kg
Power supply: 28 V DC < 350 W
Temperature range: –40 to +55°C
Altitude: up to 50,000 ft

The DV 6410 series airborne digital recorders used in the central pod of a Mirage 2000

Operational status
In production and in service for various airborne, shipborne and space applications. The DV 6411 recorder has been selected for the UK Royal Air Force Project RAPTOR (Reconnaissance Airborne Pod for Tornado). RAPTOR integrates a day-night medium level stand-off imaging capability in Tornado GR. Mk 1A and Tornado GR. Mk 4 aircraft.

Contractor
Enertec.

VERIFIED

DV 6420 ruggedised rackmount digital recorders

The DV 6420 series of ruggedised rackmount digital recorders suitable for large aircraft consists of the DV 6420 and DV 6220. Data rates are 10 to 240 Mbits/s for the DV 6421 and 5 to 120 Mbits/s for the DV 6221, when driven by the users clock (Mode A) and 0 to 240 Mbits/s for the DV 6421 and 0 to 120 Mbits/s for the DV 6221 (Mode B), when used with plug-in buffer memory. Recording time varies between 44 hours at 5 Mbits/s and 55 minutes at 240 Mbits/s.

The DV 6420 is the rack-mounted version of the DV 6410. The basic characteristics are common to both versions, but the rack-mounted version accepts D1 medium or large cassettes and has a storage capacity of 790 Gbits on a D1-L cassette.

A number of standard computer interfaces are available such as HIPPI, SCSI2 and VME64. A higher rate version, the DV 6820, provides 520 Mbits/s throughput rate capability.

Specifications
Dimensions: 483 × 311 × 593 mm
Weight: 65 kg
Power supply: 28 V DC or 115/220 V AC, 50/400 Hz, single phase, <450 W
Temperature range: –20 to +45°C
Altitude: up to 10,000 ft

Operational status
In production and operating in a number of airborne, shipborne and ground platforms. The DV6221 recorder

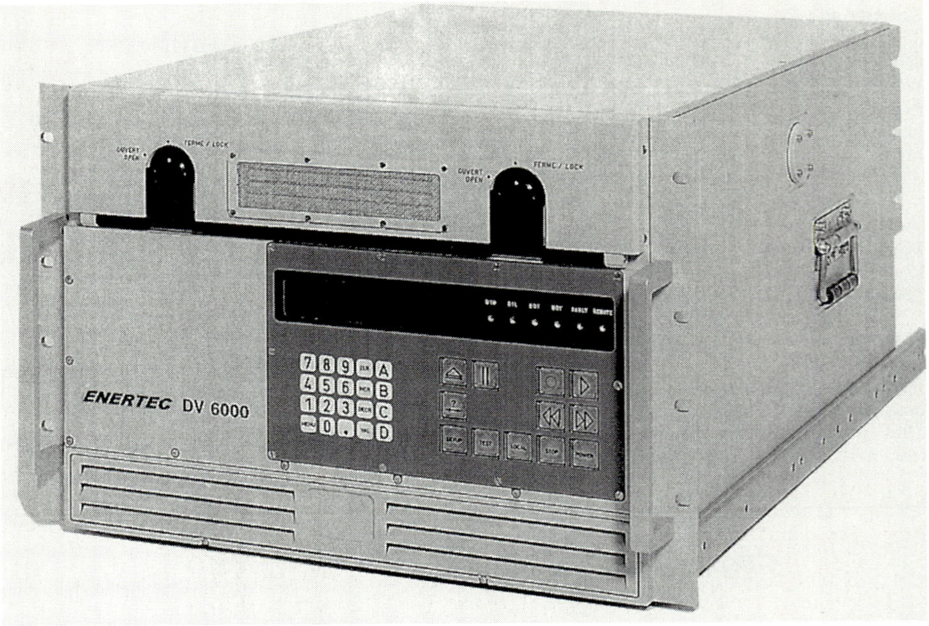

The Enertec DV6421E ruggedised rackmount digital recorder ***1999***/0051321

has been selected as the mission recorder of the acoustic sub-system of the Royal Air Force Nimrod MRA4 maritime patrol aircraft. It has also been reported that this recorder is utilised in the successful Raytheon ASTOR system.

Contractor

Enertec.

VERIFIED

ME 4110 airborne instrumentation recorder

The ME 4110 is a lightweight and compact recorder designed for use during the flight testing of military fighters and other aircraft. It uses 10 in reels of 1 in wide magnetic tape.

Specifications

Dimensions: 400 × 285 × 160 mm
Weight: 20 kg
Power supply: 22-32 V DC, 200 W
Recording time: 7 min - 16 h
Number of tracks: 14 or 28
Recording modes:
FM (up to 500 kHz)
direct (up to 2 MHz)
PCM (up to 4 Mb/s)

Operational status

In production and in service.

Contractor

Enertec.

VERIFIED

ME 4115 airborne recorder/ reproducer

The ME 4115 is an advanced, microprocessor-controlled recorder designed for use during the testing of aircraft, ships or vehicles or as an ELINT, ASW or reconnaissance mission recorder and can be configured for any application. It uses 15 in reels of 1 in wide magnetic tape and has built-in error correction electronics and read/write heads to give 14 or 28 tracks IB, WB or DD magnetic heads.

Depending on the disposition of the electronics chassis relative to the tape deck, several standard versions are proposed: 14 or 28 track record/ reproduce either for anti-vibration mount or 19 in bay installation, tape deck separated from the electronics for integration on board small fighters and 28 track record-only and monitoring.

This latest configuration is fully compliant with MIL-STD-1610 and STANAG 4238 Annex B analogue acoustic recording standards.

The ME 4115 accepts a wide range of high data rate digital formats up to 200 Mbits/s including standard interfaces for MIL-STD-1553 bus monitoring. It has been selected as a flight test recorder by most airframe manufacturers for programmes such as the Alenia/ Embraer AMX, Northrop Grumman F-14 and Airbus A340.

Specifications

Dimensions: 581 × 394 × 176 mm
Weight: 35 to 45 kg
Power supply: 22-32 V DC or 115 V AC, 300 W
Recording speed: 4.75-394 cm/s
Number of tracks: 14 or 28
Recording modes:
FM (up to 1 MHz)
direct (up to 4 MHz)
PCM (up to 8 Mb/s)
Cumulative data rate: over 200 Mbits/s

Operational status

In service on French and Italian Atlantique maritime patrol aircraft. The ME 4115 is operated by the US forces in several airborne and shipborne programmes as the AN/USH-33(V)2. More than 900 units have been produced.

Contractor

Enertec.

VERIFIED

PC 6033 general purpose digital cassette recorder

The PC 6033 records serial data on a continuous basis over a long period of time. Typical applications are performance and maintenance recording, engine health monitoring, aircraft testing and reconnaissance. The system complies with ARINC 591. Cassettes can be replayed at 80 times the recording speed.

The PC 6033 performance/maintenance recorder

Specifications

Power supply: 19-32 V DC or 115 V AC, 400 Hz
Recording capacity, data rate:
50 h, 138 Mbits
50 h at 768 bits/s
25 h at 1,536 bits/s
12.5 h at 3,072 bits/s
Number of tracks: 12
Recording code: biphase L or M
Error rate: <1 bit in 10^5

Operational status

In production for and in service with several airlines; selected for the A320.

Contractor

Enertec.

VERIFIED

PE 6010 and PE 6011 digital flight data accident recorders

The PE 6010 is a lightweight digital flight data accident recorder, in which weight saving is accomplished by a relaxation of the dynamic penetration requirement. In terms of accident protection, however, it meets all other requirements of FAA TSO C51a.

The PE 6011 is more compact, meets TSO C51a completely with only a small weight increase and has 50 per cent more capacity than the PE 6010. The new design of tape deck is simpler, with attendant gains in reliability and maintainability.

Specifications

Dimensions:
(PE 6010) ½ ATR short (length 319 mm)
(PE 6011) ½ ATR (length 296 mm)
Weight:
(PE 6010) 7.7 kg
(PE 6011) 10.2 kg
Power supply:
(PE 6010) 18-30.5 V DC
(PE 6011) 12-32 V DC
Recording capacity, data rate:
(PE 6010) 16 h, 768 bits/s
(PE 6011) 8 h, 2,308 bits/s
Replay data rate:
(PE 6010) 4,608 bits/s
(PE 6011) 18,464 bits/s
Accident survival:
(PE 6010) TSO C51a (except penetration is static)
(PE 6011) TSO C51a

Operational status

The PE 6010 is no longer in production. In service with Dassault Mirage F1 fighters flown by the air forces of Egypt, France, Greece, Iraq, Kuwait, Morocco and Spain. It has been supplied for the Indian HAL Ajeet trainer and French Navy Super Frelon helicopters.

Contractor

Enertec.

VERIFIED

PS 6024 cassette memory system

The PS 6024 is a digital magnetic cassette tape recorder/reproducer designed for operation under severe environmental conditions. It comprises a cassette drive unit, microprocessor-controlled central processing unit and electronic interface circuits. Its chief function is as an easily integrated mass memory unit for military computers with standardised interfaces.

Specifications

Dimensions: ⅜ ATR short
Interfaces: V24/RS-232, MIL-STD-1553B, Digibus HDLC as required
Capacity: 6 Mbytes formatted (1 kbyte blocks)
Data density on tape: 1,600 bits/in
Max recording rate: 2.2 kbytes/s usable data
Recording format: forward and backward serial recording in biphase code
Recording speed: 12 in/s (300 mm/s)
Rewind speed: 50 in/s (1,270 mm/s)

Max length of data blocks: 2,048 (2 kbyte) bytes
Interblock gap: 20 mm

Operational status

In service in German and Italian Panavia Tornado aircraft. No longer in production.

Contractor

Enertec.

VERIFIED

VS2100 compact airborne multichannel video digital recorder

The VS2100 compact airborne multichannel video digital recorder is a standalone, ruggedised unit, which provides for simultaneous acquisition and recording of 4 video sources (PAL or NTSC), 2 audio channels and one aircraft databus on a single, removable, digital data cartridge. It provides greater than 5 hours record time, depending on the selected configuration.

Specifications

Dimensions: 270 (H) × 160 (W) × 136 (D) mm
Weight: 5.5 kg, incl 0.5 kg DDC
Video acquisition:
coherent filtering into D1 or Half-D1 image size
max aggregate video acquisition rate:
4 x 25 images/s in PAL (7.6 Mbytes/s)
4 x 30 images/s in NTSC (7.6 Mbytes/s)
Video and audio compression: ISO MPEG2 compression algorithm
Interfaces: RS-422A, MIL-STD-1553B, MIL-STD-3910
Power: 28 V DC, < 70 W typical (120 W with pre-heaters)

Operational status

Enertec states that the VS2100 has been trialled on Mirage 2000 aircraft, and that it is in undisclosed French service. Like the DS4100 it is also said to be under evaluation by the UK Defence Evaluation Research Agency.

Contractor

Enertec

VERIFIED

The Enertec VS2100 compact airborne multichannel video digital recorder **2000**/0080288

MONITAIR flight data recorder

The MONITAIR flight data recorder is designed to record the airframe and engine parameters of turbine engined helicopters.

MONITAIR memorises: the date, time and duration of each flight (with partial flights) and makes daily reports with cumulated flight time; it records maximum values of engine pitch/torque, turbine temperature, turbine speed and rotor rpm during start sequence and throughout flight; it also records the value and duration of all turbine temperature exceedances.

MONITAIR is tamperproof, and all disconnections are indicated. It is supported by a WINDOWS-based support system.

Specifications

Storage capacity:
(single-engined helicopter) 9 h
(twin-engined helicopter) 6.30 h
Flight memory:
(single-engined helicopter) 1,400 flights
(twin-engined helicopter) 1,000 flights
Autonomous storage: 75 days
Dimensions: 260 × 155 × 90 mm
Weight: 3.2 kg
Power: 15-35 V DC, 250 mA

Operational status

The system can be installed on every type of turbine helicopter, and it has been integrated on to all Eurocopter helicopters, and a wide range of Bell and Sikorsky types.

Contractor

Monit'air.

VERIFIED

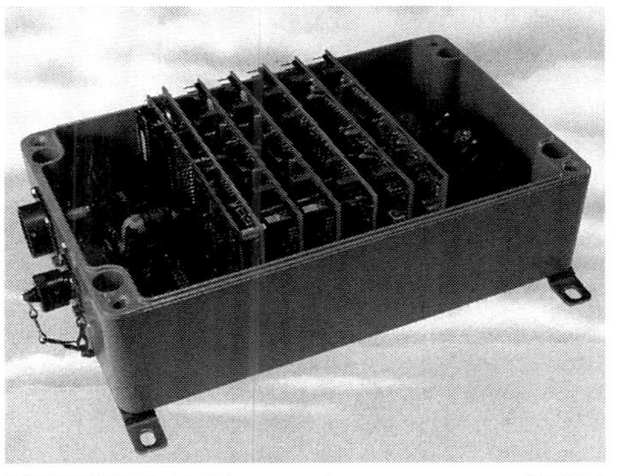

MONITAIR flight data recorder 0015301

Jumbo memory equipment

Jumbo is a compact mass memory suitable for demanding onboard data processing or exchange applications, such as: mission planning interface, digital terrain storage, mission data recording/replay and flight test instrumentation. Designed for use on airborne, naval or ground systems, Jumbo is composed of a data transfer unit and a family of data transfer cartridges.

The Jumbo data transfer unit can house different types of cartridges implementing solid-state Flash EEPROM chips or ruggedised 1.8 in magnetic disk drives, thus fulfilling a wide spectrum of environmental specifications.

A dedicated version of Jumbo for ground applications is available for connection to any standard data processing unit (PC or workstation).

As an option, an add-on board and dedicated software give Jumbo the capability of a high-performance digital map generator (Mercator equipment).

Specifications

Dimensions:
(data transfer unit) 95 × 127 × 222 mm
(cartridge) 27 × 105 × 165 mm
Weight:
(data transfer unit) 2.5 kg
Power supply: 28 V DC, 30 W
Temperature range: –35 to +71°C
Cartridge capacity: 20-2,000 Mbytes

Operational status

In production.

Contractor

SAGEM SA, Defence and Security Division.

VERIFIED

15M/125X general purpose processor

The 15M/125X has been selected as the main computer in the Dassault Aviation Atlantique 2 aircraft to process the data received by the various sensors. It is the most recent addition to the Sextant range of general purpose units and employs many internal operating procedures common to previous units. It is a 32-bit machine with up to 1 Mops capability. A 1 Mbyte memory can be addressed and versions of the processor are typically installed in a 1 ATR case. LTR real-time software support is provided and numerous applications for high-performance military equipment are anticipated.

Operational status

No longer in production. In service in two French Navy Atlantique aircraft.

Contractor

Sextant.

VERIFIED

Flight Data Acquisition Unit (FDAU)

The Sextant Flight Data Acquisition Unit (FDAU) and its associated data entry panel are compatible with the digital flight data recorder requirements of ARINC 573 and 717.

The FDAU codes aircraft parameters into digital format and transmits them to the data recorder. The data entry panel allows the crew to enter flight number, time and events into the system. It also acts as a fault monitor.

Operational status

In production.

Contractor

Sextant.

VERIFIED

Full format printer for B717/B747-400/MD-11/MD-90

The full format printer for the Boeing 717, 747-400 and MD-11/MD-90 is based on ARINC 744 recommendations and uses thermal printing technology. It features 300 DPI resolution and high buffer capacity, to interface with major onboard systems including the maintenance computer. To meet future requirements including FANS and electronic

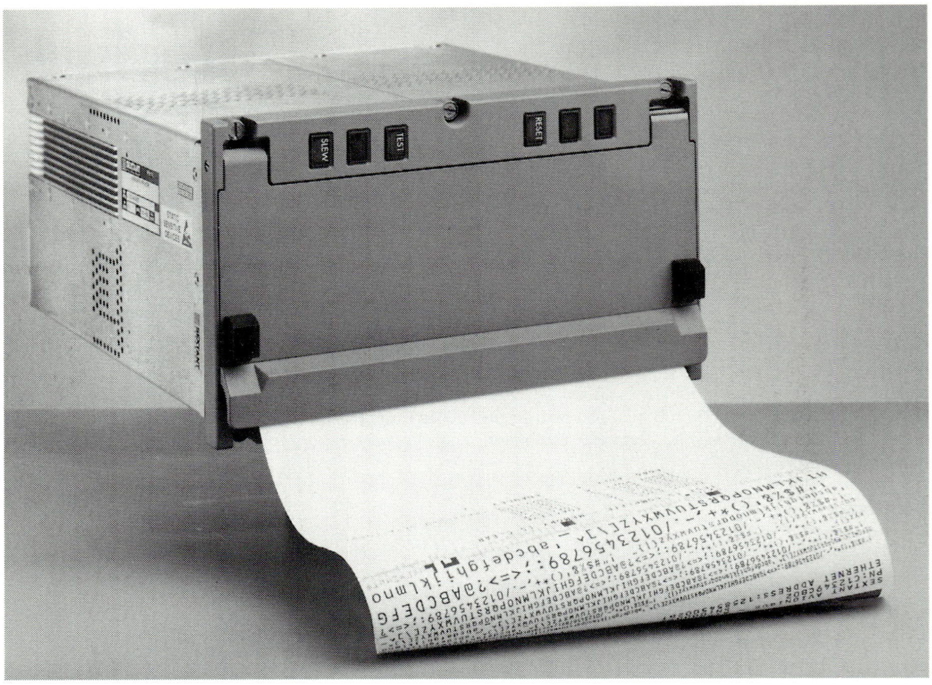

Full format printer for B717, B747-400/MD-11/MD-90 **1999**/0051322

library systems, the full format printer features a graphic interface capability associated with a high-speed datalink. It can handle a wide variety of texts and complex diagrams, making it an ideal link for the crew to provide quality hard-copy reports. The full format printer is installed in the cockpit and may also be installed in the cabin (centronics interface).

Operational status
In production.

Contractor
Sextant.

VERIFIED

Full format printer for the Airbus A330 and A340

The full format printer, as envisioned for use on A330 and A340 commercial transport aircraft, has to comply with a high-performance standard level. Primarily located in the cockpit, the full format printer provides for high-speed and high-quality printing capacities, while achieving a low-noise level and a high operational reliability.

The full format printer communicates with multiple onboard systems such as the flight management system, maintenance system, ACARS and aircraft condition and monitoring systems. It has to be able to process large quantities and various types of data transmitted through up to 12 ARINC 429 databusses.

Operational status
In production for the Airbus A330 and A340.

Contractor
Sextant.

VERIFIED

Full format printer for the Boeing 777

The full format printer for the Boeing 777 is based on thermal printing technology. It features full graphics capability and interfaces with major onboard systems, including the Aircraft Information Management System (AIMS) and the Electronic Library System (ELS).

Featuring 300 DPI resolution for quality equal to a laser printer, the full format printer handles a wide variety of texts and complex diagrams, making it an important link in the onboard data management chain.

The full format printer is installed on the central pedestal of the Boeing 777, behind the throttles.

To meet future requirements for FANS or electronic library services, the full format printer features graphic hardware interface capabilities with a high-speed datalink. A cabin installation is optional.

Operational status
In production for the Boeing 777.

Contractor
Sextant.

VERIFIED

Type 130 Air Data Computer (ADC)

The modular design Type 130 air data computer is intended for use on new-generation aircraft and for the retrofit of fighter aircraft, and has already been chosen for various Mirage retrofits.

The ADC comprises two pressure sensors for measuring static and total pressures; a CPU board including arithmetic unit; memories and sensor measurement circuits; an input/output board for all analogue and digital multiplexed bus interfacing; and a power supply board. A large number of built-in tests is available and permanent monitoring of all functional circuits takes place during flight, the results being expressed as maintenance words stored in a protected memory or dispatched on the digital lines.

Pressure sensors are the Sextant vibrating quartz blade Type 51 pressure sensors.

Specifications
Dimensions: 2 MCU
Weight: 4.6 kg
Power supply: 115 V AC, 400 Hz, single phase, 30 VA
Reliability: 10,000 h MTBF

Operational status
In production for retrofit of Mirage aircraft.

Contractor
Sextant.

VERIFIED

Type 300 Air Data Unit (ADU)

The Type 300 ADU is designed to measure static and differential (or pitot) pressures and impact temperature. From this data, the ADU computes the true airspeed and provides air data parameters through an ARINC 429 bus. Analogue outputs are available as an option.

The unit is ideally suited to provide the primary reference in helicopters and for retrofits in military aircraft.

Specifications
Weight: 1.2 kg

Operational status
In production and in service for Super Puma, Tiger and Rooivalk helicopters.

Contractor
Sextant.

VERIFIED

Type 300X Air Data Units (ADUs)

The Type 300X family of ADUs is designed to measure static and pitot pressures, static and total temperatures, and to compute and send over to the digital databus all the air data parameters.

These air data units have been miniaturised and are ideally suited as primary or secondary air data references on civil or military aircraft, helicopters and missiles.

Specifications
Dimensions: 200 × 130 × 50 mm
Weight: 0.8 kg
Power supply: 28 V DC, 5 W
Reliability:
(military) 30,000 h MTBF
(civil) 80,000 h MTBF

Operational status
The helicopter version, ADU 3000, has been selected on BK 117C2, EC 135 and EC 155. ADU 3008, the TSO C106 certified civil aircraft version, has been selected for the DASH 8-400.

Contractor
Sextant.

UPDATED

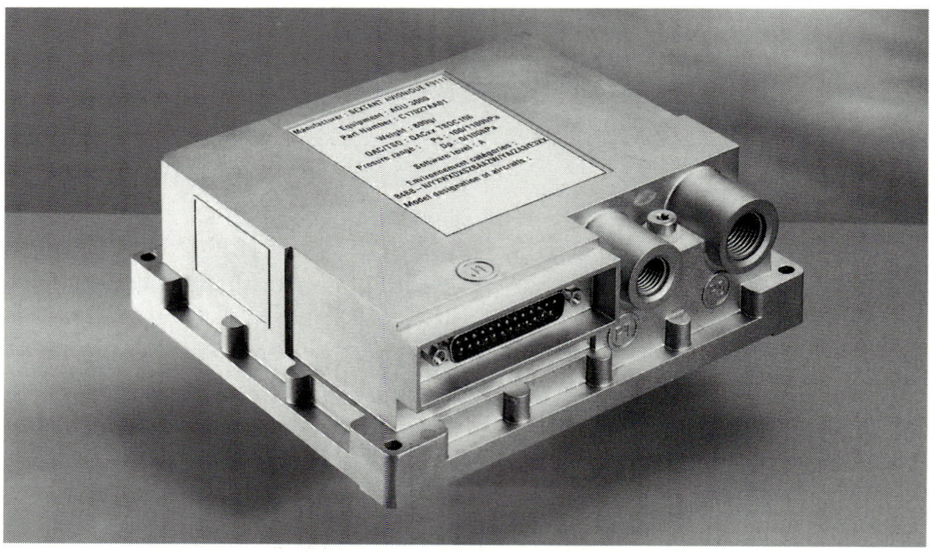

Type 3000 ADU

AC 68 multipurpose disk drive unit/airborne data loader

The multipurpose disk drive unit/airborne data loader enables the uploading and downloading of onboard computers and allows remote loading of the computer program database or program initialisation for ACMS, FMC, ACARS and TCAS, memory computer downloading and recording data reports, QAR function and other functions, from aircraft computers. The AC 68 data loader is also compatible with other data loaders in accordance with the ARINC 603 standard.

The storage medium used for transfer is a standard 3.5 in (89 mm) double-side high-density magnetic diskette. The formatted capacity is 1.44 Mbytes.

Operational status
In production for Airbus and Boeing aircraft.

Contractor
SFIM Industries (SAGEM Group).

VERIFIED

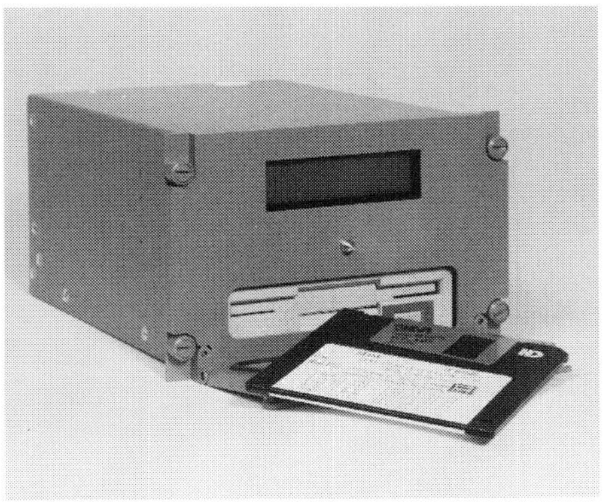

The SFIM Industries AC 68 multipurpose disk drive unit/airborne data loader

Airborne data loader ARINC 615/603

The SFIM Industries airborne data loader is compliant with ARINC 615/603 and designed to meet the requirements of all commercial aircraft.

After insertion of the floppy disk, data transfer starts immediately. A configuration file contained on the floppy disk defines the transfer characteristics: input/output links used; ARINC 429 transmission rates (Hi or Lo); and uploading/downloading actions. Exchanges are driven by the data loader of the computer and checks of the data transfer are automatic.

The front panel provides a one line, 16 character LCD.

Specifications
(cockpit installation)
Dimensions: 146 × 95.2 × 171.4 mm
Weight: 3 kg
Power: 115 V, 400 Hz, <10 VA

Contractor
SFIM Industries (SAGEM Group).

VERIFIED

Aircraft Condition Monitoring System (ACMS)

The ACMS has been designed for the acquisition of aircraft parameters in accordance with the latest airworthiness regulatory agency requirements. It processes engine, APU and aircraft monitoring, airline incident and specific investigation report data and transmits this data to a digital flight data recorder or quick access recorder. Provision is also made for customised reports. The customisation is realised through Ground Support Equipment (GSE); software is PC-compatible. The system interfaces with ACARS, multifunction Control Display Unit or CDU, onboard printer, data loader and centralised fault display system.

A new option that utilises a removable PCMCIA drive offers flight data storage equivalent to use of external recorders.

Operational status
In aircraft such as the Airbus A320, A330 and A340 and Boeing 747-400, the ACMS is a two-box configuration with separate flight data acquisition unit and data management unit. These two boxes are combined into a single unit in the Airbus A300-600 and A310, Boeing 737, 757 and 767, MD-87, -88 and -90 and the ATR 42 and 72.

Contractor
SFIM Industries (SAGEM Group).

VERIFIED

CF 368C CCD colour camera

The CF 368C onboard colour video camera is designed for use with an HUD. It is equipped with a high-resolution and high-sensitivity CCD HAD matrix sensor. The camera operates to the PAL standard.

Specifications
Dimensions: 125 × 92.5 × 81.5 mm
Weight: 0.88 kg
Power supply: 28 V DC, <10 W
Field of view: 24° (horizontal) × 18° (vertical)
Pixels: 752 (horizontal) × 582 (vertical)
Definition: >400 points per line, >350 line vertically

Contractor
SFIM Industries (SAGEM Group).

VERIFIED

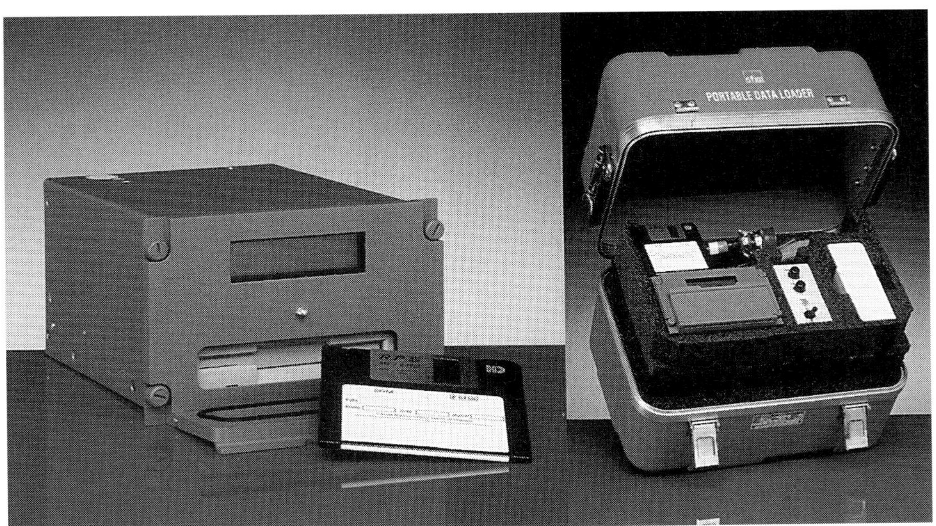

SFIM Industries airborne data loader ARINC 615/603 (left) and ground support portable data loader (right)
0002321

CF 369C CCD colour camera

The CF 369C onboard colour video camera is designed for use with an HUD. It is equipped with a high-resolution and high-sensitivity CCD HAD matrix sensor and operates on PAL or Y/C standards.

Specifications
Dimensions:
(camera head) 40 × 60 × 65 mm
(electronic control unit) 125 × 66 × 81.5 mm
Weight:
(camera head) 0.2 kg
(electronic control unit) 0.88 kg
Power supply: 28 V DC, <10 W
Field of view: 24° (horizontal) × 18° (vertical)
Pixels: 752 (horizontal) × 582 (vertical)
Definition: >400 points per line, >350 lines vertically

Contractor
SFIM Industries (SAGEM Group).

VERIFIED

CF 371 CCD colour camera

The CF 371 onboard colour video camera is designed for use with a HUD. It is equipped with a high-resolution and high-sensitivity CCD HAD matrix sensor and operates on PAL or Y/I standard specifications.

Specifications
Dimensions:
(camera head) 31 (W) × 53 (H) × 70 (D) mm
(electronic control unit) 150 (W) × 80 (H) × 100 (D) mm
Field of view: 24° (H) × 18° (V)
Power: 28 V DC

Contractor
SFIM Industries (SAGEM Group).

VERIFIED

CF 372 CCD colour camera

The CF 372 camera is a CF 371 camera with a multiplexing board allowing a temporal multiplexing of four colour or black and white videos.

Specifications
Dimensions: electronic control unit: 180 × 100 × 120 mm

Contractor
SFIM Industries (SAGEM Group).

VERIFIED

Damien 6-UAM modular mixed acquisition unit

The Damien 6-UAM system is designed for the acquisition of all types of analogue and digital parameters and may be used in combat aircraft, civil aircraft, helicopters, ships and land vehicles. It is available in several configurations either as a centralised stand-alone acquisition unit, a decentralised system comprising several units and managed by one of these units, a decentralised acquisition system comprising several units managed by a central control unit. This versatility and modularity permits the system to acquire anything from 50 to 4,000 parameters.

Damien 6-UAM is a modular system with growth capability that is suitable for both large and small systems. It features a wide selection of inputs and outputs, with the analogue parameters grouped and segregated from digital parameters. Programming is by means of software.

A Damien 6-UAM unit consists of an assembly of mechanical sections, into which printed circuit boards are plugged. Input/output data is fed through a single connector located on the front of each section. Sections are interconnected by a variable-length, flex-background circuit, which is fixed to the rear of each section. The 28 V DC power supply is fed to the central part of the unit.

Contractor
SFIM Industries (SAGEM Group).

VERIFIED

DFDAU-ACMS Digital Flight Data Acquisition Unit - Aircraft Condition Monitoring System

SFIM Industries produces a range of DFDAU-ACMS units tailored to the requirements of the following aircraft types: A300-600/A310; B737/B757/B767; MD 80 series/MD 90. All units are designed to acquire engine and aircraft data reports, including aircraft life data, weather, and other parameters by user programmable requests.

Specifications
Interfaces:
(DFDR/SSFDR recorder) ARINC 717
(MCDU) ARINC 739
(On-board printer) ARINC 740 or 744

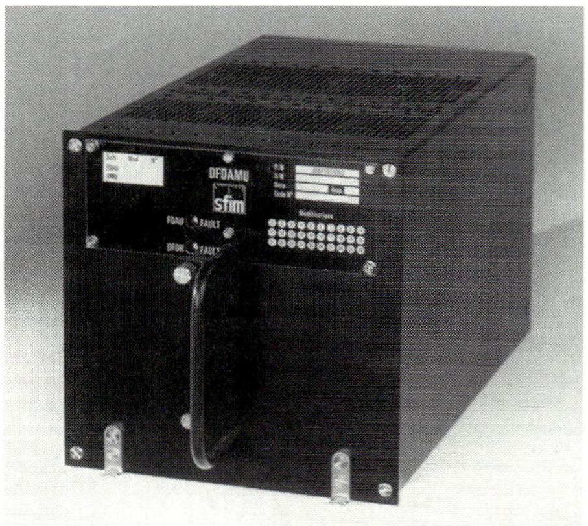

A300-600/A310 common unit DFDAU-ACMS 0015268

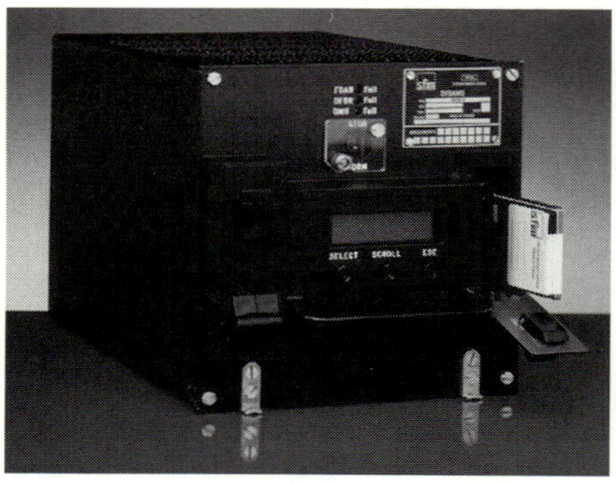

B737/B757/B767 common unit DFDAU-ACMS 0015269

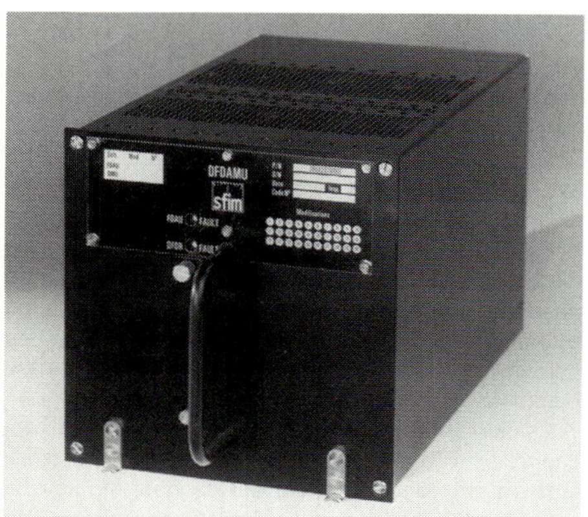

MD 80 series/MD 90 DFDAU-ACMS 0015270

Acquisition	A300-600/A310 common unit	B737/B757/B767 common unit	MD 80 series/MD 90
(analogues)	48	48	30
(discretes)	100	135	99
(ARINC 429 busses)	56	56	33

(QAR/DAR recorder) ARINC 591
(ACARS MU) ARINC 724 or 724B
(ADL/PDL) ARINC 615
Dimensions: 6 MCU
Weight: 8.5 kg
Power: 115 V AC, 400 Hz, 90 VA

Contractor
SFIM Industries (SAGEM Group).

VERIFIED

DMU-ACMS Data Management Unit - Aircraft Condition Monitoring System

The SFIM Industries DMU-ACMS is a common unit for use on the A319/A320/A321 series of commercial aircraft. The unit is designed to acquire engine and aircraft data reports, and data required for airline operational management, as may be programmed by the users.

Specifications
Interfaces:
(MCDU) ARINC 739
(on-board printer) ARINC 740 or 744
(DAR recorder) ARINC 591 (fully programmable frame)
(ACARS MU) ARINC 724B
(MDDU/PDL) ARINC 615
(CFDS (Centralised Fault Data System)) ABD 0048
Acquisition:
(discretes) 20
(ARINC 429 buses) 55
Dimensions: 3 MCU
Weight: 4.5 kg
Power: 115 V AC, 400 Hz, 40 VA

Contractor
SFIM Industries (SAGEM Group).

VERIFIED

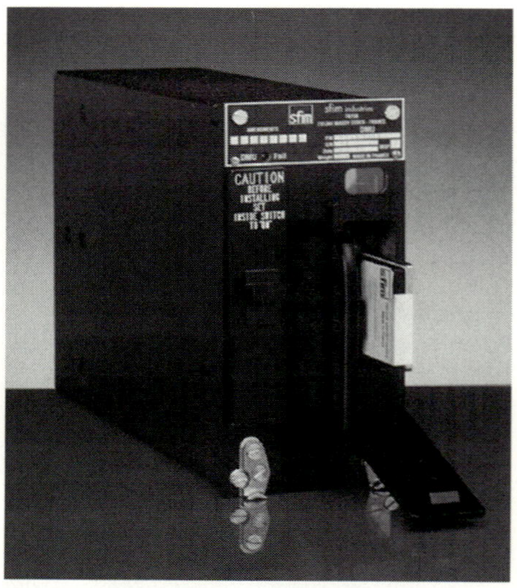

A319/A320/A321 DMU-ACMS common unit 0015271

ED 3333 data acquisition unit

For military aircraft applications, the ED 3333 is intended to acquire flight parameters; it automatically increases the flight number recorded at the beginning of each flight. The system monitors the performance of sensors and transducers and signals a warning in the event of a failure and transmits data to the crash recorder.

Specifications
Dimensions: ¼ ATR short
Weight: 3 kg
Power: 28 VA

Operational status
In production.

Contractor
SFIM Industries (SAGEM Group).

VERIFIED

ED 34XX data acquisition and processing unit

The ED 34XX has been designed for installations with weight or space constraints. It accepts the mandatory crash parameters, meets the requirements of ARINC 573 and can also take additional digital inputs from ARINC 429 databusses. After processing, information is transferred to a crash-protected recorder. The system can be expanded by adding a microprocessor and associated electronics to the same box to monitor engine and flight parameters. A printer and quick access recorder may be connected to store maintenance data.

The system can store data for between 50 and 100 flights. Information can be replayed on the ground through a low-cost commercial microcomputer system such as an IBM or Apple PC.

A new option that utilises a removable PCMCIA drive offers flight data storage equivalent to use of external recorders.

Specifications
Dimensions: ½ ATR short
Weight: 4.2 kg
Power: 30 VA

Operational status
In production for the ATR 42 and 72.

Contractor
SFIM Industries (SAGEM Group).

VERIFIED

ED 41XX, ED 44XX, ED 45XX, ED 47XX data acquisition units/data management units

The ED 41XX, ED 44XX, ED 45XX, ED 47XX are designed for the new-generation transports with digital avionics and electronic flight deck displays and are contained in ARINC 600 housing. They are designed as an expandable system by adding electronic boards. Several configurations are available: the basic FDAU for DFDR/QAR data supply; the expanded Data Management Unit (DMU) version for engine monitoring with basic engine trend monitoring reports; the expanded DMU with enhanced report monitoring for engine/APU monitoring, aircraft performance, environment reports, autoland performance as well as additional programmable reports; the expanded version with either solid-state memory for crew proficiency and flight data storage or with integrated and removable PCMCIA drive for flight data storage equivalent to external recorder.

All those units can be connected to a quick access recorder, cockpit printer, portable or onboard data loader, flight deck display (MCDU/CDU) and ACARS Management Unit for transmission of data to and from the ground.

Specifications
Dimensions:
3 MCU for DMU
6 MCU for DFDAMU
Weight: 4.5-8.5 kg (max)
Power: 40-85 W (max)

Operational status
In production for Airbus and Boeing commercial aircraft.

Contractor
SFIM Industries (SAGEM Group).

VERIFIED

ED 43XXXX flight data interface unit

The ED 43XXXX unit is intended to acquire the aircraft parameters and deliver a PCM message to the DFDR and QAR on A320, A330 and A340 aircraft.

It also interfaces with the onboard centralised maintenance system.

Specifications
Dimensions: 2 MCU
Weight: 3.7 kg
Power: 40 VA

Operational status
In production for Airbus A319, A320, A321, A330 and A340 aircraft.

Contractor
SFIM Industries (SAGEM Group).

VERIFIED

EVS 901 R videotape recorder

The EVS 901 R videotape recorder is used for storage and video restoration in harsh environments. It is equipped with a Hi-8 mm deck and features both CCIR-PAL and Y/C standards.

Specifications
Dimensions: 162.5 × 109 × 192.5 mm
Weight: 2.8 kg
Power supply: 28 V DC, 20 W
Temperature range: –30 to +55°C

Contractor
SFIM Industries (SAGEM Group).

VERIFIED

EVS 906 videotape recorder

The EVS 906 videotape recorder is a EVS 901 R with an RGB video input.

Contractor
SFIM Industries (SAGEM Group).

VERIFIED

EVS 925 videotape recorder

The EVS 925 videotape recorder is used for storage and video recovery in harsh environments. It is equipped with a Hi-8 mm deck and features both CCIR-PAL and RGB standards. The EVS 925 is controlled by a serial RS-422 link.

Specifications
Dimensions: 162.5 × 109 × 192.5 mm
Power: 28 V DC

Operational status
In production.

Contractor
SFIM Industries (SAGEM Group).

VERIFIED

EVS 1001 R videotape recorder

The EVS 1001 R videotape recorder is used for storage and video recording in harsh environments. It is equipped with a Hi-8 mm deck and complies with EIA/NTSC standards. The recording capability is 90 minutes. The EVS 1001 R provides the following functions: video and audio recording (two channels), video and audio playback, rewind, forward wind, pause, battery-driven autonomous cassette ejection.

Specifications
Dimensions: 162.5 × 109 × 192.5 mm
Weight: 2.8 kg

The SFIM FDIU is in service in the Airbus A321 ***2001***/0103889

Power supply: 28 V DC, 20 W
Temperature range: –30 to +55°C

Contractor
SFIM Industries (SAGEM Group).

VERIFIED

FDAU-ACMS Flight Data Acquisition Unit - Aircraft Condition Monitoring System

The SFIM Industries FDAU-ACMS is designed as a complete package to monitor the mandatory aircraft data parameters, together with data required for engine maintenance, on the ATR42/ATR72 commuter aircraft. The complete system comprises: the DFAU (Flight Data Acquisition Unit); the FDEP (Flight Data Entry Panel); and a three-axis accelerometer (TAA).

The FDEP displays: data, time and flight number; together with engine maintenance data.

The FDAU performs: acquisition of mandatory parameters; generation of a data frame to a DFDR/SSFDR; generation of the same parameters to the QAR and generation of time code. The system also incorporates BITE.

The ACMS function includes all reporting functions via the display memory terminal, disk and ACARS.

Specifications
Interfaces:
(DFDR/SSFDR recorder) ARINC 573
(QAR recorder) ARINC 591
objective torque output
(DMT (Display Memory Terminal)) via RS-232
(ACARS MU) ARINC 724
Dimensions:
(FDAU) ½ short ATR (127.5 × 199 × 318 mm)
(FDEP) 146 × 66.6 × 114.3 mm
(TAA) 101.6 × 91.7 × 63.5 mm
Weight:
(FDAU) 5 kg
(FDEP) 0.7 kg
(TAA) 0.4 kg
Power:
(FDAU) 28 V DC, 60 W (with FDEP)
(TAA) 28 V DC, 3.5 W

Contractor
SFIM Industries (SAGEM Group).

VERIFIED

ATR42/ATR72 FDAU-ACMS 0015272

FDIU Flight Data Interface Unit

The FDIU is designed as a common unit for the A319/A320/A321/A330/A340 aircraft. Its function is to interface with a flight data recorder to ensure recording of all mandatory aircraft parameters. It acquires both discretes and ARINC 429 bus data, and interfaces to: the DFDR/SSFDR - ARINC 717; the QAR - ARINC 591; the CFDS/OMS - ABD 0018 or 0048.

Specifications
Dimensions: 2 MCU
Weight: 2 kg
Power supply: 115 V AC, 400 Hz, 25 VA

Operational status
In production and service in a wide range of Airbus aircraft.

Contractor
SFIM Industries (SAGEM Group).

UPDATED

A319/A320/A321/A330/A340 FDIU 0015273

Mini-ESPAR 2

Designed by SFIM Industries, the Mini-ESPAR 2 has been developed for fitting as a crash recorder on helicopters or military aircraft. The Mini ESPAR 2 can acquire the basic information to cover the crash requirement.

Fitted with an entirely electronic memory, this new-generation recorder is considerably more reliable, and requires less scheduled servicing than earlier models, because of the absence of moving parts; it provides appreciable savings in weight and volume. These savings are achieved by use of a small-size static memory, made possible by the application of a specific data storage algorithm adapted to all types of aircraft and capable of recording up to 25 hours of flight data, and voice recording of up to 1 hour, in commercial aircraft. It has improved environmental protection against crush, pressure, temperature and corrosion.

During data retrieval, a recovery algorithm supplies the expected value of each parameter for comparison, to speed data analysis. A further advantage is the ability to re-read flight data without removal of the recorder. In this case, recorded data is transmitted over a high-speed line to a static data retrieval set which is then connected to the workstation.

SFIM Industries Mini-ESPAR 2 0015302

The Mini-ESPAR 2 can be adapted easily to any military or small civilian aircraft. It complies with TSO C124 and EUROCAE ED-55.

Specifications

Acquisition capacity:
(basic) 3 synchros, 15 analogues, 12 discretes, 3 frequencies
(options) 1 or 2 mixed audio channels, serial bus ARINC 429, or serial bus MIL-STD-1553B, or 2 pressure transducers
Recording duration:
(parameters) up to 24 h
(audio) up to 1 h
Frame rate: 64 or 128 words/s
Average parameter sample rate: 0.25-16 Hz, programmable
Dimensions: 183 × 148 × 273 mm
Weight: 9 kg
Power supply: 28 V DC or 115 V AC
Options: voice recording, ARINC 429 recording

Contractor

SFIM Industries (SAGEM Group).

VERIFIED

NH 90 Flight Control Computer (FCC)

SFIM Industries has been selected by Eurocopter France to design and build a flight control computer for the NH 90 helicopter.

The FCC is an essential component of the electrical flight control unit. It comprises the following two systems: the primary flight control system, which provides basic control capabilities to the helicopter and its stabilisation; and the automatic flight control system, which provides the 'hands off' high-level control modes that are required by the helicopter to fulfil its mission.

Operational status

Selected for the NH 90 helicopter.

Contractor

SFIM Industries (SAGEM Group).

UPDATED

Socrate 2/Saturne data acquisition systems

Socrate 2 is an acquisition subsystem of the Damien 6/8 family. It was designed for installation on the top of a helicopter main rotor, or in the centre of a helicopter tail rotor, so that it is able to acquire parameters from rotor blades, especially strain gauges and temperatures. It transmits up to 144 parameters on databusses through a slip ring to the cabin. The equipment is powered through the slip ring.

Saturne has the same function as Socrate 2, with a ring shape to allow the use of a mast-mounted sight on the helicopter. The acquisition capability is up to 128 parameters and the slip ring system is included in the equipment.

The equipment can also deliver the angular position of the rotor in order to correlate the position of blades with the values of parameters.

Specifications

Dimensions:
(Socrate 2) 75 to 275 mm diameter
(Saturne) 220 × 440 mm external diameter
260 mm internal diameter
Weight:
(Socrate 2) 2.6-12.5 kg
(Saturne) 30 kg

Operational status

The equipment is in production for Eurocopter Tiger helicopters.

Contractor

SFIM Industries (SAGEM Group).

VERIFIED

SSCVR Solid-State Cockpit Voice Recorder

The SSCVR has been designed in accordance with ARINC 757, TSO-C123 and ED 56(A) by SFIM Industries, Thales (formerly Thomson-CSF DETEXIS) and TEAM, to meet the requirements of all commercial aircraft for cockpit voice recording systems. The system has no moving parts, and uses compressed voice storage to provide up to 2 hours recording time.

Specifications

Recording attributes:
(method) non-sequential adaptive differential pulse code modulation and code exciter linear parameter modelling
(medium) flash memory modules
(inputs) 3 crew microphones; one area microphone; GMT (ARINC 429 format); rotor speed; flight data recorder time marker
(time) 30, 60 or 120 minutes
Dimensions: ½ ATR short - ARINC 404
Weight: 8.5 kg
Power:
(DC version) 28 V DC, 0.8 A
(AC version) 115 V AC, 400 Hz, 0.26A

Contractors

SFIM Industries (SAGEM Group).
TEAM.
Thales (formerly Thomson-CSF DETEXIS).

UPDATED

SFIM/Thales/TEAM SSCVR solid-state cockpit voice recorder 0015303

2084-XR mission computer

The 2084-XR is a general purpose digital computer designed to operate with high reliability under extremely severe environmental conditions.

Using enhanced technology such as VHSIC, ASICs and CMOS battery back-up RAMs it provides 1.2 Mips throughput with an extremely high performance/volume ratio.

Within a ½ ATR case, the 2084-XR computer comprises: an arithmetic unit; a memory of up to 512,000 18-bit words; a comprehensive input/output system (including the simultaneous management of two multiplexed buses); and a power supply.

Operational status

The 2084-XR computer is in service in all versions of the Dassault Mirage 2000 aircraft (two computers per aircraft).

Contractor

Thales (formerly Thomson-CSF DETEXIS).

UPDATED

BSDM mission data storage unit

The BSDM mission data storage unit is fitted to military aircraft like Rafale in a twin extractable cassette configuration.

Contractor

Thales (formerly Thomson-CSF DETEXIS).

UPDATED

CICS 68040 computers

The CICS 68040 computers are intended for helicopter and aircraft upgrade applications. These computers interface with the aircraft systems via two redundant 1553 buses, ARINC 429 links and other specific links.

Their modular architecture is based on a VME backplane bus and can include, other than the bus input/output and acquisition functions, several real-time 68040 central processing units programmable in Ada or in C.

Contractor

Thales (formerly Thomson-CSF DETEXIS).

UPDATED

DDVR Digital Data and Voice Recorder

Thomson-CSF DETEXIS has developed a range of new-generation flight data recorders featuring a crash-survivable Flash memory to replace the conventional on-board tape recorders.

These new recorders are available in commercial and military versions. They can be equipped with an acoustic localisation beacon and are compatible with the ground support equipment associated with conventional recorders.

Recording volumes and easy data extraction allow the new recorders to be used for flight analysis and for monitoring essential aircraft parameters.

Specifications

Dimensions: 124 × 1,567 × 272 mm
Weight: < 8 kg
Power supply: 28 V DC
Memory capacity: up to 256 Mbytes
Compliance: EUROCAE ED-55 and ED-56a

Operational status

The DDVR offers the advantages of an all-digital system. It will be certified to the latest survivability standard TSO-C124a.

A top range military version is installed on the Mirage F1, 2000, 2000-D, 2000-5 and the Rafale aircraft. On Mirage F1 and Mirage 2000 aircraft, the DDVR is interchangeable, both electrically and mechanically, with earlier tape recorders.

The earlier commercial version is certified on Airbus and Boeing 737, 757 and 767 aircraft, as well as MD-90 and IPTN 250 aircraft. It has been selected by a large number of airlines worldwide.

Contractor

Thales (formerly Thomson-CSF DETEXIS).

UPDATED

EMTI data processing modular electronics

For the future generation of avionics core systems, Thales (formerly Thomson-CSF DETEXIS) will be developing and marketing the EMTI in co-operation with Sextant.

The main characteristics are modular data processing assembly; open-ended system; adaptable to existing interfaces; simplified maintenance.

The system ensures independence between the basic software and the application programs. The programs are being developed with the ODILE software engineering environment and are defined in co-operation with Dassault Aviation. New technologies (object-oriented open environment, automatic code production) are used so that system object libraries, developed under past programs, can be re-used and the simulations and proofs can be incorporated into the environment.

Operational status

This new generation of products will be installed on the Mirage 2000-9 aircraft. The products are evaluated for use under the Rafale programme and for the retrofitting of different weapon systems.

Contractors

Thales (formerly Thomson-CSF DETEXIS).
Dassault Aviation.
Sextant.

UPDATED

Extended-storage Quick Access Recorder (EQAR)

Designed and manufactured by Thales (formerly Thomson-CSF DETEXIS), the Extended-storage Quick Access Recorder (EQAR) is jointly marketed and supported worldwide by Sextant Avionique.

EQAR is used as a quick access recorder and/or digital ACMS recorder. Based on rewritable optical disk technology, the EQAR, which is fully interchangeable with existing magnetic-tape recorders, provides increased storage capacity, enabling more parameters to be recorded at a higher rate. Recorded data can be accessed directly from DOS files read through a PC.

Designed to be fitted on civil aircraft, the EQAR system provides full-time recording throughout the flight, with maximum data integrity.

Specifications

Dimensions: ARINC 404/600 formats
Input data rate: 64-512 words/s
Storage capacity: 128, 256, or 540 Mbytes
Environmental: DO160C

Operational status

Certified and operated on most commercial aircraft including those manufactured by Airbus and Boeing, EQAR has been selected by more than 30 airlines and operates worldwide with more than 500 units ordered.

Contractor

Thales (formerly Thomson-CSF DETEXIS).

UPDATED

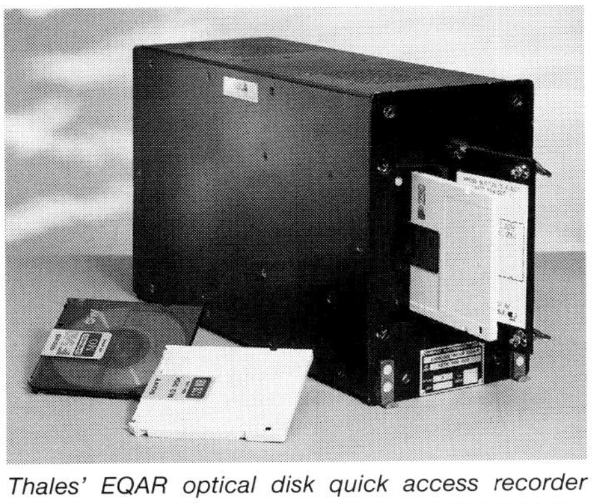

Thales' EQAR optical disk quick access recorder
1999/0051320

IMA data processing equipment

Thales (formerly Thomson-CSF DETEXIS), in co-operation with Sextant, will develop and market the data processing equipment for the new avionics core systems.

This equipment is designed to provide military aircraft with scalable computing power depending on the aircraft mission and other electronics equipment complexity. It can be used for mission control (including navigation, weapons management, electronic warfare and radar control).

The main characteristics are modular data processing; open architecture; easy adaptation to existing interfaces; independent of basic software and application programs.

The modules in production are the central processing module; graphics data processing module; mass memory module; bus management module; input/output module.

This data processing equipment is provided with a module development kit; an object oriented software environment that handles system object libraries and makes software reuse easy for other applications or simulations and automatic code production.

The data processing equipment is intended for the new military aircraft as well as for the retrofit of military aircraft already in operation.

Contractors

Thales (formerly Thomson-CSF DETEXIS).
Sextant.

UPDATED

Military real-time LANS

The DIGIBUS GAM-T-101 became a French tri-service standard multiplex databus in 1982. It is used in many military systems (Mirage F1, Mirage 2000, Atlantique 2, missiles, submarines, navy ships and land-based applications).

STANAG 3910 is a dual-speed version of the MIL-STD-1553B which is widely used in numerous onboard military systems. This new databus has been adopted for the EFA and the Rafale aircraft, it can also be used for the modernisation of any MIL-STD-1553-based system.

Operational status

Over 35,000 remote terminal bus interface units and 3,500 bus controllers have been produced.

Contractor

Thales (formerly Thomson-CSF DETEXIS).

UPDATED

Test and measurement system

Thales (formerly Thomson-CSF DETEXIS) provides a new generation of test and measurement systems for airborne applications built around an innovative architecture. Implementing the most recent technology, this set of equipment uses very high-performance components to allow more high-density circuit integration and significantly reduce size, weight and costs. It results in a very flexible system that can acquire several thousand parameters.

Core of the system is SERPAN, a digital acquisition unit. When used as a stand-alone unit, this device receives most airborne busses in full or filtered mode, including MIL-STD-1553 and ARINC 429 protocols. When equipped with PCM interface boards, it becomes a PCM concentrator, allowing connection of other SERPANs or devices with PCM outputs (Irig or Daniel).

Analogue signals are acquired by means of CCE and SAMPLE units. CCE is a miniaturised container located near the sensors. Up to 15 CCEs can be coupled to the system by a high-speed bus to allow acquisition of parameters from thermocouples, strain gauges, low-level sensors, temperature and piezoelectric probes. SAMPLE is an analogue acquisition unit. Designed for vibration acquisition, this fully automatic unit outputs a PCM stream for coupling to a SERPAN core unit.

ELEFAN is a static flash-memory recorder that is utilised instead of tape devices in severe environment. The high data storage capacity of this unit (up to 2 Gbytes) allows several hours of recording. Both ELEFAN and SERPAN functions can be mixed in the same ruggedised unit.

A standard PC workstation runs ground software for configuration, ELEFAN downloading and data analysis.

The SAMPLE unit may also be used as a stand-alone system for analogue acquisition.

SAMPLE

Although SAMPLE is a part of Thales' test and measurement system, this unit may also be used as a stand-alone unit for analogue acquisition. Fully programmable, SAMPLE is designed for vibration acquisition. Main applications are measurement during load release, dummy ejection, stress analysis and other tests in very severe environments. SAMPLE is capable of absorbing shock at 200 *g* for 11 ms. When housed in MAGISS, a patented technology from Thales for device protection, this unit is shock absorbent at 500 *g* for 6 ms.

The SAMPLE unit is equipped with analogue acquisition and memory boards. Each acquisition board can receive up to four channels (differential inputs) with programmable excitation, gain, offset, filtering and sampling rate. Storage capacity is from 64 to 384 Mbytes. SAMPLE can also acquire up to 12 digital pulses through differential inputs with opto-electronic isolation. Each input may be used to trigger the recording process with a programmable timer. Coding of signals is made on 12 bits with 0.1 per cent accuracy.

For particular applications, SAMPLE can acquire both analogue information and ARINC 429 busses. Merged data are stored into memory, when demultiplexing is made by ground workstation at the time of downloading.

SAMPLE is available in two widths, according to the customer needs. The smaller unit can support 16 channels and can be equipped with a battery for autonomous powering. The larger unit has 32-channel capability.

A standard PC workstation runs ground software for configuration, downloading and data analysis.

SAMPLE is currently in use for testing parachutes, ejection seats and released loads. Also in use for stress measurement on military carrier aircraft.

Operational status

Currently in use for flight testing of Mirage 2000, Rafale, and several foreign customers' retrofit programmes.

Contractor

Thales (formerly Thomson-CSF DETEXIS).

UPDATED

Video recording systems

Mission sight recording is accomplished by a system consisting of, an OTA 204 monochrome or OTA 300 colour video camera and an OEV 301 onboard magnetic-tape recorder. The CCD video camera replaces the earlier AA8-400 film camera. A two-unit version of the video recording system has been developed for the holographic HUDs of future combat aircraft, sights used on present and new-generation helicopters and sights on the gyrostabilised turrets of future tanks.

Operational status

The OTA 208 video camera and the OEV 301 magnetic-tape recorder were to upgrade French Navy Super Etendard aircraft.

Contractor

Thales (formerly Thomson-CSF Optronique).

UPDATED

GERMANY

Display Video Recording System (DVRS)

The Display Video Recording System (DVRS) has been designed by Bavaria Keytronic Technologie for the German Air Force Tornado. It is also suitable for army, navy and commercial applications. The system can handle up to seven video inputs with a resolution of 400 TVL/PH each, plus audio channels, event and weapon release markers, databusses, radar and FLIR.

The system consists of a HUD camera, electronics unit, videotape recorder with up to 180 minutes recording tape and a control panel.

The HUD camera consists of the video sensor head with optical components and CCD video colour sensor module, and an electronics assembly consisting of power supplies, control logic, BIT circuitry and video amplifiers. HUD symbology is combined with the picture of the outside world and transmitted in colour PAL format via the multiplexing unit to the VTR.

The electronics unit consists of an analogue-to-digital converter unit, multiplex circuitry, microprocessor, control BIT circuitry and power supplies.

The Hi-8 mm VTR is based on a TEAC V 80 recorder. The microprocessor-controlled panel gives the operator complete up-to-date information on the system, including remaining recording time. The control panel consists of the mode selection switch, video input, selection control, alphanumeric display and amplifiers.

The system provides up to 180 minutes recording time on each cassette which can be changed easily during flight.

Specifications

Dimensions:
(HUD camera) 190 × 45 × 210 mm
(electronics unit) 120 × 325 × 170 mm
(videotape recorder) 107 × 139 × 156 mm
(control panel) 55 × 145 × 127 mm

Operational status

In service in German and Italian Air Force Tornado aircraft.

Contractor

Bavaria Keytronic Technologie GmbH.

VERIFIED

The display video recording system in service in German and Italian air force Tornado aircraft, showing the electronics unit (top), video recorder (right), control panel (bottom), HUD camera (left) ***1999***/0051323

Flight Safety Recording System (FSRS)

The Flight Safety Recording System (FSRS) provides information for analysis of accidents and severe crashes, training of pilots, flight attendants and ground crew and surveillance of passenger and cargo compartments and exterior equipment. It consists of cameras, an electronic control unit and a Hi-8 mm format video recorder.

The cameras may be black and white or colour cameras installed inside or outside the aircraft. The cameras may be installed in the cockpit to record instruments and actions, in the passenger/cargo compartment to observe and record the situation and outside the aircraft to observe and record equipment and the outside world. They may be designed for low-light level operation, with infra-red illumination. The electronic control unit multiplexes all video signals, controls the recording and has an optional interface to GPS and aircraft busses to record navigation and other aircraft data.

Operational status

FSRS is designed for installation in rotary- and fixed-wing aircraft and can be tailored to different applications or customer requirements.

Contractor

Bavaria Keytronic Technologie GmbH.

VERIFIED

Data transfer system

EADS (formerly DaimlerChrysler Aerospace AG), Defense and Civil Systems' data transfer systems comprise portable solid-state data carriers as key elements with easily expandable memory capacity and associated data carrier adaptors for various ground and onboard installations. The data carrier adaptors have a similar design and are easily interchangeable by replacing the interface card which interfaces the equipment to ground computers or onboard systems. Onboard data carrier adaptors are available for MIL-STD-1553B and RS-422 interfaces. The portable and removable pocket-sized solid-state data carriers are used like a floppy disk to store any kind of data in an intelligent way.

EADS produces several types of CMOS EEPROM-based portable solid-state data carriers with storage capacities ranging from 64 kbytes to 8 Mbytes. The new generation of data carriers is equipped with a powerful built-in RISC processor combined with a specific controller ASIC providing autonomous data storage management and a high-speed serial datalink with 10 Mbits/s. The firmware routines implemented in the RISC processor provide application independent data management procedures which are directly comparable to the hard disk file system of standard DOS PCs. Programmable applications enable a large number of files of different sizes and a flexible structure of hierarchical directories limited only by the storage size of the data carrier. Up to 25 files can be opened simultaneously. For multi-user operation, a file can be opened for read access by up to 25 users simultaneously.

Specifications

Dimensions:
(Alpha Jet) 180 × 127 × 36 mm
(Tornado, C-160, EF 2000) 180 × 127 × 73 mm
(PAH-2 Tiger) 180 × 127 × 53.9 mm
(data carrier) 80 × 55 × 25 mm
Weight:
(Tornado) 1.7 kg
(Alpha Jet) 1.2 kg
(PAH-2 Tiger) 1.3 kg
(data carrier) 0.1-0.2 kg
Power supply: 28 V DC, 25 W

Operational status

Qualified for carriage on the Alpha Jet, Tornado, Eurofighter Typhoon and PAH-2 Tiger.

Contractor

EADS (formerly DaimlerChrysler Aerospace AG), Defense and Civil Systems.

UPDATED

Multiple Dislocated Flight Data Recorder system (MDFDR)

The new-generation ultra lightweight Multiple Dislocated Flight Data Recorder (MDFDR) is part of an error tolerant modular solid-state crash and cockpit voice recording system which comprises up to four MDFDR recorders providing sufficient recording capacity. Due to the modularity, small dimensions and standardised interfaces, this recording system is suited for installation into helicopters and general aviation aircraft.

The MDFDR system consists of a Multiple Recorder Data Acquisition Unit (MRDAU), one to four MDFDR, Hand-Held Terminal (HHT) and ground station, designed to collect onboard data (audio, analogue, frequency, mil-bus, HDU and discrete inputs). It filters data to avoid exceedance and logistic relevant data and stores the resulting different data groups in the FDR and SSDC.

After flight, the data can be checked on board by the HHT or can be transferred via the HHT or SSDC to the ground station for detailed checkout.

Specifications

Dimensions:
(steel ball) 55 mm diameter
Data interface: RS-485 up to 1 Mbit/s (asynchronous datalink)
Power: 8-12 V DC, 1 W
Temperature range:
(operating) –55 to +85°C
(storage) –55 to +85°C
Weight:
(1 MDFDR) 0.35 kg
(MRDAU) 1.5 kg
Memory capacity: 4-32 Mbyte, 1 to 8 tracks, min track size 2 Mbyte

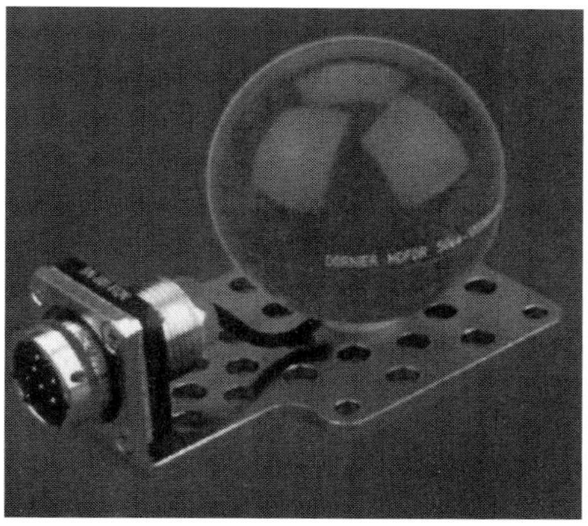

Multiple Dislocated Flight Data Recorder 0015305

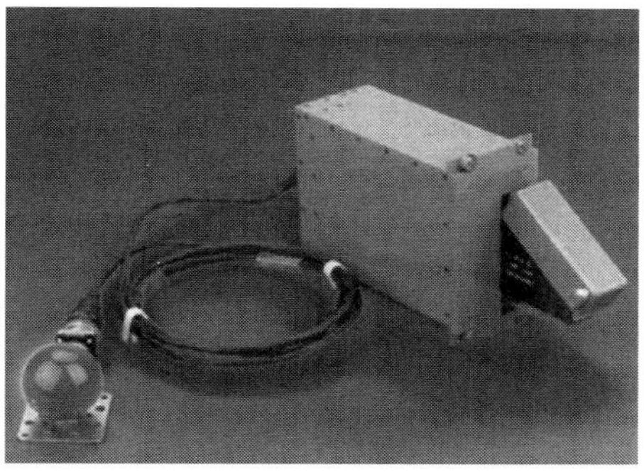

MRDAU Multiple Recorder Data Acquisition Unit 0015306

Hand-held terminal 0015308

Multiple Recorder Data Acquisition Unit (MRDAU)

The MRDAU controls and supplies up to four MDFDR. The application specific programmable data interfaces of the MRDAU provide flexibility to adapt the MDFDR system to various helicopters and general aviation aircraft. For maintenance recording functions, the MRDAU has an optional data carrier interface which can store up to 170 Mbyte of flight data.

Specifications

Dimensions:
(front panel) 146 × 76.2 × 6 mm
(case) 180 × 127 × 73 mm
Weight: 1.5 kg (excl data carrier)
Power: 28 V DC, 25 W
Temperature range:
(onboard equipment, operating) –55 to + 71°C

Optional data carrier 0015307

Hand-Held Terminal (HHT) (readout unit)

The HHT is a powerful battery-powered product to read out the MDFDR and the MRDAU. It is capable of 7 to 10 hours of continuous use. It provides a large 640 × 480 pixels, easy to read backlit display, with VGA graphics. The HHT is IBM PC/AT compatible, it is possible to connect an external keyboard, mouse, printer, monitor and floppy disk drive. Optional plug-in modules are available for SCSI, PCMCIA and other ISA bus interfaces.

Contractors

EADS (formerly DaimlerChrysler Aerospace AG), Defense and Civil Systems.

UPDATED

Flight Control Data Concentrator (FCDC)

The Flight Control Data Concentrator (FCDC) is part of the flight control system of the Airbus Industrie A319, A320 and A321 aircraft. Its main functions are to validate, concentrate and store in-flight status and failure data of the flight control system. The electrical interfaces are a number of ARINC 429 and discrete inputs and outputs. Outline and mounting are in accordance with ARINC 600.

Specifications

Dimensions: 2 MCU
Weight: 3.4 kg
Power supply: 28 V DC, 16 W
Interfaces:
(input/output) ARINC 429, SGS and SAV discrete
(input) DC analogue
Reliability: >18,000 h MTBF

Operational status

In production. The FCDC entered operational service in 1988 on board the A320, which is equipped with a dual FCDC installation. It is also in service on the A319 and A321 aircraft.

Contractor

LITEF GmbH.

VERIFIED

LAC-L LITEF avionic computer landing aid

LAC-L is used as the central processing unit of the AJAX landing aid made by ADS, St Petersburg. In the

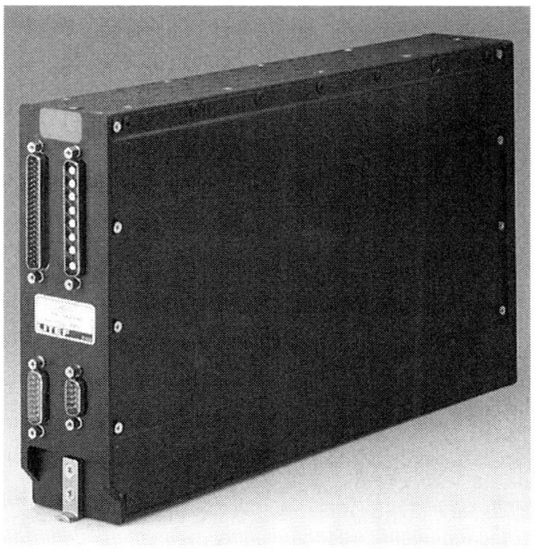

LAC-L LITEF avionic computer landing aid ***0005426***

AJAX system, LAC-L processes TV-picture data to autonomously determine relevant navigation parameters including distance to touch down, velocity and height in real time and provides it for display to the pilot.

Specifications

Dimensions: 320 × 57 × 194 mm ARINC-600-6 2MCU
Weight: 2.3 kg
Power consumption: 28 V DC, 25 W
Data input: Video from 2 CCD cameras
Data output: 2 video outputs (PAL, RGB)

Contractor

LITEF GmbH.

VERIFIED

The upgraded main computer of the Panavia Tornado

Main computer for the Tornado

The in-service main computer for navigation and weapon delivery in the Tornado has been upgraded to fulfil increased computing requirements. It is now based on a multiprocessor architecture which still uses the original LR-1432F digital airborne computer but adds one or more high-performance 68040 processor modules. This new structure enables execution of all existing software, as well as recently introduced Ada software running on the 68040 processors.

The computer is equipped with various input/output interfaces including a high number of Panavia interfaces, one or two MIL-STD-1553 bus systems, discretes and special type interfaces.

Powerful graphics capabilities have been added.

Operational status

The upgraded main computer is in service.

Contractor

LITEF GmbH.

VERIFIED

Air data computer/air data transducer for the JAS 39 Gripen

The air data computer/air data transducer for the Saab JAS 39 Gripen combine the primary air data computer with an air data transducer channel as back-up in a single electronics box. Both functions are fully independent of one another and are electrically decoupled. The air data computer contains two high-precision pressure sensors which convert the static pressure and the total pressure into an electronic frequency signal. The signal is digitised and fed to a 16-bit microprocessor. Using software algorithms, the classic air data are computed, taking into account further parameters such as barometric altitude correction and total air temperature. Air data such as speed, altitude and Mach number are passed to aircraft systems via MIL-STD-1553B interfaces.

The other part of the unit contains two additional identical pressure sensors which operate in the same way as in the air data computer. In contrast to the computer, however, the final data are not computed. Instead the static and total pressures are passed to the flight control computer via an ARINC 429 interface. Due to the complete separation of the functions of the air data computer and air data transducer, the air data transducer has its own power supply unit, a separate processor and a separate software programme. In the flight control computer, the pressure values supplied by the air data transducer are used to calculate the air data, taking into consideration the barometric altitude correction and total air temperature, and the result compared with the values from the air data computer.

Both channels of the Nord-Micro air data computer and air data transducer have a highly developed self-testing capability and are able to check their own correct function or to provide information on any faults which may have occurred. In the event of an error, the faulty function is detected. The use of hybrids, gate arrays and LCC circuits enable the units to be housed in a 4 MCU box.

Operational status
In service on the Saab JAS 39 Gripen.

Contractor
Nord-Micro AG & Co. OHG.

UPDATED

JAS 39 Gripen air data computer/air data transducer
0015274

DATaRec-A4 acoustic recording system

The DATaRec-A4 is a complete portable four-channel acoustic/analogue recording system. The system provides connectors, preamplifiers and power supplies for Bruel and Kjäer microphones. With 24-bit digital signal processing and filtering, exceptional phase error specification allows accurate correlation of data between channels. Weighting filters can be selected and, with pre-emphasis, the dynamic range will be extended to 130 dB.

A front panel LCD provides bar graph and alphanumeric displays of input/output voltage, power and spectra. The IRIG generator/decoder allows synchronisation to the internal or external IRIG source, accurately time-stamping all data. Additional data that may be recorded includes voice, RS-232 and opto-coupled rpm and speed data.

With eight different channel configurations, a patented self-calibration system, rugged but lightweight chassis and battery operation, the A4 may be used wherever portability is of importance.

Specifications
Dimensions: 270 × 88 × 265 mm
Weight: 6 kg (incl battery)
Power supply: battery operated for up to 2 h
10-36 V DC or 100-240 V AC (with adaptor)

Contractor
Racal Heim GmbH (a Thales company).

UPDATED

DATaRec-A16 analogue recording system

The DATaRec-A16 is a comprehensive analogue recording system, expandable from 16 to 80 channels. Internal signal conditioning modules allow direct connection of common transducers. With all channels sampled simultaneously, 24-bit digital signal processing and filtering, exceptional phase error and dynamic specifications are achieved. The large dynamic range leads to extremely low harmonic distortion and noise levels.

A quick-look facility for all input and output signals can display minimum, maximum and mean levels.

The IRIG generator/decoder allows accurate time-stamping of all data, using internal or external time sources. Additional data that may be recorded includes voice, RS-232 and opto-coupled rpm and speed data.

With 51 different channel configurations, patented self-calibration system and a rugged but lightweight chassis, the A16 can be used in a wide variety of applications.

Specifications
Dimensions: 325 × 133 × 320 mm
Weight: 12 kg
Power supply: 10-36 V DC or 100-240 V AC, 50 W

Contractor
Racal Heim GmbH (a Thales company).

UPDATED

DATaRec-D3 digital recording system

The DATaRec-D3 is a versatile digital recording system for PCM, MIL-STD-1553 and RS-422 data systems. An optional PCM merger card allows four independent PCM streams to be recorded. The IRIG generator/decoder provides accurate time-stamping of all data using internal or external time sources.

The MIL-STD-1553 interface card permits a dual-redundant 1553 bus to be recorded. The clock, either internal or external with 1 μs accuracy, allows the RT response and intermessage gap times to be measured and the 1553 stream to be accurately replayed.

Additional data that may be recorded includes voice, RS-232 and opto-coupled rpm and speed. Memory buffers allow bursts of higher rate data to be recorded or the 3 hour record time to be significantly extended.

The use of rugged but lightweight chassis and universal auto-sensing power supplies allow the D3 to be used in a wide variety of applications.

Specifications
Dimensions: 270 × 88 × 230 mm
Weight: 4.2 kg
Power supply: 10-36 V DC or 100-240 V AC (via adaptor), 25 W

Contractor
Racal Heim GmbH (a Thales company).

UPDATED

DATaRec-D4 digital recording system

The DATaRec-D4 digital recording system is able to record at the rate of 4 Mbits/s for 2 hours, with 1μs resolution time stamping. The DATaRec-D4 contains an IRIG-B time code generator and decoder. This locks to an external source or can be used as the source for equipment external to the D4. The time is maintained by battery back-up. The IRIG-B time code is reconstructed during replay.

The D4 will store different configurations in non-volatile memory. One of these, the default configuration, is loaded when power is applied to the D4. This configuration includes the mechanical status of the tape drive. The recorder may therefore be installed in inaccessible locations and recording is made and ended simply by operating a 28 V power switch. There are simple remote-control function switches for play, record, stop, rewind and fast forward.

There are many useful accessories for the D4, such as a hand-held remote-control unit and a cockpit mounting remote control with a sunlight readable LED display and simple control for the pilot's use. The D4 has a large amount of recording capacity. This allows the recording and playback of event markers, voice commentary, RS-232 data and rpm pulse input. Up to 32 thermocouple, or low-frequency analogue, signals may be recorded at the same time as the main PCM stream or MIL-STD-1553 data, via the RS-232 input, by connecting an optional T32 system.

The D4 is of modular construction, which provides for easy upgrade or modification. For example, an additional MIL-STD-1553 recording interface or replay module may be incorporated. There are five free slots and all optional interface boards can be mixed up to the sum bit rate of 4 Mbits/s.

The D4 operates from power supplies in the range 20 to 36 V DC and will operate without errors even when subjected to intermittent power failures in the region of 50 ms. There is also an accumulator option which saves all data during a power interruption of more than 50 ms. It is easily installed in an interface slot, like an interface card.

The construction of the D4 is extremely robust but lightweight and is designed for aerospace and automotive applications. The tape drive is mounted on built-in vibration isolators and these allow error-free operation under vibration up to MIL-STD-810C of 5 *g* and 10 *g* acceleration.

Specifications
Dimensions: 124 × 180 × 260 mm (½ ATR)
(cassette) 73 × 54 × 10.5 mm
Weight: 6 kg
Power supply: 20-36 V DC, 70 W (max)
Temperature range: –10 to +60°C

Operational status
Selected for the Royal Navy Westland Lynx Mk 8 helicopter UK.

Contractor
Racal Heim GmbH (a Thales company).

UPDATED

The DATaRec-D4 high-speed digital data recording system
0001293

DATaRec-D10/D12/D40 digital cassette tape recording system

The D10/12/40 have been designed for aerospace and defence applications and are 10, 12 and 40 Mbits/s nominal (with compression turned off) digital tape recorders. Their construction is extremely robust, yet lightweight. The heavy-duty 0.5 in Digital Linear Tape (DLT) tape drives are mounted on built-in vibration isolators which provide error-free operation in harsh environments to MIL-STD-810C (5 *g* vibration and 10 *g* acceleration).

Modular construction is provided with six free slots, which allow users to configure the recorder for particular applications and to modify it for future programs. All optional interface boards can be mixed without limitation (for example, 8 + channel analogue input and output combined with multiple MIL-STD-1553B bus recording and reconstruction). The recorders can merge and record multiple digital and analogue data streams up to combined maxima of 10, 12 or 40 Mbits/s (actual data rates due to low overheads) without the use of compression and can achieve data rates up to 150 per cent using built-in compression. Analogue bandwidths from 300 kHz (D10) to 1.2 MHz (D40) make the systems ideal for use as mission recorders particularly in sonobuoy recording applications. RS-232 serial, SCSI, and power-on remote control is available and remote contact mission configuration. The D10/12/40 have a built-in IRIG-A/B/G/J time code generator and decoder, and a serial GPS time code decoder. A high-stability oscillator (0.44 ppm) is available to improve time code accuracy.

Specifications

Recording time:
D10: 2 h, at full bit rate, 10 Mbits/s
D12: 3.6 h, at full bit rate, 12 Mbits/s
D40: 1.8 h at full bit rate, 40 Mbits/s
Recording time can be increased with lower bit rate, and with use of built-in data compression (data dependent)
Recording capacity:
10-15 Gbyte, 20 Gbyte and 35 Gbyte (uncompressed)
Dimensions: 3/4 ATR wide

Operational status

Selected by Sikorsky for the flight test certification programme for the S-92 Helibus.

DATaRec-D12 0001294

Contractor

Racal Heim GmbH (a Thales company).

UPDATED

DATaRec-E8 digital cassette tape recording system

The DATaRec-E8 is an eight-channel PCM recorder. Data is recorded on a 3.8 mm wide magnetic tape using the helical scan recording system. Digital Audio Tape (DAT) cartridges and drives are used for the storage of data. An efficient error-correction process provides data security for the system.

Incorporating the input/output filters and the cassette drive, the eight-channel PCM electronics comprise a compact unit. The inputs and outputs are connected to BNC jacks without intermediate adaptor. The number of channels can be selected to one, two, four or eight. Signal bandwidth is 5 kHz for each channel with eight-channel operation. Input and output filters can be dispensed with in many applications.

All analogue channels are sampled at the same time. The unit incorporates eight analogue/digital and eight digital/analogue converters with a resolution of 14 bits.

In additon to data, date and time are recorded continuously. A voice channel for recording comments is also provided. A built-in loudspeaker allows the voice recording to be replayed. A headphone connection and line output are provided. A standard IRIG-B time code can be recorded instead of voice recording. Index marks which were set automatically or manually during recording can be searched at high speed during replay.

The DATaRec-E8 magnetic tape system is especially suitable for mobile use. Small dimensions and low power consumption are important features. The units operate correctly even in vibration of up to 5 *g*. Three alternative power supply sources allow universal use. The 6 V, 1.8 Ah storage battery permits the PCM to operate for 2 hours. A DC voltage adaptor or a mains adaptor can be used instead of the storage battery. Voltage range is from 10 to 30 V.

Specifications

Dimensions: 235 × 92 × 176 mm
Weight: 2.8 kg
Power supply: 6 V, 1.8 Ah battery
10-30 V DC

Contractor

Racal Heim GmbH (a Thales company).

UPDATED

OptoRec Q1 and Q2 quick access recorders

The OptoRec Q1 and Q2 quick access recorders are based on Motorola 56001 DSPs. The Q1 includes a VGA module and only a keyboard and monitor need to be connected to make the recorder act as a PC. The Q2 uses a 96001 CPU, running at 33 MHz, and can be connected to a PC via the RS-232 port. Both recorders provide a standard Centronics port for a printer.

The Q1 and Q2 will record ARINC 573 and 575, with options that allow multiple ARINC 429 and analogue data streams to be recorded simultaneously. An intelligent data decoder unit automatically detects and decodes the different data types. A clock with battery back-up provides accurate date and time annotation on the recordings. An ASCII character stream can be recorded at up to 9,600 baud for report writing. Status and error conditions are indicated via a 12 digit front-panel LED display.

The standard recording medium is 3.5 in optical disk conforming to ISO/CD 10090, with an option for a DAT streamer. The disk allows up to 43 hours record time at 512 words/s and all data is recorded in DOS format files. The disk may be removed and read by any standard 3.5 in DOS-compatible optical drive. The Q1 also has an option for a hard disk drive.

Data to be recorded is buffered via a 128 kbyte non-volatile RAM and reliable recording is achieved even when subjected to power failures up to several seconds. Up to 16 Mbytes of memory are allocated separately for retaining application, or mission specific programs.

Contractor

Racal Heim GmbH (a Thales company).

UPDATED

Storehorse and Storeplex range of instrumentation recorders

The Storehorse range provides record and replay in 7, 14, 28 and 42 track configurations. Heads meeting IRIG intermediate band, wideband and double density standards are available for operation in the speed range 15/32 to 120 in/s. A wide selection of interchangeable data channels is available to provide DR, FM, dual-mode FM/DR, digital and voice recording modes up to a maximum of 4 MHz bandwidth at 120 in/s tape speed. The Storehorse features automatic calibration and equalisation, microprocessor control and other advanced techniques to ensure efficient and accurate recording. The extended bandwidth capability of Storehorse DD-4 is of particular benefit in predetection and telemetry recording.

The Storeplex range offers a new modular concept in helical scan digital recording. The tape transport module uses economical S-VHS tape, a reliable high-capacity easy-to-handle medium. Up to 64 analogue and/or digital channels are housed in a compact signals module. Channels can be recorded using any combination of bandwidths and clock rates up to the maximum system aggregate data rate of 51.2 Mbits/s. Storeplex analogue channels feature fully automatic calibration to provide a dynamic range of 96 dB and channel-to-channel phase accuracy of better than 0.5° at 45.5 kHz. Digital channels offer from 1 to 16-bit parallel input at rates between 2 kbits/s and 51.2 Mbits/s. An integrated SCSI interface allows rapid direct-to-computer download of signal data. The system offers a flexible accurate solution for acoustic, vibration and telemetry data acquisition.

Operational status

The Storehorse range is established worldwide for use in radio, radar and sonar monitoring, with many applications in aerospace development and range telemetry recording.

The Storeplex range has become a standard solution for acoustic, sonar and vibration measurement applications.

Contractor

Racal Heim GmbH (a Thales company).

UPDATED

Twenty eight Storehorse instrumentation recorders are installed on the Advanced Range Instrumentation Aircraft (ARIA)

The Racal Heim Storeplex Delta instrumentation recorder offers 51.2 Mbits/s recording capacity and variable tape speed

Modular avionic computers

A range of modular avionic computers are designed by Teldix for a variety of airborne roles.

The Missile Control Unit (MCU) is designed to control intelligent missiles fitted to aircraft such as the Panavia Tornado. The MCU serves as a computing and interface system between the aircraft's main digital computer and the relevant input/output circuits of the missiles.

The MCU is a 16-bit multimicroprocessor containing the interface circuits to convert, process and pass internally fed data to the missiles via the MIL-STD-1553B databus. The Launcher Decoder Unit (LDU) provides the interface between the MCU and individual missiles. The MCU and LDU are in production.

Operational status

In production.

Contractor

TELDIX GmbH.

VERIFIED

INTERNATIONAL

AH-64D Longbow Apache HUMS

Created jointly by Smiths Industries Aerospace, Strategic Technology Systems, Inc, and the Boeing Company, the AH-64D Longbow Apache HUMS provides comprehensive health and usage monitoring. The diagnostics for avionics, rotors, transmission, engines and airframe are integrated with the helicopter's data management and display systems. It includes exceedance monitoring and cockpit voice and flight data recording.

The system comprises the Expanded Maintenance Data Recorder (EMDR); MultiPurpose Display (MPD); PCMCIA card receptacle and portable ground station.

System features include displays, warnings and pilot initiated functions fully integrated with the Longbow data management system covering 1,800 fault codes; data download via PCMCIA card or MIL-STD-1553B databus to portable PC ground station; ground station for HUMS data management and maintenance. EMDR features include data recording based on the Longbow Integrated Maintenance Support System (LIMSS); Stewart Hughes HUMS; 28 vibration inputs; eight tachometer inputs; blade tracker (day/night); voice recorder; 80 Mbyte crash-survivable memory; up to 160 Mbyte non-crash survivable memory; MIL-STD-1553B databus terminal; RS-422, RS-232, RS-485 serial communications.

The Strategic Technology Systems maintenance data recorder forms part of the AH-64 Longbow Apache HUMS ***1999**/0051324*

Contractors

Smiths Industries Aerospace.
Strategic Technology Systems, Inc.
The Boeing Company.

VERIFIED

Avionics computers for Eurofighter Typhoon

Air data transducer for Eurofighter Typhoon

The Eurofighter Typhoon air data transducer, currently being developed by an international consortium consisting of BAE Systems, Bavaria Avionik Technologie and Tecnobit, provides high integrity and fast dynamic response air data information.

The air data transducer features a combined multifunction pitot-static and flow angle mobile vane developed by Sextant. It provides local air data parameters, such as pitot, static and differential pressures, via a dedicated MIL-STD-1553B databus to the flight control computer. Calculations of airspeed, altitude, Mach number, angle of attack and angle of sideslip are generated to support the Eurofighter Typhoon's artificial aerodynamic stabilisation.

Specifications

Dimensions: 138 × 125 × 155 mm
Weight: 3.2 kg
Power supply: (heater) 115 V AC, 450 W
±20 V DC, 15 W

Contractors

BAE Systems, UK.
Bavaria Avionik Technologie GmbH, Germany.
Tecnobit SA, Spain.

Attack Computer/Navigation Computer (AC/NC) for Eurofighter Typhoon

The main functions of the Avionic Computer (AC) and Navigation Computer (NC) are to provide data for navigation, attack and identification. The two computers, which are identical in their hardware configuration, are based on a modular multiprocessor architecture. The central processing units are equipped with Eurofighter-selected standard microprocessors MC 68020 CPU and MC 68882 floating point co-processor. In order to meet the stringent performance requirements for floating point arithmetic, the central processing units are equipped with additional customised, floating point, accelerators, which can perform arithmetic operations extremely quickly. This special purpose co-processor is based on a high performance RISC technology.

Each computer is equipped with interfaces to two STANAG 3910 high-speed fibre-optic digital data buses.

The software provided is developed mainly in Ada, applying standardised software development tools and methods. It includes built-in-test software, an adapted Ada run-time system and a set of Ada packages to provide a defined interface between target specific input/output interfaces and the application software developed by Eurofighter GmbH. (Main contract from DASA).

Eurofighter Typhoon front computer 0001276

Contractors

TELDIX GmbH, Germany (prime).
Alenia, Italy (workshare partner).
Computing Devices Company Ltd, UK (workshare partner).

Cockpit Interface Unit (CIU) for Eurofighter Typhoon

The Cockpit Interface Unit (CIU) handles and controls all cockpit-relevant data. The extensive data throughput from a STANAG 3910 high-speed fibre optic digital databus to a STANAG 3838 electrical databus

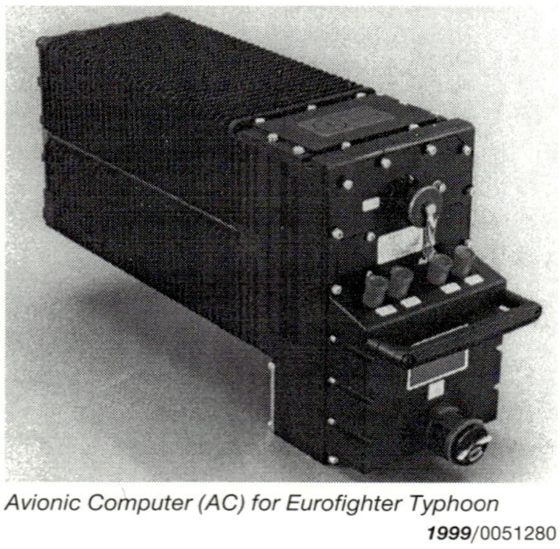

Avionic Computer (AC) for Eurofighter Typhoon ***1999**/0051280*

Cockpit Interface Unit (CIU) for Eurofighter Typhoon
1999/0051281

and vice versa is handled by a modular multiprocessor system.

The modular multiprocessor system of the CIU is based on the Eurofighter-selected MC 68020 CPU supported by MC 68882 floating point co-processor. This high data throughput is realised by a complex hardware structure and TELDIX developed ASIC's. Integrated built-in-test functions (BIT) are used.

The software provided is developed mainly in Ada, applying standardised software development tools and methods. It includes built-in-test software, an adapted Ada run-time system and a set of Ada packages to provide a defined interface between target specific input/output interfaces and the application software developed by Eurofighter GmbH. Each of two CIU's necessary per single seat aircraft has the capability to control the STANAG 3838 cockpit bus. (Main contract from BAE Systems).

Contractor

TELDIX GmbH, Germany.

Defensive Aids Computer (DAC) for Eurofighter Typhoon

The Defensive Aids Computer (DAC) performs defensive aids management functions and calculations to assist the pilot in achieving mission success.

The integrated modular microprocessor architecture system based on the Eurofighter-selected MC 68020 CPU handles the interfaces for activity request, suppression and blanking discretes and chaff/flare dispenser control.

Interface Processor Unit for Eurofighter Typhoon
1999/0051282

Information is transferred and received via one STANAG 3838 electrical databus and one STANAG 3910 high-speed fibre optic digital databus.

The software provided is developed mainly in Ada, applying standardised software development tools and methods. It includes built-in-test software, an adapted Ada run-time system and a set of Ada packages to provide a defined interface between target-specific input/output interfaces and the application software developed by Eurofighter GmbH. (Main contract from DASA).

Contractors

TELDIX GmbH, Germany (prime).
Alenia, Italy (workshare partner).

Front computer for Eurofighter Typhoon

ENOSA, leading a consortium with Alenia Difesa, Smiths Industries and VDO, has been awarded the development contract for the front computer for the Eurofighter Typhoon. The front computer forms part of the Utilities Control System (UCS) of the aircraft; it is of compact design and is fully integrated with other primary aircraft systems.

The front computer provides all the functions for control and monitoring of: the environmental and temperature control system; the life support system; the crew escape system.

Design features of the front computer are that it is fully compliant with Eurofighter Typhoon requirements and that it is databus MIL-STD-1553B compatible, allowing it to communicate with the rest of the UCS. It is based on a Motorola 68020 microprocessor.

Contractors

ENOSA, Spain.
Alenia Difesa, Avionic Systems and Equipment Division, GF-Sistemi Avionici, Italy.
Smiths Industries Aerospace, UK.
VDO, Germany.

Interface Processor Unit (IPU) for Eurofighter Typhoon

The Interface Processor Unit (IPU) is the data acquisition, management and processing unit of the integrated monitoring, test and recording system for the Eurofighter. It is based on a modular multiprocessor architecture and includes the Eurofighter-selected MC 68020 CPU standard microprocessors and MC 68882 floating point co-processor.

The IPU is equipped with interfaces to two STANAG 3910 high-speed fibre optic digital databusses and one STANAG 3838 electrical databus. Intelligent controllers are provided for the interfaces, which perform time critical functions for data transfer on the high-speed bus, the addressing and formatting of data and error checking.

In addition, the IPU also contains both electrical and fibre optic links to various aircraft subsystems. A digital signal processor handles sensor data and compresses digitised audio data, using special compression algorithms. (Main contract from BAE Systems).

Contractors

TELDIX GmbH, Germany (prime).
Alenia, Italy (workshare partner).

VERIFIED

Harpoon Airborne Command, Launch and Control System II

BAE Systems and The Boeing Company have developed a range of command, launch and control systems compatible with all standards of Harpoon and SLAM missiles.

The Harpoon Airborne Command, Launch and Control System II (HACLCS II) adds new capability at greatly reduced cost and increases the potential for Harpoon carriage on smaller aircraft, such as the IPTN-235 and Fokker 50.

The BAE Systems HACLCS II computer, as the main subsystem of HACLCS II-2, provides all the control and power signals necessary to interface with the aircraft systems, including a MIL-STD-1553B databus, and to control four or six missile interface configurations on the P-3 aircraft and on fast-jet applications.

Specifications

Dimensions: 191 × 194 × 309 mm
Weight: 9.5 kg
Power supply: 115 V AC, 2 A (max)
28 V DC, 1 A (max)

Operational status

In preproduction.

Contractors

BAE Systems.
The Boeing Company.

VERIFIED

Helicopter Flight Data Recording/Health and Usage Monitoring System (FDR/HUMS)

The HUMS part of the FDR/HUMS system combines vibration and usage monitoring of critical dynamic power-train components with techniques such as chip detection, rotor track and balance, engine power assurance, cycle counting, exceedance monitoring and oil analysis. Integrated with the FDR, the FDR/HUMS supports helicopter operations, safety and maintenance - both civil and military.

The system comprises a data retrieval unit, sensors and data sources (to customer requirements), crash survivable FDR, and a ground station for fleet data storage.

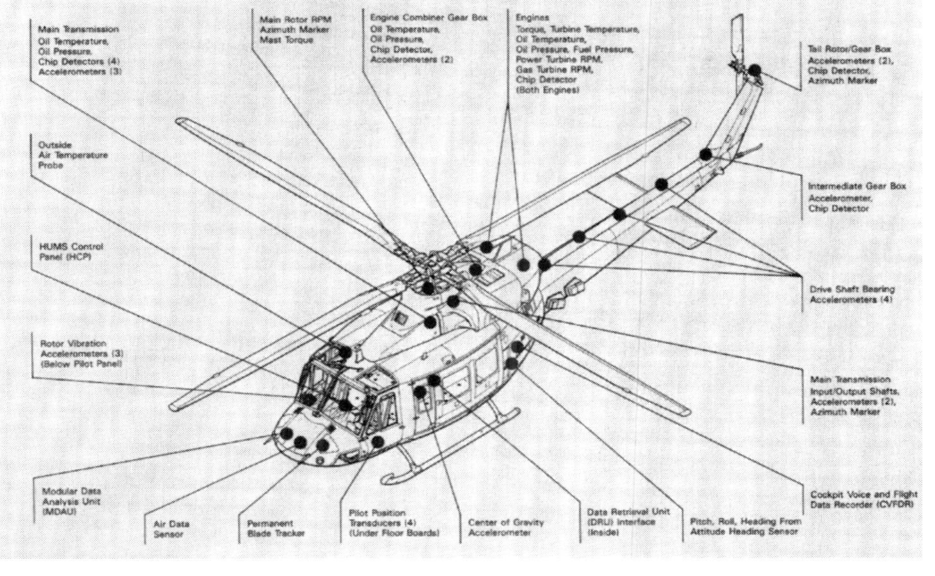

Bell 412 helicopter installation FDR/HUMS
0015311

Operational status

FDR/HUMS in the forms of North Sea HUMS, EuroHUMS™ and AHUMS™ are in service in North America, Europe, Australia and Southeast Asia. The HUMS element of the system is being supplied by Smiths Industries Aerospace. FDR/HUMS installations have been produced for Bell 412/212 and Sikorsky S-76 helicopters. Applications for these systems include North Sea HUMS: S61N, AS332 Mk 1, BV234 and S76; EuroHUMS: AS332 Mk 1, AS332 Mk 2, AS532 Mk 1 and AS532 Mk 2; AHUMS: Bell 412, CH47D and S-76.

Contractors

Teledyne Controls.
Smiths Industries Aerospace.

VERIFIED

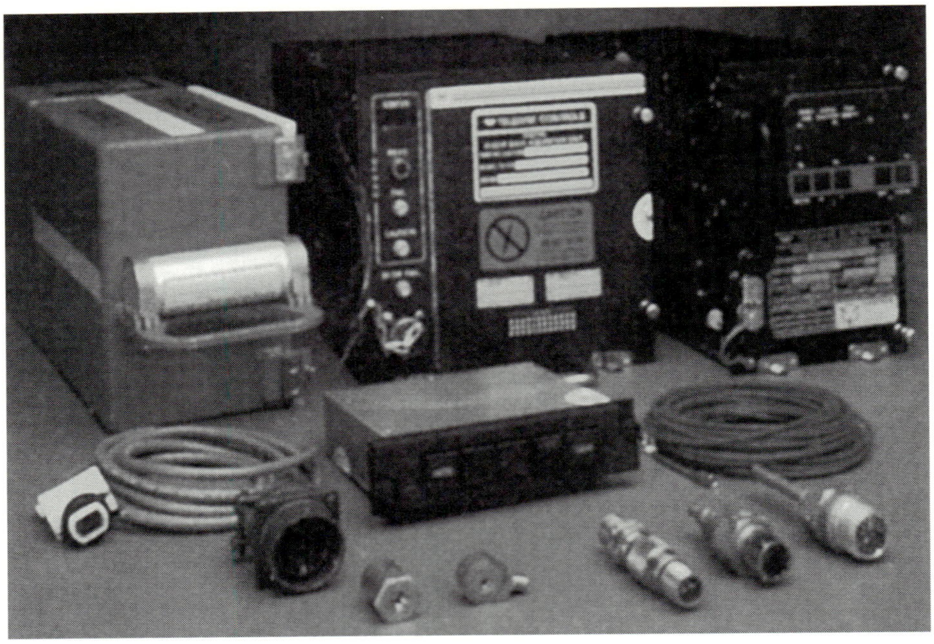

Helicopter FDR/HUMS equipment
0015310

Interface unit for the Tiger helicopter

The interface unit is a data acquisition, management and preprocessing computer for the basic avionics of the French/German Tiger helicopter. It is based on a modular multiprocessor architecture and includes the MC 68020 processor. It is equipped with input/output modules for data transfers via interfaces of different types, such as MIL-STD-1553B, ARINC 429, analogue, frequency measurement and discrete.

The equipment and the application software is developed in Ada. It includes BIT capabilities, an adapted Ada run-time kernel and flexible data transmission, handling, formatting and processing so that interface parameters are configurable via a parameter table which allows changes in acquisition rate or selection of preprocessing without any modification of the software itself.

Operational status

Qualified.

Contractors

TELDIX GmbH.
Sextant (co-op partner).
VDO Luftfahrtgeräte Werk (co-op partner).

VERIFIED

Maintenance Data Panel (MDP) for Eurofighter Typhoon

INDRA EWS has been awarded a contract for the development, production investment and series production of the MDP and Portable Maintenance Data Store (PMDS) for the Eurofighter Typhoon. The MDP and PMDS apply the most advanced technology for monitoring and recording information from the aircraft for maintenance operations.

The MDP and PMDS display the following types of data: refuelling/defuelling; weapon stores data loading; mission data loading; failure data (actual and last flight status); limit exceedances data; status of consumables; life usage data; and aircraft tail number.

Limit exceedances and fatigue data generated during the last five flights, together with data from the engine and other systems connected to the MDP (via databus) are available and recorded through the PMDS.

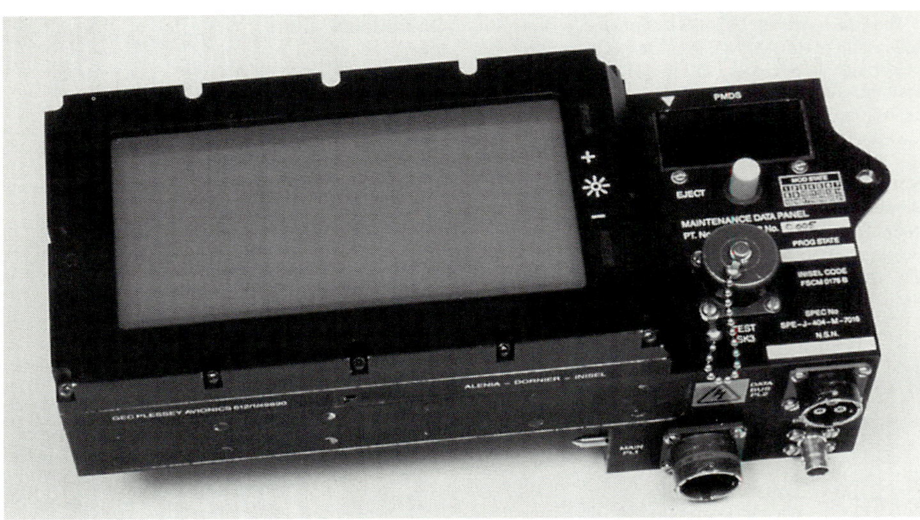

Operational status

Production investment and series production orders have been placed.

Contractors

INDRA EWS.
EADS (formerly DaimlerChrysler Aerospace AG).

UPDATED

Maintenance data panel for Eurofighter Typhoon
0001277

Pilot Maintainer Assist System PMAS™

The PMAS is being developed jointly by Avionics Specialties, Inc and Smiths Industries Aerospace.

PMAS is an evolutionary development of the Health & Usage Monitoring - Flight Data Recorder (HUM-FDR) concept. The PMAS is a computer-based maintenance management tool for data acquisition within the helicopter. It comprises a core system that is able to record and trend engine and vibration data in most types of single and light twin aircraft. The modular concept permits the operator to add functions to the core system such as rotor track and balance, gearbox and FDR similar to that of HUMS. PMAS extends the concept further by accepting maintainer assistance modules that enable the pilot to fly more safely; these include: ground proximity warning, laser tip strike and wire avoidance, carefree handling and spatial disorientation modules.

Contractors

Avionics Specialties, Inc.
Smiths Industries Aerospace.

VERIFIED

ISRAEL

ACE-5 computer

The ACE-5 is a modular high-performance airborne general purpose computer which is fully compatible with MIL-STD-1750A and MIL-STD-1553A and B. It is designed for multirole fighter mission management.

Computation performance is 1.7 Mips, with growth potential to 2.5 Mips. Some 256 kwords of memory with battery back-up are programmable at flight line level, with growth capacity to 1 Mwords. BITE provides continuous monitoring of functional integrity.

Specifications

Dimensions: 137 × 198 × 472 mm
Weight: 10 kg
Power supply: 115 V AC, 400 Hz, 3 phase
28 V DC
Environmental: MIL-E-5400 Class II
Reliability: >500 h MTBF

Operational status

In service on Israeli F-16C/D aircraft.

Contractor

Elbit Systems Ltd.

VERIFIED

Commercial Data Entry Electronics Unit (CDEEU)

In August 1998, Elbit Systems Ltd, and its wholly-owned US subsidiary EFW Inc, was contracted by Lockheed Martin Tactical Aircraft Systems (Fort Worth) to develop and deliver 25 prototypes of an enhanced data entry unit, based on advanced commercial technologies, designated the Commercial Data Entry Electronics Unit (CDEEU). The development phase is to be performed over a period of two years. Following development, Lockheed Martin plans to initiate a production order for 750 units. The CDEEU is part of the US Air Force's Common Configuration Implementation Programme (CCIP) for the F-16C/D, under which Lockheed Martin will provide significant upgrades for the Block 40 and Block 50 aircraft.

Operational status

In development.

Contractor

Elbit Systems Ltd, EFW Inc.

VERIFIED

Modular MultiRole Computer (MMRC)

The MMRC performs tasks for a full avionic suite. The MMRC is a single-multimodule computer designed to handle a broad spectrum of tasks previously performed by several individual units.

It is easily upgradeable to accept new systems and capabilities, and has built-in growth potential in processing capability, memory and interfaces.

Each module of the MMRC performs a different function: fire control; stores management; display processor (MFDs, MFCDs, HUD, HMD); display and sight helmet; integrated communication radio navigation and identification system; digital image processing and communication system; and video and VTR controller.

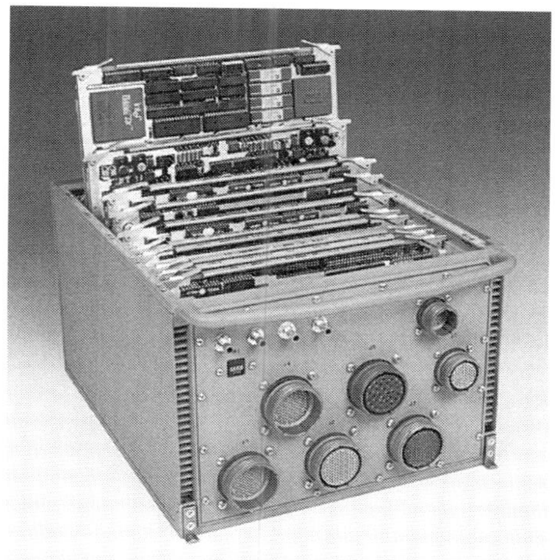

The Modular MultiRole Computer (MMRC)

The MMRC is also the central element of the HALO advanced helicopter avionics suite offered as a flexible system upgrade for a wide range of helicopters.

Operational status

In production. Selected by Embraer for fitment to AMX-T advanced trainer aircraft of the Venezuelan Air Force.

Contractor

Elbit Systems Ltd.

VERIFIED

Autonomous Combat manoeuvres Evaluation (ACE) system

The Autonomous Combat manoeuvres Evaluation (ACE) system is a flight training debriefing system aimed mainly at multiparticipant flight training. The development of ACE is the direct outcome of an operational requirement for a squadron-level replacement for the current Air Combat Manoeuvring Instrumentation (ACMI) ranges.

ACE, unlike ACMI, is not ground range dependent and this enables debriefing of any kind of flight executed anywhere, without the need for carriage of external pods or telemetry. It is installed internally on each aircraft. Installation is simple and no software or cockpit changes are required for aircraft equipped with a MIL-STD-1553 databus. The ground debriefing station is based on commercial hardware and is designed to be used at squadron level on a daily basis.

Aircraft flight data required to reconstruct manoeuvres and the operation of the avionics system is recorded independently on board each aircraft by the ACE flight monitor unit on the aircraft's VTR. The use of a C/A or P code GPS as an integral part of the ACE airborne system provides position accuracy and synchronisation, without being limited to a specific antenna range and telemetry. After landing, the cassettes of all the participants, containing the digital data and video and audio playbacks, are processed on a ground debriefing station from which a unified display file is produced and displayed graphically. Synchronised HUD video playback and audio complement and complete the debriefing material.

Rada Autonomous Combat manoeuvres Evaluation system onboard unit 0001295

Specifications

Dimensions: 70 × 165 × 320 mm
Weight: 4 kg
Power supply: 28 V DC

Operational status

In service on Israeli Air Force F-16 aircraft with an option for F-15 and F-4 2000 aircraft, to become the standard fit on all Israeli Air Force fighters.

In service on Chilean F-5E aircraft.

Contractor

Rada Electronic Industries Ltd.

VERIFIED

Data Transfer Equipment (DTE)

Rada has developed a family of data transfer systems of different memory types, sizes and interfaces. These systems are used in advanced aircraft (fixed- and rotary-wing) for data exchange between the squadron mission planning stations and the aircraft's onboard computers and systems. The system comprises a portable solid state memory cartridge on which the mission parameters are loaded in the squadron and an interface unit through which the cartridge is downloaded into the aircraft computers for system initialisation and set up.

During the flight there is a flow of information (maintenance, intelligence and so on) to and from the aircraft systems. After the mission the recorded data is downloaded into a ground station.

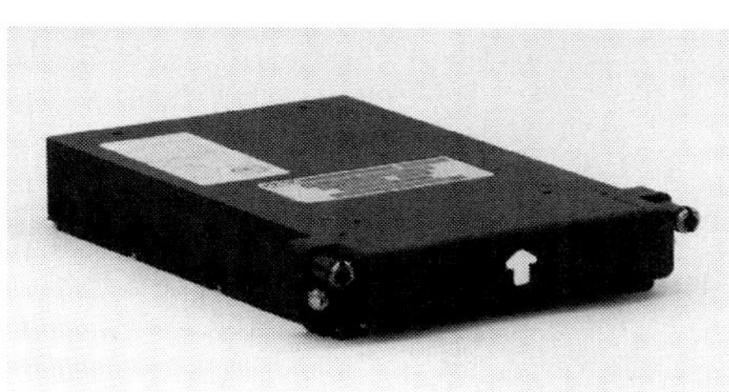

Rada Data Recorder Equipment (DRE) 0001279

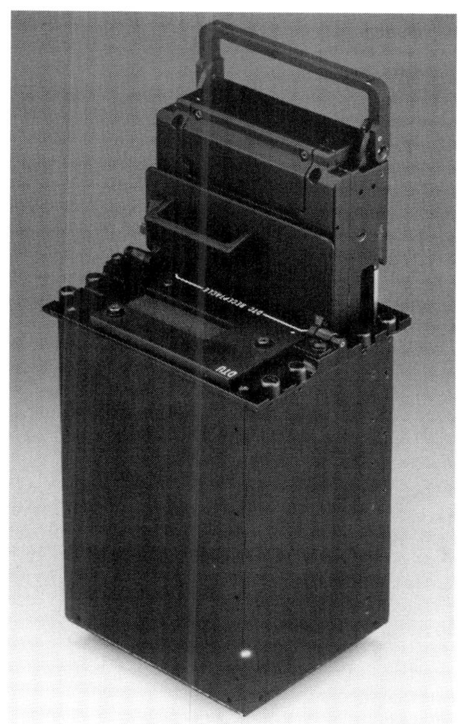

Data Transfer Equipment (DTE) 0001278

Rada's DTE are available in the following configurations:

Specifications

IDTE
Dimensions:
(DTU) 177.8 × 146 × 114.5 mm
(Cartridge) 190.5 × 120 × 41.3 mm
Weight:
(DTU) 2.5 kg
(Cartridge) 0.78-1.0 kg

Power supply: 115 V AC, 400 Hz, single phase, 17 W
Memory type: RAM/EEPROM/Flash
Memory size: 0.5 -80 Mbytes
Interface: MIL-STD-1553A/B or any tailored interface

DRE
Dimensions:
(DRU) 230 × 145 × 32 mm
(Cartridge) 97 × 110 × 20 mm
Weight: (DRU + Cartridge) 1.6 kg
Power supply: 28 V DC
Memory type: Flash
Memory size: 4 -32 Mbytes
Interface: RS-232/422

Operational status

Rada's DTE are in service on F-16A/B/C/D, F-5E, F-15 and Mirage types of aircraft.

Contractor

Rada Electronic Industries Ltd.

VERIFIED

Fatigue and Air Combat Evaluation (FACE) system

The FACE system is an airborne data acquisition system, designed to perform the two major tasks of fatigue data processing and recording and Air Combat Manoeuvring Instrumentation (ACMI). In addition, the open architecture and large storage capacity of FACE enable it to be used as an onboard integral flight test instrumentation kit.

The fatigue recorder consists of airframe fatigue recording and engine fatigue recording. In airframe fatigue recording, FACE monitors and records a set of strain gauges and multiplexer bus flight data parameters that are stored on a solid-state memory cassette. The data is downloaded into a ground station in which the aircraft fatigue index is calculated. In engine fatigue recording, FACE monitors engine parameters via the F-100 Digital Electronic Engine Controller (DEEC) and counts and records engine cycles from which engine fatigue data is extracted. Monitoring the DEEC enables the FMS to detect and record engine parameters during engine fault occurrences. This optional feature already exists in the DEEC. However, centralising this information in the FACE cassette may be an advantage.

In Autonomous Combat Evaluation (ACE), FACE monitors selected multiplexer bus and DEEC parameters and records them on a video cassette recorder. The cassettes are replayed on the ACE ground station, providing an ACMI debriefing for multi-aircraft formations.

The FACE open architecture makes it suitable for use as an onboard instrumentation kit. Programmable selected flight test data may be recorded on the system's solid-state memory cassette and on the VTR cassette, which makes it an efficient and flexible tool for flight test programmes.

Monitoring the multiplexer bus enables the FMS to record maintenance fault lists when they appear on the bus.

The FMS consists of a Flight Monitoring Unit (FMU), Data Recording Cartridge (DRC) and Video Tape Recorder (VTR). Ground equipment for debriefing purposes is also associated with the FMS.

The FMU performs processing of the aircraft video signal, originally connected to the VCR, by integrating it with the ACE and flight test data; transfer of the combined video signal to the VCR, retaining its original visual data in addition to new information; monitoring of the aircraft strains by continuous measurement of the strain gauge sensors; acquisition of engine data through a serial measurement channel connected to the DEEC; acquisition of multiplexer bus data through its interface, and monitoring of the fuel quantity by continuous measurement of the applicable signals.

The DRC stores all the data sent from the FMU. It is based on non-volatile solid-state memory modules. The DRC is used for the transfer of all the data acquired during flights to the ground station for further processing and analysis.

The VTR cassette is used to store selected flight data and engine parameters for post-mission debriefing or flight test analysis.

Contractor

Rada Electronic Industries Ltd.

VERIFIED

Helicopter Power Plant Recording and Monitoring system (PPRM)

The Power Plant Recording and Monitoring system (PPRM) is an add-on system for recording and monitoring critical helicopter power plant parameters during flight. It enables analysis of power plant performance and flight event data and provides for direct diagnostics and fault detection capability. The PPRM improves mission availability and reliability and saves time and cost in flight-line and depot maintenance.

System components are Engine History Recorder (EHR) - this airframe-mounted unit automatically records routine maintenance data and out of range events of critical parameters of aircraft power plants such as the twin/triple engine power plant of the CH-53, the twin engine power plant of the Bell 212 and the single engine power plant of the Bell Cobra AH-1; Performance Display Panel (PDP) - this cockpit-mounted panel unit continuously displays measured and calculated power plant performance parameters to the aircrew; Test and Data Extractor Unit (TDEU) - this ground crew support equipment is a hand-held computer which interfaces with the EHR and enables rapid power plant data extraction for flight line and depot-level maintenance activities.

PPRM display panel 0001408

Main features are identification and recording of specific power plant subsystem failures and fatigue cycles accumulation for 'on condition' maintenance; automatic or flight crew initiated data recording; immediate detection of power plant failures or performance degradation; continuous calculation and display of flight-critical information to the pilot including engine power and maximum hovering weight; real-time pilot warning and recording of over-limit conditions such as engine speed, temperature and torque; calculation and accumulation of Low Cycle Fatigue (LCF) and blade life.

Specifications

EHR
Dimensions: 150 × 150 × 250 mm
Weight: 3.2 kg
Supply Voltage: 18-33 V DC IAW MIL-STD-704B
Power: 10 W
Qualification:
(environmental) MIL-STD-810C
(electromagnetic) MIL-STD-461B
Communication: RS-232

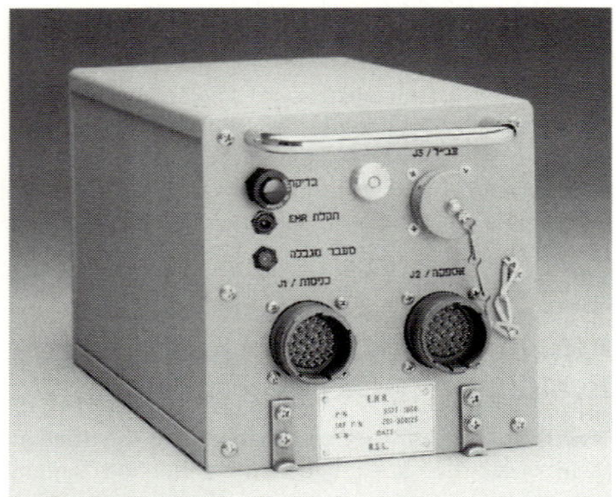

PPRM EHR unit 0001407

PDP
Display: 2 row, 16 characters NVG-compatible
Dimensions: 127 × 76 × 98 mm
Weight: 0.8 kg
Power: 6 W
Qualification:
(environmental) MIL-STD-810C
(electromagnetic) MIL-STD-461B

Operational status

In service with the Israeli Air Force.

Contractor

RSL Electronics Ltd.

VERIFIED

ITALY

Crash/Maintenance Recorder System (CMRS) for the Tornado

Alenia Difesa has developed the Crash/Maintenance Recorder System (CMRS) selected for the Tornado. The Tornado CMRS includes onboard and ground-based equipment for the collection, management and recording of flight data and subsequent computation and analysis.

The Tornado CMRS comprises an Extended Data Acquisition Unit (EDAU), Maintenance Recorder Unit (MRU), Accident Data Recorder (ADR), Mobile Quick-Look Facility (MQLF) and automatic ground station and semi-automatic test equipment.

The EDAU is based on an Alenia Difesa advanced MARA avionics computer and is used for the collection, conversion and compression of a wide range of flight signals.

The MRU records on an easily removable cassette all the EDAU processed data.

The ADR, supplied by BAE Systems, records on a crash-protected endless loop tape all the critical data for analysis in the event of an accident.

The MQLF is employed as first line ground support to perform fast analysis of the flight recorded data and to assess the aircraft serviceability status.

The CMRS has crash recording time of up to 2 hours and a maintenance recording time of 3 to 9 hours depending on the mission profile. The system records up to 125 maintenance parameters, of which up to 37 are for structural fatigue analysis and life computation, up to 32 are for engine low-cycle fatigue and up to 20 are for aircraft system monitoring. The CMRS has VDU, printer and plotter output devices and presents information in tabular or plotted form.

Operational status

In service on the Italian Air Force Tornado.

Contractor

Alenia Difesa, Avionic Systems and Equipment Division.

VERIFIED

Airborne Strain Counter (ASC)

The Airborne Strain Counter (ASC) unit, is designed to determine the different load spectra and/or local stresses that arise during operation in all aircraft structures submitted to variable loads. The unit is suitable for monitoring the fatigue life of complex structural elements, where the local stresses cannot be easily correlated to the standard usage parameters.

The ASC automatically accomplishes acquisition of a maximum of eight parameters, received from strain gauges or accelerometers positioned on the most significant points of the aircraft structure; conversion of the acquired data into digital form; processing of the data and outputting of a stress matrix; storage of the data in a non-volatile memory; interfacing with the dedicated ground support unit to allow data transfer and subsequent processing; supply of power to the sensors; and performance of self-test to check serviceability of the equipment.

The ASC features eight fully independent channels for acquisition, preprocessing and storage of data. Each channel is capable of managing the interface signals of the associated strain gauge or accelerometer. Each channel amplifies the analogue signal received from the sensor and converts it into a digital signal through an A/D converter, providing information on number of stress cycles experienced by the structure, mean values of the stress for every cycle and peak-to-peak value of the stress for every cycle.

Information is stored in a non-volatile memory on aircraft serial number, card serial number, ASC serial number, total number of stress cycles, status word, stress matrix, total number of minutes of aircraft motion and total number of minutes of operation in the presence of failures.

An RS-422 databus is used to transfer the data stored from the measurement channels to the ground support unit.

At system power on, each channel performs a self-test to check hardware and software integrity.

Operational status

In service on the Aermacchi MB-339 and the Aermacchi/Alenia/Embraer AMX.

Contractor

Logic SpA.

VERIFIED

ANV-801 computer display unit

The ANV-801 computer display unit replaces multiple control layouts with a single, easy to read unit, allowing the concentration of flight management information and data presentation. Versatile design, modular hardware and software architecture make it easy to reconfigure the ANV-801 to meet customer requirements.

A 16-bit Intel 80C186 Series CPU with 1 Mbyte memory or a 32-bit Mips R3000 Series RISC CPU with up to 8 Mbytes memory provide the ANV-801 with the computational power for solving complex tasks in real time.

The ANV-801 integrates external navigation sensors into sophisticated navigation packages providing a high-accuracy computation capability, including navigation information for the crew and steering output to flight director and autopilot. It is also available with a built-in GPS receiver sensor, providing self-contained navigation capabilities and a built-in modem for input/output datalink. The ANV-801 interfaces with a wide range of analogue and digital equipments using ARINC 429, MIL-STD-1553B, serial and modem digital interfaces and analogue interfaces. These include communications, landing and IFF equipments.

Specifications

Dimensions: 185 × 146 × 181 mm
Weight: <4.5 kg
Power supply: 28 V DC or 115 V AC, 400 Hz

Contractor

Marconi SpA.

VERIFIED

ANV-803 computer and interface unit

The ANV-803 was designed to handle real-time environments in modern avionic systems where powerful computation capability, small size and low weight are required. The powerful RISC engine and modular design of the ANV-803 meet a wide range of applications while keeping a simple monoprocessor architecture, resulting in a good price/performance ratio and maintenance cost.

System options include a complete range of input/output support modules and a comprehensive set of software development tools for Ada programming. Different interface cards are available, including MIL-STD-1553B BC/RT/BM, ARINC 429, discrete, serial and analogue.

Specifications

Dimensions: ⅜ ATR short

The Marconi ANV-803 computer and interface unit

Weight: <7.5 kg
Power supply: 28 V DC or 115 V AC, 400 Hz

Contractor

Marconi SpA.

VERIFIED

JAPAN

Altitude computer

The altitude computer receives pitot and static pressure and total temperature inputs for the calculation of aircraft altitude. It has a microprocessor and performs digital calculations. The altitude computer includes BIT and performance monitoring.

Operational status

In production for the T-2 trainer.

Contractor

Shimadzu Corporation.

VERIFIED

The altitude computer is in production for the T-2 trainer

Crash Protected Video Recorder (CPVR)

The Crash Protected Video Recorder (CPVR) is designed to provide a record of the glass cockpit data presented to the flight crew. This is a natural follow-on from the standard flight data recorder. Knowledge gained over many years in the design and manufacture of crash recorders has been utilised in the design of a prototype CPVR.

The Hi-8 mm video standard has been chosen as it is lightweight, reliable and has a bandwidth approaching 4 MHz. The CPVR provides recording times of 4 hours for NTSC and 3 hours for PAL. These two video standards, plus RS-170, provide 400 TV lines resolution. The system also has a single audio channel with a frequency response of 100 Hz to 10 kHz.

The CPVR will form part of a system utilising closed-circuit television cameras located in suitable positions both within and outside the aircraft. Camera positions will be chosen to provide the aircrew with real-time viewing of internal and external features on a suitable display. This follows recommendations from the UK Aircraft Accident Investigation Branch. The number of cameras will depend upon specific requirements, with the chosen outputs being recorded on the CPVR to enable in-depth analysis to be carried out.

Specifications

Dimensions: ¾ ATR
Power supply: 28 V DC, 15 W
Environmental: MIL-STD-810D, MIL-STD-461B, MIL-STD-462

Contractor

TEAC Corporation.

VERIFIED

V-250 series videotape recorders

The V-250 series of airborne videotape recorders are high-performance units designed for colour or monochrome recording on ¼ in video cassettes. It is ruggedised to withstand the effects of shock, vibration, high *g* loading, altitude, explosive decompression and wide temperature variations.

The V-250 AB-R has been designed as a record-only model for minimum size and weight. The V-250 AB-F both records and allows playback for in-flight and post-mission analysis. Both are available in 525- or 875-line EIA/NTSC or 625-line CCIR/PAL video formats and provide 60 minutes' (EIA/NTSC) or 51 minutes' (CCIR/PAL) recording.

Specifications

Dimensions:
(V-250 AB-R) 107 × 139 × 156.5 mm
(V-250 AB-F) 192 × 107 × 156.5 mm
Weight:
(V-250 AB-R) <3 kg
(V-250 AB-F) <4 kg
Power supply: 20-32 V DC unregulated

Contractor

TEAC Corporation.

VERIFIED

V-1000 AB-F videotape recorder

The V-1000 AB-F videotape recorder and reproducer uses high-capacity ¾ in U-matic 'S' video cassettes and is designed to record FLIR and IR linescan displays and for other high-resolution recording applications. The rugged construction provides protection against shock, vibration and low pressure.

The system is available in both 525- or 875-line EIA and 625-line CCIR video formats, providing recording times of up to 72 minutes in 525-line EIA, 60 minutes in 875-line EIA and 69 minutes in CCIR.

Specifications

Dimensions: 271.8 × 166 × 367.5 mm
Weight: <12.5 kg
Power supply: 115 V AC, 400 Hz, single phase

Operational status

Being replaced on US Navy F-14 aircraft by the TEAC America Fast Tactical Imagery (FTI) system.

Contractor

TEAC Corporation.

VERIFIED

Compact airdata transducer

Tokyo Aircraft Instrument Co Ltd has started to produce a compact airdata transducer which can be used on board as a compact airdata computer or as an airdata measurement device. The transducer has a micro-vibration pressure sensor and a semiconductor pressure sensor both of which make it possible to provide accurate airdata information at DC voltage. The transducer can be expanded to provide the interface with ARINC 429, ARINC 629, MIL-STD-1553B and other serial or parallel buses as may be required by customers.

Operational status

In production for commercial and military aircraft.

Contractor

Tokyo Aircraft Instrument Co Ltd.

VERIFIED

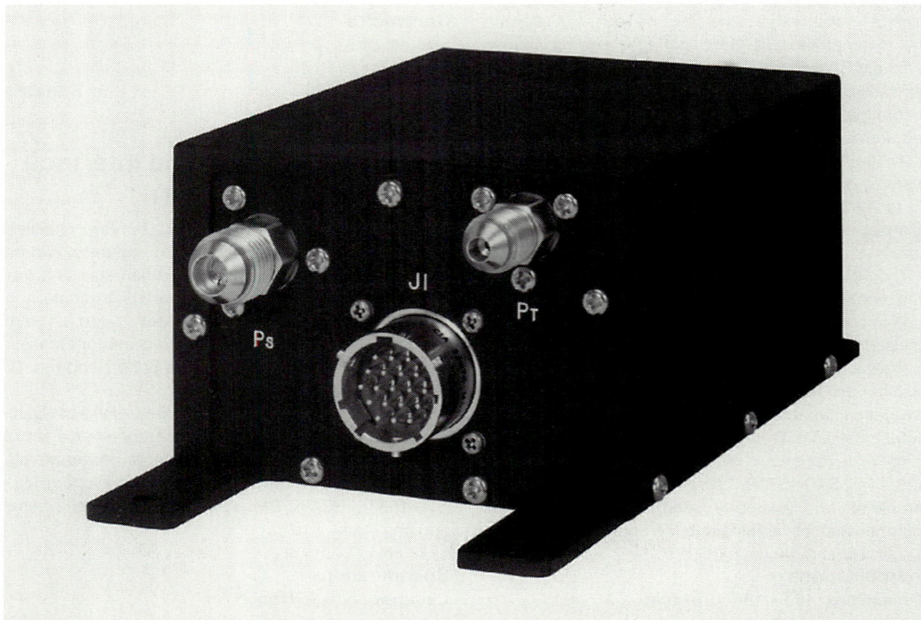

Compact airdata transducer
0001283

Truncated pyramid-shape multihole pitot probe system

This system is composed of truncated pyramid-shape multihole pitot probe tube and processing control computer (air data computer), both of which have been developed and patented by Tokyo Aircraft Instrument Co Ltd in conjunction with the National Aerospace Laboratory of Science and Technology Agency, Japan (NAL). The system measures alpha-angle, beta-angle, airspeed and altitude precisely and quickly with airflows from truncated pyramid-shape multihole pitot probe tube and provides air data information to other avionic systems via serial databusses. The system contributes to flight performance and the enhancement of STOL capability.

Operational status

In production and service in the Dornier 228 of NAL, the aerospace experimental vehicle 'ALFLEX' of NAL/NASDA.

Contractor

Tokyo Aircraft Instrument Co Ltd.

VERIFIED

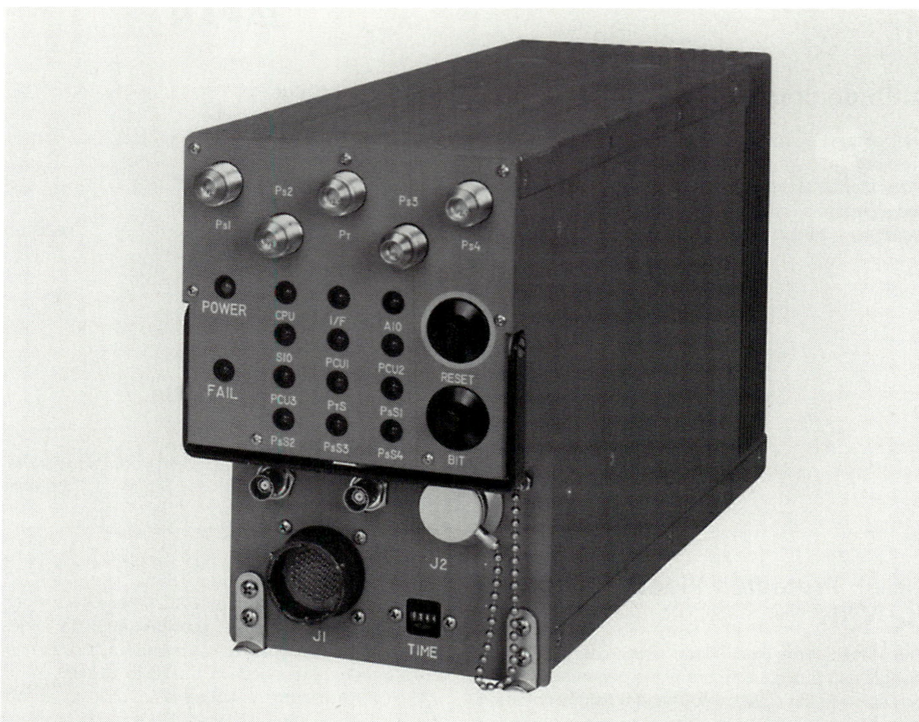

Processing control computer
0001284

NETHERLANDS

OTA series cockpit cameras

The OTA series is a range of black and white, or colour, cameras designed for the HUDs of helicopters and fixed-wing aircraft as part of a mission recording system. Used with an airborne magnetic-tape recorder, they superimpose and record HUD symbology on real-time video pictures together with flight data. The OTA cameras enable in-flight permanent recording of all the visual information received by the pilot from take-off to landing. The high-sensitivity cameras feature an automatic exposure control to provide fast adjustment to sudden changes in brightness levels. An in-flight replay capability is provided together with display options for the co-pilot.

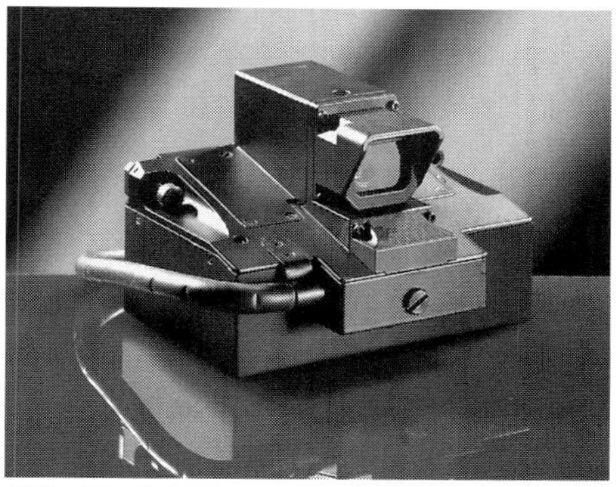

OTA 222 B/W HUD camera 0001296

Current cameras in the OTA range include the OTA-222 black/white HUD camera; and the OTA-1320 red/green/blue colour HUD camera, selected for Rafale.

Specifications

Dimensions: 75 × 65 × 95 mm
Weight: 0.65 kg
Field of view: 30° horizontal × 22.5° vertical
Line of sight: <1 mrad
CCD sensor: 756 horizontal × 575 vertical pixels
Depth of focus: 3 m to infinity
Spectral range: 0.4-0.7 μm
Bandwidth: 5.5 MHz
Dynamic range: 3-100,000 lux
Video signal: CCIR/STANAG 3350B EIA/RS-70
Synchronisation: int/ext input (RS-422)
Power: 12-60 V DC, 4 W
Temperature range: –15 to +55°C
Reliability: (operating flight hours) 5,000 h MTBF

OTA-1320 RGB HUD camera selected for Rafale 0015312

Operational status

OTA cameras are in production for the Mirage 2000-5 and Rafale fighter aircraft.

Contractor

Signaal USFA.

VERIFIED

NORWAY

EE 235 solid-state recorder

The EIDEL Eidsvoll Electronics AS EE 235 solid-state recorder is designed for use by the aerospace, military and industrial markets, and is qualified for both aircraft and missile use. Its prime purpose is telemetry data storage and features include 256 to 8,192 Mbyte data storage in non-volatile memory; serial data recording at up to 15 Mbits/s; an input data buffer for burst data; extended storage by interconnecting any number of recorders; configurable PCM encoder module for analogue and digital data collection; direct, or control box operation. The system can be configured to user requirements by adding memory and encoder modules as required.

The standard box (M2, M3, M4) is used for data storage of 32 to 1,792 Mbytes. Special systems have capabilities up to 8,192 Mbytes.

Supporting modules include: the EE 240 PCM encoder module, and the sensor signal conditioning modules.

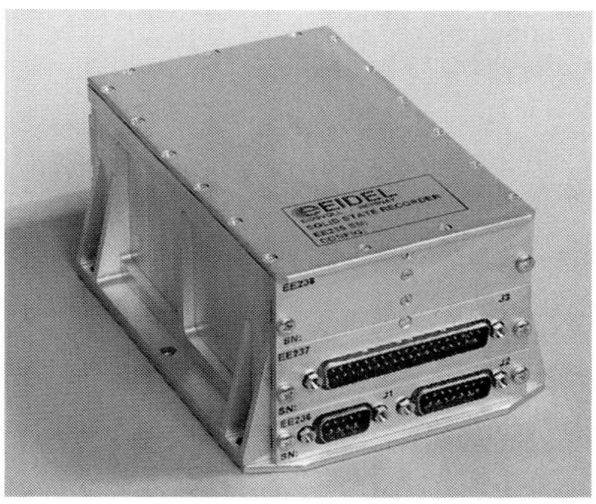

EIDEL EE 235-M3 solid-state recorder, showing EE 236 power module, EE 237 controller module, and EE 238 memory module
1999/0051325

Specifications

Dimensions:
(EE 235-M2) 105 × 143 × 49 mm
(EE 235-M3) 105 × 143 × 74 mm
(EE-235-M4) 105 × 143 × 99 mm
Weight:
(EE 235-M2) 0.8 kg
(EE 235-M3) 1.2 kg
(EE-235-M4) 1.6 kg
Input bit rate: 0-15 MHz
Playback rate: 375 kHz; 1,500 kHz, 3 MHz or external
Input: NRZ-L with CP. TTL and RS-422 level
Output: NRZ-L, CP, BiØ-L. TTL and RS-422 level

Operational status

Widely used in the aerospace, military and industrial markets.

Contractor

EIDEL Eidsvoll Electronics AS.

VERIFIED

POLAND

Solid-State Quick Access Recorder (SSQAR) family

The ATM company has designed the Solid-State Quick Access Recorder (SSQAR) with contactless data transmission between recorder and cartridge. This solution with a RAM-based removable cartridge has been developed to give the user high-quality data recording and extremely fast data replay with no pins to damage, nor tape to be mangled. The ATM-SSQAR family is also easy and flexible to operate. Its flexibility allows a wide range of different applications with most major recording systems installed on board civil and military aircraft as well as helicopters and gliders.

Programmable and able to make online analyses, the ATM-SSQAR family can display and export warning signals indicating previously programmed parameter exceedences.

To be able to meet all customer requirements ATM has designed a family of recorders consisting of: ATM-QR3 (modular, most sophisticated type); and ATM-QR4 (modular, advanced type).

Versatility is the main feature of the ATM-SSQAR family. In a single unit it combines the functions of quick access recorder with built-in real-time clock; data acquisition unit; and data management unit (programmable reports depending on installed options).

Recording time depends on the cartridge size and data format. The table below shows examples of recording time for different size ATM cartridges. Values in the table are estimated for the most popular recording standards.

Data replay from the cartridge is performed by the ATM-RD3 reader, and complete operation using a PC-compatible computer, takes up to 3 minutes, for 30 flight hours (for 64 words/s transmission speed).

The basic components include built-in real-time clock; display for recorder status and online

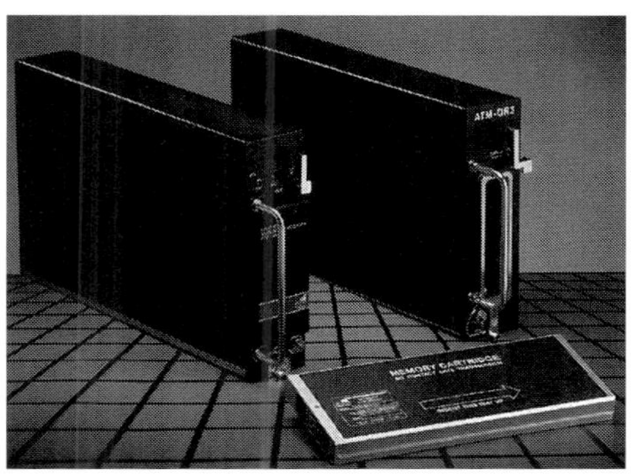

ATM's solid-state quick access recorder 0015314

Examples of recording time (hours)

Type of cartridge	**Recording system**			
	ARINC 717			**MSRP-64**
	64 words/s	128 words/s	256 words/s	
ATM-MC5/30	34	17	8	60
ATM-MC5/70	80	40	20	140
ATM-MC5/70	120	60	30	220

programmable flight data analysis (ACMS); data output for external equipment; interface for entry panel; test connector for data monitoring; and data dump.

The optional components: Depending on SSQAR type, the following optional modules are available:

ATM-QR3

Optional internal full-size modules: QR3DC (interface operating two ARINC 429 busses); QR3FT (programmable vibration spectrum analyser); QR3PE (module operating as a DAU).

Optional small size module: QR3AR (ARINC 573/717 bus interface).

ATM-QR4

Optional internal full-size modules: QR4DC (interface operating two ARINC 429 busses); QR4PE (module operating as a DAU).

Optional small size module: QR4AR (ARINC 573/717 bus interface).

Optional external modules

ATM-RT allows connection to any recorder of the ATM-SSQAR family to MSRP-64 and 256 data recording systems. ATM-DP analogue multiplexer provides the ability to process additional input signals (not available for QR2).

Specifications

Dimensions: 317 × 57 × 193 mm

Power supply: 115 V AC, 40 Hz, 25 W or 27 V DC, 20 W

Operational status

In service on the Boeing 737 and 767, Airbus A310, Antonov An-28PT, ATR 72, Ilyushin Il-62M and -76 and Tupolev Tu-134/154M/204 and Swift S-1 glider. ATM won a US$250,000 contract to fit SSQARs to all LOT Polish Airlines Boeing 737 and 767 and ATR 72 aircraft. Also fitted to military aircraft: I-22; PZL-130TC; Su-22M4; TS-11; W-3 helicopter.

Contractor

ATM Inc.

VERIFIED

ROMANIA

SAIMS data acquisition and recording system

SAIMS is designed for the acquisition and recording of the main flight parameters of all aircraft presently equipped with the SARP-12 recording systems: IAR-93, IAR-99, L-39, Mi-8, Mi-17, MiG-21, and MiG-23. It is also available for initial fit to other aircraft types.

SAIMS comprises: the data acquisition unit (UAD); protected recorder (IP); portable data copying equipment (EPPD); ground data processing equipment (ESPD) and ancillary ground equipment.

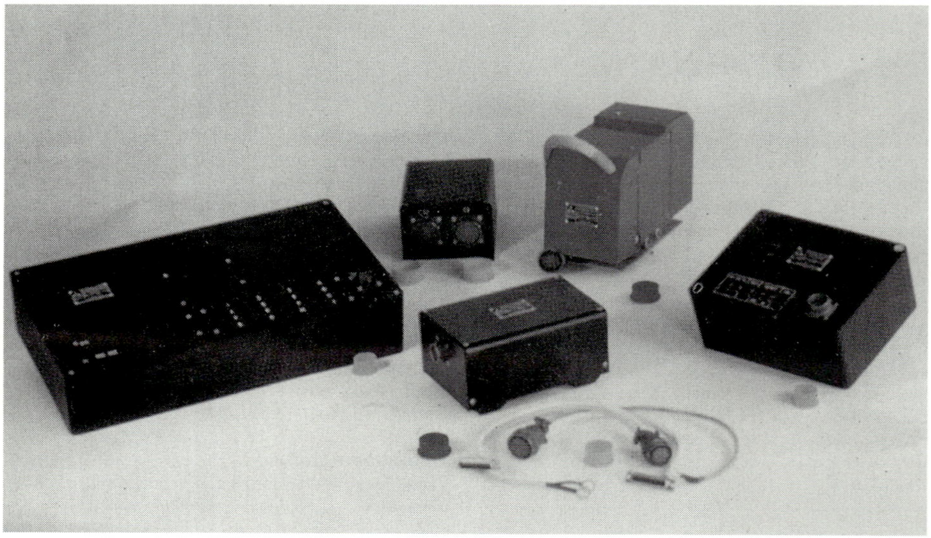

The Aerofina Avionics Enterprise SAIMS data acquisition and recording system
1999/0051326

Specifications

Signals sampled:
- 7 potentiometric signals
- 3 synchro signals, 26 V/400 Hz or 36 V/400 Hz
- 3 frequency type signals, 7-77 Hz or 36 V/400 Hz
- 1 DC voltage signal (0-32 V DC)
- 14 discrete digital signals

Sample rate: 1-10 Hz

Resolution: 8 bits

Memory capacity: 512 kbytes, 3.5 flight hours

Dimensions:
- UAD 178 × 113 × 77 mm
- IP 215 × 108 × 151 mm

Weight:
- UAD 2.5 kg
- IP: 7 kg

Temperature limits: –55 to +70°C

Contractor

Aerofina Avionics Enterprise.

VERIFIED

RUSSIAN FEDERATION AND ASSOCIATED STATES (CIS)

SVS series of digital air data computers

The SVS series of digital air data computers provide measured and computed data on: true altitude; vertical speed; Mach number; stagnation temperature; outside temperature; angle of attack; and angle of slip. Inputs include: static pressure; total pressure; stagnation temperature; local angle of attack; local angle of slip; QFE and QNH.

SVS series of digital air data computers

0018928/0015275

Specifications

	SVS-85	SVS 2Ts-00	SVS 2Ts-002	SVS PKR-1
Inputs				
(static pressure)	11.555-107.6 kPA	1.07-133.3 kPa	1.07-133.3 kPa	17.3-107.5 kPa
(total pressure)	11.55-135.4 kPa	6.66-279.9 kPa	6.66-279.9 kPa	11.5-115 kPa
(stagnation temperature)	–60 to +99°C	–60 to +350°C	–75 to +400°C	–70 to +150°C
(local angle of attack)	–60 to +60°	–60 to +60°	–60 to +60°	–60 to +60°
(QFE and QNH)	57.7-107.4 kPa	70.1-107.4 kPa	70.1-107.4 kPa	57.7-107.5 kPa
Outputs				
(true altitude)	–500 to 15,000 m	–500 to 30,000 m	–500 to 30,000 m	–500 to 12,000 m
(vertical speed)	–100 to 100 m/s	–500 to 500 m/s	–500 to 500 m/s	–60 to 60 m/s
(Mach No)	0.1-1.0	0.2-3.0	0.2-3.0	0.1-1.0
(stagnation temperature)	–60 to +99°C	–60 to +350°C	–75 to +400°C	–60 to +100°C
(outside temperature)	–99 to +60°C	–60 to +60°C	–75 to +60°C	–99 to +60°C
(angle of attack)	–60 to +60°	–60 to +60°	–60 to +60°	–60 to +60°
(angle of slip)		–30 to +30°	–30 to +30°	–30 to +30°
Power	115 V AC, 400 Hz, 50 VA	115 V AC, 400 Hz, 50 VA	115 V AC, 400 Hz, 50 VA	5 V DC, 6.5 W 15 V DC, 2 W
Weight	6.5 kg	7.0 kg	4.5 kg	0.8 kg
Dimensions	4 MCU	4 MCU	3 MCU	30 × 177 × 252 mm

Operational status

AeroPribor states that it is the only producer in Russia, and that these systems are fitted to all military and civil aircraft.

Contractor

AeroPribor-Voskhod Joint Stock Company.

VERIFIED

DAP 3-1 air data sensor

The DAP 3-1 air data sensor is designed to measure local angle of attack, pitot and static pressures.

Specifications

Airspeed range: 150 to 1,600 km/h
Altitude: up to 30,000 m (98,400 ft)
Vane deflection range: ±90°
Temperature: ±60° C
Power: 115 V AC, 400 Hz (vane heating); 27 V DC (sensor)
Weight: 5 kg

Contractor

Aviapribor Corporation.

NEW ENTRY

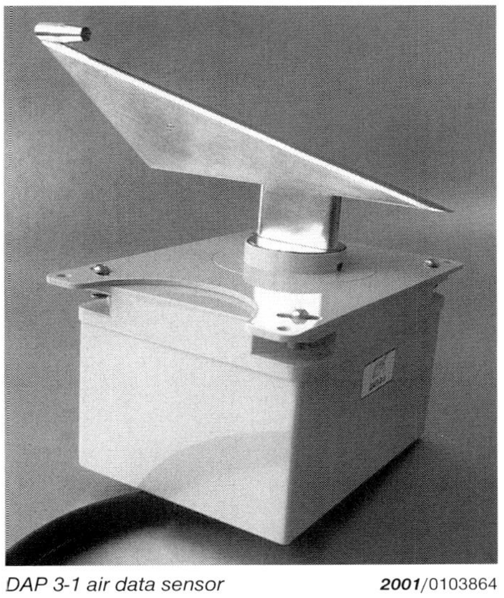

DAP 3-1 air data sensor **2001**/0103864

DAU-19 airflow direction transducer

The DAU-19 airflow direction transducer measures local angle of attack and outputs values via a proportional electrical signal. The unit is designed to be fitted to amphibians.

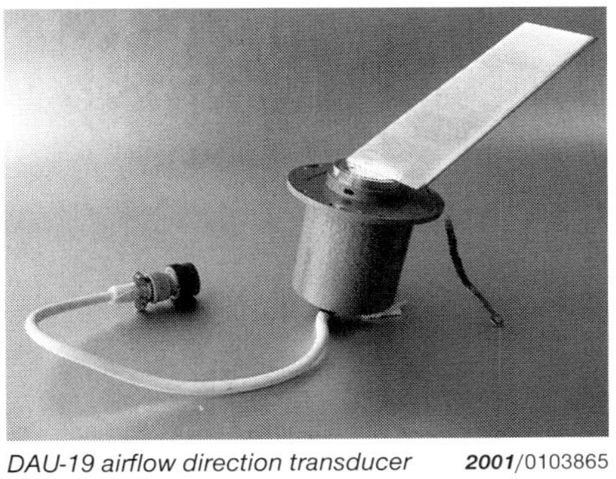

DAU-19 airflow direction transducer **2001**/0103865

Specifications

Airspeed range: up to 900 km/h
Altitude: 1,000 m (3,280 ft)
Vane deflection range: ±30°
Error: ±0.25°
Power: 27 V DC, 20 W
Weight: 0.65 kg

Contractor

Aviapribor Corporation.

NEW ENTRY

Fly-By-Wire (FBW) Container

The FBW Container represents the latest generation of integrated modular avionics. It contains flight control, autopilot and air data modules. The Container interacts with sensors and servos, performing all tasks involved with manual and automatic flight control and augmentation. The FBW Container is an open system which and can be expanded to include indication systems, GPWS or other modules, depending on customer requirements. The system is compatible with a wide variety of aircraft types/missions.

Contractor

Aviapribor Corporation.

NEW ENTRY

Fly-By-Wire Container **2001**/0103866

MVD-D1 air data module

The MVD-D1 air data module is designed to measure pitot and static pressures, calculate total air temperature (using data from a temperature sensor) and output electrical signals to onboard systems of the host aircraft utilising an ARINC-429 interface. The MVD-D1 module facilitates the calculation of altitude/airspeed parameters directly, without utilising conventional pressure systems, thereby introducing redundancy into the aircraft instrumentation system.

Specifications

Pitot pressure: 12 to 2,830 (±0.36) Hpa
Static pressure: 12 to 1,080 (±0.36) Hpa
Dimensions: 188 × 132 × 95 mm (with connector)

MVD-D1 air data module **2001**/0103868

Temperature: –60° to +240°C
Power: 27 V DC; 13 W
Weight: 0.9 kg

Contractor

Aviapribor Corporation.

NEW ENTRY

PVD-2S/PVD-K air pressure probes

The PVD-2S provides pitot and static pressure information, while the PVD-2K probe provides full pitot pressure information to air data systems.

Specifications

	PVD-2S	PVD-K
Airspeed range	Up to 1,100 km/h	Up to 450 km/h
Altitude	Up to 30,000 m	Up to 7,000 m
Pitot error	±0.02 to ±0.07 q	±0.02 q
Static error	±0.025 to ±0.058 q	±0.016 q
Electrical Power	27 V DC 140 W	27 V DC 120 W
Weight	0.4 kg	0.25 kg

Contractor

Aviapribor Corporation.

NEW ENTRY

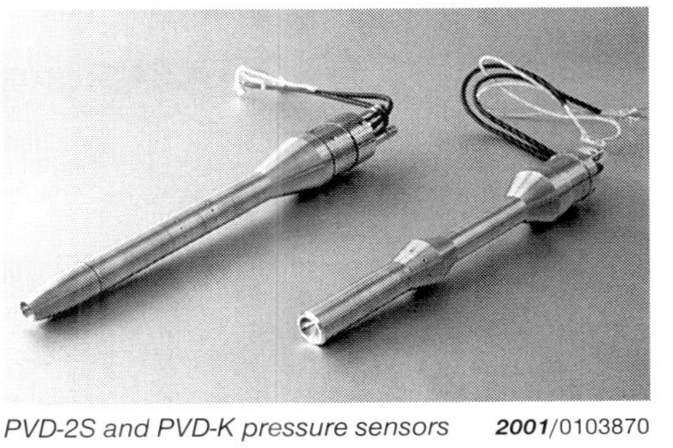

PVD-2S and PVD-K pressure sensors **2001**/0103870

SVS-V1 helicopter air data computer system

The SVS-V1 air data computer is designed for measurement of altitude and speed data on helicopters; it comprises two line-replaceable units: computer, airspeed vector transmitter.

Specifications

Airspeed longitudinal component: –90 to +450 km/h, (±3.5-5 km/h)
Airspeed lateral component: 0-90 km/h, (±3.5 km/h)
Indicated airspeed: 30-450 km/h (±2-6 km/h)
Vertical baroinertial speed: –30 to +30 m/s, ±0.3 ms
Pressure equivalent altitude: –500 to +7,000 m, ±6 m
Outside air temperature: –60 to +60°C, ±2°C
Weight: 6 kg
Power: 30 VA

Contractor

Aviapribor Corporation.

UPDATED

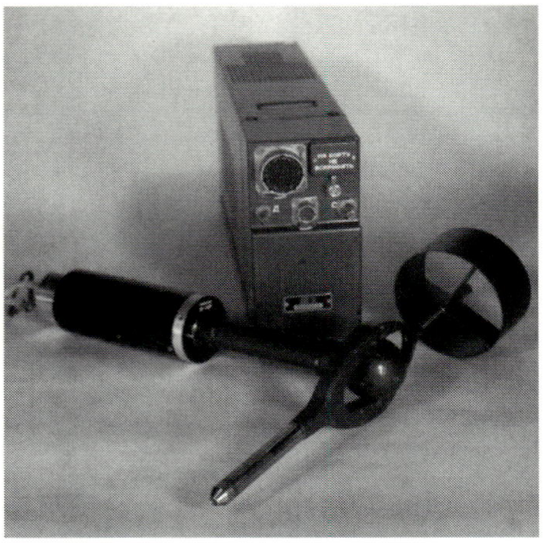

SVS-V1 helicopter air data computer system
0018240

KARAT integrated monitoring and flight data recording system

The KARAT system comprises KARAT-B the airborne monitoring and recording system and KARAT-N a portable protected computer ground data processing and analysis system.

KARAT-B comprises two line-replaceable units: Flight Data Acquisition and processing Unit (FDAU) and MultiPurpose Flight Data Recorder (MPFDR).

KARAT-B performs the following functions in the FDAU, acquisition, recording and processing of flight information from sensors and onboard systems; in-flight monitoring of onboard equipment and display on aircraft displays of relevant information; determination of maintenance requirements in the MPFDR, crash-protected storage and protection of recorded data; readout to KARAT-N of recorded data via high-speed datalink using RS-232 or RS-422 protocols.

KARAT-B employs an open architecture design that makes it suitable for all types of military and civil aircraft and helicopters. Data is protected in accordance with TSO C 124 requirements.

KARAT-N provides readout, processing and analysis of KARAT-B data; preparation of KARAT-B software and data files.

Specifications

MPFDR
Dimensions: 120 × 143 × 320 mm

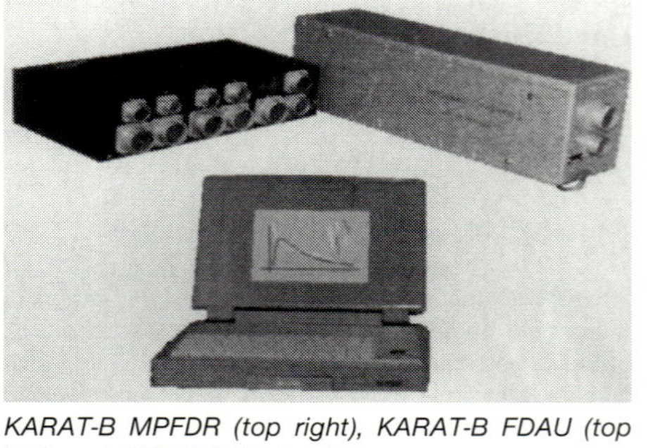

KARAT-B MPFDR (top right), KARAT-B FDAU (top left), and KARAT-N (bottom) 0015315

Weight: 10 kg
Power: 27 V DC
Memory: up to 64 Mbytes
Inputs: ARINC 717 (1 channel); ARINC 429 (1 channel)
Net interface: IOLA
Crash protection (100% at):
(impact shock) 3,400 *g* for 6.5 ms
(fire) 1,100°C for 30 min
(deep sea pressure) 20,000 ft for 1 day; 10 ft for 30 days
(pierce) 500 lb from 10 ft with 6.35 mm steel penetration pin
(static crush) 23 kN

Operational status

KARAT is installed in the upgraded MiG-21 supplied to the Indian Air Force. KARAT replaces the BASK and Tester U3L systems of the original MiG-21.

Contractors

GosNIIAS State Research Institute of Aviation Systems.
JSC Pribor Design Bureau Aviaavtomatika.

VERIFIED

C300 programmed signal processor

The C300 programmed signal processor is designed for real-time processing of radar information from the latest generation multifunction airborne radars. It comprises the following modules; signal input and buffering module; data processor and control module; computing module; video processor; non-volatile memory; input/output interfaces module; specialised communications control modules and co-processors are further options.

Specifications

Max throughput at signal processing: 3,200 Mflops
Max throughput of data processing: 64 Mips
Main memory: 48 Mbytes
Non-volatile memory: 10 Mbytes
Signal speed: 4 × 12 Mbytes/s (4 channels)
Video image resolution: 1,024 × 768 pixels
Power supply: 200 V, 400 Hz, 1,000 W
Weight: 39 kg
Software: assembler; C; parallel processing system

Contractor

Leninetz Holding Company.

VERIFIED

BTsVM-386 airborne digital computer

The BTsVM-386 airborne digital computer is a 32-bit, PC-compatible, multitask, modular, open architecture system, which can be extended by peripheral and graphics coprocessors. Three processor options are available. Program languages: C++; Modula-2; Assembler.

Specifications

Memory:
(ROM) 1.5 Mbytes (extended to 65 Mbytes)
(static RAM) 0.5 Mbytes (extended to 64 Mbytes)
(flash memory) extended to 256 Mbytes
Input/output:
(ARINC-429) 32 independent input; 16 independent output
(multiplex MIL-STD-1553B) 3 channels
(event signals) 32 input; 16 output
MTBF: 10,000 h
Power: 115 V AC; 26 V DC, 90 W
Weight: 8 kg

	i386/387-20	**i486DX-50**	**i860-25**
Word length	32-bit	32-bit	64-bit
Speed			
(fixed point)	10 Mops	50 Mops	50 Mops
(floating point)	0.7 Mops	3.2 Mops	50 Mops
(graphics)			25 k polygons/s

Contractor

Ramenskoye Design Company AO RPKB.

VERIFIED

ZBM Baget-53 integrated airborne digital computer

The Ramenskoye Design Company's ZBM Baget-53 integrated airborne digital computer employs the VME bus system and Vx Works software. The central processor module uses R3000/R3010 25 MHz microprocessors. Static RAM is not less than 1 Mbyte and 2 Mbyte of ROM is included, together with a 256 Mbyte volatile memory module and graphics processor module. The system offers 32 signal inputs and 32 signal outputs. It also complies with GOST standards for control of multifunctional control/display modules and fibre optic communication systems.

Overall dimensions are as specified by 2.5K, 3K and 4K GOST 6765.16-87 standards.

Contractor

Ramenskoye Design Company.

VERIFIED

BTsVM-386 airborne digital computer 0015276

SOUTH AFRICA

Data Transfer System (DTS)

The Data Transfer System (DTS) is a modern ruggedised solid-state non-volatile device that provides a centralised storage and archiving medium for avionics systems. It consists of the Data Transfer Unit (DTU), which is fixed in the cockpit, and a Portable Data Store (PDS). The PDS is a removable cartridge which provides for the transfer of data between a mission planning centre and the avionics host computer and records flight and maintenance data for flight evaluation and analysis.

The PDS electrical interface to the DTU takes place via the dual spring-loaded ejection mechanism, conveying both data and power. The communication between the DTU and PDS is serial, with a data transfer rate of 150,000 baud. The PDS is inserted through a slot into the DTU and is not orientation sensitive.

Specifications

Dimensions: 165 × 126 × 52 mm
Weight: <1.5 kg
Interface: RS-422, RS-485, RS-232
PDS memory: up to 4 Mbytes non-volatile RAM
Reliability: 5,000 h MTBF

Operational status

In production.

Contractor

Advanced Technologies & Engineering Co (ATE).

VERIFIED

Mission computer

The ATE mission computer is designed to meet the requirements of modern integrated avionics systems. Advanced RISC architecture and modular design provides for a high-performance generic mission computer.

A distributed processor architecture with intelligent Input/Output (I/O) peripherals allows the application software to run independently from the I/O system software. Real-time colour symbology generation/video mixing, as well as HUD stroke symbology generation, is managed by high-performance intelligent symbol generation modules. I/O interfaces include MIL-STD-1553B BC/RT, synchro, analogue, serial and discrete. A user-friendly PC-based applications software development environment is fully implemented and validated. An offline PC-based graphics generation tool allows pilot evaluation of symbology before downloading to the mission computer. Extensive BIT ensures failure detection of between 95 and 98 per cent.

The ATE mission computer

Specifications

Dimensions: 330 × 129 × 197 mm
Weight: 7 kg
Power consumption: 70 W
Reliability: 4,000 h MTBF

Operational status

Reported flight testing complete, production ready.

Contractor

Advanced Technologies & Engineering Co (ATE).

UPDATED

AM1000 distributed architecture data acquisition and processing system

The AM1000 distributed architecture data acquistion and processing system has been designed as an expandable modular COTS avionic architecture to service Health and Usage Monitoring (HUMS), Flight Data Recorder (FDR), and Cockpit Voice Recorder (CVR) systems.

Individual modules are combined, from two to six, to provide a required capability in an integrated unit. Communication between units takes place using industry standard protocols. Primary computing takes place within each decentralised functional unit. Commercially available software is utilised.

The COTS hardware is ruggedised to meet user requirements, using convection cooling and hardware enclosures to match the application concerned. Functional units include: Data Concentrator Unit (DCU) AM1-1000; Vibration Monitoring Unit (VBU) AM1-2000; Engine Monitoring Unit (EMU) AM1-3000; Engine Debris Monitoring Unit (EDMU) AM1-4000; together with supporting processor; power supply and interface units.

AM1000 distributed architecture data acquisition and processing system 0015277

Contractor

Analysis, Management & Systems (Pty) Ltd.

VERIFIED

AM3000 harsh environment computer

The AM3000 range of high-performance, high-reliability electronic modules is based on the Multibus Manufacturers Group standard for implementation of the IEEE 1296 Multibus II on MIL-STD-1389D SEM-E. This allows the performance, flexibility and reliability of Multibus II to be combined with a rugged and compact module, providing an ideal computer for harsh environments. AM3000 modules can be supplied in either military or industrial screening levels.

The military screened modules are suitable for airborne use, as well as for land and naval applications, where custom input/output modules may be added to the standard range in order to meet the requirements of new systems or mid-life upgrades to existing systems. Modules can be packaged in standard or customised enclosures, using either forced air or convection cooling. Interface standards include Multibus II, MIL-STD-1553B, RS-232, RS-422, RS-485, FDDI, ARINC 429, ARINC 404, MIL-STD-704D and MIL-STD-1275A.

Standard modules include the AM3010 32-bit RISC CPU module, AM3080 32-bit CISC CPU module, AM3030 input/output processor module, AM3020 MIL-STD-1553B communications module and the AM3060 power supply module.

Specifications

Dimensions: ¾ ATR or ½ ATR short (6-12 modules)
Weight: 7.5 kg (9 modules)
Power consumption: 80 W

Contractor

Analysis, Management & Systems (Pty) Ltd.

VERIFIED

Modules from the AM3000 harsh environment computer

Health and Usage Monitoring Systems (HUMS)

AMS Health and Usage Monitoring Systems (HUMS) is a modular system comprising data acquisition processing unit; control and display; Flight Data

Recorder (FDR) and Cockpit Voice Recorder (CVR); data transfer device; and ground replay station.

The system covers the full functional spectrum, including flight data recording, cockpit voice recording, engine health and usage monitoring, performance trending, structural fatigue monitoring, limits and exceedance monitoring, vibration monitoring and data storage and transfer.

The data acquisition unit is based on the AM3000 harsh environment avionics computer. This features IEEE 1296 Multibus II, MIL-STD-1398D SEM-E modules, Intel 80960 MC 25 MHz CPU, MIL-STD-704D power supply, multiples of 16 analogue and 32 discrete inputs, extensive BIT capability, ½ ATR short ARINC mounting or customised packaging and convection cooling.

BAE Systems Hawk Mk 100 and Analysis, Management & Systems (Pty) Ltd health and usage monitoring system
0015316

The digital flight data recorder has a solid-state memory array as the recording medium, crash protection in accordance with EUROCAE ED-55/56 and is a ½ ATR short ARINC mounting unit.

The data extraction computer is a PC-based hardware platform that can be interfaced with the Global Logistics Information System. It operates in a Microsoft Windows-based environment and can function as a data transfer device as well as a powerful analysis station.

Operational status

Variants of the system have been engineered for the Rooivalk attack helicopter HUMS, and the SMR 95 engine HUMS on Cheetah fighter aircraft.

AMS has also developed a version for the BAE Systems Hawk Mk 100 series jet trainer aircraft that will feature airframe fatigue monitoring; Engine Life Recording (ELR) for the Adour Mk 871 engines; FDR/CVR in accordance with ED55/ED56A (BASE SCR500-660); avionics equipment maintenance recording; flight line maintenance support station; fleet airframe usage management system. This system has been selected for the Hawk 100 aircraft being delivered to the Royal Australian Air Force and the NATO Flying Training Centre (NFTC) in Canada. First flight of the Hawk aircraft fitted with this AMS was in April 2000. This variant of the AMS HUMS could also be retrofitted to older Hawk aircraft.

Contractor

Analysis, Management & Systems (Pty) Ltd.

VERIFIED

GAE1200 digital audio recorder

The GAE1200 is a unit for recording and replaying audio signals, based on digital signal processing techniques. It uses a hard disk as its storage medium. This improves access time over conventional tape recorders and allows the user to move from one position in a recording to another almost instantaneously. It is thus ideally suited to communications, EW and radio monitoring applications. The GAE1200 operates either from its front panel or via a remote control.

The GAE1200 may be used with a range of SCSI hard disk sizes, yielding differing recording times. A 180 Mbyte hard drive gives a recording time of some 7 hours.

The GAE1200 was designed to take advantage of the ability to access information almost instantaneously and in virtually any order. It is also a robust replacement for a cassette recorder.

Specifications

Dimensions: 132 × 360 × 122 mm
Weight: <6 kg
Power supply: 18-36 V DC, 36 W
110/240 V AC, 50-40 Hz, 50 VA
Temperature range: 0 to +55°C
Interfaces: RS-232, parallel, transputer link

Operational status

Obsolete.

Contractor

Grintek System Technologies.

VERIFIED

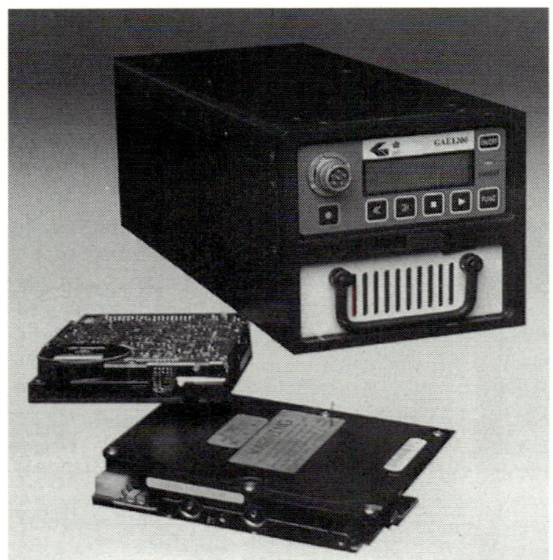

The Grintek System Technologies GAE1200 digital audio recorder

---

SWEDEN

Modular Airborne Computing System (MACS)

MACS is a standardised computer concept developed for the JAS 39 Gripen aircraft programme. It replaces the previous standard computer system - the SDS80. The system comprises the MACS computer and its software engineering environment SEEMACS.

At present, the D80E computer is used for data processing in the Gripen mission computer, radar, display and ECM systems; it uses Pascal/D0 software language. Future Gripen processing requirements will need increased processing power and use of the Ada software language. MACS is designed to fulfil these requirements, with 10-times greater processing power within similar power and volume limits. SEEMACS replaces the PUS80 support environment.

A MACS computer consists of three functional units, which communicate using VME. The processor unit handles all MACS calculations and each processor uses its own program and data memory. The communication unit facilitates communication with external devices via MIL-STD-1553B databus. The mass memory unit supports the other units.

Operational status

In development.

Contractor

Ericsson Microwave Systems AB.

VERIFIED

SDS80 standardised computing system

Designed for airborne, land and naval applications, the SDS80 is intended as a general purpose processor with cost savings extending over software development as well as hardware. The current design consists of three modules: the D80 computer, PUS80 programme development system and Pascal/D80 high-order language. By adding D80 modules the system can be expanded from a relatively simple unit to an extremely high-performance embedded computer system. Using an intermodule bus, up to 15 processing units, input/output modules and bulk storage facilities can be connected. A low-cost microprocessor complement, the D80M, has been developed, as have Ada software compilers for the D80 computer and microprocessor.

The D80 computer is the system computer for mission data, radar, display and ECM systems in the Saab JAS 39 Gripen combat aircraft. It has also been selected for the mid-life upgrade programme for the UK Royal Navy's Sea Harriers.

Specifications

Dimensions: 125 × 193 × 318 mm typical
Weight: 6.5 kg
Power supply: 56 W (AC or DC)
Word length: 32-bit
Typical speed: 1 Mips
Instruction time: from 125 μs
Interfaces: include MIL-STD-1553B
Cooling: forced air or fan

The latest, improved version of the D80 computer, retrofitted into the first batch of Gripen aircraft 0015278

Operational status
In production. In service on the JAS 39 Gripen. Also selected for the BAE Systems Sea Harrier F/A Mk 2. The latest, improved version of the D80 computer has been retrofitted into the first batch of Gripen aircraft, which originally carried an earlier version of the system.

Contractor
Ericsson Microwave Systems AB.

VERIFIED

DiRECT Digital mission RECording and data Transfer system

The DiRECT solid-state integrated mission recording system is designed to record the vast amount of information captured in the harsh environment of a modern combat aircraft. The system gives direct access to recorded information in the aircraft, or in a ground evaluation station.

The DiRECT system processes and records all video and still images, and audio, as well as radar, and bus data. By using MPEG-2 image coding, the system is well suited for environments demanding flexibility combined with ruggedness.

Specifications
Recording time: >90 minutes dynamic recording
Storage capacity: 10-25 Gbyte
Memory: non-volatile, solid-state or hard disk
Compression: MPEG-2 image compression
Coding: MPEG/Layer 2 audio coding
Environment: full military specification
Interfaces: Fibre channel, RS-170, ARINC 429, RS-422, MIL-STD-1553B and Ethernet

Operational status
In production for the JAS 39 Gripen fighter.

Contractor
Ericsson Saab Avionics AB.

UPDATED

JA 37 Viggen CD 207 mission computer

The CD 207 mission computer is based on three 486 computer module systems configured for multiprocessing. This unit replaces the predecessor CD 107 in the JA 37 Viggen Fighter as a form and fit replaceable LRU, in the stores management unit.

The CD 207 includes a MIL-STD-1553B RT/BC interface, a mass storage of 40/80 Mbyte and a master controller for the aircraft data communication to all major avionics.

Operational status
In production.

Contractor
Ericsson Saab Avionics AB.

VERIFIED

JAS 39 Gripen aircraft interface unit

The JAS 39 aircraft interface unit, which is based on a 486 computer module system, is used to interface aircraft sensors in the JAS 39 Gripen to the main electronic system. The unit contains the logic to handle warning information and operate the control panel.

Power supply control of the electronic system, normal control of the undercarriage and operation of a crash recorder are other functions of the unit. The main part of the logic in the unit is provided by software. Critical functions are implemented in dedicated logic.

The aircraft interface unit features modular design, LRU concept to facilitate maintenance, chassis integrated power supply for optimal heat dissipation, SMT for increased packing density and built-in redundancy.

Operational status
In production for the JAS 39 Gripen.

Contractor
Ericsson Saab Avionics AB.

VERIFIED

JAS 39 Gripen Environmental Control Unit (GECU)

The GECU is a computer which controls cockpit temperature, cockpit air flow, cooling pack temperature and avionics air flow. The controller receives input signals for the cockpit control panel, various sensors and trip switches, and produces output signals for the control valves and indicator lamps. Communication to and from the aircraft avionics systems is made via a redundant interface MIL-STD-1553B serial bus.

The aircraft interface unit features modular design, LRU concept to facilitate maintenance, chassis integrated power supply for optimal heat dissipation, SMT for increased packing density and built-in redundancy.

Operational status
Pre-production units delivered.

Contractor
Ericsson Saab Avionics AB.

VERIFIED

AMR 345 dual and multicommand VHF/UHF airborne radio system

The SaabTech (formerly CelsiusTech) Electronics AMR 345 dual- and multicommand system is designed for trainers, transport aircraft and helicopters. It consists of the AMR 345 VHF/UHF AM and FM speech and data transceiver and the AMR 349 control units. Performance characteristics are as for the single AMR 345 VHF/UHF transceiver described above.

It features easy handling between instructor and trainee without any priority switches, the latest command being valid, extensive non-volatile reprogrammable memories for up to 1,000 channels and BIT.

The AMR 345 is designed to be operated under the worst flying conditions with a minimum of time and attention required by the pilots. In the trainer installation the trainee has full command over the frequency selection, as in front-line aircraft, and the instructor is fully informed on what the trainee is doing and can at any time take control himself.

The AMR 345 is reprogrammable in the aircraft by means of a fill gun.

In the manual frequency mode a keyboard gives access to any frequency in the VHF or UHF bands. AM or FM is selected by a separate switch on the panel. In the preset mode, a keyboard gives direct access to 1,000 channels and can be used for immediate access to different services at all available airbases. The control features may be adjusted according to customer requirements via software modifications and change of key-tops. One version of the equipment includes VHF and UHF Guard channels.

Specifications
Dimensions:
(transceiver) 76 × 146 × 230 mm
(control unit) 76 × 146 × 40 mm
Weight:
(transceiver) 3.5 kg
(control unit) 0.65 kg
Power supply: 16-32 V DC
Power output:
(AM) 10 W
(FM) 15 W
Frequency: 104-162 and 223-408 MHz
Channel spacing: 25 kHz

Operational status
Dual and multicommand radio systems are fitted in the Pilatus PC-7 and PC-9, Raytheon Hawker 125-800, LearJet and Falcon F20.

Contractor
SaabTech Electronics AB.

UPDATED

The AMR 345 VHF/UHF transceiver dual command version

UNITED KINGDOM

FlightVu FV-0720 Aircraft Data Recorder (ADR)

The FlightVu FV-0720 ADR is a significant new additional source of information available to the air accident investigator, since it combines the Flight Data Recorder (FDR) and the Cockpit Voice Recorder (CVR) with a Flight Video Recorder (FVR) into a single unit. The provision of up to eight channels of video pictures, mounted in the cockpit, or external to the aircraft, allows investigators to fully evaluate the cockpit environment.

The solid-state digital memory is protected to the equivalent standard of ED56A, plus 1 hour fire test. In use at times other than for crash investigation, the FVR images can be viewed immediately for in-flight monitoring, or used for post flight debriefing/ maintenance activities. Additionally, data can be downloaded by 10 Mbps Ethernet connection. Remote download capability via spread spectrum microwave link is available as an option, as a replacement Quick Access Recorder (QAR).

Specifications

Recording capability:
25 h of flight data from FAR 121 parameter list
2 h voice recording of 4 channels (ED56A)
30 min video at 3 frames/s NTSC, PAL, colour or monochrome
compliant FAR 121, ED 55 and ED56A
Dimensions: 4 MCU
Weight: max 7 kg, incl underwater location beacon (excl mounting tray)

Contractor

AD Aerospace Ltd.

VERIFIED

AE3000FL series wideband analogue S-VHS data recorders

The AE3000FL series of compact ruggedised S-VHS recorders offer high performance wideband analogue data recording. Three 18 MHz models are designed for advanced communications and telemetry applications while 12 MHz, 8 MHz and 3 MHz versions are designed for long duration missions. AE3000FL series recorders are designed for intelligence gathering, communications, telemetry, radar signature and ASW applications.

For each version of the system, there are single and multichannel variants within each bandwidth range. Other input configurations can be specified as customer options. Some models can be switched between single and multichannel operation for pre- and post-detection recording from FM, linear and logarithmic receivers. Models of12 and 8 MHz can be specified to record and reproduce E1, E2 or T1 and T2 line codes and similar high rate communications data.

Each system is totally self-contained, including the power supply and a full record/reproduce capability. An optional interface is available for 12 MHz models which allows wideband analogue signals to be output in 8-bit digital format at 30 Mbit/s directly to digital analysis systems.

Operational Status

In production.

Contractor

Avalon Electronics Ltd.

NEW ENTRY

AE7000 high-performance disk recorders

Claimed to be the world's first general purpose 50 MHz (1 Gbit/s) data recorder, Avalon Electronics' AE7000 utilises the latest high-speed disk technology, offering outstanding data integrity and convenience. Achieving more than double the data rate of today's fastest conventional tape systems, the AE7000 is aimed at 'high end' data capture and simulation in intelligence gathering, electronic warfare and similar applications.

The AE7000 recorder represents a significant development in the emerging trend towards scalable, 'multitechnology' data capture, where wide bandwidth data is captured initially on fast disks before being edited for transcription onto low cost disk or tape media for analysis, archiving or onward transmission.

Data is recorded onto a group of eight 36 Gbit disks giving the system a total capacity of 288 Gbit (2.2 Terabits). The disks are housed in a shock-resistant, 'hot-swappable' crate which can be replaced in seconds or stored in a secure safe when not in use. The unit is controlled by four multifunction pushbuttons and graphical interface. Data can be output either in its original analogue or digital form, or automatically backed up to an integral Exabyte™ or AIT-2™ drive.

When integrated with a Windows NT Workstation as part of an Avalon AE7800 SIGINT Data Capture System, selected passages of data can be written directly to any networked disk or tape storage medium as computer-compatible files, which are then immediately available for analysis or onward transmission. Long-term, unattended monitoring solutions are also available.

Data interfaces include analogue (DC to 2, 20 or 50 MHz), digital (zero to 1 Gbit/s) plus a range of widely used telecommunication standards including: E1, E2, E3, T1, T2, ATM/OC3 (155 Mbit/s) and SONET/OC12 (622 Mbit/s).

Operational status

In production.

Contractor

Avalon Electronics Ltd.

NEW ENTRY

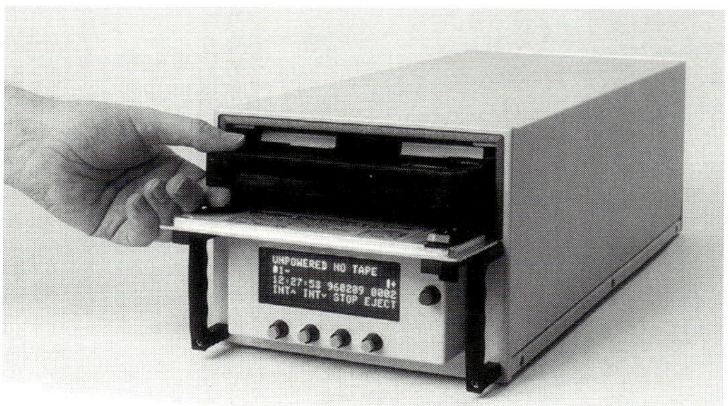

Avalon Electronics' AE3000 series wideband analogue S-VHS data recorder **2001**/0103890

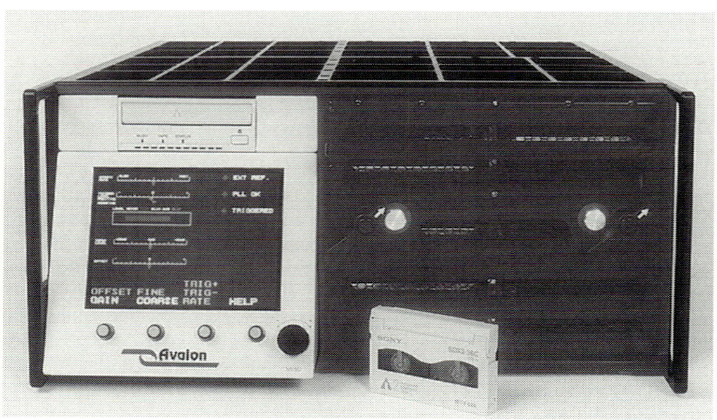

Avalon Electronics' AE7000 high-performance disk recorder **2001**/0103891

---

8 mm and Hi-8 mm sealed video recorders

Various configurations of Sealed Video Recorder (SVR) have been produced to suit different cockpit and bay mount installations. All have integral anti-vibration mounts configured for bulkhead or 5 in ARINC rack mounting, with fixed or flying connectors and local or remote controls and indicators.

Operational status

In production.

Contractor

BAE Systems.

VERIFIED

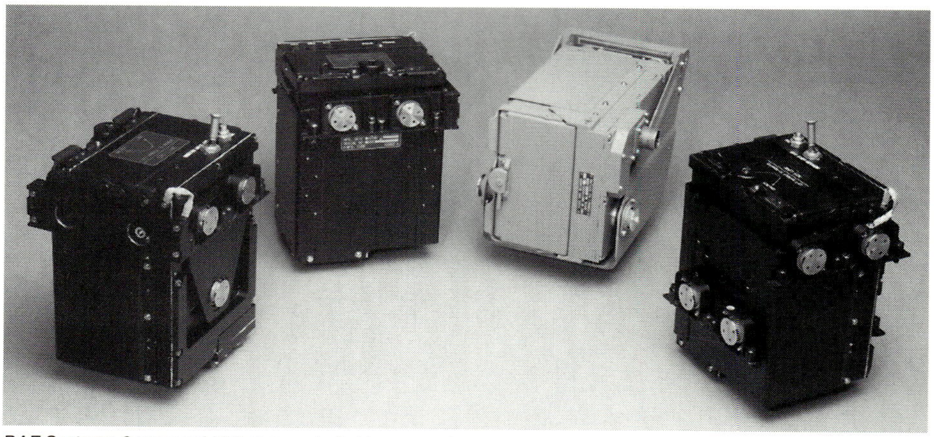

BAE Systems 8 mm and Hi-8 mm sealed video recorders 0001298

F-16 HUD video camera

This single unit high-resolution colour camera was developed to fit the F-16 C/D HUD. It replaces the existing two-unit monochrome camera.

The latest technology solid-state sensor provides high resolution, excellent sensitivity and accurate colour. Alignment accuracy is high for mounting the camera forward of the HUD. The 525-line/60 Hz NTSC or 625-line/50 Hz PAL configurations are available and the output can be Y/C or composite video.

Operational status
In production.

Contractor
BAE Systems.

VERIFIED

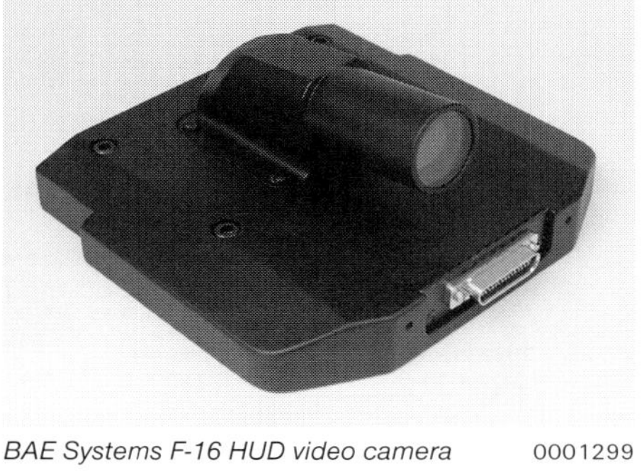

BAE Systems F-16 HUD video camera 0001299

Harrier GR. Mk 7 video recording system

The video recording system for the Night Attack Harrier GR. Mk 7 was developed as an upgrade to the GR. Mk 5 system, to incorporate recording of the FLIR video. Using two sealed video recorders, one for 'head-up' recording and one for 'head-down' recording, the system records colour HUD camera, FLIR and dual-mode tracker. Growth is built in to enable recording of all head-down displays at a later date.

Operational status
In production.

Contractor
BAE Systems.

VERIFIED

Helicopter Air Data System (HADS)

In 1979, BAE Systems designed and developed a low airspeed Helicopter Air Data System (HADS) for the AH-1S Cobra helicopter.

The system provides full three-axis, prime accuracy, air data information by utilising the Airspeed And Direction Sensor (AADS). A variant, known as the High-Integration Air Data Computer (HIADC), has been developed. It provides three-axis air data information on modern digital busses such as MIL-STD-1553B. The major parts of the system are shown below:

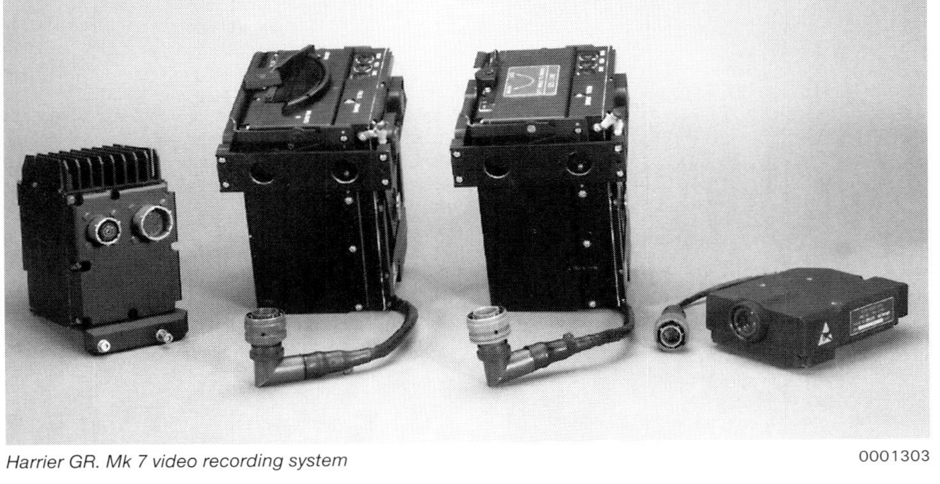

Harrier GR. Mk 7 video recording system 0001303

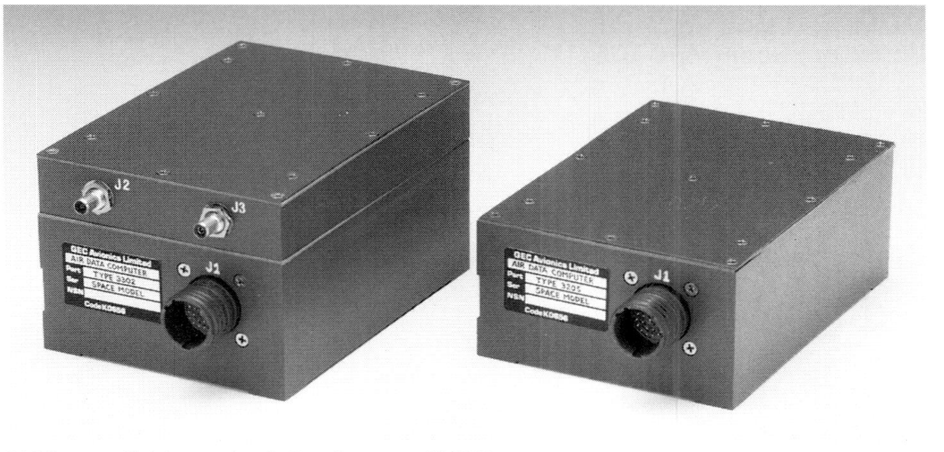

BAE Systems High-Integration Air Data Computer (HIADC) 0002346

Airspeed And Direction Sensor (AADS)
The AADS probe is mounted externally to the aircraft below the rotor and swivels to align with the local airflow. The AADS contains pitot and static pressure ports for measuring the magnitude of the local airspeed vector and a pair of resolvers for determining its direction. This enables the system to provide data for: rotor downwash velocity; ground effect; forward, rearward and lateral airspeed (to zero kt); vertical airspeed; wind direction, drift and lift margin (when integrated into an avionics suite); enhanced pilot awareness.

Installation of the system produced major improvements for operational and flight test environments in: fire control; low airspeed, low altitude manoeuvres.

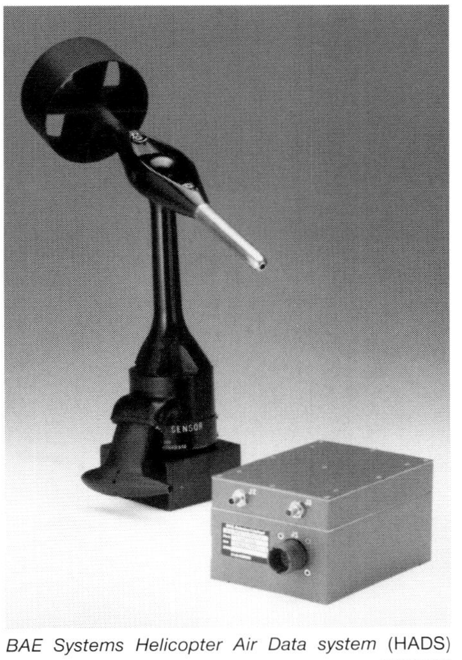

BAE Systems Helicopter Air Data system (HADS) 0001285

HIADC Interface
The HIADC interface integrates with the AADS by measuring its probe angle, air temperature and pitot and static pressures. Air data calculations are performed and the resultant parameters are made available on digital databusses.

Specifications
Airspeed and direction sensor
Dimensions: 97 × 317 × 246 mm
Weight: 1.1 kg

High integration air data computer
Dimensions: 140 × 102 × 83 mm
Weight: 1.2 kg

Operational status
Over 1,500 systems have been supplied for attack helicopters including Bell AH-1S Cobra, Agusta A-129 and Super Puma. A major order was received in 1997 for HADS in support of US Army AH-64D Longbow and British Army WAH-64 Apache helicopters.

Contractor
BAE Systems.

VERIFIED

High-Integration Air Data Computer (HIADC)

The High-Integration Air Data Computer (HIADC) utilises advanced production techniques and miniaturised air data transducers. It features low individual component count and power consumption to provide an accurate, highly reliable, compact air data computer to meet the growing market demand for distributed pressure sensing devices.

The flexible configuration concept of HIADC offers the ability to satisfy applications for both fixed-wing aircraft and helicopters.

HIADC typically interfaces to the aircraft total air temperature and angle of attack sensors; measures pitot, static and differential pressures and, having corrected for systematic error characteristics, computes a full range of accurate air data parameters. Data is digitally distributed to various aircraft systems in either ARINC 429, RS-422 or MIL-STD-1553B output formats.

HIADC has been designed as a fit-and-forget air data system, which virtually eliminates the requirements for maintenance and provides significant life cycle cost benefits.

Specifications
Dimensions: 140 × 82 × 102 mm (max)
Weight: 1.2 kg
Power supply: 28 V DC, 6 W
Reliability: >25,000 h MTBF

Operational status
In production for numerous fixed- and rotary-wing programmes worldwide.

Contractor
BAE System.

VERIFIED

Miniature Standard Central Air Data Computer (MSCADC)

The Miniature Standard Central Air Data Computer (MSCADC) is a modular digital air data computer which provides a lightweight, high-reliability system. The MSCADC can easily be configured for retrofit applications, such as for the A-4 or F-5, or new aircraft where centralised air data functions are required. MSCADC outputs include digital ARINC 429 and MIL-STD-1553B formats, analogue, synchro and discrete.

Specifications

Dimensions: 115 × 140 × 229 mm
Weight: 4.77 kg
Power: 35 W (max)
Reliability: >7,800 h MTBF

Operational status

MSCADC is in full-scale production for F-5 applications worldwide.

Contractor

BAE Systems.

VERIFIED

MOdular Data Acquisition System (MODAS)

The MOdular Data Acquisition System (MODAS) has been designed to gather and record any type of signal to be found on an aircraft during flight trials. A large range of signal input types are supported by both the Mk I and Mk II systems. Inputs include strain gauges, voltages, synchros, thermocouples, tachos, ARINC 429, MIL-STD-1553B and RS-232.

MODAS Mk I

MODAS Mk I has up to 4,096 input channels and is expandable to meet specific requirements at sampling rates of up to 128,000 samples/s. Eight separate sampling programs can be selected in flight. A pulse coded modulation digital technique is employed for recording data in either simultaneous multitrack, serial streams or IRIG-106 format. A comprehensive ground replay facility is also available and MODAS can interface with a telemetry system.

MODAS Mk II

The Mk II version of MODAS is less than half the size of the earlier model and is four times as fast. The PA 3101 processor and recorder interface unit can accept data from eight acquisition units, or up to 64 with an expansion unit. The PA3120 general purpose acquisition unit houses up to eight interfaces and two MIL-STD-1553B databusses can be monitored with the PA 3130 databus acquisition unit. Data acquisition is up to 512 k parameter samples/s.

Specifications

MODAS Mk I
Dimensions:
(acquisition/processing unit) ½ ATR
(control unit) 146 × 191 × 151 mm
(monitor unit) 146 × 191 × 166 mm
(small recorder) ¾ ATR short
(large recorder) 533 × 360 × 200 mm
Weight:
(acquisition/processing unit) 10 kg
(control unit) 3 kg
(monitor unit) 3 kg
(small recorder) 13 kg
(large recorder) 39 kg
Power supply: 115 V AC, 400 Hz, 3 phase or 28 V DC

MODAS Mk II
Dimensions:
(PA 3101, 3110 and 3130) 135 × 230 × 210 mm
(PA 3120 control and monitor unit) 66 × 146 × 132 mm
Weight:
(PA 3101, 3110, 3130) 5 kg
(PA 3120) 0.9 kg
Power supply: 28 V DC, 60 W (max)

Operational status

In service in a wide range of UK and European aircraft.

Contractor

BAE Systems.

VERIFIED

PA3520 Aircraft Integrated Monitoring System (AIMS)

The PA3520 Aircraft Integrated Monitoring System (AIMS) records accident data, engine life information and structural data. It comprises the Marconi Electronic Systems PA3521A data acquisition and processing unit and the Penny & Giles D50330 Mk 2 accident data recorder.

Accident-related data can be downloaded directly into a portable recording unit for later analysis, thus avoiding the need to remove the data recorder itself. Engine and structural data can be downloaded into a unit with a display unit, allowing instant viewing of critical data, or the information can also be stored for subsequent analysis. The PA3521A can accept discrete, shaft rotation and analogue inputs, or can interface with an MIL-STD-1553B databus.

Specifications

Dimensions: ½ ATR short
Weight: 4.55 kg
Power supply: 28 V DC, 28 W

Operational status

In production.

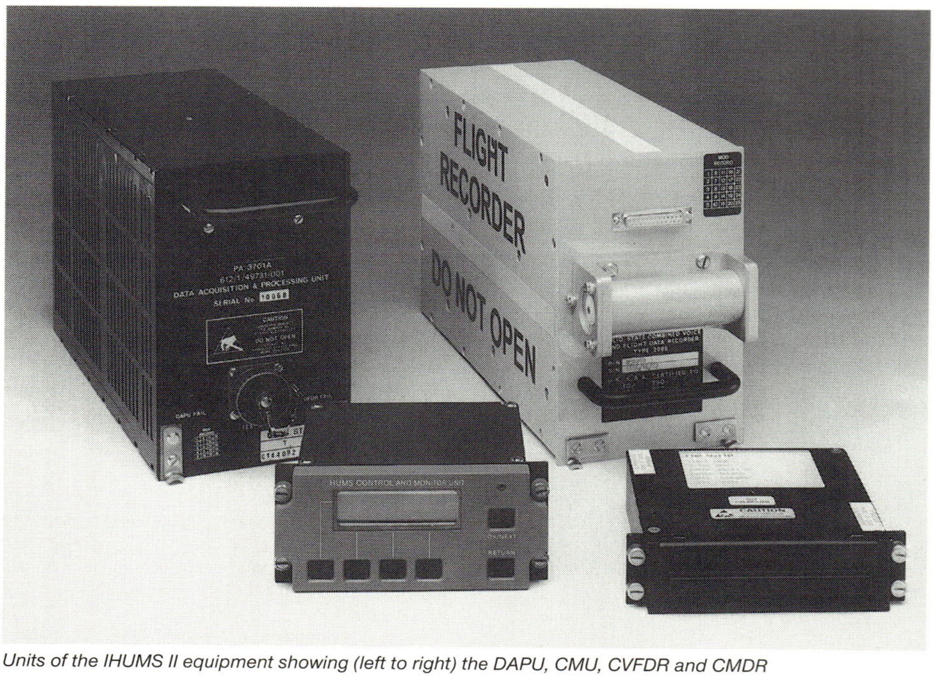

Units of the IHUMS II equipment showing (left to right) the DAPU, CMU, CVFDR and CMDR

Contractor

BAE Systems.

VERIFIED

PA3584 solid-state acquisition and recorder unit

The PA3584 solid-state acquisition and recorder unit is designed for applications where space and weight are at a premium. It offers solid-state flight data recording, cockpit video recording and data acquisition in a single ½ ATR long unit.

Suitable for helicopters and transport, business, regional and commuter aircraft, the combined unit simplifies installation and support requirements.

Equipment features include:

(1) an integral data acquisition unit for more than 33 parameters
(2) four 2-hour voice channels with an integral solid-state 25 hour duration data record capability
(3) direct recording with no data compression or decompression techniques needed
(4) full compliancy with TSOC124, EUROCAE ED-55, ED-56A and CAA Specifications 10A and 18
(5) fast data retrieval using a 386-based PC and 64, 128 and 256 words/s operating options.

Specifications

Dimensions: ½ ATR long
Weight: 13 kg typical
Power supply: 115 V AC or 28 V DC, 25 W (max)

Contractor

BAE Systems.

VERIFIED

PA3700 Integrated Health and Usage Monitoring System (IHUMS)

The PA3700 Integrated flight data recording Health and Usage Monitoring System (IHUMS) has many aircraft applications. It has been designed to fulfil the functions of a Digital Flight Data Recorder (DFDR), a Cockpit Voice Recorder (CVR) and a Health and Usage Monitoring System (HUMS). The system architecture has been optimised to provide a minimum hardware solution by the combination of these functions. Each airborne system comprises a Data Acquisition and Processing Unit (DAPU), Cockpit Voice and Flight Data Recorder (CVFDR), Card Maintenance Data Recorder (CMDR) and Control and Monitor Unit (CMU). In addition to this core system, other options are available, including a control and display unit, pilot interface panel, cockpit warning panel and quick access recorder.

The PA3701 DAPU contains all the conditioning circuitry necessary to sample and accurately monitor a wide range of different types of electrical inputs for subsequent recording, measurement or processing. The mandatory data output interfaces to a standard ARINC 573/747 Flight Data Recorder (FDR), and a standard ARINC quick access recorder. Selected mandatory data, together with raw and partially processed HUM data, is also fed to a ruggedised maintenance data recorder. Data can be displayed on the optional control and display unit as required.

There are several versions of the crash-protected recorder available to customers with differing requirements. For instance, one CVFDR provides 5 hours of continuous digital recording at a data rate of 128 12-bit words/s and 1 hour of voice on each of three separate tracks. Another FDR provides 8 to 10 hours of continuous digital data recording at a data rate of 128 12-bit words/s and is used in conjunction with a separate CVR.

The CMDR was designed to meet ARINC 615 as a high-speed data loader. The recording medium is a 2 Mbyte SRAM card. The data interface is bidirectional using RS-232C protocol. The unit is rugged, compact and the recording medium easily transportable, which makes it ideal for use as a card maintenance data recorder. The bidirectional interface to the DAPU enables upload of documentary data from a card as well as download of raw and preprocessed data for maintenance purposes.

The downloaded airborne DAPU data is supported by a comprehensive ground replay and analysis computer system.

Specifications

Dimensions:
(PA3701 DAPU) ½ ATR short
(CVFDR) 115.6 × 173.5 × 440.5 mm
(CMDR) 38.1 × 133.35 × 132.05 mm
Weight:
(PA3701) 6.2 kg
(CVFDR) 8.5 kg
(CMDR) 0.5 kg
Power supply: 28 V DC, 35 W (max)

Operational status

The system is fitted to over 120 helicopters worldwide. Systems are currently available for 14 aircraft types or variants covering the AS332L, AS332L1, AS365N, AAS365N2, Bell 212, Bell 214ST, Bell 412, S61, S76A+, S76A++, S76C, S76C+, Sea King.

The Mk II IHUMS has been selected by Sikorsky for the S-76C and S-92 production helicopters. The system architecture has been optimised to provide a minimum hardware solution and comprises a DAPU, CVFDR, CMDR and CMU.

Contractor

BAE Systems.

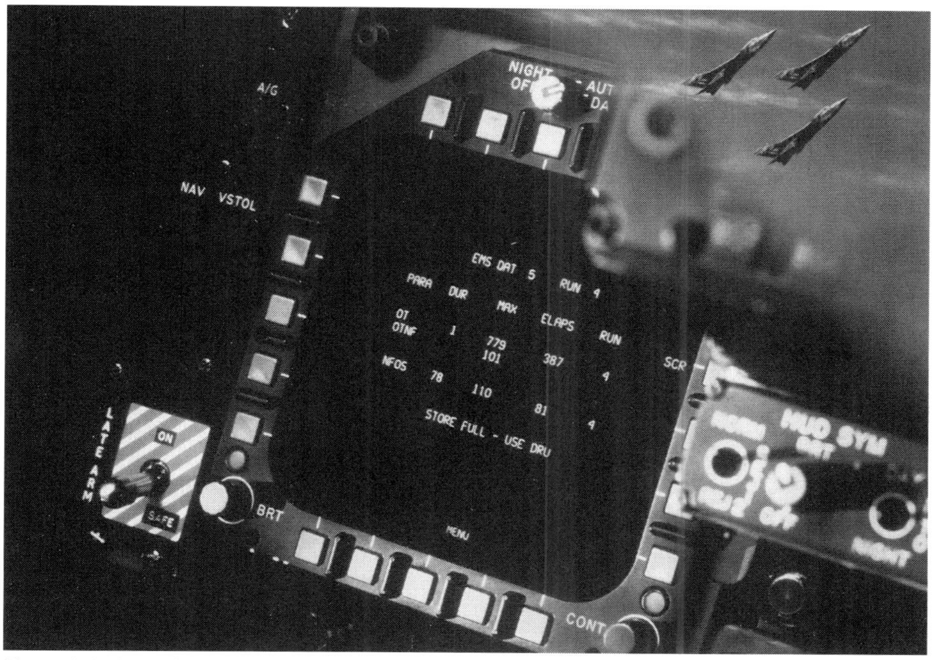

The cockpit display for the PRS2020 engine monitoring system

VERIFIED

PA3800 series flight data acquisition units

The PA3800 series, of flight data acquisition units, was developed to meet FAA and CAA requirements for flight data recording applicable from 1991. The unit samples data from a variety of input signals, which may be analogue, digital or discrete. The information is sampled in a programmable predetermined sequence and assembled into a digital data stream in a format compatible with any standard ARINC 573/717/747 digital flight data recorder. There are four or five PCB positions available for expansion of the system, perhaps taking the form of extra signal conditioners, or the unit may be expanded into an integrated microprocessor-based monitoring system to provide engine or airframe usage monitoring.

The PA3810 is for accident data acquisition, with four-card expansion available. The PA3820 is for helicopter health and usage monitoring, including accident data. The PA3830 is for fixed-wing aircraft for the Aircraft Integrated Monitoring System (AIMS), including accident data.

The latest development of the system has been designed to meet US FAA rules for August 2002, which require that a minimum of 88 parameters be recorded. This model has been selected for the Avro RJ series Regional Jet aircraft.

Specifications

Dimensions:
(PA3810) ⅜ ATR short case to ARINC 404A
Weight: 4 kg typical
Power supply: 115 V AC or 28 V DC, 15 W (max)

Operational status

In full production for both military and civil applications on fixed-wing aircraft such as the de Havilland Dash 8, Cessna Citation and Canadair Challenger and on Bell and Eurocopter helicopters.

Contractor

BAE Systems.

VERIFIED

PRS2020 engine monitoring system

Developed from the PVS1820 system, the PRS2020 is lighter, smaller, uses less power and is compatible with MIL-STD-1553 databus interfaces. It also includes a 32-bit high-speed microprocessor, plus hybrid and uncommitted logic array-based electronics.

The PRS2020 consists of several component units including:

PRS2021 engine monitoring unit: a single box data acquisition and processing unit for analogue signal conditioning, discrete digital conditioning and/or digital databus parameter recovery. The heart of this unit is a single, very large-scale integrated hybrid circuit comprising a 32-bit 19-register central processor with 4 kbytes of random access memory and 8 kbytes of read-only memory reserved for applications programs.
PRS2023 display: available for taking a quick look at data, the display can be mounted on the engine monitoring unit or remotely. The format of the display can be defined by the customer.
PRS2026 data retrieval unit: a rugged battery-powered unit for extracting data from the engine monitoring unit for subsequent analysis and for system test and calibration.

Options available include: bulk storage recorder, a portable ground replay unit and comprehensive systems software packages.

Specifications

Dimensions:
(PRS2021) 200 × 194 × 124 mm
(PRS2026) 280 × 260 × 80 mm

Operational status

Under development.

Contractor

BAE Systems.

VERIFIED

PRS3500A data/voice accident recorder

The PRS3500A consists of the BAE Systems PRS3501A data acquisition unit and the Penny & Giles D50330 accident data recorder.

The system provides 2 hours' continuous recording time, taking in up to 50 parameters at 240 words/s. The small size of the equipment makes it suitable for aircraft with limited cockpit space and digital transmission reduces the size and weight of wiring looms while increasing data integrity. Recorded data can be extracted in 6 minutes using a portable transfer device. The system comprises two units: the flight data recorder and a data acquisition unit.

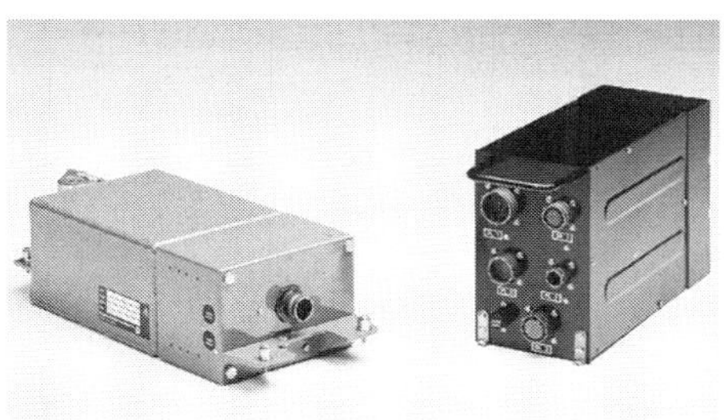

The BAE Systems PRS 3501A data acquisition unit, and Penny & Giles D50330 accident data recorder

Specifications

Dimensions:
(data acquisition unit) ½ ATR short
(flight data recorder) 115 × 172 × 440 mm
Weight:
(data acquisition unit) 4.5 kg
(flight data recorder) 8.3 kg

Operational status

In production and in service in Royal Air Force Harrier GR7 aircraft UK.

Contractors

BAE Systems.
Penny & Giles Aerospace Ltd.

VERIFIED

PV1584 data recorder

A compact unit combining recorder and acquisition electronics within a single ½ ATR case, the PV1584 is designed to replace ARINC 542 electromechanical recorders. The system is approved by all major certification authorities, including those of: Australia, Canada, Germany, UK and the USA. Its signal handling capability meets all current requirements, including those of the CAA's Specification 10 and the FAA's Part 121 section 343, with 10 per cent excess capacity.

The system can be used in conjunction with a copy recorder to extract data for analysis *in situ* without disturbing the equipment. A PV1591 flight data entry panel can be used for manual insertion of documentary information into the recording.

The PV1584 comprises a protected flight recorder at the front of the unit and a data acquisition electronics section at the rear. Design baseline is ARINC 573, input parameters being sampled at the rate of 64/s, each sample being converted into a 12-bit binary word. The resultant digital data stream is routed to the recorder section and placed on a Mylar tape loop giving 25 hours' continuous recording. The sampling rate for given parameters can be varied as required. The entire capacity of the tape can be dumped in 16 minutes by means of a Lockheed 235 data dump recorder. The unit contains an EEPROM, giving eight programs selected by changing wire links in the aircraft. The device can, therefore, be reprogrammed to meet requirements so that interchangeability could be maintained between eight aircraft types in an operator's fleet, for example. Signal types comprise 26 V synchro, DC ratio and absolute, AC ratio, potentiometer, variable resistance and frequency (tacho or pulse probe). Discretes comprise DC series and shunt and AC series and shunt.

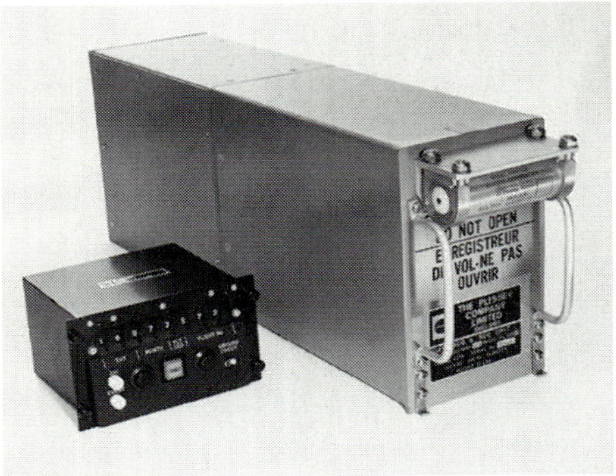

The PV1584 data recorder

Specifications

Dimensions: ½ ATR long
Weight: 11.34 kg
Power supply: 115 V AC, 400 Hz, 25 W

Operational status

No longer in production. During 1985, the PV1584 was selected to equip the UK Royal Air Force's Nimrod aircraft and the BAE Systems ATP. It is standard equipment on the BAE Systems 146 and 748, the Raytheon Hawker 800 and on the Shorts 360.

Contractor

BAE Systems.

VERIFIED

PV1591 Flight Data Entry Panel (FDEP)

The PV1591 Flight Data Entry Panel (FDEP) provides a display of the flight data recorder system status and a means of manually inserting documentary data into the recording.

Aircraft trip and date data is inserted into the system by use of eight thumbwheel switches on the front panel of the unit. The FDEP signal output represents the number indicated on one of these switches. Each of the eight switches is sampled in turn, for one data frame of the associated Flight Data Acquisition Unit/Data Acquisition and Recording Unit (FDAU/DARU). At the end of this cycle an invalid frame is output for eight frames to provide a synchronisation signal. The 16-frame sequence repeats continuously.

There are two push-buttons provided, one marked EVENT and the other marked DDI for documentary data insert, to insert markers into the recording to assist post-flight analysis of the data.

Two alternative status lights are available. One is a single indicator for use with a single box DARU. The other is a split indicator, one half marked FDAU and the other Digital Flight Data Recorder (DFDR), for use with a two-box system. Several alternative logic combinations are available, dependent on the type of FDR and the aircraft installation.

Specifications

Dimensions: 85.5 × 146 × 117 mm
Weight: 0.9 kg

Operational status

FDEPs are available for the whole range of FDAU, DARU and DFDR units.

Contractor

BAE Systems.

VERIFIED

PV1820C structural usage monitoring system

Designed to monitor airframe fatigue by continuously measuring structural loads during flight, the PV1820C is based on the company's earlier PVS1820 engine usage life monitoring system. Data and the computed results are recorded on a cassette-loaded quick access recorder for ground replay and analysis. The system comprises a PV1820C structural monitoring unit, a PV1819C control unit, quick access recorder Type 1207-003 and a transient suppression unit Type PV1845. Microprocessor operation simplifies the measurement of stress by enabling the output from a number of strain gauges (typically 16) to be monitored along with other flight parameters on a cassette recorder or solid-state data card.

Specifications

Dimensions:
(PV1820C) 94 × 194 × 394 mm
(PV1819C) 146 × 76 × 194 mm
(1207-003) 146 × 51.6 × 169 mm
(PV1845) 95 × 53 × 190 mm
Weight:
(PV1820C) 5.44 kg
(PV1819C) 1.59 kg
(1207-003) 1.77 kg
(PV1845) 1.4 kg
Sampling rate: Each parameter is defined by 10 bits of a 12-bit word, the 11th bit being a compression flag and the 12th being reserved for parity. The sampling rate for each quantity is programmable in binary steps from 1 to 128 times/s
Data compression: programmable 16:1, 8:1 or 4:1
Self-test: comprehensive automatic self-test and fault diagnosis is built in

Operational status

In production and in service.

Contractor

BAE Systems.

VERIFIED

PV1954 flight data acquisition unit

The PV1954 fulfils all of the requirements for a 32 parameter Flight Data Recorder (FDR) and provides expansion capability for additional maintenance monitoring.

Conditioning circuits enable the PV1954 to sample data from a wide variety of input signals. These inputs may be analogue, digital or discrete. The information is sampled in a predetermined sequence and assembled into a digital data stream in a format compatible with any standard ARINC 573/717/747 Digital FDR (DFDR).

A programmable read-only memory controls the input signal sampling sequence. This method of control permits the user to define the content of all the words in the frame, except synchronisation words.

The PV1954 provides an output of 64 12-bit data words/s in Harvard bi-phase format to the DFDR. As an option, this data rate may be increased to 128 or 256 words/s.

An auxiliary data output in RZ format is provided for use with an optional quick access recorder, which operates at 64 words/s. Optionally, it is possible to increase the data rate to 128 or 256 words/s.

A time synchronisation output in frequency shift key format is provided to synchronise the DFDR to the cockpit voice recorder.

Specifications

Power input: 28 V DC nominal, 15 W (max)
Dimensions: ½ ATR short case to ARINC 404A
Weight: 5 kg typical

Operational status

In service on Gulfstream IV, BAE Systems 146 Series 2 and ATP aircraft.

Contractor

BAE Systems.

VERIFIED

SCR 200 flight data/voice recorder

The SCR 200 consists of a crash-protected accident data recorder and a data acquisition unit. It has been designed for the Tornado. The accident data recorder comprises a continuous single-spool tape transport with four recording tracks. Recording rate is 128 12-bit data words/s. Total recording time is 111 minutes data and 37 minutes voice. Audio information is recorded on one track simultaneously with data on the other three.

Specifications

Dimensions:
(ADR) 425 × 172 × 81 mm
(DAU) 383 × 128 × 200 mm
Weight:
(ADR) 7.5 kg
(DAU) 7 kg
Power supply: 115 V AC, 400 Hz, 25 W
28 V DC

Operational status

In production. Over 750 systems are in service on UK Royal Air Force, Royal Saudi Air Force and Italian Air Force Tornado aircraft.

Contractor

BAE Systems.

UPDATED

SCR 300 data/voice recorder

The SCR 300 flight data recording system is designed for use on military aircraft, where space is at a premium. It is believed to be the smallest and lightest tape combined CVR/ADR system currently available.

The main function of the SCR 300 is the preservation of a flight audio and data record in the event of an accident, but it can also be used for maintenance monitoring. The system consists of a crash-protected Accident Data Recorder (ADR) and a Data Acquisition Unit (DAU).

The ADR is mechanically and thermally protected to recent FAA and CAA standards and has provided 100 per cent data and voice retrieval. The system comprises a continuous single-spool tape transport with six 30 minute tracks of information, with tracks assigned to audio and data as required. The recording rate is 128 12-bit data words/s, giving a total recording time of 3 hours. The ADR is fitted with a sonar locating beacon.

The DAU collects information from signal sources in a variety of digital, analogue and discrete forms, samples each at an appropriate rate and converts them into a stream of serial digital data which it then transmits to the ADR. The number of parameters monitored

Royal Saudi Air Force Tornado GR1 aircraft are fitted with SCR 200 flight data recorders **2001**/0103896

depends on the application. A typical system has more than 40 digital, analogue, audio and discrete signal inputs. Uniquely, the DAU stores up to 64 different aircraft frame formats within the processor. This feature greatly eases logistics in mixed aircraft fleets, where the single standard DAU is totally interchangeable among numerous aircraft types.

Specifications

Dimensions:
(ADR) 115 × 150 × 250 mm
(DAU) 89 × 125 × 200 mm
Weight:
(ADR) 6 kg
(DAU) 4 kg
Power supply: 28 V DC, 40 W

Operational status

Available against special order and in service worldwide on over 200 Jaguar aircraft, Harrier, Sea Harrier, Hawk 60/100/200, Andover, BAC 111 and Raytheon Hawker 800 fixed-wing aircraft and Chinook, Lynx, Sea King, Gazelle, Scout and Wessex helicopters.

Contractor

BAE Systems.

VERIFIED

Seven variants of the SCR 500 solid-state cockpit voice and flight data recorders, shown in front are the SCR 500 cockpit control panels (miniature and standard size options)

SCR 500 solid-state cockpit voice and flight data recorders

The SCR 500 combines the latest digital recording and flash memory techniques, to provide a lightweight cost-effective range of Cockpit Voice Recorder/Digital Flight Data Recorders (CVR/DFDRs). The SCR 500 range complies with the latest requirements of EUROCAE -55/56A, the latest TSO C123A/C124A for increased fire protection and ARINC 757/747, with recording duration to meet both current and future requirements.

The SCR 500-030 and SCR 500-120 CVRs respectively have 30 and 120 minutes' audio recording duration. The combined CVR/FDRs have a range recording duration from 30 minutes audio/10 hours data to 2 hours audio/25 hours data at 128 words/s or 50 hours at 64 words/s data rates. All SCR 500 recorders have instant real-time audio and data playback.

The SCR 500-630 and SCR 500-660 solid-state combined flight data recorders provide 30 minutes or 1 hour of four-channel audio duration, and 20/10 hours duration of 64/128 digital data words per second

The SCR 500-1530 and SCR 500-1560 solid-state CVR/FDR combined recorders provide 30 minutes or 1 hour, four-channel cockpit voice and 25 hour digital data. They are designed to provide a long duration audio/digital recorder in a single box, for large helicopters.

The SCR 500-1620 solid-state CVR/FDR combined recorder provides a full 2 hours, four-channel cockpit voice and 25 hour digital data. It is designed to replace both CVRs and FDRs in large passenger aircraft giving the safety benefits of dual redundancy. Utilising an SCR 500 Combi in place of a CVR enables regional aircraft operators to implement cost-effective FOQA procedures in conjunction with the latest analysis software package.

All models can be supplied with control panels (miniature and standard size options) and an underwater beacon. The audio provision includes four analogue channels recorded digitally and one digital data channel plus timebase. Test equipment, designed around common PC architecture, provides Downloading/Testing and Replay of all the SCR 500 family of recorders. A hand-held data downloader is also available.

Specifications

Dimensions: ½ ATR short
Weight: <7 kg for all versions
Power supply: 115 V AC, 400 Hz
28 V DC, 12 W (nominal)
Control panel dimensions:
miniature: 28.6(H) x 146(W) × 70.5(D) mm
standard: 57.2(H) × 146(W) × 70.5(D) mm

Operational status

Certified to CAA ED 55/56A, and FAA TSO C123/C124, and compliant with TSO C123A/C124A. In production and in service.

The recorders are being fitted into several major fixed-wing and helicopter programmes: prime fit for Sikorsky S92 and S76C+ (IHUMS) programmes; prime fit for Avro RJ series; prime fit for Royal Australian Air Force Hawk LIF; prime fit for Atlas Rooivalk attack helicopter; prime fit for UK MoD DHFS Bell 412s; selected for UK Royal Air Force BAe 146 and Raytheon Hawker 800 fleets.

Contractor

BAE Systems.

VERIFIED

LRUs for the range of SCADC units 0001286

Standard Central Air Data Computers (SCADC)

In 1981, BAE Systems was contracted to develop a range of Standard Central Air Data Computers (SCADCs) to upgrade US Air Force and Navy aircraft. The common standard replaces the great variety of air data systems installed in some 40 variants of 15 different types in US military service.

To satisfy these requirements, SCADC has been developed in six configurations, each having more than 80 per cent core commonality with the others. One or two special-to-type modules are included in each configuration to accommodate unique aircraft interfaces. The high commonality design has produced significant life cycle cost benefits. The equipment is designed to the latest software, computing and data transmission specifications used in US military aircraft. Over 95 per cent failure detection and isolation is achieved with the built-in test systems, and ground crew tests will enable 98 per cent of all LRU failures to be detected and isolated.

The six configurations are:

SCADC unit CPU 140/A. This unit was fitted on six US Air Force and US Navy aircraft types and includes a MIL-STD-1553B interface capability.

Specifications

Dimensions: 191 × 191 × 302 mm
Weight: 11.1 kg
Power: 72 W

SCADC unit CPU 141/A. This was fitted on the US Air Force C-5A, C-5B, C-141A, C-141B military transports. The system also includes a MIL-STD-1553B interface capability.

Specifications

Dimensions: 191 × 124 × 545 mm
Weight: 15.3 kg
Power: 102 W

SCADC unit CPU 142/A. This equipped six versions of the US Air Force General Dynamics F-111 and was designed to include MIL-STD-1553B interface capability.

Specifications

Dimensions: 191 × 356 × 483 mm
Weight: 38.1 kg
Power: 98 W

SCADC unit CPU 143/A. This computer was designed to include MIL-STD-1553B for eight versions of the F-4 Phantom.

Specifications
Dimensions: 191 × 420 × 305 mm
Weight: 20.68 kg
Power: 80 W

SCADC unit CPU 152/A. This two-channel unit was designed for all variants of the US Navy S-3 aircraft and includes dual MIL-STD-1553B navigation system and cockpit instrument interfaces.

Specifications
Dimensions: 222 × 142 × 485 mm
Weight: 15 kg
Power: 30 W

SCADC unit CPU 175/A. This computer was designed to be fitted to the US Navy Northrop Grumman F-14A, A+ and D aircraft and includes MIL-STD-1553B interface.

Specifications
Dimensions: 160 × 218 × 510 mm
Weight: 31 kg
Power: 26 W

Operational status
SCADC is no longer in production. Over 5,500 units were delivered to the US and other users worldwide.

Contractor
BAE Systems.

VERIFIED

V3500 miniature camera

The V3500 is a high-resolution solid-state sensor camera for monochrome, low-light and colour recording. It is fully automatic and has a range of auto-iris lenses available which make it well suited for head-up display recording applications as well as for helicopter sights and air vehicle stabilised platforms. The V3500 can be supplied in either a single- or two-unit configuration; the camera head can be separated from the electronics unit if required.

Specifications
Dimensions:
(camera head) 90 × 28 × 35 mm
Weight:
(camera head) 0.15 kg

Contractor
BAE Systems.

VERIFIED

ACCS 2000 and 2500 general purpose airborne computers

The ACCS 2000 (Airborne Computing and Communications System) is a family of micro programmed computers designed to provide standard low-cost data processing for a wide range of vehicles and missions; it is a derivative of the AN/AYK-14 which is the standard airborne computer for the US Navy (see entry under GD Information Systems in the USA section). The computers can be used as multiprocessor units with an extensive interface capability. There is a high degree of functional and mechanical modularity for flexible growth. The system architecture is not affected by modular configuration changes and is the feature most responsible for providing low cost and versatility.

The Single Card Processing (SCP) module, the heart of the family of computers, contains the micro programmed control, arithmetic unit, registers, micro-memory, real-time clocks, interrupt logic, bootstrap memory and bus interface.

The ACCS 2500, an upgrade of the earlier ACCS 2000, is fitted as the mission computer to all Royal Air Force Harrier GR. Mk 7 and T10 aircraft. The greater processing power of the ACCS 2500 was designed to cope with the greater demands of the night attack and poor visibility missions of the GR7. The ACCS 2500 features power conversion, SCP, memory control, two 64,000 × 18-bit word core memories, input/output unit and three MIL-STD-1553B bus controllers.

Further modular upgrades are available by supplanting the SCP with up to four Very High Speed Integrated Circuit (VHSIC) processor modules, each of which increases the processing power by between two and four times that of a single SCP. Each VHSIC processor module carries 1 Mword onboard memory and additional semiconductor memory modules are also available.

Specifications
Word length: 16-bit
Typical speed:
up to 1.2 Mips (one SCP)
up to 6 Mips (one VPM)
up to 12 Mips (two VPMs)
Input/output: 3 MIL-STD-1553B ports, up to 16 other channels
Memory:
32 kwords 18-bit core modules (900 ns access time)
32 kwords 18-bit semiconductor modules (400 ns access time)

Operational status
In service on UK Royal Air Force Harrier GR. Mk 7 aircraft. The 88 ACCS 2000 computers have been modified to ACCS 2500 standard.

Contractor
Computing Devices, a General Dynamics Company.

VERIFIED

The Computing Devices ACCS 2500 is the mission computer for the Royal Air Force Harrier GR. Mk 7
2001/0099743

ACCS 3000 mission computer

The ACCS 3000 series is a family of computers designed to meet the requirements of the next generation of military aircraft such as the Eurofighter Typhoon, or for major avionic upgrade programmes.

Eurofighter Typhoon processor modules

It uses a processor-independent multiprocessor architecture with an integrated Ada environment. Modules are available using any standard 16- or 32-bit CISC or RISC microprocessor family or with non-standard families. The processors are augmented by modules performing signal processing, display graphics processing and offering databus and standard signal interfaces, to give an integrated computing and communications system.

Specifications
Dimensions: 240 × 150 mm
Weight: 0.7 kg
Power consumption: 15 W
Processing: 2 × 68020 32-bit 20 MHz processor and floating-point co-processor, 2 Mbyte SRAM and 2 Mbyte EPROM giving 2.5 Mips performance. Software in Ada.

Operational status
The processing subsystem has been developed and selected for four main computers for the Eurofighter Typhoon: the navigation computer, attack computer, interface processing (utilities) unit and defensive aids computer.

Contractor
Computing Devices, a General Dynamics Company.

VERIFIED

ACCS 3200 fatigue monitoring and computing system

The fatigue monitoring and computing system provides real-time fatigue data gathering and analysis, via parallel processing of 16 strain gauge sensor data channels. Real-time analysis isolates structural fatigue to individual aircraft structures. A comprehensive ground analysis facility allows analysis of full sortie raw sensor data and a complex load history log by airframe, together with fleetwide data correlation.

Specifications
Dimensions:
(fatigue monitoring computer) ½ ATR rack or custom housing
(raw data recorder) 83.8 × 158 × 221.2 mm

Weight:
(fatigue monitoring computer) 9 kg
(programme and data transfer unit) <1 kg
(raw data recorder) 3 kg
Processing: 68020 20 MHz processor, 640 kbyte EEPROM, 128 k SRAM, software in Ada

Operational status
Deliveries for UK Royal Air Force Harrier GR. Mk 7 aircraft complete.

Contractor
Computing Devices, a General Dynamics Company.

VERIFIED

Digital Solid-State Recorder (DSSR)

Computing Devices' DSSR is a high performance, flexible, cost effective solid-state storage solution to meet the requirements of many data acquisition and processing applications.

The DSSR, designed to replace the Computing Devices RMS 3000 on the UK Royal Air Force Tornado GR1A, utilises many of the latest approaches to open systems and Commercial-Off-The-Shelf (COTS) design.

The DSSR provides a solid-state recording and system control functionality which can be configured for a wide range of airborne, ground based and naval applications.

The DSSR is suitable for the storage of sensor and support data from a wide range of sensors such as infra-red, electro-optical, multispectral, sonar, radar, Synthetic Aperture Radar (SAR) and other sensor types.

Optional features include: simultaneous data compression/decompression; variable ratio compression/decompression; real-time display functionality; flexible data storage options (STANAG 7023, NITFS, raw data); a detachable high storage capacity memory brick (currently up to 25 Gbytes).

Contractor
Computing Devices, a General Dynamic Company.

VERIFIED

Eurofighter Dedicated Warnings Panel (DWP)

The programmable aircraft warnings panel has been selected for the Eurofighter Typhoon

The DWP offers a high-intensity LED display which is NVG-compatible. The DWP integrates an Electronic Unit (EU) with a Display Unit (DU). The display has programmable legends which are controllable over a dual-redundant MIL-STD-1553B avionics databus. Internal redundancy is incorporated to ensure continued safe operation in the event of defects and dedicated discrete inputs are also incorporated to ensure correct warning displays should the databus sources fail.

Aircraft status, health and fault data received over the databus is categorised by the internal Ada software for display on a simple menu; amber and red characters are available, with a dedicated display area for catastrophic warnings. An interface is provided to the audio management unit to give the aircrew co-ordinated visual and audible or digitised voice warnings of any new conditions.

The DWP provides the following facilities:

(1) Three single colour warning captions for displaying high integrity or catastrophic warnings
(2) Two-colour display of system warning captions
(3) A control for the paging of warning captions
(4) Two attention getter drives
(5) Storage of up to 256 system warnings
(6) Six audio warning discretes
(7) Two modes of operation, normal and reversionary
(8) Day, dusk, night luminance modes
(9) Dual channel architecture interface with the Cockpit Interface Unit (CIU) and the Communication and Audio Management Unit (CAMU).

Specifications
Dimensions:
(display unit) 118 × 156 × 70 mm
(electronics unit) 100 × 220 × 237 mm
Weight:
(display unit) 1.7 kg
(electronics unit) 3.95 kg
Power supply: 28 V DC, 270 W (max)

Operational status
Operational on all Eurofighter Typhoon development aircraft. Production deliveries were due to commence in the fourth quarter of 2000.

Contractor
Computing Devices, a General Dynamics Company.

UPDATED

Fatigue Monitoring and Computing System (FMCS)

The Fatigue Monitoring and Computing System (FMCS) is designed to provide real-time onboard analysis of fatigue as it occurs in the aircraft structure, for the lifetime of the aircraft. The system provides the benefits of increased safety, reduced maintenance and support costs, improved availability, increased operational life and timely embodiment of structural modifications.

The FMCS onboard computer unit consists of a data processor, memory and optimally located structural sensors to capture fatigue and aircraft performance data. This data is processed in real time to produce cumulative damage profiles for each critical structural component. In addition, as a system option, the raw sensor data may be recorded to provide further detailed analysis.

The fatigue and aircraft performance data can be fully analysed on the supporting ground station to give accurate assessment of the severity and the time that damage occurred. The ground station also has database support for individual aircraft debriefing and across-fleet correlation.

Operational status
Production equipments have been delivered. Original equipment and its derivatives are now finding application for both fixed- and rotary-wing operational load monitoring and health and usage monitoring programmes for aircraft.

Contractor
Computing Devices, a General Dynamics Company.

VERIFIED

The Computing Devices fatigue monitoring system

MC50A rugged compact computer

The MC50A is based on a 1 ATR rugged chassis and provides up to five ISA/PCI slots, plus a CPU slot; in due course, the unit will support Pentium, Digital Alpha and Power PC CPUs.

The MC50A was selected as a data processing unit for the airborne Central Tactical System (CTS) for the UK Royal Air Force Nimrod MR2 Mk2 aircraft upgrade, following successful design and development of demonstration units; it will be fitted to 25 aircraft. The MC50A will function as a data processor for four low-resolution displays and one high-resolution display, as well as other computers and instruments. It will be known as the Display Drive Unit Replacement (DDUR) for the CTS Obsolete Displays Replacement (ODR) project.

Operational status

Formal airworthiness certification achieved in June 1999.

Contractor

DRS Rugged Systems (Europe) Ltd.

VERIFIED

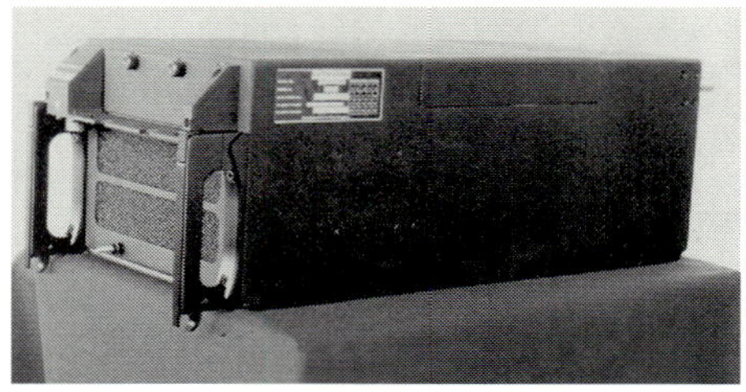

The DRS Rugged Systems (Europe) Ltd MC50A rugged compact computer **2001**/0089529

Airborne FM telemetry torquemeter

The airborne FM telemetry torquemeter provides a direct, accurate and reliable measurement of the torque in a helicopter's transmission system. The indicated torque is independent of any correction factors for air temperature, torquemeter shaft temperature, engine and gearbox life and efficiency, fuel characteristics or aircraft operating conditions.

The system consists of four component parts: rotating assembly, fixed antenna and head amplifier, signal processor and indicators. It operates without physical contact between the rotating shaft and the stationary structure and is said to have an overall accuracy of within ±1 per cent of full-scale torque. BITE is provided within the torquemeter equipment to enable system integrity and calibration accuracy to be checked and faults to be located.

Specifications

Power supply: 28 V DC, 1.5 A (max)

Contractor

GKN Westland Helicopters.

VERIFIED

Data Storage and Transfer Set (DSTS)

The solid-state data storage and transfer set is currently available with 1 Mbyte of memory capacity in a fully qualified format for installation in military aircraft.

The data transfer module has a quick-release handle mechanism for ease of operation and handling. The system is configured to be compatible with aircraft standard console mountings and includes the power supply and a databus interface for control and data transmission. The system has full read/write capability for upload and download of data.

The data transfer system can be used for upload of mission, navigation and other operational data or for the download of maintenance and mission data. Higher capacity variants, up to 400 Mbytes are in development and flight evaluation.

Operational status

In service on the UK Royal Air Force Harrier GR. Mk 7 and T. Mk 10.

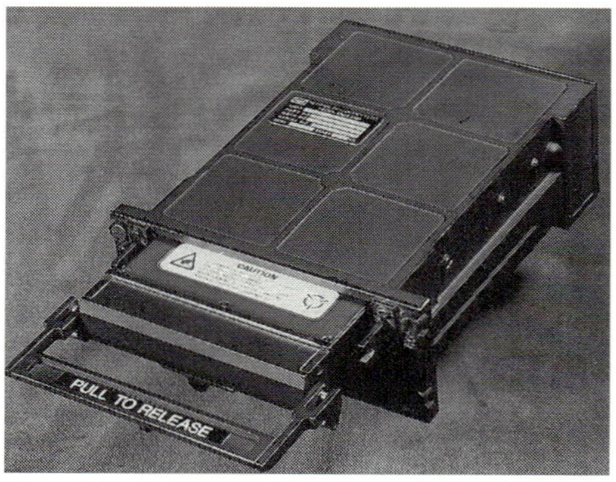

The solid-state data transfer system has been selected for UK Royal Air Force Harrier GR. Mk 7 and T. Mk 10 aircraft

Contractor

Honeywell (Normalair-Garrett Ltd).

VERIFIED

High-capacity digital data recorder

The high-capacity digital data recorder utilises the Ampex DCRSi technology, and is ruggedised for operation in severe environment applications such as helicopters.

The recorder, including interfaces to a typical ASW mission system, is contained in a 16 MCU volume envelope and has a total weight of 30 kg. The recorder is ideally suited to mission recording, as data can be recorded at any rate up to 107 Mbits/s and, because of the incremental tape motion, uses the minimum tape consistent with recording the output of the mission system. One cassette holds 385 Gbits of data, which gives over 8 hours' recording of a typical helicopter ASW mission.

The associated replay system can replay the data at up to 240 Mbits/s, and will reconstitute with high integrity an exact replica of the input signals.

Specifications

Max input data rate: 107 Mbits/s
Record time per cassette: 1 h at 107 Mbits/s; 8 h at 13 Mbits/s
Tape speed: incremental at 5.31 ips
Bit error rate (corrected): better than 1 in 10^7
Total storage per cassette: 380 Gbits
Dimensions:
(Tape Transport Unit (TTU) 11 MCU
(Data Acquisition Unit (DAU) 5 MCU
Weight: 30 kg
Power: 115 V AC, 400 Hz, less than 350 VA

Operational status

Fully qualified and in production for the UK Royal Navy EH 101 Merlin helicopter, also selected for the Nimrod MR2 Mk2 maritime patrol aircraft upgrade.

Contractor

Honeywell (Normalair-Garrett Ltd).

VERIFIED

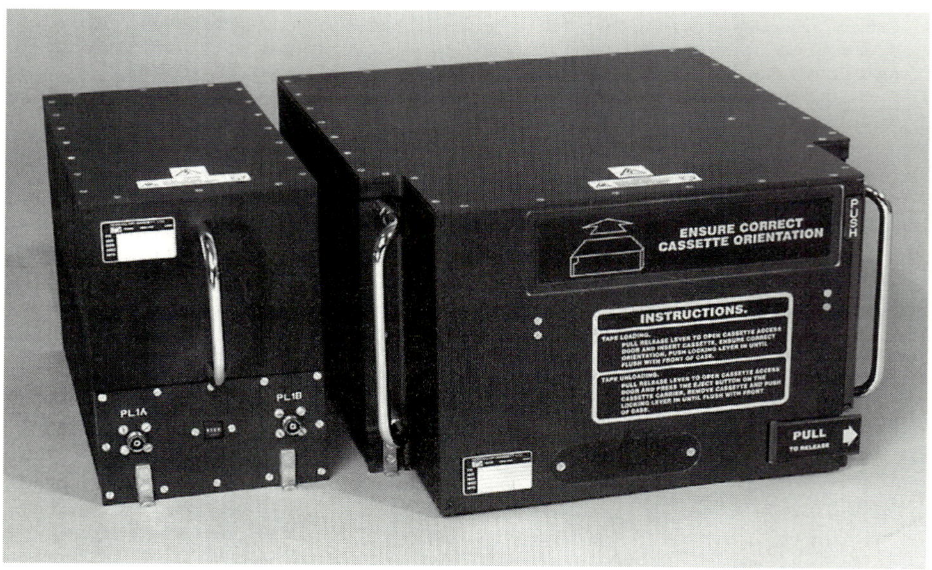

High-capacity digital data recorder showing the Tape Transport Unit (TTU) (right) and the Data Acquisition Unit (DAU) (left)

Central Maintenance Panel (CMP)

Incorporating a dual-microprocessor system, the Central Maintenance Panel (CMP) is designed to record aircraft fault data obtained from LRUs while the aircraft is either in the air or on the ground. It comprises a standard ATR enclosure constructed from light aluminium alloy.

The CMP monitors up to 192 bipolar discrete outputs from equipment or systems through remote sensors. Recorded fault data can be interrogated by ground crew for diagnostic purposes via either the operator interface or an RS-422A interface. The operator interface comprises a 32-character LCD which displays the CMP status and fault data. Interrogation push-buttons are mounted on the front face of the panel. The RS-422A interface is a serial data transmission interface enabling the operator to download fault data stored in the CMP to customer automatic test equipment. This enables data to be output to a printer to provide hard copy of the faults that have occurred in the LRU.

The CMP has six modes of operation: normal; self-test; standby; fault indication; data transmission and reprogramme. The fault indication and data transmission modes can be manually selected; all the other modes are automatically selected according to the access door position.

Specifications

Dimensions: ½ ATR
Weight: 2.91 kg
Power supply: 28 V DC, 30 W (max)
Temperature range: –54 to +71°C
Altitude: up to 45,000 ft

Operational status

Fitted to the Embraer AMX aircraft.

Contractor

Page Aerospace Ltd.

VERIFIED

2000 series Solid-State Combined Voice and Flight Data Recorder (SSCVFDR)

The 2000 series Solid-State Combined Voice and Flight Data Recorder (SSCVFDR), integrates both cockpit voice and flight data recording functions into a single unit. The recorder complies with the latest EUROCAE ED-55 and ED-56A and the FAA TSO-C123/C124 requirements for airborne recording equipment. Several options are available on the unit, such as extended audio duration, onboard maintenance system interface or integral area microphone preamplifier. The baseline unit records a minimum of 25 hours of flight data, at 128 words/s, with 30 minutes' voice on four channels and can interface to any ARINC 573 or ARINC 717 digital flight data acquisition unit or form part of a health and usage monitoring system. Continuous self-test, front panel interface, for *in situ* replay and diagnostics checks, and an underwater locator beacon are standard features.

The SSCVFDR can be hard mounted in the aircraft without shockmounts, as there are no moving parts. Because both data and audio recording is achieved in the combined crash-protected unit, the SSCVFDR is particularly suitable for airframes where weight and space are restricted. Advanced design techniques and manufacturing processes, coupled with rigorous environmental testing, mean high reliability and no periodic maintenance.

Specifications

Dimensions: 124 × 320 × 194 mm, ½ ATR Short
Weight: <9 kg
Power supply: 115 V AC, 400 Hz or 28 V DC, 20 W (max)
Temperature range: –55 to +70°C
Altitude: –15,000 to 55,000 ft
Environmental: DO-160C
Reliability: 15,000 h MTBF

Operational status

In service with military and civil operators. Current production item.

Contractor

Penny & Giles Aerospace Ltd.

VERIFIED

2000 series solid-state combined voice and flight data recorder 0001304

900 series combined voice and flight data recorders

The 900 series of flight data recorders is derived from military recorders fitted to aircraft such as the Harrier GR. Mk 7, Tucano and EH 101 Merlin. It records 5 hours of data at 128 words/s plus three channels of audio, each of 1 hour duration. An enhanced version gives 8 hours of data.

Data can be transferred at high speed for rapid extraction. The system includes an ultrasonic locator beacon with a six year battery life. The system is CAA approved and the audio section meets requirements of CAA Specification 11 and is compatible with ARINC 557; crash protection meets or exceeds TSO C51a. The 900 series has comprehensive BITE and integrity of recorded data can be confirmed on a read-after-write basis.

Specifications

Dimensions: 420 × 175 × 118 mm
Weight: 8.5 kg
Power supply: 28 V DC, 14 W

Operational status

900 series recorders have been delivered to Bristow helicopters as part of the IHUMS system on the Sikorsky S-61 helicopter. No longer in regular production, but available on special order.

Contractor

Penny & Giles Aerospace Ltd.

VERIFIED

Accident Data Recorders (ADRs)

The family of accident data recorders has been developed by Penny & Giles around a recycling mechanism using plastic-based magnetic recording tape. The track configuration is variable to meet customer needs, with up to eight tracks being currently available in various combinations of voice and data. These ADRs are fully protected according to all the major civil and military airworthiness requirements. Versions are currently in production for the Harrier GR. Mk 7, Tucano and the EH 101 Merlin.

Harrier GR. Mk 7 This unit has four data and two voice tracks arranged in two channels, each having two data and one voice track. Recording duration is 2 hours.

Tucano This has two data channels of two tracks each and three voice tracks which are recorded all the time the recorder is operating. Recording duration is 2 hours for data and 1 hour for voice.

EH 101 helicopter ADR This ADR is in production .

Operational status

In service on the Harrier GR. Mk 7, Tucano and in production for the EH 101 helicopter.

Contractor

Penny & Giles Aerospace Ltd.

VERIFIED

ADR 800 accident data recorder

The ADR 800 family of accident recorders, designed for civil and military fixed-wing aircraft and also used for helicopter flight trials, continues in service. It has a capacity of 138 Mbits and can interface either directly or via ARINC 573. Packaged into a ½ ATR short unit, with or without vibration isolators, the system is protected according to the requirements of TSO C51a. Operating temperature is between –65 and +70°C.

Operational status

In service on Concorde aircraft and some older Boeing B707 aircraft; no longer in production.

Contractor

Penny & Giles Aerospace Ltd.

VERIFIED

Cockpit Voice Recorder CVR-90

The CVR-90 is compliant with both FAA TSO C-123 and EUROCAE ED-56A survivability and environmental requirements. Voice recording on four channels is provided with a capacity of 30 minutes per channel. The CVR supports ARINC 557 and 757 compatible avionics.

Specifications

Dimensions: ½ ATR short
Colour: International orange
Operating temperature: –55 to + 70°C
Non-operating temperature: –55 to +85°C
Power: 28 V DC or 115 V AC, 400 Hz less than 15 W
Weight: 8.2 kg
MTBF: 20,000 h
Features: built-in test; fail safe erase through double electrical interlock

Contractor

Penny & Giles Aerospace Ltd.

VERIFIED

Flight Data Recorder FDR-91

FDR-91 is compliant with FAA TSO C-124 and EUROCAE ED-55 survivability and environmental requirements. A data recording capacity of 25 hours is provided. The FDR supports ARINC 573, 717 and 747 compatible avionics.

Specifications

Dimensions: ½ ATR long
Colour: International orange
Operating temperature: –55 to + 70°C

Non-operating temperature: –55 to +85°C
Power: 28 V DC or 115 V AC, 400 Hz less than 15 W
Weight: 8.4 kg
MTBF: 20,000 h
Features: built-in test; Harvard bi-phase input 64/128 bits/s; high-speed download

Contractor
Penny & Giles Aerospace Ltd.

VERIFIED

Optical Quick Access Recorder (OQAR) 0005419

MultiPurpose Flight Recorder (MPFR)

The Penny & Giles Aerospace MPFR functions as both a Cockpit Voice Recorder (CVR) and Flight Data Recorder (FDR), and it uses new materials for impact and fire protection to halve the weight of earlier generation recorders. Hardware and software variants offer wide flexibility in both functionality and recording duration.

Specifications
Cockpit Voice Recorder (CVR)
Duration: four channels × 30 minutes
option for up to 4 × 120 minutes
ATC recording: ARINC 429 digital ATC recording (low- or high-speed)
Audio bandwidth: 150 - 3,500 Hz (3 × voice channels)
150 - 6,000 Hz (area microphone channel)
Playback: real-time (read-after-write) summed output of all audio channels
high speed digital audio recovery via Ethernet into PC or ground station

Flight Data Recorder (FDR)
Duration: 25 h at up to 256 words/s
shorter duration at up to 4,096 words/s
Data acquisition: ARINC 573 or MIL-STD-1553B from data acquisition unit
Replay: ARINC 573 read-after-write
Ethernet to PC high speed
wideband datalink high speed
Fault reporting: ARINC 429 interface to OMS

Common to all configurations
Location beacon: standard Ultrasonic Locator Beacon (ULB)
Power: 28 V DC, 20 W
Cooling: convection, no forced air
Dimensions: ARINC 404A, ½ ATR short: 124 × 320 × 194 mm (W × D × H), or miniature outline: 115 × 230 × 86 mm (W × D × H) excluding ULB
MTBF:
(fixed-wing) 15,000 h
(rotary-wing) 10,000 h

Optional features
Wireless datalink system: interface
PC-card copy facility: integrated

Optional equipment
Cockpit-mounted area microphone: optional
Control unit with integral microphone: optional
Portable replay equipment: optional
Remote pre-amplifier and area microphone: optional

Operational status
Selected as a standard fit item for the Bell 609; first delivery was planned for March 2000, with flight trials planned for July to September 2000.

Contractor
Penny & Giles Aerospace Ltd.

VERIFIED

Optical Quick Access Recorder (OQAR)

Penny and Giles new high-capacity Optical Quick Access Recorder (OQAR) enables airlines to transfer data quickly from the flight line to the Flight Operations Quality Assurance (FOQA) system. The compact, ruggedised unit features a fault tolerant design with extensive internal diagnostics and provides up to 40 hours of recorded data at 1,024 words/s and 640 hours of data at 64 words/s. This data allows an airline to conduct the following types of analysis: Engine Condition Trend Monitoring (ECTM); flight operation analysis; aircraft and autopilot performance; engine exceedance reporting; automatic landing analysis and engine derate measurement.

The Optical Quick Access Recorder is designed to provide all the necessary data for effective Flight Operations Quality Assurance (FOQA) programmes. The OQAR can be programmed to record over 2,000 separate parameters, including data from the APU, brake system, and navigation computer which is typically not recorded by other flight data recorders. Airlines that implement FOQA programmes can select which parameters to record, in order to increase safety while reducing maintenance costs, to identify problems and trends more easily and to optimise crew training.

These new high-capacity data recorders document operational characteristics of the aircraft's entire flight, including take-offs and landings. Fully ruggedised, they record even when subjected to vibration or shock, so transient or abnormal events such as hard landings or fast rotations will not be missed. The unit is designed so the rewritable magneto-optical disk can be quickly removed from the aircraft during a level A check, at weekly intervals or even during a normal gate turn. As part of the FOQA programme, the flight data is then downloaded to a ground station where after de-identification, analysis for significant operational patterns can be conducted. Additional ARINC 429 inputs are provided for supplementary TCAS (Traffic Collision and Avoidance System) and GPS (Global Positioning System) recording, or for use as part of a FANS (Future Air Navigation System).

Operational status
The OQAR units are available in configurations for all new Airbus and Boeing aircraft; and are fully compatible with all digital bus aircraft including ARINC 573/717, and ACMS. They feature embedded control software, which can be upgraded while in service use. A front panel RS-232 port allows for self-diagnostic checks, relay of the stored data, or configuration changes. The recorders use industry standard, 3.5 in magneto-optical disks with MS-DOS-compatible format, which are ISO 10090 compliant.

Contractor
Penny & Giles Aerospace Ltd.

VERIFIED

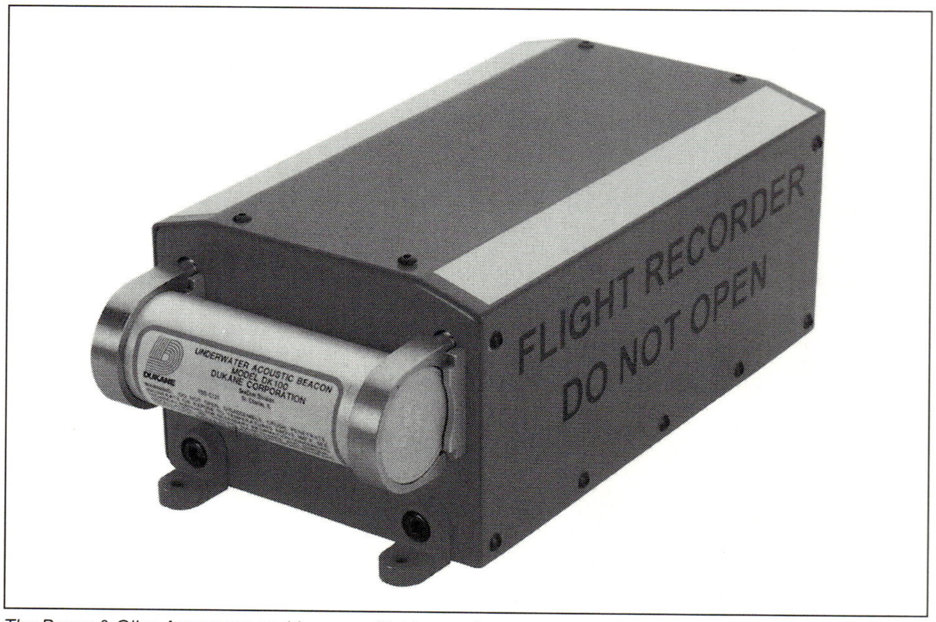

The Penny & Giles Aerospace multipurpose flight recorder **2000**/0062791

Penny & Giles airborne and ground program loaders

PLU 2000 program loader

Penny & Giles' computer program loaders are based on industry standard DC300A data cartridges and expanded capacity DC300XL units. They provide a range of features appropriate to military or commercial fixed-wing aircraft and helicopters, and land- and sea-based vehicles and vessels, in severe environments where low error rates are essential.

The PLU 2000 program loader can accommodate up to three optional modular interfaces, has a capacity of 3.2 Mbits expansible to 6.4 Mbits and can load up to 10 preselected programs. It is operated by a remote-control unit.

Operational status

In service on UK Royal Air Force Nimrod aircraft and UK Royal Navy Sea King helicopters.

Contractor

Penny & Giles Aerospace Ltd.

VERIFIED

Primary Air Data Computers (ADC)

Penny & Giles Aerospace produces primary accuracy ADCs in both miniature and ARINC 565 standard form factors. The Generic Air Data Unit (GADU) miniature ADC is a compact lightweight unit with integrated pitot/static solid-state sensors in a rugged enclosure. The GADU is available with either ARINC 429 or MIL-STD-1553 databus interfaces. A height lock function is provided on the GADU and is therefore particularly suited to rotary-wing applications.

Specifications

Miniature primary ADC GADU D60350
Dimensions: 89 × 132 × 172 mm
Weight: 1.3 kg
Power: 28 V DC, 10 W peak
Standards: RTCA DO-160C/178B
Interfaces: ARINC 429/575, airspeed and height lock, MIL-STD-1553 (optional)
MTBF: 15,000 flight hours

Generic Air Data Unit (GADU) ADC 0005416

Operational status

Fitted to a wide range of aircraft and helicopters including: Sikorsky S70/S76, Westland Lynx/Sea King, Eurocopter Dauphin, MBB BK117 helicopters; and Boeing 727/737, Lockheed Martin C-130 aircraft. Applications include: primary air data system; rotary- and fixed-wing versions for RSVM retrofit; FMS; TCAS; GPWS; windshear detection.

Contractor

Penny & Giles Aerospace Ltd.

VERIFIED

Quick access recorder

The quick access recorder is an avionic data logger of modular construction designed to provide rapid access to aircraft performance data recorded in flight. The media used is an industry standard ¼ in magnetic tape cartridge, giving low-cost data storage.

The unit fulfils the quick access recorder requirement in a civil aircraft integrated monitoring system; it also provides automatic read-after-write error detection and correction and an onboard replay facility with GMT search. The quick access recorder is microprocessor controlled and conforms to ARINC 591. It has a built-in power supply, with battery back-up for power drop-out immunity.

Specifications

Dimensions: 124 × 320 × 194 mm
Weight: <7 kg
Power supply: 115 V AC, 400 Hz, 70 VA peak
Reliability: 4,500 h MTBF

Operational status

Selected by a number of major airlines for use on a wide variety of aircraft.

Contractor

Penny & Giles Aerospace Ltd.

VERIFIED

Secondary Air Data Computers (ADC)

The Digitas and TAS/Plus ADCs provide a wide range of outputs including: altitude; airspeed; airspeed rate; baro corrected altitude; static air temperature; true airspeed, Mach and total air temperature. Depending on the model, these outputs are available on various avionic interfaces, including ARINC 429/575, MIL-STD-1553, RS 232/422 and analogue.

Specifications

Miniature primary ADC GADU D60350
Dimensions: 89 × 132 × 160 mm
Weight: 1.1 kg
Power: 28 V DC, 10 W peak
Standards: RTCA DO-160B/178B
Interfaces: ARINC 429/575, airspeed
Optional interfaces: MIL-STD-1553
MTBF: 15,000 flight hours
Applications: Secondary air data source, rotary- and fixed-wing flight data recording, HUMS

Penny & Giles Aerospace Digitas ADC 0005417

Secondary ADC TAS/Plus 90004
Dimensions: ⅜ ATR short
Weight: 2.04 kg
Power: 28 V DC, 10 W peak
Standards: RTCA DO-160C/178B, TSO-C106
Interfaces: ARINC 429/575, RS-232/-422, altitude rate/airspeed
Optional interfaces: Analogue altitude rate and Mach
MTBF: 10,000 flight hours
Applications: Business and commuter aircraft GPS, FMS, TCAS, GPWS, flight data recording data display

Operational status

Fitted to a wide range of rotary- and fixed-wing aircraft.

Contractor

Penny & Giles Aerospace Ltd.

VERIFIED

Secondary air data sources

The analogue Air Data Module (ADM) has been especially designed for GPWS (Ground Proximity Warning System) applications requiring a low-cost source of secondary air data information via an analogue channel. The ADM can be mounted in a wide variety of locations to aid retrofit of a GPWS capability.

The TP91 range of altitude and airspeed transducers provides a passive, lightweight, low-cost solution for applications such as flight data recording where a secondary or independent source of altitude or airspeed data is required.

Specifications

Air Data Module D60286
Dimensions: 89 × 132 × 159 mm
Weight: 1 kg
Power: 28 V DC, 5 W peak
Standards: RTCA DO-160C
Applications: GPWS

Altitude and Airspeed Transducer TP91
Dimensions: 83 × 90 × 82 mm
Weight: 0.8 kg
Power: 12 V DC, 4 W peak
Standards: TSO C51a
Applications: Flight data recording

Contractor

Penny & Giles Aerospace Ltd.

VERIFIED

Solid-State Quick Access Recorder (SSQAR) D51555

The Solid-State Quick Access Recorder (SSQAR) D51555 has been developed for light turbine-engined helicopters and turbofan corporate aircraft for offline trend analysis, aircraft performance monitoring and maintenance recording. It is a ruggedised, lightweight and environmentally sealed data recorder which uses industry standard solid-state memory cards. The unit acquires navigational data, typically from a GPS

The solid-state quick access recorder D51555

receiver, as well as engine and airframe parameters from various ARINC 429/573/717, RS-422, analogue and discrete sources. These inputs are sampled and decoded as necessary and the acquired data is time-stamped with a GMT value before being written to flash memory cards. Three card slots are provided, each capable of addressing up to 32 Mbytes of memory, subject to card availability of higher-capacity cards in the future.

Specifications

Power supply: 28 V DC nominal
Interfaces: ARINC 573, ARINC 429 (2), RS-422 (2), analogue (4), discrete inputs (16), run control, discrete outputs
Reliability: 15,000 h MTBF

Contractor

Penny & Giles Aerospace Ltd.

VERIFIED

Miniature primary Air Data Computer GADU D60350

The GADU miniature ADC is a compact, lightweight unit with integrated pitot/static solid-state sensors in a rugged enclosure. The GADU is available with either ARINC 429 or MIL-STD-1553 bus interfaces. A 'height-lock' function is provided on the GADU, which makes it particularly suitable for helicopter application.

Specifications

Dimensions: 89 × 132 × 172 mm
Weight: 1.3 kg
Power supply: 28 V DC, 10 W peak

Operational status

In service on a wide range of helicopters; also on Boeing 727/737 and Lockheed Martin C-130 aircraft.

Contractor

Penny & Giles Avionic Systems Ltd.

VERIFIED

0730 KEL series low-cycle fatigue counter

The 0730 KEL series calculates cumulative low-cycle fatigue and records the number of engine starts and engine hours. It also provides information on exceedances, banding of speeds, voltages and thermocouples, spool-up and spool-down times and snapshots of input data. The 0730 series records a maximum of 18 aircraft or engine input analogue parameters including up to six speed, six voltage, two thermocouple and four discrete parameters. These are all configurable in range and sensitivity to customer requirements. It provides outputs via RS-232 link to a data transfer device, printer or maintenance computer, and current and voltage sources for aircraft sensor excitation. It also provides an output to an integral LED display for manual interrogation of life usage exceedance fault codes.

Specifications

Dimensions: ¼ ATR dwarf
Weight: 2.5 kg
Speeds: pulse probe or tachogenerator 10 Hz to 25 kHz
Voltage: range selectable 0-0.5 V, 0-5 V, 0-40 V FSD
Thermocouple: 0-1,000°C Ch/Al
Discretes: 0-5 V, 0-28 V nominal

Operational status

In service with Swiss Air Force, the UK Royal Air Force Red Arrows aerobatic team and the Sultan of Oman's Air Force BAE Systems Hawks and Swiss Air Force Eurocopter Super Pumas. Selected for UK Royal Air Force Hawk aircraft, and Eurocopter Cougar helicopters.

Contractor

Smiths Industries Aerospace.

VERIFIED

0826 KEL health and usage monitor

The 0826 KEL health and usage monitor provides snapshots of input data, exceedance monitoring and incident monitoring. It also calculates cumulative low-cycle fatigue. It records up to eight speed, 32 voltage, 16 discrete and four vibration aircraft and engine parameters. Outputs are provided via RS-422 link to a data transfer device, printer or maintenance computer and the system has an integral LED display for manual interrogation of input parameter values.

Specifications

Dimensions: ⅜ ATR
Weight: 5 kg
Speed: pulse probe 10 Hz to 4 kHz
Voltage: 0-5 V FSD
Discretes: 0-28 V nominal
Vibration: 20-750 Hz buffered

Operational status

In production for and in service in the BAE Systems 146.

Contractor

Smiths Industries Aerospace.

VERIFIED

0829 KEL health and usage monitor

The 0829 KEL health and usage monitor provides snapshots of input data, exceedance monitoring and incident monitoring. It has capacity available to incorporate LCF counting. There are 30 aircraft and engine input parameters available, including up to eight speed, 15 voltage and seven discrete parameters, and the system incorporates an ARINC 429 databus. Outputs are via RS-422 link to a data transfer device, printer or maintenance computer and there is an integral LED for manual interrogation of input parameter values.

Specifications

Dimensions: ⅜ ATR
Weight: 5 kg
Speed: pulse probe 10 Hz to 4 kHz
Voltage: 0-5 V FSD
Discretes: 0-28 V nominal
Vibration: 20-750 Hz buffered

Operational status

In service with the BAE Systems ATP.

Contractor

Smiths Industries Aerospace.

VERIFIED

Digital air data computers for civil aircraft

A range of digital air data computers has been produced for both turboprop and civil jet transports. These computers meet the requirements of ARINC 706 specification, with outputs in both ARINC 429 and analogue format. They incorporate high-performance vibrating cylinder sensors and CMOS technology to minimise weight, space and power requirements.

Extensive BITE facilities are available, including a 10-flight memory to facilitate onboard checks and bench diagnostic routines.

This range of computers is demonstrating extremely high reliability on the Boeing 737 with an MTBF in excess of 30,000 flying hours.

Specifications

Dimensions: ? or ½ ATR
Weight: 4-5.9 kg
Power: 20 VA

Operational status

In production and service on the Boeing 737-300, 400 and 500, the BAE Systems ATP and Boeing 727 upgrade.

Contractor

Smiths Industries Aerospace.

VERIFIED

Digital air data computers for military aircraft

A range of digital air data computers is available for both conventional and VTOL aircraft applications. Additional channels are incorporated where increased integrity is required as in the case of VTOL aircraft.

Depending on specification, high- or low-range pressure vibrating cylinder sensors are used and outputs are provided in either analogue or MIL-STD-1553 format. Comprehensive BITE facilities are incorporated, including continuous in-flight monitoring and fault location down to module level.

Specifications

Dimensions: ⅜ ATR
Weight: 5 kg
Power: 35 VA

Operational status

In service on the BAE Systems Sea Harrier.

Contractor

Smiths Industries Aerospace.

VERIFIED

Displays and Mission Computer DMC

Smiths Industries Aerospace has developed a modular Displays and Mission Computer (DMC), which can be configured to match the exact system requirements of each application by selecting modules from an extensive library of standard electronic cards, power supplies and ATR short cases.

The standardised internal databus allows the system to be updated or expanded by inserting appropriate cards as technology evolves.

The size of the DMC depends on the number of modules needed to meet the operational requirements; ½ ATR, ¾ ATR, and 1 ATR sizes are available.

Electrical connections are via a rear panel which can be adapted to different connector configurations including: DPX and ARINC 600. The panel is detached to provide access to the backplane. Cases are cooled to ARINC 404A/600 specifications.

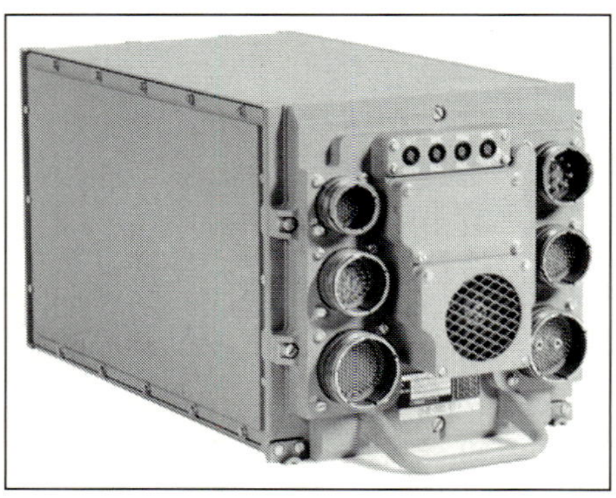

Displays and Mission Computer DMC 0015266

Operational status

The DMC is standard equipment on the next-generation BAE Systems Hawk for the Royal Australian Air Force lead-in fighter programme, and for a number of other programmes including: the BAE Systems Harrier F/A-2, Boeing AV-8B Harrier II and T-45 Goshawk.

Contractor

Smiths Industries Aerospace.

VERIFIED

Onboard Maintenance System (OMS)

The Smiths Industries' engine life computer and standard flight data recorder have been combined into a single Onboard Maintenance System (OMS). The engine life computer records running hours and engine starts, and provides real-time calculations of low-cycle fatigue and snapshots of data to highlight deviations, as well as recording excursions outside defined limits. The standard flight data recorder records structural, engine, tracking, maintenance, mishap and training or mission replay data.

The benefits of the OMS are crash survivability, crash position location and water recovery, fast accurate data transfer, management of major engine components, improved flight safety and longer life and lower cost of ownership. The system is suitable for both new build aircraft and retrofit. It consists of a Signal Acquisition Unit (SAU), Crash Survivable Memory Unit (CSMU) and engine life computer. The SAU serves as the interface to aircraft data signals and power inputs. It acquires, computes, compresses and stores data. Choosing which data to record is controlled by software and can be tailored to any application. The SAU is entirely solid state.

The CSMU can be adapted to a variety of requirements. Retention of the last 15 minutes or a full 25 hours of mishap data is available and the capability to store the last 2 hours of voice is in development.

The engine life computer processes data to compute engine life usage, thermal creep, exceedances and performance information. This is transferred into internal non-volatile memory for retrieval either manually via the integral LED display or semi-automatically via a data transfer device. The computer uses standard hardware modules and the software architecture is structured for multiple applications. External programming enables general engine algorithms, such as low-cycle fatigue and exceedance levels, to be factory programmed to suit specific applications and customer requirements. The computer includes 64 kbytes of non-volatile memory to store cumulative life usage totals, performance data and incident records. Extensive BIT is incorporated to monitor the system for correct functionality. The computer can be configured to drive annunciators for predefined input conditions, exceedances and BIT. A Deployable Flotation Unit (DFU) houses the CSMU and is equipped with a radio beacon and strobe. The buoyant DFU is released at impact or at a specified depth. The beacon and strobe provide visual and radio position location and will operate for at least 72 hours continuously.

Contractor

Smiths Industries Aerospace.

VERIFIED

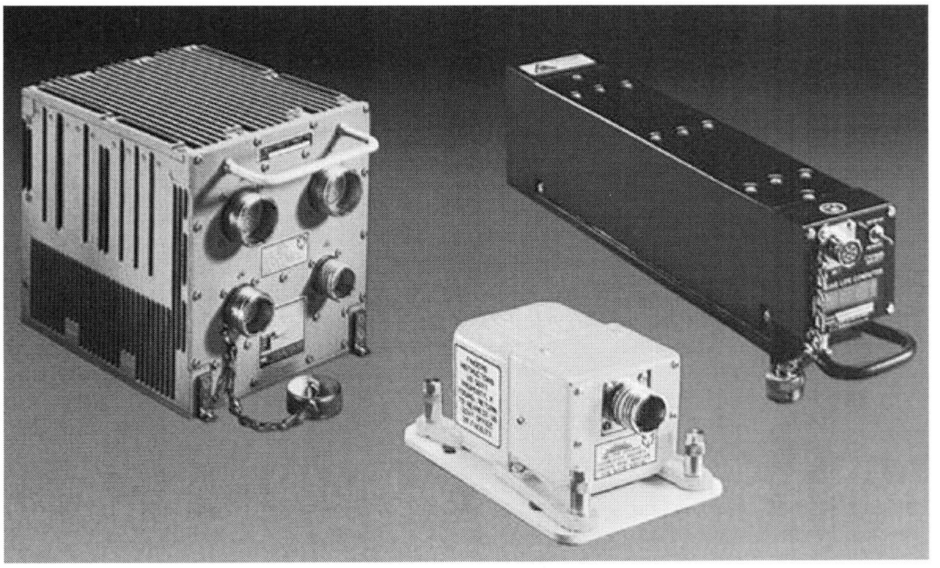

The Onboard Maintenance System consists of (left to right) the signal acquisition unit, crash survivable memory unit and engine life computer

Open Systems Architecture Mission Computer (OSAMC)

Smiths Industries Aerospace is developing a new open systems architecture mission computer including a comprehensive library of VME standard cards.

The use of the VME64 standard allows the use of only those cards which conform to this standard, including commercial-off-the-shelf (COTS) modules.

A full range of functions is available covering MIL-STD-1553B, I/O, processing, stroke and raster graphics, generators and custom interfaces.

The size of an OSAMC depends on the number of modules required and ½ ATR, ¾ ATR and 1 ATR chassis are available with five, eight and 10 card slots respectively. Forced air or integral fan cooling can be provided.

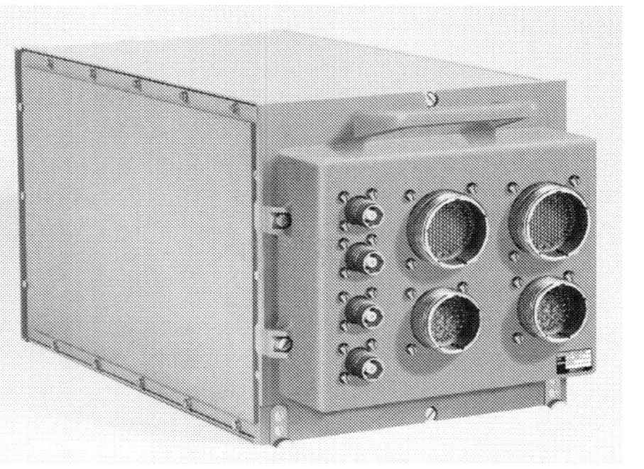

Smiths Industries open systems architecture mission computer (OSAMC) ***1999***/0051284

Operational status

Selected for the BAE Systems Harrier F/A2 (core computer) and the BAE Systems Harrier GR. Mk 7 (night attack display computer).

Contractor

Smiths Industries Aerospace.

VERIFIED

SN 170 Digital Video Recorder (DVR)

The SN 170 DVR is a hard disk digital recorder capable of recording up to eight camera inputs in a time-lapse mode. The system is intended for airborne surveillance applications with cameras mounted in the passenger cabin, or in other areas of the aircraft, as required.

The recorder may be set for continuous loop recording, with no manual intervention required. The recording time can be up to 36 hours, dependent on the recording rate selected. Recording can be viewed on the aircraft, or the disk cartridge can be removed for review at a replay station.

The replay station contains facilities for copying video sequences for evidential purposes.

Specifications

Recorder
Dimensions: ARINC 600 rack-mounted unit (3 MCU)
Weight: 3.5 kg
Power: 28 V DC
Hard disk capacity: 16 Gbytes
Video compression: 18 : 1
Recording duration: 36 h with four cameras at 1 picture/s/camera

Colour camera
Dimensions: to suit installation
Weight: 0.5 kg
Resolution: 450 TV lines
Power: 12 V DC from the recorder

Operational status

In development.

Contractor

Smiths Industries Aerospace.

VERIFIED

High-Bandwidth Video Data Recorder (HBVDR)

Data recording specialist Somerdata Ltd has teamed with Racal Defence Electronics Ltd (Wells) to claim a new concept in very wide bandwidth data capture and storage, known as HBVDR (High Bandwidth Video Data Recorder). The HBVDR has been designed initially to capture short passages of video data at bandwidths up to 30 MHz. Typical applications include the acquisition of signals from monopulse, pulse compression and passive radars.

HBVDR uses a highly flexible 'mixed media' strategy for data storage. Data is captured initially in solid-state DRAM before being transferred to an integral hard disk from which it can be sent to a networked workstation or SCSI tape drive for off-line analysis.

An important advantage of the unit over traditional data recorders is the degree of flexibility offered by its system architecture, permitting complete scaleability of the storage resources to meet current or future application needs. In its basic configuration, HBVDR can be fitted with up to 10 Gbytes of DRAM while additional memory can be accommodated in an expansion chassis if required. Current internal disk options include two 5.3 Gbyte hard disk drives while 9.1 or 18 Gbyte drives can be specified for increased on-board storage. Typical SCSI tape drive options for data export/import include JAZ®, DAT and DLT.

The system's inherent flexibility extends also to the conditioning of the input and output signals. The standard wideband video input/output modules can be upgraded to handle serial or parallel digital data at rates up 650 M/bits/s, or greater.

All records are stored as conventional binary files on disk. Removable storage uses the TAR format to facilitate data exchange with UNIX systems for off-line data analysis.

The HBVDR's 'snapshot' capability is a flexible and convenient means of capturing radar and EW signatures with high precision in their original analogue form for subsequent reply, analysis or interpretation. More than 2 minutes of continuous 30 MHz video data can be stored in the unit's 10 Gbytes of DRAM memory. Alternatively, with a suitable signal interface, the 160 MHz/1GHz IF video data output of a typical EW (electronic warfare) receiver can be captured.

The basic HBVDR is supplied as a portable or rack mounted 31 kg unit measuring approximately 381 × 534 × 500 mm and conforming to a range of typical static and mobile military platform environmental specifications.

In addition to the existing analogue radar interface, the Somerdata/Racal team anticipates the development of additional data input / output interfaces for other radar, EW and comms applications.

Contractors
Somerdata Ltd.
Racal Defence Electronics Ltd (Wells).

VERIFIED

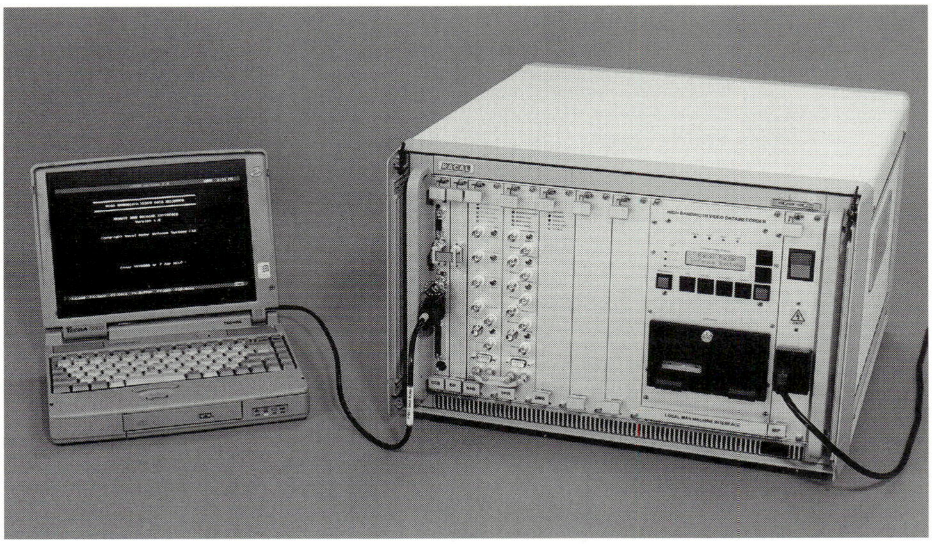

Somerdata/Racal Defence Electronics high-bandwidth video data recorder
1999/0051327

Remote Data Concentrators (RDC)

Ultra Controls is involved with the development of RDCs as an integral component of the Integrated Modular Avionics (IMA) architecture for the next generation of avionic systems. RDCs provide a localised interface for sensors, utilities and actuators in the IMA environment.

Operational status
Reported to be still under development.

Contractor
Ultra Electronics Ltd, Controls Division.

UPDATED

Series 2768 Super VHS video cassette recorder

The small lightweight Series 2768 helical scan video recorders are suitable for recording both video and data signals in hostile environments and alternative case designs enable the recorder to be located in either cockpit or equipment bay.

The recorder employs a Super VHS-C/VHS-C cassette to achieve a minimal size. Cassettes may be replayed on the ground by means of a mechanical adaptor in standard Super VHS-C and VHS video equipment, resulting in low-cost and readily available playback systems. Vinten video recorders are capable of producing recordings through high *g* aircraft manoeuvres, high vibration or gunfire.

In operation the Series 2768 recorder is sealed against sand, dust and water and contains a conditioning heater for low-temperature operation.

A significant feature of the Series 2768 is the incorporation of anti-vibration/shockmountings within the case of the recorder. This enables the recorder to be hard-mounted to the aircraft in almost any orientation and requires no sway space. The Series 2768 recorder may be used for unusually formatted signals such as those from linescan sensors.

The Airborne Recorder for IRLS and EO Sensors (ARIES) Video Cassette Recorder (VCR) is one of the Vinten Series 2768 recorders which are designed specifically for airborne environments. The recorder is available in both cockpit and bay-mounting versions.

The VCR is designed for recording linescan formatted video imagery derived from IRLS or EO sensors. It can be fitted internally or in a pod in manned aircraft or in RPVs, drones and UAVs.

Series 2768 video cassette recorder 0015318

Specifications
Dimensions: 125 × 152 × 226 mm
Weight: 3.5 kg
Recording system: rotary helical scan
Video signal system: PAL colour/CCIR monochrome 625-lines or NTSC colour/EIA monochrome 525-lines, plus other formats for non-standard signals
Recording time: up to 2 h
Event marker: visual and audio
Recording bandwidth: up to 4.8 MHz

Operational status
In production and in service.

Contractor
W Vinten Ltd.

VERIFIED

Series 3150 colour video camera

The Series 3150 colour video camera is a modular CCD camera specifically designed for airborne recording applications.

The camera is designed to provide high-resolution colour images and is readily installed as either original equipment or as an aircraft upgrading. There are two versions of the camera: the low-profile and the universal. The latter has the ability to separate the lens and sensor from the remainder of the camera by a flying lead. This new concept for an airborne colour camera enables the ideal positioning of the camera lens in a location where there may otherwise be limited room for the complete camera. Previously the only alternatives were either to produce a special camera of dedicated design or to provide an optical periscope and suffer the resulting reduction of light to the sensor.

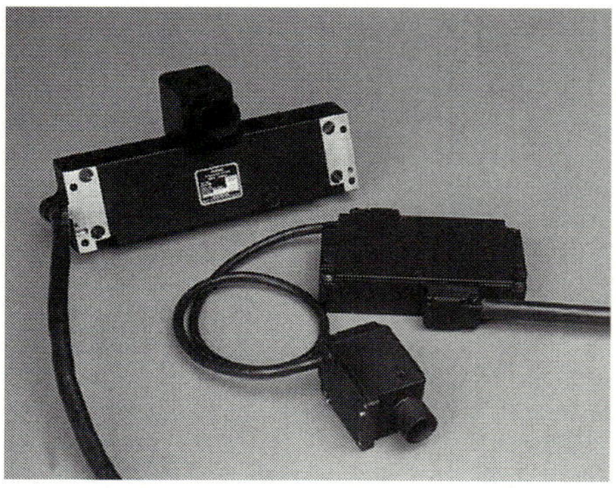

Alternative configurations of the Series 3150 colour video cameras 0015319

The camera is able to produce television pictures over a very wide illumination range and operate continuously in very demanding environmental conditions.

For cockpit installation the Series 3150 camera's standard modules are flexible enough in configuration to be orientated to suit most HUD and gunsight mounting requirements. This approach offers both an optical and cost-effective solution. The equipment can be used in association with Vinten airborne video recorders such as the Series 2768.

Specifications
Dimensions:
(camera head) 105 × 56 × 24 mm
Weight:
(camera head) 0.75 kg
Power supply: 28 V DC, 18 W nominal when used with Series 2768 VCR
Camera head sensor: solid-state imager
Field of view: 25 × 19° or 20 × 15°
Refresh rate: 50 or 60 Hz
Dynamic range: 5-170,000 lx

Operational status
In service and in production.

Contractor
W Vinten Ltd.

VERIFIED

Type 1192 recycling wire recorder

The Type 1192, based on a recording medium of stainless steel wire, preserves the last 25 hours of flight data for crash analysis. Wire has long been used as a recording medium and gives the maximum data storage for a given volume and weight. The Type 1192 mechanical transport can be used in conjunction with any data processing and normalising system providing analogue or digital information. The system is designed to withstand 100 *g*, 5 m/s shocks, the impact of a 227 kg steel bar dropped from 3 m, a static crush force of 2,270 kg, exposure to flame at 1,100°C for 5 minutes over half the outside area, and a combination of fuel, hydraulic fluid, lubricating oil, fire extinguisher fluid and salt water for 36 hours.

Specifications
Dimensions: 203 mm sphere reducing to 178 mm cylinder over central region
Weight: 7.27 kg
Power supply: 115 V AC, 400 Hz, 3 phase
28 V DC

Recording medium: 0.058 mm diameter stainless steel wire, breaking force >0.34 kg
Data transfer rate: 1,536 flux reversals/s

Operational status
In service. No longer in production.

Contractor
W Vinten Ltd.

VERIFIED

Type 1200 cockpit voice recorder/reproducer

The Type 1200 voice recorder/reproducer is designed for mounting in the confined areas typical of fighter cockpits. When connected to the communications control system, it provides intelligible speech and record/replay facilities. The system has only two controls, a selector switch for record/replay and a toggle switch for on/off, rewind and fast forward. When a microphone in the recorder detects speech it switches on automatically and continues to operate until speech stops. The recorder then continues to run for a preset period and switches itself off until the next speech sequence causes it to repeat the cycle. Cassettes may be replayed through the aircraft system or through a ground-based playback unit. The system employs standard Philips C60 and C90 audio cassettes. A version designated the Type 1200-004 can record and reproduce data as well as voice.

Specifications
Dimensions: 147 × 43 × 168 mm
Weight: 1.36 kg
Tape cassette: Philips Compact C60 (30 min audio/track) or C90 (45 min audio/track)
Tape erase: automatic before recording

Operational status
In service. No longer in production.

Contractor
W Vinten Ltd.

VERIFIED

Type 1300 pilot's display recorder

The Type 1300 camera photographically records head-up display symbology superimposed on the pilot's view of the outside world. The record can be employed for the training of combat tactics, for debriefing or for assessment of weapon aiming and target damage. Mounted on the side of the head-up display it gives the least visual obstruction to the pilot and preserves a clear rear surface on the head-up display combiner glass for control purposes. The system comprises a camera unit, film magazine and interchangeable lenses. Automatic exposure control and variable shutter speed permit the use of film emulsion speeds within a wide brightness range. The exposure is controlled by a built-in electronic sensing unit and emulsion speed is adjusted automatically by a coding feature in the film magazine that presets the appropriate controls.

Specifications
Dimensions: 210 × 190 × 45 mm
Weight: 3 kg
Power supply: 28 V DC
2 A (starting)
1 A (running)
Film type: 22 m × 16 mm × 0.1 mm double perforated cine
Framing rate: 16 or 32 frames/s
Film movement indicator: visual indication
Exposure control: integral with camera, automatic over brightness range 78-5,028 ft-lambert
Running time: 3 min at 16 frames/s
Shutter speed: 1.25-10 ms
Event marker: remote operation
Lens (with periscope):
f/2.8-22, F=25 mm, FoV 16 × 22°
f/2.8-22, F=50 mm, FoV 8 × 11°

Operational status
In service. No longer in production.

Contractor
W Vinten Ltd.

VERIFIED

Type 6051 video conversion unit

The Vinten Type 6051 video conversion unit is designed as a 525-line/60 Hz to 625-line/50 Hz video converter for use in combat aircraft. It enables 525-line/60 Hz video sources to be interconnected to 625-line/50 Hz video switching, display and recording equipment and is appropriate for use where multiple video standards exist on an aircraft. The converted video output can be synchronised to another onboard 625-line/50 Hz source.

Operational status
In production.

Contractor
W Vinten Ltd.

VERIFIED

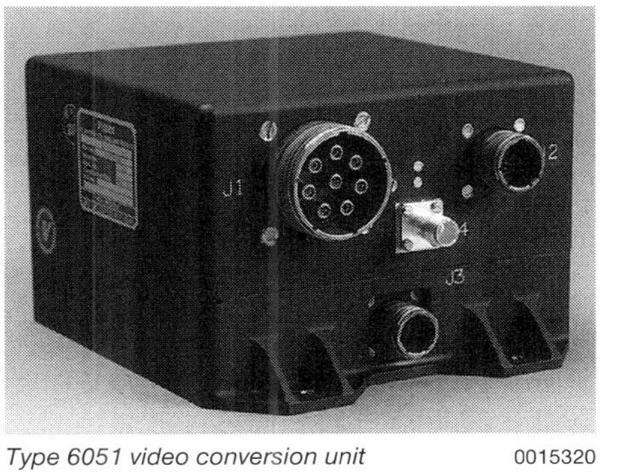

Type 6051 video conversion unit 0015320

UNITED STATES OF AMERICA

Airborne computer for the B-1B

Ametek supplies airborne computer-based systems for both military and aerospace markets. The first application of a microprocessor to an airborne product built by Ametek was the signal conditioning and distribution unit for the B-1A in the early 1970s. With the revival of the aircraft as the B-1B a redesign of the system took advantage of newer technologies which include a 16-bit I²L radiation-hardened design microprocessor and large-scale integrated components. The system interfaces with 64 engine sensors plus 12 more from the aircraft and electrical multiplex subsystem.

The B-1B system includes: a dual-redundant signal processing and power supply to enhance single-point failures; signal conditioning; sensor excitation; data conversion; linear computations; thrust computations; and periodic and initiated self-test parameter compensations for adverse temperatures to generate warning messages. Output information is transmitted via a dual-redundant digital Emux word-generated interface with aircraft avionics. The computer can perform 7 Mips.

Operational status
In service; production for the B-1B is complete.

Contractor
Ametek Aerospace Products.

VERIFIED

Data Acquisition Unit (DAU)

The Data Acquisition Unit (DAU) is designed for analogue or digital processing applications where large amounts of data need to be processed or consolidated. The system can be configured as a single- or dual-channel unit. A dual-channel configuration provides complete hardware redundancy for all parameters.

The DAU is designed using a modular approach with motherboard and plug-in function boards, which can be added as required to accommodate any aircraft sensor or communication bus. Custom boards can be designed to handle specialised sensors or communications. The system utilises advanced technology to process analogue, digital and discrete engine and airframe signals. Typical signals are filtered, converted to digital data, scaled and formatted for cockpit display or use by other aircraft signals.

In addition to analogue inputs, each DAU channel includes two ARINC 429 inputs and an RS-232 interface for communications with Sentinel displays or other aircraft systems such as the flight management system and FADECs.

All processed or stored information is transmitted via the standard ARINC 429 output bus. Optional databusses or buffered analogue and discrete outputs can be incorporated into the DAU. The built-in RS-422 port can be used to output real-time data or as an access point for maintenance interrogation.

In addition to the normal tasks of conditioning and processing signals, the DAU can perform a variety of maintenance functions. Health monitoring, exceedance recording and trending algorithms can be run in a background mode while the processor would normally be idling.

Extensive BIT isolates problems to the faulty sensor or circuit. Self-test is run continuously, with results stored in non-volatile memory for evaluation. Cross-channel communications are used to verify channel integrity.

Specifications
Weight: 4.08 kg
Power supply: 10-32 V DC or 115 V AC, 400 Hz
Temperature: –40 to +70°C

Operational status
Fitted to Bombardier Global Express aircraft.

Contractor
Ametek Aerospace Products.

VERIFIED

Engine Analyser and Synchrophase Unit (EASU)

The EASU provides continuous, real-time monitoring of engine operating parameters. It monitors engine vibration and converts the information to spectral data points, comparing it against defined limits. Exceedances are reported to the cockpit and time-stamped records are stored in non-volatile memory for downloading and analysis by maintenance staff.

For turboprop applications, the EASU functions as a propeller speed/phase synchroniser to reduce aircraft cabin noise and vibration.

Contractor
Ametek Aerospace Products.

VERIFIED

Engine data converter

The engine data converter is used for reduction of large amounts of analogue, digital and discrete data to ARINC 429 format. Some typical sources of data are the engine and transmission sensors, prop speed sensors and all the various aircraft system discretes.

The engine data converter features extremely flexible architecture, permitting maximum cost effectiveness in any application. Included on individual, easily serviced plug-in cards are all the signal conditioning and data conversion circuitry needed to interface the engine sensors with the digital databus. For use in systems where reliability is of the utmost importance, two totally independent channels are provided in each engine data converter.

For use with aircraft systems, the engine data converter offers additional optional features such as internal storage and periodic reporting of engine limit parameters, engine serial number encoding and non-volatile storage of limit exceedances, durations and engine cycle. It can also be structured to interface with other databusses, such as MIL-STD-1553B.

The design is sufficiently flexible so that it can be used with any type of engine. The software, as well as the analogue to digital circuit cards, is modular, so that changes to the engine sensors and engine performance characteristics can be readily accommodated.

One engine data converter is assigned to each engine. It is a dual-redundant device. Each half is a totally independent functioning unit having its own power supply, input converter cards, discrete units, microprocessor and digital output circuits. It shares only a common interconnect between its two halves. Isolation between the two halves is such that a failure of one half will not affect the other half's operation.

Specifications

Weight: 5.22 kg
Power consumption: 7 W per channel typical, 10 W per channel (max)
Reliability: 5,000 h MTBF

Contractor

Ametek Aerospace Products.

VERIFIED

Engine Monitoring System Computer (EMSC)

Ametek manufactures the Engine Monitoring System Computer (EMSC) for the General Electric F110 engine which equips the US Air Force F-15 and F-16C/D aircraft. The computer provides in-flight monitoring of engine exceedance, faults or trends. Engine-related signals are acquired from the engine monitoring system processor via the MIL-STD-1553B engine signal databus. Diagnostic data is retained in non-volatile memory which annunciates this data visually to the cockpit.

A secondary function downloads the data via the RS-232C series communication link to the data display and transfer unit for ground support evaluation. This link also uploads information to the EMSC such as the aircraft engine diagnostic information, time data, aircraft serialisation and life usage data.

Operational status

In service on the General Electric F110 engine on the F-15 and F-16C/D. Most of this system's modules are also used in the engine analyser unit for the Northrop Grumman E-2C's Allison T56-A-427 engine.

Contractor

Ametek Aerospace Products.

VERIFIED

Sentinel™ instrument system Data Acquisition Unit (DAU)

The DAU is used to acquire and transmit critical information from a variety of aircraft systems. It is a dual-redundant 3 MCU unit for monitoring systems such as the electronic engine controller, electrical system, air data computer, hydraulic system and a variety of other aircraft subsystems.

The DAU transmits data through both ASCB and ARINC 429 busses. Information will then be processed by the IC-800 integrated avionics computer and may be shown on the EICAS.

Operational status

The DAU has been integrated with the Honeywell Primus 2000 XP integrated avionics system for the Bombardier Global Express business jet.

Contractor

Ametek Aerospace Products.

VERIFIED

DCRsi™ 75 digital cartridge recording system

The DCRsi 75 is an inexpensive 75 Mbit/s DCRsi variant designed to fill an important niche between 'top-end' S-VHS units and Ampex's existing 107 and 240 Mbit/s DCRsi models. This recording system is targeted particularly at the growing need, particularly within the anti-submarine warfare, airborne instrumentation and telemetry areas, for a severe-environment recorder with a lower rate capability and proportionately lower cost. DCRsi 75 retains the field-proven DCRsi transverse scan recording footprint for full crossplay compatibility with other models in the range.

Operational status

Selected for the mission recorders for the Merlin EH 101 maritime reconnaissance helicopter and for the Nimrod MR2 Mk2 upgrade programme.

Contractor

Ampex Corporation Data Systems Division.

VERIFIED

DCRsi™ 107/107R Digital Cartridge Recording system

The DCRsi 107/107R rack-mount and modular ruggedised systems are 1 in transverse scan, rotary digital recorders capable of recording and reproducing at any user data rate from 0 to 13.4 Mbytes/s (0-107 Mbits/s). This capability can be sustained for over 1 hour in one tape cartridge with a total storage capacity of 48 Gbytes which is equal to four 14 in tape reels recorded on a conventional 28 track HDDR.

Capitalising on improvements in integrated circuit density and the use of ASICs has resulted in a DCRsi 107 system that is smaller, lighter and has 50 per cent less power consumption than the DCRsi system it replaces. It provides a computer-friendly mass storage data peripheral to any air, sea or land platform.

The DCRsi 107 features a format and interface that is compatible with the DCRsi and the DCRsi 240, constant packing density, 96 Mbit internal I/O buffer and data block, time code addressing and searching and automatic playback alignment. Data transfer can be continuous, in bursts or changing. A ruggedised version, the DCRsi 107R, is designed for hostile environments. The system has RS-232 and RS-422 control interfaces.

The DCRsi 107 is available in both single module, 19 in rack-mount and modular ruggedised configurations. The ruggedised DCRsi 107R is available either as a two module record only or as a three module record/reproduce configuration.

The DCRsi™ 75 digital cartridge recording system **1999**/0051319

Specifications

Dimensions:
(tape transport module) 373.4 × 274.3 × 175.3 mm
(rec/rep electronics module) 360.7 × 485.1 × 152.4 mm
(optional AC input power module) 254 × 152 × 76.2 mm
Weight:
(tape transport module) 14.75 kg
(record electronics module) 15.88 kg
(reproduce electronics module) 15.42 kg
(optional AC power module) 4.54 kg
Power supply: 28 V DC
210 W (record only)
430 W (record/reproduce)
Temperature:
–30 to +50°C (operating)
–54 to +70°C (non-operating without tape)
Altitude: up to 50,000 ft operating

Operational status

In production for German Air Force Tornado PA200 reconnaissance pods and utilised in the Eurofighter Typhoon development programme.

Contractor

Ampex Corporation Data Systems Division.

UPDATED

DCRsi™ 120 Digital Cartridge Recording system

The DCRsi 120 recorder is equipped with a fully integrated internal memory buffer which provides total isolation between the user interface and the instantaneous timing demands of the tape transport. Via this front-end buffer, the DCRsi recorder will unconditionally follow the user's data clock - any data rate from zero up to 120 Mbits/s can be recorded or played back either continuously, in bursts, or while fluctuating at any slew rate. Instant-on record capability provides for event capture without time lag associated with tape speed lock-up. No operator adjustments are required as the data rate changes. The DCRsi recorder completely emulates a solid-state FIFO memory. The system is a slave to user interface equipment, thus

DCRsi™ 120 digital cartridge recording system 0015321

simplifying the task facing the data and control interface designer.

The DCRsi 120 recorder utilises a field-proven transverse-scanning tape transport design, featuring mechanical simplicity, compact size and a short, co-planar tape path. The front loading/unloading of the tape cartridge is done instantly, simply and reliably because there are no elevators or other complex mechanisms used to transport the cartridge. The compact transverse scanner assembly contains six azimuth record/reproduce heads. The scanner assembly is self-contained and is easily replaceable in the field with a typical head life exceeding 3,000 head-to-tape hours.

The DCRsi 120 recorder is available in both a two-module and a single-unit rack-mount configuration.

The rugged 50 Gbyte-capacity DCRsi tape cartridge is made of fibre-reinforced polycarbonate. It is flame retardant, UV resistant, non-toxic, with high impact strength. Tape durability is rated at more than 200 passes.

Contractor

Ampex Corporation, Data Systems Division.

VERIFIED

DCRsi™ 240 Digital Cartridge Recording system

The DCRsi 240 rack-mount and modular system is a 1 in transverse scan, rotary digital recorder capable of recording and reproducing at any user rate from 0 to 30 Mbytes/s (0-240 Mbits/s). The byte parallel data interface used in the DCRsi, DCRsi 107 and DCRsi 240 consists of eight parallel data lines, one common clock and one enable signal. This interface compatibility, along with the common format on tape, results in full tape interchange among all three DCRsi models. Tapes recorded on the DCRsi 240 can be played on DCRsi and DCRsi 107, and vice-versa. The 240 Mbits/s transfer rate capability is accomplished by increasing the number of record/playback heads from 6 to 12 and adding a second data channel, while retaining the single channel I/O architecture intact. Data is now recorded or played back on two heads simultaneously. The data from each of the two heads is processed by separate data channels at 120 Mbits/s each and the two data streams are combined in the single 72 Mbyte data buffer which yields a sustained total throughput of 240 Mbits/s with a peak rate of up to 300 Mbits/s. This capability can be sustained over the entire tape cartridge, resulting in a storage capacity of 48 Gbytes which is equal to four 14 in tape reels recorded on conventional 28 track HDDRs.

Capitalising on improvements in integrated circuitry density and the use of ASICs has resulted in a DCRsi 240 system that is smaller, lighter and consumes less power than the previous DCRsi systems. The DCRsi 240 provides the power of a computer-friendly mass storage data peripheral to any air, sea or land platform.

The DCRsi 240 features constant packing density providing 48 Gbytes user storage per cartridge regardless of data rate, data block and time code addressing and searching and automatic playback alignment. Data transfer can be continuous, in bursts or changing. The two-module ruggedised DCRsi 240 is designed for hostile environments. The system is configured for RS-232 and RS-422 control interfaces.

Specifications

Dimensions:
(tape transport module) 373.4 × 274.3 × 175.3 mm
(rec/rep electronics module) 388.6 × 317.5 × 195.6 mm
(cartridge) 266.7 × 165.1 × 41.9 mm
(optional AC power modules) 254 × 152.4 × 76.2 mm
Weight:
(tape transport module) 14.74 kg
(rec/rep electronics module) 14.06 kg
(cartridge) 1.13 kg
(optional AC power module) 4.54 kg
Power supply: 28 V DC, 450 W typical
Temperature range:
–30 to +50°C (operating)
–54 to +70°C (non-operating without tape)
Altitude: up to 50,000 ft operating

Contractor

Ampex Corporation Data Systems Division.

VERIFIED

DCRsi Clip-On™ 1 Gbit/s airborne imagery recorder

The Ampex 1 Gbit/s airborne ultra-high rate recorder, known as the DCRsi Clip-On™, offers a 1 Gbit/s snap-shot imaging capability with instant access to cached data.

Clip-On is designed to be used in conjunction with an Ampex DCRsi™ digital cartridge recorder in an airborne image gathering role. Typically, data acquired at 240 Mbits/s or lower will be recorded on cache and tape simultaneously using the high-rate DCRsi 240. Data streams faster than 240 Mbit/s will be stored in cache and then automatically backed up to tape when the cache is nearly full or on a command from the operator. The baseline solid-state storage capacity is 5 Gbytes although larger memory sizes can be specified.

The system combines the benefits of extremely fast solid-state memory and permanent non-volatile tape storage in one inexpensive, fully integrated package. It has major operational advantages for airborne tactical reconnaissance since up to 5 Gbytes of cached imagery are always available for immediate access, giving the operator the ability to prioritise targets and downlink images almost instantaneously. The Clip-On system has already been selected for a number of classified programmes in the USA and is compatible with the whole range of Ampex ruggedised recorders including DCRsi 75, DCRsi 107 and DCRsi 240.

Operational status

Launched June 1997.

Contractor

Ampex Corporation Data Systems Division.

VERIFIED

DMS-1000 Digital Management System

The DMS-1000 includes Ampex's proprietary CSRT management system that allows capture, storage, retrieval and transmission of high-speed sensor and image data at up to 1.12 Gbits/s.

The DMS-1000 is a scalable high-speed Solid-State Memory System (SSMS) that is situated between the input/output of the source and the Ampex DCRsi™ or DIS™ recording system. The SSMS allows the input of extremely high-speed data into the data capture system without the added cost of multiple recorders.

The CSRT management system allows the operator to direct data to the recording system, into non-volatile or volatile memory, recall data for immediate display to a monitor or aircraft HUD, or direct it to a datalink system for transmission to other tactical platforms.

Specifications

Performance:
transfer rate: 0 to 1.12 Gbyte/s (140 Mbyte/s) sustained
capacity: up to 10 Gbytes in 2 Gbyte increments
memory technology: 2 Gbyte memory boards can be either high-speed DRAM or non-volatile FLASH or a mix
Interface:
data interface: 8, 16 or 32 bits, differential ECL
recorder interface: connects directly to DCRsi family of airborne recorders
control: RS-232
timecode interface: IRIG-B
Dimensions: 269.9 × 257.2 × 317.5 mm
Weight: 13.64 kg
Power: 28 V DC, 200 W

Operational status

Used for testing the Tactical Aircraft Reconnaissance Pod System (TARPS).

Contractor

Ampex Corporation Data Systems Division.

VERIFIED

DMS-1000 Digital Management System ***1999***/0051344

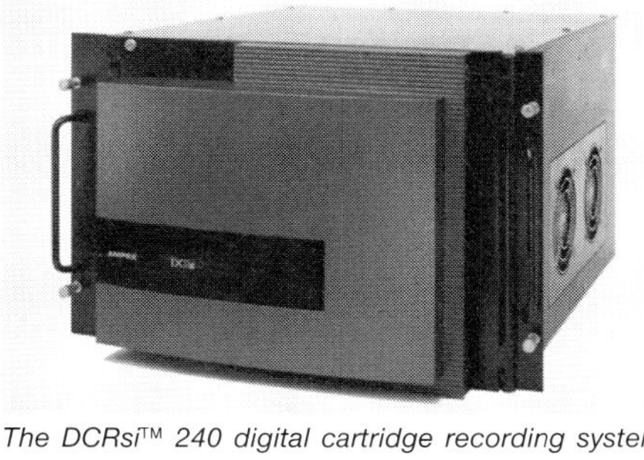

The DCRsi™ 240 digital cartridge recording system ***1999***/0051329

The Ampex Clip-On™ system offers a 1 Gbit/s snap-shot image recording and instant replay capability for airborne reconnaissance applications 0002443

Aircraft Server Unit (ASU)

The ASU provides the central processing functions associated with a network server unit and the communication link between aircraft systems and external sources. The ASU is based on an Intel class processor capable of hosting the Windows NT™ operating system and executing multiple applications. Possible applications include: digital map generators, electronic manuals, maintenance data collection, weather data, communications interface, in-flight entertainment.

Specifications

Dimensions: 4 MCU
Weight: 10 kg
Power: 115 V AC, 400 Hz, 75 W
Processor: 133 MHz Pentium, 32 bit structure
Memory: 32 Mbyte Ram (expandable), 0.5 Gbyte mass storage (expandable)
Interfaces: Ethernet, RS-232/422, ARINC 429
MTBF: > 15,000 h

Operational status

In service on the MD-11.

Contractor

Astronautics Corporation of America.

VERIFIED

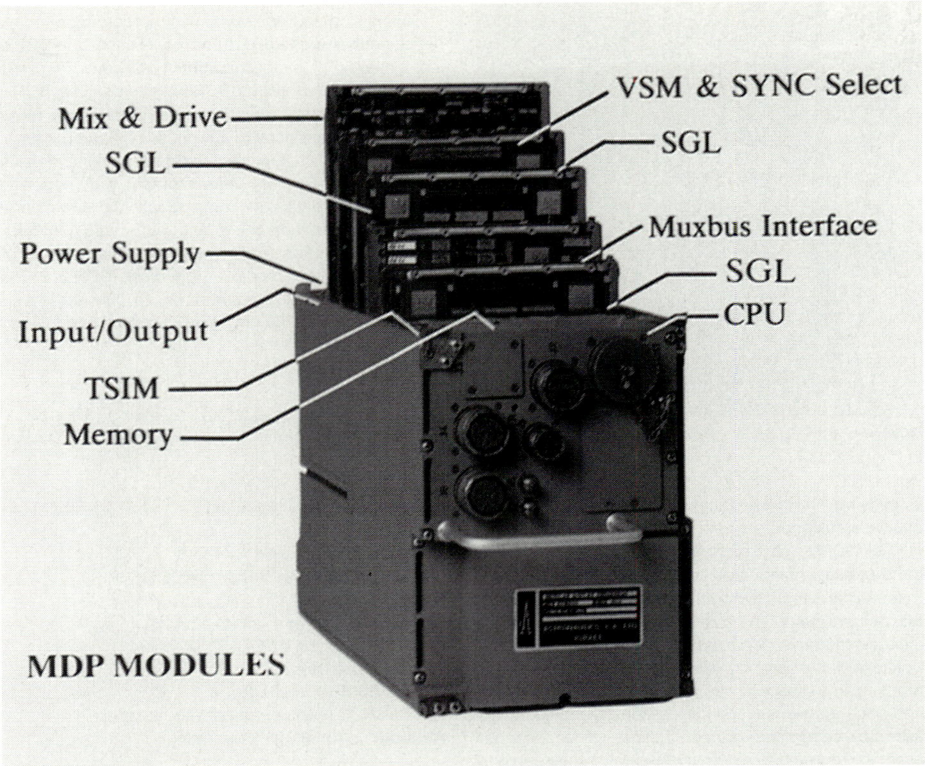

Astronautics modular mission and display processor 0015279

Digital Air Data Computer (DADC)

The digital air data computer meets full US military standards. Present aircraft applications are the A-4, F-5, F-16, L-159, Mirage and a number of additional military aircraft. It is microprocessor-based, using precision solid-state vibrating quartz pressure transducers, with analogue potentiometers and synchro outputs as well as dual-redundant serial digital databusses.

Built-in test equipment allows a high degree of self-diagnosis and the computer has considerable growth potential. It has been designed as a standard air data computer both for retrofit and new aircraft programmes.

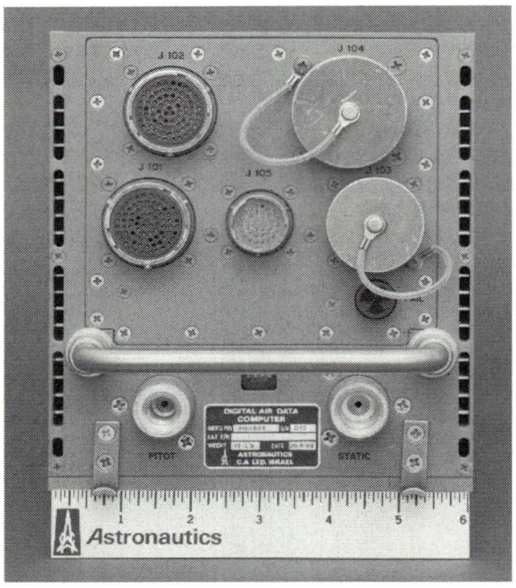

The Astronautics digital air data computer

Operational status

In production and operational in the A-4, F-5, F-16, L-159 and Dassault Mirage.

Contractor

Astronautics Corporation of America.

VERIFIED

Modular Mission and Display Processor MDP

The Astronautics modular MDP is a high-performance, advanced technology computer, using CMOS technology. It provides the basic functions required by military display and control applications, including complete weapon delivery navigation systems. The modular MDP is the heart of each one of the display, navigation and weapon control units.

The modular MDP is based on modern processor technology and distributed architecture. It is designed to drive monochrome and colour displays, HUD/HMDs and tactical displays. An integrated map module is also available. The MDP comprises five main items: the controller, which includes the CPU and the main memory; the graphics engine, a high-speed micro-programmed controller capable of driving stroke and raster displays in various combinations and formats; the video front end which mixes selected inputs with synthetic symbols; the input/output modules; power supply.

Specifications

CPU: 32-bit, 12 Mips at 25 MHz (up to 20 Mips for R-3081 CPU)
Memory: 2 Mbyte standard (4 Mbyte by upgrade)
Graphics engine: 5 million pixels/s/channel (3 graphics engines)
Graphics capability: up to 6 independent graphics channels; up to 512 graphics characters/symbols; 8 colours
ARINC buses: 429 and 629 available
Data interface: RS-232, RS-422, discretes, analogues, synchro
Number of card slots: 12
Power: 115 V AC, 400 Hz at 220 VA
Dimensions: 279.4 × 261.6 × 193 mm
Weight: 15.9 kg
MTBF: 1,200 h

Operational status

In production for A-4, C-130, F-5, Kfir, MB-339, T-38 and other aircraft.

Contractor

Astronautics Corporation of America.

VERIFIED

PAR 1000 Mini-HUMS

PAR 1000 is a compact, lightweight, Mini-HUMS system that provides automatic recording of turbine engine parameters for exceedance monitoring, health trending, and maintenance diagnostic purposes. It also monitors and records engine and airframe hours and cycle data. It can be fitted to both fixed- and rotary-wing turbine-powered aircraft.

PAR measures airframe parameters and enables health monitoring via gas path analysis. This method enables parameters such as N1, N2, torque and TOT to be recorded in the PAR computer.

Information regarding engine starts, airframe cycles, power checks, hot starts, engine cool-down times and shutdowns is monitored and stored in memory. Numbers of engine starts (cumulative), total engine time, cycles (cumulative) and flying time is available.

The standard PT6 system comprises a computer pilot's display, warning lights, associated transducers, and installation kit. Associated ATS (DOS)-based or GBS (Windows™)-based ground support systems are used to analyse the data.

Operational status

The approved application list includes a large number of Bell, Bolkow, Eurocopter and Hughes helicopter types, together with Cessna, Shorts and Raytheon fixed-wing aircraft.

Contractor

Avionics Specialties Inc.

VERIFIED

SKC-3140 bus controller and main computer for the AMX

Two SKC-3140 bus controllers and computers equip each Alenia/Embraer AMX aircraft, providing primary and back-up computations and data control.

A single ¾ ATR LRU, containing computer, bus controller and power supplies, controls data from navigation and attitude sensors, radar, head-up display, weapon aiming and communications controllers and the various control and display panels in the aircraft, and data is passed around the aircraft on a MIL-STD-1553B dual-channel databus.

Specifications
Weight: 9 kg
Power supply: 115 V AC, 400 Hz, <200 W
Reliability: > 3,600 h MTBF

Operational status
In service on the Alenia/Embraer AMX; no longer in production.

Contractor
BAE Systems Aerospace Inc, CNI Division.

UPDATED

Core Integrated Processor (CIP)

BAE Systems (formerly Lockheed Martin) Core Integrated Processor (CIP) for the C-17 Globemaster III is the central computer that controls all of the aircraft's avionics systems. The computer provides all existing mission computer functions and has growth capability to interface with, and provide data management for aircraft controls, displays and sensor data.

The CIP is a high-speed general purpose computer designed to meet the real-time processing requirements of the C-17 application. The baseline CIP is housed in a single Line Replaceable Unit (LRU) containing four subassemblies. The LRU is a full ATR cross-section, 362 mm in length, providing 10 Versa Module Eurocard (VME) module slots. The MIPS R4400 was selected as the Central Processing Unit (CPU) based on off-the-shelf multisource availability, performance and the error detection and correction capability. The VME backplane is used as the intermodule communication channel providing up to 80 Mbytes/s bus bandwidth. A fundamental requirement for the CIP design is open systems such as VME, POSIX, UNIX and VxWorks for both hardware and software. The entire software approach promotes open architectures which feature industry standards and off-the-shelf solutions for operating systems, kernels, programming languages and software development environments.

Two units will be installed on each C-17, replacing the existing three unit mission computer. The CIP will be retrofitted on existing C-17s and will be installed on all future C-17 aircraft.

Operational status
In production and in service on C-17.

Contractor
BAE Systems Controls.

UPDATED

C-17 and CIP 0001287

CP-2108A (3007A) data controller/Mission Computer (MC)

The CP-2108A (3007A) data controller/Mission Computer (MC) was designed to satisfy the combined requirements of the MC-130E Combat Talon 1, the AC-130H gunship and other programmes. Using a single part number MC suitable for multiple missions greatly reduces spares and logistics support. Level 1 test equipment requirements are also minimised by the use of extensive BIT for fault detection and isolation.

The computer employs a dual-CPU architecture. Each processor consists of a high-speed MIL-STD-1750A CPU and a dedicated local memory. The amount of local memory can be varied to suit the application.

The MC has 512 kwords of installed memory, 2.6 Mips of processing capacity and extensive I/O capability high accuracy. There is built-in memory growth to the million word limit of MIL-STD-1750A, multiplexer growth to permit a third dual-channel of MIL-STD-1553B and additional analogue and digital I/O. The unit is mechanised with several blank circuit cards that can be populated with either currently available card designs or new I/O. The 3007A MC is therefore suitable for a variety of other applications.

A plug-in CPU replacement has been developed by Marconi North America which allows one or both of the MIL-STD-1750A processors to be replaced by a 25 Mips 32-bit RISC processor. This increases total available throughput to over 50 Mips for applications such as digital terrain and digital map database systems, knowledge-based system health monitoring, threat correlation and so on. A corresponding increase in memory to 68 Mbytes can also be supported. A ½ ATR variant of the 3007A has also been developed which uses the same dual-CPU configuration.

To assist in developing and debugging software on the 3007A, a Computer Support System (CSS) has been developed. The CSS includes mainframe-based flight software development tools and utilises a micro-VAX host computer for real-time programme debug and validation. An IEEE interface is provided for memory loading and verification.

Specifications
Dimensions: 241 × 330 × 478 mm
Weight: 35.5 kg
Power supply: MIL-STD-704, 356 W (+130 W blower)
CPU: dual MIL-STD-1750A
Throughput: 2.6 Mips (DAIS), growth to 6 Mips
Memory: 512 k × 16 words with growth to 2 million words
Input/output: two MIL-STD-1553B dual-mux channels, approximately 100 discrete, various digital/DC, digital/AC, synchro/digital, digital/synchro, digital resolver and serial digital channels
Environmental: MIL-E-5400 Class 1A Category III

Operational status
In service in US Air Force AC-130H gunship and MC-130E Combat Talon I aircraft.

Contractor
BAE Systems North America.

VERIFIED

Integrated Mechanical Diagnostics - Health and Usage Management System (IMD-HUMS) for helicopters

The BFGoodrich IMD-HUMS condition assessment system reduces operation and support costs. It automatically tracks usage and limit parameters, calculates structural life usage, rotor and track balance adjustments, engine health and performance (including power and assurance checks), and assesses drive train component condition.

The system supports industry standard Cockpit Voice and Flight Data Recorder (CVR and FDR) interfaces. The accompanying Ground Station Support (GSS) package is a self-contained condition-based maintenance information management system. By continuous evaluation of component usage and condition, it can support changing the current method of inspection and automatically replacing parts solely as a function of flight hours. The current open architecture system is based upon functional and performance standards established by the US Navy for the H-60 and CH-53 and by Sikorsky Aircraft Corporation for the commercial S-76 and S-92.

IMD-HUMS acquires data from 150 input channels at a high sample rate, storing the complete data only if an anomalous condition is detected, displaying it to the pilot as a caution using a 3 in indicator. The stored data is written on to an onboard memory card for post-flight insertion into the ground station Windows-based computer to produce a series of operations, maintenance and engineering reports. The onboard system comprises: the Main Processor Unit (MPU), a ½ ATR short package; the Cockpit Display Unit (CDU), a 3 ATI five line display area multifunction key interface; and a Data Transfer Unit (DTU), containing 20 Mbytes transfer storage, typically representing 20 hours of flight data retention; and a PCMCIA Flash card. Optionally, a weight-saving Remote Data Concentrator can be used, which acquires and processes remote sensor data to reduce wire weight.

The IMD-HUMS system, showing (left to right) the data transfer unit, the cockpit display unit on the remote data concentrator, the main processor unit and the ground station **2000**/0062858

Operational status
BFGoodrich Aerospace IMD-HUMS is certified by the FAA STC for installation on the Agusta 109K2 and Eurocopter AS350 and AS355B helicopters.

In 1997, BFGoodrich was awarded a contract by the Office of the Secretary of Defense (OSD), USA Joint Dual Use Program Office to provide an open architecture HUMS for a US Navy 'lead of the Fleet' prototyping programme on six H-60 and six CH-53E model helicopters during 1999, with team members

Vibro-Meter SA of Fribourg, Switzerland and Sikorsky. The US Navy will make a production decision in 2000, based on the trial results.

BFGoodrich IMD-HUMS has also been selected as the host system for the US OSD-sponsored Joint Advanced Health and Usage Management System (JHUMS) programme. The object of the programme is to provide an opportunity for the evaluation of advanced IMD-HUMS functional modules in six US Navy H-60 and six US Army CH-47 model helicopters. Sikorsky has selected the same system as its standard HUMS for all S-76 and S-92 helicopters. Civil certification was due to begin in 1999.

In October 1999, BFGoodrich was contracted by the US Army to design and develop systems for the UH-60A and HH-60L Black Hawk helicopters; these systems were scheduled to be ready for installation and testing in early 2001.

Contractor

BFGoodrich Aerospace, Aircraft Integrated Systems.

UPDATED

RGC250 radar graphics computer

The BFGoodrich RGC250 radar graphics computer is designed to integrate all data related to storm cells, lightning strikes, intruding aircraft, and other hazards onto one multifunction weather and traffic display. Interfacing with many popular civil radars, the RGC250 permits combination of BFGoodrich's Skywatch or TCAS 1 traffic avoidance information with Stormscope (WX-1000E (429 EFIS) or WX-500) lightning detection data - all on the one display - to provide a comprehensive display of all relevant weather, traffic and terrain hazard data. The RGC250 interfaces with the indicator of the weather radar, not with the weather radar directly.

BFGoodrich RGC250 radar graphics computer
1999/0051285

Specifications

Dimensions: 37.6 (H) × 145.5 (W) × 200.9 (D) mm
Weight: 0.68 kg
Power: 18 V DC, 12 W

Operational status

US Federal Aviation Administration (FAA) Technical Standard Order (TSO) approval granted March 1999.

Contractor

BFGoodrich Aerospace Avionics Systems.

VERIFIED

VRS-3000 solid-state Vertical Reference System (VRS)

The VRS-3000 Vertical Reference System was designed as a replacement for conventional spinning mass vertical gyroscopes. The system utilises solid-state rate and level sensors and provides traditional ARINC 407-synchro information for pitch and roll attitude. The VRS-3000 also provides stable and consistent attitude, body rates, inertial pitch and roll rates and acceleration outputs in ARINC 429 digital format. It can provide attitude information to drive primary or standby flight displays and autopilots.

VRS-3000 solid-state Vertical Reference System
2001/0098572

The VRS-3000 also performs an internal system monitoring routine. This helps provide optimum levels of reliability and system confidence further enhanced by BFGoodrich Avionics & Lighting System's Standard Product Reliability Acceptance Test Program. This program includes burn-in under environmental operating conditions prior to final acceptance.

The design of the VRS-3000 uses proven solid-state components, which are currently in use for the Electronic Standby Instrument System (ESIS) model GH-3000, also developed and manufactured by BFGoodrich. The VRS-3000 can be installed in fixed-wing aircraft, helicopters, drones, and remotely piloted vehicles. The system meets or exceeds TSO-C4c, weighs only 3.5 pounds, and was designed for low power consumption.

Specifications

Dimensions: 102 × 95 × 156 mm
Weight: 1.6 kg
Power supply: 115 V AC, 400 Hz, single phase

Contractor

BFGoodrich Aerospace Avionics Systems.

NEW ENTRY

AN/ASQ-195 signal data converter set

The AN/ASQ-195 signal data converter set is a multifunctional digital processor that provides control and data interchange for radios, instrument landing systems, Tacan, automatic direction-finders, sensors such as LANTIRN and radar altimeters, air-to-air interrogation, IFF and avionics BIT. The system has been designed for the F-15E but can be adapted to other aircraft. Each of the two units serves as back-up or redundant data processor for the other unit for many of the functions.

The AN/ASQ-195 features three dual-redundant MIL-STD-1553B databus interfaces, three RS-422 databus interfaces, three MIL-STD-1750A microprocessors, six serial microcontrollers, 120 kbyte EEPROM memory, 240 programmable discrete output interfaces, 355 programmable discrete input interfaces, four synchro interfaces and 28 analogue interfaces.

Display of the INS parameters, frequencies and channel selection is provided by the up-front control panel, communicating with the AN/ASQ-195 via the RS-422 databus.

Contractor

The Boeing Company.

VERIFIED

Integrated Sensor System (ISS)

The four-year Integrated Sensor System (ISS) programme is sponsored by the US Air Force Wright Laboratory. The Boeing Company heads an industrial team which includes the Raytheon Electronic Systems and TRW. The goal of the programme is to integrate the Radio Frequency (RF) functions of Communications/Navigation/Identification (CNI), radar and electronic warfare systems in order to cut cost, size and weight of aircraft avionic systems and improve system reliability.

ISS will build on the US Air Force Pave Pillar/Pave Pace programmes which were aimed at a more integrated approach to advanced sensor processing. The keys to achieving cost and reliability improvements are the concepts of common modules, resource sharing and reconfiguration of the RF sensor systems.

The modular, open architecture design of ISS has the potential for a 50 per cent improvement in overall avionics systems costs and a 200 per cent improvement in reliability, at half the size and weight of contemporary systems. In conceptual studies, the ISS approach reduced the number of RF modules in a notional fighter/attack aircraft from 345 to 169 and the different types of modules from 70 to 33.

The team will design an ISS system architecture and then build key modules and integrate them into a system capable of demonstrating the required RF functions across a broad range of frequencies and waveforms. The team will focus on the integration of what is traditionally referred to as analogue RF support electronics units such as RF switches, converters, receivers, transmitters and other equipment.

The goal is to validate the architectural feasibility of an integrated RF system that can provide the performance and operational needs of future weapon system platforms and also be capable of insertion into current aircraft systems.

Operational status

The Boeing Company has been awarded a US$16.5 million contract by the US Air Force to develop and demonstrate an integrated sensor system.

Contractor

The Boeing Company.

VERIFIED

Rotorcraft Pilot's Associate

The Boeing Company is teamed with a consortium comprising: Honeywell, Kaiser Electronics, Lockheed Martin Advanced Technology Laboratories, Lockheed Martin Federal Systems and Raytheon Electronic Systems. Together they will develop a Rotorcraft Pilot's Associate (RPA) for the US Army's Aviation Applied Technology Directorate (AATD). The RPA programme aims to establish revolutionary improvements in combat helicopter effectiveness. This will be achieved through the application of knowledge-based systems for cognitive decision-aiding and the integration of advanced pilot facilities, acquisition, armament and fire control, communications, controls and displays, navigation, survivability and flight-control equipment.

RPA builds on advanced avionics technologies being developed by the US Army for the updated AH-64D Longbow Apache and the RAH-66 Comanche. These systems enhance the automation of such functions as flight control, information processing and weapons management. In addition to voice recognition, the RPA features an advanced helicopter pilotage system and advanced data fusion. RPA technology has applications in the civilian marketplace and offers an opportunity to aid human performance in a variety of fields.

RPA moves into the cognitive realm of data interpretation, hypothesis formulation, planning and decision making. The result is an intelligent associate that will assist the pilot in understanding the vast array

The rotorcraft pilot's associate concept demonstrator configuration (US Army) ***2000***/0048663

of battlefield information, planning the mission and managing the complex systems in modern military aircraft. Development and evaluation will occur in three stages. Initial design and assessment will be accomplished in a rapid prototyping laboratory environment. As the design matures, it will move into full mission simulation, where more rigorous and in-depth evaluations will be conducted. The third stage will install RPA in an AH-64D Longbow Apache attack helicopter, incorporating the Boeing Company's advanced digital flight control system as well as the US Army's advanced helicopter pilotage sensor system.

The Boeing Company will be responsible for system integration, architecture design, prototype development, full mission simulation, offensive systems, vehicle management and flight test. Lockheed Martin Federal Systems will provide computing technology, including the Massively Parallel Processor, and has responsibility for data distribution, mission planning, defensive systems management and external situation awareness.

Operational status

The Boeing Company and Lockheed Martin Federal Systems have received a US$70 million contract for development and demonstration of the RPA. Flight and operational demonstrations began aboard an AH-64D Longbow Apache prototype in October 1998. Testing is continuing with the current configuration understood to present data to the pilot's Helmet-Mounted Display (HMD) for use in the heads-up/eyes-out-of-the-cockpit role. Data is also presented to a second crew member both on a HMD and on three Head-Down Displays (HDDs) for mission and weapon system management.

Data sources used in the current testing are understood to include on-board radar and electro-optic surveillance sensors, as well as aircraft survivability sensors. Off-board data sources include the Tactical Receiver Intelligence eXchange System (TRIXS), Joint Surveillance and Target Attack Radar System (JSTARS), Joint Tactical Information Distribution System (JTIDS), Battlefield Combat Identification System (BCIS) and tactical command centre data.

Contractors

The Boeing Company, leading a consortium.

VERIFIED

MAXION/ATR multiprocessor system

The MAXION/ATR multiprocessor system is based on the Concurrent MAXION real-time standards-based system. It is a ruggedised system using the standard commercial-off-the-shelf MAXION board set and satisfies the requirement for a real-time multiprocessor system which is capable of being deployed in operational aircraft. It conforms to the industry standard ATR form factor. MAXION/ATR provides more than 600 Mips of power within a VME-6u double height Eurocard.

The system is based on the Mips Technologies R4400 RISC, an industry leading microprocessor. With a system bandwidth of 1.2 Gbytes/s, the system architecture is flexible and extensible. It incorporates the Scalable Multiprocessor Architecture for Real-Time (UltraSMART). This provides maximum total system performance, both in terms of computation and handling I/O data, and delivers virtually the maximum processing power possible from each multiprocessor. As requirements grow, MAXION/ATR users will be able to add new processors.

Operational status

Available in configurations of one to four processors with up to eight processor models available.

Contractor

Concurrent Computer Corporation.

VERIFIED

M362F general purpose processor

The M362F is a general purpose unit which uses parallel, binary, floating point, two's complement 16-bit processing. A typical instruction mix yields an operating speed of about 340 kips.

The M362F can be tailored to particular operations by specifying from a wide choice of memory types and standard input/output circuit modules. It is mechanised on two operating modules. The instruction repertoire can be varied by adding microprogramme memory or changing the existing microprogramme memory which consists of 10 integrated circuits. The M362F will operate with core and/or semiconductor memory. The processor provides 72 basic machine instructions, including nine special purpose types that are mechanised using the basic input/output instructions. Micro-instructions are held in a 512-word control memory, but this can be expanded to 2,048 words, providing for such operations as byte (8-bit) control jumps, skips and transfers, register variable shifts, register/register floating point arithmetic, logical immediate and register/register, and macro instructions. The last provides square root and trigonometric functions much faster than by using subroutine executions.

Development of M362F software can be accomplished on support equipment such as mini-computer directed systems, IBM 360/370 or similar facilities. Mini-computer equipment enables software development on a stand-alone basis in the laboratory.

The M362F is used as the fire-control computer in the Lockheed Martin F-16 and has a comprehensive set of Jovial or assembly coded software facilities.

Specifications

Dimensions: ½ ATR case
Weight: 6.5 kg
Power: 28 V DC
Computer type: parallel, binary, fixed and floating point, two's complement
Word length: 16-bit
Typical speed: 340 kips
Max address range: 65,536 (64 k) memory locations
Instruction set: 72 (expansible)
Input/output options
analogue/digital/analogue converter
analogue input/output multiplexer
discrete input/output (28 V)
MIL-STD-1553 bus
multipurpose serial digital processor

Operational status

In service as the fire-control computer in the F-16.

Contractor

Delco Electronics.

VERIFIED

M362S general purpose processor

Related to the M362F, the M362S is a 32-bit, high-speed general purpose processor. It uses microprogrammed, parallel, binary, fixed and floating point, two's complement operations and has a typical operating speed of 750 kips.

The processor provides 96 basic machine instructions, microprogrammed in a 512-word control memory, with possible expansion to 1,024 words. A RAM system is available and comprises semiconductor CMOS memory modules, memory controller and an error detection and correction unit. A total of 65,536 (64 k) 16-bit memory locations can be accommodated.

Packaging is either for forced-air cooling, as in aircraft bays, or a radiant cooling frame for space applications. The system has been selected for NASA's Inertial Upper Stage programme. A comprehensive set of Jovial and assembly coded software facilities is available.

Specifications

Dimensions: 365 × 365 × 152 mm
Weight: 24.5 kg
Power supply: 28 V DC, 220 W
Computer type: binary, fixed and floating point, 2's complement
Word length: 32-bit
Typical speed: 750 kips
Max address range: 65,536 (64 k) memory locations
Instruction set: 96 (expansible)
Input/output options
analogue/digital/analogue converter
analogue input/output multiplexer
discrete input/output (5 or 28 V DC)
MIL-STD-1553 bus
multipurpose serial digital processors

Operational status

In production.

Contractor

Delco Electronics.

VERIFIED

M372 general purpose processor

The M372 is a development of the M362F used as the fire-control computer for the Lockheed Martin F-16. It has been designed for applications requiring extended performance in memory, input/output, throughput and computational speed. Operating speed is typically between 570 and 720 kips, and non-volatile, electrically alterable memory of 32, 64, 128 or 256 kwords capacity can be accommodated and addressed. In addition to analogue and discrete input/output channels, up to three MIL-STD-1553 dual-redundant digital databus channels can be accommodated and the computer executes the US Air Force MIL-STD-1750 standard instruction set architecture.

The system is supported by Jovial software, and packaging dimensions can be between ½ and 1 ATR, depending on the features incorporated.

Specifications

Dimensions: 112 × 170 × 190 mm
Weight: 7.3 kg
Power supply: 28 V DC, 100 W
Computer type: binary, parallel, microprogrammed, fixed and floating point, 2's complement
Word length: 16-, 32-, or 48-bit
Typical speed: 570-720 kips
Max address range: 1,576,058 (1,024 k) words
Instruction set: MIL-STD-1750A, Notice 1, 262 instructions

Operational status

In production. The M372 computer is used as the F-16's enhanced fire-control computer, as the Lockheed Martin LANTIRN pod control computer and as the C-5B MADAR multiplexer/processor.

Contractor

Delco Electronics.

VERIFIED

Magic IV general purpose processor

The hardware used in this processor is unrelated to earlier Delco processors. Magic IV is a high-performance, all-large-scale integrated microcomputer system, which promises to provide lower cost, reduced power, weight and size and greater modularity and reliability advantages over existing machines. Typical operating speed is 250 kops. There are three basic machines: the M4116, M4124 and M4132 available with 16-, 24- and 32-bit word architectures respectively. They are seen as suitable for remote terminal or sensor orientated processor applications.

Almost exclusive use of large-scale integrated circuits has resulted in a great reduction in the number of components compared to a conventional processor. The large-scale integrated units are:

	M4116 (16-bit)	**M4124** (24-bit)	**M4132** (32-bit)
CPU control unit	1	1	1
CPU arithmetic unit	4	6	8
Input/output control unit	2	3	4
Memory controller	2	3	4
	-	-	-
Total	9	13	17

Additional large-scale integrated circuits may be needed to meet programmable communication interface and digital input requirements. MTBF estimates range from 27,000 hours for a simplex configuration to 150,000 hours for a dual-redundant system. NMOS large-scale integrated circuits are used with typical component densities of 1,000 gates and 5,000 transistors per 250 mil chip, with pair gate delays below 5 ns. Delco claims better nuclear radiation tolerance than contemporary dynamic NMOS circuits due to the static logic design, substrate bias design and exclusive use of NOR gate. Digital inputs compatible with ARINC 561, 575 and 583 can be provided.

Specifications

Dimensions: 167 × 69 × 127 mm
Weight: 1.4 kg
Power: 25.3 W
Computer type: binary, parallel, fixed point, 2's complement
Word length: 16-, 24- or 32-bit
Typical speed: 250 kops
Max address range: 32,768 (32 k) words
Instruction set: 89 (expansible)

Operational status

In production. The computer is used as the Fuel Savings Advisory and Cockpit Avionics System (FSA/CAS) computer for the US Air Force's C-135 and KC-135 aircraft. It is also used in the performance management system in the Boeing 747 and DC-10 and MD-80 series.

Contractor

Delco Electronics.

VERIFIED

Magic V general purpose processor

Magic V has been selected for engine monitor and control systems for the Boeing RAH-66 Comanche helicopter (Boeing) ***2001***/0105287

The Magic V processor is an all VLSI implementation of a CPU and memory system that executes the MIL-STD-1750A, Notice 1 instruction set architecture.

The Magic V processor utilises 3 μm bulk CMOS technology to configure a complete MIL-STD-1750A central processor in just 10 VLSI chips. An eleventh VLSI chip provides extended memory management to address up to 1 Mwords and a twelfth chip provides an IEEE-488 bus interface which enables external communication with the CPU and memory for monitoring performance and for developing software. These 12 VLSI chips are mounted on one side of a single ½ ATR size circuit card assembly. The back side of the single circuit card assembly mounts the VLSI memory controller that interfaces the CPU with the memory devices mounted on this side. Typical configurations of the M572 include up to 192 kwords of CMOS RAM plus a start-up ROM, or various combinations of RAM, EEPROM and UV PROM devices, all capable of being addressed by the programmable VLSI memory controller.

The M572 typically operates at 850,000 to 1 million operations running the DAIS instruction mix. A built-in feature of the M572 is the ability to couple multiple CPU/memory cards in a multiprocessor configuration.

Specifications

Dimensions:
(CPU and memory) 109 × 163 × 13 mm
Weight: 0.45 kg with memory side fully populated
Power:
(CPU) 2.2 W
(CPU and memory) 7.3 W
Computer type: general purpose, microprogrammed, fixed and floating point, custom CMOS VLSI, TTL compatible interfaces
Word length: 16-, 32- and 48-bit; 48-bit logic unit
Instruction set: MIL-STD-1750A, Notice 1; all specified options
Memory size: up to 192 kwords CMOS RAM on ½ ATR size circuit card; up to 256 kwords CMOS RAM on ¾ ATR size circuit card
Max address range: 1 Mwords
Addressing modes: direct, indirect, immediate, indexed, non-indexed, relative, base relative, BIT, byte
Test interface: built-in IEEE-488 interface for DMA operations, test communications, software development, and real-time performance monitoring
Input/output: all mandatory features and non-application dependent MIL-STD-1750A options implemented, discrete and digital outputs/inputs, analogue outputs/inputs, and MIL-STD-1553B bus interfaces implemented in various system level applications

Operational status

In service on C-17A and F-14D in the display systems. Selected for engine monitor and control systems on the RAH-66 Comanche helicopter.

Contractor

Delco Electronics.

UPDATED

Airborne Digital Imaging System (ADIS)

The DRS ADIS is a high-resolution, electronically shuttered, solid-state, digital imaging system in a very robust high *g* package. The system is suitable for many applications, including air superiority fighters and aircraft flight test instrumentation.

ADIS features the latest technology, including a new image sensor and internal memory module. The camera contains an ultra-high-density DRAM memory module and a new high-speed mass storage 8 Gbyte hard drive allows transfer of large amounts of digital imagery from the cameras' DRAM for storage and safe keeping.

Windows® application-based software, provides a PC access to camera set ups and the operation of up to 64 cameras. A preview mode provides an easy graphical display to review the overall set up, including pre- and post-trigger settings, plus the number of images to be captured by each camera.

DRS's ADIS is designed for high-speed imaging applications and time lapse, since it is capable of capturing one picture on demand or one pulse-one picture at a synchronous rate of up to 1,000 Hz.

With continuous Image Storing Recording (ISR) of up to 2,000 pictures using a first-in first-out principal, the ability to grab and lock the images to memory upon command is easily accomplished.

A standard feature of the system is Record On Command (ROC) and Burst Record On Command (BROC).

Recalling images from large data files is very fast, given the amount of identification information recorded with each picture. Up to four different camera recordings can be recalled simultaneously from the database and played together synchronously to analyse different views at the same time.

Airborne Digital Imaging System (ADIS) ***1999***/0051339

Specifications

Resolution: 512 × 512 pixels
Image: 24 bit colour or monochrome
Grey levels: 256 (8 bit)
Picture rates: 512 × 512, 30 to 1,000 pps (pictures per second); frame rate profiling (programmable picture rate on a frame-by-frame basis)
Recording time: 1 s @ 1,000 pps (1,000 pictures)
Image download rate: 132 Mbits/s
Picture download time: 1 picture in 16.5 ms, 1,000 pictures in <40 s
Dimensions: 87 (H) × 66 (W) × 188 (D) mm
Weight: 2.2 kg
Power: 28 V DC, 22 W

Contractor

DRS Technologies Inc, DRS Photronics, Inc.

VERIFIED

AN/AQH-9 Mission Recorder System (MRS)

The AN/AQH-9 MRS was designed to capture critical Anti-Submarine Warfare (ASW) data for contact validation, post-mission analysis and crew training. It is a lightweight, cost-effective, militarised, rotary-head integrated recording system intended to provide the tactical community with new intelligence that will improve combat effectiveness. The AN/AQH-9 mission recorder system can be readily reconfigured for the missions of rotary-wing, fighter/attack and reconnaissance aircraft, as well as Remotely Piloted Vehicles (RPVs).

Developed as the ASW mission data recorder for the US Navy's SH-60F CV inner-zone helicopter, the AN/AQH-9 is capable of recording the critical data of an entire air mission using standard VHS tape cassettes. For both recording and post-mission reconstruction, the cassettes provide ease of handling and storage of the tape media.

Weighing approximately 16.82 kg, the AN/AQH-9 mission recorder system has been built and tested to MIL-E-5400 requirements, qualifying it for the hostile environments encountered by rotary-wing aircraft.

The system can be easily tailored to satisfy the requirements of other ASW programmes, such as the SH-3H, SH-2G, SH-60B and SV-22 (the US Navy ASW variant of the V-22 Osprey) to record crucial mission data, voice and video-display information.

The AN/AQH-9, as configured on the SH-60F, records any mix of four channels of raw sonobuoy data, voice communications and time-code data, processed acoustic display data and avionics information via a dual-redundant MIL-STD-1553B data bus.

The AN/AQH-9 mission recorder system consists of four Weapon-Replaceable Assemblies (WRAs): a Mission Tape Recorder Interface Unit (MTRIU), a VHS videocassette data recorder, a Remote-Control Unit (RCU) and a video cassette. The video cassette links the airborne equipment to a ground-station playback system.

The MTRIU interfaces with various data sources on the aircraft via unique printed circuit boards. These circuit boards digitise the analogue information for combining with other digital data to form a single, serial-bit stream modified RS-170 video signal for recording.

To keep weight and volume to a minimum for critical applications, the AN/AQH-9 was configured to 'record only'. Other configurations are available for record and in-flight playback. It is shock mounted near the sonar operator's console for quick exchange of cassettes for extended recording time.

The Remote-Control Unit contains all the operating controls and indicators required to provide proper system support.

The AN/AQH-9 captures all essential mission data from the SH-60F CV helicopter for complete post-mission reconstruction and analysis.

The MTRIU contains five Shop-Replaceable Assembly (SRA) circuit boards, which configure the system to the mission avionics by controlling and formatting the data for recording. These SRAs consist of a video-acquisition module, an audio-acquisition module, a dual-redundant MIL-STD-1553B data bus interface, a Built-In-Test/Time-Code Generator (BIT/TCG) and a formatter module. In addition to the five SRAs, the MTRIU also contains a power supply module.

The audio acquisition module accepts analogue signals from sonobuoy receivers, voice communication and acoustical data from dipping sonar. The module then processes these signals by using gain-control conditioning, low-pass anti-aliasing and analogue-to-digital conversion. The processed signals are then transferred to the formatter in serial form.

The video acquisition module interfaces with processed sonar displays to record actual CRT presentation data. The analogue video data are obtained from a stroke-type CRT display and in the form of X, Y and Z signals plus cursor-gate and video-gate timing signals required for frame synchronisation. The video acquisition module performs the necessary analogue-to-digital conversion before the data reaches the formatter in serial form.

The MIL-STD-1553B interface module is a data bus operating as a remote terminal on board the aircraft. The mission tape recorder receives essential tactical, navigational and equipment-status data via the MIL-STD-1553B bus. The information from the data bus is sent to the formatter in serial form before recording. This enables the ground station playback system to recreate the tactical-situation plot as the mission develops.

The BIT/TCG module generates an IRIG 'B' time-code signal on the time-code track of the recorder. In addition, this module utilises integrated diagnostic concepts to increase the effectiveness of BIT implementation on three levels, as well as perform ORT (Operational Readiness Test) and OLSM (On-Line Status Monitoring) functions. All Remote-Control Unit functions are processed by the BIT/TCG module.

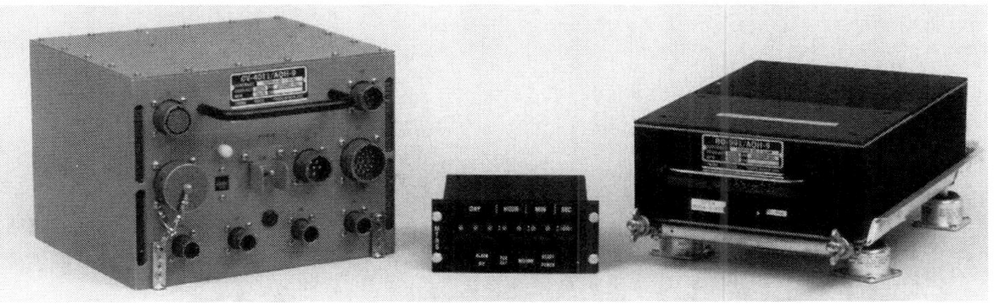

AN/AQH-9 mission data recorder system showing the interface unit (left), remote control unit (centre) and rotary-head data recorder (right) 0015323

The formatter module provides the key primary interface between the data acquisition function and the record function. The formatter combines all of the signal data from the various interface modules, except the time-code information, and formats the resulting single, serial-bit stream into a modified RS-170 data format suitable for recording. Before recording, the formatter provides the necessary Error-Correction Coding (ECC) and then geometrically scrambles the data, which allows for recovery from tape drop-outs.

The RCU is designed to be mounted in the sonar operator's console. It contains all the controls and indicators necessary to operate the MRS. These include a series of thumb-wheel switches used for setting the mission time at take-off, a time-code indicator switch to enter the appropriate mission time, a combination alarm-sensor light and a bit-function switch to indicate any malfunctions that would prevent recording, such as an unlocked cassette carrier, and to ensure proper system functionality, a record button to control tape motion, and a system power switch. All these functions are processed through the BIT/TCG module within the interface unit of the AN/AQH-9 system.

The Rotary-Head Data Recorder (RHDR) is a compact, lightweight, militarised helical-scan airborne unit, which accepts a modified digital-data stream for recording black-and-white video signals. The recorder incorporates two side tracks (time code and auxiliary voice) for expanded capabilities. The user-data rate associated with this digital video recorder using a modified RS-170 format is 3.3 Mbits/s for a 2 hour recording. The data recorder is able to accept standard VHS tape cassettes to record directly all mission data. The rotary-head data recorder has been designed and tested to meet MIL-E-5400 specifications for helicopter environments.

The RD-591 Mission Data Playback System (MDPS) is the ground element of the integrated recording system. It is capable of reproducing all the recorded data from the AN/AQH-9 MRS and can create tactical and navigational primary data plots from the MIL-STD-1553B avionics data. The AN/AQH-9 recording system contains an IRIG 'B' time-code generator, which allows the playback system to provide real-time, post-mission analysis reconstruction of mission data, as well as high-speed search for specific information.

As a counterpart reproducing system to the airborne MTRIU, the playback ground-station unit can be provided to support other weapon-system platforms, such as the SH-3H, SH-2G, SH-60B and SV-22.

Specifications

Dimensions (exclusive of handles and connectors):
Recorder: 210.8 × 289.5 × 93.9 mm
Interface unit: 256.5 × 213.3 × 193.0 mm
Remote-control unit: 146.0 × 152.4 × 66.0 mm
Weight:
Recorder: 6.6 kg
Interface unit: 9.7 kg
Remote-control unit: 0.6 kg
Power:
Recorder: 28 V DC, approximately 25 W (operation), 115 V AC, 400 Hz, approximately 250 W (includes 100 W for recorder's internal heater to extend operating temperature range)
Interface unit: 115 V, 400 Hz, approximately 300 W
Tape:
Tape cassette: Standard VHS (1/2-inch) tape format
Playing time: 2 h on a T-120 tape cassette; 2 h and 40 min on a T-160 tape cassette

Operational status

Developed as the ASW mission data recorder for the US Navy's SH-60F CV inner zone helicopter, and already fitted to approximately 140 US Navy SH-60F and 14 Taiwanese Navy SH-60F CV aircraft; it could be tailored for similar helicopter and aircraft ASW applications.

Contractor

DRS Technologies Inc, DRS Precision Echo Inc.

VERIFIED

AN/AQH-12(V) high-density digital Mission tape Recorder System (MRS)

The AN/AQH-12(V) high-density digital MRS is a high-performance, compact, lightweight, versatile record and playback unit designed specifically for military rotary- and fixed-wing airborne environments.

A flexible multiplexing system captures 2 hours of data at a user rate of 8.33 Mbps. Data integrity is high, with a robust Error Correction Code (ECC) that yields a bit error rate of less than 1×10^{-8} using certified tape. Qualified to MIL-STD-5400 Class 1b, the AN/AQH-12(V) can be configured to acquire a variety of data streams – from wideband acoustics to raster video. The AN/AQH-12(V) is derived from predecessor systems onboard the US Navy's SH-60F CV inner zone helicopter, SH-2G helicopter and the MH-53E minehunter helicopter.

Design applications for the AN/AQH-12(V) include; fixed- or rotary-wing aircraft avionics recording; instrumentation – air or ground-based; sensor

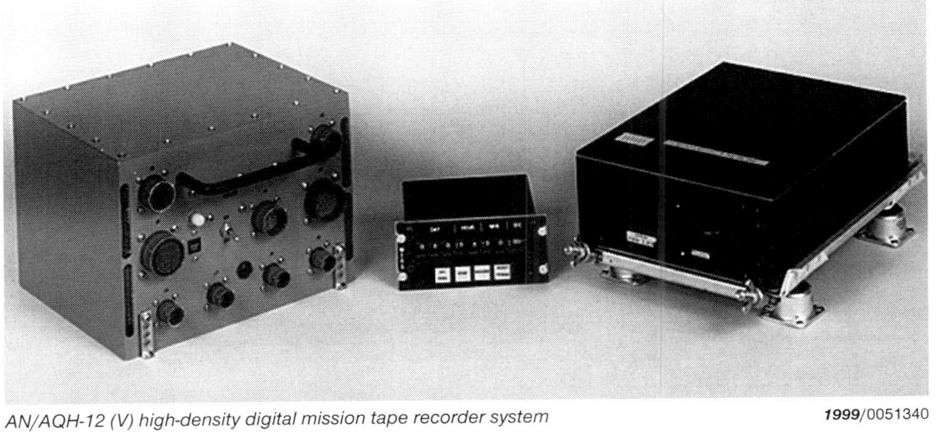

AN/AQH-12 (V) high-density digital mission tape recorder system ***1999***/0051340

recording, such as FLIR/sonar; mission analysis and maintenance analysis.

Specifications

Signal interface: RS-485 clock and data
Control interface: RS-232C 2 wire, 9,600 baud, no parity, 8 bit, 1 stop or MIL-STD-1553B
Bit error rate: 10^{-8} using certified tape
Maximum input user data rate: 8.33 Mbits/s (1 channel)
Options:
Multispeed recording: 8.33 or 16.67 Mbps software selectable, in addition to 4.17 Mbps standard
Data rate/maximum recording time (T-120 S-VHS tape):
4.17 Mbps/4 h
8.33 Mbps/2 h
16.67 Mbps/1 h
Alternative control interface: IEEE 488
Alternative data I/O: PCM, telecom
Analogue I/O mux/demux: flexible multiplexer/demultiplexer permits recording and playback of a mix of data types. For example, any mix of up to 8 DIFAR sonobuoys or 4 BARRA buoys plus full MIL-STD-1553B bus recording and digitised audio
Dimensions:
Interface unit: 256.6 (W) × 213.4 (D) × 193 1 (H) mm
Tape transport: 210.9 (W) × 289.6 (D) × 94.0 (H) mm
Weight: 19.1 kg
Power:
Interface unit: 28 V DC, 120 W
Heaters (option): 28 V DC, 100 W

Contractor

DRS Technologies Inc, DRS Precision Echo Inc.

VERIFIED

AN/USH-42 Mission Recorder/Reproducer Set (MRRS)

The AN/USH-42 is a high-performance, compact, lightweight, versatile mission recorder/reproducer set designed specifically for military aircraft. The system consists of two WRAs, the recorder/reproducer unit and the remote control and advisory panel.

The AN/USH-42 is capable of recording two channels of video data, two audio channels and digital annotation data. The two video channels to be recorded are selected from up to four video inputs. Dual, self-contained digital scan converters provide input processing of radar, FLIR, missile video and other sensors for recording on dual Hi-8 mm tape transports. Annotation data is extracted from a MIL-STD-1553B databus and recorded as closed-caption data.

The remote control and advisory panel provide control and status information to and from the recorder for each of the transports.

The AN/USH-42 is capable of airborne and ground playback with video output in standard or non-standard video formats. Ground playback is also possible using standard commercial replay equipment.

The most recent of DRS Technologies mission recorders, the AN/USH-42 is part of the US Navy's S-3B aircraft avionics system.

Specifications

Number of channels: 2 video, 2 audio
Signal acquisition interface: RS-170 raster video, radar, FLIR, other raster-type video, missile video, MIL-STD-1553, dual redundant data bus
Video resolution: 400 horizontal lines nominal
Video recording time: 2 h per channel
Signal-to-noise ratio: >40 dB
Output: RS-170 video (2 channels)

AN/USH-42( ) mission recorder/reproducer set 0015325

Dimensions:
Recorder/reproducer unit: 330.2 × 431.8 × 222.3 mm
Remote control and advisory panel: 76.2 × 254 × 146.1 mm
Weight: 18.2 kg
Power:
Recorder/reproducer unit: 28 V DC (200 W including heater)
Remote control and advisory: 5 V AC, 400 Hz

Operational status

The AN/USH-42( ) is part of the US Navy S-3B Viking avionic system; it was developed initially for use in the US Navy's A-6E Intruder aircraft.

Contractor

DRS Technologies Inc, DRS Precision Echo Inc.

VERIFIED

DCMR-24 Digital Cassette Mission Recorder

The DCMR-24 is a multifunctional digital recorder/reproducer designed around a digital 8 mm transport, utilising standard VME (6 μ) circuit cards and backplane. The digital cassette is housed in a rugged package designed for severe environmental conditions.

A three slot, industry-standard, VME backplane provides the DCMR-24 with a wide variety of recording configurations by changing a circuit card. With a simple card swap the DCMR-24 can meet changes in mission recording requirements. The DCMR-24 supports multiple analogue, digital (serial), video, and MIL-STD-1553 bus monitoring channels.

The DCMR-24 utilises AIT (Advanced Intelligent Tape) technology developed by Sony for the computer industry. With Ad-Me (Advanced Metal Evaporated) tape, the AIT provides storage capacity of up to 25 Gbytes (non-compressed) and recording speeds of up to 24 Mbyte/s. The DCMR utilises the AIT tape cassette, which incorporates a 64 kbyte memory chip for file, and header information providing fast data retrieval. This recorder can be controlled and operated with front panel switches, remotely over the MIL-STD-1553 bus, or remotely via an RS-232 port running a terminal utility program.

Specifications

Recorder type: 8 mm rotary, head helical scan with Advanced Intelligent Tape (AIT) cassette
Interface options: digital (serial), analogue, video and MIL-STD-1553 bus
Sustained data rate: 3 MB/s (up to 9 Mbyte/s with data compression)
Burst data rate: 12 MB/s (asynchronous mode enabled); 20 MB/s (synchronous mode enabled)
Buffer size: 4 MB
Record time: 2 h
Media format: AIT-1
Tape capacity: 25 GB (up to 75 GB w/compression)
Interface options:
Serial data: up to 24 I/O channels
~ (max aggregate of 24 Mb/s = 2 h record time)
impedance 120 ohm or selectable
single-ended or differential
aggregate data rate 24 Mb/s
RS-422, RS-232, HDLC, asynchronous, and synchronous
PCM NRZ, biphase, Manchester and others

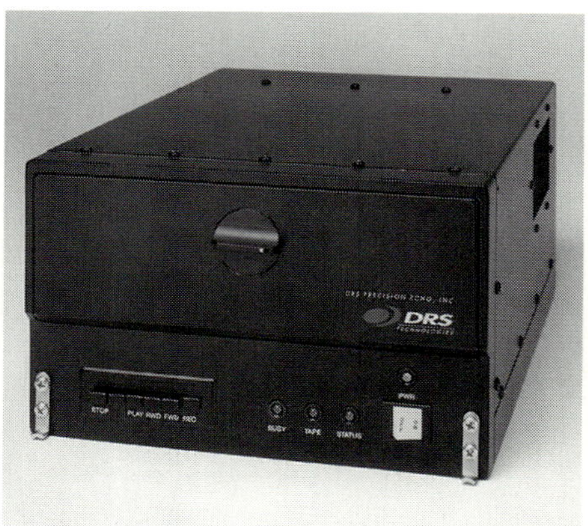

DCMR-24 Digital Cassette Mission Recorder 0015326

Video channels: 1 or 2 channels
~ colour (560 × 480 × 24 bit near full motion = 2 h record time)
~ monochrome (560 × 480 × 8 bit full motion = 2 h record time)
SCSI-2 fast/wide:
3 MB/s (up to 6-9 MB/s w/compression)
~ (24 Mb/s = 2 h record time)
12 MB/s (asynchronous mode enabled)
20 MB/s (synchronous mode enabled)
Analogue channels: up to 30 I/O channels
~ (Max 50 ksps w/16 bit resolution = 2 h record time
MIL-STD-1553B:
up to 8 dual-redundant channels
Bus interface:
single/multimode (remote terminal and/or bus monitor)
Power: 28 V DC, 120 W
Dimensions: 152.4 (H) × 247.7 (W) × 368.3 (D) mm
Weight: 9.09 kg

Contractor

DRS Technologies Inc, DRS Precision Echo Inc.

VERIFIED

DCMR-100 Digital Cassette Mission Recorder

The DCMR-100 (digital cassette mission recorder) is designed to interface with a wide variety of sensor and databus formats. The DCMR-100 also provides a common mission data storage system in a compact, rugged package.

A member of the DCMR product family, the DCMR-100 features DRS Precision Echo's Common Recorder Architecture (CRA) for the interfaces and storage device. Its VME-based design provides maximum flexibility and allows straightforward and cost-effective configuration for most imagery and instrumentation applications, often using off-the-shelf VME interfaces. Data is stored on the industry standard Digital Tape Format (DTF™) cassette allowing easy transfer of data into most workstations and data archives.

With an uncompressed tape cassette capacity of 42 Gbytes and a maximum aggregate transfer rate of 12 Mbytes the DCMR-100 is designed for recording radar, infra-red imagery, ASW acoustics and MIL-STD-1553 bus data in rugged applications where space data transfer rate and capacity are at a premium. The DCMR-100 can be controlled/operated locally by front panel switches, or remotely by RS-232 or MIL-STD-1553 bus.

Specifications

Interface options: SCSI-2 F/W, digital/serial, analogue, video, MIL-STD-1553B
Sustained data rate: 12 Mbyte/s
Burst data rate: 20 Mbyte/s
Buffer size: 32 Mbyte
Record time: 58 min
Media format: DTF™
Tape capacity: 42 Gbyte

DCMR-100 Digital Cassette Mission Recorder 0015327

Digital data:
1 to 8 I/O channels per card
Impedance 120 ohms
Differential
Maximum channel data rate:
96 Mbyte/s (synchronous)
115 kb/s (asynchronous)
RS-422, PCM NRZ, BiPhase, Manchester and others
Video channels:
1 or 2 active channels per card
Composite (NTSC, PAL), monochrome (RS-170, CCIR), B/W (Y)
Compression method: ADVS (Motion JPEG)
Colour (24 bit)
Monochrome and B/W (8-bit)
I/O Voltage: 1 V p-p
Impedance: 75 ohms
Horizontal lines – 560, 280, 140 (selectable)
Vertical lines – 480 (60 Hz), 576 (50 Hz)
Record time:
1 channel – 4.5 hours (clock rate = 20 Mbyte/s)
1 channel – 18.5 hours (clock rate = 5 Mbyte/s)
Analogue input
Single-ended or differential
Input voltage: ± 2.8 V, full scale
Input impedance: 100,000 ohms
Sampling method – Delta-Sigma (simultaneous sampling)
Resolution – up to 16 bit
Sampling frequency: 4-100 kHz
Analogue output
Single-ended
Output voltage: ± 2.8 V, full scale
Output impedance: 50 ohms
Timing: simultaneous
Resolution: up to 16 bit
Dimensions: 482.6 (H) × 393.7 (W) × 292.1 (D) mm
Weight: 25 kg
Power: 28 V DC, 250 W

Contractor
DRS Technologies Inc, DRS Precision Echo Inc.

VERIFIED

Replacement Data Storage System (RDSS)

Specifically designed to be utilised for the surveillance missions of the P-3C aircraft, DRS's Replacement Data Storage System (RDSS) collects and formats the mission event data. It is being used by the US Navy and international air forces to load the P-3C aircraft's tactical computer and signal processor mission programs, as well as to initialise the signal processor. The RDSS is designed for the P-3C aircraft, which is primarily used for forward area anti-submarine warfare surveillance missions to detect and counter subsurface threats.

The RDSS provides storage for software and data on removable, high-performance, high-capacity magneto-optical disk drives utilising laser technology. This system replaces the existing AN/ASH-33 and AN/ASH-33A magnetic tape systems, which have a limited data storage capacity of 4 Mbytes. Use of the magneto-optical disk drive technology dramatically increases the existing data storage capacity to 2.6 Gbytes. as a result, the RDSS allows data file storage on a single disk instead of the multiple magnetic tape cartridges required by older systems.

Designed for compatibility with the existing NTDS (Navy Tactical Data System) interfaces and future signal processors, the RDSS provides a file management capability for supporting multiple clients on an Ethernet interface. No change to existing operational software is needed. The RDSS physically and functionally emulates the AN/ASH-33 at the host computer interfaces, so that no changes are required to existing computer hardware, software or cabling. The increased speed provided by the magneto-optical disks reduces the time needed to load mission computers. The RDSS also provides a Local Area Network (LAN) interface, additional SCSI drive bays and VME card slots for future system growth capability. The RDSS provides added functionality, increased reliability and improved mission performance, while significantly reducing maintenance and training requirements.

The RDSS uses ruggedised Commercial-Off-The-Shelf (COTS) components to provide increased performance with significant future growth capability,

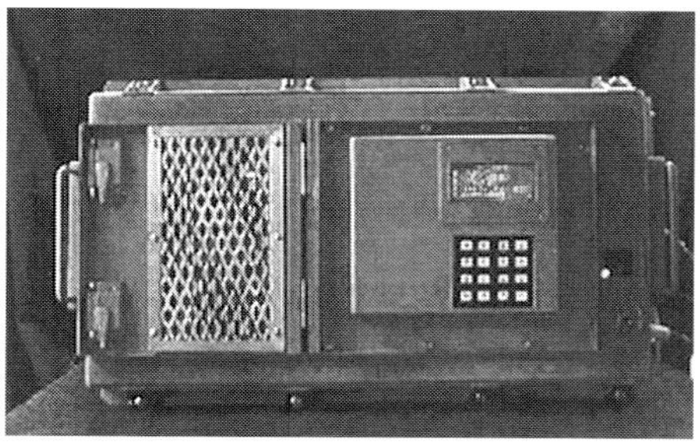

DRS Technologies replacement data storage system
***1999**/0051338*

while costing less than existing systems. Additionally, the RDSS combines the functions of the AN/ASH-33 tape drives and controller into a single enclosure, which is form/fit/function-compatible with the existing controller, reducing weight, space and power requirements on board the aircraft.

DRS's COTS-based systems meet or exceed critical performance, reliability and long-term support requirements.

Specifications
Power: 115 V AC, 400 Hz, 163 W; MIL-STD-704A compliant
Computer interfaces:
Plug-compatible with existing AN/ASH-33
32-bit MIL-STD-1397B Type C (ANEW) parallel
Proteus Digital Channel (PDC) serial
Ethernet LAN with Network File System (NFS) management
Operator interfaces:
Menu-driven plasma display
16-key front panel keyboard
Mass storage devices: accommodates up to 4 standard 5-1/4 inch half-height SCSI devices, any combination of: magneto-optical drive; hard disk drive; CD-ROM
Dimensions: Same footprint and mounting as existing AN/ASH-33 controller: 596.9 (W) × 348.0 (H) × 112.5 (D) mm
Weight: 27.3 kg

Contractor
DRS Technologies Inc, DRS Precision Echo Inc.

VERIFIED

WRR-812 airborne video tape recorder

The WRR-812 airborne video tape recorder/reproducer is designed to support airborne and ground mission applications requiring recording of up to 2 hours of video formatted data (RS-170, colour or monochrome).

This Hi-8 mm recording system delivers more than 400 lines of horizontal resolution, improved signal-to-noise ratio and a video master recording that maintains higher image quality through multiple generations of dubbing. The Audio Frequency Modulation (AFM) provides excellent fidelity and wide dynamic range with a special Voice Boost™ System to minimise background noise and enhance the recorded audio.

Applications include: airborne and ground imagery recording in hostile environments; recording of cockpit Head-Up Display (HUD) and MultiFunction Display (MFD) video on all modern fixed- or rotary-wing aircraft; recording/playback of tactical Forward Looking Infra-Red (FLIR), Low Light Level TV (LLLTV) and radar video for tactical identification, targeting and assessment; recording and in-air remotely commanded playback of Unmanned Air Vehicle (UAV), FLIR and TV for Over-The-Horizon (OTH) missions; applications requiring light weight, low power and ruggedised performance.

The WRR-812 features: light weight and small volume; a microprocessor controlled command translator interface control; Hi-8 mm or standard 8 mm format 2 hour record time per cassette; frame-by-frame advance/reverse without picture 'jitter' or video noise distortion; a linear time counter; voice boost system; auto head cleaner; optional built-in video processing with scan conversion or multiplexing capability.

Specifications
Video channel
Recording system: rotary two-head helical scanning FM system

WRR-812 airborne video tape recorder ***1999**/0051336*

Video signal: monochrome or NTSC colour, EIA standard, 1 V peak-to-peak composite video, sync negative
Input/output impedance: 75 ohms
Maximum record: SP mode: 2 h 30 min (Sony® P6-150 cassette 2 h); (Sony® P6-120 cassette)
Playback resolution: Hi-8 mm: 400 horizontal lines nominal; standard 8 mm: 240 horizontal lines nominal
Audio channel
Recording system: rotary head, FM system (2 channels)
Input/output: –7.5 dB @ 47 k ohms
Bandwidth: 20 Hz to 14 kHz
General channel
Control interface: RS-232C or RS-422A, user selectable
Dimensions: 116.8 (H) × 221.0 (W) × 241.3 (D) mm
Weight: <3.64 kg
Power: 28 V DC, 15 W nominal (plus 25 W for optional heater)

Operational status
In service on A-10 and F/A-18 aircraft for tactical data recording.

Contractor
DRS Technologies Inc, DRS Precision Echo Inc.

VERIFIED

WRR-818 airborne video tape recorder

DRS Precision Echo upgraded the WRR-818 mission recorder to produce the WRR-818 airborne video tape recorder/reproducer, designed especially to support airborne and ground mission applications requiring recording of up to 2 hours of video formatted data (composite, Y-C or RS-170). One of the features of the WRR-818 is its internal shock mount transport assembly; this eliminates extra space and cost associated with external shock mounted trays.

This Hi-8 mm recording system delivers more than 400 lines of horizontal resolution and improved signal-to-noise ratio. The Audio Frequency Modulation (AFM) provides excellent fidelity and wide dynamic range.

Applications for the WRR-818 airborne video tape recorder include: airborne and ground imagery recording; recording of cockpit Heads-Up Display (HUD) and MultiFunction Display (MFD) video on modern fixed-, or rotary-wing aircraft; recording/playback of Forward Looking Infra-Red (FLIR), Low-Light Level TV (L^3TV) and radar video for tactical identification, targeting, and assessment; recording and in-air remote command playback of Unmanned Air Vehicle (UAV) FLIR and TV for Over-The-Horizon (OTH) missions; man-portable and mobile applications requiring lightweight, low power, ruggedised performance; pilot training and post-flight debriefing; flight test instrumentation.

Features of the system include: light weight and small volume; microprocessor command translator, interface control, front panel, discrete or serial control via RS-422A; Hi-8 mm or standard 8 mm format with 2 hours recording time; high-speed Fast Forward (FF) or Fast Reverse (FR) search functions; LED status and front panel control; auto head cleaner; a three-way event-mark indicator with audio, video display and search tone.

Specifications
Video channel
Recording system: rotary two-head helical scanning FM system

WRR-818 airborne video tape recorder ***2000***/0062743

Tape format: Hi-8 mm and 8 mm
Video signal: NTSC, composite, Y-C or RS-170/A EIA standards, 1 V peak-to-peak
Input/output impedance: 75 ohms
Maximum record: 2 h (Sony® P6-120 cassette)
Playback resolution: Hi-8 mm: 400 horizontal line nominal; standard 8 mm: 240 horizontal lines nominal

Audio channel
Recording system: rotary head, FM system (2 channels optional)
Input/output: 0.9 V P-P @ 47 k ohms (output at <100 ohms)
Bandwidth: 50 Hz to 15 kHz

General
Control interface: RS-422A, discrete or manual (RS-232C optional)
Dimensions: 119.6 (H) × 147.8 (W) × 160.5 (D) mm
Weight: <2.73 kg
Power: 28 V DC, 12 W nominal (plus 25 W for optional heater).

Operational status

In service on F/A-18 aircraft and OH-58D helicopters for recording tactical data on combat missions. Selected for US Navy F/A-18E/F aircraft.

Contractor

DRS Technologies Inc, DRS Precision Echo Inc.

VERIFIED

WRR-833 tri-deck cassette video recorder/reproducer

The WRR-833 tri-deck cassette video recorder/reproducer is a high-performance, compact, lightweight, versatile video recorder and playback unit designed specifically for military fixed- or rotary-wing environments.

The system is designed for airframes where space and weight are critical factors. A switch allows recording of three channels for 2 hours or one channel for 6 hours. In case of a transport failure, the system has a graceful degradation that allows the operator to designate priorities among the data streams, so that the most important information will always be captured. Each Hi-8 mm recording system has more than 400 lines of horizontal resolution and an improved signal-to-noise ratio. An optional video/digital (internal) multiplexing system permits the recording of multichannel video or combinations of video or combinations of video and digital data on one or all three transports. The optional digital interface is designed for capturing one or more of the dual MIL-STD-1553B or asynchronous digital data streams.

Design applications for the WRR-833 tri-deck cassette video recorder/reproducer include: airborne and ground mobile imagery recording; simultaneous recording of cockpit Heads-Up Display (HUD) and MultiFunction Display (MFD) video on high-performance aircraft; mission analysis and maintenance analysis.

Specifications

Video channels
Number of channels: 3 NTSC or RS-170
Video interface: monochrome or NTSC colour, EIA standards, 1 V peak-to-peak composite video, sync negative
Input/output impedance: 75 ohms
Max record time: 2 h (Sony® P6-120 cassette) per channel; 4/6 h with sequential recording
Playback resolution: Hi-8 mm – 400 horizontal lines nominal; standard 8 mm – 230 horizontal lines nominal

Audio channels
Number of channels: 6 (2 per deck)
Input/output: –7.5 dB @ 47 k ohms
Bandwidth: 20 Hz to 14 kHz

General
Dimensions: 243.8 (W) × 149.9 (H) × 355.6 (D) mm
Weight: <9.1 kg
Control interface: RS-232C or RS-422A
Power: 28 V DC, 45 W (120 W with heater)

Contractor

DRS Technologies Inc, DRS Precision Echo Inc.

VERIFIED

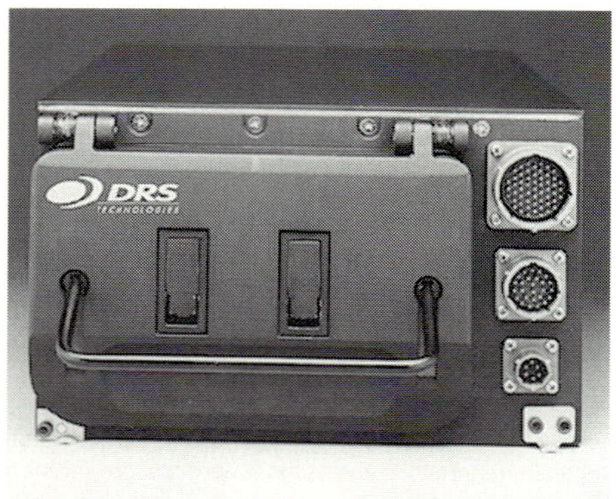

WRR-833 tri-deck cassette video recorder/reproducer ***1999***/0051337

DI-930 multichannel digital recorder

The DI-930 multichannel digital recorder uses 4 mm Digital Audio Tape (DAT) cartridges and accommodates 8 or 16 channels of audio, with up to 360 channel hours per tape.

The DI-930 unit features a backlit VGA colour display, with a solid-state trackball and Graphical Interface Unit (GIU). Playback facilities are available to the operator, via the front panel control, and front panel headphone jack or rear panel output connector. Playback of the most recent records, or analysis of older records on a previously recorded DAT tape is possible without interrupting the current recording process. Industry standard interface protocols and command sets are optionally available for remote control applications.

Specifications

Dimensions: ½ in width of 19 in rack, 5.25 in high
Ruggedised: for military air/land/sea operations

Operational status

In service with various US Army and Air Force intelligence users and with other US government agencies.

Contractor

Dynamic Instruments, Inc.

VERIFIED

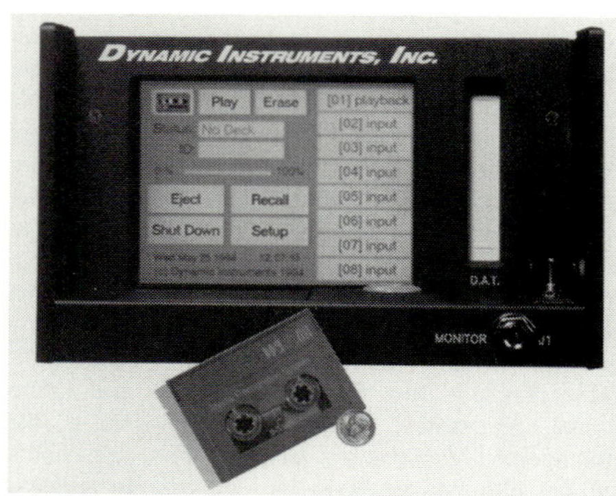

The DI-930 multichannel digital recorder ***1999***/0051330

AN/AYK-14(V) standard airborne computer

The AN/AYK-14(V) is a high-performance general purpose computer with both 16- and 32-bit processing elements. It consists of a family of interchangeable processor, memory, power, enclosure and input/output modules that can be configured to meet specific price, performance and functionality needs for a wide range of applications. The AN/AYK-14(V) computer offers high performance and high reliability with a low life cycle cost, while meeting airborne MIL-E-5400, shipboard MIL-E-16400 and land MIL-E-4158 environments. It is the US Navy's standard airborne computer and is currently being used on a wide variety of military platforms by the US and its allies.

The 16-bit version of the AN/AYK-14(V) is currently in its third generation, which is known as VHSIC AN/AYK-14(V). It provides performance of up to 20 Mips or more within a single enclosure and has a memory addressing capacity of 16 Mbytes. The VHSIC computer is fully compatible with software written for either of the earlier generations. The instruction set is compatible with that of the AN/UYK-44 and AN/UYK-20 and is supported by the MTASS software development environment. Other 16-bit modules include 128 kbyte core memory, MIL-STD-1553A/B, NTDS, RS-232, Proteus, discrete Input/Output (I/O) and application specific I/O modules. On the Boeing F/A-18, two 16-bit AN/AYK-14(V) computers are installed in a dual-redundant system configuration. One computer serves as the navigation and engine controller, while the other functions in a mission management role, handling and

The F/A-18 Hornet avionics system incorporates two AN/AYK-14(V) mission computers

processing weapons and target information. Each computer can accommodate up to 16 Mbytes of memory and provides up to 20 Mips of processing throughput.

The 32-bit version of the AN/AYK-14(V), also known as the Advanced AYK-14, is based on commercial RISC technology. Up to 225 Mips of processor performance and 180 Mbytes of memory capacity are currently available within a single enclosure. The use of commercial open system backplane and processor standards provide built-in performance and capability growth potential that can increase the performance of commercial technologies. Like the 16-bit AN/AYK-14(V) processing elements, the Advanced AYK-14 also provides modular processing and I/O components that can be configured to meet specific price and performance requirements of a range of applications. Currently, a MIPS R5271 RISC computing module processor, other processor modules, and SCSI, MIL-STD-1553A/B, RS-422, Proteus and discrete I/O modules are available.

A complete, commercially supported, Ada development environment is available for commercial workstation Sun and RS-6000 networks, providing full development and real-time debugging support. A real-time operating system with POSIX-compatible services, priority management capabilities and Rate-Monatonic scheduling support completes the Advanced AYK-14's capabilities. Ada compilation, debugging and runtime can be obtained from multiple commercial suppliers. Current implementation utilises the Rational V ADS-Advanced commercial product. Device drivers and hardware support software are available from General Dynamics Information Systems.

GD3000 Advanced Mission Computer

The General Dynamics Information Systems (GDIS) Advanced Mission Computer (AMC) is the next generation open systems processor which is deployed on a variety of platforms. The GD3000 AMC is a leading edge, flexible and rugged processing product family, which can be readily configured to meet the needs of modern military systems, from benign laboratory to harsh avionics environments. The AMC is an integrated information processing system, providing complete hardware and software solutions. It is built upon a well-defined open systems architecture allowing for rapid insertion of emerging technologies; GDIS supplies system design and integration services to ensure a precise fit to the requirements of each specific user platform.

The GD3000 AMC is a set of digital computer hardware and software that performs general purpose, I/O, video, voice and graphics processing. Communication is over multiple buses, including 1553, Fibre Channel and Local PCI and all modules integrate in an industry standard 6U VME backplane. The I/O configuration may be tailored via use of PMC mezzanine modules. The design is scaleable and expandable, with a clear and built-in path for technology upgrades and insertion. An Ethernet interface is provided to support software development and maintenance of the system.

The AMC can be used efficiently in a wide range of applications, ranging from embedded module functions to full-scale multicomputer configurations and operates reliably in extreme airborne, ground-based and shipboard conditions. The system has application in display and mission processing, information and stores management.

Specifications

Dimensions: 194 × 257 × 356 mm typical
Weight: 11-16 kg
Power: 50-300 W typical
Computer type: binary, fixed or floating point
Word length: 16- or 32-bit with double precision and floating point
Typical speed: 500 Kips to 225 Mips
Max addressing: 16 Mbytes (16-bit) to 4 Gbytes (32-bit)
Input/output options: discretes, MIL-STD-1553A or B (dual-redundant buses), NTDS fast, slow, ANEW and serial, 6 MHz Manchester, RS-232C, 85323 Proteus, RS-422, SCSI, RS-485, TM-bus

Operational status

The AN/AYK-14 is in production and selected for the Boeing F/A-18 (which has two AYK-14 mission computers), Sikorsky SH-60B Seahawk LAMPS Mk III helicopter, Northrop Grumman E-2C Hawkeye, Boeing/BAE Systems AV-8B, Northrop Grumman EA-6B Prowler, F-14D, EP-3E, ES-3A, Joint STARS, Lockheed Martin P-3C Orion and embedded modules in the AN/ALQ-149, US Air Force TAOM/MCE system. The Advanced AYK-14 is used on the Bell/Boeing V-22 Osprey in a lower cost configuration that provides 45 Mips throughput and 12 Mbytes of memory on a single board computer. The GD3000 AMC is part of the F/A-18 E/F, F-15E, AV-8B and T-45 development programs.

A modified version, designated ACCS 2500, is used on the UK Royal Air Force Harrier GR. Mk 7 aircraft (see entry under Computing Devices in the UK part of this section).

Contractor

General Dynamics Information Systems.

UPDATED

AN/UYH-15 recorder-reproducer set, sound

The AN/UYH-15 is a voice recording and reproduction system that uses modern digital speech processing technology. It is a compact system that allows operators to monitor, record and instantly recall any recorded message. Designed primarily for use with signal acquisition systems as the standard replacement for the US Army AN/UNH-17A analogue cassette recorder, the AN/UYH-15 is ideally suited for all real-time voice transcription and analysis applications. The AN/UYH-15 may be controlled by either a host computer or by one or two control display panels.

The AN/UYH-15 can record six analogue input channels simultaneously. During recording, each operator can either monitor input or play back recorded files. Operator commentary can be recorded and time-correlated to a given signal. The signal and its related commentary can be combined for output to aid in analysis.

Each recorded signal is digitally sampled and compressed before being stored on the AN/UYH-15's hard disk, which makes it possible to store 6 hours of voice input. The voice compression algorithm offers proven performance in the noisy military environment, as well as high-quality reproduction independent of the signal being reproduced.

Specifications

Dimensions:
(chassis) 133.4 × 482.6 × 412.8 mm
(control/display panel) 50.8 × 228.6 × 152.4 mm
Weight:
21.32 kg (with 2 control/display panels)
16.78 kg (without panels)
Power supply: 105-130 V AC or 208-240 V AC, single phase, 47-400 Hz or 22-30 V DC, 80 W typical
Audio channels: 6 input, 2 output
Bandwidth: 300 Hz - 4.4 kHz
Capacity: 6 h
Recording media: 170 Mbyte formatted hard disk
Interfaces: RS-232C, IEEE-488
Environmental: MIL-STD-810D, MIL-STD-461/462, TEMPEST

Contractor

General Dynamics Information Systems.

VERIFIED

Hard Disk Subsystem (HDS)

The HDS provides a highly reliable form and fit replacement for Miltope and Ampex magnetic tape transports in an airborne environment. Based on field-proven 3.5 in magnetic disk technology, the HDS provides a × 65 increase in storage capacity and reductions in weight and power. Each HDS enclosure emulates up to three tape transports with electrically isolated and redundant hardware for each magnetic tape transport equivalent set. An operator control panel provides a user-friendly interface, while extensive BIT capability allows for easy field maintenance and eliminates the need for special test equipment. The disk media is in a removable cartridge which currently uses a drive with 1 Gbyte of formatted storage.

The HDS is designed to support incorporation of future higher-capacity disk technology by simply replacing the drive in the removable cartridge with new 3.5 in disks. Capacities of 8 Gbytes are planned.

Specifications

Dimensions:
(enclosure) 469.9 × 787.4 × 314.2 mm
(1,054 Mbyte cartridge) 58.4 × 124.5 × 177.8 mm
Weight:
(enclosure) 43.1 kg
(1,054 Mbyte cartridge) 2.27 kg
Power supply: 115/200 V AC, 400 Hz, 3 phase, 165 W
Reliability: 14,500 h MTBF

Contractor

General Dynamics Information Systems.

VERIFIED

Aircraft Propulsion Data Management Computer (APDMC)

The APDMC is primarily an engine monitoring, aircraft warning and display and data transfer unit. It acquires engine data from electronic engine controls and provides engine health, status, maintenance and limit functions.

Avionics systems are connected to the warning and caution systems MIL-STD-1553B bus and ARINC 429 bus through the APDMC with 37 other aircraft LRU analogue and discrete input/output parameters. The APDMC provides aircraft stall, overspeed, take-off and horizontal stabiliser warning and display; records engine maintenance data and system flight data; and provides cockpit display data for system LRUs, fault history, propulsion and exhaust gas temperature.

The signal input/output complement includes dual-redundant MIL-STD-1553B multiplex databusses; eight receive and seven transmit ARINC 429 serial databusses; two ARINC 573-7 serial databus outputs of which one is used for an AIMS recorder output; 35 analogue inputs including a mix of RVDTs and LVDTs plus excitation; a mix of AC and DC absolute and ratiometric signals; 26 discrete inputs including series and shunt; and six discrete outputs.

The MIL-STD-1750 microprocessor and memory management unit controls an expandable 128 kbyte onboard ROM, 32 kbyte RAM and 32 kbyte NV fault storage. System BIT allows fault isolation to the offending LRU with 97.5 per cent confidence. Internal BIT provides onboard unit status. Non-volatile fault storage allows ground-based interrogation of LRU faults.

Specifications

Dimensions: 123.9 × 193 × 320 mm
Weight: 5.55 kg
Power supply: 28 V DC, 40 W
115 V AC

Operational status

In production for the US Air Force C-17.

Contractor

Hamilton Sundstrand Corporation.

VERIFIED

DFDAU 120 Digital Flight Data Acquisition Unit

Complying with ARINC 717 and performing the same functions as the FDAU for the mandatory flight recording systems on the new-generation transports built to ARINC 700, the DFDAU 120 Digital Flight Data Acquisition Unit contains a microprocessor permitting it to record some AIDS information in addition to its crash recorder functions.

Specifications

Dimensions: 6 MCU
Weight: 6.8 kg
Power: 40 W

Operational status

The DFDAU 120 is standard equipment on the Boeing 767 and 757 and is a basic option on the Airbus A310. It is also used as a building block for an expanded ACMS on the A300 and A310.

Contractor

Hamilton Sundstrand Corporation.

VERIFIED

DMU 100 and 101 Data Management Units

The DMU 100/101 Data Management Units are the brain for the ACMS system, analysing the real-time information from the FDAU, the ADAU or the DFDAU. They control the digital ACMS recorder, having program logic that determines what information to record and when to do so. They also provide data to displays, including the FDEP and airborne printer, and contain extensive built-in test equipment.

Specifications

	DMU 100	**DMU 101**
Dimensions	1 ATR long	6 MCU
Weight	16.36 kg	5.7 kg
Power	180 W	55 W

Operational status

The DMU is part of the company's Mk II and expanded Mk III ACMS system in the following aircraft:
DMU 100: Airbus A300, Boeing 747, and DC-10
DMU 101: Airbus A310 and Boeing 767.

Contractor

Hamilton Sundstrand Corporation.

VERIFIED

DMU 120 Data Management Unit

The DMU 120 is a powerful microprocessor-based data acquisition and processing tool utilised on the A320 aircraft for aircraft, engine and APU condition monitoring and for intensive aircraft systems

The DMU 120 is installed in the Airbus A320

troubleshooting. It collects comprehensive aircraft and engine data via ARINC 429 serial input ports. Real-time data processing and analysis is performed. Information in the cockpit is made available in convenient format via the MCDU and printer. Automatic, cockpit and uplink requested information supplied by the DMU is transmitted to the ground by ACARS. The DMU provides for data reduction recording to the DAR which permits economical ground-processing of airborne data.

Specifications

Dimensions: 195 × 95.5 × 287.6 mm
Weight: 4.4 kg
Power: 30 VA

Operational status

In production for the Airbus A320.

Contractor

Hamilton Sundstrand Corporation.

VERIFIED

DSS-100 Data Storage Set

The DSS is a major element of the US Navy Flight Incident Recorder and Aircraft Monitoring System (FIRAMS) on the F/A-18 aircraft. It also performs data storage functions on the A-6, F-14, AV-8B and V-22 aircraft. The DSS consists of a Data Storage Unit (DSU) and a Data Storage Unit Receptacle (DSUR). The DSUR is mounted in the cockpit. The DSU is a modular electronic memory unit which slides into the DSUR for flight and is removable for analysis off the aircraft. The DSU employs a microprocessor to extract data from a MIL-STD-1553B bus for storage in memory and vice versa. The microprocessor also expands the effective amount of internal memory by executing a data compression algorithm. The memory medium is non-volatile EEPROM installed in blocks of from 2 to 8 Mbytes. The DSUR contains no electronics or active components.

Specifications

Dimensions:
(DSU) 43 × 119 × 205 mm
(DSUR) 53 × 127 × 229 mm
(faceplate) 102 × 165 mm
Weight: 1.9 kg
Power: 12 W

Operational status

In production and service; installed in the A6, AV-8B, F-14, F/A-18.

Contractor

Hamilton Sundstrand Corporation.

UPDATED

Engine Diagnostic Unit (EDU)

Pratt and Whitney's F100-PW-220 and 229 high-performance engines are fitted with Hamilton Sundstrand's advanced engine monitoring system which includes both engine-mounted equipment and the associated ground-based diagnostic units.

An EDU is the on-engine module which acquires engine data from controls and sensors, records operating time and cycles, detects critical events and stores selected event parameters. The EDU also performs extensive self-health and data validity checks and stores this data, along with Digital Electronic Engine Control (DEEC) diagnostic data, for further analysis.

The hand-held Data Collection Unit (DCU) is used to gather data from the EDU, clear the EDU's memory and perform EDU/DEEC system diagnostics. Data from multiple aircraft may be stored in the DCU's removable memory module, providing easy data transfer between the flight line and a ground-based computer for logistics data recording.

The ground-based portable Engine Analyser Unit (EAU) interfaces with the EDU or the DEEC for data acquisition, memory examination or modification and data monitoring of the EDU/DEEC output data streams during engine operation. The EAU is also capable of performing diagnostic testing of the DEEC or EDU and exercising combined EDU/DEEC fault logic and event codes, providing a fast, accurate determination of engine problems.

Specifications

Dimensions:
(EDU) 241 × 279 × 127 mm
(DCU) 475 × 424 × 412 mm
Weight:
(EDU) 4.13 kg
(DCU) 23.6 kg

Operational status

In production.

Contractor

Hamilton Sundstrand Corporation.

VERIFIED

FDAU 100 Flight Data Acquisition Unit

Essentially a data gatherer, the FDAU 100 flight data acquisition unit contains the signal conditioning needed to rationalise the many types of signals from engine, airframe and systems sensors. Signals are multiplexed and digitised so that they can be recorded on the DFDR for accident investigation purposes. The unit complies fully with ARINC 573 and meets regulatory agency flight data acquisition requirements.

Specifications

Dimensions: ½ ATR long
Weight: 1.8 kg
Power: 70 W

Operational status

The unit is used on Boeing 727, 737, 747, DC-9 and DC-10, and Airbus A300 airliners.

Contractor

Hamilton Sundstrand Corporation.

VERIFIED

Multi-Application Control Computer (MACC)

The Multi-Application Control Computer (MACC) is a flexible computer that can be applied to a wide range of flight and subsystem control fixed-wing aircraft and helicopter applications. A key feature of the MACC architecture is that any number of MACCs can be interconnected to form a simplex, dual, triplex or quad configuration without hardware or software changes. For example, the quad configuration can be reconfigured as two separate, but linked, dual-channel computers performing different tasks. This flexibility is made possible by the Input Output Computer (IOC) which performs all Input/Output (I/O) processing in firmware, leaving the application specific processing to the general purpose processors.

The hard separation between I/O management and application processing is key to the success of the MACC architecture. Both the IOC firmware and application software can be loaded and monitored over an RS-422 link from a portable PC. The same interface can also be used for system testing and integration.

The IOC contains redundancy management, BIT and fault detection and isolation algorithms which are performed at significantly faster speeds compared to general purpose processors. Many of these algorithms are embedded in silicon microcircuits, giving the MACC system significant I/O processing bandwidth. Specifically, the IOC can perform all the I/O processing needed for 50,000 sensor sets/s. The result is a system that is robust, fault tolerant and highly flexible for meeting the requirements of most embedded control systems.

Specifications

Dimensions: 257.8 × 193.8 × 254 mm
Weight: 7.76 kg
Power supply: 115 V AC, 28 V DC, 55.5 W

Operational status

In production. Present applications include flight control, vehicle management system control, actuator and subsystem controllers.

Contractor

Hamilton Sundstrand Corporation.

VERIFIED

Advanced Recorder (AR) series Solid-State Cockpit Voice Recorder and Flight Data Recorders (SSCVR/FDRs)

The AR series of recorders is designed for use in business and general aviation aircraft and helicopters. The series comprises the following products:

(1) AR Cockpit Voice Recorder (CVR)
(2) AR Flight Data Recorder (FDR)
(3) AR Combined cockpit voice recorder/flight data recorder (AR Combi).

The newest technology, including solid-state memory, has been used in the AR series to minimise weight, reduce installation cost and delete maintenance. The new units weigh only 4 kg, require no mounting tray and need no maintenance.

The AR CVR simultaneously records up to four channels of audio for 30 or 120 minutes.

The AR FDR can record at data rates of 64, 128 or 256 words/s for 25 hours.

The AR Combi provides both AR CVR and AR FDR capabilities in one unit.

Each system provides as appropriate:

(1) One area audio channel and two or three crew audio channels
(2) One ARINC 717 or ARINC 429 data input
(3) Greenwich Mean Time (GMT) input
(4) Rotor tachometer input
(5) Two spare ARINC 429 inputs (reserved for CNS/ATM)
(6) Maintenance and status outputs.

In order to reduce weight and cockpit panel space, the microphone preamplifier is contained within the microphone unit, eliminating the need for a control panel.

Data downloading is by a hand-held download unit, connected via an RS-422 interface. Playback is on the Playback and Test Station (PATS), which can be hosted on a PC-based workstation using Windows® and Airraft Data Recovery and Analysis Software (ADRAS®).

Specifications

Dimensions: 232.4 (L) × 149.1 (W) × 142.2 (H)
Weight: 4 kg (incl underwater locating beam)
Power: 8 W nominal, 10 W max

Contractor

Honeywell Aerospace, Electronic & Avionics Lighting.

VERIFIED

The Advanced Recorder series of solid-state cockpit voice and flight data recorders ***2000***/0081866

Specifications

Dimensions: ½ ATR
Weight: 10.5 kg
Tape type, drive and speed: 0.25 in polymide-base, AC hysteresis synchronous motors, 2.75 in/s (70 mm/s)
Bandwidth: 300-5,000 Hz ±3 dB
Maintenance period: 6,000 h
Environmental:
(fire) meets requirements of TSO C84 and TSO C51a; temperature of 1,100°C over 50% of equipment surface for 30 min
(salt water immersion) 30 days
(explosion) meets or exceeds DO 160 requirements

Operational status

In service. Over 2,000 units have been built.

Contractor

Honeywell Aerospace, Electronic & Avionics Lighting.

VERIFIED

AV-557C Cockpit Voice Recorder (CVR)

The AV-557C CVR provides a crash survivable record of the last 30 minutes of flight crew conversation by recording it simultaneously on four separate tracks. The new device succeeds the earlier AV-557A and AV-557B CVRs, offering improvements in reliability and maintenance cost by incorporating integrated circuits.

The specially coated polymide-base magnetic tape recording medium is less sensitive to heat than conventional polyester film, so that data can be recovered after exposure to 240°C, compared with 110°C in conventional equipment. Recorder electronics are mounted on functional plug-in modules for ease of test and replacement. The tape has a clear window at each end so that the beginning and end of tape can be detected, capstan motor rotation electronically reversed and the record, erase and monitor heads switched for bidirectional recording.

The recorder is used in conjunction with a microphone monitor that can be installed on the flight deck. It is housed within a unit that also contains an amplifier, filter, monitor indicator, push-test and bulk-erase buttons and a headphone jack for auditing recorded information. An external microphone can also be plugged into the unit.

DL-900 data loader

Honeywell's DL-900 data loader is compatible with the NZ-600, NZ-800, NZ-900 and NZ-2000 navigation computers and the IC-615, IC-800 and IC-1080 integrated avionics computers. The system features an RS-422 interface and accepts high- and low-density diskettes. With suitable software (version 5.2 and greater) the DL-900 is capable of simultaneous NDB uploads.

Specifications

Dimensions: 146 × 57 × 210 mm
Weight: 1.27 kg
Power: 28 V DC; 2.5 W

Operational status

In production and in service with associated navigation systems.

Contractor

Honeywell Aerospace, Electronic & Avionics Lighting.

NEW ENTRY

ED-55 Solid-State Flight Data Recorder (SSFDR)

The ED-55 Solid-State Flight Data Recorder (SSFDR) is an all-solid-state implementation of a crash survivable flight data recorder. It conforms to ARINC 573, ARINC 717 and ARINC 747. The unit eliminates all moving parts and uses solid-state flash memory as the recording medium. The SSFDR comprises three shop replaceable units: the crash survivable memory, power supply and interface and control assembly.

Memory capacity allows for the last 25 hours of flight data to be stored at an input rate of 64 words/s or 128 words/s. Data is stored without the use of any data compression algorithm, which provides for maximum reliability and integrity of critical flight information. An underwater locator beacon is available as an option.

Specifications

Dimensions: ½ ATR long or short
Weight: 8.2 kg
Temperature range: –55 to +70°C
Reliability: >15,000 h MTBF

Operational status

In production and on order for Airbus aircraft, including the A319 and A320.

The first prototype of a combined SSFDR/SSCVR unit was shipped to Eurocopter for testing in the EC155 medium-lift helicopter in July 1998.

In June 1999, CASA selected the ED-55 SSFDR, as part of an extensive Honeywell Aerospace avionics package, for its new-production C-295 and CN-235 aircraft, also for retrofit installation into existing CN-235 aircraft.

Contractor

Honeywell Aerospace, Electronic & Avionics Lighting.

VERIFIED

ED-56A Solid-State Cockpit Voice Recorder (SSCVR)

The ED-56A Solid-State Cockpit Voice Recorder (SSCVR) is ARINC 757 compliant. It uses industry standard audio compression in both 30 minute and

2 hour models. Interface provisions include: one wideband and three narrowband audio channels; dedicated rotor tachometer input; FSK and GMT time recording; two ARINC 429 interfaces reserved for future ATC datalink messages; and maintenance data.

The ED-56A utilises FLASH EPROM memory and a patent Crash-Survival Memory Unit (CSMU). It is designed to replace older tape-based units.

Specifications

Dimensions: ARINC 404 ½ ATR short
Weight: <7.3 kg
Power: 35 W max

Operational status

In production and on order for Airbus aircraft, including the A319 and A320.

The first prototype of a combined SSFDR/SSCVR unit was shipped to Eurocopter for testing in the EC155 medium-lift helicopter in July 1998.

In June 1999, CASA selected the ED-56A SSCVR for its new-build C-295 and CN-235 aircraft, as part of a major avionics award to Honeywell Aerospace, Electronic & Avionics Lighting, which also includes retrofit application to its CN-235 aircraft.

Contractor

Honeywell Aerospace, Electronic & Avionics Lighting.

VERIFIED

Flight Data Acquisition Management System (FDAMS)

The Flight Data Acquisition Management System (FDAMS), combines the functionality of an ARINC 717 Digital Flight Data Acquisition Unit (DFDAU) with the real-time monitoring and troubleshooting capability of a Data Management Module (DMM) in a single 6 MCU box. It is a powerful computing machine with 2 Mbytes of solid-state mass memory.

The FDAMS provides a standard set of 64 maintenance, operational and special algorithms, four of which can be customised for troubleshooting. The airline can also enable or disable any of the 64 algorithms directly by using ground support software.

The ARINC 717 DFDAU provides the mandatory ARINC 573 flight data to the flight recorder. With its seven databases, the DFDAU meets the FAA 88-1, CAA and ICAO requirements for all newly built aircraft. To accommodate any future aircraft modifications easily, Sundstrand's databases can be reloaded from a floppy disk using an ARINC 615 data loader.

The DMM monitors a predefined set of parameters in the aircraft. It then collects and downlinks both exceedance and routine data via a cockpit printer, ACARS or data loader. The data is processed to provide both maintenance and flight operations reports necessary for monitoring the engines and other aircraft systems. Using an IBM-compatible PC, the readout programs provide information in numerical or graphical format.

Contractor

Honeywell Aerospace, Electronic & Avionics Lighting.

VERIFIED

Integrated Data Management Systems (IDMS)

The Honeywell Aerospace, Electronic & Avionics Lighting IDMS product family forms part of a much larger family of avionics systems including Ground Proximity Warning Systems (GPWS), weather radar, Communications, Navigation and Identification (CNI), Traffic alert and Collision Avoidance Systems (TCAS) and other products.

The role of the IDMS is to record voice and parameter data from these systems and to facilitate the loading and abstraction of mission management data.

The IDMS product family includes: a Solid-State Flight Data Recorder (SSFDR), a Solid-State Cockpit Voice Recorder (SSCVR), a Digital Flight Data Acquisition Unit (DFDAU), a Maintenance Data Acquisition Unit (MDAU), airborne and portable data loaders, HF data radios, Automatic Communications And Reporting System (ACARS) units and other systems' sensors and data handling units.

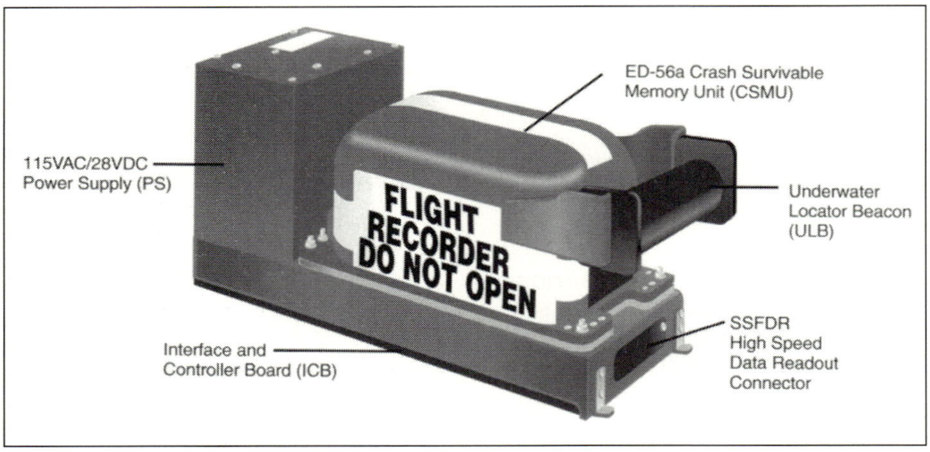

The ED-56A solid-state cockpit voice recorder **2000**/0081864

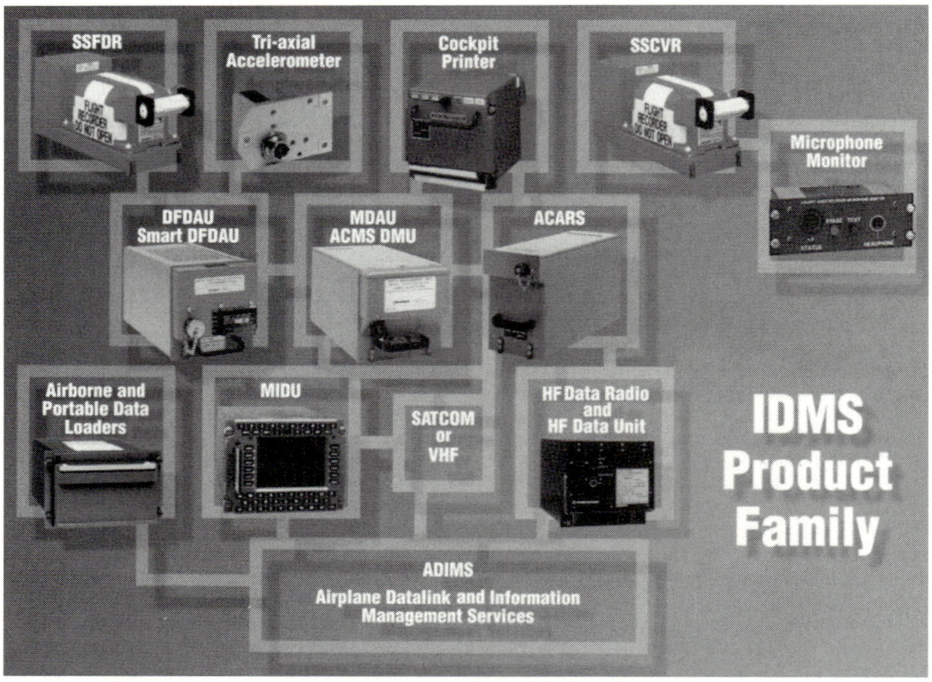

The Honeywell Aerospace, Electronic & Avionics Lighting integrated data management system **2000**/0081865

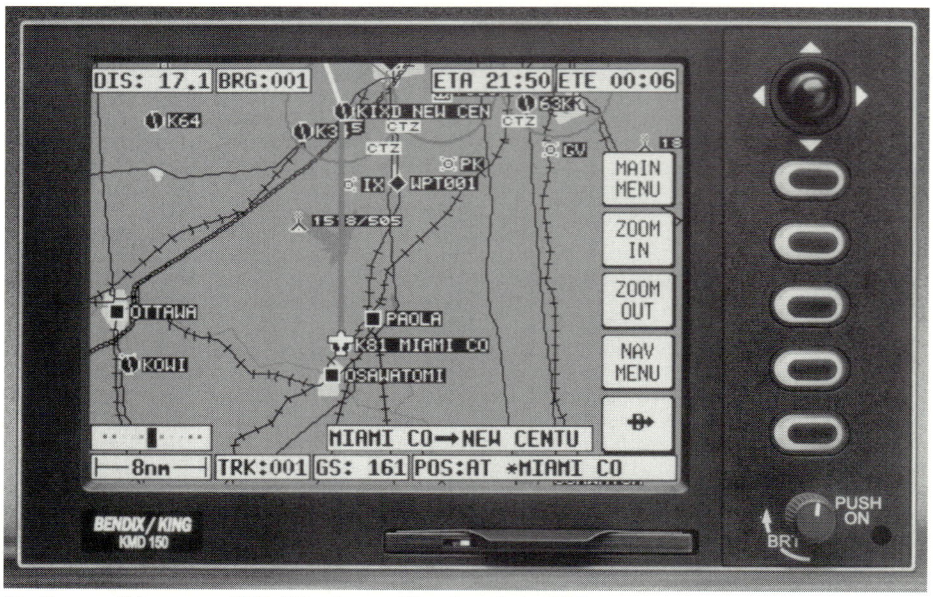

The Honeywell Aerospace KMD 150 colour multifunction display/GPS **2000**/0081844

Contractor

Honeywell Aerospace, Electronic & Avionics Lighting.

VERIFIED

KMD 150 colour MultiFunction Display MFD/GPS

The KMD 150 MFD/GPS is designed to display high-quality position information to piston-powered and light turbine aircraft users. It is part of the Bendix/King Silver Crown Plus avionics series. The KMD 150 combines aeronautical and cartographic mapping with GPS navigation information on a 5 in colour display. It also provides a range of additional display capabilities from airports, NDBs, intersections, VORs, special-use airspace, rivers, roads, lakes, cities, rail tracks and towers to WX-500 Stormscope® radar data.

The KMD 150 incorporates a built-in GPS receiver. It can also be procured without the integral GPS if that capability is already on the aircraft, or it can be used to complement Loran.

Specifications

Dimensions: 158 (W) × 175 (D) × 101 (H) mm
Weight: 1.5 kg
Screen size: 5 in diagonal, full colour
Power: 10-33 V DC, 10 W

Contractor

Honeywell Aerospace, Electronic & Avionics Lighting.

VERIFIED

Micro-Aircraft Integrated Data System (Micro-AIDS)

The Micro-Aircraft Integrated Data System (Micro-AIDS) has been developed to automate the various data logging functions on an aircraft, including the gathering and processing of data concerning engine condition and operational performance monitoring, exceedances and fuel usage and so on. A single unit is used to gather the data which is stored in solid-state memory.

The Micro-AIDS operates on the ARINC 573 data stream and recorded data can be recovered by ground crew using a portable data retrieval unit.

Specifications

Dimensions: ARINC 600 2 MCU
(3 MCU in expanded version)
Weight: 2.3 kg (3.6 kg expanded)
Power supply: 115 V AC, 400 Hz, 15 W
Capacity: 25 flights

Operational status

In production.

Contractor

Honeywell Aerospace, Electronic & Avionics Lighting.

VERIFIED

Mini-Flight Data Acquisition Unit (Mini-FDAU)

The Mini-FDAU is available in either the ARINC 404 or 600 form factor.

Output to any ARINC digital flight data recorder is generated in standard ARINC 573 format at 768 bits/s, Harvard biphase. Provisions are included to receive the DFDR BITE signal and individually display both the FDAU and DFDR status. Provisions are also included for interfacing with a Sundstrand flight data entry panel.

Input flexibility is available to permit interfacing with any aircraft configuration and to comply with diverse regulatory requirements. Data acquisition modules are available to accommodate use with analogue or digital ARINC 429 systems or with a combination of both analogue and serial databus inputs.

Input acquisition capacity is provided for recording of the FAA 17 mandatory parameters requirement. Capacity can be expanded to comply with the proposed ICAO requirements by the addition of one module.

The analogue acquisition design achieves a universal capability to acquire synchro, AC ratio, low-level DC and potentiometer signals with a single input circuit. The unit also allows automatic determination of the input signal based only on the aircraft wiring connections. This allows the Mini-FDAU to be interchanged among different aircraft without the need for reprogramming of a configuration PROM.

The Mini-FDAU is expandable to accommodate future additional input to include parameters such as those required for engine condition monitoring, a separate CPU and memory capable of screening data for storage in onboard solid-state memory. An ACARS interface expansion allows access to stored information and downlinking automatically or on demand.

Specifications

Dimensions: ARINC 600, 3 MCU
Weight: 3.63 kg (max)

Contractor

Honeywell Aerospace, Electronic & Avionics Lighting.

VERIFIED

PTA-45B airborne data printer

The PTA-45B data printer uses a high-efficiency switching power supply, which provides sufficient power to continuously print 'all-black' (all 600+ dots printing at the same time). Since normal operation runs at only 15 per cent of the machine capacity, the printer is claimed to be capable of providing up to 10 times the reliability of previously available models.

The printer permits flight crew access to hard copies of uplinked flight plans, messages and graphics, and can also print a copy of datalink information displayed on the weather radar display. The printer may be driven by a datalink system or a maintenance computer. User features include paper guides to help prevent paper jamming, a hinged spool to ease paper changing and a paper level indicator to tell how much paper is left. The unit has a MTBF in excess of 5,000 hours.

Operational status

In service on BAE Systems 146 aircraft. The PTA-45B is standard equipment on the Airbus A320 and has been selected for the Boeing 737, 747-400, 757 and 767, MD-11, MD-80 and MD-88 and Airbus A310-300, -400 and -500 airliners.

Contractor

Honeywell Aerospace, Electronic & Avionics Lighting.

VERIFIED

Units of the PTA-45B data printer family

Quick Access Recorder (QAR)

The QAR is an Aircraft Integrated Data Systems (AIDS) recorder.

Twelve-track sequential recording provides up to 50 hours of operation per cassette at an input data rate of 768 bits/s. Total storage up to 138 Mbits is available at a conservative packing density of 2,400 bits/in. The QAR can accept either Harvard biphase or bipolar RZ data input.

It features remotely selectable tape speeds to accommodate four input data rates, a ruggedised metal cassette to protect the tape in or out of the recorder, precision tape guidance within the cassette to ensure interchangeability between recorders and remote equipment status indication as a function of the BITE capability.

Specifications

Dimensions: ARINC 404 ½ ATR long
Weight: 8.62 kg
Power supply: 115 V AC, 400 Hz, 60 W (max)

Contractor

Honeywell Aerospace, Electronic & Avionics Lighting.

VERIFIED

Ruggedised Optical Disk System (RODS)

The RODS is a digital data storage and retrieval device utilising a fully erasable and rewritable 5.25 in diameter optical disk as the storage medium. RODS is designed for use in special mission aircraft and for other military applications requiring highly reliable data storage and transfer in harsh operating environments.

Typical areas of application include digital map terrain data storage, electronic document storage, signal and mission data recording, maintenance and structural data recording and general purpose data loading and data transfer.

Data capacity is 300 Mbytes user data per side and sustained transfer rate is 518 kbytes/s. Access is 100 ms maximum.

Specifications

Dimensions: 127 × 190.5 × 317.5 mm
Weight: 6.8 kg
Power supply: 115 V AC, 400 Hz or 28 V DC
Reliability: 5,000 h MTBF

Contractor

Honeywell Aerospace, Electronic & Avionics Lighting.

VERIFIED

SETS-II Severe Environment Tape System

The Severe Environment Tape System (SETS-II) comprises two modules: a tape drive system, with the recorder electronics and mechanical elements, and a tape unit containing the recording medium, magnetic recording head tape, tension device and tape-end sensor.

The system accommodates 300 ft of 0.25 in tape, with a four-track single gap read/write head, a recording density of 1,600 bits/in, transfer rate of 48,000 bits/s (phase encoded) and total capacity of 23 Mbits. The nominal error rate is 1 in 10^7 bits. Tape speed is 30 in/s write and 120 in/s search and rewind. Total system weight is 2.02 kg.

Operational status

Present status uncertain.

Contractor

Honeywell Aerospace, Electronic & Avionics Lighting.

UPDATED

Supplemental Flight Data Acquisition Unit (FDAU)

The FDAU has been designed to provide a low-cost convenient means for increasing the number of flight parameters recorded by Honeywell Aerospace's standard five-parameter Universal Flight Data Recorder (UFDR). Without replacing the UFDR, the Supplemental FDAU and recorder can together accommodate up to 11 parameters. This is accomplished by merging new parameters into the ARINC 573 data stream generated by the UFDR's internal FDAU and returning the data to the recorder section of the UFDR. The Supplemental FDAU provides a status output that is compatible with the ARINC 542 status-reporting system, causing minimal impact to existing wiring. If the Supplemental FDAU is not installed or its status is invalid, the UFDR automatically reverts to its normal five-parameter operation.

The Supplemental FDAU can accept a variety of sensor and signal types. The programmable analogue inputs accept synchro, AC ratio No 1, AC ratio No 2, 0-5 V low-level DC and potentiometer signals. Dedicated inputs include shunt and series discretes, ARINC 573 serial data from the UFDR and UFDR validity status.

The BIT capabilities of the Supplemental FDAU determine whether the recording is functioning and enunciate system status. Additionally, the Supplemental FDAU will provide validity indication that will allow faults to be isolated to the LRU.

Specifications

Dimensions: ARINC 600, 2 MCU
Weight: 2.27 kg

Contractor

Honeywell Aerospace, Electronic & Avionics Lighting.

VERIFIED

Tactical Optical Disk System (TODS)

The Tactical Optical Disk System (TODS) is a severe environment digital data storage and retrieval device utilising a fully erasable and rewritable 5.25 in diameter optical disk as the storage medium. TODS is designed for unrestricted operation in high-performance military aircraft and other platforms requiring highly reliable data storage and transfer in very severe operating environments.

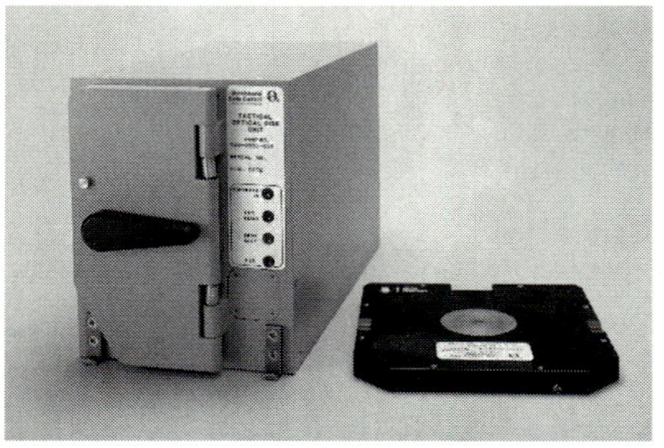

The tactical optical disk system

Areas of application include digital map terrain data storage, signal and mission data recording, maintenance and structural data recording and electronic document storage.

Data capacity is 300 Mbytes user data and sustained transfer rate is 518 kbytes/s. Access time is 120 ms maximum.

Specifications

Dimensions: 127 × 165.1 × 254 mm
Weight: 8.16 kg
Power supply: 115 V AC, 400 Hz or 28 V DC
Reliability: 5,000 h MTBF

Contractor

Honeywell Aerospace, Electronic & Avionics Lighting.

VERIFIED

Universal Flight Data Recorder (UFDR)

The 25-hour-capacity, crash-protected digital UFDR has a variety of configurations. Together with optional accessory equipment, the system can be made completely interchangeable with existing flight data recorders to ARINC 542, with digital recorders to ARINC 573/717, or expanded parameter systems intended for CAA-regulated applications.

The recorder uses Kapton tape for durability, a single motor and co-planar reel-to-reel tape transport. It can be hard mounted in the aircraft, and considerable attention has been given to reducing the number of moving (and therefore wearable) parts. High-speed data retrieval permits the entire tape content to be transferred to a copy recorder in less than 30 minutes, so that the operation may be conducted during normal turnround.

The system uses a patented feature called Checkstroke which monitors the data recorded each second and verifies its accuracy against the continuous stream of information received in the recorder's electronics. This built-in test feature ensures the validity of the recorded data.

With the release of the FAA NPRM 85-1, new versions of the UFDR have been designed. The new H model is considered a six-parameter unit (meeting that portion of the NPRM) and the new G Mod 10 model meets the expanded 11-parameter portion of the NPRM. The H Model is expansible to the G Mod 10 simply by adding one circuit card and a 5 V DC power supply.

A new feature of these models is the addition of programmable inputs which allows the user to tailor the UFDR to his particular transducer types. Another plus is the increased accuracy of the system.

The following inputs are available for the H Model: vertical acceleration, heading, pneumatic airspeed, pneumatic attitude, marker beacon, time, trip and date and two programmable inputs for altitude and airspeed.

The following inputs are available for the G Mod 10 Model: the same as for the H Model plus pitch attitude, pitch trim position, roll, flap, engines 1 to 4, total air temperature, one spare programmable and two spare 0-5 V DC.

Specifications

Dimensions: ½ ATR long per ARINC 404
Weight: 13.6 kg
Power supply: 115 V AC, 400 Hz, single phase
Compatible transducers: synchros, 0-5 V DC 3 wire, 0-18 V DC 3 wire, 0-5 V DC 2 wire, 0-15 V 2 wire, AC ratio type 1, resolvers and selsyns.
Recording capacity: 25 h at 768 bits/s
Input interface:
(digital flight data system) ARINC 573/717 data stream from remote flight data acquisition unit
(flight data recorder) vertical acceleration, altitude, airspeed, heading, marker beacon and three discrete signals (on/off, to/from and so on). Alternatively can be ARINC 573/717 input from a remote flight data acquisition unit
(expanded channel flight data recorder) same as for flight data recorder plus seven additional synchro inputs
Tape format: blocked incremental, biphase encoded, with 768 data bits, preamble, postamble and IRG
Tape type: Kapton, 0.25 in magnetic
Tape speed: recording, incremental 5 in/s = 0.414 in/s net; playback 5 in/s
Tape capacity: 450 ft for 25 h recording
Number of tracks: 8 bidirectional sequential
Playback time: 30 min with copy recorder, 2 h with sequential track transcription
Crash survivability: TSO C51a
Environmental: RTCA DO 160, TSO C51a

Operational status

In production and service for over 20 years. In US Air Force service on KC-10A Extender tankers and US Army service on Black Hawk helicopters. Also in widespread airline service.

Contractor

Honeywell Aerospace, Electronic & Avionics Lighting.

VERIFIED

Advanced Display Core Processor (ADCP)

The ADCP is a commercially-based processor, designed for the US Air Force F-15E, and sponsored by the US Defense Department's Commercial Operations and Support Savings Initiative (COSSI).

The ADCP is designed to replace two existing line replaceable units: the MultiPurpose Display Processor (MPDP) and a VHSIC (very high-speed integrated circuit) Central Computer (VCC).

Software for the ADCP has been supplied by Virtual Prototypes Inc of Montreal, Canada.

Operational status

First flight on an F-15E was in July 1998, when it was used to drive a 5 × 5 in Honeywell liquid crystal display; three more flights were planned. A preproduction ADCP was due be developed by the third quarter of 1999.

Contractor

Honeywell Defense Avionics Systems.

VERIFIED

Data storage and retrieval unit

The radiation-hardened data storage and retrieval unit has evolved from the optical disk digital memory unit developed for night attack aircraft. The 300 Mbyte unit is small and lightweight, offering a significant size advantage over larger magnetic storage systems. The unit reduces the risk of loss or compromise of vital information and makes storage of information on board more efficient.

In the Northrop B-2, two rewritable data storage and retrieval units are located in the cockpit console, providing the crew with immediate access to the mission database and the ability to record mission data and performance information in flight.

Operational status

In service on US Air force B-2 aircraft.

Contractor

Honeywell Defence Avionics Systems.

UPDATED

Air data computer for the A-10

Designed to full military environment and MIL-STD-1553 digital databus standards, the air data computer for the US Air Force Fairchild A-10A features a digital pressure transducer and a ferro-resonant power supply. Modular programming is provided and the memory has a 6 kword read-only memory and a 256 word random access memory.

Specifications

Dimensions: ½ ATR
Weight: 6.3 kg
Power: 36 W

Operational status

In service on US Air Force A-10 aircraft.

Contractor

Honeywell Engine Systems & Accessories.

VERIFIED

Digital air data computer for the AV-8B

In common with all Honeywell Engine Systems and Accessories' air data computers, this model uses a quartz pressure transducer. The unit is similar in construction to that used on the US Air Force A-10 aircraft. It features a non-volatile memory for the retention of fault messages, even if power is turned off.

Specifications

Dimensions: 129 × 193 × 355 mm
Weight: 7.2 kg
Inputs: 10
Outputs: 78

Operational status

In service on the US Marine Corps AV-8B V/STOL close support aircraft.

Contractor

Honeywell Engine Systems & Accessories.

VERIFIED

Digital air data computer for the B-1B

This computer was developed for the US Air Force B-1B and features a digital MOS large-scale integrated circuit processor and digital pressure transducers. It has an altitude reporting facility and the built-in self-test system provides continuous failure monitoring.

Specifications

Dimensions: 158 × 195 × 500 mm
Weight: 12.6 kg
Inputs: 19
Outputs: 92

Operational status

Developed for the US Air Force B-1A and in service in the B-1B.

Contractor

Honeywell Engine Systems & Accessories.

VERIFIED

Digital air data computer for the JA 37

This is an all-digital computer with similar features to those of the B-1 air data computer and has bidirectional digital communication with the aircraft subsystems. It contains both static and dynamic self-test facilities.

Specifications

Dimensions: 157 × 213 × 439 mm
Weight: 11.4 kg
Inputs: 16
Outputs: 37

Operational status
In service in the Saab JA 37 interceptor.

Contractor
Honeywell Engine Systems & Accessories.

VERIFIED

Miniature Air Data Computer (MADC)

The simplification of the interface to MIL-STD-1553B with only a few analogue inputs has made possible the Miniature Air Data Computer (MADC).

Inputs to the baseline MADC are indicated static and total pressure, total temperature (50 or 500 ohm probe), self-test and identification discrete signals with commands via the MIL-STD-1553B databus. Analogue inputs of barometric correction and angle of attack may also be provided.

In addition to the digital databus and altitude reporting code outputs, the MADC has provision for a dual-synchro drive compatible with AAU-19/A, AAU-34/A and AAU-37/A altimeters. These synchro outputs can also be software-programmed to provide other air data functions such as Mach number or true airspeed. Other miscellaneous analogue and discrete outputs tailored to a particular application can be provided. The pressure transducers are self-contained plug-in modules using fused quartz RC sensors.

Alternative mounting arrangements enhance the flexibility offered by this small package. A front panel Go/No-Go fault annunciator signals the result of the built-in test. Test points are furnished in the front panel input/output connector for fault isolation at the intermediate and depot levels of maintenance.

Specifications
Dimensions: 127 × 108 × 255 mm
Weight: 3 kg
Power: 15 W (max)

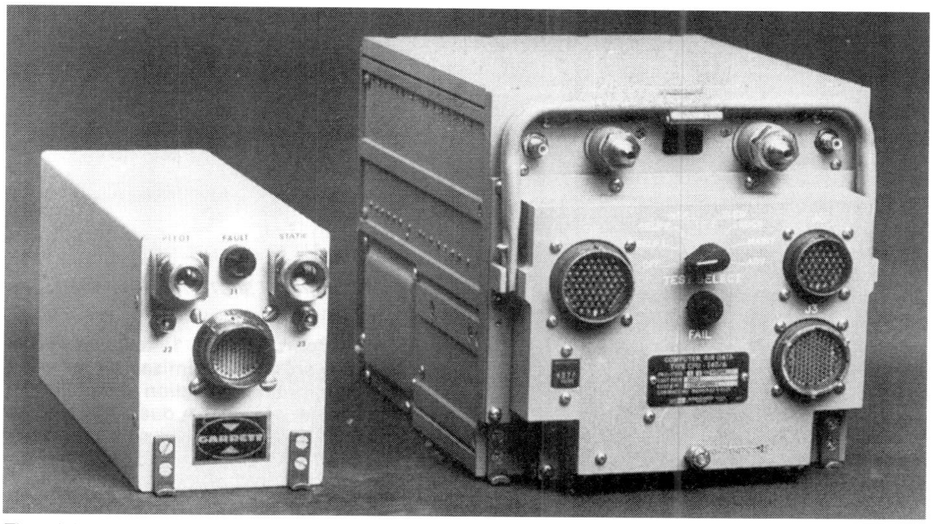

The miniature air data computer (left) is half the size and weight of a standard central air data computer (right)

Contractor
Honeywell Engine Systems & Accessories.

VERIFIED

Standard Central Air Data Computers (SCADC)

Honeywell Engine Systems & Accessories has completed contracts to supply SCADC to Lockheed Martin for the C-5B Galaxy and to Northrop Grumman for the C-2A Greyhound carrier onboard delivery aircraft.

The SCADC configuration uses a non-volatile memory for storing air data and subsystem fault information. There is a high standard of fault detection and high reliability ensures compatibility with a two-level maintenance strategy.

Specifications
CPU-140/A for the C-2A
Dimensions: 143 × 140 × 242 mm
Weight: 9.4 kg
Power supply: 115 V AC, 400 Hz, 60 W

CPU-141/A for the C-5A and C-5B
Dimensions: ½ ATR
Weight: 14.4 kg
Power supply: 115 V AC, 400 Hz, 72 W

Operational status
In service on US Air Force C-5B Galaxy and US Navy C-2A aircraft.

Contractor
Honeywell Engine Systems & Accessories.

VERIFIED

ADZ air data system

The ADZ air data system uses an AZ-241 or -242 computer, depending on whether the aircraft is a turboprop or a jet. These are now complemented in production with the AZ-600, for turboprops, and AZ-800, for jets, digital air data systems. All systems use the Honeywell patented vibrating diaphragm pressure sensor to provide outputs for altitude, airspeed, vertical speed, true airspeed, true air temperature and total air temperature information.

If required, a single computer can handle a dual-flight director installation. Consequently, no matter which director is driving the autopilot the full complement of modes is available.

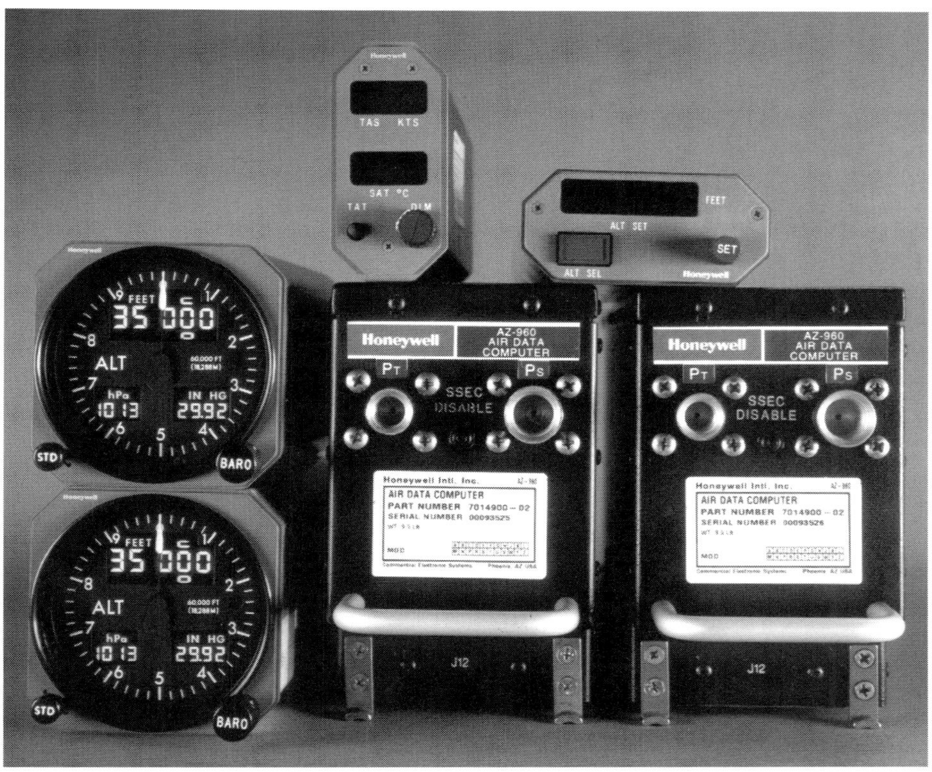

Two of Honeywell's AZ-960 advanced air data computers, shown with other elements of a complete RVSM package: left to right, dual BA-250 altimeters, a DS-125 true airspeed/total temperature indicator and an AL-861 altitude preselect controller **2001**/0106206

Specifications
AZ-241 (turboprop)
Dimensions: 193 × 79 × 361 mm
Weight: 3.9 kg

AZ-242 (jet)
Dimensions: 193 × 124 × 361 mm
Weight: 5.2 kg

AZ-600 (turboprop)
Dimensions: 193 × 92 × 362 mm
Weight: 3.7 kg

AZ-800 (jet)
Dimensions: 193 × 92 × 362 mm
Weight: 4.08 kg

Operational status
In production for a wide range of executive turboprop and jet aircraft; one of the latest is the Raytheon Hawker 1000.

Contractor
Honeywell Inc, Commercial Aviation Systems.

VERIFIED

AZ-960 Advanced Air Data Computer (AADC)

The AZ-960 Advanced Air Data Computer (AADC), designed for the Gulfstream II and IIB, is one of Honeywell's latest products in its range of air data systems. The AZ-960 AADC is fully compatible with Honeywell's legacy SP-50G and SPZ-800 autopilot systems. It enables operators to comply fully with the requirements of operations within Reduced Vertical Separation Minima (RVSM) airspace.

Operational status
In production. Honeywell has received TSO approval for the AZ-960 AADC.

Contractor
Honeywell Inc, Commercial Aviation Systems.

NEW ENTRY

Boeing 777 Air Data Module (ADM)

The Boeing 777 Air Data Module (ADM) is a pressure transducer with a single pressure input port and an ARINC 629 bus output that transmits linearised, digital serial pressure data. Each Boeing 777 aircraft is configured with six ADM units: three with pitot pressure sensors and three with static pressure sensors. The ADM transmits ARINC 629 digital serial air data to the Air Data Inertial Reference System (ADIRS) and the Secondary Attitude Air data Reference Unit (SAARU).

Specifications
Dimensions: 88.9 × 63.3 × 171.4 mm
Weight: 1 kg (max)
Power supply: 28 V DC, 10.6 W (max)
Temperature range: –15 to +70°C
Accuracy:
(100-147.4 mb) ±0.25 mb
(147.4-175.3 mb) <±0.3 mb
(175.3-1,400 mb) ±0.3 mb
Reliability: 50,00 h MTBF predicted

Operational status
In production and in service on the Boeing 777.

Contractor
Honeywell Inc, Commercial Aviation Systems.

UPDATED

DC-9 Digital Air Data Computer (DADC)

The Honeywell DC-9 Digital Air Data Computer (DADC) replaces the existing DC-9 electromechanical air data computer. The new DADC is a form, fit and functional replacement unit featuring solid-state components, including silicon chip pressure sensors that are more accurate, stable and reliable than the vibrating diaphragm predecessors.

Specifications
Dimensions: ⅜ ATR
Weight: 5.44 kg
Power supply: 115 V AC, 400 Hz, 20 W
Temperature range: –55 to +70°C
Accuracy:
(ARINC 429 altitude) ±15 ft at sea level, ±80 ft at 50,000 ft
(synchro altitude) ±20 ft at sea level, ±55 ft at 45,000 ft
(ARINC airspeed) ±4 kt at 60 kt, ±1 kt at 450 kt
Reliability: 14,000 h MTBF predicted

Operational status
In service in the DC-9.

Contractor
Honeywell Inc, Commercial Aviation Systems.

VERIFIED

Digital air data computer

Honeywell produces an all-digital ARINC 706 standard air data computer for Airbus Industrie A310 and Boeing 757 and 767 airliners. These all-solid-state systems use large- or medium-scale integrated technology, CMOS master transmitter/receiver chips and a Z8000 microprocessor. Built-in test sequences run continuously in flight to provide a 95 per cent fault-detection capability, with a 99 per cent probability of correct fault identification. Test failure information is stored in the non-volatile memory for post-flight analysis. External sensor failures can also be detected and signalled visually. The system can accommodate a 20 per cent increase in output parameters, more than 50 per cent computational growth and more than 100 per cent memory growth. The unit incorporates Honeywell patented transducers and has ARINC 429 transmitter and receiver interfaces.

Specifications
Dimensions: 4 MCU
Weight: 5.7 kg
Power: 25 W
Accuracy:
(airspeed) ±2 kt at 100 kt, ±1 kt at 450 kt
(altitude) ±15 ft at sea level, ±80 ft at 50,000 ft
(Mach) ±0.003 at 0.8-0.9 in the range 25,000-45,000 ft
Failure memory: 6 failures/10 flights
Reliability: 15,000 h MTBF

Operational status
In service.

Contractor
Honeywell Inc, Commercial Aviation Systems.

VERIFIED

HG280D80 air data computer

Standard for the DC-10 and MD-80 transport aircraft, the HG280D80 microprocessor-based ARINC 576-type digital system replaced the earlier HG280D5 computer. The transducers are new dual sensors, with high accuracy and a fast warm-up time.

The central processor memory and input/output functions have been reduced from the 360 integrated circuits on nine boards of the earlier D5 to 22 circuits on a single card. Built-in test includes both continuous monitoring and manually activated self-test. The D80's non-volatile memory automatically stores information for five flights and data may be retrieved by selecting a single switch on the front panel. If a failure has occurred either of two warning lights will illuminate.

The D80 can also accommodate limited power interruptions. A capacitatively maintained supply voltage and a CMOS memory can retain critical data during power losses of several milliseconds. When power resumes the device automatically restores all outputs based on this stored data.

Specifications
Dimensions: ½ ATR long
Weight: 5.9 kg
Power supply: 115 V AC, 400 Hz, 30 W
Temperature range: –45 to +55°C
Accuracy:
(altitude) ±15 ft at sea level, ±80 ft at 50,000 ft
(airspeed) ±5 kt at 60 kt, ±3 kt at 325 kt
Reliability: 13,300 h MTBF

Operational status
Over 1,500 units are in service. The HG280D80 is standard equipment on DC-10, KC-10 and MD-80 aircraft.

Contractor
Honeywell Inc, Commercial Aviation Systems.

VERIFIED

HG480B and HG480C digital air data computers

The HG480B and HG480C digital air data computers meet ARINC 545 and replace the analogue HG180 on Boeing 727 and 737 aircraft. The unit incorporates ARINC 429 buses.

A single-board microcomputer controls the functions and built-in test (which includes both continuous monitoring and manually activated self-tests) reduces maintenance. Ground built-in test has been expanded and includes q-pot stimulation and transmission of fixed values on all signal output lines.

The HG480 is available in several versions. The basic configuration allows for differences in aircraft wiring but is functionally identical for what are designated the B1, B2 and B3 versions. Designed to operate with conventional analogue equipment, the system also provides outputs for digital instruments and digital autopilot. The B4 version is designed primarily to interface with an all-digital flight control system but can still operate with synchro-driven instruments.

Specifications
Dimensions: ½ ATR long
Weight: 7.48 kg
Power supply: 115 V AC, 26 V AC, 400 Hz, 44 W
Temperature range: –45 to +55°C
Accuracy:
(altitude) ±15 ft at sea level, ±80 ft at 50,000 ft
(airspeed) ±2 kt at 60 kt, ±1 kt at 300 kt
Reliability: 17,600 h MTBF

Operational status
In service in the Boeing 707, 727-200/300, 737-200/300 and 747-200/300. Over 2,000 units are in service.

Contractor
Honeywell Inc, Commercial Aviation Systems.

VERIFIED

HG480E1 digital air data computer

The baseline for the HG480E1 is the Sperry ARINC 706, repackaged into a standard ARINC 404 ½ ATR case. The basic circuitry of the ARINC 706 digital air data computer remains unchanged. Two new cards were incorporated for the additional analogue outputs required for Boeing 737-300 and -400 applications.

Specifications
Dimensions: ½ ATR long
Weight: 7.48 kg
Power supply: 115 V AC, 26 V AC, 400 Hz, 35 W
Temperature range: –45 to +55°C
Accuracy:
(altitude) ±15 ft at sea level, ±80 ft at 50,000 ft
(airspeed) ±2 kt at 100 kt, ±1 kt at 300 kt
Reliability: 12,000 h MTBF

Operational status
In service on the Boeing 737-300/400.

Contractor
Honeywell Inc, Commercial Aviation Systems.

UPDATED

IC-800 Integrated Avionics Computer (IAC)

Honeywell has developed an Integrated Avionics Computer (IAC) which incorporates the functions of four major avionics subsystems into a compact lightweight package that is 75 per cent smaller than the units it replaces.

The IC-800 is a key component of Honeywell's Primus 2000 advanced avionics system. The units incorporated in the IAC include a digital automatic flight control computer which has all the features and functions of the SPZ-8000 digital autopilot, dual-display processors, a fault warning computer and a full Flight Management System (FMS) computer. The IAC, which is a single LRU, has hardware expansion capability for incorporating additional functions.

The IC-800 is in production for the Bombardier Global Express (Bombardier) **2001**/0105288

The IAC provides at least 50 per cent more processor capacity and 100 per cent more memory for each function than was previously available. Its FMS computer has a 1.9 Mbyte database. This additional memory and processor power will allow substantial growth as each function is enhanced in the future.

The IAC's reduced size and weight result from the extensive use of advanced circuit and packaging technologies. These technologies include Very Large Scale Integration (VLSI), Surface Mount Technology (SMT), Application Specific Integrated Circuits (ASIC), high-density memories, rigid flex motherboards, multilayer circuit boards and advanced thermal management. The IAC is housed in an ARINC ½ ATR short package with an integral cooling fan.

Operational status

In production for the Dornier 328 and the Bombardier Global Express business jet.

Contractor

Honeywell Inc, Commercial Aviation Systems.

UPDATED

Micro Air Data Computer (MADC)

The AZ-840 Micro Air Data Computer (MADC) provides both ARINC 429 and bidirectional Avionics Standard Communications Bus (ASCB) digital serial interfaces for the Honeywell Primus 2000, flight control and navigation systems. The MADC uses extensive surface-mount technology for minimising unit size and weight. Full air data outputs and miscellaneous inputs/outputs are provided, including ATC digitiser and input/output discretes for switching functions and aircraft identification selection. Extensive BITE monitors MADC operation and records faults on non-volatile memory for fault analysis on the ground.

The MADC supports Reduced Vertical Separation Minimum (RVSM) requirements.

Specifications

Dimensions: 106.7 × 149.9 × 157.5 mm
Weight: 2.04 kg
Power supply: 28 V DC, 16 W (max)
Accuracy:
(altitude) ±20 ft at sea level, ±150 ft at 60,000 ft
(airspeed) ±2 kt at 100 kt, ±4 kt at 400 kt
Reliability: 6,000 h MTBF predicted

Operational status

Versions of the MADC are used on the Citation III, Dornier 328 and Raytheon Hawker 1000.

Contractor

Honeywell Inc, Commercial Aviation Systems.

VERIFIED

MD-90 Central Air Data Computer (CADC)

The MD-90 Central Air Data Computer (CADC) is a derivative of the HG280D80. The CADC provides both ARINC 429 digital serial air data outputs and ARINC digital serial and DC analogue air data outputs. Two unique features of the CADC - Trigger-On-Failure (TOF) BIT and an internal temperature probe - operate in unison to record BIT failures along with corresponding CADC internal temperature data. BIT failure records and other CADC features can be accessed through the front panel connector via the RS-232 bus. Similar to its predecessor, the CADC uses two vibrating cylinder transducers for sensing pitot and static pressure inputs. The vibrating cylinder transducers can be substituted with silicon pressure sensors. Inputs are also provided from a total air temperature probe, baroset signals and programme discretes. Jointly developed by Honeywell and the Chengdu Aero Instruments Corporation.

Specifications

Dimensions: ½ ATR long
Weight: 5.35 kg
Power supply: 115 V AC, 400 Hz
Temperature range: –55 to +75°C
Accuracy:
(altitude) ±15 ft at sea level, ±80 ft at 50,000 ft
(airspeed) ±5 kt at 60 kt, ±1 kt at 450 kt
Reliability: 13,000 h MTBF predicted

Contractor

Honeywell Inc, Commercial Aviation Systems.

VERIFIED

HG1140 multirole air data computer family

Honeywell's HG1140 air data product family features a broad range of configurations from single-channel air data transducers up to complete air data computers. The product line is based on Honeywell's solid-state silicon sensors and designed to provide primary instrument quality air data information that is compliant with the most stringent Reduced Vertical Separation Minima (RVSM). With flexible input/output capabilities, configurations are easily customised to meet varying interface requirements ranging from Mil Standard 1553 to ARINC 429. The dual use nature of the product line is ideal for both commercial and military aircraft, missile and unmanned vehicle applications.

Specifications

Dimensions: 152.4 × 111.76 × 50.8 mm
Weight: 0.91 kg
Power: 28 V DC, 6 W
MTBF: 100,000+ h (AIC); 50,000 h (AUF)
Long-term stability: 0.015% F.S. over 20 years
Interface options: ARINC 429, RS422, RS485
Optional I/O: Analogue, discrete, synchro MIL-STD-1553
Qualification: RTCA DO-160 and MIL-STD-810E

HG1140 multirole air data computer 0015280

HG1141 multifunction air data computer

The HG1141 is a derivative of the HG1140 air data computer. It contains a multifunction databus so that the same configuration can be used to satisfy the Gripen aircraft's dual air data system requirements. The HG1141 provides altitude and airspeed information as well as nine other parameters in the Gripen installation.

Operational status

Fully qualified to both RTCA-DO-160 and MIL-STD 810E requirements. Fielded applications include USAF T-38C Talon Upgrade, Lockheed Joint Strike Fighter, F/A-18 E/F, JAS-39, Airbus A319-A340, Tactical Tomahawk Cruise Missile, E-6B and the Gulf Stream IV-V.

Contractor

Honeywell Inc, Sensor and Guidance Products.

UPDATED

Military air data computers

Honeywell military digital air data computers are in production and in service in the following aircraft: B-52, C-17, C-130, F-16, F-117, F/A-18, FSX and KC-135R.

The latest air data computers use patented solid-state pressure sensors and are said to be so reliable that they will never require recalibration.

Operational status

In production and service.

Contractor

Honeywell Sensor and Guidance Products.

UPDATED

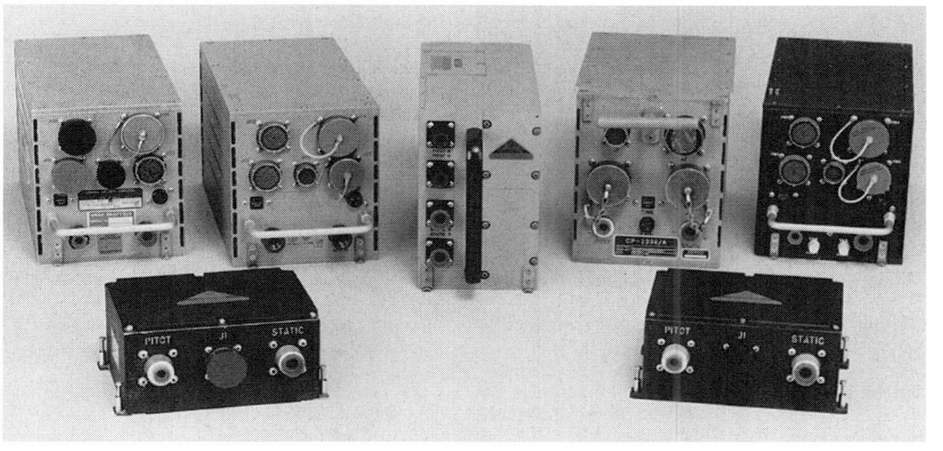

Pictured (left to right, rear) are the central air data computers for the F-15, F-16, C-17, F/A-18 and KC-135R and (front) the B-52

Sentinel® Data Transfer Systems (DTS)

The Raymond Engineering Sentinel® Data Transfer Systems (DTS) comprise a family of small, light weight, solid-state, data storage systems that have been specifically designed to provide reliable data storage in severe environment military applications.

Sentinel® Model 9410 SCSI DTS

The Sentinel Model 9410 SCSI DTS is a growth version of the Model 9510 SCSI DTS and features an industry standard SCSI-2 system interface with differential termination and dual slots which can accommodate either two type III or any combination of Sentinel PCMCIA (Personal Computer Memory Card International Association) Flash memory cards. The SCSI DTS allows for rapid insertion and removal of the Sentinel cards and features a rugged retention and ejection mechanism with a sealed field access cover.

The Sentinel SCSI DTS includes a 28 V DC power supply, dual PCMIA controller, control microprocessor and SCSI-2 interface electronics.

The system operates as a peripheral device or LRU under control of a host computer. The Sentinel SCSI

DTS is designed to accept two Sentinel PC Cards which incorporate the latest in Flash memory technology.

Sentinel cards utilise the PC-Card ATA standard command set and are compatible with PCMCIA equipped computers and laptops for programming and downloading purposes.

Specifications

Data capacity: system accepts one or two PC cards.
Interface: SCSI-2 differential.
System performance:

Data transfer rate	
to/from card	6.0 Mbytes/s burst
to/from system	1.1 Mbytes/s burst
Average access time	5.5 ms

Power: 28 V DC, <10 W
Dimensions: 163 (L) × 125 (W) × 63.5 (H) mm
Weight: 2.32 kg
MTBF: 40,000 h

Sentinel® Model 9415 SCSI DTS

The Sentinel Model 9415 is also a growth version of the Sentinel Model 9510 SCSI DTS, designed specifically for aircraft panel-mount installations.

Sentinel® Model 9422 DTS

The Sentinel Model 9422 is designed for use with an RS-422 system interface DTS, and specifically for aircraft panel-mount installations. It can accept either Type II or Type III Sentinel PCMCIA flash memory cards, and contains a 28 V DC power supply, PCMCIA controller, control microprocessor and RS-422 interface electronics that includes automatic baud rate selection.

Sentinel® Model 9450 DTS

The Sentinel Model 9450 DTS is designed for use with MIL-STD-1553A/B system interfaces, and aircraft panel-,mount installations. The system accepts two separately addressable Type II Sentinel PCMCIA flash memory cards, and contains memory control electronics, data buffer, MIL-STD-1553 interface and power supply.

Sentinel® RS-232/-422 DTS

The Sentinel RS-232/-422 has been designed for use in severe military environments. It contains a PCMCIA controller, microprocessor, RS-232/-422 interface electronics and automatic baud rate selection.

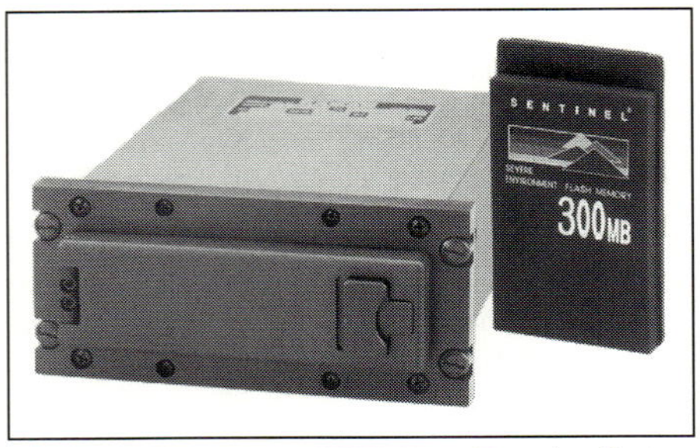

Sentinel Model 9410 SCSI Data Transfer System (DTS) ***1999***/0051291

Operational status

The Sentinel Model 9410 SCSI DTS has been selected for the C-130J of the Royal Australian Air Force and also for the SH-2G helicopter upgrade.

Contractor

Kaman Aerospace Corporation, Raymond Engineering Operations.

VERIFIED

Model A100S Solid-State Cockpit Voice Recorder (SSCVR)

The Model A100S Solid-State Cockpit Voice Recorder (SSCVR) is available for new installations or as a direct replacement for existing ARINC 557 CVRs without wiring changes to the aircraft. Existing Fairchild users installing the Model A100S SSCVR do not require new control units, microphones, or changes to existing operating procedures.

While mounted on the aircraft, recorded data cannot be extracted from the SSCVR. This protection is necessary to protect the rights of pilots and other flight crew members. The model A100S SSCVR meets the following specifications in full compliance with worldwide regulatory requirements, including EUROCAE (ED56) and CAA specifications together with FAA ISO-C123.

Specifications

Dimensions: ½ ATR short ARINC 404
Weight: 7.3 kg
Power supply: 115 V AC, 12 W

Contractor

L-3 Communications, Aviation Recorders.

VERIFIED

SSCVR open 0001306

Model A200S Solid-State Cockpit Voice Recorder (SSCVR)

The Model A200S solid-state cockpit voice recorder provides 2 hours of recording, with the last 30 minutes being redundant in a very high-quality format. This unit meets EUROCAE ED-56A and TSO C123a. Readout and copying of data is available instantly by using a hand-held downloading device (digital audio playback unit) without the need to go through a long computer conversion process. The solid-state characteristics of the Model A200S eliminate periodic maintenance requirements, and greatly increase reliability. It meets the severe vibration and environmental requirements of DO-160C without the need for a vibration-mounted tray. A built-in test function includes a tone generator that superimposes the signal on the audio channel to ensure the unit is functioning correctly. In addition, the unit has a continuous BITE that monitors the entire memory and power supply to ensure that the unit is

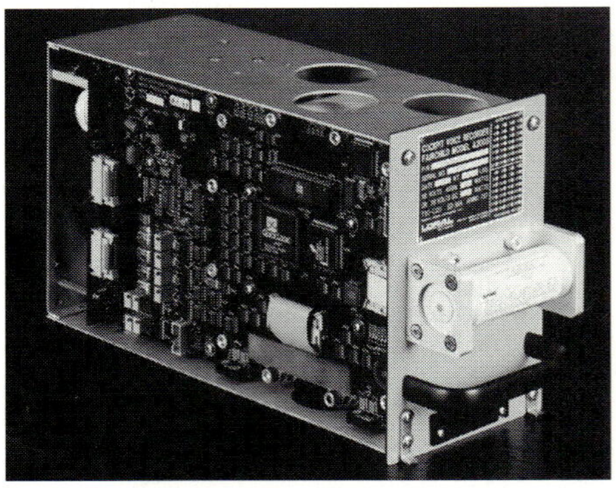

The Model A200S solid-state cockpit voice recorder

operating to the highest standards. The A200S accepts four channels of audio including pilot, co-pilot, area microphone and third crew member, PA system and data/timebase source via 429. The A200S includes a front-mounted underwater locator beacon.

With the A200S, audio retrieval and playback is instantly available by using a simple hand-held device without requiring complex computer conversion. However, when mounted on the aircraft, recorded audio data cannot be extracted from the recorder, to protect the rights of pilots and other flight crew.

Specifications

Dimensions: ARINC 404 ½ ATR short
Weight: 8.44 kg
Power supply: 115 V AC, 400 Hz
or 28 V DC, 12 W (max)
Frequency response:
150-3,500 Hz (three inputs)
150-6,000 Hz (one input)

Contractor

L-3 Communications, Aviation Recorders.

VERIFIED

Model F1000 Solid-State Flight Data Recorder SSFDR

The Model F1000 SSFDR was the first production flight data recorder to provide solid-state operation. An onboard data retrieval capability eliminates the need to remove the unit from the aircraft to recover flight data. The recorder meets or exceeds all requirements and characteristics of ARINC 542A, ARINC 573/717, ARINC 747, EUROCAE ED-55, TSO C124a and RTCA DO-160C. The F1000's solid-state design also eliminates the need for a vibration-mounted tray.

With its microprocessor-based architecture, the Model F1000 incorporates the intelligence to perform sophisticated self-tests, as well as provide fault-tolerant features. The F1000 can receive data from an ARINC 573/717 flight data acquisition unit or process raw data in the ARINC 542A mode by accepting analogue signals in synchro, DC, pneumatic, frequency and discrete formats. The recorder can accommodate parameter storage requirements of 32, 64 or 128 words/s with a recording time of 25 hours. A high-speed data dump can be accomplished in 1 minute. The Model F1000 includes a front-mounted underwater locator beacon.

Specifications

Dimensions: ARINC 404 ½ ATR long
Weight: 10.2 kg
Power supply: 115 V AC, 400 Hz
or 28 V DC, 26 W (max)
Reliability: 15,000 h MTBF

Contractor

L-3 Communications, Aviation Recorders.

VERIFIED

Model FA2100 solid-state recorder family

Basic family

Model FA2100 was originally designed by Fairchild to meet industry demands for lighter weight and higher reliability. It now includes a number of standard configurations (seven are described), together with custom configurations. Models immediately available are shown in the accompanying photograph.

Solid-State Cockpit Voice Recorder (SSCVR)

Model FA2100 SSCVR was introduced to meet the needs of commercial, regional, business and government fleets. The FA2100 meets the requirements of EUROCAE ED-56A, FAA TSO-C123a RTCA DO-160C and ARINC 757. Readout and copying (both in analogue and digital forms) data is available instantly by using a hand-held downloading device (Portable Interface - PI) without the need to go through a long computer conversion process. The solid-state characteristics of the Model FA2100 eliminate periodic maintenance requirements and greatly increase reliability. It meets the requirements of DO-160C without the need for a vibration-mounted tray. A built-in test function includes a tone generator that superimposes the signal on the audio channel to ensure the unit is functioning correctly. In addition, the unit has a continuous BITE that monitors the entire memory and power supply to ensure that the unit is operating to the highest standards. The FA2100 accepts four channels of audio including pilot, co-pilot, area microphone and third crew member, PA system, data/timebase source via RS-429 interface; rotor speed encoding is available for helicopter installations. The FA2100 includes a front-mounted underwater locator beacon.

With the FA2100, audio retrieval and playback is instantly available in real time using a simple hand-held device without requiring complex computer conversion. However, when mounted on the aircraft, recorded audio data cannot be extracted from the recorder, to protect the rights of the pilots and other flight crew.

Specifications
Dimensions: ARINC 404 ½ ATR short - footprint - height is 139.7 mm
Weight: <4.36 kg
Power supply: 115 V AC, 400 Hz or 28 V DC, 9 W (max)
Frequency response:
(three inputs) 150-3,500 Hz
(one input) 150-6,000 Hz

Solid-State Flight Data Recorder (SSFDR)

Model FA2100 SSFDR was the second generation of solid-state Flight Data Recorders (FDR) designed by Fairchild. An onboard data retrieval capability eliminates the need to remove the unit from the aircraft to play back flight data. The recorder meets or exceeds all requirements and characteristics of ARINC 573/717/747, EUROCAE ED-55, TSO-C124a and RTCA DO-160C. The FA2100 solid-state design also eliminates the need for a vibration-mounted tray.

With its microprocessor-based architecture, the Model FA2100 incorporates the intelligence to perform sophisticated self-test, as well as provide fault-tolerant features. The recorder can accommodate parameter storage requirements of 64, 128 or 256 words/s with a minimum recording time of 25 hours. Future expansion of memory to allow 512 words/s is planned. A high-speed data dump can be accomplished in approximately 2 minutes. The Model FA2100 includes a front-mounted underwater locator beacon.

The seven standard models of the FA2100 family of solid-state aviation recorders. (Top row, left to right) CVR-helicopter, HUMS configuration (programme unique); CVR-helicopter configuration (programme unique); CVR, ½ ATR short, GA model; combined FDR and CVR, ½ ATR short. (Centre) FDR, ½ ATR. (Bottom Row, left to right) CVR, ½ ATR short; FDR, ½ ATR short **1999**/0051331

Specifications
Dimensions: ARINC 404, ½ ATR - footprint - height is 139.7 mm. This flight recorder is available in both long and short versions
Weight: <4.36 kg
Power supply: 115 V AC, 400 Hz or 28 V DC, 9 W (max)

Modular Airborne Data Recording/ Acquisition System (MADRAS)

The Model FA2100 combination recorder combines the features of the SSCVR with the SSFDR to provide one combined recorder with both features. The recorder meets the characteristics of TSO-C123a/C124a, EUROCAE ED-56A/ED55, ARINC 757 and RTCA DO-160C. Readout and copying of data is available instantly by using a hand-held downloading device (Portable Interface - PI) for both flight and audio data. The solid-state characteristics of the FA2100 eliminate periodic maintenance requirements and greatly increase reliability. It meets the requirements of DA-160C without the need for a vibration-mounted tray. A built-in test function includes a tone generator that superimposes the signal on the audio channel to ensure the unit is functioning correctly. In addition, the unit has a continuous BITE that monitors the entire memory and power supply to ensure that the unit is operating to the highest standards. The FA2100 accepts four channels of audio including pilot, co-pilot, area microphone and third crew member, and PA system. The flight data is accepted via an ARINC 573/717 data stream from an onboard acquisition unit.

With the FA2100 combination recorder, the audio and data retrieval and playback is instantly available by using a simple hand-held device. However, when mounted on the aircraft, recorded audio data cannot be extracted from the recorder, to protect the rights of pilots and other flight crew.

Specifications
Dimensions: ARINC 404 ½ ATR short - footprint - height is 139.7 mm
Weight: <4.36 kg
Power supply: 115 V AC, 400 Hz or 28 V DC, 9 W (max)
Frequency response:
(three inputs) 150-3,500 Hz
(one input) 150-6,000 Hz

Recording duration:
30 min of voice/25 h of data
60 min of voice/10 h of data

Operational status

The US Federal Aviation Administration (FAA) has proposed that, by August 2000, all FDRs be upgraded in newer commercial aircraft with enough capacity to record rudder and aileron positions. Older aircraft would have to meet this requirement by August 2001. The FA2100 FDR currently meets this expanded data parameter list requirement.

The US National Transport Safety Board (NTSB) submitted an additional recommendation to the FAA that all 30-minute CVRs used in modern aircraft be replaced with 2-hour solid-state CVRs by January 2005. The FA2100 CVR already meets this recommendation.

Contractor

L-3 Communications, Aviation Recorders.

VERIFIED

Strategic/Tactical Airborne Recorder (S/TAR)

The S/TAR product line is designed to replace magnetic tape recorders in high-capacitance, high-rate, reconnaissance and instrumentation data storage applications on strategic and tactical aircraft, helicopters and UAVs.

S/TAR features include full, modular, non-volatile memory from 8 to 125 Gbytes; multiple-channel record; read-while-write to support onboard analysis and/or data downlink; built-in transparent error management.

S/TAR is a stand-alone product, but it can also be supplied as circuit cards integrated with other processing functions such as reconnaissance management systems and datalinks in a common VME chassis.

Sustained write rate: 0-400 Mbyte/s per port
Read rate: 0-400 Mbyte/s (onboard review, datalink, or ground analysis)
Sustained download rate: 640 Mbyte/s
Interfaces:
airborne data record/replay: DCRsi™ or ID-1 compatible
control channel: MIL-STD-1553 or RS-422
Power: 28 V DC, 80 W

Operational status

First production examples are available for customer evaluation. Target markets include P-3 aircraft and then fast jet types.

Contractor

L-3 Communications, Communication Systems-East.

VERIFIED

Specifications

Dimensions/weights/capacities:

Chassis style	Dimensions (mm)	Total Weight (kg)	Cartridge Weight (kg)	Capacity (Gbytes)
Removable memory	241.3 × 274.3 × 297.2	12.27	7.73	33
Removable memory	241.3 × 274.3 × 297.2	14.55	10.00	50
ARINC ATR	193.0 × 259.1 × 533.4	25.00	N/A	100
ARINC ATR	193.0 × 259.1 × 609.6	34.09	N/A	125

ADAS-7000 Airborne Data Acquisition System

L-3 Telemetry-East manufactures a wide range of telemetry and data recording systems for missile and air vehicle flight testing including the ADAS-7000, originally designed for the IAI Lavi flight test programme. This provides a master/slave recording system for aircraft use, using advanced signal conditioning and encoding hardware. A PMU-700 Series III program master unit is at the heart of the ADAS-7000; this is used as a central encoder/controller of the whole system and communicates on a MIL-STD-1553 digital data highway with a number of slave units. Over 5,000 channels can be monitored and recorded.

Contractor

L-3 Telemetry-East.

VERIFIED

ATD-800-II airborne tape system

The ATD-800-II ruggedised tape deck is a low-cost, high-performance digital data record/reproduce tape media subsystem suitable for use in the harsh environmental conditions usually associated with flight test applications.

All components of the ATD-800-II are contained in a ruggedised chassis with internal shockmounts to physically isolate all devices from externally imposed vibration. The chassis is sealed from the potential contamination of the surrounding atmosphere and an internal temperature control system is provided to maintain temperature within operating limits. The recording system is the industry recognised DLT-4000 cartridge tape mechanism by Quantam. The front panel of the ATD-800-II contains a comprehensive set of operating status indicators.

The ATD-800-II is designed to interface directly to a standard SCSI-2 controller, such as is readily available on computer systems, as well as on the L-3 Telemetry-East MiniARMOR-700 data multiplexer/demultiplexer. When used in conjunction with the MiniARMOR-700, the ATD-800-II supports recording of multiple combinations of serial and parallel data sources. For example, several channels of PCM may be combined with time, voice, MIL-STD-1553, parallel and digital and analogue inputs. Playback of the data may be accomplished by direct connection of the SCSI-2 interface to a computer system or by using a playback configuration of the MiniARMOR-700. The latter approach permits coherent reconstruction of the original data streams.

Key parameters and features include record/reproduce rates of 1.5 Mbytes/s sustained, 5.0 Mbytes/s burst, 20 Gbytes data storage capacity per cartridge, non-compressed, BER less than one error in 10E17 bits, record time of 3.6 hours at maximum rate, single-ended or differential SCSI-2 interface, tape dubbing software (requires record and playback unit) high-speed access to stored data, remote control option. Records PCM, 1553, voice, time.

Specifications

Dimensions: 387 × 173 × 136 mm
Weight: 6.8 kg
Operating temperature range: –20 to +50°C

Contractor

L-3 Telemetry-East.

VERIFIED

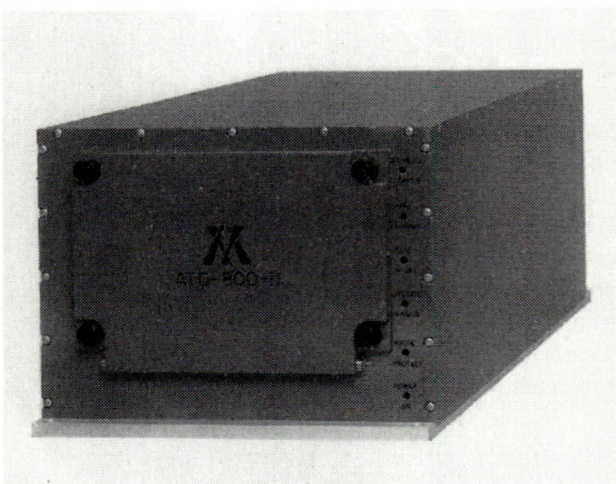

Series II ATD-800 digital data tape recorder 0002444

Common Airborne Instrumentation System (CAIS)

The CAIS was developed under the auspices of the US DoD to promote standardisation, commonality, and interoperability for flight testing.

The central characteristic of CAIS is a common suite of equipment used across service boundaries and in any airframe or weapon system testing.

CAIS products comprise airborne equipment items, that can be configured to meet project requirements, and comprehensive ground facilities to support the airborne effort:

- MDAUs: Miniature Data Acquisition Units to support both analogue and digital data acquisition
- PMU-700-C5: Programmable Master Controller Unit
- PBC-800: Programmable Bus Controller
- PCU-800C: Programmable Conditioning Unit
- MPC-800C: Miniature Programmable Conditioner
- MiniARMOR-700: High-Speed Multiplexer
- ATD-800-II: Airborne Tape Deck
- CCU-800: Cockpit Control Unit
- GSU-800: Ground Support Unit
- CBE-850: CAIS Bus Emulator
- Lab ARMOR-715: High-Speed Demultiplexer

Operational status

CAIS is used in the F/A-18E/F and F-22 aircraft programmes. L-3 Telemetry-East is the instrumentation system integrator for the F-22 programme, and is responsible for delivering a complete turnkey system.

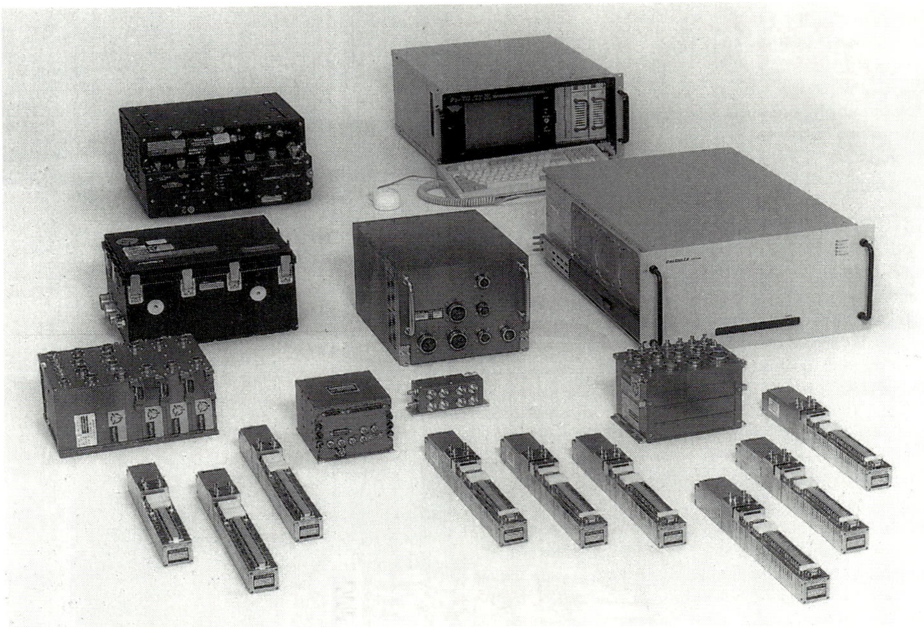

Representative CAIS system 0002349

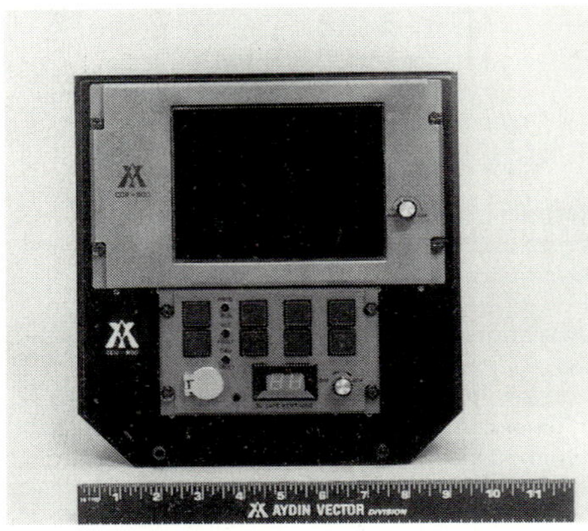

CCU-800 cockpit control unit 0002348

L-3 Telemetry-East has been selected as sub-contractor by Boeing Aircraft, Missile Systems Division to provide flight test instrumentation for its Joint Strike Fighter concept demonstration programme. L-3 Telemetry-East will also manufacture much of the flight test instrumentation equipment and integrate its Data Acquisition System for the two next-generation X-32 fighter aircraft.

Contractor

L-3 Telemetry-East.

VERIFIED

DPM-800E PCM encoder

The DPM-800E PCM encoder is a fully programmable high-performance 8- to 12-bit resolution data system for acquiring conditioned signals in severe airborne applications where flexibility and reliability are the primary requirements.

Features of the DPM-800E include EEPROM programmable via parallel IF port, 2 Mbit operation, eight programmable bit rates, user programmable for format, bits per word and gain/offset, filtered data outputs, single-ended or differential analogue inputs, single-ended discrete bilevel inputs and subcommutation and supercommutation.

The DPM-800E can be configured as a stand-alone, master or remote unit. Up to seven DPM-800Es can be configured in a master/slave or cluster configuration via the L-3 Telemetry-East standard 10-wire differential interface. This 10-wire interface can be utilised with other L-3 Telemetry-East data acquisition products such as MIL-STD-1553 bus monitors, signal conditioning units and master controllers for large distributed systems.

Specifications

Dimensions: 88.9 × 82.5 × 118.1 mm
Weight: 0.79 kg
Temperature range: –35 to +85°C

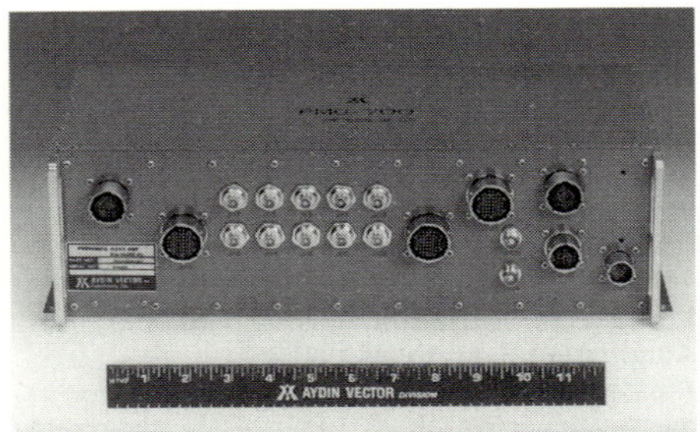

PMU-700-C5 programmable master controller unit 0002347

Contractor

L-3 Telemetry-East.

VERIFIED

MDM-700 modem

The MDM-700 modem is ruggedised for airborne use and provides a relatively inexpensive means of interfacing a computer to a duplex RF link which utilises conventional FM transmitters and receivers. The MDM-700 is a full-duplex interface. However, the modem can be used in the simplex mode without degradation and without special considerations.

The MDM-700 transmit section accepts the RS-232 signal and converts each transition into a half-sine shaped pulse. A positive transition creates a positive pulse and a negative transition creates a negative pulse. Should there be no transition activity, the modem continues to create pulses of the same polarity as the last but at a low frequency. A result of this is that a signal can be AC-coupled; DC response is not required.

The modem receive portion accepts the half-sine pulse train and recreates the original RS-232 data. Since the conversion is one of edge coding, the modem is insensitive to data rate, except for the maximum of 9,600 bps. Clock is not used nor is it required.

Specifications

Weight: 0.23 kg
Power supply: 28 V DC

Contractor

L-3 Telemetry-East.

VERIFIED

MiniARMOR-700 multiplex/demultiplex system for digital data recording

The L-3 Telemetry-East MiniARMOR-700 multiplex/demultiplex system for digital data recording is based on the design concepts of the Asynchronous Real-time Multiplex and Output Reconstructor (ARMOR) developed by Calculex, Inc.

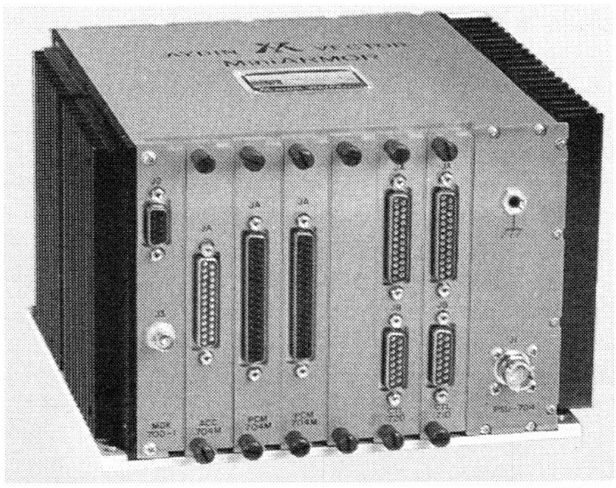

MiniARMOR-700 miniature asynchronous real-time multiplexer and output reconstructor 0002445

MiniARMOR-700 systems combine a wide variety of analogue and digital signals into a single high-speed (up to 240 Mbps) composite digital data stream for digital recording. A system configured to reconstruct the composite signals does so while maintaining inter-channel data coherency.

The MiniARMOR-700 supports two operational formats to maintain full backward compatibility with Calculex ARMOR 1 products while affording the full flexibility and performance of the advanced MiniARMOR-700 design.

The MiniARMOR-700 is a modular system with a full compliment of modules and options to support virtually any mix of analogue and digital signal types to support specific user applications. Modules may be changed in the field for user ease.

The MiniARMOR-7000 is available in a 5, 7, 9 and 11 slot configuration, with a maximum measurement of 136 × 190.5 × 261.6 mm and a maximum weight of 6.16 kg. It operates from 28 V DC, 115/220 V AC, 46-63 Hz and has an operating temperature range of –30 to +70°C.

Contractor

L-3 Telemetry-East.

VERIFIED

MME-64 Miniature Multiplexing Encoder

The MME-64 miniature multiplexing PCM encoder is a low-cost data system designed for airborne data acquisition and telemetry applications to accommodate up to 64 single-ended analogue inputs or up to 24 discrete bilevel inputs. Unfiltered PCM output is NRZ-L 0 to 5 V TTL, CMOS-compatible. Filtered PCM output is six pole Bessel response, factory-adjustable from 0±50 mV to 0±2.5 V. Bit rate can be programmed up to 1 Mbit. The frame pattern is controlled by EEPROM and is factory-programmed.

Specifications

Dimensions: 82.6 × 76.2 × 30.4 mm
Temperature range: –30 to +70°C

Contractor

L-3 Telemetry-East.

VERIFIED

MMSC-800 MicroMiniature Signal Conditioner

The MMSC-800 microminiature signal conditioner/PCM encoder combines L-3 Telemetry-East's experience of analogue and digital conditioning into a 12-bit PCM encoder. A complete family of modules provides signal conditioning for various sensors. EEPROM programmable gain/offset and sample rates provide the user with hands-off control of measurement characteristics and output data formatting. The MMSC-800 utilises thick film hybrid modular assemblies and has been qualified for missile and aerospace environments. It is available in stand-alone or master remote configurations.

The MMSC-800 features 12-bit PCM encoder with integral signal conditioning and multiplexing, EEPROM programmable gain, offset, sample rate and PCM format, ruggedised modular construction, extremely small size and RS-232 programming capability. It is military and airborne qualified and has high reliability. The system accuracy is quoted as 0.5 per cent.

Contractor

L-3 Telemetry-East.

VERIFIED

MMSC-800-RPM/E remote pressure multiplexer/encoder

The MMSC-800-RPM/E remote pressure multiplexer/encoder accepts pressure inputs directly, electronically scans the pressure inputs and conditions and encodes the resulting information. It is based on the MMSC-800 signal conditioner and PCM encoder, which has been used on many commercial test programmes and applications. Operating as a remote unit, the RPM/E can be placed in areas of the aircraft well away from the Model PCU-700 Series III central controller. Communication between the master and remote is via a 10-wire command response interface. The channel sequence and gain/offset combinations are fully programmable through the central controller. The temperature reference of the PSI scanner is electronically monitored and transmitted to the central controller to allow the user to correct for errors due to temperature changes in the scanner. Features are controlled EEPROM, programmable via an RS-232 from any personal computer terminal.

Specifications

Dimensions: 135.9 × 45.7 × 96.5 mm
Weight: 3.63 kg
Input channels: 32
Temperature range: –20 to +80°C

Contractor

L-3 Telemetry-East.

VERIFIED

MPC-800 miniature signal conditioner and encoder

The L-3 Telemetry-East miniature signal conditioner and encoder (MPC-800) is a data acquisition system component that provides all the most common required signal conditioning and encoding functions in a small modular package. The design minimises size, weight and cost, and improves survivability in extremely hostile environments. The unit is constructed from stackable conditioning and overhead modules which contain discrete, thick film hybrid and custom ASIC circuit technologies.

By utilising modular architecture, the MPC-800 can be configured to meet the exact needs of any flight test program and still allow the user the ability to reconfigure the unit in the field to adapt to changing mission requirements. Full programmability of gains, offsets, filter cutoffs, and channel sampling rates is offered through a Graphical User Interface (GUI). The MPC-800 may be operated as a master, remote, or stand-alone system.

A wide variety of standard analogue and digital signal conditioning modules are available for virtually any sensor. Custom modules can be created for unique measurement requirements.

MPC-800 miniature signal conditioner and encoder 0002350

Contractor

L-3 Telemetry-East.

VERIFIED

PCU-800 series signal conditioner and PCM encoder

The PCU-800 series signal conditioner and PCM encoder provides a ±0.5 per cent system accuracy over its operating temperature range. Available in 2, 4, 8 or 16 card slot configuration, the PCU-800 features include: EEPROM programmable format and channel parameters via RS-232; plug-in system overhead and signal conditioning cards, both digital and analogue types; universal bridge conditioning with completion, calibration and substitution; programmable balance via software control and 12-bit resolution.

The PCU-800 will operate as a stand-alone or master controller. Channel capacities vary depending on the type of signal inputs, but they can range up to more than several hundred channels per unit. Up to 40 PCUs can be clustered in a distributed data acquisition system to provide a total channel capacity of more than several thousand.

The PCU-800 is directly compatible with many other ADAS-7000 and ADAS-8000 airborne products such as the PMU-700 Series III programmable master controller, Albus-1553 all-bus monitor, MPBM-1553 microbus monitor and ATD-800 Series II ruggedised tape deck.

Contractor

L-3 Telemetry-East.

VERIFIED

SBS-500 single-rate bit synchroniser

Processing PCM data in the SBS-500 single-rate bit synchroniser suppresses receiver noise that may accompany the digital input signal and converts the input to a clean digital output. Available as a stand-alone 275 cm³ package, the SBS-500 is suitable for missiles, aircraft and space applications.

It has a clock synthesis oscillator and a data detection match filter determined by the specific bit rate. The SBS-500 can also perform code conversion from digital transmission codes and provide biphase or NRZ outputs. The SBS-500 performs well under low-input signal-to-noise conditions and is typically within 2 dB of theoretical bit detection accuracy. Acquisition time is 10 clock cycles typical and the SBS-500 has output options of TTL, HC, RS-232 and RS-422.

Specifications

Power supply: 28 V DC
Bit rate: 1-250 kHz
Temperature range: –40 to +85°C

Contractor

L-3 Telemetry-East.

VERIFIED

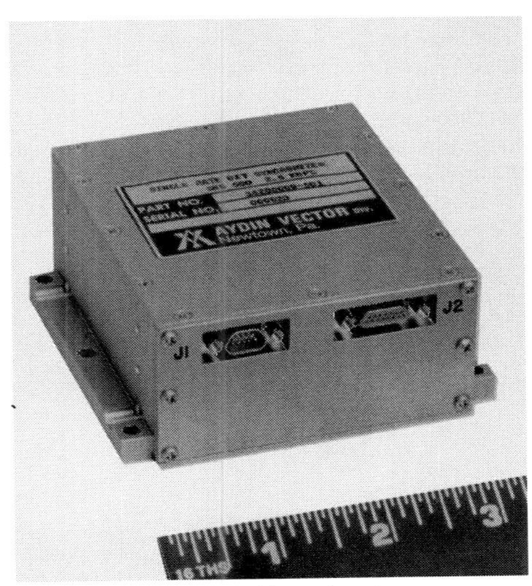

The SBS-500 single-rate bit synchroniser

C-17 Warning And Caution Computer System (WACCS)

Developed for use on the US Air Force C-17 transport aircraft, the Litton Warning And Caution Computer System (WACCS) accepts discrete aircraft data and data from the MIL-STD-1553B bus connecting specialised peripheral LRUs. It contains dual-redundant MIL-STD-1553B bus communication ports and can function as the bus controller or a remote terminal. WACCS processes signals from all sources and manipulates the inputs via Boolean logic to provide warning messages on the warning panel or by lighting annunciator lamps.

Specifications

Dimensions: 190 × 188 × 315 mm
Weight: 8.09 kg
Power supply: 28 V DC, 24 W
Reliability: 20,611 h MTBF

Operational status

In production for US Air Force C-17 aircraft.

Contractor

Litton Guidance & Control Systems.

VERIFIED

F-16 general avionics computer

The F-16 general avionics computer acts as the mission computer for the F-16C/D aircraft, providing avionics and weapons control solutions, IFF processing and navigation functions. Designed for MIL-STD-1750 ISA performance, the computer offers as standard features throughput of up to 6 Mips DIAS, 512 kbytes or greater RAM or EPROM, battery back-up MIL-STD-1553B input/output with up to four channels per module, with bus controller or remote terminal. Options include single module and custom packaging, custom input/output and BIT functions.

Specifications

Dimensions: 135 × 312 × 333 mm
Weight: 9.09 kg
Power supply: 115 V AC, 400 Hz, 70 W
Reliability: >4,000 h MTBF

Operational status

In production for the F-16C/D in service with the US Air Force and other air forces.

Contractor

Litton Guidance & Control Systems.

VERIFIED

CC-2E Data Processing System (DPS)

The CC-2E Data Processing System (DPS) is the central processor for the Boeing E-3 AWACS aircraft which provides early warning of threats and airborne control of friendly aircraft. It receives and processes data from onboard sensors, sends the completed airborne view to mission operators and links critical messages to ground commanders. The CC-2E DPS is the fourth-generation multipurpose airborne processor for the E-3.

A recent value engineering change proposal effort resulted in the development of increased memory in the CC-2E DPS, reducing the number of monolithic memory units from five to three per system.

Operational status

In service.

Contractor

Lockheed Martin Electronics Platform Integration Group.

VERIFIED

Colour Cockpit TV Sensor (CCTVS)

Lockheed Martin Fairchild Systems CCTVS updates the company's Cockpit TV System (CTVS), of which over 16,000 have already been sold to 28 countries and over 30 different aircraft types.

The CCTVS is an exact form, fit and function replacement for the current Fairchild Systems' monochrome Heads-Up Display (HUD) camera.

CCTVS cameras provide higher-resolution colour imaging sensor and recording capability, with expanded low light night performance.

Specifications

Pointing accuracy: ±0.56 mrad (at factory)
Light levels: 0.05-16,000 fL (both ALC and AEEC)
Resolution: >470 TVL/PH(); >350 TVL/PH(v)
MTBF: 32,000 h

F/A-18 low-light colour camera configuration 0015331

Operational status

Currently in production for the F-14D, F-15E, F/A-18 (3 per aircraft), MB-339 and T-45 aircraft.

Contractor

Lockheed Martin Fairchild Systems.

VERIFIED

Airborne Battlefield Command Control Centre (ABCCC III)

The Airborne Battlefield Command and Control Centre (ABCCC III) is an airborne node in the US Air Force Tactical Air Control System. It maximises the efficient use of fighter forces and other air resources by gathering real-time battlefield data from forward areas and by managing aircraft in air-to-ground operations.

Designed for the EC-130E, the ABCCC III self-contained capsule houses 15 automated workstations, allowing the battle staff to manage the tactical air assets conducting over 150 sorties/h effectively. During each 10 hour mission of the EC-130E, the ABCCC III provides a communications link to higher headquarters and co-ordinates forces in the battle area, updating aircraft on their way to the target and collecting their in-flight reports on the way out. Automated capabilities allow the battle-staff to analyse continuing combat quickly and direct offensive air support toward fast developing targets.

The ABCCC III consists of four major airborne subsystems: The Communications Subsystem (CS), the Tactical Battle Management Subsystem (TBMS), the Airborne Maintenance Subsystem (AMS) and the Capsule Subsystem (CS). There is also a major subsystem on the ground. The Mission Planning Subsystem (MPS) provides the tactical database used by the airborne maintenance technician for system initialisation. On mission completion the MPS can also be used for post-flight playback and analysis of mission data.

Once airborne, the CS and its Automated Communications and Intercom Distribution System (ACIDS) controls and secures all communications between the capsule and forward units, other aircraft and rear bases. Communications within the capsule and between the capsule and the flight crew are also handled by ACIDS.

The TBMS provides comprehensive battlefield management capabilities for up to 12 operators stationed at individual battle-staff consoles. Fast accurate access to communications, as well as tactical and map databases, enhances operator effectiveness.

The AMS provides diagnostic and fault isolation to detect hardware or software malfunctions during a mission. System initialisation and control are also integrated into AMS functions.

The Capsule Subsystem consists of the facilities providing physical, environmental and life support for operating the airborne subsystems.

Contractor

Lockheed Martin Tactical Defense Systems.

VERIFIED

AN/ASQ-212 mission processing system

The AN/ASQ-212 system consists of the CP-2044 computer and several interconnection devices which comprise a form, fit and function replacement for the AN/ASQ-114 computer, data analysis logic units and the signal data converter. The extended memory upgrade of the AN/ASQ-114 computer is transferred to the CP-2044 and used both for global and secondary memory. The CP-2044 incorporates Motorola 68030 processors to provide a throughput ranging from 10 to 25 Mips, which is 30 times greater than the current system in the P-3C Update I/III aircraft at a fraction of the current size, weight and power requirements. In the full Update III Ada implementation, less than 50 per cent of the CP-2044 minimum throughput and memory capacity is utilised.

The CP-2044 VME bus open architecture can be configured with additional processing, memory and input/output modules to meet the requirements of new subsystems such as GPS and Satcom, and of processing intensive functions such as sensor post-processing and data fusion.

Initially designed for retrofit into P-3C Update I/III aircraft, the AN/ASQ-212 can be easily tailored to the requirements of other P-3C configurations as well as new aircraft.

Operational status

The AN/ASQ-212 system was developed for the US Navy P-3C aircraft under a two-phase programme that began in September 1989. The first production systems were installed in US Navy test aircraft and training facilities beginning in May 1993, at a rate of four systems per month. Also in service with P-3C export customers.

Contractor
Lockheed Martin Tactical Defense Systems.

VERIFIED

ASPRO ASsociative PROcessor

The ASPRO Associative Processor is a small but powerful fully military-qualified 0.44 cu ft parallel processor embedded in the Northrop Grumman E-2C Hawkeye's avionics suite. It performs millions of operations per second to track potentially thousands of targets. It also performs additional display processing functions on the aircraft.

Since it was originally developed, ASPRO's processing throughput has been increased and its memory doubled, with no increase in volume. In addition, a system to meet MIL-E-16400 has been built, for use in 'Los Angeles' class submarines. Lockheed Martin has developed the ASPRO-VME, which features increased memory, system throughput, and expansion capability to accommodate new applications.

Specifications
Dimensions: 203 × 229 × 254 mm
Weight: 14.5 kg
Power: 200 W
Throughput: 50 Mops
Reliability: 4,500 h MTBF

Operational status
In service in the E-2C Hawkeye aircraft.

Contractor
Lockheed Martin Tactical Defense Systems.

VERIFIED

ASPRO-VME parallel/associative computer

The ASPRO-VME, Lockheed Martin's fourth-generation parallel processing computer, is a modular open architecture VME-compatible card set. It is capable of performing between 150 Mflops and 2.4 Gflops. The basic three-module 512 processor configuration can be expanded from 512 to 8,192 processor elements. Each of the parallel processors, called Processing Elements (PEs), contains a full 32-bit IEEE floating-point processor and a bit-serial processor.

A module occupies a single-VME card slot and consists of two printed circuit boards attached to a cold plate. The entire ASPRO-VME, in its basic configuration, requires three VME slots.

Programmable in Ada, ASPRO-VME is supported by a powerful software development tool set which allows application programmes to be easily developed. Because of ASPRO's single instruction multiple data stream architecture, modular expansion from 512 PEs to 8,192 PEs can be accomplished without rewriting software.

The parallel architecture and associative search capability produce significant gains over conventional processors for sophisticated tracking, correlation, data fusion, and situational awareness algorithms.

ASPRO-VME is available in a rugged and full Mil-Spec configuration. It can operate in a stand-alone mode or be directly embedded in commercial and rugged workstations with 6U VME slots. Applications for ASPRO-VME include command and control, correlation and tracking, data fusion, database management, signal processing, expert systems, neural networks and image processing.

Specifications
Weight: 2.59 kg
Power: 80 W

Operational status
Production units available.

Contractor
Lockheed Martin Tactical Defense Systems.

VERIFIED

Avionic common module systems

Lockheed Martin Tactical Defense Systems is the technology leader for development, application, implementation and integration of JIAWG/MASA/SHARP avionics common module processing clusters and systems. These capabilities and resources are aimed at allowing module users complete integration control from subsystem development to system platform implementation.

The foundation of common module information processing systems is extensive work in VHSIC design. A limited number of these standard VHSIC chips enables creation of a family of modules that can be used in different avionics systems. Designed in the Standard Electronic Module (SEM) E format, each module measures 149.3 × 162.6 × 14.7 mm.

The Lockheed Martin Tactical Defense Systems family of avionics common modules consists of seven processing and Input/Output (I/O) SEM-E module types, a power supply SEM-E module, backpanels, liquid or air-cooled racks and active and passive star coupler technology. Reusability and standardisation have been stressed within the design philosophy for the avionic equipment family. Lockheed Martin Tactical Defense Systems has created a generic modular, module functional design to promote reuse of ASIC devices and to support integrated diagnostics.

The Advanced General Purpose Processor Element (AGPPE) is a 32-bit RISC-based data and signal processor module ideally suited for avionic systems applications. Equipped with a Mips R3000 RISC processor, its pipelined architecture yields very high throughput while its standard SEM-E format and three input/output interfaces allow easy integration in various systems.

Advanced ceramic circuit boards and surface-mount components contribute to AGPPE's light weight, low power consumption and dense packaging. Full 32-bit operation is enhanced by a five-stage pipeline, on-chip cache control and an on-chip memory management unit. Block refilling of both instruction and data caches is supported.

The Mips R3000 processor executes instructions up to 20 times faster than the VAX 11/780. In addition the R3010 floating point co-processor chip handles floating point arithmetic compliant with the ANSI/IEEE standard.

Onboard memory resources include a 1 Mbyte SRAM which can be accessed synchronously in two CPU clock cycles. Bootstrap and debugging code can be stored in a 128 kbyte EEPROM. Separate 16 kbyte data and instruction caches, which effectively double the available cache memory bandwidth, provide instructions and operands at the CPU clock rate. Other features that enhance performance include a four-word buffer for block refill of each cache and a one-word write buffer for writes to main memory.

The module features TM bus and a Lockheed Martin designed maintenance controller ASIC with an IEEE-488 channel for console operations. A network interface controller provides a one-chip interface from the AGPPE to a data flow network, a 32-bit parallel bus.

MIL-STD-1750A processor
The MIL-STD-1750A processor module is a 3.85+ Mips VHSIC processor with 512 kwords of local SRAM, plus 8 kwords of start-up ROM and PI-bus, TM-bus and IEEE-488 standard I/O interfaces on a single-width double-sided ¾ ATR size SEM-E module. The module's ceramic printed circuit boards and surface-mount components provide a dense, lightweight, low-power, general purpose data processor module. The highly parallel pipelined architecture enables high throughput. Advanced built-in test techniques reduce life cycle costs by supporting two-level maintenance. In addition, the standard SEM-E size and three standard I/O interfaces allow easy integration into a variety of systems.

The processor module has an onboard MIL-STD-1750A maintenance controller to automate built-in test on the module, communicate with the other modules via the TM-bus and handle console operations via the IEEE-488 bus.

The MIL-STD-1750A data processor's three CMOS gate arrays, the CMOS semi-custom maintenance controller common to the Unisys common module family and the ECL clock chip are equivalent to over 160,000 gates.

High-Speed DataBus interface (HSDB)
The High-Speed DataBus (HSDB) interface module integrates the high-performance processor with a linear token-passing high-speed fibre optic databus on a single-width double-sided ¾ ATR size SEM-E module.

The HSDB module provides a dual-redundant 50 Mbit/s fibre optic system interface, processor, 256 k words of local SRAM, a subsystem interface, maintenance via the TM-bus and IEEE-488 bus I/O interfaces.

The module functions are highly integrated, using surface-mount components. The HSDB can be configured with processor/bus combinations consisting of either the high-performance 1750A processor or the Mips-based RISC processor and with either a PI or N bus subsystem interface. The HSDB interface module has a CMOS gate array, an ECL gate array and hybridised fibre optic transmitters and receivers for the HSDB interface.

DC-DC converter
The Lockheed Martin Tactical Defense Systems DC-DC converter is a single-width ¾ ATR SEM-E size power supply that provides over 200 W of regulated power from an unregulated 270 V DC bus. The converter uses a standard power supply topology with hybrids and high-frequency techniques to attain high efficiency, high reliability and small size.

Input power to the converter is +135 V DC and −135 V DC (270 V DC line-to-line) MIL-STD-704D. Output voltages are +5 V DC, −15 V DC, −5.2 V DC and +80 V DC. The efficiency of the DC-DC converter is 80 per cent at full load. Up to six modules may be paralleled to increase current capacity and/or provide redundancy.

The converter has a patented digital controller to enhance stability, improve testability and provide fault detection. The module has a slave TM-bus interface to communicate with other modules in the subsystem.

Contractor
Lockheed Martin Tactical Defense Systems.

VERIFIED

Modular Airborne Processor (MAP)

The MIL-STD-1750A Modular Airborne Processor (MAP) is a 1.5 Mips VHSIC, CMOS processor for use in UAVs such as smart weapons, RPVs, targets, and land, sea and air autonomous vehicles.

It has up to 256 kwords each of SRAM and EEPROM with 8,000 start-up memory. The basic MAP consists of two 6 × 9 in (152.4 × 228.6 mm) PCs: a processor board and an Input/Output (I/O) board. The processor board contains the 1750A chip set and resource controller ASIC along with system SRAM. The I/O board contains a dual-redundant MIL-STD-1553B bus controller interface, a MIL-STD-1553B remote terminal interface compliant with the requirements of MIL-STD-1760A, MIL-STD-1760A discretes and EEPROM memory. Two additional cards can be provided to accommodate user-defined I/O requirements.

All MAP power is provided by a pluggable PC card DC/DC converter requiring MIL-STD-704 28 V DC input. A spare slot is reserved for the optional Programme Development Card/Performance Monitor Module (PDC/PMM).

The PDC provides a standard IEEE-488 control interface for applications development and test. The PDC also provides a plug-in PMM interface to a standard VAX DRQ3B I/O channel providing enhanced Ada software development capability, including real-time interactive data collection and debugging.

The MAP is housed in a lightweight EMI compliant enclosure utilising MIL-C-38999 connectors. It is designed for conduction cooling requiring no fans or environmental controls. The MAP utilises low-cost glass-epoxy boards with pin-grid array ASICs and DIP integrated circuits.

The MAP design utilises a 3.8 Mip 1750A chip set with the RC coupled to a modified system information transfer and execution bus open architecture. This permits modular expansion on the I/O bus as required by the application. The MAP contains BIT coverage of 98 per cent as per MIL-STD-2084.

Contractor
Lockheed Martin Tactical Defense Systems.

VERIFIED

U1638A MIL-STD-1750A computer

The U1638A computer is a militarised radiation-hardened general purpose avionic computer that meets all the requirements of MIL-STD-1750A Notice 1. The U1638A computer uses the U1635 high-speed MIL-STD-1750A central processing unit. The U1638A combines this high-density high-speed microprogrammed CPU with memory, input/output interfaces and power conditioning to provide the high-speed data computation, data storage and data transfers required for avionic systems.

The U1638A computer is packaged as a MIL-STD-1788 type 10 chassis which includes 12 shop-replaceable units. One is a power supply/power conditioner and 11 are processor and Input/Output (I/O) printed circuit cards.

Using the U1635 CPU in conjunction with a high-performance cache memory, the U1638A achieves a DAIS mix performance in excess of 1.5 Mips. The main memory consists of one million words of non-volatile core memory organised as two 512,000 modules. The primary I/O interfaces are the four MIL-STD-1553B dual-redundant and multiplex databusses.

The U1638A includes a programmed I/O channel and 16 input and 16 output discretes which are programme-controlled and can be defined to meet the user's needs. Four external interrupts and five identification bits are also provided. The unit is radiation-hardened with a radiation detector, power dump circuitry, a microcontrolled main memory radiation event recovery system independent of macro software and EMP protection.

Specifications

Dimensions: 193.5 × 322.3 × 319 mm
Weight: <23 kg
Power supply:
(dual AC input) 115 V AC, 400 Hz, 3 phase, 350 W

Operational status

In service on the US Air Force B-2.

Contractor

Lockheed Martin Tactical Defense Systems.

VERIFIED

ME-1000 SSR Solid-State Recorder

The ME-1000 SSR Solid State Recorder is designed to record multiple channels of instrumentation data in the harshest of environments with the highest reliability. The SSR's small size and light-weight make it suitable for applications associated with high performance aircraft flight testing as well as for use in armored vehicles, missiles, UAVs, and other mobile platforms. A data recording capacity of up to 16 Mbits/s is available, with a a total capacity of up to 9 Gbytes.

The SSR utilises solid-state FLASH memory modules as the recording media. The ME-1000 uses commercially-available Type II or Type III PCMCIA FLASH memory modules. Type III FLASH modules are available in capacities of 110, 175, 220, and 300 Mbytes.

The SSR has slots to accept nine modules for a total recording capacity of 990 Mbytes to 9 Gbytes. Because the system uses industry-standard PCMCIA modules, future FLASH cards with even greater capacities can be installed as they become available.

Smaller Type II FLASH modules can be used for capacities less than 990 Mbytes. The reliability of data recorded in the PCMCIA modules is less than one bit error in 10^{14} bits.

Operational flexibility is provided by programmable data interfaces. Four data channels can be setup as either redundant MIL-STD-1553 data busses or PCM channels in combinations. MIL-STD-1553 data busses can be recorded in a selective mode where only programmed remote terminals and sub-addresses are captured, or in a full rate mode where complete bus traffic is captured.

The PCM channels allow interface to NRZ-L data with a bit rate clock. All interface setup is made via dip switch settings or programmable settings through an RS-232 data port using the SSRConfig Windows 95® software package.

This software package is used to set up both selective or full bus capture of MIL-STD-1553 channels and/or PCM characteristics. This information is stored in internal EEPROMs.

Two separate RS-422 asynchronous data interfaces are also standard which are designed to capture the output of a Global Positioning System (GPS) receiver to record time/space/position information. Input data rate, start/stop bits and parity are programmable. The SSR accepts an IRIG-B timecode input for time-tagging. An audio channel is also provided with the analogue voice information being sampled and stored in memory.

An event mark input is also available for storing time tags of important test points and events.

Data is not 'played back' from the SSR, but downloaded via the SCSI interface directly into a computer for analysis or archival purposes. The SSR is designed to communicate to a remote-control device through discrete and serial control and status lines. No local controls or status indications are present on the SSR.

Specifications

Data interfaces:
channels: up to 4
Types: Redundant MIL-STD-1553 busses and/or PCM (NRZ-L + Clk)
PCM: Data rate: 200bits/s to 16Mbits/s
Auxiliary interface: two RS-232/422

ME-1000 SSR solid-state recorder ***1999***/0051345

Memory capacity (PCMCIA modules):
size modules: 110, 175, 220, or 300 Mbytes (currently available)
quantity: 9
current capacity: 990 Mbytes, 1.57, 1.98, 2.7 and up to 9 Gbytes
Max recording rate: 16 Mbits/s continuous
Error rate: <1 in 10^{14} (FLASH module specification)
Power: 28 V DC, 35 W
Dimensions: 129.7 (H) × 161.3 (W) × 281.9 (D) mm
Weight: Approx 7.9 kg

Contractor

Merlin Engineering Works (a wholly owned operation of TEAC America Inc).

VERIFIED

DTR-16 wideband analogue recorder/reproducer

Metrum-Datatape's DTR-16 wideband analogue recorder/reproducer can operate in a single-channel 16 MHz or dual-channel 8 MHz mode. The recorder uses Commercial Off-The-Shelf (COTS) S-VHS tape cassettes for cost-effective operation, with a front-loading automatic cassette elevator.

The DTR-16 can be controlled via an IEEE-488 or serial RS-422/RS-232 interface; alternatively, a parallel TTL interface provides for manual remote operation.

The DTR-16 is equipped with two auxiliary longitudinal channels with a 100 Hz to 20 kHz bandwidth for time event coding data and voice annotations via a microphone input.

Specifications

Dimensions: 107.95 (H) × 279.4 (W) × 558.8 (D) mm
Weight: 11.36 kg (without cassette)
Power: 100 to 240 V AC, 47 to 400 Hz, <100 W
MTBF: 8,000 h

Operational status

In production. Suitable for ground or air benign environment application.

Contractor

Metrum-Datatape Inc.

VERIFIED

HDDR-100 data storage system

The HDDR-100 is a high performance, rugged, system which includes a removable disk storage facility ideally suited to military and aerospace applications. It includes a realtime, parity protected, Redundant Array of Independent Disks (RAID) capability.

Operating at up to 100 Mbytes/s transfer rate and 72 Gbytes of storage (expandable) it is designed for high bandwidth data storage applications, such as radar imaging and telemetry capture.

Operational status

Available.

Contractor

Metrum-Datatape Inc

UPDATED

MARS-II data recording system

MARS-II provides a systems approach to data recording. An SCSI computer enables an electronic module to utilise tape, disk or solid-state memory, with one electronic module supporting up to seven storage modules. Combined in this system is the capability to record or reproduce eight completely asynchronous, electronically configurable channels of PCM or dual-

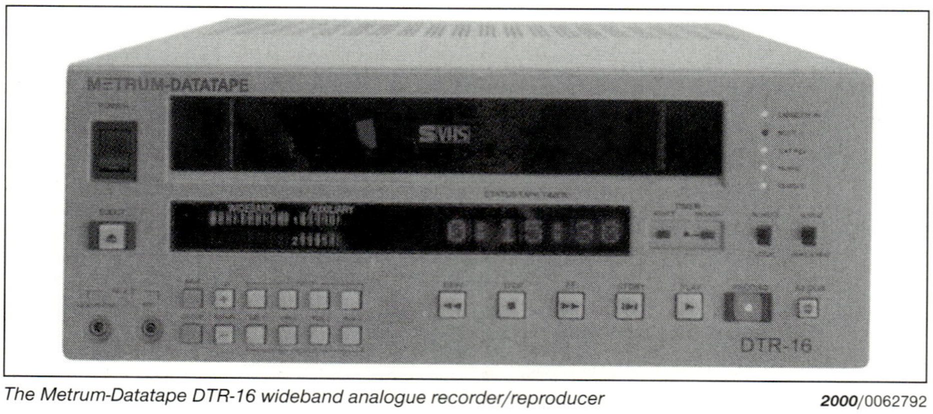

The Metrum-Datatape DTR-16 wideband analogue recorder/reproducer ***2000***/0062792

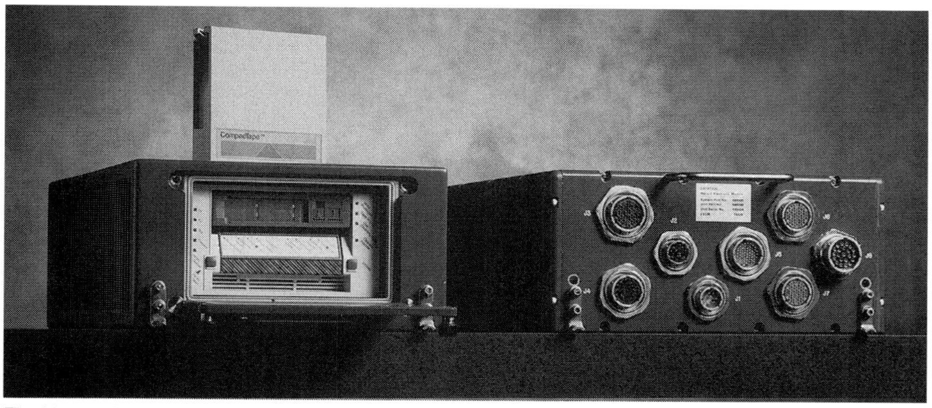

The Metrum-Datatape MARS-II consists of (left) the storage module and (right) the electronics module

redundant MIL-STD-1553B data in addition to one channel of IRIG time code data and one channel for voice annotation. In addition, MARS-II accepts NRZL, R-NRZL, biphase and analogue data formats.

Specifications

Dimensions:
(electronic module) 127 × 304.8 × 342.9 mm
(storage module) 127 × 238.1 × 342.9 mm
Weight:
(electronic module) 15.88 kg
(storage module) 6.8 kg
Power supply: 28 V DC
115/220 V AC, 47-400 Hz
Temperature range: –54 to +50°C
Altitude: up to 50,000 ft

Operational status

Variants of the MARS-II available include: Enhanced MARS-IIe (40 Mbits/s) version; MARS-II-SB (Single Box); MARS-II-L (Laboratory), and MARS-II DAS (Data Analysis System).

MARS-II-SB is specifically designed for rugged or rough-ride applications with limited or confined space; it provides two channels of standard PCM rates and one channel of high rate PCM, together with one channel of IRIG time code data and one channel of voice annotation.

Distributed in the UK and Benelux countries by Metrum Information Storage Ltd (UK).

Contractor

Metrum-Datatape Inc.

VERIFIED

Model 32HE/64HE variable speed digital recorder

The first in series of Harsh Environment (HE) recorders by Metrum-Datatape, the Model 32HE is designed, built and qualified specifically for harsh conditions. Both the Model 32HE and Model 64HE are sealed units; once the tape door is closed, the media, electronics and transport mechanism are protected from the environment. Suitable for use in fixed- and rotary-wing aircraft operations, including ASW, flight testing and other situations demanding large storage volumes.

Using the latest 100 kbpi version of Metrum-Datatape technology, the two models capture data at variable streaming rates from 0 to 32 Mbps (and 0 to 64 Mbps) and burst rates from 0 to 160 Mbpi. They record data at a density of 100,000 bpi on to broadcast-quality, high-energy, Metrum-Datatape-certified ST-160 S-VHS cassettes, that provide 13.8 Gbytes of data storage, giving 57 minutes minimum (linearly increasing as data rate reduces).

Both the Model 32HE and Model 64HE include three interfaces: a single-board mux - eight channel PCM and two channel analogue; series RS-422, TTL, ECL; parallel RS-422, TTL, ECL.

Specifications

Dimensions; 381 × 165.1 × 386.1 mm
Weight: 15.9 kg
Power: 28 V DC, 100 W (300 W with heaters operating)

Operational status

The Model 32HE was introduced in 1997. The Model 64HE, was introduced in October 1999. It provides a variable data rate up to 64 Mbits/s and stores up to 27.5 Gbytes on a single, low cost, S-VHS tape. Model 32HE recorder/reproducers can be upgraded to 64HE capability.

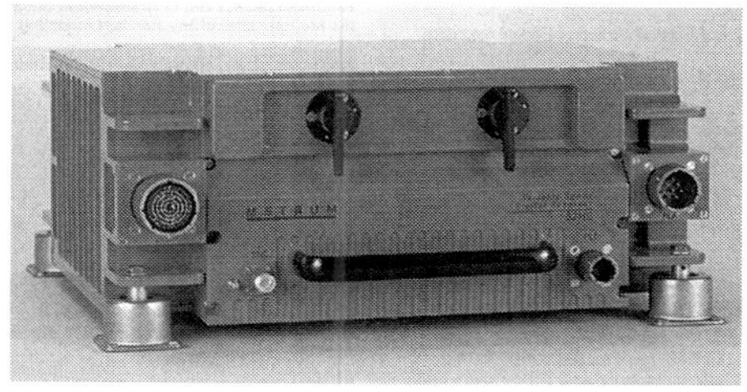

Model 32HE variable speed digital recorder 0015332

Distributed in the UK and Benelux countries by Metrum Information Storage Ltd (UK).

Contractor

Metrum-Datatape Inc.

VERIFIED

Solid-State Data Recorder (SSDR)

In September 1999, Metrum-Datatape announced their first solid-state data recorder system. Through a distribution agreement with SEAKR Engineering, Metrum-Datatape is offering the Tape And Rigid-disk Replacement System (TARRS) under the Metrum-Datatape name and SSDR model designator. The SSDR is specifically designed for applications in harsh environments where high speed recording is necessary and data integrity vital.

SSDR features include:

1. 96 Gbytes capacity in a single enclosure.
2. DCRsi compatibility.
3. Non-volatile, ECC protected, solid-state memory.
4. Modularity and expandability.
5. 100MB/s offload data transfer.
6. Data rates in excess of 30 Mbytes/s completely solid-state.

Operational status

Available.

Contractor

Metrum-Datatape Inc.

UPDATED

AN/AYK-42(V) processors

The most significant machine in the AN/AYK-42(V) range is the Norden Systems PDP-11/34M processor. All machines in the series are airborne processing units which are software and interface compatible with Digital's PDP-11 system. Fully militarised, they are suitable for applications ranging from tactical avionics to complex command and control.

The PDP-11/34M unit includes a complete processor, memory, peripheral interfaces and power supply on a single chassis. Modular unit construction ensures quick and easy replacements.

Features include: a complete PDP-11 instruction set (over 400 instructions); 1 kword cache memory option; memory expansible up to 128 kwords; 16 or 32 kword memory modules; 900 ns cycle time in core memory; memory management and protection; hardware multiply and divide; floating point processor option; hardware stack processing; and an input/output rate up to 1.1 M words/s.

Another processor in this range, the PDP-11/70M, can comprise up to four 1 ATR boxes dedicated to processing, power supply, 256 kwords memory and expander (extra input/output options) facilities respectively. This unit can perform up to 850 kips. The LSI-11M is a single-card version of the same processor design which operates at approximately 200 kips. It has 4 kwords of random access memory and can be associated with 16 or 32 kword core storage modules.

Specifications

Norden PDP-11/34M
Dimensions: 498 × 257 × 194 mm
Weight: 22.7 kg
Power: 410 W (max)
Computer type: binary, fixed and floating point
Word length: 32- or 64-bit
Instruction set: > 400 instructions (as PDP/11)
Typical speeds:
(without cache) 275 kips
(with cache) 400 kips
Input/output: 5-12 slots. Up to 1.1 M words/s
Memory:
(core) up to 128 kwords
Environmental: MIL-E-5400

Operational status

In after market support phase.

Contractor

Northrop Grumman, Electronic Sensors and Systems Sector, Norden Systems.

UPDATED

AN/ASQ-197 Sensor Control-Data Display Set (SC-DDS)

Orbital Fairchild Defense's AN/ASQ-197 Sensor Control-Data Display Set (SC-DDS), provides a link between the reconnaissance sensor suite, the reconnaissance operator and the aircraft systems providing function control, data annotation, and navigational data distribution. Originally designed for use in F/A-18D(RC)'s film system, the SC-DDS is expandable to accommodate an E-O sensor suite with an interface to an external image processing unit. This also provides an SC-DDS hybrid film and E-O sensor capability.

As presently configured, the AN/ASQ-197 controls three to five sensors. Command messages via the MIL-STD-1553 mux bus or serial interfaces to the aircraft and cockpit, determine sensor ready, run, extra picture and so on. Analogue control signals for Image Motion Compensation (IMC), altitude, and cycle rate pulses are generated for each station based on aircraft Vg/H. Sensor suite status is transmitted over the mux bus or output via discretes, providing periodic update of current system conditions.

Annotation data is formatted in both alphanumeric and MIL-STD-782 BCD formats. These data blocks are then output to appropriate recording head assembly, CRT or LED, for film annotation or for E-O video annotation to the separate image processing unit.

AN/ASQ-197 front panel controls and displays allow the operator to enter operational data, initiate system self-tests and monitor system status. Remote operate and internal BIT sections allow the operator to initiate built-in-test on any of the connected sensors, on the SC-DDS itself, or to run any of the sensors with fixed control parameters (FMC, cycle rate). Integrated AN/ASQ-197 configurations are designed for F/A-18D (RC), F-14A/D TARPS, RF-111C AUP and AV-8B ETARS aircraft.

Specifications

Data inputs: SSI, mux bus - A3818, A5232, or MIL-STD-1553B.
Sensor suite: system/sensor feedback, data demands and reference voltages.
Manual preflight: ADAS modes, fixed data film remaining, bit initiated.
Data outputs:
Aircraft mux bus: time, running, film remaining, status.
Sensor suite: Operate commands.
Filming annotation: CRT, LED matrix or strip RHA.
Power: 115 V AC, 1 phase, 400 Hz @ 100 VA max 28 V DC @ 56 W max.
Weight: (SC-DDS and tray) - 12.3 kg.
Dimensions: 225.6 (W) × 222.3 (H) × 406.4 (L) mm.

Operational status

In service in F/A-18D (RC), F-14 A/D TARPS, RF-111C (AUP), AV-8B ETARS.

Contractor

Orbital Fairchild Defense, Orbital Sciences Corp.

VERIFIED

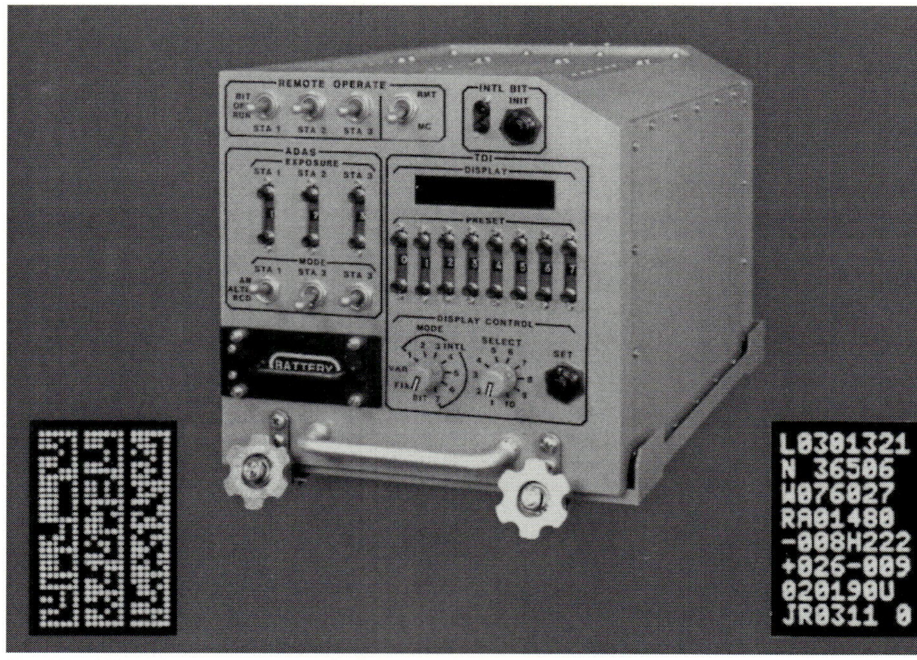

The AN/ASQ-197 Sensor Control-Data Display Set (SC-DDS) **1999**/0051289

AN/ASQ-215 Digital Data Set (DDS)

Orbital Fairchild Defense's Digital Data Set (DDS) is a high-capacity solid-state military airborne data storage and retrieval system. It provides rapid mission initialisation of the aircraft's avionics suite, as well as recording pertinent mission and maintenance data during flight; the DDS also supports in-flight processing. The AN/ASQ-215 has been adopted as the US Navy standard Digital Data Set (DDS). The DDS will support Global Positioning System data entry requirements for all MIL-STD-1553 multiplex databus-equipped US Navy aircraft. However, because many added features have been incorporated into this data management system, the DDS is suitable for a variety of tri-service military applications.

The Digital Data Set consists of two elements, a hand-carried, non-volatile solid-state data storage device known as the Data Transfer Module (DTM), and the cockpit-mounted, intelligent receptacle for the DTM, the Interface Receptacle Unit (IRU). The IRU manages and controls the data exchange between the DTM and other avionic subsystems via the aircraft's serial digital MIL-STD-1553 multiplex databus. The DTM serves as the transportable storage medium for both pre- and post-mission information exchange between ground computer stations and the airborne system.

Operating primarily as a remote terminal on the MIL-STD-1553 multiplex bus, the IRU also has the provisions to operate as a bus controller. The IRU provides an exhaustive self-test/Built-In-Test function for all circuitry, including the DTM while performing real-time bulk memory error detection and correction.

On a single circuit board, the DTM provides 2 Mbytes, expandable to 40 Mbytes with an additional card, of high-speed random access non-volatile data storage. The DTM contains only bulk memory and address decode logic to reduce overall operating expense and simplify future growth. A high degree of data integrity is assured through the use of a proven error detection and correction method.

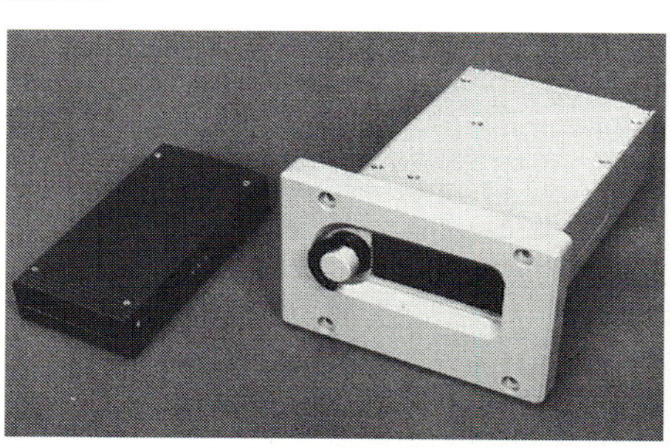

The mission data loader consists of (left) the data transfer module and (right) the interface receptacle unit

Specifications

Dimensions:
(IRU) 203 × 127 × 53 mm
(DTM) 152 × 81 × 32 mm
Weight:
(IRU) 1.50 kg
(DTM) 0.31 kg
Input power: 28 V DC at 15 W (max)

Contractor

Orbital Fairchild Defense, Orbital Sciences Corp.

VERIFIED

Data Transfer Equipment (DTE)

The Data Transfer Equipment (DTE), also known as the Data Insertion Device (DID), fulfils five main functions. These considerably reduce the workload imposed on crews and eliminate the risk of human error. The aircraft is automatically initialised and mission start procedures are launched simply by inserting a cartridge programmed during the preflight preparation. The DTE takes full control of navigation during the automatic ground-hugging phases of the flight and operates the weapons system. It also has electronic mapping and mission playback and debriefing on the various phases. The DTE reduces ground maintenance since it records and analyses the main aircraft parameters throughout the mission.

The DTE is capable of interfacing with all ground mission preparation and debriefing systems. It is operable from any allied or friendly logistic platform. The DTE also allows training in an intensive electronic warfare environment without requiring any special infrastructure.

Specifications

Dimensions:
(data transfer unit) 178 × 127 × 113 mm
(data transfer cartridge) 191 × 119 × 41 mm
Weight:
(data transfer unit) 3 kg
(data transfer cartridge) 0.7 kg

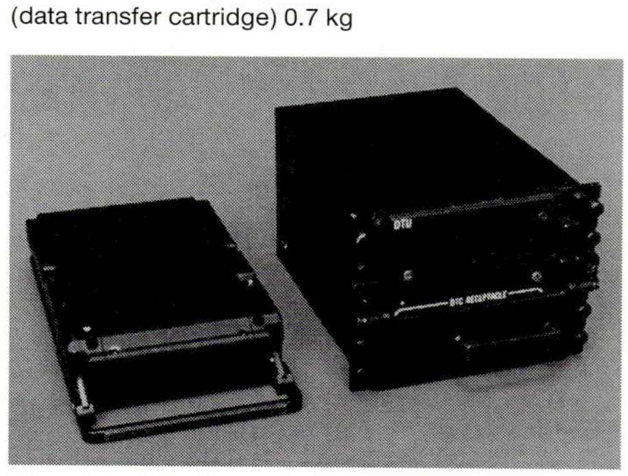

The data transfer unit (right) and data transfer cartridge (left) **1999**/0051286

Power supply: 115 V AC, 400 Hz, single phase
Cartridge capacity: 8 kwords expandable to 2 Mwords

Operational status

In service with the US Air Force, Army and Navy aboard more than 35 different aircraft, including A-6, A-10, B-1B, C-17, F-14, F-16, and SH-60. Also in service with NATO, Japan, the Republic of South Korea, and the Republic of China (Taiwan).

In late 1999, Orbital Fairchild Defense was contracted to supply an updated version of the DTE to the United Arab Emirates for use on Mirage 2000-9 aircraft.

Contractor

Orbital Fairchild Defense, Orbital Sciences Corp.

VERIFIED

Data Transfer Equipment/Mass Memory (DTE/MM)

Orbital Fairchild Defense developed the F-22 Data Transfer Equipment/Mass Memory system (DTE/MM). This system is composed of an aircraft cockpit resident Data Transfer Unit/Mass Memory (DTU/MM) and a portable Data Transfer Cartridge (DTC).

The DTC performs all data transfer functions such as mission data load, loading of operational flight programs, and in-flight data recording.

This cockpit resident DTU/MM is an intelligent mass memory system that contains up to 576 Mbytes of user non-volatile memory that provides modular growth to 2 Gbytes, The growth is obtained by the insertion of next-generation memory modules. The 100 per cent Ada operational flight program in the DTU/MM contains all functions necessary to support memory expansion including memory management up to 2 Gbytes of storage in the DTU/MM as well as the DTC.

The DTC is a portable non-volatile memory device that is used to transfer mission planning data to the aircraft for pre-mission initialisation, and is carried, typically by the pilot, from the aircraft to the mission debriefing area for unloading of in-flight recorder mission and maintenance data from the cartridge. The DTC uses a modular memory approach similar to the DTU/MM. Four memory module slots provide growth to 2 Gbytes. Present configuration provides 84 Mbytes of high-speed non-volatile random access memory. Also, this next-generation cartridge has maintained the common cartridge interface to ensure compatibility with existing ground support equipment and mission planning systems in the field.

The DTE/MM contains two memory volumes, the DTC and the Mass Memory (MM) for storage of user data files. The system maintains separate directories for each volume containing status and management data for each file present. Each volume also contains a header file used to control special functions such as automated file download for system initialisation

Data transfer equipment/mass memory for the F-22 **1999**/0051288

('boot-up' of other computers connected to the DTE/MM) and to specify bus parametrics.

File management commands allow the user to create, delete and erase files, read and write file data, copy files from DTC to MM, report file and system status information, format the volumes and perform built-in-test.

Data written into files is appended with a Reed Solomon error detection and correction code to provide robust data protection. Furthermore, if an error is detected during a data write cycle, the system provides recovery through a memory re-mapping capability, effectively enhancing system reliability.

The DTU/MM may be configured with up to nine Flash memory modules providing user memory capacity up to 576 Mbytes. This capacity may be doubled (or more) through insertion of new memory modules using next-generation memory components. Addressing capability of the DTU/MM supports up to 2 Gbytes of user memory.

The DTC may be configured with up to four SRAM memory modules providing user memory capacity up to 84 Mbytes and may also be expanded through new-generation module insertion. The DTC also supports addressing growth to 2 Gbytes. System features include full implementation of the JIAWG J88-N2 Linear Token-Passing High-Speed DataBus (HSDB) operating at 50 Mbits/s; automated download of specified system files in support of system initialisation; file management command structure; full Reed Solomon error detection and correction capability on all user memory; built-in fault log memory separate from DTU/MM and DTC memories; automatic over temperature detection with (override) system shutdown; extensive built-in-test (BIT) capability: startup BIT, initiated BIT, periodic BIT; DTU/MM operational software may be reprogrammed (updated) with system installed.

Specifications

Dimensions:
DTU/MM: 209.6 (L) × 146.1 (H) × 195.6 (W) mm
DTC: 190.5 (L) × 119.4 (W) × 40.6 (H) mm
Weight:
DTE/MM: 5.5 kg
DTC: 1.6 kg
Input power: 28 V DC, 45 W (typical)
Mass Memory data capacity: up to 576 Mbytes, growth to 2 Gbytes
Cartridge data capacity: up to 84 Mbytes, growth to 2 Gbytes
Software: 100% coded in Ada
System interface: Fibre-optics high-speed databus (JIAWG J88-N2)
Reliability (MTBF): DTE/MM system - 11,000 h

Operational status

In production for F-22.

Contractor

Orbital Fairchild Defense, Orbital Sciences Corp.

VERIFIED

Fibre Channel - Data Transfer Equipment (FC-DTE)

The FC-DTE consists of an intelligent, aircraft-mounted receptacle called a Data Transfer Receptacle (DTR) and a removable Data Transfer Cartridge (DTC).

The DTR manages and controls the data exchanged between the DTC and other avionic subsystems via the aircraft's MIL-STD-1553 and Fibre Channel busses. These busses may be used simultaneously thus allowing MIL-STD-1553 and Fibre Channel-based avionics to concurrently use the FC-DTE as a mass storage/file server device. In addition to providing the interface to the aircraft, the DTR contains significant processing capability which may be used to pre- and post-process the data which is being stored/accessed from the DTC. The DTR also contains a spare card slot which is fully wired to accept other Fairchild Defense and other third-party circuit cards. With this capability, single card functions such as a digital moving map can be added to the FC-DTE. These circuit cards have direct access to the DTC's mass storage and can interface directly to other Fibre Channel and MIL-STD-1553 devices. Thus, future aircraft upgrades can take full advantage of the FC-DTE's capabilities.

The Data Transfer Cartridge has been optimised to provide extremely high-speed access to the solid-state mass memory which it contains. Current DTC configurations allow memory densities of between 100 Mbytes and more than 1 Gbyte. The design of the DTC allows future memory upgrades and memory densities in excess of 5 Gbytes will be available in the foreseeable future.

FC-DTE features support for SCSI-3 upper level protocol with sustained write rates of more than 20 Mbytes/s and sustained read rates of more than 30 Mbytes/s; significant growth is provided via DTR spare processing throughput, DTR spare card slot, and DTC modular memory design; DTC is Air Force Mission Support System (AFMSS) compatible; the DTC is removable in explosive atmospheres; it supports NSA-approved declassification of stored data; it is qualified for multi-aircraft use including those aircraft with radiation hardness requirements; it provides 95 per cent fault detection and 95 per cent fault isolation.

Specifications

Dimensions: DTR - 208.3 (L) × 127 (W) × 152.4 (H) mm
DTC - 190.5 (L) × 119.4 (W) × 40.6 (H) mm
Weight: DTR - 2.95 kg, DTC - 1.42 kg
Power: 115 V AC, 400 Hz, single phase, 40 W maximum
Software: Ada
DTR I/F:
1 Dual-redundant MIL-STD-1553B
2 Independent FC-AL Nodes
1 RS-422
1 RS-232
DTC I/F: AFMSS compatible
MTBF: 30,000+ h with forced cooling air
23,500+ h with convection cooling

Contractor

Orbital Fairchild Defense, Orbital Sciences Corp.

VERIFIED

High-speed Solid-State Recorder (HSSR)

Orbital Fairchild Defense has completed design and development of its new High-speed Solid-State Recorder (HSSR) for initial application in airborne reconnaissance systems. This commercial-off-the-shelf (COTS) product's I/O emulates two existing standard tape recorder interfaces: Ampex DCRsi 240™ and MIL-STD-2179.

Each HSSR installation includes at least one Data Recorder Cartridge (DRC) containing non-volatile FLASH memory in configurable memory receptacles. This affords the user the ability to select only the amount of memory required, currently from 8 to 90 Gbytes. The DRCs can be removed and inserted into Fairchild's Ground Playback Unit (GPU). Alternatively, data can be electronically downloaded by connecting the Data Recording Unit (DRU) via a fibre optic connection to a RAID system in the Portable-GPU (P-GPU)

Options include voice channels, variable compression ratios, cockpit video review output, expanded memory capacity, customised data input channels. Customised configurations can be produced to meet the needs of limited space installations, such as aircraft pod installations.

Operational status

Development complete. The HSSR has been demonstrated in a US Air National Guard F-16, when it successfully recorded and reproduced digital reconnaissance imagery. The HSSR is now available for customer use in airborne reconnaissance programmes.

Contractor

Orbital Fairchild Defense, Orbital Sciences Corp.

VERIFIED

Integrated Mission Display Processor (IMDP)

The Integrated Mission Display Processor (IMDP) is an open system architecture based on VME64. It implements reliable and low-cost elements in a configurable military processor. The IMDP is designed to be applicable to airborne, shipboard and ground vehicle applications. The IMDP is configurable, using common modules, to provide mission management,

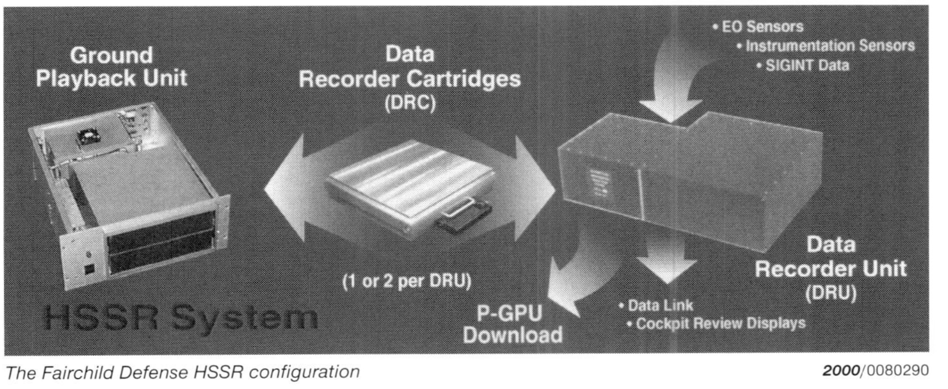

The Fairchild Defense HSSR configuration **2000**/0080290

display management, cockpit interfaces, digital map, weapons management and other functions. The IMDP provides a fully qualified, low-cost approach for new development or upgrades to existing systems.

Architecture

VME64 - Commercial standard architecture with a large supplier base, proven for cost-effective implementation of real-time military processing applications.

Interfaces

MIL-STD-1553 - Single or multiple interfaces, configurable as bus controller, remote terminal, bus monitor, or any combination.

Fibre channel - Fully compliant bidirectional interface, configurable as point-to-point or arbitrated loop.

Others - RS-422, RS-232, ARINC-429, SCSI, Ethernet, and so on.

Processing

Power PC - Fully integrated with and support for either Power PC 603 or 604.

Display processor - Single board provides video switching and outputs for two displays.

Other microprocessor options - Supports full line of Motorola, Intel, and Silicon Graphics microprocessors.

Special purpose processing - Variety of special purpose processors and functions (software) to support both graphical and numerical applications.

Memory

Mass memory module - Existing solid-state (Flash EEPROM) mass memory options (single board with up to 2 GBytes)

Embedded memory - Existing processing boards contain up to 64 Mbytes of RAM and 8 Mbytes of Flash.

Memory interfaces - Provides standard interface for other embedded memory boards (VME or PCI disk drives, and so on) or external memories.

Displays

Single board display processors. Each board provides multiple RS-170 RGB and monochrome inputs, two independent outputs, graphics rendering, and overlays. Configurable for single or multiple display processing boards.

Digital map

Existing single board map processor. Provides fully anti-aliased, RS-170 RGB, map display, threat overlay (line-of-sight), multiple scales, continuous zoom, and 3-D graphics.

Real-time operating system

VxWorks - A commercially available program which is supported throughout the VME community for real-time uni- or multi-processing applications.

Software architecture - Integrated with VxWorks to provide ease of software development or porting and integration of existing applications.

Software development

PC-based commercial-off-the-shelf (COTS) tools - Use of standard processors operating system and languages provides a fully COTS PC-based toolset.

Specifications

Operating environment: Designed and qualified for military airborne, shipboard and ground vehicle environments.

Power: 28 V DC, 270 V DC, or 115 V AC, less than 120 W.

Dimensions: ¾ ATR without cooling fan, ¾ ATR long with cooling fan.

Weight: <13.64 kg

Integrated Mission Display Processor (IMDP)

1999/0051290

Contractor

Orbital Fairchild Defense, Orbital Sciences Corp.

VERIFIED

Mega Data Transfer Cartridge with Processor (MDTC/P)

The Mega Data Transfer Cartridge with Processor (MDTC/P) stores Digital Terrain Elevation Data (DTED) and the Digital Vertical Obstruction File (DVOF) used in Terrain Referenced Navigation (TRN) and Terrain Awareness and Warning Systems (TAWS). The first application of the MDTC/P is for a Digital Terrain System (DTS) provided by BAE Systems (Operations) Limited TERPROM software for the F-16.

INS, barometric and radar altimeter data are transmitted over the aircraft databus to the MDTC/P and INS update and pilot cueing and warnings are transmitted by the MDTC/P over the bus to the INS, multifunction display, HUD and voice warning system.

Other applications for the MDTC/P include the digital moving map database for the F-22, EW simulation and the Tactical Air Combat Mission Analysis and Training System.

Operational status

Ordered by the US Air Force Reserves for use with the DTS (TERPROM) in the F-16.

Contractor

Orbital Fairchild Defense, Orbital Sciences Corp.

VERIFIED

Upgraded Data Transfer Equipment (UDTE)

Orbital Fairchild Defense's Upgraded Data Transfer Equipment (UDTE) comprises an aircraft resident receptacle called an Upgraded Data Transfer Unit (UDTU) and a removable data transfer cartridge called a Mega Data Transfer Cartridge with Processor (MDTC/P); UDTE is a form, fit, and function replacement for Orbital Fairchild Defense's Data Transfer Equipment. In addition, UDTE provides an expanded array of functions.

Data transfer capabilities include a Digital Terrain System (DTS); an Embedded Data Modem (EDM); an Embedded GPS Receiver; an Embedded Display Processor; an Embedded Training System; an Integrated Electronic Combat System; a high-performance processor to back up flight critical systems; up to 2 Gbytes of solid-state mass storage.

UDTU

Using state-of-the-art circuit design and packaging techniques, the three circuit cards in the original DTU have been reduced to a single circuit card. This card uses an industry standard 32-bit microprocessor which has enough processing throughput to retain all the original DTU's functionality while also hosting advanced software algorithms such as those required for Embedded Training and an Integrated Electronic Combat System. In addition, new capabilities may be added to the UDTU via the two unused card slots. The list of circuit cards which can populate these spare slots is continually growing and includes a ruggedised GPS receiver, an EDM and a Display Processor.

MDTC/P

The MDTC/P is a form, fit and function replacement for Orbital Fairchild Defense's standard Data Transfer Cartridges and remains fully AFMSS compatible. The MDTC/P contains up to 280 Mbytes of non-volatile, solid-state mass memory and a high-performance Digital Signal Processor (DSP) ideally suited for mathematically complex algorithms. Typical uses for the MDTC/P include mission planning, storage of Digital Map Data, storage of avionics data gathered during flight, in-flight execution of BAE Systems (Operations) Limited TERrain PROfile Matching (TERPROM) algorithm which is used as part of the Digital Terrain System and post-mission playback associated with training.

Real-Time Intelligence in the Cockpit (RTIC)

RTIC becomes a reality with the inclusion of the Embedded Data Modem (EDM) within the UDTU. Located in a spare card slot within the UDTU, the EDM takes full advantage of the UDTU's prewired backplane and connectors to bring real-time data into the aircraft. The EDM was first flown at Edwards Air Force Base in 1995 and has been proven effective with ARC-164 and ARC-210 radios, aircraft intercoms, and KY-58 crypto sets. Using standard protocols, the EDM allows real-time data, including reconnaissance photos and imagery, to be sent to and from the aircraft.

Digital moving map

The UDTU's backplane and connectors are also wired to accept a single board digital moving map. Orbital Fairchild Defense's full featured moving map provides anti-aliased RS-170 RGB outputs and has capabilities normally associated with much more expensive, stand-alone map systems. These capabilities include sun angle shading, height above threshold, threat intervisibility, continuous zoom, graphics overlays, and user-defined annotation. The moving map may also be used to display two and three-dimensional imagery.

The mass memory utilised by the moving map resides within the removable data transfer cartridge. Accordingly, the moving map electronics and mass memory are co-located within the data transfer equipment and external high-speed busses are not required at the aircraft level. As well as the obvious aircraft wiring advantages associated with this

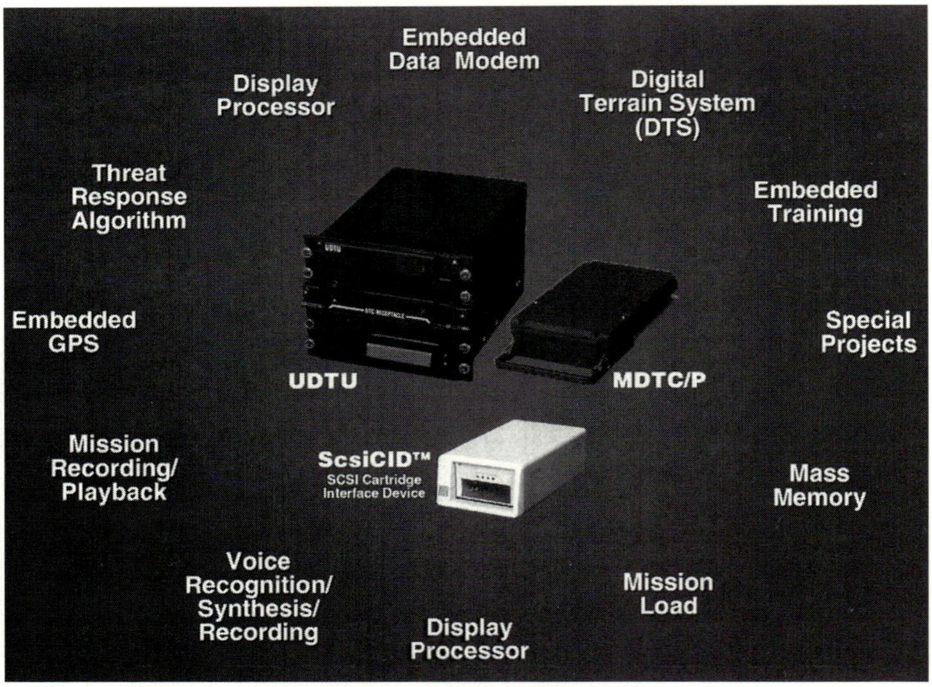

Orbital Fairchild Defense's Upgraded Data Transfer Equipment (UDTE)

0002320

approach, the use of the MDTC/P for the map storage allows standard mission planners to be used to load and prepare the map database.

Digital Terrain System (DTS)

In the Digital Terrain System (DTS), BAE Systems' TERPROM® algorithm is hosted on the digital signal processor within the MDTC/P. DTS allows aircraft integrators to incorporate significant aircraft safety of flight and lethality improvements without compromising the throughput of other aircraft systems. The numbers of aircraft using DTS continues to grow and includes platforms such as the F-16, Mirage 2000, and Jaguar.

By correlating inertial navigation and altimeter data with Digital Terrain Elevation Data (DTED) stored within the MDTC/P, precision Terrain Referenced Navigation (TRN) is accomplished. Once the aircraft is accurately located with respect to the terrain, DTS provides numerous aircraft enhancements including predictive ground collision avoidance, obstruction warning and cueing, database terrain following, and weapons ranging.

Video recording

In recent years, the amount of solid-state memory available within the MDTC/P has seen a dramatic increase. Functions previously considered to be not feasible due to the large amount of required storage are now well within the capabilities of the equipment. One such function is the recording of video data. By populating one of the UDTU's spare card slots with a video compression card, between 1.5 and 2 hours of imagery can be recorded. This video can be synchronised with other aircraft data stored in the cartridge thus improving the effectiveness of mission debriefs by eliminating the separate, non-synchronised recorders which are typically used today.

Embedded training

The advanced capabilities of the UDTE allow embedded training to be accomplished without the need to add external aircraft pods or the need to fly on an ACMI range. The use of the UDTE for embedded training was first demonstrated at Edwards AFB in 1995.

For this function, the UDTU serves as a bus monitor recording aircraft data pertinent to mission debrief, and as a bus controller able to inject synthetic threat data into the aircraft's electronic warfare systems. The data recorded during the training session is retrieved from the MDTC/P via standard mission planning ground equipment. Time tagging using GPS time can be used to synchronise several aircraft thus allowing multi-aircraft playback.

Synthetic threat data can be preplanned and stored in the MDTC/P or real-time data can be sent to the aircraft via the embedded data modem or other aircraft datalinks. By using the DTS's TRN capability, these synthetic threats are accurately positioned on the ground. In addition, DTED information stored in the MDTC/P is used to accurately simulate terrain masking. The end result is a highly accurate, life-like training scenario without the need for very expensive ranges or external equipment.

Embedded GPS

The UDTU's backplane and rear panel contain a location for the addition of an RF connector thus supporting a third party GPS card housed within one of the UDTU's spare card slots. Including the GPS card within the UDTU co-locates the GPS function with GPS almanac data stored on the data transfer cartridge. Functioning as a bus controller or remote terminal on the MIL-STD-1553 interface, the UDTU can supply the GPS information to other avionic systems.

Bus Monitor and Analysis Software (BMAS)

Avionic configurations are increasingly complex and the isolation and debug of a problem can cause extended downtime. BMAS allows the UDTE to function as MIL-STD-1553 test equipment and record user-specified MIL-STD-1553 data. All pertinent MIL-STD-1553 data can be replayed on a standard personal computer using Orbital Fairchild Defense's data logger software. At the completion of debug, the UDTE can resume normal operation.

Ground support peripherals

The MDTC/P is fully AFMSS compatible. For those applications where an AFMSS is not available, Orbital Fairchild Defense's SCSI Cartridge Interface Device (ScsiCID™) allows any personal computer or workstation to communicate with the MDTC/P over a standard SCSI-2 interface. Data rates in excess of 5 Mbytes/s allow data to be loaded rapidly into the MDTC/P's mass memory.

Specifications

Dimensions:
UDTU: 178 × 127 × 113 mm
MDTC/P: 191 × 119 × 41 mm
Weight:
UDTU: 3.0 kg
MDTC/P: 1.6 kg
Input power: 115 V AC, 400 Hz, single phase, 32 W max
Software: Ada

Contractor

Orbital Fairchild Defense, Orbital Sciences Corp.

VERIFIED

Omnidirectional Air Data System (OADS)

Developed for helicopter applications, the Omnidirectional Air Data System (OADS) consists of a mast-mounted sensor, connected cable and air data computer, and provides an airspeed output over the range 0 to 200 kt irrespective of direction. Additional outputs include forward, rearward and sideways speed components, air density, altitude, air temperature and pressure. The original Pacer OADS was used in the Bell X-22 ducted propeller tilt engine V/STOL research aircraft built in the early 1970s. The current system is in production for the US Army AH-64A Apache anti-tank and US Coast Guard Eurocopter HH-65A Dolphin search and rescue helicopters. In July 1987, a technology transfer contract was signed with the China National Aero-Technology Import and Export Corporation (CATIC). This will involve production of helicopter air data systems configured for the Z-9 helicopter, which is built in China under licence from Aerospatiale.

Specifications

Dimensions:
(sensor) 254 × 76 × 76 mm
(computer) 184 × 127 × 241 mm
Weight:
(sensor) 1 kg
(computer) 2.7 kg

Operational status

In production for the Boeing Company AH-64, Eurocopter HH-65A and Chinese Z-9 helicopters.

Contractor

Pacer Systems Inc.

VERIFIED

Series 2000 camera

The Series 2000 16 mm camera is capable of time-lapse, normal speed and up to 500 fps operation. It features interchangeable 200, 400 or 1,200 ft daylight-loading film magazines and has a synchronous phaselock option allowing synchronisation of several cameras.

The camera is extremely small when using the 200 ft magazine. Overall length and height are increased as the magazines containing greater lengths of film are installed. Magazines can be changed in a few seconds.

The Series 2000 camera is similar to the KB-21C camera system used by the US Air Force.

The Photo-Sonics Series 2000 16 mm camera

Specifications

Dimensions:
(with 200 ft magazine) 139.7 × 114.3 × 203.2 mm
Weight: 2.72 kg
Power supply: 28 V DC,12 A
115 V AC, 50-400 Hz, 3.5 A optional
Temperature range: –54 to +70°C
Acceleration: up to 25 *g*

Contractor

Photo-Sonics Inc.

VERIFIED

Super SVCR-V301 high-resolution airborne video recorder

The high-resolution Super SVCR-V301 recorder is lightweight and compact. It is designed to record video camera, infra-red sensor and multifunction displays in the stringent environment of fighter aircraft. It is designed and tested to meet MIL-STD-810C/D, including rain, sand and dust, and EMI-tested to MIL-STD-461C and -462. It is designed specifically to meet the stringent electrical, mechanical and environmental requirements encountered in modern flight test applications.

The V301 incorporates Super VHS format, rewind and playback, over 2 hours of recording, high-speed forward and reverse search, a visual event marker, comprehensive BIT, electronic frame indexing, serial and parallel interface, three audio channels and 525-, 875- and 1,023-line scan rates.

The V301's Super VHS format is not just an improvement to standard VHS. It is a distinct new format providing significantly higher picture clarity with a full 400 lines of horizontal resolution in both colour and black and white recording, providing significant improvement in line picture detail. The Super SVCR-V301 provides higher luminance signal frequency and wider frequency deviation and separates luminance and chrominance signals to minimise the degradation of image quality from cross colour and dot interference. The signal-to-noise ratio in the V301 has been significantly increased by broadening the frequency deviation from 1 to 1.6 MHz. Raising the carrier frequency also reduces interference with chrominance signal and substantially increases contrast range.

The V301 will record in both standard and Super VHS formats. This allows the use of existing VHS ground playback equipment until it is replaced with higher resolution Super VHS equipment. A host of full-function commercial ground playback equipment is currently available, all capable of playing cassettes recorded on the V301. Super VHS format tapes cannot be played back on standard VHS systems. However, they can be transferred or edited down to ¾ in Umatic or the standard VHS format.

Specifications

Dimensions: 111 × 212 × 290 mm
Weight: 7.2 kg
Power supply:
115 V AC, 400 Hz, 120 W (heater only)
22-30 V DC, 33 W

Operational status

Flight test aircraft on which the V301 has been installed include the F-15, E-2C, F/A-18, P-3 and RF-4C. Operational programmes include the Tornado GR. Mk 4, AH-1W Upgrade and F/A-18 export military sales. The V301 is under consideration for a number of other advanced operational programmes.

Contractor

Photo-Sonics Inc.

VERIFIED

F-16 modular mission computer

The computer system for the F-16 provides multiple processing functions in a single chassis. It replaces three present computers in the earlier configuration and provides processor power to support the addition of capabilities such as FLIR and digital terrain functions. SEM-E format modules provide data processing, avionic display, power supply and bus interface for the computer. Various aircraft system functions are defined by the application software operating on the computer. The primary computing module processors are based on the R3000 32-bit RISC instruction set architecture. Each processor module provides 16 VAX/Mips throughput and has 1 Mbyte of on-module memory. The computer, as configured for the F-16 with 16 digital modules installed, weighs 17.78 kg.

Operational status

Initial deliveries began in 1993.

Contractor

Raytheon Electronic Systems.

VERIFIED

Integrated Core Processor (ICP)

Raytheon Electronic Systems has been selected by Lockheed Martin Tactical Aircraft Systems to provide the ICP for its two JSF Joint Strike Fighter demonstrator aircraft.

The ICP is the central computer system for the JSF, including all the embedded computing elements for multiple subsystems. It provides the digital processing resources for sensors, communications, electronic warfare, guidance and control and cockpit displays. The ICP implements an open system architecture that maximises the use of commercially supported products and standards, to define, implement and support an affordable ICP that uses open-system concepts to enable seamless incorporation of new technologies.

Contractor

Raytheon Electronic Systems.

UPDATED

Matchwell integrated data fusion system

Matchwell is an advanced development project funded by the US Air Force Wright Laboratory. The goal of Matchwell is to provide early threat warning to aircrew. As air defence systems proceed through the targeting process, they emit sufficient information to allow an observer armed with the appropriate sensors and intelligence data to determine their location and intentions. Matchwell accomplishes this by performing fusion at Levels 1, 2 and 3 to provide data fusion and situation assessment. Use of artificial intelligence techniques allow all this to be executed in real time.

Matchwell is implemented on two DEC 3100 workstations. One hosts the environment simulator that generates an integrated set of signal reports; the other hosts the threat warning software. The workstations provide high-resolution graphic displays of the actual and inferred situation. The software is written in a combination of Ada and C to facilitate ease of transition to an embedded real-time environment.

Contractor

Raytheon Electronic Systems.

VERIFIED

Processor Interface Controller and Communication (PICC) module for the F-22

Raytheon Electronic Systems produces the MIL-STD-1750A instruction set architecture-based Processor Interface Controller and Communication (PICC) module for the F-22 aircraft. Each module provides data processing, MIL-STD-1553B bus communication, intermode communications via serial interchannel datalink and analogue and digital input/output for the vehicle management system.

Contractor

Raytheon Electronic Systems.

VERIFIED

Lockheed Martin's X-35C Naval Version of the JSF utilises Raytheon's ICP (Lockheed Martin) **2001**/0105289

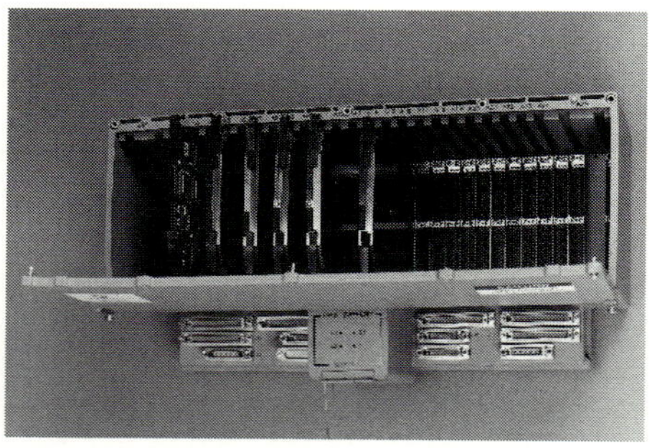

The processor interface controller and communication module for the F-22

ADS-85/86/850 digital air data systems

The ADS-85/86/850 series is similar to the obsolete ADS-82 except that it uses a solid-state piezo-resistive sensor. The ADS-85 interfaces with the Rockwell Collins electromechanical air data instruments and has an ARINC 429 bus for auxiliary systems. The ADS-86 interfaces with Rockwell Collins CRT air data instruments. The ADS-850 contains an ARINC 429 bus for flight control and attitude and heading communications.

Specifications

Dimensions: ½ ATR
Weight: 2.52 kg
Power: 18 W

Operational status

In production and service.

Contractor

Rockwell Collins.

VERIFIED

Data Management System (DMS)

The Data Management System (DMS) is a step towards achieving paperless operations, with electronic control of information. In the cockpit, approach and departure information, taxi and ground diagrams are electronically maintained and can be narrowed to the definition required by the pilot with the pan and zoom feature. Maintenance, cabin and operations data can be accessed at a single terminal, which also serves as the data load and retrieval centre. The Collins DMS is packaged in a single cabinet, using line-replaceable modules. These allow significant reduction in space and weight requirements and need no forced-air cooling.

The DMS consists of a multichannel computer, with information contained in mass storage modules for quick and efficient data access. Information is displayed on high-resolution colour flat-panel displays. The entire system is networked with a fibre optic distributed data interface.

The Rockwell Collins DMS is adaptable to virtually all aircraft types.

Operational status

Available as an option for the Boeing 777. First deliveries were made in 1995.

Contractor

Rockwell Collins.

VERIFIED

Aircraft systems processor

SCI is currently producing two distinct versions of the Aircraft Systems Processor (ASP) for the AH-64D Longbow Apache. These are the systems processor and the weapons processor, which vary in input/output and content. These processors provide all mission data, housekeeping and weapons processing and have been successfully integrated and flight tested in the first Longbow Apache prototype aircraft. Extensive use of ASICs, SMT, double-sided circuit card assemblies and controlled impedance motherboards has resulted in a full MIL-E-5400 processor that has 330 input/output channels, dual-1750A CPUs, dual-MIL-STD-1553B interfaces and extensive BIT implementation, including built-in logic analyser and console debugging functions.

The ASP features high-density input/output combined with high-computational throughput and flexibility and low weight and size. The ASP chassis is an eight-slot air-cooled aluminium housing. Five of the slots are general purpose, one slot is dedicated to the power supply module and one slot is double width to accommodate a piggyback module or other oversize assembly. Each ASP housing can contain up to 14 CCAs on six modules, plus a power supply. All ASP modules are double-sided SEM-E size with an aluminium core to provide efficient heat transfer.

Available as an option is an R3000 CPU to replace the 1750 CPU for those applications that need significant processing power above that offered by the 1750 CPU. The R3000 CPU is based around the LSI Logic LR33000, the first integrated Mips architecture microprocessor. Within a single chip, the LR33000 combines the scalar processor and cache/memory manager portions of the Mips architecture, as well as 1 kbyte of high-speed data cache memory and 8 kbytes of high-speed instruction cache memory. Operating at 30 MHz, the R3000 CPU offers performance figures of 15 Mips or greater.

Specifications

Dimensions: 190.5 × 193 × 203.2 mm
Weight: 6.35 kg
Reliability: 6,000 h MTBF

Contractor

SCI Systems Inc.

VERIFIED

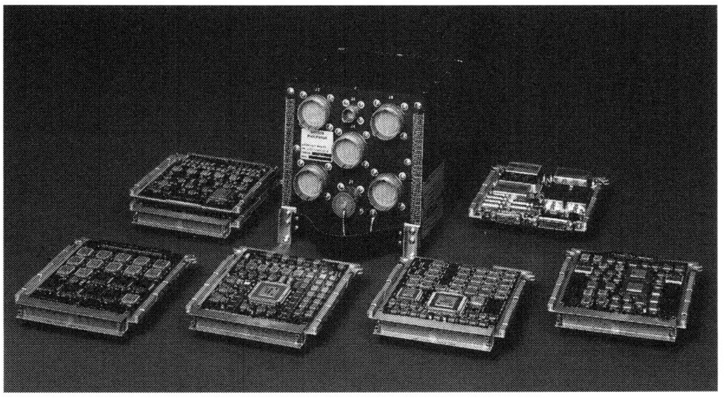

The SCI aircraft systems processor can house up to six modules

Two different versions of the ASP are being produced for the AH-64D Longbow Apache

---

Advanced Memory Unit (AMU) Model 3266N

The Smiths Industries AMU, Model 3266N, is a sophisticated data transfer and recording system, which is designed to operate on both a standard MIL-STD-1553A/B communications bus and the High-Speed Interface (HSI) Smiths Industries bus. The AMU and the associated Harris Corporation Digital Map Computer (DMC) and the W L Gore & Associates Inc High Speed Interface (HSI) cable assembly make up the new common US Navy Tactical Aircraft Moving MAp Capabiity (TAMMAC) system. Each AMU accepts up to two PCMCIA cards.

The AMU replaces earlier US Navy Data Storage (DS) and Mission Data Loader Equipment (MDLE), in addition to its digital map system.

Operational status

The TAMMAC system is scheduled to be fitted initially to five baseline US Navy TAMMAC aircraft: F/A-18, AV-8B, AH-1Z, UH-1Y, and CH-60. Other US Navy aircraft being considered for TAMMAC include the F-14, S-3, CH-53, SH-60 and P-3.

Contractor

Smiths Industries Aerospace.

VERIFIED

AN/ASH-28 signal data recorder

The AN/ASH-28 records digital data on a 25 hours capacity quickly removable cassette. This system was designed for the F-15 and is housed within a single box containing two linear accelerometers, one angular accelerometer and three gyros and interfacing with the F-15's MIL-STD-1553 digital databus.

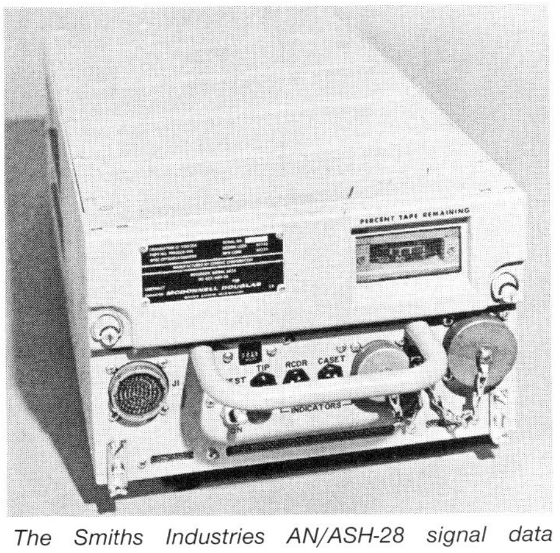

The Smiths Industries AN/ASH-28 signal data recording set

Operational status

In service in US Air Force F-15 aircraft but no longer in production.

Contractor

Smiths Industries Aerospace.

VERIFIED

Data Transfer Systems (DTS)

Data Transfer Systems (DTS) are a selectable family of automated digital data loading, recording and post-mission downloading subsystems for aircraft and surface vehicles using digital avionics. The first solid-state DTS was initiated by Smiths Industries in the 1970s to obviate digital information transfer problems encountered during the preflight loading of mission-related information. Errors and delays in aircraft readiness were being caused by the manual insertion of preflight information such as target co-ordinates and mission waypoints via aircraft system keyboards. The

The advanced memory unit model 3266N and representative-fit aircraft **1999**/0051283

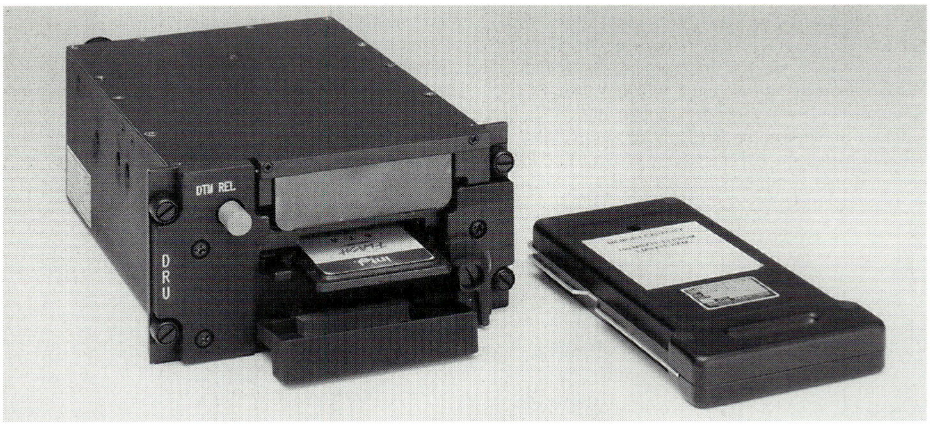

Data Transfer System 0001307

introduction of the DTS has basically eliminated preflight data entry errors while, at the same time, reducing cockpit flight deck initialisation time from about 30 minutes to a few seconds.

In addition to the loading and retrieval of normal mission data, DTS equipment is being used to initialise Joint Tactical Information Distribution System (JTIDS), Navstar GPS (Global Positioning System), missile guidance control units, digital mapping system and voice control interactive devices. DTS equipment can also provide immediate post-mission printout and analysis of operational and flight test information. It is also used within Flight Data Recorder systems as an information recording and download mechanism.

A typical DTS includes a small, portable, solid-state data transfer memory media, a receptacle into which the media is inserted and a ground-based mission data computer terminal containing an application-specific software database.

Available memory media includes a small shirt pocket-sized Data Transfer Module (DTM, or cartridge) which is offered with SRAM, Flash EPROM and combined SRAM/Flash EPROM, depending on customer preference/need. DTM memory capacities range from 16 kbytes to 150 Mbytes, with planned growth to 600 Mbytes achievable by 2000. Smiths Industries has introduced seven backward-compatible DTM improvement configurations since its initial DTM design in the 1970s. Optional PCMCIA (PC Card) portable memory media solutions are also offered for selected Smiths Industries DTS applications. PC card memory capacities currently exceed 1 Gbyte, with 6 Gbyte PC cards forecast for 2000.

The data transfer memory media receptacle, mounted in a convenient location within the aircraft, accepts the DTM and/or PC Card(s) via a safety spring-loaded door. Depending on user needs, Smiths Industries receptacles are modularly expandable to accommodate standard electronic interfaces, such as RS-422, MIL-STD-1553A/B, SCSI-2, Fibre Channel, Ethernet and combinations of these interfaces. Newer DTM/PC Card receptacles also offer a media data management microprocessor, additional memory and/or other electronics to increase system functionality. The receptacle electronics can be provided either within the receptacle or packaged remotely to conserve cockpit space. Smiths Industries cockpit receptacles can be configured to accommodate the DTM, PC Cards (PCMCIA) or a combination of DTM and PC Card memory media.

DTM/PC Card ground interface and mission planning computer terminals incorporate user-friendly software to accommodate efficient digital data loading and retrieval functions. Mission planning terminals have progressed from large expensive machines to smaller, more capable and less expensive ground computer systems. Smiths Industries offers its own tactical Mission Data Ground Terminal (MDGT) based upon personal computer (PC) technology. Also offered is a family of DTM ground interface and test equipment for use within customer designated mission planning systems. Selectable DTM ground interface and test unit/kit solutions include DTM to PC-ISA/EISA, RS-422, IEE-488 and SCSI-2 interfaces. Included within these standard mission planning systems are the Computer Aided Mission Planning System (CAMPS), Mini-CAMPS, Mission Support System-1 (MSS-1), MSS-2, MSS-2+, Air Force Mission Support System (AFMSS), Portable Mission Planning System (PMPS) and Army Aviation Mission Planning System (AMPS).

Operational status

In production. Over 80 types of fighters, multi-engine aircraft, helicopters and surface vehicles have been equipped with the Smiths Industries DTS, including over 8,500 systems to date.

Contractor

Smiths Industries Aerospace.

VERIFIED

GenHUMS

GenHUMS offers a multi-aircraft capable generic Health and Usage Monitoring System capability, using proven airborne and ground technology, based on HUMS. GenHUMS comprises the airborne GenHUMS equipment; and the HUMS Ground Station (HGS). The airborne GenHUMS components are the Data Acquisition and Processing Unit (DAPU); Cockpit Control Unit (CCU); Cockpit Interface Panel (CIP); Data Transfer System (DTS); and Optical Blade Tracker (OBT).

GenHUMS functionality implements the following health and usage monitoring functions: transmission health monitoring; engine health and usage monitoring; rotor track and balance; rotor and airframe health monitoring; aircraft usage monitoring. Acquired data is stored in crash-protected memory. Cockpit voice recording is also provided.

Operational status

Selected by the UK MoD for Chinook helicopters, with options for other UK MoD helicopters including Sea King, Puma and Lynx. On the Chinook nearly 200 parameters vital to operations are monitored. Future developments will include an upgrade with the capability to predict fatigue in real time.

Contractor

Smiths Industries Aerospace.

VERIFIED

Smiths Industries GenHUMS 0015333

Health and Usage Monitoring System (HUMS/SI HUMS)

Smiths Industries Aerospace combines the proven Cockpit Voice Recorder/Flight Data Recorder aircraft monitoring technology with proprietary Health and Usage Monitoring System (HUMS) technology to

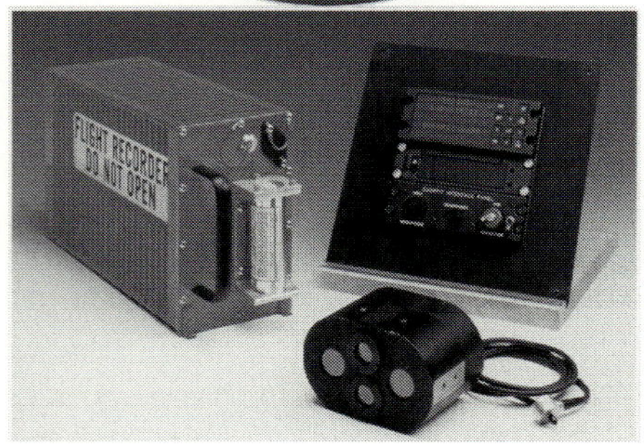

The Smiths Industries Aerospace SI HUMS
2000/0080287

provide an advanced multi-aircraft capable, 'single box' system. The HUMS has been developed using off-the-shelf technology and a proven ground support system. Latterly, this unit has carried the nomenclature SI HUMS.

Integrating the capabilities of Smiths Industries' divisions IMS, D&CS, SPS, Stewart Hughes, and MJA Dynamics with a team of leading companies in key HUMS technologies, the HUMS provides critical flexibility for cross platform and future growth requirements. The HUMS represents the logical evolution of multibox systems to a single box utilising the latest monitoring, recording, and diagnostic systems technology.

Design features include: ED55/56A automatic rotor balancing; onboard rotor track and balance; aircraft usage monitoring; rotor, engine, transmission, and gearbox health monitoring; powertrain, gearbox, and bearing diagnostics; engine and structures life usage monitoring; flight regime recognition; crash-protected voice and flight data recording, 10 to 150 Mbyte PCMCIA or ruggedised data transfer system; wide interface capability MIL-STD 1553B/RS-429/ARINC 422/717; extensive recording/monitoring/playback/analysis capability; optional onboard oil and lubrication system monitoring (Inductive Debris Monitoring system - IDM); optional engine exhaust debris monitoring (Electrostatic Engine Monitoring System - EEMS).

Contractor

Smiths Industries Aerospace.

VERIFIED

Inductive Debris Monitor (IDM)

The IDM provides real-time damage detection and is claimed to give earlier warning than any other monitoring technique for fluid wetted systems, without false alarms.

The IDM is guaranteed to detect magnetic and non-magnetic metal chips, is independent of fluid velocity and viscosity and is insensitive to foam, bubbles and dielectric properties. Metal chips are classified by type, size and rate of occurrence, detecting damage long before performance degradation and without periodic inspections, to provide up to 10 hours early warning of impending failure.

Operational status

In testing for the RAH-66 Comanche.

Contractor

Smiths Industries Aerospace.

VERIFIED

Model 3255B Integrated Data Acquisition and Recorder System (IDARS)

Smiths Industries Integrated Data Acquisition and Recorder System (IDARS) is versatile and easily adaptable to a wide variety of requirements that go beyond flight data and voice recording and into information management. Its modular design allows the IDARS to perform all the functions of a Cockpit Voice Recorder (CVR), Flight Data Recorder (FDR) and a Flight Data Acquisition Unit (FDAU) using two cards in one unit. There are three additional card slots to accommodate growth features.

IDARS supports custom tailoring of hardware and software configurations to meet specific vehicle acquisition and processing requirements.

The IDARS computes, compresses, and stores data from a wide range of analogue, digital and discrete sources. Configurable software controls the entire system and is easily modified for specific applications.

The airborne software's operational flight program characteristics can be reprogrammed and data download accomplished via a front panel connector or a readily accessible separate maintenance connector. Common ground support software tools are available to permit complete and easy access to the recorded voice and data.

Stored flight data and voice are synchronised and offer an integrated data analysis database for optional animated flight replay capability. This is supported by the Graphical Replay Animation System (GRAS) to enhance effective aircrew flight analysis and crew training. The GRAS can be hosted on a desktop or laptop to permit total or segmented flight replay.

Specifications

Dimensions: ½-ATR Short chassis per ARINC 404; 320.5 (L) × 123.95 (W) × 193.5 (H) mm
Weight: 7.2 kg, maximum without acoustic beacon
Power: +28 V DC, 40 W (maximum)
Compliant: EUROCAE ED-55 and ED-56a; FAA TSO-C123a and -C124a

Operational status

In production. Selected for the US Air Force B-1B, KC-135, T-6A II, U-2S, UH-1N;
UK MOD CH-47, Lynx, Sea King, Super Puma;
Israeli Air Force C-130, CH-53, UH-60;
Eurocopter EC 135, BK 117, BO 105.

Contractor

Smiths Industries Aerospace.

VERIFIED

Smiths Industries Model 3255B IDARS ***1999***/0051332

Removable Auxiliary Memory Set (RAMS)

RAMS includes a Data Transfer Interface Unit (DTIU), for managing data and providing system communications, and a receptacle for the associated memory module, the Data Transfer Module (DTM) or programmable cartridge. It is about the size of a cassette tape and reduces the need for large ground readout equipment. RAMS performs the functions of downloading memory, uploading operational flight plans and documentary data, formatting memory and resetting usage accumulators, flight counters and BIT history.

Data for the Standard Flight Data Recorder (SFDR) is recorded on the DTM during flight. The receptacle displays percentage of memory used and system faults. The module and receptacle perform the following SFDR functions, eliminating the need for a digital computer: load Signal Acquisition Unit (SAU) operational flight program; load documentary data into SAU non-volatile memory; format standard flight data recorder memories; clear usage accumulators and BIT records; download CSMU data; download SAU non-volatile memory; and download SAU auxiliary memory unit.

DTMs are available in 1 to 8 Mbytes battery-backed RAM capacities and 10 to 86 Mbytes EEPROM. Near-term growth to over 500 Mbytes is planned.

Specifications

Dimensions:
(module) 82.55 × 20.57 × 150.1 mm
(DTIU receptacle) 128 × 77 × 190 mm
(DTMR receptacle) 128 × 32 × 133 mm
Weight:
(module) 0.34 kg
(DTIU receptacle) 1.81 kg
(DTMR receptacle) 5 kg
Power supply: 28 V DC, 20 W

Operational status

RAMS has been installed on US Air Force B-52, F-15, KC-135 and T-38 aircraft.

Contractor

Smiths Industries Aerospace.

VERIFIED

Standard Flight Data Recorder (SFDR) system

The Smiths Industries solid-state Standard Flight Data Recorder (SFDR) system has been developed under a US tri-service specification. The solid-state SFDR can replace older oscillograph or tape recorders as well as add more comprehensive aircraft monitoring functions. It was initially introduced by the US Air Force on the Lockheed F-16 fighter.

The Series SFDR consists of the Signal Acquisition Unit (SAU) with auxiliary memory unit, the Crash Survivable Memory Unit (CSMU), the optional Flight Data Panel (FDP) and the Removable Auxiliary Memory Set (RAMS) consisting of an intelligent Data Transfer Interface Unit (DTIU) or a simple Data Transfer Module Receptacle (DTMR) and Data Transfer Module (DTM). The CSMU is also being used on Sweden's Saab JAS 39 fighter.

The SAU is capable of receiving a combination of over 600 discrete, analogue and digital multiplex bus parameters which are converted into digital information, data compressed and stored within the SAU's auxiliary memory unit, RAMS data transfer module and/or CSMU. Daily monitoring and recording of general airframe, avionics and engine health structural loads and engine low-cycle fatigue is typical. All conversion and data management functions are managed witihn the SAU as well as data monitoring and customer designated alerts. The SAU also supports rapid download to ground data logging systems and to graphical replay and analysis systems.

Selected flight data parameters are sent to a compact armoured and insulated CSMU to ensure data recovery following a flight incident. The CSMU is designed to withstand stringent mishap conditions including temperatures of up to 1,100°C, mechanical shock of 3,400 *g* and penetration. The CSMU non-volatile memory has a life expectancy of 60,000 hours which, together with the built-in test facility, permits it to be mounted in inaccessible regions of the airframe.

Routine maintenance data is retrieved from the aircraft by means of flight line Ground Readout Equipment (GRE) or an installed RAMS. Retrieval data is then transferred to a Data Recovery and Playback Evaluation System for data decompression, analysis and implementation. The STS is developing and loading Operational Flight Program (OFP) data and providing maintenance support for the SFDR.

Specifications

SAU
Dimensions: 157 × 178 × 184 mm
Weight: 6.51 kg
Memory:
(program) 572 kbytes
(scratchpad) 512 kbytes
(non-volatile) 16 kbytes
(auxiliary) up to 3 Mbytes
Environmental: MIL-E-5400 Class II
Power: <50 W

CSMU
Dimensions: 76 × 76 × 117 mm
Weight: 1.58 kg
Memory: 64 k or 256 kbytes
Min recording time:
(attack fighter/trainer aircraft) 15 min active flight, 60 min normal flight
(transport aircraft) 25 h
Environmental:
(impact) 3,400 *g* for 6 ms
(penetration) 500 lb for 15 ms
(static crush) 5,000 lb for 5 min
(fire) 1,100°C flame for 30 min, or equivalent oven test
(fluid immersion) 48 h at 1,500 ft
Power: <4 W
Locator: acoustic beacon or crash position indicator available

The Smiths Industries standard flight data recorder

Operational status

In service on both military and commercial fixed- and rotary-wing aircraft.

Contractor

Smiths Industries Aerospace.

VERIFIED

Voice And Data Recorder (VADR®)

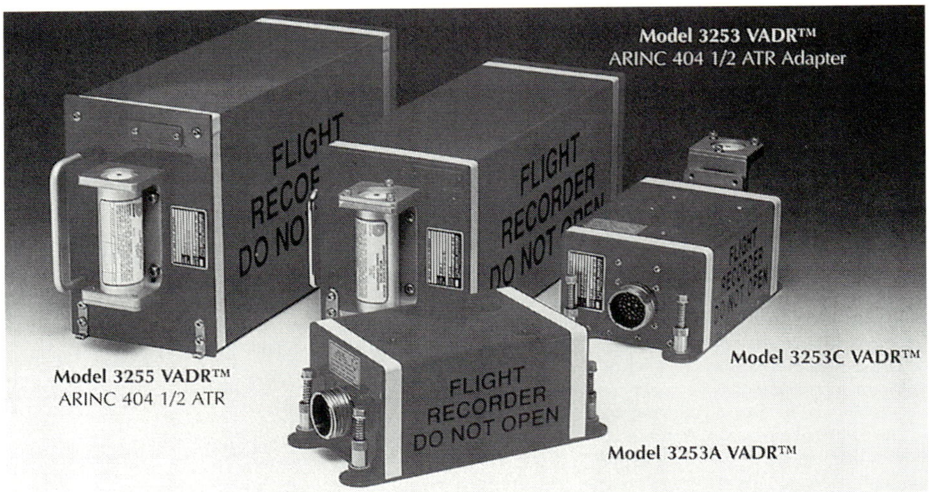

The Smiths Industries family of Voice and Data Recorders (VADR®). From left to right: Model 3255A IDARS (formerly Model 3255 VADR™), Model 3255B IDARS (formerly Model 3253 VADR™), Model 3253A VADR™ and Model 3253C VADR™
0015334

The dual-use Voice And Data Recorder (VADR®) combines the functions of a Cockpit Voice Recorder (CVR) and Flight Data Recorder (FDR) into a highly reliable, lightweight package, available for military and civil applications. The compact recorder system can be hard mounted in virtually any location or orientation, affording the original equipment manufacturer or avionics integrator increased installation flexibility. The VADR® utilises solid-state memory technology, offering increased reliability and low power consumption. The crash-protected memory packaging techniques are based on the US Air Force Standard Flight Data Recorder (SFDR) Crash Survivable Memory Unit (CSMU).

The VADR® provides data collection and mishap recording of audio data and aircraft flight and system parameters to support post-incident analysis. It can be configured as a TSO-C124a Flight Data Recorder (FDR), a TSO-C123a Cockpit Voice Recorder (CVR) or an ARINC 757 combined FDR/CVR. This family of recorders meets the survivability requirements of EUROCAE ED-55 for FDRs and ED-56/56A for CVRs for both ejectable and non-ejectable recorders. The VADR® offers mounting provisions for an underwater locator beacon.

Additional configurations of the VADR® include an ARINC 404 adaptor tray for form, fit and function replacement of existing recorder systems, a CVR/FDR combining voice and MIL-STD-1553, ARINC 429 and ARINC 717/747 serial bus interfaces, and a high-capacity data only FDR for MIL-STD-1553, ARINC 429, ARINC 717/747 or RS-422 serial data recording.

The VADR® is available in a number of models offering audio capacities from one channel × 30 minutes to four channels × 120 minutes. Data capacities from 2 to 30 hours are available depending on application.

Specifications

Dimensions vary with model
Weight varies from: 3-3.5 kg
Audio frequency response:
3 channels: 150-3,500 Hz
1 channel: 150-6,000 Hz

Operational status

In production and in service on a wide variety of fixed- and rotary-wing aircraft, including: US presidential helicopters, AV-8B, B1-B, ALX Super Tucano, AH-64A, Beech 300, C-2, C-130J, CH-47D, CL-60Y, EC-135, F/A-18C/D, F-111, HH-60J, HH-65A, JPATS, Lear 65, MH-47E, MH-60K, OH-58D, OH-X, RAH-66, SH-60J, UH-60, UP-3, US-1A, VH-3, VP-3, VH-60 and WAH-64 aircraft.

Contractor

Smiths Industries Aerospace.

VERIFIED

NuHums™ health and usage monitoring system

The NuHums is a modular, lightweight design that performs all signal processing in a single ½ ATR unit. The NuHums can be upgraded to add an optional flight data/cockpit voice recorder or third party diagnostic system.

The NuHums performs real-time onboard diagnostic processing for in-flight analysis of helicopter rotors, drivetrains, gearboxes and engines. It utilises the RADs-AT rotor track and balance system, and provides immediate feedback on any parameter exceedance.

NuHums modular architecture permits the customer to select features to match the requirement, and to add new features when required. The only essential element is the Data Acquisition Unit (DAU); sensors, cockpit displays, recorders, and other elements are options.

Specifications (DAU)

Channels: 48 analogue and 48 digital channels
Internal memory: 64 Mbytes
Dimensions: ½ ATR
Power: 28 V DC, 30 W
Weight: 4.55 kg

Operational status

In production, and in service on a number of helicopter types.

Contractor

SPS Signal Processing Systems, a division of Smiths Industries Aerospace.

VERIFIED

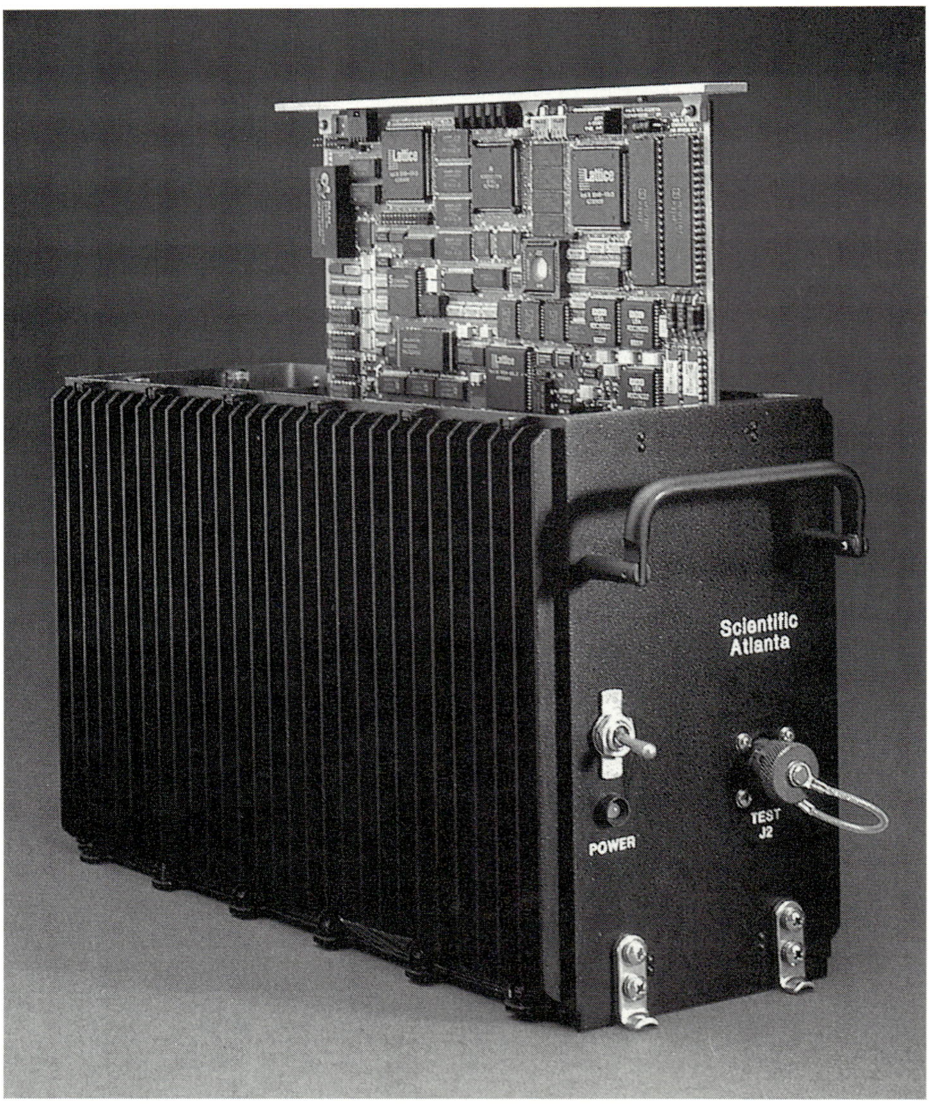

NuHums health and usage monitoring system
0015335

Interference Blanker Unit (IBU)

The Interface Blanker Unit co-ordinates the operation of the aircraft's onboard transmitters and receivers to prevent mutual interference and/or receiver damage when operating on multiple overlapping frequencies. Each aircraft transmitter issues a 'blanking request' discrete before activation of its RF output stage. The IBU detects these 'blanking request' signals and issues a discrete 'blanking output' signal to any victim receivers which are operating on overlapping frequency bands. The victim receiver uses the 'blanking output' signal to either disable its front end receiver circuitry or to filter out any input signals for the duration of the 'blanking output' signal. The IBU automatically handles any combination of simultaneous overlapping 'blanking request' signals to ensure that any victim receiver is disabled as long as any offending transmitter is active. The mapping of 'blanking request' channel to 'blanking output' channel, the 'blanking request' trigger threshold and detection characteristics, the 'blanking output' hold-off delay, pulse-width, pulse extension and pulse truncation characteristics, and the 'blanking output' voltage level are all fully programmable on a channel by channel basis.

Specifications

Dimensions: ARINC ½ ATR short
Weight: 4.55 kg
Power: 115 V AC, 400 Hz, single phase, 100 V A
MTBF: 10,000 h

Operational status

Contracted for F/A-18E/F; backward compatible to F/A-18C/D, AV-8B, AH-64D and other aircraft.

Contractor

Strategic Technology Systems Inc.

VERIFIED

Strategic Technology Systems interference blanker unit ***1999***/0051287

DV-60 AB-F Digital Airborne Video Tape Recorder (DAVTR™)

The TEAC Model DV-60AB-F is a new digital airborne video tape recorder, compatible with all V-80AB-F installations and interfaces. It provides the following capabilities: record and playback, colour; full frame clean still and event mark. It utilises DVC format small cassette (Mini DV cassettes) in the DVC tape format standard.

Specifications

Video:
Signal standard: NTSC
Resolution: 500 lines
Audio:
PCM 16
One stereo 48 kHz
Max record time: 60 min; 80 min (using AY-DVM80E), 120 min capability in development
Dimensions: 148.1 × 100.1 × 161.1 mm
Weight: 2.5 kg
Power: 28 V DC, 10 W (40 W with heater)

Contractor

TEAC America Inc.

VERIFIED

DV-60AB-F Digital Airborne Video Tape Recorder (DAVTR™) ***1999***/0051333

Fast Tactical Imagery (FTI) system

The FTI system replaces the existing TEAC V-1000 recorder fitted to F-14 aircraft. It comprises an updated TEAC Hi-8 mm recorder capable of 2 hours recording, rather than the 30 minutes of the predecessor system, together with a PhotoTelesis transceiver.

The FTI system will enable F-14 crews to record, transmit and receive still-frame images from any of the recorded inputs in the cockpit, including that of LANTIRN pods, TV camera systems, TARPS digital imagery pods, HUD video and other tactical imagery.

Images can be tagged by the F-14 backseat crew and transmitted to co-operating forces, using the AN/ARC-182 secure UHF radio for line of sight operation.

Operational status

Contracted to TEAC from Raytheon Electronic Systems. The upgraded system entered service in 1999.

Contractor

TEAC America Inc.

VERIFIED

V-80AB-F, V-82AB-F and V-83AB-F Hi-8 mm Airborne Video Tape Recorders (AVTR)®

The TEAC V-80AB-F, V-82AB-F and V-83AB-F Hi-8 mm AVTRs offer single-, dual- or triple-deck configurations respectively, that provide fully qualified, off-the-shelf, ruggedised capability, designed for HUD, FLIR, RWR, MFD recording, based on 15 years actual combat aircraft operational service.

Performance features include a 4 MHz bandwidth, 400 TV lines of resolution, full remote control and event mark generation, using standard commercial cassettes.

An optional plug-in board can be added to directly time stamp GPS, ZULU, or IRIG clocks onto the video tape. This option allows for exact synchronisation of video from multiple aircraft involved in tactical, test, or training missions. Flight crews can also select whether to write the time on the visible picture or use Vertical Interval Time Code (VITC) stamping that does not encroach on the picture space.

The V-83AB-F triple-deck requires no change to existing V-1000AB-R aircraft mounting trays or control panels.

The TEAC V-80AB-F Hi-8 mm ruggedised single-deck airborne video tape recorder ***1999***/0051334

The TEAC V-83AB-F Hi-8 mm ruggedised triple-deck airborne video tape recorder ***1999***/0051335

The TEAC Integrated Debriefing System (TIDS) ground stations can synchronise and control up to four commercial video recorders when using Hi-8 mm tape cassettes, and each TIDS can be daisy-chained to create a larger system comprising up to 32 synchronised playback decks.

Specifications

V-80AB-F:
Recording decks: one
Tape format: Hi-8 mm and 8 mm (NTSC/PAL/S-Video)
Recording time: 120 min (NTSC); 90 min (PAL)
Horizontal resolution: 400 TV lines (nominal)
Control interface: manual, discrete, RS-422
Dimensions: 148 (W) × 120 (H) × 161 (D) mm
Weight: <3 kg
Power: 28 V DC, 15 W (45 W with heater)

V-83AB-F:
Recording decks: three independent (two for V-82AB-F)
Tape format: Hi-8 mm and 8 mm (NTSC/PAL/S-Video)
Recording time: 120 min (NTSC), or 360 min in series
Horizontal resolution: 400 TV lines (nominal)
Control interface: discrete, RS-422
Dimensions: 243 (W) × 151 (H) × 330 (D) mm
Weight: <8 kg
Power: 28 V DC, 45 W (135 W with heater)

Operational status

TEAC claims that its single-, dual-, and triple-deck units are flying on over 75 per cent of the Western world's military aircraft.

In November 1999, TEAC was awarded a contract to supply the V-83AB-F triple-deck AVTR® system to the Royal Australian Air Force for the F/A-18 Cockpit Video Recording System (CVRS) programme. The system's proprietary time code insertion capability will allow each tape to be coded during recording with a time code synchronised to the Global Positioning System (GPS) universal time reference, thus permitting the synchronisation of the recordings of several aircraft on the same mission. Deliveries will begin in February 2000, with the first aircraft upgrades scheduled to be completed by October 2000.

Contractor

TEAC America Inc.

VERIFIED

VSC-80 video systems

The VSC-80 series comprises the VSC-80A and the updated VSC-80B. The VSC-80B comprises the TEAC V-80AB-F Hi-8 mm 525-line airborne video recorder with the Merlin Engineering precision scan converter, to produce a 60 per cent improvement in image quality and resolution over the existing TEAC V-1000AB-F U-matic Airborne Video Tape Recorder (AVTR) system. Both the VSC-80A and VSC-80B are direct form/fit replacements for the V-1000AB-F.

FLIR or day-TV images from video sensors can be captured at a rate of four per second and viewed immediately on the cockpit display. If desired, captured images can be selected for transmission, cropped, deleted, or saved for later transmission. Images received from ground stations, or other aircraft, can also be immediately viewed on the onboard display and retransmitted when required. Using image compression, transmission time is less than 15 seconds.

Specifications

Tape format: Hi-8 mm and 8 mm
Recording time: 120 min (NTSC) min
Function: record and playback; burst transmission
Dimensions: 271.8 (W) × 165.5 (H) × 369.5 (D) mm
Weight: <8.4 kg
Power: 115 V AC, 400 Hz, 75 V A (115 V A with heaters)

Operational status

Selected for US Army and Netherlands Army AH-64D Longbow Apache helicopters.

Contractor

TEAC America Inc.

VERIFIED

The TEAC VSC-80B video system **2000**/0081863

Combo Data Transfer Systems (DTS)

The Combo DTS is a military Data Transfer System. It loads and records data to a 3.5 in floppy disk or to dual PCMCIA flash memory cards to and from the avionics system via the MIL-STD-1553B databus. It can support two separate MIL-STD-1553B channels and function as a bus controller.

The Combo DTS acts as a remote terminal to load navigation databases and as a bus controller in the memory/loader/verifier mode to load operational programs into aircraft systems.

The floppy disk is MS-DOS compatible, readable by IBM-compatible PCs and supported by the AFMSS and PFPS mission planning systems.

65000 Combo DTS

The 65000 Combo DTS is a derivative of the current military production system.

ARINC 615 portable data loaders

Demo Systems' ARINC 615 portable data loaders are used to upload navigational databases and operational programs to commercial DTS, using 3.5 in computer disks and to download maintenance and other information.

Operational status

The Combo DTS is in use on MH-53J and HH-60 helicopters, AC-130H and AC-130U gunships, a variety of C-130 programmes including the MC-130H Combat Talon II aircraft and on RC-135 and WC-135 aircraft.

Demo Systems claim that their ARINC 615 portable data loaders support about 80 per cent of the world's airlines and are used on a wide range of systems and aircraft, including Airbus, Boeing, Lockeed Martin and Raytheon aircraft.

Contractor

Tecstar, Demo Systems Division.

VERIFIED

T'AIMS I (Teledyne Aircraft Integrated Monitoring System)

T'AIMS I

T'AIMS I evolved from the Teledyne Modular Data Analysis Unit (MDAU) and Ground Support Equipment (GSE). T'AIMS I performs exceedance monitoring based on cockpit displays, and it automatically detects any limit exceedance in the engine, rotor or transmission. It also monitors flight hours, engine hours and engine cycles, performs health checks and displays results to the flight crew. Exceedances and performance data are saved in non-crash-protected memory.

There are three system elements: the Modular Data Analysis Unit MDAU; T'AIMS I control panel; and the ground station computer - a laptop PC to which flight data can be downloaded on the flight line. The MDAU interfaces to the engine, gearbox and rotor sensors; the air data system; cockpit displays and the control panel.

T'AIMS II

T'AIMS II expands the capabilities of T'AIMS I by adding a crash-protected solid-state voice and data recorder; increased capability for aircraft performance monitoring; and increased capability for data analysis in the ground station system.

Contractor

Teledyne Controls.

VERIFIED

T'AIMS modular data analysis unit 0015336

Solid-State CVR-30A Cockpit Voice Recorder

The SSCVR-30A solid-state cockpit voice recorder provides 30 minutes of data. It accepts four channels of cockpit audio, converts the audio to digital format and stores the data in solid-state non-volatile flash memory. A fifth channel digitally records helicopter rotor speed. The recorder is fully ARINC 557 compatible.

Two control units are available. One conforms to ARINC 557 specifications, the other is a slimline unit designed for installation where cockpit panel space is at a premium and ARINC 557 specifications are not required. Both utilise an LED display for signal level.

Specifications

Dimensions:
(recorder) ½ ATR short
(ARINC control unit) 57.2 × 148 × 92.1 mm
(slimline control unit) 38.1 × 146 × 98.6 mm
Weight:
(recorder) 10.5 kg
(ARINC control unit) 0.5 kg
(slimline control unit) 0.34 kg
Power supply: 27.5 V DC, 1 A
115 V AC, 400 Hz, 0.2 A option
Reliability: 18,000 h MTBF

Operational status

In production.

Contractor

Universal Avionics Systems Corporation.

VERIFIED

Solid-State CVR-30B/120 Cockpit Voice Recorders

The SSCVR-30B/120 solid-state cockpit voice recorders are the latest Universal Avionics Systems Corporation solid-state CVRs. They feature the same capabilities as the SSCVR-30A and are fully ARINC 757/557 compatible. The SSCVR-30B records 30 minutes of data; the SSCVR-120 records 120 minutes. The recorders accept four channels of cockpit audio, convert the audio to digital format and store the data in solid-state, non-volatile flash memory, together with helicopter rotor speed and time.

Specifications

Dimensions: ½ ATR short
Weight: 5.85 kg
Power supply: 27.5 V DC, 1 A
115 V AC, 400 Hz, 0.2 A optional
Reliability: 30,000 h MTBF

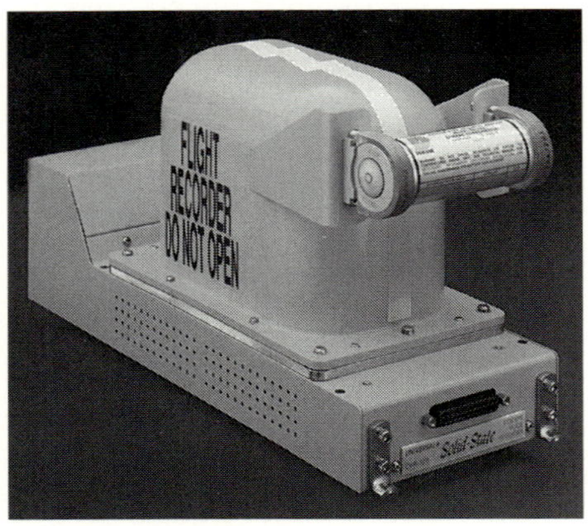

Solid-state CVR-30B/120 cockpit voice recorder 0015337

Operational status

US FAA certified TSO 123a and ED56A.

Contractor

Universal Avionics Systems Corporation.

VERIFIED

Pave Pace integrated avionics architecture

The Pave Pace programme continues the further development of integrated system avionics architectures begun by its predecessor, the Pave Pillar programme. Pave Pace will both enhance and extend the Pave Pillar architecture.

Several enhancements are planned. One will employ advanced digital multichip packages, using silicon-on-silicon technology. This approach will allow SEM-E-sized line replaceable modules, roughly 150 × 150 × 15 mm, to operate at speeds approximately one billion Floating point Operations Per Second (Flops). Substantial weight and volume reduction and concomitant reliability increases result from the need for three to four times fewer modules, the need for dramatically fewer module-to-backplane connectors and more than a three-fold decrease in solder joints. However, the densely packaged circuitry and clock rates of around 100 MHz will generate heat removal requirements in excess of 150 W per module. Pave Pace solves this problem with another Pave Pillar enhancement by using liquid flow-through modules, where the coolant, normally confined within the rack enclosure, is brought into the centre of a hollow module frame that supports circuitry on either side. Within this frame, the fluid interacts with a metallic heat exchanger, allowing a shorter thermal path between the heat source and coolant. Data shows 200 W can be removed at circuit temperatures of 83°C. Projections show that a 10,000 to 20,000 hours MTBF reliability for such modules is possible at an estimated cost of US$25,000 to US$30,000.

Another enhancement will be the use of modular high-speed optical network crossbar switches, operating at 1 to 2 Gbits/s, that will allow fault-tolerant interconnections between sensor-based signals and signal processing centres located in the racks. These same switches will be used for rack-to-rack communication and fault-tolerant rack display routeing for digitised video signals.

Extensions to the Pave Pillar architecture are the most dramatic, however. An integrated sensor system will be implemented. Here, the same Pave Pillar concepts of building a small family of modular digital building blocks to create system processing functions is used, except the modular building blocks are multifunction apertures and the family is a small number of analogue-based SEM-E-sized synthesisers, receivers, transmitters and switches. Projections show that an integrated RF system has the potential to reduce cost and weight of RF support electronics by approximately 50 per cent and improve sensor system availability by over a factor of three to four by using new circuitry and fault-tolerance techniques. Key technologies to be used include millimetre and microwave-integrated circuits, low temperature co-fired ceramic packaging, multi-arm spiral antennas for multifunction use spanning 200 MHz to 6 GHz, and a broadband active array capable of performing radar and EW functions. The modular family concept is projected to reduce the number of modules required, relative to federated sensors, from about 186 to 105, the number of module types from 40 to 22 and the number of RF apertures from 40 to 13.

The next stage in the Pave Pace programme is to develop an integrated RF sensor system. In this, the common pools of resources will be capable of performing the different functions that are needed in the RF area.

Common frequency converters, which convert the analogue signal from the antennas via an RF interconnection network to digital common IF signal, process the signal first. The objective of the frequency converter is to filter, amplify and convert the analogue signal to a common IF signal to be sent over an IF network to the pool of receiver modules.

A small number of types of receiver modules would process the signal and convert it to its digital I and Q data form. Critical technologies include programmable digital filters and wideband analogue to digital converters.

For some signals, it is still anticipated that there will be a need for digital signal preprocessing before they are sent to the common avionics system data and signal processors. However, because of the speed and throughput performance of current digital processing technology, the requirement for a dedicated preprocessing function might become obsolete.

During the Integrated Sensor System programme, the architectural features and concepts of this type of common modular design for RF resources will be demonstrated.

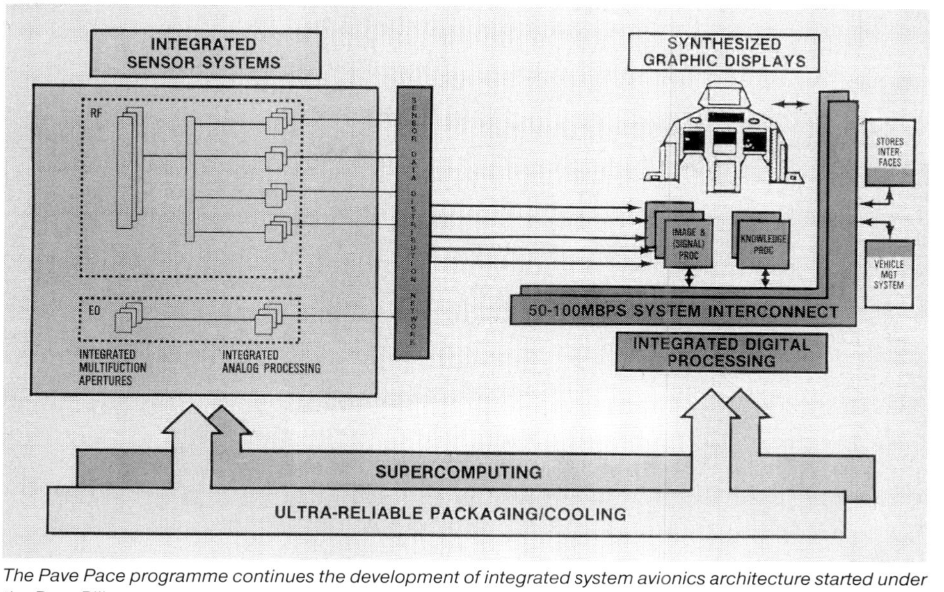

The Pave Pace programme continues the development of integrated system avionics architecture started under the Pave Pillar programme

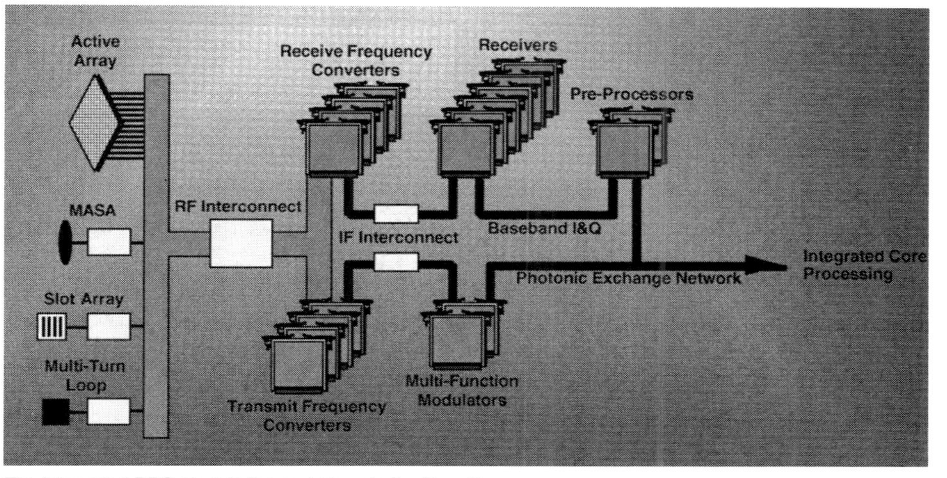

The Integrated RF System is the next stage in the Pave Pace programme

Operational status

Three functional design contracts for the Pave Pace system have been completed by Boeing and Lockheed Martin. A flow-through cooling module and rack have been built and successfully tested by AT & T. RF packaging, using LTCC and MMIC circuits, has been demonstrated by Northrop Gumman; a 2 Gbits/s optical switch and a SEM-E co-fired digital processor operating at 800 to 1,000 Mflops were started in 1992. Further system-level design work will result in both hardware and software specifications which will finalise the Pave Pace configuration. In 1994, a programme to build and integrate the Integrated Sensor System began. Real-time laboratory-based demonstrations were to be conducted during 1998-99. Present project status is not known.

The two teams currently designing the integrated RF system are Boeing/Raytheon Electronic Systems/TRW and Lockheed Martin/Boeing/Harris/Northrop Grumman.

Contractor

US Air Force Materiel Command.

VERIFIED

Modular Health Usage Monitoring System (M-HUMS™)

The Wainright Technologies Limited Liability Company (LLC) M-HUMS has been designed for the Eurocopter Dauphin 365N, -N1, -N2 helicopter. It comprises: a basic module for usage monitoring that is JAR-OPS 3 compliant; module 1 for Rotor Track & Balance (RT&B); module 2 for health monitoring; and module 3 for engine vibration monitoring.

The basic module provides for: basic aircraft parameters recording (NG, NF, T4, TRQ, NR); automatic Power Assurance Check (PAC); limitation exceedance recording; system integrity checking; electronic flight and log reports.

Module 1 provides: automatic on-board RT&B; full rotor adjustments in one flight; optionally a temporarily mounted tracker.

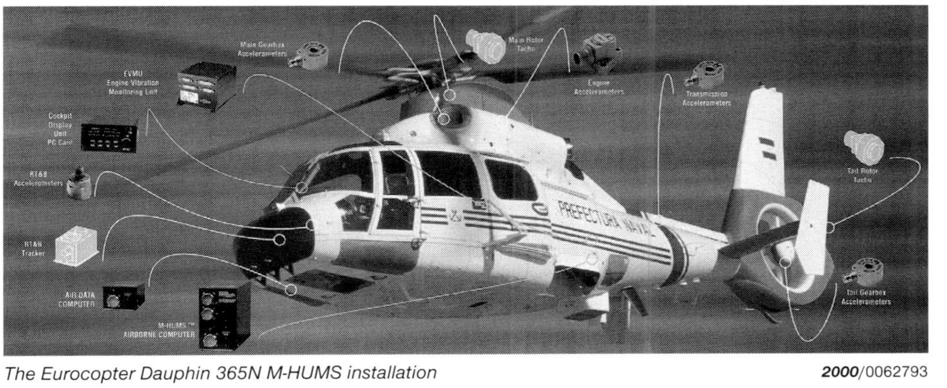

The Eurocopter Dauphin 365N M-HUMS installation **2000**/0062793

Module 2 provides: transmission and gearbox vibration monitoring; signal averaging processing techniques; Eurocopter-approved 'threshold' set-up data.

Module 3 provides: engine vibration through engine run-up; sampling of recorded data and comparison with specified threshold data; visual and aural warning for the pilot before take-off if an exceedance has occurred.

A portable ground station is provided to abstract data from the helicopter system.

Specifications

Weight: 15 kg
Certification: DGAC and CAA approved

Operational status

Standard fit for Eurocopter Dauphin 365N, -N1, and -N2 helicopters.

Contractor

Wainwright Technologies, LLC.

VERIFIED

MILITARY AVIONICS (INCLUDING PARAMILITARY)

Military/CNS, FMS, data and threat management
Military targetting, display and stores management

MILITARY CNS, FMS, DATA AND THREAT MANAGEMENT

AUSTRALIA

Marine surveillance system

The ASTA low-level airborne marine surveillance system includes a 360° Litton Canada AN/APS-504V(5) search radar, a FLIR Systems 2000G infra-red detection system with recorder and a comprehensive avionics suite with DME, Omega/VLF and an optional inertial navigation system and a two-axis autopilot. These systems are integrated and operate together to detect, track and identify targets and geographically locate them with date, time, latitude and longitude co-ordinates. The system, fitted to a suitable platform, provides a surveillance package that can carry out detection and identification, often without the crew of the target being aware of being under surveillance.

The Litton AN/APS-504(V)5 features digital display subsystems, a track-while-scan mode capable of tracking up to 20 targets simultaneously, coherent pulse compression, frequency agility and videotape recording. Detection of a 2 m^2 target in Sea State 3 conditions at ranges up to 65 km is claimed.

Signals provided to the FLIR Systems 2000G by the radar and the long-range navigation system give it the capability to identify marine targets on a 24 hour basis and record these targets on video.

Operational status

The ASTA marine surveillance system was developed for the Searchmaster N22S Series 2. This aircraft is in service with the US Customs Service in its drug interdiction programme.

The cockpit of a Searchmaster aircraft fitted with the ASTA marine surveillance system

Contractor

ASTA Components.

VERIFIED

ALR-2002 Radar Warning Receiver (RWR)

BAE Systems Australia ALR-2002 is claimed to be the first Australian Radar Warning Receiver (RWR) designed to meet the operational requirements of the Royal Australian Air Force (RAAF) F/RF-111C/6 strike, F/A-18 tactical fighter, S-70A Black Hawk and air transport aircraft.

BAE Systems Australia was initially engaged to undertake full-scale engineering development of the ALR-2002 RWR. The engineering development programme includes flight testing.

The modular design of the ALR-2002 has enabled development of a family of RWRs with common use of hardware and software models.

The system utilises a dual receiver architecture with parallel processing to maximise probability of intercept and situation awareness.

Key features of the ALR-2002 system include: near 100 per cent probability of intercept against specified emitters; full situation awareness to aircrew during high-density emitter environments; high angle of arrival accuracy over a wide RF bandwidth; the ability to detect a variety of continuous wave and pulsed emitters; software developed in Ada.

The ALR-2002 RWR has been designed to operate in conjunction with other electronic warfare systems fitted to the host platform. The system can co-ordinate the responses to specific threats from a variety of sensors in addition to its own receivers. It has also been designed to act as the bus controller in a fully integrated EW suite.

The system software is a key element of the design and will be 100 per cent developed by British Aerospace Australia. The signal processing algorithms are being produced using knowledge and experience gained through the ALR-2002 Concept Demonstrator programme which was successfully demonstrated to the Royal Australian Air Force in 1993.

ALR-2002 is an important component of the Royal Australian Air Force's Project Echidna, which will provide an integrated EW upgrade to a range of Royal Australian Air Force, Army and Navy fixed- and rotary-wing aircraft, including F-111C, C-130J, S-70A-9, CH-47D and Sea King Mk 50A. As well as ALR-2002, Project Echidna aircraft will be equipped with ECM jammers, missile approach warners and chaff/flare dispensers. Later addition of towed radar decoys and a laser warning system is also possible.

Contractor

BAE Systems Australia Ltd.

VERIFIED

CANADA

Tri-mode Synthetic Aperture Radar (TriSAR)

The TriSAR produced by Array Systems Computing is a real-time, high-resolution, airborne image processing system. It produces detailed radar cross-section images in three modes: strip map; Range Doppler Profiling/Inverse SAR (RDP/ISAR); and spotlight.

The ability to provide detailed radar cross-section data enables TriSAR to fulfil a number of specialist maritime patrol functions, including iceberg classification; tactical ice surveillance; ocean mapping; search and rescue; harbour surveillance; battlefield surveillance; land mapping.

In the spotlight mode, the antenna is directed at the target for a predetermined length of time, data is collected and processed to produce a single, static, high-resolution image of the target. The spotlight mode is able to image both stationary and moving targets. For moving targets, an adaptive sub-aperture focusing algorithm automatically constructs a finely focused image of the moving target. Aircraft motion is fully compensated. The radar illumination time for a target is substantially reduced. Spotlight SAR is the operator's preferred mode for imaging a potentially threatening maritime target. The spotlight technique is also capable of producing fine resolution images of non-moving targets

In the strip map mode, TriSAR provides an endless strip of imagery parallel to the aircraft's flight; the strip map mode delivers high-resolution imagery in real time.

The RDP/ISAR mode is similar to the traditional ISAR, in that it produces a continuous series of snapshots of moving targets, but it can also continue taking snapshots, while analysis of earlier snapshots is in progress.

Array Systems Computing provides data recording and ground processing facilities to support operational use of the TriSAR system.

Contractor

Array Systems Computing.

VERIFIED

AN/APN-208 and AN/APN-221 Doppler navigation systems

The AN/APN-221 Doppler set is supplied for night/adverse weather search and rescue helicopters. This system was derived directly from the AN/APN-208(V) helicopter Doppler navigation system developed to meet the requirements of ASW and search and rescue helicopters, where versatility in operation and interface options are essential. A recent addition has been the integration of GPS inputs into the navigation solution to provide enhanced performance.

A specific application of the APN-221 Doppler navigation set, which was developed under the sponsorship of the US Air Force Systems Command, Aeronautical Systems Division, is the Pave Low III/Sikorsky HH-53 medium-lift helicopter. The operational requirements of this aircraft called for an extension of the AN/APN-208(V)'s capability to include guidance in poor weather or darkness through the use of advanced computer techniques and cockpit displays. Improvements in the AN/APN-221 have since been incorporated into the AN/APN-208.

The four-unit APN-221 comprises a four-beam lightweight antenna, ¾ ATR short signal data converter with 16-bit microcomputer, 6-line by 12-character control and display unit and a steering/hover indicator. The displays are also available in a form that is compatible with Generation III ANVIS night vision goggles. It provides three pilot-selectable navigation co-ordinate systems, with automatic conversion from one to another. These are latitude/longitude, worldwide alphanumeric UTM and arbitrary grid. Data can be stored for up to 75 mission waypoints, 10 targets of opportunity or 25 library waypoints (including Tacan beacon locations). There are also three pilot-selectable search patterns with automatic turning point computation and navigation/guidance outputs: creeping line, expanding square and sector. The system provides aircraft velocity outputs to the flight control system, enabling coupled hover manoeuvres and automatic approaches to the hover to be conducted.

The transmitter/receiver/antenna uses a Gunn diode RF source to generate four Janus configuration beams, and FMCW modulation gives high accuracy and immunity from carrier noise, precipitation, surface spray and reflections from nearby objects such as airframe structure and sling loads.

Specifications

Dimensions:
(antenna) 439 × 439 × 113 mm
(signal data converter) 194 × 198 × 319 mm
(control/display unit) 146 × 114 × 165 mm
(steering/hover indicator) 83 × 83 × 127 mm
Weight:
(antenna) 5.59 kg
(signal data converter) 9.45 kg
(control/display unit) 3.54 kg
(steering/hover indicator) 1.14 kg
Transmission: 4 beam Janus, time-shared (200 ms/cycle), 3 × 6.7° beamwidth from Gunn diode, 13.325 GHz with FMCW modulation optimised for flight envelope
Velocity range:
(forward) −50 to 300 kt
(lateral) 100 kt
(vertical) ±5,000 ft/min
Accuracy:
(forward) 0.3% speed along velocity vector ±0.2 kt
(lateral) 0.32% speed along velocity vector ±0.2 kt
(vertical) 0.2% speed along velocity vector ±20 ft/min
Inputs: pitch/roll attitude and heading from AHARS or INS. Various optional interfaces with map displays, Tacan, sonar, radar and air data computer systems.
Microcomputer:
(architecture) word addressed, 16-bit microprocessor

Operational status

The AN/APN-208 is in service with the armed forces of many NATO and other countries. The AN/APN-221 is in service for the US Air Force Sikorsky Pave Low III HH-53 helicopter. Variants include CMA 806A/B and CMA 708C. Over 500 systems in service.

The AN/APN-208(V) Doppler system

Contractor

BAE Systems Canada Inc.

VERIFIED

AN/ASH-503 voice message system

The AN/ASH-503 voice message system is a single-box device using advanced microprocessor and memory technology to store and reproduce high-quality digitally produced speech messages.

The AN/ASH-503 Micro VMS can store 15 three-word messages or 13 seconds of continuous speech. Each system incorporates extensive self-test facilities. High-quality speech reproduction is ensured by using a 30 kHz digitising sampling rate. Any type of voice or language can be stored and reproduced. Each message can be preceded by an alerting tone and repeated as often as necessary. The messages can be sorted in order of priority so that, in the event of coincidental inputs, the most important message is reproduced first.

Specifications

Dimensions: 64 × 65 × 67 mm
Weight: 0.45 kg
Power supply: 28 V DC, 2.5 W

Operational status

In production and in service.

Contractor

BAE Systems Canada Inc.

VERIFIED

CMA-2012 Doppler navigation sensor (AN/ASN-507)

The CMA-2012 is a newly developed, single-LRU Doppler navigation sensor designed for both rotary- and fixed-wing aircraft applications where navigation aiding, back-up navigation and hover are of primary importance. Based on Canadian Marconi's experience in the design and development of Doppler radar systems and related technologies, the design parameters of the CMA-2012 were set to achieve significant reductions in size, weight and cost compared to current systems, while enhancing performance capabilities and improving reliability.

With the CMA-2012, Canadian Marconi has achieved these improvements by implementing innovative design features, including: digital signal processing for real-time signal analysis; optimised hover-hold mode for precision hover with drift rates significantly less than 1 m/minute; tactical modes include silent, horizontal beam cutoff and EW equipment compatible intermittent track.

Now available in a commercial format suitable for para-military helicopters in Search and Rescue (SAR) operations such as Heli-Dyne Systems, Sentinel Inc and Bell-412.

Specifications

Velocities (land):
Vx (forward) −50 to 250 kt ±0.3% Vt
Vy (lateral) −100 to 100 kt ±0.3% Vt
Vz (vertical) ±5,000 ft/min ±0.3% Vt
$Vt^2 = Vx^2 + Vy^2 + Vz^2$
Altitude: 2 to 15,000 ft
Power: 28 V DC/45 W max
Cooling: convection
Weight: 5.5 kg
Environment: MIL-E-5400 Class 1A/DO-160C
Reliability: 6,700 h MTBF
Dimensions: 372.6 × 345.3 × 49.5 mm
Standard interfaces: MIL-STD-1553B, ARINC-429

Operational status

The CMA-2012 has been selected for five models of scout/anti-tank helicopters, as well as a variety of other tactical helicopters in North and South America, Europe, Asia and Africa.

A derivative variant, the CMA-2012W, is being supplied to GKN Westland and Agusta for fitment to 15 Cormorant (SAR) helicopters ordered by the Canadian Armed Forces and 17 helicopters ordered for the Italian Navy's maritime patrol helicopter programme.

Contractor

BAE Systems Canada Inc.

VERIFIED

CMA-2012 Doppler navigation sensor 0015338

AN/ASA-64 Magnetic Anomaly Detector (MAD)

The AN/ASA-64 MAD identifies and marks local distortions in the earth's magnetic field induced by the presence of submarines. The operator is alerted by visual and aural alarms, thereby reducing the level of experience needed to operate the system. As the system does not require constant monitoring, the operator can devote more time to other sensors.

CAE has completed a product improvement programme in support of the AN/ASA-64 submarine anomaly detector originally built for the US Navy P-3C Orion maritime patrol aircraft. The improved version has increased processing power. A variant for helicopters is proposed.

Specifications
Dimensions:
(control unit) 102 × 146 × 90 mm
(ID-1559 processor) 229 × 150 × 153 mm
Weight:
(control unit) 0.68 kg
(ID-1559 processor) 3 kg
Power supply: 115 V AC, 20 W
Environmental: MIL-E-5400

Operational status
In service.

Contractor
CAE Electronics Ltd.

VERIFIED

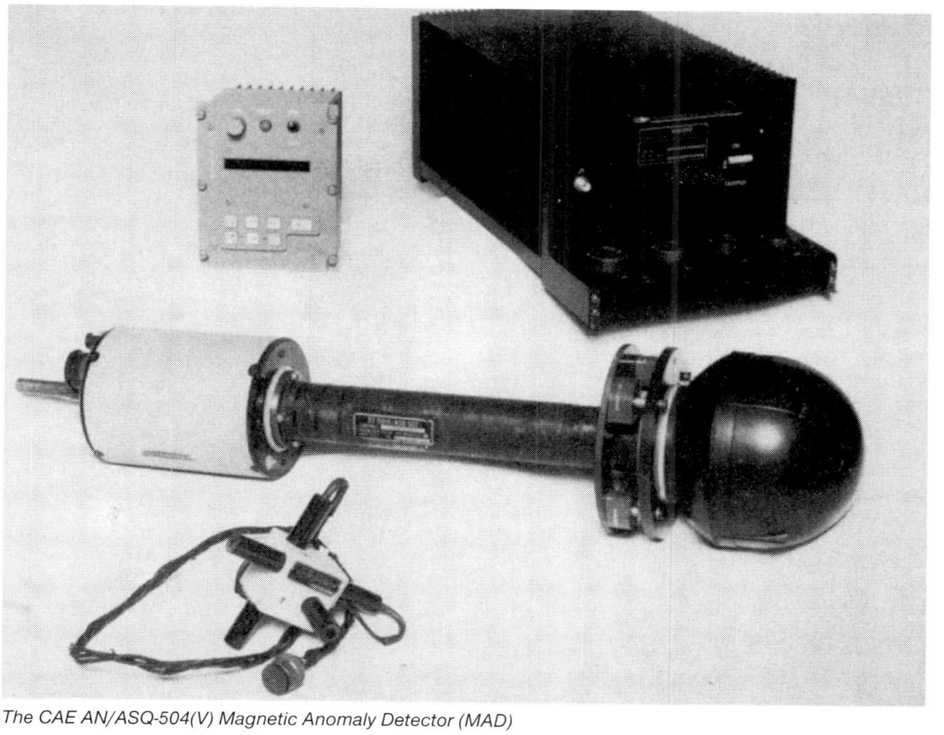

The CAE AN/ASQ-504(V) Magnetic Anomaly Detector (MAD)

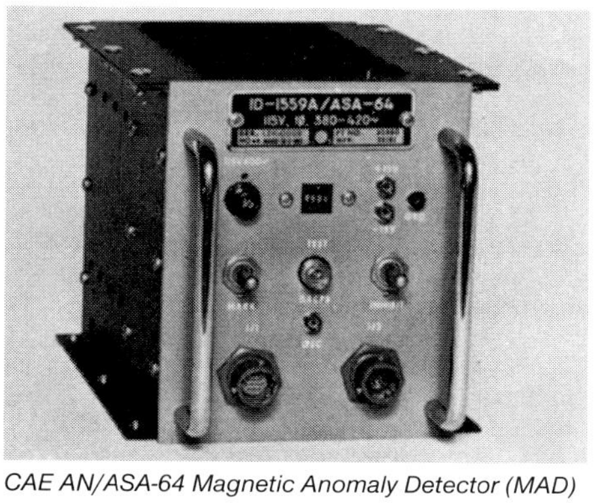

CAE AN/ASA-64 Magnetic Anomaly Detector (MAD)

AN/ASA-65(V) nine-term compensator

CAE developed the AN/ASA-65(V) semi-automatic Magnetic Anomaly Detector (MAD) compensator to improve the effectiveness of MAD on aircraft with only manual compensation for aircraft interference with the earth's magnetic field. Previously, fixed-permalloy strips and copper coils were mounted in the MAD boom to create induced and eddy-current fields equal and opposite to those caused by the aircraft. These compensators had to be custom-designed for each individual aircraft, took a long time to adjust on flight test and did not cater for the changes which take place during the aircraft's life. CAE also says that new, more sensitive MAD equipment needs greater precision than fixed compensators can provide.

The AN/ASA-65(V) compensates for permanent interference after only five minutes' flying, compared with about an hour needed for manual compensators. The system allows for manoeuvre interferences after 30 to 45 minutes, improving MAD detection range, especially when frequent manoeuvres are performed and conditions are turbulent. This compares favourably with manual compensation procedures which traditionally took 90 minutes.

The all-solid-state AN/ASA-65(V) is compatible with all current MAD systems. Internal patch connectors are used to adjust the system for the aircraft concerned.

Specifications
Dimensions:
(control indicator) 229 × 146 × 165 mm
(electronic control amplifier) 197 × 149 × 346 mm
(magnetometer assembly) 152 mm cube
(coil assembly) 89 mm cube
Total weight: 13.4 kg
Power supply: 115 V AC, 100 W
28 V DC or AC, 10 W for panel lamps
Figure of merit: <1 gamma
Max compensation field: 50 gamma on each side of aircraft
Reliability: >1,800 h MTBF

Operational status
In service. The nine-term compensator is used by the US Navy P-3C Orion and S-3A Viking ASW aircraft.

Contractor
CAE Electronics Ltd.

VERIFIED

AN/ASQ-504(V) Advanced Integrated MAD System

The AN/ASQ-504(V) Advanced Integrated MAD System (AIMS) is an inboard system for helicopters, fixed-wing aircraft and lighter-than-air platforms. This fully automatic system improves detection efficiency while reducing significantly the operator's workload. For helicopter installation, the detecting head is mounted inboard the aircraft, thus providing 'on-top' contact when over a target by eliminating the time delay inherent in a towed detecting head system.

The AN/ASQ-504(V) system combines sensitivity and accuracy with ease of operation. Its advanced signal processing capabilities eliminate aircraft-generated interference, reduce geological and solar interference and provide automatic contact alert, both visually and audibly. Detection data, via the control indicator, (an avionics bus interface or RS-343 video output), allows the operator to determine if the aircraft is within target acquisition range.

AIMS eliminates the hazard associated with towed systems in a helicopter application. It also allows surveillance and manoeuvrability at higher speed, thereby increasing patrol range and detectability and reducing the incidence of false alarms. When used with dipping sonar, transition between the systems can be performed quickly and effectively.

AIMS comprises a very high sensitivity optically pumped magnetometer, vector magnetometer, amplifier computer and control indicator. AIMS can operate independently, or it can accept and execute commands from common control/display units via a MIL-STD-1553 digital databus, while displaying latency free magnetic traces through video output.

Specifications
Dimensions:
(control indicator) 181 × 146 × 145 mm
(amplifier computer) 193 × 262 × 552 mm
(vector magnetometer) 165 × 165 × 165 mm
(detecting head) 178 × 801 mm
Weight: 30 kg
Power supply: 108/118 V AC, 380/420 Hz, single-phase, 200 VA
Sensitivity: 0.01 gamma (in flight)
Feature recognition: automatic target detection; visual and audible operator alert
Environmental: MIL-E-5400

Operational status
In service. In July 1987, CAE announced a C$38 million contract to supply 242 ASQ-504(V) AIMS systems to equip Sea King helicopters of the UK Royal Navy and Nimrod MR. Mk 2 aircraft of the UK Royal Air Force. Deliveries started in early 1989 and continued until late 1991. The system has also been ordered by several other countries for various helicopter and fixed-wing applications.

Contractor
CAE Electronics Ltd.

VERIFIED

Polaris

CAL Corporation is developing a polarimetric version of the Synthetic Aperture Fully Focused Imaging Radar Equipment (SAFFIRE) system called Polaris. The Polaris system includes a microstrip dual-polarisation antenna, with the associated waveguide switches to select horizontal or vertical transmit polarisation. This antenna is designed to provide a minimum of 20 dB isolation between orthogonal polarisation channels. An additional receiver provides the capability of simultaneously receiving two orthogonal polarisation returns. The four returns, HH, HV, VH and VV, are used to synthesise the effect of transmitting in any polarisation and receiving in any polarisation state. This polarisation preprocessing enhances the detectability of targets such as ships at sea, manmade objects in foliage, different types of terrain and vegetation and oil slicks. The parameters for synthesis are operator chosen, using a simple menu or graphic interface.

Polaris options include a high-resolution colour video display and a high-density tape recorder.

Operational status
In development.

Contractor
CAL Corporation.

UPDATED

SAFFIRE

The Synthetic Aperture Fully Focused Imaging Radar Equipment (SAFFIRE) produces map-like displays with range and azimuth resolutions of 37.5 m, in swaths of 25, 50 or 100 km and at scales of 1:1,000,000 or 1:125,000, overlaid with a latitude and longitude grid and alphanumeric annotation. A selection of single- or dual-sided operation, horizontal or vertical beam polarisation, a variable number of looks and STC (Swept Time Constant) are available.

The system has two fixed-mounted dual-polarised microstrip printed circuit antennas positioned along the underside of the aircraft fuselage, each 40 cm high and either 5 m, 2.5 m or 1.23 m in length. A high peak power, coherent-on-reception transceiver provides coherent signals to the digital processor which performs multilook registration, yaw compensation, clutter lock and autofocus processing. The radar data is recorded with a choice of dry processed paper, film or 8 mm format high-density digital tape recorder and can be sent to a video monitor unit or real-time downlink. SAFFIRE interfaces with the host aircraft inertial navigation system, GPS receiver or radar altimeter and conditioned electrical power supply.

The transmitter produces a peak power of 200 kW, a pulsewidth of 230 ns, a frequency of 9,250 MHz and a PRF between 550 and 1,100 Hz slaved to the aircraft groundspeed. The system sensitivity provides detection of a 2 m² target at the maximum range. At maximum range the synthetic aperture length, consisting of 64 elements, presumed over three to six radiated pulses, is 37.5 m, the same as the aperture spacing. The SAFFIRE integrated sidelobe ratio is −15 dB.

SAFFIRE operates at altitudes between 5,000 and 20,000 ft, at a velocity of between 150 and 300 kt and with a squint angle between ±15°.

Specifications
Dimensions:
(transmitter) 444 × 482 × 610 mm
(receiver) 266 × 482 × 610 mm
(processor) 266 × 482 × 610 mm
(display and control) 266 × 482 × 610 mm
(recorder) 310 × 482 × 508 mm
(antenna) 400 × 5,000, 2,500 or 1,250 mm
Weight:
(transmitter) 60 kg
(receiver) 15 kg
(processor) 45 kg
(display and control) 15 kg
(recorder) 57 kg
(antenna each) 20, 10 or 5 kg
Power supply:
(transmitter plus receiver) 115 V AC, 400 Hz, 3 phase, 1,200 W and 28 V DC, 6 A
(processor) 115 V AC, 400 Hz, single phase, 500 W
(display and control) 28 V DC, 2 A
(recorder) 115 V AC, 400 Hz, single phase, 225 W

Operational status
In production. During late 2000, it was reported that the United States had proposed US$100 million in aid to the Columbian National Police to purchase equipment including a Sikorsky UH-60 Blackhawk helicopter equipped with SAFFIRE.

Contractor
CAL Corporation.

UPDATED

The CAL Corporation SLAR 100 is fitted in the Boeing Canada DHC-7 Dash 7 for use over the Northern Territory in all weather conditions

SLAR 100 Side-Looking Airborne Radar

The first SLAR 100 system was installed in a modified de Havilland Dash 7 operated by the Canadian Atmospheric Environment Service, which uses the aircraft to monitor ice floes and other reconnaissance duties. The equipment entered service at the end of 1985 and has been joined by a second set, fitted to a Lockheed L188 Electra.

The basic SLAR consists of a control unit, hard-copy film recorder, transmitter/receiver, central processor unit and 5.18 m antennas. The basic SLAR can be upgraded to a synthetic aperture fixed focus radar by integrating an options processor containing digital downlink interface and/or tape drive interface, Doppler beam-sharpening processor, moving target indicator, 2.44 m antennas and constant false alarm rate.

The SLAR 100 has a maximum range of 100 km on either side of the aircraft, producing map-like displays in swaths of 25, 50 or 100 km, at scales of 1:1,000,000, 1:500,000, 1:250,000 or 1:125,000 overlaid with a latitude and longitude grid.

The system has two antennas positioned along the underside of the fuselage for the Dash 7 application, each 5.28 m long and 40 cm high; equally the antennas could be mounted in an external pod. A magnetron transmitter is used. The radar imagery is combined in a central processor with aircraft attitude and navigation data before being recorded on a roll of thermally developed black and white film. The radar data can also be displayed in the aircraft or datalinked to a ground station.

The SLAR 100 radar in the Dash 7 would normally operate at between 5,000 and 10,000 ft. Aircraft roll angle must be maintained within ±4° and yaw to within ±15°.

The transmitter operates with a peak power of 200 kW, a PRF of 800 Hz, a pulse-width of 0.23 μs and at a frequency of 9,250 MHz. In the SLAR 100 the along-track range resolution is proportional to target range, being 7.8 m/km. Across track the range resolution is constant at 37.5 m.

Specifications
Dimensions:
(transmitter) 444 × 482 × 584 mm
(central processor) 265 × 482 × 559 mm
(control unit) 265 × 482 × 406 mm
(antenna) 404 × 5,285 mm
Weight:
(transmitter) 60 kg
(central processor) 29 kg
(control unit) 22 kg
(antenna (each) 36.5 kg
(total system) 267 kg

Operational status
In service in Boeing Canada DHC-7 Dash 7 aircraft operated by the Canadian government.

Contractor
CAL Corporation.

VERIFIED

SLAR 300 Side-Looking Airborne Radar

The SLAR 300 is a real aperture side-looking non-coherent airborne imaging radar with a 100 km range on either side of the aircraft which produces a 200 km swath. The radar's high sensitivity is due to the use of a 250 kW peak power magnetron transmitter operating at I/J-band, although a G-band version can be provided. The SLAR 300 operates with a vertical or horizontal polarisation, which is selectable if a dual-polarisation antenna is used. The system works with either fixed or gimballed antennas mounted on both sides of the aircraft.

The CAL Corporation lightweight low-cost high-efficiency modular dual-polarised microstrip antenna can be used with this system, where size and weight are primary considerations. Output devices include a film recorder, downlink, digital tape recorder and video display.

SLAR 300 is the basic model of an upgradable family of radars. Options such as synthetic aperture processing, polarimetric processing and moving target indicator processing may be added after delivery to upgrade this system.

Specifications
Range: 100 km each side
Range resolution: 37.5 m
Azimuth resolution: 7.5 m/km of range
Swath width: 25, 50 or 100 km
Swath offset: 0, 25, 50 or 75 km
Aircraft groundspeed: 150-330 kt
Altitude: 5,000-20,000 ft
Max squint angle: ±15°

Contractor
CAL Corporation.

VERIFIED

AN/UYS-503 ASW acoustic processor

The AN/UYS-503 is a small, lightweight acoustic processing system designed for use in a variety of airborne, surface and subsurface surveillance platforms. It employs a processor architecture that exploits the advantages of high-density digital technology and, consequently, has remained ahead of the ever changing threat to meet the detection, localisation and attack challenges posed by the latest generation of nuclear submarines and diesel submarines operating in shallow waters. It functions as a complete system by providing all the required input signal conditioning, signal processing and analysis, post-detection processing and control and display processing.

A typical maritime patrol aircraft configuration would feature concurrent processing of 32 or 64 sonobuoys, while a helicopter configuration could include processing for eight or 16 sonobuoys and a low-frequency dipping sonar. Unique algorithms provide acoustic data fusion functions that combine available data to compute fixes and automatically track targets of interest. Other features of the AN/UYS-503 include proprietary algorithms that provide a consistent and reliable detection and localisation capability for broadband swathes and emissions and a new colour capability that is an intrinsic part of the signal processing and greatly reduces operator workload while enhancing detection and tracking performance.

Sensors processed include, analogue or digital: OMNI, DIFAR/VLAD, VLAD, DICASS, CAMBS, CODAR, BARRA, dipping sonar bathythermal, ambient noise.

Specifications

Dimensions: 262 × 396 × 246 mm
Weight: 27 kg (16 DIFAR system)
Frequency: full-band DIFAR
Input channels: any standard sonobuoy receiver
Control input: MIL-STD-1553B, RS-232C or RS-422
Tactical data output: MIL-STD-1553B, RS-232C, RS-422 or other as specified
Video output: RS-343 composite video colour or monochrome

Operational status

Currently in service with ASW aircraft of the Australian, Canadian, Swedish, UK and US defence forces. Also in service with the Japanese Maritime Self-Defence Force.

Contractor

Computing Devices Canada Ltd.

VERIFIED

ASW-503 mission data management system

The ASW-503 is a flexible mission data handling system designed to support maritime operations by airborne, surface and subsurface surveillance. It provides the necessary tactical processing, database management, control and display functions to integrate and manipulate a suite of sensor, navigation, communication and stores/armament subsystems for the effective conduct of operations.

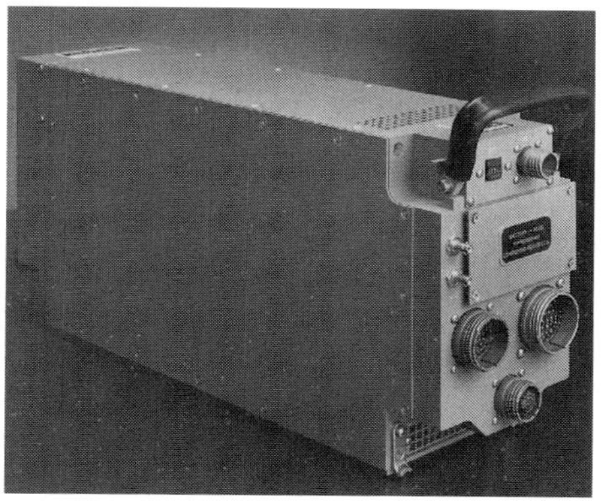

The Mission System Processor (MSP) for the ASW-503 mission data management system

In the ASW-503, individual subsystem interfaces are accommodated without modification to the subsystem. All integration activities are contained within the ASW-503, allowing the procurement of off-the-shelf systems without incurring high integration costs.

The heart of the ASW-503 is the Mission System Processor (MSP) and its associated software developed in Ada. It is a ½ ATR multiprocessor unit which includes all the electronics needed to support complete mission system functionality. Each MSP is capable of driving two independent workstations or a single workstation comprising an Integrated Control Panel (ICP) and two high-resolution colour displays. All subsystems are controlled through the programmable ICP and information is displayed on a high-resolution colour VDU with 1,280 × 1,024 pixels. Tactical data relating to the current mission is held in a central database and is used to develop a map representation referred to as the Tactical Situation Display (TSD).

In a typical MPA configuration, three MSPs are used to drive and control six displays and control panels. This results in six universal workstations, with each station being able to perform any operational role selected by the operator. In this system, one MSP is designated the master and acts as bus controller and maintains the central tactical database. Communication between the MSPs and ICPs is achieved through a high-speed serial channel.

Specifications

Dimensions:
(MSP) 132 × 262 × 546 mm
(ICP) 615 × 101 × 316 mm
Weight:
(MSP) 15.4 kg
(ICP) 12.4 kg

Contractor

Computing Devices Canada Ltd.

VERIFIED

AN/APS-140(V) radar

AN/APS-140(V) is the US designation for a configuration of the AN/APS-504(V)5 radar. This configuration includes back-to-back radar and ESM parabolic antennas, a three-axis pedestal and a MIL-STD-1553B control interface (see below for a general description of the APS-504(V)5 radar).

Operational status

In production.

Contractor

Litton Systems Canada.

VERIFIED

AN/APS-503 radar

The AN/APS-503 is a lightweight radar, with operational parameters suitable for maritime helicopters. It consists of five units: a transmitter/receiver, pedestal with parabolic antenna, PPI-type azimuth range indicator, radar control unit and range bearing unit.

Specifications

Transmitter frequency: 9.2-9.4 GHz, fixed
Power output: 50 kW peak
Pulsewidth: 0.5 µs
PRF: 400 Hz
Receiver noise: 8 dB
Antenna: 61 × 46 cm parabolic
Gain: 30 dB
Polarisation: horizontal
Stabilisation: automatic compensation for pitch and roll to ±20°
Scan rate: 30 rpm
Antenna pedestal: 2-axis
Display types: 180 mm PPI with P-7 phosphor
Range scales: 3-250 n miles
Markers: range markers; heading strobe; azimuth cursor and range dot, with digital readouts
Weight: 45 kg

Operational status

No longer in production, but still in service with Canadian Armed Forces' Sea King helicopters. There have been 50 AN/APS-503 radars produced.

Contractor

Litton Systems Canada.

VERIFIED

AN/APS-504 radar

The AN/APS-504 airborne search radar was developed for the Canadian Armed Forces' Tracker aircraft. It is a development of the AN/APS-503 radar and is designed for maritime search operations. It uses a 100 kW peak power magnetron at I/J-band with two transmitter pulsewidths: 0.5 and 2.4 µs. The AN/APS-504 includes a two-axis antenna unit with parabolic antenna, transmitter/receiver unit, analogue PPI display unit and radar control unit.

Operational status

No longer in production. Fourty AN/APS-504 radars have been produced.

Contractor

Litton Systems Canada.

VERIFIED

APS-504(V) series radar

The APS-504(V) series of airborne search radars are designed primarily for maritime patrol applications. They can be installed in either fixed- or rotary-wing aircraft. In addition to coastal and offshore surveillance missions, these radars can also be used for weather avoidance, low-resolution land mapping and navigation.

The APS-504(V)2 is a commercial version of the AN/APS-504 airborne search radar, which was developed specifically for Canadian Forces' Tracker aircraft. It uses the same 100 kW (peak power) I-band magnetron and transmitter pulse widths of 0.5 and 2-4 µs. The system consists of a two-axis antenna unit with parabolic antenna, transmitter/receiver unit, analogue PPI display and radar control unit. Fifty-three APS-504(V)2 radars have been produced.

Following the success of the APS-504(V)2, the APS-504(V)3 was developed to include an improved transmitter/receiver. It also includes a digital signal processor and scan converter that produces a ground-stabilised PPI display in a high-resolution 875-line video format. Navigation and cursor data are overlayed on the non-fading radar video, which can also be recorded and played back. Several sizes of high-performance flatplate antennas are available and are mounted on a two-axis pedestal. Twenty-five APS-504(V)3 radars have been produced.

The APS-504(V)5 radar is the most advanced of the Litton APS-504 family of airborne search radars. It employs a TWT-based transmitter with widband frequency agility, high-ratio pulse compression, scan-to-scan integration and digital signal processing to enhance the detection of sea-surface targets, including targets with radar cross-sections as small as 1 m² in Sea State 3. The APS-504(V)5 can be configured to meet various installation and performance requirements. It has been installed in aircraft ranging from small twin-engined turboprops, such as the Beech 200 to larger jet aircraft such as the Boeing 737. More than sixty-five APS-504(V)5 radars have been produced.

The APS-504(V)5 is available with a choice of two-axis or three-axis pedestals. The two-axis pedestal provides the smallest swept volume. The three-axis pedestal provides pitch-and-roll stabilisation of the antenna at scan rates up to 120 rpm, but requires a larger radome in order to accommodate the larger swept volume. A broadband antenna offers the best performance for maximum range and clutter reduction, but a selection of offset-horn parabolic antennas is also available.

Litton also offers the Tactical Data Management System, which interfaces to the APS-504(V)5 and other avionics and provides advanced tactical capabilities.

Specifications

Power: per MIL-STD-704C; 115 V AC, 400 Hz 3-phase; and 28 V DC
Control interface: MIL-STD-1553B or radar control unit
External interfaces: MIL-STD-1553B, RS-422, ARINC 407, ARINC 429, discretes
Transmitter: Travelling Wave Tube Amplifier (TWTA)
Transmit frequency: 8.9-9.4 GHz, 16 frequencies, with selectable agility patterns
Transmit power: 6.6 kW (peak)
Compressed pulsewidths: 200 ns and 32 ns
Scan rates: 7.5-120 rpm
Track-while-scan: 20 targets
Display type: RS-343 875-line video
Formats: PPI/multi,: full scan, sector scan, range delay
Resolution: 800 × 800 pixels; alphanumeric overlay: 400 × 495 pixels
Range scales: 3-200 n miles
Weight: 180 kg (with three-axis pedestal); varies with configuration

Operational status

The APS-504(V)2 and APS-504(V)3 are no longer in production. The APS-504(V)5 is still in production.

Litton's inertial referenced flight inspection system

Contractor

Litton Systems Canada.

VERIFIED

Inertial Referenced Flight Inspection System (IRFIS)

The Litton Inertial Referenced Flight Inspection System (IRFIS) is a totally self-contained en route and terminal navigation aid calibration system. The IRFIS is suitable for calibration of Cat I, II and III ILS and MLS facilities, as well as VOR, Tacan, DME, NDB and other navigation aids, in accordance with ICAO Doc 8071, Volumes I and II.

Unlike current systems, the IRFIS does not require ground personnel operating tracking equipment, nor is its operation impeded by inclement weather conditions. Automatic inspection capability is provided using updates from the Litton aircraft position sensor, multiple DME reception or pilot fixes to refine the positional data supplied by the inertial navigation system.

The IRFIS performs automatic real-time calculations, following completion of each inspection run, and instantly displays the summary of results to the operator. A large plasma monitor provides an unambiguous display in a clear format. Interactive software dialogue validates all operator inputs and, through the use of programmed function keys, requires minimal keyboard entry.

The Rolm 1866 militarised computer performs reliable and consistent data processing functions, including automatic calculation of ILS and MLS parameters and analysis of VOR/Tacan error curves, bends, scallops and roughness. A Rolm 2150 interface unit handles the receiver signal conditioning and automatic frequency tuning functions.

Inertial guidance is provided by the LTN-92 ring laser gyro INS which performs all required navigation and steering computations, provides autopilot inputs for waypoint to waypoint navigation and orbital steering about any pilot-designated point.

The Aircraft Position Sensor (APS) is a self-contained position sensor used to perform fully automatic inspection of ILS, MLS and PAR facilities. Mounted on the underside of the aircraft fuselage, the APS updates at each end of the runway are used by the software resident in the Rolm computer to improve the INS position data. The aircraft's position determining accuracy is further enhanced through the use of an 18-state Kalman filter and a Bryson-Frazier smoother error estimator.

Non-fading hard-copy calibration data is presented via an RMS GR 33-1 printer/plotter which allows the transfer of screen data to paper. A Targa data storage unit records inspection data on bubble memory for long-term storage.

The IRFIS components may be packaged in a standard side-facing console or in a compact forward-facing console. With the forward-facing style, a separate equipment rack is normally mounted opposite the console, facing the operator. The rack contains the dedicated IRFIS computer, receivers and test equipment, while all controls required for flight inspection are available at the console.

Operational status

In service with Transport Canada, the Civil Aviation Administration of China, the Japanese Air Self-Defence Force, the Netherlands Department of Civil Aviation, the Royal Air Force, the South African Department of Transport, the Republic of China (Taiwan) Air Force and the Thailand Department of Aviation.

Contractor

Litton Systems Canada.

VERIFIED

Next-generation Litton flight inspection system

Litton's next-generation flight inspection system combines the latest digital technology with carefully developed software to provide high performance and reliability with excellent operational characteristics. The common core and building block design allows each system to be customised to meet the requirements of individual users. Possible system configurations range from the basic Semi-Automatic Flight Inspection System (SAFIS) to the fully Automatic Flight Inspection System (AFIS) with a range of options. The console can be easily installed and removed to allow the aircraft to be used for other roles. Next-generation flight inspection systems are designed to accommodate all current and planned flight inspection functions in a format that is flexible, versatile, user-friendly and cost-effective. The system consists of the flight inspection processing subsystem, the position reference subsystem and optional equipment.

The flight inspection processing subsystem includes the operator's console, with computer, recording equipment and radio receivers. It features a high-resolution colour display terminal with keyboard and trackerball, a data storage unit which uses removable magneto-optical diskettes, a printer, a graphic recorder and flight inspection receivers. The VMEbus computer houses dual processors and interface modules. A 68040 processor provides high-speed data processing while an 80486 processor supports the operator interface. Software in C++ runs on DOS with Windows, providing a very user-friendly environment. Navigation information is displayed on an EHSI, while an optional EADI is also available.

The position reference subsystem provides the reference against which navaid facilities are calibrated. GPS is used for checking en route facilities but calibration of landing aids requires greater precision. For ILS, MLS and PAR, the AFIS uses a combination of differential GPS, INS and over the runway position updates to achieve a positional accuracy of better than 30 cm in three axes at the runway threshold. For the SAFIS, a telemetry tracking system is used to track the aircraft during approaches.

Optional accessories are available for the flight inspection system. These include an oscilloscope which allows selected parameters to be examined and a spectrum analyser which allows RF signals to be analysed for interference. A signal generator will allow onboard calibration of receivers under control of the flight inspection computer. Although tuning of the receivers is normally performed automatically by the computer, an optional radio tuning unit provides conventional manual tuning capability from a single universal tuning head.

Specifications

Dimensions:
(basic system) 0.4 m^3
(average system) 0.56 m^3
(fully loaded system) 0.75 m^3
Weight:
(basic system) 136 kg
(average system) 295 kg
(fully loaded system) 360 kg

Litton's next-generation flight inspection system

Operational status

In production. In service with the Directorate General of Air Communications Indonesia installed in Learjet 31A aircraft. A further system has been delivered to the Korean Ministry of Transportation for installation in a Canadair Challenger CL 603-3R.

Contractor

Litton Systems Canada.

VERIFIED

Tactical Data Management System (TDMS)

Litton's Tactical Data Management System (TDMS) and APS-504(V)5 radar together form a high-performance integrated radar sensor and tactical display system. The APS-504(V)5 is an airborne I/J-band search radar, primarily used to detect and track sea surface targets in the presence of clutter. The TDMS provides a radar display with an advanced digital map, overlay and data interface to control and enhance the operation of the radar. FLIR video can also be displayed. The TDMS/radar combination provides surveillance of maritime activity within and beyond the 370 km (200 n miles) Economic Exclusion Zone (EEZ).

The TDMS is a VME-based data management system, which contains application software for mission and flight management, provision of tactical aids and maintenance of tabular data for presentation to the operator. In addition to a high-resolution colour display, the operator interface includes a keyboard, trackerball, radar rotary controls and keypad panels. The TDMS interfaces to navigation systems such as the LTN-92. Using the TDMS to integrate radar and navigation data, the operator develops tactical and navigation plots to fit the mission requirements. From simple missions to complex tactical operations, the TDMS provides an essential aid to tactical decision making.

Ground support for the TDMS is provided by the Mission Support Facility (MSF). This PC-based facility is provided so that preflight mission planning, data entry and post-flight mission analysis independently of aircraft operations. For mission planning, the MSF is used to generate and edit operational and mission-specific TDMS parameters. It also to prepares map data for the intended patrol area. Preflight data is transferred to the TDMS on a high-capacity optical disk. Data collected and stored during flight on the optical disk can be offloaded to the MSF for post-flight mission analysis. The MSF can sort and print various reports, including mission summary, flight log and target summary reports. The operator can also reconstruct and display selected mission events, such as aircraft flight path and selected targets. Third-party database programs may be used for further statistical/historical data processing.

The APS-504(V)5 radar is the most advanced of the Litton APS-504 family of airborne search radars. It employs a TWT-based transmitter with wideband frequency agility, high-ratio pulse compression, scan-to-scan integration and digital signal processing to enhance the detection of surface targets, including targets with radar cross-sections as small as 1 m^2 in Sea State 3. The APS-504(V)5 can be configured to meet various installation and performance requirements. It has been installed in aircraft ranging from small twin-engined turboprops such as the Beech 200 to larger aircraft such as the Boeing 737.

Specifications

Weight: 105 kg, varying with configuration
Power supply: 115 V AC, 400 Hz, single phase, 28 V DC
Operator interface: colour monitor (1,280 × 1,024 pixels), alphanumeric keyboard, two programmable display keypads, trackball, read/write optical disk, radar rotary control panel
Operator display: 1,248 × 946 colour pixels

Operational status

In production.

Contractor

Litton Systems Canada.

VERIFIED

Electronic warfare training system for CT-133 aircraft

Lockheed Martin Canada is to design, develop, manufacture and integrate three electronic warfare training systems installed in Canadian Forces Challenger 600 aircraft, together with ground support facilities; the modified aircraft will be designated CT-133.

The airborne suite is designed to provide training in: threat evaluation; communications and signal detection; standoff jamming and chaff corridors; escort jamming; defensive ECM.

EW systems to be fitted to the aircraft include the following:

- ESM: AN/ALR-504 (modified AN/ALR-76)
- high-power communications jammers: AN/ALQ-504 (Zeta ZS-1920)
- high-power C, D, E, F and G-band jammers: AN/ULQ-21
- high-power (coherent) I-band jammer: AN/ULQ-21(V)
- chaff dispenser: AN/ALE-503(V)2

Operational status

Programme ongoing.

Contractor

Lockheed Martin Canada.

UPDATED

SSAR Spotlight Synthetic Aperture Radar for CP-140 aircraft

Lockheed Martin Canada's Spotlight SAR is an adaptation of the AN/APS-506 multimode radar for the Canadian Forces' CP-140 Aurora aircraft. Spotlight SAR is being developed as an affordable real-time airborne reconnaissance and surveillance imaging radar, and in particular to provide exceptionally long standoff range detection and classification of ships and submarine periscopes.

Spotlight SAR is based on the enhancement of conventional radar components: receiver, transmitter antenna and related control units to provide detailed cross-section images in three modes: Spotlight, Strip Map and Range Doppler Profiling/Inverse SAR (RDP/ISAR), whilst retaining the original conventional radar modes. In the Spotlight Mode, the aircraft's lateral motion is used to form a long synthetic aperture. Resolution (<1m) is sufficient to establish fine details of naval and ground-based targets from long standoff ranges.

The Strip Mode provides an endless strip of high-resolution, real-time imagery along the aircraft's flight path, and as much as 60° off track.

Employing enhanced ISAR techniques, Range Doppler Profiling provides a continuous series of snapshots (frames) of moving targets for analysis and classification.

Spotlight SAR has potential both for military and civilian SAR missions.

Operational status

Lockheed Martin Canada, collaborating with the Canadian Department of National Defence, has defined and designed SSAR to provide the operator interface. Lockheed Martin Canada is leading a team comprising: Applied Analytics, Array Systems Computing, the IMP Group and Raytheon Canada for development of the Spotlight SAR system, for the Canadian Forces' Bombardier CP-140 Aurora aircraft.

Contractor

Lockheed Martin Canada.

VERIFIED

SHOALS Scanning Hydrographic Operational Airborne Lidar Survey)

SHOALS is a laser-based scanning bathymeter manufactured by Optech Inc. It offers hydrographers a cost-effective, rapid means of surveying shallow coastal or inland waters, with depth and position accurate to International Hydrographic Organisation 'order 1'. In addition, SHOALS is claimed to have a unique topographic mapping capability that provides continuous coverage from shallow waters onto the adjacent shore with a single flight line.

Effective over most coastal areas of critical concern to the hydrographer, the system's speed and flexibility in deployment and its convenience in sounding dangerously shallow or highly confined and irregular areas, offer many advantages in survey planning and deployment strategy.

Airborne Lidar Bathymetry (ALB) is a technique aimed primarily at augmenting bathymetric mapping capabilities in shallow coastal waters. This well-proven technique is now being used for many mapping and surveying applications. Airborne laser bathymeters employ a technique called lidar (Light Detection and ranging) to determine depth by measuring the differential time-of-flight of two optical pulses transmitted from an aircraft. One pulse is reflected from the water surface and the other reflected from the water bottom through the air-water interface.

The Optech SHOALS in a Bell 212 helicopter

Optech's SHOALS system is designed to operate from a helicopter or a small fixed-wing aircraft. It measures water depths 400 times per second, at an altitude of 200 to 300 m and can be used by day or night.

The SHOALS system consists of an airborne component that acquires the sensor data used to determine depth and position and a ground-based component that is used for mission planning and data analysis. In operation, the pilot uses a guidance monitor to fly the aircraft along predetermined flight lines. Depths, or underwater objects, are detected in real time and displayed to the system operator, along with navigation information. The operators display also includes a real-time display of previously surveyed areas, depicted as swaths of colour-coded depths.

Specifications

(for the next-generation ALB system)
Depth measurement: 0-50 m
Depth and position accuracy: IHO Order I
Sounding frequency: 1000 soundings/s
Hydrographic altitude range: 200-400 m
Topographic vertical accuracy: 0.15 m
Topographical horizontal accuracy: altitude/1000 (such as 1 m at1000 m)
Topographical frequency: simultaneous at 1000/s
Topographical altitude range: 300-1000 m
Swath width: ½ altitude (nominal)
Size: <1.5 m^3
Weight: 170 kg
Power: 28 V DC, 70 A

Operational status

Optech produced the first operational airborne laser bathymeter for the Canadian Hydrographic Service in 1985. The system, the LARSEN 500, has surveyed worldwide and is still operational in its down-sized configuration. The SHOALS system, delivered to the US Army Corps of Engineers in 1994, has undergone recent upgrades and is also operational worldwide. Two Hawk Eye systems, largely derived from SHOALS, were delivered to the Swedish Navy and Swedish Hydrographic Department in 1994 and 1995, via a prime contractor in Sweden. One Hawk Eye system is operational throughout the Baltic region, while the other has been transferred for surveying in Indonesia.

Contractor

Optech Inc.

UPDATED

CHILE

Caiquen II radar warning receiver

The Caiquen II radar warning receiver has four independent receivers, each with its own 2 to 18 GHz wideband crystal video receiver and antenna, offering coverage through a full 360° in azimuth and ±40° in elevation. PRFs between 200 Hz and 12 kHz can be processed.

There are eight LRUs: four antennas, two RF units, a computer and a control and display unit. The cockpit display unit indicates the radar types being detected: 'A' for airborne and ground-based acquisition radars, 'V' for early warning radars, 'CW' for continuous wave fire-control radars and 'LKN' for locked on fire-control radars with pulse modulation. The display is divided into 12 sectors, indicating the bearing of the received signals.

Specifications

Volume: 0.011 m^3
Weight: 8 kg
Altitude: up to 40,000 ft

Operational status

In production, and in service. The system equips the Chilean Air Force's Hunter FGA.71 aircraft and is understood to be under development for installation in the country's Mirage 50 among others.

Contractor

DTS Ltda.

VERIFIED

Caiquen III radar warning receiver

Caiquen III is a wideband radar warning receiver operating over the 2 to 18 GHz frequency range and giving full 360° coverage in azimuth and ±40° in elevation. The system is fully automatic and programmable, provides real-time analysis of the direction and type of threats and presents the data to the pilot on a bright dot matrix LED display divided into eight sectors. The threat is classified as a surveillance, acquisition or fire-control radar in lock on mode, or as a CW radar from a missile. Relative threat priorities can also be assessed.

The Caiquen III system comprises eight LRUs: four antennas, two RF units, a computer and a display and control unit. The four orthogonal spiral antennas are each connected to a wideband crystal video receiver, which ensures the detection of pulsed and CW signals. The receiver outputs are digitised and fed into a high-speed digital microcomputer, which processes the incoming signals in real time, de-interleaves individual pulse trains and establishes type, strength and direction of the emitters based on a comparison with a preprogrammed library. This information is presented to the crew on a dot matrix LED display, together with an audio warning signal. The display is capable of indicating radar type, received signal strength and octant of arrival for up to three threats simultaneously.

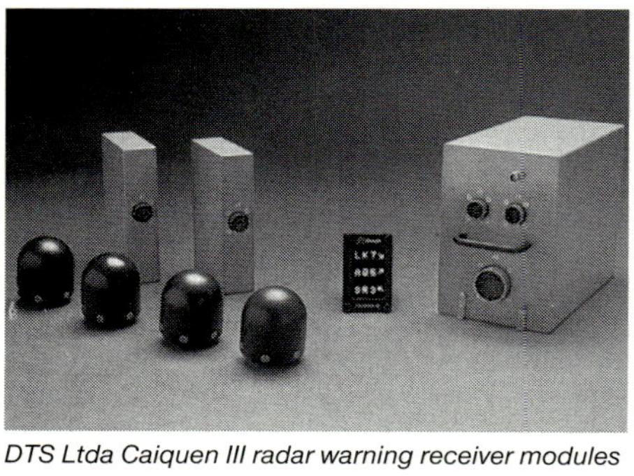

DTS Ltda Caiquen III radar warning receiver modules

Caiquen III features a compact and modular design, which allows for very low demand on aircraft space and simple retrofit installation on existing aircraft. It also features a high level of reliability and maintainability through the use of MIL-qualified components and complete BITE for online fault diagnosis. It can be interfaced with the Eclipse chaff/flare dispensing system to provide an integrated self-protection capability, as well as onboard systems for blanking of self-emitted signals.

Specifications

Weight: 14 kg
Power supply: 28 V DC, 3.5 A

Operational status

Installed on Chilean Air Force Hunter and Mirage aircraft. Development is under way for installation on the F-5 aircraft.

Contractor

DTS Ltda.

VERIFIED

DM/A-104 radar warning receiver

The DM/A-104 is a wideband radar warning receiver for helicopters and combat aircraft that provides instantaneous detection of threat radar emitters. The system consists of four orthogonal spiral antennas, each connected to wideband crystal video receiving channels that provide 360° coverage in the 2-18 GHz frequency range for the detection of most search, acquisition and fire-control radars.

A powerful digital processor and advanced de-interleaving software ensure real-time automatic radar sorting and evaluation. A cockpit display presents information on the most dangerous threats. In addition, an audio warning system is sent to the intercom system. The threat itself is classified and the pilot is informed whether his aircraft is being targeted by a surveillance, acquisition or fire-control radar in lock on mode and whether the transmission is from a continuous wave radar.

The DM/A-104 can be interfaced with the DM/A-202 chaff/flare dispensing system, to provide an automatic self-protection capability, and onboard systems for blanking of self-initiated signals. It features a compact and modular design that does not demand much aircraft space and can be retrofitted easily on any combat aircraft. It is designed for a high level of reliability and maintainability, with a complete BIT capability for online diagnosis.

Specifications

Frequency: 2-18 GHz in 4 sub-bands
C/D-band 0.7-1.3 GHz
Sensitivity: –50 dBm
Accuracy: >10° RMS

Contractor

DTS Ltda.

VERIFIED

DM/A-202 chaff and flare dispensing system

The DM/A-202 is a self-protection system for helicopters and combat aircraft that provides chaff and/or infra-red-flares to break the lock of radar and IR-guided missiles. It consists of five LRUs: the cockpit control unit and four launching units containing the chaff and flare cartridges. The control unit includes a bright display which shows the amount of chaff and flare cartridges remaining and a mode selector for selecting one of four different launching sequences.

Each launching unit comprises an easily removable magazine, allowing quick reloading of the cartridges, and the associated firing circuits. A typical configuration consists of 108 chaff cartridges and 54 flare cartridges contained in the four launching units. The system can be expanded to handle up to eight launching units.

Flexibility and ease of operation have been achieved by the use of a fast reprogrammable microprocessor that ensures adaptability of the system to changing tactical situations. Simplicity of operation has also been enhanced by the installation of control switches on the aircraft throttle and/or stick, allowing the pilot to operate the system without removing his hands from the controls. Modes of operation are manual, semi-automatic and automatic. A safety switch is incorporated in the system so that all the stores can be jettisoned in an emergency.

To avoid reducing the aircraft operational load carrying capacity, the launching units are normally attached externally to the rear fuselage or to the sides of the ventral and wing pylons. Internal or semi-recessed installation can also be adopted, depending on the aircraft configuration and available space.

The DM/A-202 also provides an interface with the DM/A-104 radar warning receiver for full aircraft self-protection.

Operational status

The DM/A-202 is operational and being offered for export.

Contractor

DTS Ltda.

VERIFIED

Eclipse chaff/flare dispensing system

Eclipse is a self-protection system for helicopters and combat aircraft that provides chaff and/or infra-red flares to break the lock of radar and infra-red-guided missiles. It consists of three types of LRUs, including the cockpit control unit, and four launching units containing the chaff and flare cartridges. The control unit features a bright display, which indicates the amount of cartridges remaining, and a mode selector on which the pilot can select one of four different launching sequences.

Each launching unit comprises an easily removable magazine, allowing quick reloading of the cartridges, as well as the associated firing circuits. The normal payload of the system consists of 64 RR-170 chaff cartridges and 34 MJU-7B flare cartridges contained in the four launching units. The system can be expanded easily to handle up to eight launching units.

Flexibility and ease of operation have been achieved by the use of a fast and reprogrammable microprocessor that guarantees adaptability of the system to changing tactical situations. Simplicity of operation has also been enhanced by the installation of

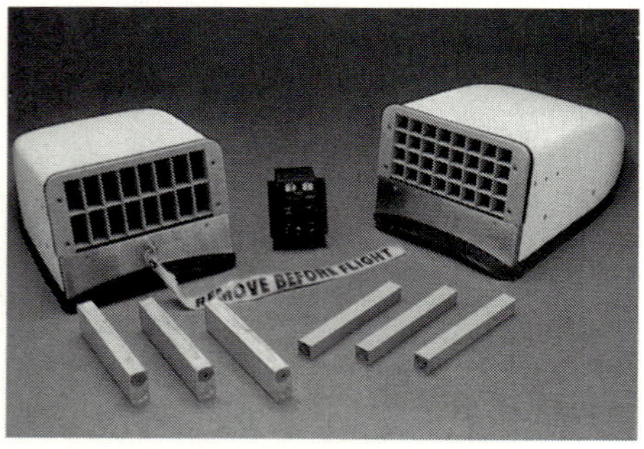

The DTS Ltda Eclipse chaff/flare dispensing system

The DTS Ltda Eclipse control unit

control switches on the aircraft throttle and stick, allowing the pilot to operate the system without taking his hands off the controls. A safety switch is incorporated in the system so that all the stores can be jettisoned in an emergency.

To avoid reducing the aircraft operational load-carrying capacity, the launching units are normally attached externally to the rear fuselage or to the sides of the ventral and wing pylons. Internal or semi-recessed installation can also be adopted, depending on the aircraft configuration. Eclipse can be interfaced with the Caiquen radar warning receiver to provide full aircraft integrated self-protection.

Operational status
Installed on Chilean Air Force Hunter aircraft. Installations for the F-5 and Mirage are under development.

Contractor
DTS Ltda.

VERIFIED

EWPS-100 EW system

The EWPS-100 has been developed to protect helicopters and combat aircraft from present and future radar-controlled weapon systems. It operates over the 0.7 to 18 GHz frequency band and provides a low-cost and effective answer to operational requirements in the air-to-air and air-to-ground roles. An integrated EW system architecture has been developed using modular techniques, proven software and hardware building blocks. The EWPS-100 integrates the DM/A-104 radar warning receiver, DM/A-202 chaff/flare dispenser and DM/A-401 self-protection jammer.

The main features of the EWPS-100 are a high probability of threat interception, high sensitivity, initiation of countermeasures, power management in time and frequency and control of chaff/flare cartridge dispensing.

Specifications
Frequency: 0.7-18 GHz
Detection: pulse, pulse Doppler, CW
Sensitivity: –50 dBm
Maximum pulse density: 500,000 pps
Display: 16 threats simultaneously

Operational status
The EWPS-100 is operational and being offered for export.

Contractor
DTS Ltda.

VERIFIED

Itata ELINT system

The Itata ELINT system is a high-sensitivity electronic intelligence-gathering system that can detect, locate and measure the parameters of emissions from search, acquisition and fire-control radars. Itata consists of a fully programmable superheterodyne receiver, a digital pulse analyser and a high-gain wideband rotating parabolic antenna which provides 360° coverage and bearing information to an accuracy of within a few degrees. Although intended primarily for light transport aircraft, Itata can also be installed in ships or ground vehicles.

The receiver operates over a frequency range of 3 MHz to 18 GHz in six bands. It can be used either in a wide open mode over the complete frequency range or in a selective mode over a single band. After detection of a transmission, the receiver locks on automatically and measures the frequency and other parameters. Digitised data of each intercepted signal can be recorded automatically for subsequent analysis.

Specifications
Frequency: 3 MHz-18 GHz in 6 bands
Azimuth coverage: 360°
Azimuth beamwidth: (E/F-band) 8°, (J-band) 1.8°
Polarisation: circular

Operational status
Operational with the Chilean Air Force on Beech 99A aircraft.

Contractor
DTS Ltda.

VERIFIED

CHINA, PEOPLE'S REPUBLIC

204 Airborne interceptor radar

It is the first I/J-band monopulse airborne interceptor radar, successfully developed by Leihua for all-weather F-8 fighters. It has search, acquisition and tracking capabilities and can be used for attack on flying targets with gun, rockets and missiles in association with an onboard fire-control computer and optical gunsight.

It provides good anti-interference, high reliability and easy maintenance.

Specifications
Detection range: 29 km
Scan range:
(azimuth) ±38°
(elevation) –12 to +24°
Volume: 0.145 m^3
Weight: 145 kg
Frequency: I/J-band

Operational status
The 204 radar was confirmed through final evaluation in 1984 and put into mass production.

Contractor
China Leihua Electronic Technology Research Institute.

VERIFIED

698 side-looking radar

The I/J-band 698 side-looking radar is designed specifically for detection of periscopes and ships. The radar features coherent moving target detection, slotted feed double parabolic reflector antenna, parametric amplifier, high-stability local oscillator, coherent receiver, IF log amplifier and digital filter. Detection ranges are quoted as 60 km against ships and 17 km against a periscope.

Specifications
Detection range: periscope 17 km; ship 60 km
Display ranges:
(transversal) 60 km (normal)
(longitudinal) 30 km (searching)
High resolution:
(transversal) 300 m (searching)
(longitudinal) 50 m
Operational altitude: 50-500 m (searching)
Volume: 0.8 m^3
Weight: 230 kg

Operational status
In service.

Contractor
China Leihua Electronic Technology Research Institute.

VERIFIED

China Leihua Electronic Technology Research Institute 204 airborne interceptor radar

CWI illuminator

The continuous wave illuminator, now used for F-8 fighters, is designed to perform semi-active radar guidance of air-to-air medium-range interception missiles when operated in combination with airborne radar.

Operational status
It is in production with an annual output of 30 to 50 sets. There are a series of CWIs, including CWI-A, CWI-B, CWI-C and CWI-D, available for different types of aircraft.

Specifications
Frequency: I/J-band
Radiation power: 200 W
FM noise: LFM<–99 dB/Hz/10 KHz
AM noise: LAM<115 dB/KHz/10 KHz
Weight: 40 kg
Volume: 0.035 m^3

Contractor
China Leihua Electronic Technology Research Institute.

VERIFIED

JL-7 fire-control radar

The multifunction JL-7 fire-control radar for the F-7C aircraft is designed to search, detect and track airborne targets and carry out air-to-ground ranging. It can be used for attack on air or ground targets using missiles, guns or bombs in association with a gunsight or HUD.

Specifications
Volume: 0.12 m^3
Weight: 115 kg
Frequency: J-band
Range:
(detection) 27.8 km
(track) 18.5 km
Coverage:
(azimuth) ±35°
(elevation) −13 to +17°
Altitude: 2,300-65,000 ft

Operational status
In production and in service with Chinese Air Force F-7C aircraft.

Contractor
China Leihua Electronic Technology Research Institute.

VERIFIED

China Leihua Electronic Technology Research Institute JL-7 fire-control radar 0001964

China Leihua Electronic Technology Research Institute JL-7 fire-control radar installed on the F-7C aircraft 0001965

JL-10A airborne interceptor radar

The JL-10A airborne radar is designed for the fighter requirements of medium-range omnidirectional attack, close-range dogfight, look-up and look-down and surface moving target attack over land and sea. It is the first airborne full-wave pulse Doppler radar, with high, medium and low PRF, produced in China. The JL-10A is a highly digitised system which uses a slotted array antenna, signal exciter and sophisticated signal processor.

Specifications
Range: 59.3 km look-up, 53.7 km look-down (5 m^2 target)
Tracking range: 29.6 km
Range resolution: 150 m
Range accuracy: 15 m
Reliability: >70 h MTBF

Operational status
Development and flight trials have been completed successfully.

Contractor
China Leihua Electronic Technology Research Institute.

VERIFIED

China Leihua Electronic Technology Research Institute JL-10A airborne interceptor radar 0001191

Shenying multimode airborne radar

The Shenying multimode airborne radar is a coherent I/J-band pulse Doppler system which will provide fighters with capabilities of medium-range dogfight, look-up and look-down weapon delivery and ground or sea moving target attack.

The antenna is a flat plate slotted array which features low sidelobes and full azimuth and elevation monopulse operation. A gridded TWT transmitter is employed, which operates at low, medium and high PRFs. The receiver, with low-noise front end, consists of two channels which provide the monopulse sum and multiplexed difference channels (azimuth and elevation). Shenying incorporates digital signal/data processors to handle radar mode control, to conduct the built-in test and to perform radar signal and data processing.

Operational status
Under development.

Specifications
Detection range:
(search) 80 km (max); 54 km (look-down)
(tracking) 40 km (look-up); 32 km (look-down)
Search angle:
(azimuth) ±60°
(elevation) ±60°
Frequency: I/J-band

Contractor
China Leihua Electronic Technology Research Institute.

VERIFIED

PL-7 fire-control radar

The PL-7 is a lightweight, monopulse, J-band fire-control radar developed for the Chinese Air Force and designed for use in fighter aircraft. It has air-to-air and air-to-ground modes and can operate in conjunction with a fire-control computer, IFF, head-up display or aiming sight.

The radar consists of 18 LRUs located in the nose and cockpit. It has five modes in air-to-air: search (from 400 m to 30 km and through ±45°); manual acquisition; boresight; attack/track (up to 15 km, through ±45° and down to 2,300 ft altitude) and transponder. The three air-to-ground modes are: slant range, attack and acquisition.

A horizontally polarised antenna is used, with a 3.4° beamwidth in azimuth, 5.6° in elevation and a 30 dB gain.

Specifications
Volume: 0.23 m^3
Weight: 115 kg
Peak power: 75 kW
Reliability: 50 h MTBF

Contractor
China National AeroTechnology Import & Export Corporation.

VERIFIED

CT-3 airborne radio

The CT-3 is an airborne AM VHF radio set covering the 100 to 150 MHz frequency range. It has 601 channels, 20 of which may be preset. Channel spacing is 83.333 kHz. When a VHF antenna with a travelling wave coefficient of more than 0.4 is used and the radio is operated at 14 W, communication range is over 120 km at a flight altitude of 3,300 ft, 230 km at 16,500 ft and 350 km at 33,000 ft. Maximum permissible altitude is 82,000 ft.

Contractor

China National Electronics Import & Export Corporation.

VERIFIED

GT-1 chaff and infra-red flare dispensing set

The GT-1 is a chaff and flare dispensing set which has been developed to provide self-protection to fixed-wing aircraft and helicopters. Chaff provides radar countermeasures in the 2 to 18 GHz frequency range and flares provide IR countermeasures in the 1 to 3 μm and 3 to 5 μm wavelengths. The complete system consists of a programme controller, an operations control, dispensers and cartridges and can be interfaced with a threat warning system to form a self-protection system. The standard configuration is 36 chaff and 18 flare cartridges. Dispensing can be manual or automatic and cartridges can be launched singly or in pairs.

The GT-1 dispensing chaff and flares from a Nanchang A-5C aircraft

Specifications

Weight: 40 kg (without cartridges and cables)
Power supply: 27 V DC, <0.7 A (static), <3 A (ignition)
380 V AC, 50 Hz, 3 phase, 2 kW
Coverage:
(azimuth) 360°
(elevation) 0 to +85°

Operational status

In production.

Contractor

China National Electronics Import and Export Corporation.

VERIFIED

BM/KG 8601 repeater jammer

The BM/KG 8601 repeater jammer operates in the E/F- and G/H-bands and is available for installation in strike and fighter/bomber aircraft to counter airborne tracking, anti-aircraft fire control and SAM guidance radars. The jammer has a high power output, minimal repeater delay time, threat management through RF channelling and wide antenna coverage. It provides multijamming techniques.

Operational status

In production.

Contractor

Southwest China Research Institute of Electronic Equipment.

VERIFIED

BM/KG 8605/8606 smart noise jammers

The BM/KG-8605 operates in the I/J-band as a smart noise jammer performing a hybrid type of jamming which incorporates some of the features of both noise and deception jamming.

The BM/KG 8606 operates in the I-band and features both orthogonal and dual circularly polarised jamming techniques.

Both the BM/KG 8605 and the BM/KG 8606 operate in conjunction with chaff/flare dispensers providing cross polarisation jamming and fast and accurate set on. They are small, lightweight and have a low power consumption.

Operational status

In production.

Contractor

Southwest China Research Institute of Electronic Equipment.

VERIFIED

BM/KJ 8602 airborne radar warning system

The BM/KJ 8602 is a radar warning receiver designed for tactical and other combat aircraft. It consists of a digital signal analyser, a CRT display unit, control box, several receivers and a number of antenna units. It features wide frequency coverage in two bands, 0.7 to 1.4 GHz and 2 to 18 GHz, and is capable of dealing with multiple threats. Automatic sorting and identification of threat emissions are provided. The system can operate in conjunction with ECM units and chaff/flare dispensers.

The BM/KJ 8602 system is compact and lightweight, making it suitable for tactical aircraft where space and weight are strictly limited.

Specifications

Weight: 20 kg
Frequency: 0.7-14 GHz and 2-18 GHz
Response time: 1 s
Capacity: 16 threats simultaneously
Coverage:
(azimuth) 360°
(elevation) –30 to +30°
Accuracy: 15° RMS

Contractor

Southwest China Research Institute of Electronic Equipment.

VERIFIED

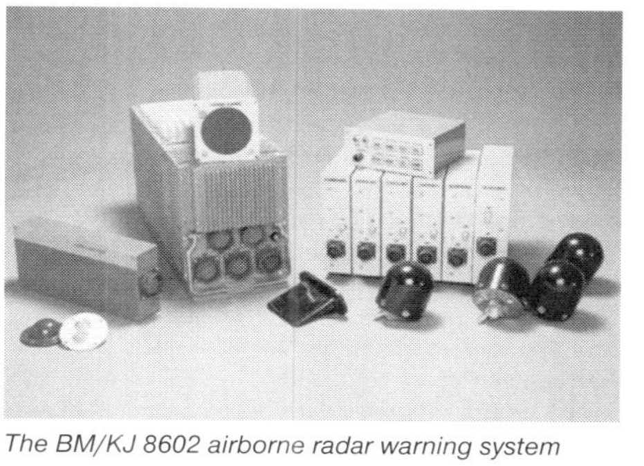

The BM/KJ 8602 airborne radar warning system

BM/KJ 8608 airborne ELINT system

The BM/KJ 8608 ELINT system detects, locates, identifies and analyses radar emitters deployed on the ground and at sea with high probability of intercept, high sensitivity and the accurate measurement of parameters. It features wide frequency coverage, high sensitivity and long operational range, automatic signal identification, emitter position fixing capability, operations in a dense RF environment and BITE.

Specifications

Power supply: 115 V AC, 400 Hz; 28 V DC
Frequency: 1-18 GHz
Frequency accuracy: 5 MHz
Coverage: (azimuth) 360°
Accuracy: (bearing) (1-8 GHz) 5°, (8-18 GHz) 3°

Contractor

Southwest China Research Institute of Electronic Equipment.

VERIFIED

CZECH REPUBLIC

SO-69/ICAO airborne transponder

The SO-69/ICAO airborne transponder is a reconstructed and updated version of the Soviet SO-69 system. It complies with ICAO design standards. The encoder unit S-ICAO and control unit O-ICAO can be modified to support IFF Mk10 modes 1 and 2.

The new transponder can easily be retrofitted in aircraft equipped with the SO-69 transponder. It weighs less than the existing SO-69 equipment and consumes less power.

Test equipment designated ZZ-69/ICAO is available to support the system.

Specifications

Dimensions/Weight:
Transmitter-receiver unit PV-ICAO:
412 × 64 × 160 mm/5.5 kg
Encoder unit S-ICAO: 436 × 86 × 160 mm/2.3 kg
Control unit O-ICAO: 71.5/98 × 112 × 59 mm/0.35 kg

Communication unit BK-ICAO: 217 × 133 × 59 mm/1.1 kg
Frame S0-69: 436 × 155 × 216 mm/2.4 kg
Receiver frequency: 1,030 MHz (bandwidth 6 MHz)
Transmitter frequency: 1,090 MHz
Transmitter power: 250 to 500 W
Modes: A, A/C, B
Power supply: 27 V DC and 115 V AC/400 Hz

Operational status

The SO-69/ICAO transponder has already been installed in MiG-23, MiG-29, Su-22 and Su-25 aircraft.

Contractor

Elektrotechnika-Tesla Kolin a.s.

VERIFIED

SO-69/ICAO airborne transponder
0002198

DENMARK

AN/ALR-DK Radar Warning Receivers (RWRs)

Royal Danish Air Force AN/ALR-69 Radar Warning Receivers (RWRs) have been reconfigured to the AN/ALR-DK, the Mini-69, for use on lightweight helicopters of the Royal Danish Army Air Corps. The lightweight, miniaturised RWR has been developed by Danish industry in co-operation with the Royal Danish Air Force to decrease logistics and software support significantly while providing helicopters with an advanced RWR capability. The basic Mini-69 weighs 13.6 kg.

The AN/ALR-DK system consists of an azimuth indicator, prime control panel, auxiliary control panel, signal processor, frequency selective receiver, C/D-band receiver and four amplifier-detectors.

The AN/ALR-69 has been modified to the AN/ALR-DK by a number of alterations to reduce size, weight and input power. The control panel has been rebuilt as a smaller LRU which optimises helicopter functions. In the CM-479, existing circuit cards have been repackaged in a smaller, lighter enclosure with the option for a second CPU/memory circuit card. The CM-479 power supply has been replaced by a new circuit card in the signal processor to match helicopter power requirements. The amplifier-detectors have been repackaged into two LRUs, each containing circuit cards from two of the AM-6639s. New antennas have been installed, but existing AN/ALR-69 antennas can be reused. The frequency selective receiver and C/D-band receiver are not used in the helicopter installation.

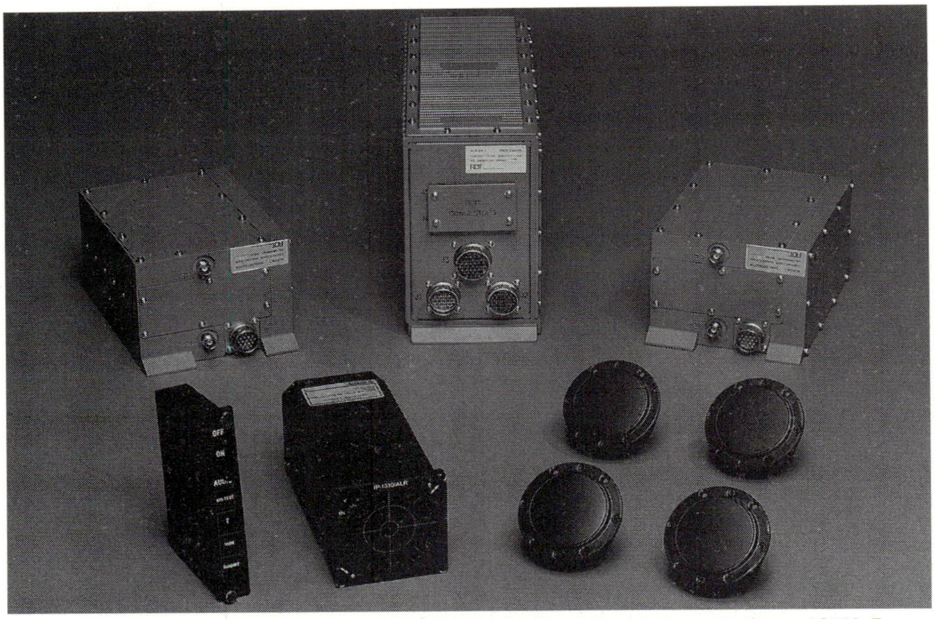

The Mini-69 has been modified from the AN/ALR-69(V) for Royal Danish Army Air Corps AS550 Fennec helicopters

Operational status

The AN/ALR-DK system is installed in Royal Danish Army Air Corps AS 550 C2 Fennec helicopters.

Contractor

Radartronic A/S.

VERIFIED

EWMS Electronic Warfare Management System AN/ALQ-213 (V)

TERMA's EWMS was originally developed for the F-16 in close co-operation with the Royal Danish Air Force. The objectives were to reduce workload and to ensure prompt and effective use of onboard EW subsystems. It was subsequently developed into a generic EWMS to fulfil the requirements of EW control for a large number of aircraft types, including: fighters, helicopters and transport aircraft.

The EWMS replaces all existing discrete EW control panels and indicators, except the Radar Warning Receiver (RWR) azimuth indicator.

EWMS functions

EWMU and ECAP. All subsystems are controlled from the Electronic Warfare Management Unit (EWMU), either by operating the software driven menus or through dynamic processing and automatic activation by the Electronic Combat Adaptive Processor (ECAP).

Modes of operation include:

(1) Manual: the pilot/operator selects and activates the countermeasure programme for the threat.

(2) Semi-automatic: the ECAP processor analyses the threat signals, computes the most effective combination of countermeasures and cues the pilot. He then initiates the computer script programme by activating a consent switch located on the control stick. The pilot keeps his hand on throttle and control stick.

(3) Automatic: analysis and selection of the threat adapted countermeasure responses take place as in semi-automatic mode, but the system will automatically initiate the response without pilot intervention. The pilot

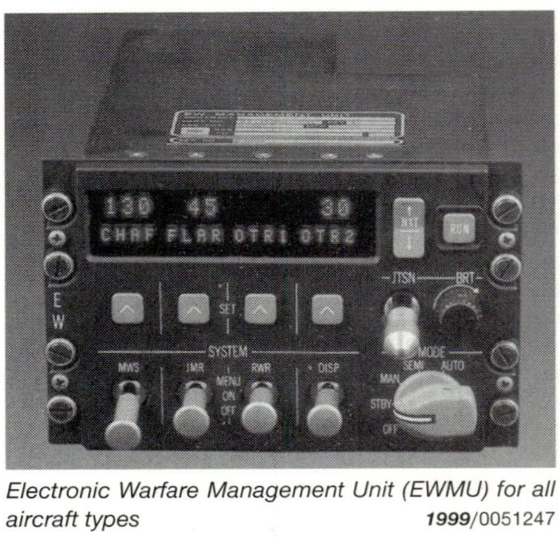

Electronic Warfare Management Unit (EWMU) for all aircraft types
1999/0051247

will be notified via synthetic voice message, display message/graphical symbols or multidimensional audio.

EWPI. Up-front control and display is effected through the Electronic Warfare Prime Indicators (EWPI) (one type for the F-16, and one type for the other aircraft). The EWPI presents the pilot with prioritised up-front information about the status of the EW subsystems, as follows:

(1) Dispenser: indication of usable decoy payloads remaining, including low level cues and indication of dispenser system activity.

(2) Jammer: operating mode/technique selection and jammer status.

(3) Radar warning: dedicated function buttons and status lights.

Electronic Warfare Aircraft ground equipment Panel (EWAP)
1999/0051249

(4) Missile warning: subsystem activation and indication of declared missile threat approach angle and time to impact, data displayed depends on the capability of the missile warning system concerned.

EWAP. System software - the Operational Flight Program (OFP) and mission data can be loaded

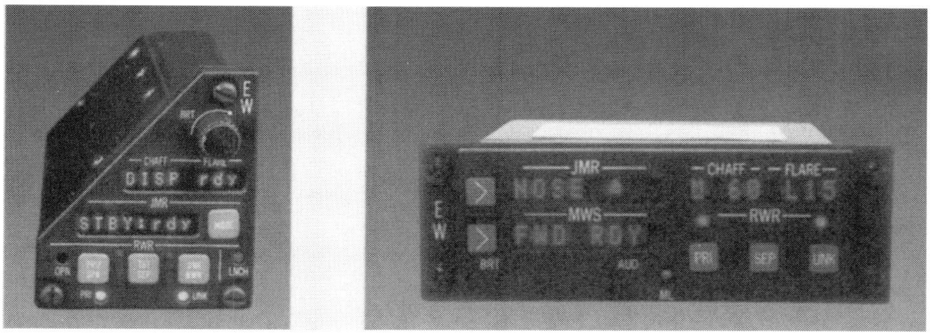

Electronic Warfare Prime Indicators (EWPI) for F-16 (left) and other aircraft (right) ***1999***/0051248

through the Electronic Warfare Aircraft ground equipment Panel (EWAP), or via the MIL-STD-1553B mux bus. The EWMS provides for multiple in-flight selections of pre-flight loaded jamming and dispensing programmes via menus.

EWMS interfaces

In addition to integrating all EW self-protection subsystems into one co-ordinated electronic warfare self-protection suite, the EWMS also interfaces the EW suite with the aircraft core avionics system.

The EWMS provides multiple electrical interface signals, including MIL-STD-1553B avionics and EW buses, RS-232, RS-422, RS-485, and PPD serial digital buses, plus a number of software programmable discrete I/O lines.

The EWMS is designed with a flexible systems architecture, which enables it to integrate and control a variety of EW subsystems. The EWMS functions are not limited to EW system control, the EWMS has also been chosen by the Danish, Belgian and US air forces to operate their F-16 Tactical Reconnaissance Systems. The EWMS provides a multiprocessor architecture, providing growth to meet new EW, and other operational requirements. Examples of EW subsystems controlled by the EWMS are:

(1) Dispensers: ACMDS, PIDS+, ALE-47, -50.

(2) RWRs: ALR-56M, -67, -68, -69, -69IV, and SPS-1000,

(3) Jammers: ALQ-119, -131, -162, -165, -176, -184.

(4) Missile Warning Systems (MWS): AAR-47, -54, -57, -60, and EL/M-2160.

EWMS upgrades

TTD. TERMA offers a Tactical Threat Display (TTD), based on multifunction display technology, for aircraft that have inadequate threat display capability. The TTD incorporates a three-dimensional audio warning for the pilot's earphones. The TTD is available in two versions: a semi-triangular version for the F-16, as a replacement for the EWPI and the RWR CRT, and a rectangular version for other types of aircraft. The TTD is a graphics colour display and it provides a number of soft keys to operate EW subsystems.

TDE. The Tactical Data Equipment (TDE) is a Commercial-Off-The-Shelf (COTS)-derived item, which associates the TTD with the Tactical Data Unit (TDU), to meet identified operational requirements for integrated EW system mission load, organic threat data recording, post-mission database reporting, and effective analysis by means of the detachable PC-compatible PCMIA Tactical Data Cartridge (TDC), plugged into the unit.

EWMS integrated self-protection suite. The EWMS integrated self-protection suite typically comprises a fully integrated EW self-protection suite, controlled by the EWMS and upgraded with the TTD and TDE. The suite contains an RWR, MWS, advanced countermeasures dispensing system, active ECM jammer and also in some examples, a reconnaissance pod.

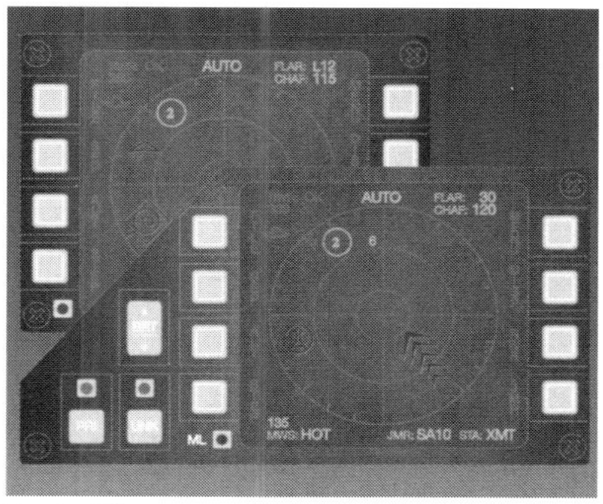

Tactical Threat Display (TTD) ***1999***/0051250

Operational status

EWMS has been selected by Belgium, Denmark, Germany, Holland, Norway, Portugal and the USA. It has been selected for the F-16 MLU by Belgium, Denmark, Holland and Norway. It is also fitted to a number of transport types, including C-130 aircraft of the Royal Danish Air Force, Royal Netherlands Air Force and the Portuguese Air Force. In US service it is designated AN/ALQ-213(V), and fitted to US Air Force HH-60Gs, US Air Force Reserve and US Air National Guard A-10s and F-16s (for tactical reconnaissance system control).

Development of the TDU, as part of the PIDS+ programme for the F-16 MLU (with the Per Udsen Company) is at the flying prototype stage.

The Royal Australian Air Force has contracted TERMA to fit the EWMS for an interim upgrade programme to the F-111C, before Project Echnida is completed. The system, designated AN/ALQ-213(V) will be used to control and operation of the countermeasures dispenser, jamming system, EW sensors and other systems, including reconnaissance pods.

Contractor

TERMA Electronik AS.

VERIFIED

C-130 self-protection concept

The Royal Danish Air Force (RDAF), the Royal Netherlands Air Force (RNLAF) and the Portuguese Air Force (POAF) decided to implement a self-protection suite for their C-130s consisting of a Radar Warning Receiver (RWR) and an Advanced CounterMeasure Dispenser System (ACMDS) of the ALE-40 type/format, together with different types of active ECM systems and Missile Approach Warning Systems (MAWS).

The concept was developed by TERMA Industries Grenaa AS and the RDAF. After extensive testing both in Denmark and in international trials like MACE and EMBOW the RDAF, RNLAF, and the POAF awarded TERMA Industries Grenaa AS production contracts. The system is operational on the three air forces' C-130 aircraft.

The installation is a symmetrical installation with eight standard ALE-40 type dispenser magazines on each side and 2 × 2 magazines in the nose gear bay (for flares only). Four magazines on each side are tilted downwards primarily for ejection of flares, but the system can be equipped with chaff and/or flares in any of the dispensers as required for the mission. The dispensers are operated by four sequencers and two EMI-filters on each side, and two sequencers and one EMI-filter in the nose gear bay.

The RWR antennas and preamplifiers are installed in wingtip pods providing perfect coverage with no shadowing and almost no cable losses. The installation can be performed by the customer or a customer-selected company with supervision by TERMA Industries Grenaa AS.

The ACMDS and wingtip pods can be delivered as a kit consisting of wingtip pods, magazine canisters, magazines, sequencer housings including sequencers and EMI-filters, fairing cable kits and the Electronic Warfare Management System (EWMS). Alternative control systems, as specified by the customer, can be used. Customer Furnished Equipment (CFE) would be the RWR, MAWS, ECM system to be installed.

One of the major benefits of the system is that there is no penetration of the pressure hull. The structural impact and requirements for reinforcements are thus minimised. The impact on other systems is limited to re-routeing of a few tubes, wires and cables. The installed weight is less than 36.4 kg.

The EWMS controls the system and provides advanced dispensing techniques for both chaff and flares. The EWMS provides control of all on-board EW-systems - and links them together for automatic response. Test flights have proved that the wingtip pods act as winglets and reduce stalling speed by a few knots as well as making the aircraft more stable at slow speed. When the dispensers are not in use, a set of cover plates can be installed to protect magazines and breech-plates from the outside environments.

Operational Status

The system has accumulated 30,000+ flight hours, on C-130 aircraft of the Royal Danish, Royal Netherlands and Portuguese air forces.

Contractor

TERMA Industries Grenaa AS.

VERIFIED

The TERMA Industries Grenaa AS C-130 self-protection concept ***1999***/0051277

F-16 PIDS+ chaff dispensers and missile warning system integrated weapon pylon

In the mid-1980s, the Royal Danish Air Force tasked TERMA Industries Grenaa AS to develop a modification for the weapon pylon of the F-16 to accommodate additional chaff dispensers and new dispenser electronics.

The F-16 PIDS (Pylon Integrated Dispenser Station) was developed and is now operational in the Belgian Air Force, Royal Danish Air Force, Royal Netherlands Air Force, US Air Force Reserve and US Air National Guard with more than 600 units having been delivered. The introduction of passive Ultra-Violet Missile Warning Systems (UV MWS) which could be accommodated in the pylon, made it possible to introduce the next-generation PIDS, designated PIDS+, which features both chaff dispensers and MWS. The PIDS+ programme is a multinational programme sponsored by the Royal Danish Air Force, Royal Norwegian Air Force and Royal Netherlands Air Force, with participation from the US Air National Guard. The programme was initiated because of the greatly increased affordability of a pylon installation compared with a fuselage installation. The PIDS+ is designed for use on F-16 stations 3 and/or 7. All conventional weapons and ECM pod capabilities are retained. The PIDS+ is integrated with, and operated from, the existing onboard control system (AN/ALE-47 control panel or TERMA Electronic Warfare Management System (EWMS)). Standard F-16 A/B/C/D pylons can be modified and the PIDS can be used on standard F-16s as standard weapon pylons. Existing PIDS can be retrofitted to PIDS+ configuration. Flexible pylon design can accommodate various MWS systems. The turbulent airflow available provides for better chaff blooming when compared with fuselage dispensers, and considerably better chaff blooming at 'dash speed'. IR flares are dispensed from the optimum location (aft fuselage), giving up to 200 per cent chaff capacity improvement. Up to 90+ per cent MWS sensor coverage is possible with six sensors.

On F-16 MLU aircraft only a cable kit has to be installed from the electronics bay to the wheel well. On non-MLU F-16 aircraft an additional harness kit is required to be installed in the wheel well. The nose and aft sections of the pylons are rebuilt to accommodate the dispenser system and the MWS. The pylons are modified at the manufacturer's facility inclusive of repaint and a complete check-out. The pylons are thus ready for installation when returned. As an option the MIL-STD 1760 harness update can be included in the modification package.

The dispenser part of the PIDS+ comprises two standard AN/ALE-40/47 ACMDS-type dispensers installed in the aft fairing interchangeable for inboard/outboard dispensing. The two dispensers are located identically to the two forward dispensers on the PIDS. The MWS installation comprises six sensors (three in each pylon) allowing for a 'clear of trailing edge' view. The systems that can be installed include the AN/AAR-54, AAR-57, AAR-60 and the Rafael Guitar MWS.

The PIDS/PIDS+ is also available with ECM systems such as the AN/ALQ-162 as an alternative to the chaff/MWS capability. The Royal Danish Air Force has 50+ pylons operational with the AN/ALQ-162 and associated antennas installed. With the pylons permanently installed on stations 3 and 7 it is possible to install an adaptor on the missile launcher which makes it possible to convert the F-16 from the ground-attack role to air-to-air role in a few minutes. The adaptor simply fits to the MAU-12 rack on the pylon. The adaptor fits both the AIM-9 Sidewinder and the AIM-120 AMRAAM. The advantages are thus that the conversion time from ground-attack to air-to-air is greatly reduced and that the wing loading is reduced compared with installation on stations 2/8.

Specifications

	Standard pylon	PIDS	PIDS+
Weight	125.5 kg	160.9 kg	170.9 kg
Length	2,515 mm	2,566 mm	2,802 mm
Width	114.3 mm	228.5 mm	256.5 mm
Height	381 mm	381 mm	381 mm

Operational Status

Over 600 PIDS in service. Flight testing of PIDS+ was planned for 1998; details of trial results are not available.

Contractor

TERMA Industries Grenaa AS.

VERIFIED

TERMA Industries Grenaa AS F-16 PIDS+ ***1999***/0051276

Twin underwing installation of MCP-7 on Royal Netherlands Air Force Fokker 60 aircraft ***1999***/0051246

The MCP-7 Modular Countermeasures Pod ***1999***/0051245

MCP-7 Modular Countermeasures Pod

The basic building block of the MCP-7 is a barrel module containing two ALE-40 or ALE-47 chaff and flare dispenser magazines and a sequencer switch. Two or four such barrels, which may be set at 15° intervals in order to dispense in the optimum direction, are attached to the strongback. The baseline system is completed by a nosecone containing a further dispenser magazine, and a tailcone accommodating a safety switch. The MCP is controlled from a TERMA EW Management System or similar installation. Options include the addition of RWRs, MAWs and towed decoys.

The MCP can be carried underwing, using standard NATO 350 mm lugs.

The designator -7 defines that the MCP-7 is able to carry a total of 7 magazines in any mix of chaff and flares (except for the forward flare only). A wide range of chaff and flare types can be dispensed including:

Flares: M206, MJO-7B
Chaff: RR-170, RR-180

Specifications

Length: 2,270 mm
Diameter: 381 mm
Height: 434 mm
Weight:
(empty) 65 kg
(with typical load) 132 kg

Operational status

Twin installation on Royal Netherlands Air Force Fokker 60 aircraft, typically carrying a load of 240 infra-red flares and 360 chaff cartridges.

Contractor

TERMA Industries Grenaa AS.

VERIFIED

FRANCE

DV400 high-resolution DV video recorder

The DV400 video recorder is an advanced DV system designed for high-resolution, direct recording from Head-Up Display (HUD) camera and multifunction displays on military aircraft. The digital video format offers up to 500 lines of horizontal resolution and high S/N ratio.

The DV400 is able to record PAL, Y/C PAL or RGB video images. A digital interface facilitates direct linking of the VCR to a PC. The video recording system comprises a two-rotary head helical scanning system.

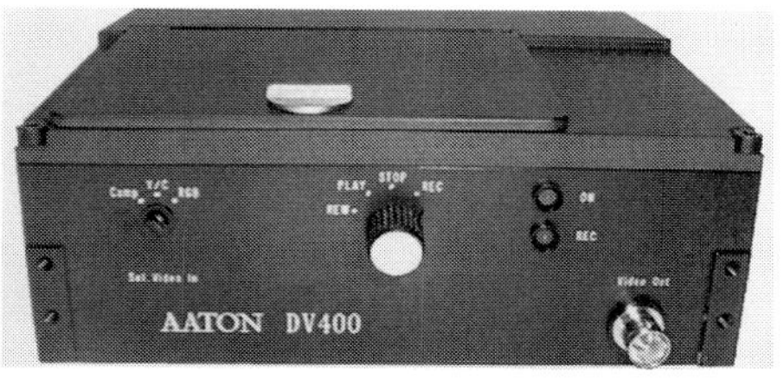

The AATON DV400 high-resolution DV video recorder
2000/0062216

Audio recording uses rotary heads and a PCM system with 12 bits (32 kHz) or 16 bits (48 kHz) operation. Tape speed can be 18.83 mm/s (SP mode) or 12.57 mm/s (LP mode); recording/playback time is 60 minutes (SP mode) or 90 minutes (LP mode).

Specifications

Dimensions: 150 (L) × 160 (W) × 70 (H) mm
Weight: < 2 kg
Power: 28 V DC (10 to 32 V), <10 W

Contractor

AATON.

VERIFIED

MV300 video selector multiplexer

The AATON MV300 video selector multiplexer provides for the multiplexing of up six synchronised video inputs and the selection of three video outputs among the video inputs; input source 1 is the master and slave sources must be synchronised. Mode control is via an RS-232 interface and three logic control inputs. The MV300 processes CCIR 50 fields/second video signals: Y/C PAL signals, PAL composite or black and white signals.

Specifications

Dimensions: 167 (L) × 110 (W) × 139 (H)
Weight: 2 kg
Power: 28 V DC, 12 W

Contractor

AATON.

VERIFIED

The AATON MV300 video selector multiplexe
2000/0062217

5000 series chaff/flare countermeasure dispensers

Alkan has developed countermeasure dispensers that can accommodate either chaff or infra-red flare cartridges or a combination of the two. The cartridges are arranged in interchangeable, easily handled magazines loaded in the modular dispenser. Typically, a single dispenser will accommodate five to seven modules. Each module houses a magazine containing, for example, 18 × 40 mm diameter chaff cartridges.

The dispenser electronic management system performs the firing sequences created by software whether it is connected to an RWR or not. It permanently manages the inventory of available cartridges and provides the necessary information to the cockpit control unit which displays the status of the complete equipment.

The CADMIR dispenser pylon using the same technologies is specifically designed for semi-conformal installation on the Mirage 2000 and 2000-5.

Operational status

The system is in service in French Air Force Jaguar aircraft. Two dispensers are fitted under the wing in a conformal installation near to the aircraft fuselage. Each dispenser contains seven modules (Alkan Type 5020).

The same system is also in service on the Dassault Mirage F1, on which it is installed on a special wing hard point.

The Alkan Type 5081 pod is designed to fit either to the JATO point of the MiG-21 or to any 14 in (356 mm) standard armament hard point. The system was in service on the French Navy Super Etendard until the type's retirement.

The same concept is used in an internal configuration on the Mirage 5. The Type 5013 dispenser which houses four standard magazines is in series production for an export customer.

Contractor

Alkan.

UPDATED

The Alkan 5020 decoy dispenser on a Jaguar aircraft

The ELIPS chaff/flare dispenser for helicopters and light aircraft

ELIPS helicopter self-protection system

The Electronic Integrated Protection Shield (ELIPS) has been designed for helicopters and light aircraft. It is a multidecoy launching system adaptable to any kind of chaff or flares, operating in automatic, semi-automatic and manual modes. Direct coupling to any type of Radar Warning Receiver (RWR) or Missile Approach Warner (MAW) for automatic sequence selection and launching is available.

The type and quantity of ammunition available, selected firing sequence and functioning status of the system are displayed either on the cockpit control unit or on a centralised multifunction ECM display. Loading of the various decoys is identified through codes allocated to magazines or through an integrated recognition system. Two, four or more magazines may be controlled without any modifications to the system.

Specifications

Weight:
(2 chassis configuration) <12 kg empty,
<23 kg loaded with IR flares
Firing interval: 25 ms-1 min
System response: <50 ms

Contractor

Alkan.

VERIFIED

SPIRIT electronic warfare system

SPIRIT is an electronic warfare system for transport aircraft. It comprises a high-capacity dispenser providing up to 84 cartridges, containing up to 376 flares, on the aircraft. It consists of a Type 5160 decoy dispenser, scabbed on to each side of the fuselage, and an NVG-compatible control unit which offers automatic, semi-automatic or manual operation.

Operational status

Fitted to French Air Force C-160/C-160NG aircraft interfaced with the Thales (formerly Thomson-CSF) Sherloc radar warning receiver and an Elta missile approach warning system.

Contractor

Alkan.

UPDATED

Corail countermeasures equipment

Corail is a radar and optronic countermeasures equipment designed for various versions of the Mirage fighter and can also be applied to transport aircraft. The system can be housed in a conformal or external pod; two pods can carry up to 256 decoy cartridges according to type - EM, IR and EO cartridges can be used.

Operational status

In production for Mirage F1 aircraft in service with the French Air Force.

Contractor

Matra BAe Dynamics SAS.

VERIFIED

Corail self-protection countermeasures system under the French Air Force Mirage F1 CT wings

DDM missile launch detector

The DDM missile launch detector provides automatic detection of a missile launch plume, locates it in flight and instantaneously transfers threat data to the aircraft ECM system. It incorporates passive infra-red detector techniques to ensure covert operation in severe ECM environments and is capable of locating missiles using any type of seeker. DDM features advanced signal processing to give extremely high probability of launch detection and a low false alarm rate. It includes a mosaic detector array to provide infra-red signature discrimination and has the capability to handle 40 tracks simultaneously.

DDM can be configured for different applications by arrangement of its component modules to meet specific size, weight and coverage requirements.

The DDM-2000 has been developed for the Dassault Aviation Mirage 2000. In this, the electro-optical head and signal processing unit are integrated into a single unit mounted at the end of the Magic AAM launcher on either side of the aircraft.

In the DDM-Prime version, which forms part of the Rafale Spectra self-protection system, the electro-optical head and signal processing unit form separate modules.

Specifications

Weight:
(electro-optical head) 5.6 kg
(signal processing unit) 3.6 kg
Coverage: 180°
Accuracy: ±2°

Operational status

Reported to be entering service with French Air Force Mirage 2000N and 2000D aircraft and Rafale.

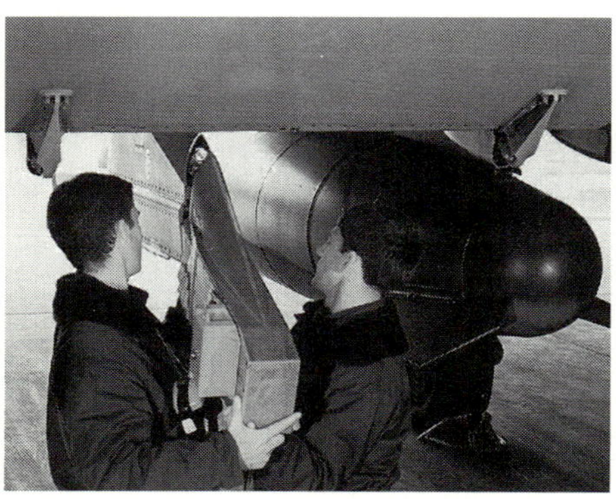

Missile launch detector (DDM) for the countermeasure equipment of a Mirage 2000 combat aircraft

Contractors

Matra BAe Dynamics SAS.
SAGEM SA, Defence and Security Division.

UPDATED

LCM (Lance-Cartouches Modulaire) modular cartridge dispenser

Developed as part of the Rafale aircraft Spectra EW system, the LCM comprises a variable number of identical cartridge dispenser modules controlled by a single computer. LCM is the modular cartridge dispenser proposed by Matra BAe Dynamics SAS for Spirale NG, Spectra, Myriad and Saphir M EW systems. The LCM system is compatible with the following standard cartridges: 1 × 1 in, 19, 40 and 60 mm diameter. It is also claimed to be compatible with future smart ammunition rounds, digitally programmable in flight. Each module will provide a load capacity of between six (LISCA-type) and 72 (Mucalir-type) cartridges.

Contractor

Matra BAe Dynamics SAS.

VERIFIED

LEA (Leurre Electromagnétique Actif) active radar decoy

LEA is being developed by Matra BAe Dynamics SAS and Thales (formerly Thomson-CSF DETEXIS). It is an expendable pseudo-repeater mini jammer that is programmable in flight before firing.

Operational status

In development.

Contractors

Matra BAe Dynamics SAS.
Thales (formerly Thomson-CSF DETEXIS).

UPDATED

LISCA (Leurre Infrarouge á Signature et Cinématique Adaptée) Smart IR decoy

LISCA is being developed by Matra BAe Dynamics SAS and SNPE. Its propulsion and aerodynamics allow its separation path from the aircraft to be optimised to deceive IR-homing seeker heads. Its ejection system is common with that of LEA, and compatible with the LCM system.

Operational status

Development.

Contractors

Matra BAe Dynamics SAS.
SNPE Societé Nationale des Poudres et Explosifs.

VERIFIED

Phimat chaff dispenser

Phimat is a light low-drag radar decoying system designed for supersonic speeds and suitable for all operations in Tornado, Harrier, Super Etendard and various types of Mirage aircraft. It can generally be adapted to all types of modern aircraft, particularly those with weapon stations capable of taking the Magic or Sidewinder missiles.

The system consists of a cylindrical dispenser pod, chaff pack tubes, ejection mechanism, associated

Phimat chaff dispenser under Super Etendard aircraft

electronic block and a control box located in the cockpit.

Specifications

Dimensions: 3,600 mm long × 180 mm diameter
Weight (loaded): 105 kg

Operational status

Operational with the French Air Force and Navy and the UK Royal Air Force. Production ceased.

Contractor

Matra BAe Dynamics SAS.

VERIFIED

Saphir chaff and flare system

The Saphir chaff and flare system is optimised for use with helicopters. Four cartridge dispensers are used and the system can be deployed manually, fully or semi-automatically or in a survival mode.

Saphir chaff and flare system operating on a Cougar helicopter ***1999***/0051251

Operational status

Saphir A was designed to equip the Super Puma helicopter. The system has been installed on Puma, Cougar, Lynx and Ecureuil helicopters. During the 1991 Gulf War, French Army Gazelles and Pumas and French Air Force Pumas were equipped with Saphir.

Saphir B was designed for light helicopters and is in service on the Gazelle and Lynx.

A version known as Saphir M, developed jointly by Buck Systems and Matra BAe Dynamics SAS, has been selected for the NH 90 TTH (Tactical Transport Helicopter) and NH 90 NFH (NATO Frigate Helicopter) and Tiger helicopters. It uses Saphir technology and Buck expendables.

Contractor

Matra BAe Dynamics SAS.

VERIFIED

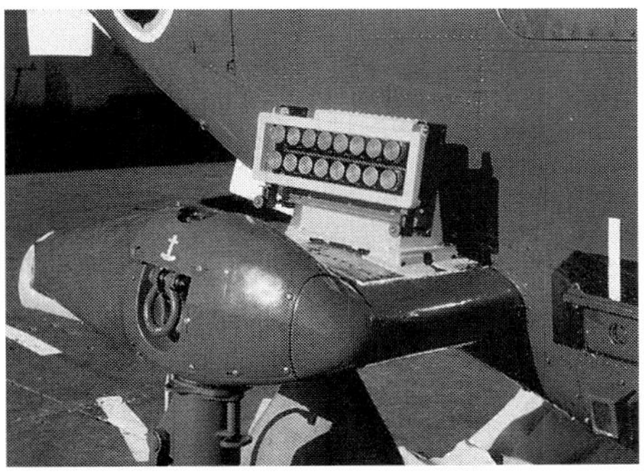

SAPHIR-B self-protection countermeasure system equipping a Lynx helicopter of the French Navy

Spirale chaff and flare system

The Spirale chaff and flare system was developed for the Dassault Mirage 2000 fighter and entered service at the end of 1987. The system comprises two cartridge dispensers located under the rear fuselage, two chaff dispensers located at each wingroot, two electronics boxes and two fixed missile detectors in the Magic missile launchers. Spirale NG is a new version being developed for the Mirage 2000D. Spirale NG will have a new-generation computer as well as being integrated with the DDM (Missile Launch Detector). With two outfits per aircraft and decoying capacity extensions - these units will be carried on six ridge-mounted cartridge dispenser modules (LCM) and four cartridge dispenser modules under the fuselage, resulting in 12 cartridge dispenser modules being available on the Mirage 2000D.

Spirale dispenser mounted on a Mirage 2000 aircraft

Operational status

Spirale is in service with several air forces on the Mirage 2000.

Contractor

Matra BAe Dynamics SAS.

VERIFIED

Sycomor chaff and flare system

The Sycomor chaff and flare system is intended for the various export versions of the Dassault Mirage F1 fighter and can be packaged either in a 2.95 m long externally mounted pod or in a 2.5 m conformal pack. Each pack has three chaff dispensing tubes and seven cartridge magazines. Each pod has the capacity for two packs.

The Sycomor ECM system on a Mirage F1 fighter

Operational status

Sycomor is operational with Mirage F1s flown by several air forces.

Contractor

Matra BAe Dynamics SAS.

VERIFIED

DF-206NF ADF radio navigation system

The DF-206NF ADF radio navigation system is the replacement system for the DF-206, of which thousands of units are in service on many types of aircraft. The DF-206NF features improved reliability, extended frequency range, ARINC 429 compatibility, light weight/small size, easy maintenance and direct retrofit of the DF-206.

The heart of the DF-206NF system is the 51Y-7NF receiver. The receiver is associated with the ANT-206NF combined sense/loop antenna, BCD control unit and an operational ARINC 429 control unit. The 51Y-7NF is a preplan product improvement to allow incorporation of additional features such as scanning on two channels and slow hop rate for ECCM compatibility.

The DF-206NF ADF radio navigation system with (left to right) the antenna, receiver and control unit

The BC-206NF control unit has been designed to provide frequency and mode control of any version of the receiver. The control unit exists in several versions, depending on voltage and panel lighting, and is also NVG compatible.

Specifications

Dimensions:
(receiver) 335 × 57 × 112 mm
(control unit) 146 × 57 × 110 mm
(antenna) 216 × 149 × 43 mm
Weight:
(receiver) 1.8 kg
(control unit) 0.8 kg
(antenna) 1.5 kg
Power supply: 27.5 V DC, 0.4 A
26 V AC, 400 Hz, 0.3 A
Frequency: 190-2,999.5 kHz
Channel spacing: 500 Hz
Preset channels: 20

Operational status

Operational on French Army Aviation Puma and Fennec helicopters, and French Air Force Tucano trainer aircraft. Selected for the Tiger European helicopter programme.

Contractor

Rockwell-Collins France, Blagnac.

VERIFIED

DF-430F tactical direction-finder

The DF-430F is Rockwell-Collins France's new generation of automatic tactical V/UHF direction-finder.

The DF-430F features an embedded synthesised receiver that covers the 30 to 400 MHz frequency band, thus enabling the DF-430F to function as a full stand-alone tactical direction-finder.

The DF-430F has been designed for simple installation on any type of aircraft (fixed- or rotary-wing). It provides rapid and accurate bearing acquisition.

The DF-430F is a three LRUs system including one DF Antenna (ANT-430F); one Receiver and Processing Unit (RPU-430F); one Control and Display Unit (BC-430F), as an option.

DF-430F main characteristics are, 30 to 400 MHz frequency range; AM or PM antenna modulation according to frequency band; fully solid-state antenna; fast bearing acquisition (50 ms burst); high bearing accuracy; dead reckoning capabilities; remote-control capability via ARINC 429 or MIL-STD-1553B interface; easy DF-301E mechanical retrofit; watertight package; and flush-mount installation.

Specifications

Typical range (up to the line of sight):
30-88 MHz: 100 n miles (20 W)
100-400 MHz: 100 n miles (5 W)
Bearing Accuracy
<3° forward/backward axis
<5° for other bearings
Dimensions;
ANT-430F: diameter: 278.5 mm; height: 107 mm
RPU-430F: ARINC 600 ¼ ATR short
BC-430F: width: 146 mm; depth: 150 mm; height: 95.5 mm
Weight:
ANT-430F: 2.6 kg
RPU-430F: 3.2 kg, ARINC 600 ¼ ATR short
BC-430F: 0.8 kg

Operational status

Selected for NH 90 multinational helicopter programme.

Contractor

Rockwell-Collins France, Blagnac.

VERIFIED

IPG-100F GPS receiver

The IPG-100F is a five-channel C/A and P/Y code single-frequency GPS receiver with selective availability and anti-spoofing capabilities.

The IPG-100F houses the receiver and its associated control panel in a single dzus-mounted box. The display includes two lines of 16 LED alphanumeric characters, each providing direct access to all GPS data, waypoints and navigational information. A keyboard with 12 keys provides access to system mode selection and data entry. Both the display and the keyboard are NVG-compatible.

The receiver is built with five independent channels providing continuous tracking of satellites in view, with permanent selection of the best four to calculate a navigation solution.

The IPG-100F includes a circular connector on the front panel to allow the operator to load crypto keys into the PPS-SM modules. The unit is still unclassified even when the crypto keys are loaded. Two installation configurations are available: Stand-alone as an autonomous RNav or remote sensor connected with the aircraft navigation computer through standard RS-422 or ARINC 429 ports.

The positions of up to 999 waypoints, airports or navaids can be loaded from a Jeppesen database on an RS-232 line either from a personal computer or magnetic card reader.

Specifications

Dimensions: 146 × 85 × 150 mm
Weight: 1.7 kg
Power supply: 10-32 V DC, 25 W
Accuracy:
3-D position error (SEP) 16 m, full dynamics
3-D velocity (RMS) 0.3 m/s, full dynamics
(time) 100 ns RMS

Operational status

In production for, and in service with, the French Army Aviation, the French Air Force, UK Royal Air Force AWACS, UK Royal Navy Sea Harrier and Polish Air Force MiG-29 aircraft.

Contractor

Rockwell-Collins France, Blagnac.

VERIFIED

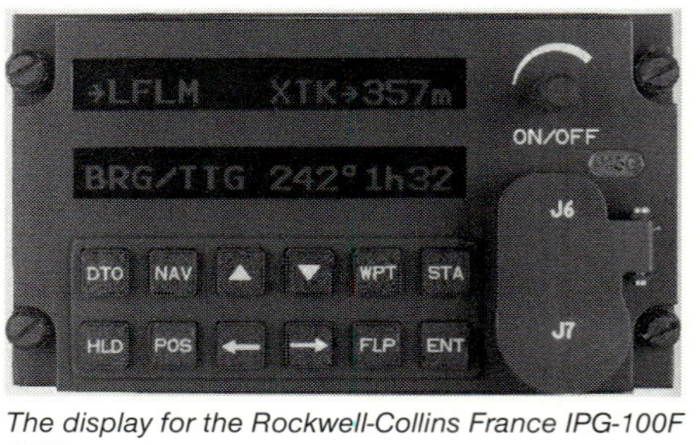

The display for the Rockwell-Collins France IPG-100F GPS

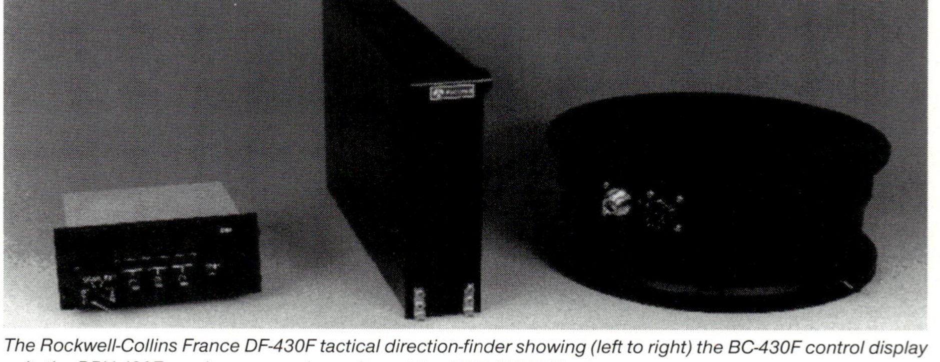

The Rockwell-Collins France DF-430F tactical direction-finder showing (left to right) the BC-430F control display unit, the RPU-430F receiver processing unit, and the ANT-430F DF antenna 0015344

IPG-120F GPS receiver

The IPG-120F GPS receiver is derived from the earlier IPG-100F system; it embeds Rockwell's latest GPS core engine. The IPG-120F is a 12-channel P/Y code L1/L2 GPS receiver. It is a form/fit replacement for the IPG-100F unit.

The IPG-120F GPS core engine provides the following features: RAIM (Receiver Autonomous Integrity Monitoring); WAGE (Wide Area GPS Enhancement); secure DGPS; fast initial acquisitions/direct Y code acquisition; high jam-resistance; all-in-view tracking and navigation, using Jeppesen database information and user waypoint data; NVG compatibility.

Specifications

Dimensions: 146 × 85 × 155 mm
Weight: 1.7 kg
Power: 10-32 V DC, 25 W
Position accuracy:
PPS: <15 m SEP
SPS: <100 m horizontal

Operational status

Upgrade in progress from IPG-100F.

Contractor

Rockwell-Collins France, Blagnac.

VERIFIED

MDF-124F direction-finder

Developed for Search and Rescue (SAR) operators, the function of the MDF-124F direction-finder is to localise the signal emitted by a distress beacon in order to rescue survivors on the ground or at sea.

The MDF-124F, with its embedded receiver, constantly scans and monitors 121.5, 243 and 406 MHz frequencies and meets the needs of SAR operators for an airborne system that reduces mission time by homing in on a localised beacon. It can also home in on a non-localised beacon because of its long-range detection capability (better than 100 nm). By combining the scanning of both 406 MHz, for long-range signal reception, and 121.5 and 243 MHz, for short-range reception, rescue time is reduced. Information provided by the COSPAS-SARSAT satellite detection system concerning position of a beacon is updated every 90 minutes.

The MDF-124F consists of the MDF-124F antenna and distress receiver unit, packaged in a single watertight unit, and the BC-124F control unit.

The antenna is a fully static rotating antenna controlled by a unique patented driver. This antenna creates an AM modulation of incoming signals in the 100 to 400 MHz frequency range and a PM modulation on a 406 MHz incoming SARSAT signal.

The MDF-124F distress receiver is made of a dual-frequency AM receiver, on 121.5 and 243 MHz, and of a 406 MHz PM receiver. The two receivers are fully compatible with all types of modulation produced by International distress beacons. The MDF-124F also includes a VHF/UHF low-noise preamplifier for use with an external receiver in the 100 to 400 MHz frequency range.

The signal processor receives the demodulated audio signal from the internal distress receiver or from the external VHF/UHF receiver. The MDF-124 is based on phase comparison performed by a microprocessor. The signal processor delivers relative bearing in ARINC 407 (three-wire synchro) and ARINC 429 formats. When the MDF-124F is on COSPAS-SARSAT frequency, the bearing is estimated during beacon silence if the aircraft heading is available.

The BC-124F control unit allows the selection of the desired distress frequency or of one external receiver out of four. The control unit also indicates the reception of a distress signal on a non-selected distress frequency. If several COSPAS-SARSAT signals are received simultaneously, the BC-124F allows the selection of the desired one.

Specifications

Dimensions:
(MDF-124F) 90 × 315 mm diameter
(BC-124F) 66.6 × 146 × 150 mm (dzus mounted)
Weight:
(MDF-124F) 3.5 kg
(BC-124F) 0.7 kg

Power supply: 27.5 V DC, 700 mA
Interfaces: bearing output and heading input, ARINC 429 and ARINC 407

Operational status
In production for the French MoD, Australia, Canada, Japan, Spain, Poland and China.

Contractor
Rockwell-Collins France, Blagnac.

VERIFIED

MDF-124F(V2) direction-finder

Recent miniaturisation of technology together with the experience gained, especially in maritime surveillance operation, have led Rockwell-Collins France to develop and release a new version of MDF-124F called MDF-124F(V2).

MDF-124(V2) is a full stand-alone SAR and tactical direction-finder; its embedded synthesised receiver allows it to operate the V/UHF 100-407 MHz frequency range without any external receiver.

MDF-124F(V2) comprises: MDF-124F(V2): antenna and receiver unit; BC-124F(V2): control and display unit.

MDF-124F(V2) features the following direction-finding capabilities: direct access of any of the three international distress frequencies or auto-scanning/alert on these frequencies; direct access to ARGOS channel; direct access to VHF-FM maritime channels 16 and 70; direct access to manual/preset user's frequencies in DF Tactical mode; combined tactical/auto alert SAR modes allowing DF operation on one user's tactical manual/preset channel while still monitoring the three international distress frequencies. Other characteristics are identical to the MDF-124F.

Specifications
Dimension, weight, interfaces and power supply are the same as for the MDF-124F.

Operational status
In production for Australia, Belgium, China and Japan.

Contractor
Rockwell-Collins France, Blagnac.

VERIFIED

RSC-125F personnel locator system

RSC-125F is a personnel locator system which provides combat SAR platforms with the advanced capabilities required for the very demanding combat SAR mission.

This equipment has been developed to recover downed pilots or aircrew members using personal locator beacons.

RSC-125F is a full stand-alone system, with embedded receiver, which can be installed on both fixed- and rotary-wing platforms. Different interfaces give RSC-125F great flexibility and allow it to be integrated within ARINC-429 or MIL-STD-1553B architectures.

RSC-125F Personnel locator system, showing BC-125F (left) LPT-125F (centre) and ANT-430F (right)

RSC-125F also includes tactical DF capabilities in the 30 to 400 MHz band and civil SAR frequencies, when operating all International distress frequencies.

RSC-125F is a three LRUs system including one DF antenna (ANT-430F); one Localisation Processor and Transmitter unit (LPT-125F); one control and display unit (BC-125F) (optional).

The main features of RSC-125F are compatibility with existing AN/PRC-112 and AN/PRC-434 Personal Locator Beacons (PLB); civil SAR (121.5-243 MHz-406.025 MHz COSPAS-SARSAT, VHF/FM maritime channels 16 and 70) and Argos capabilities; tactical DF in the 30 to 400 MHz band (25 kHz step); automatic scanning function on distress frequencies; remote-control capabilities through ARINC 429 or MIL-STD-1553B interfaces; NVG compatible; high-accuracy bearing and distance information on beacon; computed bearing during gap between the interrogations; easy flush-mount installation.

Specifications
Dimensions:
ANT-430F: diameter: 278.5 mm, height: 107 mm
LPT-125F: ARINC-600 ¼ ATR short
BC-125F: 146 × 66.6 × 150 mm
Weight:
ANT-430F: 2.6 kg
LPT-125F: 4.5 kg (ARINC 600 ¼ ATR short)
BC-125F: 0.8 kg

Operational status
In production for French military aircraft.

Contractor
Rockwell-Collins France, Blagnac.

VERIFIED

Airborne Multiservice/Multimedia Communication System (AMMCS)

The Airborne Multiservice Multimedia Communication System (AMMCS) provides voice, computer data, telex and facsimile air-to-ground communications facilities over different media such as satellite, HF, VHF or UHF radio.

Message handling facilities allow editing, retrieving, modifying and encryption/decryption of messages. The AMMCS provides automatic store-and-forward message services including logging, routeing and relaying. It supports a wide variety of message types and formats including telefax, telex, and e-mail. Multiple compression algorithms allow message transmissions to be speeded up. Message routeing is made automatically using different criteria such as urgency, medium availability, cost and priority.

The reliable transmission of long messages, such as high-quality telefax or error-free computer data files, over HF implies some specific modems, datalink protocols and transmission management protocols.

AMMCS integrates the Rockwell MDM-2501, multi-waveform adaptive high-speed modem. MDM-2501 offers 12 different standard waveforms, ensuring interoperability on different media. Among these waveforms is the MIL-STD-188-110A single-tone waveform which allows the efficient transmission of data up to 2,400 bits/s with forward error correction on HF between 300 and 3,000 Hz. When this waveform is used, the automatic multipath compensation provided by this modem overcomes the data rate limitations usually imposed on HF channels.

The AMMCS uses a multimedia FED-STD-1052 link protocol. It can operate over half- or full-duplex channels. This type of robust protocol is mandatory for HF links when propagation conditions may vary drastically and the link can be interrupted for variable periods of time. This protocol supports the automatic adaptability of system parameters to optimise the transmission performance in terms of throughput and bit error rate.

The AMMCS features functions which automatically adapt its parameters during a transmission depending on the type of link, type of service, required performance and propagation characteristic variations to optimise the performance in terms of bit error rate, required transmit power, transmission delay and throughput. These functions automatically select the best HF propagating channel and change the frequency when conditions degrade. The system uses the MIL-STD-188-141A/FED-STD-1045A ALE protocol on HF.

The AMMCS may present different levels of automation. In a fully automated system, direct dialling is offered to aircraft users for voice communications. The system selects the appropriate media and uses the dialling transcription to perform automatic selective call and link establishment procedures. A similar procedure applies to telex, data or faxes queued in the message handling system file for transmission. AMMCS uses the routeing information contained in the message envelopes to establish the appropriate link automatically and to manage the message transmission.

The AMMCS system is integrated into a high-impact magnesium alloy case laptop PC featuring 486/33 MHz Intel processor, up to 500 Mbyte hard disk, high-view angle touch panel colour screen and mouse or trackball interface. A TEMPEST version is available.

Contractor
Rockwell-Collins France, Blagnac.

VERIFIED

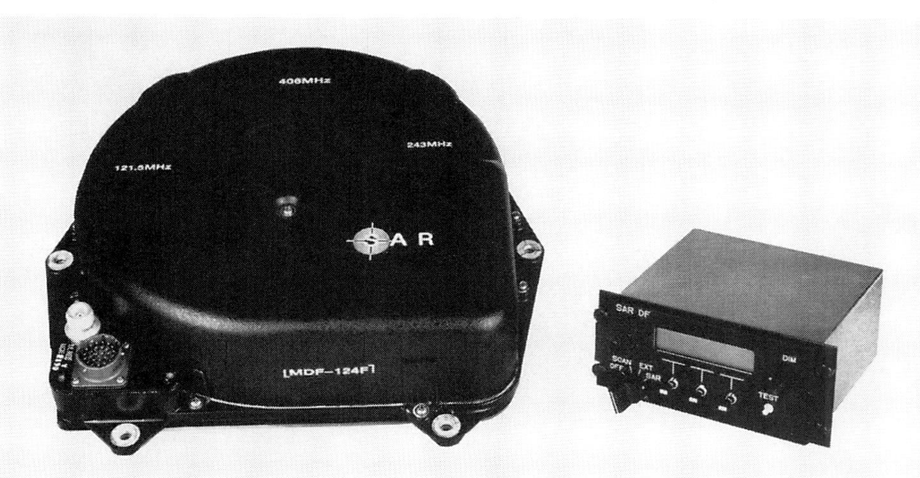

MDF-124F/MDF-124F(V2) direction-finder, showing (left) the antenna/receiver unit and (right) the control unit
0001311

DLP DataLink Processor system for NATO interoperability

The DLP (DataLink Processor) is an advanced tactical datalink server that supports TADIL-A (Link 11A), TADIL-B (Link 11B) and TADIL-J (Link 16) for NATO interoperability. It receives, transmits, processes and displays tactical data in accordance with STANAG 5511 and/or STANAG 5516 message specification.

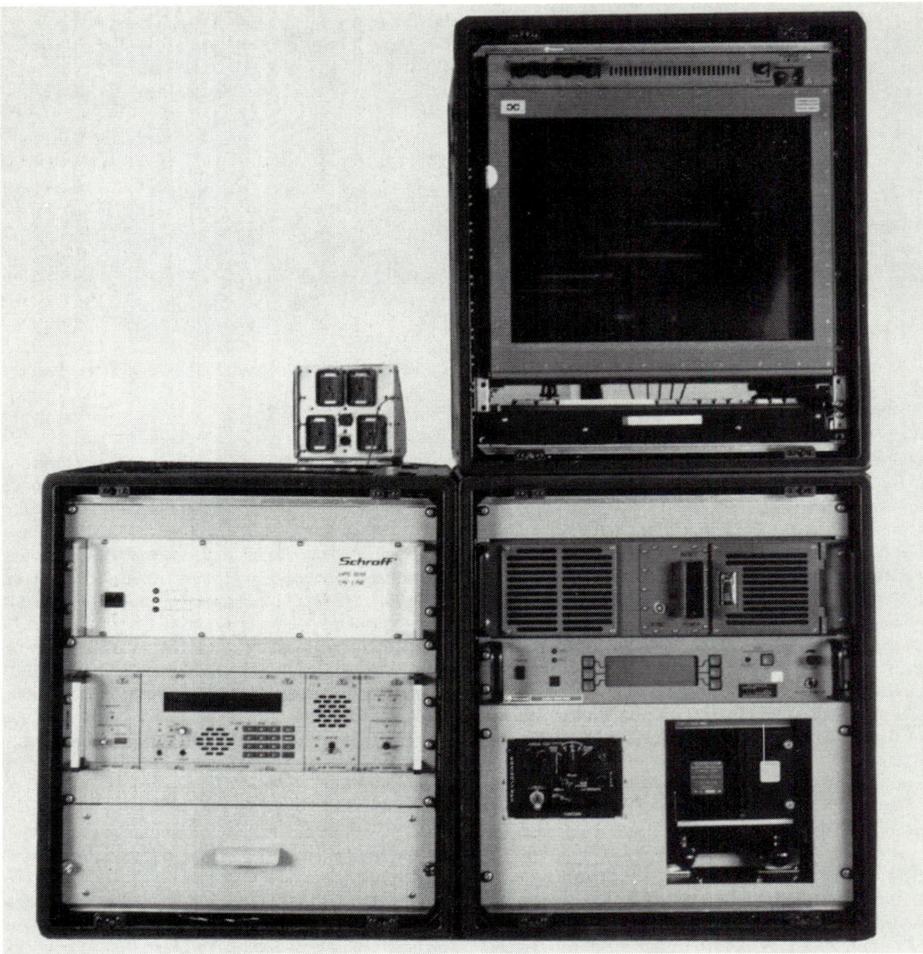

DLP DataLink Processor 0001310

The DLP has the capability to simultaneously process data provided by two or more networks (that is Link 11/Link 16, 2 × Link 11).

Depending upon the nature of the application, the DLP can be configured as a tactical datalink front-end processor only, or as a tactical datalink processor together with graphic display and keyboard/trackball.

The DLP provides the necessary external interfaces including NTDS/ATDS, MIL-STD-1553B, Ethernet and serial interfaces.

The DLP product family includes equipment designed for aircraft, ground station (fixed/transportable) and shipboard applications. It uses commercial workstations for benign environments, 19 in ruggedised VME-based equipment for ground and shipboard and ATR VME-based equipment for airborne environments.

Operational status

In production since 1996. In operation with the French Navy, French Air Force and UK Royal Navy.

ADLP-100F Airborne DataLink Processor for NATO interoperability

The ADLP-100F is part of Rockwell-Collins France DataLink Processor (DLP) product line. It is an advanced tactical datalink processor specifically designed for Link 11 airborne applications including both fixed- and rotary-wing platforms.

The ADLP supports the capability to receive, process, transmit and display tactical data in accordance with STANAG 5511 message specification for NATO interoperability. Depending upon the nature of the application, the ADLP-100F can be installed on board the platform as a front-end processor connected to the onboard mission system through a MIL-STD-1553B interface or equivalent, or as a stand-alone subsystem together with a display and keyboard/trackball.

ADLP-100F Airborne DataLink Processor 0011850

The ADLP-100F is a VME-based system packaged in an ATR format for easy installation on board the platform. It provides an ATDS interface to a KG-40(A) crypto and a 1553B interface or equivalent to the onboard mission system or sensors.

The ADLP-100F hardware is qualified for both fixed- and rotary-wing aircraft.

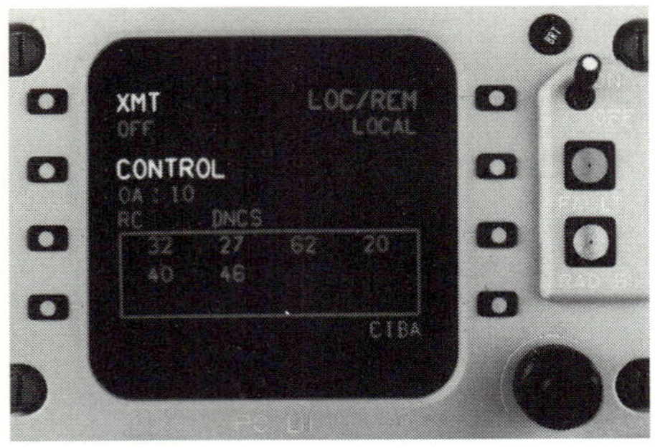

The Rockwell-Collins France MCU-2202F data terminal set control

Operational status

Selected by ECF to equip the Super Puma and Panther helicopters with Link 11 for a foreign navy; qualified for the French Navy ATL2 aircraft.

Contractor

Rockwell-Collins France, Blagnac.

VERIFIED

MCU-2202F data terminal set control

The MCU-2202F data terminal set control has been designed for the control of the Collins MDM-2202 Link 11 modem. It is a processor-controlled unit which provides configuration and control instructions to the data terminal set and provides address storage and sequencing instructions for control of the network roll call operation.

Specifications

Dimensions: 146 × 105 × 242 mm
Weight: 3.6 kg
Power supply: 28 V DC
Temperature range: –20 to +70°C
Altitude: up to 55,000 ft

Operational status

In production for aircraft of the French Navy.

Contractor

Rockwell-Collins France, Blagnac.

VERIFIED

SDF-123F sonobuoy direction-finder

The SDF-123F sonobuoy direction-finder provides relative bearing information from any sonobuoy signal in the 136-174 MHz frequency range through a companion FM receiver. It consists of two LRUs: the ANT-123F static rotating antenna and the MPU-123F

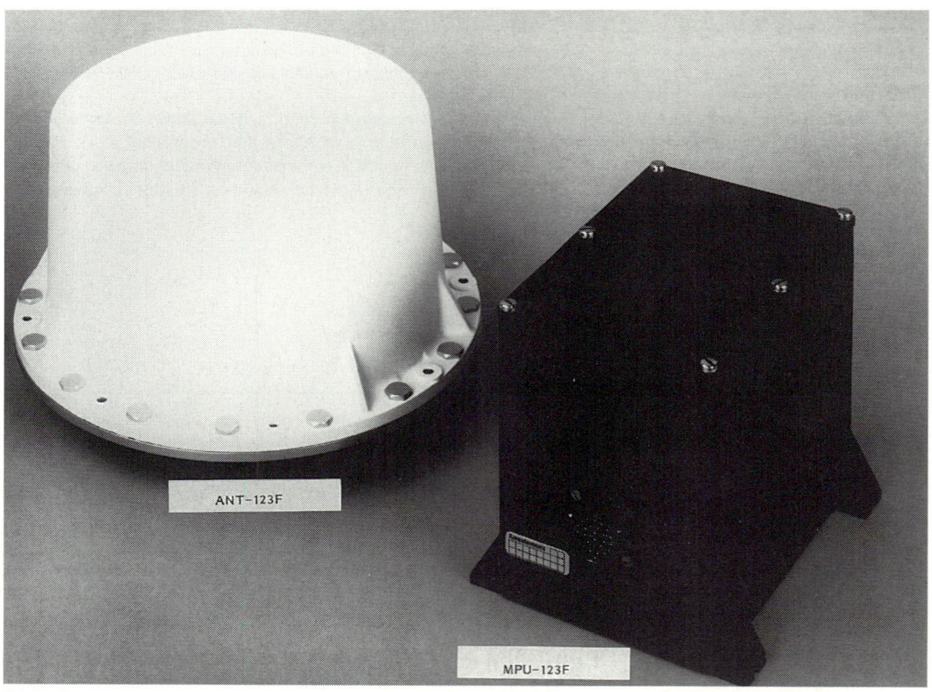

SDF-123F sonobuoy direction-finder units, showing the antenna unit ANT-123F (left) and processing unit MPU-123F (right) 0011862

monitoring and processing unit. The system is fully solid-state, uses a standard sonobuoy FM receiver and has On-Top Position Indication (OTPI).

The ANT-123F is a directional rotating antenna, made of eight elements which are switched according to a specific pattern in order to simulate a low-speed antenna rotation. This creates a frequency modulation of the incoming VHF signal, the relative phase of which contains the bearing information.

The MPU-123F processing unit is the heart of the SDF-123F system and supplies power to the system, provides signal generation for antenna control, processes the demodulated AF signal from the companion FM receiver and generates the bearing signal. Output is available in ARINC 407 (three-wire synchro) or (optionally) ARINC 429 format.

Specifications

Dimensions:
(MPU-123F) 135 × 166 × 146 mm
(ANT-123F) 130 × 250 mm diameter
Weight:
(MPU-123F) 1.5 kg
(ANT-123F) 2.1 kg
Power supply: 26 V AC, 8 VA
27.5 V DC, 1 A

Operational status

In production for Eurocopter ASW helicopters, Pakistani ATLI and Chilean Navy P3 aircraft.

Contractor

Rockwell-Collins France, Blagnac.

VERIFIED

Embedded GPS receiver

The SAGEM GPS receiver is an 'all-in-view' dual-frequency receiver, with Precise Positioning Service (PPS capability). The receiver is designed to ensure INS/GPS tight-coupling, with special attention given to data latency and time tagging.

Comprehensive integrity monitoring is performed, including a RAIM algorithm, in order to ensure the consistency of satellite signals.

Proprietary ASICs and signal processing provide: miniaturisation and reliability; fast acquisition and high resistance to Electronic Counter Measures (ECM).

The SAGEM GPS receiver is also available in Standard Positioning Service (SPS) version. Both PPS and SPS versions are capable of differential positioning through the use of standard connections.

Specifications

Positioning accuracy: in accordance with international standards

Operational status

Current applications are: French Air Force Mirage F1, the NH 90 helicopter (French Army and French Navy), and various export customers.

Contractor

SAGEM SA, Defence and Security Division, Paris.

VERIFIED

MAESTRO nav/attack system

SAGEM SA has designed the Modular Avionics Enhancement System Targeted for Retrofit Operations (MAESTRO) with flexibility so it can be easily tailored to any customer's specific requirements and offer capabilities at par with those of current front line fighters.

MAESTRO provides full inertial and GPS capability. It utilises SAGEM navigation and mission computer systems that integrate the following functions: the Embedded GPS-Inertial (EGI) concept; Terrain Contour Matching (TERCOR) to provide stealth navigation and a covert attack capability; the SAGEM high-capacity data transfer system offering digital moving map display capability; a wide field-of-view FLIR, that is compatible with the HUD, and other aspects of the glass cockpit; Hands On Throttle And Stick (HOTAS); air-to-ground and air-to-air fire control; a multimode pulse Doppler radar and/or laser range-finder; a full EW suite comprising radar warning, missile launch detection, chaff and flare self-protection and/or jamming systems; and the all digital CIRCE 2001 mission planning system.

The SAGEM MAESTRO is designed for retrofit applications

MAESTRO can be adapted to new aircraft of western or eastern origin.

Operational status

In production. Applications include the upgrade of Belgian Air Force Mirage 5, Chilean Air Force Mirage 5 and C-101 combat/trainer aircraft, Indian Air Force Jaguar, PZL Irdya Polish trainer/attack aircraft and Pakistan Air Force Mirage III and 5.

Contractor

SAGEM SA, Defence and Security Division, Paris.

VERIFIED

Mercator digital map generator

The Mercator digital map generator provides pilots of modern aircraft and helicopters with presentation of a colour moving map display.

To improve flight safety and operational effectiveness, the Mercator digital map displays terrain data in colour referenced to aircraft altitude. Mercator is derived from the Jumbo mass memory equipment and offers up to 2 Gbyte capacity, corresponding to a very large geographical area at various scales.

Key features of the Mercator digital map generator include a north-up or track-up map/DTED display; real-time smooth rotation, pan/scroll and zoom; the ability to drive up to two multifunction colour displays; up to 2 Gbyte mass storage cartridge, allowing 1,700,000 km^2 (1:100k) on line map data; standard MIL-STD-1553B/ RS-422 databus interface; simultaneous functions of mass memory server on MIL-STD-1553B (mission planning data) and digital map generation.

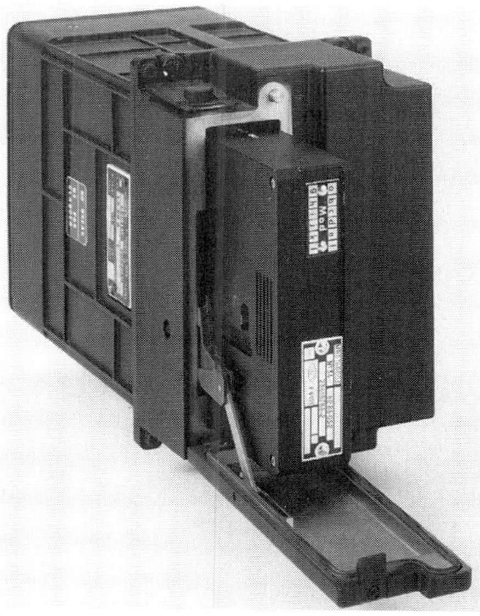

SAGEM Mercator, onboard digital map generator
0015345

Specifications

Dimensions: 95.2 × 127 × 222 mm
Weight: 2.7 kg
Power: 28 V DC, 35 W

Operational status

In production.

Contractor

SAGEM SA, Defence and Security Division, Paris.

VERIFIED

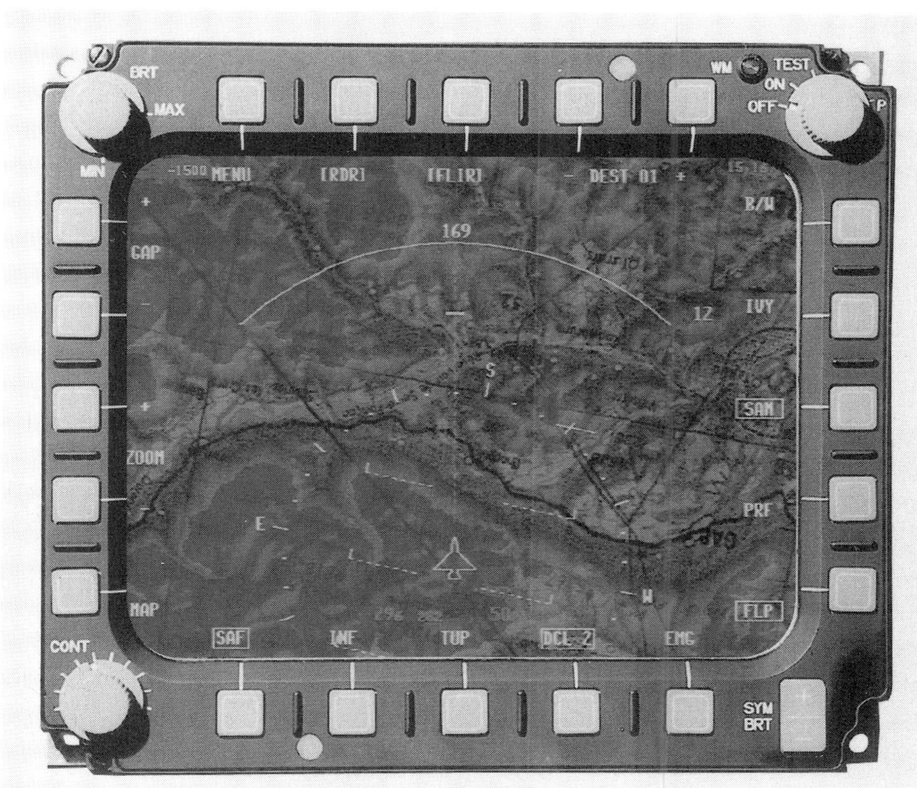

The SAGEM Mercator map display shows colour-coded altitude data ***1999***/0051581

SAMIR missile launch detector 0001258

SAMIR missile launch detector

The SAMIR (Systeme d'Alerte Missile Infra-Rouge) missile launch detector provides automatic detection of a missile launch plume, locates it in flight and instantaneously feeds threat data to the aircraft ECM system. It incorporates passive infra-red detector techniques to ensure covert operation in severe ECM environments and is capable of locating missiles using any type of seeker. SAMIR features advanced signal processing and a mosaic detector array to provide infra-red signature discrimination. It can handle 40 tracks simultaneously.

SAMIR can be configured for different applications by arrangement of its component modules to meet specific size, weight and coverage requirements. It can be installed in a variety of aircraft types such as combat aircraft, helicopters and transport aircraft.

Specifications

Weight: 10 kg
Coverage: 180°
Accuracy: ±2°

Operational status

In production for French Air Force Mirage 2000N, 2000C, 2000D and for Rafale. It is understood that the version fitted to Mirage 2000 aircraft is configured into one LRU and designated SAMIR 2000, while the version for Rafale is configured in two separate modules and designated SAMIR Rafale.

Contractor

SAGEM SA, Defence and Security Division.

VERIFIED

Sigma ring laser gyro inertial navigation systems

The Sigma family of inertial navigation systems implements a combination of high-performance ring laser gyro sensors, accelerometers and a multichannel GPS receiver. Such systems are intended for use in aircraft equipped with a multiplexed databus.

Sigma systems offer the benefits of tight hybridisation between ring laser gyros, accelerometers and a GPS receiver supported by a multisensor Kalman filter for both alignment and navigation. The synergy between these three elements brings the following advantages: reduction in size, weight and power consumption through the integration of inertial and GPS functions; short alignment time; sensor performance and integrity monitoring for GPS and INS; automatic in-flight calibration of inertial sensors; long-term stability of inertial performance, and higher resistance to jamming with improved dynamic behaviour of the GPS.

All versions provide: aircraft position, velocity and attitude information; computation of navigation and steering information to waypoints; position updating by navigation fixes; terrain reference updating; some versions combine the navigation function and weapon delivery computations, consisting of ballistics, determination of release point, ripple spacing of weapons, safety pull-up information; head-up display information for target acquisition and commands for the blind release of weapons; attack modes; air data computations and multiplex bus control.

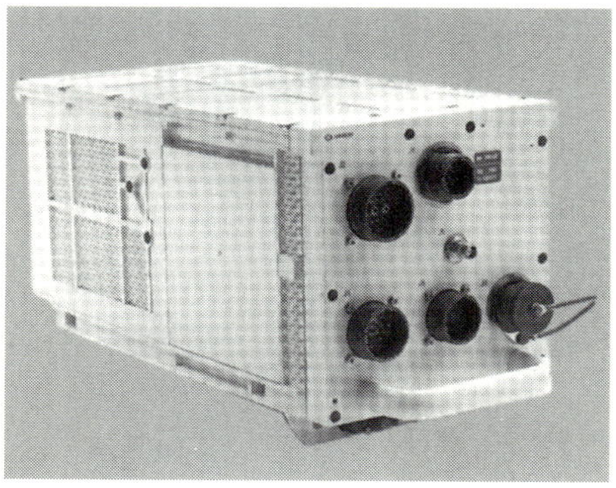

Sigma 95MF (MultiFunction) inertial nav/attack system 0015347

Sigma 95MF inertial nav/attack system

The Sigma 95MF (MultiFunction) uses three high-accuracy ring laser gyros and three accelerometers and is fitted with an embedded GPS receiver. Sigma 95MF is the heart of MAESTRO, SAGEM's avionics system. Sigma 95MF is a highly integrated system which provides a combination of high-performance, hybrid inertial navigation, attack computations and mission management tailored to advanced multirole combat or tactical transport aircraft.

Specifications

Dimensions: 209 × 200 × 406 mm
Weight: 17 kg
Power supply: 28 V DC, <90 W (115 V AC/400 Hz optional)
Accuracy: 0.6 n mile/h (inertial mode)

Operational status

In production for the Pakistan Air Force Mirage III upgrade.

Sigma 95N inertial navigation system

The Sigma 95N (Navigation) uses three high-accuracy ring laser gyros and three quartz accelerometers. It performs hybrid inertial/GPS navigation. It interfaces with the aircraft avionics systems through a MIL-STD-1553B multiplexed databus. The versatility of its interfaces enables the Sigma 95N system to be integrated easily with a wide range of carriers, avionic configurations and therefore operational situations.

Specifications

Dimensions: 209 × 200 × 385 mm
Weight: <15 kg
Power supply: 28 V DC, <45 W
Accuracy: 0.6 n mile/h (inertial mode)

Operational status

In production for the French Air Force/French Navy Rafale aircraft (under programme name RL-90) and for the Cougar helicopter.

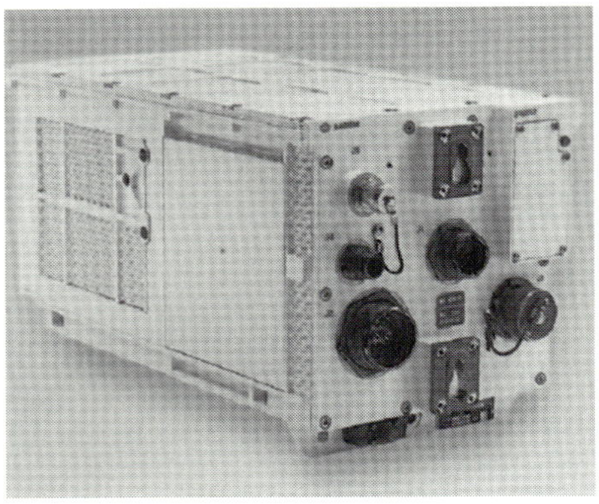

Sigma 95N (Navigation) inertial navigation system 0015348

Sigma 95L (Light) inertial navigation system 0015349

Sigma 95L inertial navigation system

The Sigma 95L (Light) is a compact inertial reference system fitted with a SAGEM embedded GPS. It has been designed for applications on helicopters and fixed-wing aircraft requiring a lightweight, small size, high-performance navigation system.

Sigma 95L can function as a self-contained attitude and heading reference system or as a full inertial navigation system. It performs hybrid inertial/GPS navigation. It interfaces with the aircraft avionics system through a MIL-STD-1553 multiplexed databus and ARINC 429.

Specifications

Dimensions: 180 × 125 × 280 mm
Weight: <8.5 kg
Power supply: 28 V DC, <35 W
Accuracy: 1 n mile/h (inertial mode)

Operational status

Off-the-shelf production. A (SAPHIR) version of Sigma 95L, which includes an integrated GPS receiver, is in production for the NH 90 helicopter programme, where it will provide the main navigation data and inertial references for the fly-by-wire system.

Contractor

SAGEM SA, Defence and Security Division.

VERIFIED

Telemir infra-red communication system

Telemir uses an infra-red beam for air-to-air, omni-directional air-to-ground, ground-to-air and ground-to-ground communications. The airborne equipment consists of an optical head (mounted on top of the tailfin) and a processing unit. It is extremely difficult to jam.

The system is used by a carrier-based aircraft for the reception of navigational updating and reference data such as attitude, location and speed from the ship's inertial navigation system for aligning its own INS. A new version is now available with a MIL bus 1553 datalink.

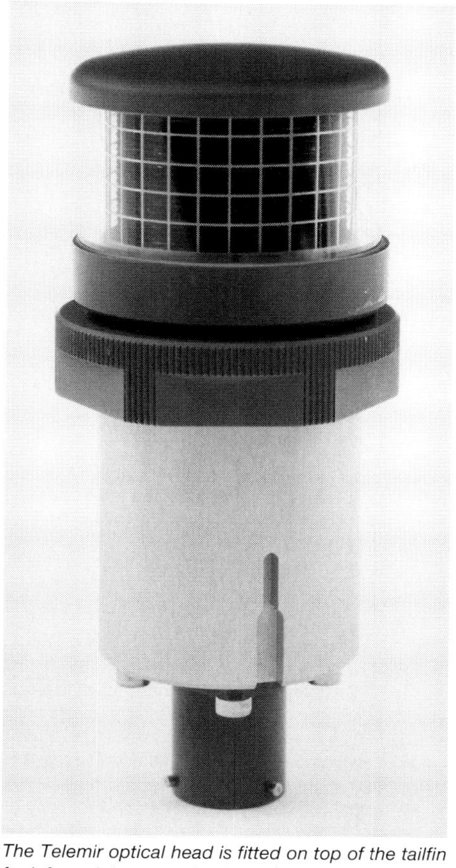

The Telemir optical head is fitted on top of the tailfin for infra-red data transfer (Rafale version) 0001173

Operational status

In service in French Navy Super Etendard aircraft and integrated into the naval version of Rafale.

Contractor

SAGEM SA, Defence and Security Division.

VERIFIED

Uliss inertial navigation and nav/attack systems

Uliss modular systems all employ high-accuracy inertial components consisting of two dynamically tuned gyros and three dry accelerometers, a microprocessor-controlled computer working at 1 Mops with EPROM memory and highly integrated and hybrid circuits. As an option, Uliss can be equipped with an embedded GPS receiver for high-performance INS/GPS coupling and a higher-performance RISC processor using Ada language.

Uliss systems fall into two categories, in both of which the main inertial navigation unit is contained within a ¾ ATR short case. In the first category are navigation versions with a position accuracy of better than 1 n mile/h. In the second category the navigation function is combined with the computation necessary for weapon delivery. Alignment time is 90 seconds for stored heading and 5 to 10 minutes for self-contained gyrocompassing. Standard interfaces permit the systems to be linked to other equipment via MIL-STD-1553B databusses or ARINC serial data lines. A failure detection system can detect faults at module level with a high level of confidence, isolate them and signal their presence on a magnetic annunciator without external test equipment.

More than 80 per cent of the components and subassemblies are common to all members of the Uliss family, the principal differences being in specific interfaces and computation functions.

In all, close to 2,000 Uliss units are in operation on 25 different types of aircraft.

Uliss 45 inertial navigation system

The Uliss 45 system has been optimised for high accuracy in long-range navigation and certain other special applications. Its interfaces comply with ARINC 561 and embody significant flexibility, for example, in order to communicate with two DME or Tacan receivers, with Kalman filtering for better accuracy. The equipment is used on long-range transport aircraft and as an accurate position and velocity reference for flight development purposes.

Specifications

Dimensions:
(navigator) 420 × 194 × 191 mm
(control/display unit) 209 × 114 × 127 mm
Weight:
(navigator) 16 kg
(control/display unit) 3 kg
Power supply: 115 V AC, 400 Hz, 250 VA
Accuracy: 1 n mile/h CEP for flights of up to 10 h duration.

Operational status

In service in Transall C-160 and Boeing KC-135 tankers of the French Air Force, and ATL-2 aircraft of the French Navy.

Uliss 52 inertial navigation system

The Uliss 52 navigation system is designed for high-performance combat aircraft. The program is based on Ada. Avionics information and commands are distributed by a digital multiplexed databus. It comprises three units: a UNI 52 inertial navigator, a PCN 52 control/display box and a PSM 52 mode selector fitted with an automatic insertion module to allow information such as flight plan, system data and maintenance information to be fed in.

Specifications

Dimensions:
(navigator) 386 × 194 × 191 mm
(control/display unit) 208 × 114 × 127 mm
Weight:
(navigator) 15 kg
(control/display unit) 3 kg
Power supply: 115 V AC, 400 Hz, 3 phase, 250 VA
Accuracy: 1 n mile/h CEP

Operational status

In production for French Air Force Dassault Mirage 2000DA, N and D aircraft and the export version Mirage 2000-5 and retrofit of Mirage 2000-5F of the French Air Force.

Uliss 92 inertial nav/attack system

The Uliss 92 combines in a single box all the functions of an inertial navigation and fire-control system. It is based on Ada programming and comprises three units: UNA 92 inertial/attack box, PCN 92 control/display unit and PSM 92 mode selector.

Specifications

Dimensions:
(navigator) 386 × 194 × 191 mm
(control/display unit) 216 × 116 × 153 mm
(mode selector) 151 × 41 × 135 mm
Weight:
(navigator) 16 kg
(control/display unit) 3.5 kg
(mode selector) 1 kg
Power supply: 115 V AC, 400 Hz, 3 phase, 220 VA
Accuracy: 1 n mile CEP position, 5 mrad weapon delivery

Operational status

In production for various Mirage upgrade programmes.

Contractor

SAGEM SA, Defence & Security Division.

VERIFIED

ADC 31XX Air Data Computers

The ADC 31XX is a new line of air data computers for military applications. They are based on a new generation of sub-miniature pressure sensors (P90 sensors), which use solid-state, vibrating beam resonators and modern microprocessor technology.

The modular design and alternate packaging arrangements available, make it easy to adapt the system to a number of primary flight needs (AFCS and fly-by-wire) and secondary or back up requirements (navigation and displays).

Specifications

Dimensions: 121 × 140 × 131 mm
Weight: 1.3 kg
Power: 28 V DC, 10 W (maximum)
Interfaces: MIL-STD-1553B, RS 422, ARINC 429 low and high speed

Operational status

Reported to be in production.

Contractor

Sextant.

UPDATED

ADU 3000/3008 Air Data Units

The ADU 3000 is a new line of air data measurement units, based on a new generation of micro-machined sub-miniature pressure sensors that produce a small, highly reliable, sensor unit. Different types of sensor can be used for differing applications:

(1) The P 90 sensor is a vibrating beam system that is used for high-measurement ranges. Two of these sensors are used in the ADU 3008 (one for pitot channel and one for the static channel). This version is optimised to the needs of transport aircraft - both civil and military.
(2) The P 92 sensor uses piezo-resistive gauges for measurement in a dual installation. This unit is optimised for low-measurement ranges and is intended for helicopter applications.

In addition to pressure measurements on pitot and static channels, the ADU 3000/3008 is used to perform computations related to air data parameters, including barometric corrections.

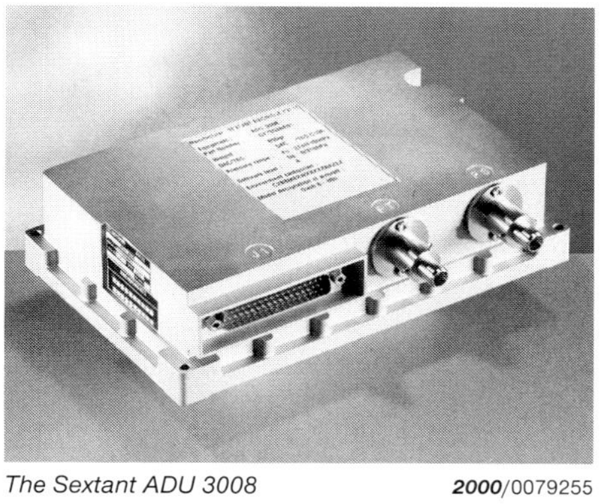

The Sextant ADU 3008 **2000**/0079255

Specifications

Dimensions:
(ADU 3000): 150 × 130 × 50 mm
(ADU 3008): 200 × 130 × 50 mm
Weight:
(ADU 3000): 0.8 kg
(ADU 3008): 1.1 kg
Power:
(ADU 3000): 28 V DC, 5 W
(ADU 3008): 28 V DC, 7.5 W
MTBF: 30,000-80,000 h

Operational status

In production for several helicopters (EC 135 and EC 155) and transport aircraft (C-130 and Dash 8-400).

Contractor

Sextant.

VERIFIED

AP 205 autopilot

The AP 205 autopilot was designed for the French version of the Sepecat Jaguar. It is a simple and rugged system allowing pilots to fly hands-off while conducting other operational tasks.

Autopilot facilities include pitch attitude and altitude hold in the pitch axis and heading hold in the roll axis. Trim functions provide automatic pitch trim and manual trim assist in roll. Further modes for navigation or guidance purposes can be added. Modular construction is claimed to minimise maintenance and the system is designed to be retrofitted into aircraft not equipped with autopilot. The first autopilot was delivered to the French Air Force in November 1981.

Specifications

Dimensions: 124 × 210 × 253 mm
Weight: 5.4 kg
Power supply:
200 V AC, 400 Hz, 3 phase, <40 VA
26 V AC, 400 Hz, 3 phase, <1 VA
28 V DC, 15 W

Operational status

In service in the French Air Force Sepecat Jaguar.

Contractor

Sextant.

VERIFIED

AP 305 autopilot/flight director system

The AP 305 autopilot and flight director system is derived from an earlier Sextant Avionique system, the Tapir, which equipped civil Fokker F27 and French Air Force Nord 262 light transports. For French Navy Nord 262s a radio altimeter low-altitude mode has been added for low-flying operations over the sea. It is an analogue system with sideslip suppression, turn co-ordination and Cat. I auto-approach capability. Flight director/autopilot facilities include altitude capture and hold, selected heading capture and hold, VOR/ILS capture and tracking, attitude hold and appropriate operation indications and alerts.

There is also a sideslip detector and four servo actuators.

Specifications

Dimensions:
(system computer) 421 × 125 × 194 mm
(control panel) 180 × 146 × 57 mm
(mode selector) 135 × 146 × 27 mm
Weight:
(excl actuators) 8.6 kg
(incl actuators) 21.6 kg
Power supply: 115 V AC, 400 Hz, single phase, <80 VA
28 V DC, <150 W

Operational status

In service in French Navy and Air Force Nord 262 aircraft.

Contractor

Sextant.

VERIFIED

AP 405 autopilot and autothrottle system

The AP 405 has been specially designed to monitor height above sea level, to meet the requirements of the French Navy for the low-altitude and high-speed operations of its carrierborne Dassault Super Etendard. Precise angle of attack holding is provided by the autothrottle, including during carrier approaches when precise control is a deciding factor in achieving consistency and safety of operations.

Autopilot functions are pitch hold, altitude capture and hold, heading hold (with control wheel steering override), semi-automatic pitch trim and the provision for instinctive disconnect to make rapid flight path changes. At very low altitudes there is also a radio altitude hold mode with continuous flight path monitoring. The autothrottle system can capture and hold a selected angle of attack.

Specifications

Dimensions:
(system computer) 409 × 198 × 94 mm
(control/indicator panel) 115 × 26 × 150 mm
(autothrottle actuator) 120 × 120 × 200 mm
(angle of attack selector panel) 37 × 45 × 50 mm
Weight: 7.9 kg
Power supply:
115 V AC, 400 Hz, single phase, <60 VA
26 V AC, 400 Hz, single phase, <1 VA
28 V DC, <60 W

Operational status

Fitted to French Navy Super Etendard aircraft.

Contractor

Sextant.

UPDATED

AP 505 autopilot

The AP 505 has been designed for M2.0+ combat aircraft. Pilots of the Dassault Mirage F1, which is equipped with the system, claim to be satisfied with its reliability and ease of use. The system can be switched on before take off, and engaged or disengaged by a handgrip trigger on the control column. In basic mode it maintains the longitudinal attitude, held when the pilot releases the stick trigger, and either the heading or bank angle, depending on whether re-engagement is effected at a bank angle of less or more than 10°. Autopilot modes provide automatic flight at a preselected altitude, heading or VOR/Tacan/ILS bearing. Limits on attitude hold facilities are ±40° in pitch and ±60° in roll.

Specifications

Dimensions:
(system computer) 190 × 202 × 522 mm
(function selector unit) 132 × 142 × 27 mm
(heading selector unit) 99 × 35 × 80 mm
Weight: 14.9 kg
Power supply:
200 V AC, 400 Hz, 3 phase, <100 VA
26 V AC, 400 Hz, single phase, <6 VA
28 V DC, <15 W

Operational status

In service in Dassault Mirage F1 aircraft. More than 600 sets have been built and supplied to eight countries.

Contractor

Sextant.

VERIFIED

AP 605 autopilot

The AP 605 is a digital flight control system developed for the Dassault Mirage 2000 supersonic combat aircraft. The Series 7000 computer allows future extensions to the basic auto-pilot, providing new modes suited to the particular missions flown by the aircraft. Flight control signals, which pass between the autopilot and the Mirage 2000 fly-by-wire flight control system, use a digibus serial data transmission system.

The AP 605 provides semi-transparent control, meaning that it is engaged or disengaged simply by releasing or taking hold of the control stick, which has a trigger switch in the handgrip. There is a high degree of internal monitoring, and computer design and organisation have been configured to reduce onboard maintenance. In basic mode the autopilot will hold pitch angle to any value within the range ±40°, or bank angle in the range ±60°. Additionally, there are altitude-capture and hold modes, preset altitude acquisition and fully automatic approach capability down to 200 ft. The Mirage 2000N version autopilot includes a terrain-following mode.

The Mirage 2000D version includes coupling with the air-to-ground fire-control system and the ability to provide very low-altitude capture and hold over the sea.

Specifications

Dimensions:
(system computer) 194 × 124 × 496 mm
(control unit) 24 × 146 × 115 mm
Weight: 12.7 kg
Power supply:
200 V AC, 400 Hz, 3 phase, <100 VA
26 V AC, 400 Hz, single phase, <1 VA
28 V DC, <30 W

Operational status

In service in Dassault Mirage 2000 aircraft.

Contractor

Sextant.

VERIFIED

AP 705 autopilot

The AP 705, equipping the Atlantique 2 ASW aircraft, is designed to provide precise flight path control and a high level of safety at very low altitudes above the sea. The system can hold a course while maintaining altitude. These levels of performance are made possible by the quality of inertial data available on the aircraft and by the versatility of the microprocessor-based computer. Automatic built-in testing enables operation of the system and its safety devices to be checked before take off and also facilitates onboard maintenance.

The three-axis autopilot includes pitch trim and can drive a flight director system. Autopilot modes permit pressure altitude hold, glide slope beam tracking (in Category I weather minima) and radio altitude hold over the sea at very low levels in reduced visibility. Lateral modes provide for holding the heading or course at the time of engagement, heading hold and homing or tracking on radio navaids, or navigation waypoints.

Specifications

Dimensions:
(system computer) 384 × 256 × 194 mm
(control unit) 200 × 164 × 67 mm
(servo-actuator) 185 × 183 × 101 mm
Weight: 23 kg
Power supply:
200 V AC, 400 Hz, 3 phase, <50 VA
28 V AC, <150 VA

Operational status

In production for the Dassault Atlantique 2 maritime patrol aircraft.

Contractor

Sextant.

VERIFIED

AP 705 autopilot

AP 2000 autopilot system

The AP 2000 autopilot system is designed specifically for VFR or IFR operation with light and medium helicopters. The system performs attitude stabilisation, basic cruise and approach modes for two or three axes of the helicopter. Built-in automatic testing enables operation of the system and its safety devices to be checked before take off and also facilitates onboard maintenance.

The AP 2000 autopilot system consists of one fully digital computer, one controller/programmer and serial and trim actuators. The AP 2000 autopilot computer drives serial actuators on the cyclic and yaw axes to perform stabilisation, attitude hold with beep trim control and heading hold. The parallel actuators are driven by the computer to achieve an automatic trim function. With the AP 2000 system, cruise and approach can be flown using standard modes such as altitude hold, heading select, IAS hold, VOR navigation, glide slope and localiser approach or back course localiser approach. Altitude and vertical speed select modes are optional.

One of the main features of the AP 2000 system is its adaptability towards the various possible helicopter configurations in terms of certification, operational or flight characteristics requirements.

The AP 2000 system also incorporates numerous features for simplified automatic line and shop maintenance.

Operational status
In development.

Contractor
Sextant.

VERIFIED

APFD 800 Autopilot/Flight Director

The APFD 800 all-digital autopilot/flight director has been developed for a wide range of aircraft, from light jet or prop fighters to medium transport, ASW or SAR aircraft.

It has a modular architecture and all the subassemblies are designed for easy installation on existing aircraft as part of the upgrade operation. It offers basic and higher modes, with many specific functions in the horizontal and vertical planes. The system is designed for use down to very low altitude, with or without autothrottle. It can be coupled with radio navigation, INS or the flight management system and is proposed with brushless DC motor actuators.

The APFD 800 provides semi-transparent control of the aircraft, as it can be instinctively engaged or disengaged hands-on-stick.

Specifications
Dimensions:
(computer) 4 MCU
(control unit) 146 × 57 × 140 mm
(mode selector) 38 × 140 × 146 mm
Weight: 7 kg total
Power supply: 28 V DC, 73 W

Operational status
Selected by the French Navy for upgrade of its Alizé maritime patrol aircraft; it is also fitted to Nimrod aircraft.

Contractor
Sextant.

VERIFIED

B 39 autocommand autopilot

Introduced into French Air Force Dassault Mirage III interceptors since 1975 to replace older autocommand systems, the B39 is available for retrofit to other versions of the same aircraft type. It is a relatively simple system, but uses modern integrated circuit techniques to confer benefits in terms of performance, safety and maintenance standards. The new autocommand computer is physically interchangeable with the original equipment.

Autopilot functions are reduced to attitude and altitude hold modes. Attitude hold includes short-term capability, stability augmentation and uniform artificial feel load against load factor irrespective of flight conditions.

Specifications
Dimensions: 264 × 200 × 140 mm
Weight: 6.5 kg
Power supply: 200 V AC, 400 Hz, 3 phase, <25 VA
28 V DC, <1 A

Operational status
In service in Dassault Mirage III aircraft.

Contractor
Sextant.

VERIFIED

Cirus attitude and heading reference system

Cirus is a hybrid inertial attitude and heading reference system. Its basic design involves the use of inertial/magnetic heading and air data hybridisation techniques. The system generates heading, attitude, angular velocities, accelerations, true airspeed, pressure altitude and temperature.

The APFD 800 autopilot/flight director has been selected for the upgrade of French Navy Alizé aircraft

As an option, it can also deliver position and groundspeed data in conjunction with a Doppler navigation radar or a GPS receiver. The groundspeed vector supplied by the GPS system, with its medium-term accuracy, can be combined with the angular and linear speed data generated by Cirus, with its excellent short-term accuracy, to form a versatile high-performance AHRS/GPS coupling system.

Specifications
Dimensions: 260 × 150 × 150 mm
Weight: 5 kg
Power supply: 28 V DC, 50 W
Alignment time: <1 min

Operational status
In production. The Cirus system is fitted to the Eurocopter Super Puma Mk 2 helicopter. It has been selected for the French Air Force C-160 Transall retrofit (Cirus 1600 and Totem 200) and maritime patrol Alizé aircraft (Cirus 1500).

Contractor
Sextant.

VERIFIED

Gemini 10 navigation and mission management computer

The Gemini 10 computer contains a Jeppesen database and, in addition to carrying out conventional FMS functions, also performs tactical military functions. Connected to the VH100-T HUD, it supports missions such as assault landings, airdrops and tactical low-altitude flights.

The computer's compact design is due to its monobloc concept, with the navigation computer and control and display unit integrated in a single item of equipment. This layout offers installation advantages of lower weight, smaller volume and higher reliability, together with easier maintenance and lower power consumption in operation. The Gemini 10's dual architecture helps increase the probability of mission success by allowing reconfiguration on the validated FMS, in case of failure of one of the two units.

The software comprises 150,000 lines in Ada language, conforming to the requirements of DO178, and allows the display of approximately 150 pages of information. The display offers 12 lines of 20 characters each and five variable label keys with functions depending on the mission.

Operational status
Production of the Gemini 10 started in 1993 for the C-160 Transall upgrade.

Contractor
Sextant.

VERIFIED

Mk 3 Magnetic Anomaly Detector

The Magnetic Anomaly Detector (MAD) Mk 3 is specifically designed for use both on fixed-wing aircraft and inboard on helicopters to detect the presence of a submersible by measuring the disturbance to the earth's magnetic field. It is an integral, digital, solid-state airborne system, which becomes operational at switch on without any warm-up time. Target parameters are automatically delivered in real time on a CRT control and display unit.

The system consists of three separate units: a detection unit, a computer and a control/display unit. A MIL-STD-1553B databus interface card is included in the system. A graphic recorder is available as an option.

The sensor operates on the nuclear magnetic resonance principle and uses the precession of protons in a liquid, the frequency of which, measured by the pick-up coils, is proportional to the magnetic field to be measured. Compensation is employed to eliminate from the received signal all disturbances created by the magnetic element in the aircraft and their movement in the earth's field. The compensator uses a 16-term model representing the magnetic components in the aircraft. The MAD Mk 3 includes rapid aircraft identification modes.

The target is detected and located automatically. This is a fundamental role, allowing the MAD system to

The Sextant MAD probe on the Dassault Aviation Atlantique 2

give a high detection probability with a very low false alarm rate in extracting the target signal from the background noise. The Sextant MAD system is based on a mathematical comparison between the current MAD signal and an analytical model of the target signals, as opposed to the more conventional use of threshold detection in several frequency bands. All computing tasks are performed by the Alpha 732 proprietary 1 Mops digital computer operating in Pascal.

Specifications

Dimensions:
(detection element) 1,250 × 125 mm diameter
(computer) ½ ATR
(control/display unit) 190 × 146 × 162 mm
Weight:
(detection element) 5.5 kg
(computer) 10 kg
(control/display unit) 3.5 kg

Operational status

In production for the French Navy Dassault Atlantique 2 maritime patrol aircraft, and in development for the Italian Navy EH 101 ASW helicopter.

Contractor

Sextant.

VERIFIED

Mk 3 towed magnetometer for helicopters

The Mk 3 towed magnetometer version of Sextant's MAD equipment is intended for ASW helicopters. It operates in exactly the same way as the system described previously and, in order to eliminate disturbances created by the helicopter, the detector probe is placed in a streamlined housing which is towed at the end of a 70 m cable. The digital computer measures the signals from the sensor and transforms them into suitable formats for the graphic recorder at the operator's station.

The housing assembly includes the detection probe and an electronic unit, combining to produce a nuclear oscillator whose Larmor frequency is a function of the magnetic field exerted on the probe. The geometry of the probe is chosen so that its position with respect to the magnetic field vector can be ignored.

Maintainability is assisted by in-flight checking of the towed magnetometer housing, computer, control unit and recorder. The ground-test points are easily accessible and the subassemblies are plug-in units.

Specifications

Background noise: typical deviation 0.006 gamma
Measurement range: 25,000-70,000 gamma
Sensitivity for relative field output: 1, 2, 5, or 10 gamma for 100 mm stylus deviation
Paper speed: 6, 75, or 300 mm/min

Streamlined body
Dimensions: (length) 1,300 mm × (diameter) 160 mm
Weight: 16 kg

Computer
Dimensions: 346 × 124 × 194 mm
Weight: 7.5 kg
Power supply: 200 V AC 400 Hz, 3 phase, 100 W
28 V DC, 5 W

Control unit
Dimensions: 165 × 150 × 57 mm
Weight: 1 kg

Recorder
Dimensions: 190 × 150 × 190 mm
Weight: 4 kg
Power supply: 115 V AC 400 Hz, 50 W
28 V DC, 20 A

Winch and cradle
Dimensions: 1,330 × 350 × 755 mm
Weight: 44 kg
Power supply: 27 V DC, 40 A

Operational status

Helicopter version in production and in service.

Contractor

Sextant.

VERIFIED

NASH-Night Attack System for Helicopters

NASH — Night Attack System for Helicopters – is designed to provide the capability required for combat helicopters to carry out attack and counter-insurgency missions at night and in all weather conditions. In addition to flight management and navigation functions, this system displays information on observation, target selection and fire control.

Observation, target detection and fire-control functions are handled by the following systems:

(1) The Victor thermal imager, integrated either in the gyrostabilised sighting system, or in a Chlio gyrostabilised platform, coupled with the existing sighting system. The FLIR image is displayed on the SMD66 smart multifunction liquid crystal display

(2) Two Topowl® binocular helmet-mounted sight/displays, which can display images from an infra-red sensor or integrated light intensifiers. Topowl® offers a very wide field-of-view, with images projected directly on to the visor.

Navigation and mission management functions are provided by a Nadir 1000 navigation system, working in conjunction with a GPS receiver, Stratus attitude and heading reference system and a Doppler radar.

All flight information for pilot and co-pilot are shown on the SMD66 head-down display and the Topowl HMS/D.

Operational status

Sextant's night navigation system has been selected for the Rostvertol Mi-35.

Contractor

Sextant.

VERIFIED

Rooivalk helicopter avionics system

In August 1994, Sextant signed a contract with the South African avionics integrator, ATE Pty Ltd to supply key avionics equipment for the Rooivalk attack helicopter being developed and manufactured by South African aircraft manufacturer, Denel Aviation.

Equipment supplied by Sextant comprises the liquid crystal displays, helmet-mounted sight/displays, ring laser gyro navigation system with embedded GPS, standby instruments and pilot handgrips.

Sextant has completed delivery of all the basic avionics and expanded basic avionics package for the development phase and has started delivery of basic avionics equipment for the first Rooivalk production models.

The basic avionics of the Rooivalk are off-the-shelf items, which were all qualified on the Tiger combat helicopter. Sextant is providing a complete, redundant navigation system, including two inertial navigation units, each based on a PIXYZ® three-axis ring laser gyro, air data equipment, two magnetometers and a GPS receiver. The navigation system is built around the

The Rooivalk helicopter carries the Sextant avionics system 0001318

inertial navigation unit, which functions as inertial sensor, navigation computer and controller for other peripherals. Three complete systems have been delivered. Flight tests have been completed on a Puma flying test bed, and are continuing on the Rooivalk EDM helicopter.

The Rooivalk helicopter will be equipped with Sextant's MultiFunction liquid crystal Display (MFD 66). These displays offer excellent readability and a good viewing angle under any lighting conditions, even direct sunlight. They are part of a complete family of LCDs developed by Sextant for both civil and military applications, including rotor-wing aircraft and are already chosen for a number of programmes, including Tiger, NH 90 and EC135. Eight displays have been delivered and are undergoing flight testing onboard the Rooivalk EDM helicopter.

As part of an expanded basic avionics package, Sextant is providing the Topowl® Helmet-Mounted Sight/Display (HMS/D) developed as part of development contracts for the Tiger programme. Topowl® is one of a new generation of biocular, wide field-of-view, day/night capable sight/display systems. It provides projection of images on the helmet visor – either video or synthetic images or from integrated image intensifiers.

Operational status

Three HMS/Ds have been delivered and have undergone flight tests on the Rooivalk EDM helicopter. In November 1998, the first production-standard unit passed its acceptance tests. The system is now in production; seven navigation, 36 MFD 66 and 10 HMS/D units had been delivered by November 1999. The first three Rooivalk helicopters had been delivered to the South African Air Force by October 1999.

Contractor

Sextant.

VERIFIED

Stratus and Totem 3000 flight systems

Sextant produces three types of Ring Laser Gyros (RLG):

(1) a three axis monolithic PIXYZ®, 14 cm path length, 0.1°/h in run stability;
(2) a three axis monolithic PIXYZ®, 22 cm path length, 0.001°/h in run stability;
(3) a single axis, 33 cm path length, 0.001°/h in run stability.

With this family of RLG, Sextant produces several inertial reference units for a large range of applications.

STRATUS

Stratus is an attitude and heading reference system that uses the RLG PIXYZ® 14 cm path length. It is used as a basic reference system for attitude and hybrid navigation (with Doppler radar and/or air data) on military helicopters, advanced trainers and missile systems.

Specifications

Attitude: 0.2°
Magnetic heading: 0.3°
Position: 1% of distance (with Doppler)
Weight: 5.2 kg
Power supply: 115 V AC, 45 VA
Interfaces: ARINC 429 and MIL-STD-1553B
Environment: as per MIL-STD-810D

Operational status

Selected for the Tiger and Rooivalk helicopters.

TOTEM 3000

Totem 3000 is an inertial/GPS navigation system for high-performance military aircraft. It utilises the PIXYZ® 22 cm path length RLG and the Sextant Topstar® 1000 GPS receiver board. The use of these two highly integrated and high-performance subassemblies leads to significant improvement of cost, size, and reliability compared with conventional single axis RLG.

All the functions are in compliance with SNU 84-1 and STANAG 4294

Specifications

Attitudes/heading: 0.05°
Position: 0.5 n miles/h CEP (inertial mode)
21 m (95%) (inertial/GPS PPS mode)

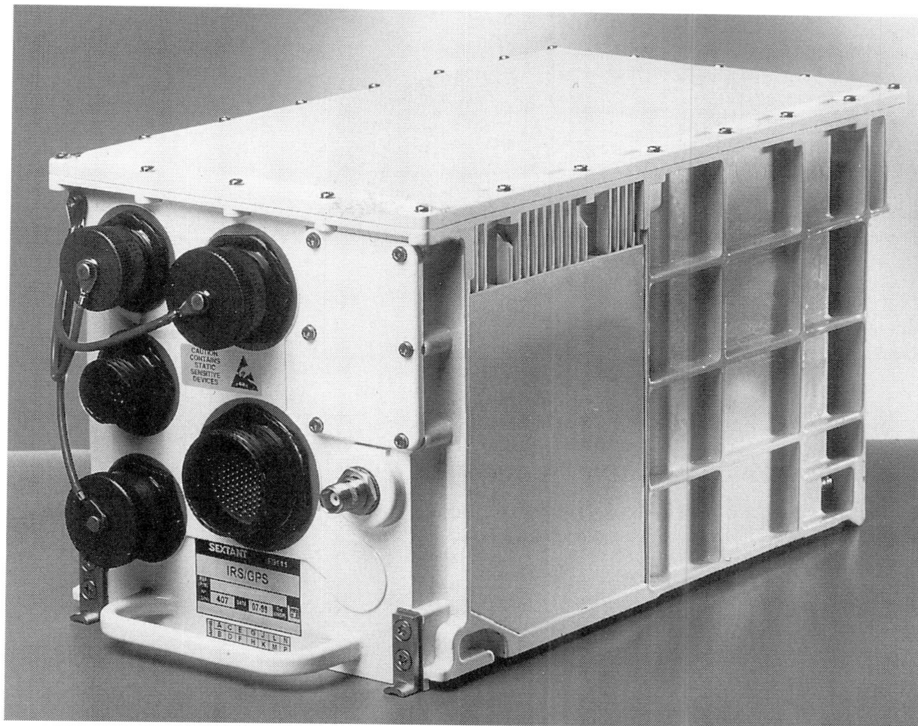

The Sextant Totem 3000 navigation and flight management system **2000**/0079262

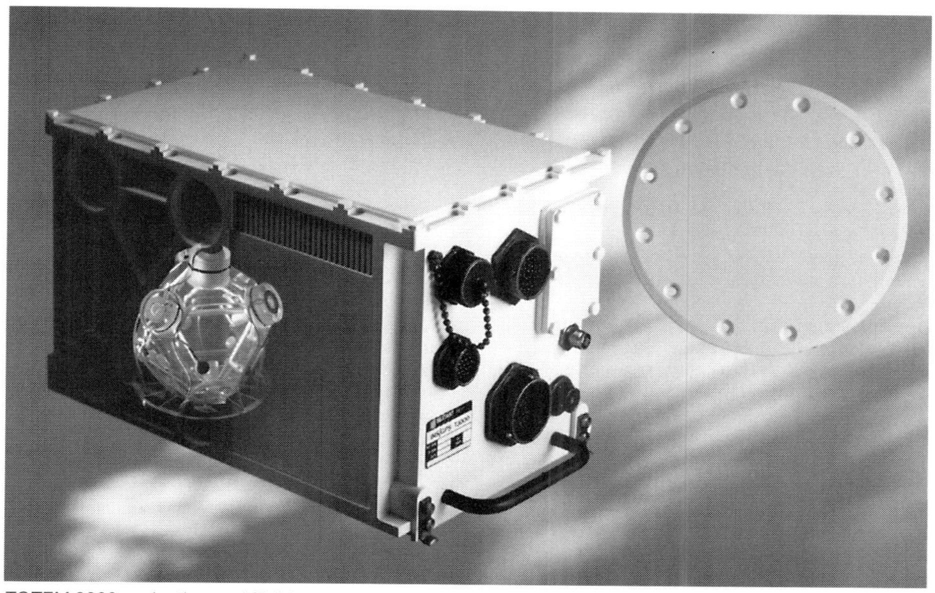

TOTEM 3000 navigation and flight management system 0001317

STRATUS attitude and heading reference system 0001315

Velocity: 0.7 m/s RMS (inertial mode)
0.1 m/s (95%) (inertial/GPS PPS mode)
Dimensions: 177.8 × 177.8 × 279.4 mm
Weight: 8 kg including GPS
Power consumption: 40 W
Interfaces: standard MIL-STD-1553 B and ARINC 429
Environment: as per MIL-STD-810 D

Operational status

Selected on Alphajet, I-22, MiG-AT and retrofit programmes including Hawk, MiG-21, MiG-29, Mirage F-1, Mirage 2000-9 and Lockheed Martin C-130 aircraft.

Contractor

Sextant.

VERIFIED

Topstar 100 0001320

Topflight® avionics suite

The Topflight® avionics suite is based on avionics designed for Rafale, and it has been designed for basic or advanced training aircraft, as well as for combat aircraft such as the Mirage F1 or Su-22, either as original equipment or for retrofit. It enhances aircraft operational capabilities in a complex environment by supporting the pilot in all navigation tasks and in both air-to-air and air-to-surface missions.

Topflight® incorporates advanced technologies, such as holography, liquid crystal displays, laser gyros, head position direction, new-generation computers, GPS, digital mapping and voice command. These are employed in a helmet-mounted display, a key part of the man/machine interface tailored to night mission capability; multimode smart head-up display; multifunction liquid crystal displays; laser gyro navigation system; digital autopilot, and the Precise Position Service (PPS) version of GPS. Topflight® is built around a modular mission computer and symbol generator.

The compact design, incorporating smart liquid crystal displays, gives Topflight® a multiple mounting capability and the modular computer provides the performance levels needed for multirole aircraft.

Operational status

Selected for the modernisation of Spain's Mirage F1, Argentina's A-4 and India's MiG-21 aircraft. It is also in production for the MiG-AT upgraded-avionics programme, in which Topflight® comprises:

1. a SHUD wide field-of-view multimode head-up display, which also controls MIL-STD-1553B databus management;
2. two reconfigurable centralised control panels;
3. five multifunction colour LCD units (two for the front seat and three for the rear seat) in a 5 × 5 in (MFD 55) format;
4. standby instruments;
5. a mission management computer and symbol generator which provide weapon systems simulation;
6. a Totem 3000 laser gyro Inverted Navigation System (INS);
7. a Topstar GPS/GLONASS stand-alone receiver;
8. a UMPT air data system.

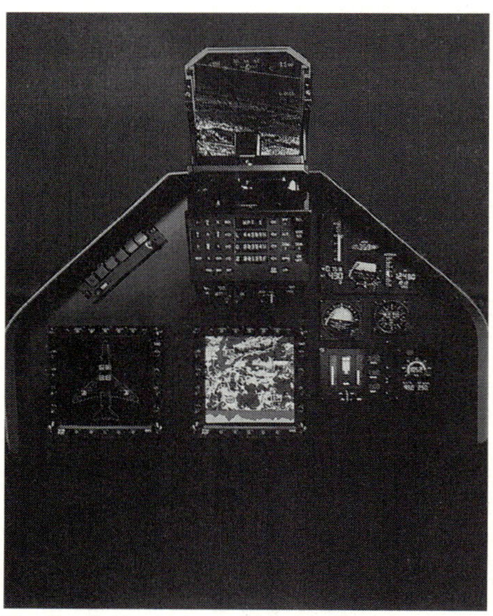

The Topflight avionics suite for training and combat aircraft

A derivative system called Smart Topflight® has been developed for the Iryda advanced trainer built by PZL Mielec. In this configuration Smart Topflight® comprises:

1. a SHUD for piloting, air-to-air and air-to-ground fire-control and MIL-STD-1553B databus management;
2. a Totem 3000 laser gyro INS for navigation, databus management in back-up modes and system interfaces;
3. two configurable centralised control panels;
4. one or more multifunction colour LCD (SMD 54) to copy the SHUD to the back seat;
5. standby instruments;
6. an UMPT 30 air data system.

Contractor

Sextant.

VERIFIED

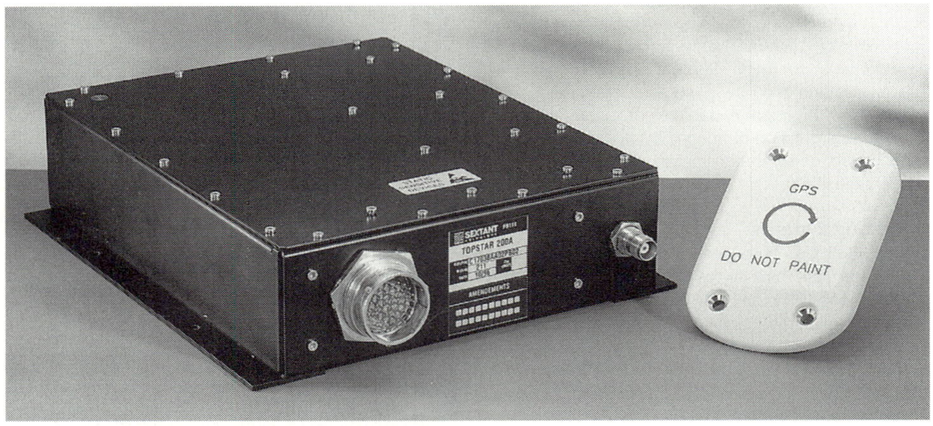

The Sextant GPS Topstar® 200 A with antenna **2000**/0079263

Topstar® family of GPS receivers

Topstar is a family of multichannel continuous tracking sensors. They provide time, three-dimensional position, and speed information at a data rate of up to 10 Hz.

They provide data for speeds up to 800 kt, and for high acceleration applications. They provide accuracy in compliance with international standards. Additional functions such as DGPS, relative navigation, combined GPS/GLONASS processing, and RAIM are offered as options.

Four products are available: three for aircraft applications, detailed below

Topstar 100

Topstar 100 is a stand-alone sensor for high-dynamic military applications in aircraft, helicopters and missiles.

GPS Topstar® 1000 and antenna **1999**/0051588

Specifications
SPS or PPS operation
8 parallel channels
Dynamic operation: 1,500 m/s - 10 *g*
Accuracy: as per STANAG 4294
Dimensions: 130 × 240 × 80 mm
Weight: 2 kg
Power supply: 28 V DC, 20 W
Interfaces: MIL-STD-1553B, ARINC 429, RS-422
Environment: as per MIL-STD 810-E

Operational status
Topstar 100 is in production for Mirage F-1, Mirage 2000, Rafale aircraft and for Apache, Cougar, Rooivalk and Tiger helicopters.

Topstar 1000
Topstar 1000 is a new version of Topstar 100, designed as a PCB module to be integrated in host equipment like IRS. Topstar 1000 interfaces with host computers through a dual-port RAM and RS-422 databus. It has growth provision for GLONASS reception and processing.

Specifications
Dimensions: 150 × 150 × 19 mm
Weight: 0.75 kg
Power construction: <12 W

Operational status
In production. Topstar 1000 is embedded in TOTEM 3000 IRS, which has been selected for several retrofit programmes, including Alpha Jet, C-130, and MiG-21. Topstar 1000 has also been selected for the Storm Shadow/SCALP EG and other missile programmes.

Topstar 200 A
Topstar 200 A is an ARINC 743A-compliant stand-alone sensor for civil aviation.

Specifications
C/A (L1) + W.A.A.S. capability
15 parallel channels
Dynamic operation: 2,000 kt, 3 *g*
Accuracy: as per ARINC 743A, RAIM GIC and differential RAIM; and SCAT 1c RTCA DO 217 optional capabilities
Dimensions: ARINC 743A 2MCU format or ARINC 743A alternate format (190 × 240 × 63 mm)
Weight: 2 kg
Power supply: 28 V DC, 15 W
Interfaces: as per ARINC 743A
Environment: as per RTCA DO 160c
TSO C 129 C1 label

Topstar 2020
Topstar 2020 is a version of Topstar 200, designed as a PCB to be integrated in host equipment such as the IRS or multimode receiver MMR.

Operational status
Selected for the Boeing MD 82 and for the Airbus family.

Contractor
Sextant.

VERIFIED

12000 VHF AM/FM radio

The 12000 is a remote-controlled VHF transceiver designed for air traffic and air-to-ship communications in the 100 to 173 MHz frequency range with a channel spacing of 12.5 kHz. The AM/FM radio can operate with direction-finding equipment and sonobuoys in anti-submarine warfare and is capable of radio relay and voice scrambling. It has basic output powers of 15 and 3 W, with a 6 mW discrete capability. An integrated Guard receiver function is incorporated. The equipment's interface could be either ARINC 410 or 429. The transceiver is suitable for combat aircraft, helicopters and maritime surveillance aircraft. The transceiver can be controlled by the TC 20 control unit.

Specifications
Dimensions:
(transceiver) 197 × 61 × 375 mm
Weight:
(transceiver) 5 kg
Power supply: 22.5-31.5 V DC
Frequency:
(transmission) 100-156.975 MHz
(reception) 100-173.5 MHz

Operational status
In service with the French Navy and Air Force.

Contractor
Thales (formerly Thomson-CSF Communications).

UPDATED

The 12000 VHF AM/FM radio

12100 VHF FM radio

The 12100 radio is a tactical VHF FM remote-controlled airborne transceiver designed specifically for air-to-ground communications between military aircraft and ground forces.

It covers 30 to 88 MHz. In addition to the standard military FM mode, it provides interoperability with security forces, with a 12.5 kHz step and lower frequency deviation mode.

The Series is capable of KY 58 cyphered voice, external homing and radio delay. Nominal power is 15 W, with a reduced power output of 3 W.

The transceiver can be controlled by the TC 20 control unit.

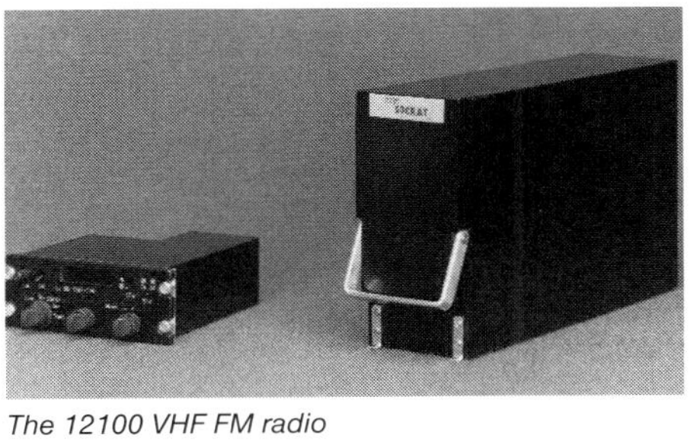

The 12100 VHF FM radio

Specifications
Dimensions:
(transceiver) 200 × 100 × 340 mm
Weight:
(transceiver) 5 kg

Operational status
In service with the French Air Force and Navy on the AS 355.

Contractor
Thales (formerly Thomson-CSF Communications).

UPDATED

3527 HF/SSB radio

The 3527 HF/SSB radio comprises the 3527F transceiver, 3527H remote control, 3596A antenna coupler, 3597A digital preselector and the 3598A FSK modem. The preselector filter enables the simultaneous operation of two 3527H units on the same aircraft, one for transmission and the other for reception.

The fully solid-state transmitter supplies a modulated peak power of 400 W. The transceiver is driven by a digital synthesiser tuned for setting to 280,000 channels spaced at 100 Hz intervals. The antenna coupler provides tuning with wire antennas from 10 to 30 m long. The FSK modem handles messages of 50 to 100 bauds.

The comprehensive fault identification BITE increases maintenance efficiency and the modular construction minimises repair time by fast access to all circuit modules.

Specifications
Power supply: 115 V AC, 400 Hz, 3 phase
Power output: 400 W PEP, 200 W average (SSB)
Modes: USB, data (USB), CW, Link 11
Frequency stability: ±5 × 10^7
Number of channels: 280,000
Operational specification: STANAG 5035, AIR 7304

Operational status
In production.

Contractor
Thales (formerly Thomson-CSF Communications).

UPDATED

Thales' 3527 HF/SSB radio

ABD 2000 jammer

The ABD 2000 is a self-protection detector/jammer designed for the export version of the Mirage 2000. This multimission aircraft is intended to operate in a wide range of theatres and its self-protection system is able to counter all types of threats.

The ABD 2000 is installed internally, thereby avoiding using up ordnance pylon points and/or limiting the aircraft flight envelope. The complete system weighs 80 kg and consists of four LRUs: a main unit containing

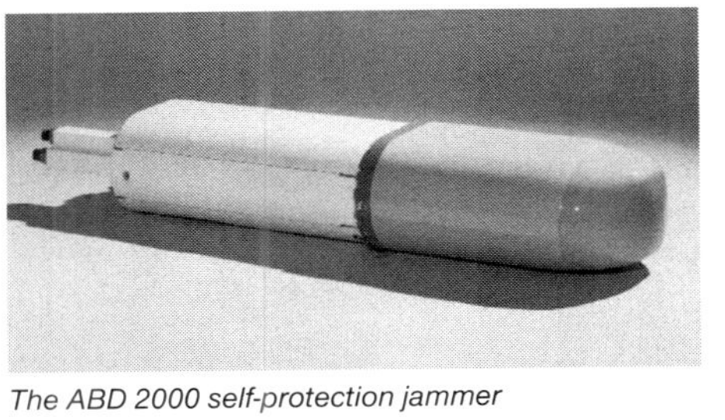

The ABD 2000 self-protection jammer

Mirage 2000 aircraft fitted with the Thales' ABD 2000 self-protection jammer, showing the rear antenna above the jetpipe

the receiver, jamming channels, transmitter and the aft antennas; a left-hand conformal unit containing the electrical power supplies for the system; a right-hand conformal unit containing the computers controlling the system operation; an antenna providing forward coverage and located on the vertical fin of the aircraft.

The ABD 2000 system is capable of detecting and identifying RF emitters selecting the more dangerous threats alerting the pilot on the countermeasures display unit; and automatically initiating jamming appropriate to the type of threat. It has a very wide frequency coverage, is entirely controlled by microprocessors, and can be programmed in accordance with user's operational scenarios. The system is operated by the countermeasures control panel fitted in the cockpit. Using this panel the pilot can select one of four positions: Off, SIL (the system detects a threat), Jam, Test.

The ABD 2000 can be integrated in a complete EW system, known as the Integrated CounterMeasures System (ICMS), which has been developed by Dassault Aviation, Thales (formerly Thomson-CSF DETEXIS) and Matra BAe Dynamics SAS.

Specifications

Frequency: H-, I- and J-bands
Coverage: all sectors
Reaction time: 0.6 s
Detector sensitivity: up to −60 dBm
Measured parameters: RF frequency, frequency agility, pulsewidth, PRI, duty cycle, sector, polarisation, level, type
Multi-jamming modes: continuous and spot noise, barrage noise, cover pulse jamming, count down, velocity deception, blinking.

Operational status

In service on the Mirage 2000 aircraft of several air forces.

Contractor

Thales (formerly Thomson-CSF DETEXIS).

UPDATED

Agrion maritime surveillance

Agrion is a member of the Iguane family of maritime surveillance radar systems and exists in several versions. It is designed primarily for use aboard helicopters or light aircraft forming a part of task forces, employed for support at sea or for coastal protection. Several types of antenna are available to meet the requirements of various aircraft. The Agrion 15 version allows the guidance of the AS 15TT Aerospatiale air-to-surface missile.

Agrion operates in the I/J-band, using pulse compression and frequency agility to ensure high performance on maritime targets in all combinations of weather, sea state and operating altitude. These same techniques also provide maximum protection against electronic countermeasures.

The system provides operational missions such as surface and anti-submarine warfare, over-the-horizon targeting for shipborne surface-to-surface missiles, search and rescue, marine environmental protection, navigation and weather avoidance.

Operational status

Agrion 15 radars are reported to be in service aboard Eurocopter AS 565SA Panther anti-submarine/anti-ship warfare helicopters operated by the Saudi Navy.

Contractor

Thales (formerly Thomson-CSF DETEXIS).

UPDATED

AHV-9 and AHV-9T radio altimeter

The AHV-9 was designed for the Mirage 2000-5, and the AHV-9T for the Panavia Tornado.

Low- and high-altitude bands can be used, the variable output being available from a microprocessor-based receiver unit. The system features CMOS integrated electronics, stripline technology antenna and built-in fault detection capability.

Specifications

Dimensions:
(indicator) 61 × 61 × 158 mm
(transmitter/receiver unit) 109 × 154 × 324 mm
(antenna) 78 × 88 × 33 mm
Weight:
(indicator) 0.9 kg
(transmitter/receiver unit) 4.5 kg
(antenna) 0.2 kg
Power supply: 115 V AC, 400 Hz, 70 VA
Altitude: up to 50,000 ft
Accuracy: 1 ft ±2% of altitude
RF power: 60 MW (FM/continuous wave)
Outputs: 5 digital, 3 analogue

Operational status

The AHV-9T is in service in German Panavia Tornados.

The AHV-9 radio altimeter, and the NC 12 Tacan receiver are fitted to Mirage 2000-5 aircraft, and will be fitted to the Mirage 2000-9.

Contractor

Thales (formerly Thomson-CSF Communications).

UPDATED

AHV-12 radio altimeter

Designed for use on the Dassault Mirage 2000 and Atlantique 2, the AHV-12 radio altimeter incorporates recent digital developments. Its main features are wide altitude range, high-accuracy and high-integrity levels.

Specifications

Typical installation
Dimensions:
(antenna) circular, rectangular or small size
(transmitter/receiver unit) 193 × 90 × 315 mm
(indicators) ARINC 429 (digital), ARINC 552 (analogue)
Weight:
(transmitter/receiver unit) 5 kg
Power supply: 115 V AC, 400 Hz, 45 VA
Altitude: up to 70,000 ft
Accuracy: 1 ft ±1% of altitude

Operational status

In production and service in Mirage 2000 variants, Atlantique 2 and US reconnaissance aircraft.

Contractor

Thales (formerly Thomson-CSF Communications).

UPDATED

Thales' AHV-12 digital radio altimeter

AHV-17 digital radar altimeter

The AHV-17 is specifically designed for Rafale. It has very low probability of intercept due to an RF management system which adjusts output power as a function of altitude and enhanced receiver sensitivity. The AHV-17 has excellent jamming detection capability, which enables it to have superior resistance even against advanced jammers.

Chin-mounted Agrion 15 radar on an AS 565SA Panther anti-submarine/anti-ship warfare helicopter

The AHV-17's modular construction, based on a series of integrated tests, allows effective diagnosis of failures and simple replacement without the need to replace or adjust the defective module. Modules can be interchanged independently.

Specifications
Volume: <3 litres
Weight: <3 kg
Power supply: 28 V DC to MIL-STD-740D, 40 W
Frequency: 4.2-4.4 GHz
Altitude: up to 30,000 ft
Accuracy: 3 ft or 1% of altitude
Reliability: >5,000 h MTBF

Operational status
In production for the Rafale aircraft.

Contractor
Thales (formerly Thomson-CSF Communications).

UPDATED

AHV-18 compact radio altimeter

The AHV-18 is the most advanced military radio altimeter of the Thomson-CSF product line. It is multimode equipment, of modular design, which can be adapted for use with fixed- and rotary-wing aircraft, missiles, RPVs and flying weapons.

It has comprehensive ECCM capabilities using power management and jamming detection modes.

Specifications
Dimensions:
(indicator) 3 ATI
(transmitter/receiver unit) 124 × 81 × 81 mm
(antenna) 105 × 90 × 38 mm
Weight:
(indicator) 1.2 kg
(transmitter/receiver unit) 1.2 kg
(antenna) 0.13 kg
Altitude: up to 5,000 ft
Accuracy: 1 ft ±2% of altitude
Reliability: >5,000 h MTBF

Operational status
In production for the Agusta A 109 helicopter and Penguin Mk 2 missile.

Contractor
Thales (formerly Thomson-CSF Communications).

UPDATED

AHV-2100 digital radar altimeter

The AHV-2100 fully digital radar altimeter has been specifically designed for modern military helicopters. The simple hardware and up-to-date technology provide high reliability and performance at low altitude and in hovering.

The AHV-2100 has a dual-processing chain and DO-178 qualified software. In addition, power management of the RF output reduces the probability of interception at low altitude over water and the combination of a narrow receiver bandwidth with high-performance digital signal processing provides resistance to jamming.

Specifications
Dimensions: 110 × 90 × 190 mm
Weight: <2.2 kg
Power supply: 28 V DC
Altitude: up to 5,000 ft
Accuracy: 3 ft ±5% of altitude
Reliability: 5,000 h MTBF
Interfaces: ARINC 429, dual-redundant MIL-STD-1553B or analogue

Contractor
Thales (formerly Thomson-CSF Communications).

UPDATED

AHV-2900 digital radar altimeter

The AHV-2900 is a fully digital radar altimeter designed for future combat aircraft and tactical fighters. It is optimised for low-altitude penetration. It includes dual-processing chains and ARINC 429 or MIL-STD-1553B interfaces.

Contractor
Thales (formerly Thomson-CSF Communications).

UPDATED

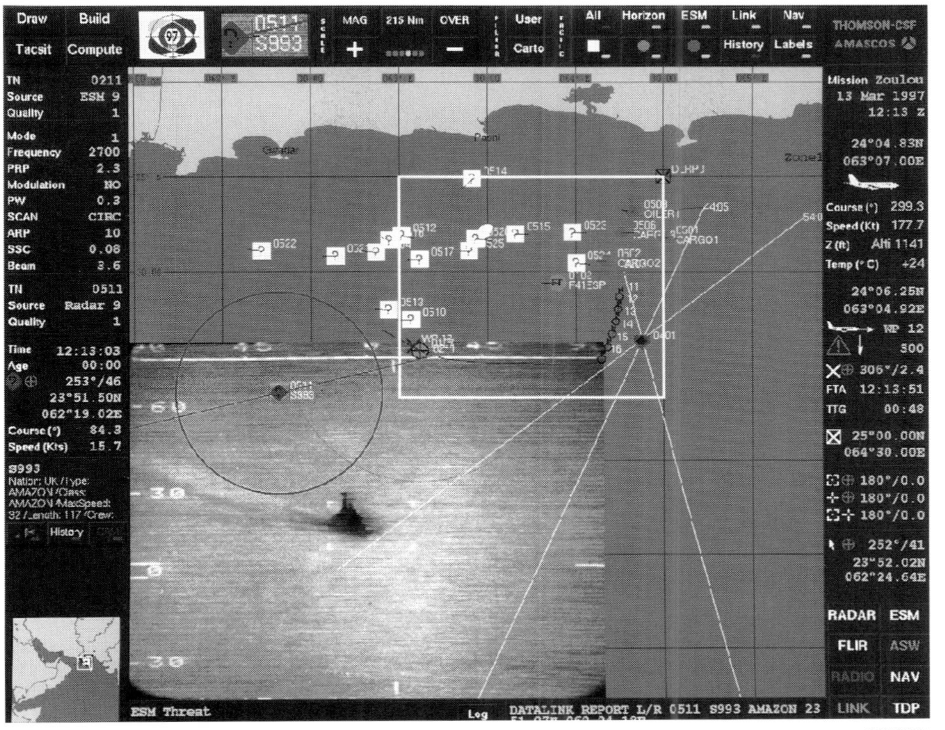

AMASCOS display 0011861

AMASCOS multisensor system

AMASCOS (Airborne Maritime Situation Control System) is designed for building up and updating tactical situations in real time and as a decision aid for operators. It is a family of maritime systems, which uses a modular approach to system design and can be integrated on any type of fixed-wing aircraft or helicopter.

The three versions of AMASCOS - AMASCOS 100, AMASCOS 200 and AMASCOS 300 - correspond to the broad categories of mission requirement ranging from simple maritime surveillance to anti-surface and anti-submarine warfare. The typical AMASCOS configuration integrates Thomson-CSF DETEXIS equipment such as radar, FLIR, sonics, MAD and communications, but its modular architecture makes it possible to tailor each system to a specific requirement.

AMASCOS 100
AMASCOS 100 is a lightweight configuration, weighing less than 250 kg. It includes radar and FLIR plus a tactical computer and is suited for a wide range of missions, such as EEZ surveillance, search and rescue and law enforcement. AMASCOS 100 is suitable for fitment to light turboprop aircraft or carrier-based helicopters with an operating crew of one to two.

AMASCOS 200
AMASCOS 200 adds ESM equipment to the AMASCOS 100. It is suitable for anti-surface warfare and can be extended to provide an anti-surface warfare capability. AMASCOS 200 is suitable for fixed- and rotary-wing aircraft of the 8 ton class, with two or three operators.

AMASCOS 300
The AMASCOS 300 is the most versatile version and is likely to comprise the Ocean Master radar, Nadir Mk II inertial GPS, Chlio FLIR, DR 3000 ESM, Link W datalink, Sadang 1000 sonobuoys, HS 312S dipping sonar and MAD Mk III. This version is suitable for both anti-surface and anti-submarine warfare missions and is also suitable for command and control assignments. The heart of the system is a dedicated tactical computer, which collates and processes data from different sensors and other onboard equipment. AMASCOS 300 is suitable for installation on any maritime patrol aircraft of over 10 tons with a crew of three or more operators.

Operational status
In 1996, the Indonesian Navy chose AMASCOS for its six NC-212 maritime patrol aircraft (Ocean Master radar and Chlio FLIR) and its three NBO 105 helicopters (Ocean Master radar). System fits also include the Sextant Gemini navigation computer.

In 1998, AMASCOS was selected by the United Arab Emirates for its IPTN CN235-220 Maritime Patrol Aircraft (MPA). This installation includes the Ocean Master radar, Thomson-CSF DETEXIS Chlio FLIR and Sextant Gemini navigation system. It is also likely to include the Thales (formerly Thomson-CSF DETEXIS) DR 3000 ESM system.

Contractor
Thales (formerly Thomson-CSF DETEXIS).

UPDATED

Anemone radar

Anemone was intended for the modernised Super Etendard.

It consists of a nosecone (antenna and circuitry), an aircraft/radar interface unit and radar controls.

The Anemone radar operates in the I/J-band with frequency agility. A wideband monopulse flat slotted array antenna with low-level sidelobes is provided and reinforced ECCM is incorporated. In the air-to-surface mode, the system enables a surface target to be detected and tracked. Targets are detected in the search mode with an elevation angle automatically adjusted as a function of the selected range angle. Following target designation and lock on, the change to track-while-scan or continuous tracking is automatic. In the air-to-air mode the radar allows linear scan, search with semi-automatic acquisition and continuous tracking.

The Anemone radar

Operational status

The Anemone was fitted to the modernised Super Etendard of the French Navy. It is assumed that production ceased when the type retired from service. It is reported that more than 40 systems were delivered.

Contractors

Thales (formerly Thomson-CSF DETEXIS).

UPDATED

Antilope V radar

The Antilope V radar was designed for the Mirage 2000N. Its basic functions are terrain-following, ground-mapping, interlace (terrain-following and ground-mapping), air-to-air and air-to-surface. Essential characteristics of Antilope V are:

(a) J-band transmission providing high ground reflectivity
(b) high-speed vertical scanning of the antenna
(c) asymmetric antenna, providing accurate localisation of obstacles in the path of the aircraft
(d) an antenna of the flat slotted-array type producing a weighted polar diagram with very low level diffuse sidelobes
(e) receiver with a wide dynamic range and image-frequency suppression
(f) image sharpening of the monopulse type with compression in the elevation plane for highly accurate determination of the height of obstacles
(g) real-time radar data processing
(h) continuous and automatic test system
(i) protection against reception by antenna sidelobes.

Radar information is displayed on a head-up display and on a three-colour multimode CRT head-down display, as well as being sent to the navigation and weapon systems. The system can provide terrain-following commands at 300 ft (91 m) and 600 kt, computing a preset obstacle clearance height and with a preselected *g* level.

Operational status

The Mirage 2000N radar has been in service since 1987. First deliveries of an upgraded version of Antilope V for the Mirage 2000D began in late 1992.

Contractors

Thales (formerly Thomson-CSF DETEXIS).

UPDATED

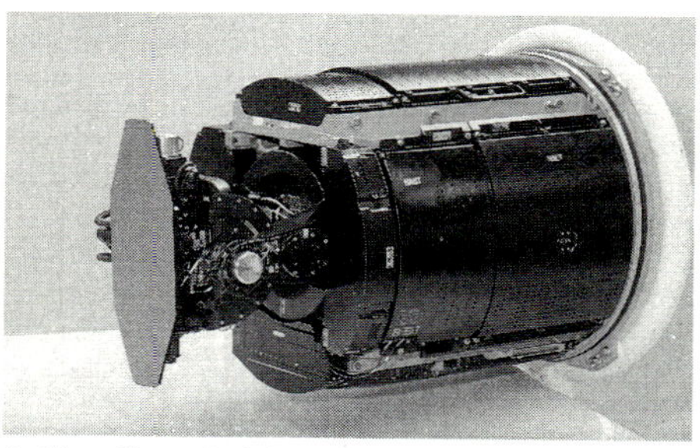

Antilope V terrain-following and navigation radar

ASTAC airborne ESM/ELINT system

ASTAC is an ESM/ELINT system intended primarily for detection and identification of enemy ground radar stations. It is suitable for medium- and high-altitude standoff reconnaissance missions, or at low altitude on the battlefield and in penetration where it can collect data for the avoidance or destruction of anti-air defences.

ASTAC airborne ESM/ELINT system

The ASTAC ESM/ELINT pod mounted, under fuselage, on a French Air Force Mirage F1-CR reconnaissance aircraft
1999

It consists of an airborne pod fitted with interferometers for high-precision direction-finding, two compressive receivers, a data recorder and a data downlink. One receiver is used to obtain a very precise measurement of radar frequency, and the two receivers working together can handle frequency-agile emitters. The pod is capable of handling up to 20 radars/s. ASTAC weighs about 400 kg and can be adapted easily to any combat or light transport aircraft equipped with accurate navigation equipment. Subsystems can also be installed on board the aircraft.

Information from the airborne sensors can be displayed on board and transmitted by secure datalink to a ground station where it is processed and the various radar parameters are displayed in alphanumeric format. These parameters include whether the radar is mobile or fixed, its type, frequency, PRF, pulsewidth, intra-pulse modulation and agility in the B- to K-bands. In addition to the alphanumeric display, a graphic display is provided which incorporates a map showing positions of detected emissions, together with known airfields, radar sites and so on. An airborne display provides real-time onboard information to the crew. Applications also exist for ground stations.

Specifications

Dimensions: 3,960 long × 406 mm diameter
Weight: 400 kg
Frequency: C-J band with A-B band and K-band as options
Accuracy: sub-degree

Operational status

In production for French Air Force Mirage F1CRs and for Japanese Air Self-Defence Force RF-4EJ aircraft. Demonstration flights have been carried out on F-16.

Contractor

Thales (formerly Thomson-CSF DETEXIS).

UPDATED

Barem jamming pod

The Barem self-protection jammer is designed to protect tactical aircraft and helicopters from radar-directed missiles and similar threats. Barem is able to jam pulse PD and CW threats. Its threat handling capability is reported to be two threats simultaneously fore or aft. Housed in a pod cleared for flight at speeds above M2, the system has front and rear receive and transmit antennas coupled to the superheterodyne receiver and TWT transmitter which work under automatic microprocessor control over the H-, I- and J-bands. Threats detected by the threat identification library and the jamming techniques generator are easily reprogrammable on the flight line. Received signals are recorded in flight for subsequent analysis and storage in the library.

Specifications

Dimensions: 3,450 (length) × 160 mm (diameter)
Weight: 85 kg
Power required: 700 VA

Operational status

In service on French Airforce and export Mirage and Jaguar aircraft.

Contractor

Thales (formerly Thomson-CSF DETEXIS).

UPDATED

BARRAGE radio jammer

BARRAGE is a tactical radio communication jammer that uses digital jamming mode generation techniques.

BARRAGE operates in the V/UHF bands and has a transmission power in excess of 100 W.

BARRAGE may be coupled with a direction-finding receiver to optimise jamming efficiency in dense environments.

Thales' Barem jamming pod can be flown on tactical aircraft at speeds above M2

Specifications
Weight: 10 kg
Power consumption: 50 VA

Operational status
BARRAGE is suitable for helicopters, light multipurpose aircraft and UAVs.

Contractor
Thales (formerly Thomson-CSF DETEXIS).

UPDATED

BER 8500/8700 transceivers family

BER 8500/8700 is a family of transceivers covering the VHF and UHF bands from 100 to 173.975 MHz and 225 to 400 MHz in A3 and A1D (A9) AM modes and in F3 and F1D (F9) FM modes. ECCM versions include Have Quick II embedded operation.

The family consists of the BER 8523 covering VHF and UHF AM/FM with Guard receiver, the BER 8524 VHF and UHF AM/FM with Guard receiver and ECCM, the BER 8751 UHF AM/FM and the BER 8752 UHF AM/FM with ECCM.

The ECCM capability is characterised by TRANSEC provided by embedded Have Quick II circuits; COMSEC provided by an external cypher unit (the family is compatible with KY 58 in both NRZ and diphase operation) and SCP 5000 compatibility through direct connection for external TDP 5000 processor.

BER 8500/8700 transceivers are controlled by one or several units of the ECCM VHF/UHF control systems, including the BCA 1217 main VHF/UHF or UHF control unit, DBC 1317 secondary VHF/UHF or UHF control unit or DMC 1317 and DMCB 1317 loading modules. These modules allow control of the transceivers, display of the mode, frequency and alarm and loading of ECCM parameters. The BER 8523 and BER 8751 can be controlled by the TC20 control unit.

Specifications
Dimensions:
(BER 85xx) 57 × 202.5 × 493 mm
(BER 87xx) 57 × 202.5 × 455 mm
Weight:
(BER 85xx) 8.5 kg
(BER 87xx) 7.5 kg
Power supply: 28 V DC or 115 V AC 400 Hz
(transmit) 180 W
(receive) 35 W
Frequency:
(BER 85xx) 100-173.975 MHz and 225-400 MHz
(BER 87xx) 225-400 MHz

Operational status
In service with the French Air Force and several other air forces.

Contractor
Thales (formerly Thomson-CSF Communications).

UPDATED

BF radar warning receiver

The Type BF radar warning receiver provides the crew with warning of most categories of airborne and surface radar threats and with an indication of their direction.

Four wideband antennas are used, linked to a video receiver and, when necessary, a synchronisation unit. In the Mirage the system control and display unit is integrated with other cockpit equipment, but it can be provided separately for other applications.

The receivers comprise photographically etched spiral antennas and microwave circuits for an RF test oscillator, limiter modulator diodes, high-pass filter, detector circuit, video modulation and preamplification. The two side-mounted antennas lie flush with the fin structure, while fore and aft antennas on the fin have conical radomes. An audio alarm is generated when threats are detected and approximate threat direction is indicated by one or more of four signal lamps. The threat is also categorised by one of three lamps which indicate conventional pulse radar, continuous wave or interrupted continuous wave radar or track-while-scan ground radar.

Specifications
Dimensions:
(flat antennas) 148 (diameter) × 53 mm (depth)
(conical antennas) 82 (diameter) × 360 mm (depth)
(synchronisation unit) 209 × 110 × 45 mm
(control box) 146 × 95 × 40 mm
(indicator unit) 68 × 61 × 61 mm
(video receiver) ¼ ATR short
Weight: 9.2 kg (total system)
Power supply: 200 V AC, 400 Hz, <500 VA

Operational status
No longer in production. In service in Mirage F-1A and F-1C, Mirage III and Royal Netherlands Air Force F-5 aircraft. In widespread use in the aircraft of a number of countries in the Middle East. A version designated TMV 008H is produced for helicopters.

Contractor
Thales (formerly Thomson-CSF DETEXIS).

UPDATED

Caiman noise/deception jamming pod

Probably developed from the earlier Alligator pod, Caiman is designed for the SEAD (Suppression of Enemy Air Defences) role. It can be installed underwing or on a fuselage pylon. The pod is self-contained, ram-air entering the unit through an annular intake to drive a power turbine and to provide cooling. Within the pod are fore and aft receiver antennas and two independent jammers each weighing 130 kg. The radiated power is 750 W in I-band and 3 kW in D-band. It can be operated in manual, semi-, or fully automatic mode. The system is combat proven.

Caiman Mk 2 is another podded Thales (formerly Thomson-CSF) jamming system which has been specifically designed for attack jamming against surveillance radars, including battlefield and AWACS type systems. It instantaneously covers two full octaves of hostile frequencies and can jam up to 20 threats simultaneously. The pod which contains TWT transmitters, superheterodyne receiver, autonomous power supplies and antennas, can be carried either under the wing or the fuselage of the aircraft. For tracking purposes, Caiman Mk 2 operates over a wider frequency range than Caiman and has an output jamming power in excess of 1,000 W. It incorporates an extensive memory capacity and modular design software for reprogramming to meet additional and changing threats. As with Caiman, the Mk 2 has manual and automatic operating modes.

Specifications
Dimensions: 5,950 (length) × 410 mm (diameter)
Weight: approx 500 kg

Operational status
In service, but no longer in production. Caiman is fitted to the Mirage F1, Mirage 2000, Jaguar and F-5 and has been supplied to export customers for several aircraft types.

Contractor
Thales (formerly Thomson-CSF DETEXIS).

UPDATED

Camel/BEL expendable jammer family

Camel/BEL is a miniaturised, expendable, electromagnetic, active, very wideband, coherent, decoy jammer for self-protection of aircraft against missiles and tracking radars. It presents a credible electromagnetic signature with regard to velocity, direction and range. Camel/BEL is, therefore, able to counter CW, pulsed and pulsed Doppler radar threats, in particular those radars used in active and semi-active missile seeker heads, either air-to-air or ground-to-air, and including radars with monopulse direction-finding or other counter-countermeasures.

Dimensions of the Camel/BEL cartridge are compatible with current-generation decoy dispensers.

Operational status
Limited-rate initial production.

Contractors
Thales (formerly Thomson-CSF DETEXIS).
Etienne LACROIX.

UPDATED

Carapace threat warning system

Carapace is a version of the passive part of the EWS-16 system for the F-16 aircraft (see later item). The main part of the system is a hybrid receiver, including IFM, crystal video and superheterodyne receivers, plus an interferometric direction-finding array.

Carapace is believed to cover the C to K frequency bands and is designed to detect, identify and localise all modern threats with great accuracy. It is able to analyse accurately, on a pulse-to-pulse basis, emitters at long range in severe ECM environments and offers an ESM capability. Accuracy of direction-finding is classified but is probably of the order of 1°.

Operational status
Carapace is in service with the Belgian Air Force F-16 aircraft.

Contractor
Thales (formerly Thomson-CSF DETEXIS).

UPDATED

Cyrano IV radars

The Cyrano IV is a development of the Cyrano series of airborne radars and is a multirole system. The Cyrano IV-M is the newest version of this series. Basic functions performed by the Cyrano radar are:
(a) air-to-air search
(b) automatic tracking
(c) interception and fire-control computations
(d) dogfight engagements
(e) home-on-jam mode
(f) ground-mapping

Additional options are:
(a) contour mapping
(b) terrain-avoidance
(c) blind let-down
(d) air-to-ground ranging.

The Cyrano IV-M model incorporates track-while-scan facilities and is also suitable for air-to-sea search and tracking roles in addition to those listed for the Cyrano IV.

Data can be presented by means of a Type 196 gunsight or CRT HUD and inputs to weapon systems are also available. Other aircraft systems providing inputs to the radar system include an inertial or gyro platform for altitude reference information and an air data computer for aircraft performance and ambient parameters.

Operational missions include interception, air superiority or interdiction using guns or missiles; all-weather penetration; air-to-ground attack with guns, bombs and rockets.

The Caiman jammer mounted under the starboard wing of a Mirage F1 fighter

Thales' Cyrano IV-M multimode radar

The Cyrano IV-M, as well as being a multifunction radar, differs from its predecessor in the embodiment of new technology which confers improved reliability and maintainability on the later model. Cyrano IVMR is reported to have been developed for use on the Mirage F1-CR-200 tactical reconnaissance aircraft. It is said to incorporate ground mapping, contour mapping, air-to-surface ranging and blind let-down operating modes. Cyrano IVM3 is noted as having been designed for use on the Mirage 50 and as a retrofit equipment for the Mirage III and V. Incorporating technology from the RDM and RDI programmes, Cyrano IVM3 is reported to offer air-to-air, air-to-surface and maritime operating modes.

Operational status

Cyrano IV series radars are reported to be in service aboard Mirage F1 aircraft operated by the air forces of Ecuador, France, Greece, Iraq, Jordan, Kuwait, Libya, Morocco, South Africa and Spain. Cyrano IVM3 radars are noted as being used in Venezuela's eight aircraft Mirage 50EV/DV upgrade programme.

Contractor

Thales (formerly Thomson-CSF DETEXIS).

UPDATED

DataLink Interface Processor (DLIP)

The DLIP is a key component for managing L1, L11, L14, L16 and L22 tactical datalinks. The DLIP is built around a modular, scalable, architecture based on COTS modules. The DLIP is able to simultaneously process messages and protocols from multiple datalinks. This allows it to integrate datalinks within command and control systems or platforms.

Operational status

The DLIP is available in air, land and seaborne versions. It is being proposed for NATO's ACCS LOC 1 air defence programme, and for the Erieye AEW&C aircraft selected by the Hellenic Air Force.

Contractor

Thales (formerly Thomson-CSF Communications).

UPDATED

DAV air-to-air warning and surveillance radar

The DAV warning and surveillance system for helicopters detects low- and medium-altitude air threats, including hovering helicopters; it provides accurate co-ordinate data at standoff range. After detection, DAV sorts the threats into aircraft and helicopter categories and identifies the helicopters by their type.

The radar is attached to the rotor, so it rotates at rotor speed, thereby delivering data at a high refresh rate over 360° in azimuth.

DAV delivers a number of data items: target designation, classification, identification, warning, target tracking and air-to-air missile fire control. It provides real-time comprehensive air-to-air situation awareness information.

DAV can provide information to an Forward Area Air Defence (FAAD) system. As an FAAD extension, it provides a capability for warning, attack and destruction of low-flying targets beyond the action ranges of the ground-to-air defence facilities.

The system has been designed for all-weather operation. Its location accuracy is such that it delivers accurate azimuth, elevation, range and radial speed data on targets for air-to-air missile engagement. It also allows very short response times for gun firing either by the carrier helicopter or other attack helicopters.

DAV features a multitargeting autotracking capability. This E/F-band radar's radiated power endows it with high-detection probability performance up to 9 km while remaining mostly covert. It is equipped with a frequency-hopping transmitter.

Radar integration into the helicopter's system is via a MIL-STD-1553B bus.

Specifications

Dimensions:
(antenna module) 420 × 900 mm diameter
Weight:
(aerial unit) 60 kg
(interface) 5 kg
Frequency: E/F-band
Transmitter: solid-state, frequency-agile
Antenna/radome: rotational velocity same as that of platform helicopter main rotor
Coverage: 360° azimuth; 24° elevation
Receiver: superheterodyne
Doppler analysis: FFT processing
Target recognition: fixed-wing and rotary-wing aircraft
Helicopter identification: from 'library' signatures
Target designation information: azimuth, elevation, range and speed

Operational status

Selected for the Tiger HAP; this radar has also been evaluated by the US Army.

Contractor

Thales (formerly Thomson-CSF DETEXIS).

UPDATED

DB-3141 noise jamming pod (low-band Remora)

The DB-3141 H- to I-band jamming pod has a single receiver, a travelling wave tube jammer and fore and aft transmitter antennas. It provides a simple active electronic warfare capability for Dassault Mirage fighters and possibly also for that company's Super Etendard. DB-3141 jams several threats simultaneously, including low-power Doppler radars. The system has a look-through capability, enabling it to discontinue jamming as soon as threat reception ceases.

Specifications

Dimensions: 3,520 (length) × 250 mm (diameter)
Weight: 175 kg
Power supply: 200 V AC, 400 Hz, 1.7 kVA

Operational status

In service on Dassault Mirage aircraft.

Contractor

Thales (formerly Thomson-CSF DETEXIS).

UPDATED

DB-3163 noise jamming pod (high-band Remora)

The DB-3163 noise jamming pod is designed to provide self-protection against both air and ground radar threats. Pulse and continuous wave emitters can be detected, identified and countered. A superheterodyne receiver performs a frequency scan search on emissions received by antennas at both ends of the pod. During preflight preparation, bands can be selected; up to three threats or groups of threats can be jammed simultaneously. An internal bootstrap air cooling system is employed and the system is energised from the aircraft's power supplies.

DB-3163 is combat proven.

Specifications

Dimensions: 3,520 (length) × 256 mm (diameter)
Weight: 175 kg
Power supply: 200 V AC, 400 Hz, 1.7 kVA
Frequency: I/J-band preprogrammed

Operational status

In service with a number of air forces.

Contractor

Thales (formerly Thomson-CSF DETEXIS).

UPDATED

DR 3000A ESM receiver

The DR 3000A is the airborne version of the DR 3000 Series designed for land, ship, submarine and aircraft operation. The 3000A is suitable for both fixed-wing aircraft and helicopters and consists of a processing unit, display unit and an antenna system and six DF and intercept aerials. It provides very high detection capability over the complete 360° combined with high sensitivity over the frequency bands from D to J. Reliable identification is based on efficient de-interleaving even in very dense electromagnetic environments, accurate parameter measurements and artificial intelligence techniques.

The DR 3000A is small, modular, flexible and can cope with the most commonly envisaged threats. It will meet all requirements for warning, surveillance, ELINT and target designation through ECM. The basic performance can be enhanced by various options to provide improved capabilities for ELINT and DF accuracy.

Thales' Remora jamming pod

DR 3000 control and display unit

Total weight of the system, including processing unit, control and display console and the antennas, is below 80 kg.

Operational status

Procured by Pakistan for installation in Atlantique maritime patrol aircraft. Selected by the Hellenic Air Force for its Erieye AEW & C aircraft.

Contractor

Thales (formerly Thomson-CSF DETEXIS).

UPDATED

DR 4000A (TMV 202) ESM suite

The DR 4000A (TMV 202) is an ESM suite designed for high-sensitivity instantaneous intercept probability and automatic processing against electromagnetic threats in the D- to J-bands. It is intended for use in fixed-wing aircraft and helicopters for ELINT, surveillance, threat detection, identification and data handling system processing and is a version of the DR 4000 Series of equipments. The basic system consists of two sets of six DF antennas, an omnidirectional antenna, modules housing RF amplifiers, an IFM processing unit and an operator console containing controls and three-colour graphic and alphanumeric displays.

With reprogrammable logic and a three-colour display, the probability of interception in both direction-finding and frequency discrimination is claimed to be 100 per cent with only a single pulse, as a result of the crystal video amplifier techniques used. The sensitivity is sufficient to intercept pulse compression signals. The system can be interfaced with any data handling system and the chaff launcher or jammer components of the ECM suite through suitable databusses or point-to-point links.

Specifications

Weight: (including antennas) 169 kg
Power: 1,800 VA

Operational status

Uncertain.

Contractor

Thales (formerly Thomson-CSF DETEXIS).

UPDATED

ERA-8500 VHF/UHF radio

The ERA-8500 transceiver covers the VHF and UHF bands from 100 to 173.975 MHz and 225 to 400 MHz in A3 and A9 AM modes and F1, F3 and F9 FM modes at 25 kHz spacing. It consists of the BER 8500 V/UHF transceiver and the BCA 11XX or BCA 12XX control unit. Its tuning time is compatible with very fast frequency-hopping modes. Power output is 15 W in AM and 20 W in FM.

The ERA-8500 is designed for use by high-performance military aircraft and is offered with a Have Quick II module incorporated in the transceiver. It operates between –55 and +90°C at altitudes up to 100,000 ft.

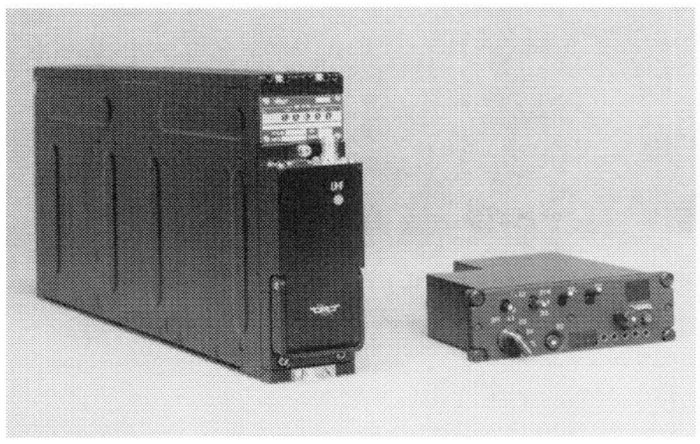

The ERA-8500 VHF/UHF radio

The ERA-8500 transceiver is coupled with the TDP-5000 radio processing unit to form the SCP-500 integrated ECCM radio communication system.

Specifications

Dimensions:
(BER 8500 V/UHF R/T unit) 60 × 202.5 × 493 mm
(BCA 11XX main control unit) 146 × 76 × 173 mm
(BCA 12XX auxiliary control unit) 146 × 47 × 173 mm
Weight:
(BER 8500 V/UHF R/T unit) 8.5 kg
(BCA 11XX main control unit) 1.3 kg
(BCA 12XX auxiliary control unit) 1 kg
Power supply: 115 V AC, 400 Hz or 28 V DC
(transmit) 180 W
(receive) 35 W

Operational status

In service.

Contractor

Thales (formerly Thomson-CSF Communications).

UPDATED

ERA-8700 UHF radio

The ERA-8700 transceiver is limited to the UHF band between 225 and 400 MHz at 25 kHz spacing. It consists of the BER 8700 UHF transceiver and the BCA 11XX or BCA 12XX control unit.

The equipment is available with the Have Quick II frequency-hopping mode.

Specifications

Dimensions:
(BER 8700 UHF R/T unit) 60 × 202.5 × 455 mm
(BCA 11XX main control unit) 146 × 76 × 173 mm
(BCA 12XX auxiliary control unit) 146 × 47 × 173 mm
Weight:
(BER 8700 UHF R/T unit) 7.5 kg
(BCA 11XX main control unit) 1.3 kg
(BCA 12XX auxiliary control unit) 1 kg

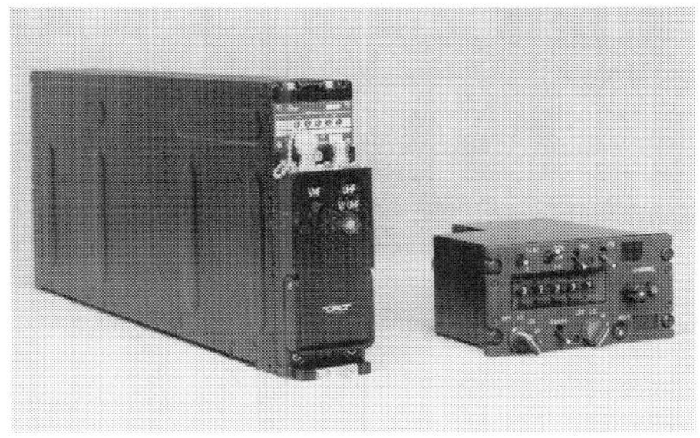

Thales' ERA-8700 UHF radio

Power supply: 115 V AC, 400 Hz or 28 V DC
(transmit) 180 W
(receive) 35 W

Operational status

In service.

Contractor

Thales (formerly Thomson-CSF Communications).

UPDATED

EWR-99/FRUIT radar warning receiver

The EWR-99 FRUIT radar warning receiver is designed for helicopters. It is a user programmable, database oriented system covering a very wide frequency range and making use of full-band Instantaneous Frequency Measurement (IFM) technology. The system is able to detect all RF signals including CW, pulse, and pulse Doppler emitters in high-density electromagnetic environments. More than a simple radar warning receiver, the EWR-99 offers smart management of onboard dispensers and can interface with a missile warning system.

Operational status

Selected to equip the French Army Aviation's entire helicopter fleet (Gazelle, Panther, Puma and Super Puma); also selected by PZL - Swidnik to equip its Sokol multirole helicopters.

Contractor

Thales (formerly Thomson-CSF DETEXIS).

UPDATED

EWS-16 self-protection system

The EWS-16 is an integrated system which consists of a threat warning system and an active jammer. The threat warning system can detect, identify and localise all

EWR-99 radar warning receiver 0010927

modern threats with great accuracy. All threats can be identified in less than 1 second without any ambiguity, even in dense electromagnetic environments. The active jammer features high radiated power and can counter pulsed and CW radars. Analysis of threats is on a pulse-by-pulse basis, with very accurate identification and priority assessment, a clear pilot interface, and full integration with the weapon system. The EWS-16 is fully programmable and makes use of a separate threat library. The ESM subsystem can accurately locate ground-based radars and can record data during flight.

Features of the EWS-16 system include: a management and compatibility unit based on a 32-bit processor; a crystal video receiver and a high-speed wideband superheterodyne receiver; an instantaneous wideband direction-finding interferometer; an IFM receiver; real-time spectral analysis processing; a plug-in Emitter Identification (EID) and mission report module; a multiple threat jammer employing a high-power transmitter; and synergy between jammer and decoy dispenser.

Operational status

A version of the passive part of the EWS-16, known as Carapace, is in service in the Belgian Air Force for its entire fleet of F-16 aircraft.

Contractor

Thales (formerly Thomson-CSF DETEXIS).

UPDATED

EWS-21 radar warning system

The EWS-21 radar warning receiver is adapted from the EWS-A (see next item) for MiG-21 aircraft upgrade applications. It is effective against all types of radar emitters, offering full instantaneous band coverage. The system uses adaptive processing to deal with high-duty cycle radars and jammers and is capable of in-flight recording for post-flight analysis. In the passive target designation role the EWS-21 provides data for the MiG-21 Kopyo radar and modernised avionics. Integration within the aircraft has been studied thoroughly with Mikoyan.

Operational status

Ready for production, probably initially aimed at the update of India's MiG-21 fleet, as part of a new avionics suite.

Contractor

Thales (formerly Thomson-CSF DETEXIS).

UPDATED

EWS-A radar warning system (AIGLE)

The EWS-A (Electronic Warning System for Aircraft) is a compact radar warning system (10 kg) designed for modernisation for aircraft self-protection systems. The system is lightweight and can detect all radar threats including CW, pulsed and pulsed Doppler emitters, even in the densest environments. EWS-A is also sometimes designated as AIGLE.

The key features of the EWS-A are a very short reaction time, high confidence of emitter identification due to Instantaneous Frequency Measurement (IFM) and a 100 per cent probability of interception. The EWS-A will also control jammers and/or chaff launchers (if fitted to the aircraft), and can be integrated to the aircraft weapon system through serial link and/or multiplex bus interface. The EWS-A is fully user programmable.

Operational status

In full-scale production for the entire French Air Force Mirage F1 fleet of aircraft (110 units). In an associated programme, Thales (formerly Thomson-CSF DETEXIS) has been contracted to provide a mid-life update for the Barax self defence jamming pods, also fitted to the French Air Force Mirage F1 fleet.

Three variants of EWS-A have been derived to meet the following aircraft upgrade programmes and have been validated during flight tests: one for Mirage III, Mirage 5 and Mirage 50 aircraft with Dassault Aviation; one for F-5 aircraft with Northrop Grumman; and one for MiG-21 aircraft with Mikoyan.

Contractor

Thales (formerly Thomson-CSF DETEXIS).

UPDATED

Gabriel SIGINT system

Thales (formerly Thomson-CSF DETEXIS) has developed complete SIGINT electronic intelligence systems for integration onboard aircraft such as the DC-8, Boeing 707, Transall and C-130. One of these, Gabriel, configures ASTAC technology for detection, analysis and localisation of radar emissions and a COMINT subsystem, provided by Thales for detection, interception, classification, listen-in, analysis and localisation of radio communications. The system offers a high degree of automation to assist the operators to accomplish all types of mission.

Operational status

Two C-160 Transall aircraft equipped with the Gabriel system are in service with the French Air Force.

Contractor

Thales (formerly Thomson-CSF DETEXIS).

UPDATED

French Air Force Transall aircraft are equipped with the Gabriel SIGINT system

Horizon battlefield surveillance system

The Horizon system (Hélicoptère d'Observation Radar et d'Investigation sur Zone) has been developed and manufactured for the French Army Aviation (ALAT) for tactical intelligence data gathering. It is a development of the earlier Orchidée concept evaluation system.

The Horizon system includes an I-band long-range, ground, surveillance system. This can detect, from a standoff position, in all weather conditions, day and night, moving objects over large areas up to 200 km distant, including wheeled and tracked vehicles, helicopters (moving and hovering), aircraft and ships. Each moving object is detected, localised and automatically analysed and classified. Up to 20,000 km² can be surveyed every 10 seconds. Airborne activities are controlled from an operator panel in the Cougar helicopter. Collected data is transmitted to a ground station over the secure, all-digital, frequency-hopping, Agatha datalink, up to a distance of 150 km.

Each Horizon system comprises one fully equipped ground station with Agatha datalink; two AS 532 UL Cougar helicopters, each equipped with Target MTI radar unit and operator console, navigation and communication equipment and secure Agatha datalink.

The Target radar is an all-digitial, frequency agile, I/J-band, Doppler MTI radar. It combines mechanical and electronic scanning and provides instantaneous panoramic surveillance of either 360° or a sector bounded by the following operating parameters: 60 and 90° sectors at scanning rates of 2, 4 or 8° per second, independently of helicopter course/heading and speed.

Each Horizon system is able to work alone, or with other Horizon systems, or as part of a larger C³I system.

The Horizon system is also proposed to NATO, within the framework of an Alliance Ground Surveillance acquisition programme.

System elements are provided by the following contractors: Eurocopter International: Cougar helicopter; Thales (formerly Thomson-CSF DETEXIS): Target MTI radar; Agatha datalink and ground station.

Specifications

Target radar
Range:
(clear weather) 200 km
(rain/cloud) 150 km
Resolution:
(range) 40 m
(velocity) 2 m/s
Scan sector: 360°
Scan rate: 2,4 or 8°/s
Agatha datalink
RF: J-band
Data rate: up to 0.5 Mbyte/s
Range: up to 150 km

Operational status

In 1991, a demonstrator system was used in Desert Storm operations by the French Daguet division and by the American XVIIIth Army Corps. During NATO operations in Kosovo in 1999, the Horizon system was used operationally.

Two systems, each comprising two helicopters and one ground station, have been delivered to the French armed forces.

Contractors

Thales (formerly Thomson-CSF DETEXIS).
Eurocopter International.

UPDATED

ICMS Integrated CounterMeasures Suite

Thales (formerly Thomson-CSF DETEXIS) and Matra BAe Dynamics SAS, have developed an integrated internally mounted EW suite for the Mirage 2000 aircraft. This is a highly sophisticated system, known as the Integrated CounterMeasures Suite (ICMS), where all parts are linked to a central interface and management unit.

The system incorporates three warning receivers designed by Thales. These receivers consist of a version of the Serval equipment (see later entry), a superheterodyne receiver to detect CW radar, pulse compression signals and low-power pulse Doppler signals, and a receiver/processor mounted in the aircraft nose to detect missile command links. The missile detector function can also incorporate an infra-red warning receiver designed by Matra BAe Dynamics SAS.

Horizon radar on AS 532 UL Cougar helicopter 0001192

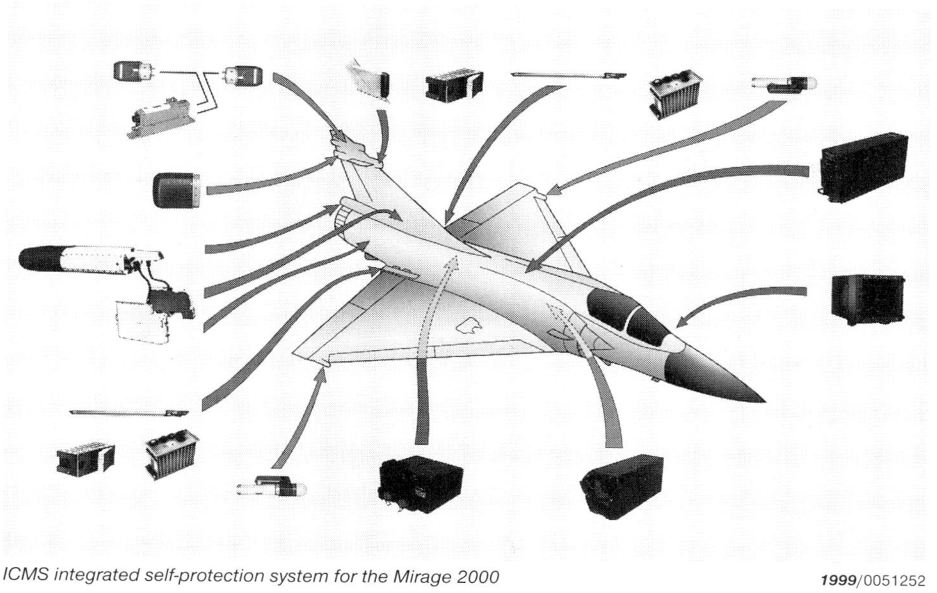

ICMS integrated self-protection system for the Mirage 2000 **1999**/0051252

Two detector-jammers are included, each with its own receiver which allows it to operate should the basic radar warning receiver be out of action. These detector-jammers consist of a high frequency sub-unit to counteract airborne and surface-to-air threats, and a low-frequency sub-unit to operate against surface-to-air threats in the lower part of the spectrum. The Spirale chaff/IR flare dispenser is also included in the overall system to provide passive countermeasures. Spirale is an internally mounted equipment which dispenses stores through openings in the aircraft structure.

Operational status
In production for the Mirage 2000-5. In service on aircraft operated by the air forces of Greece, Qatar and Taiwan. The latest development of the system is understood to be designated the ICMS Mk 2, and to incorporate the SAMIR infra-red sensor made by Matra BAe Dynamics SAS and SAGEM.

Meanwhile, a further derivative is being offered for the UAE Mirage 2000-5 upgrade (also designated Mirage 2000-5 Mk 2) and Mirage 2000-9 aircraft, designated IMEWS (Integrated Mission Electronic Warfare Suite). It is being developed to customer specification by Thales in partnership with Elettronica; this new system will incorporate the latest digital technologies for frequency measurement, high-sensitivity detection and electromagnetic jamming. Specific equipment items are reported to include an Elettronica RWR and the Matra BAe Dynamics SAS Spirale active jammer.

Contractors
Thales (formerly Thomson-CSF DETEXIS).
Matra BAe Dynamics SAS.

UPDATED

Iguane sea surveillance radar

Iguane is a maritime surveillance radar that has been produced to replace the DRAA2A sea surveillance radar fitted in the Breguet Alizé and for the Dassault Atlantique 2 long-range maritime patrol aircraft.

The system operates in the I/J-band, using pulse compression and frequency agility to ensure high performance against maritime targets in all combinations of weather, sea state and operating altitude and to provide maximum protection against ECM. Operational missions performed by Iguane include surface and anti-submarine warfare, over-the-horizon targeting for shipborne surface-to-surface missiles, environmental protection, navigation and weather avoidance.

Operational status
In service on the Breguet Alizé update programme and for the Dassault Atlantique 2. The Italian Air Force has fitted 18 Iguane systems to its fleet of Atlantic 1 aircraft.

Contractor
Thales (formerly Thomson-CSF DETEXIS).

UPDATED

MSPS EW system

The Modular Self-Protection System (MSPS) is a very lightweight, easy to install EW suite, mainly designed for retrofit in existing aircraft. It combines the Sherloc RWR with the Barem jamming pod to provide a combined crystal video/superheterodyne receiving system for accurate threat description and fast data transmission between the two equipments for speedy reaction. The Alkan 5081 decoy system is also integrated into the system.

The MSPS EW system can be installed on a large number of fixed-wing aircraft and helicopters. The jamming part can be installed internally. MSPS is fully automatic and reprogrammable, and can accept additions such as chaff and flare countermeasures, laser detection equipment and support jamming systems. The weight of the system is 100 kg when the jammer is pod-mounted and 80 kg when it is internally installed.

Operational status
Produced for the French Navy Super Etendard upgrade.

Contractor
Thales (formerly Thomson-CSF DETEXIS).

UPDATED

MWS 20/DAMIEN Missile approach Warning Systems

The MWS 20/DAMIEN family of radar-based active missile approach warning systems has been designed for self-protection of helicopters, and fixed-wing combat aircraft.

It detects approaching missiles and triggers the appropriate countermeasures at the optimum time. Its main functions are to: provide warning of all types of attacking missiles; calculate in real time the speed, range, direction of detected missiles, and time to impact; control a decoy-launcher in automatic mode in order to considerably reduce the decoy consumption.

Insensitive to weather conditions, MWS 20/DAMIEN is also not affected by infra-red decoy (flares) illumination or other heat sources, and is designed to give a low false alarm rate.

Due to the use of state-of-the-art technologies (ASIC components and Doppler solid-state transmitter), MWS 20/DAMIEN is extremely reliable. It has a BIT and modular architecture to simplify maintenance.

MWS 20/DAMIEN is highly miniaturised comprising a 10 kg airborne electronic unit and four conformal antennas.

Operational status
MWS 20 has been ordered to equip the helicopters of the French Special Forces. It is in full scale production.

Contractor
Thales (formerly Thomson-CSF DETEXIS).

UPDATED

MSPS on Super Etendard

MWS 20/DAMIEN Missile approach Warning System 0010926

Myriad radar warning system

Myriad is a warning receiver in the millimetric-wave region. It is designed for the protection of helicopters, light aircraft or armoured vehicles from smart weapons using millimetre-wave seekers. It provides immediate warning over 360° coverage and weighs approximately 8 kg.

Operational status

In development.

Contractor

Thales (formerly Thomson-CSF DETEXIS).

UPDATED

NC 12 airborne Tacan interrogator

The NC 12 is a very small and lightweight Tacan interrogator which fully complies with STANAG 5034, the improved MIL-STD-291. It is an all-digital system fitted with ARINC 429, ARINC 582 or MIL-STD-1553B standard outputs. It provides the pilot with digital distance and bearing relative to Tacan beacons on the NR 13 control box indicator. The NC 12 interrogator can be used either on new aircraft or for retrofit applications.

In air-to-air mode the pilot is supplied with distance to another aircraft fitted with an airborne beacon. A warning light indicates the presence of any jamming of the bearing and distance information.

The NC 12 interrogator is also fully adapted to advanced radio navigation through the provision of W and Z channels for DME-P compatibility.

Specifications

Dimensions:
(NR 13 control box) 144 × 63 × 57 mm
(NC 12 transmitter/receiver) 318 × 91 × 193.5 mm
Weight:
(NR 13 control box) 0.7 kg
(NC 12 transmitter/receiver) 5.5 kg
Power supply: 28 V DC, 1.4 A, 40 W
Transmission power: 350 W
Frequency:
(transmit) 1,025-1,150 MHz
(receive) 962-1,213 MHz
Number of channels: 126 X channels, 126 Y channels
Accuracy: 0.1 n miles and 1°
Altitude: up to 100,000 ft

Operational status

In production for the Mirage 2000-5 and Mirage 2000-9 aircraft. In service in the Mirage F1-CT, Super Puma, Dauphin, UH-60, CH-47 and Tucano. Selected for the Rafale.

Contractor

Thales (formerly Thomson-CSF Communications).

UPDATED

Thales' NC 12 airborne Tacan interrogator 0015352

NRAI-7(.)/SC10(.) Identification Friend-or-Foe (IFF) transponder

The NRAI-7 is a solid-state Mk XII diversity transponder which inhibits replies to interrogator sidelobe transmissions and automatically codes special replies to provide assistance in the position identification of particular aircraft and in emergencies. The diversity function is provided by a dual receiver with inputs connected to upper and lower antennas, a system for comparison of received signals and an antenna switch that directs the response to the antenna that has received the strongest interrogation signal. This allows more accurate identification, particularly during aircraft manoeuvres which can blanket or interrupt signals. The pilot may also insert codes such as radio failure alert and warning of hijackers aboard. It is available in one- or two-box housing (see NRAI-9A). The single-box version is claimed to be one of the smallest transponders in the world. A naval version is also available.

Operational status

NRAI-7 transponders are reported to have been installed on Mirage 2000, Mirage F-1, Mirage III, Mirage IV, AS 332 and AS 335, C130 Transall and various other aircraft as well as on board several ships of the French and other navies. It has also been integrated in various Polish aircraft, including MIG fighters. Over 2,000 systems have already been delivered to the French defence forces and those of other countries.

Specifications

Peak power: 500 W
Sensitivity: –77 dBm
Frequency: 1,030 MHz (receive); 1,090 MHz (transmit)
Modes available: 1, 2, 3A/C and Mode 4 capability
No of codes: 32 (Mode 1); 4,096 (Modes 2 and 3/A); 2,048 (Mode C)
Dimensions: 130 × 127 × 145 mm
Weight: 3 kg

Contractor

Thales (formerly Thomson-CSF Communications).

UPDATED

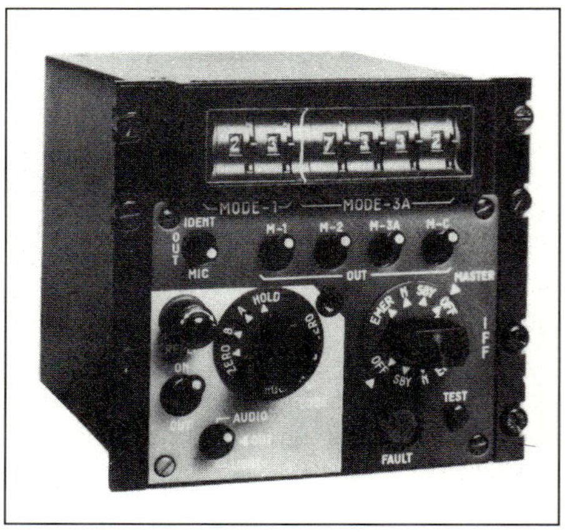

Thales' NRAI-7(.)/SC10(.) IFF transponder

NRAI-9(.)/SC15(.) Identification Friend-or-Foe (IFF) transponder

The NRAI-9A is essentially a two-box version of the NRAI-7 which incorporates a number of improvements. These include the elimination of sidelobe response, and automatic special-code referral, with positive identification permitting a ground operator to locate a particular aircraft. Special emergency codes may also be employed, chosen by the pilot, such as radio failure. Dual-receiver channels connected to upper and lower antenna and comparison circuits provide a diversity function.

Operational status

NRAI-9 is thought to be fitted to ATL2 maritime patrol aircraft.

Specifications

Frequency: 1,030 MHz (receive); 1,090 MHz (transmit)
Power output: 500 W peak
Modes available: 1, 2, 3A/C and Mode 4 capability
Codes: 32 (Mode 1); 4,096 (Modes 2 and 3/A); 2,048 (Mode C)

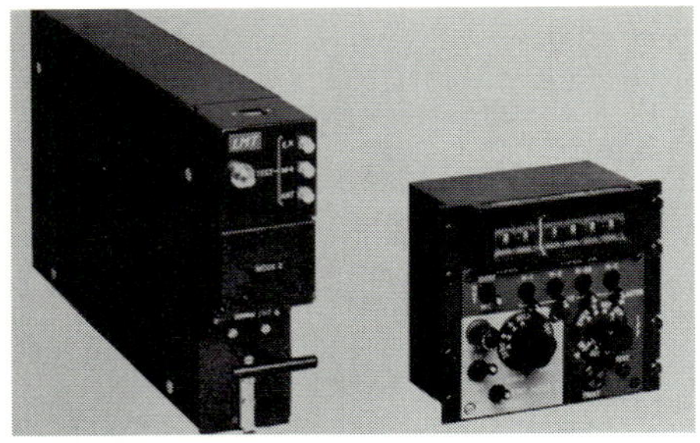

Thales' NRAI-9(.)/SC15(.) IFF transponder

Dimensions: 58 × 193 × 361 mm (transmitter/receiver); 127 × 130 × 80 mm (control unit)
Weight: 2.5 kg (transmitter/receiver); 1.4 kg (control unit)

Contractor

Thales (formerly Thomson-CSF Communications).

UPDATED

NRAI-11/IDEE 1 Mk XII interrogator-decoder

The airborne NRAI-11/IDEE 1 interrogator-decoder is used to identify a friendly target and determine its range and azimuth. The interrogator-decoder and its control box are integrated in the IFF system of Dassault Mirage 2000 aircraft. The system includes all the functions needed for IFF air-to-air identification and includes an encoder, transmitter, RF switch, two receivers, analogue processing unit, defruiter, passive decoder, evaluator, extractor, staggering circuit and automatic self-test.

Volume and power consumption have been substantially reduced through the use of a solid-state transmitter, low-power integrated and monolithic high-density circuits, hybrid and custom LSI circuits and switching power supply.

Specifications

Dimensions: 124 × 80 × 194 mm
Weight: 12 kg
Power supply: 115 V AC, 400 Hz, 3 phase, 150 VA
Transmission power: 1 kW peak
Receiver sensitivity: –79 dBm between 1,087 and 1,093 MHz
Operating modes: 1, 2, 3/A and Mode 4 capability

Operational status

In service on the Dassault Mirage 2000.

Contractor

Thales (formerly Thomson-CSF Communications).

UPDATED

Thales' NRAI-11/IDEE 1 Mk XII interrogator-decoder is in service in the Mirage 2000

ORB 32 radar systems

ORB 32 is a range of I-band airborne radars. It features very low weight and power consumption and can be installed on a wide range of helicopters and aircraft. ORB 32 is built from modular subassemblies, enabling easy extension and is designed for exclusive economic zone control, anti-surface warfare, anti-submarine warfare, active missile fire control, search and rescue, radar navigation and weather avoidance.

It features pitch and roll stabilisation, 360° azimuth and 30° elevation scanning and 60, 120, 180 and 240°

sector scans, and azimuth, bearing or true motion stabilisation. The peak power is typically 70 kW.

The ORB 32 is available in a number of versions:

The ORB 3201 and ORB 3211 are simple, compact and lightweight systems suitable for small aircraft or helicopters, specially designed for surface reconnaissance, exclusive economic zone control and search and rescue. They provide navigation information and weather avoidance.

The ORB 3202 and ORB 3212 are airborne reconnaissance and target designation radars. When integrated into a weapon system, their purpose is to detect, designate and accurately track two sea targets. Target co-ordinates may be automatically transmitted to active missiles carried by aircraft or helicopters or to a launch vessel for over-the-horizon targeting.

The ORB 3203 and ORB 3214 are one element of an anti-submarine warfare weapon system for helicopters or aircraft. They enable helicopter station holding in ASW, tactical situation information, guidance and aircraft attack on a designated target, navigation, weather and mapping. The ORB 3203 and 3214 perform both primary and secondary radar functions. Use of transponders makes identification of helicopters flying at low altitude possible, even if the primary echo is in sea clutter.

Operational status

In production for the French Navy and the armed forces of a number of other countries. Installed in Dauphin, Super Frelon, Nord 262, Super Puma and Boeing Vertol 107 helicopters.

Contractor

Thales (formerly Thomson-CSF DETEXIS).

UPDATED

ORB 37 radar system

ORB 37 has been designed to meet the French Air Force navigation requirements for the C-160 Transall transport aircraft. It carries out weather avoidance and accurate ground-mapping functions and is fitted with an interrogation facility for beacon homing.

The system consists of seven units; a slotted array flat-plate antenna, a transmitter/receiver, a power supply, a Plan Position Indicator (PPI) high-definition circular display for ground-mapping at the navigator's station, a digital PPI display on the flight deck and two control units, one for each station.

For maximum efficiency, the antenna scans at low rate for the weather mode and at high rate for the ground-mapping mode. The corresponding pulsewidths are 2.5 and 0.4 μs.

Specifications

Frequency: I-band (9,375 MHz)
Power output: 10 kW

Thales' ORB 37 weather and navigation radar

Thales' ORB 32 radar is used for surveillance and fire-control applications in maritime aircraft

Operational status

ORB 37 is reported to be in service with the French Air Force.

Contractor

Thales (formerly Thomson-CSF DETEXIS).

UPDATED

Over-The-Horizon Target (OTHT) designation system

The Over-The-Horizon Target (OTHT) designation system is designed to detect and locate targets beyond the horizon and to transmit the co-ordinates to a coastal battery or ship. The complete system consists of equipment for acquisition and transmission.

The ORB-32 panoramic radar detects, locates, identifies through the use of IFF data and automatically tracks the targets. The ORB-32-03 version for helicopters contains an antenna with IFF capability in the radome and, in the fuselage, a frequency-agile transceiver, junction box, radar control unit, scan converter, 9 in TV scope and a track-while-scan processing unit. The IFF option includes an IFF adaptation unit and IFF control unit.

The transmission of data takes place by conventional or jam-protected UHF radio over the frequency range 225 to 400 MHz. The jam-resistant version consists of a TDP-500 Series unit which allows data transmission in the frequency-hopping mode at 64 hops/s and generates system synchronisation, and an ERM-9000 transceiver equipped with a frequency-hopping unit. The conventional version consists of an MSA-300 modulator which converts the message into two LF tones for the UHF transceiver, and a CDM-6000 which receives information in the form of LF tones and converts them back to digital form.

Contractor

Thales (formerly Thomson-CSF Communications).

UPDATED

PAJ-95 detector/jammer

PAJ-95 is a pod-mounted airborne detector/jammer for self-protection of combat aircraft. It has a supersonic capability, with a diameter of 157 mm and a weight of 85 kg. Covering a very wide frequency band, and capable of multiple and independent jamming techniques, it can detect and counter all types of threat. It is fully automatic and is able to perform surveillance, alert and jamming functions over the complete frequency band. Threat and jamming techniques libraries can be reprogrammed by the user and loaded before take-off. Reaction time is very short and since it is entirely controlled by microprocessors, it can be adapted to future threat developments. PAJ-95 has the benefit of the most modern technology, through several updates.

The name Barax also appears in some literature when referring to the PAJ-95 system.

Operational status

Versions of the PAJ-95 are in service for the French Mirage III, Mirage F1, and Jaguar. A particular version has been manufactured in co-operation with Spanish industry for the Spanish Air Force.

Contractor

Thales (formerly Thomson-CSF DETEXIS).

UPDATED

PHALANGER ESM/ELINT system

PHALANGER is a new-generation ESM/ELINT payload for airborne platforms, including unmanned aerial vehicles, helicopters and light multipurpose aircraft. A pod version can be installed on combat aircraft.

Aimed at detecting, identifying, and localising ground-based radars, PHALANGER delivers radar tracks for real-time display and analysis. The battlefield tactical situation and electromagnetic order of battle (EOB) can be displayed either on board the carrier platform or in a remote ground-based STRATEGIE family ESM/ELINT station collecting the data from the payload via datalink.

The key features of PHALANGER are: high sensitivity and good direction-finding; small volume and weight (20 kg) for easy installation on UAVs and helicopters.

Operational status

PHALANGER has been validated by the French MoD during ground and flight tests.

RBE2 radar for the Dassault Rafale aircraft 0001193

Contractor
Thales (formerly Thomson-CSF DETEXIS).

UPDATED

RBE2 airborne radar

The RBE2 is the first of a new class of airborne radars using the Thales (formerly Thomson-CSF DETEXIS) electronic scanning process. It is a multirole radar with air-to-air and air-to-surface capabilities designed to meet the operational requirements of the French Air Force's Rafale B and C and also for the French Navy Rafale M.

For the air defence role the RBE2 is able to carry out all air defence functions including search in look-up and look-down modes, identification, automatic multi-target tracking and dogfight. Detection is optimised by means of automatic waveform selection of high, medium or low pulse recurrence frequencies. Multitarget tracking capability is improved by the use of a two-plane electronically scanned antenna which can track existing targets and simultaneously search for new targets to provide the RBE2 radar with its simultaneous multimode capability.

The RBE2 generates simultaneous radar-to-missile datalinks for the air-to-ground role and provides Rafale with all-weather deep strike capability. It also provides close support and battlefield interdiction, by means of automatic Terrain-Following/Terrain-Avoidance (TF/TA), high-resolution mapping and ground moving-target search and tracking. As a result of the electronic scanning antenna, the simultaneous operation of TF/TA and air-to-air modes brings vital self-defence capabilities while flying penetration missions and ground clearance capabilities in air-to-air missions.

In addition to air defence and air-to-ground roles, the RBE2 is optimised for shipping strike missions with long-range detection, multitarget tracking and target recognition capability.

The RBE2 radar is fully integrated with the EW optronics suites to provide a real-time, multisensor, weapon system.

Operational status
In production since 1997, the RBE2 will be in service on Rafale M in the French Navy by 2001 and in service with the French Air Force by year 2004.

Contractor
Thales (formerly Thomson-CSF DETEXIS).

UPDATED

The RBE2 two-plane electronically scanned multimode radar

RC 400 compact multimission multitarget radar family

The RC 400 product line is a family of modular, lightweight, compact, multimission, multitarget radars. It is composed of: the RC 400-1 with 100 W Average Transmitted Power (ATP); the RC 400-2 with 200 W ATP; and the RC 400-4 with 400 W ATP.

RC 400-4

The RC 400-4 radar has been designed for a wide range of fighters and advanced training aircraft. RC 400-4 technology is derived from the RDY radar, which equips the Mirage 2000-5. The baseline architecture comprises four Line-Replaceable Units (LRUs): the antenna unit; the exciter/receiver; the transmitter; and the processor unit. This configuration can be reconfigured as required to fit the nose of most aircraft types and only requires non-filtered air-cooling. Several antenna options are available, based on a low0inertis, flat slotted-plate design, with very low side lobes.

The RC 400-4 design features track-while-scan, smart automatic management of scanning and target prioritisation. This provides accurate, multiple fire-and-forget capability, together with datalink requirements for missile control. The radar utilises high-, medium- and low-PRFs to optimise performance in all Beyond Visual Range (BVR) interception missions, as well as combat phases of operation. Within a given mode, the PRF is automatically managed from bar to bar, with respect to attack geometry.

The RC 400-4 offers the following operational capabilities:

(1) Air-to-air: all-aspect look up down detection; automatic management of low-, medium- and high-PRFs; automatic lock-on; simultaneous multitarget engagements; IFF.
(2) Air-to-ground: improved ground mapping with azimuth beam compression; Doppler beam sharpening; ranging; high-resolution mapping; and contour mapping for low-altitude penetration.
(3) Air-to-sea: track-while-scan on two targets; sea target calibration.

Specifications
Transmitter unit: I/J-band; air cooled
Receiver unit: 2 channels; wide dynamic range
Processor unit: fully programmable signal and data processing; 1 Gflop
Antenna unit: monopulse flat slotted array; elliptical or circular; IFF interrogator
Weight: 130 kg
Power consumption: 3.6 kVA

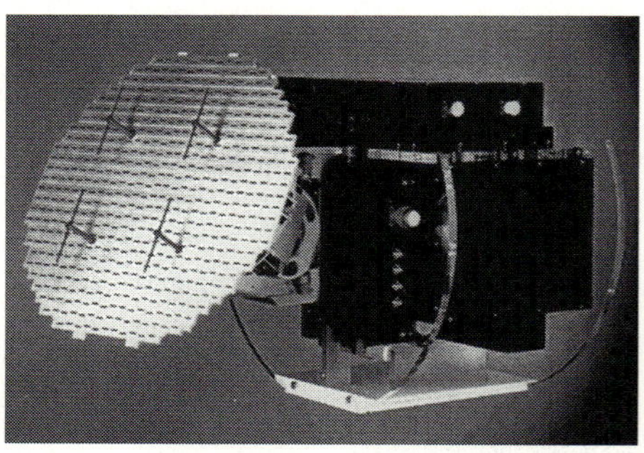

Thales' RC 400 compact multimission multitarget radar 0011895

RC 400-2

The RC 400-2 also provides air-to-air, air-to-ground and air-to-sea modes, with the following capabilities:
1. Air-to-air: all-aspect look up/look down detection; track-while-scan; automatic waveform management.
2. Air-to-ground: ground mapping with optional SAR high-resolution mapping; moving target indication and tracking of two targets; manual terrain avoidance.
3. Air-to-sea: detection and tracking; target radar cross-section assessment.

Specifications
Antenna unit: different sizes available
Weight: <110 kg
Power: 2.3 kVA

RC 400-1

The RC 400-1 performs the same roles as the RC 400-2, but it is matched to smaller combat aircraft such as the Super Etendard IV and AlphaJet.

Specifications
Antenna unit: different sizes available
Weight: <100 kg
Power: 1.5 kVA

Operational status
Flight testing in 1998. Initial deliveries planned for 2000. Marketed for the MiG-29 upgrade market and other combat aircraft.

Contractor
Thales (formerly Thomson-CSF DETEXIS).

UPDATED

RDI (Radar Doppler à Impulsions)

The RDI is one of two pulse Doppler radars (the other being the RDM) developed for France's Dassault Mirage 2000. It is intended for the all-altitude air superiority and interception version and is based on a travelling wave tube, I/J-band transmitter radiating from a flat, slotted plate antenna. The radar range capability is designed for high-performance Beyond Visual Range (BVR) interception and is commensurate with employment of the 40 km range Matra Super 530D semi-active homing air-to-air missile. The performance of the radar, and of other systems on the aircraft, benefits from the digital signal handling and transmission of information by databus. Considerable electronic countermeasures resistance is built into the equipment, which can operate in air-to-air search, long-range tracking and missile guidance, and automatic short-range tracking and identification modes. Although designed for air-to-air operation, the system incorporates ground-mapping and air-to-ground ranging modes.

The high PRF available (100 kHz+) guarantees accurate target speed assessment with a Thomson-CSF patented process to give range data at maximum range in search mode (as well as in tracking mode).

Air-to-air modes include air-to-air search; long-range TWS or continuous target tracking and missile guidance; automatic short-range tracking for missiles or guns. Air-to-ground modes include ground mapping; contour mapping; air-to-ground ranging. Total weight of the 11 line-replaceable units is 255 kg.

Operational status
In production and service with the French Air Force, this radar has been updated with enhanced capabilities.

Contractor
Thales (formerly Thomson-CSF DETEXIS).

UPDATED

Thales' RDI radar equips the Dassault Mirage 2000

RDM (Radar Doppler Multifunction)

The RDM monopulse Doppler I/J-band radar is in service with the French Dassault Mirage 2000. Whereas the RDI is designed for interception and air combat, the coherent, multimode, all-digital, frequency-agile RDM is intended largely for the multirole export version. It operates in air defence/air superiority, strike and air-to-sea modes.

In the air-to-air role, the system can look up or down, range while searching, track-while-scan, provide continuous tracking, generate aiming signals for air combat and compute attack and firing envelopes. For the strike role it provides real-beam ground-mapping, navigation updating, contour-mapping, terrain-avoidance, blind let-down, air-to-ground ranging and Ground Moving Target Indication (GMTI). In the maritime role it provides long-range search, track-while-scan and continuous tracking and can designate targets for active missiles.

For air-to-air combat, the RDM provides a 120° cone of coverage, the antenna scanning at either 50 or 100°/s, with ±60, ±30 or ±15° scan. A 5 m^2 radar cross section target can be detected at up to 111 km range. For air-to-air gun attacks, the 3.5° beam can be locked to the target at up to 19 km range, with automatic tracking within the head-up display field of view, or in a 'super-search' area, or in a vertical search mode. In look-down, air-to-air scenarios, a 5 m^2 radar cross section target can be detected at up to 46 km range.

Options include a continuous wave illuminator and Doppler beam-sharpening. Comprehensive electronic counter-countermeasures are incorporated.

Significant improvements to the radar, particularly to the look-down function, have been completed. Improvements include hardware and signal processing.

Operational status

In service. The RDM radar has equipped the first Mirage 2000 squadrons in the French Air Force, as well as the air forces of Egypt, Greece, India, Peru and the UAE.

Contractor

Thales (formerly Thomson-CSF DETEXIS).

UPDATED

RDN 2000 Doppler velocity sensor

The RDN 2000 J-band Doppler velocity sensor has been designed for light, medium and heavy helicopters, where autonomous navigation and assistance for the pilot is needed for missions of all types over land and sea. It employs FM/CW techniques and digital signal processing to provide accurate ground velocity data. The lightweight single unit includes an antenna, transmitter/receiver, power supply and signal processor.

Coupled with a computer and an AHRS or INS, the RDN 2000 provides a stealthy autonomous navigation system which is insensitive to countermeasures. In addition, it can operate with a local jammer.

The system provides high performance when flying over land or sea, especially over calm water and it has an automatic land/sea transition capability. This allows autopilot operation irrespective of flying conditions.

Specifications

Dimensions: 437 × 240 × 80 mm
Weight: 4.2 kg
Power supply: 28 V DC, 25 W
Transmit power: 50 mW
Accuracy (95%):
0.15% of Vt or 0.1 m/s (along track)
0.22 % of Vt or 0.1 m/s (across track)
0.15% of Vt or 0.2 m/s (vertical)
Reliability: 8,640 h MTBF

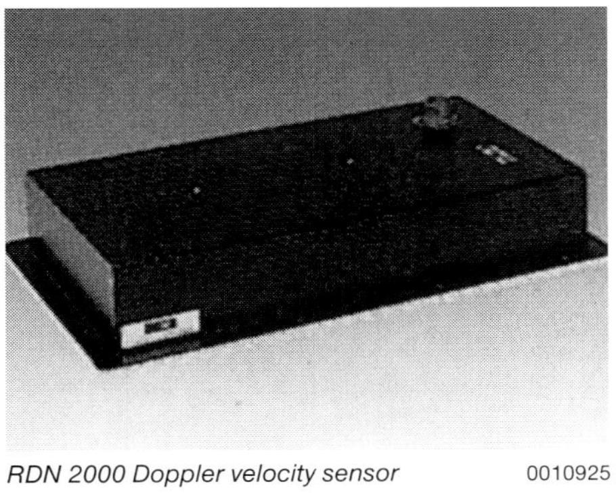

RDN 2000 Doppler velocity sensor 0010925

Operational status

The latest in Thales' (formerly Thomson-CSF DETEXIS) range of Doppler velocity sensors for civil and military helicopters; it is claimed that the RDN 2000 radar is particularly suitable for all-weather maritime missions. First deliveries began in 1998.

Contractor

Thales (formerly Thomson-CSF DETEXIS).

UPDATED

RDN 85-B Doppler velocity sensor

The RDN 85-B is a single-box Doppler velocity sensor designed for use in helicopters, said to have good operating characteristics over calm seas. The radar interfaces with an ARINC 429 digital databus and along and across track velocities are also transmitted as DC signals for display by a hover meter and for coupling to an autopilot. There is BITE for in-flight and ground system checkout.

Specifications

Dimensions: 437 × 437 × 130 mm
Weight: <10 kg
Power supply: 28 V DC, 30 W
Frequency: 13.325 ±20 MHz
Transmitter power: 30 MHz
Velocity range: −50 to +350 kt
Altitude: up to 20,000 ft
Accuracy: 0.15% of velocity or 0.12 kt
Transmitter: Gunn diode oscillator
Reliability: >3,200 h MTBF

Operational status

In production for the French Navy export search and rescue Puma and Dauphin helicopters, French Army Super Pumas and the Fennec helicopter for Singapore. More than 300 units have been ordered.

Contractor

Thales (formerly Thomson-CSF DETEXIS).

UPDATED

RDY multifunction radar

RDY is the multifunction Doppler radar for the Mirage 2000-5 aircraft. The three main modes of operation are air-to-air, air-to-ground and air-to-sea.

In air-to-air mode, the RDY detects very low- or high-altitude targets at long range, irrespective of their angle of approach. It presents the pilot with tactical situation analysis, offers multitarget tracking and IFF interrogation.

In air-to-ground mode, the RDY employs Doppler beam-sharpening to provide all-weather, low-altitude penetration capability, including terrain avoidance and ground-mapping functions.. It provides the system with air-to-ground ranging for ground attack missions.

In air-to-sea mode, the RDY is able to detect targets, even in high sea states at up to 150 km, perform multitarget tracking and target allocation for ASMs such as AM 39 or Kormoran II.

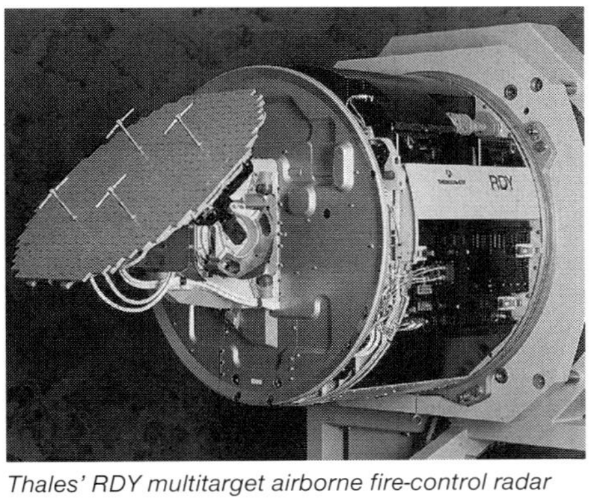

Thales' RDY multitarget airborne fire-control radar

The radar uses high-, medium- and low-PRF waveforms. High PRF is used for long-range detection of low-flying targets; low PRF for high-flying targets and a medium PRF for all-aspect medium-range detection in ground clutter.

Signal processing is carried out by a programmable signal processor with a very large computational capability: a speed of 100 Mcops is claimed. This computational throughput enables additional options such as Doppler beam-sharpening, Ground Moving Target Indicator (GMTI) and efficient track-while-scan air-to-sea modes. Target handling capability is quoted as the ability to detect 24 targets and to track-while-scan the eight which pose the greatest threat.

A 60 cm diameter flat-plate phased-array antenna is used with four integral high-frequency dipoles for IFF interrogation.

Operational status

In production since 1995. The RDY radar equips the Mirage 2000-5 for the French Air Force and export customers, reported to be Qatar (12 aircraft) and Taiwan (48 single-seat and 12 two-seat aircraft). RDY is also being retrofitted into 37 French Air Force Mirage 2000C aircraft and was reported to be in full operational service in 1997. At the Paris 1999 Air Show, Thales (formerly Thomson-CSF DETEXIS) unveiled a new SAR, high-resolution imaging function which provides the RDY radar with an all-weather reconnaissance capability. The Mirage 2000-9 aircraft for the UAE will be fitted with the latest variant of the radar – reportedly designated, RDY2.

In 1999, the government of Greece announced its decision to order 15 Mirage 2000-5 aircraft and to retrofit 10 older Mirage 2000 aircraft, making the total requirement for 25 new RDY radars.

Contractor

Thales (formerly Thomson-CSF DETEXIS).

UPDATED

Romeo II obstacle avoidance radar

Romeo is an M-band obstacle avoidance radar intended primarily for helicopter use. It employs a solid-state transmitter/receiver, operating at a frequency of 94 GHz, with high-density digital processing and a programmable computer. It has a very high detection capability on small targets such as power cables, pylons, antennas, bridges, trees and offshore rigs. The small size and weight of Romeo allows it to be fitted in a wide range of fixed-wing aircraft and helicopters.

Display presentation consists normally of a colour landscape image display, with different colours representing obstacles. Red is the danger colour and warns of close and dangerous objects. A head-up or helmet-mounted display can also be used for synthetic presentation of obstacle warning. Power output is in the region of 400 mW giving a range in the fully developed models of approximately 3,000 m.

Operational status

Romeo II is believed to be undergoing trials and a variant with a maximum detection range of 1 km is believed to have been developed. Romeo is also understood to have been tested by the US Army aboard an AH-1 attack helicopter.

Specifications

Frequency: M-band (94 GHz)
Range: up to 3,000 m in the fully developed model
Power output: 400 mW

Thales' Romeo obstacle avoidance radar

Dimensions: 145 × 110 × 70 mm (control unit); 240 × 240 × 320 mm (antenna); 315 × 190 × 190 mm (processing unit)
Weight: 1 kg (control unit); 11 kg (antenna); 12 kg (processing unit)

Contractor
Thales (formerly Thomson-CSF DETEXIS).

UPDATED

Sarigue SIGINT system

A French Air Force DC-8 aircraft has been fitted with a SIGINT suite, known as Sarigue. The size of the aircraft provides room for over 10 workstations.

In early 1993, Thales (formerly Thomson-CSF DETEXIS) was awarded a contract for the Sarigue NG (Nouvelle Génération) in which a DC-8 will be fitted with advanced ELINT and COMINT suites. The output from these will be collected to provide the SIGINT picture.

Sarigue NG is based on ASTAC equipment (see earlier item) and the Thales TRC 290/600 Series communications equipment. TRC 290 receivers operate in the VHF/UHF bands, while the TRC 600 Series is a range of receivers, analysers and DF equipment covering the frequency range 0.1 to 1,350 MHz.

Operational status
A DC-8 aircraft equipped with Sarigue is in service with the French Air Force. In May 1993, Thales was awarded a contract to install Sarigue NG in a French Air Force DC-8, which is scheduled to enter service in 2000.

Contractor
Thales (formerly Thomson-CSF DETEXIS).

UPDATED

SACRE non-comm ESM/ELINT system

The SACRE system is a lightweight fully automatic sensor and a programming/analysis workstation.

SACRE is designed for tactical situation assessment (ESM function) and EW intelligence data gathering (ELINT function).

Because of its small volume and light weight, SACRE may be integrated in light carriers such as UAVs or helicopters, as well as in highly mobile land platforms.

The data are automatically intercepted during the mission and can either be transmitted to the workstation for real-time processing by means of a datalink or stored in a mass memory.

Mission data are processed in the workstation to evaluate tactical situation assessment: radar interception, measurement, identification and localisation. Technical inter-pulse and intra-pulse analysis is also performed to refine technical data bases.

The SACRE sensor utilises interferometry techniques to provide very accurate direction-finding with a wide instantaneous bandwidth to precisely determine threat status and location.

Inter-pulse and intra-pulse (option) fine grain analysis allows precise identification of threats. A finger-print capability is a further option.

Specifications
Frequency: D to J band (C and K bands optional)
DF accuracy: 0.5° RMS
Instantaneous angular coverage: 2 × 120°
Weight: 25 kg

Operational status
In development.

Contractor
Thales (formerly Thomson-CSF DETEXIS).

UPDATED

SB25A combined interrogator-transponder

The SB25A combined interrogator-transponder includes Mode S capability and a new type of electronic scanning technology.

Operational status
SB25A equips the Rafale aircraft.

Contractor
Thales (formerly Thomson-CSF Communications).

UPDATED

SCP 5000 VHF/UHF secure radio communications system

The SCP 5000 secure radio communications system is designed for military platforms. It provides voice and data transmission services protected against the electronic warfare threat such as jamming, eavesdropping, intrusion and localisation.

The system features pseudo-random fast frequency hopping, time variable unsigned synchronisation dwell, voice and data cyphering and use of the complete UHF band. The Have Quick II mode is available as an option.

The SCP 5000 offers fixed-frequency plain analogue voice, KY 58-compatible fixed-frequency cyphering voice, frequency-hopping plain or cyphered voice, point-to-point data transmission to transmit mission data and inter-weapons system co-operation data and data transmission over a TDMA network to allow participants to exchange tactical information between aircraft or between command and control centres and aircraft.

The SCP 5000 airborne system comprises one or two BER 8500 or 8700 V/UHF or UHF transceivers, a TDP 5000 processor to control the frequency hopping, two BCA 1217 control boxes and an optional BVA 500 frequency display.

The TDP 5000 processor is connected to the aircraft system via the avionics bus or a dedicated link. It manages the frequency-hopping procedures; communicates cyphering/decyphering; manages data transmission, message formatting, link or network management and error detection and correction; and interfaces with the platform for voice digitisation and bus coupling.

Specifications
Dimensions:
(BER 8500) 57 × 202.5 × 493 mm
(BER 8700) 57 × 202.5 × 455 mm
(TDP-5000) 57 × 193.5 × 380 mm
(BCA 1217) 146 × 47.6 × 148 mm
(BVA 500) 41.5 × 41.5 × 125 mm
Weight:
(BER 8500) 8.5 kg
(BER 8700) 7.5 kg
(TDP-5000) 7 kg
(BCA 1217) 1.2 kg
(BVA 500) 0.5 kg
Power supply:
(BER 8500/8700) 28 V DC or 115 V AC, 400 Hz
(TDP-5000, BCA 1217, BVA 500) 28 V DC
Frequency: 100-156 and 225-400 MHz
Channel spacing: 25 kHz

Operational status
In production for the Mirage 2000 for the French Air Force and the export version Mirage 2000-5.

Contractor
Thales (formerly Thomson-CSF Communications).

UPDATED

Serval radar warning receiver

Fitted to French Air Force and export versions of the Dassault Mirage 2000 fighter, Serval warns the pilot when the aircraft is being illuminated by surface or airborne radars of a hostile nature; friendly/hostile discrimination is done by comparing the characteristics of the illumination energy with those of emitters held in the reprogrammable system threat library. Frequency coverage is understood to be E-J band, utilising channelised, crystal video receiver technology.

Serval uses four detection antennas mounted on the wingtips and fin feeding a hybrid analogue/digital processor. The CRT display unit shows the strength and direction of the threat emitter and whether it is ground-based or airborne. Details of several emitters can be shown simultaneously. At the same time an audio alarm sounds in the pilot's headset.

Operational status
Fitted to French Air Force and export Mirage 2000 aircraft.

Thales (formerly Thomson-CSF DETEXIS) is understood to be developing an upgraded version of Serval, designated Serval NG-D (New Generation-Distance). The 'distance' descriptor is said to be added because Serval NG-D includes new software algorithms that better determine the distance of the emitter that is illuminating the RWR.

Contractor
Thales (formerly Thomson-CSF DETEXIS).

UPDATED

Sherloc TMV 011 radar warning receiver

The Sherloc TMV 011 RWR is a system designed for fixed-wing aircraft or helicopters. It incorporates a crystal video receiver, high-speed digital processor and a radar signals library which is easily reprogrammable on the flight line. It also delivers operational flight reports. Of modular design, there are many ways to build Sherloc units into a system to meet specific requirements for extensive self-protection.

The SCP 5000 has been selected for the French Air Force Mirage 2000 and the export Mirage 2000-5

Threat data is presented to the pilot in the form of alphanumeric symbology on a colour display unit, the symbols denoting the identification of threat lethality and the position on the CRT being relative to the threat's bearing. Up to eight emitters can be presented simultaneously. Alternatively, a simple light-emitting diode display can be used, indicating threat classification and relative strength. Sherloc operates in the D- to J-bands. System weight is 11 kg for aircraft and 9.5 kg for helicopters.

The latest Sherloc version, Sherloc F, incorporates an instantaneous frequency measurement receiver.

The helicopter specific version is called Sherif.

Specifications

Weight: 13 kg
Frequency: D- to J-band; options for C- to K-band
Accuracy: >10°

Operational status

In production. Between 150 and 200 Sherloc RWRs have been sold. In service in the Mirage 50, Mirage F1, French Air Force C-135F and C-160 and Super Puma and Dauphin helicopters.

Contractor

Thales (formerly Thomson-CSF DETEXIS).

UPDATED

SICOP-500 integrated radio communication system

SICOP-500 is an integrated ECCM radio communication system designed for air force applications. It operates in the VHF and UHF bands and capabilities include plain data and voice, cipher voice, jam-resistant voice and data, ground-to-air, air-to-air and air-to-ground communications.

The jam-resistant operation is achieved by using frequency-hopping techniques for voice and data transmissions, while error correction is provided for data transmissions. For voice transmissions the ciphering can operate both in jam-resistant and fixed-frequency modes.

The SICOP-500 comprises an ERA-8500 VHF/UHF transceiver, ERA-8700 UHF transceiver (see earlier items) and TDP-500 ECCM radio processor unit. It is designed for use by high-performance military aircraft in a jamming environment. It operates between –55 and +90°C at pressure altitudes up to 96,000 ft.

Thales' SICOP-500 advanced ECCM radio system

Specifications

Dimensions:
(BER 8500 V/UHF R/T unit) 57 × 202.5 × 493 mm
(BER 8700 UHF R/T unit) 57 × 202.5 × 455 mm
(BCA 1111 main control unit) 146 × 76 × 173 mm
(BCA 1211 auxiliary control unit) 146 × 47 × 173 mm
(TDP-500 radio processing unit) 57 × 193 × 380 mm
(BCSA-500 radio processing control unit) 146 × 38 × 100 mm
(BMD-500 security unit) 80 × 60 × 160 mm
Weight:
(BER 8500 V/UHF R/T unit) 8.5 kg
(BER 8700 UHF R/T unit) 7.5 kg
(BCA 1111 main control unit) 1.3 kg
(BCA 1211 auxiliary control unit) 1 kg
(TDP-500 radio processing unit) 6 kg
(BCSA-500 radio processing control unit) 0.5 kg
(BMD-500 security unit) 0.8 kg

Operational status

In service on Dassault Mirage 2000 and various other platforms.

Contractor

Thales (formerly Thomson-CSF Communications).

UPDATED

SLAR 2000 surveillance radar

The SLAR 2000, also known as the Raphael TH, is an all-weather side-looking airborne radar employing synthetic aperture and pulse compression techniques to provide high-quality mapping. The airborne part of the system is pod-mounted on a combat aircraft. It can also be installed in the cargo bay of a commuter or transport aircraft. Radar information is transmitted via datalink to a ground station where it is displayed in real time.

Thales' SLAR 2000 is fitted on the Dassault Mirage F1 CR

The radar is highly directional and operates in the I/J-band. It has an effective beamwidth of a few mrads, providing a sharp and accurate radar map of the ground. The system also features an MTI capability. Ground echoes are processed on the aircraft, then transmitted to a ground station for automatic exploitation. This ground station is air- or ground-transportable.

Operational status

In service with the French Air Force and the air forces of several other countries. It was used operationally during the 1991 Gulf War and for Allied Force operations in the former Yugoslavia.

Contractor

Thales (formerly Thomson-CSF DETEXIS).

UPDATED

Spectra EW system

Thales (formerly Thomson-CSF DETEXIS) and Matra BAe Dynamics SAS have developed the Spectra self-protection system for the Rafale ACT/ACM aircraft. The system is the first ever in France to cover electromagnetic, laser and infra-red domains. It makes use of sophisticated techniques, such as interferometry, digital frequency memory, electronic scanning, multispectral infra-red detection, artificial intelligence, and substrate technologies (MMICS on GaAs substrates and VHS INS). Spectra includes an active phased-array transmitter.

The system is fitted internally in the Rafale, with over 10 locations distributed throughout the aircraft, and integrated through a specific EW databus and a central processor. Spectra may also be installed on the outside of the aircraft.

Thales is responsible for EW system integration, electromagnetic detection, jamming and laser warning functions. The laser warning function is provided by the DAL (Détecteur d'Alerte Laser) system. Matra BAe Dynamics SAS provides the DDM (Missile Launch Detector) (two-sensor type) and the LCM (Modular Cartridge Dispenser) (with four cartridge dispenser modules and two chaff dispenser dual tubes). Dassault Aviation is responsible for aircraft integration.

Operational status

The programme was launched in 1990 under a French MoD contract. The first prototype was delivered in 1993. First flights on Rafale occurred in 1994.

The Spectra system has been fully tested, and deliveries of the first series-production systems commenced in 1998.

Contractors

Thales (formerly Thomson-CSF DETEXIS).
Matra BAe Dynamics SAS.

UPDATED

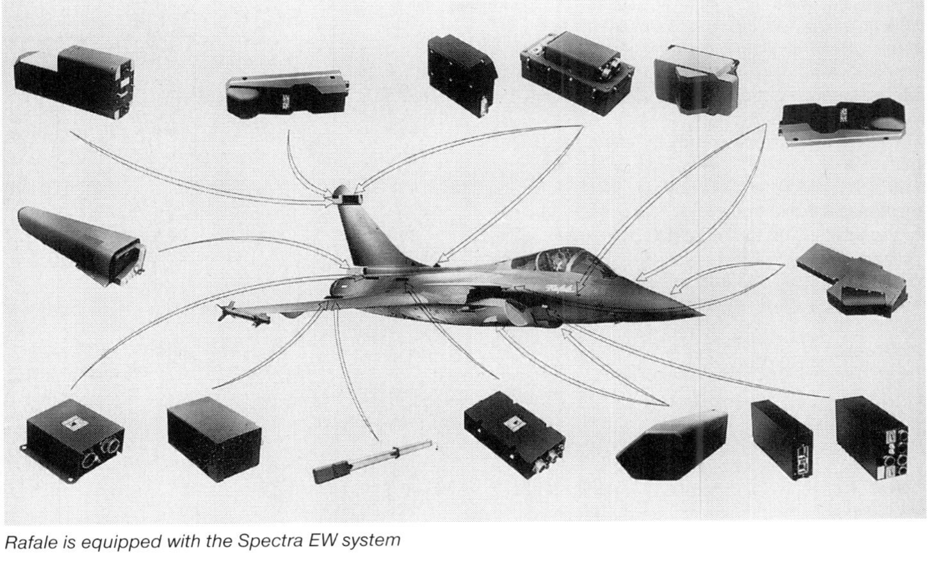

Rafale is equipped with the Spectra EW system

Spectrum Airborne Surveillance (SAS) system

The Thales (formerly thomson-CSF Comsys) SAS system is designed to intercept and localise sophisticated radio transmitters throughout the civil and military radio spectrum. The system's radio direction-finders cover frequencies from 20 kHz to 3,000 MHz, while its surveillance capability covers 300 kHz to 3,000 MHz.

Using modular architecture, the all-digital SAS system comprises a network of fixed antennas mounted within the flight deck; an interception/direction-finding antenna system mounted externally on the aircraft; one or more Windows NT™-based workstations, and recording devices. The technical analysis facilities of the SAS system are designed to enable it to identify the transmissions intercepted.

Thales has developed an entirely new network of patch antennas for the SAS system. Thales claims that this system is particularly suitable for localising radio transmissions from commercial cellular networks, that it is easy to integrate, and that it provides a high degree of stealth.

Operational status
Thales is marketing the SAS concept for installation on the F406 Vigilant or similar aircraft, but states that the concept implementation can be scaled to meet other intelligence collection requirements.

Contractor
Thales (formerly Thomson-CSF Comsys).

UPDATED

Spider active expendable jammer

Thales (formerly Thomson-CSF DETEXIS), in association with Matra BAe Dynamics SAS, is developing the Spider expendable self-protection jammer which can be fitted in current chaff and flare containers. This decoy is made of an electronic payload, using the latest technologies such as MMIC, a GaAs amplifier with a high degree of integration to fit in the limited volume, a vehicle housing the payload and a battery.

Spider is designed to counter most modern threats such as active coherent missile homing heads. It has the basic capability to defeat monopulse tracking.

Operational status
Believed to be under test.

Contractors
Thales (formerly Thomson-CSF DETEXIS).
Matra BAe Dynamics SAS.

UPDATED

SPS-H Self-Protection System for Helicopters

SPS-H is a concept that can be configured to meet specific requirements. The concept is based on a core system comprising: the EWR-99 Radar Warning Receiver (RWR) and EW system computer; the MWS 20 Missile Warning System; a selection of available off-the-shelf solutions for chaff and flare dispensing. Optional equipment includes: various available Laser Warning Receiver (LWR) systems; interface software for additional infra-red and/or radio frequency jammers to complement the chaff and flare elements of the core system.

The central computer exercises: threat data fusion; smart control of countermeasures (automatic, semi-automatic and manual modes); decoy control; electromagnetic compatibility functions; interface control via MIL-STD-1553B, ARINC or RS databus.

User self-sufficiency is provided under software control of the Operational Flight Program (OFP), clearly separated from the threat library.

Contractor
Thales (formerly Thomson-CSF DETEXIS).

UPDATED

Syracuse II airborne terminal

Thales (formerly Thomson-CSF DETEXIS) has developed an airborne terminal to work in conjunction with the Syracuse II satellite military telecommunications system. The terminal will enable an aircraft to communicate at any time with other terminals on land or at sea within the coverage of the satellite. It is designed for long-range aircraft such as the Atlantique 2 maritime patrol aircraft, KC-135 in-flight refueller and military transport aircraft.

Syracuse antennas installed on a C-160 Transall aircraft

The terminal employs two electronically scanned active antennas, each with several hundred active elements. The two transmit/receive antennas, mounted in the fuselage of the aircraft, have no significant protrusions and hence no adverse effect on drag. Faulty elements will be inhibited individually and it is estimated that satisfactory communications will be possible with up to 10 per cent of the elements defective. The system will therefore have a high tolerance to component failure, giving it a high MTBF.

Operational status
An electronically scanned active antenna was successfully flight-tested in 1991. Present status unknown.

Contractor
Thales (formerly Thomson-CSF DETEXIS).

UPDATED

Syrel ELINT pod

The Syrel pod is a fully automatic electronic reconnaissance system attached by a special centreline pylon on Dassault Mirage F1, Mirage 3 and Mirage 2000 aircraft. It can be used during medium- and high-altitude standoff missions, as well as low-altitude penetrations, automatically to acquire and record data relating to the identification and location of ground-based electronic systems. It is intended to provide reliable information on radars for early warning systems, search and acquisition, ground-control interception and fire control for anti-aircraft artillery or missiles.

The pod has two antenna sets at both front and rear, receiver units, an amplifier and recorders in its centre section. The pylon houses a cooling system which has a ram-air intake in the pylon leading edge. High-speed operation is assisted by thick-film and microwave circuit assemblies on ceramic substrates. Thales (formerly Thomson-CSF DETEXIS) also produces first and second line maintenance equipment for use with the pod.

Syrel is combat proven.

Thales' Syrel ELINT pod

Specifications
Dimensions: 3,570 length × 420 mm diameter
Weight: 265 kg

Operational status
No longer in production; believed to be in service on aircraft of the Spanish and other air forces, including Mirage III, Mirage F1 and Mirage 2000.

Contractor
Thales (formerly Thomson-CSF DETEXIS).

UPDATED

TC 20 control unit

The TC 20 control unit is a multipurpose VHF/UHF control unit which controls all the Thomson-CSF Communications fixed-frequency transceivers.

It can control two transceivers simultaneously. Programming allows the automatic adaptation of the control unit to the transceiver configuration. Mode controls provided include frequency setting from 30 to 400 MHz, 800 preset channels in four fields, Guard emergency control and Guard receiver, squelch, power and secure mode control. Preparation of the next frequency or channel to be used is possible on the second line of the display for an instantaneous transfer to the active line by a one-touch button. It is NVG-compatible.

Specifications
Dimensions: 57 × 146 × 120 mm
Weight: 1.25 kg
Power supply: 28 V DC

Operational status
In production.

Contractor
Thales (formerly Thomson-CSF Communications).

UPDATED

TDS-H radar warning receiver

The TDS-H radar warning receiver is the latest product in the Thales (formerly Thomson-CSF) family of radar warning receivers. It has a very wideband receiver, with an instantaneous frequency measuring capability. The whole receiver is miniaturised and weighs only 4 kg. It is fully programmable.

Operational status
TDS-H has been selected for the entire French Army Light Aviation (ALAT) combat helicopter force, and by the Polish armed forces for the Sokol helicopter.

Contractor
Thales (formerly Thomson-CSF DETEXIS).

UPDATED

THOMRAD 6000 V/UHF ECCM transceiver series

The THOMRAD 6000 product line is designed for voice and data communication and provides a full tri-service interoperability in ECCM mode. It covers the VHF and

UHF bands from 30 to 400 MHz and operates in AM and FM modes. All transceivers include the channel spacing at 8.33 kHz in accordance with the new ICAO requirements.

ECCM protection is provided by use of the NATO SATURN mode, or a proprietary embedded fast frequency-hopping and ciphering capability that uses modern digital modulation and synthesiser technology. The provision of embedded COMSEC and TRANSEC modules reduce weight and save installation and integration effort. The THOMRAD 6000 product line features, in addition, the NATO Have Quick II mode and can be associated with external crypto devices.

Data transmission mode is available for point-to-point links or through Time Division Multiple Access protocol. Use of the datalink mode, with fast frequency hopping and error detection and correction procedures, ensures both transmission reliability and minimal detectability. THOMRAD 6000 radios are designed to transmit and receive Link 11 and Link 22.

The THOMRAD 6000 product line includes two airborne versions (TRA 6030 and TRA 6020), ground and shipborne versions (TRG 6030), and a manpack version (TRM 6020). Together they represent a fully integrated and interoperable communication system for all air/ground, air/air and surface/surface communications.

Particular versions are available for specific requirements.

Specifications

Dimensions:
(TRA 6030) 90 × 194 × 320.5 mm
(TRA 6020) 127 × 120 × 183 mm
Weight:
(TRA 6030) 7 kg
(TRA 6020) 5 kg
Power: 28 V DC

Operational status

All versions in production. Selected for Rafale, Mirage 2000, E2C, C-130, Cougar and Panther.

In March 1999, a variant, designated TRA 6021, was selected by the UK for Royal Air Force Canberra PR.9, Hercules C-130K (C.1/3), Nimrod MR.2 and R.2, and VC-10 aircraft, and for Royal Navy Jetstream T.3 observer trainer aircraft. Aircraft installation by October 1999 was planned. The same equipment has been selected for the British Army WAH-64 Apache attack helicopter development programme, and it was evaluated during 1999 by the UK Royal Navy for its helicopter SATURN programme.

Contractor

Thales (formerly Thomson-CSF Communications).

UPDATED

TLS-2020 MultiMode Receiver (MMR)

The TLS-2020 MultiMode Receiver is a new VOR/ILS receiver designed for fighters and helicopters. It fully complies with new ICAO - Annex 10 regulations (FM immunity requirements). It also includes a VHF receiver for the DGPS datalink.

The TLS-2020 receiver can optionally be configured with Marker, MLS, DGPS and GPS functions by simply adding modules.

The TLS-2020 MMR utilises the latest technology developed both for the commercial aviation and French military programmes. The TLS-2020 belongs to the TLS-2000 product line.

Specifications

Basic functions: VOR/ILS/VHF receiver for DGPS
Options: Marker, MLS, DGPS and GPS
Dimensions: ¼ ATR short
Weight: <4 kg
Power: 28 V DC
Interfaces: MIL-STD-1553B, ARINC 429, analogue interfaces

Operational status

Selected for the Rafale and Nimrod MRA4 aircraft, and the NH 90 helicopter.

Contractor

Thales (formerly Thomson-CSF Communications).

UPDATED

Thales' TLS-2020 MultiMode Receiver (MMR) 0015353

TLS-2030 MultiMode Receiver (MMR)

The TLS-2030 MultiMode Receiver is a new VOR/ILS receiver which complies with ARINC 755.

Thales' TLS-2030 MultiMode Receiver (MMR) 0015354

The TLS-2030 is the military version of the TLS-755 MMR that has been developed for commercial aircraft. The VOR function has been added to the functions of the TLS-755.

It offers VOR/ILS capability and a VHF receiver for DGPS functions.

Optionally it can be configured with MLS, DGPS and GPS. The TLS-2030 belongs to the TLS-2000 family.

Specifications

Basic functions: VOR/ILS/VHF receiver for DGPS
Options: MLS, DGPS and GPS
Dimensions: 3 MCU
Weight: 4 kg
Power: 115 V, 400 Hz
Interfaces: ARINC 429

Contractor

Thales (formerly Thomson-CSF Communications).

UPDATED

Thales' Transceivers Airborne TRA 6020 (left) and TRA 6030 (right), of the THOMRAD 6000 V/UHF ECCM transceiver series 0011853/0011854

TRA 2020 VHF/UHF radio

The main features of the TRA 2020 are a frequency range of 100 to 400 MHz, Guard receiver, clear or KY 58 secure voice, homing/direction-finding output, discrete power and remote control. The main criteria which guided the design were reliability, maintainability, lightness, low power consumption and the ability to adapt to any military platform.

The TRA 2020 architecture is centred around two printed circuit boards using SMC technology and carrying all power supply, transmission, synthesiser and main receiver functions and optional Guard receiver and control interface functions. The easily adaptable mechanical configuration allows installation on any type of platform. The optional control interface is compatible with MIL-STD-1553 or ARINC 429 busses. In the ARINC 429 versions, it is controlled by the TC 20 control unit.

Specifications

Dimensions: 356 × 57.15 × 193 mm
Weight: <4.5 kg
Power supply: 28 V DC
(VHF) 118-156 MHz
(UHF) 225-400 MHz
Channel spacing: 12.5 kHz

Status

In production.

Contractor

Thales (formerly Thomson-CSF Communications).

UPDATED

The TRA 2020 VHF/UHF radio and its associated TC 20 control unit (right) 0011843

TRA 6020 V/UHF ECCM airborne transceiver series

The TRA 6020 series of airborne transceivers comprises the TRA 6021, which offers fixed frequency, Have Quick, operation with the NATO SATURN comsec capability, and the TRA 6025 in which the SATURN capability is replaced by a proprietary ECCM algorithm.

The equipment includes an ECCM management system and IDM/EDM compatibility. Ancillaries include a remote control unit, agile filters, multicouplers, and external booster and tuned antennas.

Specifications

Frequency bands:
30 to 87.975 MHz
108 to 173.975 MHz
225 to 399.975 MHz
Channel spacing: 8.33 and 25 kHz
Transmit power:
(AM): 10 W
(FM): 15 W
(LPI): LPI operation available
Modulation: AM, FM, MSK, FSK
Guard channels: 121.5 and 243 MHz
Functions:
(fixed frequency): clear voice, secure voice, Link 11
(ECCM) clear voice, secure voice, point-to-point data with TDMA option
Comsec: external KY 58 - KY 100, or proprietary embedded comsec
Remote control: MIL-STD-1553B, ARINC 429 or RS 422
Power: 28 V DC, 30 W receive only, 150 W transmit
MTBF: 2,000 h
Dimensions: 127 × 124 × 183 mm
Weight: 5 kg

TRA 6020 series V/UHF ECCM airborne transceiver
2000/0080280

Contractor

Thales (formerly Thomson-CSF Communications).

UPDATED

TRC 9600 VHF/FM secure radio communication system

The TRC 9600 is a VHF/FM military airborne transceiver designed for military platforms with a high ECCM protection level ensuring reliable communications in a dense electronic warfare situation. It is the airborne version of the PR4G system, fully interoperable with manpack and vehicular versions and covers the frequency range 30 to 88 MHz at 25 kHz spacing to give 2,320 channels.

The TRC 9600 provides communications protected from interception, direction-finding, jamming, listening-in and spoofing. The lightweight and compact 10 W transceiver embodies frequency hopping, free channel search and high-security digital encryption functions.

Its light weight, small size and high reliability result from the use of advanced and well-proven technology such as powerful and fast microprocessors of the new HCMOS 68000 family, VLSI circuits, wide use of surface mount components and the use of proximity filters for co-site operation.

Frequencies are generated by a digital synthesiser with an ultra-rapid acquisition time, driven by a high-

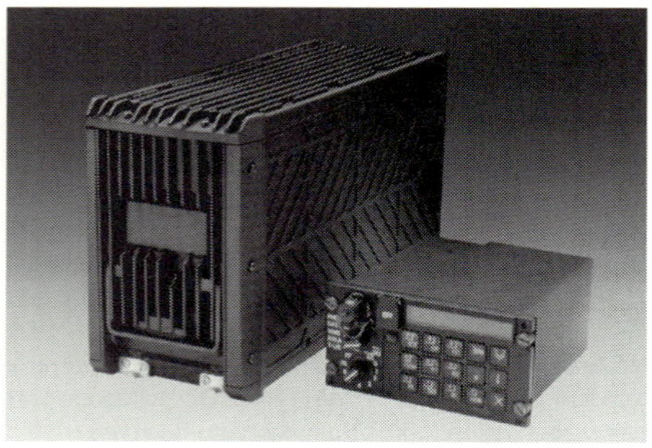

Thales' TRC 9600 VHF/FM secure radio communication system comprises the TRC 9610 transceiver unit (left) and the TRC 9620 control unit (right)

stability oscillator. Seven channels can be memorised and are stored for more than a year. The output power can be 10, 5 or 0.5 W depending on the requirement. In the analogue fixed-frequency mode, the TRC 9600 is directly interoperable with existing VHF/FM sets and is provided with a noise squelch.

The system consists of the TRC 9610 transceiver unit, TRC 9620 control unit, shockmount, antenna with logic converting unit and power supply. Accessories include a KY 58 encryption unit, relay cable, aircraft interphone system and fill device.

Specifications

Dimensions:
(control unit) 145 × 76 × 160 mm
(transceiver unit) 125 × 196 × 340 mm
Weight: 8 kg
Power supply: 28 V DC

Contractor

Thales (formerly Thomson-CSF Communications).

UPDATED

TRES Tactical Radar ESM System

TRES is a combined Tactical Radar and ESM System designed for fitment to naval helicopters to provide wide area maritime surveillance, early warning and

TRES Tactical Radar ESM System for naval helicopters 0011860

weapon control. The system integrates the Varan sea surveillance radar and the DR 2000 Dalia ESM systems.

Contractor

Thales (formerly Thomson-CSF DETEXIS).

UPDATED

TSB 2500 Combined IFF Interrogator and Transponder (CIT)

The TSB 2500 consists of two LRUs: the Combined Interrogator/Transponder (CIT) and the Antenna Control Unit (ACU) or the Antenna Adaptor Unit (AAU). The system is compatible with various types of electronically (with ACU) or mechanically (with AAU) scanned antennas. The TSB 2500 is available in both interrogator and interrogator/transponder versions and is a highly flexible system capable of meeting the needs of many military platforms. It is modular in both design and operation.

It is a Mk XII and Mode S level 2/3 transponder and a Mk XII interrogator. It also incorporates provisions for the integration of future mode 5 new-generation IFF, for both transponder and interrogator functions.

The CIT integrates all transmission, reception, signal, and data processing functions required by an interrogator/transponder. It interfaces with the host platform via a MIL-STD-1553B databus.

The ACU controls electronically scanned antennas. The AAU functions as a booster and an RF front end. The packaging of ACU and AAU in separate boxes allows their installation close to the antenna system providing minimum RF losses in the cables.

The TSB 2500 can interface with any NSA Mode 4 crypto computer or secure crypto computer (for non-NATO applications) that is interface compliant with STANAG 4193. It can also be fitted with a dual KIT/KIR appliqué crypto computer on the front panel.

Specifications

Interrogator
Power: >32 dBW
Frequency: 1.030 ±0.2 MHz
Operating modes: 1,2,3/A,C,4 (Mode S upgradable)

Transponder
Power: 500 W (±2 dB)
Frequency: 1.090 ±0.5 MHz
Operating modes: 1,2,3/A,4,S (Mode S upgradable)
Dimensions
(CIT) 228.6 × 157.2 × 193.5 mm
(ACU) 230 × 115 × 105 mm
(AAU) 32 × 193 × 290 mm
Weights
(CIT) <10 kg
(ACU) 5.5 kg
(AAU) <4 kg

Operational status

Selected for Rafale and Erieye aircraft, and the NH 90 helicopter.

Contractor

Thales (formerly Thomson-CSF Communications).

UPDATED

TSC 2000 and TSC 2050 IFF Mk XII/Mode S diversity transponders

The TSC 2000 and 2050 IFF systems are true diversity Mk XII and Mode S level 3, TCAS-compatible transponders, that are fully compliant with ICAO Annex 10, STANAG 4193 and DOD-AIMS-65-100B standards. Moreover they include all provisions for the integration of Mode 5 for a new-generation IFF, in accordance with NATO recommendations, and with any NSA Mode 4 crypto computer or secure crypto computer (for non-NATO applications) that is interface compliant with STANAG 4193. Optionally it can be provided with mechanical adaptor to fit the crypto computer as an appliqué.

The TSC 2000 was co-developed by Thomson-CSF Communications and DaimlerChrysler Aerospace AG.

The TSC 2050 has been designed to be shelf mounted in non-pressurised zones and is intended for both advanced aircraft or for retrofit purposes. It is provided with a MIL-STD-1553B bus interface, or can be operated through a control box.

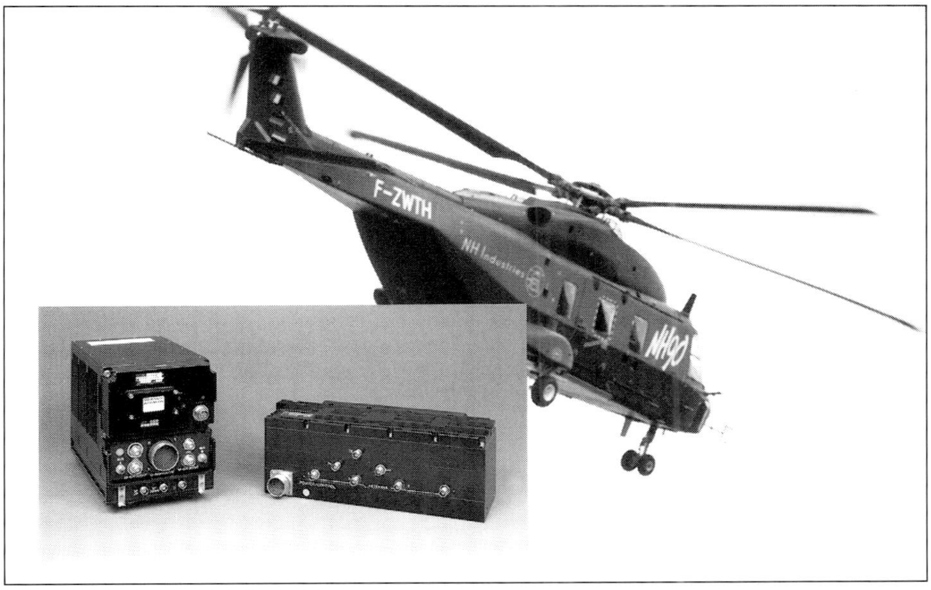

The TSX 2500 IFF interrogator equips the NH 90 NFH helicopter **2000**/0080279

Specifications

Dimensions: 136 × 124 × 212 mm
Weight: <5.2 kg
Modes: 1,2,3/A,C,4 and S level 3

Operational status

The TSC 2000 has been selected for both the Franco-German Tiger helicopter and for the TTH (Tactical Transport Helicopter) and NFH (NATO Frigate Helicopter) versions of the NH 90 by Eurocopter Deutschland. The TSC 2050 has been selected by the UK for the Nimrod maritime patrol aircraft and for global retrofit of Romanian aircraft.

Contractor

Thales (formerly Thomson-CSF Communications).

UPDATED

TSX 2500 series of IFF interrogators

The TSX 2500 has been ordered to equip the Rafale aircraft and the Erieye AEW&C system.

A derivative version has been chosen by Agusta to equip the NATO Frigate Helicopter (NFH) version of the European NH 90 military transport helicopter, where it will operate with the European Navy Radar (ENR), a mechanically scanned radar developed by the European consortium comprising Thales (formerly Thomson-CSF DETEXIS), EADS (formerly DaimlerChrysler Aerospace AG) and Alenia Difesa FIAR.

The latter contract follows the award from Eurocopter Deutschland for the supply of TSC 2000 Mode S IFF transponders for the TTH (Tactical Transport Helicopter) and NFH versions of the NH 90.

Contractor

Thales (formerly Thomson-CSF Communications).

UPDATED

Varan sea surveillance radar

Varan is essentially an Iguane radar (see earlier item) with a smaller antenna which makes it suitable for virtually all the present and planned lightweight maritime patrol aircraft and helicopters. It has been fitted to the Dassault Falcon Gardian of the French Navy and selected for the naval version of the ATR 42 transport aircraft and Eurocopter SA 365F Dauphin 2 helicopter.

The system provides real-time pollution and ice detection. Key factors are I-band operation, pulse compression over several pulsewidths, frequency agility for electronic counter-countermeasures and beacon detection. The unspecified but low-peak power level, associated with high receiver sensitivity, increases the difficulty of detection by hostile radars. Typical detection ranges, in Sea State 3 to 4 are snorkel 55 km, fast patrol boat 110 km and freighter 240 km. Total weight of the system's six units (including antenna) is 111 kg.

Operational status

In service on the French Navy's Falcon 20H Guardian aircraft. In service on Super Puma helicopters and Falcon 20 aircraft with foreign customers.

Contractor

Thales (formerly Thomson-CSF DETEXIS).

UPDATED

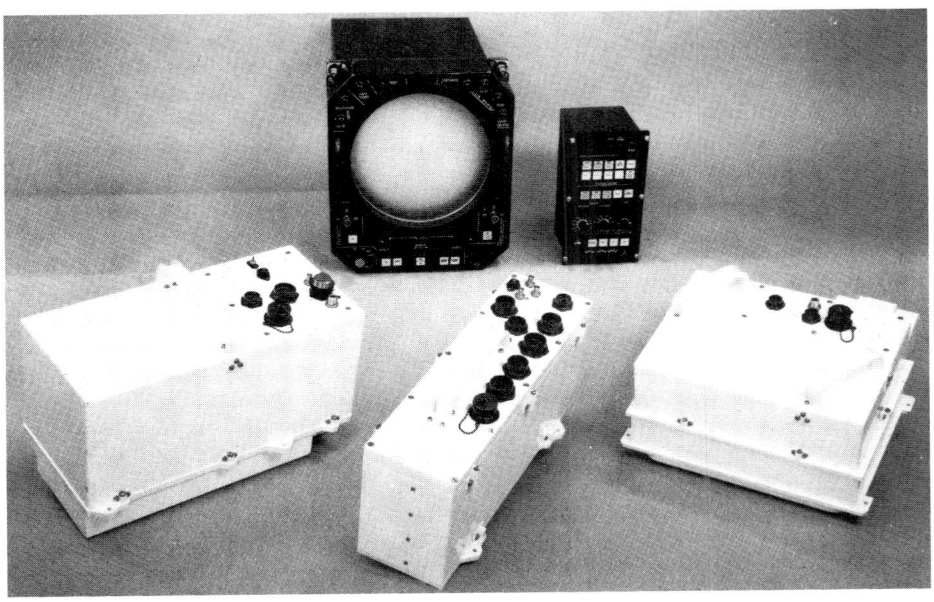

Thales' Varan surveillance radar

DataLink Interface Processor (DLIP)

The DLIP is used to manage L1, L11, L14, L16 and L22 tactical datalinks. The DLIP utilises a modular, scalable architecture based on COTS modules. The DLIP can simultaneously process messages and protocols from multiple datalinks, to provide for easy integration of data from different sources within command and control systems.

The DLIP is available in land, ship and airborne configurations.

Operational status
The DLIP has been selected for the Greek Air Force AEW aircraft, based on the Embraer EMB 145 airframe and ERIEYE radar, being supplied by the Ericsson/Thomson-CSF DETEXIS joint venture (designated Ericsson Thomson-CSF AEW Systems AB). The DLIP is also proposed for NATO's ACCS LOC1 air defence programme.

Contractor
Thomson-CSF Comsys.

VERIFIED

DUAV-4 helicopter sonar

The DUAV-4 is an active/passive directive sonar designed for submarine surveillance and location (azimuth, distance and radial speed). It is specially designed for use on board light ship-based helicopters such as the Lynx. It may also be fitted on small surface vessels.

The DUAV-4 differs from conventional sonars in its signal processing system, which is designed to give improved detection, especially in severe reverberation conditions such as in shallow waters. The sonar can be operated in either the active or passive mode. True bearing, range and radial speed are measured in the active mode; true bearing only in the passive mode.

A combined display unit permits surveillance display, for initial detection, or plotting display, for precise azimuth determination. Total weight, including the electronic rack, cables and dome, is 250 kg.

Operational status
The DUAV-4 is in service with the French Navy, Royal Netherlands Navy and several other navies. Over 90 systems have been produced. In some French Navy Lynx helicopters, the DUAV-4 will be replaced by the HS12.

Contractor
Thomson Marconi Sonar SAS.

VERIFIED

HS 12 helicopter sonar

The HS 12 is an active/passive panoramic helicopter version of the SS 12 small ship sonar and uses the same electronics as that version. It has similar capabilities for operation in shallow or noisy waters and has a system weight of 230 kg, making it suitable for installation on light helicopters such as the Lynx. The HS 12 transducer is lowered and raised by a hydraulic winch at high speed.

Operation in CW and FM modes is possible and digital signal processing is employed by the system's microprocessor. Automatic tracking of two targets and transmission of elements to an external equipment, such as a plotting table, are provided.

The system operates on 13 kHz in the active mode and in the 7 to 12 kHz range passively. A total of 12 preformed beams is employed, giving 30° sectors, with a maximum range of about 10 km. The display consists of four quadrants, these being obtained by processing adjacent beams. The operator can select a CW mode which provides target range and Doppler. FM processing can also be selected and a sector mode is provided.

Specifications
Weight: 230 kg

Operational status
In series production for French and foreign navies. Fitted to Sea King helicopters.

Contractor
Thomson Marconi Sonar SAS.

VERIFIED

HS 312 ASW system

The HS 312 is an acoustic system for helicopters, incorporating the facilities of the HS 12 system and the SADANG acoustic processor. The equipment functions in both passive and active CW and MF modes and has a longer range than the HS 12. The acoustic subsystem can be fitted to process sonobuoys simultaneously with the dipping sonar.

Only a single operator is required and the light weight and compactness of the system mean that the HS 312 can be fitted to any type of light- or medium-size helicopter. Performance of the basic components has been improved by integration of the processing units, integration of a standardised keyboard and the use of only one display screen.

Operational status
In series production for a foreign navy. Designed for Super Puma and Cougar naval helicopters.

Contractor
Thomson Marconi Sonar SAS.

VERIFIED

LAMPARO processing equipment

Derived from SADANG equipment, LAMPARO airborne sonar signal processing and display systems are designed as modern digital equipments for ASW fixed- and rotary-wing aircraft.

The processing capabilities of the systems are based on a single processing unit in a light and compact package, for use by a single ASW operator. The same wide range of facilities and processing modes is available. The system is biased towards the processing of omnidirectional passive and active buoys.

Signals are displayed on a CRT in TV format and a hard copier can be connected for permanent recording. An on-top position indicator is used in conjunction with a radio compass for homing on the sonobuoy. Weight of the complete system is 70 kg.

Operational status
In operational service.

Contractor
Thomson Marconi Sonar SAS.

VERIFIED

Upgraded DUAV-4 helicopter sonar

The upgraded DUAV-4 (DUAV-4 UPG) differs from the DUAV-4 in its acoustic processor which is now based on the SADANG new-generation acoustic processor. The system offers more powerful signal and data processing, as well as an improved man/machine interface.

The DUAV-4 UPG also offers specific software tools and operator aids designed to meet the requirement for detection in shallow, rocky bottom waters.

Operational status
The upgraded DUAV-4 sonar has been supplied to the Royal Swedish Navy.

Contractor
Thomson Marconi Sonar SAS.

VERIFIED

GERMANY

Helicopter self-defence system

The Buck helicopter self-defence system is a modular, low-weight dispenser armed with chaff and flares that are tuned to the signature of the helicopter. It consists of a transport and firing unit installed on each side of the helicopter, with an operating and indication unit in the cockpit. A combination of firing sequences can be used, such as a single round of chaff or flares, a single round of chaff and flares, a multiple sequence of flares or a multiple sequence of chaff and flares.

The device is allied with, and activated by, a radar warning receiver or missile approach warner. Activation can be completely automatic, semi-automatic or manual.

Specifications
Weight: 25 kg including cartridges
Dispenser capacity: 8 rounds each
Ammunition type: 55 × 200 mm calibre chaff and flare cartridges

Operational status
Fully developed

Contractor
Buck Systems GmbH & Co.

VERIFIED

Tornado Self-Protection Jammer (TSPJ)

The TSPJ is a third-generation self-protection jamming pod, designed by EADS (formerly DaimlerChrysler Aerospace AG) for German Air Force and German Navy Tornado aircraft.

Main design features are: it is a generic, modular, jamming system with no threat-specific hardware; has an integrated receiver/processor; has a large repertoire of jamming techniques; all functions are user-programmable via Mil-Bus; there are autonomous and radar warning controlled modes; the modular design and intelligent interface permits adaptation to other platforms.

The system includes a Digital Radio Frequency Memory (DRFM) for the production of coherent jamming signals to defeat coherent threat radars; the DRFM developed by EADS not only contains the RF and memory section to digitise and store signals precisely, but also all techniques generators needed for generation of the required jamming signals.

Operationally, the TSPJ provides a fully software programmable ECM system, capable of operation against both coherent and non-coherent threat systems. The pod can also be used for high resolution data collection, and as the basic building block of radar jammer simulator systems.

The TSPJ self-protection jammer is carried on the Tornado outer wing station ***1999***/0051253

Operational status

The first TSPJ systems were delivered to the German Air Force in early 1998. In production.

Contractor

EADS (formerly DaimlerChrysler Aerospace AG), Defense and Civil Systems.

UPDATED

AN/ALQ-119GY/ALR-68

The German Air Force has decided to update its AN/ALQ-119 ECM pods to meet the modern airborne and ground-based threat environment. EADS (formerly DaimlerChrysler Aerospace AG) has developed a modification kit consisting of RF modules, logic modules, wiring and software to fulfil the requirement. The main added features are: continuous frequency coverage; improved repeater techniques; improved receiver capability; new jamming techniques; introduction of an intelligent interface to the radar warning receiver; improved maintenance. The updated configuration is designated AN/ALQ 119GY and AN/ALQ 119GY/ALR-68, when it is integrated with the Litton Advanced Systems Division AN/ALR-68 ARWS. The AN/ALR-68 ARWS Advanced Radar Warning System is described in the Litton Advanced Systems Division section.

Operational status

The upgrade is designed for the German Air Force F-4F aircraft; the enhancement has been proven during formal testing, including tests against real threat systems.

Contractor

EADS, Defence Electronics and Telecommunications.

UPDATED

LWR Laser Warning Receiver

The EADS (formerly DaimlerChrysler Aerospace) LWR is designed to be integrated with the Threat Warning Equipment (TWE) (see entry in International section) to provide multiple threat warning capability for tactical helicopter systems.

The LWR utilises high-sensitivity near infra-red detectors to provide precise measurement of the angle of arrival of laser pulses in azimuth, with single pulse detection close to 100 per cent, and high immunity against sunlight and explosion flashes.

The LWR provides detection, analysis, identification, bearing, measurement and transmission for visual display and audio alarm of threat laser illumination from laser range-finders, illuminators and beam riders. Options include: an extension module for long wavelengths (far infra-red); a stand-alone version; Mil Bus interface; elevation angle measurement; tank and ship versions.

Specifications

Sensor coverage: 360° in azimuth, adjustable to customer requirements in elevation (up to 90°)
Azimuth accuracy: 10° rms
Reaction time: audio alarm within fractions of a second
False alarm rate (after processing): <1.5 per hour
Power consumption: 51 W (2 × 25.5 W)
Dimensions:
(height) 110 mm
(width) 129 mm
(length) 240 mm
Weight: 3.6 kg (2 × 1.8 kg)

Operational status

Designed to be part of the TWE Threat Warning Equipment, selected for the NH90 TTH (Tactical Transport Helicopter).

Contractor

EADS (formerly DaimlerChrysler Aerospace AG), Defense and Civil Systems.

UPDATED

EADS' Laser Warning Receiver 0018292

AN/ALQ 119GY ECM pod 0018297

AN/ALQ 119GY on German Air Force F-4F 0018296

Sky Buzzer towed decoy

The EADS (formerly DaimlerChrysler Aerospace) Sky Buzzer towed decoy is a high-power (100 W) radar-jamming and decoying system optimised for the protection of both fighter and transport/maritime patrol aircraft. It is effective against a wide range of aircraft including monopulse types.

Sky Buzzer's towed body contains only the jamming transmitter, receiving both the power for operating the transmitter as well as the information for the decoy RF responses via the fibre optic datalink and the high voltage conductors incorporated in the tow cable. Sky Buzzer utilises the same techniques generator as the Tornado Self-Protection Jammer (TSPJ).

Two versions of the towed decoy are available: Sky Buzzer system with deployment system, allowing recovery of the towed body by parachute; Sky Buzzer system with active winch, allowing retrieval of the towed body during flight.

The Sky Buzzer has the following capabilities: high output power (100 W); frequency coverage for all relevant threats; ECM techniques against CW, pulse, and pulse Doppler radars; special techniques to confuse radar operators; techniques to protect the decoy body; redundant decoy resources; fast and reliable deployment and stable flight behaviour.

Operational status

The Sky Buzzer system has been successfully flight tested on different aircraft, including Tornado and F-4.

The Sky Buzzer has been evaluated on Tornado aircraft in single and multiship formations, and there are no restrictions in respect of aircraft speed or manoeuvre. During system tests, at the test aircraft speed limit of Mach 1.4, the tension of the tow cable reached only 18 per cent of its specified value; 5 *g* turns induced 30 per cent of maximum tension.

Sky Buzzer is available to order.

Contractor

EADS (formerly DaimlerChrysler Aerospace AG), Defense and Civil Systems.

UPDATED

EADS' Sky Buzzer towed decoy 0018295

Tornado pod installation of the Sky Buzzer towed decoy system 0018294

STR 700 IFF transponder

The STR 700 IFF transponder is used by nearly all German armed forces aircraft. It features high reliability, small dimensions, low weight and diversity of operations.

The STR 700 meets AIMS specifications. It consists of two basic units: the receiver/transmitter and the control unit with logic section. The control unit has interfaces for the Mode 4 decoder/coder facility and for an altitude encoder. The STR 700 has been specifically designed for applications where diversity operation is required and is provided with two receiving channels. It can also, however, be equipped with a single receiving channel only. The receiver/transmitter of modular design is accommodated in a standard ½ ATR short case for both models. Usually the control unit consists of the actual control section, with switches and lamps and of the logic section for decoding the interrogation signals and for coding the reply signals. If the available space in the cockpit is too narrow, the logic section can be accommodated separately from the control section.

Reliable identification of the target aircraft depends largely on the correct action of the transponder. Consequently the STR 700 transponder is provided with a large number of test circuits, to reveal the functional condition at any time automatically by internal interrogations. For system checking on the ground, the receiver/transmitter of the two-channel version is provided with 12 LEDs to indicate proper performance of the individual subassemblies.

Specifications

Dimensions:
(receiver/transmitter) 124 × 193 × 382 mm
(control unit/logic section) 146 × 134 × 155 mm
Weight:
(receiver/transmitter) 9.8 kg (single channel)
10.8 kg (two channel)
(control unit/logic section) 3.2 kg
Power supply: 16-32 V DC, up to 70 W
Frequency:
(receive) 1,030 ±1.5 MHz
(transmit) 1,090 ±3 MHz

Operational status

In production and service in aircraft of the German armed forces.

Contractor

EADS (formerly DaimlerChrysler Aerospace AG), Defense and Civil Systems.

UPDATED

Tornado ECR system

The Tornado ECR is in service with the German and Italian air forces

The Tornado ECR is based on the sixth Tornado production batch build standard and incorporates a MIL-STD-1553B databus, upgraded radar warning and active electronic countermeasures equipment and an improved missile control unit.

The Tornado ECR features: an Emitter Location System (ELS) to pinpoint, identify and display hostile radar emitters; an imaging infra-red system for all-weather day and night reconnaissance; the Operational Data INterface (ODIN) for transmission of near-realtime reconnaissance data to following aircraft and ground centres; FLIR to enable covert low-level flight in adverse weather conditions and at night; advanced displays and powerful computers to give the crew more time for tactical decision making; the HARM anti-radiation missile; an advanced interface concept to employ existing and future smart weapons and jammers and advanced avionic architecture that can grow and adapt to the demands of the threat in the 1990s and beyond.

The Raytheon Systems Company ELS allows the passive autonomous acquisition, identification and precise location in range and angle of radiating threats. The information is displayed to both members of the crew for target selection. After selection, the co-ordinates are used for cueing the HARM missiles as well as for handover to follow-on forces or for onboard storage.

The ELS detects, identifies and locates radar emissions through the use of a high probability of intercept receiver system. It features multi-octave RF coverage, phase interferometric antenna arrays for precision direction-finding, passive ranging channelised receivers and a multiple MIL-STD-1750A digital processor. The system operates across the RF spectrum for all primary surface-to-air and airborne threats. Data acquired by the system is transferred to the tactical displays of both crew members for threat assessment. The ELS is interconnected with the aircraft avionic and defensive aids databusses, and the emitter library is loaded from the mission data transfer system.

The ELS can contribute to the ECR mission in other ways by assisting the crew during reconnaissance missions. It can be used to identify and locate mobile targets or targets of opportunity and also provide steering information for optimal sensor operation.

The fundamental Tornado ECR tasks are recce-attack or Pathfinder operations. For this the aircraft is equipped with the LITEF ODIN, a digital datalink. ODIN uses the UHF/VHF and HF frequency bands for transmitting near-realtime reconnaissance information to following aircraft and to ground command posts. As a data interface ODIN converts signals from analogue to digital and vice versa. ODIN messages received by the communications system are automatically transferred for display to the crew via the avionic databus. Message formats on the weapon system operator's screens are used for preparing transmissions. Received and transmitted data can be recorded in the mission data transfer system. Voice communication is not affected by operating the datalink facility.

The internally mounted FLIR, developed by Zeiss, provides both aircrew with navigation, reconnaissance and attack information in adverse weather and in night conditions. FLIR enhances the covert penetration and attack capabilities of the Tornado ECR.

The ECR Infra-red Imaging System (IIS) has a horizon-to-horizon capability which allows area and point reconnaissance. It provides a near-realtime onboard display of the recorded image on the weapon system operator's screens. Evaluation of the displayed image by the crew results in a reconnaissance in-flight report which can be transmitted to ground stations and follow-on forces by the ODIN datalink. The recording medium is a high-resolution dry silver film which is fully developed seconds after the target images are recorded.

The IIS consists of the IR linescanner, electronic components for power distribution, formatting, processing and amplification, electromechanical components for film recording and developing and the control panel. It provides video images to the displays via the computer symbol generator. The IIS is also used to record ELS electronic intelligence data.

Operational status

In service with the German and Italian air forces.

Contractor

EADS (formerly DaimlerChrysler Aerospace AG), Military Aircraft.

UPDATED

μ-INS/GPS

Flight control, guidance, integrated navigation, and mission management are integrated in the μ-INS/GPS navigation system HD991A1.

The core μ-INS/GPS contains the Inertial Measurement Unit (IMU) HG1700, a system/processor module, a Global Positioning System (GPS) and a power supply module.

Flexible system architecture permits use of specific application software, using Honeywell's Embedded Computer Toolbox and Operating System (ECTOS).

The IMU integrates three Honeywell GG1308 miniature Ring Laser Gyros (RLG) and three digital accelerometers as well as electronics in a strapdown configuration. It provides fully compensated inertial data on a digital, serial output.

The μ-INS/GPS system processor module is based on the Intel 80960MC microprocessor and is the processing centre for navigation processing. It also contains digital inputs/outputs and three user-configurable RS-422 serial interfaces with selectable baud rates up to 1 Mbaud.

The embedded 12-channel L1 C/A code GPSCard™ represents NOVATEL's newest generation of GPS receivers with DGPS capability. It provides excellent position and velocity performance with a very short reacquisition time even under high dynamic conditions.

Specifications

Dimensions: 210 × 120 × 125 mm
Weight: <3.0 kg
Power: 20 W, 11 to 36 V DC
Interfaces: 3 serial RS-422
Accuracy:
GPS receiver with coarse acquisition (SA on)
(position) <34 m (1σ)
(velocity) (N,E,D) <0.5 m/s (1σ)
(pitch/roll) <0.1° (1σ)
Heading:
(stationary align) <6.0°
(in-flight align) <1.0° (C/A code)

Contractor

Honeywell Regelsysteme GmbH.

VERIFIED

Global Positioning laser Inertial Navigation (GPIN) equipment for German Tornado

The Global Positioning laser Inertial Navigation (GPIN) is part of a German Tornado upgrade programme. GPIN is an advanced technology lightweight ring laser system with a fully embedded GPS (INS/GPS) based on the Litton LN-110G product family.

Operational status

Contracted for the German Tornado New Avionics Structure upgrade programme. First flights of the GPIN system occurred in 1999. Deliveries are expected to begin in 2000.

Contractor

LITEF GmbH.

VERIFIED

Inertial Measurement Unit (IMU) for Eurofighter Typhoon

The IMU is part of the Eurofighter Typhoon quadruplex fly-by-wire flight control system. It has a strapdown design and is housed in a single box which is internally separated into four channels.

Measurement of aircraft body angular rates and linear accelerations and calculation of attitude and heading angles are accomplished by the IMU. It performs autonomous heading alignment at system start up. To provide damping of air data disturbances and to provide a back-up source of air data. The IMU computes angle of attack, sideslip, TAS and altitude data based on inertial measurement and augmented by signals from the air data sensors, when available.

The IMU contains four dual-axis dynamically tuned LITEF gyroscopes and eight single-axis LITEF pendulum accelerometers mounted in a skewed-axis orientation. This combination of accelerometers and gyroscopes is about half the number of sensors used in conventional systems architecture and therefore represents a considerable weight saving.

Contractor

LITEF GmbH.

VERIFIED

LCR-88 Attitude and Heading Reference System

The LCR-88 is a strapdown Attitude and Heading Reference System (AHRS) specifically designed to meet the requirements of the commuter and general aviation market. A version with increased angular rate capability, designated LCR-88A, is available for applications in trainer aircraft. It contains a pair of two degrees-of-freedom dry-tuned gyros and three linear accelerometers. The design is based on LITEF's reliability proven LTR-81 ARINC 705 AHRS. The system operates with 28 V DC power. Several configurations are available to meet the interface requirements for integration into the available avionics packages on the market.

Specifications

Dimensions: 388 × 124 × 194 mm
Weight: 6.3 kg
Power supply: 28 V DC, 85 W
Reliability: >7,000 h MTBF

Operational status

In service in business jets, turboprops, commuter aircraft and helicopters. The LCR-88M has been selected for the German Air Force C-160 Transall.

Contractor

LITEF GmbH.

VERIFIED

Mission computer/weapon computer for the F-4F

The LITEF mission computer/weapon computer for the German Air Force Phantom F-4F features microprocessor technology based on the Motorola 68020 with 32-bit architecture, fast CMOS-SRAM/EPROM, CMOS, ASICs to reduce power, space and weight, high-density packaging with melted core multilayer and surface-mounted technology, high reliability and extensive BIT. The computer contains two spare slots for future expansion.

Specifications

Dimensions: ½ ATR
Weight: 6.5 kg
Power supply: 28 V DC, 50 W (max)

Operational status

German Air Force F-4F Improved Combat Efficiency (ICE) programme.

Contractor

LITEF GmbH.

VERIFIED

ODIN Operational Data INterface

The purpose of the Operational Data INterface (ODIN) is to establish air-to-air and air-to-ground data transfer communication via already existing voice communication channels in the HF and VHF/UHF frequency ranges. The mode of communication is half-duplex, so that ODIN can only transmit or receive

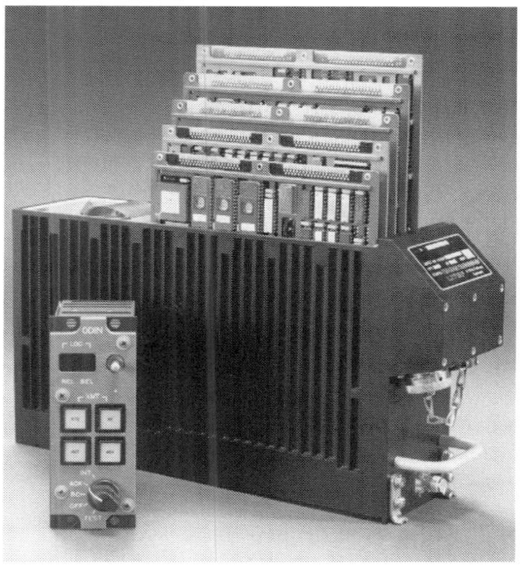

The LITEF Operational Data INterface (ODIN)

ODIN is employed in Panavia Tornado aircraft of the German and Italian air forces **2001**/0103892

information at any given time. When data transfer is not taking place normal voice communication is possible.

The ODIN system consists of two LRUs: the electronics unit and the control panel. The initiation of message transfer via a radio channel and the related transmission mode are controlled by the control panel. Acknowledge, broadcast, interrogate and automatic modes for message transfer are available.

ODIN features a 68000 communication processor and an ADSP 2100 modem processor plus a MIL-STD-1553 interface. The data transfer rate is 2,400 bauds for VHF and UHF and 300 bauds for HF.

Operational status

In service with Panavia IDS and ECR Tornado aircraft of the German and Italian air forces.

Contractor

LITEF GmbH.

UPDATED

Air data systems

Nord-Micro AG & Co. OHG manufactures a range of air data systems including Air Data Computers (ADC) and engine air intake control computers for military subsonic and supersonic aircraft. Modular design concepts are used, together with ASICS and hybrid circuits to produce lightweight, compact, low power consumption systems. For critical data, redundant dissimilar algorithms are used to avoid single point failure and improve fault tolerance. Extensive built-in test and health monitoring functions are also included.

Tornado air intake control system

The Tornado air intake control system comprises two digital air intake control units and one pilot's panel. It controls the engine intake ramp position via an electro-hydraulic actuator.

JAS39 Gripen air data computer/air data transducers (ADC/ADT)

The ADC/ADT is configured as two separate computers in one common housing. They communicate with the fly-by-wire flight control computer via ARINC 429 as well as MIL-STD-1553B databus interfaces.

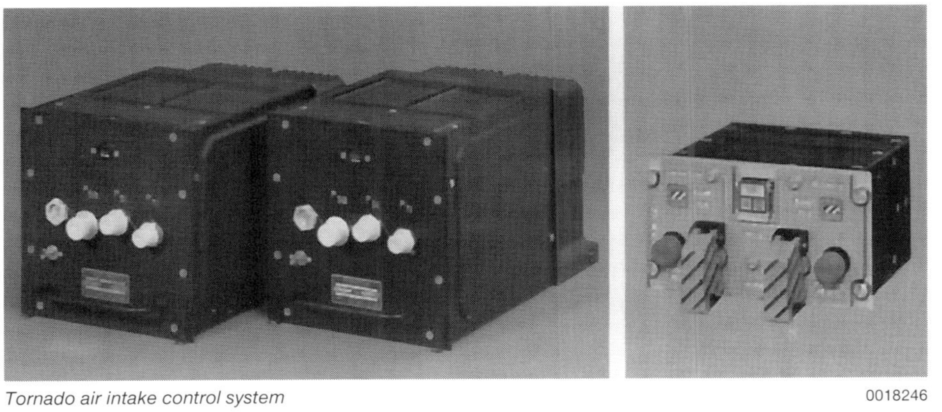

Tornado air intake control system 0018246

Eurofighter Typhoon ADCs

Two ADCs are used for calibration of the Air Data System on-board the first three Eurofighter Typhoon development aircraft.

Operational status

In service.

Contractor

Nord-Micro AG & Co. OHG.

UPDATED

610 and 620 series VHF and UHF radios

The Rohde and Schwarz 610 family of VHF and UHF radio communication systems comes in two basic versions: a single panel-mount cockpit unit or a remotely controlled transmitter/receiver with a panel-mounted control unit. The controllers for the VHF and UHF variants are electrically and mechanically identical. They permit parallel operation of both a VHF and a UHF transmitter/receiver from a single controller and/or the operation of a single transmitter receiver from two control units. All systems in the 610 series provide not only radio telephonic communication but also incorporate a 16 kbit baseband data transmission and ADF facilities.

Technical specifications of the VHF and UHF variants are virtually identical. Each type has a transmitter power output of 10 W at normal supply voltage of 28 V, or 1 W carrier wave with a reduced emergency power supply of 16 V. Frequency range of the VHF systems is from 100 to 155.975 MHz and that of the UHF equipments from 225 to 399.975 MHz. Guard receivers cover the emergency channels of 121.5 and 243 MHz respectively. Frequency setting increments are 25 kHz in each case and the channel spacing is also 25 kHz, although in the case of UHF systems the spacing is optionally adaptable to 50 kHz. Up to 30 channels, plus the Guard channel, may be preselected and a remote frequency/channel indicator is an optional accessory.

All 610 series transmitter/receivers are compatible with one another and the various modules have precisely designed interfaces to permit easy replacements to be made without need for adjustment. The systems are suitable for retrofitting and Rohde and Schwarz has produced replacement kits with tailored adaptor trays for aircraft such as The Boeing Company F-4F and RF-4E. A range of Special-to-Type Test Equipment (STTE) is available, with first line test sets which can isolate faults down to module level and full-scale automatic test equipments for base repair facilities.

Specifications

Dimensions:
(XU 610 VHF cockpit version) 127 × 124 × 165 mm
(XU 611 VHF remote-controlled) 127 × 124 × 165 mm
(XD 610 UHF cockpit version) 127 × 124 × 165 mm
(XD 611 UHF remote-controlled) 127 × 124 × 165 mm
(GB 600 remote-control unit) 146 × 76 × 110 mm
Weight:
(XU 610 VHF cockpit version) 5.2 kg
(XU 611 VHF remote-controlled) 4.6 kg
(XD 610 UHF cockpit version) 5.2 kg
(XD 611 UHF remote-controlled) 4.6 kg
(GB 600 remote-control unit) 1.9 kg

Operational status

In production.

Contractor

Rohde & Schwarz GmbH & Co KG.

VERIFIED

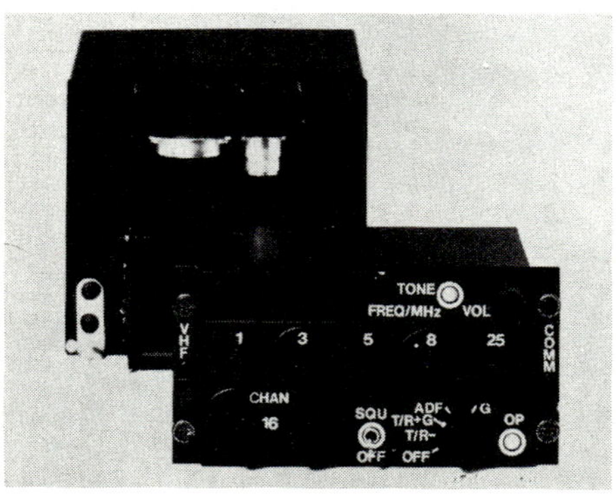

The Rohde and Schwarz 610 VHF/UHF radio

Have Quick 610/611 series VHF/UHF ECCM radios

The Have Quick airborne radio is based on the Rohde and Schwarz 610 Series of radios. As a retrofit for the German Air Force The Boeing Company F-4F and RF-4E Phantom aircraft, the radio includes UHF Have Quick ECCM and its hardware is compatible with Have Quick II. The retrofit adaptor provides an extension to a combined VHF/UHF radio. The radio complies with STANAG 4246 and is interoperable with Rohde and Schwarz Secos 400 Have Quick ground-based or shipborne radios.

Operational status

In service in German Air Force F-4F and RF-4E aircraft. The XD 610H1 with Have Quick I/II is integrated into the Tornado ECR aircraft. The XD 611 H1 is also listed as a Have Quick I/II radio suitable for the airborne search and rescue role.

Contractor

Rohde & Schwarz GmbH & Co KG.

VERIFIED

HF 510 radio system

The HF 510 is an airborne system for air-to-air and air-to-ground communication using AME and SSB to transmit voice and data during nap of the earth flying under non-line of sight and extreme environmental conditions over the whole operational area. Suitable for both helicopters and fixed-wing aircraft, the system includes types for voice and data or voice only and consists of 100 W or 400 W HF transceivers, antenna tuning units and control units.

The system features include high-density integration, a high degree of modularity and built-in speech processor, test programme and silent tuning.

The XK 510 HF data transceiver, especially developed for the German Army Eurocopter PAH-2 Tiger helicopter, has passive and active channel analysis, automatic channel selection from a preset frequency list, adaptive reaction and error detection and correction. Modes are fixed channel, adaptive or ECCM.

Specifications

Dimensions:
(GB 510 control unit) 146 × 229 × 165 mm
(XK 510 D1 transceiver) 223 × 194 × 320 mm
(XK 516 S1 transceiver) 255 × 194 × 320 mm
(FK 510 antenna tuning unit) 123 × 194 × 320 mm
(FK 511 antenna tuning unit) 255 × 194 × 320 mm
Weight:
(GB 510 control unit) 4.5 kg
(XK 510 D1 transceiver) 13.5 kg
(XK 516 S1 transceiver) 19 kg
(FK 510 antenna tuning unit) 5 kg
(FK 511 antenna tuning unit) 5 kg
Power supply:
(XK 510 and 515) 28 V DC
(XK 516) 115 V AC, 400 Hz, 3 phase
Power output:
(XK 510) 100 W PEP CW
(XK 515) 150 W PEP
(XK 516) 400 W PEP
Frequency: 2-29.999 MHz
Channel spacing: 100 kHz
Preset channels: 100 or any number with computer control or cockpit management

Operational status

In production for German Army Eurocopter PAH-2 helicopters.

Contractors

Rohde & Schwarz GmbH & Co KG.
EADS (formerly DaimlerChrysler Aerospace AG), Defense and Civil Systems.

UPDATED

Secos 610 VHF/UHF ECCM radios

Secos 610 secure radios are designed for airborne and land-mobile use and are based on the Rohde and Schwarz 610 series of radios. The Secos 610 features digital encryption, medium-speed frequency hopping with collision-free operation of a large number of networks and interoperability with the ground-based or shipboard Secos 400. The frequency range for ECCM operations is 225 to 400 MHz. There is an additional fixed-channel plain language AM option covering the 100 to 156 MHz VHF band. Channel spacing is 25 kHz.

The equipment consists of transceivers, control units and a key entry device. Control is through a compact

The Rohde and Schwarz Secos 610 UHF ECCM transceiver

cockpit transceiver; there are also remote-controlled versions. One common unit controls both the UHF and optional VHF. Power supply is 28 V DC and power output is 15 W on FM. MIL-STD-1553B-controlled versions are also available.

Operational status

In service in F/A-18 aircraft.

Contractor

Rohde & Schwarz GmbH & Co KG.

VERIFIED

Series 4400 VHF/UHF transceivers

Series 4400 transceivers are designed for land, sea and air use, in the frequency band 100 to 512 MHz (extension available from 1.5 to 512 MHz). Series 4400 systems provide voice and data transmission in fixed-frequency or ECCM mode.

Series 4400 transceivers are designed for interoperability with all currently used and planned NATO and non-NATO standards.

Operational status

In production.

Contractor

Rohde & Schwarz GmbH & Co KG.

UPDATED

Series 6000 VHF/UHF transceivers

The multiband, multimode, multifunction Series 6000 VHF/UHF transceivers are 'software radios', designed to provide UHF and VHF, AM and FM, voice and

The Series 6000 cockpit panel-mount version **2000**/0062218

The Series 6000 remote-control version **2000**/0062219

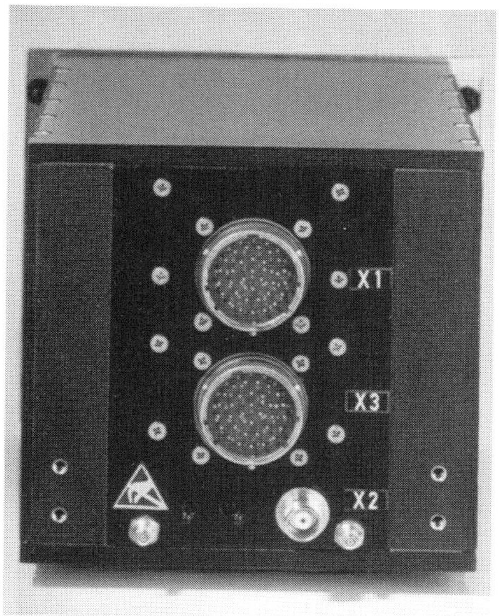

The Series 6000 MIL-STD-1553B version ***2000***/0062220

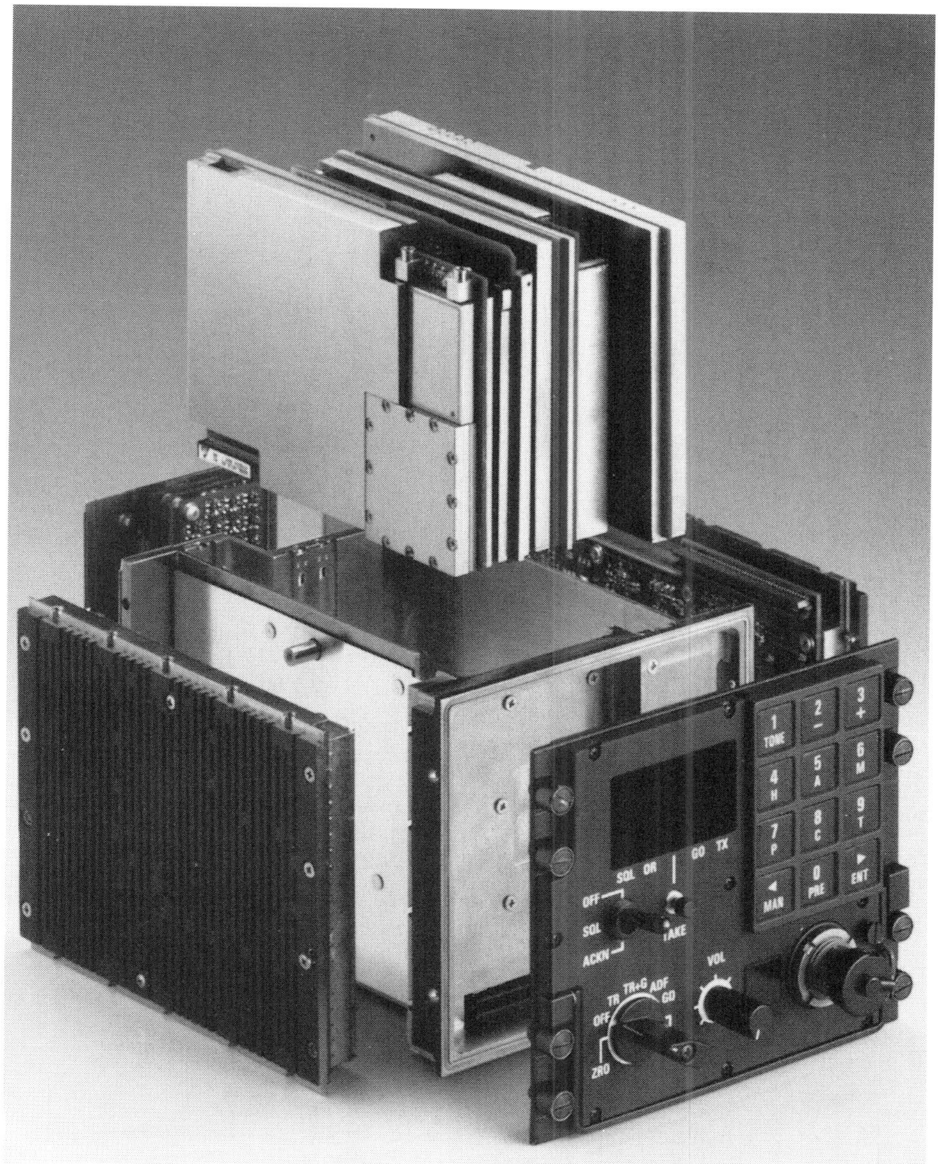

The Series 6000 cockpit panel-mount version showing construction layout ***2000***/0062221

high-rate data communications (64 kbit/s) in clear or Electronic Protective Measures (EPM) mode with embedded COMSEC and TRANSEC capability. The Series 6000 transceivers are capable of operating with several EPM techniques: Have Quick I and Have Quick II, SATURN and SECOS® (Secure ECCM Communication System). Pre-Planned Product Improvement (P^3I) features will allow upgrade to new developments in the EPM scenario simply by loading software.

Series 6000 transceivers support NATO communications standards Have Quick I, Have Quick II and SATURN, as well as the non-NATO standard SECOS, thus allowing communication in NATO's own secure network, while at the same time providing interoperability with the radio equipment of other armed services.

The VHF section supports use of both 25 and 8.33 kHz operation.

Series 6000 transceivers are available as a cockpit panel-mount version, a remote-control version, and a MIL-BUS version to provide maximum installation flexibility.

Series 6000 transceivers are form and fit replacement for a large number of earlier airborne transceivers.

Operational status

Launched in April 1999. In production.

Contractor

Rohde & Schwarz GmbH & Co KG.

VERIFIED

XK 401 HF/SSB radio

The XK 401 was developed jointly by Rohde and Schwarz with DaimlerChrysler Aerospace AG for the trinational Panavia Tornado, with the aim of providing reliable air-to-air and air-to-ground communication over long ranges.

The system covers the HF band from 2 to 29.999 MHz in 100 Hz increments, providing more than 280,000 channels, of which any 11 are preselectable. All channels are individually selectable by means of decade switches. Operational modes are upper sideband, A3J (duplex) and continuous wave. Transmitter power output is 400 W peak power for modulated transmission and 100 W in carrier wave mode transmission.

The XK 401 comprises three basic units: the XK 401 transmitter/receiver, the GB 401 remote controller and the VK 241 power amplifier. Additionally, two optional units, the FK 241 antenna tuner and a control frequency selector, are available and are recommended for optimum operation. The system is of modular construction and solid-state components are used.

The transmitter/receiver section uses digital synthesis frequency-generation techniques and the synthesiser itself is said to possess outstandingly good noise characteristics. Two intermediate frequencies, 72.03 MHz and 30 kHz, are used for both transmission and reception paths. The receiver section has automatic squelch which operates if the HF/SSB level exceeds an adjustable threshold. In the transmitter section, the lower sideband is suppressed by mechanical filters. Built-in test equipment is incorporated.

The remote controller, which may be up to 50 m from the other major units, contains all necessary system controls as well as storage facilities for the 11 preselectable channels.

Two identical amplifier modules, with a common output matching circuit, form the power amplifier section, this duplication being applied in the interests of reliability. In normal operation both modules are in use and, in the event of one module failing, the only consequence is reduction of output power rather than total power output breakdown. Thermal and mismatch overload protection circuits, which include open and short-circuit protection, are provided. Heat dissipated by the amplifier section is extracted by an external ventilator.

Use of the optional antenna tuner and the control frequency selector considerably enhances system performance. The former unit matches the base impedance of the antenna to transmitter/receiver output impedance and, during reception periods, acts as a preselector. The control frequency selector is used to control the digital tuning information. Average tuning time for the system is less than half a second, but use of the control frequency selector unit enables tuning information to be stored for all preselected channels and the tuning time is thus further reduced. No power is radiated during tuning, resulting in radio silence being maintained during such an operation.

Specifications

Dimensions:
(controller) 146 × 86 × 165 mm
(transmitter/receiver) 124 × 194 × 319 mm
(power amplifier) 257 × 194 × 319 mm
Weight:
(controller) 1.8 kg
(transmitter/receiver) 11.2 kg
(power amplifier) 17.2 kg

Operational status

In service in the Panavia Tornado.

Contractors

Rohde & Schwarz GmbH & Co KG.
EADS (formerly DaimlerChrysler Aerospace AG), Defense and Civil Systems.

UPDATED

XT 3000 VHF/UHF radio

The XT 3000 is a combined VHF/UHF air-to-air and air-to-ground transmitter/receiver covering the frequency ranges 100 to 162 MHz and 225 to 400 MHz, in both FM and AM modes for voice and data. Frequency increments are spaced at 25 kHz intervals but the channel spacing is switchable by increments of 25, 50 or 100 kHz as required. Transmitter output is 10 W. Up to 28 operational channels, together with the international distress frequencies of 121.5 and 243 MHz, may be preselected on remote-control units. The system incorporates built-in test equipment and inputs for special-to-type and automatic test equipment.

The XT 3000 comprises a single VHF/UHF transmitter/receiver, VHF and UHF amplifiers, two control units for remote operation and a channel/frequency indicator. It meets MIL-E-5400 and MIL-STD-810B, -461, -462, -463, -781B and VG 95211.

Specifications

Dimensions: ½ ATR short plus two ¼ ATR short units
Weight: 30 kg

Operational status

In production and in service.

Contractor

Rohde & Schwarz GmbH & Co KG.

VERIFIED

XT 3011 UHF radio

The XT 3011 represents what may be regarded as the UHF section of the XT 3000 system, to which it is almost identical. Exceptions are the lack of a VHF transmitter/receiver and appropriate amplifier module and the inclusion of an integrated but independent Guard receiver.

Frequency range is from 225 to 400 MHz with a frequency spacing of 25 kHz. Channel spacing is normally 50 kHz but this may be modified to give 25 kHz increments. Modulation mode is AM only.

The system is operable from either one or two remote-control positions and there are three different types of controller available. Provision is also made for a remote frequency/channel indicator.

Specifications

Dimensions:
(transceiver) ½ ATR short controllers in accordance with MIL-STD-25212
Weight: 7.5 kg

Operational status

In service.

Contractor

Rohde & Schwarz GmbH & Co KG.

VERIFIED

The Rohde and Schwarz XT 3000 VHF/UHF radio with, from left to right (top), the UHF power amplifier, transmitter/receiver and VHF power amplifier. Two control units and a remote frequency indicator are shown at the bottom

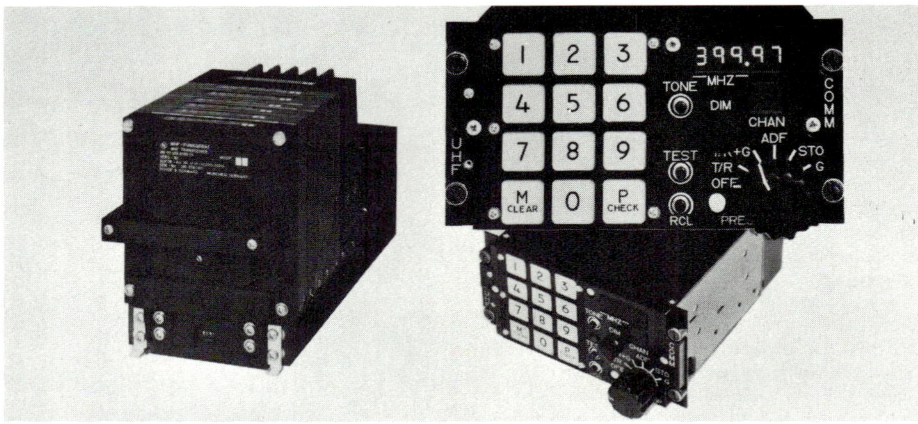

The Rohde and Schwarz XT 3011 UHF transceiver with control units

INDIA

400AM IFF transponder

The 400AM IFF Mk X operates in Modes 1, 2 and 3A/C, with 4,096 codes to full ICAO standards.

Specifications

Power supply: 27.5 V DC (nominal)
115 V AC, 400 Hz, 90 W (option)
Temperature range: –55 to +55°C
Altitude:
(pressurised) up to 70,000 ft
(unpressurised) up to 40,000 ft

Hindustan's 400 AM IFF transponder ***1999***/0044966

Contractor

Hindustan Aeronautics Ltd.

VERIFIED

405A IFF transponder

The 405A IFF transponder is all-solid-state and of modular construction. It provides automatic replies to appropriate ground or airborne interrogators operating on the Mk X IFF system, transmitting on 1,090 MHz and receiving on 1,030 MHz. It operates in Modes 1, 2, 3A and 3C on the full 4,096 codes.

Specifications

Dimensions:
(transponder) 122 × 201 × 201 mm
(control unit) 146 × 81.5 × 70 mm
Weight:
(transponder) 9 kg
(control unit) 0.6 kg
(mounting tray) 1.4 kg
Power supply: 115 V AC, 400 Hz, 150 VA (max)

Contractor

Hindustan Aeronautics Ltd.

VERIFIED

Hindustan's 405A IFF transponder ***1999***/0044967

Audio Management Unit (AMU) 1303A

AMU 1303A is designed for use in fighter and trainer aircraft. It provides management control of two transceivers, three receivers and seven audio warning signals. Additional features include: a telebrief facility; voice-operated switch for hands-free operation.

Hindustan's audio management unit 1303A
1999/0044790

Specifications
Dimensions:
(junction box) 60 × 175 × 180 mm
(station box) 67 × 146 × 180 mm
Weight:
(junction box) 1.1 kg
(station box) 1.2 kg
Power supply: 27.5 V DC, 1 A
Bandwidth: 300 Hz to 3.5 kHz

Contractor
Hindustan Aeronautics Ltd.

VERIFIED

COM 32XA HF/SSB communication system

The all-solid-state modular construction HF/SSB communication system is designed for air-to-air and air-to-ground communication on 2.06 to 29.999 MHz (COM 326A) or 2.5 to 23.5 MHz (COM 327A). Channel spacing is 100 kHz and the system has seven in-flight programmable preset channels and instant manual selection of any channel.

The salient features of the system include solid-state design with software-controlled advanced automatic antenna matching, high-stability frequency synthesiser, high-power solid-state amplifier with protection against high VSWR, high-performance receiver with front panel configurable logic, in-flight programming of channnel and modes and BITE.

Specifications
Dimensions:
(receiver/exciter) 190 × 194 × 320 mm
(power amplifier) 190 × 194 × 320 mm
(ATU for COM 327 and 325) 150 × 180 × 321 mm
(control unit) 146 × 124 × 102 mm
Weight:
(receiver/exciter) 11.5 kg
(power amplifier) 15 kg
(ATU for COM 327 and 325) 7.5 kg
(control unit) 1.5 kg
Power supply:
(COM 326A) 200 V AC, 400 Hz, 3 phase
(COM 325/327/328/329A) 115 V AC, 400 Hz, single phase
27.5 V DC
(receiver) 100 W DC
(transmitter) 850 W AC 100 W DC
Temperature range: −40 to +55°C
Altitude: up to 65,000 ft

Operational status
The COM 325A is fitted on the Hawker Siddeley 748, COM 326A on the Jaguar, COM 327A on the An-32, COM 328A on the Do 228 and the COM 329A on the Il-38.

Contractor
Hindustan Aeronautics Ltd.

VERIFIED

COM 105A VHF transceiver

The all-solid-state modular construction COM 105A VHF transceiver provides air-to-air and air-to-ground voice telephony communication in the VHF band between 118 and 136 MHz at 25 kHz spacing, giving 720 channels. Twenty preset channels are available.

Specifications
Dimensions:
(transceiver) 135 × 145 × 247 mm
(control unit) 104 × 82 × 95.5 mm
Weight:
(transceiver) 4.3 kg
(control unit) 0.5 kg
Power supply: 27.5 V DC
(transmit) 108 W
(receive) 55 W
Temperature range: −40 to +55°C
Altitude: up to 65,000 ft

Contractor
Hindustan Aeronautics Ltd.

VERIFIED

Hindustan's COM 105A VHF transceiver ***1999***/0044792

COM 150A UHF transceiver

The COM 150A UHF transceiver provides 7,000 channels at 25 kHz spacing between 225 and 399.975 MHz for radio telephony (A3) transmissions. There are four preset channels.

Hindustan's COM 32XA HF/SSB communication system ***1999***/0044791

Hindustan's COM 150A UHF transceiver ***1999***/0044793

Specifications
Dimensions: 121 × 174 × 294 mm
Weight: 6.5 kg
Power output: 5 W
Temperature range: −55 to +55°C
Altitude: up to 70,000 ft

Contractor
Hindustan Aeronautics Ltd.

VERIFIED

COM 1150A UHF communication system

The AM transceiver COM 1150A is designed with hybridised circuits for improved reliability, reduced volume and weight, providing A3 communication in VHF band.

Specifications
Dimensions:
(transceiver unit) 124 × 178 × 250 mm
(control unit) 80 × 80 × 107 mm
Weight:
(transceiver unit) 5 kg
(control unit) 0.3 kg
Operating frequency: 225-399.975 MHz
Channel spacing: 25 kHz
No of preset channels: 10
Type of transmission: A3
Receiver sensitivity: 97.0 dBm
Antenna impedance: 52±2 ohms
Audio output: 100 mW into 150/600 ohms
Power output: 5 W (nominal)
Microphone: EM type (150 ohms)

Contractor
Hindustan Aeronautics Ltd.

VERIFIED

Hindustan's COM 1150A UHF communication system
1999/0044794

IFF 1410A transponder

The IFF 1410A transponder system has been designed for operation with a stand-alone control unit, or from a centralised controller through MIL-STD-1553B bus. The equipment uses all-solid-state technology hybrid modules to achieve high reliability and ease of maintenance. Full operation in accordance with ICAO Annex 10 is provided. Extensive self-test diagnostic features help in identifying the faults to functional level. A secure mode of operation is available as an option.

Hindustan's IFF 1410A transponder 0011893

Specifications

Operation modes: 1, 2, 3/A, C, Secure
Codes: 4,096 in Modes 1, 2, 3/A
2,048 in Mode C
Indentification facility
Military emergency facility
Power supply:
(AC) 108-118 V, 400 Hz, single phase, 0.6 A (max)
(DC) 22-32 V, 0.5 A (max)
Dimensions: 310 × 100 × 194 mm (18 mm extra height with mounting tray)
Weight:
(without mounting tray) 7 kg
(with mounting tray) 8 kg

Contractor

Hindustan Aeronautics Ltd.

VERIFIED

Intercom - Audio Management Unit (AMU) 1301A

The Intercom AMU 1301A is designed for use in fighter aircraft, transports and helicopters to cater for up to five crew members in a net with call and conference facilities. It provides management control of five transceivers, three receivers and five audio warning signals. Additional features include: VOS for hands-free operation; dual redundancy.

Specifications

Dimensions:
(junction box) 120 × 120 × 250 mm
(station box) 76 × 146 × 210 mm
Weight:
(junction box) 1.4 kg
(station box) 1.5 kg
Power supply: 27.5 V DC, 2 A
Bandwidth: 200 Hz to 2.5 kHz

Contractor

Hindustan Aeronautics Ltd.

VERIFIED

RAM-700A radio altimeter

The RAM-700A radio altimeter features all-solid-state modular construction. Operating on a frequency of 4.2 to 4.4 GHz, it provides indication of height over terrain up to 5,000 ft to an accuracy of ±2 ft up to 100 ft and four per cent above that.

Specifications

Dimensions: 124 × 97.2 × 366 mm
Weight: <4 kg
Power supply: 115 V AC, 400 Hz, single phase, 80 VA

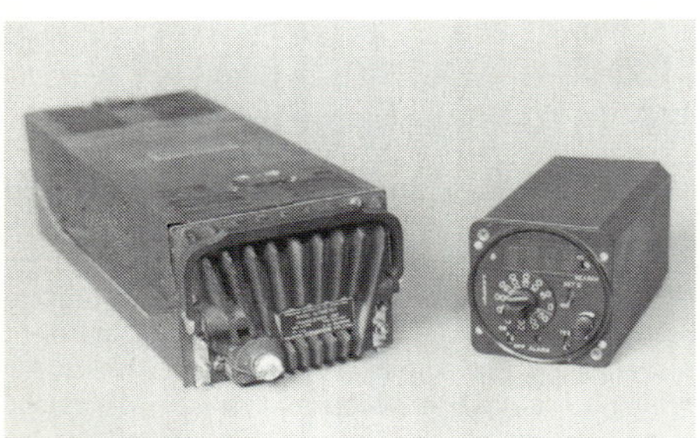

Hindustan's RAM-700A radio altimeter ***1999***/0044968

Hindustan's AMU 1301A Intercom

Contractor

Hindustan Aeronautics Ltd.

VERIFIED

RAM 1701A radio altimeter

The RAM 1701A currently being developed by HAL, Hyderabad for the LCA programme, is an upgraded version of the existing RAM-700A, incorporating state-of-the-art technology. It can be interfaced to an onboard mission computer. The equipment is modular in design, highly reliable and small in size.

Specifications

Transmitter frequency: 4,200-4,400 MHz
Power: <35 W
Power supply: 22 V DC to 29 V DC
Altitude range: 0-1,500 m
Altitude accuracy: 0-30 m (±1 m +3%)
30 to 1,500 m (±4%)

Hindustan's RAM 1701A radio altimeter ***1999***/0044969

Pitch limit: ±20°
Roll limit: ±25°
Dimensions:
(without mounting tray) 123 × 175 × 176 mm
(with mounting tray) 132 × 230 × 194 mm
Weight:
(without mounting tray) 3.5 kg
(with mounting tray) 4 kg

Operational status

Radio altimeters manufactured by Hindustan Aeronautics Ltd are fitted on An-32, MiG-21, Jaguar, Dornier aircraft and on Cheetah and ALH helicopters.

Contractor

Hindustan Aeronautics Ltd.

VERIFIED

VUC 201A VHF/UHF system

The all-solid-state modular construction VUC 201A is a standard air-to-air and air-to-ground radio, offering 2,240 channels between 100 and 155.975 MHz and 7,000 channels between 225 and 399.975 MHz, at 25 kHz spacing. The Guard channel can be tuned between 238 and 248 MHz and 19 channels can be preset. It includes BITE.

Specifications

Dimensions: 160 × 155 × 357 mm
Weight: 13 kg
Power supply: 27.5 V DC, 100 W Rx, 550 W Tx
Power output: 10 W VHF, 20 W UHF
Temperature range: –55 to +55°C
Altitude: up to 70,000 ft

Contractor

Hindustan Aeronautics Ltd.

VERIFIED

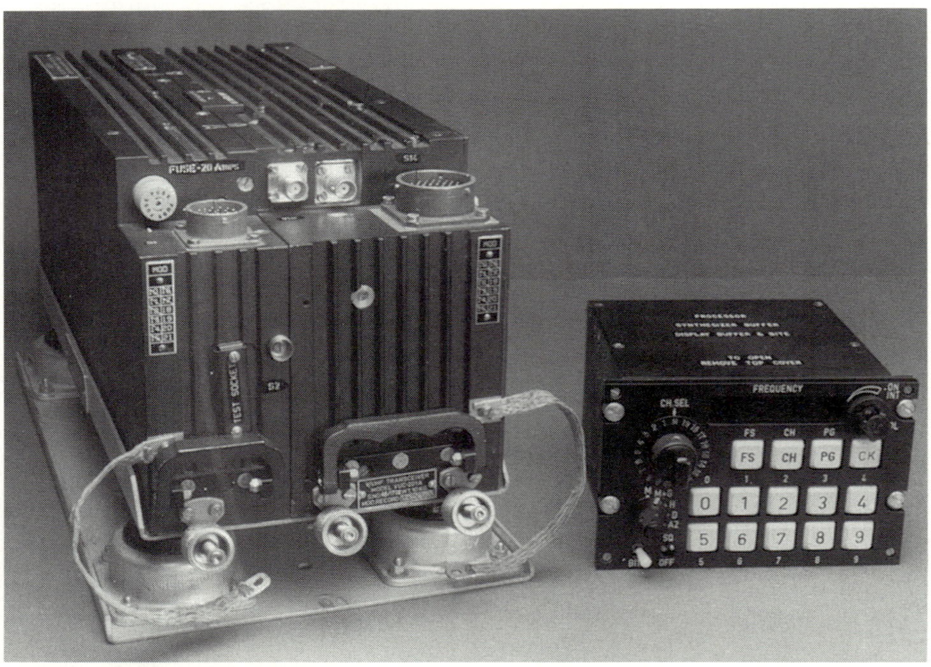

Hindustan's VUC 201A VHF/UHF system ***1999***/0044795

INTERNATIONAL

Advanced Self-Protection Integrated Suite (ASPIS)

Litton Advanced Systems Division, Raytheon Electronic Systems and BAE Systems North America have teamed to produce the Advanced Self-Protection Integrated Suites (ASPIS) which automatically detect and counter hostile threats.

ASPIS uses interchangeable subsystems, so the configuration is flexible. The suite can grow by upgrading existing subsystems, by adding subsystems or by reprogramming to counter the changing threat environment. This flexibility also facilitates installation in a wide variety of aircraft.

ASPIS consists of Litton AN/ALR-93(V)1 RWR/EWSC Raytheon ALQ-187 active jammers and BAE Systems North America countermeasures dispensing systems.

Interfaced with the AN/ALR-93(V)1, the Raytheon AN/ALQ-187 is a fully integrated, power-managed electronic countermeasures system. It can counter multiple pulse, pulse Doppler and CW threats, including SAM, AAA and air-to-air weapons. The ALQ-187 features user-programmable threat data and ECM techniques files for automatic detection of single or multiple threat radars.

The BAE Systems North America countermeasures dispenser contribution to ASPIS is the ALE-47 TACDS (Threat Adaptive Countermeasures Dispenser System). The AN/ALE-47 is a threat adaptive dispensing system that can prioritise and automatically dispense the correct type and amount of chaff and/or flares to counter single or multiple threats. Also available as part of the ASPIS system is the EADS (formerly DaimlerChrysler Aerospace AG)/Litton MILDS AN/AAR-60 Missile Launch Detection System, which utilises UV technology detection to minimise false alarms and provide accurate bearing data on approaching missile threats.

Operational status

The ASPIS is fully operational on the F-16C/D aircraft of the Hellenic Air Force.

Contractors

Litton Advanced Systems Division.
Raytheon Electronic Systems.
BAE Systems North America.
EADS (formerly DaimlerChrysler Aerospace AG), Defense and Civil Systems.

UPDATED

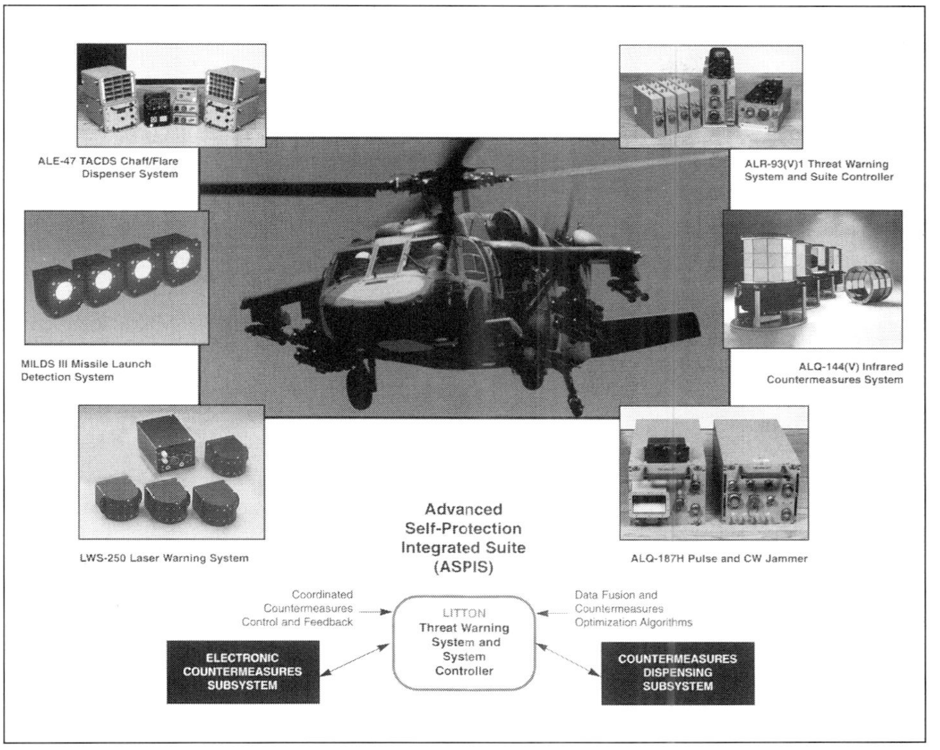

Advanced Self-Protection Integrated Suite (ASPIS) 0018268

Airborne Active Dipping Sonar

The Airborne Active Dipping Sonar (AADS) system consists of the sonar array, reeling machine, sonar transmitter/receiver and signal and display processor. It is designed to be fully compatible with all types of maritime helicopters in service today.

The sonar transducer contains the acoustic projector for sonic pulse generation and an expandable array of receiving hydrophones. The projector resonates at the frequency and duration of the pulse generated by the sonar transmitter. It is capable of producing maximum acoustic source levels at any depth below 6 m. Centre frequencies are selectable for optimal acoustic performance in different environments. The expandable receiving array provides a large acoustic aperture for increased sensitivity and directivity. The expanding mechanism is designed to fail-safe standards to ensure safe retrieval of the array.

Rapid deployment and retrieval of the sonar transducer is provided by the lightweight reeling machine. Layering and stowage of the required length of the extra strong cable has been tested with an MTBF of 4,440 cycles. Replacement of cable and assembly can be accomplished during the refuelling cycle.

The Sonar Transmitter/Receiver (STR) generates the transmit waveforms and provides the output power to drive the acoustic projector. It also processes the signal returns from the transducer receive array. The STR is located in the helicopter cabin and provides local control functions for the dipping sonar subsystem, including the reeling machine, the acoustic transducer and the interfaces between the dipping sonar system and the signal processor.

The GTDAR AMSAR future active radar for combat aircraft 0001198

Contractors

Thomson Marconi Sonar SAS.
Raytheon Electronic Systems.

VERIFIED

Airborne Mine Neutralisation System (AMNS)

Lockheed Martin Naval Electronics & Surveillance Systems and STN ATLAS Electronik are adapting the shipborne Atlas SEAFOX Mine Neutralisation System for helicopter use.

As an expendable mine neutralisation system, the AMNS is controlled from a host platform and includes an underwater vehicle that hunts for and kills mines. The vehicle, called SEAFOX, is a self-propelled, unmanned, wire-guided munition with homing capability, that expends itself during the mine destruction. Vehicle tracking is by host platform sonar activation of a dorsal-mounted transponder. Mine acquisition, homing and classification is by a sonar system capable of horizontal mechanical scanning with a resolution down to 0.9°. The control and guidance system provides control of attitude, heading, depth and altitude sensors. A television camera and headlight are used for visual identification, with data transfer via a fibre optic cable.

Contractors

Lockheed Martin Naval Electronics & Surveillance Systems.
STN ATLAS Electronik.

VERIFIED

AMSAR Airborne Multirole multifunction Solid-state Active array Radar

GEC Thomson DASA Airborne Radar (GTDAR) is a joint European Economic Interest Group (EEIG) formed by Marconi Electronic Systems (now BAE Systems), Thomson-CSF DETEXIS (now Thales) and DaimlerChrysler Aerospace AG (now EADS) to develop future radars for European combat aircraft. Target programmes are initially for upgrade of aircraft such as Eurofighter Typhoon, Rafale, Gripen, Tornado and Mirage; and subsequently initial fit for future European fighter programmes.

With the support of UK, French and German MoDs, GTDAR is executing the AMSAR programme under contract from the French STTE. The work is equally divided between the three companies in Edinburgh, Elancourt and Ulm. AMSAR also involves participation from government laboratories including the Defence Evaluation Research Agency (DERA) in the UK, the Centre Electronique de l'Armement (CELAR) and the Office National d'Etudes et de Recherches Aerospatiales (ONERA) in France and Forschungsgesellschaft für Angewandte Naturwissenschaften (FAGN) in Germany.

AMSAR will employ an Active Electronically Scanned Array (AESA) radar configuration, comprising a fixed antenna and a large number (1,000-2,000) of GaAs MMIC Transmit/Receive Modules (TRMs), to provide independent control of phase and amplitude with the ability to generate multiple beams. The beam agility available provides the following operational modes simultaneously: air-to-air, air-to-ground, terrain following and terrain avoidance. In addition, the technology produces the following ECM/ECCM capability: multiple jammer nulling; stealthiness; low probability of intercept (low sidelobes); multistatic operation and wide bandwidth.

The programme concept also involves development of multiple antenna AESA configurations to provide a panoramic field of view and unconstrained interception strategies.

The AMSAR programme defines three phases: 1993-1998 Phase 1; 1998-2005 Phase 2; 2005+ Phase 3.

Operational status

The first phase, involving testing of a 144 TRM antenna in an anechoic chamber is believed to be complete. In phase 2, a second demonstrator will be assembled with 1,000 TRMs for testing by 2002, as a Technology Demonstrator Programme (TDP); operational capabilities will be established in the latter part of phase 2; phase 3, involving product development will run in parallel with phase 2 from 2002. Programme acceleration is being evaluated.

Contractor

GTDAR, Elancourt.

Subcontractors

EADS (formerly DaimlerChrysler Aerospace AG), Defense and Civil Systems.
BAE Systems, Edinburgh.
Thales (formerly Thomson-CSF DETEXIS).

UPDATED

AN/AAQ-24(V) DIRCM Nemesis Directional Infra-Red CounterMeasures suite

Nemesis is a Directional Infra-red CounterMeasures (DIRCM) system that detects, acquires and tracks new technology IR threat missiles, and then defeats them by accurately focusing an intense beam of modulated IR energy onto the attacking missile. It is intended that the system will be updated later by the addition of a laser jammer.

Full fitting compatibility is designed into the system for a wide range of platforms from large fixed-wing transport aircraft to small helicopters.

Northrop Grumman is the team leader and overall system integrator. Northrop Grumman is also providing the passive warning element, based on its AN/AAR-54 system, and the jamming lamp and techniques generator. Rockwell is to provide the Fine Track Sensor (FTS) for the transmitter azimuth axis. BAE Systems is providing operator controls, threat assessment and the transmitter unit.

The Northrop Grumman advanced, compact, high-performance, lightweight Missile Warning System utilised in the AAQ-24 DIRCM suite is based on AN/AAR-54, also known as PMAWS (Passive Missile Approach Warning System). This passively detects missile plume energy, tracks multiple energy sources and classifies each source as a lethal missile, a non-lethal missile (not intercepting the aircraft), or clutter. Its very fine Angle-Of-Arrival (AOA) capability delivers rapid and accurate hand-off to the IRCM pointing/tracking subsystem. Fine AOA processing provides detection ranges nearly double that of existing fielded passive systems and greatly reduces false alarm rates. This provides all-weather and all-altitude operation while protecting against multiple simultaneous engagements in dense clutter environments.

The system utilises a wide field of view sensor and compact processor. From one to six sensors can be employed, providing up to full spherical coverage.

BAE Systems contribution to the AAQ-24 DIRCM programme includes:
development and assessment of system operating parameters for specific threats;
development, manufacture and qualification of prime system AC power source;
design and manufacture of operator/system interface control panels;
development of the high-performance transmitter unit for both the fixed- and rotary-wing AAQ-24 DIRCM suites.

BAE Systems will manufacture various circuit card assemblies, provide operational system support and participate in the design verification and testing of AAQ-24.

The BAE Systems transmitter unit consists of a pointing turret with a payload of a Fine Track Sensor (FTS) and an IR jammer. The transmitter unit, when cued by the Missile Warning System, acquires the incoming missile, tracks it and projects a high-intensity IR beam at the target.

The agile pointing system ensures that the AAQ-24 DIRCM transmitter unit can rapidly acquire the approaching missile, and the four-axis tracking system ensures that the maximum jamming energy is maintained on the target throughout the engagement. Each AAQ-24 DIRCM transmitter is equipped with a laser path to enable the system to be upgraded at a future date.

Rockwell's Fine Track Sensor is the eye of the high-performance transmitter, located on the azimuth axis. During a threat situation, the image of the incoming missile is electronically processed by the FTS. The electronic image is used by the AAQ-24 DIRCM system to 'close the track loop' by locking the transmitter onto and maintaining an IR beam on the incoming missile until it is defeated. The capability of the FTS to carry out the system requirements is based on proven Mercury Cadmium Telluride technology.

AAQ-24 takes advantage of the high-resolution, high-sensitivity, large area focal plane array detector based imaging system. The FTS features fast cool-down time and high sensitivity, which affords post-burnout tracking over an extended temperature range. The FTS operates in the mid-wave sector of the IR spectrum to

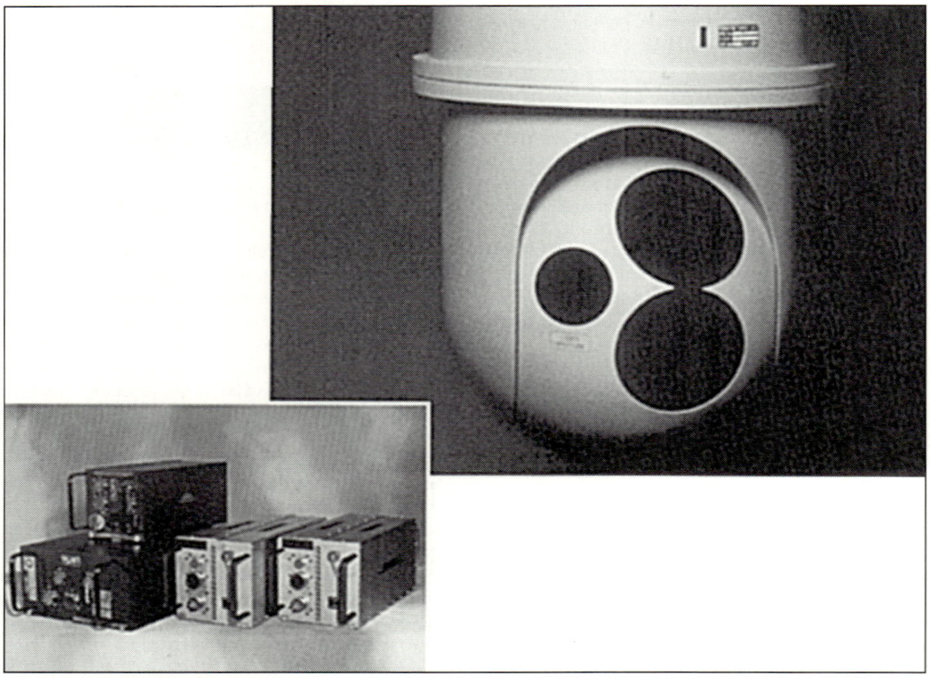

Fixed-wing AAQ-24 DIRCM suite ***1999**/0051254*

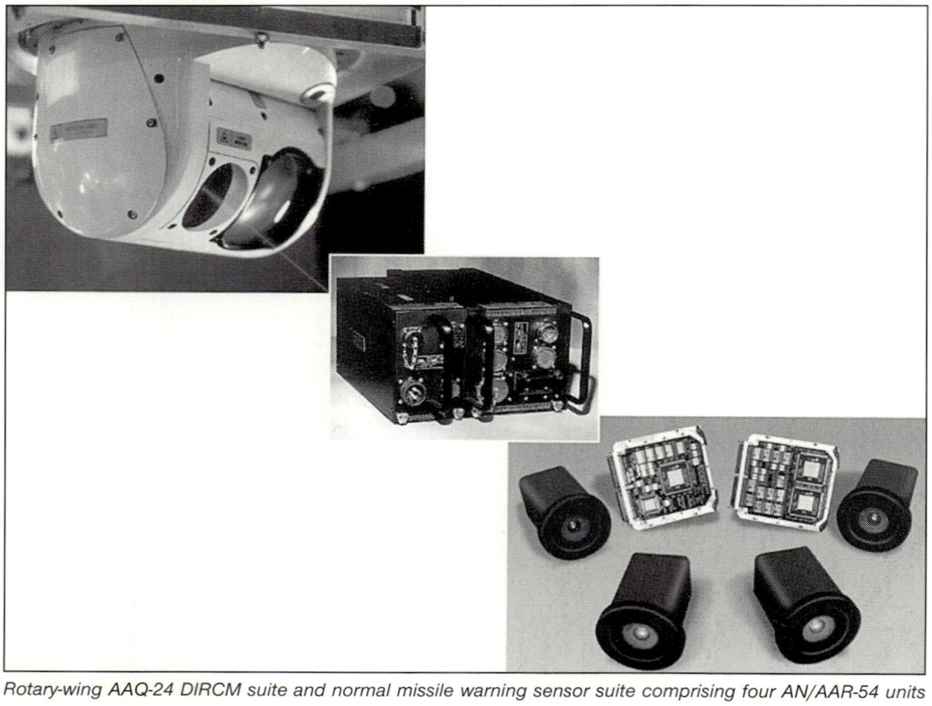

Rotary-wing AAQ-24 DIRCM suite and normal missile warning sensor suite comprising four AN/AAR-54 units ***1999**/0051255*

take advantage of the higher contrast and increased discrimination capabilities within that region.

Specifications - fixed-wing

Total weight:
(without laser) 219.3 kg
(with laser) 228.8 kg
Power (simmer):
(without laser) 3.2 kVA, 0.6 kW
(with laser) 3.2 kVA, 0.75 kW
Power (transmit):
(without laser) 19.7 kVA, 0.6 kW
(with laser) 19.7 kVA, 0.96 kW

Specifications - rotary-wing

Total weight:
(without laser) 55.7 kg
(with laser) 60.5 kg
Power (simmer):
(without laser) 0.98 kW
(with laser) 0.98 kW
Power (transmit):
(without laser) 2.85 kW
(with laser) 3.00 kW

Operational status

The Preliminary Design Review (PDR) and Critical Design Review (CDR) have been completed. Effectiveness testing and aircraft integration are under way for a range of selected fixed- and rotary-wing platforms. The system is ready for integration of the laser countermeasures element.

Initial flight trials began aboard a Sea King helicopter in October 1997 at GKN Westland Helicopters facility in Yeovil, UK, and were completed in July 1998.

Nemesis DIRCM successfully completed its first live-fire test at the US Army's White Sands Missile Range in June 1998, when the system correctly detected, declared, acquired, tracked and jammed the modern Band IV IR missile threat in a series of live-fire test engagements. Testing at White Sands continued up to November 1999, when the target used was a UH-1 helicopter suspended below a cable car.

Present plans call for the UK to instal Nemesis in 186 aircraft, including airlifter, tanker and VIP aircraft and six types of helicopter, and contracts have been placed for 131 sets. The US Air Force SOCOM plan is for fitment of 60 AAQ-24 systems to its MC-130E/H Combat Talon and AC-130H/U gunship aircraft, and production contracts have been issued worth US$130 million.

A miniaturised version of AAQ-24, called 'Wanda' is reported to incorporate a laser to produce low-band (1-2 micron) and mid-band (3-5 micron) infra-red energy for direction at the target.

Meanwhile the US Navy has initiated the Tactical Aircraft Directed InfraRed CounterMeasures (TADIRCM) requirement for fast-flying tactical aircraft. A DIRCM-derivative is being offered for this requirement which features a smaller jamming head and an infrared detector (rather than the present ultraviolet system), to better suit the high altitude requirement. This system is competing against a derivative of the Sanders Advanced Threat InfraRed CounterMeasures (ATIRCM) system, known as Agile Eye.

Contractors

Northrop Grumman Corporation, Integrated Systems and Aerostructures Sector, Rolling Meadows, USA.
Northrop Electronics & Systems, Integration Division, UK.
BAE Systems, Plymouth and Silverknowes, UK

VERIFIED

AN/AQS-22 Airborne Low-Frequency Sonar

The AN/AQS-22 Airborne Low-Frequency Sonar (ALFS) is a US Navy project for a dipping sonar which will largely replace the AN/AQS-13F. It is scheduled to be fitted to the SH-60R, as part of the update from SH-60B, SH-60F and SH-60H to SH-60R standard. No technical details have been released, except that the associated processor will be the UYS-2 manufactured by Lucid Technology (formerly AT & T). The sonar will use the expandable sonar array and reeling machine subsystem of the Thomson Marconi Sonar FLASH system.

The Tiger PAH-2 automatic flight control system ***2000***/0079239

Operational status

Trials of the FLASH array and reeling machine have been completed for the US Navy, together with acoustic in-water trials.

SH-60R technical and operational evaluation is scheduled for 2000-2001, with initial operational capability in 2002.

Contractors

Thomson Marconi Sonar SAS.
Raytheon Electronic Systems.

VERIFIED

Automatic Flight Control System (AFCS) for Eurocopter Tiger PAH-2 helicopter

The Automatic Flight Control System (AFCS) is a digital autopilot, which is superimposed on the mechanical flight control system of the helicopter. The duplex digital AFCS provides the following functions: command and stability augmentation/attitude hold, to stabilise the aircraft as a weapon platform and reduce gust sensitivity; tactical modes, including three-axis hover, a line-of-sight mode for weapon aiming/weapon delivery and gunfire compensation; standard cruise modes providing four axis control.

Each AFCS consists of two identical Flight Control Computers (FCC) and two identical Control Panels (CPs). The FCC contains five modules with the following elements and functions: a CPU Board with two microprocessors, one for system management functions and the other one for running the control laws; an Input/Output Controller (IOC) for input/output control, analogue I/O, cross-channel datalink, communication and hardware synchronisation; an input/output board for discrete I/O requirements; a Power Supply Unit (PSU) for the generation of internal voltages and LVDT/RVDT excitation; and a rear module for EMC filtering and lightning strike protection.

The AFCS can be controlled from either of the two control panels, which are installed in the front and rear cockpits. Apart from the ARINC link, communication between avionics equipment and the AFCS is established with a MIL-STD-1553B bus.

The AFCS software is programmed in Ada using state-of-the-art software development techniques.

The Tiger AFCS is a joint development by Nord-Micro (acting as prime contractor) and Sextant.

Operational status

The first flight of an AFCS-equipped Tiger helicopter was in 1993. Completion of flight testing on the Tiger prototype helicopters was planned for the end of 1999.

Contractors

Nord-Micro Elektronik Feinmechanik AG.
Sextant.

VERIFIED

Belgian Air Force C-130H avionics upgrade programme

In January 1992, the Belgian Air Force selected Honeywell Defense Avionics Systems and Sabena to design, develop and produce a state-of-the-art cockpit avionics upgrade for the C-130H. The objective for the modernisation programme was to provide additional performance functions and, at the same time, improve the aircraft's availability to carry the Belgian Air Force C-130H fleet well into the next century. In response to these objectives, the Honeywell/Sabena team orchestrated a solution that included: a precision navigation solution; a fully automated digital flight management system to reduce the crew's workload; a digital solution for an integrated autopilot/flight director; electronic flight instruments; a digital, ground map weather radar; and a two-level maintenance support system that draws heavily on the BIT functions resident within the flight management computer and the individual LRUs.

The system designed for this modernisation programme is referred to as the Integrated Vehicle Mission Management System (IVMMS). A mix of commercial-off-the-shelf and military non-development hardware was utilised to keep the IVMMS affordable. Both MIL-STD-1553B and ARINC 429 bussing is used to integrate the various military and commercial hardware. Sufficient I/O for the Flight Management computer has been included to directly integrate a myriad of existing COMM/NAV radio sensors and analogue avionics. System requirements and qualification test profiles were specifically tailored by the Belgian Air Force to reflect the existing operating conditions experienced by C-130 aircraft accomplishing a worldwide mission scenario. This foresight in tailoring the design requirements to existing conditions rather than standard MIL requirements is responsible for the cost-effective design that directly satisfies the Hercules' mission.

Belgian Air Force C-130H cockpit showing the Honeywell Defense Avionics Systems upgraded avionics management system. ***1999***/0051637

Honeywell is responsible for the avionics design, integration testing, qualification and production hardware. Sabena is responsible for the installation of the IVMMS hardware and integration of the IVMMS with the other C-130H Hercules systems.

Operational status

The first of 12 Belgian Air Force C-130Hs officially rolled out 31 May 1995. Production deliveries will be complete for the BAF C-130H fleet in 1999.

Contractors

Honeywell Defense Avionics Systems.
Sabena Technics.

VERIFIED

CLARA CO_2 Laser Radar

CLARA is an obstacle avoidance radar. Development of CLARA derives from an Anglo/French government-to-government initiative, resulting in a consortium comprising BAE Systems and Thales (formerly Thomson-CSF DETEXIS). Under the work-share arrangements, two identical demonstrator units are being produced and tested on a fixed-wing aircraft in the UK and on a helicopter in France.

CLARA is a self-contained CO_2 laser radar housed in an environmentally controlled pod and mounted on a helicopter or fixed-wing aircraft. It is designed to avoid obstacles such as cables, pylons and so on, and will also provide other functions such as terrain-following, target ranging and designation, short-range true airspeed measurements and moving target indication.

Operational status

In development under contract to the Service des Programmes Aeronautiques of France and the Defence Evaluation Research Agency in the UK, it is understood that demonstration systems have been fitted to helicopters in both countries.

Flight trials started in 1997, and they were scheduled to be completed by the end of 1998; present status uncertain.

Contractors

BAE Systems.
Thales (formerly Thomson-CSF DETEXIS).

UPDATED

Digital fly-by-wire flight control system for the Eurofighter Typhoon

BAE Systems (Operations) Limited leads a consortium comprising BGT of Germany, Alenia Difesa, Avionic Systems and Equipment Division of Italy and INDRA EWS of Spain, which is under contract to DaimlerChrysler Aerospace (DASA) to develop the Flight Control Computer (FCC) and Stick Sensor and Interface Control Assembly (SSICA) for the Eurofighter Typhoon aircraft.

The Flight Control System (FCS) provides the aircraft with a full-time fly-by-wire control system which controls the aircraft via its 11 primary and secondary flying control surfaces to give the aircraft carefree handling characteristics together with outstanding agility and manoeuvrability throughout the flight envelope. The Eurofighter Typhoon has no mechanical back-up system and is therefore totally dependent on this digital system. Redundancy of the system is ensured through replication of both the hardware and software, and should a failure occur in any one computer, this can be isolated and the aircraft will continue to fly normally.

To provide the necessary integrity and safety, the system uses a quadruplex design with each of the four FCCs containing identical hardware and software. Each computer contains Motorola 68020 32-bit microprocessors and software compiled in Ada. Comprehensive built-in test provides a continuous system monitoring for both ease of system maintenance and assurance of the continued high integrity performance. ASICs have been widely used to ensure a low component count to maximise performance and reduce system size and weight. The FCC also contains an interface to the aircraft's utility bus to provide integrated vehicle management and to a STANAG 3910 fibre optic avionics bus to allow integration with the mission avionics.

Operational status

Currently fitted to development aircraft.

Contractors

BAE Systems Operations.
Bodenseewerk Gerätetechnik GmbH/BGT.
Alenia Difesa, Avionic Systems and Equipment Division.
INDRA EWS.

UPDATED

ENR European Navy Radar for the NH 90-NFH

EADS, FIAR and Thales have signed an agreement to co-operate in the design, manufacture and marketing of an airborne sea-surveillance radar – the ENR European Navy Radar – for the naval version of the NH 90 helicopter (NH 90-NFH).

The ENR is directly derived from the APS 784 produced by Eliradar (a consortium of FIAR and Officine Galileo) and the Ocean Master radar produced by Thales (formerly Thomson-CSF DETEXIS) and EADS (formerly DaimlerChrysler Aerospace AG). It features state-of-the-art developments from these radars, including Inverse-Synthetic Aperture Radar (ISAR) processing.

Operational status

Based on the proposal submitted by the ENR partners, the radar was selected by the participating governments. The contract was placed in July 1998.

Contractors

Alenia Difesa, Avionic Systems and Equipment Division, FIAR.
EADS (formerly DaimlerChrysler Aerospace AG), Defense and Civil Systems.
Thales (formerly Thomson-CSF DETEXIS).

UPDATED

ERWE II Enhanced Radar Warning Equipment

EADS (formerly DaimlerChrysler Aerospace) and Litton Advanced Systems Division teamed to develop and produce the Enhanced Radar Warning Equipment (ERWE) and its updates. The first model produced was ERWE I. The model now in production is ERWE II. The ERWE II detects and analyses hostile radars illuminating the aircraft. It provides the crew with display and audio information, and provides other elements of the EW system with data on which to effect appropriate responses.

The main features of the ERWE II are that it is: a multi-receiver and processor architecture, with wide instantaneous bandwidth, monopulse, E-/J-band and receiver, high sensitivity narrow band receiver; special C-/D-band receiver for detecting missile guidance signals. It includes two Mil Bus interfaces; discrete interfaces for look-through management; embedded maintenance facilities; provision for the user to programme threat libraries, identification and recording requirements.

The operator is provided with alphanumeric threat displays, audio and two selectable modes of operation: terminal mode: providing smart analysis algorithms and identification/classification of threats; bypass mode:

Eurofighter Typhoon displaying the carefree handling characteristics of its quadruplex FBW system ***2001***/0103894

providing high probability of intercept of scanning emitters, real-time indication of signals and recording capability.

Operational status

In service with German Air Force and Navy Tornado PA200 aircraft.

Contractors

EADS (formerly DaimlerChrysler Aerospace AG), Defense and Civil Systems.
Litton Advanced Systems Division.

UPDATED

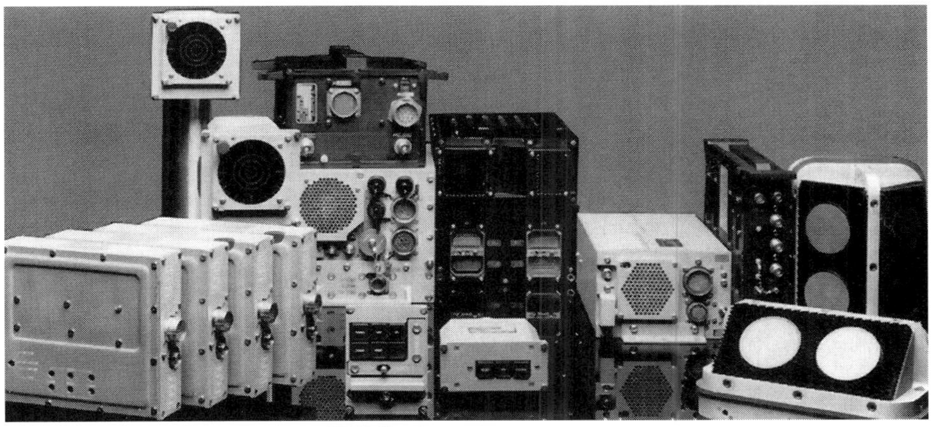

Enhanced Radar Warning Equipment ERWE II units 0018293

Eurodass Defensive Aids SubSystem (DASS)

Eurodass DASS is designed to provide Eurofighter Typhoon aircraft with spherical ESM coverage, forward and rearward missile warning, fibre optic towed decoys, active phased array jamming, and options for future growth to meet the aircraft's mission as an air superiority fighter.

The Eurodass Consortium comprises BAE Systems (Operations) Limited in the UK, Elettronica SpA in Italy and INDRA in Spain; together they will provide the Eurodass DASS for the Eurofighter Typhoon aircraft being procured for UK, Italy and Spain. German participation in DASS is subject to negotiation. Design activity for the system is led from BAE Systems (Operations) Limited at Stanmore, UK.

Operational status

The Eurodass Consortium was awarded a £200 million contract in March 1992 for the development of DASS. The work-share was initially split approximately 60 per cent to the UK, and 40 per cent to Italy. In May 1998, a £170 million contract was awarded for the Production Investment (PI) contract for the Eurodass DASS. The contract covers the establishment of processes and facilities across the Eurodass Consortium to enable the production of DASS through to 2015.

The Eurodass Consortium claims that test results have demonstrated that the requirements of the system design concept for guarantee of performance and a very high level of reliability and availability will be met, or exceeded, at entry into service.

Milestones to date include supersonic flight with a deployed BAE Systems (Operations) Limited Ariel towed decoy and first flight with an active Missile Approach Warner (MAW), as part of the DASS test programme.

Eurofighter Typhoon at takeoff; note the wingtip pods carrying DASS elements ***2001***/0103895

Contractors

Eurodass Consortium members:
BAE Systems (Operations) Limited, UK.
Elettronica SpA, Italy.
INDRA Spain.

UPDATED

Euroradar ECR 90

The multimode radar for Eurofighter Typhoon is currently in development by the Euroradar Consortium led by BAE Systems. The consortium was awarded the contract after a competitive evaluation with other systems.

ECR 90 is a third-generation coherent I/J-band multimode radar, based on the technology of Blue Vixen. It incorporates a significant increase in processing power, which will exploit fully the high information content of the advanced transmission waveform, with wideband spread-spectrum, allowing operations to be maintained in a hostile EW environment. The radar comprises six line-replaceable units; the scanner; a waveguide unit; a two-module receiver; and a two-module transmitter.

Air-to-air features of the ECR 90 radar include look-up and look-down capability; multitarget track-while-scan; target identification and prioritisation. Electronic CounterMeasures (ECM) resistance is incorporated to classify and counter jamming. Air-to-surface features include ground mapping/ranging, terrain avoidance, weapon release computation and sea surface search.

Operational status

ECR 90 radar is now in the latter stages of the development programme. Flight trials on a BAC 1-11 aircraft have shown that the radar performance is on schedule and the system is reliable. Prototype DA4 and DA5 Eurofighter Typhoon aircraft both flew with operating ECR 90 radars on their maiden flights early in 1997; both aircraft are now engaged on a full evaluation of the ECR 90's radar properties for the detection and tracking of airborne targets. It is reported that, in recent trials, the radar has detected and tracked fighter targets at over 160 km and large aircraft at up to 320 km.

BAE Systems has been awarded a contract for the ECR 90 to equip Eurofighter Typhoon. The contract covers the establishment of processes and facilities across the Euroradar Consortium to enable the production of radars through to 2015. The initial production order authorises an initial batch of 147 radars and spares to meet the production requirements for the first tranche of aircraft.

Continuing development work is concentrated in the area of countermeasures software to produce the full operational radar standard.

Contractors

BAE Systems (prime contractor).
Alenia Difesa, Avionic Systems and Equipment Division, FIAR.
EADS (formerly DaimlerChrysler Aerospace AG), Defense and Civil Systems.
INDRA EWS.

UPDATED

The Euroradar ECR 90 radar for the Eurofighter Typhoon

FLASH dipping sonar

FLASH is Thomson Marconi Sonar's most recent generation of dipping sonar, updating the earlier HS12 and HS312 (1980's technology) and DUAV-4 (1970's technology) systems.

FLASH (Folding Light Acoustic System for Helicopters) is a new low-frequency helicopter dipping sonar with sonobuoy processing capabilities developed by Thomson Marconi Sonar for a number of applications. It operates at five frequencies below 5 kHz, with 24 preformed beams. The winch is capable of 750 m immersion depth at a speed up to 10 m per second.

The system offers an FM mode to enable detection of low-speed targets in shallow waters and strong reverberation conditions. With its light weight and volume, the FLASH can be fitted on board light ASW helicopters such as the Super Lynx and on a wide variety of medium to large helicopters.

Operational features include: high vertical rejection to filter out helicopter noise; a folding low-frequency receiver array that emerges from the top portion of the body; a transmitter comprising stacked piezo-electric ceramic rings in the lower portion of the body; operator control of pulse shape and power level to suit water conditions; FM pulse compression to provide good detection and Doppler measurement capability with accurate ranging resolution; combination of FM and CW pulses into a single transmission.

Operational status

Thomson Marconi Sonar supplies FLASH (designated Sonar 2089 in the UK) for the long-range dipping sonar requirement of the Royal Navy's EH 101 Merlin helicopter. This system combines the Thomson Marconi Sonar SAS low-frequency array and winch with the Thomson Marconi Sonar Ltd acoustic processor, designated the AQS 950.

The FLASH array, reeling machine, transmitter and new acoustic processor are now being evaluated in flight trials by the French Navy for the NFH 90 programme.

Thomson Marconi Sonar has teamed with Raytheon Electronic Systems for the US Navy's (Airborne Low Frequency Sonar (ALFS) requirement, to be fitted to the SH-60R Seahawk helicopters, in which configuration it is designated AN/AQS-22. It is reported that Thomson Marconi Sonar would be providing the FLASH array and reeling machine elements of the system.

FLASH is in production for the UAE Naval Forces for the Super Puma helicopter.

Contractors

Thomson Marconi Sonar SAS.
Thomson Marconi Sonar Limited.

VERIFIED

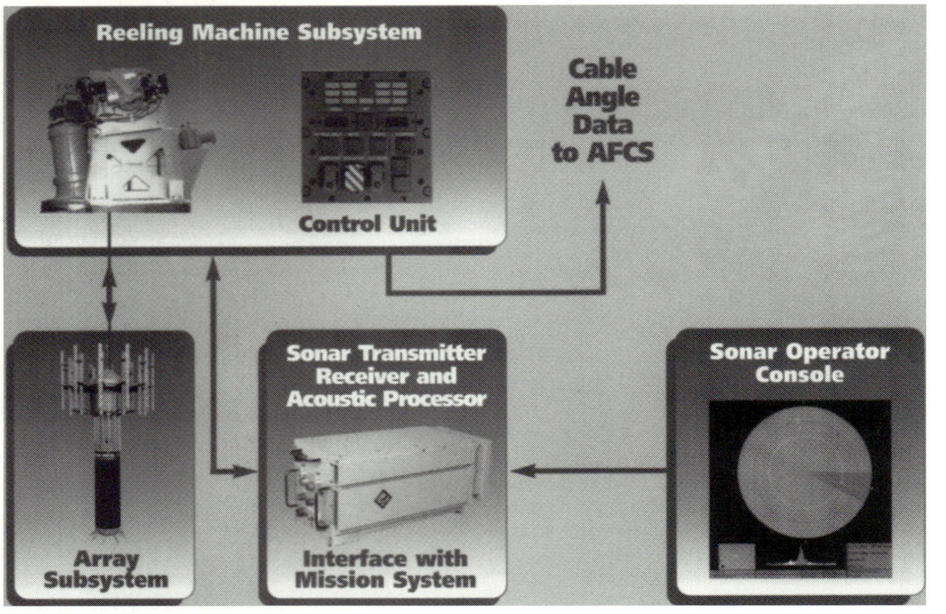

FLASH dipping sonar system components 0011863

Flight control system for the AMX

BAE Systems (Operations) Limited is co-operating with Alenia Difesa in the Fly-By-Wire (FBW) flight control system for the Italian/Brazilian AMX strike aircraft. The system comprises two dual redundant flight control computers each based on 16-bit microprocessor hardware and incorporating fail-safe software. The system commands the movement of seven control surfaces and incorporates a recently developed autopilot facility and automatic pitch, roll and yaw stabilisation. Analogue computing is used for the actuator control loops, the pilot command paths and rate damping computation. Digital computing handles gain scheduling, electronic trim and integration of the airbrake. Redundant microprocessors in the flight control computer units monitor system performance and are associated with built-in test facilities designed to provide a high confidence and rapid comprehensive system check. Testing is initiated by the pilot before flight and is conducted automatically thereafter. In addition to the flight control computers, the FBW system also includes pilot control position sensors, three-axes rate gyros and air data components.

AMX flight control computer 0001411

Operational status

In production.

Contractors

BAE Systems (Operations) Limited.
Alenia Difesa, Avionic Systems and Equipment Division.

VERIFIED

Fly-by-wire system for the Tornado

The Panavia Tornado flight control system uses triplex electronic signalling and processing, with quadruplex actuation, to provide a high degree of manoeuvrability throughout the entire flight envelope and has a mechanical back-up system for emergency control though with degraded handling qualities. The fly-by-wire processing is incorporated in the Command and Stability Augmentation System (CSAS) and the automatic flight functions, which include safety critical capability such as automatic terrain-following, are integrated with the automatic flight control system. These are separately configured but very closely allied systems. The triplex CSAS components are associated with a duplex Spin Prevention and Incidence Limiting System (SPILS) which ensures that the crew can fly the aircraft to its structural and aerodynamic limits without the risk of loss of control.

Design, development and production of the systems were a trinational venture between BAE Systems (Operations) Limited in the UK, BGT in Germany and Alenia Difesa in Italy.

Command stability and augmentation system

The CSAS is an analogue FBW manoeuvre demand system. It provides electrically signalled pitch, roll and yaw control and automatic stabilisation of aircraft response to pilot command or turbulence. Gain

Panavia Tornado SPILS computer opened up to show the four-board front connector layout

Panavia Tornado CSAS equipment includes computer units, control panels, triplex position transmitters, triple gyro packs and quadruplex first stage actuators

The Panavia Tornado benefits from fully redundant FBW, coupled with CSAS, to facilitate carefree handling at ultra-low level ***2001***/0105291

scheduling improves handling qualities and control stability over the entire flight envelope and operates in conjunction with the spin prevention and incidence limiting system. 'Carefree' manoeuvring allows the exploitation of the aircraft's full lift capability under all flight conditions without the risk of structural damage resulting from a pilot's control demand for control surface movement that would exceed the design strength of the airframe.

Individual LRUs are the CSAS pitch computer, CSAS lateral computer, CSAS control unit, pitch, roll and yaw rate gyros and pitch, roll and yaw position transmitters.

Spin prevention and incidence limiting system

SPILS is a duplex analogue system which operates in conjunction with the CSAS to achieve maximum aircraft lift in low-level flight. It limits aircraft incidence, irrespective of the pilot's demands, when maximum safe angles of attack are reached. Individual LRUs are the SPILS computer and SPILS control unit.

Automatic flight director system

The digital Autopilot and Flight Director System (AFDS) automatically controls the flight path in all modes, including terrain-following, and sends signals to the director instruments enabling the crew to monitor autopilot performance or to fly the aircraft manually. Pitch autotrim is also incorporated. The duplex self-monitoring processor configuration provides high-integrity automatic control, permitting low-altitude cruise with appropriate safety margins. The flight director remains available after most single failures. Autopilot manoeuvre demand signals are routed to the control actuators through the command and stability augmentation system. A 12-bit processor is used with 6 kbits words of stored program and 1 kbits words of data store. Typical computing speed is around 160 Kops and program cycle time is 32 ms. Individual LRUs are the two AFDS computers, AFDS control unit, autothrottle actuator and pitch and roll stick-force sensors.

Operational status

In service.

Contractors

Alenia Difesa, Avionic Systems and Equipment Division.
Bodenseewerk Gerätetechnik GmbH/BGT.
BAE Systems (Operations) Limited.

UPDATED

H-764G Embedded GPS/INS (EGI)

The H-764G EGI is a small, low weight, inertial navigation system with embedded GPS. It is based on a 32-bit Intel 80960 microprocessor, MIL-STD-1553B databus, and four RS-422 buses on one card. It is programmed in Ada and provides a triple navigation solution (pure inertial, GPS-only, and blended GPS/INS), using a Collins GEM™ GPS receiver module.

The H-764G EGI comprises the following main modules: radar altimeter, remote air data, GPS, inertial sensor, stability augmentation, processor, ARINC and discrete interfaces, and BAE Systems Terprom (optional).

Specifications

Dimensions: 177.8 × 177.8 × 248.9 mm
Weight: 8.4 kg
Pure inertial performance:
(Position) <0.8 n mile/h CEP
(Velocity) <1.0 m/s rms
Blended GPS/INS performance:
Position accuracy (SEP) <16 m
Velocity accuracy (rms) 0.1 m/s
Align time:
(Gyrocompass) 4 min
(Stored heading) 30 s
(in air align) 4 min
Power: 28 V DC, 40 W
Interfaces: MIL-STD-1553B (RT or BC) and four RS-422 serial databus (optional)
Synchro, analogue, discrete

Operational status

Installed on about 50 aircraft and helicopter types, with a total production of over 4,000 units. The H-764G EGI has been selected for the Tornado GR4/4A and F3 upgrades, with the TERPROM option, in which installation it is also known as the Tornado LINS 764GT.

Contractors

Honeywell Military Avionics.
BAE Systems.
Honeywell Regelsysteme GmbH.

VERIFIED

IFF for Eurofighter Typhoon

EADS (formerly DaimlerChrysler Aerospace AG) (Germany), MID SpA (Italy), Raytheon Systems Limited (UK) and INDRA EWS (Spain) have teamed to design, develop and produce the IFF interrogator and IFF Mode S transponder for the Eurofighter Typhoon. The system is compatible with IFF Mk XII, as defined by STANAG 4193.

EADS is also offering a derivative system for the NH 90-NFH helicopter.

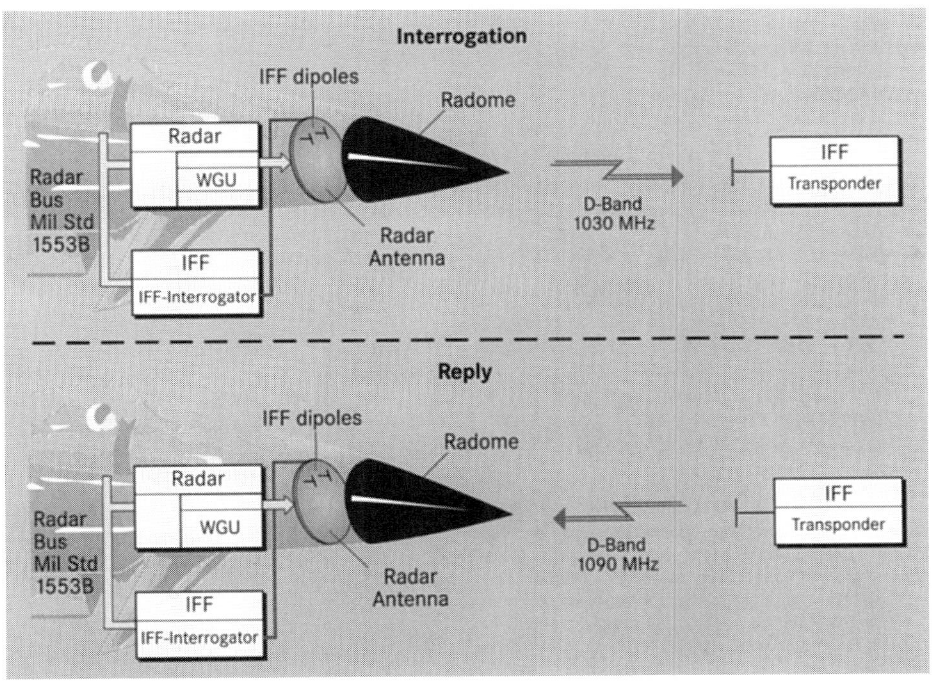

IFF for Eurofighter Typhoon 0011892

H-764G Embedded GPS/INS (EGI) ***1999***/0054178

IFF for Eurofighter Typhoon 0011891

Specifications

IFF Mode S transponder
Dimensions: ½ ATR intermediate length
Weight: 9 kg
Characteristics:
Mk XII (mode 1,2,3/A,C,4) compliant with STANAG 4193
Mode S
MIL-STD-1553B
Embedded cryptographic module
ADA software
Automatic Code Change (ACC)
Growth to Next-Generation IFF (NGIFF)
BIT modes (P-BIT, C-BIT, I-BIT)

IFF interrogator
Dimensions: ½ ATR intermediate length
Weight: 11.4 kg
Characteristics:
Mk XII (mode 1,2,3/A,C,4) compliant with STANAG 4193
MIL-STD-1553B
Embedded cryptographic module
Monopulse processing
ADA software
Automatic Code Change (ACC)
BIT modes (P-BIT, C-BIT, I-BIT)

Contractors

EADS (formerly DaimlerChrysler Aerospace AG), Defense and Civil Systems.
INDRA EWS.
MID SpA.
Raytheon Systems Limited.

UPDATED

Link 16 JTIDS (Joint Tactical Information Distribution System) airborne datalink terminals

The Link 16 JTIDS family of airborne datalink terminals evolved from early US Air Force/MITRE and US Navy studies related to Time Division Multiple Access (TDMA) communications and relative navigation. Early contractors included Hughes, ITT, Singer and Rockwell. Development of Class 1 terminals began in 1974 for E-3 AWACS; 70 were delivered. Development of Class 2/2H terminals began in 1990. Low rate initial production began in 1991. Full rate production began in 1995; over 500 Class 2 terminals have been delivered. Developments have continued to meet US and NATO requirements, and the overall range of airborne terminals now comprises the:

(1) Joint Tactical Information Distribution System (JTIDS) Class 2/2H terminals (entry under BAE Systems North America in USA section). Application: aircraft, ADGE and ship platforms.
(2) Multifunctional Information and Distribution System - Fighter DataLink (MIDS-FDL) terminal (entry under BAE Systems North America/ Rockwell Collins in USA section). Application: F-15 aircraft.
(3) Multifunctional Information and Distribution System - Low Volume Terminal (MIDS-LVT) (entry in International section). Application: fighter aircraft, ADGE and ship platforms.
(4) SHAR Link 16 low volume datalink terminal (AN/ URC-138(V)1(C) (entry under Rockwell Collins in USA section). Application: fighter aircraft and helicopters.
(5) Advanced Tactical DataLink (ATDL) (Rockwell Collins product - full details not available). Application: country-unique air/ground terminal.

Specifications

	Class 2/2H	MIDS-FDL	MIDS-LVT	SHAR	ATDL
Dimensions (mm):	RF: 396 × 259 × 193 DDP: 445 × 325 × 193	330 × 191 × 194	330 × 191 × 195	318 × 257 × 191	361 × 191 × 194
Weight (kg):	RF: 26.81, DDP: 33.18	16.36	23.18	20.00	20.45
Power (W):	200	50	200	200	200
Range (nm):	300/500	185	400	400	400
TACAN:	Yes	No	Yes	Option	Option
Voice (kbps):	2.4/9.6	No	2-channel; 2.4/16	2-channel; 2.4/16	2-channel; 4.8/16
MTBF (h):	500	N/A	N/A	>2,000	N/A
Availability:	In production	1999	2001	1999	1999

The main capabilities provided by this family of terminals includes: jam resistant, crypto secured, data and voice, line-of-sight, communication, navigation and identification, with relay for extended range, using TDMA, frequency-hopping transmissions, in the 960 to 1,215 MHz band.

Operational status

In production or selected for the following platforms: US Air Force: ABCCC, AWACS, B-1B, F-15, F-16, JSTARS, RC-135 Rivet Joint, tankers, transports. US Navy: F-14D, F/A-18, E-2C, EA-6B, S-3. US Marine Corps: AV-8B, F/A-18. NATO: AWACS, Eurofighter Typhoon, Rafale, Sea Harrier F/A-2, Sea King AEW Mk.7, Tornado F-3.

Contractors

BAE Systems North America.
Rockwell Collins.

VERIFIED

Long Star Electronic Warfare (EW) system

The Long Star EW system is a lightweight, modular architecture that offers detection, location and support jamming capabilities against radar and communications emitters. It is designed for rapid installation aboard helicopter platforms. The system comprises Electronic Support Measures (ESM) and instantaneous frequency measuring receivers, an identification processor, a power-management unit, an operator console, an ESM/instantaneous direction-finding antenna assembly and a multibeam array transmitter. The transmitter chain can be customised for specific missions.

Rafael's Long Star EW system is designed for applications such as the AH-64 battlefield attack helicopter ***1999***/0017861

Operational status

The Long Star EW system is available and has been offered to the US Army to meet a late 1990s

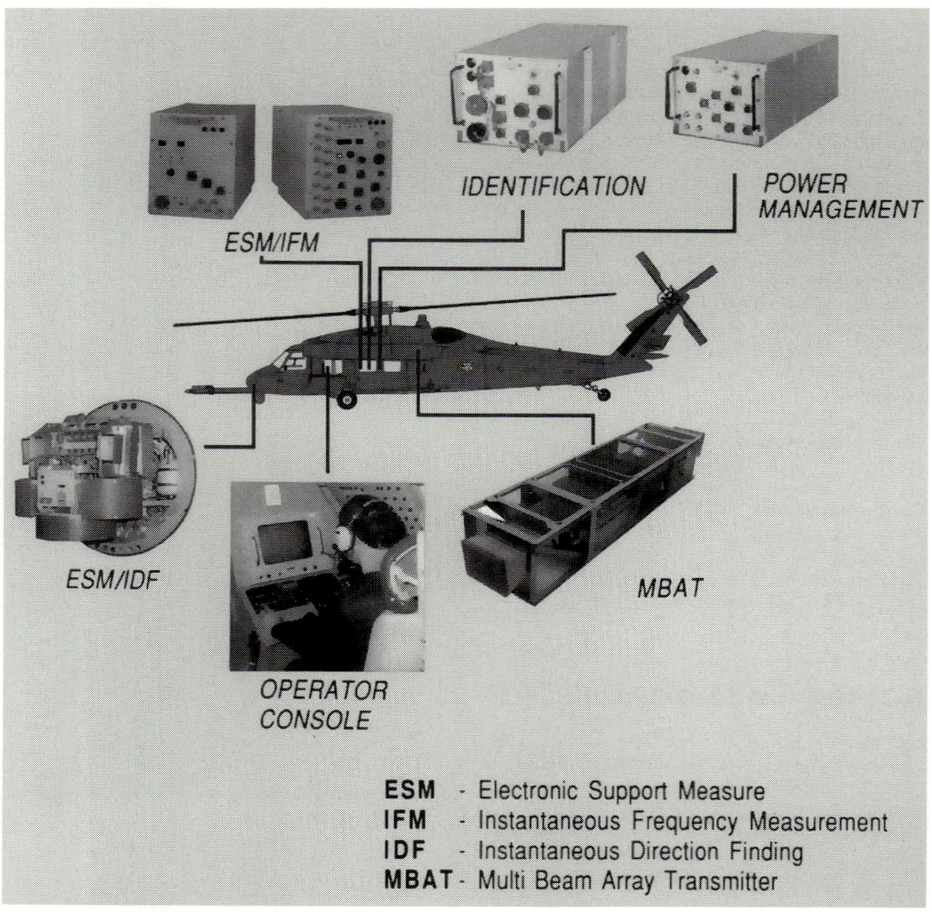

Long Star ASJ installation in UH-60A helicopter 0018289

requirement for an airborne Army Support Jamming (ASJ) system. The US Army ASJ bid was made in collaboration with Northrop Grumman.

Contractor

Rafael Electronic Systems Division.

UPDATED

MCR500 solid-state combined cockpit voice/flight data recorder

The MCR500 is a ruggedised version of the BAE Systems SCR500, specifically targeted at fighter environment applications, developed jointly by Analysis, Management & Systems (Pty) Ltd and BAE Systems. It combines the latest digital memory and material technology with special packaging to produce a CVR/FDR for the most rugged environments.

In addition, the unit provides expansion capability to integrate a custom Flight Data Acquisition (FDA) interface, thus achieving a single-box solution. All MCR500 recorders comply with the requirements of ED56A, TSO-C123A and TSO-C124A Draft.

The MCR500 makes use of standard SCR500 control panels, which include the following features: CVR erase function; CVR/FDR BIT indicator; replaceable front panels for custom installation; standard and slimline options for direct exchange fit; a built-in cockpit area microphone.

Specifications

Recording:
(audio) 4 digital audio channels of 60 minutes duration
(digital data) 64/128 words/s, 10 h duration

Dimensions: 124 (W) × 360 (L) × 188 (H) (including ULB) mm
Weight: 8 kg (including ULB)
Power: 28 V DC, 12 W
.

Operational status

Selected for the Royal Australian Air Force new Hawk Lead-in-Fighter.

Contractors

Analysis, Management & Systems (Pty) Ltd.
BAE Systems.

VERIFIED

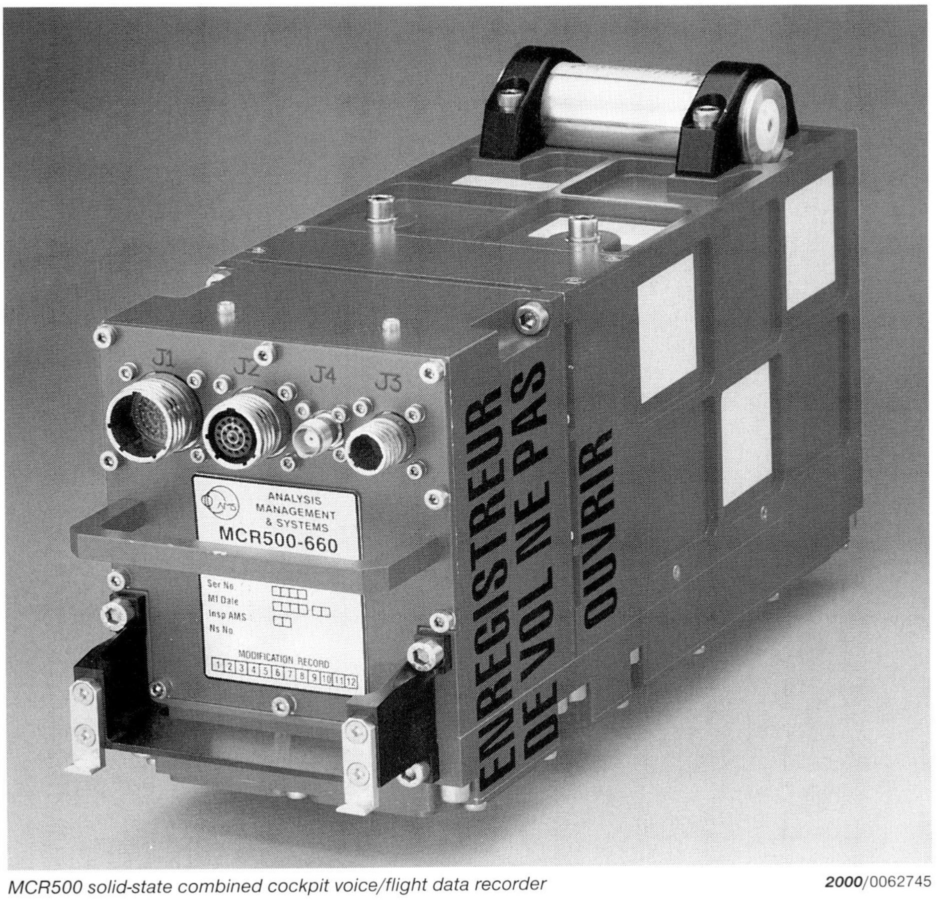

MCR500 solid-state combined cockpit voice/flight data recorder **2000**/0062745

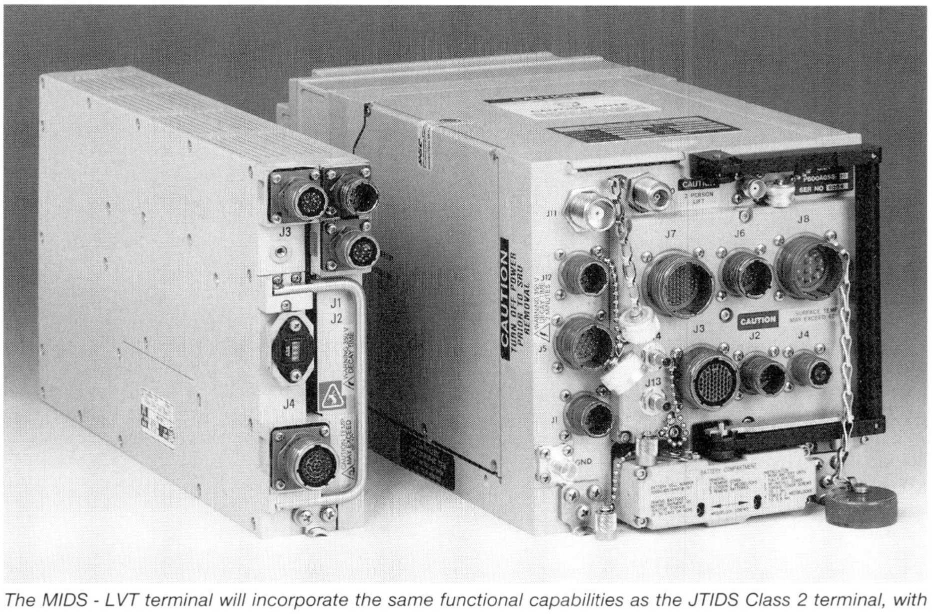

The MIDS - LVT terminal will incorporate the same functional capabilities as the JTIDS Class 2 terminal, with significant decreases in size, weight, power and cost **1999**/0044796

MIDS-LVT Multifunctional Information and Distribution System - Low Volume Terminal

The Multifunctional Information and Distribution System (MIDS-LVT) is under final development by a joint venture comprising: BAE Systems North America (USA), Thales (France), MID (Italy), EADS (Germany) and INDRA (EWS) (Spain). These companies operate through the prime contractor, MIDSCO headed in New Jersey, USA. In 1994, an Engineering Manufacturing Development (EMD) contract was awarded by the International Program Office.

MIDS is to become the standard NATO interoperable data communications system, implementing Link 16 protocols under STANAGs 4175 and 5516. The MIDS terminal will also implement the interim JTIDS message standard, to ensure compatibility with the JTIDS Class 1 terminals which are still in use. It serves a nodeless network with a fast data exchange rate. Also voice transmission capability is provided.

While maintaining full interoperability with the family of JTIDS terminals, MIDS is a new-generation design that will satisfy a broad range of Link 16 applications. To allow its use in the latest and future aircraft designs, the MIDS terminal has been reduced in size, weight, cost and power consumption to less than half that of the present JTIDS Class 2 terminal. It retains, however, all of the JTIDS capabilities, including three-dimensional receiver coverage, 200 W transmit power, complete TACAN capability, full relative/geodetic navigation, precise self-identification and powerful anti-jam capabilities. MIDS-LVT is configured in a single main terminal box and a standoff remote power supply. It will have a MIL-STD-1553 bus interface, allowing standard interoperation with designated aircraft systems, plus a high-speed optical databus (3910) and X.25 and ethernet interface capabilities. An exhaustive built-in-test (BIT) function eases the failure identification and location. Minimum maintenance effort has to be spent by the modular and interchangeable construction.

The MIDS terminal is designed around multiprocessor architecture, with industry standard VME and IEEE 486 busses offering a truly open architecture. This gives maximum flexibility and additional growth potential. Receiver/transmitter modules incorporate the latest MMIC technology.

Operational status

In the current development phase, the participating countries have funded 108 EMD terminals. MIDS customers include: Eurofighter Typhoon, F-15, F-16, F/A-18 and Rafale. International production is expected to exceed 5,000 terminals. Earlier programme deliveries include 11 MIDS simulators that are being used to integrate the capability into participating aircraft.

Contractors

BAE Systems North America, Greenlawn/Wayne.
Thales (formerly Thomson-CSF Comsys).
EADS (formerly DaimlerChrysler Aerospace AG).
MID (Marconi-Italtel-Defence).
INDRA EWS.

UPDATED

MILDS® AN/AAR-60 missile detection system

MILDS® AN/AAR-60 is a passive imaging sensing device. It is optimised to sense the solar blind spectral band of radiation that is emitted from an approaching hostile missile exhaust plume.

MILDS® AN/AAR-60 is designed to detect potential missile threats and provide maximum warning time.

The MILDS® AN/AAR-60 system comprises one to six identical sensor heads including optics, filters and imaging processing. The imaging sensor allows the system to track and classify UV sources, determine angle of attack and priority. Then the system can accurately initiate the proper countermeasures for the threat and warn the pilot, so he can begin evasive manoeuvres. The inherently high spatial resolution of MILDS® combined with advanced temporal

MILDS® six sensor LRUs functioning in master-slave configuration 0018288

processing significantly increases the probability of detection while virtually eliminating false alarms.

The sensor head is a self-contained LRU requiring no additional hardware. One of the sensor heads operates as a master while the rest are slaves. Only the master sensor head handles the communication with the EWSC (Electronic Warfare System Computer) and countermeasure dispenser; this being a purely software function the hardware of the different sensor heads is identical.

Each MILDS® AN/AAR-60 sensor outputs a fully processed signal. The system can be fully integrated into existing threat warning systems or integrated with directed IR countermeasures systems, reporting through existing display and communication paths, or it can be installed as a stand-alone system capable of driving a display and activating countermeasures.

Specifications

Field of view: 4 sensors provide 95° elevation × 360° azimuth
6 sensors provide 180° elevation × 360° azimuth
Power consumption: <14 W per sensor
Weight: < 2.0 kg/sensor
Dimensions: (L × W × H) 108 × 107 × 120 mm excluding flange/connector
Interfaces: RS-422 optional MIL-STD-1553B, discrete lines

Operational status

MILDS® AN/AAR-60 was selected for installation on the NH90 helicopter (TTH version) by Eurocopter Deutschland and NAHEMA (the NATO Helicopter Design, Development, Production and Logistic Agency). The overall EW suite for the NH-90 TTH comprises: the Threat Warning Equipment (TWE), combining a Thales (formerly Thomson-CSF DETEXIS) radar warning receiver, an EADS (formerly DaimlerChrysler Aerospace) laser warning system and the MILDS missile launch detection system, as well as a central EW processing system. MILDS® AN/AAR-60 has completed qualification for the NH90 and is ready for serial production.

It is probable that the MILDS® AN/AAR-60 will be used for the German/French TIGER programme.

Contractors

EADS (formerly DaimlerChrysler Aerospace AG), Defense and Civil Systems.
LFK - Lenkflugkörpersysteme GmbH.
Litton Advanced Systems Division.

UPDATED

Multiplexed Airborne Video Recording System (AVRS)

DRS Hadland Ltd, a subsidiary of DRS Technologies Inc of the USA, has teamed with Bavaria Keytronic Technologie GmbH (BKT) of Germany to produce the AVRS for mission performance evaluation.

The AVRS provides up to eight channels of time multiplexed display recording for military aircraft systems, including Head-Up-Display (HUD), primary radar, Radar Homing and Warning Receiver (RHWR), TV tab displays and audio.

Radar and RHWR scan converters are built-in. It features a colour Charge-Coupled Device (CCD) HUD camera with 400 line resolution. Full manual and automatic control of multiplexing sequence and timing is provided.

The AVRS comprises four units, annotated 1, 2, 3 and 4 on the photograph:

(1) The electronics unit houses the primary radar and RHWR equipment scan converters, the eight-channel multiplexer for signals at up to 9 MHz each, power supplies, weapon event mark generators, BIT circuitry and output to the recorder.
(2) The videotape recorder is a Hi-8 mm airborne videotape recorder that records up to eight video inputs, two audio channels and event marks for up to four hours.
(3) The HUD camera is a 400 TV line colour CCD unit, which records the pilot's view and HUD symbology via the combining prism or off the combining glass; electronic HUD symbols input is also available.
(4) The control panel provides mode selection, manual or automatic control of multiplexing sequences, complete AVRS status, BIT and elapsed time functions.

The Multiplexed Airborne Video Analysis System (MAVAS) is used for analysis and replay of AVRS data. MAVAS is produced by DRS Hadland Ltd, using software analysis techniques developed by the Defence Evaluation Research Agency (DERA) at Malvern in the UK.

Operational status

The AVRS has been selected to upgrade all of the German and Italian Tornado aircraft, and a variant developed by BKT and DRS Hadland Ltd is being fitted as a fit and form upgrade on 31 of the UK Tornado GR1 aircraft. It is also fitted on Jaguar GR3/T4 upgraded aircraft of the Sultan of Oman's Air Force.

The MAVAS replay system has been designed to support a wider range of video recorder systems than purely AVRS, and it is understood that it can support all current UK Royal Air Force aircraft video recorder systems, except the Tornado F3.

Contractors

DRS Hadland Ltd
Bavaria Keytronic Technologie GmbH
Defence Evaluation Research Agency (DERA).

VERIFIED

NGIFF (New-Generation IFF)

The identification system used by NATO has long ceased to satisfy military requirements. Already in the mid-1980s a standard agreement, STANAG 4162, for a new identification system designated NIS (NATO Identification System) was produced. Although several system designs, adapted to the changed threat-situation, were evaluated, no final agreement has yet been achieved in NATO. Nevertheless, the studies conducted by the five nations (FR, GE, IT, UK and US) make it possible to define and develop equipment which is fully compliant with STANAG 4193 and ICAO Annex 10 for Mode S and which can be upgraded to NGIFF when it is defined in detail.

In the meantime, the request by civil aviation authorities for the Mode S function in military aircraft has changed the scenario. Independent from NGIFF, the transponder has to perform Mode S functions from 1 January 1999 to satisfy Air Information Circular 13/92.

Operational status

France, Germany and Italy have decided to co-operate in this effort and to split the programme into two phases with the following content:

Phase 1: Development and procurement of MK X, XII, Mode S and NGIFF upgradable transponders; definition and specifications of Mode S NGIFF upgradable interrogators; studies for Crypto, STANAG and frequency supportability.

Phase 2: Development of NGIFF modules for the transponder; development and procurement of interrogators including Mode S; development and procurement of NGIFF-Crypto.

MID SpA, DaimlerChrysler Aerospace AG and Thomson-CSF have agreed to create a combined development company (EURO-ID) for perfomance of the agreed programme.

Contractors

EURO-ID, a combined development company of:
MID SpA , Italy.
EADS (formerly DaimlerChrysler Aerospace AG), Defense and Civil Systems, Germany.
Thales (formerly Thomson-CSF Communications), France.

UPDATED

Ocean Master airborne maritime patrol radar

Thales and EADS have teamed to develop the Ocean Master airborne maritime patrol radar, designed to meet the requirements of both fixed- and rotary-wing aircraft.

The two companies claim the following major features for the Ocean Master radar: outstanding radar performance in all situations; ultra-light design and simple installation to fixed- and rotary-wing aircraft; proven state-of-the-art technology.

Ocean Master is designed to fulfil both civil and military maritime missions including Economic Exclusion Zone (EEZ) surveillance; Search And Rescue (SAR); Anti-Surface Vessel Warfare (ASVW) and Anti-Submarine Warfare (ASW); air-to-air detection.

Two basic versions of the radar are offered: the Ocean Master 100 and the Ocean Master 400. These

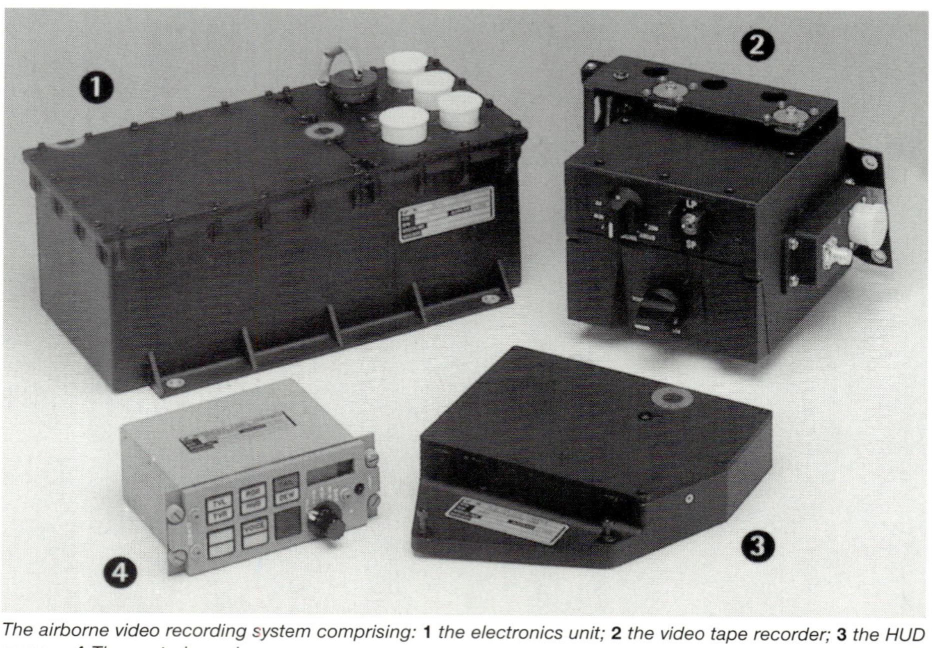

The airborne video recording system comprising: **1** *the electronics unit;* **2** *the video tape recorder;* **3** *the HUD camera;* **4** *The control panel*
2000/0064378

designations relate to the use of either a 100 W fully coherent Travelling Wave Tube (TWT) amplifier, or a 400 W fully coherent TWT amplifier. Options and growth potential are available to meet specific user requirements, including Inverse Synthetic Aperture Radar (ISAR) processing to provide ship classification capability.

The companies claim reliable range detection and tracking of all types of targets, in all sea states, due to use of the following processing characteristics: high pulse compression; frequency agility; pulse-to-pulse and scan-to-scan integration; digitial programmable processing. Ocean Master also performs the following other modes: multitarget track-while-scan; ground mapping; target classification; weather detection and beacon mode.

Ocean Master in the basic version comprises only three units: antenna unit (360° rotation and/or sector scan); transmitter unit (100 or 400 W); exciter/receiver/processor. In addition, the man/machine interface includes display and controls.

Specifications

System:
Max range: 200 n mile
Min range: 200 m
Min range resolution: 3 m
Power consumption: 115 V AC, 400 Hz; 2 kVA for 100 W model; 3.8 kVA for 400 W model
Interfaces: MIL-STD-1553B databus; ARINC 429; RS-422; video

Transmitter:
TWT type: fully coherent; 100 W or 400 W average power
Frequency: I/J-band; wideband frequency agility
PRF range: 300 Hz to 125 kHz
Dimensions: 415 × 275 × 320 mm or 566 × 208 × 260 mm
Weight: 29 kg or 39 kg

Exciter/Receiver/Processor:
Dimensions: 555 × 200 × 340 mm
Weight: 27 kg
Characteristics: pulse compression – high time-bandwidth product
automatic track-while-scan on 32 targets
pulse-to-pulse and scan-to-scan integration
digital radar map
target classification

Antenna:
Dimensions: 660 × 350 mm, or 940 × 350 mm, or 1,800 × 300 mm
Weight: 15 or 27 kg
Characteristics: stabilisation: 2-axis
tilt selectable; +4 to –29°
rotation rate; 6 to 30 rpm, automatically selected by mode
gain; 30.5 to 34 dB

Options:
Inverse Synthetic Aperture Radar (ISAR) for target classification
IFF compatibility
Interfaces with FLIR, IFF, ESM, sonics, datalink, video recorder
19 in diagonal display

Display/Control:
Dimensions: 375 × 300 × 415 mm
Weight: display 18 kg; controls 8 kg
Display unit: 14 in colour multifunction display; aircraft heading or North stabilised; 7.5 to 240 n mile scales
Controls: touch-sensitive flat panel, plus trackerball

Operational status

In production, first deliveries were made at the end of 1994. More than 30 systems have been ordered by France, Indonesia, Japan and Pakistan for use on Falcon 50, C212 maritime patrol aircraft/BO105 helicopters, US-1A aircraft and Atlantique maritime patrol aircraft respectively.

Contractors

EADS (formerly DaimlerChrysler Aerospace AG), Defense and Civil Systems.
Thales (formerly Thomson-CSF DETEXIS).

UPDATED

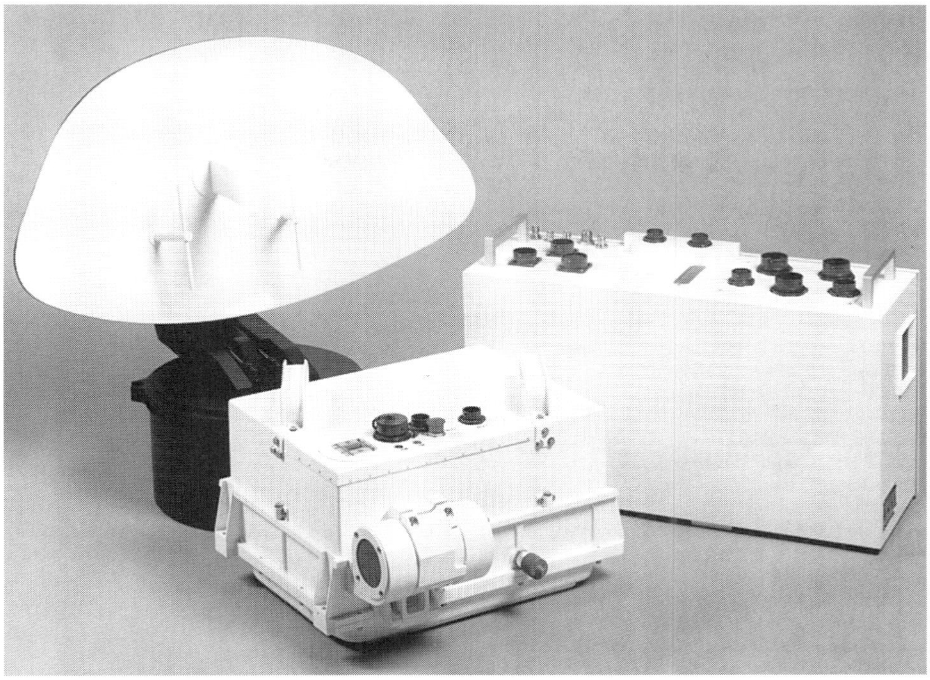

The Ocean Master airborne maritime patrol radar 0011890

Open system computers

Computing Devices in the UK, and the General Dynamics Information Systems division of its parent company in the USA, produce open system computers, featuring Commercial-Off-The-Shelf (COTS) technology, to provide integrated avionics systems for military aircraft. Typically, such integrated avionics systems combine previously discrete elements of the avionics suite, including navigation, stores management and mission processing, into a single computer using open systems architecture. This provides the user with additional functionality with significantly lower cost, weight and space requirements.

Operational status

Computing Devices provides its open system architecture computer for the UK Harrier GR7 aircraft, while General Dynamics Information Systems provides a similar solution for the US AV-8B, FA-18 and F-15 aircraft.

Contractors

Computing Devices, a General Dynamics Company.
General Dynamics Information Systems.

VERIFIED

P-3C upgrade program for RAAF

The Royal Australian Air Force (RAAF) P-3C upgrade programme comprises two major elements: Project Air 5140 and Project Air 5276. Project Air 5140 is managed by BAE Systems Australia. It comprises: replacement of the ESM kit with the Elta ALR-2001 'semi-SIGINT suite' to give the P-3C a true electronic surveillance capability. This change involves significant reallocation of the operational tasking for Sensor Station Three. The ESM and infra-red detection system operation is transferred from Sensor Station Three to a new Sensor Station Four. The new Sensor Station Four is equipped with a BARCO ruggedised RGDS 651/EX terminal unit with a 19 in monitor. About 75 per cent of the fleet has already been modified with the ALR-2001 ESM; the project was due for completion in 1998.

Project Air 5276 is managed by Raytheon Electronic Systems; it brings aircraft up to AP-3C standard and it consists of updating the avionics system and associated support facilities for the RAAF's 18 P-3C aircraft, together with structural work. Items included in the update include: Lockheed Martin's Tactical Defence Systems Data Management System (DMS-2000); Elta's EL/M-2022A(V)3 Synthetic Aperture Radar (SAR)/Inverse Synthetic Aperture Radar (ISAR); Computing Devices Canada's acoustic processor; Honeywell's fully integrated navigation suite; Raytheon Electronic Systems' ICS, HF/VHF/UHF communications system.

Contractors

BAE Systems Australia.
Raytheon Electronic Systems.
Lockheed Martin Tactical Defense Systems.

VERIFIED

The old mission computer for the Harrier GR.Mk 7 (left) and its replacement open systems computer (right)
2000/0064381

RDR-1500B multimode surveillance radar

The RDR-1500B is a lightweight airborne digital colour display 360° multimode radar designed specifically for helicopters and fixed-wing aircraft in a multitude of low- and medium-altitude missions including anti-surface vessel operations, surveillance and patrol, search and rescue, customs and fishery protection. The system is available in single- and dual-display configurations. The dual-display configuration is provided when an operator console is available on the aircraft. It consists of eight units: receiver/transmitter, interface unit, two digital colour displays, cockpit control unit, cabin/console control unit, antenna assembly with antenna drive and flat plate array and switch unit.

The transmitter/receiver operates as a short-range pulse radar for high-resolution sea search and terrain-mapping, and also as a long-range pulse radar for long-range sea search, terrain-mapping and weather avoidance. Standard radar modes include weather detection, ground-mapping, search, beacon detection and identification. Ground-stabilised, aircraft heading and north-oriented display modes are available. The RDR-1500B has the facility to offset the sweep centre to any location on the display and provide target marker capability. Information from a variety of onboard navigation sensors such as INS, Omega, VOR, DME and FLIR can be displayed independently or as overlays (except for FLIR).

Other capabilities include target position transmission via datalink. An optional video processor allows use of CCP/PAL European standard colour display and videotape recording of the images on a VHS standard video recorder.

The video processor allows the radar operator to superimpose FLIR or TV colour images automatically over the radar picture. The video processor is also used to point the stabilised gimbal of FLIR/TV sensors to the selected target. Automatic target tracking is provided by the video processor unit. The modular design of the system allows for additional growth.

Specifications

Dimensions:
(transmitter/receiver) 194.1 × 123.2 × 320.5 mm
(colour indicator) 152.4, 228.6 or 254 mm
(control panel) 133.4 × 115.5 × 168.7 mm
(interface unit) 194.1 × 189.2 × 323.9 mm
Weight: 42.4 kg
Frequency:
(transmitter/receiver for weather and search) 9,375 ±5 MHz
(receiver for beacon) 9,310 ±5 MHz
Power output: 10 kW (nominal)
PRF: 1,600, 800 or 200 Hz
Pulsewidth: 0.1, 0.5 or 2.35 μs
Beamwidth:
(azimuth) 2.6°
(elevation) 10.5°
Range: 300 km

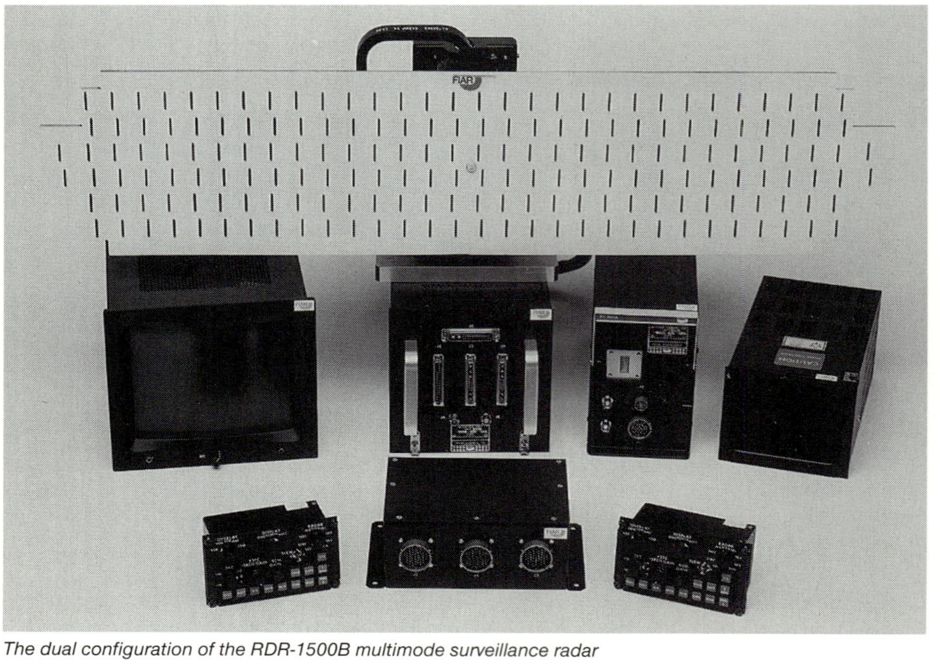

The dual configuration of the RDR-1500B multimode surveillance radar

The RDR-1500B is fitted to Agusta AB 412 helicopters

Operational status

In service. The RDR-1500B is installed on Agusta A 109 and AB 412 helicopters and the Westland Sea King and Lynx.

Contractors

Honeywell Aerospace, Electronic & Avionics Lighting.
Alenia Difesa, Avionic Systems and Equipment Division, FIAR.

VERIFIED

Silent Sentinel

Silent Sentinel is a passive payload aimed at detecting, identifying and locating ground-based radars.

Designed to be integrated on platforms such as UAV, helicopters or light aircraft, Silent Sentinel either delivers radar tracks for real-time display and analysis or uses its high-capacity recording feature for post-flight analysis.

The battlefield tactical situation can be displayed either on a compact laptop computer or on a SPARC-based ESM/ELINT workstation.

Silent Sentinel uses interferometry to give accurate direction-finding. A compressive receiver offers long-range detection, superior parameters measurement and very short acquisition times.

Specifications

Frequency range: 2-20 GHz (0.5-40 GHz optional)
Field of view: 180°
Measurement accuracy:
(frequency) 2 MHz class
(bearing) 1° class
Payload physical characteristics (typical):
(weight) 20 kg
(diameter) 200 mm
(length) 900 mm
(power consumption) <300 W

Contractors

Electronics & Space Corporation.
Thales (formerly Thomson-CSF DETEXIS).

UPDATED

SIT 421 (MM/UPX-709) transponder

The SIT 421 (MM/UPX-709) is a single-box airborne IFF transponder suitable for fitting in fixed-wing aircraft or helicopters. It operates in Modes 1, 2, 3/A, 4 and C. The receiver/transmitter includes a 500 W solid-state transmitter, dual-channel receiver and RF interface module. The first of these comprises a delay line oscillator, modulator, driver and power amplifier.

The controls for operation of the transponder, code and mode selection and so on, are mounted on the front of the equipment (which is designed for cockpit mounting) but versions are produced in which remote-control facilities are provided.

Specifications

Weight: 3.5 kg
Frequency:
(receiver) 1,030 MHz
(transmitter) 1,090 MHz
Sensitivity: –77 dBm (adjustable 69-77)
Dynamic range: 55 dB
Output power: 27 ±3 dBW at 1% duty cycle

Operational status

In production.

Contractors

BAE Systems North America.
MID SpA.

VERIFIED

SIT 432 (AN/APX-104(V)) interrogator

The SIT 432 (AN/APX-104(V)) is a lightweight airborne IFF interrogator equipment suitable for installation on helicopters or fixed-wing aircraft to provide air-to-air and air-to-ship identification facilities.

The receiver/transmitter module contains a 1,200 W transmitter, a dual-channel receiver and an RF interface module. The receiver operates at 1,090 MHz and is of a dual-channel type which, in conjunction with a dual-channel antenna, provides for receiver sidelobe suppression.

The design employs surface acoustic wave technology in the local oscillator to obtain a reliable, simple design with good stability and no field alignment requirements. The transmitter is solid-state. It accepts coded video pulse trains from an external source and the internally generated Mode 4 ISLS pulse converts the coded video pulse trains into radio frequency pulse groups for transmission as IFF interrogation.

Specifications

Weight: 6.5 kg
Frequency:
(receiver) 1,090 ± 0.2 MHz
(transmitter) 1,030 ± 0.2 MHz
Sensitivity: –83 dBm
Output power: not less than 1,200 W
Duty cycle: 1% (max)
Dynamic range: 50 dB

Contractors

BAE Systems North America.
MID SpA.

VERIFIED

SIT 434 IFF interrogator

The SIT 434 is a modular architecture IFF interrogator intended for both fixed-wing aircraft and helicopters. It has considerable growth potential. In the basic configuration, SIT 434 operates under the control of the SIT 905 control box.

SIT 434 is capable of interrogating on Modes 1, 2, 3A and 4, either separately or Modes 1, 2 or 3A interlaced with Mode 4. An associated crypto unit, with mounting and key-loading devices, is required for Mode 4.

Challenge control is possible by means of an enabling signal from the control box or radar system. Video output signals are generated by processing from the last interrogation cycle. Different symbols are generated for target, Mode 1/2/3A response and Mode 4 response.

Optional features on the SIT 434B include pulse-to-pulse defruiter to replace the internal decoding function and interface for an external active/passive decoder to replace the digital symbol generation.

For the SIT 434C, optional features include control of the interrogator and target information interface by means of an embedded dual-redundant receiver/transmitter unit designed to MIL-STD-1553B, with the ability to interrogate azimuthal sectors separately, challenge management with interface standards based on expected target position; and an antenna synchro interface to associate azimuth IFF data with the radar plot.

Specifications

Dimensions: 1 ATR
Weight: 15 kg
Peak power: 1,200 W/300 W selectable
Frequency:
(transmit) 1,030 ± 0.2 MHz
(receive) 1,090 ± 0.2 MHz
Duty cycle: 1% (max)

Contractors

EADS (formerly DaimlerChrysler Aerospace AG), Defense and Civil Systems.
MID SpA.

UPDATED

STR 2000 IFF transponder

The STR 2000 is an IFF transponder system with full capability for Mark XII, as defined in STANAG 4193, and additional capability for Mode S level 3 as defined in ICAO, Annex 10.

The system, which was developed by EADS (formerly Daimler-Chrysler Aerospace AG) and Thales (formerly Thomson-CSF), consists of a transponder and a dedicated control and display unit (CADU). For the use of Mode S level 2 and 3 with all available communication protocols, an Air DataLink Processor (ADLP) is necessary on the host aircraft for connection to the transponder's MIL-STD-1553B serial databus interface. The STR 2000 transponder then provides remote terminal functions for the databus, thereby providing the capability of bidirectional data exchange to the Mode S ADLP of the aircraft.

Without an ADLP, the STR 2000 transponder, together with the CADU, will work in elementary Mode S level 2.

The STR 2000 transponder provides all interfaces for either an external or an appliqué Mode 4 cryptographic computer as standard. An upgrade of STR 2000 provides all capabilities for TCAS (ARINC 429). The system can also be upgraded to Mode 5 (NGIFF) standard.

Specifications

Dimensions:
(Transponder) ½ ATR short (318 × 124 × 194 mm)
(CADU) 146.1 × 133.3 × 81 mm
Weight:
(Transponder) 6.8 kg (without appliqué crypto)
(CADU) 1.5 kg

Operational status

In production for several NATO countries.

Contractors

EADS (formerly DaimlerChrysler Aerospace AG).
Thales (formerly Thomson-CSF Communications).

UPDATED

Tiger combat helicopter avionics system

Sextant, TELDIX GmbH and EADS (formerly DaimlerChrysler Aerospace AG) are responsible for developing the Eurocopter Tiger combat helicopter's navigation system. It is a hybrid design, which includes a Doppler radar, radio altimeter, GES Pixyz three-axis ring laser gyros and air data modules. Together, these units provide the data required by the Automatic Flight Control System (AFCS) (developed by Nord-Micro AG & Co. OHG and Sextant), navigation, guidance, head-down display systems (Sextant and VDO-Luftfahrtgeräte Werk) and control and display unit (Rohde & Schwarz GmbH and Sextant).

Cockpit of the Eurocopter Tiger combat helicopter 0015358

TELDIX GmbH, Sextant and VDO-Luftfahrtgeräte Werk have together developed the five onboard computers: the ACSG (Armament Computer Symbol Generator); MCSG (Mission Computer Symbol Generator); BCSG (Bus Computer Symbol Generator); RTU (Remote Terminal Unit); and the CDD mission data concentrator.

The Helmet-Mounted Sight/Display (HMS/D) has been developed by Sextant and VDO-Luftfahrtgeräte Werk. It features a 40° field of view, with a thermal imager slaved to the pilot's head position, allowing the crew to fulfil day/night all-weather missions.

In parallel to the HMS/D system, BAE Systems and TELDIX GmbH have developed for the German UHT, a binocular 24-hour Mission Capable Helmet System. The system has a 40° field of view, displays pictures of integrated IITs and from a FLIR, overlay's possibilities for symbology, and has an advanced head tracking system.

Operational status

Development nearly complete. Preliminary Declaration of Design and Performance available (Pre-DDP).

Contractors

Sextant.
EADS (DaimlerChrysler Aerospace AG).
TELDIX GmbH.
Nord-Micro AG & Co. OHG.
Rohde & Schwarz GmbH & Co KG.
VDO-Luftfahrtgeräte Werk GmbH.
BAE Systems, Avionics.

UPDATED

TRA 6032/XT 621 P1 V/UHF SATURN airborne transceiver

The TRA 6032 (French identity) or XT 621 P1 (German identity) is an airborne secure radiocommunication system designed for clear and encrypted voice and data transmission, in simplex or half duplex mode in the 100 to 156 MHz and 225 to 400 MHz frequency range.

TRA 6032 uses SATURN as the main transmission security technique which provides the highest level of protection against severe jamming, direction-finding and deception. Communication security is performed with standard NATO crypto devices.

It offers downwards interoperable voice modes such as AM and FM in fixed frequency and first-generation Have Quick I and II ECCM techniques.

The TRA 6032 is capable of data transmission, in fixed frequency as well as in SATURN mode and is particularly adapted to Link 11 and NATO Improved Link Eleven (NILE) or Link 22 transmission.

The TRA 6032 is a 3 MCU transceiver compliant with ARINC 600 and is designed for the rotary-wing environment. Its Guard receiver monitors the 121.5 and 243 MHz international distress frequencies and the 156.525 MHz GMDSS frequency.

Specifications

Dimensions: 90 × 194 × 320 mm
Weight: 7.6 kg
Power supply: 28 V DC

Operational status

In series production for the Tiger helicopter. Selected for NH 90.

Contractors

Thales (formerly Thomson-CSF Communications).
Rohde & Schwarz GmbH & Co KG.

UPDATED

TWE Threat Warning Equipment

The TWE includes a laser warner, made by EADS (formerly DaimlerChrysler Aerospace), and an Instantaneous Frequency Measurement (IFM) radar warner made by Thales (formerly Thomson-CSF DETEXIS).

The TWE was designed initially for the Eurocopter Tiger helicopter. It is intended to provide warning of hostile surveillance and fire-control radars, tracking radars, CW illuminators, and missile seekers, as well as laser range-finders, laser illuminators and laser missile guidance systems.

The TWE is designed to be the core of a defensive aids suite, being linked to a chaff and flare dispenser, missile launch detector or missile approach warner, with growth to include ECM capability via MIL-STD-1553B interface.

The TWE system can be easily programmed to meet differing system and scenario requirements.

Specifications

Weight: <12 kg
Frequency:
(radar) E- to K-bands
(laser) 0.4-1.1 μm (option: 0.4-1.7 μm)
Coverage: 360° azimuth, ±45° elevation
Accuracy:
(angular) better than 10° rms
(frequency) less than 20 MHz
Resolution:
(laser) 4°
Library size: >1,000 threats
Dimensions:
(CPU) 158 × 194 × 321 mm
(antenna) Ø100 × 88 mm
Power: <170 W

Operational status

TWE has been developed for installation aboard the Tigre HAC (Hélicoptère Anti-Char)/ HAP (Hélicoptère d'Appui et de Protection) and UHU (Unterstützungs Hubschrauber) battlefield escort / anti-tank helicopters which are on order for the French and German armies respectively.

TWE threat warning equipment 0001259

TWE also forms the basis of a French - German helicopter applicable Electronic Warning System (EWS) which has been selected for the NH 90 TTH (Tactical Transport Helicopter) version. In this application the TWE radar and laser warning system is teamed with a MILDS® AN/AAR-60 passive missile approach warner and interfaced with a chaff and flare dispenser.

In the NH 90 NFH (NATO Frigate Helicopter) version, TWE is replaced by the Elettronica DETE 90 ESM, which interfaces with the Saphir-M system.

Contractors

Thales (formerly Thomson-CSF DETEXIS).
EADS (formerly DaimlerChrysler Aerospace AG), Defense and Civil Systems.

UPDATED

Utility control system for the Eurofighter Typhoon

The Eurofighter Typhoon utility control system is based on the utilities management system which was successfully demonstrated on the Experimental Aircraft Programme (EAP). The system integrates, via a MIL-STD-1553B databus, the control and monitoring of a number of aircraft functions, including environment, cabin temperature, life support, crew escape, fuel management, fuel gauging, secondary power, hydraulics and other miscellaneous systems. The main benefits derived from the system are in weight saving and reliability. In addition, there is a greater fault tolerance and better damage resistance, system health is monitored and trends can be identified, aiding planned maintenance.

Operational status

Contracts awarded for production tooling.

Contractors

Alenia Difesa, Avionics Systems and Equipment Division.
INDRA EWS.
Smiths Industries Aerospace.
VDO Luftfahrgeräte Werk Adolf Schindling GmbH.

VERIFIED

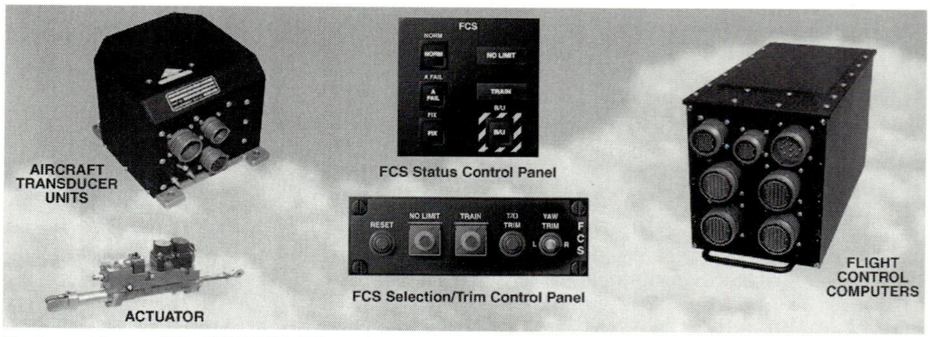

Equipment items of the YAK/AEM-130 quadruplex digital flight control system **2000**/0062222

YAK/AEM-130 quadruplex digital flight control system

The YAK/AEM-130 multirole lightweight jet trainer is specified with a quadruplex, digital, fly-by-wire, flight control system designed by Teleavio srl of Italy and BAE Systems North America. Design features of the control system include fault-tolerant redundancy management providing electrical two-fail-operate capability; a quad-channel redundant primary actuation system; multifunction air data computation and sensing; and extensive in-flight built-in testing.

Contractors

BAE Systems North America, Aircraft Controls
Teleavio srl.

VERIFIED

ISRAEL

Enhanced airborne Communication, Navigation and Identification (ECNI) system

The Enhanced airborne Communication, Navigation and Identification (ECNI) system enables integrated, centralised and computerised control and resource management of communication, radio navigation and identification devices, using avionics controls and displays as the man/machine interface. It also enables back-up control and management of vital CNI functions during failures, using a dedicated control and display. Management and control of all audio sources such as the crew, radios, alarms, weapons and EW is provided using a digital audio matrix which also provides intercom and alarm generation as by-products. The ECNI manages and controls synthetic voice and tonal alarms.

ECNI consists of the main control unit which contains all the electronics required for system implementation, an integrated back-up control box which is used as a redundant unit for control and audio during failures and additional control and audio panels for the use of other crew members. In the normal mode the ECNI is designed to be operated using the avionics controls and displays, and in the back-up mode the system is operated via the dedicated control box. The main control unit includes a digital/audio switching matrix which enables intercommunication between system subscribers according to predefined communication maps or dynamic real-time resource selection. Alarm control and management are principal features of ECNI. There are two types of alarms: synthetic voice and regular tone. The system is capable of translating existing tonal alarms into synthetic voice alarms, according to the latest human engineering requirements.

Contractor

Elbit Systems Ltd.

VERIFIED

F-16 Avionics Capabilities Enhancement (ACE)

The F-16 ACE is being developed by Elbit Systems Limited and the Lahav Division of Israel Aircraft Industries, supported by the Israel Defense Force. First flight is planned for late 2000. The upgrade is applicable to all F-16A/B aircraft and to F-16C/D aircraft up to block 50.

Cockpit enhancements include: three Astronautics/Elbit Systems 5 × 7 in multifunction liquid crystal colour displays; an Elop/Elbit Systems wide angle head-up display with LCD-equipped up-front panel; the Elbit Systems Display And Sight Helmet (DASH).

The planned sensor suite includes: the Elta E/L-2032 radar; Elisra SPS-1000 radar warning receiver and ASPS electronic warfare suite (developed for the Israeli F-15I); Rafael Litening targeting pod; Elta EL/M-2060P pod-mounted synthetic aperture radar.

Central processing is controlled by an Elbit Systems open-architecture mission and display processor system, with navigation being provided by an Elbit Systems GPS/INS interfaced to a moving map display.

Additional avionic equipment could include an Elta electronic countermeasures pod, and autonomous air combat manoeuvring instrumentation system.

Aircraft enhancements include structural upgrades, and additional fuel carrying and weapons capabilities.

Operational status

In development. First flight was planned for late 2000.

Contractors

Elbit Systems Limited.
Israel Aircraft Industries, Lahav Division.

VERIFIED

Flightsight rangeless AACMI

Flightsight is a rangeless Autonomous Air Combat Manoeuvring Instrumentation (AACMI) system that provides in-flight recording for highly effective comprehensive post-mission debriefing and analysis. It can be installed in a wide range of fighter aircraft without any expensive modifications.

Flightsight is an autonomous integral system connected to the aircraft avionics that gathers information into standard storage devices. The system offers debriefing capability with no geographical limitations. Navigation data from the aircraft arrives from the INS and GPS, giving exact position and performance information. The digital data is downloaded to the debriefing computer, together with other aircraft data. The computer can give an online display from its disks.

Flightsight provides a comprehensive review of the mission. It can be installed permanently either separately or as part of an aircraft upgrade. It does not use a missile pod and thus can be used in actual operational missions, in addition to training exercises. Voice and video are synchronised with the situation display. Flightsight interfaces with any existing avionics system based on a MIL-STD-1553 databus, and uses existing data from the navigation radar, EW and stores management systems.

Contractor

Elbit Systems Ltd.

VERIFIED

HELIA – integrated modular avionics weapon system for helicopters

HELIA is an advanced integrated modular avionics system upgrade for utility, attack and reconnaissance helicopters designed to provide: increased situational awareness and survivability; reduced crew workload; nap-of-the-earth capability in adverse weather conditions; target detection/recognition at long range with various armament configurations.

HELIA can integrate a range of avionics and armament control systems according to specific customer requirements.

The system is based on a Helicopter MultiRole Computer (HMRC) (an open architecture state-of-the-art avionics/armament computer for effective sensor management), displays, fire control, navigation and mission management. The HMRC also provides MIL-STD-1553B channels and diverse I/O capabilities.

Integrated into the system is a navigation system (embedded GPS/INS or Doppler), a modular integrated display and sight helmet system (which provides a binocular system for displays, LOS and image intensifying), a digitised moving map display, a data transfer system, a glass cockpit design supported by hands-on collectives and sticks for man/machine interface enhancement and an observation/targeting system (FLIR, CCD and laser range-finding, with provisions for a laser designator). In addition, a video/audio transmit and receive capability permits communication with other aircraft and ground stations. The armament system is based on a 20 mm turreted gun, anti-tank missiles, rockets and air-to-air missiles integrated with the avionics system.

HELIA is also supported by a map planning station for digitising different types of maps and updating information with obstacles and terrain data; a mission planning work station, for pre-mission planning and debriefing of data using a data transfer cartridge, and a debriefing station for interfacing with the aircraft's video/audio transmit and receiving capability in real time.

Operational status

HELIA has been operational in the Israeli Air Force CH-53 2000 platform since 1993. Full system capabilities (avionics and weapon) have been installed and integrated on the Romanian Air Force Puma 330 platform.

Contractor

Elbit Systems Ltd.

VERIFIED

AES-210/E ESM/ELINT RWR system

The AES-210/E is a sophisticated system, which performs ESM, ELINT and radar warning system tasks in very dense RF environments. It can be installed in small- or medium-size aircraft or ships for applications such as maritime and overland surveillance, ELINT information gathering and platform self-protection. As a programmable system designed with a modular and flexible architecture for future growth, the AES-210/E is optimised to receive all known and foreseeable radar signals, identifying highly complex signals and threats and employing an advanced combination of receiving techniques.

The AES-210/E offers very high intercept probability, sensitivity and reliability, providing very accurate direction-finding. A number of different modern DF techniques can be provided to meet operational requirements, including advanced interferometric methods to achieve the most accurate data. The system has five fast CPUs and can be operated automatically or by a single operator.

System features include:
(1) Automatic detection, measurement and identification of radars.
(2) Long-range detection together with high probability of intercept, using a combination of wide- and narrow-band receivers.
(3) Accurate direction finding implemented by amplitude comparison.
(4) DTOA and interfometric techniques.
(5) Onboard emitter library updating and mission planning.
(6) Integrated self-protection capability, including RWR display in cockpit.
(7) In-flight recording of data for further analysis.
(8) MIL-STD-1553 and ARINC 429 interfaces.
(9) Datalink communications to remote users.
(10) Windows NT® operation.

Specifications

Weight: 40 kg typical
Frequency: 0.5-18 GHz
Coverage: 360° azimuth
Accuracy: 7° RMS coarse, 3° RMS fine
Library: >1,000 emitters
Interface: RS-232, RS-422, MIL-STD-1553B, ARINC
Environmental: MIL-STD-5400T
Power: 500 W

Operational status

The AES-210/E has been installed on several types of aircraft, including: Beechcraft B200/B200T, Dornier 228, Falcon 20, Learjet 36A, M117, Sea Hawk and Sea Sprite.

The AES-210/E has been selected for the Royal Australian Navy Maritime Patrol Aircraft (MPA) helicopter fit, together with Elisra's LWS-20 laser warning system and chaff and flare dispenser system. Elisra is the main EW system integrator.

Contractor

Elisra Electronic Systems Ltd.

VERIFIED

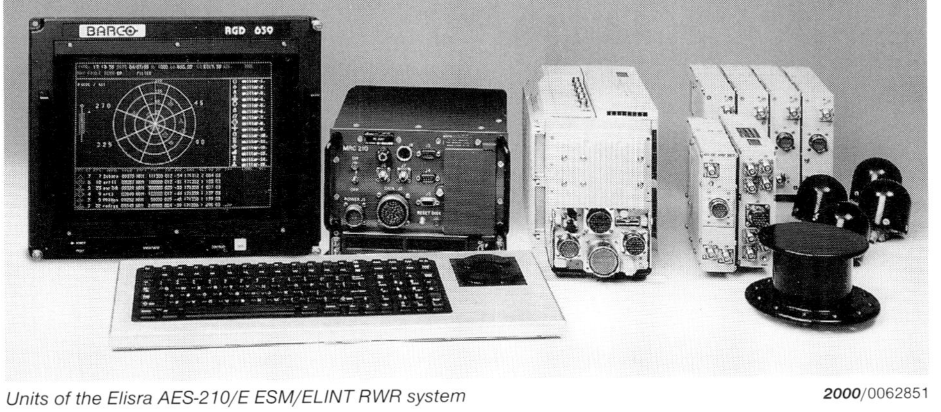

Units of the Elisra AES-210/E ESM/ELINT RWR system **2000**/0062851

ASPS EW suite for the F-15I and F-16I

The ASPS EW system is Elisra's most advanced EW suite. It is based on a combination of proven and new-generation technologies. It is in production and

The Elisra Electronic Systems ASPS EW suite for F-15I **2001**/0089535

operational aboard Israel Air Force F-15I aircraft. It has been selected for the 50 new F-16I aircraft Israel agreed to purchase in July 1999.

Contractor
Elisra Electronic Systems Ltd.

VERIFIED

Giga-Links digital communication systems

Giga-Links is a family of digital communication products for the transmission of high data rates generated by wideband imaging systems. Typical applications include airborne observation sensors such as high-resolution electro-optical cameras and thermal imaging scanners, as well as high-rate ELINT sensors.

Giga-Links operates at microwave frequencies and uses a modular concept; a single-channel implementation for rates of up to 150 Mbits/s and parallel channels for higher rates. Compression techniques are used to reduce the data rate when frequency bandwidth is constrained or when higher rate sensors are used.

The channelised airborne transmitter consists of frequency sources, QPSK modulators and solid-state power amplifiers. Error correcting codes are used to improve the link performance. A compact and lightweight design is used to meet the specific airborne requirements.

The associated ground station consists of a sensitive receiving chain with auto-tracking capability.

For those applications requiring data compression, dedicated units are added at the transmitting and receiving ends, using an adaptive compression algorithm implemented in VLSI. The sensor output data rate is reduced by up to a factor of four, while maintaining the excellent image quality required for reconnaissance applications. An extensive monitoring and test capability is included in both the aircraft and ground installation.

Contractor
Elisra Electronic Systems Ltd.

VERIFIED

LWS-20 laser warning system

Four laser sensors receive and detect laser pulsed signals and send the detected pulses to the Laser Warning Analyser (LWA). The LWA characterises each laser pulse for its angle of arrival, relative time of arrival and amplitude. It then processes the single pulses or pulse trains and identifies the threatening emitter, based on preprogrammed library information. Identified threats are displayed on the screen of the self-protection system as alphanumeric symbols. These symbols enable the crew to establish the type, azimuth and lethality of the threats. In addition, audio warnings are sent to the crew through the aircraft intercom system.

The LWA incorporates a powerful military microprocessor with EEPROM for combat software, programmable tables and in-flight recording. The LWS-20 may include a MIL-STD-1553B RWR interface or other ECM interfaces.

Specifications
Dimensions:
(analyser) 146 × 106 × 78.7 mm
(control unit) 54 × 146 × 19 mm
(display unit) 172 × 82.5 × 82.5 mm
(sensors) 150 × 640 × 860 mm
Weight:
(analyser) 1.5 kg
(control unit) 0.2 kg
(display unit) 1.2 kg
(sensors) 0.6 kg each
Power supply: 28 V DC, 50 W

Operational status
In production and in service.

Contractor
Elisra Electronic Systems Ltd.

VERIFIED

PAWS passive airborne warning system ***2000***/0062859

PAWS Passive Airborne Warning System

The Passive Airborne Warning System (PAWS) is a lightweight missile launch and approach warning system designed for fighter aircraft and attack helicopters. It was specified, characterised and integrated by Elisra. It consists of four to six infra-red staring sensors and one processor. The system may operate independently or may be integrated with an EW suite.

PAWS detects the missile heat and tracks it even in high-clutter environments. It determines when a missile threatens the aircraft, provides an accurate readout of approach direction and an estimate of the time to intercept. It can also select the appropriate narrowbeam, or flare dispenser, countermeasures and activate them automatically. The system has a very low false alarm rate even in violent manoeuvres and when operating in a high-clutter environment.

PAWS consists of a high-power processor, built by Elisra, and an IR sensor built by Elop Electronic Industries. It features missile launch and approach warning, discrimination between threatening and non-threatening missiles, multithreat warning capability, accurate approach direction and time-to-go estimation and automatic activation of countermeasures.

PAWS integrates with the SPS-65V-2 radar warning receiver, the LWS-20 laser warning receiver, and a chaff and flare dispenser. It is also able to enslave a directional infra-red countermeasures system.

Specifications
Power source: 28 V DC (MIL-STD-704D)
115 V, 400 Hz, 1Ø
Power consumption: 200 W
Dimensions:
(sensor) 120 × 120 × 200 mm
(processor) short ½ ATR (203 × 389 × 127 mm)
Weight:
(sensor) 4 × 2.5 kg
(processor) 10 kg
MIL specification: MIL-STD-5400, Class I B
Communication interfaces: RS-422, MIL-STD-1553B MUX BUS, transputerlink

Operational status
Fully developed.

Contractor
Elisra Electronic Systems Ltd.

VERIFIED

SPJ-20 self-protection jammer

The SPJ-20 is a powerful, yet compact, self-protection jamming system, designed to provide protection against ground and airborne radar-guided fire control systems. It is suitable for both fixed- and rotary-wing aircraft.

The main features include:
(1) An acquisition receiver and ECM exciter all in a single LRU - no separate RWR is required.
(2) Powerful, multiport, pulse and CW transmitters (up to two transmitters supported).
(3) Wideband operation.
(4) ECM response against pulse, CW and pulse Doppler threats.
(5) Interfaces to other onboard avionics and chaff and flare dispensers.
(6) In-flight data recording and a field programmable threat library.

The acquisition receiver is a wideband, VCO-based, fast IFM/superheterodyne receiver. It performs amplitude monopulse direction of arrival measurement threat identification and classification.

The ECM resources include: a noise transponder based on a fast VCO; pulse and CW repeater channels; multi-resource range, Doppler, noise and amplitude ECM techniques, generators and trackers.

The transmitter is a 4-tube, multimode (pulse/CW) mini TWT cluster with TWT outputs providing combined in phase and amplitude. The combined output signal is steerable to any combination of output ports on a pulse-by-pulse basis.

LWS-20 laser warning receiver 0002225

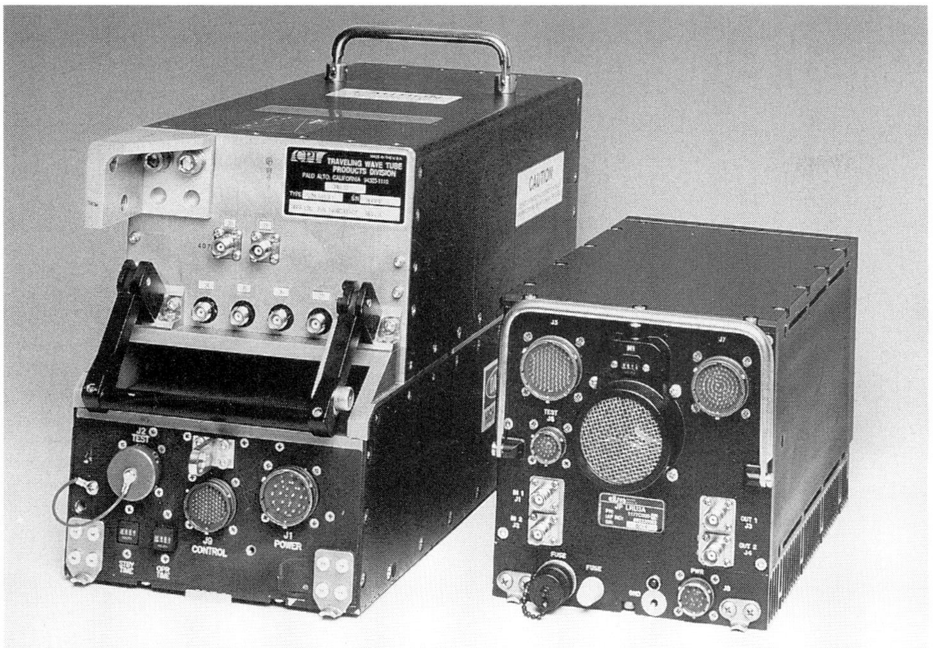

The SPJ-20 self-protection jammer ***2000***/0062852

Mission support equipment includes: a Mission Loader/Verifier (MLV) on a laptop PC; a Pre-Flight Message Generator (PFMG); and a Post-Mission Data Analyser (PMDA).

Specifications

Frequency: 6-17.5 GHz (transmit); 2-18 GHz (receive)
Interfaces: MIL-STD-1553B, RS 422, RS 232
Environmental qualification: MIL-E-5400T Class 2X
Weight: receiver/jammer: 12 kg; each transmitter 30 kg
Power: receiver/exciter: 28 V DC, 240 W; each transmitter: 115 V AC (3-phase), 3,000 W

Contractor

Elisra Electronic Systems Ltd.

VERIFIED

SPS-20(V) airborne self-protection system

The SPS-20(V) is a low-cost, low-volume and lightweight radar warning system designed to fit existing helicopters and aircraft. It detects and displays pulsed radar threats operating within the 0.7 to 18 GHz frequency range. A 3 in (76 mm) display unit provides an alphanumeric and graphic representation of the type, course, angle of arrival, relative lethality and status of the analysed radar threats. The SPS-20(V) is equipped with a fast microprocessor which executes data processing and interfacing tasks.

The system features wideband acquisition including C/D-band, optional high-sensitivity CW detection and recording of flight events and emitter parameters for playback and evaluation.

The system is fully programmable and there is an optional provision for a tie-in with the flare/chaff dispensing system.

Specifications

Dimensions:
(analyser) 155 × 118 × 120 mm
(dual-channel receiver) 178 × 66 × 96 mm
(control unit) 146 × 59 × 19 mm
(display unit) 125 × 83 × 83 mm
Weight:
(analyser) 2.5 kg
(dual-channel receiver) 1.1 kg each (2)
(control unit) 0.2 kg
(display unit) 1.2 kg
Power supply: 28 V DC (MIL-STD-704)
Environmental: MIL-STD-5400

Operational status

In production and in service.

Contractor

Elisra Electronic Systems Ltd.

VERIFIED

SPS-45(V) integrated airborne self-protection system

The SPS-45(V) is an airborne self-protection system which integrates two subsystems: the SPS-20(V) which detects pulse radars from the low band to 18 GHz and the SRS-25 which is a superheterodyne receiver detecting CW, high PRF and low ERP radars. The LWS-20 may also be incorporated as an option to detect laser pulses. Identified threats detected by the subsystem are displayed to the pilot on a 3 in unit which provides a graphical representation of type, angle of arrival, relative lethality and status.

The SPS-45(V) offers complete threat reprogrammability for new and changing environments through an extensive, easily updated emitter library. A portable memory loader/verifier enables the loading or unloading of the operational software and emitter tables in the field.

Four orthogonally dispersed cavity-backed spiral antennas receive pulsed signals in the 2 to 18 GHz frequency range and send them to the two dual receivers. The video pulse output from both receivers is fed to the analyser which tags every pulse with its angle of arrival, amplitude, pulsewidth and time of arrival.

The analyser CPU processes these characterised pulses, de-interleaves individual pulse trains and identifies the threatening emitter by comparing the pulse parameters with preprogrammed library information. Pulse signals in the low band are also detected and characterised to provide additional information and threat status identification.

The superheterodyne receiver is fed via couplers from the dual receiver.

The input circuitry enables selection of one of the four antennas. The signal detected from the selected antenna passes a preliminary filter, mixes with the output signal of a voltage-controlled oscillator and goes to an IF section. The signal is then amplified, filtered and detected. The video signal is transferred to the video circuit for conversion to digital values. The final analysis and identification of the threat is done by a microprocessor and the results are transferred to the analyser via a serial datalink.

The analyser is responsible for integrating the data from the subsystems into one common threat file.

The analysed threats are displayed on the display unit screen. In addition, audio warnings are dispatched to the pilot through the aircraft intercom system. The threat representations on the display screen allow the pilot to establish the azimuth, lethality and emitter type for up to 16 of the most lethal threats. Up to 250 different symbols can be programmed to represent specific threat types. Additional alert capability for missile launch status is provided.

The system operating mode is controlled by front panel push-buttons on a single control unit. This unit enables the pilot to power the system, activate the automatic self-test procedure, delete preprogrammed threats from the display and define modes of operation.

Each of the two subsystems incorporates its own microprocessor with EEPROM and RAM memory. Each microprocessor works autonomously using its own emitter and mission table.

The SPS-45(V) consists of SPS-20(V) and SRS-25 subsystems

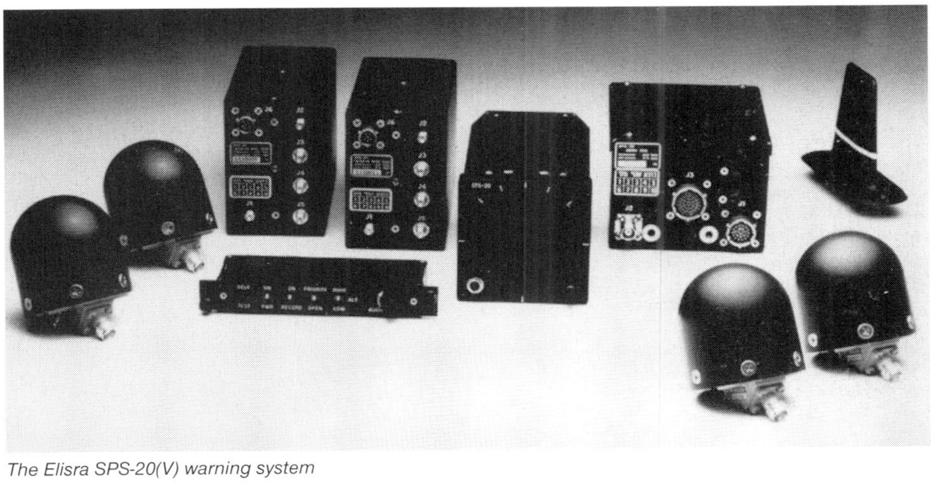

The Elisra SPS-20(V) warning system

The Elisra SPS-65(V) self-protection system **2000**/0089536

Specifications

Weight:
(SPS-20) 7.5 kg
(SRS-25) 3.8 kg
Power supply: 28 V DC (MIL-STD-704)
Frequency: low band to 18 GHz
Environmental: MIL-STD-5400, Class II

Contractor

Elisra Electronic Systems Ltd.

VERIFIED

SPS-65(V), SPS-65V and SPS-65V-2 Self-Protection Systems

The SPS-65(V) is a sophisticated lightweight system with outstanding sensitivity which is capable of intercepting and analysing a wide range of known and anticipated emissions from the battlefield, including laser radiation sources. The design of the SPS-65(V) has been enhanced by technically implementing combat experience.

The SPS-65(V) comprises three subsystems. The SPS-20(V) low-volume lightweight radar warning system detects, processes and displays radar threats operating within the 0.7 to 18 GHz frequency range. The SRS-25 which is a superheterodyne receiver detecting modern CW, high PRF and low ERP radars. The LWS-20 laser warning receiver detects, identifies and locates hostile laser-guided weapons.

A 3 in display unit provides an alphanumeric representation of the type, relative bearing, relative lethality and status of up to 16 analysed threats. In addition, audio warnings are dispatched to the pilot through the intercom system. A flare/chaff dispensing system, which can be automatic or crew-activated, is available for countermeasures capabilities.

Software and emitter tables can be loaded in the field into the emitter library in a matter of seconds. This provides complete and immediate adaptability to changing threat environments. The analyser records flight events and unloads them into a tape cassette for playback after flight.

The SPS-65V is an upgraded new generation system based on the SPS 65(V). It is a lightweight, small volume, low cost self-protection system dedicated to helicopters.

Specifications

Weight:
(SPS-20(V)) 7.5 kg
(SRS-25) 3.8 kg
(LWS-20) 2.5 kg
Power supply: 28 V DC
Frequency: 0.7-18 GHz
Environmental: MIL-STD-5400 Class II

SPS-65V-2 integrated RWR and LWS

The SPS-65V-2 is a sophisticated low-weight, low-volume system that integrates three subsystems: SPS-20V-2, SRS-25 and LWS-20. The SPS-20V-2 detects classical pulse radars within the low-band up to 18 GHz. The SRS-25 uses a superheterodyne receiver to detect modern continuous wave radars, high-PRF (Pulse Repetition Frequency) radars and low-ERP (Effective Radiated power) radars. The LWS-20 detects and analyses laser signals incident on the aircraft.

Operational status

Already in use by major defence forces worldwide, the SPS-65V has recently been fully qualified by the German military authorities and is currently installed in a CH-53G helicopter.

The SPS-65V-2 is in service.

Contractor

Elisra Electronic Systems Ltd.

VERIFIED

SPS-1000V-5 self-protection system

The SPS-1000V-5 is a compact, lightweight, state-of-the-art, airborne, self-protection system, which was specifically designed to be a Radar Warning Receiver (RWR) form-fit replacement for a wide range of front line fighter aircraft.

The system is field proven and is operational on a variety of airborne platforms. The system detects, identifies and displays modern emitters over the C- to J-bands. It can be extended to higher frequencies if required.

The SPS-1000V-5 features two integrated, independent dual receivers, which are controlled by three microprocessors interconnected by a fast internal bus, allowing for a large throughput and fast response time. The time domain receiving subsystem provides continuous and instantaneous coverage of the full frequency range and near 100 per cent probability of intercept. A multichannel superheterodyne/IFM receiver subsystem provides higher sensitivity/ detection range and selectivity for CW, high-PRF, low-ERP and pulse Doppler radars. An additional microprocessor controls the displays human and aircraft interfaces.

The Mission Support System (MSS) was designed to provide immediate response for any operationally required update. It allows for complete programmability and analysis of multiple PRF threats as well as the human machine interface of the system, such as displayed symbols (selectable from a built-in library), audio tones, missile launch warning, fade in/ fade out times and other data.

The system has growth potential to meet the requirements of different platforms and operational scenarios. The SPS-1000V-5 system contains a large array of interfaces including: discretes, MIL-STD-1553B, RS 422, RS 232, and blanking inputs and outputs.

Specifications

Weight:
(central receiver processor) 14 kg
(pilot display and control) 2.4 kg
(antennas and front end receivers) 6 kg
(RF switch) 0.8 kg
Power supply: 115 V AC, 400 Hz
28 V DC (MIL-STD-704)
Environmental: MIL-STD-5400, Class II

Operational status

Selected for the Royal Australian Air Force C-130H interim upgrade.

Contractor

Elisra Electronic Systems Ltd.

VERIFIED

The SPS-1000V-5 self-protection system **2000**/0062853

SRS-25 airborne receiver

The SRS-25 detects and analyses CW, high-PRF and medium-ERP radar signals and determines their direction of arrival. It forms part of the SPS-45(V) and SPS-65(V) integrated helicopter self-protection systems. Continuous wave or high-PRF signals in the 6.5 to 18 GHz frequency range are analysed and signals can be recorded for later playback, analysis or training. An extensive threat library, which can be updated easily between flights, is incorporated in the unit. Four cavity-backed spiral antennas are used for accurate direction-finding.

Specifications

Dimensions: 101 × 122 × 250 mm
Weight: 4 kg
Frequency: 6.5-18 GHz
Coverage: 360° azimuth,

Operational status

Available.

Contractor

Elisra Electronic Systems Ltd.

VERIFIED

ALR-2001 ESM

A sophisticated airborne ESM able to measure radar operating parameters with sufficient precision to 'fingerprint' individual radars from others of the same type.

ALR-2001 also has COMINT capabilities from V/UHF to microwave datalink.

It provides direction-finding capability, reportedly to better than 1°.

Operational status

Selected by the Royal Australian Air Force for its P-3C aircraft.

Contractor

Elta Electronics Industries Ltd.

VERIFIED

ARC-740 UHF secure radio

The ARC-740 provides UHF communication for air-to-air, air-to-ground and ground-to-ground use for plain speech and secure speech, using frequency hopping at 10 hops/s and data transmissions at 2.4 kbits/s. The frequency band from 225 to 399.975 MHz is covered by 7,000 channels at 25 kHz steps in both AM and FM modes; any channel can be manually selected. There are also ECCM-protected normal and high-powered modes. There are 99 preset channels and a continuous watch is kept on the Guard frequency.

Specifications

Dimensions:
(transceiver) 250 × 127 × 177 mm
(power amplifier) 190 × 170 × 145 mm
(control box) 143 × 146 × 124 mm
Weight:
(transceiver) 7 kg
(power amplifier) 4.9 kg
(control box) 1.8 kg

Contractor

Elta Electronics Industries Ltd.

VERIFIED

The Elta ARC-740 UHF secure radio

COMINT/ELINT system for the Arava

The COMINT/ELINT system for the Arava twin-engined light transport can be rolled on or off a multipurpose aircraft or be permanently installed. The COMINT operator uses sensors such as the Elta EL/K-1250 with omnidirectional antennas, direction-finding and signal recording. The ELINT operator has 0.5 to 18 GHz bandwidth coverage with omnidirectional antennas, signal analyser, data processor and recorder facilities. Jamming can be accomplished using antenna arrays in the aircraft tail, giving approximately 240° azimuth coverage in the lower hemisphere. Spot and band jammers of 20 to 400 W power are used.

Specifications

In production.

Contractor

Elta Electronics Industries Ltd.

VERIFIED

EL/K-1250T VHF/UHF COMINT receiver

The compact EL/K-1250T synthesised receiver operating in the 20 to 510 MHz band is used as a building block for larger COMINT or EW systems such as the Arava EW. The unit has four selectable intermediate frequency filters which demodulate AM, FM, continuous wave and single sideband signals. Intermodulation protection is claimed from RF preselection by voltage tracking filters and the fast tuning synthesiser settles within 500 ms between channel changes. Remote digital control operation is possible and the compact dimensions, low weight and power consumption are achieved by extensive use of advanced microcircuit technology.

Specifications

Dimensions: 114 × 193 × 356 mm
Weight: 7 kg
Frequency: 20-510 MHz
Frequency accuracy/stability: ±1 ppm
Synthesiser settling time: 500 µs
IF bandwidth: select 4 of 10, 20, 50, 100, 300, 600, 1,000 kHz
Noise figure:
(20-180 MHz) 12 dB
(180-510 MHz) 11 dB

Operational status

In production. The unit is in service with Israeli and other armed forces.

Contractor

Elta Electronics Industries Ltd.

VERIFIED

EL/K-7010 tactical communications jammer system

The EL/K-7010 is a modular family of communication jamming systems designed for stand-alone operation and providing for signal search and acquisition, preset channel monitoring and automatic jamming modes. It can also be integrated within a larger electronic warfare system. Computer control of power management, fast reaction jamming and multikilowatt effective radiated power provide simultaneous multiple target capability. New interception tasks are accomplished rapidly by a fast scanning receiver and automatic signal sorting in the system computer. In addition to airborne installations, ground-based systems suitable for armoured personnel carriers and air conditioned shelters are available.

Operational status

In production.

Contractor

Elta Electronics Industries Ltd.

VERIFIED

EL/K-7032 airborne COMINT system

The EL/K-7032 is designed for the surveillance and interception of radio signals in the 20 to 500 MHz frequency range, operation being largely computer-controlled.

A typical airborne system would include a supervisor's station having a computer controller, two VHF/UHF radios, a display and a data recorder. This station can work with up to four traffic collection stations, each having up to four radios and data recorders.

The EL/K-7032 has a frequency resolution of 1 kHz (with 10 Hz an option) and can intercept AM, FM, CW and HF-SSB transmissions as required.

Specifications

Frequency: 20-500 MHz
Resolution: ±1 kHz (10 Hz optional)
Modes: AM and FM (CW and SSB optional)

Operational status

In production and in service. Fitted to the Phalcon early warning aircraft.

Contractor

Elta Electronics Industries Ltd.

VERIFIED

The Elta EL/K-7032 airborne COMINT system

EL/K-7035 all-platform COMINT system

The EL/K-7035 COMINT system intercepts, monitors, locates, analyses and reports on radio communications in the 20 to 500 MHz range. It is suitable for air, sea or ground applications.

The standard EL/K-7035 system includes a supervisor console, two to five operator consoles, a system controller and mass storage, RF distribution and antenna, a DF system, plotter position and datalink. The supervisor and operator consoles would each be equipped with two to four COMINT receivers, two to four controllers, two to four dual-channel tape recorders and a ruggedised computer. As an option these consoles could also have an IF panoramic display and time code generator/reader.

Contractor

Elta Electronics Industries Ltd.

VERIFIED

The Elta EL/L-8202 advanced self-protection jammer pod mounted under the wing of an Israeli Air Force Kfir

EL/K-7200 DME/P-N airborne interrogator

Elta has been awarded a contract by the US Department of Transportation to develop DME/P-N airborne interrogators for MLS. Elta's EL/K-7200 Precision (P) and Normal (N) DME operates as a subsystem of the MLS, providing precision range and range rate information to the aircraft.

DME/P ground transponders will soon be available for installation in MLS equipment at selected major airports. In its N mode, the DME is compatible with existing transponders. Each DME/P-N includes an interrogator and controller unit for installation in the cockpit.

The EL/K-7200 utilises large-scale integration components and advanced mass production technologies. It rapidly acquires normal and precision ground transponders and remains locked-on during all flight manoeuvres including initial approach, final approach and missed approach through the implementation of proprietary algorithms and adaptive digital filters.

Operational status

FAA qualified.

Contractor

Elta Electronics Industries Ltd.

VERIFIED

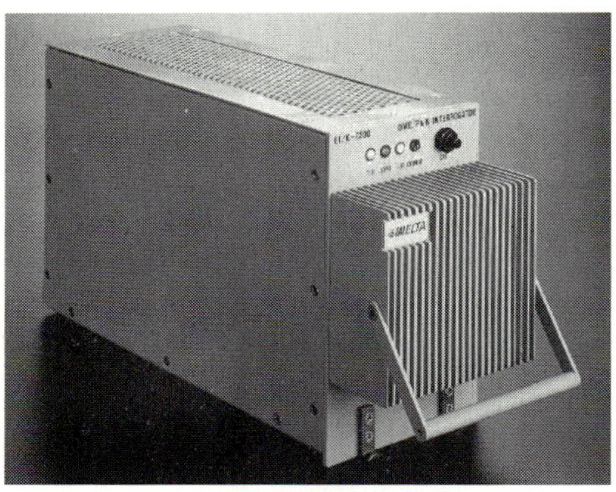

The Elta EL/K-7200 DME/P-N interrogator

EL/L-8202 advanced self-protection jamming pod

Specifically designed to counter Soviet radars operated by the Syrian forces during the 1982 Lebanon war, the EL/L-8202 combined receiver/jammer system is compatible with the IAI Kfir fighter and any similar aircraft able to accommodate the pod, which resembles that of the US Air Force AN/ALQ-131. F- to J-band (3 to 20 GHz) coverage is provided with high-power broad-band jammer outputs through either fore or aft antennas. Beam-shaping features are incorporated and the threat library plus jamming techniques and stored mission data can be reprogrammed on the flight line. There is an integral ram-air/liquid cooling system. The logic unit within the pod interfaces the jammer with remote radar warning receivers and cockpit displays. Other systems in the family are the smaller pod-mounted EL-8200 system which weighs 113 kg, but offers 80 per cent of the capability of Cerberus. Two versions are available, the EL-8222 and EL-8224, as well as the EL-8223 internal fit system.

Specifications

Dimensions: 2,900 × 260 × 390 mm
Weight: 200 kg
Power: 1.7-2.3 kVA
Frequency: F- to J-bands

Operational status

In service with the Israeli Air Force. A version known as Cerberus was developed with EADS (formerly DaimlerChrysler Aerospace) for German Air Force Tornado IDS aircraft.

Contractor

Elta Electronics Industries Ltd.

UPDATED

EL/L-8230 internal self-protection jammer

Suitable for IAI Kfir and Lockheed Martin F-16 sized aircraft, the EL/L-8230 self-protection system is designed to combat both surface and airborne threats. It operates across the G- to J-bands (4 to 20 GHz) with separate receiver and transmitter antenna groups and can generate noise or repeater jamming signals. It is designed to use standard avionics bay cooling air and, being fully contained within the aircraft, does not increase drag. The radio frequency unit, radio frequency power amplifier and logic unit are combined in one box.

Specifications

Dimensions: 250 × 240 × 600 mm
Weight: 42 kg
Power consumption: 1.8 kVA

Operational status

In service with the Israeli Air Force.

Contractor

Elta Electronics Industries Ltd.

VERIFIED

EL/L-8231 internal self-protection system

The EL/L-8231 is an internally mounted ECM set, operating in the H-, I- and J-bands, designed to protect combat aircraft or helicopters from attack by missiles with CW radar guidance. Incoming signals are automatically analysed and the appropriate jamming signal transmitted. The set also interfaces with the aircraft's RWR.

Specifications

Volume: 15 litres
Weight: 18 kg
Power consumption: (transmit) 600 VA

Operational status

In production and in service.

Contractor

Elta Electronics Industries Ltd.

VERIFIED

EL/L-8233 Integrated Self-Defence System (ISDS)

Elta's EL/L-8233 ISDS was specifically designed to meet the operational requirements of small, light fighter aircraft. Featuring a modular architecture, advanced technology, proven software and hardware, ISDS as developed for F-5E and F-5F aircraft includes: a Radar Warning Receiver (RWR); mini CW repeater jammer; and chaff/flare dispenser, all integrated through a MIL-STD-1553B databus.

The RWR provides warning of: pulse, pulse Doppler, CW and other exotic radar threats. It provides high probability of detection beyond lethal threat envelopes, in a very short cycle time, with positive identification based on threat frequency and time domain characteristics. It offers high integration with the active jamming system and automatic release of chaff and flares, together with recording of unidentified threats, and field programmability.

The CW repeater jammer provides self defence deception against enemy weapons that use pulse Doppler radar and semi-active radar missile seekers in conjunction with CW illumination. Both Velocity Gate Pull Off (VGPO), and angle deception techniques are exploited in the CW repeater. The system is programmable using ELINT data and Pre-Flight Message (PFM).

The chaff and flare dispenser is a smart, threat-adaptive design that can be tailored in capacity to meet customer requirements.

Specifications

RF coverage: 7-17 GHz (radar jammer)
ERP: 200 W (radar jammer)
Target coverage: 4 simultaneously (radar jammer)
Weight: 50 kg (radar jammer)

Operational status

Available, possibly ordered for Chilean Air Force F-5 aircraft.

Contractor

Elta Electronics Industries Ltd.

VERIFIED

EL/L-8233 Integrated Self-Defence System (ISDS)
0018251

EL/L-8240 self-protection system

The EL/L-8240 is an internally mounted self-protection system suitable for modern combat aircraft. It offers radar warning, jamming and deception over a large frequency range and wide angular coverage. It can be integrated with other avionics systems. A distributed processing system is used and the unit can be readily reprogrammed to meet changing threats.

Specifications

Weight: 136 kg
Power required: 7 kVA

Contractor

Elta Electronics Industries Ltd.

VERIFIED

EL/L-8260 COmprehensive Self-Protection Suite (COSPS)

The Elta EL/L-8260 COSPS is an active self-protection suite for helicopters. It detects, identifies and analyses threat, instantaneously alerting the air crew, to provide complete situational awareness. It integrates radar, laser and missile approach warning systems and activates appropriate countermeasures.

Contractor

Elta Electronics Industries Ltd.

VERIFIED

EL/L-8300 airborne SIGINT system

In operation in large aircraft such as the Israeli Air Force's converted Boeing 707s, the EL/L-8300 is a long-range highly sophisticated SIGINT system said to be capable of detecting communications and other electronic signals at ranges up to 450 km. Received and processed data can be transmitted to the EL/L-8353 ground command and control centre for further processing, evaluation and dissemination of data.

The EL/L-8300 incorporates the EL/K-7032 COMINT system (see earlier entry), the EL/L-8312A ELINT system (see later entry) and the EL/L-8350 command and analysis station on the aircraft. Training for operators of the EL/L-8300 can be undertaken on the L-8351 simulator and post mission analysis is done on the EL/L-8352 system.

Specifications

Frequency:
(ELINT) 0.5-18 GHz (0.03-40 GHz optional);
(COMINT) 20-1,000 MHz (2-1,500 MHz optional)
Coverage: 360° azimuth
Accuracy: ±3°
Library: 2,000 emitters (5,000 optional)

Operational status

In service in Israeli Air Force Boeing 707 aircraft, and incorporated in the Santiago Boeing 707 ECM aircraft of the Spanish Air Force. It is also reported that the South African Air Force operates at least one Boeing 707 with a version of the EL/L-8300 SIGINT system.

The EL/L-8300 is also being provided to Lockheed Martin Fairchild Systems for integration into the DASS for the UK Nimrod MRA.4; in this role it will function as an ESM.

Contractor

Elta Electronics Industries Ltd.

VERIFIED

EL/L-8303 ESM system

The EL/L-8303 is a computerised ESM system for shipborne, airborne and ground-based applications. It is designed to intercept, digitise, identify, display and record radar signals. The processed data can be examined by an operator or used to activate automatic ECM or chaff systems. The system incorporates computer control and processing features. It has an omnidirectional instantaneous direction-finding capability which is claimed to be highly sensitive and accurate with a high intercept probability across a wide bandwidth. Alphanumeric keyboard inputs and graphical display output of data are standard features and the operator can access part of the emitter library. The system also includes built-in test circuits.

Intercepted radar signals are measured and classified according to pulsewidth, frequency, azimuth, power and time of arrival and are then fed to the processor. This conducts signal sorting, recognition functions and emitter identification and produces graphical data outputs. A modular system configuration permits optimisation of the system to meet customer cost and performance requirements.

Specifications

Power supply: 115 V AC, 400 Hz, 4 kVA
Frequency: 2-18 GHz (optional down to 0.5 GHz)
Azimuth accuracy: 5° RMS (3° option)
Max pulse rate: 500,000/s
Sensitivity:
(wideband) –60 dBm
(narrowband) –80 dBm
Dynamic range: 60 dB
Reaction time: 1 μs (typical)
Installation: two 19 in (487 mm) racks
Environmental: MIL-E-5400 and MIL-E-16400

Operational status

In production.

Contractor

Elta Electronics Industries Ltd.

VERIFIED

EL/L-8222 self-protection jamming pod

The Elta EL/L-8222 radar self-protection, power-managed, jamming pod is designed for fitment to fighter aircraft. It is described as covering the RF band 6 to 18 GHz, with the ability to jam four targets simultaneously, using look-through techniques. Flightline reprogramming and control of mission data, together with, modular construction, integral ram air cooling and MIL-STD-1553B/RS-422 interface control, are all claimed to provide ease of installation and use, together with high MTBF.

Elta's EL/L-8222 self-protection jamming pod under test on an F-16 ***1999***

Specifications

RF band: 6-18 GHz
Jammer ERP: 200 W
Target coverage: 4 simultaneously
Weight: 120 kg
Dimensions: 2,540 × 254 × 254 mm

Operational status

Being marketed for fighter aircraft.

Contractor

Elta Electronics Industries Ltd

VERIFIED

EL/L-8312A ELINT/ESM system

The EL/L-8312A system forms a part of the EL/L-8300 SIGINT set (see earlier entry) or can operate alone, covering the 0.5 to 18 GHz frequency band to detect, analyse and identify signals out to the radar horizon. Direction-finding is fast and accurate and bearings are stored in the associated computer and correlated with aircraft navigational data to give a display of the actual location of selected transmissions on a colour graphic console. The EL/L-8312A incorporates the EL/L-8312R receiver, EL/L-8320 signal parameters measurement equipment and the EL/L-8610 computer.

Operational status

In production and service with the Phalcon and a variety of other aircraft. In June 1991, it was announced that Singapore had awarded a US$20 million contract for EL/L-8312A systems for four Enforcer Mk 2 maritime patrol aircraft. The EL/L-8312A has also been supplied for retrofit to Royal Australian Air Force P-3C Orions.

An enhanced version of the EL/L-8312A ESM is being offered, together with an enhanced variant of the Phalcon Airborne Early Warning (AEW) radar, by the joint Raytheon Electronic Systems/Elta Electronics

The Elta EL/L-8300 SIGINT system in operation in an Israeli Air Force Boeing 707 aircraft. Also installed on Boeing 707 ECM aircraft of the Spanish Air Force

The Elta EL/L-8300 SIGINT system in operation in an Israeli Air Force Boeing 707 aircraft. Also installed on Boeing 707 ECM aircraft of the Spanish Air Force

Industries Ltd consortium that is marketing worldwide its Airborne Early Warning and Command and Control (AEW&C) systems.

Contractor
Elta Electronics Industries Ltd.

VERIFIED

EL/M-2001B radar

The EL/M-2001B is a range-only I/J-band radar for single-seat tactical aircraft operating in air-to-air and air-to-ground modes. The target is detected visually while acquisition and tracking is accomplished automatically by the radar. The system can operate in heavy ground clutter. Information from the radar can be displayed on the head-up display or fed into the weapon control computer for weapon delivery computation. The six LRUs are based on solid-state technology, with the exception of the travelling wave tube, and have considerable reserves for future growth.

Specifications
Dimensions:
(diameter) 450 mm
(length) 790 mm
Antenna diameter: 195 mm
Weight: <50 kg
Power supply: 115 V AC, 400 Hz, 3 phase, 1 kVA, DC 30 W

Operational status
In service with IAI Kfir fighters of the Israeli Air Force.

Contractor
Elta Electronics Industries Ltd.

VERIFIED

The Elta EL/M-2001B radar on an Israeli Air Force Kfir aircraft

EL/M-2022A maritime surveillance radar

The EL/M-2022A multimode maritime surveillance radar is designed for fixed- or rotary-wing aircraft. The main missions for the radar are maritime target surveillance, ASW, search and rescue and economic zone control. Modular hardware design, software/remote control and a flexible avionic interface ensure easy installation in a variety of airborne platforms. The radar features multiple Track-While-Scan (TWS) for up to 100 targets, expand and freeze capabilities and sector or full-scan coverage. Optional features include integral IFF compatibility and air-to-air detection.

Operational modes consist of long-range sea surveillance, small target detection for periscopes based on a high-resolution waveform, high-resolution mapping based on Doppler beam-sharpening, moving target indication, navigation and weather capabilities, and SAR/ISAR/range signature classification.

The EL/M-2022A consists of an ultra-low sidelobe planar-array antenna, two-axis electric drive system and coherent TWT-based transmitter. The receiver/processor features a wide dynamic range receiver and programmable signal processor.

Three basic configurations are available: EL/M-2022A(V)1, EL/M-2022A(V)2 and EL/M-2022A(V)3.

The radar is suitable for both fixed- and rotary-wing aircraft.

EL/M-2022A(V)1 is a lightweight radar for small airborne platforms such as helicopters and UAVs, with medium-range detection and a 50 target TWS capability. Typical detection ranges in Sea State 3 are

LRUs of the EL/M-2022A(V)2 maritime surveillance radar

130 km against a small ship. Weight of the EL/M-2022A (V)1 is 65 kg.

EL/M-2022A(V)2 is a version with a long-range periscope detection capability. Typical detection ranges in Sea State 3 are 55 km on a periscope and 130 km on a small ship. The EL/M-2022A(V)2 weighs 86-95 kg.

EL/M-2022A(V)3 is a version with long-range classification, periscope detection and a 100 target TWS capability. Detection ranges in Sea State 3 are 55 km on a periscope and 130 km on a small ship. The EL/M-2022A(V)3 weighs 95-103 kg.

Operational status
The EL/M-2022A(V)3 is in production for various customers. The radar has been selected for the Australian P-3C upgrade programme.

Contractor
Elta Electronics Industries Ltd.

VERIFIED

EL/M-2032 radar

The EL/M-2032 is an advanced pulse Doppler multimode fire-control radar designed for multimission fighters, for both air-to-air and air-to-ground missions.

Modular hardware design, all-software control and flexible MIL-STD-1553B avionic interface ensure that the radar can be installed in various fighter aircraft such as the F-4, F-5, F-16, Mirage and MiG-21 and customised to meet specific requirements. Antenna size can be adapted to the space available in the aircraft nose.

In air-to-air operation the radar offers long-range target detection, automatic target acquisition in close combat situations, single-target track for weapon delivery and track-while-scan.

In air-to-ground missions the radar provides air-to-ground ranging, real-beam map, ground moving target

The Elta EL/M-2032 installed in MiG-21

indication, sea search, Doppler beam-sharpening for high-resolution mapping, terrain-avoidance and beacon modes, and a look-down/shoot-down capability.

The radar consists of the antenna, transmitter and receiver/processor.

The planar-array antenna features ultra-low sidelobes and two-axis monopulse operation.

The transmitter is a TWT coherent transmitter.

The receiver/processor includes a programmable signal processor which provides full software control of the system.

Specifications
Weight: 95-105 kg, depending on antenna
Power consumption: 2 kW
Range: 37-75 km on small fighter aircraft

Operational status
More than 250 EL/M-2032 radars have been installed or contracted as part of eight upgrade programmes, including retrofit to Chilean F-5E, Romanian MiG-21, Lancer, and Turkish F-4 Phantom 2000 aircraft. Configurations are also being developed for the Northrop Grumman F-5 Plus package, and for F-16 aircraft.

Contractor
Elta Electronics Industries Ltd.

VERIFIED

EL/M-2060P airborne Synthetic Aperture Radar (SAR) pod

The EL/M-2060P is an innovative, field-proven, high-performance, pod-mounted SAR reconnaissance system, that can be fitted to a standard stores position on the centreline station of modern fast jet aircraft such as F-16, F-18, Gripen and Tornado.

The EL/M-2060P is completely autonomous, self-contained, all-weather, day and night high-resolution reconnaissance radar system. ELTA claims that it produces images that approach photographic resolution. It is able to maintain capability in poor visibility conditions, under smoke and cloud cover and against a wide variety of man-made camouflage.

The EL/M-2060P synthetic aperture radar system concept **2000**/0064379

The system consists of a detachable, pod-mounted, externally carried, SAR; a Ground Exploitation System (GES); and an in-built bidirectional datalink.

The collected SAR imagery and data undergoes on-board, on-line, real-time processing and is transmitted to the GES for further automated interpretation. As an alternative or parallel mode, the collected data is recorded on board for retransmission or later interpretation on the ground.

Operation of the pod is highly automated, minimising the load on the pilot. In a typical fast jet installation, the EL/M-2060P system provides real-time collection and interpretation of intelligence data, with the ability to provide surveillance out to 120 km range, giving total coverage of more than 50,000 km^2/h.

Modes of operation include:

(1) Strip mode: strip mode provides fast coverage of large areas at standoff ranges, and mapping in sufficient detail for target detection and overall assessment of an entire region;

(2) Spot mode: spot mode provides detailed examination of designated areas of interest and high resolution for target classification;

(3) Strip/Ground Moving Target Indication (GMTI) mode: strip/GMTI mode highlights moving targets within the SAR strip.

The integral datalink belongs to Elta's EL/K-1850 microwave datalink family and operates in C- or I-/J-band; it provides a full duplex capability and line of sight operation.

The GES integrates Commercial-Off-The-Shelf (COTS) technology with advanced data processing software to perform computer-aided data exploitation and intelligence reporting.

Specifications

Pod weight: 590 kg
Aircraft interfaces:
(power) 4.3 kW maximum (3.5 kW typical)
(cooling) self sufficient
(video output) RS-170
(control) MIL-STD-1553

Operational status

ELTA and Lockheed Martin Tactical Aircraft Systems have signed an agreement to co-operate in the sale of the EL/M-2060P SAR in the USA and the rest of the world. The USAF has performed demonstration trials aboard an F-16 aircraft.

An upgraded version of the EL/M-2060P SAR reconnaissance pod is being offered to the US Navy for its SHAred Reconnaissance Pod (SHARP) requirement. The upgraded pod, known as ELMSAR, is intended to increase F/A-18E/F reconnaissance capability in the 2003 timeframe as the F-14 equipped with the TARPS reconnaissance pod leaves service. The ELMSAR differs from the original EL/M-2060P in replacing the Israeli-designed datalink with multiple alternative datalinks of US origin. Lockheed Martin Tactical Aircraft Systems will be responsible for the post-processing aspects of the proposed upgrade system.

Contractor

Elta Electronics Industries Ltd.

VERIFIED

EL/M-2060P podded SAR/GMTI system

The EL/M-2060P is a Synthetic Aperture Radar (SAR)/ Ground Moving Target Indicator (GMTI) podded design, intended for use on fighter aircraft in the airborne reconnaissance role.

The system is reported to offer wide area strip, high-resolution spot, and moving target indicator modes, and to be essentially self-contained, incorporating an internal Ground Position System (GPS) capability for position determination, and a datalink for mission reports.

System weight is given as 590 kg.

Elta also offer a derivative system, designated EL/ M-2055, which only weighs 50 kg. This is believed to be for operation with an unmanned air vehicle, and presumably has much reduced capability, in view of the very significant weight reduction.

Operational status

Elta and Lockheed Martin have signed a co-operation agreement to market the EL/M-2060P system.

The Elta EL/M-2060P synthetic aperture radar pod exploded view **2000**/0064380

The EL/M 2060P synthetic aperture radar pod installed on an F-16 aircraft **2001**/0067727

Contractor

Elta Electronics Industries Ltd.

VERIFIED

EL/M-2075 Phalcon AEW radar

The EL/M-2075 Phalcon is a solid-state D-band conformal array radar system for use on a Boeing 707 and other aircraft. Phalcon is intended for airborne early warning, tactical surveillance of airborne and surface targets and intelligence gathering. It will also have the command and control capabilities needed to use this information.

The system uses six panels of phased-array elements: two on each side of the fuselage, one in an enlarged nosecone and one under the tail. Each array consists of 768 solid-state transmitting and receiving elements, each of which is weighted in phase and amplitude. These elements are driven by individual modules and every eight modules are connected to a transmit/receive group. Groups of 16 of these eight module batches are linked back to what is described as a prereceive/transmit unit, and a central six-way control is used to switch the pretransmit/receive units of the different arrays on a time division basis.

Each array scans a given azimuth sector, providing a total coverage of 360°. Scanning is carried out electronically in both azimuth and elevation. Radar modes include high PRF search and full track, track-while-scan, a slow scan detection mode for hovering and low-speed helicopters (using rotor blade returns) and a low PRF ship detection mode. These modes can be interleaved to provide multimode operation in any scanning sector. Typically, 2 to 4 second scan rates are used in high-priority sectors and 10 to 12 second rates in low-priority sectors.

Operational status

Development has been completed and Phalcon has been delivered to the Chilean Air Force; by October 1998 the system is reported to have performed more than 100 missions and over 500 operational flight hours.

In October 1998, Elta Electronics Industries teamed with the Raytheon Company (Greenville) to co-operate, on a worldwide basis, in the development, production and marketing of Airborne Early Warning and Command and Control (AEW&C) systems, based on Elta's Phalcon radar and EL/L-8312A ESM expertise and Raytheon's integration, communications and mission control expertise. The aircraft selected to form

The Elta Phalcon AEW radar is designed for installation in a Boeing 707

the basis of the system is the Airbus Industrie A310-300 wide-body twin-jet.

It was also reported that an Antonov Il-76 aircraft, belonging to the People's Liberation Army Air Force (PLAAF) of China was delivered to Israel in 1999 for installation of the EL-2075 Phalcon system. It is understood that Israel subsequently withdrew from negotiations with China over production of the system, possibly under pressure from the US.

Contractor
Elta Electronics Industries Ltd.

UPDATED

EL/M-2160 missile approach warning system

The EL/M-2160 missile approach warning system provides warnings of missile attack and automatically activates chaff and flare dispensers to protect the aircraft. The EL/M-2160 is an all-solid-state pulse Doppler radar. It provides time-to-impact and direction information to enable timely and effective response.

The equipment consists of the Transceiver Processing Unit (TPU), RF head, antennas and cockpit control unit.

Specifications
Dimensions: (TPU) 508 × 254 × 216 mm
Weight: 34 kg (TPU and RF head)
Power consumption: 400 W

Operational status
The system has been designed for F-16 aircraft and adapted for transport aircraft and helicopters. The system is currently in service with several European and Asian air forces on a variety of aircraft and helicopters including C-130, C-160, Fokker, Antonov-32, Bell-212 and Mi-17. The system has been successfully tested against live missiles firing and has proved effective in protecting the platforms in several missile firing engagements.

Development of a low weight (15 kg) modern pulse Doppler radar missile approach warning system is in progress.

Contractor
Elta Electronics Industries Ltd.

VERIFIED

Elta's EL/M-2160 missile approach warning system ***1999***/0054165

NTS/NTS-A Night Targeting System

Since 1982, Tamam has been involved in upgrade programmes to provide Cobra attack helicopters with laser ranging, designation and night attack capabilities. Developed by Tamam under sole-source contract from the US Marine Corps and Israeli Air Force, NTS has accumulated considerable operational use.

Night Targeting System (NTS-A) is a new version of the NTS, with enhanced performance suitable for a wide range of different attack helicopters (including the AH-1 W/S/P Cobra). In NTS-A, the original optical tube is removed, freeing space in the helicopter cockpit, for the use of flat-panel displays, and allowing the system to be compatible with different types of attack helicopter, and with modern glass cockpit architectures, such as the US Marine Corps AH-1W 4BW upgrade programme.

NTS-A has the following capabilities:
Installation flexibility with TOW I, TOW II, Hellfire, and other weapons;
Target acquisition during day, night and limited visibility conditions;
Laser Designator and Range-finder System (LDRS) for laser-guided weapons and for measuring target ranges;
TV Tracker (TVT), providing target auto-tracking during day and night;
Guidance for all types of TOW missiles;
Fully-automatic in-flight boresight capability;
Navigation and tactical data display;

NTS-A night targeting system-A 0015363

Display of NTS-A standard video signal (both the FLIR and TV camera pictures) on a multifunction display in both the gunner and pilot cockpits;
Built-in growth potential for future integration with other systems onboard, via (2)MIL-STD-1553B databusses.

Specifications
FLIR - 2nd generation
FLIR Fields of View (FOV) and magnification (relative to display viewing angle of 40 × 25°):

Wide FOV (H × V)	18.0 × 24.0°	× 2.0
	30.0 × 40.0°	An option for pilotage
Medium FOV (H × V)	5.2 × 3.9°	× 9.0
Narrow FOV (H × V)	1.47 × 1.1°	× 32.1
Zoom NFOV (H × V)	0.73 × 0.55°	× 64.3

Automatic TV Tracker (TVT) - with prediction, adjust and offset modes:
Field of regard:
(azimuth) 90° to the left, 95° to the right
(elevation) up 30°, down 60°
Turret slew rate:
(low-angular velocity) 2°/s
(high-angular velocity) 90°/s
(acceleration) 60°/s³

TVC features:
(frame transfer Charged Couple Device (CCD)), ½ in
(number of pixels): 780 × 576
Missile interfaces available:
Hellfire missile; Rafael NT-D missile
all types of TOW missiles (TOW, TOW-1, TOW 2A, TOW 2B)
Weight: 129 kg (excluding aircraft installation kit)
Power:
(average power) 400 W
(max power) (during laser operation) 650 W

Operational status
In production. Over 300 systems have been ordered.

Contractor
Israel Aircraft Industries Ltd, Tamam Division.

VERIFIED

TN-90 compact inertial navigation system

The TN-90 is a small lightweight low-cost strapdown inertial system customised according to mechanical and electrical requirements.

Typical applications are for guided and smart missiles and bombs, unmanned air vehicles,

NTS-A mounted on Cobra AH-IS helicopter 0015362

helicopters, electro-optic and targeting systems, instrumentation pods, flight control and as a back-up for the INS.

Specifications

Dimensions: 128 × 127 × 119 mm typical
Weight: 2.4 kg
Outputs: linear acceleration, velocity and angular rates, position, azimuth and Euler angles
Interface: multiple RS-422 serial, MIL-STD-1553B mux-bus electrical
Accuracy (with CA code GPS):
(position) 100 m SEP
(velocity) 0.1 m/s
(attitude) 0.2°
(azimuth) 0.2-0.4°

Operational status

In production.

Contractors

Israel Aircraft Industries Ltd, Tamam Division.

VERIFIED

TN-90Q/G compact inertial navigation system

Tamam's TN-90Q/G is a compact lightweight unit that integrates strapdown, inertial and GPS navigation. The system incorporates into the package, a proven off-the-shelf GPS receiver tightly coupled with an inertial measurement unit.

The system provides positioning, velocity, time, attitude, heading, angular rate and acceleration.

Specifications

Dimensions: 209.4 × 114.2 × 93.9 mm
Weight: 3.5 kg
Power: 45 W typical, 28 V DC
Interfaces: RS-422 or MIL-STD-1553B mux-bus
Accuracies:

	C/A code GPS	with P(Y) code
Position	120 m	20 m
(after 5 min)	350 m	250 m
Velocity	0.6 m/s	0.1 m/s
Attitude	0.1°	0.03°
Azimuth	0.2°	0.1°

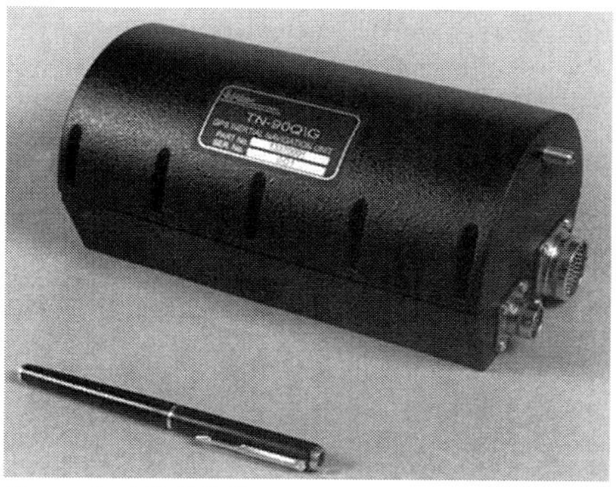

Tamam's TN-90Q/G compact inertial navigation system 0015364

Operational status

In production.

Contractors

Israel Aircraft Industries Ltd, Tamam Division.

VERIFIED

TR-90 compact inertial navigation system

Tamam has developed and produced a family of small lightweight low-cost strapdown inertial reference units, customised according to mechanical and electrical requirements.

Typical applications are for guided and smart missiles and bombs, unmanned air vehicles, optronic and targeting systems, instrumentation pods, antenna stabilisation, flight control and as a back-up navigation system.

Specifications

Dimensions: 128 × 127 × 82 mm typical
Weight: 1.8 kg typical
Outputs: velocity and angular increments, linear acceleration, angular rates
Interface: multiple RS-422 serial

Operational status

In production.

Contractor

Israel Aircraft Industries Ltd, Tamam Division.

VERIFIED

Guitar-350 passive missile warning system

The Guitar-350 is a lightweight totally passive, autonomous warning system which detects missiles by sensing the electro-optical emissions from their engines. The system provides audio and visual alarms, allowing time for the aircraft to activate its defensive systems.

Guitar-350 is based on a patented sensor and detection algorithm, which enables it to identify the type of missile, track its trajectory and transfer data on time-to-impact for automatic control of the infra-red countermeasures subsystem. It emits no electromagnetic or electro-optical signals, which might be detected. The system is designed to detect ground-to-air and air-to-air missiles and has an extremely low false alarm rate.

Guitar-350 features coverage of all possible attack angles and has high discrimination against background interference. It withstands adverse environmental conditions and is compatible with cockpit displays.

Specifications

Azimuth coverage: 360°
Elevation coverage: 120°
Warning time: 4-6 s
Weight: less than 15 kg
Power: less than 200 W
Interface: MIL-STD-1553B

Operational status

Test flown. Suitable for helicopters, fighter and transport aircraft. Guitar 350 supersedes the earlier Guitar 300 system.

Guitar 350 passive missile warning system detector ***1999***/0051256

Contractor

Rafael Electronic Systems Division.

VERIFIED

Kingfisher ESM system

Kingfisher is an ESM system which is designed to provide fast and accurate surveillance of the electronic battlefield over the 2 to 18 GHz frequency range. It may be installed in fixed-wing aircraft or helicopters. Using broadband, fast response Instantaneous Frequency Measurement (IFM) and Instantaneous Direction-Finding (IDF) receivers, Kingfisher provides almost 100 per cent probability of radar pulse intercept. It is designed with modularity, allowing for customised hardware and retrofit into existing systems, and has considerable growth potential.

Kingfisher is able to operate successfully in dense electromagnetic environments. It receives signals over

Kingfisher ECM system 0018286

a wide range and amplitude simultaneously, evaluating parameters such as frequency, direction, pulsewidth, pulse type and time of arrival. The system then activates filtering and correlation algorithms, after which the data are further processed for additional information such as

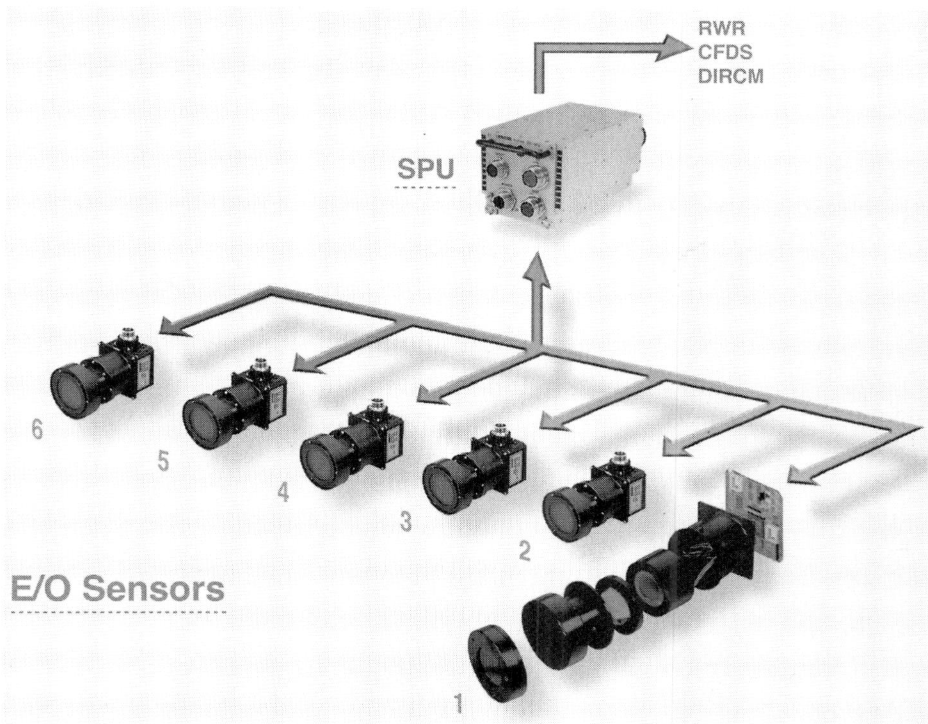

Guitar 350 system configuration ***1999***/0051257

PRI, frequency range and scan and emitter antenna scan.

If required, Kingfisher can carry out automatic activation of passive and active countermeasures to form a fully power-managed EW suite. In addition, it will support the targeting of optically or thermally guided or unguided weapons.

Specifications

Frequency: 2-18 GHz (0.5-40 GHz optional)
Coverage: 360° azimuth, 40° elevation
Accuracy:
(frequency) ±1.5 MHz RMS
(bearing) 1-2°
Sensitivity: –60 dBm

Operational status

Fully developed and tested in real scenarios, in simulators and on EW ranges.

Contractor

Rafael Electronic Systems Division.

VERIFIED

3-D GPS attitude determination receiver

The Rokar 3-D GPS attitude determination receiver is a 6 to 24 channel receiver with one to four antenna inputs, designed to provide 360° coverage in high-dynamic airborne scenarios.

Each antenna is connected to 6 to 12 channels depending upon the configuration and application.

This line of receiver products provides ready-to-use solutions for attitude determination and for antenna coverage, eliminating the need for specialised GPS antennas.

Key features are fully spherical antenna coverage with no SNR degradation; attitude determination; off-the-shelf, low-cost antennas to replace wraparound antennas; differential operation; high-dynamic scenarios operation; INS integration; multipath and spoofing rejection.

Contractor

Rokar International Ltd.

VERIFIED

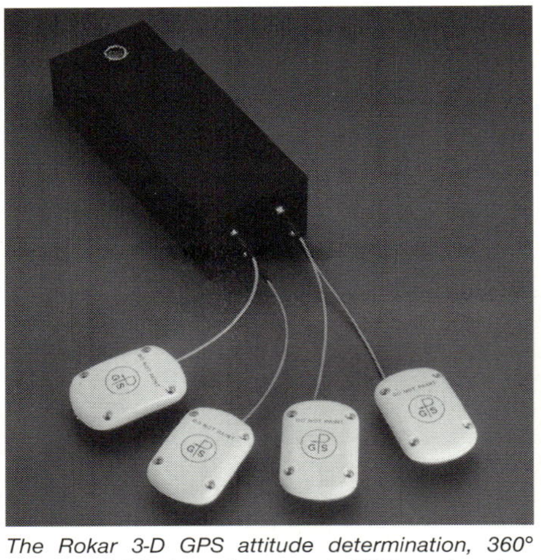

The Rokar 3-D GPS attitude determination, 360° coverage, high-dynamics scenario GPS receiver, four antenna configuration 0015359

Advanced Digital Dispensing System (ADDS)

The Advanced Digital Dispensing System (ADDS) is a computer-controlled, threat adaptive countermeasures dispensing system. It is designed to protect aircraft from both ground and air threats by dispensing decoy payloads. The system is configured for high-performance aircraft, helicopters, transport aircraft and maritime patrol aircraft.

ADDS can dispense chaff, flares and RF expendables in single shot or multiple simultaneous dispense programmes to provide multispectral responses that best counter the threat.

ADDS incorporates improvements derived from combat experience and hundreds of installations on a wide variety of aircraft, including the A-4, AH-64, F-4, F-5, F-15 and F-16.

Operational status

In service, with several ADDS delivered worldwide.

Contractor

Rokar International Ltd.

VERIFIED

GPS NAVPOD rugged receiver

The Rokar GPS NAVPOD is an 'all-in-view', parallel tracking, 12-channel GPS receiver, designed for demanding pod and combat aircraft applications.

It is a fast reacquisition C/A code receiver, which processes navigation data signals transmitted from all satellites in view. In addition to regular navigation data, the NAVPOD also outputs pseudo ranges, delta ranges, system time, measurement quality and other data required for coupling with Inertial Navigation Systems (INS).

The Rokar GPS NAVPOD high-dynamics, rugged GPS receiver 0015360

The system is capable of operating in differential navigation mode, and of receiving/producing differential GPS corrections.

The receiver incorporates the latest in digital ASIC design and Receiver-On-Chip (ROC) technology.

Contractor

Rokar International Ltd.

VERIFIED

GPS SWIFT high-velocity, high- acceleration receiver

The Rokar GPS SWIFT is an 'all-in-view', parallel tracking, 12-channel GPS receiver. The GPS SWIFT is intended for use in platforms that travel at high velocity and high accelerations.

The GPS SWIFT provides accurate position, velocity, acceleration and acceleration-rate (jerk), as well as time and other GPS data. The system is capable of operating in differential navigation mode, and of receiving/producing differential GPS corrections. It is also designed for tight integration with other sensors such as Inertial Navigating Systems (INS).

Key features are high-velocity (2,000 m/s) and high-acceleration (15 *g*) operation; provision of navigation data including acceleration and jerk; INS integration.

Contractor

Rokar International Ltd.

VERIFIED

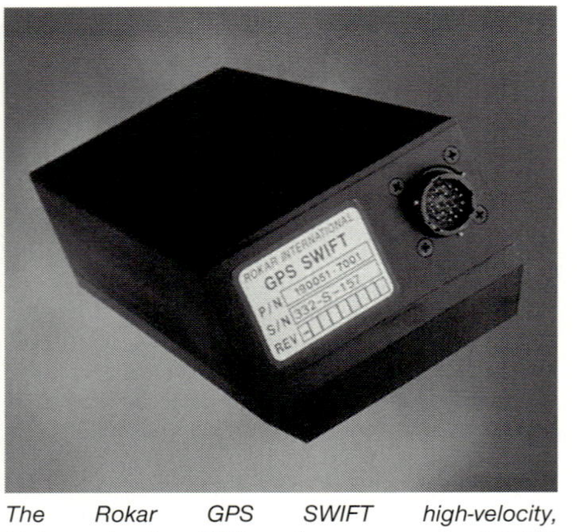

The Rokar GPS SWIFT high-velocity, high-acceleration, GPS receiver 0015361

CDF-3001 airborne V/UHF COMINT/DF system

CDF-3001 is an airborne self-contained COMINT/DF system designed to acquire, monitor and measure the Direction Of Arrival (DOA) of tactical radio transmissions. The operational system is driven by a Windows-based man/machine interface, and advanced hardware implementation provides fast scan and DF capability. The system is linked to the host aircraft navigation and command and control system via an RS-232 interface. The CDF-3001 system is suitable for both fixed- and rotary-wing application.

Specifications

Frequency coverage: 20-500 MHz (optionally 1.5-1,000 MHz)
Monitored signals: AM and FM

Contractor

Tadiran Electronic Systems Ltd.

VERIFIED

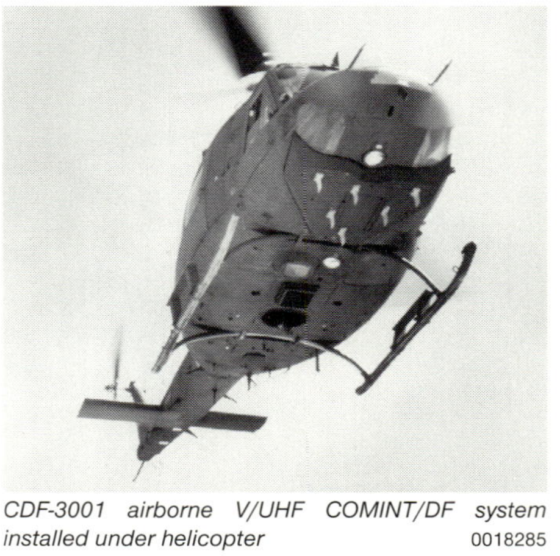

CDF-3001 airborne V/UHF COMINT/DF system installed under helicopter 0018285

TACDES SIGINT system

TACDES is an airborne SIGINT system consisting of the RAS-1B or RAS-2A ELINT system, the ACS-500 automatic VHF/UHF COMINT system and the TDF-500 automatic direction-finding unit integrated with a computerised C³I system. TACDES is installed on large transport aircraft such as the Boeing 707 or Lockheed Martin C-130 and performs real-time acquisition, location, processing and reporting of communications and radar signals between 20 MHz and 18 GHz.

Accurate instantaneous wide-angle coverage is provided by interferometric COMINT and ELINT DF systems. The system features an enhanced command and control capability and real-time secure voice and data communications.

ACS-500 Automatic VHF/UHF COMINT System

The ACS-500 is an airborne self-contained system designed for acquisition and monitoring of tactical radio nets which operate in the 20 to 500 MHz frequency band. The system features spectrum scanning in programmable frequency sectors and provides active display of up to 100 preprogrammed frequencies with identification, direction, signal level and time of reception data, and fully automatic direction-finding. It has an operator oriented hands-off monitoring capability and built-in test equipment.

Operation has been simplified by adopting a user-friendly man/machine interface.

Main functions are to provide: spectrum scanning from 20 to 500 MHz, with reporting of new active, previously unknown or unrecognised stations; scanning of known frequencies and updating operators' activity displays; and a monitor, analysis and recording facility for any active stations.

Specifications

Frequency: 20-500 MHz (1.5-1,000 MHz optional)
Preprogrammed frequencies: 100
Accuracy: 2.5° RMS typical
Modes: AM, FM

RAS-1B ELINT and ESM system

The RAS-1B airborne interferometric ELINT system, installed on large transport aircraft, covers the 0.7 to 18 GHz frequency band (0.5 to 40 GHz optional) to obtain the enemy Electronic Order of Battle (EOB). It can measure frequency, pulse repetition frequency, amplitude and Direction Of Arrival (DOA) data on an average of two emitters per second, with exceptionally high accuracy. Operation has been optimised for weak emitter recognition in high-density electromagnetic environments and an activity file is maintained which contains identification, classification and updated data on recognised threats. A combination of wide bandwidth acquisition IFM receiver and a narrowband superheterodyne analysis receiver is used to ensure 100 per cent probability of detection and high system throughput. Operation can be fully automatic or man/machine oriented in simple manual mode to accommodate up to two operators.

The RAS-1B ELINT and ESM system operators' consoles

Specifications

Frequency: 0.5-18 GHz (0.5-40 GHz optional)
Frequency resolution: 1.25 MHz
Frequency accuracy: 2.5 MHz
DOA accuracy:
(from 2-18 GHz) 0.7° (typical), 0.3° (optional)
(from 0.5-2 GHz) 1.3° (typical), 0.5° (optional)
PRI resolution: 100 ns
PRI accuracy: 200 ns
Instantaneous bandwidth:
(analysis) 40 MHz
(acquisition) 8 GHz
Sensitivity: –70 dBm typical
De-interleaving capability: Up to 8 pulse trains in the same frequency/DOA cell

RAS-2A ELINT and ESM system

The RAS-2A airborne, wingtip-mounted, interferometric tactical ESM system is deployed on small- and medium-size transport aircraft. In the 0.5 to 18 GHz frequency band it can measure frequency, PRI, amplitude and direction of arrival data with excellent accuracy, with 360° azimuth coverage. Operation has been optimised for a single operator who can both display an EOB scene and passively target an anti-radiation or anti-ship missile. The high sensitivity of the system ensures weak signal processing even at long ranges. Automatic file activity is maintained for quick identification and classification of threats.

The RAS-2MT is the configuration for maritime surveillance and targeting.

Specifications

Frequency: 0.5-18 GHz
Frequency resolution: 0.2 MHz typical
Frequency accuracy: 0.4 MHz typical
DOA accuracy:
(left/right) 0.8°
(fore/aft) 5°
Coverage:
(azimuth) 360°
(elevation) ±10°
Sensitivity: –75 dBm
Instantaneous bandwidth: 1, 5, 15, 40 MHz

TDF-500 VHF/UHF automatic direction-finding system

The TDF-500 is an airborne automatic direction-finding system using a powerful central computer, high-performance twin-channel receiver and antenna array mounted on the aircraft wings and under-surfaces. It covers the 20 to 500 MHz band and can determine direction of arrival to 1.5°. The antenna ports connect to an RF switch matrix which selects antenna pairs and feeds them to the twin-channel receivers. In each sampling period a phase measurement is performed between one antenna pair and successive phase measurements are processed to obtain direction of arrival information.

Specifications

Frequency: 20-500 MHz
Power consumption: 400 W
Response time: less than 300 ms
Modulation: AM, FSK, PSK SSB
Accuracy: 2° RMS between 30 and 500 MHz
4° RMS between 20 and 30 MHz
Coverage:
(azimuth) 360°
(elevation) –2 to –7°
Polarisation: vertical

Operational status

In production for use on large transport aircraft that are required for special surveillance tasks.

Contractor

Tadiran Electronic Systems Ltd.

VERIFIED

ASARS Airborne Search And Rescue System

The Tadiran Spectralink Airborne Search and Rescue System is used in the rescue of downed aircrew by an airborne platform, special forces pick up, or drop zone marking. It facilitates rapid rescue under adverse conditions and is designed for use both in combat and peacetime. Both covert and standard beacon mode operation are possible, as well as voice communication.

ASARS consists of the ARS-700 airborne system and the PRC-434A radio communicator. The system utilises advanced range measurement technology, enabling accurate and quick survivor location in bad weather and difficult or hostile terrain.

First pass pick up is made possible by the accurate azimuth and range measurements provided by the lightweight airborne system. A beacon mode supplements the rescue radio's special transponder capabilities, and downed aircrew survival is enhanced by the fallback voice communication capability on every channel.

Short burst-type interrogations (600 ms) and special modulation techniques eliminate the dangers of repeated or continuous transmissions being picked up by enemy forces. Ten preprogrammed frequency channels, out of 3,000 operating frequencies, further enhance security. Up to a million call codes allocated to every radio transponder enable selective interrogation, minimising the number of transmissions. Several survivors can be handled by one airborne system, with each one clearly identified. If the selective call code of the transponder radio is unknown, it can be automatically retrieved by the airborne system via secure algorithms, enabling the successful completion of the rescue mission.

The transponder radio is activated automatically upon bail-out, and it can be easily operated with one hand by a disabled survivor. The radio is exceptionally rugged, compact and lightweight, and simple to use. It is powered by a durable lithium battery that provides a long operating life.

The transponder radio is compatible with both the Cubic AN/ARS-6V and the Tadiran ARS-700, and operates over the 225 to 300 MHz frequency range, with an effective range of 200 km. The beacon mode and default voice capabilities are compatible with other rescue systems and techniques.

ASARS can be installed in all types of fixed-wing aircraft and helicopters, without the need for any aircraft modification. It is easy both to install and remove; re-installation takes less than 30 minutes. The compact control unit mounts in the standard radio slot on the instrument panel. System parameters are easily programmed using a standard PC interface.

The Tadiran Spectralink ARS 700 airborne search and rescue system

DF accuracy is ±2.5° RMS. Range measurement accuracy is 50 m.

Specifications

ARS 700 airborne system
Frequency coverage: 225-300 MHz

Effective range: 200 km
Final approach accuracy: 20 m
RF channels: 3,000
Range measurement accuracy: 50 m
DF accuracy: 5° rms
Interrogation cycle: 600 ms
Communication: 2-way voice on every channel
Dimensions:
(avionics unit) 190 × 195 × 380 mm
(control display unit) 100 × 127 × 130 mm
(antenna switching unit) 50 × 150 × 150 mm
(remote display unit) 76 × 76 × 50 mm
Weight: 18 kg
Power supply: 28 V DC, 10 A
Power output: 20 W peak (20 W, 2 W, 0.4 W PEP)

PRC 434 radio transponder
Dimensions: 186 × 74 × 38 mm
Weight: 0.9 kg
Power supply:
(lithium battery) 15 h at 1:10 T/R ratio
Power output: 1 W RMS

Operational status

Selected by the French, Israeli, Italian, Thai and Turkish air forces and the US Navy.

Contractor

Tadiran Spectralink Ltd.

VERIFIED

ASARS-G Airborne Search And Rescue System with GPS and relay

ASARS-G maintains all the existing facilities of ASARS; in addition, it includes: a complete survivor position locating system, with GPS navigation receiver, DF and full navigation support. Designated ARS-700G; it includes a secure data channel relay capability; voice channel reporting on both VHF and UHF; remote activation of embedded GPS features by the airborne units (ARS-700G and ARS-434R); a simple operator interface that turns the PRC-434 rescue radio into a complete airborne relay system, designated ARS-434R.

Specifications

(for survivor rescue radio PRC-434G)
Frequency range: 220 - 410 MHz, and 121.5 MHz
Channels: 7,000
Programmed channels: 10 + 2 guard
Position accuracy: GPS P-code or C/A code
Modulation:
(voice or swept tone) AM
(narrowband data) FSK
(transponder) OOK, PSK for ASARS
Activation: manual or automatic upon ejection
Remote operation: beacon mode; navigaton mode; data transfer mode
Transmit power:
(beacon in UHF) 2 W peak
(voice in UHF) 0.75 W CW
(121.5 MHz) 0.1 W peak
adaptable power management for data transfer

Contractor

Tadiran Spectralink Ltd.

VERIFIED

ITALY

APS-705A surveillance radar

The APS-705 I/J-band surveillance radar was designed for naval helicopters, but is equally suitable for maritime patrol aircraft and land-based helicopters. The radar's main functions are navigation, search and detection, target localisation, target tracking and designation for weapon aiming and mapping.

To allow 360° coverage, different antenna sizes are available to fit the specific requirements of the aircraft in terms of location and space. Line of sight stabilisation is provided and there are different selectable antenna rotation rates and manually controlled antenna tilt.

The radar has two I-band transmitter/receivers for frequency diversity operation at 25 kW.

The use of a modern processor enables such facilities as digital signal processing, sector transmission and blanking, built-in test, track while scan for multiple targets, dense environment tracker, freeze mode, colour raster scan display, interface and integration with weapon systems, datalink and airborne sensors such as FLIR, ESM and sonar.

The radar can be integrated also with Officine Galileo UPX-719 beacon system.

Different versions are also available; the APS-705B is a simplified configuration with MIL-STD-1553B databus interface for use on board modern utility aircraft, or as a pod-mounted configuration for retrofit applications, where it can be used with anti-surface missiles

Operational status

In service with AB 212, SH-3D and EH101 helicopters of Italian Navy and several other navies.

Contractor

Alenia Difesa, Avionic Systems and Equipment Division, Officine Galileo.

VERIFIED

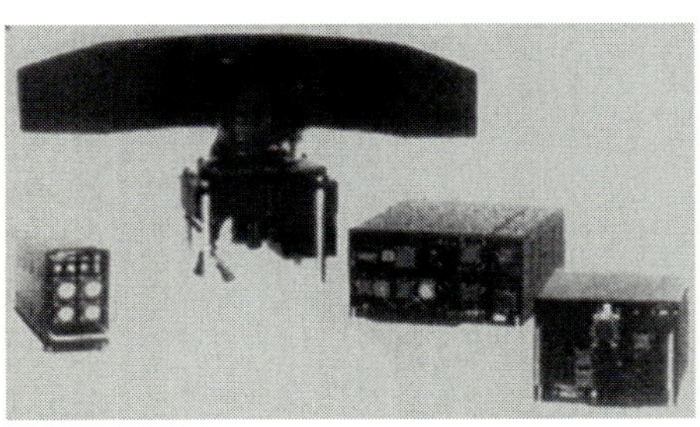

Alenia Difesa APS-705A search and rescue radar
0011886

APS-707 search and rescue radar

The APS-707 is a frequency-agile I/J-band radar, for aircraft and helicopters, designed for installation where low weight, power consumption and cost are important factors. It is a military qualified search radar featuring 360° scan for surface surveillance, small target

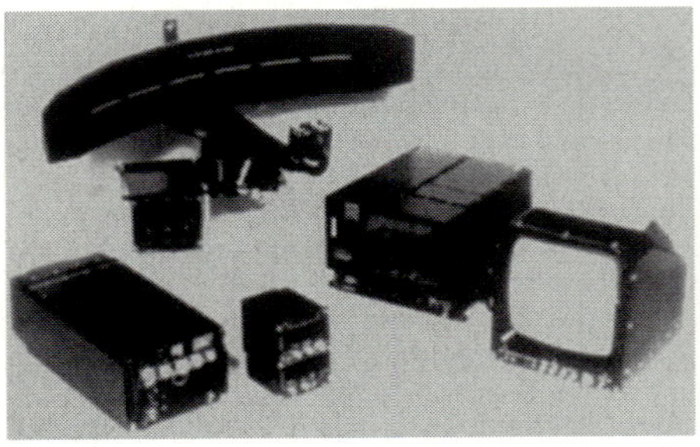

Alenia Difesa APS-707 search and rescue radar
0011885

detection for ASW operations, target designation for ASV attacks and radar mapping. The APS-707 can also display IFF, ESM, beacon and sonar data. An advanced video signal processor facilitates the presentation of integrated data combined with information from other sensors on a large TV monitor, with a cockpit repeater display as an option. Also available is a MIL-STD-1553B interface, a sonar interface and a track-while-scan facility.

Operational status

In production.

Contractor

Alenia Difesa, Avionic Systems and Equipment Division, FIAR.

VERIFIED

APS-717 search and rescue radar

The APS-717 family consists of two search and navigation radar systems which are tailored to the requirements of individual customers. It is suitable for many roles, including search and rescue, surveillance, navigation and target designation.

The APS-717(V)1 is a lightweight radar which is suitable for both fixed-wing aircraft and helicopters. It operates in the I/J-band, providing detection over 180° in azimuth with automatic stabilisation. It can be integrated with the navigation system and a FLIR sensor. Other features include Constant False Alarm Rate (CFAR), scan-to-scan integration, pulse-to-pulse integration, a freeze mode and colour display with graphics.

The APS-717(V)2 is a high-performance upgrade of the APS-717(V)1 and offers a number of additional features. These include 360° azimuth coverage, integration and automatic initialisation of FLIR and LLTV, a video recorder output and an optional track-while-scan capability covering 32 targets.

Operational status

The APS-717(V)1 is in service on HH-3F helicopters and G222 aircraft of the Italian Air Force. The APS-717(V)2

Alenia Difesa APS-717(V)1 search and rescue radar
0011884

is in service on Italian Harbour Authority AB 412 helicopters.

Contractors

Alenia Difesa, Avionic Systems and Equipment Division, Officine Galileo
FIAR.

VERIFIED

Creso airborne battlefield surveillance radar

Creso is one of the sensor systems under development as part of the Italian Army's Surveillance and Target Identification Subsystem (SORAO) CATRIN command, control, communications and intelligence system. It comprises both air and ground elements. The airborne element is installed aboard an Agusta AB 412 helicopter, and comprises the Creso battlefield surveillance radar, with ESM and FLIR sensors, together with a datalink system for air/ground data transfer.

Creso's operational roles are detection of ground moving targets beyond the FEBA (Forward Edge of the Battle Area); production of a target count in battlefield areas designated to it; high-precision localisation of designated targets.

Creso is an I-band, pulse Doppler radar; it utilises a coherent TWT transmitter and features wideband frequency agility; high-resolution pulse compression; programmable FFT processor; zoom capability; high

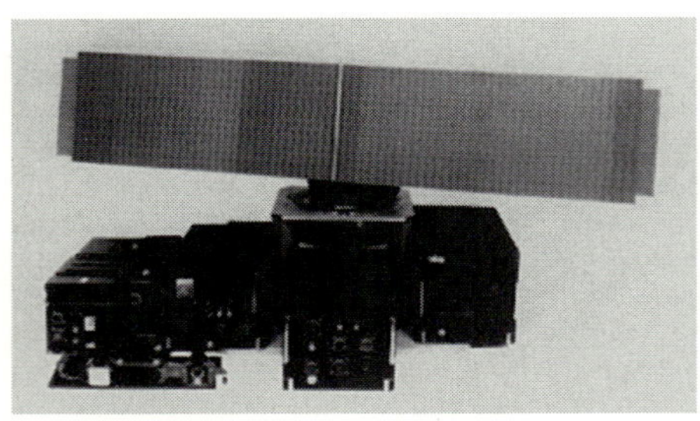

Alenia Difesa Creso airborne battlefield surveillance radar
0011889

ECCM resistance; and growth capability for air-to-air surveillance.

Operational status

It is understood that there is one flight trials model fitted to an Augusta AB 412 helicopter. The Creso system is the Italian submission to the NATO multinational Alliance Ground Surveillance (AGS) programme.

Contractors

Alenia Difesa, Avionic Systems and Equipment Division,
FIAR.

VERIFIED

Grifo multimode radar family

The Grifo multimode pulse Doppler I/J-band radar is a compact radar with a high degree of modularity and low life cycle costs designed for air superiority aircraft. It has look-up and look-down capabilities.

The radar features a monopulse flat plate antenna array, coherent TWT transmitter, pulse compression, wideband frequency agility and a programmable waveform generator and FFT processor. It is fully compatible with semi-active and active missiles and has a high immunity to ECM.

The Grifo is suitable for fitting to a number of aircraft such as the Mirage, A-4, F-5, MiG-21, Super 7 and several trainer/light attack aircraft.

Grifo 7 radar

The Grifo 7 is designed to be installed in the nose of CAC F-7 aircraft. It has full look-up and look-down air-to-air capabilities through the use of pulse Doppler and medium PRF waveform, plus an air-to-ground ranging mode to support CCIP/CCRP.

Two modes are selectable in air-to-air. Super-search is used for the acquisition and tracking of the highest priority target in the HUD field of view. The radar allows the missile seeker to be slewed to the target line of sight for offset delivery. In boresight, fixed antenna pointing is used for automatic acquisition and tracking of the nearest target.

Most of the hardware is common with the other versions of the Grifo family. Grifo 7 has compatibility with IR missiles, rockets, guns and free-fall bombs.

Specifications

Weight: 55 kg

Operational status

In production for the CAC F-7. 100 ordered by Pakistan for retrofit to the PAF F-7 fleet.

Grifo F radar

The Grifo F has been developed for retrofit in the Northrop F-5E/F. It is an I-band multimode pulse Doppler radar which offers eight air-to-air, four air combat and nine air-to-surface modes. It uses a TWT transmitter with both pulse compression and wideband frequency agility, a programmable FFT processor and a monopulse flat plate array antenna, of which two sizes are available.

Operational modes for air-to-air employment include range-while-search, velocity search, track-while-scan, single-target track, situation awareness and air combat. For air-to-ground operation the modes are real-beam mode, Doppler beam-sharpening, air-to-ground ranging, sea map, surface moving target indicator, surface moving target track and beacon.

Specifications

Weight: 80 kg

Alenia Difesa Grifo 7 radar for F-7 aircraft 0001199

Agusta AB 412 helicopter showing Creso radar antenna beneath the nose; also shown is the four-port ESM system above the radar, and the datalink below the tailboom

Operational status

In production for the Northrop F-5E/F.

Grifo L radar

Grifo L is a variant of Grifo F selected by the Czech Air Force for integration into its Aero Vodochody L-159 fighter trainer.

Grifo M3 radar

Similar to the Grifo F, the Grifo M3 has been developed for the Mirage III. It makes use of a different antenna array to fit the nose of the Mirage III. Modes are as for the Grifo F.

The Grifo M3 is compatible with a variety of weapon systems including semi-active missiles. The Grifo M21 is similar to the M3 and is under development for the MiG-21.

Operational status

In production for the Mirage III, 35 ordered by Pakistan.

Contractors

Alenia Difesa, Avionic Systems and Equipment Division,
FIAR.

VERIFIED

HEW-784 radar

The HEW-784 is a Heliborne Early Warning radar for the Airborne Early Warning (AEW) version of Italian Navy's EH101 helicopter. In addition to the detection of surface targets, the HEW-784 is designed to be capable of detecting airborne targets with extremely low radar cross section.

Specifications

Frequency: I/J-band.
Antenna: parabolic with 360° scan and sector scan capability; the antenna requires 3 m diameter radome.

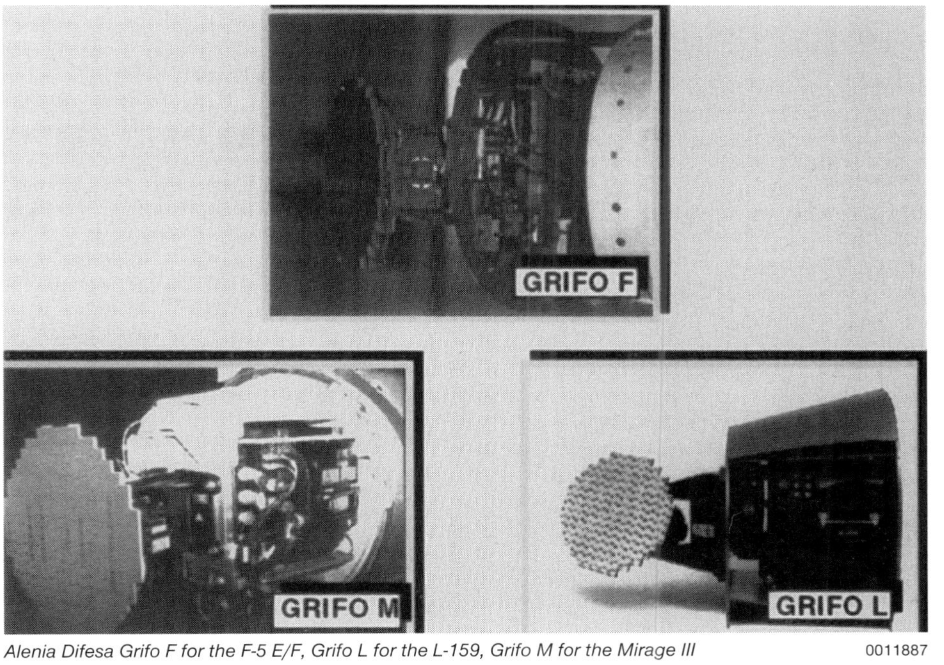

Alenia Difesa Grifo F for the F-5 E/F, Grifo L for the L-159, Grifo M for the Mirage III 0011887

Transmitter: coherent travelling wave tube with very high average power.
Features: pulse compression; pulse-to-pulse frequency agility; adaptive, multiple target look-up, look down, air-to-air, track-while-scan strategies; scan-to-scan integration; datalink to task force surface ships; Inverse Synthetic Aperture Radar (ISAR).
Display: digital multifunction displays, with two separate scan converters for video displays; different formats and orientations can be presented on each; integrated Identification Friend or Foe (IFF).
Modules: eight separate line replaceable units.
Interfaces: MIL-STD-1553B, multiple video outputs in standard format.

Operational Status
In production for the Italian Navy EH 101 helicopter

Contractor
Alenia Difesa, Avionic Systems and Equipment Division, FIAR and Officine Galileo.

VERIFIED

Maritime Patrol Mission System (MPMS)

The new Alenia Difesa Maritime Patrol Mission System is based on a company-developed core system Mission Management System (MMS) integrated with various COTS sensors, (including: 360° search radar, SLAR radar, IR/UV scanners, day/night electro-optical systems, ESM, communications and datalink) and a flexible man-machine interface, with software-based tactical solutions and pre-programmed patterns.

The MMS is based on a MIL-STD-1553B databus for data exchange and MPMS integration. It features software controlled video-routing, digital maps, data fusion, communications and datalink to/from fixed and mobile stations. Its open architecture supports system expansion for increased mission requirements and upgrades.

The MPMS can support a wide range of maritime missions, including: Exclusive Economic Zone (EEZ) operations, search and rescue, vessel identification, coastal surveillance, environmental pollution detection, and Anti-Submarine Warfare (ASW) (with additional sensors).

The modular design enables Alenia Difesa to offer a multi console design on larger regional air transport types, and a down-sized configuration on small fixed-wing and rotary-wing aircraft.

The Alenia Difesa Maritime Patrol Mission System (MPMS) multi console configuration. ***1999***/0041391

Operational status
Two customers have selected the Alenia Difesa MPMS for installation on ATR 42 aircraft.

Contractor
Alenia Difesa, Avionic Systems and Equipment Division.

VERIFIED

Scipio radar family

The Scipio family is a series of coherent, lightweight, compact frequency-agile radars. The range is designed to suit one-man operations on fighter aircraft with multi-role capabilities. It has a total weight of less than 75 kg, with a high MTBF and low MTTR. An extensive BIT system provides fault detection and location. Scipio offers an air-to-air mode with look-down capabilities, air combat mode with automatic detection, designation and tracking, sea mode for target detection and tracking of surface vessels and ground ranging and mapping.

The Alenia Difesa SCP-01 airborne radar has been designed for Brazilian Air Force AMX aircraft

The SCP-01 is the first member of the Scipio family. Its main features include I-band frequency, pulse compression and pulse Doppler techniques, frequency-agile TWT transmitter, high throughput, fully coherent, software reconfigurable signal processing, monopulse tracking, track-while-scan, MIL-STD-1553B interface and colour video output with graphics.

Operational status
The SCP-01 has been designed for the Brazilian Air Force AMX and is in production.

Contractor
Alenia Difesa, Avionic Systems and Equipment Division,
Officine Galileo.

VERIFIED

SIR SSR family

The SIR SSR family is a modular SSR system. By adding modules, SIR can be expanded from the SIR-R minimum version, through SIR-M to the most complex SIR-S version. The system is designed so that the cost of expansion is totally borne by the added unit.

The system consists of a transmitter, receiver, programmable processor/controller and power supply. The processor/controller is a firmware programmed unit which forms the extractor and provides all control signals and narrowband formatting. The processor/controller provides all the programmable functions for the transmitter and receiver.

SIR-R gives good performance and is especially well suited to areas where interference with other SSRs is not a problem. The system uses the programmable transmitter, standard SSR receiver and processor/controller. The antenna is a 14 or 28 element standard hog-trough antenna.

SIR-M has a full monopulse configuration and uses the same transmitter and processor/controller as the SIR-R. The receiver is configured in a sum and difference version and uses the same logarithmic amplifiers as the SIR-R. The antenna is an open array optimised for the monopulse capability.

SIR-S is an SIR-M system with additional modules to give a full Mode S capability. The receiver and processor/controller remain the same as for SIR-M.

Specifications
Frequency:
(transmit) 1,030 ± 0.1 MHz
(receive) 1,090 MHz
Modes: 1, 2, 3A, B, C, D (Mode S optional)
Pulse duration: 0.8 μs
Duty cycle: 1% for SIR-R/SIR-M, 10% for SIR-S

Contractors
Alenia Difesa, Avionic Systems and Equipment Division,
FIAR.

VERIFIED

SL/ALQ-34 ECM pod

The SL/ALQ-34 ECM pod is available as a self-defence aid for high-performance aircraft. The SL/ALQ-34 provides a radar warning facility and can jam anti-aircraft artillery or surface-to-air missile radars. Current versions operate across a 6 GHz bandwidth in the H- to J-band (6 to 20 GHz) but lower frequency systems at C-, D-, E- and F-bands, effective against air defence search radars, are in prospect.

The receiver processor conducts threat assessment and ranking using a stored library of threat data, and performs jamming power management tasks. Probably two travelling wave tube transmitters are used and cooled by a closed-loop liquid cooling system. Supersonic high-altitude clearance has been obtained.

Operational status
In production.

Contractor
Alenia Difesa, Avionic Systems and Equipment Division.

VERIFIED

SL/ALQ-234 self-defence pod

The SL/ALQ-234 self-defence pod fulfils the requirements for aircraft self-defence with a minimum of penalty to flight and combat capabilities. The SL/ALQ-234 is designed for standard fuselage or wing pylon installation. A small cockpit control panel provides the pilot with threat warning and jamming reaction information. Self-contained ram-air turbine power generation and cooling, as well as the modular pylon attachment technique, make the SL/ALQ-234 suitable for immediate installation on a variety of supersonic aircraft.

Real-time processing and power management are the main factors leading to the multiple threat

SL/ALQ-234 self-defence pod

self-defence capability of the SL/ALQ-234. For pulse threats the system covers the I- and J-bands and provides noise and deception jamming. CW coverage is provided in the H- to J-bands and the threat is countered by deception jamming.

The system can operate even in the complete absence of prestored threat data, through software generated self-adaptive modes. Against known threats prestored in the system memory it can perform more complex and specific reactions. Positive self and mutual screening of the aircraft is achieved against the most advanced ground-based air defence systems.

Specifications

Dimensions: 3,825 (length) × 414 mm (diameter)
Weight: 270 kg
Frequency:
(pulse) I/J-bands
(CW) H/J-bands
Power output: 7.5 kVA
Altitude: sea level to 30,000 ft
Speed: M1.1 at sea level, M1.5 at 30,000 ft

Operational status

In service with various air forces.

Contractor

Alenia Difesa, Avionic Systems and Equipment Division.

VERIFIED

UPX-719 transponder

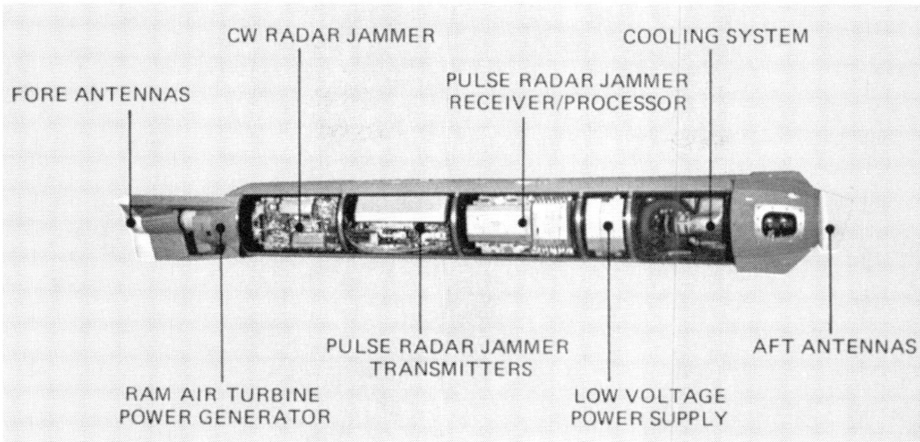

The internal arrangement of the SL/ALQ-234 self-defence pod

The Alenia Difesa UPX-719 transponder

The UPX-719 transponder forms a part of the SMA Intra (interrogator/transponder) system which links ships with their co-operating helicopters. The UPX-719 is the airborne transponder; the UPX-718 is the shipborne interrogation part of the system. The shipborne transmission to the airborne transponder can be integrated with the main radar. Ships can identify up to 10 helicopters and vice versa, due to the characteristics of the coding system. Range is over 100 km for helicopters operating at 1,500 ft.

Operational status

In production for Italian and other navies.

Contractor

Alenia Difesa, Avionic Systems and Equipment Division, FIAR.

VERIFIED

Apex jamming pod

The Apex deception jamming pod uses the existing ELT 555 pod shell and is claimed to be effective against all types of fire-control radars. Frequency coverage against both pulse and CW emitters is quoted as being extended. The pod weighs 140 kg.

Apex uses fast digital RF memory technology to facilitate coherent response techniques and the system is front-line programmable.

Operational status

Apex is suitable for aircraft such as the Hawk 100/200, MB-339, F-5 and Mirage F1.

Contractor

Elettronica SpA.

VERIFIED

Aries tactical support and training EW system

Aries can be used either for training EW specialists or for tactical ELINT/COMINT and surveillance and support ECM missions. It is designed to be housed in a small transport aircraft and typically has a supervisor's position and two operator stations. Data collected during tactical ELINT/COMINT and surveillance missions can be transmitted over a secure datalink to the ground for near-realtime evaluation.

The Aries system consists of two subsystems: Aries-A which is dedicated to all activities in the radar frequency band and Smart Guard (see later entry) which covers the communication bands.

Aries-A detects, analyses, identifies and locates emissions in the C- to J-bands and, during ECM support missions, drives a set of jammers covering the D- to J-bands.

The Elettronica Aries modular airborne EW system for tactical support and training

Smart Guard carries out similar functions in the VHF and UHF frequency ranges, with HF available as an option.

Aries is fully compatible with navigation systems such as inertial, Omega and GPS and is provided with a look-through facility to permit signal reception during jamming.

Operational status
In service since 1980. More than 60 Aries systems have been sold to a number of air forces.

Contractor
Elettronica SpA.

VERIFIED

ARWE Advanced Radar Warning Equipment

ARWE was designed for the Italian Tornado IDS aircraft. It is based on a multiprocessor architecture, with MIL-STD-1553B interface connections to other elements of the Defensive Aids SubSystem (DASS).

The ARWE system architecture comprises:
(1) four DF channels with 360° azimuth coverage;
(2) E-/J-band and millimetric frequency coverage;
(3) frequency selective receiver with: multiple IF bandwidths (wide bandwidth to optimise probability of detection and response time; medium bandwidth as a compromise to meet general processing requirements; narrow bandwidth to provide discrimination of CW signals);
(4) adaptive bandwidth selection controlled by the signal environment and the active electronic countermeasures requirements;
(5) instantaneous frequency measurement within the IF band to ensure fast emitter identification and designation of active countermeasures.

Operational capabilities include:
(1) 3 in cathode ray tube display;
(2) E-/J-band and millimetric frequency coverage;
(3) detection of Low Probability of Intercept (LPI) and MultiFunction Radar (MFR) emitters;
(4) exotic emission detection capability;
(5) compatiblity with on-board high duty factor radars and jammers;
(6) DASS management capability;
(7) flight line reprogrammable.

Operational status
Operational on Italian Air Force Tornado IDS aircraft.

Contractor
Elettronica SpA.

VERIFIED

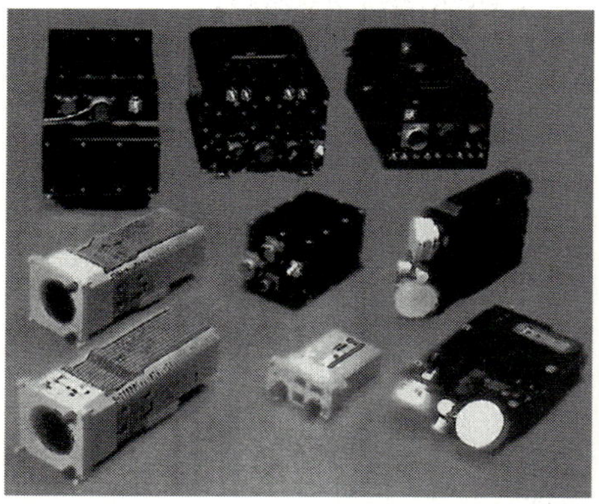

ARWE advanced radar warning equipment 0018284

Colibri integrated ESM/ECM system

The Colibri system is an integrated ESM/ECM equipment designed for installation on any type of medium-sized helicopter operating in the naval surveillance, anti-submarine and anti-ship roles, to extend the ESM/ECM range of the parent ship. The system provides a helicopter with the facilities for passive surveillance, DF analysis, identification and location of radar emitters, even when the system is operating in a dense electromagnetic environment. It provides for effective self or mutual protection against radar-guided missiles.

Colibri can be configured to meet particular operational roles by combining a variety of basic modules. The basic configuration is represented by the ELT/161 ESM and the radar warning system. A helicopter's radar warning function can then be supplemented by the ELT/562 deception jammer to obtain effective self-protection and the configuration can be further expanded by the addition of the ELT/261 for ESM capability and the ELINT role. A further unit, the ELT/361, can be added to achieve even greater protection capability.

In the basic configuration, Colibri comprises the ELT/161, consisting of receiver, antenna system, warning processor and radar warning display. The ELT/161, operating in conjunction with the ELT/261 system, gives overall capability in that field and is designed to fulfil the maritime surveillance and EW roles to extend the electronic warfare range of the parent ship. In this role, the system supplies instantaneous DF and display of intercepted emissions, frequency measurement of pulse and CW emissions, automatic analysis, identification of source and automatic tracking of selected emissions. The addition of the ELT/562 deception jammer provides the helicopter with self-protection against sea-to-air and air-to-air missiles. The ELT/361 noise jammer is added to suppress enemy surveillance radars. The system is capable of the following instantaneous surveillance coverage:

Altitude	Effective radius against radar type (in miles)	
	I-Band (8-10 GHz)	F-Band (3-4 GHz)
300 ft	26	34
1,000 ft	41	50
2,000 ft	47	63

Operational status
No longer in production. In service with helicopters of several navies.

Contractor
Elettronica SpA.

VERIFIED

ELT/156(V) radar warning receiver

Suitable for fighter, light strike/attack aircraft and helicopters, the ELT/156 series of lightweight passive detection systems provides broadband 360° azimuth coverage and can distinguish anti-aircraft artillery, surface-to-air missile and air-to-air missile illuminations in bearing and range. Miniature crystal video receiver technology and advanced signal processing, including built-in test equipment, contribute to minimise system weight. A dual-mode CRT display, on which synthetic or raw video data can be shown, is standard. An audio output is available to complement the display. Latest versions of the equipment are lighter than earlier production systems.

Specifications
Weight:
(4 antennas) 0.15 kg each
(2 RF heads) 1 kg each
(signal processor) 5.2 kg
(display unit) 1.9 kg
(control panel) 0.6 kg
(total weight) 10.3 kg

Operational status
In service.

Contractor
Elettronica SpA.

VERIFIED

ELT/156X radar warning receiver

The Elettronica ELT/156X radar warning receiver features very wideband miniature crystal video receiving systems and provides full azimuth and frequency coverage up to millimetric-waves and including distributed multiprocessing. The use of bit-slice microprocessing and EEPROM technology allows flexible and high-speed processing, giving fast reaction times and adaptability to threat evolution. Threat

The Elettronica ELT/156(V) radar warning receiver

The ELT/156X radar warning receiver showing (from left to right) antennas, processor and (front) the control panel, (rear) the display

parameter software is reprogrammable at flight line level. The dual cockpit display incorporates both a raw mode, to give an immediate picture of the surrounding environment, and a synthetic mode, to give a clear indication of threat identity. The system also includes an aural alarm.

The ELT/156X may be integrated with other onboard EW systems and sensors via two RS-422A serial links.

The system consists of two antennas/RF heads, signal processor, display and control panel. The version for helicopters, the ELT/156X(V2), employs four single-channel RF heads.

Specifications

Weight:
(antenna/RF head × 2) 2.2 kg
(signal processor) 10.5 kg
(display) 2.1 kg
(control panel) 0.7 kg
(total weight aircraft version) 17.7 kg
(total weight helicopter version) 20.1 kg

Operational status

Now in production for a number of air forces.

Selected by the Italian Navy for its AS/ASVW and AEW EH 101 helicopters, also selected for the Brazilian AMX light fighter aircraft.

Contractor

Elettronica SpA.

VERIFIED

ELT/158 radar warning receiver

The ELT/158 is designed for combat aircraft and provides high-speed detection and indication of threats through a full 360° coverage. The dual-mode cockpit display can show raw or synthesised data. The system can be integrated with a laser warning receiver, has fully automatic built-in test facilities and can be reprogrammed on the flight line.

Contractor

Elettronica SpA.

VERIFIED

ELT/263 ESM system

The ELT/263 airborne ESM is designed principally for use in maritime aircraft such as the Beech 200T but has been configured for other types such as the Guardian, CASA C212, Learjet 35A, F27 Maritime, Bandeirante, Piaggio P166 and Britten-Norman Maritime Defender. The surveillance of coastal waters is its main function. Detection, analysis and identification of emissions in the E- to J-bands is performed, in addition to bearing measurement; this permits emitter location by using a series of successive bearing measurements.

The ELT/263 equipment, in addition to its surveillance tasks, provides for emitter location by the triangulation method. Emission analysis, identification and bearing information can be transferred to external users such as ground stations, naval units and the aircraft's radar operator and tactical navigator and can also be recorded on an optional printer.

The main items of the ELT/263 are a set of four DF antennas, an omnidirectional antenna set, DF receiver, IFM receiver, ESM display and control console, radar warning processor and radar warning display unit.

Operational status

In service. The equipment is in current production for several customers.

Contractor

Elettronica SpA.

VERIFIED

ELT/457-460 supersonic noise jammer pods

A set of four related noise jammer pods is available for use on any high-performance strike/fighter or light attack aircraft, mounted on an underwing pylon. All units are self-contained with a ram-air turbine generator in the nose section and each is dedicated to a particular waveband. A heat-exchanger is situated behind the

The Elettronica ELT/460 supersonic pod

turbine and there are fore and aft antennas in all pods. The ELT/459 and ELT/460 versions have additional antennas beneath the body of the pod. Processing within the system allocates threat jamming power on a proportional basis against any type of pulsed radar threat and includes built-in test equipment and control of blanking with other aircraft systems. All units are claimed to be highly resistant to ECCM, including any type of frequency agility, and incorporate a large threat library. Maximum operating speeds are M1.1 at sea level and M1.5 at 40,000 ft.

Specifications

Dimensions: 3,120 (length) × 340 mm (diameter)
Weight: 145 kg
Interface: STANAG 3726 and MIL-A-8591D

Operational status

In service.

Contractor

Elettronica SpA.

VERIFIED

ELT/553(V)-2 airborne pulse and CW jammer

The ELT/553(V)-2 airborne deception jammer features fully automatic operation and is programmable at flight line level. It is designed to ensure maximum flexibility, operating either as a stand-alone system, or in co-operation with on board RWRs.

ELT/553(V)-2 comprises:
(1) a low-band pulse module (E- to H-band)
(2) a high-band pulse module (H- to J-band)
(3) a CW module dedicated to counter semi-active missile threats.

ELT/553(V)-2 utilises:
(1) a channelised receiving section architecture for quick and reliable lock-on and recognition, even in very dense environments;
(2) repeater jammer philosophy to increase effective jamming power and hence jamming effectiveness in the presence of ECCM techniques;
(3) dedicated CW mode to counter semi-active missile threats;
(4) effective operation against staggered and jittered PRI emitters, as well as against frequency diversity and frequency agility threats;
(5) automatic deception of locked-on radars;
(6) multi-threat capability;
(7) RS422 interfaces.

Optional capabilities include:
(1) automatic deception of pulse and CW radars, TWS included;
(2) pulse by pulse power management and optimisation across 10 channels of operation;
(3) very high ERP;
(4) flight line programmability;
(5) complete BITE.

ELT/553(V)-2 airborne pulse and CW jammer 0018283

Specifications

LRUs	quantity	weight (kg)
High-band pulse module	1	40
Low-band pulse module	1	37
CW module	1	20.4
Control panel	1	1.4
High-band antenna	2	0.2 × 2
Low-band antenna	2	0.4 × 2

Power consumption: 2.5 kVA

Operational status

Understood to be selected for the Brazilian AMX light fighter aircraft and possibly Italian Tornado aircraft.

Contractor

Elettronica SpA.

VERIFIED

ELT/554 deception jammer

The ELT/554 is specifically designed for installation in attack helicopters and operates in the J-band. It is lightweight and compact and reacts rapidly to incoming signals. The jamming programmes are intelligent and several threats can be jammed simultaneously.

Contractor

Elettronica SpA.

VERIFIED

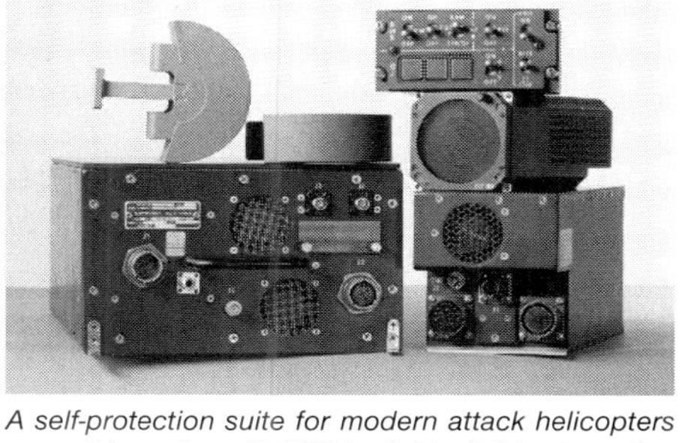
A self-protection suite for modern attack helicopters comprising the ELT/554 lightweight deception jammer (left) and the ELT/156 radar warning receiver

ELT/555 supersonic self-protection pod

The ELT/555 has a similar configuration to the ELT/457 system, with the same ram-air turbine on the nose of the pod. Internal equipment includes fore and aft facing antennas which receive pulsed and continuous wave signals and transmit pulsed responses, plus separate fore and aft facing continuous wave transmitter antennas on the undersurface. The system is sensitive to H- to J-band (6 to 20 GHz) threats and is designed to operate in dense electromagnetic environments. It has multiple target contrast capability and features BITE. Operating envelope and mechanical interface specifications are the same as those of the ELT/457. The cockpit display can be tailored to customers' requirements.

The Elettronica ELT/555 deception jammer pod on a Learjet 35A multimission trainer

Specifications

Dimensions: 3,000 (length) × 340 mm (diameter)
Weight: 150 kg

Operational status

In service.

Contractor

Elettronica SpA.

VERIFIED

ELT/558 self-protection jammer

The ELT/558 self-protection jammer operates at the lower end of the frequency spectrum and features wide-angle coverage, fully automatic responsive jamming operation, high ERP, low false alarm rate, high immunity to ECCM, very short reaction time and effective reaction against many threats. The system can be integrated with the aircraft's avionics.

Operational status

In service on the Mirage 2000.

Contractor

Elettronica SpA.

VERIFIED

The ELT/558 receiver unit (left) and ELT/558 transmitter unit (right) 0018282

ELT/562 and ELT/566 deception jammers

The ELT/562 and ELT/566 are repeater jammer systems for internal installation on aircraft and helicopters. ELT/562 is designed to counter pulse threats, while ELT/566 combats continuous wave threats. One or both jammers may be associated with an airborne ESM system such as the Elettronica Colibri. The jammers are effective against H- to J-band threats and are intended to protect aircraft and helicopters in battlefield environments.

Operational status

In production.

Contractor

Elettronica SpA.

VERIFIED

Fast Jam ECM system

Fast Jam is designed for operations over the VHF and UHF communications frequencies. It is intended mainly for airborne applications, where maximum advantage is obtained by the platform elevation and speed, although the compact and lightweight construction of the system also allows installation on land vehicles and ships. The main features of the system include modularity, coverage of the VHF and UHF bands with optional coverage of the HF band, high speed and intercept capability and multiple frequency jamming capability.

In the Fast Jam configuration the system consists of a search and analysis receiver subsystem, monitoring subsystem and jammer subsystem. The first two subsystems perform the same functions as the basic ESM system described in the later entry for Smart Guard. The addition of the jammer subsystem allows the jamming of intercepted communications channels, to disrupt enemy communications.

The jammer subsystem operates under the control of the system computer and is fully automated. By time-sharing, it is able to jam up to six channels simultaneously without performance degradation. The channels to be jammed are either preset or can be selected by the operator according to the specific operational requirements. A look-through function allows an adaptive jam/receive time management so that simultaneous search and jamming functions can be performed.

Manual control of the jammer subsystem is available as a back-up to fully automatic operation. A high-power wideband all-solid-state amplifier is used in the jammer subsystem and VSWR and thermal automatic switch off circuits are employed to avoid permanent damage to the amplifier in case of overheat or antenna failure. Various types of modulation are available from external modulation sources.

Operational status

In service. The system has been installed in a number of different types of aircraft.

Contractor

Elettronica SpA.

VERIFIED

Heloborne solid-state ESM/ECM system

Elettronica is working on a solid-state heloborne ESM/ECM system. The work is stated to be based on development work done for the Eurofighter Typhoon Eurodass, but to be optimised for attack helicopters.

The capability is stated to include pulse-by-pulse, steerable directive, multitarget operation in nanosecond timescales, using Digital Radio Frequency Memory (DRFM) technology.

The aim is to provide 360° spatial coverage, using three antennas.

Operational status

In development, production standard equipment is expected during 2001, with as yet no confirmed orders.

Contractor

Elettronica SpA.

UPDATED

Sea Petrel RQH-5(V) airborne ESM/ELINT systems

The Sea Petrel RQH-5(V) is a family of systems which can meet EW requirements ranging from threat detection and analysis to electronic intelligence. The various configurations and options allow the system to be tailored to a specific requirement. All the RQH-5(V) Series of equipments have capabilities for integration with other systems and provide target parameters and direction of arrival information for weapon systems. The small size and low weight of RQH-5(V) components make the system readily adaptable to current airborne, naval and ground installations with a minimum of effort. The system can be operated after minimal instruction and is designed for maximum operating time with minimum servicing.

The system covers frequencies from 0.65 to 18 GHz, with an optional extension to 40 GHz, and is entirely automatic. It provides real-time automatic extraction, analysis and tracking of all incoming radar signals. It also provides pulse, intrapulse and fine analysis for ELINT, including frequency fine measurement, measurement of jitter and stagger, frequency or PRI agility analysis, histogram preparation, detection and recording of antenna pattern and related amplitude histogram. The operator has only to view the system display which provides data on up to 200 emitters. The RQH-5(V) can operate without any prior knowledge of the electromagnetic scenario with no significant reduction in performance. On the other hand, the ELINT capability in terms of emitter parameter statistical analysis and pulse and intrapulse analysis, allows complete characterisation and analysis of the radar signals and keeps records of them for post-flight data collection.

The basic components of the RQH-5(V) are the antennas, direction-finder receiver, IFM receiver and data extractor.

The antenna group includes one omnidirectional antenna and four or eight DF antennas. Various DF and fine DF antenna types are available to cover different frequency ranges. The modular approach of each antenna module allows different installation configurations, to cater for any platform constraints.

Each antenna unit incorporates the associated electronic circuitry. Direction-finding is performed by a wide open omnidirectional and instantaneous receiver using amplitude comparison monopulse techniques. An eight port configuration is normally adopted, but it can also use a four element DF antenna subsystem for radar warning receiver applications.

The IFM receiver features high sensitivity and high probability of detection, wide open operations, fast response in order to operate in a multimillion pps environment and very high instantaneous dynamic range. Two versions are available – the FR-6 and FR-7 – providing different RF measurement accuracy.

The data extractor consists of a very powerful multiprocessor structure, specially adapted for real-time applications as a derivative of the AYK-204 airborne computer. The resulting automatic data extraction process has a very high acquisition speed of up to 60 new emitters every 20 ms, even in a completely unknown environment. This unit includes some standard I/O interfaces, such as serial lines, video graphic, memory expansion and MIL-STD-1553 bus developed for the Alenia Difesa AYK-204 airborne computer and its derivative versions.

The Elettronica ALR-733(V)2 ESM system for maritime patrol aircraft **2000**/0080283

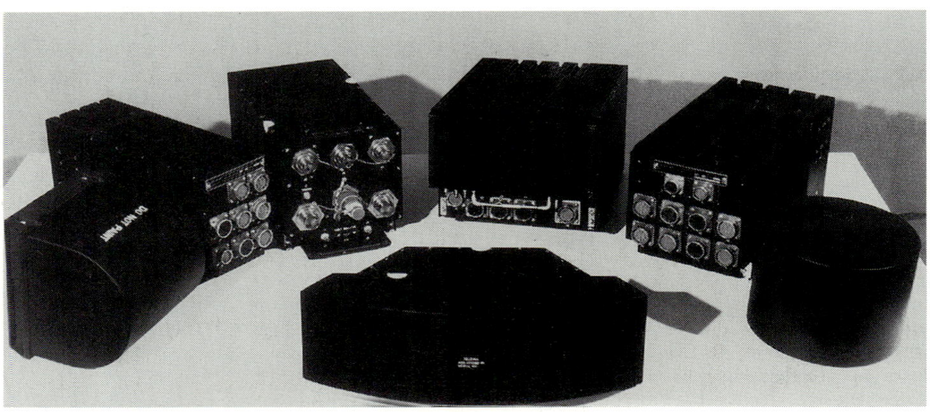

The Sea Petrel RQH-5(V) ESM/ELINT system

Several graphic and alphanumeric display modes are available to the operator. This includes frequency and direction of arrival of the signal, tabular lists, emitter characteristics, true or relative bearing, frequency-agile deviation, PRF, jitter and stagger values, emitter name and threat level and emitter scan period and type. Tactical, panoramic and geographic modes for ESM, radar and navigation are also available.

The identification library handles up to 3,000 modes of emitter parameters. The operator can store previous known data together with pre-assigned threat level and confidence level. The system compares extracted emitter data with the library and provides an immediate alert on high-interest emitters.

Several optional components are available. A high-accuracy direction-finder uses a multiple beam antenna system and a crystal video amplitude sectoral monopulse receiver. Several multibeam flat arrays, covering different frequency ranges, are available to provide different instantaneous fields of view and very high DF accuracy. A K-band high-gain steerable antenna and a formatter unit provide high-sensitivity detection and direction of the emitter signals. A fine analysis ELINT receiver gives supplementary information on the active extracted emitter, including the presence of simultaneous multiple RF or intrapulse modulations.

The following configurations are available:

The SL/ALR-730 Series offers electronic support measures and ELINT for all types of platforms including large and medium maritime patrol aircraft and large, medium and small helicopters. It is based on superheterodyne receiver technology.

The SL/ALR-740 Series offers RWR functions combined with automatic signal analysis for post-flight intelligence, and is designed for installation on small aircraft or helicopters. Average DF accuracy of 10° RMS is provided, with automatic warning and emitter parameter measurements. The ALR-741-R uses multiple-IFM receivers.

The SL/ALR-780 Series allows the integration of ECM modules in the above ESM equipment.

Specifications

Frequency: 0.6-18 GHz
Sensitivity: –60 dBm
Accuracy: ±2.5° RMS

Operational status

Stated to be in service. Current and future applications are quoted to include the AB412, EH 101 and NFH-90 helicopters.

A version of the SL/ALR-730, the ALR-735(V)3, has been selected by the Italian Navy for the EH 101 helicopter in its ASW/ASVW (Anti-Submarine Warfare/Anti-Surface Vessel Warfare) and AEW (Airborne Early Warning) roles.

The ALR-733(V)2, an ESM system for maritime patrol aircraft, is reported to have been selected by the Italian Guardia di Finanza for use on an ATR-42 aircraft. The line replaceable units appear to be different, but nonetheless the ALR-733(V)2 is believed to be part of the overall RQH-5(V) family.

The system is also said to represent the Italian contribution to a joint Italian/German programme (with EADS - formerly DaimlerChrysler Aerospace AG, Ulm), termed MPA 2000, for future maritime patrol aircraft requirements.

Contractor

Elettronica SpA.

UPDATED

Smart Guard COMINT system

The Smart Guard COMINT system is designed for intelligence monitoring of the VHF/UHF communications band. Although it is mainly intended for airborne applications, where maximum advantage is obtained by the platform height and speed, the compact and lightweight construction of the system allows installation on ground vehicles and in ships.

The basic ESM system consists of a search and analysis receiver and a monitoring subsystem. In this configuration the system allows the search and intercept of communications through a computer-controlled operation. The search is carried out over the complete bandwidth, or on specified sub-bands or channels. On operator control the receiver stops on a specific channel, so allowing the demodulation and analysis of the particular channel. A continuous monitoring of up to eight channels is provided by up to eight remotely controlled receivers and associated recorders. The frequency tuning of each monitoring receiver is set automatically by the system computer. Voice and associated data are recorded for subsequent analysis. In this configuration the system is manned by a single operator.

In the Smart Guard configuration, the basic ESM system is augmented by a DF and fixing subsystem.

The Smart Guard basic ESM subsystem

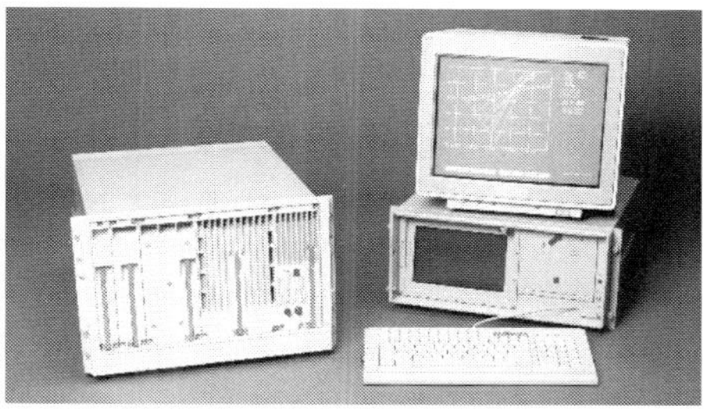

The Smart Guard DF and emitter fixing subsystem

The latter aids the capability to measure the DOA of the communications transmissions, and to determine their location. The DOA measurements are performed by a specific dual-channel superheterodyne receiver controlled by the system computer, and are obtained through an interferometric measurement. In addition to the DOA, the frequency value and signal strength are measured for each specific channel.

An interface with the navigation system provides the actual position of the platform in such a way that for each specific emission a set of data is stored. This contains DOA, frequency value, signal strength and platform position. The fixing of specific communication emission is performed by a dedicated computer and associated software algorithms by using the DOA and platform position data of a specific channel. At least two significant DOA measurements are required for a fixing computation.

In the Smart Guard configuration, two operators are required: one to control the basic ESM system and one dedicated to the fixing operation. A ground-based retrieval and analysis system is available as an option.

Integration of the Smart Guard COMINT system with the ELT/888 ELINT system makes it possible to monitor effectively the full electromagnetic environment and to identify and locate all enemy weapon systems.

Operational status

In service.

Contractor

Elettronica SpA.

VERIFIED

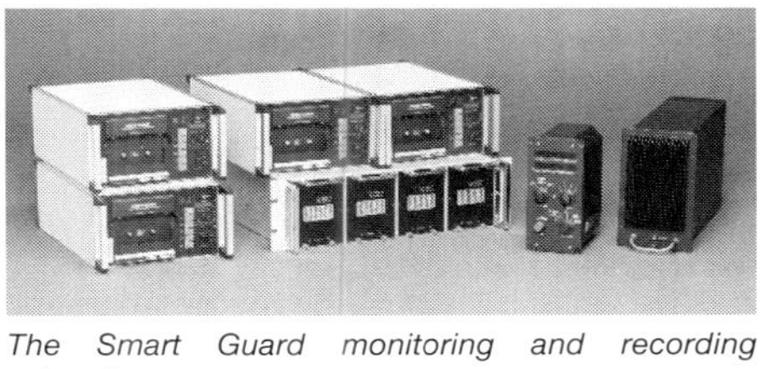

The Smart Guard monitoring and recording subsystem

APS-784 ASV/ASW radar

FIAR and Officine Galileo/SMA formed a consortium known as Eliradar to develop the APS-784 radar for the Italian Navy version of the EH 101 helicopter.

The radar provides high detection over a wide area and at long range, in adverse weather conditions. The APS-784 is able to detect and track even small targets such as liferafts, wooden boats and periscopes in rough sea states.

The APS-784 is a coherent radar characterised by pulse compression, a TWT transmitter, pulse-to-pulse frequency agility and track-while-scan capability, which is suitable for integration in sophisticated avionics suites. Main roles are for ASV, ASW and search and rescue.

The radar has 360° scan, missile launch assistance and weather detection modes.

Eliradar APS-784 ASV/ASW radar for the Italian Navy EH 101 helicopter 0011888

The radar is packaged in four LRUs and transmits in the I/J-band. It has four operational modes featuring track-while-scan with adaptive strategies: anti-surface vessel, anti-submarine warfare, weather and short range. It offers 360° surveillance with linear and circular switchable polarisation and includes an IFF Mk XII antenna array. It has two independent scan converters and scan-to-scan integration.

Target classification (profiling) is an optional capability.

Operational status

In production for the Italian Navy EH 101 helicopter.

Contractor

Eliradar (Alenia Difesa, Avionic Systems and Equipment Division, FIAR and Officine Galileo/SMA).

VERIFIED

AN/ARC-150(V) UHF AM/FM transceiver

The AN/ARC-150(V) designation represents a family of small, high-performance, lightweight, airborne UHF transceivers manufactured by Elmer and based on a Magnavox design as well as Elmer's own research and development. It has produced many versions, including the 10 W panel-mounted ARC-150(V)10, the 10 W remote-controlled ARC-150(V)2 and the 30 W remote-controlled ARC-150(V)8.

Elmer has also developed a series of control panels and frequency/channel repeaters to meet specific installation requirements on different aircraft and helicopters. A feature of this family is 'slice' assembly, which simplifies maintenance and facilitates growth.

A series of mounting adaptors has been developed and produced to allow the basic ARC-150 transceivers to replace older UHF radios such as the AN/ARC-51BX, AN/ARC-109, AN/ARC-52 and AN/ARC-552 without any mechanical or electrical modification to the aircraft.

A version of the ARC-150(V) with ECCM capabilities has also been developed using the frequency-hopping Have Quick technique. This capability is easily implemented by the substitution of the synthesiser slice and by minor changes on the control panel unit. No mechanical or electrical modifications are required on the aircraft. The modified radio retains the normal non-hopping mode.

The ARC-150(V)8 is a development of the basic 30 W ARC-150(V) transceiver incorporating the FM modulation capability by means of an additional slice. This facility makes the unit particularly suitable for use with data, frequency-shift keying and secure voice modems.

It has been demonstrated to be fully compatible with the Vinson KY-58 system both in AM and FM and in the diphase and baseband modes. It has also been successfully used as a main component of an airborne system for UHF satellite communication.

Specifications

Dimensions:
(10 W panel-mounted RT-1136) 146 × 124 × 193 mm
(10 W remote RT-1051) 127 × 120 × 183 mm
(30 W remote RT-1073) 127 × 120 × 291 mm
Weight:
(10 W panel-mounted RT-1136) 4.3 kg
(10 W remote RT-1051) 3.7 kg
(30 W remote RT-1073) 6 kg
Power supply: 28 V DC
Power output: 10 W (30 W for the AN/ARC-150(V)8 model)
Frequency: 225-400 MHz
Channel spacing: 25 kHz
Preset channels: 20 using electronic memory (MNOS)
Frequency accuracy: 2 kHz
Guard receiver: 243 MHz
Operating modes: AM voice, ADF, homer, secure voice/data, ECCM

Operational status

In production and in service with Italian and other armed forces. Over 1,000 units have been delivered and installed on a wide range of aircraft and helicopters including the Panavia Tornado, Aermacchi MB-339, Aeritalia G91Y and F-104S fixed-wing aircraft, and Agusta A 109, Agusta-Bell 212, Agusta-Sikorsky SH-3D and HH-3F and other helicopters.

Contractor

Elmer SpA (a Marconi Communications company).

VERIFIED

RV-4/213/A VHF/FM transceiver

The RV-4/213/A VHF/FM equipment is an airborne transceiver suitable for light aircraft and helicopters. It is in effect an airborne version of the Italian Army RV-3/213/V, which has been repackaged for installation in small aircraft. It is compatible with, and equivalent to, the Thomson-CSF Comsys TRAP-113 and is intended to replace equipment of the ARC-131 type. Tuning is fully automatic and digital over the frequency range 26 to 72 MHz in 50 kHz steps. Modes of operation are F3E FM voice and homing and RF power output is 20 W.

Specifications

Dimensions: 320 × 92 × 194 mm
Weight: 6 kg

SP-1450 intercommunication system ***1999***/0044797

Operational status

In production, and in service.

Contractor

Elmer SpA (a Marconi Communications company).

VERIFIED

SP-1450 intercommunication system

The SP-1450 intercommunication system is a centralised, digitally controlled, audio management system for the control and amplification of all external communications. It utilises fibre optics and appropriate wiring layouts to minimise crosstalk and electromagnetic interference to satisfy Tempest requirements. Control is effected via MIL-STD-1553B databus, or a dedicated remote-control panel to perform the switching function and via asynchronous serial databus between the user unit and central switching unit.

The system comprises:
SP-1451 switching unit;
SP-1452 headset interface unit;
SP-1453 main control panel;
SP-1454 secondary user unit;
SP-2049 direct voice output/warning tone generator.

A single switching unit can support up to eight main users and four secondary users, providing interface and control for up to:
8 transmit/receive units;
1 ANDVT narrow band crypto device;
2 wideband crypto devices;
1 Link 11 modem (HF and V/UHF);
1 sonar stereo unit;
12 navigation/mission systems;
4 telebriefs.

Operational status

Selected for: EH-101 (UK Royal Air Force SH; UK Royal Navy ASW; Italian Navy ASW) and NH-90 NFH and TTH versions.

Contractor

Elmer SpA (a Marconi Communications company).

VERIFIED

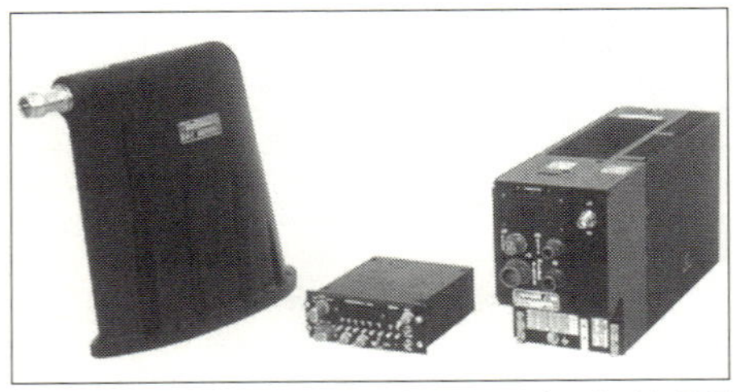

SRT-170/M (left to right: ATU, control unit, ½ ATR receiver/exciter/amplifier) ***1999***/0044798

SRT-[X]70/[X] HF transceivers

The SRT-[X]70/[X] series is an advanced family of HF transceivers providing the following features:
continuous frequency coverage between 2 to 30 MHz for both transmit and receive; voice and data (including Link 11) communication in secure and clear modes;
comprehensive modulation capabilities (USB, LSB, ISB, AM, CW, FSK, and CPFSK);
embedded selcal capability to ARINC 714;
embedded Automatic Link Establishment (ALE) to MIL-STD-188-141-2A;
embedded 2,400 bits/s modem;
remote control via MIL-STD-1553B or ARINC 429 databus, or dedicated remote control unit.

The SRT-[X]70/[X] family comprises:
SRT-170/L: 100 W power output transceiver
SRT-170/M: 175 W power output transceiver
SRT-270/L: 200 W power output transceiver
SRT-470/L: 400 W power output transceiver
SP-648 or SP-648/LA remote control panels
SP-649/L receiver-exciter, common to all versions
SP-480/L (100 W), SP-480/L1 (200 W) or SP-484/L (400 W) power amplifiers
CP-2001/[X] advanced computer-controlled remote multifunction control/display
SP-1325/L (200/100 W) or SP-1325/L1 (400 W) pre-post selectors
ATU: antenna couplers to match wire and probe (SP-1127/L for 200/100 W, ATU-1992/LM4 for 400/200 W), notch (ATU-1992/LM for 200/100 W,

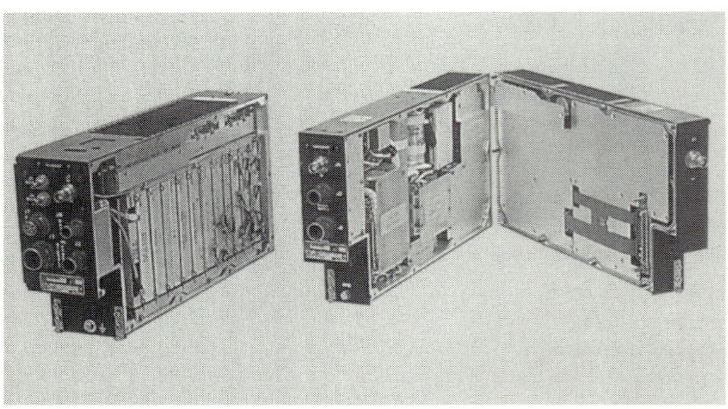

SRT-270/L internal view of 649/L receiver/exciter ***1999***/0044801

SRT-270/L and SRT-470/L with ATU for loop antenna and notch antenna ***1999***/0044800

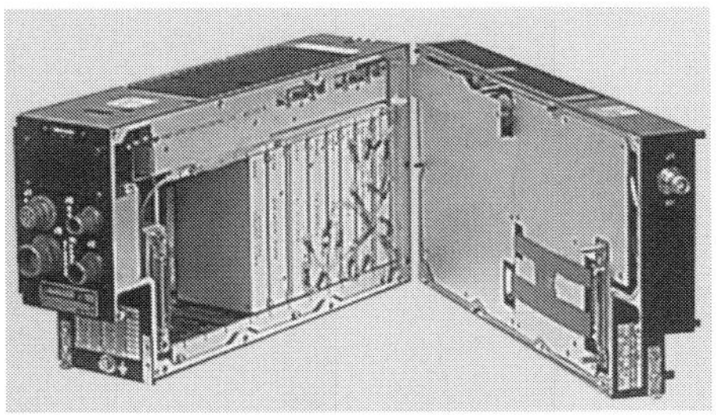

SRT-170/M internal view of receiver/exciter/amplifier unit **1999**/0044799

ATU-1992/LN for 400/200 W) and towel bar antennae (ATU-1992 for 200/100 W, ATU-1992/LN for 400/200 W).

The SRT-170/[M] is a DC power version of the SRT-[170/270]/L variants.

Design characteristics are based on modular digital radio concepts with extensive use of DSP and DDS techniques and digital processors to minimise cost of ownership and optimise flexibility and growth to meet the requirements of STANAG 4444 (ECCM), STANAG 4538 (ARCS) and STANAG 4539 (fixed frequency). The CP-2001[X] provides simultaneous management and control of all on-board communications, intercom systems and navigation systems, as well as providing management of special facilities such as ALE, data modems and satcom; it features multilayer display techniques and GEN III NVG capability.

Operational status

SRT-170/M selected for: EH 101 (UK Royal Air Force SH, UK Royal Navy Merlin and Italian Navy ASW versions); NH 90 both NFH and TTH versions.

SRT-470 fitted to Italian Air Force Tornado aircraft.

SRT-470/L selected for: UK Royal Air Force Nimrod MRA4; Australian, UK and Italian air forces C-130J.

Contractor

Elmer SpA (a Marconi Communications company).

VERIFIED

SRT-194 VHF/AM radio

Elmer's SRT-194 is a VHF/AM transmitter/receiver covering the 108 to 156 MHz band with channel spacings of 25 kHz. Up to 20 channels, plus an additional Guard channel, are preselectable. The system provides air-to-air and air-to-ground communication in AM mode, together with modulated carrier wavetone and retransmit facilities; ADF and homer functions are also optionally available by the addition of appropriate ancillary equipment. The system comprises a transmitter/receiver unit with a remote controller; an additional channel/frequency selector can also be provided.

A modular 'slice' style of construction is employed and new functions can be incorporated by simply adding appropriate slices. Construction is all-solid-state and miniaturisation techniques are extensively employed. Power consumption is low and the thermal design is claimed to be extremely efficient, so that no forced-air cooling supply is required. The system is compatible with a wide range of interphone systems and identical pin-to-pin connections with Elmer UHF systems render the SRT-194 readily interchangeable with these without change to aircraft wiring or mountings. Transmitter output power is 10 W minimum. The equipment conforms to MIL-E-5400 Class II modified specification.

Specifications

Dimensions:
(control panel) 83 × 146 × 181 mm
(transmitter/receiver) 228 × 127 × 127 mm
(channel/frequency indicator) 83 × 33 × 154 mm
Weight:
(control panel) 1.4 kg
(transmitter/receiver) 4.3 kg
(channel/frequency indicator) 0.5 kg

Operational status

In production, and in service.

Contractor

Elmer SpA (a Marconi Communications company).

VERIFIED

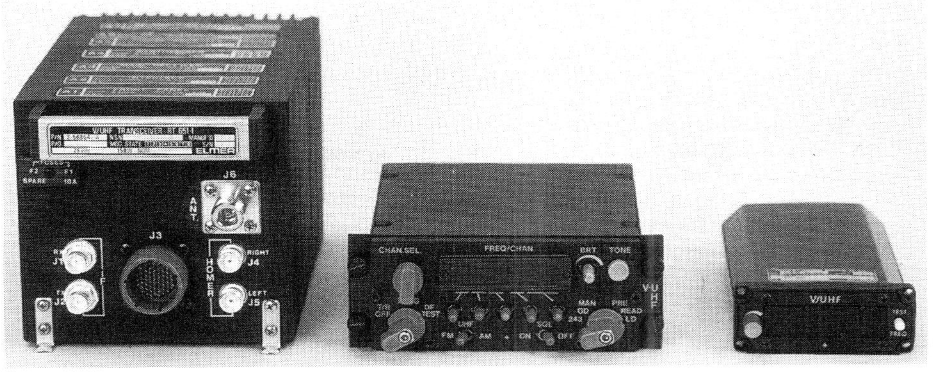

The Elmer SRT-651 VHF/UHF transceiver

SRT-651 VHF/UHF AM/FM transceiver

The SRT-651 is an all-solid-state, compact, lightweight airborne transceiver covering the 30 to 400 MHz frequency range and using the most advanced techniques in the area of large-scale integration components and microprocessors. The system comprises a CP-1200 control panel, RT-651 receiver/transmitter and optional ID-1151 channel/frequency repeater.

Conceived and designed for airborne use, the SRT-651 is also suitable for a wide range of applications where space and weight are limited. The flexible modular 'slice' design permits easy assembly and disassembly and allows expansion of the operating functions. The RF power rating can be easily upgraded to 30 W by changing the transmitter slice. Different interface standards are available, including an Elmer serial data interface using control panel CP-1200 or MIL-STD-1553B or ARINC 429.

BIT facilities can be implemented by using different plug-in cards for the interface slice. The test result is automatically monitored on the control panel display, which gives a direct identification of the faulty slice.

The SRT-651 has facilities for ECCM including frequency-hopping ('Have Quick' has been implemented) and spread spectrum pseudo-noise modulation. It can be operated with a wide variety of ancillaries including homer indicators, VHF/UHF direction-finders, UHF emergency beacons, KY-58 secure voice modems in diphase and baseband modes, Elmer SP-1212 ECCM spread spectrum voice and data modem, NATO multitone Link 11 data modem and FSK modems. Various mountings can be provided to retrofit the SRT-651 directly into existing VHF and UHF installations.

Specifications

Dimensions:
(RT-651) 126.7 × 120.6 × 224 mm
(CP-1200) 146 × 57.1 × 150 mm
Weight:
(RT-651) 3.5 kg
(CP-1200) 1.4 kg
Power supply: 28 V DC
(transmit) 150 W max
(receive) 25 W max
Power output:
(AM) 10 W
(FM) 15 W
Frequency: 30-88 MHz VHF/FM, 108-156 MHz VHF/AM, 156-174 MHz VHF/FM and 225-400 MHz UHF AM/FM
Channel spacing: 25 kHz in all bands
Guard receiver: 40.5, 121.5, 156.8 and 243 MHz automatically selected
Preselected channels: 20 using electronic memory

Operational status

Fitted to Italian Air Force Tornado aircraft.

In production for Italian, Brazilian and other armed forces and for installation on Alenia/Aermacchi/Embraer AMX aircraft.

The SRT-651/A variant is fitted to Augusta A129 Mangusta helicopters of the Italian Army.

Contractor

Elmer SpA (a Marconi Communications company).

VERIFIED

SRT-651/N V/UHF - AM/FM transceiver

The SRT-651/N transceiver is designed for military operation, to provide voice and data communication over the 30 to 400 MHz band (optionally 30 to 470 MHz). It can be configured to provide:

EPM/jamming resistance through use of frequency hopping techniques including Have Quick and Saturn (export EPM capability by use of EASY II);
Link 11 operation, in association with a Link 11 data terminal unit;
secure voice/data operation in association with a wide variety of crypto modems;
selective call and tone activated repeaters capability;
CASS/DICASS capability;
continuous coverage from 30 to 470 MHz;
70 MHz IF standard modem interface for satcom;

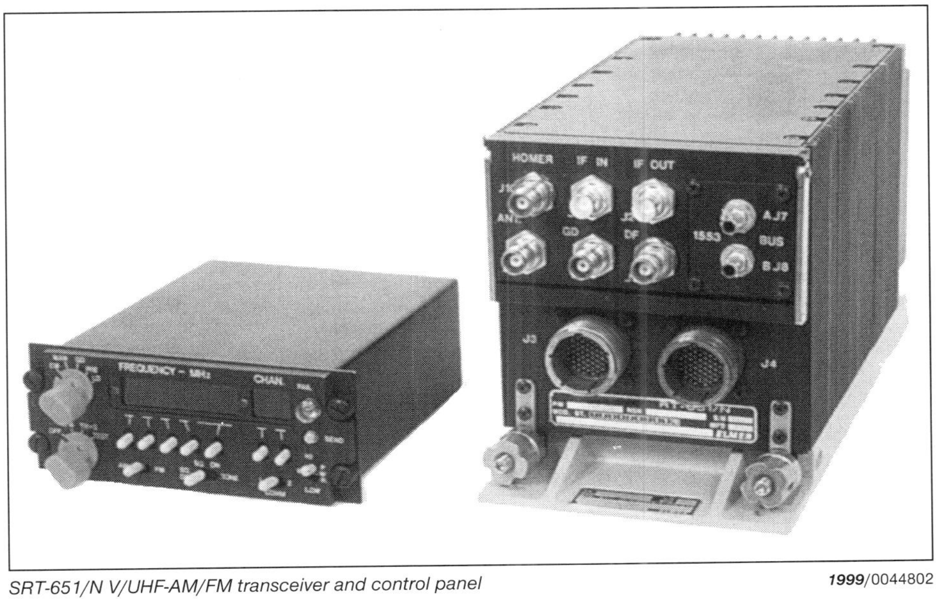

SRT-651/N V/UHF-AM/FM transceiver and control panel **1999**/0044802

databus control via MIL-STD-1553B or ARINC 429;
dual system operation.

The SRT-651/N is available in the following configurations:

standard version: RT-651/N transceiver featuring fixed frequency operation; CP-9000 cockpit control panel;
databus controlled version: RT-651/N transceiver incorporating a databus controller;
Have Quick/Saturn/Easy II version: RT-651/N transceiver featuring fixed frequency and frequency hopping operation; CP-9000 (Have Quick/Saturn/Easy II) cockpit control panel;
Have Quick databus controlled version: RT-651/N (Have Quick/Saturn/Easy II) transceiver incorporating databus controller.

Ancillary equipment can be associated with the transceiver to enhance capabilities including:
homer indicator ID 1351A (homer circuitry is built into the SRT-651/N);
direction-finders;
Link 11 data terminal set;
data modem/FSK modem;
crypto modem in diphase and baseband modes;
selcal and Tone Activated Repeater operation via CP-2001.

Operational capabilities include:
30 to 88 MHz VHF-FM operation with ground forces;
transmit inhibitor of 108 to 116 MHz VHF-AM band to avoid interference with VOR or landing systems;
118 to 156 MHz VHF-AM operation with ATC;
156 to 174 MHz VHF-FM maritime half-duplex operation;
225 to 400 MHz UHF-AM/FM air/ground operation;
400 to 470 MHz UHF-FM option for compatibility with Tone Activated Repeaters required by law enforcement agencies.

Specifications

Frequency bands:
VHF-FM: 30-88 MHz, 156-174 MHz (half duplex)
VHF-AM: 108-156 MHz
UHF-AM/FM: 225-400 MHz
UHF-FM: 400-470 MHz (option)
Preset channels: 99 (with CP-9000 or CP-2001)
Channel spacing: 25 kHz (12.5; 8.33; 5 and 2.5 kHz as options)
Guard channels: 40.5, 121.5, 156.8 and 243 MHz, automatically selected with operating band
Transmit power:
AM: 10 W
FM: 15 W
Power consumption: 28 V DC; 195 W (Tx) and 55 W (Rx)
MTBF: 2,000 h

Operational status

Selected for the following programmes: C-130J (Australian, UK and Italian air forces); Eurofighter Typhoon; EH 101 (UK Royal Air Force SH, UK Royal Navy Merlin, Italian Navy ASW); NATO AWACS upgrade; NH 90 NFH and TTH versions; Nimrod MRA4, Tornado UK Royal Air Force. Italian armed forces Sicral satellite communications.

Contractor

Elmer SpA (a Marconi Communications company).

VERIFIED

The Elmer SRT-653 VHF/UHF transceiver is installed on Italian Air Force Tornado aircraft

SRT-653 VHF/UHF transceiver

Designed to meet a requirement of the Italian Air Force for its version of the Panavia Tornado, the SRT-653 VHF/UHF transceiver comprises the RT-1051 UHF transceiver, SP-1047 VHF transceiver, SP-1203 adaptor/power supply, SP-1204 mounting tray, CP-1001 control panel and ID-1150 frequency repeater.

The RT-1051 UHF transceiver belongs to the basic ARC-150(V) family. The SP-1047 VHF transceiver has been derived by Elmer from this system and maintains its general characteristics.

Specifications

Dimensions:
(ID-1150) 86 × 50 × 159 mm
(SP-1047/RT-1051 on tray) 265 × 155 × 408 mm
(CP-1001) 146 × 95.2 × 165.5 mm
Weight:
(ID-1150) 0.35 kg
(SP-1047 and RT-1051 on tray) 13.5 kg
(CP-1001) 1.6 kg
Power supply: 115 V AC, 400 Hz, 3 phase
(transmit) 320 VA max
(receive) 140 VA max
Power output: 10 W
Frequency: 108-156 and 225-400 MHz
Channel spacing: 25 kHz
Preset channels: 17 + 2 Guard channels

Operational status

In service on Italian Air Force Panavia Tornado aircraft.

Contractor

Elmer SpA (a Marconi Communications company).

VERIFIED

LN-93EF ring laser gyro inertial navigation system

The LN-93EF is based on the LN-93 ring laser gyro standard navigator built by Litton to SNU-84 standards. However, the EF version, modified to Eurofighter Typhoon requirements, includes upgraded electronics, a fibre optic databus, and Ada software. The result is a system with the same accuracy and maturity but with a 30 per cent reduction in size and weight than the SNU-84 system.

Specifications

Dimensions: ¾ ATR (length 360 mm)
Weight: 16 kg
Power supply: 115 V AC, 400 Hz, 85 W
28 V DC back-up
Temperature range: −40 to +71°C
Align time:
(gyrocompass) 4 min
(memorised) 30 s
Accuracy:
(position) 1 n mile/h CEP
(velocity) 0.8 m/s RMS
(attitude) 0.05° RMS
(heading) 0.1° RMS

Contractor

Litton Italia SpA.

VERIFIED

ANV-301 Doppler navigation system

The ANV-301 Doppler navigation system provides a self-contained worldwide precision navigation capability, utilising Doppler-derived measurement of aircraft velocity and external input of aircraft attitude and heading. The ANV-301 design approach allows system functions to be tailored to specific customer requirements at minimum cost and the electrical interface to be tailored for specific aircraft avionics and subsystems for additional enhancement of system capability.

The system comprises four LRUs: the receiver/transmitter radar, signal data converter, control display unit and pilot steering indicator.

Three pilot-selectable navigation co-ordinate systems with automatic co-ordinate conversion are available: latitude/longitude, worldwide alphanumeric UTM and arbitrary grid.

Contractor

Marconi SpA.

VERIFIED

ANV-351 Doppler velocity sensor

The ANV-351 Doppler velocity sensor provides precision measurement of helicopter velocity components. The ANV-351 is a single-unit sensor specifically designed for integrated avionic systems and weighing less than 7 kg. It comprises two main modules: the receiver/transmitter radar and the signal data converter.

The receiver/transmitter radar module is of fixed design and all-solid-state construction. It comprises separate four-beam transmit/receive antennas, utilising printed circuit planar-array technology. The module is available in two versions, optimised for either ASW or nap of the earth operation.

The signal data converter module comprises power supply, Doppler signature processing and MIL-STD-1553B interface.

Contractor

Marconi SpA.

VERIFIED

ANV-353 Doppler velocity sensor

The ANV-353 Doppler velocity sensor provides precision measurement of helicopter velocity components. It is a single-unit low-power sensor specifically designed for integrated avionic systems and weighing less than 5 kg. It comprises two modules: a receiver/transmitter radar and signal data converter.

The receiver/transmitter radar module is of fixed design and all-solid-state construction. It comprises separate four-beam transmit/receive antennas utilising printed circuit planar-array technology. The redundant fourth beam provides high accuracy during extreme attitude manoeuvres. This module contains all RF signal generation and processing functions and receives its power input, modulation and timing signal from the signal data converter.

The signal data converter comprises power supply Doppler digital signal processing and discrete and ARINC 429 interfaces. An MIL-STD-1553B interface is an option. Individual beam frequency shifts are extracted from the spectrum received from the receiver/transmitter radar module, utilising a single time-shared IF channel and digital signal processing.

The extensive use of advanced components allows controllable RF transmitted power and automatic land/sea transition.

Contractor
Marconi SpA.

VERIFIED

MIDA airborne data terminal

The Message Interchange Distributed Application (MIDA) is a digital datalink used for data communication between manned or unmanned aircraft and the ground control station. It is a spread spectrum bidirectional microwave link structured on a time division basis.

MIDA includes a video processing section composed of an ADPCM codec for video data bandwidth compression and an automatic lock follower unit performing the automatic tracking of the target selected by the command and control station.

MIDA is able to provide both downlink and uplink facilities. Downlink includes codified monochrome video data from a 625/50 CCIR standard camera output or uncodified monochrome video data at a low frame rate or MTI, ESM, SAR surveillance radar data and telemetry and aircraft status data on altitude, airspeed, heading and failures. Uplink includes remote command data for camera platform steering commands, target designation, aircraft flight profile and mission updating; and link service data for codified or uncodified video mode selection, surveillance data retransmission requests, acknowledgement procedures and so on. DME capability is also provided by MIDA.

MIDA operates in the D- or J-bands. It consists of the airborne data terminal which has two LRUs: the spread spectrum transceiver unit and the antenna unit. Besides the microwave, IF and baseband sections, the transceiver unit contains the processing units which perform the video signal bandwidth compression and target automatic tracking. The airborne data terminal is interfaced via ARINC 429 or MIL-STD-1553 to the airborne main computer.

Contractor
Marconi SpA.

VERIFIED

The Marconi MIDA airborne data terminal

PRT 403 airborne transmitter

The PRT 403 comprises the transmitter and a high-gain rotating aerial to which the transmitter is connected by a waveguide feed. The equipment is completely solid state including the modulator, except for the power amplifier which employs a klystron tube. It is used for transmitting mid-course guidance information to a missile.

Both the logic data handling section and the RF units employ LSI, thin film and microstrip hybrid circuits, multilayer and crossover thick film hybrid circuits, allowing extremely small weights and dimensions.

The main functions of the equipment are periodically checked by BITE which is initiated by the system computer. Major defects and performance degradation are detected and displayed to the operator in real time. A major failure to the power unit and its auxiliary circuits instantaneously operates automatic protection circuits.

The equipment is designed and manufactured for operation in the MIL-E-5400 Class 1B environment.

Specifications
Dimensions: 257.1 × 193.5 × 497 mm
Weight: 25 kg
Power supply: 115 V AC, 400 Hz, 3 phase

Contractor
Marconi SpA.

VERIFIED

RALM-01 laser warning receiver

The RALM-01 laser warning receiver provides laser threat detection and classification capability. Specially designed for airborne platforms, the RALM-01 can either be integrated with other EW equipment or work as a stand-alone equipment and directly drive countermeasures systems.

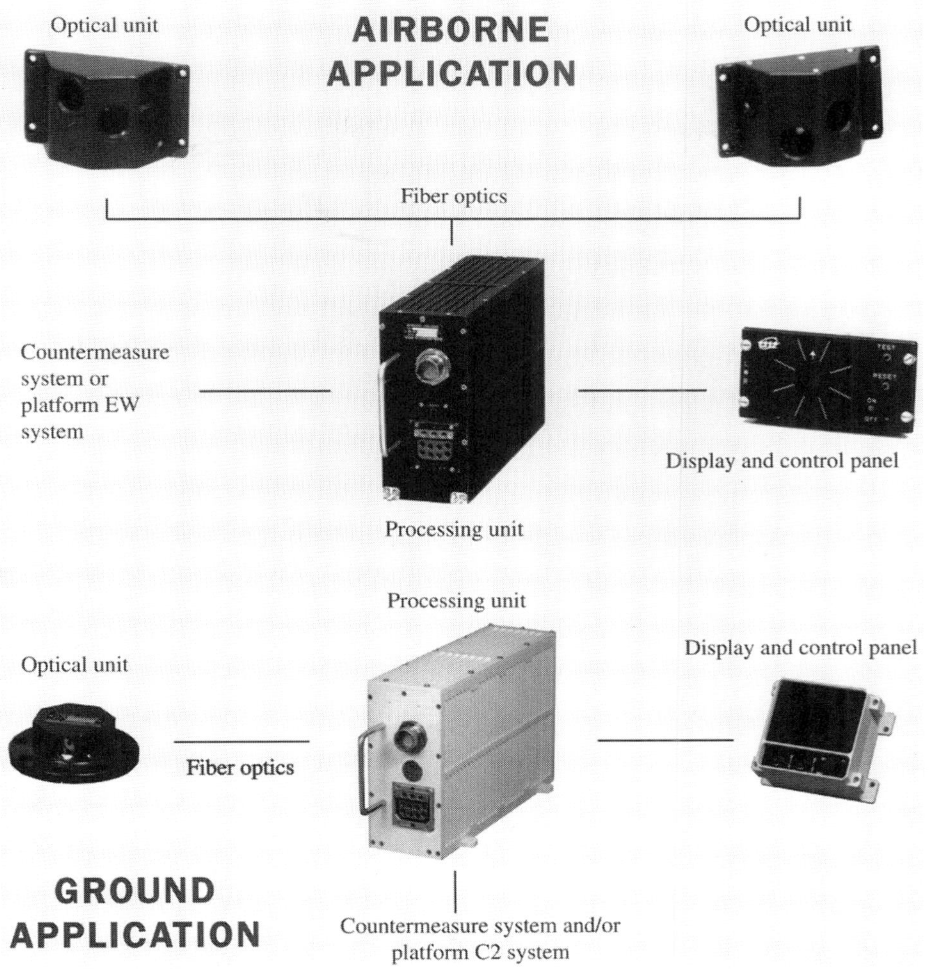

RALM-01 laser warning receiver 0018252

Based on two side-mounted head sensor units, the RALM-01 standard version provides full 360° azimuth coverage and 90° vertical coverage in the visible and near infra-red bands, while the RALM-01/1 version makes use of two additional side-mounted head sensor units to provide far infra-red band coverage.

The threat direction is shown, to an accuracy of 45° in azimuth, on the display together with the threat type. An audio alarm is also generated. Either an RS-232C or MIL-STD-1553 interface is provided for integration with other EW equipment. A blanking input signal and a countermeasures start output signal are also available.

Specifications
Azimuth coverage: 360°
Elevation coverage: 90°
Sensor band: 0.5 to 1.8 μm
Extended sensor: 8 to 12 μm
Weight: 4.5 kg
Dimensions:
(sensors (× 2)) 90 × 50 × 40 mm
(processor) ⅜ ATR short

Operational status
In production for Mangusta A129 and HH-3F SAR helicopters, and C-130J aircraft.

Contractor
Marconi SpA.

VERIFIED

SIT 421 family of Identification Friend or Foe (IFF) transponders

The SIT 421 family of airborne IFF transponders comprises the following models:
(1) SIT 421 cockpit-mounted set;
(2) SIT 421T remotely controlled set, including SIT 901 control unit for full capability (or C-6280A(P) APX control box providing reduced capability);
(3) SIT 421T-1553 bus-controlled set.

Both the SIT 421 and SIR 421T models have also been adopted by the Italian Navy for frigate and hydrofoil use, where it is given the nomenclature MM/UPX-709.

The SIT 421 provides operation in Modes 1, 2, 3/A, 4 and C. For Mode C operation, the transponder operates with an external pressure/altitude digital converter.

The receiver-transceiver comprises a 500 W solid-state transmitter, a dual-channel receiver and an RF interface module. The solid-state transmitter includes a delay line oscillator, a modulator, a driver and a power amplifier.

The dual-channel receiver handles space diversity operation to ensure reliable transponder response to the received interrogation. The receiver may also be set for operation with one antenna only (single-channel operation).

The signal processor consists of a video processor, a decoder, a coder and a video interface with the KIT-1A-/TSEC for Mode 4 operation.

The SIT 421 includes an open board slot to accommodate anti-jam circuits and other improvement circuits, as growth options.

The SIT 421 IFF cockpit-mounted transponder
2000/0062223

The SIT 421T remotely-controlled IFF transponder
2000/0062224

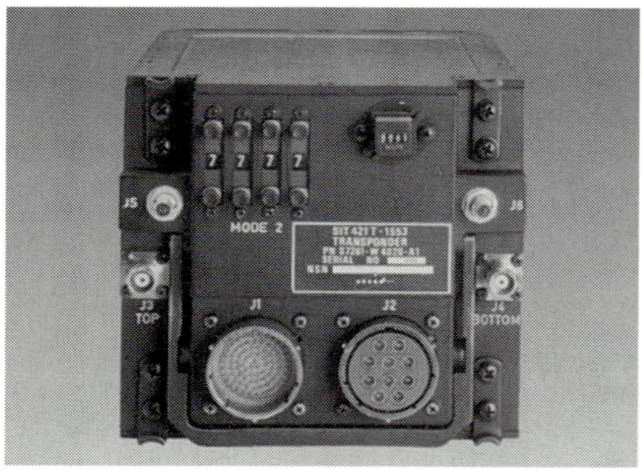

The SIT 421T-1553 bus-controlled IFF transponder
2000/0062225

SIT 421T/SIT 901 control unit

The SIT 421T directly interfaces with the SIT 901 control unit for full operational capability, or a C-6280A (P) APX control box for reduced operational capability.

SIT 421T-1553

The SIT 421T-1553 fulfils the same functions as the SIT 421T, but it is bus controlled via a MIL-STD-1553B bus.

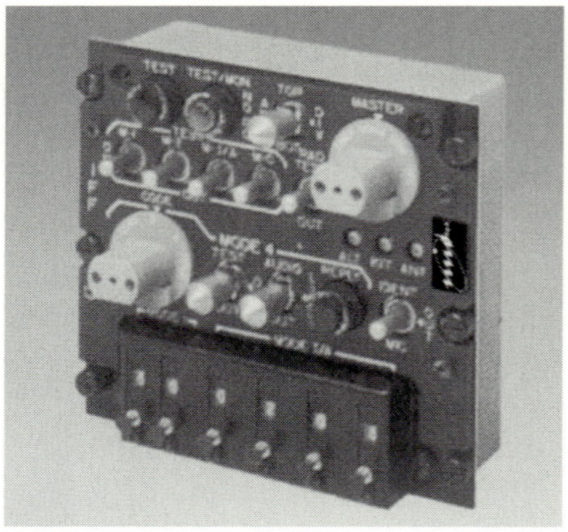

The SIT 901 IFF control unit **2000**/0062226

Being computer controlled, it includes automatic functions, including automatic code change.

Specifications

Output power: 27 ±3 dB at 1% duty cycle
Dimensions:
(SIT 421) 145 (W) × 134 (H) × 190 (D) mm
(SIT 421T) 136.5 (W) × 136.5 (H) × 213 (D) mm
(SIT 901) 133 (W) × 146 (H) × 43.2 (D) mm
(SIT 421T-1553) 146.8 (W) × 136.9 (H) × 219.8 (D) mm
Weight:
(SIT 421T) 4.1 kg
(SIT 901) 0.8 kg
(SIT 421T-1553) 5 kg
Power:
(SIT 421) 28 V DC, 50 W
(SIT 421T) 28 V DC, 50 W
(SIT 421T-1553) 28 V DC, 65 W

Contractor

MID SpA (Marconi Italia Defence).

VERIFIED

SIT 434 family of Identification Friend-or-Foe (IFF) interrogators

The SIT 434 family of IFF interrogators is designed for fixed- and rotary-wing aircraft. Modular architecture enables it to comply with the characteristics of different platforms by adding or replacing different modules, which include:

(1) SIT 434A basic airborne IFF interrogator unit, remotely controlled by the SIT 905 control unit;
(2) SIT 434B airborne IFF interrogator unit, associated with an external active/passive decoder;
(3) SIT 434C airborne IFF interrogator, with plot extractor and active/passive SIF decoding, driven by MIL-STD-1553B bus.

The SIT 434A basic airborne IFF interrogator
2000/0062227

All members of the SIT 434 interrogator family include the following features: operation in accordance with STANAG 4193 in Modes 1, 2, 3/A, 4 and C, either single or interlaced; RSLS and ISLS functions, when operated with a suitable antenna; M4 evaluation, decision threshold and anti-spoof measures; ECCM protection by means of adaptive thresholds; passive decoding - both 'bracket' and 'stretched code' are selectable for display, a dual-code capability is provided for M3 overlapping periods; digital symbol generation providing individual video outputs for SIF and M4 friendly target indication.

Optional features include:

(1) SIT 434B: pulse-to-pulse 2/2 defruiter (replacing the internal decoding), and an interface for an external active/passive decoder (replacing the digital symbol generation);
(2) SIT 434C: control of the interrogator and target information interface by means of a dual-redundant RTU per MIL-STD-1553B; challenge management including interrogation sector(s) capability (based on expected target position), and interlaced pattern and interrogation PRF optimisation; antenna synchro interface to associate azimuthal data to the SIF/M4 plot.

Associated equipment includes the SIT 905 (for SIT 434A and SIT 434B models only); an antenna subsystem (sum and difference); and a Mode 4 computer.

Specifications

Transmitter power: 1,200/300 W (selectable)
Dimensions: 1 ATR short
Weight: 15 kg
Power: 115 V AC, 400 Hz, 140 W
MTBF: 1,000 h

Contractor

MID SpA (Marconi Italia Defence).

VERIFIED

JAPAN

AFMS controller for the SH-60J

The Automatic Flight Management System (AFMS) controller is used in the automatic flight management system installed in the Mitsubishi/Sikorsky SH-60J anti-submarine helicopter. The AFMS provides a flight management integrated display, alarms and flight management. The AFMS controller is a digital computer and provides the data processing both for the automatic flight management system and for navigation equipment management.

Operational status

In service in the Mitsubishi/Sikorsky SH-60J helicopter.

Contractor

Japan Aviation Electronics Industry Ltd.

VERIFIED

Automatic flight control system for the F-1

The automatic flight control system installed on the Mitsubishi F-1 fighter uses a digital computer in a single system. The system monitors attitude, azimuth and altitude hold, reducing the pilot's workload and contributing to flight safety and mission performance. It includes fail-safe control.

Operational status

In service in the Mitsubishi F-1 aircraft.

Contractor

Japan Aviation Electronics Industry Ltd.

VERIFIED

JSN-8 strapdown laser Attitude and Heading Reference System (AHRS)

The JSN-8 high-performance laser AHRS is suitable for military and civil applications. The sensors are three Honeywell laser gyroscopes with three JAE JA-5 accelerometers. All the sensor equipment and associated processing is enclosed in a single unit and there is a separate control panel. In its current form the equipment operates solely as an attitude and heading reference system, but it can be developed to include inertial navigation functions and so has inertial navigation type interfaces.

It can operate in high angular rate conditions of up to 400°/s roll rate and does not suffer any performance degradation due to acceleration. It has a short reaction time and does not require a flux valve device. Reliability is quoted as in excess of 2,300 hours. In-flight alignment can be accomplished, aided by such systems as Doppler, Tacan or Navstar, and magnetic heading can be provided as an output.

Specifications

Dimensions:
(control panel) 100 × 146 × 76 mm
(attitude/heading reference unit) 250 × 330 × 180 mm
Weight:
(control panel) 0.8 kg
(AHRS) 19.8 kg
Power supply: 115 V AC, 400 Hz, 30 W
28 V DC, 115 W
Input/output:
(digital) ARINC 429 or MIL-STD-1553B or special proprietary

(analogue) synchro attitude signal outputs, and analogue turning rate signal
(discrete) valid signal and bit command output
Alignment time:
(normal) 2.5 min
(stored heading) 1 min
Environmental: MIL-E-5400T Class 2X
Electromagnetic interference: MIL-STD-461A, -462

Operational status
In production.

Contractor
Japan Aviation Electronics Industry Ltd.

VERIFIED

Laser gyro inertial navigation system

The strapdown ring laser gyro inertial navigation system is installed on the CH-47 transport helicopter. The use of ring laser gyros in the system permits increased reliability, longer life and so on, than a conventional platform-type inertial navigation system, considerably reducing life cycle costs, including maintenance and overhaul costs.

Operational status
In service in Kawasaki/Boeing CH-47 helicopters of the Japanese Air Self-Defence Force.

Contractor
Japan Aviation Electronics Industry Ltd.

VERIFIED

Strapdown Attitude and Heading Reference System (AHRS) for the SH-60J

The strapdown attitude and heading reference system installed on the Mitsubishi/Sikorsky SH-60J anti-submarine helicopter detects pitch, roll and azimuth angles and turn rates with reference to all three axes of the body and supplies this data to other airborne equipment. Combined with a Doppler navigation system, the strapdown attitude and heading reference system can also function as an inertial navigation system. The data is supplied to other equipment through a MIL-STD-1553B databus. The strapdown system uses a small ring laser gyro, is compact and has higher performance, reliability and life than a conventional attitude and heading reference system.

Operational status
In service in Mitsubishi/Sikorsky SH-60J helicopters of the Japanese Maritime Self-Defence Force.

Contractor
Japan Aviation Electronics Industry Ltd.

VERIFIED

Strapdown Attitude and Heading Reference System (AHRS) for the T-4 aircraft

The strapdown attitude and heading reference system for the Kawasaki T-4 is an integrated sensing system. It provides aircraft acceleration signals, angular velocities and attitude in three axes to other airborne equipment. Data transmission is via ARINC 429 databusses. Laser gyros are incorporated in the system, leading to high reliability and service life.

Operational status
In service in Kawasaki T-4 aircraft of the Japanese Air Self-Defence Force.

Contractor
Japan Aviation Electronics Industry Ltd.

VERIFIED

J/ALQ-5 ESM system

Used on ESM variants of the Kawasaki C-1 medium transport, the J/ALQ-5 ESM system receives and jams surface-to-air missile radars.

Operational status
In service on the Kawasaki C-1 ESM variant.

Contractors
Mitsubishi Electric Corporation.
NEC Corporation.

VERIFIED

J/ALQ-6 jamming system

The J/ALQ-6 jamming system is in development to provide an airborne radar jamming capability for F-4EJ and Mitsubishi F1 fighters.

Operational status
In development for F-4EJ and Mitsubishi F1 aircraft.

Contractors
Mitsubishi Electric Corporation.
NEC Corporation.

VERIFIED

J/ALQ-8 jamming system

The J/ALQ-8 is in development for the F-15J; it may be a variant of the J/ALQ-6 set. It is understood to provide countermeasures in the 1 to 4, 4 to 8 and 7.5 to 18 GHz bands. ALQ-8 jammers integrate with the J/APR-4 radar warning system.

Operational status
In development for the F-15J.

Contractors
Mitsubishi Electric Corporation.
NEC Corporation.

VERIFIED

J/APQ-1 rear warning receiver

The J/APQ-1 rear warning receiver has been developed for the F-15J aircraft. It is designed to cope with both radar and infra-red threats and will automatically activate chaff and flare countermeasures. The starboard side of the tail of the F-15J will be modified to allow the fitting of a radar antenna with a diameter of 200 mm. A visual indicator and audio alert are positioned in the cockpit.

Operational status
Developed for the F-15J. Believed to have entered service in 1992.

Contractors
Mitsubishi Electric Corporation.
NEC Corporation.

VERIFIED

Radar for the FSX

A new airborne radar for the FSX close support aircraft has been developed by Mitsubishi Electric Corporation under the management of the Japanese Defence Agency's Technical Research and Development Institute. The radar has a 66 cm diameter active phased-array antenna made up of 750 modules. It is reported to have track-while-scan facilities, with each module generating computer-controlled radar beams.

Operational status
In development.

Contractor
Mitsubishi Electric Corporation.

VERIFIED

J/APR-4/4A radar warning system

The J/APR-4 was designed for the F-15J/DJ aircraft. It is able to process multiple inputs simultaneously in a dense electromagnetic environment, and has a digital computer with a reprogrammable software package to allow reconfiguration for future requirements. The indicator provides for daylight viewing and multithreat data presentation in alphanumeric and graphic format. The system is also designed to interface with other EW equipment such as the J/ALQ-8.

The J/APR-6 is installed in Japan Air Self-Defence Force F-4EJ Kai aircraft

The J/APR-4A is the advanced model of the J/APR-4 radar warning receiver. Its specification calls for the ability to process multiple inputs simultaneously in a dense electromagnetic environment. The system incorporates a digital processor with reprogrammable software which permits reconfiguration to meet developing threats. A tactical situation CRT display presents multiple-threat data in alphanumeric and graphic form. Interfaces with other onboard electronic countermeasures systems, such as J/ALQ-8 jammers, can be accommodated.

Operational status
In production for the F-15J/DJ.

Contractor
Tokimec Inc.

VERIFIED

J/APR-5 and J/APR-6 radar warning systems

The J/APR-5 and J/APR-6 are developments of the J/APR-4 system, with additional capability designed to cope with current threats. Actual sizes and weights vary according to application.

Operational status

All RF-4E reconnaissance aircraft are fitted with the J/APR-5. The J/APR-6 equips the F-4EJ Kai aircraft.

Contractor

Tokimec Inc.

VERIFIED

The Tokimec J/APR-5 RWR equips Japan Air Self-Defence Force RF-4E aircraft

AFCS for helicopters

Production of the Automatic Flight Control System (AFCS) for helicopters has started, as well as for fixed-wing aircraft. The system consists of the FCC-20 Digital Flight Control Computer (FCC), DCP-20 Flight Control Panel and ADC-20 Air Data Computer. The AFCS is a single channel system, but the FCC in the system features not only digital auto-pilot and digital SCAS functions, but also analogue back-up SCAS functions and a dual PSU system, which contribute to flight safety.

Operational status

In production.

Contractor

Tokyo Aircraft Instrument Co Ltd.

VERIFIED

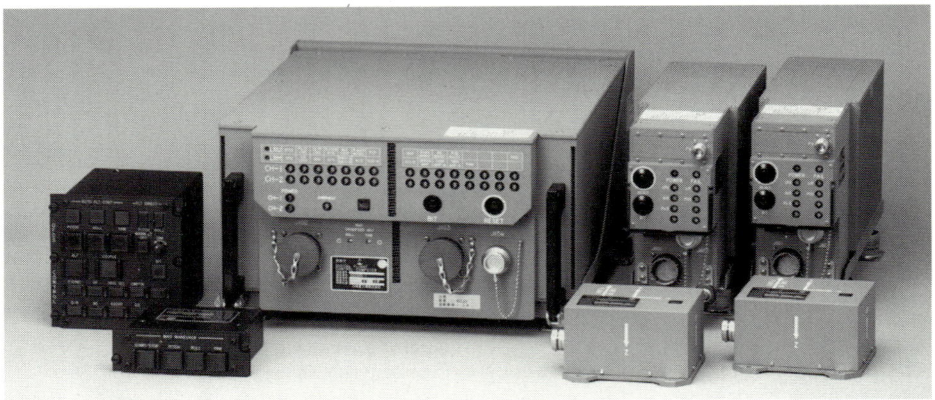

The Tokyo Aircraft Instrument Co AP-120 DFCS

AP-120 DFCS

AP-120 Digital Flight Control System (DFCS) is a high-grade comprehensive flight control system which consists of Digital Flight Control Computer (DFCC), two altitude controllers, two accelerometers, main and sub control panels. The DFCC, as a core unit, has both autopilot and flight director functions. It features dual-redundant fail-operative/fail-passive capability and is inertially smoothed to give high-accuracy stable altitude-hold capability. High maintainability is ensured through extensive use of an In-Flight Performance Monitor (IFPM), Built-In Tester (BIT) and non-volatile maintenance memories, and the use of selected high-reliability parts and advanced thermal design techniques. The system can be expanded to provide a MIL-STD-1553B databus interface and other enhancements.

Operational status

In production and service in the Lockheed/Kawasaki P-3C aircraft and its derivatives.

Contractor

Tokyo Aircraft Instrument Co Ltd.

VERIFIED

Four-axis sidestick controller

The four-axis sidestick controller has been developed by Tokyo Aircraft Instrument Company in association with Kawasaki Heavy Industries. It consists of a sensor unit and a control box. The system has quadruple redundancy in each axis.

The sensor is a force deflection right-side stick, having four-bridge solid-state strain gauges, which provide electric signals for pitch, roll, yaw and CP to a flight control computer.

Operational status

Under evaluation in the Kawasaki BK-117 fly-by-wire experimental helicopter.

Contractor

Tokyo Aircraft Instrument Co Ltd.

VERIFIED

The four-axis sidestick controller is under evaluation in a BK-117 fly-by-wire helicopter

Hybrid Inertial Sensor Unit (HISU)

The Hybrid Inertial Sensor Unit (HISU) is a core sensor of the Attitude and Heading Reference System (AHRS) now planned. The core sensor, the development of which has been completed, incorporates three Fibre Optic Gyros (FOGs) and three accelerometers, which provide three-axis angular rates and three-axis accelerations.

The sensor is to be integrated with a strapdown system (also under evaluation), into one component.

Operational status

Under evaluation.

Contractor

Tokyo Aircraft Instrument Co Ltd.

VERIFIED

Hybrid Inertial Sensor Unit (HISU) 0001333

NETHERLANDS

Link 11 DataLink Processor (DLP)

The Signaal Special Products (SSP) Link 11 DLP meets the requirements of STANAG 5511 or Opspec 411 for the operation at airborne Link 11 DLP systems in the standard Link 11 environment for all NATO countries and members of the Partners for Peace organisation.

SSP's Link 11 DLP can be fully integrated into an aircraft's command and control system, it can be operated as a stand-alone unit with a simple interface to a host system allowing track data to be passed to/from the Link 11 network. The stand-alone Link 11 DLP is equipped with a Digital Geographic Information System (DGIS), which displays track data on a map or chart of the operational area.

The most common airborne version of the Link 11 DLP is housed in a fully certified. Alternatively airworthy ¼ ATR unit, operating on 28 V DC, but other configurations are available.

Operational status

Ready for full production.

Contractor

Hollandse Signaalapparaten BV, Signaal Special Products.

VERIFIED

Link 11 datalink system

The Signaal Special Products division of Hollandse Signaalapparaten produces lightweight Link 11 systems fully qualified for all airborne environments. The system hardware was developed by Ultra Electronics Limited for a UK programme (see entries for T618 Link 11 Data Terminal Set (DTS) and T619 Link 11 DataLink Processor (DLP) in the UK section under Ultra Electronics, Sonar and Communication Systems). All software is developed by Signaal Special Products and can be tailored to meet specific requirements.

The DLP has provision for an embedded crypto unit, together with all the necessary controls, resulting in a considerable volume and weight reduction.

The DTS provides the modem and network control functions as defined in MIL-STD-188-203-1A and provides functionality for a unit to act as the Net Control Station (NCS) or 'picket' (active or passive).

The Link 11 application software resident in the DLP enables the exchange of track data, management data, status data and commands between shipborne, airborne and land-based units participating in the network.

The DTS provides the following operational capabilities, both multitone and single-tone operation; automatic or manually selected diversity reception; a modem containing link monitoring software that provides performance and reception quality analysis functions.

The DLP provides all necessary message handling, track correlation, gridlock, filters and conflict management functions. It also provides for exchange of real-time and non-realtime track data, management data such as IFF/SIF data reports and Link 11 management functions.

Specifications

Data terminal set

Interfaces:
(serial) TDS interface to DLP
(control) ARINC 429, MIL-STD-1553B or RS-232C
(audio) 600 ohm balanced interface to HF or UHF transceiver
Dimensions: ¼ ATR short
Weight: 4 kg
Power:: 115 V AC, < 30 VA
28 V DC, < 25 W

Datalink processor

Interfaces:
(serial) ATDS to crypto unit
(control) MIL-STD-1553B
Dimensions: ½ ATR short
Weight: 8.5 kg
Power: 115 V AC, < 40 VA
28 V DC, < 25 W

Contractor

Hollandse Signaalapparaten BV.

VERIFIED

Link-Y Mk 2 datalink

The Link-Y Mk 2 datalink system exchanges system track data, management data, status data and commands between shipborne, land-based and airborne units participating in the datalink network.

Link-Y Mk 2 allows the establishment of a fleetwide common tactical database through the exchange of real-time and non real-time track data. By exchanging management data, such as IFF/SIF reports, information difference reports and conflict reports, a complete common tactical picture is established.

The exchange of commands and status data, such as force disposition orders, weapon readiness reports

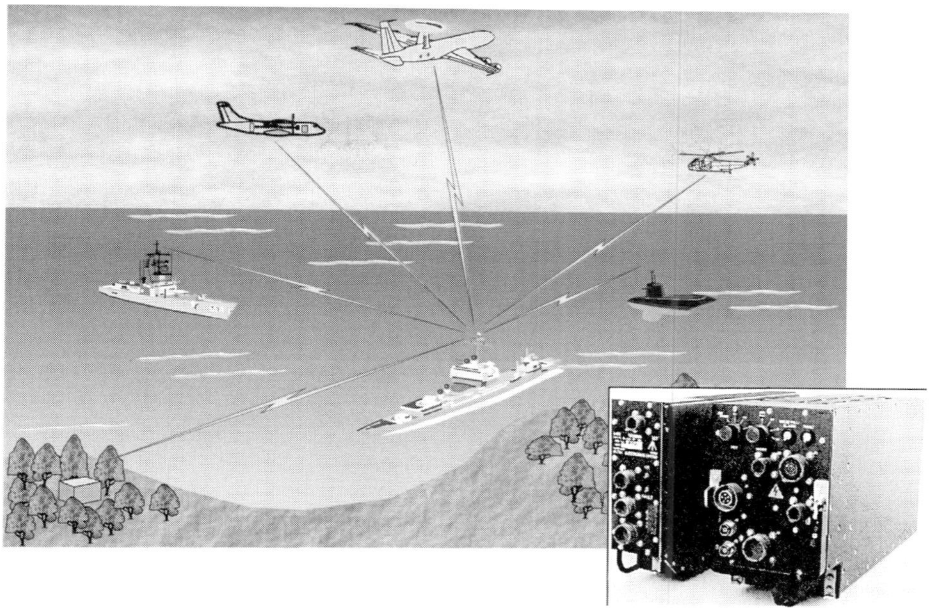

The Signaal Special Products' Link 11 datalink airborne data terminal set and datalink processor, together with diagrammatic concept of operations
2000/0064382

and force engagement status reports, is an indispensable tool for engagement planning.

Engagement execution is supported through the exchange of commands, such as weapon doctrine orders, engagement orders and hold-fire orders.

Link-Y Mk 2 allows a maximum of 31 units to participate in the network. Reporting normally takes place using a TDMA method. To each unit, one or more time slots are automatically allocated by the net control station. In radio silence mode all units maintain silence but urgent messages are transmitted by means of a single report.

The Link-Y Mk 2 terminal can be set in the Link-Y Mk 1 mode, resulting in complete interoperability with older Link-Y units which lack the automatic slot allocation and encryption facilities.

The system operates on HF, VHF and UHF frequencies. Effective range is 925 km for HF and line of sight for VHF and UHF.

Specifications

Dimensions:
(Link-Y terminal) 57.2 × 194 × 318 mm
(optional control unit) 146 × 66 × 68 mm
Weight:
(Link-Y terminal) 4 kg
(optional control unit) 0.4 kg
Power supply: 28 V DC
Data rates: 300, 600, 1,200, 2,400 and 4,800 bits/s

Operational status

In production for several navies.

Contractor

Hollandse Signaalapparaten BV, Signaal Special Products.

VERIFIED

Vesta transponder

Vesta is a landing and identification aid primarily intended for ship-based helicopters. It consists of two parts: the helicopter transponder and the ship receiver. Vesta enables accurate display and tracking of friendly helicopters on the radar display, even in heavy clutter environments. The transponder principle is based on a radar-triggered VHF reply. Operation is possible with any synchronised surveillance radar (either shipborne or shore-based) in the 1 to 10 GHz band.

For every radar pulse the transponder receives, a VHF reply pulse is transmitted, followed by a code pulse. Up to five helicopters can be identified by means of preselected codes (extension up to 64 is possible). The return signal is received and processed by the Vesta receiver in the ship, which identifies and decodes the transponder reply. Unwanted VHF reply pulses are rejected by digital filters controlled by the allocated radar on the ship.

The Vesta helicopter system consists of a fully solid-state transponder, a control unit, two radar pick-up antennas and a VHF transmitting antenna.

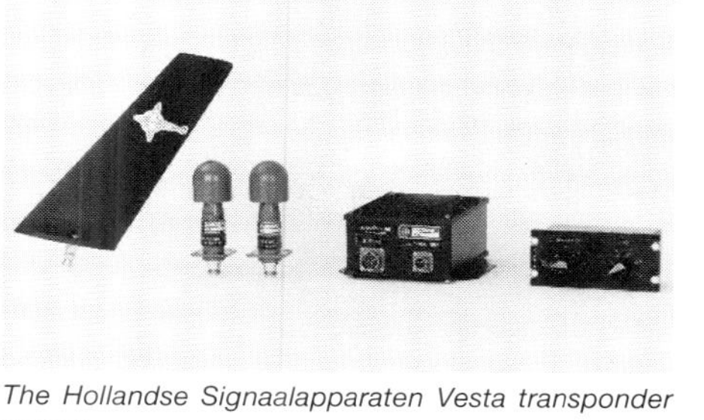

The Hollandse Signaalapparaten Vesta transponder system

The two radar pick-up antennas are used to guarantee a combined sensitivity pattern which is virtually omnidirectional. The control unit has only two switches; one for sensitivity selection and power on/off and the other for code selection.

Specifications

Dimensions:
(transponder) 167 × 85 × 194 mm
(control unit) 146 × 66 × 68 mm
(radar pick-up antenna) 45 × 106 × 48 mm
(VHF transmitting antenna) 50 × 254 × 123 mm
Weight: 3 kg
Power supply: 28 V DC, 14 W
Frequency:
(transmitter) VHF A-band
(receiver) 1-10 GHz
Transmitter peak power: 10 W
Range: 0-230 km
Pulse duration: 2.2 μs nominal
Number of helicopter codes: 5 standard
(optional extension up to 64)

Operational status

Fitted to several types of helicopters. Over 55 Vesta airborne transponders have been delivered to several navies.

Contractor

Hollandse Signaalapparaten BV, Signaal Special Products.

VERIFIED

Vesta-VC datalink

The Vesta transponder receiver system can easily be extended with a voice channel datalink. This makes it suitable for data transfer and over-the-horizon targeting for naval vessels or ground control stations by using the helicopter. The target data is received from the helicopter sensors.

The datalink is established via an existing communications voice channel. The Vesta transponder switches the available radio in the helicopter from voice to data, after which the data is transmitted via the voice channel in a frequency shift keyed signal.

A receiver extractor and datalink unit on the base station then converts this data to an 'own position' reference by triangulation of the helicopter position and target data, and interfaces the data handling system (the weapon control system) with the radar displays. Vesta is compatible with all airborne and base system interfaces.

Operational status

In production for several navies.

Contractor

Hollandse Signaalapparaten BV, Signaal Special Products.

VERIFIED

NORWAY

Samover jammer

Samover is a pod-mounted programmable responsive noise jammer operating in the 8 to 16 GHz band. It is fitted to F-5 aircraft to provide standoff jamming for F-16s launching Penguin Mk 3 anti-ship missiles.

Samover automatically identifies and responds to threats via a multiprocessor control system and provides CW, ICW and pulse signals. It is a VCO-base, liquid-cooled pod which can be integrated with a radar warning receiver and chaff and flares, if required for self-protection.

Specifications

Dimensions: 2,750 long × 254 mm diameter
Weight: 150 kg

Operational status

In service with Royal Norwegian Air Force F-5 aircraft.

Contractor

Kongsberg Defence & Aerospace AS.

VERIFIED

RUSSIAN FEDERATION AND ASSOCIATED STATES (CIS)

ASO series expendables

The ASO-2, developed from the ASO-21 fitted to the Yak-28PP 'Brewer-E' and MiG-21, was originally fitted to the Mi-24 'Hind' helicopter but proved unsatisfactory since the helicopter's slow speed and the range of the flares was insufficient to provide a reasonable separation between aircraft and flares. An improved version, believed to be the ASO-2bis, was fitted to helicopter fuselages, firing forward of the centreline in two banks of three launchers. This version has been seen on Mi-24/Mi-25 helicopters.

A further development, the ASO-3, has been seen on the Mi-8MT Hip helicopter, as well as on other combat helicopters. It features a neater installation and offers much better protection to the electronic components. The ASO-3 has a more powerful rocket discharge system and has been fitted extensively to a number of combat aircraft, including the MiG-27 and MiG-29, where it may also have the designation BVP-30-26M.

Operational status

In service.

VERIFIED

LIP missile approach warner

LIP is a microwave Doppler missile approach warning system. In its original fit, on board Mi-24 helicopters serving in Afghanistan, a single set was installed to cover the rear and underneath of the helicopter. This coverage was inadequate since it provided no warning of missiles approaching from the forward arc or fired from high ground down on to the helicopter. The most recent installation, on the Ka-29 marine assault helicopter, has a second antenna covering forward arcs.

LIP antennas have also been observed on board the Su-25 Frogfoot ground attack aircraft. In this case, a single antenna is fitted in the tail cone directly underneath the Sirena 3 radar warning system.

Operational status

In service. The L-006 variant is reported to be fitted to Su-27SK aircraft, where it is integrated with the Ekran monitoring and warning system, and APP-50 chaff and flare dispensers, in a fit with 96 (50 mm) cartridges.

VERIFIED

Russian radars

The table outlines Russian Federation radar systems and programmes. Full details are not available for all programmes, but data is provided in the subsequent equipment entries, where modern Russian source information is available.

Name	*Contractor*	*Aircraft*	*Type*
SPPZ	AeroPribor	Civil/military aircraft	Ground proximity warning system
IFF 60P	All-Russian JSC	Civil/military aircraft	IFF system
SRZO-KR	Kazan Scientific Research Institute	Military aircraft	IFF system
Berkut	Leninetz Holding Company	IL-38 (May)	ASW radar
Kinzhal-V	Leninetz Holding Company	Mi-28N, Ka-50	Attack radar. Possibly pod-mounted under stub-wing
Khishnik	Leninetz Holding Company	Su-32	Fire-control radar
Korshin	Leninetz Holding Company	Tu-142 (Bear-F), Ka-25 (Hormone)	ASW radar
NIT	Leninetz Holding Company		Side-looking airborne radar
Obzor	Leninetz Holding Company	Tu-95MS (Bear-H)/Tu-160 (Blackjack)	
PN	Leninetz Holding Company	Tu-22K, Tu-22M	
PNA-D	Leninetz Holding Company	Tu-22M3 (Backfire-C)	
Aisberg-Razrez	Leninetz Holding Company	Military/civil survey	Dual-band side-looking radar, ground/water surface surveying
Duet	Leninetz Holding Company	Civil aircraft	Dual-band weather/navigation radar
High-resolution airborne radar	Leninetz Holding Company	Proposal	Obstacle avoidance and poor visibility aid
Neva	Leninetz Holding Company	Civil aircraft. Mapping radar	Multifunction weather/obstacle/ground
Sea Dragon	Leninetz Holding Company	Tu-142MK, Il-38, Ka-28	Maritime surveillance
VID-95	Leninetz Holding Company	Civil aircraft and landing radar	Short-range, high resolution, approach
Kvant	NIIP, Vega	An-71 (Madcap), YAK-44	Airborne early warning radar
Sabla	NIIP, Vega	MiG-25RB (Foxbat-D)	Side-looking airborne reconnaissance radar
Shmel	NIIP, Vega	A-50 (Mainstay)	Airborne early warning radar
Shomol	NIIP, Vega	MiG-25RB (Foxbat-D)	Side-looking airborne reconnaissance radar
Shtyk	NIIP, Vega	SU-24MR (Fencer-E)	Side-looking airborne reconnaissance radar
N001/ N011	NIIP, Zhukovsky	Su-27M/Su-35.	I/J-band fire-control radar
N001E		Su-27SK	
N011M	NIIP, Zhukovsky	Su-30 MKI, Su-35/Su-37	Phased-array derivative of N011; I/J-band fire-control radar
N012	NIIP, Zhukovsky	Su-37	Tail radar
N014	NIIP, Zhukovsky	Su-27KUB (Su-33UB)	Fire-control radar
VEGA-M	NIIP, Zhukovsky	Tu-154M-ON	Airborne Surveillance: Open-skies
Zaslon (N007)	NIIP, Zhukovsky	MiG-31	Phased-array
Zaslon-M	NIIP, Zhukovsky	MiG-31M	Reportedly abandoned by NIIP
M002	Phazotron-NIIR	Yak-41M	N010 Zhuk development. Multifunction air-to-air/ air-to-ground/map/terrain follow-avoid
N010 Zhuk	Phazotron-NIIR	MiG-29M	Multifunction air-to-air/air-to-ground/map
Arbalet	Phazotron-NIIR	Kamov helicopters: Ka-52 Alligator	Multifunction, air-to-surface, air-to-air

Name	*Contractor*	*Aircraft*	*Type*
Gukol	Phazotron-NIIR	Marketed for light strike/attack aircraft, transports	Weather/navigation radar
Kopyo (or Komar)	Phazotron-NIIR	Mig-21-93	Multimode, multifunction, coherent PD air-to-air/ air-to-ground
Kopyo-25	Phazotron-NIIR	Su-25TM (Frogfoot-B), Mi-28N	Multimode, multifunction, coherent PD air-to-ground
Kopyo-Ph	Phazotron-NIIR		Phased-array version of Kopyo
Moskit	Phazotron-NIIR	MiG-ATC advanced combat trainer aircraft	Multimode, multifunction, coherent PD air-to-air/air-to-ground
Mosquito	Phazotron-NIIR	Marketed for Jaguar upgrade	Multimode, multifunction, coherent PD air-to-air/air-to-ground/air-to-sea maritime radar
Phathom	Phazotron-NIIR/Thomson-CSF	Marketed for SU-22 upgrade	Co-operative derivative of Kopyo
RP-21 Sapfir	Phazotron-NIIR	Many versions of MiG-21	Basic air-to-air radar
RP-22 Sapfir-21	Phazotron-NIIR	Many later versions of MiG-21	Basic air-to-air radar
S-23 Sapfir-23	Phazotron-NIIR	Many versions of MiG-23	PD air-to-air radar
RP-25 Sapfir-25	Phazotron-NIIR	MiG-25 variants	PD look-down shoot-down air-to-air radar
N019 Sapfir-29	Phazotron-NIIR	MiG-29	PD look-down shoot-down air-to-air radar
Topaz N019M	Phazotron-NIIR	MiG-29S, ME. Marketed for MiG-23, Mig-29 upgrades	Modified version of N019 Sapfir-29
N019 ME Topaz	Phazotron-NIIR	Reported fitted to Malaysian MiG-29	
N019 MF	Phazotron- NIIR	MiG-29 upgrades	Modified version of N019M, optimised for air-to-air.
N019 MP Topaz	State Enterprise V Tikhomirov Scientific Research Institute of Instrument Design NIIP	MiG-29 SMT	Variant of N019M with SAR capability
Pharaon	Phazotron-NIIR	Light fighter nose radar, or heavy fighter tail radar	Multifunction fire control, air-to-air and air-to-ground
Sokol	Phazotron-NIIR	Marketed for latest Sukhoi aircraft	Multimode, multifunction, coherent, PD
Zhuk	Phazotron-NIIR	Fitted to MiG-29M, and marketed for MiG-25, MiG-29, MiG-33 upgrades	Multimode, multifunction, coherent PD, AA/AG
Zhuk-27	Phazotron-NIIR	Marketed for SU-27, SU-30MK	Derivative of Zhuk aimed at Su-27 upgrade market
Zhuk-Ph	Phazotron-NIIR		Phased-array version of Zhuk
RP-3S5	Phazotron-NIIR	Marketed for MiG-35	Multimode, multifunction, coherent, digital PD
IFF 6201R/6202R	RadioPribor	Civil/military aircraft	IFF system
OA/ WASP	State Enterprise V Tikhomirov Scientific Research Institute of Instrument Design - NIIP	Light fighter radar	

VERIFIED

Korshin in the Ka-25 ASW helicopter

An Il-38 'May' with the Berkut search radar

The Beriev/A-50 Mainstay AWACS carries Shmel radar

Sirena-3 radar warning receiver

Variants of the Sirena-3 Radar Warning Receiver (RWR) have been installed in a wide range of aircraft types of the former Soviet Bloc, including the MIL Mi-24, Mikoyan MIG-21, 23 and 25 (Sirena 3M, 360° field of view) and the Sukhoi Su-25K.

Operational status

In service.

UPDATED

SPS series radar jammers

The SPS series of jammers includes a large number of systems, fitted to a wide variety of aircraft, in both dedicated EW roles and self-protection fits.

SPS-5-28, known as Fasol, is fitted to the specialist EW aircraft: Su-24MP 'Fencer-F' and Tu-22P 'Blinder-E'.

SPS-22-28 and SPS-44-29 are obsolescent/obsolete systems fitted to Yak-28PP 'Brewer-E' and Tu-16PP 'Badger-J' EW aircraft.

SPS-63, -66, -68 form part of the Mi-8PPA 'Hip-K' battlefield EW helicopter.

SPS-130, -140 Gvozdika jammers were first-generation systems fitted to MiG-27 and MiG-25 aircraft.

SPS-150 Lyutik self-protection jammers fitted to MiG-25RB 'Foxbat-B'.

SPS-160 Geran family of self-protection jammers fitted to Su-24 and Su-27.

SPS-170 family of self-protection jammers scheduled for installation in Su-34, -35 aircraft (may include Sorbtsiya-S).

Operational status

In service.

VERIFIED

SRO/SRZO series IFF transponder/interrogator

The SRO/SRZO series IFF transponder/interrogator is standard fit in RFAS aircraft. A large number of versions exist.

Operational status

In service.

VERIFIED

IFF 60P system

The 60P is a new-generation Russian IFF system. The following elements of the system are described:

(a) the airborne radar interrogator (designated CP3-1P (facility 6231P))

(b) the special facility 61P which is described (*verbatim*) as: 'designed for automatic ciphering interrogation-reply correspondences in the general spoofproof identification mode (mode II) of the 60P IFF system' (designated 6110P-10 airborne configuration and 6110P-21 in land-/sea-based configuration).

CP3-1P (facility 6231P)

The airborne radar interrogator CP3-1P (facility 6231P) of the 60P IFF system is designed for identification of friendly air, land and sea platforms. The 6231P facility operates on a co-operative question and answer principle and provides the following capabilities:

(a) general identification 'friend or foe' of air, overwater and ground vehicles equipped with transponders and detected by radar

(b) selective identification of air, overwater and ground vehicles on the principle 'where are you?'

(c) position location of ground vehicles.

The 6231P facility comprises 8 LRUs: unit 551 - receiver-transmitter; unit 542 - reply evaluation unit; unit 541 - video processor; unit 591 - power supply unit; unit 581-1 - controls; unit 2410 - SLS modulator; unit 526 - RF switch; unit 531-5 - auxiliary controls unit.

Specifications

Maximum range of positive identification: not less than detection range of associated radars
Angular resolution: determined by the radiation pattern of the CP3-1P interrogator antenna (autonomous or built into the radar antenna)
Transmitter power output: 1.5 to 3.5 kW peak
Receiving/decoding sensitivity: –76 dBm
Total weight: not less than 26.3 kg
Total volume: not greater than 23.3 dm³
Power consumption: 115 V AC (±5%), 380-1,050 Hz, <2.1A; +27 V DC ±10%, <0.7A; +10 V DC, <0.7A; –10 V DC, <0.2A

Operational status

Designed for both fixed- and rotary-wing aircraft.

Special facility 61P

The special facility 61P forms part of each element of the land, sea and airborne 60P system IFF installation. The description of its function is believed to represent the cryptographic function of the 60P IFF system.

The airborne equipment is designated 6110P-10. Code data is loaded using the 6110P-40 input device; data is loaded using a clockwork timer and time-set mechanism.

Special facility 61P comprises 5 units; unit 61C - Ciphering/Deciphering Unit (CDU); unit 61Y - rectifier; unit 61K - regulated power supply; unit 55 M - timer; unit 61E - interfacing unit.

Specifications

Weight: 8.95 kg
Volume: 7.2 dm³
Power consumption: 115 V AC, 400 Hz; 27 V DC; <30 W

Contractor

All-Russian JSC Nizhegorodskaya Yarmarka.

VERIFIED

CP3-1P airborne radar interrogator of the 60P IFF system
0011881

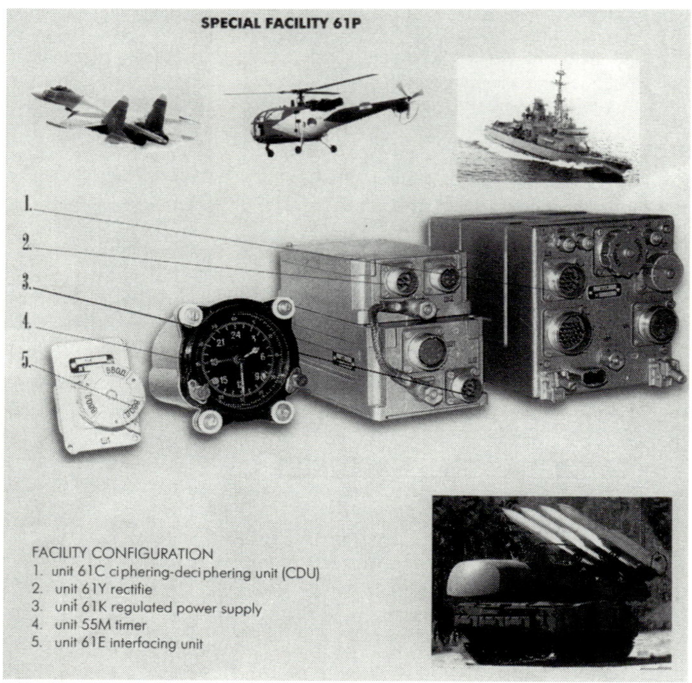

Special facility 61P cipher equipment of the 60P IFF system 0011880

20 SP M-01 airborne flare dispenser system

The 20 SP M-01 flare dispenser system provides protection against infra-red homing missiles. It has the following modes of operation:

(1) automatic: the system automatically selects the optimum flare dispense programme, responds to the commands of the EW suite and allows the aircrew to concentrate on their primary mission;

(2) manual: pre-programmed dispense programmes are available for manual selection and activation by the aircrew;

(3) accelerated: employment takes place at the maximum dispense rate, on aircrew command – emergency ejection.

The system provides multiple firing pulses to enable single, double, triple and quadruple payload dispensing when required. Emergency ejection is also available.

System architecture comprises: a program loading unit; a control unit; two switching units; and four dispenser magazines.

Specifications
Maximum payload: 120 flare cartridges
Salvo length: 1-4 flares
Salvo spacing: 0.01-10.0 s
Power: 27 V DC, 100 W

Operational status
Fitted to MiG-29 and Su-27 aircraft.

Contractors
Aviaavtomatika.
Joint Stock Company PRIBOR.

VERIFIED

20 SP M-01 airborne flare dispenser system
0018280

EDSU-77 fly-by-wire flight control system

The EDSU-77 is designed to provide aircraft stability and control in pitch, yaw and roll for the An-70 tactical transport aircraft in manual, automatic and override control modes of flight operation; it comprises the following units:

- stability and control computers: 4 units
- analogue interfaces: 2 units
- maintenance panel: 1 unit
- linear acceleration sensors: 12 units
- angular rate sensors (unified): 18 units
- position sensor (triple): 20 units
- emergency trim control: 1 unit

electromechanical actuators and sensors as required

Specifications
No of redundant channels: 6
Weight: 250 kg
Power: 115 V AC, 400 Hz, 1,000 VA

Operational status
Fitted to An-70 propfan medium range, wide bodied, tactical transport aircraft.

Contractor
AviaPribor.

UPDATED

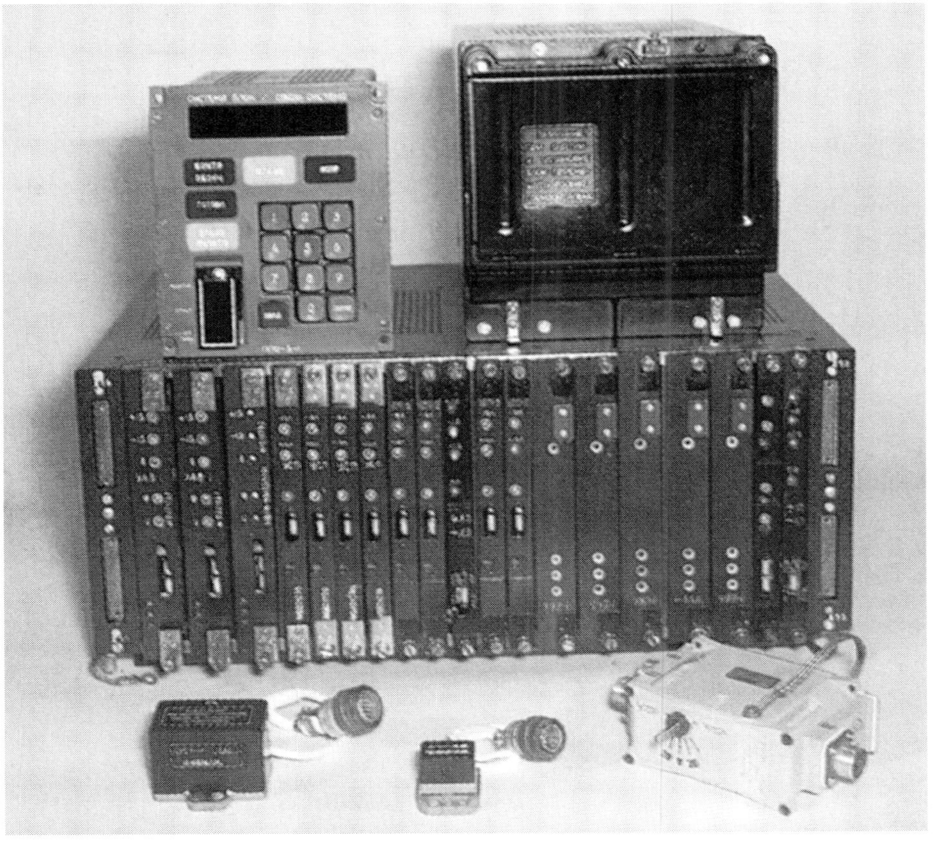

EDSU-77 fly-by-wire control system for the An-70 aircraft
0018228

Irtysh EW system on Su-39 Strike Shield

The Central Scientific Institute for Radiotechnical Measurements TSNIITI, Omsk is the system designer for the complete EW system - known as Irtysh - on the Su-39 Strike Shield aircraft (also known as Su-25TM). Data from the following elements of the system is displayed on both the Head-Down Displays (HDDs) and the Head-Up Display (HUD).

Pastil radar warning receiver
The Pastil RWR covers radio frequencies from 1.2 to 18 GHz, with the ability to intercept pulse, pulse Doppler, and continuous wave signals. It can operate in a stand-alone mode or be integrated with the electronic countermeasures system. It has antennas in the front and side of both wing tip fairings, and in the tail-sting of the aircraft.

An upgraded version, described as Pastel-K, has been reported (the difference in spelling presumably being due to transliteration and the designator 'K' possibly indicating a K-band capability).

Omul MSP-25 Electronic CounterMeasures ECM pod
The ECM system is located in the two pods on the outer weapon stations of both wings. It is said to 'cover the necessary radio frequency bands to counter expected threats', and to provide essentially 360° angular

Su-39 starboard wing view, showing the Omul ECM pod on the outer weapon station, and the Pastil RWR antennas on the forward end and side of the wing tip fairing. The port wing carries an identical installation
0018334

Su-39 Strike Shield aircraft, showing: Shkval EO sighting system in the nose, Kopyo radar pod under belly, Omul ECM pod under each wing, Pastil RWR on each wingtip 0018336

coverage, except for a 15° half angle cone each side of the normal to the aircraft centreline. The configuration comprises two identical pods, both of which have receive and transmit capability fore and aft; the operational configuration being to receive on one pod and to transmit on the other to overcome isolation problems. The system is reported to provide both noise and deception countermeasures, including range, angle and velocity gate pull-off.

Shokogruz Infra-Red CounterMeasures IRCM

The main IRCM system, known as Shokogruz, is an active jammer mounted in the tail-sting. It is a modulated IR power source, which is claimed to protect the engines at all thrust levels up to 95 per cent. Above 95 per cent thrust, flares are used to augment the system; however, a new active jammer called SNOP is being developed to cater for the higher thrust level requirement. The IR flare dispenser system, known as UV-26, holds a total of 192 flares.

Operational status

Installed in the Su-39.

Contractor

Central Scientific Institute for Radiotechnical Measurements TSNIITI, (possibly also known as Omskavtomatiki) Omsk.

VERIFIED

Su-39 tail view, showing the Shokogruz active infra-red countermeasures system, and above it the rear-facing antennas of the Pastil RWR 0018335

Airborne ARM control/ESM systems

Avtomatiki is offering the ATsU-1 and ATsU-2 pods, which detect emitter activity, and generate target designation data for Anti-Radiation Missile (ARM) launch.

The ATsU-1 pod is described as being capable of processing pulse and CW data in one system-specific frequency band, while the ATsU-2 pod is reported to handle pulse radars in one specific band, and 'quasi-CW' emitters in a second frequency band.

Both systems are quoted as being able to respond to 10 separate emitters and to cue two ARMs at a time.

The ATsU-2 pod has two antennas (one on each side of the pod), while the ATsU-1 pod has only a single antenna array.

Avtomatiki, also produces an associated wide-angle SRR system (presumably an ESM) to cue the ATsU pods. The SRR equipment is reported to cover 2 to 18 GHz and ±50°.

Specifications

Weight: each pod weighs 216 kg
Angular coverage:
(azimuth) ±32°
(elevation) 0 to –16°
Azimuth accuracy: ±1°
Ranging accuracy (in co-ordination with the aircraft navigation system): ±10%

Contractor

CKB Avtomatiki, Omsk.

VERIFIED

SPO series radar warning receivers

Variants of the SPO series of Radar Warning Receivers (RWR), from the CKB Avtomatiki Omsk design bureau are fitted to many types of combat aircraft of the former Soviet Bloc:

SPO-2

Also reported as Sirena-2, the SPO-2 RWR is installed in the Tupolev Tu-95/-95K strategic bomber/cruise missile carrier.

SPO-10

SPO-10 is reported as an H/J band RWR, installed in early production Mikoyan MIG-29 aircraft.

SPO-15

SPO-15 is the designation given to a series of RWRs fitted to a range of aircraft, including the MIL Mi-24, Mikoyan MIG-29 (SPO-15LM variant, 360° coverage), Sukhoi Su-24 and Su-27.

SPO-23

The SPO-23 is the latest development RWR for the CKB and is reported as a replacement for the SPO-15, applicable to MiG-21, -23, -29 and Su-22 aircraft, and to the Ka-50 and Ka-52 attack helicopters. SPO-23 can be configured to cover 4 to 11 GHz or 4 to 18 GHz. The associated antenna system is matched to the application: Ultra-broadband cones are used for helicopter applications, while a Broadband Beam Length (BBL) array, comprising conformal radome units, each with four or five outputs, provides adequate angular resolution to match targeting requirements for the Kh-25P and Kh-31P anti-radiation missiles. A threat library of 128 emitters is incorporated. Weight is 18 to 30 kg depending on configuration.

Operational status

In service.

Contractor

CKB Avtomatiki, Omsk.

UPDATED

Video Recorder System (VRS)

The Electroavtomatika VRS is designed to record data displayed on the Head-Up Display (HUD), as well as other data deriving from TV cameras. The VRS is a replacement system for the FKP-EU cine-camera system installed in MiG-29 and Su-27 aircraft.

The VRS comprises a memory unit with a replaceable non-volatile memory video cassette; miniature TV camera; logic and power unit; ground replay unit based on an IBM PC.

Specifications

Cassette memory: 1,024 video frames
Video data recording rate:
8 exposures/s with 2 s delay;
8 exposures/s without delay;
1 exposure/s;
exposures linked to synchronising pulses.
TV camera resolution: 50 lines minimum
Video data scanning standard: 50 interlacing fields
Dynamic range: 0.1 to 10,000 lx
Operating temperature range: –40 to +50°C
Video storage time: 30 days

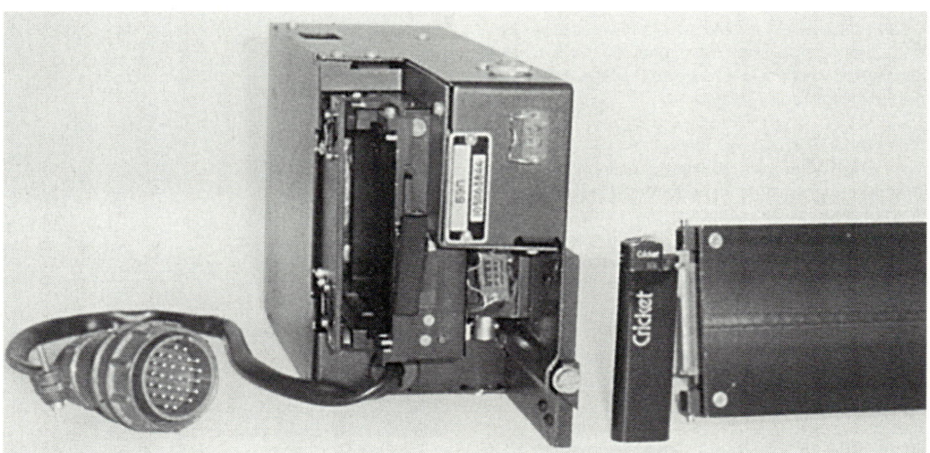

The Electroavtomatika video recorder system, showing memory unit, camera and logic/power unit ***2000***/0062228

MTBF: 80,000 cycles minimum
Power: +27 V DC, 25 W
Weight: 4.5 kg

Contractor

Electroavtomatika OKB.

VERIFIED

L-166B1A airborne fixed source IR jammer

The L-166B1A fixed-source infra-red jammer is designed for helicopter application on Mi-8MT/Mi-17, Mi-24, Mi-25 and similar foreign helicopter types. The system uses an IR radiation source and mechanical modulator. It is claimed to offer protection against IR guided missiles such as Sidewinder, Mica, Strela, Redeye and Chaparral. System weight is said to be no more than 20 kg and power consumption 2.8 kW.

Contractor

Elers-Electron Ltd.

VERIFIED

Gorizont chaff/flare dispensers

UP-P1, UV-P2 and L-028K airborne chaff dispensers fitted to aircraft such as Tu-22, Il-76, Su-24 and Su-27.

APP-50MR and APP-50MA chaff/flare decoy launchers are fitted to a variety of aircraft including: An-22, Tu-160, Su-24 and Su-27.

Operational status

In service.

Contractor

Gorizont.

VERIFIED

SRZO-KR airborne interrogator-transponder

The SRZO-KR airborne interrogator-transponder is designed as an upgrade for earlier equipment of the Kremni-2 (2 m) IFF system.

(Editor note: believed to be the system known as SRO-2)

The SRZO-KR equipment is claimed to employ the newest circuit and technical design and component configurations, including microwave transistors, bodyless elements and thinfilm technology.

Four operating modes are described:

Mode I: general identification
Mode III: individual indentification
KO: checking identification
Distress: calling

Specifications

Transmitter power: >400 W
Receiver sensitivity: –72 dBm
Weight: <12 kg
Volume: <17 dm^3
Power consumption: 27 V DC, <195 W; 115 V AC, <115 VA
MTBF: 1,500 hrs

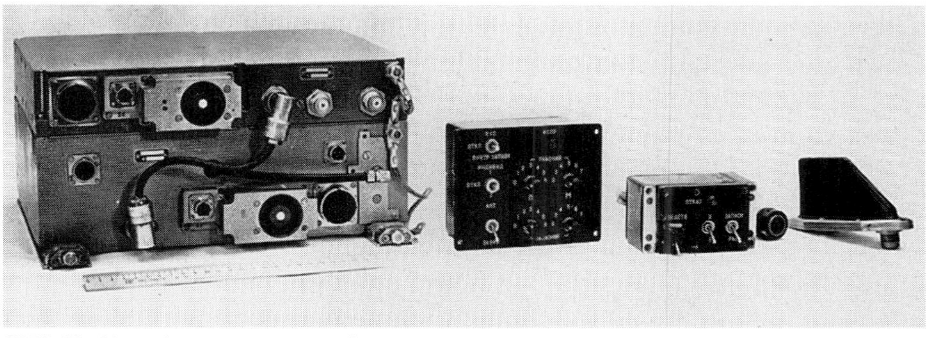

SRZO-KR airborne interrogator-transponder 0011879

Contractors

Kazan Scientific Research Institute of Radio-Electronics (product designer)
RadioPribor (producer).

VERIFIED

Aisberg-Razrez airborne dual-band side-looking radar

The Aisberg-Razrez radar is intended for observation of the Earth's surface. The radar is designed for detailed radar surveys of vast areas of ground and water surface. The system uses dual-band frequencies (centimetric and metric wavebands) to optimise efficiency regardless of weather conditions, time of day/night, and season of the year.

Radar data can be relayed over standard communication systems (including satellite datalinks) to specially equipped ground terminals to optimise operational efficiency. The system can also be integrated with other sensors (including the VIDS-95 radar) to increase operational flexibility.

Specifications

Aisberg Radar
Wavelength: 2 cm
Antenna beamwidth: 12 minutes of angle
Signal polarisation: HH, VV, HV, VH, circular
Range: 100 km
Scan strips: 64, 32, 16 km
Resolution:
(in range) 15 m
(across flight path) 12 minutes of angle
(in SAR mode) 5 m
Accuracy of map control: 30 m
Processing: non-coherent MTI, SAR
Display: TV, SVGA PC
Recording: Digital 2.5 Gbytes
Weight 320 kg
Power consumption: 2.1 kW

Razrez Radar
Wavelength: 3 m
Antenna beamwidth: 30°
Signal polarisation: HH, VV
Range: 50 km
Scan strips: 32, 16 km
Resolution:
(in range) 30 m
(along flight path) 30 m
Accuracy of map control: 30 m
Processing: SAR
Weight 120 kg
Power consumption: 1.4 kW

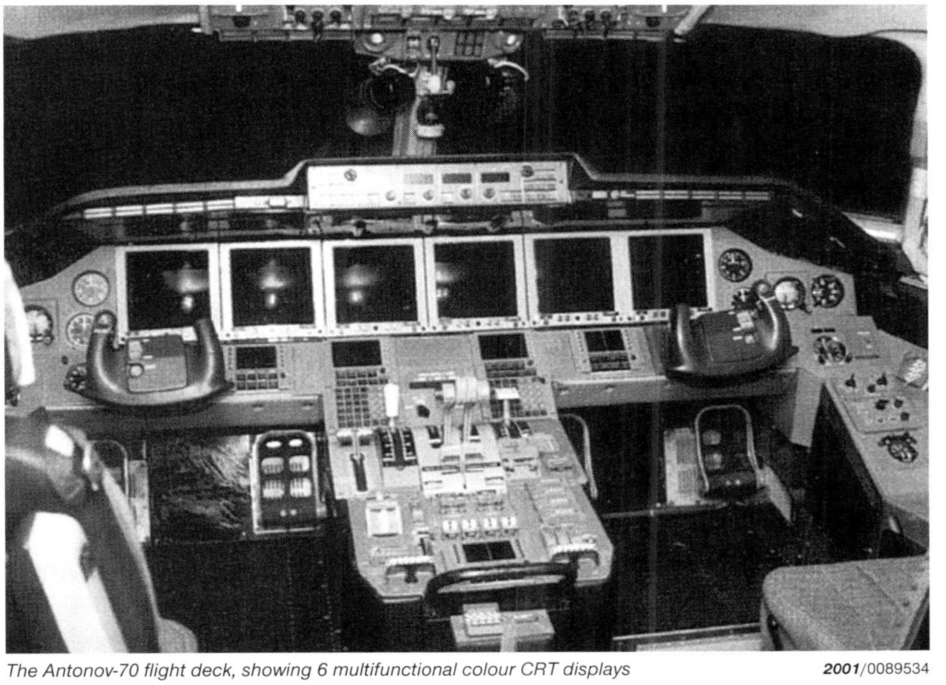

The Antonov-70 flight deck, showing 6 multifunctional colour CRT displays **2001**/0089534

Contractor

Leninetz Holding Company.

VERIFIED

Antonov-70 avionic system

AviaPribor is responsible for integration of the whole avionic suite for the An-70 aircraft. AviaPribor also developed a number of the onboard systems, including:

(1) mission management system.
(2) flight control system.
(3) fly-by-wire system.
(4) strapdown inertial system.
(5) satellite navigation system.
(6) standby instruments.
(7) MIL-STD-1553 bus control system.

The airborne information system of the An-70 aircraft is designed by the Leninetz Holding Company. It comprises: centralised firmware, unified information acquisition, processing and output to the flight crew and ground-based maintenance personnel of data on the status of the aircraft functional systems. It is designed to reduce the cost for maintenance and to permit aircraft operation at poorly equipped airfields for 30 days or 200 flight hours. It comprises two subsystems: the Information and Warning Subsystem (IWS), and the Monitoring and Maintenance Subsystem (MMS).

The IWS provides:

(1) reception of parametric and warning information from functional systems via digital datalinks and equipment sensors;
(2) preprocessing of received information;
(3) continuous monitoring of aircraft functional systems' technical status and crew actions;

The Antonov- An 70 aircraft **2001**/0089533

(4) generation and transfer of parametric information and monitoring results to the electronic display system, monitoring and maintenance subsystem and emergency warning system.

The MMS provides:

(1) reception of parametric and warning information, generated by IWS and functional systems;
(2) information processing;
(3) presentation of parametric information and processing results on the displays;
(4) recording of information in operation recorder in automatic and manual modes;
(5) output of information to the printer in automatic and manual modes;
(6) manual input of changed-over constant and other auxiliary information from the control panel;
(7) generation and record of monitoring results in non-volatile memory;
(8) presentation of information stored in non-volatile memory and operation recorder on the MMS display;
(9) initiation of aircraft systems' test through the MMS control panel.

'MONITORING' is the basic operating mode of the system, in which the monitoring of the aircraft systems' technical status and crew action, as well as the record of parametric information and monitoring results are provided.

Flight crew and maintenance personnel receive the following information:

(1) operating information on the electronic display system, emergency warning system's panel and MMS screen (it is used by the crew in-flight and ground-based services when controlling the monitored systems and for evaluating their status);
(2) report information, recorded on paper tape (it is used on the ground for post-flight evaluation of the monitored systems' technical status and the crew actions, as well as for detected failure unit removal);
(3) report information, recorded in the operation recorder, with pre- and post-histories of errors in the aircraft systems, and information about the aircraft systems failures, stored in non-volatile memory.

'TESTS' is a mode in which the IMS and MMS technical status monitoring and their failure location up to a line-replaceable unit are performed.

'OR OUTPUT' is a mode in which the readout and analysis of information, recorded in the Operation Recorder (OR), are performed directly on board the aircraft. The information processing and its subsequent displaying and printing in the form of digital information and plots are envisaged in this mode.

'FAILURES' is a mode in which the displaying and printing of the reference information, stored in non-volatile memory, are produced on the ground.

Other elements of the avionic system, including the navigation system, flight data system, computing systems and display system (6 multifunctional colour CRTs) were developed and manufactured by Elektroavtomatika.

Specifications

Information and warning subsystem:
(input analogue signals) 560
(input digital signals) 800
(ARINC-429 input channels) 20
(No of types of channels) (analogue) 12; (digital) 2
(multiplex exchange channels) 8
(total no of input parameters) 8,000
(output digital signals, +27 V) 32
(No of generated messages) 2,000
Presentation of parametric and warning information on 5 displays of the electronic display system
(ROM capacity) 300 kbyte
(main memory capacity) 80 kbyte
Monitoring and maintenance subsystem:
Ground-based display:
(screen diagonal) 23 cm
(No of lines) 28
(No of displayed character types) 121
Operation recorder:
(recorded data capacity) 300 Mbyte
(frame capacity, 16-bit word) 1,024
(continuous recording time) 24 h
accelerated playback at readout
(capacity of information, recorded in non-volatile memory) 100 kbyte
(No of input parameters) 8,000
(No of generated messages) 8,000
(ROM capacity) 600 kbyte
(Main memory capacity) 40 kbyte

Contractors

AviaPribor
Elektroavtomatika OKB
Leninetz Holding Company.

UPDATED

NIT side-looking airborne radar

NIT is a side-looking radar designed for collecting detailed data on earth surface conditions. Intended roles include survey and supervision of natural resources; ice survey; fire, flood, and disaster monitoring.

The radar provides coverage of a swathe up to 80 km wide with different polarisations. Design capabilities include on-line, real-time presentation of the radar picture on TV displays in the air; provision for digital or analogue recording; geographic referencing; a datalink to transmit information to ground and shipborne receiver/display systems.

Contractor

Leninetz Holding Company.

VERIFIED

Sea Dragon maritime surveillance mission system architecture

The Sea Dragon maritime surveillance mission system architecture is a proposal offered by the Leninetz Holding Company of St Petersburg. The roles proposed include: Anti-Submarine Warfare (ASW); maritime surveillance; Anti-Surface Warfare (ASuW); Search And Rescue (SAR) and surface environment monitoring. The mission suite comprises: radar; electro-optics, Magnetic Anomaly Detector (MAD); acoustic system; Electronic Support Measures (ESM); and Mission Control Computer Unit (MCCU).

The radar system proposed provides: long-range detection of surface vessels in rough sea states and against precipitation and jamming environments; detection of small targets and emergency beacons in high sea states; air-to-air detection against a background of sea clutter; target acquisition and tracking, with range heading and velocity data; Synthetic Aperture Radar (SAR) and Inverse SAR (ISAR) modes.

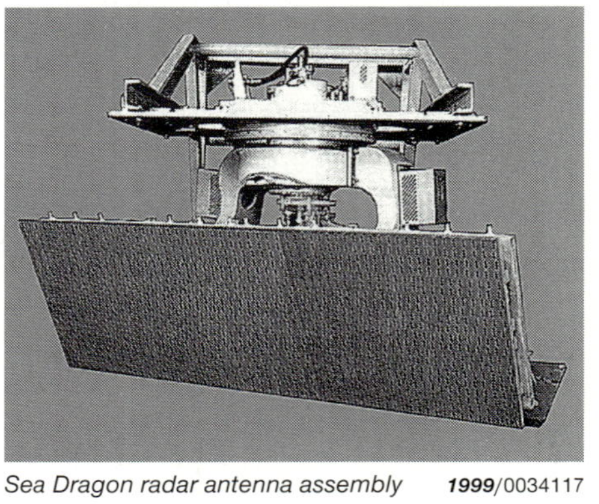

Sea Dragon radar antenna assembly **1999**/0034117

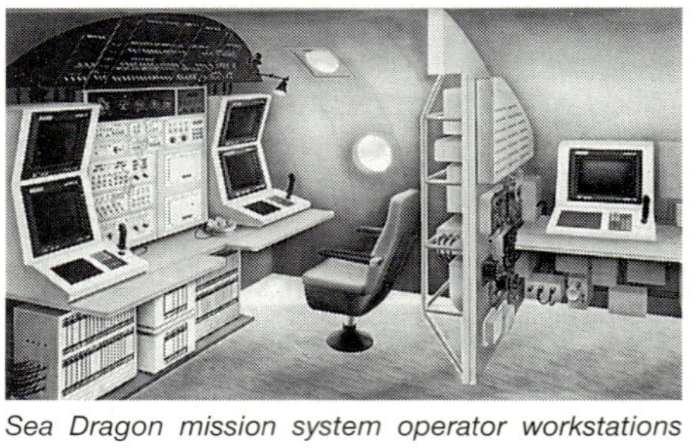

Sea Dragon mission system operator workstations **1999**/0034118

Sea Dragon ESM antenna system **2001**/0089530

The electro-optic surveillance system combines Infra-Red (IR) and Low Light Level TV (LLLTV) cameras. There are two video channels and features include: high-resolution and selectable field of view IR and TV images; 360° coverage; 3-D gyrostabilised gimbal and autotrack controlled by the Mission Control Computer Unit (MCCU).

The MAD system comprises the sensitive quantum-mechanical magnetometer and provides: real-time data processing; use of target detection and aircraft-generated interference compensation combined algorithms with a high degree of automation; high accuracy of the beam range and beam passage time determination; reliable magnetic and electromagnetic interference clutter protection.

The acoustic system provides: automatic submarine detection and localisation by means of active and passive sonobuoys with target motion parameter measurement; operation near ice edges; underwater acoustic conditions reconnaissance and contacts prediction; multichannel receiving and processing capability.

The ESM provides detection of radar emissions with multi-octave frequency coverage in dense signal environments; monopulse measurements; analysis of frequency and modulation data; emitter identification and multitarget tracking.

The Sea Dragon mission system is centrally controlled by the Mission Control Computer Unit (MCCU), which provides: three identical operator positions with access to all processing functions, via multiplex databusses. Each operator station is equipped with two multicolour 13 in Barco MPRD 134 liquid crystal displays and multifunction consoles. Each operator has a digital map display and the ability to perform data fusion from all sensors, together with access to aircraft navigation and flight plan data. Laser disk memory is used to support system requirements.

A derivative system of Sea Dragon called Stirzh has been reported that uses only the radar and electro-optical subsystems in a single operator configuration.

Contractor

Leninetz Holding Company.

UPDATED

Streege maritime surveillance mission system architecture

The Streege maritime surveillance mission system architecture is a proposal offered by the Leninetz Holding Company of St Petersburg. The roles proposed include: maritime surveillance including infra-red, optical and radio frequency bands imaging; search and rescue; surface environment monitoring; frontier and Exclusive Economic Zone (EEZ) monitoring.

The mission suite comprises: Synthetic Aperture Radar (SAR) and Inverse SAR (ISAR) radar imaging; infra-red and Low Light Level TV (LLLTV) cameras mounted in fully stabilised platforms; data processing and multifunction display systems; datalinks to co-operating ground and sea-based platforms. System growth options include: Magnetic Anomaly Detectors (MAD); gas analysing and radiation monitoring systems.

Contractor

Leninetz Holding Company.

VERIFIED

VEGA-M Open-Skies airborne surveillance system

VEGA-M is an airborne surveillance system designed for Open-Skies procedures on the Tu-154M-ON aircraft, with a mission crew of five operators.

The complete system comprises the airborne surveillance system installed on a Tu-154M-ON aircraft; an onboard digital recording system; a communications package; a ground-based data gathering and processing system.

The sensor package comprises an aerial photography system; a RONSAR side-looking synthetic aperture radar, a RADUGA line-scan IR sensor; a TV-camera system. The sensor system is supported by an onboard computer system comprising five 486DX PCs, and specialised digital recording based on the VITYAZ'-ON recorder.

VEGA-M airborne surveillance system 0011882

Specifications

Sensors	Height	Surface Resolution
Aerial camera system		
panoramic cameras	8,000 m+	0.3 m
framing mapping cameras	1,500 m+	0.3 m
framing oblique cameras	1,500 m+ (slant range)	0.3 m
TV system		
mapping cameras and oblique	500 m+	0.3 m
IR system		
low altitude cameras	1,500 m+	0.5 m
high-altitude cameras	3,000 m+	0.5 m
Side-looking SAR	500-12,000 m	3 × 3 m

Contractor

Moscow Scientific Research Institute of Instrument Engineering MNIIP.

VERIFIED

OSA/WASP light-fighter radar

OSA/WASP is an electronically scanning light fighter radar. It performs both air-to-air and air-to-surface modes, with acquisition and tracking capabilities, using adaptive beam shape and sidelobe levels. It is claimed to be able to simultaneously track eight targets, while engaging four targets, and maintaining acquisition search.

The antenna technology carries the type name SKATE within the Research Institute. It provides simultaneous shaping of a sum and two tracking patterns in a combination antenna of two arrays within a common structure for independent I-band and J-band operation within a common structure.

OSA/WASP is claimed to be compatible with a wide range of Russian infra-red, semi-active and fully active radar missiles.

Specifications

Radar frequency: I/J-band
Transmitter power: 700 W
Angular coverage: 120° cone
Detection range: 85 km for 5 m^2 target
Tracking range: 65 km
Volume: 420 dm^3
Weight: 120 kg
Power consumption: 3.6 kW at 400 Hz; 500 W at +27 V DC

Operational status

Believed to be in development; marketed for MiG-29 upgrades, also reported to be intended for combat versions of the MiG-AT and Yak-130 training aircraft.

SKATE antennas are also being proposed for Su-27 and MiG-31 upgrades.

Contractor

NIIP State Enterprise V Tikhomirov Scientific Research Institute of Instrument Design.

VERIFIED

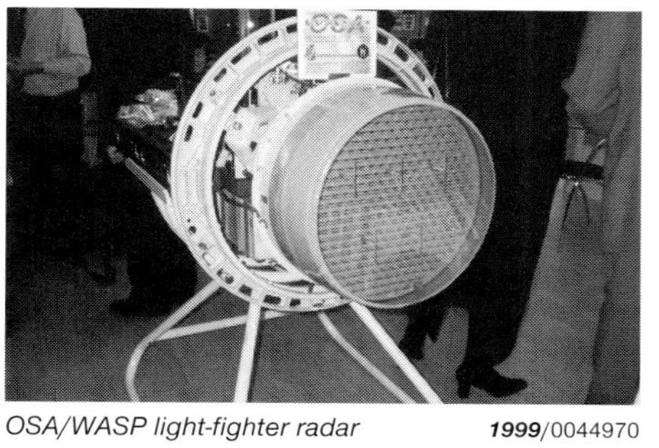

OSA/WASP light-fighter radar **1999**/0044970

Su-27SK radar system N001E

The N001E radar has been designed by the NIIP Tikhomirov Institute for the Su-27SK aircraft.

The N001E radar is a multimode, pulse Doppler radar capable of target detection and tracking against both approaching and receding targets in look-up and look-down modes. Together with a separate digital computer, the integrated system is designated as the RLPK-27 radar sighting system.

The radar can detect an approaching aircraft at 100 km and lock-on at 80 km. Track-while-scan against 10 targets can be accomplished, leading to selection of one target for interception.

The N001E radar provides all target illumination and guidance commands needed for operation of the complete range of R-27 medium range missiles (R-27R1 radar semi active missile, R-27T1 heat-seeking missile, including the R-27ER1 and R-27ET1 extended range derivatives), and R-73 short-range air-to-air missiles (R-73E heat-seeking), as well as the GSh-301 30 mm cannon.

The RLPK-27 integrated radar sighting system represents one element of the overall SUV-27 weapon control system. Other elements of the SUV-27 system include: the OEPS-27 optical electronic sighting system; the SEI-31 integrated cockpit display system; the ILS-31 head-up display; the Shchel-3UM helmet-mounted display; an IFF system and integrated computing system.

Operational status

The N001E radar is in production, and in service on Su-27SK aircraft.

Meanwhile, NIIP Tikhomirov is contracted to upgrade the N001E radar, to improve ECCM characteristics, to upgrade the air-to-air multitarget capability and to provide air-to-ground capability.

The improved air-to-air capability is predicated on use of RVV-AE medium range active radar-homing air-to-air missiles. These modifications will enable the Su-27SK to engage two air-to-air targets simultaneously.

The objective of the air-to-ground modes upgrades is to provide for mapping, surface target detection and anti-ship operations, using Kh31A active radar-homing anti-ship missiles and Kh-31P or Kh25MP anti radar missiles, together with smart bombs.

Contractor

NIIP State Enterprise V Tikhomirov Scientific Research Institute of Instrument Design.

VERIFIED

Arbalet FH01 combat helicopter multifunction radar for Kamov-52 Alligator

Arbalet is a multifunction air-to-surface and air-to-air radar proposed for the Kamov-52 Alligator attack helicopter. It provides day/night/all-weather combat capability.

Air-to-surface modes include the following over-land and over-water capabilities:

- target detection and localisation
- moving target indication and data track
- ground mapping
- terrain-following/avoidance
- air-to-surface missile and gun control

Air-to-air modes include:

- air target detection and tracking
- air-to-air missile and gun control

The Arbalet radar comprises two separate transmitter/antenna elements. The main antenna, located in the nose of the helicopter, is said to operate in Ka-band (at 8 mm wavelength), and to provide the main air-to-ground capabilities. A separate antenna is located in the radome on top of the rotor shaft; this is said to operate in D-band (at 20 cm), and to provide area surveillance.

Note that in some Phazotron literature, Arbalet is also spelled Arbalest.

Operational status

Arbalet appears to be more of a marketing proposal than a substantive system at present. It is noteworthy that the same Ka-52 Alligator airframe (061) is also (more often) shown with an electro-optic fit in the chin and no rotor mast radome.

Arbalet combat helicopter multifunction radar **2000**/0062856

Phazotron states that a prototype will be available in 2000. Meanwhile, the surveillance capability is being marketed as a possible surveillance/navigation fit for Ka-27/-28 and Ka-32 helicopters.

Contractor

Phazotron-NIIR, Scientific & Production Company.

VERIFIED

Gukol weather/navigation radars

Four variants of the Gukol radar have been marketed by Phazotron as I/J-band (8-12 GHz) weather/navigation radars for light strike/attack aircraft, helicopters and military/civil transports. Operating modes are claimed to include weather and obstacle detection/avoidance; real beam and synthetic aperture mapping; navigation and beacon tracking modes.

Gukol has been designed to meet relevant ARINC interface requirements.

The four variants have been offered with antenna sizes varying from 370 to 670 mm and equipment weights from 15 to 65 kg.

The largest variant, marketed for larger transport/tanker types, is said to have an L/M-band blind landing capability.

Operational status

The first prototype has been manufactured and the second is in construction. It is understood that orders have been received for Il-96 and Tu-204 aircraft requirements.

Contractor

Phazotron-NIIR, Scientific & Production Company.

VERIFIED

Kopyo airborne radars

Four variants of the Kopyo radar are being marketed: Kopyo; Kopyo-25; Kopyo-Ph; and Phathom. The names 'Komar' and 'Super Komar' are sometimes used in the marketing material apparently interchangeably with Kopyo.

Kopyo airborne radar

The Kopyo radar is an all-weather, coherent, multimode, multiwaveform search-and-track radar that uses digital processing to provide the features and flexibility needed for both air-to-air and air-to-surface missions. It was derived from technology developed for the Zhuk radar.

Air-to-air modes include range while search in look-up and look-down mode; single-target track; track-while-scan of eight targets and simultaneous engagement of two targets; air combat modes (vertical scan, HUD search, wide angle, boresight).

Air-to-surface modes include real beam ground map; Doppler beam sharpening to 0.45° (1:10); synthetic aperture beam sharpening to give 30 × 30 m resolution; enlargement, freezing capability; track-while-scan four targets; ground moving targets indication/track; air-to-surface ranging.

The Kopyo radar is compatible with many weapons including: Kh31A, R27R1, R27T1, R-73E, R60MK, RVV-AE, other precision weapons and iron bombs.

Specifications

Detection range:
57 km approaching targets
35 km receding targets
Angular coverage: ±10°, ±30° azimuth, 2 or 4 bars in elevation

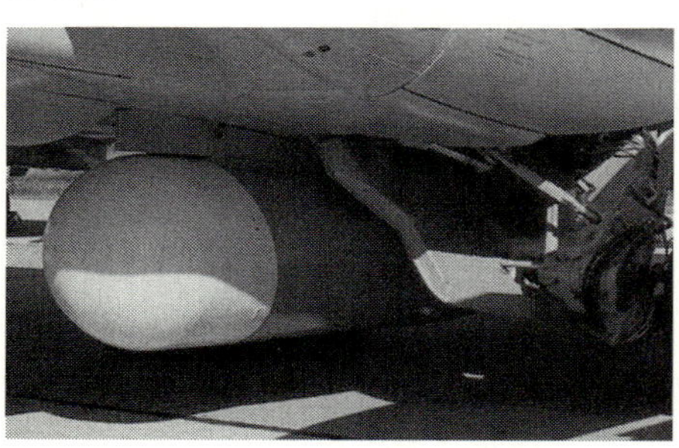

Kopyo-25 radar fitted to an SU-25TM to enable enhanced attack capability

Ka-52 Alligator helicopter showing nose-mounted radar and mast-mounted radome **2000**/0062854

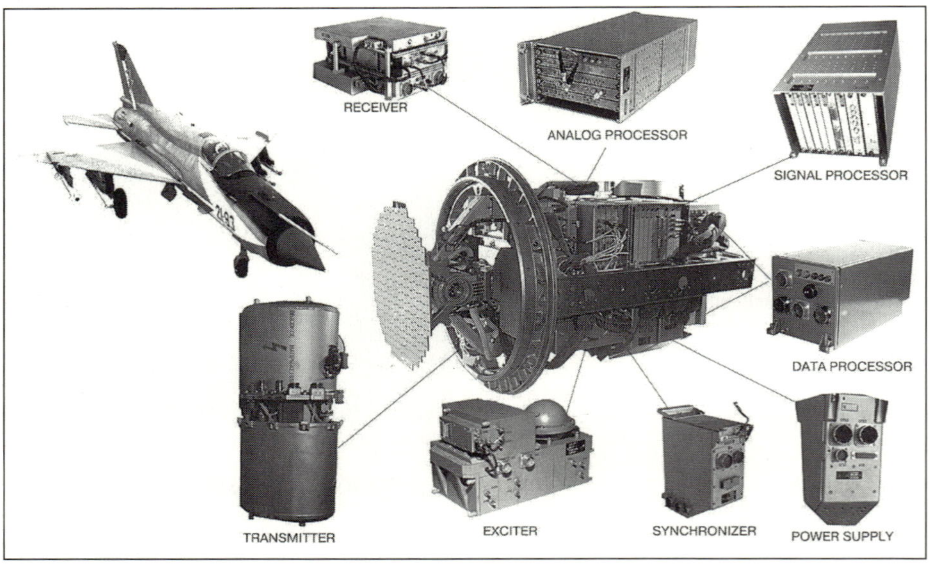

Kopyo-25 system units 0001202

Radar frequency: I/J-band
Peak power: 5 kW
Average power: 1 kW
Input power: 8.5 kVA, 400 Hz, 1 kW DC
Weight: 130 kg
Volume: 400 dm^3
Reliability: 120 h MTBF
Cooling: air, liquid

Operational status

Fitment of Kopyo is part of the Indian Air Force upgrade of 125 MiG-21bis aircraft to MiG-21-93 configuration. Proposed to China for the F-7II and A-5 aircraft, as Komar, and for the new FC-1 aircraft as Super Komar. Being widely marketed for MiG-21, MiG-23, F-5, Mirage-1 Hawk 200 upgrades.

Kopyo-25 airborne radar

Kopyo-25 is a derivative version of the Kopyo radar, designed specifically for carriage on the Su-25TM aircraft. It is carried in an under-fuselage pod. Addition of the Kopyo-25 radar to the Su-25TM gives the aircraft day/night, all-weather capabilities as well as considerably enhanced efficiency in air-to-air combat.

Operational capabilities include detect and track air targets in the automatic mode, (including targets flying at low altitude over land or sea); designate targets and engage them with radar- and IR-guided air-to-air missiles or guns; high-speed vertical search and automatic lock-on of visible targets in close combat, in association with the use of high-manoeuvrability dogfight missiles; ground mapping with real beam and Doppler sharpening and scaling-up of the chosen sector of the map.

Specifications

As for Kopyo.

Operational status

Designed for Su-25TM, also shown at Zhukovsky 97' Air Show on the Su-39 Strike Shield aircraft.

Kopyo-Ph

Kopyo-Ph is a derivative of the Kopyo radar that employs a phased-array antenna using technology derived from the N011M programme tested on Su-35.

Operational status

Development.

Phathom

Phathom is a collaborative venture between Phazotron and Thomson-CSF. From Phazotron come elements of the Kopyo radar including antenna, receiver, transmitter, and primary power supplies. Thomson provides the data and signal processing from its RDY radar. Analogue processors are a joint development.

Operational status

Development.

Contractor

Phazotron-NIIR Scientific & Production Company.

VERIFIED

Moskit/Mosquito radar

These two radars share a common name (Mosquito being the translation of Moskit), and appear to have identical specifications, despite the different roles marketed.

Moskit

The Moskit radar is a coherent, multimode, digital fire-control radar that provides weapon delivery and dogfight capabilities. It is smaller, lighter and less expensive than fighter radars of its class. Technology employed in the Moskit radar is reported to derive from the Kopyo programme. Moskit is claimed to detect and track targets at all aspects and altitudes, and to provide the following capabilities:

Air-to-air modes: range while search in look-up and look-down; eight targets track-while-scan and two target simultaneous engagement; air combat: HUD search, slewable scan, boresight, vertical scan.

Air-to-ground modes: real beam ground map; Doppler beam sharpening; synthetic aperture; enlargement; freeze; beacon; two targets track-while-scan; air-to-air ground ranging; ground moving targets track.

Moskit comprises five LRUs: a flat slot array antenna; air-cooled TWT transmitter; monopulse four-channel coherent receiver; 280 Mflops programmable signal processor; 1 Mflops effective speed, 512K static RAM, 1.5M ROM data processor. The Moskit radar is compatible with such weapons as Kh-29L, Kh-29TD, Kh-31A, Kh-31PE, Kh-38, RVV-AE, as well as KAB500KR iron bombs.

Specifications

Detection range:
25 km approaching targets
15 km receding targets
Angular coverage: ±10°, ±30° azimuth, 2 or 4 bars in elevation
Radar frequency: I/J-band
Peak power: 4 kW
Average power: 0.3 kW
Input power: 2.1 kVA, 200 V, 400 Hz; 0.2 kW 27 V DC
Weight: 70 kg
Volume: 300 dm^3
Cooling: air

Operational status

Marketed by Phazotron as a possible capability for the MiG-ATC (Advanced Trainer Combat) concept variant of the MiG-AT, if the combat capability is required by any customers for the aircraft. As yet, no orders have been placed for the combat capability.

Mosquito

Mosquito is being offered to the Indian Air Force as a maritime radar upgrade for its Jaguar aircraft. Mosquito is claimed to offer the following capabilities.

Air-to-sea modes: detection of sea targets to 100 km in Sea State 4 to 6, co-ordinates measurement accuracy of 300 m^2, and engagement using Sea Eagle air-to-surface missiles.

Air-to-surface modes: detection and co-ordinates measurement of sea ports and fleet anchorage and engagement of them using unguided missiles.

Air combat modes: detection, lock-on and tracking of air targets and engagement of them using Western and Russian guided air-to-air missiles; HUD screen; slewable scan; boresight; and vertical scan.

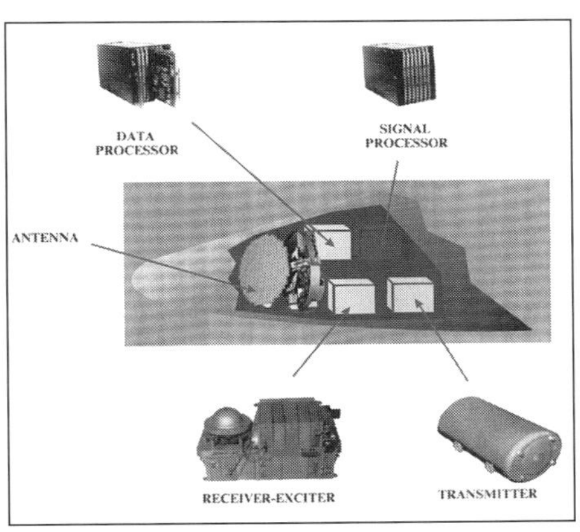

Moskit airborne radar for advanced combat trainers 0011877

Operational status

Marketed for MiG-23, MiG-23BN and MiG-27 upgrades. Also marketed as Moskit-23 for the MiG-23-98-1 and MiG-23-98-2 aircraft upgrades.

Contractor

Phazotron-NIIR Scientific & Production Company.

VERIFIED

N019M/N019ME/N019MF/N019MP Topaz multifunction fire-control radars

Topaz is a multifunction, multimode, coherent, pulse Doppler, air-to-air modernisation of the N019 Sapfir 29 radar. N019M is intended for installation in MiG-29 and MiG-23 aircraft. The modernisation is claimed to increase operational effectiveness, particularly in ECM conditions. Although designed as an air-to-air radar, Topaz can be modified to incorporate air-to-surface modes. It is compatible with a wide range of weapons, including: R27ER1, R27T1, R27ET1, RVV-AE, R73E.

Specifications

Detection range (against 3 m^2 target):
(approaching target) 80 km
(receding target) 40 km
Scan angle:
(azimuth) ±15°, ±25°, ±70°
(elevation): 4, 6 lines
MTBF: 100 h
Weight: 350 kg

Operational status

N019M is understood to be fitted to later model production MiG-29 aircraft. The 'E' designator indicates that the radar is an export variant. N019ME is reportedly fitted to Malaysian MiG-29ME aircraft.

N019MF and N019MP are being marketed for MiG-29SMT and other MiG-29 upgrades by Phazotron-NIIR and Tikhomirov-NIIP respectively. Phazotron states that the N019MF is optimised for air-to-air operation, while the N019MP has been optimised for the air-to-ground role. The name Zhemchug has also been applied to radars for MiG-29SMT but no details are available and it is not known whether the name applies to either of the N019MF or NO19MP.

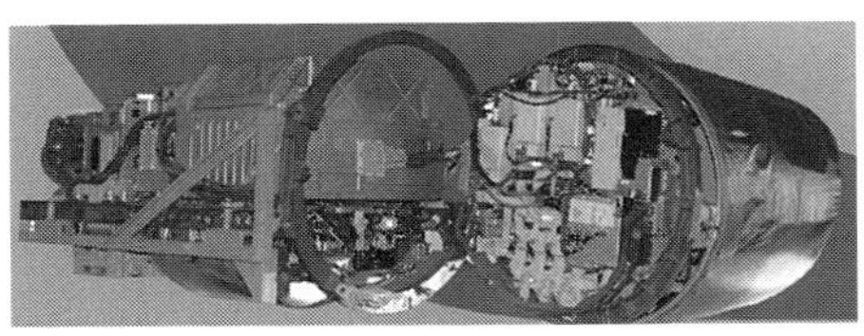

Topaz multifunction fire-control radar 0011874

Contractor

Phazotron-NIIR Scientific & Production Company.

VERIFIED

Pharaon multifunction airborne fighter radar

Pharaon is a new multifunction X-band (NATO I-/J-band) airborne fighter radar, designed to offer comprehensive air-to-air and air-to-ground capabilities. Pharaon is a relatively lightweight radar marketed as a nose fire-control radar for light fighters, or as the tail cone radar for heavy fighter aircraft.

Pharaon employs a new method of implementing a phase controlled antenna to radically reduce beam switching time and hence to increase target handling capability. Pharaon is a new radar, rather than merely a new antenna technique but, nonetheless, the antenna technology it employs can be retrofitted into other radars and Zhuk-M is cited as a prime candidate.

The antenna employs a novel, slotted waveguide concept in which the launching aperture is not the conventional series of regularly spaced slots in a horizontal waveguide, but a series of non-regularly spaced, circularly polarised ferrite phase shifters placed in the face of feeders that trace the path of a circular arc, fed from the outside of the antenna. The result of this implementation is claimed to be beam switching times reduced by a factor of 10 when compared with the conventional slotted waveguide design, and consequent increase in simultaneous target handling capability from 12 to 30.

Specifications

Detection range: 75 km against air targets
Angular coverage: +/−70° in azimuth and elevation
Peak power: 4 kW
Average power: 0.3 kW
Weight: 75 kg

Operational status

On development contract from Sukhoi. Pharaon is in laboratory test, and initial flight testing is planned for 2000.

Contractor

Phazotron-NIIR Scientific & Production Company.

VERIFIED

RP-35 multimode airborne radar

The RP-35 is a coherent, multimode, digital fire-control radar that provides a comprehensive set of all weather air-to-air and air-to-surface modes, with superior dogfight and weapons delivery capabilities. The air-to-air modes provide the capability to detect, track and

Pharaon multifunction airborne fighter radar **2000**/0067189

engage targets at all aspects, even in the presence of ground clutter. Air-to-surface modes provide extensive ground mapping, target detection, location and tracking capabilities, as well as navigation features.

The RP-35 radar is designed for use with the MiG-35 aircraft, and is compatible with a wide range of air-to-air and air-to-surface weapons, including: Kh-29T, Kh-31A, Kh-35U, Kh-38, R-27ER1, R-27ET1, R-27R1, R-27T1, R-73E, RVV-AE, KAB-500KR.

The RP-35 radar is designed for single-pilot operation. All combat-critical controls are integrated into the throttle grip and stick controller (HOTAS). Air-to-air and air-to-surface information is displayed on the Head-Up-Display as well as the Multifunction Cockpit display.

The antenna was originally designed as a phased-array with electronic scanning to provide high gain and low sidelobes at all scan angles. However, Phazotron stated, at Dubai 2000, that the phased array antenna had been deleted and a cheaper slotted array with mechanical scanning had been substituted to reduce price. Phazotron went on to state that the cheaper configuration could still achieve the basic operational requirement for engagement of two to four targets - albeit less satisfactorily - in that engagement of 4 targets is still possible if all are close to the boresight (+/−40°), but only 2 targets for widely separated angles (+/−70°). The transmitter is a liquid-cooled TWT. The receiver is a three-channel system.

The RP-35 is claimed to have the following capabilities:

Air-to-air modes: 24 target track-while-scan simultaneously; range while search; air combat - vertical scan, HUD search, boresight, wide angle, velocity search; raid cluster resolution, automatic terrain avoidance.

Air-to-surface modes: four target track-while-scan; ground moving target indication/track; air-to-ground ranging; real beam ground map; Doppler beam sharpening; synthetic aperture; enlargement; freezing; beacon.

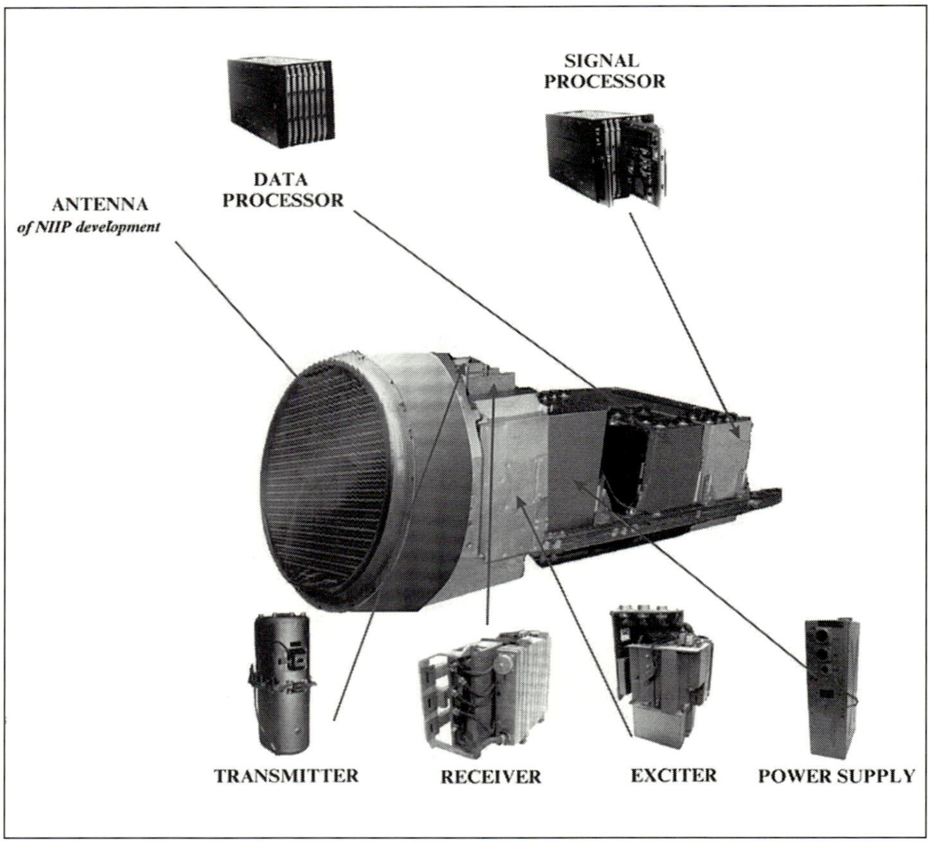

RP-35 multimode airborne radar 0011876

The MiG-25 carries the RP-25 Sapfir-25 interception radar

Specifications

Detection range:
(approaching targets) 140 km
(receding targets) 65 km
Angular coverage: ±20°, ±60° azimuth, 2 or 4 bars in elevation
Radar frequency: I/J-band
Peak power: 8 kW
Average power: 2 kW
Input power: 12 kVA, 200 V, 400 Hz; 2 kW 27 V DC
Weight: 220 kg
Volume: 500 dm^3
Cooling: air/liquid
Reliability: >120 h MTBF

Operational status

Designed to meet the requirements of the MiG-35.

Contractor

Phazotron-NIIR Scientific & Production Company.

VERIFIED

Sapfir airborne fire-control radars

The Sapfir series radars have been in use for up to 40 years on MiG-21, MiG-23, MiG-25, and MiG-29 aircraft. They were designed by predecessor organisations of Phazotron-NIIR Scientific & Production Company.

RP-21 Sapfir
RP-21 Sapfir was widely fitted to early model MiG-21 variants. It is a basic I-band air-to-air fire-control radar, credited with the following operational capabilities: detection range 20 km; tracking range 10 km; azimuth cover 60°.

RP-22 Sapfir-21
RP-22 Sapfir-21 was widely fitted to later model MiG-21 variants. It is a basic low J-band air-to-air fire-control radar, credited with the following operational capabilities: detection range 30 km; tracking range 15 km; azimuth cover 60°; elevation cover ±20°.

S-23 Sapfir-23
S-23 Sapfir-23 was widely fitted to MiG-23 variants. It is a pulse Doppler low J-band air-to-air fire-control radar, credited with the following operational capabilities: detection range 70 km; tracking range 55 km; azimuth cover 60°; elevation cover ±60°.

RP-25 Sapfir-25
RP-25 Sapfir-25 was widely fitted to MiG-25 variants. It is a look-down/shoot-down I/J-band pulse Doppler radar, credited with the following operational capabilities: detection range 100 km; tracking range 75 km; azimuth cover 112°; elevation cover ±60°.

N019 Sapfir-29
N019 Sapfir-29 was installed on initial production MiG-29 aircraft. It is reported to be a look-down/shoot-down pulse Doppler fire-control radar, which operates in I/low J-band. Credited operational capabilities include the ability to track 10 targets simultaneously; detection range 100 km; tracking range 70 km; azimuth cover 134°; elevation cover −38 to +60°.

Contractor

Phazotron-NIIR Scientific & Production Company.

VERIFIED

Sokol multimission airborne fire-control radar

Sokol is a coherent, multimission, digital fire-control radar that provides a comprehensive set of all-weather air-to-air and air-to-surface modes. The diverse operating modes required to meet the multimission design concept are achieved by employing a variety of complex and flexible waveforms involving low-, medium- and high-pulse repetition frequencies.

The Sokol radar is intended to be installed in next-generation multirole aircraft; it is compatible with the datalink requirements of a wide range of air-to-air and air-to-surface weapons, including: Kh-31A, R-27R1, R-27T1, R-73E, RVV-AE and other precision-guided weapons.

The antenna is a phased-array with electronic scanning, which provides high gain and low sidelobes at all scan angles. Originally it was to a Tikhomirov-NIIP design, but a Phazotron-NIIR design has been used to replace it. The transmitter is a liquid-cooled TWT. The receiver is a multiple-channel system, with low noise figure. The signal and data processors use flexible high-order language programming.

Sokol is claimed to have the following capabilities:

Air-to-air modes: 24 target track-while-scan and simultaneous engagement of six targets; range while search look-up/look-down; air combat manoeuvring - vertical scan, HUD search, boresight, wide angle; velocity search; raid cluster resolution, automatic terrain avoidance.

Air-to-surface modes: four target track-while-scan; ground moving target indication/track; air-to-ground ranging; real beam ground map; Doppler beam sharpening; precision velocity update; synthetic aperture; enlargement; freezing; beacon.

Specifications

Antenna diameter: 1 m
Detection range:
(approaching targets) 180 km
(receding targets) 80 km
Angular coverage: ±20°, ±60° azimuth, 2 or 4 bars in elevation
Radar frequency: I/J-band
Peak power: 8 kW
Average power: 2 kW
Input power: 12 kVA, 200 V 400 Hz; 2 kW 27 V DC
Weight: 275 kg
Volume: 600 dm^3
Cooling: air/liquid
Reliability: >120 h MTBF

Operational status
Designed to meet the requirements of the Su-37. In late 1999, the first prototype was undergoing laboratory testing. Flight testing will be on a Su-27KUB.

Contractor
Phazotron-NIIR Scientific & Production Company.

VERIFIED

Zhuk airborne radars

N010 Zhuk was designed by the NIIR Moscow (Scientific Research Institute for Radio Engineeering, Moscow), now part of Phazotron-NIIR. N010 Zhuk (together with the N011 radar that equips Su-27M) was the first multifunction Russian radar capable of tracking air as well as surface targets. The radar uses a slotted flat-plate antenna and digital computer. It was introduced in 1988 on the MiG-29M. Development has continued under Phazotron, and the following models are now being marketed:

Zhuk is the definitive type. It is a coherent, multimode, multimission, digital fire-control radar intended for MiG-29, MiG-33 and their upgrades.

Zhuk-27 is a variant marketed for Su-27 upgrades and Su-30 variants.

Zhuk-M appears to be the most advanced variant of Zhuk so far developed; it is marketed for the MiG-29SMT, MiG-29K and other MiG-29 upgrades.

Zhuk-Ph is a derivative of Zhuk that features a phased-array antenna system.

A key feature of the Zhuk radar is its programmability. The radar can be programmed to respond to new threats, to incorporate improved operating modes, to integrate new weapons and to respond to new electronic countermeasures by software change.

Zhuk is compatible with the datalink requirements of a wide range of air-to-air and air-to-surface weapons, including: Kh-31A, R-27T1, R-73E, RVV-AE and other precision-guided weapons and iron bombs.

Zhuk is claimed to have the following capabilities:

Air-to-air modes: 10 target track-while-scan and simultaneous engagement of 2 to 4 targets; range while search look-up/look-down; air combat manoeuvring - vertical scan, HUD search, boresight, wide angle, automatic terrain avoidance.

Air-to-surface modes: four target track-while-scan; ground moving target indication/track; air-to-ground ranging; real beam ground map; Doppler beam sharpening; aircraft velocity measuring for navigation system updating; synthetic aperture; enlargement and freezing capability.

Zhuk-M appears to be a very major redesign. Size and weight are significantly reduced, while power and capability have been increased; antenna design is also significantly changed. Performance figures quoted are for 20 track-while-scan targets, and 4 simultaneous engagements, at up to 120 km detection range. Zhuk-M is also claimed to be able to perform automatic terrain avoidance and hovering helicopter detection.

Specifications
Detection range:
(approaching targets) 80 km (Zhuk-M 120 km)
(receding targets) 40 km (Zhuk-M 50 km)
Angular coverage:
azimuth: ±20°, ±60°, ±90°
elevation: 2 or 4 bars (Zhuk-M +60/-40)
Radar frequency: I/J-band
Peak power: 5 kW (Zhuk-M 6 kW)
Average power: 1 kW (Zhuk-M 1.5 kW)
Input power: 8.5 kVA, 200 V 400 Hz; 1.5 kW 27 V DC (Zhuk-M 12 kVA)
Weight: 250 kg (Zhuk-M 220 kg)
Volume: 800 dm^3 (Zhuk-M 350 dm3)
Cooling: air/liquid
Reliability: 120 h MTBF (Zhuk-M 200 h)

Operational status
N010 Zhuk in service on MiG-29M; advanced versions being marketed for Russian Air Force and export aircraft. When fitted to the Chinese J-8IIM aircraft, it is called Zhuk-8II.

Contractor
Phazotron-NIIR Scientific & Production Company.

VERIFIED

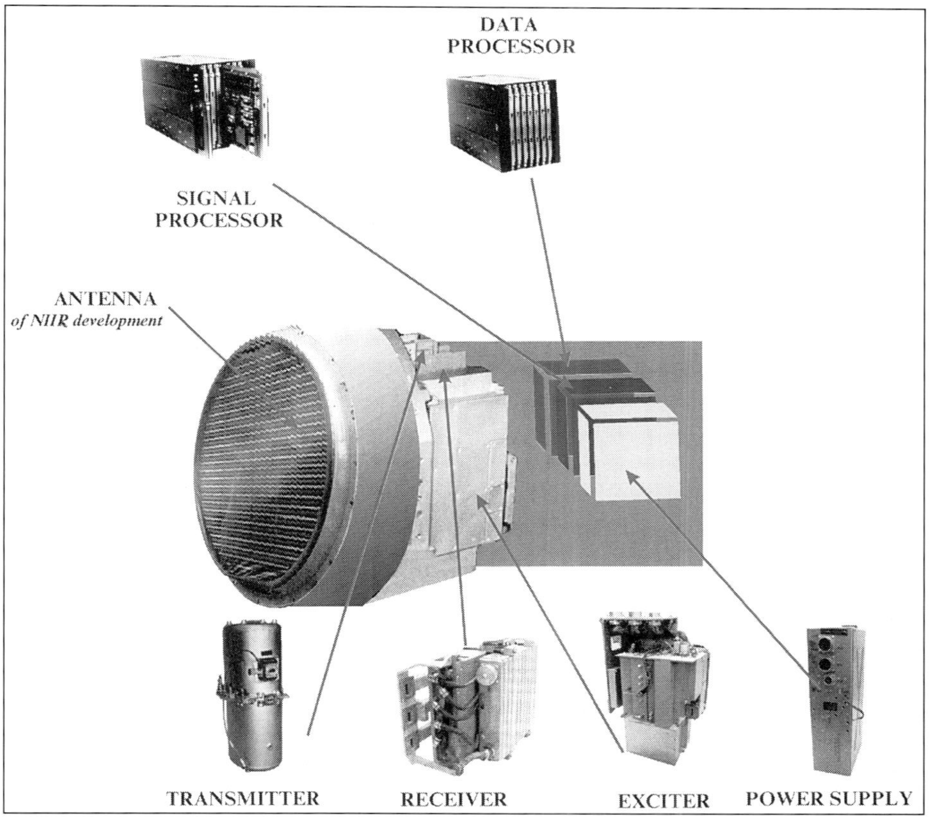

Sokol multimission airborne fire-control radar 0062855

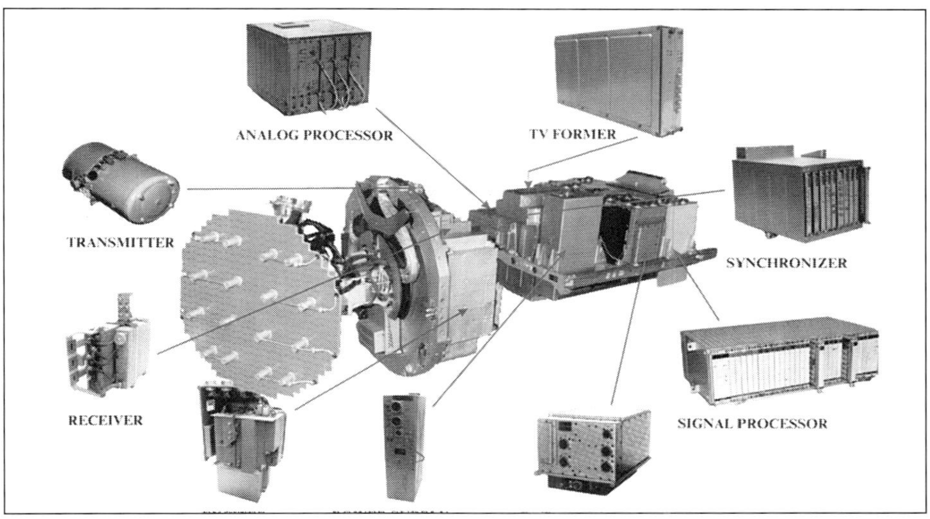

Zhuk airborne radar 0011873

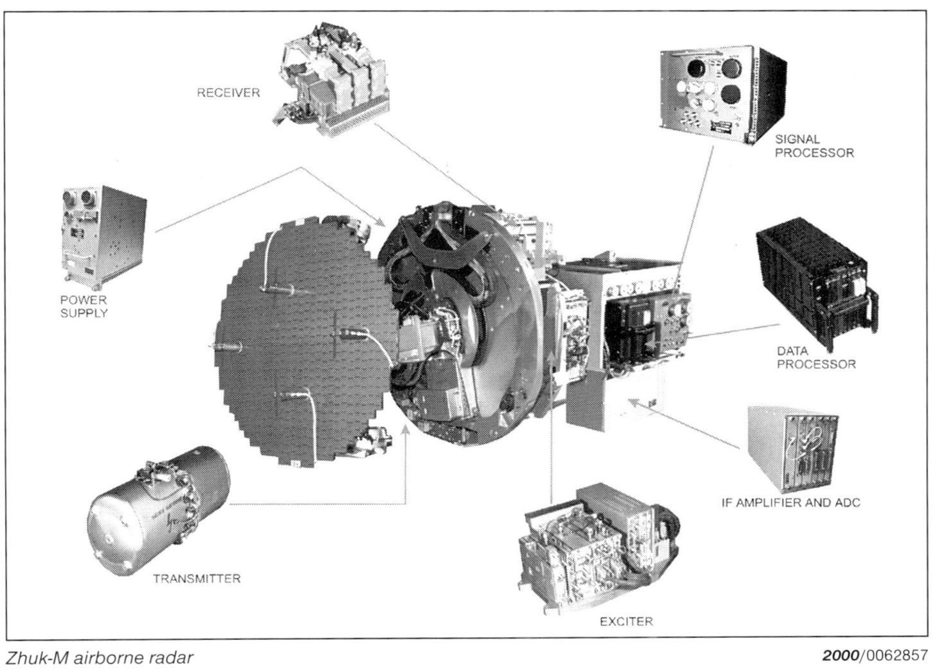

Zhuk-M airborne radar **2000**/0062857

Airborne radar jammers - Gardeniya, Schmalta and Sorbtsiya

Gardeniya, Schmalta and Sorbtsiya are all airborne jammers, that are believed to belong to the same family.

The Gardeniya family includes both pod-mounted and internal radar jammers, that can operate autonomously and automatically. Family capabilities are reported to include: self-protection; noise jamming in the 10, 20 and 70 cm wavebands; and communications jamming (Gardeniya IFUE - fitted to Mi-17P). Internally mounted versions of the Gardeniya family are reported to be fitted to MiG-29 aircraft.

Sorbtsiya (also reported as Sorbtsiya-S) wide spectrum jamming pod has been shown fitted in wing tip pods on Su-271B, naval Su-27 'Flanker', Su-34 and the Su-35 advanced 'Flanker' derivative.

Operational status

Many systems are in service; it is probable that development continues. The Su-27SK is reported to carry an optional Gardeniya installation, using two wing-tip pods.

Contractor

Pleshakov Scientific & Industrial Corp (GosCNRTI), Design Bureau, Moscow.

VERIFIED

Sorbtsiya on an Su-34

Su-30 MKI communications system

Polyot Research and Production Company identifies the communications system for the Su-30 MKI to comprise the following capabilities:

simultaneous voice and data communication between air and ground, using HF, VHF and UHF frequencies;
automatic data transfer of targeting data;
anti-jamming capabilities;
two-position control of communications and intercommunication functions;
emergency frequency monitoring;
automated BITE.

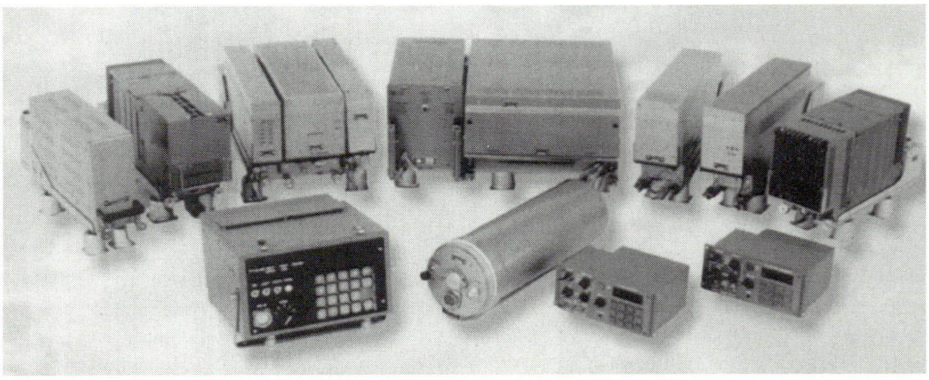

Su-30 MKI communications system (equipment details not divulged) **1999**/0044803

Specifications

Frequency: 2-18, 100-150, 220-400 MHz
Modes:
HF: AM, SSB
V/UHF: AM, FM, FT
Preset channels:
HF: 20
V/UHF: 20
data: 20
MTBF: 4,700 h

Operational status

Stated to be the fit for Su-30 MKI aircraft delivered to the Indian Air Force.

Contractor

Polyot Research and Production Company.

VERIFIED

MS communication and voice warning system

The MS communication and voice warning system is designed to meet the internal and external communication requirements of up to three aircrew, including provision of voice warnings.

System components comprise the B27-MS amplifier and switching unit and B7-MS control panels.

System capabilities include:

intercommunication facilities for up to three crew members and two ground personnel
control of four radios
monitoring of seven caution signals and two radio navigation signals
separate volume control of the intercom and radio facilities
operation of 256 voice warnings

Specifications (B27-MS)

Dimensions: 274 × 126 × 194 mm
Weight: 5 kg
Power: 18 to 31 V DC, <10 W

Operational status

The MS communication and voice warning system is fitted to the MiG-AT, MiG-29M and YAK-130 aircraft.

Contractors

PRIMA Scientific-Production Enterprise (developer).
A.S. Popov GZAS JSC (manufacturer).

VERIFIED

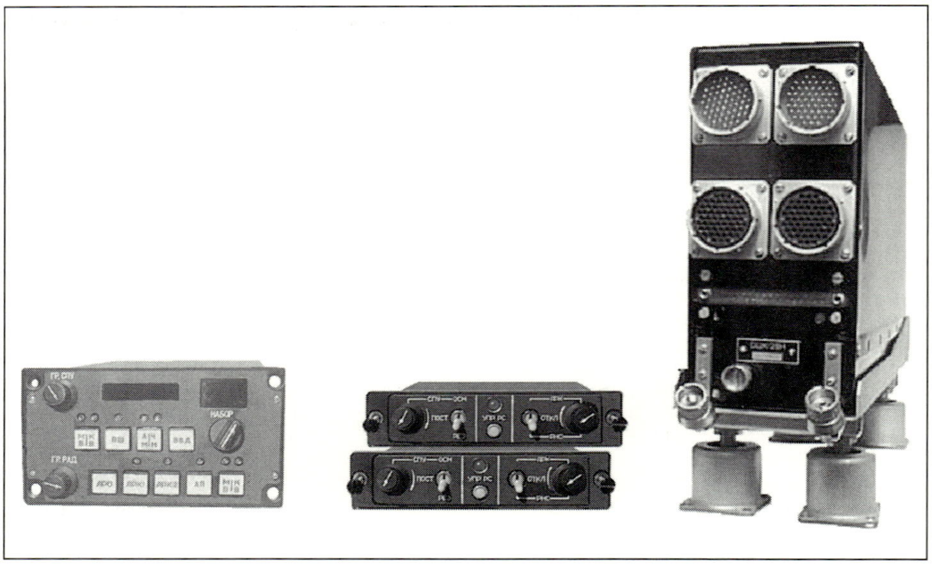

The MS communication and voice warning system **2000**/0062794

MSPD voice and data communication system

The MSPD voice and data communication system is designed to meet the internal and external voice and data communication requirements of up to two aircrew.

System capabilities include:

a) intercommunication facilities for up to two crew members and three ground personnel
b) control of four radios
c) control of data transmissions air-to-air and air-to-ground
d) monitoring of seven caution signals and two radio navigation signals
e) separate volume control of the intercom and radio facilities
f) operation of 256 voice warnings

Specifications

Data rate: 4, 75, 150, 300, 1200, 2400, 4800 9600 bit/s
Dimensions: 190 × 194 × 320 mm
Weight: 10 kg

Operational status

The MSPD voice and data communication system is fitted to the Sukhoi S-54 aircraft.

Contractors

PRIMA Scientific-Production Enterprise (developer)
A.S. Popov GZAS JSC (manufacturer).

VERIFIED

The MSPD voice and data communication system **2000**/0062795

SPGU-35 communication system

The SPGU-35 communication system is designed to meet the internal and external communication requirements of a three- or four-man crew involved in paramilitary operations such as helicopter casualty evacuation in helicopters such as the Ka-226 and fire fighting operations in the Be-200 amphibian. It also provides for crew monitoring of audio caution and radio navigation systems.

System components comprise the following:

a) one B27-35 amplifier and switching unit

b) up to three B7-35 control panels

c) one or two B7A-35 or B-7B-35 control panels

d) one GGO-3 × 2 (two channels × 3 W) address amplifier.

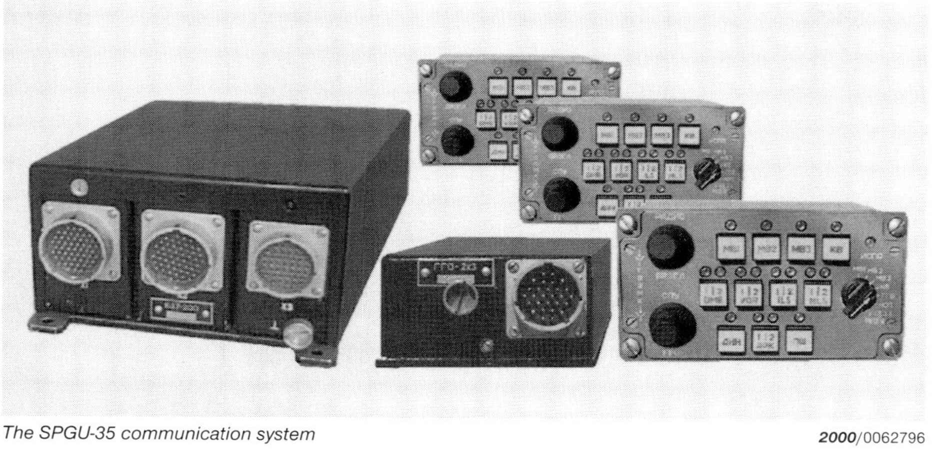

The SPGU-35 communication system **2000**/0062796

SPGU-35 capabilities include:

a) crew communication over up to five air-ground radios

b) control of up to 12 radio navigation signals

c) monitoring of up to eight caution signals

d) passenger address

e) communication for up to three crew with ground maintenance personnel

f) interface to the voice recorder and automatic test systems

g) separate volume controls for different functions

h) 6 W audio output power.

Specifications

Dimensions and weight:
(B-27-35 amplifier switching unit) 220 × 160 × 90 mm; 3 kg
(B7-35 control panel) 146 × 80 × 126 mm; 0.8 kg
(B7A-35 and B7B-35 control panels) 146 × 32 × 110 mm; 0.4 kg
(GGO-3 × 2 unit) 131 × 90 × 50 mm; 0.8 kg
Power: 18 to 31 V DC

Contractors

PRIMA Scientific-Production Enterprise (developer)
A.S. Popov GZAS JSC (manufacturer).

VERIFIED

IFF 6201R/6202R and 6231R/6232R systems

RadioPribor has manufactured airborne radio and navigation equipment for installation on all types of civilian and military aircraft and helicopters produced in Russia and the other RFAS countries.

The IFF responders 6201R and 6202R are fitted on all military and civil aircraft and helicopters. The model 6202R differs from the 6201R in that it incorporates additional signal amplifiers for use on heavy aircraft that have long SHF cable runs.

In aircraft equipped with radar, the 6231R or 6232R interrogators are installed to provide the interrogation function.

The combined system provides three modes of operation: general identification modes; individual identification modes providing 84 interrogation slots and 100,000 reply slots; identification of objects in distress (distress mode with interrogation/alarm modes without interrogation).

RadioPribor's 6201R IFF responder 0010928

Operational status

Widely deployed on civil and military aircraft and helicopters manufactured in Russia and the RFAS countries.

RadioPribor's 6231R IFF interrogator 0010929

Contractor

Production Association RadioPribor.

VERIFIED

705-6 attitude heading reference system

The 705-6 attitude heading reference system is part of the inertial navigation system of the MiG-29. It provides the following data outputs: gyroscopic heading, pitch and roll angles, and absolute orthogonal accelerations.

Specifications

Normal erection: 15 min
Fast erection: 3 min
Errors:
(normal erection) 0.15°
(fast erection) 0.7°

Operational status

Fitted to MiG-29 aircraft.

Contractor

Ramensky Instrument Engineering Plant JSC.

VERIFIED

705-6 attitude heading reference system
0018236

Zaslon fire control radar for MiG-31

Zaslon is the airborne fire-control radar fitted to MiG-31 aircraft. The designator N007 has been used in association with this radar.

Zaslon is an I-band pulse Doppler radar, with a look-down/shoot-down capability. It uses phased array radar techniques to scan a 140° conical sector, and to track up to 10 targets simultaneously. Up to four engagements can be handled simultaneously, with a maximum detection range of 200 km. Radar data can be transferred between MiG-31 aircraft by a datalink to aid raid management.

An upgraded variant of Zaslon, designated Zaslon-A, was developed for the MiG-31B aircraft, which went into production in 1990. It apparently provided for longer missile intercept range, and was integrated with new modes of operation of the aircraft datalink and upgraded navigation computers.

Meanwhile, a further derivative of the Zaslon radar, designated Zaslon-M, was developed for the MiG-31M development aircraft. It is said to have been a much more capable radar, with far greater target detection range and tracking capability, and simultaneous operation of up to six RVV-AE medium range radar active homing missiles and R-73 short range missiles. This aircraft did not go into series production.

By 1995, a multirole export derivative of the MiG-31, designated MiG-31FE, was being marketed, which included an anti-ship mode, and the ability to use antiship missiles. This aircraft did not attract any sales, but in January 1999, a Russian Air Force variant of this aircraft, designated MiG-31BM, was demonstrated at MAKS '99 and is reported to form part of the Russian Air Force's fourth generation aircraft upgrade programme, with the MiG-29SMT. This aircraft is said to carry an upgraded variant of Zaslon with increased target detection range and surface target detection capability.

Operational status

The Zaslon is in service with MiG-31 and Zaslon-A with the MiG-31B aircraft of the Russian Air Force and the Khazakhstan Air Force. Zaslon-M is associated with the MiG-31M development programme, and is not believed to have been put into active service. MiG-31BM carries the latest variant of Zaslon, which includes antiship capabilities.

Contractor

State Enterprise V Tikhomirov Scientific Research Institute of Instrument Design NIIP.

VERIFIED

Zaslon on MiG-31

YB-3A flare dispenser system

The YB-3A flare dispenser is a modular design, that offers a large number of hardware configuration options, and considerable programming flexibility.

The basic configuration comprises four electronic line replaceable units: the control panel, a programme selector (double unit), a computer unit, and a safety unit. Two types of dispenser units are available, capable of dispensing 8 × 50 mm calibre cartridges and 32 × 26 mm calibre cartridges respectively. The control system is capable of addressing up to 512 flare locations, contained in multiple dispenser unit configurations.

Control capabilities include:

(1) 50,000+ flexible random interval programming options;
(2) eight preliminary installed programs, with flexible selection in flight;
(3) five variable program parameters;
(4) 1-8 cartridges in salvo;
(5) 0.025-16 seconds interval between salvos;
(6) 18 minutes maximum program duration, or until 'all gone';
(7) 8 - 512 flare payload;
(8) one-year programmed life;
(9) emergency ejection of all flares.

The control panel provides the following capabilities: program loading, fire/stop control, payload remaining indication, rapid fire and built-in-test control, variable brightness display.

The program selector provides the following features: automatic operation, random intervalometer capability, one year programmed life.

Specifications

Dimensions and weights:
(cockpit control panel) 164 × 64 × 104 mm; 0.5 kg
(program selector) 186 × 221 × 212 mm; 5.5 kg
(computer unit) 172 × 373 × 214 mm; 9.8 kg
(safety unit) 94 × 90 × 66 mm; 0.35 kg
(dispenser units) 130 × 384 × 172 mm for the 32 × 26 mm cartridge dispenser; 7.4 kg
130 × 384 × 285 mm for the 8 × 50 mm cartridge dispenser; 8.5 kg
Power: 24-30 V DC, 150 W; 115 V AC, 400 Hz, 200 VA

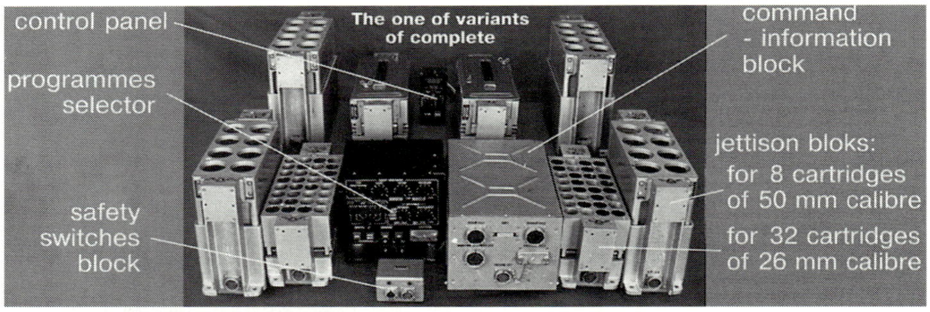

YB-3A flare dispenser - typical configuration 0018305

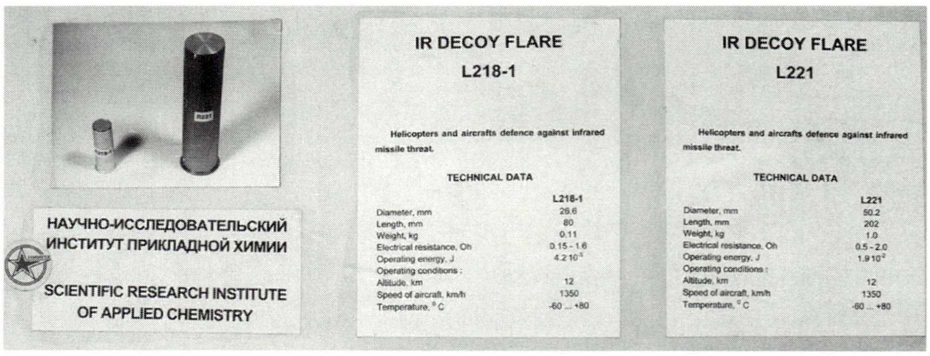

L-218-1 and L221 flare cartridges for YB-3A flare dispenser (Paul Jackson) 0018300

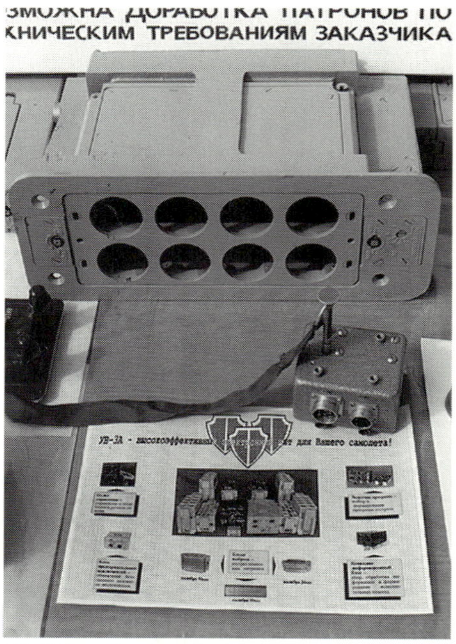

YB-3A flare dispenser 8 × 50 mm cartridge dispenser unit (Paul Jackson) 0018302

YB-3A flare dispenser control panel (Paul Jackson) 0018304

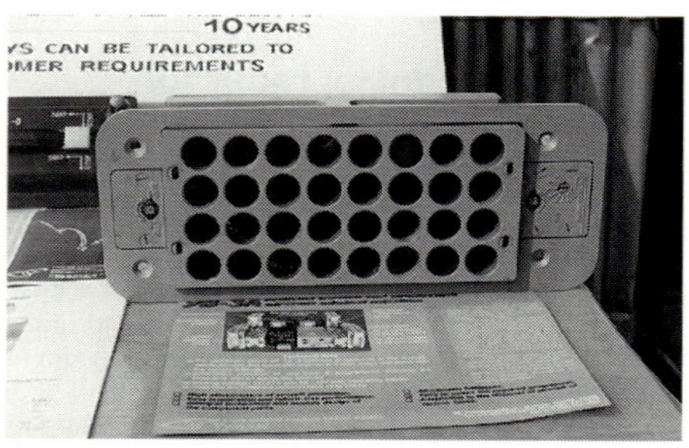

YB-3A flare dispenser 32 × 26 mm cartridge dispenser unit (Paul Jackson) 0018301

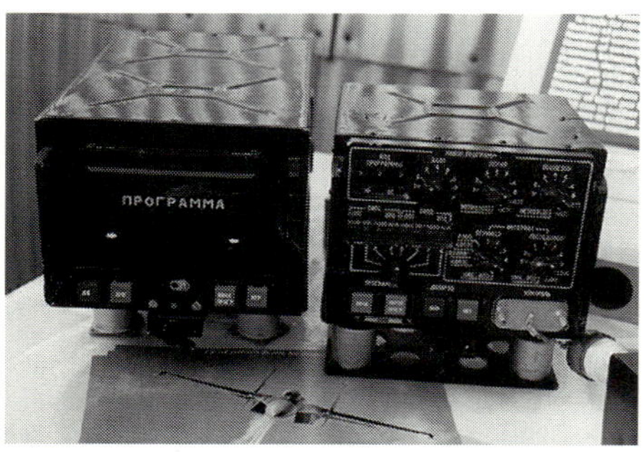

YB-3A flare dispenser double program selector unit (Paul Jackson) 0018303

Cartridges: Developed by the Scientific Research Institute of Applied Chemistry.

	L218-1	L-221
Calibre	26.6 mm	50.2 mm
Length	80 mm	202 mm
Weight	0.11 kg	1.0 kg
Aircraft speed (max)	1,350 km/h	1,350 km/h
Dispenser barrel life	700 bursts	250 bursts

Operational status

Designed for both aircraft and helicopter application; widely deployed.

Contractor

Vympel State Machine Building Design Bureau.

VERIFIED

YB-3A emergency ejection of all flares
0018299

Zenit L166VIAZ infra-red jammer

Fitted to Mi-24 helicopters since 1983. Two basic variants exist, UZV-1 and UZV-2 reported to be similar to Sanders ALQ-144.

Operational status

In service.

Contractor

Zenit Foreign Trade Firm, State Enterprises p/c SA, Zverev Krasnogorsky Zavad.

VERIFIED

SOUTH AFRICA

Pelican TACCO system

The Pelican TACCO system provides for the integration of sensor and datalink information, the compilation of tactical picture and the overall management of Maritime Patrol Aircraft (MPA) operations. It assists the Tactical Co-ordinator (TACCO) with the co-ordination and control of the aircraft's missions, including surveillance and intelligence gathering, vectored attacks on surface ships, anti-submarine warfare and search and rescue operations. The system also facilitates mission planning and debriefing. The basic configuration of the Pelican system allows integration with the following aircraft subsystems: navigation, ESM, acoustics, radar, sonobuoy launcher, and datalink controller. Other integrations can be added to match customer requirements.

Contractor

ADS Altech Defence Systems.

VERIFIED

Airborne Laser Warning System (LWS)

The airborne Laser Warning System (LWS) provides threat identification and Direction-Finding (DF) indication of laser range-finders, designators and lasers used for missile guidance purposes. The system is designed to interface with an existing onboard RWR/ ESM host system and is available with one of three LWS sensor models (LWS-200, LWS-300 and LWS-400). All the sensor models use the same Laser Warning Controller (LWC) for data processing and interfacing to the host EW system. On detection of a threat, an audio and visual (display) alarm is generated via the host EW system, or if configured as a stand-alone system, via a dedicated Threat Display and Control Unit (TDCU) developed by Avitronics. The LWS-300 and LWS-400 are form-fit compatible, and all three sensor models are electrically compatible allowing for easy and cost effective upgrading.

Broad coverage of the laser spectrum ensures detection of all known current threats. The threat library is user programmable and field loaded into the LWS via the host EW system.

The system is able to record threat parameters encountered during the mission for post analysis.

Airborne Laser Warning Controller with sensor models (from left to right) LWS-400, LWS-200 and LWS-300
1999/0051259

Specifications

LWS-200
Wavelength coverage: 0.6-1.8 μm
Threat coverage: Ruby, GaAs, NdYAG, Raman Shifted NdYAG and Erbium Glass lasers
AOA accuracy: azimuth 11° rms
Spatial coverage:
azimuth 360° (99° per sensor)
elevation 60°
Probability of intercept: >99% for a single pulse
Dimensions: 103 × 86 × 64 mm
Weight: 0.8 kg per sensor

LWS-300
Wavelength coverage: 0.5-1.8 μm
Threat coverage: the same as LWS-200 plus doubled NdYAG lasers
AOA accuracy: azimuth 15° rms
Spatial coverage:
azimuth 360° (90° per sensor)
elevation 60°
Probability of intercept: >99% for a single pulse
Dimensions: 115 × 90 × 76 mm
Weight: 1.2 kg per sensor

LWS-400
Wavelength coverage: 0.5-1.8 μm and 2-12 μm
Threat coverage: the same as LWS-300 plus CO_2 lasers
AOA accuracy: azimuth 15° rms
Spatial coverage:
azimuth 360° (90° per sensor)
elevation 60° (0.5-1.8 μm)
elevation 40° (2-12 μm)
Probability of intercept: >99% for a single pulse
Dimensions: 107 × 90 × 76 mm
Weight: 1.2 kg per sensor

Laser warning controller
Dimensions: 188 × 89 × 131 mm
Weight: 2.5 kg

Contractor

Avitronics (Pty) Ltd.

UPDATED

Electronic Surveillance Payload (ESP)

The Electronic Surveillance Payload (ESP) is a derivative of Avitronics' Emitter Location System (ELS) tailored for installation on UAVs. The system was originally designed to operate as a stand-alone ESM system integrated with the Kentron Seeker II UAV, providing information on an enemy's electronic order of battle through emitter identification and location.

The ESP consists of an acquisition and analysis receiver and controller integrated into a single unit, coupled with a nose-mounted interferometric antenna array. Emitter data is transferred via the UAV data link to a ground based REmote Terminal (RET) for threat display and control of the system.

The ESP provides the following functions:
- acquisition, analysis and precision DF of emissions from search, tracking and fire-control radars
- accurate bearings and signal parameter measurement of emitters

- gathering and recording of detailed emitter data for ESM/ELINT analysis
- high Probability Of Intercept (POI) for search radars using the wide-open acquisition radar
- onboard mini-flash data recorder
- frequency measurement for designated emitters
- an autonomous mode which provides for the detection of low POI emitters.

Specifications

Dimensions:
- controller: 193 × 127 × 343 mm

Mass:
- controller: 10 kg
- antenna array: 6 kg

Antennae: phase amplitude matched
Frequency coverage: 0.5 to 18 GHz
Frequency resolution: 1 MHz
Instantaneous bandwidth: 1 GHz or 80 MHz narrow band
Direction-finding: 1° RMS (>2 GHz); 3.5° RMS (700 MHz)
Field of view (fully calibrated): 240° in azimuth (in 3 sectors); 70° in elevation

Contractor

Avitronics (Pty) Ltd.

NEW ENTRY

Emitter Location System (ELS)

Emitter Location System (ELS) showing ELS controller (left) and 2-18 GHz antenna array ***1999***/0051258

Accurate direction-finding is important to ESM systems for the geo-location of emitters and for directing jammers and weapons. Integrated with the MSWS multisensor warning system, the ELS provides the high accuracy DF required for these tasks.

The ELS consists of an integrated receiver and controller and a number of interferometric antenna arrays, dependent on the system requirements. The main features of the system are: low mass and volume by using a single channel switched receiver; intra-pulse channel switching for a single pulse DF capability; high DF accuracy using a combination of phase and amplitude comparison technique; pulse Doppler handling capacity; high sensitivity.

The ELS is designed as an integral part of MSWS to enhance the DF capability of the radar warning function. All display and control is via the host system. The ELS functions include: acquisition and analysis of search, track and fire control radars; provision of accurate bearings for emitters designated from the host system; frequency measurement for designated emitters; detection of low probability of intercept emitters in an autonomous mode; gathering of emitter data for ESM/ELINT analysis.

The ELS is designed for installation on fixed- and rotary-wing aircraft and remotely piloted vehicles.

Specifications

Dimensions:
(controller) 193 × 127 × 343 mm
(2-18 GHz antenna array) 110 × 90 × 250 mm
(0.5-18 GHz antenna array) 180 × 80 × 500 mm
Mass:
(controller) 14 kg
(2-18 GHz antenna array) 5 kg
(0.5-18 GHz antenna array) 7 kg
Frequency coverage: 2-18 GHz (0.5-18 GHz optional)
Frequency resolution: 2 MHz in narrowband mode
Instantaneous bandwidth: 1 GHz/80 MHz
Direction-finding: 1° RMS
Field of view: 120° in azimuth per antenna array; 70° in elevation per antenna array

Contractor

Avitronics (Pty) Ltd.

UPDATED

MAW-200 Missile Approach Warning system

The MAW-200 Missile Approach Warning system operates as stand-alone equipment, or as part of an integrated Defensive Aids Sub-System (DASS). It is a totally passive ultraviolet system which provides detection and timely warning of the approach of surface-to-air and air-to-air missiles. Upon positive detection of the approaching missile, a priority interface to the chaff and flare dispensing system is activated for immediate and automatic dispensing of countermeasures against the threat. A visual Direction-Finding (DF) warning is provided to the aircrew via the display unit of an EW suite (or in the case of a stand-alone system via the Threat Display and Control Unit - TDCU) accompanied by the appropriate audio alarms.

The MAW subsystem consists of four sensors and a processing card, which resides in the Electronic Warfare Controller (EWC) of the EW suite. On the stand-alone version, processing is carried out in a dedicated processor unit. Each sensor is responsible for processing its detection algorithms. The processing card inside the EWC is responsible for the further processing of data received from the various sensors and for the built-in test control and management of the MAW sensors. The MAW-200 has a multi-threat capability and can track at least eight targets simultaneously.

Main features include: totally passive ultra-violet detection; low false alarm rate; no in-flight recalibration required; instantly online, no cooling required; comprehensive self-test routines.

Specifications

EWC dimensions: 343 × 127 × 193 mm
EWC weight: 14 kg
TDCU dimensions: 128 × 127 × 120 mm
TDCU weight: 2.2 kg
Sensor dimensions: 230 × 130 × 130 mm
Sensor weight: 3.1 kg
Detection method: passive ultra-violet
Detection range: >5 km for shoulder-launched missiles
Spatial coverage: 360° azimuth with 4 sensors
DF resolution (azimuth): better than 5°
Multithreat capability: at least 8 targets simultaneously

Operational status

In production.

Contractor

Avitronics (Pty) Ltd.

UPDATED

Multi-Sensor Warning System (MSWS)

The Multi-Sensor Warning System (MSWS) provides tactical aircraft with a complete warning capability for self-protection. The capability includes radar warning, laser warning, and missile approach warning. The architecture provides for a variety of sensors to be integrated into and managed by the system, allowing the user to upgrade the system.

Generic design and low unit count allow easy installation in aircraft ranging from helicopters to fighters. Complete spherical coverage is available and the system provides full threat identification. Threat identification parameters are user definable. The MSWS is flight line programmable and includes extensive BIT facilities.

The system includes a radar warning function for pulse Doppler and CW radars in high pulse density environments, a man/machine interface via a multifunction display and interface to and control of automatic chaff and flare dispensing systems.

The RWR features an Instantaneous Frequency Measurement (IFM) receiver, covering the 2 to 18 GHz

MAW-200 sensors with Electronic Warfare Controller (EWC) and Threat Display And Control Unit (TDCU)
2001/0103886

band in 4 GHz steps. It is reported to be able to cope with pulse densities up to 2.5 Mpps and to display worst situation threats within 500 ms, using 32-bit parallel processors.

The laser warning system is reported to cover 0.5 to 12 μm wavelengths, offering detection capability against laser range-finders, designators and missile guidance lasers, providing both threat classification and bearing.

The MAW uses ultra-violet detection techniques and is said to typically provide 5,000 m warning of shoulder-launched missiles.

The standard configuration comprises four sensor heads for each of the RWR, LWR and MAW functions. The display shows the nature and status of received signals, together with relative bearing, lethality and tabulated parametric data. Audio warning is provided to alert the crew to display data.

The system can be integrated with recording facilities for use in the Intelligence gathering role.

Growth options include an interface with an active ECM system activated automatically on threat detection, avionic system interface via a MIL-STD-1553B bus.

Specifications

Dimensions:
(EW controller) 343 × 127 × 193 mm
(threat display and control unit) 128 × 127 × 120 mm
(front end receiver × 4) 176 × 45 × 158 mm
(spiral antenna × 4) 110 × 110 × 67.5 mm
(LWS-400 sensor × 4) 107 × 90 × 76 mm
(MAW-200 sensor) 230 × 130 × 130 mm
Weight:
(EW controller) 14 kg
(threat display and control unit) 2.2 kg
(front end receiver × 4) 3 kg (per unit)
(spiral antenna × 4) 0.7 kg
(LWS-400 sensor) 1.2 kg (per sensor)
(MAW-200 sensor) 3.1 kg (per sensor)

The MultiSensor Warning System (MSWS) ***1999***/0051261

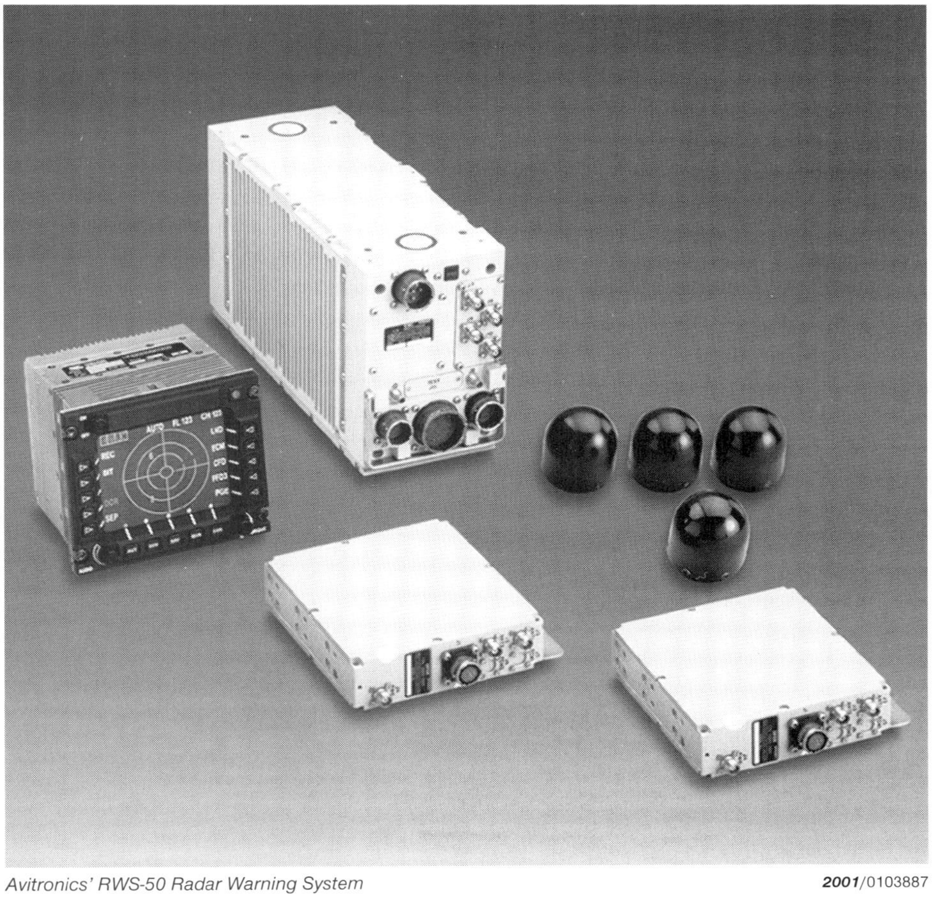

Avitronics' RWS-50 Radar Warning System ***2001***/0103887

Radar Warning System -50 (RWS-50)
Frequency coverage:
0.7-40 GHz (pulsed signals)
0.7-18 GHz (CW signals)
Direction-finding: 10-12° rms for pulsed signals in the 2-40 GHz range
Spatial coverage:
360° (azimuth or spherical)
90° (elevation or spherical)
Pulse density capability: >2.5 million pulses per second
Frequency resolution: 10 MHz

Laser Warning System-400 (LWS-400)
Wavelength coverage: 0.5-1.8 μm and 2-12 μm
Direction-finding: 15° rms (azimuth)
Spatial coverage (per sensor):
azimuth 90°
elevation 60° (0.5-1.8 μm)
elevation 40° (2-12 μm)
Laser threat coverage: doubled NdYAG, Ruby, GaAs, NdYAG, Raman Shifted NdYAG, Erbium Glass and CO_2 lasers
Laser threat types: range-finders, designators and lasers used for missile guidance (beam riders)
Probability of intercept: >99% for a single pulse

Missile Approach Warning-200 (MAW-200)
Operating frequency: solar blind UV band
Direction-finding: better than 5°
Spatial coverage: 94° conical field of view per sensor
Multithreat capability: capable of tracking at least 8 targets simultaneously

Contractor

Avitronics (Pty) Ltd.

UPDATED

RWS-50 Radar Warning System

The RWS-50 system provides tactical aircraft with a comprehensive radar warning capability for self-protection. This capability can be extended to include laser and missile approach warning.

In its basic configuration, the RWS-50 consists of four 2 to 18 GHz spiral antennas, two-dual detector amplifiers, an analyser unit and a colour multifunction display and control unit. This configuration includes an interface for automatic or manual control of a chaff and flare dispensing system.

The RWS-50 features a versatile threat library, flexible architecture, parallel processing, high sensitivity, high probability of intercept/low cycle time and low power consumption.

Upgrade options include; 0.7 to 1.4 GHz detection capability (omni or full DF); CW detection capability (omni or full DF); extended RF range 0.7 to 40 GHz (in one antenna); frequency measurements (via external superheterodyne/IFM subsystem); increased sensitivity/dynamic range; MIL-STD-1553B interface; LWR capability; MAW interface; spherical coverage.

Specifications

Dimensions:
(EW controller) 343 × 127 × 193 mm
(threat display and control unit) 128 × 127 × 120 mm
(dual-detector amplifier) 176 × 45 × 158 mm
(spiral antenna) 110 × 70 × 70 mm
Weight:
(EW controller) 14 kg
(threat display and control unit) 2.2 kg
(dual-detector amplifier × 2) = 3 kg
(spiral antenna × 4) 0.4 kg
Power supply: 28 V DC, 140 W
Frequency: 2-18 GHz
Spatial coverage:
(azimuth) 360°
(elevation) 90°
Direction-finding: 10° rms

Operational status

In production.

Contractor

Avitronics (Pty) Ltd.

UPDATED

TR 2800 airborne HF transceiver

The TR 2800 is a new-generation 100 W airborne HF transceiver. Its features include: frequency-hopping capabilities for enhanced ECCM performance, compatible with the TR 250/390 mobile base station HF transceivers; selective calling with channel enhancement and speech enhancement techniques to minimise HF noise and interference.

System control is exercised through the CU2832 control unit and the PA2810 400 W power amplifier is available, as an option, to meet high-power requirements. The CU2832 provides a 2 × 20 line alpha-numeric display. It is NVG compatible and provides links via RS 485 to the radio system and by RS 232 to the fill device.

The system can be configured with three different ATUs to meet aircraft and role requirements:

(1) The AT2820 100 W, loop, ATU has been developed to provide a high-angle antenna for near vertical incidence skywave propagation, thus eliminating the skip or dead zone, while providing communications in uneven and mountainous terrain up to a distance of 600 km. The loop antenna provides up to 6 dB gain and it is extremely fast tuning as required for the ECCM and ALE operation

(2) The AT2821 100 W wire ATU provides efficient matching to short wire antennas as installed in fixed-wing aircraft operating over long distances. Capable of storing 99 pretuned frequency settings, the ATU provides fast tuning (typically less than 10 ms) for ECCM and ALE operations

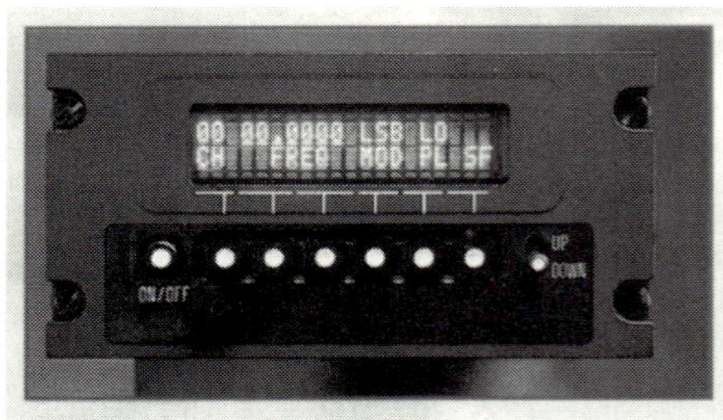

CU2832 control unit **2000**/0081870

(3) AT2822 400 W ATU provides fast tuning and high power for ECCM and ALE operations.

The TR 2800 transceiver features advanced electronic techniques to minimise HF noise and interference using a digital voice enhancement technique in accordance with CCIR445-1 and reliable DSP squelch.

Specifications

Dimensions
(transceiver): ½ ATR short, inclusive of 100 W power amplifier
(AT2820): 350 (H) × 3000 (L) × 120 (W) mm
(AT2821): ½ ATR short
(AT2822): full ATR short

Weight:
(transceiver): < 10 kg
(AT2820): < 10 kg
(AT2821): < 8 kg
(AT2822): < 15 kg

TR 2800 airborne HF transceiver 0011841

Operational status

The system is being integrated into the South African Air Force Rooivalk helicopter and into the upgraded C-130 aircraft.

Contractor

Grintek Comms.

VERIFIED

GUS 1000 Audio Management System (AMS)

The GUS 1000 AMS is a fully digital system. It is based on the GusBus (LIM), an IYU-T G703 and G704 based serial bus for distribution of audio, data, control and status information. The Gus Control Panel (GCP) and each module in the Communications Management Unit (CMU) contain a LIM.

The CMU is a configurable mechanical housing for a number of independent functional modules, which are individually integrated onto the GusBus. Functional modules provide:

(1) equipment control and status monitoring; audio management
(2) data management and networking
(3) interfaces (audio/analogue/digital/video/discretes)
(4) GPS (passive frequency hopping and aircraft back-up GPS)
(5) warning tones and synthetic speech generation
(6) gateway facilities (MIL-STD-1553B and ARINC 429)
(7) digital recording (voice and data) and voice command capability.

The GCP provides integrated control, display and audio functions. It also allows direct access to the GusBus for equipment such as radios. The GCP is NVG compatible and graphics capable.

A fillgun is provided for system application software and mission specific software. Extraction of mission data post-flight is also possible.

A full range of radio navigation and communication products covering the spectrum from 2 - 420 MHz is available to support the system.

The GUS 1000 system is available as a modular concept to meet the needs of all types of aircraft and helicopter, from basic training aircraft to multirole transport and maritime types.

Operational status

GUS 1000 is understood to have been in service on aircraft of the South African Air Force for some time. In 1998, it was selected as the AMS of choice for the Saab-BAE Systems Gripen multirole fighter aircraft in international markets.

Grintek Comms have also entered talks with Bell Helicopter Textron, Canada concerning the application of the GUS 1000 AMS for Bell commercial helicopters.

Contractor

Grintek Electronics Systems.

VERIFIED

GSY1500 VHF/UHF communications jamming system

The main purpose of the GSY1500 VHF/UHF jamming system is the disruption and/or jamming of communications channels and emissions in the 20 to 500 MHz frequency band. The GSY1500 is modular and can be mounted in airborne platfoms, such as helicopters and transport aircraft, or land-based or shipborne platforms.

Jamming features include look-through capability to monitor continued target presence, selectable optimised counter modulation types for various target signals, time-division multiplex jamming or frequency division multiplex jamming, jamming of up to 20 prioritised target frequencies, effectiveness against fixed-frequency voice and data communication links, selection of different output power settings and suppression of unwanted harmonics and spurious signals.

Intercept features include rapid scanning of the entire frequency band, noise-riding signal detection with digital signal processing techniques, high sensitivity with ranges of up to 450 km, demodulation facilities for both AM and FM modulation and rapid look-through monitoring and detection during jamming cycles.

Control features include computer control for high system flexibility, display of system status and detected activities, real-time operator interaction and short system reaction time, prediction of jamming effectiveness based on an analysis of ground and air communication links, pretasking of the system via a

GSY1500 V/UHF communications jamming system mounted on a helicopter with 100 to 500 MHz log-periodic dipole array antenna 0018278

floppy disk, programmable safeguards for own force signals, mission log and analysis on hard disk, comprehensive BIT and software flexibility to accommodate specific user requirements.

The GSY1500 consists of a wideband fast setting receiver, graphical display and computer unit with keyboard, countermeasure generator unit, fast setting RF synthesiser, power amplifier with transmit/receive switches and harmonic filters, audio intercom panel and RF filter unit.

Specifications

Dimensions:
(control console) 700 × 820 × 1,700 mm
(amplifier console) 580 × 650 × 1,110 mm
Weight: <500 kg, depending on options
Power supply: 115 V AC, 400 Hz, 3 phase
or 220 V AC, 50 Hz, single phase
28 V DC
Frequency: 20-100 MHz or 100-500 MHz or both
Output power:
(20-400 MHz) 500 W or 1,000 W options
(400-500 MHz) 400 W or 800 W options
Scanning speed: 80 channels/s
System reaction time: ≤10 ms
Number of pretasked channels: 20

Operational status

Obsolete.

Contractor

Grintek System Technologies.

VERIFIED

The GSY1501 airborne communications EW system can be expanded to accommodate up to 15 operators

GSY1501 airborne communications EW system

The GSY1501 is a comprehensive airborne communications EW system ideally suited for installation on passenger aircraft, cargo carriers, business jets or similar types. The system provides a complete capability to enable the detection, interception, direction-finding, recording and disruption by jamming or deception of enemy command and control communication networks. It provides coverage of the frequency spectrum from 20 to 1,000 MHz and can be extended to operate from 1.5 MHz. Because of its modular approach in design, the GSY1501 can be adapted to accommodate up to 15 operators and additional capabilities can be added as required.

The GSY1501 consists of a number of subsystems as detailed below.

RF reception and distribution

A 20 to 500 MHz and 500 to 1,000 MHz blade antenna with omnidirectional response in the azimuth plane provides reception over the full frequency range. An antenna distribution unit feeds received RF signals to the receivers in the system and a blanking unit protects the antennas during active transmission.

Spectrum scanning

Scanning receivers perform scanning of the spectrum in the band selected by the operators.

Direction-finding

A seven-channel interferometer direction-finder determines the bearing of a signal in the 20 to 500 MHz frequency band or, optionally, up to 1,000 MHz. The DF is fed by an antenna array with wide aperture to provide high accuracy.

System control

ESM and ECM system control units consist of a number of processors which control the scanning receivers and direction-finder, power amplifiers, synthesisers and counter modulation generators.

Operator workstations

The number of workstations is configurable, depending on the available space and platform limitations. A workstation consists of a monitoring receiver, digital voice recorder, spectral display unit, operator interface unit and intercom control unit. The operator interface is an environmentally hardened personal computer with a colour flat panel display and conventional PC keyboard.

Jamming

The system consists of 30 to 100 MHz and 100 to 500 MHz high-power antennas capable of transmitting more than 1 kW of RF power, RF jamming synthesisers each having three independent RF channels capable of AM or FM modulation, counter modulation generators which generate the baseband signals from a digital source and are used as modulation sources for the jamming signal and power amplifiers covering the 20 to 100 MHz and 100 to 500 MHz bands which are capable of generating up to 1 kW of RF power.

Specifications

Frequency:
(ESM) 20-1,000 MHz (1.5-30 MHz option)
(ECM) 30-500 MHz
Scan rate:
(1.5-30 MHz) 0.3 MHz/s
(20-1,000 MHz) 4-16 GHz/s
Accuracy: ±1.5° RMS
DF agility: 40 bearings/s

Operational status

Obsolete.

Contractor

Grintek System Technologies.

VERIFIED

ANM-90 GPS navigation system

The ANM-90 GPS navigation system has been designed to provide precise three-dimensional position, velocity and time for air, land and sea military forces. The equipment is aimed at a variety of military users, therefore, particular attention has been paid to the logistics elements, environmental conditions, volume, weight and life-cycle cost.

The ANM-90 is available in four configurations: ANM-90 manpack, ANM-90 vehicle-mounted manpack, ANV-90 integrated vehicle system and ANA-90 airborne GPS sensor. The ANA-90 operates on the ARINC 429 databus and adheres to ARINC 743 which defines standards for interfacing GPS with airborne navigation computers. The ANA-90 provides position, velocity, heading and time to a range of navigation computers such as the RNS 252 and RNav 2 Racal computers.

Specifications

Dimensions: 63.5 × 215.9 × 241.3 mm
Weight: <3.5 kg
Power supply: 10-32 V DC, <0.4 A
Temperature range: −15 to +65°C
Accuracy (with selective availability) (95%):
(2-D position) 100 m
(altitude) 160 m
(velocity) 1 kt
(differential) (2-D position) <25 m
(altitude) <25 m
Reliability: >4,000 h MTBF

Contractor

Reutech Defence Industries (RDI) (Pty) Ltd.

VERIFIED

ACA 340 V/UHF airborne radio

The ACA 340 multimode radio is an airborne tactical radio. In addition to the full-frequency coverage required by modern air forces, the ACA 340 has advanced ECCM features such as frequency hopping, data transmission and encryption. It is capable of operating in extreme environmental and electronic conditions.

The control of the radio has been designed to be integrated into existing avionics, or stand-alone controls can be supplied.

Specifications

Dimensions: 124 × 385 × 193 mm
Weight: <8.26 kg
Power supply: 28 V DC
Frequency: 30-400 MHz

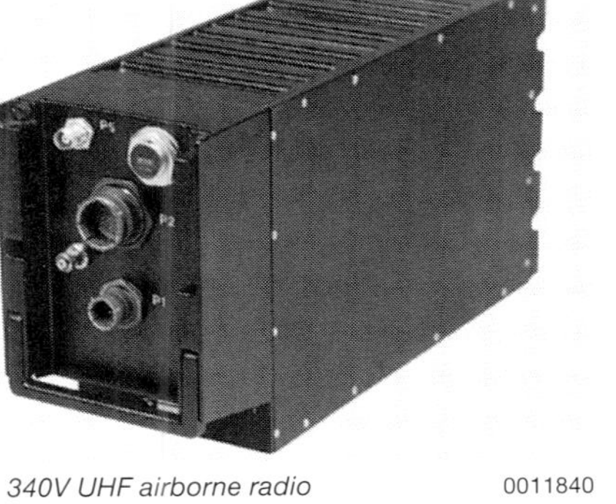

ACA 340V UHF airborne radio 0011840

Channel spacing: 25 kHz
Modulation: AM, FM, data, binary FM secure speech
RF output: 10 W AM (80% modulation) 15 W FM
Preset channels: 99
Temperature range: −40 to +71°C
Altitude: up to 70,000 ft

Contractor

Reutech Defence Industries (RDI) (Pty) Ltd.

VERIFIED

ACR 500 series air/ground V/UHF transceivers

The ACR 500 series is RDI's latest air/ground V/UHF transceiver family. The 500 system comprises two airborne transceivers; the ACR 500 V/UHF tactical airborne transceiver, and the ACR 520 VHF tactical airborne transceiver; two ground-based transceivers; the GBF 500 UHF power-agile filters transceiver and the GBR 500 V/UHF ground-based transceiver.

The ACR 500 and ACR 520 transceivers employ frequency-hopping and software encryption to ensure communications security, and very high rate direct sequence spread spectrum techniques for 'own probability of detection' communications, and to reduce the probability of interception (ACR 500 only).

High-quality vocoded speech transmission (essentially digitised speech) is used to ensure speech quality.

Other features include bandwidth efficient/4DQPSK data mode, and noise-resistant DSP squelch.

ACR 500 V/UHF transceiver

The ACR 500 is a multimode/multiband transceiver operating over the 30 to 420 MHz range. It also incorporates a fully synthesised auxiliary receiver with analogue and digital modes, to allow simultaneous voice and data reception on two frequencies, to improve situational awareness.

Specifications

(ACR 500 V/UHF transceiver)
Frequency range: 30-420 MHz
Channel spacing: 12.5 kHz; 25 kHz
Modulation formats: AM, FM (WB/NB); SSB (USB); CPFSK Data (binary); Pi/4 DQPSK
RF output, max: 32.4 W PEP AM; 20 W FM, SSB

ACR 500 V/UHF tactical airborne transceiver 0011839

Collocation: full operation 5% off frequency
ECCM: fast frequency hopping, direct sequence spread spectrum
Encryption: Cat A, built in
Audio capabilities: analogue voice, CVSD, high-quality vocoder
Data capabilities: high-speed, addressable, network function
Auxiliary receiver: built-in
Power: 28 V DC
Dimensions: 124 × 385 × 193 mm
Weight: <9.5 kg

ACR 520 VHF transceiver

The ACR 520 operates in the 30 to 88 MHz portion of the VHF band, its capability for communications with ground forces, using the GBR 500 receiver, and SSB for extended range operation.

ACR 520 VHF tactical airborne transceiver 0011838

Specifications

(ACR 520 VHF transceiver)
Frequency range: 30-88.975 MHz
Channel spacing: 12.5 kHz; 25 kHz
Modulation formats: AM, FM (WB/NB); SSB (USB); CPFSK Data (binary); Pi/4 DQPSK
RF output, max: 25 W PEP SSB; 15 W FM
ECCM: fast frequency hopping
Encryption: Cat A, built in
Audio capabilities: analogue voice, CVSD, high-quality vocoder
Data capabilities: high-speed, addressable, network function
Power: 28 V DC
Dimensions: ½ ATR short
Weight: <6 kg

Contractor

Reutech Defence Industries (RDI) (Pty) Ltd.

VERIFIED

ENAV 100 - Doppler velocity sensor

The ENAV 100 – Doppler velocity sensor is a single unit comprising a solid-state transmitter, receiver and associated microprocessor circuitry to provide complete velocity component information.

When used in conjunction with a navigation computer, such as the ENAV 150, an autonomous navigation system can be provided. The ENAV 100 has been designed specifically for helicopters and low dynamic fixed-wing aircraft.

The standard output format conforms to ARINC 429 although other outputs, for example MIL-STD-1553B or conventional pulsed output can be provided. Low-range DC analogue outputs are also available to drive conventional hovermeters.

The single transmitter is switched sequentially into each of three beams in a Janus configuration. This method allows a reduction in the number of components, weight and cost and leads to improved reliability.

Continuous wave transmission is normally used and gives better performance and freedom from height aberrations, compared to modulated systems. Other modes include interrupted CW for reduction of rain returns when flying in precipitation, and a low-power (or Stealth) mode.

Adaptor rings can be provided to retrofit older Doppler installations with the modern, lightweight ENAV 100.

Specifications

Dimensions:
(overall including flange) 117 × 403.2 × 436.4 mm
(aircraft cut-out size) 357.2 × 390.3 mm
Weight: 11 kg (max)
Microwave transmitter: J-band varactor multiplier 13.325 GHz ±20 MHz continuous wave; transmitter power 200 mW nominal (2 mW low-power mode)
Power supply: 28 V DC, 90 W (max)
Velocity range:
−50 to +300 kt, along heading, Vx
±100 kt, across heading, Vy

Operational status

Currently in service in aircraft of the South African Air Force.

Contractor

Reutech Systems.

VERIFIED

ENAV 150 - airborne navigation computer

The ENAV 150 airborne navigation computer is a low-cost, lightweight solution designed to meet the demand for an advanced navigation system that can be fitted to both fixed-wing aircraft and helicopters.

The ENAV 150 is compatible with the ENAV range of Doppler velocity sensors, to provide an autonomous Doppler navigation system. The ENAV 150 can also accept inputs from a range of inertial attitude and heading reference systems, to give a fully integrated hybrid navigation system with accuracies superior to a pure inertial navigation system. The computer will also accept position updates from radio navigation aids such as GPS, Loran C and Omega/VLF.

The variable intensity display can be clearly viewed in bright sunlight.

Up to 100 waypoints can be stored. They may be loaded manually via the keyboard, or from an optional Data Transfer Device.

The system can provide guidance to the autopilot for route and tactical steering as well as search and rescue patterns.

The ENAV 150 has synchro, analogue, discrete and ARINC 429 inputs to cope with a variety of attitude and air data signal formats. The computer's ARINC 429 input/output capability is fully compatible with a wide range of modern avionic architectures.

Specifications

Dimensions: 124 × 146 × 201 mm
Weight: 3.5 kg
Power supply: +28 V DC, 40 W
Waypoint capacity: up to 100 waypoints
Route capacity: single route of up to 20 waypoints

Operational status

Currently in service in aircraft of the South African Air Force.

Contractor

Reutech Systems.

VERIFIED

ENAV 200 - hovermeter

The primary function of the ENAV 200 hovermeter in any helicopter installation is to provide indication of low speeds either during or when entering and leaving the hover flight configuration. The indicator is designed to operate directly from the ENAV 100 Doppler velocity sensor.

The ENAV 200 contains three moving coil movements which show helicopter velocities over the following ranges:

Along heading	10 kt backward to 20 kt forward
Across heading	15 kt left to 15 kt right
Vertical	500 ft/min downwards to 500 ft/min upwards

The meter also includes warning flags for both 'memory' and for 'power off' and is internally lit.

Overall dimensions meet ARINC Specification Number 408 for three ATI size.

Specifications

Weight: 0.9 kg
Power: 28 V DC
Accuracy: the error in displayed velocity will not exceed 1 kt in the along heading and across heading directions and 40 ft/min in the vertical direction
Flight envelope:
(speed) −50 to +350 kt
(altitude) 0 to 20,000 ft

Operational status

Currently in service in aircraft of the South African Air Force.

Contractor

Reutech Systems.

VERIFIED

ENAV 300 VOR/ILS

The ENAV 300 system includes a number of VOR/ILS units which can be tailored to suit most fixed-wing and rotary-wing aircraft installations. The ENAV 300 offers interfaces to most types of RMI, HSI or PNI indicators. Microprocessor-controlled frequency generation, non-volatile memory, double frequency LC-displays for active and present channels and electro-optic friction-free frequency selection are the most innovative characteristics of the VOR/ILS/GS receiver ENAV 300 (Series).

The built-in VOR/LOC converter can drive either a standard cross pointer OBS indicator or an RMI indicator via its DC Sin/Cos output (via a TMI amplifier).

Various options are available to drive various types of indicators.

Contractor

Reutech Systems.

VERIFIED

ENAV 400 - Distance Measuring Equipment (DME)

The ENAV 400 (DME) is a modern state-of-the-art system which is built and tested to exact standards. This system offers a cost-effective solution for distance measuring equipment replacement for the upper general aviation market.

The ENAV 400 offers the following features as standard: channels; BDC or ARINC two out of five interface; Nav 1 or Nav 2 system selection; high-visibility plasma display with automatic dimming; range display up to 200 n miles; groundspeed display up to 999 kt; time to station display up to 99 minutes; AF station ident; and standard 11 to 33 DC voltage operation.

As an option the system may be configured with a slave unit or alternatively with two slave units and a remote switching unit.

The ENAV 400 (Series) consists of the DME Interrogator Type ENAV 420; Remote Master Indicator Type ENAV 430; and Remote Slave Indicator Type ENAV 435.

Contractor

Reutech Systems.

VERIFIED

PA-5429 radar altimeter

The PA-5429 pulsed airborne radar altimeter provides the height between the altimeter and the underlying terrain/surface for heights from 0 to 5,000 ft. Control/data is via ARINC 429 or analogue/discrete interfaces. The unit operates in the mid-J-band (~15 GHz) and features a single-LRU configuration, eliminating the need for separate RF feed cables and antennas. The altimeter has good ECCM performance with a Low Probability of Intercept (LPI) and comprehensive anti-jamming features, making it suitable for a wide range of applications, including high-performance and transport aircraft, helicopters and missiles.

Specifications

Dimensions: 219 × 77 × 118 mm (direct mounting to aircraft skin)
Weight: <3 kg
Power supply: 28 V DC to MIL-STD-704
Consumption: 17 W nominal, 25 W max
Height range: 0-5,000 ft
Accuracy: ±3 ft for heights 0-100 ft
±3% for heights 100-5,000 ft

Contractor

Tellumat (Pty) Limited.

VERIFIED

PT-730 airborne TACAN

The PT-730 airborne TACAN interrogator provides distance, relative bearing, range rate, morse-decoded identification and time to a selected TACAN ground station in T/R mode or airborne beacon in A/A mode. In A/A mode, the equipment also provides distance to the nearest TACAN-equipped responding aircraft, as well as replies to a maximum of five TACAN-equipped interrogating aircraft.

The equipment consists of a receiver/transmitter unit and an optional control panel and DME display. Control and data interfacing is via ARINC 429, RS-422 or MIL-STD-1553B. It features microprocessor based design, dual antenna with manual or automatic selection, solid-state 700 W transmitter, continuous, pilot-initiated and power-up self-test.

Specifications

Dimensions: ¼ ATR
Weight: <5 kg
Power supply: 28 V DC to MIL-STD-704
Consumption: 50 W nominal, 65 W max
Channels: 126-channels, with X or Y coding. May be linked to VOR/ILS or VHF COMM channel tuning
Range: limited by line of sight, instrumented to 399 n miles
Accuracy: ±0.1 n miles

Contractor

Tellumat (Pty) Limited.

VERIFIED

PT-1000 IFF transponder

The PT-1000 provides Mk XII transponder capability and is suitable for airborne and marine applications. The transponder supports South Africa's national secure mode, which is available in a country-specific export version.

The Tellumat PT-1000 IFF transponder

The transponder provides full diversity (dual antenna) decoding and replies to modes 1,2,3/A,C and the secure mode, according to STANAG 4193.

The system consists of a tray-mounted transponder and an optional internal cryptographic module and panel-mounted Control/Display Unit (CDU).

Control and status interfacing to the transponder is via ARINC 429.

Specifications

Dimensions: ⅜ ATR (transponder)
5½ in panel, 127 × 113 × 43 mm (CDU)
Weight: <5 kg inclusive of the crypto module
Power supply: 28 V DC, to MIL-STD-704
Power consumption: 35 W nominal, 55 W max

Contractor

Tellumat (Pty) Limited.

VERIFIED

PT-2000 IFF/Mode S transponder

The PT-2000 IFF/Mode S transponder is a compact (⅜ ATR) unit, with externally accessible secure mode cryptographic computer within the volume. It offers the following capabilities: full mode 1, 2, 3/A and C interrogation/reply functionality; secure mode operation (supplied with Mode 4 capability, which can include a customised national secure mode); the ability to be upgraded to the 'Successor IFF' system; Automatic Code Changing (ACC) facility for secure mode; Mode S Extended Length Message (ELM) capability by software upgrade; software/firmware upgrade and access to maintenance data via front panel connector. The PT-2000 operates off the aircraft 28 V DC supply, and it requires no special cooling. Control and data interfacing via dual-redundant MIL-STD-1553B or ARINC-429 digital databus; TCAS interface via ARINC-429 link.

Specifications

Dimensions: ⅜ ATR short (including cryptographic module)

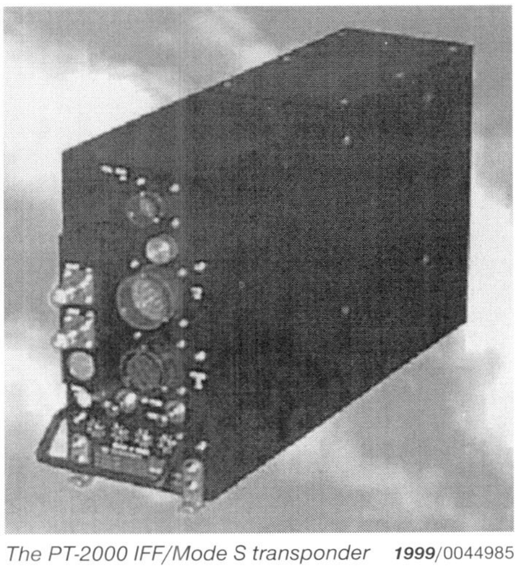

The PT-2000 IFF/Mode S transponder ***1999***/0044985

Weight: <6.5 kg
Power: 28 V DC, 60 W maximum
MTBF: 4,000 h (calculated)

Contractor

Tellumat (Pty) Limited.

VERIFIED

XBT-2000 X-band radar transponder

The XBT-2000 is a radar transponder providing encoded replies to interrogations from airborne or shipborne X-band (NATO I-band) weather radars operating in the weather band (9,200 to 9,500 MHz). The transponder provides compatibility with the DO-172 radar beacon mode standard (encoded replies), as well as fully independent digital tuning of transmit and receive frequencies for alignment with specific radars. The transponder is suitable for man-pack deployment or may be mast mounted.

Applications include demarcation of remote runways or drop zones as well as helicopter decks. Optional packaging is available for airborne use.

Specifications

Dimensions: 172 × 215 × 64 mm
Weight: <5 kg, including rechargeable battery pack and antenna
Power supply: 10.5 to 28 V external, or integral Ni/Cd battery
Power consumption: 2.5 W nominal, 5 W max
Frequency range: receiver and transmitter independently tunable between 8,500 and 9,500 MHz beacon mode 9,375 MHz receive, 9,310 MHz transmit
Set-up: stored in non-volatile memory after configuration via PC RS-232 terminal interface

Contractor

Tellumat (Pty) Limited.

VERIFIED

SPAIN

ELIOS ELINT system

ELIOS (ELINT Identification and Operating System) is a powerful tool for processing and utilising an ELINT database and is complementary to other ELINT systems. It is capable of a number of functions, ranging from gathering and processing data to mapping processes. It provides fast identification, electronic database management and library loading. The main characteristics of ELIOS are a library of threats with a high-powered capacity for response and reaction, tracking and repetition of missions on the ground, and the mapping and position-fixing of threats.

ELIOS simplifies mission preparation, records collected data and allows detected emissions to be compared to memory files. It increases the speed of management, allows the system operator to access all databases and allows tailoring of the electronic database format. High security of data is provided by double protection against loss, software inviolability and cryptographing of files.

The system is supplied in rugged hardware and can be interfaced and installed in a variety of land, sea and air platforms.

Operational status

Fully developed.

Contractor

ELT SA.

VERIFIED

NIDJAM jammer

The Navigation/Identification Deception Jammer (NIDJAM) operates over the frequency range 950 to 1,250 MHz in the band used mainly for Tacan, DME and IFF systems.

The NIDJAM superheterodyne receiver detects emissions over the band in continuous and discrete scan modes. It controls the activity in a set of previously selected frequencies by means of the discrete scan and identifies nav/ident systems and their modes of operation. Deception signals are selected from a series of preset signals or fixed patterns, although random patterns may be employed.

Scanning sensitivity is better than −88 dBm and scan speed is up to 1,000 channels/s. The radiated power in the jammer assembly exceeds 300 W CW per channel. Blade type monopole antennas are used for scanning and jamming in airborne applications.

Specifications

Frequency: 950-1,250 MHz
Sensitivity: −88 dBm
Scan speed: up to 1,000 channels/s
Radiated power: 300 W CW per channel

Operational status

Fully developed.

Contractor

ELT SA.

VERIFIED

Signal Identification Mobile System (SIMS)

The Signal Identification Mobile System (SIMS) is configured for airborne, shipborne and land-based applications. It allows detection, direction-finding, analysis and library storage over the 1 to 18 GHz frequency range and transfers all data to remotely located sites through a built-in datalink.

SIMS consists of an antenna unit, RF unit, direction-finding and panoramic signal displays, a computerised system controller, video analyser and peripherals.

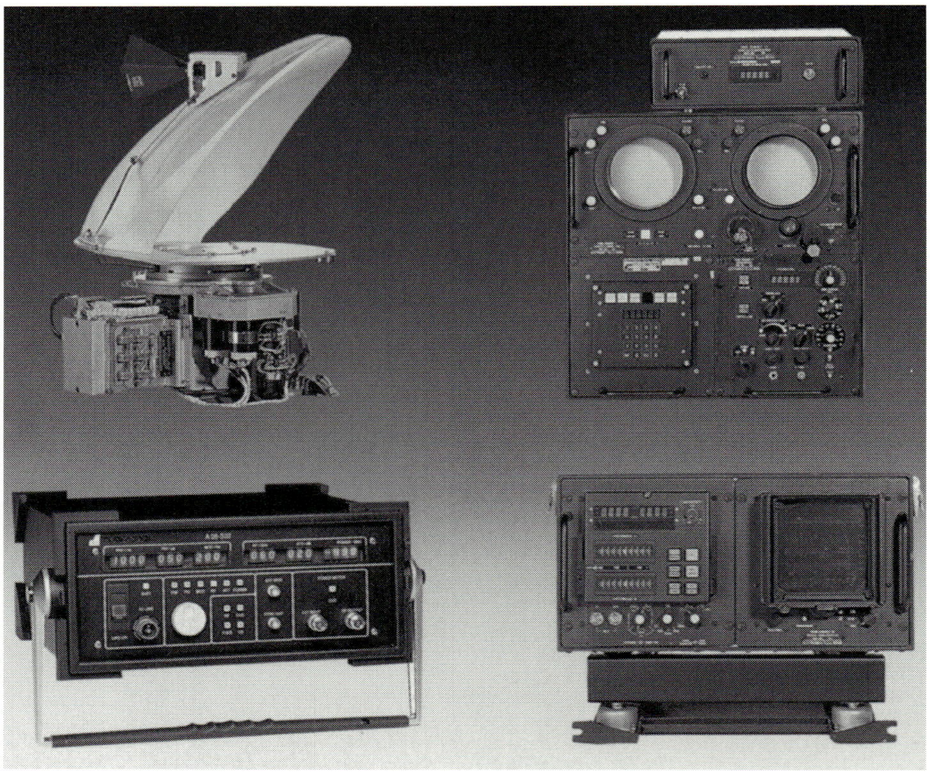

The Signal Identification Mobile System

Specifications

Power supply: 115 V AC, 400 Hz
Frequency: 1-18 GHz pulse or CW signals
Accuracy:
±4° (1-2 GHz)
±3° (2-4 GHz)
±2° (4-18 GHz)

Contractor

ELT SA.

VERIFIED

SOCCAM COMINT system

The modular communication observation and control system (SOCCAM) is configured for airborne, shipborne and land-based applications. It is a COMINT system for tactical and strategic missions and spectrum control over the frequency range 20 to 500 MHz.

SOCCAM provides functions for scanning, searching and detecting active transmissions. The system detects activity in a series of discrete bands and analyses intercepted signals to allow the operator to determine transmission characteristics and store them for subsequent analysis.

SOCCAM has two different operating modes: operation in an unknown scenario when the system searches for active transmissions in the mission area, and operation in a known scenario, where the operator has prior knowledge of threats in the area and seeks to locate and monitor them.

Specifications

Frequency: 20-500 MHz
Sensitivity: −100 dBm
Modes: AM, FM, CW and PLS
Accuracy: 4° RMS
Interfaces: RS-422, IEEE-488

Operational status

In service.

Contractor

ELT SA.

VERIFIED

Taran airborne ESM/ECM system

Taran is a modular computer-controlled airborne EW system for both ESM and ECM in the communications and tactical navigation/identification bands. The system is intended primarily for airborne applications and can be installed on a wide range of fixed-wing aircraft and helicopters.

Two different bands can be operated: low band for communication systems and high band for navigation/identification systems. The low-band subsystem consists of a search and analysis receiver, monitoring console and jamming set. The high-band subsystem comprises a search and analysis receiver, monitoring console and a deception jammer.

The main features of Taran include the ability to operate in a dense electromagnetic environment, and to resist multiple threats simultaneously. There is a high degree of automation in order to reduce reaction time, and growth potential to cater for the evolving threat. Taran has total communication capacity with the command and control system and a simple man/

machine interface and ergonomic design. The system is modular, able to adapt to the needs of different platforms and has high reliability and ease of maintenance.

The system is installed with a suitable set of consoles with interconnection routeing depending on the type of airborne platform.

Specifications

Power supply:
115 V AC, 400 Hz, 1.36 kVA (2.6 kVA with jammer operating)
28 V DC, 1.5 kW (15.2 kW with jammer operating)

Operational status

In production for the Spanish Air Force and in service on the Falcon 20.

Contractor

ELT SA.

VERIFIED

ALR-300 Radar Warning Receivers (RWRs)

The ALR-300 RWRs have been designed to detect, locate, identify and display to the pilot the existing threat environment form pulse and continuous wave emissions, regardless of illumination mode. The two systems in the series are the ALR-300/V1 and the ALR-300/V2. The main difference between them is that the ALR-300/V2 has a superheterodyne receiver/DIFM.

Operation is controlled by software based on realtime multiprocess with several 68000 microprocessors (four on the ALR-300/V1 and five on the ALR-300/V2.

The equipment consists of: four E/J-band DF spiral antennas; four E/J-band channelised crystal video receivers; a processor control unit and an azimuth indicator display with synthetic audio warning, together with a C/D-band monopole antenna and channelised receiver.

The system provides identification of all detected pulse and CW radar emitters and warns the pilot of threats by means of both an alphanumeric graphical CRT display and voice synthesised messages in accordance with the lethality of the detected emitters. The system includes the ability to record up to 100 emitters during flight.

The ALR-300 system includes full mission and maintenance hardware and software support facilities. Reprogramming on the flight line can be accomplished by use of EEPROM mission loading equipment that forms part of this support capability.

Specifications

Frequency range: 0.5 to 18 GHz
Signals detected: pulse and CW
Antennas:
C/D-band: one monopole (0.5 to 2.0 GHz)
E/J-band: 4 flat spirals (2 to 18 GHz)
Wide band receivers:
C/D-band: one channelised receiver
E/J-band: 4 quadrantal, channelised receivers
Narrow band receiver (ALR-300/V2 only): one superheterodyne receiver
Azimuth coverage: 360° azimuth
Elevation coverage: ±30°
DF accuracy: 12° RMS
Simultaneous emitter tracking: 30
Simultaneous emitter display: 16
Emitter library: 100
Standard interfaces: MIL-STD-1553B, RS 422
Dedicated interfaces: chaff/flare displensers
Weight:
(ALR-300/V1): 20 kg
(ALR-300/V2): 30 kg
Power consumption:
(ALR-300/V1): 200 W
(ALR-300/V2): 350 W

Operational status

In operation on Mirage F-1 fighter aircraft, CASA C-101 trainer aircraft, and on Chinook and Super Puma helicopters. In production for Beech aircraft.

Contractor

INDRA EWS.

VERIFIED

EN/ALR-300(V)2 Radar Warning Receiver

The EN/ALR-300(V)2 radar warning receiver upgrades the EN/ALR-300(V)1 by the addition of a superheterodyne/DFIM receiver, a power supply, and new software to improve the equipment performance.

Radar data are presented to the operator through alphanumeric symbols on a high-brightness display. The position of the symbols on the display indicates lethality and bearing of the detected signal and identification of the radar. Blinking of the symbols, together with sounding of an audio alarm indicate high-priority radar threats.

Specifications

Weight: 38 kg
Power consumption: 400 W
Frequency: C- to J-bands in five sub-bands
Coverage: 360° azimuth
Accuracy: >12° RMS

Operational status

The EN/ALR-300(V)2 is the standard equipment for Spanish combat aircraft. It is in operation on the Mirage F1 fighter aircraft, and in production for Super Puma helicopters.

Contractor

INDRA EWS.

VERIFIED

EN/ALR-310 Radar Warning Receiver

The EN/ALR-310 is a fully programmable crystal video radar warning receiver which incorporates a digital computer to provide emitter identification in complex signal environments. The system incorporates a readily programmable emitter library and provides unique identification of all detected pulsed and CW emitters by alphanumeric symbols on the CRT display. Up to 15 radars may be displayed simultaneously.

Specifications

Weight: 20 kg
Power consumption: 180 W
Frequency: 0.7-1.5 GHz and E/J-band, in 2 bands
Coverage: 360° azimuth
Accuracy: >15° RMS

Operational status

Selected for Spanish Army helicopters.

Contractor

INDRA EWS.

VERIFIED

NAT-5 tactical navigation system

The NAT-5 is a computer-based navigation system for airborne ASW tactics. It comprises a control unit, central processor unit and display unit, and accepts information from any or all of Doppler, true airspeed, AHRS, GPS, sonar and radar.

The display unit shows updated positional information plus seven predefined contacts in latitude and longitude, X-Y co-ordinates or any of four different polar co-ordinate systems, with graphic scales from 4.5 × 3 up to 288 × 192 n miles.

The control unit is a microprocessor-based control panel with a keyboard including function, numeric and control keys and a joystick for quick data entry. The control unit allows easy system operation by means of up-down menus on the screen.

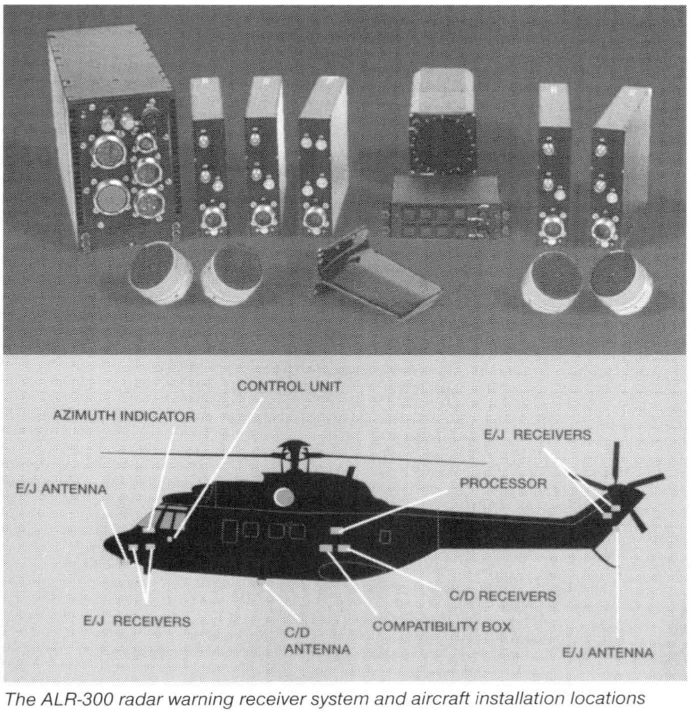

The ALR-300 radar warning receiver system and aircraft installation locations
2000/0080278

The EN/ALR-310 is fitted in Spanish Army helicopters

Graphics operation is provided by means of a powerful graphics processor with up to 100 kbytes of stored standard search patterns and a cartridge, located in the control unit, with up to 512 kbytes for predefined maps, tactics and tactical scenarios. The cartridge can also store system airborne data for debriefing purposes.

The system provides a serial output for a sensor operator control keyboard and interfaces for cockpit instruments such as the BDHI.

The operational software and the interface electronics are designed in a modular form to accommodate each future change with a minimum impact on the configuration of the system.

Optionally, the system includes ground support equipment and a navigation sensors simulator for training, testing and maintenance facilities. The ground support equipment has a special software utility which allows the user to create his own tactics and maps for loading into the cartridge.

Specifications

Dimensions:
(control unit) 135 × 146 × 181 mm
(central processor unit) 194 × 124 × 320 mm
(display unit) 265 × 325 × 333 mm
Weight:
(control unit) 2.5 kg
(central processor unit) 10 kg
(display unit) 17 kg

Operational status

In production and in service in Spanish Navy Sikorsky SH-3D/G Sea King and SH-3D Sea King AEW helicopters.

Contractor

INDRA EWS.

VERIFIED

SWEDEN

AESA Active Electronically Scanned Array antenna

Ericsson AESA (Active Electronically Scanned Array) is a new airborne radar project currently in development at Ericsson Microwave Systems.

It is intended for the next-generation Gripen aircraft as a multimode radar. The AESA technology will improve the radar's overall performance, especially its target detection and tracking capability. Beam direction can, for instance, change instantaneously, detection range will be considerably increased, and jamming suppression further improved.

The AESA radar will feature multibeam capability with all beams individually and simultaneously controlled. It can also operate simultaneously as a fire-control and obstacle warning radar, and be used both in intercept and ground attack missions. As a consequence of the very large number of transmitter and receiver modules, the radar will have a high system availability through graceful degradation.

AESA active electronically scanned array antenna 0002218

Operational status

In November 1994, Ericsson received an order from the Swedish Defence Materiel Administration (FMV) for an airborne radar study utilising a new, active phased-array technology, which uses a thousand or more transmitter and receiver modules.

In mid-1997, Ericsson commenced the first test of the AESA system. Ground tests were performed with a model having some 100 transmitters and receivers. These tests were expected to continue until mid-1998 and the test results were promising.

The next development phase, which has already started, will lead to a full-scale active antenna system for airborne trials. It is expected to start flight trials in 2002, with potential application as a mid-life update for Gripen beyond the year 2010.

For this second phase, Ericsson and FMV are looking for a partner and discussions are currently taking place with companies in the USA and Europe.

Contractor

Ericsson Microwave Systems AB.

VERIFIED

CARABAS SAR with inflatable boom antennas deployed **1999**/0044972

CARABAS surveillance radar

The Coherent All RAdio BAnd Sensing (CARABAS) is a prototype airborne low-frequency surveillance radar which can penetrate both foliage and the ground. It can be used by both military and civil authorities for tactical surveillance and remote sensing.

CARABAS is a wideband, synthetic aperture radar operating at 20 to 90 MHz frequencies necessary to penetrate the ground and foliage. The depth of ground penetration varies largely with the type of terrain, whereas foliage penetration is much less sensitive to the type of vegetation.

The CARABAS system has applications in detecting military targets such as underground bunkers and hardened shelters, as well as weapons and buildings screened by foliage. Ericsson states that the CARABAS SAR sensor imaging resolution is so good that it is only diffraction-limited.

The system uses a novel antenna design consisting of two inflatable booms which are deployed from the aircraft in flight. Each boom is 5.5 m long by 0.3 m diameter, and is attached to a bracket on the aircraft tail. The two booms are inflated by an air compressor before take-off and during flight are kept rigid by the effect of ram air. Each boom consists of several thousand small components, making it act as a dipole antenna. Power is fed to each boom and the surface radiates a signal to the ground, alternately from left and right antennas, and receives the returns.

The data requires lengthy and highly sophisticated processing and, at the current stage of development, is recorded on tape and analysed on the ground. In the long term, it is proposed that some of the processing is carried out on the aircraft.

CARABAS is also claimed to be well-suited to UAV installation.

Specifications

Frequency range: 20-90 MHz
Resolution: 2 × 1 m
Surveillance area: >1 km^2/s

Operational status

CARABAS II has replaced the CARABAS I system as testing continues.

Contractors

Ericsson Microwave Systems AB.
Swedish National Defense Research Establishment (FOA).

VERIFIED

Erieye AEW&C Airborne Early Warning & Control mission system radar

The Erieye AEW&C mission system radar features active, phased-array technology. The antenna is fixed, and the beam is electronically scanned, which Ericsson claims provides improved detection and significantly enhanced tracking performance compared with radar-dome antenna systems.

Erieye detects and tracks air and sea targets out to the horizon (and beyond due to anomalous propagation); instrumented range is 450 km. Typical detection range against fighter-sized targets is approximately 350 km, in a 150° broadside sector, both sides of the aircraft. Outside these sectors, performance is reduced in forward and aft directions.

Erieye is understood to operate as a medium- to high-PRF pulse Doppler, solid-state radar, in E/F-band (3 GHz), and to comprise 192 two-way transmit/receive modules, that produce a 1° pencil beam, steered as required within the operating 150° sector each side of the aircraft (one side at a time). It is understood that Erieye has some ability to detect aircraft in the 30° sectors fore and aft of the aircraft heading, but has no track capability in this sector. The aircraft could be manoeuvred to permit tracking of targets in these sectors.

The electronically scanned antenna is controlled by an automatic intelligent energy management system, developed to utilise the phased-array technology implemented, pulse-by-pulse, to illuminate any desired azimuth. The ability instantaneously to direct the radar energy in any wanted direction is used by the operator to optimise power management for any particular scenario by assigning priorities to areas of interest, thus optimising probabilities of detection and overall system performance.

Operational status

Erieye is in series production for the Swedish Air Force, where it is implemented in the Saab 340B aircraft. This system is understood to carry the Swedish Air Force's system designation FSR 890, and the Erieye radar to be designated PS-890. Four aircraft have been handed over to the Swedish Air Force. Two more systems were due to be delivered in 1999.

The Erieye system has been selected by the Brazilian government as the airborne element of the SIVAM system for surveillance of the Amazonas. The contract is for five Erieye systems. Delivery began in 1999 for installation in the Brazilian Embraer EMB-145 aircraft, designated EMB-145SA when converted. The EMB-145SA first flew in May 1999; the planned in service dates are given as April, July and August 2001, and February and May 2002. Equipment includes the Erieye radar, with integrated IFF system, and command, control and communication system, that will be integrated into the SIVAM command and control system.

In December 1998, Greece selected Erieye as supplier of its AEW&C requirement, based on four systems integrated on EMB 145 aircraft. A final contract was signed on 1 July 1999; deliveries are to begin in 2002.The system is to be delivered by a consortium formed by Ericsson Microwave Systems and Thomson-CSF DETEXIS (called Ericsson Thomson-CSF AEW Systems AB for the discharge of this contract). Thomson-CSF DETEXIS will provide the NATO-interoperable TSX2500 IFF identification and communication systems and the self-protection systems.

Marketing discussions are in progress with a number of other countries in Asia, Europe and the Americas.

Contractor

Ericsson Microwave Systems AB.

UPDATED

The Swedish Air Force has selected the Saab 340B as the carrier for the Erieye mission system 0010932

The Ericsson Erieye Airborne Early Warning & Control (AEW&C) system on the Embraer EMB 145 aircraft ***2000***/0079238

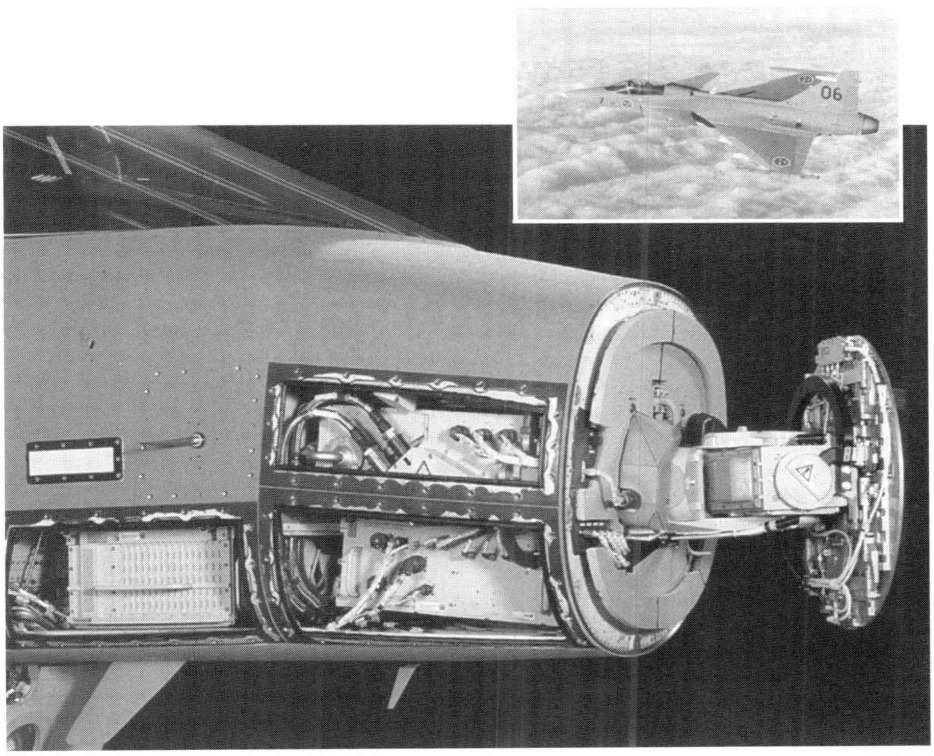

The Ericsson PS-05/A radar for the JAS 39 Gripen aircraft 0010930

PS-05/A multimode multimission radar for the JAS 39 Gripen

Ericsson Microwave Systems AB provides the multimode, multimission, I-band (9 to 10 GHz), pulse Doppler radar for the JAS 39 Gripen. Ericsson collaborated with GEC-Marconi (now Marconi) in the 1980s on the original development of the radar. The PS-05/A radar of JAS 39 Gripen, and the Blue Vixen radar installed in the UK Royal Navy Sea Harrier FA 2 aircraft are understood to still share some elements (antenna, computer, radar exciter).

The PS-05/A radar provides the following multimode capabilities: air combat, ground attack, high-resolution mapping and reconnaissance to meet the multimission operational requirements of the JAS-39 Gripen aircraft. All principal modes are software driven, and include:

(1) air-to-air operation, using high-PRF and medium-PRF pulse Doppler waveforms, to provide long-range search; auto-acquisition; multiple-target track-while-scan; multiple-priority target tracking; short-range, wide-angle search and track for air combat; high-resolution single-target tracking; raid assessment; missile mid-course update
(2) air-to-ground operation, using low-PRF pulse Doppler and frequency agility waveforms, to provide search; tracking; high-resolution mapping; air-to-surface ranging.

The radar matches the datalink requirements of the AMRAAM and MICA air-to-air missiles, and is claimed to have excellent ECCM capabiities.

The main waveform modes are:

(1) HPD - a high-PRF, pulse Doppler mode for clutter rejection, primarily designed for use against airborne approaching targets
(2) MPD - a medium-PRF, pulse Doppler mode for clutter rejection, primarily designed for use against approaching and receding targets; a special high-resolution sub-mode is designed for target tracking
(3) LPD - a mode using Doppler processing for clutter rejection designed for use against surface targets
(4) LPRF - a low-PRF mode with pulse-to-pulse frequency agility for use against surface targets and for real-beam mapping

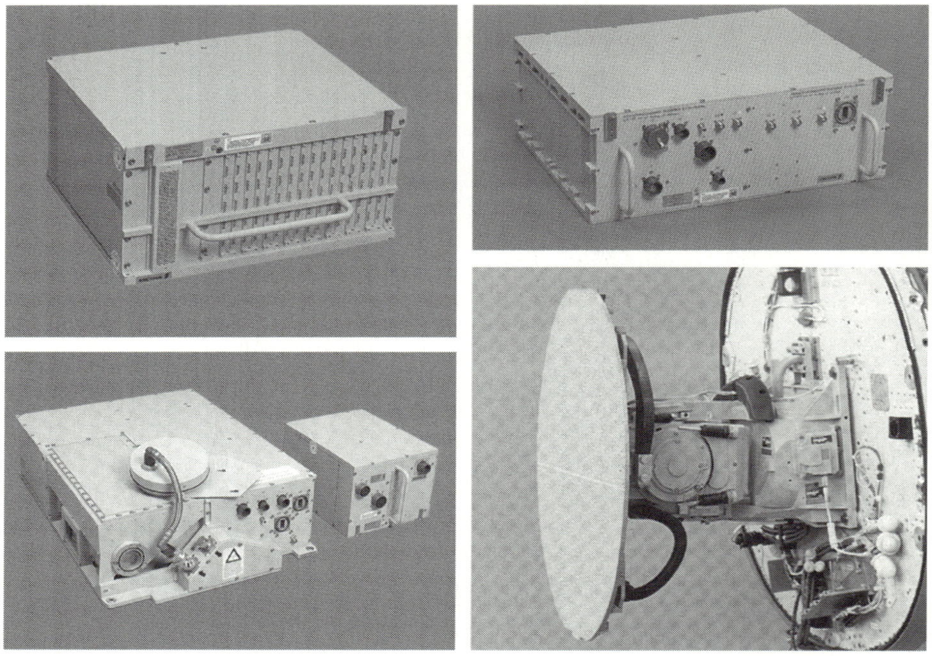

PS-05/A radar antenna unit, (bottom right) power amplifier and transmitter auxiliary units (bottom left), high-frequency unit (top right) and signal and data processor unit (top left) 0010931

Ericsson PS-46/A radar installed in the Viggen fighter

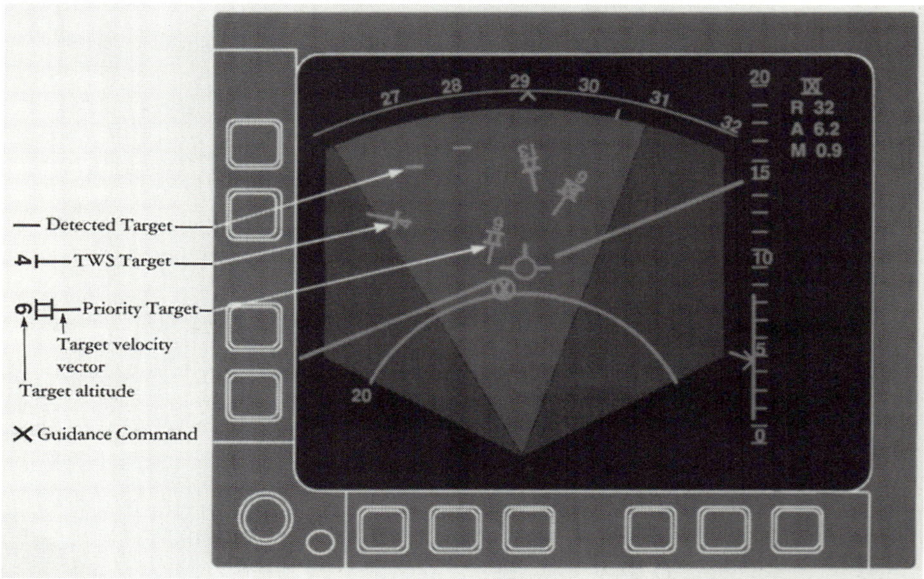

PS-05/A display with typical data format *0001209*

(5) AGR - a mode exclusively designed for ground target ranging
(6) DBS (Doppler Beam Sharpening) - a Synthetic Aperture Radar (SAR) mode utilising Doppler processing for high-resolution mapping, with high-angular coverage obtained by continuous antenna scanning
(7) SLM (SpotLight Mode) - an SAR mode utilising Doppler processing for very high-resolution mapping.

The PS-05/A radar comprises five Line-Replaceable Units (LRUs):

(1) antenna Unit: a lightweight, slotted waveguide, monopulse, planar array, featuring guard antenna, IFF dipoles and all-digital servo control, weighing 25 kg
(2) power Amplifier Unit and Transmitter Auxiliary Unit: 1,000 W average power, flexible waveform, liquid cooled, TWT system, with two LRUs together weighing 73 kg
(3) high-Frequency Unit: multiple-channel receiver, monopulse, pulse-to-pulse frequency-agile, microwave integrated circuit design, internally software controlled, weighing: 32 kg
(4) signal Data Processor: D80 multiprocessor concept fully programmable unit, employing ASIC technology, with programming in High Order Language, Pascal, weighing 23 kg.

Specifications

Weight: 156 kg
Power: 115/200 V AC, 400 Hz, 8.2 kW; 28 V DC, 250 W
Antenna: 600 mm planar-array
Interface: MIL-STD-1553B

Operational status

Some 70 units of the PS-05/A radar for the Gripen aircraft have been delivered.

Two significant improvements to radar performance are now being introduced: delivery of a new signal processor has started. It will be installed in all new production Gripen radars, and retrofitted into exisiting aircraft radars. This processor has higher capacity and improved functions in both air-to-air and air-to-ground modes. Radar system software has also been upgraded to facilitate future upgrades.

Development of a third-generation data processor, designated MACS (Modular Airborne Computer System), has also started. MACS will replace the D80 computer as the aircraft's system computer in the radar, cockpit displays and the EW system from the third batch of aircraft. Introduction of Batch 3 aircraft will also include three 6.2 × 8.3 in multifunction colour displays to replace the three existing (EP17) 5 × 6 in monochrome displays.

Contractor

Ericsson Microwave Systems AB.

VERIFIED

PS-46/A radar

The software-controlled, multimode, pulse Doppler PS-46/A radar has been developed for the JA 37 fighter versions of the Swedish Air Force's Viggen. In view of the numerically small size of Sweden's defence force, great emphasis has been placed on operational availability and readiness; all-weather capability and effectiveness in an electronic countermeasures environment are also important requirements. Designed to cope with high-performance aircraft, transports and helicopters, the system has wide-angle coverage, look-down capability and can operate at all altitudes.

The multimode requirements of the PS-46/A are air-to-air and air-to-ground. The latter is met by using conventional non-coherent pulse waveforms, but the former calls for more sophisticated techniques. The standard radar functions are controlled by a data processor that extracts information from the raw radar and transfers it to other aircraft systems. A digital bus distributes all signals within the radar itself with minimum wiring. For the guidance of semi-active homing missiles an illuminator transmits a continuous wave RF signal through the radar antenna.

Control of the system through suitable software enables parameters to be changed or optimised according to the needs of flight development programmes without time-consuming equipment changes; similar modifications can be introduced during service according to changing military requirements, and radar signatures adopted for peacetime training and exercises can be easily changed during conflict to thwart enemy intelligence and countermeasures.

Specifications

Weight: 300 kg
Power:
(coherent transmitter) 500 W average
(continuous wave illuminator) 200 W
Frequency: I/J-band, bandwidth >10% of spectrum
Performance: detection range >50 km in look-down mode
Antenna: 900 mm diameter
Processor: digital signal processor, 32-bit floating-point, 500 kword programme memory
Modes: search, acquisition (automatic via HUD, semi-automatic via HDD), tracking (track-while-scan, continuous track), target illumination, air-to-ground ranging
High-resolution ground-mapping is optional
Reliability: 180 h MTBF

Operational status

In service on the JA 37 Viggen. No longer in production.

Contractor

Ericsson Microwave Systems AB.

VERIFIED

Side-Looking Airborne Radar (SLAR)

This inexpensive, real aperture SLAR is designed exclusively for maritime patrol and surveillance (for fishery protection, oil pollution monitoring and search and rescue), with emphasis on simplicity and ease of operation. It consists of five LRUs: antenna, transceiver, digital signal processor, control unit and television display. The radar images can be recorded on standard video recorders. Digital radar video can also be recorded for subsequent computer processing. With a video link between aircraft and ground station or ship, images can be transmitted with no degradation of quality.

The digital signal processor is designed around a 1.6 Mbit television memory display. Up to 2,000 range cells can be processed to an accuracy of 6 bits (giving 64 grey-tone levels). The presentation and performance can be varied to suit customers' needs and other features include a reference grey scale, level mapping, positive or negative picture representation and automatic positioning of targets. Monochromatic television images recorded in the air can be converted to colour on the ground.

For double-sided coverage a glass fibre pod, containing two antennas, is carried under the fuselage, or antennas are carried in individual pods on each side of the aircraft.

Specifications

Weight: 70 kg
Power supply: 28 V DC, 15 W
Frequency: 9.4 GHz
Pulsewidth: 0.5 μs
Beamwidth:
(horizontal) 0.5°
(vertical) 33°
Peak output power: 10 kW
PRF: 1 kHz

Operational status

In operation in one country outside Sweden. Three systems are also fitted to CASA 212 aircraft operated by the Swedish Coastguard for oil pollution detection and ice mapping.

Contractor

Ericsson Microwave Systems AB.

VERIFIED

Three CASA 212 aircraft operated by the Swedish Coastguard are fitted with the Ericsson SLAR

Erijammer A100 (ALQ-503)

The Ericsson Saab Avionics Erijammer A100/AN/ALQ-503 pod

The Erijammer A100 jammer system is a manually or automatically computer-controlled jammer pod for tactical use and ECCM training of air defence fighters and AAA operators. The system provides Responsive Electronic Warfare Training and Support (REWTS) by giving the operator situational awareness with built-in RWR, look-through capability and a set on receiver. Over 50 smart noise, advanced range, velocity and angle deception modes and combinations of these modes are available. The pod is also capable of providing simulation of missile seeker radars. Single or multithreat capability is provided by an advanced frequency memory loop, a set on receiver and selectable bandwidths.

The pod is entirely self-contained and requires only power from the carrier aircraft. The system is controlled by an ECM operator, through a cockpit control box, and programmable EEPROM; it is 100 per cent reprogrammable in the air. The analysis and subsequent jamming of incoming signals over 360° with coverage for the selectable high- and low-gain antennas give the system high flexibility in tactical flying and training.

Specifications

Dimensions: 3,235 (length) × 426 mm (diameter)
Weight: 210 kg
Power supply: 115 V AC, 400 Hz, 3 phase, 3 kVA
28 V DC, 5 A
Frequency: H- and I/J-bands

The Erijammer A100 ECM training pod carried under a Pilatus PC-9

Output power: 350 W, ERP 1-2 or 10 kW
Coverage:
(horizontal) 360°
(vertical) ±30°
Speed: M0.2 to M1 +

Operational status

In service in the Canadian, Swedish and Swiss air forces.

Contractor

Ericsson Saab Avionics AB.

VERIFIED

Erijammer A110 EW system

The Erijammer A110 is based on the proven REWTS (Responsive Electronic Warfare Training System) concept, which traces its heritage to the Erijammer A100 (ALQ-503) system, operational with the Swedish, Swiss and Canadian air forces.

The Erijammer A110 is a complete dual-use system that may be used for training without revealing or compromising the system's tactical capability. The switch between training and tactical roles is accomplished in seconds by downloading software suitable for the planned mission via the system's interface. With the ability to operate in both autonomous and manually-controlled modes, the Erijammer A110 system provides both an effective ECM training system and also a capable tactical jammer, which in addition incorporates an ESM capability.

The Erijammer A110 comprises four main subsystems: A110 pod; Control Unit (CU); Display Unit (DU); and Mission Planning Workstation (MPW).

The Erijammer A110 pod is designed for standard NATO 14 and 30 in hard-point installations. The pod is ram-air cooled. The pod has 360° receiver and transmitter coverage, provided by three 120° antennas. The system incorporates three receivers: a Continuous Wave/High Pulse Doppler (CW/HPD) receiver; a Direction-Finding Receiver (DFR); and a Set-On Receiver (SOR). The CW/HPD receiver alerts the system to high-duty cycle emitters. The DFR identifies the sector of operation. The SOR is a narrowband receiver used in some jamming modes and in ESM modes to analyse emitters.

The jamming transmitter can be directed through any one of four antennas: three evenly spaced 120° antennas covering 360° and a high-gain antenna in the forward sector. The pod can generate spot and barrage noise, noise deception, amplitude modulation, velocity deception, coherent and FML-based range deception, radar simulation and combination modes.

Specifications

Frequency range: 6.8-10.5 GHz
Sensitivity:
(low-duty cycle) –45 dBm
(high-duty cycle) –60 dBm
ERP:
(wide-lobe antennas) 1 kW (typical at 9.5 GHz)
(high-gain antenna) 10 kW (typical at 9.5 GHz)
Antenna coverage:
(wide-lobe antennas) 3 antennas, each 120 × 40°
(high-gain antenna) 20 × 20°
(Doppler receiver) 2 antennas, each 70 × 40°, and 2 antennas, each 120 × 40°
Input power: 115 V AC, 2.5 kVA; 28 V DC, 6A
Dimensions:
(pod length) 3,233 mm
(pod diameter) 424 mm
(control unit) 250 × 145 × 103 mm
(display unit) 178 × 134 × 134 mm
Weights:
(pod) <230 kg
(control unit) <4 kg
(display unit) <4 kg
MTBF: >200 h

Operational status

Under delivery to the Swedish Air Force.

Contractor

Ericsson Saab Avionics AB.

VERIFIED

Erijammer A110 under the starboard wing of a Swedish Air Force JA 37 Viggen ***1999***/0051264

EWS 39 - Gripen integrated EW suite

Ericsson Saab Avionics is responsible for the EW suite in the Gripen fighter, designated EWS 39. EWS 39 integrates the RWR/ESM and passive missile warning system with the radar, Infra-Red Search and Track System (IRSTS), and the ground-air datalink, to provide an overall integrated defensive system.

Sensor data is compared with library data for all the sensors and systems to evaluate threat level, alert the pilot, and initiate appropriate countermeasures responses.

EWS 39 is designed to handle all types of off-board, integrated and pod-mounted jammers, chaff and flare dispensers. SaabTech Electronics AB supplies the RWR/ESM, twin BOL chaff dispensers, twin BOP B Countermeasures Dispensers (CMDS) for the deployment of BO 2D Towed Radar Decoys (TRDs) and a BOP C IR flare dispenser.

JAS 39 Gripen aircraft, fitted with built-in EWS 39 EW suite and external Erijammer pod ***1999***/0051265

Operational status

The first generation of the EWS 39 is in service with the Swedish Air Force. The second generation (Gen 2) of the system is under development, with narrowband receivers added to the wideband receivers used in the first generation of EWS 39, for increased sensitivity and selectivity. As an option, the system can be supplied with interferometers for accurate angle-of-arrival measurements, as well as several antennas for increased spatial coverage.

The Gen 2 RWR design also forms the basis of the SaabTech Electronics AB RWR selected for the German Air Force and Navy Tornado IDS and ECR update programme.

Contractor

Ericsson Saab Avionics AB.

UPDATED

NINS/NILS autonomous navigation and landing system

Saab NINS is a total navigation solution including modular hardware and software based on digital geographical databases; its main characteristics are that it utilises inertial or Doppler navigation system, radar or laser altimeter and air data system; and that it provides high real-time position, terrain and velocity accuracy, plus error awareness.

Saab NILS is a landing system that takes full advantage of the NINS concept, creating a glidepath based on NINS data; its main characteristics are that it utilises only onboard sensors, reducing cost and increasing flexibility.

The system offers an alternative navigation solution that is fully realtime and independent of ground- or space-based equipment. The software package can be integrated both with aircraft and cruise missiles, with selected hardware and software features as an 'add-on' package for existing platforms or in new designs where the full benefit of Saab NINS can be utilised. Key features of the system include high position and velocity accuracy; fully autonomous operation; awareness of surrounding terrain.

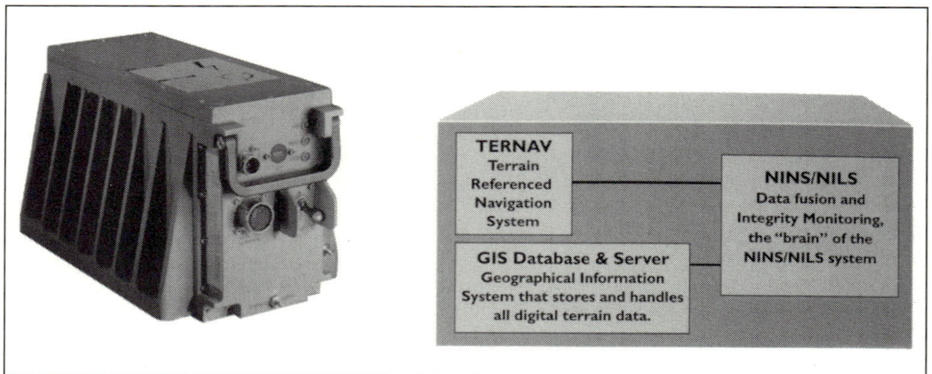

NINS/NILS system computer architecture (right) and hardware (left) ***1999***/0051597

Saab NILS is a landing system that takes full advantage of the NINS concept by creating a glidepath based on NINS navigation information. Only onboard sensors are utilised, which reduces cost and increases flexibility for operations from austere or dispersed bases with no ground equipment.

Saab TERNAV is a non-linear terrain-referenced navigation algorithm that has been developed over 20 years and achieves military GPS performance in the envelope. The system is operational in Swedish Air Force AJS 37 Viggens and in full-scale development for deployment in the JAS 39 Gripen.

A high-precision continuous position fix can be derived from the basic sensor data Inertial Navigation System (INS), Air Data Computer (ADC) and Radar Altimeter (RA) and from the Geographical Information System (GIS) database.

The GIS database and server is where all terrain databases, including those for landing, are stored and prepared for real-time access. TERNAV is one application that uses GIS, together with others such as passive terrain-following and ground-proximity warning.

The data fusion function handles all incoming and outgoing data streams from sensors and system functions. It is the decision-maker in the system and feeds data to all the other functions as well as to the cockpit displays.

At the heart of the system is a Kalman filter for optimal processing of all incoming data streams. The filter utilises all navigation information in the system to estimate the position, velocity and attitude of the platform.

The pilot's standard cockpit displays show information output from the data fusion function, including navigation, approach information and ground-proximity warnings.

Integrity monitoring is a diagnostic system for failure detection and exclusion of failed-sensor signals. With support sensors (for example GPS), it is possible to achieve integrity monitoring at an even higher level, allowing for graceful degradation while still supporting many system functions.

Specifications (dependent on sensors and terrain characteristic)
Horizontal position error: 5-50 m (CEP 50%)
Vertical position error: 2-4 m (1 sigma)
Horizontal velocity error: 0.05-0.5 m/s (CEP 50%)
Vertical velocity error: 0.02-0.2 m/s (1 sigma)
(with negligible degradation for 3 min over water)
Options:
ground proximity warning
terrain following
virtual reality input to the HUD
passive target ranging
integrity monitoring with GPS
mid-air proximity warning with LINCS

NINS/NILS display concept on HUD and head-down displays ***1999***/0051598

Operational status
NINS/NILS is operational in Swedish Air Force AJS 37 Viggen aircraft and is in full-scale development for the JAS 39 Gripen aircraft.

Contractor
Saab Dynamics AB.

VERIFIED

BO2D RF expendable decoy

BO2D is designed to provide fighter aircraft with self-protection against active and semi-active radar seeker missiles. It is a broadband repeater pyrotechnically ejected from a chaff/flare dispenser. It utilises internal battery power and communicates with the host aircraft through a towline. RF transmissions can be switched on or off and different ECM modes can be chosen while the decoy is being towed. The pilot is thus able to launch the BO2D without transmitting when closing in on a possible threat. After use, the towline is cut with a pyrotechnic cutter.

The decoy is a broadband, high-gain repeater with FM capability. It utilises all solid state advanced MIC technology. It comprises the following units: receiving antenna, signal amplifier, modulator, power amplifier and transmitting antenna.

The decoy is housed in a standard 55 mm diameter dispenser magazine, such as the BOZ/BOP types of pyrotechnic dispenser, to provide convenient and cheap installation. The cartridge, which also contains the towline and a brake mechanism, is about 100 mm longer than the standard flare cartridge. The loading and operation of the decoy are similar to those of a flare cartridge. For aircraft without a 55 mm dispenser, a small dedicated launcher is also available.

Specifications
RF coverage: H-, I- and J-bands
Weight: <2 kg (including cartridge)
Cartridge size: 55 mm diameter

Operational status
The BO2D is being developed for the Swedish Air Force. Qualification testing was conducted in 1998, with prototype and initial aircraft testing beginning in 1999. A production decision by the Swedish Air Force was expected in 2000.

Contractor
SaabTech Electronics AB.

UPDATED

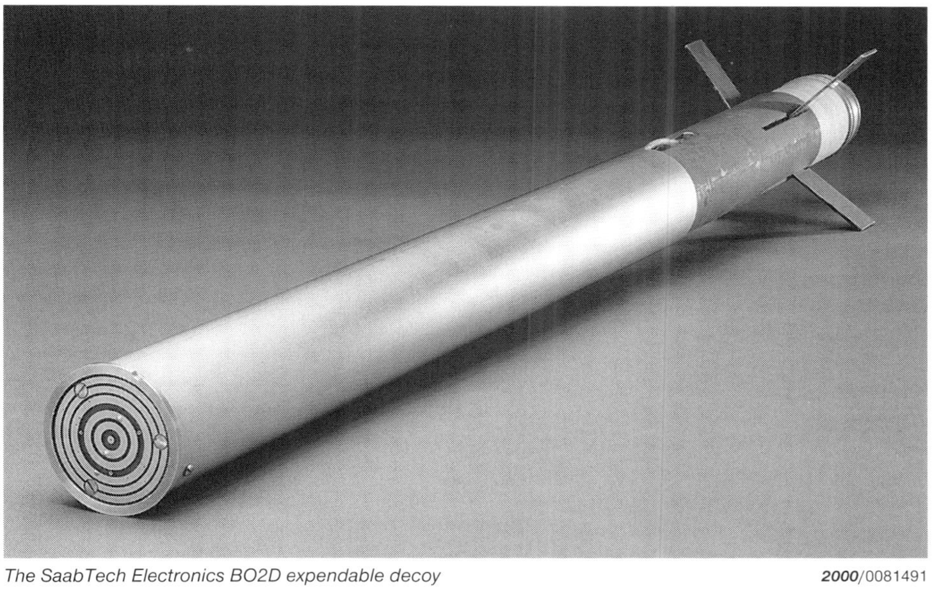

The SaabTech Electronics BO2D expendable decoy ***2000***/0081491

BO 300 passive countermeasures system

The SaabTech Electronics (formerly CelsiusTech Electronics) BO 300 passive countermeasures system consists of BOL and BOP Countermeasures Dispensing Systems (CMDS) and a BOC controller module. The BO 300 system has been developed to be applicable to a wide range of aircraft types/configurations, with particular emphasis on high

BOP/B countermeasures dispenser on Gripen aircraft 0018277

BOL chaff dispenser in LAU-7 launch rail configuration for the Royal Air Force 0018276

BOL/LAU-138 shown with AIM-9L loaded to a US Navy F-14A Tomcat **2001**/0017863

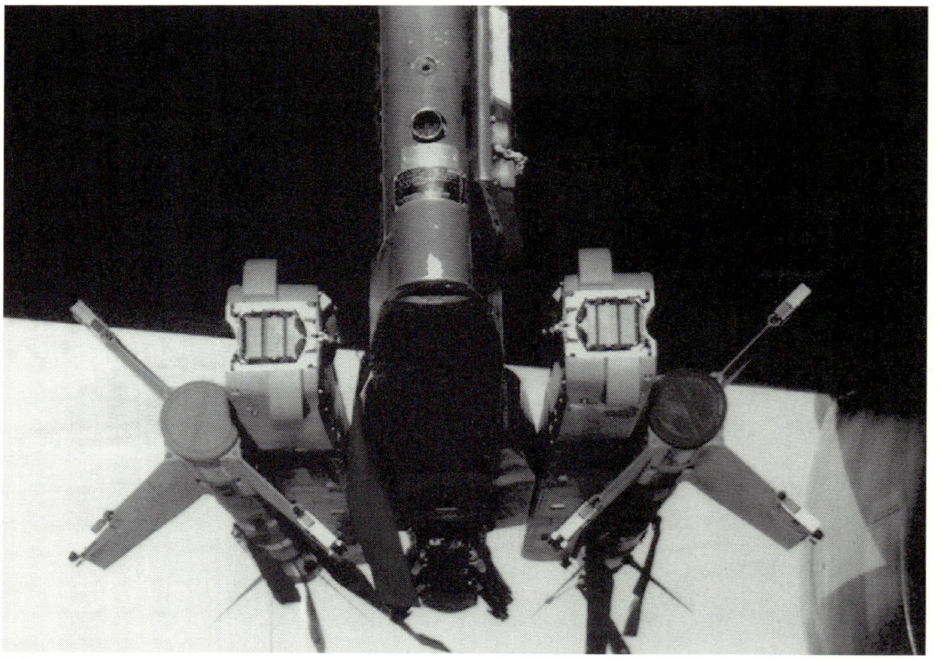

BOL/LAU-128 shown with AIM-9L **2001**/0017864

reliability and safety. Dependent on the host aircraft Electronic Countermeasures (ECM) equipment standard, the BO 300 system can operate in fully automatic, semi-automatic or manual modes.

BOL CMDS

The BOL CMDS was originally developed for insertion into the LAU-7 Sidewinder launch rail (the combined launcher/dispenser is known as LAU-138 in US service), thus facilitating the installation or enhancement of passive EW capability in an aircraft without reduction in available air-air weapon stations. In this configuration, the LAU-7 launcher has been redesigned to incorporate a BOL module in the rear half and a revised and rearranged missile seeker cooling system in the nose. The BOL system is currently compatible with a wide range of air-to-air missile launch rails which includes the LAU-127/LAU-128/LAU-129 family of AIM-120 AMRAAM and AIM-9X missile launchers, the Frazer-Nash Common Rail Launcher (CRL) and the ACMA MPRL.

The system comprises an electromechanical dispenser assembly, which can accommodate up to 160 standard packages (US nomenclature RR-184) of either chaff or IR countermeasures (CMs), mounted in plastic holding frames. With spacing controlled by an onboard countermeasures computer, the individual payloads are ejected into the turbulent airflow aft of the launcher which disperses the packages and, in the case of chaff, aids in rapid 'blooming' of the resultant cloud. If required, the BOL dispensing mechanism can be installed in a scab housing where no suitable launch rail is available. SaabTech claims that the BOL electromechanical dispensing method is more precise than pyrotechnic ejection and offers advantages in terms of safety on the ground and maintainability. BOL operates with standard dispensers such as the ALE-39, -40, -45 and -47. Loading and downloading of CM packages is achieved quickly (less than 1 minute) and easily while the dispenser is fitted to the aircraft; this can be done either by hand or with a special BOL loader. The BOL CM dispenser can be function-tested with the aid of the BOL test set.

BOP/A CMDS

The 31 kg (empty) BOP/A CMDS is made up of beam-shaped launch assemblies that accommodate six, two round capacity magazine modules and are compatible with 55 mm diameter chaff and Infra-Red (IR) decoy flare (up to three per round) cartridges. The system ejects its payload downwards and towards the rear of the host platform and is loaded from below. BOP/A is designed for ease and rapidity of installation; a typical fighter fit incorporates two launch assemblies. For larger airframes (such as transport aircraft), a four launch assembly installation is used. In the case of fighter applications, BOP/A can also be installed conformally to minimise drag. Maximum reload time is given as 60 seconds and individual BOP/A launch assemblies communicate with their host aircraft's Electronic Warfare (EW) system via an integral electronics unit and an RS-422 serial datalink. The system can be fully integrated into an onboard countermeasures suite and an optional IR sensor can be mounted in the rear of the launch assemblies to provide confirmation of IR decoy flare ignition.

BOP/AT, BOP/AX CMDS

The BOP/AT and BOP/AX CMDS make up a family of countermeasure dispensers for fighter and transport aircraft. BOP/AT is designed to be scab mounted onto the top of the fuselage of a fighter aircraft. So installed, the equipment ejects expendables upwards and forwards, an arrangement that allows IR decoy flares to develop their full spectrum characteristics in close proximity to the aircraft, thereby maximising their effectiveness. BOP/AX is designed to be scab mounted onto the side of the fuselages of a transport aircraft. The equipment can eject payloads upwards, downwards and forwards. By fitting double or quadruple installations, BOP/AX can be configured to dispense payloads in almost any direction required, thereby allowing the equipment's control circuitry to select the most effective ejection sequence for any detected threat.

BOP/AT and BOP/AX dispensers can accept all types of expendables with lengths up to 210 mm. These include standard 1 × 1, 2 × 1 and 2 × 2.5 in and 36 or 40 mm diameter cartridges, in a single-load or mixed-load configuration (for example, one type of chaff cartridge and two different types of flare cartridge). The dispenser automatically identifies each payload and the quantity of each type is reported to the control system. Ignition pulses are automatically adjusted by the dispenser software to match the particular cartridge ignition characteristics. The dispenser can carry up to nine magazines, each holding up to 24 cartridges, depending on type. The electronic control system is derived from the control system of the BOP/B and BOP/C dispensers and offers very high dispense rates and other advanced features such as automatic load identification. The electronics are fully integrated in the dispenser.

Dispensing is normally controlled via a databus and can be performed in almost any sequence. Nested sequencing and random dispensing are supported. The large capacity of the dispenser and the sophisticated control system allow preventive serial dispensing. If an expendable should fail to eject, the dispenser detects this and a new attempt is immediately initiated with another cartridge of the same type.

If an integrated MAW is used, the system can be set to automatic, manual or preprogrammed response.

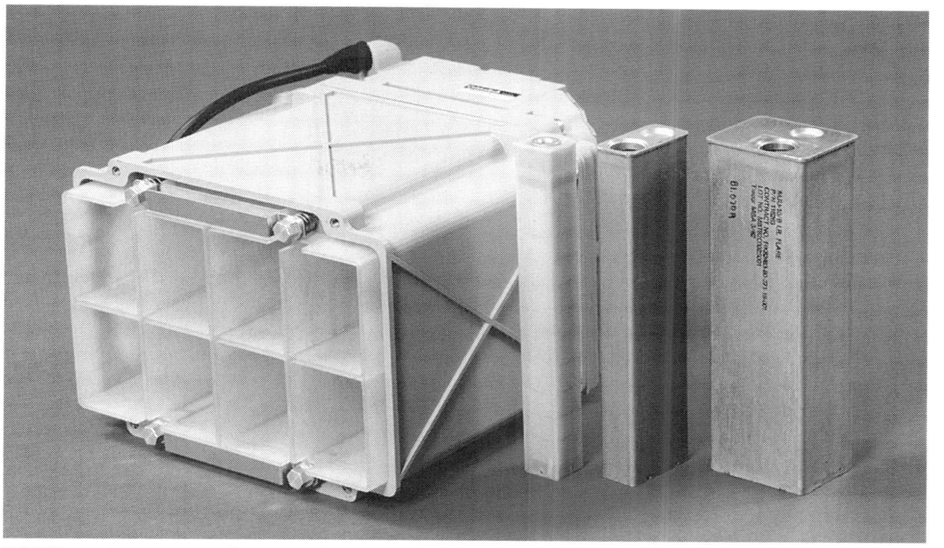

BOP/C countermeasures dispenser, shown with 1 × 1, 2 × 1 and 2 × 2.5 in flare cartridges **2000**/0081490

BOP/B CMDS

The BOP/B CMDS is designed specifically for fighter aircraft applications and, in common with the other BOP CMDS, can be loaded with chaff and/or expendable jammers in addition to flares. BOP/B is scab mounted or as a rear extension of a pylon and can accommodate 15 MJU-7, six MJU-10 or six 55 mm calibre cartridges. The system's dispenser module comprises a magazine, a breech plate and an electronics unit, and features breech plate coding pins to identify the type of payload inside a particular magazine. The ejection of ECM payloads is controlled by the BOC control unit via databus or discrete signal.

BOP/C CMDS

BOP/C, based on a dispenser developed for Sweden's JAS-39 Gripen, is a high performance countermeasures dispenser designed for fixed-wing and helicopter applications. BOP/C may be fitted either internally or externally and features interchangeable magazines, which allows for different expendable cartridge sizes, or a mix of differing cartridge sizes in the same payload. This feature enables the system to accept all types of expendables with lengths up to 210 mm, including standard 1 × 1 in, 2 × 1 in, 2 × 2.5 in, 36 mm and 40 mm diameter cartridges. In common with the other BOP dispensers, BOP/C features include payload type and quantity identification, databus/serial link ejection control, preventative serial dispensing, nested sequencing and random dispensing and automatic adjustment of ignition pulses to suit payload characteristics. In addition, BOP/C offers high dispensing rates coupled with extremely short reaction times, control of dual squib cartridges and capability to communicate with advanced expendables (allows for programming of advanced RF decoys prior to dispensing or while under tow. BOP/C is designed to be integrated with the host aircraft threat sensors and controlled via Defensive Aids Sub System (DASS) computer, or the complimentary BOC Control Unit.

BOC Control Unit

The BOC is a combined cockpit control unit and system programmer. The controller can easily be connected to different types of warners and aircraft systems by means of databusses and discrete signals. Reprogrammable non-volatile memories are used for storage of mission-specific data. BOC can control up to eight BOL or BOP dispensers simultaneously and can also control the most current pyrotechnic dispensers. The BOC controller can provide adaptive processing, to optimise response in a specific threat situation, and stores management, to keep an accurate record of the remaining load and give full flexibility in load configuration and optimum use of expendables.

Operational status

The BOL dispenser is currently in service with the UK Royal Air Force on the Harrier GR Mk 7 and the Tornado F3, the US Navy on the F-14 Tomcat and the Swedish Air Force JA-37 Viggen aircraft. It will also be integrated into the CRL for Eurofighter Typhoon and scab mounted to a weapons pylon on the JAS-39 Gripen. Flight tests of BOL integrated into the LAU-128 MRL on the F-15 began in 1997 as part of the US Department of Defense's Foreign Comparative Test (FCT) programme; additional trials leading to complete certification of BOL on the F-15 were planned for 2000. The US Navy has successfully trialled the scab mounted variant of BOL on the F/A-18C/D Hornet and S-3 Viking, with evaluation and certification work on the LAU-128/BOL variant for wingtip mounting on the F/A-18C/D ongoing.

BOP/A CMDS is installed aboard Viggen aircraft of the Swedish Air Force with BOP/B systems being fitted to J 35J Draken and JAS 39 Gripen aircraft of the same service. BOP/AT has been tested aboard a Tornado strike aircraft.

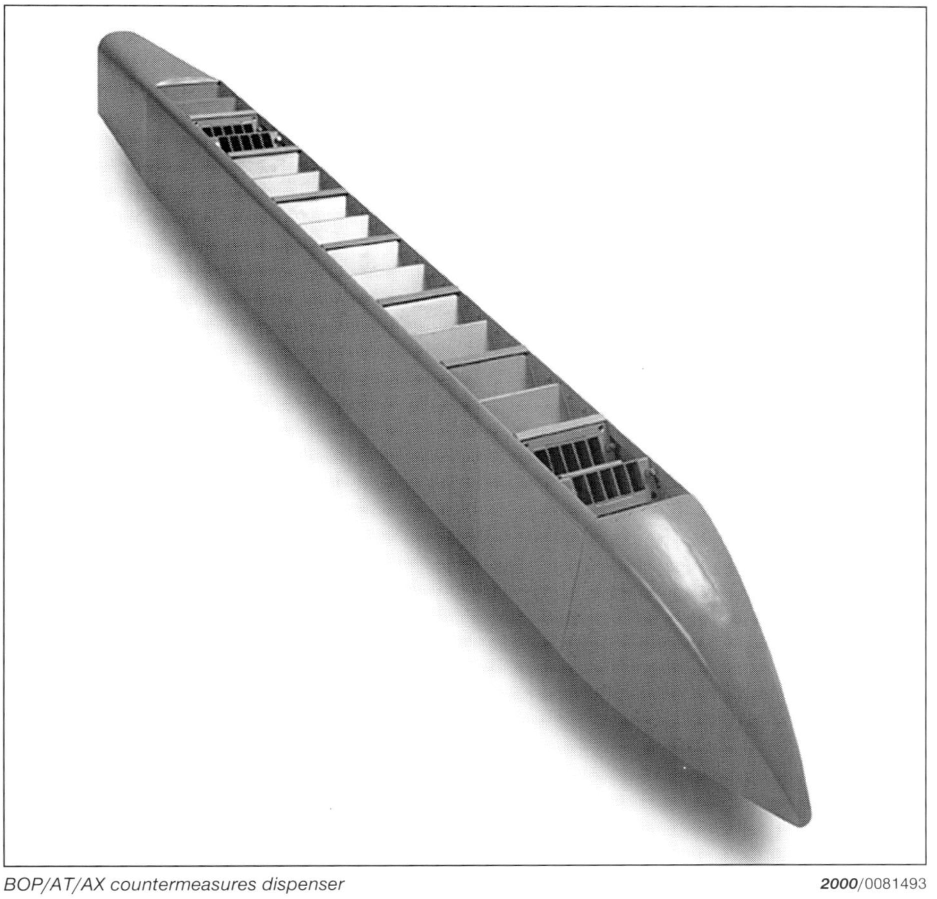

BOP/AT/AX countermeasures dispenser **2000**/0081493

Specifications

BOL CMDS

Weight (fully loaded):
(LAU-138): 56.5 kg
(scab-on fit): 24.0 kg
(dispenser only): 19.0 kg
Capacity: 160 packs
Reloading time: <1 min
Control signals: discrete input - one dispense signal MIL-STD-1553B or RS-485 serial link
Power: 115 V AC, 400 Hz

BOP/A CMDS

Chaff capacity: twelve standard 55 mm diameter cartridges
Flare capacity: twelve standard 55 mm diameter cartridges, each cartridge holds 2 or 3 flares
Reloading time: 1 minute (max)
Control signals: discrete input - one dispense signal MIL-STD-1553B or RS-422 serial link
Safety signal: MASS (+28 V)
Weight: 31 kg (without cartridges)

BOP/AT, BOP/AX CMDS

Cartridge capacity: nine magazines of:
twenty four 1 in × 1 in, twelve 2 in × 1 in, six 2 in × 2.5 in; sixteen 36 mm or ten 40 mm diameter payloads
Reloading time: 3 minutes (max)
Control signals: one dispense signal (discrete input, +28 V) MIL-STD-1553B or RS-485 (databus)
Safety signals: MASS (+28 V); safety pin (option)
Weight: 2.5 kg (single empty magazine); 40 kg (dispenser structure)
Dimensions: (L x W x H): 2,650 x 240 x 355 mm

BOP/B CMDS

Chaff capacity: six standard 55 mm diameter cartridges

Flare capacity: six standard 55 mm diameter cartridges, each cartridge holds 2 or 3 flares
Reloading time: 1 minute (max)
Control signals: discrete input - one dispense signal MIL-STD-1553B or RS-422 serial link
Safety signal: MASS (+28 V), safety pin
Weight: 11.2 kg (with empty cartridges)

BOP/C CMDS
Capacity: forty 1 × 1 in, or twenty 2 × 1 in, or eight 2 × 2.5 in cartridges
Repetition rate: 10 ms
Power supply: 115 V single phase, 400 Hz
Control signals: one dispense signal (discrete input, +28 V) MIL-STD-1553B or RS-485
Safety signal: MASS (+28 V), safety pin (option)
Weight: 7 kg (including magazine)
Dimensions: 179 (L) × 236 (W) × 281.2 (H) mm

BOC control unit
Dimensions: 76 × 147 × 165 mm

Contractor
SaabTech Electronics AB.

UPDATED

BOZ 3 training chaff dispenser

BOZ 3 is a high-capacity chaff dispenser originally developed for a Swedish Air Force requirement and since adapted for training use in view of its electromechanical design. It can be manually or automatically initiated and has both break-lock and corridor chaff modes.

Operational status
In service with several countries for training purposes.

Contractor
SaabTech Electronics AB.

UPDATED

BOZ 100 Series ECM dispenser

The SaabTech Electronics AB (formerly CelsiusTech) BOZ 100 series is an advanced ECM chaff and IR flare dispenser originally developed for the Swedish Air Force, but also supplied to overseas customers. It is capable of sophisticated break-lock and corridor operation at subsonic and supersonic speeds. The unit has a comprehensive EW system interface and is suitable for use in deep penetration, strike, reconnaissance and electronic warfare roles. It is microprocessor-controlled and has a reprogrammable program memory.

Specifications
Dimensions: 4,000 (length) × 380 mm (diameter)
Weight: 325 kg
Attachments: 14 or 30 in (356 or 762 mm) Minimum Area Crutchless Ejector (MACE) or conventional lugs

Operational status
In service with the Swedish Air Force under the designation BOX 9. About 700 pods have been supplied for the Panavia Tornado IDS programme. The German Air Force and Navy use the BOZ 101, the Italian Air Force uses the BOZ 102 and the Royal Air Force uses the BOZ 107. As a result of lessons learned during the Gulf War, installation of outboard pylons enabled BOZ 107 (coupled with the Sky Shadow ECM pod) to be fitted to Royal Air Force Tornado F3 interceptors as part of an active and passive upgrade to the aircraft ECM capability. This upgrade was carried over to Tornado F3s leased to the Italian Air Force, enabling the use of the BOZ 102.

Contractor
SaabTech Electronics AB.

UPDATED

The BOZ 3 training chaff dispenser

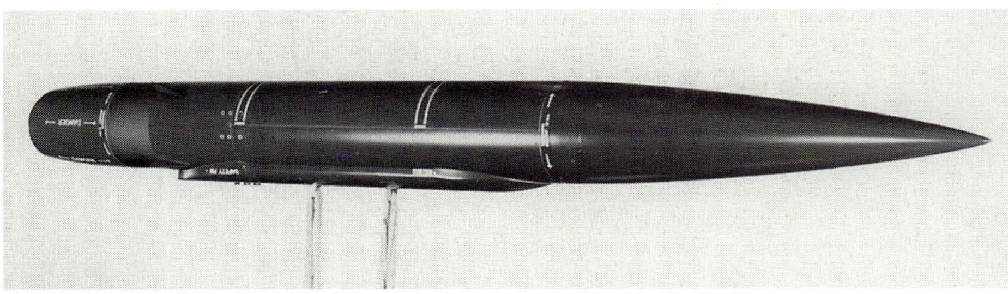

The BOZ 100 series chaff/flare dispenser for high-performance aircraft 0001266

JAS 39 Gripen radio communication system 0011837

JAS 39 Gripen radio communication system

The JAS 39 Gripen communication system is a dual-transceiver installation for simultaneous voice and data communication on the VHF and UHF frequency bands. The system can also operate as a relay link, and there is provision for adding a third Command and Control (C^2) data receiver.

The system provides both clear analogue voice and encryption modes, together with ECCM capabilities to ensure jam resistant communication for both voice and data. The Aircraft DataLink (ADL) operates in time sharing mode and enables continuous exchange of data between several aircraft, presented to the pilot via the main computer.

Functions and features included in the system comprise:

(a) The secure voice function is mechanised by use of a vocoder for narrowband digitising, and by applying encryption and error correction coding. The vocoder is specially designed for operation, including voice recognition, in noisy environments, as found in the Gripen aircraft cockpit
(b) A ground telecommunication amplifier function is incorporated to provide for analogue (voice) and digital (voice and data) telebrief operation while on the ground to eliminate the need for unnecessary free-space transmissions
(c) A tone generator and speech synthesiser function is provided to provide warning and alert signals as well as voice messages to the pilot
(d) A data transfer unit provides for loading: pre-coded frequency channels; crypto and ECCM keys; mission data; time synchronisation; and other requirements for the aircraft avionics system pre-flight briefing and post-flight analysis activities. During 1997, a more modern data transfer unit has been produced with application to other air and ground platforms, which uses 40 Mbytes of memory expansion and additional interfaces including MIL-STD-1553B and Ethernet.

Operational status
In production for JAS 39 A and B versions, and for the Saab 340 (popularly named Argus), and the S100B airborne radar aircraft. In a modified version, it forms part of the Swedish Air Force Tactical Radio System (TARAS). It is also planned as a baseline for the communication system in the JAS 39 Gripen export system.

Contractor
SaabTech Electronics AB (formerly CelsiusTech Electronics AB), Communications Systems.

UPDATED

SSC Maritime Surveillance System

The Maritime Surveillance System (MSS) was developed for the Swedish coastguard and has been in operation for 20 years. It is used for maritime surveillance tasks including Exclusive Economic Zone (EEZ) protection; oil pollution detection and assessment; fishing activities monitoring; border patrol.

The MSS comprises:
(a) Sensor data package that includes SLAR, IR/UV scanner, MicroWave Radiometer (MWR), camera(s), FLIR (optional), FLAR interface (optional), user-defined sensors (optional);
(b) Sensor Data Processor (SDP) that processes data from the selected sensor set;
(c) Data Management Unit (DMU) that provides high-performance image timing data and position information, display processing, sensor control, mission control and data reporting and datalink administration.

The current version of the MSS system is the MSS5000. MSS5000 has been improved over the previous versions of the MSS in a large number of ways, the main ones being real-time resampling of all sensor data to georeferenced image; real-time display of all sensor images together with the digital map; Side-

Looking Airborne Radar (SLAR) swath with 160 km wide coverage; new improved graphical user interface and image display windows.

The SLAR, manufactured by Ericsson, is designed exclusively for airborne maritime surveillance. The SLAR is an imaging radar which produces map-like images, suitable for superimposition on a digital map. Its main applications are surveillance of sea traffic, oil spill detection, fishery protection, search and rescue and sea ice mapping.

The IR/UV scanner, manufactured by Daedalus, is used to obtain high-resolution imagery of accident sites and is ideal for mapping oil spills. The infra-red and ultraviolet channel images are both resampled and presented in real time on top of the digital map.

The MWR, manufactured by Ericsson, is a scanning radiometer, used for detailed mapping of oil spills. It measures the thickness of the oil on the water surface and estimates the total oil volume. The MWR image can be superimposed on the other sensor images in the digital map.

The camera(s) are used for high-quality, high-resolution photographic evidence. Both hand-held and vertically mounted installations are possible. Each frame is automatically annotated with relevant time and position data.

All sensors of the MSS5000 can be operated single handed. The operator can also monitor resulting images in real time, and superimpose selected images on the digital map. Image interpretation is supported by online image analysis tools. This gives an excellent overview, as well as detail when needed.

Selected mission report text and images can be transmitted to the headquarters on the ground or to a coastguard ship to support, for instance, oil spill clean-up operations.

Operational status

MSS systems, or selected subsystems thereof, have been delivered to coastguard organisations in China, Germany, India, Netherlands, Norway, Poland, Portugal, Sweden, UK and the USA. They have been installed in aircraft such as CASA-212, Cessna 402B and 404, Fairchild Dornier Do 28, Fairchild Dornier Merlin, Turbolet L410 and Y12.

Contractor

Swedish Space Corporation.

VERIFIED

TURKEY

LN100G EGI Embedded GPS Inertial navigation system

The LN-100G EGI system is an advanced technology unit that includes a sensor assembly with three Zero-lock™ Laser Gyros (ZLGs), an A-4 accelerometer triad and, five electronics assemblies together with two spare card slots, all installed in a very small, lightweight package.

The core LN-100G INS/GPS is optimised for individual applications by appropriate additions of I/O cards and other modules installed in the spare card slots, by modification of the software for different I/O messages and formats, by modified mode control and tuning of the Kalman filter. The LN-100G is an open architecture and hardware/software flexible unit which can be adapted to various air platforms including rotary-wing, fixed-wing and unmanned air vehicles, without the need to change vehicle architecture.

The LN-100G provides three simultaneous navigation solutions; hybrid GPS/INS, free inertial and GPS only. The LN-100G optimally combines GPS and INS features to provide enhanced position, velocity, attitude and pointing performance, as well as improved acquisition and anti-jam capabilities. The embedded GPS module is an L1/L2 CA/P(Y) code unit capable of accepting RF (or IF) inputs from the GPS antenna system. Stand-alone GPS PVT data and stand-alone INS data are provided for integrity and fault monitoring purposes.

Specifications

Laser gyro features: (18 cm nondithered, Zero-Lock™ Laser Gyro and miniature accelerometer technology)
Accuracy:
(position) <10 m CEP
(velocity) <0.01 m/s
Dimensions: 280 × 180 × 180 mm
Weight: 9.8 kg
Interfaces: MIL-STD-1553B/ARINC/discrete databuses

Status

In production.

Contractor

Aselsan Inc, Microelectronics, Guidance and Electro-Optics Division.

UPDATED

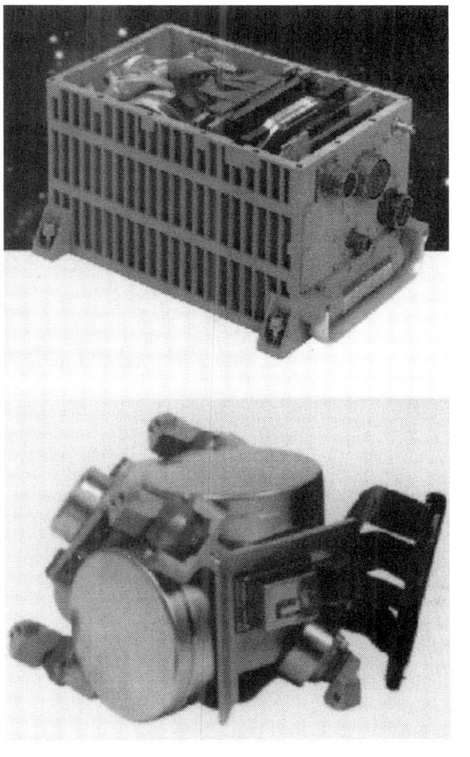

Aselsan LN100G EGI (above) and ZLG/miniature accelerometer (below)
0015372

AN/ALQ-178(V)3 and (V)5 integrated self-protection systems

The AN/ALQ-178(V)3 and (V)5 integrated self-protection systems, are manufactured by MiKES Microwave Electronic Systems Inc, for installation in Turkish Air Force aircraft. The AN/ALQ-178 is an advanced, internally mounted, self-protection system specifically designed for high performance fighter aircraft, including the F-16 and F/A-18. It has been fully operational since 1986. See also entry on AN/ALQ-178 Rapport III ECM system under Lockheed Martin Electronic Defense Systems.

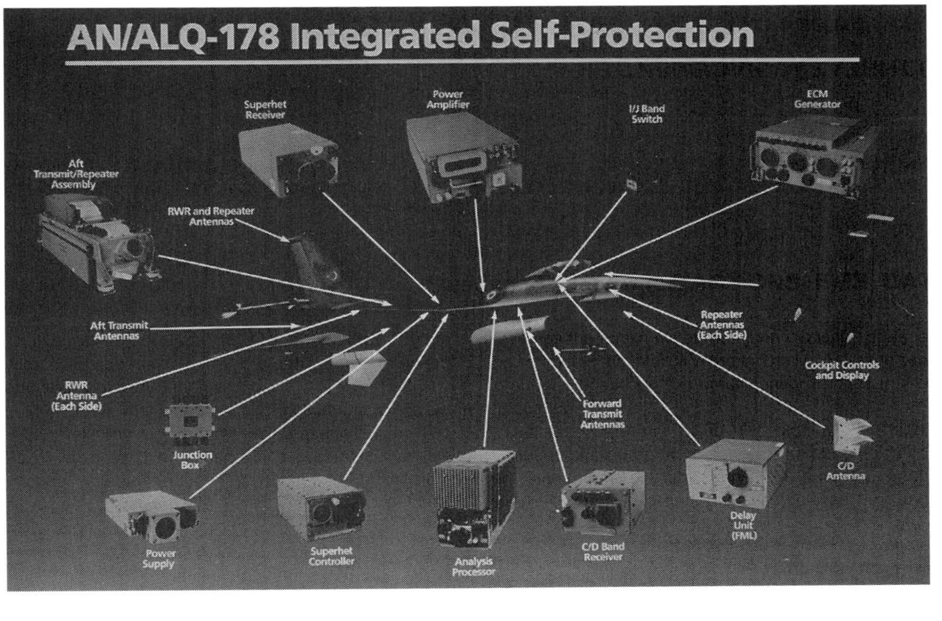

MiKES Microwave Electronics Systems Inc AN/ALQ-178 integrated self-protection system
0002228

The AN/ALQ-178 receiver utilises a superheterodyne receiver to provide precision frequency measurement and required sensitivity. The integrated architecture of the AN/ALQ-178, utilising a common receiver system, results in rapid threat identification and counter responses (both RF ECM and chaff and flare) for an exceptional level of aircraft survivability in dense environments.

Operational status

The AN/ALQ-178(V)3 (also known as SPEWS-1) is operational on the Turkish Air Force's 160 F-16 Block 30/40 aircraft. The AN/ALQ-178(V)5 has been selected for the Turkish Air Force's 80 F-16 Block 50 aircraft.

Contractor

MiKES Microwave Electronic Systems Inc.

VERIFIED

UKRAINE

Inertial Navigation System (INS)

The Central Design Office (CDO) Arsenal claims over 20 years' experience in development of monobloc laser for airborne navigation purposes. Its latest platform-independent INS product, designed for helicopter and missile applications, uses a three-axis laser gyro and pendulous accelerometers. The laser technology used employs a high-stability, two-frequency He-Ne linear laser; work has also been done with CO_2 lasers.

Specifications

Laser gyro:
angular rates measured: ±90°
zero drift (1 sigma): 0.05–0.1°/h
relative error of scale factor: 10^{-4}–10^{-5}
Accelerometer:
angular rates measured: ± 100 m/s^2
zero drift (1 sigma): 10^{-2} m/s^2
relative error of scale factor: 10^{-4}
Precision readiness time: 180 s
Volume: 10 dm^3
Weight: 12 kg
Power consumption: 70 W
Operating temperature limits: –20 to +50°C

Contractor

Arsenal Central Design Office.

VERIFIED

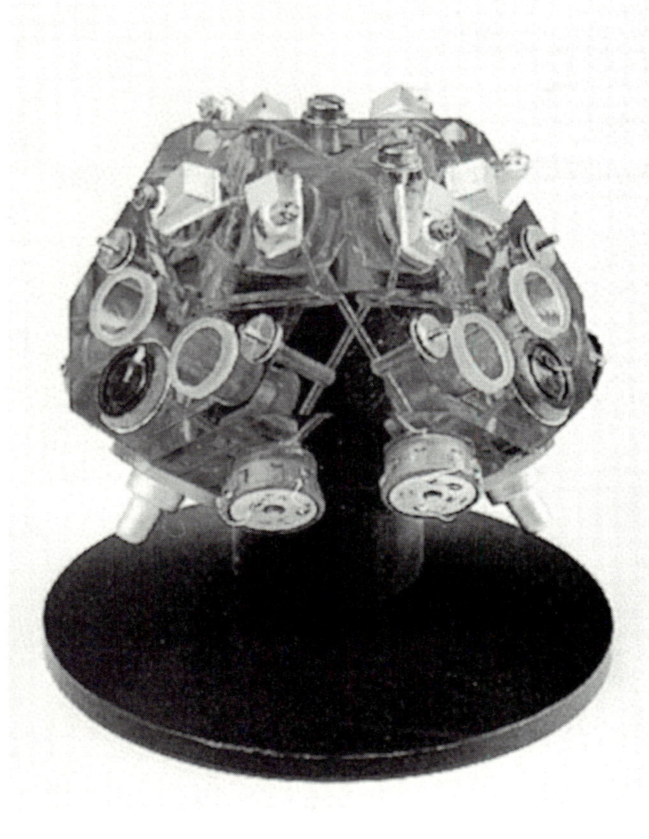

Arsenal CDO ring laser gyro system
1999/0051599

UNITED KINGDOM

Tactical Datalinks for Tanker Aircraft

Aerosystems International Tactical Communications Division has developed a range of high quality, innovative and cost effective datalink systems for a range of tactical aircraft, with emphasis on Commercial-Off-The-Shelf (COTS) software packages, facilitating short lead-times and precise tailoring of installations to meet customer requirements. One of these systems is an austere JTIDS capability for tanker aircraft, originally developed to meet an Urgent Operational Requirement (UOR) for Royal Air Force TriStar and VC10 tankers.

The operational benefits of JTIDS on Tankers have been proven during conflicts such as Bosnia and Kosovo:

- Tanker and refueling units have accurate information on relative position/location.
- Off load fuel state and Link 16 Precise Participant Location and Identification (PPLI) are transmitted by the Tanker.
- Tanker tow lines can be dynamically changed in response to tactical requirements.
- Tankers can act as relay platforms, transmitting between networks, overcoming line of sight issues.
- Tanker crews have accurate and up-to-date situation awareness, showing precise position of all the friendly and hostile forces, allowing Tankers to operate more effectively and safely.

Aerosystems International offer three installation options: Fully integrated with host aircraft avionics system, stand-alone, as an aircraft upgrade, or as a role fit for specific aircraft missions and roles.

The basic components of the system are:

- JTIDS terminal (which can use existing aircraft Tacan or IFF aerials)
- Processor control interface unit, which acts as the core of the system
- Display unit
- Keyboard
- GPS or aircraft navigation input (a lightweight GPS receiver is an option)

For a fully integrated system, an existing aircraft display unit is used, with processing software integrated with the aircraft computer system and the JTIDS terminal housed in a suitable avionics bay. For a stand-alone system, a dedicated aircraft display, processor and keyboard is provided according to aircraft layout. Navigation information can be either from the aircraft system or via an optional lightweight GPS receiver. For a role fit, the system is crated, which can then be seat-rail mounted and installed when required, typically in less than 2 hours.

The system can interface with a wide variety of terminal types; a MIDS terminal is available.

Specifications

Display (For non integrated versions):
Type: LCD active matrix
Screen dimensions: 211.2 × 158.4 mm
Colours: 262,144
Viewing angle: ±45° horizontal, +10/–30° vertical
Signal input: RGB, VGA or SVGA
Temperature range: –20 to +55°C
EMI/EMC: MIL-STD-461C, FCC-B
Dimensions: 297 × 241 × 140 mm
Power: 28 V DC, or 115 V AC/400 Hz

Processor Control Interface Unit (For non integrated versions):
Video: RGB
Interfaces: MIL-STD 1553B, ARINC-429, RS 422
Temperature range: –25 to +50°C
Cooling: Self-contained fan
EMI/EMC: MIL-STD-461C, FCC-B

Dimensions: 210 (W) × 274 (H) × 442 (D) mm
Weight: 11 kg
Power: 115 V AC/400 Hz

JTIDS Terminal (MIDS):
Interface: MIL-STD 1553B
Cooling: Conductive air
EMI/EMC: MIL-STD-461C, FCC-B
Dimensions: 193.5 × 190.5 × 343 mm
Weight: 20.5 kg
Power: 60 Hz AC or 400 Hz MIL-STD 704A
Output: 200 W

Operational status

The UK Ministry of Defence (MOD) raised an UOR in June 1999 stipulating a requirement to fit up to ten VC10 and TriStar aircraft with the Joint Tactical Information Distribution System (JTIDS). A contract to fit an initial batch of three TriStars (one KC1 and two K1s) and three VC10 MK3/4s was placed with Aerosystems International on the 11th June 1999. The fitting of JTIDS to two RAF TriStar aircraft and one VC10 was accomplished in three months. The first flight test of the system on board the aircraft was successfully completed at RAF Brize Norton in September and the RAF released the JTIDS system for operational service in early October 1999.

Contractor

Aerosystems International, Tactical Communications Division.

NEW ENTRY

Aerosystems International austere JTIDS installation in a Royal Air Force Tristar tanker aircraft (Aerosystems International) ***2001***/0099744

Advanced airborne speech recogniser for ASR 1000

The Model ASR 1000 speech recogniser was developed to meet the UK Ministry of Defence's requirements for an advanced airborne speech recogniser. Its development has followed an evolutionary path after successful flight trials by the Defence Evaluation Research Agency (DERA) of other Alenia Marconi Systems speech recognisers such as SR-128 and Macrospeak in Tornado aircraft and the Wessex helicopter. The system incorporates the advanced technology of the Macrospeak commercial recogniser with additional enhancements to meet the DERA's specific requirements for accurate performance in the noisy environment of fast jet aircraft, where lower signal-to-noise ratios are normal. A complementary ground-based speech template preparation facility uses advanced statistical techniques to produce robust templates for better performance when the pilot is under stress and other cases where the variable nature of speech has proved to be a problem in the past.

The ASR 1000 is a speaker-dependent 1,000-word continuous recogniser suitable for two-crew operation and performs its tasks in real time. It features automatic gain control with a dynamic noise mask and can implement a true finite state syntax for each crew member. It is contained in a ½ ATR case but the technology, when proved, will be customised to meet customers' specific requirements for integration in advanced avionics and other critical areas.

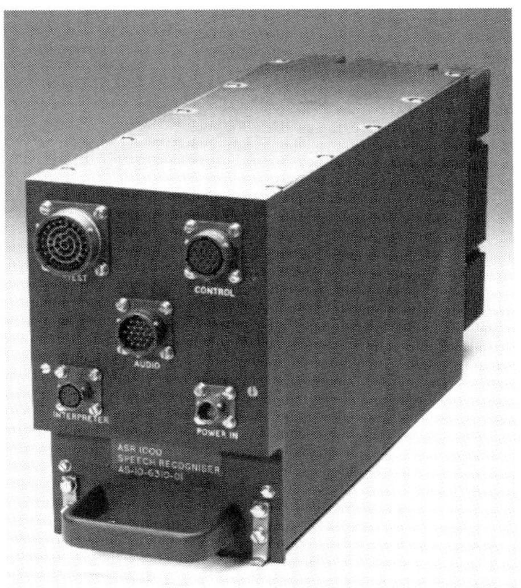

The Alenia Marconi Systems speech recognition unit

Applications will include its use by aircrew to control communications, navigation and weapon delivery systems; it will also be used for low bit-rate communications.

Specifications

Dimensions: ½ ATR short
Vocabulary size: 1,000 words

Operational status

In service.

Contractor

Alenia Marconi Systems Limited.

VERIFIED

Helicopter secure speech system

Alenia Marconi Systems has developed the helicopter secure speech system to provide high-quality secure voice encrypted communications for military helicopters, to interface with the existing avionics and be interoperable with the current secure ground-based communications system. The equipment is easily fitted for specific missions or roles, or may be used as a permanent fixed installation.

The system comprises the Digital Master Unit (DMU) with integral VHF FM ARC-340 radio and encryption equipment. Multiple user access is provided within the aircraft by means of user selector boxes over which the pilot has overall control via the Master Control Unit (MCU). A crew member may therefore operate over a secure net within the aircraft or to the ground if required. Features of the system include full air certification to military specification, transilluminated controls with a night vision facility and a rebroadcast and intercom facility.

Specifications

Dimensions:
(DMU) 420 × 540 × 237 mm
(MCU) 147 × 97 × 60 mm
(USB) 147 × 97 × 60 mm
(controller) 153 × 203 × 205 mm
Weight:
(DMU) 24.3 kg
(MCU) 0.6 kg
(USB) 1.1 kg
(controller) 2.6 kg

Operational status

In production.

Contractor

Alenia Marconi Systems Limited.

VERIFIED

MarCrypDix Air secure speech system

MarCrypDix Air is a secure speech encryption equipment developed specifically for installation in civil and military aircraft. It is designed to interface with any airborne digital communication system and provides high-grade speech and data encryption. There is storage for eight different key variables or code settings, fully automatic synchronisation with zero error extension and negligible range degradation.

The key variables are generated by a key management unit and are inserted either manually, using a hexadecimal keyboard, or may be generated automatically within the key management unit using a true random noise generator. Transfer of the key variables from the management unit to the encryption unit is accomplished by a key fill gun. This is a battery operated, pocket-sized device with an internal memory system, data interfacing being performed optically. Key variables stored within the gun can be transmitted to any number of encryption units. Provision is also made for the instant erasure of all information stored within the gun's memory. The number of key variables is 2^{128}, with a key variable length of 128 bits, plus a further 16 bits set uniquely for each customer.

The equipment is self-monitoring and will inhibit transmission should any fault occur in the equipment, thus ensuring that sensitive information cannot be transmitted with less than complete protection.

Although MarCrypDix Air has been designed for the airborne environment, the same unit, suitably mounted, can be employed in a ground role for ground-to-air or ship-to-air communication.

Specifications

Dimensions: 193.5 × 320.5 × 58 mm
Weight: 4 kg

Operational status

In production and in service.

Contractor

Alenia Marconi Systems Limited.

VERIFIED

Secure speech communication system

The secure speech communication system has been developed to enable strategic and intelligence information to be passed between aircraft and ships without risk of interception. Messages can be transmitted in the complete confidence that only the recipient can understand the contents. The system has been designed to allow integration with existing communications control equipment in naval and air force aircraft with minimum aircraft modification.

System development has involved the design and manufacture of A and B model LRUs, system test rigs and special-to-type test equipment, and commissioning of the aircraft rigs. The production programme has been built on a combination of modern control methods and advanced automatic test procedures to ensure the total set of units for each aircraft comes together as a proven system. All aspects of overall system cost, including those of ownership, installation, crew training, maintenance and support have been reduced by employing a modular concept. This concept allows system configuration to be cost-effective, giving maximum flexibility to meet customer requirements, with a built-in provision for expansion.

The system uses microprocessor technology to control the routeing of audio paths. MIL-STD-1553 and ARINC 429 data highways are used to transfer control commands between the various elements of the subsystem. The system also incorporates BIT and circuit redundancy features.

Operational status

In production for the Royal Navy EH101 Merlin helicopter.

Contractor

Alenia Marconi Systems Limited.

VERIFIED

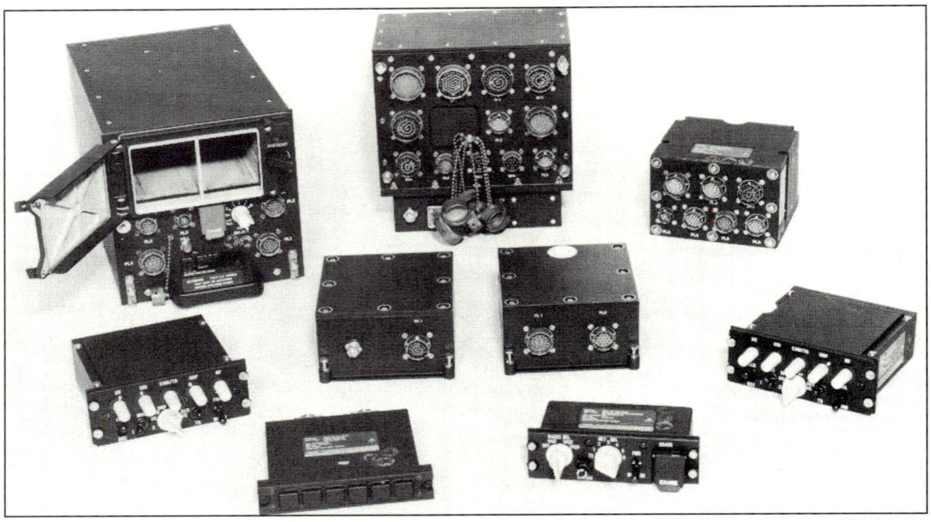

Units of the Alenia Marconi Systems secure speech communication system

The Alenia Marconi Systems four-board vocoder pack

Voice encoder/decoder

Alenia Marconi Systems has developed a channel voice encoder/decoder (vocoder) equipment designed to operate in environments of high acoustic noise and high bit error rates, such as those encountered in tactical military roles. The 2,400 bits/s vocoder incorporates a robust pitch extractor, regenerator and so on. According to Alenia Marconi Systems the system provides improved resistance to acoustic noise and transmission errors over other systems such as linear predictive vocoders. The pitch extraction technique is less sensitive to acoustic noise and forward error correction, and median smoothing/majority vote methods applied to the pitch information result in reduced sensitivity to transmission errors.

Good communication has been achieved between operators in high acoustic noise and transmission conditions. Subjective tests, using standard helmets and face-mask microphones, have been carried out in a noise chamber where operators were subjected to levels of 120 dB (typical fighter noise) and 115 dB (fighter bomber level), said to be representative of high-speed low-level flight. Communication could be maintained at these noise levels for random transmission error levels of up to 2.5 per cent.

The system is almost completely digital in character and is based upon an Intel 8085 microprocessor chip combined with a special purpose digital filter bank. It is now being manufactured as a four-board pack.

Specifications

Dimensions: 51 × 176 × 181 mm
Weight: 1 kg

Operational status

In production as a four-board pack.

Contractor

Alenia Marconi Systems Limited.

VERIFIED

1220/1223 Series laser warning receiver

The 1220 Series laser warning receiver is a modular expandable system with centralised processing and miniaturised remote sensor heads. It may be operated as a stand-alone LWR with its own display or as part of an integrated system with, for example, the BAE Systems Sky Guardian 200 RWR.

A premission programmable threat library provides the capability for selective threat identification. The panel-mounted display is NVG-compatible and provides sector indication and other threat data.

An optional module provides a record of up to several thousand date and time tagged laser detections.

Specifications

Dimensions:
(sensor) 25 (length) × 50 mm (diameter)
(electronics unit) 127 × 194 × 323 mm
(display) 80 × 80 mm
Weight:
(sensor) 0.4 kg
(electronics unit) 5 kg
(control panel) 0.4 kg
(display panel) 0.4 kg
Power supply: 24-28 V DC or 115 V AC, 400 Hz
Field of view: 360° azimuth × 45° elevation
Angular resolution:
(azimuth) ±22.5° (15 or 10° option)
Spectral range:
(basic system) 0.35-1.1 μm
1-1.8 μm and 8-11 μm options
Spectral resolution: 15-100 nm
Interfaces: RS-422 or MIL-STD-1553B

Operational status

In production for high-performance aircraft applications. The system has been sold to unspecified customers in Europe and Asia.

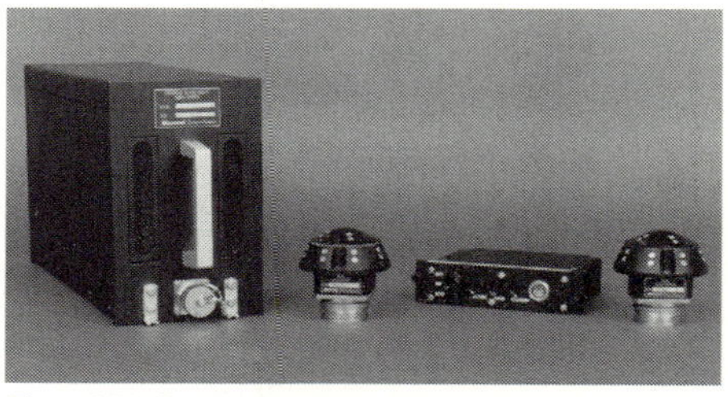

The 1220 Series laser warning receiver consists of electronic unit, control unit and sensors

An updated variant, the 1223 LWR, forms part of the BAE Systems HIDAS system selected for the British Army WAH 64 Apache helicopter. The model 1223 has very high sensitivity and extended waveband coverage, to provide detection of laser beam-riding threats at the point of launch, as well as rangefinders and target designators.

Contractor

BAE Systems.

VERIFIED

AA34030 sonobuoy command transmitter

The AA34030 sonobuoy command transmitter is a variant of the AD 3400 VHF/UHF transmitter/receiver, which has been in full production for a number of years. This variant has been developed in conjunction with the AQS-903 acoustic processing system and is currently in service with several air forces around the world.

The equipment is a UHF transmitter which has been specifically developed to provide the RF downlink between an airborne acoustic processing system and active sonobuoys. The transmitter provides AM or FM analogue or digital communications over the frequency range 282 to 292 MHz with channel increments of 25 kHz, although for normal sonobuoy operations 50 kHz is used. Control of the transmitter is via an ARINC 429 digital data highway and a number of discrete inputs. Control information is normally provided by the acoustic processing system, although a discrete control unit can be provided.

Specifications

Dimensions: 194 × 125 × 256 mm
Weight: 6 kg max

Operational status

In production for the UK Royal Navy EH 101 Merlin helicopter.

Contractor

BAE Systems.

VERIFIED

AD190 (ARC-340) VHF/FM radio

The AD190 radio, a company designation for the ARC-340 system, is a tactical VHF/FM equipment designed specifically for air-to-ground communication between military aircraft and ground forces. It is in service with the UK armed forces, in conjunction with Army Clansman VHF equipment, and is in operation in a wide range of fixed-wing aircraft and helicopters in many parts of the world.

The system provides clear, and secure, speech communication, data transmission, automatic rebroadcast for extending the range of tactical communications and homing facilities. The homing mode may be used simultaneously with a communications channel without mutual interference. The range capability of the homing facility is the same as the communications range.

The AD190 covers the VHF band from 30 to 75.975 MHz at selectable channel spacing of either 25 or 50 kHz increments. Tuning is silent and instantaneous and transmitter output power is selectable at either 1 or 20 W.

Multistation collocated operation is possible with a maximum of three systems in the same aircraft. This is subject to the proviso that antennas must be sited at least 3 ft apart and frequency separation of 3.5 per cent for three systems or 3 per cent for two systems must be maintained. The AD190 incorporates diagnostic BITE.

Specifications

Dimensions:
(controller) 146 × 67 × 85 mm
(transmitter/receiver) 385 × 197 × 125 mm
Weight:
(controller) 0.9 kg
(transmitter/receiver) 9.2 kg

Operational status

In production, and service.

Contractor

BAE Systems.

VERIFIED

AD380 and AD380S Automatic Direction-Finders (ADF)

Latest in a number of automatic direction-finding systems developed and produced by BAE Systems, the AD380 is designed to ARINC 570 and covers the frequency range 190 to 1,799.5 kHz in 0.5 kHz steps while its variant, the AD380S, covers 190 to 1,599.5 kHz. The latter additionally covers the international maritime distress frequency of 2,182 kHz with the ability to tune to ±0.5 kHz on either side of the nominal frequency, rendering it particularly suitable for search and rescue. This difference apart, both systems are designed to the same standard.

AD380 systems have automatic, crystal-controlled frequency selection and are of all-solid-state construction with instantaneous electronic tuning. Built-in test facilities are incorporated.

A range of controller options is available, permitting single, dual or programmable operation. Standard controllers come in three versions, all of which provide selection of frequency and mode, ADF, antenna, test or beat frequency oscillator, together with volume control. The decade frequency selectors display the operational frequency in 1 in (25 mm) high numerals. Frequency controls comprise three concentric knobs which operate the logic frequency circuitry.

The first type of controller is the G4032E, a single frequency version, and the second type, the G4033E, is used for tuning two receivers. Mode facilities in each case are selected by toggle switches and each controller has a test button. The third controller type is the G4034E, designed for rapid retuning of a single receiver to one of two frequencies, which can then be selected by operation of a transfer switch when a white bar is displayed across the figures of the frequency not in use. Mode facilities are selected by a rotary switch.

A programmable controller, the AA-3809, conforms to the dimensions, form factor and electrical requirements of ARINC 570. It provides a four-channel preselect facility and can control both versions of the AD380. Six push-buttons permit instantaneous selection of frequencies previously entered into the memory store, which retains information when the equipment is unpowered.

The displayed frequency is always that to which the receiver is tuned and the operator can change stored information while in flight or select the 'N' button which allows normal operation of the controller. A brightness control, which is independent of the main aircraft panel lighting system, is incorporated. Normal mode facilities are also available for pilot operation.

Specifications

Dimensions:
(receiver) ¼ ATR short
Weight:
(receiver) 4.5 kg

Operational status

Over 700 ADF 380 units have been manufactured.

Contractor

BAE Systems.

VERIFIED

AD1990 radar altimeter

The AD1990 radar altimeter is a covert radar altimeter, which was designed to meet the UK Royal Air Force's needs in the 1990s. The AD1990 radar altimeter directly replaced the initial-fit altimeter in the UK Royal Air Force Tornado aircraft, using all existing fixtures and fittings.

The digital signal processing techniques incorporated in the receiver allow the extraction and simultaneous tracking of height both above the ground and above obstacles such as trees. These two outputs enable the pilot to operate more safely when flying at low level and are also used by the Terrain Reference Navigation (TRN) system to enhance overall navigation performance. The AD1990's fast dynamic response time eliminates the need for groundspeed compensation of height data within the TRN system. Inherent in the signal processing technique is the ability to identify and reject unwanted signals from underslung stores and landing gear, a traditional problem for radar altimeters. Reliable operation is obtained from its maximum operating altitude of 5,000 ft down to ground level.

An important innovation, at the time, was that the altimeter remains covert in operation, rendering it virtually undetectable by the enemy. Such Low Probability of Intercept (LPI) is achieved by spreading the transmitted signal over a very wide bandwidth through the application of pseudo-random phase modulation and adaptive power tailoring which, in addition, gives a high resistance to jamming. AD1990 can also be applied to other modern military aircraft where the ability to remain undetected is the key to mission success.

In addition to analogue height output, the system can be configured in either Panavia or MIL-STD-1553B interfaces.

Specifications

Dimensions: 109 × 154 × 318 mm
Weight: 5.25 kg
Power supply: 28 V DC, 55 W (max)
Frequency: 4.3 GHz
Range:
(height) 0-5,000 ft
(speed) 0-800 kt
(pitch) 0 to ±60°
(roll) 0 to ±60°
Accuracy: ±3 ft or ±3% whichever is greater
Temperature: –55 to +90°C

Contractor

BAE Systems.

VERIFIED

AD2770 Tacan

The AD2770 Tacan navigation and homing aid is suitable for all types of aircraft. This system is used on the majority of front-line aircraft in service with UK forces. It provides range and bearing information from any selected ground Tacan station or from any suitably equipped aircraft and is available in a number of forms offering outputs in digital ARINC 429 or analogue form or a combination of the two. Output signals may be provided to drive range/bearing or deviation indicators on a pilot's panel or to interface directly with a computer.

The system has full 252 channel X and Y mode capability and operates up to 300 n miles range with a range accuracy of better than 0.1 n mile. Bearing accuracy is said to be better than 0.7° on normally strong input signals. It comprises two units: a transmitter/receiver hard-mounted in the avionics bay and a panel-mounted remote-control unit. A switching unit, also installed in the avionics bay is required when two antennas are fitted. A mounting tray with a cooling air blower is necessary if a cooling air supply is not available from the aircraft's own air-conditioning system.

The transmitter/receiver section with a digital interface only is contained in a ¾ ATR short case with a front doghouse. Versions with analogue outputs are accommodated in a case of longer dimensions to house the additional circuitry. Signal processing circuitry is largely digital in the interests of system reliability, and continuous integrity monitoring techniques eliminate the risk of erroneous outputs.

Range and bearing analysis and output formats are prepared in a general purpose computer module called the analyser. This allows both range and bearing signal processing to use the same circuitry, with a consequent reduction in the number of components. The range system uses a parallel search method, said to be unique, which by making use of all signal returns achieves a very rapid lock on.

The AD2770 system operates in three modes: receive (giving bearing information only); transmit and receive; and air-to-air (providing range information only). Transmitter frequency range is from 1,025 to 1,150 MHz with an output of 2.5 kW peak pulse power. Receiver frequency coverage is from 962 to 1,213 MHz. The tracking speed range is from 0 to 2,500 kt.

The design of the AD2770 system is flexible both electrically and mechanically and alternative configurations with appropriate form factors, output characteristics and mechanical and electrical interfaces can be provided for new aircraft or for retrofit.

Specifications

Dimensions:
(control unit) 57 × 146 × 83 mm
(transmitter/receiver standard unit) 194 × 191 × 380 mm
(antenna switch) 69 × 130 × 56 mm
Weight:
(control unit) 0.45 kg
(transmitter/receiver standard unit) 14 kg
(antenna switch) 0.25 kg

Operational status

In service in the Tornado GR. Mk 1 and F3, Nimrod MR. Mk 2 and Sea Harrier aircraft.

Contractor

BAE Systems.

VERIFIED

AD2780 Tacan

A follow-on from the AD2770 series supplied for the Panavia Tornado and other front-line types, the AD2780 is also proposed for military applications. The system provides slant range and relative bearing to a standard

Tacan station, range rate (which approximates to groundspeed when not used in conjunction with the company's area navigation system), time to go to waypoint or station, ARINC 429 serial data output and 252 channels in X and Y modes. Outputs of range are available in digital format to ARINC 429 and in analogue to dial and pointer displays.

Specifications

Dimensions: 127 × 153 × 318 mm
Weight: 3.5 kg
Frequency:
(transmitter) 1,025-1,150 MHz
(receiver) 962-1,213 MHz
Range rate output: 0-999 kt with accuracies ±15 kt for 0-300 kt, and ±5% for 300-999 kt
Time to station output: 0-99 min
Tracking speed: 0-1,900 kt, 0-20°/s
Memory:
(range) 10 s
(bearing) 4 s

Operational status

In production and service in the British Army Gazelle and Royal Netherlands Air Force Eurocopter BO 105 helicopters and UK Royal Air Force Shorts S312 Tucano trainers.

Contractor

BAE Systems.

VERIFIED

AD3400 multimode secure radio system

The AD3400 provides, in a single transmitter/receiver, coverage of the entire airborne line of sight frequency band from 30 to 400 MHz with channel increments of 25 kHz. Up to 20 channels may be preselected in this range.

In effect, the system provides the equivalent of four radios in a single unit since it covers the following bands and modes: VHF FM for tactical close support, the civil ATC VHF AM band, the civil and maritime VHF FM bands and the military UHF AM and FM bands. Its other major feature is that it provides secure communications for both speech and data transmission. For speech transmission in a secure mode, an encryption unit is used and for data security the appropriate modem is connected to the radio equipment.

Separate, continuously operating Guard receivers are incorporated and homing and ADF facilities are also available with suitable keying and antenna installations.

Construction of the system features 'slice' techniques and Large-Scale Integrated (LSI) circuitry. The system is based on Intel 8085 microprocessors and uses distributed processing. The extensive use of LSI and proven components provides a high measure of reliability. Serviceability is further enhanced by the comprehensive BITE facility which can provide rapid fault diagnosis, particularly since intermittent fault data is stored in a non-volatile memory for maintenance purposes.

The system is convection cooled and protected by a sensing device which progressively reduces power output at high temperatures to prevent transmitter damage. The transmitter is also protected from damage caused by short or open circuits at the transmitter output.

The AD3400 has been designed for ease of installation with particular attention to retrofit requirements in older aircraft in which it is often possible to fit two of these systems in the space formerly occupied by a single radio.

Specifications

Dimensions:
(AA34024 controller) 57 × 146 × 152 mm
(AA34001 transmitter/receiver) 194 × 125 × 256 mm
(AA34601-1 encryption unit) 194 × 58 × 320 mm
Weight:
(AA34024 control unit) 1.3 kg
(AA34001 transmitter/receiver) 6.5 kg
(AA34601-1 encryption unit) 4.5 kg

Operational status

In production.

Contractor

BAE Systems.

VERIFIED

AD3430 UHF datalink radio

The AD3430 is a UHF Link 11 compliant radio which operates in the 225 to 399.975 MHz band. It is tunable in 25 kHz increments and provides transmission and reception of clear and secure speech using amplitude or frequency modulation. Secure speech operation is provided in association with a BID 250 encryptor. The capability for providing transmission and reception of Link 11 data using frequency modulation is provided in accordance with all the general radio and UHF specific Link 11 radio requirements of MIL-STD-188-203-1A.

The equipment can deliver maximum power outputs of 20 W in AM operation and 60 W in FM operation with a transmit duty cycle of 100 per cent. The primary control of the radio is via an ARINC 429 serial data highway.

Specifications

Dimensions: 365 × 194 × 124 mm
Weight: 9.5 kg

Operational status

The AD3430 is in production for the UK Royal Navy EH 101 helicopter.

Contractor

BAE Systems.

VERIFIED

AD3500 V/UHF radio

The AD3500 was selected in 1986 as the standard tactical V/UHF radio on the UK Royal Air Force's Harrier GR. Mk 7 aircraft. The radio covers the standard VHF and UHF frequency bands, in both AM and FM, with 30 channels being preset and automatic or manual selection between AM and FM operation. The system also offers both clear and secure speech and advanced capabilities in an ECCM environment. The radio has a MIL-STD-1553 interface.

Specifications

Dimensions:
(RX/TX) 241 × 127 × 124 mm
(control unit) 152 × 146 × 57 mm
Weight:
(RX/TX) 4.5 kg
(control unit) 1.3 kg
Power supply: 28 V DC
Power output:
(AM) 10 W
(FM) 15 W

Operational status

In service in UK Royal Air Force Harrier GR. Mk 7 aircraft and Sea King Mk IV helicopters.

Contractor

BAE Systems.

VERIFIED

AMIDS Advanced Missile Detection System

BAE Systems is developing a family of advanced performance missile approach warners based on the pulsed Doppler radar detection techniques of the PVS2000. The Advanced MIssile Detection System (AMIDS) will provide extended azimuth detection coverage, detection range and warning time in comparison to present generation missile approach warning systems, and be capable of being configured for any airborne platform.

A derivative of the AMIDS system, being developed with Elettronica SpA, will form part of the Eurofighter Typhoon Defensive Aids SubSystem (DASS). The Eurofighter Typhoon missile approach warner will be configured into transmitter and receiver/processor main LRUs together with two forward and one rearward antenna assemblies. This configuration permits later integration of a passive missile launch detector, giving expanded detection performance.

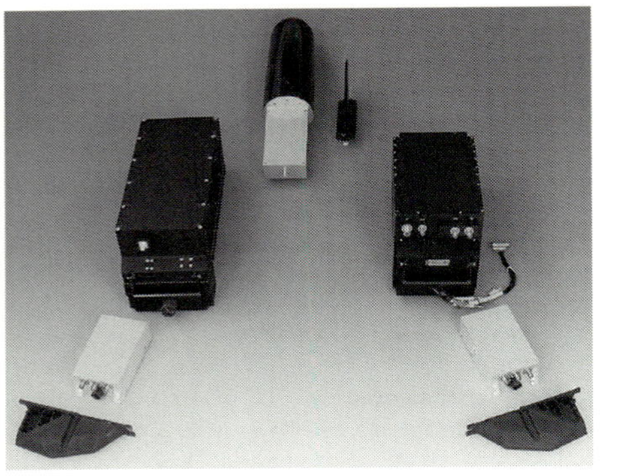

The Advanced Missile Detection System

Specifications

Dimensions:
(processor) 480 × 160 × 190 mm
(transmitter) 380 × 160 × 190 mm
Weight:
(processor) 18 kg
(antennas) 2 kg each
(transmitter) 13 kg
(low-noise amplifiers) 2 kg each

Operational status

In development for the Eurofighter Typhoon.

Contractor

BAE Systems.

VERIFIED

Apollo radar jammer

Apollo is an advanced radar jammer for aircraft self-protection. It is designed to ensure survival in hostile environments by countering a wide range of threats and is available as a pod or installed within the airframe.

Apollo comprises modules developed from in-service BAE Systems equipment. These include warning receivers, jammers and digital processors combined in a lightweight compact system. Apollo operates automatically, so that jamming management does not add to the pilot's workload.

The podded Apollo fits on a standard hardpoint and can be quickly installed and removed, giving commanders the freedom to redeploy it quickly on other aircraft.

Different performance specifications are available: from a simple repeater jammer, to a more capable system that incorporates a frequency set on receiver and noise deception capability, through to advanced single or dual DRFM options.

Apollo's jamming responses are software driven. As well as producing noise, it generates carefully modulated response patterns to confuse and deceive fire-control radars. By using various jamming techniques, Apollo is effective against pulse and CW radar signals. It is capable of dealing with multiple threats simultaneously.

Where the aircraft is fitted with a radar warning receiver such as the BAE Systems Sky Guardian, data can be passed between Apollo and the RWR to optimise intelligent interaction. This combination offers the advantages of an integrated defensive aids system and controls the decoy dispensers to fire chaff and flares at the best possible moment to co-ordinate with the jamming.

Apollo is designed to be as self-sufficient as possible. If the aircraft's own electrical system cannot generate

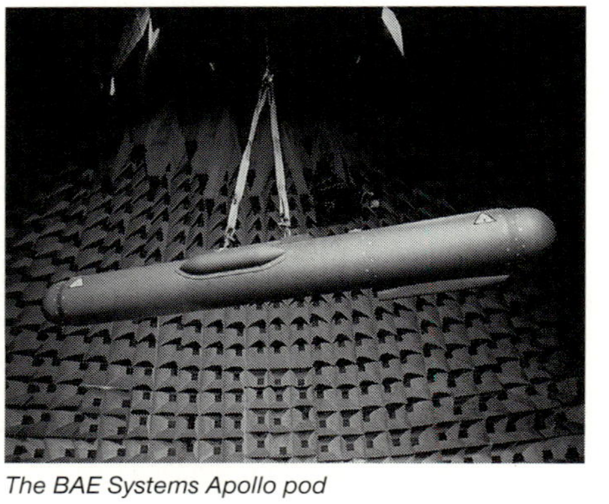

The BAE Systems Apollo pod

the power needed, a ram-air turbine is fitted to the podded variant to supply the unit.

Apollo's digital processors identify radars by comparing their parameters with the built-in EW library. The jamming technique most likely to defeat that radar is selected and used. The control system organises jamming responses in order of priority by assessing various factors. These include whether a signal is known to be a threat, whether it is locked on and its direction relative to the aircraft. Apollo can be programmed on the flight line with the latest threat intelligence.

In an aircraft fitted with a radar warning receiver, Apollo momentarily blanks out the jamming signal so that the RWR can sample the threat environment and this updated threat information is then passed back to the jammer. The operation is synchronised to maintain peak jamming performance.

Specifications

Dimensions: 2,650 (length) × 280 mm (diameter)
(length with RAT) 2,985 mm
Weight:
130 kg, 160 kg (with RAT)
40 kg (inboard)
Power supply: 115 V AC, 400 Hz, 3 phase
28 V DC
Frequency: H- to J-bands (other bands optional)

Operational status

Fully developed and flight tested.

Contractor

BAE Systems.

VERIFIED

Ariel airborne Towed Radar Decoy (TRD)

Ariel is a second generation compact, lightweight and cost-effective radar jamming system which is recoverable during or after flight for repeated operational employment on subsequent missions.

Ariel may be mounted within the airframe of the host aircraft, in a pod on a standard underwing pylon or as a scabbed-on fit to the fuselage skin. Two different types of decoy have been developed. The early first-generation device was a relatively high-power RF linked repeater decoy, which is equivalent to towing a slaved transmitter. The current second-generation version, which is now the standard operational variant in UK Royal Air Force service is a high-power datalinked decoy which can function as both a towed radar repeater and deception jammer. Both variants depend on equipment aboard the host aircraft, including the host aircraft's RWR, the EHT power supply unit and the winch control unit. The tow cable is of multi-element fibre optic construction and includes electrical conductors and strength members. The towed body includes the preamplifiers and final TWT amplifiers, the transmit antennas and the power conditioning unit. The RF-linked decoy uses the aircraft radar warners and an onboard ECM techniques generator, with the decoy interface module passing signals down an optical fibre to a photo detector in the decoy. The decoy transmits all types of ECM signals, as generated by the onboard techniques generator or the existing onboard ECM systems.

For large aircraft, the decoy is deployed and recovered using an associated onboard winch mechanism. After deployment on smaller combat aircraft, it is jettisoned before landing and may be recovered by optional parachute for reuse.

Regardless of its deployment method, the Ariel fibre optic tow cable is constructed to allow the decoy to be deployed throughout a full sortie, rather than just when a threat is detected.

Ariel is effective against monopulse, semi-active radars and home-on-jam systems. It has fully programmable countermeasures techniques and can operate either in a stand-alone mode or be integrated with an existing onboard EW suite. It has been flight proven between 150 kt and M1.2 and will be progressively cleared throughout the Eurofighter Typhoon flight envelope, as the basis of the TRD fit for Eurofighter Typhoon which is understood to include a double installation in a pod on the aircraft's starboard wingtip.

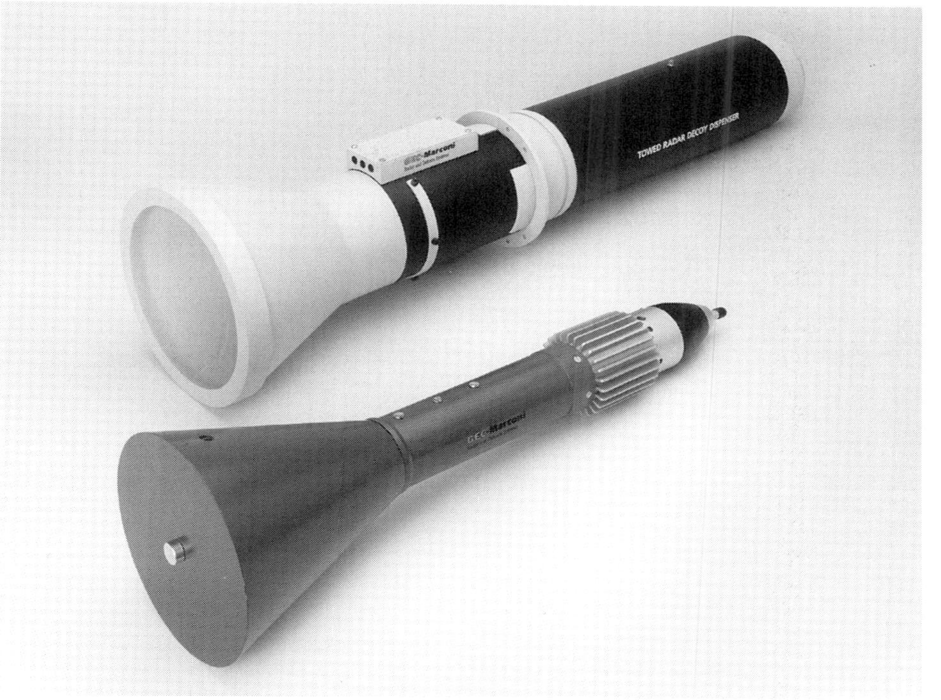

Ariel airborne towed radar decoy (below) and dispenser (above) 0018275

Specifications

Frequency: 2-18 GHz
Coverage: spherical
Techniques: noise, repeater and advanced countermeasures

Operational status

First fitted on UK Royal Air Force Nimrod MR2 aircraft in the 1991 Gulf War, Ariel has been installed on Tornado F Mk 3 aircraft, where it is fitted in a modified BOZ pod. Flight testing on a C-130 aircraft, fitted with an ALQ-131 pod as host is also reported from the US Air Force. Selected for Eurofighter Typhoon. Reports also indicate that Ariel is included in the UK Royal Air Force Jaguar '97 upgrade programme.

Contractor

BAE Systems.

VERIFIED

ASN-900 series tactical processing systems

The ASN-900 series tactical processing systems provide facilities to correlate and process data for display from the wide variety of sensors and navigation systems installed in modern maritime patrol aircraft, ASW/ASV helicopters and AEW aircraft. The system enables the tactical co-ordinator to display data in an easily assimilated form and assist in the solution of complex navigation, intercept and attack problems. ASN-902 and 924 systems, which are based on AQS-902/920 hardware, have a flexible design, which makes them readily adaptable for use as the central element in an integrated mission management system. It can replace the variety of individual sensor system

The integrated mission avionics suite in a Sea King Mk 42B

control and display units with common integrated units, providing flexibility of operation. Standard ARINC 419, 429 and MIL-STD-1553B data interfaces allow installation as original equipment or as a retrofit.

Systems integration considerably improves the efficiency and flexibility of any mission avionics suite, minimises the weight of combinations of multiple sensors and simplifies logistic and training problems with common control units and multipurpose displays. BAE Systems offers various levels of systems integration, extending to a totally integrated mission management system. The company has carried out systems integration for the advanced Sea King Mk 42B. This programme incorporates an AQS-920 Series acoustic processor and, as its core element, an ASN-902 tactical processing system.

The ASN-902 system, in service in Sea King Mk 42B helicopters, has a monochrome display, whereas the ASN-924, which is in production, has a colour display to enhance the information presented to the tactical co-ordinator. A new series, designated ASN-990, is being developed. These systems are based on power PCs and high-resolution colour displays with picture-in-picture capability.

Operational status

The ASN-902 tactical processing system is in service in the Sea King Mk 42B. The ASN-924 is in production.

Contractor

BAE Systems.

VERIFIED

Automatic Flight Control System (AFCS) for the Lynx helicopter

BAE Systems has produced several variants of the AFCS for the Lynx helicopter. The auto-stabiliser provides attitude stabilisation in the pitch and roll axes with yaw stabilisation in heading hold. These stabilisation demands are fed to limited authority series actuators, which drive the main powered flying controls. The autostabiliser is active throughout the flight envelope and operational during autopilot flying.

All autostabilisation channels are fully duplicated with monitoring facilities to indicate discrepancy between lanes. Additional pitch axis stability is provided by the inclusion of a Collective Acceleration Control (CAC) which senses normal aircraft acceleration and applies a collective pitch demand to compensate for the pitch rate divergence and instability at high forward speed and aft Centre of Gravity.

The autopilot modes of Barometric Altitude Hold, Radio Altitude, Radio Altitude Acquire and Heading Hold operate in their respective axes individually or in combination. Automatic transition and cable hold modes operate simultaneously in pitch, roll and

The control system for the Lynx helicopter includes computer, controller and drive units 0001412

collective axes. The autopilot is simplex in the pitch, roll and yaw axes but is duplicated In the collective axis. Control signals are fed into the limited authority actuators and also the full authority parallel actuators to prevent series actuator saturation during autopilot manoeuvres.

Operational status

In service with the British Army since 1975, the Automatic Flight Control System for the Lynx helicopter has been delivered to 10 different users and in all, over 600 systems have been sold.

Contractor

BAE Systems.

VERIFIED

AWARE ESM system

AWARE is a product inherited by BAE Systems from the Ferranti Company. It is not, at present, actively marketed.

The Advanced Warning of Active Radar Emissions (AWARE) range of equipments provides advanced warning of radar emissions for use in light strike and transport aircraft and helicopters or in marine craft and ground vehicles.

The system is modular in concept and design, providing variants of the basic system by simple changes of subassembly or LRUs. The expansion facilities offered provide an ESM suite and integration with ECM systems and will cover from C-band up to the millimetric region.

The flexibility of design of AWARE allows systems to be configured to match user requirements, from a basic RWR to a full ESM suite, and it is suitable for operation in a wide range of airframes, from light battlefield helicopters to high-performance aircraft. There are currently four variants in the AWARE range: AWARE-3, AWARE-4, AWARE-5 and AWARE-6. All have the same basic modules, the only difference being in the man/machine interface.

AWARE-3

AWARE-3 is the basic ESM system, covering the E- to J-bands and intended for battlefield helicopters, light tactical and logistic aircraft. The system detects and identifies pulse, pulse Doppler, CW and ICW signals. An immediate identification of the threat is provided and the particular threat type is identified.

Identification is achieved by comparison with a stored threat library. The library, threat priorities and ECM interfaces are under software control which is loaded before a mission, allowing rapid updating of threats and priorities on a mission by mission basis. Additionally, security of system and intelligence data is enhanced by the ability to clear the library and software. The system operates in peak pulse densities of several hundred thousand pulses/s, radiated by many simultaneous emitters.

The AWARE display is heading-compensated with computer-controlled symbology. An extensive symbol library can be programmed with symbols of the user's choice.

AWARE-3 consists of four planar spiral antennas, a hand-portable programme loading unit, two dual-channel crystal video receivers, a fast Instantaneous Frequency Measurement (IFM) receiver and a powerful signal processor. Control of the system can be achieved either as an autonomous system, where the control unit and a 3 in display are employed, or as an integral part of the mission management system, with control being exercised via a MIL-STD-1553B standard interface. All LRUs are convection cooled.

The four antennas are mounted in mutually orthogonal azimuth directions. The signals received are passed to the two dual-channel crystal video receivers. These receivers provide the preamplification, detection and compression of the video signals for input to the signal processor. They also provide controlled RF outputs to the IFM receiver. The output of the IFM receiver is presented to the signal processor as a digital number.

The signal processor determines the direction of arrival and time of arrival of the signals. The results of these operations, when combined with output of the IFM receiver, allows de-interleaving of the incoming signals to be performed. Once de-interleaved and characterised, the signals are classified by comparing them with a stored library of known emitters. Following classification, the source of emission and its range and bearing may be displayed in simplified plan form on the monitor or passed to the aircraft mission management system via a monitor or via a MIL-STD-1553 databus.

AWARE-4

AWARE-4 is an ESM system for maritime helicopter applications. The right-hand screen retains the PPI presentation of emitters to maintain the radar warning role, while the left-hand screen is a tote for the display of emitter parameters.

AWARE-5

AWARE-5 retains the AWARE-3 control unit but has no display. In this variant, the data are displayed on either the aircraft HUD or the AI radar display, via the MIL-STD-1553B or ARINC 429 databusses. AWARE-5 is for use in light strike aircraft.

AWARE-6

AWARE-6 is similar to AWARE-4, but the display unit is replaced by two 10 in colour monitors. AWARE-6 is for use in maritime patrol aircraft operations where more space is available and a larger tote that can display the whole track table is required.

Specifications

System weight:
(AWARE-3) 13 kg
(AWARE-4) 15 kg
Power supply: 28 V DC
150 W (AWARE-3)
180 W (AWARE-4)
Frequency: 2-18 GHz
Accuracy: better than 10° RMS

Operational status

In production for customers in the UK and overseas. AWARE-3, under the designation ARI 23491 (Rewarder), equips Lynx and Gazelle helicopters of the British Army. AWARE-3 also equips Lynx helicopters of the Royal Netherlands Navy. Systems are being used for both airborne and shipborne applications.

Contractor

BAE Systems.

VERIFIED

Blue Fox interception radar

Blue Fox is a lightweight radar designed to fulfil the dual roles of airborne interception and air-to-surface search and strike. It was developed to form part of a fully integrated weapon system.

Blue Fox operates in the I-band and uses frequency agility to enhance the radar's immunity to ECM and improve its ability to detect small targets in bad weather or rough sea states.

For air-to-air interception, Blue Fox can be used for lead-pursuit or chase attacks and incorporates a transponder mode for identifying friendly aircraft or ships.

Blue Fox is built on the LRU principle. Each component part of the radar, such as the transmitter,

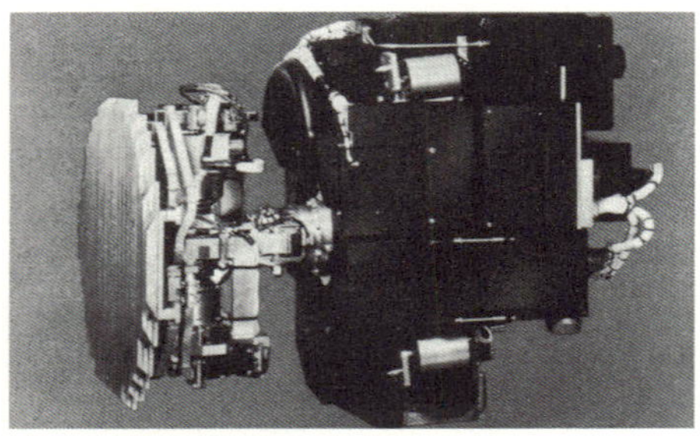

The BAE Systems Blue Fox radar

The AWARE ESM system is in service in Lynx and Gazelle helicopters

receiver, processor and amplifier, can be checked easily or removed independently for servicing. The antenna is a flat aperture slotted array, stabilised in pitch and roll.

The radar display provides a bright digital scan-converted picture. Superimposed on the display are the flight symbols showing aircraft altitude, speed, heading and so on, so that the pilot can monitor and control the aircraft's manoeuvres while using the radar.

Operational status

In service in the BAE Systems Sea Harrier FRS.1 of the UK Royal Navy and the Indian Navy. Deliveries are complete.

Contractor

BAE Systems.

VERIFIED

Blue Hawk radar

The Blue Hawk is a new I-band lightweight, coherent, multimode pulse Doppler radar designed for both new lightweight fighters and for the upgrade market. It employs low-, medium- and high-PRF waveforms and offers multiple functions for air interception, close air combat, air-to-ground and anti-ship operation. Aimed at today's typical threats, Blue Hawk has substantial processing capacity to allow it to be matched to a customer's particular requirements, and to adapt to future changes in the threat.

Blue Hawk weighs 107 kg and consists of four main elements; transmitter, receiver/exciter, combined display and data/signal processor, and the antenna. The modular design allows flexibility of installation, especially in platforms where space may be restricted. Blue Hawk is fully compatible with MIL-STD-1553B databus and may be integrated readily with a total avionic system. It is compatible with a wide range of current weapons and ordnance. A major feature of the design philosophy is the wide use of standard components and conventional manufacturing technologies. Blue Hawk employs a fully programmable digital unit comprising signal, data and display processing, with 50 per cent spare throughput and 50 per cent spare memory capacity for growth. It is optimised for HOTAS control. A number of inherent design features minimise the effects of ECM.

Specifications

Frequency: I-band (9.6-9.9 GHz) frequency agile over multiple channels
Antenna: planar-array mechanically scanned +60° in azimuth and +60° in elevation. Antenna size can be varied to suit installation
Receiver/exciter: 2-channel receiver plus frequency generator
PRF: wide range from (low PRF) 800 Hz to (high PRF) 120 kHz
Power output: 8 kW peak; 160 W mean, a 400 W option is available
Detection performance: 51 n miles in look-up mode; 30 n miles in look-down mode; out to 80 n miles in ground-mapping mode
CW illuminator: can be offered as an integrated option for MRAAM control
Weight: 107 kg
Power requirements: 2.5 kW, 400 Hz, 3 phase

Operational status

Blue Hawk has completed the ground proving stage of development and has undergone a flight trials programme within the UK.

Contractor

BAE Systems.

VERIFIED

Blue Kestrel maritime surveillance radars

The Blue Kestrel surveillance radar family consists of the Blue Kestrel 5000, 6000 and 7000. They are allied to the Seaspray family of radars, using common modules throughout the range. This minimises cost and provides natural growth paths for the customer.

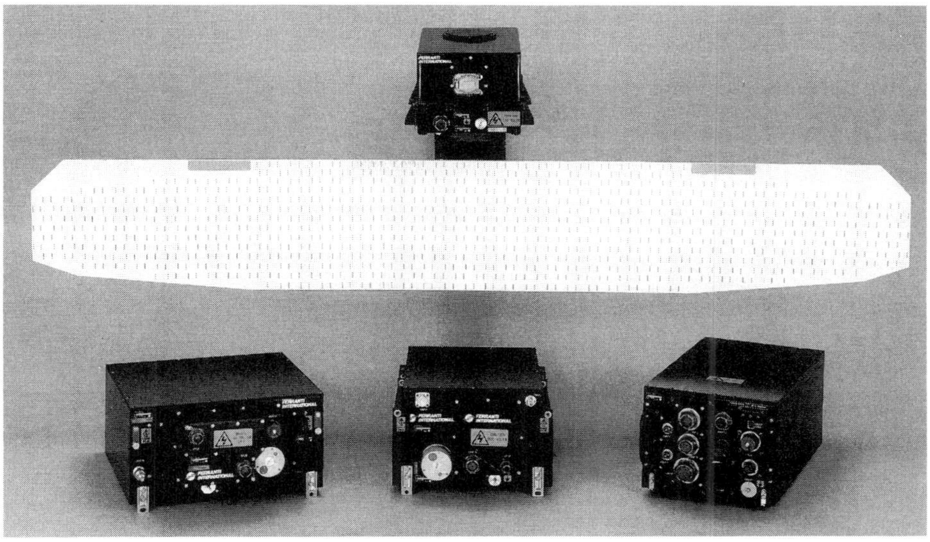

The main units of the BAE Systems Blue Kestrel 5000 radar

The EH 101 Merlin helicopter is being fitted with the BAE Systems Blue Kestrel 5000 radar, and may receive the Blue Kestrel 7000 in future capability upgrade

Blue Kestrel 5000 maritime surveillance radar

Blue Kestrel 5000 is a pulse compression radar sensor developed for the UK Royal Navy EH 101 Merlin helicopter. It consists of a large flat plate antenna and Travelling Wave Tube (TWT) transmitter to provide the Merlin with the optimum power aperture product, combined with a high-gain receiver and digital processor. The radar sensor is interfaced with and controlled by the platform's mission avionics management system via the MIL-STD-1553B databus.

This highly integrated sensor approach to the radar results in a high-performance pulse compression surveillance radar in a particularly compact and lightweight four-line replacement unit configuration. It is suitable for application in a range of sophisticated naval helicopters or maritime patrol aircraft.

Specifications

Scanner type: planar-array
Frequency: I-band
Transmitter: low-peak power, high-mean power TWT
Pulsewidths: selectable
PRF: selectable
Coverage: 360°
System weight: 102 kg
Features: pulse compression, CFAR, multiple TWS and operator-selectable scan-to-scan integration

Operational status

Blue Kestrel 5000 has completed development, and production deliveries have commenced. The development phase included the building of 10 systems, all to full flying standard, and the completion of several thousand hours running time in ground-based rigs, two specially converted Sea King flying testbeds, and the EH 101 Merlin prototype. Lockheed Martin ASIC, the Merlin programme prime contractor, placed the production contract for Blue Kestrel 5000 in October 1992.

Blue Kestrel 6000 maritime surveillance radar

Blue Kestrel 6000 is a coherent pulse Doppler radar sensor. It is conceived as a modular upgrade to the Blue Kestrel 5000 sensor, where the only major change required is to insert the pulse Doppler upgrade module. It brings together BAE Systems experience with the pulse compression Blue Kestrel 5000, the multimode pulse Doppler Blue Vixen, and Inverse Synthetic Aperture Radar (ISAR) work, which has been jointly funded by the UK Defence Evaluation Research Agency (DERA) and BAE Systems.

Flight trials of a modified Blue Kestrel 5000 in a DERA Sea King testbed have been successfully completed, producing real-time images of co-operative and opportunity surface contacts from a helicopter. The resulting radar sensor provides enhanced air-to-surface and air-to-air detection, standoff classification capability, and a superior anti-submarine warfare capability with sub-clutter target detection. Additional options include moving target detection and high-resolution synthetic aperture radar ground mapping.

Specifications

Frequency: I-band
Type: coherent multimode
Transmitter: TWT fixed frequency/frequency agile
Scanner type: planar-array
Coverage: 360°
LRUs: 4
System weight: 125 kg

Features: CFAR, enhanced TWT, ISAR classification, ASW and air-to-air modes
Options: MTI and SAR ground mapping

Operational status

Blue Kestrel 6000 development is complete.

Blue Kestrel 7000 maritime surveillance radar

Meanwhile, a more comprehensive upgrade, designated Blue Kestrel 7000, is nearing development completion. Blue Kestrel 7000 will have a full pulse Doppler, multifunction, COTS-based, open architecture, reprogrammable capability (the same capability can also be accommodated within the smaller Seaspray radar, where it will be designated Seaspray 7000). Blue Kestrel 7000 is planned for a future EH 101 capability upgrade programme.

Contractor

BAE Systems.

VERIFIED

Blue Vixen radar

Blue Vixen is a lightweight multimode, coherent, pulse Doppler airborne interception radar operating in I-band. It is a true multimode radar, maintaining full power in all modes, with all-weather operation in look-down and look-up modes and over-sea and over-land detection of targets. It incorporates high-resolution ranging in air-to-air as well as its land and sea search modes.

Blue Vixen operates in low-, medium- and high-PRF modes. Selection of the appropriate PRF for optimum detection is automatic, and depends on background clutter and target density. Low PRF is used in the look-down mode to provide accurate range and velocity with all-aspect detection. High PRF provides for the look-down detection of targets approaching at high speed in a high-clutter environment.

Automatic track-while-scan and single-target track are available in air-to-air modes with electronic counter-countermeasures, and air combat modes. It is claimed to be the first AI radar in the world to be designed from the outset with full AMRAAM compatibility. It is also compatible with other MRAAM, Sea Eagle and Sidewinder.

Blue Vixen has been designed with a flexible LRU configuration. The antenna, receiver and transmitter are co-located in the nose bay while the radar power unit and the processor can be conveniently located nearby. Weight of the complete system is 145 kg.

Operational status

Currently in production for, and in service on, the Royal Navy Sea Harrier F/A Mk 2 aircraft.

A successful programme of trial firings of AMRAAM from a Sea Harrier fitted with Blue Vixen took place in 1993-94 in the USA. Blue Vixen enables up to four AMRAAM missiles to be ripple-fired, maintaining guidance datalink while simultaneously tracking the targets.

Digital flight computer for the F-14 0001410

In November 1998, the UK MoD contracted BAE Systems to implement JTIDS on Sea Harrier F/A Mk 2 Blue Vixen radar; development and integration work is planned for the period 1998-2004.

Contractor

BAE Systems.

VERIFIED

Digital Flight Control System (DFCS) for the F-14

A digital version of the existing Flight Control System (FCS) has been developed by BAE Systems for the F-14 Tomcat under the US Foreign Comparative Test Program.

This new digital system will replace the analogue version in form, fit and function but with an improved set of high angle of attack and power approach configuration flight control laws combined with a redundancy management scheme which provides a fail operational/fail safe capability.

The control laws for the power approach configuration are designed so that there is virtually no requirement for the pilot to provide co-ordinating rudder pedal inputs in response to lateral stick deflections and the redundancy management scheme has the aim of providing a near fail operational architecture.

Each of the three DFCSs are of similar design but each contains specific aircraft interface and software functions appropriate to its function. The DFCSs fulfil the existing FCS functions and contain growth provision to accommodate additional specified functions, if required later.

Operational status

On 28 March 1996, the US Navy signed a not to exceed price for the first phase of the contract to complete the development and manufacture of 93 ship-sets of digital flight control computers and control panels. Further phases of the contract are expected to cover the manufacture of an additional 151 ship-sets for other variants of the F-14 aircraft, and also to provide US support for the programme.

Evaluation and flight test programme continues.

Contractor

BAE Systems.

VERIFIED

FIN 1000 series inertial navigation systems

FIN 1000 is the designation of a family of inertial navigation systems based on the BAE Systems gimballed inertial platform that use floated rate integrating gyros and precision force feedback accelerometers. The group includes particular systems optimised for long-term high accuracy and for rapid reaction alignment. Versions include:

FIN 1010 Developed for the Panavia Tornado, the FIN 1010 has all-digital interfaces and an accuracy of better than 1 n mile/h. Another version of the Tornado system, with analogue interfaces and comprehensive route navigation, is fitted to the Mitsubishi F-1 advanced trainer.

FIN 1012 Fitted to UK Royal Air Force Nimrod MR. Mk 2 aircraft, the FIN 1012 is optimised for long-term accuracy.

FIN 1031 The FIN 1031 Navigation Heading and Attitude Reference System (NavHARS) is fitted in UK Royal Navy Sea Harrier F/A-Mk 2 aircraft. The inertial platform is stabilised by twin two-axis ruggedised oscillogyros developed by BAE Systems. NavHARS utilises other aircraft sensors, which provide Doppler radar velocities, true airspeed and flux valve magnetic heading.

A 2-minute alignment can be achieved on land or sea with further refinements when the aircraft is airborne.

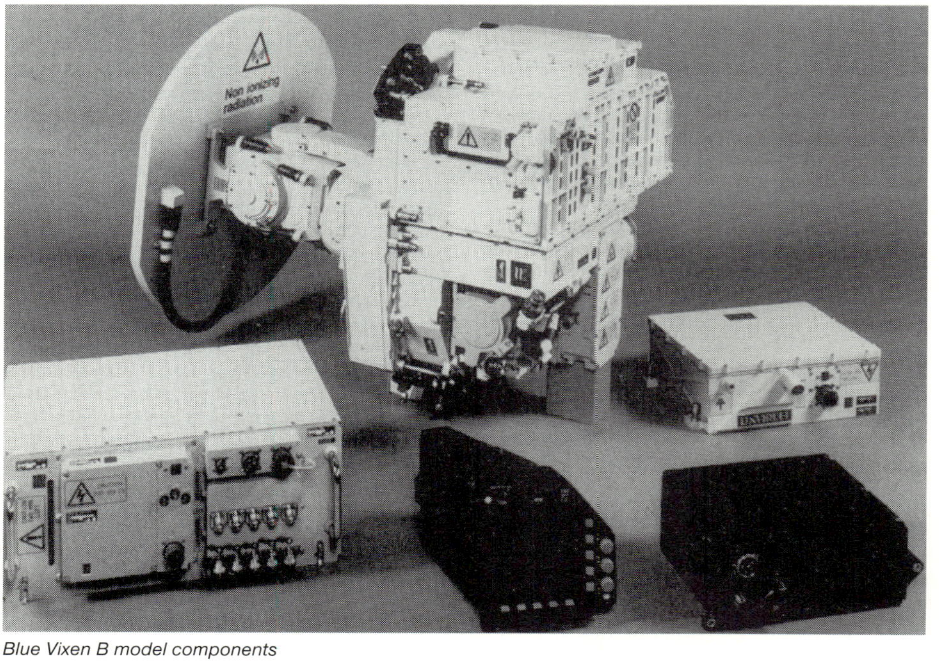

Blue Vixen B model components

The FIN 1031B NavHARS has been fitted as part of the mid-life update installation in the UK Royal Navy Sea Harrier F/A Mk 2. It is similar to the FIN 1031 system but has two independent dual-redundant MIL-STD-1553B databusses. The addition of a twin MIL-STD-1553B databus is primarily to facilitate interfacing with the Blue Vixen radar, AMRAAM and revised avionics such as the new bus control interface unit.

Specifications

Inertial platform unit
Dimensions: 212 × 215 × 332 mm
Weight: 11.9 kg

Processor unit
Dimensions: 261 × 199 × 381 mm
Weight: 13.64 kg

Control/display unit
Dimensions: 147 × 152 × 139 mm
Weight: 2.5 kg
Power supply: 200 V AC, 400 Hz, 3 phase
28 V DC for switching and lighting

FIN 1064 FIN 1064 is an integrated navigation and attack system, providing inertial navigation and a wide range of weapon delivery modes. The system has been in service since the early 1980s in the UK Royal Air Force Jaguar ground attack/reconnaissance aircraft, and is also fitted to Jaguars of the Omani and Ecuadorian air forces. Navigation data is provided by a FIN 1000 series gimballed inertial platform. The system was fitted in a mid-term upgrade of the aircraft systems and provides a suite of analogue and digital interfaces to the aircraft sensors and control/display systems; it also provides the capability to upgrade the system software at LRU level using a portable, solid-state, programme loader. Weapon aiming computation is provided for both air-to-air and air-to-ground modes. More recently, FIN 1064 has been updated to incorporate a MIL-STD-1553B databus, integration of inertial and GPS navigation data, and integration with the TIALD (Thermal Imaging and Laser Designator) pod. The mission and weapon-delivery capability of the system has been significantly enhanced by in-service software upgrades, in line with the expanding role of the Jaguar aircraft.

FIN 1075 The FIN 1075 inertial navigation system was selected for the Harrier GR. Mk 5 and GR. Mk 7 and is currently in service with the UK Royal Air Force. The system is form, fit and function interchangeable with the AN/ASN-130 navigation system in the US Marine Corps AV-8B. The inertial platform interfaces with the Harrier GR. Mk 7 databus, avoiding the need for a dedicated control/display unit.

The inertial platform can be aligned on land, at sea or in flight. The ground alignment mode has a wander azimuth and does not require an initial heading input.

Specifications

Dimensions: 193 × 286 × 356 mm
Weight: 20 kg
Reaction time: <3 min
Gyrocompass align: 7 min for 0.8 n mile CEP
Accuracy:
(navigation) 0.8 n mile CEP
(heading) ±0.1°
(attitude) ±0.1°
Reliability: 1,500 h MTBF

FIN 1075G The FIN 1075G is a variant of the FIN 1075 which is designed to operate in conjunction with a stand-alone GPS receiver to provide a continuous, precision navigation solution under high dynamic conditions with significant periods of GPS outage.

FIN 1075G has been evaluated at Boscombe Down and has been used extensively in overseas Harrier GR. Mk 7 operations.

The same system is also capable of a GPS aided moving base alignment and is currently being assessed to provide the Harrier GR. Mk 7 with an 'at sea' alignment capability.

Contractor

BAE Systems.

VERIFIED

FIN 1110 two gimbal inertial navigation system

The FIN 1110 is a lightweight and compact inertial navigation system for helicopters and maritime and transport aircraft. The system is contained in a ½ ATR short box weighing less than 10 kg and costing about one third the price of a full inertial navigation system. Complexity is avoided by the use of only two gimbals (hence the designation two gimbal INS) in place of the four needed in aerobatic aircraft. At the same time the system has a strapdown azimuth sensor gyro. The adoption of a gimballed platform rather than a strapdown system for this purpose has several advantages. The most significant is the rapid and independent alignment capability, the system finding true north to an accuracy of 0.4° within 4 minutes of powering up. The system is thus freed from the inherent errors caused by variations in the earth's magnetic field. Continuing alignment in the air permits operation from moving platforms such as ships.

A 15-state Kalman filter is used providing a much greater degree of accuracy than could be obtained by either Doppler or INS in isolation.

FIN 1110s operate as vertical and heading reference systems in conjunction with the Racal Defence Electronics Searchwater radar which equips the Sea King AEW helicopters of the UK Royal Navy.

Specifications

Dimensions: 334 × 125 × 194 mm
Weight: 9.8 kg
Alignment time: 2-5 min

Operational status

No longer in production. In service in UK Royal Navy Sea King AEW helicopters to stabilise the radar antenna and to improve navigation performance and in the Swedish Air Force Super Puma.

Contractor

BAE Systems.

VERIFIED

FIN 3110G ring laser gyro INS/GPS

The FIN 3110 INS/GPS is designed to meet the requirements of military aircraft, helicopter, self-propelled howitzers, artillery, land vehicles and marine craft.

The FIN 3110 is an Integrated Navigation System (INS) consisting of a ring laser gyro inertial sensor and an embedded Global Positioning System (GPS) receiver module. This system is capable of providing precise and continuous outputs of navigation heading and attitude data to the weapon and flight control systems.

The FIN 3110 is a small (177.8 × 177.8 × 279.4 mm), lightweight (10.5 kg) unit consuming just 55 W from a 28 V DC power source. The INS and GPS are closely integrated in a Kalman filter, which combines the high position accuracy of the GPS receiver with the angular rates and linear acceleration of the INS on a 1553B databus. Other interfaces could be available if required. The applications software is written in Ada and is executed on a Motorola 68040 processor.

The FIN 3110 was the culmination of 15 years of research and development of ring laser gyro technology. The advent of the single module GPS receiver and processor in 1993 allowed the integrated navigation system to be manufactured for the UK MoD and export programmes.

Specifications

Alignment times:
(gyrocompass) 4 min
(rapid reaction) 30 s
Inertial performance:
(position) <0.8 n mile/h CEP
(velocity) (N,E) <2.5 ft/s RMS
(velocity) (vertical) <2 ft/s RMS
GPS performance:
(position) (spherical error) <16 m SEP
(velocity) (per axis) <0.1 m/s RMS
MTBF: >5,000 h
Power: 28 V DC, <55 W
Dimensions: 177.8 × 177.8 × 279.4 mm
Weight: 10.5 kg

Environmental requirements:
MIL-E-5400T;
Temperature altitude operation to Class 2 (optionally Class 2X for forced cooling);
Tested to MIL-STD-810E (temp, altitude, vibration), MIL-STD-461C (EMC), MIL-STD-704E (power supply)
Interfaces:
1 or 2 dual-redundant MIL-STD-1553B (RT or bus control); PTTI/Have Quick; RF for GPS antenna; Dual RS-422 instrumentation; ARINC 429; Synchro/analogue; Panlink

Operational status

In production.

Contractor

BAE Systems.

VERIFIED

FIN 3110 GTI

The FIN 3110 GTI (GPS, Terrain, Inertial) is a variant of the FIN 3110 navigation system designed for standoff missiles and aircraft.

The system is based on a Ring Laser Gyroscope (RLG) inertial system capable of autonomous operation to 0.8 n mile/h and includes two sophisticated Kalman filters for the provision of horizontal and vertical integrated navigation solutions using additional sensor data. In addition to the basic inertial system the unit has the capacity to be equipped simultaneously with an embedded military GPS receiver, a Digital Terrain System (DTS) and special-to-type analogue and/or digital interfaces.

The GPS receiver can be either military code or civil C/A code.

The DTS module provides both terrain referenced navigation and terrain following functions and is comprised of: terrain elevation data storage memory (EEPROM); mission specific digital terrain elevation data; data processing hardware; terrain-referenced navigation update algorithms; and terrain-following algorithms providing steering commands.

A feature of the FIN 3110 GTI is the direct use of GPS and TRN measurement data in the same integrated navigation Kalman filter to give improved performance and increased robustness through improved sensor cross-mounting.

The overall size of the FIN 3110 GTI is the same as the basic FIN 3110.

Contractor

BAE Systems.

VERIFIED

Fly-by-wire control stick assemblies

BAE Systems has produced high-integrity position sensors for a wide range of aircraft including the pilot's stick sensors and the rudder pedal position sensors for the Tornado aircraft; pilot position sensors for the AM-X; and 'relais' jack sensors for the Concorde AFCS. This range of sensors has now been extended to include fly-by-wire passive control sticks for the Eurofighter Typhoon and F-22 in centre and side console configurations respectively.

Eurofighter Typhoon centre stick controller

The 'passive' unit utilises springs and dampers to provide feel to the pilot and contains high integrity electrical position sensors. They are configured as two axis displacement units utilising multiredundant position sensing to provide high integrity input demands to the flight control system. No mechanical outputs are provided from these passive units and the force and damping characteristics are fixed. Mass balancing can also be specified which minimises stick movement caused by aircraft acceleration (see also International Section for details of consortium partners).

F-22 side stick controller

The modular construction provides easy maintenance and also enables the force/displacement characteristics of the units to be tailored to meet the particular aircraft requirements. All of the units are high integrity, flight safety critical items manufactured to the highest standards using class 1 components and manufacturing techniques. Careful choice of materials ensures the strength-to-weight ratio is optimised to meet aircraft weight constraints, provide a robustness capable of withstanding an applied pilot load in excess of 200 lb and still give reliable precision manoeuvring throughout the flight envelope and life of the aircraft.

Further developments

A design currently under development is an electrically back-driven active pilot controller which will allow adjustable spring rates, breakout forces, detents and damping ratios to provide situational awareness to the pilots at all times. With feel characteristics capable of being tailored to suit individual applications, it will also permit electrical connection between two sticks such that if one stick is moved, the other will follow.

Active throttle units using the same technology and possessing a high degree of commonality with the active stick unit are also under development.

In aircraft with a requirement for dual sticks, such as a civil aircraft, helicopter or military trainer, and fitted with a full-time fly-by-wire system, linked active control sticks can be provided, packaged to suit the individual requirements.

Operational status

Stick assemblies are being manufactured and supplied as part of the Eurofighter Typhoon and the F-22 programmes.

The most recent development is for the supply of active stick units for the US Joint Strike Fighter programmes for both Lockheed Martin and Boeing.

Contractor

BAE Systems.

VERIFIED

F-22 side-stick controller 0001413

Foxhunter airborne interception radar

Foxhunter is the multimode airborne interception radar for Tornado F Mk 3. A substantial part of the signal processing is performed digitally in addition to digital radar data handling. The equipment design anticipated trends in offensive tactics, such as low-level penetration and use of ECM, and the latest improvements give growth potential into supporting active missiles. Additionally, Foxhunter has the flexibility to operate as part of ground- or AEW-based control environments while retaining the ability to perform autonomously.

Foxhunter is designed to detect and track subsonic and supersonic targets at ranges in excess of 185 km at both low and high level.

Foxhunter operates in I-band (3 cm), and in its primary mode uses the pulse Doppler Frequency Modulated Interrupted Continuous Wave (FMICW) technique. At the heart of the radar is a master timing and synchronising unit, employing phase-locked loop techniques, which generates the complex waveforms for amplification by the transmitter, and the accurately related reference signals and precise timing pulses employed within the receiver and signal processing circuits.

Compact and lightweight surface acoustic wave devices, provide signal waveforms for the pulse compression modes of the radar. The antenna uses the 'Elliot' twist-reflecting cassegrain principle which combines rigidity and light weight with extremely low levels of spurious radiation lobes, a key feature in the rejection of ground clutter and jamming signals. This type of antenna has been developed to a very advanced level of performance. The scanner employs a hydraulic drive mechanism and very high-grade servos to achieve the speed and precision of beam pointing and stabilisation demanded especially during manoeuvre.

The TWT transmitter has the inherent flexibility to handle the differing waveforms used in the various modes of operation. The heart of the signal processing system is a digital processor employing Fast Fourier Transform techniques of frequency analysis to filter the signal returns into narrow frequency channels. Target echoes are segregated from clutter returns and a particularly uniform detection threshold is achieved against all target velocities.

Foxhunter also provides target illumination for the Sky Flash medium-range semi-active radar air-to-air missiles and, since a significant ECM operating environment can be expected, the radar system incorporates many EPM features.

Product Improvement Programme

Since Foxhunter was introduced into service, the product improvement programme has been defined in two main stages. Stage 1 introduced numerous modifications and permitted radar functions that were closely tied to HOTAS modes. The TWS and EPM algorithms were also refined. The Stage 2 standard introduces a number of major improvements, including a completely redesigned data processor and supporting software, as well as significant modification to both the transmitter and receiver paths. The resultant effect is a much higher quality of TWS performance in a high RF noise environment. The new processor also enables sophisticated automatic and manual scan management techniques to be employed in TWS; it is also capable of handling a significantly greater number of tracks. Additionally, it provides scope for significant growth potential beyond Stage 2, including the ability to support active radar missiles.

The latest installed standard in British and Italian Tornado F3 aircraft is Stage 2G+. Stage 2G+ was developed as a result of lessons learned in the 1991 Gulf War, and included much more automation of search techniques. Royal Saudi Air Force aircraft remain at Stage 1 standard.

The latest development standard is designated Stage 2H. Development of Stage 2H is complete, and deployment in all British and Italian Tornado F Mk 3 aircraft was planned to begin at the end of 1999. Stage 2H incorporates maintenance and support improvements and mechanical interfaces for further pre-planned hardware product improvements.

Operational status

Foxhunter is in operational service with the UK Royal Air Force, the Royal Saudi Air Force and the Italian Air Force. Stage 1 was introduced into service in 1989. The latest development standard, Stage 2H is entering

BAE Systems Foxhunter radar installed in the nose of a UK Royal Air Force Tornado F Mk 3 aircraft

service with the UK Royal Air Force and the Italian Air Force.

BAE Systems is working on a concept proving programme to provide the UK Royal Air Force with Augmented Logistic Support (ALS) for the Foxhunter radar.

Contractor

BAE Systems.

VERIFIED

Hermes ESM system

Hermes is an EW surveillance system for the interception and analysis of radar signals, providing data for planning operations and for defence. It can form part of an air defence network, both in the maritime surveillance role and from land-based sites.

In the airborne mode, Hermes can be configured from a basic radar warning receiver to a complete ESM system for the AEW role and reconnaissance. It can also be fitted into UAVs.

Hermes intercepts radar signals at long range using wideband superheterodyne receivers that give unrivalled handling of continuous wave and high-duty cycle signals. The modular design means that Hermes can be installed in a wide variety of platforms and can be fitted inboard on helicopters and medium-size aircraft or in a pod on fast jets.

Hermes is fitted with interfaces for navigation, data exchange and the automatic triggering of electronic countermeasures. These interfaces can be tailored to specific platform requirements.

Specifications

Weight: 110 kg approx
Frequency: C- to J-bands
Signal types: pulse, CW, ICW, RF agile, PRI agile, jammers
Coverage:
(azimuth) 360°
(elevation) ±45°
Emitter library: >1,000 modes

Operational status

The system is in service with the Indian Navy on its Westland Mk 42B Sea King helicopters.

Contractor

BAE Systems.

VERIFIED

HIDAS Helicopter Integrated Defensive Aids System

Founded on the needs of the demanding attack helicopter role - where short engagement ranges demand rapid reaction times - HIDAS components are designed to operate together to provide radar, laser and missile threat warning, together with advanced radar and IR countermeasures and off-board decoys.

Component systems of HIDAS are as follows:

(1) RWR radar warning receiver: Sky Guardian 2000 RWR from BAE Systems; Sky Guardian 2000 covers the E-J band, with optional frequency extensions downwards to C/D bands, and upwards to K band.
(2) LWR laser warning receiver: Series 1223 LWR from BAE Systems; with full beam-rider detection capability.
(3) CMWS common missile warning system: AN/AAR-57(V) from Sanders, a Lockheed Martin Company; the AN/AAR-57(V) is the Common Missile Warning System (CMWS) of the ATIRCM/CMWS; each CMWS set provided for the HIDAS system will include an electronic control unit (ECU), which processes threat data, and four passive electro-optic missile sensors, being produced by Lockheed Martin Infra-red Imaging Systems.
(4) RF radar frequency jammer: Apollo RF jammer from BAE Systems; covering the H-J band, with optional extensions down to E-G bands, and upwards to K band.
(5) CMDS CounterMeasures Dispensing System: Vicon 78 series 455 from W Vinten Ltd; comprising three dispensers, one for chaff and two for flares. The chaff will be fired upwards towards the tail rotor. Flares will be fired from dispensers on the rear fuselage. Marconi Electronic Systems envisages that HIDAS will employ 'smart flares' emerging from current R&D programmes to improve IRCM capability.
(6) DIRCM directional infra-red countermeasures from the current NEMESIS international programme to provide bands I-IV IR band jamming.

To achieve the high degree of integration essential to timely reaction, the DAS management module is incorporated within the Sky Guardian 2000 RWR. This enables data from all the DAS sensors (RWR, LWR, CMWS) to be correlated against the large threat library. The DAS manager then determines and automatically responds with the optimum mix of radar, IR and off-board countermeasures. The DAS communicates with the aircraft mission system over a MIL-STD-1553B databus and can display data or system status reports on the aircraft multifunction displays or on dedicated EW displays.

By combining and correlating data from all its sensors, HIDAS is able to present to the crew a comprehensive picture of the surrounding environment. Thus, crews have the option to take pre-emptive action; for instance, the HIDAS detection can be used to cue the on-board weapon system.

HIDAS is configured to ensure radar, laser and missile threats are detected, identified, declared to the crew and, where appropriate, countermeasures automatically instigated, in under two seconds - typically one second from threat detection.

HIDAS automatically selects the appropriate form or combination of radar jamming, IR jamming, chaff or flare dispensing. Where the crew require it, the countermeasures can be restricted to semi-automatic operations and a manual override is available at all times.

HIDAS is fully user reprogrammable at the flight line and provision is made for a mission library of up to 4,000 threat modes. Individual modes can be linked readily to enable the positive identification of weapon systems.

A 'smartcard' PCM port can be provided on the control unit to facilitate the loading of pre-flight messages. The large capacity of these cards allows the same device to be used for recording in-flight mission data. Software loading and recording functions can also be carried out by aircraft systems over the MIL-STD-1553B Bus.

The BAE Systems MERLIN is a PC-based software support system that enables the user to quickly create their own mission libraries, specify the required countermeasures and carry out post flight analysis of mission data.

Control of HIDAS is exercised either through a simple dedicated control unit with associated display, or through 'soft' controls on multifunction displays via the aircraft mission computer.

HIDAS supports both MIL-STD-1553B and RS-422 interfaces, and provides a combined situation picture and status report for display to the crew on the aircraft multipurpose display.

Operational status

Selected for the British Army WAH-64 Apache attack helicopter programme. Successful flight trials of the HIDAS aboard the WAH-64 were conducted in mid-2000.

Contractor

BAE Systems.

UPDATED

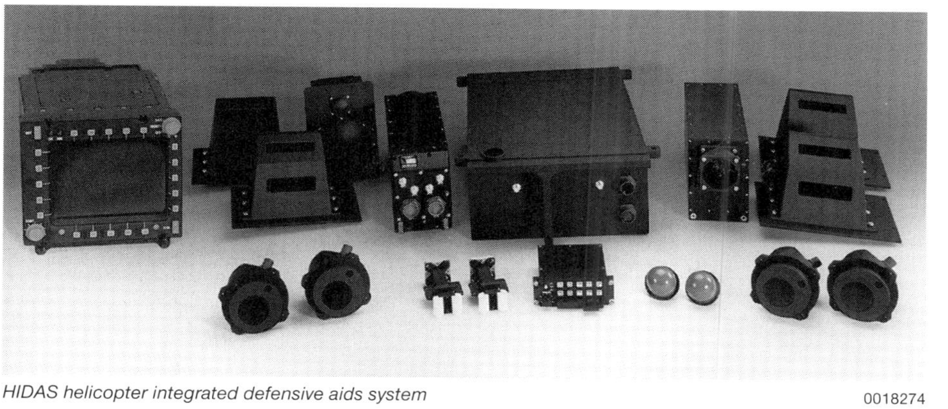

HIDAS helicopter integrated defensive aids system 0018274

Infra-red jammer

The BAE Systems infra-red jammer is designed to provide protection against missiles using an infra-red seeker head. It is intended to be mounted externally on a helicopter and operates by radiating modulated energy from an infra-red source. The output from the jammer is modulated mechanically by means of shutters. The output signal enters the missile seeker head, impairing its ability to track. The IR source is a graphite radiating element which is hermetically sealed within a sapphire envelope.

The optical assembly, which surrounds the lamp, rotating at high speed, passes between the lamp and slots in the outer drum. The drive to the optical assembly comes directly from a brushless DC motor integrally contained within the jammer mounting structure.

Surrounding the optical system is a cylinder with 16 longitudinal slots spaced around its circumference. The slots are covered by a window, which provides environmental protection to the working parts and also serves to transmit IR wavelength radiation and to block visible emissions. The jammer is therefore covert in the visible spectrum.

The modulated IR signal emitted by the jammer has the effect of degrading the tracking ability of IR-guided missiles beyond the point at which they can be effective. The jammer radiates adequate power in the spectral band used by the seeker head to protect most helicopter types.

Field of view of 360° in azimuth and ±30° in elevation can be provided. These apply to the jammer prior to its installation on a helicopter. The pilot requires only a simple on/off control; two signals are provided to monitor lamp and motor operation. Weight of the complete system is 18 kg.

Operational status

In service.

Contractor

BAE Systems.

VERIFIED

LINS 300 Laser Inertial Navigation System

BAE Systems manufactures a family of LINS 300 SNU 84-1 inertial navigation equipment which meets the requirements of many different types of aircraft. LINS 300 provides vehicle acceleration, velocity, position, heading and attitude. Outputs are available in MIL-STD-1553B, ARINC 429 and synchro formats. For the EH 101 Merlin, LINS can align when airborne using a Kalman filter to interface with a Doppler velocity sensor or GPS; this also enables shipborne alignments. An ARINC 429 interface and Doppler/GPS air alignment mode are unique among SNU 84-1 format systems.

LINS 300-10 is the baseline SNU-84 system used on fixed-wing aircraft and helicopters. LINS 300-20 is the baseline ARINC 429 system with air alignment.

Specifications

Dimensions: 460 × 191 × 194 mm
Weight: 20 kg
Alignment time:
(gyrocompass) 8 min
(stored heading) 1.5 min
Accuracy:
(position) 0.8 n mile/h CEP

(velocity) (horizontal) 2.5 ft/s
(vertical) 2 ft/s
(rates) 400°/s

Operational status

In 1990, LINS 300-11 was selected as baseline equipment for the Hawk 100 and 200 series aircraft, and is in service with three air forces. In 1995, LINS 200-21 was also selected for the UK Royal Air Force utility variant of the EH 101 helicopter.

Contractor

BAE Systems.

VERIFIED

PA5000 series radar altimeters

The PA5000 is a software-controlled altimeter operating in the J-band using advanced microwave and signal processing surface mount VLSI integration techniques.

The transmit and receive antennas are both included within the unit outline and thus the PA5000 is fuselage mounted, requiring only a single fuselag e cutout. RF feeders are not required. Operating in the J-band, the PA5000 series radar altimeter provides a covert system and gives precision accuracy and resolution for all high-performance fixed-wing, helicopter, UMA and RPV applications. In helicopter applications, excellent hover and nap of the earth performances are available.

PA5000 series radar altimeters offer height ranges up to 5,000 ft with 2 per cent accuracy, are designed to be compatible with most types of aircraft digital and analogue interfaces and are qualified to MIL-STD-461C and MIL-STD-810D.

Specifications

Dimensions: 218 × 76 × 138.6 mm
Weight: 3.2 kg
Power Supply: 19-32 V DC, 25 W (max) at 28 V
Altitude: 0-5,000 ft options
Accuracy: ±2% height + 2 ft
Reliability: 5,000 h MTBF

Operational status

Installed in Agusta/Sikorsky HH-3F and the NH500 helicopters.

Contractor

BAE Systems.

VERIFIED

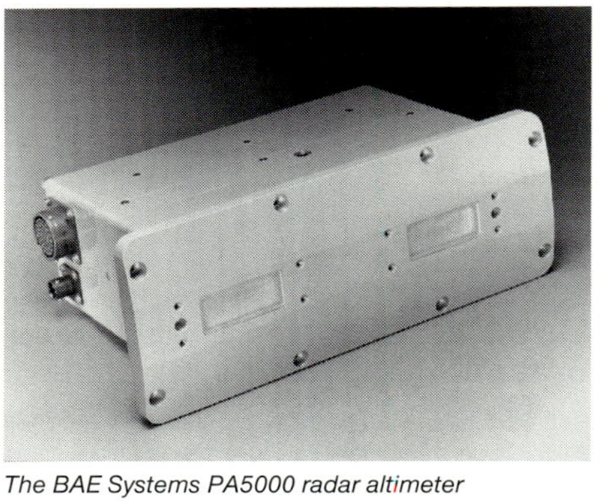

The BAE Systems PA5000 radar altimeter

PA5200 series radar altimeters

The PA5200 series radar altimeter operates in the mid-J-band using microwave Field Effect Transistor (FET) technology. Software-controlled signal processing techniques are used to enable reliable performance to be achieved up to 5,000 ft with a transmitter power of only 0.5 W. Surface-mount technology is used to give a low-volume, high-reliability package which includes the antenna.

The PA5200 radar altimeter uses a dual leading-edge tracker to ensure tracking of the nearest object. Continuous automatic monitoring of the system ensures high reliability down to ground level.

Specifications

Dimensions: 225 × 76 × 117 mm
Weight: 3 kg
Power supply: 28 V DC, 26 W (max)
Altitude: up to 5,000 ft (can be extended)
Accuracy: ±3% height + 3 ft
Temperature range: –40 to +50°C
Reliability: 5,000 h MTBF

Contractor

BAE Systems.

VERIFIED

PA5495 radar altimeter

The PA5495 radar altimeter operates in mid-J-band using microwave Field Effect Transistor (FET) technology. Software-controlled signal processing techniques are used to enable reliable performance to be achieved up to 5,000 ft with a transmitter power of only 1 W. Surface-mount technology is used to give low volume. Separate antennas are provided to be compatible with existing C-band installations.

The altimeter uses a dual leading-edge tracker to ensure tracking of the nearest object. Continuous automatic monitoring of the system ensures high reliability with accurate height indication right down to ground level.

Specifications

Dimensions: 140 × 220 × 85 mm
Weight: 4 kg
Power supply: 28 V DC, 26 W
Altitude: 0-5,000 ft (can be extended)
Accuracy: ±3% height + 3 ft
Temperature range: –40 to +70°C
Reliability: 5,000 h MTBF

Contractor

BAE Systems.

VERIFIED

PA6150 airborne IFF/SSR transponder

The PA6150 Mk XII IFF and Mode S SSR transponder is a remote terminal on a MIL-STD-1553B databus. A second, autonomous version uses a separate transponder control and display unit interfaced to the transponder LRI over RS-422A serial interfaces.

The transponder system is specified to meet the requirements of STANAG 4193, Part 1 for Mk XII IFF operation and ICAO Annex 10, Vol 1, Part 1, 4th Edition incorporating Amendment 68 for Mode S Level 3 operation. For the Mode S datalink, it will support Communications Capability Levels 1 to 3. Reply diversity operation is based on received signal level and timing at the two independent receiver channels. Provision is also being included for expansion to NG IFF at a later date.

The transponder system will accept the following Mk XII IFF interrogations and challenges: Modes 1, 2, 3A, 4 and C; automatic code changes for Modes 1 and 3A. Mode 2 will be set by switches mounted on the transponder LRI which are accessible when the LRI is installed in the aircraft.

For Mode S, the transponder will accept the following interrogations/challenges: Intermode A/C/S all-call; intermode A/C only all-call; short air-to-air surveillance; altitude request surveillance; identity request surveillance; Mode S only all-call; comm-A, altitude request; comm-A, identity request and comm-C, ELM.

The transponder system will have comprehensive BIT facilities, including power-up BIT, continuous BIT and initiated BIT.

Specifications

Dimensions:
(transponder) 194 × 124 × 318 mm
(control unit) 95 × 146 × 100 mm
Weight:
(transponder) 9.5 kg
(control unit) 2.2 kg
Power supply: 28 V DC
Frequency:
(transmitter) 1,090 ± 0.5 MHz
(receiver) 1,030 MHz

Contractor

BAE Systems.

VERIFIED

PA7010 Expendable System Programmer (ESP)

The PA7010 Expendable System Programmer (ESP) is designed to interface with radar warning receivers, jamming systems and chaff and flare dispensers, and to integrate the various EW sensors for the maximum operational effectiveness. The interface is produced via digital datalinks and discrete signals; crew monitoring and intervention is embodied in the design.

On detection of a threat, ESP is alerted to command the release of expendables. An easily reconfigurable stores library determines the appropriate release pattern and sequence. ESP then commands the response from one or more dispensers.

ESP has been designed for military combat aircraft and has a predicted reliability in excess of 1,000 hours MTBF. Modular construction and comprehensive BIT simplify maintenance.

Specifications

Dimensions: 127 × 127 × 190.5 mm
Weight: 2.7 kg
Power supply: 28 V DC, 10 W

Operational status

In production. The system is fitted in the UK Royal Air Force Harrier GR. Mk 7, in which it interfaces with the Zeus ECM system and onboard countermeasures dispensers. It has also been selected for the Eurofighter Typhoon.

Contractor

BAE Systems.

VERIFIED

PA7030 laser warning equipment

The PA7030 laser warning equipment offers both instantaneous warning and bearing information on radiation by ruby and neodymium lasers. The system consists of a direct detector head unit and up to four indirect detectors mounted externally on the aircraft. A compact display and control unit is mounted in the cockpit.

The indirect detectors are sensitive to pulsed laser radiation scattered by the aircraft and so can detect designators and range-finders even when these do not directly illuminate the direct detector. Electronic filtering discriminates against glints and flashes to give an extremely low false alarm rate. The direct sensor gives instantaneous bearing on a radiation source to within 15°.

Specifications

Dimensions:
(PA7031 direct detector) 152 × 48 mm
(PA7032 scatter detector) 105 × 80 × 130 mm
(PA7034 signal conditioner) 260 × 160 × 87 mm
(PA7033 display) ¼ ATR short
Weight:
(direct detector) 2 kg
(scatter detector) 1.5 kg
(signal conditioner) 2 kg
(display) 2 kg
Power supply: 28 V DC, <12 W
Coverage:
(direct detector) 360° (azimuth), –45 to +10° (elevation)
(scatter detector) as required

Operational status

Development equipment has had trials in helicopters and fixed-wing aircraft in the UK, USA and Canada.

Contractor

BAE Systems.

VERIFIED

PA9052SM GPS receiver

The PA9052SM GPS receiver provides full Precise Positioning Service (PPS) capabilities and includes all the interfaces expected of a military GPS receiver including MIL-STD-1553B, ARINC 429/575, PTTI and RS-422. Standard Positioning Service (SPS) variants are also available. As well as generating a basic navigation solution, the PA9052SM can provide area navigation facilities and is suitable for use in an integrated navigation system.

BAE Systems is also providing a five/six-channel GPS receiver module for the Eurofighter Typhoon project. This module also provides full PPS capability and is suitable for use in embedded applications.

Specifications

Dimensions:
(PA9052 receiver) 216 × 194 × 90 mm
(PA9915 antenna) 89 mm diameter
Weight: 4.3 kg

Operational status

Deliveries of GPS receivers to customers worldwide started in 1989. They were selected for the development phase of the AH-64D Apache Longbow helicopter and are currently in service on UK Royal Air Force Tornado, Jaguar, Harrier and Nimrod aircraft.

Contractor

BAE Systems.

VERIFIED

PA9360 GPS modules

The PA9360 family of GPS modules provides a flexible solution for applications requiring an embedded GPS receiver capability. The PA9361 module is the first in the family and is designed for applications in the airborne, naval and land environments. Advanced ECL and VLSI ASIC technology has been used to achieve full six-channel, dual-frequency capability within a single module.

The PA9361 module uses software from the PA9000 Series of military GPS receivers and is suitable for embedded use in a range of navigation equipments. Installation and support are simplified by the incorporation of a PPS-SM device which ensures that the module is unclassified, even when loaded with encryption keys, and the ability to power feed a remote preamplifier via the single-cable RF input. Performance is maximised by the availability of 10 Hz GPS measurements and full navigation capability in high dynamic environments, even in the unaided mode. Extensive BITE and a high inherent reliability combine to lower the customer's support tests.

Specifications

Dimensions: 150 × 150 × 25 mm
Weight: 1.4 kg
Power supply: 5 V DC, 15 V DC or battery
Temperature range: –54 to +71°C
Accuracy:
(position) 16 m (SEP)
(velocity) 0.2 m/s (95%)
(time) 100 ns (1σ)
Reliability: >10,000 h MTBF calculated

Contractor

BAE Systems.

VERIFIED

PTR283 Mk1/PVS1280 Mk II IFF interrogators

The PTR283 and PVS1280 interrogator systems have been designed to meet the requirements of in-flight secondary radar interrogators. The transmitter/receiver uses pulsed oscillator techniques employing automatic frequency control and a logarithmic receiver using silicon integrated circuits. The equipment interrogates on Modes 1, 2 and 3A, the pulses driving the modulator being generated by the encoder/decoder.

The PTR283 Mk1 equipment consists of a lightweight D-band transmitter/receiver unit, an associated encoder/decoder and a control unit. Other units associated with the system are an antenna switch, dual-antenna system and L-trace radar display.

The PVS1280 Mk II system consists of a D-band lightweight transmitter/receiver and encoder/decoder, which offers the facility of active decoding and defruiting. This equipment is designed to integrate into an airborne primary radar system. Control of the PVS1280 is performed by the radar controller. It offers ISLS operation and an Interrogation SideLobe Suppression (SLS) switch is available to enable the transmitted power to be distributed equally to the two antennas and the two antennas to be fed alternately in phase and anti-phase for ISLS operation.

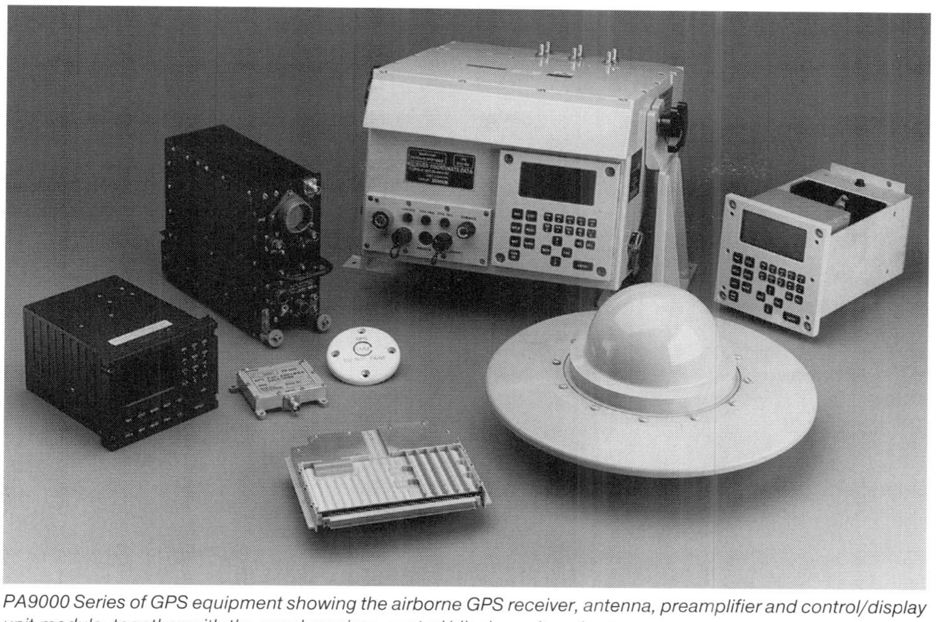

PA9000 Series of GPS equipment showing the airborne GPS receiver, antenna, preamplifier and control/display unit module, together with the naval receiver, control/display unit and antenna

Specifications

Frequency:
(transmit) 1,030 ± 0.5 MHz
(receive) 1,090 ± 0.2 MHz
Pulse length: 0.8 ± 0.2 μs
Duty cycle: 0.11%
Decoder: 496 codes

Contractor

BAE Systems.

VERIFIED

PTR446A transponder

The PTR446A lightweight transponder identifies aircraft in response to secondary radar interrogation and covers civil and military modes. Emphasis in design was placed on reliability combined with small size and low weight and these qualities have been achieved by the use of specially designed micro-electronic circuits. A digital shift register replaces conventional delay lines in the decoder/encoder circuits, so providing time delays independent of temperature. Integrated circuits are used for the logic and video processing circuits and the logarithmic response intermediate frequency amplifier. Decoder, encoder and associated switches in the control unit reduce the number of interconnecting wires to five and substantially cut down the installation weight. The transmitter/receiver houses the pulse selection and power-supply modules. Three-pulse sidelobe suppression is incorporated.

Two control units are available for use with the transmitter/receiver. The smaller of the two is the PV447, of which there are six versions with the following capabilities:

PV447 - Mode 1 or 3A/B and Mode C or off
PV447A - Mode 1 or 2 and Mode C or off
PV447B - Mode 2 or 3A/B and Mode C or off
PV447C - Mode 1 or 3A and Mode 2 or off
PV447D - Mode A or B and Mode C or off
PV447E - Mode A/B or off and Mode C or off.

An alternative to the PV447 is the PV1447 control unit, which meets the requirements of NATO STANAG 4193 for IFF Mk XA and provides Modes 1, 2, 3/A and C. Automatic code changing is provided on Modes 1 and 3A with storage capacity for 48 codes in each mode. Manual code entry for these modes is via a front-panel keypad. Mode 2 codes are entered through screwdriver adjusted switches reached through the top cover of the unit.

The transponder can be used with either control unit without modification to the transmitter/receiver. Comprehensive self-test is incorporated in all units.

Specifications

Dimensions:
(transponder) 57 × 127 × 254 mm
(PV447 control unit) 146 × 57 × 102 mm
(PV1447 control unit) 146 × 95 × 165.1 mm
Weight:
(transponder) 1.7 kg
(PV447) 0.48 kg
(PV1447) 1.6 kg
Power output: 24.7 dBW
Pulse rate: 1,200 replies/s, each containing up to 14 reply pulses
Triggering sensitivity: –72 to –80 dBm

Operational status

In production and in service with helicopters of the UK Royal Navy, Army and Royal Air Force, and on the BAE Systems Hawk aircraft.

Contractor

BAE Systems.

VERIFIED

PTR 1721 V/UHF radio

The PTR 1721, a combined VHF and UHF radio covering the UHF band from 225 to 400 MHz and the VHF band from 100 to 156 MHz, is designed for all types of military aircraft. The ability to communicate on VHF in addition to UHF offers increased flexibility to air forces which, on occasion, may need to operate from civil airfields equipped with VHF facilities only. The PTR 1721 has been chosen for the RAF Panavia Tornado aircraft.

The PTR 1721 offers up to 9,240 channels, 2,240 on VHF and 7,000 on UHF. Channel spacing in the VHF band is at 25 kHz and standard spacing in UHF is at 50 kHz although an optional interval of 25 kHz spacing is available. Frequencies are selectable directly from the remote-control unit which also provides preselection of up to 17 channels with the addition of the UHF and VHF Guards at the international distress frequencies of 243 and 121.5 MHz respectively. Frequency synthesis techniques ensure good frequency stability and channel selection characteristics.

The system is entirely solid-state in construction and, wherever possible, employs conventional technology in the interests of reliability enhancement. The receiver is varactor-tuned and the frequency synthesiser is compared against a single reference oscillator.

A sealed case houses the transmitter/receiver unit and access to the modules is gained by removal of the sides of the case, to which the modules themselves are attached. Heat is conducted from the modules via the chassis, which acts as a heatsink, to the sides of the case and hence to external air. Forced-air cooling is directed through the mounting tray installation which also acts as a vibration insulator. Forced-air cooling may be dispensed with in less demanding aircraft environments.

Switching for dual-control operation is external to the system and in two-seat aircraft identical control units are installed at each crew position. Options include an antenna lobe switch for azimuth homing requirements and a UHF antenna switch for automatic direction-finder operation.

Recently, the company has introduced an optional modification which is available for both in-service and new radios. This modification provides a frequency-hopping capability which gives a significant improvement in communications performance when the radio is operated in a jamming environment.

The environmental temperature range extends from –40 to +70° C ambient and the normal operational altitude is to a maximum of 50,000 ft, although operation is possible for short periods at altitudes of up to 70,000 ft.

Specifications

Dimensions:
(controller) 94 × 145 × 175 mm
(transmitter/receiver) 125 × 194 × 339 mm
Weight:
(controller) 1.8 kg
(transmitter/receiver) 11.25 kg

Operational status

In service in UK Royal Air Force Tornado aircraft.

Contractor

BAE Systems.

VERIFIED

PTR 1741 VHF/AM radio

The PTR 1741 equipment may be regarded as the VHF counterpart of the PTR 1751 system, both units being members of the same family of military airborne radios. Covering the frequency range from 100 to 155.975 MHz, the PTR 1741 provides 2,240 channels at 25 kHz separation. If desired, however, channel spacing can be set at steps of 50 kHz to ease interface with older equipment, which does not have the same frequency stability as the new-generation systems. As in the case of the PTR 1751, the PTR 1741 is available with a choice of manual or manual/preset control unit; however, only a single power output level of 10 W is available.

The PTR 1741 shares a number of common features with the PTR 1751 with regard to build standard, operating range, options and so on. Both are suitable for dual UHF/VHF installation, their electrical interfaces and mechanical dimensions being identical. A single controller may be used to operate the two transceivers in such a dual installation.

Specifications

Dimensions:
(manual controller) 146 × 48 × 108 mm
(preset controller) 146 × 95 × 108 mm
(transmitter/receiver) ½ ATR short × 160 mm
Weight:
(manual controller) 0.7 kg
(preset controller) 1.4 kg
(transmitter/receiver) 5 kg
(Guard receiver module) 0.3 kg

Operational status

No longer in production but still in service.

Contractor

BAE Systems.

VERIFIED

PTR 1751 UHF/AM radio

The PTR 1751 is a lightweight UHF/AM radio designed for all types of military fixed-wing aircraft and helicopters. It provides 7,000 channels in the frequency band 225 to 399.975 MHz at a channel spacing of 25 kHz and is available with either 10 or 20 W transmitter outputs. Channel spacing of 50 kHz is available as an option.

Both versions comprise a single transmitter/receiver unit with either a manual controller or an optional manual and preset controller. Each controller provides full selection of the range of 7,000 channels together with control of the built-in test functions; the manual/preset unit additionally permits preselection of up to 30 channels, plus Guard frequency, through incorporation of a non-volatile memory store. A remote frequency and channel indicator is also available.

Options include continuous monitoring on the 243 MHz international distress frequency via a separate Guard receiver module which plugs directly into the main transmitter/receiver chassis, an external homing unit and a wideband secure speech facility which requires no additional interface equipment.

Recently the company has introduced an optional modification which is available for both in-service and new radios. This modification provides a frequency-hopping capability which gives a significant improvement in communications performance when the radio is operated in a jamming environment.

The PTR 1751 conforms generally to DEFSTAN-07-55 and operates satisfactorily over a temperature range from –35 to +70° C. It is of all-solid-state modular construction with high reliability as a principal design aim, and a claimed MTBF of 800 hours.

Specifications

Dimensions:
(transmitter/receiver)
(10 W version) ½ ATR short × 160 mm high
(20 W version) ½ ATR medium × 160 mm high
(manual controller) 146 × 48 × 108 mm
(preset controller) 146 × 95 × 108 mm
Weight:
(transmitter/receiver)
(10 W version) 5 kg
(20 W version) 6.7 kg
(preset controller) 1.4 kg
(manual controller) 0.7 kg
(Guard receiver module) 0.3 kg

Operational status

In production and service.

Contractor

BAE Systems.

VERIFIED

PVS1712 radar altimeter

The PVS1712 J-band pulse radar altimeter is designed for applications ranging from helicopters to high-performance fixed-wing aircraft. The use of a high operating frequency has enabled the design of a single-unit system, containing transmit, receive and electronic functions. The unit uses a pulse leading-edge tracking technique to measure precisely the time interval for radar pulse travel. It is an all-solid-state design and requires no warm-up. Very short 4 ns pulses are used. The system includes automatic error correcting circuitry which fully compensates for errors due to internal delays and their time/temperature drift.

The transmitter is a 5 W Gunn diode oscillator and a pseudo-homodyne receiver is used with a low-noise video preamplifier and main amplifier, the gain of the latter being geared to expected signal level at the range tracking point.

Specifications

Typical installation
Dimensions: 218 × 116 × 76 mm
Weight: 2 kg
Altitude: 0-1,000 ft
Accuracy: ±2% height + 2 ft at 500 ft,
±5% height + 2 ft above 500 ft

Operational status

In service in British Army Gazelle helicopters and towed aerial targets.

Contractor

BAE Systems.

VERIFIED

PVS2000 Missile Approach Warner (MAW)

PVS2000 is a radar-based Missile Approach Warning (MAW) equipment designed to equip the UK Royal Air Force's Harrier aircraft. The MAW uses a low-powered pulse Doppler to detect approach of missiles, and automatically initiates countermeasures. The system is optimised to provide protection against air-to-air and ground-to-air heat-seeking missiles.

The MAW comprises: a transmitter/receiver unit, lightweight antenna, signal processing unit and cockpit control unit.

PVS2000 family of equipment

The Transmitter/Receiver Unit (TRU) contains a low-power solid-state transmitter and receiver, analogue processing and digital conversion circuitry. It is designed for installation in an unconditioned area of the aircraft tail within 1 m of the antenna.

The remote lightweight antenna is connected to the TRU, using low-loss Gore cable, or may be attached directly to the TRU. The antenna must have a clear view aft.

The Signal Processing Unit (SPU) houses the digital signal processing necessary for detection and assessment of signal returns. All interfaces with other aircraft avionics systems, BITE monitoring and module level diagnostics are handled by this unit. The SPU is installed in a conditioned avionics bay.

The optional Cockpit Control Unit (CCU) provides on/off control and incorporates a BITE status indicator lamp. A small auxiliary blade antenna is mounted on any suitable site with a relatively clear view forward.

The radar is mounted in the tail of the aircraft and should be linked to an expendables dispenser for the automatic deployment of flares and chaff.

Specifications

Dimensions:
(transmitter/receiver unit) 156 (length) × 250 mm (diameter)
(signal processing unit) ½ ATR short
(optional cockpit control unit) 38 × 146 × 75 mm
(main antenna/radome) 100 (length) × 260 mm (diameter)
(auxiliary antenna) 150 × 100 × 260 mm
Weight:
(transmitter/receiver unit) 5.5 kg
(signal processing unit) 7 kg
(cockpit control unit) 1 kg
(main antenna/radome) 0.7 kg
(auxiliary antenna) 0.25 kg

Operational status

In service on UK Royal Air Force Harrier GR. Mk 7 aircraft.

Contractor

BAE Systems.

VERIFIED

Radar Homing and Warning Receiver (RHWR)

The Radar Homing and Warning Receiver (RHWR) is produced in two versions: an amplitude comparison version for strike aircraft and a high-accuracy version for air defence aircraft. Both versions are effective against radar threats likely to be encountered and are combat proven. Frequency coverage is available from C- to J-bands. RHWR can detect and identify threats up to and beyond the radar horizon. The information obtained is correlated with the aircraft interception radar tracks to allow for the classification and allocation of target priorities. High-accuracy DF is obtained by using interferometers. RHWR uses a powerful digital processor, which accepts the parametric data from the receivers, including frequency, PRI, pulsewidth and scan characteristics. The information is compared with a comprehensive emitter library. The identified emitters are passed to the cockpit displays which provide the crew with both tabular and graphical presentation of threats, as well as the overall radar scenario.

The RHWR has been upgraded by BAE Systems to meet the requirements of UK MoD Air Staff Requirement 907. Increased processing power and additional memory has been added and the capability for the RHWR to exchange data with the Sky Shadow ECM pod has been provided. Sky Shadow has also been upgraded under the same programme. While each system retains the ability to operate independently, under normal conditions the RHWR acts as the defensive aids system controller of an integrated system.

Operational status

In service in the Tornado F Mk 3 of the UK Royal Air Force and Italian Air Force, and with the Tornado ADV of the Royal Saudi Air Force.

Contractor

BAE Systems.

VERIFIED

Rangeless Airborne Instrumentation Debriefing System (RAIDS)

RAIDS eliminates the need for deployment to a fixed air combat manoeuvring range, and permits air-to-air and air-to-ground weapon simulation, together with electronic warfare training as required, and safety alerts.

RAIDS is housed in a P4B pod with integral inertial measurement unit (Litton LN-200), GPS receiver (Rockwell Collins GNP-1®) and datalink. No aircraft modifications are required, providing AIM-9 and MIL-STD-1553 interfaces are available. A full debriefing system is provided.

Specifications

Datalink:
(frequency range): 2.2 - 2.4 GHz
(frequencies available): 32 within 100 MHz
(simultaneous handling): 200 aircraft, without pre-booking
(peak output power): 26 - 28 W
(range): 60 n miles (120 with relay)
GPS receiver: 10 channel, P/Y or C/A code, 24 state Kalman filter, WAGE
Inertial measurement unit: FOG, 1°/h attitude accuracy
Data storage: 240 MB PCMCIA card, upgradeable
System accuracy (1σ):
(position X and Y): 5 m
(position Z): 8 m
(no drop bomb score): 15 m
(velocity): 0.1 m/s
(pitch/roll/heading): ±1.0°
Ground station:
mission planning
3D debrief
plan, angular and cockpit views
background mapping
digital terrain elevation data
synchronised with aircraft video recordings
DGPS capability
COTS workstation

Operational status

RAIDS has been chosen for seven air forces. It is flying with the UK Royal Air Force on Tornado F-3 aircraft for EW training, and with the Operational Evaluation Unit at Boscombe Down on Harrier GR Mk 7, Jaguar, and Tornado GR Mk 4 aircraft. BAE Systems is bidding to the UK Royal Air Force for a buy of 158 units.

Contractor

BAE Systems.

VERIFIED

Seaspray 2000 airborne radar

Seaspray 2000 is an I-band maritime surveillance radar which has been optimised for civil, Economic Exclusion Zone (EEZ) surveillance operations. A range of different sized, full-colour combined control and display units provides the man/machine interface. Control of the highly automated radar is via simple on-screen menus. Output ports are provided for hard-copy records, and video and one or more displays can be provided as control stations or repeaters. Superior contact detection in adverse conditions is provided by the high-peak power, frequency-agile transceiver, Constant False Alarm Rate (CFAR) processing, operator-selectable scan-to-scan integration, and a range of optional features. Options include an operator-selectable, side-looking array, radar for ground-mapping and pollution control, operator-selectable circular polarisation for enhanced performance in precipitation and multiple target track-while-scan. The display presentation is enhanced by a comprehensive set of synthetic overlay facilities, including digital coastlining, customer-defined waypoints and operator-defined variable waypoints.

The BAE Systems rangeless airborne instrumented debriefing system installation ***2000***/0080284

Specifications

Frequency: I-band
Peak power: 90 kW
Transmitter: frequency-agile magnetron
Pulsewidths: 2 selectable
PRF: 4 selectable
Scanner type: front-fed, elliptical section, double curved paraboloid
Azimuth scan: 360° continuous, 180 and 60° sector scan selectable
LRUs: 4
System weight: 80 kg
Features: CFAR and operator-selectable scan-to-scan integration
Options: operator-selectable SLAR, circular polarisation, multiple TWS and coastlining, a range of antenna sizes (up to 1 m diameter) and one or more of a range of display sizes (up to 0.35 m diagonal)

Operational status

Seaspray 2000 is in service in Reims F 406 Caravan II Vigilante aircraft with the UK government's Scottish Fisheries Protection Agency. It is in service in the Dornier 228 maritime patrol aircraft with the Finnish Frontier Guard, and also in the Dornier 228 with another UK government agency. In the Dornier 228 applications it is fully integrated with the BAE Systems multirole turret system FLIR.

Contractor

BAE Systems.

VERIFIED

Seaspray 3000 airborne radar

Seaspray 3000 (previously known as Seaspray Mk 3) introduces significant advances over Seaspray Mk 1, most notably through the provision of a new digital processor and full 360° scan. The system is designed primarily for operation in light, agile, shipborne naval helicopters to provide detection and tracking of small targets in adverse conditions for the Sea Skua missile. For this purpose it retains the combat proven monopulse lock-follow target illumination of its predecessor.

Consisting of six LRUs, Seaspray 3000 is configured to operate on a MIL-STD-1553B databus, with multiple additional standard interfaces being provided. The resultant lightweight, compact and flexible system is applicable to both the retrofit and new aircraft. Operating in the I-band, Seaspray 3000 uses a magnetron to provide high transmitted power and very high-speed agility to provide these platforms with high-detection performance in sea and weather clutter and in ECM conditions. The digital processor provides advanced features, including constant false alarm rate and track-while-scan facilities to optimise target detection and reduce operator work load.

The operator is provided with a comprehensive tactical situation display of scan-converted television format, in monochrome or colour. In addition, the output of other sensors (FLIR, ESM, Datalink and so on) may be displayed. The control unit employs a menu structure with soft key options.

Specifications

Frequency: I-band
Peak power: 90 kW
Transmitter: high-speed, spin-tuned magnetron
Pulsewidths: 2 selectable

BAE Systems Seaspray 3000 radar units

PRF: 4 selectable
Coverage: 360°
LRUs: 6
System weight: 90 kg
Target illumination: monopulse lock-follow
Features: CFAR and operator-selectable scan-to-scan integration
Options: operator-selectable circular polarisation, a range of antenna sizes (up to 1 m diameter), and a range of display options

Operational status

Seaspray 3000 is in service with the Republic of Korea Navy and Brazilian Navy in GKN Westland Super Lynx, in the Turkish Navy in Agusta Bell 212, and in the German Navy in GKN Westland Sea King helicopters. The German Navy has also ordered Seaspray 3000 for its new Super Lynx Mk 88A helicopters and the Royal Navy is upgrading its Seaspray Mk 1 radars to Seaspray 3000 standard in its Lynx Mk 1 helicopters. Seaspray 3000 has also seen service in a Fokker F27 maritime patrol aircraft and has been proven with Sea Skua in land-based coastal battery and fast patrol boat applications.

Contractor

BAE Systems.

VERIFIED

Seaspray 7000 airborne radar

Seaspray 7000 is a pulse Doppler, pulse compression, multimode, COTS-based, open architecture, programmable airborne maritime surveillance radar. It can be operated as a stand-alone radar or as the heart of a fully integrated avionics suite. It is the resultant system from combining the proven man/machine interface and processing of Seaspray 3000 with BAE Systems' Travelling Wave Tube-based (TWT-based) pulse compression technology, developed in conjunction with the Blue Kestrel 7000 radar.

Comprising six LRUs, which are form/fit compatible with the Seaspray 3000, Seaspray 7000 provides the optimum performance for medium-size maritime patrol aircraft and naval helicopters. Processing capabilities include: pulse compression, Synthetic Aperture Radar (SAR), Inverse SAR (ISAR), and Moving Target Indication (MTI).

Specifications

Frequency: I-band
Transmitter: Ring bar TWT power amplifier
Pulsewidths: selectable
PRF: selectable
Coverage: 360°
LRUs: 6
System weights: total 62.2 kg
(scanner): 10.2 kg
(transmitter): 22.5 kg
(receiver): 12.0 kg
(processor): 17.5 kg
Dimensions:
(scanner): 737 mm swept dimension
(transmitter): 145 (W) × 381 (D) × 243 (H) mm
(receiver): 207 (W) × 381 (D) × 225 (H) mm
(processor): 260 (W) × 475 (D) × 200 (H) mm
Features: pulse compression, CFAR, multiple TWS, operator-selectable scan-to-scan integration, SAR and ISAR
Options: operator-selectable circular polarisation, range of antennas and displays

Operational status

Seaspray 7000 has been designed for, and is in service in the Super Lynx 300 radome. The numbering gaps in the Seaspray range (4000, 5000 and 6000) are explained as follows: Seaspray 4000 was a development project that has matured into the Seaspray 7000; the designators 5000 and 6000 have been reserved for the Blue Kestrel range of radars.

Contractor

BAE Systems.

VERIFIED

Seaspray Mk 1 maritime surveillance radar

Seaspray is a family of maritime surveillance and targeting radar systems. Since the 1970s this family has

Westland Super Lynx helicopters of the South Korean Navy are equipped with the Seaspray 3000

The Seaspray chin radome on the Super Lynx 300 helicopter **2000**/0080286

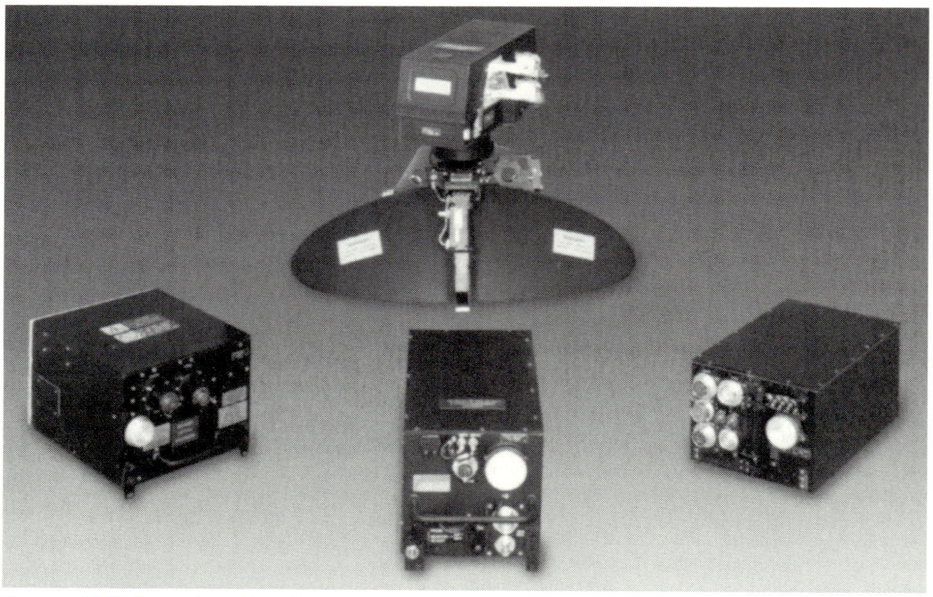

Seaspray 7000 line replaceable units **2000**/0080285

The BAE Systems Seaspray Mk 1 in the nose of a Westland Lynx

been developed to include Seaspray Mk 1, Seaspray 2000, Seaspray 3000 and Seaspray 7000, over 500 of these radars have been ordered.

Allied with the Seaspray family of radars is the Blue Kestrel range of maritime surveillance radars. Commonality of modules is used throughout both the Seaspray and Blue Kestrel ranges. This minimises cost and provides natural growth paths for the customer. Details of the Seaspray range are given below. The Blue Kestrel range is described in a previous entry.

Seaspray Mk 1 is an I-band airborne maritime surveillance and targeting radar, which was developed in the early 1970s as an integral element of the solution to a UK Royal Navy Staff Requirement. This requirement was raised as a result of the 1967 sinking of the Israeli destroyer *Eilat* by two Styx missile-armed fast patrol boats. The requirement demanded a light, agile, shipborne helicopter, which would be able to detect and neutralise long-range surface-to-surface missile-armed fast patrol boats while keeping them outside the engagement range of the helicopter's mother ship. The solution was the Westland Lynx helicopter armed with BAE Systems Sea Skua missiles. To minimise the weight of the missiles and to enable four of them to be carried, semi-active radar homing was selected for their guidance. The link was the helicopters' prime sensor and target illuminator, the Seaspray Mk 1. This lightweight radar provides the high-performance, frequency-agile detection of small targets in adverse conditions and tenacious monopulse lock-follow target illumination for the Sea Skua.

Specifications

Frequency: I-band
Peak power: 90 kW
Transmitter: high-speed, spin-tuned magnetron
Pulsewidths: 2 selectable
PRF: 3 selectable
Coverage: 180°
LRUs: 5
System weight: 75 kg
Target illumination: monopulse lock-follow

Operational status

Over 300 Seaspray Mk 1 radars have been delivered in GKN Westland Lynx helicopters to Brazil, Denmark, Germany, Netherlands, Norway and the UK. The Royal Navy also operates land-based coastal-sited versions of Seaspray Mk 1 as Checksite to monitor the effectiveness of shipborne EW equipment. Seaspray Mk 1 has also been delivered to the Republic of Korea for land-based trials, and to Pakistan in ex-Royal Navy Lynx helicopters.

Contractor

BAE Systems.

VERIFIED

Sky Guardian 200 radar warning receiver

Sky Guardian 200 uses crystal video receivers and the latest digital processors to give analysis of threats and high probability of intercept of signals. Information is presented to the crew by alphanumeric presentation on either an indicator bearing unit, a multifunction display or a HUD. Sky Guardian 200 will control a range of countermeasures, including jamming and chaff and infra-red flare dispersion.

The Sky Guardian 200 radar warning receiver

Specifications

Weight: 24 kg
Frequency: E-J in 3 bands (C/D- and K-bands optional)
Frequency measurement: pulse, pulse Doppler, CW, ICW
Coverage: 360° azimuth instantaneous
Accuracy: better than 10°
Response time: <1 s
Emitter library: 400 emitters with 2,500 modes

Operational status

In production. Sky Guardian 200 is in service on many fixed-wing aircraft and helicopters, including: UK Royal Air Force Jaguar GR. Mk 1B (Jaguar '96 and '97) and VC10, UK Royal Navy Sea Harrier F Mk 2, Matadors of the Spanish Navy, Omani Air Force Hawk 200 aircraft, J350 aircraft of the Austrian Air Force and other air forces. It has been selected for the Czech L-159 aircraft.

Contractor

BAE Systems.

VERIFIED

Sky Guardian 300 and 350 ESM systems

Sky Guardian 300 is an advanced airborne electronic support measures system, which detects radar emitter signals from outside their own detection range. It can recognise the bearing and mode of operation of an emitter, provide fine DF and prioritise threats. The system includes a library capable of recognising up to 4,000 different signals, a central processor providing identification times of less than a second and programmable display with a variety of graphic and tabular formats on full colour raster scan monitor.

Sky Guardian 300 allows the operator to prioritise and select relevant data and also has the ability to compile and update emitter libraries at both first line and national level. The system will integrate with other onboard sensors to provide a comprehensive picture. It will provide automatic warnings and initiate countermeasures.

A variant of the Sky Guardian 300, the Sky Guardian 350, includes a rotating antenna and special receiver facility to provide long-range detections and raid assessment. It offers improved signal measurement and high accuracy DF.

Specifications

Weight: 28 kg plus display
Frequency: E- to J-bands (C/D-band optional)
Frequency measurement: DIFM
Coverage: 360° instantaneous
Accuracy: better than 2.5°
Response time: <1 s
Library: over 4,000 emitters

Operational status

Fully developed and on order for specified customers.

Contractor

BAE Systems.

VERIFIED

Sky Guardian 2000 radar warning receiver

Sky Guardian 2000 is configured for rapid response in a high-density electromagnetic environment and has the processing capability to display multiple threat signals in less than a second. The system operates either as a stand-alone radar warner or integrated with an ECM system such as the Apollo radar jammer or other countermeasures. Sky Guardian 2000 will accept data from the Type 1220 laser warning receiver and display the threats on a common indicator bearing unit.

Sky Guardian will provide long-range detection of airborne and surface threat radars, emitter identification and built-in recording for post-flight analysis. Other key features are display persistence for fleeting intercepts, low-band targeting options and the ability to integrate into a multifunction cockpit control and display. The system is extendable to meet future threats.

Sky Guardian 2000 employs advanced signal processing algorithms and provides high sensitivity as well as accurate RF measurement. An emitter library of 4,000 emitter descriptions can be loaded by a PCMCIA smart card incorporated into the control unit. Sky Guardian 2000 has been designed from the outset as the core of a Defensive Aids System (DAS) and includes DAS control functions.

Specifications

Weight: 13 kg
Frequency: E- to J-bands (C/D- and K-bands optional)
Frequency measurement: pulse, pulse Doppler, CW and ICW
Coverage: 360° instantaneous
Accuracy: better than 10°

Operational status

Fully developed and on order for specified customers. It is understood to have been ordered by the Austrian Air Force for J350E Draken fighters, and it has been

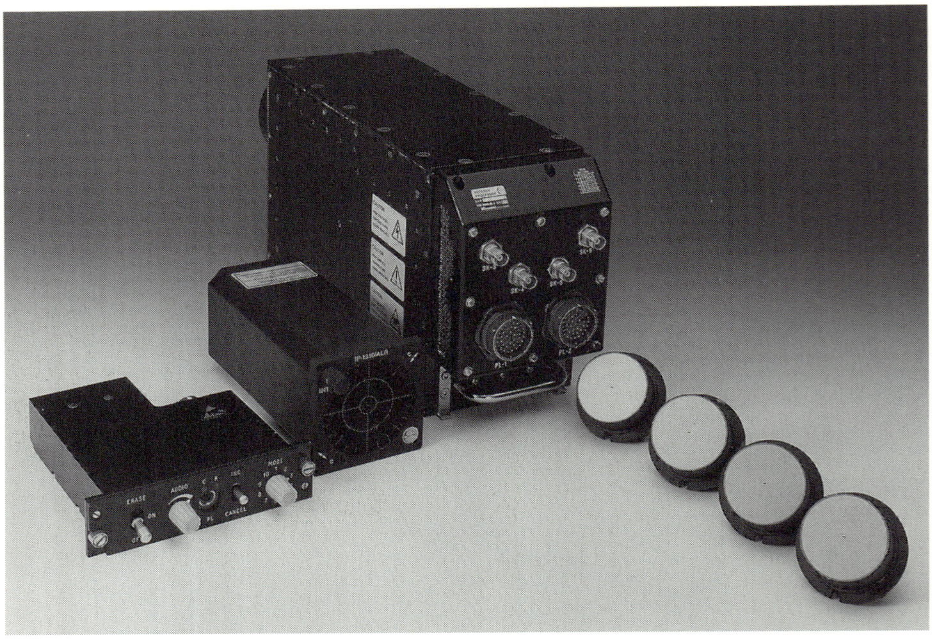

The Sky Guardian 2000 radar warning receiver

selected for UK Royal Air Force Merlin HC.3 support helicopters. Sky Guardian 2000 forms part of the Helicopter Integrated Defensive Aids Suite (HIDAS) for the British Army WAH-64D Apache Longbow helicopter.

Contractor
BAE Systems.

VERIFIED

Sky Shadow ECM pod

BAE Systems was the prime contractor in the production of the ARI 23246/1 Sky Shadow ECM pod developed for use with the UK Royal Air Force Panavia Tornado. Other components of the system are provided by Racal Defence Electronics and BAE Systems.

A high-power travelling wave tube amplifier is used with a dual-mode capability for deceptive and continuous wave jamming. A voltage-controlled oscillator uses a varactor-tuned Gunn diode to cover the full frequency band. The set-on receiver is a Racal design, and signal processing uses BAE Systems hardware.

The pod incorporates both active and passive electronic warfare systems and includes an integral transmitter/receiver, processor and cooling system. It has radomes at both ends and is stated to be capable of countering multiple ground and air threats, including surveillance, missile and airborne radars. Power management is automatic and modular construction allows the system to be adapted to differing operational missions.

An enhanced version of Sky Shadow has been developed. This features a high-performance data processor running on Ada software. The range of jamming activities has been extended, with greater flexibility for front line reprogramming.

Sky Shadow has been upgraded, with the Tornado RHWR, as part of the UK MoD Air Staff Requirement 907. The improved Sky Shadow is able to exchange data with the RHWR. When operating in the linked mode, RHWR acts as a Defensive Aids System (DAS) controller and both equipments offer the synergistic benefits of an integrated system.

A Royal Air Force Tornado carrying the ARI 23246/1 Sky Shadow ECM pod on each outer pylon

Specifications
Dimensions: 3,350 (length) × 380 mm (diameter)

Operational status
BAE Systems was awarded a contract to upgrade the Sky Shadow ECM pod for the UK Tornado GR.4 aircraft. The upgraded Sky Shadow system is in production.

Contractor
BAE Systems.

VERIFIED

Skyranger airborne radar

Skyranger is a lightweight airborne weapon control radar developed for light fighter and light attack aircraft in the retrofit market. It consists of three main units; antenna, transmitter/receiver and signal processor/power supply, and since the amount of space available for retrofit programmes can often be limited and irregular in shape, the modularity of Skyranger has been established at printed circuit card level. The individual cards can, therefore, be packaged into housings designed for the space available. The equipment has been designed as part of an integrated avionics suite, the other system being an air data computer, radar altimeter, head-up display, weapon aiming computer and secure communications.

Skyranger accepts discrete digital commands from a cockpit-mounted control panel and provides output data in the form of a digital serial link (ARINC 429) to a HUD and other weapon aiming systems. It has two main modes, guns and missiles, the former having a shorter range wide-angle beam. In the missile mode the radar energy is fed from the feed horn and reflected back from the parabolic antenna in a 6° beam with a maximum range of 15 km. For gun attacks, the radar energy is fed out directly from the antenna through a polarised window. This results in an 18° beamwidth and a range of 5 km. Minimum range is 300 m for guns and 150 m for missiles, with a ranging accuracy in the order of ±15 m below 3 km and ±30 m above. Target relative velocities from –500 to +1,000 m/s may be handled.

The equipment operates in I-band and has a 5 per cent pulse-to-pulse agility. MTBF is given as 200 hours and the equipment contains built-in test systems. The current version has a fixed antenna.

Specifications
Frequency: I-band
Range: 15 km for missiles, 5 km for gun operation
Range resolution: 150 m
Pulse-to-pulse agility: 5%
Power supply: 27 V DC, <50 W; 115 V 400 Hz, single phase, <400 VA
Weight:
(antenna) 4 kg
(transmitter/receiver) 25 kg
(signal processor/power supply) 8 kg
(total installed weight) 40 kg

Operational status
In service on the F-7 fighter aircraft of the People's Republic of China, as part of the BAE Systems integrated avionics system. More than 300 systems have been produced.

Contractor
BAE Systems.

VERIFIED

Super Skyranger airborne radar

Super Skyranger is a low-cost multimode radar for light fighter and light attack aircraft. It is based on the Skyranger and is a direct replacement for Skyranger in the improved version of the Chinese F-7 fighter (designated F-7MG) and as an upgrade in the MiG-21 airframe.

Super Skyranger has a full look-down shoot-down capability using a planar-array antenna which can scan to ±30° dependent on the aircraft installation. It can provide target range, range rate and line of sight data to the aircraft's avionic system via ARINC 429 serial link and has retained the excellent ECCM features of the original Skyranger.

Contractor
BAE Systems.

VERIFIED

Super Skyranger airborne radar implemented on the Chinese F-7MG fighter

0011866/0011868/0011867

TERPROM®/TERPROM®II

TERPROM® is a digital terrain system developed by BAE Systems. Its capabilities include high accuracy, drift-free navigation, predictive ground proximity warning system (GPWS), obstruction warning and cueing, database Terrain Following (TF) and enhanced weapon aiming. The system is entirely passive and produces no forward emissions.

TERPROM® uses standard digital terrain elevation data and vertical obstruction information stored in on board computer mass memory. A Kalman filter compares this data with information from the radar altimeter, allowing in-flight calibration of errors in the aircraft inertial navigation system.

Terrain referenced navigation accuracies of 30 m CEP (horizontal) and 3 m LEP (vertical) are achievable over terrain roughnesses as low as 2 per cent. System robustness is such that accurate navigation can be maintained for long periods when flying over water or during total radio silence.

The recently launched TERPROM®II product has dramatically improved navigation performance which, combined with the use of optional GPS velocity data, eliminates any nuisance ground proximity warnings during tactical flying.

TERPROM® generates 'predictive' ground proximity warnings which are not reliant on current radar altimeter inputs but are based on the output from the navigation capability. Warnings may therefore be provided even when the aircraft is approaching the terrain inverted.

The predictive algorithm is tuned to match the performance of each specific aircraft platform. This, together with the high accuracy and terrain referenced nature of the system's navigation solution, allows timely and accurate warnings to be provided even when the aircraft is being flown aggressively as low as 250 ft above ground level.

TERPROM® provides precursory directional warnings of man made obstructions (for example radio masts) in the aircraft's flight path. This allows the pilot to visually locate and fly around rather than over obstructions, reducing his chance of being detected.

TERPROM® generates terrain following signals, which may be displayed to the pilot. Unlike conventional terrain following radar systems, TERPROM® achieves total freedom of manoeuvre by the use of its onboard map. Through its constant awareness of the shape of the terrain beyond the immediate horizon, TERPROM® enables aircraft to hug contours more closely, reducing exposure to attack.

TERPROM® passive ranging allows undetected approach and accurate weapon aiming accuracy. With its knowledge of terrain elevation and true aircraft height, TERPROM® enables height above target, and hence impact location, to be computed continuously and accurately.

A single shot look aside ranging mode is also available, providing a 'passive laser range-finder' capability.

Inputs to TERPROM® include the: inertial navigation system data; baro/inertial altitude; radar altimeter; Pilot selections (TF and GPW heights); GPS data (optional).

Outputs from TERPROM® are as follows: inertial navigation system corrections; ground proximity warnings; obstruction warnings and cues; terrain following marker; ranging data.

TERPROM®/TERPROM® II is available in the:BAE Systems Modular Navigation System; Honeywell H764G embedded INS/GPS; OSC Fairchild Defence Mega Data Transfer Cartridge with Processor (MDTC/P); AYK14 and Open Systems mission computers from Computing Devices.

Operational status

TERPROM® was originally developed by BAE Systems for low flying fast jet aircraft, and has been selected for, or is in service on, Eurofighter Typhoon, F-16, Harrier GR. Mk 7, Jaguar, Mirage 2000 and Tornado GR. Mk 4.

The transport aircraft variant of TERPROM, known as the Terrain Awareness Warning System (TAWS), has been selected for the US Air Force's C-17 fleet.

A version of TERPROM for rotary-wing aircraft has been developed, and was successfully trialled by the British Army in September 1999 on a Lynx helicopter. The purpose of the trial was primarily to demonstrate TERPROM's capabilities on helicopters flying transit and terrain missions. Dynamic terrain advisory cues combine with a map of the surrounding terrain enable the pilot to assess his course of action while remaining at low level, with terrain cues being provided down to 100 ft. Further evaluation with the UK military, as well as trials with a US helicopter manufacturer are planned for 2000.

Contractor

BAE Systems.

VERIFIED

Terrain Reference Navigation (TRN)

Terrain elevation data can be exploited to achieve automatic micro-navigation and covert terrain-following of the necessary high integrity. The micro-navigation employs modern radar altimeters incorporating variable output power and spread spectrum techniques to provide low probability of detection. In Spartan, the radar altimeter is used to map the vertical cross-section of the terrain beneath the aircraft for matching within the database. This Terrain Referenced Navigation (TRN) produces fixes every 2 seconds. The fix matching algorithm is robust, recovering quickly from the larger inertial errors likely to be met after a prolonged water crossing or from any local errors in the terrain data.

The TRN fixes are used to Kalman filter the INS outputs and thus, even with medium-grade INS, the aircraft position is known with extremely good accuracy and high confidence. With this knowledge, the database can be scanned over the projected flight path to predict the ground profile ahead. The TF algorithm is therefore able to define and demand a totally safe kinematic flight path that ensures the closest maintenance of clearance level, even in manoeuvring flight. The sudden unmasking of close terrain and ballooning over hill crests that can occur with conventional TF radars is avoided. The system also monitors achieved and forecast ground clearances against those required and gives ground proximity warning where necessary. Additionally, obstacles are included in the database and their location and indicated height projected on to the pilot's HUD or helmet-mounted display for obstacle cueing.

Specifications

Dimensions: ¾ ATR short
Weight: 15 kg
Power: 200 W
Coverage: 230,000 sq miles
Database programming: 13 min
Environmental: MIL-STD-810, MIL-STD-461 and 462
Reliability: 3,000 h MTBF

Digital colour map

Advances in techniques for lossless compression, storage and recovery of digital data, combined with developments in cockpit colour displays, have been exploited to produce two- and three-dimensional digital map presentations with wide area coverage. Topographical and cultural data can be stored in pixel or vector formats and recalled for colour video display as realistic reproductions of paper maps or as terrain representations. Additional switchable colour palettes are available to suit different applications, for example to suit cockpit lighting conditions for night operations. Depending upon how the database is compiled, the resulting display can be decluttered of unwanted detail. The viewing area can be rotated in orientation, swiftly and smoothly manoeuvred about the stored area and viewed at any one of the standard aeronautical map scales or with a zoom capability. Additionally, on a sortie-to-sortie basis a mission routing and intelligence overlay can be generated and loaded into the database for use in flight.

The addition of terrain elevation information to the database generates a wide range of enhanced tactical head-down display options to overlay the basic map. For example, dynamic relative height shading provides terrain-avoidance assistance, while ground-to-air intervisibility displays, with ground threat positions and effective ranges, aid threat-avoidance manoeuvring. Conversely, dynamic air-to-ground intervisibility displays demonstrate the achievement of terrain-masking and provide a pseudo radar display to ensure that any mapping radar transmissions are initiated only when the target or fix point is in radar view. Head-up display enhancement is achieved by the generation and precisely placed display of ridge lines to give added pilot confidence in poor visibility or when limited to flat contrast FLIR pictures.

Operational status

In service in Harrier II, Jaguar, Tornado and C-130J aircraft.

Contractor

BAE Systems.

VERIFIED

Type 453 laser warning receiver

The Type 453 laser warning receiver is designed for use in fixed-wing aircraft, helicopters and armoured fighting vehicles. It provides a countermeasure to laser-guided weapon systems and provides an audible alarm and visual display showing the direction from which the threat has originated. Output from the receiver can be combined with radar warning equipment to give an integrated battlefield threat warning system or with a defensive aids system.

The system uses a number of dispersed sensors to detect incident laser radiation. This counters the problem of detecting a very narrow beam which at any instant would be illuminating only a small part of the airframe. The sensor head configuration can be tailored to provide spherical or hemispherical cover and presents the laser threat bearing to the crew as sectoral information. The number of sensors required is tailored to the aircraft type. The sensor protrudes through the aircraft skin as a 25 mm hemisphere and occupies a space of 50 mm depth behind the skin.

The sensor heads are completely passive. Laser radiation is routed to the central processing unit by fibre optic cables. This feature eliminates risks of false alarms being generated by radio frequency interference.

Operational status

In advanced development for the Eurofighter Typhoon.

Contractor

BAE Systems.

VERIFIED

Zeus ECM system

Zeus is a compact multipurpose EW integrated defensive aids system for combat aircraft. It provides a

The BAE Systems Zeus internal ECM system

complete defensive system consisting of a radar warning receiver and a jammer. The RWR is the latest in a series of passive intercept systems and is based on IFM and fast superheterodyne techniques. It is able to intercept and measure the characteristics of all radar-controlled systems which may threaten the aircraft. The RWR will measure parameters including direction of arrival, time of arrival, frequency, PRI, pulsewidth, amplitude, scan interval and scan rate. Signals are passed to the Zeus digital processor which identifies radar type, displays it and gives audio warning. The processor also controls the jammer and other means of countermeasures such as chaff, flares and decoys. The transmitter will jam both pulse and CW radars, while control features ensure that home-on-jam radars are not given sufficient time to lock on to the aircraft. Different jamming modes are available with Zeus for the operational requirements of individual customers.

Specifications

Weight: 118 kg
Frequency: C- to J-bands
Coverage:
(azimuth) 360°
(elevation) ±45°
Response: 1 s typically
Jamming: noise, VGPO, deception, co-operative

Operational status

Zeus is in service with the UK Royal Air Force on the Harrier GR. Mk 7 aircraft. Installations have been designed for other aircraft, including the JAS 39 Gripen, AV-8B, F-16 and F/A-18. Zeus has successfully completed technical and operational evaluations in France, the UK and USA.

Contractor

BAE Systems.

VERIFIED

---

7 series homing systems

The 7 series is a versatile building block homing system, capable of providing both broadband and emergency guard channel homing. In its simplest form, a 7 series installation comprises a homing indicator, an antenna feed unit, an antenna system and the aircraft receiver. When an independent self-contained emergency guard channel homing system is required, the 7 series uses a Chelton 7-28 Series two-channel receiver in place of the aircraft receiver. It is also possible to interface both the aircraft receiver and the 7-28 Series receiver with the 7 series, thus providing broadband plus guard channel homing. The 7 series interfaces with all AM receivers, including the ARC116, ARC159, ARC164, ARC182, ARC186, PTR1751 and VHF20, and also with those FM equipments having AM facilities.

The 7 series comprises a number of sub-units which include:

Series 7-24 homing indicator units. These are self-contained homing directors incorporating all necessary electronic control circuits, voltage regulators and computer reference amplifiers, packaged within an 89 mm housing. The indicators have a variable sensitivity control located on the front panel and are edge-lit.

Series 7-27 indicator feed units. These comprise the electronic circuitry otherwise packaged within the 7-24 homing indicator units and are intended to feed existing aircraft navigation indicators or flight directors.

Series 7-25 antenna feed units. These contain the necessary antenna phasing and switching circuits to couple antennas into the homing system. Unlike the sub-units listed above, the choice of antenna feed unit is dependent upon the frequency band required.

Series 7-28 twin-channel emergency guard receivers. These small receivers are designed to provide a completely self-contained homing system at VHF and UHF distress frequencies.

Series 7-60 complete homing adaptors. These are combinations of Series 7-27 indicator feed units and Series 7-25 antenna feed units packaged in one small module.

Specifications

Weight: less than 1 kg
Power: 28 V DC, 500 mA max

Operational status

More than 700 systems are in service with military and commercial operators worldwide.

Contractor

Chelton (Electrostatics) Ltd.

VERIFIED

700 Series sonobuoy and VHF homing receivers

700 Series receivers are frequency synthesised, dual-bandwidth VHF receivers designed to form part of a system to provide port and starboard and fore and aft VHF homing with automatic on-top indication. The dual bandwidth permits reception of 25 kHz spaced signals when in the narrowband mode, or 375 MHz spaced sonobuoy signals in the wideband mode. An audio output is available to provide aural port-starboard homing and recognition of amplitude-modulated signals.

The 700 Series was developed from and embodies technology and subsystems used in the Chelton VHF/UHF homing systems, of which more than 700 are in service for worldwide military and civil operations. A complete system comprises four LRUs plus antennas, and features frequency-synthesised dual-bandwidth operation, up to 99 preset channels with full OTPI interface, low weight and small size.

Specifications

Type 700 receiver
Weight: 1.325 kg
Power supply: 28 V DC, 500 mA
Frequency:
(Type 700-1) 100.00-173.500 MHz,
(Type 700-2) 117.975-173.975 MHz

Type 710 control unit
Weight: 0.98 kg
Power supply: 22-31.5 V DC, < 1.0 A
Frequency:
(Type 710-1) 100.00-159.975 MHz,
(Type 710-2) 117.975-173.975 MHz
Channel mode frequency:
(Type 710-1) 136.00-173.500 MHz,
(Type 710-2) 136.00-173.500 MHz

Operational status

Chelton 700 Series sonobuoy and VHF homing receivers have been ordered by Bristow Helicopters, the Helicopter Services (Navy), the UK Royal Navy, the Royal Norwegian Navy and the Royal Swedish Navy.

Contractor

Chelton (Electrostatics) Ltd.

VERIFIED

730/930 Series sonobuoy homing and channel occupancy system

The 730 series 99-channel sonobuoy homing and channel occupancy system operates over the frequency range 136.000 to 173.500 MHz and has homing or channel occupancy independent modes of operation.

In the homing mode, the system supplies azimuthal information so that the aircraft may be flown overhead a sonobuoy. Additional information indicates whether the sonobuoy is ahead or astern of the homing aircraft. This information is further decoded to give an indication of the instant the aircraft passes overhead the sonobuoy.

In the channel occupancy mode, the receiver measures the received signal strength to show when a selected channel is clear for the reception of sonobuoy signals.

The system comprises eight LRUs: the Type 730-1 radio receiver, Type 731-1 homing/channel occupancy switch, Type 732-1 azimuth switch coupler, Type 733-1 fore and aft coupler, Type 734-1 receiver mounting tray, azimuth antenna assembly, fore and aft antenna assembly and channel occupancy antenna assembly.

The heart of the system is the radio receiver which interfaces the system to the aircraft via a dual-redundant MIL-STD-1553B databus.

In the channel occupancy mode, the radio receiver provides the necessary control signals to disable the azimuth and fore and aft switch couplers. The receiver cycles through channels as designated by the mission computer and provides information on the databus channel status, together with received signal strength.

In the homing mode, the radio receiver provides the necessary control signals of the azimuth and fore and aft switch couplers to switch the RF input of the radio receiver between the antenna assemblies.

BITE is capable of diagnosing 95 per cent of all defects to LRU level.

Chelton's experience with the 730 Series system has led to the development of the 930 Series VHF/UHF direction-finding systems. These provide 360° bearing information in analogue or digital format to accuracies better than 5° and use high reliability, stepper motor driven, rotating antennas and enhanced software.

The Chelton 730 Series sonobuoy homing and channel occupancy system comprises (left, top to bottom) the Type 731-1 homing/channel occupancy switch, the Type 733-1 fore and aft switch coupler, the Type 732-1 azimuth switch coupler, (centre) the Type 730-1 radio receiver and (right) the Type 734-1 receiver mounting tray

Specifications
Weight:
(all LRUs) 8 kg

Operational status
The 730 Series sonobuoy homing and channel occupancy system has been selected for the Westland/Agusta EH 101 Merlin ASW helicopter.

Contractor
Chelton (Electrostatics) Ltd.

VERIFIED

ACCS 3100 audio management unit

The ACCS 3100 Audio Management Unit (AMU) provides complete integration of audio selection, aural warnings generation and speech recognition functions within a single unit. Selections and control are normally provided via the MIL-STD-1553B avionics databus from an integrated avionics suite, although a dedicated control panel can also be offered.

The AMU implements all the intercommunications functions required on the next generation of military aircraft such as the Eurofighter Typhoon or for avionic upgrades to existing aircraft. Intercommunication is provided between radios, navaids, pilot, co-pilot, ground crew, recorders and the avionics system status outputs which activate aural warnings. Digital signal processing techniques effect noise tracking voice-operated switching for hands-free intercom, and noise filtering and voice-operated gain adjustment devices ensure audio signals routed to transmitters are of optimum clarity and modulation.

Aural warnings include voice warnings as well as attentions and tones, as commanded by external status inputs from dedicated discretes or via a MIL-STD-1553B databus. Voice warnings storage for up to 200 messages is currently provided. Digitised voice output of status and informatory data can also be activated by direct voice input commands from the aircrew.

Direct voice input uses speech recognition algorithms within the AMU to ensure accurate performance despite ambient cockpit noise and stress-induced voice variations. A programmable 200 word vocabulary, together with a dynamic syntax pointer responsive to avionics system status, gives sufficient flexibility to meet most demanding applications.

The AMU uses a powerful Ada-programmed 68020 processor, running BIT as well as normal operating programs; a non-volatile maintenance memory is available to store fault data. The AMU databus interface is compatible with MIL-STD-1553B, STANAG 3838 and STANAG 3910 fibre optic databus protocols.

Specifications
Dimensions: 380 × 200 × 124 mm
Weight: 7.7 kg
Power: 100 W
Processing: 30 Mips

Operational status
In service. Also selected as the communications and audio management unit for the Eurofighter Typhoon.

Contractor
Computing Devices, a General Dynamics Company.

VERIFIED

ACCS 3300 digital map generator

Computing Devices has designed rugged map display equipment intended for installation in harsh airborne

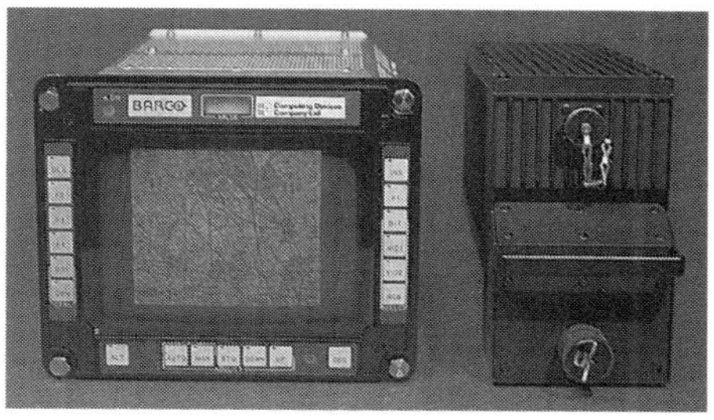

The ACCS 3300 digital map generator for the EH 101 Merlin helicopter with a Barco display

environments. The ACCS 3300 digital map generator has been selected for the UK Royal Navy Merlin helicopter. This design has a scalable architecture in order to meet a range of performance criteria, from static north-up maps to full rotating tracking maps with overlays. This will allow presentation of both raster and true vector data with efficient man/machine interface to facilitate zoom and a list of additional functions. The growth capability enables additional functions such as intervisibility, two- and three-dimensional displays. Map data is optimally compressed and stored to provide the operator with control of the map, to enable display of key map and overlay information.

The complexity of overlays can include text with sensor data, providing the ability to perform sensor and map fusion and object correlation presented on a range of low- to high-resolution displays.

Map database expertise has been achieved in compressing raster data and, more significantly where a structured dataset is required, in automatically generating vector data from a raster map using its own Unix-compatible knowledge system. The advantage of utilising vector map data is its easy interpretation, enhancement and attribution. The vector dataset and its associated overlays are more easily maintained and updated than raster.

Specifications
Dimensions: ½ ATR
Power supply: 115 V AC, 400 Hz, 110 W
28 V DC optional
Outputs: RS-170, RGB 625-line CCIRR
Environmental: MIL-STD-810E
Reliability: >2,000 h MTBF

Operational status
In service in the UK Royal Navy EH 101 Merlin helicopter.

Contractor
Computing Devices, a General Dynamics Company.

VERIFIED

Eurofighter Typhoon cockpit Video and Voice Recorder (VVR)

The Computing Devices' Eurofighter Typhoon cockpit Video and Voice Recorder (VVR) integrates a Cockpit and Video Interface (CVI) with an Off-The-Shelf (OTS) TEAC V-80AB-F Hi-8 mm Airborne Video Tape Recorder (AVTR).

A video demultiplexing system allows post mission replay of recorded (video, MIL-STD-1553B mission data) data on four standard video monitors using a range of search facilities.

The Eurofighter Typhoon VVR is capable of recording colour and/or monochrome video, audio data and the indication of the occurrence of event marking signals.

The Eurofighter Typhoon VVR interfaces with: two Computer Symbol Generators (CSGs) for the recording of video; the Computing Devices' Communications and Audio Management Unit (CAMU) for the recording of audio; dual redundant MIL-STD-1553B terminals to accept event marking, input selection and mode control signals and to output status data; two discrete output VVR status signals; one discrete input signal to facilitate manual mode control.

The two CSGs output multiplexed video display data to the VVR in RGB format. The minimum record time is 90 minutes.

The VCI comprises four functional blocks: Control Mode and Monitoring module (CMM); Video, Audio and Data module (VAD); Power Supply Module (PSM); Case Electrical Assembly (CEA) including motherboard and Elapsed Time Indicator (ETI).

The CMM module incorporates a microcontroller, MIL-STD-1553B chipset, control interface, data encoder and test bus interface. The VAD module incorporates the discrete interfaces, video interfacing and channel controls and audio conditioning. The PSM accepts +28 V power from the aircraft power systems.

Operational status
Development systems are flying on Eurofighter Typhoon development aircraft; production deliveries are due in 2000.

Contractor
Computing Devices, a General Dynamics Company.

VERIFIED

Eurofighter Typhoon Cockpit Video and Voice Recorder (VVR) **1999**/0051343

Pod SAR

The UK Defence Evaluation Research Agency (DERA) is leading a UK MoD applied research programme for a podded SAR (Synthetic Aperture Radar) and G-MTI (Ground-Moving Target Indicator) mode radar to provide detailed battlefield all-weather day and night penetrating reconnaissance capabilities for Tornado and Eurofighter Typhoon in areas where standoff systems and other more vulnerable systems cannot see or be employed.

DERA has awarded Racal Defence Electronics the single concept study to design the technology demonstrator. The demonstration radar will have a high-resolution SAR output of both spot and strip modes, together with G-MTI capabilities.

Pod SAR will complement other assets, such as ASTOR. The radar will be installed in the DERASmart Pod, which can be configured for a variety of sensor fits.

Operational status
Technology demonstrator programme. Intended to enter service in about 2004.

Contractors
Defence Evaluation Research Agency (DERA)
Racal Defence Electronics.

VERIFIED

Terrain referenced mission systems

Research is continuing into the exploitation of terrain referencing principles and database technology used in conjunction with Electro-Optic (E-O) sensor systems. Given precise knowledge at all times of aircraft spatial position and attitude and access to a digital topographical database, it is possible to synthesise real-time perspective imagery of the terrain scene around the aircraft. This image, when properly fused with information from an E-O sensor such as a FLIR, can form the basis of a low-flying aid in situations where information from the FLIR is less than adequate. The concept is readily extended to portraying other database information such as targets, flight path obstructions and power cables. This last application is particularly attractive and is perceived as offering a major contribution to flight safety during fast jet and helicopter operations.

Terrain and obstacle scene generation combined with FLIR has recently been shown to be a viable concept. The basic philosophy is that of enhancing the FLIR image by cueing features that are not otherwise clearly visible due to the prevailing thermal conditions. Development of the concept is continuing using ground rig facilities together with flight trials on Andover and Hunter laboratory aircraft equipped with the necessary enabling technologies. On the Hunter the system is implemented as part of the PENETRATE demonstrator programme.

PENETRATE (Passive Enhanced Navigation with Terrain Referenced Avionics) is a venture to develop an integrated covert mission system exploiting terrain reference technology. The system, mounted in a small pod, comprises a mass data store and a computing suite undertaking such functions as terrain referenced navigation, terrain/obstacle scene generation, tactical routeing and mission management. The mass data store is a 400 Mbyte capacity optical disk system developed for use in a combat aircraft environment.

Operational status
Continuing research and development.

Contractor
Defence Evaluation and Research Agency, Farnborough.

UPDATED

Visually Coupled System (VCS)

A Visually Coupled System (VCS), currently under development at the Defence Evaluation and Research Agency, Farnborough, will enable research into Helmet-Mounted Display (HMD) flight symbology, platform characteristics and the human factors of VCS operation at night and in adverse weather conditions. Navigation accuracies of better than 20 m are anticipated from an integrated INS/GPS/terrain referenced navigation suite.

A Lynx helicopter has been extensively modified to accept a chin-mounted pan and tilt platform containing a thermal imaging sensor. The platform moves in synchronisation with head movements which are detected by DC head trackers, and a stabilised 1:1 thermal image of the outside world is displayed on a 53° field of view HMD, together with essential flight symbology. The symbology can be rapidly reprogrammed in order that optimised displays for specific flight conditions can be designed and proven.

The scope of the research will shortly be expanded to investigate the potential of obstacle warning devices, enabling operation at low level or close to obstructions with increased confidence.

The benefits of this research may be enjoyed by military and civil helicopter operators alike. Nap of the earth flight at night or in adverse weather allows military helicopters to operate covertly on the battlefield, increasing survivability and mission effectiveness. There are advantages for search and rescue and offshore rig helicopter operators where knowledge of the presence of sometimes highly dynamic obstructions, such as ship superstructures, is essential.

Contractor
Defence Evaluation and Research Agency, Farnborough.

UPDATED

Genesis SR rugged multiplatform computers

The Genesis SR (Short Rack) is an adaptable, rugged, rackmount computer designed for installation in wheeled or tracked vehicles, military aircraft, surface ships and submarines. Typical applications include tactical communications systems, mission planning, combat and logistics support.

The unit can accommodate different processor engines (Intel Pentium™, Power PC™, Compaq Alpha™, Sun UltraSPARC™), it has configurable I/O panels, and can be operated from interchangeable AC, DC or AC/DC power supplies. Optionally it can be supplied to Tempest BTR 01/210.

Specifications
CPU options: Intel Pentium™, Power PC™, Compaq Alpha™, Sun UltraSPARC™
Internal drives: Will support 'hot swap' and disk monitoring (RAID)
Input/Output: Configurable I/O panel (optionally: 104-key hinged removable keyboard with integrated trackball
Display: DRS Rugged Systems' FPR rugged flat panels from 12 to 20 in
Power: 110-230 V AC (option 400 Hz and 19-32 V DC)
Dimensions: height 4U, depth 400 mm
Weight: 19.5 kg

Operational status
A variant system, designated Genesis Ultra, is being supplied as the Link 11 driver for UK Royal Air Force Nimrod MR2 Mk2 aircraft.

Contractor
DRS Rugged Systems (Europe) Ltd.

VERIFIED

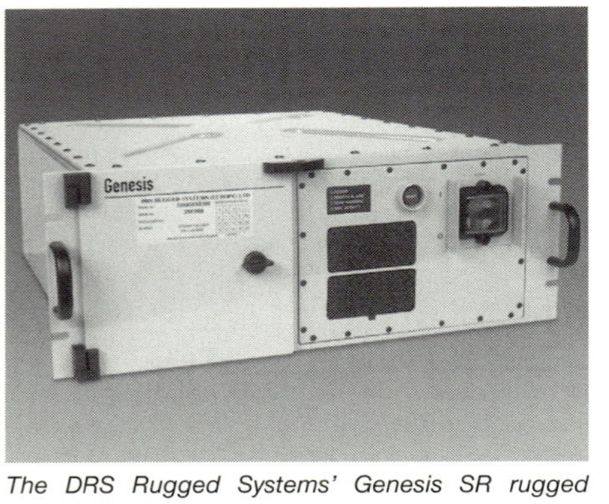

The DRS Rugged Systems' Genesis SR rugged multiplatform computer **2000**/0062797

Mission recording system

The Honeywell (Normalair-Garrett Ltd) recording system is designed for use in ASW operations. It uses Ampex Corporation DCRSi digital cassette recording technology.

Operational status
In production for the EH 101 Merlin HM Mk 1 ASW helicopter.

Contractor
Honeywell,
(Normalair-Garrett Ltd.)

VERIFIED

NGL sonobuoy dispenser

The Honeywell (Normalair_Garrett Ltd.)NGL sonobuoy dispensing system

The sonobuoy dispenser forms part of an automatic system for the deployment of sonobuoys during ASW operations.

The Honeywell (Normalair-Garrett Ltd) sonobuoy dispenser is a 10-station rotary device for the carriage and release, by gravity, of various sonobuoys in any combination of the standard 'A', 'F' or 'G' sizes. It can be remotely controlled from a tactical mission console or the stores management system and can be used to release a predetermined pattern of sonobuoys.

The dispenser, although primarily designed for a helicopter application, can also be used for fixed-wing operations, where the addition of a pressurisation valve and associated safety interlocks are required to ensure that sonobuoys cannot be dropped above aircraft pressurisation levels.

The composite dispenser is supported by a base structure which is attached to the floor of the aircraft by means of receptacles in the aircraft seat rails. Additionally, the dispenser is braced at the top by a top-steady arrangement which attaches to a spigot on the top cover.

Specifications
Weight: 34.5 kg, excluding the aircraft/dispenser interface unit and top steady bracket.
Power consumption: Drive mechanism: 115/200 V, 3 phase, 400 Hz AC, 300 W-28 V DC, 8 A.
Release solenoid: 28 V DC-13 A for 160 ms

Operational status
In production for EH 101 Merlin HM Mk 1 ASW helicopter. Selected for Nimrod MRA4.

Contractor
Honeywell,
(Normalair-Garrett Ltd.)

VERIFIED

Advanced Digital Radio ADR + VHF radio

The ITT Defence Ltd software programmable ADR + VHF radio has been evolved from the proven Single Channel Ground Airborne Radio System (SINCGARS); it uses the proven functionality with the benefit of an all-digitial signal processing architecture and features:

(a) Embedded GPS position and user information in data transmissions, enabling accurate monitoring of friendly force positions for situational awareness

(b) New frequency-hopping packet data waveform with forward error correction, reduced on-air transmit time, reduced transmission overhead, and improved message throughput

(c) Improved channel access algorithm which combines voice and packet data on a common net, with packet throughput rates and minimal impact on voice communication

(d) Bowman Internet Communications Controller (BICC) with a programmable interface which converts protocols, providing a seamless flow of data across the battlefield and facilitates horizontal and vertical integration of command and control.

Waveform improvements have been implemented to improve simultaneous voice and data transmission.

The BICC has been developed to interface both the UK Bowman battlefield communications system and the US Force XXI digitisation programme.

The ADR + VHF radio interfaces directly with the main battlefield Bowman radios and the compatible ITT-enhanced portable VHF radio.

Specifications

Frequency range: 30.00-87.975 MHz
Modulation: Gaussian Minimum Shift Keying (GMSK)
Channels: 2,320
Channel spacing: 25 kHz
Modes of operation: fixed frequency; frequency hopping; free channel search
Preset channel operation: 6 single-channel; 6 frequency hopping; 6 free channel search
Data rate: 30 kbps
Transmit power: 16 W (adjustable)
Power: 28 V DC
Dimensions: 139.7 × 195 × 104 mm
Weight: 3.73 kg

Advanced Digital Radio ADR + VHF radio 0002126

Contractor

ITT Defence Ltd.

VERIFIED

ARI 5983 I-band transponder

The ARI 5983 I-band transponder provides a means of locating, identifying and providing navigational assistance to aircraft outside normal radar coverage and range. It is interrogated by a primary radar and gives an edge-of-band response. The response codes are selected on a simple control unit from which a comprehensive BIT routine can be initiated. The control unit also allows either manual or electronic switching between the transponder's two antennas to ensure optimum coverage.

The transponder receives interrogation signals, via the antenna, from pulse radars at any frequency in two bands 100 MHz wide. When interrogated, the transponder will respond with either a single-RF pulse, which provides enhancement of the radar return, or a coded group of up to six pulses, as selected on the control unit, which allows identification. There are 16 different reply codes available.

The transmitted power can be reduced by approximately 11 dB via another switch on the control unit. The transponder output will automatically be suppressed during the operation of other I-band equipment in the aircraft. Similarly, a pulse is supplied by the transponder to allow suppression of other equipment in the aircraft operating in the same frequency band when the transmitter is operating. The BIT self-test facility generates an interrogate signal which is fed into the transponder input. A green LED on the control unit indicates correct transponder operation.

Options include double- or multiple-pulse interrogation to minimise false triggering when several I-band radars are transmitting in the same area.

Both the transponder and control unit are fully NATO codified.

Specifications

Dimensions:
(transponder) 160 × 217 × 87 mm
(control unit) 147 × 117 × 48 mm
Weight:
(transponder) 2.7 kg
(control unit) 0.45 kg
Power supply: 28 V DC, 40 W (max)
Frequency:
(receive) 9,190-9,290 MHz and 9,360-9,460 MHz
(transmit) 9,310 ±7 MHz
Bandwidth: ±50 MHz
Sensitivity: –93 dBW
Output power: 135 W (min) to 300 W (max) peak
Pulse duration: 0.45 μs ±0.1 μs

The M/A COM ARI 5983 IFF transponder

Reply code: 6-pulse code, 16 settings single-pulse reply capability
Pulse spacing: 2.9 μs nominal
Duty cycle: 0.005 (max)

Operational status

In service with the UK Royal Navy on the EH 101 Merlin, Lynx, Sea King and Sea Harrier FRS.1 and F/A-2. Also exported for Dauphin and Lynx helicopters and the Do-228.

Contractor

M/A COM Ltd.

VERIFIED

ML3500 radar transponder

To supplement the more sophisticated ARI5983 I-band transponder, M/A COM has introduced the ML3500 radar transponder. The ML3500 provides an edge-of-band response when interrogated by an I-band radar, and is ideal for enhancing the radar echoing area of small air and surface targets. Although in essence an active corner reflector, the transponder's response can be simply coded to aid target identification and, being edge-of-band, reduces primary plot clutter on radars with tunable receivers.

The ML3500 is lightweight and fully weatherproof.

Specifications

Dimensions: 120 × 170 × 55 mm
Weight: 1 kg
Power supply: 12 or 24 V DC, 10 W
Frequency:
(receiver) 9,000-9,600 MHz
(transmitter) 9,200-9,400 MHz (factory set)
Sensitivity: –43 dBm (min)
Interrogate pulsewidth: 0.15 to 1.5 μs
Stability: ±10 MHz
Pulse duration: 0.2 to 1 μs
Reply code: 5 output pulses - one for range mark, four customer settable identifiers
Duty cycle: 1% (max)
Temperature range: –20 to +50°C

Contractor

M/A COM Ltd.

VERIFIED

Hostile Fire Indicator (HOFIN)

The Hostile Fire Indicator (HOFIN) uses shockwave sensors to detect gunfire aimed at the host aircraft, and is a passive warning system suitable for helicopter use. It uses a five-armed sensor (four arranged in one plane at 90° to each other and one perpendicularly) which is mounted beneath the helicopter. Shockwaves generated by projectiles are converted to electrical signals and then processed in an electronic unit which occupies a ⅜ ATR short box. This produces outputs which can be used to generate audio or visual warning of nearby arms fire. An audio warning lasts for approximately one second after a shockwave has been detected and a visual warning is presented on a 4 ATI size unit which has eight 45° wide segments. Four segments which indicate the approximate direction from which the shock is detected will illuminate for about five seconds.

Specifications
Dimensions:
(sensor array) 305 × 305 × 195 mm
(computer) 94 × 418 × 228 mm
(indicator) 106 × 106 × 125 mm
Weight:
(sensor array) 1.93 kg
(computer) 2.72 kg
(indicator) 1 kg
Power supply: 22-28.5 V DC, 60 W
Sensitivity: responsive to supersonic projectiles at 20 m
Temperature range: –20 to +50° C
Humidity limit: 95% non-condensing

Operational status
In service with British, Italian and Canadian forces.

Contractor
MS Instruments plc.

VERIFIED

DVR2000B airborne Digital Video Recorder

The Prostab DVR2000B airborne video recorder uses a compact digital tape to record video pictures in real time. It provides up to three hours recording time. Prostab claims that use of the latest DVCAM digital format provides high-quality pictures that are superior to those obtained from current analogue tape recorders, while enabling the recorder itself to be compact and lightweight. The DVR2000B is designed for harsh military environments and can be qualified to MIL-STD-810D and MIL-STD-461C.

The DVR2000B has been developed from the Prostab S-VHS airborne video recorder and it offers the advantage of digital recording for very high resolution with low noise levels. It can use either a mini-cassette to provide 60 minutes of recording, or a standard cassette to provide three hours of recording. Units can be cascaded to provide longer duration continuous recording.

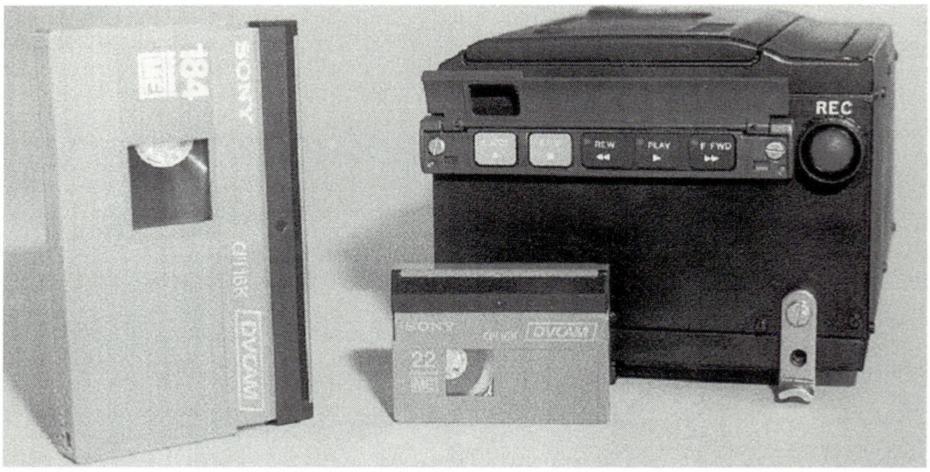

The Prostab DVR2000B airborne digital video recorder **2000**/0064372

The DVR2000B can be built to be form/fit/function compatible with existing analogue systems to enable existing installations to be upgraded economically to the new digital format.

The DVR2000B can be equipped with remote controls for user functions (record, stop and others) which can be either hard wired or via serial interface. Other options include time/date/text insertion onto the recording.

Tapes can be replayed using a desktop DVCAM playback station and copied on to other formats as required.

Specifications
Recording system: DVCAM digital tape
System format: 4:2:2 digital
Recording duration:
(mini-cassette) up to 60 minutes
(standard cassette) up to 180 minutes
Video bandwidth: 25 Hz to 5.5 MHz; +1/–2 dB
Video S/N: 55 dB minimum
Audio input: 2 channel
Video input: PAL composite or S-video digital camera input optional
Power: 28 V DC, 12 W
Operating temperature: –40 to +50°C
Dimensions: 200 × 180 × 100 mm maximum
Weight: 5 kg

Contractor
Prostab, A Smiths Industries Company.

VERIFIED

AMS 2000 multifunction Control Display Navigation Unit (CDNU)

The AMS 2000 multifunction CDNU is a flexible navigation and management system with embedded P(Y) code GPS that is easily configurable to suit the customer's requirements and the host airframe. Typically Search and Rescue (SAR) specific functions are added for aircraft/helicopters, having a SAR role.

This CDNU is configured as a navigation computer using GPS and, if available, a combination of sensors including IN, Doppler and Air Data. From this baseline, the CDNU capability may be increased to meet additional customer requirements - thanks to the modular approach adopted for hardware and software. The CDNU can provide a full mission management facility in single or dual configuration using MIL-STD-1553B and ARINC 429 databus.

The AMS 2000 CDNU provides centralised control and display of the chosen avionics suite. Data and instructions are inserted manually via the keyboard or Data Transfer Device (DTD). The DTD is used for flight planning and for post-flight data retrieval. A customised mission planning station is available.

The CDNU software is written in Ada. The non-volatile Flash memory is reprogrammable via the DTD. The CDNU's large memory capability allows a wide range of equipment interfaces to be supported - in addition to navigation sensors.

The AMS 2000 multifunction CDNU is compatible with GNSS and provides a control and display function for the Racal Avionics Satellite TRansceiver (STR) system. The CDNU may be used as a cockpit mission system or a tactical system, without any additional interfacing hardware being required.

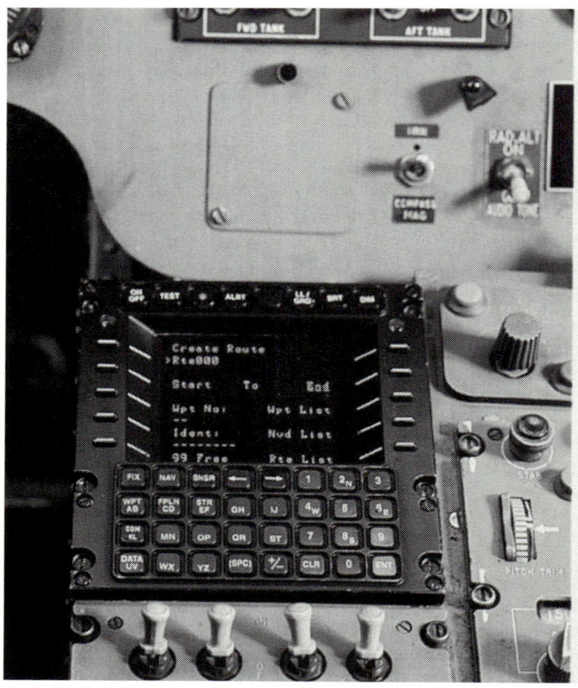

AMS 2000 CDNU 0002442

Specifications
Size: 161.43 × 145.5 × 216.5 mm
Weight: 5 kg (max)
Power: 28 V DC, 45 W (max - for full module complement), typically 36 W
Cooling: convection - no forced air required
Environmental: MIL-STD-810E; MIL-STD-461C/D RTCA/DO-160C
Memory: 4 Mbyte Flash PROM
I/O: MIL-STD-1553B; ARINC 419/429 (Hi/Lo) RS-232, -422; DC, AC, synchro, discretes
GPS: embedded C/A or P(Y) 5 channel to STANAG 4294. Includes Have Quick, BID 250 and KYK 13 interface

Operational status
Under contract to UK MoD for; Lynx HMA Mk 8, Merlin HM Mk 1, Merlin SH Mk 3, Nimrod MR2 Mk2, Nimrod MRA 4, and Sea King Mk 3, 4, 5 and 6.

Contractor
Racal Avionics Ltd.

VERIFIED

CDU/IN/GPS

Racal Avionics is responsible for full integration of an Inertial Navigation and Global Positioning System (IN/GPS) and a Control Display Unit (CDU) to interface with the Central Tactical System of the UK Nimrod MRA4 aircraft. The CDU concerned is a derivative of the AMS 2000 Control Display and Navigation Unit, made by Racal Avionics, and the combined unit is to be known as the CDU/IN/GPS.

Operational status

In development for Nimrod MRA4.

Contractor

Racal Avionics Ltd.

VERIFIED

Doppler velocity sensors

There are five models in the Racal Doppler velocity sensor family: the Doppler 71 and 72 antenna units for helicopters and fixed-wing aircraft respectively; the Doppler 80 for helicopters and light aircraft; and Doppler 91 and 92 antenna units for helicopters and fixed-wing aircraft respectively.

The Doppler 91 and 92 units use the best features of the previous models and incorporates microprocessor technology and new manufacturing techniques. These units use waveguide antennas and varactor multiplier transmitters for high accuracy at greater altitudes.

The Doppler 80, with printed antennas Gunn diode RF source and switched beams, is for low-level helicopters operations where low weight is particularly important.

Antenna units for helicopters have a speed range of −50 to +300 kt forward and 100 kt laterally. The fixed-wing sensors have a corresponding speed range of −50 to +1,000 kt and 200 kt laterally. Velocity data can be provided in either analogue or ARINC 429 digital format and MIL-STD-1553 databus may be specified. The microwave signals produced by the Doppler 90 series are specially tailored to reduce errors created by heavy rain, snow and hail.

The transmission characteristics of the Doppler 91 and 92 units can be remotely controlled. A low-power stealth mode can be selected for minimum detectability, or the transmitter switched off if the beam goes above the horizon, for example when the aircraft is banking. They transmit information to other aircraft systems in ARINC 429 serial digital data form, but the MIL-STD-1553B remote terminal format can also be supplied.

The Doppler 91 can be supplied in a configuration that is optimised for rotary wing over-water operations, where transition to hover and auto-hover are autopilot functions dependent upon the maximum amount of continuously available velocity data.

Specifications

Dimensions:
(Doppler 71/72) 406 × 406 × 127 mm
(Doppler 80) 356 × 381 × 80 mm
(Doppler 91/92) 358 × 391 × 118 mm
Weight:
(Doppler 71/72) 16.5 kg
(Doppler 80) 8.6 kg
(Doppler 91/92) 11 kg
Power supply:
(Doppler 71/72) 115 V AC, 400 Hz
(Doppler 80 and 91/92) 28 V DC

Operational status

In production. The Doppler 91 has been selected for the EH 101, Sea King, and Lynx helicopters. The Doppler 71 is installed on UK Royal Air Force Aerospatiale/Westland Puma, Sikorsky/Agusta SH-3D, Westland Sea King and UK Royal Navy and British Army Lynx helicopters and Belgian Air Force Sea King Mk 48s. Over 3,500 Doppler 71s are in service.

Contractor

Racal Avionics Ltd.

VERIFIED

LWCCU-LightWeight Common Control Unit

The LWCCU is based on the AMS 2000 CDNU (Control Display and Navigation Unit). The LWCCU is the main pilot/co-pilot interface with the aircraft's mission, tactical, navigation and communications systems. It provides significant airframe weight saving (9.8 kg), as well as increased reliability and maintainability over earlier systems.

Operational status

Selected by the UK MoD for the UK Royal Navy Merlin Mk 1 helicopters and the UK Royal Air Force EH 101 support helicopters. Four LWCCUs will be fitted in each of the Royal Navy's 44 Merlin Mk 1 helicopters: two for the pilot/co-pilot and one each for the mission equipment operators in the cabin. Each of the Royal Air Force's 22 Merlin Mk 3 helicopters will carry only one LWCCU, in the cabin, for use in a maintenance role.

Racal Avionics will install its LightWeight Common Control Unit into UK Royal Navy Merlin and UK Royal Air Force EH 101 support helicopters 0015374

Selected by the Italian MoD for its EH 101 ASW and utility helicopters.

Contractor

Racal Avionics Ltd.

VERIFIED

RA 800 Communications Control Systems (CCS)

The RA 800 series of CCS provides full compatibility with aircraft warning and management systems, and is designed to interface with Health and Usage Monitoring Systems (HUMS) to enhance overall maintainability and flight safety.

Two variants of the RA 800 CCS are available, the RA 800 and the RA 800 Light.

The RA 800 Light offers improved ElectroMagnetic Compatibility (EMC) and reduced weight by use of fibre optic instead of copper routing - a 90 per cent in wiring weight is claimed. The RA 800 Light is a fully digitised system incorporating advanced digital signal processing to ensure high signal quality and integrity, with system flexibility, making it ideal for helicopter installations.

Operational status

The RA 800 CCS is in service on BAE Systems Hawk aircraft in several countries, and is being delivered to the UK Royal Air Force for fitting on EH 101 and Chinook helicopters following selection for the UK Royal Air Force's Support Helicopter (SH) programme.

In September 1999, the RA 800 CCS was selected to equip the Hawk 115 trainer aircraft of the NATO Flying Training Centre in Canada, and the RA 800 CCS Light system was selected for the Canadian EH 101 Cormorant Search And Rescue (SAR) helicopter programme.

Contractor

Racal Avionics Ltd.

VERIFIED

Racal Avionics Management System (RAMS)

The RAMS is a family of avionics management systems that can be configured to meet the operational requirements of military and commercial operators for helicopters and fixed-wing aircraft. The purpose of the equipment is to reduce the aircrew workload and so enhance flight safety and the chance of mission success.

The RAMS family starts with the basic RAMS 1000 navigation management system. The RAMS 3000 is similar, but configured around a MIL-STD-1553B databus. The RAMS 2000 is the non-MIL-STD-1553B equivalent of the RAMS 4000.

The RAMS family has a flexible hardware and software architecture for the user-friendly control and display of avionics systems, as well as accomplishing a variety of interfacing and processing tasks. Simplex and duplex system configurations can be provided, depending on the level of redundancy, interfacing and processing required. Management of the Communication, Navigation and Identification (CNI) subsystems can be provided, as well as mission management for the sensors and subsystems specific to the operational role of the aircraft. Other functions provided are performance management, including fuel management and Health and Usage Monitoring and Sensing (HUMS) of engine and transmission parameters.

The interface between the aircrew and the avionics systems is one or more Control and Display Units (CDUs) with alphanumeric keyboard and dedicated function keys; the CDUs are night vision goggle compatible. Associated with the CDU are one or more Processor Interface Units (PIU), commonly called the mission computer, with a hardware and software configuration specific to the operational requirement.

MIL-STD-1553B is the primary interface between the PIU, CDU and other compatible sensors and subsystems, with the bus control provided by the PIU; other interfacing can be analogue or digital, with a variety of industry standards such as ARINC 429, RS-232 and RS-422. A pocket book-sized Data Transfer Device (DTD) with solid-state memory is used to enter information and instructions from a DTD ground loader into the various battery-supported RAMS databases maintained in the PIU.

A variety of liquid crystal display remote indicators, monochrome and colour raster and stroke displays with associated symbol generators, stiff stick controllers and auxiliary control panels are available to make up any required configuration.

A Racal Avionics RAMS 4000 forms the heart of the UK Royal Navy Lynx Central Tactical System (CTS), with two specially extended control and display units and a central tactical situation display. Additional equipment includes a dual-stores management unit for weapons control and two processor interface units to link RAMS to existing navigation, communications and other systems. A data transfer device allows mission data to be loaded quickly into the system before flight. The high flexibility and wide range of enhancements of RAMS allows it to be tailored for present day and future requirements.

Specifications

Type 5401 processor interface unit
Dimensions: 129 × 194 × 414 mm
Weight: 5.9 kg (depending on number of interface modules)

Type 5407 control and display unit
Dimensions: 146 × 229 × 248 mm
Weight: 4.5 kg

The integrated crew station on the Boeing 530MG helicopter with the RAMS multifunction display (top) and control/display unit

Type 5404/5 series data transfer device with receptacle
Dimensions: 146 × 38 × 171 mm
Weight: 0.6 kg

Operational status

In production although superseded by the AMU 2000 CDNU. Notable applications of RAMS are the Central Tactical System (CTS) for the UK Royal Navy Lynx helicopter; the tactical data system in the Royal Danish Navy Lynx; and the Boeing 530MG Defender multirole helicopter.

RAMS has also been fitted in Royal Swedish Air Force Super Puma rescue helicopters and a RAMS version with multifunction displays has undergone trials on German Army PAH-1 anti-tank helicopters. It has been configured for use in maritime patrol fixed-wing aircraft and helicopters.

Contractor

Racal Avionics Ltd.

VERIFIED

RNav 2 navigation management system

Certificated by the UK Civil Aviation Authority in 1984, and now TSO C129 A/C115 B compliant, the RNav 2 navigation management system is in widespread use by helicopter operators supporting the North Sea oil industry and search and rescue, corporate and special mission operators worldwide. It is also in use in military rotary- and fixed-wing aircraft where the operating environment does not justify the expense of extended military environmental specifications.

The equipment comprises a Control and Display Unit (CDU) and a Navigation Computer Unit (NCU), and can accept inputs from GPS, Decca, VOR/DME, Loran C and Doppler. Four separate navigation plots are maintained within the system, any of which may be selected for guidance. In the latest systems any navigation plot with temporarily invalid sensor input is able to revert to Doppler updating. With certain sensor combinations a total of five input sensors is therefore possible.

The computer provides a variety of display and guidance outputs. Analogue outputs are available for the retrofit installation while digital databusses permit interfacing with EFIS and digital AFCS. Inputs of fuel flow can be accepted from a variety of flowmeter types. Fuel computations include range, endurance, fuel remaining over each waypoint and at destination and estimates of these values when a helicopter is in the hover.

A choice of CDU types is available with letters first or numbers first alphanumeric keys, Gen II or Gen III NVG capability and compatibility with Health and Usage Monitoring Systems (HUMS).

NCU hardware variants provide for interfacing with a variety of VOR/DME types, HSIs and RMIs. Software variants offer options such as grid navigation, vertical navigation, transition down and compatibility with the Eurocopter Super Puma Mk II integrated flight data system.

RNav 2 is operating worldwide using GPS as one of its input sensors. It will interface with GPS receivers

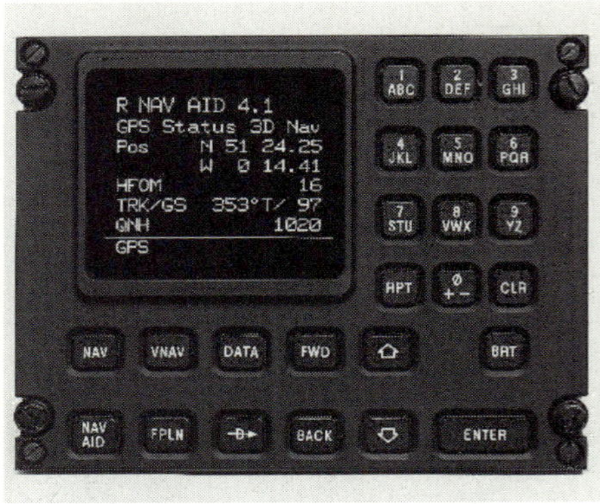

The display for the Racal Avionics RNav 2 area navigation system

which conform to ARINC 743. It forms the core of search and rescue systems certified by a growing number of airworthiness authorities for IFR use in helicopters such as the Eurocopter Super Puma and the Sikorsky S-76. It is also in demand by fishery and environmental protection agencies in a variety of fixed-wing aircraft.

Specifications

Dimensions:
(control/display unit) 146 × 114 × 208 mm
(navigation computer unit) 124 × 184 × 324 mm
Weight:
(control/display unit) 2.7 kg
(navigation computer unit) 5.5 kg

Operational status

In production. Used in large numbers for offshore oil support. Military users include the UK MoD, United Arab Emirates AB-412 SAR and Royal Norwegian Air Force Sea King Mk 43 update.

Also fitted to fixed-wing aircraft such as Dornier 228s of the UK Ministry of Agriculture, Food and Fisheries, Cessna Caravan IIs of the Scottish Department of Agriculture, Food and Fisheries, Dutch police Turbine Islanders and British Army Defenders.

Contractor

Racal Avionics Ltd.

VERIFIED

RNS 252 navigation system

RNS 252 is a single-unit panel-mounted navigation computer which can accept inputs from Doppler and one additional sensor such as GPS or Loran C. Sensor control is exercised through the computer keyboard and sensor data is accessed on the computer's dot matrix display.

The system accommodates 200 waypoints which are numbered but may also carry a five-character ident. Any or all waypoints may be vectored if required. Waypoints may be loaded manually through the keyboard or automatically through a data transfer device. A variety of steering modes is available. Steering guidance is available in the basic form of steering arrows on the computer display, but outputs are available for driving instrumentation and both AC and DC analogue autopilots. A weather radar output provides a navigational overlay for use with digital colour radars.

RNS 252 is compatible with Racal Avionics Type 70, 80 and 90 Doppler sensors and with GPS sensors which conform to ARINC 743.

Specifications

Dimensions:
(standard version) 124 × 146 × 201 mm
(Supertans version) 161 × 146 × 240 mm
Weight:
(standard version) 3.5 kg
(Supertans version) 4.2 kg
Power supply: 28 V DC, 40 W

Operational status

In production. In service in fixed- and rotary-wing aircraft worldwide including British Antarctic Survey Twin Otters, Indonesian Army BO 105s, Norsk Luftanbulanse BK 117s and Royal Moroccan Air Force Puma and Gazelle helicopters. Versions with GPS are in service in UK Royal Air Force Chinook and Puma, UK Royal Navy Sea King and British Army Lynx helicopters.

Contractor

Racal Avionics Ltd.

VERIFIED

Supertans integrated Doppler/ GPS navigation system

The Supertans integrated Doppler/GPS navigation system is a replacement for the TANS series of equipments. Based on the RNS 252 (see previous item), it combines a six-channel GPS receiver with any of the Racal family of Dopplers. The package has been designed as a drop-in replacement for current TANS equipments.

Supertans uses all existing connectors and cables, with only minimal additional cabling required for the GPS installation. The system maintains and enhances all the present capabilities of the TANS/Doppler system while adding the precise navigational accuracy of GPS.

Specifications

Dimensions:
(Supertans) 240 × 161 × 146 mm
(GPS receiver) 368 × 197 × 57 mm
(antenna) 102 × 10 × 95 mm
Weight:
(Supertans) 4.2 kg
(GPS receiver) 2.75 kg
(antenna) 0.5 kg
Power supply: 28 V DC, 40 W

Operational status

In series production and in operational use in UK Royal Air Force Chinook, UK Royal Navy Sea King and British Army Lynx AH. Mk 7 helicopters.

Contractor

Racal Avionics Ltd.

VERIFIED

AN/ULQ-19(V)3 ECM system (RACJAM AIR)

The AN/ULQ-19(V)3 (RACJAM AIR) system is a helicopter-mounted VHF responsive jammer which provides an airborne communications ECM capability without the need for the aircraft to be assigned permanently to the jamming role. In US service it is known as the AN/ULQ-19(V)3; in service with UK forces it is known as RACJAM AIR.

The system consists of an ESM subsystem used to find targets for the jammer, the RJS3100 100 W VHF responsive jammer and a blade antenna fitted to the aircraft.

The jamming signal is radiated through an aerodynamically efficient blade antenna mounted on a suitable inspection panel on the underside of the aircraft, with a coaxial feeder routed into the cargo bay. There is no requirement to remove this antenna when the aircraft is not being used for jamming. The antenna is tuned rapidly to each target frequency as it is attacked, thus ensuring maximum transmitted power and effective range. The antenna tuning rate is compatible with the rapid retuning capability of the jammer system.

The rest of the system is contained within two easily deployed, shockmounted containers, each weighing less than 85 kg, secured to the aircraft standard cargo lashing points. It requires no special wiring or modifications to the existing power or RF distribution systems and can be installed in less than 15 minutes. Once installed, the system is connected to the aircraft power supply and intercom system and is then ready for operation.

Operational status

In service. It has been certified on the UH-1H and UH-1N helicopters. No longer in production.

Contractors

Racal Communications Inc.
Racal Defence Electronics Ltd.

VERIFIED

ARI 5955/5954 radar system

The ARI 5955/5954 radar system is designed for ASV, ASW and search and rescue roles, specifically for helicopters. ARI 5955 is the radar sensor and processing system and ARI 5954 the IFF transponder providing identification of friendly aircraft and surface craft.

The system operates in the I-band and has an antenna that can be gyrostabilised to compensate for aircraft motion. Antenna tilt is adjustable from the radar operator's position both above and below a horizontal datum.

In 1992, a contract was awarded to upgrade the ARI 5955 UK Royal Air Force Sea King search and rescue radars. This was achieved by incorporating the Super Searcher radar signal processing unit and displays. The upgraded system is designated ARI 5955/2.

Operational status

No longer in production, but still in service in a number of countries. However, the upgraded system using the ARI 5955 front end and the Super Searcher processor is still available as a low-cost high-performance option. Subsequent updates have been carried out on Royal Australian Navy Sea King Mk 50s. This system is a variant of the 5955 radar designated AW391(A).

Contractor

Racal Defence Electronics Limited.

VERIFIED

ARI 5980 searchwater radar

Searchwater is the commercial name for the ARI 5980 radar which is standard equipment on UK Royal Air Force Nimrod MR. Mk 2 maritime reconnaissance aircraft; it was designed to replace the ASV Mark 21 radar. The system formed part of a major mid-life refit and 31 Nimrod MR. Mk 1s were converted to Nimrod MR. Mk 2 configuration by the addition of Searchwater and other improvements.

Searchwater is designed for all-weather, day or night operation outside the defensive range of potential targets.

The system comprises a frequency-agile radar which uses pulse compression techniques and a pitch and roll stabilised scanning antenna with controllable tilt and automatic sector scan. IFF equipment is included to interrogate surface vessels and aircraft.

The signal processor enhances the detection of surface targets (including submarine periscopes) in high sea states. An integrating digital scan converter permits plan-corrected presentation and classification of target and transponder returns. The single radar observer in the aircraft is presented with bright, flicker-free television-type PPI, B and A scope displays in a variety of interactive operating modes. Weather radar and navigation facilities are provided within the system.

A real-time dedicated digital computer relieves the radar operator of many routine tasks, while continuously and automatically tracking, storing and analysing data to provide position information and automatic classification for a number of ship targets at the same time. Built-in test facilities provide for automatic detection and diagnosis of faults.

The facilities offered by Searchwater reduce the vulnerability of the host aircraft by permitting operation in a standoff mode, avoiding the need to fly over the target for visual identification. Over-the-horizon targeting is also provided for such missiles as Sea Eagle and Harpoon.

The system is entirely modular, with the interfaces and mechanical construction designed for ease of fault location and replacement. Major units are functionally self-contained as far as possible with a minimum of interconnections. Extensive use is made of hybrid and integrated circuit techniques. The transmitter uses solid-state frequency generators and mixers, followed by two cascaded travelling wave tubes. A fluorocarbon liquid cooling system is employed. The scanner, which both transmits and receives radar and IFF signals, uses a reflector of lightweight construction based on resin-bonded carbon fibre.

Operational status

Searchwater entered service aboard UK Royal Air Force Nimrod MR. Mk 2s in 1979.

The Racal Defence Electronics Searchwater radar display in a UK Royal Navy Sea King Mk 2 AEW helicopter

The Racal Defence Electronics Searchwater antenna on a UK Royal Navy Sea King Mk 2 AEW helicopter. When in use the radome is inflated. The system rotates clockwise through 90° to give clearance for landing 0001211

Searchwater in the Sea King

As a result of demands from the UK Royal Navy for more effective radar sensors for organic fleet defensive surveillance following operations in the Falklands, Searchwater was modified for use aboard converted Westland Sea King anti-submarine helicopters, redesignated Sea King AEW Mk 2. Ten Searchwater systems were acquired to permit 24 hour AEW coverage for the UK Royal Navy.

The system was also supplied to the Spanish Navy for AEW helicopters under a £13 million deal announced in September 1984. Deliveries began in October 1986; the system entered service in mid-1987 and is operational on three Spanish Navy Sikorsky SH-3D helicopters.

Operational status

The first deliveries of Sea King AEW Mk 2s, fitted with modified Searchwater radars, to the UK Royal Navy were in 1985. Deliveries to the Spanish Navy began in 1986. The UK Royal Navy bought two more radars in mid-1991.

Searchwater 2000AEW

In February 1997, the UK MoD appointed Racal Defence Electronics Limited as prime contractor for the radar and mission system upgrade of the UK Royal Navy's Sea King AEW Mk 2 helicopters. Acting as prime contractor, Racal Defence Electronics Limited is responsible for all aspects of the upgrade, including equipment supply, installation, aircraft modification and certification, together with provision of logistic support.

Racal is supplying its new-generation Searchwater 2000 AEW radar for the upgrade. This radar variant retains the high-performance maritime surveillance features of the existing in-service Searchwater radar, but offers considerably improved AEW performance using new high-power transmitters and advanced pulse Doppler signal processing techniques. The new radar was under development as a private venture for five years, and provides an advanced high-technology hardware and software solution with considerable weight saving and improved reliability. Searchwater 2000MR (for Nimrod MRA4) and Searchwater 2000

AEW (for Sea King AEW) have a high degree of commonality, which will provide operational and logistic standardisation benefits to both the UK Royal Navy and the UK Royal Air Force.

The updated helicopter is designated the Sea King Mk 7. The 10 in-service Sea King Mk 2 helicopters are being updated over seven years (1997-2003). The contract includes an extensive logistic support package covering mission and maintenance trainers and workshop support down to second line, including the provision of support on board aircraft carriers at sea. Since contract award, Racal has received an order for three additional systems to enable expansion of the AEW fleet to 13 aircraft. The additional three helicopters will be conversions from Sea King Mk 5/6 AEW.

Searchwater 2000MR

Racal Defence Electronics Limited was contracted by Boeing Space & Defense, as a result of its contract from BAE Systems for supply of the Nimrod MRA4 mission system, to supply 21 Searchwater 2000MR radars over eight years (1997-2004), a date that has slipped as a result of overall project delay. The contract includes provision of integrated logistic support.

Searchwater 2000MR and Searchwater 2000AEW (see above) have a high degree of commonality, and airborne system tests were conducted on an Air Atlantic DC 3 aircraft in 1998. Array Systems Computing Inc of Canada is supplying radar imaging software for the Searchwater 2000MR from its TriSAR programme.

Contractor

Racal Defence Electronics Limited.

VERIFIED

BCC306 helicopter VHF/FM transmitter/receiver

The BCC306 is a development of the Clansman RT351 VHF receiver/transmitter in service with the British Army for vehicle and manpack use and is intended for tactical use between airborne and ground units. The BCC306 provides air-to-ground and air-to-air voice communications over the 30 to 76 MHz frequency range, giving 1,841 channels at 25 kHz spacing. The equipment functions in the single-frequency simplex mode and uses F3E narrowband modulation. It is simple to operate with channel selection and all other control functions via a remote-control unit.

The system consists of the transmitter/receiver, remote-control unit, power supply unit, antenna tuning unit and an aircraft interface unit, with an overall weight of 10.37 kg. The transmitter/receiver is an RT351 equipment from which all the controls have been removed and resited on a remote-control unit near the pilot.

The tuning unit is capable of tuning the 1.7 m whip antenna over the complete frequency range. The unit does not require frequency setting or antenna impedance information, antenna matching and tuning being carried out automatically following operation of the appropriate switch on the control unit.

Operational status

In operational service with the UK and other armed forces. Although developed originally for installation in helicopters, especially the Westland Wessex, the equipment has also been fitted in various types of light aircraft. No longer in production.

Contractor

Racal Defence Electronics Ltd.

VERIFIED

Corvus ELINT system

Corvus is a self-contained ELINT receiving system which includes signals analysis and magnetic recording. The equipment provides a complete turnkey system requiring only an antenna and turntable system.

A major feature of the system is its ability to record and analyse intrapulse modulation on wideband frequency-agile emitters. This modulation is measured at up to 200 MHz sampling rate and the measurements are correlated with the standard pulse descriptors.

Analysis of the live or recorded data can be performed by the integral pulse analyser. Alternatively,

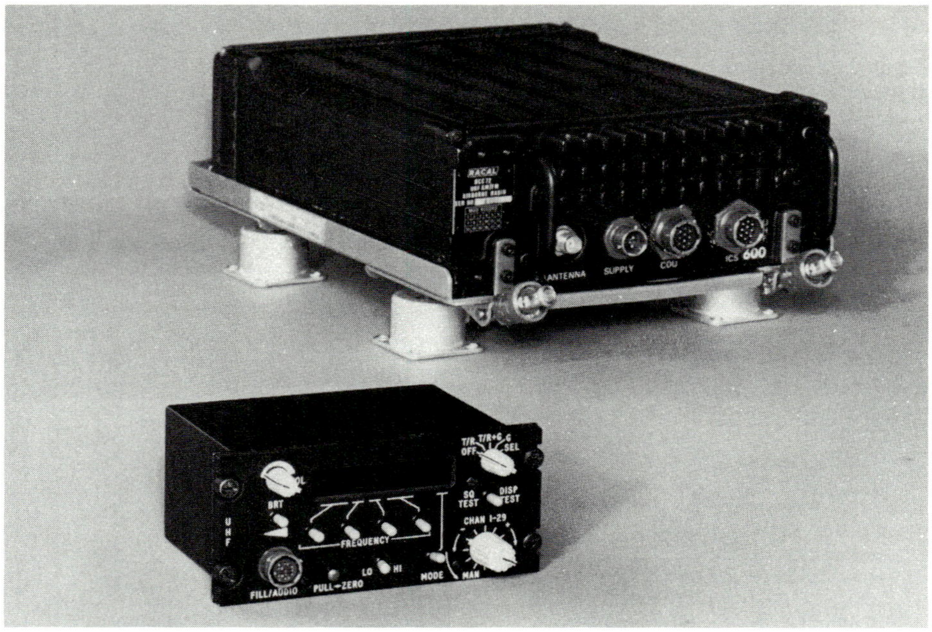

The Racal-Radio BCC 72 transceiver and BCC 584B control

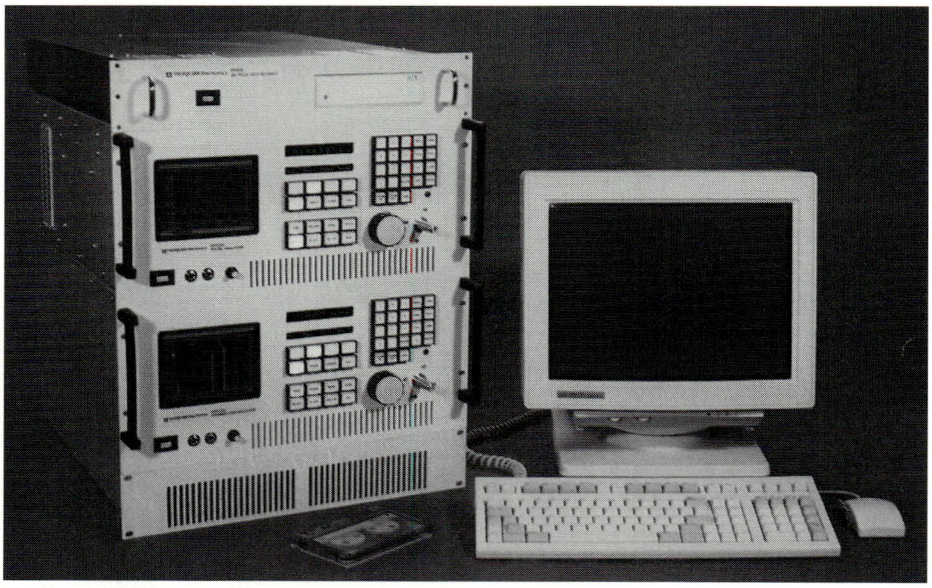

The Corvus ELINT system

the EP4240 separate off-line analysis system is available. This allows a more detailed analysis to be made and ELINT libraries to be compiled, with rapid retrieval of data.

The system covers the frequency range of 0.5 to 18 GHz and options are available for wider ranges. The instantaneous bandwidth of the channelised receiver is 1 GHz.

Modern radars employ modulation during each pulse in order to improve range resolution and produce profiles of targets. Some radars have unintentional modulation during each pulse and this can be used to provide unique identification.

Corvus can record every pulse of a radar over many antenna scans, complete with intrapulse data. This is measured by sampling every 5 ns and measuring both instantaneous amplitude and instantaneous frequency. These samples are then digitised and recorded directly on to a magnetic tape recorder.

The sampling and digitising technique produces a recording bandwidth of up to 50 MHz and this is essential for observing the intrapulse modulation of modern complex military radars without distortion.

Corvus links the recording technique to the channelised receiver to enable every pulse of a frequency-hopping radar to be reproduced, even if the hop range is as wide as 1 GHz.

A 2 GHz bandwidth channelised receiver is now in development.

Contractor

Racal Defence Electronics Ltd, Wells.

VERIFIED

DRFM TG Digital Radio Frequency Memory Techniques Generator

Racal Defence Electronics (RDE) has been contracted by Lockheed Martin Fairchild Systems to provide its DRFM TG for integration into the Nimrod MRA4 Defensive Aids Sub-System (DASS).

System architecture provides for the Lockheed Martin Fairchild Systems AN/ALR-56M(V) RWR to cue the DRFM TG, which is then used to drive the Fibre-Optic Towed Decoy (FOTD) version of the Raytheon Systems Company AN/ALE-50 towed decoy system.

The DRFM TG has been developed by RRDS in its work on advanced jamming system technology in association with the UK Defence Evaluation Research Agency (DERA). It is understood that a twin-DRFM installation is to be used to meet the high data throughput specification for the Nimrod MRA4 system.

Operational status

All 21 Nimrod MRA4 aircraft are due to be fitted with the system.

Contractor

Racal Defence Electronics Ltd.

VERIFIED

EP4000 Series of EW modules

A range of advanced modules for electronic warfare has been developed by Racal. The range includes the EP4410 pulse analyser and the EP4220 channelised receiver.

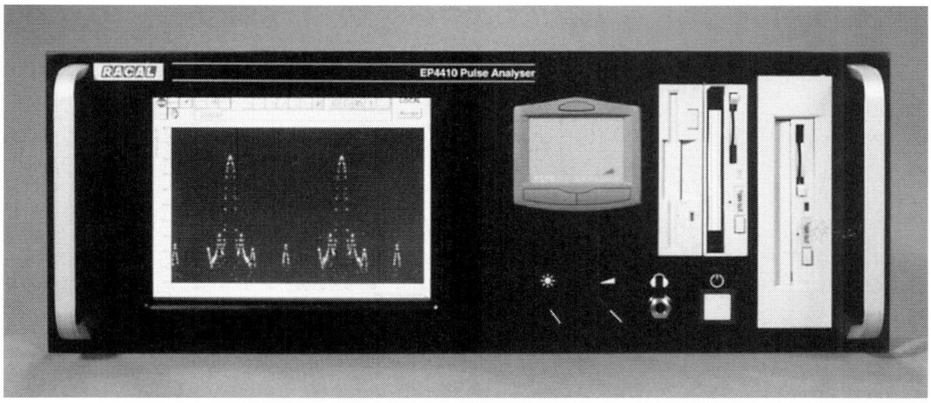

The EP4410 Pulse Analyser 0001269

The EP4410 is designed to interface with the EP4220 or conventional swept superheterodyne receivers. It is an advanced pulse analyser and uses qualification on descriptor parameters to enable signals to be selected for analysis and displayed on the high resolution colour graphics display. In addition to pulse descriptor information, the VME processor-based analyser measures intrapulse frequency and amplitude information by frequent sampling of the signals through each pulse. The EP4410 has the capacity to perform an analysis on up to 64,000 pulses, with intrapulse information automatically correlated with pulse descriptor data.

A digital receiver option is available for users wishing to analyse amplitude, frequency and phase intrapulse characteristics with minimum receiver colouration effects.

The EP4220 is a wideband channelised receiver and has been specifically designed for ELINT, ESM and signal monitoring applications which require high sensitivity and wide instantaneous bandwidth. Operating in the 0.4 to 18 GHz frequency band, it offers very high performance, detecting complex signals in a dense electromagnetic environment. Its wide 1 GHz bandwidth enables full examination of frequency-agile transmitters. Within each channel, the detection bandwidth is sufficiently narrow to give measurements comparable with narrowband ELINT receivers.

A 2 GHz bandwidth channelised receiver is now in development.

Specifications

EP4410 pulse analyser
Dimensions: 483 × 178 × 575 mm
Weight: 20 kg
Power supply: 115/240 V AC, 40-60 Hz, 300 W

EP4220 channelised receiver
Dimensions: 483 × 310 × 580 mm
Weight: 50 kg
Power supply: 115/240 V AC, 40-60 Hz, 400 W
Frequency: 0.5-18 GHz

Contractor

Racal Defence Electronics Ltd, Wells.

VERIFIED

Griffin Radar Warning Receiver (RWR/ESM)

Griffin is designed for fixed- and rotary-wing aircraft. It provides full surveillance, identification and threat warning in dense signal environments. It uses advanced SADIE customised Very Large Scale Integration (VLSI) ESM processors to give continuous radar identification within one second from a mission library of 3,500 radar modes. It provides automatic data recording of intercepts for post-mission analysis using the programmer/loader unit (PLU). The same PLU is used for loading pre-flight generated ESM library and mission data. Griffin uses advanced instantaneous frequency measurement (IFM) receivers, with improved fidelity of measurement.

Griffin detects pulse, pulse Doppler and continuous wave radars beyond lethal range, and provides prioritised audio and visual cues in the primary threat bands in multi-emitter and high pulse density areas.

System elements include:
(a) four circularly polarised, planar cavity backed spiral antennas, designed for surface mounting;
(b) Dual Receiver Units (DRU); each pair of antennas feeds into a DRU; the DRU supplies both RF and video to the signal processor unit;
(c) a Signal Processor Unit (SPU); which contains the IFM that accurately measures the RF on a pulse-by-pulse basis, or repetitively sampled CW; and the SADIE processor;
(d) Display and Control Unit (DCU): a dual control CRT unit that provides threat warning and tote separately; the DCU is NVG compatible; display options include a single CRT indicator with separate control, and remote display via MIL-STD-1553B databus;
(e) programme loader unit (PRU).

Specifications

RF range: 2-18 GHz; options for C/D, K and M (millimetric) bands
Azimuth coverage: 360°
Elevation coverage: ±35°
Sensitivity:
(pulse) 45 dBmi (typical)
(CW) 55 dBmi (typical)
POI/Pulse data rate: 100%; >500 kpps
Bearing accuracy: better than 8° RMS
Frequency agiles; simple/agile/hopper
PRI: fixed/stagger/jitter/grouped
Scan types:
circular/sector/conscan/lock-on/unknown
Pulses to form track: 6
Number of tracks: 200
Threat report time: 1 s

Contractor

Racal Defence Electronics Ltd.

VERIFIED

Jaguar-U (BCC 72) UHF radio

The Racal Jaguar-U system is a range of radio communications equipment operating in the UHF band, available for airborne, shipborne and land use. The airborne version, BCC72, is designed to operate over the frequency range 225 to 400 MHz and provides fixed frequency or frequency-hopping FM with selectable encryption, and fixed frequency AM compatible with current operational systems. It provides 7,000 channels at 25 kHz spacing, with 30 programmable channels (including a Guard channel on the distress frequency of 243 MHz) with flexibility to select clear, secure, fixed and hopping modes.

Design of Jaguar-U is based on experience gained in the Jaguar-V systems and employs the same method of medium-speed frequency hopping to protect against interception, direction-finding and jamming. To simplify frequency management, the 225 to 400 MHz band has been divided into 13 sub-bands, allowing the co-sited operation of multiple radio systems without interference. However, the radio can also be programmed to use any one of three larger hop bands. In either mode, orthogonal hop sets are available to assist frequency management. Large numbers of nets can operate in the same frequency bands and individual bands can be barred to avoid jamming or to protect other fixed frequency stations. Selective communication can be performed within an individual net or, conversely, a radio can be selectively barred from a net should it be captured. Once the radios have been programmed with hop codes and frequencies they synchronise automatically without the need for time of day input. Communications security is also enhanced by the use of either a built-in encryption unit using a second keystream generator or any external 16 kbits/s system in fixed or frequency-hopping modes

The airborne BCC 72 system consists of a rack-mounted transceiver and a panel-mounted control/display unit. The transceiver provides output powers of either 10 mW or 15 W on FM and 40 mW or 40 W PEP on AM, output level being selected on the controller. Two control/display units are available, depending on the aircraft requirements. The BCC 584B controller gives full manual selection of any of the 7,000 channels plus selection of the 30 programmable channels, hopping or fixed mode selection, low- or high-power selection and frequency/mode display by light-emitting diodes.

Specifications

Dimensions:
(transceiver) 230 × 90 × 350 mm
(controller BCC 584B) 145 × 66 × 104 mm
Weight:
(transceiver) 6.5 kg
(controller BCC 584B) 0.8 kg

Operational status

In production and in service with a number of users.

Contractor

Racal Defence Electronics Ltd.

VERIFIED

Griffin radar warning receiver on Hawk aircraft 0001267

Kestrel ESM system

An ESM system in production for ELINT, surveillance and reconnaissance tasks, Kestrel has a near 100 per cent probability of intercept of any radar emission within its frequency range. Kestrel also utilises the latest processing techniques to give the operator virtually real-time information. In service with UK forces, it is known as Orange Reaper.

The system receives and processes radar emissions over 0.6 to 18 GHz. A six-port amplitude comparison system measures the bearing of the emissions and provides instantaneous digital information over 360° in azimuth. At the same time, a frequency measurement receiver provides an instantaneous frequency indication.

Digitised information on all threats, as accumulated pulse by pulse, is passed to the processor. In this equipment overlapping pulse trains from different radars are derived and their pulse repetition frequency and frequency agility characteristics determined. The digitised data is then passed into the main processor where long-term information is extracted. Radar identification is made by comparing measured and derived radiation parameters with those stored in a library of known emitters. Full information about the radar signal environment is then presented to the system operator on an ordered tabular or tactical display or, alternatively, the data can be transferred to remote displays using a standard data highway.

Orange Reaper has been selected for the Westland/Agusta Royal Navy EH 101 Merlin helicopter.

Specifications

Weight: 55 kg (max)
Frequency: 0.6-18 GHz (optionally up to 40 GHz)
Frequency accuracy: 5 MHz RMS
Accuracy: ±3.5° RMS
Radar storage capacity: 2,000 modes (min)
Processing time: 1 s (max)

Operational status

In production and operational on Lynx helicopters of the Royal Danish Navy, with Orange Reaper versions ordered for 44 EH 101 for the Royal Navy under a £30 million contract.

An export version of Kestrel, known as Kestrel II, is being evaluated by a number of countries.

Contractor

Racal Defence Electronics Ltd.

VERIFIED

The Racal Kestrel ESM system is fitted to the Royal Navy EH 101 Merlin helicopter

MAREC II Maritime Reconnaissance radar

The MAREC II (MAritime REConnaissance) radar was developed to meet the requirements for a small, efficient, low-cost system to conduct maritime patrol, including coastguard surveillance, search and rescue, fishery and oil rig protection and pollution control. Suitable for fixed-wing aircraft and helicopters, it has a true motion plotting table display covering 360° in azimuth and a range of 460 km, together with a pilot's display.

Specifications

Frequency: 9,345 MHz
Power output: 80 kW
Pulsewidth: 0.4 and 2.5 μs
PRF: 200 and 400 Hz

Operational status

The system has flown in the BAE Systems Coastguarder demonstrator - a version of the HS748 - the prototype Airship Industries Skyship 500 airship and in Dornier 128 Skyservants for maritime patrol operations. Three MAREC II radars were supplied to the Indian coast-guard; these have now been upgraded to Super MAREC standard.

Contractor

Racal Defence Electronics Ltd.

VERIFIED

Microwave Aircraft Digital Guidance Equipment (MADGE)

The Microwave Aircraft Digital Guidance Equipment (MADGE) is a military landing system, which can be used to provide night and all-weather landing guidance and terminal area navigation information, for all types of fixed-wing aircraft and helicopters under the strictest emission control conditions, to reduce detection by the enemy. The system consists of ground and airborne equipment. The ground equipment may be installed on board a ship, in addition to its use for tactical operations in the field.

MADGE can be used for multiple simultaneous approaches and is passive unless correctly interrogated. It includes a two-way datalink between aircraft and ship or ground and selective aircraft identification. Navigation information is provided to an aircraft out to a range of 30 n miles (55.6 km) and landing guidance is provided out to 15 n miles (27.8 km) on ILS or Automatic Carrier Landing System (ACLS) indicators.

Aircraft equipment consists of the pilot's controller, logic unit, transmitter/receiver, two antennas and an antenna switching unit, plus the use of the aircraft normal instrument display. All equipment operates from 28 V DC.

Specifications

Dimensions:
(controller) 65.5 × 146 × 84 mm
(logic unit) 194 × 91.9 × 384 mm
(transmitter/receiver) 198.9 × 91.4 × 383.5 mm
Weight:
(controller) 0.8 kg
(logic unit) 4.9 kg
(transmitter/receiver) 5.7 kg

Operational status

In service in the UK Royal Navy for use with BAE Systems Sea Harrier aircraft.

Contractor

Racal Defence Electronics Ltd.

VERIFIED

MIR-2 ESM system

The MIR-2 airborne ESM system, also known as Orange Crop, combines airborne warning system and search receiver functions using an advanced digital receiver served by a fully solid-state wideband antenna system covering the 0.6 to 18 GHz radar frequencies. Six

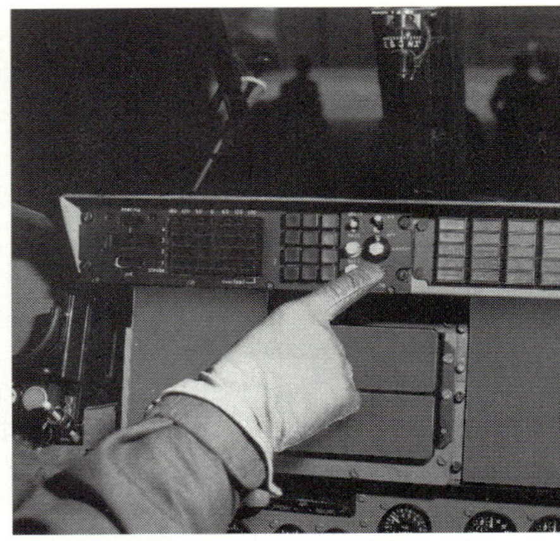

The Racal Defence Electronics MIR-2 ESM cockpit control indicator unit

antennas monitor a full 360° in azimuth and provide a very high intercept probability, improving the crew's location and identification of friendly and enemy radars. The control-indicator unit, on the cockpit coaming, is a compact light-emitting diode display which indicates frequency band, amplitude and relative bearing on the main display. On an associated fine bearing unit the true bearing and radar PRF are indicated to a high degree of accuracy. The complete system of eight units comprises six antennas, a processor and the control/indicator unit.

Specifications

Weight: 47.7 kg
Frequency: 0.5-18 GHz
Pulsewidth: 0.15-10 μs
PRF: 0.1-10 kHz
Coverage: 360°

Operational status

In production and in service with UK Royal Navy Lynx and Sea King helicopters and UK Royal Air Force C-130 Hercules aircraft. In August 1994, Racal announced a US$10 million order for MIR-2 systems for Brazilian Navy Lynx helicopters. Sales now exceed 300 sets.

Contractor

Racal Defence Electronics Limited.

VERIFIED

Prophet radar warning receiver

Prophet is an operationally proven radar warning system designed to reduce combat aircraft vulnerability to radar associated threats on the battlefield. It provides the crew with timely and unambiguous threat warning, permitting appropriate countermeasures, such as manoeuvres or chaff discharge, to be employed. System design has taken account of the likely need for frequency extensions in order to increase the survivability of the battlefield helicopter.

A UK Royal Navy Westland Lynx is equipped with the Racal MIR-2 warning receiver

The Racal Prophet radar warning receiver

The receiver detects signals over 2 to 18 GHz and, by means of a processor with a programmable threat library, alerts the pilot to imminent hostile action. The system is designed to operate in a dense RF environment. A four-port antenna system provides direction-finding, using signal amplitude comparison to derive bearing. The system has an extremely low false alarm rate and threats are recognisable with a high degree of confidence. An audible warning is provided by a tone generator and fed to the aircraft internal communications system.

The Prophet display uses light-emitting diodes and has three lines, each dedicated to a threat. Each line can show an arrow indicating threat direction and a three-character alphanumeric identifier. Display colour and brightness is such that it can be viewed in bright sunlight and through night vision goggles. Also provided, as an option for fixed-wing aircraft, is a full threat display on the HUD and multifunction displays.

The system can be readily reprogrammed and has high in-service reliability.

Specifications

Weight: 10 kg
Power supply: 28 V DC, 80 W
Frequency: 2-18 GHz (pulse and CW); options: millimetric wave and laser threat warning
Coverage: more than 100 radar modes
(azimuth) 360°
(elevation) ±30°
DF accuracy: 10° RMS
Detection time: <1 s

Operational status

In service with UK Royal Navy Sea King HC4s and British Army Gazelles and other air forces. Prophet is in production for the BAE Systems Hawk 100 and 200.

Contractor

Racal Defence Electronics Ltd.

VERIFIED

Super MAREC radar

Super MAREC (MAritime REConnaissance) is an upgrade of the MAREC II involving improved software and the replacement of the MAREC II's 430 mm plotting display with the 356 mm colour television-type display developed for Super Searcher. This saves weight and space and increases the tactical navigation

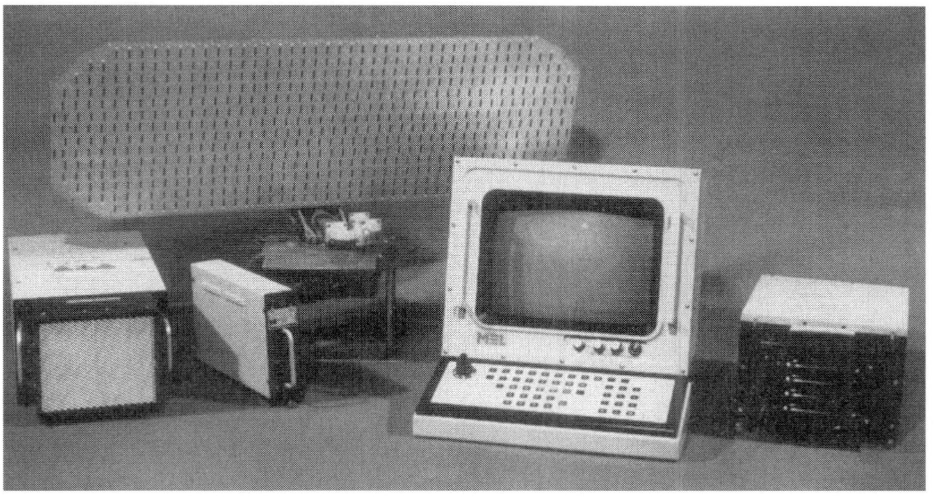

The Racal Defence Electronics Super Searcher radar

facilities available to the operator. Additional facilities include multiple target track-while-scan and navigation overlay data.

Enhancements to Super MAREC include the addition of a full ISO contoured weather avoidance mode, a ground-mapping mode and an extra radar mode to give improved close-range target detection. Digital map overlays are also provided. A smaller display and joystick option is offered as a replacement for the large display and keyboard to permit radar operation from the cockpit by the second pilot.

Operational status

In production and service with the Indian coastguard, which has a requirement for 36 radars.

Contractor

Racal Defence Electronics Ltd.

VERIFIED

Super Searcher airborne radar

Super Searcher is a development of the Sea Searcher and Super MAREC radars and is fitted to UK Royal Navy Sea King helicopters. Designed for multithreat maritime operations, it is a lightweight I-band command and control radar with a high-definition display on a range of colour raster scan CRTs that can show true motion or centre PPI with variable sector scan.

By comparison with Sea Searcher and Super MAREC, the Super Searcher has a greater detection, target tracking and guidance performance. The system has an inbuilt guidance capability, which can be adapted for all fire-and-forget, anti-surface vessel, sea-skimming missiles.

The system incorporates three selectable pulsewidths, one of which provides high definition of small targets in bad weather. Contact recognition is also improved by the use of the latest microprocessor techniques and signal processing algorithms.

The colour CRT facilities include freeze-frame and memory storage with graphical overlays of tactical and navigational symbology. The display can show either true or relative motion with or without offsets. Data is displayed either in latitude and longitude or as a grid reference. The system has multiple track-while-scan capability and provides a range of navigational facilities including waypoint markers. As an option, digitised maps can be loaded prior to the mission, allowing the use of map overlays on the radar image. A comprehensive library of symbols, that can be vectored as required, eases the operator's task, particularly in intelligence storage and extraction. The system is compatible with IFF/SSR interrogators and can be interfaced with a wide range of other sensors such as FLIR, ESM and sonar.

Super Searcher Ground-Mapping Radar (GMR) is a development of the Super Searcher radar and is a further addition to the Racal Defence Electronics Limited family of lightweight airborne radars. The radar uses a three-axis cosecant^2 antenna combined with high-resolution displays and upgraded processing to enhance ground features. The low-cost system can be used for high-altitude radar fixing, ground-mapping and ground target attack training.

Operational status

In production. Super Searcher is the principal sensor on board the 20 Sea King Mk 42 ASW helicopters ordered by the Indian Navy. It has also been installed on the eight Sikorsky S-70B helicopters of the Royal Australian Navy. Brazil has been supplied with 20 Super Searchers, for fitting to Embraer EMB-111 maritime patrol aircraft. Further systems have been fitted to UK Royal Air Force Dominie aircraft for navigation training. Over 100 Super Searchers have been sold worldwide.

Contractor

Racal Defence Electronics Limited.

VERIFIED

ASTOR airborne standoff radar

ASTOR is a UK MoD sponsored project designed to provide detailed over-the-border surveillance of the land battle and major hostile ground forces. It fulfils UK MoD Staff Requirement (Land/Air) 925, valued at £800 million.

The objective is to provide high-resolution static imagery with the ability to detect moving targets. The tactical objective is to provide 24-hour observation and targeting intelligence of enemy first and second echelon forces and to support peacekeeping operations. A secure datalink to mobile ground forces and ground-based interpretation facilities also forms part of the system.

The primary sensor radar will be a Synthetic Aperture Radar with Moving Target Indication (SAR/MTI). The contract award is for five aircraft, plus eight supporting ground stations, to enter service from 2005.

Raytheon Systems Limited was awarded the prime contract, covering Development, Production and Support, in December 1999. The System is based on the Bombardier Global Express aircraft, with 25 per

The Raytheon Systems Limited ASTOR design using the Bombardier Global Express aircraft **2000**/0062744

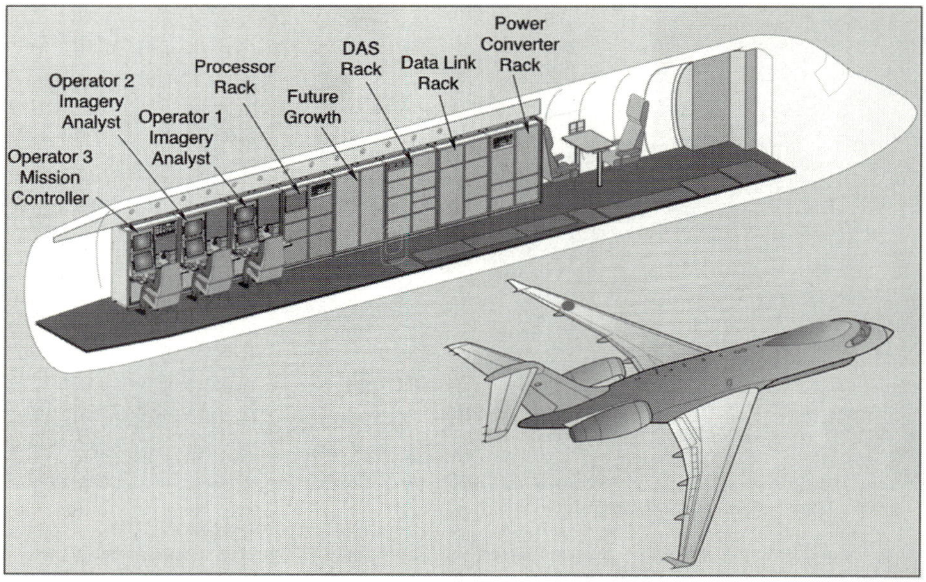

The interior design of the Raytheon Systems Limited ASTOR airborne system **2000**/0062746

cent input from Bombardier's subsidiary Short Brothers in Belfast.

The radar system is to be a developed version of the ASARS 2 radar currently in use in US Air Force U-2R aircraft, with developments incorporated from the ASARS Improvement Programme, together with a completely rebuilt antenna, to be provided by BAE Systems, who will also provide the defensive aids suite. The Defence Evaluation Research Agency (DERA) Malvern will provide design consultancy in motion compensation and autofocus. The radar architecture is designed to allow for later upgrades as new radar technology becomes available through the in-service life of ASTOR.

Ultra Electronics, with Cubic Defence Systems, is to provide the secure broadcast datalink. The ground stations will be provided by Motorola, in conjunction with Raytheon Systems Limited, Marshal! Specialist Vehicles and DERA. Six of the ground stations will be mobile variants, mounted on Steyr Pinzgauer 6 × 6 trucks; the remaining two will be built as deployable shelter-based versions.

Operational status

The UK MoD and Raytheon Systems Limited signed the UK ASTOR contract in December 1999.

Contractor

Raytheon Systems Limited.

UPDATED

Cossor Interrogation and Reply Cryptographic Equipment (CIRCE)

Raytheon Systems Limited supplies a range of modules for all types of IFF applications to facilitate a nationally secure IFF operation in Secure mode which is similar to, but not interoperable with, NATO Mode 4. The modules can be added to any IFF system that is compatible with the NATO Mk XII standard (STANAG 4193).

Two key components of CIRCE are the programmer and the fill gun. The programmer is used to hold the key variable necessary to operate the CIRCE units. It is held in a secure location and the electronic fill gun is then used to transfer the key variable data from the programmer to the various IFF systems. The key variable data can be changed whenever required using the fill gun.

Raytheon Systems Limited CIRCE - IFFcryptographic unit 0011871

Specifications

Dimensions: 123 × 50 × 129 mm
Weight: 1.42 kg

Operational status

In service with three non-NATO countries.

Contractor

Raytheon Systems Limited.

VERIFIED

GAS-1 GPS adaptive antenna system

The GAS-1 system protects GPS systems from deliberate countermeasures and radiated interference. Using a Controlled Reception Pattern Antenna (CRPA) and the associated Antenna Electronics (AE), the system employs a null steering technique to provide protection against six jamming signal sources. The speed of null steering is greater than the dynamics of any known production or developmental aircraft platform, ensuring that the tracking of jamming sources is not disrupted by aircraft manoeuvres.

The system provides protection for multiple GPS based navigational/targeting aids including inboard stores and weapons. All up system weight is less than 9.1 kg.

Operational status

The GAS-1 system is a derivative of equipment already in service with air forces in the UK and Australia. GAS-1 entered service with the US Air Force in 1998, when production deliveries commenced.

The GAS-1 system replaces the earlier STR-2200 CRPA and STR-2400 ACU models, and is compatible with both RF and IF variant GPS receivers.

For smaller aircraft, UAVs, land vehicles and ships, a compact version called PAGAN is available which provides anti-jam protection to the GPS receiver against 3 simultaneous interfering signals using a 4-element, 7 in, CRPA and an AE4 form factor Antenna Electronics (AE).

Contractor

Raytheon Systems Limited.

VERIFIED

IFF 2720 transponder

IFF 2720 is a microminiature IFF Mk10A identification friend or foe/secondary surveillance radar transponder for use on all types of military aircraft and helicopters. On Modes 1, 2, 3/A and B the full 4,096 codes are available and 2,048 codes are available on Mode C for altitude reporting. In addition there are circuits for the identification facility (SPI or I/P) and military emergency. The system comes in two units: a transmitter/receiver and a controller, each of which features easily accessible circuit boards.

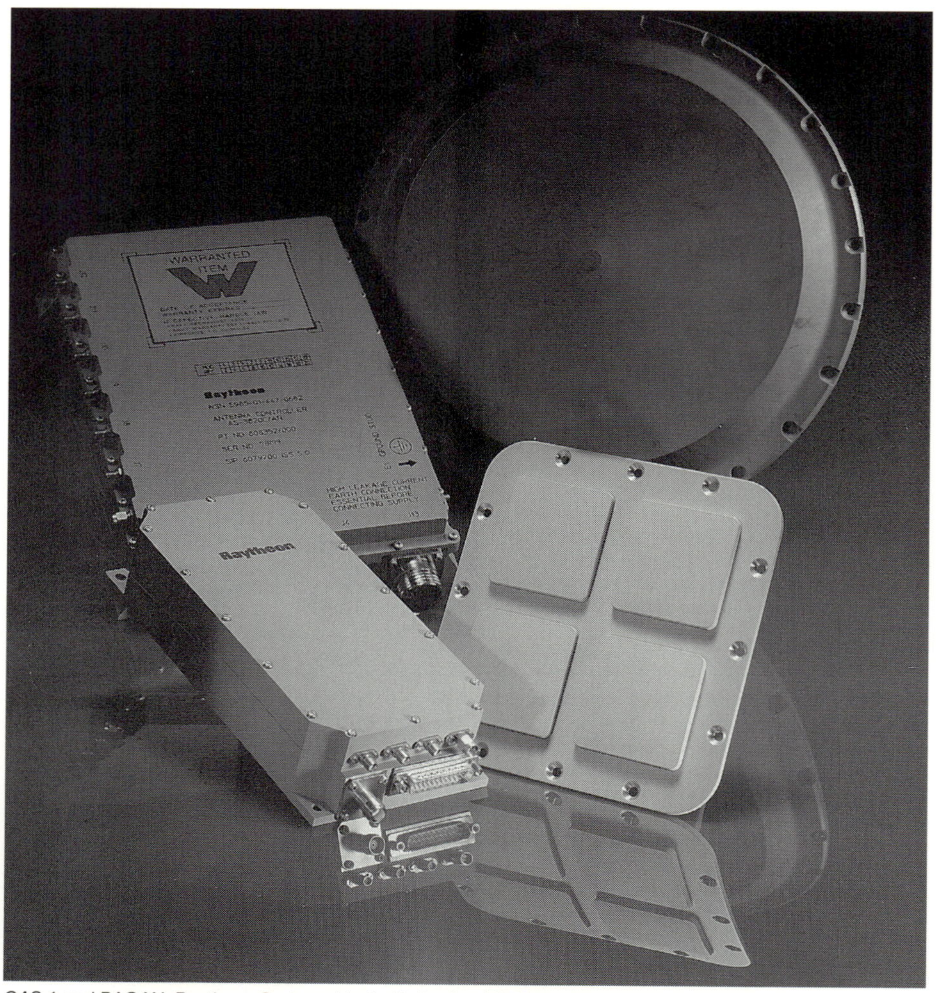

GAS-1 and PAGAN, Raytheon Systems Limited's family of GPS anti-jam antenna solutions **2000**/0062848

Early deliveries of Indonesia's BAE Systems Hawks have the IFF 2720. Later Hawks have the IFF 4720 transponder

Electronic warfare provisions include resistance to continuous wave, modulated continuous wave and pulse jamming, sidelobe rate limiting, short pulse and spurious interference protection, single-pulse rejection and long-pulse discrimination.

The microprocessor-based control unit for use with this system is designated IFF 2743. It has a light-emitting diode display which can be fully dimmed.

Specifications

Dimensions:
(IFF 2720 transponder) ⅜ ATR short
90 × 194 × 314 mm
(IFF 2743 full facility control unit) 146 × 57 × 98 mm
Weight:
(IFF 2720 transponder) 4.6 kg
(IFF 2743 full facility control unit) 0.7 kg
Frequency:
(transmitter) 1,090 MHz
(receiver) 1,030 MHz
Power output: 27 dBW (500 W) (min)
Receiver sensitivity: −76 dBm
Dynamic range: >50 dB
Sidelobe suppression: 3-pulse
Qualification: STANAG 5017 Edition 2, ICAO Annex 10

Operational status

IFF 2720 is in production and equips many strike, transport, combat and trainer aircraft; it is fitted to early export versions of the BAE Systems Hawk.More than 1,000 systems have been delivered to over 20 countries.

Contractor

Raytheon Systems Limited.

VERIFIED

IFF 3100 transponder

The IFF 3100 is a single-package transponder tailored to the Royal Air Force Panavia Tornado aircraft.

Claimed advantages over previous systems are small size, lower weight and simpler installation. Although the component density is high, reliability is ensured by the use of high-grade, close-tolerance circuits and a four-port circulator protects the output stages from the effects of any antenna mismatch. Open or short-circuit conditions at the antenna do not damage the transponder.

Extensive integrity monitoring is incorporated during operation and when the test button is pressed. Checks cover receiver sensitivity, receiver centre frequency, mode decoding, aircraft reply coding and transmitter power level.

Interrogation Modes are 1, 2, 3/A, B and C and the reply capability covers 4,096 codes for Modes 1, 2, 3/A and B. Provision for the use of the X-pulse is included and there are 2,048 codes for Mode C.

Specifications

Dimensions: 146 × 132 × 165 mm
Weight: 5.3 kg
Qualification: ICAO Annex 10, STANAG 5017 Edition 2

Operational status

In service with Panavia Tornado aircraft with UK equipment fit including Saudi Arabian aircraft.

Contractor

Raytheon Systems Limited.

VERIFIED

IFF 3500 interrogator

The IFF 3500 airborne interrogator employs monopulse techniques to achieve high accuracy in the measurement of target bearing. The IFF 3502 variant incorporates an automatic code-changing system to enhance security and eliminate the possibility of incorrect code setting.

The transmitter employs P2 emphasis to provide antenna beam-sharpening. P1 and P3 are transmitted on the antenna sum channel and P2 on the difference channel. Selectable 3 or 6 dB of P2 emphasis is available. Advanced video processing circuits for degarbling, defruiting, decoding and for echo and multipath suppression are contained within a single unit. Passive and active decoding are provided and two channels of passive decoding enable comparison during the overlap period between code changes. Active decoding provides serial readout of the 4,096 reply codes.

Manual and continuous automatic built-in test circuitry checks transmitter power, interrogation coding, receiver sensitivity, defruiting/decoding, bearing accuracy and integrity of the transmission feeders.

Specifications

Dimensions: 1 ATR short case to ARINC 404A
Weight: 20.7 kg
Power supply: 115 V AC, 400 Hz, single phase
28 V DC
Frequency:
(transmitter) 1,030 MHz
(receiver) 1,090 MHz
Power output:
(P1, P3) 30.5 dBW
(P2) 0, +3 or +6 dBW above P1 power
Spurious outputs >76 dB below 1 W
Sensitivity (decoding): −80.5 dBm
Dynamic range: 60 dB
Spurious responses: 60 dB down outside pass-band
Bearing resolution: dependent on antenna configuration, but around 5% of angle between intersection points of control and interrogate patterns
Qualification: compatible with NATO STANAG 5017 Edition 3

Operational status

In service with the Royal Air Force Tornado F2 and F3 aircraft and Royal Navy Sea King AEW helicopters. Also fitted to Tornado aircraft delivered to Saudi Arabia.

The equipment will also be fitted during an extensive upgrade of Boeing 737 Surveiller aircraft which are used in a maritime patrol role by the Indonesian Air Force.

Contractor

Raytheon Systems Limited.

VERIFIED

IFF 4500 interrogator

IFF 4500, which is a monopulse interrogator, is suitable for a wide variety of applications. Tactical fighter and maritime patrol aircraft surveillance platforms are already projected.

As well as operating in Modes 1, 2, 3A, C and 4, IFF 4500 has built-in growth potential for upgrade to include Mode 5 system, as specified in STANAG 4193 Part V.ACC (Automatic Code Change) and Mode S can also be added as customer options.

Control of the interrogator from the host primary radar can be via a MIL-STD-1553B databus or via a discrete control alternative. Target reports are fed again via a MIL-STD-1553B databus for display alongside the target information generated by the host primary radar.

Mode 4 or CIRCE cryptograhic units are built in to the IFF 4500, but these 'add-ons' can be removed for IFF Mk10A only applications.

Specifications

Dimensions: 1 ATR short or ½ ATR medium (depending upon requirement)
Cooling: 1 ATR short-convection
½ ATR medium forced air
Peak output power: 1 kW or 2 kW options ±2 dB

Operational status

IFF 4500 is just entering production for Royal Air Force Tornado F3 aircraft.

Contractor

Raytheon Systems Limited.

VERIFIED

IFF 4700 series transponders

There are two members of the Raytheon Systems Limited 4700 Series of IFF transponders which are designed to meet the NATO IFF Mk XII specification (NATO STANAG 4193); both versions use a common set of modules so they can be supported by a common spares inventory.

IFF 4720 series transponder is designed for a wide range of applications, and is a plug-in replacement for the IFF 2720 system (see previous entry). The IFF 4720 has full dual-redundant decoders on each receiver channel to provide optimum anti-jamming performance and the system can be operated by a dedicated remote-control panel, or via a MIL-STD-1553 databus. Full standard coverage of Modes 1, 2, 3/A, 4 and C is offered and the IFF 4720 can operate up to 70,000 ft altitude.

Specifications

Dimensions: 90 × 194 × 314 mm
Weight: 4.5 kg

Operational status

In production. The IFF 4720 is standard equipment on the BAE Systems Hawk 100 and 200; also fitted to some MiG-29 aircraft and several Naval IFF applications.

IFF 4760 transponder is a NATO Mk XII-compatible remotely controlled IFF designed also to act as a MIL-STD-1553 bus controller. The unit provides an alternative form factor to that of the IFF 4720. All 4700 series transponders also offer tighter frequency tolerances, all-solid-state construction and VLSI processing to enhance performance; operation is identical to the IFF 4720.

Specifications

Dimensions: 136 × 136 × 213 mm
Weight: 4.5 kg

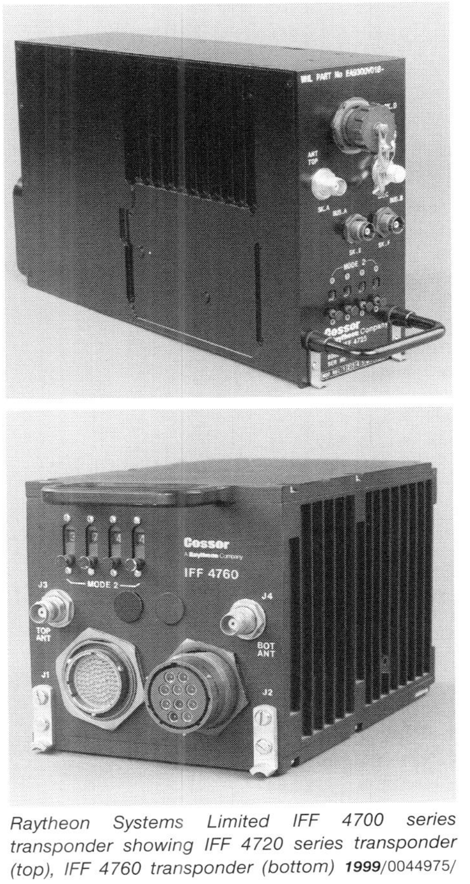

Raytheon Systems Limited IFF 4700 series transponder showing IFF 4720 series transponder (top), IFF 4760 transponder (bottom) **1999**/0044975/0044976

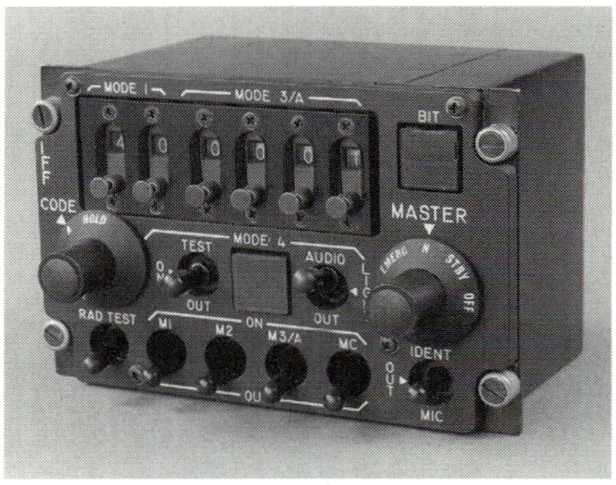

IFF 4770 control unit ***1999***/0044973

Operational status

In production and in service. Over 250 sets have been purchased by the UK MoD.

IFF 4770 control unit is designed to operate with IFF 4720 and IFF 4760 transponders. It is an NVG-compatible package which is intended to occupy minimum cockpit space, while at the same time having good ergonomics.

Operational status

In service in CN-235, Hawk 100, Merlin and MiG-29.

Contractor

Raytheon Systems Limited.

VERIFIED

IFF 4800 transponder

As well as operating in Modes 1, 2, 3A, C and 4, the IFF 4800 operates in Mode S Level 3 which is the latest civil aviation selective address SSR system. By using Mode S, military aircraft will be able to continue to use civil airspace when conventional SSR modes are phased out in the early years of the next century.

Additionally, IFF 4800 incorporates the following facilities:

(a) Interface for Mode S Air Link Data processor
(b) Airborne Collision Avoidance System (ACAS) interface

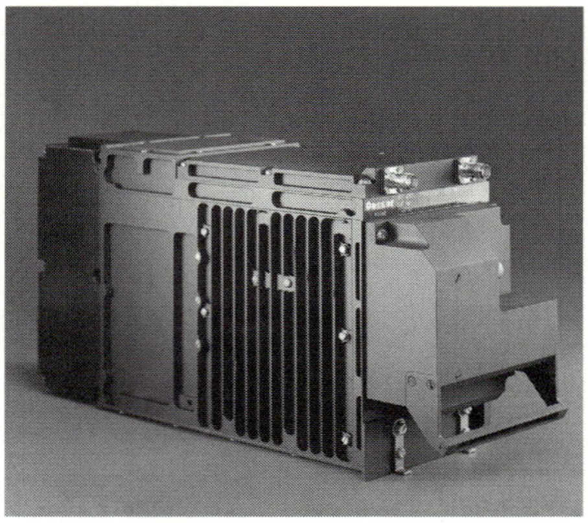

Raytheon Systems Limited IFF 4800 transponder ***1999***/0044977

(c) GPS position reporting interface
(d) Built-in Mode 4 or CIRCE cryptographic computer
(e) Provision for the incorporation for growth to Mode 5 as specified in STANAG 4193 Part V
(f) Automatic code change on Modes 1 and 3A

The IFF 4870 control unit is used together with the IFF 4800. For unified avionics control systems, a MIL-STD-1553D databus version of IFF 4800 is available.

Specifications

Dimensions: 1 ATR short
Weight: 8.5 kg
Peak output power: 500 W ±2 dB
Cooling: convection cooled

Operational status

Just entering production. Specified for Royal Australian Air Force Hawk, Royal Air Force Tornado F3 CSP (Capability Sustainment Programme) and IPTN CN 235 maritime patrol aircraft.

In May 2000, the IFF 4810 Transponder completed a successful first flight trial, installed in a Tornado F3 aircraft. Raytheon claim that the 4810 transponder will be the first Successor IFF to be supplied to the UK armed forces.

Contractor

Raytheon Systems Limited.

UPDATED

STR 2515 series receiver processor unit

The STR 2515 series receiver processor units are high-dynamic, all in view, ¼ ATR short units. The RPU provides Standard Position Service (SPS) C/A code and, to US DoD authorised users, the very accurate Precise Position Service (PPS) P/Y code. A major advantage of the STR 2515 Series is that the C/A code version of the RPU can be simply upgraded at the factory to P/Y code once the customer has an agreement with the US DoD.

The STR 2515 Series features separate L1 and L2 channels, for higher resistance to jamming. It is capable of handling all selective availability and anti-spoofing aspects required by US DoD authorised users and is the first European GPS RPU to be unclassified when keyed. It has parallel operation of MIL-STD-1553B, ARINC 429, ARINC 572 Gray coded baro altitude, RS-422 and PTTI/Have Quick interfaces and operates with either FRPA or CRPA systems without modification.

A Kalman filter is available which, when embedded in the host GPS, helps to overcome the problems associated with close coupled navigation solutions and the security implications associated with the export of GPS corrected pseudo-range and range rate data. Kalman filter solutions provide a flexible method of calibrating inertial, altitude and heading reference systems and Doppler, or any combination of all three.

Recent developments have added RAIM, WAGE, relative differential and KCPT to the STR 2515, which now forms part of the UK MoD's 'HIGGER' system that offers compatibility with the US DoD JPALS requirements.

Specifications

Dimensions: 320 × 60 × 194 mm
Weight: 4 kg
Power supply: 115 V AC, 400 Hz or 28 V DC, 40 W
Reliability: >7,000 h MTBF

Operational status

Deployed internationally on many fixed- and rotary-wing aircraft, and on ships.

Contractor

Raytheon Systems Limited.

VERIFIED

AN/APN-194 radar altimeter

The AN/APN-194 is a low-profile, lightweight, all-solid-state radar altimeter which provides analogue and digital outputs to interface with an automatic flight control system. The unit conforms with the high environmental specifications of most military aircraft.

Specifications

Dimensions: 79.4 × 97.3 × 188.2 mm
Weight: 2 kg
Power supply: 115 V AC, 400 Hz, 28 V DC
Frequency: 4.3 GHz
Pulse repetition frequency: 20 kHz
Radiated power: 5 W peak
Altitude: 0-5,000 ft standard
Output signals: digital and analogue
Accuracy: ±3 ft or 4%
Manoeuvrability:
(typical) pitch and roll ±45° within stated accuracy
Temperature range:
(operating) −55 to +95°C
(non-operating) −65 to +125°C

Operational status

In production.

Contractor

Smiths Industries Aerospace.

VERIFIED

Electrical Power Management System (EPMS) for the AH-64C and D Apache

The EPMS for the AH-64C and D Apache helicopters provides switching, monitoring and protection of aircraft primary and secondary distribution of AC, DC and battery power. The main functions of the system are to provide automatic load switching, power source and load distribution protection, as well as electrical system diagnosis. Key benefits are reduced aircraft weight, build time reduction and improved BIT and maintainability.

Operational status

In production for AH-64D.

Contractor

Smiths Industries Aerospace.

VERIFIED

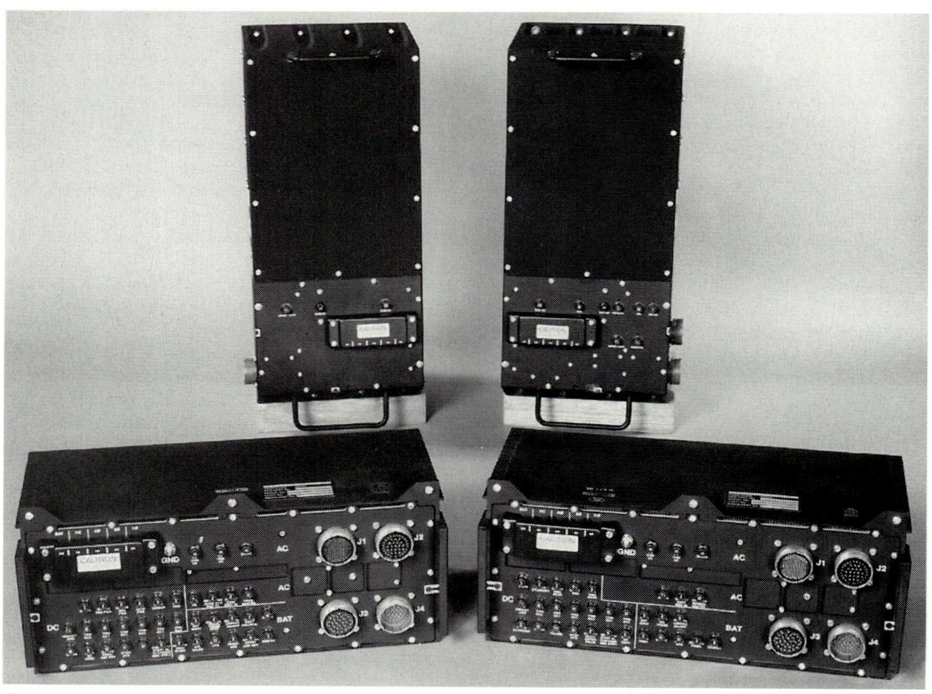

Electrical Power Management System for the AH-64D Longbow Apache 0001395

Eurofighter Typhoon voice control system

Smiths Industries Aerospace has been selected by Computing Devices, Hastings Ltd and endorsed by Eurofighter GmbH to provide the direct voice input system for the Eurofighter Typhoon programme. A design and development contract has been awarded to the company's Defence Systems division, Cheltenham, covering a Speech Recognition Module (SRM) for installation in the communications audio management unit to be supplied for Eurofighter Typhoon by Computing Devices.

It is claimed that Eurofighter Typhoon will be the first production aircraft to incorporate interactive voice technology as a standard fit. Some 25 different cockpit functions will be controlled by voice command through a system containing a 200 word vocabulary developed by Smiths Industries Aerospace. Activities such as calling up screen displays, selecting radio channels and frequencies and other similar functions will be controlled simply though verbal instructions from each individual pilot, even when flying under the most extreme operating conditions.

Each pilot will be required to enrol his voice in a ground-based PC station in advance of each flight. The system is capable of recognising that pilot's individual voice patterns at all times and can cope with physical strains which are placed on the vocal chords through the effects of *g* forces and when flying at high speeds or avoiding enemy aircraft under combat conditions.

Leading up to the Eurofighter Typhoon application, prototype voice systems developed by Smiths Industries were successfully trialled in a number of aircraft including the F-16, F/A-18 and the AV-8B/ Harrier. These led to the development of an interactive voice module designed specifically for the AV-8B/ Harrier mission computer. The trials proved the potential of such systems to reduce pilot workload, enhance situation awareness and improve aircraft safety.

The development of voice technology has benefited from the R&D at the Speech Research Unit of the Defence Evaluation Research Agency (DERA) Malvern, and application development in the DERA Speech Systems Group, Farnborough.

As a result of this co-operation, a contract has also been signed for Smiths Industries to equip a GKN Westland Lynx helicopter with an SRM to evaluate performance in rotary-wing aircraft.

Other land/sea applications of this technology is being investigated.

Operational status

Smiths Industries' SRM contract for Eurofighter Typhoon includes the provision of hardware to support the aircraft development programme, with options covering the future production phase.

Contractor

Smiths Industries Aerospace.

VERIFIED

Nimrod MRA. 4 - avionics systems

Smiths Industries Aerospace will supply and support four major elements of the new avionics suite for Nimrod MRA. 4 - the UK's updated maritime patrol aircraft. They comprise the Navigation and Flight Management System (NFMS), based on the latest FMS for the Boeing 737-600/700/800, with added military capabilities; the Utility Systems Management System (USMS) controlling utilities such as hydraulics and environmental systems; the Stores Management System (SMS) handling weapons delivery; and the Full Authority Digital Engine Control (FADEC) system for the four BR710 engines.

Operational status

In development. Equipment for the first two development aircraft has been delivered. Production deliveries commence in April 2001.

Contractor

Smiths Industries Aerospace.

VERIFIED

Self-contained attitude indicators

Smiths Industries Aerospace manufacture a number of self-contained attitude indicators which feature electrical erection and lockable mechanical caging. These high quality units can be supplied with features such as synchro outputs, pitch trim, various sphere markings, case mountings and lighting configuration including NVG-compatibility.

Specifications

Accuracy: ±0.5°
Range:
(roll) ±360°
(pitch) ±85°
Power supply: 28 V DC
Installation: Case ARINC 3ATI, 0-20° panel
Weight: 1.7 kg

Operational status

In production.

Contractor

Smiths Industries Aerospace.

VERIFIED

SN500 Automatic Flight Control System (AFCS)

The SN500 is a high-integrity AFCS designed for the most demanding applications. The system can be integrated both with existing and future facilities, such as MLS, and is suited for all Sea King roles including ASW, OTHT, SAR, AEW and utility. It features duplex stabilisation, attitude and heading hold, fly-through manoeuvring, automatic trimming, barometric height hold, airspeed hold, heading acquire, RNav, radar altitude hold, transition up and down, hover with hover trim and overfly.

The analogue autostabiliser provides tight hands-off control combined with flexible hands-on manoeuvring. The duplex lanes of computing are integrated with existing simplex series actuators and, in the cyclic axes, with existing hydraulic beepers. The system is designed to provide a level of safety suited to demanding environmental conditions. The computing circuits are modular and use separate plug-in boards in each lane. Each board is self-contained with its own stabilised power supply. Lane 1 and Lane 2 circuits are separated by a partition and use separate wiring. BITE facilitates ground testing and preflight check-out provides continuous interlane monitoring.

The digital autopilot uses the latest technology to provide coupled control of a wide range of modes. The use of digital techniques allows the specific requirements of each mode, both standard and special to role, to be taken into account. The use of extensive self-testing facilities and performance monitoring within the simplex microprocessor give unambiguous indication to the crew of any detected failures and, where safety requires it, initiates a fly-up procedure. In all axes but collective, autopilot outputs are summed into the duplex lanes of the analogue autostabiliser. In the collective axis, functions are monitored using analogue techniques.

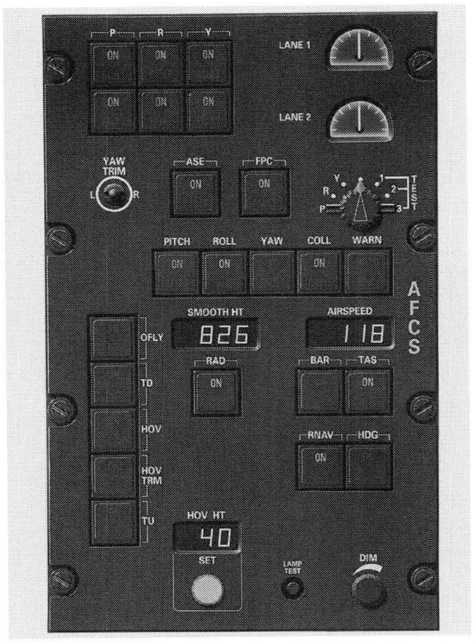

The pilot's controller for the SN500 flight control system has engagement buttons for the duplex autostabiliser and autopilot modes

Specifications

Dimensions:
(sensor unit) 65 × 130 × 212 mm
(autostabiliser computer unit) 194 × 125 × 398 mm
(pilot's controller) 245 × 146 × 213 mm
(autopilot computer unit) 194 × 125 × 398 mm
(hover trim controller) 86 × 146 × 277 mm
Weight:
(sensor unit) 1.9 kg
(autostabiliser computer unit) 5 kg
(pilot's controller) 5 kg
(autopilot computer unit) 5 kg
(hover trim controller) 1 kg
Power supply: 115 V AC, 400 Hz
28 V DC

Operational status

Developed as a retrofit for the SH-3 Sea King helicopter. Now in service on UK Sea Kings.

A variant of the SN500, designated SN501, has been certified for the Sikorsky S61 helicopter.

Contractor

Smiths Industries Aerospace.

VERIFIED

Solid-State power distribution system for F-22

The Smiths Industries family of power management and distribution products includes the F-22 solid-state power distribution system, which controls and distributes secondary 28 V DC and 270 V DC power to aircraft loads. This advanced technology solid-state switching system is housed in four aircooled line replaceable Power Distribution Centres (PDCs), which also control the distribution of 115 V AC over airframe located circuit breakers and relays.

The modular design of the F-22 power distribution system allows greater flexibility of load management than conventional power distribution systems. Each PDC uses a common footprint Solid-State Power Controller (SSPC) design approach. The 28 V and 270 V SSPCs are the same size and are available in four amperage ratings: 5, 7.5, 12.5 and 22.5 A.

The SSPC modules are double-sided, surface mount boards which incorporate Smiths Industries' patented integral buried bus bar. This common design for both 28 V and 270 V SSPC modules further enhances design flexibility and cost-effectiveness.

Contractor

Smiths Industries Aerospace.

VERIFIED

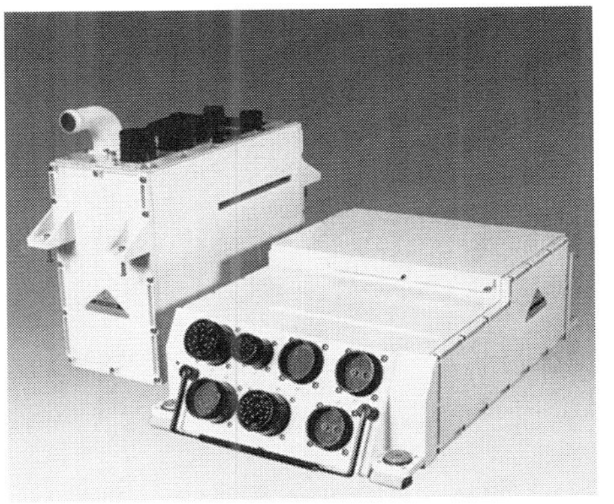

Two of the four power distribution centres for the F-22.
1999/0051646

Airborne ASW crew trainer

Thomson Marconi Sonar Limited produces Anti-Submarine Warfare (ASW) Air Crew Trainers (ACTs). These provide fully dynamic, real-time ASW training for maritime patrol aircraft and ASW helicopter crews. ACT systems offer comprehensive acoustic, tactical and crew co-operation and co-ordination training by simulating sonobuoy, dipping sonar, target and ocean environmental data. Data can be controlled, processed and displayed on the aircraft's acoustic and tactical systems in the same way as real data. Training is therefore achieved without the need to deploy sonobuoys or dipping sonar, or use co-operating submarines and surface vessels, thus substantially reducing costs and increasing training opportunities. There are two basic variants of ACT.

ACT 1 is a software-based system developed specifically for the Thomson Marconi Sonar AQS 901 acoustic processing systems now in service on UK Royal Air Force Nimrod MR.2 Mk 2 and Royal Australian Air Force P-3 Orion maritime patrol aircraft.

ACT 2 is a hardware-based system that can be interfaced with any acoustic processor resident in the platform or, by means of a VHF radio link, with an ACT ground station. ACT 2 can simulate the characteristics of any sonobuoy, dipping sonar, target and ocean environment specified by the user. It comprises an Acoustic Simulation Unit (ASU) and an Exercise Control Unit (ECU). These may be physically separate devices, or integrated into one device depending upon the application. The ASU executes the ASW scenarios and generates synthetic passive and active acoustic data, it can be configured to simulate any number of independent sonobuoy and dipping sonar channels. The ECU is an intelligent control terminal with an integral keyboard and display. The ECU is also used for the temporary storage of exercise scenario data. ACT 2 systems currently available can simulate up to 64 independent data channels from any combination of LOFAR, DIFAR, Barra, Ranger, CAMBS, DICASS and bathythermal sonobuoys and can generate simultaneously the acoustic signature of up to six independent targets. Simulation of specific dipping sonar and other types of sonobuoy can be easily accommodated within the existing hardware. There are no limitations on the number and variety of exercise scenarios for training. These are created and maintained by each customer, using an Exercise Support System (ESS). The ECU and ESS functions can be performed using a single/common IBM PC-compatible computer, which can also be used as a classroom trainer and for post-flight debriefs. ACT 2 systems have been successfully integrated with aircraft navigation systems and communications systems to provide multi-aircraft co-ordinated ASW exercises. The systems have also been integrated as part of a full mission system simulator. Variants of the ACT 2 system are also offered as embedded trainers within Thomson Marconi Sonar's new generation of acoustic processing systems including TMS2000.

Operational status

ACT 1 is in service with the UK Royal Air Force and Royal Australian Air Force. ACT 2 is in service with several customers, including the UK Royal Navy.

Contractor

Thomson Marconi Sonar Limited.

VERIFIED

Airborne ASW crew trainer **1999**/0044963

The AQS 901 colour display in a Nimrod MR.2 Mk 2 **1999**/0044962

AQS 901 acoustic processing systems

The Thomson Marconi Sonar AQS 901 acoustic processing system is installed in both the UK Royal Air Force Nimrod MR.2 Mk 2 and the Royal Australian Air Force Lockheed P-3C Orion maritime patrol aircraft and has been in service since 1979. The AQS 901 is able to handle data from all types of sonobuoys in the NATO inventory. This includes advanced sonobuoys such as Barra, CAMBS and DIFAR.

Every aircraft contains two AQS 901 systems, each with two CRT displays and two electrographic chart recorders (hard-copy display). A fifth display, for system management activities, shows processed and tabular data from either system. Control and interrogation of processed data is achieved using a mix of keyboard, on-screen menus, rollerball and cross-wire cursor.

A series of software and hardware updates over the past 12 years has enabled the AQS 901 to continue to fully satisfy current ASW requirements.

In 1995, the UK Royal Air Force's Nimrod fleet was further upgraded when its AQS 901 systems were updated to include advanced colour processing.

Operational status

In service. AQS 901 equips over 30 UK Royal Air Force Nimrod MR.2 Mk 2s and 20 Royal Australian Air Force P-3C Orions. Deliveries were completed in August 1986.

Contractor

Thomson Marconi Sonar Limited.

VERIFIED

AQS 902/AQS 920 series acoustic processors

The AQS 902 and its export variants, the AQS 920 series, have been designed for ASW helicopters, maritime patrol aircraft and small ships. This range of lightweight and versatile systems will handle data from all current and projected NATO inventory sonobuoys and will interface with a wide range of 31 or 99 RF channel sonobuoy receivers. An AQS 902/920 series can process and display up to eight buoys or a combination of sonobuoys and dipping sonar simultaneously. The modular structure of the AQS 902/920 series allows systems to be tailored to specific customer requirements. The signal processor unit converts received sonobuoy signals into digital form, carries out the necessary filtering and analysis and processes the data into a suitable form for display. It is packaged in a 1 ATR(S) box. Acoustic data is displayed on a CRT or electrographic paper chart recorder. Either or both may be specified, but the CRT display requires a post-processor unit to prepare data for presentation and provides many additional operator aiding facilities. The operator uses a key panel and rollerball or stiff stick cursor control device to specify processing modes and

The AQS 902G/DS installation in a UK Royal Navy Sea King Mk 6

display formats, measure and extract information and access the wide range of automatic and semi-automatic system facilities.

The AQS 902/920 offers numerous processing and display facilities including simultaneous passive and active processing, wide and narrowband spectral analysis, broadband correlation, CW and FM active modes, passive auto alerts, Doppler fixing, calculation of target range and speed, bathythermal processing and ambient noise measurement. Additionally, the system provides automatic tracking and calculation of line frequency, target bearing and range and signal-to-noise ratio.

Thomson Marconi Sonar claims that a unique feature of these systems is the acoustic localisation plot which presents the operator with a geographical plot of sonobuoy bearing and range information.

Operational status

Configurations of this system are in service with the UK Royal Navy, the Royal Swedish Navy, the Pakistan, Indian and Italian navies. Further variants have been sold to Northrop Grumman for S-2(T) Turbo Tracker aircraft.

The AQS 902G-DS is in service with the UK Royal Navy Sea King Mk 6 ASW helicopter, 150 systems having been delivered and over 80 export variants are in service throughout the world.

Contractor

Thomson Marconi Sonar Limited.

VERIFIED

AQS 903/AQS 930 series acoustic processing system

The AQS 903 acoustic processor and its new compact, powerful, lightweight derivatives can process and display data from all current and NATO-projected sonobuoys as well as dipping sonars. AQS 903 is fitted to the UK Royal Navy's EH101 Merlin and is currently being updated to improve the passive processing capability and reduce weight. Derivatives are being considered for several maritime helicopter and patrol aircraft.

The systems use distributed processing and the latest COTS technology to achieve optimum performance. The modular design of the system architecture allows variants to be configured to meet customer requirements from 8 to 32 processing channels.

Operator aids ensure that the ever-increasing mass of information available to the operator is exploited fully and effectively to achieve mission success. Systems can interface to other avionics systems for sonobuoy launching, navigation, tactical tracking and weapon release.

Operational status

Selected for the UK Royal Navy EH 101 Merlin helicopter. The production order for 44 systems is nearing completion. The upgrade programme is now under way.

Contractor

Thomson Marconi Sonar Limited.

VERIFIED

The EH 101 Merlin ASW helicopter

AQS 940 acoustic processor

AQS 940 is an advanced, powerful airborne acoustic processor which uses a cost-effective and totally open architecture based on rugged state-of-the-art, off-the-shelf hardware. The existing and proven algorithms used in the AQS 903/930 series have been re-hosted and enhanced with new colour broadband and data fusion techniques. The system can also be provided with the latest processing and display techniques for multistatics, providing increased area coverage in littoral waters.

The system is designed for use by one to four operators, with each operator having access to all the processed data as graphical data fusion plots, localisation plots, acoustic formats or tabular information.

All current and projected NATO sonobuoy types can be processed; sonobuoy data may be received from either traditional analogue receivers or from new digital receivers.

Operator workload is reduced by the user-friendly interface and, more importantly, the wealth of operator aids. In particular, the data fusion display focuses the operator's attention on areas of acoustic interest.

The system can be easily interfaced to other avionics systems via any industry standard.

Contractors

Thomson Marconi Sonar Limited.

VERIFIED

CRISP Compact Reconfigurable Interactive Signal Processor

Thomson Marconi Sonar Limited has developed 'CRISP', a Compact, Reconfigurable Interactive Signal Processor which is based on many years experience of providing powerful signal processors to navies around the world.

Several versions of CRISP are now available to support the surface, subsurface and airborne processing community.

A high-resolution (1,024 × 1,024 pixels) colour graphics display (available as a flat screen) with pop-up menus and trackermouse roll and click selection allows interactive and user-friendly operation.

Using modular boards already in operational naval service, CRISP brings the following advances to Mission Support Systems (MSS): simultaneous broadband and narrowband analysis; digital DIFAR processing; frequency range from infrasonic to intercept; very fast time replay; short-term event capture and analysis; rapid software reconfigurability; commonality with other UK Royal Navy and Royal Air Force equipment.

Operational status

In service with the UK Royal Navy, UK Royal Air Force and Canadian Forces.

Contractor

Thomson Marconi Sonar Limited.

VERIFIED

Operator Aiding SYStem (OASYS)

The performance of the current generation of ASW sonar systems depends upon the successful fusion of the skills of a highly trained operator and the use of sophisticated computer equipment. A balance must be maintained between the need to present the operator with the widest choice of detailed information and the ability to view, comprehend and evaluate such large volumes of data.

The OASYS provides the acoustic system operator with a toolset of functions and graphical displays which enable the operator to make the most effective use of the acoustic processor and available sonobuoy stores. The OASYS includes a set of database functions for providing environmental, target, sonobuoy and processor information. By making use of all this information, OASYS can provide tools for sonobuoy stores management, tactical advice and processing recommendations. Although OASYS has been designed primarily as an operational system management aid, its portability and extensive toolset allows it be used as a mission planning tool within a ground-based PC environment.

Contractor

Thomson Marconi Sonar Limited.

VERIFIED

Position Estimation by TRack Association (PETRA)

With the ever increasing levels of information available from an acoustic processor, it has become more important to provide the operator with tools which sort the data into a manageable form. The Position Estimation by TRack Association (PETRA) subsystem has been designed to meet this task by estimating the source positions directly from sets of frequency lines which have been associated across sonobuoys. The resulting data can be presented as an overlay to a Tactical Situation Display (TSD) or used as the raw input data for a classification system.

Contractor

Thomson Marconi Sonar Limited.

VERIFIED

TMS 2000 series acoustic processing system

The TMS 2000 is an advanced and powerful airborne acoustic processor. It uses a cost-effective and totally open architecture based on rugged, state-of-the-art,

TMS 2000 series acoustic processing system (typical passive display) 0001187

off-the-shelf hardware. The TMS 2000 uses existing and proven algorithms which have been developed as part of the AQS 900 and SADANG series of processors. They have been rehosted and enhanced with new colour broadband, data fusion, and multistatic processing techniques.

The system is designed for use by one to four operators, with each operator having access to all the processed data as graphical data fusion plots, localisation plots, acoustic formats or tabular information. The system can support multiple high-resolution colour monitors.

All current and projected NATO sonobuoy types are processed and sonobuoy data may be received from either traditional analogue receivers or from new digital receivers.

System control uses the latest techniques of pull-down or pop-up menus, augmented by on-screen editing and fast dedicated keys.

The system can be easily interfaced to other avionics systems via any industry standard or MIL-STD-1553B interface.

Passive functions

The TMS 2000 provides a range of advanced processing detection modes, which cover a full range of target noise characteristics including narrowband frequency lines, broadband signals, swaths, transients and DEMON signals. Each channel has two main processes, plus up to six auxiliary processes which are tuned to the target characteristics and environmental conditions. Computer assistance is provided to link database information to observed signal data to aid target classification. DOUBLE DEMON processing ensures full coverage of both the carrier noise bandwidth and modulating spectrum to achieve permanent detection of transient cavitation phenomena.

Active functions

The TMS 2000 has the capability to process up to 16 DICASS sonobuoys simultaneously in the multistatic mode as well as the monostatic mode, and so satisfies typical brown shallow water operational requirements. Typical coverage is 200 square nautical miles with 16 DICASS sonobuoys.

Specifications

Sonobuoy capability: 8, 16, or 32 sonobuoy channels; compatible with all NATO sonobuoys
Passive processing: up to 64 DIFAR per unit; two main and up to six auxiliary processes for each channel; manual/automatic noise nulling; cardioid processing; computer-assisted classification
Active processing: 16 DICASS multistatic/monostatic modes; ping-to-ping integration, auto-handling of pulse sequences; reverberation suppression and stationary artefact removal
Displays: passive and active, with colour enhancement; passive and active geographic energy plots

Operational status

Ordered by the UK Royal Navy for the EH 101 Merlin helicopter, and by the Spanish Air Force and UAE Navy.

Contractor

Thomson Marconi Sonar Limited.

VERIFIED

Type 2069 sonar

The Type 2069 sonar helicopter dunking system is installed in Sea King helicopters. It provides full 360° coverage and is understood to be effective at ranges in excess of 7,000 m. Full azimuth coverage is provided in stepped fashion over 90° arcs progressively, but manual control allows the operator to concentrate on any particular sector. The system is programmed to undertake an automatic search. The operator is provided with audio, visual Doppler and visual sector sonar information, and close contact maintenance is provided for the tracking of nearby targets and those at greater depths.

The sonar may be employed for either surveillance or attack control, or for both simultaneously. Pulse length and detection range settings are operator selectable to optimise the system according to sea conditions and the tactical situation.

The Type 2069 sonar was derived from the 195M employing new solid-state transmitters and with a longer cable. The sonar transducer has been re-engineered to provide a greater operating depth and is now integrated with the AQS 902 G/DS acoustic system.

The total system provides simultaneous control, processing and display of sonobuoys and sonar. It is installed in the UK Royal Navy Sea King Mk 6.

The Type 2069 dunking sonar sensor

Operational status

Type 2069 is now in service with the UK Royal Navy and several other navies.

Contractor

Thomson Marconi Sonar Limited.

VERIFIED

Landing gear control for Eurofighter Typhoon

Landing gear control modules control the operation of the landing gear system, monitor various test signals, provide the power supply for the braking system and communicate with other avionic systems via a STANAG databus.

Operational status

In production for the Eurofighter Typhoon.

Contractor

Ultra Electronics Ltd, Controls Division.

VERIFIED

A628 speech pre-processor

High extraneous noise levels at the transmitter reduce the quality of speech received over RF links, vocoders and internal communication systems. This is a particular problem over digitised systems as the transmitted signal is often unusable at the receiver.

The A628 speech pre-processor provides clear recognisable speech over communication links by reducing the effects of background noise and enhancing speech quality. It automatically adjusts to the acoustic environment, and is particularly suitable for digital systems.

The A628 is able to separate noise and voice signals and, using active techniques, cancels noise picked up by the microphone before it reaches the transmitter. Not only does the speech pre-processor reduce noise from the transmitted audio signal but it can also enhance the speech waveform at the receiver. This ensures that received speech is much easier to understand. The A628 is of particular importance to operators of HF radio systems, which are typically noisier than other RF links. A further feature of the A628 is its ability to adapt to the acoustic environment, providing the precise amount of processing to ensure that the speech content can be recognised.

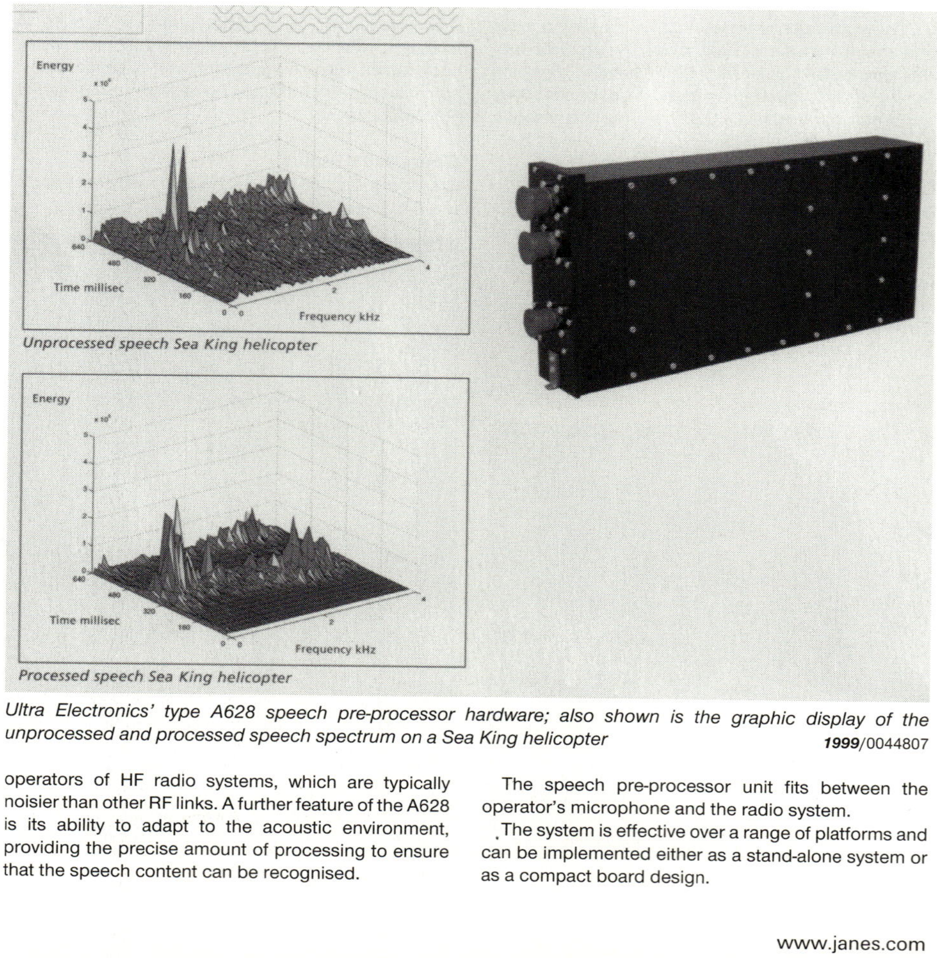

Ultra Electronics' type A628 speech pre-processor hardware; also shown is the graphic display of the unprocessed and processed speech spectrum on a Sea King helicopter **1999**/0044807

The speech pre-processor unit fits between the operator's microphone and the radio system.

The system is effective over a range of platforms and can be implemented either as a stand-alone system or as a compact board design.

Contractor

Ultra Electronics Limited, Sonar and Communication Systems.

VERIFIED

AQS 970 acoustic processor

The Ultra AQS 970 airborne acoustic processor has been selected for the Nimrod MRA4 programme and is being developed through collaboration between Ultra Electronics and Computing Devices of Canada. The CDC UYS-503 system is already in service with the defence forces of Australia, Canada, Sweden and the US. The AQS 970 interfaces with the Tactical Command System (TCS) and the Sonobuoy Receiving System (SRS) in fixed-wing Maritime Patrol Aircraft (MPA) and Anti-Submarine Warfare (ASW) helicopters, providing signal conditioning, signal and data processing, display generation and control functions.

The Ultra AQS 970 is designed to meet the detection, localisation and attack requirements posed by the latest-generation nuclear and diesel submarines operating in open ocean and littoral water environments. The modular architecture of the AQS 970 and its ample processing power and memory capacity enable it to accommodate readily new buoy types and processing techniques without hardware modification.

The system employs a scalable hardware and software architecture, configurable as 16, 32 or 64 identical processing channels, providing full 64-buoy processing capability (subject to buoy RF channel limitations) for all NATO variants of the following existing or planned sonobuoy types:

Passive search and localisation
- DIFAR
- LOFAR
- VLAD
- CODAR (DIFAR/omnipairs)
- HIDAR (digital DIFAR)
- BARRA

Active search and localisation
- CAMBS
- DICASS
- RO
- SR(SA)903 Active sonobuoy search system

Special purpose
- Ambient Noise Monitoring (ANM)
- BATHY
- Built-in Acoustic Test Signal Generator (ATSG) and Tactical Acoustic Trainer (TAT)

Processing and display capabilities include bearing coherent processing and effective colour-coded displays; energy map and acoustic situation displays; narrowband, broadband/swath, diesel, DEMON, and transient detection processes; variable band CODAR (using DIFAR buoys); adaptive interference rejection/anti-jam processing (DIFAR, BARRA, CAMBS)

Operator aids include Gram markers and harmonic dividers, CPA analysis, Lloyd's mirror analysis, display magnify/freeze/zoom/pan, cross-hair and hook, adjustable thresholds and integration times, concurrent multiresolution processing, rapidly selectable operator-defined display formats and combinations.

General aids provided include frequency/bearing/range trackers, synchronised audio, range prediction, data fusion and contact tracking, auto alerts.

Operational status

Ultra Electronics is providing the integrated acoustic system for Nimrod MRA4.

Contractor

Ultra Electronics, Sonar and Communication Systems.

VERIFIED

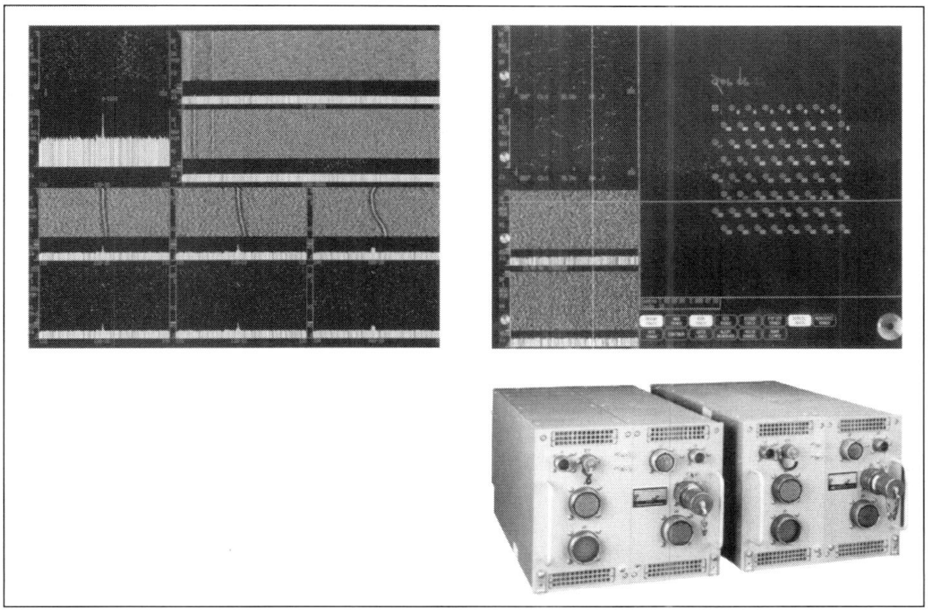

Ultra Electronics Limited, Sonar and Communication Systems AQS 970 acoustic processor hardware and two display formats ***1999***/0044961

ARR 970 sonobuoy telemetry receiver

The ARR 970 sonobuoy receiver system has been designed to satisfy the requirements of current Anti-Submarine Warfare (ASW) aircraft, while providing additional facilities to meet future requirements. It amplifies and demodulates FM sonobuoy transmissions and provides the output to an acoustic processor for analysis and display. The ARR 970 provides excellent RF and acoustic performance as well as lightweight and compact size.

Two models are available. A multichannel version offers 16-, 20- and 32-channel options, while a single channel version provides only the 32-channel capability. Both models include options for an On-Top Position Indicator (OTPI) function, channel scanning and a sonobuoy positioning system.

Specifications

16-channel system
Dimensions: 545 × 215 × 345 mm
Weight: 33 kg
Power supply: 115 V, 3 phase, 400 Hz, MIL-STD-704A: 300 VA
RF channels:
99+ test (375 kHz spacing, standard)
505 (75 kHz spacing, optional)
Audio interfaces: balanced, 2 Vrms for 75 kHz peak deviation
IF bandwidth: 240 kHz, standard
Control interfaces: MIL-STD-1553B, RS-422

32-channel system
Dimensions: 545 × 328 × 345 mm
Weight: 39 kg
Power supply: 115 V, 3 phase, 400 Hz, MIL-STD-704A: 450 VA
RF channels:
99+ test (375 kHz spacing, standard)
505 (75 kHz spacing, optional)
Audio interfaces: balanced, 2 Vrms for 75 kHz peak deviation
IF bandwidth: 240 kHz, standard
Control interfaces: MIL-STD-1553B, RS-422

Operational status

Selected for use on Nimrod MRA4.

Contractor

Ultra Electronics, Sonar and Communication Systems.

VERIFIED

R605 sonobuoy receiving set

The R605 sonobuoy receiver is in use on the Royal Navy EH 101 Merlin ASW helicopter. The receiver amplifies and demodulates four simultaneous FM sonobuoy transmissions and provides the output to the acoustic processor for analysis and display. It provides coverage of 99 RF channels per receiver module and is mechanically interchangeable and connector compatible as a direct replacement for the AN/AAR-75.

For applications where the receiving sets must be mounted at a distance from the receiving antenna, a remote preamplifier is provided. MIL-STD-1553B control is available. Control functions include RF channel selection, sonobuoy type for optimal bandwidth control, RF level reporting and self-test initiation and reporting.

The R605 receiver features a large RF dynamic range and high sensitivity over a broad base bandwidth. In addition, the receiver provides a wide, linear phase,

Ultra Electronics ARR 970 sonobuoy telemetry receiver multichannel (16-, 20 and 32-channels) (left) and single-channel (32- channels) (right) configurations 0011858

audio output with close uniformity between each of the four audio channels. Digital sonobuoy data undergoes matched filter detection and clock regeneration in the receiver. Differentially driven clock and data signals are available for each of the four receiver outputs at any of the data rates and code functions.

Specifications

Dimensions: 159 × 351 × 232 mm
Weight: 10 kg
Power supply: 115 V AC, 400 Hz, 3 phase, 100 VA max
Frequency: 136-173.5 MHz
Channels: 99 RF, 4 acoustic
Interfaces: MIL-STD-1553, RS-422, ARINC 429
Reliability: >2,000 h MTBF

Operational status

In serial production. In service on UK Royal Air Force Nimrod MR.2 Mk 2 and Royal Australian Air Force P-3C Orion aircraft, and selected for the UK Royal Navy EH 101 Merlin helicopter.

Contractor

Ultra Electronics, Sonar and Communication Systems.

VERIFIED

The Ultra R605 sonobuoy receiver

T618 Link 11 Data Terminal Set

The T618 Data Terminal Set (DTS) is a small lightweight unit suitable for helicopter, fixed-wing and other applications. It provides all Link 11 modem and network control functions defined by MIL-STD-188-203-1A, STANAG 5511 and ADat P-11 in picket and network control modes.

As a modem, the DTS converts the Link 11 data into audio tones suitable for transmission over radio circuits. As a network controller, it provides error corrections in USB, LSB and Diversity modes and roll call management functions in the net.

The T618 includes features which are of great importance to Link 11 operators. It provides an outlet to display the performance of every station in the network in real time; the link monitor displays the received signal quality on both upper and lower sidebands. The percentage of time each picket responds to interrogation and the analysis of the data being received is also displayed for each picket unit.

For the system maintainer, the T618 contains a BIT which isolates malfunctions to card level. In addition, various loopback modes provide signal paths which help to isolate difficulties in system configuration, including multiple stations. The combination of these controls and the link quality assessment provide a high visibility of the system operation.

The single-tone Link 11 system enhancement is available with the T618. Provided as a switchable option, it enables the DTS to be compatible with both single tone and conventional Link 11 modes.

The DTS is also programmable for operations in combinations of TDMA network protocols and frequency hopping.

The T618 DTS can be upgraded easily to provide NATO Improved Link 11 (NILE) performance.

Specifications

Dimensions: ¼ ATR short
Weight: 3.6 kg
Power supply: 115 V AC, <30 W
Interfaces: serial data ATDS, radio interfaces as defined by MIL-STD-188-203-1A
Computer control: ARINC 429, MIL-STD-1553B, RS-232C as options
Temperature range: –40 to +55°C
Reliability: 7,000 h MTBF (ARW environment)

Ultra Electronics Link 11 datalink processor (left) and Data Terminal Set (right) 0011828

Operational status

In serial production. Developed for, and in use on, the Merlin ASW helicopter; selected for the Nimrod MRA.4 aircraft.

Contractor

Ultra Electronics, Sonar and Communication Systems.

VERIFIED

T619 Link 11 datalink processor

The T619 datalink processor acts as a Link 11 processor. It uses up-to-date technology to provide a small lightweight unit suitable for helicopter, fixed-wing and other applications.

The datalink processor interfaces with both STANAG 5511 and OPSEC 411. It provides database management, track correlation, gridlock, message assembly/disassembly, transmit and receive filtering, operator control and monitoring. All functions are automated where practical to reduce the operator workload to a minimum.

The unit consists of a single processor card and an interface card, both of which are VME-based. A spare slot is available to double processor throughput and memory capability or to provide additional functions such as to integrate a single card cryptographic device. For this purpose the datalink processor has the appropriate qualifications and is equipped with all the required controls on its front panel.

Also available is the A628 speech processor, which allows secure speech with active noise reduction and can be applied to LPC-10 vocoders.

Specifications

Dimensions: ½ ATR short
Weight: <10 kg
Power supply: 115 V AC, 40 Hz, <40 W
Interface: MIL-STD-1553B
Temperature range: –40 to +55°C
Reliability: 6,000 h MTBF (ARW environment)

Operational status

In serial production. In use on the Merlin ASW helicopter; selected for the Nimrod MRA.4 aircraft.

Contractor

Ultra Electronics, Sonar and Communication Systems.

VERIFIED

Vicon 70 countermeasures pod

Vicon 70 is a modular, lightweight dispensing pod which uses components of the Vicon 78 Series 455 system to provide self-protection against radar threats and heat-seeking missiles. It is designed for helicopters and fixed-wing aircraft where structural or other constraints preclude the use of airframe-mounted dispensers.

Vicon 70 can be configured for two dispensers, or for up to eight dispensers by adding modules to the pod. Modules can also be used to house radar warning receiver and missile approach warner LRUs. It can be fitted to any fixed-wing aircraft or helicopter type equipped with weapons racks or fuselage or wing pylons with 14 in NATO attachments. The firing trajectory of the dispenser modules can be preset to optimise the effectiveness of the chaff and flares. Manual or fully automatic versions of Vicon 70 are available. The automatic version interfaces directly with the radar warning receiver via a serial datalink or databus.

Unlike the internal CMDS installation, the podded solution requires no major airframe modification and provides a greatly enhanced chaff dispensing capability. The additional benefit of the pod is that countermeasures dispensing may be carried as role equipment and can be downloaded should self-protection not be required.

Specifications

Dimensions: 2,705 length × 356 mm diameter
Weight: 210 kg fully loaded with 8 dispensers

Contractor

W Vinten Ltd.

VERIFIED

Vicon 78 airborne decoy dispensing systems

Vicon 78 is a family of advanced lightweight chaff and IR decoy dispensing equipment in production for high-performance combat, tactical transport and maritime patrol aircraft, helicopters and RPVs. Vicon 78 systems can be fitted as original equipment or upgrade existing systems. It is a defensive aids system which provides self-protection by passive ECM against radar-guided and IR-seeking air- and ground-launched missiles and radar-directed anti-aircraft artillery fire.

Vicon 78 systems provide manual or fully automatic threat adaptive capability. Fully automatic operation, greatly reducing aircrew workload, is achieved by interfacing the dispensing system to a threat detection sensor. In automatic mode the programme to be dispensed is downloaded from the Radar Warning Receiver (RWR) or Missile Approach Warner (MAW). In semi-automatic mode the programme to be dispensed is downloaded from the RWR and initiated by the crew. In manual mode the crew can dispense preset chaff or IR flare programmes.

Systems can be offered with multidispenser configurations to cater for large transport and maritime patrol aircraft down to a single lightweight dispenser for smaller aircraft, each of which can accept interchangeable chaff or flare magazines containing up to 64 cartridges. Dispensers can also be configured to dispense expendable jammers, towed decoys and other format expendables as required. The dispensers may be internally mounted, semi-recessed with fairings or carried in a lightweight pod.

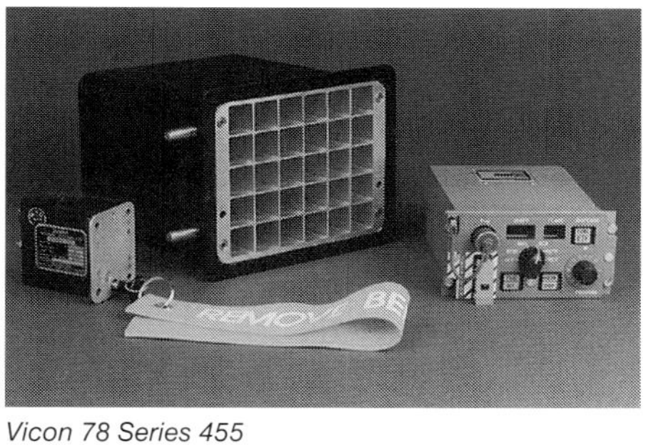

Vicon 78 Series 455

Rear view of a Hawk showing the Vicon 78 Series 300 countermeasures dispenser

Vicon 78 Series 200

Vicon 78 Series 200 equipment has been supplied for the Sea Harrier programme. The Series 200 CounterMeasures Dispensing System (CMDS) is a two-dispenser system, controlled by a salvo control processor unit and manually operated by the pilot. Vicon 78 Series 200 equipment has successfully undergone EMC testing at Boscombe Down and has been selected by the UK MoD. The Series 205 CMDS has the ability to dispense ALE-39 and ALE-40 format rounds from a common dispenser.

Vicon 78 Series 203 equipment was developed and supplied to CASA of Spain for an export CASA 101 programme. It is a four-dispenser system controlled by a two-box processor with a fully automatic mode of operation via an interface to the aircraft Litton AN/ALR-80(V) RWR. The Series 203 system also equips the CASA CN-235 tactical transport aircraft, interfacing to the Litton AN/ALR-85 RWR.

The Vicon 78 Series 210 was selected by the Royal Air Force for operations by the Tornado F3 in the Gulf War. It utilises a new format dispenser that fires a 55 mm two shot infra-red decoy flare. After the 1990-91 Gulf War, the Series 210 was incorporated as a fully approved modification on the Tornado F3 aircraft.

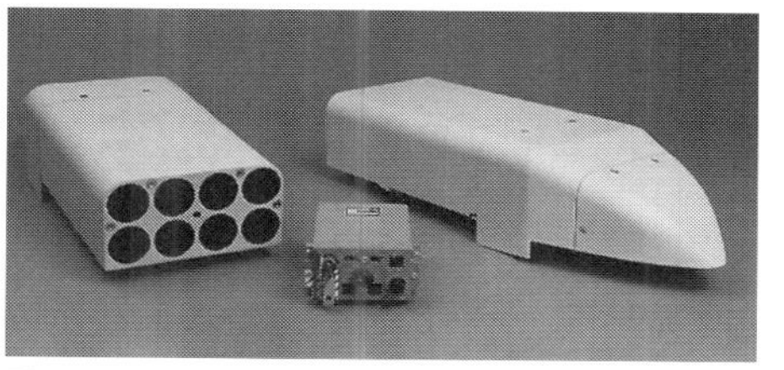

The Vicon 78 Series 400 is fitted to UK and Italian Tornado F Mk3 aircraft

Vicon 78 Series 300

The Vicon 78 Series 300 equipment was developed for the BAE Systems Hawk 100 and 200 Series aircraft. In the Hawk, the Series 300 CMDS is a digitally controlled,

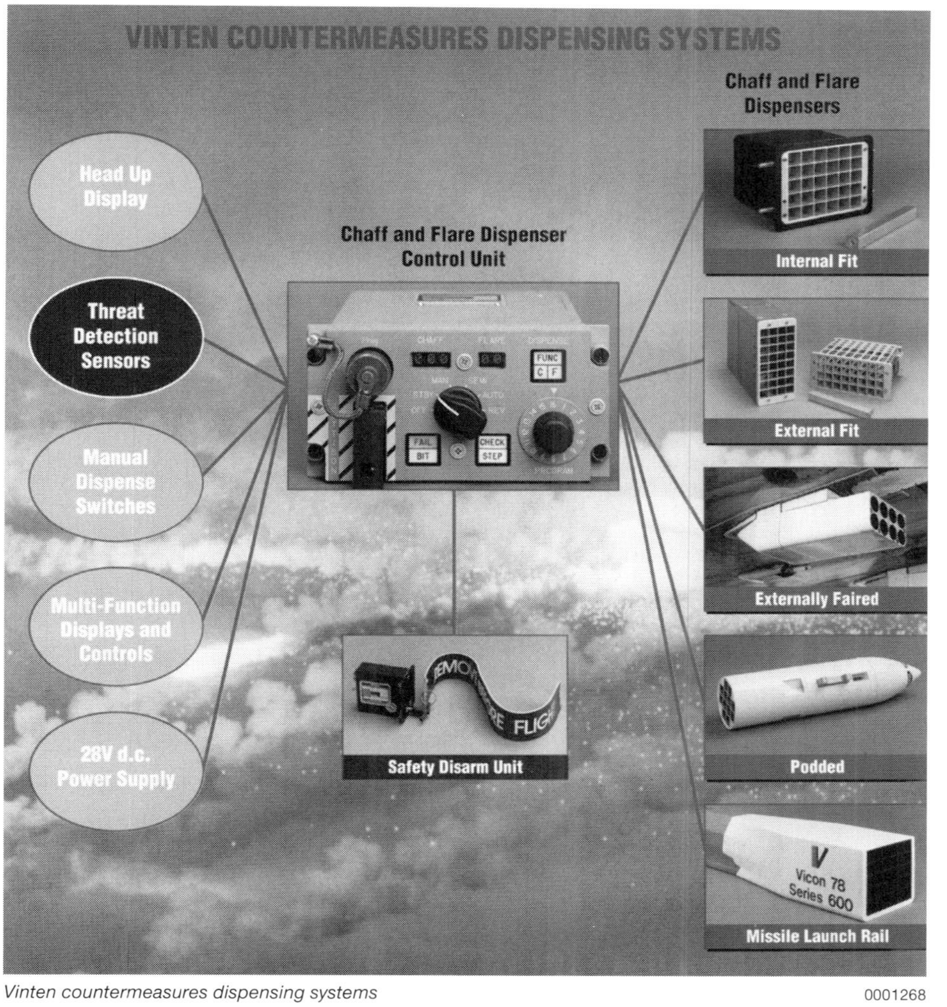

Vinten countermeasures dispensing systems 0001268

The VICON 78 Series 455 countermeasures dispensing system selected for the UK WAH-64 Apache HIDAS **2000**/0062752

fully automatic two-dispenser system that interfaces via an RS-422 link to the Racal Prophet or Marconi Sky Guardian RWRs. The Series 300 system has been designed with the capability to control up to 16 dispensers, with each dispenser having the capability to be configured to hold up to 64 chaff or flare cartridges. Each dispenser can accept interchangeable chaff and flare magazines. When power is applied, the system will identify the number and type of expendables in each dispenser and display the count of chaff and flare cartridges on the cockpit control unit.

A continuous BIT facility is executed by the system; in the event of failure, fail indications are displayed on the cockpit control unit. The BIT is also designed as a first line maintenance facility to identify faults and so speed rectification.

The Vicon 78 Series 300 CMDS has been designed to interface with any RWR or ESM system that has a serial datalink interface capability. The RWR/CMDS interface has been successfully achieved with Litton, BAE Systems, Avionics and Racal. Agreements are also in place with other RWR manufacturers. Vicon 78 CMDSs have the ability to interface to MAWs and missile launch detectors.

Vicon 78 Series 400

The Vicon Series 400 incorporates the advanced electronics of the Series 300, which have been improved and miniaturised, with the eight cartridge 16 shot 55 mm dispenser format of the Series 210. The system utilises a single miniature LRU Chaff and Flare Dispenser Control Unit (CFDCU), which can be either cockpit- or remote-mounted, and provides the system control, processing and programme loading functions.

The Vicon 78 Series 400 has been selected for a fleetwide upgrade of Royal Air Force Tornado F Mk 3 aircraft.

Vicon 78 Series 420

The Series 420 is based on the 400 system electronics, with the CounterMeasures Dispensers (CMD) being reconfigured to a 1 × 1 in payload, 8 × format suitable for helicopter platforms. Each 8 × 4 CMD can carry 32 expendables and the system can be configured for up to four dispensers. The 420 system is currently equipping British Army Air Corps Lynx and Gazelle helicopters.

Vicon 78 Series 455

The Series 455 CMDS is the latest advanced lightweight version of the VICON 78 series suitable for helicopters, fast jet, Maritime Patrol Aircraft (MPA) and transport aircraft. Due to its low mass and reduced number of LRUs, the Series 455 CMDS is ideal for retrofit and upgrade programmes, while allowing ease of integration into new build aircraft. The system can control up to 24 countermeasures dispensers and up to two BOL chaff dispensers. Several dispenser formats are available, including variants that are interchangeable with ALE-40/-47 LRUs. A wide selection of payload options is supported including standard, dual-shot and modular expendable block decoys. In addition, all Series 455 dispensers are able to accommodate up to six different types of payload simultaneously. The system can be expanded to suit large, multidispenser installations by the addition of further dispensers without the need for sequencer LRUs. The VICON 78 Series 455 has been selected as the CounterMeasures Dispensing System (CMDS) element of the UK WAH-64 Apache attack Helicopter Integrated Defensive Aids System (HIDAS), and forms part of the Nimrod MRA4 defensive system. It has also been configured for Hawk 100/200 upgrades.

The Vicon 78 Series 420 is currently equipping British Army Corps Lynx and Gazelle helicopters 0001270

Vicon 78 Series 500

The Vicon 78 Series 500 lightweight stand-alone dispenser utilises the electronics of the Vicon 78 Series 300 system reconfigured to provide a simple and cost-effective dispensing solution for light aircraft, helicopters and RPVs. The 12-shot dispenser has a simple mechanical and electrical interface and can be operated from the cockpit or, in the case of an RPV, via an RF link. The system can be configured to interface to an MAW.

Vicon 78 Series 600

The Vicon 78 Series 600 is a chaff and flare dispenser which can be mounted on the end of a missile launch rail. The system allows additional countermeasures to be carried without requiring major airframe modifications and with only a small weight and drag penalty.

Operational status

Vicon 78 dispenser applications are as follows: Vicon 78 Series 200 Sea Harrier FRS Mk 51 aircraft; Vicon 78 Series 203 export-CASA 101 (interfacing with AN/ALR-80(V) RWR) and CN235 (interfacing with the AN/ALR-085 RWR) programmes; Vicon 78 Series 210 fully approved modification for Royal Air Force Tornado interceptors post 1990/1991; Vicon 78 Series 300 all export Hawk training/light strike aircraft; Vicon 78 Series 400 UK and Italian Tornado F Mk 3 aircraft; Vicon 78 Series 420 British Army Air Corps Gazelle and Lynx helicopters; Vicon 78 Series 455 Royal Air Force Nimrod MR2 Mk2 aircraft, Royal Air Force Nimrod MRA4, export Hawk training/light strike aircraft, Aerovodochody L159 light multirole combat aircraft, export C-130 aircraft, and several rotary-wing aircraft including the UK WAH-64 Apache HIDAS, Agusta A109 and MI-24.

Contractor

W Vinten Ltd.

VERIFIED

UNITED STATES OF AMERICA

LoFLYTE™

LoFLYTE™ was developed under funding from the US Air Force and the National Aeronautics and Space Administration, using technology previously developed under contracts with the US Navy, the National Science Foundation, and the National Aerospace Plane Joint Program Office. The 'waverider' shape, so named because it is designed to ride on the shockwave created at hypersonic speeds, was proved to be aerodynamically sound through extensive wind tunnel testing at NASA Langley Research Center. In August 1996, LoFLYTE™ was introduced at the Experimental Aircraft Association Fly-in Convention in Oshkosh, Wisconsin. LoFLYTE™ is currently undergoing flight testing at Edwards Air Force Base in California.

LoFLYTE™ incorporates several important innovations in its design, including the M5 Waverider design, neural air data 'smart skin' technology, neural network inner and outer loop flight controls, neural network engine controls, advanced real-time data acquisition and control system, adaptive 'fly-by-light' communications, neural network fault diagnosis, hypersonic flowpath demonstration and nozzle concepts, tiperons, 'pilot induced oscillation' detection and compensation, and small-scale jet-powered aircraft prototyping. Once flight testing of the current 100 in model is completed, the next phase of the project is to build a 23 ft version of the design.

The LoFLYTE™ Waverider flight model 0003301

At the heart of LoFLYTE™'s innovative control system is Accurate Automation's Neural Network Processor (NNP®). The NNP®, developed under SBIR funding from the US Navy, is the only multiple instruction/multiple data neural processor currently on the market. Artificial neural networks are a class of important computational methods that loosely emulate biological neural assemblages of neurons and their dendritic connections that form the brain. Artificial neural networks work like the brain to recognise patterns, make decisions, and perform many other common tasks. They exploit massively parallel processing and distributed memory in ways similar to the brain.

A chief goal of the LoFLYTE™ programme is to demonstrate that neural networks can safely and efficiently control an aircraft during flight. This neural network design adapts and learns during flight. As the aircraft is flown, the controller adapts its actions to varying flight conditions. When it later encounters the same conditions, it recalls the best control strategy. Adaptation and learning are achieved within the scope of a carefully constructed proof of stability. The learning consistently results in improved performance.

Operational status

Development.

Contractor

Accurate Automation Corporation.

VERIFIED

LoFLYTE™ neutral network processor 0003302

AN/ALQ-161 ECM system for the B-1B

AIL Systems is prime contractor for the AN/ALQ-161 suite for the US Air Force Rockwell B-1B. This was launched in 1972 under AIL Systems leadership for the Rockwell B-1A bomber, but the effort came to a halt when the aircraft was cancelled in June 1977. The programme was, however, reinstated in October 1981. AIL Systems accordingly resumed its task as ALQ-161 project leader, having participated in limited flight trials with one aircraft since 1979. The current equipment was developed under restart and initial production contracts, worth more than US$1,700 million, which were awarded to AIL Systems and its subcontractors. Each aircraft set is understood to have cost around US$20 million, approximately one-tenth the cost of a complete B-1B, and comprises no fewer than 108 LRUs, with a total weight of 2,300 kg exclusive of cabling, displays and controls.

The operating frequency range is approximately 0.5 to 10 GHz, to cover early warning, ground-controlled interception, surface-to-air missile and interceptor radar frequencies. Jamming signals in the higher regions of the ECM spectrum are emitted from three electronically steerable, phased-array antennas: one in each wing-glove leading-edge and the third in the fuselage tailcone. Each antenna provides 120° azimuth and 90° elevation coverage. Lower frequency signals are emitted from quadrantal horn antennas mounted alongside the high-frequency equipment.

Major subcontractors in the AN/ALQ-161 programme were Northrop Grumman, Litton Industries and Sedco Systems. All companies served as subsystem managers and have responsibilities for receivers, data processing and jamming techniques. Northrop Grumman provides the low-frequency jamming antennas and Sedco is supplying the phased-array antennas.

AIL Systems was responsible for system integration and for the LRUs, including 51 unique designs. Most of the LRUs are about 0.03 to 0.06 m^3 in volume and weigh between 18 and 36 kg. The majority are readily accessible and can be removed easily or installed by one or two people. Total power required is about 120 kW.

Improvements introduced during the programme included: a frequency extension into the K-band to improve the warning and jamming performance; extension into the low-frequency domain below 200 MHz to improve the response of the warning system; introduction of a digital radio frequency memory to permit the deception jamming of more advanced radars, notably those employing pulse Doppler techniques; and introduction of a new tail warning system. AIL Systems was awarded a US$9.1 million contract in September 1983 to extend the ALQ-161 system to include tail warning. An IBM 101D computer functions as the central processor. The computer uses software based on Jovial to identify the function of every hostile radar, assess its potential threat and assign a jamming priority. Three sets of phased-array antennas mounted in the wing leading-edges and tail provide full 360° coverage in Bands 6, 7 and 8, antennas for lower frequencies being located fore and aft in the airframe.

Two complete systems were delivered for continuing trials with a modified Rockwell B-1A in the Summer of 1984 and for the first production B-1B; the latter made its initial flight in October 1984. Production rate was built up to four systems a month by May 1986. In August 1985, AIL Systems was awarded a US$1,800 million contract for the final 92 ALQ-161 shipsets.

Problems in development were reported in mid-1986 leading to the first 22 B-1Bs not being equipped with the tail warning system which had been scheduled to enter service in mid-1987. Problems with the tail warning system and with repeatability of results from the system as a whole were a part of a list of deficiencies in the B-1B's performance made public at the end of 1986. Of a US$600 million request made by the US Department of Defense at that time to extend the capabilities of the B-1B, almost US$100 million was allocated to bring the ALQ-161 up to its original specification and solve other problems (according to the DoD), while the bulk of a US$131 million element within the US$600 million total requested for expenditure in 1988/89 was to upgrade the defensive avionics suite to meet capabilities which have emerged since the original baseline was set in 1982.

The AIL AN/ALQ-161 system for the B-1B

Operational status

Production of 100 systems is complete and the system is installed in the B-1B. In February 1991, AIL was awarded a US$16.5 million contract for AN/ALQ-161A systems for the Northrop B-2 and an additional US$5.491 million to upgrade systems for the B-1B.

The B-1B Defensive Systems Upgrade Program (DSUP) now reportedly being applied to the whole B-1B fleet incudes Block B, C, D and E enhancements to the ALQ-161 system. The details of Block B upgrades are not available. Upgrades designated Block C to E are understood to involve: Block C improved direction finding and sector blanking (initiated in 1996); Block D improved jamming capability (to begin in 1999); Block E improved emitter detection (later). Presumably, one or other of these upgrades will interface the ALQ-161 DSUP to the ALE-50 towed decoy system also being procured for the B-1B. The B-1B ALE-50 installation is understood to include an eight-unit fit in two quadruple launchers.

Contractor

AIL Systems Inc.

VERIFIED

AN/APS-144 airborne surveillance radar

The AN/APS-144 is a modular radar developed for over land and over water wide area surveillance for tactical, border and interdiction applications. Operating modes of the system are surface moving target indication, SAR imaging and airborne intercept. Its primary use is surface surveillance from manned aircraft, helicopters and UAVs. Typical targets are ground vehicles and personnel, watercraft and low-flying fixed-wing aircraft and helicopters. A cueing feature is included in the system for aiming electro-optical sensors for close examination and identification of moving targets. Target classification features are incorporated in the radar processor.

The modular configuration of the AN/APS-144 allows it to be configured for specific applications. For the airborne intercept mode, a pod-mounted configuration is utilised which is adaptable to light aircraft. An I-band version of the radar has been developed for long-range, wide area surveillance. The potential for D-, E-, F- and H-band versions is inherent in the modular design of the series.

The AN/APS-144 consists of a receiver/exciter, power amplifier, antenna feed, pedestal control, digital control interface and control/display unit.

Specifications

Weight: 55 kg (manned aircraft)
Power supply: 28 V, 660 W
Frequency: J-band coherent, frequency agile, pulse Doppler
PRF: 3, 4, 5, 6.25 kHz
Coverage: 360° or sector (azimuth)
Range: 30 km
Antenna rotation: 11 and 18°/s

Operational status

A prototype was tested by the US Army in 1991 on a UH-60 helicopter and successfully demonstrated the capability to detect groups of personnel. Production is for use in drug interdiction operations.

Contractor

AIL Systems Inc.

VERIFIED

AN/ALE-43 chaff cutter/dispenser pod

The AN/ALE-43 is a high-capacity chaff system which holds rolls of chaff material and cuts it to the appropriate dipole length during operation. Lundy claims to have overcome the difficulties associated with past chaff cutters by developing a unique chaff roving supply system.

The system has a chaff roving hopper behind which is the chaff cutter. A remote processor controls operation of the cutter.

Chaff is drawn simultaneously from up to nine chaff roving packages in the hopper. Each roving package passes through a guide tube which terminates at draw rollers and a cutting roller. As dipole lengths are cut, they are discharged into a turbulent airflow and distributed efficiently behind the pod. Each cutter assembly consists of a drive motor, clutch/brake unit and three indexing cutting rollers, the latter embodying blades which yield specific combinations of dipole lengths.

The system can be podded or mounted internally. It is frequently used as a training aid, but its primary wartime functions would be for anti-ship missile defence, area saturation operations over battlefields, corridor seeding and aircraft self-protection.

Specifications

Podded version
Dimensions: 3,370 (length) × 480 mm (diameter)
Weight:
(empty) 139 kg
(loaded) 284 kg
Chaff payload: 8 × RR-179 roving packages
Max dispensing rate: 7.2 × 10^6 dipole in/s

Internal version
Weight:
(basic) 37 kg
(max loaded) 191 kg
Chaff payload: 9 × RR-179 roving packages
Max dispensing rate: 8.1 × 10^6 dipole in/s

Common characteristics
Power supply: 115 V AC, 400 Hz, 1.7 kVA
28 V DC, 2.5 A
Max continuous dispense time: 11 min
On-time: select 1-9 s in 1 s steps or continuous
Off-time: select 1-9 s in 1 s steps
Altitude: 0-50,000 ft

Operational status

In service. The podded version can fit a variety of standard pylon attachments and is used on the B-52, F-4 Phantom, EA-4A Skyhawk and EA-6B Prowler. Internally mounted versions have been used on the ERA-3B Skywarrior and NKC-135.

Contractor

Alliant Defense Electronics Systems Inc, Lundy Products Group.

VERIFIED

AN/ALE-43(V)1/(V)3 chaff countermeasures dispenser set

The AN/ALE-43(V)1/(V)3 chaff countermeasures dispensing set is designed for two types of installation: as a pod (mounting per MIL-A-8591 both 14 and 30 in mounting), or internal mount. In the pod installation, the chaff cutter assembly is mounted on a structural bulkhead at the rear of the pod centre section. This assembly consists of a drive motor, clutch/brake unit, cutting mechanism and supports. The cutting mechanism comprises a rubber platen roller and three cutter rollers.

Specifications

AN/ALE-43(V)1 Pod Version
Dimensions: 3,370 (length) × 482 mm (diameter)
Weight (empty): 154 kg
Power requirements: 115/208 V, 400 Hz, 3-phase 1,700 VA and 28 V DC, 2.5 A
Chaff supply: standard RR-179/AL roving bundle 18 kg
Chaff payload: 144 kg (8 bundles)
Dispensing rate: 216 g/s per 8 roving bundles
Modes (2):
(pulse): 1-9 s in 1 s intervals
(continuous): 660 s

AN/ALE-43(V)3 Internal Version
Dimensions: 408 × 338 × 310 mm
Weight (empty): 36 kg
Power requirements: 115/208 V, 400 Hz, 3-phase 1700 VA and 28 V DC, 2.5 A
Chaff supply: standard RR-179/AL or special roving bundle
Chaff payload: up to 225 kg (up to 9 roving bundles)
Dispensing rate: 27 g/s per roving bundle
Modes (2):
(pulse): 1-9 s in 1 s intervals
(continuous): 880 s

Operational status

The AN/ALE-43(V) pod version has been proven on SH-3 helicopter, EA-6A, EA-7A, F-4, F-16, F-18, P-3, Learjet 36 and other aircraft. Internal installation versions are operational on such aircraft as: EW Falcon, EW Challenger, ERA-3B, NKC-135 and LearJet 35.

Contractor

Alliant Defense Electronics Systems Inc, Lundy Products Group.

VERIFIED

AN/ALQ-176(V) ECM pod

The AN/ALQ-176(V) is a pod-mounted jamming system designed, developed and tested to fulfil aircraft mission requirements for ECM support, standoff jamming and combat evaluation and training. The pod system comprises transmitter modules and a complementary antenna housed in canisters to form a slim pod with a 10 in (25 cm) diameter. The AN/ALQ-176(V) is mounted on standard aircraft wing or fuselage stores/munitions stations. The pod offers the option of using aircraft internal power or a ram-air turbine.

The modular design of the AN/ALQ-176(V) pod allows flexibility in ECM configurations. A two- or three-canister arrangement is available. The AN/ALQ-176(V)1 two-canister configuration houses up to three voltage-tuned magnetron transmitters. The AN/ALQ-176(V)2 three-canister version provides housing for five transmitters. Dependent on frequency band and tube selection, each transmitter produces 150 to 400 W CW (up to 30 per cent bandwidth) at efficiencies of greater than 50 per cent. The transmitters provide ECM across the specified threat frequencies. Transmitter protection circuits safeguard the system and provide fault status to the cockpit control panel.

The AN/ALQ-176(V) has been specifically engineered to allow rapid turnround in meeting the demands of new threat frequencies. It can accommodate new demands through selection of different stored parameters or through easy replacement of the transmitter tube or antenna. Transmitter spares, maintenance costs, times and training skills required are greatly reduced by the use of standard transmitter design. In addition the design incorporates built-in test and provides for parameter/mode selection at the cockpit control panel.

The flexibility in ECM pod configuration and standard aircraft stores/munitions stations installation enables the AN/ALQ-176(V) ECM pod to be used on a wide variety of tactical and support aircraft.

Specifications

Dimensions: 250 mm diameter
(AN/ALQ-176(V)1) 1,990 mm length
(AN/ALQ-176(V)2) 2,590 mm length
Weight:
(transmitter module) 17 kg
(AN/ALQ-176(V)1 pod) 112 kg
(AN/ALQ-176(V)2 pod) 196 kg
Input/output power:
(AN/ALQ-176(V)1) 2.6 kVA/1.2 kW
(AN/ALQ-176(V)2) 4.5 kVA/2 kW
Frequency: 0.8-10.5 GHz
Transmitter power output: 150-400 W per tube CW (dependent on frequency and tube selection)
Modulation: noise (various)
Control: ALQ-10725 and C-9492 aircraft ECM cockpit control panel

Operational status

The US Air Force, the Royal Norwegian Air Force, the Royal Thai Air Force, and the Canadian Ministry of Defence have the AN/ALQ-176(V) in their inventories.

Contractor

Alliant Defense Electronics Systems Inc, a wholly owned subsidiary of Alliant Techsystems Inc.

VERIFIED

AN/ALR-75(V) surveillance receiver

The AN/ALR-75(V) is a surveillance receiver providing coverage over the frequency range 0.1 to 18 GHz. It consists of an aircraft style set of control panels and a number of remote-control units in ATR configurations. The system includes complete digital control, computer compatibility, scratchpad memory and digital refreshed displays. The AN/ALR-75(V), also known as the SCR-2100, is one of a number of equipments in the SCR series of receiving equipments.

Operational status

In operational service in the ERA-3B, NKC-135A and

EC-24A aircraft, and also as a ground-based equipment in AN/ULQ 13(V)1 vans.

Contractor

Andrew SciComm.

VERIFIED

SCR5101-B1 VHF/UHF receiver

The SCR5101-B1 module is a VHF/UHF receiver consisting of a high performance synthesiser, RF converter and DSP board in a single slot VXI module. The receiver optimises rejection of unwanted signals in narrow, medium and wideband applications through use of three selectable IF roofing filters. RF preselection is provided over the full 20 to 3,400 MHz input range.

The final IF, centred at 21.4 MHz, is available on the front panel in analogue and high speed digital parallel form. The Digital Down-Converter (DDC) accomplishes selectable filtering, resulting in improved output signal quality versus analogue filtering. There are 32,768 different ouput bandwidths, from 250 Hz to 250 kHz, available through the C4x Communication port.

The receiver's two TMS320C44 digital processors provide substantial Digital Signal Processing (DSP) capabilities.

Dual channel audio output is available in a number of modes. The receiver can also perform FIR low pass, band pass, or high pass filtering to remove random audio noise bursts or undesired signals.

The receiver can be controlled via VXI's standard register or message-based protocols, through the exposed C4x communications ports or RS-232. Virtual VXI control panel software and drivers are available.

Specifications

Frequency range: 20-3,400 MHz
Frequency resolution: 1 Hz
Digital output: 12-bit parallel ADC
32-bit C4x Comm Port
– Demodulated Data
– Decimated I/Q Data
21.4 MHz analogue output: pre-detected IF from RF converter
Bandwidth: 8 MHz (max)
Demodulations: AM, FM, USB, LSB, ISB, CW
Dimensions: 30.5 × 228.6 × 330.2 mm
Weight: 4.55 kg

Contractor

Andrew SciComm.

VERIFIED

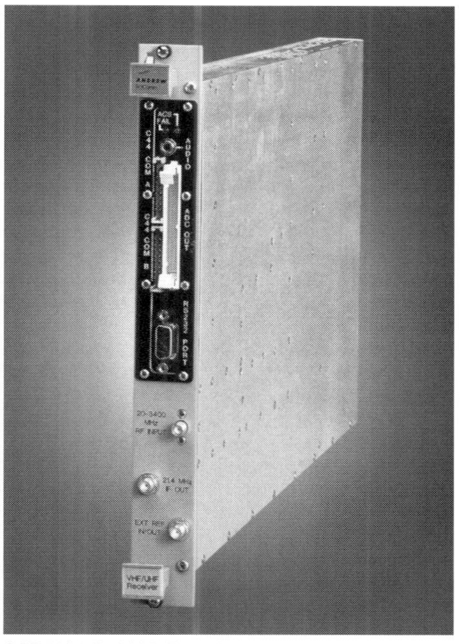

SCR5101-B1 V/UHF receiver ***1999***/0044814

AN/ALR-52 ECM receiver

The AN/ALR-52 is a multiband Instantaneous Frequency Measuring (IFM) receiver for airborne or land applications. A naval variant is also produced, as the AN/WLR-11. Typical systems cover the microwave frequency band from 0.5 to 18 GHz, using receiver modules that cover octave bandwidths. Additional capabilities include: provision for selection of a particular signal or frequency band; blanking of signals which are of no further interest; separation of interleaved pulse trains; measurement of radar parameters; analysis of CW in addition to pulse trains; and direction of arrival measurement. The complete equipment provides for two operator positions, each equipped with a control unit and a display. The two IFM control units are capable of complete control over any band selected. If the same band is selected by both units, the video processing unit gives priority control to one position only. All intercepted signals, however, are presented on the displays at both positions.

A modified version of the ALR-52 has been developed and deployed, with one of the operator positions replaced by an interface unit to link the IFM receiver with a digital computer. This system employs a 1 to 18 GHz DF antenna system. The computer uses the data to tag every received pulse with its frequency, pulsewidth, time of arrival and azimuth. This information is stored and used to develop a library of emitter parameters.

Operational status

In operational service. The AN/ALR-52 is fitted to the EP-3E ELINT aircraft operated by the US Navy.

Contractor

ARGOSystems (a Condor Systems business unit).

VERIFIED

AN/AYR-1 ESM system

The AN/AYR-1 is an advanced ESM system which is part of the upgrade of US Air Force and NATO E-3 AWACS aircraft. It will carry out passive interception, identification and analysis of radar and radio signals. The antennas are believed to be installed in a blister on either side of the aircraft, just aft of the cockpit, and in nose and tail arrays, to provide 360° azimuth coverage. The extra processing units required to analyse signals quickly and classify radar returns, plus associated equipment, are expected to add some 855 kg to the overall aircraft weight.

Operational status

In production for the US Air Force and NATO. A January 1997 firm fixed-price contract award to Boeing Defense and Space Group provided for integration of four ESM systems to E-3F AWACS aircraft at a value of US$32.4 million. Contracting authority is Electronic Systems Center, Hanscom AFB.

Contractor

ARGOSystems (a Condor Systems business unit).

VERIFIED

AR-700 ESM/DF system

The AR-700 ESM/DF system is a passive electronic warfare system designed for tactical operation in a dense signal environment. It automatically receives and processes signals in the frequency range 2 to 10 GHz to identify and determine threat potential and the bearing of intercepted signals. The information is presented to the operator on alphanumeric and tactical graphical displays. Alarms warn the operator if specific threats are detected. The processed information can be logged on a printer and can be routed to the datalink processing system.

A cassette recorder is used to load emitter and platform libraries into the signal processor. The emitter library can be created, expanded or changed by keyboard entry at the operator console. The keyboard is also used to operate the system and select various display formats through the use of special function keys and alphanumeric inputs. Additional special function keys are located on the operator control unit.

A special interface in the advanced signal processor sends emitter reports to the datalink interface. The system also receives aircraft heading from the inertial navigation system in the bearing processor unit.

The AR-700A is a higher performance version of the AR-700, with missile configurations for use on a variety of platforms. It has wide open 0.5 to 18 GHz frequency coverage, a near 100 per cent probability of intercept, a 1 second reaction time and excellent DF accuracy.

Specifications

Frequency: 2-8 GHz, 8-18 GHz
Frequency accuracy:
3 MHz (2-8 GHz)
5 MHz (8-18 GHz)
PRF: 123 to 20,000 pps
PRF resolution: 0.1 Hz
Accuracy:
(coarse) 5° (70% angular coverage),
8° (95% angular coverage)
(fine) 2.3° (70% angular coverage),
4° (95% angular coverage)
Library: 990 emitters
Emitters tracked: 200

Operational status

In production.

Contractor

ARGOSystems (a Condor Systems business unit).

VERIFIED

AR-730 ESM/DF system

The AR-730 ESM/DF system is a modular equipment designed for maritime patrol aircraft and anti-submarine warfare applications. It provides surveillance and threat warning over the frequency range 0.5 to 18 GHz with automatic signal and data processing and precision DF. The system consists of antenna assemblies, an RF processing assembly and a control/processing assembly.

The antenna assemblies are designed for simple installation on any aircraft and consist of an omnidirectional antenna channel for signal acquisition, rotating DF antenna for direction-finding and optional coarse DF antennas and crystal video receivers. For the AR-730, maximum weight savings are realised by mounting the high-gain DF antenna on the back of the radar antenna. Alternative configurations include separate antenna/radome mountings on the underside of the aircraft.

The RF processing assembly contains the receivers. A DIFM receiver provides high probability of intercept. A superhet provides added sensitivity for finding low-power emitters, allowing searching of dense portions of the frequency spectrum and the 0.5 to 18 GHz band.

The control/processor assembly includes a signal processor and optional cassette recorder, printer and operator display/keyboard. Automated acquisition and identification of intercept is provided, with prompt notification to the avionics computer system. Information provided to the avionic system includes intercept reports, intercept updates, threat alerts and systems alarms. The only information required for processing is a threat/identification database.

Contractor

ARGOSystems (a Condor Systems business unit).

VERIFIED

AR-900 ESM system

The AR-900 is a high-performance, fully automatic, passive surveillance system that has been designed for both Electronic Warfare (EW) and ELINT applications. It provides both 100 per cent Probability Of Intercept (POI) and full 2 to 18 GHz frequency coverage. The system comprises three major assemblies: the antenna assembly, the receiver/processor unit, and the operator's workstation. The antenna assembly uses a wideband omnidirectional antenna and a wide open amplitude monopulse DF subsystem that provides better than 3° RMS bearing accuracy.

The receiver/processor unit comprises: two Digital Instantaneous Frequency Measurement (DIFM) receivers, an amplitude-monopulse bearing processor, and a signal processor.

This equipment can process 1,000,000 pps. The signal processor receives digitised high-resolution

Specifications for AR-900 ESM system

Frequency range	**2 to 6 GHz**	**6 to 18 GHz**
System sensitivity	–65 dBm	–65 dBm
Azimuth coverage, instantaneous	360°	360°
DF measurement accuracy, rms	3.5°	2°
Displayed resolution	1.0°	1.0°
Dynamic range	>70 dB	
Signal types received	Conventional pulse trains, agile frequency (±10%), staggered PRI (2-16 positions), jittered PRI (±10%), frequency/phase/amplitude or pulse, CW, FMCW, pulse Doppler	
Frequency measurement		
(displayed resolution)	1 MHz	1 MHz
(accuracy, rms)	2 MHz	3 MHz
Pulsewidth measurement		
(range)	0.1-230 μs (0.05-230 μs with 2 dB reduction in sensitivity)	
(resolution)	0.05 μs	
Amplitude measurement		
(range)	>60 dB	
(resolution)	1 dB	
PRI measurement		
(range)	2-20,000 μs	
(resolution)	0.1 μs	
(display)	PRI (μs) or PRF (Hz)	
Scan types	Circular, conical, bidirectional, unidirectional	
Polarisations received	Horizontal, vertical, slant linear, circular	
System alarms	Threat, steady illumination, CW, amplitude	
Threat library capacity	10,000 emitter modes	
Number of signals tracked	500	
System reaction time	1 s max	
Pulse density capacity	1,000,000 pps	
Options		
(RF extensions)	0.5-2.0 and 18-40 GHz	
(superheterodyne receiver)	0.5-18 GHz	
(fine DF)	1.0°	

frequency measurements for each received Radio Frequency (RF) pulse and Continuous-Wave (CW) signal from the DIFM receivers, while the amplitude-monopulse bearing processor provides direction of arrival and amplitude of each pulse. The signal processor processes this information in parallel, compares the pattern with those in a signal library to identify the emitter and its platform, and alerts the operator to a high-threat signal within one second of acquiring the signal. Emitter libraries can be programmed with up to 10,000 emitter modes. The operator's workstation is made up of a keyboard with integrated trackball; a high-resolution colour display; a 3.5 in floppy disk drive; CD-ROM; printer; and an embedded computer.

The AR-900's operator is provided tactical and intelligence information through simple, comprehensive displays. The activity, tactical graphics, and tactical summary pages give the operator threat-warning information, while the frequency × azimuth, frequency × PRI, and frequency × amplitude pages are provided for analysis. Various displays aid in system setup, and an intercept report generator is provided to prepare intelligence reports.

Contractor

ARGOSystems (a Condor Systems business unit).

VERIFIED

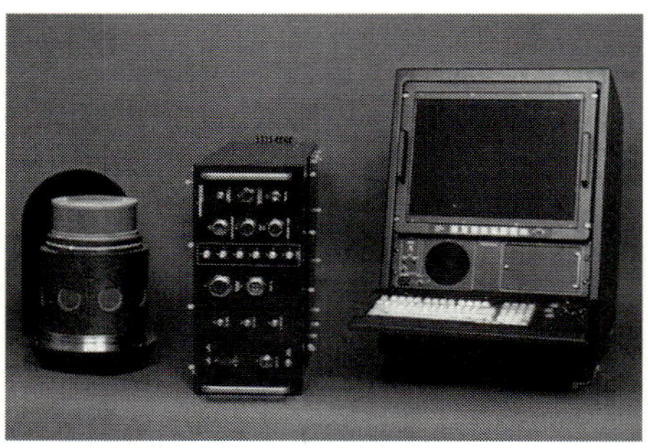

AR-900 ESM system 0018250

AR-7000 airborne SIGINT system

The AR-7000 is designed to provide airborne reconnaissance, direction-finding and geolocation of emitters in all bands from 20 MHz to 18 GHz. Three subsystems are used to provide this full band coverage of ELINT, COMINT and ESM.

The ELINT subsystem provides for operator controlled signal search, analysis and recording of all pulsed signals from 0.1 to 18 GHz. Three superheterodyne tuners are provided for the 0.5 to 18 GHz tuning range; a fourth heterodyne tuner covers the 0.1 to 1 GHz range. A pulse analyser performs detailed numeric emitter parameter measurements.

The ESM subsystem provides instantaneous wideband intercept, automatic detection, identification and DF on all signals from 0.5 to 18 GHz. The wideband instantaneous coverage in a dense complex environment is provided by three DIFM receivers and by crystal video receivers which furnish monopulse DF data. Additional DF capability is given by a fine DF subsystem which uses phase interferometers for high-accuracy DF measurement on a tasked basis. An emitter library of up to 5,000 emitters may be tracked in the active emitter file.

The COMINT subsystem provides up to 20 receivers from as many as five operators to use in monitoring the 20 MHz to 1 GHz band. In addition, a DF subsystem furnishes DF on communication signals from 20 to 500 MHz. The DF subsystem uses two phase-matched receivers and two antenna arrays to perform highly accurate multiple differential phase measurements on signals from 20 to 500 MHz. The multiple differential phase measurements are converted to DF estimates by the subsystem processor.

Operational status

The AR-7000 forms the basis of the Fokker 50 SIGINT Black Crow aircraft proposal.

Contractor

ARGOSystems (a Condor Systems business unit).

VERIFIED

MPAvionics system

ARGOSystems' MPAvionics is an adaptable, readily customised MPA mission system that integrates operational control and display of information from the wide range of sensors, communications and navigation equipment used in maritime patrol missions. In addition to MPA, the MPAvionics system can serve as the basis of an integrated SIGINT system, Airborne Early Warning (AEW) system, ground- or ship-based support station, or any other complex sensor information collection system with selection of the required sensors.

The heart of the MPAvionics system is the ruggedised Distributed Processing and Display System (DPDS). The DPDS is a streamlined mission command and control centre that processes and displays integrated subsystem data and controls the mission surveillance radar, IFF interrogator, Infra-Red (IR)

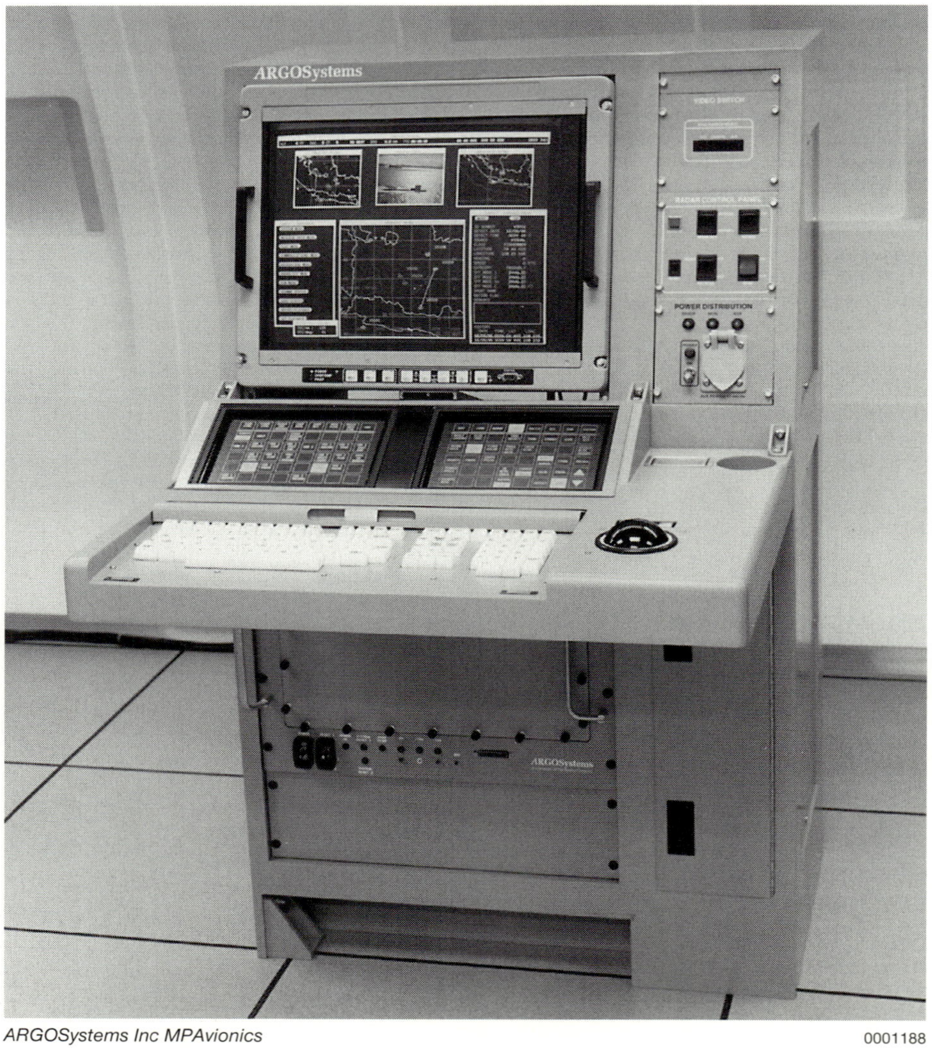

ARGOSystems Inc MPAvionics 0001188

sensor, ESM, acoustic, navigation, GPS, communications and datalink subsystems. It replaces dedicated sensor control and display units with integrated controls to minimise the number of operator actions required to control individual sensors. The DPDS provides a complete picture of the real-time tactical situation during MPA missions on the tactical plot, a map overlaid with radar, video, and contact information.

The DPDS records detailed aircraft, Built-In Test (BIT) status and contact data for post-mission analysis; enables premission data preparation; and integrates with voice and datalinks to transmit critical information to supporting platforms or to a ground-based MPA operations centre. The DPDS also provides aircraft navigation references for mission sensors and displays.

The DPDS consists of one or more workstations (consoles), linked by an Ethernet Local Area Network (LAN). Consoles are modular in design and have an open and distributed VME bus architecture that allows the user to specify a desired number of consoles and to allocate subsystem display and control among consoles to meet mission and geographic needs. This approach ensures that technological advances in both hardware and software can be readily integrated into the system as the state of the art evolves.

Each DPDS console comprises RISC-based display and input/output processors and multifunctional displays, a keyboard and trackball and two electroluminescent Programmable Entry Panels or PEPs, displaying touchscreen buttons. The touchscreen buttons map to customisable menus and submenus, designed specifically to control the sensors and subsystems in the system and to control the tactical plot and manage content and status information.

The DPDS has 11 multiwindow displays. Up to four windows can be used for real-time video displays or playback of prerecorded video. The display format permits operators to simultaneously display multiple windows of the tactical situation plot, contact and status information, and radar, IR, and ESM videos. All windows can be moved, sized, opened, or closed as desired, enabling the operator to optimise the displays to meet the precise demands of any tactical situation. Display data can be captured on a video recorder and added to data stored on the mission loader/recorder to provide a complete record of the mission which can be used as documentary evidence to establish legal liability for violations of territorial waters. All console displays are loaded from a master mission database maintained in the input/output processor. Removal of the mission loader/recorder enhances security for sensitive mission data and permits easy exchange of data files at a ground station.

A typical DPDS configuration allocates mission responsibilities to functional operators via consoles for a Tactical Controller (TACCO), one or more Sensor Operators (SENSOs) and an ASW operator. Each operator is assigned primary responsibility for the operation and control of certain sensor subsystems, aircraft systems and other control functions, including tactics and weapons. Each console in the DPDS provides integrated graphical user interface controls for all subsystem control and display functions and can be configured for individual operator responsibility or particular mission. All DPDS consoles are essentially identical in hardware and software, except for placement of manual equipment controls, which may vary by console.

The DPDS uses a real-time operating system, based on commercial-off-the-shelf standard and an Ada development tool set to provide a flexible mission application development environment and enable successful performance of time-critical operations, such as sensor interfacing and co-ordination. DPDS software can be executed on multiple consoles in the aircraft or on a ground-based console to provide simulation and debugging capabilities and permit development and testing of mission data on any operator or ground-based workstation. The modular structure allows for reuse of existing software when the system must be modified to address evolving geographic or mission needs.

A ground-based MPA Operations Centre supports operational and organisational activities by providing the means to classify, store and disseminate data for mission analysis and maintenance of national database resources. ARGOSystems provides integrated logistics support, including software maintenance and development training programs.

In fielded MPAvionics systems, the DPDS has been interfaced with the following sensors and aircraft subsystems:

Litton Systems Canada AN/APS-504 (V)5 search radar;
Raytheon Electronic Systems AN/APS-134(LW) search radar;
Raytheon Systems Limited Model 3500 IFF interrogator;
BAE Systems TICM II MRT-S infra-red detector;
FSI-FLIR-2000 infra-red detector;
ARGOSystems AR-900 ESM system;
Litton LTN-92 Inertial Navigation System (INS);
Sextant AOC-31 Air Data Computer (ADC).

Operational status

The MPAvionics mission system is a derivative of deployed systems on the Boeing 737 Surveiller and other operational platforms. It is currently in service with the Indonesian Air Force in the 737 Surveiller MPA. The DPDS has been demonstrated worldwide on the IPTN CN-235 MPA under MPA mission conditions.

Contractor

ARGOSystems Inc (a Condor Systems business unit).

VERIFIED

Flight director computers

Astronautics manufactures both analogue and digital flight director computers for most military fixed-wing aircraft and helicopters. The flight director computer provides both two- and three-cue steering information to the pilot which may be presented on either cross-pointers or a combined single indicator.

A flight director computer which is compatible with GPS is currently being manufactured for the US Army Sikorsky UH-60A helicopter. The flight director computer accepts digital signals directly from the GPS receiver and other radio navigation aids and converts them to analogue signals for display on the flight instrument. The navigation information received from the GPS receiver provides horizontal deviation, vertical deviation and glide slope angle.

Operational status

In production.

Contractor

Astronautics Corporation of America.

VERIFIED

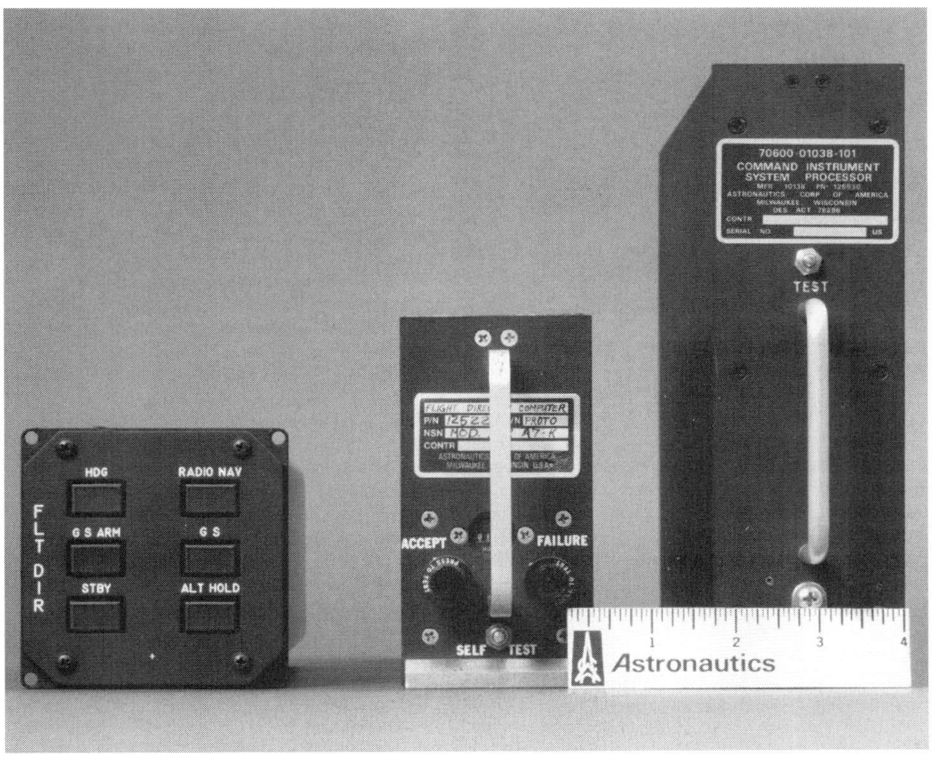

The Astronautics Macchi MB-339 flight director computer (left), A-10 flight director computer (centre) and Black Hawk command instrument system processor (right)

Modular Mission and Display Processor (MMDP)

The MMDP is a mission computer and multichannel display processor. Its high throughput CPU (12-20 mIPS) and large memory (up to 4 Mbyte), provides the capability to perform as a primary mission computer, as well as display driver, data interface/converter, graphics processor, bus controller, and weapons delivery/stores management system for military aircraft.

The MMDP is a modular flexible design, providing an easy path to application-specific software and system enhancement. Software language can be Ada, 'C' or JOVIAL.

Specifications

Design qualifications: ARINC 429 and ARINC 629; MIL-STD-1553B (BC or RT); MIL-STD-5400; RS 232/422
Processor: 32 bit RISC R-3081/R-3000 CPU
Input/output:
(input video): 8 channels
(colour channels): 6
(monochrome output channels): 6
(discrete): I/O
(analogue): inputs
(HUD stroke/raster drive): yes
(colour map): yes

Operational status

The MMDP forms part of the US Air Force T-38 Avionics Improvement Program (AIP); it is also being used for an A-4 aircraft upgrade programme in Asia and for the Italian MB-339.

Contractor

Astronautics Corporation of America.

VERIFIED

AN/APR-44(V) radar warning receiver

The AN/APR-44(V) has been adopted by the US Army and Navy.

The system consists of three units. The Model AS-3266 antenna is a monopole unit which provides omnidirectional, vertically polarised coverage within a 50° elevation sector and is designed to be mounted on a flat horizontal surface. The second unit, the receiver, is a radio frequency chopped, crystal video type with a bandpass filter followed by a radio frequency switch/detector module, a linear video amplifier and processing circuitry. The video amplifier output is routed to a comparator before being compared with a radio frequency chopping signal in the processing logic. This procedure detects any continuous wave signal and triggers a 3 kHz audio output, a logic output, an alert lamp drive and a 2 Hz logic blink signal. The third unit is a control panel which includes the alert light indicator. Three versions of the system are available. The 44(V)1 incorporates the R-2097 receiver for coverage of the H/I-bands, the 44(V)2 uses the R-2098 J-band receiver and the 44(V)3 covers H-, I- and J-bands.

Specifications

Dimensions: 165 × 94 × 229 mm
Weight: 0.85 kg
Frequency: selectable
Bandwidth: selectable
Sensitivity:
(min) –45 dBm
Max RF input level:
(integral limiter) 1 W CW

Operational status

In service in different versions with various US Army and Marine Corps helicopters.

Contractor

BAE SYSTEMS Aerospace Electronics, Inc.

UPDATED

CrossJam 2000 training and testing jammers

CrossJam 2000 is a family of modern jamming systems designed to be versatile in meeting the requirements for ECCM training or radar testing. It is designed to simulate numerous self-protection and escort jamming systems and can be configured from a simple noise jammer to a sophisticated deception system capable of providing ECCM training for the most advanced fighters or AEW and GCI radars.

CrossJam 2000 is available in bands covering 0.5 to 18 GHz. It can be configured with various high-power amplifiers ranging from 100 to 500 W and is available in an AN/ALQ-167 pod or custom mounted internally in a business jet, such as Lear, Falcon or Challenger, or a fighter such as the F-5.

The full system is capable of generating all significant ECM techniques in use. Available in the CrossJam 2000 are noise, range, set-on and Doppler techniques.

Operational status

CrossJam 2000 is used to train NATO fighter pilots. The Canadian Department of National Defence has selected the system as an integral component of the Electronic Support and Training Challenger aircraft.

CrossJam 2000 is an integral component in the Canadian Forces Challenger EW training aircraft

Contractor

BAE SYSTEMS Aerospace Electronics, Inc.

UPDATED

Threat Emitter Simulator System (TESS)

The Threat Emitter Simulator System (TESS) is a programmable threat simulator that is tunable in flight. It can be used to simulate threat aircraft radar and air-to-surface emissions and is used for training weapon systems operators.

TESS can be housed in either a standard AN/ALQ-167 pod shell for external carriage or EIA standard 19 in racks for internal carriage. The major assemblies are the transmitter assembly, magnetrons, antennas, associated RF components, interface assembly and low-voltage power supply. Both configurations have identical assemblies. For the pod configuration, the assemblies are mounted on an electronics tray assembly which slides into the pod.

In flight, the operator can choose from 15 programmed scenarios which vary frequency and scan patterns. Closed-loop remote magnetron tuning is aided by a frequency measurement device which prevents excessive drift.

TESS has been designed for flexibility, while requiring a minimum of component replacement between bands. For the lower bands of 7.8 to 8.5 GHz and 8.5 to 9.6 GHz, only a change of magnetron is required. For the upper bands of 12 to 13.2 GHz and 14 to 15.5 GHz, the magnetron and associated RF components are replaced. Mechanical packaging of TESS allows rapid replacement of components for band change or maintenance functions.

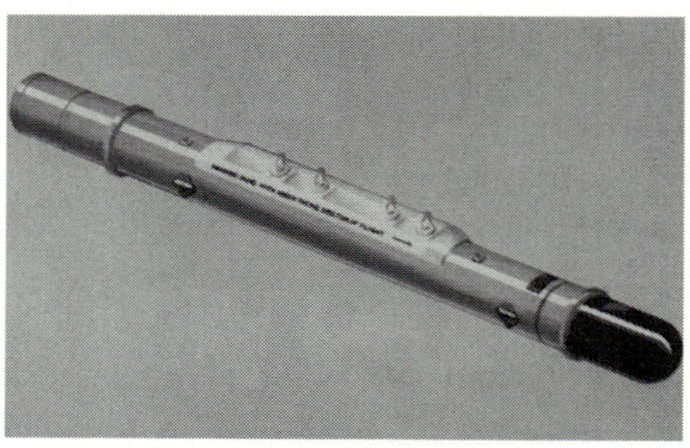

The Threat Emitter Simulator System pod

The TESS pod has been designed and tested to meet military environmental requirements and is qualified for carriage on commercial and military aircraft at speeds of up to 750 KCAS or M1.3.

Specifications

Dimensions:
(pod) 3,284 length × 254 mm diameter
(internal) 482.6 × 488.9 × 609.6 mm
Weight:
(pod) 170.1 kg
(internal) 79.38 kg
Power supply: 115 V AC, 400 Hz, 3 phase
Frequency:
7.8-8.5 GHz, 8.5-9.6 GHz
12-13.2 GHz, 14-15.2 GHz
Environmental: MIL-E-5400 Class 1A

Operational status

In service on the CT-133 aircraft, with the Canadian Forces.

Contractor

BAE SYSTEMS Aerospace Electronics, Inc.

UPDATED

ULQ-23 FutureJam countermeasures

The ULQ-23 electronic countermeasures set, known as FutureJam, was developed by the US Navy. Future applications are being investigated, including airborne rack-mounted systems and increased technique capabilities.

FutureJam provides programmable ECM at a lower power level for jamming a radar system. Coupled with an external amplifier and antenna, the system can be used as a complete airborne jammer. An enhanced jamming system can be configured with the addition of a real-time controller and an external set on receiver.

FutureJam is capable of various noise jamming techniques, including barrage, narrowband, spot and swept. Various deception techniques are also available, either alone or in combinations, including AM, swept AM, RGPO, random Doppler, multiple frequency repeater, velocity gate stealing and narrowband repeater noise. With the addition of a real-time computer, multiple technique scenarios can be generated.

Contractor

BAE SYSTEMS Aerospace Electronics, Inc.

UPDATED

AN/ASN-128B Doppler/GPS navigation system

The AN/ASN-128B is the US Army's standard lightweight helicopter airborne Doppler/GPS navigator and comprises three units: a receiver/transmitter/antenna, signal data converter and computer/display unit. A steering hover indicator can also be included as an option. With inputs from heading and vertical references, the system provides aircraft velocity, present position and steering information from ground level to above 10,000 ft.

BAE SYSTEMS, under contract to the US Army, embedded a military code GPS receiver into the AN/ASN-128 GPS receiver. The PCY code receiver was integrated into the signal data converter and the computer display unit software was modified to display both Doppler navigation and GPS data. Over 2,100 AN/ASN-128 systems were modified and are currently installed in UH-60A/L Black Hawk and CH-47D Chinook helicopters. These modifications are applicable to most other AN/ASN-128 systems.

BAE Systems North America has also developed a field kit enabling the AN/ASN-128 Doppler navigator to interface with a stand-alone Trimble GPS receiver, for continuous update of Doppler-derived present position with valid GPS data.

Specifications

Volume: 20,724 cm^3
Weight: 13.61 kg
(hover indicator) 0.9 kg
Propagation: 4-beam configuration operating FM/continuous wave transmissions in K-band. Beam shaping eliminates the need for a land/sea switch. The single transmit/receive antenna uses full aperture in both modes to minimise beamwidth and reduce fluctuation noise
Number of waypoints: >100
Self-test: localisation of faults at LRU level by BITE
Reliability:
(complete system) >2,100 h MTBF

Operational status

In production. Many thousands of AN/ASN-128 and AN/ASN-128B sets have been manufactured.

The AN/ASN-128B is in service in the Sikorsky UH-60A/L and CH-47D Chinook. The AN/ASN-128 is in service with US Army Bell AH-1F and Boeing AH-64A helicopters; also in Royal Australian Air Force Bell UH-1H, UH-60 and CH-47D, Hellenic Air Force UH-1H, Jordanian Army AH-1S, Pakistan AH-1S, South Korean CH-47, UH-60 and AH-1S, Spanish Army Eurocopter BO 105 and CH-47B, Taiwan Army CH-47B and Turkish

UH-60. The system, with hover indicator, has been provided for the German Army Eurocopter BO 105, PAH-1, UH-1D and VBH helicopters. Also built under licence in Japan for the JASDF AH-1S and CH-47D. In use in Austria, Bahrain, Brunei, China, Denmark, Dubai, Egypt, Greece, Netherlands, Singapore, Spain, Taiwan and Thailand. Version kit deliveries started in the second quarter of 1996.

Contractor
BAE SYSTEMS Aerospace Inc, CNI Division.

UPDATED

AN/ASN-137 Doppler navigation system

The AN/ASN-137 is the multiplexed version of the AN/ASN-128. It is compatible with the MIL-STD-1553B databus and has ARINC 575 or 429 outputs.

The AN/ASN-137 uses the same receiver/transmitter/antenna as the ASN-128 and has an optional control/display unit of the same size, with the same front panel as the ASN-128. Additional features of the AN/ASN-137 include hover bias correction for precision hovering, 12-point magnetic deviation entry and the addition of latitude and longitude and UTM grid zone outputs in MIL-STD-1553 output.

Specifications
Weight: 11.7 kg
Power supply: 28 V DC, 87 W
Reliability: 2,800 h MTBF

Operational status
In production and in service. The AN/ASN-137 is installed on the US Army's Bell OH-58D and the US Navy's Sikorsky VH-3D and VH-60 helicopters. It is also currently installed on AH-64A helicopters, on the US Army MH-47E and MH-60K helicopters and the US Air Force MH-60G Pave Hawk helicopter.

Contractor
BAE SYSTEMS Aerospace Inc, CNI Division.

UPDATED

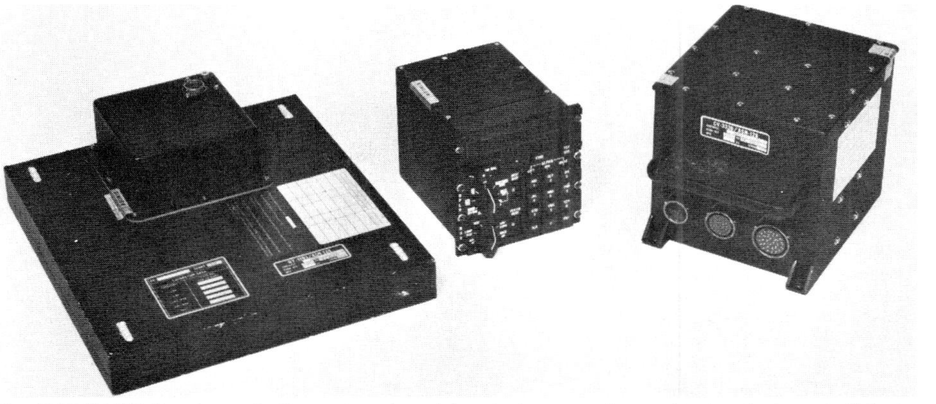

The AN/ASN-128 Doppler navigation system: from left, velocity sensor (transmitter/receiver), control/display unit and signal data converter

AN/ASN-157 Doppler navigation set

The AN/ASN-157 is a Doppler navigation set in a single LRU, emulating the two-unit AN/ASN-137. As well as navigation data, the system provides velocity data in all three axes, including computations for hovering and translational flight in helicopters.

Specifications
Dimensions: 370 × 342 × 57 mm
Weight: 5.7 kg
Reliability: 7,300 h MTBF

Operational status
In production, and in service. Installed on the Agusta A 109 helicopter for the Belgian Army, the CH-47D glass cockpit Chinook cargo helicopter for the Netherlands Air Force and the AH-64D Longbow Apache attack helicopter for the US Army. Selected for the WAH-64D Longbow Apache attack helicopter for the British Army.

Contractor
BAE SYSTEMS Aerospace Inc, CNI Division.

UPDATED

AN/ASW-38 automatic flight control set for the F-15

The ASW-38 provides the control functions for the F-15A, B, C and D versions. It provides three-axis command augmentation to improve the fighter's handling qualities over the very wide speed/height envelope. The analogue system incorporates two channels in each of the three axes, also supplied by dual sensors. Three-axis fail-safe trim control and stall inhibit functions are also provided. The set comprises two computers (one for pitch, the other for roll and yaw), a three-axis dual-channel rate sensor box, two-axis dual-channel accelerometer box, dual-channel pressure sensor unit, stick force sensor and pilot's controller.

Specifications
Dimensions:
(computers) 152 × 152 × 381 mm
(rate sensor) 114 × 114 × 178 mm
(acceleration sensor) 102 × 57 × 102 mm
(pressure sensor) 114 × 76 × 127 mm
Weight:
(total set) 17.8 kg
Power: 265 W
Reliability: 1,100 h MTBF over 10,000 flight h

Operational status
In service in the F-15A, B, C and D. No longer in production.

Contractor
BAE Systems Controls.

UPDATED

Digital flight control system for the F/A-18

The F/A-18 flight control system is a digital four-channel Fly-By-Wire (FBW) system operating the aileron, stabilator and rudder primary flying controls, leading-edge and trailing-edge flaps and nosewheel steering. The system incorporates 32 servo loops to drive and control these functions. At the same time the conventionally mounted control column, in contrast to the F-16's sidestick controller, has a mechanical link to the ailerons for reversionary pitch and roll control. All FBW computations are accomplished by four digital computers operating in parallel, accepting inputs from linear variable differential transformers sensing control column and rudder pedal movement and analogue motion sensors, and formulating commands to the redundant electrohydraulic servo-actuators driving the control surfaces. The system therefore remains operational after two failures. It has a high degree of integration with other aircraft equipment, communication being accomplished by means of a MIL-STD-1553 databus.

The flight control electronics set for the F/A-18 Hornet

Two special display modes are used in conjunction with the system. The first is a flight control failure matrix, the second provides recovery guidance during spins and both use the same CRT unit. In the first mode, the display is selected to show details of the fault, after the pilot's attention has been drawn to its presence by an indication on the central warning system. The display signals 'X' at the position of the fault on a schematic diagram of the system painted on the screen. In the second mode, the display shows the position of the control column needed to recover from a spin, assumed to exist when yaw rates greater than 15°/s occur simultaneously with speeds of 125 kt or below. This mode is selected automatically, having absolute priority over other displays when this combination of speed and yaw rate occurs. The two modes were originally provided for flight test purposes, but have been retained in production aircraft following recommendations from the Naval Air Test Center and test pilots.

The system comprises two flight control computers, each incorporating two microprogrammable digital processors specially designed for flight control applications, two rate sensor boxes, two acceleration sensor boxes, an air data sensor box, rudder pedal force sensor and pilot's control panel.

Operational status
In service on the F/A-18.

Contractor
BAE Systems Controls.

UPDATED

Electronic Flight Control System for the C-17

The C-17 transport is designed to carry men and equipment into austere airfields and drop zones near the front line. Operation of a large aircraft at low speed needs a flight control system that provides stable yet responsive control, augments the basic stability of the aircraft at low speeds and is extremely reliable. These requirements are being met through development of a quadruple redundant digital fly-by-wire control system for both the primary and secondary flight controls of the C-17. The Electronic Flight Control System (EFCS) controls the actuators that drive the movable ailerons, flaps, leading-edge slats, spoilers, horizontal stabilisers, elevators and rudders.

All flight controls are electrically commanded in their normal mode of operation. A mechanical system provides back-up control between pilot controls and the elevator, aileron and lower rudder.

The EFCS is a full-time, full-authority fly-by-wire control system with stability augmentation in all axes. Electronic flight control functions are provided by quadruple redundant sensor, computation and actuation channels. In the fly-by-wire control mode, the sensors relay pilot force signals to the four Flight Control Computers (FCC). The FCCs are programmed, using other sensor information and inputs, to generate flight control signals that control flight path and attitude. In addition, the EFCS provides automatic aileron and stabiliser trim, autopilot, flight director and autothrottle functions. An angle-of-attack limiting system is provided to prevent stall during high-lift configuration.

The EFCS is designed to provide continued operation following both failures and battle damage. This capability is achieved through four-channel architecture which can continue full-function operation on two channels. The channels are separated on the aircraft and the components are located so as to minimise the risk of losing multiple channels through battle damage.

Operational status

In service on the C-17.

Contractor

BAE Systems Controls.

UPDATED

Flight Control Electronics Assembly-Upgraded for JAS 39 Gripen

The JAS 39 Flight Control Electronics Assembly-Upgraded (FCEA-U) by Lockheed Martin is a digital, high performance triplex flight control computer which provides full authority, fly-by-wire control of seven primary control surfaces, four secondary control systems, and interfaces with a variety of sensors and I/O devices. The FCEA-U system is provided to Saab Scania AB, Saab Military Aircraft for the Swedish Air Force.

Within each channel of electronics, the FCEA-U contains two high performance 32-bit processors. A Motorola 68040 performs primary flight control law processing and is interfaced via dual port RAM to a Texas Instruments TMS320C30 processor which acts as the I/O processor in addition to computing back-up control laws. A 10 MHz serial cross-channel datalink with built-in error checking is closely coupled with a data acquisition system that can exchange data between channels with less than 150 μs latency.

The electronics are housed in a single, air-cooled, lightweight chassis which is designed with an integrated flexible harness/mother-board and removeable circuit cards for easy maintenance. Reliability of the FCEA-U exceeds 1,500 h.

Components for the C-17 electronic flight control system

Specifications

Dimensions: 358 × 454 × 192 mm.

Operational status

In production for and in service in the JAS 39 Gripen aircraft.

Contractor

BAE Systems Controls.

UPDATED

Flight Control System for the V-22 Osprey

The V-22 Osprey Flight Control System (FCS) is a fly-by-wire system composed of triple, dual Primary Flight Control System (PFCS) processors and triple Automatic Flight Control System (AFCS) processors. The Flight Control Computers (FCCs) provide interfaces for the swashplate, elevator, rudder, flap and pylon primary actuators. The FCCs also provide interfaces for the cockpit control force/driver actuator, cockpit control thrust driver actuator and nosewheel steering secondary actuators. The FCC interfaces with the MIL-STD-1553B avionics multiplex bus and also incorporates a dedicated flight control system multiplex bus in each channel. The system is two fail-operational with respect to PFCS functions and one fail-operational with respect to AFCS functions.

Each FCC incorporates dual 1750A processors for the PFCS control functions and one 1750A processor for the AFCS control function. An input/output processor is used for data management and built-in test functions. The PFCS and AFCS processors interface with the I/O bus via a dual-port memory to enhance computerised throughput.

The PFCS electronically connects the pilot's controls to the various control surface actuators for safe operation of the aircraft. The automatic flight control system provides the necessary level of control augmentation required for reliable mission performance.

The PFCS provides basic control of the aircraft by connecting and mixing pilot control inputs during the helicopter, transition and aeroplane modes of operation. All actuators are commanded from the PFCS processors and each dual PFCS processor has failure detection and shutdown logic so that either processor in each channel can independently shutdown the actuator channel controlled by that channel.

The automatic flight control system provides stability and control augmentation and mission related selectable modes of the flight control system. It is a triplex cross-channel monitored system which provides operation after first failure in most triplex sensor paths. On a second failure, the function is shut down in a fail-safe manner.

Automatic flight control system inputs to the primary flight control system are voted at the input interface of the PFCS. The selected inputs are further restricted by rate and authority limiting in the control laws of the primary flight control system.

A non-redundant, dual-computation analogue back-up computer, in conjunction with analogue elements of the PFCS, provides for continued control of the aircraft in the event of a total loss of the digital processing functions. It also directly controls the back-up modes of the pylon actuator and electronic engine controls.

The flight control computer processor monitors the back-up computers continuously in normal operation. Built-in test for the back-up is controlled by the I/O processor during preflight and maintenance modes of operation.

The system consists of:

Flight control panel: this provides pilot/co-pilot control for the automatic flight control system.

Engine control panel: lever type engine condition controls are provided for the V-22's two engines. Illuminated push-button switches are used to select and indicate which engine controller is active. A third lever controls the rotor brakes.

Cockpit interface unit: the Cockpit Interface Unit (CIU) passes cockpit discrete data to the dedicated flight control system multiplexer bus. The CIU also contains a roll rate sensor.

Flight control computers: one flight control computer using 1750A architecture is provided for each channel. Primary and automatic flight control system functions each incorporate separate processing. Within the computer, the systems communicate via a common dual-port memory designed so that the hardware failures on the AFCS bus or memory side do not cause failures of the PFCS functions.

Flight Control Electronics Assembly-Upgraded (FCEA-U) for JAS 39 Gripen (above)

0001443/0001442

Analogue back-up computer: a non-redundant, dual computation channel analogue back-up computer provides for continued control of the aircraft in the event of a system failure of the digital processing functions.
Conversion actuator electronic interface unit: this provides control of the electric back-up pylon actuator motor.
Flap panel: this allows the pilot/co-pilot to select the operation of the flaps.

Operational status
In production for the V-22 Osprey.

Contractor
BAE Systems Controls.

UPDATED

Stability augmentation system for the A-10

Stability augmentation for the US Air Force A-10 in pitch and roll is provided by a two-axis, two-channel system that not only improves handling qualities but makes for more accurate weapon delivery. Each channel in a given axis is independent of the other, except for monitoring cross links, and both are disengaged if significant desynchronisation occurs. Separate engage switches permit single-channel operation, however.

The system also provides turn co-ordination, compensation for the pitching moment caused by operation of the speed brake or recoil from the GAU-8 30 mm cannon. The pitch and yaw monitors can also be self-tested.

The system comprises a single-box computer and pilot's control panel. The computer contains six rate sensors, four of which are solid-state devices manufactured by Lockheed Martin.

Specifications
Dimensions:
(computer) 155 × 196 × 318 mm
(control panel) 146 × 86 × 165 mm
Weight:
(total) 7 kg
Power: 73 W

Operational status
No longer in production. In service in the US Air Force A-10 aircraft.

Contractor
BAE Systems Controls.

UPDATED

Standard Automatic Flight Control System (SAFCS) for the EA-6B

The Standard Automatic Flight Control System (SAFCS) for the EA-6B Prowler is a standard computer design suitable for application to a variety of Navy aircraft. This capability is made possible by the inclusion of additional input/output circuitry and multiple memory options. Growth in redundancy up to triplex is possible through the inclusion of a cross-channel databus for exchanging information between computers.

The EA-6B SAFCS currently interfaces with analogue sensors and a MIL-STD-1553B avionics databus and drives electrohydraulic and trim actuators. Command augmentation, attitude and directional control servo-loops provide aircraft stability and control.

An extensive built-in test is provided together with non-volatile RAM for in-flight fault data storage. Maintenance information displayed in the front panel simplifies testing and logistic support. It also provides a quick turnaround resulting in higher availability. Use of hybrid circuits and low-power electronics minimises space, power and weight considerations while maximising reliability.

Specifications
Dimensions: 178 × 145 × 267 mm
Weight: 9.1 kg
Power supply: 115 V AC and 28 V DC
Power dissipation: 42 W
Cooling: convection, no forced cooling air
CPU throughput: 429 kips, DAIS mix
Interfaces: RS-232, MIL-STD-1553B
I/O: 96 analogue inputs, 12 analogue inputs, 128 discrete I/O, 1553 and cross-channel datalink. (Above I/O is not a limitation, but current capability.)
Programs: Jovial and Assembly

Operational status
In service in the EA-6B Prowler. The system has also been adapted for US Air Force A-10 and TR-1 aircraft and US Army AH-64 helicopters.

Contractor
BAE Systems Controls.

UPDATED

Analogue fully fly-by-wire system for the Pre-Block 40 F-16

The system is entirely analogue in nature, with four channels in each of the three axes providing the necessary redundancy to meet the failure criterion of being able to continue flight after two failures. At the heart of the system is a 20 kg box containing the four flight control computers, each rigorously isolated from electrical contact with any other. Each has its own power supply and individual input signals from the pilot's sidestick controller, sensors and air data system. Each computer processes its signals and provides independent output commands to the servo actuators that move the flying control surfaces.

Fault isolation within the computer box, and subsequent reconfiguration of the computers to compensate for the new situation, are achieved by the use of analogue output voters that interface sensors and computations with movement of the servo-actuators. A 63-function self-test schedule is run automatically as part of preflight checks.

Specifications
Dimensions: 457 × 254 × 203 mm
Weight: 20 kg
Power: 150 W
Digital channels: 4
Independent back-up: 1
Fault tolerance: 2 fail-operate
Analogue inputs: 88
Memory:
(PROM) 16 k
(RAM) 2 k
Language: LSI HOL
Reliability: 2,181 h MTBF

Operational status
In service in F-16A and pre-Block 40 C single-seat and F-16B and D two-seat versions.

Contractor
BAE Systems North America, Aircraft Controls.

VERIFIED

Digital Flight Control System (DFCS) for the F-15E

Whereas the standard F-15 has a Lockheed Martin analogue flight control system, the additional requirements of the F-15E for deep penetration with emphasis on terrain-following and terrain-avoidance, allied with new sensors, called for a digital triplex automatic system of greater performance and reliability. Digital technology permits control laws to be optimised both for air combat and for high-speed low-level flight. Lear Astronics is also responsible for the digital pressure sensors used to provide speed and aerodynamic pressure sensing for the system. The system is based on MIL-STD-1750 instruction set architecture and uses Lockheed Martin Fairchild Systems 9450 microprocessors.

Operational status
In service on F-15E.

Contractor
BAE Systems North America, Aircraft Controls.

VERIFIED

F-22A - VMS/IVSC: Vehicle Management System/Integrated Vehicle Subsystem Control

The VMS combines flight and propulsion controls, while the IVSC controls the aircraft utilities. The VMS utilises a triplex digital flight control system, with no electrical or mechanical back-up, to provide full carefree handling and enhanced manoeuvrability, with a sidestick fly-by-wire controller.

The IVSC modules control the following services: electric power, hydraulics, fuel systems, integrated warning/caution/advisory functions, diagnostics and health monitoring, auxiliary power, environmental and life support functions.

Equipment modules are included in the avionics common module racks. Raytheon Electronic Systems provides common 1750A processor modules for these systems.

Operational status
Installed in F-22A EMD aircraft.

Contractor
BAE Systems North America, Aircraft Controls.

UPDATED

Flight control augmentation system for the KC-135R

The flight control augmentation system for US Air Force Boeing KC-135 tankers retrofitted with General Electric/SNECMA CFM56 turbofan engines is essentially a high-authority single-channel yaw damper, based on a rate gyro, that monitors engine rpm differences on the outboard engines and provides a correction signal to the rudder servo if the difference exceeds a given threshold value. The system is nuclear hardened.

Operational status
In service on the Boeing KC-135R.

Contractor
BAE Systems North America, Aircraft Controls.

VERIFIED

Flight control computer for the RAH-66

In 1991, Lear Astronics was selected by Boeing Helicopters to be the flight control computer subsystem contractor for the RAH-66 Comanche helicopter, which is being developed in association with Boeing Defense and Space Group. Lear Astronics (now BAE Systems North America) has overall responsibility for delivery of an integrated subsystem and for complete systems integration of the triplex computers. The company is also developing the controller, flux valve, engineering control panel, flight control panel and back-up computer.

Operational status
In development for the RAH-66 Comanche helicopter.

Contractor
BAE Systems North America, Aircraft Controls.

UPDATED

Advanced Digital Dispenser System

The Advanced Digital Dispenser System (ADDS) is a microprocessor, computer-controlled, threat adaptive countermeasures dispensing system providing aircraft self-protection against current and future ground and airborne threats. ADDS utilises user programmable dispensing programmes, adapted for the specific threat and optimum engagement parameters, to control and dispense chaff, including dual chaff cartridges, flares, RF and future types of offboard expendable decoys.

With simple software modifications, aircraft configurations can be tailored for high-performance aircraft, helicopters and transport and maritime patrol aircraft. ADDS incorporates improvements derived from combat experience and hundreds of installations on a wide variety of aircraft including the A-4, F-4, F-5, F-15, F-16 and others.

ADDS features include automatic, semi-automatic and manual operational modes; payload inventory; discrete inputs from aircraft MIL-STD-1553B and RS-422 databusses to interface with onboard RWR, MLD and TWS warning sensors; solid-state sequencer switch providing multiple firing pulses with low burst intervals and automatic and crew initiated BIT.

Operational status

In production for many air forces. Installed as standard equipment on a number of aircraft types.

Contractor

BAE Systems North America, Austin.

VERIFIED

ALE-47H Threat Adaptive Countermeasures Dispenser System (TACDS)

ALE-47H TACDS is a lightweight form and fit replacement for the M-130 helicopter dispensing system using solid-state microprocessor technology to perform threat adaptive, automatic programming for dispensing chaff, flare and advanced expendables as they become available. The system interfaces with other onboard EW and avionics systems such as radar warning receivers, missile warning systems, laser warning systems and the aircraft avionics, compares these inputs against the threat response files and selects the optimum dispense program.

ALE-47H TACDS provides three in-flight selectable modes of operation: manual whereby four preprogrammed dispense programs can be activated from the cockpit control unit; semi-automatic where the system prompts the aircrew that a threat exists and selects the optimum expendable/sequence for aircrew activation; automatic where the threat analysis (based on sensor and avionics inputs), dispense programs and activation performed by TACDS are fully automatic without aircrew action.

ALE-47H TACDS can be easily retrofitted into any aircraft equipped with the M-130 system or as new installations. ALE-47H TACDS has been designed to communicate over several communications busses such as the MIL-STD-1553A/B and RS-422. Built-In Test (BIT) is included with both continuous automatic and aircrew-initiated BIT providing fault isolation to the system replaceable unit.

Capabilities designed into ALE-47H TACDS provide offboard countermeasures to counter existing IR and RF threats with growth provisions for future threats. Flexible load/mission mixes are available with five payload types per magazine, 15 magazine mixes, multiple simultaneous firing pulses and extremely short burst intervals provided by the solid-state sequencer switch. Automatic misfire detection and correction eliminates the M-130's unreliable flare detector to ensure that decoys are dispensed at the critical moment. Threat response files and mission-specific dispense programs can be developed by the user and uploaded into the system on the flight line.

Operational status

ALE-47H TACDS has been developed and is in production for international users.

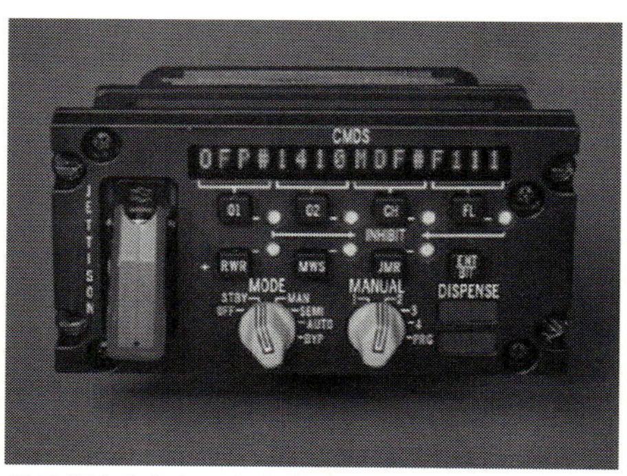

ALE-47H digital cockpit display unit **1999**/0051272

Contractor

BAE Systems North America, Austin.

VERIFIED

AN/ALE-38/41 dispenser system

The AN/ALE-38 (US Air Force) and AN/ALE-41 (US Navy) equipments are high-capacity bulk chaff dispensers. They employ dispensing techniques which provide continuous dipole dispersal and instantaneous bloom for laying chaff corridors. They can also be used for aircraft self-protection and can be turned on automatically by the radar warning receiver. Precut dipoles are sandwiched between two wraps of mylar film with six 22.7 kg rolls of this composite being carried in each pod.

Operational status

ALE-38 and -41 systems are understood to be in US Air Force and US Navy service respectively.

Contractor

BAE Systems North America, Austin.

VERIFIED

AN/ALE-39 series ECM dispensers

The AN/ALE-39 is a first-generation system for the protection of tactical aircraft from ground- and air-launched missiles and radar-directed anti-aircraft guns. It represents the culmination of over 10 years of dispenser system refinement by the US Navy and is designed with flexibility of response to meet present and future needs in a changing threat environment.

The system is capable of accommodating up to three types of expendable payloads loaded in any combination of multiples of 11. Chaff, IR flares or expendable jammers can all be dispensed manually or automatically in accordance with preset programmes. The dispensing function can be initiated by the pilot, or crew member in the case of the F-14. The system is also capable of accepting dispense commands from aircraft warning receivers.

The programme flexibility, payload loading versatility, operational modes and wide selection of payload dispensing options and sequences, including simultaneous multiple flare ejection, random chaff ejection and rapid dump of flares, combine to make the AN/ALE-39 one of the better first-generation analogue-programmable countermeasures-dispensing systems in the US inventory.

ALE-39B is an upgrade of ALE-39 and is designed to protect tactical aircraft from missile and radar-directed anti-aircraft gun threats. It is capable of accommodating up to three types of expendable payloads (chaff, Infra-Red (IR) flares and expendable jammers), loaded in any combination of multiples of 10. All three types of payload can be dispensed manually (single payload) or automatically in accordance with preset programmes. The dispensing function can be initiated by the pilot (or by the weapon systems operator in the case of the F-14). The system is also capable of accepting dispense commands from aircraft warning receivers.

Operational status

Obsolete ALE-39 series dispensers are reported to have been installed on A-6, AV-8B, EA-6B, ES-3A, F-14, F/A-18, S-3B and SH-2F/G aircraft of the US Navy, US Marine Corps and other air forces/naval air arms. More than 1,000 systems have been delivered since the initial production release in October 1973. ALE-39 is no longer in production where it has been replaced by AN/ALE-47.

The AN/ALE-40 chaff/flare dispenser for the US Air Force F-16 fighter

Contractor

BAE Systems North America, Austin.

VERIFIED

AN/ALE-40(V) dispenser system

The AN/ALE-40 offboard, expendable countermeasures, dispenser system was the precursor to the AN/ALE-47. This intervalometer and stepper-switch-based system was developed in the 1960s and has been widely installed. AN/ALE-40 is being upgraded or replaced with the form-fit-compatible AN/ALE-47 LRUs.

Operational status

Obsolete. Widely deployed in F-16 export aircraft. Upgrade or replacement required to support operational readiness.

Contractor

BAE Systems North America, Austin.

VERIFIED

AN/ALE-45 countermeasures dispenser

The AN/ALE-45 is a microprocessor-controlled countermeasures dispenser system developed for the F-15 aircraft. It responds automatically to threat notification from the radar warning receiver, tail warning set or pilot.

The system consists of a programmer assembly and four dispensing switch assemblies. Eight magazines containing expendable stores are attached to the switch assemblies. The programmer assembly houses the central processor and the input/output circuitry. An operational flight program which contains the dispenser programs, and other preprogrammed data defining the functional operation of the system, is changed easily through a front panel connector. The software is designed to provide automatic priority selection of dispensing program based on computations of threat sources from the radar warning receiver or the pilot and aircraft flight data. The dispensing program parameters are selectable, including payload class, burst count and interval and payload count and interval.

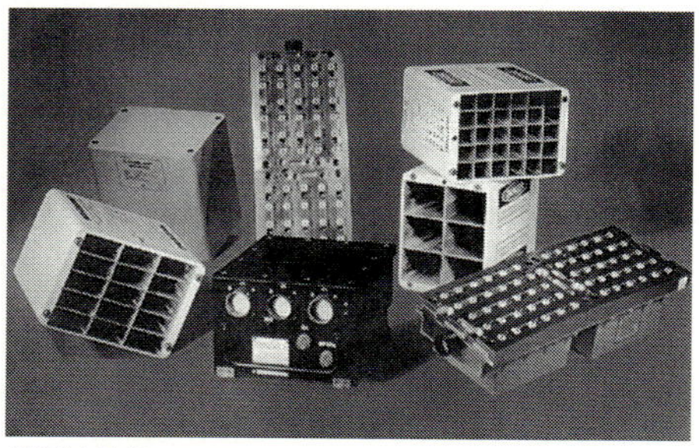

AN/ALE-45 countermeasures dispenser **1999**/0051273

The four dispensing switch assemblies are identical and interchangeable. Their primary function is the firing of payloads.

The AN/ALE-45 has the capability to dispense RR-170 and RR-180 chaff cartridges, MJU-7 and MJU-10 flare cartridges and other payload types of similar form factor.

Operational status

In service on F-15. Becoming obsolete.

Contractor

BAE Systems North America, Austin.

VERIFIED

AN/ALE-47 countermeasures dispenser system

The AN/ALE-47 is an offboard expendable countermeasures dispenser system that is intended to replace the AN/ALE-40 and AN/ALE-39 on currently equipped aircraft and to be fully integrated into numerous new aircraft.

The system interfaces with other onboard EW and avionics systems through two MIL-STD-1553 databusses, two RS-422 busses or discrete lines. The AN/ALE-47 uses threat data received over these interfaces to assess the threat situation and determine the appropriate countermeasures response. Dispense routines tailored to the immediate aircraft and threat environment may be automatically, semi-automatically or manually dispensed, depending on the mode selected by the user.

Threat response determination is embedded in the system operational software. The software consists of the Operational Flight Program (OFP) and the Mission Data File (MDF). The OFP processes input threat and aircraft position data to compute the optimum offboard response. The MDF contains the aircraft and mission specific parameters used by the OFP and is fully flight line reprogrammable.

Other features include modular automated test equipment; a Consolidated Automatic Support System (CASS) compatibility; Built-In Test (BIT) to the Shop Repairable Unit (SRU); a Shop Replaceable Assembly (SRA); hardware/software growth provisions; reprogrammable manual dispense routines; ALE-40 form-fit compatibility and ALE-39 Sequencer form-fit compatibility.

The AN/ALE-47 programmer LRU has been enhanced and reduced to a single card processor (SCP) to facilitate advanced avionics architecture and simplify aircraft integration. The SCP hosts all versions of the AN/ALE-47 OFP and provides all electronic warfare suite interfaces developed for the AN/ALE-47 system. The SCP provides the NDI solution for the next generation CMDS for all future aircraft.

Operational status

In service on the F-16 and F/A-18, C-5, C-17, C-130, C-141, HH-60, MH-47, MH-60 and the P-3 and slated for service on the E-8C, V-22, VH-60 and the VH-3. A combined ALE-47/AAR-47 system, called the Airlifter Defense System, (ADS) is fitted to the C-17.

Contractor

BAE Systems North America, Austin.

VERIFIED

AN/ALE-47 countermeasures dispenser system ***1999***/0051274

AN/ALE-47(V) TACDS ***1999***/0051271

AN/ALE-47(V) Threat Adaptive Countermeasures Dispenser System (TACDS)

The AN/ALE-47(V) TACDS provides an integrated, threat-adaptive, reprogrammable, computer-controlled capability for dispensing expendable decoys (chaff, flares and others) to enhance aircraft survivability in sophisticated threat environments. The threat-adaptive features of the TACDS are provided by the customer-defined Table-Look-Up Mission Data File (MDF). The TACDS Operational Flight Program (OFP) derives all (fixed and adaptive) dispensing programmes based entirely upon the customer-developed MDF. The system interfaces with other onboard EW and avionics systems through two MIL-STD-1553 databusses, two RS-422 busses or discrete lines.

The following operating modes are available: automatic, semi-automatic, manual and bypass. Mode selection is performed by the crew using a Cockpit Display Unit (CDU). An emergency jettison mode is available to dispense all expendables on command. In automatic, the CMDS evaluates the threat data from on-board EW sensors and integrates it with stored threat information to determine the optimum counter-measures dispense program; the system dispanses expendables without operator action. In semi-automatic mode, the CMDS determines the dispense program as in the automatic mode, but signals the operator to manually activate the dispense. In manual mode, the operator manually selects and depresses one of six pre-determined programs. In bypass mode, the operator manually dispenses expendables in degraded mode after a partial system failure by reconfiguring the system in flight.

Other features include: Built-In Test (BIT) to the Shop Repairable Unit (SRU); a Shop Replaceable Assembly (SRA); hardware/software growth provisions; reprogrammable manual dispense routines; ALE-40 form-fit compatibility and ALE-39 Sequencer form-fit compatibility. Support equipment available for the TACDS includes the Tracor Mission Planning System/ Threat Matrix Generator (TRACOMPS/TMG) that permits efficient development of MDFs customised to the customer's threat environments; Memory Loader Verifier (MLV) for loading MDFs into the system; and a Maintenance Test Station (MTS) to provide depot level repair capability.

Operational status

Currently in service for fighter, cargo and rotary-wing aircraft for several international users. Contracts for the US Air Force continue until 2002.

Contractor

BAE Systems North America, Austin.

VERIFIED

M-130 Dispenser

The M-130 aircraft general purpose dispenser is a lightweight countermeasures dispenser system developed for the US Army, using AN/ALE-40 technology. The system is designed for employment on tactical helicopters or fixed-wing aircraft, such as the AH-1, CH-47, OH-58, UH-1, RU-21 and OV-1D, to provide self-protection from radar-directed weapons by dispensing chaff and from infra-red homing weapons by dispensing flares.

The M-130 system is of modular design to allow flexibility of operational configuration. The modules consist of a cockpit control unit, electronics module, one or more dispensers, cabling and aircraft adaptors. The cabling and adaptors are unique to the aircraft type, whereas the other modules are common among all aircraft. The M-130 uses the M-1 countermeasures chaff and the M-206 countermeasures flare as payloads. Each payload module accommodates 30 chaff units or flares of a single type. The system configuration may be either a single dispenser of

The M-130 (with the cover removed) on a UK Royal Air Force Chinook helicopter (Paul Jackson)

30 capacity or a double dispenser of 60 capacity. The system weight, with 60 chaff units or flares, is 21.8 kg.

The M-130 will function with input from the infra-red threat warning device directly to the electronic module, obviating the need for pilot action.

The chaff units and flares have a nominal 25 by 25 mm square configuration and are 210 mm in length. The form factor is identical to the RR-170A/AL chaff used in the AN/ALE-40 and is functionally interchangeable.

Operational status

In service on a number of aircraft types.

Contractor

BAE Systems North America, Austin.

VERIFIED

AN/APX-111(V) Combined Interrogator/Transponder (CIT)

The AN/APX-111(V) CIT consists of an interrogator, a transponder and two associated cryptographic computers in a single, small, lightweight unit. For retrofit applications the CIT can replace six or seven boxes making up the existing interrogator and transponder, yielding reduced weight and freeing space for other avionics equipment.

The AN/APX-111(V)'s architecture enables it to support Mk X (SIF), Mk XII (Mode 4) or custom crypto systems. The crypto module is integral but removable from the front panel. Mode S transponder capability and growth to Mk XV (NIS) are also part of the system architecture.

The CIT meets international standards for IFF and ATC including US-DoD AIMS 65-1000B and NATO STANAG 4193. AN/APX-111(V) utilises miniature low-profile fuselage-mounted electronically scanned antenna arrays. It provides full interrogator range capability without the need for external amplifiers. The modular design is all-solid-state with a unique approach to thermally efficient cooling. A MIL-STD-1750 processor and MIL-STD-1553B databus are included.

The latest IFF techniques are provided, including digital target reports for a clear operator display, monopulse processing for accurate target azimuth, a statistical reply evaluator for high-confidence identification, a defruiter for dense environments and cryptographic coding for security. Both continuous and operator-initiated built-in tests are provided, with reporting via the MIL-STD-1553B databus.

Specifications

Combined interrogator/transponder RT-1679(V): 208 × 146 × 320 mm
Beam forming network C-12222(V): 178 × 249 × 76 mm

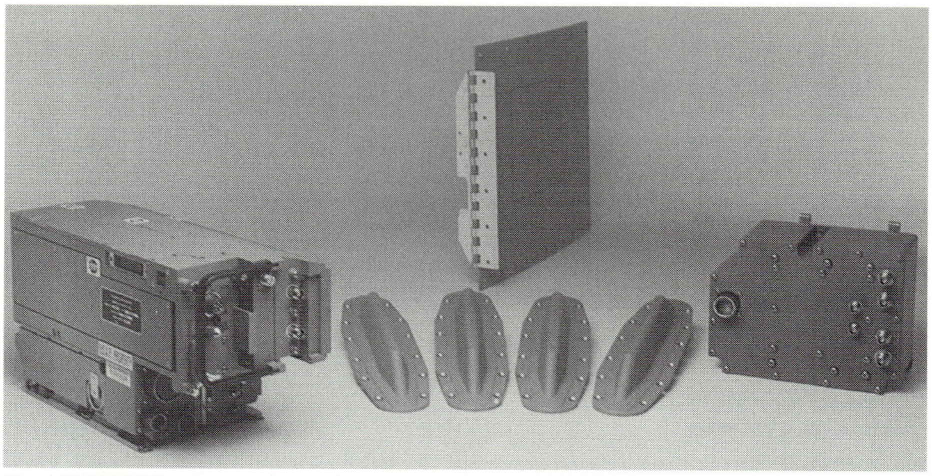

The AN/APX-113(V) showing the combined interrogator/transponder (left), the lower interrogator antenna (centre, rear), the fuselage-mounted antenna elements (centre, front) and the beam-forming network (right)

Fuselage-mounted antenna elements AS-4267(V): 29 × 55 × 237 mm
Weights:
RT-1679(V): 15 kg
C-12222(V): 5.4 kg
AS-4267(V): 0.2 kg

Operational status

Fitted in F/A-18 FMS and F-16 MLU programmes, and Greek and Turkish air force F-16 aircraft.

Contractor

BAE Systems North America, Greenlawn.

VERIFIED

AN/APX-113(V) Combined Interrogator/Transponder (CIT)

The AN/APX-113(V) Combined Interrogator/Transponder (CIT) is a complete Mk XII identification system which includes crypto computers. It consists of one unit and incorporates growth for the next generation of IFF and combat aircraft identification equipment. The AN/APX-113(V) provides both interrogation and IFF responses on IFF Mode XII Modes 1, 2, 3/A, 4, C and S Level 3.

The multiple antenna configurations feature electronic or mechanical scan. The system features Ada software and a MIL-STD-1553 bus interface.

Specifications

Dimensions:
(combined interrogator/transponder) 209.8 × 152.4 × 368.3 mm
(beam-forming network) 165.1 × 212.9 × 101.6 mm
(fuselage-mounted antenna elements) 39.4 × 82.6 × 332.7 mm
(lower interrogator antenna) 15.2 × 431.8 × 355.6 m
Weight:
(combined interrogator/transponder) 14.52 kg
(beam-forming network) 4.54 kg
(fuselage-mounted antenna elements) 0.23 kg
Power supply: 28 V DC, 200 W
Range: 185 km
Coverage:
(azimuth) ±60°
(elevation) ±60°
Accuracy:
(range) 500 ft
(azimuth) ±2°
In-beam targets: 32
Reliability: 1,600 h MTBF

Operational status

Developed specifically for F-16 Falcon. Fitted to F-16 Block 16 A/B MLU, Block 20 A/B. Also fitted to ASW/surveillance helicopters and the Japanese FS-X fighter.

The US Air Force plans to retrofit the AN/APX-113 to its F-16 Block 50/52 aircraft. This installation is designated Advanced Identification Friend or Foe (AIFF). AIFF flight trials are due to commence in June 2001, with aircraft installation starting in July 2002.

BAE Systems (Rochester, UK) has been selected to supply Successor Identification Fried or Foe (SIFF) systems for retrofit on UK Royal Navy Sea Harrier F/A Mk 2 aircraft. The capability required is to upgrade the present IFF Mk Xa to Mk XII Modes 1,2 3/A, C and S Level 2. Development work, based on the AN/APX-113 system, is to be completed by March 2000, with full-scale production to begin in mid-2000.

Contractor

BAE Systems North America, Greenlawn.

VERIFIED

AN/ARN-155 Precision Landing System Receiver (PLSR)

The PLSR is a multimode precision landing system receiver designed to replace existing DoD navaid avionics. PLSR is interoperable with both civilian (ICAO/FAA compliant) and military Cat.I and II landing aids, including Instrument Landing System (ILS), Microwave Landing System (MLS) and Differential Global Positioning System (DGPS). PLSR meets the new international FM interference requirements (a critical requirement in Europe for ILS operation) and also incorporates the latest GPS JPO GRAM compliant, all-in-view, P/Y GPS receiver technology. In addition to the embedded VHF datalink for local area DGPS augmentation, PLSR also contains VOR capability. A growth version adds the J-band Pulse Coded Scanning Beam (PSCB) capability for US Navy carrier and marine operations.

The Receiver Processor Unit (RPU) is capable of being controlled via a MIL-STD-1553B digital databus or ARINC 429 interface, or with an optional Control Display Unit (CDU). The RPU also contains discrete inputs and outputs for compatibility with older aircraft.

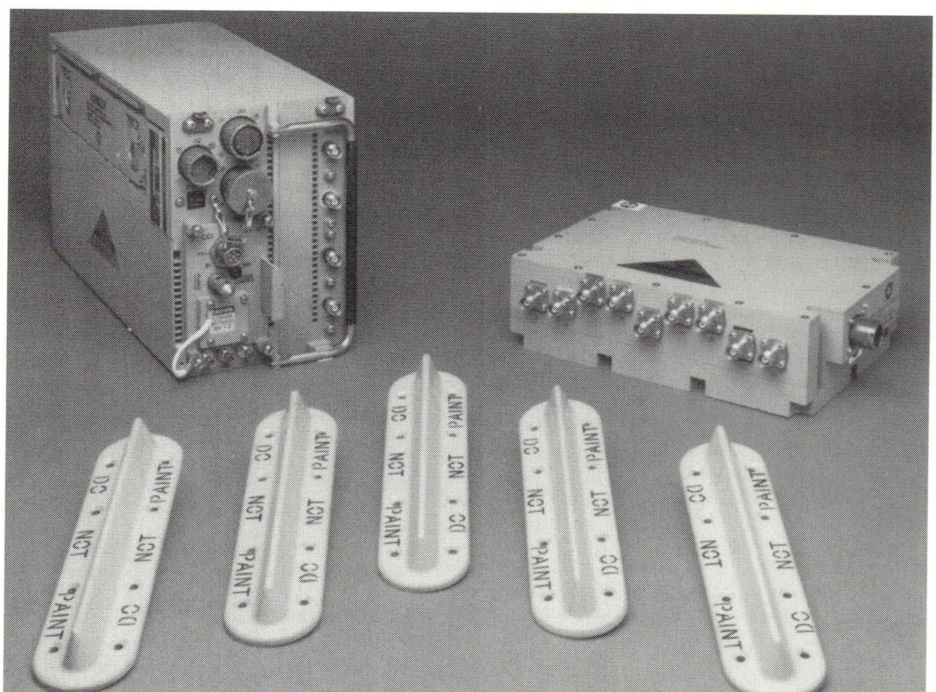

AN/APX-111(V) combined interrogator/transponder ***1999***/0044984

Precision landing system receiver 0003107

The AN/ARN-155 comprises two units: the R-2556/ARN-155 Receiver Processor Unit (RPU) and the C-12377/ARN-155 radio and Control and Display Unit (CDU). It accepts DME, TACAN and GPS ranging information from external avionics for computed mode MSL approaches and accommodates both split site offset and collocated landing site scenarios.

Specifications

Dimensions:
(RPU) 258 × 127 × 91 mm
(CDU) 146 × 121 × 38 mm
Weight:
(RPU) 4.49 kg
(CDU) 1.22 kg
Power supply: 28 V DC, <30 W
Interfaces: MIL-STD-1553, ARINC 429, discretes
Environmental: MIL-E-5400 Class 2
Reliability: >10,500 h MTBF per MIL-HDBK-217E

Operational status

BAE Systems North America was selected by the US Air Force in June 1993 to build and test 30 preproduction ILS/MLS 2-Band PLSR systems, including a full qualification programme. In August 1996, the US Air Force included an Engineering Change Proposal (ECP) contract award to add DGPS with a 12-channel, P/Y code, all-in-view GPS receiver, thereby providing an ILS/MLS/DGPS three-band PLSR system. In December 1996, the US Air Force selected PLSR for the C-17 fleet, adding VOR as a new mode and initiating a production programme with deliveries in late 1997. Follow-on full-scale production is believed to have started in 1998.

The AN/ARN-155 PLSR is a leading candidate for future landing avionics in the tri-service US DoD's Joint Precision Approach and Landing System (JPALS) study.

Contractor

BAE Systems North America, Greenlawn.

VERIFIED

AN/ARR-78(V) Advanced Sonobuoy Communication Link (ASCL)

ASCL is the US Navy standard sonobuoy receiver and is deployed on the Orion P-3C Update III and Viking S-3B ASW aircraft. The P-3C Update III employs two ASCLs in a 32-acoustic-channel expansion configuration (CHEX). ASCL operates on both types of aircraft in conjunction with standard VHF antennas. The following units comprise a typical ASCL configuration.

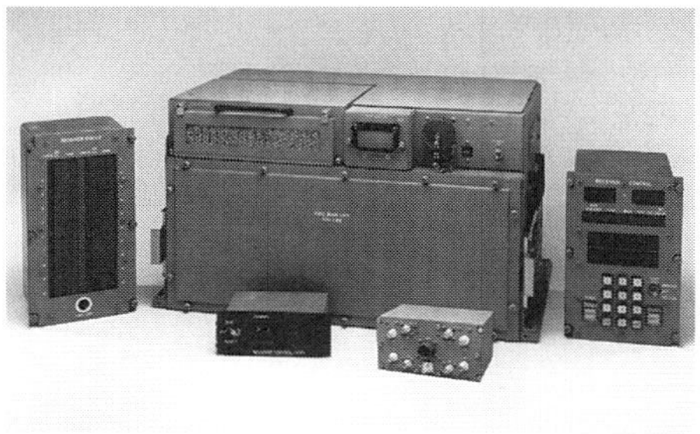

AN/ARR-78(V) *Advanced Sonobuoy Communication Link* (ASCL) 0011864

RF preamplifier

The AM-6875 is an optimised high-performance, low-noise VHF preamplifier that provides amplification and prefiltering of received RF signals.

Radio receiver

The R-2033 (P-3C, Update III) or the R-2066 (S-3B) contains 20 fully synthesised receiver modules (16 acoustic and four auxiliary) and one each of the following modules: RF/ADF amplifier multicoupler, reference oscillator, I/O Proteus digital channel (P-3C), I/O Manchester digital channel (S-3B), I/O processor, clock generator, BITE and DC power supply module. Each single-conversion module includes mixer conversion, frequency-synthesised local oscillator, demodulator, and output interface circuits.

Each of the acoustic receiver modules processes FM/analogue signals at any of the 99 channels in the extended VHF band. Each of the four auxiliary receiver modules processes FM/analogue signals at any of the 99 channels and provides one channel for selection and processing of the On-Top Position Indicator (OTPI) signals, two channels for the operator to monitor acoustic information and one channel to monitor the RF signal level in any of the RF channels. Common receiver modules are interchangeable.

BITE circuits provides comprehensive end-to-end evaluation of each receiver from the VHF preamplifiers to the receiver output interface circuits. BITE is initiated automatically by the computer (such as Proteus) and/or by the operator through the Indicator Control Unit (ICU). Performance status is displayed on the ICU and routed to the computer.

Indicator Control Unit (ICU)

The C-10126 ICU provides the operator with a means for manual control of each receiver channel frequency assignment, receiving mode and self-test. It also displays status of the receiving set and operator entry information.

Receiver Indicator (Receiver Status)

The ID-2086 Indicator continuously displays the control mode setting, the RF channel number, and the received signal level for each receiver.

Receiver Control (OTPI)

The C-10127 Receiver Control Unit provides the operator with control over the OTPI.

Specifications

Frequency: extended VHF
Receiver: 20 (16 acoustic/4 auxiliary)
Channels: 99 per receiver
Audio Output:
(analogue) 2 V rms balanced
(monitor) 50 mW, 300 ohms
Power: 115 V AC±10%, 380 to 440 Hz, 3-phase, 500 W; 18 to 32 V DC, 7 W; 26.5 V AC±10%, 400 Hz, 50 W

Operational status

Installed in US Navy P-3C Orion and S-3B Viking aircraft.

Contractor

BAE Systems North America, Greenlawn.

VERIFIED

AN/VRC-99 (A) communications system

The AN/VRC-99 (A) is a programmable, wideband, secure, open architecture communication system. It provides virtual circuit and datagram service guaranteeing reliable, simultaneous, multichannel voice, data, imagery, and video transmission.

The AN/VRC-99 (A) is available in ground vehicular, shipboard and airborne configurations. Digital signal processing provides flexibility and operational simplicity in end-to-end communication connectivity, packet formatting, and packet switching protocols. Use of spread spectrum modulation (LPI/AJ), growth for transmit power controls (LPD), and embedded encryption provides for the security and integrity necessary in modern strategic and tactical networks. The basic system provides coverage from 1,300 to 1,500 MHz.

The AN/VRC-99 (A) can provide wireless LAN service for both Line-Of-Sight (LOS) and Beyond-Line-Of-Sight (BLOS) data and/or voice links. Adaptive routing provides for network survivability in dynamic mobile subscriber applications as required by the Tactical Internet/Warfighter Information Network (WIN) Architecture.

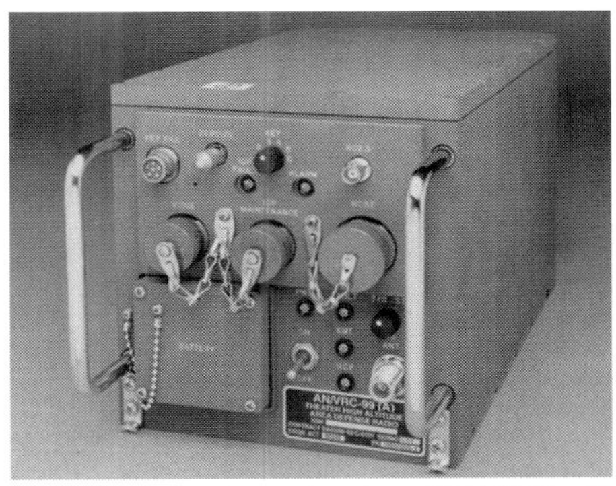

BAE Systems North America AN/VRC-99 (A) communications system ***1999***/0044845

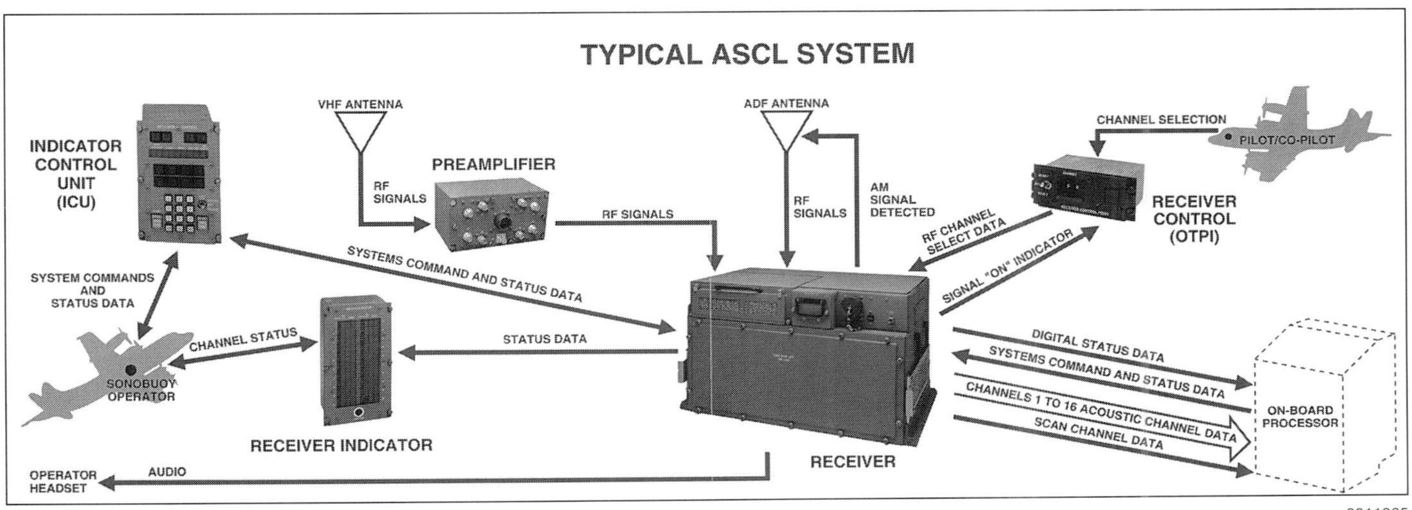

Typical ASCL system 0011865

System features include:
(a) digitised voice/data/secondary imagery/video
(b) user selectable burst rate of 156 kbps to 10 Mbps with growth option for fully adaptive operation*
(c) low probability of intercept and antijam (LPI/AJ) utilising specialised spread spectrum techniques
(d) low probability of detection (LPD) via transmit power control (ground options)
(e) embedded type 1 encryption
(f) EKMS/AKMS compliant
(g) forward error correction
(h) frequency-hopping mode
(i)automatic initialisation/network entry at power on – no user setup required
(j)asynchronous/synchronous interfaces; RS-232; RS-422 with X.25 LAPB link layer; Ethernet 802.3
(k)TEMPEST-tested COMSEC module with networking efficiency design
(l) interoperable with MSE Voice Circuit and Packet Data Switches
(m) US Army Technical Architecture (ATA)/Joint Technical Architecture (JTA) compliant.

Specifications

Frequency band: 1,300 to 1,500 MHz (19 channels)
Frequency agility: selection by software controls
Power output: 10 W (optional 50 W with external PA)
Channel access: CDMA/FDMA/TDMA/SDMA/ALOHA/hybrids
Synchronous data: up to 10 Mbps for SDLC/HDLC; 10 Mbps for Ethernet
Asynchronous: up to 100 Kbps
Voice: 4-wire conditioned diphase (4WCDP)
Dimensions: ¾ ATR 193.3 × 190.5 × 318.8 mm
Weight: 11.36 kg
Power: 28 V DC
MTBF: 4,000 hours w/PA

Contractor

BAE Systems North America, Greenlawn.

VERIFIED

GRA-2000 Low Probability of Intercept (LPI) altimeter

The GRA-2000 LPI altimeter, has been selected by the US Joint Services Program Office to replace the AN/APN-194, -171, -209 and -232 series altimeters on the majority of tactical jet, helicopter and transport aircraft employed by the US Department of Defense.

The design is based on a very simple durable design employing a single I/F down-convert and specialised algorithms to provide exceptional stealth and jam resistance; it is an outgrowth of the AD-1990 SARA system being procured for Tornado aircraft in the UK.

Specifications

Dimensions: 185.4 × 97.3 × 77.5 mm
Weight: < 3.2 kg
Reliability: 8,000+ h
Accuracy: ±2 ft or 2% (1σ)
Performance: 0 to 35,000 ft AGL

Operational status

In development against the F/A-18E/F AN/APN-194 configuration.

Contractor

BAE Systems North America, Greenlawn.

VERIFIED

AN/AAR-57 Common Missile Warning System (CMWS)

The AN/AAR-57 CMWS provides missile warning for rotary-wing, transport and tactical aircraft in all military services. It is claimed to have a very low false alarm rate and to exceed NATO staff requirements for a missile warning system in this respect. It is also able to detect and declare missile identity before missile burnout. A very high detection-processing speed and self-contained image-stabilisation assembly ensure high performance on manoeuvrable fighter aircraft.

The CMWS consists of an Electronics Control Unit (ECU) developed by Sanders (now BAE Systems, following acquisition in 1999) and up to six Electro-Optical Multiple Sensors (EOMS), designed and developed by Lockheed Martin Infra-red Imaging Systems (LMIRIS) of Lexington, Massachusetts. The ECU processes threat data and provides cueing for expendables and directable infra-red jammer elements of the Advanced Threat Infrared CounterMeasures (ATIRCM) system. The EOMS employs ultra-violet detector technology, to detect radiation from the plume of the attacking missile.

Specifications

Weight:
(AN/AAR-57, including one ECU and six EOMS): 14.44 kg
(EOMS): 1.18 kg
(ECU): 7.36 kg
Power:
(AN/AAR-57, including one ECU and six EOMS): 449 W (353 W without anti-icing)
(EOMS): 12 W without anti-icing, 28 W with anti-icing
(ECU): 281 W

Operational status

The US Army, PEO Aviation, is managing the tri-service CMWS programme as part of the ATIRCM programme, in Redstone Arsenal, Alabama.

In August and September 1999, the US Army and Sanders conducted 12 flight tests of the CMWS system (with a single EOMS) aboard a US Army UH-60 Blackhawk helicopter. The system was exposed to both simulated and actual industrial and military ultra-violet sources of radiation at varying altitudes in varying weather conditions.

Customers include the US Army, Navy, Marine Corps and Air Force for A-10, Apache, Blackhawk, C-5, C- Prospective US military customers for the CMWS include the C17, C-130, C-141, Chinook, CV-22, F-15, F-16, Kiowa and RC-135. Other national armed forces service programmes, include the UK WAH-64 Apache and UK Nimrod MRA4.

Contractor

BAE Systems North America, Information and Electronic Warfare Systems

UPDATED

Marconi CNI Division GRA-2000 LPI altimeter 0002221

AN/ALQ-126B Deception Electronic CounterMeasures (DECM) system

AN/ALQ-126 is a DECM system developed by the US Naval Air Systems Command for its Tactical Air Electronic Warfare Programme. It provides wider frequency coverage than the preceding AN/ALQ-100 system and was initiated in response to new threats from anti-aircraft, surface-to-air missile and airborne interception systems. The initial production model (ALQ-126A) was widely used by the US Navy aboard aircraft such as the A-4, A-6, A-7 and F-4 aircraft.

An improved version, the ALQ-126B, offers increased frequency coverage, and incorporates a digital instantaneous frequency measurement receiver, improved deception techniques and more modem construction, packaging and cooling arrangements. ALQ-126B includes a distributed microprocessor control system to enable the system to be reprogrammed on the flight line and improve signal processing. Covering the 2 to 18 GHz band frequency band, the system is capable of generating a variety of jamming modulations including inverse conical scanning, range gate pull off, swept square wave and main lobe blanking.

ALQ-126B operates either autonomously or as part of an integrated weapon system comprising the AN/ALR-45F, APR-43 or ALR-67 radar warning equipment, the AN/ALQ-162 continuous wave radar jamming system, and the HARM, Sparrow, Phoenix and AIM-120 missiles. ALQ-126B is compatible with the A-4, A-6, A-7, F-4, F-14 and F/A-18 aircraft. ALQ-126B also forms part of the ALQ-164 ECM pod (see later entry).

Specifications

ALQ-126B
Dimensions: 411 × 270 × 609 mm
Weight: 86.3 kg

Operational status

The US Navy has awarded BAE Systems North America (formerly Sanders) contracts to the value of nearly US$500 million for the production of the ALQ-126B with production and deliveries continuing. The equipment is also installed on Australian F/A-18, Kuwaiti F/A-18, Malaysian F/A-18, Canadian CF-18 and Spanish EF-18 aircraft. Deliveries to these countries are complete.

Contractor

BAE Systems North America, Information and Electronic Warfare Systems.

UPDATED

AN/ALQ-144A(V)

BAE Systems' (formerly Sanders) ALQ-144A(V) countermeasures system provides protection against a wide spectrum of infra-red threats. Its small size, light weight and low power consumption make the AN/ALQ-144A(V) ideal for helicopter installation. It is an omnidirectional system consisting of a cylindrical source surrounded by a modulation system to confuse the seeker of the incoming missile. Phase Lock (PL) installations of two, or more, systems permit protection of larger helicopters or fixed-wing aircraft. The units may be mounted on either the top or bottom of the airframe or, in the case of the PL configuration, both. The new phase lock configuration (V)5 is a pair of ALQ-144A transmitters that are electrically phased

BAE Systems AN/ALQ-144A(V) transmitter and control unit **1999**/0051269

through a junction box. They are controlled by a single Operator Control Unit (OCU), which has the capability to select from nine different jamming options.

Specifications

Weight: 12 kg (transmitter); 0.45 kg (control unit)
Dimensions: transmitters: 320 mm (diameter) × 340 mm (height)
control units: 28 × 147 × 134 mm
Field of view: 360° azimuth
MTBF: 300 h

Operational status

More than 5,000 systems have been delivered to the US Army CECOM for fielding in all US armed services and to a dozen export customers, for the following applications:

ALQ-144A(V)1 - Utility, fitted to: HH-60, MH-60, Puma, UH-1H and UH-60B helicopters;
ALQ-144A(V)1 - Observation, fitted to: Lynx, OH-58D helicopters and OV-10 aircraft;
ALQ-144A(V)3 - Attack, fitted to: A-109, A-129, AH-1F, AH-1W, AH-64 and Tiger helicopters;
ALQ-144A(V)5 - Phased Lock Special Mission, fitted to: SH-2F/G, SH-3D, SH-60B/R, Super Puma and VH-60 helicopters.

In February 2000 an order for the Turkish government was completed - for a large number of AN/ALQ-144A(V)1 units and a smaller number of AN/ALQ-144A(V)5.

Contractor

BAE Systems North America, Information and Electronic Warfare Systems.

UPDATED

AN/ALQ-156 Missile Approach Warner (MAW)

The AN/ALQ-156 is an airborne, internally mounted, pulse Doppler missile-warning system, originally developed for slow- and low-flying helicopters and fixed-wing aircraft, but later modified for use in other types of aircraft. The system is solid-state, incorporating four doughnut-shaped antennas, a transceiver and a control unit. The system recognises missile threats by comparing closure rates and other ballistic parameters, warns the pilot aurally and triggers the countermeasures command to dispense flares, chaff or other expendables. When coupled with radar and laser warning sensors, the ALQ-156 can initiate automatic selection of the appropriate countermeasure.

The AN/ALQ-156(V) is a 360° pulse Doppler radar missile detector that illuminates an incoming missile, detects the RF reflection and measures the missile's range and velocity. It then accurately determines time-to-go and provides the optimum triggering of an expendable to protect the host platform. The system can be integrated with a range of countermeasures dispensers such as the ALE-40, ALE-47 and M-130 systems to provide infra-red protection and with a range of RWRs, such as the APR-39, ALQ-56, ALR-56 and MK-20. Integrated it can also improve situational awareness.

A later model, the AN/ALQ-156A, was developed for the US Navy and was designed for use in tactical aircraft. It integrates an RWR and deceptive ECM data to allow smart control of decoys and flares and has a much longer detection range.

Operational status

The AN/ALQ-156 is fitted to a variety of aircraft, including the US Army CH-47, EH-1H, EH-60B, BV-10, REC-12D/H/K, RU-21A/B/C/D and RV-1D platforms.

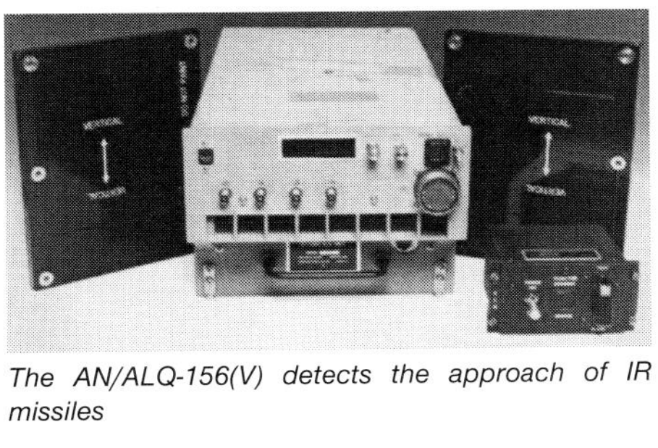

The AN/ALQ-156(V) detects the approach of IR missiles

The F-22 Raptor aircraft ***2000***/0079236

The US Air Force uses the system on its C-130 transports and the Royal Air Force on its Chinook helicopters. Several other nations use the system on C-130 aircraft and Chinook helicopters. Nearly 1,000 systems have been delivered. An external pod-mounted configuration has been flight tested successfully.

In January 2000, the US Army Communications and Electronics Command (CECOM) contracted Sanders (now BAE Systems North America) to supply 17 systems (designated AN/ALQI-156(V)2/(V)3T) for the Hellenic Army's CH-47D helicopters. This version of the AN/ALQ-156(V) was developed as part of the NATO C-130 Apollo programme. It incorporates improved false alarm rejection capability.

Contractor

BAE Systems North America, Information and Electronic Warfare Systems.

UPDATED

AN/ALQ-164 ECM pod

The AN/ALQ-164 is an airborne pod-mounted jammer for both pulse and CW modes. The pod incorporates unmodified ALQ-126B and ALQ-162 jammers. It can operate fully autonomously or integrated with an RWR and other avionics on board the aircraft. It is reprogrammable to provide for threat changes. The ALQ-164 is intended for fitting on the centreline pylon of the AV-8B aircraft and has applications for a wide variety of tactical and transport aircraft.

Specifications

Dimensions: 2,160 × 445 × 406 mm
Weight: 188 kg
Power supply: 115 V AC, 400 Hz, 3 kVA
Frequency: G- to J-bands

Operational status

In service with the US Marine Corps AV-8B. The Spanish Navy has ordered the ALQ-164 for the EAV-8 Harrier. It is also reported that ALQ-164 has been ordered by the Italian Navy for its AV-8B+ aircraft.

Contractor

BAE Systems North America, Information and Electronic Warfare Systems.

UPDATED

AN/ALR-94 electronic warfare system

The AN/ALR-94 electronic warfare system is an integral element of the F-22 Raptor aircraft' sensor suite. It is reported to provide RF warning and countermeasure functions, together with missile-launch warning, and to be integrated with the AN/ALE-52 flare dispenser.

Operational status

In February 1999, Sanders (now BAE Systems North America) completed production and delivery of the first 11 AN/ALR-94 suites for the F-22 EW Engineering and Manufacturing Development (EMD) programme.

Units shipped from BAE Systems to the F-22 Avionics Integration Laboratory (AIL) in Seattle, include: advanced apertures and associated electronics, a Remote Aperture Interface Unit (RAIU) and RF unit, as well as aircraft power supplies, test stands and workstations. BAE Systems is also providing, among other items, the overall Communications, Navigation and Identification (CNI) antennas for the F-22.

Contractor

BAE Systems North America, Information and Electronic Warfare Systems.

UPDATED

AN/USQ-113 communications jammer

The AN/USQ-113 provides the EA-6B Prowler aircraft with its communications jamming capability. To meet the expanded role of the aircraft now that it has subsumed the US Air Force support jamming tasks (after retirement of the EF-111A) as well as those of the US Navy, an upgrade programme has been initiated.

Operational Status

In September 1996, Sanders (now BAE Systems North America) were contracted to develop three preproduction systems during a non-recurring Engineering and Manufacturing Development (EMD) phase. The contract also called for the upgrade of 30 existing USQ-113 systems, beginning in July 1997, scheduled for completion before the end of 1998. Under another contract issued in October 1998, Sanders were contracted to provide 33 AN/USQ-113(V)2 Phase III systems and two improved operator panels. The Phase III systems consist of: new receivers, power amplifiers and transmitters developed to extend the frequency range of the system and to replace the operator display. The new display has an improved Liquid Crystal Display (LCD) that runs Windows™-based software. Additionally, a signal recognition device is being developed for insertion in a new system controller. With the device, aircrews will be able to analyse the signals they receive. This upgrade forms part of the ICAP III (Improved Capabilities III Warfighter Upgrade System) upgrade to the EA-6B Prowler aircraft being managed for the US Navy by Northrop Grumman Corporation.

Contractor

BAE Systems North America, Information and Electronic Warfare Systems.

UPDATED

ATIRCM/CMWS AN/ALQ-212(V)

The AN/ALQ-212(V) Advanced Threat IR CounterMeasures (ATIRCM) system, combined with the AN/AAR-57 Common Missile Warning System (CMWS), is in development by BAE Systems North

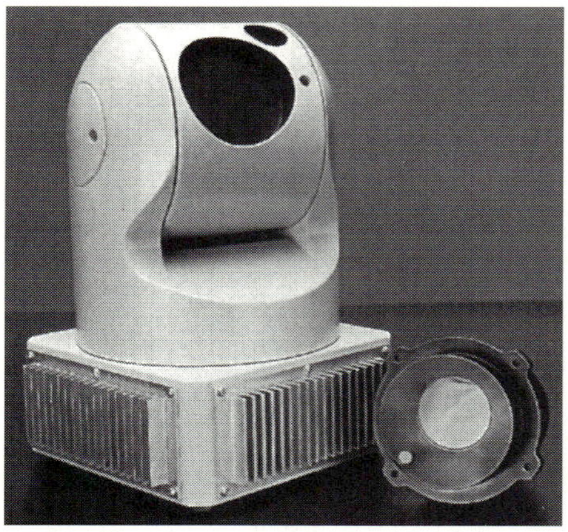

ATIRCM/CMWS counters advanced threats using the Common Missile Warning System to cue directable jammers and dispensers 0018254

America (formerly Sanders and Lockheed Martin Infrared Imaging Systems). The system combines the missile warning (CMWS), with an advanced jammer that integrates laser and xenon lamp technology, and an enhanced dispenser to counter current threats. The design also includes pre-planned product improvement plans for countering future threat systems. When the missile warning sensors detect a threat, the ATIRCM system will slew one or two jamming heads in the direction of the missile. A missile tracker on the jammer head will further refine the direction of the jamming transmission, ensuring that the main beam remains on the target. The system will also cue the release of flares from the dispenser.

BAE Systems' directable IRCM suite provides comprehensive protection against a multi-tier array of infra-red threats. The system detects, acquires, tracks and then jams these missiles with a focused beam of infra-red energy. The AN/ALQ-212(V) incorporates both a robust laser and an arc lamp to provide protection in all IR threat bands and has proven its ability to defeat modern IR missiles in live-fire tests.

The ATIRCM/CMWS suite functions as a stand-alone warning system (the AN/ALQ-57(V)), using common missile warning sensors and electronics coupled to expendables, or as a complete jamming suite (the AN/ALQ212(V)), incorporating one or more jam heads.

The ATIRCM system is claimed to provide over 100 times the jamming power, at less weight, than traditional IRCM systems.

Operational status

The US Army is the main customer for the AN/ALQ-212(V) and has issued BAE Systems North America with a contract to integrate the system onto the AH-64D Longbow Apache. The system is also slated for other US services and other countries are considering application to their platforms, or have already selected it. It had been expected that the system would be fitted to about 25 different types of US Army, Navy, Marine Corps and Air Force rotary-wing, transport and fighter aircraft. However, in April 2000, the US Air Force announced a reduction in its projected requirement from 853 systems to 362 systems, and at the same time the US Navy reduced its projected requirement from 665 to 264 systems.

Contractors

BAE Systems North America Information and Electronic Warfare Systems.

UPDATED

Improved Self-Defense System (ISDS) Infra-Red CounterMeasures (IRCM)

Developed as an improvement on Sanders' Self-Defense System, ISDS is a fuselage- or pylon-mounted IR countermeasures system which is designed to provide protection against IR homing anti-aircraft missiles. The system comprises a new multiband IRCM transmitter, an electronic control unit and an operational control unit. System configuration varies according to aircraft type with a typical installation incorporating an electronic control unit/transmitter package for each engine. A single, cockpit-mounted operation controller can control up to four transmitters. ISDS has also been flight tested in a ram-air turbine-

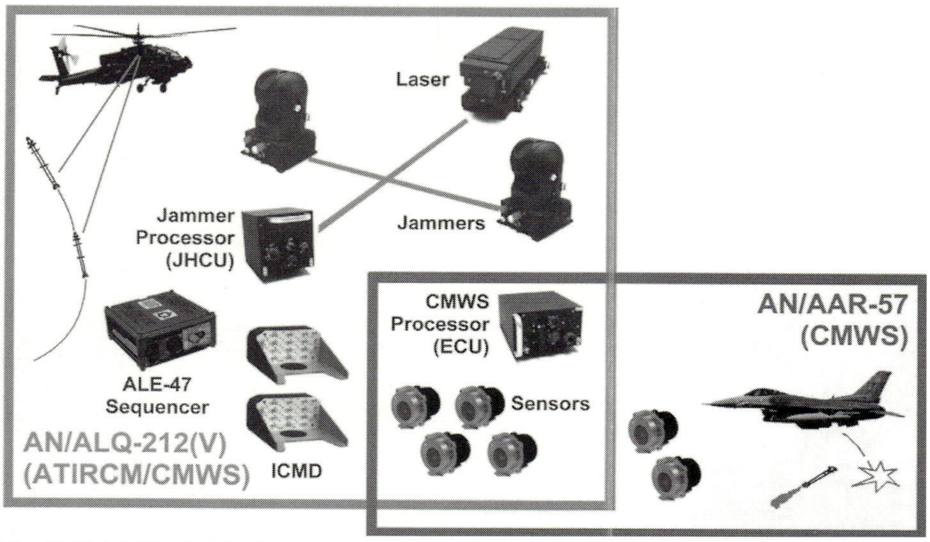

The AN/ALQ-212(V) ATIRCM/CMWS configuration ***1999***/0051270

powered pod configuration. Within the system as a whole, the individual transmitters weigh 29.5 kg; the electronic control unit, 2.3 kg and the 146 × 125 × 57 mm operator control unit, 1.56 kg.

Operational status

ISDS is in production and in service on both turboprop and jet-powered aircraft. More than 150 ISDS systems are reported to have been delivered.

Contractor

BAE Systems North America, Information and Electronic Warfare Systems.

UPDATED

Joint SIGINT Avionics Family (JSAF) Low-Band SubSystem (LBSS)

In February 1997, Sanders, (now BAE Systems North America), was selected by the US DoD Joint Airborne Signals Intelligence (SIGINT) Programme Office for development and demonstration of the Joint SIGINT Avionics Family (JSAF) Low-Band SubSystem (LBSS). The LBSS is a key element in the planned upgrade of US Army, Navy and Air Force airborne SIGINT capabilities.

BAE Systems North America will serve as prime contractor and system integrator and will also develop signal processing hardware and software for the system. Major subcontractors include: Radix Technologies Inc of Mountain View, California; Applied Signal Technologies (APSG) of Sunnyvale, California; and TRW System Integration Group, also of Sunnyvale. Radix will provide radio frequency (RF) and digital signal processing subsystems; APSG will develop special signal processing subsystems; and TRW will be responsible for high-speed networking and computing subsystems.

The JSAF low-band subsystem is a platform-independent, modular, reconfigurable suite of hardware and software that can address multiple mission scenarios aboard a variety of aircraft. It will significantly enhance the ability of reconnaissance platforms to detect the locations of modern enemy communications systems and to provide targeting accuracy on them, thereby providing real-time intelligence on enemy intentions and capabilities to the warfighter.

Operational status

Under the development contract, the consortium will design, develop, integrate and flight test three demonstration units. Delivery of the first development unit was scheduled for 2000. The programme is administered by the US Air Force's Aeronautical Systems Center, Wright-Patterson AFB, Ohio.

Initially, JSAF LBSS will be deployed on US Air Force's and RC-135 Rivet Joint aircraft and U2 aircraft, as the US Navy's EP-3 aircraft. JSAF LBSS will also be capable of deployment on unmanned aerial vehicles (UAVs) in the future.

Contractor

BAE Systems North America, Information and Electronic Warfare Systems.

UPDATED

AN/AAR-44 missile warning receiver

The AAR-44 infra-red passive airborne warning receiver continually searches a hemisphere while tracking and verifying missile launches. It warns the crew of missile

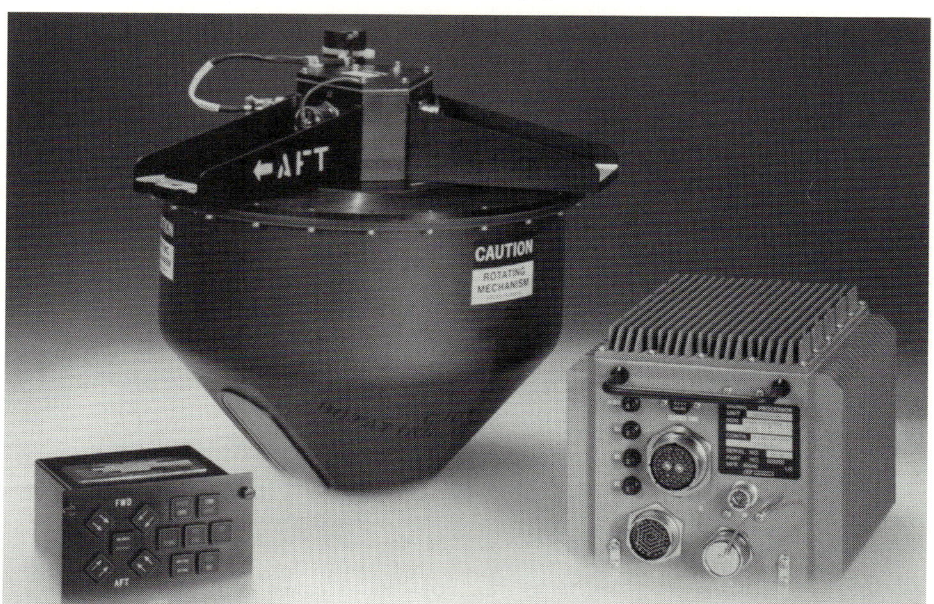

AN/AAR-44 missile warning system with (left to right) control display unit, conical detector head and processor

position and automatically controls countermeasures to neutralise threats and enhance survival.

Design features of the equipment include: automatic pilot warning and countermeasures command, continuous track-while-search processing, multiple missile threat capability and countermeasures discrimination capability. Multidiscrimination modes against solar radiation and terrain and water reflection, minimise false alarms. MIL-STD-1553B bus interface capability is incorporated. A number of sensor unit configurations are available to fulfil specific field of view requirements.

Specifications

Control and display unit
Dimensions: 102 × 142 × 150 mm
Weight: 1.12 kg

Processor
Dimensions: 200 × 212 × 252 mm
Weight: 8.82 kg

Sensor/Dimensions: 400 × 369 mm (diameter)
Weight: 17.91 kg

Displays: sector threat indicators; external countermeasures command and audio tone

Operational status

The AAR-44 is thought to be in service aboard US Air Force C-130 aircraft. AAR-44 equipment operated by the US Air Force's Special Operations Command is understood to have been the subject of a processor upgrade programme. The AAR-44 is also reported to be in Belgian Air Force service.

An improved version, designated AN/AAR-44A, has been announced that includes upgraded technology and a DIRCM interface.

Contractor

BAE Systems North America, Mason.

VERIFIED

AN/AAR-44(V) advanced missile warning system

The AN/AAR-44(V) advanced missile warning system is a collaborative development by BAE Systems North America (Mason) and Raytheon Electronic Systems (Goleta)of the earlier AN/AAR-44 equipment.

The AAR-44(V) features include: passive multicolour infra-red detection for positive missile warning, with minimum false alarm rate, using colour, size, intensity, and trajectory discriminants; multiple simultaneous threat detection, not confused by flares or multiple objects; wide volume coverage (±135 × 360°); accurate angular detection (better than 1°), with laser-pointing growth potential.

Designed for use on virtually any airborne platform, the AAR-44(V) can be mounted internally, via a 4 in diameter opening in the aircraft skin, externally into a weapon-carrying pylon, or into the underside of the lower gondola of the ALQ-184 ECM pod.

Specifications

Dimensions:
(baseline) 101.6 × 152.4 × 330.2 mm + 152.4 mm dome
(alternate) 101.6 × 152.4 × 406.4 mm
Weight: 9.09 kg
Detection range: beyond lethal range
Altitude: to 45,000 ft
False alarm rate: claimed to be insignificant (classified)
Coverage (per sensor head): ±135 × 360°
Cueing accuracy (AOA): better than 1°
Multiple threats: virtually unlimited
Warning time: classified
Reliability: over 1,000 h
Interfaces: MIL-STD-1553B, RS-422, RS-232
Power: 28 V DC, 1.1 A; 115 V AC, 400 Hz, 1.0 A

Operational status

Extensively proven in field tests; not yet known to be in operation.

Contractors

BAE Systems North America, Mason.
Raytheon Electronic Systems, Goleta.

VERIFIED

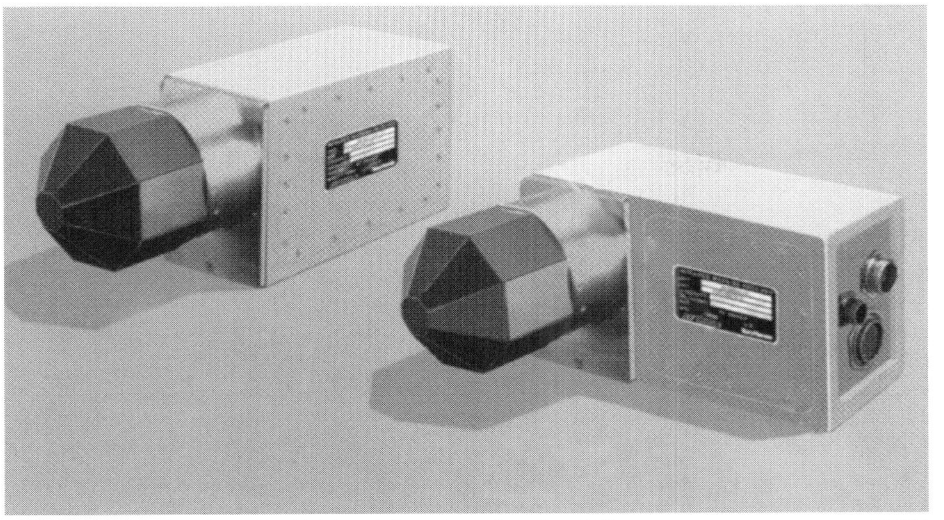

AAR-44(V) baseline configuration (right) and alternate configuration (left) 0002111

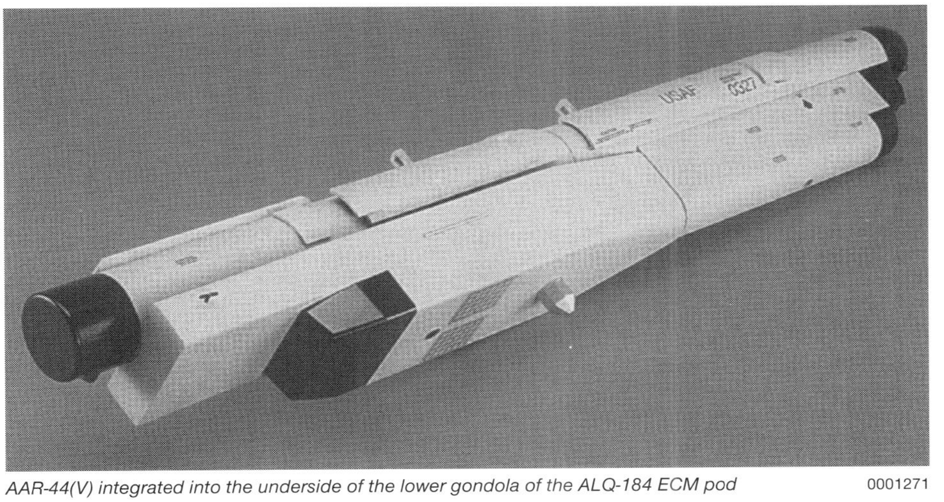

AAR-44(V) integrated into the underside of the lower gondola of the ALQ-184 ECM pod 0001271

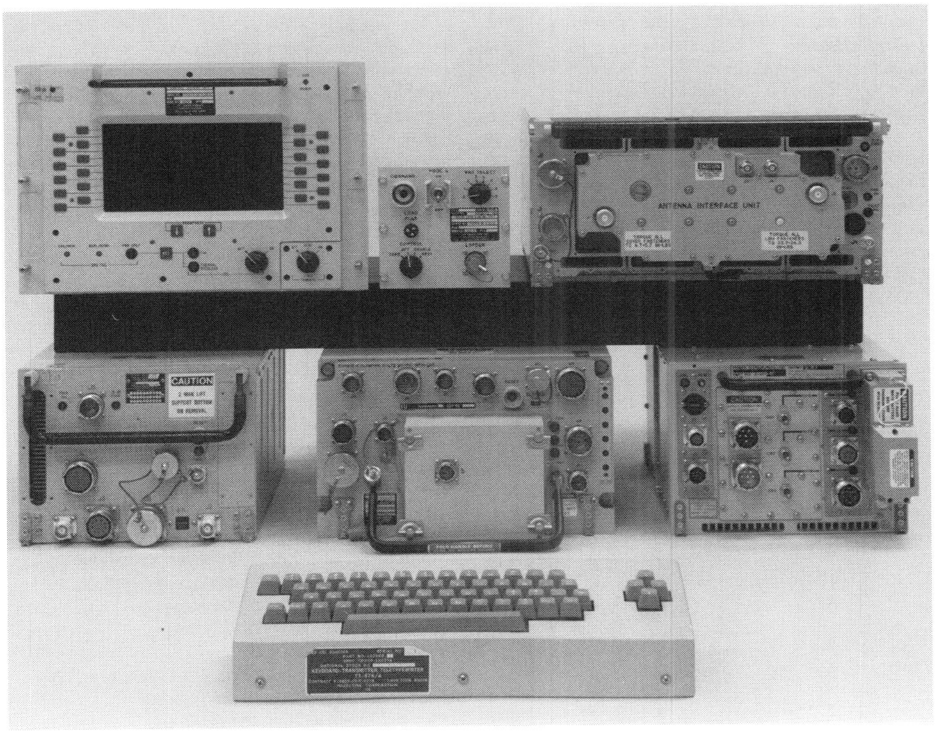

Class 2H JTIDS equipment

Joint Tactical Information Distribution System (JTIDS)

The Joint Tactical Information Distribution System (JTIDS) is a US joint service command and control support system providing secure jam-resistant communications and embedded navigation and identification for land, sea and air platforms. Using Time Division Multiple Access (TDMA) technology, it provides high-capacity networking among diverse airborne and surface users. It allows all stations to share an integrated awareness of the combat situation for friendly forces as well as detected threats in real time, using the tri-service multinational message catalogue TADIL-J. It can thus provide a language-independent method to ensure co-ordinated operations on a multinational battlefield. The data received can be displayed in both symbology and language of the host nation's platform. Thus, data received by a UK system would use English as the language for a given message; an Italian system could display the same message in Italian.

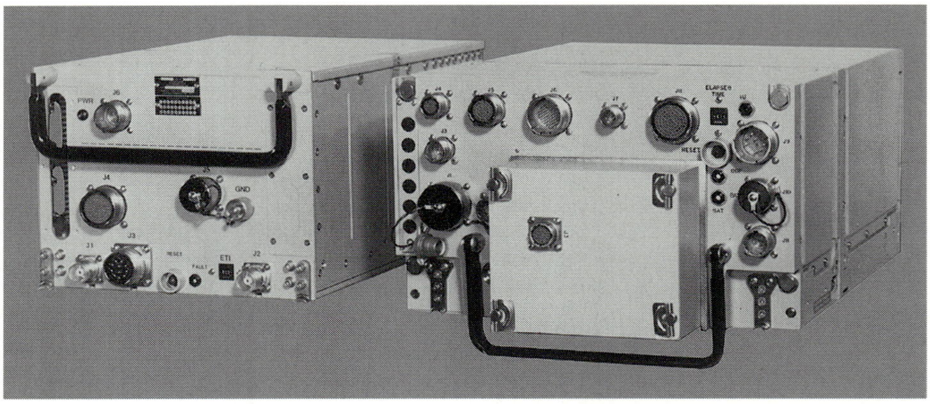

The Class 2 JTIDS terminal for the US Air Force F-15 with the receiver/transmitter (left) and the data processor group (right)

JTIDS operates on 51 frequencies in the 960 to 1,215 MHz frequency band, sharing with Tacan but strictly avoiding the IFF transponder frequencies which are also in the band. The JTIDS TDMA scheme breaks up time into 7.8125 ms time slots and allocates these slots to users, based on projected traffic demand. Users employ their time slots to transmit while listening during all other times, thus enabling relays of opportunity, to maximise the probability of message reception. Participants routinely inject information into the network through their regular broadcast slots without necessarily knowing who needs the information, using a broadcast-oriented architecture. These broadcast slots might include identification and location data, as well as host processor-generated traffic such as target acquisition tracks and platform status information on weapons and fuel status or equipment readiness.

JTIDS is designed to survive the highest levels of enemy radio electronic combat. The links are encrypted with the latest approved crypto devices. Jam resistance is achieved by multiple techniques through fast hopping over all frequencies, direct sequence spread spectrum of the waveform and Reed-Solomon forward error correction. The result of the signal processing gains from this combination of techniques means that the JTIDS omnidirectional radiation pattern can offer as much signal improvement as if it were being transmitted by a highly directional antenna (such as parabolic dish), without the difficulties of beam pointing or the limitations of single path links.

The initial emphasis in the development of JTIDS hardware was on Class 1 terminals developed from 1974. First deliveries took place in 1977 and resulted in terminals on board the Boeing E-3A AWACS aircraft, in ground terminals in the UK Royal Air Force IUKADGE system and in transportable shelters designated as Adaptive Surface Interface Terminals (ASIT) used by the US Army and US Air Force ground control facilities. The terminal is also used in the NATO Air Defense Ground Environment. The Class 1 terminal comprises rack-mounted components occupying almost 0.26 m³ and weighing approximately 192 kg.

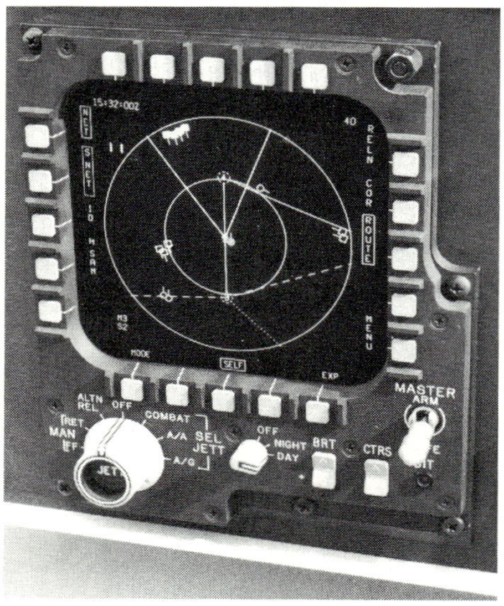

JTIDS information shown on the colour display of an F-15. The aircraft is shown at the centre of the display, concentric circles indicate distance in nautical miles. Round symbols indicate friendly aircraft, rectangles are unknowns and triangles are enemy aircraft

Operational status

The operational fielding of JTIDS began, primarily with navy and army (FAAD) platforms. Multiservice demonstrations, notably the All-Services Combat Identification Evaluation Team (ASCIET), exercises, and combat operations demonstrated the capabilities of Link 16 systems such as JTIDS.

Exercises and combat experience have proved interoperability among US, UK and French JTIDS equipment. The Japanese AWACS aircraft are reported as being equipped with JTIDS capability.

JTIDS Class 2 terminal

The Class 2 terminal, for fighter aircraft installation, began its development early in 1980. It was designed by Singer (now BAE Systems North America) in a leader-follower arrangement, with Rockwell Collins as the subcontractor providing RF design expertise.

The Class 2 terminal, designated by the USA as the AN/URC-107(V)1, AN/URC-107(V)6, AN/URC-107(V)8, and AN/URC-107(V)10, provides the basic JTIDS functions of high-capacity data communications, embedded TACAN functionality, dual-channel/dual-mode integral secure voice, dual-grid relative and geodetic navigation, GPS interface, PPLI self-identification messages, crypto-secure and jam-resistant connectivity and system status monitoring and reporting.

Since JTIDS operates in the TACAN and IFF frequency band, extensive interference protection circuits are built into the equipment to prevent inadvertent interference with those systems. Circuits continuously monitor the JTIDS radio operation and can shut down transmission in the event of any out of specification emission. During peacetime activities, the terminal operation maintains restricted duty cycles, further limiting the potential for interference with civilian facilities. The terminal incorporates SRU/LRU BIT hardware and software, achieving 98 per cent fault detection and 95 per cent fault isolation.

The Class 2 terminal comprises two boxes occupying approximately 0.045 m³ and weighing 57 kg. One box houses the receiver/transmitter circuits; the other is a dual LRU component. The dual LRUs are the digital processor section, standard for all terminals and applications, and the interface unit which is customised for each host platform. The digital processor section carries out all signal processing and digital computing functions, as well as advanced position location and tactical air navigation. The key output of the system is the display of information passed over the JTIDS network, allowing network users to share a common situational awareness. In the fighter terminal, the JTIDS interface feeds the multifunction display, which is capable of selective control by the pilot and automatic display under instructions in preloaded mission files. Selected platforms can receive precise direction from a central control point or can support local decision making based on the shared awareness. In other installations, the JTIDS data can be displayed on host processor displays, providing real-time depiction of the combat situation.

Specifications

Dimensions:
(data processing group) 324 × 193 × 485 mm
(receiver/transmitter) 257 × 193 × 395 mm
Weight:
(data processing group) 36 kg
(receiver/transmitter) 25.8 kg
Power supply: 120/208 V AC, 50/60/400 Hz or 240/280 V DC, 1,400 W
Power output:
(TDMA) 200 W
(Tacan) 500 W
Data rate: 238 kbits/s
Range: (normal) 557 km, (available) 928 km

Operational status

The final Class 2 production contract is expected to continue until at least the year 2001.

Terminals are currently being integrated into US Air Force aircraft: F-15, Rivet Joint, Joint STARS, and Airborne Battlefield Command/Control Centers (ABCCC) and the ground-based Modular Control Equipment (MCE). They are also being fitted into the US Navy's F-14 aircraft and the US Marine Corps' Air Defense Command Post (ADCP). They are also operational in many UK Tornado aircraft. Successful testing of Class 2 terminals on US Navy terminals on US Navy submarines has shown the usefulness of JTIDS for the entire fleet.

JTIDS Class 2H terminal

The Class 2H terminal is derived from the Class 2 terminal, to replace the Class 1 terminal. It occupies 0.15 m³ and weighs 155 kg, while giving the additional capabilities of increased throughput (115 kbits/s v 28.8 kbits/s), relative navigation and TACAN functionality, interoperability with TADIL-J message protocols and increased system functionality such as an increased quantity of crypto variables and over-the-air initialisation and rekey.

The Class 2H terminal is the high-power 1,000 W output transceiver version of the basic Class 2 terminal. It currently has three configurations. The first (AN/URC-107(V)4 and AN/URC-107(V)5) is an airborne system with receiver/transmitter, data processor group, high-power amplifier/antenna interface group, control monitor set and power conditioner and crypto loading devices. The other two configurations are for land-based use (AN/URC-107(V)9) and for surface ships (AN/URC-107(V)7). The Class 2H terminal possesses all the functional characteristics of the basic Class 2, and it is totally interoperable.

Specifications

Volume: 0.15 m³
Weight: 155 kg
Power supply: 120/208 V AC, 50/60/400 Hz, 3 phase
Power output:
(TDMA high power) 950 W
(TDMA low power) 235 W
(Tacan) 500 W
Data rate: up to 238 kbits/s
Range: (normal) 557 km, (available) 928 km
Message types: TADIL-J or interim JTIDS message standard
Interface: MIL-STD-1553B

Operational status

Class 2H terminals are currently being integrated into US Naval surface ships and E-2C Hawkeye aircraft, US Marine Corps Tactical Air Operations Modules (TAOM), and into upgrades to US Air Force E-3 AWACS aircraft. They are also fitted to UK and French AWACS aircraft.

JTIDS Class 2M terminals have been ordered by the US Air Force for JTIDS operations.

JTIDS Class 2M terminal, designated AN/GSQ-240, is a single-box variant of the basic JTIDS terminal, designed for land-based operations.

Contractors

BAE Systems North America, Wayne.
Rockwell Collins.

VERIFIED

MIDS-FDL Multifunctional Information and Distribution System - Fighter DataLink

BAE Systems North America and Rockwell Collins have entered into a Limited Liability Company (LLC) to qualify, build and sell Link 16 Multifunctional Information Distribution System (MIDS) Fighter DataLink (FDL) terminals for the US Air Force F-15 aircraft. Features include: full Link 16 interoperability -TADIL-J and IJMS; 50 W transmit power and >1 W LPI mode; dual antenna transmit and receive to provide the following capabilities: position, location and identification messages; relative and geodetic navigation; adaptable host interface processing.

Specifications

Performance characteristics:
Pseudo random frequency hopping
51 frequencies - 969 to 1,206 MHz
Frequency hop rate - 13us
238 kbps Data Rate/expandable to 2 Mbps
Jam resistant and crypto secure
128 possible nets
Dimensions: ¾ ATR form factor
Weight: Maximum 20 kg (option dependant)
Power supply: 115 V AC, 400 Hz, 350 W average

Operational status

Contracts have been awarded by the US Air Force to integrate the MIDS-FDL onto F-15 C/D and F-15E aircraft.

Contractors

BAE Systems North America, Wayne.
Rockwell Collins.

VERIFIED

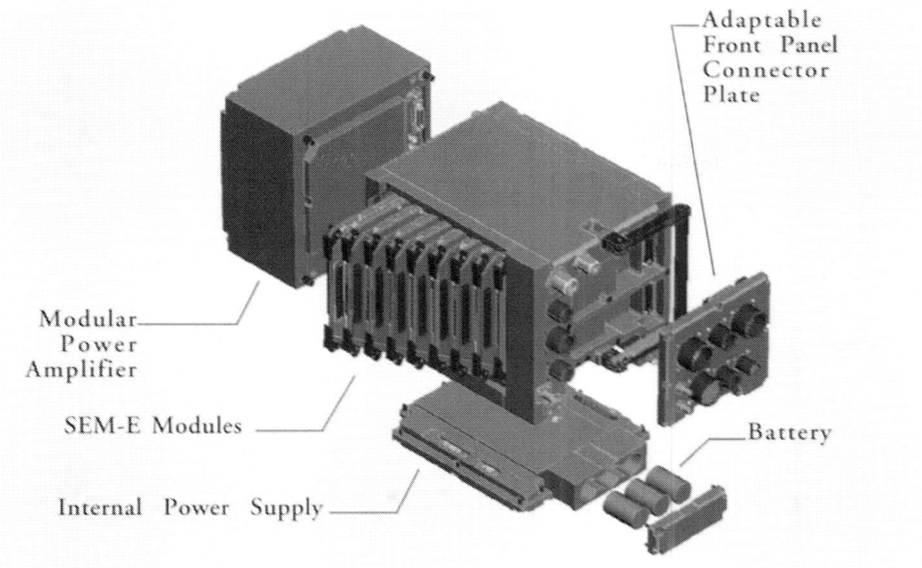

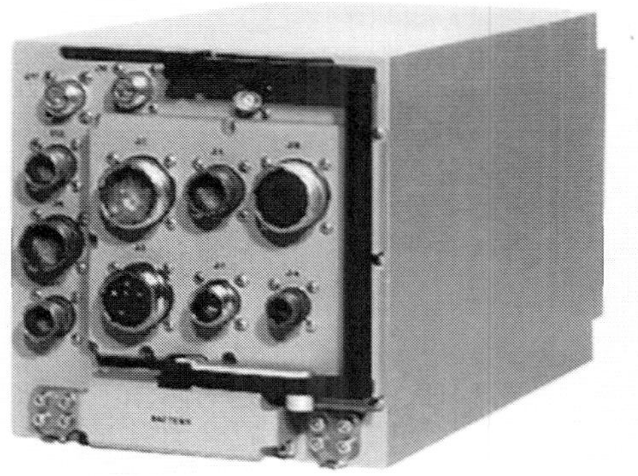

MIDS-FDL Fighter DataLink
0011842

AN/ARN-154(V) Tacan

The AN/ARN-154(V) lightweight airborne Tacan is a remotely controlled multistation tracking system utilising large-scale integrated circuits and CMOS technology. The system consists of the RT-1634 receiver/transmitter, ID-2472 indicator unit, MT-6734 Tacan control unit mounting base and antenna.

The AN/ARN-154(V) has the capability to track up to four ground stations simultaneously in range and two in bearing. Tracking velocity is up to 1,800 kt. It incorporates a sine/cosine bearing output, along with an ARINC 547 CDI interface and ARINC 547/579 low- and high-level flags. Bearing information is pilot selectable from either of the two tracking channels. The system has both X and Y mode MIL-STD-291B air-to-air ranging, plus antenna switching to allow the aircraft to be configured for dual antennas.

The AN/ARN-154(V) provides an ARINC 568 digital range output from either of the two tracking channels for display on remote EFIS or HSI displays.

The RT-1643 is offered in a number of configurations to interface to most navigation flight management systems. The AN/ARN-154(V) can be used as a pilot-controlled positioning system or as a blind navigation sensor controlled by one or more long-range navigation systems.

An ARINC 429 version is available as an option. This version will output range on three separate channels and bearing on two. The input tuning is standard ARINC 429 bus. This unit is ideal for updating VLF/Omega or inertial navigation systems in helicopters and fixed-wing aircraft, or as a stand-alone Tacan system.

The AN/ARN-154(V) system can utilise an ID-2472 indicator unit which displays distance, Tacan radial or bearing, decoded station ident, groundspeed and time to station. Either or both Tacan stations may be displayed simultaneously. A second ID-2472 can be installed to provide independent cockpit instrumentation.

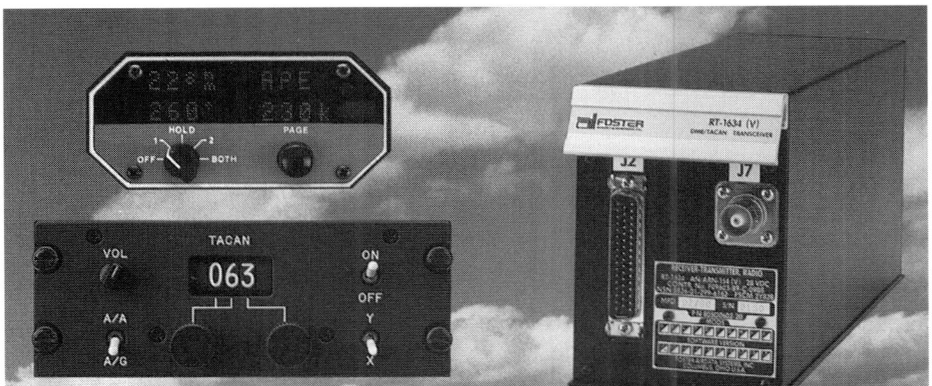

The AN/ARN-154(V) Tacan showing (top left) the ID-2472 display unit, (bottom left) a control unit and (right) the RT-1634 receiver/transmitter

Specifications

Dimensions:
(display unit) 3ATI × 175.3 mm
(receiver/transmitter) 290.1 × 87.4 × 127.8 mm
(control units) 50.8-56.9 × 71.1-146 mm
Weight:
(display unit) 0.48 kg
(receiver/transmitter) 2.81 kg
(mounting base) 0.2 kg
(control units) 0.57-0.91 kg
Power supply: 18-32 V DC,1.25 A (max)
Frequency:
(receive) 962-1,213 MHz
(transmit) 1,025-1,150 MHz
Range: up to 400 n miles
Accuracy:
(range) ±0.1 n mile
(bearing) ±0.5°
Temperature range: −54 to +71°C
Altitude: up to 70,000 ft

Contractor

BFGoodrich Aerospace Avionics Systems.

VERIFIED

Digital Fuel Quantity System upgrade FQS/(DFQS) for P-3 aircraft

The technology utilised in this capacitance-type DFQS offers P-3 users performance benefits over the original equipment system. Solid-state, micro-processor-based DFQS architecture incorporates robust Built-In-Test (BIT) and pre-calibrated components which guarantee system mean time between failure rates of 5,000 h minimum. The DFQS consists of six flight station indicators, five repeater indicators, 23 DC full-height compensated fuel tank units, four attitude compensation units, one low-level warning control unit, and all associated wiring. The DFQS provides accuracy which meets MIL-G-26988C Class II, and will operate with a range of fuel including JP-4, JP-5, JP-8 and NATO F-35.

Operational status

Selected by the US Navy in 1994 for the P-3C sustained readiness programme.

Contractor
BFGoodrich Aerospace Aircraft Integrated Systems.

VERIFIED

Fire/Overheat Detection System (F/ODS)

The Fire and Overheat Detection System (F/ODS) provides continuous fire and overheat detection in engine nacelles and the auxiliary power unit compartment. The system also detects hot air leakage from engine bleed air ducting and identifies the location of the overheat. On detection of an overheat, the system sends commands to the bleed air isolation valves. The F/ODS reports this information to the mission computer via the MIL-STD-1553 databus.

The system consists of a dual channel controller that monitors 44 sensor loops and operates as a master and slave with cross-channel fault checking to achieve high-integrity fire detection.

Based on this technology, a system can be designed and customised to meet the fire detection needs of most aircraft. The system may also be incorporated into an integrated utility system where several aircraft subsystems are combined in a single package.

Specifications
Dimensions: 318 × 191 × 194 mm, ARINC 404 ¾ ATR Short
Weight: 6.9 kg

Contractor
BFGoodrich Aerospace Aircraft Integrated Systems.

VERIFIED

Integrated Utilities Management System

The Integrated Utilities Management System (IUMS) is a system that allows for the functional and physical integration of utilities subsystems, and the effective sharing of resources. What traditionally required 20 to 30 stand-alone black boxes can now be configured in a flexible, lightweight, small network of enclosures, each sharing common electronics, processors and power supplies. For logistics support, the departure from traditional subsystems will result in greater system fault isolation, allow a two-level maintenance structure and reduce maintenance time and cost.

Operational status
The system is being provided for the F-22 aircraft.

Contractor
BFGoodrich Aerospace Aircraft Integrated Systems.

VERIFIED

Vibration, Structural Life and Engine Diagnostic system (VSLED)

The Vibration, Structural Life and Engine Diagnostic System (VSLED) is a health monitoring and diagnostic system designed for the V-22 engines and transmission units.

The airborne unit performs data acquisition and signal processing in real time on the aircraft and also stores data for later analysis on the ground. This system may be adapted for application on other aerospace platforms.

The design of the VSLED airborne unit is modular in both hardware and software so that the system can be easily modified for a different interface complement.

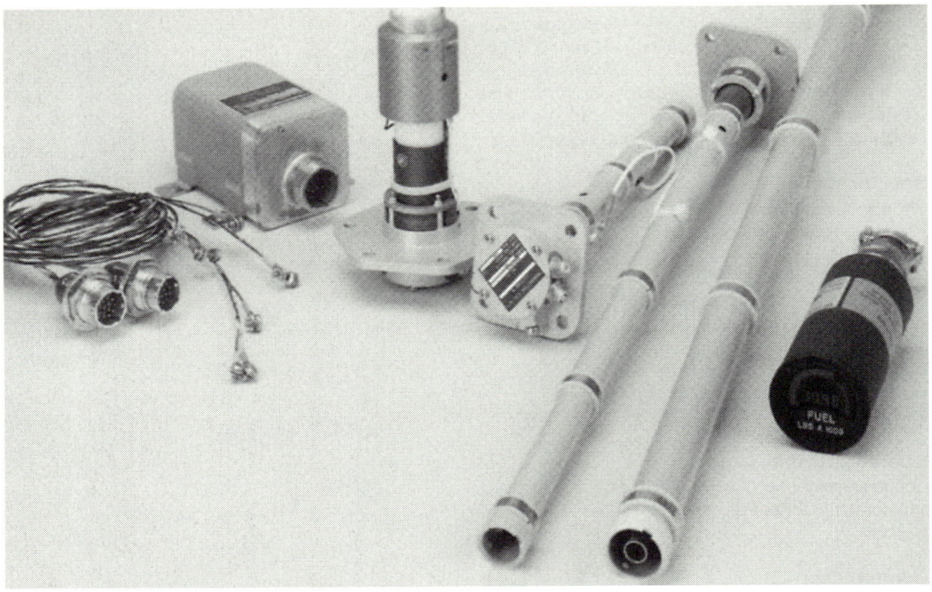

Digital fuel quantity system for P-3 aircraft upgrade 0001431

The BFGoodrich Integrated Utilities Management System is being fitted to the F-22

Specifications
Dimensions: 229 × 203 × 203 mm
Weight: 9.1 kg

Contractor
BFGoodrich Aerospace Aircraft Integrated Systems.

VERIFIED

ACTIVE - Advanced Control Technology for Integrated Vehicles

ACTIVE - Advanced Control Technology for Integrated Vehicles is a joint NASA, US Air Force, Boeing and Pratt & Whitney programme. ACTIVE is assessing the use of engine thrust vectoring for improving the range, fuel economy, manoeuvrability and survivability of advanced aircraft.

The test results from the 60 planned ACTIVE flights over the next few years will provide useful information for improving the performance of a variety of advanced civil and military aircraft.

The use of thrust vectoring for pitch and yaw control reduced the need for aerosurface controls, which add drag and weight to an aircraft and work well only within limited flight regimes. Future ACTIVE flight tests could include testing the F-15's manoeuvrability with some, or all, of its vertical tail area removed.

The F-15 demonstrator is equipped with Pratt & Whitney axisymmetric thrust vectoring nozzles attached to two high performance F100-PW-229 engines. As prime contractor for the ACTIVE programme, The Boeing Company is responsible for integrating the nozzles and engines with the F-15 demonstrator.

Contractor
The Boeing Company.

VERIFIED

F-15 Precision Direction-Finding (PDF) system for SEAD missions

The F-15 PDF is designed to detect and accurately locate enemy radars and other RF transmitters. The system consists of a receiver and digital processor that is connected to conformal antenna arrays developed by Boeing. The system was jointly developed with Litton Advanced Systems Division and TRW.

For demonstration, the PDF receiver processor has been installed in the F-15's 20 mm cannon ammunition bay. The conformal antennas are positioned around the nose of the aircraft, aft of the radome.

By providing aircrews with precise targeting, situation awareness and threat information, the PDF will add a key sensor to the existing suite of integrated avionics.

When coupled with the JTIDS/Link-16 system, the PDF could be used to help direct attacks against SAM sites and other defence weapon systems, or to warn other aircraft of their presence.

Operational status
Boeing began a series of flight tests in August 1996 to demonstrate the use of a new Precision Direction-Finding (PDF) system on the F-15 Eagle for conducting air defence suppression missions.

Contractors
The Boeing Company.
Litton Advanced Systems Division.
TRW.

VERIFIED

Reuseable navigation software module

Boeing demonstrated its new reuseable navigation software module in an F-15 Eagle equipped with a commercial MIPS R4400 SC processor. The single processor card exceeds the processing power of the IO cards in the computer. The same navigation software module was successfully flown in an AV-8B Harrier II equipped with a commercial power PC processor. The reuseable software has also been trialled in an F/A-18 again equipped with a power PC processor.

The reuseable software project is part of a Boeing Company affordability initiative to develop aircraft with more flexible, open architectures. The approach is based on a modular object-orientated design, as well as the commercial POSIX standard and hardware independent high-order languages. This can cut the costs of software development and maintenance in half. The use of commercial processors can also significantly reduce recurring and non-recurring hardware costs.

Contractor
The Boeing Company.

VERIFIED

E-3 Airborne Warning And Control System (AWACS)

See the entry for AN/APY-2 radar for the E-3 AWACS under Northrop Grumman in this section for a full description of the radar system.

Contractor
The Boeing Company, Boeing Information, Space & Defense Systems.

VERIFIED

F-15 ACTIVE test aircraft 0001444

Reusable navigation software module demonstrated in an F-15 Eagle equipped with a MIPS R4400 SC processor 0001339

The Boeing Company E-3 Airborne Warning And Control System **1999**/0044983

AN/ALQ-167(V) Yellow Veil Jamming Pod

The AN/ALQ-167(V) Yellow Veil dual-band airborne jammer provides jamming against surface and airborne sensors. Antenna coverage is both forward and aft. The pod is suitable for both helicopters and fixed-wing aircraft and can be used for tactical self-protection or standoff escort jamming roles or for training.

Cockpit selectable frequency and jamming mode provisions allow manual updates in flight or the jammer can be interfaced with ESM equipment. A broad range of jamming modes is available in the standard configuration and special jamming requirements can be accommodated.

Pod-mounting is the standard configuration, but the electronics can also be configured for internal carriage or for drone or RPV requirements.

Specifications

Dimensions: 3,700 (length) × 23 mm (diameter)
Weight: 100-150 kg
Output power: 100-200 W, ERP 1 kW (min)
Frequency: 0.1-18 GHz in customer specified bands

Operational status

Yellow Veil is pod-mounted on UK Royal Navy Lynx and Sea King helicopters.

Contractor

Condor Systems Inc.

VERIFIED

AN/ALR-81(V)1 and ALR-81(V)3 ESM/ELINT systems

The AN/ALR-81(V)1 is a fully synthesised microprocessor-controlled receiving system designed for general search, collection, signal analysis and recording. It consists of two units: a tuner and a control/display/demodulator. The tuner is a fully synthesised microwave tuner able to cover the 0.5 to 40 GHz frequency range. It features low-phase noise combined with instantaneous IF bandwidth. The second unit includes a multitrace digitally refreshed spectrum/IF pan display, plus a complete analysis demodulator. It is a single compact unit and includes MIL-STD-1553B remote-control interface.

The AN/ALR-81(V)1 employs high-resolution, digitally refreshed RF and IF pan displays. Spectrum information is provided over the entire frequency display in the form of combined multitrace band scan and IF pan displays with alphanumeric annotation. In the manual mode of operation, a split screen display is provided to allow the operator an overview of the frequency band and a high-resolution IF pan output.

The AN/ALR-81(V)3 is an integrated ELINT system. It features a broadband spinning DF antenna, multiple low-phase noise, wide instantaneous bandwidth superheterodyne receivers and an operator control and display station. The system includes a millimetre-wave antenna and an integral down converter, packaged on the DF antenna.

The operator control unit allows control of both antenna and receiver parameters. It features a high-resolution vector graphics display with alphanumerics. Additional features include multitrace, digitally refreshed spectrum/IF pan displays, multimode DF displays and a complete analysis demodulator. The millimetre-wave downconverter features automatic threshold detection. This allows the tuner to scan bands of interest and then switch to millimetre-wave bands when activity is detected. The AN/ALR-81(V)3 receiver/DF system can be remotely controlled via a MIL-STD-1553B interface bus.

Operational status

In production and in service.

Contractor

Condor Systems Inc.

VERIFIED

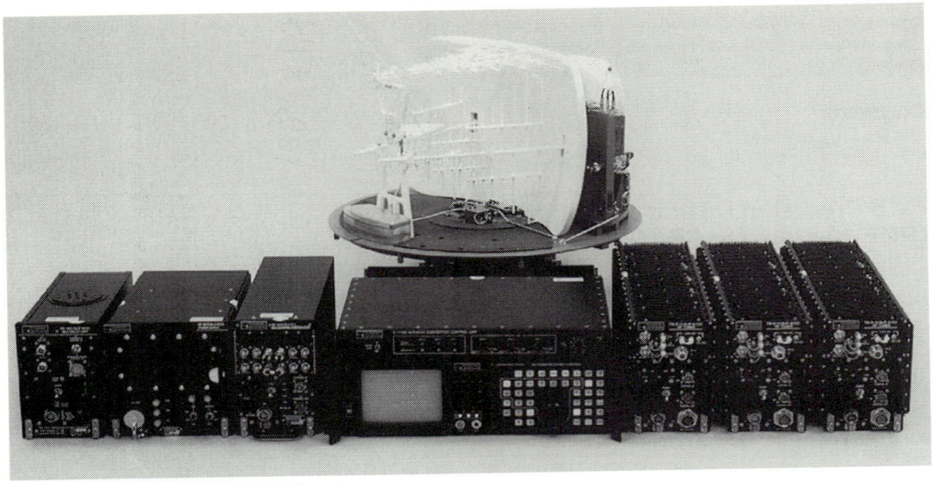

The Condor AN/ALR-81(V)3 ELINT system

The Condor ALR-801 ESM/ELINT system

AN/ALR-801 ESM/ELINT system

The ALR-801 is a multipurpose maritime patrol ESM/ELINT system for airborne applications, even in smaller aircraft. It features NDI subsystems and can receive over the frequency range 0.1 to 18 GHz and provide automatic DF from 0.4 to 18 GHz. Both the receive and auto-DF frequency range can be extended to 40 GHz with NDI hardware.

Using a high-gain shaped beam-spinning DF antenna and associated pedestal control unit, a preamplifier, TN-618 superheterodyne tuner (also noted as using IFM receiver technology) and the SP-103 signal processor, the ALR-801 searches the environment, de-interleaves multiple pulse trains, identifies emitters using a user-supplied library and automatically performs direction-finding on each signal.

The combination of a high-gain antenna, tuner with 500 MHz instantaneous bandwidth and the intelligent search strategies resident in the signal processor results in near 100 per cent probability of interception. Radar parameter measurements are made with extreme precision to eliminate ambiguities in signal identification.

The ALR-801 weighs less than 100 kg and consumes less than 1 kW of prime power. Growth potential includes interactive controls and displays to place a human operator in the loop, the CS-D-250 high-speed digitiser for broadband digital recording of radar signals and the Signal Processing Tools (SPT) software for post-flight analysis of signals, and digital receivers for FMCW and LPI radar detection.

Operational status

Designed for maritime patrol aircraft; available.

Contractor

Condor Systems Inc.

VERIFIED

AN/APR-46 receiving system

The AN/APR-46 receiving system, also known under the company designation WJ-1840, is a rapid scanning highly sensitive heterodyne receiver, able to detect threat warning radars at very long range. The receiver is also equipped for emitter bearing determination, as well as signal parameter extraction and identification. It can be interfaced to onboard avionics through a dual-redundant MIL-STD-1553B interface.

The intercept portion of the AN/APR-46 provides frequency coverage between 30 MHz and 18 GHz. Direction-finding capability is available between 0.5 and 18 GHz. Antenna installations containing the omnidirectional and DF antennas can be mounted on both the top and bottom of the aircraft. A single operator can select either or both and configure the receiver scan strategy to optimise coverage against expected threats. Fully automated search, identification and reporting functions are possible for stand-alone systems.

Threat identification is provided to assist the operator by analysing the radio frequency, pulse repetition frequency and pulsewidth of the intercepted signal. These values are then compared with parameters contained in an emitter library.

Specifications

Frequency: 0.5-18 GHz
Emitter PRF: CW to 100 kHz
Emitter pulsewidth: 200 ns to CW
Probability of intercept: 100% within bandwidth of receiver
Accuracy:
10° RMS (0.5-2 GHz)
5° RMS (2-18 GHz)

Operational status
In production for a US Air Force airborne application. Conversion to monopulse DF capability is in progress.

Contractor
Condor Systems Inc.

VERIFIED

AN/APR-46A and AN/APR-46A (V)1 ESM systems

Both the AN/APR-46A and AN/APR-46A(V)1 systems are designed for the panoramic threat avoidance role. They are fitted with highly sensitive and selective superheterodyne receivers covering the frequency band 30 MHz to 18 GHz, utilising one or two channels of microwave tuners.

Digital scan control with multimode operation and an eight-trace digitally refreshed pan display is implemented to provide band scan, sector scan, and manual modes. AN/APR-46A(V)1 adds monopulse direction-finding to the AN/APR-46A capability, with DF accuracy of 5° rms over the 0.5 to 18 GHz band.

Contractor
Condor Systems Inc.

VERIFIED

CS-2010 HAWK Receiver/DF system

The CS-2010 is a multiple operator, pooled resource ELINT/ESM system covering the frequency range 20 MHz to 40 GHz, designed for airborne, shipborne or land-based applications. The family of equipments includes microwave tuners, demodulators, VHF/UHF up converters, 18 to 40 GHz down converters and spinning DF antenna subsystems. System elements, or resources, are maintained by the system as a pool of resources and are shared by the operators. Operators acquire resources from the pool as required and release them back to the pool when no longer needed. Operators interface to the CS-2010 through a resources control unit, a keyboard and a pair of high-resolution raster scan RGB display units. The RCC provides the operator with control of the pooled resource elements and provides the operator with real-time displays of both antenna and receiver functional outputs.

The CS-2010 features a variety of operational modes, including common band scan, analysis and DF mode. Common band scan is a display synthesised from the RF scan data produced by all unused tuners in the system resource pool - that is, the display is produced by parallel scanning all tuners which have not been acquired by an operator for use in analysis of DF operations. The normal and expanded analysis displays provide the operator with RF, IF pan displays and LIFM for receiver channels acquired by the operator from the resource pool. The operator has control over five independent traces representing up to five independent digitally generated polar, inverse polar and rising raster rectilinear, and supports the manual DEF capabilities of the system.

The system is built around VersaNet, a time domain multiple access bus. VersaNet distributes control information between system elements and carries digitised display and audio information gathered in the system elements to the RCC. VersaNet enables additional resources such as tuners, demodulators, resource control units and pedestal electronic units to be added to the system as required. The principle tuner in the family is the TN-618 microwave tuner which features very low phase, 500 MHz wide IF and 10 kHz step size over 0.5 to 18 GHz. Block up converters and down converters extend the frequency coverage. The tuner also includes a built-in single bandwidth demodulator. An external full-function demodulator with eight IF bandwidths, IF pan and Log/Lin/FM video inputs is also available.

Operational status
In production.

Contractor
Condor Systems Inc.

VERIFIED

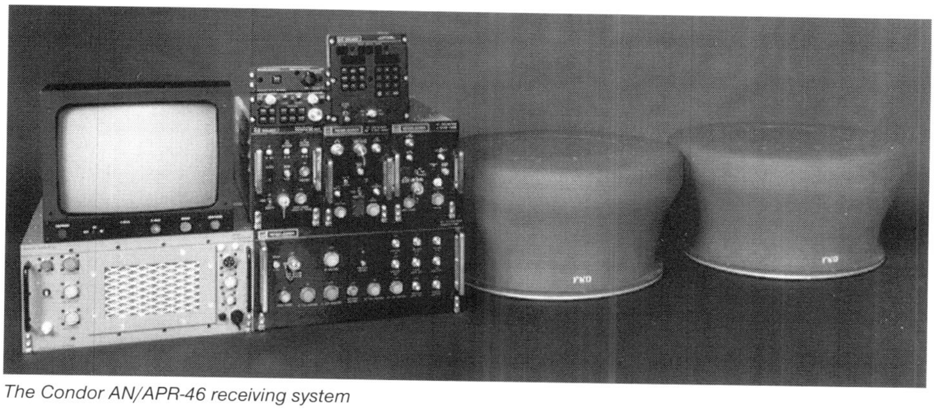

The Condor AN/APR-46 receiving system

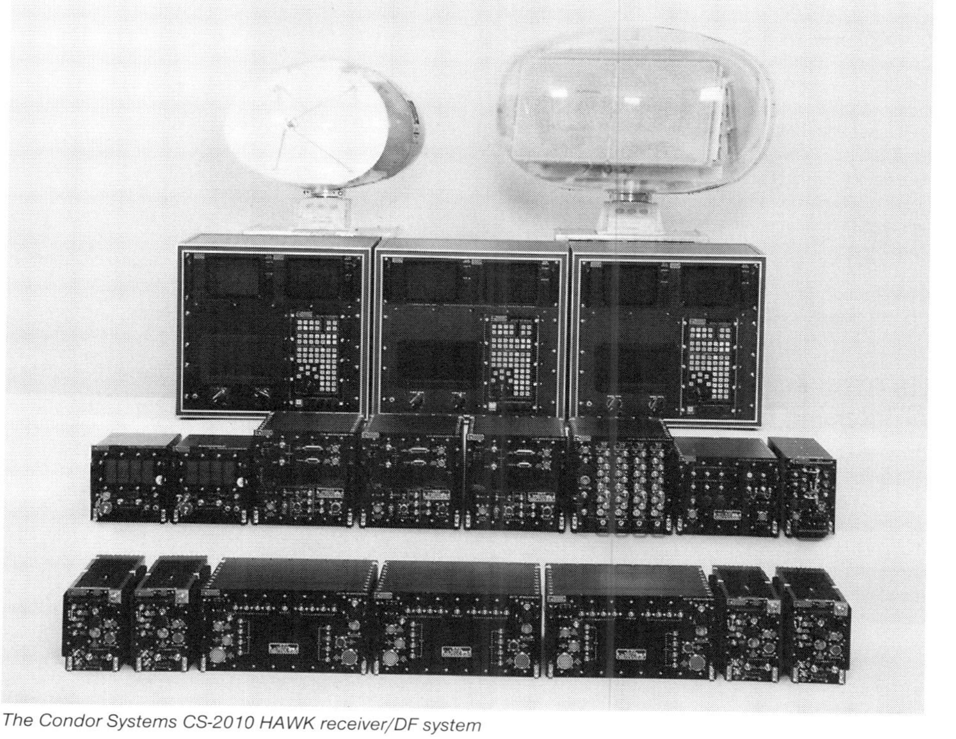

The Condor Systems CS-2010 HAWK receiver/DF system

CS-3360 lightweight ESM system

The CS-3360 is a complete ESM system with antennas, receivers, pulse processor and multifunctional display. It combines the functions of radar warning, tactical surveillance and ELINT collection. It is designed for installation on aircraft, ships and land vehicles, where space is limited but high performance is required. Emitters are detected, identified and displayed with direction of arrival.

The SP-2300 processor can process normal, stagger, jitter, multipulse, CW and complex radar signals. De-interleaving software is able to separate pulse trains in dense signal environments and measure the parameters of individual signals. Precision pulse interval measurements, with sub-nanosecond accuracy, are achieved with adaptive pulse tracking algorithms and an internal rubidium timing standard.

The SP-2300 can quickly identify signals using a built-in signal identification library with space for over

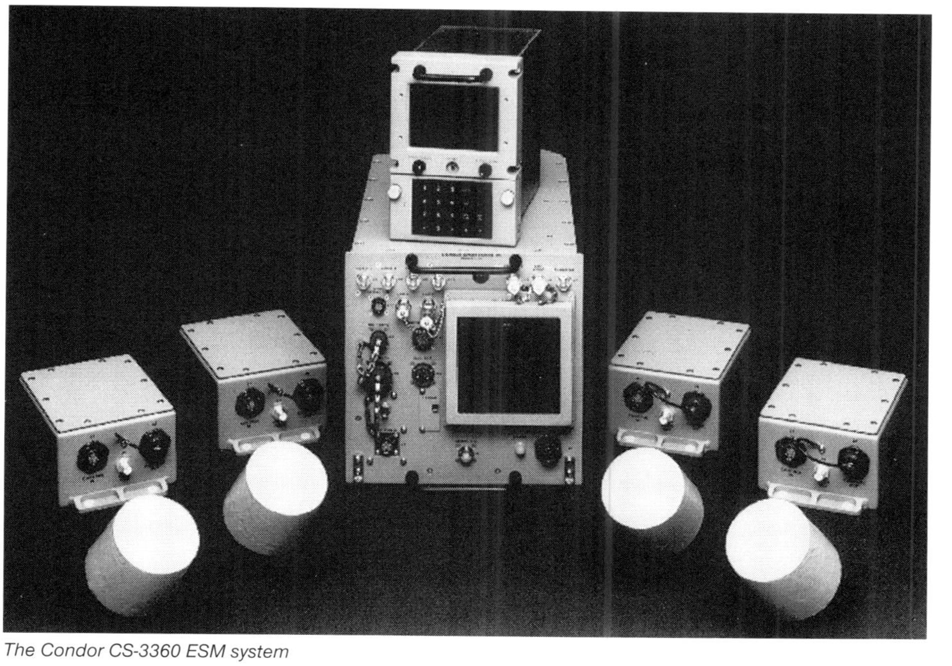

The Condor CS-3360 ESM system

2,000 emitter modes. An internal editor allows field updates to library entries. Each library entry allows for information on signal priority, high/low frequency, high/low PRF/PRI, high/low pulsewidth, library entry number, signal identification, scan type, scan period, illumination time, function of emitter, emitter mode for a set of parameters, up to two lines of text information, power in kW and antenna gain in dB.

Features of the CS-3360 also include 0.5 to 18 GHz frequency coverage, IFM receiver, high sensitivity and probability of intercept, high-speed pulse DF processing, instantaneous azimuth coverage, library loading from floppy disk and interfaces to time and navigation, external computer and SIGINT equipment. It weighs less than 30.4 kg.

Other systems in the series include: CS-3300 rapid deployment systems designed for applications requiring maximum capability, minimum weight and volume, and rapid set-up time. This includes the CS-3350 and CS-3360 systems. The latter is a lightweight, multiplatform equipment. Both systems operate over the 0.5 to 18 GHz band; CS-3500 tactical signal detection and classification systems; CS-3700 wideband ELINT intercept systems, latest version listed as CS-3701; CS-3900 remotely controlled ELINT receiver systems which can be used in remote fixed sites or unmanned aerial vehicles. They operate automatically or manually allowing remote control of frequency and pulse processing modes.

Contractor
Condor Systems Inc.

VERIFIED

CS-6700 ACES automatic processing ESM systems

CS-6700 ACES is a high-performance ESM/ELINT system for use on airborne, ground-mobile, shipborne or fixed-site installations. Frequency coverage is 0.5 to 18 GHz (expandable to 40 GHz). High-sensitivity, wide band superheterodyne and IFM receivers are implemented, together with automatic and manual radar signal analysis facilities. A Windows™-based HMI includes: scan control, library editing, and graphical signal analysis.

Contractor
Condor Systems Inc.

VERIFIED

CS-36500 Signal Intelligence Receiving System (SIRS)

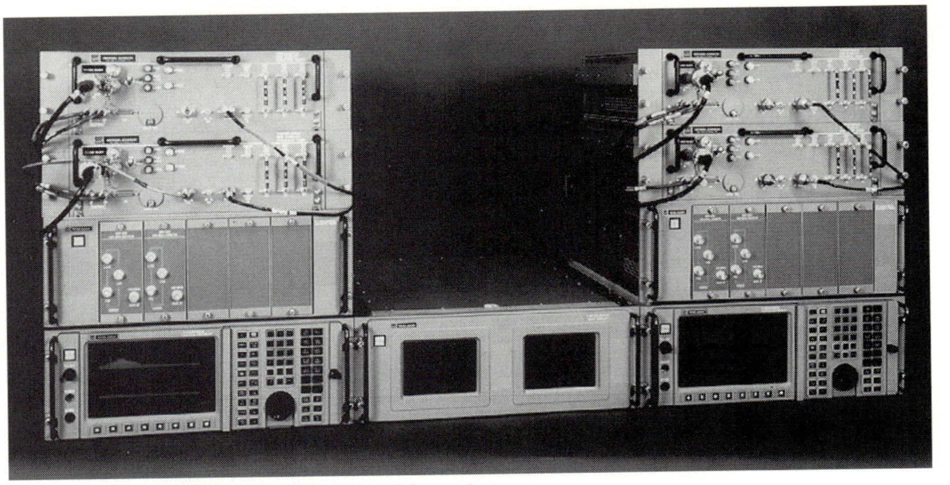

The Condor CS-36500 signal intelligence receiving system

The CS-36500 Signals Intelligence Receiving System (SIRS) is a 0.03 to 40 GHz ESM/ELINT/COMINT receiver which maximises current capability and future adaptability. It can be configured as a single/independent or as multiple/interactive operator positions. This modularly designed superheterodyne receiver consists of a C-100 control/display unit; an optional C-200 scan display; a CY-100 equipment frame with one or more demodulator plug-ins, depending upon desired IF bandwidth and video performance; and one or more octave or multi-octave tuners, or the FXT-1XX millimetre extensions. Narrowband and wideband (to 500 MHz) tuners and demodulators are available.

The C-100 controls SIRS operations; displays operating parameters, status, IF Pan and analysis mode video; and accepts configuration programming. This unit has internal removable mass memory and supports multiple standard interfaces.

The C-100 panel includes eight soft keys for maximum versatility. Controller panels are backlit and may be NVG-compatible. Parameter values are entered via three methods; numeric keypad, cursor increment (arrow keys) and slew knob (shaft encoder). Operation is user-friendly with automatic prompting. An optional external keyboard is available for touch-typist execution.

The C-100 ElectroLuminescent (EL) display provides operation, configuration, and diagnostic reports; BITE status; flexible RF, IF and time spectrum display with a selectable refresh/decay rate; as well as an AM/FM display mode for accurate measurement of broadband emitter's frequency excursions. SIRS may also include an audio alarm which alerts the operator to important but infrequent events such as detection of energy above 18 GHz.

The C-200 scan display is capable of RF panoramic display for up to eight additional tuners.

The SIRS controller design incorporates features which facilitate acquisition of difficult-to-detect emitters, such as those with a low duty cycle. For example, scan coverages and sweep rates may be optimised to enhance probability of intercept for specific signals. In addition, a scan priority feature allows the operator to choose the number of scan repetitions over each frequency sector relative to the other sectors. Mission profiles, called Receiver Instruction Sets (RISs), can be created online (or offline on a personal computer) and stored in the C-100. They include receiver control parameters such as mode, frequency scan limits, frequency markers, frequency lock-out sectors, priority, dwell times, attenuation, sweep rate, threshold levels, video select, IF pan settings, and IF bandwidths.

The CY-100 equipment frame, containing internal power supplies and video switching, accepts plug-ins including five demodulator families, various switching matrices, and custom units.

The CS-36500's architecture provides the user with the flexibility necessary to meet varying mission requirements and facilitate future system expansion. Its modular design permits adjustment of the receiver's frequency coverage and configuration. Single- and multiple-operator configurations with narrowband and/or wideband analysis ability are standard. Future expansion is accomplished by connecting additional units. The system executive will automatically recognise the additions. Controllers in multi-operator systems are interactive and may hand-off tasking to each other, or borrow currently unused assets.

The C-100 uses multiple microprocessors and incorporates MIL-STD-1553B digital bus architecture. The system can include any combination of octave and multi-octave-band tuners. Retrofits may utilise existing assets, either analogue or digitally tuned. The basic system is able to control 15 tuners and demodulators. Tuners offering special characteristics, such as phase-locked scan, are exploited by the system software.

Contractor
Condor Systems Inc.

VERIFIED

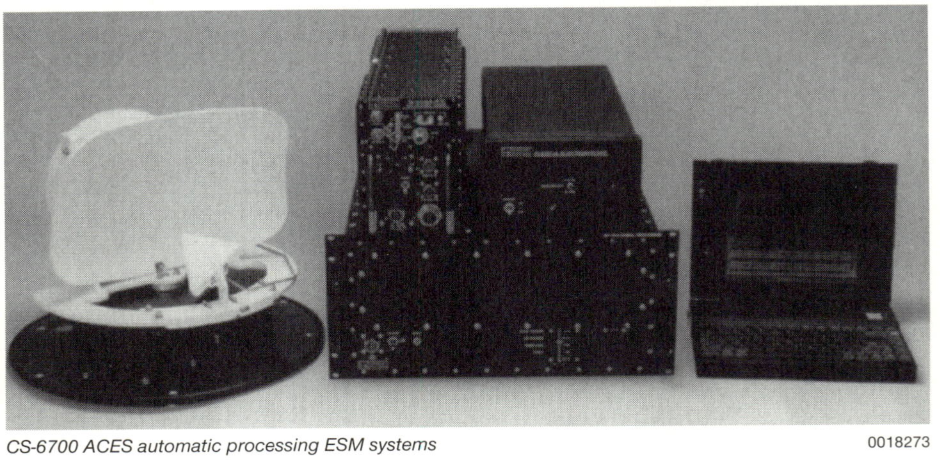

CS-6700 ACES automatic processing ESM systems 0018273

SP-2060 pulse processor

The SP-2060 pulse processor measures radar signal parameters, providing fast, accurate results. Operation is automatic, giving the operator a display of signal activity with identification of specific radar types. A continuous log of activity is stored in memory for post-mission analysis. Requiring only a single coaxial connection, it is compatible with any type of radar receiver. Two modes of control/display units are available. The EI-5100 control/display unit has a 9 in (178.6 mm) display and mounts in a standard 19 in rack. The EI-1400 control/display with a 5 in (127 mm) display is available for aircraft installations where space and weight are limited. Both models have built-in floppy disk drives.

The SP-2060 processes normal, stagger, jitter, CW and complex radar signals. De-interleaving software is able to separate pulse trains in dense signal environments and measure the parameters of individual signals. A high-speed digitiser and a maths co-processor allow processing of high pulse rate radars. Precision pulse interval measurements with sub-nanosecond accuracy are achieved with an internal rubidium timing standard and adaptive pulse tracking algorithms.

A time of arrival phase tracking algorithm is able to separate and measure two emitters with identical pulse repetition intervals. The algorithm remembers the time of arrival phase from scan to scan and does not require continuous signal reception.

The SP-2060 can quickly identify specific signal types using a built-in signal library with space for over 2,000 emitter modes. An internal text editor allows field updates to library entries. Entries are loaded from, and stored on, floppy disks.

The disk recording capability enhances the utility of the SP-2060. A single disk can store multiple libraries, enabling quick updates to match the local threat environment. Digitised pulse-by-pulse descriptors can be recorded for post-mission analysis. Recording of the intercept file on disk ensures a continuous log of mission signal activity.

Contractor
Condor Systems Inc.

VERIFIED

Wideband secure voice and data equipment

Condor Electronic Systems Division (previously Whittaker Electronic Systems) wideband secure communications equipment consists of the Voice Processor Unit (VPU) or Voice Processor Unit with Data mode (VPUD), Remote-Control Unit (RCU), Key Fill Device (KFD) and Key Transfer Device (KTD). The equipment is easily installed in aircraft, ground stations and ships and readily interfaces with existing UHF/VHF AM/FM military transceivers with X-mode capability. Modifications are available for operation with frequency-hopping radios.

Electrical and mechanical interfaces ensure direct replacement for US KY-58 crypto. The operator can erase cryptographic keys from memory, and an automatic alarm security feature continuously tests for proper secure operation. Key usage accountability is monitored by a non-resettable usage counter in the KFD. Key memory retention is provided during power-off or transient conditions. Resynchronisation occurs within half a second in late entry or temporary loss situations. A remote over-the-air rekey option is available and there is a retransmission capability for relay link operation.

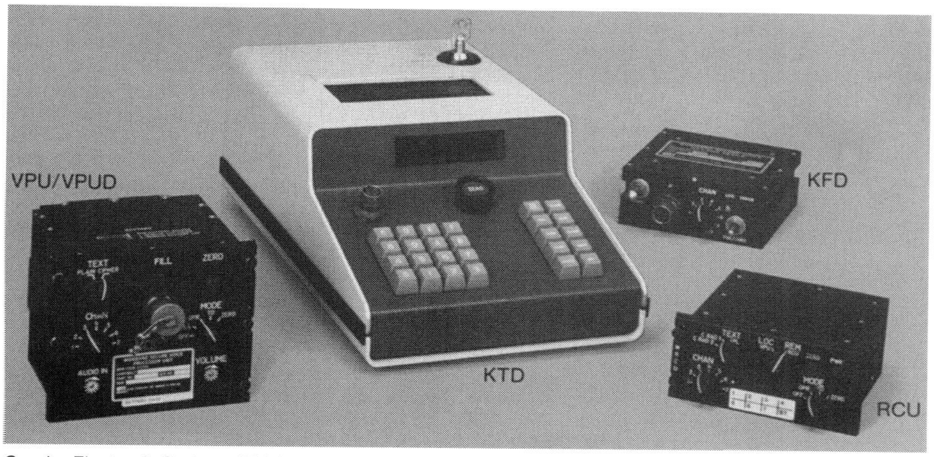

Condor Electronic Systems Division wideband secure communications equipment consists of (left to right) the VPU/VPUD, KTD, KFD and RCU

Specifications

Dimensions:
(VPU/VPUD) 146.1 × 124 × 121.9 mm
(RCU) 146.1 × 66.8 × 121.9 mm
(KFD) 129.5 × 58.4 × 149.9 mm
(KTD) 254 × 152.4 × 350.5 mm

Weight:
(VPU/VPUD) 2.36 kg
(RCU) 1.09 kg
(KFD) 1.18 kg
(KTD) 8.92 kg

Power supply:
(VPU/VPUD, RCU) DC per MIL-STD-704A
(KFD) battery self-power
(KTD) 115/230 V AC, 50-400 Hz

Temperature range:
(KTD) –40 to +55°C
(VPU/VPUD, RCU, KFD) –54 to +71°C

Altitude:
(KTD) up to 15,000 ft
(VPU/VPUD, RCU, KFD) up to 70,000 ft

Reliability:
(RCU) 60,000 h MTBF
(VPU/VPUD) 7,000 h MTBF
(KFD) 38,000 h MTBF
(KTD) 19,000 h MTBF

Contractor

Condor Electronic Systems Division.

VERIFIED

ASPJ AN/ALQ-165 Airborne Self-Protection Jammer

The AN/ALQ-165 Airborne Self-Protection Jammer (ASPJ) is designed for F-14D, F-16C, F-18C/D, A-6E and AV-8B aircraft and can be installed internally or in a pod. The equipment incorporates the latest technology in TWTs, microwave components and packaging. It can be electrically reprogrammed on the flight line and the BIT and modular plug-in design permits rapid replacement of assemblies at the operational level without external test equipment. Weight of the system is between 91 and 150 kg, depending on configuration; it occupies a space of 0.6443 m^3.

The AN/ALQ-165 has the ability to select automatically the best jamming techniques to use against any given threat, based on the jamming system's own computer data and real-time data of the threat signal from the receiver/processor. The computer software can be modified to accommodate new threats as they arise. The equipment covers the frequency range in two bands and is technically expansible to cover a greater frequency range if required.

The transmitters within the system can jam a large number of threats simultaneously over various ranges and in different modes. The computer selects the power and duty cycle criteria based on threat parameters detected and processed by an advanced receiver subsystem. An augmented version with an additional transmitter power booster is also available.

AN/ALQ-165(V)

Following renewed enthusiasm for the ASPJ system in the US, the US Naval Air Systems Command in December 1997 issued a US$45 million firm fixed price contract to procure: 36 AN/ALQ-165(V) ASPJ systems for US Navy F/A-18 C/D and F-14D aircraft; 15 AN/ALQ-165(V) ASPJ systems as US Navy spares; 10 AN/ALQ-165(V) ASPJ spares for the US Air Force; and data services support for the US Navy ASPJ programme. Contract fulfilment to be by December 1999. The nature of the upgrade represented by the addition of (V) to the nomenclature is not known.

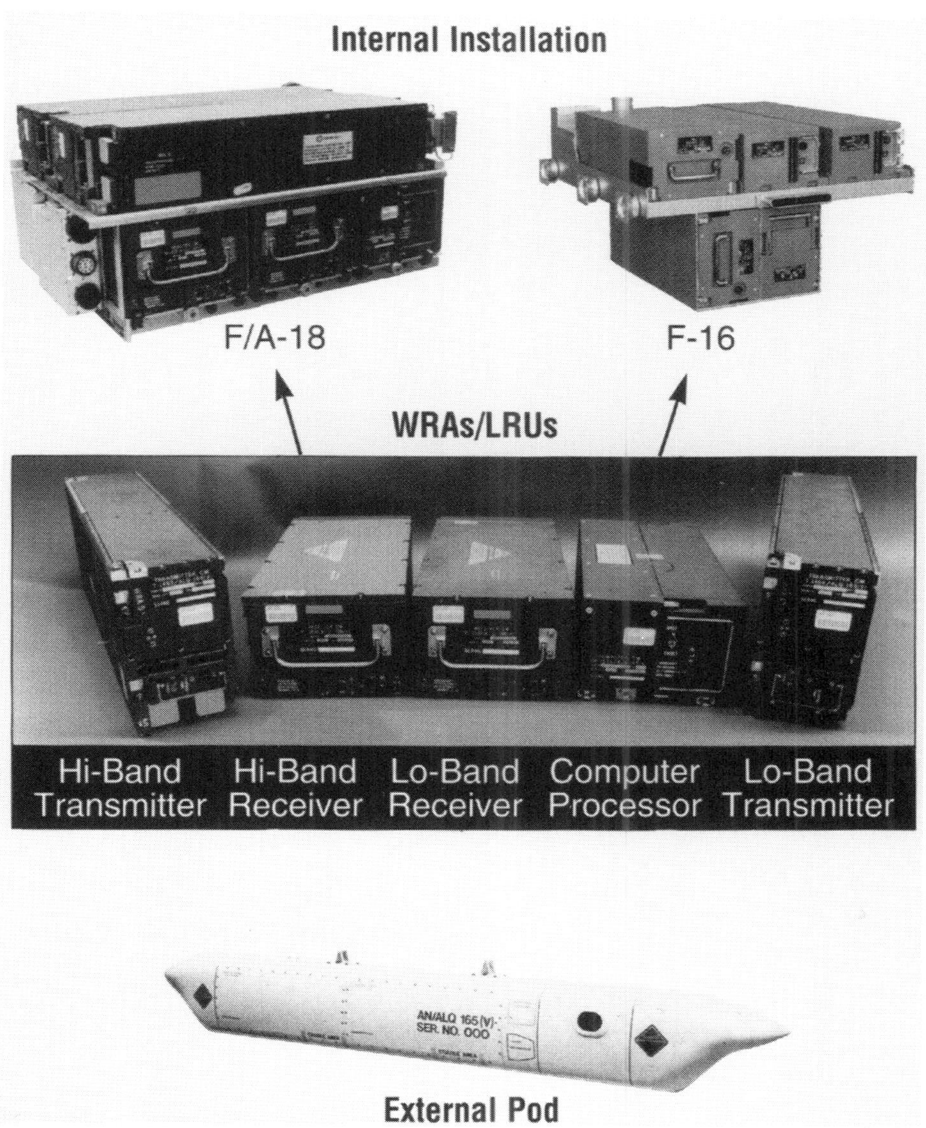

ASPJ AN/ALQ-165 pod installation (below) and internal fit configuration (above) 0018272

Specifications

Dimensions:
(processor) 141.7 × 199.4 × 403.9 mm
(low-band receiver) 141.7 × 199.4 × 400.6 mm
(high-band receiver) 141.7 × 199.4 × 400.6 mm
(low-band transmitter) 121.4 × 208 × 644.7 mm
(high-band transmitter) 121.4 × 208 × 644.7 mm

Weight:
(processor) 17.69 kg
(low-band receiver) 16.78 kg
(high-band receiver) 16.78 kg
(low-band transmitter) 30.84 kg
(high-band transmitter) 29.48 kg

Operational status

September 1994, procured by Finland for its F/A-18 aircraft.

December 1994, procured by Switzerland for its F/A-18 aircraft.

July 1995, deployed on US Navy and US Marine Corps F/A-18 C/D aircraft.

December 1996, procured by the Republic of Korea for its F-16 aircraft; first modified aircraft delivered to ASPJ standard in Feb 1999.

December 1997, new ASPJ systems procured by the US Navy (as above).

Contractor

Consolidated Electronics Countermeasures. (An ITT Industries, Avionics Division/Northrop Grumman Corporation, Electronic Sensors and Systems Sector joint venture.)

VERIFIED

ASPJ AN/ALQ-165 internal fit on F-14D
0018271

CDR-3580 VHF/UHF DSP receiver

The CDR-3580 is a VHF/UHF DSP (Digital Signal Processing) half-rack receiver; optional side-by-side or full 19 in rack versions are available.

Specifications

Frequency range: 20-1,200 MHz, 10 Hz tuning resolution
Detection modes: LSB, USB, CW, AM, FM
Sweep and scan: 100 channels per second
Channels: 250 programmable channels
IF bandwidths: 17 synthesised digital IF filters (1 kHz to 240 kHz)
Remote control: RS-232, RS-422 or IEEE-488
Dimensions: 19 in half-rack, 3.5 in high

Operational status

In production.

Contractor

Cubic Communications Inc.

UPDATED

Cubic Communications' CDR-3580 DSP receiver
1999/0044818

LCR-2000 and LCR-3000 series DSP receivers

The LCR (Low Cost Receiver) range of DSP (Digital Signal Processing) receivers comprises: the LCR-2000 single rack-unit (no display or speaker) and LCR-2400 three rack-unit (with display and speaker) LF-HF DSP receivers; the LCR-3000 single rack-unit (no display or speaker) and LCR-3400 three rack-unit (with display and speaker) V/UHF DSP receivers.

Specifications

LCR/SMR-2000 and LCR/SMR-2400
Frequency range: 10 kHz to 30 MHz, 1 Hz tuning resolution
Detection modes: LSB, USB, ISB, CW, AM, FM, PM, FSK
Sweep and scan: 100 channels per second
Remote control: RS-232 or RS-422
Dimensions: 19 in rack unit

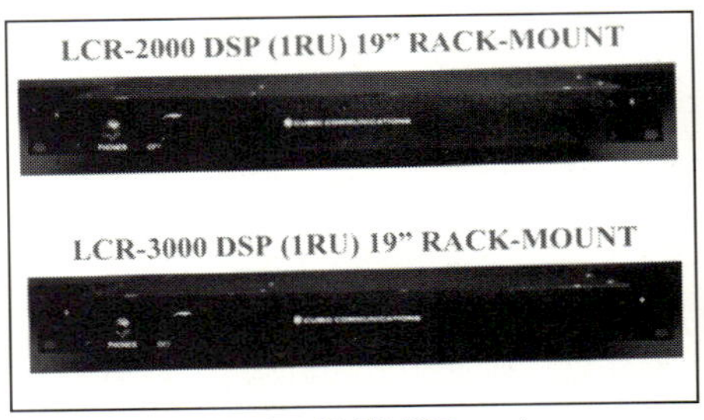

Cubic Communications' LCR DSP receivers
1999/0044819

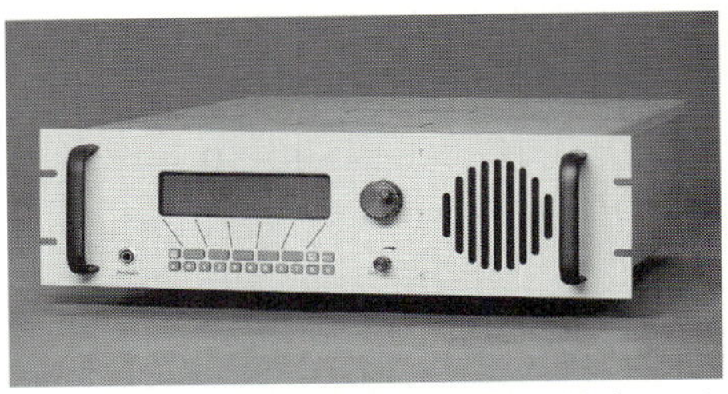

LCR-2400 and LCR-3400 DSP(3RU) 19 in rack-mounted units share similar enclosures ***2000***/0062849

LCR-3000 and LCR-3400
Frequency range: 20-3,000 MHz, 10 Hz tuning resolution
Detection modes: LSB, USB, CW, AM, FM, PM
Sweep and scan: 100 channels per second
Remote control: RS-232 or RS-422
Dimensions: 19 in rack unit

Operational status

In production.

Contractor

Cubic Communications Inc.

UPDATED

PA-5050A 1 kW power amplifier

The model PA-5050A solid-state amplifier features continuous-duty operation and is designed to operate with the T-4180 HF DSP exciter, capable of LSB, USB, ISB, AM, FM, CW and FMfax operating modes. Included in the PA-5050A is the PS-7130A power supply consisting of five power supply modules. This integrated design concept minimises the effect of single point failures by providing gradual degradation of system power output.

PA-5050A is compliant with MIL-STD-188-141A; remote serial control is possible by RS-232, RS-422 or RS-485.

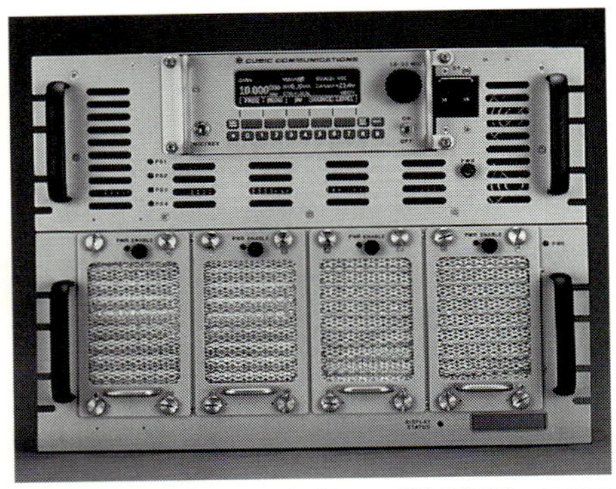

The Cubic Communications PA-5050A power amplifier (lower unit) 0011823

Specifications

Frequency range: 1.6-30 MHz
Power output: 1 kW peak
Dimensions: 197 x 483 x 538 mm
Weight: 11.3 kg

Operational status

This product is currently in production.

Contractor

Cubic Communications, Inc.

UPDATED

T-4180 LF-HF Digital Signal Processing (DSP) exciter

The T-4180 DSP exciter incorporates digital signal processing microchip technology designed to provide greater linearity and spectral purity. It is intended for use with the PA-5050A 1 kW power amplifier, and operates at 1.6 to 30 MHz with 1 Hz tuning resolution.

The T-4180 exciter is available in a half-rack chassis. The unit is capable of local or remote-control operation. In addition, operation is menu-driven and little or no

The Cubic Communications T-4180 DSP exciter
0011822

operator training is required. The exciter has a digitally tuned IF filter and 250 programmable channels. Operating modes include LSB, USB, ISB, AM, AM, CW, FSK and FMfax. BITE constantly troubleshoots the receiver and, when a problem is detected, a message on the unit's display screen identifies the module which should be removed and replaced. Remote control via RS-232 or RS-422.

Operational status
This product is currently in production.

Contractor
Cubic Communications, Inc.

UPDATED

VXI-3000 series DSP receivers

The VXI-3000 range of DSP (Digital Signal Processing) receivers comprises: the VXI-3250A HF DSP receiver; VXI-3550A DSP VHF/UHF DSP receiver; VXI-3570A VHF/UHF DSP receiver. Supporting systems comprise: the VXI-4400 DF processor; VXI-6180 HF distribution unit; VXI-6581 LO distribution unit; VXI-6582 LO distribution module (clam); VXI-8110 reference frequency source.

Specifications
VXI-3250A
Frequency range: 10 kHz to 30 MHz, 1 Hz tuning resolution
Detection modes: LSB, USB, ISB, CW, AM, FM
Sweep and scan: 100 channels per second
Programmable channels: 250
Standard bandwidths: 51 bandwidths from 100 Hz to 16 kHz
Optional interface: C40

VXI-3550A
Frequency range: 20-1,200 MHz
Detection modes: LSB, USB, CW, AM, FM
Sweep and scan: 100 channels per second
Standard bandwidths: 18 bandwidths from 1 to 240 kHz
Optional interface: C40

VXI-3570A
Frequency range: 20-3,000 MHz, 10 Hz tuning resolution
Detection modes: LSB, USB, CW, AM, FM
Sweep and scan: 100 channels per second
Programmable channels: 250
Standard bandwidths: 18 bandwidths from 1 to 240 kHz
Optional interface: C40

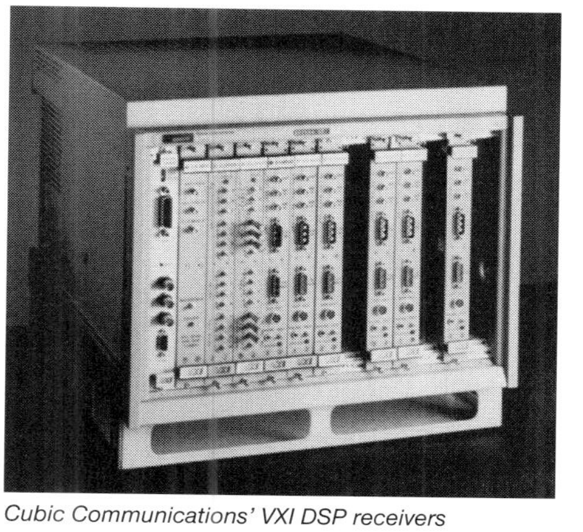

Cubic Communications' VXI DSP receivers
1999/0044817

Operational status
In production.

Contractor
Cubic Communications, Inc.

UPDATED

AN/ARS-3 sonobuoy reference system

The AN/ARS-3 is a major part of the Update II improvements to the US Navy's P-3C ASW aircraft. Comprising 10 blade antennas and a receiver-converter, the system provides for the passive detection and location of sonobuoys transmitting on 31 standard channels. The system uses the P-3C's Univac CP-901 computer for processing and display. It uses the aircraft's inertial navigation system to determine the geographical position of the buoys. In conjunction with the CP-901, the AN/ARS-3 can locate a faulty module 95 per cent of the time. Unlike the AN/ARS-2, the system needs no cooling air.

Specifications
Dimensions: 279 × 546 × 311 mm
Weight:
(including antennas) 35.8 kg
(receiver-converter only) 24.5 kg
Power supply: 115 V AC, 150 W
Temperature range: –54 to +71°C
Cooling air: none
Vibration: MIL-T-5422E
Environmental: MIL-E-5400
Reliability: 1,000 h MTBF

Operational status
In service on P-3C Orions flown by the US Navy, Australia, Japan and Netherlands.

Contractor
Cubic Defense Systems.

VERIFIED

AN/ARS-4 sonobuoy reference system

The AN/ARS-4 is a modified version of the AN/ARS-2 designed for the S-3B ASW aircraft. It is retrofitted into all S-3 aircraft in the US Navy inventory. The unit's updated electronics provide improved performance, reliability and maintainability, as well as the ability to locate sonobuoys operating on the 99 sonobuoy channels.

Specifications
Dimensions: 222 × 229 × 483 mm
Weight: (receiver-converter only) 15.9 kg
Power supply: 115 V AC, 150 W
Temperature range: –54 to +71°C
Cooling air: 0.24 kg/m
Vibration: MIL-T-5422E
Environmental: MIL-E-5400
Reliability: (SBR only) 2,862 h MTBF

Operational status
In service in US Navy S-3 aircraft.

US Navy S-3B Viking ASW aircraft are equipped with Cubic AN/ARS-4 sonobuoy reference systems

Contractor
Cubic Defense Systems.

VERIFIED

AN/ARS-5 sonobuoy reference system

Designed for the P-3C Update III, Cubic's AN/ARS-5 is an improved version of the AN/ARS-3. It has 99-channel capability as well as advanced design and built-in expansibility for such additional capabilities as anti-jamming, communications intelligence and search and rescue. The AN/ARS-5 incorporates the latest advances in self-test antenna radiation and built-in test equipment; the key difference between the ARS-5 and the other systems in the series is that it incorporates an embedded computer, while the others rely on computing within the aircraft.

Specifications
Dimensions: 279 × 546 × 311 mm
Weight:
(including antennas) 35.8 kg
(receiver-converter only) 24.5 kg
Power supply: 115 V AC, 150 W
Temperature range: –54 to +71°C
Cooling air: none
Vibration: MIL-T-5422E
Environmental: MIL-E-5400
Reliability: 1,000 h MTBF

Operational status
In production for the US Navy's P-3C Update III programme.

Contractor
Cubic Defense Systems.

VERIFIED

AN/ARS-6(V) Personnel Locator System (PLS)

The personnel locator system consists of the AN/ARS-6(V) guidance system manufactured by Cubic Defense Systems and the Motorola AN/PRC-112 survival radio. The survival radio remains silent until interrogated by the airborne guidance system, when it responds with a coded 300 ms pseudo-noise burst transmission. The AN/ARS-6(V) guidance system then displays range and steering directions in the rescue aircraft.

The AN/ARS-6(V) includes a compact radio transceiver, high-gain antenna set, control unit and instrument panel display.

Light, compact and modular, the system is easily transportable. It can be installed in a UH-1H or UH-60 in less than 30 minutes, and the system's simple controls and logical sequence make it extremely easy to operate.

The AN/ARS-6(V) is operated from the Control/Display Unit (CDU). The operator programmes Channel A and B frequencies from the 3,000 channels between 225 and 300 MHz or selects 243 and 282.8 MHz on preset channels. Steering commands and ranging information are displayed on a small liquid crystal Remote Display Unit (RDU) which is easily mounted on the aircraft instrument panel. Range to target is measured first in miles, then in feet. Steering commands are presented on a bar graph display that indicates the position of the target relative to the heading of the aircraft. Both the RDU and the CDU have sunlight-readability and full NVG compatibility.

The Receiver/Transmitter (RT) is a fully synthesised UHF transceiver optimised to operate with the AN/PRC-112 survival radio. It weighs less than 10 kg and is packaged in a ¾ ATR case. The shock-isolated RT mounting base is adaptable to either temporary installation on the aircraft floor or permanent mounting. The RT's 1553B bus interface can configure the AN/ARS-6(V) as a remote terminal on the databus, integrating control and display with existing aircraft functions and eliminating the need for the RDU and CDU.

The antenna group consists of twin blade antennas on the aircraft ventral surface. By mounting the antennas on the aircraft centreline, signal scattering effects from aircraft structures are minimised to achieve the specified accuracy.

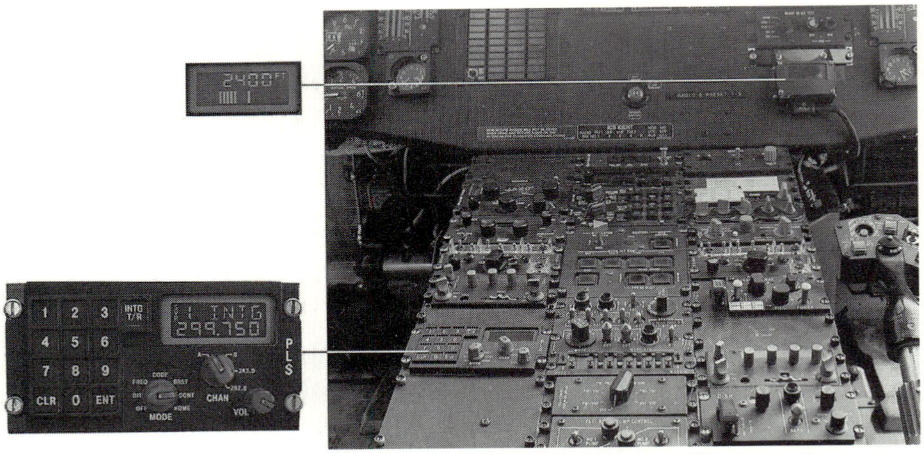

The RDU shows range and steering directions in a daylight-readable, NVG-compatible display. (Display shows survivor is 2,400 ft away, to the left of aircraft heading) ***1999***/0051607

Specifications

Dimensions:
(antenna set) 228.6 × 116.8 × 50.8 mm
(receiver/transmitter) 317.5 × 193 × 198.1 mm
(antenna switching unit) 144.8 × 193 × 33 mm
(control display unit) 101.6 × 146 × 76.2 mm
(remote display unit) 38.1 × 76.2 × 48.3 mm
Weight:
(antenna set) 0.59 kg
(receiver/transmitter) 9.89 kg
(antenna switching unit) 1.04 kg
(control display unit) 1.3 kg
(remote display unit) 0.27 kg
Power supply: 28 V DC, 4.5 A (max)
Power output: 10 W
Frequency: 225-300 MHz
Available channels: 3,000 (25 kHz steps)
Reliability: 2,650 h MTBF

Operational status

In production since 1988 for the US Army, Navy and Air Force and other customers.

Contractor

Cubic Defense Systems.

VERIFIED

AN/AYD-23 GPWS: Ground Proximity Warning System

Adaptable to all classes of rotary-wing aircraft, GPWS can be installed as a ¼ ATR (short) avionics package interfaced with discrete sensors, or via the ARINC 429/ MIL-STD-1553B databusses. It can also be embedded in computers already on board the aircraft.

The GPWS continuously monitors important aircraft parameters in real time:

(1) Attitude (roll and pitch) relationship to terrain (altitude and closure rate acceleration).
(2) Terrain type and terrain trends (rising and falling).
(3) Validity and reliability of sensor inputs.
(4) Response capabilities of the pilot and the aircraft.

These inputs are used to determine the dynamic state of the aircraft. From this data, the GPWS algorithm computes the following:

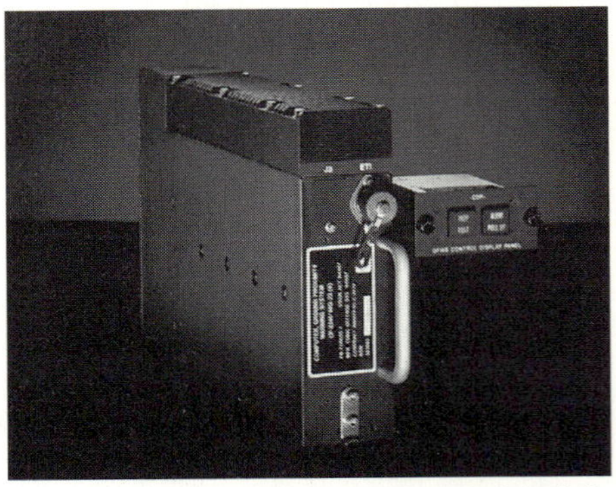

Cubic Defense Systems AN/AYD-23 ground proximity warning system 0018926

(1) The predicted altitude loss due to pilot response time.
(2) The predicted altitude loss due to roll recovery.
(3) The predicted altitude loss during the actual recovery.
(4) The predicted altitude loss due to terrain change.

The sum of these altitude loss quantities is used by the GPWS to compute the need for warnings to be issued.

The GPWS is designed to use existing sensors, including radar altimeter, barometric altimeter, attitude gyros, accelerometers, air data sensors. Growth capability to use forward-looking warning radar systems and windshear data is included in the design.

The GPWS algorithm continually assesses the validity of input data and is optimised for highly dynamic tactical flight environments.

Operational status

In service in Canadian Forces C-130 cargo aircraft and on order for US Navy H-46 and H-53 helicopters, with delivery scheduled for early 1999 - assumed to have started.

Contractor

Cubic Defense Systems.

VERIFIED

AN/URQ-34 anti-jam tactical datalink

The AN/URQ-34 is an anti-jam tactical datalink transmission system operating in the Ku (NATO J-band) frequency band, built to support US Army and US Air Force real-time combat sensors. It was originally developed for the StandOff Target Acquisition System (SOTAS) and was delivered and deployed for special projects after SOTAS was discontinued.

The AN/URQ-34 uses fast frequency-hopping technology to defeat the hostile ECM threat specified for the Modular Integrated Communications and Navigation System (MICNS) and has a demonstrated range of 240 km. Input of digital data can be in clear or encrypted and the data rate is 25 to 100 kbytes nominal. It meets all applicable MIL, MICNS and SOTAS standards for environment, nuclear hardening, ballistic hardening, MTBF and MTTR. The MTBF is given as between 480 and 1,680 hours. The AN/URQ-34 makes extensive use of BITE, giving fault location to module level.

Operational status

In service with the US Army and US Air Force as the Joint Surveillance Target Attack Radar Systems (J-STARS) datalink.

Using a combination of US Army and company records and funds, Cubic has begun a System Improvement Program (SIP) to reduce Size, Weight And Power (SWAP) and cost of the AN/URQ-34 by 50 per cent.

Contractor

Cubic Defense Systems, Inc.

VERIFIED

SCDL Surveillance and Control DataLink

Joint STARS (Surveillance Target Attack Radar System) radar data, comprising wide-area search/ moving target indicators, synthetic aperture radar/fixed target indicators, low-reflectivity indicators and sector search information, is broadcast in real time, in continuous communication through Cubic Defense Systems SCDL to the Army Ground Station Modules (GSMs).

SCDL is a time division, multiple access datalink with flexible frequency management capability. The link provides reliable and secure performance in high jamming environments. The link establishes a network consisting of the E-8 aircraft and ground stations.

The main features of the network are: broadcast radar messages to all GSMs within line of sight; GSM access to the aircraft on a time-ordered manner; relay of messages from one GSM to another; automatic acknowledgement of error-free receipt of messages; up to eight multiple networks operating simultaneously and within the same geographical area; all data encrypted.

SCDL uses very wide band fast frequency hopping, coding and data diversity to achieve robust jamming resistance. Novel techniques are used to achieve link acquisition and re-acquisition. The uplink messages are implemented with a unique modulation approach which provides accurate determination of path delay

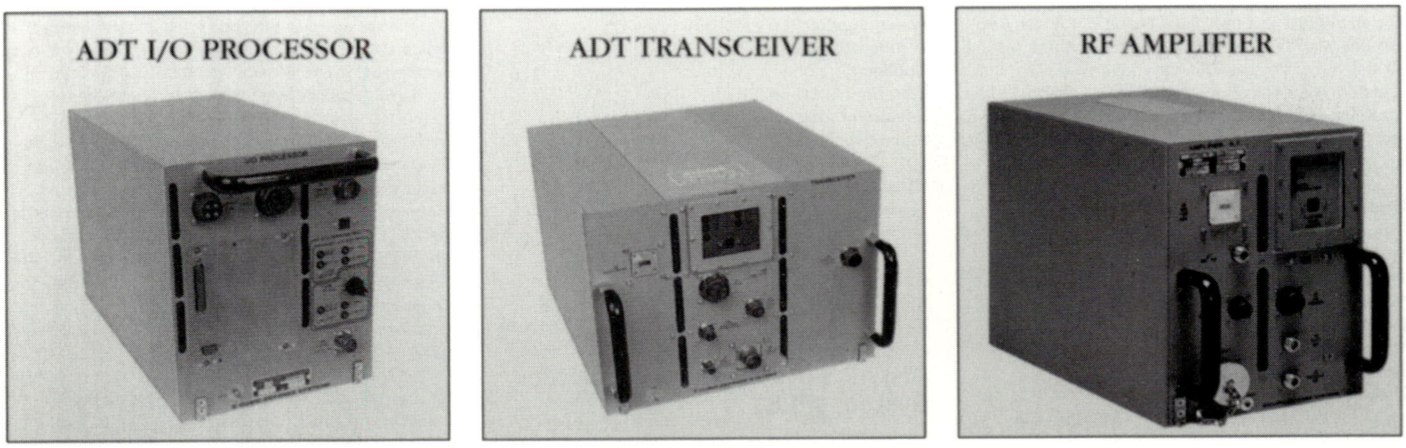

Cubic Defense Systems SCDL airborne data terminal ***1999***/0044820

between the E-8 aircraft and the GSM. This provides the capability for a highly reliable, short duration, one-time uplink message to achieve low probability of enemy detection of GSM transmissions.

Airborne Data Terminal (ADT)
The Airborne Data Terminal (ADT) comprises three line replaceable units: an Input/Output Processor (IOP); a transceiver and an RF amplifier. The IOP provides the primary interface with the aircraft through a MIL-STD-1553B databus. It prioritises, encrypts and forwards radar messages for transmission over the datalink, accepts uplink messages, and provides overall control of the datalink, including BIT.

The transceiver contains the downlink transmitter and uplink receiver, which operate with the antenna mounted on the bottom of the E-8 fuselage. All of the datalink functions of modulation, coding, frequency spreading and power amplification are contained in this unit.

The RF amplifier is slaved to the transceiver and provides the capability to transmit and receive data from the top-mounted antenna. The dual antenna installation provides full and continuous ground coverage at all aircraft attitudes, including turns.

Ground Data Terminal (GDT)
The Ground Data Terminal (GDT) comprises four units: the nuclear and ballistically hardened antenna unit; the Lower Control Unit, which performs data coding, decoding, and datalink timing; the Joint STARS Interface Unit (JSIU) provides the interface with the GSM using a MIL-STD-1553B databus and provides encryption and decryption of all data, it also controls the datalink and BIT; the AC-AC converter converts 50-60 Hz prime power to 400 Hz for use by the datalink.

System Improvement Program (SIP)
The aim of the SIP is to reduce the Size Weight And Power (SWAP) of the GDT by 50 per cent, to make space in the GSM for other C3SR systems.

MR (MultiRate) SCDL
At present, SCDL functions at 64 kbit rate. The MR SCDL upgrade will provide for the ability to function at rates from 64 k to 1.544 Mbits, with no degradation in anti-jamming performance. The US Department of Defense is understood to have approved funding for this programme in August 1998.

Broadcast Moving target indicator DataLink (BMDL)
BMDL is being proposed by Raytheon for their ASTOR bid to the UK government. BMDL is virtually identical to SCDL used by US JSTARS and would provide inter-operability between the two aircraft types. BMDL comprises the Air Data Terminal – the transceiver and I/O processor in the aircraft, and the Ground Data Terminal – the masthead antenna assembly, interface unit and control unit, housed in a vehicle or ground station.

Operational status
Operational in Joint STARS E-8 aircraft.

In May 2000, Cubic Defense Systems was awarded a US$6.9 million contract to design, build and test a prototype and feasibility demonstration of a High Capacity DataLink (HCDL) development of the SCDL. The task is to be completed by May 2001.

Contractor
Cubic Defense Systems.

VERIFIED

Stand-alone Sonobuoy Reference System (SRS)

Under an independent research and development programme, Cubic is developing a stand-alone Sonobuoy Reference System (SRS) that can allow virtually any fixed-wing aircraft or helicopter to have full SRS capabilities. This unit will incorporate such additional features as anti-jamming, DME, COMINT and search and rescue. Other features include self-calibration and built-in fault isolation for easier maintenance.

This stand-alone capability can be incorporated into either the AN/ARS-4 or the AN/ARS-5 chassis and can interface with any aircraft databus. Using its own embedded computer system, the system can operate in both active and passive modes with sonobuoys operating on up to 99 channels.

Performance specifications for the stand-alone SRS are identical to those for the AN/ARS-5 system.

Contractor
Cubic Defense Systems.

VERIFIED

AN/URC-138(V)1(C) SHAR Link 16 Low Volume Terminal (LVT)

The AN/URC-138(V)1(C) Information Distribution System provides anti-jam protected, encrypted, high throughput data distribution. Because it is low cost, it makes JTIDS participation affordable. The small size and low weight of the AN/URC-138(V)1(C) terminal makes it suitable for a broad variety of tri-service platforms.

The AN/URC-138(V)1(C) terminal is waveform, message format and network compatible with existing JTIDS (TADIL J) Link 16 systems. The terminal provides JTIDS interoperability between the US tri-services and NATO forces.

JTIDS provides situation awareness by providing threat, target and friendly ID, position and status information among participating platforms in near-real-time, with anti-jam security without any voice communication.

It provides full stacked net capacity, up to 128, and full JTIDS data throughput. The system can automatically exchange information from a variety of platform sensors. This can include functions such as IR and optics scan, target identification and steering commands. Real-time data updates can also be used to provide landing cues.

In addition to robust data communication, the-AN/URC-138(V)1(C) terminal also provides two voice ports to enable secure voice communication in a jamming environment.

AN/URC 138(V)1(C) SHAR Link 16 Low Volume Terminal (LVT) 0001183

Specifications
Operational characteristics
Net participation:
TDMA
128 nets maximum
128 time slots/sec/net
Message catalogue: J-Series messages as defined in STANAG 5516
Frequency operating range: 969-1206 MHz
Informational data rates: 28.8 to 238 kbs
Anti-jam: Frequency hopping, forward error detection/correction
Range (max): 400 n miles

Electrical characteristics
RF power output (max): 200 W
I/O data/voice:
Data: MIL-STD-1553
Voice port 1: LPC-10 (2.4 kbs)
Voice port 2: CVSD (16 kbs)
Antenna: JTIDS compatible
Primary power: 120 V AC 3ø 400 Hz, 750 W

Physical characteristics
Dimensions: 318 × 256.3 × 191 mm
Weight (max): 20 kg
Packaging: SEM-E module form factor cooling, forced air mounting

Operational status
Selected for the UK Sea Harrier F/A Mk 2, and Sea King AEW. The contract, awarded in November 1995, called for the design, development, qualification and production of up to 65 production Link 16 terminals. Thirteen terminals, used for qualification testing and integration, were shipped in 1998. Production readiness was achieved in mid 1999, and terminal delivery is planned for March 2000 to the end of 2001.

Contractor
Data Link Solutions (DLS), a limited liability company of Rockwell Collins.
Marconi CNI Division, a BAE Systems company.

VERIFIED

Carousel 400 series inertial systems

The Carousel 400 series of inertial reference systems is designed to meet the requirements of advanced aircraft and provide an upgrade for older aircraft. It consists of the Carousel 411, 424, 444 and 448 systems. These systems are based on the hemispherical resonator gyro which has no moving parts and no known wear-out mechanism.

Carousel 411. The Carousel 411 inertial navigation system is a direct replacement for ARINC 561 systems and no operational, electrical or mechanical changes are required. It provides the advantages of a new system, while maintaining all the current interfaces. Any combination of Carousel IV and 411 systems can be installed on the aircraft. The Carousel 411 provides a completely self-contained worldwide navigation capability. In addition, it provides digital outputs in both ARINC 561 and 704 formats. The Carousel 411 can make use of the existing Carousel IV CDU or an optional RNav CDU with no wiring changes. The optional RNav CDU has full alphanumeric capability and, along with an optional navigation database and autotune circuit card, makes the system a lateral flight management system. Also available as an option is an embedded GNSS receiver/processor.

Carousel 424. The Carousel 424 MG small inertial navigation system consists of a strapdown platform, digital computer and associated input/output hardware. It is designed for the military environment and has provision for an embedded GPS receiver. The Carousel 424 also provides an extra card slot for unique aircraft functions or interfaces such as flight control, radar or air data. It can make use of existing CDUs for independent display of navigation data. An optional CDU is available.

Carousel 444. The Carousel 444 is a 4 MCU ARINC 704 inertial reference system. It is designed for use in new production aircraft. The reference unit consists of a strapdown platform, digital computer and associated input/output hardware. Use of this solid-state instrument, combined with supporting electronics, makes accuracy and reliability an inherent feature of the reference unit. An inertial system display unit is available to provide mode selection and insertion of initial position. The Carousel 444 can be used in 10 MCU configurations, with the addition of an adaptor tray. An embedded GNSS receiver/processor is available as an option.

Carousel 448. The Carousel 448 is a 4 MCU inertial reference system and air data computer which meets the ARINC 738 ADIRS specification. In addition to all the information available from the Carousel 444, the Carousel 448 provides full air data outputs including true airspeed, Mach number, air temperature and angle of attack. An embedded GNSS receiver/processor is also available for the Carousel 448.

Contractor
Delco Electronics.

VERIFIED

Carousel IV Inertial Navigation System (INS) (AN/ASN-119)

During the late 1960s the Carousel IV inertial navigation system was the subject of the largest ever single military procurement of such equipment, when it was chosen by the US Air Force for its fleet of C-5A Galaxy and C-141 StarLifter transports and KC-135 tankers. It had earlier been chosen as standard fit for the Boeing 747, when it was fitted as a three-system installation to become the first certified commercial inertial navigation system. A guaranteed MTBF of greater than 1,250 hours was an important factor in Boeing's choice of the Carousel IV. Some 7,000 sets have now been delivered to support 30 military programmes and more than 60 airlines. The Carousel IV has a demonstrated MTBF of more than 3,000 hours on commercial aircraft.

Improved versions of the Carousel IV have been built for the Boeing E-3A Sentry AWACS under the designation AN/ASN-119, the Titan III ICBM missile fleet, numerous helicopter applications and the airborne element of the US Army's Guardrail intelligence programme. The AWACS installation comprises two ASN-119 platforms operating in conjunction with a single Northrop Grumman AN/ARN-129 Omega system and a Litton AN/ADN-213 Doppler velocity sensor. The combination provides an accuracy of better than 1 n mile on a 10-hour mission.

Each system comprises three elements: an inertial navigation unit with gyros, accelerometers and computing functions; a control/display unit; and a mode selector unit. A battery unit to maintain operation during power transients is optional.

Specifications
Dimensions:
(inertial navigation unit) 215.9 × 259.1 × 510.5 mm
(control/display unit) 114.3 × 146.1 × 152.4 mm
(mode selector unit) 146.1 × 38.1 × 50.8 mm
Weight:
(inertial navigation unit) 25 kg
(control/display unit) 2.1 kg
(mode selector unit) 0.45 kg

Operational status
In service. Replaced in the C-141 and KC-135 by the Rockwell Collins PACER CRAG avionics refit.

Contractor
Delco Electronics.

VERIFIED

Fuel Savings Advisory and Cockpit Avionics System (FSA/CAS)

Delco provides the Fuel Savings Advisory and Cockpit Avionics System (FSA/CAS) for the upgrading of the current Boeing C-135 and KC-135 fleet of about 700 aircraft.

In this system, a CRT control and display unit replaces one of the standard inertial navigation system units to simplify crew management tasks; in particular, all basic data for flight use can be entered at this one point. In addition to the usual engine and navigation information, a new fuel management panel and centre of gravity display have been installed for inputs of fuel state, disposition and usage. This latter function is invaluable in tanker operations and ensures efficient fuel allocation and usage. The panel is linked to other avionic systems by a MIL-STD-1553B databus.

The FSA provides commands to the pilots during climb, cruise and descent using flight path optimisation algorithms and flight manual data for lift, drag and thrust. An additional mode computes all required take-off and landing parameters based on crew inputs for present conditions.

Many KC-135s are being refitted with GE/SNECMA CFM56 engines, which provide extra thrust over earlier types. FSA/CAS is integrated with this engine, offering more economic operation.

Specifications
Dimensions:
(FSA computer) ½ ATR long
(control/display unit) 146 × 181 mm
(bus controller) 207 × 131 mm
(fuel panel) 457 × 123 mm
(fuel management computer) 235 × 249 mm
(remote display unit) 83 × 83 mm
Weight:
(FSA computer) 11.9 kg
(control/display unit) 4.5 kg
(bus controller) 4.1 kg
(fuel panel) 12.5 kg
(fuel management computer) 21.8 kg
(remote display unit) 0.95 kg
Power:
(FSA computer) 95 W
(control/display unit) 44 W
(bus controller) 34 W
(fuel panel) 92.8 W
(fuel management computer) 139.5 W
(remote display unit) 10 W

Operational status
In service on US Air Force Boeing C-135 and KC-135 aircraft.

Contractor
Delco Electronics.

VERIFIED

Low-Cost Inertial Navigation System (LCINS)

The Low-Cost Inertial Navigation System (LCINS) development programme extends Delco's Carousel class systems' activity into applications that require somewhat lower performance at substantially lower cost. Typical navigation accuracy rating for the LCINS is 2 to 4 n miles/h. The LCINS is a strapdown configuration, utilising Incosym Inc two-degree-of-freedom gyros. The entire inertial reference assembly is substantially reduced in size. A digital microprocessor performs all the measurement data processing, instrument torquing computation, scaling, attitude and navigation functions. Steering commands and other autopilot interfaces are provided.

Specifications
Dimensions: 152 × 152 × 215 mm
Weight: 3.0 kg
Power supply: ±15 and ±5 V DC, 35 W

Operational status
The LCINS has been selected as the Three Axis Inertial Measurement System (TAIMS) for a classified military aircraft programme and as the inertial sensor assembly for a second classified military programme.

Contractor
Delco Electronics.

VERIFIED

Mission computer and display system for the C-17

In September 1986, Delco Electronics was awarded a contract for the development of a prototype mission computer and electronic display system for the C-17.

Although the computer is similar to a flight management system, it features a number of other functions including parachute cargo extraction guidance and automatic flight control system coupling for station-keeping in formation flying.

Operational status
In service on US Air Force C-17 transport

Contractor
Delco Electronics.

VERIFIED

AN/USQ-130(V) MX-512PA Link-11/TADIL-A data terminal

The AN/USQ-130(V) Link-11/TADIL-A data terminal set is designed to provide all required modem and network control functions in a Link-11/TADIL-A system using either HF or UHF radio equipment. The equipment meets the data terminal set requirements of MIL-STD-188-203-1A and may be operated as a picket or net control station in a TADIL-A net. As a net control station, the AN/USQ-130(V) accepts addresses from the tactical data computer or from a separate control panel.

The equipment provides all the modes of Link-11/TADIL-A systems including net control or picket, high- and low-data rate, net test, net synchronisation, short broadcast, long broadcast and full-duplex (for single station system tests and sidetone verification). Doppler correction circuits which operate independently on both sidebands are operator-selectable. The AN/USQ-130(V) also operates in the Improved Link-11 Waveform (ILEW) mode.

The AN/USQ-130(V) can be externally controlled by a computer over a MIL-STD-188-114, RS-232C-compatible asynchronous control interface, or 1553 databus. The AN/USQ-130(V) may also be controlled from a separate remote-control panel using menus standard to the MX-512P DTS family.

The set is programmable. All modem, network control and link monitoring functions are performed digitally in microprocessors using a modular, multiprocessor architecture. Selection of the conventional Link 11 or ILEW is made over the remote-control interface.

The single-tone waveform for Link-11 provides improved performance in HF Link-11 networks on an SSB HF channel. Single-tone Link-11 uses an eight-phase modulated 1,800 Hz tone. Adaptive equalisation is used to demodulate the signal under the severe

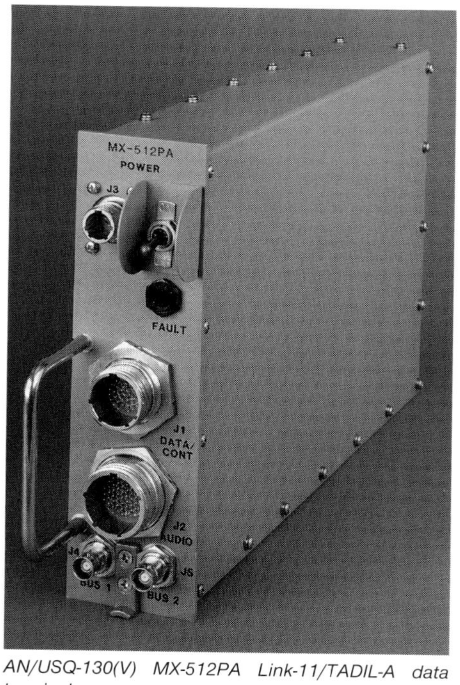

AN/USQ-130(V) MX-512PA Link-11/TADIL-A data terminal 0002119

multipath conditions typical of HF propagation paths. Error detection and correction codes are used to provide enhanced message throughput.

The AN/USQ-130(V) provides, as an option, a 2,400 bits/s, full-duplex, RS-232C satellite-wireline interface which transmits and receives compatible Link-11 data in digital form. Link-11 data may be sent over satellite, wireline, or other tactical circuits. The AN/USQ-130(V) can be operated in either the digital mode, the conventional mode, or in a mixed mode (gateway), where some pickets operate in the digital transmission mode and some in the conventional HF or UHF mode.

The unit provides, as another option, link quality analysis indicators which include multipath spread, fading bandwidth, net cycle time since last reply, and tone power spectrum for each participating unit in the network. Using these indicators, an operator can troubleshoot equipment failures and configuration set up problems in the net and determine when HF propagation problems require a change of radio frequency. BITE provisions in the AN/USQ-130(V) include loop-back functions which verify operation of the system.

The data terminal set is compatible with ATR short, measures 193 × 57 × 32 mm, and weighs 4.6 kg. It is powered from 24 to 32 V DC and meets certain requirements of MIL-HDBK-217E, MIL-E-5400T, MIL-STD-188 and MIL-STD-1553B.

Specifications

Dimensions:
(size): ¼ ATR-short
(height): 19.3 cm
(width): 5.7 cm
(depth): 32.0 cm
(weight): 4.6 kg

Power:
24 to 32 V DC
28 W

Reliability:
MTBF: Over 10,000 h per MIL-HDBK-217F at 50°C AIC

Contractor

DRS Communications.

UPDATED

GA-540 TADIL-A/Link-11 Serial DataLink Translator (SDLT)

The GA-540 Serial DataLink Translator (SDLT) provides the interface in a TADIL-A/Link-11 system between a serial encryption device and a tactical data system processor. The GA-540 serial interface to the encryption device meets the Airborne Tactical Data System (ATDS) interface requirements of MIL-STD-188-203-1A (Appendix D-2). All data buffering, timing and ATDS handshaking are performed automatically. The GA-540 connects to the tactical data system processor using one of the following interfaces: RS-422/RS-423 conditioned diphase; transformer-coupled conditioned diphase; RS-422/RS-423 synchronous; RS-423 asynchronous; VME bus.

The GA-540 is offered as a 19 in rack-mountable unit, as a ¼-ATR-short airborne unit, or as a 6U-VME card.

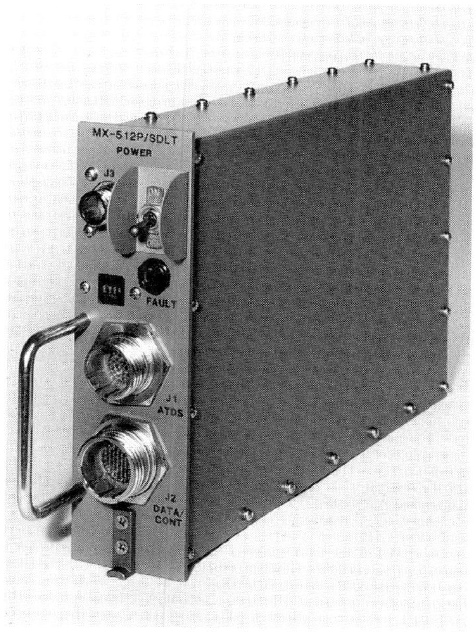

GA-540 TADIL-A/Link-11 Serial DataLink Translator (SDLT) 0002120

The rack-mounted and ATR units include a power supply, BIT indicator, and panel-mounted connectors. The VME card set can be hosted in any VME operating environment.

Specifications

Dimensions:
(size): ¼ ATR-short
(height): 19.3 cm
(width): 5.7 cm
(depth): 32.0 cm
Power:
24 to 32 V DC
15 W
Reliability: 78,000 h

Contractor

DRS Communications.

UPDATED

Ice detection system

DNE Technologies has developed an ice detection probe which employs what is claimed to be the most sensitive method of ice accretion detection currently available. The probe operates in a cyclic fashion using the thermal characteristics of the ice which forms on it combined with the heat of fusion effect as the ice is formed. The signal produced may be used to initiate a warning signal or to activate the aircraft's de-icing system automatically. Between detection cycles, the probe is cleared of ice by an integral heater circuit. It is claimed that an ice accretion thickness of 0.12 mm can be detected within 5 seconds. A test facility for airborne confidence checking and for ground servicing purposes is incorporated. The standard ice detector is designed to sense ice formation on fuselage and nacelle air intakes for turbine engines, but other versions are available for carburettor ice detection on piston-engined aircraft and for other ice sensitive areas.

Operational status

DNE Technologies ice detectors are in service on the B-1B, B-2, the F-16C/D and F-117. An ice detector is under development for the F/A-18E/F and a flush-mounted ice detector is also being developed.

Contractor

DNE Technologies Inc.

VERIFIED

Digital Terrain System (DTS)

As an enhancement to the present F-16 Data Transfer Equipment, Fairchild has integrated BAE Systems' TERPROM® algorithm with a Data Transfer Cartridge mass memory and a high performance processor. The resulting Digital Terrain System (DTS) capability is contained within the form factor of the original Data Transfer Cartridge (DTC). The Data Transfer Unit (DTU), the cockpit-resident, intelligent receptacle, remains unmodified resulting in straight forward retrofit or insertion into existing platforms.

The Mega Data Transfer Cartridge with Processor (MDTC/P) performs all existing data transfer functions of mission data load and in-flight data recording on a non-interfering basis while concurrently running DTS algorithms. The cartridge interface has been preserved to ensure compatibility with existing ground support equipment and mission planning sytems. The original DTC and the MDTC/P cartridges can be interchanged on aircraft without affecting normal data transfer operations.

The BAE Systems' TERPROM® terrain correlation algorithm has been successfully implemented in numerous aircraft, including the F-16, Jaguar and Harrier. Its unique attribute is its ability to precisely locate the aircraft with respect to its on-board terrain data base under adverse conditions – thus providing stealthy, all weather, night operation.

The MDTC/P hosts the Digital Terrain Elevation Data (DTED) database in a non-volatile, solid-state mass memory array. tHis memory was developed and qualified by Fairchild as part of continuing memory system development. This memory is actively being employed on multiple, severe environment, military platforms and requires no special conditions for operation or application.

The cartridge embedded processor works in conjunction with the cartridge mass memory providing an efficient method of processing DTS algorithms. Direct processor access to the terrain database obviates the transmission of bulk data over the MIL-STD-1553 data bus simplifying system integration, reducing bus bandwidth and simplifying bus control.

Through the precise location of the aircraft with respect to terrain and knowledge of the real-time kinematic model of the aircraft, the TERPROM® algorithm can compute ground intercept and provide warning cues to the pilot in advance of entering a dynamically unrecoverable state.

Specifications

Dimensions:
DTU – 178 × 127 × 113 mm
MDTC/P – 191 × 127 × 113 mm
Weight:
DTU – 3.0 kg
MDTC/P – 1.6 kg
Power: 115 V AC, 400 Hz, single phase, 32 W
Cartridge data capacity: Typical configuration uses 72 Mbytes. Growth to over 1 Gbyte
MDTC/P software: coded in Ada
System interface: dual redundant, MIL-STD-1553B
Reliability (MTBF): DTS system – 12,000 h.

Contractor

Fairchild Defense Division of Orbital Sciences Corporation.

VERIFIED

AN/AKT-22(V)4 telemetry data transmitting set

The AN/AKT-22(V)4 system relays up to eight channels of sonobuoy data from an ASW helicopter to a ship. It comprises the T-1220B transmitter-multiplexer, the C-8988A control indicator, an AS-3033 antenna and a TG-229 actuator. The sonobuoy signals are received by dual AN/ARR-75 radio receiving sets and passed into the transmitter-multiplexer. The control indicator has four trigger switches, each of which disables two data channels. Composite trigger tones are brought into the multiplexer separately via the control indicator and combined with the sonic data channels and the single voice channel. The resulting FM signal is used to modulate the transmitter.

The data transmitting set has a DIFAR operating mode in which the extra voice channel is inoperative and the composite FM modulating signal is disconnected from the transmitter input. Two DIFAR sonobuoy transmissions enter the transmitter-multiplexer on dedicated channels. After conditioning in an amplifier-adaptor, the DIFAR signal is split into two components: DIFAR A and DIFAR B. The A signal is conditioned in a low-pass filter, while the B signal drives a variable-cycle oscillator centred at 70 kHz and then passes through a bandpass filter. The two filter outputs are combined linearly and the resulting composite modulation signal drives the transmitter. A switch on the controller indicator controls whether the normal sonic or composite DIFAR signals are transmitted.

The AS-3033 antenna has two sections: a VHF element for receiving the sonobuoy signals and a UHF part which sends the multiplexed data down to the ship. The ship receives the information on an AN/SKR telemetric data receiving set.

Specifications

Power supply: 115 V AC, 3-phase, at 0.85, 1.5, and 1 A respectively
Warm-up time: <1 min in standby mode < 15 min under environmental extremes
Operating stability: > 100 h for continuous or intermittent operation
Frequency: 2,200-2,290 MHz, 1 of 20 switch-selectable E-L-band channels
Multiplexer inputs:
(a) 8 sonar data channels (7 with 10-2,000 Hz bandwidth; 1 with 10-2,800 Hz), at 0.16-16 V
(b) 4 sonar trigger channels, 26-38 kHz, at 1-3 V
(c) 1 voice channel, 300-2,000 Hz bandwidth, at 0-0.25 V
(d) 2 composite DIFAR channels, 10-2,000 Hz bandwidth at 3-6 V or 10.6 V
Channel phase correlation: difference in phase delay between any 2 passive data channels < 1° (10-500 Hz)

Operational status

In service with US Navy.

Contractor

Flightline Electronics Inc.

VERIFIED

AN/ARN-146 On-Top Position Indicator (OTPI)

The AN/ARN-146 radio receiving set is a 99-RF channel On-Top Position Indicator (OTPI) which is used on board ASW rotary- and fixed-wing aircraft to provide bearing and on-top position indication of deployed sonobuoys. When used in conjunction with a suitable ADF system, the AN/ARN-146 enables an operator to locate and verify the position of sonobuoys which are operating on any of 99 RF channels.

The AN/ARN-146 has a modular solid-state design for use with computer control, either via RS-422 directly or when connected with the AN/ARR-84 sonobuoy receiver. The AN/ARN-146 is form and fit interchangeable with the R-1651/ARA and R-1047 A/A OTPI receivers. It is compatible with ARA-25, ARA-50 and OA-8697/ARD ADF antenna systems.

The AN/ARN-146 set consists of the R-2330/ARN-146 receiver and the C-11699/ARN-146 radio set control. The R-2330/ARN-146 is a VHF AM receiver, which receives signals from sonobuoys operating on any of 99 RF channels in the frequency range from 136 to 174 MHz. This receiver houses all the electronics, including internal transfer circuits for the switching of RF, baseband, power and phase compensation circuits for sharing the DF system between the OTPI and an associated UHF receiver system for the ADF. The R-2330/ARN-146 also contains a PLL synthesised local oscillator, digital address decoder, voltage tuned BP filters, AGC and BIT circuitry.

The C-11699/ARN-146 radio set control is an optional manual control box which is used in place of the RS-422 bus or the MIL-STD-1553B dual bus of the AN/ARR-84. This control box provides the capability to apply power to the AN/ARN-146 system to select any one of 99 RF channels, to activate the BIT circuitry and to display adequate signal strength and results of the BIT via the adequate signal strength indicator.

Specifications

Dimensions:
(radio receiver) 125.8 × 134.6 × 77.47 mm
(radio set control) 57.15 × 127 × 146 mm
(mounting plate) 15.2 × 144.8 × 77.47 mm
Weight:
(radio receiver) 2.15 kg
(radio set control) 0.57 kg
(mounting plate) 0.27 kg
Power supply: 115 V AC, 400 Hz, single phase, 15 VA
28 V DC, 20 W
Frequency: 136-173.5 MHz (99 channels)
Sonobuoy compatibility: DIFAR, LOFAR, Ranger, BT, CASS, DICASS, VLAD, CAMBS, Barra, ERAPS, HLA, ATAC, SAR
Reliability: 3,000 h MTBF

Operational status

In production for aircraft of the US Navy. Fitted to SH-60B helicopters.

Contractor

Flightline Electronics Inc.

VERIFIED

AN/ARR-72 sonobuoy receiver system

The AN/ARR-72 sonobuoy receiver system is used on the P-3C patrol aircraft, in conjunction with acoustic signal processors and a digital computer. The AN/ARR-72 system receives, amplifies and demodulates FM signals transmitted by deployed sonobuoys in the 162.25 to 173.5 MHz VHF band. The AN/ARR-72 is compatible with the AN/SSQ-36, 41, 50, 53 and 62 sonobuoys. The receiver is in five parts: AM-4966 preamplifier, CH-169 receiver, SA-1065 audio assembly, C-7617 control indicator and SG-791 Acoustic Sensor Signal Generator (ASSG) which performs diagnostic functions. The AN/ARR-72 system is compatible with LOFAR, CODAR, BT, RO, CASS and DICASS equipment.

The AN/ARR-72 is a dual-conversion superheterodyne VHF receiver system. Radio frequency signals are received at the dual-aircraft VHF blade antennas, amplified by the system's AM-4966 dual RF amplifiers and passed on to the CH-169 31-channel receiver assembly. Within this assembly, a multicoupler distributes the preamplified RF to the 31 fixed tuned receivers, where it is further amplified and demodulated to provide baseband audio and RF level signals.

Each of the 31 receiver channels contains a channelised first converter, an IF filter, a second converter and a discriminator/amplifier. The first converter contains a crystal-controlled local oscillator and mixed/IF circuit. The plug-in discriminator/amplifier provides RF level and FM signal detection. The optional phaselock discriminator is directly interchangeable with the discriminator/amplifier module. The first and second converter and discriminator/amplifier assemblies are plug-in units and, excepting the local oscillator crystals, are identical for the 31 channels.

The SA-1605 audio assembly includes 19 audio switching and amplifier cards and two audio power supply regulators. The audio assembly accepts the baseband and RF level outputs of the 31 receivers and outputs them to the computer and processing equipment. Each of the 19 audio channels contains a 31 by 1 switching matrix to select the output of a given receiver. This receiver signal is then amplified and provided through two individually buffered outputs to the processing equipment. Selection of a particular receiver channel may be accomplished by the digital computer or the C-7617 dual-channel control indicator. RF level for the selected receiver is displayed on the corresponding control-indicator meter.

The SG-791 ASSG is the BIT for the receiver system. RF test signals are generated for application to the preamplifier, the multicoupler or to a radiating antenna. Internal circuits generate simulated signals for the testing of normal LOFAR, extended LOFAR, range only and BT processing equipment. External modulation inputs are provided to accept modulation from devices such as the target generator in the demultiplexer of the AN/AQA-7(V) DIFAR equipment, thus allowing end-to-end checks of sophisticated ASW equipment.

The ASSG contains a redesigned multifrequency oscillator to provide for generation of the 31 individual RF frequencies. This replaces a previously utilised crystal turret. This synthesiser reduces the channelling time from 10 seconds to less than a second.

Specifications

Weight:
(preamplifier) 0.9 kg
(receiver) 26.53 kg
(audio) 17.69 kg
(ASSG) 8.71 kg
(8 control indicators) 14.51 kg

US Navy SH-60 LAMPS Mk III helicopters have Flightline Electronics AN/ARR-72 sonobuoy receivers

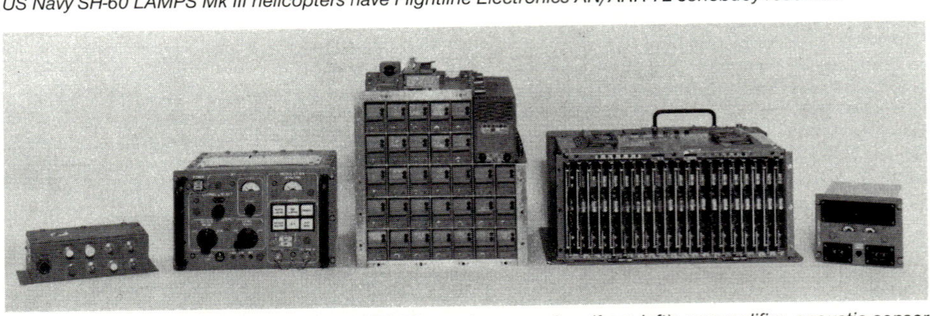

Modules of the Flightline Electronics AN/ARR-72 sonobuoy receiver (from left): preamplifier, acoustic sensor signal generator, receiver, audio assembly and control indicator

Total weight: 68.36 kg
Power supply: 115 V AC, 400 Hz, single phase, 300 W
28 V DC, 280 mA for panel lighting
18 V DC, 250 mA for ASSG annunciators
Frequency: (31 channels) 162.25-173.5 MHz at 0.375 MHz spacing
Noise figure: 5 dB max (3.5 dB is typical)
IF rejection: 66 dB min (>100 dB typical)
Image rejection: 66 dB min (>100 dB typical)
High audio level: 16 VRMS at ±75 kHz deviation
Standard audio level: 2 VRMS at ±75 kHz deviation
Crosstalk: >54 dB min
Output isolation: 60 dB min
Audio frequency response: ±1 dB from −20 Hz to 20 kHz; ±6 dB from 5 Hz to 40 kHz
Audio distortion: −20 Hz to 5 Hz <3% 5-18 kHz <5%
Controlled audio phase characteristics: −100 Hz to 3 kHz ±5°
Audio noise: 10-300 Hz 2 mV
301-2,400 Hz 3 mV
Operational stability: 500 h
Operating life: 20,000 h
Reliability: 500 h MTBF including ASSG and 10 dual-channel control indicators

Operational status

In service with P-3C aircraft and Sikorsky SH-60 LAMPS Mk III helicopters.

Contractor

Flightline Electronics Inc.

VERIFIED

AN/ARR-75 sonobuoy receiving set

The AN/ARR-75 sonobuoy receiving set is a 31-RF channel FM receiver designed for ASW fixed-wing aircraft and shipboard applications. Independent receiver modules provide four simultaneous demodulated audio outputs, each capable of selecting one of 31 RF input channels.

The AN/ARR-75 receiving set is composed of two units. The first of these units is the OR-75/ARR-75 receiver group assembly, which consists of the PP-6551/ARR-75 power supply, four R-1717/ARR-75 receiver modules and the CH-670/ARR-75 chassis. The receiver group assembly contains the majority of the system's electronics.

The second receiving set unit is the C-8658/ARR-75 or C-10429/ARR-75 radio set control. This control unit provides independent RF channel selection and signal strength monitoring of any one of the 31 RF channels for each of the four receiver modules.

Specifications

Dimensions:
(receiver) 203 × 186 × 305 mm
(controller) 122 × 145 × 71 mm
Weight:
(receiver) 9.8 kg
(controller) 1.1 kg
Power supply: 115 V AC, 400 Hz, 3 phase, 75 VA max
27 V (C-8658) or 5 V (C-10429), 5.6 W for lighting at either 27 V (C-8658) or 5 V (C-10429)
Frequency: 162.25-173.5 MHz (31 channels)
Noise figure: 5 dB at 50 ohms
Specifications: MIL-STD-461, 462, 463, 781, MIL-E-5400, MIL-R-81681, AR-5, 8, 10, 34
Operating life: 20,000 h
Reliability: 1,500 h MTBF

Operational status

In service on the LAMPS Mk I, LAMPS Mk III and SH-3H helicopters.

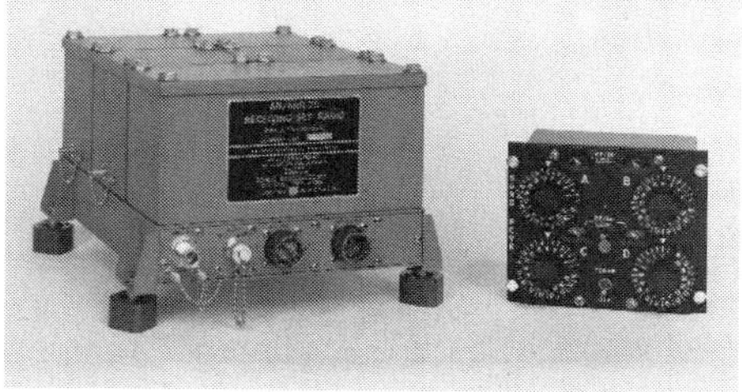

The Flightline Electronics AN/ARR-75 sonobuoy receiver

Contractor

Flightline Electronics Inc.

VERIFIED

AN/ARR-84 sonobuoy receiver

The AN/ARR-84 is designed to receive signals from all current and planned future US and allied sonobuoys. This receiver has four acoustic channels capable of receiving signals from up to four deployed sonobuoys simultaneously on any of 99 RF channels. The AN/ARR-84 is installed on a variety of ASW platforms, including rotary- and fixed-wing aircraft as well as large surface combatants and fast patrol craft. This receiver is a form and fit replacement for its predecessor, the AN/ARR-75 31-channel sonobuoy receiver.

The AN/ARR-84's major assemblies include a power supply, four identical receiver modules and the electrical equipment chassis. The power supply contains the input/output assembly for MIL-STD-1553B and the FSK printed circuit board, allowing digital sonobuoy data extraction.

Additional features of the AN/ARR-84 include dual-selectable IF bandwidths for optimum performance with a wide variety of sonobuoy types. Outstanding AM rejection and mechanical vibration immunity severely reduce spurious returns due to propeller/rotor multipath and platform vibration. Low susceptibility to conducted and radiated energy prevents desensitisation and allows the AN/ARR-84 to be utilised near strong onboard emitters and shipboard search radars.

Flightline also provides an optional radio set control, which is intended for use when an online computer is not available and which provides power control and BIT operation for the receiver group. RF channel selection, sonobuoy type selection and RF level readout is provided independently for each of the four receivers in the group. Operator input is effected through a multifunction keypad, with signal strength provided as a histogram display and entry readback/failure data provided by a 16-character message display.

Specifications

Dimensions: 190.5 × 381 × 254 mm
Weight:
(receiver) 11.34 kg
(optional control box) 1.81 kg
Power supply: 115 V AC, 400 Hz, 3 phase, 100 VA
Frequency: 136-173.5 Hz (99 channels)
Sonobuoy compatibility: DIFAR, LOFAR, Ranger, BT, DICASS, VLAD, CAMBS, Barra, ERAPS, HLA, ATAC, SAR
Environmental: MIL-E-5400 Class 1B
Reliability: 1,500 h MTBF

The Flightline Electronics AN/ARR-84 99-RFchannel sonobuoy receiver shown with the AN/ARN-146 OTPI (on the right)

Operational status

In service with aircraft of the US Navy. Fitted to SH-60B helicopters.

Contractor

Flightline Electronics Inc.

VERIFIED

AN/ARR-502 sonobuoy receiver

The AN/ARR-502 sonobuoy receiver is intended for a wide variety of maritime patrol aircraft and ASW helicopters. The standard configuration system provides for 16 acoustic receiver channels and an OTPI function in a lightweight package. However, the system's modularity allows it to be easily reconfigured to any specific application. The receiver module is a plug-in assembly containing advanced circuitry and software selectable tuning flexibility, allowing for reception of thousands of RF channels down to 5 kHz channel spacing with the current standard being 99. This hybridised receiver module weighs only 0.36 kg and also contains AM demodulation capability.

The AN/ARR-502 features extremely low acoustic receiver baseband noise floor, active mixer technology for unsurpassed third order intermodulation and spurious response performance, threshold extending FM detector with programmable characteristics for sonobuoy type optimisation, progressive AGC with programmable time constants to improve multipath, propeller artefact and countermeasure rejection. It also has latest-generation surface acoustic wave IF filtering for improved phase and delay response to provide the best bearing accuracy for multiplexed buoys and best data error rate for digital buoys, high-resolution self-calibrating signal strength indication and high-resolution local oscillator tuning which can be used to optimise performance with off-tuned sonobuoys and for operating in a countermeasure environment.

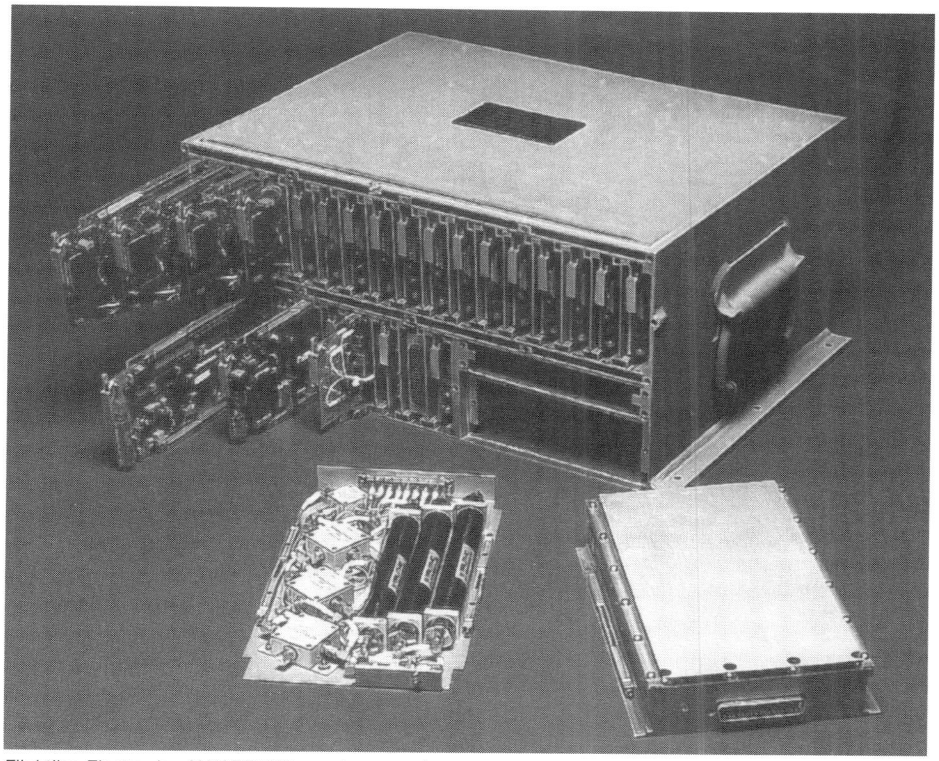

Flightline Electronics AN/ARR-502 sonobuoy receiver system **1999**/0044964

The system is capable of providing acoustic, On-Top Position Indicator (OTPI) and SRS receivers in the same package and has optional dual-antenna input with each receiver capable of independently using either antenna through autonomous or external control. The OTPI receiver is compatible with AN/ARA-25, AN/ARA-48/50 or DF-301E antenna systems. Independent or simultaneous operation from MIL-STD-1553B databus or manual control units is possible, as is operation with either internal or antenna collocated preamplifiers. Variable receiver configurations allow 16, 20 or 40 acoustic channel packaging options in a single package.

Reliability and maintainability features are fully solid-state with no mechanical relays and self-test hardware and software perform complete sensitivity and signal-to-noise ratio tests on receiver modules. All self-test failure data is logged in non-volatile memory and no special test equipment is needed. The system has full ATE compatibility, with test connectors on modules. Receiver module design is common to any type of analogue, digital, OTPI, scanning, aural monitoring or SRS receiver.

Specifications

Dimensions: 406.4 × 304.8 × 203.2 mm
Weight: 21.77 kg
Reliability: >1,000 h MTBF

Contractor

Flightline Electronics Inc.

VERIFIED

R-1651/ARA On-Top Position Indicator (OTPI)

The solid-state modular construction R-1651 radio receiving set is a 31-channel On-Top Position Indicator (OTPI) which is being used on board ASW rotary- and fixed-wing aircraft to provide bearing information and on-top position indication of deployed sonobuoys. This OTPI, when used in conjunction with a suitable ADF system, enables the operator to locate and verify the position of deployed sonobuoys operating on any of 31 VHF RF channels.

The R-1651/ARA is form and fit interchangeable with the R-1047 A/A OTPI. It is compatible with the ARA-25 and ARA-50 antennas and also available in a modified configuration which is compatible with the DF-3010E antenna system. The R-1651/ARA is controlled by the optional C-3840/A control box.

The R-1651/ARA is a single-conversion, 31-channel superheterodyne VHF receiver which is designed to receive signals from deployed sonobuoys over the frequency range of 162.25 to 173.5 MHz.

The antenna input is switched by a relay to the receiver or the UHF receiver, on external command. The RF signals pass through bandpass filters, are amplified and converted to an IF of 25 MHz. After IF amplification and detection, the resultant audio signal is amplified to the level required by the ADF system. The AGC level is developed at the detector, amplified and used for gain control of the IF and RF stages and to operate the adequate signal strength circuits.

The local oscillator provides one of 31 possible LO signals. The same number of crystals is appropriately selected using external controlled code lines containing a 6-bit binary code. Line receivers, decoding gates and matrix drivers connect one of the 31 crystals to the local oscillator circuits.

The optional C-3840/A radio receiver control is used to switch the antenna system, by energising an external coaxial relay, from the UHF/ADF mode of operation to the sonobuoy OTPI mode of operation to select any one of the 31 RF channels and to indicate adequate signal strength when the R-1651/ARA receives an RF signal of –86 dBm or greater.

Specifications

Dimensions: 76.2 × 124.46 × 134.6 mm
Weight: 1.59 kg
Power supply: 115 V AC, single phase, 15 VA
28 V DC, 20 W
Frequency: 162.25-173.5 MHz (31 channels)
Altitude: up to 35,000 ft
Temperature range: –54 to +55°C
Operating life: 5,000 h
Reliability: 1,000 h MTBF

Operational status

In service in the US Navy P-3C Updates I and III and the Sikorsky SH-60 LAMPS Mk III.

Contractor

Flightline Electronics Inc.

VERIFIED

Data Loader Recorder (DLR) RD-664/ASH

The Data Loader Recorder (DLR) provides the capability to store and retrieve digital data on a non-volatile, transportable media. The Removable Media Cartridge (RMC) is based on field-proven, high-capacity 3.5-inch (9 Gigabytes) Winchester disk products currently used on many shipboard, airborne, ground-mobile, and ground-fixed platforms. The DLR contains one 9 Gigabyte RMC, a Power Converter Assembly (PCA), and a SCSI Differential Converter (SDC). The system is designed to facilitate operational use and maintenance; it does not require any periodic maintenance. The DLR has a computer system interface that meets SCSI-2 and SCSI-3 differential. The DLR operates from 28 VDC input through an EMI filter and the PCA converts to +5 VDC and + 12 VDC to power the RMC and the SDC. When the hinged access door to the RMC is opened, power is removed from the RMC. The DLR can operate through significant shock and vibration levels through the use of a shock/vibration isolation frame for the RMC. The operating temperature is -20°C to +55°C with an MTBF of 15,000 hours. The DLR performs self-diagnostics on power up and when commanded over the SCSI bus. A fault indicator and a disk activity indicator are provided on the front panel.

Specifications

Dimensions: 146.1 × 228.6 × 279.4 mm
Weight: 7.26 kg
Formatted capacity: 9 Gbytes
Power supply: 28 V DC
Temperature range: -20°C to +55°C
Shock: 20 g 5.5 ms
Altitude: 35,000 ft
Reliability: 15,000 h

Contractor

General Dynamics Information Systems.

NEW ENTRY

Laser transmitter

General Dynamics Information Systems is developing an improved airborne laser transmitter for aircraft-to-submarine communications, under contract to the US Navy. Experiments conducted during the past few years with a blue-green laser system have demonstrated the feasibility of communicating through cloud with a submerged submarine at significant depths, and the existing high-performance equipment is being upgraded to operate with a high area coverage rate in various weather conditions and water types. Laser communication systems offer a variety of benefits for use with submerged submarines compared with conventional radio communications, including less distortion as the beam travels through water, higher data rates, greater security and less susceptibility to jamming.

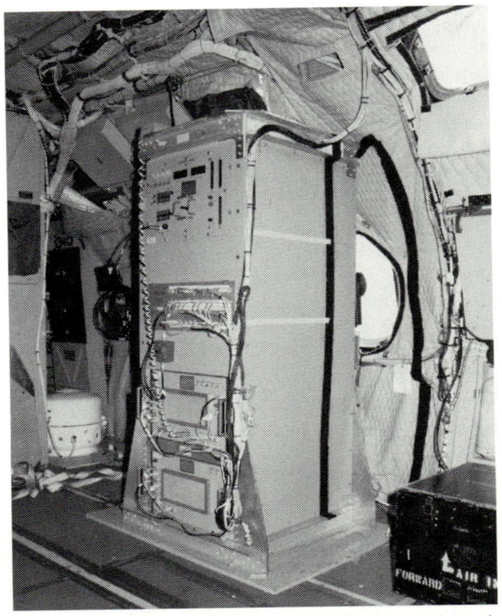

The General Dynamics Information Systems air-to-air laser communication system

It is intended that the airborne transmitter will provide nearly instantaneous communication between carrier-based aircraft and submerged submarines in tactical situations. In addition, advanced versions of the transmitter would enable a space platform to communicate both tactical and strategic messages over large areas.

Lasers are also being developed for air-to-air applications under the Have Lace Laser Airborne Communications Experimental programme. GTE is producing two terminals, mainly from off-the-shelf equipment, which include laser transmitters and receivers as well as acquisition and tracking equipment.

Operational status

In development. The blue-green laser transmitter was mounted in a P-3C Orion aircraft for experimental flights. The Have Lace terminals were evaluated on board a Boeing KC-135 testbed aircraft. GTE is also developing submarine optical receivers based on atomic resonance filter techniques.

Contractor

General Dynamics Information Systems.

UPDATED

Removable Memory Systems

General Dynamics Information Systems produces a range of Removable Memory Systems (RMS) intended for applications requiring removable ruggedised data storage. The series provides compact, cost effective, and upgradable memory solutions.

RMS-1018

The RMS-1018 is based on field-proven, high-performance 2.5 in Winchester disk technology. The RMS-l 018 consists of an environmentally sealed disk with over 18 Gbytes of formatted data storage. It has internal mechanical isolation for shock and vibration and a standard SCSI-2 interface. An internal heater and heater control for extended temperature operation is optional. Compatibility with previous RMS series products is maintained.

Specifications

Dimensions: 58.4 × 124.5 × 177.8 mm
Weight: 1.54 kg
Formatted capacity: 18 Gbytes
Power supply: +5, +12 or +28 V DC
Temperature range: –51 to +71°C
Shock: 30 g 11 ms and 125 g 2 ms
Altitude: 50,000 ft
Reliability: 130,000 h

RMS-1072

The RMS-1072 is based on 3.5 in Winchester magnetic disk technology. The system consists of an environmentally sealed disk with 72 Gbytes of formatted data storage and either a standard Ultra SCSI or a standard Fibre Channel Arbitrated Loop (FC-AL) interface. An internal heater and controls for extended temperature operation is optional. The heater requires 30 watts of either + 12 V DC or +28 V DC.

Specifications

Dimensions: 58.4 × 124.5 × 177.8 mm
Weight: 2 kg
Formatted capacity: 72 Gbytes
Power supply: 115 V AC, 60-400 Hz
Temperature range: –40 to +71°C
Shock: 15 g 11 ms
Altitude: 60,000 ft
Reliability: 130,000 h

RMS-1910

Similarly to the RMS-1072, the RMS-1910 is based on 3.5 in Winchester magnetic disk technology. The RMS-1910 consists of an environmentally sealed disk with 9.1 Gbytes of formatted data storage vibration and a standard SCSI-2 interface. An internal heater and controls for extended temperature operation is also optional. The heater requires 30 watts of either +12 V DC or +28 V DC.

Specifications

Dimensions: 58.4 × 124.5 × 177.8 mm
Weight: 2 kg
Formatted capacity: 9.1 Gbytes
Power supply: 115 V AC, 60-400 Hz
Temperature range: –40 to +71°C
Shock: 15 g 11 ms
Altitude: 60,000 ft
Reliability: 130,000 h

RMS-1648

The RMS-1648 is based on 2.5 in Winchester disk technology. The RMS-1648 consists of an environmentally sealed disk with over 6.4 Gbytes of formatted data storage. It has internal mechanical isolation for shock and vibration and a standard SCSI-2 interface. An internal heater and heater control for extended temperature operation is optional. Compatibility with previous RMS series products is maintained.

Specifications

Dimensions: 58.4 × 124.5 × 177.8 mm
Weight: 1.54 kg
Formatted capacity: 6.4 Gbytes
Power supply: +5, +12 or +28 V DC
Temperature range: –51 to +71°C
Shock: 30 g 11 ms and 125 g 2 ms
Altitude: 50,000 ft
Reliability: 130,000 h

Contractor

General Dynamics Information Systems.

NEW ENTRY

Tactical Disk Recording System - LANTIRN (TDRS-L)

The TDRS-L provides the capability to compress and record digital video from the LANTIRN FLIR simultaneously with other aircraft navigation and sensor data. It is configured for installation in the standard LANTIRN targeting pod. The unit consists of a chassis with liquid coolant cold plate, a Recorder Electronics Unit (REU), a Removable Memory System (RMS) and a Power Converter Unit (PCU). Interfaces to the TDRS-L include the FLIR video, one RS-232 port for control and status, a second RS-232 port for maintenance, an Ethernet port and an external SCSI port.

The digital video is compressed by a jpeg module through the selection of one to five predefined quantisation tables and formatted by the processor into an industry-standard graphics file format (AVI). All of the data is recorded on a removable disk cartridge, model RMS-1910 (> 90 minutes recording time, typical). The digital video files can be replayed on a PC or workstation with a standard CODEC and standard multimedia software. The aircraft data is recorded in separate files on the removable disk. This data is synchronised to the digital video by frame number.

Specifications

Dimensions: 255.3 × 262.9 × 152.4 mm
Weight: 11.35 kg
Formatted capacity: 9 Gbytes
Power: 115 V AC, 60-400 Hz
MTBF: 5,000 h

Contractor

General Dynamics Information Systems.

VERIFIED

Tactical Disk Recording System - RAID (TDRS-R)

The Tactical Disk Recording System-RAID (TDRS-R) provides a high-performance data storage system suitable for use in harsh environments. Minimum sustained transfer rates of over 90 Mbytes/second and burst transfers of 100 Mbytes/second are possible. The TDRS-R is housed in two ½ ATR-style chassis; the Storage Array Module (SAM) and the electronics unit. The SAM is mounted in a SAM Isolation Frame (SIF) that provides shock and vibration isolators for the disk array and can be easily inserted and removed without the use of tools. The electronics unit, which can be configured with various I/O interfaces, contains the RAID controller and an optional general processor. The standard I/O interface for TDRS-R is Fibre Channel Arbitrated Loop (FC-AL). The SAM currently has a total formatted capacity of 288 Gbytes when used in a RAID configuration, or 360 Gbytes raw capacity. SAM modules can be daisy-chained for greater capacities.

Specifications

Storage Array Module (SAM)
Dimensions: 238.8 × 132.1 × 208.3 mm
Weight: 6.4 kg
Formatted capacity: 288 Gbytes with redundancy
Power supply: +5 V DC and +12 V DC
Temperature range: -40°C to +71°C
Shock: 15 g 11 ms
Linear acceleration: ±9 g all axes
Altitude: 50,000 ft
Reliability: 26,000 hours

Electronics Unit
Dimensions: 355.6 × 124.5 × 198.8 mm
Weight: 5.9 kg without cards installed
Power supply: + 115 V AC single phase or +28 V DC
Temperature range: -40°C to +71°C
Shock: 15 g 11 ms
Linear acceleration: ±12 g all axes
Altitude: 50,000 ft
Reliability: 18,000 hours (includes electronics)

Contractor

General Dynamics Information Systems.

UPDATED

FCC100 automatic flight control system

The FCC100 system was designed specifically for the US Army Sikorsky UH-60 Black Hawk helicopter. It meets requirements calling for quick and easy field maintenance and repair and is claimed to have been the first application of a dual-digital processor-based stability augmentation system for a production helicopter. Features include three-axis stability augmentation with turn co-ordination, pitch, roll and yaw attitude and altitude/airspeed hold modes.

Specifications

Dimensions: 356 × 305 × 217 mm
Weight: 8 kg
Power: 75 W

Operational status

In service on the Sikorsky UH-60 Black Hawk helicopter.

Contractor

Hamilton Sundstrand Corporation.

VERIFIED

FCC105 automatic flight control system

The FCC105 automatic flight control system introduced modern digital technology features in Sikorsky HH-60, CH-53E, SH-60 and MH-53E helicopters. It is a dual-redundant system which provides three-axis stability augmentation, turn co-ordination and hands-off flight. It can also include a stick-force feel system and has four degrees of freedom facilities and hover augmentation. The most recent version, on the Sikorsky MH-53E helicopter, includes MIL-STD-1553B digital data interfaces. A non-redundant derivative is utilised on the SH-60B. A further derivative of this, with the nomenclature 'general computer', has been developed for use on H-60 helicopters other than the SH-60B.

The baseline 53E computer used in the Sikorsky SH-60B includes automatic approach to hover, three-axis Doppler and radio altitude coupling, with provision for tactical navigation computer coupling.

Specifications

Dimensions: 190 × 432 × 266 mm
Weight: 11.5-13.5 kg
Power: up to 83 W

Operational status

In service on Sikorsky CH-53E, MH-53E, HH-60 and SH-60 helicopters.

Contractor

Hamilton Sundstrand Corporation.

VERIFIED

AN/ASW-54(X) Link 16 Interoperable Tactical DataLink (LITDL)

The AN/ASW-54(X) Link 16 Interoperable Tactical DataLink (LITDL) multinet TDMA architecture supports mixed force operations and provides a real-time tactical situational awareness capability to tactical forces.

Multilink flexibility ensures connectivity to existing datalink systems and an innovative gateway systems architecture interoperates with the broad area command and control JTIDS net.

LITDL contains an airborne datalink for tactical aircraft and UAVs. The transponder mode, for operation with a command and control terminal, also contains a fighter-to-fighter situational awareness mode. The integrated self-contained unit consists of a UHF transceiver, programmable modem, MIL-STD-1553 interfaces and a military power supply. BIT is incorporated in the unit.

Specifications

Dimensions: 128.3 × 180.6 × 273 mm
Weight: 6.23 kg
Power supply: 28 V DC
Frequency: 300-324.9 MHz
Channels:
(FSK) 250 100 kHz channels,
(anti-jamming) 250 orthogonal channels

Contractor

Harris Corporation.

VERIFIED

Digital Map Generator (DMG)

The Digital Map Generator (DMG) uses stored terrain data to generate moving terrain map displays, to eliminate the need for paper maps in the cockpit and to improve the pilot's situational awareness. These map displays can either be digitised representations of standard aeronautical charts or topographical displays created from digitised elevation and cultural feature data. Symbology overlays are added to show mission features such as waypoints, flight paths and threat and target positions, and assist the pilot with terrain-aided navigation and ground collision avoidance.

Harris is also providing the high-speed databus interface for the data transfer unit, a mass memory storage device which stores digital map and mission planning data.

Operational status
In development for the RAH-66 Comanche helicopter.

Contractor
Harris Corporation.

VERIFIED

Digital Terrain Management and Display (DTM/D)

The Digital Terrain Management and Display (DTM/D) has been test flown on the US Air Force Advanced Fighter Technology Integration (AFTI)/F-16 aircraft. It combines a Harris digital map generation system with a colour multifunction display from Honeywell Aerospace, and is expected to provide a navigation system for combat aircraft 10 times more accurate than conventional systems but with only minimal electromagnetic emissions, at the same time leaving the aircraft's radar free for other purposes.

The DTM/D correlates radio altimeter information with a digitally stored map, providing for covert navigation which will, according to the US Air Force, enhance terrain-masking tactics and give the pilot the ability to look beyond hills in front of and to the side of the aircraft with a passive ranging capability. This will allow the pilot to work out the best route well in advance and be able constantly to relate his position to threats and targets.

Operational status
Flight testing in the AFTI/F-16 has been completed. Harris worked under a US$5.9 million contract jointly funded by the US Army and Air Force.

Contractor
Harris Corporation.

VERIFIED

AN/USQ-86(V) datalink

The AN/USQ-86(V) is a modular integrated communications and navigation system which provides a nuclear and ballistic-hardened anti-jam datalink for air-to-ground applications. Modules of the system, which has been developed under the auspices of the US Army's Electronics Research and Development Command, can be used with a variety of manned aircraft and remotely piloted vehicles.

In the manned aircraft role the system can be used as part of a standoff target radar system wideband sensor configuration for remote-control airborne intercept, to compromise enemy communications and to locate hostile command and control facilities.

The key element in all these facilities is the secure air-to-ground datalink. This is achieved by a combination of spread spectrum modulation techniques, message coding, signal processing and advanced antenna techniques.

Operational status
In late development.

Contractor
Harris Corporation, Electronic Systems Sector.

VERIFIED

AN/ZSW-1 weapon control datalink

The AN/ZSW-1 weapon control datalink is flight qualified on the F-15E for use with the GBU-15 and AGM-130 standoff guided weapons. It provides real-time target acquisition and battle damage information at long standoff range and is resistant to jamming in all RF environments. The AN/ZSW-1 is fully MIL-qualified and is adaptable to other aircraft and missile configurations.

Operational status
In production.

Contractors
Harris Corporation, Electronic Systems Sector.
Raytheon Electronic Systems.

VERIFIED

Multifunction/multiband antenna subsystem

The tactical airborne antenna array with multiband and multifunction antennas is integrated into a single aperture structure and configured so that the entire structure can be conformally mounted on the aircraft fuselage. An additional feature is the electronics system which is needed to interface the aperture to radios and also provides anti-jam adaptive processing. It covers VHF, UHF and L-, S- and C-bands (NATO D, E/F and G/H bands).

The switching electronics is a highly integrated array of RF switches, power dividers and receive amplifiers which provide multi-use selection of each major antenna band. The resource manager is a general purpose processor with interfaces to a radio via mission avionics bus and intercoms. Its function is to manage antenna resources in order to optimise coverage and minimise interference.

Specifications
Weight: 58.1 kg
Reliability: 1,800 h MTBF

Contractor
Harris Corporation, Electronic Systems Sector.

VERIFIED

TCDL Tactical Common DataLink programme

The US Defense Airborne Reconnaissance Office (DARO), in conjunction with the US Defense Advanced Research Projects Agency (DARPA), who are the contracting and technical agency, has selected three contractor teams for Phase I of the TCDL programme.

The TCDL programme is a multiple phase, multiple award programme to develop a family of CDL interoperable digital links to support both unmanned and manned airborne reconnaissance platforms including: Outrider, Predator, Reef Point, Rivet Joint, Joint STARS, Airborne Reconnaissance Low (ARL) and others. The TCDL will support air-to-surface transmission of radar, imagery, video and other sensor information at ranges up to 200 km. The TCDL will operate with existing CDL systems operating at the 10.71 Mbps return link and the 200 kbps command link data rates.

Operational status
Phase I, during which the teams will develop designs for airborne and surface TCDL terminals for tactical unmanned vehicle applications, this will last six months. DARO and DARPA will then select two teams to continue into Phase II, an 18 month prototype development and demonstration effort.

Contractor teams
US DARO and US DARPA contracting three teams.

Team 1:Harris Corporation, Melbourne; BAE Systems North America, Fort Wayne; TSI Telsyn Inc, Columbia.

Team 2: Lockheed Martin, Salt Lake City; Rockwell Collins, Cedar Rapids.

Team 3: Motorola, Scotsdale; Raytheon Electronic Systems, Falls Church; Cubic Defense Systems, San Diego.

VERIFIED

AN/APN-215(V) radar

The AN/APN-215(V) colour radar is a weather, surface search and precision terrain-mapping system derived from the successful and widely used RDR-1300 commercial system. It is designed for heavy twins, turboprops and transport helicopters. Low weight and a 445 km range suit it to utility and reconnaissance aircraft and it was chosen for the US Army versions of the Beech King Air, the U-21 and the RU-21.

In conjunction with other equipment the APN-215 can display navigation pictorial information overlaid on the weather map, together with pilot-programmable pages of checklist information such as en route navigation data and emergency procedures.

The system comprises three units: a 305 mm pitch and roll-stabilised antenna, transmitter/receiver and colour control/display unit.

Operational status
In service with the US Army and Coast Guard.

Contractor
Honeywell Aerospace, Electronic & Avionics Lighting.

VERIFIED

AN/APN-234 multimode radar

The AN/APN-234 is a lightweight airborne digital colour display multimode radar designed to provide sea search weather detection and terrain-mapping for a variety of military aircraft including rotary- and fixed-wing types ranging from light to heavy twins. The system consists of a receiver/transmitter, combined colour display/control unit, stabilised antenna and optional interface unit.

The AN/APN-234 is identical to the AN/APN-215 except for the addition of the sea search function.

Operational status
In service with the US Navy on the Northrop Grumman C-2A Greyhound and the Lockheed Martin EP-3E aircraft.

Contractor
Honeywell Aerospace, Electronic & Avionics Lighting.

VERIFIED

AN/APS-133 radar

The AN/APS-133 digital colour radar is a high-performance weather, beacon-homing and terrain-mapping system designed for large commercial and military transports. The RDR-1FB (Type 1) was originally launched for the retrofit of the US Air Force's C-141 Starlifter fleet and has since been installed on the C-5, E-3 and KC-10 aircraft among others.

The multicolour display can be used in conjunction with other equipment to show programmable checklists or to superimpose navigation or other information on the weather map. The system employs digital processing and microcomputer techniques, as well as a solid-state modulator.

In November 1984, the company delivered to the US Navy and Marine Corps the first units of the RDR-1FB (M) (Type 2) improved land-mapping version for its fleet of KC-130 tankers and C-130 transports; the unit now equips the entire fleet of 70 aircraft. It was specially suited to US Marine Corps requirements, with a high PRF, short pulsewidth, enhanced digital processor and selectable sector-scan antenna to improve radar navigation at low level.

The US Air Force has selected the Type 2 for its E-4 NEACP and VC-25A (Air Force 1 Boeing 747) aircraft and the US Navy uses the unit in its EA-6A Intruder aircraft fleet.

Significant landmarks and continental shorelines up to 555 km away can be portrayed in the ground-mapping mode by using the high-power output concentrated into a pencil beam. At the same time, discrete details such as lakes, rivers, bridges, runways and runway approach reflectors readily show up on the colour display. To improve range resolution at short ranges the system operates with 0.4 µs (RDR-1FB(M)) or 0.5 µs (RDR-1FB) pulses in contrast to the 5 µs pulses used for long-range ground- and weather-mapping.

In the air-to-air mode, the APS-133(V) detects and tracks other aircraft during rendezvous, formation and air refuelling. Aircraft of C-130/C-141 size can be tracked to 56 km, but may still be resolvable at ranges as little as 550 m depending on relative bearing, aspect and altitude.

To provide long-range homing to remote ground destinations or tanker aircraft, the APS-133(V) operates at I-band frequencies - 9,375 MHz for beacon interrogation and 9,310 MHz for beacon reception. The identification of closely spaced pulse reply codes at long ranges is made possible by the marker and delay modes of the radar indicator. In the marker mode a variable marker is positioned on the screen just in front of the beacon reply. When switched to the delay mode the display presentation starts at the marker range. The range switch can then be moved to select a shorter range scale, yielding an expanded view of the area containing the beacon reply.

Derived from the Bendix/King RDR-1F used on many hundreds of airliners, the AN/APS-133(V) comprises five LRUs: a 762 mm fully stabilised split-axis parabolic antenna that provides specially shaped search or fan beams for terrain-mapping and skin painting, a transmitter/receiver, a colour display, a radar control unit and an antenna sector-scan control unit (RDR-1FB (M) only).

Specifications

Weight:
(antenna) 15.8 kg
(transmitter/receiver) 22.2 kg
(sector-scan control unit) 1.2 kg
(colour indicator) 6.3 kg
Frequency: I-band (9,375 MHz transmit, 9,310 MHz receive)
Power output: 65 kW
PRF:
200 pps (Type 1 system)
200 and 800 pps (Type 2 system)
Pulsewidth:
(weather) 5 μs
(beacon) 2.35 μs
(mapping) 0.5 μs (or 0.4 μs Type 2, selectable)

Operational status

In service with US Air Force transport aircraft, notably C-5A, C-17 and C-141 and KC-10 Extenders, also E-4A, E-3A, VC-25A and in US Navy/USMC aircraft such as C/KC-130, EA-6A, E-6A, A-3 and YP-3C.

Contractor

Honeywell Aerospace, Electronic & Avionics Lighting.

VERIFIED

AN/APS-133(TTR-SS) multimode radar system

The AN/APS-133(TTR-SS) multimode radar is a coherent pulse Doppler system. It is designed for military tanker and transport aircraft.

The APS-133's advanced digital signal processing architecture provides operational advantages like frequency agility, pulse compression and Doppler beam-sharpening with monopulse resolution enhancement. The system has a precision ground-mapping capability and the capability to detect and display weather and turbulence. To aid in differentiating between mountain shadows and low reflectors such as lakes, and to allow detection of ridge lines, the system utilises a selectable fast time constant. When selected, this gives the appearance of a three-dimensional picture. Both the receiver/transmitter and the digital processor provide fault isolation to the LRM level, and fault isolation and storage to the LRU level.

The APS-133(TTR-SS) features calibrated turbulence detection and display on all range settings, signal attenuation compensation for more accurate weather display, solid-state digital design and independent roll axis stabilisation. There is automatic beacon decoding for APX-78 and APN-69 beacons, monopulse operation for separating closely spaced targets, and a freeze-frame facility. The system has a MIL-STD-1553B dual-digital bus for FMS and INS interface, antenna stabilisation amplifiers for pitch, roll and tilt and multishade monochrome display capability with colour enhancement. It can interface with and display station-keeping data from the AN/APN-169C and IFF data from the AN/APX-76.

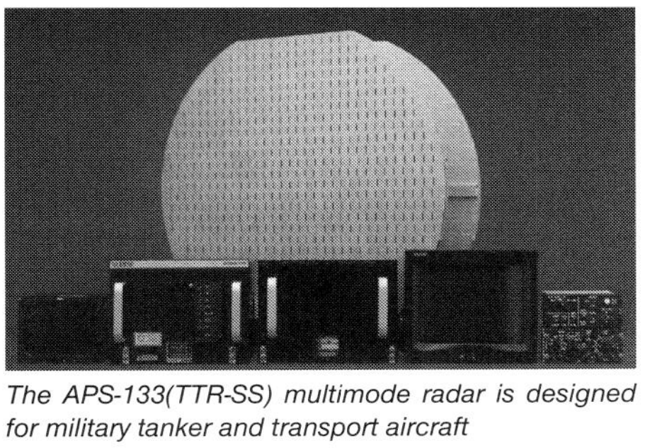

The APS-133(TTR-SS) multimode radar is designed for military tanker and transport aircraft

Specifications

Weight:
(total system) 62 kg
Power supply: 115 V AC, 275 W
5 V AC, 10 W (panel lighting)
28 V DC, 100 W

Contractor

Honeywell Aerospace, Electronic & Avionics Lighting.

VERIFIED

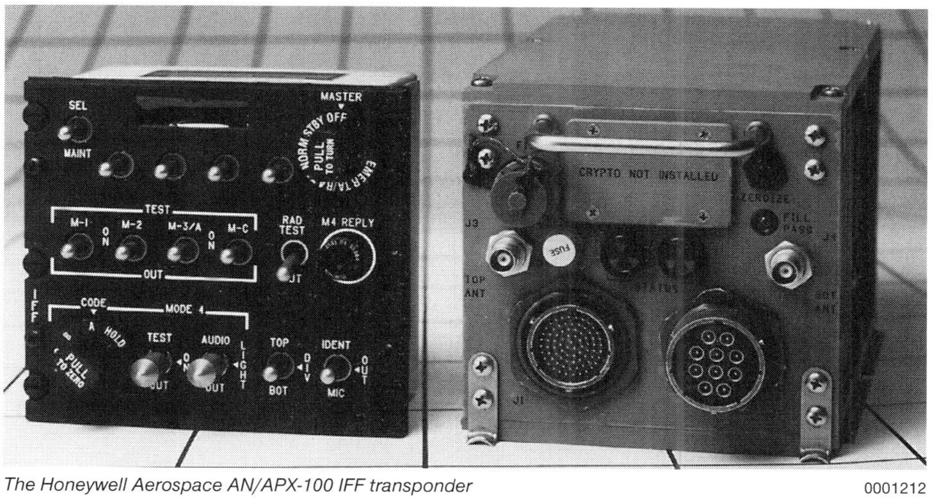

The Honeywell Aerospace AN/APX-100 IFF transponder 0001212

AN/APX-100 IFF / Mode S transponder

The AN/APX-100 IFF mode S transponder is a panel-mounted IFF transponder using microminiature technology in both digital and RF circuitry. It is in production for a number of US military aircraft. The system is a completely solid-state, modular constructed equipment with a complete dual-channel diversity system, comprehensive BIT, digital coding and encoding and a high anti-jamming capability. Two antennas form part of the equipment and the diversity system receives signals from each and switches the transmitter output to the antenna which received the stronger signal. This is designed to cure the problems of poor coverage with a single antenna.

The transmitter is all-solid-state with a 500 W peak power output obtained from four parallel microwave transistors. Two additional transistors complete the transmitter oscillator and driver stages. The diversity system provides improved antenna coverage and allows improvement in performance in overloaded, jamming and multipath environments. Automatic overload control and anti-jamming features are also incorporated.

For aircraft with very limited cockpit space, an equipment bay-mounted configuration, the RT-1157/APX-100(V), is available. In addition, the RT-1471/APX-100(V), a databus version operating in accordance with MIL-STD-155B, is available.

The latest in the APX-100 family includes Mk XII and embedded Mode 4 integrated crypto, full Mode S Level 3 and GPS position reporting for advanced air traffic control systems and TCAS systems. There are also NVG-compatible units available which operate at low light levels.

Specifications

Dimensions: 136.5 × 136.5 × 212.7 mm
Weight: 4.53 kg
Frequency:
(transmitter) 1,090 ± 0.5 MHz
(receiver) 1,030 ± 0.5 MHz
Peak power: 500 W ± 3 dB

Operational status

The APX-100 replaced the AN/APX-72 as the standard US military ATC transponder. Over 13,000 units have been built to date.

In production for all new US Navy, Army and Air Force aircraft including the AF-1, AH-1S/T, AH-64, AV-8B, C-5, C-12, C-17, C-20, C-21, C-23A, CH-47, C-130, EC-2C, EC-130, F-14D, F-18, F-22, HH-60, HH-65, LAMPS, MH-47, MH-60, OH-58, OV-10, RAH-66, SH-60, T-45A, UH-60, V-22 and VC-6.

The USA is delivering 113 APX-100 Mk XII airborne transponders to Hungary under a US$12.7 million contract.

The equipment is also being produced under licence in Japan by Toyocom.

Contractor

Honeywell Aerospace, Electronic & Avionics Lighting.

VERIFIED

AN/ASK-7 data transfer unit

The AN/ASK-7 data transfer device, originally designed for use with the Boeing AGM-86 air-launched cruise missile, has now been adapted for the US Air Force B-52 and B-1B aircraft.

The unit transfers large quantities of digital information from ground-based master systems to aircraft equipment. It will also monitor performance parameters as part of the central integrated test system on the B-1B. In the B-52, it forms a small part of the current Offensive Avionics System (OAS) update programme to improve the effectiveness of the aircraft.

Major features of the AN/ASK-7 include full record/replay capability, MIL-STD-1553B interface, error correction code, remote electronics, extensive built-in test and nuclear hardening. The system comprises a control unit, containing most of the recorder electronics, and an instrument-mounted cassette drive unit accommodating two cassettes at a time.

Specifications

Weight:
(cassettes) 3.1 kg
(cassette drive unit) 3.6 kg
(control unit) 6.8 kg
Power supply: 115 V AC, 400 Hz, 20 W

Operational status

In service.

Contractor

Honeywell Aerospace, Electronic & Avionics Lighting.

VERIFIED

Digital fully fly-by-wire system for the F-16C/D

The system has evolved from the equipment developed by Honeywell Aerospace for the US Air Force Advanced Fighter Technology Integration (AFTI) programme based on an F-16 airframe. A fundamental change from the flight control system in the AFTI aircraft is, however, the progression from three digital plus one analogue channels per axis to four digital channels, the configuration favoured for operational reliability (the system has to remain operational after any two failures). The four digital channels use MIL-STD-1750A architecture utilising Jovial high-order language and each has a processor, memory, input/output functions, discrete failure logic, MIL-STD-1553B digital databus and serial link to other channels. The computer is designed to remain fully operational following any two consecutive failures within the quadruplex part of the system. The single unit, weighing 22.7 kg and housed in a 1 ATR long box, has the equivalent function of the four separate units on the AFTI aircraft and occupies about half the space. In all other respects the system meets the form, fit and function requirements that permit it to replace directly the earlier analogue system.

Computing and processing technology is based on large-scale and very large-scale integration and gate arrays. It is designed to accommodate further growth, incorporating VHSIC components, without the need for new software. Flight critical functions are hard-wired into the system for the highest integrity, but less critical signals communicate with other equipment via the digital databus. The system has 48 kwords of PROM memory, 2 kwords of RAM scratchpad memory, 8 kwords of input/output scratchpad memory and a 2 kword memory to record faults.

Benefits of the all-digital system over the previous analogue one are given as better reliability, lower power density, smaller number of components, greater ease of tuning to meet changes and lower life cycle costs.

Honeywell Aerospace considers this flight control system to be an important step forward, since it can be seen as the basis for future integrated flight-critical full authority flight control systems incorporating thrust/side-force vectoring, terrain-following and fire control.

Specifications

Dimensions: 514 × 273 × 222 mm
Weight: 22.7 kg
Power: 200 W
Reliability: >2,150 h demonstrated MTBF

Operational status

Installed in later F-16C and F-16D single-seat and two-seat fighters.

Contractor

Honeywell Aerospace, Electronic & Avionics Lighting.

VERIFIED

The RN-262B (AN/ARN-127) VOR/ILS system

RN-262B (AN/ARN-127) VOR/ILS system

The RN-262B VOR/ILS system comprises a compact, solid-state 200-channel VOR/Loc remote-mounted receiver, a 40-channel glide slope receiver and marker beacon receiver, and a panel-mounted control/display unit. All three receivers operate independently.

Specifications

Dimensions:
(receiver) 183 × 130 × 319 mm
(control unit) 146.1 × 66.7 × 114.3 mm
Weight:
(receiver) 4.5 kg
(control unit) 1 kg
Power supply: 26 V AC, 400 Hz, 25 VA
28 V DC, 1.5 A

Operational status

In service. The system has been adopted by the US Air Force, Army and Coast Guard.

Contractor

Honeywell Aerospace, Electronic & Avionics Lighting.

VERIFIED

AMS 2000 Avionics Management System

Honeywell's AMS 2000 is a fully integrated modular avionics concept/system designed to meet the needs of military tanker/transport/maritime patrol aircraft. It integrates flight management systems, flight controls, cockpit instruments and cockpit controls. Modular architecture is built around the system processor with growth to TCAS, MLS, HUD, digital map, NVIS-compatibility, DGPS and CNS/ATM.

Operational status

In service with C-130 and P-3 aircraft of various countries.

Contractor

Honeywell Defense Avionics Systems.

VERIFIED

ARINC 758 Communications Management Unit (CMU)

The ARINC 758 CMU is an airborne communications router that supports datalink service access between aircraft datalink applications, such as AOC/AFIS, CPDLC and ADS and their corresponding service providers; it includes an Aircraft Personality Module (APM).

The CMU initially provides an ARINC 724B-compatible datalink router is through which all character-oriented data transmitted to and from the ground-based Aircraft Communications Addressing and Reporting System (ACARS) network. This CMU is designed to ARINC Characteristic 758 and can be upgraded, by software, to an Aeronautical Telecommunications Network (ATN) router when protocols and application infrastructure are available to support CNS/ATM datalink applications.

The initial CMU provides Level 0.1A functional capability, with growth via software updates to Level 2-D, as application functionality becomes available. Level 0.1A functionality is operationally equivalent to an ARINC 724B ACARS management unit. This includes the use of a VHF Data Radio (VDR) operating with ARINC 618 protocols and modulation interfacing to the CMU via an ARINC 429 interface. Level 2-D functionality is operational over the ATN utilising high-speed services and incorporating Air Traffic Services (ATS) applications.

The CMU MMI is provided via an ARINC 739-compatible, Multipurpose Control and Display Unit (MCDU) interface to display datalink status, entry of text messages for transmission and the display of received text messages for review. A hard copy of datalink messages is provided via an interface with an ARINC 740/744 airborne printer.

Access to the ground network is provided via several ACARS air-to-ground sub-networks.

The CMU interfaces to other onboard systems such as the flight management computer and maintenance data computer utilising ARINC 749 interfaces.

The ARINC 758 CMU can be customised to meet operational needs, including future requirements associated with the implementation of the CNS/ATM and ATN systems. The ability to support dual-CMU installations will be available from the third quarter of 2000.

Specifications

Dimensions: 124 (W) × 193.5 (H) × 324 (L) mm
Weight: <5.5 kg

Operational status

Available January 2000.

Contractor

Honeywell Inc Defense Avionics Systems.

VERIFIED

Avionics Control and Management System for the CH-47D (ACMS)

The Avionics Control and Management System (ACMS) for the CH-47D helicopter provides the primary interface for the pilot and co-pilot to aircraft pilotage, mission and air vehicle systems. Although some components are essential to more than one function, the ACMS can be divided into the following general functional areas: pilotage subsystem, mission subsystem and air vehicle subsystem.

The pilotage subsystem provides control and display of aircraft sensor data, selection of navigation sources, and other functions essential to flying the helicopter. The EFIS, a primary element of the pilotage subsystem, replaces traditional mechanical gauges, indicators and control heads with electronically generated displays, allowing the operator to select and tailor presentations to provide the optimum level of information during each mission phase.

The mission subsystem provides mission planning utilities, storage and retrieval of navigation reference points, and control of communication and navigation radio lists. Pre-mission planning is normally accomplished on a ground-based workstation and downloaded to the aircraft via a data transfer cartridge. The air vehicle subsystem allows continuous performance monitoring of all major aircraft systems including warning/caution/advisory annunciations and the capability to interface to the health and usage monitoring system.

The air vehicle subsystem monitors engines and transmissions, fuel and hydraulic systems, flight controls and actuators, the electrical system, and other aircraft systems and equipment. Critical information about the engines and rotors is continuously available on torque/cruise/RPM displays.

System design incorporates redundancy to minimise vulnerability to failures and damage and to enhance the accuracy of navigation equipment. The ACMS interfaces with other aircraft equipment via two dual-redundant MIL-STD-1553B multiplex databusses, special purpose high- and low-speed ARINC 429 buses, and aircraft discrete, synchro and proportional analogue I/O interfaces. The ACMS also receives video signals from the weather radar system for display in the cockpit. A digital map function provides real-time tracking of helicopter position overlaid on paper chart or digital terrain-elevation data displays. Processing and subsystem interfaces are managed by four integrated system processors (two mission, two air vehicle). The primary crew interface is supported by four multifunction displays (two mission, two air vehicle) and two control/display units. Additional information is displayed on two electronic flight instrumentation systems, each comprising of an electronic attitude/direction indicator and an electronic horizontal situation indicator. Two optical display assemblies, which mount onto the pilots' night vision goggles, provide flight symbology during night-time operation. Navigation redundancy is accomplished through use of a Kalman filter to blend navigation sensor data into the most accurate solution of aircraft position and motion, even during transient periods or loss of individual navigation sensor inputs.

The ACMS provides comprehensive integration and control of mission management, air vehicle management and pilotage subsystems. Its flexible architecture will accommodate growth well into the next century. Central computers provide control and data handling capability for a wide variety of aircraft systems and equipment. This control and computation ability decreases crew workload and enhances navigation accuracy and mission safety. Overall, the ACMS provides the pilot and co-pilot with added functional capabilities, which contribute to a more effective and efficient mission.

Operational status

In service with the Royal Netherlands Air Force. Two additional export customers have also placed contracts.

Contractor

Honeywell Defense Avionics Systems.

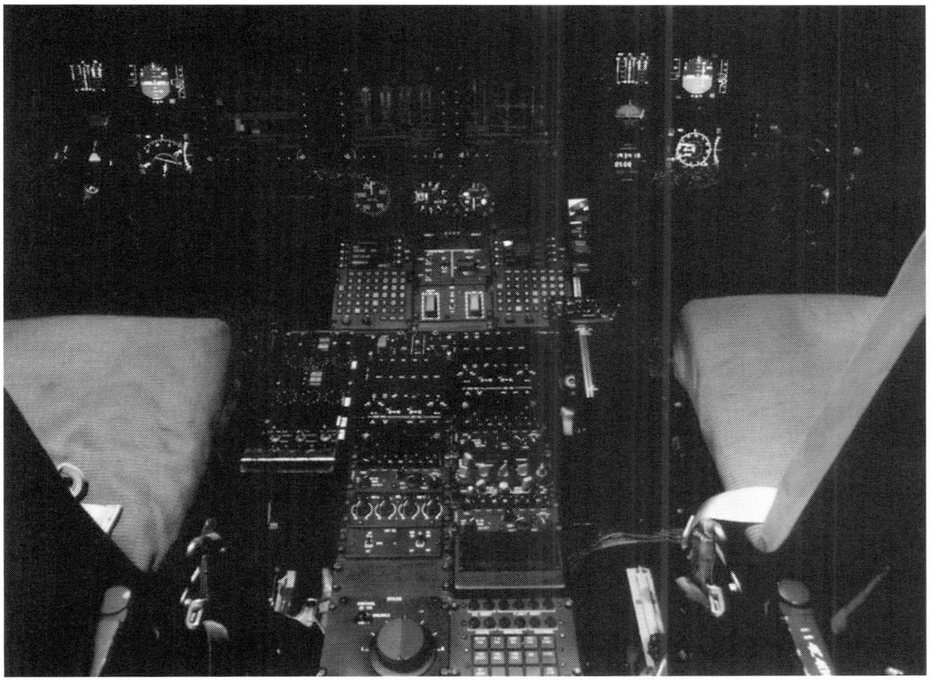

Royal Netherlands Air Force (RNLAF) CH-47 upgraded cockpit

VERIFIED

C-130J Digital Autopilot/Flight Director (DA/FD)

The system consists of dual, Automatic Flight Control Processors (AFCP), each with two central processor units for redundant control law computation and mode logic implementation for: autopilot, flight director and autothrottle performance. System inputs to the control surfaces (ailerons, rudder, elevators) are made through three, individually housed, servo drive units containing: servo motor, tachometer, synchro, solenoid engage clutch, drive spline and electrical connector.

Honeywell's digital autopilot/flight director provides a proven solution for the C-130J programme. Early versions of the system flew on C-130E/Hs with the US (RAMTIP and PRAM), Royal Australian and Royal New Zealand Air Forces. An identical hardware version is flying on the Belgian Air Force C-130Hs. Aerodynamic control laws and flight director functions have been thoroughly tested in a series of development programmes for the US Air Force and for foreign governments.

The C-130J version of this autopilot is qualified to both MIL-STD and Federal Aviation Administration requirements in accordance with MIL-STD-810, MIL-STD-454, MIL-STD-461, AC 25.1309-1A, DO-178A and DO-160C. The autopilot interfaces with C-130J sensors and avionics via a MIL-STD-1553B databus. The interface to the three servo drive units and the autothrottle is via direct wire, to ensure adequate response time.

Key design features include: digital electronics programmed in Ada; fail-passive operation; dual flight control computers, each with two central processor units for redundant control law and mode logic computation (internal comparison of LRU outputs verify each LRU operation); flight director system internal to the autopilot computer; comprehensive built-in test covering 98 per cent of circuitry (power-up, continuous, maintenance diagnostics); parameter modification permitting selected control law gains to be adjusted under flight test-only conditions; autothrottle control for precise speed control; each LRU provides all of the main computing and power amplification functions for the digital autopilot/flight director system; comparison of servo outputs for validity and tolerance (disagreement results in system disengagement).

Automatic flight control processor **1999**/0051638

Operational modes for the autopilot include: heading and altitude select; coupled navigation (global positioning system, inertial navigation, Doppler); radio navigation (VOR/LOC, back-course, TACAN, GS); altitude hold; turn/pitch knob adjust; coupled flight management system operation; pitch/roll go-around; autothrottle; heading-hold; speed-hold; vertical navigation; yaw damper; turn co-ordination; and speed select.

Specifications

Dimensions: 292 × 203 × 505 mm
Weight: 15.8 kg

Operational status

In production.

Contractor

Honeywell Defense Avionics Systems.

VERIFIED

Digital Automatic Flight Control System (DAFCS) upgrade for the B-52, C (KC)-135 and C-130

The DAFCS provided for the B-52 G/H, C-135 A/E/R and C-130 have much in common. Although the total number of LRUs comprising the DAFCS differs depending on the type of aircraft involved, there is a very significant reduction in the total number of LRUs when compared to the analogue system being replaced. In addition, the DAFCS reliability has significantly increased the availability and performance of its host platform. The primary common LRUs comprising the DAFCS are the Flight Control Set Processor (FCSP), the Air Control Unit Sensor (ACUS) and the Status Test Panel (STP).

The FCSP controls the digital autopilot system; it contains a dual Z8001 processor which can be upgraded to a MIL-STD-1750 capability. This LRU contains a plug-in slot for a MIL-STD-1553 circuit card assembly that can be installed in the future to support aircraft growth requirements.

The ACUS provides air data sensor information to the FCSP to support autopilot modes: Mach hold, altitude hold, attitude hold, and so on. The STP provides the man/machine interface to assist maintenance diagnostic testing and provides accessibility to specific failure code readouts.

The DAFCS was delivered in sufficient quantities to allow an upgrade of up to 262 B-52s, 736 C (KC)-135s and a single prototype for evaluation purposes on the C-130 aircraft. A modified version of the DAFCS also has been incorporated into the US Air Force Chief-of-Staff's C-135C aircraft, commonly referred to as the Speckled Trout aircraft.

Operational status

Currently installed on all in-service B-52 and C (KC)-135 aircraft as well as on the Speckled Trout aircraft.

Contractor

Honeywell Defense Avionics Systems.

VERIFIED

Digital map system

Using the technology and experience gained from the US Navy's AV-8B and F/A-18 night attack programme as a baseline, Honeywell has developed an advanced digital imaging system.

The capabilities of the base system have been expanded to include terrain reference navigation,

Honeywell Defense Avionics Systems DAFCS showing left to right: status test panel, air control unit sensor, flight control set processor, cockpit control panel **1999**/0051640

ground collision avoidance, perspective view, terrain-following computations, threat intervisibility and sensor blending for advanced avionics systems.

Digitised aeronautical charts and Defense Mapping Agency Digital LandMass System (DLMS) Level 1 and 2 terrain and feature data are stored and recalled from the digital memory unit to generate a full-colour moving map which provides precise navigation and route selection to and from the point of target acquisition and weapons delivery.

The digital memory unit is a stand-alone mass memory unit designed to provide increased storage to military systems requiring high reliability and quick access times. The unit contains a flight-proven fully militarised optical disk capable of storing 520 Mbytes of data. Information is written to the disk by a laser diode in the head assembly. A beam of light is focused on the medium to produce a hole. During the read operation, the hole is recognised as a bit of data. Unlike magnetic media, which can be scratched by a head crash, data stored on the optical disk is written to a thin alloy encased in a protective substrate and a clear protective cover. Digitised reconnaissance photos, radar data, Landsat data, emergency procedures, let-down and approach plates, mission data and flight plans can be stored with digitised aeronautical charts and DLMS data for a variety of missions.

The digital map computer performs all the airborne map generation. The configuration on board the US Navy AV-8B night attack aircraft contains 11 circuit cards, a motherboard, a power supply and additional space to accommodate functions such as terrain reference navigation, terrain-following, ground collision avoidance, target hand-off, in-flight route planning, ridgelines, perspective view, sensor blending and threat intervisibility. The computer receives operational parameters, which include aircraft state vector, map scale, zoom factor, north up/track up, database type, declutter select/deselect and video output mode, from the mission computer via a MIL-STD-1553B multiplex bus. Data requests are then sent to the digital memory unit via a fibre optic link.

DLMS elevation data is compared to absolute elevation limits or limits relative to the aircraft altitude, resulting in a clearance band display. Colour is assigned, based on the band and sun angle shading. Variably spaced contour lines can be added. The overlay selection includes linear, area and point feature data such as roads, rivers and towers in the DLMS mode, and flight path, waypoint symbols, text and threat symbols in the DLMS and digitised chart modes. The video generator combines the background scene memory with the overlay memory to provide a composite video output. The system supports two red/green/blue video outputs and two monochrome video outputs at either 525/60 Hz or 625/50 Hz line rates.

Processing is distributed across several high-performance microprocessors rather than a single unit. Data flow is pipelined through the digital map computer and parallel processing is performed wherever possible. The use of parallel processing provides a level of fault tolerance, since a failure in the map generation does not prohibit the graphics overlay operation.

Operational status

Production deliveries for the US Marine Corps AV-8B began in June 1989. Honeywell is also under contract to provide the digital video mapping system for the V-22 Osprey tiltrotor aircraft.

Contractor

Honeywell Defence Avionics Systems.

UPDATED

HP-700 satcom High Power Amplifier (HPA)

The HP-700 satcom HPA is an 8MCU unit designed for installation in an equipment bay, or near the satcom antenna. The HP-700 is a class AB linear amplifier, which provides RF amplification of the L-band (NATO D-band) signals generated by the SD-700 Satellite Data Unit.

It is designed to operate with the Inmarsat 3rd generation spot beam system.

Contractor

Honeywell Inc Defense Avionics Systems.

VERIFIED

SD-700 satcom Satellite Data Unit (SDU)

The SD-700 satcom SDU is a 6 MCU unit, which is intended for both the military and commercial aircraft markets. It is designed to operate with the Inmarsat 3rd generation spot beam satellite system.

The SD-700 is configured to provide four channels (three voice and one data). As an option, growth to seven channels (an extra three voice channels) can be provided by the use of an additional circuit card. Maximum data rates are 600, 1,200 and 21,000 bits l/s.

Contractor

Honeywell Inc Defense Avionics Systems.

VERIFIED

Stability Augmentation/Attitude Hold System (SAAHS) for the AV-8B

The SAAHS was the first digital system to be applied to any of the Harrier series of aircraft. It comprises a limited authority stability augmentation system with some conventional autopilot functions to reduce pilot workload. The single-channel system operates on all three axes and is accompanied by a mechanical back-up for reversionary operation. Extensive self-monitoring circuitry switches out the system in the event of a fault, so that it fails passive. Monitoring is accomplished in three ways: equipment, software monitoring of equipment and software monitoring of performance. In addition a comprehensive preflight schedule is provided.

Stability augmentation modes comprise three-axis rate damping in both vertical and cruise flight and in transition between these phases. It also includes rudder/aileron and stabiliser/aileron interconnects to improve turn co-ordination. Autopilot modes include altitude, attitude and heading hold, automatic trim, control column steering and airspeed limit schedules. A rudder pedal shaker alerts the pilot to the onset of potentially dangerous sideslip during the hover.

Principal element of SAAHS is a 6.1 kg digital flight control computer with 2901-bit slice, 16-bit processor working at 470 Kips, with UVEPROM. The flight control programme occupies 20 kwords of memory of which 9 kwords are taken by the control law code and 11 kwords by the built-in test schedule. The other units comprise a three-axis rate sensor, lateral accelerometer, stick sensor and forward pitch amplifier. In December 1982, the system demonstrated a completely hands-off automatic vertical landing.

Specifications

Weight: 12.3 kg total

Operational status

In service on the US Marine Corps AV-8B.

Contractor

Honeywell Defense Avionics Systems.

VERIFIED

AN/APN-59E(V) Search Radar

The APN-59E(V) was developed in the late 1970s to replace the earlier AN/APN-59B with improved performance and reliability for a variety of retrofits. The I/J-band radar was tailored to a closely defined mission profile with unusually stringent quality assurance demands. For example, component selection was made on the basis of a number of engineering and data-bank recommendations such as the Government/Industry Data Exchange Programme and the resulting system has been verified by rigorous testing to Advisory Group on Reliability of Electronic Equipment (AGREE) type testing. MTBF is given as 219 hours.

The principal modes are search, navigation, weather-mapping and beacon homing. To accommodate all these, the operator can choose pencil or fan beam with a variety of pulse lengths and repetition rates; the system can be set up for angle sector or 360° scan.

All LRUs are interchangeable with those of the AN/APN-59B so that separate stocks of spares are not needed for flight line support and gradual upgrading of a system can be accomplished over a period of time and without aircraft stand-down.

Alternative configurations range from single azimuth/range displays, driven by the radar as an independent system, to more complex installations with up to three displays. Where requirements are particularly critical the system can be connected to a compass and a dead-reckoning computer for the most accurate navigation fixes. The weight of a typical configuration is about 84 kg.

Operational status

Retrofitted to US C-130, C/KC-135 and RC-135 fleets, and C-130 fleets of various air forces.

This radar was replaced on US Air National Guard C/KC-135 aircraft by the Rockwell Collins FMR-200X radar as part of the PACER CRAG upgrade programme. The system is no longer in production.

Contractor

Honeywell Sensor and Guidance Products.

UPDATED

AN/APN-171 radar altimeter

For over 25 years, US Navy helicopters have been flying with the APN-171 (HG9000) as their standard altimeter. Many functions are available with this system, including low-altitude warning, radar altitude warning set input, aircraft rate, aircraft altitude errors and landing gear warnings.

Three basic Honeywell AN/APN-171 radar altimeter systems are currently available: the HG9010 is a 0 to 1,000 ft system, the HG9025 is a 0 to 2,500 ft system and the HG9050 is a 0 to 5,000 ft system. All AN/APN-171 systems are available with standard or special output signals to represent a particular altitude range.

Specifications

Dimensions: 337 × 194 × 125 mm
Weight: 6.1 kg
Power supply: 115 V AC, 400 Hz, 10 VA
Altitude:
(HG9010) 0-1,000 ft
(HG9025) 0-2,500 ft
(HG9050) 0-5,000 ft
Accuracy: ±5 ft ±3%

Operational status

In service as standard on all US Navy helicopters. A modification kit is available to upgrade existing systems.

Contractor

Honeywell Inc, Sensor and Guidance Products.

VERIFIED

AN/APN-194 radar altimeter

The AN/APN-194 radar altimeter is standard on all navy fixed-wing and high-performance aircraft including the F-14 and F/A-18. Functions include low-altitude warning, radar altitude warning set input, aircraft rate, aircraft altitude errors, landing gear warnings and both analogue and digital outputs.

Several height indicators interface with the APN-194. The AN/APN-194 was introduced in the early 1970s as a form, fit and function replacement for the AN/APN-141 and Honeywell has now produced over 8,000 of these altimeters. Since then, the AN/APN-194 has been upgraded twice, incorporating a solid-state transmitter and production/installation enhancements.

Options include transmitter power management. The AN/APN-194 can interface with up to four height indicators.

Specifications

Dimensions: 185 × 97 × 82 mm
Weight:
(radar altimeter) 2 kg
(height indicator) 0.7 kg
Power supply: 115 V AC, 400 Hz, 25 W
Frequency: FCC approved 4.3 GHz
Transmit power: 5 W
PRF: 20 kHz
Pulsewidth: 0.02 or 0.20 μs
Altitude: 0-5,000 ft
Accuracy: ±3 ft ±4%

Operational status

In service as standard on all US Navy fixed-wing high-performance aircraft.

Contractor

Honeywell Sensor and Guidance Products.

UPDATED

AN/APN-209 radar altimeter

The AN/APN-209 radar altimeter is standard on all US Army helicopters. Functions include transmitter power management, low- and high-altitude warnings, NVG compatibility, analogue and digital outputs and integration of indicator, receiver and transmitter. For more installation flexibility, a version of the AN/APN-209 with the transmitter/receiver separate from the indicator is available.

Honeywell has produced over 9,000 AN/APN-209 altimeters. It was introduced in the early 1970s and, since then, has been upgraded twice, incorporating a solid-state transmitter and producibility enhancements.

Options include MIL-STD-1553B databus, voice warning and LPI enhancements.

Specifications

Dimensions:
(receiver/transmitter) 145 × 83 × 83 mm
(indicator/receiver/transmitter) 199 × 83 × 83 mm
Weight:
(receiver/transmitter) 1.4 kg
(indicator/receiver/transmitter) 1.9 kg
Power supply: 28 V DC, 25 W
Altitude: 0-1,500 ft
Accuracy: ±3 ft ±3%

Operational status

In production and in service as standard on all US Army helicopters.

Contractor

Honeywell Sensor and Guidance Products.

UPDATED

AN/APN-224 radar altimeter

The APN-224 was developed specifically for the Boeing B-52 and meets the nuclear hardening and high-reliability specifications of that aircraft. The system's performance and ability to withstand severe environments led to its selection by the US Air Force for the B-1B. The APN-224 has also been selected for the US Air Force's A-10 and the US Air National Guard's F-16.

Functions of the AN/APN-224 include nuclear hardness, aircraft rate, blanking pulse and both digital and analogue outputs.

A modified version of the AN/APN-224, which has a MIL-STD-1553B databus, has been developed for the F-16 LANTIRN aircraft.

Specifications

Dimensions:
(B-52) 213 × 127 × 86 mm
(F-16) 213 × 160 × 102 mm
Weight:
(B-52) 2.1 kg
(F-16) 3.6 kg
Power supply: 115 V AC, 400 Hz, 45 VA
Altitude:
(B-52) 0-5,000 ft
(F-16) 0-10,000 ft
Accuracy: ±5 ft ±4%

Operational status

In service on US Air Force/US Air National Guard B-52, B-1B and F-16 LANTIRN aircraft.

Contractor

Honeywell Sensor and Guidance Products.

UPDATED

AN/ASN-131 SPN/GEANS precision inertial system

The Standard Precision Navigator/Gimballed Electrostatic Aircraft Navigation System (SPN/GEANS), designated AN/ASN-131, was developed primarily under sponsorship of the US Air Force Avionics Laboratory at Wright-Patterson Air Force Base, Ohio.

The basic SPN/GEANS system consists of an Inertial Measurement Unit (IMU), an Interface Electronics Unit (IEU) and a complete software library. For a stand-alone system capability, these two units are supplemented with a Digital Computer Unit (DCU). The IMU contains, in addition to the Velocity Measuring Unit (VMU) and two ESGs, temperature control electronics, accelerometer pulse rebalance and V readout electronics, precision timing reference, gimbal control electronics and Built-In Test Equipment (BITE) functions, as well as serial digital databus communication electronics. The IEU provides power conversion, control and sequencing electronics, additional BITE circuits and a common serial databus interface with other units of the inertial system and other subsystems.

The ESG has only one moving part, a suspended hollow beryllium ball, which is combined with two optical pick-offs to give error and timing signals. These in turn are used to drive the IMU platform gimbals to maintain a stable reference base for the accelerometers. Three highly accurate, single-axis accelerometers (contrasted with the two-degree-of-freedom ESGs) are used within the VMU which is mounted on the stable platform inner element. These accelerometers, oriented in an orthogonal triad configuration, measure accelerations directly and provide the incremental velocity pulses to the computer which uses them in its software algorithms to calculate velocity and position parameters.

The ESG system requires little or no reliance on other navigation aids for most aircraft applications and thus can be described as self-contained.

In addition to the traditional function of position determination or basic navigation, the higher accuracy outputs of velocity and attitude data have opened up a new realm of possibilities for stabilisation and/or motion compensation for other non-inertial sensors. These include high-precision radars, sonars, lasers, optical and electro-optical devices.

SPN/GEANS is being deployed throughout the entire US Air Force B-52 strategic aircraft and F-117A stealth fighter fleets. Widespread use is also expected for long-range reconnaissance and patrol missions, specialised cargo and transport usage in both military and civil applications and tactical military aircraft.

Operational status

In service with the US Air Force on the B-52 bomber and F-117A stealth fighter fleets.

Contractor

Honeywell Inc, Sensor and Guidance Products.

VERIFIED

Automatic Flight Control System (AFCS) for the C-5B

Honeywell provided the flight control system for the US Air Force C-5A Galaxy. It was claimed to be the first operational flight control system to provide complete Cat. III automatic landing and roll-out guidance with fail-safe performance. Honeywell provides an improved (though still analogue) five-box form, fit and function version of the original flight control system for the C-5B; the computer boxes fit into the same racking as in the C-5A, have the same connectors and are functionally interchangeable with the flight control boxes in the earlier aircraft. Commonality of the two analogue systems, in order to exploit the heavy investment in ground test and avionics maintenance equipment, was a major factor in the decision to go for a minimum change system. The principal purpose of its upgrading is to improve reliability and ease of maintenance. The system incorporates three-axis stability augmentation and autothrottle.

Operational status

In service in the C-5B Galaxy. Depot repair modification for C-5A.

Contractor

Honeywell Sensor and Guidance Products.

UPDATED

H-423 ring Laser Inertial Navigation System (LINS)

The H-423 Laser Inertial Navigation System (LINS) was developed according to the US Air Force SNU-84-1 specification which was the RLG version, an update to ENAC77-1. The system is a self-contained unit comprising three Honeywell GG1342 RLGs, three solid-state Sundstrand QA2000 accelerometers and associated electronics along with a dual-redundant MIL-STD-1553B databus and a complete MIL-STD-1750A navigation processing package. It also contains built-in test circuitry to achieve a greater than 95 per cent fault detection.

Under the US Air Force contract, Honeywell is guaranteeing the H-423 will achieve 2,000 hours MTBF in a fighter/helicopter environment and 4,000 hours MTBF in a transport environment. Honeywell claims the H-423 has the highest reliability and maintainability and lowest life-cycle costs ever achieved by a military aircraft inertial navigation system, allowing the US Air Force to go from a three-level to two-level maintenance system. These performance and reliability advantages have been the primary drivers for the US Defense Services now adopting RLG technology for all future aircraft inertial navigation systems.

Specifications

Dimensions: 459.7 × 193 × 200 mm
Weight: 22 kg
Power supply: AC 140 VA, DC 125 W
Interface: dual 1553B digital databus
Accuracy: <0.8 n miles/h with full performance from 22 stored heading alignment
Specification: SNU-84-1, FNU-85-1

Operational status

In August 1985, the US Air Force selected the H-423 as the standard inertial navigation system for its C-130, a number of its fixed- and rotary-wing aircraft. Up to the end of 1989 more than 1,000 military systems had been delivered.

In November 1990, the system was selected for Royal Australian Air Force F-111 aircraft in a US$5 million contract and for US Air Force F-16 upgrade programmes. Since then, H-423 systems have been selected for upgrades of Belgian and Danish F-16s and for the Swedish JAS 39, Taiwanese IDF and the Indian Light Combat Aircraft.

In September 1992, the H-423E, a high-accuracy version of the H-423, was selected to upgrade the F-117A fleet. The high standard of accuracy is achieved through the use of an enhanced specialised version of the H-423 standard software.

Contractor

Honeywell Inc, Sensor and Guidance Products.

VERIFIED

H-770 ring laser gyro inertial navigation system

Honeywell developed the H-770 ring laser gyro inertial navigation system for the F-15.

The H-770 uses the same ring laser gyro and accelerometers as the H-423 (see earlier entry), but contains two separate databusses for the F-15 applications. The system can determine for itself in which variant of the F-15 it has been installed, and perform accordingly. The H-770 has a reliability 20 times better than the AN/ASN-109 system that it is replacing in the F-15.

Specifications

Dimensions: 381 × 213 × 330 mm
Weight: 26.8 kg
Power required: AC 140 VA, DC 130 W
Accuracy: 0.4 n mile/h CEP
Alignment:
(gyrocompassed) 4 min
(stored heading) 30 s
Reliability: 2,000 h MTBF

Operational status
Honeywell had been contracted to retrofit the US Air Force fleet of F-15A, B, C, D and E variants.

Contractor
Honeywell Inc, Sensor and Guidance Products.

VERIFIED

HG7170 radar altimeter

The HG7170 radar altimeter is a dual-redundant altimeter system. Functions of the HG7170 include MIL-STD-1553B databus, aircraft rate, transmitter power management, full BIT monitoring, continuous BIT, isolated grounding, nuclear event detector and nuclear hardness.

Specifications
Dimensions: 338 × 193 × 155 mm
Weight: 5.7 kg
Power supply: 28 V DC, 38 W
Altitude: 0-5,000 ft
Accuracy: ±2 ft ±4%

Contractor
Honeywell Inc, Sensor and Guidance Products.

VERIFIED

HG7700 Series radar altimeters

The HG7700 is specifically designed for the low-cost high-performance requirements of tactical manned and unmanned vehicles. The HG7700 features automatic test capability, noise immune tracker and low power consumption. Available options include digital output, transmitter power management and integrated antenna.

Specifications
Dimensions: 194 × 95 × 72 mm
Weight: 1.55 kg
Power supply: 28 V DC, 12 W
Altitude:
(HG7702) 0-2,500 ft
(HG7705) 0-5,000 ft
Accuracy: ±5 ft ±3%

Operational status
In production for fighter aircraft, as well as RPVs and tactical weapon systems. The HG7705 is fitted to the Saab JAS 39 Gripen.

Contractor
Honeywell Sensor and Guidance Products.

UPDATED

HG7800 radar altimeter module

Honeywell has introduced the new, low-cost HG7800 altimeter module. A unique approach, achieved through the application of ASIC and MIMIC chip technology, delivers the entire radar altimeter on a single, SEM-E short-circuit card. The single-card altimeter is optimised for use as an embedded function and can be added to an available slot within an existing system, such as Honeywell's H-764-G Embedded GPS/INS (EGI) or other integrated systems. The HG7800 is flying in US Navy AH-12 Cobra and UH12 helicopters, USAF T -33, T -38, F-5 fixed-wing aircraft and in a number of US Navy unmanned air vehicles.

HG 7800 options include 5,000 and 8,000 ft variants, and all versions are available as stand-alone LRUs with their own chassis. The HG7800 can be customised for unmanned air vehicles, fixed-wing and rotary-winged platforms.

Specifications
Dimensions: SEM-E Short (5.88 × 6.06 × 0.50 in)
Weight: 0.34 kg
Altitude accuracy: ±2 ft or ±2%
Altitude range: 0 to 8,000 ft
PRF: 35 kHz, PRN coded
Pulsewidth: 30 to 256 ns
Frequency: FCC approved, 4.3 GHz
Transmit power: power managed, 1.0 W peak
Power supply: +5/−15 V DC
Interface: RS 422/485

Operational status
In service in US Navy and Air Force aircraft.

Contractor
Honeywell Sensor and Guidance Products.

NEW ENTRY

HG7808 radar altimeter module

The HG7808 provides an entire low-cost radar altimeter on a single (SEM-E short) circuit card. The module is highly integrated through the use of ASIC technology and optimised for use as an embedded function added to open/expansion card slots of existing systems such as Honeywell's H-764-G Embedded GPS/INS (EGI) or other integrated systems. The standard circuit card configuration and flexible I/O facilitate integration of the module into other customer-defined installations/applications. The system is also available as a stand-alone LRU with its own chassis.

Specifications
Dimensions: SEM-E Short (5.88 × 6.06 in)
Weight: < 0.45 kg
Altitude accuracy: ±2 ft or ±2%
Altitude range: 0 to 8,000 ft
Manoeuvre angles: ±45° up to 5,000 ft, ±30° up to 8,000 ft (nominal antennae, typical terrain roughness, -3 dB cable loss)
Frequency: C-band, 4.3 GHz
RF power: power managed, +30 dBm peak

Operational status
Honeywell's HG7808 has been selected for the US Air Force T-38 Avionics Upgrade program, various F-5 upgrades worldwide, AH-1W, H1 and the tactical Tomahawk Cruise missile.

Contractor
Honeywell Sensor and Guidance Products.

NEW ENTRY

HG8500 Series radar altimeters

The HG8500 is specifically designed for the low-cost, high-performance tactical requirements of manned and unmanned vehicles. The result of Honeywell's independent development efforts, the HG8500 series is a solid-state altimeter utilising Honeywell MMIC technology. Available models feature analogue or digital, or both analogue and digital output, with altitude ranges of 0 to1,000 ft or 0 to 10,000 ft. The HG8500 offers pulsed leading edge tracking with power management for low detectability. The HG8500 series altimeters are installed in several US Navy missiles and target vehicles.

Specifications
Dimensions: 86 × 86 × 142 mm (3.4 × 3.4 × 5.6 in)
Weight: 1.4 kg
Altitude accuracy: ±2 ft or ±2% (dependent on specific model)
Altitude range: 0 to 10,000 ft (0 to 2,000 ft and 0 to 5,000 ft most common)
Frequency: 4.3 GHz
Track rate: 2,000 ft/s (minimum)
PRF: 25 kHz
Power: +28 (±4.0) V DC, 16 W Max

Operational status
Operational on US navy land attack missiles and remotely controlled target vehicles.

Contractor
Honeywell Sensor and Guidance Products.

NEW ENTRY

HG9550 LPI radar altimeter system

Honeywell claims that the HG9550 represents a quantum leap in altimeter capabilities over its earlier products in reliability, covertness, size, weight and cost.

The system was developed by Honeywell and jointly qualified with the US Air Force. The HG9550 was designed as a form, fit and function replacement for MIL-STD-1553B versions of the US Air Force Combined Altitude Radar Altimeter (CARA).

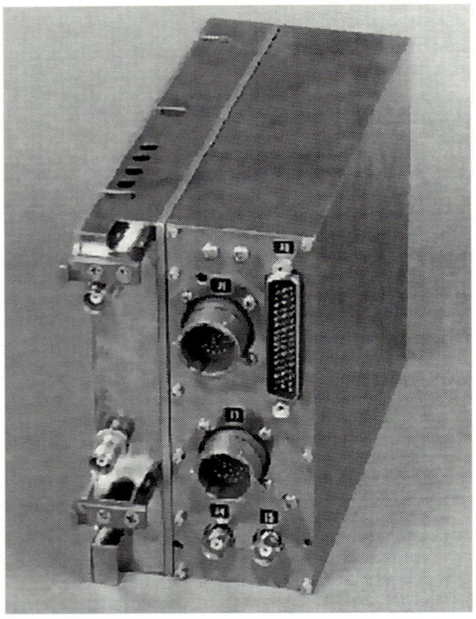

Honeywell HG9550 LPI radar altimeter system 0018070

The HG9550 is designed to provide the high accuracy and programmable features of pulsed altimeter designs with the sensitivity and Low Probability of Intercept (LPI) advantages of coherent designs; with less than 1 W of power it is virtually undetectable. A microprocessor permits system characteristics such as track rate and ECCM response to be varied as a function of real-time inputs, or to be preprogrammed according to mission requirements. It is designed to be an element of GCAS systems.

Specifications
Dimensions: 90 × 60 × 222 mm
Weight: 4.43 kg
Frequency: 4.3 GHz
Range and track rate: 0 to 50,000 ft ±2,000 ft/s (minimum)
Altitude accuracy:
(analogue) ±4 ft or ±4% (whichever is greater)
(digital) ±2 ft (0 to 100 ft); ±2% (100 to 50,000 ft)
Manoeuvre angles: ±60° up to 3,000 ft; ±45° up to 5,000 ft; ±10° up to 50,000 ft
RF power: power managed: controlled at 10 dB above track threshold, with less than 1 W transmit power
Input power: 28 V DC (MIL-STD-704), 28 W nominal (35 W maximum)
Programmable features: track rate, ECCM response, sensitivity, altitude range, output formats
LPI features: frequency agility, power management, jittered code and PRF

Operational status
The HG9550 is an off-the-shelf, fully qualified system, currently in production for US Air Force C-130J, UK C-130J, Argentine A-4 upgrade, Boeing Joint Strike Fighter and the Lockheed Martin Joint Strike Fighter. The HG9550 has recently completed an extensive flight test qualification for all blocks of F-16 and was due to be integrated onto the USAF C-17 Globemaster during the fourth quarter of 2000.

Contractor
Honeywell Sensor and Guidance Products.

UPDATED

LG81xx radar altimeter antennae

Honeywell's LG81 series of radar altimeter antennae (RAAs) have been in continuous service for more than 30 years on air vehicles flown by the US Army, Navy and Air Force as well as NASA and commercial applications. Over 30,000 RAAs in more than 40 configurations have been fielded.

The RAAs are ruggedised, low-profile, conformal micro strip designs, suitable for a wide range of platforms including fighters, bombers, cruise missiles, smart munitions and UAVs. Honeywell produces the US Navy's standard fixed-wing RAA, the US Army's standard helicopter RAA, and numerous specially adapted antennae for unique applications, including antennae for the Space Shuttle and Mars Lander missions and low-RCS designs for covert applications.

Specifications

RAA Part No.	LG81T	LG81K	LG81AT	LG81V
Dimensions (mm)	89 × 89 × 2.3	108 × 88 × 2.8	146 dia × 2.8	178 × 254 × 5.6
Weight	46 g	122 g	251 g	637 g
Gain	10.0 dBi	10.0 dBi	11.0 dBi	15.0 dBi
E Plane × H Plane Beamwidth	52° × 50°	55° × 50°	42° × 51°	23° × 25°
Application	General	US Navy fixed-wing (standard)	US Army helicopter (standard)	high altitude

Operational status

In service in all US Navy fixed-wing aircraft, US Army helicopters, F-16, Saab JAS-39 Gripen, Lockheed C-130 aircraft, Tomahawk Cruise Missile and AGM-130 weapon systems and the Space Shuttle.

Contractor

Honeywell Sensor and Guidance Products.

NEW ENTRY

AN/ARC-201 VHF/FM transceiver (SINCGARS-V)

The AN/ARC-201 SINCGARS-V is an airborne VHF/FM frequency-hopping radio and is an all-solid-state equipment for use in helicopters, light observation aircraft and Airborne Command and Control (ACC) aircraft like the C-130. It provides single-channel and frequency-hopping modes. A six-channel non-volatile preset memory is incorporated for single-channel and ECCM modes. An interface and controls for the AM-7189A/ARC 50 W amplifier are incorporated.

The AN/ARC-201 is only available in the MIL-STD-1553B multiplex bus-remote configuration. It is interoperable with the current VHF/FM radios in the single-channel mode, and with the Single Channel Ground and Airborne Radio Subsystem (SINCGARS) VRC-87 to VRC-92 and manpack PRC-119 ground radios in the frequency-hopping mode. Electroluminescent lighting is provided on the front panel, compatible with the use of NVGs.

The radio has module and component commonality with the ground SINCGARS communication equipment and makes extensive use of LSI circuitry and DSP microprocessors for high reliability. A BIT function isolates faults to the module level with 90 per cent confidence. A data-rate adaptor is integrated in the radio to interface with data devices for communication. An automatic single-channel cueing capability in the ECCM modes allows a single-channel user to alert members of an ECCM net. Internal and external COMSEC can be used to provide secure communications in voice and data modes.

In 1998, the AN/ARC-201D radio was introduced as a MIL-STD-1553 bus-controlled unit. The aircraft integrator supplies the audio intercom, antennas, mounting and cabling. The integration of COMSEC and the Data Rate Adaptor combines three line replaceable units into one and reduces overall weight of the aircraft. Additional features such as improved error correction, enhanced data modes (including packet data), more flexible remote control, and GPS capability enable the radio to assume a number of roles supporting future digital battlefield requirements. The ARC-201D is backward compatible with existing SINCGARS data modes.

The ITT Aerospace/Communications Division RT-1478D/ARC-201D VHF/FM transceiver **2000**/0079235

Specifications

Dimensions:
(C-11466) 146 × 76 × 132 mm
(RT-1478) 127 × 102 × 259 mm
Weight: (RT-1478) 3.1 kg
Power output: 10 W
Frequency: 30-87.975 MHz
Channel spacing: 25 kHz
Channels: 2,320

Operational status

In production. Contracted requirements for the US Army continue until 2002. Ordered for the UK WAH-64 Apache helicopter.

Contractor

ITT Aerospace/Communications Division.

VERIFIED

AN/ALQ-136 radar jamming system

The AN/ALQ-136 family of airborne jamming systems is designed to protect rotary- and fixed-wing aircraft from radar-guided weapons. The AN/ALQ-136 enables the host aircraft to carry out a mission from a low-level approach, pop up and complete its mission before the hostile radar can initiate a fire sequence.

The system consists of three main units: a control unit, two spiral antennas - one each for transmit and receive - and a transmitter/receiver. When the aircraft is illuminated by a hostile radar, the jammer automatically analyses the received pulses, compares them with its threat library, assigns a priority and then provides the most appropriate countermeasure response.

The AN/ALQ-136 is software reprogrammable to allow adaptation to the swiftly changing threat scenarios and can simultaneously handle multiple threats. The system is internally mounted and integrated on numerous US Army helicopters.

Operational status

In service. More than 1,400 systems have been delivered to the US Army and are deployed on AH-64 Apache and AH-1 Cobra attack helicopters. Other platform applications include the MH-53J Pave Low, MH-60 Black Hawk, EH-60 Quick Fix, MH-47E Chinook and AH-1W SuperCobra helicopters and the A-10 and RC-12/RU-21 fixed-wing aircraft.

The AN/ALQ-136(V)1/5 jamming system is carried by US Army AH-64 Apache attack helicopters

AN/ALQ-136(V)2

The AN/ALQ-136(V)2 is an enhanced version with increased receiver sensitivity, frequency range and threat-handling capability. The system design employs extensive use of Thick Film Hybrid technology (TFH) and multilayered PCBs. A MIL-STD-1553B databus is used to facilitate system integration.

Specifications

Dimensions:
(receiver/transmitter) 177.8 × 330.2 × 408 mm
Weight: 31.8 kg

Operational status

200 systems have been delivered to the US Army.

Contractor

ITT Industries, Avionics Division.

VERIFIED

AN/ALQ-172 Electronic CounterMeasures system

The AN/ALQ-172 is an advanced ECM system which was combat proven during Operation Desert Storm. It is installed in US Air Force B-52 strategic bombers and Special Operations MC-130E/H Combat Talon I and II and AC-130U/H Gunships.

System design features of the AN/ALQ-172 include full automation, multiband coverage, simultaneous multiple threat recognition and jamming, digital computer control, advanced jamming techniques, high effective radiated power, threat reprogrammability, high-gain array antenna, threat warning display, dual MIL-STD-1553B databus interface and extensive BIT.

Reportedly there are three upgraded versions of the system: the AN/ALQ-172(V)1, AN/ALQ-172(V)2 (which incorporates a phased-array antenna) and the AN/ALQ-172(V)3.

Operational status

In service in US Air Force B-52 and Special Operations MC-130E/H Combat Talon I and II, AC-130H and AC-130U/H gunship aircraft. System upgrades under way include expansion of AN/ALQ-172 system memory and extended frequency coverage designated AN/ALQ-172(V)3 for the AC-130H upgrade programme.

In 1998, the US Special Operations Command contracted ITT to upgrade its C-130 fleet to the AN/ALQ-172(V)1 (ECP-93) standard at a cost of US$18 million. Procurement options permit increase to US$44.6 million. This upgrade provides expanded system ECM and OFP memory capacity.

Contractor

ITT Industries, Avionics Division.

VERIFIED

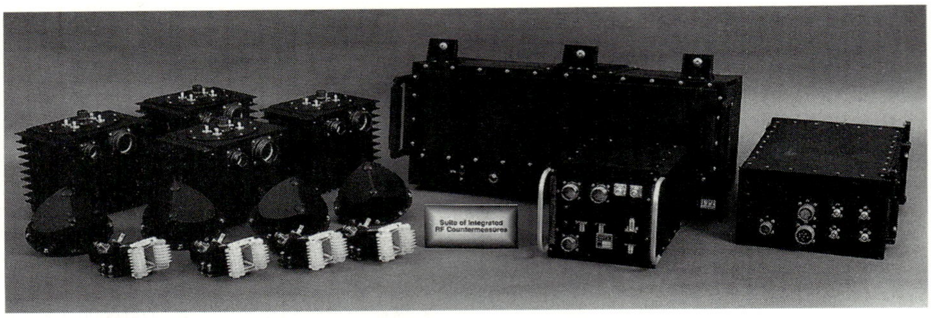

Suite of Integrated RF Countermeasures (SIRFC)

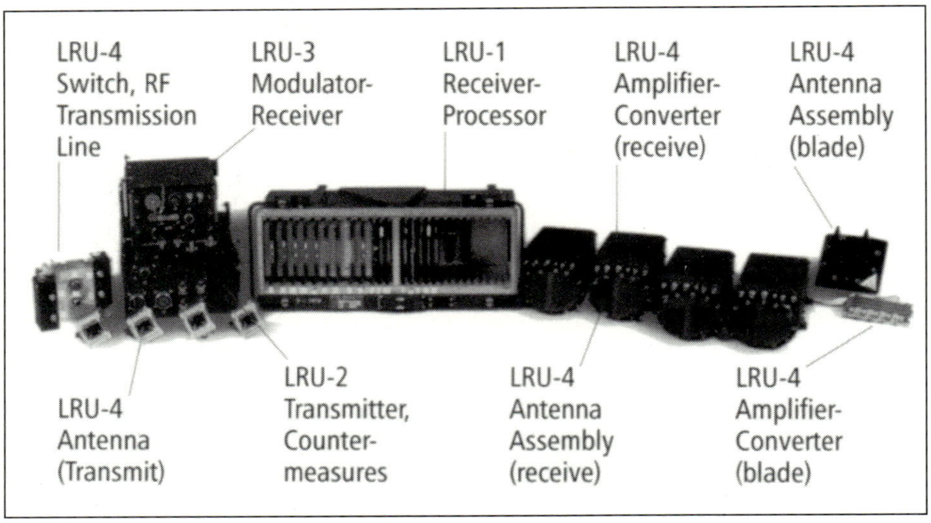

Suite of Integrated RF Countermeasures (SIRFC) **2000**/0081869

AN/ALQ-211 Suite of Integrated RF Countermeasures (SIRFC)

The AN/ALQ-211 SIRFC system is a fully integrated airborne electronic combat system that provides real-time battlefield situational awareness, radar warning and self-protection capabilities to a multitude of aircraft.

SIRFC counters multiple, simultaneous pulse, pulse Doppler and continuous wave threats. It intercepts, analyses, passively ranges and geolocates multiple air- and ground-based RF threat signals and co-ordinates appropriate countermeasures or weapons cueing.

High-power pulse and continuous wave ECM transmitters provide protection against coherent and non-coherent radars. Through its advanced countermeasures module, SIRFC is claimed to provide robust countermeasure techniques against monopulse radars.

The SIRFC team comprises: ITT Avionics as team leader, with BAE Systems North America (formerly Tracor Electronic Systems) providing the interferometer and preamplifiers and EMS Technologies responsible for the radar jammer techniques generator.

The AN/ALQ-211 SIRFC system provides an embedded capability which directly supports the battlefield digitisation initiatives of the US Army. SIRFC's real-time situational awareness reports and updates can be disseminated horizontally and vertically to aircrews and battlefield commanders. The location of radar-directed threats can be transmitted from the aircraft to the ground commander over the tactical datalink and displayed on appliqué hardware. The situational awareness and onboard countermeasures capabilities of the AN/ALQ-211 SIRFC system allows aircrews to evade or defeat a diversity of airborne and ground-based threats, significantly enhancing survivability.

Operational status

In June 1996, the AN/ALQ-211 Suite of Integrated RF Countermeasures (SIRFC) programme successfully completed the Critical Design Review (CDR). (Note: The AN/ALQ-211 SIRFC system was formerly called the Advanced Threat Radar Jammer (ATRJ). In June 1996, the system received its official nomenclature of AN/ALQ-211 SIRFC.)

ITT Industries' Avionics Division delivered the first AN/ALQ-211 Suite of Integrated RF Countermeasures (SIRFC) Engineering and Manufacturing Development (EMD) system to the US Army in August 1998, followed by the second system delivery in October 1998.

The first SIRFC EMD system was delivered to the US Army in August 1998 after completion of government-witnessed acceptance testing at ITT's SIRFC laboratory in Clifton, New Jersey. The SIRFC EMD hardware also successfully completed preliminary integration and installation testing at Boeing's AH-64D Longbow Apache Aircraft Integration Laboratory (AIL) in Mesa, Arizona. Among the hardware and software delivered to the US Army were the SIRFC LRU-1: receiver/processor; LRU-3: advanced countermeasures module; LRU-4: antenna group, and software to support the system's sensor fusion processing functions for the AH-64D Longbow Apache.

Delivery of the second SIRFC EMD system to the US Army was achieved in October 1998. System #2 includes the full complement of equipment as delivered in System #1 in addition to LRU-2: transmitter. The system underwent 'Safety of flight' tests at ITT's facility before delivery to Boeing in January 1999. Contractor flight testing of the of the SIRFC system started in March 1999, and continued throughout 1999. US Army Operational Test & Evaluation (OT&E) is due to start early 2000. Production of the ALQ-211 SIRFC system is planned for late 2000.

The US Army's current Operational Requirements Document (ORD) calls for the installation of the ALQ-211 SIRFC system on the AH-64D Longbow Apache, Special Operations MH-47E and MH-60K helicopters and designated UH-60 and CH-47 aircraft. ITT Industries is also under contract with Bell Textron for the development of SIRFC for use on US Air Force Special Operations CV-22 tilt-rotor aircraft.

The AN/ALQ-211, together with Sanders AN/ALQ-212 Suite of Integrated IR Countermeasures (SIIRCM) and Raytheon's AN/AVR-2A(V) form the designated helicopter Defensive Aids Sub-System (DASS) for US Army helicopters, but the US Army is deferring production decisions, pending operational decision.

Contractor

ITT Industries, Avionics Division.

VERIFIED

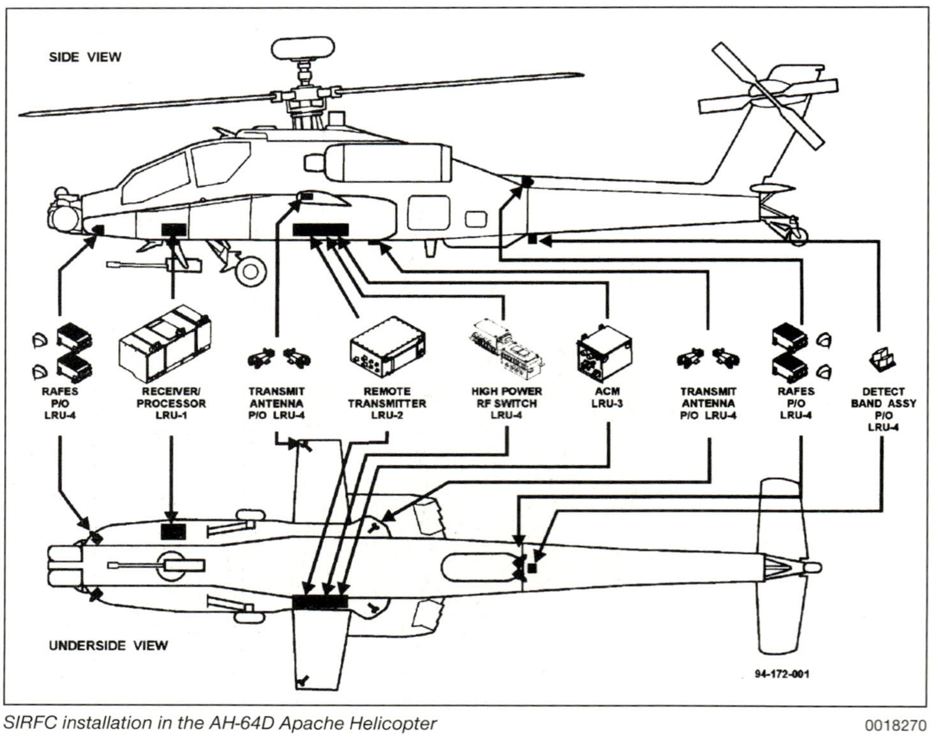

SIRFC installation in the AH-64D Apache Helicopter 0018270

Magic Lantern for SH-2G helicopter

Magic Lantern uses a blue-green laser and camera array to scan the water below the helicopter from surface level down to keel depth. The system correlates multiple scans to identify mines. It is able to sweep the entire 'upper column' at and below the ocean surface, and is cleared for day and night operations. Accurate navigation data from the GPS enables the SH-2G crew with Magic Lantern to locate mines precisely.

The podded Magic Lantern system replaces the Magnetic Anomaly Detector (MAD) on a strengthened hardpoint on the right side of the helicopter. The Magic Lantern system display in the aft cabin of the SH-2G shows the Sensor Operator (SENSO) mine detection symbology and real-time video imagery of suspected mine contacts. The ASN-150 Tactical Navigation system (TACNAV) provides signals to the HSI in the cockpit enabling the pilots to fly predetermined search patterns. A Tactical Decision Aid (TDA) has been developed to aid aircrews in determining search patterns and analysing post-mission data. A miniaturised airborne GPS is part of the SH-2G modifications to provide EOD crews with precise position data.

The SH-2G Super Seasprite advanced glass cockpit developed by Kaman Aerospace and LittonGuidance & Control Systems 0002197

Operational status

Kaman Aerospace is contracted to modify eight helicopters for the US Navy Reserve.

Contractor

Kaman Aerospace Corporation.

VERIFIED

SH-2G Super Seasprite Integrated Tactical Avionics System (ITAS)

Working with Litton Guidance & Control Systems, Kaman has created an advanced glass cockpit which contains a highly automated Integrated Tactical Avionics System (ITAS), enabling a crew of two to fly the aircraft and manage its multimission equipment suite. The ITAS is a low-risk avionics system designed specifically to meet the requirements of the Royal Australian Navy.

The Kaman-Litton ITAS, driven by two mission data processors, integrates the input of radar, thermal imager and electronic protection measures for manageable cockpit presentations. It enables the Super Seasprite crew to attack targets with a variety of anti-ship missiles. Electronic Flight Instrumentation System (EFIS), engine and transmission data, tactical plots, and sensor imagery are posted on any of the SH-2G's four colour multifunction displays.

The new glass cockpit retains the under-glare shield caution/advisory panels introduced originally on the SH-2G, but eliminates all electromechanical instruments except for back-up airspeed and altitude gauges and a standby compass.

Operational status

Selected by the Royal Australian Navy for its fleet of 11 helicopters.

Contractors

Kaman Aerospace Corporation.
Litton Guidance & Control Systems.

VERIFIED

KN-4060 series Ring Laser Gyro/ Inertial Navigation Systems (RLG/ INS)

The KN-4060 family of navigation systems offers a low-cost solution for a wide variety of fixed-wing aircraft and helicopter and UAV applications. The systems feature the Kearfott ring laser gyro currently under contract to the US Navy and US Army.

The KN-4060 is designed to operate in conjunction with embedded or federated GPS receivers for enhanced navigation performance and advanced satellite reacquisition. The model KN-4068GC includes an Embedded GPS Receiver (EGR). The system's modular architecture allows for alternative navigation aiding inputs, including Doppler radar, baro-altimeter and airspeed or remote located GPS receivers via MIL-STD-1553 or RS-422 data protocols. The KN-4060 family provides navigation functions in digital formats including MIL-STD-1553B, RS-422 and ARINC 429. As a full featured navigation system, the KN-4060 provides precision position and velocity over extended periods, as well as the attitude and heading output of a Standard Attitude Heading Reference System (SAHRS). The small, lightweight, rugged chassis can accommodate field replaceable sensor assemblies of three levels of performance to meet customer requirements.

The Kearfott Guidance & Navigation Corporation KN-4065 improved standard attitude heading reference system. Note the KN-4060 series and KN-4068GC ring laser gyro inertial navigation systems utilise the same box enclosure **2000**/0062845

Specifications

Dimensions: 177.8 × 177.8 × 279.4 mm (NB: The KN-4060 uses the same box enclosure as the KN-4065)
Weight: 9.5 kg
Power supply: 28 V DC, 35 W or 115 V AC, 400 Hz back-up battery capability
Outputs:
(digital) MIL-STD-1553B, RS-422, ARINC 429
(analogue) synchro, discrete, Magnetic Azimuth Detector (MAD)
GPS receiver:
(operating frequencies) L1/L2, L1
(antispoof/enhanced antijam) P(Y), C/A code
Channels: 5
MTBF: >6,000 h

Operational status

The system has been flight-tested by A & AEE Boscombe Down, UK; Holloman AFB, New Mexico; and Patuxent River, Maryland.

Contractor

Kearfott Guidance & Navigation Corporation.

VERIFIED

KN-4065 Improved Standard Attitude Heading Reference System (ISAHRS)

The KN-4065 Improved Standard Attitude Heading Reference System (ISAHRS) provides both attitude heading reference and full navigation outputs. ISAHRS features Monolithic Ring Laser Gyro (MRLG) technology for high reliability, low cost and weight and low power comsumption.

Items in this family of Kearfott ring laser gyro systems include:
The Improved Standard Attitude Heading Reference System (ISAHRS);
The Modular Azimuth Position System (MAPS);
The Low cost Attitude Heading Reference System (LAHRS);
The Attitude Motion Sensor Set (AMSS).

Specifications

Dimensions: 279.4 × 177.8 × 177.8 mm
Weight: 9.3 kg
MTBF: 6,750 h
Outputs: MIL-STD-1553B; Magnetic Azimuth Detector (MAD) interface; Mode Control Unit (MCU) interface; synchro outputs for pitch, roll and heading
Performance:
(heading) 0.5° (1σ) RMS (8 min alignment time)
(pitch/roll) <0.1° (1σ) RMS
(dynamic range) 400°/s max (all axes) ± 10 *g* acceleration
(back-up navigation) 2 n miles/h

Operational status

The KN-4065 has been certified by the US Department of Defense for use in military aircraft and is in production for several types of US Navy aircraft.

Contractor

Kearfott Guidance & Navigation Corporation.

VERIFIED

KN-4068GC Ring Laser Gyro/ Inertial Navigation System (RLG/ INS) with Embedded GPS Receiver (EGR)

The KN-4068GC system is designed for fixed- and rotary-wing aircraft. It contains RLG technology and a C/A code embedded GPS receiver for enhanced, closely coupled navigation performance, and it accepts other navigation aiding inputs via MIL-STD-1553 or RS-422 databuses or analogue navigation aids (flux valve). During GPS availability, blended INS/GPS performance can provide position accuracies of 300 ft.

Specifications

Dimensions/weight: as for KN-4060 and KN-4065
GPS receiver:
(operating frequency): L1
(antispoof/enhanced antijam): C/A code
(channels): 5

Contractor

Kearfott Guidance & Navigation Corporation.

VERIFIED

KN-4070 Monolithic Ring Laser Gyro/Inertial Navigation System (MRLG/INS)

The KN-4070 is designed for aircraft and UAV operation. It is an MRLG-based INS, that is designed to operate in conjunction with an embedded P(Y) or C/A code GPS receiver for enhanced navigation performance and faster satellite acquisition. The KN-4070 provides navigation and autopilot functions in digital formats including MIL-STD-1553B, RS-422 and RS-232.

Specifications

Dimensions: 231.1 × 137.2 × 152.4 mm
Weight: < 5 kg
Power: 28 V DC, 35 W
GPS receiver:
(operating frequencies): L1/L2
(antispoof/enhanced antijam): P(Y), C/A code
MTBF: > 6,000 h

The Kearfott Guidance & Navigation Corporation KN-4070 MRLG/INS **2000**/0062844

Contractor

Kearfott Guidance & Navigation Corporation.

UPDATED

KN-4071 Attitude Heading Reference System (AHRS)

Kearfott's KN-4071 AHRS uses a patented Monolithic Ring Laser Gyro (MRLG) and a triad of Kearfott MOD VIIA force rebalance accelerometers as the inertial reference. The system provides analogue outputs of heading and attitude for cockpit, flight director or weapon computer input.

The unit includes a self-contained GPS card to improve accuracy of altitude and heading over standard electromechanical AHRS. Growth to full navigation capability is optional.

The AHRS comprises three LRUs: the KN-4072 , a Signal Conversion Unit (SCU) and the Mode Select Unit (MSU).

Specifications

Compass accuracy: +/– 0.7° typical
Slaved mode:
(heading): +/– 0.5° of MAD
(attitude): +/– 0.25°
Gyro mode:
(heading): +/– 0.5°/h
(attitude): +/– 0.25°
Enhanced mode:
(heading): +/– 0.25°
(attitude): +/– 0.15°
Analogue outputs: four heading; three pitch; three roll
Optional digital outputs: MIL-STD-1553; RS-422, RS-232
MTBF: 2,000 h

Contractor

Kearfott Guidance & Navigation Corporation.

VERIFIED

KN-4072 Digital Attitude Heading Reference System (AHRS) GPS/ INS

The KN-4072 digital AHRS GPS/INS is designed for a wide variety of aircraft, UAV and missile applications. It features Kearfott's Monolithic Ring Laser Gyro (MRLG) operating in concert with an embedded P(Y) or C/A code GPS receiver for enhanced navigation performance and faster satellite acquisition. The KN-4072 provides navigation, heading, attitude, velocity, position, together with angle and velocity rate data and autopilot functions in digital formats including MIL-STD-1553B, RS-422 and RS-232.

Specifications

Dimensions: 231.2 × 137.2 × 152.4 mm
Weight: < 5 kg
Power: 28 V DC, 35 W
GPS receiver:
(operating frequencies): L1/L2
(antispoof/enhanced antijam): P(Y), C/A code
MTBF: > 6,000 h

Contratcor

Kearfott Guidance & Navigation Corporation.

VERIFIED

The Kearfott Guidance & Navigation Corporation KN-4072 digital AHRS GPS/INS **2000**/0062843

The Kearfott Guidance & Navigation Corporation KN-4071 attitude heading reference system **2000**/0062846

ATX-2740(V) airborne datalink system

The ATX-2740(V) is a lightweight, exportable, ruggedised COTS version of the US Common DataLink (CDL) satisfying STANAG 7085 for digital, point-to-point, datalinks. It provides real-time full duplex sensor data and voice communications between aircraft and ground stations at selectable wide bandwidth data rates.

The ATX-2740(V) system comprises three units: Airborne Modem Assembly (AMA); Radio Frequency Electronics (RFE); one or more fixed or steerable antennas.

Specifications

Frequency: X-band or Ku-band (NATO I/J-band or J-band)

Tuning: 5 MHz steps
Data rates:
uplink: up to 200 kb/s aggregate data rates (option to 10.71 Mp/s)
downlink: selectable aggregate data rates to 274 Mb/s
Dimensions (W × H × D):
AMA 259.1 × 218.4 × 505.5 mm
RFE 241.3 × 101.6 × 431.8 mm
directional antenna 228.6 mm dia
Weight: AMA 15.9 kg
RFE 6.8 kg
antenna 3.2 kg

Contractor

L-3 Communications, Communications Systems - West.

VERIFIED

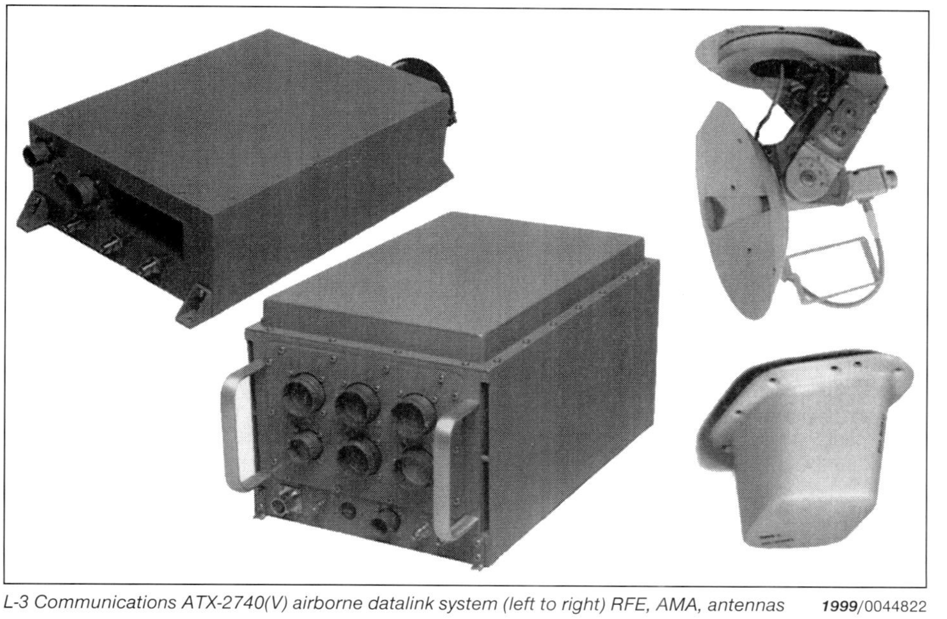

L-3 Communications ATX-2740(V) airborne datalink system (left to right) RFE, AMA, antennas ***1999**/0044822*

Tactical Common DataLink (TCDL) airborne datalink

The TCDL programme is a co-operative effort between the US government, L-3 Communications and Rockwell Collins to develop a communications architecture that supports current and future requirements. Its design emphasises Common DataLink (CDL) interoperability, small size, low weight, affordability, and an open, modular and scalable architecture using COTS technology.

The TCDL airborne system is built in a compact, light weight form factor, specifically for UAVs and manned, non-fighter environments. The data rates, modulation techniques, and transmission frequencies are fully interoperable with CDL systems. The programmable modem offers a variety of other waveforms and data rates that can be selected during flight. The Ku-band (NATO J-band) operating frequency is readily changed by replacing the RF converter and the antenna. Options include bulk encryption and programmable uplink and downlink data rates up to 45 Mbps. Full duplex voice and data modes are available. Either omni or directional antennas can be employed and the system can be upgraded to air-to-air operation.

Specifications

Frequency: uplink 15.15-15.35 GHz
downlink 14.40-14.83 GHz
optional X-band (NATO I/J-band)
tuning 5 MHz steps
Data rates: variable up to 45 Mbps; CDL interoperable at 10.71 Mbps; full duplex symmetric or asymmetric
RF power: 2 W (higher power optional)
Bit error rate: 10^{-6} with COMSEC (10^{-8} without COMSEC)
MTBF: >3,800 h
Interface: RS-422
Dimensions (H × W × D): link interface assembly: 76.2 × 171.5 × 254.0 mm; microwave modem assembly: 76.2 × 304.8 × 254.0 mm
Weight: 7.05 kg (including omni antenna)

Operational status

Designed by L-3 Communications, to be brought into production by Rockwell Collins, Government Systems. The TCDL entered flight testing in February 1999.

Contractors

L-3 Communications, Communications Systems - West.
Rockwell Collins.

VERIFIED

Tactical common datalink airborne data terminal ***1999**/0044823*

AN/AQS-13 sonar for helicopters

The AN/AQS-13 is a helicopter dunking system, the -13B and -13F models being in production. It is one of a series of equipments which began with the AN/AQS-10 in 1985. The AN/AQS-13B is a long-range active scanning sonar which detects and maintains contact with underwater targets through a transducer lowered into the water from a hovering helicopter. Opening or closing rates can be accurately determined and the system also provides target classification information.

The AN/AQS-13B has significant advantages in operation and maintenance over earlier systems. To aid the operator, some electronic functions have been automated. Maintenance has been simplified by eliminating all internal adjustments and adding BITE circuits. These advantages were brought about by the use of the latest electronic circuits and packaging techniques, which also reduced system size and weight.

An Adaptive Processor Sonar (APS) has been developed for the system, to enhance detection capability in shallow water and reverberation-limited conditions, while eliminating false alarms from the video display. The APS is a completely digital processor employing fast Fourier transform techniques to provide narrowband analysis of the uniquely shaped CW pulse transmitted in the APS mode. The display retains the familiar PPI readout of target range and bearing, but APS adds precise digital readout of the radial component of target Doppler. With APS, processing gains of greater than 20 dB with zero false alarm rates have been measured for target Dopplers under 0.5 kt.

The AN/AQS-13E was the first system to integrate APS and sonobuoy processing in a common processor, the sonar data computer. Improvements to this system led to the AN/AQS-13F.

The higher energy transmitted with the longer pulse APS mode, combined with the narrowband analysis, also substantially improves the figure of merit in the non-reverberant conditions typical of deep water operations. Measured processing gains for APS under ambient wideband noise limited conditions exceed 7 dB.

The AN/AQS-13F has been designed to provide rapid tactical response against the most advanced submarine threats. It is a sister equipment to the AN/AQS-18 (see next item) and is identical in many respects. A new transducer, when lowered to depths down to 450 m, permits instantaneous range improvements of over 100 per cent compared to previous systems. Very high-speed reeling allows a dip to maximum depth to be completed in approximately 3 minutes. The powerful omnidirectional transducer providing 216±1 dB source level is integrated with a sensitive directional receiver array providing azimuth resolution in a small rugged unit. The sonar data computer offers digital matched filter processing for 200 and 700 ms sonar pulses, as well as sonobuoy control and processing. The azimuth and range indicator and receiver provides a video display for the operator.

Specifications

Weight:
(13A) 373 kg
(13B) 282 kg
(13F) 280 kg
Frequency: 9.25 to 10.75 kHz
Sound pressure level:
(13B) 113 dB
(13F) 216 dB
Range scales: 1, 3, 5, 8, 12, 20 n miles (0.9, 2.7, 4.6, 7.3, 11, 18.3 km)
Operational modes:
(13A) active 3.5 or 35 ms, MTI, APS, passive, voice communications, key communications
(13F) active 3.5 or 35 ms rectangular pulse, 200 or 700 ms shaped pulse, MTI, passive 500 Hz (bandwidth 9 to 11 kHz), SSB voice communications on 8 kHz
Visual outputs:
(13A) range and bearing
(13B) range, range rate, bearing and operator verification
Audio output:
(13A) single channel with gain control
(13B) dual channel with gain control plus constant level to aircraft intercom
Recorder operation: bathythermograph, range, aspect, MAD self-test
Operating depth:
(13F) 1,450 ft at 50 ft hover

Operational status

The AN/AQS-13 is widely used by US forces and 1,000 sets have been ordered or supplied for the helicopters of 15 foreign navies in Asia, Europe, the Middle East and South America. The AN/AQS-13F was selected by the US Navy and is now in operation on the SH-60F carrier-based ASW helicopter.

Contractor

L-3 Communications Ocean Systems Division.

VERIFIED

AN/AQS-18 sonar for helicopters

The AN/AQS-18 is a helicopter long-range active scanning sonar. The system detects and maintains contact with underwater targets through a transducer lowered into the water from a hovering helicopter. Active echo-ranging determines a target's range and bearing and opening or closing rate relative to the aircraft. Target identification information is also provided.

The AN/AQS-18 is an advanced version of earlier dunking sonars made by AlliedSignal and includes digital technology, improved signal processing and improved operator displays. The system consists of a small high-density transducer with a high sink and retrieval rate, a built-in multiplex system to permit use of a single conductor cable, a 330 m cable and compatible reeling machine and a lightweight transmitter built into the transducer package. The Adaptive Processor Sonar (APS), which provides enhanced performance in shallow water areas, is an integral part of the system.

The AN/AQS-18 offers a number of improvements over earlier dipping sonars. These include increased transmitter power output to give longer range, high-speed dip cycle time and reductions in weight of all units.

The APS increases detection capability in shallow water and limited reverberation conditions, while eliminating false alarms from the video display. The APS is a digital processor which uses fast Fourier transform techniques to provide narrowband analysis of the uniquely shaped CW pulse transmitted in the APS mode. The PPI display retains the normal readout of target range and bearing.

The APS processing gain improvement over the normal AN/AQS-18 analogue processing is 20 dB for a 2 kt target and 15 dB for a 5 kt Doppler target. The higher energy transmitted with the longer pulse APS mode, combined with the narrowband analysis, also improves operation in the non-reverberant conditions more typical of deep water. In general, the gain improvement above a speed of 10 kt exceeds 7 dB.

The latest version is the AN/AQS-18(V) which is available with both 300 and 450 m length cables.

Specifications

Weight: 252 kg plus 13.3 kg for APS
Frequency: 9.23, 10, 10.77 kHz
Sound pressure level: 217 dB/μPa/yd (0.9 m)
Range scales: 1,000, 3,000, 5,000, 8,000, 12,000, 20,000 yds
Modes: 3.5 or 35 ms pulse (energy detection) and 200 or 700 ms pulse (narrowband analysis)
Visual outputs: range, range rate, bearing, operator verification
Audio output: dual channel with gain control plus constant level to aircraft intercom
Recorder operation: bathythermograph, range, ASPECT, MAD, BITE
Operating depth: 330 m

Operational status

In production and in service with the German, Greek, Italian, Japanese, Portuguese, Spanish, Taiwanese and US navies on Lynx, SH-3, SH-60J and S-70C(M)-1 helicopters. Selected for the Egyptian SH-2G(E).

Contractor

L-3 Communications Ocean Systems Division.

VERIFIED

AN/AQS-18(V) dipping sonar system

The AN/AQS-18(V) is the export version of the AN/AQS-13F helicopter dipping sonar. These systems employ a transducer which is lowered into the water from a hovering helicopter to detect and maintain contact with underwater targets. Active echo-ranging determines target range and bearing and opening and closing rate relative to the helicopter. Target classification indications are also provided.

The AN/AQS-13F/18(V) series of dipping sonars is specifically designed for ASW helicopters where great mobility is required for fast reaction. AN/AQS-13F/18(V) equipped helicopters are well-suited for redetection of contacts, target localisation and weapon delivery against shallow and deep water threats. ASW helicopters are often required to search areas which are difficult for other sensor platforms, including shallow water with high noise areas, coastal regions, constrained passages, high-density shipping lanes and areas of concentrated naval activity.

The AN/AQS-18(V) deployed from a German Navy Westland Lynx Mk 88 helicopter

Features of the series include high source level to provide long-range, shallow water signal processing for high reverberation areas, capability to control, process and display sonobuoys in a single integrated system, high-speed reeling machine to achieve maximum depth and retrieval within 3 minutes and adaptable reeling machine designs compatible with a wide range of helicopters.

AN/AQS-13F/18(V) systems include digital technology, improved signal processing, extensive use of hybrid integrated circuits and improved operator displays.

AN/AQS-13G/18A

The AN/AQS-13G/18A is the latest stage in evolutionary development of the system. The improved 'dry end' of the system provides 14 dB improvement over earlier versions, thus providing search rates which are more than four times greater.

Specifications

Weight: 275 kg total
Frequency: 9.23, 10.003 and 10.744 kHz
Operating depth: 440 m
Sound pressure level: 217 ±1 dB/μPa/yd (0.9 m)
Range scales: 0.9, 2.7, 4.6, 7.3, 11, 18 km
Modes:
(active) 3.5, 35 ms pulses, 200, 700 ms shaped pulses (passive), (communicate) SSB at 8 kHz
Raise speed: 6.7 m/s average
Lower speed: 4.9 m/s
Water exit speed: 1.5 m/s

Operational status

Users include Greece, South Korea, Spain and Taiwan.

Contractor

L-3 Communications Ocean Systems Division.

UPDATED

AQS-18A dipping sonar system

The AQS-18A dipping sonar system represents a development of the medium frequency AN/AQS-18(V) dipping sonar. The dome control, reeling machine and transducer of the AN/AQS-18(V) have been interfaced with a powerful digital processor, control unit and colour display. The advanced processing brings greater performance through high-resolution digital processing, greater contact memory space and the flexibility to increase the number of sonar beams, type and length of pulses and menus of operator displays.

The AQS-18A has additional pulse lengths of 1.6, 3.2 and 4 seconds. The longer pulses put more energy on the target and provide higher Doppler resolution for maximum performance in high reverberation shallow water conditions. An FM mode is available for extremely low Doppler target detection and maximum range resolution. The total ASW system improvement of the AQS-18A is 14 dB over current systems and can provide more than four times the area search rate of the AN/AQS-18(V).

The AQS-18A has spare processing power and space for additional processing features such as computer-aided detection and classification, multisensor target fusion, embedded training and performance prediction, based on environmental data collected during past or current missions.

MIL-STD-1553 databus protocol facilitates integration with other aircraft subsystems and components. BIT eases support and boosts availability. The new weapon-replaceable assemblies have a significantly higher MTBF than current systems.

Specifications

Weight: 265 kg total
Frequency: 9.23, 10.003, 10.774 kHz
Sound pressure level: 217 ±1 dB/μPa/yd (0.9 m)
Range scales: 0.9, 2.7, 4.6, 7.3, 11, 18, 29 km
Operating depth: 440 m
Modes:
(active) 3.5, 35 ms pulses, 0.2, 0.7, 1.6, 3.2, 4 s shaped; 0.625 s FMs (passive), (obstacle avoidance), (communicate) SSB at 8 kHz

Raise speed: 6.7 m/s average
Lower speed: 4.9 m/s
Water exit speed: 1.5 m/s

Operational status
The AQS-18A is in production and has been delivered to the Egyptian and Italian Navies.

Contractor
L-3 Communications Ocean Systems Division.

UPDATED

HELRAS Active Dipping Sonar

HELRAS, previously known as LFADS (Low Frequency Active Dipping Sonar) is similar in size and weight to the mid-frequency AN/AQS-18(V). HELRAS has been demonstrated on SH-3 Sea King, EH 101 and SH-60 helicopters. The cable interface, sonar processor, sonar control unit, display, reeling machine and control devices are common between the 10 kHz AQS-18 and the 1.38 kHz HELRAS systems.

HELRAS is capable of depths down to 440 m and has figure-of-merit sufficient to achieve convergence zone detections in deep water, and transmission/receive characteristics optimised for extremely long ranges in shallow water. The low-frequency capability designed into HELRAS using proprietary transducer and beam-forming technology allows multiple boundary interactions and reduced reverberation contamination of the received signals. Use of high-resolution Doppler processing and shaped pulses achieves detection of targets even at speeds as low as 1 kt. Extended duration FM pulses are available to detect the near zero Doppler target as well.

Specifications
Operating depth: 440 m
Operating frequencies: 1.33-1.43 kHz
Source level: 220 dB/μPa/yd
Range scales: 1, 3, 5, 10, 15, 20, 30, 40, 60, 80, 100, 120 kYd (thousands of yards); 0.9, 2.7, 4.6, 9, 14, 18, 27, 37, 55, 73, 91, 110 km
Operational modes: Active-CW to 10 s pulses (FM to 5 s pulses); passive; UQC; raytrace
Raise speed (avg): 4.3 m/s
Lower speed (max): 4.9 m/s
Water exit speed: 1.5 m/s
Seating speed: 0.6 m/s
System weight: (stand-alone)
(Processor and Display) 59 kg
(R/M and Cable) 95 kg
(Transducer) 141 kg
(Total weight) 295 kg

Operational status
HELRAS has been demonstrated in differing water conditions, including the Timor and Mediterranean Seas, Vestfjorden Fjord and against varying targets, including diesel-electric submarines.

It is reported that HELRAS is being ordered by the Italian Navy for its EH 101 fleet, with a first delivery in February 2001. Other potential sales include the NFH 90 helicopter.

Contractor
L-3 Communications Ocean Systems Division.

UPDATED

GPS Receiver Application Module (GRAM)

The L-3/Interstate Electronics Corporation GPS receiver provides highly accurate position, velocity and time information, using the NAVSTAR system. It operates effectively in high jamming and spoofing environments.

Operational status
Used in the navigation systems of AH-64, AV-8B, B-1B, F-117 and F/A-18 aircraft, among others.

Contractor
L-3/Interstate Electronics Corporation.

VERIFIED

RCC-210 series Command Control Receivers

The RCC-210 series UHF command control receivers are solid-state FM receivers designed for operation in aerospace environments where size, weight and power consumption are critical. Models are available with input frequency ranges of 1,400 to 1,550 MHz, 1,700 to 1,850 MHz and 2,200 to 2,400 MHz.

Specifications
Weight: 0.48 kg
Power supply: 28 V DC, 220 mA

Contractor
L-3 Telemetry-East.

VERIFIED

RCC-223-SYN series airborne Command Control Receiver

The RCC-223-SYN series airborne command control receiver is designed to meet missile, UAV and aircraft datalink telemetry requirements. It is a dual conversion superheterodyne UHF receiver capable of operating in S-band with a 200 MHz tuning range and 1 MHz steps. It has dual IF bandwidths that are locally selectable.

Specifications
Dimensions: 152.4 × 82.6 × 52.5 mm
Weight: 1.19 kg

Operational status
Available.

Contractor
L-3 Telemetry-East.

VERIFIED

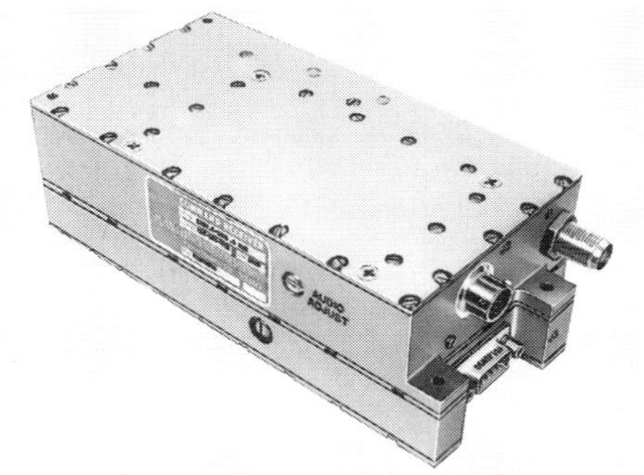

L-3 Telemetry-East RCC-223-SYN S-band 200 channel airborne command control receiver **1999**/0044815

ST-800S/L series wideband microwave transmitter

L-3 Telemetry-East's ST-800S/L series microwave transmitters are designed for highly reliable operation in the severe environmental flight conditions of missiles, space vehicles and aircraft, where size and weight efficiency are critical. These solid-state, crystal stabilised, true FM telemetry transmitters can accommodate various modulation formats such as standard analogue pre-emphasised video, TTL (Transistor Transistor Logic), differential TTL and fully isolated differential TTL (opto-coupled).

Available in 2, 5 and 10 W minimum power output, the ST-800S/L measures 50.8 × 76.2 × 20.3 mm, excluding connectors and weighs 0.2 kg. It operates in the frequency range of 2,200 to 2,400 MHz and 1,435 to 1,540 MHz.

Contractor
L-3 Telemetry-East.

VERIFIED

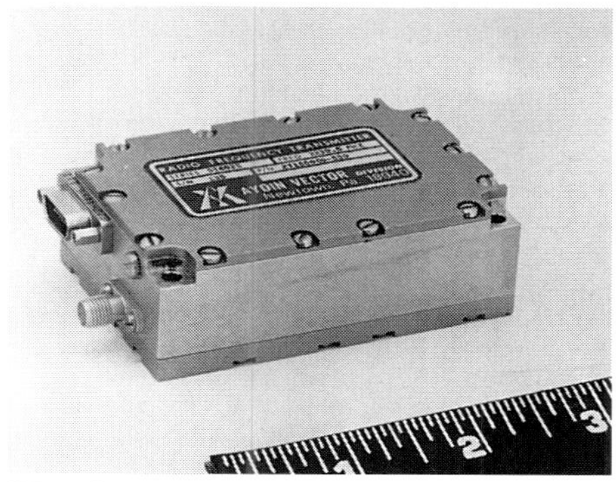

ST-800S/L series wideband microwave transmitter 0011827

T-300 series airborne UHF transmitter

The T-300 series is a subminiature solid-state crystal-stabilised UHF/FM transmitter capable of transmitting wideband telemetry and digital multiplex signals. It is designed for extremely reliable operation in the severe environmental flight conditions associated with missiles, space vehicles or aircraft.

The T-300 operates at 2,200 to 2,400 MHz, with a frequency stability of ±0.002 per cent. Power output is 5 W. The T-300 series meets IRIG-106-93 standards.

Specifications
Dimensions: 63.5 × 38.1 × 19 mm
Power supply: 28 V DC ±4 V
Temperature range: –20 to +70°C

Contractor
L-3 Telemetry-East.

VERIFIED

AN/ARC-51 UHF radio

The Lapointe AN/ARC-51 is a military airborne radio covering the UHF band from 225 to 400 MHz in which it provides 3,500 channels. The system, intended mainly for high-performance aircraft, is available in a number of versions designated ARC-51A, ARC-51AX, ARC-51B and ARC-51BX. These, together with a wide range of controllers, give users the flexibility to assemble a number of configurations to suit particular installations. ARC-51 combinations are also used in surface vehicles and in other land applications.

All variants operate in AM/DSB mode and have a transmitted power output of 20 W. They provide azimuth homing facilities, when used in conjunction with suitable indicator equipment, and can be used for automatic rebroadcast purposes.

Although many different types of controller are available, a typical ARC-51 installation would provide preselection of up to 20 channels together with manual selection of any of the total of 3,500 frequencies covered. An independent Guard receiver provides simultaneous continuous monitoring of the international UHF distress frequency of 243 MHz. Other installations permit the dual control of one or more transmitter/receivers from more than one crew position, and miniature frequency indicator displays are available for use in situations where space is limited.

A feature of this system is the hermetically sealed pressurised container which allows full operational performance in unpressurised avionics bays at aircraft altitudes up to 70,000 ft when used in conjunction with a forced-air cooling supply. This sealed case assembly also renders the system particularly suitable for operation in situations where dust or water contamination could otherwise be expected, such as in desert vehicles or high-humidity tropical environments. The system's normal operating temperature range is from –54 to +71°C.

Specifications

Dimensions: 429 × 222 × 171 mm
Weight: 14.5 kg

Operational status

Variants of the ARC-51 are in service with the Belgian, Canadian, German, Indian, Italian, Thai and US armed forces.

Contractor

Lapointe Industries.

VERIFIED

AN/ARC-73A VHF nav/com radio

Lapointe's AN/ARC-73A is an AM VHF combined navigation and communications transmitter/receiver system for light military aircraft.

The transmitter section covers the frequency band 116 to 149.95 MHz in which it provides 680 channels. The receiver, which can drive ILS and VOR indicators, covers a wider band to receive ground-based navigation aid signals. This band extends from 108 to 151.95 MHz and provides 880 channels. Channel spacing is at 50 kHz increments in both transmitter and receiver sections. Transmitter power output is 20 W. The system is remotely controlled.

Specifications

Dimensions:
(transmitter) 401 × 89 × 193 mm
(receiver) 318 × 89 × 191 mm
(controller) 160 × 145 × 56 mm
Weight:
(transmitter) 6.7 kg
(receiver) 4.7 kg
(controller) 0.76 kg

Operational status

In service.

Contractor

Lapointe Industries.

VERIFIED

ASCOT Aerial Survey COntrol Tool

The Aerial Survey COntrol Tool (ASCOT) is a GPS-supported navigation system for use in aerial survey. It provides interface to airborne equipment such as the GPS receiver, aircraft navigation system, cameras and mission recording. It offers interactive mission planning, programming of photo annotation, better survey flight navigation, automatic triggering of the camera, accurate overlaps and sidelaps, reductions in aircrew from, typically, three to two, the capability to repeat a specific portion of a flight, in-flight annotation of photos and availability of all key mission data in digital form for graphical processing, documentation and further processing.

ASCOT consists of a control computer, control station and pilot display. In addition, there are software modules for mission planning, flight execution, post flight analysis and utility programs.

The control computer is based on a powerful industrial computer designed for use under extremely demanding conditions and suited to the airborne environment. The choice of control computer takes into account aspects such as system and function reliability, safety and electromagnetic tolerance. A silicon disk has been selected as the standard data storage device, mainly because of its reliability under turbulent air conditions and in unpressurised survey aircraft at high altitudes. The architecture of the computer is designed around an Intel processor for fast real-time computations. The operating system is MS-DOS. Depending on the type of application, alternative RAM memory sizes and clock frequencies are available.

The control station is used for general management and aircraft guidance during survey missions. It can be mounted on the NSF3 navigation sight of a Wild camera system and allows supervision and control of the survey flight from the usual camera operator's seat. It can also be mounted in any other suitable location. The navigation station combines a monitor and alphanumeric control keyboard. The standard display consists of a large, flat-panel electroluminescent display with a high resolution. The bright flicker-free image is especially contrasted and clear. A liquid crystal colour display is also available. The alphanumeric keyboard is used for the control of the system, as well as for mission comments. With this set-up, the navigator can access all the functions of the system during flight.

The graphic guidance display for the pilot consists of either a liquid crystal display or an electroluminescent monitor. For aircraft with a pressurised cabin, a CRT display is available.

Contractor

LH Systems, LLC.

UPDATED

AN/USC-42 (V)3 UHF satcom and line of sight communication set

The AN/USC-42 (V)3 Miniaturised Demand Assigned Multiple Access (Mini-DAMA) set is a downsized member of the TD-1271 terminal family. It achieves interoperability with the US Navy's TD-1271B/U multiplexer and AN/WSC-3 and the AFSATCOM system. The Mini-DAMA will function in nine operational modes. Among them is 25 kHz Satcom; here the system will support navy TDMA-1 network operations and non-TDMA communications. On 25 kHz line of sight channels, it will support short-range tactical communications. On 5 kHz UHF Satcom channels, it will interoperate with navy non-TDMA communications, US Air Force DAMA network operations, US Air Force non-TDMA communications and AFSATCOM network operations. Product improvement growth paths exist for embedding AFSATCOM IIR and Have Quick IIA capabilities.

The US Navy's FLTBDCST, CUDIXS/NAVMACS, SSIXS, OTCIXS, secure voice TACINTEL, TADIXS A and ORESTES will use Mini-DAMA for data exchange. Mini-DAMA modem/receiver/transmitters will come in two configurations: 483 mm rack for ship and shore installations and as a 1 ATR-long package for aircraft.

Principal components include an integrated modem/Receiver/Transmitter (R/T), a separately housed power amplifier and an external Display Entry Panel (DEP). The airborne version will contain a remote operation display/entry panel. Operations will be either half- or full-duplex through the Mini-DAMA embedded radio or through a 70 MHz IF interface to an external receiver/transmitter. The AN/USC-42(V)3-Mini-DAMA is configured for airborne platforms, while the AN/USC-42 (V)2 is the equivalent submarine, ship and shore-based version.

Specifications

Frequency: 225-399.995 MHz
Channel spacing: 5 kHz or multiples of 5 kHz
Power output: 100 W
Temperature range: –32 to +55°C
Reliability: 2,000-4,000 h MTBF

Operational status

In production, and in service. Preplanned product improvement (P³I) growth paths exist for embedding COMSEC, vocoding, AFSATCOM IIR, Have Quick IIA and SAFEVET. OTCIXS II is embedded in all terminals.

Contractor

Linkabit, a Titan company.

VERIFIED

AN/USC-42(V)3 Mini-DAMA Satcom terminal ***1999***/0044838

LSM-1000/MD-1333/A UHF DAMA satcom modem

The LSM-1000 is a federated (free-standing) modem, which is capable of implementing Demand Assigned Multiple Access (DAMA) MIL-STD-188-181, -182, and -183 waveforms when used in conjunction with UHF Satcom radios. It can be operated in either full- or half-duplex applications and includes an embedded CTIC chip to implement the encrypted Orderwire Channel operation. The LSM-1000 is configured as a ½ ATR extra-short, low-cost modem in a rugged airborne chassis, providing growth capability for embedding Vocoder and COMSEC functions plus two spare SEM-E card slots. The user configures the modem using a selected platform control device.

Control of the modem may be accomplished via a MIL-STD-1553B interface, a standard RS-232 or RS-422 computer serial interface, or with a dedicated remote Control Indicator (CI). Key loading is accomplished with DS-102-compatible fill devices using the front panel keyfill interface.

The LSM-1000 is compatible with external KG-84A and KY-57-58 COMSEC equipment and Advanced Narrowband Digital Voice Terminals (ANDVTs). The LSM-1000 provides various serial and discrete interfaces including radio control and status to support system applications. It is also available as SEM-E modules for embedded modem applications. The LSM-1000 is prewired to accept an embedded SEM-E Vocoder/COMSEC module available as a plug-in option. A companion mounting tray is also available.

Operational status

Currently available.

Contractor

Linkabit, a Titan company.

VERIFIED

MD-1035B/A UHF dual voice/data modem

The MD-1035B/A UHF dual modem was designed to meet a US Air Force Electronic Systems Division requirement for communication over both current and anticipated satellite systems with widely varying characteristics. It provides a variety of modulation/demodulation, convolutional or block-coding error control and multiple access options in a single package. Communication and network control functions are performed by a flexible, multistack microcomputer which permits demodulation of virtually any digital signalling scheme through software changes. It can interface with a number of RF systems including those operating in frequency bands above the nominal UHF range.

Features include AFSAT 1 and AFSAT 2 (US Air Force Satellite) modulation schemes, dedicated interleaving, error control coding/decoding, input/output and network control and BIT facilities which permit fault identification down to card level. Incorporated firmware changes provide an additional channel for use with the Single Channel Transponder (SCT), 2,400 bits/s for secure voice, coding and interleaving to reduce scintillation mitigation, probing for SCT report-back, 1,200 bits/s and demultiplexing for fleet broadcast and 2,400 bits/s data for tactical operations.

The dual modem upgrade programme is extending the equipment capability further, to operate in a MILSTAR UHF network and to receive the SCT AFSAT 1 and 2 type downlinks. The upgraded modem will receive modified versions of the AFSAT 2 signalling from the SDS and DSCS 3 SCT and AFSAT 1 channel 1.5 from the DSCS 3 SCT. This programme includes development of a MILSTAR payload and command post simulator to allow DMU testing before the availability of the MILSTAR system.

Specifications

Dimensions:
(control indicator) 173 × 147 × 175 mm
(telegraph modem) 198 × 127 × 362 mm
(electrical equipment) 98 × 135 × 435 mm
Weight:
(control indicator) 2.54 kg
(telegraph modem) 8.63 kg
(electrical equipment) 1.59 kg

Operational status

In production and in service. The MD-1035B/A is installed in strategic force elements of the US Air Force in aircraft such as the B-52, B-1B, KC-10, EC-135G and RC-135G. The system is the basis for the Titan Linkabit airborne command post and US Navy attack submarine modem/processor.

The Titan Linkabit MD-1035B/A UHF dual modem

Contractor

Linkabit, a Titan company.

VERIFIED

AN/APR-39A(V)1, AN/APR-39A (V)2, 'AN/APR-39A(V)3 and AN/APR-39B(V)1/3 threat warning systems

Note: The AN/APR-39(V) was an analogue system made by E-Systems. The AN/APR-39A(V) was a digital form-fit replacement. AN/APR-39A(V)1 and AN/APR-39A(V)3 are made by Lockheed Martin Fairchild Systems, with Litton Advanced Systems Division as a second source. AN/APR-39A(V)2 and AN/APR-39B (V)1/3 are made by Litton Advanced Systems Division.

AN/APR-39A(V)1

Designed for helicopters and other light aircraft operating at low level and in Nap-Of-the-Earth (NOE) mission environments, the digital AN/APR-39A(V)1 provides audio and visual (NVG compatible) warning. LRUs include: one digital signal processor; 2 H/M band crystal video receivers; four H/M band spiral antennas; one C/D band receiver; one C/D band blade antenna; one display unit and one control unit.

Specifications

Frequency range: H/M and C/D bands
Weight: 7.05 kg
Power supply: 28 V DC, 58 W
Interfaces: RS-422 databus, MIL-STD-1553B databus optional, E-O warning systems, missile launch detectors, radar jammers, CW warning receivers

AN/APR-39A(V)2

The AN/APR-39A(V)2 is designed for use on Special Electronic Mission Aircraft (SEMA), helicopters and non-high-performance fixed-wing aircraft such as the C-130 and CV-22. In addition to threat warning, the system acts as a controller in an integrated, digitally controlled, electronic warfare survivability suite. The AN/APR-39A(V)2 is integrated with laser warning, missile warning systems and countermeasures dispenser systems. Major LRUs include: four quadrant antennas; two dual channel receivers; one central processor/receiver; one display unit (NVG compatible); one control unit; and one blade antenna.

Specifications

Frequency range: E - K and C/D bands
Weight: 15.8 kg
Power supply: 28 V DC, 200 W
Interfaces: RS-422 databus, MIL-STD-1553B databus optional, E-O warning systems, missile launch detectors, radar jammers, CW warning receivers, and mission data recorders

AN/APR-39A(V)3

Designed for helicopters and other small aircraft operating at low level and NOE missions, the AN/APR-39A(V)3 provides C/D and E to J band coverage with audio and visual warning (NVG compatible).

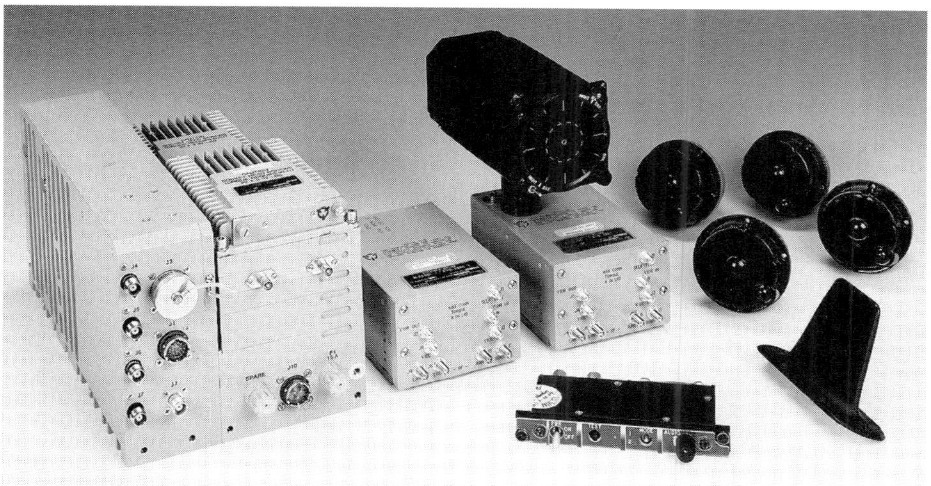

The Litton AN/APR-39A(V)2 threat warning system

The AN/APR-39A(V)1 threat warning system **2000**/0062865

The AN/APR-39A(V)3 threat warning system **2000**/0062864

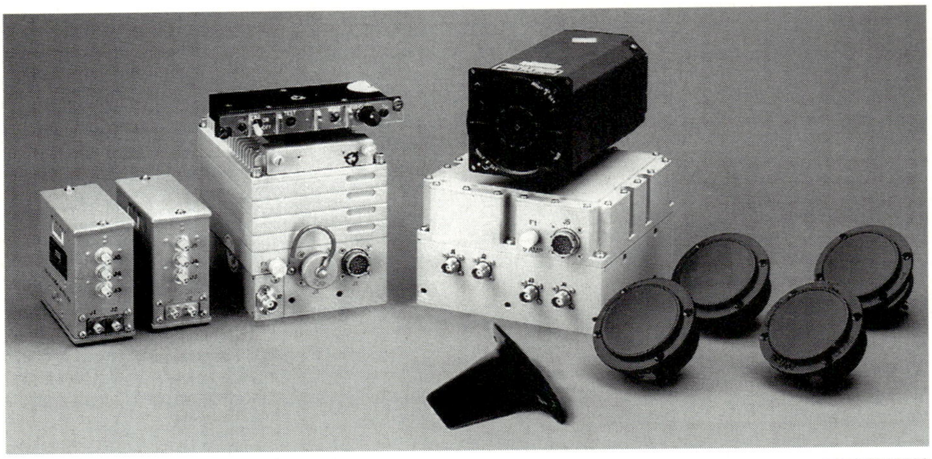

The APR-39B(V)1/3 threat warning system **2000**/0062866

It comprises 10 LRUs, which are form/fit compatible with the analogue APR-39(V)1: one digital signal processor; 2 crystal video receivers; 4 E/J band antennas; one C/D band antenna; one display unit; one control unit.

Specifications

Frequency range: E - J and C/D bands
Weight: 8 kg
Power supply: 28 V DC, 58 W
Interfaces: RS-422 databus, MIL-STD-1553B databus optional, E-O warning systems, missile launch detectors, radar jammers, CW warning receivers

AN/APR-39B(V)1/3

The AN/APR-39B(V)1/3 utilises APR-39A(V)2 technology and includes dual-channel crystal video receivers, plus a Tuned RF (TRF) receiver, to provide enhanced direction-finding of CW and pulse Doppler radars, enhanced frequency measurement for improved ambiguity resolution, and better high-pulse density performance. The system is fully programmable. AN/APR-39A(V)1 and AN/APR-39A (V)3 can be upgraded to this new RF threat warning configuration. The upgrade adds one LRU to the AN/ APR-39A(V)1 configuration at a weight penalty of 4.9 kg, and with no change within the cockpit.

AN/APR-39A(VE) threat warning system

The AN/APR-39A(VE) is aversion of the AN/APR-39A (V)1/3 used by the Royal Norwegian Air Force. These systems have now been upgraded using the AN/ APR-39B(V)1/3 upgrade kits and are now known as the Viking threat warning system.

Operational status

The AN/APR-39A(V)1, AN/APR-39(V)2, and AN/ APR-39A(V)3 are operational in US and foreign military service. The AN/APR-39B(V)1/3, or Viking, is operational in Norwegian service.

Sales continue, and in June 1999, Litton Advanced Systems Division announced a US$70 million order for 350 AN/APR-39A(V)2 systems for US combat helicopter and fixed-wing platforms.

Contractor

Litton Advanced Systems Division
Lockheed Martin Fairchild Systems.

UPDATED

Airborne Digital Automatic Collection System (ADACS)

The Airborne Digital Automatic Collection System (ADACS), an ESM and data collection suite, is an advanced configuration of the AN/ALR-66B(V)3 ESM system. It encompasses integrated ESM capabilities in conjunction with precision emitter parameter collection and measurement. The system handles modern threats including CW, interrupted CW, pulse Doppler, low probability of intercept, jitter/stagger/agile, pulse compression and frequency-agile radars, and performs precision measurements on all intercepted emitters. The design is flexible for easy installation in both fixed- and rotary-wing aircraft.

The system provides full sensitivity and over-the-horizon detection, covering a full 360° in azimuth and ±45° in elevation in the C- to J-bands in high-density EW scenarios. ADACS has automatic operation of both ESM and data collection with precision parameter measurement and recording of all intercepted emitters. The virtually infinite memory capacity permits full mission recording. Removable data record and library modules ensure data security.

Operational status

Lockheed Martin ordered eight systems for the P-3C for a foreign customer in March 1991.

Contractor

Litton Advanced Systems, Inc.

UPDATED

AN/ALR-66(V)4 high-sensitivity ESM system

The AN/ALR-66(V)4 countermeasures receiving set is an airborne electronic system which detects and identifies radars in the C- to J-bands. It is currently in operation on US Navy Boeing E-6A aircraft.

Key features of the system include ultra-high RWR sensitivity, excellent DF accuracy, positive emitter identification in high-density environments, self-protection for the aircraft, automatic operation and complete BIT capability. The system is designed with EEPROM data memory so that the library can be reprogrammed to recognise the identifying characteristics of radar emitters other than those contained in the original threat library. Utilising the computer memory loader, the system can be totally reprogrammed on the flight line. As the system is software intensive, the life-cycle costs are extremely low.

Operational status

In service on US Navy E-6A aircraft.

Contractor

Litton Advanced Systems, Inc.

UPDATED

AN/ALR-66B(V)3 surveillance and targeting system

The AN/ALR-66B(V)3 was developed for the US Navy as an enhancement in capability for its multimission aircraft. The ALR-66B(V)3 surveillance and targeting system is the successor to the US Navy's AN/ALR-66A (V)3 system.

Integrated with the aircraft's radar antenna, the system provides ultra-high system sensitivity and precision DF accuracy. Interfaced with other aircraft sensors, the AN/ALR-66B(V)3 provides operation on a non-interfering basis, as well as interfacing with aircraft navigation systems, central computer and display.

Radar antenna modification techniques provide C- to J-band signal reception and precision direction-finding capabilities required for over-the-horizon targeting. Simultaneous operation of the radar and ESM surveillance and targeting functions is allowed. Alternatively, a dedicated spinning antenna may be used for 360° C- to J-band precision measurements.

Advanced signal processing techniques for instantaneous, positive emitter identification in high-density environments are used. A multimode emitter library, which uses EEPROM technology, permits rapid

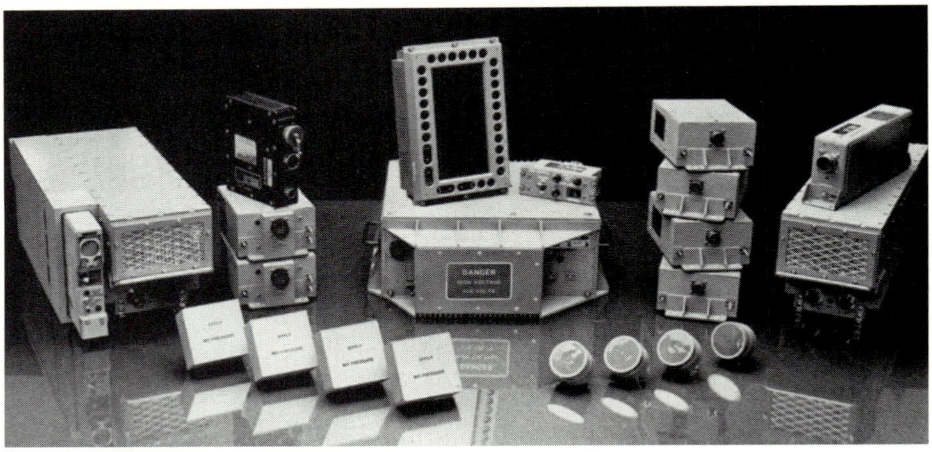

The AN/ALR-66B(V)3 surveillance and targeting system

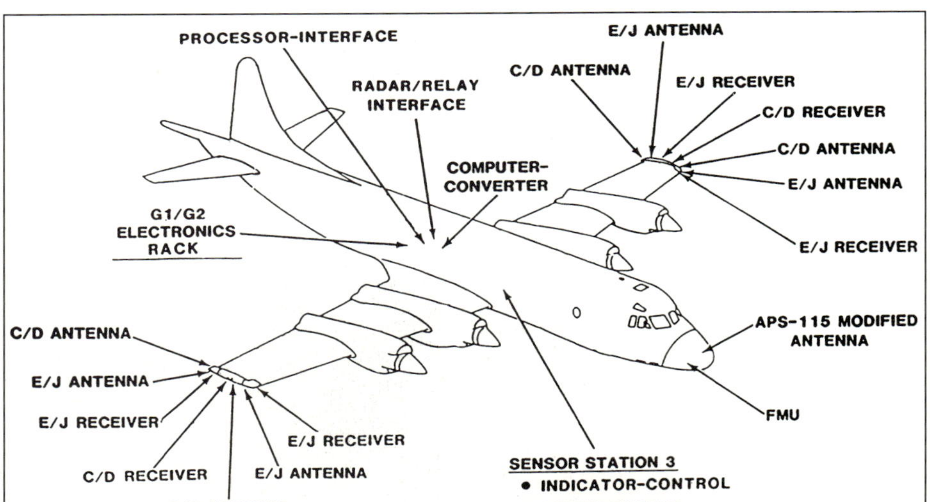

The AN/ALR-66B(V)3 layout in the Lockheed P-3C Orion ASW aircraft

reprogramming of total library scenarios and individual emitters. Emitters not found in the library are displayed by their generic characteristics with precise parameter measurements. The use of EEPROM technology for the AN/ALR-66B(V)3 eliminates the need for hardware modification when changes to the memory are made.

The basic modes of operation are:

Surveillance mode: the highest priority emitters are presented on the plasma display in positions corresponding to their range and bearing, using unique symbology which is either emitter or platform related.

Targeting mode: precise targeting data on any emitter is presented on the plasma display. This operating mode provides the OTH targeting data required for weapon activation.

Emitter waveform analysis: video is accepted from the computer-converter and data is presented directly on the plasma display, allowing the operator to analyse emitter waveforms.

Specifications

Weight: 87.3 kg
Receiver: crystal video
Frequency: contiguous over the C- to J-bands
Warning and identification: all pulsed radars including pulse Doppler, CW, ICW, LPI, 3-D, jitter/stagger, pulse compression and agile
Radar storage: >2,000 emitter modes in removable library storage module
Symbology: 1, 2 or 3 symbols per emitter as desired, programmable

Operational status

The AN/ALR-66(V)3 has been procured by the US Navy and other international customers for Lockheed Martin P-3B and P-3C aircraft.

The AN/ALR-66(V)3 has been upgraded to the AN/ALR-66B(V)3 for the US Navy.

Contractor

Litton Advanced Systems, Inc.

UPDATED

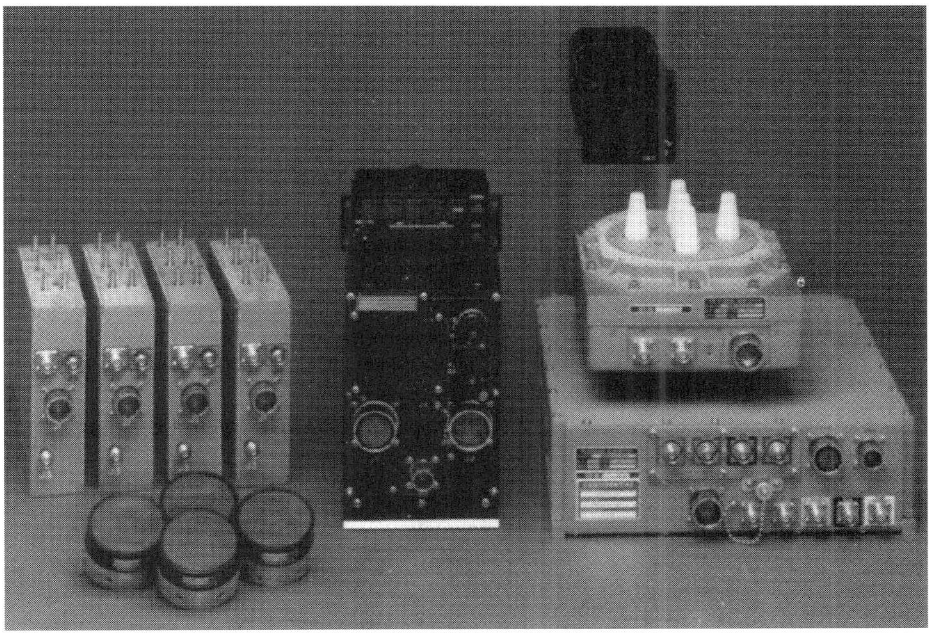

Litton Advanced Systems' AN/ALR-67(V)2 countermeasures warning and control system 0018269

AN/ALR-67(V) countermeasures warning and control system

The AN/ALR-67 countermeasures warning and control system is the successor to the US Navy's AN/ALR-45 system installed in A-6E and F-14A tactical aircraft. It is the standard threat warning system for tactical aircraft and was specifically designed for the A-6E/SWIP, AV-8B, F-14B, F-14D and F/A-18. The system detects, identifies and displays radars and radar-guided weapon systems in the C to J frequency range. The system also co-ordinates its operation with onboard fire-control radars, datalinks, jammers, missile detection systems and anti-radiation missiles. The dispensing of expendables, such as chaff, flares and decoys, is controlled by the AN/ALR-67(V)2.

The AN/ALR-67(V)2 consists of broadband crystal video receivers, a superheterodyne receiver, an integrated low-band receiver, an antenna array and an alphanumeric azimuth indicator. The system is field-programmable and includes provision for software updates at squadron level. The system is fully compatible with MIL-STD-1553B databus requirements and features interface control of such systems as HARM, ALQ-126A/B, ALE-47, ALQ-162 and ALQ-165.

The AN/ALR-67(V)2 comprises the following units:

(1) four small spiral high-band antennas to provide 360° azimuth RF coverage
(2) four wideband, high-band quadrant receivers
(3) a low-band array plus receiver to provide 360° azimuth low-band coverage
(4) a narrowband superheterodyne receiver for signal analysis functions
(5) twin CPU
(6) threat display
(7) control unit.

The AN/ALR-67(V)2 in turn has been given a significant enhancement in capability, through Engineering Change Procedure ECP-510 to the AN/ALR-67E(V)2 standard.

AN/ALR-67E(V)2

ECP-510 provided a card-for-card upgrade of the AN/ALR-67(V)2 to the AN/ALR-67E(V)2 standard. It provides a significant increase in system sensitivity in the presence of strong signals and offers a significant increase in computer pulse processing capability. The upgrade also features the capability to detect and exploit unique signals for improved tactical awareness. To better manage high density signal environments, the AN/ALR-67E(V)2 incorporates Application-Specific Integrated Circuit (ASIC) chips that increase the system's processing power five-fold. The AN/ALR-67E (V)2 provides additional enhancements including a 10-fold improvement in detection ranges when in the presence of a wingman's radar signals; it also incorporates INS stabilisation for accurate display in high 'G' manoeuvres.

The AN/ALR-67E(V)2 modular architecture permits further enhancement, without change to the aircraft system, to provide yet more computer power, increased detection range, and improved target discrimination.

AN/ALR-67(V)3/4

For clarity, note that the AN/ALR-67(V)3/4 developments of the AN/ALR-67(V) are made not by Litton, but by the Raytheon Systems Company.

Operational status

In production. Over 1,600 AN/ALR-67(V) and AN/ALR-67(V)2 systems have been sold. The AN/ALR-67(V) has been supplied for the AV-8B, F-14 and F/A-18 to the US Navy and US Marine Corps, Australia, Canada, Finland, Kuwait, Malaysia, Spain and Switzerland.

Contractor

Litton Advanced Systems, Inc.

UPDATED

AN/ALR-68A(V)3 advanced radar warning system

The AN/ALR-68A(V)3 advanced radar warning system is now deployed operationally by the German Air Force. It is a digital, radio frequency, threat warning receiver system that replaced the AN/ALR-46.

The ALR-68A(V)3 is a broadband crystal video, field-programmable system which provides in-cockpit threat parameter programming and hand-off to tactical ECM systems. It uses a digital threat processor containing software provisions unique to the German threat scenario.

Operational status

Manufactured in association with DaimlerChrysler Aerospace AG, the AN/ALR-68A(V)3 is installed in C-160, F-4F and RF-4E aircraft of the German Air Force.

Contractor

Litton Advanced Systems, Inc.

UPDATED

AN/ALR-69 radar warning receiver

An outgrowth of the ALR-46(V) radar warning receiver, the AN/ALR-69 employs a Frequency Selective Receiver System (FSRS) and a low-band launch alert receiver (Compass Sail) added to the basic ALR-46. Compass Sail detects and analyses SAM guidance beams to warn the pilot of missiles tracking toward the aircraft. The system is designed to activate ECM resources automatically. The signal processor provides executive control for the FSRS. It accepts video inputs from five receivers and processed information from the FSRS, sorts and analyses the data, identifies and labels the received radar signals, tracks the status of these radar signals and generates signals to provide threat warning to the operator. Functions performed by the FSRS include warning and direction-finding on CW signals, accurate frequency measurements on pulse signals for ambiguity resolution, threat antenna scan type and rate analysis and jammer frequency set on for jammer power management and blanking functions.

The Litton Advanced Systems AN/ALR-68A(V)3 advanced radar warning receiver **2000**/0062862

Operational status

More than 3,500 systems have been delivered to the US Air Force and to international air forces. The AN/ALR-69 is used on the A-10, C-130, F-4D and F-16A/B aircraft. No longer in production.

Contractor

Litton Advanced Systems, Inc.

UPDATED

AN/ALR-73 detection system

The AN/ALR-73 passive detection system is an airborne ESM equipment developed for the US Navy's E-2C airborne early warning aircraft. It is an improved version and successor to the AN/ALR-59 and, because of the extensive update, has been given its own designation. The ALR-73 is intended to augment the AEW, surface, subsurface and command and control functions of the E-2C by enhancing the threat detection and identification capabilities of the aircraft. It is a completely automatic, computer-controlled, superheterodyne receiver processing system that communicates directly with the E-2C command and control central processor. The design of the system was motivated by four major considerations: very high probability of intercept in dense environments; automatic system operation; high reliability; and ease of maintenance.

Features of the ALR-73, which are related to its intercept probability performance, include 360° antenna coverage, four independently controlled receivers, dual-processor channels and a digital closed-loop rapid tuned antenna. Features concerned with automatic operation are low false alarm rate, automatic overload logic, an AYK-14 computer that adaptively controls hardware and degraded mode operation.

The system uses 52 grouped antennas in four sets, one for each of several wavebands. Each complete set is positioned to look at a 90° sector. The forward and aft antennas are in the fuselage extremities and the sideways-looking aerials are in the tailplane tips. All receiver sets are under separate control, so wavebands are scanned independently and simultaneously in all sectors. Each antenna has dual-processing channels and uses digital closed-loop rapid tuned local oscillators. The latter provide instantaneous frequency measurement with fast time response and high-accuracy frequency determination.

Receiver outputs are collected at a signal preprocessor unit which performs pulse train separation, direction-finding correlation, band tuning and timing and built-in test equipment tasks. Data is then in a form suitable for the general purpose digital computer, which has overall control of electronic surveillance measures operations and will vary frequency coverage, dwell time and processing time according to prescribed procedures. Control of these parameters is aimed at maximising the probability of intercepting signals on particular missions. Other onboard sensor data and crew inputs will determine the technique adopted. Data such as signal direction of arrival, frequency, pulsewidth, pulse repetition frequency, pulse amplitude and special tags are sorted by the computer and transmitted to the E-2C central processor.

The ALR-73 can measure direction of arrival, frequency, pulsewidth and amplitude and PRI simultaneously. Scan rate information is also available if called for by the central processor. Special emitter tags can be provided. The ALR-73 detects and analyses electromagnetic radiation within the microwave portion of the spectrum and sends emitter reports of pulsewidth, PRI, direction of arrival, frequency, pulse amplitude and special tag to the E-2C's central processor via its own data processor. The ALR-73 immediately reports new emitters to the central processor, which performs the identification function. It eliminates redundant data on emitters for a programmable period of time, thus significantly reducing the data rate to the central processor. The ALR-73, a multiband, parallel scan, mission programmable system covers the frequency range in four bands through step sweeping. Programmable frequency bands and dwell time permit very rapid surveillance of priority threat bands. Non-priority bands are also monitored, but at a reduced rate. Probability of intercept is increased without sacrificing sensitivity through the detection of both real and image sidebands.

The US Navy E-2C Hawkeye carries the Litton AN/ALR-73 ESM system

Operational status

In production, and in service on the US Navy Northrop Grumman E-2C and with the forces of France, Japan, Singapore and Taiwan.

The AN/ALR-73 has also been successfully installed on C-130 aircraft.

Contractor

Litton Advanced Systems, Inc.

UPDATED

AN/ALR-87 radar warning system

The AN/ALR-87 advanced threat warning system is designed with all the necessary performance features to meet identified and postulated threat environments. It is a replacement for the AN/ALR-46(V) family of radar warning receivers. The AN/ALR-87 is based on a digital IFM receiver and is controlled by MIL-STD-1750A microprocessors. The system is fully programmable with in-flight recording capability, operates in the C- to J-bands to provide rapid emitter identification in complex and dense threat environments and interfaces with active and passive onboard systems.

It consists of four high-band antennas, four quadrant receivers, a low-band antenna, low-band receiver and power supply, processor/receiver, azimuth indicator and indicator control unit.

Specifications

Dimensions:
(high-band antenna × 4) 50.8 × 70.1 mm diameter
(quadrant receiver × 4) 175.3 × 154.9 × 40.6 mm
(low-band antenna) 101.6 × 127 × 76.2 mm
(low-band receiver/power supply)
254 × 152.4 × 101.6 mm
(processor/receiver) 127 × 193 × 370.8 mm
(azimuth indicator) 81.5 × 81.5 × 222.1 mm
(indicator control unit) 45.7 × 129.5 × 109.2 mm
Weight:
(high-band antenna × 4) 0.45 kg
(quadrant receiver × 4) 10 kg
(low-band antenna) 0.13 kg
(low-band receiver/power supply) 5.45 kg
(processor/receiver) 11.82 kg
(azimuth indicator) 1.36 kg
(indicator control unit) 0.68 kg
Frequency: C- to J-bands
Power: 580 W
Processors: three MIL-STD-1750A microprocessors
Memory: 128 k EEPROM, 162 k RAM
Mission data recorder: 1 M × 17 dynamic RAM
Interfaces: MIL-STD-1553B AV and EW bus, RS-422, RS-232

Operational status

200 systems have been procured by the Swiss Air Force for Northrop Grumman F-5E and F and Dassault Aviation Mirage III aircraft. No longer in production.

Contractor

Litton Advanced Systems, Inc.

UPDATED

AN/ALR-91(V)3 radar warning system

The AN/ALR-91(V)3 radar warning system is an upgrade of earlier Litton AN/ALR-46(V) radar warning receivers. The AN/ALR-91(V)3 is the designation given to the AN/ALR-46 RWR upgraded with Litton's CM-518 digital processor replacing the CM-442A processor, and the AM-9101 Single Channel Amplifier Detectors (SCADs) replacing the AM-6639 high-band receivers.

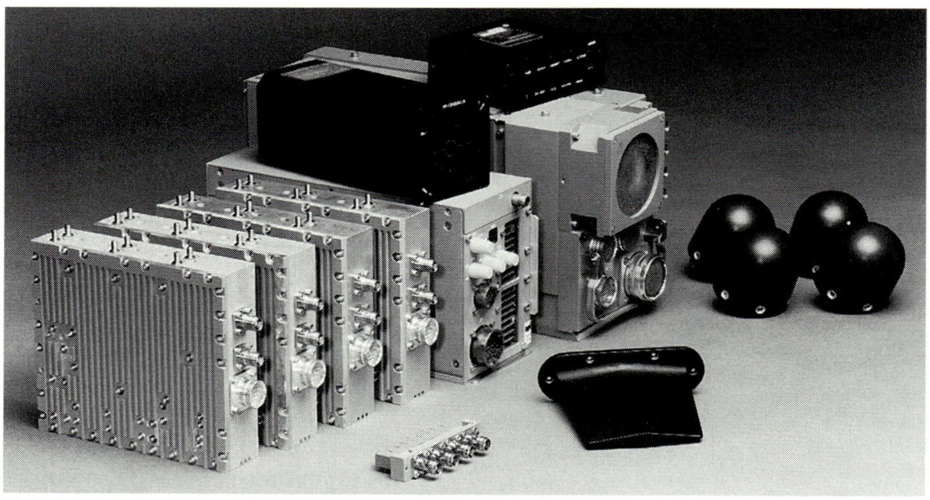

The Litton Advanced Systems AN/LAR-87 radar warning system **2000**/0062869

Litton Advanced Systems' AN/ALR-91(V)3 radar warning system **2000**/0062868

The updated system offers the following improvements, relative to the ALR-46 RWR: up to 10 times the detection range; operation in pulse densities five times greater; detection and DF of pulse, PD and CW emitters.

A recording capability for mission playback and training is also provided, together with a flight line reprogramming capability

EW system integration can be accomplished using MIL-STD-1553B and RS-422 interfaces provided.

Specifications

RF coverage: 5-18 GHz
Spatial coverage: 360° azimuth; ±60° elevation
DF: 15° RMS
MTBF: > 600 h

Operational status

Operational on F-5E/F, F-4E and RF-4E aircraft.

Contractor

Litton Advanced Systems, Inc.

UPDATED

AN/ALR-93(V)1 Radar Warning Receiver/Electronic Warfare Suite Controller (RWR/EWSC)

Litton Advanced Systems' AN/ALR-93(V)1 radar warning receiver and electronic warfare suite controller
2000/0062867

The AN/ALR-93(V)1 is a computer-controlled RWR that provides automatic detection and display of RF signals in the C/D and E- to J-bands.

The AN/ALR-93(V)1 is designed to operate in dense, complex RF threat emitter environments with near 100 per cent Probability of Intercept (POI). By combining the wideband acquisition capability of an amplified Crystal Video Receiver (CVR) with the fast frequency measurement of an Instantaneous Frequency Measurement (IFM) receiver and the selectivity of a SuperHeterodyne Receiver (SHR), the AN/ALR-93(V)1 is able to reduce threat ambiguities, increase detection range and still achieve a high POI.

The AN/ALR-93(V)1 provides threat warning to the aircrew both visually and aurally; display of warnings can be integrated with other aircraft warning systems.

The AN/ALR-93(V)1 also provides a flexible mission recording and software support system, providing for post-mission analysis and mission optimisation.

Specifications

Weight: <27.2 kg
Frequency coverage: C/D, E to J-band
Receiver types: amplified CVR, IFM, SHR
Radar types: all pulsed radars including pulse Doppler, CW, ICW, LPI, jitter/stagger, pulse compression, frequency-agile, PRI agile and agile-agile
DF accuracy: 15° rms (E- to J-bands), omni directional (C/D-band)
Emitter library storage: 1,800 modes
Reliability: 525 h MTBF
Interface options: 2 dual redundant MIL-STD-1553B, RS-232C, RS-422, discretes
Warnings: visual and aural
Programming: both OFP and emitter library are contained in EEPROM
Software: 'C' higher order language.

Operational status

Operational on 80 F-16C/D aircraft for a NATO customer believed to be part of the ASPIS EW system ordered by Greece. Also provided for fit to CN-253 and A-4M aircraft for international customers.

Contractor

Litton Advanced Systems, Inc.

UPDATED

AN/ALR-606(V)1 series radar warning receivers

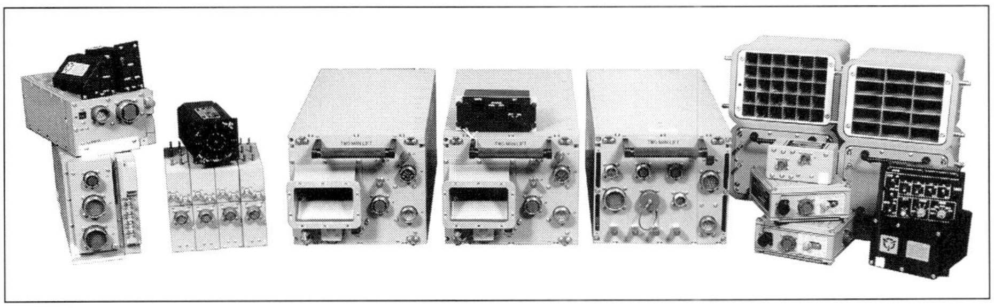

Aspis EW system
2000/0062861

The AN/ALR-606(V)1 ESM system is a fully digital radar warning receiver. It is based on the AN/ALR-66A(V)1, with which it shares the same hardware configuration.

The ALR-606(V)1 is designed especially for the export market by offering item for item replacement of older analogue and digital RWR systems in previous generation maritime patrol aircraft such as the S-2 Tracker, TA-7 and C-130. It also provides full digital signal processing, high-intensity CRT, E- to J-band coverage and alphanumeric symbology.

Use of EEPROM allows on-site reprogramming capability for display symbology and program parameters through the use of a separate Computer Memory Loader (CML).

The ALR-606(V)1 has recently been upgraded to an ALR-606A(V)1 configuration which incorporates new higher-sensitivity receivers, a liquid crystal display, increased EEPROM memory, enhanced processing and allows complete OFP and emitter library reprogramming through the use of a CML.

Specifications

Weight: 28 kg
Frequency: E- to J-band continuous
Warning I/D: all pulse types; CW, ICW and pulse Doppler, LPI, jitter/stagger, pulse compression, and frequency-agile radars
Receiver type: crystal video
Azimuth: 360° coverage
Symbology: up to 3 symbols per emitter, programmable by customer
Interface: RS-232 or MIL-STD-1553
Reliability: >630 h MTBF demonstrated (MIL-STD-781)

Operational status

No longer in regular production, having been replaced by the newer ALR-606A(V)1, the ALR-606(V)1 is in worldwide use in numerous types of aircraft.

Contractor

Litton Advanced Systems, Inc.

UPDATED

AN/ALR-606(V)2 series surveillance and direction-finding systems

The AN/ALR-606(V)2 surveillance and direction-finding system is specifically designed for use in maritime patrol aircraft, offering coverage in the C- to J-bands and over-the-horizon emitter location. It was derived as an export version of the AN/ALR-66A(V)3 system. The equipment provides advanced capabilities in such areas as precision DF accuracy, high sensitivity for over-the-horizon detection, precise frequency measurement, advanced signal processing coupled

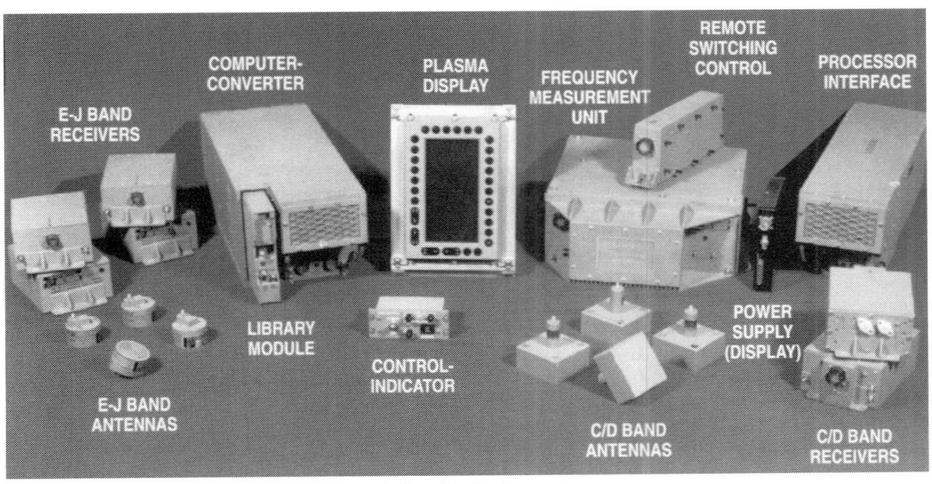

Units of the AN/ALR-606(V)2 surveillance and direction-finding system

with expanded data memory, multimode operator interactive display and controls, precision emitter parameter measurements, integration with other aircraft primary sensors and EEPROM flight line reprogramming. Upgrades to the ALR-606A(V)2 can be accomplished with minimal or no aircraft wiring change. Upgrade includes higher sensitivity E- to J-band receivers, expanded emitter library capacity, improved processing and in-flight signal-of-interest programming. Additional upgrades to ALR-606B(V)2 have been developed which include incorporation of a dedicated, controllable DF antenna and the addition of a data recorder to support ground analysis of ESM intercepts.

Operational status

The AN/ALR-606(V)2 has been provided for the Northrop Grumman S-2 maritime patrol aircraft, Sikorsky S-70 helicopter and for numerous international customers. The ALR-606(V) as well as the upgraded ALR-606A(V)2 and ALR-606B(V)2 systems are in production.

Contractor

Litton Advanced Systems, Inc.

UPDATED

AN/APR-39A(V)1,3,4 family of Radar Warning Receivers (RWR)

The Litton Advanced Systems' AN/APR-39A(V)1/3/4 are a family of low-cost RWRs, combining lightweight, small size and low-power consumption. The A(V)1/3/4 provides rapid 360 degree detection and identification of RF threats. The APR-39A(V)1 is the standard US Army RWR and is also widely utilised in USAF, USN and USMC aircraft. The APR-39A(V)3 is generally for export and is identical to the A(V)1 except for frequency coverage. The APR39A(V)4 is identical to the A(V)1 with the addition of a 1553B interface card and is used specifically on the AH-64D .

Operational status

There are in excess of 5,000 APR-39A(V)1/3/4 systems fielded worldwide; this includes all 4 US services and approximately 15 international users. Current installations for the system include the AH-1G/S/W, AH-64A/D, CH-47D, MH-4, UH-1H and A-109 helicopters, the F-5A/B, L-35, L-100 and DH-7/8 fixed-wing aircraft, together with various shipboard applications. The APR-39(V)1 and APR-39A(V)1 systems in the USMC/USN are being replace by the new APR-39A(V)2 and APR-39B(V)2.

Contractor

Litton Advanced Systems.

Specifications for AN/APR-39A(V)1,3,4 family of Radar Warning Receivers (RWR)

	AN/APR-39A(V)1	AN/APR-39A(V)3	AN/APR-39A(V)4
Frequency Range:	E-M and C/D bands	E-K and C/D bands	E-M and C/D bands
Weight:	6.8 kg (15 lb)	6.8 kg (15 lb)	7.0 kg (15.5 lb)
Power:	100 W, 28 V DC	100 W, 28 V DC	110 W, 28 V DC
Display:	NVG Compatible CRT	NVG Compatible CRT	MFD/Mission Computer
Interfaces:	RS-422	RS-422	Dual 1553B (RT) RS-422
Integration:	AAR-47, AVR-2, AVR-2A, PC-M/LV	AAR-47, AVR-2, AVR-2A, PC-M/LV	AAR-47, AVR-2, AVR-2A, PC-M/LV

NEW ENTRY

AN/APR-39A/B(V)2 Radar Warning Receiver (RWR)/ Electronic Warfare Management System (EWMS)

The Litton Advanced Systems' AN/APR-39A(V)2 and AN/APR-39B(V)2 are the newest RWR/EWMS in the US government inventory, with a requirement in excess of 1,000 units. This low-cost system provides state-of-the-art technology, combining light weight, small size and low-power consumption. The APR-39A/B(V)2's highly sensitive multireceiver architecture provides rapid 360° detection and identification of RF threats in the E-K Bands and is the EWMS for the USMC/USN Suite of Integrated Sensors and Countermeasures (SISCM). SISCM is comprised of an EWMS, RWR, Missile Warning System (MWS), Laser Warning System (LWS) and Countermeasures Dispensing System (CMDS), providing multispectral threat warning and semi-automated and fully automated countermeasures deployment. An RF Jamming capability is currently under consideration.

The AN/APR-39B(V)2 is the newest variant and differs from the A(V)2 in several significant areas. The primary area is an Electronic Warfare Management System (EWMS) card, which is added to the system. The EWMS has its own stand alone 603R processor specifically for EW management, integration and situational awareness applications. The EWMS card was originally designed to integrate with new glass cockpit aircraft such as the AH-1Z, UH-1Y, KC-130J and MV-22. The system has a Duplex, Dual 15538 capability, allowing it to be a bus controller on the EW bus and an RT on the avionics bus. This serves to improve situational awareness by allowing the sharing of EW sensor information, EOB data and onboard navigation data that can all be overlaid on a cockpit digital map. The EWMS card also provides for mission data recording and playback.

Specifications for AN/APR-39A/B(V)2 Radar Warning Receiver (RWR)/Electronic Warfare Management System (EWMS)

	AN/APR-39A(V)2	AN/APR-39B(V)2
Frequency Range:	E-K and C/D bands	E-K and C/D bands
Weight:	15.8 kg (35 lb)	15.8 kg (35 lb)
Volume:	12,710 cc (775 cu in)	12,710 cc (775 cu in)
Power:	200 W, 28 V DC	200 W, 28 V DC
Display:	NVG Compatible CRT	MFD/Mission Computer
Mission Data Recorder:	N/A	2-8 h recording time
Interfaces:	BC Dual 1553B RS-422	Dual 1553B (BC) Dual 1553B (RT) RS-422 RS-232
Integration:	AAR-47, AVR-2, AVR-2A, AAR-60, ALE-47, USQ-131	AAR-47, AAR-47(V)1, AAR-47(V)2, AVR-2, AVR-2A, AAR-60, ALE-47, USQ-131, PC-M/LV, AH-1Z/UH-1Y

Operational status

The system entered service in November 2000 and is to be installed in a wide range of USMC and USN rotary- and fixed-wing aircraft, including the AH-1W/Z, UH-1Y/N, CH/MH-53, HH-60, MV-22 and KC-130. Additional customers for the system include Greece, New Zealand and Taiwan.

Contractor

Litton Advanced Systems.

NEW ENTRY

AN/APR-39B(V)4 Radar Warning Receiver (RWR)

Litton Advanced Systems' AN/APR-39B(V)4 is a low-cost system that provides state-of-the-art technology in RWRs, combining lightweight, small size and low-power consumption. The APR39B(V)4's highly sensitive multireceiver architecture provides rapid 360 degree detection and identification of RF threats in the E-K Bands. The -39B(V)4 can be purchased as a new stand-alone system or as an upgrade kit to the current AN/APR39A(V)1/3 RWR. The AN/APR-39B(V)4 upgrade kit consists of one new small LRU (RF Deck), four new plug and play CCAs and new quadrant receivers and antennae. In most cases, the old A(V)1/3 wiring harness can be retained and only RF cables would need to be added.

Improved capabilities over the old AN/APR-39A (V)1/3 system include:

- greatly improved ambiguity resolution
- much higher sensitivity
- higher pulse density handling capability
- AI/PD handling
- complex/modern threat detection
- directional CW detection
- increased processing capability
- dual 1553B databus
- built-in growth capability

Specifications

Frequency range: E-K and C/D bands
Weight: 11.8 kg (26 lb)
Power: 150 W, 28 V DC
Display: NVG compatible CRT or MFD interface
Interfaces: Dual 1553B (RT), RS-422
Integration: AAR-47, AVR-2, AVR-2A, PC-M/LV

Contractor

Litton Advanced Systems.

NEW ENTRY

LR-100 Warning and Surveillance Receiver

The LR-100 is a lightweight radar signal receiver covering 2 to 18 GHz. Options are available for 70 to 200 MHz and 18 to 40 GHz. This receiver has been built to provide precision RWR, ESM and ELINT measurements with a total installed weight of less than 23 kg, including receiver, antennas, cables and brackets. The LR-100 system is a complete, two channel interferometer receiver system with a 500 MHz bandwidth and VME-based processor.

Phase and amplitude calibration signals are injected at the antenna to achieve precision angle and location measurements.

Emitter identification and location data are passed digitally for warning, display and analysis and recording. The only support equipment needed for the LR-100 is a Windows™-compatible PC. Help and maintenance manuals are built into Windows™-based software. User defined receiver function and identification parameters are programmed with the same software tool. The Litton software tool can graphically display this information on the vehicle or via datalink. The LR-100 can be modified in real time for special receiver modes or directed tuning.

Specifications

Dimensions:
(receiver) 254 × 177.8 × 127 mm
(array) 304.8 × 304.8 × 90 mm
Weight:
(receiver) 11 kg
(array) 1.8 kg
Field of view: ±65° (azimuth and elevation)

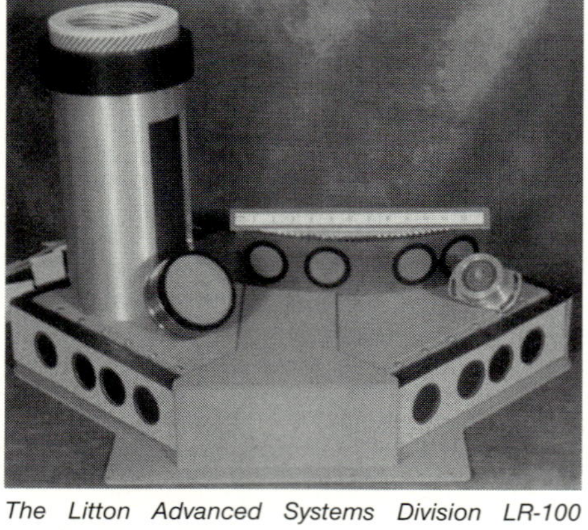

The Litton Advanced Systems Division LR-100 warning and surveillance receiver **2000**/0062860

Accuracy:
amplitude DF: < 15°
interferometer DF: 0.78° rms

Operational status

Built as a COTS product. Proposed to the United Arab Emirates as part of both the Lockheed Martin F-16 bid. It has been selected for the Royal New Zealand Navy Kaman Aerospace SH-2G Super Seasprite multimission helicopters.

Contractor

Litton Advanced Systems, Inc.

UPDATED

LR-4500 microwave collection system

The LR-4500 microwave collection system was designed to perform ESM and ELINT functions over the frequency range of 0.5 to 18 GHz. Information is gathered by onboard antenna and receiver/processor equipment, which may be installed on aircraft, on ships or at ground sites. Aircraft installations may be either internal or external in a pod. Collected data may be processed locally or transmitted to an operations centre by datalink. Information may also be recorded on magnetic tape for post-mission analysis. The airborne system can be installed in either manned or unmanned vehicles.

The complete system includes airborne, ground processing and maintenance support equipment. The airborne component of the system weighs 102 kg.

Operational status

Equipment has been delivered to an unnamed customer.

Contractor

Litton Advanced Systems, Inc.

UPDATED

LT-500 Emitter Targeting System

LT-500 is a passive precision interferometric RF Emitter Targeting System (ETS). The fully militarised LT-500 ETS is a recent joint development of Litton Advanced Systems Division and TRW Avionics. Designed to achieve the goal of balancing performance, cost, and reliability in a single receiver/processor electronics package, its primary application is passive precision RF targeting of modern air-ground weapons via retrofit and upgrade of existing tactical fighter aircraft.

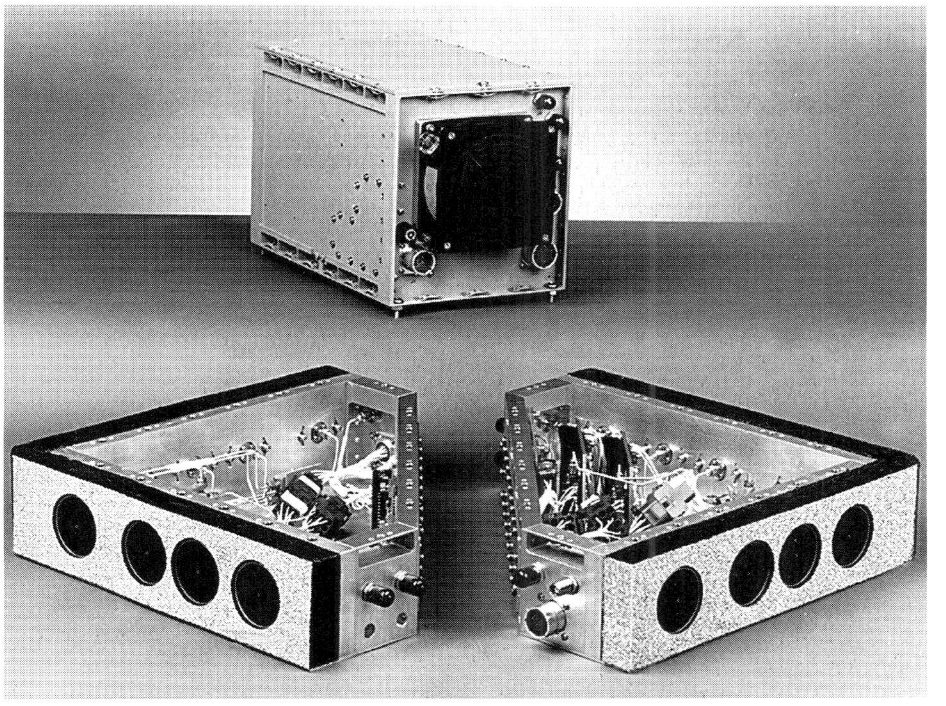

LR-100 array open **2000**/0062870

This cost-effective implementation consists of a totally modular 'building block' architecture (SEM-E size electronics modules weighing 34 kg total without antenna) that may be easily tailored in capabilities to match specific emitter tracking/targeting mission requirements. The LT-500 ETS may be installed with a variety of interferometer antenna array configurations and is operated via a MIL-STD-1553 control/display interfaces. An embedded, 32-bit JIAWG compliant, RISC processor hosts the Ada language 'Operational Flight Program'.

Employing Litton-patented interferometric direction-finding and ranging techniques, the LT-500 achieves very rapid situation awareness, emitter identification and precision geolocation in dense electromagnetic environments. The LT-500 ETS is compatible with either internal or pod installation on board the F-15, F-16, F/A-18 or similar tactical aircraft. Because of its modularity, it may be easily adapted to a wide variety of other airborne, shipboard, or ground-based platforms.

Contractor

Litton Advanced Systems, Inc.

UPDATED

AN/APN-217 Doppler Radar Navigation Systems

The AN/APN-217 Radar Navigation System (RNS) is a lightweight, low-power, self-contained Doppler radar velocity sensor. The RNS unit detects and processes Doppler-shifted frequency returns from continuous wave, time-multiplexed radar beams to determine three orthogonal velocities in aircraft heading (Vh), drift (Vd) and vertical (Vz) co-ordinates. The AN/ARN-217 is available in -217, -217(V)2, -217(V)3, -217(V)5 and -217(V)6 models and can accommodate output formats of ARINC or MIL-STD-1553 and provides DC analogue voltages for driving hover indicators and automated flight control systems. Applications encompass helicopters and medium-performance fixed-wing aircraft.

Specifications

Dimensions: 164 × 423 × 408 mm
Weight: 12.72 kg
Power supply: 28 V DC, 48 W
Velocity range:
(heading) −40-350 kt
(drift) −100-100 kt
(vertical) −4,500-4,500 ft/min
Accuracy:
(heading) <0.3% over land, <0.4% over water
(drift and vertical) <0.2%
Reliability: 15,257 h MTBF

Operational status

The AN/APN-217 is the standard Doppler for US Navy and US Marine Corps aircraft.

Contractor

Litton Guidance & Control Systems.

VERIFIED

AN/APN-218 Doppler Velocity Sensor (DVS)

The AN/APN-218 Doppler Velocity Sensor (DVS) is a reliable, high-performance, nuclear-hardened Doppler radar for fixed-wing aircraft. Usually referred to as the Common Strategic Doppler (CSD), it is combined with a GroundSpeed Drift Indicator (GSDI) to process digital velocity data from the radar and display the groundspeed and drift angle. The DVS is an LRU consisting of 10 Shop Replaceable Units (SRUs). The GSDI is also an LRU and consists of three SRUs.

The AN/APN-218 continuously measures three velocities: heading (Vh), drift (Vd) and vertical (Vz). These provide accurate in-flight data to navigation equipment. It can output data for all three co-ordinates from the microcontroller and process it into either MIL-STD-1553A or ARINC 575 formats.

Specifications

Dimensions:
(sensor) 636 × 708 × 168 mm
(GSDI) 146 × 76 × 155 mm
(CDU) 146 × 152 × 165 mm
Weight:
(sensor) 33.18 kg
(GSDI) 1.5 kg
(CDU) 3.9 kg
Power supply: 115 V AC, 400 Hz, 10 VA
Velocity range:
(heading) 96-1,800 kt
(drift) ±200 kt
(altitude) up to 70,000 ft
Reliability: >9,740 h MTBF

Operational status

In production. AN/APN-218 is in service in US Air Force B-52, C-130, KC-135 and MC-130 aircraft and on the C-130J.

Contractor

Litton Guidance & Control Systems.

VERIFIED

AN/APN-233 Doppler velocity sensor

The AN/APN-233 can be used either as a single unit velocity sensor providing outputs to other aircraft systems or with a Control/Display Unit (CDU) and HSI as a self-contained navigation facility. It was designed for applications in which size, weight, performance and reliability are critical factors.

The APN-220 family evolved from a small, lightweight Doppler sensor originally designed for the US Army and was subsequently qualified by that service and by the US Air Force and the German Air Force.

An optimum velocity range and near-zone rejection are offered for each application.

The CDU combines the functions of navigation computer and control/display unit and contains an incandescent alphanumeric display panel and a keyboard for entering data and selecting operational modes. Up to 10 waypoints can be accommodated, and a non-volatile scratchpad memory holds critical information during power transients or interruptions.

Specifications

Typical fixed-wing applications
Dimensions:
(sensor) 426 × 291 × 113 mm
(CDU) 152 × 146 × 165 mm
Weight:
(sensor) 9.66 kg
(CDU) 3.86 kg
Power supply:
(sensor) 28 V DC, 28 W
(CDU) 28 V DC, 30 W

Output: heading, vertical velocity and groundspeed/drift to aircraft systems, for example AFCS, or to CP-1251 and HSI
Number of waypoints: 10 entered via front panel keyboard
Velocity range (typical system):
(speed) –40 to 600 kt
(drift) ±150 kt
(vertical) ±5,000 ft/min
Altitude: up to 50,000 ft
Accuracy:
(over land) 0.25% + 0.2 kt
(over sea) 0.3% + 0.2 kt
Self-test: BITE diagnostic programme locates faults at first-line level to 95% confidence
Reliability: 2,600 h MTBF demonstrated in the Alpha Jet

Operational status
In production. The system has been produced for the US Navy C-2A Greyhound and US Marine Corps OV-10D Bronco observation post aircraft. Additional production for the S-2, DHC-5 and CH-47 aircraft has been completed.

The system has been chosen by the German Air Force for its Dassault/Dornier Alpha Jet strike/trainers as a velocity sensor to provide data to the navigation and weapons delivery systems. Versions of the equipment have also flown on RPVs and helicopters.

Contractor
Litton Guidance & Control Systems.

VERIFIED

AN/APX-( )MAT IFF transponder

The AN/APX-( )MAT contains, in a single unit, all Mk XII, Mode S and cryptographic IFF transponder functions with either a MIL-STD-1553B multiplex bus or a discrete control panel interface. It operates on Modes 1, 2, 3A, C, 4 and S. It is compatible with AIMS and STANAG requirements and uses advanced microwave packaging techniques to minimise space and weight.

Features include a KIV-2 COMSEC appliqué that uses an electronic key fill loader to eliminate requirements for a computer and mechanical code loader. Mode S Level 1 is provided as a standard feature, with optional growth to Levels 2 and 3.

Specifications
Dimensions: 136.5 × 136.5 × 212.6 mm
Weight: 4.55 kg
Frequency:
(receive) 1,030 ± 0.5 MHz
(transmit) 1,090 ± 0.5 MHz
Power output: 27 dBW ± 2 dB at 1.2% duty cycle

Contractor
Litton Guidance & Control Systems.

VERIFIED

AN/APX-92 IFF transponder

One of a family of IFF systems produced by Litton, the APX-92 is suitable for all types of aircraft, including helicopters. It offers response in Modes 1, 2, 3/C, 4 and C. In Mode 4 it operates in conjunction with a Mode 4 computer to give complete recognition and monitoring capability, while for Mode C it interfaces with an appropriate encoding altimeter. The 500 W transmitter and the receiver are both all-solid-state and the signal processor uses linear and digital circuitry.

Specifications
Dimensions: 133 × 146 × 254 mm
Weight: <4.54 kg
Power supply: 28 V DC, 55 W

Operational status
In service.

Contractor
Litton Guidance & Control Systems.

VERIFIED

AN/APX-101(V) IFF transponder

The AN/APX-101(V) diversity IFF transponder is all-solid-state and consists entirely of replaceable modules. The transponder is housed in a single LRU and is mounted in the airframe without the use of shock-mounts. Crystal-controlled pulsewidth, discrimination, decoding and encoding ensure accurate response to interrogations. BIT circuits monitor the critical parameters of the transponder and provide an immediate status indication.

The transponder operates in Mk XII Modes 1, 2, 3A, C and 4. It receives RF interrogations from two antenna systems, decodes the interrogation into the proper mode, encodes the selected reply and transmits the coded RF reply through the correct antenna.

Specifications
Dimensions: 152.4 × 127.3 × 278 mm
Weight: 6.53 kg
Power supply: 28 V DC, 65 W
Frequency:
(receiver) 1,030 MHz
(transmitter) 1,090 ± 1.5 MHz
Temperature range: –54 to +70°C
Altitude: up to 100,000 ft

Operational status
Standard equipment on A-10, E-3A, F-5E, F-15 and F-16 aircraft. Over 3,000 units have been ordered.

Contractor
Litton Guidance & Control Systems.

VERIFIED

AN/APX-108 IFF transponder

The AN/APX-108 is a single unit, reduced size Mk XII diversity IFF transponder. Using advanced microwave packaging, high-speed analogue/digital converters, advanced digital signal processing and CMOS LSI gate array circuitry, the component count, volume, weight and power dissipation of the AN/APX-108 has been minimised. It operates in Modes 1, 2, 3A, C and 4.

Features include a COMSEC appliqué that uses electronic key fill to eliminate the requirement for a KIT-1A computer. For interface flexibility, the AN/APX-108 can be interfaced with either a MIL-STD-1553B multiplex bus or a C-6280 control box.

Specifications
Dimensions: 152.4 × 162.5 × 229 mm
Weight: 6.4 kg
Power supply: 28 V DC, 55 W nominal
Frequency:
(receiver) 1,030 ± 0.5 MHz
(transmitter) 1,090 ± 0.5 MHz
Output power: 500 W min at 1% duty cycle
Temperature range: –40 to +71°C
Altitude: up to 70,000 ft
Reliability: 2,500 h MTBF

Operational status
In production and in service.

Contractor
Litton Guidance & Control Systems.

VERIFIED

AN/APX-109(V)3 Mk XII IFF combined interrogator/transponder

The Litton AN/APX-109(V)3 represents a major advance in IFF system design. AIMS and STANAG requirements are both provided in hardware design and system implementation utilises a fully integrated system approach, making practical use of advanced technologies.

The AN/APX-109(V)3 is an efficient solid-state design that reduces the component count, weight and volume of the IFF system significantly. Employing advanced signal processing, the AN/APX-109(V)3 offers a prioritised four-channel operation; adaptive thresholding of the received video to improve performance in high-noise and jamming environments, and a monopulse processing of received target video for enhanced azimuth accuracy.

Specifications
Dimensions: 152 × 213 × 368 mm
Weight: 15.5 kg (including COMSEC appliqué)

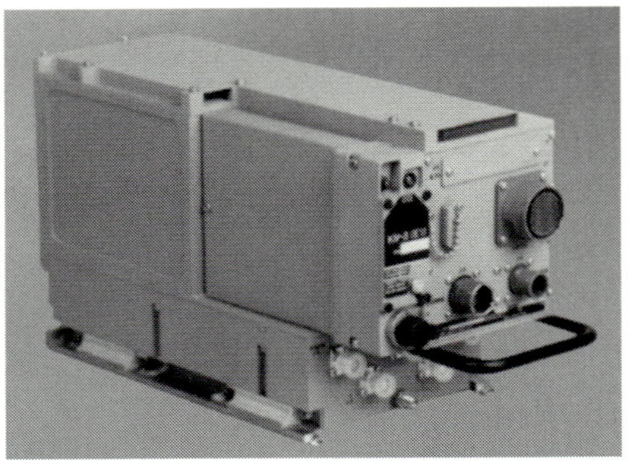

Litton AN/APX-109(V)3 Mk XII IFF combined interrogator/transponder **1999**/0044986

Power supply: 28 V DC, 150 W
Frequency:
(receive) 1,030 ± 0.2 MHz
(transmit) 1,090 ± 0.2 MHz
Temperature range: –40 to +71°C
Altitude: up to 70,000 ft
Reliability: >2,000 h MTBF

Operational status
In production.

Contractor
Litton Guidance & Control Systems.

VERIFIED

AN/ASN-130A (LN-38A) carrier aircraft inertial navigation system

The AN/ASA-130A, designated LN-38A by Litton, is a complete system contained within a single weapons replaceable assembly, the principal elements being a P-1000C platform, G-1200 gyros, A-1000 accelerometers and LC-4516C digital computer. The P-1000C provides attitude, heading and velocity information, while the LC-4516C computer is a 16-bit general purpose device with 30 kbyte memory operating at 238 Kops. It has a 10 MHz serial digital databus in MIL-STD-1553 format and synchro, analogue and discrete signal input/output capabilities.

Specifications
Dimensions: 427 × 287 × 190 mm
Weight: 15.9 kg
Power supply: 115 V AC, 400 Hz, 3 phase
(start) 800 W
(run) 185 W
Accuracy: 1 n mile/h CEP
Reliability: 1,950 h predicted MTBF

Operational status
No longer in production. About 2,000 of these systems have been produced. The system is used on US Navy and US Marine Corps AV-8B, A-6, EA-6B, F/A-18A and F-14D aircraft. Also in service in Australia, Canada and Spain in the F/A-18 and in the UK in the Harrier GR. Mk 7.

Contractor
Litton Guidance & Control Systems.

VERIFIED

AN/ASN-139 (LN-92) carrier-based aircraft inertial navigation system

The Litton LN-92 is being produced as the US Navy AN/ASN-139 Carrier Aircraft Inertial Navigation System (CAINS II) ring laser gyro INS as a form, fit and function replacement for the AN/ASN-130A. It is designed for operation by high-performance carrier-based aircraft. The accuracy specifications are better than 1 n mile/h CEP and 3 ft/s velocity with a 4 minute reaction time.

The gyro in the LN-92 is the Litton LG-9028 28 cm ring laser gyro used in conjunction with the A-4 accelerometer triad. This system incorporates MIL-STD-1750 processors for navigation and signal data processing. The reliability of the unit permits the navy to employ a two-level maintenance concept.

Specifications
Dimensions: 427 × 287 × 190 mm
Weight: 21.46 kg
Reliability: 3,992 h MTBF

Operational status

In production and in service with US Navy AV-8B, C-2A, E-2C, EA-6B, F-14D and F/A-18 aircraft. The system has been selected by Finland, Kuwait, Malaysia and Switzerland for the F/A-18 and has been flight-tested in the UK Royal Air Force Harrier GR. Mk 7.

Contractor

Litton Guidance & Control Systems.

VERIFIED

AN/ASN-141 (LN-39) inertial navigation unit

The AN/ASN-141, designated LN-39 by Litton, is the sensing and data processing device that was chosen by the US Air Force as the basis for the standard inertial navigation system on the A-10 and F-16.

The unit contains a P-1000 platform, stabilised by two G-1200 gyros and mounting three A-1000 accelerometers, and an LC-4516C general-purpose computer.

Particular emphasis has been placed on reliability. This is accomplished partly by the use of large/medium-scale integrated circuits and hybrid components, allowing a significant parts count reduction.

Specifications

Dimensions:
(inertial navigation unit) 191 × 193 × 460 mm
(control/display unit) 146 × 152 × 185 mm
Weight:
(inertial navigation unit) 17.36 kg
(control/display unit) 3.73 kg
Power:
(starting) 550 W
(running) 180 W
Accuracy (position, velocity):
(gyrocompass) 0.8 n mile/h CEP, 2.5 ft/s
(stored heading) 3 n miles/h CEP, 3 ft/s
Align time:
(gyrocompass) 8 min at 21°C
(stored heading) 1.5 min at 21°C
LC-4516C computer: 16-bit single or 32-bit double precision, 65,536 words, semiconductor RAM/ROM or EPROM 24 k words total
Reliability: 740 h MTBF specified goal over full military environment

Operational status

In service in A-10 and F-16 aircraft. The system is also in service in the Brazilian/Italian AMX, Japan Air Self-Defence Force F-4EJ, Thai Air Force F-4 and F-16s in Bahrain, Egypt, Korea and Turkey. Over 3,000 AN/ASN-141s have been ordered.

Contractor

Litton Guidance & Control Systems.

VERIFIED

AN/ASN-142/143/145 (LR-80) strapdown Attitude and Heading Reference System (AHRS)

The AN/ASN-142/143/145, designated LR-80 by Litton, was developed as a flexible system to meet the growing operational requirement for an all-attitude dynamically accurate attitude reference for air, land and sea applications. It is a lightweight self-contained LRU that provides heading, pitch and roll and angular rate reference information for a wide variety of military applications including cockpit display, sensor stabilisation, fire control and autopilot reference.

The equipment senses vehicle angular rates and translational accelerations by means of two Litton G-7 gyroscopes and three Litton A-4 accelerometers.

Specifications

Dimensions: 194 × 192 × 259 mm
Weight: 8.1 kg
Power supply: 28 V DC, 60 W nominal
115 V AC, 60 W (620 W warm-up)
Accuracy:
(heading) 0.5° RMS
(pitch and roll) <0.25° RMS
(angular rate) 0.25°/s RMS

The Litton AN/ASN-139 INS is installed in the Northrop Grumman F-14D

Operational status

In production. Over 1,000 systems have been delivered to several programmes for US and overseas services, primarily for the Boeing AH-64 Apache, Bell OH-58D Kiowa Warrior and Sikorsky UH-60.

Contractor

Litton Guidance & Control Systems.

VERIFIED

AN/ASN-150(V) central tactical system

The AN/ASN-150(V) central tactical system controls and displays navigation, communication and armament data for a wide variety of fixed-wing aircraft and helicopters. It includes two dual-redundant tactical data processors, communications and armament system controllers, multifunction displays and various display and intercom system control panels. Operator interface with the system is provided by four control/display units, each with an alphanumeric keyboard and LED display.

The system interfaces with other avionics by a dual-redundant MIL-STD-1553B databus and by discrete interconnections.

The display system features a high-resolution multifunction display with either 6 × 9 or 19 in diagonal sizes. Up to three live video windows can be presented including radar, acoustic data, FLIR, monochrome or video camera. Window size is under operator control and can present comprehensive graphic, symbol and moving map presentations for tactical data transfer to other land, sea or air platforms.

Specifications

Dimensions:
(tactical data processor) 257 × 193 × 387 mm
(communications system controller) 356 × 178 × 443 mm
(armament system controller) 191 × 152 × 262 mm
(multifunction display) 267 × 211 × 381 mm
(control/display unit) 116 × 170 × 178 mm
(intercom system) 127 × 38 × 46 mm
(display and control panel) 86 × 147 × 76 mm
Weight: 123 kg
Reliability: 250 h MTBF

Operational status

In production and in service with US Navy HH-60H and SH-60F, US Naval Reserve SH-2G and US Coast Guard HH-60J. The system is also in use internationally in S-70 variants procured or selected by Greece, Kuwait, China and Thailand. The central tactical system has been modified for use in fixed-wing aircraft such as the China S-2T.

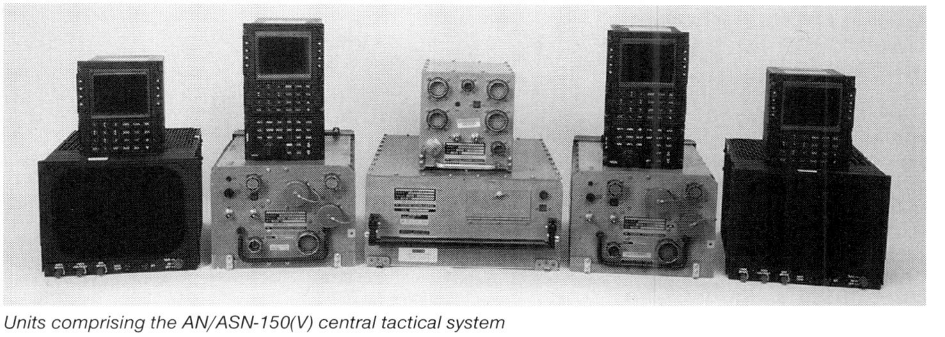

Units comprising the AN/ASN-150(V) central tactical system

Contractor

Litton Guidance & Control Systems.

VERIFIED

CN-1655/ASN (LN-94) ring laser gyro INU

The CN-1655/ASN, designated LN-94 by Litton, is a variant of the CN-1656/ASN (LN-93) developed as a ring laser gyro form, fit and function replacement for the Litton LN-31 conventional gyro INS previously fitted in the F-15. The system uses the same inertial assembly and most of the electronic circuits of the LN-93, but is packaged in the LN-31 chassis and has input/output circuitry that is specific to the F-15.

Operational status

Orders have been placed for retrofit into all US Air Force F-15A, B, C and D aircraft, as well as for Israel and Japan. The LN-94 is also qualified for the F-15E.

Contractor

Litton Guidance & Control Systems.

VERIFIED

CN-1656/ASN (LN-93) ring laser gyro INU

The CN-1656/ASN, designated LN-93 by Litton, is the US Air Force standard ring laser gyro INU. This unit is now in production, employing 28 cm path length ring laser gyros. Following a contract awarded in 1985, units have been widely fitted to US Air Force aircraft and to aircraft in many other countries.

The LN-93G is form, fit and functionally interchangeable with the LN-39 and LN-93 standard INUs, as well as any other systems that are in compliance with SNU-84-1 used in numerous types of US and foreign tactical aircraft. The LN-93G ring laser gyro INU/GPS is in a single box which contains the US Air Force standard navigation unit integrated with a Collins militarised six-channel P code GPS receiver module.

Specifications

Dimensions: 521 × 204 × 193 mm
Weight: 22 kg

Operational status

In production. The LN-93G is in development and a large number of units has been ordered for US fighter aircraft.

Contractor

Litton Guidance & Control Systems.

VERIFIED

Communication Control Group (CCG)

The Communication Control Group (CCG) was developed for the US Navy S-3B ASW aircraft. It provides intercom system control between crew stations, air-to-air and air-to-ground communications control, data communications via datalink and TEMPEST secure/clear isolation. The system comprises a Communication System Controller (CSC), two control indicators and four system control indicators.

The CSC provides the interface between the radios and the intercom system, as well as other audio and digital system interfaces. In response to display units, the CSC connects and configures radios and other communications equipment. The CSC maintains communication configuration and control menus for the display units, provides control of all modes of frequency selection for the HF and VHF/UHF radios in addition to the sonobuoy receiver and the On-Top Position Indicator (OTPI). Complete communication control is provided to multiple crew stations as well as access to all radios. The CSC provides control of UHF radio channelisation frequencies and the antenna switching unit. It can also control switching for secure voice equipment.

The control indicators work in conjunction with the CSC to provide management of communications configuration and mode control, provision of navigation initialisation and update capability, provision of armament system status and mode control, provision of acoustic sensor status and updating capability and display of mission avionics diagnostics and BIT information.

Specifications

Dimensions:
(CSC) 178 × 350 × 477 mm
(control indicator) 122 × 146 × 240 mm
(radio system control indicator) 169 × 146 × 2.1 mm
Weight:
(CSC) 23.63 kg
(control indicator × 2) 4.09 kg
(radio system control indicator × 4) 6.4 kg
Power supply:
(CSC) 115 V AC, 400 Hz, 150 W 28 V DC, 15 W
(control indicator) 115 V AC, 400 Hz, 30 W,
5 V AC, 400 Hz, 15 W, 28 V DC, 24 W
Reliability:
(CSC) 2,837 h MTBF
(control indicator) 10,501 h MTBF
(radio system control indicator) 12,039 h MTBF

Operational status

In production for the US Navy S-3B ASW aircraft.

Contractor

Litton Guidance & Control Systems.

VERIFIED

GPS Guidance Package (GGP)

The GPS Guidance Package (GGP) is an ARPA-sponsored programme to develop a family of low-cost, modular, miniature GPS-based guidance systems that can easily be configured to support a broad spectrum of US military platforms. The GGP wil consist of the Litton Miniature Inertial Measurement Unit (MIMU) and the Rockwell Collins Miniature 12-channel GPS Receiver (MGR).

Specifications

Dimensions: <1,630 cm³
Weight: 3.19 kg
Accuracy: <16 m SEP with GPS
Output: position, velocity, attitude, heading, angular rate and acceleration

Operational status

In development. GGP phase 1 was initiated in 1995 with two prototype demonstration units. Phase 2, beginning in 2000, will address improved performance, flight management and control, size and weight reduction, and production.

Contractor

Litton Guidance & Control Systems.

VERIFIED

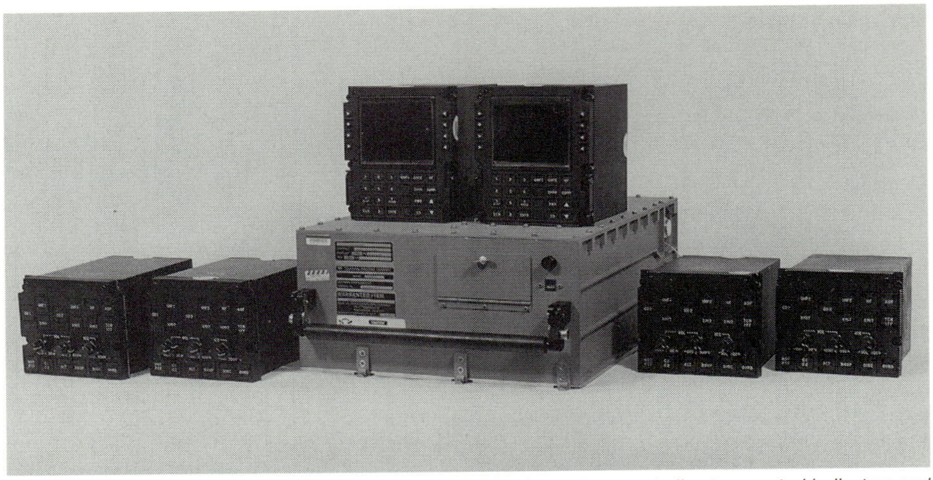

The Communication Control Group showing the communication system controller, two control indicators and four radio system control indicators

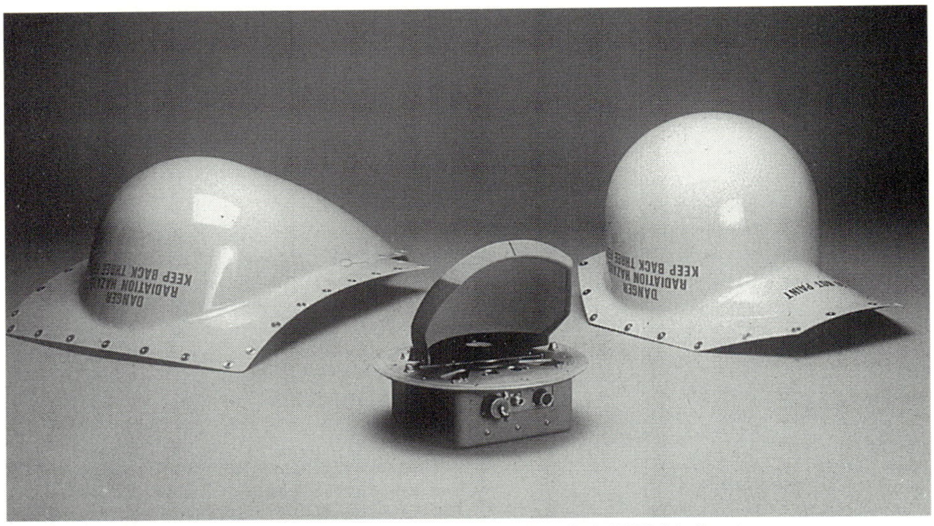

The Litton AN/ARQ-44 datalink antenna and radomes for the SH-60B LAMPS II helicopter

LN-100 advanced navigation system

In October 1989, Litton was awarded a contract to develop a low-cost inertial system featuring low power, weight and volume. The LN-100 employs non-dithered Litton zero-lock laser gyro and A-4 accelerometer instrument technologies, together with a 22-state Kalman filter which integrates Doppler velocity and GPS position and velocity inputs, to serve as a dual-flight control reference and high-accuracy navigation system for the AH-64D Longbow Apache programme.

In April 1991, the LN-100 INU was selected by Boeing to provide the Navigation Quality Inertial Sensor (NQIS) for the RAH-66 Comanche helicopter inertial navigation system. The LN-100 is also used as the inertial reference system for the Lockheed/Boeing F-22 advanced tactical fighter.

Specifications

Dimensions: 241.3 × 177.8 × 177.8 mm
Weight: 8.6 kg
Power supply: 28 V DC, 25 W
Accuracy: 0.8 n mile/h
Reliability: 5,000 h MTBF

Operational status

Selected for the AH-64D Longbow Apache and RAH-66 Comanche helicopters and the F-22A Raptor.

Contractor

Litton Guidance & Control Systems.

VERIFIED

LN-100G navigation system

The LN-100G has evolved from the proven LN-100 product line. All LN-100 systems use common hardware and software elements, affording economies of scale from high-rate production. Major US DoD programmes that require lightweight inertial navigation systems have contracted for the LN-100. These include the RAH-66 Comanche helicopter and the F-22. The

The LN-100 has been selected for the AH-64D Longbow Apache, RAH-66 Comanche and F-22

LN-100G, with an embedded GPS receiver, has been selected by the US EGI tri-service programme office for the F-18 and EA-6B upgrades, by the US Navy for the T-45A Cockpit 21 programme and by Boeing for the Japanese 767 AWACS.

The LN-100G provides three simultaneous navigation solutions; hybrid GPS/INS, free-inertial and GPS only. The processing heart of the LN-100 product line is the 32-bit PowerPC Motorola microprocessor; the software is Ada.

The LN-100G optimally combines GPS and INS features to provide enhanced position, velocity, attitude and pointing performance as well as improved acquisition and anti-jam capabilities.

The addition of a GPS receiver package on a single circuit card provides a complete hybrid navigation unit with very low size and volume. Two spare card slots allow for the addition of analogue I/O modules, ARINC interfaces, Low Probability of Interception (LPI) radar altimeter, air data and other expansion modules.

The embedded GPS module is an L1/L2 CA/P(Y) code unit capable of accepting RF (or IF) inputs from the GPS antenna system. Stand-alone GPS PVT data and stand-alone INS data are provided for integrity and fault monitoring purposes.

Contractor
Litton Guidance & Control Systems.

VERIFIED

LN-200 inertial measurement unit

The LN-200 is a three-axis strapdown inertial measurement unit designed for tactical missile and unmanned vehicle guidance, flight control and sensor stabilisation applications. It is very small and light and employs fibre optic gyros and silicon chip accelerometers for extreme ruggedness and high reliability.

A version of the LTN-200, identified as the LN-201, has been selected as the flight control quality inertial system for the US Army Boeing/Sikorsky RAH-66 Comanche helicopter.

Litton, teamed with Rockwell Collins, has a contract with the US Advanced Research Projects Agency (ARPA) for an integrated inertial/GPS system. The system, known as GPS Guidance Package and identified as the LN-250, has a goal of a navigation grade inertial system with full GPS capability in a 0.0016 m^3 package.

Operational status
In development.

Contractor
Litton Guidance & Control Systems.

VERIFIED

Model 3044 radar altimeter

The Model 3044 radar altimeter was designed and developed for manned aircraft applications and operates from 0 to 10,000 ft. It employs hybrids, application-specific integrated circuits and MIL-STD-1553B interface to provide a highly reliable lightweight production unit. The altimeter transmits and receives a very low-RF signal that is phase shift-key modulated using a pseudo-random code. The phase shift-key technique provides the equipment with the capability to measure over the full altitude range using a very low-level RF output.

The altimeter comprises two subassemblies. The microwave subassembly contains a filter board assembly and four microwave hybrids: the transmitter, receiver, dielectric resonant oscillator and video amplifier. The microwave hybrids provide increased reliability at reduced part count. The electronic subassembly contains an MIL-STD-1553B interface, signal processor and power supply. The application-specific integrated circuit incorporated into the signal processor enables digital signal processing at a greatly reduced part count.

Contractor
Litton Guidance & Control Systems.

VERIFIED

RINU(G) navigation system

Litton's Guidance & Control Systems division has been awarded a contract by the US Navy to provide replacement navigation systems for all of the US Navy's P-3 maritime and C-130 patrol and cargo-type aircraft.

Production options for up to 784 systems to be delivered up to 2001 could bring the total contract value to Litton to more than US$50 million.

The systems will be direct replacements for existing Litton units aboard the aircraft since the late 1970s. Two systems are installed in each aircraft.

The new systems, called RINU(G), combine in a single unit the continuous, long-term accuracy of the latest laser gyroscope navigation technology with the geographical precision of a global positioning system satellite signal receiver.

The P-3/C-130 navigation units will be based on Litton Guidance & Control Systems division's production LN-100G system.

Operational status
In production.

Contractor
Litton Guidance & Control Systems.

VERIFIED

TEC-60i combined interrogator/transponder

The airborne IFF TEC-60i combined interrogator/transponder, when functioning as an interrogator, challenges and identifies co-operative active targets. These targets are suitably equipped with compatible transponders and operate within effective IFF range. The unit also functions as a transponder responding to valid interrogations. The unit is a compact, lightweight system installed in the equipment bay, with a control panel mounted in the aircraft cockpit. Peak output power is 1 kW.

Operational status
In production and in service.

Contractor
Litton Guidance & Control Systems.

VERIFIED

AN/APN-59F(V) search radar

The AN/APN-59F(V) search radar is an upgrade of the AN/APN-59E(V). Principal modes are search, skin painting, ground-mapping/navigation, weather mapping and beacon homing. Pencil or fan beams with a variety of pulse lengths and PRFs are selectable. The antenna can scan a full 360° or in selected sectors. The LRUs are interchangeable with those of earlier AN/APN-59 systems and the system can be gradually upgraded without aircraft stand-down.

Many configurations are possible, ranging from single azimuth/range displays to more complex installations with up to three displays.

The APN-59(X) search and navigation radar is the ultimate upgrade of the AN/APN-59 series. It has been designed for simple direct backfit on a box-by-box basis for all C-130/135-type aircraft. All performance features of the older systems are retained or enhanced including the addition of multifunction colour displays. System MTBFs greater than 1,000 hours have been achieved by power reduction and the elimination of problem components. The system has been designated the AN/APN-242 under a US Air Force contract.

Specifications
Weight: 84 kg (typical system)
Frequency:
(transmitter/receiver) 9,375 ± 40 MHz
(beacon) 9,310 MHz
Peak power: 70 kW
Beamwidth: 3°
Pulsewidth: 0.35-4.5 μs variable

Contractor
Litton Marine Systems (Sperry Marine).

VERIFIED

AN/APN-242(V) search/navigation radar

The AN/APN-242, formerly the AN/APN-59(X), is a long short-range colour weather and navigation radar with integrated navigation and air data overlays via the AN/ASN-165 Display Group. Principal features include weather detection and display in colour, black/white, or green for night vision goggle compatibility; enhanced terrain-mapping with navigation using a latitude/longitude-stabilised electronic cursor; aircraft detection/skin painting concurrent with other operating modes; beacon interrogation and display; and optionally, IFF display.

The APN-242 was designed by Litton Marine Systems as a follow-on to the AN/APN-59 now in use worldwide on over 1,500 C-130 and C-135 aircraft. The radar is suitable both for new installations and as a form, fit and function replacement for the APN-59. Performance has been improved in every operating mode by increasing range, resolution and accuracy. By redesigning high failure rate components, the APN-242 achieves a system Mean Time Between Failure (MTBF) rate greater than 1,000 h.

The APN-242 antenna subsystem is a flat plate array which is stabilised to aircraft attitude reference systems and which, through the elimination of all gears, achieves reliabilities approaching 7,000 h. The array rotates 360° or can sector-scan. The antenna beam can be tilted vertically and the pattern can be instantaneously switched between pencil and fan to achieve the desired illumination. Pulsewidths and PRFs are operator selectable. The low-noise receiver and lower-power transmitter improve APN-59 performance and the long life 10,000 h magnetron has the power to skin paint fighter aircraft at extended ranges through intervening rain showers. The AN/ASN-165 is the standard display group.

When replacing the APN-59, the APN-242 does not require an aircraft modification for installation and either the antenna or receiver/transmitter (R/T) subsystems can be dropped in as direct replacements for APN-59 subsystems. On these aircraft, the system can be installed on the flight line in a day by unit-level maintenance using existing radar cabling, connections and mounting brackets.

Specifications
Frequencies:
(radar operation) I-band; 9,375 ±10 MHz
(beacon reception) 9,310 MHz
Transmitted power: 25 kW nominal peak (new high-reliability magnetron)
Ranges: 2.5 to 20 (2.5 n mile increments), 25, 30, 50, 100, and 240 n miles
Pulse length: multiple lengths (0.20, 0.8, 2.35, and 4.5 ms) automatically selected for different ranges and functions
Scanning features: 360°
Scan rates: 12 rpm on long-range functions; 45 rpm on short-range functions
Sector scan: (basic) approx 80°, centred about forward position, variable sector with alternate control
Antenna beam selection: pencil or equal energy (fan) beam; both with 3° azimuth beamwidth, instantaneous electronic switching
Antenna stabilisation: stabilised to existing aircraft reference throughout a range of ±15° pitch and ±30° roll
Controls: independent navigator and pilot controls
System components/weight:
(basic) 7 components/70 kg
Power:
(basic) 115 V ± 5% (380 to 420 Hz), 800 VA average

Operational status
The AN/APN-242 is in production for a US Air Force customer.

Contractor
Litton Marine Systems (Sperry Marine).

VERIFIED

ADS-18 Advanced UHF Radar (AURA) C-130J-30 AEW&C

ADS-18 AURA is the advanced UHF radar, designed by Northrop Grumman Integrated Systems and Aeronautical Sector, selected by Lockheed Martin for its C-130J-30 aircraft Airborne Early Warning & Control (AEW&C) proposals. No details of the radar design are yet available.

Operational status

Marketing proposal.

Contractors

Lockheed Martin Aeronautical Systems.

Northrop Grumman Corporation, Electronic Integrated Systems and Aeronautical Sector.

VERIFIED

Apollo

Apollo is the name for a US Air Force foreign military sales programme that provides a Defensive Aids Suite (DAS) for C-130 aircraft.

Apollo comprises the AN/ALQ-156 pulse Doppler MAW, the AN/AAR-44 and AN/AAR-47 electro-optic MAWs and an AN/ALE-40 chaff/flare dispenser. Alternative versions may also include the AN/ALR-56M RWR and the AN/ALE-47 smart chaff/IRCM dispenser.

Operational status

The US government initiated a 'quick-fix' programme to support aircraft involved in UN peacekeeping duties, and Apollo installations have now been fitted to C-130 aircraft from a number of foreign air forces, including the Australian, Norwegian and Swedish air forces.

Contractor

Lockheed Martin Aeronautical Systems.

VERIFIED

A modified T-4 UK Royal Navy Harrier is used to flight test the side-stick control of the aircraft while performing various STOVL tasks. The testing is being conducted on behalf of Lockheed Martin's JSF programme.
1999/0051641

Compass Call electronic warfare aircraft

The EC-130H Compass Call version of the Hercules transport is designed to identify and disrupt enemy radars and communications.

The Compass Call aircraft was a response to an urgent request from Tactical Air Command for a communication jammer. Development of equipment and operational tactics are the responsibility of the Tactical Air Warfare Center. The aircraft entered service in mid-1982 with the 41st Electronic Combat Squadron. It is distinguished from all other Hercules C-130 variants by the large vertical aerial forward of the vertical fin, and by two more underwing antennas.

Compass Call has been modified and upgraded over the past decade, mainly to improve its jamming power and characteristics, its capability in multithreat situations and the overall software and system management. It is understood that a US Air Force programme is under way to enable the system to jam more signals simultaneously and at longer ranges.

Systems on the aircraft include the Raytheon Electronic Systems ALQ-173, ALQ-175 and ALQ-198 high-band systems with digitally tuned receivers.

Operational status

The standard configuration in operational service is designated Block 20. A Block 30 upgrade, aimed at improving reprogrammability, signal acquisition, threat identification and threat location, is under way; three aircraft were delivered in 1997; the remainder will be delivered by mid-1999. US Air Force plans call for six Block 20 aircraft to be updated to Block 35 standard starting in FY2000, and subsequently for the rest of the force to be upgraded to Block 40 standard. It is reported that Block 35 modifications will replace remaining analogue technology systems, and instal the Tactical RAdio Countermeasures System (TRACS) and digital compressor omitted from Block 30. Block 40 requirements have yet to be finalised.

Contractor

Lockheed Martin Aeronautical Systems.

VERIFIED

SAMSON system

SAMSON (Special Avionics Mission Strap On Now) is a quick turn around kit facilitating reconfiguration of C-130 aircraft for special missions such as search and rescue, photo reconnaissance, drug interdiction, electronic surveillance/countermeasures, air sampling and radio relay. The kit comprises an equipment pod which replaces the C-130E/H external fuel tank, a cargo pallet-mounted operators' console, interface wiring, and any necessary additional external components. The kit may be installed or removed within one work shift by a crew of four to six persons and is accomplished with no permanent modification to the aircraft.

Operational status

The system prototype has undergone numerous field tests by US and foreign C-130 users. A production version of the system is currently in service in a reconnaissance role.

Contractor

Lockheed Martin Aeronautical Systems.

VERIFIED

SATIN EW system

SATIN (Survivability Augmentation for Transport Installation) is a minimum modification approach to equip C-130 aircraft with self-protection sensors and countermeasures. The installations are designed to be installed by a small team in a few days. All external or cockpit components are provided with replacement cover panels, so that the systems can be removed when not required, or used on other aircraft.

SATIN components are selected from the standard military inventories of the particular user to minimise logistics problems. Systems which have been installed for various users include the AN/ALR-69 Radar Warning Receiver, APR-39A(V)1 Radar Signals Receiving Set, AN/ALQ-156 ESM system, AN/AAR-47 Missile Warning Set, AN/ALQ-157 Self-Protection System and the AN/ALE-39, ALE-40 and ALE-47 Countermeasures Dispenser Systems. Other systems can be incorporated, as dictated by user requirements.

Operational status

SATIN has been installed on various versions of C-130 aircraft for several users and the basic installation approach has been incorporated into production C-130 aircraft.

Contractor

Lockheed Martin Aeronautical Systems.

VERIFIED

Vectored-thrust Aircraft Advanced Control (VAAC)

In late 1998, Lockheed Martin Aeronautical Systems completed 20 hours of flight testing in the United Kingdom to support development of the STOVL (Short Take Off Vertical Landing) variant of the UK Royal Navy Joint Strike Fighter.

The flights were conducted with the UK's Defence Evaluation Research Agency (DERA) Vectored-thrust Aircraft Advanced Control (VAAC) vehicle, a modified Harrier. The evaluation programme tested side-stick control of the aircraft in various STOVL tasks at flight speeds from hover to 200 kt. The testing confirmed that a side stick provides satisfactory control of a STOVL aircraft at the low speeds where the aircraft is airborne.

A total of 36 flights were conducted at the flight test facility at DERA Boscombe Down in the United Kingdom. Test pilots from Lockheed Martin, BAE Systems (Operations) Limited, the US Marine Corps and the UK Royal Air Force participated in the evaluation.

The test was divided into two phases: a calibration phase where test control laws and stick characteristics were validated followed by an evaluation phase. Evaluation tasks included pattern work, approach and final transition tasks, and precision and aggressive hover tasks.

DERA, provided the aircraft for the flight test. The VAAC Harrier is a two seat T-4 trainer fitted out with a fully digital flight control system for advanced STOVL control research. VAAC is equipped with an integrated sensor suite and an advanced flight control computer that can be easily altered to provide varied levels of augmentation for STOVL control law research. Basic flight safety is provided by the VAAC's front seat pilot, who can take control of the aircraft at any point during a test should the need arise.

The VAAC was modified for this test to provide the functionality of Lockheed Martin's JSF STOVL aircraft. Modifications included installation of a side-stick controller in the rear cockpit; alteration of the rear cockpit throttle to provide HOTAS – Hands-On-Throttle And Stick – functionality; and modification to VAAC's fly-by-wire controls to provide JSF STOVL aircraft-type responses.

The side-stick system used in the evaluation was manufactured in Buffalo, New York, by Calspan, an Operation of Veridian. The variable-feel side stick is powered by the aircraft's hydraulic system and controlled by a dedicated Calspan computer located in the cockpit. Characteristics of the stick, such as break-out force, force and deflection gradients, and dynamic attributes, can be varied in real time, in flight, by the VAAC's flight control computer.

A major portion of the flight test evaluation was dedicated to providing variations in side-stick characteristics to gather an assessment of side-stick control for STOVL flight, and to gather data for the development of the side-stick controller for the JSF Concept Demonstrator Aircraft.

Contractor

Lockheed Martin Aeronautical Systems.

VERIFIED

EC-130E airborne TV and radio broadcasting station

The latest version of psychological operations airborne TV and radio broadcasting EC-130E aircraft is capable of broadcasting TV in full colour in any format worldwide. Aircraft modification kits and installations include VHF and UHF antennas housed in 6 ft diameter, 23 ft long (1.829 × 7.01 m) pods under each wing dedicated to higher-frequency TV and rotatable to alter polarities; VHF low-frequency TV antennas mounted on each side of the tailplane; upgraded transmitters; equipment for formatting TV signal to worldwide standards and increased output power from 1 to 10 kW. Two retractable trailing-wire antennas provide both HF and VHF AM omnidirectional radio broadcast coverage.

The aircraft are upgraded versions of EC-130E Volant Solo aircraft.

Operational status

In service on US Air National Guard EC-130E aircraft.

Contractor

Lockheed Martin Aircraft and Logistics Centers.

VERIFIED

Electronic Flight Control Set (EFCS) for the B-2

The digital fly-by-wire Electronic Flight Control Set (EFCS) consists of two parts: a quadruple redundant set of Flight Control Computers (FCCs) and quadruple redundant pilot sensor assemblies. Each FCC is a single-channel, real-time digital computer, connected via a MIL-STD-1553 bus controller to the avionics sytems, remote terminals, pilot sensor assemblies, cockpit panels, nosewheel steering and hydraulic system. The computer includes a PACE 1750A processor with a throughput of 2 Mips.

There are three pilot sensor assemblies: yaw pedal, pitch stick and roll stick. The sensor assemblies transform pilot input to an electronic signal which communicates with the FCC.

Specifications

Dimensions:
(FCC) 388.4 × 345.9 × 193.5 mm
(pilot sensor assembly) 215.9 × 66 × 91.4 mm

Weight:
(FCC) 18.6 kg
(pilot sensor assembly) 1.81 kg

Operational status

In service in the Northrop Grumman B-2.

Contractor

Lockheed Martin Control Systems.

VERIFIED

AN/APG-78 Longbow radar

A Lockheed Martin/Northrop Grumman joint venture company (Longbow Limited Liability Company LLLC) is working with The Boeing Company to supply the Longbow radar and RF Hellfire fire-and-forget missile to the AH-64D Apache helicopter for the US Army. The joint venture is also working with Westland Helicopter Ltd to supply the radar and missile for the WAH-64 Apache for the UK. A lightweight variant is expected to be integrated into the US Army's RAH-66 Comanche reconnaissance and light attack helicopter.

The AN/APG-78 Longbow radar system comprises a mast-mounted, millimetric-wavelength (35 GHz), fire-control radar, which allows the crew to search, detect, locate, classify and prioritise up to 256 ground targets (tanks, air defence units, trucks and so on) in adverse weather. Up to 16 of these 256 prioritised targets can be displayed simultaneously, at ranges up to 8 km for moving targets. An integrated, Radar Frequency Interferometer (RFI), is located below the radar. It performs passive seeking, to provide directional cues for the fire-and-forget capability for the Hellfire missile. The RFI is designated AN/APR-48A. It is supplied by Lockheed Martin. The RFI can perform triangulation calculations, to give target position, using information datalinked from a second co-operating helicopter.

Operational status

Following a successful proof-of-principle programme, the joint venture received a US$314.9 million contract in 1991 for full-scale development of the radar. The

AN/APG-78 Longbow mast-mounted radar dome ***1999***/0044987

The AH-64D Longbow Apache helicopter with mast-mounted Longbow radar 0018072

Longbow system completed Initial Operational Test and Evaluation (IOT&E) in 1995. Contracts for the first two lots comprising a total of 20 fire-control radars were awarded in March 1996 and February 1997.

Meanwhile, in May 1996, contracts were awarded to Longbow International (LBI), the joint venture's international marketing unit, to supply Longbow systems to the British Army for all 67 of its Westland Attack Helicopter WAH-64. The first Apache Longbow radar was delivered to Westland GKN in June 1998, two months ahead of schedule.

In November 1997, the US Army awarded the joint venture a five-year, US$565 million contract to build 207 Longbow radars. The US Army plans that only a proportion of its fleet of helicopters will carry the Longbow radar/RFI capability. Similarly, the Royal Netherlands Army procurement for AH-64D includes only a limited fit of the Longbow radar/RFI systems. Of the other countries so far planning procurement of the AH-64D, none include either the Longbow radar or the RFI system.

Contractors

Lockheed Martin Electronics & Missiles.
Northrop Grumman Corporation Electronic Sensors and Systems Sector.

VERIFIED

AN/ALR-47 radar homing and warning system

Developed for the US Navy S-3A Viking ASW carrier-borne aircraft, the AN/ALR-47 is a comprehensive passive electronic warfare system which uses four cavity-backed planar spiral antennas in each wingtip. The aerials are orthogonally directed to enhance monopulse direction-finding, ensuring that threat direction is measured very accurately. Associated with them are twin, highly sensitive, narrowband receivers and a comprehensive processor. Manual or automatic system operation is possible, control being exercised over frequency band limits, speed of tuning and signal selection. The processor indicates the frequency scanning limits, scan speed, pulse length, pulse repetition frequency and bearing limits of any detected radar transmissions.

Operational status

No longer in production. In service on Canadian Forces CP-140 Aurora aircraft, in which application it is designated ALR-502, and formerly in the S-3A. In the S-3A it has been replaced by the AN/ALR-76 as part of the avionics upgrade which denotes the S-3B version.

Contractor

Lockheed Martin Electronics Platform Integration Group.

VERIFIED

AN/ALR-76 ESM system

The AN/ALR-76 is a form, fit and function replacement for the ALR-47 and is part of the Weapons Systems Improvement Program (WSIP) for the S-3B ASW aircraft. Compared with the ALR-47 the ALR-76 has an extended frequency range (reportedly including microwave coverage) and improved capabilities which include the automatic classification and location of emitters. The ALR-76 utilises VLSI technology and is of modular construction to facilitate future growth. A standard fit is understood to include eight planar antennas - presumably two sets of four.

Specifications

Weight: 61 kg

Operational status

The AN/ALR-76 is in full production for the US Navy S-3B weapons system improvement programme, as well as the EP-3E and ES-3A electronic reconnaissance aircraft. It is also installed on the Canadian EST Challenger aircraft.

Contractor

Lockheed Martin Electronics Platform Integration Group.

VERIFIED

AN/APR-48A Radar Frequency Interferometer (RFI)

The AN/APR-48A RFI is standard fit on the AH-64D Longbow Apache, but at present it is not fitted on AH-64D helicopters that do not carry the Longbow radar. It is designed to minimise target acquisition and platform exposure times and operates in conjunction with radar and electro-optical sensors to extend the useful range, decrease acquisition time and provide positive identification. It provides high sensitivity and

The AN/APR-48A radar frequency interferometer ***2000***/0081867

precision sensor cueing in a lightweight and modular configuration suitable for both airborne and ground platforms.

The system is fitted under the radar dome. The total weight of the system is 13.4 kg, including the antenna (3.6 kg), the receiver (3.9 kg) and the processor (5.5 kg). The system can be packaged as a single LRU or multiple LRUs to meet platform requirements. Expansion modules will allow growth to millimetric-wave capabilities.

The RFI system design incorporates a four-element interferometer coupled with a three-channel phase receiver. A four-element coarse DF array is used for initial signal acquisition. When a fine DF measurement is required, a four-element, long baseline interferometer is used which is said to provide a DF accuracy of better than 1°. The receiver is a full four-channel amplitude and three-channel phase measuring unit.

The signal parameter measurement function includes analogue video and IF circuits for measuring the signal amplitudes, radio frequency and interferometer phases. The digital processing is carried out in the processor which sorts the incoming pulses, computes DF, characterises and identifies emitters and determines threat priority.

It is understood that an improved development of the AN/APR-48A will be available soon, based on use of a higher throughput Motorola/IBM Power PC603e giving much higher throughput, lower cost and lower weight.

Operational status

The AN/APR-48A has been developed for the US Army AH-64D Longbow Apache and OH-58D Kiowa helicopters. 222 systems are in production for the US Army AH-64D Longbow Apache. Deliveries began in November 1998 and are projected to continue until February 2003. In addition 62 sets are on order for UK WAH-64 Longbow Apache helicopters. Deliveries are scheduled from December 1998 to February 2003.

Contractor

Lockheed Martin Electronics Platform Integration Group.

VERIFIED

AN/UYS-1 signal processor

The AN/UYS-1 is a fully programmable, high-performance signal processor capable of processing a wide variety of sensor data from analogue or digital inputs. Programmable software and modular hardware features have enabled the UYS-1 to satisfy processing requirements for 16 different US weapon systems.

A total of 160 S-3A Viking aircraft were modified, beginning in 1987, as part of the programme that denotes the S-3B standard. The UYS-1 also equips the P-3C Update III standard aircraft, deliveries of which to the US Navy began in 1984: this standard is now in fleet use and it is planned that 167 Orions will be modified.

The same processor has also been installed in the US Navy's SH-60B Lamps III helicopters. It is to be replaced by the AN/UYS-2 signal processor in the SH-60R.

Specifications

Dimensions: 1,455 × 594 × 284 mm
Weight: 109 kg

Operational status

In service with the US Navy.

Contractor

Lockheed Martin Electronics Platform Integration Group.

VERIFIED

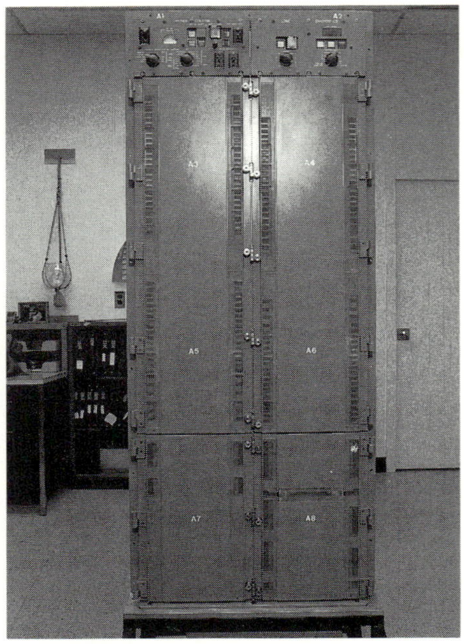

The Lockheed Martin AN/UYS-1 signal processor

IFPS Intra-Formation Positioning System

Lockheed Martin is developing an IFPS Intra-Formation Positioning System for use by US Air Force Operations Command on its MH-53J helicopters and MC-130H tanker aircraft during covert tanker operations.

IFPS is the result of a competitive contract awarded to Lockheed Martin by the Wright Laboratory/Flight Dynamics Directorate, Wright Patterson Air Force Base. IFPS provides added capability to fly multiship formation operations in day/night adverse weather. Through co-ordinated situational awareness,

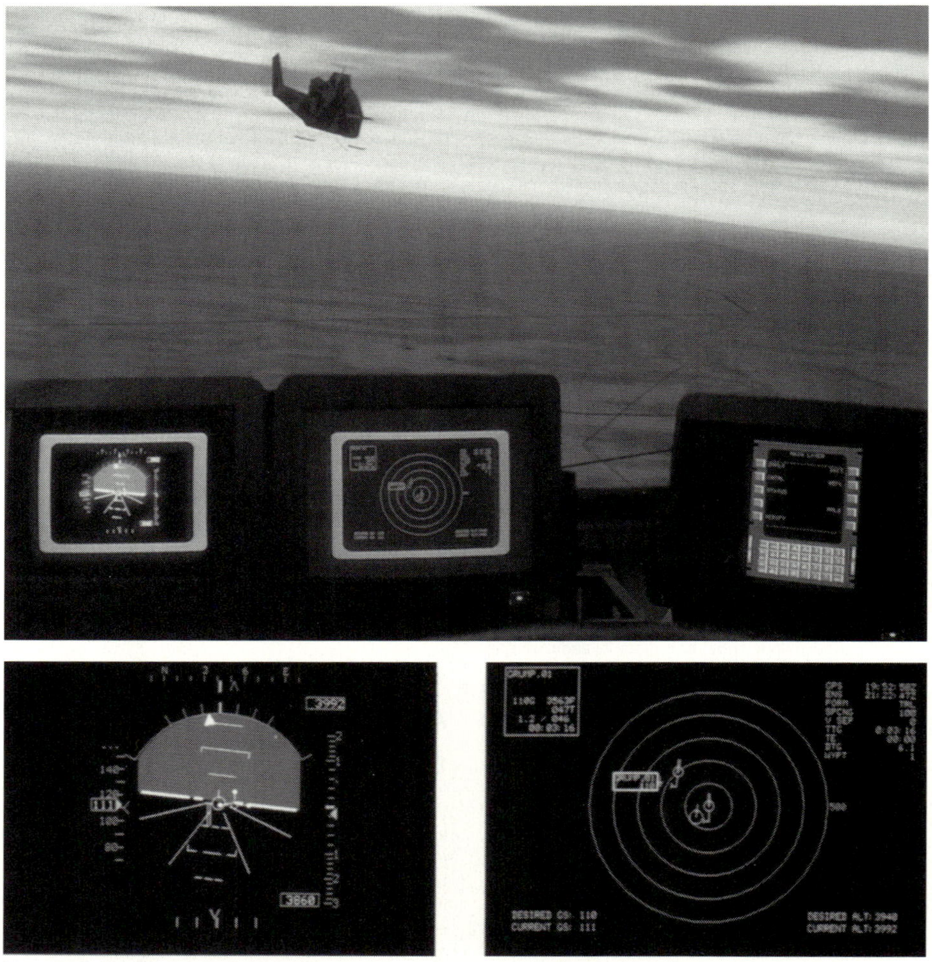

Top: *Both in-cockpit and out-of-cockpit flight cues may be viewed through the flight director, situational awareness and control display unit displays. Visual simulation of the aircraft joining the formation is also available to the pilot.* ***Bottom left:*** *The IFPS flight director ensures safe, low-stress, rendezvous or formation flight under any visibility.* ***Bottom right:*** *Situational awareness display shows position of other aircraft in formation* ***1999***/0044824

IFPS further enhances the pilot's ability to execute rendezvous, join-up, ingress/egress, formation station keeping and manoeuvring, collision detection/ avoidance manoeuvres and co-ordinated assault missions.

IFPS was originally developed for Special Operations helicopters and tanker transport aircraft. However, increased capability and safety in virtually every aspect of formation operation in adverse weather is also provided.

IFPS provides spherical situational awareness and allows highly accurate formation positioning of large and small flights of aircraft, while allowing the lead aircraft flexibility to manoeuvre as required in three dimensions.

Relative Global Positioning System (RGPS) provides the high level of accuracy required for safe formation flight in all types of weather. Furthermore, a Low Probability of Interception (LPI)/Low Probability of Detection (LPD) datalink allows the formation to operate undetected.

Unique guidance algorithms provide formation cues for any standard type of formation as well as automatic cueing for changes in formation type, lead changes, spacing changes, formation split and join-up. Rendezvous algorithms safely place separate formations within visual range and provide cues to ensure both formations arrive at the air refuelling control point at a designated time.

The Collision Warning/Collision Avoidance (CW/CA) algorithm projects the path of all participating aircraft to detect potential collisions. If a potential collision is identified, collision detection and three-dimensional collision/avoidance steering cues are provided to each interfering aircraft, while CW/CA situational awareness is displayed in participating aircraft.

IFPS uses a Have Quick II derivative radio datalink incorporating special LPD features. The Have Quick II derivative radio is selectable between IFPS or a full AN/ ARC-164 compatible mode. In the IFPS mode, the Have Quick II datalink provides simultaneous digital data and voice communications between intra- and interformation aircraft.

The system architecture for IFPS uses existing avionic suites. Current processors, displays, Inertial Navigation Systems and Global Positioning Systems are used to develop RGPS solutions. Existing AN/ ARC-164s are upgraded to Have Quick II derivative radio via a field retrofit kit. IFPS software can either be embedded into the existing host aircraft mission processor or added as part of a MIL-STD-1553B compatible IFPS processor unit.

IFPS provides a situational awareness graphics display format on standard aircraft multifunction displays. display formats can also be overlaid on existing radar Plan Position Indicator (PPI) formats. The IFPS situational awareness display provides pilots with a picture of where their own aircraft is in relation to others in the same or other participating formations. It also indicates where their aircraft is relative to the desired formation position. IFPS provides coverage up to 10 aircraft per formation for a maximum of 20 independent formations.

Flight director cues are provided by IFPS on existing flight director displays. Through cues overlaid onto the flight director, the pilot is provided with guidance for proper aircraft pitch, roll and power inputs. Following these cues maintains the aircraft in its desired formation and ensures accurate time on target or air refuelling control times.

The IntraFormation Positioning System is highly flexible and, with the addition of a portable ground terminal, can provide drop zone and landing zone control, survivor/team location, LPD/LPI communications, target verification/identification, friendly air defence penetration and tactical precision landing.

Four separate types of message will be used on the AN/ARC-164 low-probability of intercept (LPI) radio to correlate position with: aircraft in the same formation (A-Net datalink); aircraft in other formations (E-Net datalink); refuelling rendezvous (R-Net datalink); secure voice communications.

Operational status

The system is in US Air Force service.

Contractor

Lockheed Martin Electronics Platform Integration Group.

VERIFIED

AN/ALQ-78 ESM system

The AN/ALQ-78 ESM system is used aboard the Lockheed Martin P-3C Orion maritime patrol aircraft as its electronic support measures sensor. The antenna is carried on a pylon under the inner wing. It automatically detects and measures the characteristics and bearings of intercepted radar signals of anti-submarine and electronic warfare interest. The measured parameters and bearing of the intercepted signals are supplied to the aircraft central data processing system for evaluation, recording and presentation on the aircraft displays.

The system uses a high-speed rotating antenna and a scanning, superheterodyne receiver for acquisition of signals in specific frequency bands of particular interest to the P-3C. Operation is mostly automatic, based on parametric data. The countermeasures set normally operates in an omnidirectional search mode. When a radar signal of interest is acquired and analysed, ALQ-78 automatically initiates a direction-finding routine. The signal data is processed by the central data computer and formatted for readout on a multipurpose display.

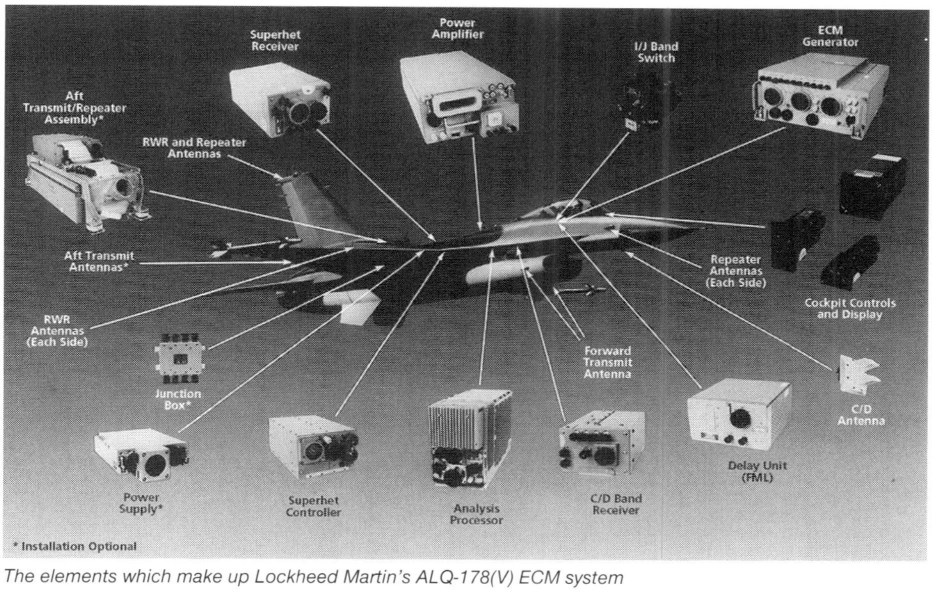

The elements which make up Lockheed Martin's ALQ-178(V) ECM system

Operational status

ALQ-78 is understood to have been in service aboard Lockheed Martin P-3C Orion patrol aircraft since 1969. As part of the US Navy's Orion update effort, ALQ-78 is noted as being scheduled for replacement by the AN/ALR-66(V)5 system. ALQ-78 is also known to have been built under licence by Mitsubishi for use on board Japanese P-3C aircraft. In US service, ALQ-78 has been upgraded to ALQ-78A standard.

Contractor

Lockheed Martin Fairchild Systems.

VERIFIED

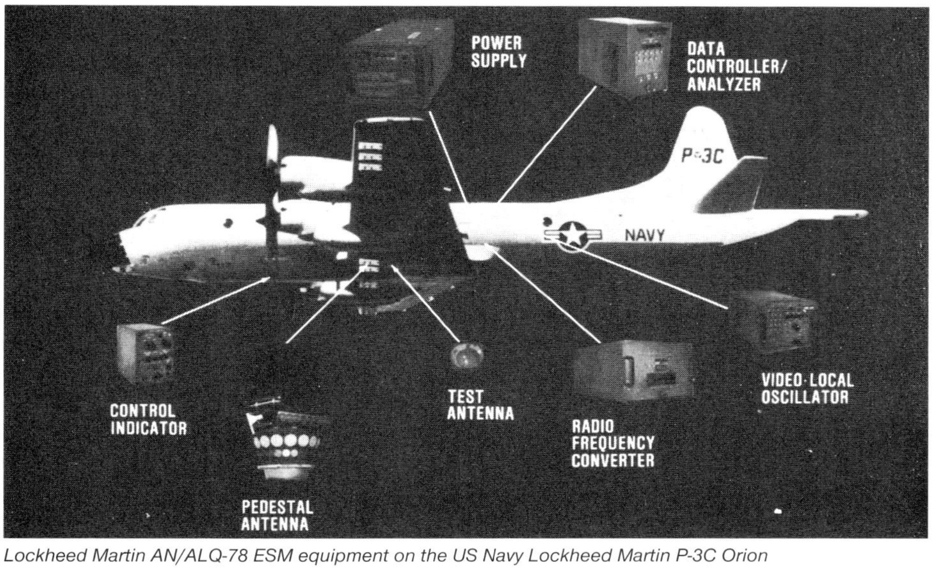

Lockheed Martin AN/ALQ-78 ESM equipment on the US Navy Lockheed Martin P-3C Orion

AN/ALQ-178(V) radar warning and Electronic CounterMeasures (ECM) suite

AN/ALQ-178(V) is an integrated radar warning and active countermeasures suite for tactical fighter aircraft, such as the F-16. An earlier version was known as RAPPORT III. The system utilises a central programmable computer for data analysis and system control, with independent microprocessors to direct Radar Warning Receiver (RWR), display, jamming and countermeasures dispensing functions. The wideband RWR continuously scans the threat radar environment. Detected signals are de-interleaved and identified by radar type and displayed to the pilot on a cathode ray tube in legible, unambiguous format. The power management algorithm optionally matches the countermeasures to the RWR's constantly changing threat picture. Separate forward and aft jammers are used for maximum spatial coverage. The jammers are accurately set for maximum power output at each victim radar's frequency to maximise effective radiated power and jammer effectiveness. The jammer frequency range provides coverage of the entire threat band while the jammer power is sufficient to counter these radars throughout their lethal area. The countermeasures dispenser is automatically controlled, including the pilot's display and cues, to provide optimum response to the threat between active jamming and chaff dispensing.

Although the ALQ-178(V) consists of an RWR, an active jammer and countermeasures control, the RWR can be installed as an independent threat warning system (controlling a countermeasures dispensing system), with the jammer added later.

The latest derivative of the ALQ-178(V) family incorporates technological advancements and packaging to increase capabilities significantly. Enhanced performance is provided by Digital RF Memory (DRFM) technology, agile jamming channels, distributed high speed/capacity processors and precision direction-finding capabilities.

A follow-on to ALQ-178(V), combined with the AN/ALR-56M or other RWR, is known as the AN/ALQ-202(V) autonomous jammer.

Specifications

RF coverage: E-J band
Targets: 32 targets simultaneously
Weight: 203 kg
Power: 2.5 kW
Interfaces: MIL-STD-1553B and RS-232

Operational status

ALQ-178(V)3 was selected by Turkey in January 1989 for its F-16 aircraft and, in October 1989, it was announced that Turkey had awarded a contract worth US$325 million for supply of systems, test and support, and ancillary equipment. Deliveries ran from October 1991 to early 1996. Lockheed Martin and the Turkish company Kavala have formed a joint venture known as MiKES to produce the ALQ-178(V)3 in Turkey after initial deliveries from the USA.

ALQ-178(V)1 is fitted to F-16s of the Israeli Air Force with the threat warning subsystem. A portion of the system was built under licence in Israel by Elisra. The Israeli application may cover aircraft up to batch 3 F-16C/Ds but excludes the country's F-16D-30 aircraft. ALQ-178(V)1 incorporates a set of crystal video (DF) receivers which have been eliminated from the ALQ-178(V)3.

Contractor

Lockheed Martin Fairchild Systems.

VERIFIED

AN/ALQ-202 autonomous jammer

The AN/ALQ-202 autonomous jammer is an advanced capability ECM jamming system for internal installation on F-16, F/A-18, and other aircraft. The ALQ-202 provides automatic and prioritised ECM responses to the full spectrum of radar threats, including advanced airborne interceptors and surface-to-air missiles. The ALQ-202 is fully integrated with onboard avionic systems including Radar Warning Receiver (RWR), Fire-Control Radar (FCR), chaff/flare dispenser and mission computer. System interfaces are designed for non-interference, allowing fully compatible operational performance of all onboard avionic systems. The ALQ-202 integration with the CounterMeasures Dispenser (CMD) system enables fully automated/co-ordinated electronic and dispensed countermeasures operations. The ALQ-202 is designed to operate independently, or interfaced with any RWR, including the ALR-56M Advanced Radar Warning Receiver (ARWR), and earlier ALR-67 and ALR-69 systems.

The ALQ-202 incorporates an internal Jammer Support Receiver (JSR). The JSR automatically scans through a programmed frequency range of interest, providing a rapid and independent threat detection capability against both pulsed and CW threat signals. The ALQ-202 further determines the direction of arrival of each identified threat radar, and selects from a prioritised multilevel list of ECM techniques to automatically counter threat radars with an optimised directional response.

The ALQ-202 continually searches for new threats and updates system information on those previously detected. Jamming is both transponder and repeater type, and includes continuous and time-gated responses. The system utilises an extensive library of complex deception and denial jamming techniques, with proven effectiveness.

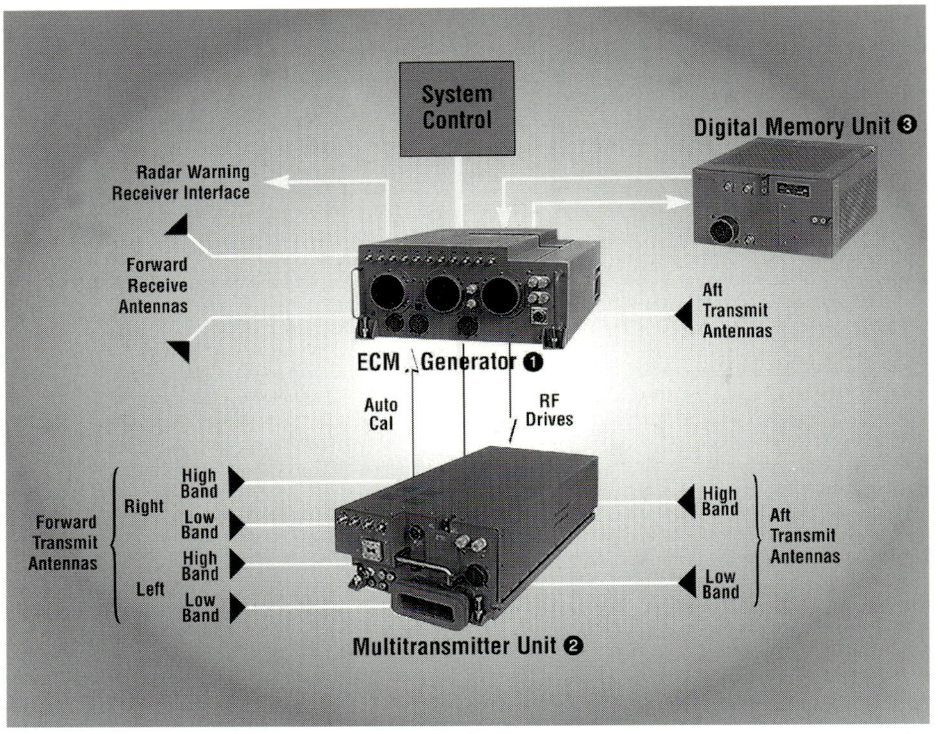

Lockheed Martin AN/ALQ-202 autonomous jammer units 0009915

The ALQ-202 system comprises an ECM generator, multitransmitter unit(s), Digital RF Memory (DRFM) and system control unit. The ECM generator contains the jammer support receiver and ECM technique generating functions. The DRFM is provided for additional jamming techniques capability. Multitransmitter units utilise new technology travelling wave tubes for improved efficiency and reliability, with critical low (non-interfering) thermal noise level, to assure compatible operation with other aircraft systems.

Contractor

Lockheed Martin Fairchild Systems.

VERIFIED

AN/ALR-56A/C Radar Warning Receiver (RWR)

The ALR-56 RWR is used with the AN/ALQ-135 Internal Countermeasures Set jamming equipment to form the Tactical Electronic Warfare System (TEWS) of the F-15 fighter aircraft.

The basic ALR-56 system incorporates the R-1867 processor/low-band receiver, the R-1866 high-band receiver, the IP-1164 display, the C-9429 immediate action control unit, the PP-6968 power supply, a TEWS controller and an antenna array. In more detail, the processor and low-band receiver unit contains three major sections: a single-channel low-band superheterodyne receiver, a dual-channel Intermediate Frequency (IF) section, and a processor.

The low-band receiver is electronically tuned under control of the processor. The dual-channel IF section operates with either the high-band dual-channel receiver or the low-band single-channel receiver. Selection of the receiver to operate with the dual-channel IF section is under the control of the processor.

The processor contains a preprocessor and a general purpose digital computer. The preprocessor contains all the video circuits for analysing intercepted signals. It also provides digital outputs to the computer which represent the measured signal parameters. On the basis of these measurements, a digital output is provided to the display and an audio signal generated. The computer also controls all ALR-56 system functions and is software programmable for the expected threat environment.

The solid-state, digitally controlled, dual-channel high-band receiver is capable of scanning an extremely large portion of the electromagnetic spectrum. Its single conversion superheterodyne design provides high sensitivity and the use of dual yttrium indium garnet Radio Frequency (RF) preselection is claimed to afford excellent selectivity and spurious rejection. High frequency accuracy is obtained through the use of a frequency synthesised local oscillator.

The RF input ports are configured to accept two main antenna inputs to each channel and two additional RF inputs which may be designated to either channel. All switching functions are provided internally.

The receiver is capable of receiving signals over a large dynamic range for analysis and Direction-Finding (DF) measurements. A self-contained precision signal source is provided for calibration and/or built-in test over the entire tuning range. Provision is also made for multiplexing out important receiver functions.

All receiver functions are controlled by serially generated NRZ Manchester coded data. The precise data rate also serves as the reference clock for the receiver frequency synthesiser.

On the countermeasures display, rapid threat evaluation for aircraft defensive manoeuvres is accomplished through automatic intensity control utilising ambient light sensing techniques. Sharp, high-contrast, unambiguous alphanumerics and special symbology allow immediate assessment of the overall tactical threat, without requiring pilot display adjustments as external ambient light fluctuates. Phosphor screen-optical filter matching provides contrast enhancement to overcome direct sunlight contrast degradation.

A range-bearing data presentation is provided with special clutter elimination programmes as well as special threat status. Built-in test circuits determine display system malfunction and automatically indicate GO conditions. An integral lighting panel provides control illumination for night-time viewing.

The antenna system consists of four circularly polarised spiral antenna assemblies (each within its own radome) and a blade antenna. Collectively, this antenna system provides omnidirectional acquisition and direction-finding over the operating frequency range of the RWR. The four spiral antennas, which cover the high-band frequency range, are mounted in a manner that provides 360° azimuth coverage. DF operation is accomplished under computer control. The blade antenna provides omnidirectional coverage of the low-band frequency range.

The TEWS control unit provides for control of the RWR as well as other countermeasures subsystems. A single switch controls the entire RWR on/off function, and separate switches and relays are provided for the subsystems under RWR control. Audio tone cues are controlled from this unit also.

The other control panel is for activating in-flight functions that require instant pilot reaction. In addition to countermeasures subsystem mode control, interrogated data readout on this unit can be called up

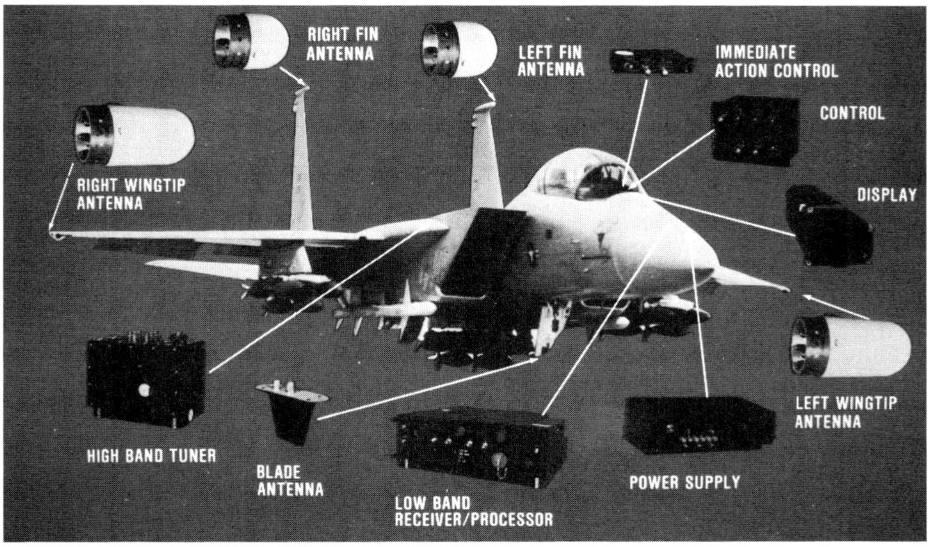

Lockheed Martin AN/ALR-56C radar warning equipment on the US Air Force F-15 Eagle

from the subsystems by means of push-button operation.

A major update known as the ALR-56C has now been completed with improvements in the processor to handle new threats, greater signal densities and other changes. ALR-56C is a digitally controlled, dual-channel superheterodyne receiver covering the E- to J-bands, and is capable of sorting and identifying all current and projected threats to the aircraft. RF input ports are configured to accept two main antenna inputs to each channel and two additional RF inputs that may be designated to either channel, with all switching functions provided internally.

ALR-56C was designed to improve on the ALR-56A signal acquisition capability to include the latest threat parameters and to improve on situational awareness and terminal radar direction. The system maintains the ALR-56A capabilities in the areas of analysis, establishment of priorities, jammer management, passive countermeasures management, and visible and audio displays and warnings.

Operational status

ALR-56 is in service aboard US Air Force and Saudi Arabian F-15 aircraft. ALR-56C equipments are reported to be installed on F-15E strike aircraft and as being retrofitted to F-15Cs and Ds as part of the two types' Multinational Staged Improvement Programme. ALR-56C is reported to be replacing ALR-56A throughout the F-15 inventory as well as being the standard installation on all new build aircraft.

Contractor

Lockheed Martin Fairchild Systems.

VERIFIED

AN/ALR-56M(V) Radar Warning Receiver (RWR) system

The AN/ALR-56M(V) has become the US Air Force's standard RWR and is designed to meet the 21st century threat environment and operational requirements. ALR-56M(V) was selected as the replacement for the AN/ALR-69(V) RWR in a US Air Force competitive fly-off programme in 1988. ALR-56M (V) has successfully completed all required US government testing, including US DoD operational test and evaluation and has been in production since 1989. Over 900 ALR-56M(V) systems are known to be in service with the US Air Force or export customers.

ALR-56M(V) is stated to have an internal jammer hardware interface; sensitivity greater than maximum threat engagement ranges; adaptive selectivity options to provide threat sorting in dense environments; frequency agility to allow rapid adaptive RF scanning; measurement ability to sort and identify time of arrival, pulse repetition frequency, signal strength, pulsewidth, angle of arrival and RF frequency on a monopulse basis; throughput to handle high pulse density within the response time requirement; and an interface capability to all aircraft avionics via dual redundant MIL-STD-1553B busses - one for aircraft avionics (inertial navigation system, fire-control radar and control/display systems) and one for electronic warfare system integration (AN/ALE-47 countermeasures dispenser, internal or external jammer, flight test instrumentation and memory loader/verifier unit). ALR-56M(V) also provides the ability to store multiple, in-flight selectable mission data tables. The operational flight programme and the mission data tables are completely separate entities, allowing for easy reprogramming of parametric data. ALR-56M(V) supports two-level maintenance, enhanced by extensive built-in test.

The system consists of four direction-finding receivers, each connected to one of the four high-band quadrant antennas, a C/D-band receiver/power supply, connected to a low-band antenna assembly, a superheterodyne receiver, a superheterodyne controller and an analysis processor which communicates with the control panel and azimuth indicator and provides two MIL-STD-1553B busses. ALR-56M(V) is effectively a time-shared, rapid, agile single-channel superheterodyne receiver for processing the signals intercepted by the four high-band antennas and the low-band antenna assembly. Wideband, fast, frequency-agile, sensitive receivers provide intra-pulse parametric measurement capability. A computer-controlled tunable radio frequency notch filter, adaptive digital signal processing and parametric screening provide the ability to control the input to the high-speed control processor unit resulting in an unambiguous rapid threat warning, even when operating in the densest, most sophisticated threat environment.

Operational status

ALR-56M(V) entered worldwide operation in the US Air Force in 1992. ALR-56M(V) has been selected for overseas application on F-16 and C-130 aircraft, including Foreign Military Sales contracts for South Korea and Taiwan.

ALR-56M(V) is reported to be mandated for all US Air Force AC-130U, CV-22 and Block 50/52 F-16C/D aircraft. The system is also a retrofit item for US Air Force Block 40/42 F-16 aircraft and under consideration for use in B-1B strategic bombers. In addition, it is the recommended replacement for the AN/ALR-69(V) on the F-4, MH-53J, C-130, Joint STARS, A-10, MC-130E, EC-130 and C-141. It is the baseline offer for C-130J aircraft.

A January 1998 US Air Force contract provided for delivery of 490 replacement computer kits. More than 1,000 units have been ordered.

Contractor

Lockheed Martin Fairchild Systems.

VERIFIED

AN/APR-39(V) Radar Warning Receiver (RWR)

The AN/APR-39(V)1 was designed and developed by E-Systems (now Raytheon Electronic Systems). The APR-39(V)2 processor was designed and developed by

Units of AN/APR-39(V)1 RWR system

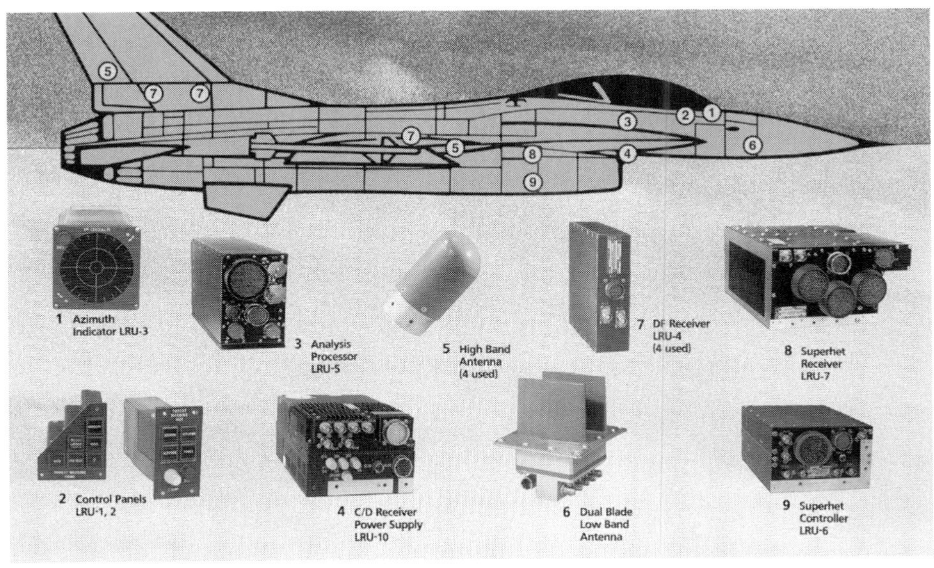

The elements making up ALR-56M(V) as applied to the F-16

Lockheed Martin under contract to the US Army Electronic Command. A third version, the AN/APR-39A (V)1 is also a Lockheed Martin product. This version incorporates a number of updates, including the provision of millimetric-wave threat warning.

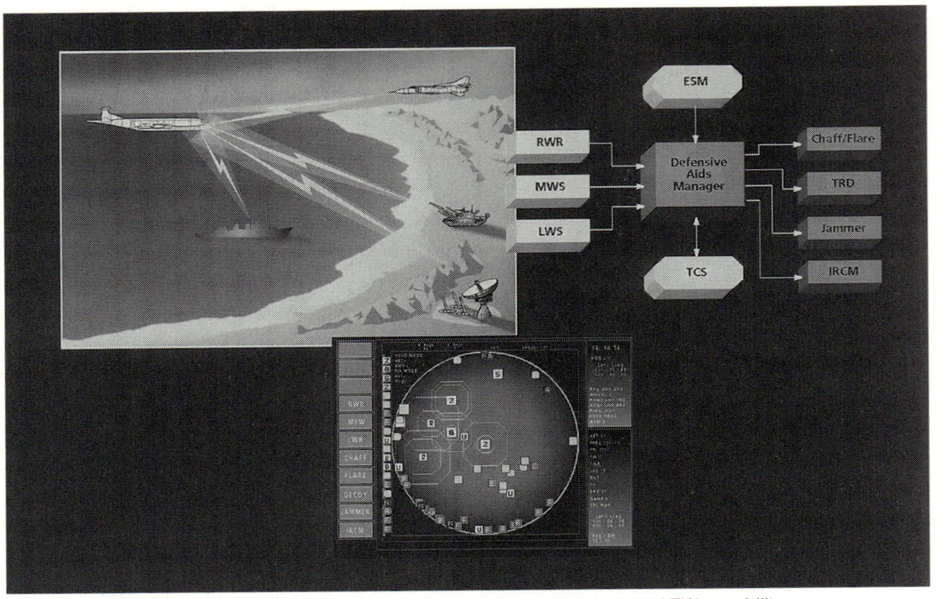

DASS 2000 provides maritime patrol with an effective multispectral, integrated EW capability

Description

The APR-39(V)1 RWR equipment provides automatic warning of emitters in the E, F, G, H, I and most of J radar bands, as well as the appropriate portions of C- and D-bands. It is intended for use on either fixed-wing aircraft or helicopters. The equipment provides indications of bearing, identity and the mode of operation of detected signals with the acquired data being displayed on a cockpit indicator. Proportional pulse repetition frequency of displayed signals and alarm tones are presented to the crew via an integral audio warning subsystem.

The basic APR-39(V)1 system comprises two dual-video receivers, four spiral cavity-backed antennas, one blade antenna, an indicator unit, a comparator and a control unit. Without cables and brackets, this weighs 3.63 kg.

An updated version, APR-39(V)2 has been produced by Lockheed Martin. In this variant, the comparator is replaced by a Lockheed Martin CM-480/APR-39(V) digital processor. This performs signal sorting, identification of emitters, bearing computation and character generation for the presentation of threat details in alphanumeric form on the cockpit display. The unit weighs 6.5 kg and incorporates an adaptive noise threshold and angle-gate, programmable pulse repetition interval filters, and coded emitter outputs. A 19,000-word programmable read-only memory/ random access memory is provided.

Operational status

APR-39(V) series RWRs are reported to have been installed in 15 types of aircraft and patrol/fast attack craft. Several thousand APR-39(V) systems are reported to have been produced, a figure which includes 230 for installation aboard German PAH-1 anti-tank helicopters.

Contractors

Lockheed Martin Fairchild Systems.

VERIFIED

AN/APR-39A(V) threat warning systems

AN/APR-39A(V)1

The AN/APR-39A(V)1 is an upgrade of the earlier analogue APR-39(V)1 radar warning system. It is a lightweight system designed for helicopters and light fixed-wing aircraft. It provides visual and aural warning of hostile radar energy incident on the host aircraft over a very wide frequency range. The system displays multiple threats, (with the highest priority indications highlighted) while keeping track of other emitters in the environment.

A digital display clearly identifies the threat type and azimuth from the aircraft. It also indicates if the threat is searching, locked and tracking, or when the radar lock is broken. The alphanumeric display will automatically separate symbols that are grouped too closely together. The aural warning is by synthetic voice.

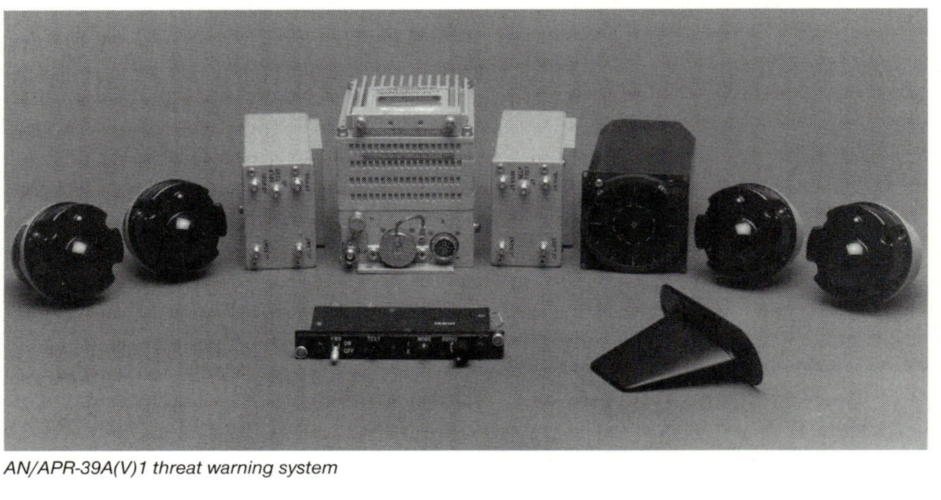
AN/APR-39A(V)1 threat warning system

AN/APR-39A(V)3

AN/APR-39A(V)3 uses APR-39A(V)1 technology and provides continuous coverage through dual-channel crystal video receivers. This system is deployed on US Army and NATO aircraft. It is specifically designed for helicopters and other light aircraft operating at very low levels. It consists of 10 line-replaceable units which are form-fit compatible with the analogue APR-39(V)1. Frequency range covered is E/J and C/D.

Operational status

In service and, in the case of the AN/APR-39A(V)3, also in production. (See also Litton Advanced Systems Division entry for AN/APR-39A(V)2 and 'AN/APR-39A (V)3 with CW' and threat warning systems.)

Contractor

Lockheed Martin Fairchild Systems.

VERIFIED

AN/TLQ-17A countermeasures set

The modular AN/TLQ-17A is an advanced tactical communications countermeasures set developed for the US Army. It is configurable to meet user requirements for a number of host platforms including helicopters.

The AN/TLQ-17A incorporates computer-controlled search, surveillance and jamming in the HF and VHF ranges, microprocessor technology for advanced EW applications, BITE and modularity for ease of maintenance.

Specifications

Frequency:
(band 1) 1.5-20 MHz
(band 2) 20-80 MHz
Operating modes: search/lockout, priority search/ lock on, monitor/automatic and scan band sectors
Effective radiated power: 10-550 W
Receiver tuning time: 1 ms
Preselected frequencies: 255
Reliability: 400 h MTBF

Contractor

Lockheed Martin Fairchild Systems.

VERIFIED

DASS 2000 Electronic Warfare (EW) suite

The Lockheed Martin DASS 2000 electronic warfare suite is a turnkey, fully integrated defensive aids suite which can be configured to meet specific customer requirements and is applicable to a wide range of platform types. The suite's long-range, multispectral threat detection and identification capabilities provide the aircrew with full situation awareness through sensor fusion, and ensure the application of optimum self-protection countermeasures. These co-ordinated capabilities ensure platform survivability in all phases of mission prosecution.

DASS 2000 resources typically include a radar warning receiver, missile warning system, laser warning system, chaff/flare dispensers, electronic countermeasures jammer, towed radar decoy and infra-red countermeasures. Lockheed Martin will adapt the configuration to user requirements, integrate it into the platform ensuring full compatibility, and provide a full range of training and logistics support.

DASS 2000 features interactive functions and a full colour display that provides the EW operator with tactical awareness. The display enables threat assessment as well as automatic countermeasures response by using the defensive aids resources.

A DASS 2000 based system is on order by the UK MOD, via British Aerospace, as the Nimrod MRA.4 DASS.

Operational status

A system based on DASS 2000 has been selected for the UK Nimrod MRA.4.

Nimrod MRA.4 Defensive-Aids SubSystem (DASS)

The Nimrod MRA.4 DASS is not yet fully defined, but the following elements are understood to be included: the Lockheed Martin Fairchild Systems AN/ALE-56M radar warning receiver; the Raytheon Electronic Systems ALE-50 towed radar decoy; the Racal Defence Electronics Digital Radar Frequency Memory Techniques Generator (DRFMTG).

System architecture is based on receipt of threat warning via the AN/ALE-56M and/or an as yet unspecified missile approach warning receiver; production of a suitable threat countermeasure in the techniques generator, and its transfer via EO cable to the AN/ALE-50 towed radar decoy.

The Racal techniques generator is based on advanced twin Digital Radio Frequency Memory (DRFM) units, which are used to classify the threat and customise the most coherent jamming response.

Operational status
In development.

Contractor
Lockheed Martin Fairchild Systems.

VERIFIED

EW-1017 surveillance system

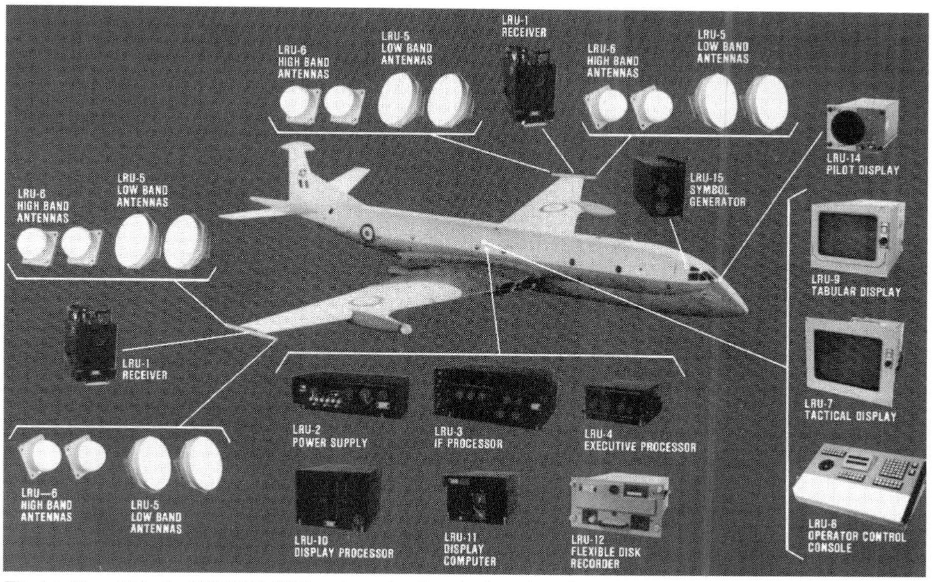

The Lockheed Martin EW-1017 ESM system showing the location of units on the Nimrod MR2 Mk2 aircraft

As an airborne electronic surveillance system, EW-1017 automatically acquires and identifies emissions within the C- to J-bands. The system is also designed to receive and identify all those emissions illuminating the aircraft (including short bursts) in particular when it is operating in very dense signal environments. The warning of possible danger is given both visually and aurally on a display unit while preferential scan is used to ensure immediate recognition of possible lethal threats.

EW-1017 consists of antenna arrays, receiver, a processor system, a pilot display, and ESM operator interactive display subsystem. Broadband spiral antennas are used to provide omnidirectional coverage which, together with their separate multiband receivers, are mounted in pods on each wingtip. This location drastically limits aircraft 'shadowing' and the proximity of the receiver cuts signal losses to a minimum. Angular bearing of the emissions is determined by using selected pairs of antennas.

The hybrid superheterodyne receiver combines high acquisition probability, high sensitivity, frequency accuracy and a high degree of frequency selection and selectivity. A broad bandwidth is used in the acquisition mode to obtain the initial intercept with a narrow bandwidth then being used for accurate bearing measurement and analysis. To ensure processing capability in highly dense signal conditions a high-speed digital computer performs the data processing functions, supplemented by microprocessors. This enables the receivers to scan the frequency band continuously on a reprogrammable basis, so that conventional, continuous wave and agile signals are processed for identification. For special applications, the receiver can be interfaced via a smart post processor unit to a centralised tactical display and control system.

An interactive display subsystem provides a full range of operator facilities to manage and optimise collection. It also provides a readily accessible real-time emitter and platform library storage and analysis capability, facilitated by a modern data management system. A control/display unit allows the operator to monitor and control the automatic surveillance function to resolve possible ambiguities and evaluate and use to the best advantage the data displayed.

Operational status
EW-1017 is in service with the German Naval Air Arm (Atlantique maritime patrol aircraft) and the UK Royal Air Force (Nimrod MR2 Mk 2 maritime patrol and E-3D Sentry airborne warning and control system aircraft). In UK service, the system is designated as ARI-18240 Yellowgate.

In April 1999, Racal Defence Electronics Limited was contracted by the UK Royal Air Force to upgrade the processing system and threat management software for the UK E-3D aircraft using Commercial-Off-The-Shelf (COTS) processors.

Contractor
Lockheed Martin Fairchild Systems.

VERIFIED

Intelligence and Electronic Warfare Common Sensor (IEWCS)

Lockheed Martin Federal Systems is contracted to provide the US Army's IEWCS under a five-year contract worth US$276.5 million - initially integrate on HMMWV trucks and EH-60L helicopters.

The EH-60L configuration known as Advanced Quick Fix (AQF), and AN/ALQ-151(V)3, is an update to be applied to 32 of the original 66 helicopters that are variously described by the terms: the EH-60A Quick Fix 11B; and AN/ALQ-151(V)2.

The Quick Fix 11B configuration was integrated by BAE Systems North America, with the main AN/ALQ-151(V)2 installation coming from TRW Electronics Systems; delivery of all 66 systems was completed in 1989. Quick Fix 11B is understood to provide an intercept, D/F and location capability from 20 MHz in the HF band to 80 MHz in the VHF band.

The Advanced Quick Fix, AN/ALQ-151(V)3, configuration is understood to be designed to provide a more comprehensive capability over a much wider bandwidth. It is being integrated by Lockheed Martin Federal Systems, using the Lockheed Martin Federal Systems Communications High Accuracy Location System - Exploitable (CHALS-X), and the Sanders TACJAM-A equipment, to provide detection, D/F, identification and location across a wide band of HF, VHF, UHF and SHF communications emitters. Development and operational testing of the AQF configuration was completed in 1999.

Contractors
Lockheed Martin Federal Systems.
Sanders, a Lockheed Martin Company.
BAE SYSTEMS Aerospace Electronics, Inc.
TRW Electronics Systems.

UPDATED

The EH-60 L Advanced Quick Fix

AN/AAR-47 missile warning set

The AN/AAR-47 missile warning set detects the plume of approaching surface-to-air missiles and gives the pilot an indication of range and bearing. Decoy systems, such as the AN/ALE-39 flare/chaff dispenser, can be automatically triggered and the system is also compatible with the AN/APR-39A radar warning receiver.

The feasibility of such a warning system was first demonstrated by Loral in December 1977 and, following development of the AAR-46 system in 1979, full-scale development of the AAR-47 started in March 1983. The AAR-47 is small, needs no cryogenic cooling and has an MTBF of 1,500 hours.

The system consists of six sensors, a central processor and a control indicator. The sensors are based on ultra-violet technology. They are hard-mounted on the skin of the aircraft and provide full coverage, with overlap to protect against blanking. The complete system weighs only 18.2 kg. Within 30 seconds of electrical power a self-initiated BIT programme is completed and the system is operational with no in-flight down time for recalibration. The processor analyses the data from each sensor independently and as a group, and automatically deploys the appropriate countermeasures. In the case

The AN/AAR-47 missile warning set

The Lockheed Martin AN/ALQ-204 mounted on the fuselage of a Royal Air Force VC10 (Paul Jackson)

The Lockheed Martin AN/ALQ-204 integrated into the tail fairing of a Gulfstream G-IV (Edward Downs)
2001/0087831

of a flare failure, the AAR-47 automatically commands a second flare.

Multidirectional threats are automatically analysed and prioritised for countermeasures sequencing. The control indicator displays the incoming direction of the highest priority threat for tactical manoeuvres.

A shorter sensor has been developed for use in locations, such as the F-16 fuselage, where depth is limited.

AN/AAR-47 technology is used as the basis of the Common Missile Warning System (CMWS) element of the Advanced Threat Infra-Red CounterMeasures (ATIRCM) and Common Missile Warning System (CMWS) being developed by Sanders and Lockheed Martin IR Imaging Systems.

Specifications

Dimensions:
(sensor) 120 × 200 mm
(processor) 203 × 257 × 204 mm
Weight:
(4 sensor system) 14 kg
(sensor) each 1.5 kg
(processor) 7.9 kg
Power supply:
(4 sensor system) 28 V DC, 75 W
(4 W per sensor, 59 W for processor)
Coverage: 360° azimuth (given by 6 sensors)

Operational status

AN/AAR-47 systems are installed on transport aircraft and helicopters operated by Australia, Canada, UK and USA.

The US Naval Air Systems Command (NAVAIR) has issued a Request For Proposals (RFP) for an upgrade to the AN/AAR-47 sensors and software. The plan is to add laser warning and to increase the sensitivity of the optics.

Contractors

Lockheed Martin IR Imaging Systems.

VERIFIED

AN/ALQ-157 infra-red countermeasures system

The AN/ALQ-157 provides multiple simultaneous protection for large heavy-lift helicopters and medium-size fixed-wing aircraft against SAM and AAM threats. The system employs advanced components and microprocessor technology to allow operator jamming code selection and reprogrammability for future threats. The two fuselage-mounted synchronised jammer assemblies provide continuous 360° protection against threats launched from any direction. The power module, line filter and pilot control indicator can be placed anywhere within the aircraft.

Specifications

Weight: 99.8 kg
Power supply: 115 V AC, 3 phase, 400 Hz
28 V DC

Operational status

Lockheed Martin received the initial production contract in December 1983 for systems for the US Marine Corps' CH-46 helicopter. US and international aircraft equipped with the AN/ALQ-157 include the SH-3, CH-46, CH-47D, H-53, Lynx, C-130 and P-3C. Nearly 700 systems have been sold worldwide.

Contractor

Lockheed Martin IR Imaging Systems.

VERIFIED

AN/ALQ-204 Matador IR countermeasures system

The AN/ALQ-204 Matador is a family of infra-red countermeasures systems which has 11 different configurations. The system is suitable for all types of large transport aircraft with unsuppressed engines, one transmitter per engine being recommended for maximum protection. The basic system consists of transmitters, a Controller Unit (CU), Power Supply Unit (PSU) and an operator's controller. Transmitters are electronically synchronised by the CU which controls and monitors one or two transmitters. The operator's controller is common to all configurations, controls from one to seven transmitters and incorporates a system status display.

Each transmitter contains an IR source which emits pulsed radiation to combat an IR missile. Preprogrammed multithreat jamming codes are selectable on the operator's control unit and all new codes can be entered as required to cope with new threats. Each transmitter weighs 22 kg nominal and can be pod mounted for aircraft attachment.

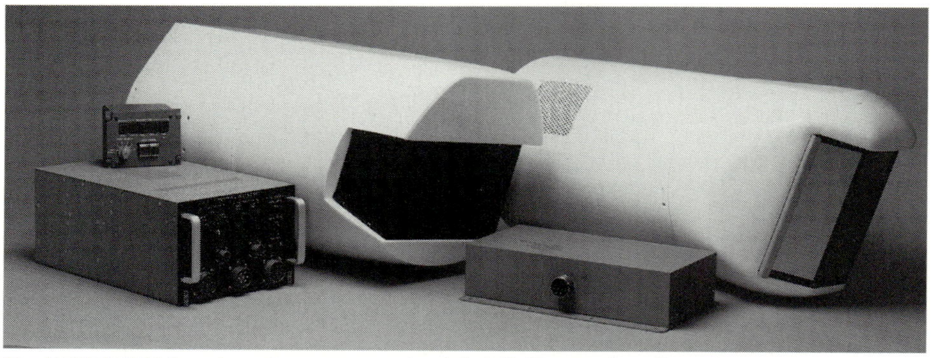

The AN/ALQ-204 infra-red countermeasures system is installed in a number of large fixed-wing aircraft

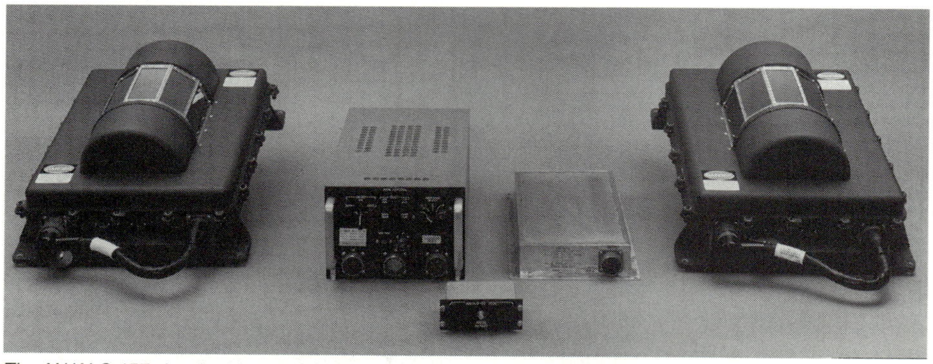

The AN/ALQ-157 showing the two fuselage-mounted jammer assemblies (left and right) and, in the centre, the power module/pilot control indicator, line filter

Operational status

In service on the Boeing 747 and 707 (derivatives), Lockheed Martin L-1011 and BAE SYSTEMS VC10, 146, Andover, Airbus -340 and Gulfstream G-IV aircraft.

All current applications are FAA certified for operation on Head of State, commercial and civilian aircraft.

Contractor

Lockheed Martin IR Imaging Systems.

UPDATED

Challenger IR jammer

A small omnidirectional IR jammer, especially suitable for helicopters and light aircraft, Challenger can be installed in fixed aircraft apertures or deployed on retractable platforms. The jammer head weighs about 9 kg and power consumption is within the range 1 to 4 kW. The jammer assembly is driven by a compact controller in the aircraft. Single or dual transmitter configuration can be selected to provide 360° azimuth protection. Challenger, an upgraded AN/ALQ-157, was designed to be flexible and to offer a wide variety of installation options.

Operational status

In production and in service on Lynx and UH-1 helicopters.

Contractor

Lockheed Martin IR Imaging Systems.

VERIFIED

Defendir IR jammer

Defendir is a family of systems which can be used for small and large helicopters, as well as turbojet transport fixed-wing aircraft. It can be supplied in various configurations, such as an aft-mounted jammer for aft protection, an aft-mounted jammer with side windows for aft and side protection, a nose-mounted jammer for forward protection or aft and nose jammers for all-round protection.

Operational status

In development.

Contractor

Lockheed Martin IR Imaging Systems.

VERIFIED

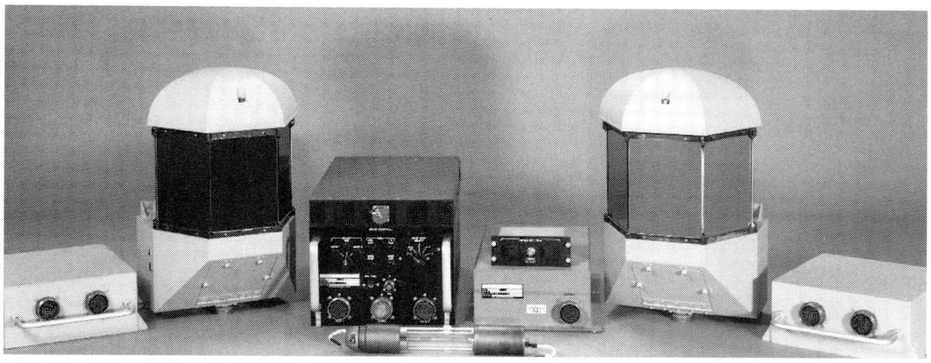

The Lockheed Martin Challenger infra-red jammer

Passive missile warning set

The passive missile warning set is designed for the C-130 to detect potential missile threats. It discriminates against false targets, declares the approaching missile threat when this is within the optimum countermeasure interval and sends a signal to the countermeasures dispenser. The system is totally passive, requiring only the E-O signature from the missile plume. Non-threat false alarms are eliminated by E-O spectral selection and signal processing algorithms.

The four sensors are hard-mounted on the aircraft skin to provide 360° horizontal coverage. These are wide field of view staring E-O receivers that collect in-band radiation and convert it to electrical signals. Background clutter is rejected spectrally. The signal processor receives the signals from the four sensors and uses temporal algorithms to distinguish threats from non-threat or false signals. There are no moving parts in the sensors or the signal processor and no cooling air is required. The pilot control indicator contains BIT initiation plus a quadrant threat indicator.

Contractor

Lockheed Martin IR Imaging Systems.

VERIFIED

AN/APG-67(F) attack radar

The AN/APG-67(F) has been developed for retrofitting into Northrop F-5 aircraft and other fighter aircraft.

A significant characteristic of the APG-67, stemming from its Modular Survivable Radar (MSR) ancestry, is the choice and layout of circuitry to enable the system to be updated for new technologies as they appear, such as very large-scale integration, without having to redesign the LRUs. Growth features can also be added in this way.

In air-to-air missions the system searches for, automatically acquires and tracks targets in both look-up and look-down situations. In its air-to-surface role the radar provides real-beam mapping, high-resolution Doppler beam-sharpened ground-mapping (40:1), and air-to-ground ranging. Variants of the radar can include synthetic aperture imaging.

The heart of the processor contains 16 Shark 2106D processors. Twelve perform signal processing functions and the other four perform Hardware I/O, radar mode control and aircraft mission computer interface functions, together with self-test radar processing associated with motion compensation, target tracking, antenna motion control and raster scan output to the cockpit display.

The processor integrates raw target data from the antenna so that reliable reports can be established. The unit incorporates a programmable signal processor and its use is considered to be the key to what is claimed to be this unit's exceptional performance. The processor performs a variety of functions, including fast Fourier transformation, moving target indication, pulse compression and motion compensation, and can operate with variable waveforms and bandwidth. The transmitter produces coherent I/J-band radiation from a travelling wave tube and can operate at low, medium and high PRFs, with variable power outputs and pulsewidths. The antenna is a flat plate slotted array with low-sidelobe sensitivity and with ±60° scan in azimuth and elevation. Scan width is selectable and elevation search can be accomplished in 1, 2 or 4 bar modes. The receiver has a low-noise front-end to maximise detection range.

The AN/APG-67(F) version for the new aircraft and retrofit market offers a full range of air-to-air capabilities including range-while-search out to 148 km, three air combat modes and single target track. It also provides a complementary set of air-to-ground modes including map, expand, Doppler beam-sharpening, freeze and surface moving target indication and track over ground or rough seas.

It consists of three LRUs and incorporates a MIL-STD-1553B database interface and MIL-STD-1750A computer architecture. There is extensive built-in test.

Lockheed Martin has collaborated with Geophysical and Environmental Research Corp to develop GSAR, a variant of the AN/APG-67 radar to equip business aircraft for the SAR role. GSAR has a range of 150 km, with swath widths up to 40 km wide. Resolution is reported to be 3 m, improving to 1 m in spotlight mode.

Specifications - APG-67(F)

Volume: 0.049 ms
Weight: 74 kg
Frequency: I/J-band
Power: 2,600 VA

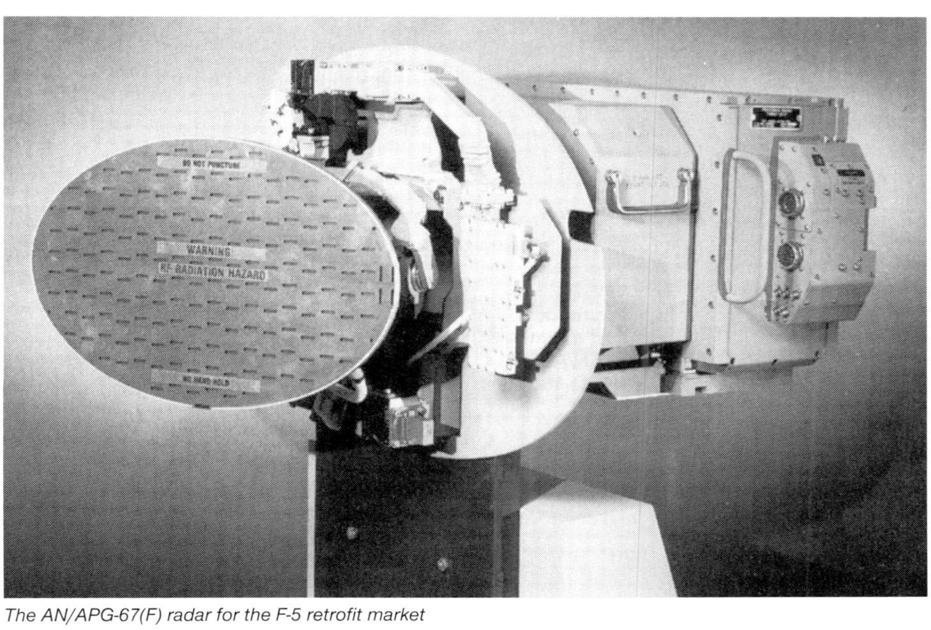

The AN/APG-67(F) radar for the F-5 retrofit market

Transmit power: 350 W
Antenna size: 279 × 711 mm
Detection range (fighter-size targets):
(look-down) 39 km
(look-up) 72 km
(sea targets, 50 m^2) 57 km Sea State 1
(max range, all modes) 148 km
Accuracy (air-to-ground): 15 m or 0.5% of range
Reliability: design 235 h MTBF

Operational status

In production. Taiwan has selected the APG-67 for its KTX-II Indigenous Defence Fighter project.

Contractor

Lockheed Martin Naval Electronics & Surveillance Systems.

VERIFIED

AN/APS-125/-138/-139/-145 airborne early warning radars

The AN/APS-125/-138/-139/-145 series of airborne early warning radars have all been developed for the Northrop Grumman E-2C Hawkeye aircraft.

AN/APS-125

The AN/APS-125 radar was designed for the airborne early warning role, and operates in the UHF band. It was reported to have been the original fit in US Navy E-2C Group 0 aircraft, and in the E-2C Group 0 aircraft exported to Egypt, Israel and Japan, during the late 1970s and early 1980s.

AN/APS-138

The AN/APS-138 was an update of the AN/APS-125 during the mid-1980s. It reportedly included minor circuit changes and a low-sidelobe antenna, designated the Total Radiation Aperture Control Antenna (TRAC-A). AN/APS-138 radars were reportedly in production between 1983 and 1987, and the US Navy is understood to have updated AN/APS-125 radars to the AN/APS-138 standard. AN/APS-138 radars were also reportedly sold to Singapore on its E-2C aircraft.

AN/APS-139

The AN/APS-139 radar is a development of the AN/APS-138 that improves ECCM performance and surface surveillance capabilities, reportedly in production from 1987 to 1989. Reportedly no radars have been updated from AN/APS-138 to AN/APS-139 standard.

AN/APS-145

Latest in the series is the AN/APS-145. It is an update of the AN/APS-138 and AN/APS-139 radars, and has been retrofitted to all US Navy E-2C aircraft from 1990. The AN/APS-145 development reportedly specifically addresses the problem of overland clutter and provides fully automatic overland targeting and tracking capability, an improved IFF system, expanded processing and new colour displays. The present radar is said to perform very well over sea and desert, but to degrade rapidly when terrain becomes more rugged. To reduce the false alarm rate, a feature known as 'environmental processing' is being developed. This adjusts the sensitivity of the radar cell by cell, according to the clutter and traffic in each cell. To enable large aircraft to be detected at long range (up to 400 n miles), a new lower PRF is used and, to match this development, the E-2C's rotordome rotation rate is slowed from 6 to 5 rpm. A third PRF is also introduced allowing the radar to operate with different PRFs during scanning, to eliminate blind speed problems caused by single PRF operation. The AN/APS-145 entered service in 1992. It is fitted to US Navy E-2C Group II aircraft and to Taiwanese E-2C and to French E-2C aircraft.

Lockheed Martin and Northrop Grumman have signed an MoU to collaborate in supplying airborne early warning and control (AEW&C) systems to overseas customers, based on Lockheed Martin supplying the AN/APS-145 radar and Northrop Grumman being responsible for integration of the mission system. This agreement applies to ADE&C variants of the Lockheed Martin C-130J and to other platforms, including the Northrop Grumman E-2C, but excluding the Boeing E-3 AWACS.

Operational status

Aircraft fits are reported to be as follows: E-2C Group 0 aircraft fitted with AN/APS-125/-138; E-2C Group I aircraft fitted with AN/APS-139; E-2C Group II aircraft fitted with AN/APS-145; all US Navy aircraft upgraded to AN/APS-145 configuration. Four US Coast Guard P-3 AEW&C aircraft are reportedly fitted with the AN/APS-138 radar.

Because the E-2C will serve the US Navy for many years, more capable Hawkeyes are on order. Northrop Grumman restarted its Hawkeye production line in 1994, after the US Navy ordered the first four of an expected 36 new Group II E-2Cs. In April 1996, the eighth and ninth new Hawkeyes were initiated under an advance procurement contract. In addition, Northrop Grumman is updating Group I Hawkeyes to Group II configuration for the US Navy. Modified aircraft are fitted with the AN/APS-145 radar system with fully automatic overland targeting and tracking capability, an improved IFF system, a 40 per cent increase in radar and IFF ranges, expanded processing capacity, new high-target-capacity colour displays, JTIDS for improved secure, anti-jam voice and data communications and GPS-based navigation capability.

The AN/APS-145 surveillance radar has been retrofitted to US Navy E-2C Hawkeye AEW aircraft

E-2C Group II radar display 0018073

Beginning in 2001, all US Navy production deliveries will be Hawkeye 2000s; 21 have been ordered so far. All international customer deliveries would be standardised on this configuration. Japan has already begun a phased upgrade of its E-2Cs to incorporate the Hawkeye 2000 technology standard.

France has ordered one and the Republic of China two. Meanwhile, Northrop Grumman is marketing the concept of an 'Advanced Hawkeye' for the period after 2010 based on use of an electronically scanned radar. The development radar will be incorporated into a Lockheed Martin C-130 AEW aircraft for flight testing.

Contractor

Lockheed Martin Naval Electronics & Surveillance Systems.

VERIFIED

AN/AYA-8C data processing system

The AN/AYA-8C data processing system for the US Navy P-3C Orion anti-submarine warfare aircraft provides the interface between the AN/ASQ-212 general purpose digital computer and the aircraft systems and so constitutes a major part of the P-3C's mission avionics.

The system is connected to all the crew stations on the aircraft: tactical co-ordinator, non-acoustic sensor, navigation/communications, acoustic sensor and flight deck. In addition, it communicates directly with the radar interface unit, armament/ordnance system, navigation systems, sonar receiver, magnetic anomaly detector, infra-red detection system, electronic support measures, sonobuoy reference system and Omega.

The P-3C's data processing activities are divided into four logic units, forming separate boxes in which all electronic operations are conducted in a combination of digital and analogue formats. Various keysets and panels complete the system hardware.

Logic Unit 1 interfaces between the central computer and four types of peripheral information system: manual data entry, system status, sonobuoy receiver and auxiliary readout display.

The manual entry subsystem provides the communication between the various operator stations and the central computer. Each operator has a panel of illuminated switches and indicators by which he communicates with the central computer. System status for the navigation and magnetic anomaly detector is received and stored by part of Logic Unit 1, then transmitted to the central computer. The computer receives the status information on demand, or when any status changes. Finally, the auxiliary readout display logic interfaces between the central computer and the auxiliary displays at the tactical co-ordinator and nav/com stations. The radar, sonar antenna, infra-red detection system and electronic support measures interfaces are also achieved by Logic Unit 1.

Logic Unit 2 is concerned with communicating between the central computer and the navigation and armament/ordnance systems. This logic unit transmits Doppler and inertial navigation data to the computer, and instructions to launch search-and-kill stores.

Logic Unit 3 controls the CRT displays provided for the tactical co-ordinator, sensor operators and pilot. The tactical co-ordinator and pilot displays can generate characters, vectors and conics, while the sensor displays use characters and vectors only.

Logic Unit 4 is mainly an expansion unit comprising two items: the Data Multiplexer Sub-unit (DMS) and the Drum Auxiliary Memory Sub-unit (DAMS). These provide extra input/output capacity and memory capacity respectively.

The DMS can service four input and output peripherals, as selected by the central computer. One output channel presents characters, vectors and conics for the auxiliary sensor display and one input/output channel is used for the aircraft's Omega navigation system, a command launch system for the Harpoon anti-ship missile and sonobuoy reference system.

The DAMS was incorporated to give an additional 393,216 words of memory to the computer, so that the operational program could be expanded to accommodate extra functions and equipment.

Various keysets and control panels allow access to the central computer via the data processing system. A universal keyset allows the transfer of information between the computer and the nav/com operator. The pilot uses his own keyset for controlling the information presented on his CRT display for entering navigation stabilisation data, dropping weapons and flares and entering information on visual contacts. The ordnance panel displays the commands which the computer has given to the ordnance operator concerning status and position of the search stores, such as sonobuoys, which are available for deployment. Finally, there is an armament/ordnance test panel which monitors the output from the data processing system Logic Unit 2 to those systems.

Operational status

In service on P-3C Orion aircraft.

Contractor

Lockheed Martin Naval Electronics & Surveillance Systems.

VERIFIED

HIgh Performance Active Sonar (HIPAS)

Lockheed Martin is developing the HIgh Performance Active Sonar (HIPAS) to meet the US Navy's Airborne Low Frequency Sonar (ALFS) requirements. HIPAS is a heliborne dipping system that provides both active and passive detection and tracking of submarines.

Designed for installation on the SH-60F carrier-based inner zone and SH-60B LAMPS III helicopters, as well as the HV-22 Osprey tilt-rotor aircraft, HIPAS fits in the same space, weight and power envelope as present sonar systems.

Operational status

Under development.

Contractor

Lockheed Martin Naval Electronics & Surveillance Systems.

VERIFIED

HOWLS experimental airborne radar

The HOstile Weapons Location System (HOWLS) radar is an extremely versatile coherent J-band phased-array radar, designed to investigate techniques from which solutions may be derived for detecting and locating fixed tactical targets in a clutter environment. It has been developed by Lockheed Martin under the sponsorship of the US Advanced Research Projects Agency and management of MIT's Lincoln Laboratory.

The key to the radar system is a lightweight, phased-array antenna. The antenna system utilises a low-cost, highly reproducible phased-array element that combines the radiating element and PIN diode phase shifter in a single integrated structure to minimise weight and cost.

HOWLS has six basic modes of operation. These are contextual ground map, fixed-target detection, ground moving target detection, low-level Doppler spectrum signature analysis, projectile detection and tracking, and Doppler beam-sharpening. Through microprocessor control, the radar has mode parameter flexibility and is capable of interleaved multimode operation. Selectable parameters include antenna scanning pattern and rate, PRF, data sampling window position and data sampling pattern. The radar is capable of switching among parameters in an orderly manner so that up to four operating modes can be interleaved.

Results of the experimental HOWLS programme will provide the technology base from which radars for airborne platforms can be configured to meet a variety of missions.

Contractor

Lockheed Martin Naval Electronics & Surveillance Systems.

VERIFIED

F-16C/D Block 40/42 and Block 50/52 Common Configuration Implementation Program (CCIP)

The F-16 CCIP will configure nearly 700 US Air Force Block 50/52 and Block 40/42 F-16 aircraft with common avionics. The total programme is valued at approximately US$1.6 billion.

The purpose of the F-16 CCIP is to provide: state-of-the-art capability; expand avionics growth potential (primarily through software upgrades); reduce potential for diminishing manufacturing sources; provide commonality for the Block 50/52 and Block 40/42 fleets.

The CCIP configuration includes the following systems: Modular Mission Computer (MMC); colour multifunction display set; common data entry electronics unit; electronic horizontal situation indicator; Joint Helmet Mounted Cueing System (JHMCS made by Vision Systems International); Link 16 Multifunction Information Distribution System Low Volume Terminal (MIDS LVT) with TACAN and new weapons such as the AIM-9X air-to-air missile and the AGM-158 Joint Air-to-Surface Standoff Missile (JASSM).

In addition, integration of an Advanced Identification Friend or Foe (AIFF) with air-to-air interrogation capability is being included for the 251 Block 50/52 aircraft under a separate contract action authorised in August 1999. The system selected is the BAE Systems North America AN/APX-113 Advanced Identification Friend or Foe (AIFF).

After the CCIP modifications, the Block 50/52 and Block 40/42 versions will be essentially interchangeable in terms of cockpit, avionics and weapons capability. Residual systems differences in the two blocks will be in the pilots' head-up display unit version, the fire-control radar version, the AIFF and the engines.

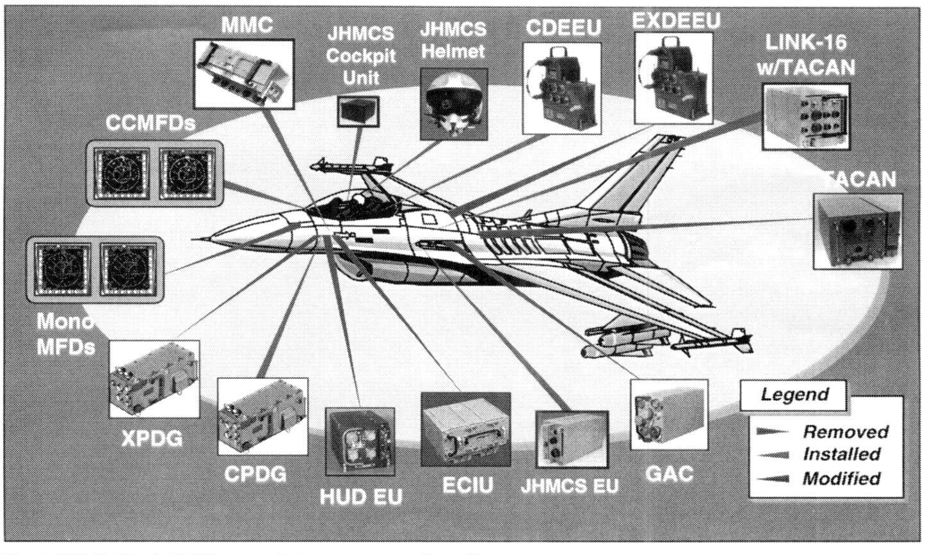

The F-16C/D Block 40/42 upgraded common configuration **1999**/0054219

The CCIP core avionics and cockpit will provide commonality with the F-16A/B Mid-Life Update aircraft and the Block 50 aircraft still being purchased by the US Air Force and international customers.

Operational status

Lockheed Martin Aeronautical Company began hardware development for CCIP under a contract in May 1998. Initial flight testing began in June 1999. Associated software development began in November 1998 and full integration flight testing will begin in June 2001.

Lockheed Martin will provide the production modification kits, beginning in July 2001. In September 1999, Lockheed Martin received the first increment of a contract worth US$108 million to begin kit production activities for the upgrade of 64 Block 50/52 aircraft. The aircraft will be modified at the USAF Ogden Air Logistics Center at Hill Air Force Base, Utah. The first aircraft will complete modification in early 2002. The Block 50 aircraft will have priority and will be completed between 2002 and 2006. Block 40 aircraft will be modified between 2005 and 2009.

In May 2000, Lockheed Martin received a 15 month extension of the Engineering Manufacturing and Development (EMD) phase to March 2003, to incorporate hardware design modifications to facilitate the HARM Targeting System capability, and to ensure radio frequency compatibility with the Link 16 datalink system.

Contractor

Lockheed Martin Tactical Aircraft Systems.

UPDATED

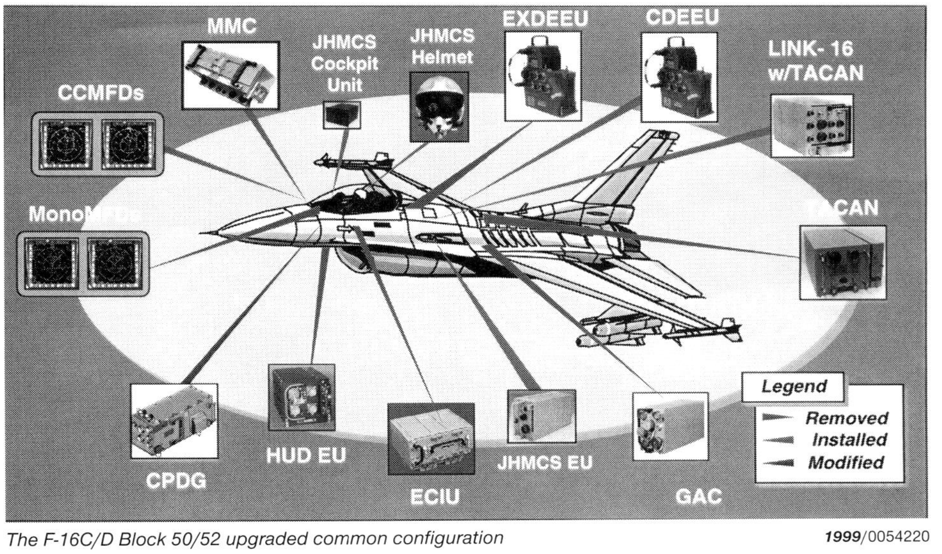

The F-16C/D Block 50/52 upgraded common configuration **1999**/0054220

A US Air Force Block 50 F-16C aircraft on a defence suppression mission - one of the aircraft destined for the F-16 Common Configuration Implementation Program **2000**/0081868

AN/ASQ-212 mission processing system

The AN/ASQ-212 system consists of the CP-2044 computer and several interconnection devices which comprise a form, fit and function replacement for the AN/ASQ-114 computer, data analysis logic units and the signal data converter. The extended memory upgrade of the AN/ASQ-114 computer is transferred to the CP-2044 and used both for global and secondary memory. The CP-2044 incorporates Motorola 680030 processors to provide a throughput ranging from 10 to 25 Mips, which is 30 times greater than the current system in the P-3C Update I/III aircraft at a fraction of the current size, weight and power requirements. In the full Update III Ada implementation, less than 50 per cent of the CP-2044 minimum throughput and memory capacity is utilised.

The CP-2044 VME bus open architecture can be configured with additional processing, memory and input/output modules to meet the requirements of new subsystems such as GPS and Satcom and of processing intensive functions such as sensor post-processing and data fusion.

Initially designed for retrofit into P-3C Update I/III aircraft, the AN/ASQ-212 can be easily tailored to the requirements of other P-3C configurations as well as new aircraft.

Operational status
The AN/ASQ-212 system has been developed for the US Navy under a two-phase programme that began in September 1989. The first production systems were installed in US Navy test aircraft and training facilities beginning in May 1993, at a rate of four systems per month. Plans called for 150 operational systems to be installed beginning in October 1993. In addition, two foreign P-3C operators took delivery of the AN/ASQ-212 beginning in 1993, and the system was supplied for Royal Norwegian Air Force P-3C aircraft between 1997 and 1999.

Contractor
Lockheed Martin Tactical Defense Systems.

VERIFIED

AN/UPD-4 reconnaissance system

The AN/UPD-4 was developed for the US Air Force as a successor to the AN/APD-102A. It has since been enhanced and designated AN/UPD-8. The AN/UPD-4 side-looking radar is an all-weather, high-resolution reconnaissance sensor system for the airborne collection of tactical and strategic intelligence information. Utilising I/J-band radar energy to illuminate selected terrain swathes, it operates equally well in daylight, darkness and adverse weather.

The AN/UPD-4 consists of an AN/APD-10 synthetic aperture radar mounted in the aircraft, a ground-based correlator/processor set and test consoles for the maintenance of airborne and ground equipment. Airborne datalink transmitters and ground receiving and processing systems are optional additions to provide radar information to ground commanders more rapidly than the basic system.

The airborne system includes multi-element, phased, linear waveguide antennas individually gimballed for left- or right-side operation. Look-angle of the antennas is maintained in flight by a control system that receives error signals from the aircraft inertial navigation system and gimbal-mounted gyros and accelerometers.

The radar mapping recorder contains the optical, electronic and electromechanical assemblies required to record the radar video data and associated coded data on photographic film. This information is displayed on two 127 mm CRTs as four intensity-modulated traces that are transferred by two mirror assemblies and four recording lenses to focus the images on the film. Mode strips that convey essential operational information and multi-element blocks containing information pertinent to the mission are recorded on the film. For datalink operation, an identical recorder is used at the correlator/processor on the ground to establish a continuous recording/correlation path. Film speed is carefully controlled in either application to maintain a fixed relationship to aircraft groundspeed.

The equipment has several modes of operation, providing a variety of standoff distances and altitudes and the option of recording only Fixed Target Indication (FTI) or both FTI and Moving Target Indication (MTI). Imaging of the terrain at either side can be obtained at the discretion of the operator. With datalink-equipped systems, the information may be transmitted to a ground station in real time for recording and processing. The final imagery is recorded on 241 mm wide film in four channels at a scale of 1:1 million for all modes. In the along-track direction, targets and terrain features are imaged in terms of distance travelled and represent true ground separation. In the across-track direction, the imagery is recorded in slant range or the distance from the aircraft to the target.

Specifications
Volume: 0.57 m^3
Weight: 282 kg
Frequency: I/J-band
Swath width: 18.5 km
Range: 55 km
Resolution: 3 m

Operational status
In service with RF-4EJs of the Japanese Air Self-Defence Force.

Contractor
Lockheed Martin Tactical Defense Systems.

VERIFIED

AN/UPD-8/-9 and AN/APD-14 side-looking reconnaissance systems

AN/UPD-8/-9
The AN/UPD-8 and AN/UPD-9 are updated and improved versions of the AN/UPD-4, with much greater range. The AN/UPD-8 is a side-looking, synthetic aperture airborne reconnaissance radar system for the US Air Force. The system provides an all-weather, day/night, standoff, tactical and strategic reconnaissance capability and, with its extended-range antenna pods and airborne datalink electronics, is the latest in a sequence of Lockheed Martin airborne reconnaissance radars.

Operating in the I/J-band, the system records with equal effectiveness during daylight and darkness and through periods of adverse weather that would render other types of reconnaissance systems ineffective. The airborne radar system's standoff capability, which is increased by 60 per cent in the new equipment, allows coverage of border regions, harbours and coastal areas without violating another nation's airspace or national waters. Maximum range, limited by power, is about 130 km. At such distances, resolution is 6.1 m in azimuth and 4.6 m in range but resolution increases at shorter ranges. At 50 to 90 km the radar resolution is 3 m, allowing large aircraft to be identified and showing clear images of vehicle formations.

Lockheed Martin has also developed a pod-mounted version of the UPD-8. Similar in size to a 1,250 litre fuel tank and weighing less than 682 kg, the pod includes the main antenna, support electronics and a datalink. A version of this pod has been flight tested on an RF-4B and is to be supplied for US Marine Corps F/A-18(RC) reconnaissance aircraft. The podded system can also be installed on an executive jet aircraft.

Another upgrade of the UPD-4 is the AN/UPD-9 developed for the US Marine Corps. This is similar to the UPD-8, with a datalink, but does not have the long-range features. Basic standoff range is about 80 km.

The airborne sensor part of the overall system, the AN/APD-12, operates in eight modes selectable by the operator according to aircraft altitude and the distance to the target area. The system can record to left or right of the aircraft track with the normal antennas or with the extended-range antennas which are mounted in a centreline pod. Target information can be transmitted to the ground in real time through a datalink or can be recorded in flight for later processing on the ground. The former facility enables a ground commander, perhaps hundreds of kilometres away, to evaluate targets while the aircraft continues its mission.

Specifications
Volume:
(UPD-8) 1.305 m^3
(UPD-9) 0.58 m^3
Weight:
(UPD-8) 93.5 kg
(UPD-9) 321 kg
Frequency: I/J-band
Swath width: 18.5 km
Range: (UPD-8) 93 km, (UPD-9) 55 km
Resolution: 3 m

Operational status
In operational service with Israeli Air Force RF-4B aircraft.

AN/APD-14
Sandia National Laboratory has modified the AN/UPD-8 design to form the AN/APD-14 SAR for Open Skies (SAROS) set. The treaty on Open Skies specifies SAR resolution no better than 3 m resolution. SAROS, installed aboard an OC-135 aircraft, has a centre frequency of 9.6 GHz and a mapping swath of 10 n miles.

Contractor
Lockheed Martin Tactical Defense Systems.

VERIFIED

LAIRS Lockheed Martin Advanced Imaging Radar System

The current LAIRS is the fourth generation of upgraded Synthetic Aperture Radar (SAR) electronics based upon the original LAIRS that was developed from the Advanced Synthetic Aperture Radar System (ASARS-1). Although ASARS-1 was developed for the SR-71 Mach 3 environment, LAIRS has been specifically built and tailored to operate in the medium altitude, subsonic through transonic speed regimes. LAIRS uses the latest in SAR algorithms and is composed of both an airborne segment and a ground segment.

The airborne segment is built in a modular design and can be easily modified to accommodate specific customer requirements. The system incorporates commercial-off-the-shelf (COTS) technology in many of the components, including the airborne or ground processing system. The collected data from the airborne sensor can be transmitted directly to the ground via a datalink as phase history data, recorded on the aircraft for later processing on the ground or processed in real time on the aircraft and transmitted to the ground as imagery data. The datalink also has a feature for both downlink and uplink of data.

LAIRS is capable of providing both fixed target imagery (FTI) and moving target imagery (MTI). The FTI modes are swath mode and spotlight modes. The swath mode provides a wide area (normally 19 km wide) image of the earth or sea for surveillance and provides medium-resolution imagery (normally 3 m resolution). The swath can be continuous in either a parallel path or an oblique path to the flight vector. Wider swaths and finer resolution swaths are available, and are easily incorporated into the modular design of LAIRS. The second FTI mode is spotlight mode. LAIRS has two spotlight modes, one for classification and a higher-resolution mode for identification of targets. Spotlight modes can image targets either forward or aft of the aircraft and within the squint limits of the antenna. LAIRS also has an MTI mode for imaging moving targets. The mode typically uses a sector scan image out to 185 km and can image moving targets within a 50 km sector that can be positioned anywhere from 35 to 185 km away from the aircraft.

The LAIRS ground station uses state-of-the-art computer equipment and software to store, retrieve, combine, exploit and report. Typically, LAIRS is

Specifications for LAIRS

	Swath mode	Spotlight mode	MTI mode
Min slant range	<19 km	<19 km	35 km
Max slant range	>185 km	>185 km	185 km
Resolution	<3 m	fine and ultra-fine	<3 m

The antenna/gimbal assembly used in STacSAR™

The STacSAR™ radome configuration as used in a light aircraft application

displayed on soft copy so that the user can optimise data analysis from imagery produced by the LAIRS system. The number of workstations, the management of imagery software, and hard copy production and display equipment can be tailored to the specific requirements of the user.

Operational status
LAIRS radar equipment is reported to be operational aboard the US Customs Service P-3 aircraft, and will be fielded in an executive jet for export in 1999.

Contractor
Lockheed Martin Tactical Defense Systems

VERIFIED

P-3C upgrade programmes

US Navy P-3C Anti-surface warfare Improvement Program (AIP)
The AIP effort involves system upgrades to the P-3C Update III aircraft to improve its capability to support a range of naval operations including: ASuW; over-the-horizon targeting and command, control, communications and intelligence activities.

Major upgrades include non-acoustic sensor enhancements (radar, ESM, IR detection system); survivability improvements; data management system upgrades and a communications upgrade (ICS, VHF, tactical communications, UHF, SATCOM).

Equipment to be fitted in the programme includes: the AN/APS-137B(V)5; the AN/AVX-1 Cluster Ranger electro-optical sensor and a missile approach warner. Provision for Link 16 to be added is a preplanned product improvement.

Operational status
The exact funding status of the programme is uncertain, but it could be applied to any number required of the 146 aircraft that are already modified to Update III standard. It is understood that the P3-AIP fleet update is required by 2005 and that it is expected to be in use until the new Large Land-Based Aircraft (LLBA) becomes available between, approximately, the years 2015 and 2020.

Royal Australian Air Force (RAAF) P-3C Upgrade Program
See International section for details of this programme.

Royal Norwegian Air Force (RNoAF) P-3C Upgrade Improvement Program (UIP)
The RNoAF P-3C UIP will upgrade four aircraft with a variant of the US Navy's AIP tailored for Norwegian requirements; upgrades include improvements to: non-acoustic sensors (IRDS, new ESM, APS-137 ISAR/SAR radar); SATCOM and new ICS; new DMS.

Operational status
Modification of the four aircraft began during 1997, with final aircraft delivery in 1999.

Contractor
Lockheed Martin Tactical Defense Systems.

VERIFIED

STacSAR™ Small Tactical Synthetic Aperture Radar

The Small Tactical Synthetic Aperture Radar (STacSAR™) is a lightweight (less than 29.5 kg) J-band SAR system produced for use on unmanned aerial vehicles or light manned aircraft. It uses the latest radar technology and processing techniques. STacSAR™ provides fixed target imagery in both spotlight and search modes, as well as moving target indications. Onboard processing is included, allowing real-time exploitation. STacSAR™ is a modular design with inherent flexibility. System parameters are programmable under software control. SAR modes include both a strip map search mode and a higher-resolution spotlight mode along with a moving target indicator mode. The user can specify mode parameters in selecting range and azimuth resolutions and swath widths. Transmitter pulsewidth and repetition frequency can also be varied to achieve maximum average power for selected modes. The system is designed for datalink operation where imagery can be downlinked in real time to user ground stations. STacSAR™ consists of two subsystems: an antenna/gimbal assembly and an electronics package.

The radar electronics package used in STacSAR™

Operational status
The system is performing fight demonstrations aboard manned light aircraft.

Specifications
Wavelength: J-band (17 GHz)
Resolution: 0.5-50 m programmable
Swath: 500-5,000 m programmable
Slant range: 10 km (with 12 W transmitter)
Weight: < 29.5 kg

Contractor
Lockheed Martin Tactical Defense Systems.

VERIFIED

AN/ALQ-122 ECM system

The AN/ALQ-122 is the power-managed radar jamming system in use on the B-52 bomber. The equipment searches for, acquires and tracks radar threat signals, and generates narrowband, low-duty cycle ECM signals to deny range and azimuth to enemy radars. Signals are linear amplified by solid-state amplifiers in the aircraft.

Operational status
In operational use on B-52 aircraft. An advanced mixed-mode deception jammer is reported to be in development by Motorola for the US Air Force, based on work on this equipment and the Northrop Grumman Corporation, Electronic Sensors and Systems Sector AN/ALQ-131(V) ECM pod.

Contractor
Motorola Inc, Government & Systems Technology Group.

VERIFIED

AN/APS-131 / APS-135 side-looking radars

The AN/APS-131 is a Side-Looking Airborne Radar (SLAR) used for the detection of ships and boats, for search and rescue, and for the detection of oil pollution. The aircraft installation consists of six main subassemblies: antenna, receiver/transmitter, synchroniser, amplifier, recorder/processor/viewer and control unit.

An area of up to 200 km on either side of the aircraft can be mapped when both arrays are in use and one can be selected if mapping of only one side is required. A range control determines the width of the target area to be mapped and presented on the photo-radar map. This control has four settings corresponding to 25, 50, 100 and 200 km wide scans by each antenna. When used in conjunction with the antenna switch to select either left, right or both arrays, maps corresponding to four standard scales can be presented on the display at 1:250,000, 1:500,000, 1:1 million and 1:2 million. Radar and aircraft operational data is annotated on the film. This data, together with the latitude and longitude printed on the film, helps the measurement of map co-ordinates for any feature observed on the radar image.

Operational status
The AN/APS-131 was developed under contract to the US Coast Guard and is in service on HU-25A aircraft. A similar radar, the AN/APS-135 is used on the US Coast Guard HC-130.

Contractor
Motorola Inc, Government & Systems Technology Group.

VERIFIED

Side-Looking Airborne Modular Multimission Radar (SLAMMR)

A version of the AN/APS-131/135, the SLAMMR, is offered internationally for a wide variety of maritime surveillance and border patrol applications. SLAMMR has been delivered, in a maritime surveillance configuration on the Boeing 737-200 aircraft, to Indonesia, and also to another country in a border patrol configuration on a C-130H aircraft.

The radar is designed for a variety of applications such as ice patrol, fishery protection, anti-drug smuggling, mapping and search and rescue. In particular, its performance over the sea is sufficiently good to reveal details of oil spills from tankers, and provide maritime reconnaissance data to a maximum range of 185 km on either or both sides of the aircraft. The sensor uses a planar array 2.5 or 5 m long, mounted below the aircraft or along the side of the fuselage, feeding an operator's control/display console. The two antennas are arranged to point perpendicular to the aircraft longitudinal axis. From this position the two antennas can survey the sea or ground on either side of the aircraft out to the horizon.

A notable application of SLAMMR is the patrol maintained by the US Coast Guard to provide iceberg warning to shipping. Before the advent of side-looking radar, iceberg observations were conducted visually and therefore could only be performed in good visibility. Now with SLAMMR, the US Coast Guard HC-130 Hercules can maintain a much more consistent monitoring.

Operational status
No longer in production. In service on Lockheed Martin C-130 and Boeing 737 aircraft.

Contractor
Motorola Inc, Government & Systems Technology Group.

VERIFIED

AN/APN-232 Combined Altitude Radar Altimeter

The Combined Altitude Radar Altimeter (CARA) is an all-solid-state 0 to 50,000 ft FM/CW radar altimeter system operating at a nominal frequency of 4.3 GHz. It consists of a receiver/transmitter, antennas and indicators and offers inherent low probability of intercept and anti-jam capability as well as conventional analogue and digital outputs for aircraft avionic systems.

Low Probability of Intercept (LPI) performance is achieved by control of the power output so that the transmitted power is the least amount required for signal acquisition and tracking. The system mechanisation automatically adjusts the required transmitter power to maintain normal system operation over varying terrain, aircraft altitude and attitudes. The FM/CW technology, operating over a 100 MHz bandwidth, provides inherent spread-spectrum capability which further reduces detectability.

The AN/APN-232 is easily maintained and offers low life cycle cost. The equipment incorporates highly modular packaging. The assemblies, subassemblies and components are 100 per cent screened and tested to stringent military standards. Ease of maintenance is achieved by using proven microprocessor-based digital display fault monitoring. These elements combine to produce a predicted system MTBF of greater than 2,000 hours. Low life cycle cost is obtained from high system MTBF, automatic self-test with fault isolation to both the line- and shop-replaceable units and a two echelon maintenance concept which maximises the use of shop-replaceable units that may be discarded rather than repaired.

Specifications
Dimensions:
(receiver/transmitter) 88.9 × 160 × 222.2 mm
(indicator) 82.55 × 82.55 × 104.1 mm
(fixed-wing antenna) 101.6 × 101.6 × 5.3 mm
(rotary-wing antenna) 142.2 × 158.7 × 5.3 mm
Weight:
(receiver/transmitter) 4.76 kg
(indicator) 1.13 kg
(fixed-wing antenna) 0.23 kg
(rotary-wing antenna) 0.34 kg
Power supply: 28 V DC, 100 W max
Accuracy:
(analogue) ±2 ft ±2%
(digital) ±2 ft (0-100 ft), ±2% (100-5,000 ft), ±100 ft (5,000-10,000 ft), ±1% (>10,000 ft)

Operational status
In production, with over 10,000 APN-232s delivered. CARA is the US Air Force's standard altimeter.

Further technology enhancements are in development as kit module replacements to the basic system. Among these is a tested capability to track terrain and features simultaneously.

Contractor
NavCom Defense Electronics Inc.

VERIFIED

The Navcom AN/APN-232(V) CARA system

AN/ARN-84 Tacan

Operational features of the ARN-84 Tacan include X-Y position fixing, air-to-ground link with any standard Tacan station, air-to-air bilateral ranging, receive only for bearing and beacon identification, radio frequency datalink, air-to-air bearing and inverse mode. The system provides automatic antenna switching together with digital and analogue range and bearing outputs to aircraft instruments and computers. The solid-state design gives a level of reliability previously unmatched according to NavCom. The range is 300 n miles.

Operational status
In service. The system is fitted to 15 types of US military aircraft including the A-4M, A-7E, C-130, E-2C, F-14A, S-3A, AH-1T and SH-3H. Over 3,200 ARN-84s have been produced. The AN/ARN-84 is no longer in production.

Contractor
NavCom Defense Electronics Inc.

VERIFIED

AN/AAR-54(V) Passive Missile Approach Warning System (PMAWS)

AN/AAR-54(V) installation on the pylon integrated dispenser system (PIDS+) of an F-16 MLU aircraft
1999/0051266

Northrop Grumman and the US Department of Defense (DoD) have developed and tested a PMAWS which can be used on a wide variety of aircraft and ground vehicles. The system is available to provide internal advanced missile warning for tactical and transport aircraft, helicopters and armoured fighting vehicles. It is also used to provide identification, missile tracking information and target cueing to Directed Infra-Red CounterMeasures (DIRCM) systems. Because of the adaptive design of AAR-54(V), all applications can use common hardware and software.

The fine 1° angle of arrival discrimination capability of AAR-54(V) provides greatly reduced false alarm rates as well as detection ranges nearly double that of existing ultra-violet systems. Passive time to intercept is another system feature providing optimum cueing to

The AN/AAR-54(V) Passive Missile Approach Warning System (PMAWS) consists of up to six sensors and an electronic unit

countermeasures dispensers. AAR-54(V) provides all-weather, all-altitude operation while protecting against multiple simultaneous engagements in dense clutter environments.

The system consists of wide field of view, high-resolution ultra-violet sensors and a modular electronics unit. From one to six sensors can be utilised, providing up to full spherical coverage. Full in-flight built-in test and fault isolation to a single sensor or electronics line-replaceable unit provides level-2 maintenance.

Operational status

Northrop Grumman, the US DoD and the UK Ministry of Defence are believed to have completed AAR-54(V) design verification testing. This trials programme included several live fire demonstration programmes with the warner installed on QF-4 drones, a cable car test rig and ground vehicles. The system has also been integrated into and demonstrated with a DIRCM system and the AN/ALQ-131 electronic countermeasures system. In April 1995, the UK's Ministry of Defence and the US Special Operations Command awarded Northrop Grumman a US$35 million engineering, manufacturing and development/production contract for the AN/AAQ-24(V) DIRCM system which includes AAR-54(V). In this application, the equipment provides missile detection, lethal missile declarations and fine angle of arrival hand-off to the DIRCM. AAQ-24(V) is scheduled to be installed on 14 types of UK and US fixed-wing aircraft and helicopters.

Outside the AAQ-24(V) application, AAR-54(V) has been selected by the Portuguese Air Force for use on a percentage of its C-130H fleet. Here, a six-sensor AAR-54(V) configuration is teamed with the BAE Systems North America Inc AN/ALE-40 dispenser system and TERMA's Electronic Warfare Management Unit to create a defensive aids suite.

The air forces of Denmark, the Netherlands and Norway have selected the AN/AAR-54(V) PMAWS for their F-16 Mid-Life Update (MLU) programme. The configuration is controlled by a Terma Electronik Electronic Warfare Management System (EWMS) and installed into TERMA Industries Gvenaa AS Pylon Integrated Dispenser System (PIDS+). The F-16 MLU installation utilises a six-sensor PMAWS configuration: three sensors in each of two pylons. This solution requires no costly aircraft wing-to-wing wiring modifications and, with this implementation, the pylons can be used, mission by mission, on any F-16 MLU aircraft, without fleetwide modifications.

Northrop Grumman supported integration activities and a total of 20 flights to prove the installation, was completed in September 1998. Testing was undertaken in association with the three procuring air forces; in addition, the Belgian Air Force participated in an observer role. Missile threat data was displayed on the AN/ALR-69 RWR display. PMAWS alert information was routed to the onboard AN/ALE-40 countermeasures dispenser.

The AN/AAR-54(V) system has also been selected, with the AN/ALE-47 dispenser, for the Royal Australian Navy upgraded S-70B Seahawk helicopters.

Contractor

Northrop Grumman Corporation, Electronic Sensors and Systems Sector.

UPDATED

AN/ALQ-119 noise/deception jamming pod

Initiated as project QRC-522 in 1970 and one of the most numerous jamming pods in service with the US Air Force, the AN/ALQ-119 was one of the first dual-mode noise and deception jammers to appear. It was used initially on the F-4 Phantom, but was subsequently adopted for the Fairchild A-10 and Lockheed Martin F-16 as well as the F-111 and F-15. The system has a three-band frequency range transmitter which covers the terminal threat range. Both noise and deception jamming modes can be employed. Each pod has dual-mode travelling wave tube emitter elements. The pod has the gondola cross-section introduced by Northrop Grumman on the AN/ALQ-101 pod.

The AN/ALQ-119(V)15 model is recognisable alongside earlier versions of the pod by the addition of a radome below the front end of the gondola portion. Many earlier versions of the pod with the US Air Force and other air forces were upgraded to this standard. Features of the (V)15 version include automatic control of power radiated, frequency selection and signal type. A shorter-body version, the AN/ALQ-119(V)17, is also operated by the US Air Force.

With the appearance of new air-to-air and ground-to-air threats, the US Air Force, in the early 1980s, instituted a major improvement programme, designated AN/ALQ-119A (Seek Ice). Raytheon was appointed to strip out and replace much of the existing electronics and incorporate current Rotman lens technology. The changes facilitate reprogramming and make for greater reliability and ease of servicing. The upgraded pod is designated AN/ALQ-184.

EADS (formerly DaimlerChrysler Aerospace AG) has developed the AN/ALQ-119 pod for use by German Air Force F-4F Phantom aircraft; this variant is designated AN/ALQ-119GY (see EADS AG entry).

Operational status

In service with F-4, F-16 and A-10 aircraft. The total production run exceeds 1,600 units. Sets are operated by Germany and Israel. Superseded in US Air Force service by the AN/ALQ-131 system.

Contractor

Northrop Grumman Corporation, Electronic Sensors and Systems Sector.

UPDATED

AN/ALQ-131 noise/deception jamming pod

The AN/ALQ-131 is an automatic, highly reliable modular self-protection system. It was designed to provide advanced broadband coverage against all types of modern radar-guided weapons. The ALQ-131 is carried externally on a variety of front-line, high-performance aircraft. It is certified on front-line aircraft such as the A-7, A-10, C-130, F-4, F-15, F-16, and F-111.

The modular design of the pod structure and electronic assemblies, plus its central computer software architecture, enable the ALQ-131 system to adapt quickly to a broad spectrum of EW applications. This feature proved valuable in the Gulf War, where ALQ-131s provided over 48 per cent of the US Air Force tactical aircraft EW self-protection. The Block II version experienced no combat losses in over 12,000 combat sorties. AN/ALQ-131 pods also provided protection during Bosnian operations, on US Air Force F-16, A-10, C-130 and Royal Netherlands Air Force F-16.

Increased effectiveness can be achieved by incorporating various mission modules, including missile warning systems, offboard and towed countermeasures dispensers and advanced technique generators. This capability allows maximum flexibility in countering all threat types.

The basic structure elements of the pod are modular canisters that provide structural support, cooling and environmental protection. Each canister is an I-beam structure which also serves as a cold plate. Both sides of the I-beam form equipment bays into which functional equipment modules are mounted. The modules can also be mounted in lower equipment bays on the bottom surface of the I-beam. Using common mounting techniques, each canister can accommodate several equipment modules which can be removed directly from the bays without disassembly of the pod. The system is 2.83 m long and its weight ranges from 260 to 324 kg.

The functional organisation of the system is centred around the Interface and Control (I/C) module which contains a programmable digital computer as the system controller. The modules required for a given configuration are connected to the I/C by a digibus that carries all sensor and control data. A memory loader/verifier allows operational flight and mission-specific program software to be loaded into the pod on the flight line in less than 15 minutes.

The I/C module also contains a digital waveform generator that can permit up to 48 simultaneous waveforms for deception modulation. When any ECM technique requires a deception waveform, the latter's values are transmitted to the onboard equipment via a waveform distribution bus.

Maintenance of the ALQ-131 is based on the pod's Centrally Integrated Test System (CITS) which provides a comprehensive functional check of system operation, both in flight and on the ground. During flight, the CITS continuously monitors the operational status of the equipment, including repeater channel modulation, high voltage of the TWTs, noise power output, primary bus voltage and the integrity of the computer memory.

The F-16 carries the AN/ALQ-131 pod under the fuselage

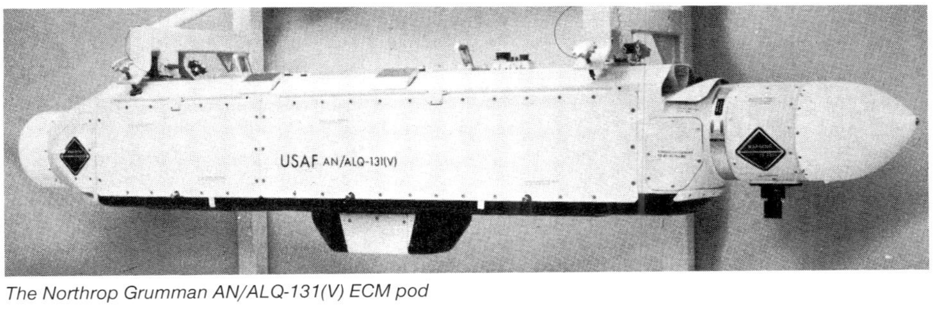

The Northrop Grumman AN/ALQ-131(V) ECM pod

The pod is a software reprogrammable system which allows a tactical commander to tailor the ALQ-131's responses for the mission requirements. Utilising the ALQ-131's ability to be flight line reprogrammed, mission-specific data can be created in response to threat changes and, by using a memory loader verifier, changes loaded into the pod's digital computer.

Another capability of the system is its self-contained power management feature. This is included in a receiver/processor module which detects radar threats, measures their key parameters and performs weapon type and operational mode identification. This information is then used by the ALQ-131's I/C module to select optimum jamming techniques automatically and tailor their parameters for countering all detected threats. Numerous threats can be countered simultaneously, each with independent techniques, using the receiver/processor's PRI tracking capability.

Specifications

Dimensions:
(one band shallow) 2,210 × 297.2 × 533.4 mm
(two band shallow) 2,819.4 × 297.2 × 533.4 mm
(three band deep) 2,819.4 × 297.2 × 635 mm
Weight:
(one band shallow) 175.54 kg
(two band shallow) 263.1 kg
(three band deep) 298.92 kg
Power supply: 115 V AC, 400 Hz
Reliability: 125 h MTBF

Operational status

More than 560 Block I and 460 Block II pods have been delivered to the US Air Force. A production programme to update the original Block I configuration to the latest Block II version is under way. Over 370 pods have been procured by the air forces of Bahrain, Belgium, Egypt, Israel, Japan, Netherlands, Norway, Pakistan, Portugal, Singapore and Thailand. International deliveries are continuing.

In October 1993, the US Air Force completed EMI/EMC testing of the ALQ-131 modified with an AN/ALQ-153 active missile warning system and AN/ALE-47 countermeasures dispenser on an F-16 aircraft. In April 1994, the ALQ-131 was demonstrated with a fully integrated AN/AAR-54 passive missile warning system and AN/ALE-47. In August 1998, the ALQ-131 was demonstrated with a high-power fibre optic Towed Radar Decoy (TRD) on a C-130 aircraft.

A self-contained external ECM suite configuration of ECM pod, missile warning system and countermeasures dispenser is being considered, to provide tactical aircraft with protection against IR-guided missiles while eliminating Group A aircraft modifications.

The pod has been combat proven in Bosnian operations on US Air Force A-10, C-130 and F-16 aircraft and Royal Netherlands Air Force F-16s.

New contracts for the AN/ALQ-131 system continue to be issued by the US Air Force, both for continued customer support and for system upgrades for both US Air Force and Foreign Military Sales (FMS) requirements. Recent awards for Block II conversion kits have been made on behalf of the Royal Norwegian Air Force (16 kits) in June 1999, and the Egyptian Air Force (39 kits) in January 2000.

In 1999, Northrop Grumman was contracted by Mitsubishi Electric Corporation to provide ALQ-131 production kits for local assembly, test and delivery to the Japanese Air Self-Defence Force.

Contractor

Northrop Grumman Corporation, Electronic Sensors and Systems Sector.

UPDATED

AN/ALQ-135 jamming system

The AN/ALQ-135 Internal Countermeasures Set (ICS) is a component of the Tactical Electronic Warfare System (TEWS) for the US Air Force F-15; it is installed in various configurations in the F-15A, C, and E variants and operates with the AN/ALR-56 radar warning system and the AN/ALE-45 countermeasures dispenser.

The AN/ALQ-135 is an advanced jamming system which uses high-powered transmitters. All equipment is mounted internally and jamming system management is self-contained in the F-15A, C and E.

The Northrop Grumman AN/ALQ-135 jamming system is carried internally in the F-15

The basic ALQ-135 (Band 1.5 and Band 3 variants) consists of seven LRUs plus appropriate waveguides and antennas. Two LRUs are receivers, four are amplifiers and one is a preamplifier. Over the years the system has continued to evolve along with the capabilities of the aircraft and changes in the threat. While maintaining commonality with the original system and support electronics, the AN/ALQ-135 has been updated to include full band coverage and extremely effective jamming techniques flexibility.

For installation in the two-seat F-15E, the frequency range of the system has been expanded by the addition of a Band 3 transmitter/receiver/processor and power amplifier for the aft radiating antenna. Bands 1 and 2 have individual jammers in older F-15s, but in the F-15E variant they are being combined into a Band 1.5 low-band jammer that is half the size of the bands 1 and 2 jammer. From a logistical standpoint, the Band 1.5 system is more than 70 per cent common with the existing Band 3 system, thereby maximising commonality and minimising support costs. Major features of the AN/ALQ-135 system are:

Receiver: channelised, multimode for fast response and activity detection throughout the band, variable bandwidth superheterodyne with IFM and sophisticated blanking/lookthrough.

Control: 20 microprocessors in a federated architecture with parallel processing for fast system response and flexibility/reprogrammability.

Techniques: demonstrated capability for emerging/developing radar threats with advanced techniques and coherent response.

Operational status

In service with F-15. Northrop Grumman has produced more than 1,750 AN/ALR-135 systems.

The band 1.5 subsystem underwent flight tests in 1998. Approval for upgrade of 170 AN/ALQ-135 systems to Band 1.5 configuration was given by the US Congress in November 1998; this indicates that not all the US Air Force's 220 F-15Es will receive the upgrade.

The Band 1.5 version is now in full production, against an order for 17 Band 1.5 kits issued by the US Air Force in late 2000, with a requirement for work to be completed by December 2001.

As of January 2001, US Airforce technicians, assisted by field engineers from Northrop Grumman Corporation's Electronic Sensors and Systems Sector, had successfully completed the first operational installation of the Band 1.5 ECM subsystem in an F-15E aircraft. Following installation, the AN/ALQ-135 Band 1.5 was fully field-tested to verify successful threat countering using Northrop Grumman's Virtual System Analyser (VSA), which was designed and developed specifically for the ALQ-13 system. In addition to testing all of the electronic components, the VSA is used to inject threats into the system to ensure operational viability. The project is managed by the F-15 Development System Program Office at Wright-Patterson AFB, Ohio. To date, the AN/ALQ-135 has operated in more than 6,600 hours of combat, beginning with Desert Storm and continuing through recent operations in the Balkans and Kosovo without a single aircraft loss due to threats covered by the ECM system.

Contractor

Northrop Grumman Corporation, Electronic Sensors and Systems Sector.

UPDATED

AN/ALQ-153 threat warning system

The AN/ALQ-153 threat warning system is a frequency/PRI-agile pulse Doppler radar that detects and discriminates between approaching missiles and aircraft. It is a range-gated Doppler system capable of detecting the approach of the entire spectrum of hostile aircraft and land-, sea- and air-launched missiles. Range and time-to-impact are computed automatically, updated continuously and compared with in-flight selectable timing criteria so that automatic countermeasures dispensing and aircraft manoeuvres can be generated. The system continuously displays the most imminent threat while tracking multiple missile engagements. Both 360° all-aspect and tail-only coverage versions are available.

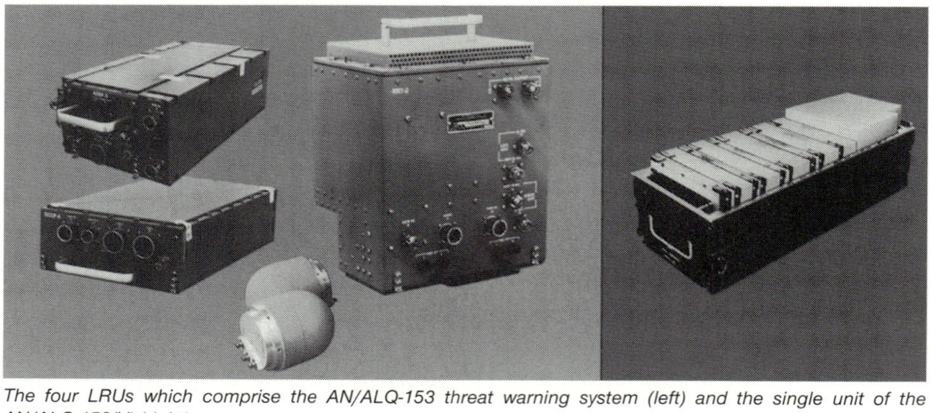

The four LRUs which comprise the AN/ALQ-153 threat warning system (left) and the single unit of the AN/ALQ-153(V) (right)

Specifications

AN/ALQ-153

Dimensions:

(radar receiver transmitter) 408.9 × 251.5 × 342.9 mm
(analogue data signal processor) 457.2 × 254 × 160 mm
(digital data signal processor) 457.2 × 279.4 × 121.9 mm
(antennas) 177.8 length × 152.4 mm diameter

Weight:

(radar receiver transmitter) 10.34 kg
(analogue data signal processor) 21.55 kg
(digital data signal processor) 16.78 kg
(antennas) 1.59 kg

AN/ALQ-153(V)

Dimensions: 660.4 × 203.2 × 152.4 mm
Weight: 31.75 kg

Operational status

Production is complete with the delivery of 321 systems for US Air Force B-52G/H aircraft. Spare LRU production was completed in 1991, bringing the total number of systems up to 400.

The 360° all-aspect AN/ALQ-153(V) has undergone extensive flight testing on tactical aircraft, while the tail-only system has successfully demonstrated in-flight missile detection, deception and low false alarm performance on the B-1B and B-52 in low-level, high-clutter environments. Using modern components, options are being considered to improve reliability and reduce volume to approximately 0.023 m^3 and weight to less than 32 kg.

In October 1993, the US Air Force completed EMI/EMC testing of the ALQ-153 installed in an ALQ-131 ECM pod on an F-16 aircraft.

Contractor

Northrop Grumman Corporation, Electronic Sensors and Systems Sector.

UPDATED

AN/ALQ-155(V) B-52 power management system

The AN/ALQ-155(V) countermeasures power management system forms part of the reliability and maintainability programme to improve the defensive avionics of the B-52 aircraft. It provides integral set on receivers for each jamming transmitter, plus increased effective radiated power density through accurate frequency set on. The system is a power management evolution for the ALT-28(V) active ECM set providing automated hand-off from the ALR-46 radar warning receiver with near-instantaneous jammer response. It is computer managed and field programmable. The system contains automatic frequency control in all modes and a wide variety of ECM techniques that are automated, semi-automated or manual. A 12-transmitter upload capability is provided.

A variety of improvements is incorporated in the ALQ-155(V) including frequency agility against multiple threats, pulse repetition interval trackers, cover pulse jamming techniques, false target generation through pseudo-random noise, coherent and incoherent jamming and downlink jamming. It has a hybrid IFM receiver and central receiver capability, programmable noise optimisation, increased pulse-up power for CW to pulse operations and compatibility with the AN/ALQ-172 deception I/J-band jammer.

Northrop Grumman has been placed on contract for a reliability and maintainability upgrade to solve obsolete component issues within the programme while improving performance through the use of non-developmental items from Northrop Grumman Corporation's most advanced countermeasures systems.

Operational status

Production complete. About 300 ALQ-155s were produced.

Contractor

Northrop Grumman Corporation, Electronic Sensors and Systems Sector.

UPDATED

AN/ALQ-162 countermeasures set

Development of the ALQ-162 was started in 1979 under contract to the US Navy; the US Army later joined the programme. It is a small, reprogrammable radar jamming system which can be supplied with its own receiver/ESM management processor, or made compatible with many existing types of radar warning receiver processor systems. The system is software-programmable to meet new threats, and can be installed in pod, pylon or internal fit configurations.

ALQ-162 is fully integrated with the following systems: radar warning receivers (AN/APR-39, -43, -45, -46, -66, -67 and -69); pulse jammers (AN/ALQ-126 (A and B) and AN/ALQ-136); CM dispensers (AN/ALE-36, -40 and M-130).

The AN/ALQ-162 Pulse Doppler (PD) upgrade, also known as Shadowbox II, has advanced pulse Doppler capabilities to counter emerging threats and provide increased aircraft survivability. No changes to form, fit, or aircraft wiring is required to implement this upgrade.

Specifications

Dimensions: 161 × 184 × 420 mm
Weight: 18 kg
Power supply: 115 V AC, 400 Hz, 3 phase, 650 W
Heat dissipated: 480 W
Reliability: >300 h MTBF demonstrated

Operational status

ALQ-162 installations include: US Navy - A-4M, A-7E, AV-8B, F-4S and RF-4B; US Army - EH-1, EH-60, OV-1D, RC-12D, RU-21, RV-1D; US Army Special Operations Command (ASOC) - MH-47D/E and MH-60K/L; NATO - C-130, F-16 and Saab F-35 Draken, more than 650 installations. Enhancements planned by Northrop Grumman for the ALQ-162 include:

(1) Power Plus: utilising advanced microwave power technology, which will be available in conjunction with the Shadowbox II PD upgrade, the effective radiated power will more than double;
(2) Increased Threat Handling: the current ALQ-162 defeats multiple threats in the same band simultaneously; incorporating new technology will allow the system to defeat multiple threats in different bands simultaneously;
(3) Additional Pulse Capability: the current Shadowbox II defeats CW and PD threats. When combined with available technologies currently used in other Northrop Grumman EW products - Tactical Radar Jammer (TRJ) and AN/ALQ-135 - the ALQ-162 will be a more capable system able to defeat CW, PD and pulse threats;
(4) Potential Growth: other growth areas include integration with towed radar decoys and integration with DIRCM.

Contractor

Northrop Grumman Corporation, Electronic Sensors and Systems Sector.

VERIFIED

AN/APG-66(V) Series radars

The AN/APG-66(V) Series radars are multimode, versatile systems installed on 16 different airborne platforms operating in 20 countries. The APG-66(V) Series provides improved performance, functionality, reliability and maintainability over its predecessor, the AN/APG-66 which was designed for the Lockheed F-16. Over 2,300 of the original version of the radar have been produced and deployed worldwide.

The AN/APG-66(V) Series radar is based on the multinational F-16 Mid-Life Update (MLU) APG-66(V)2 radar. Improvements made to incorporate the latest technology include a newly developed signal data processor, higher-power transmitter, low-power RF speed and sensitivity and faster antenna phase shifting.

Consolidating the functions of the radar computer and digital signal processor reduces system size, weight, power and cooling requirements, while providing seven times greater processing speed and 20 times greater non-volatile memory than the AN/APG-66 system. A Doppler correlator reduces the false alarm rate by classifying radar returns as either ground vehicles, weather, mutual interference or sidelobe returns. The processing and fully programmable graphics capability in the signal data processor provides numerous options for radar operational mode growth, colour display support and growth capabilities to perform combined multisensor processing functions such as for FLIR or missile warning.

Significant advantages over the original AN/APG-66 have been realised, including twice the operational range, 40 per cent higher reliability and 16 per cent reduction in system weight. A four to one improved ground-map resolution has also been achieved.

AN/ALQ-162 installed in the outboard pylon on F/A-18 0018248

Greater situation awareness is provided by Track-While-Scan (TWS), Situation Awareness Mode (SAM), multitarget track and improved protection against electromagnetic interference. Support for multiple employment of up to six missiles is provided and the radar is compatible with a variety of Beyond Visual Range (BVR) air-to-air missiles and anti-ship missiles.

The radar provides air-to-air and air-to-surface modes in all weathers against all-aspect targets in look-up/shoot-up and look-down/shoot-down engagements, including high-clutter environments. The flexible multi-mode fire-control sensor is designed to provide integrated target and navigation data. Air-to-air modes provide the capability to detect and track multiple targets in the presence of ground clutter and electromagnetic interference. Air-to-surface modes provide extensive mapping for navigation, target detection, tracking and ranging.

The upgrade of the air-to-air capability of both individual aircraft and the F-16 units provides a significant air superiority margin that will also include a degree of air space management, incorporating both control aircraft and ground station datalinks providing real-time target information. Operating modes include 10 target TWS, Range-While-Search (RWS) look-up and look-down, six shot AMRAAM against six targets, medium-resolution DBS (Doppler Beam-Sharpening) and enhanced ECCM capabilities.

Key radar features in air-to-air are search and track mode incorporating TWS, RWS and SAM functions. SAM offers a better track quality than TWS on multitarget track, plus the capability to track one or two targets while maintaining RWS scan volume. The TWS mode can, in addition to providing information on up to 10 targets, provide search information on up to 64 other targets.

There is a multitargeting capability for up to 10 target tracks, two target SAM tracks, single target track and BVR engagements. Inclusion of a BVR capability is a key point in order to provide MLU-modified aircraft with the ability to use medium-range radar-guided missiles such as Sparrow, AMRAAM and MICA.

Air Combat Manoeuvre (ACM) features automatic acquisition of targets at short range in four predetermined scan patterns. Other information available includes the search altitude display which provides a reference to the altitude of all radar search returns and an indication of target relative groundspeed as well as an indication of the target aspect.

In the air-to-surface mode the APG-66(V)2 radar provides a significant improvement in both ground attack and anti-ship roles, especially in conditions of extreme clutter. Functions exist for the provision of a radar map video for target detection and navigation as well as providing an improved capability against shadowed or hidden targets. Using the ground map as a base, the APG-66(V)2 is also able to provide an overlay of navigation and reference points especially for rendezvous locations, weapon release points and waypoints. Operating modes include real-beam map, enhanced ground map, medium and coarse Doppler beam-sharpening, fixed target track, ground moving target indication and SAM in ground map interleaved mode.

Continuous updating of the navigation system is also enhanced through air-to-ground ranging which, combined with mission data information, provides confirmation of reference points along a proposed mission route and information regarding terrain-avoidance for obstacles in either overfly or avoidance modes that can be preprogrammed into the mission data package.

Additional features include moving ground target tracking that can also be superimposed on an improved definition ground map and, in both ground and sea modes, the ability to detect and fix on targets in open country or at sea. The sea mode provides a capability for the accurate detection and tracking of targets in both low and high sea states. The long-range sea mode provides a detection range reported to be as much as 148 km.

The radar may be readily adapted to a platform through expeditious software modifications supported by the fast Express software development environment and by selecting one of the nine antennas available or by resizing the array as necessary.

Features providing for integration flexibility include NTSC and PAL/CCIR video formats, 480 × 480 colour video resolution with 64 grey shades or 256 colours, fully programmable multifunction displays, MIL-STD-1553B remote terminal or bus controller, discrete input/output for HOTAS, CW illumination LRU and RAM or ECS cooling air capability.

Possible growth capabilities include weather awareness, blind let-down, wide swath map at medium altitude, terrain-following/terrain-avoidance, synthetic aperture radar, internal FLIR, missile warning and ESM processing.

Specifications

Volume:
(APG-66(V)1) 0.102 m^3
(APG-66(V)2) 0.097 m^3
(APG-66H) 0.082 m^3
(APG-66T) 0.08 m^3
Weight:
(APG-66(V)1) 134.3 kg
(APG-66(V)2) 115.9 kg
(APG-66H) 107.7 kg
(APG-66T) 98.4 kg
Power supply: 115 V AC, 3 phase, 400 Hz, 3,209 VA
28 V DC, 115 W
Frequency:
(APG-66(V)1) 6.2-10.9 GHz
(APG-66(V)2) 8-10 GHz
Search angle: 120° (azimuth and elevation)
Azimuth scan: ±10°, ±30°, ±60°
Elevation coverage: 1, 2 or 4 bar
Range scale: 10, 20, 40, 80 n miles (19, 37, 74, 148 km)
Electronic protection: multiple features, EMI pulse editor, fast phase shifting, frequency agility.
Maintainability: 5 min MTTR
Reliability: >210 h MTBF (over 300 h MTBF demonstrated)

Operational status

The initial variants in the AN/APG-66 Series, the APG-66T and APG-66H, have been deployed in the AT-3 trainer and BAE Systems Hawk 200 aircraft respectively. The APG-66H is currently in production.

The APG-66(V)1 was developed for the US Air Force. The contract for 270 systems was completed in 1991.

Development of the APG-66(V)2 for the European F-16A/B Mid-Life Update (MLU) programme was initiated in 1992. The programme will update 48 Belgian, 61 Danish, 136 Netherlands and 56 Norwegian air force AN/APG-66 radars to AN/APG-66(V)2 standard. DT&E was conducted at Edwards AFB in May 1995. Test missions were flown by the Netherlands Air Force in Autumn 1996. Production radar software was completed in January 1997.

Flight trials for both the APG-66(V)2 and (V)3 radars were conducted in 1994. These successfully demonstrated doubled operational detection range; false alarms reduced by a factor of 10 even in the presence of multiradars and severe ECM; simulated six-shot AMRAAM capability and ground-mapping range improved to 148 km.

The APG-66(V)3 is in production for the Taiwanese Air Force F-16 and 157 systems are currently under contract. Deliveries began in November 1995. This variant has a CW illumination capability for the AIM-7 Sparrow missile.

Northrop Grumman is also under contract to deliver 30 ARG-1 radars – a version of the APG-66(V) customised for the A-4.

In May 1999, US FMS ordered a further 49 AN/APG-66(V)2 upgrade kits, 24 for Belgium and 25 for Portugal.

Northrop Grumman has marketed Commercial Off-The-Shelf (COTS) derivatives of the APG-66(V)2 and APG-66(V)3 under the names APG-66(V)X and APG-NU.

Contractor

Northrop Grumman Corporation, Electronic Sensors and Systems Sector.

VERIFIED

AN/APG-68 radar for F-16 aircraft

The AN/APG-68 is a coherent, multimode, digital fire-control sensor that provides a complement of 25 all-weather air-to-air and air-to-surface modes with superior dogfight and weapons-delivery capabilities. The air-to-air modes provide the capability to detect, track and engage targets at all aspects and at all altitudes even in the presence of ground clutter. Target information in the air-to-air modes is presented by a synthetic video on a 'clean scope' display. Air-to-surface modes provide extensive mapping, target detection, location and tracking, as well as navigation capability.

The AN/APG-68 radar fitted to US Air Force F-16C/D aircraft 0018074

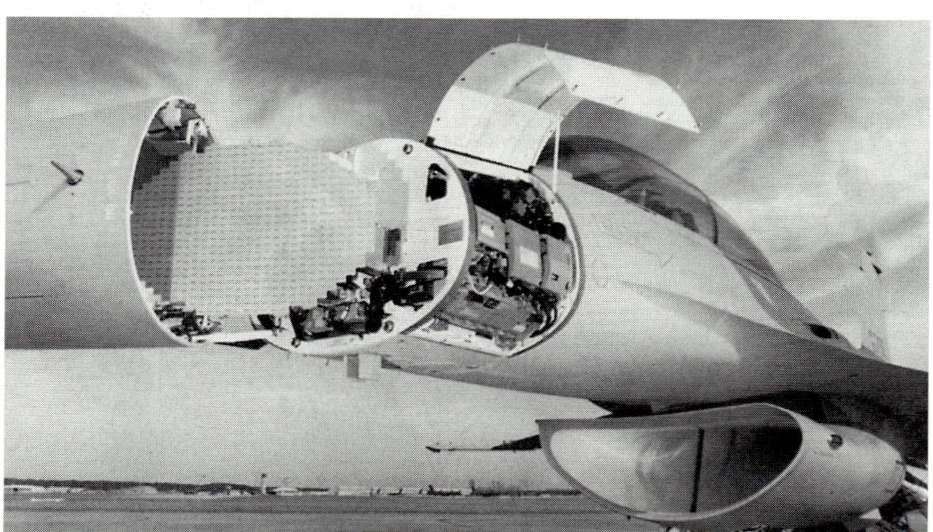

AN/APG-66(V)2 for European participating governments' F-16A/B mid-life upgrade programme 0001217

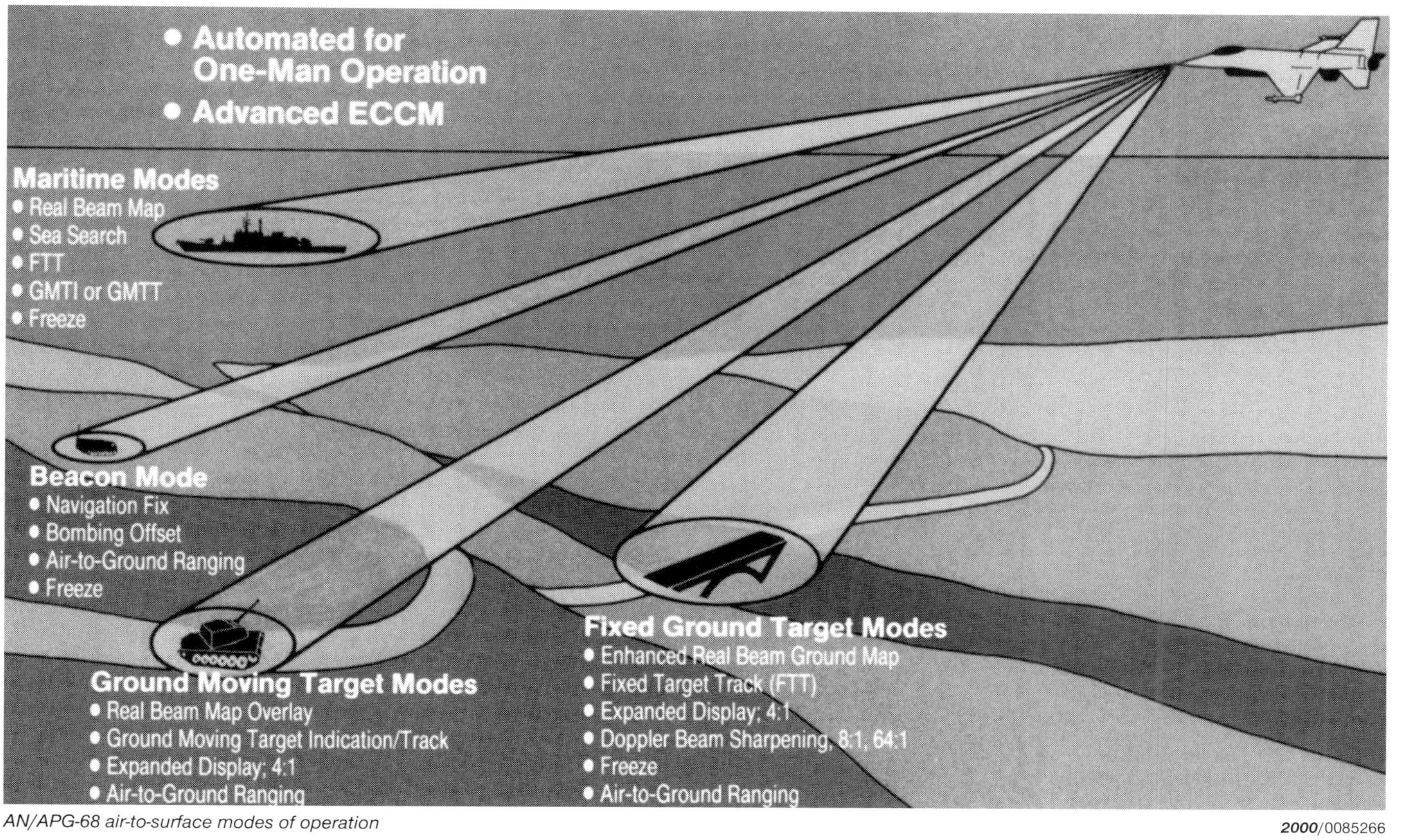

AN/APG-68 air-to-surface modes of operation **2000**/0085266

The AN/APG-68 features a highly programmable modular architecture derived from the Northrop Grumman AN/APG-66. The major operational improvements incorporated into the AN/APG-68 include extended detection range and multiple target tracking, beyond visual range utilisation of AMRAAM (Advanced Medium-Range Air-to-Air Missile) and Sparrow (AIM-7F/M) missiles and enhanced ECCM (electronic counter-countermeasures).

Since its initial deployment, the APG-68 has undergone regular improvements to performance and reliability.

The AN/APG-68 features an electrically driven antenna, a Modular Low-Power Radar Frequency (MLPRF) unit that features plug-in modules for growth and ease of maintenance, and an air-cooled Dual Mode Transmitter (DMT) that permits the radar to operate using low-, medium- and/or high-pulse repetition frequencies (PRFs). This hardware capability and flexibility allow the radar to be optimised easily for any air-to-air or air-to-ground search, track or mapping scenario.

The Northrop Grumman AN/APG-68 multimode high-performance fire-control radar for the Block 30/40 F-16C/D aircraft has demonstrated a Mean Time Between Failures (MTBF) of more than 150 hours in the field. For Block 50, the AN/APG-68 has been upgraded with a more advanced airborne signal processor (APSP), which has increased the system reliability to over 300 hours MTBF while enhancing system performance and growth potential of the radar, at a reduced cost. It provides 25 separate modes of operation, including long-range, all-aspect detection and tracking, simultaneous multiple target tracking, and high-resolution ground mapping.

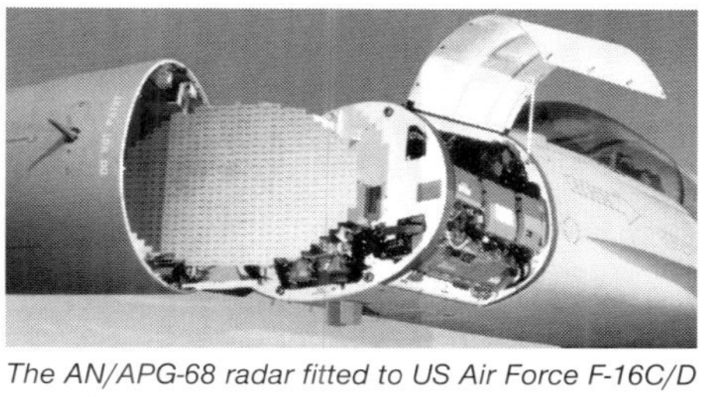

The AN/APG-68 radar fitted to US Air Force F-16C/D aircraft **2000**/0085265

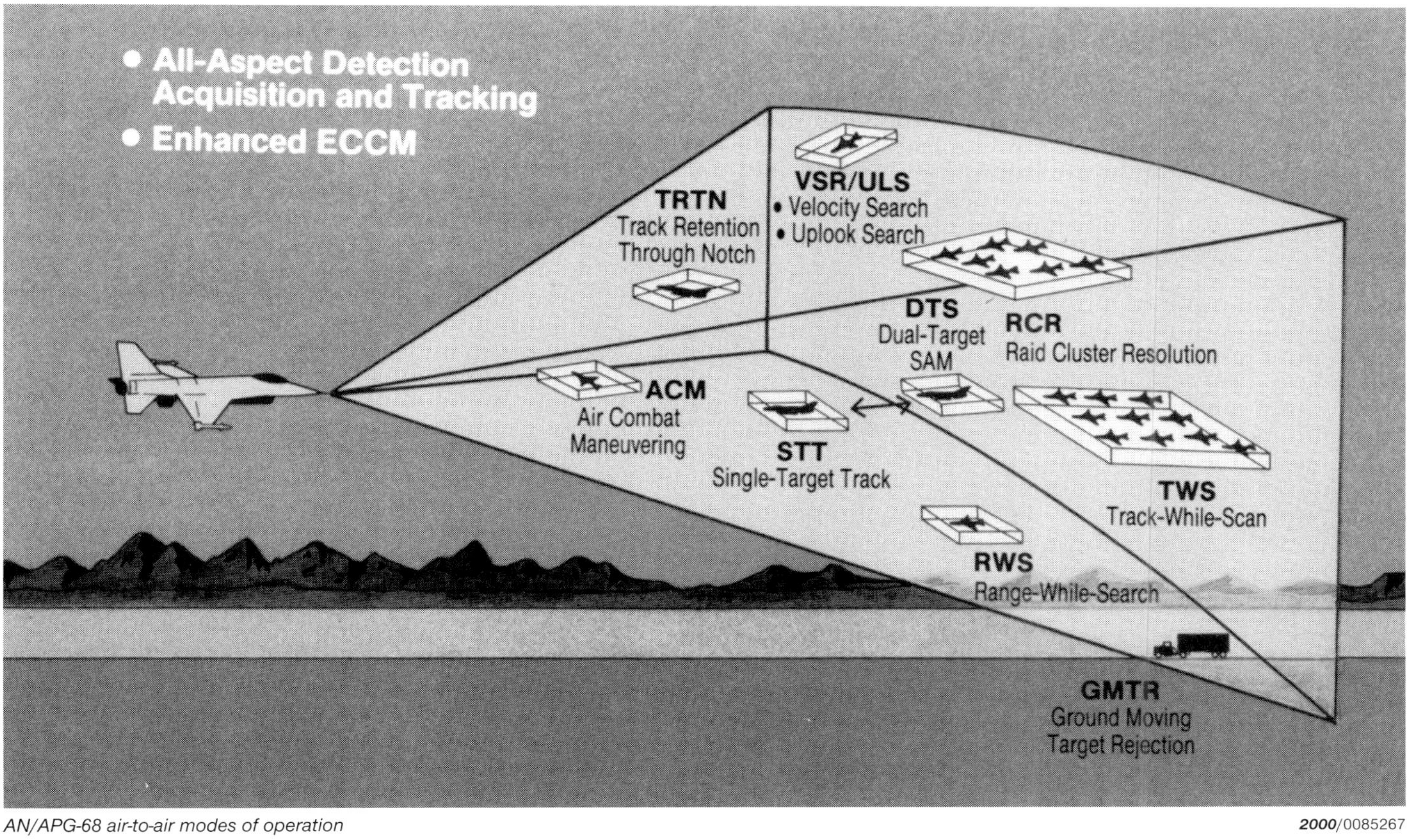

AN/APG-68 air-to-air modes of operation **2000**/0085267

Meanwhile, a software upgrade for the US Air Force fleet of over 400 Block 40/42 F-16 C/Ds, released in September 1998, was termed 40T5. It makes the Block 40/42 aircraft more capable in the Close Air Support (CAS) role and in other missions. Related hardware modifications will be installed at US Air Force depot Ogden, Utah, starting in early 2000. Software 40T6 is in development and will be fielded in 2001, together with further hardware upgrades. Major enhancements will include common missile warning system, additional smart weapons, and an upgrade to the digital terrain database. The next step after tape 40T6 in the Block 40/42 evolution comes as part of the US Air Force's F-16 Common Configuration Implementation Program that harmonises Block 40/42 and 50/52 versions into a common configuration of core avionics and software; planned fielding date is 2004.

AN/APG-68(V)X and AN/APG-68ABR

Northrop Grumman is developing an improved version of the AN/APG-68, called the AN/APG-68(V)X, through the use of low-risk technology insertions. Areas of improved performance are reported to include a modular receiver/exciter, a COTS processor, and a synthetic aperture ground mapping capability. The AN/APG-68(V)X is targeted at the F-16 upgrade market. An initial availability date of 2003 is mentioned.

In parallel, Northrop Grumman is also developing the AN/APG-68ABR (Agile Beam Radar) as a more radical development path, based on the use of a new-technology Active Electronically Scanned Array (AESA) antenna. The AN/APG-68ABR is targeted principally at the new-build aircraft market. Flight tests of the AESA concept are understood to have been conducted by Northrop Grumman in 1999, aboard its BAC 1-11 test aircraft. A possible date of 2004/5 is suggested for initial availability.

Specifications

Volume: 0.13 m^3
Weight: 172.3 kg
Frequency: 8-12 GHz
Transmitter: gridded, multiple peak power travelling wave tube
Antenna: planar array, 740 × 480 mm
Search range:
(air-to-air) 160 n miles (296 km)
(air-to-ground) 80 n miles (148 km)
Range scales: 10, 20, 40, 80, 160 n miles
Azimuth scan: ±10°, ±25°, ±30°, ±60 °
Elevation coverage: 1, 2, 3 or 4 bar
Number of units: 4 LRUs, all in the nose of the aircraft
Maintenance: 30 min MTTR
Reliability: 300 h MTBF

Operational status

More than 6,500 copies of the AN/APG-68 and its predecessor, the AN/APG-66, have been produced. Besides the US Air Force, the AN/APG-68 is operated by the air forces of Bahrain, Egypt, Greece, Israel, Korea and Turkey.

The designation of the radars delivered to Bahrain was AN/APG-68(V)5.

Radar designations for the new F-16C/D Block 50+ aircraft ordered by the Hellenic Air Force are reportedly either the APG-68(V)7M or APG-68(V)XM.

The designation of twenty-four new radars, to be provided for Egyptian F-16 aircraft by the end of 2001 is AN/APG-68(V)8.

Radars offered to Israel, associated with a possible sale of F-16C/D Block 50/52 aircraft, are reported to carry the designations APG-68(V)7 or APG-68(V)X.

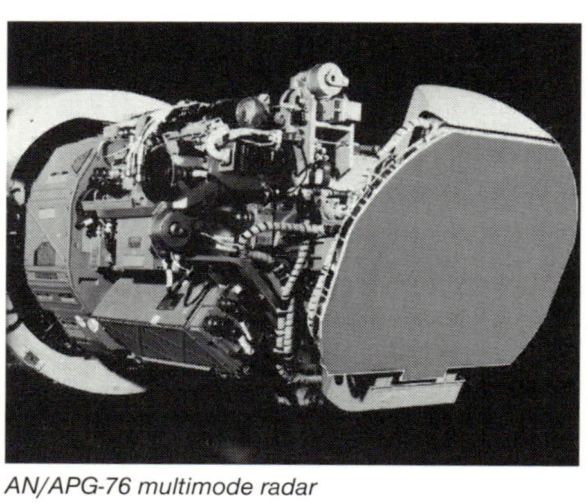

AN/APG-76 multimode radar

The F-22 Raptor carries the AN/APG-77 multimode radar **2000**/0085268

It is reported that the sale of 80 Block 60 F-16 aircraft to the United Arab Emirates is for the APG-68ABR (Agile Beam Radar) derivative.

Northrop Grumman is also marketing the APG-68(V)X, and/or the APG-68ABR, to the US Air Force to meet part of the F-16C/D Multi-Staged Improvement Program (MSIP-IV).

Contractor

Northrop Grumman Corporation, Electronic Sensors and Systems Sector.

VERIFIED

AN/APG-76 MultiMode Radar System (MMRS)

APG-76 MMRS is designed to provide the F-4 Phantom with enhanced air-to-air and air-to-ground capabilities. Air-to-air capabilities include look-up, look-down and beacon mode, as well as air track and air combat modes. Air-to-ground performance is enhanced through the inclusion of real-beam ground map, high-resolution Synthetic Aperture Radar (SAR) and Doppler beam-sharpening, ground mapping with simultaneous ground moving target indication, and beacon modes. A powerful clutter suppression interferometer provides a clutter-free resolution SAR map, as well as multitarget tracking capability. The radar's air-to-ground capabilities have also been used in the Gray Wolf technology demonstration (see operational status). Here, the radar is mounted in a pod and is described as offering a number of operating modes including real beam, Doppler sharpening and spotlight. In real beam, the Gray Wolf application is noted as having a maximum detection range in excess of 185 km while the Doppler beam-sharpening capability allows 26 × 26 km sectors to be scanned with high-resolution values. The equipment's spotlight mode is thought to incorporate three submodes, the most sensitive of which provides 0.3 m resolution on targets at ranges of up to 130 km. The Gray Wolf application is also noted as being able to track up to 75 moving surface targets simultaneously.

Operational status

A podded APG-76 codenamed Gray Wolf is noted as having been tested as a theatre ballistic missile launcher detection and ground surveillance sensor aboard a USAF F-16 and a US Navy S-3B.

Contractor

Northrop Grumman Corporation, Electronic Sensors and Systems Sector, Norden Systems.

UPDATED

AN/APG-77 multimode radar

A joint venture of Northrop Grumman's Electronic Sensors and Systems Sector and Raytheon Electronic Systems is developing the advanced AN/APG-77 radar for the US Air Force F-22. The radar's range will give an F-22 pilot unprecedented capability in air-to-air combat, allowing the pilot to track, target and shoot at multiple threat aircraft.

The first developmental F-22 radar has begun system-level integration and testing at the ESSS facility in Linthicum, Maryland, US. The radar system is currently in a testing laboratory, where hardware and software integration is taking place as part of system-level tests. The radar is the first of 11 systems to be delivered by the Northrop Grumman/Raytheon Electronic Systems team as part of the Engineering and Manufacturing Development (EMD) phase of the programme.

The radar employs a low observable active aperture, electronically scanned array that features a separate transmitter and receiver for each of the antenna's radiating elements. This type of antenna provides the agility, low radar cross-section and wide bandwidth necessary to support the F-22's air superiority mission, providing long-range, multitarget, all-weather capability. It will support multiple missiles launched against enemy aircraft. The radar is expected to have mean-time-between-maintenance actions of more than 400 hours.

Systems integration and testing of the first F-22 radar is expected to extend over approximately 18 months. Major elements of the first radar, including active electronically scanned antenna array, supporting electronics and search/track software were delivered in April 1998 to The Boeing Company's F-22 Avionics Integration Laboratory in Seattle, Washington, where engineers are integrating the radar with other F-22 avionics. Additional radar systems will be flight-tested on a Boeing 757 testbed aircraft and on F-22 flight test aircraft. In December 1998, Northrop Grumman delivered to Boeing the second of three planned software blocks – 2 months ahead of contract. Delivery of Block 2 radar software, which ultimately will be integrated with other F-22 avionics systems, follows successful stand-alone qualification testing of software in Baltimore.

F-22 first flight took place in September 1997. Preproduction verification continues.

Specifications

Frequency: 8-12 GHz
Power per antenna module: 10 W
Reliability: 2,000 h MTBF estimated

Contractors

Northrop Grumman Corporation, Electronic Sensors and Systems Sector.
Raytheon Electronic Systems.

VERIFIED

AN/APN-241 airborne radar

The AN/APN-241 is a lightweight, fully coherent pulse Doppler radar, based on the Northrop Grumman AN/APG-66/68 Series of fire-control radars. It has been developed to provide precision airdrop and navigation

AN/APN-241 Radar Antenna **2001**/0092870

radar capabilities for military tanker and transport aircraft, and in addition to provide a long-range weather mode, including detection of turbulence and windshear.

The AN/APN-241 has five radar modes: weather/turbulence, predictive windshear, high-resolution ground map, skin paint and beacon. The system has three display modes: station-keeping, flight plan and Traffic Collision Avoidance System (TCAS). The system can interleave radar modes, allowing the crew to view and control separate modes simultaneously while, at the same time, overlaying any of the three display modes. It is capable of accommodating a crew of two or three. The open systems architecture of the AN/APN-241 will allow advanced functions such as Synthetic Aperture Radar (SAR), terrain-following/terrain-avoidance, maritime surveillance and autonomous landing capability to be added without significant development costs. A SAR upgrade was successfully flight tested during 1999.

APN-241NT

Northrop Grumman has been contracted by BAE Systems Australia to develop and supply seven examples of an updated variant of the APN-241, known as APN-241NT, using an antenna that allows it to equip medium-sized transport aircraft as part of the upgrade of Royal Australian Air Force HS 748 navigator trainers.

Specifications

Antenna size: 610 × 890 mm
Weight:
(antenna) 31.8 kg
(electronics unit) 28 kg
Frequency: 9.3-9.4 GHz
Power output: 146 W peak, 11 W average
Range: 515 km
Coverage:
(azimuth) ±135°
(elevation) −15 to +10°

Operational status

The AN/APN-241 is in service on C-130H aircraft of the US Air Force, Royal Australian Air Force and Portuguese Air Force. It is also a candidate for upgrades to numerous tanker and transport aircraft worldwide. More than 230 radars of this type have been delivered worldwide. The AN/APN-241 is the baseline fit for the C-130J and the C-27J Spartan Lockheed Martin Alenia Tactical Transport Systems' medium tactical airlifter. In 1999, a variant of the AN/APN-241, designated AN/APN-241B, was proposed to the US Air Force to replace the existing AN/APS-133(V) systems on the C-17 Globemaster III transport aircraft. The most significant difference between the two variants was the replacement of the 36 inch aperture radar used on the basic model with one of 22 inch aperture to fit into the C-17.

Contractor

Northrop Grumman Corporation, Electronic Sensors and Systems Sector.

VERIFIED

AN/APQ-156/-148/APS-130 series multimode radars

AN/APQ-156/-148/APS-130 series radars are Ku-band airborne multimode equipments specially developed to combine the functions of two radars previously required by US Navy A-6A all-weather attack aircraft in a single radar. The APQ-156 system is a modification of the APQ-148 to accommodate the addition of a forward-looking infra-red sensor/laser target recognition attack multisensor system to the A-6E aircraft. APS-130 is the system derivative fitted to some US Navy EA-6B electronic warfare aircraft. Functions performed by the APQ-156/-148/APS-130 include:
(a) search
(b) ground mapping
(c) tracking and ranging of fixed or moving targets
(d) terrain-avoidance or terrain-following
(e) beacon detection and tracking.

A track-while-scan capability provides simultaneous range, azimuth and elevation data for weapon delivery. As in other systems, range and azimuth markers must be placed on the target, but elevation data are available on a continuous basis and are derived from a separate phase interferometer array carried below the main scanner dish. The latter has a width of about 1 m and is illuminated by a conventional horn feed to produce a very narrow beam in azimuth.

The beam has a cosec^2 profile in elevation and this, with the interferometer elevation data provided, eliminates the need for mechanical scanning in the elevation plane. The interferometer array consists of two adjacent rows of 32 horns and moves with the main dish. Energy reflected from ground targets arrives at the upper and lower rows with a time difference, which is measured by phase comparison techniques and translated into angular information.

There are two cockpit displays in the A-6E APQ-148 installation: a 13 cm storage tube unit for the pilot, and an 18 cm Direct View Radar Indicator (DVRI) for the bombardier/navigator. In the case of the APQ-156, only the DVRI is provided. Terrain data from the radar system are also presented on a vertical display for the pilot. The system incorporates comprehensive built-in test facilities. System weight is about 227 kg.

Contractor

Northrop Grumman Corporation, Electronic Sensors and Systems Sector, Norden Systems.

UPDATED

AN/APQ-164 multimode radar

The AN/APQ-164 is the radar installed in the US Air Force B-1B aircraft. This radar combines technology from the F-16 AN/APG-68 radar and the Electronically Agile Radar (EAR) programme of the US Air Force.

The B-1B radar generates data for navigation, penetration, weapon delivery, and for certain other functions such as air refuelling. There are four modes in the AN/APQ-164 system that provide the navigation capability. The primary mode is a high-resolution synthetic aperture radar mapping mode, backed up by a monopulse-enhanced real beam ground-mapping mode. The system also detects weather ahead and can display ground beacon returns over a real beam image. The penetration functions of the radar include automatic terrain-following and terrain-avoidance. For weapon delivery the radar provides four different functions. The first is a velocity update mode, similar to a Doppler navigator, which generates velocity information for the inertial navigation system. Coupled with an accurate Global Positioning System receiver in the avionics system, velocity update produces a dynamic, precision antenna calibration correction. Second, there is a ground moving-target detection and tracking capability for both fast- and slow-moving vehicles. Third is a high-altitude altimeter function that provides a very accurate measure of local height above the ground. Fourth is a monopulse targeting mode that provides accurate height to the on-scene selected fixed target.

The synthetic aperture mode provides the operator with a high-resolution image of an area of ground that can be chosen by the avionics system or the operator. Long-range maps can be made and five different map scales displayed. The synthetic aperture mapping mode accepts the co-ordinates of a waypoint from the avionics system and makes a map centered on that point. To make an image, the antenna is electronically scanned to the waypoint location. The radar transmits a train of pulses, gathers data for the image, and then switches itself off. At the same time, the image is stored in the radar and presented on the display in a rectangular, ground co-ordinate display.

The radar provides the basic data required for automatic terrain-following. It scans the ground in front of the aircraft and measures the terrain in a range versus height profile out to 19 km and stores that data in the computer. The profile data is sent across the multiplex bus to the terrain-following control unit where the data is used to generate climb/dive commands. This flight profile is then automatically fed into the pilot's flight control system. Since the radar is not continuously scanning in terrain-following, a very low

APS-130 is the system derivative fitted to some US Navy EA-6B electronic warfare aircraft **2001**/0106205

update is used, helping to reduce the risk of detection. This rate is variable and depends on aircraft altitude, manoeuvres, groundspeed and terrain roughness. Under normal conditions updates are made at 3 to 6 second intervals. However, if the terrain demands it, data can be gathered continuously.

The AN/APQ-164 in the B-1B is a dual-redundant system, with two complete and independent sets of Line-Replaceable Units (LRUs), except for the phased-array antenna. This was the first airborne application of this technology for combat aircraft. Only one set of LRUs is used at a time, the other being maintained on standby.

The phased-array is an outgrowth of the antenna developed on the EAR programme. It contains 1,526 phase control modules and allows virtually instantaneous beam movement to any point in the antenna field of regard. When the radar mission requires a forward, right or left region of regard, the antenna is physically movable to three different positions on a roll detent mount. The radar can, therefore, look off to either side of the aircraft or forward by rolling the antenna about an axis. The normal antenna position is looking forward. However, when the antenna is rolled to one side, the field of view extends from the aircraft nose back to about 115°, permitting a look off to the side of interest without having to change aircraft heading. Once physically moved to one of the three available positions, the antenna is locked into a detent. From the fixed spot, it can be scanned electronically ±60° in azimuth and elevation by means of a unit on the antenna called the beam-steering controller, which controls all 1,526 phase control modules.

Specifications

Frequency: I-band
Transmitter: gridded, multiple, peak-power TWT (similar to the AN/APG-68 transmitter)
Antenna: phased-array electronically scanned, 1,118 × 559 mm
Operating modes: (air-to-ground) high-resolution mapping, monopulse enhanced real beam mapping, automatic terrain-following, manual terrain-avoidance, velocity update, ground moving target detection and track, high-altitude calibrate, ground beacon; (air-to-air) weather mapping, air-to-air beacon, rendezvous mode. Growth for full conventional standoff capability and a full air-to-air mode complement is provided.
Weight: 570 kg

Operational status

AN/APQ-164 is installed aboard US Air Force B-1B Lancer bomber aircraft.

Contractor

Northrop Grumman Corporation, Electronic Sensors and Systems Sector.

VERIFIED

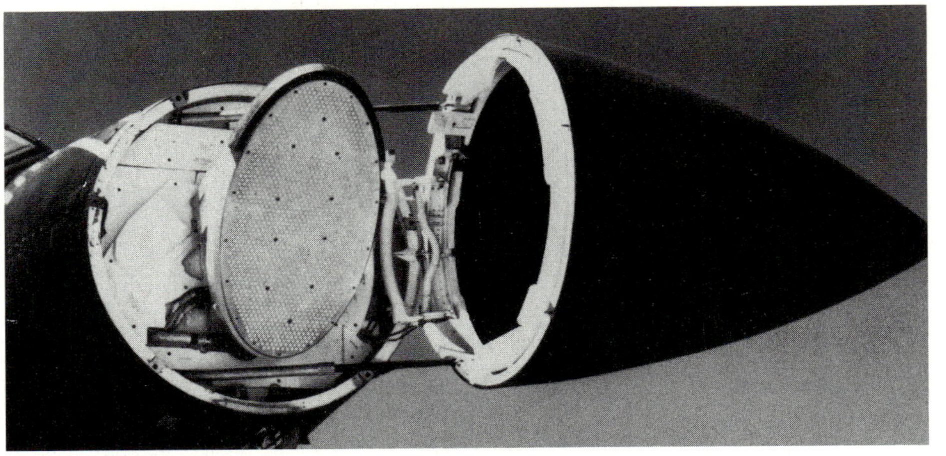

AN/APQ-164 airborne radar

AN/APY-2 radar for the E-3 AWACS

The Boeing Company is prime contractor to the US Air Force for the E-3 programme. Northrop Grumman's Electronic Sensors and Systems Sector, based in Baltimore, Maryland, builds the production AN/APY-2 surveillance radar, which is installed in a Boeing 707-320 aircraft modified with a large radome for the radar and other antennas. The AN/APY-2 surveillance radar is also installed on the Boeing 767 AWACS aircraft. The radar transmitter and receiver, communications gear, and other command and control systems are located inside the aircraft.

The E-3 radar provides full, long-range surveillance of high or low-flying aircraft. A maritime capability provides surveillance of moving or stationary ships. The E-3 aircraft operates during all kinds of weather and above all types of terrain.

The E-3 surveillance system detects and tracks both enemy and friendly aircraft in a large volume of air space. Low-flying aircraft, which can escape detection by ground-based radar, are detected by the E-3 aircraft.

The AN/APY-2 high-PRF pulse Doppler radar with its digital signal processing provides the look-down surveillance capability that is the key to the E-3 airborne warning and control system.

The AN/APY-2 radar looks down at the ground, distinguishing between ground reflections (clutter) and radar returns from aircraft hugging the ground to escape detection. In addition to detecting either high- or low-flying aircraft and ships out to the aircraft horizon, the E-3 radar also offers long-range aircraft surveillance above the horizon. For an E-3 aircraft altitude of 9,000 m (30,000 ft), the radar look-down range to the horizon is about 400 km (245 miles).

The E-3 aircraft provides a real-time assessment of both enemy action and friendly resources. With an E-3 aircraft, the airborne commander has available the information that is needed to detect, assess and counter an enemy threat.

High-PRF pulse Doppler radar - concept: The basic advantage of high-PRF Doppler radar is that it provides better Doppler separation of moving targets from ground clutter. Additional radar techniques minimise ground clutter and maintain high sensitivity. Some of these are an extremely low-sidelobe antenna, ultrastable frequency generation and digital processing techniques.

An inherent feature of the E-3 radar design is its flexibility to accommodate future growth through software control. Many signal processing functions are accomplished with programmed instructions rather than the specific hard-wired arrangements of circuitry so that functions or processes can be updated by altering programme instructions. These are programmable in flight; thus tactical programmes can be adjusted to respond to changes in the tactical situation.

Radar operating modes: The target-handling capability of the E-3 radar is enhanced by its operation in various modes, depending on the nature of the tactical situation. Each 360° radar scan can be divided into as many as 24 azimuth sectors and each sector can be operated with its own set of operation modes. Available radar modes are as follows:

(1) pulse Doppler non-elevation mode scans with high-PRF Doppler to provide look-down surveillance of aircraft out to the radar's horizon, but it does not measure target elevation.

(2) pulse Doppler elevation scan is similar to the previous mode, but includes an electronic vertical scan of the radar beam to provide target elevation.

(3) beyond-the-horizon mode of operation is used for long-range surveillance of medium- and high-altitude aircraft. Since the radar beam is above the horizon, there is no ground clutter and a low-PRF radar pulse is used to obtain the range and azimuth of target aircraft.

(4) passive scanning mode operates with the radar transmitter 'off' and the receiver 'on' to obtain ECM information, such as the locations of enemy jammers.

(5) maritime mode uses a very short radar pulse to provide the high resolution required to detect moving and anchored surface ships.

Multimode operation and sectoring permits maximum potential of the radar to be concentrated in sectors where the need is greatest. The pulse Doppler mode and beyond-the-horizon mode can be used simultaneously in an interleaved manner, as can the maritime and pulse Doppler non-elevation modes.

Technological features: To permit cancellation of the main-beam clutter in the radar receiver, extremely stable signal generation and advanced signal processing are used. The improvements in signal generation capability necessary for the E-3 radar were made through advances in circuit components, including oscillator crystals and transmitter tubes.

The signal processing advances were made possible by the application of digital signal processing and by analogue-to-digital converter developments for the E-3 application. Radar control is accomplished by operating commands from the central computer.

Much of the high performance achieved by the E-3 pulse Doppler radar derives from the antenna and rotodome design. Low antenna sidelobes are necessary to minimise sidelobe clutter return and avoid reduction in detection performance for target returns that do fall within the sidelobe clutter region. The low sidelobe patterns that have been achieved in the E-3 radar antenna and rotodome represent a significant aspect in antenna design. The advance was made possible by the application of digital computing techniques to the design of the antenna, and by the use of high-precision digital techniques in manufacturing the antenna. The Boeing rotodome design enabled the low-sidelobe characteristics of the antenna to be maintained when radiating through the rotodome.

In addition to minimising sidelobe clutter, the low-sidelobe antenna design is also a major contributor to the radar's resistance to jamming. The highly directional nature of the antenna when receiving, rejects a jamming signal except when the antenna is pointed directly at the jamming sources. Hence, jamming signals can be easily identified and located.

E-3 radar Electronic Counter-CounterMeasures (ECCM) capabilities: The E-3 airborne surveillance radar system, with its inherent ECCM capabilities, can provide effective command and control under jamming conditions.

The rotodome of the Boeing E-3A Sentry houses the AN/APY-2 surveillance radar scanner

The slotted planar-array antenna of the AN/APY-1/2 radar 0018075

The E-3 ECCM features, such as low antenna sidelobes and inherent chaff rejection, are an integral part of the radar design, rather than add-on 'fixes'. This results from the pulse Doppler radar design to eliminate the effect of ground returns (clutter) and the requirement for the E-3 to operate effectively against a variety of electronic countermeasures.

Filtering is used in the pulse Doppler radar receiver to remove mainbeam clutter. This filter also eliminates the radar signals received from stationary or slowly moving targets, such as chaff.

Another form of ECCM available is the ability to switch to a radar frequency which is not being jammed. This negates the jammer effect even though the jammer remains within the radar line of sight.

The primary E-3 radar ECCM capability is derived from the very low antenna sidelobes. Flight tests of the E-3 radar, integrated with other E-3 subsystems such as a computer and tracking display consoles, have shown that the E-3 can operate successfully against powerful airborne and ground jammers.

The E-3 radar also has the ability to determine the relative bearing of a jammer with the radar in either an active or passive mode. The mobility of the E-3 enables it to perform self-triangulation. The system computer indicates the position data on an operator's console and computes the intercept path of the fighter aircraft designated to attack the jammer.

E-3 mobility also permits the use of jammer avoidance tactics to reduce or eliminate the jammer effects. The E-3 aircraft can drop below the radar horizon in relation to the jammers, thereby eliminating reception of the jamming energy. This tactic would result in some loss of low-altitude coverage but will enable target detection above the horizon line of sight to continue unaffected by the jammers.

Simultaneous operation of two or more E-3 aircraft enhances the overall system effectiveness against jammers and non-jamming targets. The E-3 has the ability to transmit information via a datalink system to other E-3 aircraft or to ground stations, producing a synergistic effect and improving the surveillance and battle management functions.

E-3 radar design: Radars for E-3 aircraft consist of three major subsystems: the slotted planar-array antenna located in the rotodome on top of the aircraft, the radar receivers and processors located in the centre of the aircraft cabin, and the radar transmitter located in the lower cargo bay. Total weight of the radar system is 3,742 kg.

The slotted planar-array antenna rotates with the aircraft rotodome at six rpm to provide horizontal radar scanning. The antenna face consists of 30 slotted waveguide sticks and measures 7.3 m × 1.5 m.

Vertical scanning and height-finding are performed by electronic scanning techniques using ferrite phase shifters. The phase shifters, phase-control electronics, receiver protectors and receiver paramplifiers are mounted on the back of the antenna. The phase shifters are located on one side of the antenna and the electronics on the other side for weight balance and easy access during maintenance. Northrop Grumman-developed receiver protectors use a radioactive igniter power source for long life and fail-safe operation even during radar shutdown.

The radar transmitter consists of eight pressurised vessels located in the lower cargo bay. An overhead rail system permits easy removal of the transmitter units through a lower hatch.

The high-power transmitter chain is completely redundant with an inflight switchover capability should a malfunction occur. This transmitter redundancy, along with extensive redundancy in other parts of the E-3 system, assures high radar reliability and high probability of E-3 mission success.

Only two tubes are used in the transmitter chain: a high-power Klystron and a travelling wave tube driver. All other elements of the transmitter are solid state.

The radar receiver and digital signal processor are located in a single cabinet in the centre of the E-3 aircraft. Digital printed circuit boards and receiver electronics are accessed through hinged doors. Critical circuitry is backed up by redundant circuit boards.

A radar-dedicated digital computer and high-speed digital data processor control and monitor radar operation, reject ground clutter from radar returns, perform frequency analysis on signal returns, correlate radar returns to determine presence of legitimate targets, and digitally format the output target reports in range, velocity, azimuth and elevation.

Digitised radar information is passed in near-real-time to the E-3 central processor and displays. The central processor correlates radar reports over successive scans to form target tracks. Navigation inputs are supplied to the radar computer to adjust for aircraft motion and altitude.

Digital radar signal processing provides significant advantages over conventional analogue processing in terms of cost, weight, complexity, reliability, maintainability and operability. Since the data processor is software programmable, system flexibility and room for potential system growth to meet changing environments also are advantages.

Use of high-reliability components throughout the radar, digital technology, use of integrated circuits and functional groupings of circuits, redundancy of critical circuitry with automatic switchover, built-in-test and fault isolation to an individual circuit board, and inflight maintenance capability all add up to make the E-3 radar highly reliable and easy to maintain.

Built-in test constantly monitors radar operation under control of radar computer software. Normal system operation is interspersed with fault detection tests. A 98.5 per cent probability of detecting online faults was demonstrated by the radar. In the event of a malfunction, the radar computer automatically reconfigures the system using available redundant circuitry. After correction, the results are displayed to the radar maintenance technician.

Spares for most non-redundant portions of the E-3 radar are carried on the aircraft and substituted while in the air. This attention to radar reliability and easy maintenance assures a high probability of mission success for the E-3 aircraft. Operational availability has consistently been greater than 95 per cent in the AWACS worldwide operation.

AWACS Radar System Improvement Programme (RSIP): Northrop Grumman's Electronic Sensors and Systems Sector (ESSS), under a contract worth more than US$350 million from the US Air Force Materiel Command's Electronic Systems Center, undertook the Engineering Manufacturing Development (EMD) phase of an E-3 AWACS Radar System Improvement Programme (RSIP). RSIP is a joint Air Force/NATO programme.

The RSIP contract includes the design, development and flight test of improvements to the AWACS AN/APY-1 and -2 radars to maintain operational capability against the growing threat from smaller radar cross-section targets, cruise missiles and electronic countermeasures. The contract also includes significant improvements in the man-machine interface and reliability and maintainability. RSIP represents the most significant upgrade to the E-3 radar since its development in the early 1970s.

The major portion of RSIP is devoted to increasing radar sensitivity against small targets through replacement of the digital Doppler processor and radar data correlator with a state-of-the-art Surveillance Radar Computer (SRC), and translation of the associated software into Ada language. In addition to handling the upgraded radar processing load, the SRC will contain adequate growth reserves to accommodate further radar upgrades.

Improvement in the man-machine interface will result from modification of the radar control and maintenance panel by incorporation of a spectrum analyser, special test equipment and new displays for monitoring the surveillance environment as well as the maintenance status of the radar system.

The RSIP EMD programme included fabrication of the first five modification kits for the US Air Force and one modification kit for NATO.

Under the initial production contract, Boeing and Northrop Grumman (ESSS) were to build 18 modification kits for NATO, four for the US Air Force, seven for the UK. Follow-on options have been exercised so that RSIP production for 43 of the total production of 70 aircraft has been approved. All 43 systems are scheduled to be operational by 2002.

RSIP kit installation is by US Air Force at Tinker AFB for US Air Force; DaimlerChrysler Aerospace AG for NATO; BAE Systems (Operations) Limited for UK.

Operational status

The E-3 AWACS became operational with the US Air Force in early 1978. To date, 34 E-3s have been delivered to the US Air Force, 18 to NATO, and five to the Royal Saudi Air Force. Production of seven E-3 radars for the United Kingdom and four E-3 radars for France has also been completed. Additionally, the first 4 radars for the Japanese 767 AWACS have been delivered, with 2 of the additional Japanese 767 systems becoming operational in early 1999.

The RSIP has been installed on all 18 NATO aircraft, all seven UK Royal Air Force aircraft and three US Air Force aircraft.

In May 2000, the US Air Force awarded Raytheon Electronic Systems a contract to incorporate 435 Common Large Area Display Sets (CLADS) in the US Air Force AWACS aircraft.

Contractor

Northrop Grumman Corporation, Electronic Sensors and Systems Sector.

UPDATED

AN/APY-6 airborne surveillance radar

The AN/APY-6 is a high-performance airborne Synthetic Aperture Radar (SAR) that provides Ground Moving Target Indication (GMTI) and sea surveillance capability. It has been developed as a technology demonstration system for the US Office of Naval Research. It had its first flight in the first quarter of 1999.

The AN-APY-6 radar uses a Travelling Wave Tube (TWT) power amplifier and a flexible planar-array antenna that can be configured for forward looking, side looking or 360° cover in nose or belly-mounting locations. The receiver is a 4-channel unit that provides three MTI channels and one SAR channel.

Precision location of fixed and moving targets is achieved, with slow speed, clutter suppression MTI, to provide single pass two- or three-dimensional strip map data at 0.3 m resolution.

System architecture is based on an Ethernet with fibre channel interconnectivity and industry-standard programming in 'C'.

The AN/APY-6 system is air-cooled and weighs less than 205 kg.

The technology applied to this project is reported to complement Northrop Grumman work on the Joint Strike Fighter (JSF) Multifunction Integrated Radio Frequency System (MIRFS) technology demonstrator.

The project name Gray Wolf has been linked to this technology demonstrator programme.

Contractor

Northrop Grumman Corporation, Electronic Sensors and Systems Sector, Norden Systems.

VERIFIED

AN/AQS-14/14A/14A(V1) sonar

The AN/AQS-14 sonar equips the US Navy Sikorsky MH-53E helicopter; first deliveries took place in 1984 and the system has been in service since June 1986. To date, 32 systems have been delivered. Northrop Grumman has delivered additional AN/AQS-14A systems with a new airborne electronics console which greatly improved system performance. The AQS-14 saw extensive use in the Red Sea during 1984 and in the Gulf during 1987/88 and 1990/91 during Operations Desert Shield and Desert Storm.

Used for minehunting duties, towed from helicopters, hovercraft or small surface vessels, the AQS-14 is a side-looking multibeam sonar with electronic beam-forming, all-range focusing and an adaptive processor. The underwater vehicle is 3 m long and has an active control system which allows it to be towed at a selected distance above the seabed or under the surface. The vehicle is controlled by an operator in the helicopter and it is connected by a non-magnetic cable. The operator, assisted by computer-aided detection and classification, has a real-time sonar display on which he can mark targets of interest. A tape recorder allows the recording, classification, position logging and review of data concerning mines and similar objects.

The latest variant of the system is the AN/AQS-14A (V1). In May 1999, Northrop Grumman was awarded a one year contract, valued at US$3.2 million, for development, integration and test of an engineering design model to incorporate laser line scanning technology. The addition of this, laser-driven, electro-optical capability will enable operators to actually identify mines. The upgrade will also provide an increase in side-scan sonar resolution of × 4 the current resolution at close range and × 2 the current resolution at long range. Testing of the system was completed in November 2000.

Operational status

In production and in service with the US Navy on MH-53E helicopters, small craft, remotely controlled vehicles and multimission air cushion vehicles.

Contractor

Northrop Grumman Corporation, Electronic Sensors and Systems Sector.

UPDATED

The antenna for the ASSR-1000 surveillance radar is 7 m wide and just under 4 m high

ASSR-1000 surveillance radar sensor

The ASSR-1000 is a radar sensor designed for the long-range detection of both air and maritime targets. It provides detection over rough terrain and high sea states. It is lightweight and adaptable for installation on a variety of platforms, including the Sentinel 1000 and 1200 airships.

The ASSR-1000 utilises technology from the Northrop Grumman Low-Altitude Surveillance System (LASS), with a solid-state transmitter, signal processor and post-processor. The lightweight radar has been carefully adapted for the airship airborne application and is planned to fly at altitudes up to 15,000 ft, providing a line of sight of 222 km. It is a D-band fully solid-state coherent pulse compression system, featuring an ultra-lightweight antenna, fully solid-state transmitter, frequency diversity and automatic target detection and tracking. It is capable of interfacing with operational displays, navigation equipment and datalinks. The ASSR-1000 includes an integrated AN/TPX-54 beacon interrogator.

It has a low probability of false alarm with a high probability of detection on small fluctuating targets, with a field-proven 60 dB moving target improvement factor over terrain and sea conditions. The antenna is a circular reflector, stabilised in both pitch and roll, with a wideband feed horn used for both the search and the beacon radars. The signal processor has special features that allow it to detect small targets in the surrounding environments. These features include full range I and Q processing, pulse compression, six pulse canceller with variable time periods between pulses, constant false alarm rate processing and dual-frequency transmission selectable from 93 frequencies.

Specifications

Dimensions:
(antenna) 4,572 mm diameter
Weight: 850 kg nominal
Power: 30 kW
Frequency: 1,215 to 1,401 MHz
Pulsewidth: pulse pairs of 48 and 51 μs separated by 60 MHz
PRF: 375 pps average
Range:
(instrumented) 300 km
(2 m² air target) 278 km in clutter
(4 m² marine target) out to horizon in Sea State 3
Range resolution: 0.3 km
Azimuth coverage: 360°
Azimuth accuracy: 0.25°
Data refresh rate: 12 s
Reliability: >1,500 h MTBCF

Operational status

In production and in service with the US Customs Service and several overseas customers.

Contractor

Northrop Grumman Corporation, Electronic Sensors and Systems Sector.

UPDATED

Channelised Programmable Digital Radio (PDR)

The modular Programmable Digital Radio (PDR) can implement virtually any narrow or wideband function from 1.5 MHz to 2 GHz for comms, nav, ident, landing systems or datalinks. A common module transceiver pair of receiver/processor and exciter/processor is the basic core of the system with RF input and digital data output. The common modules can be programmed to provide any function at system turn-on or during a mission or flight segment. Multisystem simultaneous operation is achieved by adding one or more receiver modules and exciter modules to the system. The architecture associated with the PDR offers simplistic system fault detection and isolation. In addition, system functional redundancy is a part of the basic design via software reconfiguration of a lower priority function automatically or the automatic reprogramming of a spare module. The PDR is designed for both military and commercial applications.

Specifications

Frequency bands: (extendable to higher bands of operation)
Exciter: 2 MHz to 2 GHz continuous
Receiver: 100 kHz to 2 GHz continuous
Instantaneous bandwidth: 14 MHz
Modulation: AM/FM/PM/FHSS/DSSS
Voice/data interface: JTA compliant; supports TCP/IP from hosts
Encryption:
Embedded AN/CSZ-12 demonstrated
External interconnection of Type 1 INFOSEC: KY-57, KG-84, KY-99 demonstrated
Other embedded options available
Modem: Programmable narrowband DSP/wideband FPGA based design
BIT: GO/NO GO automatic with power up; fault isolation to LRU module level
MTTR: 15 min LRU replacement
MTBF: 5,500 h
Environmental : MIL-STD-810E
Supply voltage: MIL-STD-704: 28 V DC

Operational status

The PDR is currently in development. The core common modules have been developed and were demonstrated in 1995. The existing form factor utilises an ARINC standard packaging. The PDR has been successfully demonstrated during Exercise ACT II 97 with the 8th Airborne Corps of the USAF on an EC-135 aircraft. The PDR is currently in use by the US Army Battle Command Battle Lab in their Warfighter Information Network - Terrestrial (WIN-T) Proof-of-Concept (WIN-POC) Testbed.

Contractors

Northrop Grumman Corporation, Electronic Sensors and Systems Sector.

UPDATED

Joint STARS AN/APY-3 Joint Surveillance Target Attack Radar System

The Joint Surveillance Target Attack Radar System (Joint STARS) designated AN/APY-3, is a long-range air-to-ground surveillance and battle management system. It is capable of looking deep behind hostile borders to detect and track ground movements, in both forward and rear echelon areas and to detect helicopter and fixed-wing aircraft. Joint STARS provides air and ground commanders with the intelligence and targeting data for management of their war-fighting assets.

Joint STARS is a complex of systems. It comprises an airborne platform, four major subsystems and a ground station module that receives, in near-realtime, radar data processed in the aircraft. The four major subsystems, all integrated on the airborne platform, consist of an advanced radar; internal and external communications including UHF, VHF and HF voice links, the Joint Tactical Information Distribution System (JTIDS) and a newly developed surveillance and control datalink; operations and control including advanced computers and 18 operator display stations which perform the data processing and display functions for tens of thousands of targets and C^3I operation; and a self-defence suite which is in the process of being specified. Joint STARS is a joint US Air Force and US Army development, with the air force being responsible for the airborne segment and the army for the ground segment.

Joint STARS detects, locates, classifies, tracks and targets potentially hostile ground movements in virtually all weather. It operates in near-realtime and in constant communication through a secure datalink with army mobile ground stations that, in turn, can use TACFIRE and the advanced field artillery tactical data system to talk to artillery for fire support or to the all-source analysis system using US message format. Joint STARS will also maintain constant communication with air force tactical command posts via JTIDS. The platform for Joint STARS is a modified and militarised version of the Boeing 707-300 Series aircraft.

The technology employed in the radar was initially demonstrated as part of the Air Force/DARPA Pave Mower programme in the late 1970s. Major technological achievements of Joint STARS include the software-intensive displaced phased-centre slotted array antenna radar with several concurrent operating modes; the unique 8 m antenna mounted under the fuselage of the aircraft developed by Norden Division of United Technologies; very high-speed processors, each capable of over 600 Mops; high-resolution colour graphic and touchscreen tabular displays; the wideband surveillance and control datalink and over one million lines of integrated software code. The radar operates as either a side-looking synthetic aperture

The JSTARS radar antenna unit 0018077

The JSTARS aircraft, showing the JSTARS radar under the forward fuselage 0018076

radar for the detection of fixed or stationary targets, or a Doppler radar to track slow-moving targets such as tanks or troop platforms.

In September 1987, the air force, following the August Joint Requirements Oversight Committee review, increased the number of E-8 aircraft from 10 to 22. In April 1988, the Conventional Systems Committee, after reviewing the Joint STARS airborne segment, concluded that the airborne platform should be changed from used aircraft to new 707s. However, in October 1989, the air force returned to the plan to continue the airborne portion of the programme with used 707 aircraft, on the grounds of the higher cost of new aircraft, to meet an initial operational capability in 1997.

In September and October 1990, operational field demonstrations in a dense electromagnetic environment and poor weather were completed in Europe. After six weeks of flying demonstrations, consisting of 25 missions and 110 flying hours, Joint STARS concluded the demonstration of the system's capabilities and data gathering for the continuing development of the programme. US Air Force and Army officials assessed the demonstration as a complete success.

In November 1990, Northrop Grumman received a contract for the development of the third Joint STARS full-scale development aircraft and follow-on full-scale development work. Meanwhile, the two E-8A prototype aircraft (T-1 and T-2) achieved outstanding success in the 1991 Gulf War, and provided strong impetus for full development of the E-8C production configuration.

Specifications

Antenna: 7.3 m long, side-looking, phased-array housed in canoe-shaped radome under forward fuselage aft of nose landing gear; scanned electronically in azimuth, steered mechanically in elevation from either side of aircraft.

Operating modes: Wide area surveillance; fixed target indication; Synthetic Aperture Radar (SAR); moving target indicator; target classification.

Workstations: 17 identical workstations for system operators; one navigation/self-defence workstation; each operator can perform: flight path planning and monitoring; generation and display of cartographic and hydrographic map data; radar management, surveillance and threat analysis; radar; radar data review; time of arrival calculation; jammer location; pairing of weapons and targets, and other functions.

Communications: Surveillance and Control DataLink (SCDL) for transmission to ground stations; JTIDS; TActical Data Information Link-J (TADIL-J); Satcom; two encrypted HF radios; three encrypted VHF radios with provision for SINgle Channel Ground and Airborne Radio System (SINCGARS).

Operational status

The two prototype E-8A aircraft were designated T-1 and T-2. The third aircraft - the first designated E-8C, is the preproduction Model (T-3). The first production E-8C (designated P-1), was handed over to Air Combat Command in June 1996. Joint STARS was approved in September 1996 for full-rate production for a planned buy of 19 production aircraft for US Air Force service, since reduced to 13 aircraft, in full production. The current configuration is termed Block 0, essentially that of the test and evaluation aircraft.

Initial Operational Capability (IOC) was achieved in December 1997, after delivery of the third production aircraft to the US Air Force in November 1997. Meanwhile, in June 1997, the US Air Force awarded Northrop Grumman two contracts for a computer replacement programme to install modern commercial-off-the-shelf (COTS) processor technology and fibre-optic links to the upgraded operator consoles.

Based on the success of the E-8A aircraft in the Gulf War, NATO expressed interest in procurement of the JSTARS system for its NATO Airborne Ground Surveillance (AGS) programme and Northrop Grumman has appointed DaimlerChrysler Aerospace AG to be lead European contractor for its bid for the NATO AGS aircraft.

New higher-resolution SAR modes offered to NATO include Enhanced (SAR) (ESAR); Inverse (SAR) (ISAR); and SAR Swath.

In November 1998, the US Air Force awarded Northrop Grumman Electronic Sensors and Systems Sector (ESSS) a Pre-Planned Product Improvement (P3I) contract to develop the next-generation advances to JSTARS. The P3I contract, which could have a value of US$776 million, is the first element of the Radar Technology Insertion Program (RTIP). Including production, the total value of RTIP is expected to exceed US$1.3 billion More recently, Northrop Grumman received a separate Indefinite Delivery/ Indefinite Quantity (ID/IQ) contract with a potential value of US$1.2 billion for upgrade work on JSTARS. The contract extends to March 2005. The first project in the ID/IQ contract is development of an integrated satellite communication system for JSTARS, valued at approximately US$40 million.

The main elements of the RTIP are a new 2-D electronically scanned array radar, a Commercial-Off-The-Shelf (COTS) technology signal processor and improved software for operation and control of the workstations inside the JSTARS platform.

Northrop Grumman ESSS and Raytheon Electronic Systems have agreed a '50-50' workshare on the radar portion of the RTIP.

Northrop Grumman Integrated Systems and Aerostructures (ISA) Sector will continue as the RTIP prime contractor with Raytheon Electronic Systems as a subcontractor to Northrop Grumman ESSS.

In March 2000, the US Air Force received its seventh Joint STARS aircraft. An eighth aircraft is due to be delivered in 2000, and the contract calls for four more to be produced by December 2001. The US Air Force is reported to want to increase its fleet of JSTARS aircraft above the present plan for 13 aircraft.

Contractor

Northrop Grumman Corporation, Electronic Sensors and Systems Sector.

UPDATED

Multifunction Integrated Radio Frequency System (MIRFS) for Joint Strike Fighter (JSF)

A team led by Northrop Grumman's Electronic Sensors and Systems Sector (ESSS) is working under a US$48.5 million contract from the Joint Strike Fighter (JSF) programme office to design, build and flight test the Multifunction Integrated Radio Frequency System (MIRFS).

Under the contract, the ESSS-led team is designing a new active electronically scanned array, a MultiFunction Array (MFA), which will help to reduce the future JSF aircraft's avionics system cost by 30 per cent and its weight by 50 per cent. Other team members are Litton Advanced Systems Division, Raytheon Electronic Systems, BAE Systems North America and Harris.

The Northrop Grumman antenna will perform radar, high-gain electronic support measures, and Communication/Navigation/Identification (CNI) functions. The antenna will be flight-tested using existing radar support electronics from Northrop Grumman's APG-77 radar being developed for the F-22 aircraft, and a Commercial-Off-The-Shelf (COTS) emulator of the Integrated Core Processor (ICP). Software will be reused from the F-22 radar and from Northrop Grumman Norden Systems' AN/APG-76 radar. Northrop Grumman will flight-test the new MFA in the company's BAC 1-11 testbed aircraft against various airborne and ground-based targets, such as the SCUD missiles used during the Gulf War. The radar is reported to use pulse-to-pulse MTI and SAR technology to provide two-dimensional imaging of the target.

In parallel, Northrop Grumman is developing a Common Integrated RF (CIRF) concept that will combine radar and electronic warfare equipment into an advanced open architecture RF environment. The CIRF will replace the radar support electronics used in the initial flight tests and along with the advanced antenna and ICP will form a complete flying prototype of the MIRFS.

Operational status
Development.

Contractor
Northrop Grumman Corporation, Electronic Sensors and Systems Sector.

VERIFIED

The Boeing 737 AEW&C aircraft, showing the MESA radar antenna on the top of the fuselage **2000**/0079022

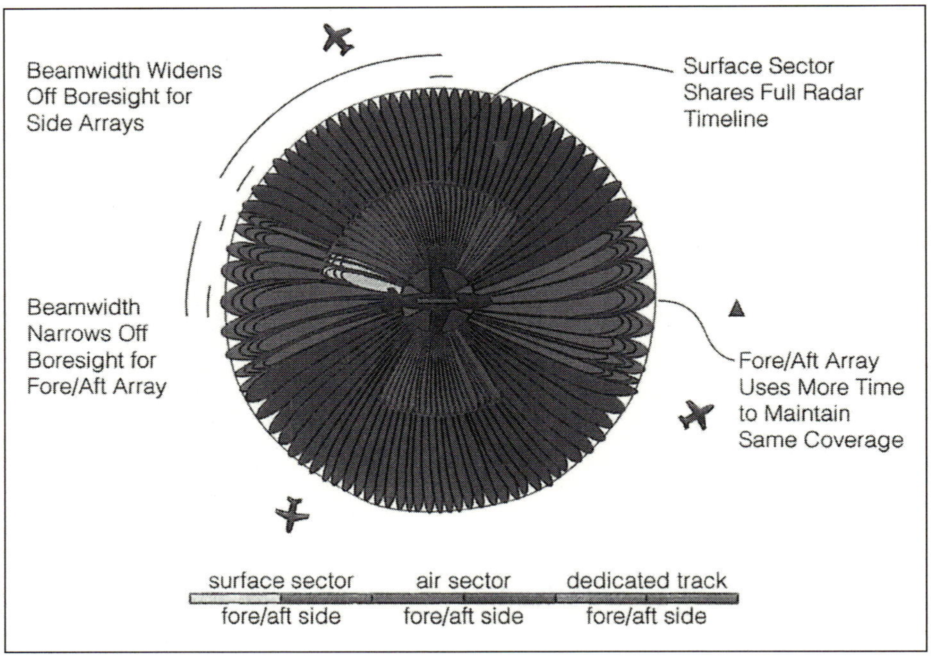

Boeing 737 AEW&C typical radar scan pattern **2000**/0079024

Multirole Electronically Scanning Aircraft (MESA) system

The MESA system is a low-cost, high-performance modular airborne surveillance radar system, installed on a slightly modified Boeing 737-700 aircraft, as the main sensor of the Boeing 737 Airborne Early Warning & Control (AEW&C) system.

Northrop Grumman Electronic Sensors and Systems Sector (ESSS) is responsible for the MESA radar; Boeing Information & Surveillance Systems is performing all aircraft modification and systems integration work.

The MESA radar supports a variety of air surveillance missions including airborne, ground, maritime, environmental, drug interdiction and border patrol missions. The system provides for the detection of low-flying aircraft, helicopters, tactical ballistic missiles and both stationary and moving targets in a clutter environment.

Other sensor avionics that can be integrated with MESA to create a comprehensive AEW&C capability include IFF, ESM, FLIR and LLTV, all coupled to a suite of workstations and voice and data communications facilities.

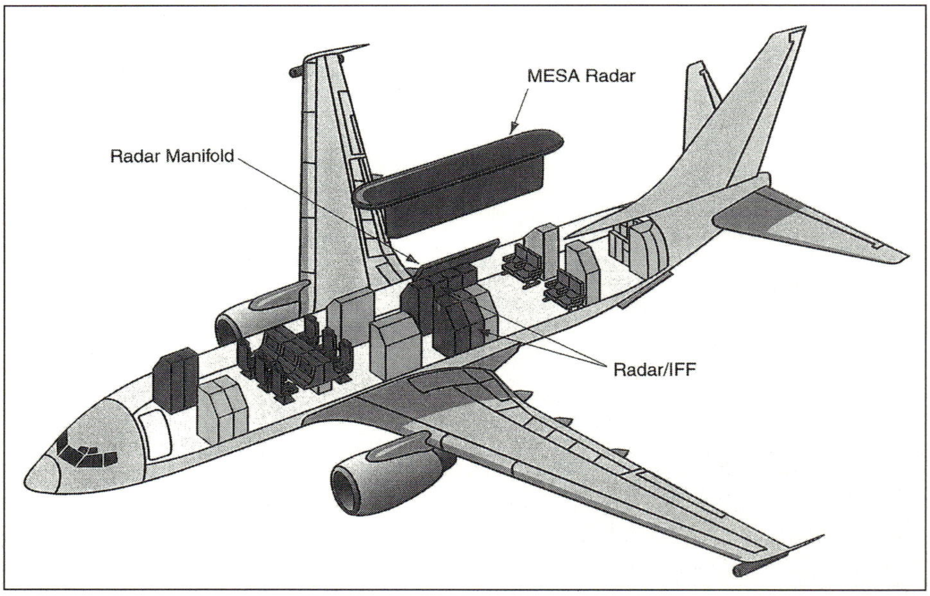

The Boeing 737 AEW&C equipment configuration **2000**/0079023

The MESA radar utilises a static, electronically scanned antenna on the top of the aft main fuselage. There are three antenna arrays, arranged as a dorsal fin with an array on each side, and a 'top hat' antenna array atop the dorsal fin.

The dorsal fin arrays provide coverage +/–60° from the broadside axis, the 'top hat' array covers +/–30° from the fore and aft axis (with overlaps). The operating frequency is 1,200 to 1,400 MHz, and the antenna arrays comprise 'hundreds' of silicon carbide solid-state transmit/receive elements.

The MESA radar provides 360° horizontal coverage to 190 n miles. It functions in three different modes: low Pulse Repetition Frequency (PRF) for ground and maritime detection; high PRF for air-to-air detection; and spot mode for detailed tracking of high-priority targets and enhanced long-range performance. 360° scan time can be varied by the operator, between 3 and 40 seconds, and it can be optimised between the three available modes during every rotation. Sector width is constrained by beamwidth, which varies between 2 and 8°. Increased range and higher probability of detection is provided in the spot mode by use of an increased beam dwell period to increase average power on target.

The three antenna arrays provide Identification Friend or Foe (IFF) capability, as well as active radar. The arrays cannot function simultaneously in both modes, the 'top hat' section may function in IFF-mode, while the dorsal fin arrays are in active radar mode, and vice versa.

The mission system, implemented by Boeing Information & Surveillance Systems, is an open system architecture derived from experience gained in the AirBorne Laser and Nimrod MRA 4 mission system programmes. Up to 14 workstations can be fitted. All workstations are similarly configured. All positions can be used for the full airborne early warning and control function, including management of communications. Radar and communications management functions have been automated to the point where no specialist operators are required for these activities, and no specialist airborne maintenance position is planned. The workstations are configured against the skin of the aircraft to maximise space and minimise weight.

The open systems architecture permits integration of other onboard sensors, including infra-red, television and electronic support measures systems. It can also interface datalink systems to co-operating air or ground platforms for data transfer. Boeing says that the overall Human Machine Interface (HMI) will be compatible with the NATO AWACS mid-term upgrade programme.

The 737-700 aircraft is fitted with additional fuel tanks, and the strengthened wing and landing gear of the 737-800. An additional distinguishing feature is a fin below the rear fuselage added for aerodynamic stability. At present it does not contain any electronic components, but its location appears to be suitable for self defence equipment.

The flight deck is standard 737-700, but with the addition of a sixth flat screen display, between the pilots, for presentation of mission tactical data and threat scenario information. The Marconi Electronic Systems HUD 2020 Visual Guidance System (VGS), certified on the 737-800, is also available, and forms a part of the Wedgetail installation.

Mission ceiling is 41,000 ft, with normal operations between 29,000 and 40,000 ft. Time on task at 300 n miles is 8 to 8.5 hours, without flight refuelling. A flight refuelling capability is an option.

Operational status

A radar proof-of-concept flight demonstration was completed in 1994, and flight testing continued until 1996.

Boeing Information & Surveillance Systems incorporated the MESA radar in its Boeing 737 Airborne Warning and Control (AEW&C) proposal for the Royal Australian Air Force (RAAF) Australian Project Air 5077 Wedgetail requirement. A contract award for seven aircraft was made in 1999, with first aircraft delivery planned for 2004. It was reported in May 2000, that the contract had been revised down to six aircraft with an option for the seventh aircraft. The total contract cost is estimated to be US$1.18 billion.

Contractor

Northrop Grumman Corporation, Electronic Sensors and Systems Sector.

UPDATED

OASYS Obstacle Avoidance SYStem

Northrop Grumman Corporation, in conjunction with the Night Vision and Electronic Sensors Directorate (NVESD) of the US Army Communications and Electronics Command (CECOM), has developed an Obstacle Avoidance SYStem (OASYS). It provides obstacle detection and situational awareness for helicopter low-altitude flight operations, enabling military helicopter pilots to fly at very low altitudes and avoid obstacles such as wires, cables, towers, trees and terrain features.

The OASYS uses a solid-state laser diode radar to detect wires, towers, poles and antennas. The laser, along with a rotating holographic scanner, generates a wide-area scan in front of the helicopter. OASYS can detect 25 mm wires at ranges up to 400 m in poor weather (2 km visibility).

The system comprises a sensor, processor and pilot control panel. Design criteria includes high-detection probability, low cost, weight and volume and high reliability. Information is presented on a HUD or can be fed to a helmet-mounted display.

Operational status

Under development. OASYS has undergone successful trials on a US Army JAH-IS helicopter and on the NUH-60 STAR (Systems Test bed for Avionics Research) helicopter as part of the Automated Nap Of the Earth (ANOE) programme being conducted by NASA's Ames Research Centre. The UK Defence Evaluation and Research Agency (DERA) is currently conducting ground trials of the prototype system prior to a potential future flight test.

Contractor

Northrop Grumman Corporation, Electronic Sensors and Systems Sector.

NEW ENTRY

AN/AAQ-8(V) (QRC 84-02) infra-red countermeasures pod

The AN/AAQ-8(V) is a multithreat infra-red countermeasures system capable of operating in a supersonic environment. This pod is a second-generation system updated to meet new and continuing threats, and has been extensively deployed. The system is mounted in an aerodynamically faired pod and can be configured with a ram-air turbine, allowing protection independent of aircraft power and cooling resources.

Specifications

Dimensions: 2,290 (length) × 254 mm (diameter)
Weight:
(AAQ-8(V1)) 107 kg
(AAQ-8(V2)) 120 kg
Power supply: 115 V AC, 400 Hz, 3 phase, 4 kVA
28 V DC, 20 W

Operational status

In service on fixed-wing combat and transport aircraft including A-7, C-130, F-4, F-5E and Mirage F-1C.

Contractor

Northrop Grumman Corporation, Integrated Systems Sector.

UPDATED

AN/ALQ-99F Tactical Jamming System (TJS)

The AN/ALQ-99 is a large and sophisticated ECM jammer designed for the SEAD Suppression of Enemy Air Defences role for the US Navy EA-6B and US Air Force EF-111A aircraft.

It has been operational for over 20 years, and has undergone numerous upgrades to meet changing operational requirements and to take advantage of technical advances. During this time, it has progressed through nomenclature changes up to the current configuration, which is believed to be known as AN/ALQ-99F(V).

It was originally designed in two configurations: as a full internal fit for the EF-111A; and as a five external underwing/fuselage pod-fit for the EA-6B.

Since the EF-111A has now been withdrawn from US Air Force service, the EA-6B aircraft, and with it the AN/ALQ-99 ECM system is being further upgraded to meet both the US Navy and the US Air Force SEAD requirements, and many more years' operational service is expected from the EA-6B/ALQ-99 system.

The EA-6B configuration comprises either two or three operator positions, from which specialist crew members exercise automatic/semi-automatic/manual control over the System Integrated Receiver (SIR) situated in the aircraft tail fin, and the five underwing/fuselage pods. Each pod includes transmitter elements and high-gain electronically steerable antennas (in both nose and tail), together with ram-air turbines that provide the requisite electrical power.

The ALQ-99 was designed from the outset to meet specific threat requirements with very high effective radiated power. To meet this design aim, the pod has always been fitted with high-gain, directional antennas, and specific limited bandwidth transmitters. The original version only covered four bands, but additional bands have been added through the years to meet new operational requirements, and coverage is now understood to be from VHF up to I/J-band in separate frequency-specific bands: designated from band 1 up to band 9/10.

Recent upgrades to the ALQ-99 pod have been integrated with EA-6B aircraft upgrades. In 1976, ICAP 1 (Increased CAPability) improvements were introduced to reduce processing time, add new displays, and improve communication, navigation and IFF functions. Then, in 1982, ADVCAP (ADVanced CAPability) was implemented to update the OR-262 receiver/processor and ALR-73 passive detection system.

In the 1980s, ICAP 2 configurations, designated -82, -86 and -89 for the years of their introduction were implemented. Current ICAP 2 Prowlers carry five integrally powered pods with 10 very high-power jamming transmitters. Each pod contains an exciter that optimises the jamming signals for transmission by two powerful transmitters. The ICAP 2 Prowler carries exciters that generate signals in any of seven frequency bands. Each pod can jam in different bands simultaneously. In addition, the Prowler can carry any mix of pods, depending on mission requirements.

The latest defined configuration is designated ICAP 2 Block 89A. The ICAP 2 Block 89A configuration was flown for the first time in June 1997. US Naval approval of the Block 89A configuration was originally planned for 1999, with approval for a fleet fit of 104 aircraft.

ICAP 2 Block 89A is understood to include the following upgrades to the ALQ-99 system: an upgrade to the universal exciter, low- and high-band transmitters; and the AN/USQ-113(V)2 Phase III communications jammer. It is also understood to include the following avionics upgrades: new radios (AN/ARC-210); an embedded inertial navigation

The EA-6B aircraft fitted with AN/ALQ-99 pods on both outer wing stations and under the belly **2000**/0085270

system; a global positioning system, as well as hardware and software enhancements to the navy's standard mission computer AN/AYK-14; new instrument landing system and COTS electronic flight instrumentation system.

In January 1998, Tracor Aerospace Electronic Systems Inc (now part of BAE Systems North America) was awarded a US$60 million contract to manufacture 120 AN/ALQ-99 Band 9/10 transmitters for the US Navy Air Systems Command (with options for 61 additional transmitters and spares). A further contract for the fabrication of 84 Band 9/10 transmitters was issued in 1999, value US$42 million. The company is also involved with production of AN/ALQ-99 Bands 1 and 2, as well as development and production of the Low-Band Transmitter.

After a common configuration is achieved for 104 aircraft in mid-2000, ICAP 3 will be launched; it will add a reactive jamming capability, which is understood to include a receiver upgrade, upgrades to the HARM targeting system, and improved narrowband jamming capabilities against modern SAM systems. ICAP 3 will also include a variety of communication datalinks and displays for enhanced situational awareness.

The ICAP III programme is also understood to include the requirement for the wiring and interface for the MIDS (Multifunction Information Distribution System) to enable the EA-6B to transmit data to, and receive from, other US military airborne and ground locators, as well as NATO platforms. ICAP 3 Initial Operational Capability (IOC) is slated for 2004.

The latest variant of the ALQ-99 is designated the ALQ-99F.

Contractors

Northrop Grumman Corporation, Integrated Systems Sector

Many contractors have been involved with the AN/ALQ-99/EA-6B system through the years. It is understood that the current ICAP 2 Block 89A team is led by the Northrop Grumman Corporation, Integrated Systems Sector, with Sanders, Litton Advanced Systems Division, and PRB Associates.

In early 1998, it was announced that the ICAP 3 programme valued at US$150 million had been awarded to Northrop Grumman, Bethpage. Initial operating capability is planned for early 2004, and the aircraft are expected to remain in operational service until at least 2015. Subcontractors involved in the project include:

(1) Litton Advanced Systems Division for the passive receiver technology;
(2) Sanders for upgrades to the AN/USQ-113 tactical communications jamming system, and for new tactical displays;
(3) PRB Associates for an updated mission computer and new display software;
(4) COMPTEK Amherst Systems Inc for advanced jamming software.

UPDATED

AN/APR-50 defensive management suite

AN/APR-50 is the US Air Force designation for the ZSR-63 ESM suite and radar warning receiver for the B-2 aircraft. The system is classified and very few technical details have been released. It would appear, however, that the complete installation consists of a number of antennas feeding nine radio frequency front ends to detect and analyse a wide variety of signals. It is assumed that each of the front ends is tuned to a different part of the frequency spectrum. Five receivers receive the outputs of the front ends and pass these to the processor. The APR-50 is believed to be linked to the ZSR-62 system, a classified part of the B-2 defensive management suite in development by Northrop Grumman which employs advanced concepts in EW technology.

Operational status

In January 1993, Northrop Grumman was awarded a US$117 million contract to continue development of the AN/APR-50. Northrop Grumman was also awarded US$53.9 million to carry on with ESM development, including extension of the frequency range. It is believed that this originally was for Band 2 and the extension was to cover Band 4 from 500 MHz to 1 GHz.

Contractor

Northrop Grumman Corporation, Integrated Systems Sector.

UPDATED

The MIRTS installed on a UK Royal Air Force Raytheon Hawker 800 (Paul Jackson)

Modularised Infra-Red Transmitting System (MIRTS)

The Modularised Infra-Red Transmitting System (MIRTS) is a derivative of the AN/AAQ-8(V). It is an advanced subsonic, infra-red countermeasures system for deployment in a wide range of aircraft, including helicopters, which can be carried internally or pod mounted. MIRTS utilises advanced jammer technologies including a variable optics/reflector design to provide optimum aircraft infra-red signature coverage, combined with advanced digital electronics and mode-switching power supplies to enhance system reliability, maintainability and versatility.

Specifications

Transmitter/receiver
Dimensions: 228 × 240 × 635 mm
Weight: 23.6 kg
Power supply: 115 V AC, 400 Hz, 3 phase, 3.3 kVA
28 V DC, 5.6 W

Operator Control Unit
Dimensions: 146 × 57 × 127 mm
Weight: 0.68 kg
Power supply: 5 V AC, 400 Hz, 3 phase, 28 V DC, 10 VA

Electronics control unit
Dimensions: 190 × 259 × 318 mm
Weight: 6.6 kg
Power supply: 115 V AC, 400 Hz, 3 phase, 308 VA
28 V DC, 22.4 W

Operational status

In service. The system has been tested on the rear fuselage of a UK Royal Air Force VC10. Aircraft installations include the Boeing 707, 747 and DC-8, Eurocopter Puma; Falcon 20, Fokker 27, and Raytheon Hawker 800.

Contractor

Northrop Grumman Corporation, Integrated Systems Sector.

UPDATED

Starfire self-protection suite

The Starfire laser-based Infra-Red CounterMeasures (IRCM) self-protection suite has been developed as the ATIRCM solution. Its jamming effectiveness is increased by the use of superior sightline stabilisation in the most demanding flight environment. The heart of the system is an advanced and combat proven pointing and tracking system, keeping Starfire's laser accurately on the target. Starfire represents an integrated and reliable IR defensive self-protection suite that can be fitted on military and commercial aircraft.

The Starfire suite provides fast accurate threat missile location in a combined missile approach and warning surveillance system. The high-power jammer acts only on demand from an alert from an approaching missile. The suite tracks missiles in all modes of operation. All-aspect self-protection is provided with a power-managed architecture. Full fitting compatibility is designed into the system for a wide range of helicopters and fixed-wing aircraft.

Contractor

Northrop Grumman Corporation, Integrated Systems Sector.

UPDATED

Tiger IV Avionics

Northrop Grumman has integrated digital weapons systems avionics into an F-5E Tiger IV under a co-operative research and development agreement with the US Air Force. The enhanced F-5 features improved pilot situational awareness display, advanced multimode radar capabilities and a ring laser/GPS navigation unit, integrated via a MIL-STD-1553B databus.

Participating suppliers include Honeywell Aerospace, standard air data computer and inertial navigation system; Strategic Technology Systems Inc, stores management system; Litton Systems Inc, Guidance and Control Division, airborne video camera, mission computer; Northrop Grumman ESSS, AN/APG-66 radar; Fairchild Defense OSC, data transfer system; TEAC America Inc, airborne video tape recorders, Thomson-CSF DETEXIS EWS-A RWR.

Operational status

Tiger IV Avionics has completed flight test evaluation.

Contractor

Northrop Grumman Corporation, Integrated Systems Sector.

UPDATED

AN/AIC-28(V)1, 2, 3 and 4 audio distribution systems

The AN/AIC-28(V)1 audio distribution system provides COMSEC-compatible internal communications among aircrew and direct access communications between them and personnel in support of aircraft mission and maintenance activities. The Audio Distribution System (ADS) consists of up to 27 subscriber control panels located at the various crew stations, two central switching units to route audio traffic and a programming, display and test panel to control, monitor and test the ADS.

The key to achieving ADS secure communication performance lies in the microminiature LSI design which provides extremely low levels of cross-talk (greater than 100 dB between communication channels). Interlock circuitry prevents any subscriber from transmitting simultaneously on a clear and a secure or classified channel. The interlock also inhibits all direct access transmissions from a subscriber station initiating a public address announcement.

Any subscriber control panel or station can signal any other subscriber station for a private conversation via a four-channel selective intercom. Other subscribers can be added to form progressive conferences.

Access to radios by subscribers is accomplished remotely from the programming, display and test panel. Subscribers are provided with panel controls for selection of monitor and transmit functions and controls for adjusting the receive volume level for up to four radios at a time. The ADS retains all selective interphone and radio access operations in the event of an aircraft power failure.

The programming display and test panel enables a control operator to programme selectively the direct access radio and intercom nets to the various subscribers. It also displays the current programme status by visual display of the channels assigned to each subscriber. The programming display and test system isolates faults down to a replaceable primary unit level.

The central switching unit interfaces directly with the system's subscriber stations and serves as an audio distribution matrix. It consists of an array of audio multiplexer devices with appropriate station and channel input/output audio buffers and station data buffers. All audio switching and mixing is performed within the central switching unit so that, irrespective of the total number of channels a subscriber chooses to monitor simultaneously, composite earphone audio is conveyed from the central switching unit to the subscriber station via a single twisted cable pair. Conversely, when a subscriber transmits, the microphone audio is conveyed on a single twisted pair to the central switching unit which distributes it to other subscriber stations and to radios on the channels on which he is transmitting. The two central switching units operate synchronously in a master/slave relationship.

Four separate types of subscriber station panels can operate with the system: a flight deck audio panel, a mission audio panel, a special audio channel and maintenance audio panel.

All flight deck and mission audio panels can initiate public address announcements which are broadcast over all headsets and loudspeakers in the aircraft. When in public address mode, all direct access radio transmission from the originating subscriber station is inhibited, and when originating from the flight deck, the speaker nearest the originating station is muted. All-weather 127 and 203 mm speakers are provided. The public address amplifier is packaged for standard air transport racking. The package contains controls and an output power meter for testing and adjusting the 120 W main amplifier, a 16 W auxiliary amplifier and the speakers. Provision is made for the incorporation of recorders to log all two-way external communications.

The system has a built-in self-test capability at both system level and at subscriber station level. Replacement of primary units is possible during mission operations. A semi-automatic test station is available to facilitate ground maintenance and repair of primary units.

System features include the automatic assignment of nets on the selective intercom together with logic to decode call digits and determine which nets are available to the respective subscribers; capability of subscribers to transmit on more than one communication net; direct access to as many as 22 radios with additional access from flight crew stations to seven navigation monitors, and capability for distributing audible alarm signals to subscribers in response to sensors and switches external to the ADS.

The AN/AIC-28(V)2 and AN/AIC-28(V)3 audio distribution systems are identical to the AN/AIC-28(V)1 system except for an increased number of subscriber audio stations and direct access radios available for mission operations. These changes required expansion of the programming display and test panel and the central switching unit.

The AN/AIC-28(V)4 tanker audio distribution system provides secure communications between crew members and radio equipment. The system consists primarily of sunlight-readable refuelling audio panels located at stations for the boom operator, observers and loadmaster; flight deck audio panels and their associated clear-secure panels, located at the aircraft commander's, pilot's, flight engineer's and navigator's stations; central switching units to route the audio traffic; a radio select panel to programme access to radios, and an ADS test panel to test and monitor operation of the system.

The system audio traffic is divided into clear and secure communications. The central switching unit interlock circuitry prevents compromise of secure communications, such as transmitting simultaneously on a clear and secure channel. Any subscriber panel can be used to signal other subscribers for private or conference conversations, or can be used to make public address announcements, which are broadcast over the headsets and loudspeakers.

Operational status

In service. The AN/AIC-28(V)1, 2 and 3 are no longer in production.

Contractor

Palomar Products Inc.

VERIFIED

AN/AIC-29(V)1 intercommunication system

The AN/AIC-29(V)1 intercommunication system provides secure internal communications between helicopter crew members and direct access between crew members and radios and/or security equipment for external communications. The system can accommodate up to 15 transmit and 18 receive radio channels and provides greater than 100 dB cross-talk isolation between a transmit and any other transmit or receive channel. The system consists of six crew station units, a single maintenance station unit and a communications switching unit. The crew station units are NVG-compatible. The communications switching unit performs switching and mixing of audio channels in accordance with digital data multiplexed from each crew station unit. The switching unit's response to the multiplexed data depends on the communication plan programmed into the system.

The system provides emergency back-up intercom and radio communication selection which bypasses the communications switching unit for audio transmit and receive functions.

System features include the capability for distributing audible alarms to crew stations in response to sensors/switches external to the intercommunication system. The system also provides interface with the MIL-STD-1553 databus, two-way chime call capability and built-in test circuits.

Operational status

In production for the US Customs Service's fleet of P-3 aircraft for anti-drugs operations. Six shipsets were ordered in September 1999 for installation by Lockheed-Martin Aeronautical Systems in Greenville in 2000.

Contractor

Palomar Products Inc.

VERIFIED

AN/AIC-30(V)1 and (V)2 intercommunication systems

The AN/AIC-30(V)1 intercommunication system provides communications among helicopter crew members and direct access to radio communications in support of mission activities. The system includes a microprocessor-controlled communications switching unit which is controlled by multiplexed data from six crew station units during normal system operation. Interlock circuitry prevents a crew member from transmitting on a clear channel while receiving on a secure channel. Cross-talk isolation between any two channels is greater than 126 dB at 1 kHz.

An intercom back-up call channel, which bypasses the switching unit matrix, may be activated at the crew station units. This back-up channel may be constantly monitored using the crew station unit master volume control. In addition, two cockpit crew station units are provided with a direct back-up interface to two radio transceivers. The intercom back-up call channel is also used for extended ground communication operations with the switching unit power removed. Two audio frequency amplifiers provide volume control for communications at the maintenance stations.

The AN/AIC-30(V)2 intercommunication system is an enhancement to the AN/AIC-30(V)1 intercommunication system for use in larger aircraft. The AN/AIC-30(V)2 system increases the number of crew stations and audio frequency amplifiers, modifies the communications switching unit, expands the radio channels and hard-wired back-up channels and provides dual-headset and microphone capability at each crew station unit.

The communications switching unit modification includes added circuit cards and wiring for multiplexing control and audio access transmit and receive. A fourth connector is included on the front panel of the switching unit to accommodate the additional audio traffic.

The crew station unit front panel changes reflect the additional radios and navaids utilised for the system. Each crew station unit provides dual-headset capability. The cross-talk isolation between transmit and receive channels is greater than 126 dB at 1 kHz. The crew station unit also provides intercom back-up circuits and is NVG-compatible. The three audio frequency amplifiers provide headset volume control at the maintenance or remote crew stations.

Operational status

In service. The AN/AIC-30(V)1 is no longer in production.

Contractor

Palomar Products Inc.

VERIFIED

AN/AIC-32(V)1 intercommunication system

The AN/AIC-32(V)1 intercommunication system provides secure internal communications between crew members and direct access to mission radios and security equipment for external communications. The system's primary units consist of a communication control unit, four flight deck crew station units, five mission area crew station units, eight maintenance station units and one maintenance control unit.

Audio traffic from all station units is controlled by the communication control unit during normal system operations. A back-up operating mode, initiated from selected crew station units, bypasses the communication control unit to provide hard-wired access to a set of predetermined radios. A back-up intercom network, integrated with the call function on all station units, is also provided for emergency intercommunication between crew members.

A test switch on each crew station unit permits preflight verification of audio and lamp indicator circuitry and digital interface with the communication control unit. A public address system, accessed from the flight crew station, allows announcements over the loudspeakers and headsets. Auxiliary control units, located at selected mission area crew station units, expands the total system direct access capability to 30 transmit and 36 receive channels.

Operational status
In production and in service.

Contractor
Palomar Products Inc.

VERIFIED

AN/AIC-34(V)1 and (V)2 intercommunication systems

The AN/AIC-34(V)1 secure intercommunication system is a programmable microprocessor-controlled modular audio and digital communication distribution system designed for use on board fixed- and rotary-wing aircraft or in ground-based C³ shelters. The AN/AIC-34(V)1 provides internal communications between crew members as well as crew member access to mission radios and communication security equipment for external communications. Designed in accordance with MIL-E-5400, it is qualified to MIL-STD-810 for air transport, MIL-STD-461 for EMI and NACSIM 5100 for TEMPEST compliance.

The AN/AIC-34(V)1 intercommunication system comprises a Communication Control Unit (CCU), five Crew Station Units (CSU), 16 Control Display Units (CDU), an Emergency Audio Panel (EAP), three Maintenance Station Units (MSU) and 26 jack boxes.

The CCU is a microprocessor-controlled modular audio switching and control unit that provides up to 32 crew stations with access to as many as 48 receive channels and 30 transmit channels. It also accepts up to 70 binary discrete inputs and provides up to 122 switched relay outputs for discrete control and crypto switching.

The CSUs and CDUs are the crew member interfaces to the communication control unit and communication assets. The functions of channel select switches and volume controls on the CSU front panel are firmware dependent and are reconfigurable to meet the needs of specific missions or system requirements. The CDU allows operator assignment of communication assets and displays communication system operational and BIT status on an integral eight-colour CRT display. Both units are capable of operating with two headsets when operated in a monaural mode or can operate with one headset when operated in a binaural mode. Monaural or binaural operating modes are configured by jumpers at the unit connectors in the aircraft wiring.

The MSU provides intercom access for maintenance personnel and ground crew. Each communication control, control display and maintenance station unit is connected to a jack box. This allows switching of multiple microphone inputs and provides for switching of audio from an auxiliary source or attached unit to various headset or speaker interfaces.

The EAP provides an operator with the capability to select multiple levels of degraded or emergency back-up modes of operation to provide for continued operation in the event of hardware failures or battle damage.

The AN/AIC-34(V)2 secure intercommunication set is a subset of the AN/AIC-34(V)1 system, and is used in airborne military applications where the crew complement is significantly less than that of the AIC-34(V)1 but the requirement to access numerous radio channels and the need for the ability to reconfigure the operating modes of the onboard communication assets are similar. The AN/AIC-34(V)2 intercommunication set is designed to be compliant with the requirements of MIL-E-5400, MIL-STD-810, MIL-STD-461 and NACSIM 5100.

The set comprises a CCU, five CSUs, two CDUs, an EAP, three MSUs and eight jack boxes.

The CCU is the same as that used in the AN/AIC-34(V)1 except that it contains an additional set of relay circuit cards to provide for greater flexibility and expanded switching modes of the communication assets. The firmware resident in this CCU is also unique to the requirements of the AN/AIC-34(V)2.

The CSU, CDU, MSU, EAP and jack boxes are identical to those used in the AN/AIC-34(V)1 except for the CSU front panel switch legends.

Specifications
Dimensions:
(CCU) 497.6 × 257 × 193.7 mm
(CSU) 165 × 146 × 152.4 mm
(CDU) 165.1 × 146 × 152.4 mm
(EAP) 106.7 × 146 × 47.8 mm
(MSU) 78.7 × 160 × 107.9 mm
(jack box) 150.5 × 127.5 × 43.2 mm
Weight:
(CCU) 19.8 kg
(CSU) 2.7 kg
(CDU) 3.6 kg
(EAP) 0.45 kg
(MSU) 1.03 kg
(jack box) 0.34 kg

Operational status
In service with the US Navy.

Contractor
Palomar Products Inc.

VERIFIED

AN/AIC-38(V)1 and AN/AIC-40(V)1 intercommunication systems

The AN/AIC-38(V)1 and AN/AIC-40(V)1 secure intercommunication systems are programmable microprocessor-controlled modular audio and digital communication distribution systems designed for use in both fixed- and rotary-wing aircraft or in ground-based C³ shelters. The AN/AIC-38(V)1 and AN/AIC-40(V)1 provide internal communications between crew members as well as access to mission radios and communications security equipment for external communications. Designed in accordance with MIL-E-5400, the systems are qualified to MIL-STD-810D for air transport, MIL-STD-461 for EMI and NACSIM 5100 for TEMPEST.

The systems consist of a Communication Control Unit (CCU), Crew Station Units (CSU), a Digital Switch Unit (DSU) and Emergency Audio Panel (EAP).

The CCU is a microprocessor-controlled modular audio switching and control unit that provides up to 32 crew stations with access to up to 48 receive channels and 30 transmit channels. It also accepts up to 70 binary discrete inputs and provides up to 122 switched relay outputs for discrete control and crypto switching. The communications connectivity plan resident within the CCU may be reconfigured in real time by a higher-order controller via MIL-STD-1553B databus, or by operator control via a CSU.

The CSU provides crew member interface to communication assets. Channel select keyswitches and volume controls on the CSU are firmware controlled and are reconfigurable to meet the needs of specific missions or system requirements.

The DSU, under CCU control, routes eight bidirectional channels of digital data and control lines to data terminals and radios or communication security equipment. Using the EAP, multiple levels of emergency back-up operation are operator selectable for continued operation after hardware failures or battle damage.

Operational status
In production and in service.

Contractor
Palomar Products Inc.

VERIFIED

AN/AIC-39(V)1 intercommunication system

The AN/AIC-39(V)1 secure intercommunication system is a programmable microprocessor-controlled modular audio and digital communication distribution system designed for use on board both fixed-wing aircraft and helicopters or in ground-based C³ shelters. The AN/AIC-39(V)1 provides internal communications between crew members as well as crew member access to mission radios and communications security equipment for external communications. Designed in accordance with MIL-E-5400, the AN/AIC-39(V)1 is qualified to MIL-T-5422 and MIL-STD-810D for air transport, MIL-STD-461 for EMI and NACSIM 5100 for TEMPEST.

The system consists of a Communication Control Unit (CCU), Crew Station Units (CSU), Dual Crew Station Units (DCSU), Audio Amplifier Units (AAU), Auxiliary Control Units (ACU), an Emergency Audio Panel (EAP) and secure jack box.

Crew station unit for the AN/AIC-34, AN/AIC-38, AN/AIC-39 and AN/AIC-40 intercommunication sets

The CCU is a microprocessor-controlled modular audio switching and control unit that provides up to 32 crew stations with access to up to 48 receive channels and 30 transmit channels. It also accepts up to 70 binary discrete inputs and provides up to 122 switched relay outputs for discrete control and crypto switching. The communications connectivity plan resident within the CCU may be reconfigured in real time by a higher-order controller, via MIL-STD-1553B databus, or by operator control via a CSU or DCSU.

The CSU and DCSU provide crew member interface to communication assets. Channel select keyswitched and volume controls on the CSU and DCSU are firmware controlled and are reconfigurable to meet the needs of specific missions or system requirements. DCSUs are identical to CSUs except that the DCSUs are equipped with additional audio circuitry to support a subordinate ACU.

The ACU, used in conjunction with the DCSU, provides limited access to communication assets at remote crew positions. The ACU operates with a subset of the functions provided by the DCSU.

The AAUs provide a binaural headset and microphone interface to the CCU, but have no annunciators or indicators.

The EAP allows operator selection of multiple levels of emergency back-up operation. The back-up capability permits continued operation after hardware failures or battle damage.

Operational status
In production and in service.

Contractor
Palomar Products Inc.

VERIFIED

CCS-2100 Communications Control Systems

The Palomar Products CCS-2100 Communications Control Systems provide intercommunication solutions for secure applications. The secure facilities and extremely high channel-to-channel isolation, provide for multilevel security, private radio channels and private intercom channels.

The integrated, microprocessor-controlled, modular distribution system is designed for real-time adjustment of communication assets to meet the operational requirements of each mission.

The modular design permits systems to be designed to fulfil simple internal/external communication requirements or to provide full command and control systems that include crypto and datalink assignments, clear/secure radio relay, clear/secure simultaneous broadcast, unlimited conference networks and selective dial facilities. In addition, systems can be integrated with a host computer via a MIL-STD-1553B interface or other network links.

A typical CCS-2100 architecture provides a full TEMPEST-compliant communications control capability to flight deck crews and mission personnel that enables them to exercise control over the aircraft's crypto units, radio transceivers, mission receivers and navigation receivers.

Individual control units that comprise the CCS-2100 include Integrated Panel Crew Station Units (IPCSUs); Liquid Crystal Control Display Units (LCCDUs); Crew Station Units (CSUs); a Communication Control Unit (CCU) (full ATR or half ATR sizes available); Maintenance Station Unit (MSU); and Emergency Audio Panel (EAP).

Features include simultaneous clear and secure transmissions; TEMPEST compliance; ElectroMagnetic Pulse (EMP) hardened; voice and data switching with binaural audio; redundant and emergency back-up operation; recorder control/playback.

Operational status

Palomar Products' CCS-2100 systems are widely used in US military aircraft and in the military aircraft of other nations.

Contractor

Palomar Products Inc, Integrated Communication Systems.

VERIFIED

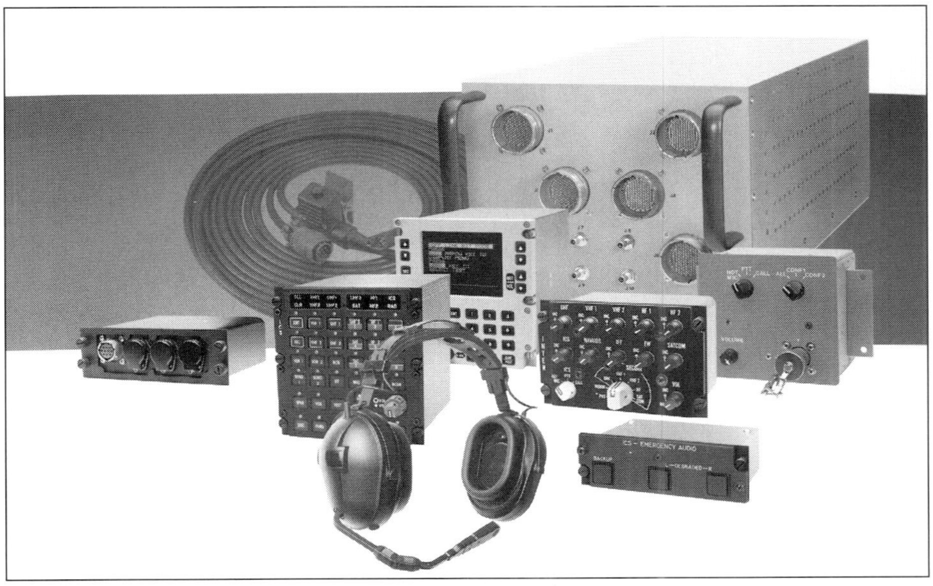

Palomar Products' CCS-2100 communications control system units **2000**/0062230

DSS-100 digital intercom system

The DSS-100 is Palomar's newest digital intercommunication system. It uses Commercial-Off-The-Shelf (COTS) components in an open VME architecture, providing a complete solution for systems demanding high-quality digital voice/data switching with combined system management.

Features include a distributive architecture, which can readily be tailored to customer requirements. The system is a secure digital switch with physically separate 'red' and 'black' buses. The red and black bus structure is maintained throughout all components of the communications suite, including the digital operator positions.

The DSS-100 provides complete management of the communications suite and co-ordinates modes of transceivers, cryptos, data modems and host computers with actions and indications at the operator positions.

The DSS-100 features include up to 2,048 channels; selectable bandwidths; direct signal processing voice algorithms; binaural audio (spatial optional); unlimited conferencing; a variety of colour liquid crystal touchscreen operator display panel styles.

Contractor

Palomar Products Inc, Integrated Communication Systems.

UPDATED

Palomar Products' DSS-100 digital intercom system components **2000**/0064384

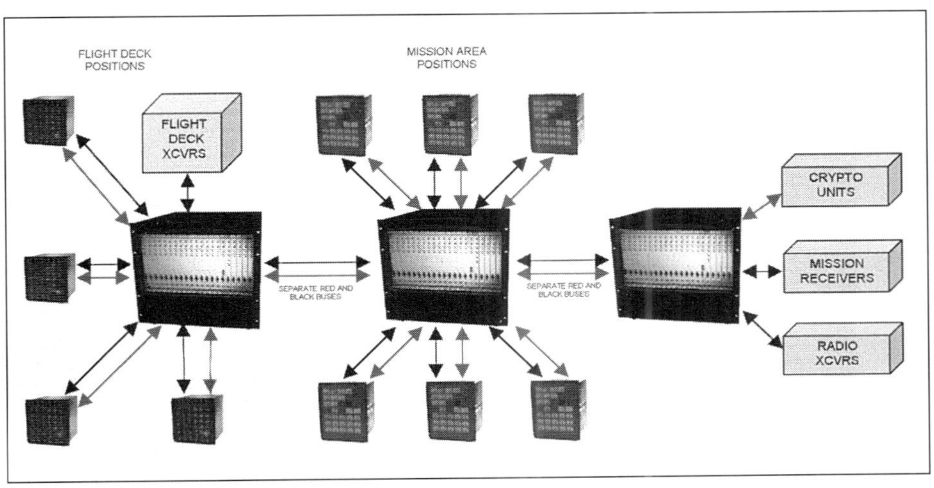

A representative system configuration for the Palomar Products DSS-100 digital intercom system **2000**/0064383

AAR-58 missile warning system

The AAR-58 is reported to be a private venture concept development by Raytheon Electronic Systems to produce a two-colour, scanning, infra-red warning system to overcome the deficiencies of ultra-violet warning systems in sunlight at high altitudes.

The scanning technique is being investigated to provide greater angular coverage than existing staring array systems and hence less stringent siting requirements on the airframe. It is anticipated that the AAS-58 would form part of a more comprehensive warning capability.

Contractor

Raytheon Electronic Systems.

VERIFIED

Advanced Narrowband Digital Voice Terminal (ANDVT)

The Advanced Narrowband Digital Voice Terminal (ANDVT) provides half-duplex secure voice and data communications for a variety of military tactical applications, including shipboard, land-based and airborne. Typical user modes include secure voice, data and signalling, point-to-point and modem processor only.

The CV-3591 Basic Terminal Unit (BTU) provides voice and modem processing by using two similar signal processors. Communications security is achieved by encoding and decoding the digital data to and from the voice processor external data device.

In the standard terminal configuration, the KYV-5 COMSEC Module (CM) is a front panel plug-in unit.

Specifications

Dimensions:
(BTU) 193.8 × 124.7 × 337.8 mm
(CM) 158.75 × 123.95 × 76.2 mm
(MPU/VPU) 157.5 × 124.7 × 75.4 mm
(interface unit) 146 × 69.85 × 285.75 mm
Weight:
(BTU) 9.89 kg
(CM) 1.63 kg
(MPU/VPU) 1.27 kg
(interface unit) 2.13 kg
Temperature range: –46 to +95°C
Altitude: up to 70,000 ft
Reliability: 2,000 h MTBF

Contractor

Raytheon Electronic Systems.

VERIFIED

Advanced Tactical Targeting Technology (AT3)

In October 1998, the US Air Force Research Laboratory, Wright-Patterson AFB awarded the Raytheon Electronic Systems a US$6.575 million cost-plus-fixed-fee contract to design, develop, and ground and flight demonstrate a passive tactical targeting system for the lethal Suppression of Enemy Air Defences (SEAD), known as the Advanced Tactical Targeting Technology (AT3) Programme. The expected project completion date is October 2002.

Contractor

Raytheon Electronic Systems.

VERIFIED

Airborne Integrated Terminal Group (AITG)

The AITG is being designed by Raytheon Electronic Systems, St Petersburg, Florida on contract to the US Air Force Electronic Security Command, Hanscom AFB.

The AITG is designed to provide the following operational links between high value US Air Force assets, such as E-3A AWACS and other US forces:

UHF satcom (DAMA) 240 to 320 MHz
UHF LOS 225 to 400 MHz (Have Quick II)
VHF 30 to 90 MHz Army (SINCGARS)
VHF 118 to 156 MHz ATC (8.33 kHz)
VHF 130 to 156 MHz LOS (land mobile)
UHF 405 to 512 MHz LOS (growth)

The AITG comprises: the AIT receiver/transmitter/modem family; the Remote-Control Unit (RCU) (also referred to as the enhanced remote-control unit); the LNA diplexer and system software.

The AIT receiver/transmitter/modem family comprises: the 30 to 512 MHz receiver/modem; 225 to 400 MHz UHF 100 W transmitter; AIT 30 to 400 MHz VHF/UHF 100 W transceiver; and personality modules to adapt the radio to platform AC or DC power and to provide baseband interfaces.

The 30 to 512 MHz receiver modem and 225 to 400 MHz modules can be configured to produce: a 30

Airborne integrated terminal group, (left to right) remote-control unit, receiver/transmitter/modem, DNA diplexer
1999/0044825

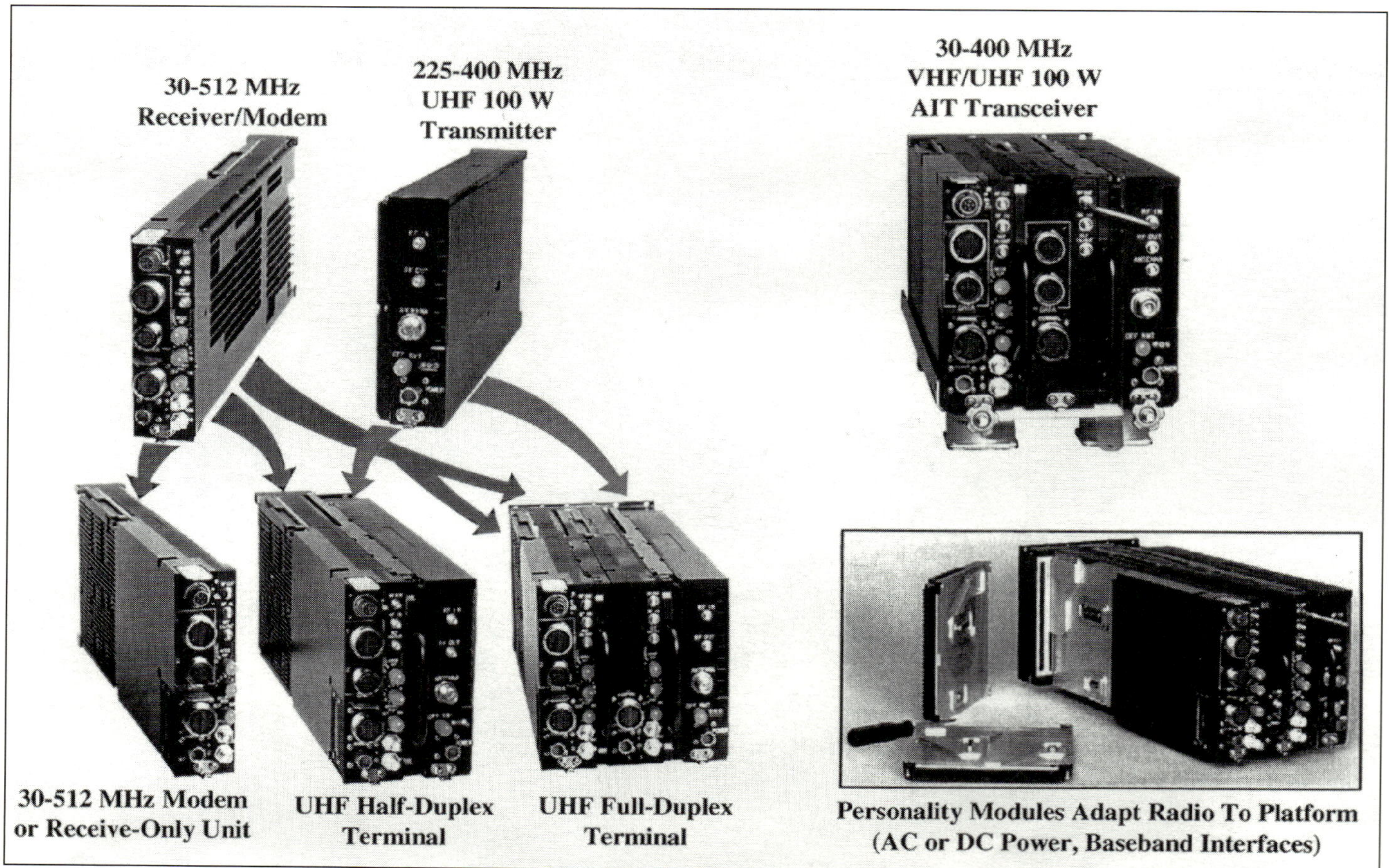

Airborne integrated terminal family
1999/0044826

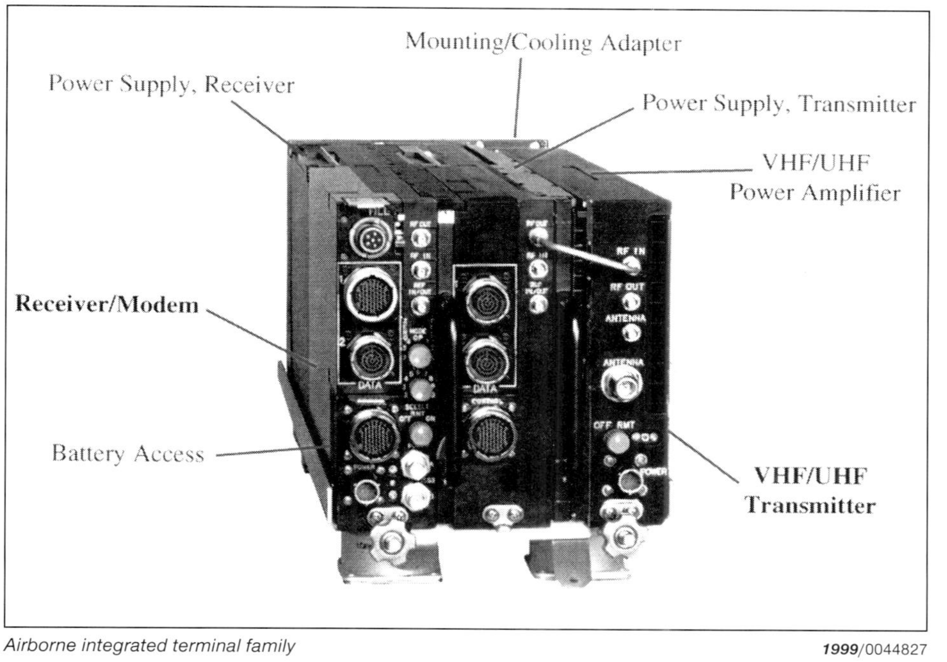

Airborne integrated terminal family ***1999***/0044827

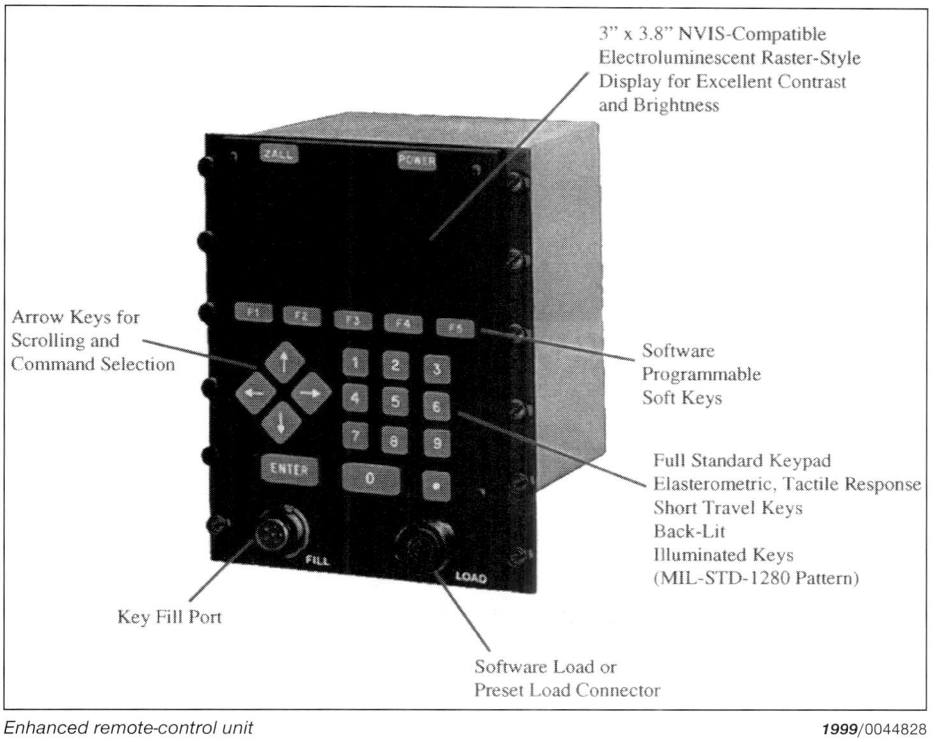

Enhanced remote-control unit ***1999***/0044828

to 512 MHz modem or receive-only unit; a UHF half-duplex terminal; or a UHF full-duplex terminal.

The modem functions at rates varying between 75 bps and 64 kbps (including 75, 300, 600, 1,200 and 2,400 bps).

The AITG will be delivered with the new MIL-STD-188-181B UHF satcom high-speed waveform: 9.6 kbps in 5 kHz-channel (2.4) ; and 48.0 kbps in 25 kHz-channel (16) to increase the UHF satellite capacity from 347 to 1,018 kbps.

Crypto functions include: KG-84; KY-58; KYV-5; KGV-10 and KGV-11.

The RCU is used to exercise network management and downloading of the mission plan. The RCU uses 3 × 3.8 in, Windows-style, NVIS-compatible screens; together with a full standard keypad, arrow keys, and software programmable soft keys; it can be upgraded to faster COTS processors. The design concept is that future capabilities should be added by software packages, to minimise cost of change.

Operational status
A total of 447 units being produced for US Air Mobility Command and Air Combat Command.

Contractor
Raytheon Electronic Systems.

VERIFIED

Airborne remote-controlled ESM system

The airborne remote-controlled ESM system consists of a number of radio receivers and antennas operated by remote radio control from a ground facility which is equipped with the necessary processing and display equipment. It is intended to be carried in small, relatively inexpensive aircraft and the system configuration can be tailored to the particular aircraft type. Onboard equipment would normally consist of radar detection units covering the frequency range 0.5 to 18 GHz, radio monitoring receivers covering 1.5 MHz to 2 GHz and radio direction-finding systems covering 20 to 1,200 MHz. The datalink is a full-duplex highly directional system.

Operational status
Fully developed as a private venture.

Contractor
Raytheon Electronic Systems.

VERIFIED

Airborne Shared Aperture Programme (ASAP)

The ASAP is sponsored by the US Naval Air Warfare Center and is intended to demonstrate a multifunction, broadband, shared aperture system for use on future air and sea systems. Wideband individual transmit and receive modules would make up the ASAP sensor aperture, providing the capability to transmit and receive radar, communications, ECM and ESM signals over a broad frequency band.

Operational status
A sole-source contract has been awarded to demonstrate ASAP technology.

Contractor
Raytheon Electronic Systems.

VERIFIED

AN/AAS-44(V) thermal imaging/laser designation system

The AN/AAS-44(V) is a multipurpose thermal imaging and laser designation system. It provides long-range surveillance, target acquisition, tracking, range-finding and laser designation for the Hellfire missile and for all US tri-service and NATO laser-guided munitions.

Operational status
The AN/AAS-44(V) is deployed on US Navy HH-60H helicopters and on the US Navy Light Airborne MultiPurpose System (LAMPS) MK III SH-60B armed helicopters as part of the Block I upgrade programme. Production continues for both programmes. By mid-1999, 69 systems were in US Navy service with a further 35 on order.

Contractor
Raytheon Electronic Systems.

UPDATED

AN/ALE-50 Towed Decoy System (TDS)

The ALE-50 TDS provides protection against RF threats. When deployed, the decoy seduces RF guided missiles away from the host aircraft. This stand-alone system requires no threat-specific software and communicates health and status to its host aircraft over a standard databus.

The ALE-50 TDS consists of a launch controller, launcher and towed decoy. The decoy control/monitor electronics and power supply are contained in the launch controller. The launcher, which holds the decoy magazine, can be customised to fit any candidate aircraft. The decoy has a 10-year shelf life and is packaged in a sealed canister, which also contains the pay-out reel.

The AN/ALE-50 launcher is 'dual compatible' for launching both AN/ALE-50 and AN/ALE-55 decoys.

Operational status
The ALE-50 programme was a joint US Navy/US Air Force/Raytheon Electronic Systems development. The MultiPlatform Launch Controller (MPLC) is the standard launch controller for all installations.

US Air Force F-16 aircraft were the first operational users, for which decoy production began in early 1997.

The F-16 ALE-50 installation is fitted in a specially adapted AMRAAM pylon manufactured by Lockheed Martin Tactical Aircraft Systems (Fort Worth). Each F-16 carries two pylons, Lockheed Martin is on contract to deliver 413 ALE-50 capable pylons by June 2000, to begin equipping the US Air Force's fleet of Block 40 and 50 F-16s. The US Air Force has a stated requirement for a further 548 pylons. Eventually, the US Air Force hopes to equip its Block 25, 30 and 32 with ALE-50.

Other users, for which launcher-specific installations have been developed, are the US Air Force B-1B (a 2 × quadruple launcher installation is being fitted as part of the Defensive Systems Upgrade Programme) and US Navy F/A-18E/F.

An installation has been developed for integration with the AN/ALQ-184 ECM pod, designated AN/ALQ-184(V)9 (see separate entry).The US Air Force

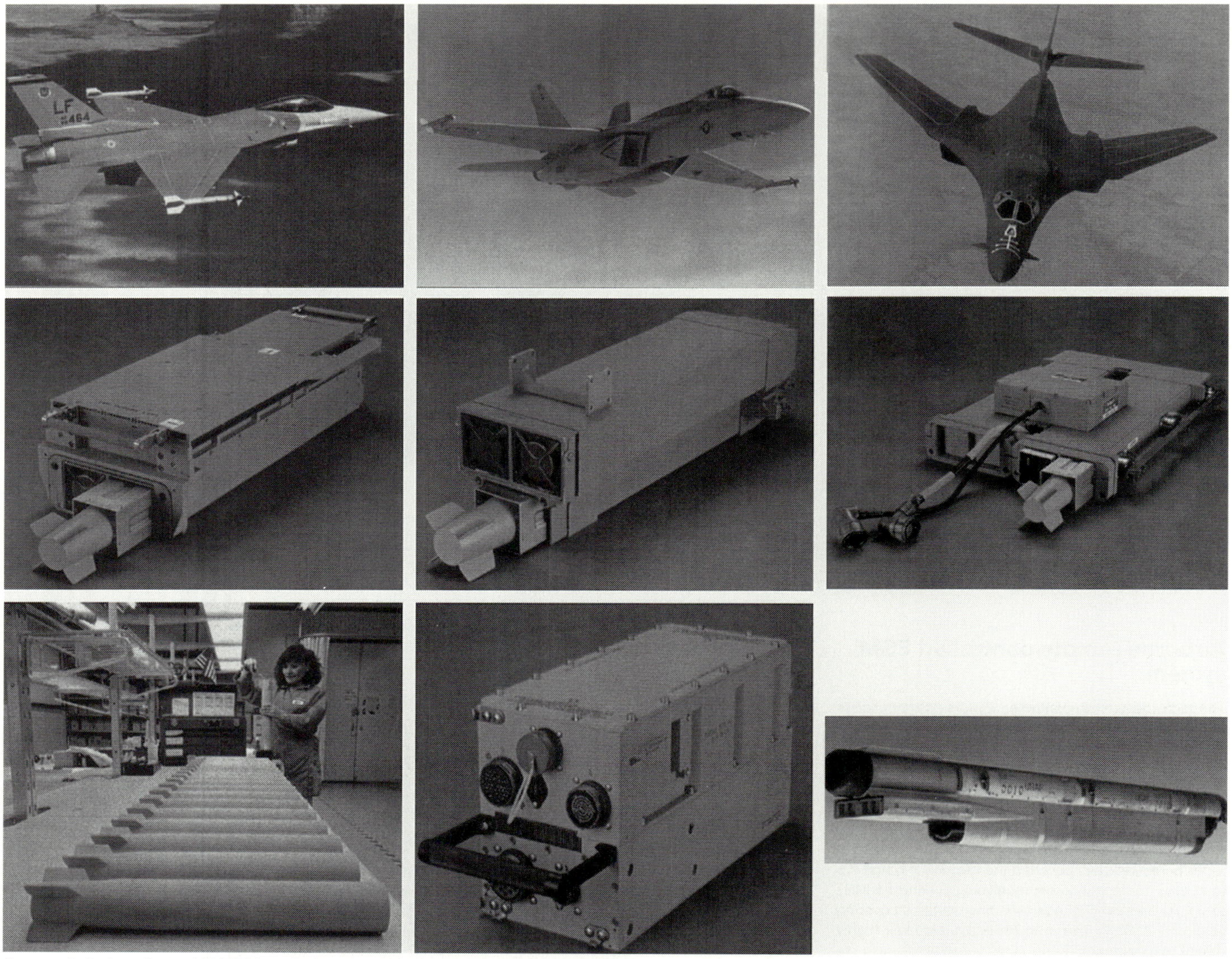

Composite 9-view photo of ALE-50 installation, showing: F-16 Launch/Launch Controller (left); F/A-18E/F T-3 Launcher (centre); B-1B 1 × 4 Launcher (right); MultiPlatform Launch Controller (bottom centre); ALQ-184(V)9 pod with 1 × 4 launcher (bottom right); and ALE-50 production configuration (bottom left) 0018259

approved full-rate production of the modified AN/ALR-184(V)9 version in October 1999.

Raytheon Electronic Systems, using IR&D funds, is currently developing an infra-red towed decoy to expand the capability of the ALE-50 system to protect against both RF and IR threats.

A derivative variant of the ALE-50 is also being developed for the UK Royal Air Force Nimrod MRA4 maritime patrol aircraft, where it will be linked to the AN/ALR-56M RWR for threat cueing and to a techniques generator being developed by Racal Defence Electronics based on DRFM technology.

In December 1997, Raytheon Electronic Systems was awarded a US$35.5 million contract to provide 1,522 towed decoy rounds (1,372 for the US Air Force and 150 for the US Navy) applicable to the AN/ALE-50(V) countermeasures system on the F-16 and F/A-18 aircraft. In 1998, the US Air Force awarded production orders for B-1B and F-15 aircraft. By the end of 1999, more than 4,000 units had been delivered to the US Air Force and the US Navy.

Contractor

Raytheon Electronic Systems.

UPDATED

AN/ALQ-108 IFF jamming pod

The AN/ALQ-108 jamming pod is used in the US Navy's Northrop Grumman E-2C Hawkeye, Lockheed Martin EP-3A Aries and S-3A Viking types, and some German Air Force F-4 Phantoms, to improve survivability in ASW and ELINT operations by jamming IFF transmissions.

ALE-50 pylon on Block 50 F-16 (small underwing pylon outboard of larger wing weapons pylon) ***1999***/0051267

Operational status
Production of about 300 sets is reported.

Contractor
Raytheon Electronic Systems.

VERIFIED

AN/ALQ-128 threat warning receiver

The AN/ALQ-128 is the standard threat warning receiver on the US Air Force's F-15 aircraft, forming part of the Tactical Electronic Warning System (TEWS) with the ALR-56, ALE-45 and ALQ-135 systems. The ALQ-128 has been in production since 1980. Little is known about this system's performance; it possibly provides coverage of the higher-frequency bands above the J-band limit of the ALR-56.

Operational status
In service on the US Air Force F-15.

Contractor
Raytheon Electronic Systems.

VERIFIED

AN/ALQ-184(V) self-protection jamming pod

The AN/ALQ-184(V) ECM pod is designed to provide effective countermeasures against surface-to-air missiles, radar-directed gun systems and airborne interceptors by selectively directing high-power jamming against multiple emitters. An upgrade of the AN/ALQ-119(V) pod, the AN/ALQ-184(V) functions as a repeater, transponder or noise jammer. Multibeam architecture based on proven Rotman lens antenna technology provides substantially reduced countermeasures response time with a tenfold increase in ERP at 100 per cent duty factor (a figure of 9 kW is quoted in some sources). High reliability is inherent in the architecture due to redundant mini-TWTs, reduced operating voltages and lower operating temperatures. Maintainability has been improved by the use of digital circuitry, automatic gain optimisation, elimination of manual adjustments and extensive computer-driven built-in test.

Specifications
Length: 3,957 mm
Weight: 289 kg

Operational status
Raytheon Electronic Systems has delivered more than 880 pods to the US Air Force since 1989. The ALQ-184 operates on US Air Force A-10, F-4, F-15 and F-16 aircraft. AN/ALQ-184(V) has also been selected by the Taiwanese government for its F-16 aircraft, and deliveries began in 1997. Sale of an additional 48 pods to Taiwan was authorised by the US government in June 2000.

In 1996, the US Air Force awarded Raytheon Systems Company a contract to upgrade the ALQ-184. Under the contract, Raytheon Electronic Systems will build 225 reprogrammable low-band kits, which will provide the pods with rapid flight line programming, improved life cycle costs and additional modulation jamming techniques. The contract also includes options for upgrade kits to all ALQ-184 ECM pods produced to date.

Contractor
Raytheon Electronic Systems.

UPDATED

AN/ALQ-184(V)9 combined ALQ-184 ECM pod and ALE-50 towed decoy system

The AN/ALQ-184(V)9 is a development of the AN/ALQ-184 ECM pod, in which a scab-fit unit of four ALE-50 towed decoy system units is added to the aft end of the pod.

The ALE-50 system consists of a launch controller, launcher and towed decoy. The decoy protects the host aircraft against radar-guided missiles by providing a more attractive target and seducing them away from the aircraft.

Technique co-ordination between the two systems will be managed by an Advanced Correlation Processor (ACP). The ACP in the (V)9 pod has been tested by the US Air Force Air Warfare Center.

The ACP will make the decision between the ALQ-184 and ALE-50 threat responses in order to employ the most effective counter to the threat.

Raytheon Electronic Systems is developing an infra-red towed decoy to expand the capability of both ALE-50 and ALQ-184(V)9 systems to provide equally effective protection against both RF and IR threats.

Integration of ALE-50 into ALQ-184 also results in significant improvement of basic ALQ-184 capability. To provide space for the ALE-50 launch controller and four decoy launcher, the current low-band controller is modernised and made field programmable by conversion of 12 1970s vintage circuit cards into two 1990s technology circuit cards. The two-card low-band modification improves MTBF to the degree that the addition of ALE-50 LRUs are completely offset and the resultant ALQ-184(V)9 MTBF is better than the basic pod.

ALQ-184(V)9 will have an improved ability to communicate with its host aircraft and other onboard systems, such as missile warning receivers and radar warning systems, via newly installed dual redundant MIL-STD-1553B interfaces.

As with the basic ALE-50, ALQ-184(V)9 will offer growth to fibre optic decoy capability using proven technique generation in the ALQ-184 pod.

Operational status
The US Air Force is reported to have approved full rate production of the AN/ALQ-184(V)9, but details of an actual build programme are not clear, because it is reported that there is uncertainty about funding.

Contractor
Raytheon Electronic Systems.

VERIFIED

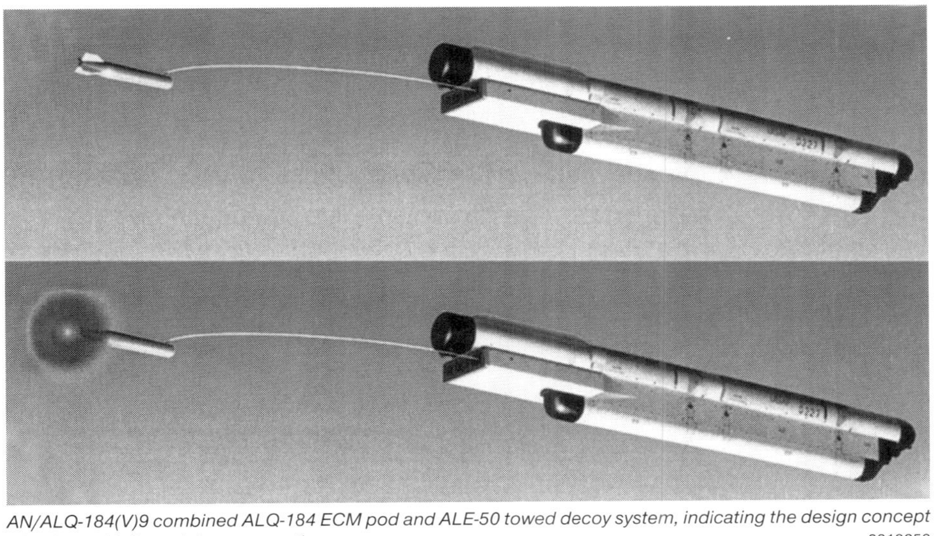

AN/ALQ-184(V)9 combined ALQ-184 ECM pod and ALE-50 towed decoy system, indicating the design concept of radar and infra-red decoy operation 0018256

AN/ALQ-184(V) self-protection jamming pod 0018258

AN/ALQ-184(V)9 configuration, showing 1 × 4 ALE-50 launcher unit at centre rear 0018257

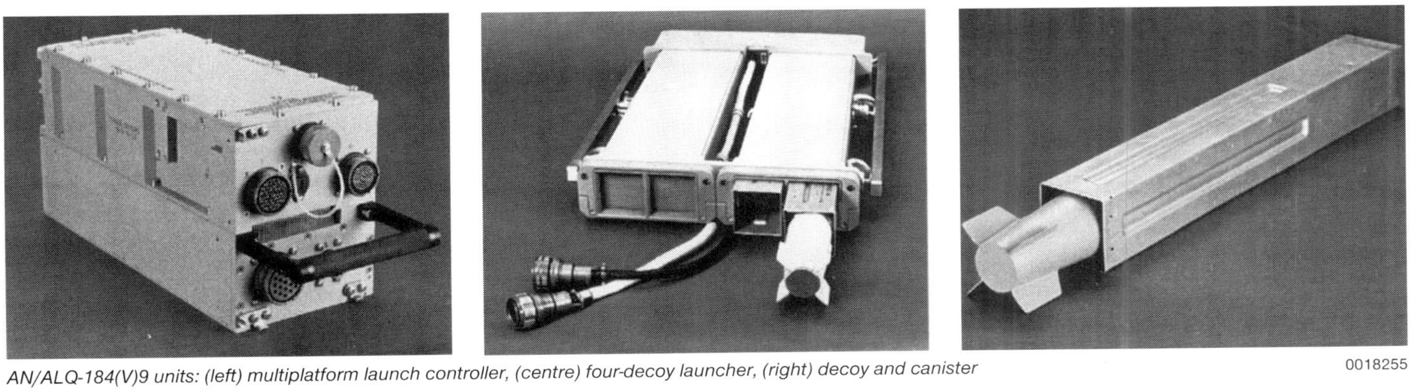

AN/ALQ-184(V)9 units: (left) multiplatform launch controller, (centre) four-decoy launcher, (right) decoy and canister 0018255

AN/ALQ-187 internal countermeasures system

The AN/ALQ-187 is an internally housed, fully automatic jammer system integrated with radar warning and flare/chaff systems for tactical aircraft self-protection. The only fully integrated jammer on the F-16, it is also readily adaptable to installation in the F-4, F-18, A-7 and Mirage 2000 aircraft. The system detects and defends against surface-to-air missiles, anti-aircraft artillery and air-to-air interceptor weapon systems. It can interface with the AN/ALE-39 or AN/ALE-40 chaff/flare dispensers and the AN/ALR-66, AN/ALR-69 or AN/ALR-74 radar warning receivers.

The fully software programmable system automatically detects single and multiple-threat radars of the same or different technologies and selects the most effective of 13 Electronic CounterMeasures (ECM) techniques to counter pulse, Pulse Doppler (PD) or Continuous Wave (CW) ground-based, shipborne or airborne emitters. ECM techniques include Range Gate Pull Off (RGPO), Range on False Targets, Inverse Gain, Velocity Gate Pull-Off (VGPO), Doppler False Target, Swept Spot Noise and False Scan/False Lobe. The system is flight line reprogrammable, allowing immediate incorporation of the latest intelligence threat data and ECM techniques. Advanced power management ensures maximum jamming effectiveness. A variant of the AN/ALQ-187 jammer, designated the AN/ALQ-187H, is used as part of the Litton Applied Technology ASPIS Advanced Self-Protection Integrated Suites system.

Specifications

RF coverage: H-J bands
Dimensions:
(control unit) 86 × 146 × 83 mm
(forward transmitter unit) 222 × 336 × 533 mm
(aft transmitter) 256 × 222 × 533 mm
(processor) 203 × 188 × 458 mm
Weight:
(control unit) 1.35 kg
(forward transmitter) 43.1 kg
(aft transmitter) 29.5 kg
(processor) 14.1 kg
Frequency: 6.5-18 GHz

Operational status

The AN/ALQ-187 is operational with the A-7, F-4 and RF-4 aircraft of a NATO country.

Contractor

Raytheon Electronic Systems.

UPDATED

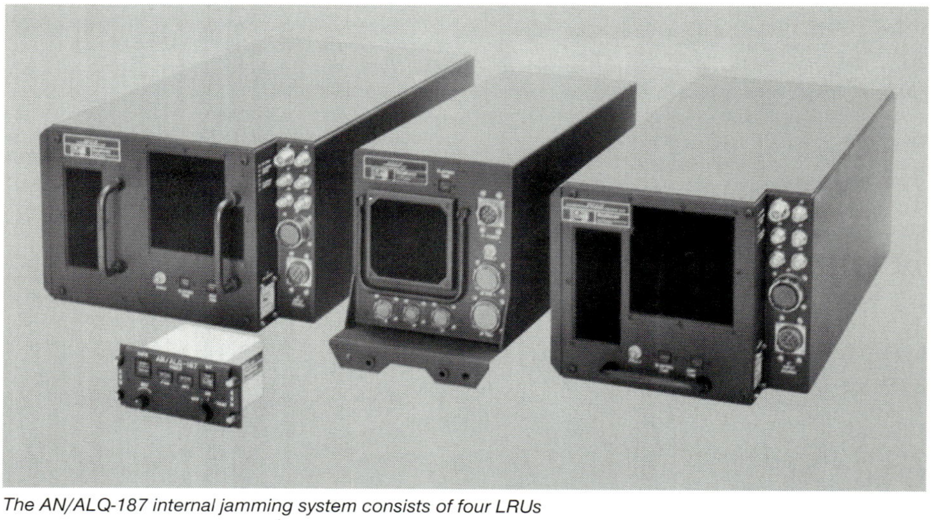

The AN/ALQ-187 internal jamming system consists of four LRUs

AN/ALR-50 radar warning receiver

The AN/ALR-50 was part of a very substantial US Navy programme throughout the early 1970s.

Operational status

At least 1,300 sets were delivered to the US Navy and used on A-4 Skyhawk, A-7 Corsair, EA-6A Intruder and EA-6B Prowler, F-8J/RF-8G Crusader, F-14 Tomcat, RA-5G Vigilante and RF-4B/F-4N Phantom. The AN/ALR-50 is no longer in production.

Contractor

Raytheon Electronic Systems.

VERIFIED

AN/ALR-67(V)3 and 4 countermeasures receiving set

The AN/ALR-67(V)3 and 4 is commonly known as the Advanced Special Receiver (ASR) set. It has been designed to be the standard US Navy Radar Warning Receiver (RWR) for carrier-based tactical aircraft such

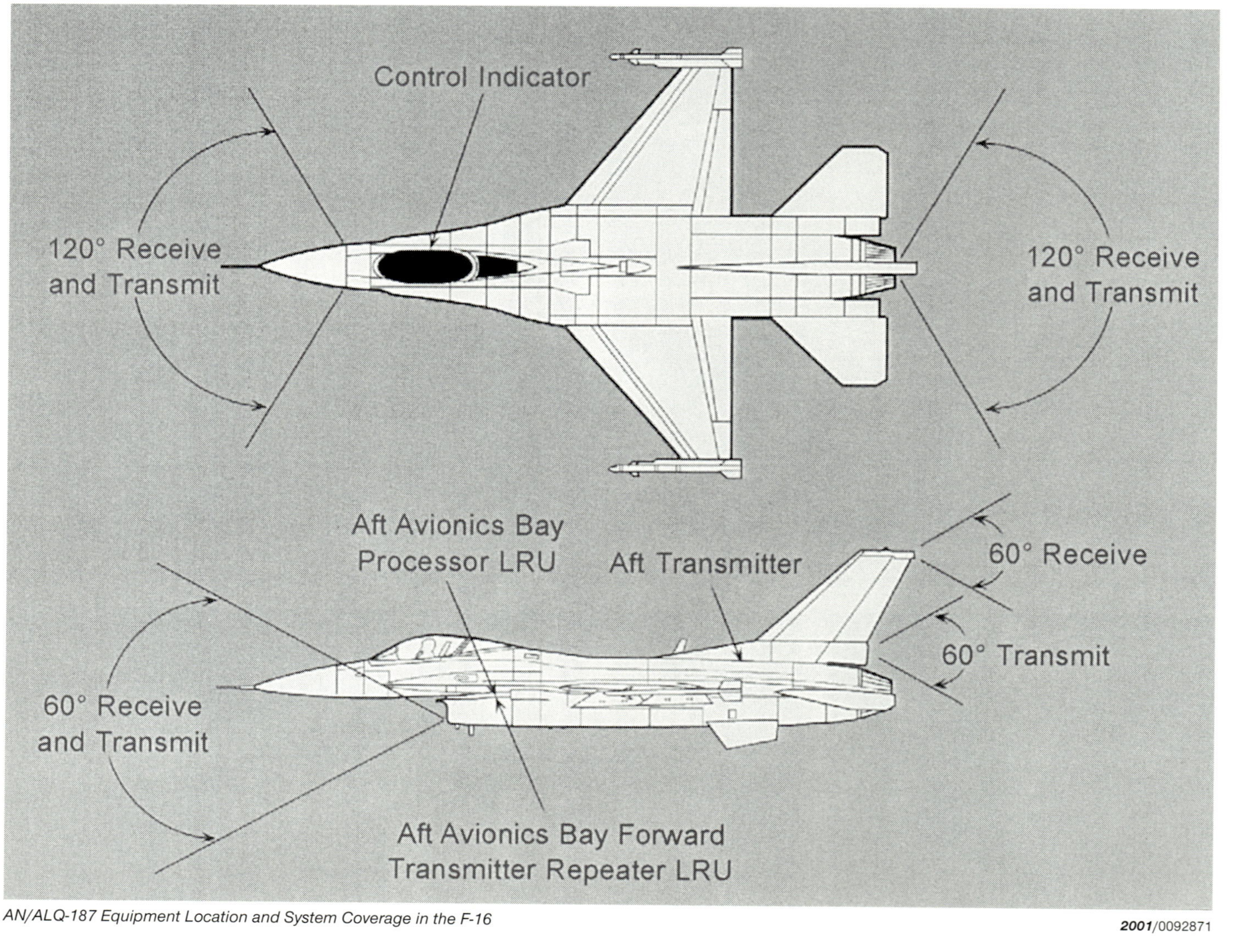

AN/ALQ-187 Equipment Location and System Coverage in the F-16

2001/0092871

as the F/A-18C/D, F/A-18E/F, F-14C/D, A-6E and AV-8B. Actual procurement plans appear to be heavily constrained by funding limitations.

The system comprises seven Weapon Replaceable Assembly (WRA) types to give a total of 13 WRAs. These consist of four integrated antenna detectors, four quadrant receivers, countermeasures receiver, countermeasures computer, low-band integrated antenna, control status unit and azimuth display indicator.

Each Integrated Antenna Detector (IAD) includes a dual-polarised microwave antenna, MMW antenna and supporting electronics. For ease of installation in various aircraft types, the IAD is available in three different housings.

Each IAD feeds signals into an associated Quadrant Receiver (QR) which conditions the received energy for processing. Conditioning involves filtering, amplification and frequency conversion to IF. Band structure has been optimised to minimise interference from onboard jammers.

The Countermeasures Receiver (CR) accepts preconditioned signals from the QR and low-band WRAs and generates digital words describing the parameters of the pulsed and CW radar waveforms detected. Parameters include amplitude, angle of arrival, time of arrival, frequency, pulsewidth and modulation.

The Countermeasures Computer (CC) incorporates an Ada programmable 32-bit JIAWG-compatible computer. Software in the CC processes the pulse and CW data to characterise, identify and prioritise intercepted threats based on their potential lethality. The CC manages all external interfaces, including two MIL-STD-1553B busses and special interfaces to various jammers and missile launch computers.

The low-band integrated antenna receives and conditions signals in the lower frequency region and is intended to counter those elements of the SAM threat that utilise lower frequencies.

The Control Status Unit (CSU) is a push-button WRA (carried over from the AN/ALR-67(V)2), fitted in the cockpit that enables the pilot or maintenance technician to issue commands to the system.

The Azimuth Display Indicator (ADI) is a 3 in (76.2 mm) diameter CRT cockpit display, also carried over from the AN/ALR-67(V)2, used to show intercepted threats. The ADI may not be required in the F/A-18E/F, AV-8B or F-14D programmes.

The AN/ALR-67(V)3 and 4 is understood to form part of the IDECM programme, and to include an interface to the AN/ALE-50 towed decoy system.

Operational status

Production of the AN/ALR-67(V)3 is planned for the F/A-18C/D and F/A-E/F, the F-14C/D and the AV-8B. In August 1999, Raytheon was awarded a contract for full-rate production of the AN/ALR-63(V)3 for the US Navy F/A-18E/F. The contract is for 34 complete installations, together with 40 spare quadrant receivers and five countermeasures receivers. These are to be delivered between April 2001 and January 2002. Meanwhile, the US Navy has contracted for 20 AN/ALR-67(V)3 systems for use on F/A-18 C/D aircraft in war zones. These are to be used as a cross-deck upgrade to their existing AN/ALR-67(V)2 systems (manufactured by Litton Advanced Systems Division), but the remainder of the F/A-18 C/D force will continue to use the AN/ALR-67(V)2 for normal training missions.

It is understood that US Marine Corps AV-8B aircraft will continue to use an updated AN/ALR-67(V)3, rather than retrofitting the AN/ALR-67(V)4. Raytheon is reported to be seeking funding for development of a HARM Targeting System (HTS)interface for the AN/ALR-67(V)3 and 4.

Contractor

Raytheon Electronic Systems.

VERIFIED

AN/ALR-89(V) integrated self-protection system

The AN/ALR-89(V) radar and laser warning system incorporates three subsystems: the AN/ALR-90(V) RWR which detects pulsed radars in the C- to J-bands, the AN/APR-49(V) RWR which is a superheterodyne receiver for detecting modern pulse and non-pulse radars and the AN/AVR-3(V) laser system which detects laser emissions.

The AN/ALR-67(V)3 and 4 operational architecture ***1999***/0051268

Identified threats which are detected by the three subsystems are presented to the pilot on a 3 in (76.2 mm) display that provides an alphanumeric representation of the type, angle of arrival, relative lethality and status of the threats.

The ALR-89 offers complete threat reprogrammability for new and changing environments through an extensive, easily updated emitter library file. A portable loader unit enables loading of operational software and emitter tables and downloading of recorded data in the field. The integrated ALR-89 or one of its subsystems is suitable for the self-protection of helicopters or fixed-wing aircraft regardless of their size. Electrical design and mechanical configuration both adapt to new installations and provide an upgrade of AN/APR-39(V)1 installations. This upgrade can be implemented in various levels of performance using the basic APR-39 system and selected LRUs from the ALR-89 system.

Specifications

Weight:
(AN/ALR-90(V)) 5.76 kg
(AN/APR-49(V)) 4.04 kg
(AN/AVR-3(V)) 3.86 kg
Power supply: 28 V DC, 210 W (MIL-STD-704)
Frequency: C- to J-bands, 2 laser bands
Coverage: 360° azimuth
Environmental: MIL-E-5400, Class II

Operational status

In production. It is understood that the AN/ALR-89(V) is part of the Defensive Aids System (DAS) of the Northrop Grumman RQ-4A Global Hawk.

Contractor

Raytheon Electronic Systems.

UPDATED

AN/ALR-90(V) self-protection system

The AN/ALR-90(V) is a microprocessor-controlled airborne self-protection system which employs a crystal video detector system to detect pulsed radars in the C- to J-bands. Detected radars, which are analysed and identified as threats, are presented on a 3 in (76.2 mm) display that provides an alphanumeric representation of the type, angle of arrival, lethality and status of threats.

The AN/ALR-90(V) offers complete threat programmability for new and changing environments. This is accomplished with an extensive, readily updated emitter library file. A portable loading unit enables field loading of the operational software and emitter tables and downloading of recorded threat data.

The system has provisions for integration with various avionics systems, such as radar, for blanking, and other digital data and ECM sensors. It can also be interfaced with the AN/APR-49(V) radar warning receiver and the AN/AVR-3(V) laser warning system.

Specifications

Weight: 5.77 kg
Frequency: C- to J-bands in three bands
Signal types: pulsed radars
Coverage: 360° azimuth

Contractor

Raytheon Electronic Systems.

VERIFIED

AN/APG-63(V) fire-control radar

The AN/APG-63 is the principal sensor for the F-15 Eagle. The system was designed around three main objectives: capability, reliability and maintainability. Capability is aimed at providing one-man operation for the tracking of hostile aircraft at long range and close in, in a look-down situation; a clutter-free display with all appropriate information, including guidance and steering information, on a head-up display; simplified controls and co-ordination with weapons, and a secondary air-to-ground facility.

The requirement for the AN/APG-63 called for an MTBF of 60 hours, a level never before approached for this class of equipment. As a corollary, maintenance time would be reduced to about a quarter of that of previous systems. High reliability and relatively simple maintenance schedules reduce the turnround time. This is reflected in the smaller numbers of flight line and

An F-15A Eagle fighter with nose radome swung aside to give access to the AN/APG-63 radar

An F-15C Eagle fighter with nose radome swung aside to show the AESA upgrade (Boeing) **2001**/0103897

maintenance personnel needed. This is largely due to the use of hybrid and integrated circuits to reduce the number of components and to the use of only four basic module sizes in the entire system.

The system comprises nine LRUs: exciter, transmitter, antenna, receiver, analogue processor, digital processor, power supply, radar data processor and control unit. It has a wide look-angle and its antenna is gimballed in all three axes to hold target lock on during roll manoeuvres. The clutter-free head-down radar display gives a clear look-down view of target aircraft silhouetted against the ground, even in the presence of heavy clutter from ground returns. This look-down, shoot-down ability is achieved by using both high and medium PRFs, by digital data processing and by using Kalman filtering in the tracking loops. The system's gridded travelling wave tube permits variation of the waveform to suit the tactical situation.

False alarms are eliminated, regardless of aircraft altitude and antenna look-angle, by a low-sidelobe antenna and frequency rejection of both ground clutter and vehicles moving on the ground, so that only real targets are displayed.

Primary controls for the multimode, pulse Doppler I-/J-band radar are located on the control column, allowing the pilot to keep his head up during fast-moving situations. Three special modes are provided for close in combat: supersearch, vertical scan and boresight. These enable automatic acquisition of, and lock on to, targets within 18.5 km. In the supersearch mode the radar locks on to the first target entering the head-up display field of view. In the vertical scan mode the radar locks on to the first target that enters an elevation scan pattern at right-angles to the aircraft lateral axis. In the boresight mode the antenna is directed straight ahead and the radar locks on to the nearest target within the beam. In all tracking modes the target position is displayed on the HUD if the target is within the field of view. This greatly increases the range of visual detection.

The use of interleaved high- and medium-PRF waveforms provides look-down, shoot-down capability for both approaching and retreating targets. The technology used is a digital signal processor. Incoming signals are sampled and their frequency content analysed by Fourier transforms on individual samples.

Although designed specifically for the air-to-air role, the F-15 has emerged as a potent ground attack fighter, and the APG-63 has target ranging for automatic bomb release for a visual attack, a mapping mode for navigation and a velocity update for the inertial navigation system.

The APG-63 is compatible with AIM-12, AMRAAM, AIM-7F Sparrow and AIM-9L Sidewinder air-to-air missiles. Antenna search patterns and radar display presentations are selected automatically by means of a three-position switch on the throttle for the type of weapon to be used (medium-range radar homing, short-range infra-red missiles or M61 cannon).

While the eventual reliability goal of 60 hours specified by MIL-STD-1781 was met consistently on bench tests, in the field, service equipment is currently giving about 30 to 35 hours MTBF.

All APG-63 radars produced since mid-1980 incorporate a programmable signal processor and a high-speed digital computer that enables the system to respond quickly to new tactics or weapons by changes to software rather than extensive hardware modifications. This change was introduced with the improved F-15C and F-15D models. Aircraft produced before the processor was introduced were scheduled to receive it as a retrofit item.

In December 1983, the US Air Force awarded a $274.4 million contract to upgrade the control computer and armament control system as part of the MultiStaged Improvement Programme (MSIP) launched in February of that year. MSIP arose out of a McDonnell Douglas study, beginning in June 1982, which showed the need to improve the avionics system, and specifically the radar, in order to maintain reasonable combat superiority over likely adversaries. Radar improvements include a memory increase to 1 M words, a radar data processor speed tripled to 1.4 Mops and improvements to both transmitter and receiver. The outcome of this is the AN/APG-70.

In March 1984, the F-15 was selected as the US Air Force's new dual-role air control and interdiction fighter. The upgraded radars for the two-seat F-15E version incorporate very high-speed integrated circuit technology to increase computational speed and improve ECM performance.

Specifications

Volume: 0.25 m^3
Weight: 221 kg total
Number of LRUs: 9
Frequency: I/J-band (selectable)
Transmit power: 12.975 kW
Reliability: 60 h MTBF

Operational status

In service with the F-15 Eagle. The last system was delivered in September 1986, by which time some 1,000 sets had been built, including co-production in Japan. Four P-3A Orions, on loan from the US Navy, have been modified to include an AN/APG-63 radar mounted in the aircraft nose, for anti-drug-smuggling operations with the US Customs Service.

The US Air Force has upgraded the AN/APG-63 radars (upgrade designation AN/APG-63(V)1) in some of its F-15C/D aircraft.

The AN/APG-63(V)1 programme began in October 1994 with the award of an Engineering and Manufacturing Development (EMD) contract valued at US$175 million. This phase was essentially complete by the end of 1999. Meanwhile, performance testing began in 1997 for Low Rate Initial Production (LRIP), funded under contracts totaling US$96 million. In July 1999, a contract for the AN/APG-63(V)1 Production 1 lot of 22 radars was awarded for aircraft retrofit by November 2001. A further production order for 25 sets was placed in June 2000 for delivery by December 2002. A production run of five consecutive Fiscal Years (FY99-03) is expected for 162 radar systems, with possible follow-on buys and the prospect of export sales.

The upgraded AN/APG-63(V)1 radar utilises software from the later model APG-70 radar to provide a substantial increase in computing power and electronic counter-countermeasures (ECCM) capability, while improving the air-to-ground mode.

During the last quarter of 1999, the US Air Force placed a contract to equip a limited number (18) of its F-15C fleet with a further upgrade of the AN/APG-63 radar, to be designated AN/APG-63(V)2. The Boeing Phantom Works unit led a team that received the US$250 million contract to install Raytheon's Active Electronically Scanned Array (AESA) radar. The aim of the upgrade was to improve situational awareness and to make better use of the AIM-120 missile capability by increasing air-to-air detection and tracking range. The upgrade also included an Advanced Identification Friend or Foe (AIFF) system from BAE Systems North America. The program was used to provide experience of the technology that will underpin the US Air Force F-22 and Joint Strike Fighter aircraft systems. The Air Force F-15 System Program Office's Projects Team at Wright-Patterson Air Force Base, Ohio, managed the program for the US government.

In an AESA system, the traditional mechanically scanning radar dish is replaced by a stationary panel covered with an array of hundreds of small transmitter-receiver modules. Unlike a radar dish, these modules have more combined power and can perform different detection, tracking, communication and jamming functions in multiple directions simultaneously. An AESA offers greater precision to detect, track and eliminate multiple threats more quickly and effectively than traditional radar. Further, because the AESA eliminates the hydraulic and electrical systems associated with mechanically operated radars, its reliability and maintainability are dramatically improved.

The final three upgraded F-15C aircraft were delivered during December 2000. In addition to the F-15C AESA, Raytheon is also developing AESAs for the F/A-18E/F Super Hornet for the US Navy and the X-32 A/B Joint Strike Fighter Boeing will build, if it wins the competition for that program.

Contractor

Raytheon Electronic Systems.

UPDATED

The AV-8B Harrier II Plus is equipped with the AN/APG-65 radar

AN/APG-65 multimission radar

In the air-to-air role the APG-65 radar incorporates the complete range of search, track and combat modes, including several previously unavailable in an operational radar. Specifically these modes are gun acquisition (the system scans the head-up display field of view and locks on to the first target it sees within a given range); vertical acquisition (the radar scans a vertical slice of airspace and automatically acquires the first target it sees, again in a given range); and boresight (the radar acquires the target after the pilot has pointed the aircraft at it). In the raid assessment mode the pilot can expand the region around a single target that is being tracked, giving increased resolution around it and permitting separation of closely spaced targets. As a gunsight, the radar operates as a short-range tracking and lead-computation device using frequency agility to reduce errors due to target scintillation. All the pilot has to do is to put the gunsight pipper on the target and press the firing button.

Other modes are long-range velocity search (using a high-PRF waveform to detect oncoming aircraft at high relative velocities); range while search (high- and medium-PRF waveforms interleaved to detect all-aspect targets, not only head-on but at any line of sight crossing angle), and track-while-scan which can track up to 10 targets simultaneously and display eight. When combined with autonomous missiles such as AMRAAM, this mode confers a launch-and-leave capability and the simultaneous engagement of multiple targets.

In the F/A-18's air-to-ground role, the APG-65 has six modes: terrain-avoidance, for low-level penetration of hostile airspace; precision velocity update, when the radar provides Doppler signals to update or align the aircraft's inertial navigation system; tracking of fixed and moving ground targets; surface vessel detection, in which the system suppresses sea clutter by a sampling technique; air-to-surface ranging on designated targets; and ground-mapping. Two Doppler beam-sharpening modes are provided for these air-to-ground modes.

During tests in early 1983, the system demonstrated (a year ahead of schedule) the 106 laboratory test hours MTBF required by contract. Two systems chosen at random for a reliability demonstration operated under test for a total of 149 hours without a failure.

All fault-finding is conducted with Built-In Test (BIT) which currently operates at module LRU level. The biggest improvement in BIT technology with the AN/APG-65 has been the reduction in false alarm rate. The US Navy requirement is to be able to detect and locate 98 per cent of all radar faults by BIT. The US Navy also specifies 12 minutes MTTR, which involves running a BIT test, locating the fault, removing and replacing the faulty line-replaceable unit and running a BIT check on the new unit.

In its first six months of operation, the initial Marine Air Group II at El Toro in California had only six radar engineers to support a 12 aircraft unit flying 30 hours a month.

Under the terms of an Australian Industrial Participation (AIP) programme signed in December 1981, Philips Electronics Systems in Australia is engaged in final assembly and test of the radar and co-produces the data processor. In February 1984, it was announced that Marconi Española SA had been licensed to build low-voltage power supply modules for the radars being produced for Spain's F/A-18s.

German Air Force F-4F ICE programme

In May 1985, the German Ministry of Defence chose the AN/APG-65 as a major element in the Improved Combat Efficiency (ICE) programme for the German Air Force F-4F Phantom. This radar is also designated AN/APG-65GY.

DaimlerChrysler Aerospace AG built the radars under licence and ICE includes a Honeywell laser inertial navigation system, Marconi Electronic Systems air data computer, DaimlerChrysler radar display and Raytheon Electronic Systems cockpit displays. DaimlerChrysler Aerospace AG is also contracted to provide the ICE upgrade to the Hellenic Air Force as part of its F-4E upgrade.

Radar for AV-8B

A variant of the APG-65 entered service in the AV-8B Harrier II Plus with the US Marine Corps and Italian and Spanish navies during 1993.

Specifications

(APG-65 in F/A-18)
Volume: <0.126 m^3 excluding antenna
Weight: 154.6 kg
Number of LRUs: 5
Frequency: 8-12 GHz
Antenna: low-sidelobe planar-array with fully balanced direct electric drive replacing hydraulics and mechanical locks of previous systems
Transmitter: liquid-cooled, contains software-programmable gridded travelling wave tube amplifier
Receiver: contains A-D converter
Radar-data processor: general purpose with 250 K 16-bit word bulk-storage disk memory
Signal processor: fully software-programmable, runs at 7.2 Mops
Reliability: 120 h MTBF. Built-in test equipment detects 98% of faults and isolates them to single replaceable assemblies that can be changed in 12 min without adjustments or setting up

Operational status

In service with US Navy, US Marine Corps, Canadian Forces, Royal Australian Air Force, Spanish Air Force and Kuwaiti Air Force F/A-18 aircraft; US Marine Corps, Spanish Navy and Italian Navy AV-8B Harrier II Plus aircraft and the German Air Force F-4F ICE programme.

In production for AV-8B Harrier II Plus aircraft of the US Marine Corps, Italian Navy and Spanish Navy. Contracted for the Hellenic Air Force F-4E upgrade in the ICE configuration.

Contractor

Raytheon Electronic Systems.

VERIFIED

The US Navy F/A-18 Hornet is equipped with the AN/APG-65 radar

AN/APG-70 radar for the F-15E

The radar for the two-seat F-15E combat aircraft is a substantially improved version of the APG-63, designated AN/APG-70. It is also installed in a limited number of F-15C/D aircraft.

By comparison with the earlier system, the APG-70 has a far greater RF bandwidth, a larger look-down target detection range, a one-third increase in MTBF and is packaged in eight units instead of nine. Four of the LRUs (radar data processor, programmable signal processor, analogue signal converter and receiver/exciter) are completely new, while the transmitter and control unit have been modified; only the power supply and the antenna remain unchanged. The radar uses VLSI technology, making the programmable signal processor five times faster than previous designs while having three times the memory.

Expanded built-in test is provided, giving unambiguous fault detection and isolation and the ECCM capability is improved to combat new and more advanced threats. The system is compatible with existing and new missiles such as AIM-7F/M Sparrow, AIM-9 Sidewinder and AIM-120 Advanced Medium-Range Air-to-Air Missile (AMRAAM), and with 20 mm cannon.

In the air-to-air role the radar has five search modes: range while scan with high PRF, medium PRF or interleaved PRF, range-gated high PRF and velocity search. The radar also has single-target track, track-while-scan and raid assessment track modes and vertical search, super search, boresight and auto-guns target acquisition modes.

In the air-to-ground role the radar can produce a high-resolution or a real-beam ground map; the specification calls for 2.6 m resolution at 75 km range. There is also a precision velocity update mode and air-to-ground ranging. For the future, terrain-following/terrain-avoidance, ground moving-target track indicator and fixed-target track modes are planned.

In December 1987, Hughes (now Raytheon Electronic Systems) announced a US$58 million contract to supply a variant of the APG-70 radar as the fire-control system in the US Air Force Special Operations Force's AC-130U aircraft. This aircraft, of which 12 have been ordered, has a 105 mm howitzer and 25 mm and 40 mm cannon. Five new air-to-ground modes were developed for this application: fixed-target track, ground moving target indication and track, projectile impact point position, beacon track and a weather mode. The existing antenna and signal processor were modified and a digital scan converter added to the system. The first system was delivered at the end of 1988.

Specifications

Volume: 0.25 m^3
Weight: 251 kg
Frequency: 8-20 GHz selectable
PRF: multiple
Number of LRUs: 8
Range:
(air-to-air) 185 km
(automatic acquisition) 500 ft-20 n miles
(ground map) 50 n miles+
Resolution:
(ground map) 2.6 m at 75 km
Reliability: 80 h MTBF

Operational status

In service in the F-15E and late model F-15C/D aircraft and on the F-15S (Saudi Arabia) and F-15I (Israel).

Contractor

Raytheon Electronic Systems.

UPDATED

AN/APG-71 fire-control radar for the F-14D

AN/APG-71 is an enhancement of the AN/AWG-9 radar originally fitted to F-14 aircraft.

Compared with the AWG-9, the APG-71 offers better overland performance, expanded velocity search capability, a larger target engagement zone, a raid assessment mode and programmable electronic countermeasures and clutter control features.

The APG-71 is essentially a digital version of the radar section of the AWG-9, with greatly improved ECM performance acknowledging the new and vastly more sophisticated jamming technologies that have appeared since design of the F-14 was frozen. New modes include medium-PRF all-aspect capability, monopulse angle tracking, digital scan control, target identification and raid assessment, but the number of boxes comes down from 26 to 14. The system also employs some of the elements developed by Hughes for the new F-15 radar, the AN/APG-70; for example the APG-71 has a signal processor which is 86 per cent common in modules with that in the APG-70. The AWG-9's transmitter, power supply and aft cockpit tactical information display are retained for the APG-71.

The APG-71 also incorporates non-co-operative target identification, by which radar contacts may be identified as friendly or hostile, at beyond visual ranges, through close examination at high resolution, of the returns; this technique obviates deficiencies and ambiguities in IFF equipment.

The APG-71's antenna retains the gimbal system of the AWG-9 and adds a new array with low sidelobes and a guard channel to eliminate sidelobe penetration of ground clutter and electronic warfare interference. The APG-71 also provides an improved radar master oscillator which significantly increases the number of radar channels and provides frequency-agile operation, with low sidelobes.

The signal processor has four processing elements as against the three found in the APG-70, giving an operating speed of 40 Mcops (million complex operations per second). The radar data processor also has a large degree of commonality with that in the APG-70, differing only in the interface cards, and operates at 3.2 Mips (million instructions per second).

The aft cockpit digital display for the APG-71 was originally developed for the AWG-9 but never put into production, while the tactical information display, originally designed for the F-106 aircraft, remains largely unchanged from the version in the F-14A. Significantly, the APG-71's software is written in Jovial.

The AN/APG-70 in an F-15C aircraft

The AN/APG-71 radar in an F-14D

Specifications

Volume: 0.78 m^3
Weight: 590 kg

Operational status

In service on F-14D aircraft.

Contractor

Raytheon Electronic Systems.

VERIFIED

AN/APG-73 radar for the F/A-18

The AN/APG-73 radar is an all-weather, coherent, multimode, multiwaveform search-and-track radar for upgraded versions of the F/A-18A, later versions of the F/A-18C/D and the F/A-18E/F aircraft. It is based on the AN/APG-65 fitted to earlier versions of the F/A-18, but uses new signal and data processors and a revised receiver/exciter to perform both air-to-air and air-to-surface missions. Raytheon Electronic Systems claims this will give the new radar more than three times the speed and memory of the AN/APG-65 and will make it compatible with an Active Electronically Scanned Antenna (AESA).

The new data processor unit is a multifunction processor consisting of a signal processor function and a data processor function. The signal processor has four million words of bulk memory capacity and its throughput has been increased to 60 million complex operations per second by use of multichip modules. The data processor function is a general purpose, dual-1750A computer that provides software loads to the signal processor and controls to the rest of the radar.

Phase II of the AN/APG-73 upgrade incorporates a Motion-Sensing Subsystem (MSS) unit; a stretch waveform generator module; and a Special test equipment, Instrumentation and Reconnaissance (SIR) module, as well as reconnaissance software. These enhancements give the F/A-18 the hardware capability to produce high-resolution ground maps (comparable with those of the F-15E and U-2). It also allows precision strike missions using advanced image-correlation algorithms, further improving weapon designation accuracy.

Air-to-air modes include: high PRF; velocity search; high/medium PRF range-while-search; four short range automatic acquisition modes; track-while-scan; gun director; raid assessment/situation awareness; single target track.

Air-to-surface modes include: Doppler beam-sharpened sector and patch mapping; medium resolution SAR; radar navigation ground mapping; real beam ground mapping; fixed and moving ground target indication/track; air-to-surface ranging; terrain avoidance; precision velocity update; inverse range angle; and sea surface search with clutter suppression.

Reconnaissance modes are strip map and spotlight map.

Specifications

Volume: (excluding antenna) 0.126 m^3
Weight: 154 kg
Frequency: 8-12 GHz
Number of LRUs: 5 (transmitter, MSS, power supply, data processor, receiver) plus antenna

Operational status

First operational units of the AN/APG-73 were delivered in mid-1994.

The AN/APG-73 is in production for the US Navy F/A-18E/F aircraft. It is also in production for the US Marine Corps F/A-18A upgrade, designated ECP-583.

The APG-73 was also specified for F/A-18C/Ds ordered by Finland, Malaysia, Switzerland and Thailand.

In January 2000 Raytheon was awarded a contract US$200 million to supply 71 AN/APG-73 radar units for the Royal Australian Air Force F/A-18 A/B aircraft upgrade (presently fitted with the APG-65), together with 40 radar kits for the US Navy (19) and US Marine Corps (21). A further award, valued at US$8.8 million, was made in May 2000 for nine updated receivers to support the Royal Australian Air Force requirement for completion in May 2002. The APG-73 radar, similarly, forms part of the projected Canadian F/A-18 upgrade programme.

Latest development models for the F/A-18E/F have been reported to use the designator AN/APG-73(V).

The exact relationship between the AN/APG-73 and the Active Electronically Scanned Array (AESA) radar, selected by Boeing to provide a state-of-the-art radar for future US Navy F/A-18E/F Super Hornet aircraft, is unclear. In November 1999, under an Advance Agreement between Boeing and the US Navy, Raytheon is to develop an integrated AESA radar prototype. The next step is expected to be a US Navy contract award in early 2001, for Engineering and Manufacturing Development (EMD). If the contract is awarded, deliveries of the AESA radar could begin in 2004. The AESA development is intended to improve the air-to-air detection and tracking range; add higher resolution air-to-ground mapping modes at longer ranges; improve situational awareness and reduce support costs.

Contractor

Raytheon Electronic Systems.

UPDATED

AN/APQ-122(V) radar

The AN/APQ-122(V) is a dual-frequency nose radar developed for use in the US Air Force Adverse Weather Aerial Delivery System (AWADS) programme for installation in C-130E transport aircraft. This long-range navigation sensor is used for weather avoidance and navigation in supply dropping missions. The equipment provides ground-mapping out to more than 385 km, weather information up to 278 km and beacon interrogation up to 444 km when using the I-band frequency radar. J-band frequencies are used when short-range high-resolution performance and target location are required. In the J-band mode the radar provides a high-resolution ground map display to permit target identification and location for position fixing and aerial delivery missions. In this mode the radar will detect and display targets with a radar cross-section of 50 m^2 while operating in rainfall of 4 mm/h.

In addition to the dual-frequency system designed for AWADS, designated AN/APQ-122(V)1, three other configurations have been developed. The AN/APQ-122(V)5 is a single-frequency I-band radar which has been developed as a direct replacement for the AN/APQ-59 radar used in C-130 and E-4B aircraft. Facilities include long-range mapping, weather evaluation and avoidance and rendezvous. A navigation training version of the AN/APQ-122(V)5, the AN/APQ-122(V)7, has been designed for use in the T-43A aircraft. Another dual-frequency radar, the AN/APQ-122(V)8, incorporates a terrain-following capability and is used on Combat Talon 1 MC-130 aircraft.

Operational status

The AN/APQ-122(V) has been supplied to the US Air Force and the air forces of Argentina, Australia, Bolivia, Cameroon, Congo, Denmark, Ecuador, Egypt, Gabon, Greece, Indonesia, Iran, Israel, Italy, Jordan, Libya, Malaysia, Morocco, Nassau, New Zealand, Niger, Nigeria, Oman, Philippines, Portugal, Saudi Arabia, Singapore, Spain, Sudan, Thailand, Venezuela and Zaïre. The radar has been installed in C-130H, RC-130A, KC-135A, RC-135A, RC-135C and E-4B aircraft.

Contractor

Raytheon Electronic Systems.

VERIFIED

AN/APQ-126(V) terrain-following radar

The AN/APQ-126(V) is a forward-looking variable configuration airborne navigation and attack radar which was produced for the US Navy A-7E and US Air Force A-7D aircraft. It operates in the J-band and its primary functions are ground-mapping, air-to-ground ranging and terrain-following/terrain-avoidance. The radar also features adverse weather look-through using selectable polarisation, slaved antenna pointing in air-to-ground ranging and variable tilt control which allows the pilot to optimise ground map displays and highlight points of interest.

Operational status

The AN/APQ-126 radar equips the A-7H aircraft of the Greek Air Force and the A-7P and TA-7P aircraft of the Portuguese Air Force. A variant, known as the AN/APQ-158, has been developed for the HH-53 helicopter. Production is now complete with over 1,000 units delivered.

Contractor

Raytheon Electronic Systems.

UPDATED

The AN/APG-73 radar for later versions of the F/A-18

AN/APQ-158 radar

The AN/APQ-158 is a multimode forward-looking radar used primarily for terrain-following/terrain-avoidance at low altitudes in the Pave Low III night/adverse search and rescue helicopter, the Sikorsky MH-53J. The equipment is similar to the AN/APQ-126 but is modified for compatibility with the unique helicopter characteristics and the Pave Low III mission requirements. The radar contains 15 LRUs which provide the same basic modes of operation as the AN/APQ-126.

System upgrades have provided this radar with the ability to supply updates in all modes except terrain-following, and to perform terrain-following missions over very high clutter areas such as cities.

Operational status

In service in MH-53J helicopters.

Contractor

Raytheon Electronic Systems.

VERIFIED

The AN/APQ-174 MultiMode Radar (MMR) provides the ability to operate at altitudes down to 100 ft in adverse weather by day or night

AN/APQ-168 multimode radar

The Sikorsky HH-60D Night Hawk helicopter is designed to penetrate hostile territory during darkness to rescue downed aircrews or deliver and retrieve special operations teams. The task calls for long-distance nap of the earth flying and accurate navigation.

The radar can operate in terrain-clearance, terrain-avoidance, air-to-air ranging and cross-scan modes, the latter combining ground-mapping or terrain-avoidance with terrain-following. A terrain storage facility permits the radar to have a reduced duty cycle thereby reducing the probability of detection by enemy ESM equipment.

The system has increased electronic countermeasures resistance, improved weather penetration, better guidance in turning flight, a power management function for semi-covert operation and low-beam reflectivity. Extensive BITE provides a high degree of fault isolation and detection. The system is carried in a pod in the nose of the aircraft.

Specifications

Dimensions: 1,420 mm long × 330 mm diameter
Weight: 113 kg
Power supply: 115 V AC, 3 phase, 200 VA
28 V DC
Reliability: 144 h specified MTBF

Operational status

In service with the US Air Force Sikorsky HH-60D Night Hawk helicopter.

Contractor

Raytheon Electronic Systems.

VERIFIED

AN/APQ-174 MultiMode Radar (MMR)

The AN/APQ-174 multimode radar has been developed for US Army, Navy and Air Force combat rescue and special operations missions, for use on aircraft such as the HH/MH-60, CH/MH-47, HH-53 and the V-22. The radar is a derivative of the LANTIRN terrain-following radar and the AN/APQ-168 multimode radar and maintains commonality with five of the six LANTIRN LRUs. The system will enable an aircraft to perform special operations and search and rescue missions at night, in adverse weather conditions and in a high-threat environment.

AN/APQ-174 modes include normal, power management and weather, terrain-following, terrain-avoidance, ground-ranging, beacon and weather. Set clearances are 100, 150, 200, 300 and 500 ft. Weather performance is enhanced by the use of selectable circular polarisation and operation in 10 mm/h of rain is claimed. The system includes extensive internal monitoring, periodic and manually initiated BIT and end-to-end test.

The AN/APQ-174 allows operations at low altitudes, down to 100 ft above the ground by day or night. MFR improvements include the addition of weather detection and beacon interrogation modes to help in navigation and rendezvous. Other upgrades to the radar include expansion of software memory, conversion to electrically erasable memory and addition of an obstacle warning signal to existing video displays. Another upgrade, adding a new set clearance altitude to the terrain-following mode, will allow the aircrew to train at a safer altitude.

Specifications

Dimensions:
(pod) 330 mm diameter × 1,090 mm
(radar interface unit) 760 × 330 × 480 mm
Weight: 114 kg
Reliability: 144 h MTBF specified

Operational status

The AN/APQ-174 MMR is deployed on the US Army Special Operations Aircraft (SOA) MH-60K and MH-47E.

Contractor

Raytheon Electronic Systems.

VERIFIED

The AN/APQ-168 radar for the Sikorsky HH-60D Night Hawk helicopter

AN/APQ-181 radar for the B-2

The AN/APQ-181, designed specifically for the Northrop Grumman B-2 Spirit stealth bomber, operates in the J-band using 21 separate modes for terrain-following and terrain-avoidance; navigation system updates; target search, location, identification and acquisition; and weapon delivery.

The radar is a completely redundant modular system which employs two electronically scanned antennas, sophisticated software modes and advanced low probability of intercept techniques that match the aircraft's overall stealth qualities.

To meet reliability specifications and provide for operational redundancy, the AN/APQ-181 provides two separate radar sets, each consisting of five Line Replaceable Units (LRUs): antenna, transmitter, signal processor, data processor and receiver/exciter, with all but the antennas able to function for either or both radar units. These LRUs weigh 953 kg and have a volume of 1.47 m³. Six LRUs are symmetrically positioned on the sidewalls of the nose-wheel well and the two radar data processors are stacked vertically in the aft wall of the well.

Each 260 kg antenna is mounted in a cavity behind a large radome some 8 ft outboard of the aircraft centreline, just below the flying wing's leading edges. Antenna locations are marked by large, slightly darker, rectangular patches visible on the underside of the aircraft. The antennas look down and outward. They are electronically steered in two dimensions and feature a monopulse feed design that is claimed to enable fractional beamwidth angular resolution. A beam-

The Northrop Grumman B-2 bomber carries the Raytheon Electronic Systems AN/APQ-181 radar
2001/0103893

steering computer establishes phase shifter settings based on pointing direction commands from the radar data processor. A Smiths Industries' motion sensor subsystem fitted to the antenna uses a modified strapdown inertial platform to measure antenna motion. This allows the radar to compensate for motion during synthetic aperture radar mode operation. The antenna design features its own power supplies, liquid cooling and line-replaceable modules.

The remaining LRUs in the radar are derived from other Raytheon products used in aircraft such as the F-15C/D/E and F/A-18. All radar units communicate over a dual-redundant MIL-STD-1553 databus, and are hardened to withstand transient radiation and electromagnetic pulse effects. The radar was designed for very stringent environmental requirements exceeding those of other radars. This is because the vibration experienced by a B-2 operating at low altitude is considered to be especially severe for equipment because of the aircraft's stiffer, less flexible structure.

Operational status
In service.

Contractor
Raytheon Electronic Systems.

UPDATED

AN/APR-49(V) self-protection system

The AN/APR-49(V) is a microprocessor-controlled airborne self-protection system which uses a superheterodyne receiver to detect pulsed and CW signals in the H- to J-bands. An integral high-speed microprocessor controls the receiver, executes data processing and performs the required interfacing tasks. The receiver scans the programmed frequency ranges, measures frequency of detected signals, calculates the angle of arrival, analyses emitter parameters, identifies threat radars and provides data for display. The processed threat data are presented on a 3 in (76.2 mm) display that provides an alphanumeric representation of the type, angle of arrival, relative lethality and status of threats.

Other features of the AN/APR-49(V) include high-resolution emitter separation and identification; the ability to record emitter parameters in flight for post-flight playback, signal analysis and training; simple and effective interface with laser and other pulse radar warning receivers; field loading of the emitter library and downloading of the system recorder; and provisions for interface with chaff/flare dispensing, ECM and missile warning systems.

The AN/APR-49(V) superheterodyne receiver is connected to an array of four spiral antennas. The receiver input enables selection of any four antennas in a controlled sequence to enable angle of arrival determination. The optional remote RF preamplifier enables selection and amplification of signals detected by the remote antennas. An integral signal processor not only determines DF but also characterises each signal, including frequency, and identifies a threat emitter by comparing the signal parameters with preprogrammed library information.

The digital signal analyser in the receiver incorporates a computer featuring a 32-bit microprocessor with EEPROM for combat software and emitter tables and RAM for temporary storage. When the APR-49(V) is configured with a wideband radar warning receiver it is capable of effective and accurate determination of radar warning for self-protection.

Specifications
Weight: 6 kg with indicator, remote preamplifier and 4 antennas
Frequency: H- to J-bands
Signal types: high-PRF and low-ERP pulsed signals and CW signals
Coverage: 360° azimuth or selected coverage

Contractor
Raytheon Electronic Systems.

VERIFIED

AN/APS-115 radar

The AN/APS-115 is one of the Raytheon Electronic Systems family of airborne search radars and is an I-band frequency-agile system of modular design. It is employed principally for ASW and maritime roles. The AN/APS-115 is a dual system to provide 360° coverage for the P-3C Orion land-based ASW aircraft. One antenna is mounted in the nose and the other in the rear. In addition to the underslung stabilised antenna assemblies, the equipment includes dual receiver/transmitters, an antenna position programmer, dual radar set controls and a common antenna control unit.

Operational status
No longer in production, but still in operational service.

Contractor
Raytheon Electronic Systems.

VERIFIED

AN/APS-124 search radar

The AN/APS-124 search radar was specially designed to be part of the comprehensive avionics suite for the US Navy Sikorsky SH-60B Seahawk ASW helicopter built to satisfy the Light Airborne MultiPurpose System (LAMPS) Mk III requirement. One of the problems associated with the operation of these medium-size helicopters from the 'Spruance' class destroyers on which they serve is that of stowage, particularly the height limitation. The APS-124 is therefore designed around a low-profile antenna and radome and consists of six LRUs.

Optimum detection of surface targets in rough sea is accomplished by several unique features including a fast-scan antenna and an interface with the companion OU-103/A digital scan converter to achieve scan-to-scan integration. The system is associated with a multipurpose display and with the LAMPS datalink so that radar video signals generated aboard the aircraft can be displayed on LAMPS-equipped ships.

The system operates in three modes covering long- and medium-range search and navigation and fast scan surveillance. The display ranges are selectable up to 74 km and the false alarm rate is adjustable to suit conditions.

The system is designed around the MIL-STD-1553 digital databus to communicate with other aircraft equipment and the modular design facilitates installation on other aircraft.

Specifications
Weight: 95 kg
Coverage: 360° azimuth
Display range: out to 160 n miles (selectable)
Pulse length:
(long range) 2 μs
(medium range) 1 μs
(short range) 0.5 μs
PRF:
(long range) 470 pps
(medium range) 940 pps
(short range) 1,880 pps
Scan rate:
(long range) 6 rpm
(medium range) 12 rpm
(short range) 120 rpm

Operational status
Service deployment of the LAMPS III Seahawk helicopter began in 1983. More than 300 systems are in service.

Contractor
Raytheon Electronic Systems.

VERIFIED

AN/APS-134(V) radars

The APS-134(V) anti-submarine warfare and maritime surveillance radar is the international successor to the US Navy's AN/APS-116 periscope detection radar. The APS-134(V) incorporates all the features of the former system while improving performance and adding capabilities, including a new surveillance mode.

The heart of the radar is a fast-scan antenna and associated digital signal processing which, says Raytheon Systems Company, form the only proven and effective means of eliminating sea clutter. This technique is used in two of the three operating modes, the third being a conventional slow scan for long-range mapping and navigation. The transmitter power is 500 kW.

In Mode 1, periscope detection in sea clutter, high-resolution pulse compression is employed with a high PRF and a fast-scan antenna, actual values being 0.46 m, 2,000 pps and 150 rpm. Display ranges are selectable to 59 km. There is an adjustable false alarm rate to set the prevailing sea conditions and scan-to-scan processing is employed.

Mode 2, long-range search and navigation, operates at medium resolution and with a low PRF, low scan and display ranges selectable to 278 km. Actual values are 500 pps and 6 rpm.

Mode 3 operates, again at high resolution, for maritime surveillance. A low PRF (500 pps) is used in conjunction with an intermediate scan speed of 40 rpm. Display ranges are selectable to 278 km and an adjustable false alarm rate is used together with scan-to-scan processing.

The system is also available in an offline configuration, with its own 10 × 10 in (254 × 254 mm) CRT control/display unit. Online operation linked in with other aircraft systems is accomplished via a MIL-STD-1553 digital databus, with the digital scan converter providing raster scan video for other aircraft displays. The weight of the entire APS-134(V), including the waveguide pressurisation unit, is 237 kg. The equipment is compatible with the inverse synthetic aperture radar techniques developed by Raytheon Systems Company for long-range ship classification.

Operational status
In service with Dornier/Dassault Aviation Atlantique ASW aircraft of the German Navy, Republic of Korea Navy and Pakistan Navy P-3C aircraft, Portuguese Air Force P-3P aircraft and as part of the Royal New Zealand Air Force Lockheed Martin P-3B Orion update programme.

The APS-134(V)6 radars on the Republic of Korea Navy P-3C aircraft are to be updated by the addition of ISAR (Inverse Synthetic Aperture Radar) capability. Some AN/APS-134(V) radars are also being upgraded to a so-called AN/1APS-137(V)6 standard.

AN/APS-134(V)7 radar
The AN/APS-134(V)7 offers technological improvements over its predecessor, the AN/APS-134(V). It features the developments of the US Navy's AN/APS-116 family of periscope detecting radar systems.

The AN/APS-134(V)7 radar system includes the periscope detection, long-range maritime surveillance and navigation modes from the AN/APS-134 radar. To that are added capabilities for improved periscope detection, advanced digital signal processing, multiple track-while-scan, dual-channel digital scan conversion, ESM countermeasures and 0-level built-in radar system diagnostics. There is also an optional capability to record radar and FLIR video.

The system is designed to detect small targets in high sea states at long range. Long-range performance is achieved by using a 500 kW high-power transmitter, 35 dB high-gain antenna and custom-developed low-noise preamplifier of less than 3 dB noise figure. These features provide the signal-to-noise ratio necessary to achieve long-range capability. Detecting small targets in the sea clutter environment is an inherent problem of maritime surveillance radar. To overcome this limitation, pulse compression and scan-to-scan processing are employed in the AN/APS-134(V)7.

The AN/APS-134(V)7 features simultaneous 32-target high-resolution tracking, 360° capable PPI coverage with sector scan, picture within picture PPI and B scan display format, dual-channel multilevel digital scan conversion and advanced digital signal processing. Multiple radar configurations are available for offline, MIL-STD-1553B and ANEW databusses. The system is adaptable to multiple radar video display configurations. The AN/APS-134(V)7 interfaces with other aircraft systems such as IFF, FLIR and ESM.

Operational status
In production for the Fokker Maritime Enforcer Mk 2 patrol aircraft.

Contractor
Raytheon Electronic Systems.

VERIFIED

AN/APS-137(V) inverse synthetic aperture radar

The APS-137(V) is an improved version of the AN/APS-116 periscope detection radar which is standard on the US Navy's S-3 aircraft. Over 200 units of the AN/APS-116 I/J-band radar have been produced and the APS-137(V) introduces an Inverse Synthetic Aperture Radar (ISAR) mode. Funding for this development, which increases radar processing and introduces a standard surveillance and automatic classification capability, started in 1982 and was scheduled to continue into 1994 as part of the avionics upgrade which denotes the S-3B version.

The APS-137(V) offers long-range detection and classification of ships, the radar producing a recognisable image of the target vessel. The image is derived from the Doppler shifts of the returns, compared with the reference level. Using pulse compression and fast scan processing to eliminate sea clutter, the APS-137(V) provides improved periscope detection and high-altitude maritime surveillance. Multiple target tracking (track-while-scan) is also available.

The APS-137(V) is compatible with the seekers in Harpoon, Tomahawk and other missiles, and can interface directly into weapons system computers. By comparing images before and after an attack, target battle damage can be assessed.

Operational status

Full production for the S-3B retrofit programme commenced in 1987. In January 1987, Raytheon Electronic Systems received a contract to supply AN/APS-137(V) radars to equip US Navy S-3B and P-3C and US Coast Guard C-130 aircraft, with deliveries extending to 1992. This was extended at the end of 1987 to include a further nine sets and long lead items for 16, to equip the S-3B. A further contract was awarded in August 1991 for 24 AN/APS-137(V) systems.

The AN/APS-137B(V)5 radar forms part of the P-3C Anti-surface warfare Improvement Programme (AIP) scheduled for 146 US Navy P-3C Update III-equipped aircraft. The modification programme is under way with an order for 11 modified radars to be completed by September 2000.

Contractor

Raytheon Electronic Systems.

VERIFIED

AN/AQA-7 sonobuoy processor

The AN/AQA-7 is the standard processor aboard the US Navy's P-3A/B and C (Update II) ASW aircraft. The system works in conjunction with the AN/SSQ-53B DIFAR sonobuoy, the SSQ-62B DICASS and SSQ-77A VLAD buoys. A new development programme, Improved Processing And Display System (IPADS), will significantly enhance the capabilities of the sensor station operators and the tactical co-ordinators on P-3A and P-3B aircraft.

Operational status

In service in the P-3. The system is in service with five foreign navies and some 1,325 AN/AQA-7s have been delivered.

Contractor

Raytheon Electronic Systems.

VERIFIED

AN/ARC-114A VHF radio

The AN/ARC-114A is a military airborne radio which covers the VHF band from 30 to 75.95 MHz and provides 920 channels at a spacing of 50 kHz. The system also incorporates an independent Guard receiver to monitor a single, pretuned Guard channel in the frequency range 40 to 42 MHz.

A compact lightweight single container unit, the ARC-114A is designed primarily for helicopters, although its high-altitude performance up to 50,000 ft means it is suitable for higher-flying fixed-wing aircraft.

The ARC-114A operates in FM mode and can transmit either voice or data signals. Transmitter power output is 10 W. Facilities include azimuth homing and a rebroadcast capability is also available when the system is used in conjunction with a suitable transmitter/receiver.

The system is of all-solid-state construction and employs crystal-controlled digital synthesis techniques for frequency generation. In receive mode, varactor and band selection tuning methods are employed and for transmit mode the frequency selection is accomplished by indirect synthesis and digital tuning techniques.

Specifications

Dimensions: 105 × 146 × 215 mm
Weight: 3.1 kg
Temperature range: –32 to +55°C

Operational status

No longer in production. In service in a number of forces around the world. The ARC-114A has been produced in quantity for the US Army which procured over 5,000 systems for service in its Bell UH-1H and Sikorsky UH-60 utility helicopters.

Contractor

Raytheon Electronic Systems.

VERIFIED

AN/ARC-115A VHF radio

The AN/ARC-115A broadly has the same range of facilities as the company's AN/ARC-114A system but provides, within the frequency band 116 to 149.975 MHz, 1,360 channels at a spacing of 25 kHz. Like the ARC-114A, it has a rebroadcast facility when operated in conjunction with suitable equipment. It also has an azimuth homing capability.

Specifications

Dimensions: 124 × 146 × 235 mm
Weight: 3.26 kg

Operational status

No longer in production. In service with the UH-1H helicopters of the US Army.

Contractor

Raytheon Electronic Systems.

VERIFIED

AN/ARC-164 UHF radio

The ARC-164 is the basic member of a family of radio communications equipment and subvariants, each designed for particular applications yet with a high degree of commonality.

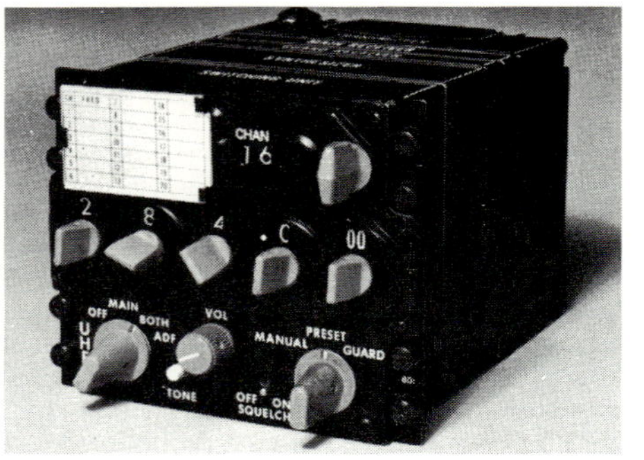

The ARC-164 UHF radio

The basic ARC-164 covers the UHF band, providing 7,000 channels over the range 225 to 400 MHz in 25 kHz increments. Any 20 channels may be preselected.

A fully solid-state system, the ARC-164 is distinguished by its 'slice' module construction in which a series of modules, connected by a flexible harness, are simply bolted together to form the desired electronic configuration. A typical simple system would comprise transmitter, receiver, guard receiver and synthesiser. The control unit may either form part of this consolidated package or be remotely located. The modular approach adopted allows growth capability, extra modules being added as required. A range of optional facilities, such as data transmission, secure speech and ECCM capability, is available by the addition of the appropriate slices.

A number of directly connected or remote-control units are produced for the ARC-164. These include: a simple frequency-selection controller; a 32-channel preset control with LED readout of the selected channel; a 20-channel preset unit with provision for two-cockpit take-control; a microprocessor controller with 400 UHF and VHF, AM or FM, preset channels, liquid crystal channel and frequency readout display and the capability of controlling up to four systems simultaneously. Additional remote frequency/channel indicators are available. Also available is a variety of mounting trays to suit differing installations for new types of aircraft and for the updating of older aircraft equipment.

A remote ARC-164 radio compatible with MIL-STD-1553B databus operation has been developed under contract to the US Army and is in production and service. Panel-mounted radios and certain controls can be furnished with ANVIS Green A lighting compatible with NVG in accordance with MIL-STD-85762. These features can also be obtained by retrofitting appropriate radios and controls. A Low Probability of Intercept and Detection (LPI/LPD) version of the ARC-164, which allows covert UHF communications, is known as StealthComm.

StealthComm waveform features include a hybrid of several LPI/LPD techniques: hybrid direct sequence and frequency hopping; feature suppression modulation techniques; receive sensitivity enhancements; 60 dB of power control.

StealthComm has enhanced data capability from 16 to 80 to 120 kbytes.

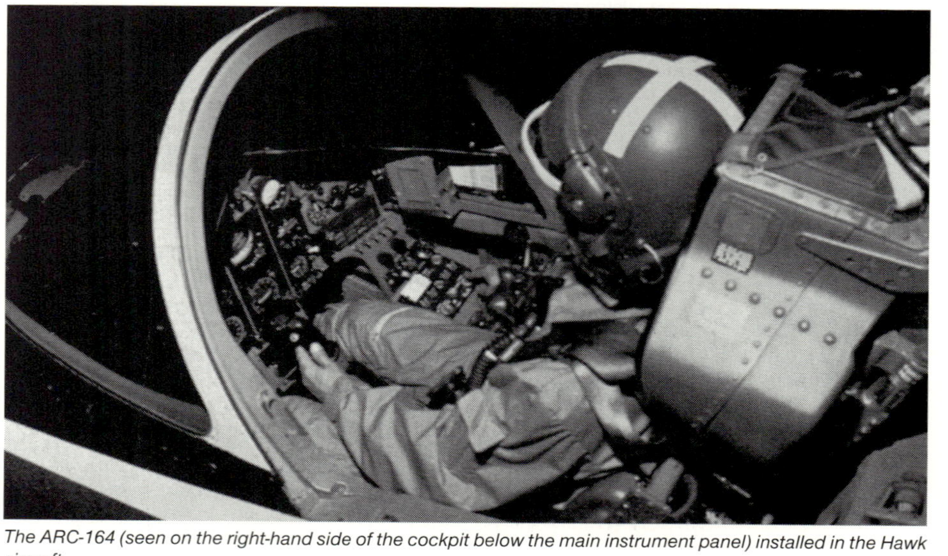
The ARC-164 (seen on the right-hand side of the cockpit below the main instrument panel) installed in the Hawk aircraft

All StealthComm features can be incorporated in existing ARC-164 radios through mod kits; or new ARC-164/LPD radios can be installed as direct replacements form-fit for existing ARC-164s.

Specifications

Dimensions:
(transmitter/receiver)
(10 W version) ½ ATR × 178 mm
(30 W version) ½ ATR × 374 mm
(controller) ½ ATR × 83 mm
Weight:
(transmitter/receiver)
(10 W version) 3.7 kg
(30 W version) 6.8 kg
(controller) 2 kg
Frequency: 225-400 MHz
Channels: 7,000
Channel spacing: 25 kHz
Power output: 10 W (standard), 30 W (uprated)
Reliability: 2,000 h MTBF demonstrated

Operational status

In production and service. More than 65,000 ARC-164s have been produced to date.

The system has become standard fit for a wide range of US Air Force aircraft including the F-15 and F-16 fighter aircraft.

The ARC-164 equips the Royal Navy's Sea King helicopters and Sea Harrier aircraft and the Hawk, Jaguar and other Royal Air Force aircraft. Further Hawk aircraft, notably those delivered to Kenya, are also fitted with the ARC-164, as are some Strikemaster strike/trainers.

Contractor

Raytheon Electronic Systems.

VERIFIED

AN/ARC-181 TDMA radio terminal

The AN/ARC-181 Time-Division Multiple Access (TDMA) terminal is a secure, jam-resistant radio terminal for airborne surveillance, command and control centres developed for US and NATO E-3A AWACS aircraft, Hawk missile batteries and NATO Air Defence Ground Environment centres under the Airborne Early Warning/Ground Environment Integration Segment (AEGIS) programme. The terminal comprises a communications processor, transmitter, receiver, high-power amplifier and control and display panel.

Part of the JTIDS programme, the terminals provide the channel for continuous communications exchange, resulting in a constantly updated information pool which is available to all network members. Spread spectrum, data interleaving and frequency-hopping techniques give enhanced data and jam-resistance capabilities.

Operational status

First production terminals went into service with NATO Boeing E-3A aircraft and ground stations in 1983.

Contractor

Raytheon Electronic Systems.

VERIFIED

AN/ARC-187 UHF radio

The ARC-187 is a further development of the company's ARC-164 US Air Force standard system which has been adapted to meet a US Navy requirement for a low-cost terminal designed to operate with communication satellites, principally the US Navy's FltSatCom. It covers the UHF band from 225 to 400 MHz, in which range it provides 7,000 channels at increments of 25 kHz. Up to 20 channels may be preselected.

Operating modes include AM with a secure speech facility and FM and FSK data transmission in both analogue and digital form. ECCM capability is incorporated internally in the receiver/transmitter.

The system, which is remotely controlled, uses standard ARC-164 'slice' modules including a modified synthesiser section designed for compatibility with communications satellite data rate requirements.

A new control for the ARC-187 has been developed, which provides for compatibility with Satcom and MIL-STD-1553B databus modes of operation. This control incorporates ANVIS Green A lighting in accordance with MIL-STD-85762 for compatibility with Gen III NVG. A Satcom modem for the ARC-187 is under development.

Specifications

Dimensions:
(controller) 132 × 147 × 124 mm
(transmitter/receiver) 440 × 153 × 143 mm
Weight:
(controller) 2 kg
(transmitter/receiver) 7.4 kg
Frequency: 225-400 MHz
Channels: 7,000
Channel spacing: 25 kHz
Power output: 30 W (AM), 100 W (FM/FSK)

Operational status

In service in US Navy P-3C aircraft.

Contractor

Raytheon Electronic Systems.

VERIFIED

AN/ARC-195 VHF radio

The AN/ARC-195 radio may be considered as a variant of the UHF AN/ARC-164, with which it has a component commonality of 93 per cent. It is available in 10 and 30 W output versions. The major differences between the two systems consist of some component value changes and the substitution in the AN/ARC-195 of a synthesiser with a frequency standard appropriate to the VHF section of the radio frequency spectrum.

The AN/ARC-195 covers the VHF band from 116 to 156 MHz, providing 1,750 channels at a frequency separation of 25 kHz. In other respects it is almost identical to its UHF counterpart. Control units are also almost identical and certain units from the AN/ARC-164 range of controllers may be used for combined UHF/VHF operation.

Operational status

In production and in service.

Contractor

Raytheon Electronic Systems.

VERIFIED

AN/ARC-222 SINCGARS radio

The AN/ARC-222 SINCGARS radio is the replacement for the AN/ARC-186 and is designed for air-to-ground and air-to-air communications. It includes SINCGARS-V capability and covers the frequency ranges 30 to 87.975 MHz in VHF FM, 108 to 151.975 MHz in VHF AM (108 to 115.975 MHz receive only) and 152 to 174 MHz FM for the International Maritime Band.

It consists of a control unit and a receiver/transmitter.

Specifications

Dimensions:
(control unit) 57.4 × 127 × 95.8 mm
(receiver/transmitter) 120.6 × 127 × 174.6 mm
Weight:
(control unit) 0.73 kg
(receiver/transmitter) 4.67 kg
Power supply: 28 V DC, 4.5 A max
Frequency:
30-87.975 MHz (FM)
108-151.975 MHz (AM) (108-115.975 MHz receive only)
152-174 MHz (International Maritime Band)

Operational status

Deliveries to the US Air Force began in April 1997. It is part of the baseline fit for the C-130J aircraft, and is listed for the Apache helicopter. 8.33 kHz channel spacing capability is reportedly being added.

Contractor

Raytheon Electronic Systems.

VERIFIED

AN/ASB-19(V) Angle Rate Bombing Set (ARBS)

The AN/ASB-19(V) Angle Rate Bombing Set (ARBS) was designed for US Marine Corps aircraft to improve day and night bombing accuracy when operating in the close support role using unguided weapons. The system provides accurate delivery, irrespective of target velocity, wind velocity, target elevation or dive angle. It is compatible with guided ordnance and can be used to direct gunfire and air-to-ground rockets. ARBS was originally designed for application to the US Marine Corps A-4M and is also the primary weapons delivery system for the US Marine Corps and Spanish Navy AV-8B and the Royal Air Force Harrier GR.Mk 7 aircraft.

ARBS comprises three main subsystems: a dual-mode tracker, weapon delivery computer and control unit. The tracker includes a laser and pilot-controlled television tracking equipment, both using a common optical system. The dual-mode tracker automatically tracks targets designated by a laser (either ground-based or by an accompanying aircraft) or targets which are television-designated by the pilot of the ARBS-equipped aircraft itself. The tracking system's common optics enable transition from laser to television tracking mode to be accomplished without losing the target.

Tracking information is passed from the dual-mode tracker to the weapon delivery computer. The weapon delivery computer performs computations for weapon trajectory and fire control, position control of the dual-mode tracker during target acquisition, digital filtering of the dual-mode tracker angular rate signal outputs and automatic fire or weapon release signals to the armament and other systems.

The weapon delivery computer receives aircraft to target line of sight angle and angle rate data from the tracker when that unit has achieved target lock-on. This data, combined with true airspeed and altitude information from the air data computer, when processed by the weapon delivery computer, provides the weapon delivery solution. Target position, weapon release and azimuth steering information are generated and presented to the pilot via a head-up display.

The function of the ARBS control unit is to interface with the weapon delivery computer and to provide the pilot with control of the tracker modes and entry of display, target, navigation and maintenance information.

When operating in the laser mode, the sensor automatically acquires the target, which is illuminated from an external laser source. The sensor, via the weapon delivery computer, presents steering signals to the pilot on the head-up display. In television mode, visible light from the common optical system is directed to form an image on the television display. Tracking of small low-contrast poorly defined moving targets is said to be possible even with changes in aspect ratio, in the presence of competing clutter or when the target is partially lost to view behind ground obstructions.

Following head-up acquisition in the television mode, the pilot is shown a × 7 magnified image of the target on the cockpit television monitor display. He may at this stage use a hand control to slew the tracker gate on to a new track point or on to a nearby alternative target.

Weapon release in either laser or television tracking mode may be made automatically or manually, with a weapon release time and a continuously computed impact point being generated by the weapon delivery computer. A weapons data insert panel provides entry into the weapon delivery computer of weapon characteristics and rack type for those stores being carried for a specific mission.

ARBS permits delivery of any type of weapon at any attitude and airspeed combination at any dive angle or in level flight. The system accuracy is said to be sufficient for first-pass precision delivery in close air support or interdiction missions against land or seaborne targets, but navigation steering commands are also generated to provide second passes against hardened targets. This re-attack facility provides the pilot with head-up steering information for return to a designated target, the location of which is retained in the system memory until a new target is designated by the pilot.

The ARBS in the nose of a US Marine Corps AV-8B

The system's television tracking element also provides a limited air-to-air capability for daylight operations, its × 7 magnification being particularly useful for visual identification and tracking of airborne targets.

Angular coverage of ARBS extends from ±35° in azimuth and from +15 to –70° in elevation at roll angles of ±450°. These limits are sufficiently wide to allow extensive flexibility in weapon selection and delivery profiles and permit multiple roll manoeuvres to take place on the attack approach run without loss of target lock on. The tracking equipment's laser is NATO coded.

The ARBS weapon delivery computer processor on the A-4M is a 16 kword capacity system extensible to 32 kwords. With 16 kwords, the system has far more capability than is required for weapon release and steering command computation and may be used for additional purposes such as the pilot-initiated self-test of the ARBS system which performs operational readiness testing and fault isolation functions. By modification of the computer input/output scaling factor, the ARBS can be configured to interface with aircraft avionics other than those of the A-4M and the Harrier AV-8B. Three basic aircraft interfaces are required: a vertical reference to provide pitch and roll data, true airspeed and a servoed optical sight or head-up display on which to present azimuth steering commands, continuously computed impact point and bomb release information.

ARBS systems can be configured, in either internally or pod-mounted versions, to suit the requirements of other aircraft. Several pod configurations have been formulated for candidate aircraft with attachment hooks at 356 and 762 mm centres, compatible with standard pylon or fuselage centreline mountings. Where existing cockpit controls cannot be used, two small ARBS control units may be fitted at any accessible position. The A-4M control units are configured with standard 146 mm widths adaptable to most cockpit consoles.

Specifications

Volume: 0.07 m^3
Weight: 58.18 kg

Operational status

In service in US Marine Corps and Spanish Navy AV-8B aircraft and Royal Air Force Harrier GR. Mk 7s.

Contractor

Raytheon Electronic Systems.

VERIFIED

AN/ASQ-208 magnetic anomaly detection system

The AN/ASQ-208 is a derivative of the AN/ASQ-81 magnetometer that operates on the atomic properties of optically-pumped metastable helium atoms to detect variations in total magnetic field intensity. The AN/ASQ-208 is a digital system which incorporates microprocessor technology to achieve aircraft compensation, multiple channel filtered display and threshold processing. Two system configurations are available: one for inboard installations and one for towed installations.

The system features a high-sensitivity helium sensor and choice of inboard or towed configuration. Control implementation is by MIL-STD-1553B or Manchester databus for online operation or dedicated offline control unit.

Enhanced signal recognition is provided by three-channel filtered data for optimum operator display and automatic detection with range/confidence estimate. Automatic aircraft compensation is provided by electronic compensation of aircraft interfering terms and an automatic figure of merit estimator.

The AN/ASQ-208(V) is configured for minimum impact on existing aircraft or new aircraft installations and utilises existing AN/ASQ-81(V) aircraft wiring. It requires a single cable change for the vector sensor.

System performance enhancements include shallow water detection, elimination of dedicated Magnetic Anomaly Detection (MAD) compensation flights and automatic detection.

Operational status

In initial production, and in service with Sikorsky S-76N.

Contractor

Raytheon Electronic Systems.

VERIFIED

AN/ASQ-213 HARM Targeting System (HTS)

The AN/ASQ-213 HTS was designed to be fitted to US Air Force F-16C/D Block 50D aircraft and provide targeting information for the AN/AGM-88A HARM missile in the manned SEAD role. US Air Force aircraft in the HTS role are designated F-16CJ.

The HTS detects, identifies, and locates hostile radars and provides data needed by HARM for

The Raytheon HARM Targeting System (E) **2000**/0079234

US Air Force F-16CJ carrying HTS on the starboard chin station

calculation of targeting and launch parameters. In its most effective mode ('range known'), the HTS displays the target location to the pilot for HARM designation and firing.

HTS/HARM-equipped F-16 aircraft often operate with RC-135 Rivet Joint or EA-6B Prowler aircraft to optimise effectiveness.

The HTS system is carried on the starboard chin station of the F-16 aircraft; it weighs 40 kg.

Operational status

The HTS system is in service on US Air Force F-16CJ aircraft. The HTS pod is to be upgraded to provide improved range-finding algorithms that will increase accuracy against radiating targets. The R7 system upgrade, planned for 2005, will enable the HTS to use targeting data that has been transmitted to the F-16CJ SEAD aircraft over the JTIDS (Joint Tactical Information Distribution System) Link 16 terminal, that will be fitted to F-16 Block 50 aircraft starting in 2004.

Raytheon is said to be discussing an HTS integration on F/A-18C/D with the US Navy.

HARM Targeting System (E)

Raytheon is now marketing a variant of the HTS system under the designation HARM Targeting System (E) (believed to be Export). The capability offered includes the 'range known' mode. Dimensions are 200 × 1420 mm (diameter and length), with a weight of 41 kg.

Contractor

Raytheon Electronic Systems.

UPDATED

The AN/AVR-2A laser detecting set consists of four sensor units and an interface unit comparator

AN/ASQ-81(V) Magnetic Anomaly Detection system (MAD)

The AN/ASQ-81(V) Magnetic Anomaly Detection (MAD) system was developed for US Navy use in the detection of submarines from an airborne platform. The system operates on the atomic properties of optically pumped metastable helium atoms to detect variations of intensity in the local magnetic field. The Larmor frequency of the sensing elements is converted to an analogue voltage which is processed by bandpass filters before it is displayed to the operator.

Four configurations of the AN/ASQ-81(V) are available: two for use within an airframe and two for towing behind an aircraft. The US Navy uses the AN/ASQ-81(V)1 in the land-based P-3C Orion, where it is housed in a tail sting. The AN/ASQ-81(V)3 is installed in the carrier-based S-3 Viking aircraft, where it is extended on a boom.

The AN/ASQ-81(V)2 is a towed version employed by the US Navy on Sikorsky SH-3H and Kaman SH-2D helicopters. It is also in service with other countries, including the Netherlands for use on the Westland Lynx, Japan for use on the Mitsubishi HSS-2, and with forces employing the Hughes 500D helicopter.

The second towed version, the AN/ASQ-81(V)4, is used by the US Navy on the Sikorsky SH-60B LAMPS III helicopter.

All versions of the AN/ASQ-81(V) have the same C-6983 detecting set control, AM-4535 amplifier and power supply unit. The AN/ASQ-81(V)1 and 3 use a DT-323 magnetic detector, while the AN/ASQ-81(V)2 and 4 have a TB-623 magnetic detecting towed body. The towed version is controlled by the C-6984 reel control, which works the RL-305 magnetic detector launching and reeling machine.

Operational status

In production, and in service.

Contractor

Raytheon Electronic Systems.

VERIFIED

The Royal Netherlands Navy P-3C Orion uses the AN/ASQ-81(V)1

AN/AVR-2A(V) laser detecting set

The AN/AVR-2A(V) laser detecting set detects, identifies and characterises optical signals and provides audible and visible warning of laser threats to the aircrew. It consists of four staring array SU-130A(V) sensor units and a CM-493A interface unit comparator. The AN/AVR-2A(V) interfaces with all variants of the AN/APR-39(V) Series radar signal detecting set to function as an integrated radar and laser warning receiver system.

The AN/AVR-2A(V) system provides 360° coverage and can identify laser range-finders, designators and beamriders. It contains a reprogrammable, removable, user data module. P³I expansion includes a MIL-STD-1553B interface.

Operational status

In production for the US Army, Navy, Marine Corps and Special Forces for the AH-1F Cobra, AH-64 Apache, AH-64D Longbow Apache, AH-1W SuperCobra, MH-60K Black Hawk, MH-47E Chinook, OH-58D Kiowa Warrior, HH-60H Combat Rescue, UH-1N Huey, and V-22 Osprey helicopters. Also selected for the UK EH 101SH and WAH-64 helicopters. Under development for: RAH-66 Comanche, C-130 Hercules, SH-60B LAMPS, and SH-60F CV helo aircraft.

Contractor

Raytheon Electronic Systems.

VERIFIED

AN/AVR-3(V) airborne laser warning system

The AN/AVR-3(V) is an airborne microprocessor-controlled warning system for detecting laser emissions in two threat bands, with provision for a third. Identified threats, which are detected and analysed, are presented to the pilot on a 3 in (76.2 mm) CRT display that provides an alphanumeric representation of the type of laser and the angle of arrival.

The AVR-3 offers complete threat reprogrammability for new and changing environments as laser weapons increase and mature. This is accomplished by an extensive and easily updated emitter library file. A portable loader unit enables the loading of operational software and emitter tables and the downloading of recorded signal data in the field.

The AVR-3 is suitable for the self-protection of helicopters and fixed-wing aircraft regardless of size. Electrical design and mechanical configuration provide for the use of up to eight laser sensors to ensure complete spatial coverage for all aircraft.

The AVR-3 features detection of modern pulsed lasers, direction-finding of received signals, high-resolution emitter separation, sophisticated signal processing, alphanumeric azimuth threat display and full system BITE. The system is able to record emitter parameters during flight for post-flight playback and signal analysis. It has provision for a tie-in with flare and chaff dispensers, ECM and missile warning systems.

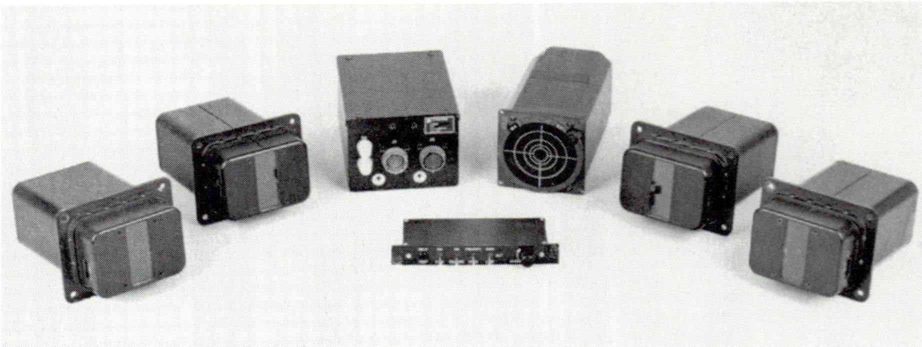

The AN/AVR-3(V) airborne laser warning system consists of a number of sensors (left and right), a laser analyser (centre left), a display unit (centre right) and a control unit (front centre)

Specifications

Dimensions:
(laser sensor) 63.5 × 85.1 × 149.9 mm
(laser analyser) 146 × 106.2 × 78.7 mm
(control unit) 54.1 × 146 × 19 mm
(display unit) 158 × 80.5 × 80.5 mm
Weight:
(laser sensor) 0.59 kg
(laser analyser) 1.5 kg
(control unit) 0.2 kg
(display unit) 1.2 kg
Power supply: 28 V DC, 90 W (max) (MIL-STD-704)
Frequency: 2 laser bands, provision for a third
Coverage: 360° azimuth
Environmental: MIL-E-5400

Contractor

Raytheon Electronic Systems.

VERIFIED

AN/AWG-9 weapon control system for the F-14A/B

The AN/AWG-9 is the aircraft element of the Phoenix AIM-54 air-to-air missile weapon control system in the early model Northrop Grumman F-14 aircraft. It comprises a fire-control radar and target illuminating radar, digital computer and displays. Provision is also made for the automatic exchange of datalink information between the AWG-9 and the Naval Tactical Data System (NTDS) or Airborne Tactical Data System (ATDS) for target designation and other functions. For onboard target acquisition, the AN/AWG-9 includes a long-range high-power pulse Doppler radar. This system has a look-down capability which enables it to pick up moving targets in the ground clutter that normally obscures targets from a conventional radar.

In addition to its long range, the AN/AWG-9 introduced, for the first time in an aircraft, the ability to track many targets at once with computer-aided selection of target priority. The system is designed for consecutive launch and simultaneous guidance of up to six AIM-54 missiles against separate targets, at ranges in excess of 110 n miles, using time-sharing techniques for mid-course guidance.

In addition to the AIM-54, the AN/AWG-9 can be used with the AIM-7 Sparrow and AIM-9 Sidewinder, and for controlling the M-61 Vulcan 20 mm cannon.

The radar antenna is a planar slotted-plate array with a high-aperture efficiency for increased radar range and, when it was introduced, represented a significant advance in design.

The pulse Doppler radar represents several years of development in transmitter tubes, crystal filters and planar-array techniques to enable the long-range requirements of the AIM-54 to be met. A low-noise parametric amplifier in the receiver section contributes to the long-range capability. Advanced Doppler techniques make look-down target acquisition possible. A flexible high-capacity computer permits simultaneous tracking of a large number of targets and aids the radar intercept officer in the assignment of priorities and in missile firing.

Target detection, tracking and ranging functions for all F-14 air-to-air weapon configurations are handled by the AN/AWG-9 radar and it can operate in either pulse Doppler or conventional pulse modes. A separate TWT provides CW illuminating energy for the semi-active homing AIM-7 Sparrow.

Data processing in the AN/AWG-9 is performed by a general purpose digital computer that features high-speed operation and a large memory capacity in an extremely compact package. The central computer keeps track of targets detected by the radar while the radar continues to search. Based on a preprogrammed logic, the computer evaluates threats, generates steering information for the pilot and paints a complete tactical situation for the radar intercept officer in standard symbology, based on data generated either internally or obtained through external links.

Information from the AN/AWG-9 is displayed on two CRTs. A 127 mm diameter unit is used as a multimode display for the presentation of raw radar target information and IFF returns. A larger 254 mm unit is used to display processed data. This includes target track information and alphanumeric and symbolic data obtained via datalink from other units. This display is also used as a computer readout device for the presentation of computer-generated missile and target assignments.

Specifications

Volume: 0.78 m^3
Weight: 590 kg

Operational status

The AN/AWG-9 radar is operational in US Navy F-14A/B aircraft. The last AN/AWG-9 weapon control system was delivered in August 1988; 695 were produced. Production of spares continued until

The AN/AXQ-14 datalink pod mounted inboard of a GBU-15 glide bomb, under the fuselage of a US Air Force F-15

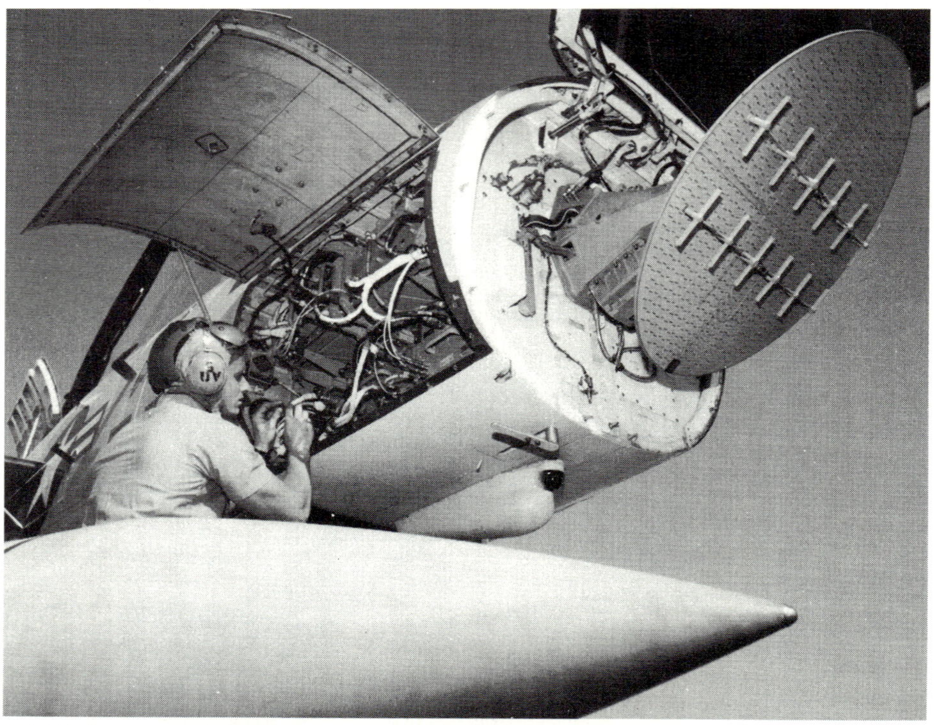

AN/AWG-9 weapon control system undergoing maintenance on the Northrop Grumman F-14. Note the IFF array on the front face of the flat-face antenna

mid-1989. The weapon control system for the F-14D Super Tomcat is designated AN/APG-71, which retains the vital features of the AN/AWG-9, but includes upgraded processing speed, flexibility and signal/data processor capacity.

Contractor

Raytheon Electronic Systems.

UPDATED

AN/AXQ-14 datalink

The AN/AXQ-14 is a two-way communication datalink to guide the GBU-15 glide bomb. It provides a video and command link between the command aircraft and the weapon, enabling the systems operator to remain in the control loop while the weapon is being directed to its target. In effect, the datalink permits a command authority similar to a fly-by-wire system, in which the operator can transmit guidance instruction from launch to impact. Alternatively, he may select any one of a number of autonomous weapon control modes, including an override mode which permits target updating or redesignation as required.

The extended weapon control capability conferred by the datalink contributes to weapon system performance in terms of standoff range and operational utility. Target acquisition is deferred until the weapon, rather than the command aircraft, is closer to the target. Tactically, the aircraft can leave the target zone immediately after launch.

The AN/AXQ-14 system comprises three major elements: a datalink pod mounted on the command aircraft, a datalink control panel used in conjunction with an existing display within the aircraft and a weapon datalink module mounted on the rear of the weapon itself.

The pod is an aerodynamically shaped container mounted on a standard stores carriage strongpoint on the fuselage centreline or on an underwing station, according to aircraft type. It contains four LRUs comprising:

1. An electronics section incorporating all radio frequency generating and receiving equipment, a demultiplexer to decode all aircraft command and pod control signals, an encoder and antenna controls.
2. A phase-scanned array for weapon tracking in normal operation.
3. A forward horn antenna, that provides additional coverage.
4. A mission tape recorder, which maintains a permanent record of weapon video data.

The pod is suitable for high-performance aircraft, is certified for operation at speeds in excess of M1.0 and is also compatible with high- and low-altitude operations. There is said to be no compromise of aircraft performance attributable to carriage of the pod.

Used in conjunction with an existing display system, the aircraft control panel acts as the interface between the weapon system operator and the weapon guidance system. The panel accepts signal inputs from the aircraft as well as from its own controls, and formats these into discrete commands as required via the datalink. Although the panels are tailored to the individual requirements of the aircraft type and intended customer usage, each unit accepts the standard configurations of the GBU-15/AGM-130 datalink and the pod.

Attached to the aft of the GBU-15/AGM-130 weapon is the ultimate component in the datalink chain, the weapon datalink module. This simultaneously transmits video from the weapon's seeker-head and processes incoming command signals from the aircraft to the weapon. Heading changes during the weapon's flight are effected through discrete command signals. Dual-analogue command channels enable the operator to slew the weapon in pitch and yaw during approach to the target.

Digital techniques are employed in the AN/AXQ-14 system and the transmitter is of all-solid-state construction. The system's electronically phase-scanned antenna array provides the datalink with high-rate tactical manoeuvring capability. A comprehensive range of test equipment is provided, including:

1. A flight checkout unit for testing aircraft cables from the pod connection point.
2. An aircraft simulator unit, which permits functional checks of the control panel.
3. A weapon simulator unit for test of the aircraft pod and isolating faults down to LRU level.

Used together, these two latter units permit full system functional checkout.

Two primary launch modes are envisaged for operation of the GBU-15 weapon and AN/AXQ-14 control combination: low-altitude penetration and high-altitude standoff. It is claimed that use of the datalink has improved weapon delivery accuracy over non-link weaponry in various profiles from airborne platforms such as the US Air Force's B-52, F-4 and F-15 aircraft; and US FMS F-16 aircraft. The system is also said to be compatible with the A-4, A-7 and F/A-18 aircraft. Potential weapon applications include Harpoon, Maverick and cruise missiles.

Operational status

In production, and in service.

Contractor

Raytheon Electronic Systems.

VERIFIED

AN/URQ-33(V) Class 1 JTIDS terminal

The Joint Tactical Information Distribution System (JTIDS) uses frequency-hopping, spread spectrum, automatic relay and other high-technology techniques to provide data and voice communications which are highly resistant to jamming.

The AN/URQ-33(V) Class 1 JTIDS terminal is used on board the US Air Force and NATO E-3 AWACS and for a number of ground-based applications for communicating information on command and control, surveillance, intelligence, force status, target assignments, warnings and alerts, weather and logistics.

JTIDS uses a computer-controlled Time Division Multiple Access (TDMA) technique in which information is transmitted in short bursts lasting only a fraction of a second. Bursts are synchronised by computer with bursts from other users, to allow simultaneous transmission on the network without causing interference. The JTIDS burst is spread in frequency, encoded and hopped across a number of frequencies in a split second, making it hard to intercept and almost impossible to jam. The receiver selects pertinent data by means of software filtering.

JTIDS communications are automatically relayed by other terminals. This extends the range beyond the line of sight, and it provides another layer of defence against jammers. It also enables terminals to provide users with position and navigation data, without the need for extra equipment, by means of the highly accurate message time of arrival measurement, which can be converted to range between the transmitter and the receiver.

JTIDS is broadcast in the 960 to 1,215 MHz frequency range. The system consists of a radio set control, transceiver processor unit, high-power amplifier, high-power amplifier power supply, low-power amplifier power supply and antenna coupler. It also includes a general-purpose digital computer, programmed to perform most of the communications tasks and interface with the host platform's computer.

Contractor

Raytheon Electronic Systems.

VERIFIED

ASARS-2 Advanced Synthetic Aperture Radar System

The high-resolution ASARS-2 side-looking radar was designed for the US Air Force U-2 high-altitude battlefield surveillance aircraft, and satisfies a US Air Force/Army requirement for a standoff intelligence gathering system. ASARS-2 was launched in 1977 under the designation AN/UPD-X and first test flown on board a U-2R reconnaissance aircraft in 1981. It is likely that the resolution approaches limits set by altitude, residual airframe vibration and atmospheric turbulence. The antenna is V-shaped to enable the ground on either side of the aircraft's track to be surveyed without the aircraft having to manoeuvre. Guidance for the ASARS-2 antennas is provided in part through several other sensors and information is transmitted via a datalink to a special ASARS-2 deployable processing station on the ground. Here, the signals are converted into strip-maps and spotlights and are available within minutes for air force and army commanders.

ASARS-2 consists of a search mode for wide area ground coverage, featuring both a Moving Target Indicator (MTI) and a fixed target indicator and a spotlight mode for high-resolution coverage of smaller areas which can detect fixed targets only.

In 1988, Raytheon Electronic Systems was awarded US$20 million to develop an Enhanced Moving Target Indicator (EMTI) which, in part, is designed to add MTI to the spotlight mode. This entailed software modifications to the ground-based processing component and the addition of components in the airborne ASARS-2 receiver/exciter and processor control unit.

Operational status

In service with US Air Force U-2 high-altitude reconnaissance aircraft.

The ASARS-2 Improvement Program, which began in June 1996, is divided into several phases. The initial phase (completed at the end of 1998), known as the contingency phase, consisted of integrating the legacy ASARS-2 electronic scanning antenna, transmitter, and low-voltage power supply, with two new units based on COTS hardware, the Receiver/Exciter/Controller (REC) and the onboard processor. Flight testing, completed at the end of 1998, tested and evaluated the new Asynchronous Transfer Mode (ATM) interfaces and the hardware/software of the onboard signal/data processor and the REC. The ATM interfaces connect the radar to the air/ground datalink and provide a COTS architecture.

Contractor

Raytheon Electronic Systems.

VERIFIED

C-1282AG remote-control unit

The C-1282AG Remote-Control Unit (RCU) provides the capability to control and monitor remotely the operation of the RT-1273AG Multimission UHF Satcom Transceiver (MUST) or the MD-1269A MultiPurpose Modem (MPM) over an asynchronous MIL-STD-188-114A balanced interface. Frequency and mode selection for the AN/ARC-171 and preset selection for the AN/WSC-3 are provided by the RCU via the MPM.

The C-1282AG RCU allows full control of all MUST and MPM radio and modem features, including frequency, data rate, modulation type, modem emulation mode, transmit power level and BIT. Menu and Arrow keys allow the operator to scroll through various menus to view or update radio or modem configurations. The Preset switch allows access to each of the eight radio and modem preset configurations. The volume knob may be used to attenuate plain text receive audio out of the radio or modem.

The C-1282AG RCU has an illuminated panel designed to MIL-P-7788E which provides a menu-driven keypad/display interface. Display intensity is adjustable and tracks the externally controlled edge-lit panel brightness. The panel and display are designed for upgrade to NVG compatibility.

The Raytheon Electronic Systems C-1282AG remote-control unit

Specifications

Dimensions: 146 × 76.2 × 133.35 mm
Weight: 0.91 kg
Power supply: 19-30 V DC, 0.5 A max
Temperature range: –45 to +55°C
Altitude: up to 30,000 ft

Contractor

Raytheon Electronic Systems.

VERIFIED

CA-657 VHF/AM radio

The CA-657 is another VHF variant of the ARC-164 UHF system and its derivative, the VHF ARC-195. Its frequency band coverage, however, is somewhat broader than the ARC-195 since it covers from 100 to 159.975 MHz, in which range it provides 2,400 channels. Microprocessor memory control systems used with the CA-657 permit alternative preselection of up to either 20 or 30 channels. Channel separation is at intervals of 25 kHz and the transmitter output power is 10 W.

Construction of the system is based upon that of the ARC-164 and a high degree of component commonality exists.

Operational status

In production, and in service.

Contractor

Raytheon Electronic Systems.

VERIFIED

Commanders' Tactical Terminal

Commanders' Tactical Terminal/Hybrid-Receive only (CTT/H-R)

The Commanders' Tactical Terminal/Hybrid-Receive only (CTT/H-R) is a multichannel, multifunction terminal used to receive real-time intelligence reports from a variety of sources. It allows the tactical user to receive data from the Tactical Reconnaissance Intelligence Exchange Service (TRIXS) or the Tactical Information Broadcast Service (TIBS), while simultaneously receiving the Tactical Receive equipment and related APplications (TRAP) and Tactical Data Information eXchange System-Broadcast (TADIXS-B). This capability provides the user with a significant increase in access to real-time intelligence data and offers the flexibility to utilise whatever resources are available in the theatre.

CTT/H-R is the result of an evolutionary development that began with the full-duplex CTT field terminals and has continued with the development of a CTT/TEREC terminal, a multichannel CONSTANT SOURCE receiver system, a TIBS interface unit and a TIBS receive only unit. The CTT/H-R combines the functions of all these systems into a single receive-only terminal.

The CTT/H-R is packaged into a single unit and uses open architecture VME technology. Some of the key VME modules used include embedded COMSEC modules based on CTIC and Ricebird crypto chips, multiple 680X0 microprocessor modules and a MIL-STD-1553 interface. Additional modules include a single-board UHF receiver/synthesiser and a single-board modem using TMS320 processors.

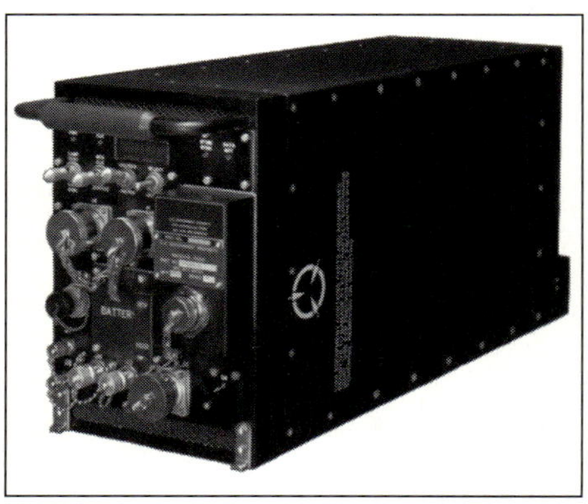

The Raytheon Electronic Systems commanders' tactical terminal/hybrid – receive only (CTT/H-R) **1999**/0044829

The Raytheon Electronic Systems commanders' tactical terminal 3-channel (CTT(3)) **1999**/0044846

The CTTR/H-R is qualified for installation in fixed-wing aircraft and helicopters, as well as vehicles and ships. It uses Ada software and has a download capability for changes or upgrades.

Specifications

Dimensions: 190.5 × 266.7 × 520.7 mm
Weight: 19.5 kg
Power supply: 100-132 V AC, 47-140 Hz, 180 W or 200-264 V AC, 47-66 Hz, 180 W
Frequency: 225-400 Hz
Channel spacing: 5 and 25 kHz
Temperature range: –20 to +50°C
Operating modes: TIBS, TRIXS, TDDS (TRAP), TADIXS-B-Tops and OBP
Demodulation: SBPSK or BPSK 2,400, 4,800, 9,600 and 19,200 bps
SOQPSK 4,800 and 19,200 bps
QFSK 32 kbps
Altitude: up to 30,000 ft

Operational status

In production for aircraft, helicopters, tracked and wheeled vehicles, ships, submarines and ground shelters.

Commanders' Tactical Terminal 3-channel (CCT(3))

The CTT (3) is central to the US Army's Integrated Battlefield Targeting Architecture (IBTA) and a vital link in the US Army's modernisation objectives. It supplies the critical datalink to battle managers, intelligence centres, air defence, fire support and aviation nodes across all service lines.

CTT (3) allows users to exploit all modes and rates of intelligence broadcast networks: Tactical Receive equipment and related APplications (TRAP), TActical Data Information eXchange System-B (TADIXS-B), Tactical Information Broadcast System (TIBS) and Tactical Reconnaissance Intelligence eXchange Service (TRIXS).

It provides tactical intelligence data transmission and network control in the TRIXS and TIBS networks; C4I on the move and general situational awareness; support of sensor queuing, Secondary Imagery Dissemination (SID); over-the- horizon targeting and dynamic tasking of forces. CTT (3) enhances survivability through passive tracking, increases response times through early warning, provides data accuracy sufficient for a fire solution and accommodates collateral or SCI traffic.

Configurations available are:

CTT/H3: full duplex in TIBS, TRIXS and GPL networks while receiving two additional channels of intelligence broadcasts or GPL UHF

CTT/H-R3: receives three simultaneous channels of intelligence broadcasts or GPL.

Specifications

Frequency range: 225 to 400 MHz
Channelisation: 5 and 25 kHz
Operating modes: full- or half-duplex
Primary power: 100-132 V AC, 47-440 Hz, single phase or 200-264 V AC, 47-66 Hz single phase battery-backed key storage and classified software
MTBF: 3,000 hours
Modulation: FM analogue and Secure Voice, FSK, QFSK (32 kbps), BPSK, SBPSK 975, 300, 1,200, 4,800, 9,600, 19,200 and 38,400 bps)
Encoding/decoding: selectable differential or non-differential, convolutional encoding and Viterbi decoding
Signal acquisition: ±900 Hz offset at 40 Hz per sec
Interfaces: KG-84, KY-57/58, KGR-96, RS-232/422, MIL-STD-1553B, headset, intercom, maintenance
ECCM features: Have Quick II, Adaptive Array Processing (null-steering option), Quadrature Diversity Receiver/Combiner (option)
Transmitter: output power adjustable 11-20 dBW
Dimensions: RRT, RBP, R3, all full ATR tall-long
Weight: RRT, RBP, R3, all 29.55 kg
Cooling: self-contained forced air fans
Temperature: –45 to +55° C operating
–50 to +70° C nonoperating
Altitude: 30,000 ft operating
Vibration/shock: tracked and wheeled vehicles, helicopter, jet and propeller aircraft, shipboard
EMI/EMC: MIL-STD-461

Operational status

In production. The two-channel versions have been fielded to all US Services.

Contractor

Raytheon Electronic Systems.

VERIFIED

CV-3670/A digital speech processor

The CV-3670/A is an airborne digital, linear predictive data converter which provides digitised speech output at a data rate of 2.4 kbits/s, permitting narrowband system operation on standard voice quality circuits. The output can be multiplexed with other databit streams to allow simultaneous voice and data transmissions. Independent data clocks, provided by the data terminal to the transmitter analyser and receiver synthesiser of the speech processor, are used to synchronise the equipment to the receiver and transmitter of the data terminal.

The CV-3670/A is a remotely operated unit packaged in a ½ ATR box, with voice input/output via the aircraft intercommunication system. It is operated by use of the C-10085/A controller/indicator.

Contractor

Raytheon Electronic Systems.

VERIFIED

Direction-finding system

Received RF signals between 20-1,500 MHz route to distribution and switching circuitry contained in a modular RF unit. This unit provides 14 reconfigurable HF/VHF/UHF RF inputs and is designed to be cascadable, allowing for additional antenna elements or arrays, simply by adding other RF units. The RF unit also includes a switchable dual-channel HF upconverter for coverage of the 2-20 MHz frequency range.

The IF outputs of the two Nanomin receivers are each routed into coherently clocked A/D converters and then into a two-channel DDC. The DDC digitally fine tunes to the specified frequency, generates in-phase and quadrature components of each channel, down-converts the desired portion of the IF bandwidth to baseband, and digitally filters the signal with linear phase Finite Impulse Response (FIR) filters. The DDC also performs basic preprocessing and controls data flow over the VME bus to the Motorola 68040 processor.

Switchable IF filters operate digitally within the DDC. The set of IF bandwidths can be changed easily by downloading a different set of coefficients. The digital decimation filter uses FIR filter designs that exhibit ideal phase characteristics as well as excellent roll-off features. This digital demodulation concept offers unique flexibility for tailoring the DF receiver to specific user needs.

Accurate angle of arrival measurements can be obtained for both data and voice signals using virtually any type of modulation, including AM, FM, continuous wave, single sideband, and independent sideband. Precision results on digital signals can also be achieved using frequency-shift keying or pulse-shift keying modulation.

The DF processor is a single board computer using the Motorola 68040 processor, with its floating point math co-processor running at 25 MHz. Along with 4 Mbyte of random access memory (RAM), a 16 Mbyte electrically erasable programmable read-only memory (EEPROM) board is included for non-volatile storage of calibration data. Interface boards are included for Ethernet, IEEE-488 and navigation systems (ARINC and Synchros).

The interior of the EC-24A electronic warfare training aircraft

Angle-of-arrival and position information feed into the DF processor for line-of-bearing (LOB) calculation.

The system handles pulse-type signals with pulse-widths down to 0.4 microsecond and with duty cycles as low as 0.0005.

Specifications

Frequency range: 2-1,500 MHz
DF Accuracy: <1° RMS estimated for fine DF using typical airborne VHF/UHF arrays
DF Sensitivity: −116 dBm (10 dB S/N in 6.4 kHz bandwidth)
DF Output: LOB, quality factor, frequency, time, in/out of geosort limits, emitter location (calculated in workstation)
Options:
(search): 100 freq/s (software upgrade only)
(classification): AM, FM, FDM, FSK, PSK
(copy): digital demod-AM, FM, SSB

Contractor

Raytheon Electronic Systems.

VERIFIED

E-SAT 300A satellite receiver

The E-SAT 300A is an all-digital satellite communication system designed for aircraft applications. Airborne telephone calls and data messages are automatically transmitted via the INMARSAT network to ground stations. Two telephone calls can be made simultaneously, while a third channel is available for relaying data messages. The system is designed to accommodate growth to eight simultaneous telephone channels.

The steered high-gain single-helix antenna features small size and weight and minimises the number of LRUs. The design eliminates the keyhole and is mechanically steerable through 360° in azimuth and −30 to +90° in elevation.

Specifications

Dimensions:
(satellite data unit) 9 MCU
(radio frequency unit) 12 MCU
(Class A high-power amplifier) 8 MCU
(antenna control unit) 6 MCU
(system power supply) 8 MCU
(radome) 374.7 × 431.8 × 2,794 mm
Weight: (total system) 99.79 kg
Power: 1.3 kVA
Frequency:
(receive) 1,530-1,559 MHz
(transmit) 1,626.5-1,660.5 MHz
Reliability: 31,500 h MTBF

Contractor

Raytheon Electronic Systems.

VERIFIED

The Raytheon Electronic Systems direction-finding system 0001273

EC-24A electronic warfare training aircraft

The former Chrysler Technologies Airborne Systems (CTAS) converted a DC-8-54 as an electronic warfare aircraft for use by the US Navy's Fleet Tactical Readiness Group (FTRG) which acts as the Orange Force in exercises. The EC-24A, which is the platform for the Orange Force Commander, is based at Waco and operated and maintained by the civilian crews of CTAS.

The EC-24A has a crew of 10, including three flight crew, six systems operators and the Orange Force commander. There are seats for a further 20 people on the aircraft.

The aircraft is equipped with: two AN/ALT-40 radar jamming systems with steerable antennas; two Rockwell Collins AN/ASQ-113 communications systems jammers; the Rockwell Collins AN/ASQ-191 communications jammer system; two AN/ALE-43 chaff dispensers effective over the A/J-bands; two Scientific Communications AN/ALR-75 ESM receiver systems with pulse analysers to give onboard signal identification capability; six Rockwell Collins AN/ARC-159 UHF; four Rockwell Collins AN/ARC-190 HF and two Rockwell Collins AN/ARC-186 VHF transceivers; two OE-320 direction-finding systems; two Hewlett-Packard HP-322 computers; one KY-58 secure communications system and one AN/PSC-3(V)1 Satcom radio.

Raytheon also operates and maintains two US Navy NKC-135A aircraft in support of FTRG, in a similar role to the EC-24A.

AN/ASQ-191 system

The AN/ASQ-191 has been developed by Rockwell Collins for use in airborne, fixed or mobile ground applications and covers the frequency range 225 to 400 MHz. It is utilised in the EC-24A. The AN/ASQ-191 employs two basic modes of operation: a communications mode for voice comms or deception, and a command, control and communications countermeasures mode for monitoring and jamming. It can operate against fixed-frequency or slow frequency-hopping radios. The AN/ASQ-191 incorporates a number of automatic scanning, monitoring and jamming functions which cover the complete frequency range, parts of it or selected frequencies. The jammer has an output of about 100 W and the jamming sequence includes a 6 ms look-through. A 400 W jamming variant is available.

Contractor

Raytheon Electronic Systems.

VERIFIED

Emitter Location System (ELS)

Raytheon Electronic Systems, under contract to DaimlerChrysler Aerospace AG on behalf of Panavia, developed an Emitter Location System (ELS) for use on the Tornado ECR aircraft. The ELS detects, identifies and locates radar emissions through the use of a high probability of intercept system. It features multi-octave frequency coverage, phase interferometric antenna arrays for precision direction-finding and passive ranging, channelised receivers and multiple 1750A digital processors for operation in dense signal environments.

The ELS operates across the frequency spectrum for all primary surface-to-air and air-to-air threats. Data acquired by the system is transmitted, via a MIL-STD-1553B databus, to the tactical displays of both crew members, who can then make threat assessments and activate the appropriate countermeasures. Threats that have been located by the ELS can also be communicated to follow-on forces through the Operational Data INterface (ODIN). This data can then be used for threat avoidance or suppression operations.

The design uses a SAW channeliser in a cued analysis receiver configuration. In this configuration the delay lines allow the channeliser to measure the frequency of the incoming signal and then, using a fast-settling local oscillator, tune and cue up the narrowband receivers for subsequent analysis and pulse report generation. The channeliser provides the cued analysis contiguous high signal-selectivity across a wide instantaneous bandwidth. The basic channeliser design consists of multiple parallel SAW filter banks, each bank having multiple signal channels.

High-speed real-time emitter analysis is accurately processed for positive threat identification. In addition to this, accurate DF and passive ranging is computed within the processor, correctly locating and displaying large numbers of emitters. The processed information can then be used to cue weapons, and could be configured to steer jamming pods.

The configuration of the ELS in the Tornado ECR consists of 10 individual LRUs. The two antenna/RF converter units are mounted conformally with the aircraft's left and right wing nibs near the base. The remaining units are distributed within the wing shoulder and gun bays. An operator control is provided in the ELS control panel and the Tornado TV tab displays.

Operational status

In service in German and Italian Air Force Tornado ECR aircraft.

Contractor

Raytheon Electronic Systems.

VERIFIED

Escort ESM system

Escort is a family of electronic support measures for maritime patrol aircraft, small ships and ground-based vehicles. It is generally applicable to adapted versions of business aircraft. Escort provides signal detection, sorting, analysis, identification and reporting of radar frequency emitters. Wide instantaneous spatial and frequency coverage, combined with sophisticated processing techniques, provides high probability of intercept. All mission data can be recorded on tape cartridge for post-mission analysis. The ELINT configuration provides precision direction-finding by using phase interferometer antennas in conjunction with the basic receiver processor and display system.

The basic system operates over the 2 to 18 GHz frequency band with growth options available. It uses two antenna units, each consisting of a four-element antenna array, a high-speed RF switch and various control and amplification circuitry. The receiver unit accepts RF from port and starboard antenna units and performs frequency, time of arrival, amplitude and angle of arrival measurements on each received pulse. The processor consists of a 16-bit general purpose computer with 128,000 16-bit words of memory and a variety of interface circuits. The search and dwell process used by the receiver as it steps between sub-bands to intercept and collect emitter data is controlled by the processor software. Because of the wide bandwidth of the receiver, the search cycle over the entire frequency/azimuth domains is typically accomplished within one main beam illumination of a radar, ensuring high probability of intercept and a very low intercept time for new emitters.

The ESM operational software program maintains an extensive emitter data file, which includes the measured parameters as well as time of initial and most recent intercepts. Formats of CRT displayed outputs vary according to user requirements.

Specifications

Volume: 0.112 m^3
Weight: 91 kg
Frequency: 2-18 GHz
Processing: 100 emitters
Accuracy: 8° DF, 5 MHz frequency, 1 μs PRI

Operational status

Fully developed. The system is believed to form part of the RC-350 Guardian airborne reconnaissance system.

Contractor

Raytheon Electronic Systems.

VERIFIED

GEN-X expendable decoy

Raytheon Systems Company has developed the generic expendable, GEN-X, as a follow-on to the AM-6988A POET system. GEN-X is an active transponder which emits a signal, after being ejected from its aircraft-mounted AN/ALR-39 or AN/ALE-47 dispenser, to decoy radar-guided missiles. GEN-X has a much greater frequency band than POET and can be programmed with its operating frequency before flight or in flight by information from the RWR.

Internally the system is made up of four primary electronic modules which make extensive use of monolithic, hybrid and silicon integrated circuitry. Individual rounds are fired from their sleeve by a gas squib and deploy three spring-loaded fins for flight stability.

An ignition primer activates an onboard battery and, once powered up, the round begins a sequential search of its predefined frequency bands until a high-priority threat signal is identified. The unit's transmitter is then activated, to provide a highly attractive radar target for the incoming missile.

Operational status

Advanced development was completed in 1986 and full-scale development in 1990. The first production contract, worth US$67.8 million, for 7,000 GEN-X decoys was awarded in May 1992. Production deliveries for US Navy and Marine Corps aircraft began in 1994, since when there have been further contract

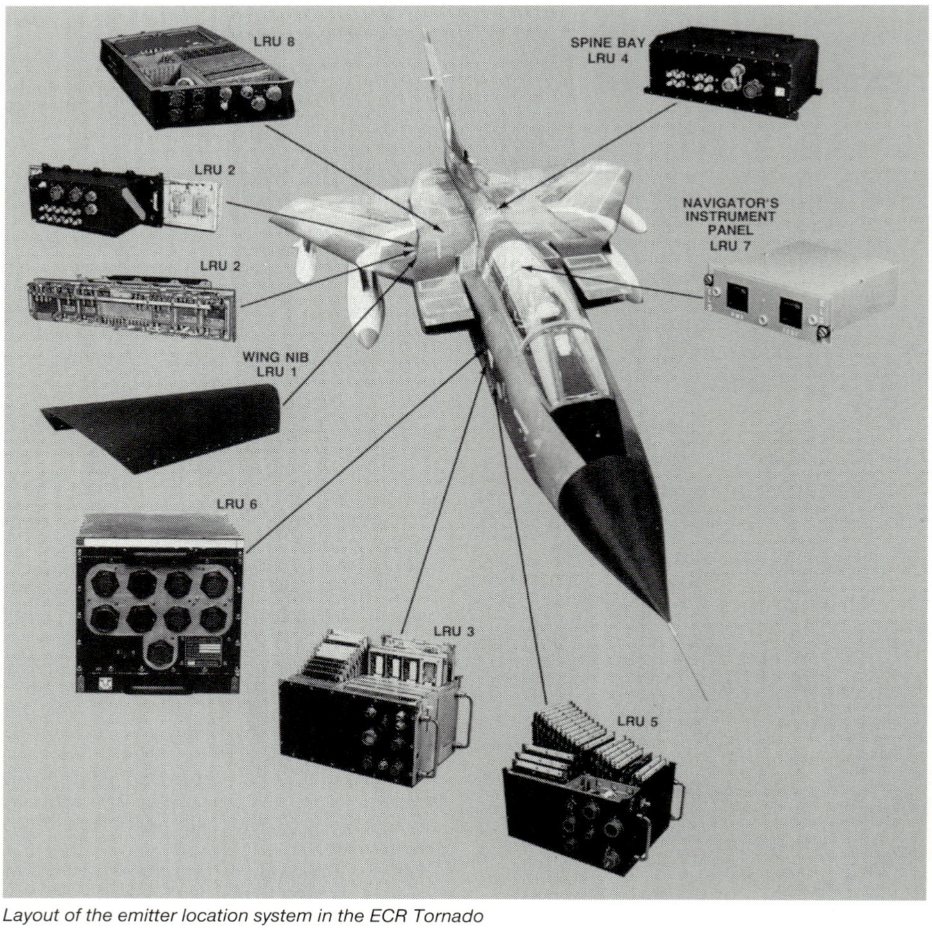

Layout of the emitter location system in the ECR Tornado

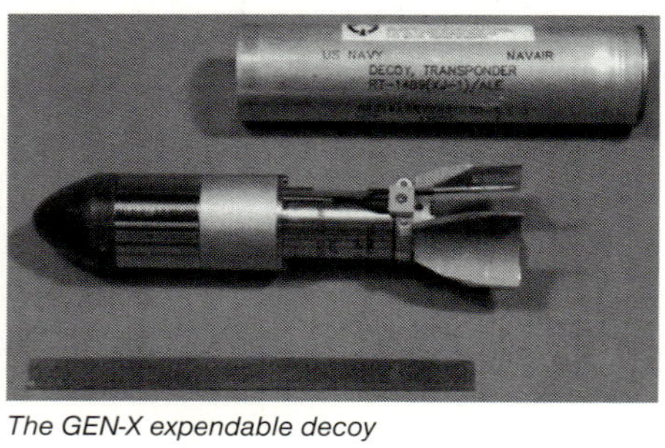

The GEN-X expendable decoy

awards for the US Navy. It is reported that the UK Royal Navy has bought GEN-X for the Sea Harrrier F/A Mk2 aircraft.

Contractor

Raytheon Electronic Systems.

VERIFIED

GPS-Aided Targeting System (GATS)

The US Air Force B-2 bomber is equipped with the GPS-Aided Targeting System to reduce target location error. GATS uses the B-2s GPS navigation system and the aircraft's synthetic aperture radar to provide accurate target location data, that is then passed to the GPS-Aided Munitions (GAM); together they form the GATS/GAM targeting system of the B-2.

Operational status

In service.

Contractor

Raytheon Electronic Systems.

VERIFIED

Ground-mapping and terrain-following radar for the Tornado

The Tornado IDS radar has no official designation, but is often referred to as the Tornado Nose Radar (TNR). The TNR is also used in the ECR version of the Tornado.

The all-weather day or night radar comprises two essentially separate systems that share a common mounting, power supply and computer/processor. They are the Terrain-Following Radar (TFR) and the Ground-Mapping Radar (GMR). The first is used for automatic high-speed, low-level approach to the target and escape after an attack. The second is the primary attack sensor for the IDS Tornado and operates in air-to-ground and air-to-air modes to provide high-resolution mapping for navigation updating, target identification and fire control.

The radar enables the crew to fix the aircraft position by updating the Doppler-monitored inertial navigation system, provides range and tracking information for offensive or defensive weapon delivery, and commands, via the autopilot, a contour-hugging flight profile that reduces the chance of detection by hostile air-defence radars. ECCM, dramatically enhanced with the introduction of the Phase 1 upgrade improvement, is used to provide relative immunity from interference in severe ECM environments.

The three units comprising the system are the radar sensor (transmitter/receiver package for TFR and GMR), a digital scan converter and a radar display unit in which a moving-map image can be superimposed on a radar picture for navigation updating and target identification.

The GMR operates in the J-band frequency band with nine modes: readiness, test, ground-mapping, boresight contour mapping, height-finding, air-to-ground ranging, air-to-air tracking, land/sea target lock on and beacon homing.

The TFR operates at J-band frequencies and has four modes: standby, ground standby, test and terrain-following. In the latter mode the aircraft can be flown automatically or manually through head-up display steering information. The pilot can also select ride comfort (for a given speed, the closer the allowable ground clearance, the less comfortable the ride owing to the greater *g* levels needed to stay on the commanded flight profile).

Both systems have extensive built-in test features to ensure a high degree of fault isolation and comprehensive reversionary modes.

Operational status

In service on Tornado IDS and ECR aircraft.

Contractor

Raytheon Electronic Systems.

VERIFIED

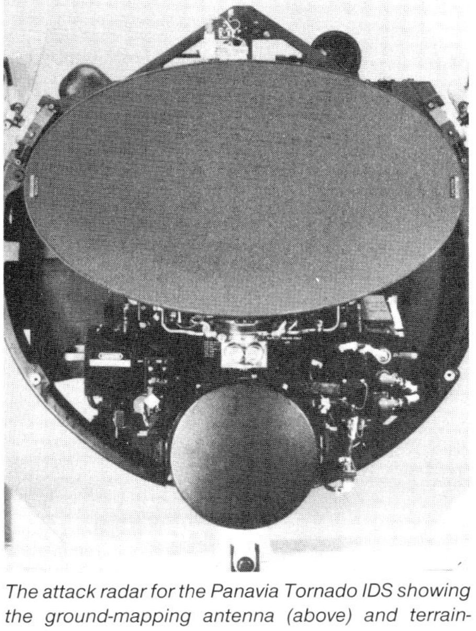

The attack radar for the Panavia Tornado IDS showing the ground-mapping antenna (above) and terrain-following antenna (below)

Have Quick system

Have Quick provides the user with an effective air-to-air, air-to-ground and ground-to-air jam-resistant UHF voice communication capability for operations in jamming environments.

The Have Quick is an applique system. It consists of an ECCM modification to selected airborne and ground-based radios, which gives them a frequency-hopping capability. Part of the strength of the system comes from the use of channels in an apparently random manner so that no pattern is evident to the external observer. Jamming is consequently more difficult.

The frequency-hopping scheme is implemented by storing a pattern of the frequencies to be used for a given day within every Have Quick radio and utilising this pattern according to the time of day. For every time slot in the day, where each time slot is a small part of a second, there is a specific frequency which must be used for a given communications net, whether it is transmitting or receiving. This frequency changes pseudorandomly from one time slot to the next. Thus, Have Quick terminals require some means to store the frequency pattern for channel use on a given day and also an accurate clock to control the times at which the pattern is consulted.

The Have Quick radio retains the normal non-hopping mode where it uses any one of the 7,000 channels available in the 225 to 400 MHz UHF communication band. The use and operation of the radio in normal mode is essentially unchanged from present-day procedures.

If jamming is encountered, the Have Quick radios can switch over to the ECCM mode and continue their communication. In order to permit this switch over, the radios must be suitably primed so that they will be synchronised in the ECCM mode; this is usually done before take-off.

The Have Quick system also has a capability termed 'multichannel' or 'break-in' operation. This permits a Have Quick radio to receive two simultaneous transmissions on the same net while avoiding the beat note which typically prevents the listener from understanding either transmission. The technique is implemented automatically in the transmitter where it is recognised whether or not the net is already in use. If so, the transmitter side-steps by one 25 kHz channel. Since the receiver is set to wideband mode, the second signal is received in addition to the original. The multichannel operating capability can be selected by the operator, whenever the system is in ECCM mode.

Operational status

Have Quick has been defined at two levels of complexity - Have Quick 1 and Have Quick 2. It is in service in a large number of NATO military aircraft, and Have Quick applique systems have been developed by a number of NATO radio manufacturers. AN/ARC-164 is perhaps the most common host radio for the Have Quick system.

Contractor

Raytheon Electronic Systems.

VERIFIED

HISAR® Integrated Surveillance And Reconnaissance system

HISAR is an I/J-band Synthetic Aperture Radar (SAR) mapping and surveillance system. It is designed for border surveillance, remote sensing, and specialist monitoring of all types (including oil pollution, deforestation and economic zones).

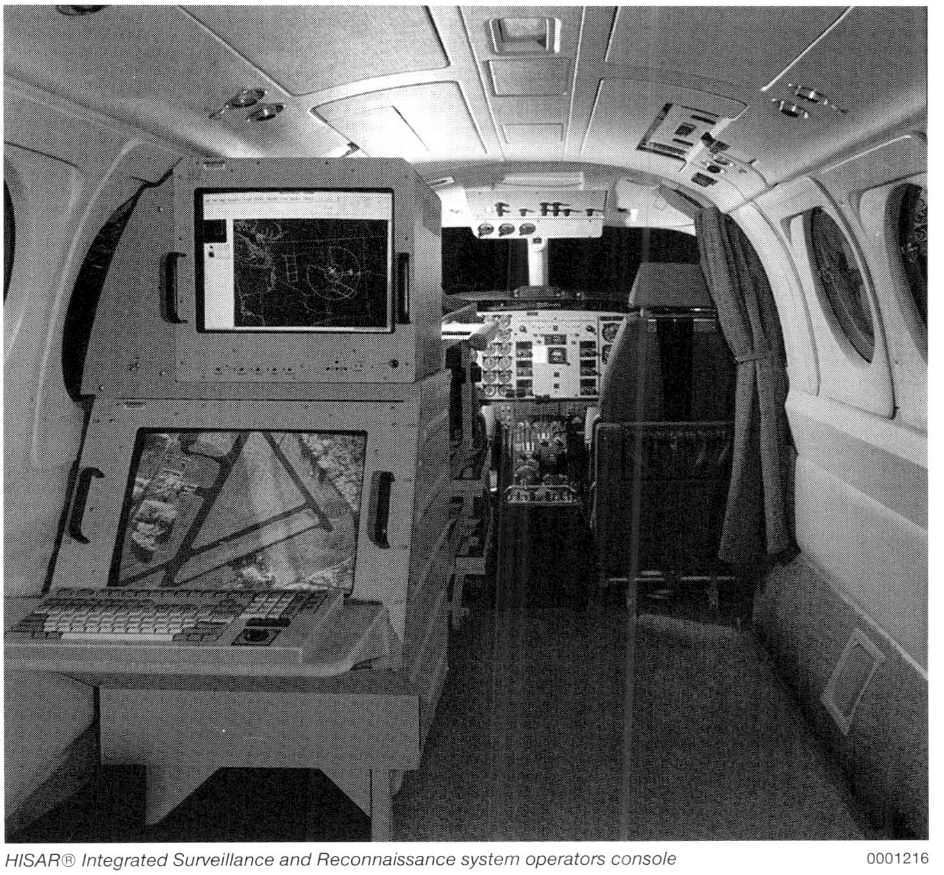

HISAR® Integrated Surveillance and Reconnaissance system operators console 0001216

HISAR® synthetic aperture radar installed aboard the Beech King Air 200T. 0018071

HISAR's development is based on experience with ASARS-2 and the AN/APQ-181 radar that equips the B-2 bomber.

HISAR features three imaging modes:

1. Wide-area search with a resolution of 20 m.
2. A combined SAR/MTI Strip mode, with 6 m resolution at a range of 20 to 110 km.
3. A SAR Spot mode with 1.8 m resolution.

Moving Target Indication (MTI) is a key feature of the HISAR radar that detects and tracks moving land and sea targets.

HISAR is supported by a self-contained, compact mission-planning system, that is used for pre-flight planning, and post-mission analysis.

Specifications

Dimensions: 0.83 m^3
Weight: 245 kg (full system)
Power consumption: 4,930 W

Operational status

Raytheon states that HISAR is installed on five different platforms, including the US Army Airborne Reconnaissance Low-Multifunction (ARL-M), the Beechcraft King Air 200T and the DARPA Global Hawk high-altitude endurance unmanned aerial vehicle.

HISAR is also being marketed internationally for civil applications. These include environmental monitoring, border surveillance and maritime patrol. The system is marketed as a complete integrated reconnaissance and surveillance suite, in which form additional sensors are added, typically, the Raytheon Electronic Systems DB 110 electro-optical sensor and the AN/AAQ-16 infra-red sensor.

Contractor

Raytheon Electronic Systems.

VERIFIED

Joint Tactical Radio System (JTRS)

The JTRS is a US Department of Defense (DoD) sponsored project, intended to provide all US armed services with a common family of software programmable radios. The US Army is lead-Service for the JTRS project.

Operational status

A consortium, lead by Raytheon Electronic Systems, with ITT Aerospace Communications and Marconi Aerospace Electronic Systems, was awarded a US$21.7 million contract in October 1999 to define a JTRS architecture and to construct four prototype radios. It is reported that the consortium is also to contribute US$9.6 million to the definition project, which is due to be completed by November 2000.

Contractors

Raytheon Electronic Systems (consortium leaders).
ITT Aerospace Communications (partner company).
Marconi Aerospace Electronic Systems, a BAE Systems company (partner company).

VERIFIED

Joint Tactical Terminal/Common Integrated Broadcast Service-Modules (JTT/CIBS-M)

JTT/CIBS-M is a high performance, software programmable radio (derivative of the Commanders' Tactical Terminals), providing plug-and-play modular functionality that is backward and forward compatible with the Integrated Broadcast Service (IBS).

The Raytheon Electronic Systems Joint Tactical Terminal/Common Integrated Broadcast Service-Modules (JTT/CIBS-M) ***1999***/0044847

JTT/CIBS-M incorporates the Army Technical Architecture (ATA), Joint Technical Architecture (JTA), Defense Information Infrastructure Common Operating Environment (DII COE), Technical Architecture for Information Management (TAFIM) and Joint Vision 2010.

JTT/CIBS-M provides critical datalinks to battle managers, intelligence centres, air defenders, fire support elements and aviation nodes across all services and aboard airborne, sea-going, subsurface and ground mobile mission platforms.

JTT/CIBS-M allow the warfighting Commanders in Chief (CINCs), US Army, US Air Force, US Navy, US Marine Corps, US Special Operations Forces (SOF) and other agency users to exploit the IBS intelligence networks: Tactical Reconnaissance Intelligence eXchange System (TRIXS), Tactical Information Broadcast Service (TIBS), Tactical Related APplications (TRAP) Data Dissemination System (TDDS) and Tactical Data Information eXchange System-B (TADIXS-B-TOPS and OBP).

In addition, JTT/CIBS-M will support the evolving IBS broadcast architecture, including changes to message formats and transmission protocols and use of different portions of the RF spectrum.

JTT/CIBS-M provides an architecture that supports multiple terminal configurations, emerging technology insertion, P3I, and module integration into other terminals and processors.

Specifications

Operating modes: TIBS, TRIXS, TDDS (TRAP), TADIXS-B-TOPS and OBP, 5 and 25 kHz DAMA, SID and GPL
Interfaces: LAN: 10/100 MHz Ethernet, RS-232/422, MIL-STD-188-114, MIL-STD-1553B
Frequency: 225-400 MHz
Operating modes: full- or half-duplex
Transmit power: adjustable 11-20 dBW
ECCM: Have Quick II
EMI/EMC TEMPEST: MIL-STD-461/462, NACSIM 5100
Prime power: 100-132 V AC, 47-400 Hz, single phase, or 200-264 V AC, 47-66 Hz, single phase; battery-backed key storage and classified software
Dimensions: JTT-T/R (12 RX/4 TX; 1 full ATR long and 4X 1/4 ATR long); JTT-R (1 full ATR long)
Weight: 28.2 kg per LRU maximum
Cooling: self-contained or platform-provided forced air
Temperature: –32 to +43° C (operating)
–47 to +71° C (nonoperating)
Altitude: 60,000 ft
Vibration/shock: tracked and wheeled vehicles, helicopter, jet and propeller aircraft, shipboard and submarine
MTBF: 6,500 h minimum

Contractor

Raytheon Electronic Systems.

VERIFIED

MD-1269A multipurpose modem

The MD-1269A multipurpose modem is an advanced digital signal processor-based, full-duplex 70 MHz modem for UHF Satcom and line of sight communications links. It simplifies existing communications systems and terminals by combining the important modulation modes of six currently used modems into a single unit.

Digital data shaping provides spectral containment and permits 2,400 bps BPSK and 4,800 bps OQPSK operation over a 5 kHz satellite channel. The MD-1269A also includes baseband interfaces with audio equipment, data terminals or encryption devices to provide narrowband AM and FM voice, TADIL A (link 11), TADIL B, TADIL C (Link 4), KY-57/58 (Vinson AM and FM), KY-65/75 (Parkhill), KG-84 and ANDVT compatibility.

The MD-1269A has a standard 70 MHz RF interface and can be used with any 70 MHz compatible receiver/transmitter in the non-DAMA modes. The 70 MHz interface is synthesised to provide 5 kHz channelisation capability with the R/T operation on 25 kHz channels. For DAMA operation, the R/T must be compatible with the stricter requirements for DAMA operation. The modem is compatible with the US Navy 25 kHz TDMA-1 DAMA and the US Air Force 5 kHz USTS DAMA. All non-DAMA capabilities are preserved in the DAMA version of the MD-1269A for backward compatibility.

The MD-1269A includes synchronous and asynchronous control/data ports for various local or remote-control options. All control and data ports can be configured as balanced or unbalanced MIL-STD-188-114A interfaces to a host terminal controller or C-1282AG control head. The entire modem is contained in a ½ ATR airborne package and is ruggedised to withstand airborne, shipboard or tactical environments.

Specifications

Dimensions: 123.95 × 193.55 × 497.84 mm
Weight: 11.34 kg
Power supply: 115 or 230 V AC, 47-440 Hz, 50 W max
Temperature range: –40 to +55°C
Altitude: up to 30,000 ft

Contractor

Raytheon Electronic Systems.

VERIFIED

Model 960 extended environment COTS computer

The model 960 extended environment COTS computer is a repackaged implementation of the Digital Equipment Corporation 500 MHz AS/500 Alpha-based workstation.

Operational status

Raytheon is contracted to provide Boeing with 174 workstation processors and 55 input/output processors for incorporation into the Boeing-supplied Tactical Command System for the UK Royal Air Force Nimrod MRA4 aircraft.

The Tactical Command System integrates and displays the output from the primary sensors, the defensive aids subsystem, armament control system and communications suite. Each aircraft carries seven workstations located in the aircraft's rear cabin which are capable of being reconfigured, control and display mission system elements. Two workstations are dedicated to the acoustic system, and the remaining five can be reconfigured into any sensor display format to meet mission requirements.

Contractor

Raytheon Electronic Systems.

VERIFIED

MX 42000 UHF Satcom datalink

The MX 42000 UHF Satcom datalink is suitable for a wide range of manned and unmannned airborne applications, as well as ground-based ones. It has been adapted from the MXF-420 which has been selected by the US Army for the Joint Services Multiband AN/PCS-5 programme.

The MX 42000 features embedded, variable ratio video/image compression, an embedded ViaSat DAMA modem to provide full 5 and 25 kHz capability and embedded COMSEC. It is also available with multiband VHF/UHF capability in addition to its UHF Satcom capability and can be operated as a voice/data communications relay.

Specifications

Volume: 0.0112 m³
Weight: 9.98 kg
Power output: 300 W (with 100 W amplifier)

Contractor

Raytheon Electronic Systems.

VERIFIED

MX8000 series military GPS receivers

The MX8000 series are single card, high-performance military GPS receivers designed for embedded applications. The MX8100 provides six parallel continuous tracking, switchable channels, each independently tracking code and carrier measurements for smooth navigation results, high accuracy and greater data availability. Designed to the requirements of GPS-EGR-600, both hardware and software interfaces provide quick and easy mating to military inertial instruments so that initial communication is quickly established.

MX8000 receivers feature a flexible modular architecture, intended for growth and adaptation to changing military requirements. High computer throughput, memory margins and software modularity allow new requirements to be added with minimum impact on the weapon system. The field reprogrammability feature allows mission software to be changed as required to meet specific tactical requirements. The receivers can be cold keyed using standard techniques. Receiver default condition is the Y-code, allowing direct P/Y code acquisition in time critical applications.

Specifications

Dimensions: 149.4 × 148.6 × 16.5 mm
Weight: 0.27 kg
Power supply: 5 V DC, 7.5 W
Accuracy:
(position) 15 m SEP, 2-3 m with DGPS
(velocity) 0.1 m/s
(time) 100 ns
Temperature range: –40 to +85°C

Contractor

Raytheon Electronic Systems.

VERIFIED

MXF-400 Series V/UHF communication system

The MXF 400 family of VHF/UHF multiband multifunction communication systems uses a modular technology common to ground and airborne systems.

The MXF-400 modules are adapted to manpack, vehicular and airborne applications through the use of interfaces proven on the PACER SPEAK, ARC-164, ARC-187 and ARC-222.

Raytheon Systems Company is developing the MXF-430 and AN/PSC-5 receiver-transmitters into a form, fit and function replacement for the RT-1319 in the VRC-83 and GRC-206. The MXF-440 will provide a complete UHF/VHF LOS/UHF Satcom communication system, including Have Quick II, COMSEC and DAMA, within the footprint of the ARC-187. The MXF-450 provides UHF/VHF LOS communication with Have Quick II and COMSEC in a shell that is form, fit and functionally interchangeable with an RT-1504/ARC-164 or RT-1614/ARC-164.

Raytheon Electronic Systems is also evaluating the following developments for the MXF-400 family:

1. SATURN UHF ECCM for NATO applications as an option on current contracts.
2. 8.33 kHz channel spacing in VHF-AM for ICAO ATC requirements in European airspace.
3. TDMA in VHF-AM for FAA ATC requirements in US airspace.
4. Link 4 and Link 11 datalink capability.
5. SINCGARS operation.
6. Expansion of frequency band coverage to 512 MHz, with continuous tone code subaudible system.
7. Video compression imagery capability.

Operational status

In development.

Contractor

Raytheon Electronic Systems.

VERIFIED

The NKC-135A US Navy electronic warfare aircraft

NKC-135A electronic warfare training aircraft

A Boeing KC-135A aircraft has been converted into an NKC-135A electronic warfare aircraft for use by the US Navy, NAVSEA and other research and development organisations.

The aircraft is fitted with a comprehensive suite of EW equipment which includes two ALT-40 radar jamming systems covering the A/B- to I/J-bands, an ALR-75 ESM system, an OE-320 DF system interfaced to the ALR-75, a Fleet Tactical Readiness Group Airborne EW System equipped with HP 9826 computers and a colour graphic display, two ALE-43 chaff systems and a USQ-113 comms jamming system covering VHF and UHF bands. Two pylons have been fitted, each capable of carrying up to 1,000 kg of external stores in pods. Additional power for the EW systems is provided by the installation of four 90 kVA generators.

Operational status

In service with the US Navy.

Contractor

Raytheon Electronic Systems.

VERIFIED

RAYFLIR-49 and -49(LG) Day/Night Airborne Thermal Sensor (DNATS)

The Raytheon Electronic Systems Forward-Looking Infra-Red (FLIR) RAYFLIR-49 is a low-weight, multiple-purpose, Focal Plane Array (FPA) thermal imaging sensor for navigation, surveillance, maritime, search and rescue and troop transport missions.

The RAYFLIR-49 includes two Weapon Replaceable Assemblies (WRA): Turret Unit WRA-49 (TU) WRA-1 and Electronics Unit WRA-49 (EU). The RAYFLIR-49(LG) combines a three-chip, colour TV, with continuous zoom and an eye-safe laser range-finder with the FPA FLIR.

Features of the RAYFLIR-49 system include multiple field of view; electronic image stabilisation; Local Area Processing (LAP); an adaptable interface and multimode video tracker; MIL-STD-1553, ARINC 429 and RS-422 controls; 525 or 625 line video standard.

Specifications

Fields of view:
(wide) 22.5 × 30°
(medium) 5 × 6.67°
(narrow) 1.3 × 1.7°
Electronic zoom: 2:1 and 4:1
Gimbal angular coverage:
(azimuth) 360° continuous
(elevation) +40 to –150°

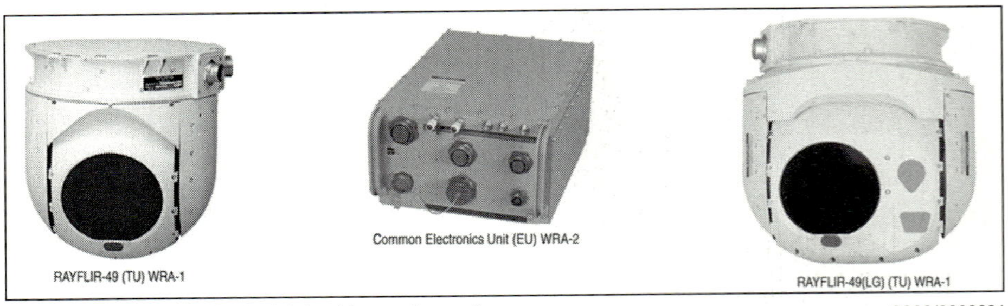

RAYFLIR-49 and -49(LG) day/night airborne thermal sensor ***2000***/0062231

Gimbal slew rate: 3 rads/s
Gimbal stability: 4 axis < 25 μrad
Gimbal angular resolution: <100 μrad
Interface slaving: radar, INS, mission or weapon computers
Control interfaces: MIL-STD-1553, ARINC 429, RS-422
Video outputs: 525 RS-170, 625 CCIR
Max airspeed: > 300 kt
Dimensions:
(turret unit) 323.85 diameter × 375.92 height mm
(electronics unit) 306.32 (W) × 413.52 (L) × 199.14 (H) mm
Power: standard aircraft power
Options: cameras, lasers, illuminators, displays, recorders, spot trackers and other avionics
/SP

Contractor
Raytheon Electronic Systems.

UPDATED

RC-135 Rivet Joint

Rivet Joint is the programme name given to the 14 US Air Force RC-135V and RC-135W SIGINT aircraft. Details of the fit are classified, but they are reported to have up to 17 operator positions, with one automated ELINT, two manual ELINT, and up to 13 COMINT stations.

A further two RC-135U aircraft (Combat Sent) are employed on technical SIGINT tasks.

All these aircraft are derived from the C-135, itself a derivative of the Boeing 707.

Operational status
All aircraft are in active service, and form an integral part of all US Air Force airborne intelligence collection and targeting plans. More details on the Joint SIGINT Avionics Family (JSAF) Low-Band Sub System (LBSS), destined to be fitted first into Rivet Joint, is given in the EW/USA listing for the Sanders company.

Contractor
Raytheon Electronic Systems.

VERIFIED

RT-1273AG DAMA UHF Satcom transceiver

The RT-1273AG UHF Satcom transceiver is an advanced full-duplex transceiver which upgrades and simplifies existing communications systems by combining modem and transceiver functions into one ATR unit. It incorporates the functionality of the Digital Signal Processor (DSP) based on the MD-1269A multipurpose modem which provides interoperability with all existing and planned UHF satellite communication equipment. The DSP architecture is designed to allow new waveforms to be programmed and incorporated as software updates. This integrated Satcom system reduces cabling, hardware size, weight and power consumption.

The RT-1273AG transceiver section features an integral 100 W transmitter and a highly sensitive receiver for UHF line of sight or satellite communication. Transmitter power output is variable from 1 to 100 W for optimum link performance when using either low- or high-gain Satcom antennas. A high-performance synthesiser provides independent 5 kHz channel selection of receive and transmit frequencies from 225 to 400 MHz for full-duplex operation.

The RT-1273AG modem section is based on the MD-1269A multipurpose modem. The modem section features a DSP-based architecture that offers multiple modulation modes and data rates. DSP-generated AM and FM is provided for low-distortion line of sight communications. Data rates from 75 to 38.4 bits/s are selectable for binary FSK, BPSK, Shaped BPSK, QPSK, Offset QPSK, or Shaped Offset QPSK operation. Doppler frequency offsets up to 1 kHz can be tracked by the modem to allow airborne operation. To maximise data throughput while minimising adjacent channel interference, sidelobes of the transmit signal are significantly reduced by advanced digital signal processing techniques that shape the modulated waveform. This permits 2.4 kbps BPSK and 4.8 kbps OQPSK operation over a 5 kHz satellite channel in compliance with JCS requirements.

The Demand Assigned Multiple Access (DAMA) waveforms are processed internally, allowing the RT-1273AG to participate in either 5 or 25 kHz DAMA networks with no additional equipment. Two simultaneous full-duplex baseband ports are available in the DAMA modes. Provisions for adding embedded COMSEC and digital voice have been designed into the equipment.

The RT-1273AG includes synchronous and asynchronous control/data ports for various local and remote-control options. Both ports can be configured as balanced or unbalanced MIL-STD-188-114A interfaces to a host terminal controller or control head.

Specifications
Dimensions: 256.5 × 193 × 497.8 mm
Weight: 22.68 kg max
Power supply: 115 or 230 V AC, 50/60/400 Hz, single phase
115 V AC, 400 Hz, 3 phase
28 V DC
Frequency: 225-399.995 MHz
Temperature range: –45 to +55°C
Altitude: up to 30,000 ft

Contractor
Raytheon Electronic Systems.

VERIFIED

SADL Situational Awareness DataLink

The SADL Situational Awareness DataLink integrates air force close air support aircraft with the digitised battlefield via the US Army Enhanced Position Location Reporting System (EPLRS). SADL provides fighter-to-fighter, air-to-ground and ground-to-air data communications that are robust, secure, jam-resistant, and contention free. With its inherent automatic and on-demand position and status reporting for situational awareness, SADL provides an effective solution to the air-to-ground combat identification problem.

Fighter-to-Fighter Operation
The SADL radio is integrated with aircraft avionics over the 1553 multiplex databus, providing the pilot with cockpit displays of data from other SADL-equipped aircraft as well as EPLRS-equipped aircraft and ground units. SADL is capable of fighter-to-fighter network operation without reliance on ground-based EPLRS network control. Fighter positions, radar targets, and ground target positions are shared relative to the fighter's own inertial navigation system or global positioning system. Based on the number of fighters on the network, the data capacity of the network can be customised from the cockpit. Automated fighter-to-fighter relay and adaptive power control capabilities ensure connectivity, jam resistance, and reduced probability of undesired detection.

Air-to-Ground Mode
In the air-to-ground mode, the pilot commands the SADL radio to synchronise with a specific ground network based on encryption keys. The fighter's radio then returns to sharing fighter-to-fighter data while recording ground positions from the ground EPLRS network. EPLRS tracks the fighter and provides the fighter position and altitude for ground-to-air combat identification. At the beginning of an air-to-ground attack, the pilot uses a switch on the control stick to request a view of the five closest friendly EPLRS-netted positions shown with Xs on both the head-up display and the multifunction displays. The pilot decides whether to fire based on the proximity of friendly positions to the target for effective air-to-ground combat ID. Friendly air defence units also have a picture of friendly air tracks to minimise fratricide.

Operational status
The Joint Interoperability Test Command at Fort Huachuca, in conjunction with the USAF 422nd Test and Evaluation Squadron, successfully conducted gateway interoperability tests connecting the Joint Tactical Information Distribution System (JTIDS) to SADL. Tactical awareness displays were reliably exchanged in both directions between two SADL-equipped F-16s and a JTIDS Class 2-equipped F-15 aircraft. SADL's update rates and message structure permitted exchange of TADIL-J message sets for displays of relative fighter positions, radar targets, aircraft fuel status, weapon loads and additional information. In addition to the 1,816 EPLRS radios fielded to divisions of the Contingency Force, EPLRS received a full-scale production award in 1997 for a multiyear 1997-98 buy of 2,100 radios. A further 535 SADL radios have been ordered by the US Air National Guard for F-16 Block 25/30 and A-10 aircraft assigned to the close air support role.

Contractor
Raytheon Electronic Systems.

VERIFIED

Sea Vue (SV) surveillance radar

Raytheon Electronic Systems Sea Vue (SV) radars are modern, lightweight, high-performance systems. They employ proven long-range surface detection techniques that support the entire range of surveillance missions. Inverse Synthetic Aperture Radar (ISAR) and SAR modes provide real-time imaging of maritime, land-based and coastline targets. The APS-134 (LW) was the initial product of the Sea Vue (SV) radar family.

Radar features include Sea Vizion™, Raytheon's unique combination of system parameters and target signal processing; fully coherent ISAR, SAR and MTI processing; pulse compression, pulse-to-pulse frequency agility and PRF jitter.

Radar modes available include surface surveillance and small target detection; navigation and mapping; weather avoidance; ISAR and SAR processing. Available enhancements include moving target discriminator; ISAR classification aids; Doppler Beam Sharpening (DBS); coherent look-down air target detection and tracking; control and display options.

Specifications
Detection performance:
(tanker) 230 n miles
(patrol boat) 95 n miles
(life raft) 30 n miles
Transmitter: coherent, solid-state, TWT; 8, 15 or 50 kW peak power options; 9.5-10 GHz; linear FM, fixed, frequency agile, or biphase coded waveforms
Antenna: parabolic or flat plate; stabilised, 360° scan; 6, 60, 120 rpm; sector and searchlight search, IFF capability
Receiver/Exciter/Synchroniser/Processor (RESP): linear FM pulse compression; digital pulse compression; Sensitivity Time Control (STC); Automatic Gain Contol (AGC)
MTBF: >500 h
Power: 28 V DC; 115 V AC, 400 Hz

Operational status

Demonstrated on PC-12 and IPTN CN-235 aircraft. Operational on Royal Australian Air Force Coastalwatch Dash-8; Japanese Self Defense Force U-125 SAR; and Italian Customs ATR-42 aircraft.

Contractor

Raytheon Electronic Systems.

VERIFIED

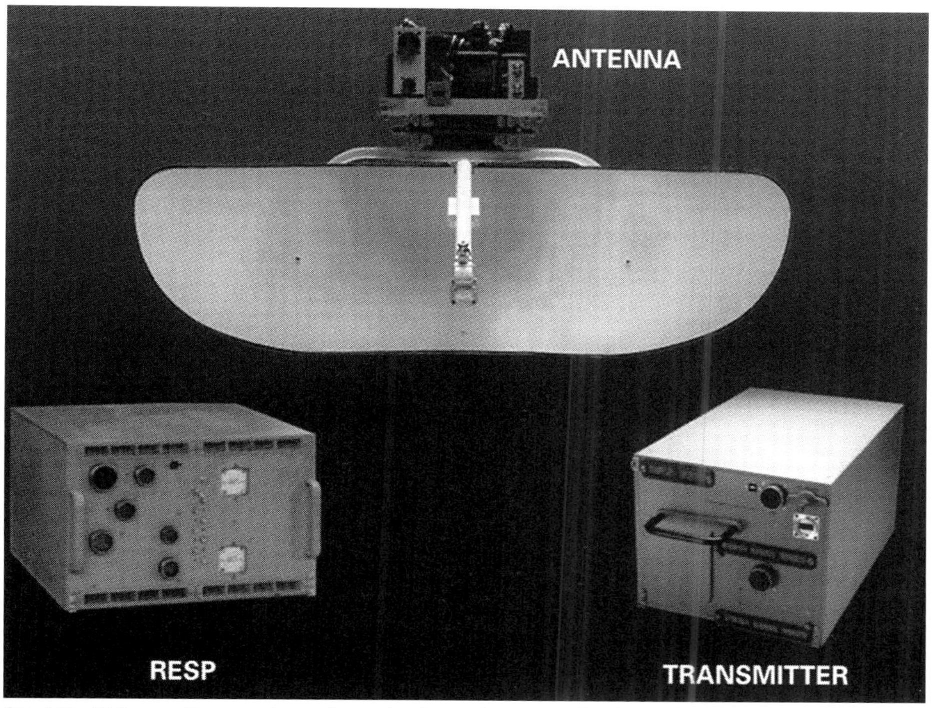

Sea Vue (SV) surveillance radar system, showing antenna, receiver/exciter/synchroniser/processor, and transmitter units
0018078

Dimensions	Width	Height	Depth	Weight
antenna	1,240 mm	650 mm	n/a	23 kg
transmitter	330 mm	279 mm	498 mm	30 kg
RESP	391 mm	257 mm	498 mm	37 kg

Terrain-Following Radar (TFR) for LANTIRN

The LANTIRN TFR retains the basic facilities available in the Tornado, including the ability to operate under enemy jamming in bad weather and at very low level. In addition, system modifications permit the host aircraft to maintain terrain-following capability while executing turns at a maximum rate of 5.5°/s compared with the Tornado's 2°/s, and at angles of bank up to 60°. Whereas the Tornado TFR, and previous terrain-following radars, employed magnetrons, the LANTIRN TFR has a travelling wave tube, giving considerably better performance, particularly in an ECM environment, and higher reliability. The pilot can control the aircraft's flight path by selecting terrain-clearance heights of from 100 to 1,000 ft.

The LANTIRN TFR subsystem began testing in a Lockheed Martin F-16 during 1983 and quickly demonstrated the ability to fly down to 100 ft above the terrain.

Operational status

In service on US Air Force F-15E and F-16D, Israeli Air Force F-15I and Royal Saudi Air Force F-15S aircraft.

Contractor

Raytheon Electronic Systems.

VERIFIED

LANTIRN under F-16

Airborne PC card receptacle DR-200

The Rockwell Collins airborne (PC card) receptacle is an intelligent microprocessor-based unit that permits access to an aeronautical worldwide database and mission-specific data from two PC card sockets and communications via the MIL-STD-1553B databus. The unit accepts standardised memory-type PCMCIA PC cards of type-I, II or III sizes. The unit provides connector, pin-out and software compatibility with the Collins DR-902(v) data receptacle and cartridges. When used in conjunction with a Digital Aeronautical Flight Information File (DAFIF) or Jeppesen formattted PC card, the unit will provide instant retrieval of all stored airport, heliport, communication and navigation information. Multiple data files and subdirectories can be accessed using DOS-like commands via the databus. The PC cards can be programmed with

Airborne PC Card receptacle DR-200 and mission planning station
0001338

the off-the-shelf desktop and laptop computers using DOS file protocol and plugged directly into the DR-200 receptacle.

All DR-200 operations are controlled by the aircraft bus-controller using MIL-STD-1553B messages. The unit provides status, type and directory information for each inserted card. Data file access and control flexibility is provided through DOS-like commands. When given an ICAO lookup command, the unit performs data retrieval and display formatting for the requested worldwide airport, heliport, navaid or intersection. Companion ground station software, designed to operate on a desktop or laptop computer, provides transfer and filtering of DAFIF information from CD-ROM to PCMCIA cards. The unit has the capability to continously search for and monitor the worldwide database for the 12 nearest airports, navaids or waypoints. The processor module is designed around a high-performance 80C186 microprocessor operating at 12 MHz. This module contains 64 kbyte RAM, 64 kbyte programme memory, 64 kbyte non-volatile memory. Interfaces include: MIL-STD-1553B, discrete I/O, RS-422 and ARINC 429.

Specifications
Size: 57 × 146 × 153 mm
Weight: (max) 1 kg
Power: (max) 5 W

Contractor
Rockwell Collins.

VERIFIED

AN/ARC-22XX HF radios

In June 1999, the UK MoD awarded Rockwell Collins (UK) Ltd a firm-price contract of US$17 million for the replacement of the UK Royal Air Force's E-3D AWACS aircraft HF radiosystems. The AN/ARC22XX triple-fit radios will be supplied by Rockwell Collins Government Systems and installed by Marshall Aerospace.

The AN/ARC22XX radios will feature Automatic Link Establishment (ALE), an embedded multimode modem with advanced waveforms, and custom frequency management/conflict software to enhance mission performance and reduce operator workload.

Each aircraft system will comprise triple-fit, 400 W, HF transceivers, controlled over a MIL-STD-1553B serial bush using Control Display Units (CDUs).

Operational status
The first trial installation is to be completed in the last quarter of 2000, with all seven aircraft to be modified and accepted into service by mid-2001.

Contractor
Rockwell Collins.

VERIFIED

AN/ARC-182 VHF/UHF radio

The Rockwell Collins ARC-182 is a combined VHF/UHF military communications system designed for all types of fixed-wing aircraft and helicopters. Small and light enough to be especially attractive for installation in the lighter aircraft classes, it covers the frequency bands from 30 to 88 MHz in FM, 116 to 156 MHz in AM, 156 to 174 MHz in FM and for the UHF band 225 to 400 MHz in both AM and FM modes. Additionally, a receive-only facility covering the band 108 to 116 MHz is provided for navigation purposes. Channel spacing throughout the range is at 25 kHz intervals. A total of 11,960 channels, up to 28 of which may be preselected, is available and Guard channel coverage on the 243 MHz emergency frequency is also provided. A Guard precedence mode is activated by single-switch selection.

Developed for a US Navy requirement, the ARC-182 is also in service with the US Air Force. The provision of a comprehensive communication system, which also covers the marine band, renders the equipment useful in a number of tactical roles, especially those associated with extended economic zone maritime patrol. Later additions include a facility to scan up to five channels, allowing monitoring of multiple communication nets, and an added satellite communications ability which allows transoceanic flights with line of sight radio.

Specifications
Dimensions: 165 × 146 × 124 mm
Weight: 4.54 kg

Operational status
Rockwell Collins has been awarded annual US Navy contracts with options that, to date, total more than 7,000 sets. These contracts include receiver-transmitters and associated mounts, controllers, filters, high-power amplifiers and so on for US Navy fixed-wing aircraft and helicopters and ground-transportable applications. The contracts also contain foreign military sales arrangements for aircraft destined for Australia, Norway, Spain, and other countries. The AN/ARC-182 is no longer in production.

Contractor
Rockwell Collins.

UPDATED

AN/ARC-182(V) high-power/frequency-agile transceiver system

A variant of the AN/ARC-182, the AN/ARC-182(V) is a frequency-agile system that is designed for simultaneous operation of several transceivers in airborne applications. The system permits operation of transmitters and receivers frequency spaced as close as 5 per cent, with antenna isolation as small as 20 dB in the 225 to 400 MHz band in both single channel and frequency-hopping modes.

In panel-mount configuration, the equipment consists of the RT-1324A transceiver, the 835Z-1 filter and the 938K-2A antenna. In remote-mount configuration, the equipment consists of the RT-1250A transceiver, the 835Z-1 filter, the 938K-2A antenna, the MT-4934 mounting tray and the C-10319B controller.

The system provides enhanced AN/ARC-182 performance in UHF at 30 W while permitting standard ARC-182 operation in the VHF band. The high-power amplifier is wideband, while the electronically tuned filter used in both transmit and receive allows medium-speed frequency hopping. In addition, a synchronous filter capable of providing high-energy pulsed emitter protection is provided.

Specifications
Dimensions:
Transceiver: 146 × 165 × 124 mm
Controller: 146 × 140 × 57 mm
Indicator: 84 × 153 × 34 mm

Weight:
Transceiver: 4.54 kg
Controller: 1.02 kg
Indicator: 0.49 kg

Operational status
The AN/ARC-182(V) is in service in 42 countries on more than 120 airborne and ground-based installations. Modification kits are available to add NVG compatibility and HQ or HQII capability. The AN/ARC-182(V) is no longer recommended for new platform integration or system designs. The system remains a special order product item, replaced by the AN/ARC-210(V).

Contractor
Rockwell Collins.

UPDATED

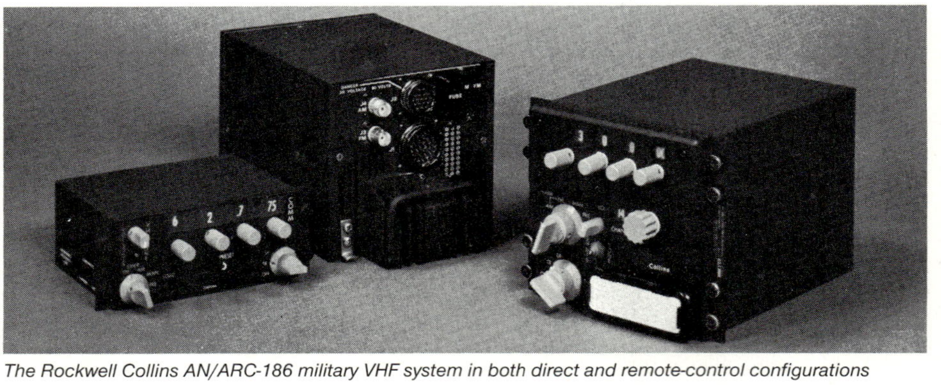

The Rockwell Collins AN/ARC-186 military VHF system in both direct and remote-control configurations

AN/ARC-186(V)/VHF-186 VHF AM/FM radio

The Rockwell Collins AN/ARC-186(V)/VHF-186 is a tactical VHF AM/FM radio communications system designed for all types of military aircraft.

The basic ARC-186(V) is a solid-state 10 W system of modular construction which provides 4,080 channels at 25 kHz spacing. The 1,760 FM channels are contained in the 30 to 88 MHz band and 2,320 AM channels within the range 108 to 152 MHz. A secure speech facility can be used in both AM and FM modes and the equipment is compatible with either 16 or 18 kbit secure systems in diphase and baseband operation.

Up to 20 channels may be programmed for preselection on the ground or in the air. Preselection is accomplished through incorporation of a non-volatile NMOS memory which continues to retain data in the event of a loss of power supply. Two dedicated channel selector switch positions cover the FM and AM emergency channel frequencies of 40.5 and 121.5 MHz respectively.

Either panel mounting with direct control through an integral controller or remote mounting with an identical control panel presentation is possible. A half-size remote controller which contains the same control functions is also available and a typical configuration in a two-seat aircraft would comprise a full panel mount in the pilot's cockpit with a half-size controller at the crew position. In these dual-control configurations, a manual take-control switch provides full communications control for either crew member.

Conversion from panel to remote control is made by removing the panel controller and replacing it with a plug-in serial control receiver module. A typical conversion is said to require less than 5 minutes. Frequency displays on both types of controller are immune to fade-out during periods of low voltage.

Circuitry of the ARC-186(V) is of modular design. Seven module cards are held in place by the body chassis or card cage and are electrically interconnected by a planar card in which all hard wiring has been virtually eliminated. All radio frequency lines in the interconnecting planar card have been buried to minimise electromagnetic interference. Individual module cards are readily removable and may be replaced in the field to reduce fault finding and repair time.

Current options for the ARC-186(V) include AM/FM homing facility, but growth capability has been designed into the equipment from the outset and possible future developments could include SELCAL, burst data and target hand-off. One present simple modification, carried out by replacement of the decoder module in the remote transceiver, permits the radio to be directly connected to a MIL-STD-1553 digital databus and it is claimed that the system will be equally compatible with avionics suites of future generation equipment. An additional possibility is the uprating of transmitter output power.

US Air Force testing has demonstrated a MTBF in excess of 9,000 hours.

A principal design objective for the ARC-186 series equipment was that it should be capable of easy retrofit in existing installations. Since the system is considerably smaller than the equipment it is designed to replace, this is accomplished by use of plug-in adaptor trays which permit rapid replacement without disturbance to existing aircraft wiring harnesses.

Specifications

Dimensions:
(remote-mounted transmitter/receiver)
127 × 165 × 123 mm
(panel-mounted transmitter/receiver)
146 × 165 × 123 mm
(half-size remote control) 146 × 95 × 57 mm
Weight:
(transmitter/receiver) 2.95 kg
(remote-control) 0.79 kg
(FM homing module) 0.45 kg

Operational status

In production and in service. More than 28,000 sets have been delivered worldwide. It has been selected by the US Air Force as the standard equipment for all aircraft requiring VHF AM/FM capability.

Rockwell Collins is to upgrade 183 AN/ARC-186 transceivers to VHF-186R configuration for the Royal Netherlands Air Force; deliveries are to begin in late 2000. The upgraded transceivers will meet International Civil Aviation Organisation (ICAO) requirements for VHF radios to operate with 8.33 kHz channel separation and FM immunity in European airspace. The upgraded radios will be fitted to the F-16 aircraft, and Cougar and CH-47D helicopters.

Contractor

Rockwell Collins.

UPDATED

AN/ARC-190(V)/HF-190 HF airborne communication systems

AN/ARC-190(V) aircraft communications system

The Rockwell Collins AN/ARC-190(V) is a military HF transmitter/receiver designed as a replacement for a number of earlier HF systems in a US Air Force modernisation programme. The system is therefore particularly suited to retrofit applications as well as for installation as original equipment in a wide range of aircraft. The AN/ARC-190 is the mainstay of HF communications in the US Air Force, having been installed in a large variety of fixed- and rotary-wing aircraft, such as the C-130, KC-135, C-141, C-5, C-9, KC-10, B-1, B-52, C-17, F-15, F-16, H-60 and S-2T. As well as retrofit kits for 618T systems, a MIL-STD-1553 system is available. The system can be used in either 1553 or non-1553 applications. A selective calling (SELCAL) AM detector is available in some versions of the ARC-190.

The ARC-190 is automatically tuned in both receive and transmit modes. Built-In Test Equipment (BITE) and modular construction provide for rapid fault isolation to the box and module level for quick repair. The unit covers the 2 to 30 MHz band, providing 280,000 channels, any 30 of which are preselectable, in incremental steps of 100 kHz. Operational modes include USB, LSB, AME and CW. Data transmission facilities are also available in USB and LSB modes and in these modes the system is able to operate with audio-frequency shift keying or multitone modems.

The system is remotely controlled and dual control of the radio is possible from two crew stations. Serial data control is applied between each of the major units in the system and this is said to render it adaptable to future requirements such as SELCAL and remote frequency management. Transmitter power output is 400 W. The ARC-190 is supplied with AC or DC interface power, as well as MIL-STD-1553 control scheme.

Construction of the ARC-190(V) is all-solid-state. The full system includes an antenna coupler, which ensures compatibility with military cap, shunt, wire, whip or probe antenna systems. An F-1535 bandpass filter unit is available to provide added selectivity and overload protection for improved receiver performance in a strong signal environment. In the transmit mode it provides additional filtering to the exciter RF output. The system operates to pressure altitudes up to 70,000 ft over temperatures ranging from –55 to +71°C.

Rockwell Collins supplies the US Air Force with an automatic communications processor for the ARC-190(V) which automatically selects the optimum HF frequency after the operator has selected the station to be called. The processor also features anti-jam modes. After test and evaluation, production of the processor commenced in 1988.

The Rockwell Collins AN/ARC-190(V) HF radio

HF-190 aircraft communications system

The HF-190 is a derivative of the ARC-190. It has the same features and capabilities as the ARC-190, plus operation on 176 preprogrammed maritime frequencies and two international distress frequencies. The system is TSO approved and is authorised by the FCC for use on aeronautical (Part 87) and mobile maritime (Part 83) channels.

High Frequency Data Link (HFDL) upgrade

The HFDL mode provides worldwide automated-data communications directly between aircraft systems and the ARINC ground network. HFDL is part of the Aeronautical Telecommunication Network (ATN) and supports a wide variety of data applications. The HFDL configuration of the AN/ARC-190 consists of an RT-1341(V)8 receiver transmitter and a CP-2024(MIL Nomenclature TBD) Automatic Communications Processor (ACP). Upgrade kits are available for both the RT and processor to upgrade prior versions of these equipments to the HFDL configuration.

Specifications

Dimensions:
(controller) 114 × 146 × 67 mm
(transmitter/receiver) 484 × 257 × 194 mm
(antenna coupler) 545 × 211 × 188 mm
Weight:
(controller) 0.68 kg
(transmitter/receiver) 23.60 kg
(antenna coupler) 10.89 kg

Operational status

In production and in service in a wide variety of US Air Force and US Navy aircraft.

Contractor

Rockwell Collins.

UPDATED

AN/ARC-210(V) multimode integrated communications system

The Rockwell Collins AN/ARC-210(V) multimode integrated communications system was derived from the AN/ARC-182 system to provide multimode voice and data communications in either normal or-jam-resistant modes through software reconfiguration. The RT-1556 transceiver is capable of establishing two-way communication links over the 30 to 400 MHz frequency range within tactical aircraft environments. There are currently nine different variants of the transceiver, but a great deal of commonality is retained between these.

The ARC-210 meets the 8.33 kHz European ATC channel spacing requirements.

Rockwell Collins has produced the RT-1794(C)/ARC-210 which incorporates all the features of the basic ARC-210, plus embedded Satcom/DAMA, COMSEC (KG-84, KY-58, ANDVT, RGV-11), CTIC,-MIL-STD-188-220A JVMFa, Link 4A and is software reprogrammable via the Advanced Memory Loader Verifier (AMLV) feature, also embedded.

The US Navy's F/A-18C/D and E/F, MV-22, C-2 and EA-6B, and the US Air Forces' C-17A, CV-22 and HH-60G are on contract to receive this unit. A fully functional auxiliary remote control, the C-12571/ARC-210 is available, or the system may be integrated into aircraft Control Display Units (CDUs) via-MIL-STD-1553B databus.

The transceiver is the nucleus of the multimode communication system which includes an appliqué for Have Quick, Have Quick II and SINCGARS-V waveforms. In addition, the AN/ARC-210 has been demonstrated to provide Have Quick IIA ECCM and for Link 4A and Link II data communications. The system will also provide Satcom wideband and narrowband operation. Integrated with the AM-7525/-7526 UHF high-power amplifier and MX-11641 low-noise amplifier/diplexer, the ARC-210 provides a flexible satcom terminal and complies with MIL-STD-188-181/182/183 DAMA requirements. Maritime, land-mobile, ATC and ADF are included modes of operation. The system can be controlled by a MIL-STD-1553B databus and includes two types of remote controller for manual operation (C-1189XA/ARC and C-12419A/AERC half-size units with non-embedded DAMA RTs and the full-size C-12561/ARC remote control unit, with RT-1795©/ARC interfaces). A remote indicator and a family of broadband and electronically tunable antennas are also included.

Rockwell Collins was placed on contract on 16 August 1995 to incorporate the Secretary of Defence's acquisition streamlining initiative into the ARC-210 communications system.

Planned growth capabilities include: Have Quick IIA and SATURN capability, SINCGARS Improvement Program (SIP) – data rate adaptor, USA ATC DAMA (digital voice) and Differential GPS datalink.

Specifications

Dimensions: 127 × 142.2 × 248.9 mm
Weight: 5.44 kg

Operational status

In production for US Navy (F/A-18 C/D, AH-1W, AV-8B, C-5, C-17, CH-46E, C/MH-53E, KC-130 F/R/T, UH-1N, EA-6B), US Air Force, US Marine Corps, US Army and US Air National Guard platforms plus Australian, Canadian, Finnish, Italian, Spanish, Swiss and Taiwanese forces.

The AN/ARC-210 forms part of the US Air Force C/KC-135 Global Air Traffic Management (GATM) programme. In February 2000, the US Air Force contracted Rockwell Collins to demonstrate complete end-to-end commercial Airline Operational Control (AOC) datalink and Very High Frequency DataLink (VDL) Mode 2 capability on C/KC-135 aircraft as part of the GATM upgrade programme. System demonstration and functional testing will be completed during KC-135 GATM flight tests scheduled for 2001. The AOC demonstration will show how military flight operations can be enhanced through the use of realtime commercial datalink by modifying the standard airline AOC messaging applications which are resident in the Rockwell Collins CMU-900 Communications Management Unit. In addition, standard formats for maintenance, fuel logs and flight management datalink applications will be provided.

The Rockwell Collins AN/ARC-210(V) communications system

The VDL Mode 2 enhancement supports the transition from voice-based to digital communications by using a high-speed line of sight Aircraft Communications Addressing and Reporting System (ACARS). This capability also establishes the baseline for connection to the future Aeronautical Telecommunications Network. The VDL enhancement will be provided by adding a datalink module to the existing Rockwell Collins AN/ARC-210 V/UHF radio and by upgrading the CMU-900 software. The datalink module will initially provide VDL Mode 2 functionality and ARINC 750 compliant interfaces to the dual CMU-900s. The added module also provides the necessary hardware to support VDL Mode 3 via a future software upgrade.

Over 6,000 units had been sold by mid-1999.

Contractor

Rockwell Collins.

UPDATED

AN/ARC-217(V) HF system

The AN/ARC-217(V) is a lightweight tactical HF airborne transceiver system designed for both rotary- and fixed-wing applications. When installed with a simple retrofit kit, the AN/ARC-217(V) is a direct replacement for the AN/ARC-199, but provides the additional features of embedded ECCM, automatic link establishment and data transmission. It also has a-MIL-STD-1553B control interface.

The primary requirement was that the system should establish communication links 24 hours a day with ground stations on 90 per cent of the attempts and that 90 per cent of each message be accurately transmitted and received.

Operational status

The AN/ARC-217(V) has a wide application for a variety of US military combat helicopters and fixed-wing aircraft. Also procured for the presidential helicopter fleet.

Contractor

Rockwell Collins.

VERIFIED

AN/ARC-220 HF tactical communications system

Rockwell Collins' AN/ARC-220 is a multifunction, fully Digital Signal Processing (DSP) high frequency radio intended for airborne applications. Advanced communications features made possible by DSP technology include embedded Automatic Link Establishment (ALE), Serial Tone Data Modem and Anti-jam (ECCM) functions.

The ARC-220 Advanced HF Aircraft Communications System is suitable for a variety of tactical rotary-wing and fixed-wing airborne applications. In addition to offering enhanced voice communications capabilities, the AN/ARC-220 is an advanced data communications system capable of providing reliable digital connectivity.

The ARC-220 allows communications on any one of 280,000 frequencies in the High Frequency (HF) band from 2.0000 to 29.9999 MHz. Upper SideBand (USB) voice, Lower SideBand (LSB) voice, Amplitude Modulation Equivalent (AME), Continuous Wave (CW), USB data and LSB data emission operating modes are provided. Communication is possible using either simplex or half-duplex operation. The AN/ARC-220 also includes an integrated data modem, which enables communication in noisy environments where voice communications are often not possible. Up to 25 free text data messages can be preprogrammed via data fill or created/edited in real time. Received data messages can be stored for later viewing and retransmission if desired. Built-in integration with external GPS units allow position data reports to be sent with the push of a button. Long-range wireless e-mail and Internet connectivity is possible when used in conjunction with Rockwell Collins HF Messenger™ software.

In addition to conventional HF communications, the AN/ARC-220 provides MIL-STD-148-141A Automatic Link Establishment (ALE) capability. ALE maintains a database of channel signal conditions and automatically establishes a two-way communication link on the best available frequency. After establishing a link, the operator is alerted and communication can begin. ALE Linking Protection (LP) is also provided, which prevents unauthorised link establishment and ALE jamming. The AN/ARC-220 is also claimed to be the first radio to be fielded featuring MIL-STD-188-141B Appendix A Alternative Quick Call (AQC) ALE. Rockwell Collins asserts that this advanced protocol/waveform has the potential to reduce ALE over-the-air call time by up to 50 per cent over conventional second-generation ALE.

Electronic Counter CounterMeasures (ECCM) is a frequency method used to combat the effects of communication jammers and direction finding attempts. The ARC-220 provides ECCM in accordance with MIL-STD-188-148A and CR-CX-0218-001 (Army Enhanced). ALE for both forms of ECCM is provided.

The R/T also supports the modem requirements necessary for the MIL-STD-188-141A ALE, MIL-STD ECCM and Army Enhanced ECCM digital message protocols.

System description

The AN/ARC-220 provides three transmit power output settings: low (10 W pep and average), medium (50 W pep and average) and high (175 W pep; 100 W average). Transmit tune time is typically 1 second in manual frequency selection. Transmit tune time is typically 35 ms after a frequency is tuned and tuning data stored.

The AN/ARC-220 system is composed of three Line Replaceable Units (LRU) for non-MIL-STD-1553B bus controlled platforms and two LRUs for a MIL-STD-1553B bus controlled platform. The RT-1749/URC receiver transmitter is the heart of the system with the AM-7531/URC power amplifier/coupler providing antenna matching and RF power amplification. The C-12436/URC control allows the user to control system operation.

Radio set control C-12436/URC provides the AN/ARC-220 operator interface. A six-line alphanumeric display provides system status and advisory data. The display includes variable backlighting and is NVG compatible. The radio set control internal power supply operates from the +28 V DC aircraft power.

A radio receiver-transmitter RT-1749/URC provides the electrical interface with other AN/ARC-220 LRUs and associated aircraft systems such as interphone, GPS and secure voice systems. The receiver-transmitter provides EIA RS-232C and MIL-STD-1553B serial data interface ports to accommodate various system configurations.

The receiver-transmitter uses a microprocessor to perform all ALE, ECCM (optional) and modem functions. The internal power supply operates from +28 V DC aircraft power. An external 6 to 24 V DC power source can be connected to the receiver-transmitter, which will enable retention of clock and data message information when main power is removed from the system.

The RT-1749A/URC version of the receiver transmitter does not have ECCM or ALE LP 3 capability and is intended for foreign military sales.

A power amplifier/coupler AM-7531/URC provides an electrical interface with other AN/ARC-220 LRUs and the antenna system. The PA coupler provides three output power levels: high (100 W average; 175 W PEP), medium (50 W PEP and average) and low (10 W PEP and average). Antenna coupler circuits provide impedance matching between the power amplifier output and the transmit antenna which enables maximum power transfer to the antenna. The PA coupler is digitally tuned under control of the receiver-transmitter microprocessor. First time tuning is usually completed in 1 second. The receiver-transmitter stores tuning data for previously tuned frequencies. Stored

The AN/ARC-217(V) HF transceiver is designed for nap of the earth communications. It consists of (left to right): control unit, receiver/transmitter and antenna coupler

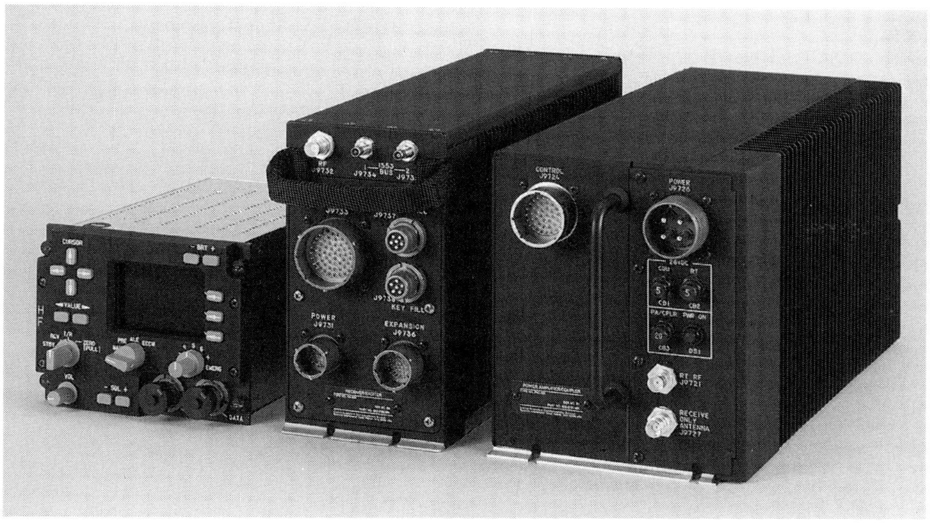

The AN/ARC-220 is an advanced HF aircraft communications system

tuning data reduces future tuning time (to typically 35 ms) when a frequency is used again. The PA coupler internal power supply operates from +28 V DC aircraft power.

Specifications

Frequency: 2.0000 to 29.9999 MHz in 100 Hz steps
Channels: 20 user programmable simplex or half-duplex
20 programmable simplex or half-duplex
20 programmable automatic link establishment (ALE) scan lists
12 programmable ECCM hopsets (with ALE capability)
Modes: USB and LSB voice and data, CW and AME
Power requirements: 28 V DC per MIL-STD-704
Usage: (RT-1749/URC) 50 W
(AM-7531/URC) 450W (transmit)
(C-12436/URC) 50 W
MTBF: 1,000 h min
Fault isolation/detection: 98% fault isolation (LRU level)
Dimensions (W × D × H): (RT-1749/URC) 105.4 × 355.6 × 194.3 mm
(AM-7531/URC) 160.3 × 428.0 × 184.9 mm
(C-12436/URC) 146 x 203.7 x 114.3 mm
Weight: (RT-1749/URC) 5.9 kg
(AM-7531/URC) 8.5 kg
(C-12436/URC) 2.3 kg
NVG compatibility: IAW MIL-L-85762F
Temperature: –40 to +55°C
Altitude: 25,000 ft max
Humidity: 95%
Vibration: 2.5 *g*

Operational status

In production and in service with US Army helicopters.

Contractor

Rockwell Collins.

UPDATED

AN/ARC-230/HF-121C high-performance radio system

The Rockwell Collins AN/ARC-230 (company designation HF-121C) high-performance radio system is designed for military voice and data HF applications requiring 400 W operation. Compliant with the requirements of MIL-STD-188-203-1A for Link 11/TADIL A and MIL-E-5400, this 400 W PEP/and average power radio provides maximum performance for airborne applications. The AN/ARC-230 has been optimised for tactical digital data communications and SIMultaneous OPeration (SIMOP) of multiple radio sets with minimum frequency and antenna separation. Embedded Automatic Link Establishment (ALE) and ECCM capabilities are also available. For a full description of the system and capabilities overview, see separate entry for Rockwell Collins' HF-121A/B/C HF radios.

Operational status

In production and operation in commercial and military aircraft in the USA and other countries.

Contractor

Rockwell Collins.

UPDATED

AN/ARN-118(V) Tacan

The AN/ARN-118(V) Tacan provides the crew and flight control systems with reliable navigation data. Advanced digital circuitry is employed to achieve space and weight savings. All-solid-state, with the exception of the transmitter tubes, the AN/ARN-118(V) unit is packaged in a ¾ ATR short configuration. Both X and Y channels are standard on the system and a self-contained automatic antenna switch enables the unit to be plugged into aircraft with either single or dual Tacan aerial installations. With full digital readout, distance and bearing can be acquired in less than 1 and 3 seconds respectively. Input and output hardware is designed to accommodate complete digital serial-controlled Tacan navigation functions, including automatic tuning by RNav or air navigation systems. In addition, the system's output is compatible with INS, RNav and automatic navigation systems to provide full Rho-Theta real-time update information to these systems. Range rate information is available for area navigation.

Specifications

Dimensions:
(control) 127 × 76.2 × 81.3 mm
(receiver/transmitter) 190.5 × 172.7 × 317.5 mm
(adapter) 43.2 × 172.7 × 317.5 mm
Weight:
(control) 0.77 kg
(receiver/transmitter) 12.23 kg
(adapter) 1.81 kg
Power supply: 115 V AC, 400 Hz, 125 W

Operational status

In production. A recent application is the US Air Force KC-10 Extender. The original US Air Force production contract for AN/ARN-118(V) units was placed in 1975 and the system is standard equipment with that service and with the US Coast Guard. It is also becoming standard for the US Army and Navy and has been chosen by over 35 countries, some 32,000 sets having been built. The ARN-118(V) will remain in production for some years.

Contractor

Rockwell Collins.

VERIFIED

AN/ARN-139(V) Tacan

The AN/ARN-139(V) transforms an aircraft into a flying Tacan station by providing air-to-air bearing and distance information. This capability is added to all the standard features of the AN/ARN-118(V) Tacan which provides standard air-to-ground and air-to-air Tacan information. Rendezvous with an ARN-139(V)-equipped aircraft or ship simplifies critical missions by providing a reliable readout of bearing and distance during the approach.

The ARN-139(V) is derived from the ARN-118(V), of which more than 32,000 units have been built. The ARN-139(V) provides distance and bearing transmission and inverse Tacan operation. The latter function allows a tanker aircraft, for example, to read the bearing and distance to a Tacan-equipped aircraft in need of air refuelling. Inverse Tacan also enables a pilot to determine the bearing to a DME ground station. The selectable range ratio capability permits the pilot, by means of a switch, to limit all replies to within four times the range of the nearest aircraft, or to concentrate on aircraft more than 30 times the distance of the nearest aircraft.

Specifications

Dimensions:
(AN/ARN-139(V)) 254 × 196 × 494 mm
(C-10059/A and C-10994 control units) 127 × 76 × 81 mm
Weight:
(AN/ARN-139(V)) 31.3 kg
(control unit) 0.9 kg
Frequency:
(transmitter) 1,025-1,150 MHz
(receiver) 962-1,213 MHz
Transmitter power:
(min) 500 W
(typical) 750 W
Receiver sensitivity: –92 dBm
Modes: X/Y channels, inverse beacon, inverse air-to-air, inverse transmit/receive, inverse receive
Range: 390 n miles
Track rate: 3,600 kt, 20°/s
Accuracy:
(distance) (digital) 0.1 n mile
(analogue) 0.2 n miles
(bearing) (digital) 1°
(analogue) 1.5°
Reliability: 1,000 h design MTBF

Operational status

In service.

Contractor

Rockwell Collins.

VERIFIED

AN/ARN-144(V) VOR/ILS receiver

The ARN-144(V) VOR/ILS receiver is said to be the first such system to be compatible with the MIL-STD-1553B digital databus. All the standard VOR, localiser, glide slope and ILS beacon facilities are available with 160 VOR channels and 40 localiser/glide slope channels being selectable at 50 kHz spacing.

A number of configurations is produced to meet specific military applications. The R-5094/ARN-514, a version of the ARN-144, is standard on F/A-18. A number of different control panels is also available.

Specifications

Dimensions: 104 × 127 × 304 mm
Weight: 3.6 kg
Power supply: 28 V DC, 25 W

Operational status

In production and in service.

Contractor

Rockwell Collins.

VERIFIED

AN/ARN-147(V) VOR/ILS receiver

The AN/ARN-147(V) is fully compatible with the MIL-STD-1553B databus. This receiver combines all VOR/ILS functions such as VOR and ILS localiser, glide slope and marker beacon in one compact system.

In late 1997, Rockwell Collins released both a new model of the AN/ARN-147(V) and a modification kit for existing receivers that provides FM interference immunity in accordance with ICAO Annex 10.

All-solid-state modular construction makes the ARN-147 a reliable receiver for either new or retrofit applications on fixed- and rotary-wing aircraft.

Rotor modulation suppression circuitry is pin-selectable in the ARN-147 for reliable operation in rotary-wing aircraft. FAA split-channel requirements are met by providing 50 kHz spacing for 160 VOR and 40 localiser/glide slope channels.

Digital and analogue outputs are compatible with the latest high-performance flight control systems, digital indicators and analogue instruments. In addition, high- and low-level deviation and flag outputs for VOR, Loc and glide slope are provided.

The ARN-147 meets the following US military standards and specifications: MIL-E-5400 Class II environment; MIL-STD-810 vibration, including gunfire vibration; MIL-STD-461/462 electromagnetic interference, and MIL-STD-704 power characteristics. An FM immunity upgrade is now available.

Specifications

Dimensions: 104 × 127 × 304 mm
Weight:
(with MIL-STD-1553B) 3.6 kg
(without MIL-STD-1553B) 3.4 kg
Power supply:
28 V DC, 25 W with MIL-STD-1553B
28 V DC, 20 W without MIL-STD-1553B

Operational status

In production and service as the standard VOR/ILS receiver for the US Air Force and currently installed on aircraft such as the C-130, C-5, C-141, CH-53, UH-1H and UH-60. In mid-1985, Rockwell Collins received a contract to supply the US Air Force with over 25,000 AN/ARN-147(V) sets for use on several types of transport aircraft.

The system has also been selected for the V-22 Osprey, US Navy T-45 and US Air Force C-17. It is being retrofitted to the US Air Force T-38 aircraft.

Contractor

Rockwell Collins.

VERIFIED

AN/ARN-149(V) (DF-206A) Automatic Direction-Finder (ADF)

The AN/ARN-149(V) was the first low-frequency automatic direction-finder to provide an internal field upgradable MIL-STD-1553B digital multiplex bus capability. This system, consisting of a receiver, control, antenna and mount, provides a low-frequency automatic direction-finding function in a lightweight easily installed set. The all-solid-state receiver eliminates all moving parts such as goniometers, synchros and mechanical tuners. Quadrantal Error Correction (QEC) is set by aircraft connector strapping, eliminating corrector modules and airframe specific internal adjustments. The antenna combines the loop and sense antennas and preamplifiers in one compact housing, eliminating expensive sense panels, couplers and special prefabricated cable assemblies. A dual version of the same antenna is also available in one aerodynamic package in either white or black colours. The receiver is controlled by a four-wire serial bus from the control. The system meets MIL-E-5400 and has an MTBF of 4,000 hours.

Specifications

Dimensions:
(receiver) 79 × 127 × 279 mm
(control) 146 × 57 × 96 mm
(antenna) 216 × 43 × 419 mm
Weight:
(receiver) 2.5 kg with MIL-STD-1553B capability
(control) 0.7 kg
(antenna) 1.4 kg
Power supply:
26 V AC, 7.8 W
28 V DC, 18 W

Operational status

In production.

Contractor

Rockwell Collins.

VERIFIED

The AN/ASC-15B command and control console is designed for helicopters such as the UH-60 Black Hawk

AN/ARN-153(V) Tacan

The AN/ARN-153(V) consists of one or two antennas, a cockpit control, the receiver/transmitter and a mounting tray.

Key features of the system are digital outputs for both distance and bearing (with optional analogue outputs), input and output MIL-STD-1553 bus capability, microprocessor-based design, dual-antenna ports, inverse bearing capability in certain applications with the 938Y-1 antenna, pilot-selectable air-to-air range ratio capability, 390 n miles range, signal-controlled search, solid-state 500 W transmitter, compatibility with ARINC 568, 582 and 429, enhanced BIT and W and Z channels for MLS compatibility.

The AN/ARN-153(V) has four basic modes of operation: receive, transmit/receive, air-to-air receive and air-to-air transmit/receive. When used in conjunction with the optional 938Y-1 rotating antenna and an optional control unit, the AN/ARN-153(V) can also provide bearing in certain applications to an air-to-air Tacan if it can transmit unmodulated squitter, as well as bearing to any DME-only ground station.

Specifications

Dimensions:
(control) 144.8 × 76.2 × 99.1 mm
(receiver/transmitter) 104.6 × 172.2 × 304.8 mm
(mounting tray) 114.3 × 50.8 × 312.4 mm
Weight:
(control) 0.9 kg
(receiver/transmitter) 6.49 kg
(mounting tray) 0.54 kg
Power supply: 28 V DC, 1.5 A nominal
Frequency:
(transmitter) 1,025-1,150 MHz
(receiver) 962-1,213 MHz
Number of channels: 126 X and 126 Y
Provision for W and Z
Range: up to 390 n miles
Accuracy:
(distance) (digital) ±0.1 n mile,
(analogue) ±0.2 n miles,
(bearing) (digital) ±0.5°,
(analogue) ±1.5°

Contractor

Rockwell Collins.

VERIFIED

AN/ASC-15B communications central

The AN/ASC-15B communications central, referred to as a command console, functions as an airborne and ground command post, providing tactical voice communications in both secure and non-secure modes. This highly mobile communications combat command centre provides NATO and US tri-service forces interoperability during all types of military operations and special missions.

The AN/ASC-15B can be operated from a UH-60A or UH-1H helicopter, or removed and configured for ground operation. It provides HF plus VHF and UHF communications in AM and FM modes, channel scanning of four V/UHF preset channels in each-AN/ARC-182 radio, automatic retransmission in VHF and UHF bands and UHF satellite communications.

The AN/ASC-15B consists of an AN/ARC-174 HF transceiver, three AN/ARC-182 V/UHF transceivers, two AM-7189A IFM power amplifiers, an MX-931B/URC repeater, an AM-7402 Satcom power amplifier and two C-11128 ECCM (HQ) controls.

The ASC-15B has been modified with three ARC-210(V) radios to replace the ARC-182s presently installed. It was given the nomenclature AN/ASC-15C after radio upgrading.

Specifications

Weight: 129.28 kg
Power supply: 28 V DC
Power output:
(2-30 MHz) 100 W PEP
(30-400 MHz) 15 W FM, 10 W AM
Frequency: 2-30 MHz and 30-400 MHz
Modes:
(2-30 MHz) HF/SSB, AME and CW
(30-400 MHz) V/UHF AM and FM

Operational status

In service with the US Army.

Contractor

Rockwell Collins.

VERIFIED

AN/USQ-146 CCW system

The Rockwell Collins USQ-146 is a reliable, versatile, low-cost, off-the-shelf CCW system featuring dual receiver-transmitters, providing conventional AM/FM voice and data communications, multiband surveillance scanning, target detection, identification, and analysis and multiwaveform jamming of targeted signals. A variety of power amplifiers and antennas are available for customised use. Other features of this PC-compatible system are:

- Windows-based operating system provides user-friendly graphical user interface
- A wide frequency range of 20 to 2500 MHz, supporting all missions with only the addition of application-specific power amplifiers and antennas
- Low-cost microwave translator module makes it affordable to increase frequency range up to 18 GHz
- System supports multiple operational uses - communications, surveillance (scanning and data collection), Electronic Support (ES) (surveillance) and Electronic Attack (EA) (active jamming)
- Built-In Test (BIT) capability verifies performance and identifies failed modules quickly
- Systems with dual AN/ARC-210 RT-1747 receiver-transmitters permit failed unit to be 'bypassed' while good unit takes over, thus eliminating single-point failures
- Wideband receive and transmit modes
- Simultaneous transmission in two bands
- Modular software is field upgradeable
- Uses Collins AN/ARC-210 RT-1747B/C receiver-transmitter to provide basic radio functions (modulation, demodulation) and frequency translation between 20 and 400 MHz
- Uses Collins CV-2100 block converter to cover 400 to 2,500 MHz band
- Separate PA Case Assembly permits custom configuration of power amplifiers without changing the configuration of the RT Case Assembly.

The USQ-146 is intended for installations including fixed station, shipboard, mobile and airborne. The system can operate autonomously or can be operated from the provided Windows control program from RS-232 or Ethernet serial interfaces.

Modes of operation

All of the operational modes and features of the USQ-146 are accessible using a Windows 95® or Windows NT® compatible graphical user interface. The program is designed for fast and efficient operation from either the keyboard or mouse. On-line help facilitates training and operation of the system.

Communications mode

In the communications mode, the USQ-146 performs receive and transmit functions for voice and data traffic between two or more stations. The following capabilities are provided:

- Simplex or half-duplex fixed-frequency operation
- HAVE QUICK, HAVE QUICK II and future SATURN ECCM modes
- AM/FM, voice/data, clear/cipher modes
- Compatible with encryption devices such as Voice Privacy™ and ANDVT
- Compatible with single- and multitone data modems such as the Collins MDM-3001
- Manual, channelised operation.

AN/USQ-146 power amplifier case assembly
***2001**/0054163*

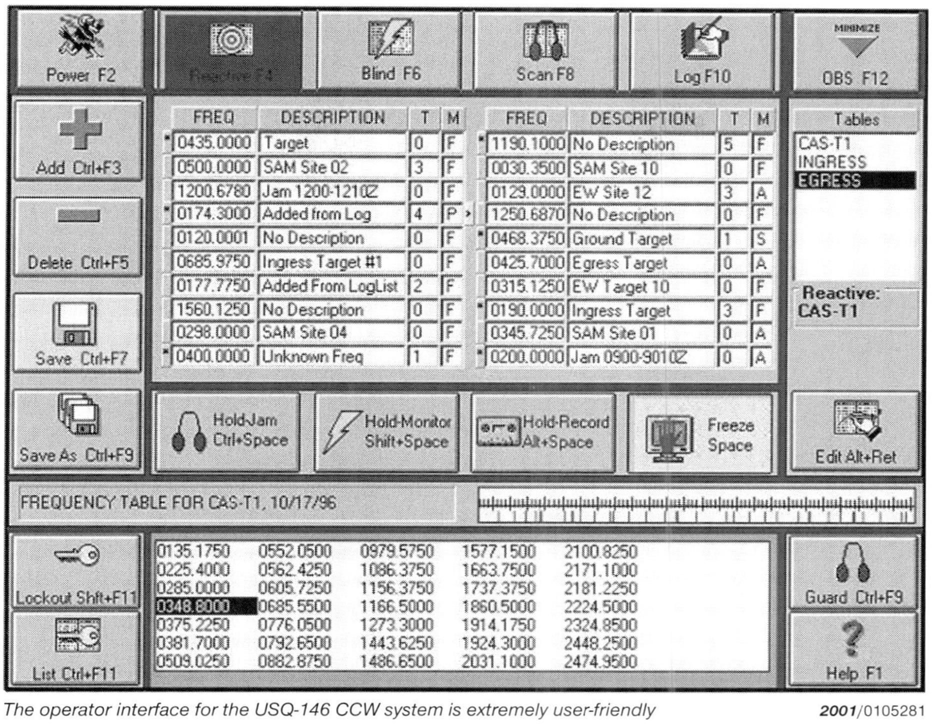

The operator interface for the USQ-146 CCW system is extremely user-friendly ***2001**/0105281*

Electronic Support (ES) (Surveillance) mode

The USQ-146 provides Electronic Support (ES) modes for detection and verification of target emitters.

- Acquisition mode - the system detects signals at widely varying energy levels while avoiding false responses due to noise
- Scan mode - the system initially scans the selected band using a 6 MHz bandwidth filter for maximum scan rate. When energy is detected, the system automatically rescans at reduced bandwidths and records the frequency and amplitude of each target emitter encountered.

Electronic Attack (EA) (jamming) modes

The USQ-146 system employs innovative jamming algorithms and features that allow it to effectively deny many potential target networks:

- Jamming mode - when a target emitter is identified in scan mode, the system consults the frequency tables to determine what action to take. The frequency tables identify which frequencies are to be jammed or protected (i.e. not jammed)
- Jamming parameters - the target parameter table defines acquisition parameters, jammer timing, modulation mode, jamming waveform and power level to be used with a particular target emitter
- Normal mode - the system scans from the selected minimum to maximum frequency, looking for target emitters
- Selective mode - the system scans only frequencies within the minimum/maximum range that are identified in the Target Preset Table or the Sector Table
- Priority mode - the system scans only frequencies which are listed in both the Target Preset Table and the Sector Table. If multiple target emitters are found, only the highest priority emitters are jammed. Lower priority targets will be jammed only if high priority targets are inactive

Specifications

Frequency range:	20 to 2,500 MHz
Frequency resolution:	100 Hz (400 to 2500 MHz), 1.25 kHz (20 to 400 MHz)
Tuning time:	500 ms
Frequency accuracy:	± 2.5 PPM, 0 to 50°C
Noise figure:	
20 to 400 MHz antenna	15 dB (typical)
400 to 2,500 MHz antenna	13 dB (typical)
Modulation modes:	CW, AM, FM, WBFM
Receive sensitivity:	–95 dBm during scan modes
Jamming modulation:	programmable wave file and/or internal generators
Elementary jamming duration:	programmable, 1 to 9,999 ms
Jamming between look through	programmable, 1 to 9,999 ms
Look-through duration:	less than 1 ms
Scan rate:	12 MHz per ms
Target emitter frequency measurement accuracy:	100 Hz
Power Input	
C2W System (Dual RT)	115 V AC, 47-440 Hz, 125 W (receive), 325 W (transmit)
Amplifier (Lucas-Zeta)	115 V AC, 400 Hz, 3-phase, 3,145 W (max)
Amplifier (ComTech PST)	115 V AC, 50-60 Hz, 3,500 W (max)
PA Transit Case	
Size	83 × 151 × 96 mm (W × D × H)
Weight	141 kg (with dual power supplies)
RT Transit Case	
Size	83 × 127 × 96 mm (W × D × H)
Weight	75 kg (with dual power supplies)
Environmental	
Temperature-operating	–20 to +50°C (except laptop computer)
Temperature-standby	–57 to +85°C
Altitude	up to 15,000 ft
Humidity	up to 100%

- Blind mode - the system jams the frequencies within the min/max range that are identified in the Target Preset Table or the Sector Table whether a target emitter is active or not
- Hybrid mode - the Hybrid mode consists of alternating between Priority mode and Normal mode
- Directed mode - the Directed mode consists of a user-defined combination of the above EA (jamming modes) as well as ES modes.

Operational status
In production.

Contractor
Rockwell Collins.

NEW ENTRY

APR-4000 GPS approach sensor

The APR-4000 combines essential navigation functions with critical landing functions into a single, integrated GPS sensor that supports en route, terminal area, precision approach and non-precision approach operations. The sensor uses uplinked final approach waypoints and differential corrections of raw GPS satellite signals to determine the aircraft's precise lateral and vertical position in relation to the associated approach path. The information presented to the flight crew is ILS look-alike deviation data.

The APR-4000 GPS approach sensor utilises a 12-channel GPS receiver and incorporates Receiver Autonomous Integrity Monitoring (RAIM) and predictive RAIM that improves fault tolerance and the integrity of the computed solution by verifying and anticipating the availability of GPS signals throughout the flight plan. The sensor is engineered to provide growth to local area augmentation system specifications for Cat I and Cat II GPS precision approaches as well as growth to regional area augmentation systems such as WAAS, EGNOS and MT-SAT.

Operational status
Available.

Contractor
Rockwell Collins.

VERIFIED

AN/ARC-186, ARC-186R and VHF-186R series VHF AM/FM digital transceivers

The Rockwell Collins ARC-186R and VHF-186R series of VHF transceivers offers digital, form/fit/function replacements for the AN/ARC-186 multiband airborne radios. The Digital Signal Processing (DSP) software-based ARC-186R transceiver provides dual-band VHF operational capability in a single, miniature package. The ARC-186R and VHF-186R transceivers comply with the latest European 8.33 kHz and FM interference ATC standards. Both combine VHF/AM and VHF/FM transmission and reception capabilities into a single compact unit, covering the 30-88 and 108-152 MHz bands. 8.33 kHz spacing will be provided in the 116-137 MHz ATC band. Standard voice and data communications in either band are possible in accordance with international standards and conventions. The radios are panel mounted (RT-1002), MIL-STD-1553 (RT-1001), or RS-422 serial remote (RT-1000). The optional half-size remote control (RC-1000) may be used with a remoted RT-1000 radio.

The pilot may select a channel in either band manually or as a preset without regard to the band. Optionally, the radios may be programmed to restrict access to allowed bands or presets.

The ARC-186R features Digital Signal Processing (DSP) technology, allowing the operational capabilities of the radio to be updated or modified by downloading software without disassembling the radio.

The ARC-186R is an all solid-state VHF radio transceiver covering the 30 to 88 MHz and 108 to 152 MHz bands.

The single-piece transceiver is designed for airborne use, either in a panel mounted or remote-controlled configuration. The ARC-186R is a direct form/fit/function retrofit for the proven ARC-186 transceiver.

The ARC-186R will comply with the current European standards for FM overload.

The transmitter utilises a proven, on-channel linear AM modulator for distortion-free, highly intelligible AM transmissions. For FM transmissions, the ARC-186R utilises the new Rockwell Collins Direct Conversion Transmitter (DCT) technique, employing a special vector modulation integrated circuit to synthesise the carrier-frequency FM waveform from the DSP digital inputs. The software-based transmit modes/waveforms/modulations are also field programmable.

The panel mounted ARC-186R (RT-1002) features a digital display (night vision compatible version available) for frequency and radio functions, but retains the proven ARC-186 basic layout, controls and functionality familiar to aircrews worldwide. Similar control layout and functionality minimises operator training needed to transition from the ARC-186 to the ARC-186R. No installation training is required because the ARC-186R directly retrofits the ARC-186.

An optional half-size remote-control head (RC-1000) provides functionality identical to the full-size ARC-186R front panel in half the panel space.

Presets and field programming of options are accessed via a front panel fill port from any RS-232 equipped computer and a special programming cable.

Specifications
Frequency range: 30-87.975 MHz FM
108-115.99167 MHz AM (receive only)
116-151.99167 MHz AM
Channel spacing: 25 kHz (30-87.975 MHz)
8.33, 25 kHz (108-152 MHz) (field programmable)
Emergency presets: 40.5 MHz FM; 121.5 MHz AM (customer programmable)
Channel presets: 100 non-volatile (customer programmable)
Carrier power: 10 W min, 16 W max (AM/FM, 52 ohms)
Dimensions (D × W × H): Panel RT 241.3 × 146.1 × 123.8 mm
Remote RT 241.3 × 127.0 × 120.7 mm
Remote control 126.5 × 146.1 × 57.2 mm
Weight: Panel RT 3.29 kg max
Remote RT 3.18 kg max
Remote control 1.14 kg

Operational status
In December 1999, Rockwell Collins was contracted to upgrade 183 AN/ARC-186 transceivers to the VHF-186R standard for the Royal Netherlands Air Force F-16 aircraft and cougar and Chinook helicopters.

Contractor
Rockwell Collins.

UPDATED

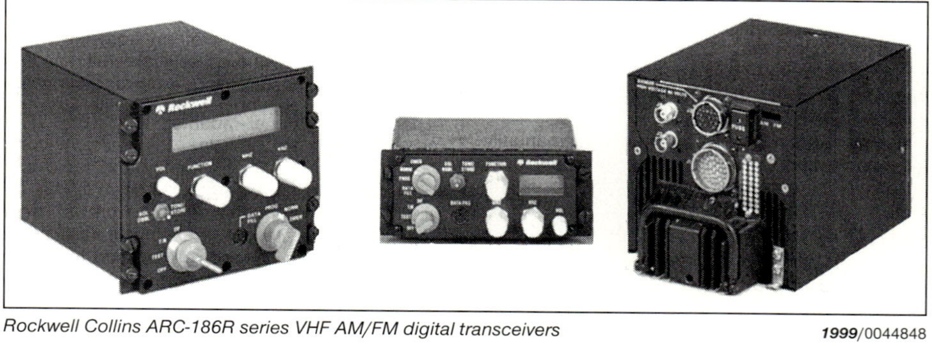

Rockwell Collins ARC-186R series VHF AM/FM digital transceivers ***1999***/0044848

Automatic communications processor ARC-190

The automatic communications processor and associated ARC-190 400 W airborne radio provide a system that automatically scans multiple frequencies, selects the best frequency on which to make a call and automatically repeats the call until contact is confirmed. The system also provides an anti-jam frequency-hopping capability for effective ECCM.

The CP-2024 Automatic Communications Processor (ACP) and the C-11814/ARC-190(V) Automatic Communication Processor Control (ACPC) operate together as a microprocessor-based remote-control subsystem which can be added to existing-AN/ARC-190(V) radio systems to automate and simplify HF radio operation. These units are completely interoperable with MIL-STD-188-141A ALE Rockwell Collins Selscan commercial air and ground units and FED-STD-1045.

The ACP combines receive scanning and selective calling under microprocessor control to monitor up to 100 preset channels for incoming ALE calls. Link quality analysis circuits measure and store signal-to-noise and bit error rate characteristics of received ALE signals for use by automatic frequency selection algorithms.

Selective calling addresses and preset channels may be programmed by the user from the ACPC front panel or from a remote ASCII terminal. All presets are stored in non-volatile memory for power-off retention.

The ACP provides frequency control of the associated ARC-190(V) HF radio in order to monitor multiple frequencies by scanning multiple preset channels chosen from a total of up to 100 stored simplex or half-duplex preset channels. Incoming ALE calls are answered automatically and the calling station's address is displayed to the user. Positive squelch is automatically broken whenever contact is established in response to an incoming call or as a result of an outgoing call. The system also provides a standard selective calling (SELCAL) capability when used with the AN/ARC-190 RT-1341(V)6, RT-1341(V)7 or RT-1341(V)8 radios.

Outgoing calls can be initiated on a station-to-station or net broadcast basis.

Specifications
Dimensions:
(ACP) 198.6 × 121.9 × 495.8 mm
(ACPC) 66.5 × 146 × 106.7 mm
Weight:
(ACP) 9.53 kg
(ACPC) 1.81 kg
Power supply: 115 V AC, 400 Hz, 110 W, 28 V DC, 25 W

Operational status
In service on C-5, C-20, C-25, C-27, C-130, C-141, KC-10 and VC-135 aircraft.

Contractor
Rockwell Collins.

VERIFIED

Automatic Link Establishment (ALE) for HF

The 309M-1 Automatic Link Establishment (ALE) processor and 514A-12 control set operate together as a microprocessor-based remote-control subsystem which can be added to existing Rockwell Collins HF radio systems to automate and simplify operation.

The 309M-1 ALE processor combines receive scanning and selective calling under microprocessor control to monitor up to 100 preset channels for incoming ALE calls. Link quality analysis circuits measure and store signal-to-noise and bit error rate characteristics of received ALE signals for use by automatic frequency selection algorithms. The ALE processor automatically mutes received audio output from the HF radio while scanning to eliminate distracting HF background noise and irrelevant channel activity.

When an automatic call is placed, the operator selects the preset ALE address of the individual station or net to be contacted and initiates the call. Automatic channel selection algorithms choose the calling

Rockwell Collins is supplying communications, navigation and identification modules for the F-22

channel from the list of channels currently being scanned. Automatic channel selections are made according to the order in which the candidate channels are ranked. The system also features a wire data messaging capability via an MS-141A AMD internal data modem.

Channels scanned, together with the choice of which of the multiple self-addresses are valid at any time, are determined by the scan list or lists selected. Multiple scan lists may be selected simultaneously, resulting in a combined list of channels for scanning purposes. The unique flexibility provided by selectable scan lists allows the 309M-1 to participate in multiple networks simultaneously.

Specifications

Compliance: FED-STD-1045, MIL-STD-188-141A (MS-141A) ALE
Associated equipment: HF-9000, 718U/ARC-174, ARC-190, HF-80
Dimensions:
(309M-1) 198.1 × 124.5 × 320 mm
(514A-12) 66.5 × 146 × 106.7 mm
Weight:
(309M-1) 5.5 kg
(514A-12) 1.36 kg
Power supply: 28 V DC
(309M-1) 30 W
(514A-12) 25 W

Contractor

Rockwell Collins.

UPDATED

CMS-80 cockpit management system

The CMS-80 cockpit management system family consists of hardware and software building blocks. The baseline CMS-80 consists of a full complement of Rockwell Collins avionics plus selected equipments from other manufacturers.

The CMS-80 unclutters the cockpit by removing individual controls which normally crowd the panel. These are replaced with centrally located control and display units which use a standard screen layout and human interface to control all equipment. System interconnection is via dual-redundant MIL-STD-1553B multiplex cables. The CMS-80 can also provide mission computing and navigation integration.

A CMS-80 option is weapons and sight integration and control. This feature has been implemented on the B-406CS Combat Scout demonstrator and the MD-530 NOTAR helicopter. A single keystroke shifts the CDU from avionics to weapons control and all options for guns, rockets and missiles can be selected and managed. Weapons selection and firing can also be accomplished from the handgrips, allowing the pilot to keep his attention focused outside the cockpit.

Various versions of the system have been chosen for the Sikorsky HH-65A and HU-25A helicopters and Lockheed C-130 for the US Coast Guard. It has also been selected for the US Air Force Fairchild A-10A Thunderbolt aircraft.

In December 1981, Delco Electronics ordered the CMS-80 under a US$15 million contract to support its commitment to provide 300 sets of Fuel Saving Advisory System (FSAS) equipment to the US Air Force, part of a programme to upgrade the fleet of Boeing KC-135s. Deliveries for this application began in January 1983 and ended in 1986. The FSAS system on the KC-135 is expected to save between 2 and 4 per cent of the fuel used by advising the crew of the most efficient speed, engine pressure ratio, altitude and descent profile.

In 1988, the system provided the baseline for the next-generation FMS-800 flight management system for the German Air Force C-160 Transall autonomous navigation system. In this application the CDU provides crew interface for communications, navigation and IFF, as well as flight management functions.

Operational status

In service. The US Army selected the CMS-80 for Special Operations (SO) and Special Electronic Mission Aircraft (SEMA). In addition, the CMS-80 has been chosen for the AH-64A Apache and for three US Navy/Marine Corps helicopters including the AH-1W, CH-46 and UH-1N.

Contractor

Rockwell Collins.

VERIFIED

Communications, Navigation and Identification (CNI) modules for the F-22

Under the CNI concept for the F-22, SEM-E modules are placed in an integrated avionics rack. CNI represents a modular avionics approach, as opposed to LRUs. These CNI modules perform avionics tasks that are accomplished today by individual LRUs. The programme is designed to provide reliable advanced avionics in a much smaller space than that occupied by current units.

Modules developed by Rockwell Collins include the VHF/UHF receiver, VHF/UHF transmitter, GPS receiver, GPS antenna electronics, single and five-channel L-band receiver and antenna interface unit. In addition to hardware modules, Rockwell Collins is also supplying the software for GPS, VHF/UHF and Have Quick II.

Contractor

Rockwell Collins.

VERIFIED

CP-1516/ASQ (ATHS) and CP-2228/ASQ (ATHS II) Automatic Target Hand-off Systems

Rockwell Collins' Tactical Data Manager (TDM) family of battlefield mission computers is used in conjunction with standard communication transceivers to provide a digital communications network. The C³I network enables command and firing element crews to manage resources and exchange target and other mission essential information, using a short data burst rather than voice communications. Data-burst transmissions minimise the possibility of jamming and lessen the probability of detection, while increasing the transfer rate of accurate battle information.

The TDM units are designed for use with the Rockwell Collins' CMS-80 and FMS-800 Avionics Flight Management Systems, which are capable of integrating with complete avionics systems, including COMM/NAV and weapons control.

CP-1516/ASQ Automatic Target Hand-off System (ATHS)

The CP-1516/ASQ Automatic Target Hand-off System (ATHS) is a battlefield mission management system. It is used in conjunction with a control and display unit and up to four standard HF, VHF or UHF radios to provide a tactical Command Control, Communications and Information (C³I) network. The digital communication network can provide for stores management, target handovers and other similar functions to be passed to airborne, artillery and ground forces.

The CP-1516 features a recall capability for 12 previously received messages and allows the transmitting of preformatted messages or free-text messages using an alphanumeric keyboard. Non-volatile memory in the unit retains all critical information in the event of a power loss. In addition, the CP-1516 maintains the current status of up to 10 active airborne missions and two preplanned missions.

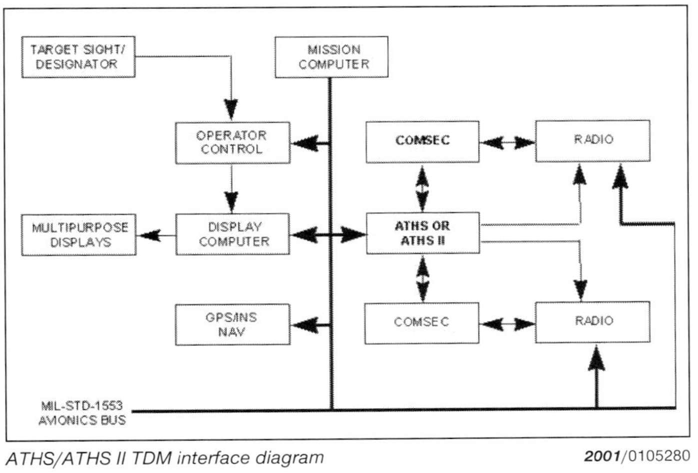

ATHS/ATHS II TDM interface diagram **2001**/0105280

The Rockwell Collins CP-1516/ASQ automatic target hand-off system and control/display unit

ATHS and ATHS II Features Comparison and Specifications

Operational features	CP-1516/ASQ (ATHS)	CP-2228/ASQ (ATHS II)
Modulation format	FSK	Same plus digital baseband/diphase, and DCT FSK tone sets
Transmission rates	75, 150, 300, 600, 1200 b/s	Same plus 5k, 8k, 9.6k, and 16 kb/s digital
Radio interfaces	4 ports, 1 modem	4 ports, 2 to 4 modems*
TEMPEST	Yes	Yes
Host vehicle interface	MIL-STD-1553B	Same plus MIL-STD-1553A
Programming language	PLM	C
Reprogramming method	Depot, via card edge	On-aircraft, via 1553 bus
Physical characteristics		
Power	35 W	17 W
Weight	4.54 kg	4.54 kg max
Size	137.2 × 167.6 × 203.2 mm	Same
Mounting	Hard mount	Same
Cooling	Convection	Same
Spare card slots	None	Four (If 4 modems implemented)

Various control/display unit options are available for data entry and display. The CP-1516 is fully compatible with the AH-64 Apache data entry panel and TADS/PNVS display, the Bell OH-58D control/display and mast-mounted sight display and combat helicopter control/display unit.

Modern electronic battlefield systems including SINCGARS, E-PLRS/JTIDS hybrid (PJH), Tacfire communications, COMSEC and all MIL-STD-1553 avionics, including digitally generated map displays, are completely compatible with the CP-1516.

The computer within the CP-1516 incorporates 8 kbytes of RAM, 2 kbytes of EPROM and 196 kwords of program memory. The system is compatible with MIL-STD-1553A and B databusses.

CP-2228/ASQ Tactical Data Modem (ATHS II)

The CP-2228/ASQ Tactical Data Modem (ATHS II, also known as the TDM-200) is an upgraded and improved version of the CP-1516/ASQ ATHS. ATHS II was developed to provide additional capabilities to meet the more stringent environmental requirements of fighter and close air support aircraft, while also meeting the needs for future datalink applications.

ATHS II is capable of transmitting and receiving FSK from baud rates of 75 to 1,220 and digital data from 75 to 16,000 bits/s. This higher frequency operation dramatically reduces transmission time, thus making it more difficult to detect and jam. It has four ports and up to four modems, which can simultaneously transmit or receive messages.

The computer within ATHS II is a Z80180 with 256 kwords program memory with potential growth to 1 Mwords. The computer also incorporates 64 kbyte RAM and 8 kbyte non-volatile memory. Mission data and operational flight program data may be programmed via the MIL-STD-1553 databus or a digital data loader.

Over 750 ATHSs are installed on various platforms which include the AH-64, OH-58D, JOH-58, ANG F-16, MH-60 A/K, MH-47 D/ E, RAN SH-70B and Belgium Aeromobility A-109 aircraft.

The newest member of the TDM family, the ATHS II entered production in 1997. It is fit and form backward compatible with earlier CP-1516/ASQ models.

*ATHS II standard configuration consists of two modems. Up to two additional modems can be added in prewired card slots.

Operational status

ATHS is in production and in service on a wide variety of US and foreign aircraft.

Contractor

Rockwell Collins.

UPDATED

DLM-700B datalink system

The DLM-700B datalink management unit is designed in accordance with ARINC 724B for all analogue and digital aircraft, including the Airbus A320, Boeing-747-400 and MD-11. The unit is part of the Rockwell Collins Airline Communications And Reporting Systems (ACARS) product line. Growth capacity is provided for additional memory or ARINC 429 interfaces. Compatibility with the Collins DL-700 and DL-500 systems can be added.

Message decoding in the DLM-700B is enhanced through the use of a Collins MSK demodulator which removes distortion from the VHF communications transceiver. A frequency management function provides automatic frequency changes without any pilot action. A channel management function, for use with Category B service providers, allows the DLM-700B to select the optimum ground station. This automatically provides fast message response time, as well as reduced message congestion in ground networks.

A communications statistics package monitors air-to-ground link performance; data is gathered on every flight leg. A message store and forward function allows flight plans, weather and ATIS information to be held for display upon demand. A flight profile report provides the crew with details on departure and arrival times, cruise times, engine data and other items. Custom data, such as frequency tables, delay codes and pilot alterable messages, can be stored in the system. Data can be loaded on the bench or in the aircraft. An optional floating point maths package is available for data processing. Using this package, complex engine reports from any analogue ARINC 717 or 573 equipped aircraft can be prepared.

Specifications

Dimensions: 125 × 193 × 325 mm
Weight: 5 kg
Power supply: 115 V AC, 400 Hz, single phase, 80 W optional 28 V DC battery, 75 mA max
Temperature range: −20 to +70°C
Altitude: up to 55,000 ft

Operational status

More than 2,300 units in service.

Contractor

Rockwell Collins.

VERIFIED

FCS-110 autostabilisation system for AH-64

The FCS-110 is a dual-redundant autostabilisation system for the US Army AH-64 Apache helicopter. The system comprises two rate gyros and two analogue computer units. Airspeed sensors and the gyros provide inputs to the computers which calculate the appropriate stabiliser angle for a mission configuration, taking account of longitudinal stability demands. Other features include pitch axis control augmentation and trim features which reduce pilot workload and improve control characteristics during low-altitude high-speed combat operations.

Specifications

Dimensions:
(one unit) 178 × 203 × 279 mm
Weight:
(one unit) 3.75 kg
Power: 40 W per unit

Operational status

In service on the AH-64 helicopter.

Contractor

Rockwell Collins.

VERIFIED

FMS-800 flight management system

The FMS-800 flight management system integrates the functions of communications, navigation and IFF control, GPS/INS navigation, flight instruments and controls, autopilot, stores and radar. It is intended for use in transport, tanker, trainer and utility aircraft. The system automates many of the functions normally carried out by the navigator. It also simplifies the complex tasks of the pilot and co-pilot, permitting them to concentrate on mission planning and execution.

The FMS-800 outputs dynamic data to the flight instruments and automatic flight control system using MIL-STD-1553B, ARINC 429/561 or analogue synchro signals. The system is compatible with existing mechanical flight instruments and analogue autopilots as well as digital systems such as the Rockwell Collins CDU-900 Control Display Unit, FDS-255 flight display system and APS-85 autopilot. The FMS-800 integrates GPS/INS navigation using a 12 state Kalman filter for airborne alignment and continued high accuracy. GPS and INS stand-alone navigation are also provided.

Workload is reduced and flight accuracy is improved by the automated processing features of the FMS-800. These features include tactical airdrops, intercepts, raster towlines, orbit/rendezvous, search patterns, VNav, FMS non-precision approach, coupled flight director/autopilot guidance and speed commands for precise time of arrival. The FMS-800 also includes embedded dynamic simulation software for ground mission rehearsal and training.

The standard system consists of dual-control display units, dual mission computers which also provide for the interface of non-MIL-STD-1553B avionics, dual-remote readout units and a data loader.

The FMS-800 flight management system is supported by a Windows 95 Mission Planning Station (MPS), which inputs data to the aircraft system via the DR-200 Airborne PC Card Receptacle. The system allows for integration of aircraft mission planning data with data from Jeppesen and Aeronautical Flight Information File (DAFIF) sources.

Operational status

The FMS-800 is operational in many international aircraft, including the German C-160 Transall and Canadian and Royal Danish Air Force C-130 aircraft. US

The FMS-800 flight management system is fitted in the E-8C Joint STARS aircraft

Air Force and US Army platforms using the FMS-800 include the B-1, C-9, C-12, E-3, E-5, E-8, KC-10 and KC-135 (PACER CRAG) aircraft. The FMS-800 has received FAA TSO approvals for installation on US Army C-12 and US Air Force C-9 and KC-10 aircraft.

Contractor

Rockwell Collins.

VERIFIED

GEM II/III/IV embedded GPS modules

The GPS Embedded Module (GEM) family represents a flexible approach to modular GPS. By offering a family of modules that share the same form factor and interface, the user is offered the ability to select various GPS performance parameters without the added expense of additional integration. To simplify the basic software integration, the GEM family's primary interface is via Dual Port RAM utilising the ICD-GPS-059 format. In addition, HAVE QUICK, KYK-13, RS-422 and 1 pulse/s input/output are also included as standard interfaces.

The GEM series offers full P/Y code performance in a standard module; current applications include manned and unmanned airborne platforms, ground vehicles and missiles. The ability to operate either as a stand-alone GPS receiver or integrated with Inertial or Doppler sensors, Mission Computers and Flight Management systems provides maximum flexibility for platform installation.

GEM II

The GEM II receiver offers full military performance to Miniature Airborne GPS Receiver (MAGR) specifications with an IF interface for retrofit of RCVR 3A, RCVR UH and RCVR OH installations.

GEM III

The GEM III receiver also offers full military performance to MAGR specifications (in the PPS configuration) with an RF interface direct to an antenna. This receiver also supports an RF preamplifier for longer cable runs while still utilising a single cable interface to the antenna.

GEM IV

The GEM IV was developed as a System Specification Equivalent (SSE) replacement for the GEM III and is form, fit and functionally compatible. GEM IV is designed to be a backward compatible replacement of Rockwell Collins' other GEM series modules and allows for an evolutionary path to GPS Receiver Application Module (GRAM) performance. Incorporating the Nighthawk Digital Signal Processor, GEM IV is a 12 channel receiver with a five channel interface, coupled with a dual-frequency L1/L2, P(Y)-code GPS receiver with Selective Availability and Anti-Spoofing (SAAS) capabilities.

Operational status

The system was selected for the US DoD's first embedded GPS and inertial programme (GINA) for the Navy T-45 trainer aircraft, as well as for the tri-service Embedded GPS/IVS (EGI) programme. The GEM IV now offers a Fast Acquisition Direct Y Capability, which utilises Rockwell Collins' High Accuracy, Low Power Time Source (HAL) and the Acquisition Correlation Engine (ACE). The ACE Digital Signal Processor interfaces with the Nighthawk ACE Digital Signal Processor and provides a large correlation bank in order to enhance the acquisition. The HAL device improves the powered-down time-keeping accuracy by more than two orders of magnitude. Combined, these two components provide superior Direct Y acquisition performance to counter the vulnerabilities of C/A code threats. Planned Enhancements to GEM IV include the following:

- all in view tracking and navigation
- DO-229 RAIM/FDE
- Carrier Phase measurements
- Standard Positioning Service (SPS)
- ICD-GPS-155 compatibility

The GEM family offers highly accurate position, velocity and time information over the full range of dynamics encountered in manned and unmanned airborne, ground and munitions applications. GEM receivers are standard fit in several inertial navigation systems as well as Flight Management (FM) and Mission Computer (MC) systems. Due to market place demands and parts obsolescence, Rockwell Collins issued a Last Time Buy (LTB) notification on the GEM II and GEM III, accepting orders up to 1 October 2000 for deliveries to be completed by 30 September 2001.

Contractor

Rockwell Collins.

UPDATED

Specifications

Performance

	C/A code without SA	C/A code with SA
Position accuracy	10 m horizontal	100 m horizontal
Velocity accuracy 3D	0.2 m/s	0.7 m/s
Time accuracy	70 ns	270 ns

Dynamic conditions: 1,200 m/s, (min) satellite signal power level, and 31 dB jammer signal level. SA degradation is assumed at the current nominal SPS level of 100 m.

Physical

	GEM II	GEM III	GEM IV
Number of channels	5	5	12
P/Y code capable	Yes	Yes	Yes
Frequency	L1/L2	L1/L2	L1/L2
Antenna interface	IF	RF	RF
Power (typical), W	8.9	6.4	6.5
Input	±5 V, -15, +21.4 V DC	±5 V DC	+5 V DC
Temp range	–54 to +85°C	–54 to +85°C	–54 to +85°C
DPRAM I/0	ICD-059	ICD-059	ICD-059
Dimensions, mm	145.35 × 144.78 × 14.73	145.35 × 144.78 × 14.73	145.35 × 144.78 × 14.73
Weight	0.45 kg	0.45 kg	0.45 kg

HFCS-800 helicopter flight control system

The HFCS-800 helicopter flight control system can be configured for single- or dual-pilot operation and offers what the company calls total mission IFR capability. Operating modes, covering operations from start-up to shutdown, have been incorporated, with special emphasis given to autopilot assistance and control at low speeds and in hover.

Two separate subsystems are used, one for Automatic Flight Control Systems (AFCS) and the other for Flight Director Systems (FDS). The AFCS provides automatic stabilisation and control, while the FDS computes automatic path steering for any desired manoeuvre. The subsystems may be purchased jointly or separately and they meet all the requirements of FAA TSO C9A and C52A.

The following list of operating modes provides an indication of the system's performance:

Airspeed hold: maintains indicated airspeed; the pilot may 'beep' airspeed after initial selection. May be used if airspeed is above 35 kt.
Vertical speed hold: holds barometric vertical speed if airspeed is above 60 kt.
Altitude hold: holds barometric altitude if airspeed is above 60 kt.
Heading select: captures selected HSI heading if airspeed is above 35 kt.
Navigation: captures and tracks selected VOR, Loc, Doppler, Omega or RNav course if airspeed is above 35 kt.
Approach: captures and tracks course, providing three-dimensional control from VOR, ILS, Loc (back course), MLS or RNav sensors. May be used above 35 kt airspeed and up to 12° glide slope.
Airspeed/vertical speed: holds both parameters if airspeed is above 35 kt.
Navigation transfer: in two-pilot operation, transfers control to co-pilot's side.
Hover augmentation: holds hover position and radio altitude.
Transition to hover: decelerates aircraft to hover at 50 ft radio altitude.
Go-around: accelerates to climb at 70 kt. May be used after take-off for departure.

Specifications

Weights:

Automatic flight control system
(AFCS computer) 6.43 kg
(AFCS panel) 1.5 kg
(servos) 0.96 kg each (three or four installed)
(feel/trim units) 1.36 kg each (two installed)
(yaw servo) 2.1 kg
(collective servo) 2.1 kg
(vertical gyros) 3.1 kg each (two installed)
(airspeed sensor) 0.5 kg
(yaw rate gyro) 0.9 kg

Flight director system
(flight director computer) 4.9 kg
(flight director panel) 0.6 kg
(attitude control) 1.5.kg
(attitude director indicator) 3.1 kg each (two installed)

Operational status

In production. The system has been selected by the US Coast Guard for installation in the Eurocopter HH-65A (AS 336G1) Dolphin helicopter.

Contractor

Rockwell Collins.

VERIFIED

ICS-150 intercommunications set

The Rockwell Collins ICS-150 intercommunications set is a fully militarised aircraft audio system which provides selectable channels of communications between aircraft crew stations. It also provides communications between each crew station and various transceivers, receivers and warning systems.

This set is particularly applicable for aircraft with multiple crew stations, a variety of communication, navigation and warning receivers or transceivers and stringent requirements for cross-talk isolation, electromagnetic interference and nuclear hardening.

The set delivered to the US Air Force for its B-1B aircraft provides an audio interface for up to eight crew stations, five ground crew/maintenance stations, eight separate avionics transceivers and 10 receivers.

An InterCommunication Set (ICS) is made up of one central control unit, up to eight crew station units and up to five maintenance station units.

Each crew station unit provides 10 receive-monitor control functions such as on/off and volume and transmit selection control of ICS, plus up to six other transmit functions. The crew station also provides master volume, hot mic, all call and LRU test facilities.

The central control unit is equipped with secure interlock capability to prevent secure communications from being heard on non-secure transmissions.

This system has high channel isolation and provision has been made for the future incorporation of a COMSEC switch capability. The COMSEC switch will permit a single speech encryption device to be switched between several radios.

The ICS-150 was designed to meet all military requirements, from parts utilisation to qualification testing.

A central mixing architecture requires very few interconnect lines between each crew station and the central control unit. Three twisted/shielded pairs of wires are used for microphone audio, headset audio and serial control data.

The ICS-150 central control unit has redundant input regulators and separate line regulators in each module to prevent any single point failure from causing system failure. An additional back-up mode is provided in case there is structural damage to the central control unit. This back-up mode provides for a direct connection between the pilot and co-pilot and two separate transceivers.

Specifications

Dimensions:
(central control unit) 124 × 193 × 497 mm
(crew station unit) 146 × 95 × 112 mm
(maintenance station unit) 117 × 91 × 99 mm
Weight:
(central control unit) 8.3 kg
(crew station unit) 0.9 kg
(maintenance station unit) 0.23 kg
Power supply: 28 V DC

Operational status

In production and in service. Applications include US Air Force B-1B, B-2 and C-135C, US Army Special Operations Force aircraft and Royal Australian Navy helicopters.

Contractor

Rockwell Collins.

VERIFIED

KC-135 Global Air Traffic Management (GATM)

In October 1999, the US Air Force selected Rockwell Collins to upgrade 544 C/KC-135 aircraft to meet GATM requirements. Rockwell Collins' solution is based on the Collins Flight2 System. This open-architecture system improves situational awareness by integrating flight operations with navigation and guidance functions. It also provides the interactive services required for the future GATM environment.

The upgrade will provide a COTS solution based on the PACER CRAG avionics baseline, while ensuring unrestricted access to civil airspace and providing enhanced military mission functionality. Rockwell Collins avionics supporting the GATM programme include: the CMU-800 for datalinks; the SAT-2000 Satcom radio; datalink modifications to the AN/ARC-190 HF and AN/ARC-210 V/UHF radios; and the GNLU-945 MultiMode Receiver (MMR).

Operational status

The initial contract award for US$39.2 million is for the Engineering Manufacturing Development (EMD) phase. This phase includes design, development, integration and testing of the completed GATM suite, which will comprise the Collins Flight2 System and is scheduled for completion in 2002. A follow-on option for eight prototype A and B modification kits was issued in December 1999, for completion by the end of August 2003. The installation of 540 production kits and 20 simulators is planned, for installation by 2008.

Raytheon Electronic Systems will perform aircraft installations at its facility in Greenville, Texas.

Contractor

Rockwell Collins.

VERIFIED

MAGR receivers

The MAGR military GPS receivers provide the user with a wide variety of input, selective availability/anti-spoofing and power supply options. All of these receivers provide MIL-STD-1553, RS-422, ARINC 429 and Have Quick interfaces.

The ability to process selective availability/anti-spoofing is controlled by the National Security Agency and the approved Rockwell Collins precise position service-security module allows selected users to process GPS crypto-variables. These crypto-variables give the user the maximum accuracy possible at all times.

The Miniaturised Airborne GPS Receiver (MAGR) is a five-channel receiver that fits within a 3/8 ATR form factor. It provides high levels of GPS performance and excellent anti-jamming capabilities on high-dynamic aircraft, including fighters and bombers. MAGR is designed to be a stand-alone GPS receiver which can operate with a MIL-STD-1553 or ARINC 429 interface or with a stand-alone control display unit. It is offered in versions that operate on 115 V AC or 28 V DC. The receiver can also be ordered with an intermediate frequency or radio frequency interface.

MAGR provides the maximum levels of navigation accuracy in the highest levels of jamming and dynamics. Operating in the Precise Positioning Service (PPS) with velocity to within 0.1 m/s, MAGR also supports accurate time information for radio system co-ordination and synchronisation.

Specifications

Dimensions: 82 × 172 × 305 mm
Weight: 5.6 kg (IF); 5.5 kg (RF)
Power: 115 V/400 Hz AC or 28 V DC
Number of channels: 5
Interface: MIL-STD-1553B, ARINC 429, RS-422, Have Quick/1-PPS IF/RF, KYK-13/KOI-18
Accuracy:
(position) <15 m SEP
(velocity) <0.1 m/s
(time) <100 ns
Reliability: >5,000 h MTBF

Operational status

The MAGR is a standard US DoD airborne GPS system. It has been chosen by all four US military services.

The receiver provides full accuracy, P/Y-code GPS capability to a wide variety of platforms, including F/A-18, AV-8B, F-16, F-117A, B-2 and S-3B. MAGR is fitted to several international aircraft, as well as selected special use platforms. It can accommodate many interfaces, including SA/AS with PPS-SM and can be integrated with existing INS and Doppler systems.

In November 1998, the US Air Force Space and Missile Command awarded the Raytheon Systems Company a US$167 million to provide a maximum of 6,700 MAGRs, for use by the US Air Force and possible foreign military sales.

As of this edition, the MAGR remains in production, but Rockwell Collins has advised that, due to the age of the design, increasing numbers of critical components are becoming non-procurable, therefore limited availability should be expected.

Contractors

Rockwell Collins.
Raytheon Systems Company.

UPDATED

AN/ARR-85 Miniature Receiver Terminal (MRT)

The Miniature Receiver Terminal (MRT) forms part of the Minimum Essential Emergency Communications Network (MEECN) that provides secure VLF links between the National Command Authority and B-52H and B-1B bombers. It includes the Datametrics quarter-page black and white printer.

Operational status

The MRT was developed for the B-52H and is in service on the B-1B.

Contractor

Rockwell Collins.

UPDATED

Modified Miniature Receiver Terminal (MRT)

The Rockwell Collins Modified Miniature Receiver Terminal (MMRT) provides a VLF/LF communication link from the National Command Authority (NCA) to the E-4B National Airborne Operations Centre and the E-6B TACAMO aircraft via the Minimum Essential Emergency Communications Network (MEECN).

The MMRT is an enhanced version of the USAF AN/ARR-85 miniature receiver terminal. It has a new High Data Rate (HIDAR) mode, improved command post operator control capabilities and advanced packaging to enhance MEECN mission performance. Like the ARR-85, the MMRT ensures automatic reception and processing of secure, long-haul Emergency Action Messages (EAM) in benign, hostile and nuclear-stressed environments.

Collins' MMRT is a common VLF/LF receive solution for USAF and USN land, sea and airborne communication platforms. It is a self-contained VLF/LF receiver/demodulator with automatic message processing and embedded cryptographic equipment that allows for on-aircraft updates. It incorporates high-performance, adaptive signal processing, a wide dynamic range front end, programmable operational modes and high-reliability advanced modular design and packaging.

The main benefits of the MMRT and improvements over the MRT are:

- Global interoperability for all MEECN/VERDIN command post platforms
- Enhanced EAM processing
- Improved anti-jam performance
- Flexible programmability to meet changing mission roles
- Affordable upgrading and capabilities expansion
- Increased operator control and management
- Lower life-cycle support costs
- Improved reliability and lower mean time to repair.

The Collins MMRT provides high-performance EAM reception to globally deployed strategic forces in support of post-cold war scenarios. The architecture includes digital signal processing enhancements to provide high data rate reception and increased anti-jam capabilities. A MIL-STD-1553B interface provides real-time system control. The MMRT uses Ada software and has built-in flexibility for future enhancements.

The main features of the system are:

- Three-channel or single-channel operation
- Interoperability with MEECN modes 15, 9 and 9 MMPM, HIDAR and VERDIN modes 22 and 23
- High dynamic range for simultaneous transmitter operation
- Three-channel TE/TM spatial diversity adaptive combining
- Anti-jam protection via multichannel null steering, NBIS and single-channel, non-linear adaptive processing
- Time diversity, three-channel reception/message combining
- Automatic dual mode search (one normal/one special)
- Frequency scanning from one to five transmitters (operator programmable), five-day mission plan
- Smaller size and lower power consumption and weight
- Low life-cycle cost through receiver hardware commonality
- Additional nuclear hardening protection.

The MMRT has optional ancillary equipment to enhance operation and adaptability for various mission roles and platforms. They include:

- An antenna coupler for platform antenna interface matching
- A radio set control for mission data set creation and receiver control/status display

Operational status

The MMRT is in service on airborne communications platforms of the US forces.

Contractor

Rockwell Collins.

NEW ENTRY

OG-187/ART-54 VLF/LF transmitter

The OG-187/ART-54 is a self-contained 200 kW VLF/LF transmitter designed to meet transmission requirements over the 17 to 60 kHz frequency range in 10 kHz increments.

The OG-187/ART-54 is an all solid-state transmitter designed to operate in the minimum shift keying, frequency shift keying, continuous shift keying and frequency shift continuous wave modes. The transmitter can be controlled locally or remotely via a standard serial bus. Automatic tuning is achieved in 10 seconds maximum. Constant surveillance tuning ensures matching antenna impedances regardless of the environmental effects after deployment.

Specifications

Dimensions: 1,874 × 1,990 × 1,450 mm
Weight: 1,545 kg
Power supply: 115/200 V AC, 400 Hz, 3 phase, 252 kVA, 28 V DC, 8 A

Contractor

Rockwell Collins.

VERIFIED

OG-188/ARC-96A VLF/LF transmitter

The OG-188/ARC-96A is a self-contained 100 kW VLF/LF transmitter designed to meet the requirements for the US Air Force World Wide Airborne Command Post mission. Operating over the 17 to 60 kHz frequency range in 10 kHz increments, it provides 100 kW to the dual trailing-wire antenna aboard the EC-135 aircraft.

The OG-188/ARC-69A is an all solid-state transmitter designed to operate in the minimum shift keying, frequency shift keying, continuous shift keying and frequency shift continuous keying modes. The transmitter can be controlled locally or remotely via a standard serial bus. Constant surveillance tuning ensures matching antenna impedances regardless of the environmental effects after deployment.

Specifications

Dimensions: 1,750 × 1,340 × 1,570 mm
Weight: 1,091 kg
Power supply: 115/200 V AC, 400 Hz, 3 phase, 119 kVA, 28 V DC, 8 A

Contractor

Rockwell Collins.

VERIFIED

PACER CRAG C/KC-135 avionics upgrade

Rockwell Collins is the prime contractor for the C/KC-135 PACER CRAG upgrade. This programme is an avionics-driven upgrade for the aircraft fleet and includes the integration and installation of the FMS-800 flight management system, FDS 255 colour flat-panel flight display system, a WXR-700X forward-looking weather radar replacement system and an embedded INS/GPS navigation system to replace the existing compass. An open systems architecture has been utilised with installed growth and functional capabilities to ensure continued interoperability with current and planned Global Air Traffic Management (GATM) requirements.

The PACER CRAG integration and upgrade is aimed at modernising the C/KC-135 cockpits and, through human factors techniques, reducing crew workload and automating many routine cockpit functions.

Rockwell Collins is prime contractor for the programme and has responsibility for design, test and installation of the flight management system, displays, weather radar and embedded INS/GPS for compass replacement.

The radar provides weather, windshear and skin paint functions. The skin paint capability allows radar detection, identification and separation maintenance of aircraft during refuelling operations.

Additionally, as part of the PACER CRAG modification, an integrated Traffic alert Collision Avoidance System (TCAS) and Enhanced Ground Proximity Warning System (EGPWS) are being installed.

Rockwell Collins is also installing a digital interphone system to replace the existing AN/AIC-10 system.

PACER CRAG flight deck **1999**/0051642

Operational status

In July 1996, Rockwell Collins delivered the first C/KC-135 PACER CRAG (Compass, Radar And GPS) aircraft with avionics upgrade to the US Air Force, signifying the formal release of the aircraft into the Qualification Test and Evaluation (QT&E) phase of the programme. Ultimately, more than 600 C/KC-135 aircraft will undergo the upgrade.

The PACER CRAG system was declared operationally suitable in October 1997. In October 1998, the US Air Force certified that the PACER CRAG system provided adequate reduction in workload to permit safe operation without a navigator. The first production aircraft was delivered in June 1998. By 30 November 1999, the fifth production option had been awarded, bringing the total shipset quantity to 505.

Contractor

Rockwell Collins.

VERIFIED

RT-1379A/ASW transmitter/ receiver/processor

The RT-1379A/ASW is an AN/ARC-182 derivative design providing a 5 kbits/s half-duplex or simplex RF datalink using the TADIL-C message protocol and modulation. As currently configured, the radio covers the 225 to 400 MHz UHF band with 25 kHz channel spacing and is compatible with a number of US Navy data systems. These include the naval tactical data system, airborne tactical data system, AN/SPN 10/42 automatic carrier landing system, AN/TPQ-10/27 precise course direction system and the inertial navigation system.

The RT-1379A/ASW interfaces with the mission computer on either of two (redundant) MIL-STD-1553B multiplex buses. Jumpers in the aircraft wiring harness determine the unique multiplex address assigned to the radio.

TADIL-C address assignment is via five octal encoded switches under a front protective cover. The radio's address is normally selected on the flight line before a mission. The last three (least significant) octal address digits can be changed by commands from the mission computer at any time, causing the radio to assume a new TADIL-C address.

Among the types of information which can be handled are two-way transfer of target information, aircraft vectoring data, INS update data, landing system data and general data reporting of aircraft status.

In general, any data which is available on the aircraft multiplex bus can be transmitted by the radio. It can be modified to communicate using a message protocol other than TADIL-C format and can accommodate other data rates up to 16 kbits/s. The radio can also be made to operate on any channelised frequency between 30 and 400 MHz, and voice communications capability can be added.

Specifications

Dimensions: 135.9 × 127 × 270.5 mm
Weight: 4.9 kg

Operational status

The RT-1379A/ASW radio is in operational service on US Navy F/A-18 Hornet aircraft.

Contractor

Rockwell Collins.

VERIFIED

Tactical Data System (TDS)

The Rockwell Collins Tactical Data System (TDS) is installed on the Royal Australian Navy's Role Adaptable Weapons System (RAN/RAWS), the Sikorsky S-70B-2 Seahawk helicopter. With this system, the Seahawk can perform ASW missions with three crew members, do all sensor processing on board the aircraft, locate, classify and prosecute targets autonomously and transfer target data automatically via digital datalink.

The TDS consists of a tactical display unit, a horizontal situation video display, a multifunction keyboard for communication navigation and identification management, an advanced mission adaptable computer, a data loader that simplifies the preflight mission preparation and a datalink to transfer target and mission data.

The S-70B-2 with the TDS has the capability to assist the crew in making tactical decisions autonomously and each crew member has access to virtually all tactical data. That implies greater situational awareness, so that the entire mission can be flown by three crewmen. System functions include navigation, flight plan management, datalink, tactical display and databus controller.

The tactical display unit is a multifunction display which provides an integrated pictorial representation of processed sensor information to the tactical co-ordinator and sensor operator. As it is raster driven, there is no limit to the amount of information that can be displayed. Two identical, yet independent, 8 × 8 in (203 × 203 mm) CRTs display standard naval combat data system symbology and various special display icons in the presentation of the tactical plot. Crew members use multifunction slew controllers to select symbols to provide access to the various display functions and to designate displayed data to the system. There are no complicated or extensive sets of

The Rockwell Collins Tactical Data System (TDS) is in service on the Royal Australian Navy's Sikorsky S-70B-2 Seahawk helicopters

buttons and switches to master and during a mission, the crew can concentrate on the tactical display, not on managing a host of individual subsystems.

A powerful mission adaptive computer links all the aircraft sensors and displays via a MIL-STD-1553B multiplex databus management system. A very efficient functionally redundant integrated design places specific emphasis on maintaining a consistently high mission success rate through graceful degradation.

Operational status

In service on the Royal Australian Navy Sikorsky S-70B-2 Seahawk helicopter.

Contractor

Rockwell Collins.

VERIFIED

TACAMO II communications system

The TAke Charge And Move Out (TACAMO) II system provides airborne VLF communications links with the US Navy strategic submarine fleet. The system is a manned communications relay link to strategic forces, normally passing messages one way from the national command to submarines and other strategic forces.

At present, a complete communications centre in the Boeing E-6A TACAMO II aircraft allows simultaneous receive and transmit throughout the frequency range VLF to UHF. The system receives multiple frequency low-level signals, while simultaneously transmitting at high power in a stressed environment. The VLF power amplifier provides amplification of the signal to 200 kW power and automatic tuning of the signal to the dual trailing-wire antenna system. This latter consists of two antennas, one nearly 1,500 m long and the other more than 8,500 m. Only the short wire is charged, the energy reradiating off the longer wire, the length of which varies with the frequency in use. The transmitted signal to the submarine is vertically polarised, with the E-6A aircraft flying in a continuous tight turn. This allows most of the antenna system to hang vertically from the aircraft.

Operational status

In service on E-6A TACAMO II aircraft of the US Navy.

Contractor

Rockwell Collins.

VERIFIED

UST-107 C2W system

The UST-107 is Rockwell Collins' latest product for the C2W market. The UST-107 is a reliable, versatile, low-cost, off-the-shelf CCW system, featuring a user-friendly interface and is PC compatible. Key features of the system are:

- Digital Signal Processing for virtually unlimited modulation/demodulation and jamming waveform modes
- Direct Conversion Technology eliminates spurious signal mechanisms and excess circuitry associated with dual and triple conversion schemes
- Windows-based operating system provides user-friendly graphical user interface
- A wide frequency range of 5 kHz to 2000 MHz supports all missions with only the addition of application-specific power amplifiers and antennas
- Low-cost dedicated receiver and exciter modules facilitate affordable upgrades to performance by adding additional modules
- System supports multiple operational uses - communications, surveillance (scanning and data collection) and active jamming
- Built-In Test (BIT) capability verifies performance and identifies failed modules quickly
- Systems with multiple receivers and exciters permit failed modules to be 'bypassed' while good modules take over, thus eliminating single-point failures
- Using the external transmit RF switch matrix, multiple exciters may be routed to a single power amplifier if mission needs require it
- Modular software is field upgradeable.

The UST-107 C2W system provides reliable, low-cost Electronic Support (ES), Electronic Attack (EA) and radio communications in the 0.5 to 2000 MHz frequency range. It is intended for installations including fixed station, shipboard, mobile and airborne. The system can operate autonomously or can be operated from the provided Windows control program from RS-232, Ethernet, or MIL-STD-1553B serial interfaces.

Functional Description of the UST-107

The UST-107 C2W system consists of multiple surveillance receivers and jamming exciters under control of a user-friendly man-machine interface (Windows NT™) running on a laptop computer. The receivers and exciters are installed in a ruggedised VME chassis which permits the equipment configuration (number of receivers and exciters) to be easily tailored to the particular mission needs and upgraded as new technology becomes available. Scanning algorithms are contained in flash memory on the VME CPU card and are initiated and monitored by the Windows program.

Modes of Operation

Communications Mode

In the communications mode, the UST-107 performs receive and transmit functions for voice and data traffic between two or more stations. The following capabilities are provided:

- Simplex, half-duplex, full-duplex operation
- Single-channel or 2-channel (ISB)
- Voice, data (in selectable bandwidths)
- Compatible with encryption devices such as Voice Privacy™ and ANDVT
- Compatible with single- and multitone data modems such as the MDM-3001
- AM, FM, SSB, ISB modulation/demodulation modes
- Other modulation/demodulation modes available as options
- Manual, channelised operation
- Optional ALE operation
- Fixed frequency or frequency hopping.

Surveillance mode

The UST-107 provides ES modes for detection and verification of target emitters.

- Acquisition mode - the system detects signals at widely varying energy levels while avoiding false responses due to noise
- Scan mode - the system records the time on and time off for each emitter and records a sample of the received audio.

Reactive jamming mode

The UST-107 system employs the following jamming algorithms and features that allow it to effectively deny many potential target networks:

- Programmable jammer acquisition, timing, waveforms and other system parameters for each target frequency or frequency band
- Optimisation prevents excessive jamming of one target when there are other targets that could also be jammed
- The Collins system also allows target signals to be 'prioritised', causing lower-priority targets to be jammed only if high-priority targets are inactive.

Pre-emptive jamming mode

The UST-107 provides a pre-emptive (blind) jamming mode, which is designed to jam weak signals or to deny an entire frequency band by time multiplexing the jamming signal.

Operational status

In production.

Contractor

Rockwell Collins.

UPDATED

Specifications

Electronic Surveillance Modes

Frequency range:	.005 to 2,000 MHz
Frequency resolution:	1 Hz
Tuning time:	500 μ s (typical), with AGC disabled and VME control
Frequency stability:	± 0.1 PPM, using electronic adjustment of the internal standard
Frequency accuracy:	± 1 PPM, 0 to 50°C
Noise Figure:	
0 to 30 MHz antenna	14 dB (typical)
20 to 1,000 MHz antenna	12 dB (typical)
Modulation modes:	CW, SSB, ISB, AM, FM
Demodulation modes:	CW, SSB, ISB, AM, FM, I/Q

Electronic attack

Frequency range:	.005 to 1,000 MHz
Jamming modulation:	programmable wave file
Amplifier output power:	100 W
Frequency switching:	500 μ sec.
Elementary jamming duration:	selectable from 1 ms to 2 min
Jamming between look through:	selectable from 1 ms to 2 min
Look through duration:	< 1 ms for two (2) antenna systems

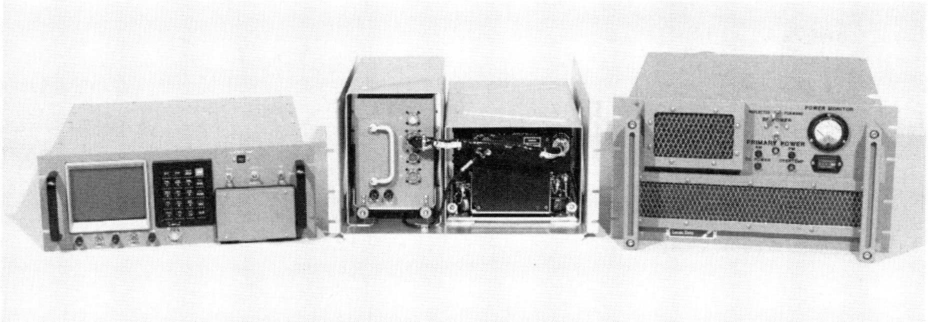

The Rockwell Collins UST-104 radio countermeasures set

UST radio countermeasures sets

The Rockwell Collins UST radio countermeasures equipment provides reliable HF, VHF and UHF ESM and ECM over the frequency range 5 kHz to 2,500 MHz. It is designed for fixed station, shipboard, mobile and airborne installations.

The range of UST equipment includes:

UST-103, for large aircraft, covering the frequency range 225 to 400 MHz with a power output of 100 W.

UST-104A, for large aircraft and ships, covering the frequency range 20 to 500 MHz with a power output of 400 W.

UST-104B, for tactical aircraft, covering the frequency range 100 to 500 MHz with a power output of 400 W.

UST-105, supporting multiple operational uses, covering the frequency range 2 to 30 MHz

USQ-146, also supporting multiple uses, including communications, surveillance and jamming, covering the range 20 to 2,500 MHz (see separate entry).

UST-107, covering the frequency band 0.005 to 2,000 MHz, is designed for fixed station, shipboard, mobile and airborne installation. The system can operate autonomously, or be operated from a PC using a Windows™ control programme provided (see separate entry).

Lancer is a spectrum surveillance and jamming system covering the frequency band 0.1 to 2,000 MHz.

Rockwell Collins' radio countermeasures sets use the 95V-1 Direct Signal Processing (DSP) receiver.

Operational status

The UST-103, UST-104A, UST-104B and UST-105 are in service but no longer in production. The superseding UST-107 and USQ-146 are in production.

Contractor

Rockwell Collins.

UPDATED

VLF/LF High-Power Transmit Set (HPTS)

The HPTS system consists of a Very Low Frequency/ Low Frequency (VLF/LF) 200 kW solid-state power amplifier and dual-trailing wire antenna system. It is designed to improve the reliability of systems that provide survivable communications links from the US Navy's E-6A TACAMO aircraft to the US strategic forces.

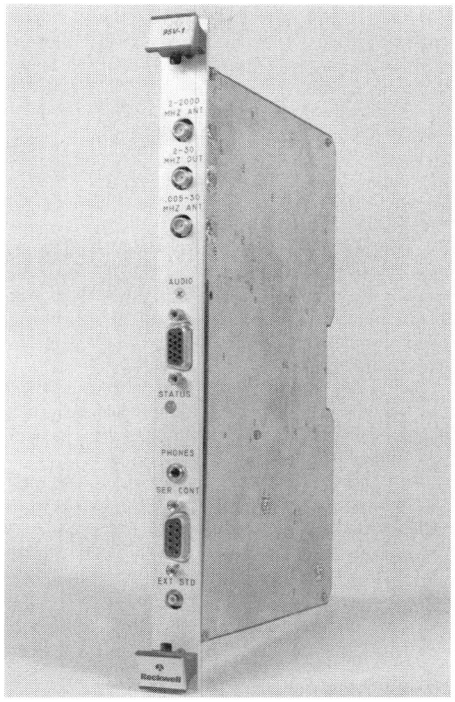

Rockwell-Collins 95V-1 DSP receiver **1999**/0054164

Operational status

In service and in production for the US Navy E-6A TACAMO aircraft.

Contractor

Rockwell Collins.

VERIFIED

AN/ALQ-167 ECCM/ECM jammer pod

The AN/ALQ-167 pod generates a wide selection of noise and deception jamming. The jamming system is modular and is based on building blocks; as a result, frequency range and performance characteristics can be tailored to a particular need. The system is designed for use on manned aircraft or aerial targets, or for laboratory applications for radar and missile system evaluation.

The AN/ALQ-167 is digitally controlled using a cockpit-mounted control box. The parameters and features that can be selected include operating bandwidth, jamming techniques, set on receiver, forward and aft radiation and power level. Up to 24 noise and deception modes are currently available.

Specifications

Dimensions: 2,300 × 3,700 mm
Weight: 140-180 kg
Output power: 100-200 W, ERP 1 kW min
Frequency: 0.85-18 GHz in customer-specified bands

Operational status

In service.

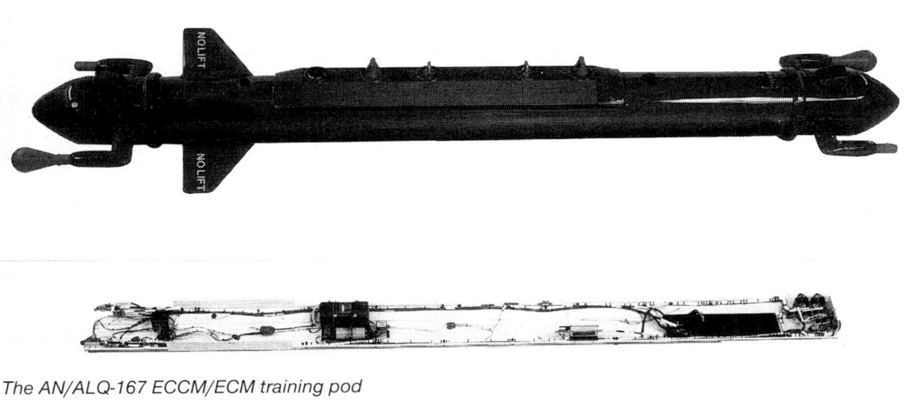

The AN/ALQ-167 ECCM/ECM training pod

Contractors

Rodale Electronics Inc.

VERIFIED

Digital Airborne Radar Threat Simulator (DARTS)

The Digital Airborne Radar Threat Simulator (DARTS) is capable of providing radar signal emission simulation of airborne search radars, targeting radars, terrain-following radars and missile seeker radars. DARTS provides for in-flight selection of emitter frequency, pulse repetition frequency, pulsewidth, stable, stagger or jitter PRF mode selection and antenna scan simulation.

DARTS is a band reconfigurable magnetron-based emitter system employing an FET modulator. This enhancement permits the selection of transmitted pulsewidth by the operator over the specified 0.1 to 2 μs range with a 0.1 μs resolution. Remote magnetron frequency tuning from the control indicator is also provided.

DARTS is housed in a standard AN/ALQ-167 pod.

Specifications

Dimensions: 3,100 × 250 mm
Weight: 147 kg
Power supply: 115 V AC, 400 Hz, 3 phase
Output power: 12.5 MW typical
Frequency: H-, I- and J-bands

Operational status

In service.

Contractors

Rodale Electronics Inc.

VERIFIED

REWTS Turbo Crow

The Turbo Crow Responsive Electronic Warfare Training System (REWTS) consists of a high-power broadband radar jammer and a high-power radar simulator combined with a sophisticated monitoring and control system. Turbo Crow is typically installed in a commercial jet (Learjet, for example) or similar military aircraft and is a dual-role system - used in both training and tactical applications.

Jamming system

The jammer provides high-power broadband jamming in user-selected bands. Jamming techniques include noise, deception, cover pulses, false targets, Doppler and combinations thereof. The system is capable of simultaneous operation in multiple bands. The jammer is modular and can be configured to suit a particular aircraft's physical characteristics. Modularity also results in significant maintenance and repair advantages. The system's operating architecture and mechanical design readily accept changes and enhancements.

Radar simulator

The radar simulator provides the Turbo Crow with the capabilities of simulating a wide variety of airborne and ground-based radars including search and tracking, missile guidance and fire control. The installation provides for simultaneous operation of jammers and radar simulator. The programmable aspects of the Digital Airborne Radar Threat Simulator (DARTS) brings a new level of sophistication to EW training and test of shipboard/ground-based ESM systems.

Monitoring and control system

The monitoring and control system is the heart of the REWTS system providing the operator with full control over the engagement scenario - whether it be training or tactical. He or she knows where the target is; can monitor the target's response to the jamming - both ECCM and manoeuvres; is able to adjust the jamming to suit the situation; and can also record any aspect of the engagement.

Optionally the Turbo Crow can be enhanced for ELINT/ESM missions with capabilities to detect, analyse and classify radar data as well as passive ranging/positioning of radars.

The system can either be carried on board or in wing-mounted pods using a hybrid configuration with the high-power amplifiers and antennas housed in pods carried on wing hard points. This configuration results in significant savings in installation costs, better aerodynamics, less onboard high-power equipment to cool, and more flexibility in aircraft utilisation.

Specifications

RF range:
(jammer) 0.85-18 GHz (3 bands)
or 0.2-18 GHz (4 bands)
(DARTS) 7.8-8.5, 8.6-9.6, 12-13.2, 14-15.2 GHz
(ELINT) 0.4-18 GHz (optional 40 GHz)
Jammer output power: 200 or 400 W typical
DARTS output power: 100 kW (min)
ELINT sensitivity: –70 dBm
DF accuracy: 3-5°
Jamming modes: noise, smart noise, deception, cover pulses, multiple false targets, Doppler, co-ordinated and combined modes
DARTS modes:
(scan) circular, steady, sector centered, sector off centre
(PRF) stable, jitter, stagger
(adjustable) pulsewidth, PRF, delay, frequency
Monitoring and control: situational awareness, measurement, control, recording and databank system
Pods: ALQ-167 or ALQ-503

Contractor

Rodale Electronics Inc.

VERIFIED

S200 self-defence jammer

S200 is a small cost-effective modular pod-mounted countermeasure system capable of generating ECM techniques with radiated power of 1 kW ERP typical. The system uses flight proven hardware from both the ERIJAMMER A100 and the Turbo Crow system.

The S200 is housed in a standard ALQ-167 pod shell and uses a standard ALQ-167 tray and proven radomes and antennas.

Specifications

Frequency range: 7.5-18 GHz
(optional) 5.3-10.5 GHz; 2.5-6 GHz; 0.85-1.4 GHz
Sensitivity:
(low duty signals) –42 dBm
(high duty signals) –42 dBm
(ERP) 1 kW typical
(antenna coverage) 2 × 110° azimuth, 40° elevation
Range deception:
(incoming pulse length) (min) 150 ns
(min delay) (max) 150 ns
(memory time) up to 10 μs or more (automatic longer memory time for pulse compression radars)
Velocity deception:
(frequency translation) up to ±127.5 kHz
(gain (including antennas)) up to 70 dB (installation dependent)
Amplitude modulation:
(swept square wave frequency) 0.01-1,000 Hz
(duty cycle) 5-95%
(sweep time) 0, 5-10 s
(modulation depth) (min) 50 dB
(input power) 200/115 V, 400 Hz, 3 phase, 28 V DC, 10A, <3 kW
Physical characteristics:
(pod length) (min) 3,200 mm
(pod diameter) 254 mm
(pod weight) (max) 180 kg

Contractor

Rodale Electronics Inc.

VERIFIED

The Rodale Smart Crow ECCM/ECM jammer is carried in the Learjet 35/36

Smart Crow ECCM/training jammer

The Smart Crow jammer is digitally controlled from a control box and a computer and is designed for an onboard EW station with jammers mounted inside the aircraft. Its modularity enables a distributed packaging configuration to fit particular limitations of the specific aircraft. Digital spot, barrage and swept noise bandwidth and velocity and range deception parameters are all digitally controlled. Frequency can be set digitally or by using the set on receiver. The system has built-in test capabilities that provide both fault isolation and calibration.

The system provides up to 25 jamming modes with all standard noise and deception techniques. A built-in set on receiver connected to antennas featuring integral amplification is used against any frequency-agile radar in the threat band. Different antenna spatial coverages can be selected. The system is modular, allowing easy installation and maintenance.

Specifications

Power supply: 115 V AC, 400 Hz, 3 phase, 5 kVA 28 V DC, 1 A
Output power: 200 W typical, 1 kW ERP (min)
Frequency: 0.5-18 GHz in customer-specified bands

Operational status

In service but no longer in production. Being replaced by REWTS Turbo Crow.

Contractor

Rodale Electronics Inc.

VERIFIED

Integrated Defensive Electronic CounterMeasures (IDECM) Radio Frequency CounterMeasures (RFCM) system AN/ALQ-214(V)

The AN/ALQ-214(V) IDECM RFCM system is a joint US Navy/US Air Force programme, lead by the US Navy. It is intended to provide a range of aircraft types with next-generation protection from RF threats. The system's primary application is the F/A-18E/F carrier-borne, multirole combat aircraft where it is integrated with the AN/ALR-67(V)3 radar warning receiver, the AN/AAR-57 Common Missile Warning System (CMWS) and the AN/ALE-55 Fibre Optic Towed Decoy (FOTD).

System design objectives include improved situational awareness through use of:

1. the Common Missile Warning System (CMWS);
2. real-time information fusion;
3. IR threat countermeasures based on the kinetic ASTE round and 'smart' dispensing routines;
4. Radio Frequency CounterMeasures (RFCM) to defeat radar threats, that incorporates features such as monopulse angle tracking, signal coherency and manual tracking techniques.

The IDECM RFCM system comprises a Techniques Generator (TG), an Independent WideBand Repeater (IWBR) and the ALE-55 Fibre Optic Towed Decoys (FOTD). Both the TG and the IWBR interface with the FOTDs through the decoy dispensing system. Electronic CounterMeasures (ECM) techniques are synthesised in the TG and transduced to optical frequencies for transmission to the FOTD via its tow line. Within the FOTD, the optical data is converted back to RF format for amplification and transmission. The IWBR provides an alternate source of ECM by passing the threat signal (as received on the host aircraft) to the FOTD.

The AN/ALQ-214(V) is understood to be capable of both independent repeater mode, and coherent digital RF memory mode operation, based on use of a coherent Digital RadioFrequency Memory (DRFM), and interchangeable RF transmitters.

The IDECM RFCM system can be tailored to customer applications by use of any combination of the elements described above. It is mandated for the US Navy F/A-18E/F and US Air Force B-1B, F-15C/E and U-2 aircraft, but could meet the future needs of a wide variety of other aircraft.

Specifications (predicted)

Total weight: 71.9 kg
Total power consumption: 1,397 W
MTBF: 600 h

Operational status

In November 1995, the contractor team of Sanders and ITT Avionics announced that Sanders (as prime) had been awarded a US$26.8 million, five-year duration, IDECM RFCM system Engineering and Manufacturing Development contract. The contractor team is required to deliver and integrate five TGs, 150 FOTDs and 100 FOTD mass models for use in development flight testing. Initial flight testing started in June 1999, with testing of the IDECM RFCM system with an operational FOTD at the Naval Air Warfare Center - Weapons Division at China Lake, California. This was part of the Engineering and Manufacturing Development (EMD) phase of the IDECM RFCM programme. The FOTD was successfully deployed and the IDECM system correctly detected, identified and radiated - via the FOTD - against prioritised threats.

The system's primary application is the US Navy F/A-18E/F aircraft. Under a contract option, work is also being done on a common, high-powered Towed Decoy System design, development and test effort for the US Air Force F-15 and B-1B aircraft, together with a B-1B architecture study to determine how the IDECM RFCM system can support the B-1B's Defensive System Upgrade Program (DSUP). Acceptance testing of the techniques generator (receiver, modulator and processor), the signal conditioning assembly and equipment rack was completed at Sanders plant in October 1998. Electronic countermeasures data generated by the on-board system is carried via fibre optic cable to the towed decoy, which converts and amplifies the signal.

Following further integration and testing at Boeing's systems integration laboratory in St Louis, the US Navy's Air Warfare Center at Point Mugu and at ITT and Sanders factories, the US Naval Air (NAVAIR) Systems Command announced its intention, in early 2000, to award Sanders a contract for the low-rate initial production of the IDECM RFCM. The requirement is expected to be for up to 80 RFCM systems and up to 340 FOTDs to support US Navy, US Air Force and foreign customer requirements. Initial deliveries for the F/A-18E/F are forecast for 2003.

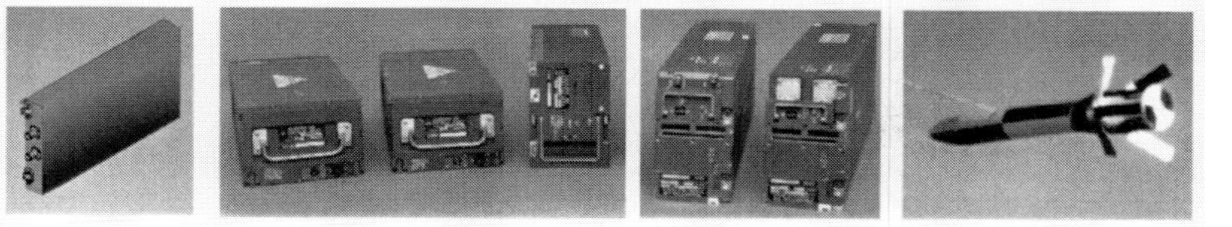

IDECM RFCM LRUs, showing, from left: the signal conditioning assembly; the receiver/processor/techniques generator; the onboard transmitters (optional); the fibre optic towed decoy 0001274

Contractors
Sanders, a Lockheed Martin Company.
ITT Defense and Electronics Avionics.

VERIFIED

Joint Strike Fighter (JSF) electronic warfare equipment

In March 1999, Lockheed Martin Tactical Aircraft Systems selected the team of Sanders and Litton Advanced Systems Division to be the electronic warfare equipment supplier for the Lockheed Martin aircraft proposal. Sanders will provide the infra-red and radio frequency countermeasures elements, while Litton (of College Park, Maryland) will be responsible for the electronic support measures aspects.

Contractors
Sanders, a Lockheed Martin Company
Litton Advanced Systems Division

VERIFIED

The Lockheed Martin JSF aircraft concept, US Air Force variant ***2000***/0079237

Auxiliary Communications, Navigation and Identification Panel for the AV-8B

The Auxiliary Communications, Navigation and Identification Panel (ACNIP) is an integral part of the communications, navigation and identification system used on the AV-8B and TAV-8B Harrier.

When interfaced with the other components of the aircraft communications system, the ACNIP performs audio amplification, control inhibit and distribution functions, generates audio warning messages in response to discrete serial and/or analogue inputs, provides code, mode, remote variable load, baseband/diphase, and control zeroing functions for two KY-58 secure communications units and controls functions of the identification system such as zeroisation and emergency operation. Additionally, the ACNIP provides logic-controlled push-to-talk switch closures and switch functions for landline telephone communications, control for ground crew communications and a hot mic capability for the operator. BIT circuitry detects 98 per cent of all electrical component failures.

The ACNIP controls a non-volatile EEPROM for the storage of the code and mode operating parameters of the secure speech units. On initial power-up, the ACNIP will update the KY-58 units to the operating modes as selected before power-down.

An LCD module, backlit and with variable illumination level located on the front panel of the ACNIP, provides a visual readout of the functional status of the KY-58 units, the operating mode, code and type of cypher used by each unit being displayed.

Operational status
In service with the AV-8B.

Contractor
SCI Systems Inc.

VERIFIED

GPS Navstar navigation systems

SCI produces two- and five-channel GPS receivers in multiple configurations; the two-channel UH receiver for helicopters, the five-channel 3A receiver for aircraft and the five-channel 3S for shipboard use. The receivers are fully qualified for operation on highly dynamic platforms in a high-jamming environment. As multichannel receivers, they are capable of processing signals from multiple satellites simultaneously, expediting initial acquisition times. The receivers are capable of providing accuracies of better than 16 m SEP in position, 0.1 m/s in velocity and 100 ns in time, even in the presence of the DoD selective availability and anti-spoofing environment.

Specifications
Dimensions:
(3A receiver) 193 × 191 × 484.8 mm
(UH receiver) 193.6 × 191.3 × 382.6 mm
Weight:
(3A receiver) 18.2 kg
(UH receiver) 11.5 kg
Power supply: 120 V AC, 400 Hz
160 V DC
Interfaces:
(3A receiver) ARINC 429 and 575, MIL-STD-1553, PTTI, KYK-13
(UH receiver) ARINC 561, 572, 575, 582, Have Quick, RS-422, KYK-13
Reliability:
(3A receiver) >1,350 h MTBF
(UH receiver) >160 h MTBF

Operational status
In service.

Contractor
SCI Systems Inc.

VERIFIED

Intercommunications set for the V-22 Osprey

The full V-22 intercommunications set consists of a Communications Switching Unit (CSU) with up to six Intercom Set Control (ISC) stations, two Audio Frequency Amplifier (AFA) assemblies, a Cabin Public Address (CPA) amplifier and four cabin PA speakers.

It provides simultaneous intercom for multiple crew stations, five channels of radio transmission and reception for each ISC, and reception of four composite navaids, five warning tones, one IFF and one radar warning receiver on two ISC panels. All audio switching and mixing functions are software-controlled and interfaces to clear and secure communications equipment are provided. There is also a digital message device I/O channel. The electroluminescent panel provides high-contrast NVG-compatible lighting.

The CSU contains the switching circuitry, logic circuit and a large portion of the audio circuitry required for ICS operation. In addition, it is the interconnect unit for the ICS system components and peripheral devices.

The intercommunications set for the V-22 Osprey

The CSU provides impedance matching, audio reporting and push-to-talk functions between the crew station ICS panels and the aircraft radios. Radio selection status is provided to the aircraft control and display subsystem through the CSU.

Each intercommunication set is essentially the individual crew station control panel, on which the frequencies, sources and signal levels are selected for monitoring and/or transmission. The audio frequency amplifier permits the selection of intercom facilities only.

The system is based on a low-power CMOS microprocessor and large-scale integration circuitry. Extensive built-in testing has been designed in for simplified maintenance. Options with the system include a digitised message card, RS-232, ARINC 429 and MIL-STD-1553 interfaces and an incandescent light panel.

Operational status

In service and in production for the V-22 Osprey.

Contractor

SCI Systems Inc.

VERIFIED

AN/APN-169/240/243/243A(V) Intraformation Positioning System (IFPS)

Sierra Research developed and manufactures the AN/APN 243 and 243A(V), a wideband datalink with multiple uses, including network datalinking and Station-Keeping Equipment (SKE). As an SKE it allows aircraft to fly fully instrumented formation flying, and to conduct aerial deliveries or instrumented approaches when using the ground-based zone marker sensor (AN/TPN 27B). The new version of Sierra's formation positioning system, AN/APN 243A(V) operates between 3,300 and 3,600 MHz. It is LPI/LPD power-programmable, uses direct sequence spread spectrum, and retains full interoperability with existing AN/APN-169, -240 and -243 systems already installed on over 750 aircraft.

The AN/APN 169, 240 and 243 SKE systems are limited to 18.5 km in-flight range and 36 participants on each of four channels. The AN/APN 243A(V) increases the internetted range to about 160 km. With the system's virtual channels, 100 or more aircraft and/or surface contacts can be accommodated in the network.

System accuracies will exceed those of the AN/APN-243 specification and not be reliant upon global positioning systems for operation, thereby assuring continued system use should GPS be denied.

Operational status

In 1970, US Air Force C-130 tactical aircraft were equipped with the original AN/APN-169; over 750 C-1, C-17, C-130, and C-141 aircraft are now equipped with the SKE system. Sierra delivered its first AN/APN 243 (SKE-2000) for the C-17 and C-130J in 1996, and is currently under contract to develop the AN/APN-243A for the C-17. The system's smaller size, lower weight, lower cost, higher MTBF, and use of military frequency band make it ideally suited for fixed-wing and rotary-wing aircraft, surface ships, vehicles, and fixed position sensors. The robust secure adaptive datalink enhanced the system for multiple uses.

Contractor

Sierra Research, a division of SierraTech, Inc.

VERIFIED

AN/ARN-136A(V) TACAN

The AN/ARN-136A(V) lightweight airborne TACAN is a remotely controlled system utilising large-scale integrated circuits and CMOS technology. The system consists of the RT-1321A/ARN-136A radio receiver/transmitter ranging unit and the CP-1398/ARN-136 azimuth computer with bearing unit. The units contain no moving parts and through exclusive use of CMOS circuits, diodes and LSI chips are rated for continuous operation at +70°.

The units meet or exceed all FAA TSO C66a requirements for high-altitude operation up to 70,000 ft. The TACAN system has been designed to work with all TACAN or VORTAC stations meeting MIL-STD-291B and FAA selection order for the US National Aviation Standard 1010.55 and ICAO Annex 10.

The AN/ARN-136A(V) (ruggedised EMI lightweight airborne TACAN) has been designed to meet ElectroMagnetic Compatibility (EMC) requirements of MIL-STD-461B Methods CE03, CS01, CS02, CS03, CS04, CS05, CS06, RE02, RS02 and RS03. In addition, RTCA Do-160B A2E1/A/MNO/XXXXXX2BABA and induced signal susceptibility category A and Z have been met, together with Shock and Random Vibration per MIL-STD-810C Method 514.2 procedure 1A and Method 516.2 procedure I30Gs. This enhancement is in use by the US Navy.

The AN/ARN-136A(V) system utilises an ID-2218/ARN-136 range indicator which displays distances up to 399 n miles, groundspeed up to 999 kt and time to TACAN station up to 99 minutes. To minimise the number of wires between the cockpit and avionics bay, where the remote boxes are installed, the AN/ARN-136A(V) TACAN incorporates a three-wire serial databus that is used for both range and tuning.

The radio receiver/transmitter utilises two LSI circuits to provide digital computations of distance, groundspeed and time to station. All tuning is performed electronically, using a digital frequency synthesiser. The unit provides 63 channels of X mode air-to-air ranging. Channels are selected by the control head (P/N7801-9000-2).

The bearing unit accepts receiver video, pulse-pair decoding information and DC power from the unit. The bearing unit derives aircraft bearing angle, with respect to local magnetic north, to or from the VORTAC or TACAN ground beacon.

The derived bearing information is then converted into several bearing data formats to enable proper interfacing with a variety of aircraft instruments. The unit is designed with two multilayer printed circuit cards, divided so that one card derives the bearing signal digitally and the second interfaces with the aircraft instruments. The bearing unit provides a bearing accuracy of ±0.5°.

Specifications

Dimensions:
(receiver/transmitter) 63.5 × 133.3 × 298.4 mm
(azimuth computer) 63.5 × 133.3 × 298.4 mm
Weight: 3.13 kg
Power supply: 11-33 V DC, 15 W

Contractor

Sierra Research, a division of SierraTech, Inc.

VERIFIED

LAMPS MK III datalink (Hawk Link)

The LAMPS MK III datalink system provides full duplex, secure and highly reliable communications between airborne and shipboard platforms. The system, better known as the Hawk Link, was designed specifically to enable the sensors and weapons of the SH-60B Seahawk, LAMPS MK III helicopter to function as integrated subsystems of the US Navy's Light Airborne MultiPurpose System (LAMPS) MK III weapon system. The Hawk Link multiplies SH-60B based processing of airborne sensor information by the power of shipboard processing capabilities through parallel operations, increasing the speed and accuracy of target/threat classification and/or localisation. Airborne sensors become, through the Hawk Link, an extension of the ship's sensor suite.

The Hawk Link system consists of two subsystems, the AN/ARQ-44 Airborne Data Terminal and AN/SRQ-4 Shipboard Data Terminal. The AN/ARQ-44 is functionally responsible for communicating SH-60B Seahawk sensor data to its parent ship over the wide data bandwidth digital downlink and receiving command and control data from the parent ship over the narrower data bandwidth digital uplink. The AN/SRQ-4 is linked with shipboard computers via an NTDS slow interface, hands-off Anti-Ship Surveillance and Targeting (ASST) sensor information (radar video and tactical plot data) to shipboard mission control consoles and hands-off Anti-Submarine Warfare (ASW) sensor information to the ship's sonar signal processing system.

Hawk Link features include: full duplex, digital communication; wide-bandwidth downlink; downlink error correction; high fade margins; up to four wide plus four narrow acoustic channels; communication security (KG-45); full Mil design with online BIT; directional (jam resistant and LPI); narrow-bandwidth uplink; uplink error detection; high-intelligibility voice com; search and track radar video; advanced sensor interfaces, interoperability; LPI upgrades under consideration.

AN/ARQ-44 Airborne Data Terminal Specification:

Multiplexer/demultiplexer, TD-1254/ARQ-44, analogue and digital sensor interfaces, 1553A interface, 18 kg.
Radio, receiver/transmitter, RT-1275/ARQ-44, FM FSK modulation, 30 kg.
Directional antenna (2 each), AS-3273/ARQ-44, Azimuth steerable, 3.9 kg each (with radome).

Operational status

The Hawk Link is operational in the US Navy's SH-60B Seahawk helicopters, and AN/SQQ-89(V) Surface ASW Combat System equipped ships including DDG-51 'Arleigh Burke' Class and DD-963 'Spruance' Class Destroyers, CG-47 'Ticonderoga' Class Cruisers, FFG-7 'Oliver Hazard Perry' Class Frigates. In addition, the Spanish Navy selected the Hawk Link to support its SH-60B Seahawks and LAMPS MK III Frigates.

Contractor

Sierra Research, a division of SierraTech, Inc.

VERIFIED

AN/ASN-162 Attitude and Heading Reference System (AHRS)

The AN/ASN-162 AHRS is currently in development as the US Air Force and Navy standard compass/attitude and heading reference system. Smiths Industries interferometric fibre optic gyros are incorporated into the system which is the first military AHRS to use these as inertial sensors.

The AN/ASN-162 will provide multiple synchro outputs of roll, pitch and heading for all aircraft attitudes. Body axis rate and acceleration data will be available via a MIL-STD-1553B databus. The inclusion of an improved Magnetic Azimuth Detector (MAD) will provide magnetic compass information to the system. The MAD and AHRS will be calibrated automatically for ease of maintenance.

The AN/ASN-162 consists of four units: the inertial reference unit, control indicator, flight data memory unit and the MAD. It is being designed for a wide range of aircraft applications. Specific aircraft interface parameters are stored in the flight data memory unit. This allows a single inertial reference unit configuration to be used in all aircraft configurations.

Operational status

In engineering development.

Contractor

Smiths Industries Aerospace.

VERIFIED

C-10382/A Communication System Control (CSC) set

The primary function of the Communication System Control (CSC) set is to provide the pilot with integrated, centralised control of data transferring capability, power switching, mode selection, operating frequencies, interconnections and signal flow routes of the aircraft's CNI equipment. The CSC provides for highly efficient operation of communications by integrating these primary CNI systems controls into the aircraft's advanced avionics architecture and also into a single convenient easy-to-operate pilot's control panel.

On the US Navy's F/A-18 aircraft, the controls and displays of the cockpit control panel are engineered to optimise pilot control of the CNI equipment. The control panel is positioned to allow the pilot to keep his eyes focused straight ahead, with only the pertinent information and controls he needs presented in his field of view.

A redundant MIL-STD-1553 multiplex bus provides connection between the CSC and the AN/AYK-14(V) mission computer for the flow of information and control. The mission computer provides CNI control signals, BIT commands and information for the control panel's alphanumeric display. In return, the CSC transmits equipment status, received CNI data, operating options and BIT response to the mission computer. Dedicated serial digital lines interface the control panel with the CSC.

To process data to and from the CSC, mission computer, control panel and CNI equipment, the CSC interfaces serial digital signals, discrete signals, analogue signals, synchro signals, avionic multiplex bus signals and audio signals. As the CSC microcomputer processes at least 1,300 parameters/s, it controls and processes the data required, leaving 40 per cent of real time available for growth.

Operational status

In service on the US Navy F/A-18 aircraft.

Contractor

Smiths Industries Aerospace.

VERIFIED

Fuel Savings Advisory System (FSAS)

The Fuel Savings Advisory System (FSAS) is intended for military transport aircraft. The system includes a Fuel Savings Computer (FSC), Control and Display Unit (CDU) and a Display Interface Control Unit (DICU). The computer has software for both the Lockheed Martin C-141 and C-5 supported in a single computer program. Identical hardware is used in both aircraft.

An aircraft wiring jumper selects aircraft type and version when the computer is installed. A Smiths Industries installation kit is used for rapidity and ease of aircraft modification. Software is written in high-level language with full documentation and support tools. The system has a specified MTBF of 2,000 operating hours.

Operational aspects include climb, cruise, descent and transition coupling to the autothrottle and autopilot. The system also provides advisory information on an alphanumeric CDU that interfaces with the aircraft's INS for enhanced navigation.

The FSAS features flight management, navigation, fuel savings, Tacan and waypoint database, take-off and landing computations, colour graphic tactical aids, moving map displays, enhanced airdrop capability, time of arrival control, altitude alerting, windshear prediction and detection and audio alert tones.

Tactical functions are provided to enhance the mission performance of the aircraft and crew. Performance advisory aids include maximum climb, obstacle clearance, VMO/MMO, buffet margins, rapid descent and optimum approach data, all of which help to improve tactical efficiency and reduce crew workload.

A built-in test system verifies overall system operational status to the flight crew and offers a method of isolating faulty LRUs without support equipment.

Operational status

In service in Lockheed Martin C-141 and C-5 aircraft.

Contractor

Smiths Industries Aerospace.

VERIFIED

Model 6000 Attitude and Heading Reference System (AHRS)

The Model 6000 represents a series of AHRS designed and manufactured by Smiths Industries for a large variety of aircraft applications. Licence production was also undertaken by Smiths Industries Newmark (UK) and KKK (Japan). Over 3,200 examples were produced between 1974 and 1990.

The AHRS provides multiple synchro outputs of roll, pitch and heading through all aircraft attitudes, including loops and spins. Heading rate and validity outputs are also provided. The system is tolerant of power interruptions of up to 30 seconds and can accommodate the body rates attainable in any aircraft type.

The Model 6000 AHRS consists of three units: an all-attitude twin gyro platform Displacement Gyroscope Assembly (DGA), an Electronic Control Amplifier (ECA) and a Compass System Controller (CSC).

Specifications

Dimensions:
(DGA) 150 × 140 × 260 mm
(ECA) 155 × 188 × 231 mm
(CSC) 146 × 38 × 102 mm
Weight: 14.1 kg incl mounting bases
Power supply: 115 V AC, 400 Hz, single phase, 90 W
Gyro drift:
(vertical) 3°/h
(directional) 0.5°/h
Environmental: MIL-E-5400 Class II

Operational status

No longer in production. Aircraft applications include the A-10A, T-45A, T-46A, Alpha Jet, Hawk, S.211, AT-3, S-70 and the IA-63. Retrofit applications include the F-106, C-141, B-52G/H, J-35 Draken, HSS-2B, PS-1, A-4 and F-5A/B.

Contractor

Smiths Industries Aerospace.

VERIFIED

Navigation Attack System (NAS)

The core of the Navigation Attack System (NAS) is the Computer Interface Unit (CIU), a specially developed computer based on dual MIL-STD-1750A processors which combines mission and navigation management with stores management and weapon delivery and which contains the main controller for the MIL-STD-1553 databus. It also generates symbols for the HUD and radar display. Some of the CIU functions, such as bus control and navigation management, are backed up by the display processor. The CIU and display processor are designed to permit easy expansion to meet future needs.

The NAS includes two MultiFunction Displays (MFDs). Normally, the right-hand MFD displays radar data, with navigation, stores or other information on the left screen. As with other modern avionics, a primary mode switch on the throttle puts the HOTAS controls in navigation, air-to-air or air-to-ground mode.

The SNU-84-1 ring laser gyro inertial navigation unit is a standard component. NAS offers the same choice of radar modes as the standard F-16 radar, selectable through the HOTAS controls.

The NAS includes a Sidewinder control system for enhanced AIM-9 off-boresight capability. Another feature of the NAS is a Smiths Industries Data Transfer Module (DTM), a read-write device using removable memory cartridges, located on the right side of the cockpit. The DTM is used in conjunction with a Mission Data Ground Terminal (MDGT). The MDGT can also be linked to a map and digitiser pad to allow the direct entry of navigation data or used to download post-flight and maintenance data.

Operational status

In service in New Zealand Air Force A-4 Skyhawk aircraft.

Contractor

Smiths Industries Aerospace.

VERIFIED

Self-Contained Navigation System (SCNS) for the C-130

The Self-Contained Navigation System (SCNS) is a fully integrated navigation and communications management system designed to enhance mission performance. Independent navigation capability is achieved through integration with the ring laser gyro INS, GPS and a Doppler velocity sensor. SCNS hardware is interchangeable on C-130 models and is easily adapted to other military aircraft.

The pilot, co-pilot and navigator each have independent control through the use of an Integrated Control and Display Unit (ICDU) which can control up to 12 radios. The ICDU communicates with the other boxes in the system via a MIL-STD-1553B databus and features a 10-row by 24-character display which is compatible with night vision goggles.

The main functions of the SCNS include navigation, flight planning, aircraft guidance and steering and control of radio and navigation systems. As well as conventional navigation facilities such as time and distance to waypoint, the SCNS also displays information regarding eight different types of airdrop.

The system is standard equipment on US Air Force C-130E and H aircraft.

Operational status

In service in C-130 aircraft.

Contractor

Smiths Industries Aerospace.

VERIFIED

The Smiths Industries Self-Contained Navigation System for the US Air Force C-130

Improved Data Modem (IDM) MD-1295/A

The design aim of the IDM is to act as an international digital communications translator, so that all elements in a multinational force can intercommunicate effectively and efficiently share sensor and tactical data without regard to nationality, language or platform.

By transmitting in a digital format, each participant's platform can receive, process and handle the data in such a manner as to facilitate the operators' specific mission requirements. However, in order that the mission processor can utilise the data, it must first go through a digital translation process to convert the data into a format that the mission processor can use. The IDM is designed to perform that function.

The IDM contains four Standard Electronic Module format-E (SEM-E) modules mounted in a chassis: two Digital Signal Processors (DSPs) as modems, a Generic Interface Processor (GIP) for link and message processing, and a power converter for power adaptation for the IDM's internal assemblies. Three additional SEM-E slots are available within the housing for expansion of the functional capability of the IDM. They could include the following modules, which are under development: an internal GPS; a multiband, multimode, transceiver module; a video card; a new cryptographic module to address TRAP and TADIXS-B; or additional DSP modules for additional channels.

The DSP module uses a TMS320C30 DSP to modulate and demodulate messages at 33 MHz clock rate/6 Mips. Each DSP can modulate or demodulate two half-duplex channels simultaneously at data rates up to 16 kbps. The DSP module includes all discrete signals required to interface a variety of radios.

The GIP module uses an Intel 80960MC RISC microprocessor capable of simultaneously processing four half-duplex channels of AFAPD, MTS, TACFIRE or MIL-STD-188-220 messages at 16 MHz clock rate/6 Mips. It also receives and transmits messages and configuration commands via the MIL-STD-1553 databus, processes data and processes configuration commands. It has 6 Mbytes of storage for warehousing messages.

The IDM was designed around the concept that changes in functionality should be accomplished through software changes inside the IDM, and not via more expensive changes to the air/ground platform Operational Flight Program (OFP). In fulfilling this design aim, software 'Radio Lookup Tables' facilitate the easy interconnection to a variety of receiver/transmitters.

Specifications

Interoperability:
AFAPD (US Air Force)
TACFIRE (US Army)
MTS (US Navy/US Marine Corps)
MIL-STD-188-220/VMF (US DoD common)
ATM (commercial)
NITF (imagery)
TADIL-J (future)
IDM radio interfaces:
ARC-164 (Have Quick II)
ARC-182
ARC-186
ARC-201 (SINCGARS)
ARC-210
ARC-222 (IDM software provides DRA function)
SINCGARS SIP
KY-58 and Whittaker
Radio Look-Up Tables provide flexibility to add other receiver/transmitter software interfaces
Analogue port (CPFSK, duobinary FSK):
(data rate*) 2,400; 1,200; 600; 300; 150; 75 bps
(tone pairs) 2,400/1,200; 2,100/1,300; 1,700/1,300
Digital port (ASK):
(data rate*) 16,000; 9,600; 8,000; 4,800; 2,400; 1,200; 600; 300; 150; 75 bps
(channel bandwidth) 20 kHz
(input signal level) 0.1 to 25 Vpp
(output signal level) 0.1 to 12 Vpp
KY-59 port:
(data rate) 16,000; 8,000; 1,200; 600; 300; 150; 75 bps
*data rates are limited only by the receiver/transmitter connected; the IDM has achieved 400,000 bps
Dimensions: 230.1 (D) × 190.5 (H) × 136.5 (W) mm
Weight: 6.8 kg

IDM hardware **2000**/0062233

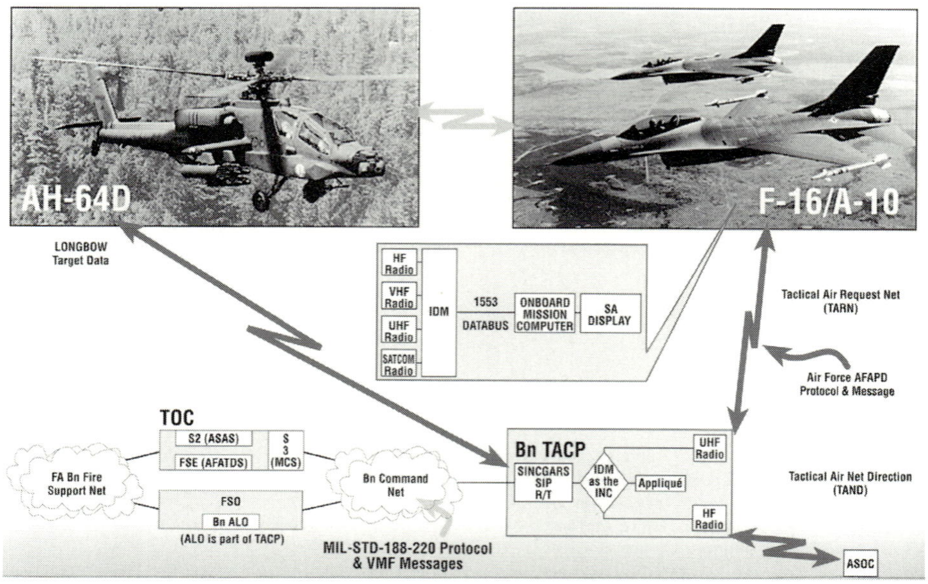

The IDM in typical tactical situation, sharing data between an AH-64D and F-16 **2000**/0062234

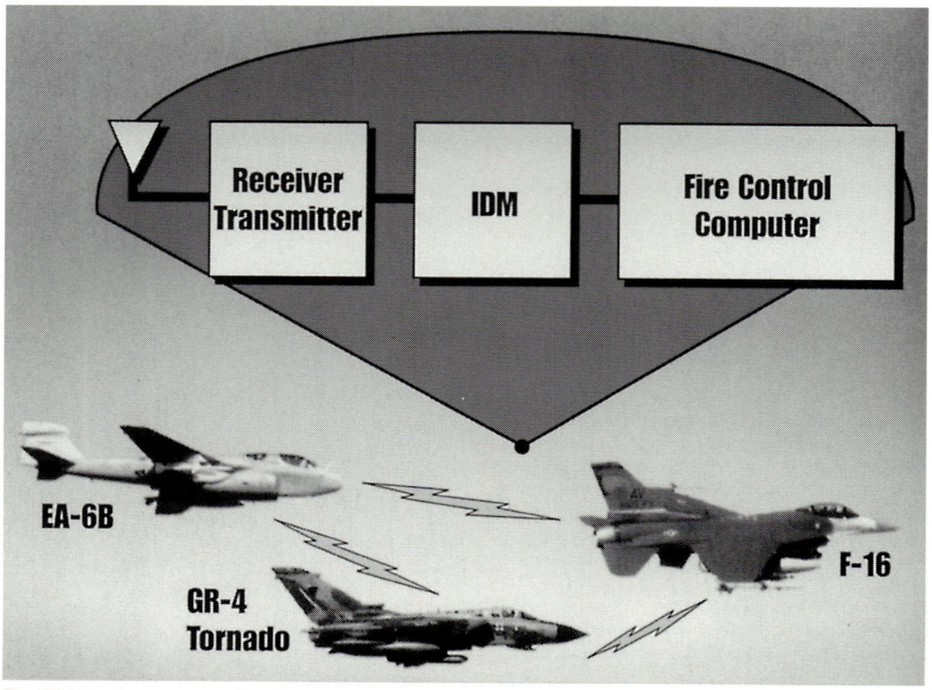

The IDM functions as a translator between different transceivers and fire-control computers on different types of aircraft **2000**/0062232

Operational status

Under a multiservice programme directed by the US Air Force F-16 System Program Office (SPO), the US Naval Research Laboratory (NRL) developed 25 First Service Units and 200 Low-Rate Initial Production (LRIP) units. Full-rate production is in progress at Symetrics Industries Inc for a projected total of over 4,000 IDMs.

Initially, the IDM was developed to provide a datalink capability for the F-16 Suppression of Enemy Air Defences (SEAD) mission. In that role, the IDM automatically hands off High-speed Anti-Radar Missile (HARM) targeting data and transmits navigational data directly from one F-16 to another.

The IDM is currently in operational service with the US Air Force on F-16 Block 40 and Block 50 aircraft. IDMs are also installed on a variety of international and US platforms, including other F-16 aircraft, the Longbow AH-64D and WAH-64, EA-6B, OH-58D, JSTARS, A2C2S, UH-60Q and E-2C.

Together with SEAD, the IDM supports other missions, including Close Air Support (CAS), Forward Air Control (FAC), air combat, joint air attack, fire support, intra-flight datalink, situational awareness datalink and command and control.

At present, US military services have taken different approaches to software support of the IDM, which has created functional differences. Nonetheless, the essential aim of full data exchange has been retained.

Contractor

Symetrics Industries Inc.

VERIFIED

AN/APQ-153 fire-control radar

The AN/APQ-153 airborne search target attack radar is a lightweight search and range tracking radar used on F-5E aircraft. This system operates in the I-band frequency range and provides stabilised search, automatic acquisition and target illumination, and automatic target ranging with boresight steering in the missile mode for head-down launch of Sidewinder AIM-9 missiles. In the gunnery modes, the radar automatically provides range and range rate outputs for targets within the sight lead-angle computation envelope.

The system consists of five main units: antenna assembly, receiver/transmitter, radar processor, indicator and set control. Combined weight is 50 kg. The antenna is a parabolic 300 × 400 mm dish with horizontal polarisation. The indicator uses a 127 mm direct view storage tube and provides a B-scope search display and a lock on missile mode display. Built-in test is provided for radar performance verification.

The APQ-153 has the following principal modes of operation:

Search mode

In search operation, a 7° elevation beamwidth is stepped up 3° at the right azimuth limit and down 3° at the left azimuth limit, providing a two-bar scan coverage of 10° in elevation. This two-bar 10° coverage can be adjusted ±45° in elevation by the pilot. Azimuth search coverage is ±45°. The search pattern is space-stabilised for aircraft pitch and roll motion to allow a given volume of space to be searched to prevent loss of the target and/or smearing of the display. Range of search coverage extends to 37 km.

Boresight missile mode

The missile mode enables the pilot to lock on to targets out to 19 km and provides aircraft steering information to align the acquisition envelope of the AIM-9 missile with the target. Once a target has been acquired, a lock on occurs and azimuth and elevation steering data are provided on the indicator display in the form of a target steering bar. The aircraft is flown towards the bar to align it within the allowable AIM error circle scribed on the overlay. Missile acquisition occurs when the steering bar is within the allowable AIM error circle and the in-range indicator is illuminated.

Air-to-air gunnery modes

There are two air-to-air gunnery modes in the APQ-153: dogfight and AA1/AA2. Both are external commands to the radar. The gunnery modes are head-up and the radar automatically provides range and range rate information to the sight. Activation of a gunnery mode causes the antenna to align to the boresight in azimuth and 2.2° down in elevation. The range gate will automatically slew from 500 to 6,000 ft at 22,000 ft/s. Acquisition is automatic for the first target encountered.

Operational status

In operational use aboard Northrop F-5E and F-5F aircraft. Over 1,400 sets have been produced. No longer in production, and is considered obsolete, having been superseded by the AN/APQ-159(V).

Contractor

Systems & Electronics Inc.

VERIFIED

AN/APQ-157 fire-control radar

The AN/APQ-157 is a dual-seat version of the AN/APQ-153 system. It consists of eight units: antenna assembly, receiver/transmitter, radar processor, front indicator, rear indicator, coupler power supply and two control units. Total weight is 64 kg. The indicators use a 127 mm direct view storage tube and provide a B-scope search display and a lock on missile display. Video trim is provided for the harmonisation of video displays. Each display can be adjusted to suit the individual operational preference and will correlate information within 1°. The control units are transferable for individual cockpit control.

All other details and modes of operation are as for the AN/APQ-153.

Operational status

No longer in production but still in service aboard Northrop Grumman F-5F aircraft. The system is considered obsolete, having been superseded by a dual-seat version of the AN/APQ-159(V).

Contractor

Systems & Electronics Inc.

VERIFIED

AN/APQ-159(V) radar

The AN/APQ-159(V) was designed in the mid-1970s as a technological upgrade of the AN/APQ-153/157 systems. It provides increased range, off-boresight lock on and angle tracking, frequency agility and a higher-gain lower-sidelobe planar-array. The AN/APQ-159(V)-1 and -2 incorporate a scan converter and television display compatible with the Maverick missile for service aboard single-seat F-5E and dual-seat F-5F, respectively. The AN/APQ-159(V)-3 and -4 versions, again for F-5E and F respectively, incorporate a Direct View Storage Tube (DVST) display for radar-only air-to-air applications.

The latest derivative is the AN/APQ-159(V)-5. The AN/APQ-159(V)-5 includes off-boresight acquisition and angle track capability to improve search and acquisition performance, and offers 100 per cent increase in reliability over earlier variants.

The AN/APQ-159(V)-5 is a lightweight forward-looking I/J-band radar designed for air-to-air detection and tracking and for installation in operational fighter aircraft. This version was developed with the intention of providing a cost-effective radar upgrade for the F-5E aircraft currently equipped with the AN/APQ-153 or -157 systems and the AN/APQ-159(V)-1 through -4 systems. The criteria for the new radar were minimum aircraft modification and no impact on weight and balance, while providing improved performance, reliability and supportability.

To meet these criteria, analysis of the earlier F-5 radar resulted in the selection of higher-reliability components, redesign of the processor and servo amplifier, and the incorporation of a gallium arsenide FET low-noise receiver. The MTBF figure was projected to improve by 100 per cent from 62 hours for the AN/APQ-153 to 127 hours for the AN/APQ-159(V)-5. Actual field experience exceeds 150 hours. In the process of improving reliability, the upgraded (V)-5 system extends target detection range to more than double that of the (V)-1 through -4, with comparable improvements also in the other modes.

The AN/APQ-159(V)-7, in addition to improved reliability and performance, was to incorporate monopulse tracking and air-to-ground ranging to provide better angle tracking and optimum weapon delivery. This version did not go beyond the development stage.

Operational status

The AN/APQ-159(V)-5 is currently in production and is operational with the US Air Force Aggressor and US Navy Adversary units. It became operational in the US Navy Adversary unit in mid-1989. It is also in service with several other air forces, including a variant which uses a television display. The (V)-5 is being actively considered by several additional countries.

The AN/APQ-159(V)-1 to -4 versions remain in serice, having been installed in over 550 aircraft in 15 different countries.

Contractor

Systems & Electronics Inc.

VERIFIED

AN/APQ-170(V)1 multimode radar

Systems and Electronics Inc developed the AN/APQ-170 radar to equip 28 MC-130H Combat Talon II aircraft procured for the US Air Force Special Operations Forces. The radar has terrain-follow/avoidance capabilities, as well as ground map, weather and beacon modes. The AN/APQ-170(V)1 is a dual radar for redundancy and best performance in each mode. The system has high reliability and, therefore, a high probability of mission success. Delivery of the first AN/APQ-170-equipped aircraft took place in 1989.

The Combat Talon II aircraft incorporate the avionics suite developed by Lockheed Martin, including the Raytheon Systems Company AN/AAQ-45 infra-red system and a four-CRT cockpit display system.

Operational status

The current version of the radar in service on the 28 MC-130H Combat Talon II aircraft is the AN/APQ-170(V)1. A programme to improve mission effectiveness and reliability and to reduce maintenance costs was awarded to Lockheed Martin Federal Systems (Owego) and Systems and Electronics Inc (SEI) as its subcontractor, in January 1997 for development, testing and installation for all 28 aircraft by November 1999. The latest upgraded/modified configuration is designated the AN/APQ-170(V)1-425 Mod.

Contractor

Systems & Electronics Inc.

VERIFIED

AN/APQ-175 multimode radar

The AN/APQ-175 radar has been developed to replace the AN/APQ-122(V) for the US Air Force in the Adverse Weather Aerial Delivery System (AWADS) C-130E aircraft. It is designed to enable the C-130 to airdrop and air land personnel and equipment during poor weather conditions. Features of the equipment will include long- and short-range precision ground-mapping, weather detection, station-keeping, and beacon integration and reception. Independent multi-purpose displays are used for both pilots and the navigator so that each can independently observe certain different radar functions. Interface to the aircraft is via MIL-STD-1553 bus to facilitate radar data distribution and interaction with other aircraft systems.

The AN/APQ-175(V)X provides the C-130 with a modern navigation and weather radar. It is a dual-frequency system which, together with frequency agility, selectable beam shaping and selectable polarisation, provides optimum performance for the various modes and conditions. Very high resolution and antenna stabilisation are provided in order to achieve the needed accuracies at all ranges for autonomous air-drop and landing. Other features include echo contouring, rainfall rate determination, false return rejection, aim point offset, and built-in test.

Operational status

Delivery of production units began in 1990. The current version is in service with the US Air Force. A very high reliability is experienced, and a programme for further mission effectiveness and reliability improvement is in progress. The AN/APM-464 depot-level system test equipment, provided by Systems & Electronics Inc, is also in service with the US Air Force.

Contractor

Systems & Electronics Inc.

VERIFIED

TC-510/AC Series HF/VHF/UHF direction-finders

The TC-510/AC Series of airborne HF/VHF/UHF direction-finder systems is designed for installation on small and large military or commercial aircraft. The system utilises a single small DF antenna, flush-mounted or externally mounted on the fuselage. Single or simultaneous independent multiple direction-of-arrival information is provided to the operator for received signals in the 1.5 to 1,300 MHz frequency range.

Variants of the system provide acquisition and direction-finder information on signals up to 2.6 GHz, including INMARSAT and new communication frequency allocations.

Operational status

In production and in service with various military and civilian agencies. TC-510/AC systems have been installed on both fixed-wing aircraft and helicopters.

Contractor

Tech-Comm Inc.

VERIFIED

Amplifier Control Indicator (ACI)

Designed for the US Navy F/A-18 E/F, the Telephonics ACI provides secure communications for onboard radios and the Mutlifunctional Information Distribution System (MIDS). The ACI controls the radio relay, IFF, ILS and cryptographic functions.

The ACI includes a voice activated switch (VOX) for crew interphone communications and extensive built-in testing that automatically checks system circuits and radio lines.

Operational status

Selected for the US Navy F/A-18 E/F programme, and for the Australian Hornet upgrade programme. Deliveries were planned to begin in late 1999.

Contractor

Telephonics Corporation.

VERIFIED

AN/APS-128 Model D surveillance radar

An upgraded version of the AN/APS-128, the Model D is an all-digital radar with a scan converter to present information in television raster format. It contains target enhancement and clutter reduction circuits for frequency agility, sensitivity time control, constant false alarm rate and scan-to-scan integration.

The system comprises a flat-plate antenna and pedestal, transmitter/receiver, radar control unit, digital scan converter, a trackerball or joystick cursor and bright display. A dual display for cockpit weather presentation is available, as well as cabin displays. A programmable microprocessor provides the alphanumerics and graphics to meet various mission requirements, while the antenna size can be altered to suit aircraft and operational needs. An alternative parabolic antenna with dual polarisation provides a pencil beam for sea search and a shaped beam for ground-mapping. The system has a 30 target track-while-scan facility and narrow/wideband datalinks. In addition, the APS-128D has successfully passed ground integration tests with the BAe Sea Skua missile.

Claimed target detection ranges are:
Snorkel or fishing vessel, Sea State 3, 55 km
Trawler, Sea State 5, 111 km
Freighter, Sea State 5, 185 km
Tanker, Sea State 5, 222 km

When functioning as a weather radar the range is 370 km.

Specifications

Weight: 91.8 kg
Frequency: 9.375 GHz
Frequency agility: 85 MHz peak-to-peak
Power output: 100 kW peak
Pulsewidth: 2.4 and 0.5 μs (0.1 μs pulse compression optional)
PRF: 1,400, 1,200 and 1,600 Hz

Operational status

In service but no longer in production. Over 90 systems have been sold worldwide. Versions of the APS-128D are in service with the Puerto Rican police and NASA, in the Falcon 900 for the Japanese Maritime Safety Agency, in CASA 212s for the Venezuelan Navy, Portuguese Air Force, Spanish Customs Service and Air Force, as well as the Argentine Coastguard and in the Skyvan for the Singaporean Air Force.

Older units are being replaced with the modern, more advanced APS-143B(V)3 OceanEye radar.

Contractor

Telephonics Corporation.

VERIFIED

AN/APS-128A/B/C surveillance radar

The AN/APS-128A/B/C is one of the most widely used maritime patrol radars. It has come into its own largely as a result of the international 200 n mile (370 km) fishing boundaries which need constant patrolling, but by low-cost aircraft to be economically effective.

The system comprises a rectangular flat-plate antenna and pedestal, transmitter/receiver, radar control module, range and bearing control module and azimuth/range indicator.

Target detection ranges are:
Fishing vessel, in Sea State 3, 46 km
Trawler, Sea State 5, 93 km
Freighter, Sea State 5, 148 km
Tanker, Sea State 5, 185 km

The system also functions as a weather radar with a range of 370 km.

Specifications

Weight: 79.1 kg
Frequency: 9,375 MHz
Frequency agility: 85 MHz peak-to-peak
Power output: 100 kW peak
Pulsewidth: 2.4 and 0.5 μs
PRF: 267, 400, 1,200, and 1,600 Hz
Antenna rotation rate: 15 and 60 rpm
Antenna stabilisation: automatic compensation for pitch and roll up to ±20°, with tilt to ±15°
Azimuth/range indicator: PPI, P-7 phosphor, 178 mm CRT, north- or aircraft heading-orientated, range scales 46, 93, 231 km

Operational status

In service. No longer in production. The AN/APS-128 is installed on all the Beech 200T maritime patrol aircraft operated by the Japanese Maritime Safety Agency and is operational with the Uruguayan Navy. The Brazilian Air Force bought 16 sets of equipment for its Embraer EMB-111 Bandeirante MR aircraft and Gabon and Chile also chose the system for EMB-111s. The Indonesian Air Force and the Royal Malaysian Air Force have installed the AN/APS-128 in their C-130 Hercules aircraft. It is also installed on the CASA 212 SAR aircraft of the Spanish Air Force.

Maritime patrol Beech Super King Air with the radome for the Telephonics AN/APS-128surveillance radar on the underside of the fuselage

Older units are being replaced with the modern, more advanced APS-143B(V)3 OceanEye Radar.

Contractor

Telephonics Corporation.

VERIFIED

AN/APS-143(V)1 and (V)2 radars

The AN/APS-143 radar is designed to provide aircraft and aerostats with good detection capability against small targets in very high sea states. It is a lightweight, travelling wave tube radar with pulse compression. This provides better detection, resolution and clutter rejection, higher average power and greater receiver/transmitter reliability. The system includes a signal processor incorporating track-while-scan, scan conversion and databus interfaces.

Specifications

Weight:
(inc display) <110 kg
Frequency: 9.3 to 9.5 GHz
Agility: fixed, or can operate with 11 agility steps of around 20 MHz over the band
Peak power: 8 kW min, 10 kW nominal
Effective peak power: 700 kW up to 93 km range; 2.4 MW above
Compression ratio: 70:1 at ranges up to 93 km; 240:1 above
Pulsewidth: 5.0 and 17.0 μs uncompressed; 0.1 μs compressed (weighted)
PRF: 2,500, 1,510, 750 or 390 Hz (range dependent)

Operational status

In production. In operation with US government aerostats and US Air Force de Havilland DHC-8 aircraft used for test range surveillance. The Malaysian Air Force has taken delivery of the APS-143 for Beech 200T aircraft, as has the Japanese Maritime Safety Agency on the Saab 340B.

Contractor

Telephonics Corporation.

VERIFIED

AN/APS-143(V)3 and AN/APS-143B(V)3 sea surveillance radars

The AN/APS-143(V)3 is a maritime surveillance and tracking radar designed for installation in a variety of fixed-wing aircraft and helicopters. It is also known as OceanEye. The system uses frequency agility and pulse compression techniques and consists of three units: an antenna, receiver/transmitter and signal processor. Options available include air search mode with MTI customised for the individual platform, air-to-surface missile guidance, integrated ESM interface,

The APS-143B(V)3 radar system ***1999***/0044997

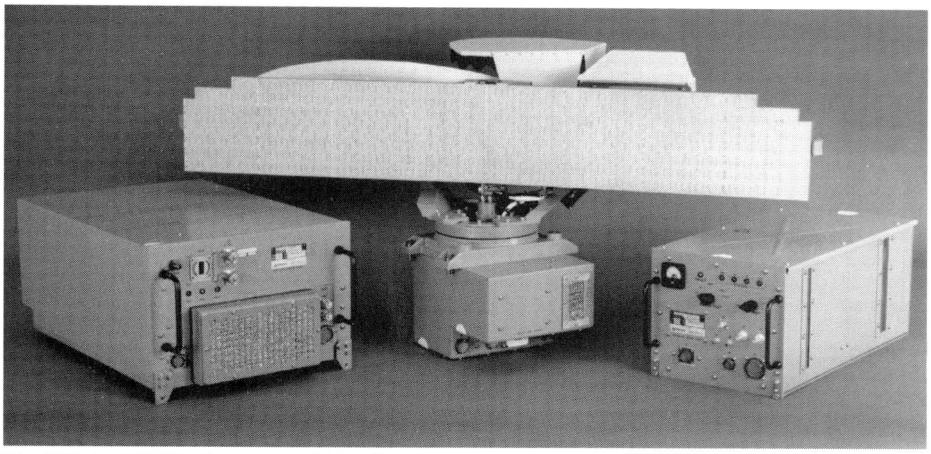

The AN/APS-143(V)3 is in service in S-70 helicopters, shown with integrated ESM antennas

integrated IFF interface and a dedicated control unit, with colour display. The latest variant, APS-143B(V)3, can be upgraded with a complete imaging capability: range profiling, ISAR, spotlight SAR, strip-map SAR.

The flat-plate antenna array can be fitted into any radome, or alternatively a parabolic type is available. It is stabilised for ±20° in pitch and roll. The transmitter is a TWT type with a peak power output of 8 kW. Options include integrated IFF.

Specifications

Weight: 81.8 kg
Frequency: 9.2-9.7 GHz
Frequency agility: 200 MHz (450 MHz for B(V)3)
Peak power: 8 kW
Pulsewidth: 5 or 17 μs to 40 μs for B(V)3, compressed width 100 ns (0.1 μs); and 23.4 μs compressed to 6.5 ns for B(V)3 imaging.
PRF: 2,500, 1,500, 800 or 400 Hz
Max range: 370 km
Compression ratio: 50:1/170:1 (1:1 to 3,600:1 for B(V)3)
Range resolution: 15 m
Azimuth accuracy: 0.5°
Image resolution: down to 1 m

Operational status

In service with Republic of China Navy, the Hellenic Navy and the Royal Thai Navy S-70C helicopters, and ordered by the Israeli Navy, the New Zealand Navy, and the Turkish Navy. Also delivered to the Ecuadorean Navy for its Beech 200T aircraft.

An advanced version, the APS-143B(V)3 OceanEye, with ISAR and SAR, has been ordered by Kaman Aerospace for the SH-2G helicopters for the Australian Navy.

The APS-143(V)3 forms part of the equipment kit of the four SH-2G (NZ) helicopters to be delivered to the New Zealand Navy by Kaman Aerospace in 2000. Pending installations of the APS-143B(V)3 include P-3 Orion, Beech 350 and Saab-340 aircraft, together with the S-70 helicopter.

Contractor

Telephonics Corporation.

VERIFIED

AN/APS-147 multimode airborne radar

The AN/APS-147 multimode radar, designed for the US Navy SH-60R helicopter, is an Inverse Synthetic Aperture Radar (ISAR) which uses the latest in high-throughput signal and data processing. Flexibility through programmability provides a product optimised for the maritime surveillance mission.

Advanced processing allows the AN/APS-147 to use a variety of waveforms to perform its mission at an output power substantially lower than traditional counterparts in maritime surveillance radars. This results in a radar with an extremely Low Probability of Intercept (LPI). Using a low peak power waveform with frequency agility, the radar can detect medium- to long-range targets without the threat of ESM interception.

Radar modes include target imaging, small target and periscope detection, long-range surveillance, weather detection and avoidance, all-weather navigation, short-range search and rescue, enhanced LPI search and target designation.

The AN/APS-147 features a flexible modular design which can be tailored to meet specific requirements and can be easily upgraded, LPI, high-resolution images for rapid classification, lightweight construction through the use of composite materials, low-input power, simple design for high reliability and maintainability, fully programmable signal processor with multiple waveform exciter and high-throughput rates and integrated IFF and SAR option.

Operational status

Telephonics Corporation delivered its first AN/APS-147 radar to Lockheed Martin Federal Systems, Owego for integration and test aboard a US Navy SH-60R helicopter in June 1999.

Contractor

Telephonics Corporation.

VERIFIED

AN/APX-103, AN/APX-103B, AN/APX-103C

The AN/APX-103 IFF interrogator system was developed for the Boeing E-3 Sentry AWACS which is operational with the US Air Force, NATO, UK, France and Saudi Arabia and will shortly be operational with Japan. The system utilises the features and functions of the co-operative beacon system commonly known as Mark XII Identification Friend or Foe (IFF). The AN/APX-103 provides state-of-the-art Air Traffic Control as well as Selective Identification and Friendly (SIF) identification of military aircraft for the airborne early warning mission. The output of the AN/APX-103 is digital target reports which are inputted to the E-3's central command and control computer for correlation of the IFF and primary radar information. The APX-103's digital processor operates accurately in dense target environments.

Operational status

The AN/APX-103 has completed a monopulse processing technology insertion upgrade and is now in service and available under the nomenclature AN/APX-103B (monopulse processing) and AN/APX-103C (monopulse with additional processing improvements). The AN/APX-103B/C technology improvements provide the operators with state-of-the-art resolution, accuracy and code correction performance, resulting in improved target handling capacity and reduced workload. Additional improvements are underway including a Mode S development for NATO and a retrofit technology insertion upgrade of the APX-103 receiver transmitter.

Contractor

Telephonics Corporation.

VERIFIED

The Boeing E-3 Sentry AWACS is equipped with the Telephonics AN/APX-103 IFF interrogator, with the antenna mounted back-to-back with the primary radar antenna in the rotating radome

AN/ARA-63 microwave landing system

The ARA-63 is the airborne portion of the US Navy's standard aircraft approach control system for landing on aircraft carriers and equips aircraft such as the A-6E, F-14A, S-3A and F/A-18. Known as a radio receiving decoder set, it works in conjunction with the AN/SPN-41 on board ships and the AN/TRN-28 transmitters at naval air stations. Pulse-coded microwave transmissions are received by the ARA-63, decoded and displayed on a standard cross-bars indicator in the cockpit. There are 20 channels in the range 14.688 to 15.512 GHz. The system has three LRUs: receiver, pulse decoder and control unit.

Operational status

In service in US Navy aircraft such as the A-6E, F-14A, S-3A and F/A-18 and also overseas. Over 3,000 units have been produced. Development and early production was by Telephonics. Follow-on production by Stewart-Warner.

Contractor

Telephonics Corporation.

VERIFIED

C-11746(V) communication system control unit

The C-11746 communication system control unit is designed for secure TEMPEST crew intercommunication and radio transmit and receive control in high-noise airborne applications.

The unit provides individual on/off and receive level control of radios and navigation receivers, voice-operated switching for hands-free intercom control and a remote select capability for HOTAS/HOCAS operation. A single unit can handle five transmit/receive radios, six navigation receivers, four auxiliary inputs and two intercom buses. Units are available with MIL-L-85762A NVG-compatible or standard edge-lit front panels.

Specifications

Dimensions: 66.675 × 127 mm
Weight: 1.18 kg
Power supply: 28 V DC, 7 W
Environmental: MIL-C-58111
TEMPEST: NACSIM 5100

Operational status

In service on the AH-64 Apache helicopter.

Contractor

Telephonics Corporation.

VERIFIED

Digital Communication Management System (DCMS)

The Digital Communication Management System (DCMS) family of intercommunication systems (AN/AIC-37 and AN/AIC-41) is a digital, distributive, TEMPEST certified intercommunication system designed to manage the communications assets of airborne platforms. It integrates and manages all the navigation aids and alert/warning tones, as well as providing for the real-time control and switching of modems and encryption devices between transceivers. Both frequency management and a MIL-STD-1553B interface are offered. In addition, the AN/AIC-37(V) has no single point failure, contains BIT to card level (eliminating the need for intermediate level maintenance) and is field reconfigurable and expandable. A voice-activated switch, capable of operating in high-noise environments, and 25 intercom nets are standard features. Individual volume control for all assets is provided, plus a variety of front panel configurations ranging from push-buttons to a full display.

The AN/AIC-37(V) operates on a redundant 10 mbits/s databus. This permits up to 84 non-blocking channels of communication between crew stations and external radio systems, and a means to interface with an external computer through a redundant

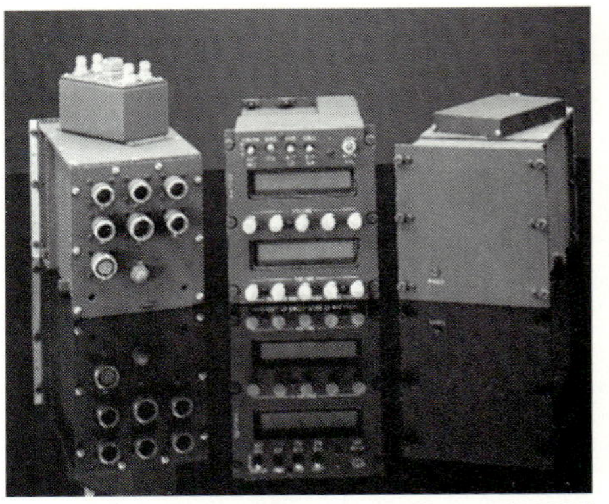

Digital Communication Management System (DCMS)
1999/0044836

MIL-STD-1553B bus. External voice and data transmissions can be made on encrypted or plain networks. The system can deny access to any crew position not authorised to receive secure information.

When operators communicate, the crew terminal digitises the audio into an allocated data time slot with a destination address to the selected interface. The destination unit then converts the digital information back into baseband audio for application to the specific asset. Similarly, when a radio or remote communication device receives audio, it is digitised and made available to any of the operators that have been enabled to receive the audio. The entire system is digitally reconfigurable, enabling asset assignment based on the operational scenario. All channels within the system can be selected, combined and monitored.

Specifications

Dimensions:
(DCI) 171.5 × 127 × 204.7 mm
(DAI) 152.4 × 127 × 242.8 mm
(DBI) 152.4 × 127 × 242.8 mm
(DII) 123.9 × 127 × 86.1 mm
(DUCK) 19.1 × 80.8 × 132.1 mm
(DBC) 57.1 × 111 × 57.1 mm
Weight:
(DCI) 4.11 kg, (DAI) 4.14 kg
(DBI) 3.64 kg, (DII) 1.02 kg
(DUCK) 0.25 kg, (DBC) 0.34 kg
Power supply: 28 V DC
Temperature range: –54 to +55°C
Altitude: up to 50,000 ft

Operational status

In production and operational on aircraft such as US Navy EP-3, ES-3 and P-3C, US Air Force E-8 JSTARS and the US President's Air Force 1. Also selected for the Norwegian Air Force P-3 upgrades and the German ATL-1 upgrade.

Contractor

Telephonics Corporation.

VERIFIED

Integrated Radio Management System (IRMS)

The Integrated Radio Management System (IRMS) is a communications management system providing control of aircraft radios, radar transponders and intercom. It provides total communications back-up in the event of battle damage through redundant control panels and centralised control units. The system comprises two Communication equipment Control Units (CCU), two Communication/Navigation equipment Controls (CNC), seven Intercommunication Set control Units (ISU), three Public Address set Controls (PAC), six Interphone Receptacle Panels (IRP) and nine Headset Receptacle Panels (HRP).

The CCU is the centralised 1750 processor-based MIL-STD-1553B bus controller unit which provides control of all audio, digital and analogue signal processing. It interfaces to the CNC, ICS and various radio equipments, via a MIL-STD-1553B interface bus. The CNC is used to tune navigation and communication radios, display radio frequencies and modes, IFF modes and VOR/ILS course selection. ICS provides microphone, PTT, VOX, audio and volume selection for all audio sources. ISU enables intercom/radio, talk/listen or PA selection. PAC selects speakers and volume levels. IRP and HRP provide microphone preamplification and headset impedance matching.

Specifications

Dimensions:
(CCU × 7) 198 × 191 × 498 mm
(CNC × 2) 95 × 184 × 178 mm
(ICS × 7) 124 × 146 × 152 mm
(ISU × 2) 67 × 146 × 89 mm
(PAC × 3) 38 × 136 × 102 mm
(IRP × 6) 64 × 102 × 38 mm
(HRP × 9) 38 × 127 × 102 mm
Weight:
(CCU) 13.8 kg, (CNC) 3.14 kg
(ICS) 2.3 kg, (PAC) 0.41 kg
(IRP) 0.23 kg, (HRP) 0.34 kg
Power supply:
(CNC, ICS, PAC) 28 V DC
(ISU, IRP, HRP) 16 V DC
Temperature range: –54 to +71°C
Altitude: up to 50,000 ft
Reliability: 3,300 h MTBF

Operational status

In production and in service. The IRMS is installed and fully operational on the US Air Force C-17A transport aircraft.

The Telephonics Integrated Radio Management System (IRMS)

Telephonics also provides the SURE-COMM loadmaster wireless communications system for the US Air Force C-17A, and the company has been contracted by the US Air Force to incorporate Global Air Traffic Management (GATM) requirements into the C-17A communications suite.

Contractor
Telephonics Corporation.

VERIFIED

Joint STARS Interior Communications System (ICS)

The Interior Communications System (ICS) is utilised on the E-8A aircraft as the communication system for the US Air Force/US Army Joint STARS. It is a fully distributed digital communications system, supporting 96 full-duplex channels for simultaneous non-blocking secure and non-secure operations. The ICS interfaces with various receiver/transmitters, secure speech devices, communication, navigation and identification receivers via General Interface Terminals (GIT), as well as supporting the interface with all mission crew members and flight deck personnel.

The ICS is fully modular and other crew members and GITs may be added as the system grows or the mission requirements change, with no changes to the existing system hardware. At the present time, there are six Flight Deck Terminals (FDT), 18 Crew member Terminals (CT) and five GITs with five channels each. Each of the mission operator CTs has a single display and keyboard. The keyboard provides radio selection or selection of the mission net, preset conference net, progressive/selective conference net, call and a telephone dialling net. The FDTs are identical to the CTs except that they contain an additional display and control keys for radio frequency and parameter selection. The GITs are digitally configurable for input and output levels, thereby allowing one design to accommodate many different peripheral devices without changing design or adjustments. Each GIT contains the input/output and control for five full-duplex audio channels. Once the system has been configured, all information within all of the units is retained in-non-volatile memory storage.

When operators communicate, the terminals digitise audio in an allocated data time slot with a destination address to the selected interface, which then converts the digital information back into baseband audio for application to the specific asset. Likewise, when a radio or remote communication device receives audio, this audio is digitised and made available to any of the operators that have been enabled to receive the audio. The entire system is digitally reconfigurable, enabling asset assignment based on the operational scenario. All channels within the system can be selected, combined, monitored and individually volume-controlled by any operator.

Battle damage that renders a single CT, FDT or GIT terminal, or multiple terminals, inoperable does not impact on system operation or affect the operation of the other system components. Each of the units is self-contained and utilises microprocessors that transfer function and allow continued operation.

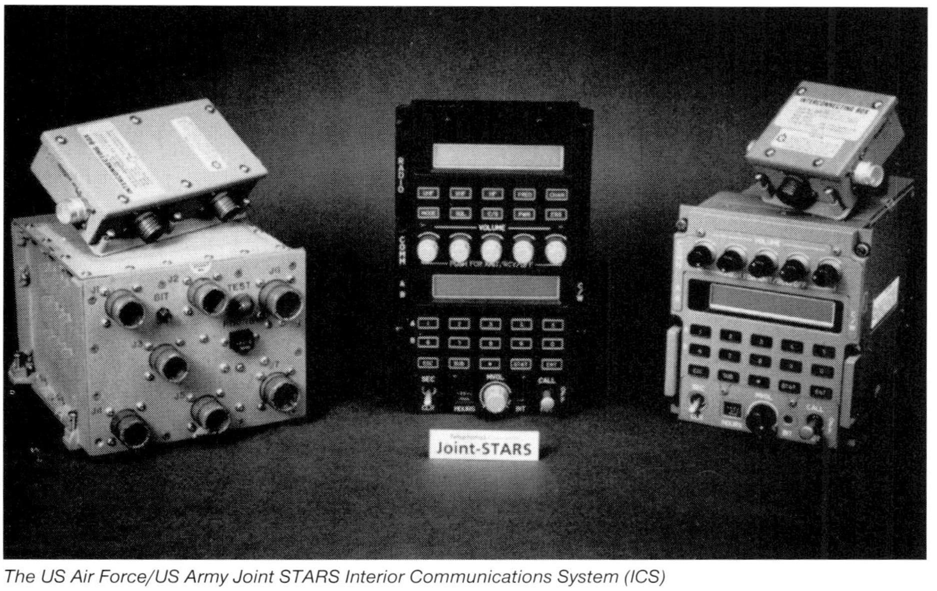

The US Air Force/US Army Joint STARS Interior Communications System (ICS)

Specifications
Dimensions:
(CT × 18) 171.5 × 127 × 177.8 mm
(FDT × 6) 247.7 × 127 × 177.8 mm
(GIT × 5) 142.7 × 171.5 × 190.5 mm
(single TAP) 47.8 × 85.1 × 85.1 mm
(dual TAP) 47.8 × 161.3 × 85.1 mm
Weight:
(CT) 3.98 kg
(FDT) 4.45 kg, (GIT) 4.05 kg
(single TAP) 0.25 kg, (dual TAP) 0.43 kg
Power supply: 28 V DC
Temperature range: –54 to +71°C
Altitude: up to 50,000 ft

Operational status
In service on the E-8A Joint STARS aircraft.

Contractor
Telephonics Corporation.

VERIFIED

OK-374/ASC Communication System Control Group (CSCG)

The Communication System Control Group (CSCG) is an airborne communication and control system utilised on the LAMPS III SH-60B helicopter. It provides centralised microprocessor-regulated access and control of the various internal and external communication, navigation, voice encryptors and antenna selections via the Relay Assembly (RA) for the four main crew members, while providing for two additional maintenance-only positions utilising the Interconnecting Box (IB).

The CSCG is designed to meet the requirements-of NACSIM 5100, MIL-E-5400, MIL-STD-461, MIL-STD-1553A and other applicable standards for airborne communication systems operating in severe helicopter environments. It provides control of communications and intercom configurations based on manual control or serial digital data instructions from an AN/AYK-14 airborne general purpose digital computer. The manual control panel switch settings of the four Remote Switching Controls (RSC) and the Control Indicator (CI) are transferred over a serial digital datalink to the central Audio Converter Processor (ACP). The ACP then performs the required functions such as tuning the radios, selecting audio for presentation to the crew members or selecting ICOM sets.

The CSCG provides integrated control via the CI of the HF radio, two UHF radios, datalink, IFF interrogator and transponder, direction-finder group, voice encryptors and antenna selections. It also provides external aircraft computer control of sonobuoy receivers, sonobuoy command via UHF radio, radio modes and frequencies and other supervisory functions, plus external radio communications access via the two UHF clear or secure radios, the clear or secure HF radio and the secure datalink. Internal ICOM communications are provided over a common ICS net and two separate conference nets: one for the pilot and co-pilot and the other for the sensor operator and the observer. Both PTT and VOX are provided for the ICOM nets and common ICS net access is also provided at two other positions called hoist and maintenance. The system distributes various warning tones such as stabilator warning, radar altimeter warning, helo threat warning and IFF Mode 4.

The Communication System Control Group (CSCG) for the LAMPS III SH-60B helicopter

A manual back-up mode is provided, enabling selected crew members to have access to radios and ICOM in the event of system failure.

Specifications
Dimensions:
(ACP) 185 × 334 × 524 mm
(CI) 372 × 146 × 165 mm
(RSC × 4) 124 × 146 × 165 mm
(RA) 81 × 147 × 112 mm
(IB) 47 × 94 × 54 mm
Weight:
(ACP) 22.3 kg
(CI) 6.4 kg, (RSC) 7.9 kg
(RA) 0.9 kg, (IB) 0.2 kg
Power supply: 115/200 V AC, 400 Hz, 3 phase, 200 VA
Temperature range: –40 to +71°C
Altitude: up to 15,000 ft
Reliability: 1,200 h MTBF

Operational status
In service in the LAMPS III SH-60B helicopter.

Contractor
Telephonics Corporation.

VERIFIED

STARCOM intercommunication system

STARCOM is a high-intelligibility audio communication system that meets the intercom needs of a wide variety of airborne and ground-based applications. The system is designed to provide secure communications capability in high-noise environments.

The baseline STARCOM system consists of Communications System Controls (CSC) and an Audio Distribution Unit (ADU). It can accommodate five transmit/receive radios, six navigation receivers, two intercom channels and controlled and uncontrolled audio warning signals. Radio and nav receive channels can be individually monitored and controlled for level. Voice-operated switching, with an adjustable threshold for hands-free intercom control, and a remote select capability for HOTAS/HOCAS operation are standard features. Up to 10 CSCs can be interconnected through the ADU. MIL-L-85762A NVG-compatible or standard front panel lighting is available.

Various additions to the baseline system can be installed to support specific applications.

Operational status
STARCOM has been installed on AH-1W, AH-6,-AS 565 MA, CE-144A, CH-47D ACMS, CH-146, MH-47D, MH-47E, MH-60G, MH-60K, MH-60L, OH-58D and UH-1N helicopters, and C-130 and P-3C aircraft. Also selected for the LCAC upgrade, SH-2G(A), SH-2G(NZ), S-70B (International Sea Hawk), and UH-60Q.

Telephonics STARCOM intercommunication system ***1999***/0044835

Contractor
Telephonics Corporation.

VERIFIED

TCOMSS-2000 + DAS Telephonics COmmunications Management System and Digital Audio System

TCOMSS-2000 Telephonics Communications Management System
The Telephonics TCOMSS-2000 is a fibre optic, open architecture, VME-based system that uses COTS hardware to provide secure digital audio and full digital control from all operator positions. TCOMSS-2000 operates on a 100 Mbps fibre optic network that supports full-duplex, non-blocking communication/management of up to 738 audio sources (crew members, external transceivers, navigation aids and encryption devices). Selectable bandwidths support audio and digital data distribution from narrow to wideband signals. System features include: unlimited conferencing, individual volume control, binaural (dichotic) audio, SIMOP, VOX, radio relay, frequency management, data transmission and real-time modem and encryption device switching. Because TCOMSS-2000 is modular, the system can be upgraded with options: auditory localisation (3-D audio), multilevel security, combined or separate red/black busses as well as interfaces with FDDI, ATM, Fibre Channel, MIL-STD-1553, RS-232, and RS-422.

TCOMSS-2000 consists of three module types: the Audio Control Subsystem (ACS); Audio Control Panel (ACP); and the jack box. The ACS provides the main interface for the TCOMSS-2000. It links the fibre optic rings with operators and communications assets. A plug-in module interfaces with the platform databus. ACS modules can be arranged in single or mulitple configurations to support varied platforms.

The ACP is the interface between the operator and the system. Using an RS-422 interface, the ACP enables the operator to manage all assets assigned to a particular crew position. This includes: independent radio transmit and receive selection and control for all assigned radios; programmable intercom/conference net selection and control; VOX; selectable dichotic capability; clear/secure selection; radio relay; and crypto and modem control.

The jack box is available for certain applications, typically larger platforms, to provide digital audio out to the headset jack. The jack box provides the proper controls and interfaces to support an airborne system. In smaller TCOMSS-2000 applications, the ACS incorporates the jack box functions.

When operators communicate, the audio is digitised and placed into an allocated time slot with a destination address within the ACS. The ACS then routes the audio to the appropriate address; either another jack box for interphone audio or for conversion to analogue for radio transmission. Similarly, when a radio or remote communication device receives audio, it is digitised and made available to any of the operators that have been enabled to receive the audio. The entire system is digitally reconfigurable, enabling asset assignment based on the operational scenario. All channels within the system can be selected, combined and monitored.

DAS Digital Audio System
DAS is to be a modified version of TCOMSS-2000 to meet the requirements of the NATO AWACS E-3A Mid-Term Modernisation Programme.

Specifications
Dimensions:
ACS 190.5 × 194.0 × 376.7 mm
ACP 146.0 × 171.4 × 127.0 mm
J-Box 165.1 × 50.8 × 127.0 mm
Weight:
ACS 9.98 kg
ACP 2.04 kg
J-Box .91 kg

Operational status
TCOMSS-2000 has been selected for the UK Nimrod MRA4 programme. Telephonics Corporation has also been selected by Lockheed Martin Federal Systems to provide a variant of the TCOMSS system for the US Navy SH-60R multimission helicopter and the CH-60 utility helicopter common cockpit programme.

DAS has been contracted by Alcatel Bell Space and Defense of Antwerp to Telephonics Corporation. Both companies will, together, conduct an engineering and manufacturing development phase to provide three laboratory systems and one flight system. The production phase will be for 17 aircraft and two simulator systems.

Contractor
Telephonics Corporation.

VERIFIED

TILS II Landing System receiver for the JAS 39 Gripen

Telephonics is producing the Tactical Instrument Landing System receiver (TILS II) for the JAS 39 Gripen. The airborne unit weighs less than 5 kg and is approximately 120 × 280 × 125 mm in size. It provides outputs to a MIL-STD-1553B databus.

The system works in conjunction with the TILS ground equipment developed by Telephonics and currently deployed throughout Sweden and Finland. This receiver and ground system are currently in production.

A variation of the receiver with DME has also been supplied for the Space Shuttle programme.

A multimode version incorporating ILS, GPS and J-band MLS is in development. This version will provide users with the flexibility to land at airfields equipped with ILS as well as Tactical Instrument Landing System (TILS) equipped roadways. The GPS feature can be used for en route navigation. All three functions will be contained in the same form factor as the current TILS II version.

Telephonics TCOMSS-2000 communication management system ***1999***/0044837

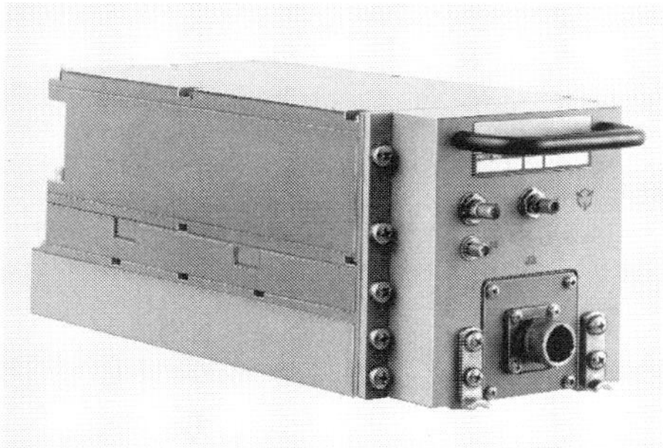

The TILS II receiver for the JAS 39 Gripen

Operational status

In service on JAS 39 Gripen. The predecessor TILS I receiver remains operational on Saab Viggen and Draken aircraft.

Contractor

Telephonics Corporation.

VERIFIED

Wireless Communications and Control System (WCCS)

The Wireless Communications and Control System (WCCS) is a wireless FM communications system which provides voice-operated hands-free full-duplex party line operation for up to six cordless headset users. The system provides highly intelligible communications in a 115 dBSPL environment through the use of a special enhanced noise-cancelling microphone and high-noise attenuation headset. Communication between users occurs within an aircraft through the use of an internally installed leaky coaxial line antenna or externally through a UHF blade antenna. Additional aircraft communications flexibility is provided by a hard-wired two-way audio interface to the aircraft interphone system. In addition to the communications capability which the system provides, it also supports a wireless hand-held remote Control Transmitter (CT) which is used to control cargo and retrieval winches.

The WCCS comprises a Receiver/Transmitter radio (RT), remote-control cargo winch controller, headset rack, up to six cordless headsets and the battery charger.

The RT, which is housed within the headset rack, is the repeater which provides full-duplex operation between users. Remote operation is achieved by using the RT in the portable battery-powered mode. Direct communications between cordless headset users can be accomplished by operating the system in the half-duplex mode, independent of the RT. The simultaneous use of multiple WCCS systems within the same operating range of each other without cross-talk interference is achieved by the use of seven frequency groups and 32-tone squelch codes.

The CT provides variable speed bidirectional cargo winch capability and two single-speed bidirectional retrieval winch capabilities.

Specifications

Dimensions:
(RT) 191 × 203 × 152 mm
(CT) 217 × 84 × 121 mm
(headset) 254 × 226 × 216 mm
(charger) 255 × 229 × 84 mm
(rack) 1,118 × 483 (stowed), 978 (extended) × 178 mm
Weight:
(RT) 2 kg
(CT) 0.55 kg, (headset) 1.23 kg
(charger) 1.9 kg, (rack) 7.73 kg
Power supply: 28 V DC, 28 W
Frequency: 410-420 MHz
Temperature range: −20 to +71°C
Altitude: up to 50,000 ft
Reliability: 5,165 h MTBF

Operational status

In production and in service in the US Air Force C-17A transport aircraft.

Contractor

Telephonics Corporation.

VERIFIED

AN/ASN-175 Cargo Utility GPS Receiver (CUGR)

Trimble's CUGR receiver is a dzus-mount P(Y) GPS navigation system for worldwide military aviation operations. It is a PPS version of Trimble's 2101 I/O Approach and meets the performance standards for Instrument Flight Rules (IFR) for en route, terminal and non-precision approach phases of flight as specified in the US Army Aircraft GPS Integration Guide (GIG). The unit continuously checks accuracy using Receiver Autonomous Integrity Monitoring (RAIM) and predicts RAIM conditions for approach. Trimble's CUGR can automatically sequence up to 20 flight plans with 20 waypoints each, showing nearest airport, planning vertical descents, displaying minimum safe altitudes as well as many other features. The receiver incorporates Jeppesen's NavData card including airports, approaches, SIDS, STARS, VORs, intersections, NDBs, airspace boundaries and MEAs. The unit can be intefaced with HSI, CDI, autopilot or air data computers to maximise the flexibility of the system, all within a single unit. The receiver is also AN/AVS-6 compatible.

Operational status

In October 1996, Trimble was awarded a contract from the US Army Communications-Electronics Command (CECOM) to provide Cargo Utility GPS Receivers (CUGRs) totalling over US$12 million. The contract provides for an initial procurement with two, one year options; over the life-time of the contract, CECOM may procure as many as 3,600 receivers. The CUGR system will be installed on US military helicopters including selected UH-1, UH-60, OH-58 and CH-47 aircraft.

Specifications

Receiver: 6-channel P(Y) code, continuous tracking, digital receiver; L1, L2 capable; 12-channel upgrade planned.
GPS antenna: L1/L2 flat panel
Dimensions/weight:
(receiver) 146 × 197 × 76 mm; 1.38 kg
(amplifier) 115 × 89 × 31 mm; 0.27 kg
(antenna) 96 × 102 × 26 mm; 0.32 kg
Performance:
Autonomous: PPS position accuracy: 3-D 16 m SEP; PPS velocity accuracy: 0.2 m/s
DGPS: position accuracy: 5 m; velocity accuracy: 0.2 m/s RMS

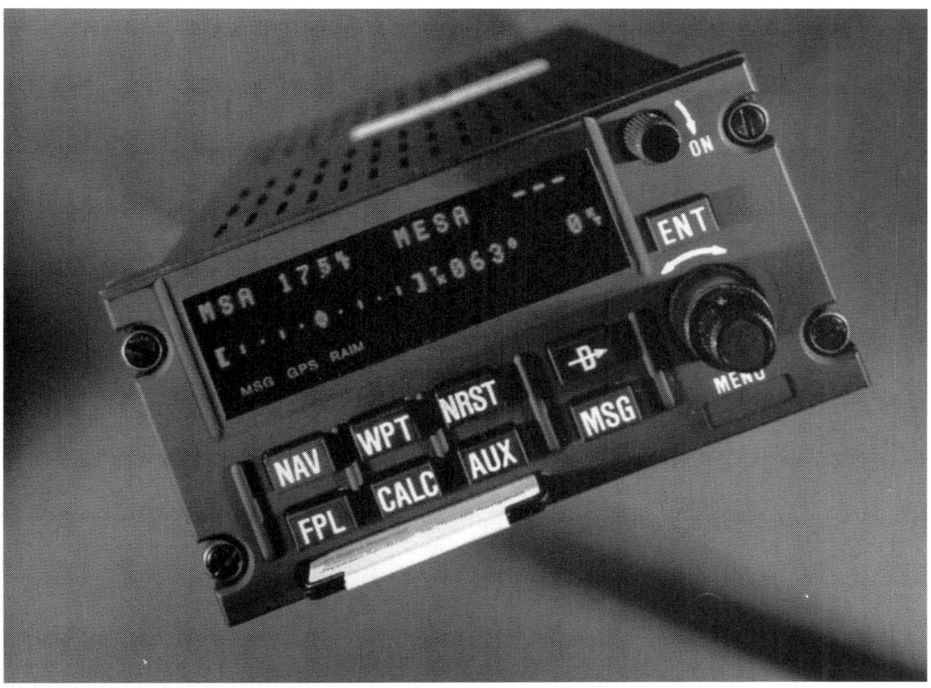

AN/ASN-175 Cargo Utility GPS Receiver (CUGR) 0015392

Contractor

Trimble Navigation Ltd, Aerospace Products.

VERIFIED

AN/PSN-10(V) Trimpack GPS receiver

The AN/PSN-10(V) Trimpack small lightweight GPS receiver is a completely self-contained navigation system. It can pinpoint position to within a few metres anywhere on earth and is built to operate under battlefield conditions. Its rugged electronics are sealed in a metal impregnated high-impact case that also houses a built-in unbreakable antenna. It is suitable for helicopters and medium-performance fixed-wing aircraft, as well as land applications.

Trimpack can operate in temperatures ranging from −30 to +65°C. Solid construction, coupled with electronics tested to MIL-STD-810D, results in a calculated MTBF of over 15,000 hours.

All controls are designed for easy operation. A rotary switch selects a wide variety of navigation functions including position and altitude, range and bearing, cross-track error, velocity and time to go. Up to 26 waypoints can be stored and then transferred from set to set over a datalink interface. No initialisation for position or time is ever required.

Trimpack automatically selects the optimal mix of satellites and when only three are visible it automatically operates in an altitude-hold mode.

With three independent channels of GPS, Trimpack can navigate under a wide range of dynamic conditions, including speeds up to 600 kt. Its multichannel design also allows precise velocity measurement and guarantees quicker acquisition of satellites. It can go from a cold start to a three-dimensional fix in 2½ minutes and can calculate a fresh position every second thereafter.

Specifications

Dimensions: 198 × 223 × 63 mm
Weight: 1.9 kg
Power supply: 2 lithium batteries, rechargeable Ni/Cd battery pack or 8 alkaline AA cells
9-32 V DC, 5 W external power option

Contractor

Trimble Navigation Ltd, Aerospace Products.

VERIFIED

Mobile Reporting Unit (MRU)

The Mobile Reporting Unit (MRU) is the mobile segment of the Trimble Mobile Automated Command and Control System and its Advanced Range Tracking System. The MRU is a fully integrated, stand-alone unit designed to withstand MIL-STD shock, vibration and temperature extremes and to maximise reliability. It may be installed on fixed-wing aircraft and helicopters, as well as on land vehicles and naval vessels.

The MRU provides real-time unit identification, precise time-tag, course, speed and autonomous or differential GPS positioning with regular telemetry updates at the rate of one per second. System Operational Status information and two-way messaging are also available when appropriate system options are specified. When equipped with the standard low-noise C/A code GPS receiver, the MRU delivers differential GPS position and velocity.

While the standard low-noise receiver is adequate for most high-performance applications, the MRU accommodates other Trimble GPS receivers for special applications. Options include full military P(Y), standard navigation and other special purpose Trimble receivers.

Specifications

Dimensions:
(1 ATR) 318 × 257 × 191 mm
(½ ATR) 318 × 129 × 191 mm
Weight:
(1 ATR) 7.5 kg
(½ ATR) 6 kg
Frequency: 403-430 MHz, 450-470 MHz
Temperature range: −30 to +60°C
Altitude: −100-50,000 ft
Accuracy - autonomous GPS:
(position) 15 m
(velocity) 1 ft/s
(time) 0.1 μs
Accuracy - differential GPS:
(position) 2.5 m
(velocity) 0.3 ft/s
(time) 0.1 μs

Contractor

Trimble Navigation Ltd, Aerospace Products.

VERIFIED

TASMAN®-ARINC-TA-12 GPS sensor

TASMAN-ARINC-TA-12 is a 12-channel, Precise Positioning Service (PPS) GPS receiver that uses both the L1 and L2 frequencies. It corrects for Selective Availability (S/A), and incorporates Receiver Autonomous Integrity Monitoring (RAIM), Fault Detection Exclusion (FDE), and it protects against spoofing. It has interface capabilities to dual FMS/IRS and air data installations using ARINC 743 standards.

The TASMAN®-ARINC-TA-12 GPS sensor
2000/0080281

It meets US government certification requirements in both PPS and SPS modes, including FDE in accordance with DO-229. It provides users with an upgrade path to WAAS as a software-only upgrade. Initialisation is not required and a three-dimensional fix is calculated nominally within 1 minute.

It provides positioning accuracies of better than 16 m and also supplies velocity and time for navigation and tracking applications.

Specifications

Dimensions:
(receiver) 127 (W) × 207 (D) × 56 (H) mm
(antenna) 96 (W) × 102 (D) × 19 (H) mm
Weight:
(receiver) 2.2 kg
(antenna) 0.6 kg
Power supply: 9-32 V DC, <5 W
Accuracy - autonomous GPS:
(position 3-D) 16 m SEP
(velocity) 0.2 m/s RMS
(time) 100 ns RMS
Accuracy - differential GPS:
(position) 5 m SEP
(velocity) 0.2 m/s RMS
(time) 100 ns RMS

Operational status

In September 1999, TASMAN®-ARINC-TA-12 received US Federal Aviation Administration (FAA) certification C-129a Class B1/C1, and therefore now provides a standalone P(Y) code GPS sensor capability that fulfils both military Precision Positioning Service (PPS) and commercial Standard Positioning Service (SPS) requirements.

Selected for the US Air Force 1 aircraft.

Contractor

Trimble Navigation Ltd, Aerospace Products.

VERIFIED

Low-observable, conformal, antennas for avionics

For more than 30 years, TRW has developed communications antennas and radio frequency (RF) components. Today, TRW is applying that knowledge to the development of low-observable, conformal aircraft antennas for a wide range of applications in integrated Communications, Navigation and Identification (CNI) avionics, Electronic Warfare (EW) systems and Unmanned Aerial Vehicles (UAVs).

These low-observable, structurally integrated antennas are designed to provide efficient radiation characteristics that meet system performance requirements in avionics. They are capable of receiving and/or transmitting signals from High-Frequency (HF) through millimetre waves and support such functions as CNI, SATellite COMmunications (SATCOM), combat ID and situation awareness.

TRW has pioneered the development of 'Smartskin' antennas. The surfaces, or 'apertures', of these antennas serve as an integral part of an aircraft's outer surface without degrading the low observability of the vehicle. Smartskin antennas provide the bandwidth and efficiency associated with antennas having frequency-independent performance. While flown on a NASA Dryden F/A-18 test aircraft, TRW's VHF 'Endcap' antenna provided complete spherical coverage with a gain improvement of between 15 and 20 dB over the combined existing upper and lower blade antennas.

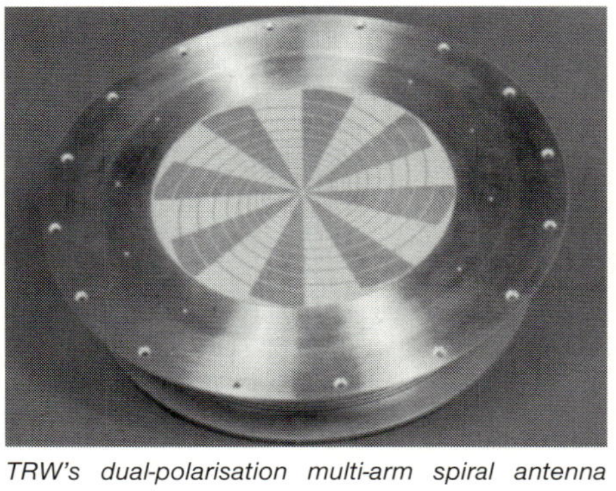

TRW's dual-polarisation multi-arm spiral antenna
1999/0044851

TRW is also studying the application of Smartskin antenna technology to the RAH-66 Comanche helicopter. This extension of the F/A-18 Endcap antenna provides simultaneous VHF and UHF communication from a single, low-observable structurally integrated antenna, meeting mission performance requirements for communication and navigation. This antenna technology enables the Identification of Friend or Foe (IFF), Global Positioning System (GPS) and HF functions.

In addition to Comanche, TRW's CNI antenna designs are being developed for new aircraft such as the F/A-18 E/F.

TRW's situation awareness antennas provide precise direction-finding capability from a single, efficient, extremely broadband, low-observable aperture. This unique antenna has applications to low-observable fighters, bombers, surveillance aircraft and ships in such functions as electronic warfare, Signal intelligence (Sigint), combat ID and other special applications.

TRW has also developed a high-gain, wide-bandwidth phased-array antenna for various applications. This was accomplished by designing, fabricating and testing a low-observable phased array possessing greater than 1,000 elements for simultaneous operation – from S-band through X-band (a five to one bandwidth). This phased-array antenna is highly efficient, since the radar-absorbing material is minimised while the very stringent low-observability requirements are met.

Finally, TRW is developing a similar phased-array antenna operating from C-band through Ka-band applicable to both military low-observable vehicles and

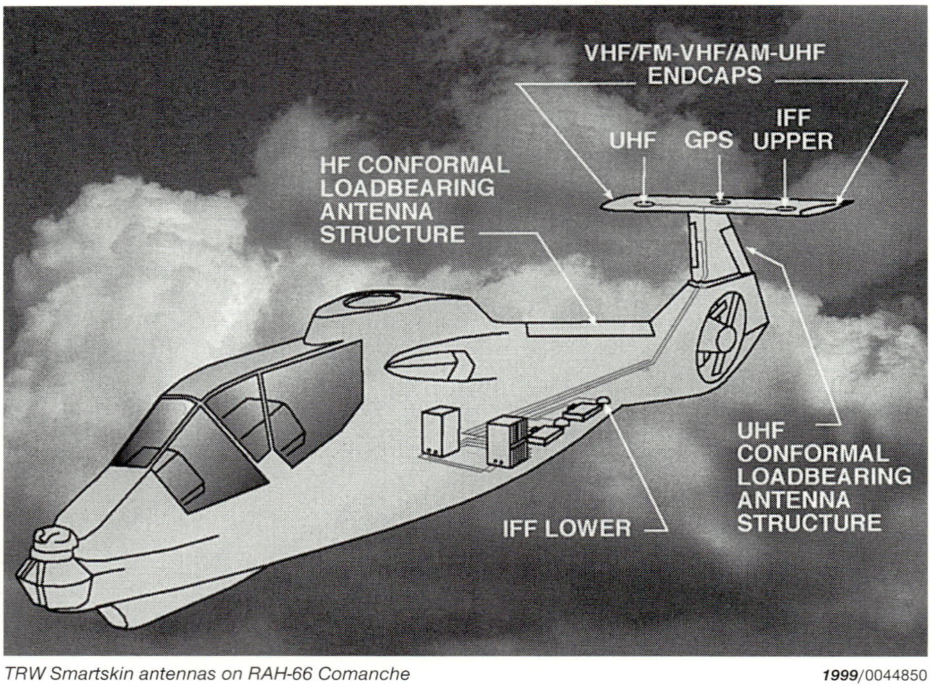

TRW Smartskin antennas on RAH-66 Comanche
1999/0044850

TRW Smartskin antennas on NASA Dryden F/A-18 **1999**/0044849

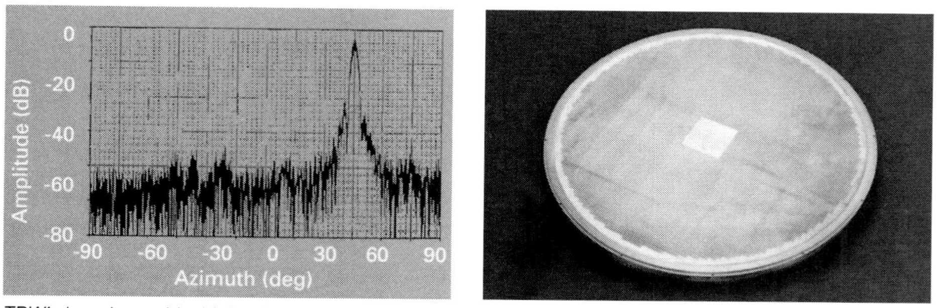

TRW's low-observable, high-gain, wide bandwidth (C- to Ka- band), phased-array antenna and test results **1999**/0044852

commercial vehicles. This design, measuring approximately 3 ft in diameter and having greater than 5,000 elements, will provide military and commercial multiSatcom functions from a common aperture that eliminates the need for multiple antennas – reducing cost, maintainability and installation issues.

Contractor
TRW Avionics Systems Division.

VERIFIED

STAR receiver system

The Special Threat Analysis and Recognition (STAR) precision direction-finding receiver system is intended for augmentation of ESM or radar warning receivers. The system uses a building-block architecture based on the INEWS (INtegrated Electronic Warfare System) heritage. The main part of the system is a dual-path, self-calibrating superhet receiver using a fast tuning phase-locked synthesiser and encoders for front-end sorting of radar parameters. An encoder chip digitises the input signal and controls phase interferometry measurements for determination of angle of arrival of the signal. Features include a programmable screening pulse environment and adaptable threshold. Also included in the system is a data processor, based on an Intel 80186 machine, which executes Pulse Recognition Interval De-interleaving (PRIDE) algorithms and formats contact reports for transmission to the host system.

The processing and control architecture of the STAR system allow it to accept high-level commands or to operate in a slave mode in the structure of a complex ESM system. Demonstrations of the system have shown an angular accuracy of 1°. It is claimed that pulses as narrow as 100 ns can be detected and between 30 and 100 targets can be tracked simultaneously.

In operation, the STAR system executes a search and tracking strategy defined by a host computer or system operator. The system scans the frequency rapidly, detects signals of interest, analyses them and calculates the angle of arrival, pulse amplitude and pulsewidth on a monopulse basis.

Specifications
Volume: 0.0141 m³
Weight: <8.2 kg

Frequency: 0.5-18 GHz
Pulsewidth detection: <100 ns without DF, 600 ns with DF
Target tracks: 30-100 simultaneously
Dynamic range: >50 dB
Accuracy: 1° RMS

Operational status
Flight tests have been carried out.

Contractor
TRW Avionics Systems Division.

VERIFIED

Target Locating Radar

The Target Locating Radar (TLR) is a self-contained radar detection, location, classification and tracking system. It can be used in aircraft or missiles or UAVs and is totally reconfigurable externally. TLR is available with a variety of wideband antennas. It searches for targets of interest and detected pulses of interest are processed to determine signal characteristics and angle of arrival. These data are reported over the bus interface for multiple targets. TLR can provide data for use by other sensors, the crew or as a direct input to a guidance system or can be used as a mid-course or terminal sensor. Other features of the TLR include light weight, small size and high accuracy.

Specifications
Volume: 0.0142 m³
Weight: 8.2 kg
Frequency: 2-18 GHz
Pulsewidth detection: 100 ns without DF, 600 ns with DF
Target tracks: 30 simultaneously
Accuracy: 1° RMS

Contractor
TRW Avionics Systems Division.

VERIFIED

ES4000 SIGINT system

The ES4000 is a complete airborne reconnaissance and surveillance system that consists of a mobile ground control and processing centre and modular surveillance and electronic warfare payloads. The system is intended primarily for light aircraft. It has 360° azimuth field of view and provides broad area surveillance with a communications signal intercept payload. Frequency range is 20 to 500 MHz, which can be extended to 1,200 MHz. Sensors are remotely controlled from a small, mobile shelter on the ground and SIGINT information is transmitted to the ground station in real time.

In addition to the signal intercept package, the aircraft can be fitted with a video and infra-red camera.

Specifications
Frequency: 20-500 MHz (1,200 MHz optional)
Frequency resolution: 100 Hz
Signal types: AM, FM, CW, USB, LSB
Accuracy: 2° RMS
Bearing files: 2,000

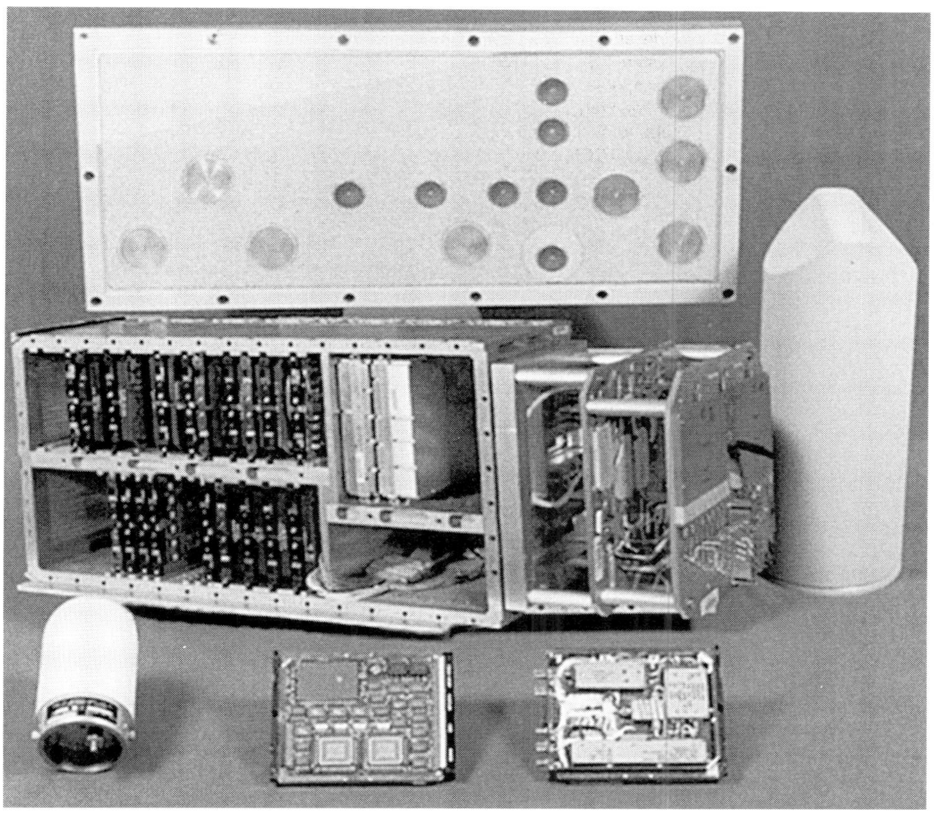

TRW's STAR precision DF system **1999**/0051275

Operational status
The ES4000 is installed in the Dornier Speed Canard SCO1-B optionally piloted vehicle.

Contractor
TRW Systems & Information Technology Group.

VERIFIED

ES5000 COMINT/ELINT system

The ES5000 is a combined communication and electronic intelligence system which can be installed in most types of aircraft. It performs signal interception, direction-finding and analysis, data collection and location of communication emitters from 2 to 1,200 MHz and non-communication emitters from 0.02 to 18 GHz. Information is transmitted by microwave link to a ground station. The airborne SIGINT equipment can be controlled either by operators on board the aircraft or remotely from the ground analysis centre.

On a typical mission, one or two airborne collection centres communicate by microwave link with the ground centre. One ES5000 can perform signal intercept, direction-finding and emitter location, producing emitter bearing data. A two-aircraft configuration can perform instantaneous emitter location on short-burst transmissions. Both systems pass data to the ground analysis centre.

Specifications
COMINT
Frequency: 2-1,200 MHz
Instantaneous coverage: 500 MHz
DF rate: 12/s coarse, 4/s fine
Field of view:
120° each side (primary)
±30° front and rear (secondary)
Elevation: −10 to +2°
Accuracy:
<2° RMS (primary)
<3° RMS (secondary)

ELINT
Frequency: 0.02-18 GHz
Coverage: 360° azimuth
Accuracy: 1-3° RMS
Measurements: PRI/PRF, pulsewidth, amplitude, scan time, DF, time of arrival, intrapulse
PRI range: 2-10,000 μs
Radar types: pulsed, 16 level stagger, CW, interval pulse, complex
Emitter library: 10,000 modes
Active emitters: 1,024

Microwave link
Frequency: E/F-band
Range: 350 km at 10,000 ft altitude

Operational status
In production and in service.

Contractor
TRW Systems & Information Technology Group.

VERIFIED

GuardRail/Common Sensor (GR/CS)

The US Army's GR/CS is an airborne SIGINT system that intercepts, locates and classifies target systems and transmits data to ground processors to provide real-time intelligence information. The GR/CS suite consists of a transportable ground-based system and RC-12 aircraft carrying remotely controlled mission equipment.

The Guardrail system is primarily for COMINT, but an ELINT capability has been added to later aircraft. The concept behind Guardrail is that data from multiple receivers are microwave downlinked to a ground-based Integrated Processing Facility (IPF).

The airborne relay facilities part of the Guardrail system is the airborne communications intelligence payload installed in electronic mission RC-12K, N and P aircraft. They contain intelligence intercept, signal classification and direction-finding equipment, and include provisions and interfaces for other signal intelligence subsystems, mission datalinks and aircraft navigation and communications systems. Collected data is datalinked directly to the IPF, or via an RC-12Q mothership.

Through a series of demonstrations, known as Precision SIGINT Targeting System (PSTS), Guardrail has proven its ability to work with national resources to provide targeting data for cross-programs operations. Guardrail has participated in the US Office of Naval Research (ONR)-sponsored Advanced Concept Technology Demonstration (ACTD), and is claimed to be already compliant with the US Joint Avionics SIGINT Architecture (JASA), with Guardrail/Common Sensor System 2 being the first operational implementation.

For the 21st century, GR/CS will be part of the US Army's Aerial Common Sensor system, which will also use the Direct Air-to-Satellite Relay (DASR) operational with Guardrail.

Operational status
Guardrail has been in operational service with the US Army for a number of years on RU-21H and RC-12D aircraft. The original Guardrail II was updated to IIA and was later updated to Guardrail IV. Further updating provided the Guardrail V under the nomenclature AN/USD-9(V)2. The US Army has developed a modification to Guardrail V that includes the Advanced Quick Look ELINT system produced by Electronics & Space Corporation and the Communications High-Accuracy Airborne Location System (CHAALS) provided by Lockheed Martin. This integrated system, known as GuardRail/Common Sensor (GR/CS), was deployed in the RC-12K.

Latest versions RC-12K, N, P and Q are fitted with upgraded CHAALS equipment and have access to the Defense Satellite Communications System (DSCS).

The US Army Communications Electronics Command (CECOM) began fielding the GR/CS Interoperability Subsystem (GRIS) in October 1998.

The GRIS is a real-time sensor management system that will provide the GR/CS with the ability to collect and analyse SIGINT data collected by the US Air Force within line of sight, using the Interoperable DataLink (IDL).

GRIS-capable variants of the aircraft include: RC-12 K, N, P and Q.

The IDL design is based on, and interoperable with, the US Air Force Contingency Airborne Reconnaissance System/Deployable Ground Station (CARS/DGS).

Contractor
TRW Systems & Information Technology Group.

VERIFIED

WTSS Wideband Tactical Surveillance System

The TRW WTSS is designed to be fitted in the Gulfstream IV-SP aircraft by means of an optional cargo door fit. The WTSS is based on COTS surveillance equipment to minimise risk.

Specifications
RF coverage: 20-2,000 MHz; 2 MHz and 40 GHz extension options
Signal search rate: 15-45 GHz/s
DF accuracy: <1° RMS
SIGINT sensors: fast signal search; D/F-emitter location; pool of fast set-on receivers; digital and audio playback; co-channel mitigation; ELINT option
Communications: air to ground datalink - G.703 compatible - 2 Mbps
Operator stations: 4-6 positions; Sun workstations; C and Ada software

Operational status
Installation proven.

Contractor
TRW Systems & Information Technology Group.

VERIFIED

TRW wideband tactical surveillance system 0018253

Advanced laser electro-optic/infra-red countermeasures

The US Air Force Wright Laboratory's Avionics Directorate is sponsoring the research and development of advanced technology for laser-based infra-red and electro-optic IR/EO countermeasures to fill operational requirements to defeat IR missiles and EO/IR tracking systems. This technology is being structured to meet both near-term and long-term demands driven by the continuing advancement and proliferation of EO/IR missiles and tracking systems.

Near-term programme
The near-term aim is to demonstrate and evolve the technology necessary for a threat-adaptable IR jamming suite capable of protecting large and high-value aircraft against the proliferated threat of surface-launched IR missiles. This IRCM suite would include provision for missile warning, fine tracking, missile identification or classification and laser jamming. The technology and performance requirement derived from this Advanced Technology Demonstration (ATD) programme would feed an offshoot development that could upgrade the specifications of current system developments, for example, the US Army Advanced Tactical Infra-Red CounterMeasures (ATIRCM) or USSOCOM DIRCM to meet US Air Force requirements. Initial Operational Capability for such a system is planned for 2000. One major aim is to satisfy user needs for data on technology needed to meet overall requirements at minimal cost. The programme will address the capability of basic open-loop jammers as well as more complex threat adaptive closed-loop techniques, signature suppression and missile warning system requirements. Wright Laboratory is currently under contract with Lockheed Martin Defense Systems at Akron, Ohio to develop the demonstration IRCM testbed to be used in field testing. The development plans for the IRCM testbed will require interface for multiple candidate missile warners and several candidate fine tracking/CM laser systems. Current plans call for using mature developmental systems such as the ATIRCM and the SOCOM/UK DIRCM as the baseline for upgrading to a higher level of capability. The programme will use IRCM lasers being developed by ARPA for future IRCM systems but is also seeking other laser source technologies, for example, the diode pumped approaches from Massachusetts Institute of Technology.

Long-term programme
In the longer term, the emerging focal plane array imaging seekers and laser-guided threats are advancing countermeasure technology. Also driving technology is the need to improve high-performance combat aircraft survivability against a threat base that is turning towards EO/IR to overcome the limitation caused by application of stealth radio frequency (RF) signature suppression techniques. The longer-term programme is researching multiple approaches to defeat robust threats while meeting the integration and operating requirements of high-performance combat aircraft. The philosophy of the programme is to pursue an integrated modular avionics capability in which the multiple functionalities offered by solid-state optical systems, high-power multiband lasers and advanced tracking and imaging recognition algorithms would provide a cost-effective compact system. Such a system may be able to provide the warning, tracking and identification functions against missiles, aircraft and ground EO/IR threats, and apply multiband countermeasures techniques against these threats. The countermeasures techniques may include both jamming and damaging options offered through application of directed energy technologies.

IR Decoys

Wright Laboratory is continuing to research expendable decoys which may be the only IRCM option for some aircraft in the future. As the US Air Force's Advanced Strategic and Tactical Infra-red Expendables (ASTE) programme/effort has entered into Engineering and Manufacturing Development and will be fielding the next-generation expendables, Wright Laboratory is funding technology programmes to support product improvements needed to meet third-generation IR threats that are included in the intelligence order of battle estimates. The Special Material Decoy (SMD) project will search for answers to two unique decoy requirements while the more mature mixed expendables programme will move toward injecting advanced decoy technology into operational use.

Contractor

US Air Force Materiel Command, Wright Laboratory.

VERIFIED

Advanced Tactical Targeting Technology (AT³)

Advance Tactical Targeting Technology (AT³) is a US Air Force Materiel Command, Wright Laboratory and the US Defense Advanced Research Projects Agency (DARPA) project intended to produce the US Air Force's next-generation lethal Suppression of Enemy Air Defenses (SEAD) system. Its aim is to produce an RF targeting system capable of reducing the missile battery targeting time from minutes to seconds. Under this concept, multiple aircraft fitted with ESM receivers will be networked to provide real-time targeting of hostile emitters within a CEP (circular error probable) of 15 to 50 m, thus enabling use of standoff GPS-guided weapons, such as the Joint StandOff Weapon.

In June 1998, three industrial teams responded to the research and development requirement: Boeing teamed with Litton Advanced Systems Division, Raytheon Electronic Systems; and Lockheed Martin Federal Systems (Owego).

Contracting agencies

US Air Force Materiel Command, Wright Laboratory.
DARPA.

VERIFIED

Ballistic Winds programme

The Ballistic Winds programme is advancing airborne light detection and ranging (lidar) wind profiler technology to provide real-time 3-D wind information to improve delivery of various projectiles from high altitude. Wright Laboratory has been developing and field testing numerous lidar systems for several years. The first systems were ground based, and were used mainly for technology advancements, measuring and quantifying atmospheric backscatter, and supporting field tests. Field tests included measuring winds to correct for parachute drift, quantifying wind vortices from large aircraft, and base lining newer airborne lidar systems. Lidar technology advancements have permitted the transition from ground-based to airborne systems. The first airborne wind profiling system was demonstrated in 1995 on a C-141 aircraft. That system uses an eye-safe 2 μm laser source which shoots a laser pulse from the aircraft through the atmosphere. The returned laser light is Doppler shifted (wind velocity component) by the dust and water vapour particles that are traveling with the wind. Then, through a sequence of laser scans in a conical manner, the system can determine the directional component. The range-resolved component is derived by the short pulse length of the laser and provides range slices of the atmosphere. As a result, the system can provide 3-D wind information from altitude to the ground with range slices of hundreds of metres and velocity accuracy of 0.5 m/s. This information can then be used to negate wind effects on projectile delivery.

The airborne laser wind profiler allows the navigator to incorporate real-time wind conditions from the aircraft to the ground and calculate a release point above 10,000 ft with high delivery accuracy. Wright Laboratory is planning to take the latest laser transceiver (smaller and more powerful than the original test transceiver) and install it in the forward section of a standard C-130 external fuel tank while keeping about 75 per cent of the fuel-carrying capacity intact. The pod-mounted approach offers the ability to mount the lidar in any C-130 without permanent modifications to the airframe and do this in one day using standard flight line equipment.

Operational status

In development.

Contractor

US Air Force Materiel Command, Wright Laboratory.

VERIFIED

Integrated Communications Navigation Identification Avionics (ICNIA)

The Integrated Communications Navigation Identification Avionics (ICNIA) radio terminal concept, relied on the use of rapidly emerging advanced Radio Frequency (RF) and digital technologies, to provide pilots with more Communications, Navigation, Identification (CNI) availability and flexibility than ever before. The ICNIA effort demonstrated that the pilot can do the same CNI functions as individual radios with the same performance, but in a smaller integrated terminal; a terminal capable of maintaining operation through multiple failures or battle damage while responding to the pilot's needs for different priorities of the CNI functions throughout the mission. Savings were projected for a 45 to 50 per cent reduction in size and weight with an accompanying 50 to 55 per cent reduction in life-cycle cost. The ICNIA programme was a US tri-service programme, air force led, with the army and navy as participating services.

The functions implemented by ICNIA were in the 2 MHz to 2 GHz range and included:

Joint Tactical Information Distribution System (JTIDS)
Enhanced Position Locating and Reporting System (EPLRS)
Have Quick
Single Channel Ground-to-Air Radio System (SINCGARS)
Global Positioning System (GPS)
Tactical Air Navigation (TACAN)
Mark 12 Interrogate
Mark 12 Transpond
Microwave Landing System (MLS)
Instrument Landing System (ILS)
VHF Omnidirectional Range (VOR)
Traffic Collision Avoidance System (TCAS)
HF
UHF
VHF
Mode S
Link 4
Link 11
Fleet Satellite Communications (FLTSATCOM).

The ICNIA terminal configurations were defined as the army terminal (ADM-1), the air force terminal (ADM-2) and the full-function terminal (ADM-3 and ADM-4). The ADM-1 terminal provided reception, transmission, processing and control of a tailored set of CNI functions for laboratory demonstration and was later flight demonstrated on an army UH-60 Black Hawk helicopter. ADM-2 was used to test Ada software activities to support the Advanced Tactical Fighter (ATF), and was delivered in place at the contractor's facility as an RF test bench and to support module repair. The ADM 3/4 terminals were delivered to Wright Laboratory's Avionics Directorate for further laboratory testing, demonstration and system development. The ADM-4 terminal was also used to demonstrate the navy's functions of Link 4, Link 11 and FLTSATCOM.

ICNIA made use of redundancy, resource-sharing, extensive Built-In Test (BIT), and Very High-Speed Integrated Circuit (VHSIC) technology to produce an architecture capable of detecting/isolating fault to the LRM level and dynamically reconfiguring to improve operation availability. The ICNIA software also employed a modular concept. As new functions or requirements became known, new software could be added to the ICNIA system to implement the new function or requirements, that is, ICNIA was software reprogrammable. This allowed flexibility to the ICNIA system that could not be realised with a federated or black box implementation. The implication is that new functions can be added to a weapon system that has an ICNIA system via a software upgrade and no new hardware modifications. This saves significant cost in Group A modifications (the cost of installing new equipment) to the weapon system. This was demonstrated by the addition of the navy functions.

ICNIA technology was transitioned to the F-22 Advanced Tactical Fighter and formed the foundation the F-22's CNI suite.

Operational status

With ICNIA technology transitioned to the F-22, the next step is to transition this technology to the retrofit aircraft market, such as the F-15 and F-16. A new programme, called the Integrated CNI Subsystem (ICNIS) programme, is planned to address technical and risk issues involved with transitioning this technology to retrofit platforms.

Contractor

US Air Force Materiel Command, Wright Laboratory.

VERIFIED

Moving and Stationary Target Acquisition and Recognition (MSTAR)

The objective of the MSTAR programme is to design, construct, and demonstrate in the laboratory an accurate and robust Automatic Target Recognition (ATR) system capable of locating and recognising time-critical targets in air-to-ground 2-D Synthetic Aperture Radar (SAR) imagery. The MSTAR programme emphasises an applied systems engineering approach that integrates advanced ATR algorithm modules to achieve accurate and robust performance in highly unconstrained ground-based image analysis scenarios. The scenarios focus on shallow hidden targets obscured by layover, partial masking, Cover, Camouflage and Deception (CC&D). The programme relaxes the constraints typically imposed by existing ATR development programmes on target, sensor, and background variants referred to as the MSTAR Extended Operating Conditions (EOCs).

The MSTAR EOCs include a minimum of 20 target classes at:

arbitrary viewing angles
radar squints off broadside at ±35°
radar depression angles up to 40°
modest levels of target camouflage and deception
30 per cent target obscuration levels in both flat and hilly terrains
multiple target configurations and articulations.

MSTAR algorithm design strategies employ a model-based or model-driven approach in which targets, backgrounds, target/background interactions, and the uncertainty with which they can be measured, extracted, and/or predicted, are each modelled to account explicitly for obscuration, clutter layover, camouflage, diffuse scattering, multipath reflections, and other operational conditions. A central tenet of this approach is that the large number of target, sensor, and background combinations found in realistic scenarios precludes the use of ATR algorithms that relies solely on precomputed templates, and which requires large amounts of measured data. MSTAR will significantly advance the state of the art of existing SAR model-based algorithms such as the Automatic Radar Air-to-Ground Target Acquisition Programme (ARAGTAP) conceived and developed by Wright Laboratory (WL) Avionics Directorate.

An associated team of module developers and a system integrator will develop the MSTAR system under a distributed, collaborative development strategy. The MSTAR modules are:

Focus of Attention - identifies potential targets and eliminates non-target areas in SAR imagery; uses contextual cues, map products, scene segmentation algorithms, map/image, and image/image registration algorithms, collateral information, image derived and externally provided area delimitation techniques and advanced target detection algorithms.

Indexing - coarsely classifies target detections and hypothesises target orientation and background effects such as occlusion and layover.

Feature Prediction - uses phenomenon-based target and background models to predict the appearance of significant, discriminating SAR image features, along with appropriate probability distributions characterising feature attribute uncertainty.

Feature Extraction - excises and characterises stable, phenomenon-based target and scene features from image regions of interest.

Matching - compares corresponding predicted and measured target and/or background features to yield a probabilistic measure or score of their 'closeness'; refines estimates of key target parameters such as configuration, articulation, and orientation.

Search Management - formally accrues evidence and dynamically controls the overall hypothesis refinement process; controls the sequence of feature prediction and matching processes; assesses the outcome of prediction and matching; terminates the hypothesis refinement process.

Operational status

Early development.

Contractor

US Air Force Materiel Command, Wright Laboratory.

VERIFIED

Radar detection of concealed time critical targets (RADCON)

RADCON is a joint effort between Wright Laboratory (WL) and the Defense Advanced Research Projects Agency (DARPA). The objective is to extend and demonstrate foliage penetration (FOPEN) technology. RADCON will develop automatic target detection algorithms to enhance the performance of a radar sensor operating from an airborne platform to detect and classify time-critical targets concealed by foliage and/or camouflage. RADCON will use real radar data to enhance the performance of target/clutter discrimination algorithms to achieve a useful probability of detection and false alarm rate at real-time operating speeds; develop a real-time signal processing capability; develop a concept of operation for a radar sensor capable of performing foliage penetration from an airborne platform and develop a foliage penetration radar system specification for an airborne platform consistent with the developed concept of operation and foliage penetration capability.

The RADCON Automatic Target Detection/ Classification (ATD/C) algorithm development began with the proven Concealed Target Detection (CTD) ATD/C algorithm as a baseline and will iteratively improve its performance. During the three years' duration of the programme, radar data will be used to evaluate and enhance the performance of automatic concealed target detection algorithms. Target versus clutter discriminating techniques including, but not limited to, polarisation diversity, angle discrimination, wavelength, target groupings as well as other discriminants, are being developed, tested and evaluated to enhance the performance of the concealed target detection algorithms. These discriminants have been tested to ensure compatibility with the modelled data used in CTD. The performance of these algorithms will be evaluated by performing analysis of the real radar data containing targets concealed under foliage and/or camouflage. Once developed, RADCON's algorithms will be placed in the Khorus image processing environment and will be delivered as Khoros Glyphs as part of a Khoros toolbox to maximise development transportability and to minimise duplication of effort.

In order properly to evaluate the concealed target detection algorithms and validate the selection of target/clutter discriminants, RADCON has developed an advanced ground-based signal processing system. This demonstration system has been implemented from a Mercury i860 parallel processing system and will process data at real-time rates. The system is a partially populated VME 9U chassis with 104 i860 processors with 16 Mbyte of memory per node. These processors are on ruggedised 9U VME cards with 16 processors per card. The benchmark performance of this system ranges from 4 to 8 Gflops depending on instruction mix. The system has a total of 1.664 Gbyte of memory and will process a single polarisation of radar data (IFP and RFI) at the real-time rate of 7.4 s/km^2 and 3.3 s for ATD/C. This system is available for further system trade studies and ATD/C algorithm development and performance analyses.

System, algorithm, and hardware trade studies have been performed to determine the ultimate system specification and concept of operation for real-time performance of IFP, RFI, and ATD/C processing of FOPEN data from an airborne platform.

Operational status

RADCON is a risk reduction programme and is in its last year of development at which time it will have automatic detection algorithms capable of integration into an airborne demonstration of a FOPEN radar capable of detecting time-critical targets concealed under foliage and/or camouflage.

Contractor

US Air Force Materiel Command, Wright Laboratory.

VERIFIED

WJ-8721 digital HF receiver

The WJ-8721 airborne digital HF receiver is one of a family of air and ground-based HF receivers that provide surveillance and monitoring of RF communications from 5 kHz to 30 MHz with 1 Hz tuning resolution. By combining analogue and Digital Signal Processing (DSP), these receivers achieve high performance at low cost in a number of packages.

Contractor

Watkins-Johnson Telecommunications Group.

VERIFIED

WJ-9104A multichannel digital tuner

The WJ-9104A tuner provides up to eight RF channels, with a frequency range of 25 to 2,000 MHz (tunable 10 to 2,600 MHz), and a maximum tuning speed of 60 µs.

An operator can configure the tuner channels via remote control or tune independently, or in a phase-coherent direction-finding mode where channels share common local oscillators. In either mode, each tuner channel provides a digitised 10 MHz instantaneous bandwidth with 12 bits of precision.

A high speed SCSI-2 handles remote operation and control of the tuner. A direct-tuning control interface allows precision triggering and timing.

Contractor

Watkins-Johnson Telecommunications Group.

VERIFIED

AN/ALQ-504 airborne VHF/UHF intercept, DF, jamming and deception systems

The AN/ALQ-504 is a small, lightweight airborne integrated intercept, DF, jamming and deception system, that can be installed in a wide range of fixed- and rotary-wing aircraft. The system can also be adapted for installation in wheeled or tracked vehicles, or on board ships.

The system performs manual or fully automatic search and detection of communications signals in the VHF and UHF bands between 20 and 400 MHz. Frequency extensions up to 2,500 MHz are available.

The AN/ALQ-504 system measures DF lines of bearing, and computes the location of intercepted signals. It enables the operator to jam selected signals or to transmit false messages for deception or training purposes. Signal jamming can also be linked to DF bearing, providing jamming of signals within a selected azimuth sector.

Jamming power output is 400 W to allow operation at a considerable standoff distance. Up to four active signals from a prioritised list of up to 10 frequencies can be jammed simultaneously using various strategies. The operator can seek and identify hostile signals, and can either jam each selected signal with one of a variety of computer-generated modulations, or implement the system's deception capabilities. The operator can use pre-recorded or processed messages as deception tools.

Contractor

ZETA, an Integrated Defense Technologies Company.

VERIFIED

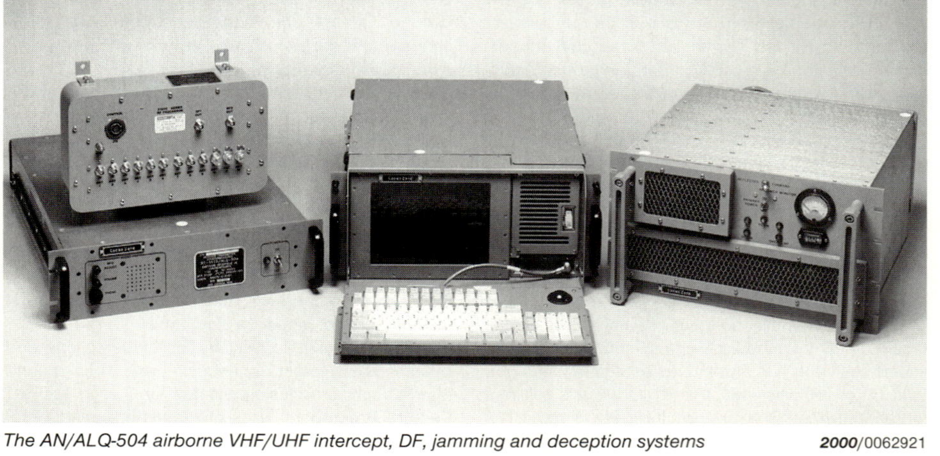

The AN/ALQ-504 airborne VHF/UHF intercept, DF, jamming and deception systems **2000**/0062921

ZS-1915 airborne communication signals intercept and DF system

The ZS-1915 system provides for the intercept, detection, monitoring, DF and location of communication signals in the 20 to 500 MHz frequency range, with optional extensions to 1,000 or 2,700 MHz. The system is designed for installation in a wide variety of aircraft, from small single-engined turboprop aircraft upwards. The ZS-1915 DF antenna array comprises a set of standard blade antennas or a radome assembly mounted on the aircraft underside. DF accuracy depends on the aircraft installation, but it is normally better than 3° RMS. The ZS-1915 system provides for continuous, or discrete, high-speed frequency search with simultaneous DF measurement on active signals. Active signals can be handed off to one or more monitor receivers, whose audio outputs can be recorded on an integrated Digital Audio Record & Playback (DARP) subsystem.

An interface to read the aircraft navigation/GPS system is provided to automatically establish emitter location from signal intercepts. An integrated Geographic Information System (GIS) accepts digital maps in a variety of formats. The location of an emitter is automatically calculated from a series of lines of bearing taken over a period of time as the aircraft flies past the emitter. Both lines of bearing and the calculated emitter location are displayed as overlays on the digital map.

Emitters operating in a network can be independently identified and located. All data obtained during a flight mission are automatically entered and stored in the system database, including flight path, intercepts, emitter locations and audio recordings. This data can be transferred by removable disk to a ground analysis station, where the complete mission can be 're-flown' and analysed.

The ZS-1915 system operator uses a rugged PC-based workstation with colour monitor and keyboard/trackerball together with the Microsoft Windows NT® operating system to provide a familiar, easy to use operating environment. A wide variety of display modes and operating screens is available to suit different operational missions. Additional monitor receivers, an increased channel capacity DARP, and a signal analysis capability are available as options.

RC 232C remote-control and network (LAN) interfaces to other airborne or ground-based systems are available as options. The client/server-based software design, using standard TCP/IP protocols, permits easy expansion to a more extensive COMINT/DF system.

Contractor

ZETA, an Integrated Defense Technologies Company.

VERIFIED

ZS-3015/4015 ultrascan intercept and direction finding systems

The ZS-3015/4015 ultrascan intercept and direction finding systems are designed for tactical and strategic signal collection requirements demanding high performance in local or remote fixed-site, vehicle, airborne or shipboard applications.

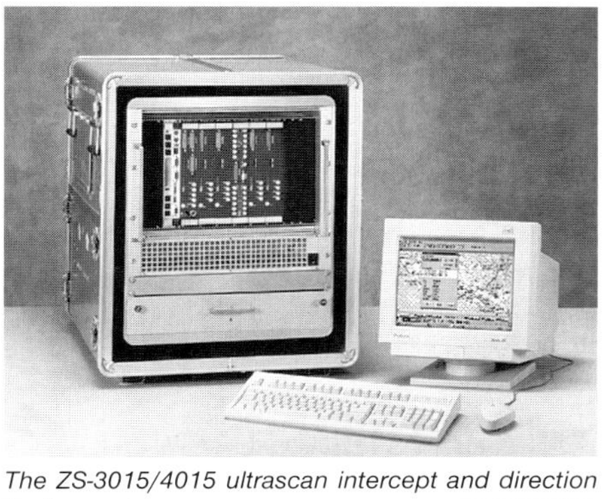

The ZS-3015/4015 ultrascan intercept and direction finding systems **2000**/0062919

The ZS-3015 (2-channel) and ZS-4015 (5-channel) systems use wideband digital receivers and Digital Signal Processing (DSP) techniques to achieve high-performance signal intercept and direction finding over the 20 to 3,000 MHz frequency range.

Precision DF measurements result from the use of a proprietary, proven, correlation interferometry DF technique. Very high spectrum scan speeds are achieved through use of a 6.4 MHz wideband multichannel receiver with Fast Fourier Transform (FFT) processing using parallel Digital Signal Processors (DSPs). A comprehensive Windows NT® Graphic User Interface (GUI) enables the operator to select a wide variety of automatic and manual operating modes.

SQL database management supports building Signals Of Interest (SOI) files with all pertinent signal parameters and mission intercept files. An integrated Geographic Information System (GIS) accepts a variety of digital maps, and it displays lines of bearing and emitter locations as overlays on the map.

Other system features include Digital Audio Record and Playback (DARP) of received audio, with correlated operator comments of signal intercept events, local or remote control, built-in networking capabilities between systems and open client-server architectures via LAN/WAN using RF, fibre-optic or wire links.

Contractor

ZETA, an Integrated Defense Technologies Company.

VERIFIED

MILITARY DISPLAY AND TARGETING SYSTEMS

AUSTRALIA

Long-Range Tactical Surveillance (LRTS) sensor

BAE SYSTEMS Australia's LRTS thermal imaging sensor is part of a family of thermal imaging sensors for land, sea and air use. It is optimised for the maritime environment. Notable features are a fourth-generation, high-definition 640 × 486 pixel staring array detector, with high-performance optics and two or three fields of view; exceptional range performance in humid conditions.

Specifications

Detector: Platinum Silicide (PtSi): 640 × 486 pixels
Spectral response: 3-5 microns
Fields of view (degrees):
(narrow) 2.9 × 2.2
(wide) 11.3 × 8.65
Weight: 1.8 kg
Interfaces: RS422/232
Video format (frame rate): CCIR (25 Hz) or RS-170 (30 Hz); programmable 640 × 486 composite monochrome or RGB graphic overlay

Operational status

In service with the Royal Australian Navy.

Contractor

BAE SYSTEMS Australia Ltd.

VERIFIED

Ranger 600 eye-safe laser range-finder

The Ranger 600 is one of a family of BAE SYSTEMS Australia's tactical sensor solutions for naval, land and airborne applications. The Ranger 600 eye-safe laser range-finder is designed for a maritime environment and provides excellent sensitivity and range performance for a laser range-finder that is safe to the unprotected eye.

Ranger 600 uses the latest high-reliability, solid-state, diode-pumped laser technology that is OPO frequency shifted to eye-safe wavelengths. Notable features include: programmable internal false alarm and probability of detection settings; high-performance transmit and receive optics, with transit alignment to within 0.2 mrad and field of view less than 1.5 mrad; hardware-interlocked safety override; synchronisation of the laser ranging function to an external imaging sensor to optimise tracking accuracy.

Specifications

Detector: avalanche photodiode
Range: 30 km
Accuracy: ±2 m
Boresight: 0.2 mrad
Output power: 7.8 mJ per pulse (min)
PRF: up to 20 Hz
Fields of View (FoV):
(receiver) 1.5 mrad
(transmitter) 0.5 mrad
Weight: 10.9 kg
Power: 24-32 V DC
Interfaces: RS-422 (optional RS-232)
Cooling: air cooled

Operational status

In production and in service with the Royal Australian Navy.

Contractor

BAE SYSTEMS Australia Ltd.

UPDATED

BELGIUM

BARCO avionics family concept

The BARCO avionics family concept focuses on three major parts of the cockpit man/machine interfacing: the Control Display and Management System (CDMS), the intelligent MultiFunction Display (MFD) and the dumb Cockpit Head-Down Display (CHDD).

The architecture is built around an open and modular architecture integrated into the general purpose avionics processing unit, allowing BARCO to adapt the different products to the individual customer requirements. Commonality of processing unit for all the Control Display Units (CDUs) and the MultiFunction Displays (MFD) yields a powerful and flexible hardware/software architecture.

The avionics processing unit houses all the option boards such as high-end graphics generators, ARINC 429, ARINC 629, MIL-STD-1553B synchro/discrete/analogue interfacing, video RS-170 and PCMCIA dataloader. All the option boards are identical for use in both the CDMS and MFD. Several sizes for the CDU keyboard panel are available. The panel sizes BARCO offers are 3 × 4, 4 × 5 and 6 × 8 in, portrait or landscape.

BARCO's CDMS control and display management system **1999**/0051650

Specifications

MFD6.8/1
Screen: 6 × 8 in – colour LCD
Power: Std 28 V DC (MIL-704)
Intelligence: flexible
Environment: military and civil

CHDD 5.4/1
Screen: 5 × 4 in – colour LCD
Input: STANAG 3350 B – RS-170 – PAL/NTSC
Output: STANAG 3350 B – PAL – RGB
Environment: military and civil

CDMS 3.4/1
Screen: 3 × 4 in – colour LCD
Modular architecture Control Display Unit (CDU)
MIL-STD-1553 remote terminal or ARINC 739 terminal
Environment: military and civil

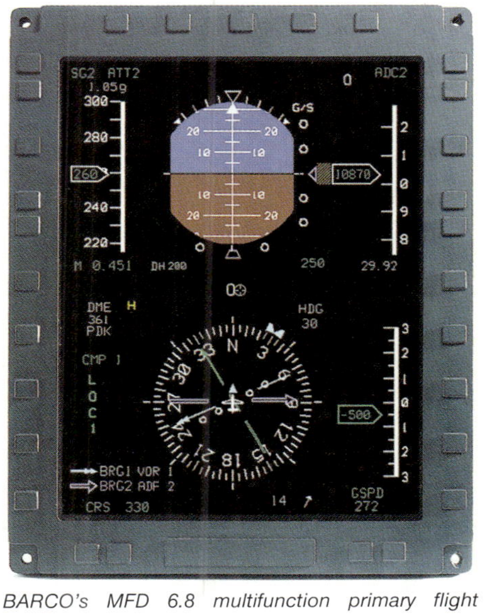

BARCO's MFD 6.8 multifunction primary flight instrumentation display **1999**/0051651

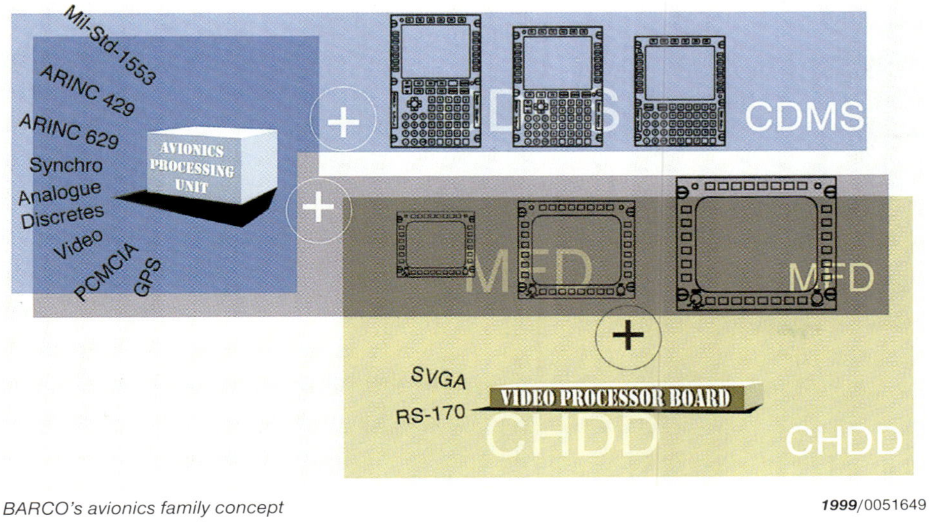

BARCO's avionics family concept **1999**/0051649

Operational status

MFD6.8/1 selected for T-33 HSI/ADI and RC-135 cockpit modernisation; CHDD 5.4/1 selected for the colour head-down display for the Boeing Night-hawk Targeting System being installed on the Eurocopter BO-105 helicopter; CDMS 3.4/1 selected for Eurocopter Tiger, C-5 cockpit upgrade programme, and for the C-27J. All the display systems described became available in 1999.

Contractor

BARCO n.v., Display Systems - Avionics Division.

UPDATED

FlexVision video products family

FlexVision is a family of hardware and software products that provide exceptional flexibility and capability in digitally mixing graphics from an independent workstation with video, radar and other graphics. FlexVision is available as a stand-alone box, or as a set of FlexVision VME or PCI cards.

The host-independent product family mixes a wide variety of sources into a single RGB video output. The video sources appear as windows within the master RGB video output, and the operator can easily position, scale or merge each window independently. There is no degradation in graphics performance when displaying in this multiwindow environment. Boards are also available for conduction-cooled environments.

Specifications

Boards available:
video mixer
colour video frame grabber mezzanine

FlexVision hardware **1999**/0051653

RGB video frame grabber mezzanine
Vxbus extension
Resolution: up to 1,600 × 1,200
Video standards: NTSC, PAL, S-video, HDTV, FLIR, PPI radar, PC video
Operating systems: HP-UX, Solaris 2.5, DEC, UNIX, VX Works, Windows NT

Operational status

FlexVision hardware and software has been selected for the Tactical Command System (TCS) display of the UK Nimrod MRA 4. It will be embedded in the Raytheon workstations and used to display the output from: the primary sensors radar, acoustics, Electronic Support Measures (ESM), Magnetic Anomaly Detector (MAD) and electro-optical surveillance and detection system; the Defensive Aids Sub-System (DASS); the armaments control system and the mission communications suite.

BARCO Display Systems has been selected by Raytheon Computer Products to supply conduction-cooled digital video mixer subsystems for the US Air Force AirBorne Laser (ABL) programme. BARCO's FlexVision video windowing subsystem will be embedded in a Raytheon Model 960 computer. BARCO has already been chosen by The Boeing Company to supply 20.1 in colour flat-panel displays for the ABL programme. Boeing will integrate both the Raytheon 960 computer with the embedded BARCO Flexvision subsystem and the BARCO 20.1 in FD 251 display into the modular console for the ABL surveillance, battle management, command, control, communications, computers and intelligence (BMC4I) system.

Contractor

BARCO n.v., Display Systems - Avionics Division.

UPDATED

The BARCO Display Systems MFD-5.4/1 multifunction primary flight instrumentation display **2001**/0089538

MFD-5.4/1 primary flight display/ engine instrument display

The MFD-5.4/1 is a multifunction primary flight instrumentation display with a sunlight-readable 5 × 4 in Active Matrix Liquid Crystal Display (AMLCD).

Specifications

Screen: 5 × 4 in AMLCD
Resolution: 640 × 480 pixels
Power: Std 28 V DC(MIL-STD-704)
Optional: night vision goggles compatible

Contractor

BARCO n.v., Display Systems - Avionics Division

NEW ENTRY

MHDD 3.3 Monochrome Head-Down Display

The MHDD 3.3 is a compact, high-resolution 2.7 in (68 mm) circular avionics display used in airborne applications. It is a monochrome, sunlight-readable

Example FlexVision display **1999**/0051652

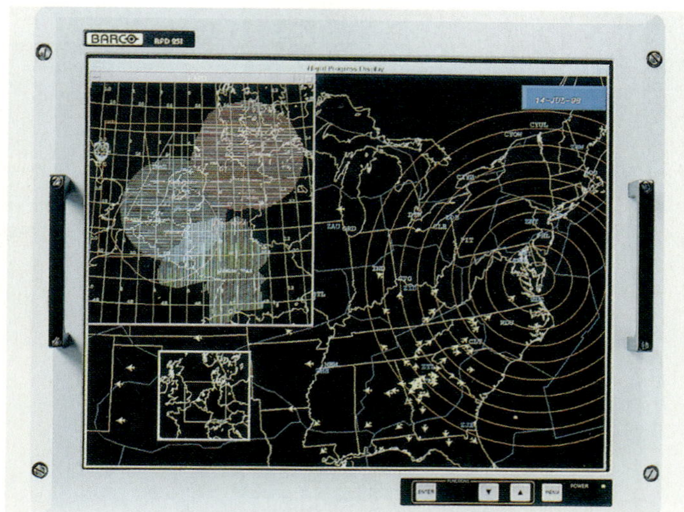

The BARCO Display Systems RFD 251 20.1 inch flat-panel display **2001**/0089537

MHDD 3.3 monochrome circular head-down display
0018219

display, with a raster scanning system suitable for fighter aircraft environmental requirements.

The MHDD 3.3 offers a high-brightness display, with very low power consumption, compact, lightweight design and brightness control located on the front of the display. It has full MIL specifications and is convection cooled, requiring no forced air.

Specifications

Dimensions: 82.6 × 82.6 × 176 mm
Weight: 17 kg
Power supply: ± 15 V DC, 20 W (max.)
Temperature range: −40 to +55°C
Altitude: up to 80,000 ft
Image dimensions: 68 mm (2.7 in) circular
Resolution: 384 × 384 pixels (cross)

Operational status

In production, utilised for international airforce programmes such as Belgian Air Force F-16 aircraft on the Carapace passive ECM subsystem. Also selected for several helicopters of the ALAT (France) as the display for the EWR-99/Fruit radar warning receiver display.

Contractor

BARCO n.v., Display Systems - Avionics Division.

VERIFIED

Boeing's V-22 Osprey tiltrotor aircraft equipped with BARCO's SFD 4.3 avionics standby flight display
2001/0089527

SFD 4.3 avionics standby flight display

The SFD 4.3 avionics standby flight display uses Active Matrix Liquid Crystal Display (AMLCD) technology to provide high quality optical performance in terms of luminance, contrast and reflectivity. It is based on a Commercial-Off-The-Shelf (COTS) design using open and modular architecture.

Operational status

Selected by Boeing for the US Marine Corps V-22 Osprey tiltrotor aircraft. Deliveries will begin in June 2001

Contractor

BARCO n.v., Display Systems - Avionics Division.

NEW ENTRY

PC16 pod control unit

The PC16 pod control unit monitors and controls aircraft gunpod functions. The PC16 features a burst controller which limits the number of rounds fired in a burst. Once the preset figure is reached, firing is automatically interrupted until the trigger is released. It includes a bar graph indicator which shows, on two 10 LED linear indicators, the total ammunition content of each pod, with each LED representing 10 per cent of the full load. When ammunition reaches less than 10 per cent of the full load, amber spare ammunition indicator lamps illuminate.

The PC16 incorporates a number of safety features. These include a non-reversible sequencing selector to prevent wrong or dangerous use of the system, a positive locking selector to prevent the gun being fired during the landing sequence, a control test which tests the system during the pre-take-off checks, a firing auto control to allow the gun to be fired only when certain conditions are met and auto-recocking to bring the gun back into operational condition after a misfire.

The system has a number of dimmable status lights. Information from these can, as an option, be fed into the aircraft intercom system to provide audio signals.

Contractor

FN HERSTAL.

VERIFIED

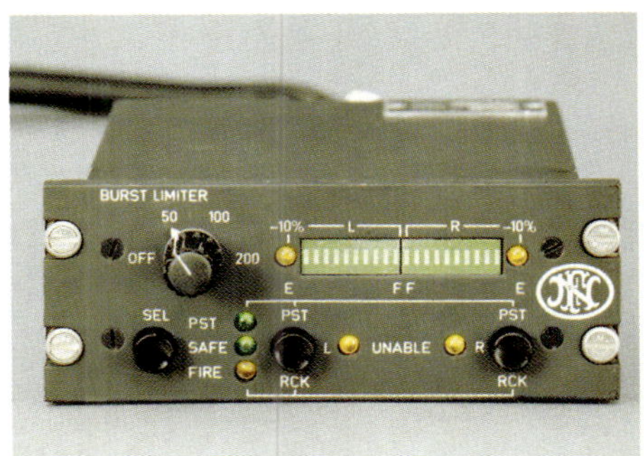

The FN HERSTAL PC16 pod control unit 0106365

HELIMUN helicopter pilots' night vision system

OIP Sensor Systems' HELIMUN stereoscopic night vision system is the latest addition to the LORIS family of night vision systems, available with 40° and ultra-wide 60° field of view.

HELIMUN has been specifically designed to fit easily on to a pilot's helmet. Principal enhancements are the new SuperGen image intensifiers, individual interpupillary adjustment, larger eye relief and exit pupil eyepieces, and a newly designed power module that is built on to the back of the helmet to increase the user's comfort. The mounting of HELIMUN can be easily adapted to fit all kinds of pilots' helmets.

The new SuperGen image intensifiers provide a high-resolution image. In addition, these new image intensifiers have higher photocathode sensitivity, signal-to-noise ratio, a more stringent blemish specification and a higher burn-in resistance compared to standard intensifiers.

HELIMUN provides a very high image quality over a much greater range of eye positions than the standard 15 mm eye relief eyepieces, ensuring that all users have the full field of view (40° or 60°) including those with deeper set eyes and those who wear corrective eye glasses. HELIMUN greatly simplifies the task of centring an eyepiece directly in front of each eye. Further, the new eye mechanism reduces monocular wobble.

Specifications

	40° FOV	**60° FOV**
Magnification	× 1	× 1
Scene illumination	min 0.1 mlux	min 0.1 mlux
Field of view	40°	60°
Spectral responses (nm=nanometre)	350 to 880 nm	350 to 880 nm
Resolution at 1 mlux (with SuperGen tube) (USAF target 85% contrast)	1.7 mrad/lp	1.9 mrad/lp
Gain	min 3.000 cd/m² lux	min 3.000 cd/m² lux
Dioptre adjustment	− 6 to + 2 dioptre	− 6 to + 2 dioptre
Interpupillary adjustment	52 to 74 mm	52 to 74 mm
Tilt adjustment	min 10°	min 10°
Exit pupil	10 mm	10 mm
Eye relief	28 mm	20 mm
Focal range	24 cm to ∞	24 cm to ∞
Flip-up/Flip-down	standard	Standard
Battery type/life	2 pieces AA-size + set of back-up batteries Alkaline: ≥ 75 h + 75 h	

Contractor
OIP Sensor Systems.

VERIFIED

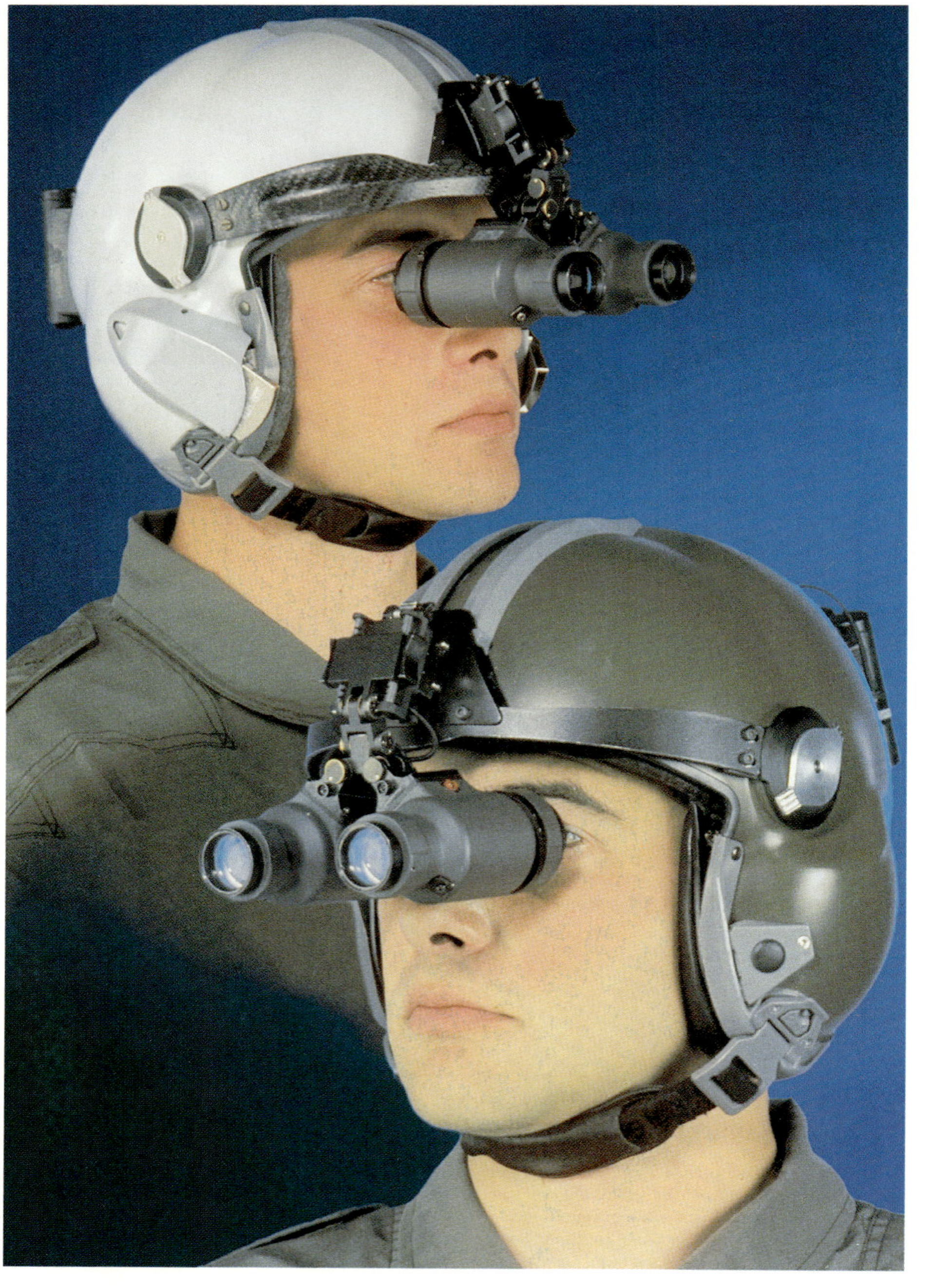

HELIMUN helicopter pilots' night vision system with 60° field-of-view
1999/0051710

CANADA

Peripheral vision display

The peripheral vision display is designed to prevent pilot spatial disorientation and improve flight safety.

The system presents the pilot with a line of light projected across the instrument panel. This laser-produced line moves parallel to the earth's horizon as the aircraft changes its attitude. As this line can be seen by the pilot's peripheral vision, he will be aware of changes in aircraft attitude even while his attention is diverted outside or to other cockpit tasks.

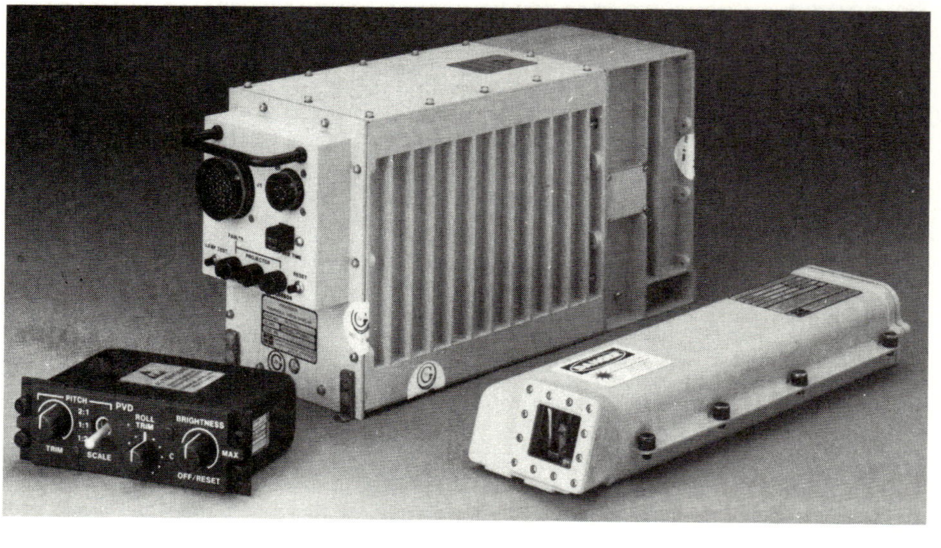

Specifications
Dimensions:
(control panel) 146 × 48 × 123 mm
(processor) 127 × 51 × 318 mm
(laser projector) 102 × 51 × 318 mm
Weight: <9 kg

Operational status
In operational service on a classified US Air Force programme. Under evaluation at the Defence Evaluation Research Agency, Farnborough on a Sea King helicopter. Also under evaluation on board the US Air Force F-16, the Canadian Forces CH-113 and the US Navy SH-2F ship motion simulator and vertifuge.

Contractor
Honeywell Aerospace Canada.

VERIFIED

The Honeywell Aerospace Canada peripheral vision display system

Aviator's Night Vision Imaging System (ANVIS)

The Aviator's Night Vision Imaging System (ANVIS) is a helmet-mounted unity power image intensifier binocular which allows low or contour flying in fixed-wing aircraft or helicopters, at night. The system is compatible with either second- or third-generation image intensifier tubes and can be used in conjunction with FLIR systems. The binocular is lightweight, can be fitted to any flying helmet and provides full peripheral vision. The complete ANVIS consists of the binocular assembly, image intensifier tubes, helmet visor interface and battery power pack.

Specifications

Weight:
(binocular) 0.463 kg
Field of view: 40°

Operational status

In service with the US Army.

Contractor

Hughes Elcan Optical Technologies Ltd.

VERIFIED

Avionic Keyboard Unit (AKU)

The Avionic Keyboard Unit (AKU), originally designed and developed for the AH-64C Apache helicopter, is available as a durable intelligent keyboard for crewstation applications.

The AKU can be incorporated in any digital cockpit or mission architecture and is fully qualified as a MIL-E-5400 Class II device for airborne applications. It is a lightweight, versatile general purpose keyboard, with MIL-STD-1553B serial digital interface or optional ARINC 429 or RS-422 serial interfaces.

Specifications

Dimensions: 124 × 155 × 63.3 mm
Weight: <1.4 kg
Power supply: 28 V DC
Display: 22 characters in 1 line
Temperature range: –54 to +55°C
Environmental: MIL-STD-810D
Reliability: >7,500 h MTBF

Contractor

Litton Systems Canada.

VERIFIED

Comanche colour and mono MultiFunction Displays (MFDs)

The MultiFunction Displays (MFDs) for the RAH-66 Comanche comply with MIL-L-85763 for the NVIS Class A mode of operation, using advanced techniques to achieve the Comanche optical specification. There are four MFDs per aircraft, with two displays providing high-resolution monochrome for sensor information and two displays that provide colour.

Specifications

Dimensions: 203.2 × 241.3 × 218.4 mm
Weight: 5.72 kg
Power: 130 W excl heaters
Display area: 152.4 × 203.2, 480 × 640 RGBG stripe

Operational status

In development.

Contractor

Litton Systems Canada.

VERIFIED

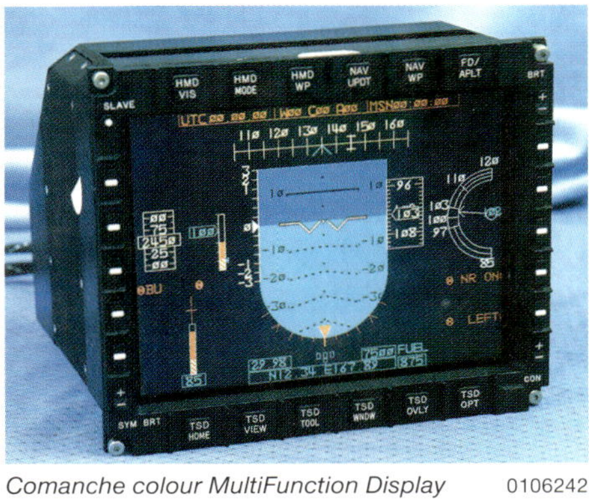

Comanche colour MultiFunction Display 0106242

Comanche MultiPurpose Display (MPD)

Litton Systems Canada has designed, developed and produced the monochrome MultiPurpose Display (MPD) to provide graphics for the Comanche aircraft and weapon functions. There are two MPDs located in each of the two cockpits, making four displays in all in each Comanche.

Litton Comanche MultiPurpose Display 0106243

Specifications

Dimensions: 129.5 × 200.7 × 149.9 mm
Weight: 2.99 kg
Power: 50 W excl heaters
Display: 91.4 × 170.2 mm, 288 × 544 pixels

Operational status

In development.

Contractor

Litton Systems Canada.

VERIFIED

Data Entry Display (DED) system

The data entry display set consists of two units: a Data Entry Display (DED) assembly and a power supply. The DED assembly is a sunlight-readable upfront display developed under the F-16 multinational staged improvement programme. The display unit, although graphics capable, usually provides five lines of 24 characters and is used for presenting communications, navigation and IFF data. The DED set has been under high-rate production since 1981.

The purpose of the DED is to take in serial data in a raster scan format and display it on a 76 mm wide by 25 mm high screen. The display screen is composed of three LED display modules, each of which provides an array of 64 × 64 pixels on a 25 × 25 mm viewing surface. The DED is designed to store a serial bit stream of data and display it in raster scan format on a screen 192 pixels wide by 64 pixels high, consisting of three modules. In order to keep the information updating process as independent from internal timing constraints as possible, two identical memories are provided in the DED. When one memory is used to refresh the screen the other memory is available for updating. When the memory which is being updated becomes completely filled, the memory functions are exchanged so that the memory with the latest complete set of data is used to refresh the screen and the other memory is available for updating. Each memory has one bit of storage available for each pixel on the screen.

In addition to the circuitry required to provide the display, the DED also contains independent fault detection and temperature alert discretes and a thermostat to enable in-flight non-interruptive monitoring.

Operational status

In service on the Lockheed Martin F-16C/D.

Contractor

Litton Systems Canada.

VERIFIED

EH 101 Merlin mission displays

Litton Systems Canada is under contract to supply 625-line CCIR-compatible Active Matrix Liquid Crystal Displays (AMLCDs) to meet the mission video display requirements for the UK Royal Navy EH 101 Merlin cabin and cockpit. A Merlin set consists of four cabin and two cockpit units.

Specifications

Dimensions:
(cockpit) 254 × 241.3 × 175.3 mm
(cabin) 396.2 × 299.7 × 182.9 mm
Weight:
(cockpit) 7.48 kg
(cabin) 10.43 kg
Power supply:
(cockpit) 130 W
(cabin) 110 W
Display:
(cockpit) 147.3 × 195.6 mm, 100 pixels/in
(cabin) 205.7 × 274.3 mm, 576 × 768 Quad RGBG

Operational status

In production.

Contractor

Litton Systems Canada.

VERIFIED

General Purpose Control Display Unit (GPCDU)

The General Purpose Control Display Unit (GPCDU) has been developed as a low-cost, rugged avionic user interface for data input and display, control and management of aircraft systems. The GPCDU incorporates a Litton/LED, 10-row by 24-character dot matrix display and a large number of interface options.

The GPCDU can also be supplied as a SNU-84-1 compatible keyboard entry and display system or tailored to specific sensor requirements.

Operational status

The GPCDU has been selected for US Air Force RC-135, RF-4C, TR-1A and U-2 aircraft and a number of other airborne platforms, including the AWACS and RF-5E, for other air forces.

Contractor

Litton Systems Canada.

VERIFIED

Up-Front Display (UFD)

The Up-Front Display (UFD) is a military qualified alphanumeric display which employs Litton's matrix LED technology. The UFD is designed to operate in the most severe airborne environments. In its standard configuration, the UFD is equipped with dual-redundant RS-422 serial ports, but can be supplied with MIL-STD-1553B or ARINC 429 interfaces.

The UFD is a compact and versatile display device, with built-in scroll and function keys. The flexible UFD can easily be incorporated in an avionic architecture requiring a display repeater and/or data display capability.

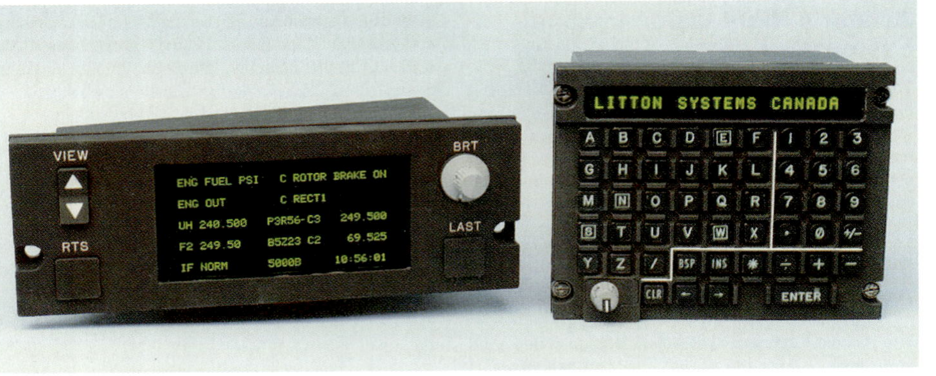

The Litton Up-Front Display (left) and avionics keyboard (right) have been selected for AH-64C and AH-64D Apache helicopters 0106244

Specifications

Dimensions: 83 × 230 × 165 mm
Weight: <2.35 kg
Power supply: 28 V DC, <57 W
Display: 10 lines × 35 characters
Temperature range: –54 to +55°C
Environmental: MIL-STD-810D

Operational status

The UFD has been selected for the AH-64C and AH-64D Apache helicopters.

Contractor

Litton Systems Canada.

VERIFIED

The Litton warning and caution annunciator is in production for the C-17 0106245

Warning and caution annunciator display

The warning and caution annunciator display consists of three LED dot matrix annunciation display zones, an LED dot matrix resettable cue display, an edge-lit information panel with key switches and a dual-multiplex databus receiver/transmitter.

The left and right LED modules and the cue display are ANVIS yellow for the display of cautionary messages. The upper zone of the centre LED module is ANVIS green B to display advisory messages and the lower zone is ANVIS red for the presentation of warning messages. The edge-lit information panel has seven push-button control switches for scrolling, display brightness, cue reset and self-test.

Specifications

Dimensions: 115 × 396 × 224 mm
Weight: 6.58 kg
Power supply: 28 V DC, 180 W, 5 V AC

Operational status

In production for US Air Force C-17 aircraft.

Contractor

Litton Systems Canada.

VERIFIED

Advanced Helmet-Mounted Display (HMD)

An advanced helmet-mounted visually coupled display system has been developed by a team of Canadian industrial and research partners including the National Research Council (NRC) in Ottawa. The HMD being developed and tested at the Council's Institute for Aerospace Research follows pilot head movement in three axes. This more accurate reproduction of pilot head movement gives pilots a performance edge.

The goal of the HMD research programme is to improve pilot performance by providing an accurate picture of what is outside the helicopter, no matter what the external conditions, using an enhanced and synthetic vision system. This system, to be demonstrated in 2000, will combine a computer-generated terrain image with an enhanced sensor image, such as infra-red or low-light level video.

The current research programme examines basic pilot performance associated with use of HMD systems. Work under way focuses on how much sensor platform roll movement is required to reduce workload and minimise spatial disorientation for pilots wearing an HMD.

Helmet-mounted displays in current use dispense with sensor platform roll movements, which means that visual input is missing information which helps a pilot to maintain spatial orientation. The NRC HMD system uses a sensor platform that rotates in yaw, pitch and roll, to improve helicopter pilot performance by more accurately reproducing pilot head movements.

The roll compensation study arose from observations of pilot behaviour during a previous HMD experiment. Head-tracking data revealed that pilots roll their heads during Visual Flight Rules (VFR) turns, opposite to the direction of aircraft bank. NRC researchers believe they do this to maintain a level horizon on their retinas, and that this is important for self-orientation.

This effect has been seen by other researchers in a fighter simulator, and in jet fighters. However, the NRC researchers believe that this is the first time the effect has been noted in helicopter operations. NRC researchers can set the platform to have no roll movement, some roll, or more roll than a person is physically capable of performing. Results show that only a small amount of platform roll is needed to make a large difference in system handling qualities. Research shows that if a sensor platform has as little as 6° of roll from the horizontal, it appears to improve greatly the pilot's ability to fly the aircraft. Pilot workload and disorientation are measurably reduced.

Operational status

Research and development.

Contractors

National Research Council Canada, Flight Research Laboratory, Institute for Aerospace Research.

VERIFIED

WESCAM® 20 for the P-3C Advanced Imaging Multi-spectral System (AIMS)

The WESCAM® 20 was developed specifically for long-range surveillance applications and combines high stabilisation with multiple, high-magnification, day and night vision cameras. Based on the demonstrated performance of the Model 20, Wescam has been selected by Lockheed Martin Tactical Defense Systems, Eagan, to supply the Advanced Imaging Multi-spectral System (AIMS) for the US Navy P-3C Upgrade III Anti-Surface Warfare (ASuW) Improvement Program (AIP).

The AIMS replaces the capabilities of the current AN/AAS-36 Infra-Red Detection Set (IRDS) and Electro-Optical Sensor (EOS) AN/AVX-1(V) in a turret system with full 360° azimuth field of regard. The AIMS will be located at the existing IRDS station.

Operational status

In October 1997, Wescam was contracted to provide two systems, with options for follow-on orders of up to 160 systems. The WESCAM® 20 AIMS variant is intended to be fielded on all AIP aircraft and is compatible with non-AIP P-3C aircraft which currently incorporate the AN/AAS-36 IRDS. It can also be installed as a stand-alone installation, or integrated with other navigation and radar systems.

In July and October 1998, Wescam announced that Lockheed Martin Tactical Defense Systems had exercised two options for US$24 million and US$8.4 million respectively, for delivery between March 1999 and early 2000 for the P-3C AIMS programme.

In June 1998, the United States Coast Guard contracted Wescam to supply a full turnkey surveillance and communication system for installation in Coast Guard HC-130H aircraft. The system comprises the Wescam Model 20 AIMS surveillance sensors, airborne workstations and a military satellite communications system. Contract value is US$15.6 million with delivery from early 1999 to early 2000. The contract includes an option to purchase additional systems worth a further US$20.4 million.

In July 1999, Lockheed Martin Tactical Defense Systems exercised a third set of options, valued at C$18.3 million, to purchase additional variants of the WESCAM® 20 multisensor camera systems for the US Navy P-3C AIP; this order is due to be delivered between the third quarter of 2000 and the first quarter of 2001.

Contractor

Wescam Inc.

VERIFIED

WESCAM® 20 Advanced Imaging Multi-spectral System (AIMS) US Navy P-3C Update III Anti-Surface Warfare (ASuW) Improvement Program (AIP) installation 0018368

DENMARK

Airborne surveillance system

The TERMA surveillance system is designed for detection of oil spills, identification and documentation of fishing violations, performance of search and rescue missions and ice-mapping by day and night and during periods of poor visibility. It integrates a wide variety of surveillance sensors, navigation equipment and video systems.

A Side-Looking Airborne Radar (SLAR) has become the primary long-range sensor for oil pollution surveillance, typically covering a 37 km swath from preferred search altitudes. An oil slick is detected by variation in reflected radar signals between oil-covered water and normal seawater. In applications like ice-mapping and ship surveillance, the SLAR covers a 74 km swath.

An Infra-red/UltraViolet (IR/UV) scanner is provided for close-range imagery and allows a rough area estimation to be made, as the aircraft passes overhead, of the oil slick detected by the SLAR. The IR system can be operated by both day and night. It provides information on the spreading oil and indicates the relative thickness within the oil slick. The ultraviolet sensor is only used during daylight. It maps the complete area covered with oil, irrespective of thickness.

A scanning radiometer system is provided for oil thickness measurements and quantification, enabling clean-up operations to attack the worst part of the spill first. The MicroWave Radiometer (MWR) measures microwaves originating from the sea surface at I/J- and K-band wavelengths.

Video cameras are used to secure evidence of oil pollution, fishery violation and other illegalities. Information can be recorded on videotape or stored as still photographs in the computer. Real-time navigation data is integrated into the picture. The video can be normal colour, highly sensitive low-light level TV or IR. A hand-held camera with a real-time data annotation capability, can be integrated into the system.

Data downlink equipment is used for transmission of real-time or stored data to a ground- or ship-based station.

Information from microwave and optical sensors can be recorded either on a standard VTR or on a high-resolution digital tape recorder.

The 355 mm Sensor Image Display (SID) provides the operator with sensor information. The SID presents the current sensor image whether it is the SLAR image, the IR/UV scanner image or the radiometer image that is selected. Real-time navigational data is integrated into the bottom of the SID format. Information on aircraft position, heading, speed and altitude, as well as date and time is presented to the operator. By means of the trackerball, the operator can move a cursor on the SID and the target position is then annotated with real time.

The 254 mm colour map display provides the operator with an outlined map of the area under surveillance. The map display is integrated with the video system. The operator can select map information, video information or both simultaneously. Map information is available from customised map data. The map can be zoomed in close steps and the operator can insert symbols at any position.

The 254 mm control panel display facilitates the operation of all surveillance sensors, back-up stores and video systems. All possibilities in these systems are pre-arranged in the control panel display as logical menus, sectioned into two or three levels. Functions within the menu are accessed by use of a two-stroke keypress or the trackerball.

Specifications

Dimensions:
(operator console) 1,150 × 575 × 1,400 mm
(observer console) 1,150 × 775 × 1,190 mm
Weight:
(operator console) 145 kg
(observer console) 150 kg
Power supply: 28 V DC, 900 VA (max)

Contractor

TERMA Elektronik A/S.

VERIFIED

Modular Reconnaissance Pod (MRP)

In 1994, the Royal Danish Air Force (RDAF) contracted TERMA Industries Grenaa AS to develop an all-new reconnaissance pod to be sufficiently flexible to accommodate current and future sensor systems. The MRP comprises three parts: a common pod structure, a sensor-particular part, and a platform-particular part.

The common pod structure includes the pod body, strongback, and electronic control system. The sensor-particular part is fitted with fixings for LRUs and sensors. The platform-particular part includes pylon attachments and electrical/mechanical interfaces.

The RDAF has selected the TERMA EWMS to control the MRP on its F-16MLU (Mid-Life Update) aircraft.

Payloads to be integrated are reported to include the Recon/Optical CA-261; EO sensors; Vinten wet film and EO sensors; Elop LAEO (Low-Altitude EO) and MAEO (Medium-Altitude EO) sensors; Lockheed Martin ATARS sensors.

TERMA Industries Grenaa AS Modular Reconnaissance Pod (MRP) 0106246

The sensor layout of the Modular Reconnaissance Pod ***1999***/0044998

Specifications

Dimensions: 4,496 × 762 × 610 mm
Weight:
(empty) 227 kg
(loaded) 544 kg
Flight envelope: +9 *g*
Data interfaces: tape recorders and datalink

Operational status

The MRP has been certified by the Royal Danish Air Force, the Royal Netherlands Air Force and the US Air Force Air National Guard. More than 20 pods are reported to be on order by the following four air forces: Royal Danish Air Force, Belgian Air Force, Royal Netherlands Air Force, and US Air Force Air National Guard. It is understood that TERMA Industries Grenaa AS, TERMA Elektronik AS and Recon/Optical may be teaming to fulfil these operational requirements.

Contractor

TERMA Industries Grenaa AS.

VERIFIED

FRANCE

TIM laser range-finders

TIM laser range-finders are high-repetition rate eye-safe laser range-finders designed to be integrated into an airborne fire control system to measure the distance to a ground-based, aerial or naval target.

They consist of a 1.54 μm transmitter integrating a 1.06 μm laser, a Raman conversion cell, a receiver, the power supplies, an interface chronometry/control and serial link RS-422 with the host system.

Specifications

Dimensions: 300 × 180 × 150 mm
Weight: 8.5 kg
Power supply: 220 V AC, 3 phase or 115 V AC, 400 Hz

Operational status

TIM laser range-finders have been selected for the Rafale aircraft and the Tiger helicopter.

Contractor

Compagnie Industrielle des Lasers (CILAS).

VERIFIED

TMS303 laser range-finder

The TMS303 laser range-finder is intended to be integrated into short- and medium-range weapon systems for range measurement of land and airborne targets, such as the SFIM Viviane sight for HOT missiles on the Gazelle helicopter.

The sight consists of a 1.54 μm transmitter integrating a 1.06 μm laser, a Raman conversion cell and a power supply; an interface/control/range processing card; a receiver and a low-voltage converter.

To remain compatible with existing systems, the TMY303, a 1.06 μm version has been developed. This could easily be upgraded to the 1.54 μm eye-safe configuration.

Specifications

Dimensions: 190 × 119 × 78.5 mm
Weight: 2.2 kg
Power supply: 28 V DC

Operational status

TMS303 and TMY303 are in series production.

Contractor

Compagnie Industrielle des Lasers (CILAS).

VERIFIED

TMS312 laser range-finder

The TMS312 laser range-finder is a version of the TMS303 (see separate entry) which is designed for medium-range applications such as air-to-ground fire-control systems. It offers higher repetition rates of 3 Hz continuous wave or 6 to 8 Hz in short bursts.

Contractor

Compagnie Industrielle des Lasers (CILAS).

VERIFIED

AIRLYNX air-to-ground surveillance and identification system

The AIRLYNX air-to-ground surveillance and identification system includes a complete suite of airborne sensors, air-to-ground datalinks and ground stations of all types, from man-portable systems to large, static ground stations. AIRLYNX is designed to fulfil day and night observation requirements for military, paramilitary and civilian users and can be fitted to a wide range of fixed- and rotary-wing aircraft.

Typically, the airborne sensor suite comprises single, dual, triple or quadruple electro-optic day/night sensors, mounted on a gyrostabilised platform, controlled by an onboard control suite that is linked, via onboard datalinks, to ground stations. Preferred sensors offered include the following range of Wescam® electro-optical systems:

1. The Wescam® 12 DS dual sensor, with 3-5 μm two-field of view thermal imager and colour daylight CCD camera.
2. The Wescam® 14TS/QS triple/quad sensor, with 3-5 μm six-field of view thermal imager, long-range colour daylight CCD camera, colour daylight CCD camera with X10 zoom and eyesafe laser range-finder.
3. The Wescam® 16DS/16DS Synergi dual sensor, with 3-5 μm three-field of view thermal imager, or 8-12 μm two-field of view thermal imager and a colour daylight 3CCD camera with × 18 zoom lens.
4. The Wescam® 20TS and 20QS tri and quad sensors, featuring TV, FLIR and laser range-finder.

System control onboard the aircraft or helicopter is exercised using the AOC 930/950 control console. This system comprises a high-resolution 9 in colour monitor (AOC 930), or a 10 in high-resolution LCD monitor (AOC 950), together with MCU 930 microwave and radio control unit, microphone and system hand controller. Optionally, a video recorder can be added.

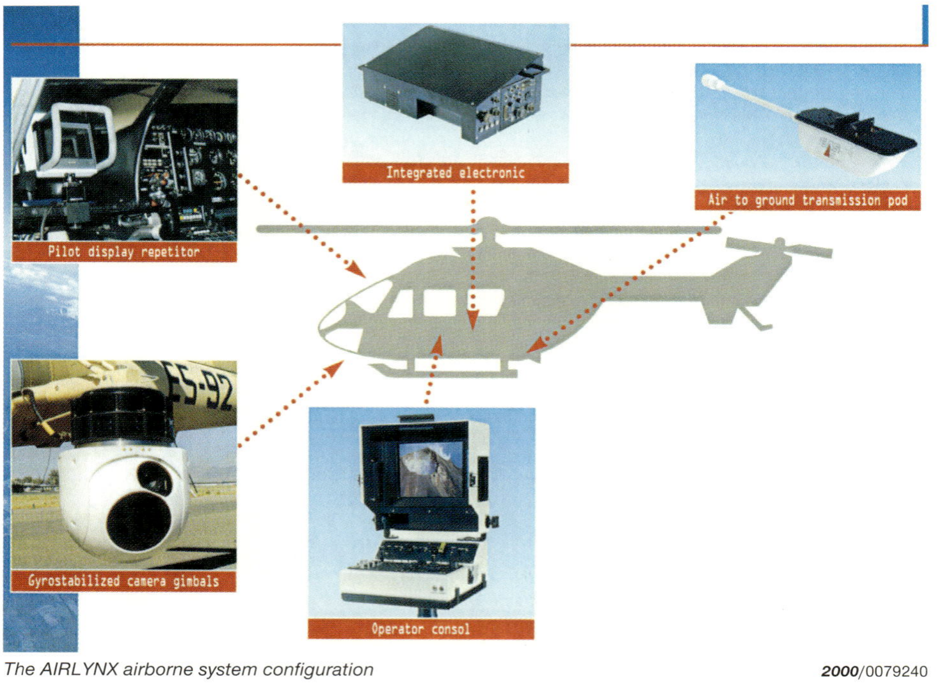

The AIRLYNX airborne system configuration **2000**/0079240

The ATU 930 transmission pod system provides the microwave link (video and audio) in both analogue or digital data formats. The pod operates at S-, C-, or X-band and offers multiplexing and encryption of both video and audio signals. It provides up to 10 km range to man-portable ground stations (designated ORS 950), 50 km to vehicle mounted ground stations (designated MRS 930V/TRS 930T) and 160 km to ground stations with large static antenna systems (designated FRS 930LR).

Contractor

EAS Europe Aero Surveillance, a subsidiary of Thomson-CSF Optrosys.

VERIFIED

Cyclope 2000 infra-red linescan sensor

The Cyclope 2000 infra-red linescan sensor is designed for airborne applications including: reconnaissance, and battlefield observation and surveillance (on aircraft, helicopters or UAVs); as a navigation aid; the monitoring and surveillance of sensitive areas; forest fire detection, pollution detection (oil on the sea's surface) and mine detection. It provides day and night infra-red images of the 8 to 12 μm spectral bandwidth to a thermal sensitivity of 0.1°C, with an angular resolution of 1 mrad. The field of view can be adapted from 60 to 120°.

The basic modular configuration can be adapted to various aircraft. Improved versions are offered with a different spectral bandwidth, multispectral detection or stereoscopy.

Specifications

Dimensions: 170 × 170 × 200 mm
Weight: <6 kg
Power supply: 28 V DC, 70 W

Contractor

SAGEM SA, Defence and Security Division.

VERIFIED

IRIS new generation FLIR

IRIS is a high-sensitivity modular thermal imager with high resolution and image quality. IRIS's main features include: up to three switchable fields of view, automatic gain and offset control, polarity selection; × 2 zoom, athermalised focusing, extended BITE, boresight alignment; digital image enhancement, hot point detection and tracking and low power consumption.

The IRIS thermal imager 0106247

The latest enhancement given to SAGEM's nav/attack system (MAESTRO) has been the night operation capability, based on integration of the IRIS internal FLIR. The IRIS second-generation IRCCD camera provides a one-to-one infra-red image, superimposed on the external world in the head-up display, in combination with the normal symbology. A narrow field of view image is also available to be displayed in the head-down display, for easy and precise target designation.

This new configuration has recently been qualified in-flight onboard a Mirage III and Mirage 2000.

The compact design of IRIS allows for internal installation leaving all hardpoints free for weaponry.

Specifications

Wavelength: 8-12 μm
Detection module: integrated detector/dewar microcooler device
288 × 4 elements IRCCD focal plane array Cd Hg Te closed-cycle Stirling microcooler
NETD: T < 0.02°C
Video output: CCIR

Operational status

In production for Mirage 2000, Mirage III and UAV programmes. IRIS is also fitted into the main sight of several attack helicopters (French and export), including the Tiger HAP helicopter where it is part of the STRIX sight, and the Rooivalk sighting system. A variant called Condor 1 is used for the sight of the Tiger HAC helicopter, and Condor 2 is integrated in the Tiger HAC navigation FLIR.

Contractor

SAGEM SA, Defence and Security Division.

UPDATED

CTH 3022 head-up display

The CTH 3022 head-up display for the Rafale aircraft is a holographic 30 × 22° field of view HUD that displays computer-generated symbology and FLIR imagery. Information is received from an associated symbol generator unit.

Operational status

Developed for the Rafale aircraft.

Contractor

Sextant.

VERIFIED

Electronic Head-Up Display (HUD) for the A330/A340

The A330/A340 HUD comprises an Optical Display Unit (ODU) and a Head-Up Display Computer (HUDC).

The ODU is installed in the overhead cockpit panel and is stowable behind the pilot's head when not in use. It moves automatically along the mounting tray to the operating position. The combiner glass is collapsible and can move forward in a crash situation. The ODU consists of optical lenses and a combiner which presents collimated symbology superimposed on the outside world, a miniature high-brightness CRT and an automatic brightness control system. The HUD controls are located in the ODU.

The HUDC is located in the electronic bay. It is connected on one side to the display management computer and on the other to the ODU. The HUDC is designed around VLSI circuits. Input parameters received from two distinct channels are monitored so that false information is instantaneously detected. The HUDC includes BITE. The symbol generator is capable of driving two optical display units.

The HUD includes growth potential for enhanced vision systems.

Specifications

Weight:
(ODU) 10 kg
(HUDC) 5 kg
Power supply: 115 V AC, 400 Hz
Field of view: 30 × 24°

Operational status

In production.

Contractor

Sextant.

VERIFIED

The HUD for the A330/A340 0106367

FCD 34 colour displays

The FCD 34 is a high-resolution shadow-mask full-colour multimode CRT display, using video scanning or stroke and raster images to show information on flight and engine control, navigation and weapon status. With a 3.5 × 4.5 in (89 × 114 mm) screen, the FCD 34 compact units adapt to the instrument panels of combat or training aircraft, either as original equipment or as a retrofit.

Operational status

The FCD 34 has been selected for the French Air Force Mirage 2000D and the Mirage 2000-5 and Mirage 2000 Strike export versions.

Contractor

Sextant.

VERIFIED

Head-Up Flight Display Systems (HFDS)

Sextant has pioneered Head-Up Displays (HUD) for commercial use since 1975. Its latest development is the Head-Up Flight Display System (HFDS), which is a

HFDS shipset 0106248

Sextant's Head-Up Flight Display System for Cat IIIb operation on Aerospatiale's B737-300 0001450

multipurpose HUD system that can be fitted to any cockpit, whether analogue or digital.

The system aim is to provide assistance to take-off and landing, to improve safety at the minima. The HFDS has been developed as a series:

Series 100, called Integrated HFDS
Series 200, called Manual HFDS
Series 300, called Hybrid HFDS

The Integrated HFDS is intended for aircraft already designed with a 'Fail Op' autopilot.

The Manual HFDS is intended for aircraft with 'fail passive' autopilots in order to reach Cat IIIa minima on manual landing with 50 ft decision height. Bombardier offers this system on Global Express aircraft.

The Hybrid HFDS is already in operation on Aerospatiale's Boeing 737-300 and on the MD 82 of Alitalia to enhance the landing capabilities to Cat IIIb standard with respective decision heights of 35 ft and 20 ft. The system allows a 75 m Runway Visual Range (RVR) for take-off.

Display symbology is matched to the series type and operational configuration.

The HFDS is also able to support Enhanced Vision System (EVS) procedures.

The HFDS comprises 4 LRUs: the projector located in the cockpit above the pilot's head; the holographic combiner with a full 40 × 28° field of view; the control panel; the computer.

Operational status

In production. Selected as optional on Global Express business aircraft. Certification was planned for 1999.

Installed as part of the C-130 Topdeck® avionics upgrade being integrated by Marshall Aerospace for the South African Air Force. C-130 aircraft.

Contractor

Sextant.

VERIFIED

Helmet-mounted sight for helicopters

The helmet-mounted sight for helicopters features visor projection for pilot comfort and safety, accurate fully qualified and flight test proven, head position sensing system, modular concept with a single-size customised helmet, NBC kit compatibility and eye protection against laser threats. It is used in target acquisition and designation and the firing of guns, rockets and air-to-air missiles.

It is a monocular system with a 6° field of view and 600 mm eye relief. The display capability includes a sight reticle and fixed-mode symbology. The head supported weight is 1.5 kg and the helmet has an optimised centre of gravity location. It is NVG compatible.

Operational status

Deliveries for Eurocopter Tiger helicopter.

Contractor

Sextant.

VERIFIED

Low-airspeed system

Sextant has acquired the exclusive VIMI licence and has developed the CLASS low-airspeed system as an answer to the problem of low-airspeed measurements in helicopters for flight management, navigation and weapon firing requirements.

Using the helicopter's flight characteristics and, in particular, the measurement of cyclic pitch controls and attitudes, CLASS allows the combination of low airspeed computed data with conventional air data measurements to provide airspeed information. This information is valid throughout the flight range and requires no additional external mobile probes.

Operational status

In production for the Eurocopter Tiger helicopter.

Contractor

Sextant.

VERIFIED

MFD 54 colour displays

The MFD 54 is a high-resolution, multifunction display using a 5 × 4 in active matrix LCD, providing excellent performance in all light conditions.

Operational status

The MFD 54 has been selected for the Mirage 2000-9.

Contractor

Sextant.

VERIFIED

MFD55/MFD66/MFD88 liquid crystal multifunction displays

The MFD55/MFD66/MFD88 multifunction displays are full-colour multimode displays using 5 × 5 in (125 × 125 mm), 6.13 × 6.13 in (155.7 × 155.7 mm) and 8 × 8 in (203.2 × 203.2 mm) active matrix liquid crystal displays in the MFD55, MFD66 and MFD88 respectively to provide good visibility under any ambient light conditions.

The displays are high-resolution units enabling display of any kind of video image overlaid with synthetic symbols. The units are fitted with surrounding soft keys.

Specifications

Dimensions:
(MFD55) 160 × 160 × 232 mm
(MFD66) 196 × 192 × 238 mm
Weight: <7 kg
Power supply: 115 V AC, 400 Hz, or 28 V DC, 130 W

Operational status

The MFD55 is in production for the LCA, MiG AT, Mirage 2000 and Sukhoi 30. The MFD66 is in production for the Tiger and Rooivalk helicopters and for Jaguar, Nimrod MRA4 and Sukhoi 30 MKI. The MFD88 is utilised in a 5-display fit in the NH 90 prototype helicopters.

Contractor

Sextant.

VERIFIED

SMD 88 smart multifunction display **2000**/0079265

Para-visual display

The para-visual display has been developed as optional equipment for the A320. Mounted on the glareshield, the instrument guides the pilot along the runway centreline during roll-out in poor visibility. Guidance is provided by vertical white and black strips, which move to the left or right, according to the deviation from runway centreline.

The display uses liquid crystal and is within the pilot's field of view when looking outside. It is connected to the flight guidance system and to the flight warning computer of the aircraft, or equivalent sources. The system comprises the autoland warning and the picture generator.

Specifications

Dimensions: 150 × 155 × 36 mm
Weight: 1 kg
Interface: ARINC 429 HS discrete inputs (autoland warning control)
Angle of view: 60° under day or night conditions

Operational status

Optional equipment for the Airbus A320.

Contractor

Sextant.

UPDATED

Rafale aircraft displays

Sextant has developed avionics for the Rafale multirole fighter comprising:

(1) Two 5 × 5 in SLCD55 colour, multifunction side-mounted displays, with touch-sensitive screens, to provide pilots with information on aircraft systems.
(2) TMC 2020 head-level display, providing colour, high-resolution synthetic images collimated for infinity, designed to enhance the pilot's situation awareness.
(3) VEH 3022 multimode head-up display, providing 'short-term' information (symbology, FLIR images) in a wide 30 × 22° field-of-view.
(4) The Topsight® helmet-mounted display, which increases the visual range for target acquisition and designation.

These displays incorporate the following interface technologies:

(1) The SLCD liquid crystal display has its own integral symbol generator, a MIL-STD-1553B bus and a touch-sensitive screen.
(2) The TMC 2020 uses LCD technology and an optical device to collimate images to infinity.
(3) The VEH 3022 incorporates developments in holography to provide a wider field-of-view.
(4) The Topsight® HMD features advanced electromagnetic head position sensing, and symbology projection on the visor.

Completing the man/machine interfaces for Rafale are a control and display unit, combined standby instrument and engine and fuel indicator.

Two computers manage the overall display system:

(1) The Modular Data Processing Unit (MDPU), developed by Sextant with Thomson-CSF DETEXIS, gathers all data processing in the aircraft (symbol generation for the HUD and HLD units, digital mapping functions, mission data, flight data and tactical situation) into one unit.

The Sextant range of active matrix liquid crystal displays **2000**/0079264

Cockpit of the Rafale aircraft (Sextant) ***1999***/0051660

Sextant T100 helicopter head-up display 0018160

Sextant T200 helicopter head-up display 0018159

(2) An alternative symbol generator for the standby instruments.

Sextant is also working on a voice recognition and control system, intended to equip a future version of Rafale. Also in development for Rafale is the CET flightpath computer. The CET prefigures the 'electronic co-pilot' which will allow combat aircraft pilots to concentrate on the tactical situation, rather than flight control management.

Sextant also supplies the following air data and system management equipment: four multifunction probes; three UMPT33 air data systems; one Topstar® precision positioning GPS receiver; one static magnetometer; a data concentrator to control interface with the control and display unit; andsensors which monitor internal pressure, temperature and other parameters.

Contractor
Sextant.

VERIFIED

Smart Head-Up Display (SHUD)

The SHUD is a fully integrated head-up display with a 24° circular field of view. Connected to the MIL-STD-1553B databus, the SHUD is able to compute and draw symbology and to present simultaneously FLIR video and flight information. The SHUD also incorporates sufficient processing power for weapon system computations. It is particularly well adapted to combat aircraft upgrading.

Sextant SHUD Smart Head-Up Display for the MiG-AT ***1999***/0051711

Operational status
Selected by Lockheed Martin for the Argentine Air Force A-4M avionics upgrade programme. Other orders include: the Mikoyan MiG-AT trainers, Spanish Air Force Mirage F1 modernisation, Iryda I-22 trainers of the Polish Air Force. Over 100 SHUDs have been produced.

Contractor
Sextant.

VERIFIED

T100 and T200 weapons sight for helicopters

The T100 and T200 electromechanical head-up weapons sights show a collimated reticle which can be moved between −10 and +7° in elevation and ±6° in azimuth, angles compatible with the requirements for air-to-air and air-to-ground weapons launch. The device is mounted on the canopy frame of the helicopter, weighs 2 kg and has a field of view of 7.5°.

The sighting system is based on a modular concept. The basic component is the T100 single- or T200 dual-axis monocular sight head which can be used for both day and night weapon firing in conjunction with third-generation microchannel NVG. Sight recording by CCD camera has been validated for both training and operational firing and is available as an option. The physical features of the sight head, such as its low weight, small size, simple and strong high-performance optical system of Angenieux lens, diode array on micro-electronic support, give it the capacity to produce a remarkably high-quality image, both by day and night.

The second main component of the sighting system is a control unit combining command and calculation functions. This unit includes the necessary controls for moving the sight head, symbology animation controls, firing tables stored in its internal memory, weapon firing computation system for guns and air-to-air missiles and sighting telescope interface. The links with the missile system and telescope are designed in particular for export requirements and include coupling with Sextant's Nadir computer for target designation. The system is equipped with a built-in automatic testing facility, with status information being displayed on the sight head.

To meet the weaponry requirements of the French Army Gazelles, which are equipped with guns and Mistral missiles, Sextant proposed a multipurpose sighting and fire-control system comprising the above-mentioned components that have already been qualified for the Gazelle. With this system, the pilot can fire the various weapons, including rockets, without any manipulation other than manual selection of functions on the front panel of the control unit.

Operational status
In production. The system has been installed on a wide variety of helicopters including the Boeing 500/530, Eurocopter BO 105 and BK 117, Sikorsky S-76 and Black Hawk, Bell 206, 406 and 412, Westland WS 70. It has also been selected to equip the Gazelle helicopters of the French Army Light Aviation Corps and the Ecureuil helicopters of the French Air Force, and by Eurocopter for export versions of the Dauphin, Ecureuil, Gazelle and Puma.

Contractor
Sextant.

VERIFIED

TMV 980A head-up/head-down display

The TMV 980A is an integrated head-up/head-down display system for the Dassault Mirage 2000 multirole fighter. It comprises a digital computer and processor to generate the display symbology and help with flight and weapon aiming computation and three display units: a VE 130 CRT head-up display, a VMC 180 interactive multifunction head-down colour display and a VCM 65 complementary monochrome CRT for electronic support measures information.

The head-up display has a high-resolution, high-brightness CRT with a collimating optical system based on a 130 mm lens providing a wide total field of view. The instantaneous binocular field of view is increased in elevation by the use of a twin-glass combiner, which transmits 80 per cent of the light incident upon it. Automatic brightness control, with manual adjustment, permits symbols to be read in an ambient illumination of 100,000 lx. The system provides continuous computation of tracer line in the air-to-air mode and impact and release points in the air-to-ground mode.

The main head-down display presents, in red, green and amber on a 127 × 127 mm CRT radar display, information such as a radar map, synthetic tactical situation and range scales, raster images from television or FLIR sensors and tactical data from the system itself or from an external source.

A helmet-mounted sight may be integrated into the TMV 980A unit to improve target discrimination and off-boresight target designation. Another option is the substitution of the VE 130 head-up display by a VEM 130 system.

Specifications

Weight:
(electronic unit) 9 kg
(head-up display) 13 kg
(VMC 180 head-down display) 14 kg
(VCM 65) 4 kg

Operational status

In service in the Dassault Mirage 2000. No longer in production.

Contractor

Sextant.

VERIFIED

TMV 1451 electronic head-up display for commercial aircraft

The TMV 1451 head-up display comprises the Optical Head Unit (OHU) and the Head-Up Display Computer (HUDC). The OHU is installed in the glareshield panel either behind the glareshield front panel or in the operational or pull-out position with the combiner appearing in the forward field of vision of the pilot. The HUDC is located in the electronics bay. It is linked on one side to the display management computer and on the other side to the OHU itself. The HUD's control panel is included in the OHU.

The OHU is an electronic head-up display system designed specifically to be installed in the glareshield of commercial aircraft and particularly in the Airbus A320. The head-up display, linked to the existing aircraft systems, can be used for roll-out guidance, visual approach guidance and monitoring of automatic approach and flare. The OHU is linked to the HUDC, which generates the symbology to be displayed, according to the flight phase.

The OHU consists of optical lenses and combiner which present collimated symbology to the pilot superimposed on the outside world, a miniature high-brightness CRT, an automatic brightness control which adjusts the symbol brightness to the required level and a control panel. Of limited volume, the OHU is designed and installed in such a way that the pilot's lower field of vision is not interrupted.

The head-up display computer is designed around VLSI circuits already in use in the display systems of A310 and A320 aircraft. Input parameters received from two distinct channels are monitored so that false information is instantaneously detected. The HUDC includes BITE. The symbol generator function of the computer is capable of driving two optical head-up displays.

Specifications

Weight:
(OHU) 10.5 kg
(HUDC) 5 kg
Power supply: 115 V AC, 400 Hz
Field of view: 24 × 15°
Reliability:
(OHU) 5,000 h MTBF
(HUDC) 13,000 h MTBF

Operational status

In production.

Contractor

Sextant.

UPDATED

Topdeck® avionics for military transport aircraft

Sextant markets the Topdeck® concept as a modular upgrade for military transport and special mission aircraft. Principal components include:

(1) Holographic Head-up Flight Display Systems (HFDS);
(2) LCD 68 full-colour 6 × 8 in LCDs for Primary Flight Display (PFD), Navigation Display (ND), Engine Indication and Crew Alerting System (EICAS), weather radar display;
(3) Multifunction Control Display Units (MCDU);
(4) Totem 3000 INS/GPS.

Operational status

Selected for the South African Air Force C-130 upgrade and for the CASA CN-235-300 and CN-295 military transport aircraft, where certification is planned for late 2000 with deliveries to begin in 2001.

Contractor

Sextant.

VERIFIED

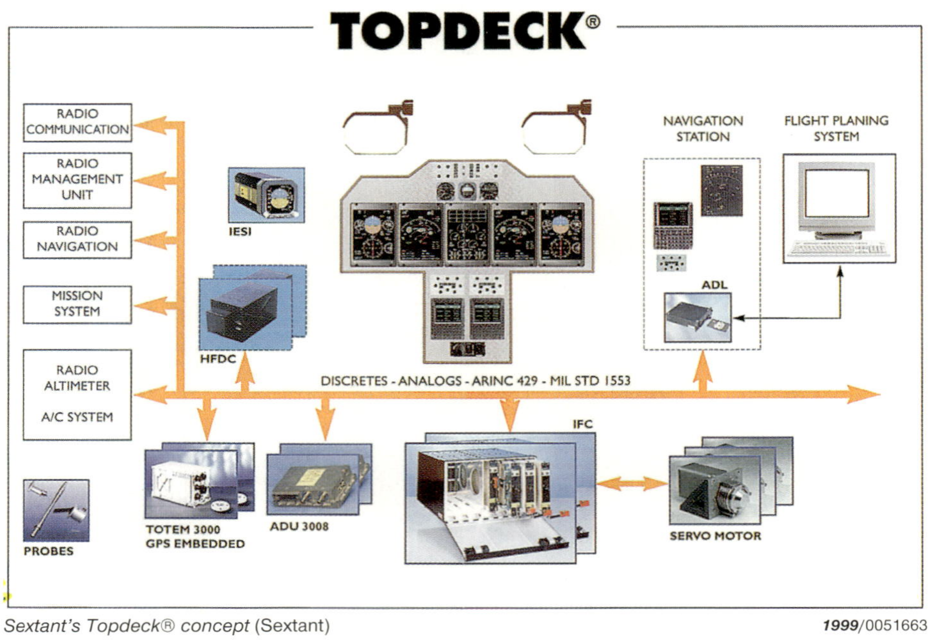

Sextant's Topdeck® concept (Sextant) **1999**/0051663

The Sextant Topdeck® installation (Sextant) **1999**/0051664

Topowl® day/night Helmet-Mounted Sight Display (HMS/D)

Sextant's Topowl® binocular HMS/D for helicopters is based on a new modular design, with optimised head-supported mass, for day or night missions. The basic helmet provides conventional physiological protection and communications, while the display module projects imagery on the visor, using integrated night vision sensors (image intensifier tubes) or FLIR video and/or synthetic symbology.

The HMS/D is a purpose-built, state-of-the-art system, utilising proven European technology. Technical and operational features include the following:

(1) Enhanced situational awareness and safety due to visor-projection technology.
(2) Binocular, wide field of view in a lightweight, well balanced, helmet.
(3) Modular design, with a basic helmet, specially adapted to the needs of each pilot and a display module that remains in the helicopter.
(4) Accurate head-position sensing (tracking) technology, which is fully qualified and flight proven in the helicopter environment.
(5) NBC compatibility, head-in displays and aircrew equipment kits.
(6) Image intensifier tubes optically integrated in the display module, giving safe dual-sensor night-mission capability.
(7) Integrated cursive symbol generator providing safe and clearly readable symbology even in full daylight.

Operational status

Deliveries of the Topowl® HMS/D have been as follows:

(1) In early 1999, for the Tiger PT2R and PT3R development helicopters.
(2) In March 1999, two production standard HMS/D units were delivered for flight testing of the NH-90 PT4 prototype.
(3) Delivery of two shipsets per month for the South African Rooivalk programme.

Contractor

Sextant.

VERIFIED

Topsight® Helmet-Mounted Sight/Displays HMS/D

Sextant has developed the Topsight® family of Helmet-Mounted Sight/Displays (HMS/D) to meet the requirements of modern combat aircraft pilots.

The first two products in this family are a fully-integrated Topsight® HMS/D for France's new Rafale multirole fighter, and the modular Topsight® E, specially designed for export markets.

The Topsight® E comprises a basic helmet with a removable display module that projects symbology on the visor. Depending on the assigned mission, the display module can be replaced with a double visor module (like a conventional helmet), or an ejection-compatible night vision module. In all cases, the pilot keeps his own oxygen mask.

Featuring an optimised centre of gravity and head-supported weight of less then 1.5 kg, Topsight® E ensures effective target designation and acquisition for all missions. A camera integrated in the display module also provides full mission replay capability.

The Topsight® E is designed from the outset for excellent physiological protection against shock, punctures, fire and facial injury, as well as safe ejection at speeds up to 625 kts.

Using the highly accurate electromagnetic head tracking system, the pilot can perform off-boresight target designation and weapon firing, enabling him to maintain optimum tactical position and/or aspect relative to any engagement/attack. Once the target is acquired, the pilot's line-of-sight is then transmitted to the aircraft's other systems, including sensors, avionics and weapon systems, for lock-on.

Sextant's Topsight® helmet-mounted sight/display family also allows reverse cueing from the aircraft systems to Topsight itself, which means the pilot's eyes are guided to the target tracked by aircraft sensors.

The Topsight® family maximises the pilot's situational awareness, by displaying flight and navigation data, weapons system status, warnings and

The Topowl® day/night helmet-mounted sight and display for the Eurocopter Tiger helicopter
***1999**/0051712*

other vital information directly on the visor. Information displays can be customised to meet specific needs, ensuring superiority in both aerial combat and ground attack missions.

Operational status

Since May 1997, as part of the overall Rafale development programme, Sextant has been carrying out a complete flight test evaluation of the Topsight® HMS/D for the French Air Force and Navy. Topsight® E flight evaluations started in early 1999.

Contractor

Sextant.

VERIFIED

VE 110, VE 120 and VE 130 head-up displays for combat aircraft

Sextant has developed a full family of head-up displays for tactical aircraft of various sizes and sophistication. The displays include advanced software for each phase of a flight, allowing accurate attacks as well as navigation and piloting.

The VE 120, chronologically the first of the family, was configured to the Dassault Super Etendard and Mirage F1.

The VE 110, especially designed for small cockpit tactical or trainer aircraft such as the most recent version of the Dassault Alpha Jet, has the most compact Pilot's Display Unit (PDU).

The VE 130, with a larger PDU, is fitted to versions of the Mirage 2000.

The instantaneous fields of view of the dual-combiner displays are typically about 20° in azimuth and 15 to 18.5° in elevation, depending on aircraft type.

Specifications

VE110
Dimensions:
(PDU) 427 × 123 × 125 mm
(EU) ½ ATR
Weight:
(PDU) 8-10 kg
(EU) 8-9 kg

Operational status

No longer in production.

Contractor

Sextant.

VERIFIED

The Topsight® E helmet-mounted sight and display for export markets ***1999**/0051713*

VEM 130 combined head-up and head-level displays

The VEM 130 is a derivative of the VE 130, which includes an additional capability for raster FLIR image presentation at head level. This display gives a field of view of 14 × 10° and the green phosphor raster display can be set at 525, 625 or 675 lines, 50 or 60 Hz.

Specifications

VEM 130
Dimensions: 645 × 378 × 145 mm (plus combiner glass)
Weight: 23 kg

Operational status

The VEM 130 has been selected for the Dassault Mirage 2000-5.

Contractor

Sextant.

VERIFIED

All-attitude indicators

SFIM Industries has developed a range of spherical indicators on which attitude information is displayed to give the pilot heading, roll and pitch information on a single dial without freedom limits around the three axes.

Some versions are fitted with command bars to display signal information from navigation aids or landing systems. A helicopter version is designated 11-2 and features a manual pitch setting mode of ±10°. The various features of the series are:

Specifications

Type	810	811	816
Roll	Yes	Yes	Yes
Pitch	Yes	Yes	Yes
Heading	Yes	Yes	Yes
ILS/VOR/ Tacan	Yes	No	Option
To-from	Yes	No	*
Beacon	Yes	No	Option
Sideslip	Yes	Yes	Yes
Failure detection	Yes	Yes	Yes
Warning flag	Yes	Yes	Yes

* separate unit, on option

Type 810 and 811
Dimensions: 97 × 97 × 203 mm
Power supply: 26 or 115 V AC, 400 Hz
28 V DC

Type 816
Dimensions: 81 × 81 × 208 mm
Power supply: 26 V AC, 400 Hz
28 V DC

Operational status

In production. Four types of instruments are available and equip BAE Systems Harrier GR. Mk 7, Dassault Mirage F1 and 2000, Sepecat Jaguar and Dassault Super Etendard, French Navy Westland Lynx helicopters and Soko Galeb, VTI/CIAR Orao and FMA Pucara aircraft.

Contractor

SFIM Industries (SAGEM Group).

VERIFIED

SFIM IS 816-1 attitude indicator 0106368

APX M 334 sights for helicopters

The designation APX M 334 covers a family of roof-mounted sighting systems for helicopters. All are single eyepiece devices and have magnifications of × 3 and × 10 with corresponding 300 and 90 mrad fields of view.

The principal requirement of the M 334 series is to provide an operational system able to detect a target at a range of about 10 km and recognise a tank at 5 km. An optional special TV camera arm allows the recording of observed images by a video recorder.

Variants of the basic system have specific capabilities:

M 334: observation and manually guided missile firing
M 334-04 Athos: observation and firing of manually guided missiles or seeker-head missiles
M 334-25: observation sight with range-finding and target localisation. When used in conjunction with a navigation system it constitutes the Osloh III system giving target absolute position. When used in connection with an aiming collimator it constitutes the Oshat system for air-to-ground gun and rocket firing
M 397 HOT: observation sight able to fire HOT automatic infra-red guided missiles produced by the Franco-German Euromissile partnership.

The SFIM Industries APX M 397 sight

Operational status

All members of the M 334 family are currently in production and service. A total of 1,500 systems are in operation in 30 countries.

The sight has been fitted to such helicopters as: the Eurocopter Gazelle 341 and 342 and Super Frélon SA 321; Westland Wasp, Lynx and WG 13; Bell 204, 205, 206/OH-58 and 212; Sikorsky SH-3D, Boeing 500 and 530, Agusta A 109, Eurocopter BO 105 and BK 117.

Contractor

SFIM Industries (SAGEM Group).

VERIFIED

CN2H-AA Night Vision Goggles (NVGs)

The CN2H-AA NVGs are a development of the CN2H. They are designed for aircrew in high-performance military aircraft and can be rapidly released with either hand before ejection. The NVGs can be equipped with Gen II and III image intensifiers, for high resolution and high sensitivity and can be mounted on any type of aircrew helmet.

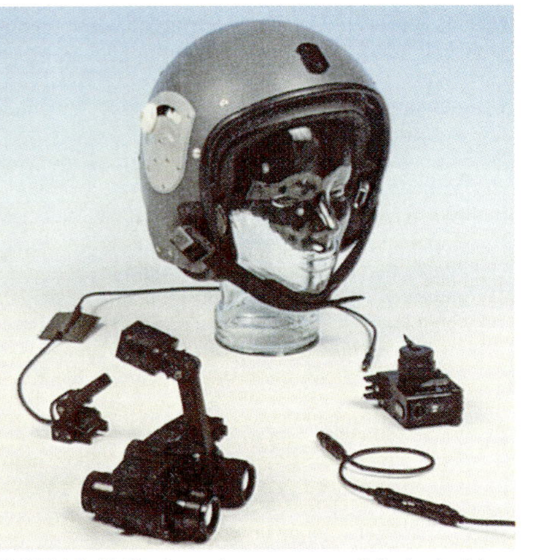

CN2H-AA NVGs have been successfully tested in a Mirage 2000 0106249

Specifications

Weight: 0.59 kg
Power supply: 28 V DC or PS31 battery
Field of view: 40 or 50°
Magnification: × 1

Operational status

In production and in service with the French Air Force.

Contractor

SFIM Industries (SAGEM Group).

VERIFIED

CN2H Night Vision Goggles (NVGs)

The CN2H NVGs are part of a range of night vision systems produced for military applications; the goggles are specifically designed for use in helicopters and fixed-wing aircraft for night piloting in tactical situations. They are fixed to the helmet, with a power pack on the back and a specially designed support bracket enables them to be immediately discarded in an emergency. Focusing and positional adjustments are available to suit the wearer. The goggles are compatible with a wide range of French, British and American helmets and they incorporate second- or third-generation image intensifiers according to requirements.

Specifications

Weight: 0.95 kg incl battery pack
Power supply: 28 V DC or 3.5 V PS 31 battery
Battery life: 20 h
Field of view: 40 or 50°
Magnification: × 1

Operational status

In production. Adopted by the French Army Aviation (ALAT), Air Force and Navy and by export customers.

Contractor

SFIM Industries (SAGEM Group).

VERIFIED

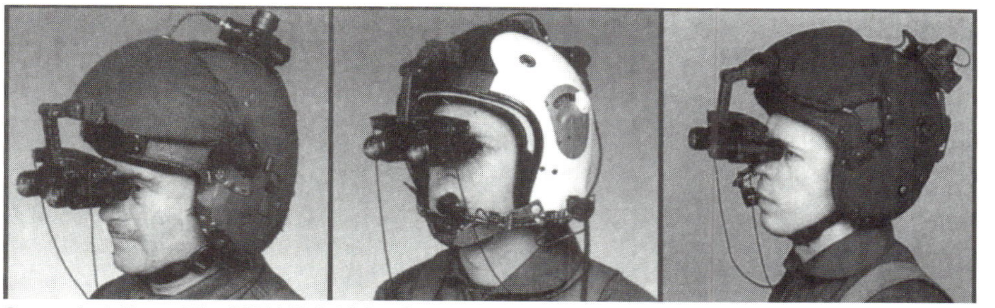

Some adaptations of the CN2H Night Vision Goggles already in service 0018367

JADE night vision goggles

JADE is a twin-channel NVG system designed for combat aircraft operations. It is compatible with high-*g* manoeuvre and emergency ejection. The 'direct' channel permits the pilot to monitor both the cockpit displays and the situation outside the cockpit, while simultaneously, the 'intensified' channel provides night vision of the same external situation by superposition.

Operational status

JADE has passed the 'windblast' qualification testing at the Centre d'Essais Aeronautiques of Toulouse (CEAT) and flight testing in the Centre d'Essais en Vol (CEV). JADE has successfully completed operational flight tests carried out by French Air Force at the Centre d'Experiences Aériennes Militaires (CEAM).

Contractor

SFIM Industries (SAGEM Group).

VERIFIED

JADE night vision goggles 0018366

The SFIM Industries Strix roof-mounted targeting system for the HAP (multirole support) Tiger helicopter 0018158

Strix day/night sights

SFIM Industries has developed a family of day/night nose- and roof-mounted sights for helicopters.

The Strix sight is fitted with: an IRCCD thermal imager; direct view channel; TV channel; high-rate laser range-finder; and micromonitor. The thermal and TV video images and firing symbols are projected onto the sight eyepiece.

The weapon system uses Strix as a data reference system so as to determine its exact line of sight in space. Sighting precision is ensured by automatic target tracking on the TV and thermal channels. The sight is designed to maintain stabilisation and precision of the line of sight throughout the entire operating range of the helicopter and is MIL-STD-1553B compatible to interface with the associated weapon systems. Composite materials are used for the structure so as to obtain the best trade-off between the mass of the integrated sensors and its own mass.

Operational status

Strix is in production for the French ALAT to be used in the HAP Tiger helicopter. A nose-mounted version of Strix is in production for the South African Rooivalk attack helicopter programme.

Contractor

SFIM Industries (SAGEM Group).

VERIFIED

Viviane day/night sight

Developed for Euromissile, Viviane is a day/night sight used on French ALAT Gazelle SA 342M helicopters for operation of the HOT weapon system. Viviane provides a full day/night firing capability.

Operational status

In service in the French ALAT on the Gazelle SA 342M helicopter.

Contractor

SFIM Industries (SAGEM Group).

VERIFIED

The SFIM Industries Viviane day/night sight is fitted on the French ALAT Eurocopter SA 342M Gazelle
1999/0106370

Up-Front Control Panel (UFCP) for Head-Up Display (HUD)

Developed by TEAM and Sextant, the UFCP enables the pilot to interface with the HUD and mission computer via a number of illuminated keys, programmable displays, variable controls and discrete interfaces.

The UFCP communicates with the HUD via an RS-422 databus. The UFCP transmits coded control command data and status information in all operating modes and additionally, transmits 'key pressed' data in backup mode. In backup mode, the UFCP receives, decodes and displays data on the UFCP programmable keys and displays.

The UFCP communicates similarly with the mission computer via an RS-422 databus.

The UFCP provides background panel illumination and background legend illumination. The UFCP also decodes data received on the RS-422, lines and displays the data on the programmable keys and displays. In normal mode, the UFCP data is received from the mission computer, in backup mode it is received directly from the HUD.

The UFCP provides the following panel configuration: 14 fixed legend illuminated keys; five fixed legend illuminated keys with key selected interior; 22 programmable legend keys; four programmable LED displays; two potentiometers; two switched potentiometers; three bi-stable switches.

Two sizes of UFCP are available. The picture below shows the smaller-sized model, the larger model has an extra line of six illuminated keys.

Operational status

Supplied for various military aircraft upgrade programmes, including the A-4M (Argentina), Iryda (Poland), Mirage F-1 (Spain) and Alphajet (Belgium).

Contractors

TEAM
Sextant.

UPDATED

The up-front control panel produced by TEAM and Sextant **2000**/0062240

Airborne reconnaissance and surveillance system

The airborne reconnaissance and surveillance system provides photographic reconnaissance by day at low and medium altitudes for surveillance missions at sea through an assembly at the front or rear of the aircraft. The dropping of charges can also be photographed through the assembly at the rear of the aircraft.

The system consists of an AA3-35-100 camera, BC3-135-2 camera, BF3-135-2 computer and P11-1 vacuum pump. Three lens cones are available: the E3-150, E3-75 and E3-300.

Specifications

Dimensions (with E3-150 lens cone): 300 × 240 × 340 mm
Weight: 16 kg
Power supply: 200 V AC, 400 Hz, 3 phase
Format: 114 × 114 mm

Operational status

In production since 1988 for the French Navy Atlantique 2.

Contractor

Thales Optronique (formerly Thomson-CSF Optronique).

UPDATED

The airborne reconnaissance and surveillance photographic system for surveillance missions at sea showing (left to right) the three lenses, control unit and computer 0106333

AP 40 panoramic film camera

The AP 40 panoramic film camera is designed for medium-, low- and very low-altitude reconnaissance missions, at very high penetration speed.

Specifications

Lens focal length: 75 mm
Frame rate: 2 to 10 frames/s
Magazine: 75 m of 70 mm film; 300 exposures with standard film
Panoramic image: 180°
Dimensions: 386 × 316 × 197 mm
Weight: 19 kg
Optional fit:
(magazine) large capacity 150 m of 15 mm film; 600 exposures with standard film
(weight) 29 kg

Operational status

Current programme: upgraded Super Etendard.

Contractor

Thales Optronique (formerly Thomson-CSF Optronique).

UPDATED

AP 40 panoramic film camera under Super Etendard 0018365

ATLIS II laser designator/ranger pod

ATLIS II is a pod-mounted laser targeting system with an automatic TV tracker, laser designator/ranger, tape recorder and interface electronics. The automatic TV tracker and laser stabilisation system reduce pilot workload, allowing tracking and designation in a single-seat aircraft. ATLIS II can be installed on a wide variety of strike or close support aircraft.

The system has an accuracy of 1 m on a target at an average firing range of 10 km and provides a high rate of success for low- and medium-level standoff attacks. It provides high image quality due to the dual-mode tracker which covers both the visible and near infra-red portions of the spectrum. ATLIS has a long-range target acquisition capability even in the most adverse weather conditions. It features a high degree of magnification

The ATLIS II pod installed on an F-16 aircraft 0018364

The Mirage 2000 with the Thomson-CSF Optronique ATLIS II laser designator pod mounted under the starboard engine
0106250

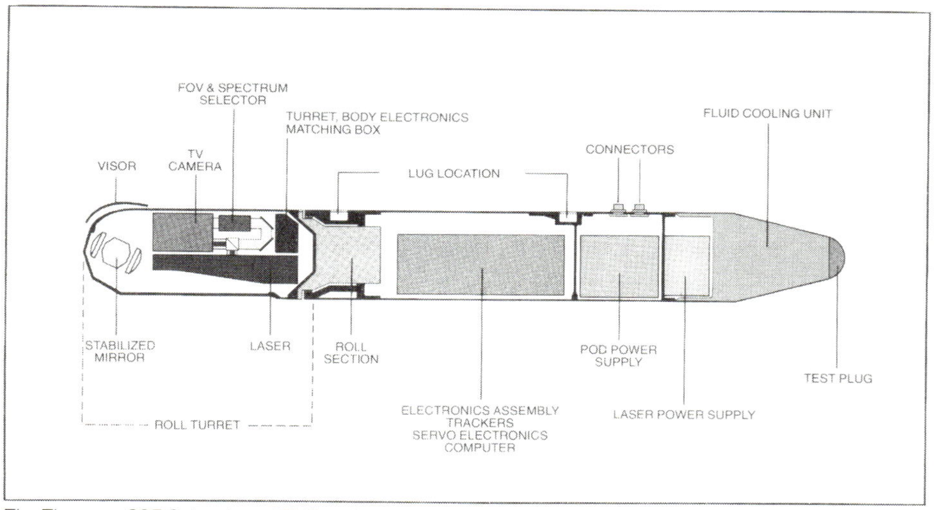

The Thomson-CSF Optronique ATLIS pod

(up to × 20), damage assessment capability, video recording and a reconnaissance mode.

Specifications

Dimensions: 2,520 (length) × 305 mm (diameter)
Weight: 170 kg
Power supply: 115 V AC, 400 Hz, 2.3 kW
Wavelength: 1.06 μm
Max operating range: dependent on the weapon selected
Angle of regard:
(roll) unlimited
(pitch) −160 to +15°
Attachments: standard 760 mm NATO bomb rack

Operational status

In production. The ATLIS 2 pod equips French Air Force Jaguars and has been ordered for export to equip Mirage 2000 and F-16 aircraft. Integration of the ATLIS II targeting pod and Aerospatiale AS-30L laser-guided missile forms the third part (Standard 3) of a four-part programme of work for the French Aéronavale Dassault Super Etendard Modernisé (SEM) that is due to be completed in 2004, for service until 2010.

Contractor

Thales Optronique (formerly Thomson-CSF Optronique).

UPDATED

Chlio-WS multisensor airborne FLIR

Thomson-CSF has developed a version of the Chlio thermal imaging system, Chlio-WS, which incorporates a second-generation detector. Chlio-WS is intended for search- and rescue-operations and is installed on helicopters.

The system uses the Wescam Synergi thermal imaging modules. The detector is a 288 × 4 cadmium mercury telluride focal plane array, operating in the spectral band 8-12 μm.

The system has four fields of view and is mounted on a gyrostabilised turret which is stabilised in two axes electro-mechanically and two axes electro-optically.

Specifications

Thermal imager
Spectral range: 8-12 μm
Fields Of View (FOV) (H × V):
(wide): 24.0 × 18.0°
(intermediate): 12.0 × 9.0°
(narrow): 3.0 × 2.2°
(very narrow): 1.5 × 1.1°
FOV switch time: 0.5 s
Features: remote switching of FOV; × 2 electronic zoom; polarity switching; automatic and manual gain; edge enhancement; symbology; NVG compatible

Colour camera
Camera type: 3CCD Sony XC-003
Lens: 14 × 10.5 Fujinon zoom lens
FOV (H × V):
(wide): 13.0 × 9.8° 21 mm
(narrow): 0.93 × 0.70° 294 mm

Gimbal
Line of sight (LOS) pan range: 360° continuous
LOS tilt range: +90 to −180°
Slew rate: 60°/s typical

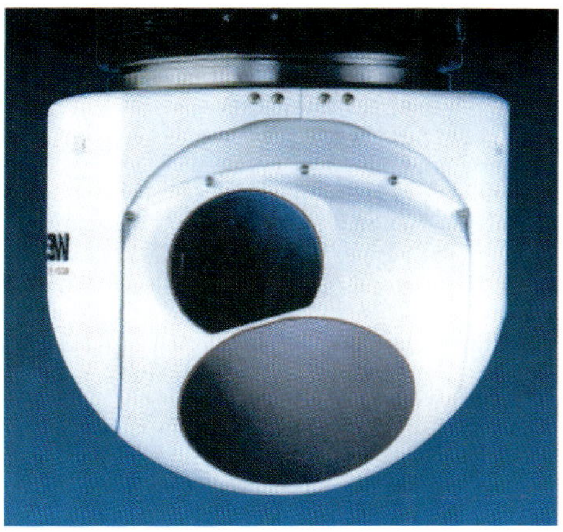

The Chlio-WS multisensor airborne sensor system **2000**/0079245

Weights
(gimbal): 40 kg
(IR electronics unit): 5.6 kg
(system control unit): 1.5 kg
(hand controller): 0.7 kg
Dimensions
(gimbal): 400 mm (diameter) × 510 mm (H)
(IR electronics unit): 245 × 181 × 160 mm

Operational status

Chlio-WS is in production for C160, Alizé and French Navy Falcon 50 Surmar aircraft. It is in service on French Air Force Super Puma helicopters and with export customers. The Wescam Synergi thermal imager is also used in the latest variant of the Thomson-CSF Convertible Laser Designator Pod (CLDP).

Contractor

Thales Optronique (formerly Thomson-CSF Optronique).

UPDATED

Convertible Laser Designator Pod (CLDP)

The Convertible Laser Designator Pod (CLDP) has been developed from Thomson-CSF Optronique's ATLIS pod. Flight tests of the CLDP/TV (TeleVision) began in mid-1986. Flight tests of the CLDP/CT (Camera Thermal) (note that in the accompanying photograph, camera thermal has been anglicised to Thermal Camera (TC)) version began in January 1988 using a Jaguar aircraft.

In September 1988, at the Landes flight test centre, a Jaguar successfully launched an AS 30 missile at a speed of 470 kt from an altitude of 213 m at a range of 8 km. The CLDP had acquired the target at a range of 13 km.

In CLDP, the laser designator is supplemented by either a TV (CLDP/TV) or a thermal camera (CLDP/CT). The CLDP features a common body with laser transceiver, electronic assembly and environmental control system and two separate nose sections which can be changed in 2 hours.

The TV head features a TV camera, gimballed mirror and a roll-stabilisation device. It also has a Field Of View (FOV) selector and a visible or near infra-red spectrum selector. This has four magnifications and corresponding fields of view.

The thermal imaging head features a gimballed optical head with a laser and thermal imager optics and

A close-up of the Convertible Laser Designation Pod with thermal camera (CLDP/CT), mounted on a Mirage 2000 0018363

CLDP/CT with thermal camera on a Mirage 2000D 0106252

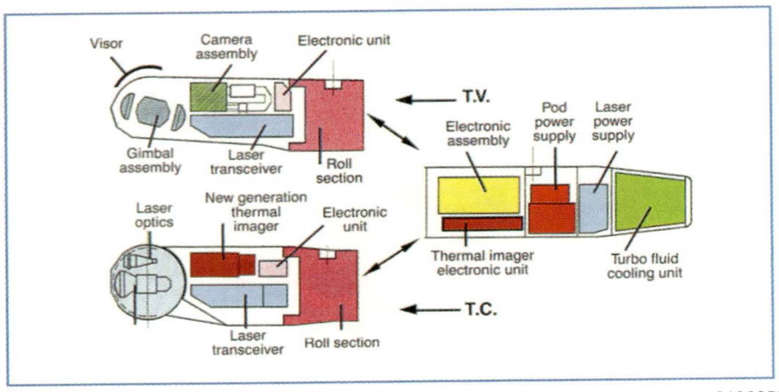

Schematic of the CLDP/TV and CT pods 0106253

roll-stabilisation device. There are four FOVs: 12/6° for navigation and 4/2° for target acquisition and tracking. The thermal imager is based upon the SMT modules.

A new version CLDP/CTS, based on a Wescam Synergi thermal imager using 288 × 4 cadmium mercury telluride focal plane array, is in production for French forces CLDP installations.

Operational status

In production. CLDP has been ordered by Abu Dhabi for Mirage 2000 and Saudi Arabia for Tornado IDS. It is already fitted to the Italian Air Force Tornado IDS. In addition, Alenia is working on a future AMX upgrade featuring the CLDP. Elsewhere, CLDP is being offered as part of the Sukhoi Su-25M5 upgrade. CLDP/CTS equips the French Air Force Mirage 2000-D

Specifications

Weight: 290 kg
Length: 2.85 m
TV
Spectral band: 0.7-0.9 μm
Magnification: × 2.5, × 5, × 10, × 20
Fields of view: 0.75, 1.5, 3 and 6°
Laser
Wavelength: 1.06 μm

Contractor

Thales Optronique (formerly Thomson-CSF Optronique).

UPDATED

Damocles multimode multifunction laser designator pod

Damocles is a multimode, multifunction laser designator pod that incorporates a third-generation thermal imager. The thermal imager operates in the 3 to 5 μm waveband and uses a staring focal plane array. The pod also includes a CCD TV camera and laser spot tracker.

A French Air Force Mirage F1 CR carrying the pod-mounted MDS 610 electro-optical sensor 0106421

The modular pod is designed primarily for laser designation but can also be used for navigation, air-to-air identification and reconnaissance roles.

Specifications

Thermal imager
Spectral band: 3-5 μm
Fields of View (FoV)
1.0 × 0.75°
4.0 × 3.0°
Electronic magnification: × 2

Laser spot tracker Laser range-finder
Wavelength: 1.06-1.54 μm
Dimensions (l × diameter): 2.5 m × 370 mm
Weight: 250 kg

Operational status

Damocles is under development for the Mirage 2000-9 and Rafale aircraft. The system will enter flight test in mid-2000. Damocles is also marketed as a combined system with NAVFLIR when it occupies only a single weapon station. As marketed to the United Arab Emirates, this dual configuration is designated SHAHAR (Damocles) and NAHAR (NAVFLIR).

Contractor

Thales Optronique (formerly Thomson-CSF Optronique).

UPDATED

The Thomson-CSF Optronique Damocles multimode, multifunction laser designator pod shown under the NAVFLIR navigation and attack system integrated into the pylon **2000**/0079243

MDS 610 MultiDistance Sensor

The MDS 610 is a passive electro-optical airborne reconnaissance sensor designed for the following daytime intelligence gathering missions: low-, medium- and high-altitude tactical reconnaissance; standoff oblique and vertical reconnaissance; fixed- or moving-target localisation and identification.

MDS 610 is gyrostabilised in two axes and can be pod or fuselage-mounted on combat, reconnaissance and surveillance aircraft. The sensor has both pushbroom and panoramic scanning modes. It can be manually or automatically controlled with real-time image display in the cockpit and is compatible with real-time transmission via datalink. Automatic control is achieved by preprogramming according to the mission.

MDS 610 is of modular design with an electro-optical CCD detector (or film), giving unlimited mission duration, stereo viewing of small surface areas and in-flight recording and replay capability.

Operational status

MDS 610 has been fitted in the Thomson-CSF Desire reconnaissance pod demonstrator which has been tested on a Mirage F1 CR aircraft. It will be carried in the Presto pod, based on the Desire demonstrator. Seven Presto pods have been ordered by the French Air Force. They are scheduled to equip Mirage F1 CR aircraft from 1999.

Specifications

Lens focal length: 610 mm – f/4
CCD detector: 10,000 pixels (0.4-1.1 μm)
Field of view: 11°
Lateral coverage: ±110°

Thomson-CSF Optronique MultiDistance Sensor (MDS) 610 0106420

Longitudinal coverage: ±20°
Weight: 120 kg
Dimensions: (d × l) 350 × 1,500 mm

Contractor

Thales Optronique (formerly Thomson-CSF Optronique).

UPDATED

NAVFLIR navigation and attack system

Thomson-CSF has developed the NAVFLIR airborne Forward-Looking Infra-Red (FLIR) navigation and attack system. This provides assistance for low-altitude flight at night and medium-range targeting in day or night-time conditions. Aircraft installation can be on the nose of a standard pylon or on a chin pod. The FLIR has a × 2 electronic zoom.

NAVFLIR can detect a target out to 20 km and perform reconnaissance at ranges of up to 10 km.

Specifications

Spectral band: 3-5 μm
Field of view
Wide: 18 × 24°, or matched to the HUD FOV
Narrow: 6°
Electronic zoom: × 2
Performance:
(for a 20 × 20 m target): 10 km
(for a 100 × 100 m target): 50 km

Operational status

Available. NAVFLIR is also marketed, as a combined installation on a single weapon station with the Damocles multimode pod. As marketed to the United Arab Emirates, this combination is designated SHEHAB (Damocles) and NAVHAR (NAVFLIR).

Contractor

Thales Optronique (formerly Thomson-CSF Optronique).

UPDATED

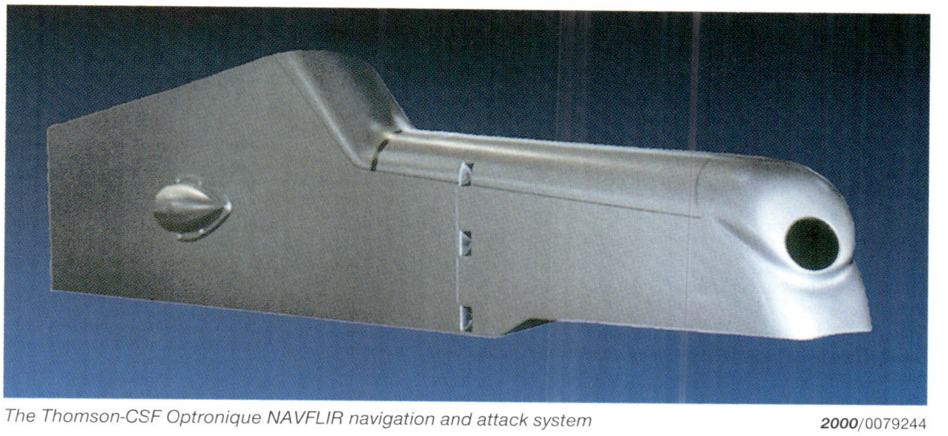

The Thomson-CSF Optronique NAVFLIR navigation and attack system **2000**/0079244

The Thomson-CSF Optronique Presto electro-optical reconnaissance pod **2000**/0079241

Optronique Secteur Frontal (OSF) for the Rafale aircraft

Thomson-CSF Optronique and SAGEM SA are co-operating for the development and manufacture of the optronic, visual and infra-red search and tracking system for the Dassault Aviation Rafale ACT and ACM.

The OSF (Optronique Secteur Frontal) is designed to aid covert missions, firing under jamming, visual identification and damage assessment in air-to-air, air-to-ground and air-to-sea operations and to provide navigation/piloting assistance. Key features include infra-red passive detection, very low false alarm rates, high-definition CCD imagery, an eye-safe laser range-finder, very large field of regard and two optical heads to ensure simultaneous search/identification/telemetry functions.

The Optronique Secteur Frontal (OSF) shown with the Rafale aircraft 0106254

Operational status

In continuing development for the Rafale aircraft. First prototypes were available for flight test at the end of 1997.

Contractors

Thales Optronique (formerly Thomson-CSF Optronique).
SAGEM SA, Defence and Security Division.

UPDATED

Presto/Desire reconnaissance pod

The Desire reconnaissance pod was a demonstrator built by Thomson-CSF Optronique under a French Ministry of Defence programme. This led to the development of the Presto pod for standoff or high-speed penetration missions. Presto provides real-time imaging, digital recording and transmission to ground processing stations.

The system includes the Thomson-CSF Optronique MDS 610 multidistance sensor, a reconnaissance management system, a high-speed digital recorder and an associated ground station. Presto is capable of operating in both pushbroom and panoramic modes, and it can function either under direct pilot control, or autonomously in flight under pre-programmed control.

Operational status

The French Air Force has placed an order for Presto pods for Mirage F1 CR and Mirage 2000D aircraft. Presto entered the in-flight development stage at the Cazaux flight test establishment in December 1999. Operational evaluation will be at French Air Force's Mont-de-Marsan flight test centre, and production deliveries are scheduled for mid-2000.

Contractor

Thales Optronique (formerly Thomson-CSF Optronique).

UPDATED

SDS 250 electro-optical reconnaissance sensor

The SDS 250 is a compact passive electro-optical airborne reconnaissance sensor designed to perform the following daytime intelligence gathering missions: low- and medium-altitude tactical reconnaissance,

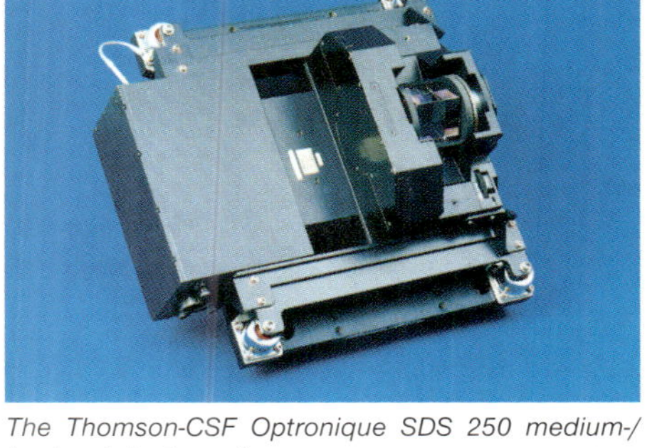

The Thomson-CSF Optronique SDS 250 medium-/low-level electro-optic sensor **2000**/0079242

vertical reconnaissance and up to 20 km standoff oblique reconnaissance.

The sensor has pushbroom scanning mode, electronic roll stabilisation and selectable operating spectral band - either visible or near-infra-red or both. The system provides real-time image display in the cockpit and line of sight position is selected from the cockpit. The system is compatible with real-time image transmission via datalink. The compactness of the sensor allows fuselage installation.

Specifications

Lens focal length: 250 mm – f/5.6
CCD detector: 6,000 pixels (0.4-1.1 μm)
Field of view: 14°
Lateral coverage: ±85°
Weight: 30 kg
Dimensions: 410 × 400 × 230 mm

Operational status

SDS 250 has been selected by the French Air Force and by the French Aeronavale for upgraded Super Etendard aircraft.

Contractor

Thales Optronique (formerly Thomson-CSF Optronique).

UPDATED

Tango thermal imager

The Tango thermal imager is a modular thermal imaging system for long-range maritime patrol aircraft. It is an 8 to 12 μm CMT thermal imager with three fields of view. The system was developed for the Dassault Aviation Atlantique 2 maritime surveillance aircraft. Key features include: a large aperture for very long range imaging; high resolution; fine image stabilisation; automatic aiming towards designated targets; aircraft databus coupling; and line-to-line integration. It is incorporated in a Sere-Bezu gyrostabilised platform, fixed under the nose of the Dassault Aviation Atlantique 2.

Day and night missions include passive detection of ships and snorkels, long-range identification of surface vessels, reconnaissance and search and rescue.

Tango 2G modular thermal imaging system for long-range maritime patrol aircraft

The Thomson-CSF Optronique TMV 630 airborne laser range-finder 0106335

Tango 2G is a second-generation thermal imager based on the Synergi thermal imaging modules developed by Thomson-CSF, Pilkington Optronics and Zeiss-Eltro Optronic. Synergi uses a 288 × 4 IRCCD detector developed by Sofradir. Tango 2G has four fields of view. It is a multisensor integration and incorporates a CCD detector for the visible and near-infra-red channel and has an auto-search mode. It is being developed for the Atlantique third-generation, the aircraft ATL3G.

Operational status

Tango is in production for French Navy Atlantique 2 aircraft.

Specifications

Tango modular thermal imaging system
Dimensions: 600 mm turret diameter
Turret weight: 85 kg
System weight: 120 kg
Gyrostabilised field of view
(Azimuth) ±110°
(Elevation) +15 to -60°
Tracking speed with speed/accuracy optimisation: 1 rad/s
Infra-red channel
Detector: CMT
Cooling: Stirling engine
Spectral band: 8-12 μm
Fields of view
(Wide) 6.45 × 4.30°
(Medium) 2.15 × 1.43°
(Narrow) 1.07 × 0.7°
Tango 2G
Dimensions: 600 mm turret diameter
Turret weight: 75 kg
System weight: 98 kg
Gyrostabilised field of view
Azimuth: ±360°
Elevation: +15 to -93°
Tracking speed with speed/accuracy optimisation, auto-search mode: 1 rad/s
Infra-red channel
Detector: 288 × 4 IRCCD CMT focal plane array
Spectral band: 8-12 μm
Fields of view: 15 × 11.2°
7.5 × 5.6°
1.5 × 1.1°
0.75 × 0.6°
Visible and near infra-red channel
Detector: CCD
Field of view: 1.5 × 1.1°

Contractor

Thales Optronique (formerly Thomson-CSF Optronique).

UPDATED

The Thomson-CSF Optronique Tango thermal imager mounted on the nose of the Dassault Aviation Atlantique 2 aircraft 0106255

TMV 630 airborne laser range-finder

The TMV 630 airborne laser range-finder equipment has been designed to meet single unit, small installation requirements and can be fitted easily to a wide range of aircraft. It provides high-precision aircraft-to-target range measurement and is claimed to increase considerably the performance of conventional weapon-aiming systems. The large field of view provided is compatible with all head-up displays and the electrical interfaces are compatible with almost all aircraft types. The high-speed and accurate laser beam-steering is specifically adapted for continuously computed impact point attacks irrespective of terrain or the nature of the weapons or aircraft altitude.

Specifications

Dimensions: 190 × 190 × 520 mm
Weight: 15 kg
Power supply: 28 V DC, 12 A
Wavelength: 1.06 μm
Range: up to 19 km
Accuracy: better than 1 mrad

Operational status

In production as part of nav/attack system on Dassault Mirage and Dassault Aviation/Dornier Alpha Jet aircraft.

Contractor

Thales Optronique (formerly Thomson-CSF Optronique).

UPDATED

TMV 632 airborne laser spot tracker and range-finder

The TMV 632 ground attack laser range-finder was developed at the request of the French DGA (Delégation Générale de l'Armament) and combines eye-safe laser ranging and angular tracking functions into a single compact monobloc system, for use on tactical ground support and training aircraft. It provides the weapon systems with extremely accurate fire control data, for both laser-guided and conventional munitions. Several variants are available, with different digital interfaces. The dual-function TMV 632 offers the same level of performance and accuracy as single-function systems.

The TMV 632 is mounted in the airframe (embedded) or fitted in a mini-pod, or inside a store-carrying pylon. The TMV 632 detects and identifies the laser spot on illuminated ground-based targets. Acquisition and tracking are automatic. A sighting reticle in the head-up display allows the pilot to aim at the target. The laser beam of the range-finder is locked to the position of the tracker, making accurate measurement of the aircraft-to-target distance possible. There are two independent safety interlocks in the TMV 632. If required, cooling air is supplied by the aircraft.

The TMV 632 laser spot tracker and range-finder in underbelly housing 0106256

The TMV 632 airborne laser spot tracker and range-finder 0018355

Operational status

Flight tests have validated rangefinding performance, and tests are underway on the angular tracking capability. The first three series-production units for Mirage F1CT aircraft are due to be delivered by the end of 2000, with shipping of the remaining 17 on order for the French Air Force scheduled for 2001.

Specifications

Overall Dimensions (L × W × H): 530 × 170 × 190 mm
Total weight: 18 kg
Wavelength
Laser: 1.54 μm
Spot tracker: 1.06 μm
Range: up to 20 km in range-finding mode and 15 km in tracking mode
Field of regard: 40° azimuth, 20° elevation
Interfaces: TM632: digibus, TM632A: ARINC, TM632B: bus 1553
Power: 28 V DC, 8 A

Contractor

Thales Optronique (formerly Thomson-CSF Optronique).

UPDATED

Victor thermal camera

The Victor thermal camera is designed to be connected to Viviane and Strix gyrostabilised sights, for SA342 Gazelle HOT and HAP Tiger escort helicopters. It can also be fitted on AS 365M Panther, BO 105 or other types of helicopters.

Victor converts the thermal radiation of landscape and objects into a visible image at TV standard and is suitable for air-to-ground and air-to-air gun or missile firing, rocket firing and as a flying aid by day and night and under adverse weather conditions.

The system displays symbols in the eyepiece of the sight or on a TV monitor and has a magnifying function of × 2 to enlarge the image. There is a processing board option for improved performance. Initial sight stabilisation is by the platform and electronic fine stabilisation is by the imager. There is a specific video output for tracking.

Specifications

Weight:
(total) 25 kg
(on roof) 17 kg
Power supply: 20-32 V DC, 140 W
Wavelength: 8-13 μm
Trifocal lens: 30 × 20°, 6 × 4°, 2.4 × 1.6°
Range: up to 4,000 m

Operational status

Victor entered service in 1988.

Contractor

Thales Optronique (formerly Thomson-CSF Optronique).

UPDATED

The Thomson-CSF Optronique Victor camera on a French Army Gazelle helicopter 0106257

GERMANY

Hellas helicopter Obstacle Warning System (OWS)

Hellas is designed to provide warning of wires and other similar obstacles to helicopters flying nap-of-the-earth and other low-altitude missions, at up to 1,000 m range, both for military and civil operations.

Obstacle detection is performed with an imaging eye-safe ladar, which generates images of the scene in front of the helicopter, while range data processing is performed in the processing unit. Processing results can be configured for display as warning information for the pilot. The first processing action provides presentation of the colour-coded, real-time image on the HMI. Wire obstacles contrast with the background to become clearly visible and recognisable.

The imaging laser radar scanning architecture is designed to ensure the generation of range images which enhance the detection of wires and wire-like objects. The scanner, designed by Dornier, provides fast line scanning. Continuous column scanning is produced with an oscillating mirror. The range imaging system is capable of producing up to 100,000 pixels per second, allowing within the image generation for configuration of line and frame rates. The fibre optic scanner is based on two nutating mirrors on a single shaft. Laser pulses injected into a single fibre are imaged via the nutating mirror onto a circular fibre array. This array transforms into a linear array, positioned in the focal plane of the field optics. During development, the system demonstrated the ability to detect 5 mm diameter wires at several hundred metres.

Hellas helicopter obstacle warning system ***1999***/0054090

Specifications

Sensor: range imaging laser radar
Wavelength: solid-state eye-safe at 1.54 micrometres
Laser pulse power: 4 kW
Receiver: InGaAs APD hybrid (Avalance Photo Diode)
Scanning: 2 axes. horizontal: fibre optic; vertical: oscillating mirror
Image repetition frequency: 2 Hz, 4 Hz possible
Field of view: 32 × 32°
Range:
>1,000 m (extended area objects, in good visibility)
>400 m (extended area objects, poor weather)
>500 m (wires >10 mm, good visibility)
>300 m (wires >10 mm, poor weather, oblique incidence)

Range resolution: <1 m
Angular resolution:
<0.35° horizontal
<0.2° vertical
Pixels:
95 horizontal
200 vertical
Volume: <36 litres
Weight: <30 kg

Operational status

Development contract awarded by the Federal Office of Defense Technology and Procurement. The system has been successfully tested on BK-117, CH-53 and UH-ID helicopters. The German Federal Border Guard has ordered 25 units for its new EC-135 and EC-155 helicopters. Initial operational deployment was scheduled for late 2000, with all helicopters to be fitted by 2002.

Contractor

EADS Defence and Civil Systems (Formerly DaimlerChrysler Aerospace, AG).

UPDATED

TV tabular display unit

The TV tabular display is capable of displaying various sensors such as a TV camera, FLIR, low-light level TV and reconnaissance cameras and alphanumeric characters. By pressing related buttons on the keyboard the operator can select the preferred sensor out of three which is then internally mixed and displayed with alphanumerics overlaid. The display information on the Display Unit (DU) is also output for recording purposes for future reference or for post-mission analysis.

The DU also generates permanent data output by two 32-bit data words that are transferred to the main computer for selection and control of a computer program. The data words are output in both true and complement form and clocked out by a continuous train of 64 kHz data synch pulses which is also presented in both true and complement form.

The display unit brightness is controlled by an automatic brightness control circuit which guarantees readability of the displayed information during changing light levels, from darkness up to 10^5 lux ambient illumination. In addition to this, the operator can, within certain limits, override the automatic control system by manually setting a brightness and contrast level.

The TV tab DU consists of modules interchangeable between like assemblies which guarantees good serviceability and low maintenance costs.

Specifications

Dimensions: 450 × 210 × 210 mm
Weight: 11.6 kg (max)
Power supply: 200 V AC, 400 Hz, 3 phase, 80 VA
28 V DC

Operational status

In service on the Panavia Tornado.

Contractor

ESW-Extel Systems Wedel.

UPDATED

TV tabular displays fitted in the rear cockpit of a UK Royal Air Force Tornado GR4 (E L Downs) ***2001***/0103860

TV tabular display for the Panavia Tornado 0106371

Infra-Red Linescanner Systems (IRLS)

The Honeywell Regelsysteme IRLS is designed to meet the German Air Force Tornado reconnaissance pod requirement. It comprises the following line replaceable units:

Scanner Receiver Unit (SRU), supplied by Lockheed Martin IR Imaging Systems;

Reconnaissance Management Unit (RMU);

Digital Tape Recorder (DTR), supplied by Ampex Inc;

Reconnaissance Control Panel (RCP), supplied by Computing Devices UK;

Reconnaissance Power Supply (RPS).

The SRU, RMU, DTR, RPS and two aerial daylight cameras will be installed in the new modular pod designed by DaimlerChrysler Aerospace AG.

The IRLS is optimised for operation between the altitudes of 200 and 2,000 ft, and at speeds ranging from 300 to 600 kt. Operation at up to 8,000 ft altitude will be achieved by software change, implemented in 1999.

The SRU senses infra-red terrain radiation and converts it to 12 parallel video channels; an array of 34 detector elements is scanned across track. The SRU also contains patented scan compensation systems.

Operational status

The IRLS is in production for the German Air Force Tornado reconnaissance pod. The Scanner Receiver Unit (SRU) was designed and built for the Infra-red Imaging System (IIS) on the Tornado ECR (Electronic Combat and Reconnaissance) aircraft.

Contractor

Honeywell Regelsysteme GmbH

VERIFIED

GB 609 Control and Display Unit (CDU)

The GB 609 is a Control and Display Unit (CDU) specially devised for avionics systems for entering and displaying different types of data. Its operating concept is determined by software, so that operating strategies can be implemented by the user.

The GB 609 is a smart CDU exhibiting high intelligence and computing capability as a result of its hardware concept, which can manage complex systems in the role of a central computer, for example as a bus control as defined in MIL-STD-1553B. An arithmetic co-processor can be incorporated as an option, accelerating substantially the execution of extensive arithmetic operations.

The panel-integrated, solid-state electroluminescent display has full graphic capability, call up of prestored masks, 14 lines × 40 characters in alphanumeric mode and three character size standards. It features high contrast and is viewable in direct sunlight. Coloured green, the display is NVG-compatible. The reliability of the ultra-flat and lightweight display, which is insensitive to vibration and shock, is unequalled by any other display technology. Unlike a CRT, it does not have to be changed at regular intervals.

Specifications
Dimensions: 146 × 217 × 185 mm
Weight: 4.5 kg
Power supply: 16-31 V DC, 45 W
Temperature range: −40 to +55°C
Program memory: 0.5 Mbytes EPROM
Data memory: 16 kbytes EEPROM, 64 kbytes RAM

Contractor
Rohde & Schwarz and Co GmbH.

VERIFIED

GB 609 control and display unit

KRb 8/24 F reconnaissance camera

The KRb 8/24 F camera achieves wide-angle panoramic coverage with undistorted framing camera geometry on a single 9.5 in (240 mm) wide film. The camera performs at highly survivable parameters for low- to medium-altitude reconnaissance missions. Each exposure covers 143° across track, with true angle forward motion compensation across the entire format. This format affords sequential along-track stereoscopic coverage. Images are without the cylindrical distortion inherent in panoramic cameras. Special Zeiss optics provide performance into the near infra-red, allowing the use of all aerial film types, including colour and camouflage detection, without refocusing. Small size and low weight permit easy installation in RPVs, pods and aircraft.

Specifications
Dimensions: 356 × 374 × 311 mm
Weight: 22 kg
Power supply: 28 V DC, 250 VA
Focal length: 80 mm
Aperture range: f/2.56-f/16

Contractor
Z/I Imaging GmbH.

VERIFIED

KS-153 modular camera system

The KS-153 is a modular camera system consisting of three different focal length configurations: The Pentalens 57, Trilens 80 and Telelens 80. The Pentalens configuration is the latest design development in the system. These three configurations have a common camera body, film cassettes and film cassette holder. The desired focal length lens and shutter assembly with format mask can easily be attached to the camera body to accommodate mission requirements. The high parts commonality provides a benefit to multiple-type camera users.

The modular KS-153 is a fully electronic, microprocessor-controlled design and uses Carl Zeiss optics optimised for high resolution. The camera system is built and tested to meet and exceed reliability and maintainability criteria for use in modern military or commercial aircraft, pods and UAVs.

The KS-153 Pentalens 57 is a high cycle rate, pulse operated, sequential frame camera designed for low- to medium-altitude, wide-angle, photography from high-performance aircraft.

The camera combines wide-angle panoramic camera coverage with framing camera geometry on a single 240 mm (9.5 in) wide film. The wide-angle coverage is provided by an optical assembly of five TOPAR A2 2/57 lenses having an effective maximum aperture of f/2.7. The 182.7° lateral coverage is displayed by five across-track images per frame, each frame covering an angular field of view of 18.27° across-track by 47.4° along-track. The field view of the side lenses are deflected by front-mounted prisms.

The KS-153 Trilens 80 is a high cycle rate, pulse operated, sequential frame camera designed for low- to medium-altitude, wide-angle photography from high-performance reconnaissance aircraft. The camera accommodates both 4 mm standard base and 2.5 mm thin base roll film in any panchromatic, infra-red or colour emulsion. Each frame covers an angular field of view of 143.5° across-track and 48.5° along-track. The wide lateral coverage is provided by the optical assembly of three S-TOPAR A1 2/80 lenses having an effective maximum aperture of f/2.56 and the field of view of the side lenses deflected by front-mounted prisms. Major camera features are true angle corrected Forward Motion Compensation (FMC), constant velocity focal plane shutter, integral intervalometer, Automatic Exposure Control (AEC), easily interchangeable interface card and continuously monitoring BITE. This BITE feature is further enhanced by an integral non-volatile BITE memory to enable post-flight analysis of in-flight transient failures.

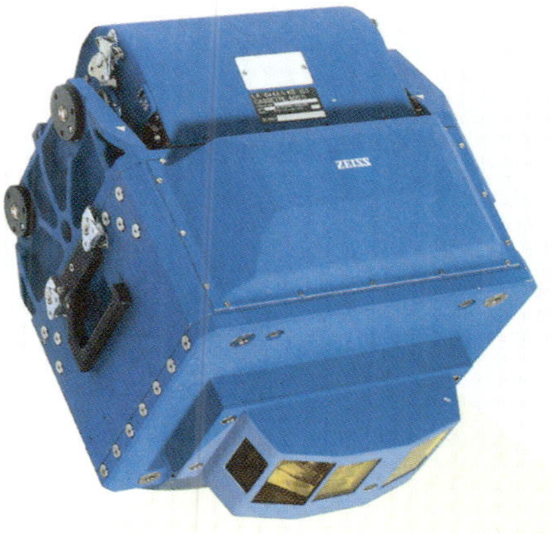

Z/I Imaging KS-153 Pentalens 57 camera
1999/0051001

The KS-153 Telelens 610 is a pulse-operated, sequential frame camera designed for medium- to high-altitude oblique photography from high-performance reconnaissance aircraft. The camera accommodates both 4 mm standard base and 2.5 mm thin base roll film in any panchromatic, infra-red or colour emulsion. Each frame covers an angular field of view of 21.4° across-track by 10.7° along-track. The camera mounts in an integral ring bearing, which allows in-flight rotation under electronic control to any desired oblique angle. The Telelens 610 has the same major camera features as the Trilens version of the KS-153 camera.

Stereoscopic viewing greatly increases intelligence gained from aerial photography and the KS-153 images appear side-by-side to provide maximum convenience for direct stereoscopic viewing without cutting the film.

Z/I Imaging KS-153 modular camera system **1999**/0044999

Z/I imaging KS-153 Trilens 80 camera **1999**/0051002

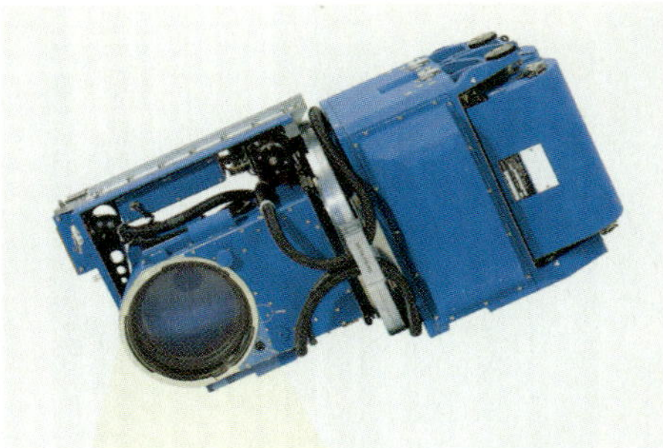

Z/I Imaging KS-153 Telelens 610 camera ***1999***/0051003

Specifications

Dimensions:
(Pentalens 57) 439 × 467 × 502 mm
(Trilens 80) 439 × 467 × 506 mm
(Telelens 610) 423 × 809 × 470 mm
Weight:
(Pentalens 57) 59 kg
(Trilens 80) 57 kg with 500 ft of film
(Telelens 610) 110 kg with 200 ft of film
Power supply: 115 V AC, 400 Hz, 3 phase
Focal length:
(Pentalens 57) 57 mm
(Trilens 80) 3.5 in (80 mm)
(Telelens 610) 24 in (610 mm)
Cycle rate:
(Pentalens 57) 10/s (max)
(Trilens 80) 10/s (max)
(Telelens 610) 4/s (max)
Shutter speed: 1/2,000 to 1/150 s
Aperture range:
(Pentalens 57) f/2.56 to f/16
(Trilens 80) f/2.56 to f/16
(Telelens 610) f/4 to f/16
Format:
(Pentalens 57) 50 × 212 mm
(Trilens 80) 72 × 222 mm
(Telelens 610) 230 × 115 mm
Film length:
(Pentalens 57) 500 ft
(Trilens 80) 200 or 500 ft
(Telelens 610) 200 ft

Contractor

Z/I Imaging GmbH.

VERIFIED

Z/I Imaging RMK TOP15 configuration showing the TOP15 camera (centre), T-TL central control unit (left) and T-CU central computer unit (right) ***1999***/0051004

RMK TOP aerial survey camera system

The RMK TOP is an aerial photography system used for survey and cartographic purposes. It provides enhanced image quality with minimum distortion and extensive image motion compensation by FMC and a gyrostabilised suspension mount. System controls and functions are monitored by a compact computer and microprocessor and a pulsed rotating disk shutter with a constant access time of 50 ms. It is particularly suitable for use with GPS-controlled navigation systems.

The basic components of an RMK TOP camera are the RMK TOP15 camera body with wide-angle PLEOGON A3 4/153 lens (or RMK TOP30 camera body with standard TOPAR A3 5.6/305 lens), T-TL central control unit, T-CU central computer unit, T-MC film magazine with FMC and T-AS suspension mount gyrostabilised in three axes.

The following options are available for navigation and system control: T-FLIGHT GPS-supported photoflight management system for flight planning and mission documentation, T-NT visual navigation telescope, T-NA automatic navigation meter for automatic V/H measurement and interfaces for aircraft specific navigation systems.

Specifications

Weight:
(RMK TOP15) 176.3 kg total
(RMK TOP30) 169.6 kg
Focal length:
(RMK TOP15) 6 in (153 mm)
(RMK TOP30) 12 in (305 mm)
Aperture:
(RMK TOP15) f/4 to f/22
(RMK TOP30) f/5.6 to f/22
Exposure time: 1/50 to 1/500 s

Contractor

Z/I Imaging GmbH.

VERIFIED

VOS digital video colour camera system

The VOS digital video colour camera system is designed to provide high-resolution airborne photographic surveillance.

VOS uses a high-resolution, multi-spectral detector. It consists of three parallel photodiode line arrays, each with 6,000 pixels for the colours red, green and blue, covering a spectral band from 350 to 1,050 nm. A cut-off filter in front of the lens limits the spectral response to 650 nm, generating a true colour image.

The line array of the camera system operates in a push-broom mode and a continuous image is generated by forward motion of the aircraft. The resolution in the forward direction is directly dependent on the aircraft speed. In order to obtain square image pixels, the control unit automatically calculates the correct line rate dependent on the altitude, speed and focal length. Post-processing is used to optimise the output of the camera, recording and processing functions.

The SMV-1 stabilised platform is used as a mount for the VOS camera system. It employs gyro technology, active control components and passive vibration damping to stabilise the camera body.

Specifications

VOS electro-optical sensor
Detector: Kodak, colour, 3 × 6,000 pixels (red, green, blue)
Pixel size: 12 × 12 μm², 12 μm pitch
Line space: 96 μm
Line rate: 1.6 kHz, maximum
Spectral response: 350-1,050 nm; colour with 650 nm filter
Data rate: 230 Mbits/s, maximum
Lens: Carl Zeiss Planar®, FL 80 mm, f/2.8
Field-of-view: 48.5° across flight path
Dimensions: 154 × 154 × 210 mm
Weight: 3.5 kg
Power: 28 V DC, 28 W

Sensor control unit
Dimensions: 483 × 133 × 390 mm
Weight: approximately 15 kg
Power: 28 V DC, 125 W

Sensor control panel
Dimensions: 150 × 200 × 105 mm
Weight: approximately 1.5 kg
Power: supplied by control unit

SMV-1 stabilised platform
Gimbals: pitch and roll axes stabilised
Gimbal freedom: ±10° each axis
Stabilisation: <70 μrad rms
Positioning accuracy: <0.5° (1σ)
Weight: approximately 10 kg (excluding camera)
Dimensions: 460 × 310 × 220 mm
Power: 28 V DC, 50 W

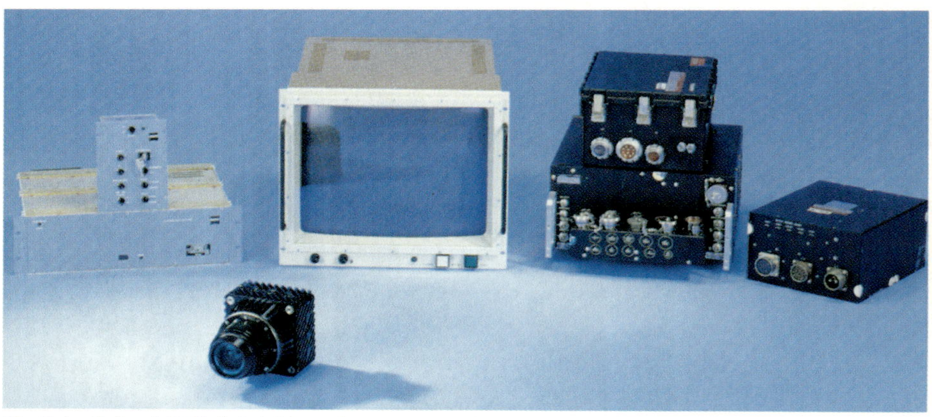

The Zeiss Optronik VOS digital video colour camera system ***2000***/0079246

Operational status

The VOS system is operated by the German 'Open Skies' force, in a three camera fit. One system is vertically installed in the observation aircraft, the other two systems in oblique positions on the left and right side, with overlapping FOVs for wide area coverage.

Contractor

Zeiss Optronik GmbH.

VERIFIED

INTERNATIONAL

AN/AVS-7 ANVIS/HUD system

The AN/AVS-7 ANVIS/HUD system projects flight data into the views of the pilots NVGs. By eliminating the need to shift attention and focus to the cockpit interior, the AN/AVS-7 Aviator's Night Vision System/Head-Up Display (ANVIS/HUD) enhances flight safety and operational effectiveness. Pilots can fly head-up, viewing the situation outside the cockpit, while at the same time receiving all essential flight data. The AN/AVS-7 features a real-time, high resolution, lightweight display unit which is easily mounted on NVGs. It is adaptable to any aircraft platform and has four independent display modes, each with declutter facilities. The pilot and co-pilot can independently select symbols and display modes from controls on the centre console and collectives, using ANVIS compatible control panel and display facilities. BIT provides 95 per cent failure detection and fault isolation.

Day-HUD optics interface directly with the AVS-7 harness to provide day symbology, including expanded symbology for day/night pilot operations, plus maintenance and training.

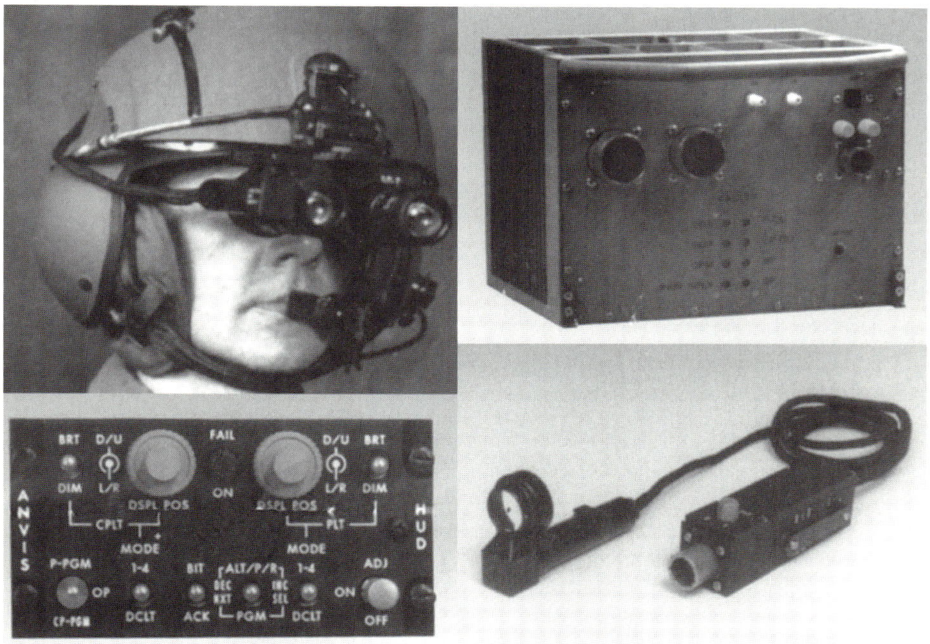

AN/AVS-7 ANVIS/HUD showing: display unit (top left), signal data converter (top right), converter control (bottom left) and sensors and harness (bottom right)
0018157

Specifications

Field of View (FoV): 34°
Display type: stroke (highly stable)
Resolution: 512 × 512 pixels
Interfaces: MIL-STD-1553B, ARINC 429, RS-422, analogue, discrete and synchro
Reliability: >1,000 h
Power: 28 V DC, <5A

Avionics B kit
Dimensions:
(converter control (CC)): 63.5 × 139.7 × 76.2 mm
(display unit (DU)): 88.9 × 38.1 × 39.1 mm
(signal data converter (SDC)): 274.3 × 190.5 × 198.1 mm
Aircraft installation A kit: sensors and harness, as required

Operational status

AN/AVS-7 is built by BAE Systems North America in collaboration with Elbit Systems Ltd. AN/AVS-7 is installed on many Israeli Air Force, US Army, US Marine Corps and US Navy rotary- and fixed-wing aircraft. AN/AVS-7 adapts to the following aircraft types: CH-46E, CH-47D, CH-53E, HH-60H, KC-130T, MH-47E, MH-60K, MV-22, UH-1N and UH60A/L. A total of 3,500 systems are in use Worldwide.

A derivative system (Gideon) has been ordered by the US Marine Corps, and by unspecified other forces for Bell AH-1W helicopters.

BAE Systems North America and Elbit Systems Ltd have together been contracted to upgrade the computer element of 1,100 AN/AVS-7 systems for the US Army. The upgrade will double the number of inputs and outputs to transfer data to the Crash Survival Memory Unit (CSMU), increase CPU speed, and facilitate programming via a front panel connector.

Contractors

Elbit Systems Ltd.
BAE Systems North America.

VERIFIED

Cockpit Displays for the A400M Military Transport

The Airbus Military Company was formed in January 1999 to manage the A400M military transport aircraft project, formerly known as the FLA (Future Large Aircraft). The industrial partners involved in the programme are BAE Systems, EADS (comprising Aerospatiale Matra, CASA and DASA), Alenia, FLABEL and TAI. Airbus Industrie is the majority shareholder in the company, bringing to the project valuable experience of building large commercial air transport aircraft and managing complex international industrial programmes.

The cockpit of the A400M will be based on the familiar and proven layout used in Airbus' large commercial aircraft, such as the A340-300/500/600. The cockpit features side stick controllers linked to the fully digital fly-by-wire system, wide angle Head-Up Displays (HUDs), large scale portrait primary flight Displays (PFDs) and Navigation Displays (NDs), coupled with large format head-down MultiFunction Displays (MFDs) linked to the aircraft Flight Management System (FMS). The cockpit, in common with all Airbus commercial designs, will be designed for two pilots, with monitoring of aircraft systems and fault warning carried out by an Electronic Centralised Aircraft Monitor (ECAM). Cockpit seating and HUD installation will cater for both pilots wearing Night Vision Goggles (NVG), such as the AN/AVS-9 (see separate entry). Cockpit displays will include elements of existing Airbus commercial aircraft displays and benefit from ongoing work for the A340-500/600 and A380 development programmes.

Operational status

Under development.

Contractor

Airbus Military Company.

NEW ENTRY

Artist's impression of the A400M cockpit, showing Head-Up-Displays for each pilot and large scale portrait primary flight and navigation displays
2001/0106208

Crusader helmet display system

BAE Systems in association with Gentex and Pilkington Optronics, is developing the Crusader helmet system for both fixed- and rotary-wing aircraft. The team is taking a fully integrated approach and is concentrating on visor projection for fighter/attack applications with full day and night capability. An optical head tracker is used for maximum accuracy and minimum latency. A 40° visor-protected field of view has been demonstrated and is being further developed.

Crusader employs Pilkington Safe-Lite helmet display optics.

Operational status

Crusader was a helmet display system technology demonstrator programme, which has lead into the Eurofighter Typhoon Helmet-Mounted Display (HMD) system.

Contractors

BAE Systems.
Gentex Corporation.
Pilkington Optronics Ltd.

VERIFIED

The Crusader helmet display system is under development
0106258

Eurofighter Typhoon Head Equipment Assembly (HEA)

The Eurofighter Typhoon HEA fulfils all the requirements of the pilot for a fully supported conventional helmet, together with full Helmet-Mounted Display (HMD) functions. It provides 24-hour, all-weather, capability to track targets using the visor display, and to designate them for attack using the microphone to communicate via the aircraft Direct Voice Input (DVI) system. Information is displayed wherever aircrew look as the HEA includes a high-speed optical tracking system to determine which way the helmet is pointing. This allows all displayed information to exactly overlay the real-world view seen by the aircrew through the visor. HEA capabilities provided include:

(1) life support.
(2) oxygen mask including positive pressure breathing for high 'G' manoeuvres;
(3) communications;
(4) microphone with Direct Voice Input (DVI) compatibility;
(5) earphones with high attenuation of external noise;
(6) aircrew protection;
(7) high speed ejection capability;
(8) nuclear Biological Chemical (NBC) hood option;
(9) anti-laser visor;
(10) air-ventilated temperature control;
(11) display of target and flight information;
(12) symbol/image generation;
(13) binocular fully overlapped 40° field of view;
(14) capture and display of the Night Vision Enhanced (NVE) outside world picture;
(15) twin NVE cameras with 40° field of view;

The HEA is based on a new Basic Mechanical Helmet (BMH) sourced from Gentex. The BMH is a two-part helmet made up of an inner helmet (custom fitted to the pilot), and outer helmet assembly (which incorporates the visor assemblies) and a lightweight oxygen mask. The electronics are integrated to the BMH. Alenia Difesa provides the display generator system, based on its work on the Eurofighter Typhoon Enhanced Computer Signal Generator (ECSG), and the HEA Processor Unit (HEAPU).

The Eurofighter Typhoon head equipment assembly **2001**/0089539

Operational status

The Eurofighter Typhoon HEA is a twin-track programme. The BMH will enter service with the first batch of Eurofighter Typhoon aircraft in 2002. The full avionics HEA will enter service with the second batch of Eurofighter aircraft in 2003. All BMHs can be upgraded to full display helmets. The full avionics HEA is due to start flight test in 2001

Contractors

BAE Systems (Operations) Limited, leading a consortium comprising Alenia Difesa, Avionic Systems and Equipment Division Gentex.

NEW ENTRY

EuroGrid European geographic information display system

The digital map generator of EuroGrid comprises the following main functional modules:

(1) The main processor for the control of the EuroGrid system with graphic-overlay management including thematic classes, control and performance of the Man-Machine Interface (MMI), and an Internal Fast Serial Link (FSL).
(2) The geographic processor for 2-D map and graphic overlay presentation, output via video link to the displays in the cockpit with the following functions: presentation of basic maps on different scales: 1:100,000, 1:250,000, 1:500,000; presentation of real moving map: translation and rotation in real time; presentation of maps with different zooming factors.
(3) A geographic memory for the storage of all necessary map data and the symbol library.
(4) The communication processor for the control of two different radio links via software, with the following main tasks: receipt and transmission of tactical messages; management of transmission and reception; administration of datalink participants; control of datalink channels (VHF/FM and HF); failure management. Communication messages are as follows: tactical, navigation and threat overlays; target hand-over; position requests and answers; text messages.
(5) A video memory (video interface) for acquisition and storage of IR and TV images. Presentation on the selected monitors beside the respective map section in which the lines of sight of the image are represented.
(6) A elevation processor for the vector map image presentation with the features: shading with or without hypsometric colours; ground avoidance presentation by coloured safe flight areas; terrain profile, presentation of terrain on the helicopter track.

Specifications

Type of system: real-time map/graphic elevation processing; 32 bit processing
Architecture: solid-state
Memory technology: raster, vector and charts
Type of map data in main map storage: raster, vector, let down, plates, charts and tactical data
Dimensions: ARINC 600, 6 MCU, 1 cockpit unit
Mass: 10.1 up to 13.1 kg for EuroGrid LRU set (incl MDE/MDT) max
Power: 130 VA for EuroGrid LRU set max
MTBF: 1000 h

Operational status

Selected for both German and French versions of the Tiger helicopter; also planned for use in the NATO NH 90 helicopter.

Contractors

DaimlerChrysler Aerospace AG, Defense and Civil Systems.
Sextant.

VERIFIED

Guardian family of display helmet systems

The Guardian family of display helmets is a range of systems that have evolved from the strategic teaming of Pilkington Optronics and Cumulus. The result is a fully integrated, lightweight, balanced, comfortable, high-performance range of display helmets for fixed- and rotary-wing applications.

The Guardian concept is based on use of an inner helmet that is individually matched as a personal-issue item to each aircrew user. The inner helmet carries all life support elements, including communications and oxygen mask. Active noise reduction and chemical defence are offered as options.

The inner helmet can support a variety of different outer helmets, which carry the Electro-Optic (EO) platform's visors, optics, display media and other electronics.

Sight Helmet

The Guardian Sight Helmet is designed to allow fighter and light attack aircraft operators to slew off-boresight missiles, thereby greatly increasing the operational effectiveness of the aircraft. The display medium is a red LED, which provides the symbols for off-boresight aiming and, in conjunction with a radar, cueing arrows to allow fast acquisition of the target. The red symbology is reflected off the visor to the pilot's eye.

With the Guardian optical head tracker constantly providing head angles and position to the weapon system, all the pilot needs do is align the LED cross with the target and press the missile designate button. The missile head then moves to the target angles and locks-on. On receipt of the missile tone, the pilot fires the missile.

As there is no dynamic symbology and minimal software in the system, minimising cockpit and aircraft integration time and cost. The Guardian Sight Helmet is in production.

Monocular display

The monocular display is similar to the Sight Helmet design, but the LED is replaced with a miniature high-brightness CRT and the drive electronics are increased in capability to match the potential of the CRT. Full symbology is available, giving the wearer full information on aircraft parameters, navigation data, weapon status and fuel. The symbology is fully programmable to meet user requirements. A feature of the monocular display is the ability to provide the pilot with raster imagery from an external sensor such as a targeting pod or infra-red search and track system. The Guardian monocular display is in production.

The Pilkington Optronics/Cumulus Guardian Sight Helmet **2000**/0079249

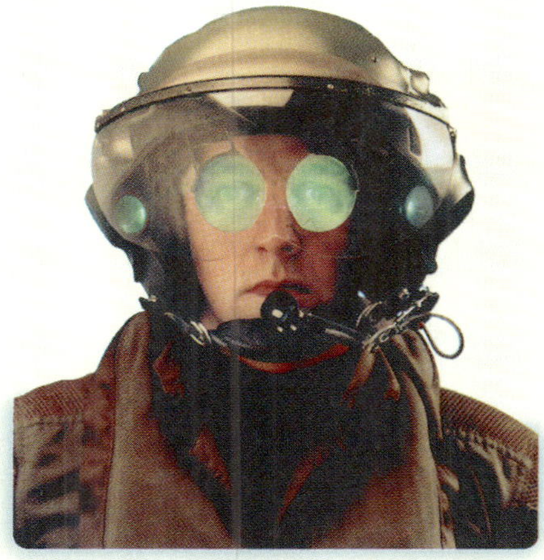

The Pilkington Optronics/Cumulus Guardian integrated systems **2000**/0079252

The Pilkington Optronics/Cumulus Guardian binocular display **2000**/0079251

The Pilkington Optronics/Cumulus Guardian monocular display **2000**/0079250

Binocular display

The first Guardian binocular helmet developed is aimed at the helicopter market and has a wide field of view display (up to 60°), which can project night vision, or helmet-mounted night vision camera imagery. The night vision cameras form a key component of the system and are designed to replace the Night Vision Goggle (NVG), or folded NVG system. Pilkington claim that the use of night vision cameras, rather than NVGs, is advantageous in reducing helmet bulk and providing a better growth path to high-definition TV-standard imagery and improved brightness control. The first binocular display system flew in April 1997.

Head tracking system

The optical head tracker is an innovative, high-performance, proven system, which is now in production. Infra-red LEDs are positioned on the helmet and are tracked by cockpit-mounted miniature cameras. The system is immune to cockpit metalwork, 100 per cent sunlight compatible and highly accurate. Performance accuracies better than 4 mrad have been demonstrated. The head tracking system is in production.

Drive electronics

The display helmet drive box provides aircraft interfaces (MIL-STD-1553B, ARINC 422/429, analogues, discretes and synchros), processing, symbol generation, display processing, deflection amplifiers and power to the system. It is able to drive all the Guardian family of sights and displays. The drive electronics are in production.

Integrated systems

Noctua is the Pilkington Optronics/Cumulus fully integrated, turnkey, pilot night vision system for attack and utility helicopter applications. It is designed to optimise conventional utility helicopter operations such as combat search and rescue, reconnaissance, scout and special forces insertion missions. Matching the display quality of the Guardian helmet is a second-generation thermal imager in a head-steered turret. The pilot can select the optimum sensor from either the turret (thermal camera), or helmet (integrated image-intensified cameras). The system is offered for new or retrofit application's.

Specifications

Accuracy: better than 5 mrad
Measuring envelope
Azimuth: –180 to +180°
Elevation: –90 to +90°
Roll: –90 to +90°
HMD imagery
Angular dimensions: 20° circular (day); 40° circular (night)
Exit pupil: 16 mm
Colour: green
Focused at infinity
Weight
Helmet (without NVGs): <1.6 kg
Total system weight: <15 kg
Power consumption: <120 W

Operational status

In June 1999, Pilkington Optronics delivered the first of two Guardian monocular display helmets to the Aircraft Systems Directorate of Defence Materiel Administration in Sweden for fitting to the Gripen simulator. The first flight of the system was conducted during 2000.

Pilkington Optronics has worked closely with Guardian team member Cumulus, in optimising the system for the Gripen aircraft. The optical head tracker installation has been adjusted to match the Gripen' cockpit layout.

Pilkington Optronics and Cumulus have developed both monocular and binocular versions of the Guardian display helmet.

Contractors

Pilkington Optronics Ltd.
Cumulus, a business unit of Denel (Pty) Ltd.

UPDATED

Head-Up Display (HUD) for the Eurofighter Typhoon

BAE Systems, leading a consortium with TELDIX, Alenia Difesa and INDRA EWS, has been awarded the development contract for the head-up display the Eurofighter Typhoon. The HUD forms part of the aircraft displays and controls subsystem and provides the primary display of flight information to the pilot.

The Pilot's Display Unit (PDU) uses diffractive optics featuring a single-element holographic combiner which consists of two glass elements bonded to produce a flat parallel-sided assembly. An optically powered hologram is recorded on photosensitised gelatine on the spherical interface sandwiched in the assembly and acts as the collimating combiner. The resulting advanced optical system, manufactured using computer-generated holographic techniques, provides new levels of display capability. Additionally, the uncluttered simplicity of the combiner support structure allows virtually a clear out-of-cockpit field of regard. The total field of view is 30° azimuth by 25° elevation and the instantaneous field of view is 30 × 20°. The display is required to operate in three modes; cursive, raster and raster/cursive.

The optical module brightness levels are optimised to operate in a very high ambient light environment, whilst minimising solar reflection and maximising outside world transmission and display uniformity from within the large eye motion box.

Associated with the PDU is a complex HUD control panel, developed by TELDIX, which is attached to the aft face of the unit and incorporates LED technology for multifunctional displays.

Operational status

In development.

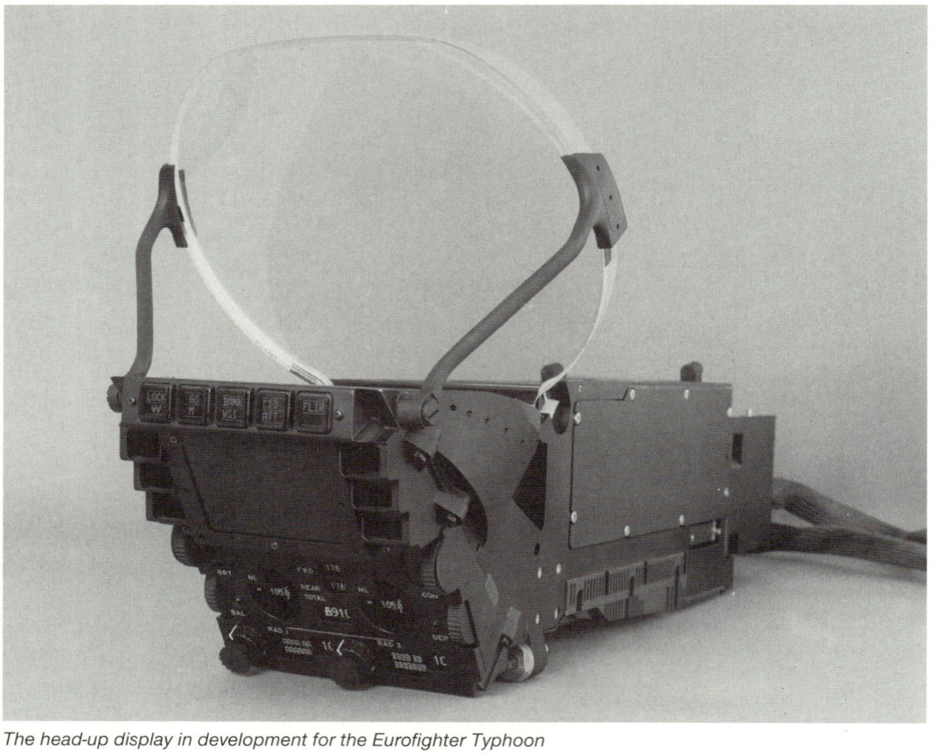

The head-up display in development for the Eurofighter Typhoon

Contractors

BAE SYSTEMS Avionics.
Alenia Difesa, Avionic Systems and Equipment Division.
INDRA EWS.
TELDIX GmbH.

VERIFIED

Head-Up Display (HUD) for the Tornado

TELDIX (Germany) is jointly responsible with Smiths Industries Aerospace (UK) and Alenia Difesa (Italy) for the head-up display system for the Panavia Tornado.

The system consists of a Pilot's Display Unit (PDU) and an Electronics Unit (EU). The PDU incorporates a 5 in (127 mm) CRT with deflection amplifiers and an extra high-tension power supply. It also includes the reflector, lens assembly and combining glass. The control panel for the display is on the PDU. The optical system incorporates a standby sight in case of HUD failure and a mounting support for a recording camera.

The EU includes a digital computer for symbol generation, symbol motion and fire-control functions and is programmable to customer requirements. The electronically generated symbols are automatically adjusted by a photoelectric cell to a preset brightness level, to cater for conditions ranging from bright daylight to non-dazzle viewing at night. The BITE circuits allow for the direct detection of a defective assembly.

Operational status

In service.

Contractors

Alenia Difesa, Avionic Systems and Equipment Division.
Smiths Industries Aerospace.
TELDIX GmbH.

VERIFIED

The head-up display for the Panavia Tornado

Helmet-Mounted Cueing System (HMCS)

The HMCS design aim is to provide the pilot with the 'first-shot' advantage. The HMCS features: integrated electronics including sight, display and symbol generation, and auto-brightness sensor; also integrated is a colour camera and Honeywell's advanced magnetic head-tracker system. The symbol-set is fully programmable.

The HMCS is suitable for single or dual cockpit use, with HGU-86, HGU-55 and HGU-53 helmet shell options.

Specifications

Symbols: stroke/raster or mixed operation
Field of view: 20°
Exit pupil: >25 mm
Eye relief: 69 mm
Accuracy: <2 mrad RMS accuracy
Weight: 1.27 kg

Operational status

The HMCS completed an intensive flight demonstration aboard a Royal Netherlands Air Force F-16 MLU aircraft in summer 1998; test pilots of the four participating F-16 MLU nations: Belgium, Denmark, Netherlands and Norway took part in the trial.

Contractors

Honeywell Inc, Sensor and Guidance Products.
BAE SYSTEMS, Operations.

UPDATED

Helmet-mounted cueing system (HMCS)
0018152

Integrated display/processing systems

Litton Guidance & Control Systems teamed with TELDIX GmbH, are developing full colour, high brightness, MultiFunction Display (MFD) systems, that include state-of-the-art thermal control for cold temperature operation and high MTBF.

The Litton/TELDIX display system includes an embedded CPU and high performance digital and video graphics processing. The multimedia display system is capable of producing the real-time 2-D and 3-D graphics required for modern digital map systems. Map memory storage can be internal, or accessed via the SCSI-2 interface.

The video processor is capable of receiving sensor inputs via RS-170, RS-343 and PAL systems, and resizing to meet the selected screen resolution. Video inputs can be overlaid, or viewed on independent windows.

Specifications

Display type: colour AMLCD
Display area: 5 × 5 in, and 6.25 × 6.25 in
Brightness: >250 Fl
Colours: 262,144
Resolution: 480 × 480, 512 × 512
Dimensions:
5 × 5 in display: 177.8 (W) × 190.5 (H) × 355.6 (D) mm
6.25 × 6.25 in display: 210.8 (W) × 215.9 (H) × 266.7 (D) mm
Weight:
5 × 5 in display: 6.8 kg
6.25 × 6.25 in display: 8.2 kg

Contractors

Litton Guidance & Control Systems.
TELDIX GmbH.

VERIFIED

Litton/TELDIX integrated display/processing systems
1999/0051666

Joint Helmet-Mounted Cueing System (JHMCS)

Vision Systems International LLC (VSI) was established by Elbit Systems Ltd of Israel, through its US subsidiary EFW Inc of Fort Worth, Texas and Kaiser Aerospace & Electronics Corporation of Foster City, California, specifically to pursue Helmet-Mounted Display (HMD) systems opportunities worldwide.

VSI was selected by Boeing and Lockheed Martin Tactical Aircraft Systems (jointly) to develop the JHMCS to equip US Air Force and US Navy fighters (F-15, F-16, F/A-18 and F-22) that the two companies produce.

The JHMCS is primarily designed to provide first shot, high off-boresight air-air weapons engagement under high-*g* conditions, allowing pilots to lock on and fire at enemy aircraft without having to manoeuvre their own aircraft into position; in combination with the AIM-9X, the JHMCS is known as HOBS - the High Off-Boresight Seeker, with the pilot utilising the helmet system to 'look and lock' to exploit the high angle-off capability of the missile.

The JHMCS, combining a magnetic head tracker with a display projected onto the pilot's visor, allows the pilot to aim sensors, air-air and air-ground weapons and view flight parameters without recourse to aircraft instruments or HUD. The helmet portion (either HGU-55/P or HGU-68/P) is common, with the removable display module replaced by Night Vision Goggles (NVG) for night flying duties, thus obviating the need for two separate helmets and facilitating pilot transition from day into night during a mission. The overall system is designed to have low-HMD weight and optimised CG for the demands of fast-jet operations.

Specifications

Field of regard: unlimited
Field of view: 20° monocular, right eye
Exit pupil: 18 mm on - 16 mm off axis
Enhanced cueing: 2 LED reticles
Eye relief: 50 mm on axis
Compatible helmets:
US Air Force: HGU-55/P
US Navy: HGU-68/P
HMD/DU weight: 1.82 kg (with mask)

The VSI joint helmet-mounted cueing system, shown fitted with the removable display module **2001**/0099195

MTBF: 1,000 h
Electronics dimensions: 228.6 × 177.8 × 127.0 mm
Electronics weight: 6.82 kg

Operational status

First flights of JHMCS were performed on US Air Force F-15 and US Navy F/A-18 aircraft in January 1998. Initial low-rate production was approved by the US Navy during July 2000 and a $10.4 million contract, to begin production of JHMCS for the F/A-18E/F Super Hornet aircraft, was received by Boeing during August 2000. The Navy will receive 37 of the systems by 2002.

Vision Systems International has been selected by Lockheed Martin to supply the Helmet-Mounted

Display (HMD) for its Joint Strike Fighter (JSF) aircraft demonstrator programme.

Lockheed Martin Tactical Aircraft Systems has been contracted by the US Air Force to provide for development and trial verification of a modification kit to incorporate the JHMCS and the Link 16 datalink into the F-16 aircraft, to support foreign military sales to Belgium, Denmark, The Netherlands and Norway. Expected contract completion date is October 2003.

Contractor
Vision Systems International.

UPDATED

Litening airborne laser target designator and navigation pod

Litening is a multipurpose, day/night, precision laser targeting and navigation pod. It contains FLIR, CCD-TV, laser spot target tracker/range-finder, laser marker and multifunctional tracker (area, EO correlation and INS) sensor for use with either conventional or laser-guided bombs. An on-gimbal Inertial Navigation Sensor (INS) has a stabilised line of sight and automatic boresight capability. The INS and software design is optimised to permit easy integration in a wide variety of modern military aircraft such as the AV-8B, F-4, F-5, F-15 and F-16, as well as the Jaguar, Mirage 2000 and Tornado.

The current build standard is designated Litening II. System components are designed to provide the following capabilities:

1. The Forward Looking Infra-Red (FLIR) is a high-resolution thermal imaging system, providing Litening II with day/night and adverse weather attack capability.
2. The Charge Coupled Device - TeleVision (CCD-TV) is designed to enhance the stand-off capability of the system during daylight operations.
3. The FLIR's wide field of view provides a HUD-compatible display for low-altitude navigation.
4. The laser spot tracker provides Litening II with the capability to track laser spots from a second laser source, while the laser marker allows Litening II to illuminate targets for co-operative attack by forces equipped with Night Vision Goggles (NVGs).

Rafael Missile Division and Northrop Grumman Corporation are working together under an agreement involving sale and production of Rafael's Litening II airborne laser target designator and navigation pod. On the International market, Rafael is prime contractor, with Northrop Grumman as subcontractor. For the Litening II system on US-built military aircraft sold to the US Government, or to allied nations through the US Foreign Military Sales (FMS) programme, Northrop Grumman is the prime contractor, with Rafael as its subcontractor.

Northrop Grumman is supplying the FLIR system for both US and international sales, except where other special arrangements are made.

PrePlanned Product Improvements (P^3I) for Litening II inlcude:

1. Provision of a 512 × 512 element mid-band infra-red (3-5 micron) (FLIR) staring-array sensor, to improve target tracking and recognition ranges by 30 per cent.
2. Sensor fusion of the CCD-TV and FLIR images.
3. Provision of a diode-pumped Nd:YAG dual-wavelength (eyesafe) laser to increase laser energy and improve designation range, while also providing for safe training.
4. Improved reconnaissance capability, as a secondary mission during air-to-ground operations.
5. Improved built-in-testing capability.

Specifications (Litening II)

Dimensions: 2,200 (length) × 406 mm (diameter)
Weight: 200 kg
FLIR TDI (spectral band 8-12 micron)
picture elements: 708 × 240
(narrow) 1.5 × 1.5°
(medium) 5.4 × 5.4°
(wide) 18.4 × 24.1°
FLIR FPA (spectral band)
(narrow) 1.0 × 1.0°
(medium) 4.0 × 4.0°
(wide) 18.4 × 24.1°
CCD camera fields of view:
(narrow) 1.0 × 1.0°
(wide) 3.5 × 3.5°
picture elements: 768 × 494
Laser designator and range-finder:
energy: 100 MJ per pulse
Trackers: advanced correlator/inertial laser spot search and track
Gimbals:
(fields of regard) +45 to –150° pitch; ±400° roll
(stabilisation) 30 microradians

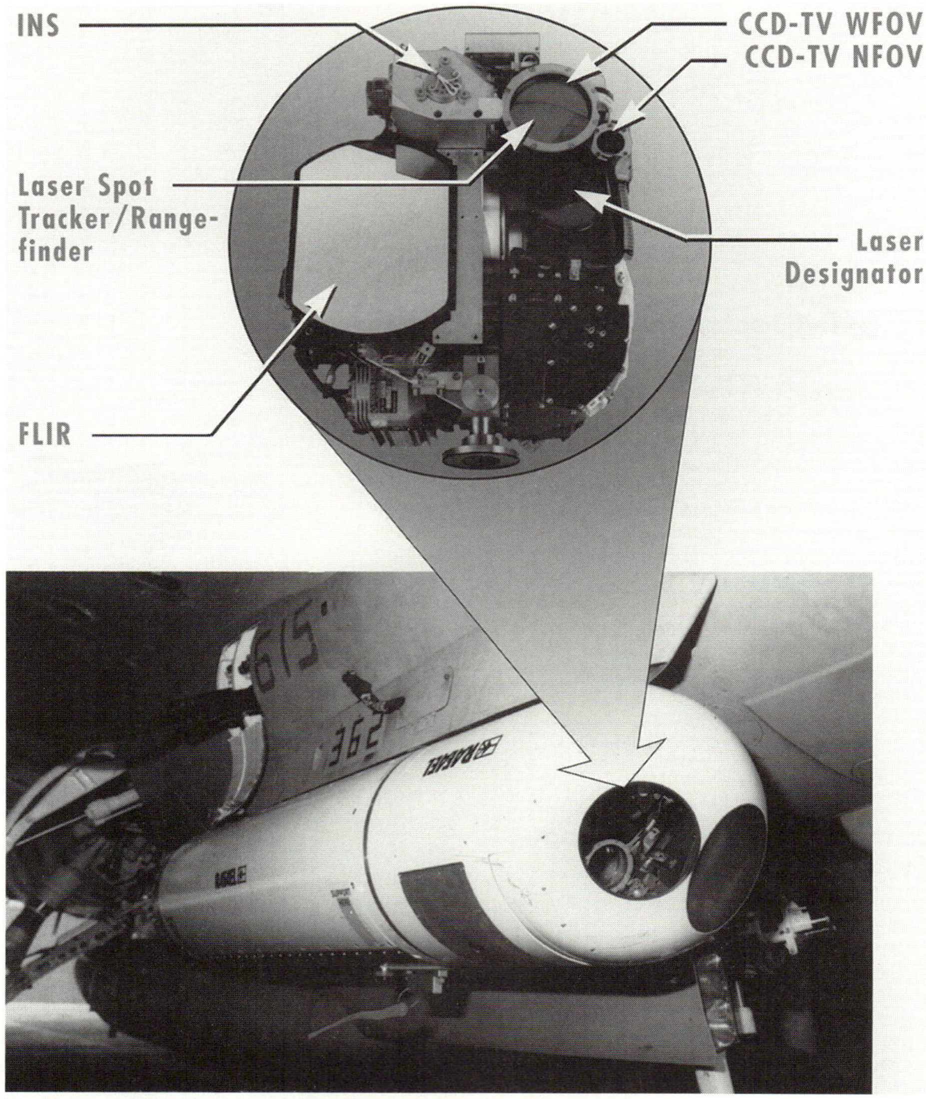

The Litening airborne laser target designator and navigation pod 0018352

The Litening II day and night navigation and targeting pod **2000**/0079248

Flight envelope:
(at low altitude) 1.2 Mach
(manoeuvre) 9G

Operational status
In August 1998, Northrop Grumman and Rafael were awarded an initial contract worth nearly US$18 million

to supply Litening II targeting pods to the US Air Force Reserve Command and the US Air National Guard for use on F-16 aircraft, where it will be designated Litening Precision Attack Targeting System (PATS). The initial contract was for eight pods to achieve initial operating capability by March 2000. Qualification tests were flown at Edwards AFB California in late 1999, with operational test and evaluation occurring in February 2000. As a result of successful testing, four F-16 equipped US Air Force Reserve Command and seventeen F-16 equipped Air National Guard squadrons will eventually be equipped with the Litening II system, a total of 192 pods.

Meanwhile the US Marine Corps has a requirement for 56 pods for the AV-8B, to be delivered by 2002/2003. In April 2000, Rafael and Northrop Grumman announced that they had received a contract to supply 9 Litening II pods to the US Marine Corps for installation on the AV-8B, and that Italy and Spain were to buy 4 pods and 2 pods respectively under the same contract award.

Rafael has also supplied the Litening pod to several nations and integrated it in the following systems:

(1) Germany: IDS Tornado, as part of the Mid-Life Update (MLU) programme being carried out by DaimlerChrysler Aerospace AG for the German Air Force and German Navy. In this instance, Litening will be fitted with electro-optical systems provided by Zeiss Optronik GmbH
(2) India: Jaguar and Mirage 2000 aircraft
(3) Israel: F-16C/D Block 30 and Block 40 aircraft
(4) Romania: MiG-21, as part of the Lancer upgrade
(5) Venezuela: F-16A/B Block 15 aircraft.

Contractors

Rafael Missile Division.

Northrop Grumman, Integrated Systems and Aerostructures Sector.

UPDATED

Multifunction Control Unit (MCU)

The Litton Systems Canada/Harris Multifunction Control Unit (MCU) is a full-capability CDU for controlling and managing a diverse range of avionic equipment, including navigation sensors, communication systems and stores. The MCU is available for custom application development with a bundled Ada software toolset.

The MCU includes a host of built-in input/output capabilities, including MIL-STD-1553B, ARINC 429 and RS-422, with other optional digital interfaces available. The MCU incorporates Litton's LED technology, on a 10 line by 24 dot matrix, for readability in high-ambient lighting environments, with optional NVIS-B compliant optics available.

The MCU incorporates a 32-bit, 24 MHz processor/co-processor and a 2 Mbyte memory as standard. Options include a 33 MHz processor and up to 6 Mbytes of flash EEPROM. A full US Air Force approved alphanumeric keyboard is standard, including 22 customisable keys for dedicated functions.

The MCU is supplied with sample application code, providing GPS flight management, planning and control, plus inertial navigation update capability. This application provides for a Trimble GPS as the sole means of navigation and includes a 200 waypoint database and other features.

Specifications

Dimensions: 184.1 × 146 × 165.1 mm
Weight: <4.54 kg
Power supply: 28 V DC to MIL-STD-704A
Reliability: >4,000 h MTBF

Operational status

The MCU has been fitted to the A-10, Tornado and various military helicopter applications.

Contractors

Harris Corporation Electronic Systems Sector.
Litton Systems Canada Ltd.

VERIFIED

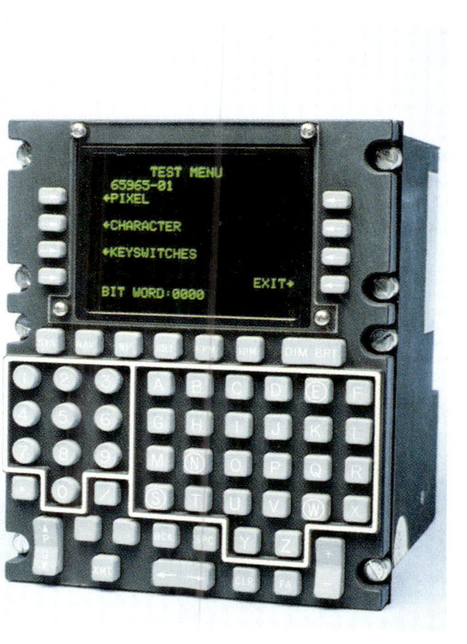

Harris/Litton Canada Multifunction Control Unit
0106259

Multifunction Head-Down Displays (MHDD) for Eurofighter Typhoon

Three Multifunction Head-Down Displays (MHDDs) are installed in the cockpit of the Eurofighter Typhoon, with six in the two-seat trainer version. The MHDD provides flight, tactical situation and sensor data, as well as vital onboard systems information, on a 6.25 × 6.25 in (158.75 × 158.75 mm) usable screen area Active Matrix Liquid Crystal Display (AMLCD). This combines exceptional brightness and high resolution to give full legibility in full sunlight of raster scan images overlaid with fine stroke graphics. Interface with other aircraft systems is provided by 17 programmable push-button key displays on the unit's bezel. The equipment is a major development of the MPCD fitted in the AV-8B/Harrier GR. Mk 7.

Smiths Industries is the project leader and design authority for the programme.

Specifications

AMLCD display area: 6.25 × 6.25 in (158.75 × 158.75 mm)
RGGB pixels: 1,024 × 1,024
Viewing angles:
(horizontal): ±35°
(vertical): -5 to +35°
Contrast: > 30:1 throughout the viewing envelope
AMLCD dimming range: > 30,000:1
Night vision compatible
Power: 165 W
Weight: < 9 kg

Operational status

In prototype production for the Eurofighter Typhoon.

Contractors

Smiths Industries Aerospace.
Alenia Difesa, Avionic Systems and Equipment Division.
VDO Luftfahgerate Werk.

VERIFIED

The Eurofighter Typhoon multifunction head-down display
2000/0080275

OSIRIS: Tiger anti-tank helicopter mast-mounted sight

The OSIRIS sight is a day and night sight which will detect and transfer potential target information to the missile seekerhead. OSIRIS is fitted with high-definition infra-red and TV cameras integrated with cockpit displays that can be operated in bad atmospheric conditions and throughout the helicopter's operating envelope.

Developed as part of the third-generation European anti-tank missile programme, SFIM Industries is the prime contractor for this sight.

OSIRIS embodies multispectral optics (bands 0.5 to 0.7, 0.7 to 1, and 8 to 12 μm) and meets stringent stabilisation specifications.

Operational status

OSIRIS production started in June 1999 for the French Army HAC version of the Tiger helicopter and for the German Army UHT version of the Tiger helicopter.

Contractors

SFIM Industries (SAGEM Group).
ESW - Extel Systems Wedel.
BAE Systems.

VERIFIED

The OSIRIS mast-mounted sight on the rotor head of the Tiger anti-tank helicopter 0018351

PIRATE Infra-Red Search and Track (IRST) system

FIAR is the leader of the Eurofirst Consortium, also comprising Pilkington Optronics, which is developing its Passive Infra-Red Airborne Track Equipment (PIRATE) for the Eurofighter Typhoon.

PIRATE will detect the infra-red signature of aircraft at very long range, over a wide field of view under conditions of poor visibility. Being a passive sensor, it enables the aircraft to gather early intelligence of threats and to manoeuvre stealthily into an advantageous position without being detected by hostile EW systems.

The system will accurately track multiple high-speed targets, prioritise them and provide high-resolution images for visual identification. It provides highly reliable information for air-to-air and air-to-ground use. The system uses proven signal-processing technology derived from the Pilkington Optronics Air Defence Alerting Device (ADAD), which demonstrates a very high suppression rate of false alarms.

PIRATE will be integrated with other onboard sensor systems for maximum sensor fusion effectiveness.

PIRATE will also locate low-level targets and provide cueing information. It provides data and imagery to head-up or multifunction head-down displays, facilitating navigation and terrain-avoidance in adverse weather conditions.

Operational status

A development contract to supply equipment for the Eurofighter Typhoon aircraft was awarded in 1992. Flight trials were conducted during 1999 using pre-production hardware.

Contractors

The Eurofirst Consortium, comprising:
Alenia Difesa, Avionic Systems and Equipment Division, FIAR.
Pilkington Optronics Ltd.

VERIFIED

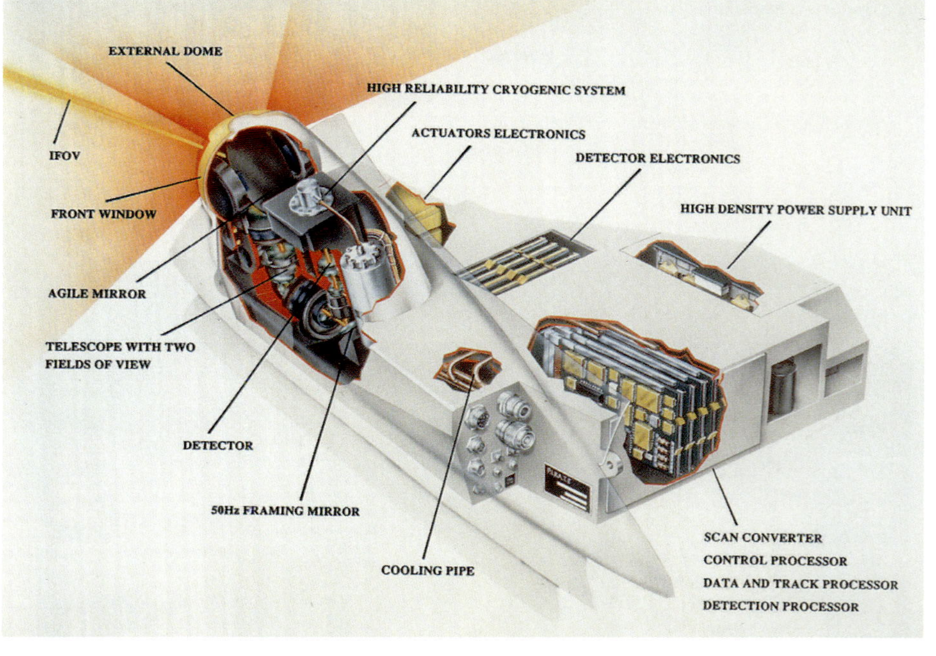

The PIRATE infra-red search and track system
0106260

Terminator infra-red targeting system

Raytheon Electronic Systems leads a private venture team that is developing the Terminator infra-red targeting system for the high-performance stand-off targeting role. Terminator is claimed to be a new, third-generation, design incorporating advanced technology with better performance and reliability than existing first-generation targeting pods, at lower cost.

BAE Systems, Avionics has contributed its third-generation Mid-Wave Infra-Red (MWIR) navigation FLIR, Laser Spot Tracker (LST), roll drive unit and aircraft-specific adapters to the design, while Raytheon has contributed its experience of mid-wave, large format, staring array technology.

The Terminator infra-red targeting pod
2000/0079253

Specifications

Focal plane: 640 × 480 InSb
Spectral band: 3.7-5.0 μm
Fields of view: 5.0 and 0.83°
Recognition range: 10 n miles
MTBF: >300 h

Operational status

Development ongoing.

Contractor

Raytheon Electronic Systems.
BAE Systems, Avionics.
Fairchild Controls.

UPDATED

Tiger helicopter Helmet-Mounted Display (HMD)

Currently under development by BAE SYSTEMS (who also call the system Knighthelm) and TELDIX, this binocular 40° field of view system incorporates a one-piece display module for both day and night mission requirements. For low-visibility and night missions, the system displays both intensified images from third-generation image intensifiers and imagery from a Forward-Looking Infra-Red (FLIR) sensor. This together with flight and weapon symbology information, is projected onto clear combiners placed in front of each of the pilots' eyes. The combiners may be rapidly flipped up out of the line of sight if required. Pilot comfort and mission capability requirements have been fully integrated and all helmet controls for tactical sensors and weapons may be controlled by Hands on Collective and Stick (HOCAS) or via a control panel. Sensors and weapons are interfaced to the helmet via BAE SYSTEMS' advanced helmet tracking system.

Specifications

Weight: 2.2 kg
Field of view: 40°
Eye relief: 30 mm

Operational status

Selected for the development phase of the Integrated Helmet System for the German Tiger helicopter. Development of the HMD is finalised with qualification complete. Integration into Tiger was conducted during 1998/9; series production contracts are under negotiation.

Contractor

BAE SYSTEMS, Avionics.
TELDIX GmbH.

UPDATED

The Tiger integrated helmet-mounted display
0106261

VH 100/130 Head-Up Display (HUD)

Sextant and Hamilton Sundstrand have produced a family of head-up displays as a part of advanced weapon control systems in helicopters, in particular for the air-to-ground firing of Stinger or Matra Mistral missiles.

The VH 100/130 is designed to provide comprehensive navigation and weapon aiming information to the helicopter crew, while being small enough to obscure the crew's vision to the minimum. In weapon aiming it can offer air-to-air rockets and guns and air-to-ground guns and missile symbology and for navigation the display is compatible with NVG. The system comprises a pilot's display unit, an electronics unit and a control panel. The total field of view is 20°.

The system consists of a Pilot Display Unit (PDU), Electronic Unit (EU) and Optional Control Unit (OCU).

The PDU utilises a miniature CRT, which can generate symbology visible in a high-brightness environment Optical lenses and a combiner are used to to present superimposed symbology over the external scene.

The EU provides power supply and signal processing for the CRT, generation of synthetic symbology, electrical interfacing with weapons system sensors, airframe and fire-control computations and aircraft sensor signal processing.

The CU controls weapons mode selection, symbol luminance adjustment, firing distance selection and weapon elevation offset.

The VH 100/130 has been flight-test fitted to a Gazelle in France, OH-58 and Bell 406 in the USA and a BO 105 in Germany. It has been selected for the Transall, the Tiger and the US Army's OH-58 helicopter Stinger missile sight subsystem programme. Initial flight trials were completed in Autumn 1986.

Specifications

Weight:
(pilot display unit) 2.99 kg
(electronics unit) 1.99 kg
Power supply: 28 V DC, 2.5 A
Field of view: 20°
Reliability: >3,600 h MTBF

Operational status

In service in US Army OH-58 helicopters. Twin VH 100/130 HUDs have been installed in the C-160 Transall upgrade. VH100/130 has been selected for Tiger.

Contractors

Hamilton Sundstrand Corporation, a United Technologies Company.
Sextant.

VERIFIED

Viper Helmet-Mounted Display (HMD) systems

The Viper family of helmet mounted displays provides a low-cost, lightweight visor-projected display capability by addition of a display module to almost any type of flying helmet. The Viper family comprises four variants; Viper I, Viper II and Viper III (now redesignated Night Viper) and Viper IV.

Delft Sensor Systems supplies the optical modules for Viper I and II to BAE Systems. Viper III (Night Viper) was developed by Delft Sensor Systems in co-operation with BAE Systems. Viper IV is a BAE Systems development

Viper I

The Viper I helmet mounted display is a monocular system providing a field of view suitable for off-boresight missile aiming.

The visor reflects dynamic flight and weapon aiming data to the pilot from a high-efficiency miniature CRT

Viper I HMD 0106263

Viper II HMD 0106262

display projected via an optical relay assembly. The optical design allows the use of a standard aircrew visor with the addition only of a neutral density reflection coating. This technique enables a high outside world transmission without colouration. The design also ensures that the displayed image is stable and accurate even when the visor is raised.

Although primarily configured for daytime use, the Viper I system is capable of displaying video from a sensor, providing the pilot with an enhanced cueing system after dark or in bad weather.

Specifications

Viper I
Weight: 1.72 kg (on a US standard flying helmet)
Eye relief: >80 mm
Eye relief: >70 mm
Exit pupil: 15 mm
Real world transmission: > 70%

Viper II

The Viper II helmet mounted display provides a binocular, fully overlapped, 40° field of view capability. The system is able to receive sensor video (for example, from a Forward Looking Infra-Red (FLIR) sensor), and display this together with overlaid flight and weapon aiming symbology.

The Viper II has a similar optical concept to the Viper I, and by use of a spherical visor, maintains a stable image within the large exit pupil regardless of visor position. Full interpupillary adjustment is also provided.

Specifications

Viper II
Weight: 1.54 kg (on a US standard flying helmet)
Field of view: 40°
Eye relief: 81 mm to visor
Exit pupil: >15 mm
Real world transmission: 88%

Other common Viper I and II features:

a) both models fit neatly onto all sizes of standard UK, US Air Force and US Navy fixed-wing aircraft helmets;
b) the HMD incorporates provision for a lightweight video camera head. When the output of this camera is mixed with scan converted helmet-mounted display symbology, a video output is available for recording purposes and subsequent use on the ground for mission review and training;
c) they provide full space stabilised symbology when coupled to a suitable Head Tracking System (HTS), such as the advanced DC system developed by BAE Systems;
d) the Viper helmet mounted display system includes an Electronics Unit (EU) and a Cockpit Unit (CU). These units can be configured to provide a variety of functions, depending on user requirements and which Viper variant is required. These functions include: display processor, video processing, MIL-STD-1553 bus interface, symbol generation, CRT display drive, high voltage power supply, display control and head tracker interface; further improvements to the Viper family of HMDs are currently in development.

Night Viper (Viper III)

Night Viper is a day and night helmet with integrated image intensifier tubes and a binocular visor projected helmet mounted display. The lightweight concept is based on the successfully flight proven Viper I and Viper II concepts. The visor combines and reflects night vision images to the pilot, which are collected by the helmet-mounted objectives. Use of a spherical visor means display accuracy is insensitive to visor rotation. The product provides both see-through vision and an extremely high degree of ejection safety. Applications include both helicopter and fixed wing aircraft without any update or modification of any electronic system. The Night Viper module can be fitted to all sizes of HGU55/P helmets. The optical relay system can be adjusted for individual interpupillary settings with no changes or modifications to the visor. Since there is no interface to the aircraft, an electronics box is not part of the system.

Specifications

Field of View: full binocluar 45°
night image: 40°
overlapping L + R: >30° horizontal
see-through image: 180 × 110°
Exit pupil: 15 mm on axis
Image intensifier tubes: standard ANVIS 18 mm inverting SuperGen or Gen3
Resolution @ 30 mlux with Omnibus IV Gen3 tubes: better than 1 mrad
Dioptre adjustment: not applicable compatible with eye glasses eyerelief > 75 mm
Interpupillary distance: 55 to 74 mm with centred exit pupil
Transmission: real world 35 % - other values upon request
Weight on HGU55/P:
(incl oxygen mask, earphones, liners, and so on): <2.3 kg
Centre of gravity: in 'Knox box' (COG displacement <0.89 inches)

Viper IV

Viper IV is a visor projected HMD system, offering a 40° field of view, using monocular or binocular display on a spherical visor. Viper IV provides a large (>20mm) exit pupil. The system is claimed to provide excellent stability with no loss of display at high G. It has been windblast tested to 550 kt, and flight proven on an F-16 aircraft, using an off-canopy tracker installation.

Night Viper pilot's day and night helmet ***1999***/0051715

Viper IV HMD ***2000***/0080272

Viper IV provides off-boresight missile designation with reverse cueing and multiple target designation and tracking in the air-to-air scenario. For air-to-ground operations, it provides target designation with CCIP/ CCRP symbology. Head line of sight is computed by an electromagnetic head tracker system, which is also used to space-stabilise the symbology, and to slew missile seeker heads or other sensors.

Viper IV HMD system provides stroke symbology over sensor video. It uses a high-brightness daytime mode. At night, video from head-steered sensors, such as FLIR or night vision cameras can be overlaid.

Operational status

The operational status of the various Viper HMD products is understood to be as follows:
Viper I has been trialled as part of the European governments F-16 Mid-Life Update (MLU) programme, and it is also involved in the X-31 concept programme;
Viper II has been trialled by DASA in the German Air Force Tornado turreted FLIR programme;
Night Viper is under trial by the Royal Netherlands Air Force.

Contractor

BAE Systems.
Delft Sensor Systems.

VERIFIED

ISRAEL

DSP-1 Dual Sensor Payload

DSP-1 is a compact day/night observation system designed for use on light reconnaissance aircraft and helicopters (as well as UAVs and patrol boats). It is a four-gimbal system, gyrostabilised in azimuth and elevation. It uses two channels: a third-generation focal plane array InSb FLIR night sensor with a continuous (× 22.5) zoom lens; and a high resolution colour Charge Coupled Device (CCD) daylight channel equipped with a (× 20) zoom lens.

Options include an Intensified Charge Coupled Device (ICCD), laser pointer; 8-12 μm FLIR (first or second generation), video tracker; radar designated pointing, MIL-STD-1553B databus and GPS interface.

Specifications

FLIR sensor:
Spectral range: 3-5 μm
Detector: InSb FPA 256 × 256 InSb
Lens: × 22.5 continuous optical zoom
FoV:
0.98 × 0.92° (narrow)
21.7 × 20.6° (wide)
IFoV: 67 μrad
NEDT: 0.02 K
Video output: PAL or NTSC
Cooler: closed cycle
Daylight camera:
Type: high resolution colour CCD 768 (H) × 494 (V) pixels
Lens: × 20 zoom
FoV:
0.92 × 0.7° (narrow)
18.6 × 13.9° (wide)
Acquisition range:
Truck detection: 25 km (daylight camera), 25 km (FLIR)
Truck recognition: 10 km (daylight), 7.5 km (FLIR)
Electro-mechanical:
Type: 4-gimbal
FoR:
+10° to −110° (elevation)
360° continuous (azimuth)
Stabilisation: better than 25 micro radians RMS
Pointing accuracy: 0.7°
Power: 28 V DC, 110 W
Dimensions: (height × diameter) 500 × 320 mm
Weight: 26 kg

Contractor

Controp Precision Technologies Ltd.

UPDATED

DSP-1 dual sensor payload **2001**/0099742

ESP-1H Light Weight Observation System

The ESP-1H is a light weight observation system designed for light aircraft and UAVs, capable of detecting vehicular sized targets at ranges up to 20 km and identifying them at up to 7 km.

The system features a three-gimbal mount, which is gyrostabilised in azimuth and elevation, and is equipped with two daylight TV cameras; a wide angle camera with a 16 mm fixed focal length and a high resolution black and white or colour Charge Coupled Device (CCD) camera with a 6 × (50 to 300 mm) zoom lens.

Specifications

Camera #1
Type: High resolution black/white or colour CCD
FoV: 1.5 × 26°
Lens: 6 × zoom

Camera #2
Type: Wide angle
Lens: 16 mm fixed focal length

Gimbal:
FoR:
azimuth: 360° continuous
elevation: +10 to −110°
Stabilisation: > 50 mrad
Pointing accuracy: 0.7°
Angular velocity: 0-50°/s
Dimensions: 283 (D) × 415 mm (H)
Weight: 8 kg

Operational status

Operational on various (unspecified) platforms.

Contractor

Controp Precision Technologies Ltd.

NEW ENTRY

ESP-600C high-resolution colour observation payload

ESP-600C is a very high-resolution, lightweight, stabilised daylight observation system, designed primarily for scout helicopters, light reconnaissance aircraft and UAVs (including IAI Scout and Searcher). It is a three gimbal system, gyrostabilised in azimuth and elevation. ESP-600C carries two high-resolution colour CCD cameras: one with a very wide fixed field of view for target detection and one with a × 15 zoom lens for target recognition and identification. Options include 3-CCD cameras, extended focal length for longer acquisition ranges.

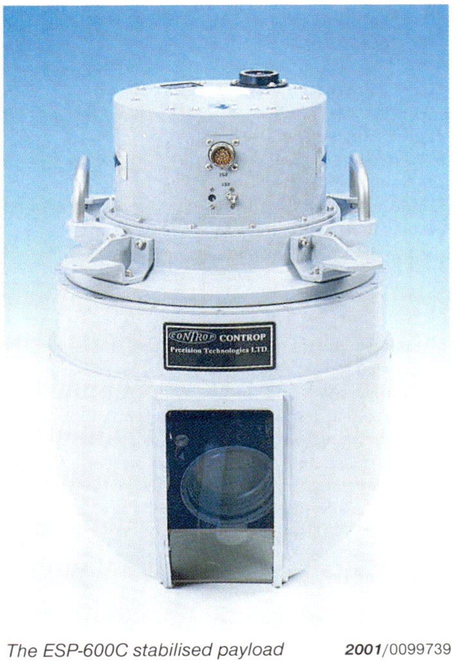

The ESP-600C stabilised payload **2001**/0099739

Specifications

Electro-Optical:
Zoom camera: High-resolution. Colour CCD (PAL or NTSC)
Lens: 15 × zoom
FoV: 0.75 to 11.5° (diagonal) (option 1.5 to 22.9°)
WFOV camera: High-resolution. Colour CCD (PAL or NTSC)
FOV: 22.6 × 17°
Acquisition ranges:
Vehicle recognition: 12 km
Vehicle detection: 30 km
Electromechanical:
FoR: azimuth: 360° (xn) continuous
elevation: +10 to −110°
Angular velocity: 0 to 40°/s
Stabilisation: better than 15 μrad
Pointing accuracy: 0.7°
Power: 28 V DC, 40 W
Dimensions: 300 (D) × 435 mm (H)
Weight: 12.3 kg

Contractor

Controp Precision Technologies Ltd.

UPDATED

FSP-1 FLIR Stabilised Payload

FSP-1 is a high-resolution, stabilised, 8-12 μm FLIR sensor with a three Field of View (FOV) telescope, designed primarily for scout helicopters, light reconnaissance aircraft, UAVs and marine patrol boats. FSP-1 features a four-gimbal system, gyrostabilised in azimuth and elevation and protected against weather effects and contamination by a rotating dome with built-in optical window.

Options include third-generation 3-5 μm Focal Plane Array (FPA) or second-generation 8-12 μm FLIR camera, video tracker, extended environmental conditions and laser pointer.

Specifications

FLIR sensor:
Spectral range: 8-12 μm
Detector: Cadmium Mercury Telluride (CMT)
FoV:
2 × 1.5° (narrow)
7 × 5.3° (medium)
24 × 18° (wide)
Resolution: 0.07 × 0.11 mrad
Sensitivity: 0.8° C (MRTD @ 7 cy/mrad)
Cooler: closed-cycle split-Stirling cycle
Acquisition ranges:
Vehicle recognition: 4 to 5 km
Vehicle detection: 12 to 20 km

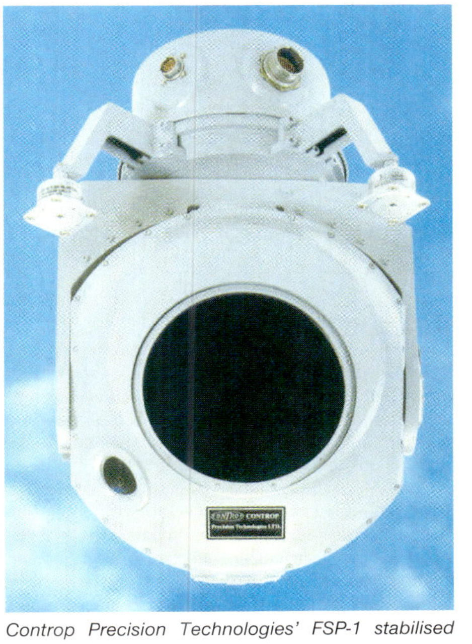

Controp Precision Technologies' FSP-1 stabilised payload **2001**/0099740

Electromechanical:
Type: 4 gimbal gyrostabilisation system
FoR:
azimuth: 360° continuous
elevation: +10 to –105°
Gimbal angular rate:
azimuth: up to 45°/s
elevation: up to 32°/s
Stabilisation: better than 25 μrad rms
Pointing accuracy: 0.7°
Power: 28 V DC, 125 W
Serial Connection: RS-422A (bi-directional)
Altitude: Up to 6,100 m (20,000 ft)
Acceleration: Up to 5 g
Temperature:
In operation: -20 to +50°C
In storage: -32 to +71°C
Dimensions: 320 (D) × 500 mm (H)
Weight: 28 kg

Operational status

The FSP-1 is operational on various platforms including the IAI Searcher UAV.

Contractor

Controp Precision Technologies Ltd.

UPDATED

MSSP-1 Multi-Sensor Stabilised Payload

Controp Precision Technologies' MSSP-1 is a rugged day/night surveillance system especially configured for use on attack helicopters (as well as multiwheeled terrestrial vehicles and marine patrol boats). It is a four gimbal system, gyrostabilised in azimuth and elevation, and equipped with three sensors - a high-resolution 8-12 μm FLIR sensor, a high-resolution Charge Coupled Device (CCD) daylight camera and a laser range-finder.

Options include an Intensified Charge Coupled Device (ICCD), 3-5 μm InSb Focal Plane Array (FPA) FLIR camera, eyesafe Laser Range-Finder (LRF), extended environmental conditions capability, video tracker, radar designated pointing, MIL-STD-1553 and a GPS interface.

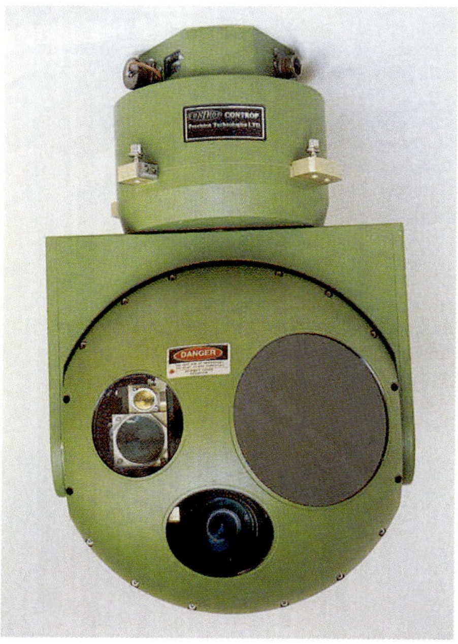

Controp Precision Technologies' MSSP-1 multi sensor stabilised payload **2001**/0099741

Specifications

FLIR sensor:
Spectral range: 8-12 μm
Detector: Cadmium Mercury Telluride (CMT)
FoV:
2 × 1.5° (narrow)
7 × 5.3° (medium)
24 × 18° (wide)
Cooler: closed cycle
Daylight camera:
Type: High resolution black/white CCD
Lens: 15 × zoom
FoV:
1.2 × 0.9° (narrow)
18 × 13.7° (wide)
Laser range-finder:
Wavelength: 1.06 μm
Range resolution: 5 m
Repetition rate: 1 pps
Electromechanical:
Type: 4 gimbal
FoR:
azimuth: 360° continuous
elevation: +25 to –110°
Stabilisation: better than 25 μrad
Power: 28 V DC, 300 W
Dimensions: 400 (diameter) × 650 mm (height)
Weight: 41 kg

Contractor

Controp Precision Technologies Ltd.

UPDATED

MSSP-3 MultiSensor Stabilised Payload

Controp Precision Technologies' MSSP-3 is a day/night observation system especially designed for Maritime Patrol applications on board aircraft, helicopters and patrol boats. In common with the MSSP-1, it is a four-gimbal system, gyrostabilised in azimuth and elevation and equipped with three sensors - a high-resolution 3rd generation 3 to 5 μm InSb Focal Plane Array FLIR camera with a dual FoV lens, a high-performance black and white/colour CCD camera with a × 15 zoom lens and an optional eyesafe Laser Range-Finder (LRF).

Options include an Intensified Charge Coupled Device (ICCD), colour CCD, a × 22.5 continuous optical FLIR zoom lens for the InSb thermal imager, a 2nd generation 8 to 12 μm thermal imager, interface to GPS and MIL-STD-1553.

Specifications

Physical Weight:
Turret: 38 kg
Control unit: 4.5 kg
Joystick: 1.5 kg

Turret Dimensions: 400 × 570 mm (D × H)

Electro-Mechanical
Field of Regard (FoR):
Azimuth: 360° continuous
Elevation: +35° to –110°
Stabilisation: better than 25 μrad RMS
Angular velocity:
Azimuth: 60°/s (max)
Elevation: 50°/s (max)

FLIR Sensor
Wavelength: 3 to 5μm
Detector: InSb 320 × 240 FPA
FoV:
2.2 × 1.65 (narrow)
11.0 × 8.2 (wide)
Cooler: closed cycle cooler
Gain Control: automatic/manual

Daylight Camera
Camera: high-resolution B+W CCD
Resolution: 550 TV lines
Lens: × 15 zoom
FoV:
1.2° × 0.9° (narrow)
18.0° × 13.7° (wide)
Gain Control: automatic/manual

LRF (optional)
Wavelength: 1.54μm
Range: 20,000 m (limited by atmospheric attenuation)
Accuracy: ±5 m
Repetition: 10 ppm (1 pps burst)

Environmental
Temperature: –20 to +50C
Vibration: 2.5 grms, 5 to 2,000 Hz
Shock: 20 g/11 ms duration
Speed: up to 300 kt

Electrical
Power: 28 V DC, 200 W
Video: CCIR or RS-170
Interface: RS-422

Contractor

Controp Precision Technologies Ltd.

NEW ENTRY

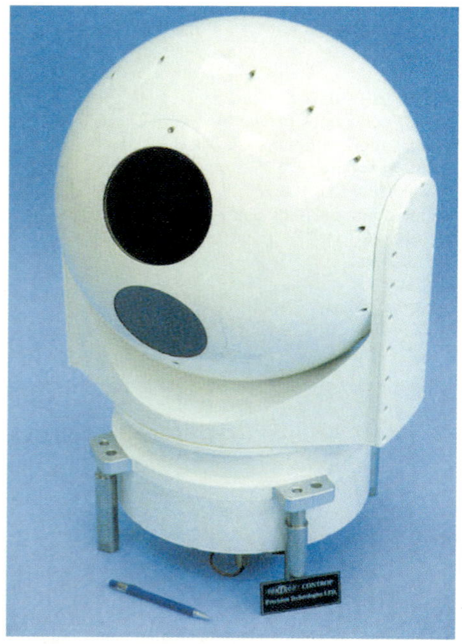

Controp Precision Technologies' MSSP-3 MultiSensor Stabilised Payload **2001**/0077827

All-Light Levels TV

The All-Light Levels TV (ALLTV) is based on an advanced high-resolution gated ICCD sensor with laser illumination for day and night surveillance. It is suitable for airborne, shipborne or land-based applications.

ALLTV comprises three subsystems: an electro-optical head, a motorised pan and tilt platform mount and a portable control unit that includes joystick, operating panel and high-resolution display. Its features include a dynamic range of 10, very high sensitivity, to allow usable pictures to be obtained at any light level, even during the darkest nights and two simultaneous electro-optical channels: wide field of view for orientation and narrow field of view for long-range observation and identification.

ALLTV creates and transmits real-time TV pictures simultaneously to multiple users via wire or radio. The system's fast real-time image processor is designed for picture enhancement and reduction of quantum noise in night-time imagery. ALLTV also allows for recording of real-time TV pictures for analysis and debriefing purposes.

Contractor

Elbit Systems Ltd.

VERIFIED

Cockpit Integration Unit (CIU)

The CIU replaces multiple control panels with a single panel-mounted control display unit. Cockpit management tasks executed via the CIU include radio communication, navigation, stores management and text messages terminal.

Specifications

Keyboard:
Full alphanumeric keyboard with function and soft keys
Sealed for sand/dust environments
Display:
High contrast, sunlight readable

Cockpit Integration Unit 0106264

Processor:
32 BIT 386 sx/387 sx instruction set
512 kword memory upgradable to 2 Mword
Interfaces:
MIL-STD-1553 bus controller or remote terminal
RS-422 half or full duplex communication channels
ARINC 575/429 communication channels
Input/Output discretes
Weight: 5 kg
Dimensions: 181 × 146 × 156 mm
Power: 28 V DC 40 W

Contractor
Elbit Systems Ltd.

VERIFIED

Colour Liquid Crystal Display (LCD)

Elbit Systems' colour LCD represents the latest generation of airborne displays. The flat panel display packages high optical performance and reliability into a compact unit with low power consumption. It is suitable for ruggedised airborne and ground applications.

It features a full-colour active matrix LCD; high-resolution graphic display; wide viewing angle; slim design (only 3 in (76 mm) thick); and high contrast in day and night operation.

Specifications
Display size: 97 × 127 mm
Resolution:
640 × 600 pixels (384,000 over display area)
640 × 200 colour triads over display area
Brightness: 50 ft-lamberts (white)
Contrast: 40 (white)
Colours: 8
Viewing angle: ±45° horizontally and vertically for a contrast of 10
Interface: RGB at TTL levels
Power supply: 28 V DC in range of 20-36 V
Power consumption: 20 W
Dimensions: 190 × 190 × 70 mm

Operational status
Operational on MiG-21 and other aircraft. Selected by Boeing to be producer and supplier for the V-22 Osprey multifunction display upgrade programme, via its wholly owned US subsidiary, EFW Inc. The V-22 Osprey installation uses a four-MFD configuration to display primary flight and navigation data, together with video imagery and digital map information.

Contractor
Elbit Systems Ltd.

VERIFIED

CTC-1/CTC-2 airborne colour TV camera

Elbit Systems' colour TV camera offers the user outstanding colour fidelity and full-band response for high-resolution colour applications. The camera is based on a 1.25 cm interline transfer CCD and micro-electronic assemblies.

The camera also incorporates a unique and specially designed Automatic Electronic Shutter (AES) function, which adjusts the shutter speed automatically to maintain a wide dynamic range of light levels (0.5 to 20,000 foot candles) all day long. The AES function also ranges from a half hour before sunrise to a half hour after sunset. The AES feature eliminates the need for a moving mechanical iris, thus improving camera reliability.

The interline transfer technique significantly increases the camera performance when compared with the old frame transfer technique.

The CTC appears in two mechanical configurations: CTC-1 (housed) in one module, CTC-2 (housed) in two modules.

The electrical and mechanical interface is compatible with BAE Systems avionics HUDs. It features high-resolution (460 horizontal TV lines); high-sensitivity complementary colour filter; NTSC/PAL and Y/C (S-VHS) output; 1/60 to 1/31,000 second Automatic Electronic Shutter (AES): wide dynamic range and excellent S/N ratio; AWB memory or auto-tracking (optional); low power consumption; military qualification; single or double housing configuration and very fast response time.

Specifications
Sensor type: 1.25 cm interline CCD (6.4 × 4.8 mm)
Picture elements: 768 (H) × 494 (V)
Colour filter: Cy, Ye, Mg, G complementary filter
TV resolution: 460 (H) × 400 (V) lines
Spectral response: 460 to 630 nm
Field of view (FoV): 17 × 22° (optional other FoVs)
Response time: approx 1 s
Automatic Electronic Shutter (AES): 1/60 s to 1/31,000 s
Power source: 115 V AC, 400 Hz
Power consumption: <10 W

Contractor
Elbit Systems Ltd.

VERIFIED

Day-HUD

Safe, daytime nap-of-the-earth flight demands that both pilots pay full attention to the external world. The ability to present comprehensive cockpit and navigational information directly in the pilot's field of view dramatically increases flight safety and reduces crew workload. The Day-HUD enables both pilots to fly head-out and receive all vital data, including altitude, height, speed and aircraft condition, flight and platform warnings at eye level. The Day-HUD head assembly adapts to ANVIS-AVS-6 or similar NVG-mountings. It comprises a miniature display source, and an optical combiner, which interfaces with the AN/AVS-7 system.

Specifications
Field of view:
(horizontal) 28°
(vertical) 20 °
Exit pupil: 10 to 12 mm
Eye relief: 25 to 40 mm
Weight:
(helmet-mounted element) 0.27 kg
(interface unit) 0.29 kg

Operational status
In production for the US Army.

Contractors
Elbit Systems Ltd.

VERIFIED

Display and Sight Helmet (DASH)

The Elbit Systems' Display and Sight Helmet (DASH) allows the crew of a combat aircraft to direct missiles or sensors on to targets or points of interest by simply looking at that point. Head position, and hence sightline, is computed. DASH slaves all armament systems to the pilot's line of sight. The pilot directs missiles, radar and INS to specific targets by looking at them and receives feedback on his visor on the target that has been acquired. The pilot can also point out a target to a second crew member by looking at it.

DASH presents the head-up display information directly on the pilot's visor so that he is always aware of flight conditions. The position of a target seen by one crew member can be electronically cued in the visor of any other person linked into the system. DASH incorporates the Nightsight integrated avionics system.

Nightsight is an integrated avionics system designed to ensure increased safety and efficiency in tactical night missions. The system's major capabilities and advantages include low-altitude navigation in dusk and darkness, target location, sensor/pilot slaving and cueing target attack, formation flying, flying in adverse weather conditions, map and aerial photo display and full head-out mission capability.

Nightsight's integrated system concept combines the Night Vision Goggles/Head-Up Display (NVG/HUD) and the Line Of Sight (LOS) tracker. Wearing the NVG/HUD, the pilot sees flight symbology collimated with the external view and can fly head-out, realising an added dimension of mission safety and operational capabilities.

The LOS tracker provides active measurement of pilot line of sight and helmet position with high accuracy in all dynamic environments. With pilot LOS slaved to aircraft sensors and vice versa, the result is

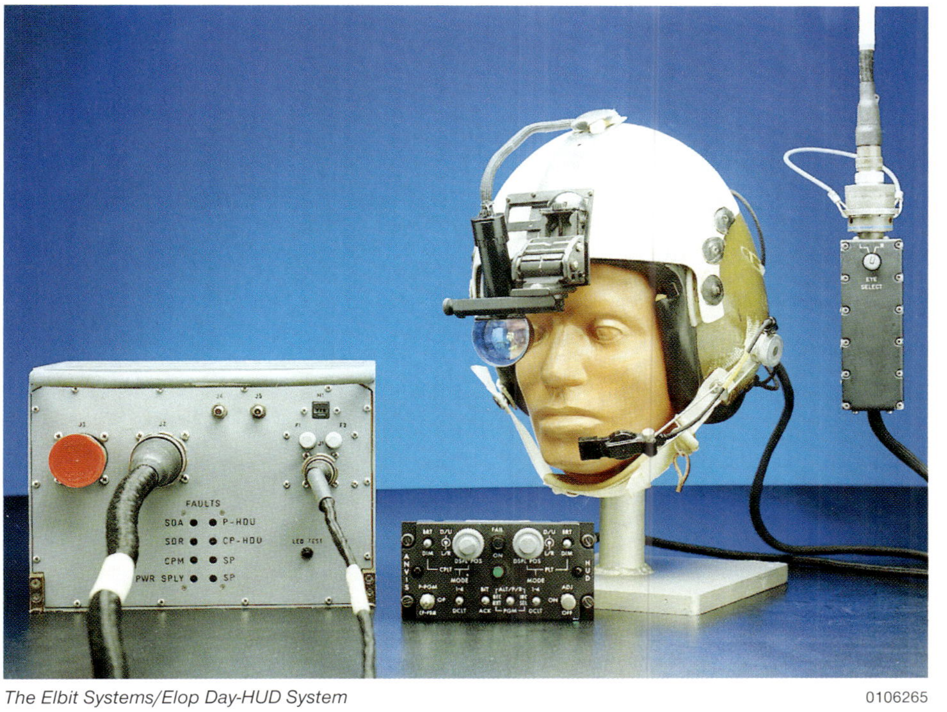

The Elbit Systems/Elop Day-HUD System 0106265

Units of Elbit Systems' DASH helmet-mounted sight 0106375

improved air-to-ground target detection and improved air-to-air interception capability.

Integrating the FLIR pod, NVG/HUD and pilot LOS measurement is a powerful connection, resulting in enhanced overall system performance. The system is designed so that the same hardware drives both day and night displays.

Specifications

Weight: 1.8 kg (total system: 13.1 kg)
Field of view: 22°
Coverage:
(azimuth) ±160°
(elevation) ±70°
(roll) ±60°
Accuracy:
Forward Cone (20°) 6 mrad
Rest of Envelope: 10 mrad
Reliability: 2,000 h MTBF

Operational status

DASH is in service on Israeli Air Force F-4, F-15, F-16 aircraft, and with foreign customers on F-5, F-16 and MiG-21.

A third-generation helmet display, more compact and lighter than the two previous versions, was reported to be undergoing flight tests during 1999.

Contractor

Elbit Systems Ltd.

VERIFIED

Engine Performance Indicator (EPI)

Elbit Systems' EPI is an advanced cockpit display that replaces nine electromechanical switches with one electronic display. The display conserves valuable space on the instrument panel for additional displays (MFCDs) required for comprehensive F-5 upgrades. The EPI is based on LCD technology and emphasises human engineering in its use of colour and combination of analogue and digital symbology.

Specifications

Display size: 110 × 58 mm area
Display technology: direct drive LCD with 4 colours
Power supply: 28 V DC, 40 W
(80 W while heater operates)
Dimensions: 86 × 125 × 273 mm
Weight: 3 kg
Interface: NVG-compatible

Operational status

Operational.

Contractor

Elbit Systems Ltd.

VERIFIED

Map Display Generator (MDG)

Designed for two-seat fixed-wing aircraft or helicopters, the Map Display Generator (MDG) has two channels capable of producing a variety of images as selected by the pilot or co-pilot. The system uses two 6 × 6 in (152 × 152 mm) multifunction colour displays. Control of the various map modes is via a keyboard mounted on the display. The MDG is equipped with an optical disk drive and removable optical disk for storing map databases on board the aircraft. Two powerful graphics units simultaneously draw raster and vector maps.

Using the mission planning ground station, the pilot can plan the intended mission in advance, co-ordinating all maps, tactical mission data, flight plans, threats, obstacles and communications. Accumulated data is then downloaded onto the optical disk which the pilot transfers to the MDG disk drive for use during the mission.

The MDG displays paper charts, vectorial maps, digital terrain elevation data and SPOT or Landsat satellite photos. With continuous improvement offline at the ground station and online at the MDG, the image presented on the multifunction colour display is clear, crisp and easily discernible. The system also provides the user with zoom, scale, plan and freeze facilities.

The MDG's computing powers are capable of calculating the line of sight between any two points, point altitude, radar threat area coverage and obstacle clear flight and landing corridors.

Operational status

Selected by the Israeli Air Force for the modernised CH-53 helicopter.

Contractor

Elbit Systems Ltd.

VERIFIED

MIDASH Modular Integrated Display And Sight Helmet

MIDASH is Elbit's latest addition to its line of head-mounted systems, developed specifically for attack and reconnaissance helicopters.

MIDASH allows pilots to fly head out day and night and integrate weapon and targeting systems using the pilot's helmet line of sight and dedicated flight and targeting symbology.

MIDASH draws on Elbit's experience in production of head-mounted systems, among them, DASH, JHMCS and ANVIS HUD, designed for both fixed and rotary wing platforms.

MIDASH will provide attack and reconnaissance helicopter pilots with wide Field Of View (FOV), see-through binocular night imagery, symbology and Line-of-Sight (LOS) cueing for both day and night operation. It comprises a standard shell helmet and a dark visor with personal fitting device, Helicopter Retained Units (HRU) and a small quick disconnect Vest Mounted Unit (VMU). The HRU and VMU are standard aircraft units attached to any pilot by snap-connectors. In addition, a stand-alone mode enables the pilot to operate the image intensified channels outside the helicopter.

The Image Intensified Display (IID) is based on a pair of SuperGen 98 (or Gen III) tubes used for night operation and a single CRT overlaid symbology channel. Both of these are projected on designated combiners.

Specifications

Night image FOV: 50° (H) × 40° (V)
Unobstructed FOV: 120° (H) × 70° (V)
Day/night symbology FOV: 30° circular
Total weight night operation: 2.2 kg
Eye relief: >50 mm
Adjustments: performed once on personal fitting device; no pre-flight adjustments necessary
Symbology contrast for day: >1:1.4 @ full sunlight (10,000 fc ambient)
See-through transmission: >50%
Cueing: electromagnetic sensor

Operational status

Elbit Systems Ltd has signed a teaming agreement with Sikorsky Aircraft Corporation for the upgrading and conversion of Black Hawk helicopters to the armed reconnaissance and attack role. The new version of

Elbit's MIDASH HUD **1999**/0051714

The Battle Hawk targeting configuration **2000**/0080291

Black Hawk, designated Battle Hawk, will incorporate the following Elbit Systems items of equipment:
MIDASH to provide flight and weapons symbology for day and night operations;
Toplite II targeting sensor, equipped with FLIR for night operations, TV for day operations, and a laser designator and rangefinder to cue weapons' operations.

Contractor
Elbit Systems Ltd.

VERIFIED

MultiFunction Display (MFD)

The multifunction display is essentially a monochrome armament control and display panel for combat aircraft with limited cockpit space, showing video information from a variety of sensors on a 5 × 5 in (127 × 127 mm) or 4 × 4 in (102 × 102 mm) CRT. The display CRT is surrounded on the face of the panel by 28 push-buttons for the selection of different functions.

Specifications
Power supply: 28 V DC per MIL-STD-704A at 40 W
Dimensions: 140 × 138 × 344 mm
Weight: 6 kg
Display size: 4 × 4 in (102 × 102 mm) or 5 × 5 in (127 × 127 mm)
Phosphor type: P43 (green)
Raster video input: 525/60, 625/50 or 875/60 lines per RS-170, RS-343 or CCIR standards
Brightness: 200 ft-lamberts (685.2 cd/m^2) min
Contrast ratio: 9:1 at 10,000 ft candles sunlight
Video bandwidth: 30 Hz-20 MHz to 3 dB points
Reliability: 1,500 h MTBF per MIL-HDBK-271A

Operational status
In production for Israeli Air Force F-4 Phantoms, IAI Kfir fighters, F-5, T-45 and MiG-21.

Contractor
Elbit Systems Ltd.

VERIFIED

Nightwatch

Nightwatch is a modular concept designed for helicopter and transport aircraft. It assures reduced crew workload, increased accuracy and avoidance of obstacles during night-time low-level flight. Elbit systems that can be integrated into the Nightwatch concept include: NVGs, NVD/HUD, HOCAS, IR filtered landing lights, IR imaging systems, moving map and cockpit mission management systems using Doppler/GPS/INS.

Operational status
Nightwatch is flying on CH-53, CH-47, CH-46, UH-60, HH-60, OH-58, UH-1, A 109, Gazelle, AS 350, AS 565, BO 105, AH-1, V-22 and C-130.

Contractor
Elbit Systems Ltd.

VERIFIED

Stores management system

Elbit Systems' stores management system is computerised, flexible, easy to manage and has dual-mode operation for maximum integrity against the accidental release of stores. Computed information is shown on an armament control and display panel which is essentially a multifunction CRT unit. The pilot prepares several weapon release schedules according to his preflight briefing so that, during the attack phase, he merely has to select the appropriate one for the task. The system comprises four units: an armament control and display panel, stores management computer, video electronic unit and an optional video recorder.

The stores management system replaces the switches and selectors which would normally control guns, weapons and fuzes, weapon data, stores release programming, weapon data insertion panel and stores emergency jettison. Preflight selections comprise: master mode for the aircraft's weapon delivery and navigation system, which takes in air-to-air combat, air-to-ground missiles, gun firing, selective jettison and navigation; CCIP, dive/toss, toss and direct weapon delivery mode; type of weapon to be used; single weapon, pair or salvo release mode; release interval and nose, tail or nose and tail fuze mode.

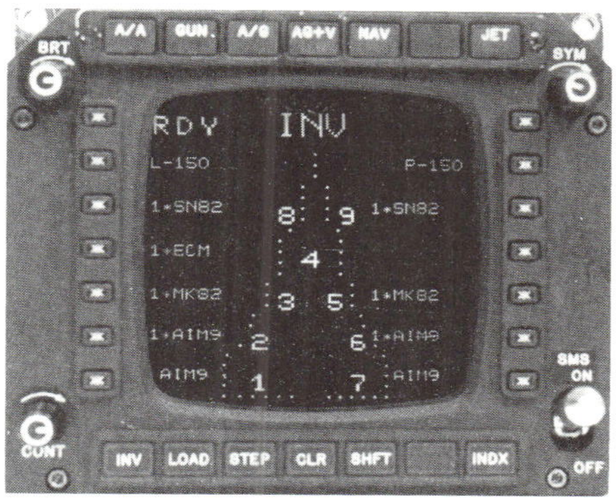

Elbit Systems' stores management system display showing the INV (inventory) mode, with store types and positions on the aircraft. The pattern indicates that four AIM-9 Sidewinder missiles and four Mk 82 and SN 82 bombs are carried on wing pylons and an ECM pod on the fuselage centreline station

The CRT continuously displays the number, location and type of stores being carried on the mission and is updated as stores are released or jettisoned. It also shows the information appropriate to the particular release programme and warns of incorrect programming, equipment failure or beyond limits conditions. The system computer can also control a videotape recorder for post-mission analysis or debriefing after a training flight.

Specifications
Computer
Dimensions: 321 × 194 × 172 mm
Weight: 13.5 kg
Redundancy: two independent computing systems, dual power supply
Memory: 58 kbyte EPROM, 3 kbyte RAM, 1 kbyte non-volatile RAM
Input/output: 160 discrete inputs, 120 discrete outputs

Armament control and display panel
Dimensions: 172 × 147 × 267 mm
Weight: 7 kg
Scanning: 525 or 875 lines
Bandwidth: 20 MHz
Contrast ratio: 7:1

Operational status
In production and in service in the Kfir, MiG-21, A-4 Skyhawk, F-4, F-5, and in other aircraft exported from Israel. It has also been selected by Embraer for the Venezuelan Air Force AMX-T advanced trainer aircraft.

Contractor
Elbit Systems Ltd.

VERIFIED

Aerial Camera Electro-optical Magazine (ACEM)

Elop's ACEM replaces the film magazine with an electro-optical magazine featuring a linear array of butted CCD detectors in the focal plane.

The retrofitted camera employs the existing optics and structural mechanics of the original aerial film camera, the ACEM and modified scanning and control units.

Real-time collected image data can be recorded in-flight and/or transmitted in real-time to the ground station, providing rapid intelligence data.

Specifications
Output line rate: 2,000 lines/s
Focal plane array: 9,200 pixels
Spectral region: 0.5-0.9 μm
Output data dynamic range: 50 dB
Total weight: 15 kg
Power requirements: 28 V DC, MIL-STD-704C

Operational status
The systems have been integrated in Israeli Air Force fighters.

Contractor
Elop Electro-Optics Industries, an Elbit Systems Ltd company.

VERIFIED

Elop's Aerial Camera Electro-optical Magazine (ACEM) 0106266

Airborne Laser Range-finder And Designator (ALRAD)

The Airborne Laser Range-finder And Designator (ALRAD) was developed to serve as the major component for a variety of high-performance airborne designation systems. The stringent requirements resulted in a compact, lightweight and easily maintainable and adaptable designator.

The airborne designation system increases the pilot's bombing effectiveness by marking the target with a laser beam and directing a laser-guided bomb on to the target. The system can be integrated into fighter aircraft as well as general utility aircraft. For this application it may be installed externally on any suitable aircraft station. The system is available in two basic configurations: either as a range-finder, with a steering mirror for high-repetition precise ranging within a cone of 20 to 30° and slaved to the HUD pipper like a radar ranger; or as a designator/ranger with steering mirrors for hemispherical coverage and slaved to the mission computer as well as to a manual control.

The ALRAD comprises the laser transceiver, a cooling unit, electronic unit and beam-steering unit. These four units can be combined modularly in accordance with system requirements. The laser transceiver and cooling unit are common to all systems while minor modifications to the electronic unit provide the flexibility to engineer the system adaptively for different requirements. Consequently, the system is well suited for upgrades, with rapid, low-cost designs for the customer's applications.

Specifications

Weight: 7.5 kg nominal
Wavelength: 1.06 μm
Energy: 80 mJ
Pulse rate: up to 20 pps, codable
Beam divergence: 0.4 mrad
Interfaces: EIA-RS-429 or MIL-STD-1553B

Contractor

Elop Electro-Optics Industries, an Elbit Systems Ltd company.

VERIFIED

Elop's Airborne Laser Range-finder and Designator ***1999***/0051005

COMPASS: COmpact MultiPurpose Advanced Stabilised System

COMPASS is designed to provide day/night search-and-track of land and sea targets, onboard weapon control and day/night navigation. It is designed for land, sea, or airborne application and is an adaptable, integrated concept, featuring modular components, add-on modules and interface/plug-in capability. It can be installed as a stand-alone system, or as part of a larger weapons system. It provides multi-operator control capability and, when installed in a helicopter, can be operated from the pilot and/or navigator/weapons operator positions.

COMPASS incorporates three sensors that can all be brought to bear simultaneously. The combination of Forward-Looking Infra-Red (FLIR), colour or black and white zoom CCD camera and Laser Range-Finder/Designator (LRFD), together with an improved tracker, enables manual or automatic search and track operations against naval, aerial and ground targets, as well as weapons guidance and day/night navigation. All of these operations may be carried out concurrently, using the combined sensor system.

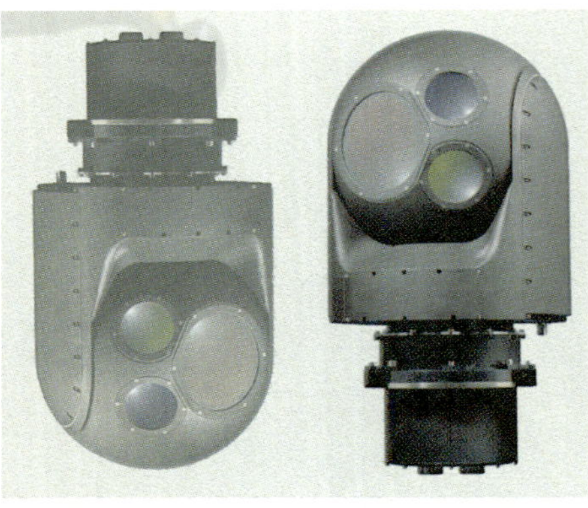

COMPASS: COmpact MultiPurpose Advanced Stabilised System for helicopters ***2001***/0092869

Modular in design, the COMPASS concept system can be readily customised to individual user requirements. System integration is facilitated by provision of a MIL-STD-1553B databus interface and RS422 avionics interface.

Specifications

FLIR (interchangeable): 1st and 2nd generation FLIR, or mini-FLIR 8-12 microns, or matrix FLIR 3-5 microns
Angular coverage: azimuth: 360°; elevation: +35 to −85° (-35° to +85°)
Stabilisation: <20 μRad
Maximum flight speed: operating: 300 kt; endurance: 400 kt
Temperature: operating: -40° to +50°C; endurance: -45° to +71°C
Weight: 31 kg (without laser); 34 kg (with laser)

Contractor

Elop Electro-Optics Industries, an Elbit Systems Ltd company.

UPDATED

COTIM Compact Thermal Imaging Module

COTIM is a very compact and lightweight thermal imaging sensor. It is designed for installation in miniature stabilised payloads for Remotely Piloted Vehicles (RPVs), light helicopters and small naval vessels.

It features: advanced focal plane detector technology with integral closed-cycle cooler; control via serial communication, with dual field of view (FoV). (Triple FoV or continuous zoom are optional.)

Specifications

Spectral region: 8-12 μm
Aperture: 100 mm
Fields of view:
(narrow) 2 × 1.5°
(wide) 7.1 × 5.3°

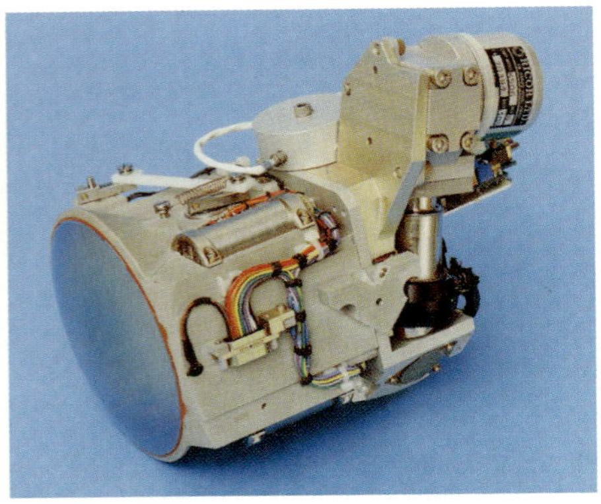

COTIM Compact Thermal Imaging Module 0106267

Instantaneous FoV:
(narrow) 0.11 mrad
Detector: MCT
Power: 40 W, 28 V DC
Weight: 2.9 kg

Operational status

In serial production.

Contractor

Elop Electro-Optics Industries, an Elbit Systems Ltd company.

VERIFIED

EO-LOROPS Electro-Optic Long-Range Oblique Photography System

EO-LOROPS Electro-Optic Long-Range Oblique Photography System outfits high-performance aircraft with an electro-optic, standoff reconnaissance system incorporating real-time data transmission or real-time data record/transmit options, in pod- or nose-mounted configurations.

The reconnaissance pod houses the camera, comprising the Cassegrain Ritchey-Chretien mirror telescope with a linear array of butted CCD detectors in the focal plane, the video processing unit and the scanning mirror.

The peripheral units, such as the datalink transmitter, digital VTR, air conditioning unit, power supply and reconnaissance management unit, also reside in the pod. Alternatively to the podded installation, Elop offers an internal installation in which the reconnaissance equipment is mounted in the nose section of the aircraft.

System control is performed from the cockpit. Real-time collected image data can be recorded in flight and/or transmitted in real time to the ground station, providing timely intelligence. The ground station incorporates the tracking antenna, datalink receiver,

Electro-Optic Long-Range Oblique Photography System (EO-LOROPS) 0106268

image enhancement and archiving capability and hard-copy and soft-copy displays.

Specifications
Camera type: E-O visible
Focal plane array: 10,000 pixels
Spectral region: 0.55-0.9 μm
Resolution:
57-70 cm/lp from 50 km, 40,000 ft
175/250 cm/lp from 90 km, 40,000 ft
Camera weight: 120 kg
Power requirements: 115 V AC, 400 Hz 3Ø, 28 V DC

Operational status
In January 2000, the EO-LOROP system was selected by the Turkish Air Force Command for the upgrade of 54 F-4E aircraft to the Phantom 2000 configuration.

Contractor
Elop Electro-Optics Industries, an Elbit Systems Ltd company.

VERIFIED

Helmet-Mounted Display Device (HMDD)

The HMDD consists of a head assembly featuring add-on capability to the ANVIS/6 or similar NVG adaptor. It contains a miniature display source, an optical subassembly and a lower electronic adaptor unit which interfaces with the aircraft symbol generator.

The HMDD features a stroke and/or raster display to comply with the existing display source and is adjustable by controls similar to those incorporated in the NVG. It is removable from the helmet, can be folded while on the helmet and is compatible with the aviator's eyeglasses and gas mask. The system is transferable to either eye, to allow use of the dominant eye, and the use of two HMDDs as a binocular display is an option.

Specifications
Weight:
(helmet element) 110 gm
(interface unit) 290 gm
Field of view: (horizontal) 28° × (vertical) 20°
Exit pupil: 15 mm
Eye relief: 25 mm

Contractor
Elop Electro-Optics Industries, an Elbit Systems Ltd company.

UPDATED

The helmet-mounted display device can be used with either eye 0106269

Laser Range-Finder Designator (LRFD) systems

Elop manufactures a range of Laser Range-Finder Designator (LRFD) systems, which have been selected for a number of modern aircraft and helicopter weapon systems, including:

(a) the Laser Range-Finder Designator (LRFD) for the RAH-66 Comanche reconnaissance attack helicopter of the US Army;

Elop's Laser Range-Finder Designator (LRFD) for the RAH-66 Comanche helicopter Electro-Optic Sensor System (EOSS) 0018349

Elop's Laser Designator for the F/A-18 NITE Hawk pod 0018348

(b) the Range-Finder Target Designator Laser (RFTDL) of the Night Targeting System A (NTS A), integrated by TAMAM Division, Electronics Group, Israel Aircraft Industries and Kollsman, is fitted to AH-1W Cobra helicopters of the US Marine Corps;

(c) the laser designator element of the F/A-18 NITE Hawk pod requirement for the US Navy, integrated by Lockheed Martin Electronics and Missiles Division where, following evaluation trials by the US Navy, Elop has been selected as a second source supplier;

(d) the Laser Tracker Receiver (LTR) widely fitted to Apache attack helicopters;

(e) the Laser Designator of the Litening pod, manufactured by Rafael Electronic Systems Division, for the air forces of Israel, Germany, Romania, Venezuela and the US Air National Guard;

(f) the Laser Range-Finder Designator (SELRD) for the Kiowa Warrior (OH-58) helicopter of the US Army, where it will be incorporated into the mast-mounted sight for aiming laser-guided weapons, replacing the present laser system, with improvements in operational performance, reliability and training flexibility;

(g) the laser Range-Finder Designator (dual mode) with diode pumped and OPO technology for the Apache (AH-64D) helicopter of the US Army, where it will be incorporated into the TADS sight for aiming laser-guided weapons, replacing the present laser system, with improvements in operational performance, reliability and training flexibility.

Laser Range-Finder Designator (LRFD) for RAH-66 Comanche helicopter
A new LRFD developed by Elop has been selected to form the LRFD element of the Electro-Optic Sensor System (EOSS), being integrated by Lockheed Martin

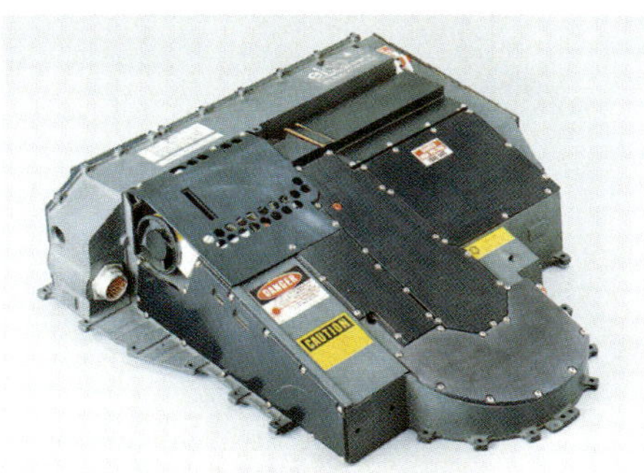

RFTDL for the Cobra helicopter 0106270

Elop's Laser Range-Finder Designator (SELRD) for the Kiowa Warrior (OH-58) helicopter (MMS)
1999/0051006

Electronics & Missiles Division for the US Army's RAH-66 Comanche reconnaissance attack helicopter.

The new Elop LRFD operates on two different spectral wavelengths, within the single system, to provide optimum performance for both the laser target designation/guidance function (1.06 micrometres), and the laser ranging fuction (1.54 micrometres) simultaneously.

The system is based on use of innovative diode-pumped technology, which improves both operation and reduces power consumption. A critical aspect of the design is that the basic wavelength is converted, through use of solid-state devices and advanced optics, to produce an eye-safe sytem, thus enhancing realistic crew training and training flexibility.

Range-Finder Target Designator Laser (RFTDL)
The RFTDL is a high-repetition rate laser designator which forms part of the Night Targeting system A (NTS A) of the US Marine Corps' Cobra AH-1W helicopter. Integration of the RFTDL into the NTS A system is by TAMAM Division of Israel Aircraft Industries.

Contractor
Elop Electro-Optics Industries, an Elbit Systems Ltd company.

VERIFIED

Model 849A Head-Up Display (HUD)

The Model 849A is designed for small fixed-wing aircraft and helicopters such as the A-4, L-39, Tucano and Pampa. Ideal for avionics upgrades, it is a compact, low-cost, lightweight, easy to maintain HUD which is capable of displaying cursive (stroke) symbology.

Specifications
Modes of operation:
stroke writing
Total field of view: 20°
Stroke brightness/contrast:
Full readability in 10,000 fL ambient light
Power supply: 115 V AC, 400 Hz, 3-phase; 28 V DC
Power consumption: 80 W
Weight:
(PDU) 4 kg
(PSVS) 7 kg
Dimensions: 102 × 100 × 285 mm
Environmental: MIL-E-5400T

Operational status
In service in the A-4 Skyhawk, L-39 and IA 63 Pampa.

Contractor
Elop Electro-Optics Industries, an Elbit Systems Ltd company.

VERIFIED

Model 921 Head-Up Display (HUD)

The HUD 921 is capable of displaying high brightness cursive (stroke) symbology with an option of overlaying it on raster display.

Versions of this HUD differ in their combiner-design to accommodate different types of cockpits. Target markets include upgrades for: F-5, IAR 99/109, Kfir,

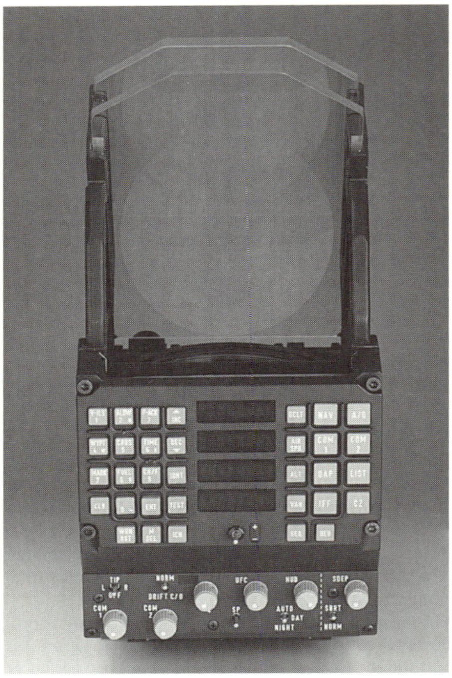

The Model 921 HUD is designed for medium and small cockpits 0018155

L-39/59, MiG 21/23/27, Mirage, T-38 and Tucano ALX aircraft.

The HUD comprises a customer tailored Up Front Control Panel (UFCP) and can accommodate a high-resolution Colour Cockpit TV Sensor (CCTVS) for post flight debriefing.

Specifications

Modes of operation:
stroke writing
stroke on raster (optional)
Total field of view: 24°
Stroke brightness/contrast:
full readability in 10,000 fL ambient light
automatic brightness control
Power supply: 115 V AC, 400 Hz, 3-phase; 28 V DC
Power consumption: 97 W
Weight: 17 kg
Dimensions: 167 × 190 × 560 mm
Environmental: MIL-E-5400T

Operational status

Selected by the US Air Force for the T-38 upgrade, as well as by countries that fly F-5, Kfir, MiG-21, Mirage, Tucano and other aircraft.

Contractor

Elop Electro-Optics Industries, an Elbit Systems Ltd company.

VERIFIED

Model 959 Head-Up Display (HUD)

The Model 959 HUD comprises a customer-tailored upfront control panel and an integral colour cockpit TV sensor for post-flight debriefing. It is designed for medium and small cockpits in aircraft such as the F-5, F-16, Kfir, MiG-21, MiG-23, Mirage and trainers.

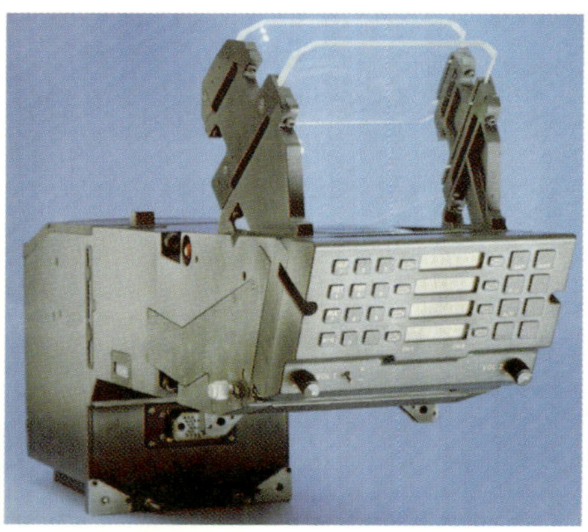

The Model 959 HUD is designed for medium and small cockpits 0106271

The HUD 959 is capable of displaying high-brightness cursive symbology with an option to overlay it on a raster display. Its shape results from a repackaging of modules from other Elop HUDs in order to maximise the field of view in a given volume. The Model 959 HUD can function as the mount for a multifunction display or other flight instrument.

Specifications

Modes of operation:
stroke writing
stroke on raster (optional)
Total field of view: 24°
Stroke brightness/contrast:
full readability in 10,000 fL ambient light
automatic brightness control
Power supply: 115 V AC, 400 Hz, 3-phase; 28 V DC
Power consumption: 100 W
Weight: 23 kg
Dimensions: 180 × 129.5/250 × 712 mm
Environmental: MIL-E-5400T

Contractor

Elop Electro-Optics Industries, an Elbit Systems Ltd company.

UPDATED

Model 967 Head-Up Display (HUD)

The Model 967 HUD is a wide field-of-view HUD. It consists of a customer-tailored Up Front Control Panel (UFCP) and includes provision for a high-resolution Colour Cockpit TV Sensor (CCTVS) for post flight debrief and rear cockpit repeater. It is designed for high-performance aircraft with medium or large cockpits such as the F-4, F-5, F-15, F-16, F/A-18, MiG-29 and Su-27/30/35.

The Model 967 is capable of displaying high-brightness cursive (stroke) symbology and raster with or without stroke symbology overlaid on it. The Model 967 is a high-reliability lightweight HUD which is third-generation NVG-compatible.

Specifications

Modes of operation:
stroke writing
raster
stroke on raster
Optical system: Single or double combiner
Total field of view: 28°

The Model 967 HUD is third-generation NVG compatible 0018154

Stroke brightness/contrast:
2,500 fL
automatic brightness control
Power supply: 115 V AC, 400 Hz, 3-phase; 28 V DC, 5 V AC
System interface: RS-422 Serial link
Weight: 21 kg
Dimensions: 170 x 168 x 656 mm
Environmental: MIL-E-5400T

Operational status

Successfully flown on Israeli Air Force F-16. Selected by a few countries for their F-4 and Su-30 upgrade programmes.

Contractor

Elop Electro-Optics Industries, an Elbit Systems Ltd company.

UPDATED

Model 979 Head-Up Display (HUD)

The Model 979 HUD comprises a customer-tailored upfront control panel and an integral high-resolution colour cockpit TV sensor for post-flight debriefing. It is designed for medium and small cockpits for aircraft such as the F-5, F-15, F-16, IAR 99 and 109, L-39, L-59, Kfir, MiG-21, MiG-23, Mirage, MiG-29.

The Model 979 is capable of displaying high-brightness cursive (stroke) symbology with an option of overlaying it on raster display.

Specifications

Modes of operation:
stroke writing
stroke on raster (optional)

The Elop Model 979 HUD **2001**/0087829

Total field of view: 24°
Stroke brightness/contrast:
Full readability in 10,000 fL ambient light
automatic brightness control
Power supply: 115 V AC, 400 Hz, 3-phase; 28 V DC
Power consumption: 100 W
Weight: 18 kg
Dimensions: 174 × 185 × 596 mm
Environmental: MIL-E-5400T

Operational status
In service in F-16 type aircraft.

Contractor
Elop Electro-Optics Industries, an Elbit Systems Ltd company.

UPDATED

The Elop MSIS installed on a helicopter 0106273

Model 981 Head-Up Display (HUD)

The Elop Model 981 HUD consists of a Pilot's Display Unit, Display Processor Unit, cockpit TV sensor and power supply unit, all combined into a single package. It features a raster and stroke display and air-to-air and air-to-ground modes. Interface to other equipment is via a MIL-STD-1553B multiplexer bus or analogue, synchro or discrete signals. The integral BIT is automatic or manually initiated.

Specifications
Weight: 43 kg
Power supply: 115 V AC, 400 Hz, 3 phase, 530 W
28 V DC, 30 W
Field of view: (horizontal) 30° × (vertical) 21°

Operational status
In service in the Israeli Air Force Phantom 2000.

Contractor
Elop Electro-Optics Industries, an Elbit Systems Ltd company.

VERIFIED

Model 982/3 Head-Up Displays (HUDs)

The Model 982/3 HUDs comprise a customer-tailored upfront control panel and an integral high-resolution colour cockpit TV sensor for post-flight debriefing. The HUDs are designed for medium and small cockpits in aircraft such as the F-5, IAR 99 and 109, Kfir, L-39, L-59, MiG-21/23/27 and Mirage.

The Model 982 and 983 HUDs are two models which differ in their combiner design to accommodate different types of cockpits. This results in the reduction of cost and weight, while maintaining high performance. Both HUDs are capable of displaying high-brightness cursive symbology with an option to overlay it on a raster display.

Specifications
Weight: 19 kg
Power supply: 115 V AC, 400 Hz, 3 phase
28 V DC

The Elop Model 981 HUD is in service in Israeli Air Force Phantom 2000 aircraft 0106377

Field of view: 24°
Environmental: MIL-E-5400T

Operational status
In service in the Kfir and F-5.

Contractor
Elop Electro-Optics Industries, an Elbit Systems Ltd company.

VERIFIED

Model 989 Head-Up Display (HUD)

The Elop Model 989 HUD consists of a Pilot's Display Unit (PDU) with integral power supply unit and an Electronics Unit (EU). The Up Front Control Panel (UFCP) may be customer tailored and the HUD includes an integral high-resolution Colour Cockpit TV Sensor (CCTVS) for post flight debriefing. The unit is is designed for medium-size cockpits in aircraft such as the AMX, MiG-21, MiG-23 and trainers.

The Model 989 features stroke, raster and stroke on raster symbology, providing for navigation, attack and

The Elop Model 989 HUD is installed in the AMX 0106272

test display modes. Interface to other equipment is via a MIL-STD-1553B multiplexer bus or through logic discretes. The integral BIT is automatic or manually initiated.

Specifications
Modes of operation:
stroke writing
raster
stroke on raster
Total field of view: 24°
Stroke brightness/contrast:
full readability in 10,000 fL ambient light
automatic brightness control
Weight: 18 kg
Dimensions: 155 x 162 x 675 mm
Power supply: 115 V AC, 400 Hz, 3 phase
28 V DC
Power consumption: 80 W
Environmental: MIL-E-5400T

Operational status
In service in the AMX.

Contractor
Elop Electro-Optics Industries, an Elbit Systems Ltd company.

UPDATED

Multisensor Stabilised Integrated System (MSIS)

The Multisensor Stabilised Integrated System (MSIS) is a lightweight fully stabilised electro-optical system, for day and night passive surveillance and tracking of surface and airborne targets. The system is designed to detect, recognise and track the complete range of manoeuvring targets from a rubber dinghy to a supertanker, as well as all types of helicopters, fixed-wing aircraft and sea-skimming missiles.

MSIS integrates three sensors: a CCD TV camera, a 6 in thermal imager and a laser rangefinder. It has a built-in automatic target-tracking computer. Optionally, it can also be equipped with a laser pointer. The stabilised ball turret provides a low wind resistance and the system obtains a picture with good resolution, high tracking accuracy and very good recognition ranges.

Operational status
In production.

Contractor
Elop Electro-Optics Industries, an Elbit Systems Ltd company.

VERIFIED

Up-Front Control Panel (UFCP)

The Elop Up-Front Control Panel (UFCP) is the pilot's input and control terminal for activating and operating the various avionic systems such as the Weapon Delivery and Navigation System (WDNS), stores management system, navigation equipment, communications equipment, IFF and so on, in advanced aircraft. It constitutes the main man/machine interface between the pilot and the aircraft avionic systems.

The UFCP displays pilot-initiated or automatic alphanumeric information received from the aircraft

Elop Up-Front Control Panel shown fitted on a HUD
0106274

avionics. Communication with the avionics is through the UFCP's digital push-buttons and information is displayed either on the UFCP or on the HUD.

The UFCP is located on the front of the pilot display unit. Electrical interface with the WDNS/MC is via MIL-STD-1553A or B, RS-422A or as specified by the customer. A microcontroller manages the UFCP functional display, illumination, validation, analogue to digital converter and communication with the WDNS/MC. Management is performed in real time.

The UFCP is part of every Elop HUD.

Contractor

Elop Electro-Optics Industries, an Elbit Systems Ltd company.

VERIFIED

Very Light Laser Range-finder (VLLR)

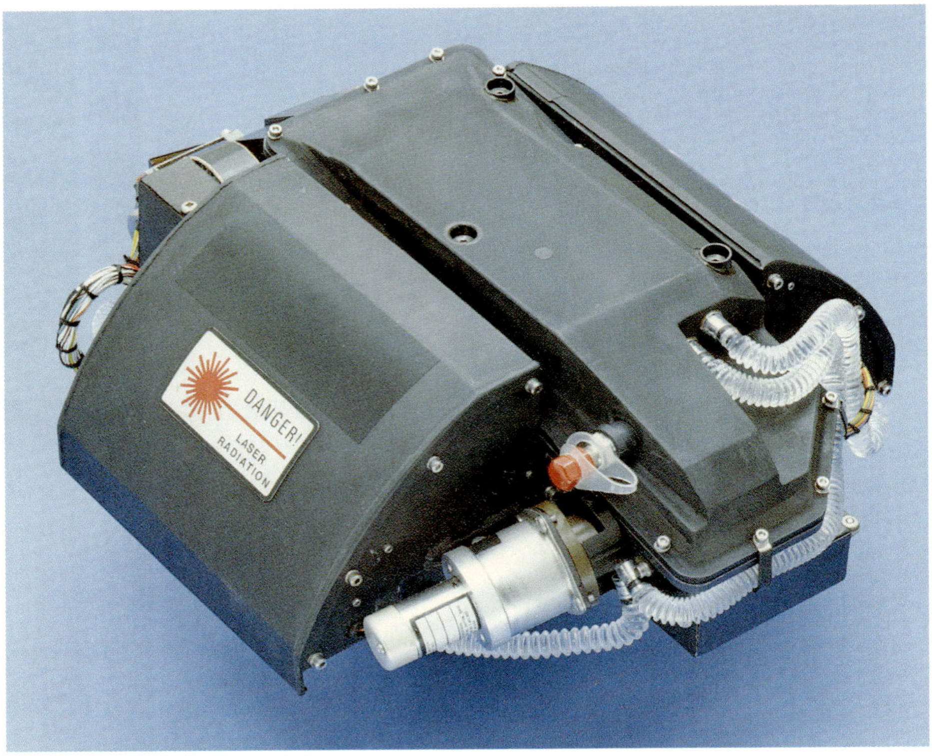

VLLR Very Light Laser Rangefinder
0106275

The VLLR is designed to respond directly to the need for greater accuracy, particularly when aiming at moving targets. It upgrades existing platforms and pods and enhances the capabilities of new systems. It is suitable for a variety of airborne systems, including RPVs, as well as seaborne and ground platforms. The VLLR is compact, integrating easily into a variety of systems; modular, being adaptable to different systems, high-speed, sending up to six laser pulses per second; ruggedised, to meet severe environmental conditions, and has trouble-free maintenance, with malfunction indicator, external test pins and plug-in PC boards. The housing is customised according to customer requirements.

The VLLR features high-efficiency, low-heat dissipation to the pod and low divergence of the laser beam.

Specifications

Weight: <3.8 kg
Power consumption: 220 W (max)
Wavelength: 1.064 μm
Output energy: 80 mJ
PRF: single shot or up to 6 pps
Beam divergence: less than 0.4 mrad
Pulsewidth: 15 ns (max)
Range: 200-9,995 m
Range accuracy: ±5 m

Operational status

In production.

Contractor

Elop Electro-Optics Industries, an Elbit Systems Ltd company.

VERIFIED

Cockpit Laser Designation System (CLDS)

The CLDS is a lightweight, small-size designator specially designed for light attack two-seater aircraft. It is mounted in the rear cockpit of the aircraft which commands the tactical operation. The CLDS can designate targets for other attacking aircraft equipped with laser-guided bombs and provide the ability to deliver the laser-guided bomb in a wide variety of modes, altitudes and ranges. It is designed for operation in close air support, battlefield air interdiction, deep strike or by a forward air controller using the magnifying sight for observation only and for sea surveillance.

The CLDS includes a TV camera and auto-tracker. This closed loop hands-off tracking system reduces the pilot's workload during the critical phase of the mission and significantly improves the tracking accuracy and manoeuvrability of the designating aircraft in the target area. The video camera, interfaced to an aircraft VTR, is also utilised for debriefing and intelligence purposes.

The CLDS can be interfaced to the aircraft Weapon Delivery and Navigation System (WDNS), enabling the WDNS to direct the CLDS line of sight towards a chosen target as a highly accurate sensor for marking targets and providing position updates to the WDNS.

The Israel Aircraft Industries cockpit laser designation system
0106336

Specifications

Dimensions: 560 × 420 × 360 mm
Weight: 40 kg
Field of view:
(wide × 4 magnification) 12.5°
(narrow × 10 magnification) 5°
Field of regard:
(forward) 45°
(backward) 25°
(elevation) ±35°
Accuracy: 0.25 mrad
Range: 8-10 km typical

Operational status

In production and sold to several countries. The CLDS can be installed in the Israel Aircraft Industries Kfir, Dassault Aviation Mirage III and 5B, Northrop Grumman F-5F and Lockheed Martin F-16B aircraft.

Contractor

Israel Aircraft Industries Ltd, Tamam Division.

VERIFIED

Multimission Optronic Stabilised Payload (MOSP)

Multimission Optronic Stabilised Payload (MOSP) is a lightweight dual or triple sensor, target acquisition and range-finder/pointing payload.

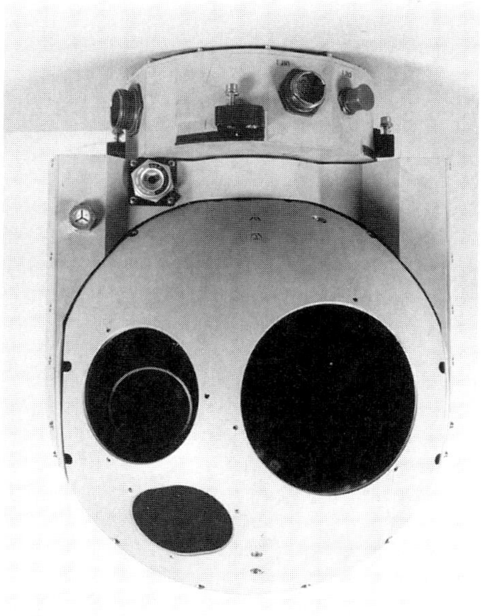

Tamam helicopter Multimission Optronic Stabilised Payload 0018347

Tamam MOSP on Eurocopter SA 342 Gazelle helicopter 0018346

MOSP has the capability to provide a stabilised view throughout the lower hemisphere, its coverage including the nadir point.

Sensor package options include: single monochrome (or colour CCD or triple colour CCD) day channel; 8 - 12 μm first-/second-generation and a 3 to 5 μm third-generation Focal Plane Array (FPA) FLIRs; laser range-finder of up to six pulses/s, or laser target illuminator.

Specifications

Dimensions: 354 (diameter) × 548 mm (height)
Weight: 26-36 kg (varies with sensors carried)
Spatial coverage:
(elevation) +15 to −105°
(azimuth) n × 360° (unlimited)
Fields of view:
TV channel: monochrome, or one- or three-colour CCD:
(narrow) 0.37° (for the long-range MOSP)
(zoom) 1.3 to 18.2°
FLIR channel: 3-5 or 8-12 μm
(narrow) 2.4°
(medium) 8.2°
(wide) 29.2°
Power consumption: 280 W (day/night), 28 V DC

Operational status

Over 300 MOSP systems have been delivered for varying types of air, sea and land-based platforms.

Contractor

Israel Aircraft Industries Ltd, Tamam Division.

UPDATED

Plug-in Optronic Payload (POP)

The Plug-in Optronic Payload (POP) is a modular, compact, lightweight electro-optical payload, designed for day/night surveillance, target acquisition, identification and location. POP is designed for light aircraft, helicopters and UAVs. It consists of a 260 mm diameter stabilised platform with a replaceable plug-in sensor unit, which may be configured to meet the customer's requirements.

The sensors include: a focal plane array infra-red sensor (3-5 μm); a long-range colour CCD TV; or a combination of both. The sensor unit can be replaced in the field within minutes. The payload includes an automatic tracker and is controlled via a parallel or serial communication channel.

Specifications

Dimensions: 260 (diameter) × 380 mm (height)
Weight (typical): payload 15 kg; control unit 0.7 kg
Fields of regard:
(azimuth) 360°
(elevation) +40 to −110°

POP installed on helicopter 0018344

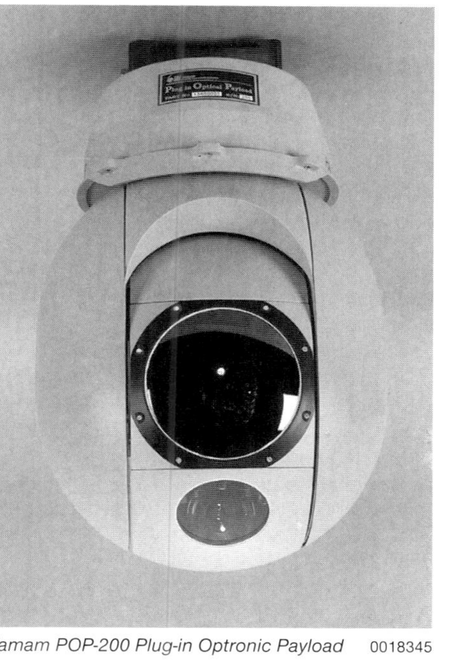

Tamam POP-200 Plug-in Optronic Payload 0018345

Fields of view:
focal plane array: 5 × 25° or 4 × 12°
TV: various zooms
TV and focal plane array: 4 × 3° to 16 × 12°

Operational status

In production.

Contractor

Israel Aircraft Industries Ltd, Tamam Division.

UPDATED

High-performance FLIR

The high-performance FLIR is a compact parallel-scan thermal imaging system which operates in the 8 to 12 μm range. It is designed to provide maximum sensitivity and range performance with limited space and optical aperture, to give a serial scan-like image quality.

Combining a Newtonian-type telescope of relatively high magnification with a small element detector array with a high spatial resolution, the high-performance FLIR provides a fast compact optical design through a 1 W bidirectional scanner. The fast optics also guarantee high-efficiency cold shielding. Image quality is ensured by full DC restoration in conjunction with a very accurate channel-to-channel responsivity equalisation scheme. Other image enhancement features include dynamic range compression from 12 to 6 bits, electronic boost, interframe averaging, automatic gain and level and optional autofocus. An increase in range performance of approximately 20 per cent is achieved by a special electronic zoom underscan mode which also provides enhanced sensitivity and improved spatial resolution in both scan and cross-scan directions.

Utilisation of gate arrays and full custom VLSI designs have further reduced the size, weight and cost of the high-performance FLIR. All the preamplifiers, together with the electronic multiplexer, are housed in the Dewar package, while the rest of the electronics are included in the PC boards. Serially multiplexed detector outputs are provided for gimballed applications.

The combination of compactness and performance make the high-performance FLIR particularly suitable for long-range surveillance, target acquisition and tracking, and navigation. Typical applications include helicopter fire-control sights, high-performance aircraft targeting and navigation sights, long-range detection of slow-moving aircraft and stationary concealed personnel, and observation of small fast-moving sea craft.

Contractor

Rafael Missile Division.

VERIFIED

Lilliput-1 observation and target acquisition system

Lilliput-1 is a miniature day observation and target acquisition system for helicopters, RPVs, ground and naval platforms. It consists of a gimballed module complete with CCD sensor and line of sight stabilisation, packaged in a composite shroud. The compact sensor has four fields of view. The narrowest, with 500 mm focal length and 20 μrad resolution, achieves exceptional performance in dynamic conditions. The CCD sensor can be replaced by an ICCD sensor for low-light conditions.

The system is lightweight, compact and self-contained and features very high resolution, line of sight stabilisation and low power consumption.

Specifications

Dimensions: 190 mm sphere
Weight: 1.45 kg (including laser range-finder)
Power supply: 28 V DC, <15 W
Field of view: 20°, 7.5°, 2.7° and 1°
Coverage:
(azimuth) 360°
(elevation) −85 to +10°

Contractor

Rafael Missile Division.

VERIFIED

Modular Thermal Imaging System (MTIS)

The MTIS is a high-performance, low-cost compact thermal imaging system in the 8 to 12 μm spectral range, consisting of a sensor and an electronic unit. It is suitable for night and adverse weather navigation, detection and recognition, search and rescue missions and surveillance.

MTIS can be adapted to a variety of platforms, such as helicopters, RPVs or land and sea platforms. It is available as an independent unit, or as part of a stabilised turret containing other sensors such as TV, laser range-finder or designator and daytime optics. Integrated in such a multisensor, MTIS becomes the core of a 24 hour fire-control system.

Specifications

Weight:
(sensor) 6 kg
(electronic unit) 7 kg

Field of view:
(wide) 24.5 × 18.4°
(medium) 7 × 5.2°
(narrow) 2 × 1.5°

Contractor
Rafael Missile Division.

VERIFIED

TAWS-05 altitude warning sensor

TAWS-05 is an electro-optical sensor for monitoring the tail altitude of civilian and military helicopters. TAWS-05 uses a gallium-arsenide (GaAs) laser diode for high-frequency accurate measurement of the distance to the ground and warns the pilot when the limits of the safe range have been reached.

TAWS-05 represents a major improvement in safety measures for helicopters landing in poor visibility or under adverse atmospheric conditions. It interfaces with standard altitude warning systems and can be integrated into operational aircraft voice communication systems.

Specifications
Dimensions: 64 × 150 × 34 mm
Weight: 0.5 kg
Power supply: 22-32 V DC, <1.5 VA
PRF: up to 10 kHz
Range: 0.5-15 m
Temperature range: –40 to +70°C

Operational status
TAWS-05 has been successfully tested on an Israel Defence Force Sikorsky CH-53 in severe flight conditions.

Contractor
Rafael Missile Division.

VERIFIED

Topaz electro-optical targeting and surveillance system

Topaz is a targeting and surveillance system for day/night and adverse weather. Topaz incorporates FLIR, CCD, Laser Range-Finder (LRF) and automatic tracking capability. It is optimised for naval applications and adaptable to a wide variety of marine platforms and other platforms, such as UAVs and helicopters. Topaz meets all relevant Mil-Specs and is recommended for fire-control, reconnaissance and search and rescue missions, coastal patrols and shore surveillance.

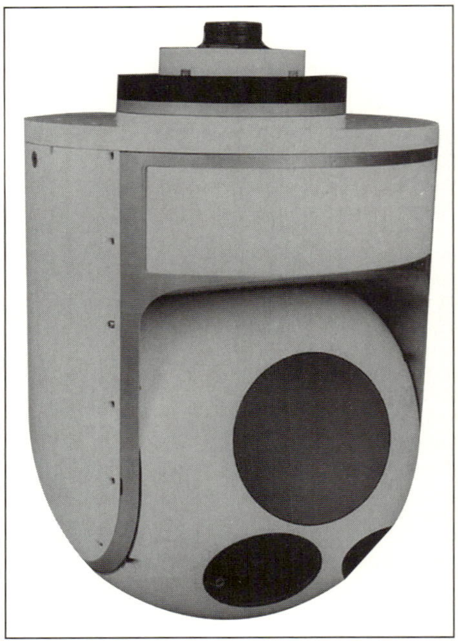

Topaz electro-optical targeting and surveillance system 0015290

Operational Status
Rafael has supplied the Topaz targeting and surveillance system to several overseas customers.

Contractor
Rafael Missile Division.

VERIFIED

Toplite multisensor payload

Toplite is a compact multisensor payload which can be integrated into various weapon systems and is adaptable to a wide range of platforms. Toplite includes CCD, FLIR, LRF (Laser Range-Finder) and their respective electronics, all packaged in a turret. It can be used for observation, tracking and pointing, slaved to either the platform's navigation, radar, or modular weapon sight, or to a helmet-mounted sight. Toplite can be integrated into helicopters and marine platforms. The lightweight and sealed payload can be installed on boats and ships of all sizes. It can be upgraded to include a laser designator, laser spot tracker, laser illuminator and in-flight boresighting.

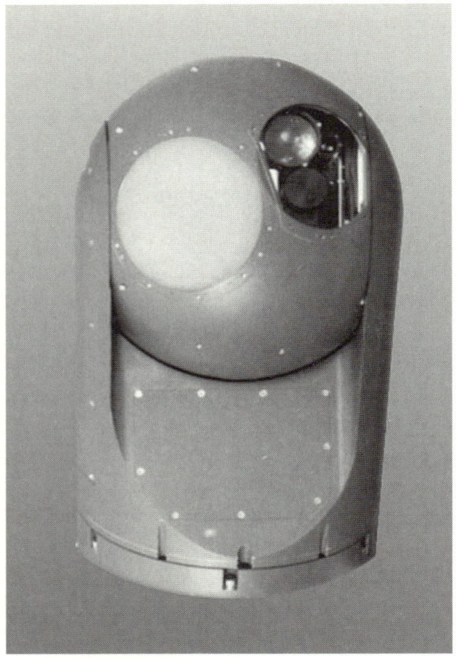

Toplite multisensor payload 0015291

Operational status
The Toplite Multisensor Payload has been procured by several countries.

Rafael and Elbit Systems have entered into a joint venture to integrate Toplite in Puma helicopters. This is part of an upgrading programme for an overseas customer.

Contractor
Rafael Missile Division.

VERIFIED

ITALY

GaliFLIR ASTRO electro-optic multisensor system

GaliFLIR ASTRO is a multisensor electro-optic system which is available in a number of configurations and is mainly intended for avionic applications such as navigation, surveillance, observation, reconnaissance, aiming and targeting.

GaliFLIR comprises a stabilised sensor platform and sensor pack. The sensor pack can include FLIR, laser designator, day/night TV cameras to suit customer requirements. The FLIR has been designed on a modular basis to improve its flexibility. The modules for series parallel scanning have been designed for applications in fixed-wing aircraft, helicopters and RPVs.

The FLIR with dual field of view optics is mounted on a high-accuracy stabilised platform. A day and night TV camera can be mounted as an option on the same platform.

GaliFLIR undergoing operational testing on an Agusta A109 helicopter 0106337

The Alenia Difesa GaliFLIR ASTRO electro-optic multisensor system **2000**/0080282

Specifications

Volume: 10 litres
Weight: 32 kg
Power supply: 28 V DC, <100 W
Wavelength: 8-12 μm
Field of view:
(scanner) 40 × 27°
(wide) 16 × 10.8°
(narrow) 4 × 2.7°
Resolution: 0.15 mrad (narrow FOV)
Detector: Sprite, CMT 8 elements
IR lines: 512
Display: TV monitor (standard CCIR 625/50)

Operational status

In production. GaliFLIR ASTRO has been selected by the Italian Navy and Coast Guard. It is installed on Agusta AB-212 and AB-412 helicopters and on EH 101 and SH-3D helicopters.

Contractor

Alenia Difesa, Avionic Systems and Equipment Division, Officine Galileo.

VERIFIED

Head-Up Display (HUD) for the AMX

Alenia Difesa produces the Type 35 HUD for the AMX. OMI designed the Pilot's Display Unit (PDU) and Alenia was responsible for the Symbol Generator Unit (SGU). The symbols generated can be changed by software and a display recorder, using either tape or film, can be fitted to the display.

Specifications

Dimensions:
(PDU) 136 × 350 × 650 mm
(SGU) ½ ATR short
Weight:
(PDU) 13.75 kg
(SGU) 8.5 kg
Power supply: 115 V AC, 400 Hz

Operational status

In production for the AMX.

Contractor

Alenia Difesa, Avionic Systems and Equipment Division.

VERIFIED

P0705 HELL laser range-finder

P0705 HELL is an airborne Nd:YAG laser range-finder designed for integration in (helicopter) stabilised electro-optic sight units with the functions of target ranging, gun and missile pointing and navigation fixing.

The fully qualified HELL system improves helicopter attack capabilities by providing extended reconnaissance and target detection ranges, increasing the probability of a kill, reducing flight times and enhancing flight safety.

PULSE (Precision Up-shot Laser Steerable Equipment): steerable laser range-finder for airborne applications 0106276

Specifications

Wavelength: 1.064 μm
Pulse power: 4 MW
PRF: 2 Hz
Range: 300-10,000 m
Interface: RS-422 serial

Operational status

In production.

Contractor

Alenia Difesa, Avionic Systems and Equipment Division, FIAR.

VERIFIED

P0708 PULSE airborne steerable laser range-finder

P0708 PULSE is an airborne steerable Nd:YAG laser range-finder.

PULSE offers superior range performance, both in CCIP and CCRP attack modes and in navigation fixing, high-pulse power and repetition frequency, good precision and accuracy and an optimised configuration for easy installation on a wide variety of platforms. It is suitable for retrofitting to aircraft such as the A-4 Skyhawk, AMX, Mirage, F-5 and MiG-21.

Specifications

Wavelength: 1.064 μm
Steering angle: within 20° pointing cone
PRF: 10 Hz
Interface: MIL-STD-1553

Operational status

In production.

Contractor

Alenia Difesa, Avionic Systems and Equipment Division, FIAR.

VERIFIED

The Alenia Difesa head-up display for the AMX

Pilot Aid and Close-In Surveillance (PACIS) FLIR

The PACIS FLIR is a thermal imaging system in the 8 to 12 μm range, designed to be installed on helicopters and fixed-wing aircraft in order to provide them with increased capability by day, night and in adverse weather operations. It creates a TV-compatible video signal for viewing in the cockpit on a standard display.

PACIS is provided with a telescope with two switchable fields of view and is steerable by a position control grip. It can be used for navigation, day/night surveillance, border patrol, search and rescue, remote sensing and monitoring or as a take-off and landing aid. PACIS can be interfaced with the aircraft avionic system.

The system is composed of a steerable platform, electronic unit and FLIR control grip connected by a cable to the electronic unit. The platform aims the FLIR optical axis in azimuth and elevation. The FLIR is equipped with a two field of view telescope: the wide field of view is used for navigation and surveillance, while the narrow is used to identify and track targets.

Specifications

Dimensions:
(steerable platform) 300 × 511 × 300 mm
(electronic control unit) 170 × 225 × 390 mm
(control panel) 146 × 124 × 165 mm
Weight:
(steerable platform) 23 kg
(electronic control unit) 8.5 kg
(control panel) 1.5 kg

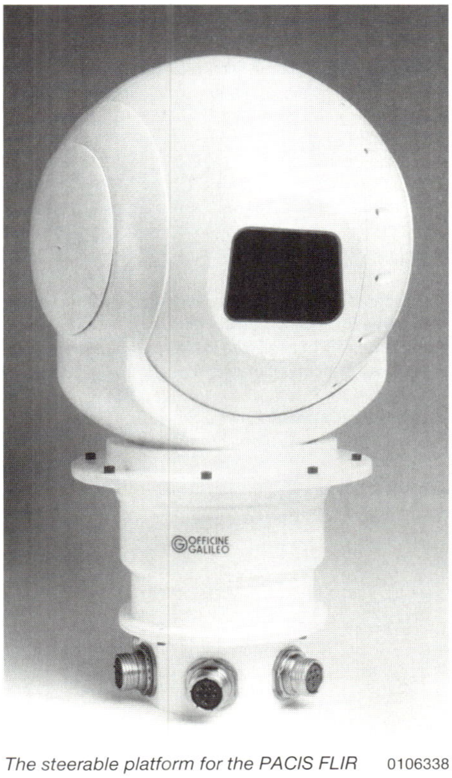

The steerable platform for the PACIS FLIR 0106338

Power supply: 28 V DC, 140 W (average), 450 W (peak)
Wavelength: 8-12 μm
Field of view:
(wide) 40 × 26.7°
(narrow × 4 magnification) 10 × 6.6°
Field of regard:
±170° azimuth,
+45 to −70° elevation
Video format: CCIR 625 lines at 50 Hz
Interface: RS-422

Contractor
Alenia Difesa, Avionic Systems and Equipment Division, Officine Galileo.

VERIFIED

PILOT infra-red integrated pod

The Marconi Pod-Integrated Localisation, Observation, Transmission (PILOT) is an electro-optical pod including a stabilised infra-red sensor, automatic target detection/localisation/tracking unit and a bidirectional ECCM video and datalink.

The PILOT infra-red sensor is a high-quality stabilised passive 24-hour all-weather infra-red imager which can be steered in order to observe targets and background in front of and below the aircraft. The video output is a standard CCIR television signal which can be displayed on the aircraft head-up display or airborne monitors, or can be video recorded.

By using image processing techniques PILOT allows automatic target detection, prioritisation and localisation with the presentation on the infra-red image of the target position, priority and spatial co-ordinates.

PILOT provides the capability of automatic tracking of the target and can be controlled both by a local operator on board the aircraft and by a remote operator through a video and datalink. The video and datalink is frequency-hopping spread spectrum and offers robust ECCM performance. This link allows infra-red image transmission to a remote station and bidirectional data communication between the pod and a remote station.

The Marconi PILOT infra-red pod 0106399

PILOT can be provided with the standard RS-422, ARINC 429 and MIL-STD-1553B avionics interfaces, which allow data communication with the airborne radar, the navigation system and the weapons management system.

Specifications
Dimensions: 2,062.9 (length) × 419.3 mm (diameter)
Weight: <128 kg
Power supply: 28 V DC, <750 W

Contractor
Marconi SpA.

VERIFIED

JAPAN

FLIR Pod for the FS-X

The Japan Defence Agency Technical Research and Development Institute is teamed with Mitsubishi Electric and Fujitsu in the development of a FLIR for the FS-X, to supplement the fire-control system for low-altitude night navigation.

The FLIR will have both air-to-air and air-to-ground applications and is specifically intended to enhance low-altitude night navigation. The sensor will have infra-red CCD detectors operating in the 3-5 μm and 8-12 μm wavelengths.

Operational status
Full-scale development for installation on the FSX aircraft. Development completion is planned for 2001.

Contractors
Japan Defence Agency Technical Research and Development Institute.
Mitsubishi Electric Corporation
Fujitsu General Ltd.

VERIFIED

Head-Up Displays (HUDs)

The Head-Up Display (HUD) provides information such as altitude, airspeed, magnetic heading, attitude, angle of attack and aiming reticle to the pilot during the mission. The information is presented as symbology with a collimated image overlaid in the pilot's view. The HUD consists of the display unit and a symbol generator.

Operational status
In production for, and in service with Japanese Self Defense Force AH-1S, F-4EJ Kai, F-15J/DJ, T-4 and US-1A aircraft. Also in limited production for the CCV-T2 and C-1 QSOL experimental aircraft.

Contractor
Shimadzu Corporation.

VERIFIED

The range of HUDs made by Shimadzu Corporation 0106277

Radar display

The radar display provides radar information to the crew in the form of symbology and/or video. The radar display consists of a forward indicator unit, aft indicator unit and electronic unit.

Operational status
In service on the F-4EJ Kai aircraft.

Contractor
Shimadzu Corporation.

VERIFIED

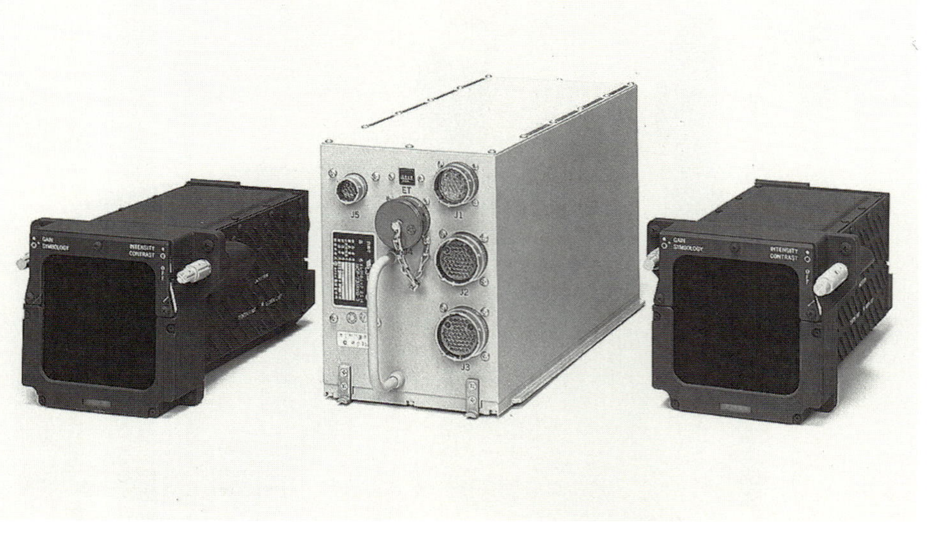

The Shimadzu radar display consists of (left to right) the forward indicator unit, the electronic unit and the aft indicator unit

KOREA, SOUTH

Electro-optical tracking system

Samsung Defense Systems produces an airborne image surveillance system to detect and autotrack targets. It has two image sensors, a CCD camera and a 2nd/3rd generation thermal imager for day and night use. The system is stabilised in four axis and is suitable for use in helicopter or UAV applications.

Contractor
Samsung Defense Systems

VERIFIED

The Samsung Defense Systems electro-optical tracking system
2000/0085272

RUSSIAN FEDERATION AND ASSOCIATED STATES (CIS)

Missile Launch and Trajectory Control System (MLTCS)

The MLTCS has been designed by a consortium to provide control of the RVV-AE and R-27 missiles on Russian and foreign aircraft. The MLTCS ensures single- and multi-target engagement using integrated data from the radar, electro-optical and helmet-mounted display sensors.

Functional capabilities include: generation of target co-ordinates; slaving of the selected missile to its allocated target; generation of launch zone data; missile launch control; target illumination control for semi-active radar missiles; transmission of mid-course guidance commands to the missiles.

Contractors
Aircraft Scientific Industries Complex MIG.
State Research Institute of Aviation Systems.
Design Bureau Aviaavtomatika.
State Scientific and Research Instrument-Building Institute.
Scientific and Research Institute for Instrument Engineering.
The State Machine-Building Design Bureau Vympel.

VERIFIED

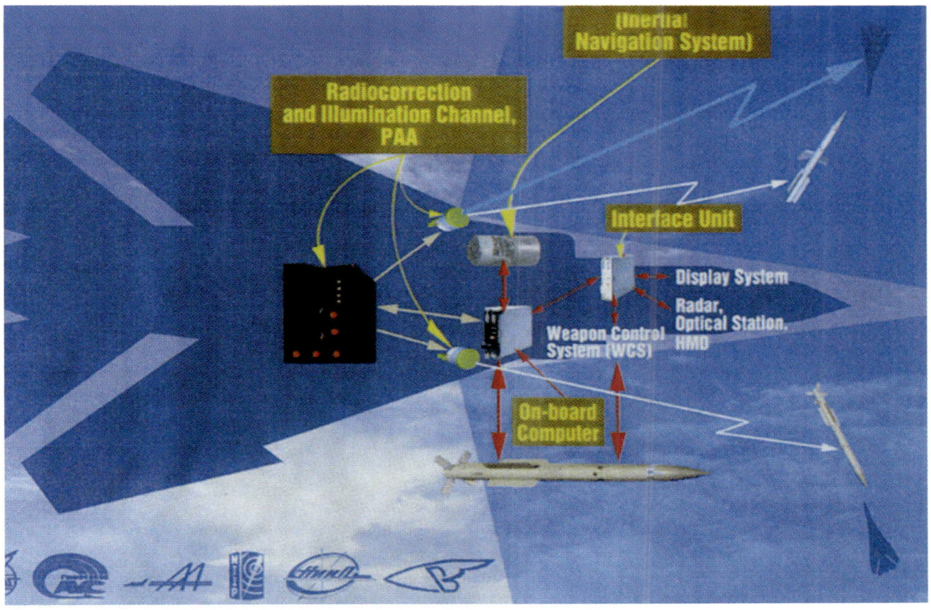

Missile launch and trajectory control system ***1999***/0051720

Head-Up display (HUD)

The Electroavtomatika HUD is designed for the generation of collimated symbolic flight/navigation and special information with a green colour image, superimposed on the scene outside the cockpit. The HUD comprises: the WCS-3 wide-angle collimating system; projection CRT control unit; built-in symbology information generation module; built-in computer module; built-in video camera for CRT image transfer.

The HUD provides the following functional capabilities: operation with MIL-STD-1553 bus systems; display of TV images exchanged in accordance with STANAG 3350B; wide-angle holographic data presentation; use of a light emitting backlight.

Specifications

Total field of view:
(vertical) 20°
(azimuthal) 30°
Instantaneous field-of-view:
(at 400 mm) not less than 20 × 30°
(at 500 mm) not less than 18 × 24°
Max viewing brightness: 30,000 cd/m^2
HUD display error not greater than:
(10° diameter field of view) 7 minutes
(20° diameter field-of-view) 10 minutes
(at wider angles) 15 minutes
TV image display characteristics:
(refresh frequency) 50 Hz
(lines) 625
(active lines) 575
(scan) left to right, top to bottom
(display levels) not less than 8 for up to 50,000 lx; not less than 5 for 75,000 lx
(image brightness) automatic or manual
(front panel controls) symbology brightness; TV brightness; TV contrast; reticule brightness; BIT control

The Electroavtomatika head-up display
2000/0062241

Operating temperature: –40 to +55°C
Weight: 20 kg max
Power: less than 80 W

Contractor
Electroavtomatika OKB.

VERIFIED

Helmet-mounted aiming system (HSTs-T)

The Electroavtomatika HSTs-T helmet-mounted aiming system is designed to compute and display line of sight angular co-ordinates of the visual target as cued by the pilot's head turn; generate collimated images of the sighting marker and launching marker in the field-of-view of the pilot's right eye.

System components comprise an easily removable helmet-mounted optical assembly with reference points device and sighting parameters display system; two pilot's head position sensors (trackers) installed in the cockpit and, as a rule, installed with a collimator-type indicator used as a mounting base; control and interface unit.

The helmet-mounted sight system provides for data exchange over MIL-STD-1553B or ARINC-429 data standards; autonomous computation of sighting angles; high-accuracy target designation for effective employment of modern weapons. Data is presented to the pilot in the form of sighting and launching markers. These markers are presented at a luminance of 35,000 cd/m^2 and their light intensification is controlled automatically.

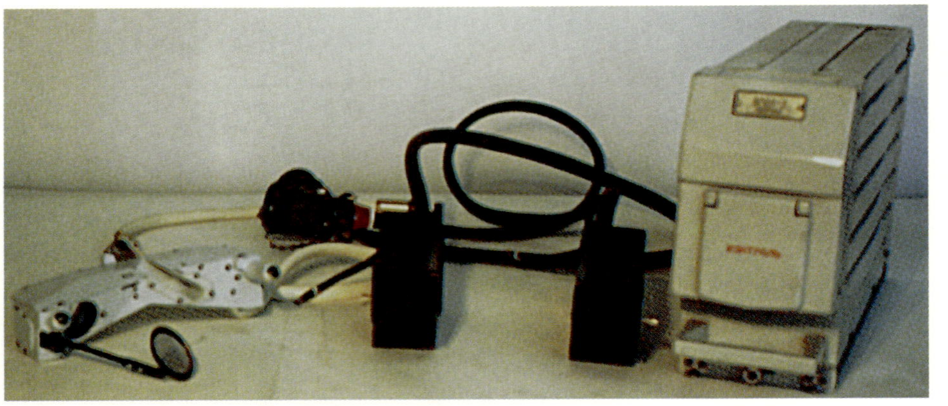

The Electroavtomatika helmet-mounted aiming system **2000**/0062242

Specifications
Angular coverage:
(azimuth:)±60°
(elevation) –15 to +60°
Max aiming error:
in the central aiming area lying between ±35° azimuth and –15 to +60° elevation: 20′ of arc outside the central aiming area: 35′ of arc
Data refresh rate: 100 Hz
System weight:
(helmet-mounted assembly) 0.3 kg
(head position trackers) 2 × 0.5 kg
(control and interface unit) 6 kg
MTBF: 2,000 flight hours
Operating temperature range: –40 to +55°C
Power: 80 VA maximum

Contractor
Electroavtomatika OKB.

VERIFIED

GEO-NVG-III Night Vision Goggles

Geophizika claims that its GEO-NVG-III (Russian designation reported to be GEO-ONV (Ochki Nochnogo Videniya)) night vision goggles employ third-generation GaAs photocathode technology to provide high responsivity in starlight/overcast conditions. The goggles are ruggedised to meet Russian military requirements for low-altitude helicopter combat, reconnaissance, and search and rescue operations.

Features include: full 40° field of view, F/1.1 at 43 mm eye relief and 10 mm exit pupil; full binocular night vision; full peripheral vision; automatic brightness control; internal power supply; quick disconnect.

Specifications
Illumination: 10^{-5} to 10 lx
Magnification: × 1
Field of view: 40°
Exit pupil and eye relief: 10 and 43 mm
Objective lens: fixed focus 25 mm, F/1.1
Focus range: 300 mm to 00
Eyepiece lens: 25 mm
Weight: 0.78 kg
Voltage required: 3 V DC, 50 mA (2 × AA batteries)
Mechanical adjustment:
(vertical) 20 mm
(fore and aft) 24 mm
(tilt) 15°
(interpupillary) 56-73 mm
(eyepiece dioptre) +4 to –4 dioptres
Photocathode: GaAs
Sensitivity:
luminous 2,856 K: 1,200 uA/1m
radiant (830 nm): 120 mA/W
equivalent brightness input (at 10^{-4} lx): 2.5 × 10^{-7} lx
S/N (at 10^{-4} lx): 15
centre resolution: 32 mm
useful cathode diameter: 17.5 mm

Operational status
Claimed to be in widespread use on Russian military helicopters.

Contractor
Geophizika-NV.

VERIFIED

GEO-NV-III-TV day/night tracking system

The GEO-NV-III-TV image-intensified, solid-state, charge-coupled device (CCD) camera is a versatile day/night tracking system, mounted on a gyrostabilised platform, that incorporates three separate channels: a CCD day sensor; an image intensifier; and a CCD night channel.

Specifications
Intensified CCD night channel
Scene illumination: 10^{-5} to 10^{-1} lux
Sensor instantaneous field of view: 10°
Combined sensor/platform field of view: +30 to –50°
Focus range: 0.3 m to ∞
Objective lens: f=75 mm; F/1.5
Camera slew rate: 20°/s
System acc
uracy: 0.5°
Dimensions: 200 (length) × 60 mm (diameter)
Weight: 0.65 kg

CCD day channel
Pixels: 512 (H) × 582 (V)
Pixel size: 7.6 × 6.3 mm
Active imaging cell size: 4.6 (H) × 3.5 mm (V)
Resolution: 480 TV lines
Field of view: 13.5°
Objective lens: f=25 mm; F/1.8
Grey scales: 10
Dimensions: 50 (height) × 60 mm (diameter)
weight: 0.1 kg

Image intensifier
Photocathode: GaAs

Operational status
Geophizika states that the GEO-NV-III-TV system is fitted to Kamov Ka-50 helicopters and to a wide range of Mil helicopters, including Mi-17, Mi-24, Mi-26 and Mi-28. Geophizika has also proposed the system for the Ka-52 helicopter.

Contractor
Geophizika-NV.

VERIFIED

Airborne Laser Radar Landing System (LRLS)

GosNIIAS has been involved with research in the area of laser radar technology for miitary aircraft surveillance and targeting. In 1994, work was started to convert this experience into production of a laser radar landing system for civil aircraft.

The LRLS is designed to enable general purpose aircraft and airliners to land safely in non-standard situations caused by adverse weather conditions, en-route obstacles, or the need to land on non-instrumented runways in emergency situations.

The LRLS provides day/night search, detection and auto-tracking of runways defined by optical reference marks in clear and adverse weather. The system determines aircraft angular and linear co-ordinate data, relative to the selected runway, and provides automatic steering commands to the flight control system, together with display of relevant information on the flight and navigation displays. The system is intended to support landings in Cat II and III conditions.

The system hardware comprises: the laser sighting unit; a computing unit; interfaces to the aircraft central computer and display systems.

Specifications
Runway detection range: over 5 km (in meteorological visibility range of not less than 1 km)
Detection range of objects on the runway and in the air: over 2 km
Weight (with two-axis stabilisation): 23 kg
Power consumption: 240 W

Operational status
Technical documentation and production models of all system components are available.

Contractor
GosNIIAS State Research Institute of Aviation Systems.

VERIFIED

Airborne Multifunction Optical Radar System (AMORS)

Since 1982, GosNIIAS has been developing a generation of Airborne Multifunction Optical Radar Systems (AMORS), designed to provide automatic object and obstacle detection and recognition in poor visibility conditions. The goals of this work were:

(a) to develop technologies, methods and algorithms as well as software and hardware for automatic scene analysis during complex processing using optical sensors with differing physical natures and data characteristics;
(b) the development of high-performance image sensors, including Doppler laser radar, uncooled IIR systems, and pulse-illumination TV sensors;
(c) the operational development and full-scale testing of optical radar prototypes.

AMORS is intended for air and ground monitoring, object detection, recognition and localisation against complex dynamic scenes. The system can detect and recognise various small-size objects, including vehicles and electrical power lines. Integration of AMORS into avionics systems enhances navigation, safety in poor weather conditions, operation at poorly equipped airfields, and rescue and reconnaissance operations.

AMORS is based on range, velocity and image intensity movement, integrated image processing, and automatic object processing algorithms. The system features high jam resistance and is intended for day/night all-weather operation. The system can comprise different laser radar, imaging infra-red and TV sensor combinations, as well as signal processing software/hardware, object recognition modules, and a TV display control panel.

Specifications

Range: up to 10 km
Resolution:
(angle) 0.4 min
(range) 1 m
(velocity) 0.3 m/s
Signal processing time for automatic object recognition: <1s
Weight: <50 kg

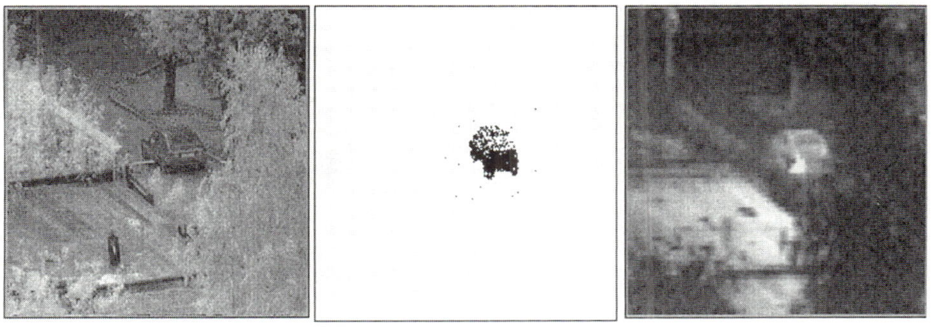

Examples of TV, laser radar and IIR images generated by the AMORS system 0018343

Operational status

Experimental prototypes have been developed and tested. Pre production development is ongoing.

Contractor

GosNIIAS State Research Institute of Aviation Systems.

VERIFIED

Laser range-finder/target illuminators

Klyon

The Klyon laser range-finder/target illuminator was developed for MiG-27M, Su-22, Su-25 and their variants.

Specifications

Ranging capability: up to 10 km; accuracy 5 m
Illumination range: up to 7 km
Weight: < 82 kg

Prichal

The Prichal laser range-finder/target illuminator is employed on Su-25TK aircraft and on Ka-50 and Mi-28 helicopters. It is used for navigation, particularly cross-country over rough terrain, and for target ranging/illumination for laser-guided bombs and missiles.

Specifications

Ranging accuracy: 5 m
Weight: 46 kg

Operational status

In service.

Contractor

Production Association Urals Optical and Mechanical Plant (PA UOMZ).

VERIFIED

Technical specifications of laser emitters for Klyon and Prichal systems are as follows:

	Lasers				
	N1	**N2**	**N3**	**N4**	**N5**
Impulse energy (J)	0.07	0.18	0.38	0.4	0.25
Pulse repetition frequency (Hz)	up to 25	up to 25	25	10	25
Cooling	fluid cooling for all systems				
Weight (kg) (incl cooling system)	2.5	6.5	7.3	3.2	6
Overall dimensions(mm) (incl cooling system)	240 × 188 × 103	420 × 180 × 160	300 × 155 × 110	40 × 390	400 × 125 × 120
Power supply	115 V; 400 Hz, 3-phase for all systems				
Power supply weight (kg)	6.5	6.5	10.5	20	9.8

Klyon laser range-finder/target illuminator 0106278

Optical-electronic sight systems

OEPS-29

The OEPS-29 optical-electronic sight system is installed on MiG-29 aircraft. It provides: search, detection and tracking of targets at all altitudes, in free airspace, on the earth background, in daytime and at night and range-finding for air and ground targets.

To increase battle effectiveness, the sighting system is integrated with the pilot's helmet-mounted display.

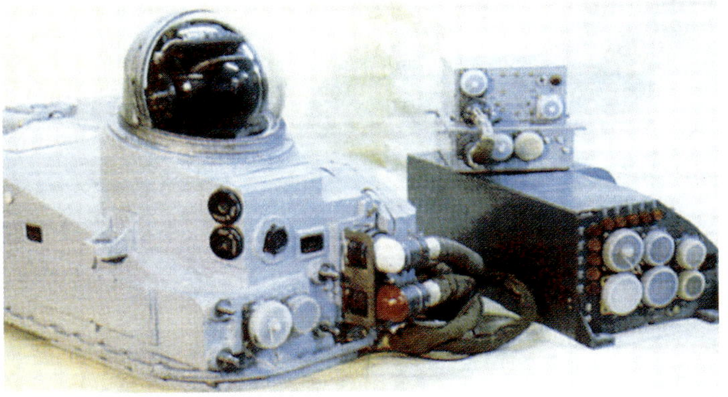

OEPS-29 optical-electronic sight system 0106279

OEPS-27 optical-electronic sight system 0106281

OEPS-27

The OEPS-27 optical-electronic sight system is installed in Su-27 variants. It provides the same functions as the OEPS-29, but it differs by providing a greater viewing angle and greater range for air target detection.

Specifications

Technical specifications of laser emitters for OEPS-27 and OEPS-29 sight systems are given in the previous entry, headed 'Laser range-finder/target illuminator'.

Operational status

In service.

Contractor

Production Association Urals Optical and Mechanical Plant (PA UOMZ).

VERIFIED

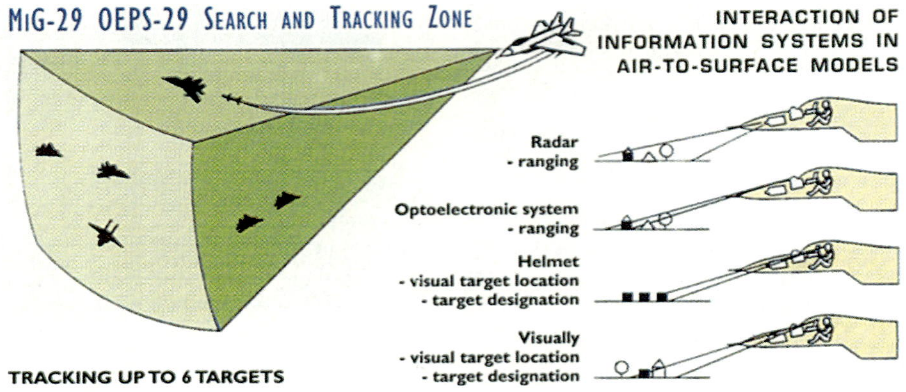

MiG-29 OEPS search and tracking zones 0106280

YOM3 Gyrostabilised Optical Electronic System (GOES)

The YOM3 GOES platforms are designed to carry thermal imagers, day and low light-level TV cameras, cine- and video-cameras, laser range-finders, and similar equipment. The first four models produced were designated GOES-1/-2/-3/-4. They were all designed to carry electro-optical payloads varying from 16 to 100 kg. The systems were suitable for both civil and military use, and payload specification is by customer choice.

A new generation of GOES platforms, designated GOES-310/-320/-330 has now been developed and put into production. The designations represent single-, double- and triple-channel systems respectively. All are designed for detection and recognition of objects in a broad range of vision angles in severe rolling and vibration conditions in both civil and military environments. Sensors available for fitment to the GOES-310/-320/-330 series include: thermal imagers, cine and video cameras, laser range finders, and

Specifications

Platform	**GOES-1**	**GOES-2**	**GOES-3**	**GOES-4**
Payload weight	100 kg	16 kg	30 kg	65 kg
Payload volume	85 dm³	13.5 dm³	20 dm³	55 dm³
Weight of optical turret	140 kg	25.5 kg	20 kg	85 kg
Dimensions of optical turret	Ø720 × 980 mm	Ø340 × 552 mm	Ø460 × 613 mm	Ø640 × 850 mm
Look angles				
(azimuth)	±170°	±170°	±235°	±135°
(elevation)	+80 to –40°	+10 to –30°	+45 to –115°	+10 to –30°
Stabilisation (micro radians):	50	70	50	50

YOM3 mounted below Shkval on Ka-50N Black Shark helicopter (Paul Jackson) 0018342

YOM3 mounted above the cockpit on Ka-52 Alligator helicopter (Paul Jackson) 0018341

The GOES-310/-320/-330 gyrostabilised optical electronic systems (from top to bottom) **2000**/0079247

YOM3 on Ka-52 Alligator helicopter (Paul Jackson) 0018340

infrared sensors in any combination of fits. System options include: compatibility with GPS, RS 232 interface, video monitor, VHS/SVHS videotape recording, an air-ground microwave datalink.

GOES-310/-320/-330
Platform
Stabilisation: 5-axis
Pointing error: <50 μrad RMS
Field of regard:
(azimuth): ±230°
(elevation): + 30 to −110°
Maximum angular rate: 60°/s

Dimensions:
(opto-mechanical unit): Ø460 × 613 mm
(electronic unit): 330 × 485 × 225 mm
(control unit): 225 × 50 × 57 mm
Weight:
(opto-mechanical unit): 55 kg
(electronic unit): 20 kg
(control unit): 0.43 kg
Power: 27 V DC, 500 W; 115/200 V AC, 3-phase, 400 Hz, 250 VA

Thermal imager: AGEMA THV-1000
receiver: 5-bar SPRITE focal plane
spectral range: 8-12 μm
cooling: integral Stirling cooler
pixels: 580 × 386
field of view (H&V):
(narrow) 5.0 × 3.3°
(wide) 20.0 × 13.3°
NETD: 0.18°C

TV system: CCD Sony EVI-331 colour
pixels: 752 × 582
TV lines: 480
focal range: f=5.4 to 64.8 mm with × 12 optical zoom
field of view: 48.8 × 37.6° to 4.4 × 3.3°

Laser range-finder (made by Production Association Urals Optical and Mechanical Plant (PA UOMZ):
wavelength: 1.54 μm
energy: 0.01-0.07 J
beam divergence: 2-3 minutes of angle
PRF: 1 Hz
Maximum range: 10 km
Measurement error: <5 m

Operational status

The company has stated that orders were placed for the GOES-320 system at the MAKS-97 air show. The GOES-320 unit was shown on the 'Hind' night vision proposal (Mi-24VN, also designated Mi-35O) mock-up shown at MAKS'99.

Variant YOM3 systems are also fitted to the Kamov Ka-29, Ka-50N and Ka-52, and Mi-24 and -35 helicopters. The accompanying photographs taken at MAKS-97 show the YOM3 system mounted below Shkval on the Ka-50N Black Shark, although it has also been shown mounted above Shkval on the same aircraft at other times. On the Ka-52 Alligator, it is mounted above the cockpit.

Contractor

Production Association Urals Optical and Mechanical Plant (PA UOMZ).

UPDATED

Airborne laser target designation pod

The Urals Optical-Mechanical Production Association is offering an airborne laser target designation pod to complement its existing range of electro-optical target tracking and designation systems.

In addition to the electro-optical components, the pod carries integral power supply, thermal regulation and control systems.

Specifications

Laser wavelength: 0.6 to 0.8 μm
Measurement range: 20 km
RMS stabilisation: 3 mrad
Field of view:
(elevation) +10 to −160°
(azimuth) ±10°
(roll) ±150°
Sight line tracking:
(angular rate) 55°/s
(angular acceleration) 60°/s²
Dimensions: 360 × 3,000 mm
Weight: 250 kg

Contractor

Urals Optical-Mechanical Production Association.

VERIFIED

A-84 panoramic aerial camera

The A-84 panoramic aerial camera is designed to provide wide-area photographic coverage of the earth's surface during daylight conditions, from medium and high altitudes. It can be set to operate automatically by onboard control system, or be controlled manually from its control panel.

The A-84 camera is equipped with image motion compensation and automatic exposure control. The camera film records navigation data from the aircraft navigation system.

Specifications

Focal length: 300 mm
Aperture: f/4.5

The Zenit A-84 panoramic aerial camera 0018339

Frame size: 118 × 748 mm
Film size: 480 m (length) × 130 mm (width)
Nominal overlap in centre frame: 25%
Along track filming distance: 160 × aircraft altitude
Linear resolution: 0.4 m at range = 2 × altitude; 0.8 m at range = 6 × altitude
Power: 27 V DC, less than 300 W; 115 V AC, 400 Hz, less than 900 V A
Weight: 160 kg

Operational status

Fitted to Tu-22 medium bomber (presumably the Tu-22MR variant), and to the Tu-154 medium transport aircraft for 'Open Skies' operations. Also reportedly fitted to the M-17 high-altitude reconnaissance and research aircraft.

Contractor

Zenit Foreign Trade Firm, State Enterprise P/C S.A. Zverev Krasnogorsky Zavod.

VERIFIED

AC-707 spectrozonal aerial camera

The AC-707 aerial camera has four separate photographic channels, four separate lenses, four filters and four automatic exposure control systems to provide photography in four predetermined spectral bands. Each lens is corrected for focal length and distortion. The camera contains image motion compensation, vacuum back film platen, range focus and temperature focus compensation. It records navigation information obtained from the aircraft navigation system, and it can be fitted with gyrostabilisation. It can be controlled automatically or manually from its control panel.

Specifications

Focal length: 140 mm
Aperture: f/2.8-f/22
Spectral zones:
Channel 1 (blue) 400-500 nm
Channel 2 (green) 480-600 nm
Channel 3 (red) 580-700 nm
Channel 4 (infra-red) 700-860 nm
Frame size: 180 × 180 mm
Sub picture size: 70 × 70 mm
Film type: MLU-4, 240 m (length) × 190 mm (width)
Operating altitude: over 50 m
Exposure time: 1/20-1/300 s
Along track filming distance: 400 × altitude
Across track coverage: 0.5 × altitude
Power: 27 V DC, less than 30 W; 115 V AC, 400 Hz, less than 5 VA; 36 V AC, 400 Hz, less than 6 VA
Weight: 212.9 kg

Operational status

Installed on Mi-8 helicopters.

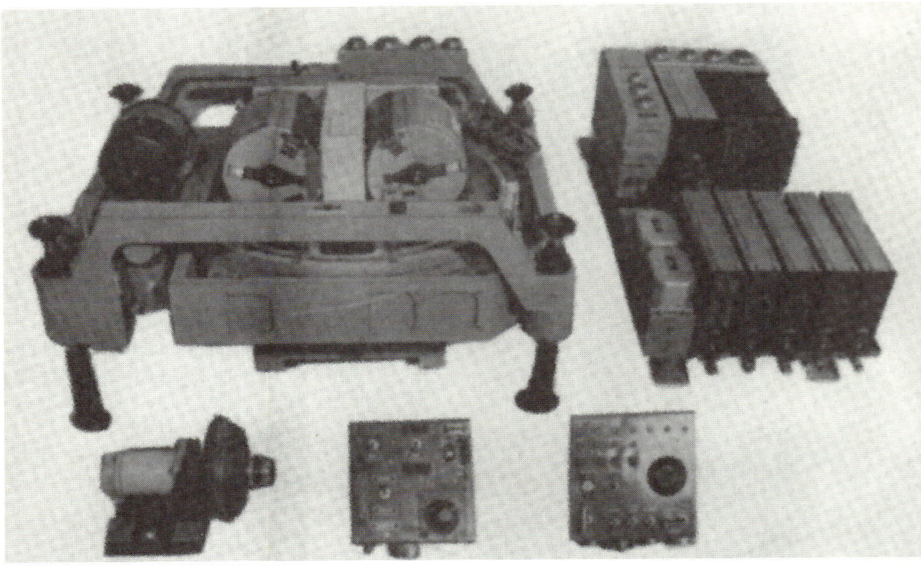

The Zenit AC-707 spectrozonal aerial camera 0018338

Su-39 Strike Shield aircraft, showing Shkval EO sighting system in the nose, Kopyo radar pod under belly, OMUL ECM pod under each wing, Pastil RWR on each wingtip 0018336

Contractor

Zenit Foreign Trade Firm, State Enterprise P/C S.A. Zverev Krasnogorsky Zavod.

VERIFIED

AK-108Ph vertical and oblique aerial camera

The AK-108Ph aerial camera is designed for simultaneous vertical and oblique photography using three across-track width options (described as Routes 1, 2 and 3), together with three vertical/oblique angular options (defined as 0° (vertical), 75° and 80°). The camera head mirror and magazine can be rotated to provide the required coverage angle.

Camera features include: automatic exposure control; linear compensation of image shift during exposure; mirror stabilisation; temperature and pressure focus compensation and aircraft navigation recording.

Specifications

Focal length: 1.8 m
Aperture: f/5
Frame size: 180 × 180 mm
Angular field of view: 6°
Film size: 240 m (length) × 190 mm (width)
Route options: 1, 2 and 3
Photographic angle options: 0°, 75° and 80° from vertical

Angle of Photography	Across track coverage, relative to height		
	Route 1	Route 2	Route 3
0°	0.1	0.18	0.26
75°	1.7	3.2	5.3
80°	3.8	8.6	21.4

Overlap, nominal: 20%
Power: 27 V DC, less than 300 W; 115 V AC, 400 Hz, less than 900 VA; 36 V AC, 400 Hz, less than 100 VA
Dimensions: 3.1 m × 600 mm diameter
Weight: 600 kg

The Zenit AK-108Ph vertical and oblique aerial camera 0018337

Shkval window (above YOM3 'ball') on Ka-50N Black Shark helicopter (Paul Jackson) 0018333

Operational status

Fitted to Su-24MR reconnaissance aircraft.

Contractor

Zenit Foreign Trade Firm, State Enterprise P/C S.A. Zverev Krasnogorsky Zavod.

VERIFIED

Shkval sighting system

The Shkval sighting system is designed as a comprehensive anti-tank electro-optical fire-control system, incorporating: TV sighting sensor, laser range-finder and target designator, and laser beam-rider. It is also fitted with a three-axis field of view stabilisation system and an automatic image correlator to ensure tracking against ground, sea or sky backgrounds. There is × 23 magnification to extend detection/tracking ranges. Tracking angles are quoted as +15 to −80° in elevation and ±35° in azimuth. Laser guidance system accuracy is quoted as 0.6 m.

Operational status

Shkval is part of the weapon system on the Ka-50 Black Shark and Ka-52 Alligator helicopters, and on the Su-24T and Su-39 multimission attack aircraft.

Contractor

Zenit Foreign Trade Firm, State Enterprise P/C S.A. Zverev Krasnogorsky Zavod.

VERIFIED

SOUTH AFRICA

LEO airborne observation systems

LEO*II* is a range of gyrostabilised airborne observation systems specifically developed for the paramilitary security market. Suitable for both rotary- and fixed-wing aircraft operating by day or night, LEO*II* is a powerful force multiplier in applications ranging from law enforcement through search and rescue to border patrolling.

Headed by the latest model combining Forward Looking Infra-Red (FLIR) with broadcast-quality colour TeleVision (TV), the range comprises various single-, dual- and triple-sensor models fitted with a selection of FLIRs, colour TV cameras and laser range-finders.

All models are based on the same compact, high stabilisation, 400 mm platform, which utilises an open modular architecture, to guarantee future upgrade

The operator's console and monitor for the LEO observation system mounted in a Eurocopter BO 105 helicopter 0106282

LEOII-A2 with 8-12µ FLIR and broadcast-quality TV camera ***2000***/0065938

LEO-II airborne observation system ***1999***/0051007

capability. The systems carry appropriate civil aviation approvals and are built to ISO 9002 and MIL-STD-10C/C/E standards. A full suite of options is available to enhance system operation.

Specifications

Thermal imagers: various 3-5µm and 8-12µm FLIRs
Colour TV cameras: various 1-CCD and 3-CCD broadcast colour cameras
Laser range-finders: various
Stabilisation: <2µrad
Platform size: 400 mm
Basic system mass: 34 kg
Options: GPS position display; searchlight/FLIR slaving units; auto-trackers; microwave downlinks.

Operational status

As of September 1999, LEO*II* was in service in South Africa and 25 other countries, including 90 per cent of UK police air support units.

Contractor

Cumulus, Business Unit of Denel (Pty) Ltd.

VERIFIED

SPAIN

SCAR armament control system

The SCAR (Sistema de Control de Armamento) armament control system provides full weapon selection and control facilities for up to six underwing carriage stations and a centrally located gun pod. The system's functions include selection of weapon station for use in a mission, selection of weapon type fitted at a station, mode selection for bomb arming, time interval selection for bomb release in ripple mode, air-to-air and air-to-ground gunnery selection, emergency mode selective jettison, weapon availability and status, stores selected and stores remaining indication and status and gun ammunition counter indication.

Designed principally for the CASA C-101, the SCAR system provides weapon selection and control facilities for bombs, rockets, guns and combined stores dispenser/launcher units and also supplies signals for control of weapon sighting and gun camera systems.

The equipment comprises a central processor, usually located in the aircraft's nose bay, a main unit, monitoring unit and emergency unit installed in the cockpit. In the case of trainer aircraft, additional monitoring and emergency units, identical to those in the attack aircraft, are mounted in the student's cockpit.

The main unit is used for all control and selection functions, with the exception of emergency jettison for which the emergency unit is used. Monitoring units indicate the status and type of stores available and the quantities of stores or ammunition remaining or expended. In the training role, the functions of the student's portion of the system can be completely inhibited by the instructor pressing a button. An error light indicates a system malfunction or a mis-selection

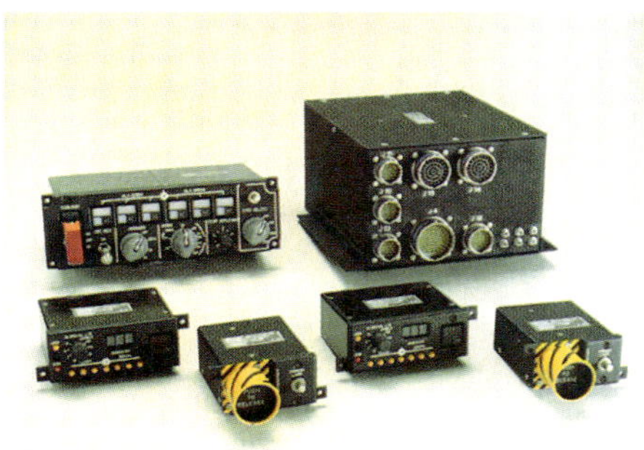

The SCAR armament control system with (rear left) the main unit, (rear right) the central processor and (front) two monitor units and two emergency units 0106378

of a store or weapon. In the case of mis-selection, the system, with the exception of emergency release, is automatically rendered inoperative. The emergency unit is hard-wired to ensure complete reliability and contains a test-lamp push-button for status checking of all indicator lamps in the system.

SCAR is a microprocessor-controlled system employing digital techniques and TTL integrated circuitry. Operational parameters can be modified via software changes and an EAROM memory ensures that stored data is retained even when the system is unpowered. The system is of modular design and conforms to the recommendations of MIL-STD-1472 and MIL-HDBK-244. It can be readily adapted to different types of aircraft by changing module input/output interfaces. An interface to a centralised WAC, MFD, HUD control/display system has been incorporated via discrete disturbance-free output lines. Compatibility with previous self-supporting systems is maximised as the new interface is provided through the central processor connector previously dedicated to the front cockpit monitoring unit and redundant in the new configuration.

The new interface lines provide not only all status information included in the substituted monitoring unit, including gun rounds, but also ripple interval status and certain control capabilities for remote gunnery selection (HOTAS interface). Power requirements are 2.5 A maximum for an illuminated panel system or 1.6 A maximum for a non-illuminated system, both at 28 V DC.

Specifications

Dimensions:
(central processor) 180 × 210 × 120 mm
(main unit) 200 × 70 × 101 mm
(monitor unit) 100 × 42 × 75 mm
(emergency unit) 70 × 40 × 81 mm
Weight:
(central processor) 3.3 kg
(main unit) 1.62 kg
(monitor unit) 0.46 kg
(emergency unit) 0.28 kg

Operational status

In service.

Contractor

Tecnobit SA.

VERIFIED

SWEDEN

AddVisor Head-Mounted Display

Specifications

Image source: 2 flat-screen panel displays
Field of view: 40° diagonal with 100% overlap
Video input: dual SXGA
Resolution: 1,280 × 1,024 full-colour pixels (SXGA)
Brightness: 200 cd/m^2
Focus: 0.7-25 m (factory adjustment)
Eye relief: eyeglass compatible
Tracker: compatible with Ascension, Polhemus, Interscene
Weight: < 0.65 kg

Operational status

Available.

Contractor

Ericsson Saab Avionics AB.

VERIFIED

The cockpit display for the JAS 39 Gripen comprises a wide-angle HUD and three head-down displays

1999/0051667

EP-12 display system for the JA 37 Viggen

The EP-12 integrated display system for the interceptor version of the Saab JA 37 Viggen collects, processes and displays all flight, navigation, radar and tactical data. It comprises five main units: head-up, radar and tactical displays, waveform generator and power supply. The three displays utilise CRTs with contrast enhancement techniques to provide clear symbology even in bright sunlight. The radar display supports all weather surveillance and interception, while the tactical display shows a track oriented moving map superimposed with navigation and tactical data.

Also associated with the EP-12 system is a cassette recorder to sample data at 4 Hz for the purpose of debriefing and training.

Operational status

In service in the Saab JA 37 Viggen. No longer in production. The tactical display is being updated in two squadrons of the JA 37 Viggen by a variant of the JAS 39 Gripen head-down colour display, as part of the Viggen Mod D upgrade programme (see MFID 68 display entry).

Contractor

Ericsson Saab Avionics AB.

VERIFIED

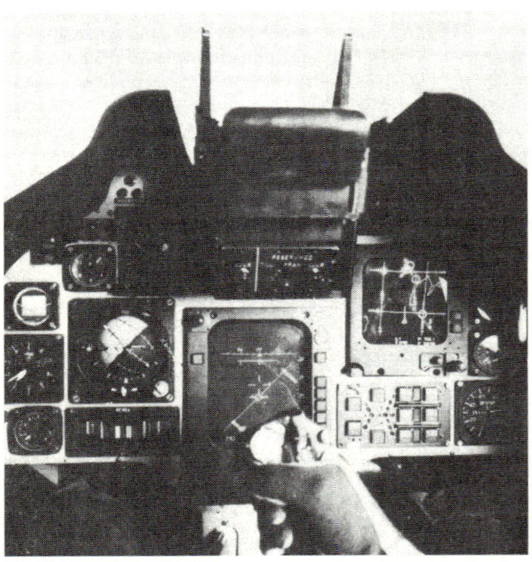

The Ericsson Saab Avionics radar display (centre) and tactical display (right) in a Saab JA 37 Viggen

0106379

EP-17 display system for the JAS 39 Gripen

The cockpit of the Saab JAS 39 Gripen features advanced electronic information presentation on four display units: three head-down displays and one head-up display. All four displays are computer-controlled, permitting the presentation to be tailored to every type of mission and flight mode. It also provides considerable flexibility and redundancy under emergency conditions.

The three head-down units are the Flight Data Display (FDD), the Horizontal Situation Display (HSD) and the MultiSensor Display (MSD). The FDD provides the pilot with flight, systems and weapons data as well as HUD back-up. The HSD employs a digital moving map showing geographical features and obstacles, such as radio towers and masts, with the tactical situation superimposed giving excellent situation awareness. Map scale and displayed information are selected manually and automatically to suit different phases of a mission. The MSD presents a computer processed radar picture in air-to-air and air-to-surface modes. The digital map can be overlaid on the radar picture. The flight data display and the multisensor display can also present imagery from other sources, such as IR or TV sensors.

The EP-17 has a display processor, divided into two functional chains each driving two displays and communicating with other aircraft systems by means of MIL-STD-1553B databusses. Each functional chain contains an Ericsson D-80 computer, the standard computing element in the aircraft. Specially developed high-performance graphics processors provide excellent dynamics to the graphics. The head-down displays employ raster-generated symbology, while stroke writing is used for the head-up display.

The lot 1 and 2 version of the Gripen aircraft employs three 6 × 4.7 in (152 × 120 mm) monochrome CRT multifunction head-down displays. The displays are dimensionally identical (the same part number). They use raster to present information and feature three selectable video systems: 525, 675 and 875 lines. Conventional mechanical standby instruments are used as back-ups for some of the more critical parameters. The head-up display uses diffraction optics and presents a large bright picture to the pilot. The field-of-view is 20 × 28°. The normal display mode is stroke, but raster generated imagery can also be presented.

The Swedish Air Force lot 3 and the export version of the Gripen aircraft employs three 6.2 × 8.3 in (158 × 211 mm) flat panel multifunction colour liquid crystal displays. The displays are dimensionally identical (the same part number). They feature high brightness, high-resolution (SVGA 600 × 800 colour pixels) liquid crystal glass. The displays are NVG compatible. The instrument panel has no space left for back-up instruments. Consequently, back-up functions are included in the multifunction displays. Each display interfaces redundant data sensors and can be supplied from the aircraft battery.

Display processing features full software control in computers and symbol generators, anti-aliasing, full colour, digital map storage and presentation in several scales and radar scan conversion. The lot 3 version will include an upgraded computer using the PowerPC chip. Even more functions will also be moved to loadable software to ensure affordable upgrades and future growth.

The multifunction colour display also comes in a smart version designated MFID 68. This display will be used in the rear seat to facilitate special operator display functions. This smart display is ideal for retrofit and integration in the JA 37 Viggen aircraft is also planned.

Recording capabilities included, with processing in the display processor, are multiplexed sensor video, MIL-STD-1553B bus data and audio using a Hi-8 mm videocassette recorder. The lot 3 version of the system features a solid-state digital mass memory (DiRECT) instead of the videotape. Several sensor systems sources, video information, bus data and audio will be digitally mixed on the mission recorder. Standard MPEG-2 coding is employed for the imagery. Both systems feature an onboard replay facility for quick evaluation.

Operational status

The Swedish Air Force has ordered 204 Gripen aircraft in three lots of which approximately 75 had been delivered by August 1999. In June 1997, Ericsson Saab Avionics was contracted for development and production of the lot 3 version featuring up-graded display processor, colour displays, head-up display and digital recording.

First delivery of upgraded aircraft is scheduled for 2001.

A variant of the EP-17 display, the MFID 68, is also being retrofitted into two squadrons of JA 37 Viggen aircraft, as part of the Viggen Mod D upgrade programme. Each JA 37 Viggen aircraft has only one multifunction head-down display, while the Gripen has three.

Contractor

Ericsson Saab Avionics AB.

VERIFIED

JAS 39 Gripen stores management unit

The stores management unit distributes and interfaces signals between the various weapons and tactical systems in the aircraft. It is based on a 486 computer module system. The unit is partly operated from the main mode selector, status panel and target acquisition panel. The unit also handles the preparation of stores, functions for firing and releasing stores in co-operation with the system computer, stores monitoring, control of the gun and control of countermeasures.

The stores management unit features modular design, LRU concept to facilitate maintenance, chassis integrated power supply for optimal heat dissipation, SMT for increased packing density and built-in redundancy.

Operational status

The MIL-STD-1553B variant of the JAS 39 Gripen stores management unit is now in production for Lot 1 and Lot 2 aircraft. The system is being updated to MIL-STD-1760 for Lot 3 aircraft, with pre-production units delivered.

Contractor

Ericsson Saab Avionics AB.

VERIFIED

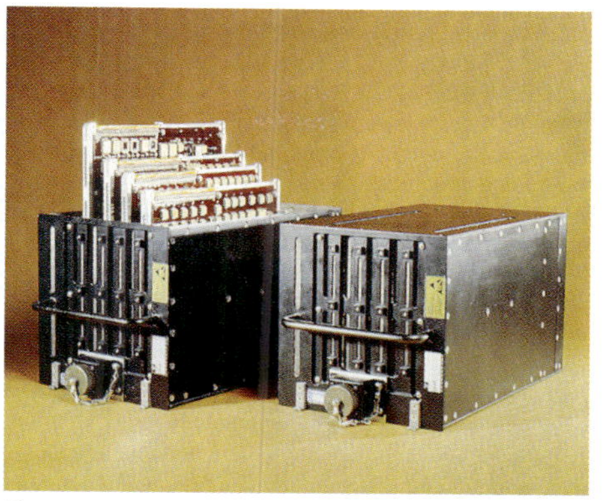

The stores management unit (left) and aircraft interface unit (right) for the JAS 39 Gripen 0106283

ARGUS 350

ARGUS 350 is a lightweight two-axis stabilised sensor platform, designed for border surveillance (coastguard and customs, drug interdiction); general law enforcement and police operations; search and rescue; electrical powerline and pipeline inspections; environmental inspection (oil spill, ecological surveys, wildlife); and fire management.

Options available include tracking systems and microwave downlink.

Specifications

Sensors:
(thermal imager) AGEMA Thermovision®1000
(daylight video) Sony EVI 311 single CCD colour
Performance:
(elevation) +10 to −120°
(azimuth) ±140°
(max air speed) 140 kt
(voltage) 28 V DC
(power consumption) 50-200 W
(weight including sensors) 23 kg

Contractor

Polytech AB.

VERIFIED

Airborne laser range-finder

The airborne laser range-finder is a very compact unit, which is easy to integrate into existing navigation and weapon delivery systems.

This high repetition-rate laser range-finder uses a simple, modular design consisting of transmitter, receiver, range counter and a deflection unit for the optical axis.

The laser is aimed at the target by slaving the deflection unit to the aircraft sighting system.

Specifications

Transmitter:
(laser type) Nd:YAG
(wavelength) 1.06 μm
(pulse energy) 20 mJ
(pulse length) approx 10 ns
(pulse repetition frequency) 1-10 Hz in bursts
(beam divergence) 0.7 mrad
Receiver:
(detector) Silicon avalanche diode
(field of view) 0.5 mrad
Range Counter:
(range, max) 20,000 m
(range, min) 200 m
(range resolution) 5 m
Range:
(typical range) ≥10 km at optical visibility >20 km
Deflection Unit:
(azimuth travel) ±10°
(elevation travel) ±10°
(accuracy) <1 mrad
(slow rate) >60°/s
Interface:
(digital bus) ARINC 429
Weight:
(total weight) approx 14 kg
Dimensions: approx 500 × 160 × 160 mm
Power:
(power supply) 28 V DC
(power consumption) 225 W

Contractor

Saab Dynamics AB.

VERIFIED

IR-OTIS Infra-red Optronic Tracking and Identification System

Saab Dynamics is currently working on design and test of a prototype of IR-OTIS, a multifunctional Infra-Red Search and Track (IRST) system, intended to provide passive situation awareness for the JAS 39 Gripen

Saab Dynamics IR-OTIS 0106415

The Saab Dynamics airborne laser range-finder **2000**/0106284

aircraft at long range, during day and night operations, against air and ground targets. The system can operate both as an IRST and a FLIR with a large scanning field. In the IRST mode, the system scans an assigned part of the airspace, detects targets automatically and tracks them while continuing to scan, thus giving the system a multifunction track-while-scan capability. In this mode, the system can scan with a narrow field of view to give long-range detection or with a wide field of view to cover a large sector in a short time. In the FLIR mode, the system generates an image covering the field of view in the assigned direction.

In air-to-air combat, the IRST mode is used for automatic search, including track-while-scan, or FLIR mode in a designated direction with automatic lock-on and continuous tracking.

In air-to-ground combat, the FLIR mode is used for observation and continuous tracking.

Specifications

Sensor wavelength: 8-12 μm
Sensor elements: 1,100-1,200 elements
Field of view: several, selectable
Field of regard: more than one hemisphere, but limited by aircraft installation
Sensor unit: 30 kg and 30 litres volume
Signal processing unit: 10 kg and 10 litres volume

Operational status

In development for the Swedish Defence Material Administration for JAS 39 Gripen; the first development model was flight tested on a JAS37 Viggen aircraft.

Since the sensor system is designed to be fitted on the Gripen and requires relatively little space, integration in other aircraft should be possible either in a new build arrangement or as retrofit equipment.

Contractor

Saab Dynamics AB.

UPDATED

SEOS 200 helicopter observation system

The Saab SEOS, stabilised electro-optical multisensor system, is designed for a variety of helicopter applications.

The system is designed to give high probability of detection and it is stabilised to match weapon targeting requirements.

SEOS consists of a stabilised sensor platform, a conditioning unit for climate control, a processing unit for image enhancement and fusion of sensor signals, and a control unit with a high-resolution head-in display.

Mounted on the stabilised platform, in the basic system, are an 8 to 12 μm second-generation thermal imager, laser range-finder and TV cameras. Optional sensors/designators include a laser designator, missile tracker/beam-rider unit, low-light-level TV camera, 3 to 5 μm thermal imager.

Depending on sensor choice, SEOS can be integrated with various weapon interfaces, including optically controlled missiles, laser-guided munitions, turreted or fixed guns, rockets and artillery fire-control systems. The sensor head can be either mast or pedestal mounted.

The image processing system features aided target tracking and detection. Image freeze/store, area tracking, thermal cueing, electronic magnification, image integration, graphics generation and overlay are also included.

Specifications

Weight: 70 kg
Dimensions: 0.5 (height) × 0.6 m (diameter)
Azimuth coverage: ±200° (optional 360°)
Elevation coverage: –30 to +85°
Fields of view (horizontal):
(narrow) 1.5°
(medium) 4.5°
(wide) 18.0°
LOS jitter: <15 mrad rms
Interface: MIL-STD-1553B, RS-422 or others

Operational status

Under development.

Contractor

Saab Dynamics AB.

VERIFIED

TURKEY

ASELFLIR-200 second-generation gyrostabilised airborne FLIR

The Aselsan ASELFLIR-200 forward-looking infra-red (FLIR) system is a light-weight, multipurpose, thermal imaging sensor for pilotage/navigation, surveillance, search and rescue, automatic tracking, target classification and targeting. The ASELFLIR-200 is an open architecture and hardware/software flexible unit which can be adapted to various air platforms, including rotary-wing, fixed-wing and unmanned air vehicles.

Key features of ASELFLIR-200 include Electronic Image Stabilisation (EIS), Local Area Processing (LAP) for image enhancement, MultiMode Tracking (MMT), analogue and digital video outputs for transmission and/or recording, MIL-STD 1553/ARINC and other discrete databusses to interface with onboard avionics, such as radar, navigation and weapon systems. The ASELFLIR-200 has three fields of view: Narrow Field of View (NFoV) for recognition and identification, Medium Field of View (MFoV) for detection and a unity Field of View (FoV) for navigation and pilotage.

The system has single, dual and triple-sensor configurations as following:

- FLIR only
- FLIR+colour CCD or FLIR + eye-safe Laser Range-Finder (LRF)
- FLIR+colour CCD + eye-safe LRF

ASELFLIR-200 is in full production and installed on various rotary-wing and fixed-wing platforms. It incorporates a second-generation 4 × 240 Focal-Plane Array (FPA) detector that operates in the 8 to 12μm band. The key features of the system provide greatly improved range performance over conventional first-generation linear array detectors and improve mission capability.

There are two weapon replaceable assemblies: a turret unit, WRA-1 and electronics unit, WRA-2. Options include a laser range-finder and/or a CCD day TV camera.

Specifications

Field of View (FoV):
(wide) 22.5 × 30°
(medium) 5 × 6.67°
(narrow) 1.3 × 1.7°
Parallel detector channels: 240 × 4 FPA
Spectral Band: 8 to 12 μm
Electronic zoom: 2:1 and 4:1
Gimbal angular coverage:
(azimuth) 360° continuous
(elevation) 40° up; 105° down

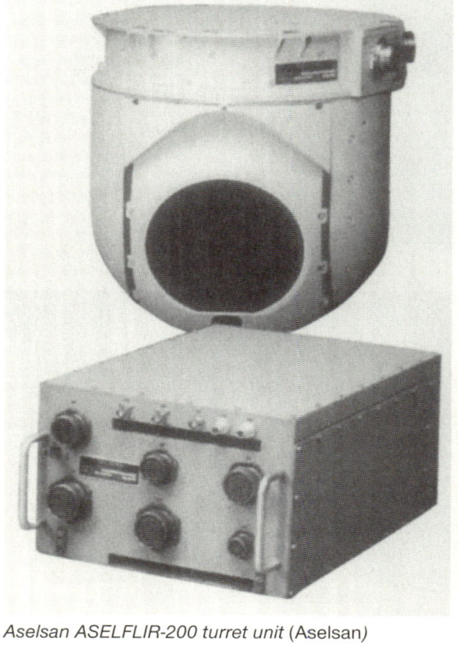

Aselsan ASELFLIR-200 turret unit (Aselsan)
2001/0018331

Aselsan M930 aviators' Night Vision Goggles, note offset mount for Cobra attack helicopter application (Aselsan)
2001/0103875

Gimbal acceleration: head steering compatible at aircraft speeds
Laser range-finder: optional
Day TV: optional
Environmental: MIL-STD-E-5400
Video outputs: analogue and digital video outputs are provided
Cooling: self-contained
Weight:
(turret unit) <31.8 kg
(electronics unit) <22.73 kg
Dimensions:
(turret unit) 323.85 (diameter) × 372.87 mm (height)
(electronics unit) 306.3 (width) × 413.5 (length) × 199.1 mm (height)
Power: standard aircraft power

Operational status
In production, and installed on various rotary- and fixed-wing platforms.

Contractor
Aselsan Inc, Microelectronics, Guidance and Electro-Optics Division.

UPDATED

MFD-268E MultiFunction Display

The ASELSAN MFD-268E MultiFunction Display is a colour flat panel display using Active Matrix Liquid Crystal Display (AMLCD) technology, with an active display area of 6 × 8 in. The display contains internal graphics, video and input/output processing capabilities to support the generation of display formats, using data obtained from system interfaces to the MFD. The MFD-268E is capable of displaying video, graphics, video with graphics overlay, split screen video/graphics and split screen graphics/graphics.

The ASELSAN MFD-268E has analogue and digital input/output interfaces and a MIL-STD-1553 interface. The -1553 interface allows operational software modification without removal of the display from the aircraft. The MFD is bus programmable in the Ada computing language for customer display format generation.

Specifications
Display type: Colour AMLCD
Display area: 152 × 203 mm (6 × 8 in)
Resolution: 1,024 × 768
NVIS compatibility: MIL-L-8562A, Class A and B
Dimensions: 208 × 259 × 254 mm (8.2 × 10.2 × 10 in)
Video input: digital, monochrome and colour
Interface: dual MIL-STD-1553B databus, RS-232, ARINC 429, analogue, synchro and discrete
Weight: 9.3 kg (20.5 lb)
Power: 22.5 to 32 V DC operational in accordance with MIL-STD-704A

Contractor
Aselsan Inc, Microelectronics, Guidance and Electro-Optics Division.

NEW ENTRY

M929/M930 aviator's Night Vision Goggles (NVG)

The M929 and M930 Aviator's Goggles offer substantial improvements over earlier night vision systems. They can be fitted with third-generation Omnibus-III or Omnibus-IV image intensifier tubes. Along with increased sensitivity, the ANVIS goggle offers improved flash response over earlier models. The M929 and M930 are identical except for their respective mounts - the M929 features a standard mount assembly, while the M930 features an offset mount assembly, for application in the Cobra attack helicopter.

Specifications
FoV: 40°, circular
System gain:
2,500 (Omnibus III, Gen III)
5,000 (Omnibus IV, Gen III+)
Resolution (on axis):
1.01 lp/mR (Omnibus III, Gen III)
1.28 lp/mR (Omnibus IV, Gen III+)
F-number: F/1.2
Weight: 590 g

Operational status
In production.

Contractor
Aselsan Inc, Microelectronics, Guidance and Electro-Optics Division.

NEW ENTRY

UKRAINE

SURA Helmet-Mounted Target Designation System (HMTDS)

The SURA Helmet-Mounted Target Designation System (HMTDS) is designed to be used either independently or in combination with other aircraft systems, concerned with the operation and aiming of guided weapon and gun systems.

The HMTDS generates target designation signals for weapons in proportion to the angles of turn of an operator's (pilot's) head, as well as collimating the image of an aiming mark and initiating one-time commands to his field of view.

The HMTDS helmet position sighting system is implemented using small IR emitting diodes mounted on the pilot's helmet, together with two scanning units (designated scanning unit A and scanning unit B) mounted one each side of the Head-Up Display (HUD). This concept facilitates use of the system in a variety of aircraft with minimum requirement for aircraft modification. It also meets safety requirements for pilot ejection and emergency evacuation, and makes it possible to install the system on three standard sizes of pilot's helmet. Units of the HMTDS comprise:

(1) the helmet-mounted sighting device, attached to the pilot's helmet, using a special bracket
(2) scanning units A and B, mounted each side of the HUD
(3) an electronics unit that processes sensor information, and interfaces it to other aircraft weapon and communications systems.

SURA scanning unit installation, either side of the HUD 0022252

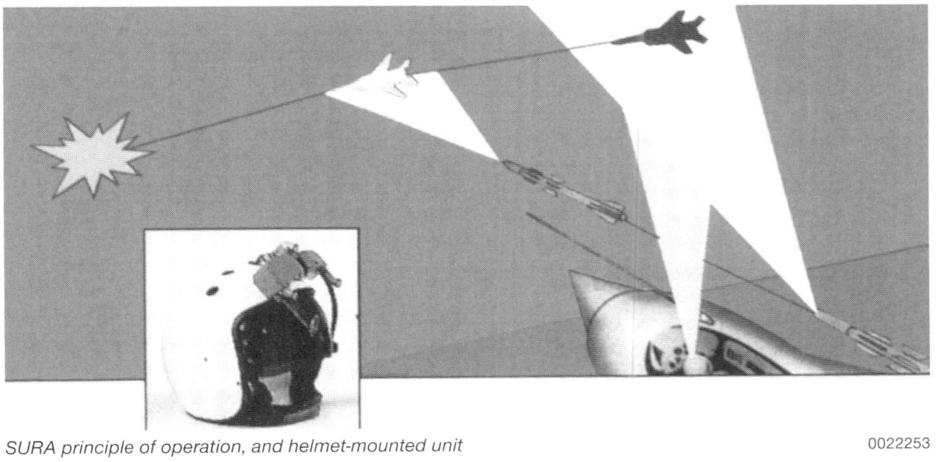

SURA principle of operation, and helmet-mounted unit 0022253

Specifications
Target designation angles:
(horizontal) +70 to –70°
(vertical) +60 to –25°
Target designation accuracy (1σ): <3 mrad
Sensor data format: asynchronous binary numerical code
Power: 115 V AC, 400 Hz, <150 VA
Weights:
(helmet-mounted sighting device) 0.36 kg
(scanning units) 0.8 kg
(electronics unit) 7.0 kg
Temperature limits: –54 to +60°C
Warm up time of HMTDS: <1 min
Continuous operation: limited to 3 h, followed by 25 min cycle time

Operational status
SURA HMTDS has been developed as a successor to the SHCH-3UM helmet-mounted target designation

systems used on MiG-29 and Su-27 aircraft, based on 15 years' design experience and the production of over 500 HMTDS.

Contractor

Arsenal Central Design Office.

VERIFIED

SKI-77 Head-Up-Display (HUD)

The SKI-77 HUD is designed for use by military fighters, ground-attack, transport aircraft and helicopters. It provides a collimated display of sensor data, together with appropriate computing capacity. It includes automatic and manual control of image brightness to match background illumination conditions, together with automatic changes in data formatting to match the different stages of the mission profile.

Specifications

Instantaneous field of view: >20° (V) × 30° (H)
Exit pupil size: 45 × 90 mm
Data display colour: green monochromatic
Weight: <31.5 kg
Power: 115/200 V AC, 400 Hz, <300 W 27 V DC, <5 W 6 V AC, 400 Hz, < 5 VA
Temperature limits: – 60 to + 60° C

SURA electronic unit and scanning units 0022251

Operational status

In development for the An-70 tactical transport aircraft.

Contractor

Arsenal Central Design Office.

VERIFIED

UNITED KINGDOM

RGS Series weapon aiming systems

The RGS Series is a comprehensive range of airborne weapon aiming systems, from the simplest of aiming devices to the more sophisticated type of lead computing HUDs. The compact size of the RGS Series allows easy installation in the smallest of cockpits, making it suitable for small turboprop and jet trainers. When cockpit space is particularly restricted, low-profile versions of the sight heads can be provided. All variants of the Series utilise the same sight head/cockpit interface.

Although the primary role of the RGS Series is for use in trainer aircraft, the more sophisticated derivatives can be utilised as alternatives to full HUDs. The Series includes the RGS1 universal manually operated fixed-type sight, RGS2 basic lead computing rate gyro sighting system, RGS2A compact gyro gunsight and weapon delivery computer for enhanced ground attack capability and RGS2R laser-augmented system based on a microprocessor for high-speed low-level attacks.

The RGS1 is a fixed optical sight head which provides a basic weapons delivery system. Incorporated in the sight head are a combining glass, fixed-plane mirror, movable spherical mirror and injected reticle system. Utilising a 100 mm wide aperture, the RGS1 provides the pilot with binocular viewing, enhancing efficiency even in the most turbulent conditions when otherwise unavoidable head movements would cause loss of the aiming mark. For ease of viewing, the RGS1 displays an aiming mark which is projected at infinity. Compensations for ordnance characteristics, mode of attack and crosswind are made by deflecting the aiming mark in azimuth and elevation. These deflections are made manually using control knobs on either side of the sight head and are displayed, in mils, on indicator scales facing the pilot. A brightness control allows the aiming mark illumination to be varied according to prevailing light conditions.

The RGS2 is a lead computing optical gunsight which comprises an optical sight head, computer and

The twin RGS2 sight head installation in a Saab 105 side-by-side two-seat trainer 0106380

gyro unit and a control unit. The sight head comprises a combining glass, fixed-plane mirror and movable spherical mirror which is suspended in gimbals, controlled by torque motors and displacement transducers. An injected reticle system, with two reticle lamps, is also included. The mirror deflection is very accurately determined by the controlling signals and does not suffer from the temperature dependence and long-term drift encountered in the frequently used open-loop system. The computer and gyro unit houses the system electronic circuitry and a gyro pack. The electronics comprise circuits for weapon aiming calculations and the system power supply. The circuits are built of electronic components assembled on plug-in printed circuit cards. The gyro pack consists of three miniature high-precision rate gyros measuring aircraft pitch, yaw and roll rates. Output signals from these gyros control the deflections of the aiming mark in the sight head. The control unit operates the system and incorporates a built-in check function. A facility to set the aiming mark manually is also provided by this unit.

The RGS2A has all the features of the RGS2 but utilises a more powerful weapons aiming computer. This computer provides improved delivery of bombs and rockets by incorporating a CCIP function.

The RGS2R has all the features of the RGS2A but with the integration of a separate laser range-finder.

The RGS2 Series can be used in a twin-sight head arrangement for training purposes in which the instructor's and trainee's aiming marks are deflected equally and in exact synchronisation. For light attack aircraft, the same principle can be used for slaving a laser range-finder or an IR missile or radar homing device to the optical sightline of the sight head.

Specifications

Dimensions:
(control unit) 38 × 146 × 84 mm
(computer and gyro unit) 114 × 175 × 275 mm
Weight:
(RGS1 sight head) 2.5 kg
(RGS2 sight head) 3.1 kg
(control unit) 0.8 kg
(computer and gyro unit) 4.5 kg
Power supply:
(RGS1) 28 V DC, 30 W (max)
(RGS2) 115 V AC, 400 Hz, single phase, 70 VA (max)
28 V DC, 60 W (max)

Operational status

The RGS Series has been in operational service since 1982. RGS systems have been fitted to the BAE Systems Hawk, CASA Aviojet C-101, Aermacchi MB-326 and MB-339, Agusta (SIAI-Marchetti) S.211, Northrop Grumman F-5, Fiat G-91R3, Saab 105, MB Vectro 2, Dassault Super Mystère, Pilatus PC-7 and PC-9, Shorts Tucano, Embraer Tucano, IA 63 Pampa, Pucara and A-4 Skyhawk.

Contractor

Avimo Ltd.

VERIFIED

4500 Series Head-Up Displays (HUDs)

The 4500 Series head-up display system is primarily designed for light attack aircraft and trainers. Designed from the outset to be a dual-mode cursive and raster display, the HUD is modular allowing the mechanical outline to be optimised to fit most aircraft installations.

The display gives a 24° total field of view with a large instantaneous field of view available from the 4.5 in (114 mm) exit lens. P53 phosphor is used on the CRT, giving a very bright display. The upfront control panel (UFCP) gives complete control over the rest of the system and a colour TV camera fitted to record the pilot's view.

The interface unit or Head-up display Electronics Unit (HEU) is a ¾ ATR box containing multiple analogue and discrete synchro interfaces, 1553B R/T or bus control, 68020 or higher-performance processors and symbol generators and graphics processors. Full weapon aiming, mission computations and head-up and head-down display symbol generation are available.

The Type 4510 has both cursive and raster displays, the latter being selected by a switch on the UFCP. The 4500 and 4510 are physically identical and were based on the optical and symbol generation technology previously used on the COMED system. The extensive production runs of COMED had removed any design problems, while the weapon aiming, interfacing and air data techniques were derived from a series of complete weapon systems designed and built by BAE Systems.

The 4510 head-up display provides a full suite of navigation symbology with steering and location cues available at all times and generates automatic or selected weapon aiming symbology for all known air-to-air and air-to-ground weapons. The control panel allows the pilot to set up the navigation system, select and display modes during flights and then change to a raster display when mission sensors, such as FLIR, necessitate.

Since the introduction of the 4500 Series HUD, systems have undergone rigorous flight trials on the Buccaneer and Harrier Nightbird aircraft at the Defence Evaluation Research Agency, Farnborough. A UK Royal Air Force Jaguar was fitted out with a BAE Systems system and flew early in 1989. The success of these trials was due in part to the flexibility of the highly programmable software in the HEU and the performance of the HUD in both cursive and raster modes.

The HUD is an integral part of various total system options at present being considered in the worldwide retrofit market. The unit can, because of its compact size, meet the installation requirements of many types of attack aircraft.

Specifications

Weight:
(HUD) 12 kg
(HEU) 14 kg
Power supply: 115 V AC, 400 Hz

Operational status

In production and in service for UK Royal Air Force Jaguar aircraft and for the A-4, C-101, F-5, Mirage III and Mirage 5 aircraft retrofits.

Contractor

BAE Systems.

VERIFIED

All Light Level TV (ALLTV) system

The ALLTV system consists of low-light level television cameras and laser systems mounted on a steerable stabilised platform based on an existing family of platforms in production for civil and military applications. The ALLTV system enables an operator to detect, identify and track targets in daylight and at night, to direct aircraft guns automatically on to a target.

Operational status

The ALLTV system was selected for installation in US Air Force AC-130U Gunship aircraft.

Contractor

BAE Systems.

VERIFIED

AN/AVQ-29 Head-Up Display (HUD) for the A-7D and A-7K

The AN/AVQ-29 HUD was developed from the HUD which equips the F-16C/D in order to give the A-7 a full night attack capability when used in conjunction with an electro-optical sensor.

Operational status

Production complete.

Contractor

BAE Systems.

VERIFIED

ATLANTIC podded FLIR system

The Airborne Targeting Low-Altitude Navigation Thermal Imaging and Cueing (ATLANTIC) pod-mounted FLIR system is designed to give ground attack aircraft night and poor weather capability on high-speed low-level missions.

The BAE Systems ATLANTIC FLIR installed under the port wing of a UK Royal Air Force Jaguar T Mk 2A

The Atlantic pod employs the BAE Systems modular FLIR system as used in the Harrier GR Mk 7 and it can also include an advanced thermal cueing system for early target detection and a laser spot tracker. The system, with multitarget compatibility, has a MIL-STD-1553B databus interface and can be integrated with existing weapons and avionics systems.

The telescope is selected to match the FLIR image to the aircraft HUD field of view. Pod cooling is by means of a self-contained environmental conditioning unit and the detector is cooled by a closed-cycle cooling engine. Time to readiness for the system is 3½ minutes typically from 20°C.

Specifications

Dimensions: 2,413 × 254 mm diameter
Weight: 100 kg nominal
Power supply: 28 V DC nominal, 210 W typical
115/200 V AC, 400 Hz, 146 W typical
Max speed: Mach 1.2 at 40,000 ft
Temperature range: –40 to +70°C
Video output: 525-line, 60 Hz or 625-line, 50 Hz

Operational status

In production for UK Royal Air Force Jaguar T Mk 2A aircraft. On order for the Royal Netherlands Air Force F-16 MLU aircraft; with 60 pods expected to be delivered between 1999 and 2001.

Contractor

BAE Systems.

VERIFIED

Cats Eyes Night Vision Goggles (NVGs)

BAE Systems Cats Eyes NVGs system allows for the combination of both a direct visual and an intensified image to be presented to the pilot's eyes. The two images are combined in a 1:1 relationship and complement each other. The benefits of the system have been extensively proven in low-level night attack

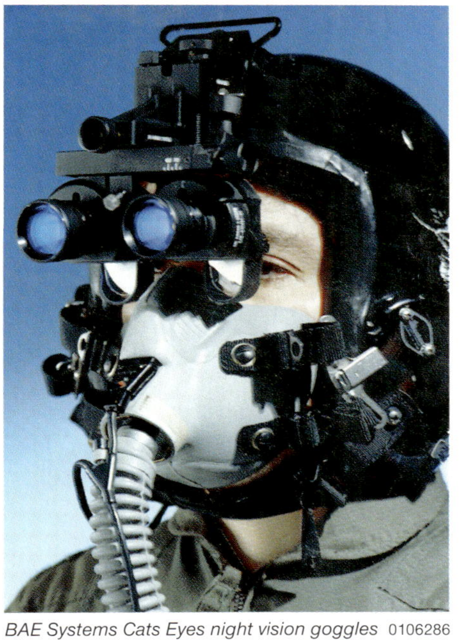

BAE Systems Cats Eyes night vision goggles 0106286

flying trials, which used a fully integrated NVG-compatible cockpit and Forward Looking Infra-Red (FLIR) generated head-up display imagery, together with a head-down multifunction display. They also performed well in combat during the 1990-91 Gulf War.

The head-up display is seen through a direct visual path, and it is not degraded by unnecessary image intensification as it would be with conventional NVG systems. Additionally, the direct vision path through the optical combiner arrangement makes monitoring of cockpit displays and instruments considerably easier while the ability to scan either side of the combiners enhances peripheral vision and ensures better spatial awareness. The direct vision path also removes problems normally associated with light to dark transitions as the intensified image becomes progressively more noticeable as the direct visual image fades. The system is compact and rugged and the restrictions on head mobility imposed by the depth of conventional NVG systems is avoided. While the system incorporates a single handed quick release mechanism for the helmet interface, it can be configured to include an automatic separation system on ejection and designed growth will enable it to accept the latest image intensifier technology as it becomes available.

Specifications

Weight: (including mount and helmet bracket) 0.82 kg
Field of view: 30°
Magnification: Unity
Eye Relief: 1.0 in (25 mm)
Exit Pupil: 0.4 in (10 mm)

Operational status

In production. Selected as the standard NVG system for the US Navy and Marine Corps fixed-wing tactical aircraft. Over 800 systems are now in service with the US Navy. They are fully qualified to aircraft carrier EMC and environmental requirements.

Contractor

BAE Systems.

VERIFIED

CGI-3 compass gyro indicator

The CGI-3 is a panel-mounted instrument containing all the elements, apart from the flux valve and a small annunciator, of a gyromagnetic compass and an RMI. Although it is ideal for installation in helicopters and executive aircraft, the CGI-3 has been fully type tested and cleared for operation in a severe military environment.

The gyro uses a wheel and gimbal assembly and the readout of heading is by means of a rotating dial with 5° graduations which moves in the conventional sense with changes in aircraft heading.

The transistorised slaving amplifier contained within the instrument case provides the necessary energising voltage for a flux valve magnetic detector unit and then amplifies and discriminates the signals received from this unit in order to maintain the indicated heading in accordance with the direction of the earth's magnetic field.

Specifications

Dimensions: 76 × 76 mm
Weight: 2.5 kg

BAE Systems GM 9 gyro magnetic compass system

Operational status
In service.

Contractor
BAE Systems.

VERIFIED

CNI/control and display unit

Developed for Hawk 100/200 aircraft, the CNI/control and display unit provides the common point for the control and data information display for the various radio, navigation and IFF systems. The unit has the primary role of controlling and monitoring the radio systems via a MIL-STD-1553B databus. It has a secondary role, upon request, to perform a reversionary bus controller function without affecting its primary role. The unit can, however, be reconfigured for other applications through software changes via an RS-424 serial datalink.

Operational status
In production.

Contractor
BAE Systems.

VERIFIED

The CNI/control and display unit has been developed for the Hawk 100/200 0106287

Combined Map and Electronic Display (COMED)

The Combined Map and Electronic Display (COMED) system map is electronically annotated with intelligence or navigation data appropriate to the particular mission. COMED provides a colour topographical map display annotated with dynamic navigation information as required. This can include aircraft track and commanded track, present position, locations of known hostile detection or anti-aircraft devices and tactical information such as the delineation of forward edge of the battle area. In addition, the system can print out alphanumeric information such as time to go to fix point or target.

COMED's CRT and projection facilities permit it to perform other operational tasks. These can include the display of high-resolution high-contrast symbology in raster form from radar, low-light television or FLIR sensors. Electronic countermeasure threats can also be shown. Tabular displays of weapons status, destination co-ordinates, or other tactical information can be shown on tabular 'forms' projected from images stored on the film. The system can be programmed with a library of aircraft and engine checklists. COMED can also act as a back-up primary flight information display, particularly for the horizontal situation and attitude director indicators.

COMED interfaces with the main aircraft navigation computer via a MIL-STD-1553 serial digital datalink. From a knowledge of film strip layout, aircraft present position and demanded scale, the main computer calculates the appropriate map drive words and transmits them to COMED by the datalink. Within the display the information is converted into a form suitable for driving the map servo resolvers.

COMED uses standard 35 mm colour film up to 57 ft (17 m) long. Typically, this provides coverage of an area of 10,000 sq n miles at a scale of 1:250,000, plus selected target areas at 1:50,000, and sufficient film frames for tabulated displays. Film replacement can be achieved through a side access panel in under a minute.

COMED units form the Projected Map Assembly (PMA) element of the AlliedSignal horizontal situation indicator, a combined map and electronic display for that aircraft. It projects a coloured topographical moving map image superimposed with data from a CRT image.

Specifications
Dimensions:
(CRT face) 139.7 × 139.7 mm
(unit) 180.3 × 217.9 × 604 mm

Operational status
No longer in production. The first production COMED system was delivered during early 1983. In May 1986 CEDAM (the latest variant of COMED) was selected for the Tornado ECR version ordered by the German Air Force.

More than 1,000 systems have been supplied for the F/A-18 Hornets with the US Navy and the armed forces of Australia, Canada and Spain.

The system has also been installed by the Indian Air Force in its Sepecat Jaguars made under licence by Hindustan Aeronautics.

Contractor
BAE Systems.

VERIFIED

DC electromagnetic helmet tracking system

The BAE Systems electromagnetic helmet tracking system has been designed for both fixed- and rotary-wing applications and determines the helmet position and orientation by measuring magnetic fields which it generates in the cockpit. This data can be used to position aircraft sensors and weapon seekers and for referencing Helmet Mounted Display (HMD) systems.

The system has been developed to overcome many of the disadvantages of competing electromagnetic tracker systems while maintaining simplicity of installation. A number of prototype systems have been delivered to US and European customers as part of fully integrated HMD systems.

A small transmitter containing three orthogonal coils is rigidly mounted within the cockpit. The coils are pulsed with currents, which generate known magnetic fields. A small magnetic sensor, again containing three orthogonal coils, is rigidly mounted to the helmet. These coils sense the direction and magnitude of the generated magnetic fields and, using this data, a processor calculates the helmet position and orientation. The pulsed nature of the transmitted signal leads to this type of system being termed a 'DC Electromagnetic Tracker' to differentiate it from the older 'AC electromagnetic' systems.

The key functional elements are the transmitter, the sensor and an electronics module, which is normally installed within the HMD system electronics unit.

This system provides full spherical coverage within the aircraft cockpit offering high accuracy and good dynamic performance. It does not affect and is not affected by other cockpit functions and is significantly less susceptible to cockpit metal than traditional systems, requiring a single generic mapping for all aircraft of a particular type.

Operational status
This system is in production.

Contractor
BAE Systems.

VERIFIED

CEDAM, the latest model COMED for the German Air Force ECR Tornado 0106381

DGI 3 - directional gyro indicator

DGI 3 is a panel-mounted directional gyroscope. Normally used as a standby heading unit, it contains a directional gyro geared to a rotating compass card, which provides a heading readout. The compass card is graduated in 5° divisions and is read against a white lubber line at the top of the unit. A yellow 'Set Heading' marker is also provided. Drift rate is a nominal 10°/h, after making allowance for the effect of the Earth's rotation.

Specifications

Dimensions: 83 mm square case, 190 mm length
Weight: 2.5 kg
Power supply: 28 V DC plus 5 V AC/DC for lighting

Operational status

Unit in production. Applications include the Hawk aircraft.

Contractor

BAE Systems.

VERIFIED

Digital Map Generator (DMG)

The BAE Systems Digital Map Generator (DMG) has been developed from COMED/CEDAM experience, to meet the mission requirements of the next generation of combat aircraft. Housed in a 3/4 ATR box, the DMG is suitable not only for these new aircraft but also for a large number of retrofit applications.

The DMG hardware and software can handle both true digital maps and digitised chart information. The digital database can be held in solid-state or in optical disk. Both of these options have been developed. Features on the presentation can be selectively displayed or erased. Hazards such as terrain and obstacles above aircraft altitude can be made to stand out in contrasting colour. Safe areas occasioned by terrain-masking can be depicted and areas may be viewed from different angles and altitudes. Data updating is rapid and simple. The map is displayed on a colour electronic display, which incorporates a multifunction keyboard and electronically generated symbology.

The map can be displayed either north up or track up, with the aircraft present position centred or de-centred on the display. Multiple map scales can be accommodated and zoom and declutter facilities are available. Map stabilised overlays for routes and navigational information can be displayed.

Operational status

In production for the UK Royal Air Force Jaguar GR. Mk 1B, Tornado GR. Mk 4 and Lockheed Martin C-130J aircraft.

Contractor

BAE Systems.

VERIFIED

Display processors/graphics generators

The advent of glass cockpits demands that the display symbol generators have a capability which matches display performance and operational requirements. BAE Systems has adopted a modular approach to graphics generation and can supply boxes which drive either a single display surface in monochrome or a full display suite of multifunction colour displays and a dual-mode head-up display.

A variety of interfaces is offered which can be modified, as necessary, to integrate with existing or new aircraft systems. As well as dedicated display generation, the boxes can have weapon aiming and mission computer functions added easily.

Operational status

In production for the UK Royal Air Force Tornado GR. Mk 4.

Contractor

BAE Systems.

VERIFIED

Engine monitor panel

The engine monitor panel for the Hawk 100 and 200 displays all necessary fuel and engine information as well as indications of any associated malfunction. The display technology combines the latest LCD technology and LED arrays to provide a flat panel which can be mounted in any cockpit position and still maintain high readability over all ambient light conditions.

The display technology is adaptable to cockpit environments where integration of analogue instruments is required to allow more use of cockpit space.

The panel incorporates a fuel 'bingo' facility which allows the pilot to set minimum fuel levels at which an audio warning will be generated. An option to this panel is the incorporation of engine low-cycle fatigue recording for post-flight evaluation.

Operational status

In production.

Contractor

BAE Systems.

VERIFIED

The engine monitor panel for the BAE Systems Hawk 100 and 200

Head-down displays

BAE Systems has developed a range of multisensor, multifunction television tabulator, E-scope and other displays. The systems comprise a CRT driven by a waveform generator embodying raster scan techniques. The waveform generator produces synthetic symbols, by what the company calls a time-shared digital technique, to give high accuracy and resolution of modulated video signals with a minimum of components; a notable feature is the elimination of the staircase effect in raster graphics. The synthetic video is directly compatible with standard television signals from a variety of sensors, so that video signals can be mixed to produce an overlay and symbology on a pictorial display.

The waveform generator is completely digital in operation and receives data in standard serial digital form. By storing this information it can synthesise continuous video signals to drive one or more displays. The signal range can be modified or extended to meet individual requirements, particularly those suited to raster applications. The display unit provides a high-contrast television picture of sensor and computed data and has a multifunction keyboard permitting the operator to communicate with the aircraft computer. A filter on the face of the CRT improves the contrast under high-ambient illumination and photosensors adjust the brightness of the display in accordance with the ambient lighting.

Operational status

No longer in production. In addition to the Tornado and Nimrod MR 2, other applications include the gunship version of the Lockheed Martin C-130 Hercules.

Contractor

BAE Systems.

UPDATED

Head-Up Display and Weapon Aiming Computer (HUDWAC) for the F-5

The HUDWAC is an avionics upgrade, for the Northrop-Grumman F-5E that allows the performance and manoeuvrability of the F-5 to be combined with the advantages of head-up flight.

The F-5 HUD utilises technology and hardware from other BAE Systems HUD programmes and performs air-to-air and air-to-surface weapon aiming calculations as well as delivering symbology for flight and navigation modes. The HUD comprises a Pilot Display Unit (PDU), an Electronics Unit (EU) and a Weapon Data Input Panel (WDIP).

The PDU fits in an existing mounting tray. The PDU's 25° total and 15.75° vertical by 16.9° azimuth instantaneous fields of view are achieved without impinging on the F-5's ejection plane. A 16 mm cinematic camera is included in the PDU, but a TV camera is available as an option. The PDU control panel includes switches for controlling symbol brightness, standby sight selection and symbol declutter. Data entry is via a keypad on the same control panel.

The EU will fit in all existing configurations of the F-5E in the same location as the LCOS gyro lead computer which is removed for HUDWAC installation. It contains a power supply, integral cooling fans and the circuit cards for processing, symbol generation and interfacing with the F-5's avionics systems.

The HUDWAC in its baseline configuration interfaces with the F-5's power supply, CADC, AHRS, fire-control system and AN/APQ-153, 157 or 159 radar. The system has flown with the LN-93, with the H423 laser inertial navigation system and also with the LN-39. Optionally, the HUDWAC will interface with the AN/APG-66, 67 or other pulse Doppler radar.

The system software provides weapon aiming calculations for weapons certified for F-5 carriage and is written to display symbology which complies with MIL-STD-1787. The system has the capability to expand and handle the AIM-9P-4 missile line of sight and off-boresight aiming.

The Marconi Electronic Systems displays which equip the Panavia Tornado. Left to right, the pilot's electronic head-down display (GR4/4A), the E-Scope and the Navigator's TV tab display unit

The system operates in three modes; navigation, surface attack and air combat.

In navigation and landing submode the display symbology shows speed, altitude, heading, velocity vector, pitch, bank, Mach number and *g*. Inertial navigation information, when available, is also displayed.

Surface attack modes include Continuously Computed Impact Point (CCIP) and Continuously Computed Release Point (CCRP) displays.

There are three air-to-air modes: missiles; Lead Computing Optical Sight (LCOS) for tracking 20 mm cannon attacks and dogfight, which combines missiles and guns; LCOS and snap-shoot or Continuously Computed Impact Line (CCIL) on the same display.

The system has built-in test facilities to isolate faults to a particular LRU. Predicted MTBF for the HUDWAC is greater than 1,400 hours. The HUDWAC's support equipment includes a semi-automatic system test set which uses a 16-bit microprocessor.

Operational status

In production.

Contractor

BAE Systems.

VERIFIED

Head-Up Display (HUD) for the C-17

The HUD for the C-17 is claimed to be the world's first HUD designed as a critical flight instrument. The unit has a single box, the electronics being built into the optical unit rather than being separate. It has a 30° azimuth by 24° elevation field of view and has twin integral MIL-STD-1750A processors. Two HUDs are fitted to the aircraft, one each for pilot and co-pilot. When not required, the HUD combiner can be folded away below the line of sight.

Specifications

Power supply: 115 V AC, 400 Hz, single phase, 100 W
Reliability: 5,000 h MTBF

Operational status

In production.

Contractor

BAE Systems.

VERIFIED

BAE Systems HUD for the F-5E

Head-Up Display (HUD) for the F-16 A/B and A-10

Notwithstanding the capability and growth potential of the standard Lockheed Martin F-16 head-up display, the US Air Force in the mid-1970s issued requirements to industry for a more advanced system. This, in conjunction with improvements in other areas, provided the basis for a substantial upgrading of the F-16s under the designation Multinational Staged Improvement Programme (MSIP). The specific improvement sought was the adoption of the Lockheed Martin LANTIRN to permit the F-16s to operate in all weathers and at night and the provision of a HUD which meets the qualification requirements, including wind blast and birdstrike.

It was soon realised that the field of view needed for this would be far greater than that available with existing head-up display technology. Conventional head-up displays with lateral and vertical fields of about 13.5 and 9° respectively could be expanded to 20 and 15° using standard optics, but this would still be less than the field size the US Air Force had set as its ultimate objective.

BAE Systems started developing a head-up display based on holographic techniques using the principles of diffractive optics and was awarded a development contract in July 1980. The system combines the wide-angle display geometry made possible by using hologram technology with a company-developed method of combining raster and cursive symbol writing that greatly reduces the amount of equipment needed for day/night head-up display. The optical train uses three combiner glasses, each of which is a sandwich, the hologram being imprinted on to a gelatin filling. It provides a field of view of 30° laterally and 18° vertically.

Installation of the BAE Systems holographic head-up display unit in the F-16 cockpit as part of the LANTIRN system
0106384

On the electronics side, it incorporates MIL-STD-1589B high-order language, MIL-STD-1750A airborne instruction set architecture and MIL-STD-1553B digital databus standards.

Operational status

In service in the F-16 with over 900 systems delivered. The entry for the F-16C/D HUD defines current upgrade work for the F-16 HUD Electronics Unit.

Contractor

BAE Systems.

VERIFIED

Head-Up Display (HUD) for the F-16C/D

In March 1983, BAE Systems announced a US$50 million order to begin production of a new, wide-angle non-holographic head-up display for the US Air Force F-16C/D fighter programme. This head-up display was based on development work undertaken for the US Air Force's Advanced Fighter Technology Integration (AFTI) programme.

The head-up display provides electronically generated symbols thrown up on a total field of view of 25° which was much wider than that attained with previous head-up displays. The instantaneous field of view (that is, the field seen by the pilot without moving his head) is 21° in azimuth and 13.5° vertical. The system uses the same electronics unit as that developed for the LANTIRN system. The symbology and raster scan pictures are particularly suited to guidance and target acquisition at night or in poor weather. The system was claimed to represent the first applications of MIL-STD-1750A processor architecture, MIL-STD-1553B digital data transmission and Jovial 73 MIL-STD-1589B high-order language.

Specifications

Pilot's display unit
Dimensions: 635 × 163 × 170 mm
Weight: 21.6 kg
Power: 98 W (including 25 W for the standby sight)
Predicted MTBF: >2,000 h

Electronics unit
Dimensions: 337 × 180 × 191 mm
Weight: 14.1 kg
Addressable memory: 64 kwords (48 kbytes EPROM, 16 kbytes RAM)
Predicted MTBF: >1,000 h

Operational status

Still in current production for the US Air Force F-16C/D, with approximately 6,000 units delivered.

In September 1999, BAE Systems was awarded a contract for development of a Commercial-Off-The-Shelf (COTS) replacement Head-Up Display Electronics Unit (HUD EU) for F-16 aircraft. The contract covers development, flight qualification and demonstration within 20 months. The new HUD EU will incorporate a MIL-STD-1750 processor emulator and other COTS components that will allow continued use of existing Operational Flight Programme (OFP) software without modification. The lighter form, fit, function replacement EU incorporates a PowerPC microprocessor and enhanced symbol generator that offers significant improvements in reliability, maintainability and support costs. The new EU will also provide significant room for future growth and functionality expansion using vacant module slots and the existing power supply. In addition, the new HUD EU will be capable of rapidly updating software over a MIL-STD-1553 data bus in seconds, rather than the 4-6 hours required with the present system.

Contractor

BAE Systems.

VERIFIED

Head-Up Display (HUD) for the F-22

BAE System was selected to develop the HUD for the Lockheed/Boeing F-22 advanced tactical fighter. This will feature a single flat combiner, which will ensure maximum clarity while preserving the degree of head freedom to perform operations in a modern fighter aircraft.

The Smart HUD is a critical flight instrument with the display processors and drivers in a single LRU. It uses differactive optics featuring a single element holographic combiner, which consists of two glass elements bonded to produce a flat parallel-sided assembly. An optically powered hologram is recorded on photosensitised gelatine on the spherical interface sandwiched in the assembly and acts as the collimating combiner. The resulting advanced optical system, manufactured using computer-generated holographic techniques, provides new levels of display capability. Additionally, the uncluttered simplicity of the combiner support structure allows virtually a clear out-of-cockpit field of regard. The total field of view is 30° azimuth by 20° elevation and the instantaneous field of view is 24 × 20°.

The optical module brightness levels are optimised to operate in a very high ambient light environment, while minimising solar reflection and maximising outside world transmission and display uniformity from within the large eye motion box.

Associated with the PDU is a complex HUD control panel attached to the aft face of the unit which incorporates LED technology and a colour camera system.

Operational status

In development. BAE Systems is also supplying the HUD for the Boeing X-32 Joint Strike Fighter (JSF) demonstrator aircraft.

Contractor

BAE Systems.

VERIFIED

Helmet-Mounted Display System (HMDS) for AH-1Z

The HMDS selected by the US Marine Corps for the AH-1Z helicopter utilises a visor projected display module fitted to the Knighthelm lightweight basic helmet. It provides 24-hour day/night capability using detachable intensified night vision cameras, and a binocular visor projected display with 40° field of view. The display symbology can be overlaid on night camera video for night operations, and the display offers resolution comparable with current night vision goggles.

A high accuracy, proven d.c. electromagnetic head tracker system is used from the Knighthelm programme.

Contractor

BAE Systems.

VERIFIED

F-16Cs and Ds have BAE Systems avionics wide-angle head-up displays

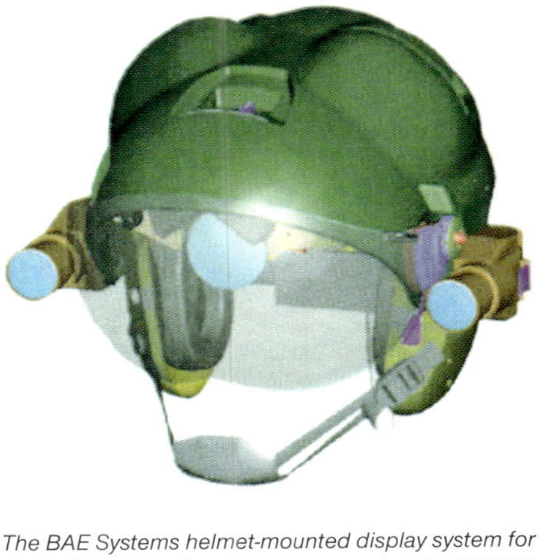

The BAE Systems helmet-mounted display system for the Bell AH-1Z helicopter **2000**/0080271

Helmet-Mounted Sighting System (HMSS)

The HMSS is a visor-projected sighting system which comprises an optical subassembly mounted to a virtually unmodified standard aviator's helmet (such as the Helmet-Integrated Systems Limited Mk 10B) with the associated electronics remotely located in the aircraft's avionics bay or similar. Sighting and cueing information is presented to the pilot by means of a high-brightness LED reticle, relayed by a prism and reflected into the pilot's eye by a dichroic patch coating on the inner surface of the clear visor.

The HMSS, originally developed in conjunction with the UK Defence Evaluation Research Agency (DERA), has now undergone various stages of upgrade to provide maximum optical performance with minimum obscuration. It comprises the BAE Systems Striker helmet-mounted sight in conjunction with Honeywell's Advanced Metal Tolerant Tracking (AMTT) technology.

The sight enables the pilot to perform off-boresight missile target acquisition, and to fully exploit the advantages of modern missile seeker head capabilities. It includes a proven ASRAAM interface. Aircraft sensors can also be pointed using the sight, and with the reticle slaved to the sensor, to cue the pilot. In a typical air-to-air situation, the HMSS and missile combination is used to acquire and lock on to a target, up to 30° off boresight.

Specifications

Weight: 0.15 kg
Field of view (circular): 40°
Exit pupil: 16 mm at centre field of view
Eye relief: Compatible with prescription spectacles

Operational status

A contract was awarded for pre-production units for use on the UK Royal Air Force Jaguar GR. Mk 1B/T2B as part of the Jaguar '97 upgrade. The production standard HMSS was flown for the first time on a UK Royal Air Force Jaguar GR. Mk 3A aircraft at DERA, Boscombe Down, in early 1999. The system is now in production for Jaguar aircraft of the UK Royal Air Force and the Omani Air Force.

Contractor

BAE Systems.

VERIFIED

The helmet-mounted sighting system 0106288

HGU B9, B19 horizon gyro units

The HGUs Type B9 are electrically operated gyroscopic flight instruments which present the pilot with continuous indication of the aircraft attitude in pitch and roll with respect to the natural horizon.

Movement of the aircraft causes the instrument, with its miniature aircraft and roll angle scale, to move in relation to the stabilised horizon bar and roll angle pointer. Thus the horizon gyro unit indicates the degree of roll by means of the pointer and scale, and the attitude of the aircraft with respect to the natural horizon by means of the horizon bar and miniature aircraft.

HGU Type B9 must be used in conjunction with a junction box which provides a junction point for the AC supply to the respective HGU and the DC supply for the potentiometer pick-offs.

HGU Type B19 is an improved form of the electrically operated vertical gyro designed to meet the requirements for attitude information to an autopilot, reduced pendulosity and reduced erection rates.

Specifications

Dimensions:
(mounting face to rear of case) 195 mm
(case diameter) 112 mm
Weight: 2.7 kg
Power supply: 115 V, 400 Hz, 3 phase AC

Operational status

In service.

Contractor

BAE Systems.

VERIFIED

HL8, HL9, HL11, HL12 Horizon Gyro Units (HGU)

The HGU is an electromechanical flight instrument, designed to provide the pilot with a constant visual indication of the pitch and roll attitude of the aircraft relative to the natural horizon.

Pitch attitude is indicated by a gyro-stabilised horizon bar registering against a fixed 'gull wing' datum engraved on the bezel glass; roll attitude is indicated by a pointer fixed on the sky-plate registering against a roll angle scale. To maintain the gyro axis in the vertical, a mechanical erection system is incorporated. The erection system is gravity-controlled and automatically responds to any deviation of the gyro from the vertical axis.

The instrument also incorporates a power failure indicator which carries a warning flag marked off. The flag is visible in the frame located in the upper half of the sky-plate. When the instrument is not operating, the flag indicates off. When the instrument is switched on the flag clears and during normal operation the flag remains in the clear position.

Specifications

Dimensions: 82 × 82 × 216 mm
Weight: 2.15 kg
Power supply: 115 V or 55 V, 400 Hz, single phase.

Operational status

Instrument fitted in Gazelle, Sioux, Whirlwind, Wasp helicopters as well as VC10 and Dove aircraft.

Contractor

BAE Systems.

VERIFIED

ISIS weapon aiming sights

The Integrated Strike and Interception System (ISIS) has been designed for the aiming of guns, rockets and bombs in the close support role and for guns and missiles in air combat.

The ISIS sight can also use range information from a laser range-finder and can serve as a low-cost alternative to a CRT head-up display by displaying pointing information from a laser seeker such as the Lockheed Martin Pave Penny or BAE Systems laser ranger marked target seeker. The company claims that adequate accuracy is coupled with low-maintenance cost and high reliability.

Operation is essentially the same for all members of the ISIS family. A two-axis rate gyro generates a lead-angle proportional to the rate of turn and a mirror attached to the rotor reflects the image of an illuminated reticle pattern through a collimating lens onto a combining glass and into the pilot's field of view.

In the ground attack role the reticle is depressed below the armament datum through an angle, which compensates for the combined effects of the weapon's ejection velocity, angle of attack, the gravity drop of the weapon during its flight and the surface wind, if known. These angles may be set up by the pilot to suit the intended attack speed, dive angle and slant range at weapon release, or he may prefer to use a predetermined set of parameters for each type of weapon carried.

In air combat, the unit measures the rate of turn of the sightline and scales it to target range, thus producing a first order lead-angle solution.

To assist steering and enable the pilot to make allowance for sudden manoeuvres by the target, most sights include a second fixed reticle representing the gun axis.

Specifications

ISIS D-195R Mk 3
Dimensions:
(sight head) 243 × 251 × 132 mm
Weight:
(sight head) 4.1 kg
(control unit) 1.36 kg
(gyro interface unit) 1.63 kg
(throttle unit) 0.2 kg

Operational status

In service in the BAE System Hawk strike/trainer. The ISIS first entered service in 1968 and current users include the air forces of Argentina, Australia, Austria, Canada, India, Indonesia, Italy, Kenya, Norway, Singapore, UK, Venezuela and the US Marine Corps. ISIS has been fitted to 25 different aircraft types including the Saab 105 (ISIS F-126), Fiat G91Y (ISIS B), Aermacchi MB-326 (ISIS F-126), Boeing A-4S (ISIS D-101), Northrop Grumman CF-5A (ISIS N), Northrop Grumman NF-5A (ISIS F-195R), BAE Systems Hawk (ISIS D-195R), HAL Ajeet (ISIS F-195R), HAL HF24 (ISIS F-124), MiG-21FL, Mostar Galeb and LAS A-4C (ISIS D-126R).

Contractor

BAE Systems.

VERIFIED

Laser Ranger and Marked Target Seeker (LRMTS)

The LRMTS is a dual-purpose unit which can be used as a self-contained laser ranger or as a target seeker with simultaneous range-finding. In the target seeking role it can be used to detect and attack any target designated by ground troops with a compatible laser, enhancing the effectiveness of battlefield close air support.

The LRMTS is an Nd:YAG laser mounted in a stabilised cage, which allows beam-pointing and stabilisation against aircraft movement. The seeker can detect marked targets outside the head movement limits. It operates at a relatively high pulse repetition frequency, thus allowing continuous updating of range information during ground attack. As range is a crucially important parameter for accurate weapon delivery, and yet virtually unobtainable on a non-laser-equipped aircraft, the LRMTS is a vital additional sensor.

In a typical operation, a forward air controller with pulsed laser target-designation equipment directs the aircraft to a location within laser detection range before switching on the ground marker equipment. Radio communications between the forward air controller and aircraft crew are minimised and positive identification of even small, hidden or camouflaged targets is assured.

Once the LRMTS has detected the laser energy reflected from the target it provides steering commands to the pilot on the head-up display. Ranging data is also shown and fed directly into weapon aiming computations for the accurate and automatic release of weapons.

The LRMTS is easy to install and harmonise with other aircraft systems and is said to be more effective than any alternative sensor during operations at the grazing angles used in low-level ground attack. By improving weapon delivery accuracy, the sensor ensures a high probability of success in single-pass, high-speed attacks.

Associated with the LRMTS head is an electronics unit which contains power supplies and ranging and seeker processing. The laser needs a transparent window and in the Jaguar is mounted behind a chisel-shaped nose, with two sloping panels. For the Harrier,

LRMTS optics in the nose of the Harrier GR. Mk 3

where there is more chance of debris accumulation during VTOL operations, the optics are protected by retractable eyelid shutters.

A specially designed installation for the UK Royal Air Force Panavia Tornado has incorporated the LRMTS into an underbelly blister. In addition to the systems in service with the UK Royal Air Force, the equipment is also used by three other air forces.

An eye-safe version of LRMTS has been developed and successful trials have been completed. Eye-safe operation has been achieved by altering the output of the laser with a Raman cell from 1.06 μm to 1.54 μm which is in the eye-safe region of the spectrum. Existing LRMTS systems can be retrofitted with this modification which will significantly reduce safety restrictions for training.

Specifications

Dimensions:
(LRMTS head) 300 × 269 × 607 mm
(electronics unit) 330 × 127 × 432 mm
Weight:
(LRMTS head) 21.5 kg
(electronics unit) 14.5 kg
Power supply: 200 V AC, 400 Hz, 3 phase, 700 VA
28 V DC, 1 A
Wavelength: 1.06 or 1.54 μm
PRF: 10 pps
Angular coverage:
(elevation) +3 to −20°
(azimuth) ±12°
Roll stabilisation: ±90°
Detection angle: ±18° from aircraft heading
Max range: >9 km

Operational status

In service. Development of the LRMTS began in 1968 under a government contract and prototype units were first flown in 1974. Deliveries to the UK Royal Air Force, accounting for over 200 units, for installation in nose housings of the BAE Systems Harrier and Sepecat Jaguar aircraft were completed in 1984. Over 800 LRMTS have been delivered to the Tornado, Jaguar and Harrier programmes, in the UK and overseas.

Contractor

BAE Systems.

UPDATED

MED 2060 series monochrome head-down displays

MED 2060 monochrome head-down displays are small high-brightness raster presentation devices designed for use where space is at a premium. Units are currently available with screen diagonals ranging from 105 to 280 mm. Different aspect ratios are selectable and 525- or 625-line variants are available. The displays can be supplied with or without a passive contrast enhancement filter matched to the CRT phosphor. The filter can be either bonded to the CRT or mounted away from its face to provide optimum visibility in specific conditions.

The display can be used as full multifunction displays for FLIR, radar or stores status, or as simple head-up display repeaters in the rear cockpit of tandem-seat trainers.

Additionally, it has facilities to switch automatically between 1:1 and 4:3 aspect ratios, catering for a range of presentations from maps and engine/systems status data to the display of video from the pilot's head-up display and FLIR night vision system. The unit is also fully compatible with night vision goggles.

MED 2060 series monochrome head-down displays

0106387

Specifications

Weight: 7.5 kg
Power supply: 115 V AC, 400 Hz or 28 V DC, 50 W

Operational status

In production. The Type MED 2060 series has been selected for the Sea Harrier F/A-Mk 2 port and starboard displays, for the BAE Systems Hawk 100/200, a variety of Mirage III and 5 aircraft, Danish SAR S-61 helicopters and for an Asian A-4 Skyhawk retrofit contract.

Contractor

BAE Systems.

VERIFIED

Miniature LCD Tacan controller for the Shorts Tucano

The miniature LCD on the Shorts Tucano is used in conjunction with the lightweight Tacan navigation equipment. Fitted in each cockpit of the Tucano, the LCD will provide normal Tacan control and display two Tacan channels, the one in use and the standby channel. Either unit is able to take control of the Tacan.

Occupying only a 2.5 × 2.25 in (63.5 × 57.1 mm) space, the panel-mounted miniature LCD controller is particularly suited to fixed-wing aircraft and helicopters where panel space is at a premium. The controller features the latest industry standard liquid crystal display technology and is compatible with most modern avionic systems.

Operational status

In service in the UK Royal Air Force Shorts Tucano.

Contractor

BAE Systems.

VERIFIED

Miniature LCD Tacan controller for the UK Royal Air Force Shorts Tucano

Modular FLIR

The modular FLIR provides a passive solution to the demanding requirements of night navigation and target acquisition. The FLIR may be installed either in a pod or integrated directly into the aircraft, with the optics looking forward through a small window. The system includes a lightweight miniaturised scanner and advanced signal processing to satisfy the demanding space and performance requirements of the airborne role. Hands-off fully automatic operation minimises aircrew workload and enhances combat survivability. The modular design of the FLIR permits simple reconfiguration to meet the space constraints of aircraft such as the Hawk 100. A variant of the production equipment for the Hawk 100 comprises two LRUs: the sensor head and the electronics unit. The electronics unit includes space provision for future growth in performance and capability. The configuration for the Harrier and AV-8B can be fitted within a 254 mm diameter pod.

By projecting a high-resolution image of the terrain ahead on the HUD, the FLIR permits the pilot to carry out aggressive manoeuvres at low altitude. In addition, an integrated thermal cuer detects hot objects, which may be potential targets, within the scene and marks them on the HUD.

The telescope is selected to match the FLIR image to the aircraft HUD field of view. The detector consists of eight parallel CMT TEDs, cooled by a closed-cycle cooling engine. Time to readiness is typically 3½ minutes at 20°C.

Other versions are available for tactical transport aircraft, such as the C-130 Hercules, or as enhanced vision systems on civil aircraft.

Specifications

Power supply: 28 V DC nominal, 200 W typical
115/200 V AC, 400 Hz, 146 W typical
Temperature range: −40 to +70°C
Video output: 625-line, 50 Hz or 525-line, 60 Hz
Interface: MIL-STD-1553 and/or discrete hard-wired

BAE Systems FLIR image on a pilot's head-up display

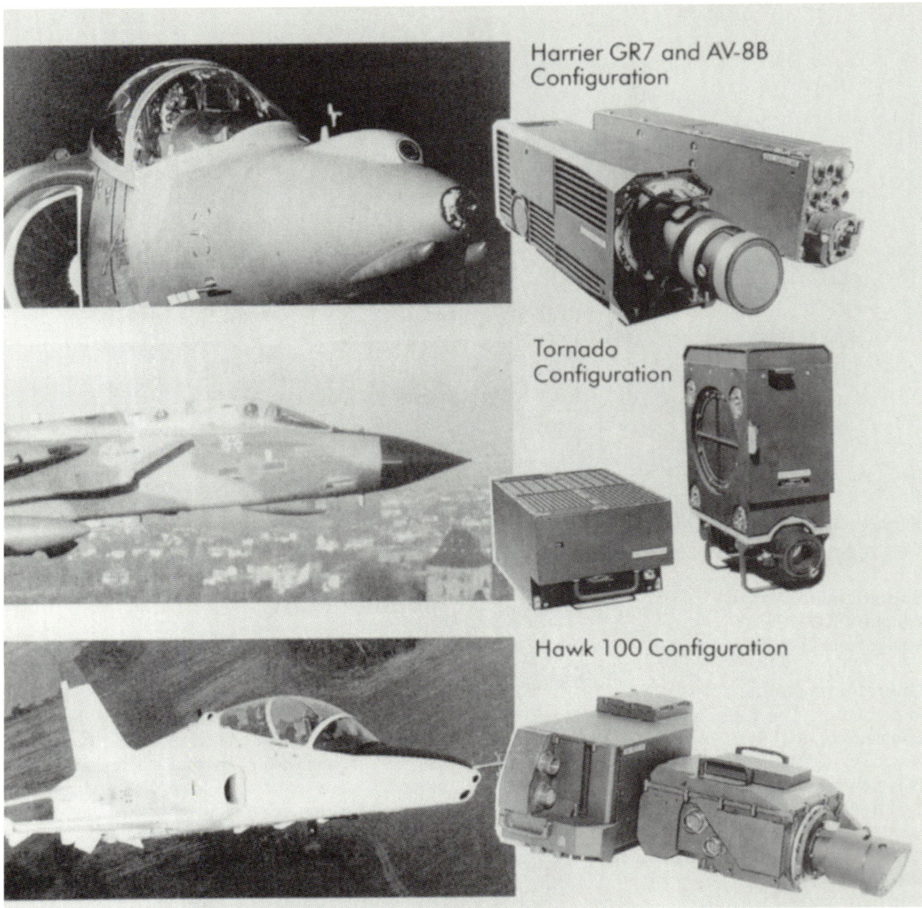

Configurations of the BAE Systems modular FLIR system

Operational status

In production for the UK Royal Air Force Tornado GR Mk 4 and Harrier GR Mk 7, AV-8B for the US Marine Corps, Spanish and Italian navies and the Hawk 100.

Contractor

BAE Systems.

VERIFIED

Monocular Head-Up Display (MONOHUD) for helicopters

Designed for installation in existing cockpits, the MONOHUD is a fully capable system in miniature. It offers flight and navigation modes to provide the pilot with full flight parameters head-up during low-level flight and landing, as well as computed weapon aiming solutions when used as a sighting system on military aircraft. It has been designed for applications in which size and weight constraints play a major role in the choice of systems. The MONOHUD system consists of a Pilot Display Unit (PDU), an Electronics Unit (EU), a High-Voltage Power Supply Unit (HVPSU) and a control panel.

The PDU can be fitted with minimal modification of the airframe. It allows an unobstructed view of instrumentation and the outside world in both the stowed and operational positions. In the operational position the PDU is situated approximately 8 cm in front of the pilot's eye and thus a 30 × 24° field of view is achieved. The PDU control panel includes manual controls for symbol brilliance and declutter.

The EU is compatible with a comprehensive range of discrete, analogue, digital, synchro, ARINC 429 and MIL-STD-1553B inputs. In addition to the input interface, processor and symbol generator, the ½ ATR electronics unit contains the low-voltage power supply and BIT. Built-in test eliminates the need for external support equipment at organisational and intermediate levels.

Specifications

Dimensions:
(PDU) 76.2 × 76.2 × 190 mm
(EU) 320 × 124 × 193 mm
(HVPSU) 178 × 153 × 102 mm
(control panel) 57 × 146 × 79 mm
Weight:
(total system) 12.7 kg
Field of view: (vertical) 30° × (azimuth) 24°
Contrast ratio: 1.2:1 against 10,000 ft-lamberts
Environmental: MIL-E-5400T, Class 1B
Reliability: 2,500 h MTBF

Operational status

In production.

Contractor

BAE Systems.

VERIFIED

MultiSensor Turret System (MST-S)

The MST-S is a versatile compact lightweight thermal imaging system suitable for helicopter and subsonic fixed-wing aircraft operations. It is NVG-compatible.

The BAE Systems multisensor turret fitted under the starboard wing of a Nimrod MR. Mk 2

The gyrostabilised turret platform provides 360° steerable field of regard in both azimuth and elevation and is fully operational at airspeeds up to 300 kt.

The standard thermal imaging payload utilises UK TICM II modules, a closed-cycle cooling engine and a continuous zoom telescope with magnification of × 2.5 to × 10. Alternative payloads with a variety of IR and TV sensors and telescopes are available.

The basic system consists of two LRUs: the turret and a control switch joystick unit. This can be expanded by the addition of a third unit to cater for the range of options which provides full integration with other aircraft systems. The options currently available include MIL-STD-1553B, ARINC 429 and RS-422 databus interfaces, video tracker, electronic magnification, radar-designated target handover and a variety of control units.

Specifications

Weight:
(turret) 40 kg
(control switch joystick unit) 2 kg
(optional control electronics unit) 12-20 kg
Power supply: 28 V DC nominal, 340 W typical
Wavelength: 8-13 μm, 3-5 μm as an option
Video output: 625-line, 50 Hz or 525-line, 60 Hz

Operational status

In production for S-61, S-76, Lynx and Sea King helicopters and Fokker F50, Dornier 228 and CN-235 maritime patrol aircraft. A quantity of MSTs has also been fitted to Nimrod MR. Mk 2. Ordered for UK Royal Air Force EH 101 Support Helicopters and as part of a surveillance system being installed on Agusta-Bell 412 EP helicopters by an export customer.

Contractor

BAE Systems.

VERIFIED

Nightbird Night Vision Goggles (NVG)

Designed specifically for fighter aircraft applications, the Nightbird NVG system has all the attributes of the NITE-OP system with the addition of full head-up display compatibility and the enhancement of pilot safety on ejection by means of automatic NVG detachment.

Utilising a special filter, the NVG optics are designed to permit viewing of HUD symbology directly, allowing the pilot to view Forward Looking Infra-Red (FLIR) raster imagery combined with HUD stroke symbology. Nightbird NVGs are designed to be employed with either Generation II or Generation III image intensifier tubes, permitting growth/enhancement of the system as Gen III technology becomes more widely available.

The goggle auto-detach mechanism, activated by movement of the ejection seat and employing a small pyrotechnic charge to operate the goggle detach lever, minimises risk of injury to the pilot from the goggles on

Nightbird night vision goggles for fast-jet aircraft 0106289

ejection by releasing them approximately 4 ms after ejection initiation. The auto-detach system has been extensively tested and is fully qualified in all applicable RAF aircraft.

Specifications

Goggle
Weight: 0.8 kg (including helmet mounting plate)
Eye relief: 30 mm (nominal)
Exit pupil: 10 mm at 300 mm eye relief
Field of view (circular): 40° (+6°, −0°) at 30 mm eye relief
Magnification: 1.0 ± 5%
Aperture: f1.2

Auto Separation System
PS/BIT unit weight: 0.5 kg
Dimensions: 153 × 64 × 48 mm
Power supply: 6 V, 2 cell $LiMnO_2$

Operational status

No longer in production. In service with the UK Royal Air Force Harrier GR Mk 7, Tornado GR Mk 1/4 and Jaguar GR. Mk 1/3.

Contractor

BAE Systems.

UPDATED

NITE-OP Night Vision Goggles (NVG)

The NITE-OP NVG are specifically designed for helicopter aircrew, enabling them to fly visually at night.

NITE-OP NVGs have a fully circular 40° field of view at nominal eye relief. The optical design provides large eye relief and exit pupil permitting the use of an eye protection visor or NBC protective mask. Manufactured using the latest composite materials, the goggles are lightweight and robust. Electrical configuration ensures high reliability and redundancy by powering each image intensifier tube separately by batteries integral to the NVGs. With no external wires or connectors required the system is completely self-contained and portable. Either Generation 2 or Generation 3 image intensifier tubes may be fitted.

Specifications

Weight: 0.8 kg
Power supply: 3.5 V batteries (independent for each channel)
15 h endurance at 0°C
Field of view (circular): 40°

Operational status

No longer in production. In service with UK armed forces and several overseas customers.

Contractor

BAE Systems.

UPDATED

A pilot of a Royal Air Force Tornado GR1 wearing Nightbird night vision goggles **2001**/0106207

Optical helmet tracking system

The BAE Systems optical helmet tracking system is designed to satisfy the requirement for a tracking system to interface with helmet-mounted displays and sights that is fast, accurate and remains uninfluenced by metal structure within the aircraft cockpit.

One or more optical sensors located on the aircraft structure detects the position of several helmet-mounted light emitting diodes (LEDs). A multiplexer energises each LED in turn and the position of the helmet in the cockpit is determined from the sensed relative position of the LED from each sensor. The aircraft-mounted electronics unit can be provided in stand-alone form or as a card set for integration into an existing unit. The system does not need to be aligned or boresighted after initial installation.

Operational status

Under development.

Contractor

BAE Systems.

VERIFIED

Nite-Op Night Vision Goggles 0106290

RAI 4 remote attitude indicator

The RAI 4 remote attitude indicator works on a synchro output from a vertical gyro. The indication is essentially a sphere, which moves relative to a fixed aircraft pitch symbol and to a peripheral bank scale for roll. The sphere has 360° of freedom in roll and ±100° in pitch.

Specifications

Dimensions: 4 ATI × 210 mm
Weight: 1.82 kg
Power supply: 115 V AC, 400 Hz, 11 VA
28 V DC plus 5 V AC/DC for lighting

Operational status

In service.

Contractor

BAE Systems.

VERIFIED

RL8 - heading repeater

RL8 is a heading repeater series that provides heading information displayed on a servo-driven moving card, with settable markers for desired heading and wind direction. The 3.25 in (83 mm) diameter instrument accepts heading information in standard three-wire synchro form. The rotating card display is calibrated through 360° in 5° increments and the compass cardinal points are designated. Wind direction is indicated by a green cursor, set by pulling out and rotating the knob on the front bezel. Required heading is indicated by a white pointer, which is set by pushing in and rotating the knob. Once set, both reference indicators rotate with the card display.

Operational status

Unit in service. Applications include Lynx, Sea King and Tornado ADV.

Contractor

BAE Systems.

VERIFIED

Sea Owl passive identification device

The Sea Owl system provides a day and night long-range target detection and identification capability for helicopters. It uses a BAE Systems thermal imaging sensor with high-magnification optics mounted in a highly stabilised platform sited on the nose of the helicopter. The high-resolution thermal image, automatically optimised for maximum picture quality, is displayed in the cockpit. The employment of advanced signal processing provides tracking and target cueing automatically, further reducing aircrew workload.

The BAE Systems' Sea Owl passive identification device on a Lynx helicopter

The system is completely integrated with the helicopter central tactical system and other avionics systems. It provides both automatic search and acquisition modes of operation, and manual control via a joystick unit. The long-range standoff identification capability of this equipment has been proved around the world in a wide range of atmospheric conditions.

Specifications

Weight:
(turret) 64 kg
(signal processor) 17 kg
(tracking unit) 13 kg
(compressor) 10 kg
Power supply: 28 V DC nominal, 570 W typical
115/200 V AC, 400 Hz, 115 W typical
Wavelength: 8-13 μm
Detector: 8 parallel CMT TED
Magnification: × 5 to × 30 switched zoom
Field of regard:
(elevation) +20 to −30°
(azimuth) +120 to −120°
Video output: 625-line, 50 Hz or 525-line, 60 Hz

Operational status

In service on the UK Royal Navy Lynx.

Contractor

BAE Systems.

VERIFIED

Striker display helmet

The Striker display helmet is a 2-part system comprising flexible, fitted, inner helmet and rigid, protective outer helmet. The visor projection provides 40° field of view. Integrated vision cameras can be

The BAE Systems Striker helmet-mounted sight as used in Jaguar aircraft **2000**/0080274

quickly removed for day use. The electronic symbol system is night vision compatible, and the optical tracking system is compatible with a magnetic environment. The Striker helmet includes an integral, pivoted, laser protective visor, and full NBC compatibility.

The Striker helmet forms part of the Jaguar Helmet-Mounted Sighting System (HMSS) production system, and the Eurofighter Typhoon Helmet-Mounted Display (HMD) development system.

In May 1999, the Striker helmet was successfully tested in a 600 kt election from a specially modified YF-4J Phantom aircraft at China Lake US Naval Air Warfare Center, as part of an assessment for a revised parachute canopy for the F/A-18 aircraft.

The Striker display helmet is also suitable for rotary-wing applications.

Operational status

In production for UK Royal Air Force and Omani Air Force Jaguar aircraft, and in development for the Eurofighter Typhoon aircraft.

Contractor

BAE Systems.

VERIFIED

Thermal Imaging Common Modules (TICM)

TICMs have been configured into a wide variety of thermal imaging systems. They are built to full military standard and use advanced optical and electronic components, including the high-performance Transferred Electron Device (TED) detector, to give high resolution and sensitivity even at long range. Telescopes and displays can be selected by system designers to meet precise operational requirements.

These indirect view systems produce a high-quality video image which can be displayed on television monitors or head-up or head-down displays. They are fully automatic and minimise operator workload. The modular basis gives flexibility in system design, enabling application-specific sensor architectures to be configured by system integrators. The ease of maintenance and repair allows cost-effective use both in new weapons and for retrofit.

TICM II

TICM II thermal imaging modules are built to full military standard and use advanced optical and electronic components, including the Marconi Electronic Systems' SPRITE detector, to give high resolution and sensitivity even at long range. Optical and electronic processing modules are based on the coaxial scanning polygon which requires only one drive motor for both azimuth and elevation scans. The modular basis gives flexibility in system design.

Images from these indirect view systems can be displayed on television monitors or head-up and head-down displays. They are fully automatic in operation and allow module replacement without any adjustment or set up time.

Specifications

Spectral band: 8-13 μm
Field of view: 625-line 50 Hz: 60 × 40°
525-line 60 Hz: 48.4 × 32.5°
Resolution: 2.27 mrad (60° field of view)
Pupil diameter: 10 mm
Min resolvable temperature difference: typically better than 0.1° C
Detector: 8 parallel CMT SPRITE
Video output: CCIR Systems I/EIA-RS-170, compatible 625/525-line 50/60 Hz

The BAE Systems Striker display helmet as developed for Eurofighter Typhoon aircraft **2000**/0080273

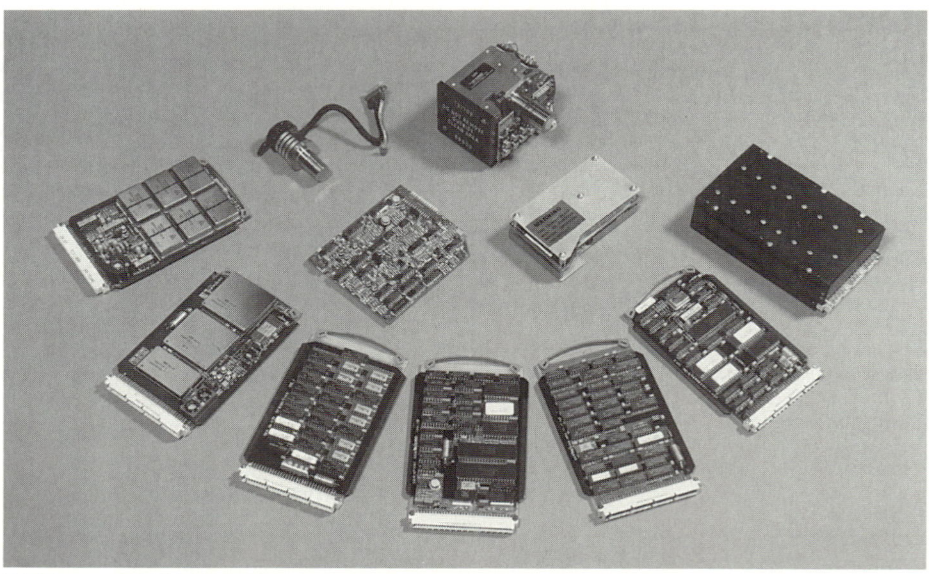

BAE Systems TICM II thermal imaging common modules **1999**/0005494

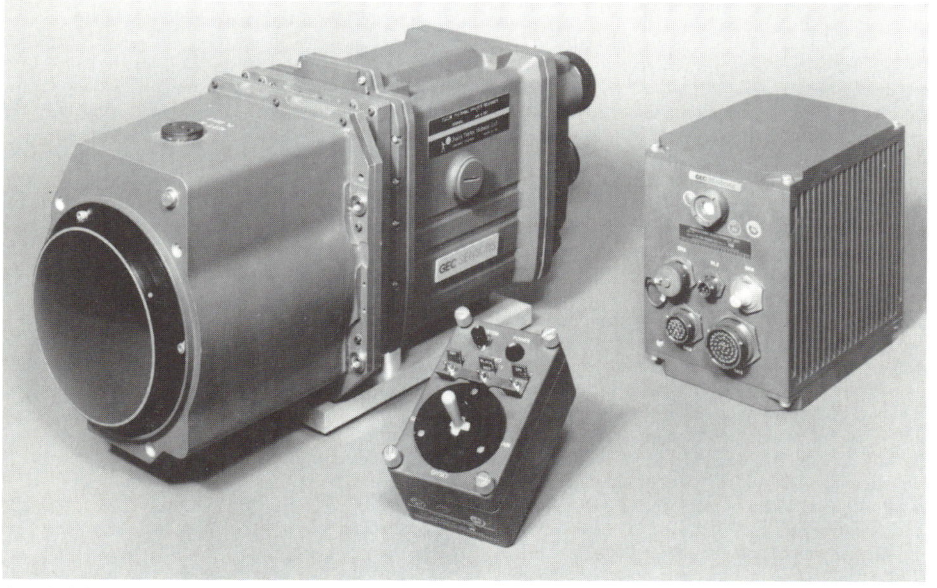

A typical thermal imaging system configured from the TICM II modules

Weights (modules in enclosures):
Scanner head: 9.5 kg
Processing electronics: 6.5 kg
System control panel: 1.03 kg

Operational status

In large-scale production for the UK MoD. Over 2,000 sets of modules have been supplied to the UK armed forces and to customers throughout the world.

Contractor

BAE Systems.

VERIFIED

TIALD laser designation pod

The Thermal Imaging/Airborne Laser Designator (TIALD) pod provides military aircraft with automatic tracking and laser designation of targets, day or night, for the delivery of laser guided and ballistic weapons. The pod carries three electro-optic (EO) sensors - a thermal imager, TV and a laser designator, which all share a common optical aperture. This unique packaging solution results in the smallest diameter pod of this type and facilitates the maintenance of an accurate common boresight for all sensors. The large sightline field of regard (FoR) afforded by the pod roll and gimbal arrangement facilitates designation without flying directly towards the target. In many cases, the full FoR of the pod is restricted in part by the airframe of the carriage aircraft, therefore, in order to prevent the laser impacting the airframe during manoeuvres, specific obscuration profiles are stored within the pod. The infra-red (IR) and TV sensors operate simultaneously with either image displayed by single switch selection. This allows the operator to quickly compare IR and TV performance in the target area and choose the optimum sensor for the attack. The IR sensor has two fields of view (FoV), wide and narrow, while the TV sensor operates only in the narrow field of view (NFoV). In NFoV, an electronic zoom facility facilitates maximum target standoff for avoidance of co-located defences.

Apart from the cockpit display and controls the pod is self-contained, taking its power from the aircraft primary supplies. Interface with the avionics is via a MIL-STD-1553B databus.

The forward section of the pod contains the thermal imager and TV sensors, the telescope and the laser designator transceiver unit. The laser, TV and thermal imager optical paths are combined within the telescope and steered over a wide angle of regard by the pod roll and gimbal arrangement. The combined optical path is stabilised against aircraft movement and pod vibration by a stabilised mirror

The static rear sections of the pod contain the ram-air cooler, electronics units and power supplies. The automatic video tracker, produced by BAE Systems, Plymouth, is included within the electronics units.

In operation, the sensor sightline would be directed on to the target area, primarily by commands from the aircraft nav/attack system, or, in reversionary modes or during unplanned attacks, by the crew. After assessment of the target area in wide field of view (WFoV), the crew would acquire and identify the target and specific desired mean point of impact (DMPI) in narrow field of view (NFoV). After confirmation of target identification on the video display, the video tracker may be engaged and locked to the target. Once the video tracker has locked onto the target, the aircraft may be manoeuvred within pod gimbal limits, with the system automatically keeping the sensor sightline locked onto the target. Depending on weapon type and delivery platform, the laser would be fired at a predetermined time to enable terminal guidance and ensure optimum weapon impact parameters.

Although used primarily to designate targets for laser-guided weapons, TIALD may also be used as an integral aircraft sensor, used to update the nav/attack system before to release of ballistic weapons, thus ensuring the most accurate release solution possible. TIALD has also been used extensively in a reconnaissance/surveillance mode at medium level, tracking and pinpointing the positions of points of interest and recording video footage for post-mission analysis.

Specifications

Dimensions: (length) 2,900 × (diameter) 305 mm
Weight: 210 kg
Sensors:
TV 0.7-1.0 μm
Laser 1.06 μm
Thermal imager 8-12 μm
Field of regard: (Tornado GR1)
Ball + 30 to -155°
Roll continuous
Field of view:
(wide) 10 × 6.7°
(narrow) 3.6 × 2.4°
Electronic zoom: × 2, × 4
Power supply: 200 V AC, 400 Hz, 3 phase, 3.3 kW (max), 1.2 kW (average)

Operational status

In production and in service with the RAF on Tornado GR. Mk 1, Jaguar GR. Mk 1B/3 and Harrier GR. Mk 7. TIALD is expected to be in service on Tornado GR. Mk 4 by early 2001. TIALD Pod development has been ongoing with a series of enhancements, including more powerful head motors and the integration of VITS automatic tracking systems (see separate entry). The latest variant of TIALD is the -400 Series Pod, which is understood to be the final version of the system.

Contractor

BAE Systems.

UPDATED

Type 105 laser ranger

The Type 105 is a high-repetition rate, Nd:YAG steerable laser range-finder developed privately by BAE Systems to provide compact low-cost accurate target ranging sensors for ground attack aircraft.

The 105 Series includes several variants to suit different aircraft installation and avionics requirements. The latest variant has been specifically designed as a two-box system for ease of installation in a wide range of aircraft. The 105 is now MIL-STD-1553B databus-compatible and is particularly suitable for aircraft updates such as the F-5E, A-4, A-10, Hawk and AMX, as well as new ground attack aircraft.

Target range is measured to an accuracy of 3.5 m, standard deviation to a range of 10 km, effectively removing the largest source of error in air-to-surface weapon delivery.

The BAE Systems Type 105 laser ranger is fitted in the Hawk 100 0106348

An RAF Harrier GR Mk 7 after return to Gioia del Colle after a mission during Operation Allied Force in 1999. The aircraft is carrying a TIALD pod on the port shoulder pylon and 4 BOL countermeasures dispensers. The empty wing pylons were probably loaded with two Paveway II LGBs for the mission (Crown Copyright) ***2001***/0087830

Low power consumption, ease of integration with aircraft avionic systems, flexible configuration, small size and frontal area are seen to be important factors in the suitability of the Type 105 for fit or retrofit in ground attack aircraft.

Specifications

Dimensions: 190 × 200 × 370 mm
Weight:
(laser) 12.5 kg
(electronics unit) 4.5 kg
Power supply: 28 V DC, 300 W
Wavelength: 1.06 μm
PRF: 10 pps
Angular coverage: within 10° semi-apex angle cone
Range: 10 km
Accuracy: 3.5 m standard deviation
Reliability: >1,000 h MTBF

Operational status

In service.

Contractor

BAE Systems.

VERIFIED

Type 118 lightweight laser designator/ranger

The Type 118 laser transceiver is suitable for integration with the visual optics of future or existing helicopter sights to designate targets for spot-tracker or laser-guided weapon applications. It is a lightweight Nd:YAG device, originally developed and produced for the mast-mounted sight on the US Army OH-58D Army Helicopter Improvement Programme (AHIP).

Specifications

Dimensions: 153 × 337 × 163 mm
Wavelength: 1.06 μm
Output energy: 110 mJ
PRF: 20 pps (max)
Range: 300 m-10 km
Resolution: 5 m

Operational status

In production.

Contractor

BAE Systems.

VERIFIED

Type 126 laser designator/ranger

The Type 126 is a high-energy Nd:YAG range-finder and designator system developed for the UK MoD Thermal Imaging Airborne Laser Designation (TIALD) pod. The Type 126, comprising separate transmitter unit and power supply/control unit for ease of pod or aircraft installation, provides very high-output energies at pulse repetition rates of up to 20 Hz in a temperate environment ranging from –54 to +71° C.

Operational status

In production and in service in the TIALD pod.

Contractor

BAE Systems.

VERIFIED

Type 221 thermal imaging surveillance system

The Type 221 thermal imaging surveillance system is designed for service with military helicopters. Developed by BAE Systems in conjunction with Pilkington Optronics, the system incorporates an IR18 thermal imager and telescope by the latter company, with sightline stabilisation steering provided by a stabilised mirror. The assembly is contained in a pod which either can be mounted beneath the nose of a helicopter or can project through an aperture in the aircraft floor.

The IR18 imager unit provides a normal field of view of 38° in azimuth and 25.5° in elevation. In the Type 221 application, users can choose a telescope magnification of either × 2.5 or × 9, with corresponding wide or narrow fields of view. The wider field of view (15.2° azimuth by 10.2° elevation) would be used for general surveillance, target acquisition or navigation; the narrow field of view (4.2° azimuth by 2.6° elevation) permits detailed observation for target identification or engagement of targets detected in the wide field of view mode. In the pod installation the system has fields of regard of +15 to –30° in elevation and ±178° in azimuth. The entire sensor system is vertically mounted above the mirror which is angled, periscope fashion, at 45° to the horizontal to provide views in the horizontal plane.

The sensor system employs SPRITE detector units cooled by a Joule-Thompson minicooler supplied with high-pressure compressed air. The air source is a bottle, mounted on the equipment and charged immediately before flight. This has a capacity of 1 litre and provides a system operation time of approximately 2½ hours. If greater endurance is required other cooling options, involving the use of mini-compressors permanently connected to the equipment, are available. The system operates in the 8 to 13 μm band and has a sensitivity of between 0.17 and 0.35° C to target background and surroundings.

The display may be presented on either 525- or 625-line television monitors. The output is either in CCIR or EIA composite video formats, as required. This television-compatible output may be displayed on one or more monitors throughout the aircraft to aid winching operations.

The Type 221 system is intended for use in medium to large helicopters and roles envisaged include maritime reconnaissance, search and rescue and integration with onboard weapon systems to improve all-weather capability.

Specifications

Dimensions:
(pod unit) 865 long × 420 mm (max) diameter
(electronics unit) 127 × 432 × 330 mm (max)
Weight:
(pod unit) 75 kg
(electronics unit) 8 kg

Operational status

In service. The Type 221 thermal imager has been fitted to a number of Aerospatiale/Westland Puma helicopters operated by the UK Royal Air Force.

Contractors

BAE Systems.
Pilkington Optronics Ltd.

VERIFIED

Type 239 pilot's night vision system

The Type 239 is a compact steerable platform for the pilot's night vision system on the Agusta A129 helicopter.

The platform, on which is mounted a Lockheed Martin thermal imager, has a two-axis movement, with ±130° range in azimuth and +75 to –60° in elevation. The platform's acceleration and slew rates for both axes are compatible with helmet tracking systems to enable the platform to function with a helmet display as a fully visual coupled system.

The platform is suitable for mounting on almost any helicopter, due to its small size and low weight, and can be fitted with a range of compact imaging systems to meet customer requirements.

Operational status

In service on the Agusta A129 and Italian HH3 helicopters.

Contractor

BAE Systems.

VERIFIED

Type 9000 Head-Up Display (HUD)

To cater for larger attack/fighter aircraft or those aircraft with less installation constraints, the Type 9000 wide-angle conventional HUD uses a 6.5 in (165 mm) exit lens. It has self-contained high- and low-voltage power supplies and is capable of cursive and raster display with cursive-in-raster flyback. High-brightness combiners give daylight viewability of the raster FLIR image. A high-resolution colour camera is fitted to record the pilot's view.

The HUD is driven by a Computer Symbol Generator (CSG), which offers full mission computing and faster/cursive symbol generation capability.

The CSG can be fitted with multiple processor cards to ensure all customer systems moding, control and operational requirements can be easily accommodated. The CSG is fully programmable using the ADA software language.

The CSG provides video routeing and mixing of external video with internally generated symbology and can operate as either a bus controller or an R/T on the aircraft 1553 B databus.

The CSG is of modular construction enabling various combination of symbol generation and aircraft interface to be readily configurable. A typical CSG

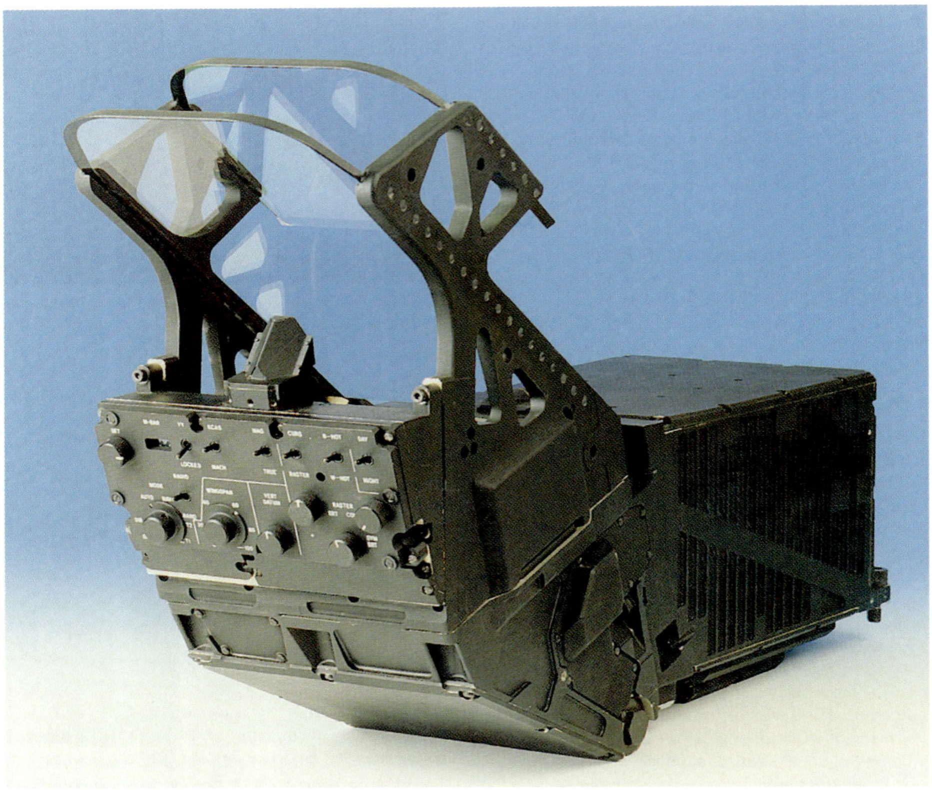

Type 9000 HUD **2001**/0106291

application utilises 12 cards in a 19 slot box providing expansion capability to add extra functions to the unit. Functions which may be added include digital mapping, terrain following and intelligent ground proximity warning.

Operational status

In production and in service in the UK Royal Air Force Tornado GR. Mk 4.

Contractor

BAE Systems.

UPDATED

VITS automatic tracking systems

VITS is a family of high-performance products which employs an expanding architecture. They provide superior tracking capability, fast processing time and reduced operator workload.

VITS is fully compliant with the complete range of military specifications and provides both VME and RS-422/RS-232 serial interfaces. The family embodies Ada software, custom ASICs and extensive BIT facilities.

The VITS family consists of VITS1000, VITS1500, VITS2000 and VITS3000. The basic VITS1000 features dual-standard video input, high-precision centroid and correlation tracking, low data latency for high system bandwidth, automatic selection of optimum tracking mode, track quality measures to aid system performance, an aimpoint refinement and boresight alignment facility. Tracking performance is maintained in high-clutter backgrounds.

VITS2000 includes all these features plus multiple object tracking, target cueing and prioritisation, automatic target acquisition and robust tracking despite decoys and obscurants.

The top-of-the-range VITS3000 will include all the features of the other VITS models plus automatic target classification using neural networks and enhanced rejection of false alarms.

The VITS1000 tracking system has been integrated with the BAE Systems TIALD (Thermal Imaging Airborne Laser Designator) pod and, following extensive user evaluation, initial production has started to refit the TIALD fleet. Trials programmes are now continuing with the integration of VITS2000 into the pod. VITS1000 is also being supplied to a variety of UK and offshore prime contractors and defence establishments.

The VITS systems are available both as off-the-shelf products and as application-tailored derivatives. One example of a tailored product has resulted in the VITS1500 system – a multiple target detection and tracking system but without the sophisticated correlation tracking subsystem.

Type 9000 HUD installed in a Tornado GR. Mk 4A (E L Downs) ***2001***/0106203

Operational status

VITS1000 is in production; VITS1500, 2000 and 3000 are under evaluation.

Contractor

BAE Systems.

VERIFIED

RMS 3000 reconnaissance management system

The RMS 3000 Series combines scan conversion and image processing of electro-optic sensor imagery with in-flight display of the terrain overflown. The RMS 3000 Series system is fitted to UK Royal Air Force Tornado GR. Mk 1A reconnaissance aircraft and provides for the simultaneous recording and display of the imagery from multiple infra-red sensors. Other system features include real-time and near-realtime display of imagery, a wide range of rolling and updating display modes, display facilities including slow speed replay, magnification and processing to optimise the displayed image and aspect correction and rectilinearisation of displayed imagery to facilitate in-flight exploitation.

The system design is optimised to ensure that the ground exploitation time is minimised. Relevant features include combined video and aircraft data annotation recording on videotape, in-flight editing of event imagery and rapid removal of video cassettes from the aircraft on return to base.

Computing Devices is developing a Digital Solid State Recorder (DSSR) LRI to replace the existing analogue videotape recorder. The DSSR will provide fast data access with greater system reliability.

Infra-red images produced by a Computing Devices RMS 3000 being displayed in the rear cockpit of a Tornado GR. Mk 1A

Operational status

In service in UK Royal Air Force and Royal Saudi Air Force Tornado GR. Mk 1A aircraft.

Contractor

Computing Devices, a General Dynamics Company.

VERIFIED

RMS 4000 reconnaissance management system

The RMS 4000 Series is an all-digital system which is designed to be common across several platforms. Key features which have been used to enhance the system performance and increase the mission success rate are use of Ada high-order language, flexible modular design with high configuration commonality across platforms, high reliability and high level of functional redundancy to ensure no single defect can cause mission critical failure, design for two-level maintenance and growth potential to accommodate new sensors, image and target processes.

The RMS 4000 provides a selection of sensors (EO, IR and SAR) with built-in growth potential, image data management, compression and decompression and routeing, multiple display formats for the selected sensor, recall of imagery for review and edit prior to data transmission, datalinking of key target imagery to provide real-time record and availability of archive data as soon as the aircraft lands.

Computing Devices is integrating Solid State Recording (SSR) technology into the RMS 4000 to provide customers with significant system cost reductions and performance enhancements.

The RMS 4000 is available for nose-pallet and pod-mounted installation.

Operational status

In production. Variants of the RMS 4000 are in production (under contracts to Lockheed Martin) for the US Marine Corps F-18 Advanced Tactical Airborne Reconnaissance System (ATARS).

Contractor

Computing Devices, a General Dynamics Company.

VERIFIED

RMS 5000 reconnaissance management system

The RMS 5000 represents the latest modular design to accommodate multiple sensor types using standard interfaces. The function of the system can be chosen by selecting the configuration from a standard set of modules that provide for multiple sensor types including E-O, IR, SAR and MMWR. The system incorporates image enhancement to correct sensor and aircraft abnormalities, which is combined with sensor correlation and automatic target recognition. Imagery data undergoes high-order data compression for in-flight display and datalinking for real-time exploitation. The system contains massive onboard data storage for in-flight and in-platform ground exploitation and a key feature is autonomous operation for single-pilot or unmanned aircraft. The RMS 5000 is available for built-in or podded use.

The Computing Devices RMS 5000 reconnaissance management system can be installed internally, or in a pod
0106350

Operational status

In development.

Contractor

Computing Devices, a General Dynamics Company.

VERIFIED

AF500 Series roof observation sights

The AF532 roof-mounted helicopter sight superseded the AF120 sight introduced in 1970. The new sight is half the weight of its predecessor, but confers greatly improved optical performance.

This gyrostabilised, monocular, periscopic telescope is designed for the gunner/observer in reconnaissance helicopters, particularly when scouting targets for anti-tank helicopters.

The device has a built-in interface for a laser designator and range-finder. It can also be adapted for night vision equipment, helmet sights and weapons, and there are facilities for attaching a television recording camera for training or intelligence gathering. The design of the optical system is such that the varying eye positions in different helicopter installations can be easily accommodated.

The gyrostabilised head protrudes above the roof forward of the rotor mast, while the down-tube and eyepiece extend downwards from the roof so that the eyepiece falls into a comfortable viewing position. The down-tube is adjustable in height and retracts sideways, locking close to the roof when not in use. A control handle is extended by the operator and adjusted in tilt so that it can be used with the right forearm resting on the knee. A horizontal thumbstick is used to steer the sightline and a direction indicator, to show its direction relative to aircraft heading, is mounted on the glareshield in front of the pilot.

The sight provides a stabilised image of the chosen field of × 2.5 magnification for search and × 10 for identification and laser operation. The sightline may be steered through ±30° and ±120° in pitch and yaw planes respectively.

AF580 systems, fitted with BAE Systems, Avionics laser ranger and target designators and integrated with a Rockwell Collins Automatic Target Handover System have been evaluated by the US Army in Bell OH-58C Kiowa helicopters. This combination of systems permits the range and bearing of the target to be determined by the observation helicopter and transmitted by datalink to an attack helicopter. In a further development, a thermal image from a separate FLIR was been injected into the sight to permit night observations. The FLIR and the sight were steered via the same controller.

Operational status

In service in British Army Air Corps' Westland Gazelle helicopters. Laser Target Designation and Range Finder (LRTDRF) equipment has been fitted to a number of systems and is now in service.

Contractor

Ferranti Technologies Limited.

VERIFIED

The AF532 sight unit installed on the roof of a British Army Westland Gazelle helicopter

Standby Attitude, Heading and rate of turn Indicating System 0018194

SAHIS Standby Attitude, Heading and rate of turn Indicating System

SAHIS is a Standby Attitude, Heading and rate of turn Indicating System designed primarily for use as an emergency back up in the event of failure of aircraft generated power and/or loss of primary aircraft attitude and heading information.

The system consists of a three-axis spherical indicator, mounted in the aircraft's instrument panel, and a gyro unit, mounted remotely. The gyro unit contains a vertical gyro for attitude reference and a directional gyro for heading reference.

The indicator displays pitch, roll and heading information by means of a sphere moving behind a fixed aircraft symbol. There is a roll scale below the sphere and slip and rate of turn indications below the roll scale.

All signals and power inputs to the indicator (except slip) are derived from or via the gyro unit.

The system can operate in either of two modes whereby heading information is derived from the aircraft's inertial navigation system or from the unit's own directional gyro.

Specifications

Freedom:
(heading) unlimited
(roll) unlimited
(pitch) ± 85°
Accuracy:
(heading) ±1°
(roll) ±1°
(pitch) ±1°
Rate of turn: up to 380°/min
Dimensions:
(indicator) 81 × 203 × 81 mm
(gyro unit) 110 × 265 × 170 mm
Weight:
(indicator) 1.8 kg
(gyro unit) 5.6 kg

Operational status

SAHIS is in production to meet UK Royal Air Force requirements on Harrier GR. Mk 7 and Harrier T Mk 10 aircraft.

Contractor

Ferranti Technologies Limited.

VERIFIED

Rugged colour Active Matrix Liquid Crystal Display (AMLCD)

In August 1999, Litton Data Systems (erstwhile Science Applications International Corporation) announced a rugged, colour, flat panel AMLCD that has been specifically designed for wide-angle viewing while maintaining visual clarity even when used in direct sunlight or during night operations. The display module is a COTS 20.1 in SXGA TFT-LCD unit with 1,280 × 1,024 resolution. It includes an EMI protective screen. Varying display sizes can be provided to meet workstation requirements.

The target application for this series of rugged display modules is for radar, sonar and GPS display.

Contractor

Litton UK Ltd, Data Systems Division

VERIFIED

DEWD Dedicated Electronic Warfare Display

Bell Helicopter Textron has selected Meggitt Avionics Inc (an operating division of Meggitt Avionics), to develop and produce the Dedicated Electronic Warfare Display (DEWD) for the US Air Force Special Operations CV-22 Osprey tiltrotor aircraft.

The DEWD is a 3 in Air Transport Indicator (ATI) size active matrix liquid crystal display that interfaces with the Suite of Integrated Radio Frequency Countermeasures (SIRFC) system to provide electronic warfare situation awareness to CV-22 operators.

The DEWD is a derivative of the Meggitt Avionics Secondary Flight Display System (SFDS).

Contractor

Meggitt Avionics.

VERIFIED

ADEPT automatic video tracker systems

ADEPT 30

The ADEPT 30 tracker is the latest in the range of Octec video trackers and is the core tracker used in both the CAATS and CAATS 2 systems. It comprises a well integrated software design on a single VME card designed for use in the Octec airborne CAATS systems, and easy integration into other manufacturers' avionics VME electronics.

The ADEPT 30 unit has a great deal of flexibility in its design, allowing closed loop control of a platform and easy integration to control panels.

A range of daughter boards gives even greater flexibility and performance, allowing input of high-speed digital data, further signal processing, picture stabilisation, improved detection capability or additional tracking channels – this configuration is known as ADEPT 30+ (see side heading below).

A range of enclosures is available from a commercial 19 in rack unit to fully ruggedised avionic enclosures. Build standards are available to meet different standards from commercial up to full MIL Spec.

Octec has integrated its trackers with over 100 different platforms, including most widely used helicopter sensor platforms.

Specifications

Dimensions: 233.4 × 160 mm Double Euro
Power: +5 V 3 A, +12 V 0.2 A, –12 V 0.2 A
Video input: Composite video 625/525 line CCIR or RS-170
No of video inputs: 2
Track modes: Centroid, Correlation, Edge, Multiple target track, Scenelock
Video output: 1 with symbology overlay
Interfaces available: VME, RS-232/422, Analogue, Discretes
Automatic video detection: variable from 2-90% of FoV

ADEPT 30+

A range of daughter boards is available which allow the ADEPT 30+ to carry a wide range of additional functions, including:

(1) input of high-speed digital data from modern sensors
(2) provision of a second tracking channel
(3) additional preprocessing
(4) electronic picture stabilisation
(5) improved detection capability

ADEPT 33 single height VME bus automatic video tracker ***1999***/0051011

(6) multimode tracking
(7) enhanced filtering
(8) colour processing.

ADEPT 33 automatic video tracker

The ADEPT 33 tracker is the latest in the range of Octec video trackers and is a new tracker with smaller form factor for requirements where size and weight are more critical. It comprises an integrated software design on a single Eurocard for easy integration into other manufacturers electronics.

The ADEPT 33 unit has a great deal of flexibility in its design, allowing closed loop control of a platform and easy integration to control panels. Built-in interfaces include RS-232/422 and the video output includes video with symbology overlay. Build standards are available to meet different standards from commercial up to full MIL specification.

ADEPT 30+ VME-based automatic video tracking system ***1999***/0051010

Specifications

Mechanical: single height VME board; forced air or conduction cooled variants available
Dimensions: 100 × 160 mm single Euro (3U) format
Power: +5 V 2.5A, +12 V 0.2 A, –12 V 0.2 A
Video input: composite video 625/525 line CCIR or RS-170
No of Video Inputs: 2
Track modes: centroid, correlation, edge, scenelock, multiple target track (optional)
Video output: 1 with symbology overlay
Interfaces available: RS-232/422
Automatic video detection: variable from 2 to 90% of FoV

Operational status

Some 500 ADEPT 30 trackers have been shipped to customers worldwide and it continues in full production. ADEPT 33 is in full production for a NATO UAV.

VSG 30 image processing system

The VSG 30 image processing system comprises a board, using the VMEbus format, which provides a range of complementary facilities to the ADEPT tracker range. The board provides for three video inputs in either monochrome or colour. One or more of these can be passed through the image processing modules, which provide a range of facilities including electronic zoom, freeze frame, frame to frame integration and filtering. Some of these processing facilities may be combined to provide more complex filtering.

A full colour symbology generator is supplied, operating in either monochrome, or in up to 15 selectable colours.

VSG 30 may be supplied with a separate single-board computer to provide additional facilities such as frame grabbing and storage in a modular system. Selection of inputs, outputs and filtering can all be remotely controlled over datalinks.

The VSG 30 is built in a range of environmental specifications to match the ADEPT 30.

Specifications

Video Inputs: 3
Format: Composite Video 1.0 V p-p
Lines: 625/525 Line
Fields: 50/60 Hz

Standards: PAL, NTSC, Y/C
Video Output: 6 max
Format: Composite (3), Y/C (3), RGB (1)
Level: 1.0 V p-p
Mechanical: VMEbus
Dimensions: 233.4 × 160 mm double Euro

Operational status
VSG 30 entered production in the third quarter of 1998.

Contractor
Octec Ltd.

VERIFIED

CAATS Compact Airborne Automatic video Tracker

CAATS is a variant of the Octec ADEPT series video trackers, and has been configured specifically as commercial ruggedised equipment for use in aircraft and helicopters. It is extremely light in weight and of small dimensions, thus making retrofitting a straightforward task.

The interfaces and control techniques available in the standard unit allow for easy integration into existing airborne electro-optical surveillance systems.

The introduction of automatic tracking for airborne systems has proved to be a great advantage in reducing operator fatigue and increasing general tracking capability. Centroid and screen lock modes are provided, the latter particularly useful in locking on to a scene in the field of view.

Specifications
Weight: 3 kg
Dimensions:
(tracking unit) 250 × 300 × 82 mm
Power: 28 V DC, 2 A (max) or 115/230 V AC
Video input: 50-60 Hz, RS-170 or CCIR
No of video inputs: 2
Tracking modes: centroid and screenlock. Automatic target detection selectable over 90% of FoV.

Operational status
Selected by UK government agencies and US DoD.

Contractor
Octec Ltd.

VERIFIED

Compact Airborne Automatic video Tracking System (CAATS) 0106292

CAATS 2 - Compact Airborne Automatic video Tracker

This new unit offers significant advantages over the earlier CAATS system, which remains in production.

CAATS 2 provides the same or better tracking performance but is lighter, is less than half the size, and uses less power than the original design. It is a sealed unit using conduction cooling. Several versions are available to meet different environmental scenarios. CAATS 2 also has the capability, using daughter boards, to enhance performance including electronic image stabilisation and input of digital video.

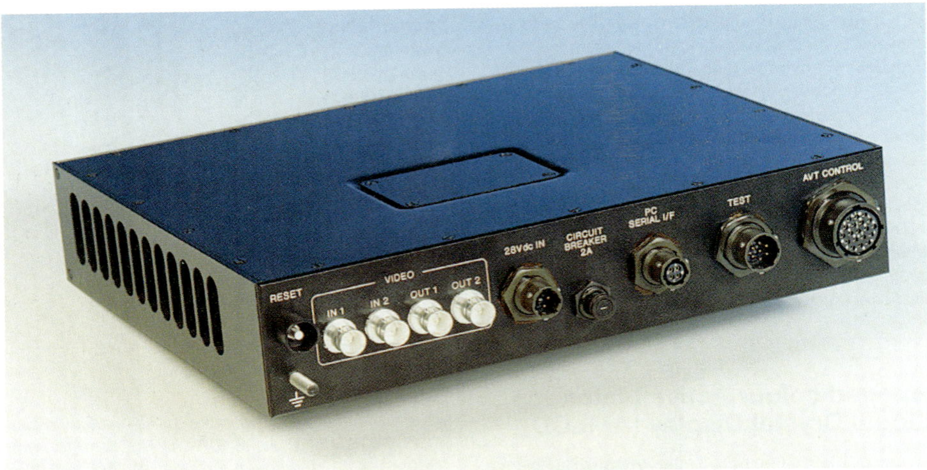

CAATS 2 automatic video tracker 0106414

The CAATS 'Scene Lock' tracking feature is retained and performance improved. This is a particularly important feature for the helicopter fit. Centroid tracking is also provided together with a range of submodes, including automatic cueing, which allows optimisation for particular target scenarios. A video symbology generator is integrated to allow on-screen data to be shown and recorded.

Specifications
Weight: 2.8 kg
Dimensions: 250 (W) × 210 (D) × 50 mm (H)
Power: 28 V DC, 1.0 A mean or 115/220V AC using adaptor
Video input: RS-170 or CCIR video input 50/60 Hz
No of video inputs: 2
Tracking modes: centroid, scenelock (other modes optional)
Video output: 1 with symbology overlay.
Automatic video detection: variable from 2-90% of FoV

Operational status
First shipments March 1997.

Contractor
Octec Ltd.

VERIFIED

Integrated Video System (IVS)

IVS is a fully integrated video system which provides distribution for a number of video sources, extensive image processing, the addition of colour symbology, tracking of a number of targets and distribution of video to workstations around the aircraft.

The equipment is packaged to meet airborne environmental and EMC standards. The units include several standard Octec assemblies, including the full range of Octec image processing cards, as follows:

1. VSG 30; an image-processing system.
2. VDB 30, a 16 × 16 video routing switcher, which can be programmed to select any of the 16 video input signals. 8 TTL inputs can also be programmed independently from 8 TTL inputs.
3. VSG 15, a compact (PMC) video routing switcher providing four video outputs, which can be programmed from four video input signals. It also provides a monochrome symbology facility.
4. IMP 15, a PMC card offering a range of complimentary facilities similar to that of the VSG 30, but in a PMC format.

Both the IMP 15 and VSG 15 can be hosted as a mezzanine board on any board capable of hosting PMC modules.

The Integrated Video System is a system tailored to meet individual airborne requirements using a range of standard modules.

Specifications
Mechanical: aluminium alloy chassis, covers and panels
Dimensions (H × W × D): 133 × 486.2 × 263 mm
Weight: 6 kg
Power: dependent on specific application

Operational status
In production for a NATO ATR 42 maritime patrol aircraft.

Contractor
Octec Ltd.

VERIFIED

SLAved Searchlight System for helicopters (SLASS)

The SLASS SLAved Searchlight System converts a standard remotely controlled searchlight to a servo-controlled system. When installed, the system provides the means to slave the searchlight to an E-O gimbal system, so that it automatically follows and points in the same direction as the gimbal. It consists of a servo control module and a remote-control unit which is used in conjunction with an E-O gimbal system. It operates with Spectrolab and other searchlight systems.

SLASS increases the performance of existing searchlight/E-O gimbal systems by reducing operator fatigue and minimising operator disorientation when his view changes to and from the video monitor scene and the outside illuminated scene.

Specifications
Dimensions (W × D × H): 127 × 165 × 38 mm
Weight: 1.0 kg
Power: 28 V DC 0.5 A mean
115/230 V with AC power converter
Environmental: temperature operating: –40 to +70° C
humidity: 10 to 80% non-condensing

Operational status
SLASS is in production for US and European markets.

Contractor
Octec Ltd.

VERIFIED

EPIC identification thermal imaging

The EPIC thermal imaging system is constructed from the UK Defence Evaluation and Research Agency (DERA) ruggedised STAIRS C modules, to provide thermal imaging for all land, sea and air platforms. EPIC offers greatly improved target recognition and identification ranges as a result of the extended spatial resolution, twice that of many first-generation systems. EPIC uses high-resolution SVGA and standard TV display formats. The module interfaces have been defined to ease system integration, maintenance and testing.

Specifications
Field of view:
(single) 5 × 3°
(dual) 5 × 3° and 17.5 × 10.5°
Optical aperture: 110 mm (telescope)
Sightline stability: ≤0.1 mrad

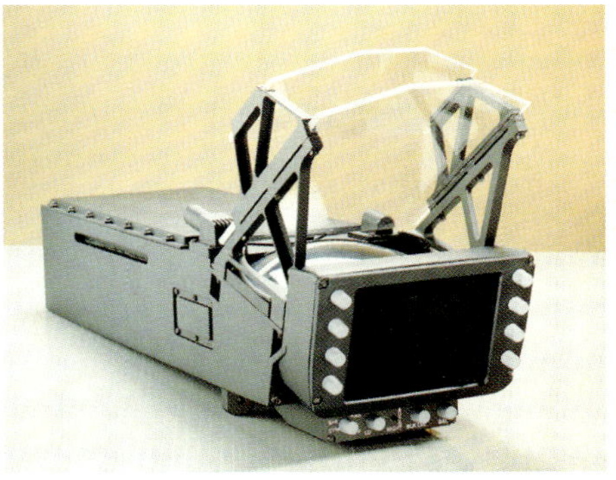

The low-profile HUD frees valuable panel space
0106294

The Smiths Industries' head-up display for the US Navy T-45C Goshawk 0018153

Operational status

Under development for new aircraft or retrofit applications in high-performance military aircraft.

Contractor

Smiths Industries Aerospace.

VERIFIED

Training Head-Up Display (HUD) system

The training head-up display is a compact low-weight system suitable for a wide range of light attack and training aircraft. Proven off-the-shelf equipment has been combined with significant growth provision for additional computing, interfacing and graphics generation.

The HUD has a total field of view of 25°. A display electronics unit responds to analogue, synchro, discrete and databus interfaces to generate HUD symbology for navigation, flight and weapon aiming data. In addition, its expansion capability has been exploited in order to provide the raster graphics for four head-down displays. A data entry panel is integral with the HUD for system control and input. A video recording system records information for post-flight debriefing. The baseline configuration features F/A-18 style symbology.

Operational status

In production for the Cockpit 21 installation of the US Navy T-45C Goshawk.

Contractor

Smiths Industries Aerospace.

VERIFIED

Type 1502 Head-Up Display (HUD)

The Type 1502 HUD is of modular construction. The unit is highly reliable and is easy to maintain; cost, volume and weight have been reduced without any compromise over optical accuracy and capability. The Type 1502 is suitable for installation in many new build attack aircraft or for retrofit, particularly where space or volume is limited.

The total field of view is 25°, achieved by using a 140 mm diameter exit lens, truncated fore and aft, and dual combiner glasses. Stroke symbology, raster video imagery or hybrid formats can be displayed. A video camera and an electronic variable standby sight can be incorporated as options. A customised data entry panel is an integral part of the HUD and enables the pilot to control the aircraft's nav/attack system and HUD mode. The Type 1502 is precision hard-mounted on the aircraft and requires no on-aircraft harmonisation.

Operational status

In service/production for the F-5E/F and Hawk 100 and 200.

Contractor

Smiths Industries Aerospace.

VERIFIED

Type 3000 integrated colour display

The Type 3000 is a self-contained colour multipurpose display designed to be integrated into aircraft navigation and attack systems. All primary flight and navigation data is displayed, replacing conventional attitude and horizontal situation indicators, plus a route map representation. In addition, flight plan, system status including fuel information and engine data, radar warning receiver and maintenance information can be displayed under the control of the 17 soft keys around the front panel.

The unit combines a graphics generator, deflection amplifiers, high-voltage power supplies and a full shadow-mask CRT into a single display unit. The display data and control functions are transmitted via the aircraft MIL-STD-1553B databus. The latest large-scale integration and hybrid packaging techniques have resulted in a compact self-contained unit.

The Smiths Industries' Type 1502 head-up display
0106396

Specifications

Dimensions: 162 × 162 × 317 mm
Weight: 11 kg
Power supply: 115 V AC, 400 Hz, 3 phase, 200 VA
Usable screen area: 127 × 127 mm
Reliability: 1,500 h MTBF

Operational status

In production for the BAE Systems Hawk 100 and 200.

Contractor

Smiths Industries Aerospace.

VERIFIED

70 mm film format cassette-loaded panoramic cameras

Over 10 variants of the cassette-loaded 70 mm film format panoramic camera have been produced since the camera was first introduced into service in the early 1970s. The camera is fitted with a high-resolution 76 mm focal length lens with a cycling rate of 2 to 10 frames/s. It is normally operated in the autocycle mode but some variants may be fired at intervals in the pulse mode when interfaced with a suitable control. The cameras are inherently very reliable, due to the design concept.

The camera is suitable for low and medium reconnaissance by day or as a medium- to high-altitude tracking camera operated in the pulse mode in co-ordination with other sensors. It may be fitted internally or in a pod on a wide range of fixed-wing aircraft, helicopters, RPVs, drones or UMA.

Principal features of the cameras include a 76 mm high-resolution lens, autocycle or pulse control, optional remote control and optional data head.

The most common types in service are the 751, 751A-E, 752, 753, 755, 910 and 914 A/B/C.

Operational status

Large numbers of the cameras are in service worldwide. Most variants are available for production. The Type 914A is in service on combat aircraft and drones.

Contractor

W Vinten Ltd.

VERIFIED

70 mm film format framing cameras

More than 20 variants of the W Vinten Ltd 70 mm film format framing camera have been produced over the past 30 years. Thousands of the cameras, which are simple to operate and require minimum maintenance, are in service worldwide. The compact cameras are suitable for low- and medium-altitude reconnaissance by day and may be fitted in a wide range of fixed-wing aircraft, helicopters, RPVs, drones and UMA, internally or in a pod.

Principal features of the 70 mm film format framing camera are magazine loading, a range of interchangeable lenses from 38 up to 280 mm, high-speed focal plane shutters and optional data head.

The most common types in service are the F95 Mks 1-10 and the Types 360, 362, 512, 518, 547, 591, 618, 880 and 881.

Operational status

Most camera variants are available for production.

Contractor

W Vinten Ltd.

VERIFIED

70 mm film format magazine-loaded panoramic cameras

The Type 900 magazine-loaded panoramic camera entered production in 1983. It is fitted with a high-resolution 76 mm focal length lens with a cycling rate of 2-17 frames/s. The Type 900A operates in the autocycle mode only, whereas the Type 900B, introduced into service in 1993, operates in either pulse or autocycle modes. The magazine accepts up to 750 ft of standard 0.004 in thick film or 1,000 ft of thin-based film.

Roles for the Type 900 panoramic camera include low- and medium-altitude reconnaissance by day, as a tracking camera for use with other sensor systems and area coverage. The camera is normally fitted inboard in aircraft or in a pod.

Principal features of the Type 900 include a 76 mm high-resolution lens, magazine loading, autocycle or pulse control (Type 900B) and a data head as standard.

Operational status

In production. Types 900A, 900B and F152 are in service.

Contractor

W Vinten Ltd.

VERIFIED

IRLS 4000 infra-red linescan sensor

The IRLS 4000 infra-red linescan sensor has been designed to meet the exacting requirements for day and night reconnaissance by high-speed aircraft and may be installed internally in aircraft, RPVs, drones or UMA or in podded installations. The sensor is fitted on the Royal Air Force Tornado GR. Mk 1A reconnaissance aircraft and has been proved in combat. The IRLS 4000 is a high-performance day and night airborne sensor operating in the 8 to 14 μm waveband with horizon-to-horizon across-track coverage. The system comprises a scanning head and an electronics unit. The linescan is cooled by a continuous rated split-Stirling closed-cycle cooling engine.

Imagery is recorded on an S-VHS airborne video cassette recorder and displayed on an Imagery Display Processor (IDP). The IDP may be located on the aircraft, providing onboard real-time display, or on the ground for post-flight exploitation. The imagery may also be transmitted to the ground when suitable datalinking equipment is fitted to the aircraft.

Specifications

Dimensions:
(sensor) 309 × 254 × 247 mm
(electronics unit) 302 × 337 × 202 mm
Weight:
(sensor) 10.5 kg
(electronics unit) 12.5 kg
Power supply: 200 V AC, 400 Hz, 3 phase, 280 W
28 V DC, 115 W
Resolution: <1 mrad
Sensitivity: <0.2° NET
Field of view: 190° scanned, 180° displayed
Stabilisation: ±30° (electronically roll stabilised)
Outputs: Video, line sync, sample clock, BITE, system ready

Operational status

In service with the UK Royal Air Force Tornado GR. Mk1A and Royal Saudi Air Force Tornado IDS. Other export sales have also been achieved.

Contractor

W Vinten Ltd.

VERIFIED

Tornado Infra-Red Reconnaissance System (TIRRS)

The TIRRS for the Tornado GR.Mk1A incorporates two Side-Looking Infra-Red (SLIR) sensors and an IRLS 4000 Infra-Red Linescan manufactured by W Vinten Ltd. The SLIR sensors embody elements of the UK TICM SPRITE detector programme.

TIRRS is the first system of its type was first employed operationally during the 1990-91 Gulf War. The system is mounted internally in the fuselage and provides horizon-to-horizon across-track coverage with roll stabilisation and gives a real-time display in the cockpit.

The output from the sensors is recorded on videotape and the operator in the rear cockpit can monitor the imagery while the sortie is under way, directly from the sensors or by replaying from the videotape recorders. The system offers a high-definition thermal picture which can be magnified or enhanced in flight.

Operational status

In service in UK Royal Air Force Tornado GR. Mk1A and Royal Saudi Air Force Tornado IDS reconnaissance aircraft.

Contractor

W Vinten Ltd.

VERIFIED

Type 214 infra-red linescan sensor

Linescan 214 is a high-performance day and night sensor operating in the 8-14 μm waveband. It has been designed for fixed-wing aircraft or helicopter airborne surveillance, with provision for in-flight imagery recording on film and, with additional equipment, real-time TV imagery viewing and recording.

The infra-red signals from scanned terrain are recorded on 70 mm film contained within the unit. The sensor also produces a video output which is recorded on videotape in a linescan format video cassette recorder. The imagery is displayed on a TV monitor via a suitable scan converter such as the W Vinten Ltd Imagery Display Processor (IDP). Imagery is normally displayed on the IDP at a ground station. However, where a suitable installation can be provided on the aircraft, imagery may be displayed in real time, recorded and/or transmitted to the ground station via a datalink.

The Linescan 214 sensor may be installed in the aircraft fuselage or mounted in a pod under the fuselage or wing. The sensor and the cooling pack are low volume, lightweight and have low power consumption. A separate control panel is supplied for remote operation.

Specifications

Dimensions:
(linescan sensor) 258 × 437 × 317 mm
(cooling pack) 272 × 140 × 105 mm
(control unit) 113 × 146 × 113 mm
Weight:
(linescan sensor) 12 kg
(cooling pack) 1.7 kg
(control unit) 1.1 kg
Power supply: 28 V DC, 2.5 A

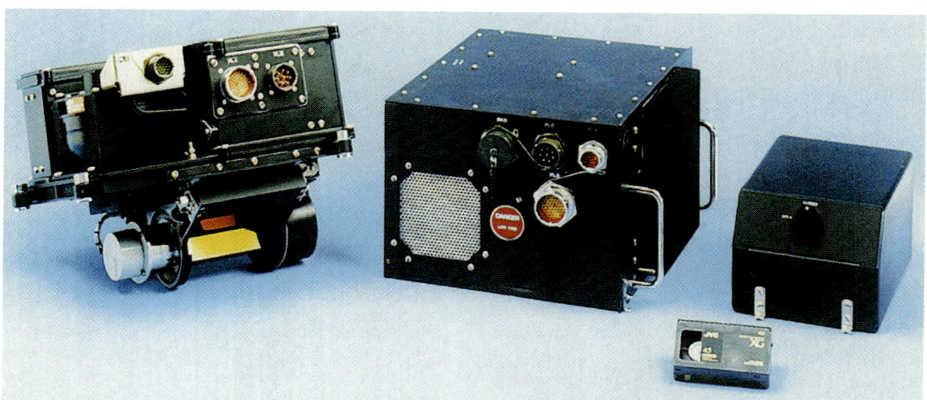

Type 4000 IRLS sensor system including cassette recorder 0106295

Resolution: 1.5 mrad
Sensitivity: 0.25° NET
V/H range: 0.2-0.7 rad/s
Field of view: 120° transverse

Operational status

No longer in production. Several hundred 201 systems have been delivered and are in service in a number of countries. The 201 system is, for example, the standard equipment aboard the Canadair CL-89 surveillance RPV. Another member of this series is the Linescan 212 system designed for use in light aircraft and helicopters. A number of these systems is also in service throughout the world.

Contractor

W Vinten Ltd.

VERIFIED

Type 401 infra-red linescan sensor

Linescan 401 is a high-performance day and night sensor operating in the 8-14 μm waveband. Infra-red signals from scanned terrain are recorded on 70 mm film contained within the unit.

The Linescan 401 sensor may be installed within the aircraft structure or in a pod under the fuselage or wing. The sensor is roll-stabilised and may be slewed 30° either side for scanning from the nadir out to the horizon.

Linescan 401 has been specifically designed for low-level day and night high-speed tactical airborne reconnaissance at speeds up to 600 kt and altitudes down to 200 ft.

Specifications

Dimensions:
(linescan) 604 × 320 × 280 mm
(roll swept diameter) 366 mm
(cooling pack) 390 × 230 × 120 mm
Weight:
(linescan) 34 kg
(cooling pack) 9 kg
Power supply: 200 V AC, 400 Hz, 3 phase, 600 W
28 V DC, 2 A
Resolution: >1 mrad
Sensitivity: >0.2° NET
V/H range: 0.025-5 rad/s
Field of view: 120° (30° left/right offset)
Reliability: >200 h MTBF

Operational status

In production and service. Seven countries have Linescan 401s in service and it is fitted to the UK Royal Air Force's Jaguar reconnaissance aircraft.

Contractor

W Vinten Ltd.

VERIFIED

Type 690 (126 mm) film format framing camera

The Type 690 (126 mm) film format framing camera entered production in 1976 and has been in continuous production with many enhancements since then. It is designed for day low-, medium- and high-altitude reconnaissance in the tactical standoff and LOng Range Oblique Photography (LOROP) roles and is fitted internally in fixed-wing aircraft and helicopters and in podded systems.

The camera was first introduced with 18 in (457 mm) and 36 in (914 mm) lenses, to be followed by 75 mm and 150 mm lenses in 1983. A high-resolution 900 mm lens replaced the 36 in lens in 1984 and a 300 mm lens was introduced in 1992. A 450 mm lens was added to the range in 1994. Both the 900 mm and 450 mm lenses are temperature and pressure controlled with a focus-for-range capability.

Type 690 cameras are most frequently fitted in pods, while others are fitted in fixed installations internally on the aircraft. In podded applications, the camera is fitted into a rotating nosecone assembly imaging through a mirror-box, enabling it to be rotated from vertical to port or starboard. On take-off and landing the camera window is rotated to point upwards to reduce contamination. The camera and rotating nose mechanism are controlled automatically through a dedicated sensor interface unit that forms part of the reconnaissance management system.

Some cameras are operated in helicopters on special anti-vibration mounts.

Operational status

In production. The Types 690, 690A and F144 are in service. A new ultra-high-resolution 900 mm lens is also in production.

Contractor

W Vinten Ltd.

VERIFIED

Type 8040B EO sensor mounted in standoff module **2000**/0062748

Type 8040B electro-optical sensor

The Type 8040B utilises the same technology as the Type 8010, but provides a replacement for the 126 mm film format Type 690 camera.

The Type 8040B sensor has an 8 μm 12,288 element charge-coupled device mounted in the focal plane. The sensor electronics record the imagery on to either a digital or analogue output.

The pixel size within the focal plane array gives 'photographic quality' resolution in the imagery. When associated with the advanced technology 450 mm lens, which offers a corresponding high-modulation transfer function, the 8040B can offer long-range resolution previously only associated with longer focal length systems.

The Type 8040B EO sensor is designed to operate between 200 and 40,000 ft at slant ranges of 300 m to 40 km. Although the spatial resolution is comparable with that of film from the Type 690 film camera, the Type 8040B provides a significant advantage over film against low-contrast targets, especially when imaged through a hazy atmosphere.

Specifications

Dimensions: depending on podded or internal installation
Weight: depending on podded or internal installation
Lenses: 450 mm
Fields of view: 12.4°, 6.2°

Contractor

W Vinten Ltd.

VERIFIED

Type 950/955 cameras

The Type 950 and 955 panoramic cameras are designed to meet operational requirements for low-, medium- and high-altitude reconnaissance. The cameras can be fitted on a variety of airborne platforms including aircraft and UMA. Both the Type 950 and Type 955 are of compact design with a high degree of commonality which enables the cameras to be installed as role change units. The design enables the cameras to be fitted in small diameter reconnaissance pods in keeping with the dimensions of other sensors including IRLS, FLIR and EO sensors.

Both camera types are digitally controlled for operations on aircraft equipped with a MIL-STD-1553 databus or with conventional aircraft avionics systems. The camera control systems allow in-flight selectable across-track coverage angles from wide to narrow field of view.

The lenses have been designed specifically to meet the resolution requirements for a range of operational requirements at various altitudes. The lenses are very high resolution and are prefocused for optimum performance over the specified range of pressure, temperature and altitude without the need to refocus the lens while airborne. The cameras record on 5 in wide standard or thin-based film.

The Type 950 camera features a high-resolution 150 mm lens for low- and medium-altitude reconnaissance. The Type 955 camera features a 300 mm lens for medium- and high-altitude reconnaissance.

Specifications

Weight:
(Type 950A) 34 kg
(Type 950C) 41 kg
(Type 955B) 50 kg
Power supply: 28 V DC

Contractor

W Vinten Ltd.

VERIFIED

Type 8040B E-O sensor image 0106297

Type 8010 electro-optical sensor

The Type 8010 electro-optical sensor is a form and fit replacement for the F95 and other 70 mm film format framing cameras such as the Types 360, 518 and 544 manufactured by W Vinten Ltd, and can therefore be retrofitted into existing reconnaissance systems as well as integrated into new installations. The primary role for the sensor is day low- and medium-altitude tactical reconnaissance and surveillance. A secondary role is day low- and medium-altitude surveillance to aid various government agencies, including the police, coastguard, drug enforcement and fishery protection agencies.

The Type 8010 sensor may be fitted internally in a wide range of aircraft, RPVs, drones and UMA or in podded systems.

Principal features of the sensor include high spatial resolution, contrast stretch for low light and haze penetration, imagery recorded on S-VHS videotape, capability for onboard real-time display and in-flight data transmission and imagery exploitation on video monitors.

Specifications

Dimensions:
226 × 181 × 264 mm (1.5 and 3 in lenses)
226 × 181 × 258 mm (6 in lens)

Type 8010 E-O sensor compatible with 152 mm lens 0106296

Resolution:
(A version): 12 μm, 4,096 elements
(B version): 8 μm, 6,144 elements
Weight: 7.8 kg plus lens
Lenses: 152 mm, 76 mm, 38 mm
Field of view: 18.3°, 35.7°, 65.6°

Operational status

In production and in service on a variety of aircraft types.

Contractor

W Vinten Ltd.

VERIFIED

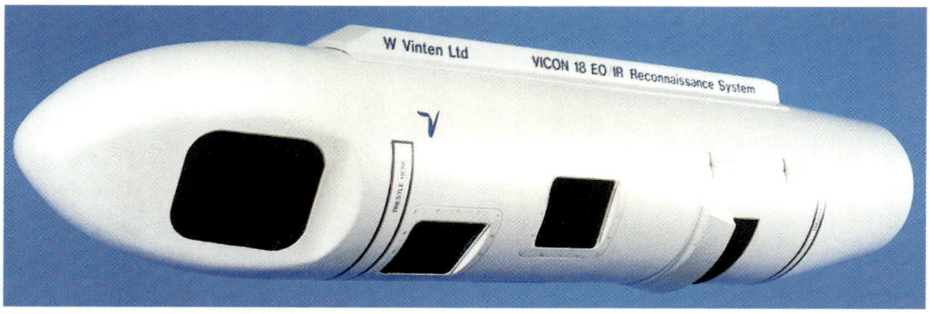

VICON 18 electro-optical reconnaissance system **2000**/0062749

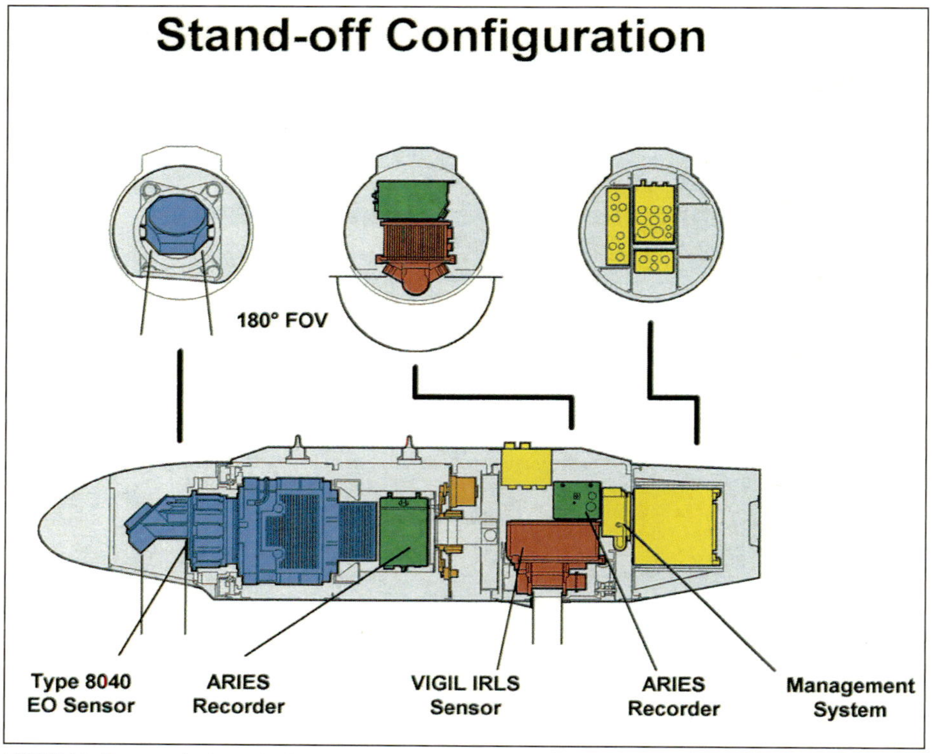

VICON 18 electro-optical stand-off configuration **2000**/0062750

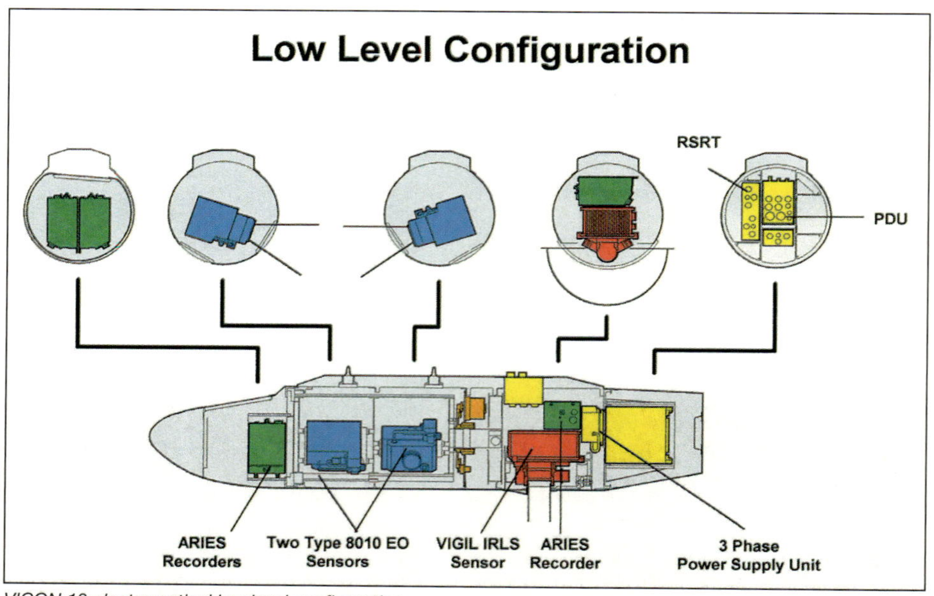

VICON 18 electro-optical low-level configuration **2000**/0062751

VICON 18 Series 601 reconnaissance system

The VICON 18 Series 601 pod is one of the wide range of VICON 18 reconnaissance pods designed for a variety of operational roles. Specifically, the Series 601 provides day time reconnaissance from a low-cost, lightweight pod for use at low, medium and high altitudes. Implicit in the pod design is the ability to integrate the pod with a variety of airframes. The pod is currently fully flight-cleared on the BAE Systems Tornado, Jaguar, Harrier and Hawk aircraft. VICON 18 Series pods are also operational on F-5, MiG, Learjet and many other aircraft types.

Film variant

The VICON 18 Series 601 pod contains two sensors: Type 690 (F144) 126 mm film format framing camera with a 450 mm focal length high-resolution lens which is mounted in a rotatable nose cone;

Type 900B 70 mm film format panoramic camera of 76 mm focal length which is mounted in the rear centre section.

The sensors are driven by a VICON 2000 Reconnaissance Management System under the control of the pilot or systems operator. The VICON 2000 System can be interfaced to a wide range of aircraft avionics systems including the MIL-STD-1553B databus. Navigational data can be imprinted on the imagery of both sensors.

The pod provides capabilities for the following operational roles: tactical standoff photography; long-range oblique photography; low-level tactical reconnaissance; area coverage.

Specifications

Dimensions:
(overall length) 2,250 mm
(standard diameter) 457.2 mm
(depth over saddle) 508 mm
(weight (estimated)) 254 kg
(centre of gravity) midway between the 14 in or 30 in attachments
Performance:
(sea level): up to 730 kt (M1.1) at 36,000 ft (11,000 m): up to 1,033 kt (M1.8)
Symmetrical normal (accelerations): +ve 9.0 *g*
–ve 2.0 *g*
Asymmetrical normal (acceleration): +ve 4.0 *g*
–ve 2.0 *g*
Rate of roll: 150°/s

Operational status

In service and in production.

VICON 18 Series 601 electro-optical/infra-red reconnaissance system

Electro-optical and infra-red configurations of the VICON 18 Series 601 have been developed and are now in series production.

Standoff configuration

The standoff configuration is fitted with the Type 8040B electro-optical sensor and the VIGIL Infra-Red LineScan (IRLS). The electro-optical sensor incorporates an 8μm 12,288 pixel Charge Coupled Device (CCD) linear array and an ultra-high-resolution 450 mm focal length lens. The lens focuses automatically from 150 m to infinity, providing an excellent standoff capability. This sensor, which has a field-of-view of 12.4°, is fitted in the rotating nose assembly of the pod giving a field-of-regard of 180° from horizon-to-horizon across the track of the aircraft. The VIGIL IRLS, which is fitted in the rear of the pod, provides high resolution day and night horizon-to-horizon reconnaissance imagery at low and medium altitude.

Low-level configuration

The low-level configuration is equipped with two Type 8010 electro-optical sensors and a VIGIL IRLS. The Type 8010 sensors, which are fitted with a 12μm 4,096 pixel CCD linear array, may be fitted with a range of lenses (38, 76 or 152 mm local length). These electro-optical sensors can be configured in a number of oblique or vertical split pair configurations, which in combination with the IR sensor, meet the requirements for a wide range of tactical reconnaissance roles.

Operational status

The VICON 18 Series 601 GP (1) Electro-Optical (EO) pod and Ground Imagery Exploitation System (GIES) have been selected by the UK Royal Air Force to replace the Jaguar aircraft's previous generation reconnaissance assets. The pod has a role change capability providing either day E-O short to medium range standoff, or low-level reconnaissance, together with day/night Infra-Red (IR) capability.

The Jaguar Replacement Reconnaissance Pod (JRRP) programme has completed low-level sensor performance trials, designed to evaluate the performance of the JRRP EO and IR linescan sensors in both day and night conditions.

VICON 70 pod 0106299

Cessna Citation fitted with VICON 70 pod 0106298

Avionics integration of the JRRP to the Jaguar GR3a (Jaguar '97) is being conducted by the Defence Evaluation Research Agency (DERA).

The JRRP sensors are controlled automatically by DERA-designed aircraft software, via the Vinten advanced reconnaissance management unit located in the pod. This minimises pilot workload and introduces sensor pointing directly from the Helmet-Mounted System (HMS). Flight testing has been successful and continues.

Contractor
W Vinten Ltd.

VERIFIED

VICON 70 general purpose modular pod

Since the introduction of the VICON 70 Series of general purpose modular pods in the early 1980s over 50 different variants have been designed or built to meet a wide range of reconnaissance, surveillance, night illumination, Forward Looking Infra-Red (FLIR) and Instrumentation requirements.

Variants have been supplied for operations on helicopters, light aircraft, civil aircraft and a range of fighter bombers flying at speeds up to M2.2.

Most variants have been designed to meet Vinten requirements but many have been built for other Original Equipment Manufacturers (OEM) particularly for research and development purposes.

Reconnaissance and Surveillance
The VICON 70 pod will accept a wide range of reconnaissance and surveillance sensors including film framing and panoramic cameras recording on 70 mm film, Infra-red Linescan (IRLS) and Electro-Optical (E-O) sensors recording on either 70 mm film or on video-tape and night illumination systems operating with either white light or IR flash.

Systems are designed to meet customers' operational requirements including the integration of CCD TV video sighting devices with onboard real-time display.

Combinations of day and night sensors have been supplied in VICON 70 pods from low V/H applications on helicopters to V/H 5 on fighter-bomber aircraft.

FLIR
A number of different FLIRs including GEC TICM II, HUGHES, and IR-18 have been integrated into VICON 70 pods for various OEMs and customers for a number of applications including navigation and attack.

Vinten-supplied FLIR pods are designed for reconnaissance and surveillance operations and are fitted with the GEC TICM II FLIR sensor, the associated processing electronics, and detector cooling system.

The imagery is recorded on videotape and in some applications the imagery is displayed in real time on board the aircraft on a video monitor.

Instrumentation pods
Variants of the VICON 70 pod have been designed to accept Photo-Sonics high-speed instrumentation cameras for recording release of weapons and other stores from combat aircraft. The pods will accept the 16 mm IVN or 16 mm IPL cameras operated from either aircraft or battery power.

Specifications
Performance: Variants cleared up to M 2.2
Attachment lugs:
Twin suspension. NATO standard 14 in or to customer specification
Connectors: Location and type to customer specification
Standard diameter: 355 mm (14 in)
Length: from 1,447 mm (57 in) minimum up to 2,300 mm (100 in) maximum approx.

Operational status
In service and in production.

Saab-British Aerospace has selected a variant of the VICON 70, designated VICON 70 Series 72c, for the Gripen aircraft. The pod will be fitted on the underfuselage pylon. A variety of electro-optic and infra-red sensors may be installed which, when combined with onboard sensors such as the radar, will enable Gripen to undertake the full spectrum of reconnaissance missions.

Contractor
W Vinten Ltd.

VERIFIED

VICON 2000 digital reconnaissance management system

The VICON 2000 modular digital reconnaissance management system entered service in 1986. It is now in service on a wide variety of combat aircraft types, interfacing with 64 K NAVHARS, MIL-STD-1553B databus and ARINC 407A, 429/10, 561 and 571. Some podded reconnaissance systems operate on multiple aircraft types, with minimal changes required to move the system from one aircraft type to another.

The VICON 2000 system is designed for use with all types of reconnaissance sensors fitted either internally or in podded systems. The system normally comprises a navigational interface unit, systems management unit and Sensor Interface Units (SIU). New sensors incorporate the SIU, reducing the number of LRUs in the system. The VICON 2000 may be controlled by a conventional cockpit control unit or via the aircraft computer system.

Components of the VICON 2000 are linked by RS-422A serial datalinks over two twisted pairs.

Operational status
In service in a wide variety of combat aircraft types.

Contractor
W Vinten Ltd.

VERIFIED

VIGIL Infra-Red Linescan Sensor (IRLS)

The VIGIL IRLS is a single LRU sensor which incorporates the electronics and the control interface

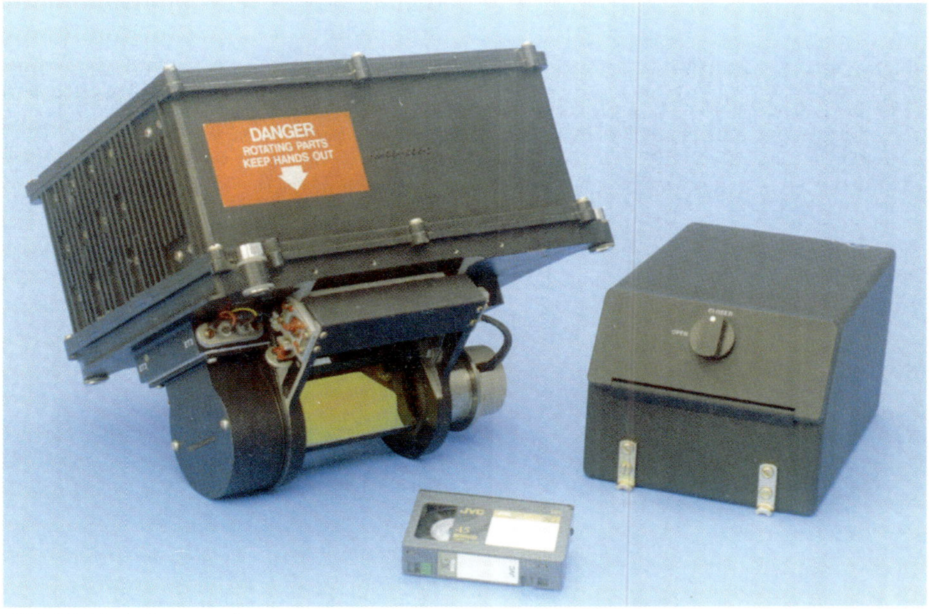

VIGIL IRLS including recorder 0106302

VIGIL IRLS flown at 1,000 ft AGL-day 0106300

VIGIL IRLS flown at 1,000 ft AGL-night 0106301

which may be either single discrete signals or RS-422 serial command link for interfacing to advanced avionic databusses.

VIGIL is designed for installation in RPVs, drones, UMA, helicopters and medium-performance fixed-wing aircraft up to a velocity/height ratio of 2.5.

It is a high-performance day and night airborne sensor with horizon-to-horizon coverage operating in the 8-14 μm waveband. The linescanner detector is a single element SPRITE CMT which is cooled by a continuous-rated Stirling closed-cycle cooling engine.

The infra-red radiation from the terrain is scanned by the sensor, producing continuous along-track imagery. VIGIL produces an analogue signal containing both line sync and video. This signal is suitable for direct interfacing with either video cassette recorder for airborne or ground recording, datalink transmitter for transmission to a ground station or an imagery display processor for scan conversion and image manipulation for display on a conventional video monitor.

Specifications

Dimensions: 309 × 254 × 247 mm
Weight: 10.5 kg
Stabilisation: Mode + 30° roll (electronically roll stabilised)
Mode 2 locked to airframe
Environmental: MIL-STD-810 and DEF STAN 07-55

Operational status

In full production and operational service.

Contractor

W Vinten Ltd.

VERIFIED

UNITED STATES OF AMERICA

AeroVision™ 2000 HMD (Helmet Mounted Display)

The *AeroVision 2000 HMD* is designed to meet the needs of law enforcement, paramilitary and military organisations. It is designed to interchange directly with the US Army Aviators' Night Vision System (ANVIS) helmet mountings and is compatible with the Gentex helmet. It is also able to display high-resolution images from external video sensors, while operating heads-up and out of the cockpit.

The Liquid Crystal Display (LCD) and associated optics provide full colour, high resolution, high brightness, wide field of view and 100% stereo overlap in a small lightweight configuration - total weight less than 0.9kg.

The design allows flight crew to use one helmet for all missions; the ANVIS goggles can be removed and rapidly substituted by the *AeroVision 2000 HMD.*

Contractor

AeroSolutions.

VERIFIED

AH-64 caution and warning system

This system provides caution and warning annunciation for the AH-64 Apache helicopter. It consists of a 60 station pilot panel, 24 station co-pilot panel and master caution warning panel. Subsystem inputs are monitored and fault information displayed as applicable on any or all three units. The legend modules are fully sunlight-readable and incorporate front lamp replacement.

Specifications

Dimensions:
(pilot panel) 146 × 206.3 × 177.8 mm
(co-pilot panel) 146 × 95.2 × 177.8 mm
(master caution) 193.5 × 25.4 × 69.8 mm
Weight:
(pilot panel) 4.4 kg
(co-pilot panel) 2.36 kg
(master caution) 0.73 kg

Operational status

In service in the AH-64 Apache helicopter.

Contractor

Aerospace Avionics.

VERIFIED

Caution and warning advisory panel and cockpit lighting controller

The all-solid-state caution and warning advisory panel continuously monitors specific subsystem inputs, displays fault information on an integral 30-channel annunciator panel and drives external warning caution advisory displays. Twenty annunciators are yellow and 10 are red. The unit also provides power to 112 separate external 5.1 V constant current cockpit annunciators. The panel interfaces with an external dimming control to provide discrete dim and a continuously variable lamp power for the annunciators.

The panel is sunlight-readable in 10,000 ft candle diffuse ambient light and is EMP- and nuclear-hardened.

Specifications

Dimensions: 200 × 72.4 × 279.4 mm
Weight: 3.4 kg
Power supply: 22-29 V DC and 6.5 V DC
Temperature range: –40 to +74°C
Reliability: 15,000 h MTBF

Contractor

Aerospace Avionics.

VERIFIED

Cockpit control panels (C-130J)

These cockpit control panels provide the functional interface between the operator and the applicable subsystems. They provide data, system status, controls and information displays.

Cockpit control panels (C-130J) 0106419

Interfaces are available through MIL-STD-1553B databus, ARINC 429 databus or via discrete wire signals. The system is available with Night Vision Imaging Systems (NVIS), non NVIS-compatible lighting, and with or without edge lighting panels.

Examples of current control panels designs include: aerial delivery; air conditioning; automatic flight control system; bleed air; caution and warning; cursor control; fire handles; flight select; fuel management; heading/course select; hoist and winch control; ice protection; landing gear; lighting control; pressurisation; radar control; and reference set/mode select.

Operational status
In production for C-130J.

Contractor
Aerospace Avionics.

VERIFIED

CV-Helo fuel management panel

The microprocessor-based fuel management panel provides fuel management and control during refuelling and fuel transfer operations on board the SH-60F CV-Helo. Fuel flow activity of the auxiliary fuel system is monitored via inputs from system sensors and displayed with indicator lights. The unit automatically controls sequential refuelling and fuel transfer functions.

Specifications
Dimensions: 146 × 120.6 × 165.1 mm
Weight: 2.72 kg
Power supply: 28 V DC
5 V AC for panel edge lighting
Processor: 2 MHz Z80 microprocessor, 1,024 bytes RAM, 4,096 bytes PROM
Reliability: 23,000 h MTBF

Operational status
In service in the US Navy Sikorsky SH-60F helicopter.

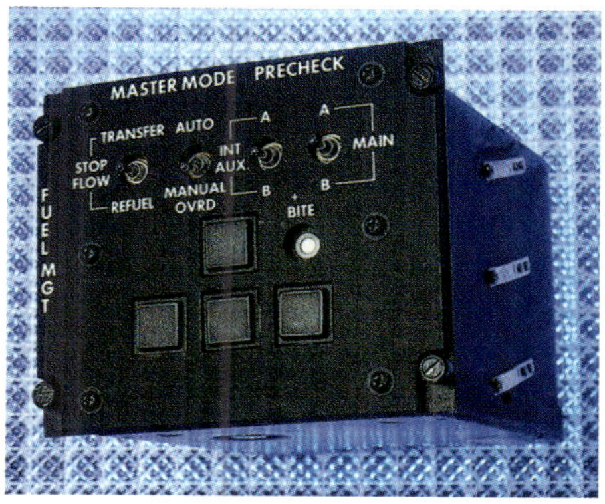

Fuel management panel in the US Navy Sikorsky SH-60F helicopter 0106397

Contractor
Aerospace Avionics.

VERIFIED

ACA Attitude Director Indicators (ADIs)

ACA 135070 ADI
Operating on the F-5 and F-16, the 3 in (76 mm), two-axis ACA 135070 ADI provides aircraft attitude information. In the flight director mode it also provides pitch and roll steering information. Glide slope information is presented on a separate pointer. When the flight director is not used, the roll steering pointer displays localiser information. The unit is military qualified and hermetically sealed.

Operational status
In service in the F-5 and F-16 of the Israel Defence Force.

ACA 129060 ADI
Presently used on the Black Hawk UH-60A, the 5 in (127 mm) ACA 129060 hermetically sealed military qualified ADI provides command as well as raw data information. The command pitch, roll and collective information is provided by Astronautics 3 cue flight director computer. Eyebrow annunciator lights provide go-around, decision height and marker beacon status. Glide slope, localiser and turn and slip information is also provided.

A similar unit used in the SH-60B Sea Hawk - the ACA 126370 - has demonstrated a MTBF in excess of 5,000 hours.

Operational status
In production and in service in the Sikorsky UH-60A Black Hawk.

ACA 131070 ADI
The ACA 131070 4 in (102 mm), two-axis unit is fitted in the Cobra AH-1S. It has cyclic pitch and roll and collective pitch steering pointers. In addition, glide slope, localiser and turn and slip information is presented. The hermetically sealed unit is military qualified.

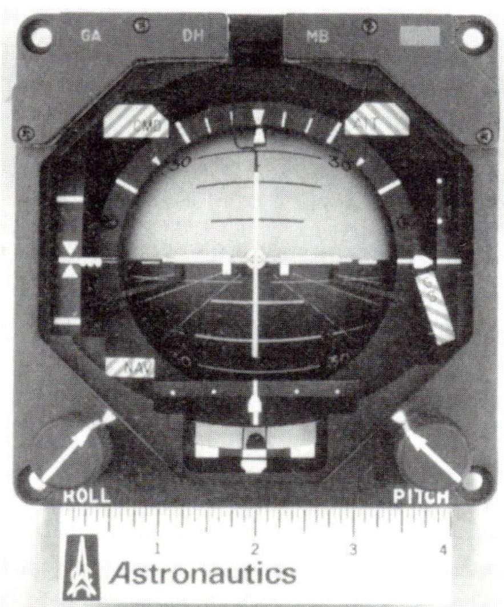

The ACA 129060 ADI as fitted in the Black Hawk UH-60A helicopter 0106398

Operational status
In service in the AH-1S Cobra helicopter.

ACA 135160 ADI
Presently being built for the T-45, the ACA 135160 is a 4 in (102 mm), three-axis ADI. It provides pitch, roll, heading and turn and slip information. Localiser and glide slope information is also provided. The instrument is military qualified.

ACA 137100 ADI
The ACA 137100 3 in (76 mm), three-axis ADI is presently in use on the F-16 and the AMX. It provides pitch, roll and heading as well as turn and slip information. Vertical and horizontal steering pointers provide glide slope and localiser information. The servos are failure monitored and the unit is hermetically sealed and military qualified.

Operational status
In service in the F-16 and Alenia/Embraer AMX.

Contractor
Astronautics Corporation of America.

VERIFIED

ACA Horizontal Situation Indicators (HSIs)

4 in (102 mm) HSI for the Hawk
British Aerospace is the customer for the 4 in (102 mm) HSI which is fitted to export versions of its Hawk light strike/trainer aircraft. Major features include a course bar indicator, to/from indicator, glide slope pointer, two bearing pointers, course set knob, course selection window and digital readout of range which is compatible with ARINC 568 digital input. The instrument meets full MIL-SPEC standards.

Operational status
In production and in service in the BAE Systems (Operations) Hawk.

ACA 113515 HSI for helicopters
The ACA 113515 HSI has proved to be a popular instrument for helicopters and has course bar indicator, to/from flag, glide slope pointer, two bearing pointers and course set knob with associated course selection window.

Operational status
In production and operational on Bell 212, 214, 412, Sikorsky S-76, S-61, Eurocopter Super Puma and Agusta AB 212, AB 412 helicopters.

ACA 123790 HSI
The ACA 123790 is a 4 in (102 mm) HSI to military specifications fitted in the Bell AH-1S Cobra helicopter. It has similar features to other members of the company's family of instruments, including a range readout compatible with ARINC 582.

The Astronautics 4 in (102 mm) horizontal situation indicator for the BAE System (Operations) Hawk 0106399

The instrument will also accept direct digital input from Doppler navigation.

Operational status
In production and in service in the Bell AH-1S helicopter.

ACA 126370 HSI for the SH-60B
The ACA 126370 HSI is a 5 in (127 mm) instrument featuring standard ARINC 407 inputs and outputs with digital interface and extensive built-in test equipment. It was designed for the US Navy Sikorsky Sea Hawk SH-60B LAMPS helicopter programme and is currently in production and operational. An MTBF in excess of 5,000 hours has been demonstrated from operational service.

Operational status
In service in the US Navy Sikorsky SH-60B LAMPS helicopter.

ACA 126460 HSI
A 3 in (76 mm) HSI, the ACA 126460 has been supplied for the F-5 and F-16 as well as the B-1B. A slightly different version is used on the AV-8C and a type suitable for use in simulators is also available.

Operational status
Operational in the F-5, F-16 and B-1B.

ACA 130500 HSI
Similar in presentation to other company HSIs, the 3 in (76 mm) ACA 130500 is supplied for the US Marine Corps AV-8B. The instrument will accept direct digital input from a Tacan.

Operational status
In service in the US Marine Corps AV-8B Harrier.

Contractor
Astronautics Corporation of America.

VERIFIED

Airborne 19 in colour display

Astronautics has designed and produced a large colour display for airborne use. This display is a full militarised high-resolution colour monitor. It has a usable diagonal screen size of 19 in (482.6 mm) and a pitch dot size of 0.31 mm, a feature which contributes to the exceptional resolution. It is equipped with extensive self-test diagnostics for fault isolation and reporting.

The colour display has numerous applications such as for airborne command, warning and control centres, JTIDS operations and control centres and ASW operations. It can easily be adapted to other environments.

Operational status

In production. Qualified for C-130 aircraft and surface ships and submarines.

Contractor

Astronautics Corporation of America.

VERIFIED

Airborne multifunction CRT displays

Astronautics' airborne multifunction (CRT) displays feature:

(1) High-resolution, high-brightness displays viewable in 10,000 ft lambert ambient light
(2) Automatic or manual brightness and contrast control
(3) Raster-scanned stroke and stroke/raster CRT operation
(4) Contrast enhancement filters
(5) Comprehensive - BIT circuitry, and compatibility with Night Vision Goggles (NVG).

Operational status

In service.

Contractor

Astronautics Corporation of America.

VERIFIED

Display processor for the 530MG helicopter

The display processor designed for the Boeing Helicopters 530MG Defender generates independently programmable symbology for the multifunction displays, control/display unit and the TOW missile display. Heart of the display system is a 16-bit high-speed general purpose processor programmed in high-order language that manages all video, digital and analogue inputs and outputs, performs real-time computations and drives two independent symbol generators each controlling raster and stroke displays.

Specifications

Dimensions: 240 × 145 × 270 mm
Weight: 8 kg
Inputs: (digital) high-speed parallel, (DC analogue) symbol brightness, 525- and/or 625-line video per EIA standard
Outputs: 4 raster video: 1 at 525/875-line, 2 at 525 per RS-330, 1 at 875 per RS-343
Symbology: azimuth and target, direction scales, altitude and rate of climb, pitch, roll, TOW firing symbology, navigation and multifunction display switch annotation.

Operational status

In production and in service in the Boeing 530MG helicopter.

Contractor

Astronautics Corporation of America.

VERIFIED

E-Scope radar repeater display for the Tornado IDS

The E-scope radar repeater display portrays video signals from the Raytheon Systems Company terrain-following and attack radar in the Panavia Tornado IDS variant. In the terrain-following mode it shows a hyperbolic shaped symbol representing the ground ahead, above which the aircraft symbol remains if the autopilot is functioning correctly. Topographical information can be shown in the ground-mapping mode. For check purposes, an operator initiated test pattern can also be brought up on the screen.

Operational status

In service in the Panavia Tornado IDS.

Contractor

Astronautics Corporation of America.

VERIFIED

Engine Performance Indicator (EPI)

The Astronautics family of Engine Performance Indicators (EPI) are either active matrix or dichroic liquid crystal display monitoring and displaying five critical engine parameters. It is designed to achieve full dual redundancy, be of minimum weight, volume and power, easily maintainable and have high reliability. The EPI also functions as an EICAS display for use in aircraft with and without FADECs.

The EPI is designed to operate over a temperature range of –55 to +85°C and within an altitude band from sea level up to 70,000 ft. It meets the requirements of MIL-E-5400T and MIL-STD-810C.

Operational status

In production for LC-200 and UH-1 aircraft.

Contractor

Astronautics Corporation of America.

VERIFIED

F-16 aft seat HUD monitor

The Astronautics Corporation of America's display unit is a compact high-resolution, high-brightness, lightweight, ruggedised monochromatic raster display presently in use in the rear cockpit of the two-seat F-16B and D aircraft.

The display is configured as a single scan rate unit of 525 lines at 30 Hz frame rate with single aspect ratio of 4:3 on a 4.125 × 5.5 in (105 × 140 mm) active screen size.

The display unit is modular. Each module can be replaced independently without adjustments to the display (including the CRT assembly replacement). Integral BIT continually checks the display circuitry, providing go/no go indications and immediate failure isolation. It also has an internal video test pattern generator.

The display has an illuminated control panel assembly and automatic brightness control circuitry to compensate for cockpit ambient light levels.

Operational status

In production and in service in F-16B and D aircraft.

Contractor

Astronautics Corporation of America.

VERIFIED

Integrated coloured moving map

The helicopter integrated coloured moving map is a high-performance low-cost system. It comprises the Astronautics modular display processor and multifunction display and provides the pilot with a coloured moving map, night and all-weather capability, nap of the earth capability and workload reduction.

The coloured moving map is stored in the modular display processor and has multiple layers including topography, tactical data, navigation data, flight guidance and obstacles. It has track up and north up modes and a zoom capability from 1:50,000 to 1:1 million.

Options include a helmet-mounted display, data gathering from sensors such as FLIR or CCD and Hands On Collective And Stick (HOCAS).

Contractor

Astronautics Corporation of America.

VERIFIED

Video display for the AH-64A helicopter

The video display for the AH-64A helicopter is a high-resolution, high-brightness CRT for displaying flight, navigation and weapon data. Information is passed to the visual display unit in the form of analogue composite video. Pitch and roll trim controls are provided on the front panel, together with a turn and slip indicator. Built-in test circuitry provides go/no-go status for the VDU and assists in checkout and fault location. The display is in use on the US Army AH-64A Apache helicopter.

Specifications

Dimensions: 186 × 152 × 318 mm
Weight: 6.59 kg

Operational status

In service in the AH-64A Apache helicopter.

Contractor

Astronautics Corporation of America.

VERIFIED

Specifications for Airborne multifunction CRT displays

	P/N 603737	P/N 602700	P/N 133000	P/N 133970	P/N 142000	P/N 133400	P/N 119450
Dimensions	138 × 155 × 475 mm	172 × 147 × 285 mm	168.4 × 200 × 305 mm	176.3 × 190.5 × 336.5 mm	217 × 292.2 × 319.4 mm	251.5 × 228.6 × 330.2 mm	254 × 342.9 × 457.2 mm
Weight	7.0 kg	6.36 kg	6.82 kg	6.82 kg	14.54 kg	18.18 kg	21.82 kg
Power	35 W/28 V DC MIL-STD-704D	30-45 W/28 V DC MIL-STD-704D	30-45 W/28 V DC MIL-STD-704D	30-45 W/28 V DC MIL-STD-704D	175 W/115 V DC 400 Hz	250 W/110 V DC 400 Hz	150 W/115 V DC 400 Hz
Display size	4.0 × 4.0 in	5.0 × 5.0 in	5.0 × 5.0 in	3.75 × 5.00 in	7.0 × 7.0 in	7.0 × 7.0 in	7.0 × 9.0 in
Brightness	200 fl	220 fl	220 fl	220 fl	160 fl	100 fl	120 fl
Scanning line or writing rate	raster 525 or 875 lines	raster 525 or 875 lines	raster 525 or 875 lines	raster 875 Lines	stroke 70,000 in/s	stroke/raster 200,000 in/s	stroke 50,000 in/s
MTBF	<3,000 h	2,000 h	3,000 h	3,500 h	2,700 h	1,000 h	853 h

Rugged 20 in AMLCD flat-panel display

The Aydin Displays rugged 20 in Active Matrix Liquid Crystal Display (AMLCD) provides 160° viewing angle, both horizontally and vertically, with resolution of 1,280 × 1,024 pixels. It is designed for use in airborne, ship and ground mobile application.

Contractor

Aydin Displays.

VERIFIED

ATD-111 LIDAR

The ATD-111 is an airborne LIDAR (Light Detection And Ranging) system developed for the US Navy to detect and classify underwater threats, primarily submarines and mines. The system is a stabilised scanning laser unit mounted aboard a Navy SH-60 Seahawk helicopter.

The ATD-11 employs a solid-state laser self-contained in a pod. The onboard operator control unit provides detection information, images and maps in GPS coordinates on a flat panel colour display.

Operational status

In development. The ATD-111 is currently undergoing upgrades for future testing to explore the potential of active laser-based technology to achieve superior mission capability in shallow waters.

Contractor

BAE Systems North America, Information and Electronic Warfare Systems.

UPDATED

Command and control display system

The command and control display system provides the man/machine interface between the mission crew and the sensors and communications systems in the Boeing E-3A AWACS. It maintains the display database, filters and positions the selected data and generates all graphics, alphanumerics and sensor target reports with real-time responses. Situation Display Consoles (SDCs), each comprising a 19 in (482.6 mm) CRT MIL-SPEC colour monitor, data entry, filtering and control panels, trackerball and keyboard with associated processing electronics, provide the mission crew with all display and control features required to carry out surveillance, weapons direction and battle staff functions. Data Display Indicators (DDIs) with monochrome monitors support the communications, maintenance and data processing functions of the mission crew. The E-3A has 14 SDCs and 2 DDIs.

The SDC presents the appropriate colour pictorial representation of the situation required to support the function assigned to the SDC by the operator. Using the high electro-optical qualities of the NDI monitor to achieve high legibility of dense data presentations under all operating conditions, the pictorial contents range from individual symbols indicating only sensor type and target positions to a combination of symbols and tabular notes that display such information as target type, speed, direction of flight, bearing, mission and altitude. Supporting tabular data, also in colour, is presented in the lower 20 per cent of the display surface. From this data, and from background pictorial information such as maps, landmarks and unsafe areas, the SDC operator can determine the appropriate responses to developing situations. The mission crew can also configure the SDCs in flight to serve as battle staff, surveillance or weapons consoles.

Specifications

Power supply: 115 V AC, 400 Hz, 3 phase
Temperature range: −54 to +55°C
Altitude: up to 40,000 ft
Environmental: MIL-E-5400 Class 1

Operational status

In service in US Air Force Boeing E-3A AWACS and Royal Saudi Air Force aircraft, and through a technology transfer on the NATO E-3 AWACS. Selected for the Japanese Air Self-Defense Force Boeing 767 AWACS fleet.

The E-3A AWACS command and control display system has 14 SDCs and 2 DDIs 0106303

Contractor

BAE Systems North America, Greenlawn/Wayne.

VERIFIED

Display and control system for the F-22

BAE Systems North America (formerly Sanders) is responsible for the display suite for the F-22. The F-22 display and control system consists of one primary MultiFunction Display (MFD), three secondary MFDs and two upfront displays. BAE Systems also provides the Graphics Processor Video Interface (GPVI), Airborne Video Tape Recorder (AVTR) and Operational Debrief Station (ODS).

The primary MFD presents a situation display of the air and ground situation.

The secondary MFDs present attack and defensive data, together with stores management information. The upfront display presents CNI data, critical flight data and warnings.

The GPVI resides in the Common Integrated Processor (CIP) and hosts software that generates the tactical situation displays. It both generates display formats and mixes these formats with video from other aircraft sources. The GPVI includes the i960MX processor and two graphics ASICs developed by BAE Systems. The Graphics Drawing Processor (GDP) ASIC is the high-performance anti-aliasing drawing processor. Using subpixel addressing and advanced anti-aliasing algorithms, the GDP generates complex display formats at better than a 30 Hz rate. The anti-aliasing provides for smooth display dynamics, without artifacts such as stair-stepped lines.

The video merge ASIC controls the merging of external video with the locally generated format and transmits the result over the fibre optic interface to the MFDs. The MFDs also utilise the GDP ASIC to enable local display generation. Thus the MFDs can either display video from the CIP or locally generated formats derived from information received over dual MIL-STD-1553 interfaces.

The GDP is a single 3,000 Mops ASIC which accomplishes on-chip translation, rotation, scaling, anti-aliasing drawing, filling and micro-positioning to better than one-sixteenth of a pixel for smooth instrument movement on an active matrix LCD medium. F-22 systems engineering trade-offs have resulted in design advances such as dual-brightness sensors for automatic compensation for cockpit ambient light level, expanded luminance ranging from 220 ft-lamberts down to 0.1 ft-lamberts for enhanced night operations with minimum canopy reflections, lamps designed to outlast the life of the aircraft and comprehensive BIT for improved safety and maintainability.

Specifications

Dimensions:
(primary MFD) 257.8 × 257.8 × 190.5 mm
(secondary MFD) 213.4 × 213.4 × 190.5 mm
(up front display) 156.2 × 130.8 × 218.4 mm
Weight:
(primary MFD) 7.03 kg
(secondary MFD) 6.12 kg
(up front display) 3.3 kg
Display size:
(primary MFD) 203.2 × 203.2 mm
(secondary MFD) 158.7 × 158.7 mm
(up front display) 76.2 × 101.6 mm

Operational status

In flight testing on the US Air Force F-22 advanced tactical fighter.

Contractor

BAE Systems North America, Information and Electronic Warfare Systems.

UPDATED

Electronic Display Units (EDU) for the B-1B

BAE Systems North America (formerly Sanders) has supplied a number of Electronic Display Units (EDUs) for the Rockwell B-1B. The EDU is a random position display capable of presenting graphic information in response to X and Y deflection of signals and an unblank signal Z.

The indicator uses a 12 in (305 mm) diagonal, square CRT with a usable display area of 8 × 8 in (203 × 203 mm). The deflection system is capable of linear response at writing rates of up to 125,000 in/s. Small signal deflection bandwidth is greater than 3 MHz. Positioning moves can be made and settled to within 0.05 per cent of full-screen in less than 35 µs. These displays depict threat and panoramic information which enables the B-1B's defensive systems operator to analyse threat situations quickly and assign countermeasures.

Operational status

No longer in production. In service on the US Air Force Rockwell B-1B.

Contractor

BAE Systems North America, Information and Electronic Warfare Systems.

UPDATED

Miligraphic display system for the P-3 AEW

BAE Systems North America (formerly Sanders) has supplied a display system for the Lockheed Martin P-3 AEW. It is a colour graphics terminal with the control electronics and the raster display integrated into a single 19 in (483 mm) rack-mounted unit. The control electronics use dual 68020 microprocessors to generate real-time displays of radar data overlaid on situation maps. The operator keyboard and the display touch control provide interactive operator control of the airborne early warning system. The display uses a high-resolution 19 in (483 mm) shadow-mask colour CRT that has been specially designed to meet the airborne

The F-22 display and control system consists of (left to right) the upfront display, secondary MFD and primary MFD 0106330

The miligraphic display system for the Lockheed Martin P-3 AEW aircraft 0106411

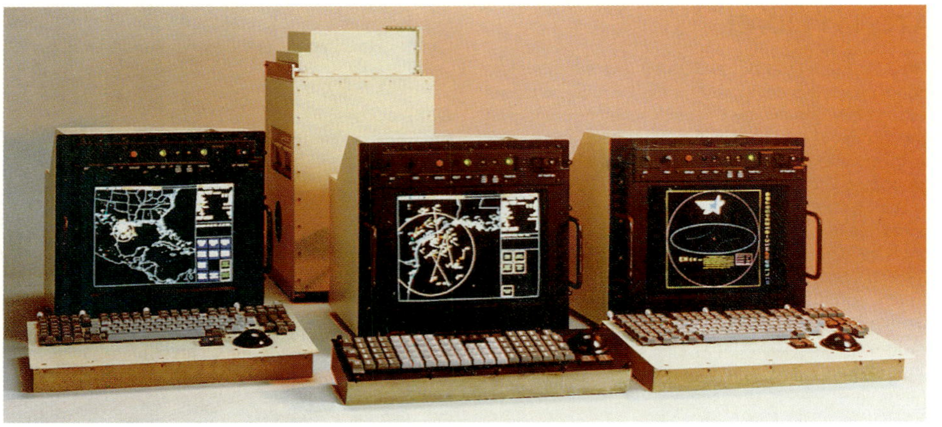

BAE Systems' control and display system for the Lockheed Martin ES-3A 0106412

environment and to protect it from varying magnetic fields. The miligraphic display software uses a multitasking executive with multiple windows that is designed to support application programming of real-time display tasks.

Operational status

Some 300 systems have been sold to the US Customs Service, US Navy and foreign customers.

Contractor

BAE Systems North America, Information and Electronic Warfare Systems.

UPDATED

Model 2799 E-HUD system

The Model 2799 E-HUD system injects critical flight data into the views of the pilot and co-pilot for heads up and out operation through NVG goggles. By eliminating the need to shift attention and focus to the cockpit interior, the Model 2799 E-HUD aviators Night Vision System/Head-Up Display enhances flight safety and operational effectiveness. Pilots can fly head-up and out to view the situation outside the cockpit, at the same time receiving all the essential flight data that would normally require a heads down function. This data can also be viewed independently from flight support personnel in the back of the aircraft via use of repeater ports. This allows the use of critical flight information for further enhancing situation awareness and mission success. The Model 2799 E-HUD system features lightweight, commercial-off-the-shelf packaging utilising flat panel AMEL display technology. The cable connecting the HUD to the symbol generator is an IEEE standard high-speed digital data connection. The optics are integrated into the eyepiece of the NVG using refractive optics thus making the Head-Up Display (HUD) a one piece assembly. The E-HUD can adapt to any aircraft platform and has four independent display modes each with declutter facilities. The pilots can independently select symbols and display modes from controls on the centre console and collectives, using ANVIS compatible control panel and display facilities. BIT isolation provides 95 per cent failure detection and fault isolation.

This same type of display technology is also being developed for Day HUD applications. The interface is picked off the MIL-STD-1553B or other bus structures that reside on the platforms. Software for symbol generation is adaptable via the MIL-STD-1553B or RS-232 interface.

Specifications

Field of view (FOV): 34°
Display type: AMEL (raster)
Resolution: 640 × 480 pixels
Interface: MIL-STD-1553B, ARINC 429, 561, RS-422 Analogue, discrete synchro
Reliability: > 10,000 h
Power: 28 V DC, <2A

Avionics
Dimensions
Data Converter Unit: 165.1 (L) × 63.5 (H) × 165.1 (W) mm
E-HUD display size: 17.8 × 17.8 mm
Aircraft installation A kit: sensors harness as required

Operational status

The model 2799 E-HUD is built by BAE Ssytems North America. This system is a direct replacement for older technology/hardware supported by the military forces. This system is replacing the hardware on C-130, HH-60, MH-53, C-141 and other special purpose platforms.

Contractor

BAE Systems North America, Austin.

VERIFIED

MultiFunctional Display (MFD)

The MultiFunctional Display (MFD) is the latest addition to the Miligraphic product line. Using the Miligraphic operating system, the 14 in (355.6 mm) flat tension mask CRT display with the 0.28 mm pitch shadow-mask is capable of displaying high-contrast high-visibility red/green/blue colour images. Three multifunctional displays are used on the ES-3A aircraft: one of these is in the cockpit and two are at crew stations. Each MFD is supported by an individual controller housed in a separate common chassis rack. This enables installation of the MFD monitors in limited access areas, with remote installation of the control electronics.

The MFD features contrast of 4:1 with a visor in 8,000 ft-candles ambient light, 91 lines/in resolution and compact construction. The monitor weighs 41.28 kg and the control unit 42.64 kg. The MFD meets MIL-E-5400F, MIL-T-5422F and MIL-STD-461.

Operational status

The MFD is used in the Lockheed Martin ES-3A.

Contractor

BAE Systems North America, Information and Electronic Warfare Systems.

UPDATED

Day and night ruggedised CCD cameras

In developing a complete line of high-resolution, day and night, ruggedised CCD cameras, Ball altered its successful line of military programme Mil-Spec cameras, focusing on design flexibility to accommodate each user's unique requirements. Several standard package configurations are available, and custom packaging allows the cameras to fit into existing enclosures. Each camera comes in either the 'L' or 'square' housing.

The large format CCD provides high resolution and sensitivity, and it allows the cameras to be used as direct replacements for 1 in vidicon-tube cameras.

A number of options are available, including: 18 or 25 mm Gen-III image intensifiers to increase night camera resolution; modified intensifier photocathodes to vary the spectral reponse; high-resolution image intensifiers with resolution greater than 45 lp/mm; custom interface electronics; timing in accordance with STANAG-3350A video format; video conditioning (log amplifier, auto black); and full military qualification. Cameras are available in three standard line rates:
525-line rate, 60 Hz, RS-170A;
625-line rate, 50 Hz, CCIR;
875-line rate, 60 Hz, RS-343A.

Contractor

Ball Aerospace and Technologies Corp.

VERIFIED

Intensified high-definition television camera

Ball Telecommunication Products Division has developed an intensified, high-definition television camera.

The camera uses a frame transfer, charge-coupled device paired with a high-resolution, 25 mm, Gen-III intensifier to obtain the absolute best in sensitivity and resolution.

The combination of an advanced automatic light control, based on intensifier gating, and an enhanced automatic gain control, ensures operation from starlight to full sunlight without the need for a mechanical iris or display brightness adjustments. The camera's 16:9 (H × V) aspect ratio, together with its outstanding hands-off performance, makes it an ideal sensor for helicopter operations including wide area, all-light-level surveillance or low-light-level microscopy.

Contractor

Ball Aerospace and Technologies Corp.

VERIFIED

TAMMAC Tactical Aircraft Moving MAp Capability

The new US Navy Tactical Aircraft Moving MAp Capability (TAMMAC) is fitted with the Smiths Industries Aerospace Advanced Memory Unit (AMU) which will replace earlier US Navy data storage and mission data loader equipment.

Operational status

The TAMMAC system is scheduled to be fitted initially to five baseline US Navy TAMMAC aircraft: F/A-18, AV-8B, AH-1W, UH-1N and CH-60, with the following aircraft being considered for TAMMAC later: F-14, S-3, CH-53, SH-60 and P-3.

Contractor

Boeing Company.

VERIFIED

Cockpit 21 for T-45C Goshawk

Cockpit 21 was developed to replace the T-45A's analogue displays with digital displays similar to those found in the US Navy's F/A-18, AV-8B Harrier II and other advanced carrier-based jets. Aircraft equipped with Cockpit 21 are designated T-45C. Since students transitioning to these aircraft from Cockpit 21 will have already mastered cockpit information management skills and situational awareness, they can concentrate on the primary mission of learning how to perform key tactical manoeuvres.

Smiths Industries Aerospace has performed the full systems integration on Cockpit 21 and supplies the Head-Up Display. Cockpit 21 comprises a Display Processor Unit (DPU), Pilot-Display Unit (PDU), and Data Entry Panel (DEP). The DEP drives five display surfaces including the PDU and four raster Head-Down Displays (HDD) (monochrome multifunction displays sourced from Elbit Systems Ltd). The new cockpit provides navigation, weapons delivery, aircraft performance and communication data to both stations in the two-seat cockpit. The cockpit also includes a Global Positioning System/Inertial Navigation Assembly and a multiplex databus that will allow expansion of cockpit capabilities to accommodate changing training requirements.

Operational status

The first Cockpit 21 aircraft was delivered in October 1997 for testing of the production configuration at US Naval Air Station Patuxent River. The second T-45C delivered in December 1997 went to US Naval Air Station Meridian for Training Wing. US Navy plans call for the existing 72 T-45A Goshawk aircraft with analogue cockpits to be upgraded to the T-45C Cockpit 21 digital configuration. Smiths Industries will provide Boeing with 103 new digital Cockpit 21 assemblies by 2004. Retrofit kits for 84 T-45As are planned.

Contractor

The Boeing Company.

VERIFIED

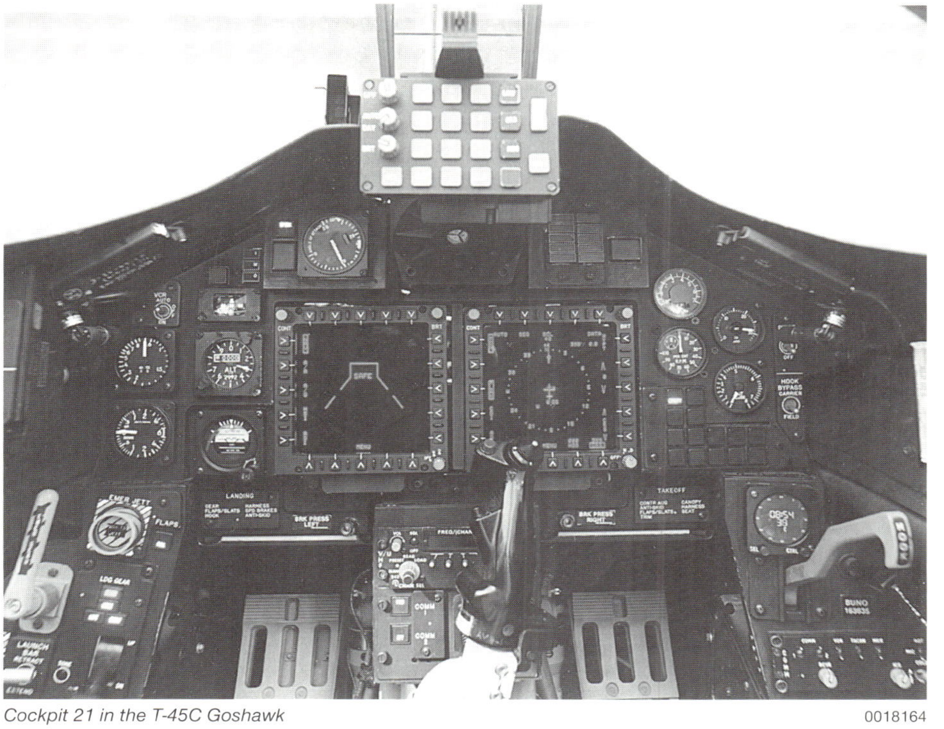

Cockpit 21 in the T-45C Goshawk 0018164

Head-Up Display (HUD) for the F-15A/B/C/D

The F-15A/B/C/D HUD consists of a display unit and signal data processor. The HUD accepts inputs from various sensors and processes and formats the display.

The HUD subsystem is being upgraded as part of the F-15 cockpit reliability and supportability enhancement programme. The display will have increased field of view, better combiner optics, higher writing speed and enhanced BIT. The signal data processor will have increased symbology capability, a helmet-mounted tracker option and enhanced BIT.

The F-15 improved HUD subsystem provides the pilot with navigation, attack and targeting and primary flight information necessary for effective flight control and weapons management. When complete, the upgraded system will be offered as an option for aircraft modification and upgrades for the F-15A/B/C/D.

Operational status

Over 1,300 HUDs have been delivered for F-15A, B, C and D aircraft over a period of 17 years. The Boeing Company has been awarded a US$3 million contract for the full-scale design and development of the improved HUD subsystem.

Contractor

The Boeing Company.

VERIFIED

The HUD subsystem in the F-15A/B/C/D is being upgraded 0106401

Integrated Conventional Stores Management system/Global Positioning System (ICSM/GPS) for the B-52H

In early 1993, The Boeing Company, Product Support Division in Wichita was awarded a US$25.1 million contract by the US Air Force Aeronautical Systems Division for work on the B-52H bomber ICSM/GPS. The Conventional Enhancement Modification (CEM) programme provides standard electrical and software interfaces for all future weapons specified for the

B-52H, using the MIL-STD-1760 protocol. The system permits the crew to load smart missiles with details of route and target, and also allows the weapon load of a B-52 to be changed from one flight to the next with only a software change loaded into the offensive avionics suite using a tape cassette. Integration of GPS provides precision navigational capability which greatly improves weapon delivery accuracy.

Operational status

Initial operational capability for the B-52H was achieved in December 1994.

Plans to add the ICSM/GPS capability for additional B-52Hs are in hand.

Contractor

The Boeing Company.

VERIFIED

Mast-Mounted Sight (MMS) for the AHIP helicopter

In October 1984, the US Army ordered the MMS for its Army Helicopter Improvement Programme (AHIP) Bell OH-58D helicopters. Mounted on the rotor mast above the rotor itself, the sight unit acts like the periscope of a submarine, permitting the aircraft to conceal itself by using natural cover but allowing the crew to peer over the cover to observe enemy dispositions and forces.

The MMS contractor, integrates stabilisation platform and electrical systems with a sensor suite purchased from Northrop Grumman. The sensors include a low-light television, a thermal imaging sensor and laser range-finder/designator for Copperhead shells and Hellfire anti-tank missiles. A feature of AHIP is the Automatic Target Handover System by which targeting information from one aircraft can be transmitted to another or to ground-based weapons. AHIP itself is the precursor to the much more ambitious RAH-66 light battlefield helicopter system.

Mounted over the main rotor drive shaft, the performance of such a system would be severely degraded by the high vibration levels in this area were it not for the soft mount devised by The Boeing Company which provides a high degree of isolation. Performance of the anti-vibration system is such that target bearing can be measured to within 20 mrad. The overriding need to keep down weight above the rotor has produced a sensor package weighing only 73 kg. Equipment bay systems add another 41 kg.

The television camera has a silicon-vidicon dawn to dusk capability with an 8° field of view for target acquisition and a 2° field for recognition. The FLIR sensor has a 120 element common module detector array and two fields of view of 10 and 3°. Television and laser systems share a common optical path to minimise the number of components. A video tracker and digital scan converter in the data processor together permit the incorporation of other features such as autotrack, frame-freeze and point-track.

In the cockpit are two multifunction displays for the presentation of video from the MMS and flight guidance and communication and navigation information. The sensors themselves are contained within a 650 mm carbon-epoxy sphere with a sightline 810 mm above the plane of rotor rotation. Data is channelled into the cockpit area via cables through a 23 mm tube running inside the drive shaft.

The MMS is applicable to a variety of helicopter platforms, including the Boeing 500 Series, the Sikorsky H-76, the Bell 406, the Agusta A 129 and the Eurocopter BK 117. The MMS can be used to provide these platforms with both AHIP scout features and light attack capability, using Hellfire and/or TOW missiles. The Apache helicopter which does not have a mast-mounted system, the Lockheed Martin TADS/PNVS system being nose-mounted, is regarded as a potential future application.

Operational status

In service in OH-58D helicopters.

Contractor

The Boeing Company.

VERIFIED

The mast-mounted sight on an OH-58D helicopter 0106403

The mast-mounted sight display is on the left of this Bell OH-58D helicopter cockpit. The view is adjusted by controls on the stick on the pilot's right-hand side 0106402

Rotorcraft Pilot's Associate (RPA)

The RPA is a display system concept, developed by a team led by Boeing, under contract to the US Army's Aviation Applied Technology Directorate (AATD). The objective is to use artificial intelligence to help pilots exploit the full potential of onboard and offboard sensors and other sources of data input to optimise mission effectiveness and flexibility.

Operational status

The RPA concept is being developed by a team led by Boeing (Mesa) and including Associate Systems Inc (Atlanta), Honeywell (Minneapolis and Teterboro), Kaiser Electronics (San Jose), Lockheed Martin Advanced Technology Laboratories (Camden), Lockheed Martin Federal Systems (Owego) and the Raytheon Electronic Systems (Dallas). First flight of a specially adapted AH-64D Longbow Apache occurred in October 1998. Testing is continuing. The current configuration is understood to present data to a pilot helmet-mounted display for use in the heads-up/eyes-out-of-the-cockpit role, and to a second crew member both on helmet-mounted display and three 10 × 8 inch head-down displays for mission and weapon system management.

Data sources used in the current testing are understood to include onboard navigation, radar and electro-optic surveillance sensors, as well as aircraft survivability sensors. Offboard data sources include the Tactical Receiver Intelligence eXchange System (TRIXS), Joint Surveillance and Target Attack Radar

System (JSTARS), Joint Tactical Information Distribution System (JTIDS), Battlefield Combat Identification System (BCIS) and tactical command centre data, integrated via the Improved Data Modem (IDM) and JTIDS.

RPA presents the pilot with a number of planning systems optimised for attack planning and sensor planning. The pilot can accept, reject, modify or override RPA suggestions.

Concept planning includes the use of RPA in other attack helicopters, in Unmanned Combat Air Vehicles (UCAVs) and possibly in fixed-wing aircraft in the Joint Strike Fighter (JSF) timeframe.

Contractors

Boeing, leading a consortium.

VERIFIED

The rotorcraft pilot's associate concept demonstrator configuration
2000/0048663

AirBorne Laser (ABL)

In November 1996, the US Air Force awarded Boeing Information & Communications Systems a US$1.1 billion Programme Definition and Risk Reduction (PDRR) contract to develop the ABL system, also known as the YAL-1A Attack Laser. The Boeing team includes TRW Space & Electronics, Redondo Beach and Lockheed Martin Missiles & Space, Sunnyvale.

The ABL is a theatre ballistic missile defence system that will consist of a laser and a beam control apparatus fitted on a 747-400F freighter aircraft.

Under the terms of the PDRR contract, Boeing's ABL team over the next five years will demonstrate that relevant ABL technologies can be integrated onto an airborne platform to shoot down theatre ballistic missiles at ranges measured in hundreds of kilometres.

Following successful completion of the PDRR phase, the US Air Force plans for an engineering and manufacturing development contract for a fully operational system, leading to a fleet of seven ABL aircraft.

Four adjunct missions will be studied during the PDRR: self-protection; protection of other high value assets; the role of the ABL in cruise missile defence; the role ABL can play in airborne surveillance.

Specifications

Laser: Chemical Oxygen Iodine Laser (COIL)
Laser power: Multimegawatt

Operational status

In April 1998, the ABL programme passed a key laboratory test, when a scaled beam control system demonstrated the laser pointing and focusing performance required for the ABL mission. A 'first light' test of the flight-weighted module, a kilowatt-class COIL was conducted at TRW's test centre in June 1998; and higher-power tests have since been conducted, with a full-power test in late 1999. The configuration of the ABL was locked during a final design review in April 2000, initiating PDRR Phase II is due to continue to 2002. An Engineering and Manufacturing Development (EMD) could begin in 2004. Production is currently forecast for 2006 to 2008 for seven aircraft.

Redefinition of the EMD phase during the April 2000 final design review formalised a cost increase from US$2.714 billion to US$3.545 billion and introduced a two year delay in the first lethality demonstration from 2003 to 2005.

Airborne Tactical Laser (ATL)

A scaled-down derivative of the ABL, designated the ATL, is being tested by Boeing Rocketdyne for use against cruise missiles, artillery projectiles and surface-to-air-missiles. The objective is to achieve a system weight of between 4,500 - 6,750 kg, so that the system could be fitted to airborne platforms such as the V-22 Osprey tiltrotor, CH-47 Chinook helicopter and C-130 or P-3 aircraft. A projected in-service date similar to that of the ABL is being suggested.

The ABL concept being developed by the US Air Force, Boeing, Lockheed Martin and TRW ***1999***/0051013

Contractors

Boeing Information and Communication Systems.
Lockheed Martin Missiles & Space.
TRW Space & Electronics.
US Air Force Phillips Laboratories.

VERIFIED

Airborne Surveillance Testbed (AST)

The AST project is a technology demonstration programme that supports development and evaluation of defensive systems to counter InterContinental and Theatre Ballistic Missiles (ICBMs and TBMs) and their warheads.

As an airborne platform, the AST also supports risk reduction testing and the evaluation of developing technologies.

Formerly known as the Airborne Optical Adjunct (AOA), AST is a key element of the US Department of Defense's Ballistic Missile Defense Organisation (BMDO).

Initial research was aimed at evaluating whether an airborne infra-red (IR) sensor could reliably provide early warning using detection, tracking and target discrimination methods, as well as provide ICBM tracking information to ground radar. The programme was later renamed AST, and the original support mission was changed to data-gathering for ballistic missile defence development and resolving issues associated with target characteristics.

Under contract with the US Army's Space & Strategic Defense Command, Boeing has modified a 767 commercial jet aircraft to carry a large multicolour IR sensor housed in a cupola atop the 767's fuselage.

IR sensors detect the comparative heat of objects. The AST sensor, comprising more than 38,000 detector elements, is sensitive enough to detect the heat of a human body at a distance of more than 1,000 miles against the cold background of space. The AST sensor has demonstrated the capability to detect, track and discriminate warheads from missile components, debris and decoys, both from the ambient temperature of an object before launch and from the heat generated during re-entry into Earth's atmosphere.

Boeing was awarded the original AOA contract in July 1984. As the integration contractor, Boeing's responsibilities included procuring the data processor, as well as all recording, communications and support equipment. Hughes Aircraft Electro-Optical Data Systems Group designed, built and maintains the IR sensor.

The AST and its mission system equipment were initially used to gather data that confirmed the system's ability to acquire, track, discriminate warheads from decoys, and provide track information to ground units. Originally, the emphasis was on ICBMs. At the time of the 1990-91 Gulf War national emphasis shifted to defending military installations and troops from TBMs. To support the US Army's studies on TMD, the AST system was used to gather data applicable to TBMs, as well as ICBMs.

AST has successfully participated in 65 missions, including eight operational exercises with real-time tactical communication links to US Army, Navy and Air Force elements to demonstrate the utility of an airborne IR platform in a TMD role.

Information acquired during tests at national missile ranges is added to a database that is used in computer-simulation models, and to confirm the system's capability to discriminate between objects. Data acquired during flights has validated onboard software, computers, and communication equipment necessary for real-time processing of tracking data and transmission of that data to ground-based defensive units.

The AST frequently operates at the Western Test Range, off the coast of California, and observes missiles launched from: Vandenburg Air Force Base, California; the US Army's Kwajalein Missile Range in the central Pacific Ocean; Pacific Missile Range Facility, Hawaii; White Sands Missile Range, New Mexico; Eastern Test Range in Florida; and Wallops Flight Facility, Virginia.

Boeing Airborne Surveillance Testbed ***2001***/0106305

During a typical strategic test mission, an ICBM launched from Vandenberg Air Force Base, for example, sends one or more ballistic missile re-entry vehicles into the Kwajalein Missile Range. Depending on the objectives assigned to AST, the aircraft will be positioned 150 to 300 miles to the side of the missile's trajectory or down range of the impact point to acquire data during boost, exoatmospheric flight, or re-entry.

For short-range TBMs, AST can observe the entire trajectory. The data acquired by the sensor is processed in real time to generate a track that may be transmitted to the ground after each target-sighting update - about every 1.5 seconds. The computers on board AST extrapolate the track forward to predict an impact point, and can backtrack to estimate the launch point.

All sensor and tracking data generated by the computers is recorded on board AST for post-test analysis and system evaluation, which takes place at a laboratory in Kent, Washington. This information is also provided to the US Army's data facility in Huntsville, Alabama. It is available to all Department of Defense services and defence contractors engaged in ICBM and TBM defence projects to enhance their knowledge of missile threats and warheads.

The AST aircraft is based in Seattle, Washington, and can deploy to any national test range within one day. It carries a flight crew of 15 for typical missions, with room for observers from various government agencies and contractors. Missions last approximately 6 to 8 hours, and the aircraft usually flies at an altitude above 42,000 feet, somewhat higher than commercial aircraft.

The technology developed for AST is applicable to many sectors of the defence community. AST personnel currently are supporting the US Navy Theatre Wide Captive Carry (NTWCC) Standard Missile (SM)-3 and AirBorne Laser (ABL) programmes.

For the NTWCC SM-3 programme, Boeing has completed aircraft, integration and two successful flight tests of the sensor for system risk reduction.

Operational status

Successfully completed 65 data collection missions. Present development is funded by a 3-year, 10-month follow-on contract dated September 1995 at a value of US$19.5 million a year. Studies are underway to assess the benefits of installing multiple sensors and experiments on the AST aircraft.

Contractor

The Boeing Company, Boeing Space and Communications Systems.

VERIFIED

IP-5110 high-resolution Digital Radar Display

The IP-5110 Digital Radar Display (DRD) converts radar into high-resolution raster scan format. Compatible with a wide variety of radars, the DRD has applications in airborne, shipboard and ground installations. Incorporating two digitisers, the DRD can provide two independent display presentations from the same radar. Graphic displays allow the overlay of ESM and other platform sensor data with radar displays, including ASW and ESM.

Input radar video is digitised at a 20 MHz rate. Overlay graphics are stored in four display planes with 16 levels of intensity. Output video resolution is selectable between 1,024 by 832 lines or 1,024 by 486 lines. Internal video matrix provides switching to video recorders or external monitors.

A trackerball allows the operator to designate radar returns quickly, to determine location in latitude and longitude and to measure range and bearing to or between multiple targets. Target types can be designated with Navy Tactical Data Systems (NTDS) symbols. Up to 64 targets can be designated.

Features include improved radar performance with digital processing; two independent digital scan converters; trackerball control; operation with any radar with pulse rates up to 10 kHz and antenna rotation up to 200 rpm; advanced measurement and display marking up to 64 concurrent symbol markers with independent target ring generation; ruggedised and MIL qualified composite video outputs selectable between high and low resolution. The IP-5110 also features programmable overlay graphics; four display planes with independent 16-level intensity control; display ranges from 1.5 n miles to 200 n miles; selectable aircraft, waypoint and ground display stabilisation; freeze and zoom functions; video outputs compatible with VHS video recorders; averaging; peak detecting; weather and fast time constant processing.

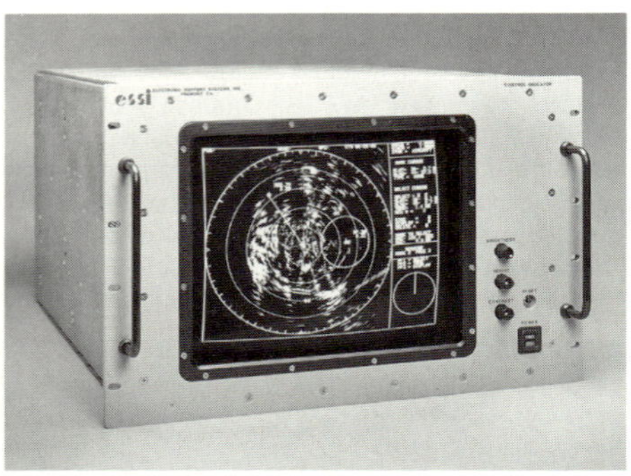

The IP-5110 high-resolution digital radar display 0106306

Contractor

Condor Systems Inc.

VERIFIED

IR/UV maritime surveillance scanner

The AADS1221 IR/UV is a sensor subsystem of the Swedish Space Corporation Maritime Surveillance System. It is specifically designed for use in maritime surveillance applications.

The IR/UV scanner, operating in the 8.5 to 12.5 μm range for IR and the 0.32 to 0.38 μm range for UV, is used at low altitude to obtain high-resolution imagery of oil spills. IR data can be obtained by both day and night, providing information on the spreading of oil and also indicating the relative oil thickness within the oil slick. Usually 90 per cent of the oil is concentrated within less than 10 per cent of the visual slick. By using IR information, clean-up operations can be directed for maximum efficiency.

UV data is obtained during daylight and maps the entire extent of the slick, irrespective of thickness. The UV data adds confidence to the IR registration by distinguishing natural thermal phenomena, such as cold upwelling water, from suspected oil pollution.

In the Maritime Surveillance System, the IR/UV video outputs are presented in real time on a split screen format colour TV monitor, with IR on the left and UV on the right, with false colour coding for image enhancement.

Specifications

Dimensions:
(scan head) 380 × 380 × 380 mm
(operator console)180 × 490 × 340 mm
Weight:
(scan head) 17 kg
(operator console) 12 kg
Power supply: 28 V DC, 15 A
Field of view: 5 mrad instantaneous, 87° total
Scan rate: 160 scans/s

Contractor

Daedalus Enterprises Inc.

VERIFIED

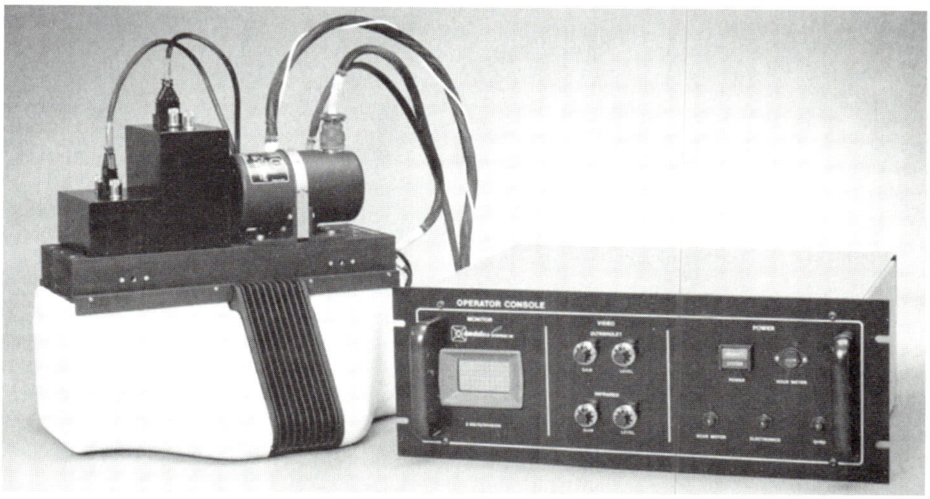

The IR/UV maritime surveillance scanner consists of the scan head (left) and an operator's console (right)
0106307

Eagle-5 colour AMLCD for military avionics

The Eagle-5 is a 5 × 5 in square, Active Matrix Liquid Crystal Display (AMLCD) designed to provide situational awareness in the most demanding visual and environmental conditions. Incorporating a special high-definition, quad-green pixel design, this advanced flat-panel display provides sharp, crisp imagery even in bright sunlight, darkness and extremes of temperature. With wide viewing angle and fast response times for real-time video performance, the Eagle-5 can be customised for thermal management, mechanical fit and assembly, and other programme-specific requirements.

Specifications

Display area: 5 × 5 in (127 × 127 mm)
Pixel resolution: 480 × 480 colour groups (96 colour groups per in)
Pixel configuration: Quad-green (RGGB; greens can be addressed independently)
Colour bit depth: 24-bit (255 grey levels)
Contrast ratio: >100:1, on axis
Reflectance: <0.1%, diffuse; <1.0%, specular
Data input: Digital

Contractors

dpiX, Inc.
Planar Systems.

VERIFIED

Eagle-6 colour AMLCD for military avionics

The Eagle-6 is a 6.25 × 6.25 in square, Active Matrix Liquid Crystal Display (AMLCD) designed to provide situational awareness in the most demanding visual and environmental conditions. Incorporating a special high-definition, quad-green pixel design, this advanced flat-panel display provides sharp, crisp imagery even in bright sunlight, darkness, and extremes of temperature. With wide viewing angle and fast response times for real-time video performance, the Eagle-6 can be customised for thermal management, mechanical fit and assembly, and other programme-specific requirements.

Specifications

Display area: 6.25 × 6.25 in (159 × 159 mm)
Pixel resolution: 512 × 512 colour groups (82 colour groups per in)
Pixel configuration: Quad-green (RGGB; greens can be addressed independently)
Colour bit depth: 24-bit (255 grey levels)
Contrast ratio: >100:1, on axis
Reflectance: <0.1%, diffuse; <1.0%, specular
Data input: Digital

Contractors

dpiX, Inc.
Planar Systems.

VERIFIED

The dpiX Eagle-19 colour AMLCD **1999**/0051694

Eagle-19 colour AMLCD

The Eagle-19 active matrix liquid crystal display (AMLCD) is designed to replace 17 to 21 in monitors based on CRT technology. It provides flat-panel display solutions suitable for mission-critical crew-station applications in platforms such as AWACS and JSTARS. With full 24-bit colour and a proprietary optical stack that enables wide-angle viewing, the Eagle-19 supports full-motion video without ghosting or smearing and is intended for display of radar, FLIR, ASW applications.

Specifications

Display area: 19.0 in diagonal; 14.8 × 11.9 in
Pixels: 1,280 × 1,024
Colour bit depth: 24 bits per pixel
Contrast ratio: >100:1 on axis
Viewing angle: ±65° horizontal; ±55° vertical
Response time: <12 ms (typical)
Data input: digital; application-specific analogue interface

Contractor

dpiX, Inc.

VERIFIED

AN/UYQ-70 Advanced Control Indicator Set (ACIS)

Designed for ship, submarine and airborne use, the US Navy has selected the AN/UYQ-70 display for its E-2C aircraft, for system compatibility. Its role is to display command and control data, and to provide a radar scan converter interface to the AN/APS-145 airborne radar. The display is a rugged full-colour 20.1 in AMLCD with 1,280 × 1,024 resolution.

The AN/UYQ-70 incorporates Q-70 high-performance commercial-off-the-shelf (COTS) modules, which features a 100 MHz (165 MHz option) VME-based processor, 9 Gbyte removable hard drive and high-power graphics.

Operational status

DRS Technologies was contracted by Lockheed Martin Naval Electronics & Surveillance Systems (Eagan), on behalf of the US Navy, in 1994, to supply AN/UYQ-70 systems for its 'Aegis'-class ships, 'Trident'-class submarines, NSSN 'Virginia' class attack submarines and E-2C aircraft, as part of the Lockheed Martin-DRS Technologies-Raytheon UYQ-70 team. The system is also planned for retrofit on the 'Ticonderoga', 'Nimitz' and 'San Antonio' class ships. Additional orders for the system continue, most recently in March 2000.

Contractors

DRS Technologies, Inc.
Lockheed Martin Control Systems.
Raytheon Electronic Systems.

VERIFIED

AN/UYQ-70 E-2C advanced control indicator set
2000/0062747

AN/ASQ-165 Armament Control Indicator Set (ACIS)

The AN/ASQ-165 ACIS is an airborne ordnance stores management system developed for the US Navy Sikorsky SH-60B LAMPS III helicopter. The system controls one or two Mk 46 Mod 0/1/2 torpedoes and up to 25 air-launched sonobuoys. Interfaces for control functions and for the ACIS microprocessor-control system are provided for two BRU-14A bomb racks, the sonobuoy launcher and a MIL-STD-1553 databus which interfaces in turn with an AN/AYK-14 airborne computer.

The system consists of a C-10488/ASQ-165 Armament Control Indicator (ACI) for the airborne tactical operator or the helicopter pilot and a CV-3531/ASQ-165 Armament Signal Data Converter (ASDC) which contains logic, interlock, driver and relay circuits. The microprocessor controls all tasks such as inventory usage, functional status and built-in self-test, with the exception of the jettison function which is hard-wired. Jettison functions are redundant and the circuitry ensures that no single failure will cause inadvertent release or inability to jettison. Communication with all parts of the system apart from the jettison function is provided by the databus.

The ACI control panel functions include: master arm voltage switching for positive armament safety; torpedo selection, arming and launch; torpedo search depth, mode, ceiling and course programming; three torpedo programme status indicators; sonobuoy selection, arming and launch; and jettison left store, right store, all sonobuoys or all ordnance.

The AN/ASQ-165 precludes unintentional launching from single point failures. The system features low-power Schottky TTL 54-LS series logic and tri-state, sunlight-readable switches and indicators. With the exception of some launch and jettison circuits, the self-test facility is quoted as being able to detect more than 98 per cent of faults. The system is compatible with automatic test equipment at both unit and module level. Failure rates for the ACI and the ASDC are stated as better than one in 25,700 hours and 3,850 hours respectively.

Specifications

Dimensions:
(ACI) 219 × 146 × 170 mm
(ASDC) 193 × 149 × 457 mm
Weight:
(ACI) 2.04 kg
(ASDC) 11.05 kg

Operational status

No longer in production. In service in US Navy Sikorsky SH-60B LAMPS III helicopters.

Contractor

Fairchild Defense Division of Orbital Sciences Corporation.

VERIFIED

AN/AWG-15 Armament Control System (ACS) for the F-14A

The AN/AWG-15 ACS for the US Navy F-14A/A+ was designed and developed for weapon stores control to be implemented from a control indicator by an aircrew member or in conjunction with an Airborne Missile Control System (AMCS) computer. The system comprises a Control Indicator (CI), a Power Switching Unit (PSU) and a set of Signal Command Decoders (SCDs), one of which is located at each weapon station.

The ACS provides the preparation and control functions before releasing Phoenix, Sparrow and Sidewinder missiles or bombs, rockets and flares. It also controls and fires the M61-A1 gun.

For air-to-air functions, the system provides missile options, control/speed gate selection, inventory/ready control, weapon/station control and release by the pilot or navigator. For air-to-ground functions, using bombs, rockets or flares, it provides weapon and station selection, inventory/ready control, attack, delivery, fuzing and weapon mix mode selection and release by the pilot. It also provides jettison command functions. Protective logic circuitry is incorporated to prevent inadvertent weapon firing or release.

The system has a high level of inbuilt redundancy with BIT for 90 per cent fault detection and automatic isolation of faulty circuits, sub-units or components.

Specifications

Dimensions:
(CI) 318 × 292 × 254 mm
(PSU) 203 × 178 × 419 mm
(SCDA) 64 × 102 × 203 mm
(SCDB) 51 × 102 × 152 mm
Weight:
(CI) 6.13 kg
(PSU) 9.09 kg
(SCDA) 0.85 kg
(SCDB) 0.5 kg

The US Navy F-14 has the AN/AWG-15 armament control system 0106404

Operational status

In September 1986, Fairchild Defense was awarded US$23.1 million by the US Naval Air Systems Command to modify the Navy's AN/AWG-15E weapon fire-control system in its F-14 aircraft. The upgraded system is designated AN/AWG-15F and over 470 aircraft were modified; electronics in the control indicator and part of the power switching units were replaced by items of more modern technology.

Earlier, in May 1985, Fairchild Defense was awarded a full-scale development contract by Northrop Grumman for the AN/AWG-15 Technology Improvement Programme (TIP) and a prototype of the new system was flight-tested in August 1986.

The first upgraded units were delivered to the navy at the end of 1987, and deliveries completed in 1989.

No longer in production. In service in US Navy F-14A and F-14A+ Tomcats.

Contractor

Fairchild Defense Division of Orbital Sciences Corporation.

VERIFIED

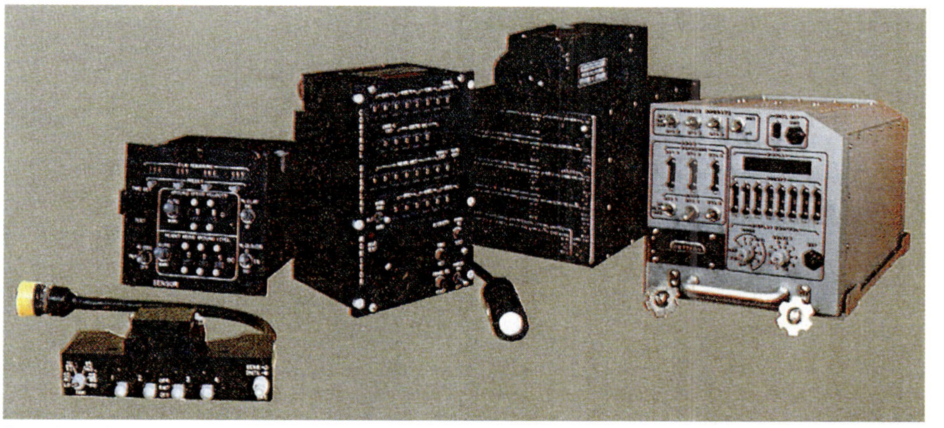

Fairchild Defense Reconnaissance Management Systems (RMS) ***1999***/0051647

Reconnaissance Management Systems (RMS)

Reconnaissance Management Systems (RMS) are Fairchild products which link aircraft avionics to unique sensor interfaces to control, monitor and annotate multisensor film and/or EO reconnaissance sensor suites.

RMS provide a single interface between aircraft and sensor suite. Evolving from early Airborne Data Annotation Systems (ADAS), today's Control Processor and Annotation (CPA) products include mux bus interfaces and microprocessor controlled, low power, high density electronics which control and report sensor operation while retaining the film annotation function. Image identification and management are greatly enhanced with machine readable, MIL-STD-782 code matrix or alphanumeric data blocks. Datalink output of this reconnaissance information is also available for correlation with images on the ground by image interpreters.

Recording Head Assemblies (RHAs) provide the means to annotate film with all mission parameters at the instant a sensor collects image data.

With the transition from film sensors to Electro-Optical (EO) sensors (analogue and digital) for usable images in near-real time, the airborne video processing function has been added to the basic control and annotation requirements. This addition converts analogue or digital imagery data from EO sensors and incorporates annotation data into the reformatted video.

Fairchild's Code Matrix Reader (CMR) is used to read the annotated films' MIL-STD-782 code block for efficient management of film results and automated search for archival retrieval of data. CMR includes a film gate through which the film is passed at a high rate. The code block data is then transmitted to an electronics unit for processing and I/O control by a host computer.

RMS combines four functions:

- Aircraft interface: The aircraft interface provides primary systems and displays interface via an existing standard digital bus: MIL-STD-1553, ARINC standards, or through analogue or discrete interface as required.
- Sensor control interface: Interfaces to sensors are implemented with the latest analogue and discrete techniques providing operator confidence and reliability, that is, synchronising, level shifting, impedance matching, on-off-ready, special interest, angular velocity stabilisation, BIT initiate, preflight testing. Special temperature sensing data, control of doors, or Environmental Control System (ECS) have been incorporated.
- Annotation: High or low speed (for LOROP) data annotation provides the capability to annotate data in cameras, infra-red sets and radar systems via CRT, LED or other special devices during flight. Aircraft parametric data can also be output in a serial digital form for flight recording or datalinking so that all information is synchronised with images on the ground with the same time and position data. EO video is annotated by recording data along with video. This is done in an auxiliary audio track on video tape or per new JSIPS format.
- Video processing: Charge Coupled Device (CCD) output data of EO sensors and/or IRLS output data are converted for compatibility with airborne monitors, video tape recorders and/or datalink formats in Video Processor Units (VPUs). Further image processing functions: video compression, enhancements, are optional. These video functions are usually provided in a unit separate from the control unit to isolate high frequency video signals in the RMS.

Operational status

RECCE aircraft	RMS systems
RF-4	AN/ASQ-90
OV-1D	AN/AYA-10
F-14 TARPS	AN/ASQ-172 or AN/ASQ-197
CP-140	RDAS
RF-5E	PSCS, A/A24Q-1 (V)
F/A-18D (RC)	AN/ASQ-197 (X)
RF-111C	AN/ASQ-197A
AV-8B	AN/ASQ-197 (X)

Contractor

Fairchild Defense Division of Orbital Sciences Corporation.

VERIFIED

Field Emission Displays (FEDs)

FED Corporation teamed with Honeywell, Harris, and Astronautics Corporation of America for the US Air Force Wright Laboratory research and development contract to develop field emitter displays for military aircraft. The heart of the military display is FED Corporation's recently developed Field Emission Display (FED), essentially a flat CRT in which the three colour electron guns are replaced with an array of emitters each responsible for a single pixel on the screen.

FED technology offers the avionics systems providers significantly easier integration issues compared with AMLCDs. The ability of field emitter displays to operate over wide temperature ranges results in less sensitivity to thermal issues of the emitter display. This translates into lower system design cost and manufacturing cost for integration companies.

Operational status

Development.

Contractor

FED Corporation (consortium leader).

VERIFIED

FV-3000 Modular Mission Display Processor (MMDP)

The FV-3000 MMDP forms the core of modern retrofit or new aircraft avionics packages offered by Flight Visions Inc. It can drive Head-Up Displays (HUDs), Helmet-Mounted Displays (HMDs) and MultiFunction Displays (MFDs) while providing spare power for mission processing.

Containing two 50 MHz 68040 microprocessors and a dual-redundant MIL-STD-1553B databus, the FV-3000 MMDP generates both flight and military displays. It has the capability to input RS-170 video, synchronise it with stroke-generated graphics and output the combined image for display on a HUD. The FV-3000 MMDP also offers air-to-ground weapon delivery processing, air-to-air gunnery and missile processing, primary navigation processing, primary MIL-STD-1553B databus, primary system control, upfront control panel switch processing, HOTAS switch processing.

The FV-3000 MMDP meets both military and civilian specifications while designed to meet the high demands of fighter aircraft. It has the capability to detect and isolate faults through a combination of operational and intermediate-level testing. In accordance with MIL-STD-810E, the FV-3000 can be fitted in jet fighter aircraft.

Specifications

Dimensions: 152.4 × 228.6 × 406.4 mm
Weight: 7.3 kg
Power supply: +28 V DC MIL-STD-704D 200 W

Operational status

The FV-3000 MMDP forms part of the US Navy F-14B integrated cockpit technologies upgrade, with the

FV-3000 Modular Mission Display Processor (MMDP) 0106308

Sparrow Hawk HUD. In this configuration, it doubles the field of view of the existing F-14B installation, and provides a very large improvement in MTBF. Further, because of the Sparrow Hawk's integral combiner, the US Navy can also replace the windscreen on the F-14Bs with the night vision-compatible version used in the F-14D.

The FV-3000 MMDP has also been selected, with the Night Hawk HUD, for the Lockheed Martin Joint Strike Fighter demonstration aircraft.

Contractor

Flight Visions Inc.

VERIFIED

Night Hawk HUD with FV-3000 modular mission display processor ***1999***/0051717

Night Hawk Head-Up Display (HUD)

Night Hawk, developed as Flight Visions' most advanced HUD, has a wide 30° total field of view and a 6in aperture, aiding high weapon delivery and navigation accuracy, together with greater flight safety. An automatic control adjusts brightness in stroke mode and brightness and contrast in raster mode. It is compatible with Gen 2 and Gen 3 night vision goggles. An optional video camera and recorder is capable of recording a high-quality image of what the pilot sees. The unit has dimensions of 135.1 × 33.9 × 586.7 mm, and a weight of 12.73 kg.

Operational status

Night Hawk has been selected by Lockheed Martin for its X-35 Joint Strike Fighter (JSF) demonstrator.

Contractor

Flight Visions Inc.

VERIFIED

Night Hawk HUD for the Lockheed Martin X-35 concept demonstrator aircraft ***1999***/0051716

Sparrow Hawk™ Weapon Delivery and Navigation System (WDNS)

Sparrow Hawk is a complete weapon delivery and navigation system providing a powerful and cost-effective solution for aircraft that perform a broad range of missions. Sparrow Hawk is designed for new fighter aircraft and for upgrading existing aircraft, and it can be customised to provide fleet commonality or to ease pilot training and/or transition across aircraft types.

The Sparrow Hawk's Head-Up Display (HUD) provides attack flexibility on a sharp and crisp stroke on raster head-up display with electronic boresight and 25° wide field of view. An automatic control adjusts brightness in the raster mode. An optional Up-Front Control Panel (UFCP) allows the pilot to select the system mode and to enter numerical data manually. It also includes an LED character display that presents alphanumeric entries used by the HUD computer. A HUD camera (optional) and video recording system can be integrated with the optical system. Two different symbol generators can drive Sparrow Hawk. In the event of failure of one, the pilot can switch to a back-up.

Air-to-air functions include LCOS, hotshoot and AIM-9; air-to-ground functions include gun and rocket dive attack, dive and level bombing with both instantaneous and delayed release, combination rocket or gun and bomb attacks in the dive toss mode and toss bombing.

The Sparrow Hawk HUD has been designed to provide a fully customised instrument interface, with symbology and upfront controls. At a total system weight of 7.8 kg, it is one of the lightest weapon delivery systems available. It can provide weapon delivery with a wide range of sensors, from those having only basic attitude heading reference systems and no radar to full inertial reference and radar. A wide range of interfaces is available, from analogue and synchro to ARINC 429 to MIL-STD-1553B. Sparrow Hawk features extensive BIT for easy maintainability and most faults can be diagnosed in the aircraft. Internal data logging and continuous in-flight testing keeps track of system health. Sparrow Hawk is Gen 2 and Gen 3 NVG-compatible and features raster display capability, integrated colour HUD camera and built-in standard training features.

Operational status

Sparrow Hawk has been selected for the US Navy F-14B upgrade, when it is used with the FV-3000 Modular Mission Display Processor (MMDP). Sparrow Hawk is also in service on many different aircraft types in a number of air forces.

Contractor

Flight Visions Inc.

VERIFIED

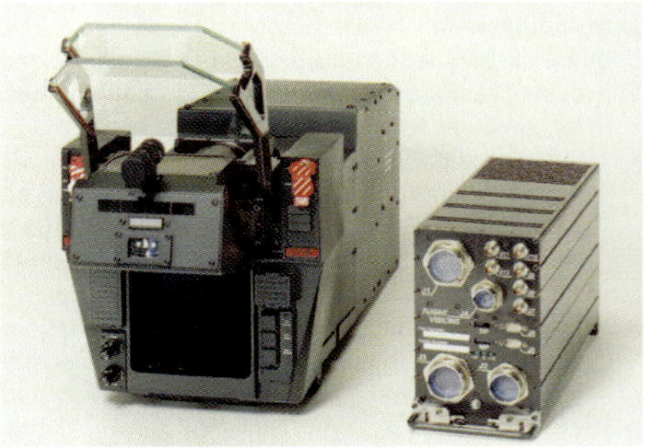

Sparrow Hawk HUD with the FV-3000 MMDP as used for the US Navy F-14B upgrade ***1999***/0051718

Flight Visions' FV-2000/Sparrow Hawk Primary Flight Display HUD is mounted at the top of the instrument panel in the Pilatus PC-7 Mk II turboprop military trainer

---

SAFIRE™ thermal imaging system AN/AAQ-22

Military-qualified as the AN/AAQ-22 thermal imaging system, the SAFIRE™ is a digital, high-resolution 8 to 12 µm wavelength system with three-axis gyrostabilisation and full 360l turret rotation. SAFIRE's wide field of view (28l) is designed for navigation and area searches. Its narrow field of view (5°) is supplemented by electronic zoom and freeze-frame features. The rugged turret has full performance at all pointing angles, including straight down. The system includes a hand-held controller and flight-hardened video display that supports recording for evidence and analysis. Options include compatibility with night vision goggles, autotracker, laser illuminator, laser range-finder, searchlight slaving, and radar and navigation interfaces. The hermetically sealed SAFIRE™ turret is flight qualified for airspeeds up to 405 kt. SAFIRE™ systems are certified on more than 20 fixed-wing and rotary aircraft.

Operational status

In service worldwide in 50 nations. Fielded by all branches of the US military. Other customers include the Royal Danish Navy, the Japanese Maritime Safety Agency, the Royal Netherlands Navy, and the Royal Saudi Naval Forces.

Contractor

FLIR Systems Inc.

VERIFIED

SAFIRE™ thermal imaging system AN/AAQ-22
0015295

Series 2000™ thermal imaging system AN/AAQ-21

Military-qualified as the AN/AAQ-21 thermal imaging system, the Series 2000™ is used widely for surveillance, narcotics interdiction, police pursuit, search and rescue and environmental patrols. The Series 2000™ is a rate-stabilised, high-resolution analogue system that operates in the 8 to 12 μm wavelength range. The gimballed turret provides full hemispheric sensor pointing. A hand-held controller directs the imager. Thermal images are displayed on a flight-hardened display unit. Images can also be recorded to a VCR for evidence or analysis. Wide field of view is 28°; narrow field of view is 5°, and is available in either × 7.5 or × 10.5 magnification. Other options include a CCD colour camera and laser pointing system.

Operational status

In service worldwide. Customers include law enforcement agencies throughout the United States, coastal patrols in Italy, and the US Air Force for use on UH-1N helicopters.

Contractor

FLIR Systems Inc.

VERIFIED

Series 2000™ thermal imaging system AN/AAQ-21 0015292

Star SAFIRE™ airborne thermal imaging system AN/AAQ-22

Military-qualified as the AN/AAQ-22 thermal imaging system, the Star SAFIRE™ is a digital, high-resolution system enhanced for imaging over water and in humid atmospheric conditions. Star SAFIRE™ is a military-qualified COTS airborne infra-red system equipped with a third-generation 3 to 5 μm indium antimonide (InSb) focal plane array detector (320 × 240). It offers three fields of view (1.4l narrow, 5.6l medium and 30° wide) for extended detection range capability and for surveillance, SAR, and flight safety. In addition to the thermal imager, Star SAFIRE™ has slots for three optional payloads (including CCD colour camera with × 2 extender, eye-safe laser range-finder, or laser illuminator). Star SAFIRE™ is a three-axis gyrostabilised system with full 360° turret rotation, including straight-down. The system's footprint and quick-release mounting match that of the SAFIRE to give fleet users platform flexibility. The system includes a hand-held controller and flight-hardened video display that supports recording for evidence and analysis. System options include navigation and radar interfaces, autotracker, video downlinks, and video recorders. The hermetically sealed Star SAFIRE™ turret is flight qualified for airspeeds up to 405 kt and is designed for the marine environment.

Specifications

Azimuth coverage: 360° continuous
Elevation: +30 to −120°
Fields-of-view:
narrow: 1.4 × 0.98°
medium: 6 × 4.4°
wide: 30 × 22.5°
Weight: 38.1 kg
Interfaces: MIL-STD-1553B, RS-170, CCIR

Operational status

Star SAFIRE™ systems are certified on more than 20 fixed-wing and rotary aircraft. Orders include those of the US Marine Corps for its fleet of UH-1N Huey helicopters in the navigational safety, search and rescue and surveillance roles; the US Navy for P-3 Orion aircraft for navigational safety, maritime patrol and surveillance missions; and the New South Wales police for law enforcement.

Contractor

FLIR Systems Inc.

VERIFIED

Star SAFIRE™ equipment components ***1999***/0051014

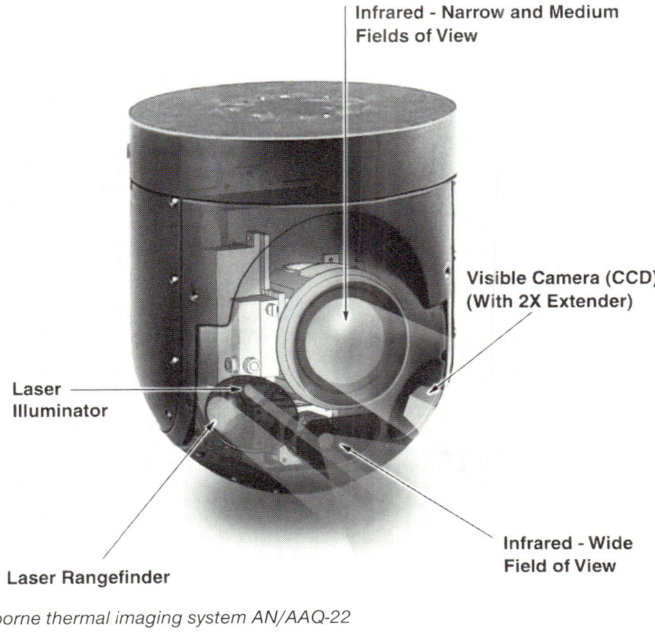

Star SAFIRE™ airborne thermal imaging system AN/AAQ-22 0018313

Thermovision 1000 series thermal imaging systems

Within the FLIR Systems portfolio, the AGEMA range of thermal imaging systems is optimised to different requirements, including Thermovision 1000; Thermovision 1000LR (Long Range); Thermovision

FLIR Systems AGEMA Thermovision 1000 thermal imager 0106309

The Thermovision 1000 CLR fits inside the LEO-400-LSPIR/SPTV platform from Irenco 0018330

1000CLR (Compact Long Range). High-resolution real-time images are obtained with excellent thermal sensitivity using the LK-4 scanning module, multi-element SPRITE detector and integral Stirling cooler, with 12-bit digital image processing for accurate image capture. Features include automatic brightness and contrast adjustment; dual field of view lens for panoramic surveillance or close inspection (telescope option for optimum resolution); image freezing and electronic zoom; user-defined alarm levels using isotherms or pre-set zone conditions; other user-defined options available.

The Thermovision systems can be integrated into the gyrostabilised four-axis platform from gimbal specialists, Irenco. Called the LEO-400-LSPIR/SPTV, this compact and rugged platform fits neatly underneath, or on the side, or on the nose of a wide variety of helicopters and fixed-wing aircraft. The possibility to include a three-chip CCD broadcast quality daylight camera in the same gimbal provides the flexibility for day/night surveillance in the same mount.

Specifications
Dimensions: 310 × 164 × 221 mm
Weight: 8 kg
Power supply: 28 V DC, 55 W
Detector: Multi-element MCT SPRITE
Wavelength: 8-12 μm
Performance: <0.1°C
Optics:
Field of View (FoV) no lens 33° (H) × 21.6° (V)
Field of View (FoV) dual lens:
(wide FoV) 20° (H) × 12° (V); IFOV 0.6 × 0.6 mrad
(narrow FoV) 5° (H) × 3.0° (V); IFOV 0.15 × 0.15 mrad
Detection capability: man: 5 km; fast boat: 18 km; light aircraft: 20 km

Operational status
Widely used by the police in the UK and Europe.

Contractor
FLIR Systems Inc.

VERIFIED

ULTRA 4000™ dual-camera 24 hour imaging system

The ULTRA 4000™ combines a three-chip CCD colour camera for daylight surveillance and pursuit with the military-qualified SAFIRE thermal imaging system, for a complete, high-resolution package that is appropriate for policing and Electronic News Gathering (ENG), regardless of light levels. The system is gyrostabilised on three axes and offers 360° turret rotation, including full look-down capability. The SAFIRE is an 8 to 12 μm thermal imager with a 28° wide field of view for navigation and area searches. Its 5° narrow field of view is supplemented by electronic zoom and freeze-frame features. The colour camera has a 16:1 continuous zoom lens with a × 2 extender to capture close-up details. One hand-held controller operates both the visible light camera and the thermal imaging system. Video images are viewed on a flight-hardened display. Images can be fed live to ground units or a television station via an optional microwave downlink, or video can be recorded to an optional VCR. Systems can be configured for compatibility with night-vision goggles. Other options include autotracker, searchlight slaving, and radar and navigation interfaces.

Operational status
In service worldwide. Customers include several UK police forces; the Indian Dept of Revenue through Pawan Hans Helicopters (India); and several television stations in the United States.

Contractor
FLIR Systems Inc.

VERIFIED

ULTRA 4000™ dual-camera 24 hour imaging system 0015293

ULTRA 6000™ dual-camera 24 hour imaging system

The ULTRA 6000™ is a lightweight, four-axis gyrostabilised system designed to meet the operational demands of law enforcement. The system includes both daylight video and thermal imaging cameras to enhance an air support unit's search, surveillance and apprehension capabilities. The ULTRA 6000™ is fitted with an advanced indium antimonide (InSb) 3 to 5 μm thermal imager, for sharp images over a range of operating temperatures and exceptional performance in the cooler temperatures encountered as a result of altitude, flight speed and seasonal weather. The CCD camera uses a × 15 continuous zoom lens for high-resolution colour video. The thermal imager is equipped with four Fields of View (FoV). For maximum imaging flexibility, the two optical FoVs (20° wide and 4° narrow) can be electronically extended for two additional FoVs (10° wide and 2° narrow). Loaded with both cameras, the gimbal (diameter: 279.4 mm; height: 381 mm) weighs less than 20 kg.

Specifications
Gimbal
Active gyrostabilisation: 4-axis stabilisation
2-axis inner (pitch/yaw)
2-axis outer (az/el)
6-axis passive stabilisation
Azimuth/elevation slew rate: >60°/s (90° max)
Dimensions: 279 × 381 mm
Weight: 20 kg
Power: 28 V DC @ 150 W

Thermal imager
Sensor: InSb
Resolution: 256 × 256 pixels
Cooling: Stirling closed-cycle cooler
FoV switch over time: <1.0 s
Focal length: 21.6 to 110 mm

ULTRA 6000™ dual-camera 24 hour imaging system on a Eurocopter AS-350 0015296

A typical installation for the UltraForce® high-resolution thermal/visual imaging system ***2000***/0085263

FoV: 20, 10, 4 and 2°* (*fields of view using × 2 electronic zoom)
Format: NTSC or PAL

Visual imager
Camera: Sekai
Pick up device: ½ in Hyper HAD 1 CCD
Resolution: 470 NTSC, 460 PAL
Gain: Auto/Manual
Video zoom: × 15 Canon continuous zoom
Focal length: 16 to 240 mm
NFOV: 1.5° (H) × 1.1° (V)
WFOV: 17° (H) × 13° (V)

Options
Autotracker, searchlight slave, LCD flat panel, GPS, VCR, microwave downlink, RS-422 interface.

Operational status
Development complete.

Contractor
FLIR Systems Inc.

VERIFIED

UltraForce - LE™ (Law Enforcement) airborne surveillance system

The UltraForce is an affordable, high-performance airborne imaging system designed for law enforcement, Electronic News Gathering (ENG), and paramilitary surveillance applications. This system provides instant-on imaging in a flexible, lightweight, cost-effective package. The system can be configured for dual-sensor use with a choice of state-of-the-art uncooled infra-red imagers, CCD daylight cameras, and high-magnification video cameras. The infra-red camera uses an infra-red Focal Plane Array (FPA) to offer dual fields of view, and the video camera provides a continuous (× 18) zoom mode. Both allow the operator to see an entire area of interest or close-up detail for easy identification of targets.

Changing payloads is a simple three-step process allowing the user easily to tailor the UltraForce system mission by mission, without the expense of two separate camera systems. The system's low weight and compact size allow for longer flight times and nose mounting on a variety of small and weight-restricted aircraft.

Stability for both cameras is assured through the use of a four-axis gyrostabilised gimbal providing stable images through 360° of rotation, even when using full zoom. These images can be recorded directly to tape or, with an optional microwave downlink, transmitted live to ground operations.

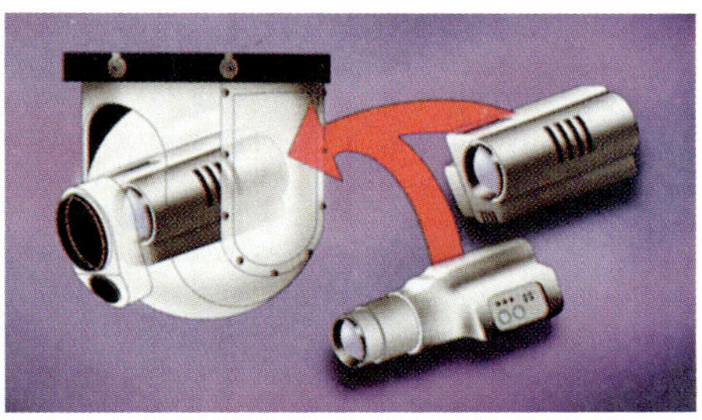

UltraForce - LE™ (Law Enforcement) airborne surveillance system ***1999***/0051015

The UltraForce® high-resolution thermal/visual imaging turret ***2000***/0085262

Specifications
Gimbal
Active gyrostabilisation: 4-axis
Pointing accuracy: <100 μrad RMS
Azimuth/Elevation slew rate: 60°/s
Power: 28 V DC @ 10 A
Weight: 3.2 kg
Dimensions: 280 × 381 mm

Thermal imager
Sensor: Microbolometer
Resolution: 320 × 240 pixels
Cooling: Uncooled
Wavelength: 7.5-12 μm
FOV switch over time: <1.0 s
Focal length: 38 to 152 mm
FOV: 6 × 4.5°
24 × 18°
Format: NTSC/PAL
Electronic Zoom: × 4
Palettes: Black hot, white hot, rainbow, iron

Built-in visual camera
Camera: Auto focus, ⅓ in CCD
Sensitivity: 6 lx
Zoom: × 8

Optional visual camera
Camera: Sony DCR-VX1000 - 3-chip ⅓in CCD
Sensitivity: 4 lx
Zoom: × 10 optical, × 20 electronic

Operational status
The system is in production.

Contractor
FLIR Systems Inc.

VERIFIED

UltraMedia™ 5-axis gyrostabilised camera system

The UltraMedia™ aerial camera system is designed to provide the highest quality long-distance surveillance capability for policing, Electronic News Gathering (ENG) and long-range sports coverage. UltraMedia™ systems have captured images from 'standoff' distances of more than two miles. The system's small size and weight make it ideal for nose mounting on many helicopters, including new fourth-generation models. The system uses a colour CCD camera and a continuous zoom lens with a × 2 extender for up to × 72 magnification. Intuitive controls can be adjusted to suit a range of user preferences. Five-axis gyrostabilisation ensures steady images, even in tightly banked turns. The gyros cannot be unseated by extreme manoeuvres and require no daily balancing or set up. The system's low profile (diameter: 381 mm; height: 444.5 mm)

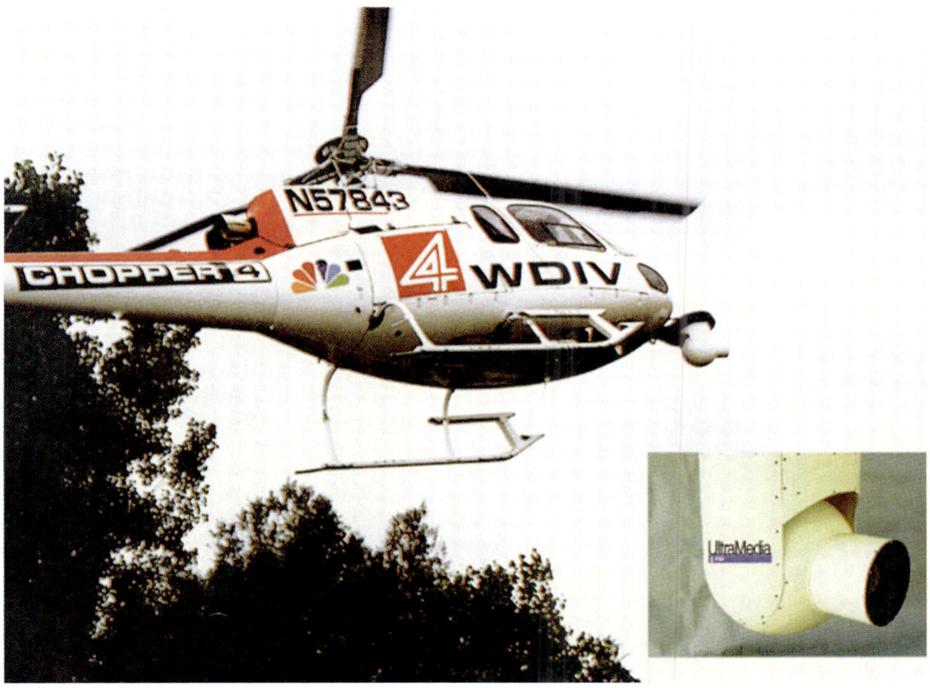

UltraMedia™ 5-axis gyrostabilised camera on a Eurocopter AS-350 ***1999***/0051016

minimises drag to improve fuel efficiency and increase flight time. Complete with camera and lens, the gimbal weighs 35.9 kg. Video images can be recorded to an onboard VCR for evidence gathering.

Specifications

Camera

UltraMedia™ is available with a broadcast-quality ⅔in colour camera and 36:1 or 33:1 lens with motorised × 2 extender. Other lens and camera configurations are available.

Gimbal

Stabilisation: 5-axis gyrostabilisation
Field of Regard:
 Elevation: +20 to −105°
 Azimuth: 360° continuous
Maximum slew rate: 60°/s
Dimensions:
 Diameter: 381 mm
 Height: 444 mm
Weight (w/camera and lens): 39.5 kg
Power: 22-29 V DC
Global power:
 quiescent: 75 W max
 continuous: 160 W max
 transient: 300 W max

Operational status

Customers include more than 50 television stations in the United States representing all four major networks (CBS, NBC, ABC, Fox) and sports and news coverage in Australia, Brazil, Japan and New Zealand. More than 50 UltraMedia™ systems are in use around the world.

Contractor

FLIR Systems Inc.

VERIFIED

UltraMedia LE® TV-only surveillance camera

The UltraMedia LE® is a TV-only system, designed for the Law Enforcement (LE) market. It is equipped with long-range optics and a super-low-light Hitachi 3-CCD camera. This system permits operation at extreme standoff distances, even under low light conditions. The camera has 46 per cent longer focal length than earlier UltraMedia systems to provide positive identification of personnel and vehicles.

Equipment features include:

1. Low light operation down to 0.15 Lux to provide full-colour video imagery during night-time operations.
2. A lightweight 5-axis gyrostabilised gimbal design offering 10-microradian performance.
3. Digital control and on-screen symbology.

Contractor

FLIR Systems Inc

VERIFIED

UltraMedia-RS™ 5-axis gyrostabilised compact system

Weighing only 15.9 kg, complete with camera and lens, the new UltraMedia-RS™ is designed for daylight surveillance and Electronic News Gathering (ENG) from small helicopters and light aircraft. The system uses a three-chip CCD colour camera and continuous zoom lens that provides up to × 40 magnification. The system's five-axis gyrostabilisation and advanced optics make it possible to capture details of unfolding events, even at considerable distance from the scene. The gimbal features 360° rotation and full look-down capability. The camera is operated by a laptop panel in the cockpit. A flight-hardened monitor provides instantaneous visual confirmation. The system's digital signal can be recorded to VCR for evidence or later analysis. An optional microwave downlink sends the signal live to the ground for real-time viewing or broadcast. The gimbal's small profile (diameter: 279.4 mm; height: 360.7 mm) minimises drag during flight. The UltraMedia-RS™ can be nose-mounted on many small aircraft.

Operational status

The UltraMedia-RS™ is used for news coverage by several television stations in the United States and for law enforcement surveillance.

Contractor

FLIR Systems Inc.

VERIFIED

The UltraMedia LE® long-range, low-light TV system mounted on a Cessna aircraft **2000**/0085264

UltraMedia-RS™ 5-axis gyrostabilised compact system 0015294

AMLCD MultiPurpose Displays (MPDs)

4 × 4 in Colour AMLCD MultiPurpose Display (MPD)

The 4 × 4 in MPD offers a complete range of formats, and is ideal for new build and retrofit requirements in fighter aircraft. High resolution, low reflectivity and excellent chromaticity is provided. VAPS software is programmable to meet customer requirements, with AMLCD technology. Simultaneous video and graphics capability is offered, with 15 grey shades (option 64), MIL-STD-1553B interfaces, NTSC output, and NVIS Class B compliance.

Specifications

Dimensions: 141 × 141 × 275 mm
Weight: 4.8 kg
Power Supply: 28 V, 95 W

6.25 × 6.25 in Colour AMLCD MultiPurpose Display (MPD)

The 6.25 × 6.25 AMLCD has been developed for the US Army Longbow Apache helicopter, where the display system comprises four 6.25 × 6.25 colour AMLCD MPDs and two Colour Display Processors (CDPs). The CDPs can be populated selectively to a maximum of four completely independent channels, using a MIL-STD-1553 bus if desired. Full feature digital map system can be embedded within the CDP. The MPD provides 64 grey shades (growth to 256). Resolution is 512 × 512 colour pixels in Quad RGGB Colour Pixel arrangement. The MPD is NVIS Class B compliant, and tested with Class A goggles in mono green mode.

Four of the Honeywell Aerospace 6 × 8 in AMLCD units (two pilot and two co-pilot) in the US Air Force C-141 cockpit configuration 0018181

4 × 4 in colour AMLCD multipurpose display 0018182

8 × 10 in AMLCD display unit 0018184

Specifications

Dimensions: 216 × 216 × 184 mm
Weight: 5.8 kg
Power supply: 115 V 3-phase 400 Hz, 115 W

Operational status

In full production for the US Army Longbow Apaches and standard for all Netherlands and UK Longbow Apaches. A total or more than 4,000 MPDs will be required for the Longbow Apache programme including remanufacture of all US Army AH-64s.

6 × 8 in AMLCD Display Unit
The 6 × 8 in AMLCD has been designed for use on the C-141 aircraft upgrade, where it acts as either a primary (ADI, HSI, Airspeed, Altitude display) or secondary (Heading, Waypoint, Weather Map) flight display. A high performance graphics processor is used with a 486DX2 general purpose processor; the Virtual Application Prototyping System (VAPS) permits the drawing and simulation of formats in near-realtime for format changes. Resolution is 480 × 640 colour pixels (RGGB Quad) with 80 colour groups/in. The display is NVIS Class B compliant, with NVIS Class A option. Fitment of 6 × 8 in AMLCDs forms part of the US Air Force C-130 and C-141 integrated avionics upgrade programme. The 6 × 8 in display is also used

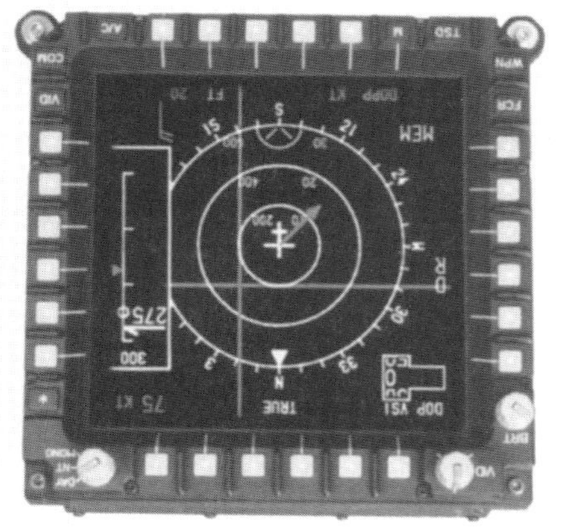

6.25 × 6.25 in colour AMLCD multipurpose display 0018180

in the Spanish Air Force C-130 modernisation programme.

Specifications

Dimensions: 196 × 246 × 136 mm
Weight: 8.2 kg
Power supply: 28 V, 117 W

8 × 10 in AMLCD unit
This is currently Honeywell Aerospace's largest ruggedised military and avionic crew station AMLCD MPD. It has been supplied for the US Army's Rotary Pilots' Associate (RPA) next-generation cockpit.

Specifications

Grey shades: 64
Resolution: 1,024 × 768 (XGA)
NVG: NVIS Class B (option)
Dimensions: 292 × 241 × 76 mm
Weight: 6.8 kg
Power: 28 V DC, 75 W

Contractor

Honeywell Aerospace, Electronic & Avionics Lighting.

VERIFIED

Moving map display for the F-15

The F-15 remote map reader provides a continuous colour or monochromatic high-resolution video signal to a multifunction display. It uses existing 35 mm film strips housed in self-indexing, interchangeable cassettes which automatically engage and align when inserted into the unit. The system can be mounted in the equipment bay, so reducing pressure on cockpit space. The map images appearing on the screen are therefore synthetic, as opposed to the real images projected on to the screen of conventional moving map displays.

The centre of the map image is automatically aligned with the aircraft present position using data from the navigation system. The unit incorporates a built-in test facility.

Specifications

Dimensions: 193 × 191 × 319 mm
Weight: 12.3 kg

Operational status

In service.

Contractor

Honeywell Aerospace, Electronic & Avionics Lighting.

VERIFIED

V-22 Osprey advanced colour display system

The system comprises four 6 × 6 in (152 × 152 mm) high-resolution colour displays driven by a pair of three-channel programmable display processors. These use a MIL-STD-1750A computer and feature triple redundant display processing.

Operational status

In production for the V-22 Osprey.

Contractor

Honeywell Aerospace, Electronic & Avionics Lighting.

VERIFIED

Colour multifunction display for C-17

Honeywell's 6 × 6 in (152 × 152 mm) colour multifunction display is a high-resolution, shadow-mask CRT display developed for the C-17 military air transport. Honeywell's newest family of full-colour, military-qualified displays includes a number of advanced technologies to provide significant performance advantages. Superior vibration tolerance, sunlight viewability and high-resolution graphics are inherent in the design of the display.

The multifunction display can be used to present primary flight and navigation information, colour weather radar, digital map information, engine instrumentation and tactical display formats. The unit presents stroke, raster or hybrid formats in 16 colours. Raster images are driven from sensor inputs with an RS-170 or RS-343 interface.

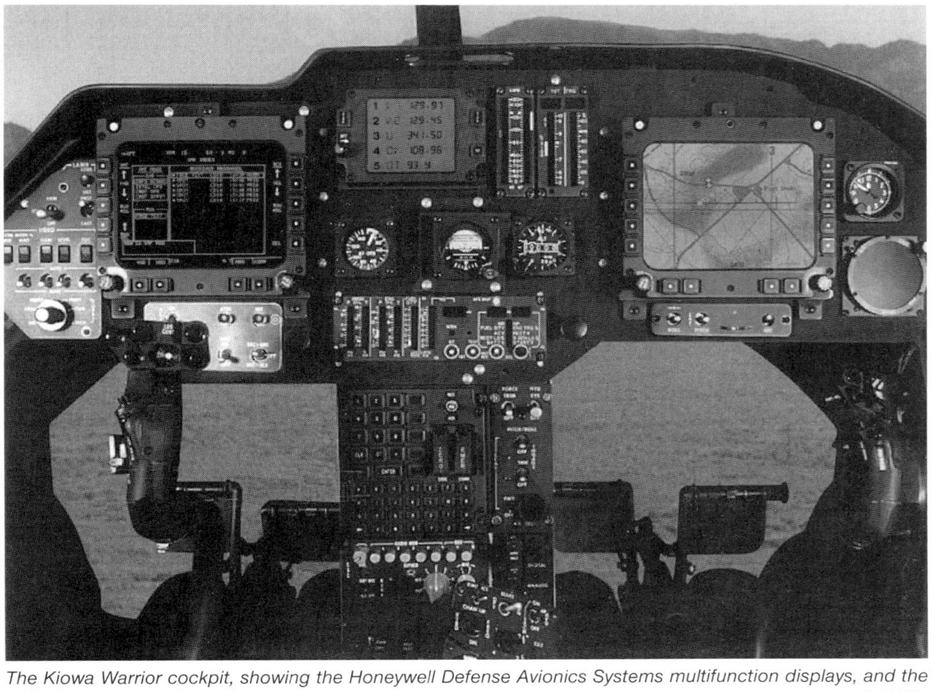

The Kiowa Warrior cockpit, showing the Honeywell Defense Avionics Systems multifunction displays, and the radio frequency display 0018170

Honeywell's multifunction display has been designed as a stand-alone 'smart' unit. The display contains a MIL-STD-1750A processor to convert stored modes into formats using aircraft parameters received over the MIL-STD-1553B bus. An internal vector generator is used to draw the formats on the screen. The unit also includes an expanded built-in test capability and can report failures to the host computer via the MIL-STD-1553B bus.

A separate display processor is not necessary when linking the 'smart' multifunction display directly to the aircraft's mission computer via the 1553B bus. This configuration reduces system weight, reduces potential interface problems and increases system reliability.

Specifications

Dimensions: 203 × 203 × 381 mm
Weight: 16.8 kg
Resolution: 799 × 820 pixels

Operational status

The new multifunction display will be introduced into the 71st of 120 production C-17s. MFD production in excess of 200 units was scheduled to begin in October 2000, with final delivery in 2003.

Contractor

Honeywell Defense Avionics Systems.

VERIFIED

Control Display System (CDS) for OH-58D Kiowa Warrior

Honeywell's Control Display System (CDS) provides embedded controls and displays for communication (ETICS), navigation, engine/power-train, Mast-Mounted Sight (MMS) and weapons, as well as Rotorcraft Map System (RMS - digital map) and Video Image cross-Link (VIXL).

The ETICS (Embedded Tactical Information Control System) (see Communications section), is part of the Honeywell CDS, and it embeds integrated Task Force XXI Variable Message Format (VMF) capabilities for command and control, fire support and situational awareness.

The rotorcraft mapping system provides Kiowa Warrior crews with situational awareness by translating data from ETICS and the Honeywell Embedded Global positioning system/Inertial navigation system (EGI) into icons that report the aircraft's position in relation to the terrain and other air and ground vehicles, both friend and foe.

The VIXL transmits near-realtime reconnaissance video images from the Kiowa Warriors mast-mounted sight over standard combat radio links to other Kiowa Warriors or ground command and control centres. In addition, Honeywell is also providing the SINCGARS/SIP (Single-Channel Ground-Air Radio System/System Improvement Program) radio, and EGI system as part of the overall Kiowa Warrior upgrade. The display system itself comprises two monochrome multifunction displays (one for each pilot), and the radio frequency display.

Contractor
Honeywell Defense Avionics Systems.

VERIFIED

MultiPurpose Colour Display (MPCD) for the F-15C/D MSIP and F-15E

The MultiPurpose Colour Display (MPCD), developed for the F-15 fighter as part of a continuing Multinational Staged Improvement Programme (MSIP) enhancement for the C and D versions, is a very high-resolution 5 × 5 in (127 by 127 mm) shadow-mask system with stroke, raster, or combined stroke/raster writing in any of 16 colours as determined by software in the symbol generator. Raster symbology is a full-colour representation of a colour sensor output. Functions include built-in test for a number of aircraft systems, graphic representation of aircraft stores configuration (replacing the electromechanical armament control panel on earlier F-15s), display of video from electro-optic sensors and weapons and interface for secure tactical information from a JTIDS communications system.

Honeywell has been contracted for a second version of the colour display for the F-15E programme. The F-15E fleet, which incorporates air-to-ground and air-to-air mission capabilities, is equipped with three Honeywell MPCDs on each aircraft. Honeywell is also under contract for the display processor for F-15E aircraft. The equipment not only powers the aircraft's three colour displays, but all the displays on the aircraft.

The display's delta gun shadow-mask tube uses dynamic and static convergence for maximum symbol fidelity. A bandpass filter and 17,000 in/s writing speed facilitates viewing in high ambient light levels, or even direct sunlight. Two CRT versions are available, one having 0.008 in phosphor dot spacing said, by Honeywell, to be only two-thirds the dot size and spacing of other shadow-mask CRTs designed for airborne applications.

The Honeywell programmable signal data processor interfaces with other equipment and sensors, generating monochrome and colour symbology for other displays. It comprises two elements: a three-channel symbol generator and a general purpose processor. Self-test circuits monitor all critical signals continuously and failures are flagged on a built-in test indicator.

Specifications
Dimensions: 172 × 180 × 387 mm
Weight: 11.8 kg
Screen size: 127 × 127 mm
Linewidth: 0.016 in typical
Video bandwidth: DC 10 MHz
Television scan resolution: 70 line pairs/in

Operational status
In service with F-15C/D MSIP and F-15E fighters. A version of this display is in the F-117A aircraft.

Contractor
Honeywell Defense Avionics Systems.

VERIFIED

Radar Control Display Unit (RCDU) for UK E-3D

Honeywell Defense Avionics Systems will develop the colour Radar Control Display Unit (RCDU) for the United Kingdom's fleet of seven E-3D Airborne Warning And Control System (AWACS) aircraft under a subcontract to Northrop Grumman Electronic Sensors and Systems Division of Baltimore, Maryland as part of the Boeing Company's United Kingdom AWACS Radar System Improvement Programme (RSIP).

The RCDU adapts proven Commercial-Off-The-Shelf (COTS) hardware Honeywell Air Transport Systems originally developed and integrated on the Boeing 777, incorporating Honeywell's flat panel, active matrix liquid crystal display technology.

Operational status
In development for the UK E-3D AWACS aircraft.

Contractor
Honeywell Defense Avionics Systems.

VERIFIED

Remote frequency display

The operational status of each of the five radios carried aboard OH-58D helicopters is displayed by means of the remote frequency presentation system. It shows, on five lines of liquid crystal readout, the radio number, operator in control, frequency, whether coded or in plain language and two alphanumeric characters for channel identification.

The display receives its information through a serial digital interface with either of the two master controller processor units, the choice of unit being left to the operator. Data for each window remains displayed until updated. If the system does not receive an update within two seconds, the screen goes blank as a warning to the operator. The screen is designed for reflective daylight viewing and is back-lit for night operation. Characters are painted in white on a black background.

Specifications
Dimensions: 146 × 99 × 102 mm
Weight: 10.8 kg
Power:
3 W normal
53 W with heater
Contrast:
20:1 in direct sunlight
16:1 when back-lit
Back-lighting: electroluminescent

Operational status
In production for and in service on the OH-58D.

Contractor
Honeywell Defense Avionics Systems.

VERIFIED

Integrated Helmet and Display Sighting System (IHADSS)

Both the pilot and co-pilot/gunner are provided with helmet units and controls to allow independent and co-operative use of the system. The IHADSS sight component provides off-boresight line of sight information to the fire-control computer for slaving weapon and sensor to the pilot's head movements. Real-world sized video imagery from the slaved and gimballed infra-red sensor is overlaid with targeting as well as flight information symbology and projected on a combiner glass immediately in front of the pilot's eye. The IHADSS allows night nap-of-the-earth flight at below treetop altitudes, without reference to cockpit instruments, and rapid target engagement.

Specifications
Dimensions:
(sensor surveying unit × 4) 111.76 × 129.54 × 78.74 mm
(sight electronics unit) 177.8 × 177.8 × 289.56 mm
(display electronics unit) 139.7 × 177.8 × 289.56 mm
(display adjust panel × 2) 152.4 × 182.88 × 76.2 mm
Weight:
(helmet × 2) 1.4 kg each
(sensor surveying unit × 4) 0.57 kg each
(sight electronics unit) 7.26 kg
(boresight reticle unit × 2) 0.23 kg each
(helmet display unit × 2) 0.57 kg each
(display electronics unit) 6.35 kg
(display adjust panel × 2) 1.58 kg each
(total weight) 23.45 kg
Power supply: 115 V AC, 400 Hz, 3 phase, 460 W
Field of view: (horizontal) 40° × (vertical) 30°
Coverage:
(azimuth) ±180°
(elevation) ±90°
Accuracy: 3 to 10 mrad RMS
Slew rate: 120°/s

Operational status
Currently in service on the Boeing AH-64A Apache and Agusta A 129 helicopters.

Contractor
Honeywell Sensor and Guidance Products.

UPDATED

Honeywell IHADSS helmet sight and display 0106407

C-130 air data and multifunction engine/oil/fuel display system

The ISS air data and multifunction engine/oil/fuel system is a quadruple-redundant solid-state system display. The air data system uses four part numbers to replace 19 part units and eliminates 11 Line Replaceable Units (LRUs). Instead of 32 electromechanical engine instruments composed of eight part numbers, the ISS solid-state system consists of 24 indicators composed of just two part numbers (Type I and Type II instrument displays).

System accuracy and redundancy fulfils requirements for RVSM operation in the North Atlantic Free Flight area and metre/foot switching facilities provides for operation in Eastern Europe and other metric airspace areas.

Air data system units are:

One Central Air Data Computer (CADC), which replaces the existing electromechanical True Air Speed (TAS) computer and 12 autopilot and airspeed sensors. It communicates via a dual MIL-STD-1553B databus and provides height accuracy of 10 ft at 50,000 ft.

Three solid-state airspeed indicators, which provide accuracy of ± 3 kt at > 100 kt. Optional features include: windshear alert with a trend display; take-off monitor; angle of attack display.

Three solid-state altimeters, which are RVSM compliant with accuracy of 30 ft at 30,000 ft.

Three voice annunciating combined altitude alerting units, which provide voice and tone warnings.

The Type I engine display automatically changes function to serve as an RPM indicator, torque indicator, temperature indicator and fuel-flow indicator. The Type II indicator displays five independent oil parameters including pressure, temperature, quantity and cooler flap settings; the same unit can also display hydraulic indications.

Operational status

Selected by Lockheed Martin for the C-130H aircraft. The system is catalogued in the US Air Force inventory. It is available for retrofit to older C-130 aircraft.

Contractor

ISS Innovative Solutions & Support Inc.

VERIFIED

C-130 aircraft cockpit showing updated ISS multifunction engine/oil/fuel displays **2000**/0081871

MERLIN night vision goggles

The Modular Ejection-Rated Low-profile Imaging for Night (MERLIN) night vision goggles have been developed for high-performance aircraft. The goggles incorporate a lower-cost Generation III tube and have a 35° field of view. Each of the two interchangeable monoculars weighs 0.25 kg and provides 20 mm of eye relief.

The company claims that MERLIN attached to a standard HGU-55 helmet extends the centre of gravity of the pilot's head forwards by only about 3 mm, allowing safe ejection. The identical halves of the goggles are attached to the helmet's visor mountings and a new visor is hung from a beak on the helmet. All optics, apart from the objective lenses, are placed behind the visor, further aiding ejection.

Specifications

Weight: 0.8 kg
Power supply: 3 V DC batteries (one per monocular)
Field of view: 35° circular
Focus range: 5 m to infinity
Magnification: ×1

Operational status

Under development.

Contractor

ITT Industries, Electro-Optical Products Division.

VERIFIED

Agile Eye™ Plus helmet

Kaiser Electronics has developed Agile Eye™ Plus, a second-generation helmet-mounted display that provides significantly greater capability than the original Agile Eye™. The field of view of Agile Eye™ Plus has been increased from 12 to 18° to improve situational awareness, and a raster display capability has been added to the stroke display mode so that information from TV cameras and other sources such as FLIR can be projected. A retractable visor replaces the clip-in visor of Agile Eye™.

Agile Eye™ Plus projects head-up display information on to the visor of the pilot's helmet superimposing, in a monocular presentation, information over the field of view without obscuring vision. The system includes two declutter modes so that the pilot can include only those items of information he needs to see. A magnetic head tracker, mounted inside the canopy, tracks the pilot's head movements and continuously updates display data to correspond to the direction of his head. Mounted in the helmet are a 0.5 in diameter CRT; a small high-voltage power supply; a system of mirrors for projecting the image on to the visor. The Agile Eye™ Plus helmet weighs just over 1 kg. The system also includes a display driver unit in the cockpit and a display processor/tracker unit in the avionics bay.

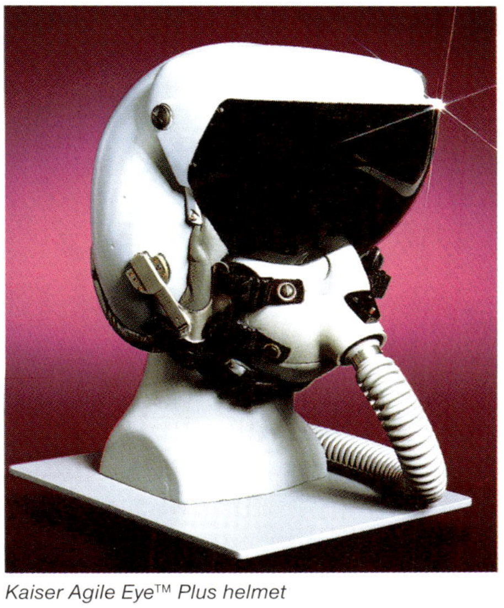

Kaiser Agile Eye™ Plus helmet

The third generation of Agile Eye™, Agile Eye™ Mk III, has now been released. This features a tracking sensor, detachable display unit, interchangeable day/night visor, visor/mask seal adjustment, high-voltage quick disconnect and 600 kt windblast protection.

Specifications

Agile Eye Mk III
Weight:
(helmet-mounted display) 1.45 kg
(cockpit control panel) 2 kg
Power supply: 115 V AC, 400 Hz
5 V AC, 400 Hz
Field of view: 20° monocular, right eye
Field of regard: unlimited
Reliability: >2,000 h MTBF

Contractor

Kaiser Electronics.

VERIFIED

Colour helmet-mounted display

The colour helmet-mounted display provides a high-quality, high-brightness, wide field of view, full-colour helmet display in a system configuration which is easy to use and compatible with aircraft and simulator installations. The system is a suitable flight research tool for experimentation in guidance, control and display flight research.

Specifications

Weight:
(control panel) 0.45 kg
(display electronics unit) 17.8 kg
(helmet-mounted display) 2.04 kg
Power supply: 115 V AC
Field of view: 40° (vertical) by 60° (horizontal)
Colours: red, green and blue
Modes of operation: day, night, boresight

Contractor

Kaiser Electronics.

VERIFIED

Display suite for the F-14D

In 1985, Kaiser was awarded the contract for the entire display suite for the US Navy's F-14D update programme. The multifunction panel display specified by the US Navy was Kaiser's Multipurpose Display Repeater Indicator (MDRI) already present in the US Navy's inventory for the AV-8B and F-16 aircraft.

In addition to the MDRI, the display suite consists of an advanced Display Processor (DP) and a wide-angle, low-profile HUD. The F-14D complement of equipment includes one HUD, two DPs and three MDRIs.

The F-14D head-up display features an advanced, extremely compact optics design that provides a 30° total field of view and dual flat holographic combiners for increased brightness, better see-through vision and raster and cursive capability.

The Display Processor (DP) uses custom gate arrays and a single-card MIL-STD-1750A processor. It can simultaneously drive up to six cockpit displays, including HUDs, multifunction displays, helmet displays and a videotape recorder, all under Mission Computer (MC) control through the primary interface, a dual MIL-STD-1553B bus. The generated functions include alphanumerics, geometric shapes, over 160 prestored formats and new formats definable over the bus. These can be mixed with 525- or 875-line raster video selected from eight possible inputs. In addition, any of the three selected raster formats may be augmented by stroke symbology written in retrace. Stroke only presentations can be in red, yellow and green on display devices utilising the Kaiser Kroma liquid crystal shutter technology.

Four processors handle the input/output and display functions of the DP. Two MIL-STD-1750A CPUs provide system function and display function processing, while a Motorola 658000 microprocessor provides the primary interface with the MC and an Intel 8031 microprocessor provides input/output functions for nine RS-422 serial channels. The software, coded in CMS-2 high-order language, is contained in EEPROMs that are reprogrammable in the aircraft through the IEEE-488 maintenance bus or MIL-STD-1553B bus.

Specifications (DP)

Dimensions: 498.3 × 190.5 × 193.5 mm
Weight: 20.38 kg
Power supply: 115 V AC, 400 Hz, 3 phase, 195 W
Inputs: 8 video, 8 analogue, 52 discrete
Outputs:
(raster) 6 display ports, 1 VTR port, 525/875-line rates, 1:1 or 4:3 aspect ratio
(stroke) 6 independent sets
(hybrid) raster graphics/stroke/video to any display
Interface: dual MIL-STD-1553B
Reliability: 1,700 h MTBF

Operational status
In service in F-14D.

Contractor
Kaiser Electronics.

VERIFIED

Display system for the F-15E

Kaiser produced the head-up/head-down displays for the F-15E aircraft. Each aircraft has two 6 × 6 in (152 × 152 mm) CRT head-down MultiFunction Displays (MFDs) in each cockpit and one holographic, wide field of view Head-Up Display (HUD).

By presenting both high-resolution, high-contrast video and fine line, fast writing stroke symbology, the MFD has the capability to fulfil a wide variety of requirements and is compatible with almost any system configuration. High-resolution video from E-O sensors, radar and missiles in either 525- or 875-line rates is fully readable in the high ambient light of the tactical cockpit. Pure stroke modes provide the high information content necessary for such displays as tactical situation displays and JTIDS. The hybrid stroke during retrace mode allows high-resolution stroke symbology to be placed on top of raster displays.

The MFD has nine plug-in subassemblies, all replaceable without harmonisation. It is fully equipped with continuous and initiated BIT. Software cueing on the display surface provides in-flight programmability through bezel-mounted push-buttons.

The HUD employs both raster scan and stroke written symbology to accommodate the FLIR imagery from the Lockheed Martin LANTIRN system and the Raytheon Electronic Systems AN/APG-70 radar. The Kaiser wide field of view head-up display employs a holographic single combiner glass, permitting a smaller and lighter support structure with less obscuration of forward view.

The holograms for this HUD are made by an associate company, Kaiser Optical Systems Inc.

Specifications
MFD
Dimensions: 345.9 × 190.5 × 196.8 mm
Weight: 10.42 kg
Power supply: 115 V AC, 400 Hz, 3 phase, 180 W
Display: 152 × 152 mm
Resolution: 100 line pairs/in
Reliability: 3,000 h MTBF

Operational status
In service on the F-15E.

Contractor
Kaiser Electronics.

VERIFIED

Head-Up Display (HUD) and weapon aiming system for the MB-339C

Kaiser supplies the head-up display and weapon aiming system for the MB-339C. The three-unit system comprises a computer/symbol generator, sweep driver unit and pilot's display unit. All or part of the same system is used in the MT-4, F-4EJ, C-1 and CCV programmes in Japan and the Taiwanese AT-3. The Kaiser HUD fitted to the Italian MB-339C, known in this application as Sabre, has dual-flat holographic combiners and a flexible upfront control panel.

Operational status
In service in the MB-339C.

Contractor
Kaiser Electronics.

VERIFIED

Head-up Display (HUD) for the A-10

The updated version of the Kaiser HUD for the US Air Force A-10 incorporates inertial navigation data from the A-10's inertial navigation system. The HUD contains a MIL-STD-1553 multiplex databus which can compute total velocity vectors and other algorithms.

Operational status
In service in the US Air Force A-10.

Contractor
Kaiser Electronics.

VERIFIED

Head-Up Display (HUD) for the AH-1S

The HUD for the US Army Bell Helicopter AH-1S Cobra helicopter is a lightweight low-profile conventional CRT that superimposes aiming information for the multibarrelled gun and TOW anti-tank missile on to the pilot's forward field of view. The programming flexibility and spare capacity of the microprocessor-controlled symbol generator permits other functions to be incorporated, such as the derivation of flight commands for nap-of-the-earth flying and laser tracking and pointing information.

Operational status
In service in the US Army Bell AH-1S helicopter. Over 1,000 HUDs have been delivered.

Contractor
Kaiser Electronics.

VERIFIED

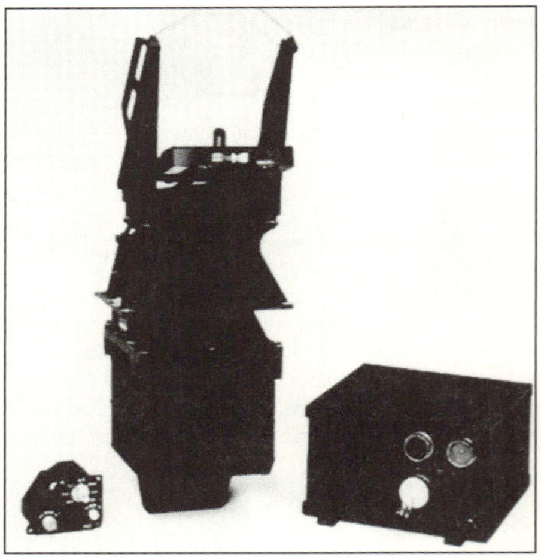

The pilot's display unit for the Kaiser Electronics head-up display in the AH-1S Cobra helicopter
0106409

The Kaiser Electronics symbol generator, combined head-up display/control unit and sweep driver for the MB-339C
0106408

Head-Up Display (HUD) for the AH-1W

Both the control/display subsystem and Full-Function Signal Processor (FFSP) are produced by Kaiser. The HUD is identical to that of the US Army's Bell AH-1S, which Kaiser also supplies. In contrast, the FFSP is an entirely new design, which features dual 68000 processors and software designed to DoD-STD-1679A requirements. The processor is programmable and includes the capabilities for vectors, circles, arcs and rotation.

Operational status
The US Marine Corps has ordered 44 Bell AH-1Ts and is expected to retrofit its fleet to the AH-1W configuration.

Contractor
Kaiser Electronics.

VERIFIED

Helmet Integrated Display Sight System (HIDSS)

The HIDSS is a second-generation binocular, wide field of view helmet-mounted display system designed for the RAH-66 Comanche helicopter. Critical for night pilotage, accurate delivery of weapons and improved situational awareness, the lightweight, high-resolution HIDSS utilises a two-piece modular helmet design, advanced optics and precision magnetic tracking to provide head-up, eyes-out operation. Driven by the high-performance SEM-E expanded display unit, HIDSS combines Gen II FLIR video, raster graphics imagery and growth to stroke symbology for a day and night, all-weather helmet-integrated display system that is adaptable to a variety of helicopter platforms and missions.

Specifications
Weight:
(control panel) 0.5 kg
(helmet-mounted display) 2 kg
(expanded display electronics unit) 14.97 kg
Power supply: 270 V DC, 335 W
Field of view: 35 × 22° (18° overlap)
Field of regard:
(azimuth) ±180°
(elevation) ±90°
(roll) ±180°
Reliability: 1,000 h MTBF

Operational status
Under development for the US Army RAH-66 Comanche helicopter.

Contractor
Kaiser Electronics.

VERIFIED

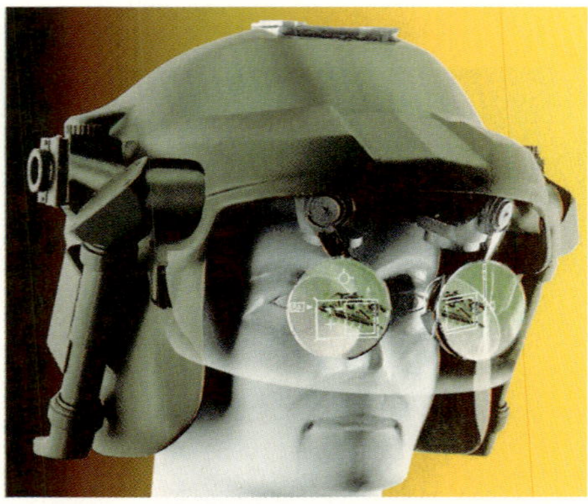

HIDSS is being developed for the RAH-66 Comanche helicopter
0106311

Integrated cockpit display for the F/A-18C/D

The integrated cockpit display in the F/A-18C/D includes the Multifunction Display Indicator (MDI) and Multipurpose Display Repeater Indicator (MDRI).

The F/A-18C/D 5 × 5 in (127 × 127 mm) NVG-compatible colour MDI and symbol generator includes an integral display processor. The processor consists of three Motorola 68020 microprocessors that handle input/output via a MIL-STD-1553 bus and provide display functions to the mission computer. The display functions include a variety of alphanumeric symbols, geometric shapes and large format macros such as a compass rose or pitch ladder. Three fully independent symbol generators are included, allowing the MDI to drive four additional displays with stroke or video symbology, or both.

The MDI suports a variety of formats, including 525-line raster in either 1:1 or 4:3 aspect ratio, 675-line square format, arc scan and wedge for display of high-resolution ground mapping radar and 875- or 1,224-line formats for sensor video. The display generators can handle three independent simultaneous video outputs and can overlay stroke-in-retrace on all of them.

Kroma liquid crystal shutter technology is incorporated into these displays, allowing stroke symbols to be presented in either red, yellow or green. The use of colour improves pilot situation awareness by providing quick differentiation between friend, enemy and unknown in a cluttered tactical situation without sacrificing resolution and brightness. Colour signals are available externally so that each MDI can drive similar remote colour displays, as in the F/A-18 night attack aircraft.

Originally designed for the night attack version of the F/A-18, the 5 × 5 in (127 × 127 mm) MDRI raster/stroke display provides night vision-compatible Kroma colour stroke and monochrome raster displays in a wide variety of formats.

While air-to-air displays are presented in colour stroke, with red, green and yellow symbology available, air-to-ground views are matched to the AN/APG-65 radar by means of special arc-scan, DBS patch and rotating raster forms which preserve in the display the inherently high resolution available from that radar. Sensor video, such as FLIR, can also be accommodated with normal rasters in 25-, 675- or 875-line format, in either 1:1 or 4:3 aspect ratios. All of these rasters may be augmented by sharp, clear, fine line stroke symbology written in retrace.

Dependent on an external symbol generator for raster sweeps and stroke symbols, the resultant small size and 'cathedral' top allow installation high in the instrument panel either on the left or right sides or, as in the case of the F/A-18, on both sides. Push-buttons around the display surface are software cued to provide in-flight programming.

Specifications

Dimensions:
(MDI) 658.4 × 170.2 × 179.1 mm
(MDRI) 397 × 170.2 × 179.1 mm
Weight:
(MDI) 18 kg
(MDRI) 9.7 kg
Power supply: 115 V AC, 400 Hz, 3 phase
Display: 127 × 127 mm
Resolution: 120 lines/in
Reliability:
(MDI) 1,500 h MTBF
(MDRI) 2,600 h MTBF

Operational status

In service on the US Navy and US Marines F/A-18C/D aircraft and on export F/A-18C/D aircraft.

Contractor

Kaiser Electronics.

VERIFIED

Head-Up Display (HUD) for the JAS 39 Gripen

Sweden awarded a production contract for a HUD incorporating diffraction-optics technology, because the advantages of this technology are claimed to be a key factor in providing the JAS 39 with its capability in the air-to-air, air-to-ground and reconnaissance roles.

Diffractive-optics HUDs have two principal advantages. First, by comparison with conventional systems, they have a much wider field of view, typically 30 × 20° compared with 20 × 15°, and are thus more suited to the new generation of combat aircraft in which weapon aiming symbology can make large angles with the flight vector. The wide field will also be useful in night operations to display data from electro-optical sensors such as FLIR.

Secondly, the combiner glass on which the symbology is superimposed on the outside world acts as a mirror reflecting only a narrow band of light. The transmission index is about 85 per cent compared with 50 to 70 per cent for refractive HUDs. The symbology is also bright enough to stand out in direct sunlight without having to operate the CRT at such high-power levels that its life is shortened.

The system also provides resistance to glare, reflections and spurious sun images. The latter is particularly important since bright sunlight can create hot spots on the display that prevent the pilot from seeing the symbology. The design, based on proprietary technology using holography and lasers, employs a single-combiner glass, eliminating the bulky support structure necessary in HUDs that support two or more.

Operational status

In production and in service in the JAS 39 Gripen.

Contractor

Kaiser Electronics.

VERIFIED

Holographic Head-Up Display (HHUD) for the F-4E

A Wide Field Of View (WFOV) HHUD is now in service on the F-4E aircraft under a contract from an international customer. The F-4E HHUD features a single curved holographic combiner which offers a 20 × 30° total and instantaneous field of view. The HHUD operates in raster and/or cursive modes and the display processor, which utilises a MIL-STD-1750A processor, is packaged with the HUD.

Operational status

Some 100 systems have been produced for the F-4E.

Contractor

Kaiser Electronics.

VERIFIED

Lite Eye™ Helmet-Mounted Display (HMD)

The Lite Eye™ Helmet-Mounted Display (HMD) is a compact, lightweight, low-cost monocular display system designed for helicopters and combat support aircraft. Specifically designed to attach to the standard ANVIS Night Vision Goggle (NVG) system, Lite Eye™ HMD provides crew members basic HUD capability (aircraft flight, engine performance and weapons symbology) in both day and night operations. Designed to accommodate a wide range of aircraft configurations and mission requirements, the core Lite Eye™ HMD system is easily expanded to include capabilities such as head tracking (magnetic or non-magnetic) and symbol generation. Also designed to accommodate a wide variety of users, Lite Eye™ HMD can be used with all military aviator helmets and provides full ranges of InterPupillary Distance (IPD), fore/aft and vertical adjustments.

Specifications

Field Of View: 20 × 15°
Exit pupil: 15 mm
Eye relief: 22 mm
Contrast ratio:
1.4 @ 10,000 fL
1.7 @ 5,000 fL
4.5 @ 1,000 fL
Resolution: 0.9 cy/mr
Transmission: 30%; 448 to 650 nm
Adjustments: IPD; fore/aft; vertical
Optical focus: Infinity
Display
Type: Active Matrix ElectroLuminescence (AMEL)
VGA: 640 × 480
Colour: monochromatic yellow
Dimming range: 100:1
Interface: VESA FPDI-1 (display controller input interface)
Average power: 5 W (5 V DC)

Contractor

Kaiser Electronics.

VERIFIED

Low-profile Head-Up Display (HUD)

Designed to deliver a 23.5 × 30° wide picture in the demanding tactical cockpit environment, the low-profile HUD for the F-14D uses modern holographic combiners to provide the highest see-through transmission and contrast. The high contrast is made even more valuable through the inclusion of a circularly polarised filter which suppresses the reflections so common in many HUDs. Used in the F-14D, the low-profile HUD provides crisp stroke symbology to display navigation, steering, flight situation and attack cues in the brightest daylight or, at the flick of the day/night switch, at subdued brightness to allow smooth adjustment of the critical night-time brightness levels.

This performance is packed into a unit of cross-section 102 × 152 mm, enabling cockpit designers to preserve forward visibility in instrument panel designs with large displays underneath and alongside the HUD. A TV recording camera is chin-mounted to view the same scene as the pilot, while eliminating the need for scan conversion.

The unit is completely self-contained. All nine shop-replaceable assemblies, including the combiner, are replaceable without harmonisation. BIT isolates 90 per cent of possible failures to these subassemblies, including the low- and high-power supplies.

Specifications

Dimensions: 674.9 × 165.1 × 274.8 mm
Weight: 17.5 kg
Power supply: 115 V AC, 400 Hz, 3 phase, 81 W
Field of view: 23.5° (elevation) × 30° (azimuth)
Reliability: 3,300 h MTBF

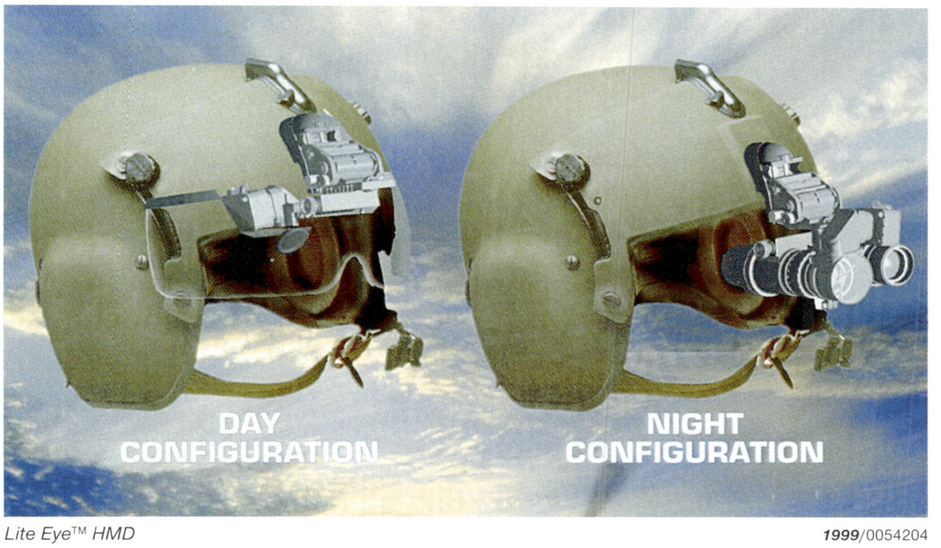

Lite Eye™ HMD **1999**/0054204

Operational status
In service with the Northrop Grumman F-14D.

Contractor
Kaiser Electronics.

VERIFIED

MultiPurpose Colour Display (MPCD) for the F/A-18E/F

The MultiPurpose Colour Display (MPCD) is a smart full colour LCD with a 6.25 × 6.25 in (158.7 × 158.7 mm) quad RGGB colour pixel resolution. Incorporating high brightness, the MPCD provides full-colour map video or high-resolution monochrome sensor video overlaid with full-colour graphics symbology. Colour improves pilot situational awareness by providing improved contrast ratio and improved resolution in full sunlight, enhancing the pilot/aircraft interface and reducing pilot workload.

A variety of video formats, including 525-line mono or GRB rasters in either 1:1 or 4:3 aspect ratios, 675-line square format for display of high-resolution ground mapping radar and 875 or 1,224-line formats for sensor video, is standard.

The MPCD architecture includes two fully independent processors which drive the MPCD as well as the Up-Front Control Display (UFCD), a separate monochrome 3.9 × 4.9 (99.1 × 124.5 mm) LCD with an infra-red touchscreen for pilot interactive control. Each display processor can provide formats using external video, internally generated graphics symbology or both, and comprises an 80960 microprocessor, high-performance graphics generator ASICs and high-speed video digitising and formatting. Primary graphics and control instructions are received via a MIL-STD-1553B bus from the aircraft mission computer. Dual-display generators allow any two independent video inputs, overlaid with internally generated symbology, to be displayed on the MPCD and UFCD.

Display functions include an extensive library of alphanumeric symbols, geometric shapes and large format macros such as a compass rose or pitch ladder. Priority information is colour-coded into symbology, alerts, messages and push-button labels.

A third 80960 processor, embodied in a standard HAC-32 MCM, provides UFCD touchscreen processing and communication, navigation and identification format processing. The HAC-32 processor has significant reserve throughput and memory to accommodate expanded future general purpose processing functions.

Specifications
Dimensions: 196.8 × 196.8 × 424.2 mm
Weight: 11.79 kg
Power supply: 115 V AC, 400 Hz, 3 phase, 246 W
Display: 158.7 × 158.7 mm
Resolution: 82 pixels/in
Viewing angle:
(horizontal) ±20°
(vertical) +35 to –5°
Interface: MIL-STD-1553B, discrete, video out
Reliability: 3,000 h MTBF

Operational status
Selected for the F/A-18E/F aircraft.

Contractor
Kaiser Electronics.

VERIFIED

Wide Eye helmet integrated display

Wide Eye is a fully integrated binocular helmet display system with retractable combiners for day and night use by helicopters at low level in all weather. It incorporates dual 1 in CRTs with stroke/raster and hybrid capabilities and image intensifiers. The optical subsystem is detachable and remains with the aircraft. The system consists of the headgear, display electronics unit, tracker and boresight reticle control unit.

Operational status
Weight:
(helmet) 1.8 kg
(system) 10.25 kg
Power supply: 115 V AC, 200 W
28 V DC, 20 W
Alignment time: 5 s
Field of view: 52 × 35°
Accuracy: <4.9 mrad

Contractor
Kaiser Electronics.

VERIFIED

All Weather Window® Enhanced Vision System (EVS)

The Kollsman All Weather Window® is an infra-red Enhanced Vision System (EVS) comprising a specially designed infra-red sensor suite and an infra-red window in the radome. It is connected through an RS-170 video interface to a stroke/raster Head-Up Display (HUD). The HUD displays the infra-red image to the pilot conformally with the real world outside the cockpit window. The All Weather Window® improves safety through greater pilot situational awareness, since it significantly enhances night/low-visibility capability. With the All Weather Window® installed, pilots with CAT I aircraft will be able to land in CAT III-visibility conditions at all CAT I, II and III equipped runways using CAT I approach procedures.

Specifications
Sensor wavelength: 3-5 μm
Field of view: 30 × 22.5°
Weight:
  nose-bay equipment: 15.9 kg
  avionics-bay equipment: 4.55 kg

Operational status
Kollsman is working with Gulfstream to provide the All Weather Window® as an option for Gulfstream IV and V aircraft. The infra-red image will be viewed by the pilots via a Honeywell/BAE Systems (Operations) HUD 2020 Visual Guidance System (VGS). The real-world image will be presented on the HUD combiner-glass and overlies the landing environment so that the pilot sees an infra-red image that is conformal to the visual landing scene.

Kollsman and Gulfstream started flight tests in October 1998; US FAA certification was planned for 1999 on the Gulfstream V, with the Gulfstream IVSP to follow, and Gulfstream IV installations to be offered as a retrofit.

Contractor
Kollsman Inc.

VERIFIED

The Kollsman Night Targeting System Advanced (NTS-A) 0106312

Night Targeting System (NTS)

The Night Targeting System (NTS) is an airborne electro-optical fire-control system designed to provide a night fighting capability for the AH-1 Cobra attack helicopter. It provides the capability to detect, acquire, track, designate and attack tactical targets at night and in limited visibility or adverse weather conditions. The system is operational in the US Marine Corps.

The NTS consists of a modified M-65 telescopic sight unit, a laser designator/range-finder, a CCD TV sensor and a FLIR sensor to provide autonomous delivery of TOW and Hellfire missiles. A video cassette recorder is also included to provide intelligence gathering and training capability. An automatic target tracker is integrated to improve tracking accuracy for weapon delivery. An upgraded configuration, the NTS-Advanced (NTS-A), eliminates the Optical Relay Tube (ORT), incorporates reliability and maintainability improvements, and integrates a TOW 2A thermal tracker. The NTS-A is currently undergoing operational testing.

Operational status
In production for variants of the Bell AH-1 Cobra helicopter in service with the US Marine Corps, and foreign users.

Contractor
Kollsman Inc.

VERIFIED

Cockpit Display for the Apache AH-64

The Apache display is a rugged, high-resolution, 5 × 5 in Active Matrix Liquid Crystal flat-panel Display (AMLCD). The display has a wide dimming range, providing true sunlight readability and night vision compatibility. The AH-64 Apache display is for weapons targeting and night vision system imagery.

Operational status

Currently in production.

Contractor

L-3 Communications Display Systems.

UPDATED

Enhanced Main Display Unit for the E-2C

The Enhanced Main Display Unit (EMDU) is the primary information display for the crew of the E-2C Hawkeye AEW aircraft. Each aircraft carries a complement of three EMDUs. The full-colour EMDU displays more than 2,000 radar tracks on an 11 × 11 in (279.4 × 279.4 mm) screen. The EMDU also allows the operators to overlay a map of the search area over their track files and display up to three separate windows containing additional information.

The EMDU utilises beam index CRT technology to provide sharp high-resolution colour images. In addition to the CRT, the EMDU contains processing subsystems that perform graphics generation, radar scan conversion and data input/output functions. The EMDU performs all display oriented processing internally, thus offloading that requirement from the E-2C's L304 central computer.

Specifications

Weight: 65.77 kg
Resolution: 900 × 900 pixels
Reliability: 1,100 h MTBF

Operational status

In service in US Navy E-2C aircraft.

Contractor

L-3 Communications Display Systems.

UPDATED

The L-3 Display Systems enhanced main display unit for the E-2C aircraft 0106313

S-3B ASA-82 pilot/co-pilot replacement display upgrade

The S-3B ASA-82 pilot/co-pilot replacement display suite comprises one 10.4 in pilot display and one 13.3 in co-pilot display per shipset. The displays are ruggedised to meet the carrier-based aircraft environment. The S-3B utilises the displays for precision targeting, tactical plots, and viewing of sensor data.

Operational status

In production.

Contractor

L-3 Communications Display Systems.

UPDATED

Actiview flat-panel AMLCD

The Actiview product line provides ruggedised, flat panel displays to meet a multitude of system architectures and installation requirements. Available with screen sizes ranging from 5 × 5 in to 21.3 in diagonal, each may be acquired as stand-alone or as integrated smart displays, and are available as COTS, rugged, full military specification or customised.

Features include embedded graphics generation, sensor video processing, multiple line rate video inputs and a full complement of interfaces. Actiview displays offer superior image quality combined with true sunlight readability.

Specifications of Actiview 20

Display area: 399.4 × 319.5 mm
Resolution: 1,280 × 1,024
Luminance: 50 fL
Colours: 16,000,000
Grey shades: 256 levels
Viewing angle: 120° cone (+60° horizontal and vertical)
Weight: 15.9 kg
Power input: 110 V AC, 60 Hz
Reliability: 20,000 h

L-3 Display Systems Actiview 20i flat-panel AMLCD **1999**/0054304

Operational status

In production.

Contractor

L-3 Display Systems.

UPDATED

Programmable Tactical Information Display (PTID) for F-14 aircraft

The PTID is an 8 × 8 in monochrome CRT display. It features high brightness and contrast; hybrid stroke on raster picture; embedded display processor with dual 1750A CPUs at 30 MHz using ADA software (MIL-STD-2167A); video select mixer and 1553B/F-14 unique I/O. It is now integrated with LANTIRN.

Operational status

In service in US Navy F-14 aircraft.

Contractor

L-3 Display Systems.

UPDATED

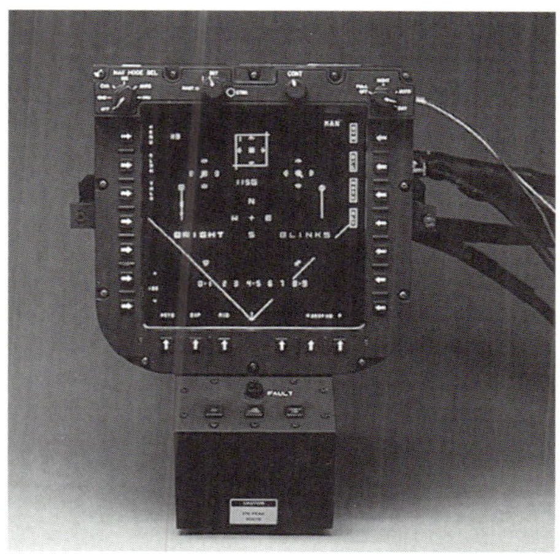

L-3 Display Systems programmable tactical information display for the F-14 aircraft 0018165

M-927/929 aviator's night vision goggles

The M-927/929 aviator's night vision goggles are based on the ANVIS design. The single point helmet-mounted system permits practically normal peripheral vision. A flip-up feature allows the pilot to pivot the assembly up out of the way whenever he wants to use unaided vision. The weight of the binocular is counterbalanced by a dual battery pack which mounts to the rear of the helmet.

The weight of the M-927/929 is 0.456 kg for the binocular assembly and 0.2 kg for the visor mount assembly, magnification is unity and the field of view is 40°. Brightness gains are 1,200 for the Generation II Plus M-927 and 2,000 for the Generation III M-929.

Operational status
In production to order.

Contractor
Litton Electron Devices.

VERIFIED

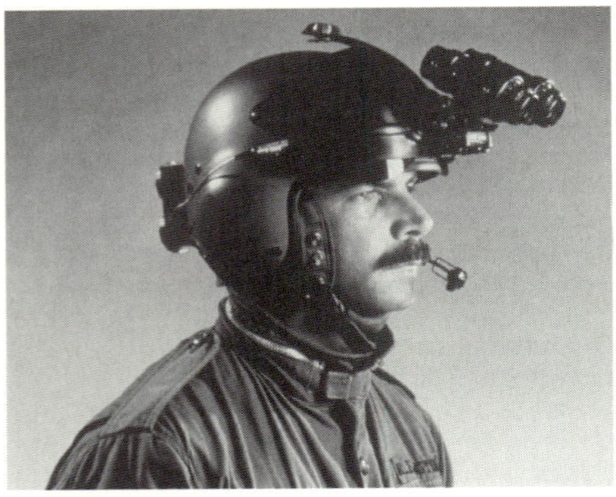

Litton M-927/929 aviator's night vision goggles

C-17 Mission computer/display unit

The C-17 computer/display unit, in conjunction with the C-17 multifunction control panel (see next item) and other keyboard units, acts as the single point of control for all primary navigation system modes and sensors to control sensors, display sensor and mission computer data and for input/output control of the mission computer. Display data is available in either 5 × 7 in (127 × 177.8 mm) or 7 × 9 in (177.8 × 228.6 mm) alpha and numeric forms.

Specifications
Dimensions: 122 × 146 × 238 mm
Weight: 4.32 kg
Power supply: 115 V AC, 400 Hz, 84 W
Reliability: >10,000 h MTBF

Operational status
In production for the C-17 aircraft.

Contractor
Litton Guidance & Control Systems.

VERIFIED

C-17 Multifunction control panel

The C-17 multifunction control panel serves as the control for data displayed on the multifunction displays and the HUDs. It is used to command the operating modes and display formats, as well as for range selection. It also controls and displays the altitude set function.

Specifications
Dimensions: 160 × 146 × 194 mm
Weight: 3.98 kg
Power supply: 115 V AC, 400 Hz, 35.8 W
Reliability: >10,000 h MTBF

Operational status
In production for the C-17 aircraft.

Contractor
Litton Guidance & Control Systems.

VERIFIED

The multifunction control panel for the C-17 0106314

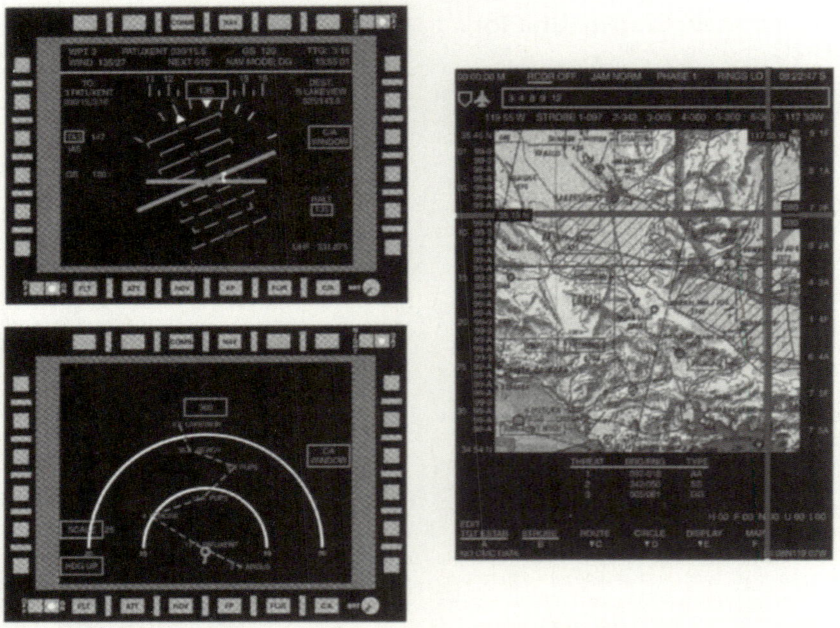

Litton Guidance & Control Systems full function AMLCD display/processing systems 0018168

EA-6B digital display group

The EA-6B digital display group is a derivative of the AN/ASN-123 tactical navigation set developed for ASW. It is employed on the EA-6B aircraft to display electronic warfare situation data and is the primary method of EW system control and operation.

The system comprises one Display Processor (DP), two Digital Display Indicators (DDIs), two Digital Display Indicator Control (DDIC) units and an Interface Computer Unit (ICU). The additional capability to drive these displays and development of a full-colour capability have been completed recently.

The DP interfaces with the aircraft master computer via a MIL-STD-1553 interface. It also interfaces with all the other components of the digital display group and master control panels via discrete lines.

The DDI provides the high-resolution, high-contrast display capability of the EA-6B avionics suite. When used in its three modes of stroke, raster and hybrid, it can display tactical navigation, situation plots and sensor image data. The modes and command data are generated by the DP, while data entry and display control are accomplished through the DDIC.

Operational status
In service on the EA-6B.

Contractor
Litton Guidance & Control Systems.

VERIFIED

Full function AMLCD display/ processing systems

Litton Guidance & Control Systems multifunction and full function display systems combine commercial and military components to provide cost-effective design. Features include thermal control for cold temperature operation and a backlight of proprietary design that provides very high output.

The full function display system is capable of real-time 2-D and 3-D graphics required for modern digital map system display. Map memory storage can be internal to the full function display system or accessed via the SCSI-2 interface.

The video processor is capable of receiving various sensor inputs (RS-170, RS-343 and PAL) and performing the appropriate resizing to meet the selected resolution. These video inputs can have digital data overlaid, or be viewed in two independent windows. Various chassis sizes are available to incorporate custom input/output channels and to accommodate mechanical constraints for cockpit installation in existing aircraft.

Specifications
Display type: colour active matrix liquid crystal display
Display area: 10, 12 or 14 in diagonal, others available
Luminance: 0.05-350 fL
Colours: 262,144
Pixel resolution: 640 × 480; 600 × 800; 1,024 × 768
NVIS: Class B
MTBF: 3,500 h
Dimensions: 252 × 211 × 152-381 mm (various models)
Weight: <11.4 kg
Power: 28 V DC, <75 W

Contractor
Litton Guidance & Control Systems.

VERIFIED

Model 1040C Doppler indicator

The Model 1040C Doppler indicator is a pilot-operated control display unit that provides functional control of the Litton AN/APN-217 and digitally displays the Doppler-provided readout of aircraft groundspeed and drift angle.

Major control functions of the system include Doppler On/Off and Land/Sea modes. Test capabilities

The Model 1040C Doppler indicator shows groundspeed (left) and drift angle (right) 0106315

include the ability to execute and report the result of the Doppler BIT and Doppler self-test velocity. Doppler BIT failures are reported by the malfunction light, while self-test velocity results are indicated on the digital display. Other Doppler indications are memory condition, indicated by a panel light, and invalid data, by dashes on the display. The Model 1040C also performs its own BIT which blanks the display in the event of a failure.

Specifications

Dimensions: 125 × 146 × 75 mm
Weight: 2.2 kg
Power supply: 115 V AC, 400 Hz
28 V DC

Operational status

In production.

Contractor

Litton Guidance & Control Systems.

VERIFIED

Model 1046 multipurpose indicator

The Model 1046 multipurpose indicator is a solid-state instrument which is compatible with third-generation night vision goggles. It uses six digital displays and two graphic displays to provide the pilot with steering and hover information, making it an ideal instrument for rotary-wing as well as fixed-wing applications.

In the navigation mode the indicator provides a fly-to graphic display as well as command heading and range to checkpoint, groundspeed and estimated time en route. In the hover mode the graphic displays show heading, drift and vertical velocities. Both modes include a system status display.

Specifications

Dimensions: 205 × 83 × 83 mm
Weight: 1 kg
Input signal: ARINC 419-compatible 32-bit data

Operational status

Installed in the US Marine Corps' OV-10 and US Navy UH-1N and CH-46 aircraft.

Contractor

Litton Guidance & Control Systems.

VERIFIED

Smart Display Unit (SDU)

The SDU combines the functions of a powerful mission computer, a versatile colour graphics processor and a colour display and control panel in a single lightweight package. The SDU includes a high-resolution 4 × 4 in (101.6 × 101.6 mm) viewable area, active matrix liquid crystal display with integral controller and back-lighting. A 32-bit CPU with 8 Mbytes of memory provides the SDU with the computational power for solving complex tasks in real time.

The SDU is suited for applications such as avionics management, navigation, ECM, tactical datalink, fire control and stores/weapons management, GPS navigation and control, internal digital map and communications control.

The display unit contains a full alphanumeric keyboard, function keys, trackerball and colour active matrix LCD which can be used to superimpose targeting symbology over a FLIR/TV video image or display graphics of system status for the operator. With snapshot of video datalink module installed, the operator can freeze the screen and transfer the whole or a portion of the image to another similarly equipped platform via standard VHF/UHF radios and crypto equipment. Multiple images can be stored for later transmission.

The SDU also provides an independent external display output and inputs for a standard keyboard and trackball which allows it to function as a powerful tactical workstation processor. Full video windowing is provided on the 800 × 600 external display output, which means two video sources can be displayed simultaneously along with a full-colour digital map with tactical symbology.

Specifications

Dimensions: 146 × 171.5 × 180.3 mm
Weight: <3.63 kg
Power supply: 28 V DC, <40 W
Temperature range: −40 to +55°C
Environmental: MIL-E-5400 Class 1A
Display: 101.6 × 101.6 mm, 512 × 512 pixels
Interfaces: MIL-STD-1553B, RS-232, RS-422, RS-170A video, discretes

Operational status

Prototype units have been built and demonstrated.

Contractor

Litton Guidance & Control Systems.

VERIFIED

Dark Star laser designator

The Dark Star laser designator is designed to provide accurate, first-strike capability, with all types of laser-guided weapons. The two-assembly system has been optimised for low volume and weight, and includes transmitter and high energy converter line-replaceable modules. Incorporating a newly-developed high-brightness resonator, this Nd:YAG (1.064 μm) air-cooled, equipment provides state-of-the-art beam divergence at high efficiency.

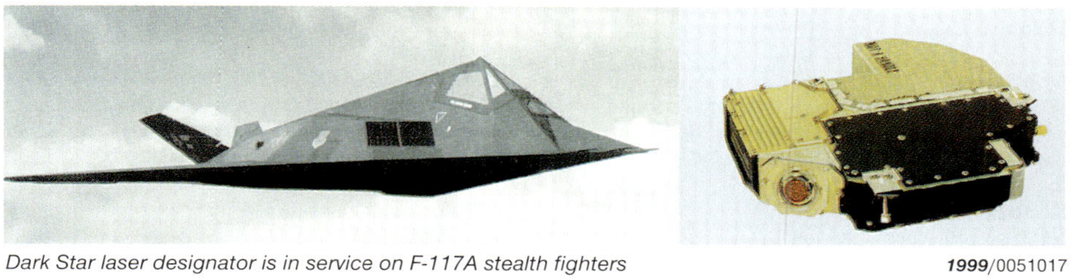

Dark Star laser designator is in service on F-117A stealth fighters ***1999***/0051017

Specifications

Wavelength: 1.064 μm
Power: 115 V AC, 3-phase, 400 Hz, 250 W; ±15 V DC, 20 W
Dimensions:
  transmitter: 292 × 292 × 61 mm
  HEC: 216 × 145 × 112 mm
Weight:
  transmitter: 5.0 kg
  HEC: 2.0 kg

Operational status

In service on the F-117A stealth fighter.

Contractor

Litton Laser Systems Division.

VERIFIED

F/A-18 C/D Laser Target Designator/Range-finder (LTD/R)

The LTD/R is integrated into the AAS-38 Night Hawk FLIR pod, the LTD/R provides the F/A-18 C/D Hornet, flown by the US Navy and Marine Corps, with advanced war fighting capabilities through improved target ranging and designation.

The LTD/R is in service on F/A-18 C/D aircraft ***1999***/0051018

Specifications

Wavelength: 1.064 μm
Range limits: 315-24,966 m
Beam diameter: 0.28 in
Beam alignment: 0.3 milliradians
Pulse repetition rate: Band I and Band II PRF codes
Weight: LTR: 5.9 kg
  LPS: 4.9 kg

Operational status

In production and in service in US Navy and US Marine Corps F/A-18C/D aircraft.

Contractor

Litton Laser Systems Division.

VERIFIED

LAMPS Mk III laser designator

The LAMPS Mk III laser designator is designed for use with all types of laser guided weapons. The laser designator system has been optimised for low volume and weight. Built-in test verifies critical aspects of system performance. LAMPS Mk III is a fully qualified military laser designator.

Specifications

Wavelength: 1.064 μm
Weight: 5.0 kg
Dimensions: 292 × 292 × 61 mm
Power: 115 V AC, 400 Hz, 3-phase, 250 W; ±15 V DC, 20 W

Operational status

In production and in service in US Navy SH-60 helicopters.

Contractor

Litton Laser Systems Division.

VERIFIED

LAMPS Mk III laser designator in service on US Navy SH-60 helicopters ***1999***/0051019

LANTIRN Laser Designator Ranger (LDR)

LANTIRN (Low-Altitude Navigation and Targeting Infra-Red System for Night) provides strike aircraft with first pass attack capability at greatly enhanced standoff ranges. The LANTIRN LDR is part of the LANTIRN targeting pod produced by Lockheed Martin. The targeting pod provides 1.06 micron laser energy for ranging targets, for the effective delivery/guidance of laser guided weapons. The LDR can also be switched to operation at 1.54 micron.

The LANTIRN LDR consists of three replaceable units: a laser transmitter/receiver, an electronics unit, and a high-voltage power supply.

The LDR provides tactical forces with day/night under-the-weather attack capability.

Specifications

Wavelength: 1.064 or 1.54 μm, pulsed switchable
Pulse Repetition Frequency (PRF): Band I, Band II and PIM
Beam diameter: 0.25 in
Beam alignment: 2.5 milliradians
Maximum range readout: 24,500 m
Power: 115 V AC, 400 Hz, 3-phase; +5 V DC; ±15 V DC; +28 V DC
Weight: 10.9 kg

Operational status

In production and in service on F-14, F-15 and F-16 aircraft of many nations. A modified version is being developed for the AH-1Z attack helicopter.

Contractor

Litton Laser Systems Division.

VERIFIED

TADS is in service on AH-64A Apache helicopters ***1999***/0051021

Mast-Mounted Sight Laser Range-Finder/Designator (MMS-LRF/D)

Litton Laser Systems produces the MMS-LRF/D which equips the US Army's OH-58D helicopters. Use of advanced packaging techniques enables the electronics, high-voltage power supplies, cooling system, range receiver and laser transmitter to be housed in a single LRU. The unit includes an asynchronous digital interface for precise communication with the MMS computer internally. All subassemblies interface with a common bus under control of a main processor which maintains system timing through software control. Built-in test circuitry periodically monitors system operation, allowing the processor to compensate automatically for component degradation.

Specifications

Wavelength: 1.064 μm
Pulse Repetition Frequency: (PRF): Band I, Band II, tri-service codes
Beam diameter: 0.52 in
Weight: 6.4 kg
Dimensions: 143 × 210 × 271 mm
Power: 28 V DC, + 15 V DC, 8.5 V DC

The MMS-LRF/D is in service on the OH-58D Aeroscout helicopter ***1999***/0051022

Operational status

In production and in service. A suitable eye-safe tactical version has been developed. A modified unit has been integrated into the all-light-level television system in the Lockheed Martin AC-130U.

Contractor

Litton Laser Systems Division.

VERIFIED

Target Acquisition Designation Sight (TADS)

Built for the Apache AH-64A attack helicopter, the TADS laser range-finder target designator and tracker enables the Hellfire system and other laser-guided munitions to lock-on to targets.

Specifications

Wavelength: 1.064 μm
Weight:
  laser targeting unit: 6.6 kg
  laser electronics unit: 8.2 kg
Power: 115 V AC, 400 Hz, 3-phase

Operational status

In production and in service on AH-64 Apache helicopters of many nations, including the UK WAH-64 programme.

Contractor

Litton Laser Systems Division.

VERIFIED

AN/AAS-35(V) Pave Penny laser tracker

The AN/AAS-35(V) is a miniaturised day and night laser-based target identification set. Used in conjunction with a laser designation system, either ground-based or in a co-operating aircraft, targets can be recognised and identified rapidly, and accurate steering data provided to ensure quick pilot reaction and accurate delivery of weapons.

A silicon pin diode detector head is used, with full lower forward hemisphere coverage plus some look-up capability. Pilot's controls permit selection of several seeker scanning patterns to improve early designator recognition. It can be used to improve the accuracy of conventional weapon delivery, or to lock up laser-guided munitions. The system is contained in a relatively small pod, which is either fuselage- or pylon-mounted to allow easy harmonisation with other onboard sensors. An aircraft adaptor module is used to integrate the sensor data with onboard processors and a pilot's control panel provides for easy use and built-in test operations.

Specifications

Dimensions: 833 (length) × 200 mm (diameter)
Weight: 14.5 kg
Power supply: 28 V DC, <10 A
Wavelength: 1.064 μm
Scan coverage:
(elevation) –90 to +15°
(azimuth) –90 to +90°
Selectable scan patterns: wide, narrow, depressed, offset
Output: direction cosines of line of sight

Operational status

In service. The first operational Pave Penny system for the A-10 was delivered in March 1977 and the US Air Force has equipped its entire fleet originally numbering 733 aircraft. Up to 380 US Air Force A-7D Corsairs also had Pave Penny installed. Starting in 1995, Pave Penny was also installed on US Air Force F-16 aircraft.

Contractor

Lockheed Martin Electronics & Missiles.

VERIFIED

AN/AAS-42 Infra-Red Search and Track system (IRST)

The Infra-red Search and Track System is designed to permit the multiple tracking of thermal energy emitting targets at extremely long range to augment information supplied by conventional tactical radars. The system enhances performance against low radar cross-section targets while providing immunity to electronic detection and RF countermeasures. High-resolution IRST provides dramatically improved raid cell count at maximum declaration ranges - information that can stand alone or be fused with other sensor data to enhance situational awareness.

The IRSTS consists of a sensor head mounted beneath the nose of the F-14D and an electronics unit just aft of the cockpit. The system is integrated with the F-14's central computer system and complements the AN/APG-71 radar providing the aircrew both target track data and infra-red imagery displays. The AN/AAS-42 operates in six discrete modes, with selectable and individually controlled scan volumes in azimuth (±80°) and elevation (±70°). The system is suitable for installation on multiple tactical platforms as either an internal fit or in a self-contained podded configuration.

Specifications

Dimensions:
(sensor head) 914.4 × 228.6 mm diameter
(electronics unit) 190.5 × 190.5 × 482.6 mm
Weight:
(sensor head) 41.28 kg
(electronics unit) 16.78 kg

Operational status

The system has been fitted to US Navy F-14D aircraft since 1994.

Contractor

Lockheed Martin Electronics & Missiles.

VERIFIED

AN/ASQ-145 low-light television system

The AN/ASQ-145 is a low-light television system designed to provide fire-control facilities for the US Air Force AC-130 Hercules gunship aircraft, on which it is a primary sensor. The system is unusual in that it uses a dual camera installation in order to provide both wide and narrow fields of view simultaneously.

Operational status

In production and service, currently being updated.

Contractor

Lockheed Martin Electronics & Missiles.

VERIFIED

AN/ASQ-173 Laser Spot Tracker/ Strike CAMera (LST/SCAM)

Developed specifically for the US Navy F/A-18 Hornet, the LST/SCAM permits crews to identify laser designated targets illuminated by either ground forces or co-operative aircraft, and to achieve more accurate tracking and weapon release performance than non-laser equipped aircraft. Target position data from the laser detector tracker is fed directly to the F/A-18 mission computer and used to provide weapon aiming and ordnance release information.

The laser spot-tracker optics are stabilised using attitude data passed to the unit from the aircraft inertial navigator. The detector is a four-quadrant photodiode type, mounted behind a hemispherical dome. Laser pulse decoder electronics are also contained in the detector.

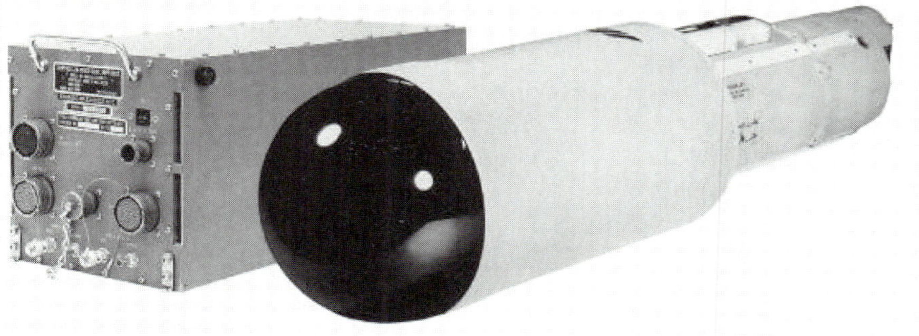

The AN/AAS-42 infra-red search and track system for the US Navy F-14D

Located in the same pod, for which Lockheed Martin is prime contractor, is a strike camera. This unit is in the aft section and has a wide field of view in the lower hemisphere. It can be slaved to the laser spot-tracker or independently controlled by the mission computer. Photographic data from this unit permits rapid assessment of damage after aircraft attacks.

For strike operations the laser spot-tracker/strike camera on the F/A-18 is mounted on the starboard fuselage side and an infra-red target acquisition/ tracker pod is carried on the port fuselage side.

Specifications

Dimensions: 2,290 (length) × 200 mm (diameter)
Weight: 73 kg
Wavelength: 1.064 μm
Scan patterns: preprogrammable/pilot selectable

Operational status

Production completed. In service in the US Navy F/A-18.

Contractor

Lockheed Martin Electronics & Missiles.

VERIFIED

EOSS Electro-Optic Sensor System for the RAH-66 Comanche

The Lockheed Martin Electro-Optic Sensor System (EOSS) uses advanced focal plane array and digital imaging technologies to provide night navigation and targeting capabilities for the US Army RAH-66 Comanche helicopter. The EOSS is designed to permit the Comanche crew to fly safely at extremely low altitudes, even in darkness and poor weather, and detect targets faster and at greater distances than currently fielded EO systems.

The Electro-Optical Target Acquisition/Designation System (EOTADS) consists of a second-generation FLIR sensor, high-resolution TV camera and laser range-finder/designator. The system uses a unique two bar scan technique to search for airborne and ground targets. Target imagery is prioritised according to threat potential and stored in computer memory, which permits the crew to remask while deciding combat tactics.

The Night Vision Pilotage System (NVPS) will permit the pilot to manoeuvre safely at nap of the earth altitudes around the clock and in poor weather or limited visibility caused by smoke, dust or haze. The system is coupled to the pilot's line of sight and

Elop Laser Designator/Range-finder element of EOSS 0106317

RAH-66 Comanche helicopter 0106316

provides high-resolution day-like imagery at extended ranges using advanced focal plane array infra-red technology.

The laser system has been developed by Elop Electro-Optical Industries Ltd of Israel. It operates at two different wavelengths within a single unit. This enables target designation to be achieved using the appropriate weapon-compatible wavelength, while ranging is achieved using an eye-safe wavelength, with obvious benefits in training operations. A diode-pumped laser and wavelength translation technology is used to produce the eye-safe wavelength which minimise size and weight.

Specifications

Weight: 204.1 kg
Power supply: 270 V DC, 3.4 kW
Temperature range: –46 to +52°C
Altitude: up to 9,800 ft
Wavelength:
(solid-state TV) 0.65 to 1 μm
(FLIR) 8-12 μm
(laser) 1.06 μm (ranging/designation)
1.54 μm (ranging only)
Reliability:
(NVPS sensor) 270 h MTBF
(EOTADS sensor) 125 h MTBF

Operational status

In September 1991, Lockheed Martin received a contract from Boeing's Helicopter Division to build five EOSS systems for a demonstration/validation prototype phase. The first EOSS FLIR was completed and began testing in 1995. The programme is being restructured; meanwhile, the FLIR components are being evaluated in a dynamic flight environment aboard an Apache helicopter.

Contractor

Lockheed Martin Electronics & Missiles.

VERIFIED

LANTIRN 2000/2000+

LANTIRN 2000

Lockheed Martin is developing enhancements to the basic LANTIRN system known as LANTIRN 2000.
The enhancements are:

(1) A quantum well, third-generation Forward Looking Infra-Red (FLIR) sensor, operating in the 8 to 12μm band. The employment of long wavelength Quantum Well Infra-red Photodetectors (QWIPs) has facilitated the construction of extremely dense detector arrays which provide for dramatic improvements in detectivity and, thus, detection range, enabling target acquisition at the extended ranges demanded by weapons such as the GBU-24 series LGB. Further, as quantum well technology is largely solid state, the FLIR sensor in LANTIRN 2000 is claimed to be 23 per cent more reliable.

(2) A solid-state diode-pumped laser; which, as a result of lower beam divergence and greater pointing accuracy, enables operation of the targeting pod at greater range and altitude (up to 40,000 ft is claimed). An eye-safe training laser with tactical performance and range is integrated into the system, which is claimed to be 17 per cent more reliable, thanks to an improved power supply, fewer parts and a cooler operating temperature.

(3) An enhanced computer system. The LANTIRN 2000 computer is smaller, weighs half as much, and uses half as much power as the computer it replaces. Throughput, memory and reliability are optimised; software, cabling and interfaces remain the same.

LANTIRN 2000+

A collateral programme designated LANTIRN 2000+, offers US customers further options:

(1) An Automatic Target Recognition (ATR) system to reduce pilot workload in the acquisition and identification of targets.

(2) A Laser Spot Tracker (LST), to further aid in target acquisition and identification via third party marking.

(3) A digital disk recorder for use in Battle Damage Assessment (BDA) of target strikes and reconnaissance mission support in surveillance modes.

(4) A TV sensor, which has been successfully tested and flown, which will provide added capability in the event of poor IR sensor performance due to atmospheric conditions.

While significantly enhancing LANTIRN's traditional mission - day/night strike interdiction - LANTIRN 2000+ will facilitate a wider range of aircraft mission capabilities to include air-to-air tracking, theatre missile defence and battle damage assessment. LANTIRN 2000+ refinements provide for greater targeting flexibility and enhanced reconnaissance capability.

LANTIRN 2000 technologies are being developed under Lockheed Martin's internal research and development programme, supported by several US Air Force programmes and the Air National Guard.

Operational status

FLIR performance has been evaluated by the US Air Force, as part of long-range staring array and ATR capabilities for the theatre defence mission. It is understood that LANTIRN 2000 is close to a production capability.

Contractor

Lockheed Martin Electronics & Missiles.

VERIFIED

LANTIRN system

The Low Altitude Navigation and Targeting Infra-red for Night (LANTIRN) system consists of two pods: the AN/AAQ-13 navigation pod and the AN/AAQ-14 targeting pod.

The LANTIRN system enables fast jet aircraft, such as the F-15E and F-16C/D, to penetrate hostile airspace at extremely low altitude and high speed, acquire their targets and deliver weapons round the clock. The initial application was the two-seat F-15E fielded by the US Air Force in 1989. LANTIRN also equips the F-16C/D. Compatibility of the LANTIRN system with single-seat navigation and targeting operations has been demonstrated on the F-16. In 1995, the US Navy selected the LANTIRN targeting pod for the F-14 Precision Strike Programme.

The system

The LANTIRN system consists of two separate sets of equipment each contained within its own pod, suitable for underwing or underfuselage attachment. Either or both pods may be carried, depending on mission requirements. This option enhances flexibility and reduces support demands. The equipment within the pods is supplied by a number of manufacturers, but overall responsibility rests with Lockheed Martin as the prime contractor.

The AN/AAQ-13 navigation pod contains a wide Field of View (FoV) (21 × 28°) single cadmium mercury telluride array FLIR unit and a J-band terrain-following radar, together with the associated power supply and pod control computer. FLIR imagery from the pod is displayed on a wide field of view holographic head-up display developed by BAE Systems (formerly Marconi Electronic Systems). This provides the pilot with night vision for safe flight at low level. The Raytheon Electronic Systems Ku-band terrain-following radar permits operation at very low altitudes with en route weather penetration and blind let down capability.

Both the F-15E and F-16 have fully automatic terrain-following with inputs from the navigation pods to the digital flight control system in the aircraft. Additionally, terrain-following may be accomplished manually by means of directive symbology presented to the pilot on the HUD.

The AN/AAQ-14 targeting pod contains a stabilised wide and narrow fields of view targeting FLIR and a Litton Systems Inc, Laser Systems Division laser designator/range-finder in a rotating nose section whose total Field of Regard (FoR), limited only by pod characteristics and airframe configuration, is of the order of ± 150°. The centre section houses the Maverick hand-off unit boresight correlator, a multimode tracker, pod-control computer, power supply and provision for Automatic Target Recognition (ATR) equipment. In certain derivatives, global positioning and inertial navigation systems are also integrated into the AN/AAQ-14. The targeting pod interfaces with the aircraft controls and displays, including the Stores Management System (SMS), to permit low-level day and night manual target acquisition and weapon delivery of guided and unguided weapons. It may be configured as a laser designator-only pod, for use with laser-guided munitions and conventional weapons, by deleting the Maverick hand-off subsystem. This configuration of the LANTIRN targeting pod has been redesignated as Sharpshooter.

Both pods have environmental control units in their tail sections to ensure that their systems will function satisfactorily over a wide range of temperatures and

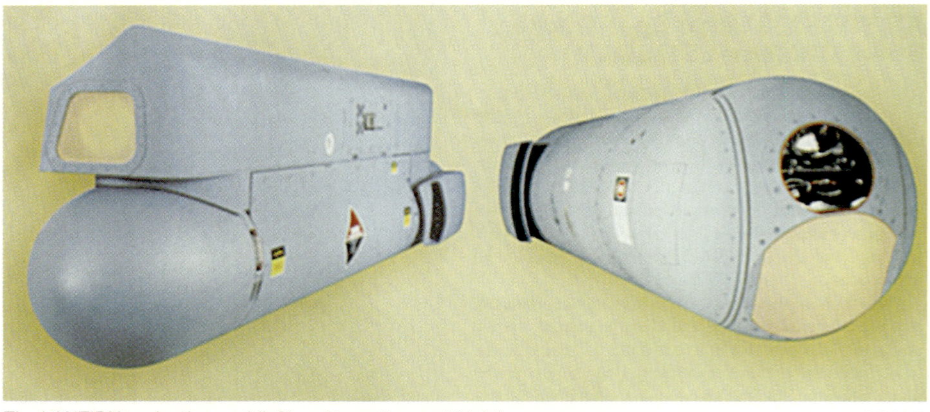

The LANTIRN navigation pod (left) and targeting pod (right) 0106318

This Lockheed Martin F-16D carries both navigation and targeting pods of the Lockheed Martin LANTIRN system under its fuselage

flight conditions. Aircraft interfaces include a MIL-STD-1553 multiplex databus, video channels and power supplies. Service ground support employs typical three-level maintenance: organisational, where no special test equipment is required; intermediate, which employs automatic test equipment, and depot servicing.

In operational use, the pilot will acquire a target using the aircraft (or pod mounted) nav/attack system to cue the line-of-sight (LOS) of the AN/AAQ-14 to the desired area. Using one of his head down displays, he will initially utilise wide FoV to confirm the target area, acquire his Desired Mean Point of Impact (DMPI) and then switch to narrow field of view to further refine and positively identify the correct aim point. The pilot will then engage the automatic target tracker. After the tracker is locked on, the pilot is able to manoeuvre his aircraft, limited only by the need to maintain line-of-sight between the pod and the target, taking care not to interrupt the LOS with his own airframe. With a stable target lock, the subsequent actions of the pilot are dictated by the type of weapon and attack profile: If the pilot is employing unguided weapons, he may simply wish to employ the laser range-finder to facilitate a highly accurate release solution. If laser guided weapons are to be used, the pilot is able to designate the target for his own weapons, or those of another aircraft. For a Maverick missile, the pod automatically hands the target off to the missile for launch with pilot consent.

Specifications

AN/AAQ-13 Navigation Pod
Dimensions: 1,985 × 310 mm
Weight: 205 kg
FLIR FoV: 21 × 28°

AN/AAQ-14 Targeting Pod
Dimensions: 2,510 × 381 mm
Weight: 241 kg
FLIR Total FoR: ± 150°
FLIR WFoV: 6 × 6°
FLIR NFoV: 1.7 × 1.7°

Operational status

In service on the US Air Force F-15E and F-16C/D and with several other air forces. A modified version of the AN/AAQ-14 targeting pod (as fitted to F-15E and F-16C/D), has been supplied to the US Navy for fitment to upgraded F-14 aircraft. It features a GPS/inertial navigation subsystem, which, coupled with the navy's limited requirement for automatic terrain following, obviates the need for a dedicated navigation pod. Production continues with the US Navy ordering a further 26 targeting pods in June 1998 for F-14 aircraft.

The first production navigation pod was formally accepted by the US Air Force in April 1987, with the final item in the contract delivered in March 1992. US Air Force targeting pod deliveries were completed in April 1994.

As of January 1999, Lockheed Martin had received orders for 763 navigation pods, 579 targeting pods, 12 Pathfinder pods and 184 Sharpshooter pods.

LANTIRN and LANTIRN derivatives (including Pathfinder and Sharpshooter) are either operational or on order with the US Air Force, US Navy and eleven other air forces: Bahrain, Denmark, Egypt, Greece, Israel, Netherlands, Saudi Arabia, Singapore, South Korea, Taiwan and Turkey.

In April 2000, Lockheed Martin was contracted by the US Air Force for Engineering, Manufacture and Development (EMD) of a modified version of LANTIRN. The modification integrates a radiometer and digital recorder into LANTIRN to enhance bomb impact assessment capability of the system. The date for completion of the EMD contract is June 2001.

In April 2000, Northrop Grumman was contracted to provide 15 LANTIRN II systems, nine for the US Marine Corps, four for Italy and two for Spain, all to be fitted to AV-8B aircraft.

Contractor

Lockheed Martin Electronics & Missiles.

VERIFIED

Low-Light-Level TeleVision system (LLLTV)

The Lockheed Martin Low-Light-Level TeleVision (LLLTV) is a general purpose multirole system. It is designed for maritime surveillance missions such as search and rescue, monitoring of territorial waters and policing offshore environmental laws. Principal features are high resolution and sensitivity at low-light levels and a small (16 mm diagonal) format. It is claimed to be capable of resolution densities in excess of 30 lines/mm and to have a wide dynamic range, which provides useful imagery around brightly lit parts of the area surveyed.

The camera head can be mounted in any attitude and is designed for hands-off operation. It uses a 16 mm vidicon tube and an 18 mm second-generation hybrid photocathode intensifier, which has extended sensitivity at the red end of the spectrum.

The system's electronic unit has a switchable line rate of either 525 or 875 lines a frame as standard but other line/frame rates are optionally available. Normal aspect ratio is 4:3 but a version with an aspect ratio of 1:1 is also available. Frame rate is 30 Hz. Resolution is 600 television lines horizontal and the dynamic range permits operation at face illumination levels from 0.1 ft candle down to starlight conditions.

The LLLTV has been chosen by the Spanish Navy for a shipboard fire-control application. The US Coast Guard has also selected an active gated version for possible use with its Dassault Aviation HU-25 Guardian aircraft, the application being to monitor and police territorial waters. In this configuration the equipment is designated AN/ASQ-174 Active Gated TeleVision (AGTV) and was flight-tested during 1984/85.

Specifications

Dimensions:
(camera head) 76 × 228 mm
(electronics unit) 241 × 184 × 165 mm
Weight:
(camera head) 1.36 kg
(electronics unit) 4.54 kg
Power supply: 28 V DC, 30 W

Operational status

In service. The LLLTV system also forms part of the US Air Force AC-130H Spectre Gunship, in which role it enables the crew to covertly illuminate targets with the aid of a separate flashlight laser. In January 1999, the US Air Force Special Operations Command contracted Lockheed Martin to produce an upgraded capability with a larger aperture lens giving improved light-gathering capability and increased magnification, an intensified CCD camera, an ambient cooled laser illumination and improved servo system.

Contractor

Lockheed Martin Electronics & Missiles.

VERIFIED

NITE Hawk targeting FLIR

Development of NITE Hawk began in March 1978 with the requirement to provide the F/A-18 Hornet with a day and night strike capability. The first systems, designated AN/AAS-38, were delivered to the US Navy in 1983.

The NITE Hawk (AN/AAS-38B) carried by the F/A-18 aircraft 0106319

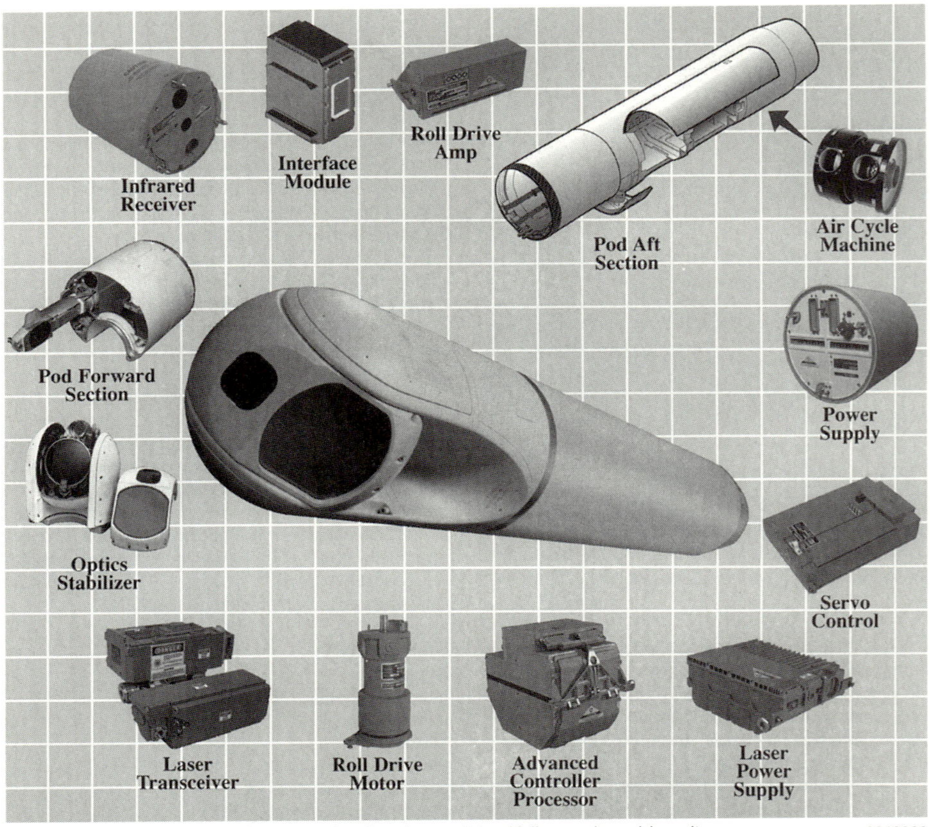

NITE Hawk SC targeting FLIR system showing the modular 12 line replaceable units 0018326

The NITE Hawk presents the pilot with real-time passive thermal imagery in a television formatted display to assist in the location, identification and tracking of targets. The system provides the aircraft mission computer with accurate target line of sight pointing angles and angle rates. Automatic in-flight boresight compensation is used to correct for dynamic flex and thus maintain accurate pointing angles throughout the aircraft's full performance envelope. Increased weapon delivery accuracy is provided through a Laser Target Designator/Ranger (LTD/R) feature (AN/AAS-38A). The LTD/R provides precise target range information and designation capability for precision-guided munitions.

Additional operational capabilities were incorporated into the NITE Hawk AN/AAS-38B, which has been in production since 1992. A laser spot tracker has been included which allows the system to search for, acquire and track targets, which have been laser designated by ground or other airborne sources. A multifunction autotracker also provides scene track, intensity centroid and geometric tracker algorithms for improved acquisition and maintenance of target lock on during attack of air-to-ground targets. Enhancements were also made to the air-to-air detection and tracking capabilities of the system.

Two further versions of the system have been developed: the AN/AAS-46 configuration for the F/A-18 E/F aircraft, and the NITE Hawk SC (self-cooled) system.

The NITE Hawk SC is designed to perform in a supersonic flight environment, providing single or multiseat aircraft with 24-hour strike capability against land- or sea-based targets. NITE Hawk SC interfaces with other aircraft avionics over a MIL-STD-1553B multiplex databus.

In some Lockheed Martin literature, the latest variant of NITE Hawk is also referred to as NITE Hawk Block III. The Block III capability includes a new third-generation FLIR to replace the first-generation FLIR in the original variant. The modification much improves detection and identification ranges and reduces maintenance costs. Installation of the new FLIR is said to involve very minor modifications to the NITE Hawk system.

Specifications

Dimensions:
(AN/AAS-38/38A/38B) 1,840 × 330 mm diameter
(NITE Hawk (SC)) 2,440 × 330 mm diameter
Weight:
(AN/AAS-38) 158 kg
(AN/AAS-38A/38B) 168 kg
(NITE Hawk (SC)) 195 kg
Field of view:
(wide) 12 × 12°
(narrow) 3 × 3°
Field of regard:
(pitch) +30 to −150°
(roll) ±540°
Stabilisation: 35 μrad
Tracking: 230 μrad
Pointing: 400 μrad

Operational status

The first production NITE Hawk system was delivered to the US Navy in December 1983. Systems have been delivered to, or are on order for: Australia, Canada, Kuwait, Malaysia, Spain, Switzerland and Thailand.

NITE Hawk SC has been flight tested on the F-14, F-15, F-16 and AV-8B. The NITE Hawk SC system is the baseline laser/designator pod for the Spanish Eurofighter Typhoon.

Lockheed Martin completed development of the NITE Hawk AN/AAS-46 configuration for the F/A-18E/F at the end of 1996.

Over 500 NITE Hawk systems have been delivered, and production into the 21st century is planned.

Contractor

Lockheed Martin Electronics & Missiles.

VERIFIED

Pathfinder navigation/attack system

Pathfinder is an international derivative of the LANTIRN navigation system.

The Pathfinder navigation/attack system uses hardware derived from the LANTIRN night navigation system (see earlier item). It consists of three LRUs which can be integrated into a pod or embedded in the aircraft fuselage or pylon. These LRUs are an infra-red sensor, power supply and environmental control unit. The first two of these LRUs are derived directly from LANTIRN but the environmental control unit is smaller and lighter, with reduced power requirements.

As with the LANTIRN navigation pod, Pathfinder imagery may be presented on a HUD or any other cockpit video display. Unlike the navigation pod, in addition to a × 1 wide field of view it also contains a × 3 magnified slewable field of view for standoff target acquisition.

Pathfinder has been demonstrated in a pod on the F-16, integrated into a pylon on the A-7 and embedded in the B-1B. It is suitable for installation on a wide variety of additional aircraft including the F-5, A-10, C-130, Dassault/Dornier Alpha Jet, Dassault Rafale, British Aerospace Hawk and Panavia Tornado.

Specifications

Dimensions: 1,950 (length) × 248 mm (diameter)
Weight: 90.5 kg
Power supply: 115 V AC, 400 Hz, 3 phase, 2.8 kW
28 V DC, 2 A
Field of regard: 77 × 84°
Field of view:
(wide) 21 × 28°
(narrow) 7 × 9°
Temperature range: −40 to +90°C
Reliability: 539 h MTBF

Operational status

Egypt has ordered 12 Pathfinder pods for the F-16. The US Air Force ordered 20 Pathfinder (and 20 Sharpshooter) pods in 1998 for F-16 aircraft for Taiwan. A further 39 Pathfinder/Sharpshooter pods were ordered by the US Air Force for Taiwan in June 2000, also for fitment to F-16 aircraft.

Contractor

Lockheed Martin Electronics & Missiles.

VERIFIED

Sharpshooter targeting pod

Sharpshooter is an international derivative of the targeting pod used in LANTIRN. Sharpshooter carries the designator AAQ-14(V1) to differentiate it from the AN/AAQ-14 pod of the LANTIRN system. It can be integrated with the Pathfinder navigation pod to permit pilots to fly at low altitudes in total darkness to the target area.

Sharpshooter provides round-the-clock targeting capabilities through the use of a 200 mm aperture infra-red sensor that has a 1.7° narrow field of view and a 6° wide field of view. A stabilised line of sight includes a 150° look-back angle and continuous roll tracker, which can track either stationary or moving targets or track a scene using area correlation. The laser designator and ranger is boresighted to the centre of the field of view of the targeting pod and is programmable for coding. Sharpshooter includes an eye-safe laser for training.

Operational status

Orders include: three for Bahrain; 12 by Egypt for installation on F-16; Greece has ordered 16; Israel has ordered 30 systems for installation on F-16 and a further 10 for F-15I; and Saudi Arabia has 48 pods on order for F-15S. The US Air Force ordered 20 Sharpshooter (and 20 Pathfinder) pods in 1998 for F-16 aircraft for Taiwan. The US Air Force ordered a further 39 Pathfinder/

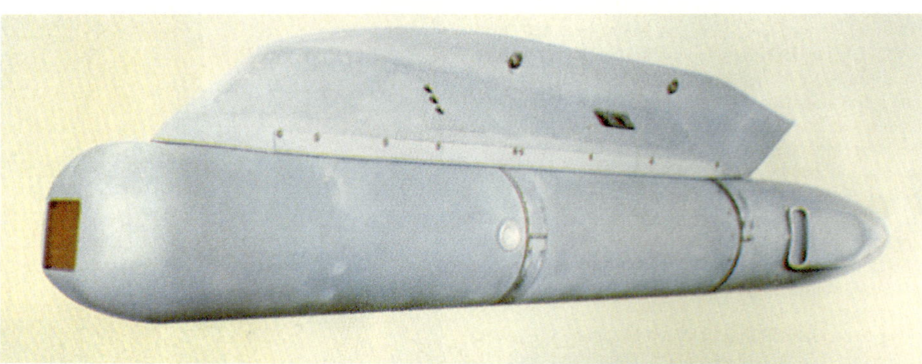

The Lockheed Martin Pathfinder is a FLIR sensor pod derived from the LANTIRN navigation pod 0106320

Sharpshooter pods for Taiwan in June 2000, also for fitment to F-16 aircraft.

Contractor

Lockheed Martin Electronics & Missiles.

VERIFIED

Sniper/PANTERA targeting FLIR

Sniper is a mature, long-range precision targeting system in a single, lightweight pod that Lockheed Martin is promoting for US applications.

PANTERA {Precision Attack Navigation and Targeting with Extended Range Acquisition) is a derivative of Sniper that is being offered for export.

Sniper provides positive identification, automatic tracking and laser designation of tactical size targets via real-time imagery presented on cockpit displays.

Sniper features include: high-resolution, third-generation, mid-wave, targeting FLIR; passive air-to-air target detection and tracking; tactical, eye-safe, diode-pumped laser spot tracker; modular, design for podded, conformal or internal fit; supersonic, low-observable configuration. The same FLIR has been adopted by the US Marine Corps as part of the Target Sight System (TSS), under development for its AH-1Z Cobra attack helicopters.

Lockheed Martin's Sniper targeting FLIR under lower fuselage of an F-16 aircraft ***1999***/0051023

Specifications

Field of view:
- **wide:** 4°
- **medium:** 1°
- **narrow:** 0.5°

Field of regard:
- **pitch:** +35 to –155°
- **roll:** continuous

Dimensions: 300 mm diameter; 2,300 mm long
Weight: 177 kg
Field MTBF: 196 h
Predicted MTBF: 614 h

Operational status

Flight tested. Indicated to be suitable for F-16 and similar aircraft.

Contractor

Lockheed Martin Electronics & Missiles.

VERIFIED

Target Acquisition Designation Sight/Pilot Night Vision Sensor (TADS/PNVS)

An AH-64A Apache with the Lockheed Martin TADS/PNVS system mounted on the nose

Lockheed Martin's Target Acquisition Designation Sight/Pilot Night Vision Sensor (TADS/PNVS) is designed to provide day, night and limited adverse weather target information and navigation capability. The TADS/PNVS system comprises two independently functioning subsystems known as TADS and PNVS.

TADS provides the co-pilot/gunner with search, detection and recognition capability by means of direct view optics, TV or FLIR sighting systems which may be used singly or in combinations according to tactical, weather or visibility conditions. The PNVS subsystem provides the pilot with flight information symbology which permits nap of the earth contour, on low-level flight to, from and in the combat area at altitudes low enough to prevent or delay detection by enemy forces.

TADS consists of a rotating turret, mounted on the nose of the helicopter and containing the sensor subsystems, an optical relay tube located at the co-pilot/gunner station, three electronic units in the avionics bay and cockpit controls and displays. TADS turret sensors have a field of regard covering ±120° in azimuth and from +30 to –60° in elevation.

By day, either direct vision or television viewing may be used. The direct vision system has a narrow field of view, 3.5° at × 18.2 magnification, and a wide field of view, 18° at × 3.5 magnification. The television system provides a narrow field of view of 0.9° and a wide field of view of 4°. For night operations the FLIR sensor has three fields of view: narrow (3.1°), medium (10.2°) and wide (50°).

Once acquired, targets can be tracked manually or automatically for autonomous attack with guns, rockets or Hellfire anti-tank missiles. A laser may also be used to designate targets for attack by other helicopters or by artillery units firing the laser-guided anti-armour Copperhead shell.

PNVS consists of a forward-looking infra-red sensor system packaged in a rotating turret mounted above the TADS, an electronics unit located in the avionics bay and the pilot's display and controls. The system covers a field of ±90° in azimuth and from –45 to +20° in elevation. Field of view is 30 × 40°.

TADS is designed to provide a backup PNVS capability for the pilot in the event of the latter system failing. The pilot or the co-pilot/gunner can view, on his own display, the video output from either TADS or PNVS, raising the probability of mission success. Although designed primarily for combat helicopters flying nap of the earth missions, PNVS may also be used as a single entity in tactical transport and cargo helicopters.

Special attention has been paid to reliability, as has the ability to remove and replace units easily and rapidly on the flight line.

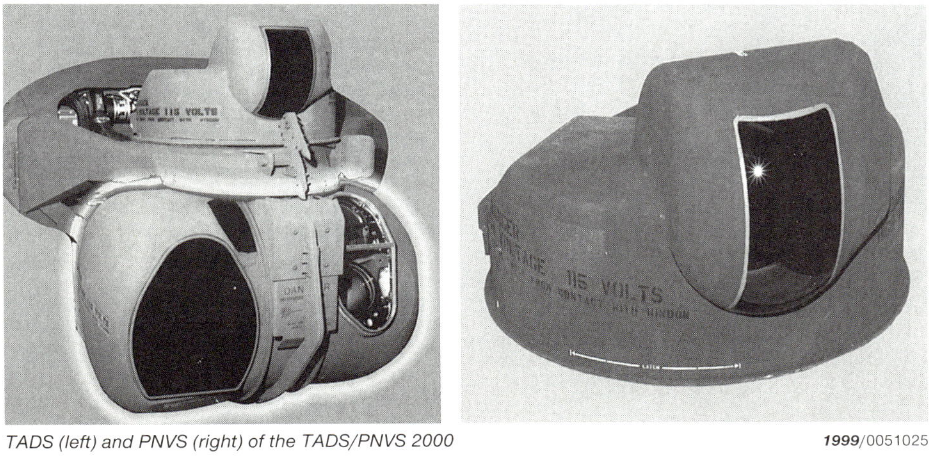

TADS (left) and PNVS (right) of the TADS/PNVS 2000 ***1999***/0051025

TADS/PNVS 2000

The US Army is initiating a major system upgrade to the TADS/PNVS that will improve reliability by more than 130 per cent and increase performance by more than 40 per cent, through addition of Comanche EOSS FLIR technology.

The lower turret will mount an advanced FLIR capable of 40 per cent greater targeting range and an improved CCD camera for day TV viewing. Improvements to the laser adopt Comanche technology for Apache. Direct-view optics are eliminated, increasing survivable crew space and enhancing display resolution with a large flat-panel targeting display.

The upgraded PNVS will provide an advanced FLIR capability with provisions for image intensification and image fusion. Coupled with the improved IHADSS, the pilot will have greatly enhanced nap of the earth (NOE) capability with built-in growth for use of the Comanche helmet display.

Operational status

TADS/PNVS has been fielded with the AH-64 since 1981. More than 900 systems have been delivered. In addition to the US Army, international customers include Egypt, Greece, Israel, Netherlands, Saudi Arabia, the United Arab Emirates and the United Kingdom.

Meanwhile in November 1996, Lockheed Martin was awarded a US Army contract to upgrade the system with 600 Charge Coupled Device (CCD) solid-state cameras, to upgrade reliability and maintainability. The US Army is also studying several other upgrades, including introduction of a second-generation FLIR system for improved range and image definition.

The first TADS/PNVS for the UK WAH-64 Apache was delivered in May 1998, by Lockheed Martin working with a team of British companies to build an internationally manufactured TADS/PNVS system.

Contractor

Lockheed Martin Electronics & Missiles.

VERIFIED

AN/AAQ-23 Electro-Optical/Infra-red Viewing System (EOIVS)

The AN/AAQ-23 Electro-Optical/Infra-red Viewing System (EOIVS) provides high-performance day or night target acquisition and navigation imagery. The system consists of a mid-wave staring array Forward-Looking Infra-Red (FLIR) sensor, sensor support electronics and video post-processor. Images are displayed on an 875-line display with TV compatibility in the cockpit. The EOIVS is based on an advanced development staring infra-red focal plane array which uses a platinum silicide substrate. The focal plane array, measuring 640 × 480 pixels, has two fields of view, with the narrower field used for target acquisition.

Specifications

Weight: 43 kg without gimbal
Power supply: 115 V AC, 400 Hz, 3 phase, 208 W
Optics: Dual field of view f/1.6 fixed focus

Operational status

The EOIVS is designed to improve the reliability and maintainability of navigation systems for the B-52 fleet. The system could also be integrated on the B-1B. A derivative of the sensor is available for other military and commercial aircraft. Production deliveries began in March 1996.

Contractor

Lockheed Martin Fairchild Systems.

VERIFIED

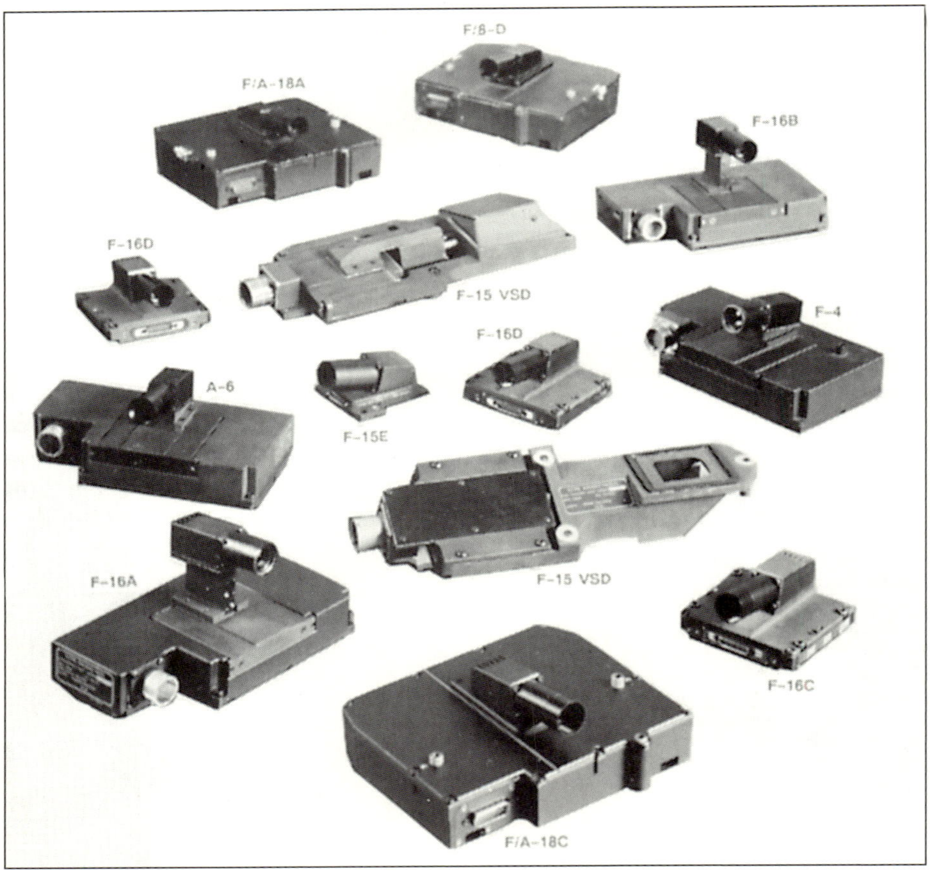

The Lockheed Martin Fairchild Systems AN/AXQ-16(V) cockpit television sensor is available in a large number of configurations 0106323

AN/AVD-5 (LOROPS) electro-optical reconnaissance sensor system

The AN/AVD-5 is a high-resolution long-range reconnaissance system designed to provide identification of tactical targets at ranges beyond 75 km. The sensor has a 1,676 mm focal length, f/5.5 reflective lens coupled with a 12,000 pixel array which has sufficient sensitivity to capture near-realtime imagery and provide haze penetration. The AN/AVD-5 has five modes of operation which can be optimised for coverage, resolution or stereo performance, and used in both manual or automatic scan.

The AN/AVD-5 is modular in design, providing flexibility in installation in a variety of aircraft or in a reconnaissance pod on fighter aircraft. The system consists of an imaging LRU, sensor control unit, power amplifier unit, junction box and interconnecting cables. Additionally, it utilises a reconnaissance management system, a switchable main electronics unit and a digital recorder, all of which are common to the ATARS suite. The digital imagery from the sensor is capable of transmission for processing in a ground exploitation system.

A dual-band (EO/IR) sensor is in development.

Specifications

Dimensions:
(imaging unit) 130.1 × 50.8 × 50.8 mm
Weight: 310.7 kg
Power supply: 115 V AC, 400 Hz, 3 phase, 362 W
28 V DC, 799 W
Range: 4.8-80.5 km

Operational status

Production systems are being shipped under a US Air Force contract.

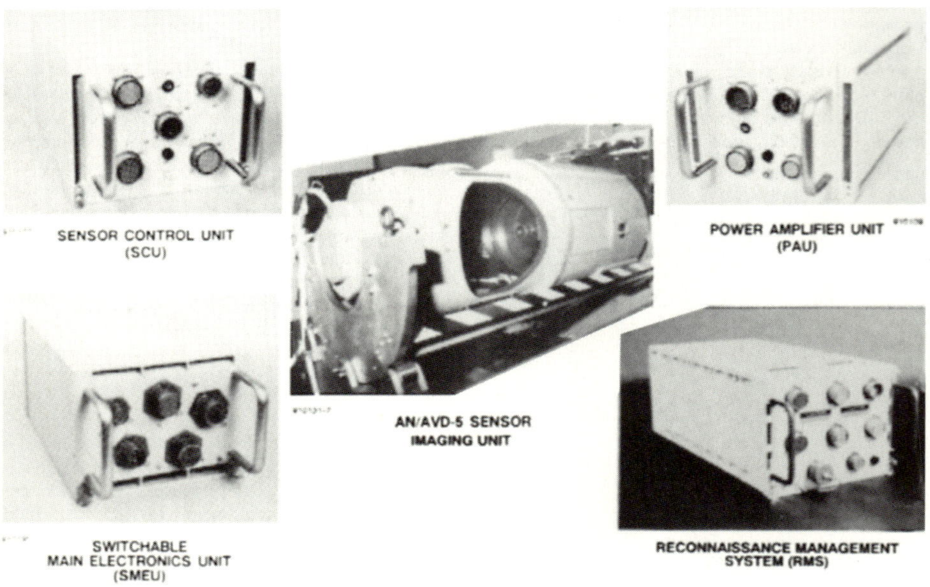

The AN/AVD-5 (LOROPS) electro-optical reconnaissance sensor system 0106321

Contractor

Lockheed Martin Fairchild Systems.

UPDATED

AN/AXQ-16(V) Cockpit TeleVision Sensor (CTVS)

The AN/AXQ-16(V) CTVS allows recording of real-time gunsight, HUD symbology, instrument panels and audio in fighters, tactical strike aircraft and commercial aircraft. The camera can also drive a monitor to provide real-time displays for the second crew member in a two-seat aircraft. It is a small solid-state TV camera configured to meet customer requirements.

The camera uses a charge-coupled device for high sensitivity and dynamic range. The principal optical element is a 17 mm, f/1.2 lens with automatic iris control capable of operating at low-light levels.

The system can replace existing film cameras or interface with any HUD. It is a form, fit, function replacement for existing monochrome cameras. Event mark, BIT, configuration control and environmental requirements have been incorporated into the design.

Specifications

Dimensions: specific to requirements
Weight: <1.36 kg
Power supply: 115 V AC, 400 Hz, 3 phase, 20 W
Lens: various
Sensor: 768 × 494 pixels
Refresh rate: 30 frames/s
Line rate: 2:1 interface
Reliability: 32,000 h MTBF

Operational status

In production. The television sensor has been tested and fielded in several US Air Force, US Navy and foreign military tactical aircraft and is specified for the F-15 and F/A-18.

Contractor

Lockheed Martin Fairchild Systems.

VERIFIED

ATARS Advanced Tactical Airborne Reconnaissance System

The Advanced Tactical Airborne Reconnaissance System (ATARS), under development by the US Navy and US Marine Corps, will provide high-resolution imagery for real-time or near-realtime reconnaissance. The Boeing Company is the prime contractor, and Lockheed Martin Fairchild Systems is the reconnaissance systems integrator.

F/A-18 (RC) aircraft equipped with the Low Altitude Electro-Optical (LAEO) and Medium Altitude Electro-Optical (MAEO) sensors for daylight operations and the Infra-Red LineScanner (IRLS) for day and night operations will be capable of satisfying the requirement for deep penetration, under the weather, time-critical reconnaissance.

The MAEO sensor allows the aircrew to obtain high-resolution imagery from three to five miles standoff, without direct overflight of targets in high threat areas.

The Reconnaissance Management System (RMS) provides control of the sensors, recorders and datalink, manages the flow of data from the sensors to the digital recorder, and from the datalink to the ground station and to the cockpit displays. The system can store 12 preplanned point, strip or area targets and 20 targets of opportunity.

Specifications

Low Altitude Electro-Optical (LAEO) sensor:
200-3,000 ft above ground level
140° field of view
vertical or forward oblique
high resolution, dawn to dusk below the weather, high speed sensor
Medium Altitude Electro-Optical (MAEO) sensor
2,000-25,000 ft above ground level
22° field of view
horizon to horizon field of regard
high resolution, medium range standoff sensor for daylight operations in high threat environments
Infra-Red LineScanner (IRLS) sensor:
200-10,000 ft above ground level
140° or 70° field of view
8-12 micron waveband
high resolution, low to medium altitude day/night reconnaissance capability

Operational status

Originally in design as an internal fit for the F/A-18 (RC) and RF-4C aircraft, ATARS can be configured for internal or pod fit, and current production is for internal fit to US Marine Corps F/A-18D aircraft, and pod fit for US Air Force F-16 aircraft (when it is designated TARS - Theatre Airborne Reconnaissance System). ATARS is also being considered by the US Navy for the F/A-18F aircraft.

Contractor

Lockheed Martin Fairchild Systems.

VERIFIED

SU-172/ZSD-1(V) Medium-Altitude Electro-Optical sensor (MAEO)

The Medium-Altitude Electro-Optical sensor operates from 2,000 to 25,000 ft with a 22° narrow field of view, 12 in (304.8 mm) focal length, f/5.6 lens for daylight operations. It is designed for medium-altitude tactical airborne reconnaissance penetration and standoff missions. High-resolution imagery is captured on a 12,000, 10 × 10 μm pixel CCD.

The SU-172/ZSD-1(V) is designed with emphasis on an integrated system approach to reduce life cycle costs. The sensor is automated to reduce aircrew workload and for use in UAVs. Stabilisation allows for target tracking in the presence of aircraft manoeuvres.

Specifications

Dimensions:
(imaging unit) 511 × 381 × 330.2 mm
Weight: <62.14 kg
Power supply: 28 V DC, 440 W
Altitude: 2,000-25,000 ft

Operational status

In production for the US Marine Corps.

Contractor

Lockheed Martin Fairchild Systems.

VERIFIED

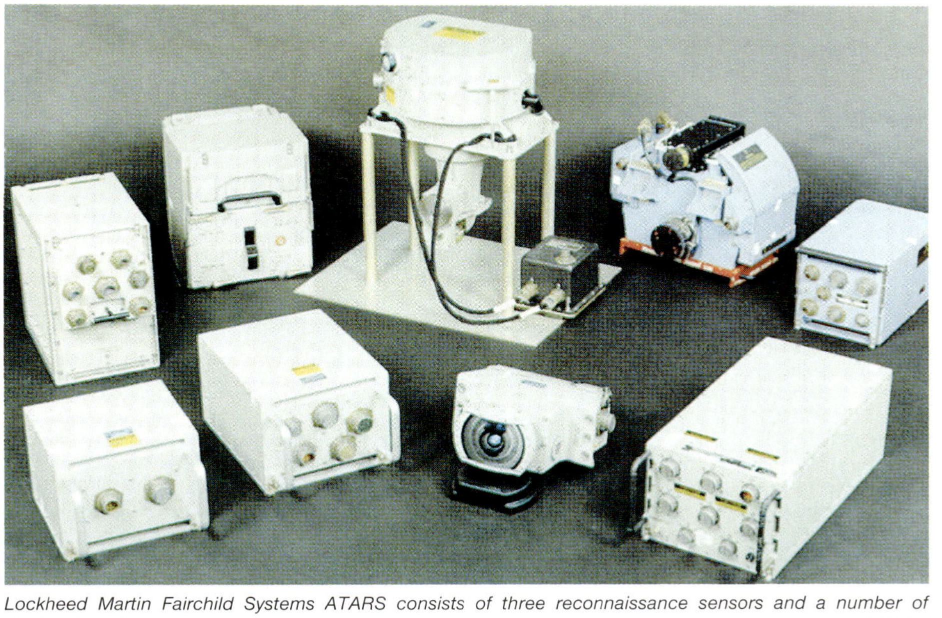

Lockheed Martin Fairchild Systems ATARS consists of three reconnaissance sensors and a number of electronic units
0106418

SU-173/ZSD-1(V) Low-Altitude Electro-Optical Sensor (LAEO)

The SU-173/ZSD-1(V) Low-Altitude Electro-Optical (LAEO) sensor operates from 200 to 3,000 ft with a low distortion 140° wide field of view and vertical or forward oblique fixed focus. The sensor provides high-resolution visible spectrum imagery for high-speed, low-altitude area coverage on tactical reconnaissance missions.

The SU-173/ZSD-1(V) consists of a sensor and electronics unit that collect imagery, provide roll correction of imagery, perform preliminary image processing and output digital image data. The low-altitude sensor system has been designed to be flexible and responsive to both mission and platform tailoring. The system may be internally mounted in manned aircraft such as the RF-4C/E or F/A-18(RC) or pod-mounted in the F-14, F-16, Tornado, Mirage, Jaguar and JAS 39, in addition to being configured for unmanned aerial vehicles.

Specifications

Dimensions:
(imaging unit) 284.5 × 121.9 × 363.2 mm
Weight: <24.5 kg
Power supply: 28 V DC, 231 W
Altitude: 200-3,000 ft

Operational status

In production for the US Navy and US Marine Corps.

Contractor

Lockheed Martin Fairchild Systems.

VERIFIED

AN/AAD-5 Infra-red reconnaissance set

The AN/AAD-5 is a dual field of view infra-red reconnaissance system which scans the terrain beneath an aircraft's flight path. The system consists of seven LRUs: the receiver, recorder, film magazine, control indicator, infra-red performance analyser, cooler and power supply.

The receiver includes the scanning optics, cooler cryostat, two 12-element detector arrays, 24 preamplifiers and the associated buffer electronics. The detector arrays are Mercury Cadmium Telluride (MCT) photoconductors which are sensitive to infra-red radiation in the 8 to 14 μm waveband. One array is used for the wide field of view and the other for narrow field of view.

The recorder consists of a CRT, recording optical components, video and sweep electronic cards, digital timing circuits, a high-voltage power supply, film speed control circuits and an auxiliary data annotation set.

An improved version, the AN/AAD-5(RC), is derived from the AN/AAD-5. The AN/AAD-5(RC) imaging process is identical to that of the AN/AAD-5. The receiver is two-thirds the size of the AN/AAD-5 receiver and uses an integral 0.25 W split-Stirling miniature cryogenic refrigerator.

Various electro-optical output options were available for both AN/AAD-5 and AN/AAD-5(RC) equipped systems. Upgrades using the Lockheed Martin IR linescanner real-time display system can add a cockpit real-time display with onboard video recording, with an optional datalink to ground stations, while maintaining the current film system operation.

Specifications

Dimensions:
(receiver) 460 × 380 × 330 mm (AN/AAD-5)
430 × 345 × 279 mm (AN/AAD-5(RC))
(recorder) 410 × 580 × 230 mm
(magazine) 380 × 180 × 250 mm
(control indicator) 150 × 100 × 80 mm
(cooler) 410 × 20 × 300 mm
(analyser) 150 × 50 × 300 mm
(power supply) 250 × 430 × 280 mm
Weight:
(AN/AAD-5) 130 kg
(AN/AAD-5(RC)) 63.2 kg
Power supply: 115 V AC, 400 Hz, 725 VA nominal
28 V DC, 118 W nominal
Magazine capacity:
106 m (conventional film)
213 m (thin-based film)

Operational status

In service with the US Air National Guard in the RF-4C and the US Navy in the F-14 TARPS. Also used by a number of air forces including Australia for the RF-111 and Germany, Greece, Korea, Spain and Turkey for the RF-4. Over 600 AN/AAD-5 systems and 168 AN/AAD-5(RC) systems have been delivered.

Contractor

Lockheed Martin IR Imaging Systems.

VERIFIED

D-500 Infra-Red LineScanner (IRLS)

The D-500 Infra-Red LineScanner (IRLS) is a later version of the AN/AAD-5 IRLS in service with the US military and several other armed forces. The D-500 is designed for pod, small aircraft and UAV applications and can fit within a 15 in diameter envelope so that it can be installed in all but the smallest pods. Its performance is the same as that of the AN/AAD-5. In addition to meeting the AN/AAD-5 film performance specification, the D-500 incorporates electronic oblique standoff viewing, BIT, videotape recording, real-time display and a datalink capability.

The D-500 consists of four basic LRUs: the receiver in which the IR image is formed at the Cadmium Mercury Telluride (CMT) detector array, the film recorder on which a visible image is formed through a CRT trace from the detector signal, the film magazine and a power supply. The intensity modulated CRT trace is coupled by a fibre optic faceplate to the recording film on which the image is composed on a line-by-line basis. The detector is composed of two linear arrays of 12 detectors, each aligned in the direction of flight. One to 12 detectors are used for recording, depending on the Velocity/Height (V/H) profile of the aircraft. The signal from the receiver is used to drive a 12-beam CRT which sweeps the signal on to film, mimicking the fashion in which the signals are received through the scan mirror at the detector. The scene sensed by the instrument can have a large intrinsic dynamic range which is first limited by the detectors to 12 bits. The range of levels displayed on the film is reduced below this by both the limited dynamic range of the CRT and by the film itself. This range is typically in the order of five bits or 32 grey levels in the film image.

The only major differences, other than size, between the AN/AAD-5 and the D-500 in the imaging process are the CRT/film coupling and the video electronics package.

Both optical efficiency and bulk have been improved by replacing the AN/AAD-5 CRT relay lens with the fibre optic faceplate CRT. The option for EO image recording was accommodated by integrating the video electronics with the receiver LRU. The D-500 fibre optic faceplate CRT eliminates the need for the relay lens.

In addition to these improvements, the D-500 has been developed for rapid insertion into tactical reconnaissance aircraft by remaining highly compatible with the AN/AAD-5(RC). Externally, the LRUs have been modified, yet internally some 80 per cent of components are identical. The D-500 receiver has been reduced in size and weight, compared to the AN/AAD-5, while the original image quality specifications have been maintained. The recorder and film magazine sizes have been reduced to complement the smaller receiver. The preamplifiers have monolithic hybrid circuitry.

The D-500 can be operated as an EO videotape recording and/or film image generating system. The recorder and film magazine can also be remotely located in a ground station to provide the mission hard-copy database. Options for the D-500 include videotape recording, on aircraft real-time display and datalink capability.

The D-500 is electronically roll corrected. Up to ±20° of roll can be corrected in both the wide and narrow modes. The automatic roll correction is accomplished by shifting a display unblank pulse within the 180° timing window provided by each facet of the scan mirror. The display unblank pulse gates out the proper section of data from the number of degrees of video collected during a scan line.

Contractor
Lockheed Martin IR Imaging Systems.

VERIFIED

Helicopter Infra-Red Navigation System (HIRNS)

The Helicopter Infra-Red Navigation System (HIRNS) is a low-weight television-compatible FLIR designed to aid navigation by day and night. The sensor head is integrated with a Ferranti steerable two-axis unstabilised platform.

A serial scanning thermal imager is used because balancing of amplifiers is not required to provide a high-resolution display. It also has the advantage of simpler processing electronics, while DC restoration ensures a bloom-resistant image and clear horizon definition. Energy from the scene is scanned horizontally with a continuously rotating eight-faceted mirror, while a flat nodding mirror carries out vertical scanning. The output signal from the SPRITE detectors is amplified and processed to provide gain and level control and is then displayed in a standard TV format.

The system may be linked with an Integrated Helmet And Display Sighting System (IHADSS) to allow the crew to perform nap of the earth missions.

Specifications
Weight:
(FLIR) 6.35 kg,
Field of regard:
(bearing) ±130°
(elevation) +20 to -60°

Operational status
In service on the Agusta A 129 helicopter.

Contractor
Lockheed Martin IR Imaging Systems.

VERIFIED

Infra-red Imaging System (IIS)

The Infra-red Imaging System (IIS) is a near-realtime, 180° field of view, high-resolution IR linescanner system developed for the German Air Force Tornado ECR aircraft. The system consists of five interactive units. Two of these units, the Recorder/Film Processor Unit (R/FPU) and the Scanner Receiver Unit (SRU) are produced by Lockheed Martin. These two units provide the infra-red detection and recording functions of the system. The other three units, the control, display and interface units, are produced by Honeywell Sondertechnik.

The SRU, used to detect horizon-to-horizon infra-red radiation emanating from the terrain, senses medium band infra-red radiation and converts it to 12 parallel video signal channels. SRU electronics provide a constant resolution footprint of the target, regardless of scan angle, throughout the widest practical ground coverage. These electrical signals are then sent to the R/FPU. The SRU is controlled by digital commands from the control unit, which are sent via the aircraft databus.

The SRU can be broken down into eight functional groups. The spin mirror group performs optical scanning of the IR scene. The IR sensor and amplifier group collect the IR energy of the scanned scene. The channel processor group performs the initial amplification and processing of the 34 preamplified detector signals. The mapping group processes signals from the channel processor group and combines them into 12 parallel video signals. The delay buffer and timing group provides proper timing for the 12 video channels and generates the file video output. The gyro and electronics group provides the roll compensation for maintaining the stability of the scanned imagery. The BIT and control group provide the token ring bus interface, central processing for the SRU and BIT circuitry. The video group provides the video output.

Only 12 infra-red signal channels, video sync signals and roll sensor data are passed directly between the SRU and the R/FPU. The modular design concept is carried out within each unit in order to simplify maintenance and logistics.

The function of the R/FPU is to receive and process the video signals from the SRU, provide the aircraft crew with the ability to view the imagery in near-realtime and produce a permanent film record of the IR imagery scanned by the SRU.

A precision CRT converts the amplitude modulated video signals to intensity modulation of the CRT beam as it sweeps across the tube face in synchronism with the rotation of the spin mirror in the SRU. An autofocus detector and focusing circuit measure the intensity of the beam and automatically correct the focus. The film is exposed by the variable intensity CRT beam as it moves in front of the faceplate. Annotation data, obtained from the aircraft via the databus, is applied to the film at proper time intervals.

The dry process film is developed within the R/FPU for onboard viewing. The film moves, at constant speed, between a rotating drum and a heated shoe. A servo-controlled heater inside the shoe maintains the temperature at 130°C, the level required for the dry process chemicals on the film to be activated by the heat, thus developing the film on board the aircraft.

Processed film is temporarily stored on the film manipulator. The film storage mechanism consists of two concentric powered drums on which the film is wound or unwound as required, to place the desired portion of the film in the viewing area. Each drum is controlled separately so that newly processed film can be wound on to the manipulator as it accumulates, while previously stored film is being wound back on to the manipulator as the operator commands viewing of earlier images.

As the film leaves the manipulator, it passes over a light table. This electroluminescent panel causes the image on the film directly above it to be projected on to the lens of a TV camera. The camera output is displayed on the operator's console. The operator can view any part of the exposed film. When previously exposed film is to be viewed, it is rerolled from its storage spool, moved back across the light table for viewing and temporarily stored on one of the two film manipulator drums. This action does not prevent newly exposed film from being stored on the other drum.

Selectable lenses give the TV camera three fields of view, effectively a zoom capability as the film is being viewed. Areas of interest on the film may be marked during viewing so that they may be quickly located at a later time.

A 107 mm reel of unexposed dry process film is available to the R/FPU. The film is provided at the correct time and speed for exposure by the CRT. The exposed film is stored on a separately controlled take-up reel.

Specifications
Dimensions:
(SRU) 520.7 × 381 × 509.8 mm
(R/FPU) 600 × 640 × 476 mm
Weight:
(SRU) 53 ±2 kg
(R/FPU) 87 kg

Operational status
In service on the German Air Force Tornado ECR.

Contractor
Lockheed Martin IR Imaging Systems.

VERIFIED

Stabilised Thermal Imaging System (STIS)

The low-cost, high-performance Stabilised Thermal Imaging System (STIS) satisfies the surveillance and navigation requirements of fixed-wing aircraft, helicopters and UAVs.

STIS features the Lockheed Martin Mini-FLIR which, with two bar SPRITE detector, is a serial scanning IR sensor incorporating DC-restored electronics for a crisp, bloom-free display. STIS is designed in modular style. The standard modules consist of the miniature FLIR scanner, the split-Stirling cooler, the focal plane/preamplifier assembly and the monolithic electronics.

The Mini-FLIR provides a TV-compatible output, making it suitable for a wide variety of displays. The stabilised thermal imaging system incorporates a three field of view telescope and is mounted in a stabilised inner gimbal capable of providing smear-free imagery in a dynamic environment.

Available options for STIS include an autotracker, laser range-finder, low-light-level TV camera and an alphanumeric display.

Specifications
Dimensions: 508 × 355 mm diameter
Weight: less than 34 kg
Waveband: 8-12 μm nominal
Field of view:
(wide) 20 × 30° (× 1.2)
(medium) 7.2 × 10.8° (× 3.3)
(narrow) 2 × 3° (× 12)
Video format: RS-170, TV-compatible, CCIR option

Operational status
In service on Dornier 228 aircraft, Eurocopter Super Puma helicopters, Skyeye RPVs and various other helicopters and RPVs.

Contractor
Lockheed Martin IR Imaging Systems.

VERIFIED

AN/AAS-40 Seehawk FLIR

The AN/AAS-40 Seehawk is a thermal imaging system using US Department of Defense common module FLIR components to provide high-resolution imagery. Designed for an aircraft or surface vessel, the system's current principal application is aboard a US Coast Guard Sikorsky HH-52A helicopter serving the primary role of search and rescue, law enforcement, maritime environmental control, marine and border control and navigational assistance. It is also installed on the Northrop Grumman S-2(T) ASW aircraft.

The Northrop Grumman AN/AAS-40 Seehawk lightweight FLIR system consists of a turret assembly, the control electronics unit and the power supply unit. FLIR imagery is fed to the cockpit-mounted display. An automatic scan capability provides constant search coverage in elevation and azimuth. The autosearch mode is enhanced by inclusion of automatic lock on which reacts to either large or small targets, as selected by the operator.

Operational status

The Seehawk system has successfully completed two years' service aboard a Beech 200T aircraft. The system was fully operational in conjunction with the other onboard avionics systems, demonstrating maritime patrol, surveillance and reconnaissance applications. The latest-generation AN/AAS-40 Seehawk was installed aboard a modified Sikorsky S-76 helicopter used as a demonstrator in support of the US Army RAH-66 Comanche programme. In production for the US Coast Guard HH-65A Dolphin helicopter and the Northrop Grumman S-2(T) ASW aircraft.

Contractor

Northrop Grumman Corporation, Electronic Sensors and Systems Sector.

VERIFIED

The AN/AAS-40 Seehawk FLIR on the US Coast Guard HH-52A helicopter

Contractor

Northrop Grumman Corporation, Electronic Sensors and Systems Sector.

VERIFIED

AN/ASQ-153 Pave Spike laser designator/ranger

The Pave Spike development programme was initiated in 1971 and delivery of 156 pod sets to the US Air Force was completed by August 1977, by which time 327 F-4D Phantoms had been converted to accept the system. A further 82 sets for foreign use were delivered up to September 1979, including a substantial number for the UK Royal Air Force and some for the Turkish Air Force.

The system is contained within an externally mounted pod, the nose section of which revolves about the pod axis to provide roll stabilisation, and a cylindrical forward portion which rotates in pitch to provide elevation stabilisation. Virtually complete lower hemisphere coverage is thus provided in a relatively compact and light arrangement. The nose section is sealed and pressurised with nitrogen and maintained at a constant temperature for optimum sensor performance. The centre section provides umbilical connections between the nose and rotating sections, the aircraft and the aft electronics system. In the aft section is a cold plate on to which are mounted the electronic LRUs. These comprise a low-voltage power supply and pod control, servo drivers, laser control, laser power supply and interfaces. The pod contains a television tracking sensor and laser designator/ranger. The television sensor can be used for target acquisition and the designator permits accurate delivery of laser-guided munitions. Laser ranging can be used to improve the delivery accuracy of conventional weapons.

The overall AN/ASQ-153 system comprises the AN/AVQ-23 pod and several system components in the aircraft. These include a line of sight indicator, control panel, range indicator, modified radar control handle and weapon release computer. The system can be used with Paveway laser-guided bombs and several other laser-guided munitions.

Specifications

Dimensions: 3,660 (length) × 205 mm (diameter)
Weight: 193 kg
Wavelength: 1.06 μm

Operational status

In service. Initially the unit was procured for use only on US Air Force F-4D and F-4E Phantoms, but it is now also employed on several other types.

AN/AVQ-27 LTDS

The AN/AVQ-27 Laser Target Designator Set (LTDS) is for installation on the canopy rail of the rear cockpit of the Northrop Grumman F-5F. The LTDS maybe installed in most two-seat fighter or fighter-bomber aircraft with only minor wiring modification. The LTDS allows the aircraft to be equipped with a removable target designation capability as required by the mission.

The system's principal elements are a stabilised operator sight unit with a retractable telescopic periscope, a laser transmitter and a 16 mm mission recording camera, all of which are bore sighted through common optics in a single self-contained package. The stabilised sighting assembly includes a rate-stabilised mirror which is controlled by a two-axis hand controller and provides two tracking modes (rate and rate-aided), a retractable periscope and associated optics. The laser transmitter consists of a single package that houses a hermetically sealed laser and its associated optics and electronics. The 16 mm mission camera integrated into the L TDS is a version of the same KB-25A/KB-26A standard gun camera used in the Northrop Grumman F-5E/F and other aircraft.

In operation the viewfinder-sight swings across into the rear seat pilot's field of view so that the crewman can track targets. Tracking may be accomplished manually, using a two-axis hand controller, or with rate or rate-aided tracking modes. The front cockpit is fitted with sight and canopy markings which allow the front seat pilot to assist the crewman in initial target acquisition and then to maintain the target within the LTDS field of view. Normally, attacks are co-operative, that is one aircraft is used for designation while accompanying aircraft will deliver ordnance.

Operational status

In production and in service on foreign F-5B and F-5F aircraft.

Contractor

Northrop Grumman Corporation, Electronic Sensors and Systems Sector.

NEW ENTRY

Electro-Optical Surveillance and Detection Systems (EOSDS)

EOSDS is a technology upgrade to an electro-optical system, developed by Northrop Grumman, called Night Giant, which comprises a forward-looking infra-red sensor and television camera. The system will perform a variety of surveillance and detection missions as part of the UK Royal Air Force Nimrod MRA4 mission system.

Operational status

Northrop Grumman will deliver 21 EOSDS systems to The Boeing Company, as mission system integrators. The mission system is in development.

Contractor

Northrop Grumman Corporation, Electronic Sensors and Systems Sector.

VERIFIED

Helicopter Night Vision System (HNVS)

Helicopter Night Vision System (HNVS) is installed in CH-53 Super Stallion transport helicopters for use on low-level tactical missions in adverse weather conditions. The system is based on equipment similar to that installed by Northrop Grumman on 10 AH-1S

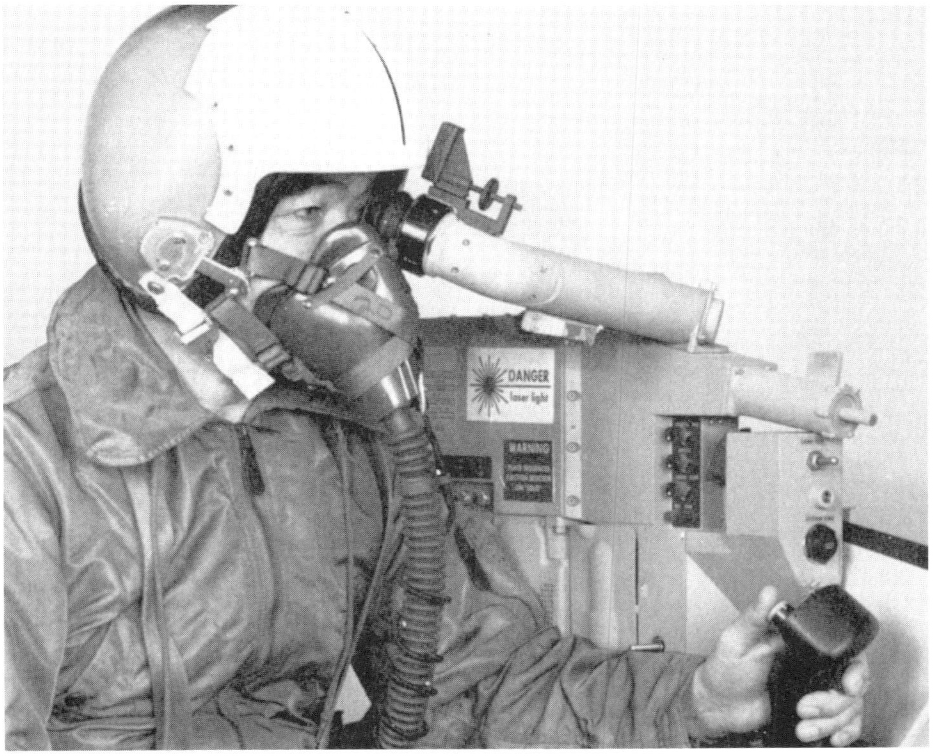

AN/AVQ-27 laser target designator set **2001**

An artist's impression of the F-16 Block 60 for the United Arab Emirates ***2000***/0085269

helicopters which were used as surrogate trainers for AH-64 pilots.

The system includes a data entry panel, master control assembly, the Honeywell Integrated Helmet And Display SubSystem (IHADSS) and control panel, video monitor/recorder and control panel, power distribution unit, heater/filter assembly, system control electronics, vapour cycle unit, symbology generator, multiplexing remote terminal, control grips and the Lockheed Martin Pilot Night Vision System (PNVS).

The PNVS is a thermal imaging system which optically senses the heat emitted by objects and converts it into video image for display to the pilot and co-pilot. The PNVS, mounted as a turret on the chin of the CH-53E, is controlled by the IHADSS worn by the pilot or co-pilot. The IHADSS monitors head position and converts head movement into azimuth and elevation commands which are sent to the PNVS. Video from the turret is displayed and flight symbology is overlaid on the cockpit displays, allowing either the pilot or the co-pilot to fly the aircraft in a head-up attitude.

Operational status

In service.

Contractor

Northrop Grumman Corporation, Electronic Sensors and Systems Sector.

VERIFIED

Integrated MultiSensor System (IMSS)

The Integrated MultiSensor System (IMSS) was initially configured specifically to perform day and night all-weather air interdiction and maritime patrol as part of the anti-drug efforts of the US Customs Service, a role in which it has been exceptionally successful. By combining and integrating a high-performance multimode radar and an infra-red imaging system, IMSS is also effective in performing reconnaissance, surveillance and search and rescue missions.

The IMSS consists of the AN/APG-66 radar fully integrated with an infra-red detection set such as the Northrop Grumman WF-360, controls and displays and an inertial navigation system.

The AN/APG-66 is a digital, coherent, multimode radar system developed to serve both the strike and fighter demands of the F-16 aircraft. It provides long-range detection and acquisition of targets at all altitudes and aspect angles in the presence of heavy background clutter. The current version of the APG-66 incorporates a new signal data processor which replaces several radar units and eliminates the need for special interface hardware when used in IMSS. The result is significant improvement in system weight, cooling, power, capability and reliability.

The primary function of the infra-red system is short-range tracking and observation of airborne, maritime and ground targets. As integrated in the IMSS, the infra-red capability has, as a passive system, proved exceptionally effective in the covert tracking of aircraft. The infra-red system can be applied independently to track a designated target by locking on to the heat differential generated by the target; it can be slaved to the radar so that the infra-red line of sight follows radar-tracked airborne targets and it can be directed to acquire and track ground targets automatically, using data from the INS. Additional capabilities include a TV camera, laser range-finder and video recorder.

The IMSS controls and displays are integrated with the radar, infra-red and INS through the radar signal data processor. A hand control unit provides the sensor operator with single-hand slew control of both sensors, including antenna elevation, radar cursor in azimuth and range, target designation and infra-red line of sight. Separate displays are provided for the radar and the infra-red, and these displays can be duplicated at several stations within the aircraft.

The INS provides the inertial references required by the radar and infra-red systems, such as roll, pitch, yaw, heading and velocities in three axes.

The IMSS incorporates continuous self-testing and BIT to isolate a fault down to an easily replaceable unit.

Contractor

Northrop Grumman Corporation, Electronic Sensors and Systems Sector.

VERIFIED

Internal FLIR Targeting System (IFTS)

The IFTS, together with the AN/APG-68ABR, forms the major part of the primary sensor suite of the Lockheed Martin F-16 Block 60 aircraft for the United Arab Emirates.

The IFTS is said to derive from LANTIRN-type technology, and to be configured into FLIR apertures fitted above and below the radome. Mid-wave FLIR sensors are proposed as best matching the high-humidity conditions of the Gulf. A wide field of view navigation and target acquisition FLIR is installed in the upper FLIR aperture, and a targeting FLIR with built-in laser/designator ranger in the lower aperture.

Contractor

Northrop Grumman Corporation, Electronic Sensors and Systems Sector

VERIFIED

Mast-mounted sight sensor suite

The sensor suite for the mast-mounted sight of the US Army's Bell OH-58D helicopters consists of a FLIR, a day television sensor and an automatic boresight device.

Operational status

In service on OH-58D Kiowa helicopters of the US Army and the armed forces of Taiwan.

Contractor

Northrop Grumman Corporation, Electronic Sensors and Systems Sector.

UPDATED

Multifunction Infra-red Distributed Aperture System (MIDAS)

In late 1999, Northrop Grumman, Linthicum Heights was awarded a US$9.9 million contract by the US Navy for research and development of the MIDAS Advanced Technology Demonstrator (ATD) programme.

MIDAS continues directly from the Distributed Aperture Infra-Red Sensor (DAIRS) technology development programme.

The aim of the MIDAS ATD programme is to reduce the risk associated with the DAIRS concept for the Joint Strike Fighter (JSF), and to allow an acceptable entry to an Engineering & Manufacturing Development (E&MD) implementation of DAIRS.

DAIRS consists of multiple IR cameras providing 360° spherical coverage for pilot's situational awareness. Additional functions to be performed by the

The Northrop Grumman sensor suite installed in the mast-mounted sight of a US Army OH-58D helicopter

system include navigation, missile warning, and infra-red search and track.

Work will be undertaken in the Northrop Grumman Linthicum Heights, Maryland plant, and it is to be completed by January 2002.

More widely, Northrop Grumman Electronic Sensors and Systems Sector (ES^3) has teamed with Lockheed Martin Electronics and Missiles, Orlando, Florida to compete to provide electro-optical systems for the JSF. Northrop Grumman has the lead for the DAIRS sensor and Lockheed Martin has the lead for the electro-optical targeting system.

Northrop Grumman ES^3 has also included The Netherlands Organisation for Applied Scientific Research (TNO) Physics and Electronics Laboratory into the team. The two organisations will collaborate on the development of advanced digital signal conditioning algorithms for DAIRS. The objective of this part of the programme is to provide adaptive, real-time, scene-based non-uniformity correction; resolution enhancement; and sensitivity enhancement. The first demonstration of these new signal-conditioning algorithms took place at TNO in November 1999.

Contractor
Northrop Grumman Corporation, Electronic Sensors and Systems Sector

VERIFIED

The Northrop Grumman WF-360 surveillance and tracking infra-red system 0106356

Contractor
Northrop Grumman Corporation, Electronic Sensors and Systems Sector.

VERIFIED

Northrop Grumman LSF

The Northrop Grumman Light weight surveillance FLIR (LSF) is a modular airborne electro-optical system for detecting targets and is designed for high-performance aircraft.

The FLIR uses a platinum silicide staring array operating in the 3.4 to 4.8 μm band that eliminates optical scanning mechanisms, thus reducing weight. It also produces a detector with improved uniformity between the elements for low noise/low NETD performance.

The small sensor head may be adapted to a number of aircraft installations and contains the system optics and IR sensors. Line of sight is controlled by stabilised gimbals which may be adjusted by the operator. There are two Fields of View (FoV), wide FoV for search and narrow FoV for target identification and recognition.

Processor electronics control the LSF functions and video processing to RS-170 standard. Optional extras include automatic tracking, inertial or external control of the line of sight and MIL-STD-1553 bus communications. The control unit is used for the selection of FoV, commands from the command menu, calibration and built-in test functions.

Specifications
Dimensions: Ø 171 mm
Detector: PtSi staring array
Band: 3.4 to 4.8 μm
FoV: 3 × 4° (narrow), 15 × 20° (wide)
FoR: 120° (azimuth), 0 to 60° (elevation)
Line of sight stabilisation: <100 μrad
Weight: 29.5 kg (total)
Power supply: 22 to 29 V DC

Operational status
In service. No longer in production.

Contractor
Northrop Grumman Corporation, Electronic Sensors and Systems Sector.

NEW ENTRY

WF-360 surveillance and tracking infra-red system

The WF-360 system features a Forward-Looking Infra-Red (FLIR) sensor operating in the 8 to 12 μm long-wave spectrum, with options for adding a high-resolution day TV camera and an eye-safe laser range-finder. These sensors are boresighted together on an optical bed housed in a stabilised gimbal designed for aircraft applications. The system provides high-resolution day and night detection and tracking of airborne, maritime and land-based targets. System capabilities include passive search and track in both air-to-air and air-to-surface modes. ARINC and MIL-STD-1553 bus interfaces are provided to communicate with other aircraft sensors such as the AN/APG-66/68 radar, flight management systems and inertial navigation systems. A high-throughput processor provides cueing of the system to radar targets or navigation waypoints, real-time display of the track point co-ordinates and fire-control solutions.

The WF-360 infra-red sensor marries US Army common modules with electronics providing DC restoration, automatic gain and level control and digital scan conversion for a standard US RS-170 525 line video output. These features, together with an advanced digital noise reducer and image enhancer, provide superior video imagery. Virtually all FLIR systems in the US DoD inventory use common module parallel scan long-wave technology because of its capability for widely varying operational scenarios throughout the world.

The aerodynamically streamlined turret utilises a two-axis stabilised platform to stabilise the optical lines of sight and point them throughout the entire lower hemisphere with look-up limited only by the aircraft structure. A broadband servo system featuring solid-state rate sensors and an inner acceleration loop provides stabilisation better than 35 μrad in typical aircraft environments ranging from helicopters to the US Air Force A-10. The turret is environmentally self-contained, employing an internal liquid-to-air heat exchanger to allow proper operation of the turret without the need for bleed or cabin air.

The WF-360's astronomical telescope provides two fields of view: a 4.5° narrow field of view and an 18.5° wide field of view. The TV sensor uses a × 6 zoom lens which can match either FLIR field of view or zoom continuously.

In addition to the TV and laser range-finder, other optional plug-in modules provide capabilities including FLIR/TV image fusion, wide area correlation video tracking and covert laser illumination for night TV

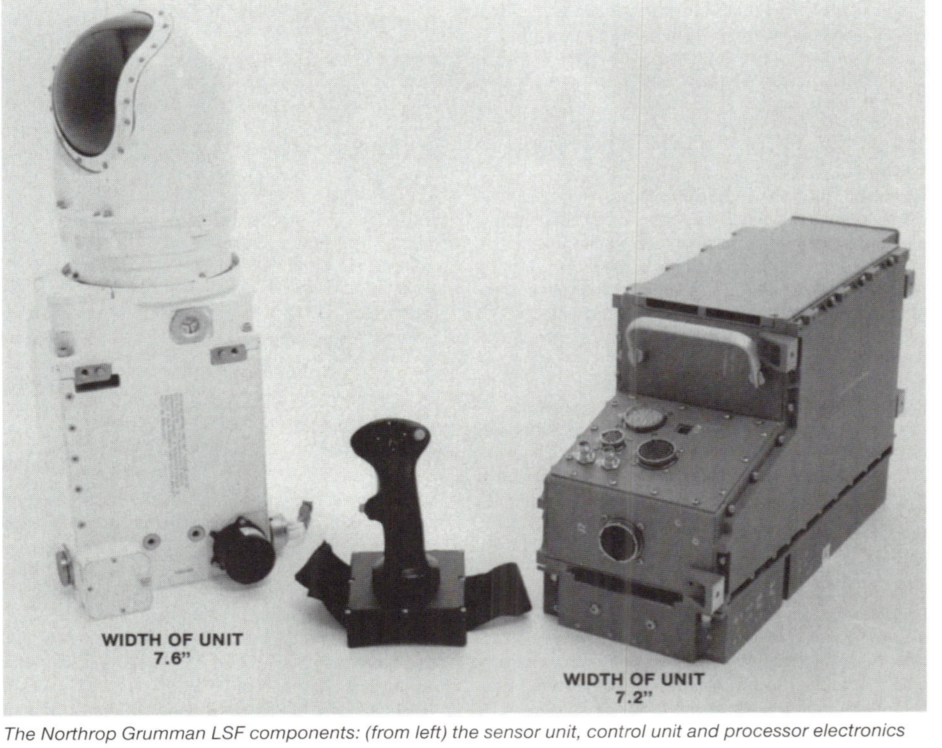

The Northrop Grumman LSF components: (from left) the sensor unit, control unit and processor electronics **2001**

RISTA Reconnaissance, Infra-red, Surveillance, Target Acquisition

RISTA has been derived from a US Army airborne mine detection system; it provides a FLIR for battle damage assessment and an Infra-Red LineScan (IRLS) for reconnaissance.

Operational Status
Ordered by the Royal Danish Air Force for its F-16 aircraft.

imagery. The computer-aided track mode, in conjunction with the automatic video trackers, provides an excellent coast mode in the event of short-term target obscuration.

Operational status

Used extensively for surveillance and drug interdiction by the US Coast Guard in the HU-25 Falcon jet and RU-38, the US Air Force in the C-26 and the US Army in the DH-7.

An improved system, featuring the second-generation common module 480 × 4 scanned focal plane array is available.

Contractor

Northrop Grumman Corporation, Electronic Sensors and Systems Sector.

VERIFIED

Colour video HUD cameras

Photo-Sonics manufactures a family of colour video HUD (Head-Up Display) cameras for use on the A-4M, A-10 and F-16C/D aircraft; detail specification changes match the cameras to each aircraft type. All cameras utilise a mix of MIL-STD and COTS components; they incorporate extensive filtering to provide noise-free, high-quality imagery, with automatic exposure control. The camera assembly, comprising periscope and colour camera, is mounted on the HUD to record what the pilot is viewing and the symbology on the HUD. Data is output to an onboard video cassette recorder.

Specifications

Horizontal resolution: (NTSC) 470 TV lines
(PAL) 460 TV lines
Picture elements: (NTSC) 768 (H) × 494 (V)
(PAL) 752 (H) × 582 (V)
Power: 115 V AC 47-440 Hz
Weight: 1.9 kg

Contractor

Photo-Sonics Inc.

VERIFIED

Optical specifications

	Lens, focal length	FOV degrees		FOV milliradians	
		H	V	H	V
A-4M	16.0 mm, f1.4	22.6	17.1	395	298
A-10	25.0 mm, f2.5	14.6	11.0	255	191
	16.0 mm, f1.4	22.6	17.1	395	298
	15.0 mm, f1.4	24.1	18.2	420	317
F-16C/D	16.2 mm, f1.4	22.4	17.1	390	294.1

High-speed camera 16 mm-1PL

The Photo-Sonics 16 mm-1PL camera provides variable speeds from 10 to 500 frames/s. The 16 mm-1PL features interchangeable magazines of 200, 400 and 1,200 ft and synchronous phase-lock plug-in operation. Utilising the magazine load concept, the 16 mm-1PL can be loaded within seconds without disturbing the camera, the optical alignment or the electrical connection.

The camera features automatic exposure control, and timing pulse marks at 10, 100 and 1,000/s. The 70 series miniaturised Film Data Recording System (FDRS) can be attached to the side of the 16 mm-1PL camera, and it can record numerical and BCD information on 16, 35 and 70 mm film on every frame up to 1,000 frames per second.

Specifications

Shutter angle: 9 settings varying from 7.5 to 160°, set by external knob.
Power: 28 V DC or 115 V AC
Weights:
camera body: 2.73 kg
magazines: 200 ft, 2.27 kg; 400 ft, 3.64 kg; 1,200 ft, 6.82 kg

Operational status

Fitted on the wing-tip of Canadian Forces CF-18 aircraft.

The Photo-Sonics 16 mm-1PL high-speed camera with film data recorder attached **1999**/0051026

Contractor

Photo-Sonics Inc.

VERIFIED

T8653 and T8660 taut shadow mask colour CRTs

A patented taut shadow-mask design for this full colour line of CRTs offers very high purity and brightness. Excellent thermal stability means that high beam currents can be achieved without loss of colour purity. This results in a rugged CRT for military aircraft applications that offers a sharp, highly readable display usable in brilliant cockpit sunlight or under demanding night vision conditions.

Specifications

Model:	T8660	T8653
Dimensions:	6 × 6 in	5 × 5 in
maximum weight:	5.0 kg	3.86 kg
maximum length:	372.1 mm	326.4 mm
minimum viewing area:	6.0 × 6.0 in 152.4 × 152.4 mm	5.0 × 5.0 in 127.0 × 127.0 mm
outer width/ height:	7.3 × 7.3 in 185.4 × 185.4 mm	6.2 × 6.2 in 157.5 × 157.5 mm

Contractor

Planar Advance Inc.

VERIFIED

Advanced helmet-mounted sight

Polhemus designs, develops and manufactures helmet-mounted sights specifically for tactical military aircraft, based on a company-patented method of measuring sight angles by the use of magnetic sensors. A fourth-generation system, marketed under the name Magnetrak, is now in production (see next entry).

A source mounted in the cockpit generates a magnetic field some distance around it, which is sensed by a sensor installed under the pilot's helmet visor. Signals are processed by a computer in an associated electronics unit and converted into angles representing the direction of the pilot's line of sight in relation to a particular datum, usually the aircraft's reference frame.

Coincidence between the helmet aiming axis and the pilot's line of sight is attained by a small helmet-mounted optical generator which projects a virtual image located at infinity so that the pilot does not have to refocus his eyes on to the helmet visor. The pilot then aligns the aiming reticle with the target. The reticle can also be used to provide cueing information which permits the pilot to be signalled or directed from an external source of information, for example, from a radar, FLIR or radar warning device or by a weapons systems operator. The same system, in conjunction with a more elaborate display, can provide the pilot with flight director symbology or imagery and information from other aircraft sensors. The complete system, irrespective of complexity, comprises a magnetic field source, helmet-mounted sensor, memory unit, electronics unit and helmet visor display unit.

Contractor

Polhemus Inc.

VERIFIED

Magnetrak helmet-mounted sight

The line of sight data acquired by the sensor/source arrangement is passed into the aircraft weapon aiming system using a MIL-STD-1553B databus link. Conversely, target data detected by one of the aircraft's sensors such as the radar or thermal imager, can be relayed to the Magnetrak system, to give target cueing information on the pilot's visor. Thus the pilot's own line of sight can be directed to a possible target for visual identification before attack.

The components of the Magnetrak system include a small three-axis sensor and source, system electronics unit and memory unit, the latter supplying mapping data to the former. There is also a visor display consisting of a parabolic visor, (and cover) and an integral LED reticle generator which projects a collimated cross-hair image on to the visor, allowing the pilot's eye to focus on target and reticle simultaneously. Discrete dots around the cross-hair give a cueing facility. The visor fits any standard military helmet without modification to the shell. The electronics unit is ½ ATR in format.

The Magnetrak offers a resolution of 0.1° and covers 360° movement in azimuth and roll axes and ±90° in elevation. The motion box, inside which the head must remain for accurate system performance, is 410 × 254 × 150 mm either side of a central point.

Specifications

Dimensions:
(source) 61 × 35 × 35 mm
(sensor) 28 × 23 × 18 mm
(electronics unit) 124 × 174 × 283 mm
(memory unit) 109 × 33 mm
Weight:
(source) 0.16 kg
(sensor) 19 g
(electronics unit) 5.13 kg
(memory unit) 0.17 kg
Power supply: 115 V AC, 400 Hz, single phase, 0.7 A

Operational status
In production.

Contractor
Polhemus Inc.

VERIFIED

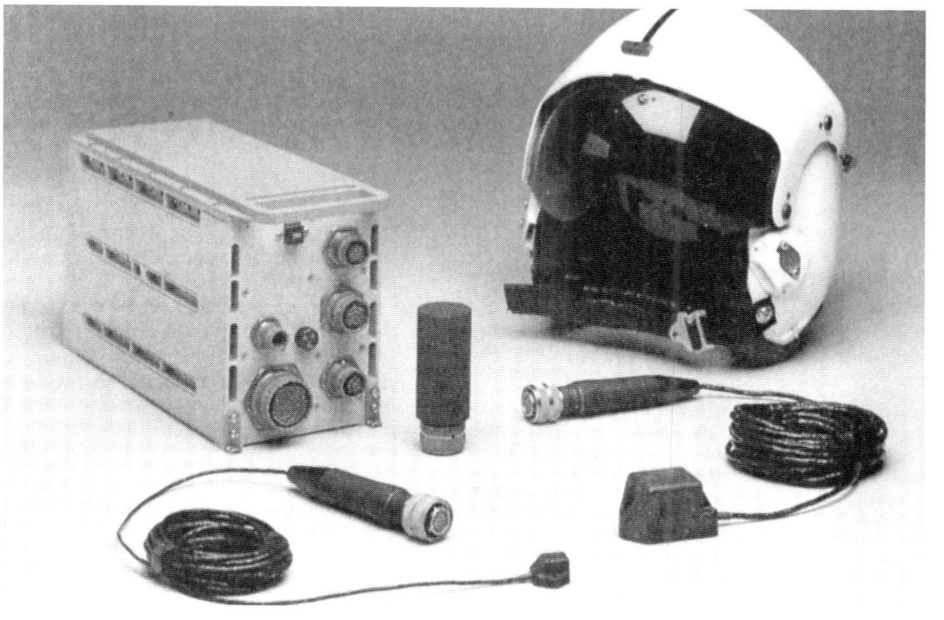

The Polhemus Magnetrak helmet sight. Left to right: system electronics unit and connecting cables, aircraft memory unit, magnetic field sensor, helmet with visor display and source unit
0106410

Airborne Electro-Optical Special Operations Payload (AESOP)

AESOP, designated the AN/AAQ-16D, combines the AN/AAQ-16 Helicopter Night Vision System (HNVS) (see following item) with a three field of view telescope and a laser designator/range-finder. The Raytheon Systems Company recently flight-tested an upgrade to the HNVS, called Hi-Mag, which provides both the high magnification capability required for target identification and the wide field of view needed for safe night and low-visibility pilotage.

The AESOP system is based on the AN/AAQ-16B Hi-Mag design, which upgraded the HNVS with a telescope featuring wide, medium and narrow fields of view, an enhanced autotracker and improved stabilisation. It also incorporates a lightweight laser target designator. Prototype systems will be used to demonstrate the ability to detect, recognise, track and direct Hellfire missiles to tactical targets. The system is designed for compatibility with the newer helicopters, in addition to test aircraft.

Operational status
In July 1994, a contract was awarded by the US Army for 15 systems, with options for a further 25 for installation on MH-60L Black Hawk and AH-6J Little Bird helicopters.

Contractor
Raytheon Electronic Systems.

VERIFIED

AN/AAQ-15 infra-red detection and tracking set

The AN/AAQ-15 is a small lightweight infra-red tracking set that has been developed under contract to the USAF for fitting to the HH-60 helicopter and the MC-130H Combat Talon II aircraft.

The system uses the infra-red receiver unit of the AN/AAS-36 with new electronics and lightweight gimbals. It has three fields of view, covers all of the lower hemisphere and has a 15° unobstructed look-up capability.

Operational status
In service.

Contractor
Raytheon Electronic Systems.

VERIFIED

AN/AAQ-16 night vision system

The AN/AAQ-16 infra-red imaging system entered initial production for the US Department of Defense in 1984. It has been selected by the US Army, Navy, Marine Corps and Air Force and international customers for a variety of helicopters.

The system provides 24-hour mission capability during night or degraded weather conditions in support of special operations, air assault, search and rescue, and anti-surface warfare.

An advanced thermal imaging system, AN/AAQ-16's high-resolution, TV-like imagery provides for low-level pilotage and navigation in the wide field of view, and in the narrow field of view, provides long-range target detection and identification. The AN/AAQ-16 combines automatic FLIR performance optimisation, and DC restoration to respond to various environmental conditions, with hands-off features of automatic gain, level, and focus for a high-quality image and eased operator workload. Either a black-hot or white-hot image can be selected by the operator. Additional features which enhance mission success include an autotracker and operator-controlled autoscan.

Variants of the AN/AAQ-16 system include: the AN/AAQ-16B which utilises the 8 to 12 μm waveband and two fields of view (30 × 40° and 5 × 6.7°); the AN/AAQ-16C which also utilises the 8 to 12 μm waveband and three fields of view (30 × 40°, 5 × 6.7° and 1.9 × 2.5°) and a dual-mode tracker.

Specifications
Dimensions:
(FLIR turret) 356 (length) × 305 mm (diameter)
(electronics unit): 305 × 200 × 412 mm
(multifunction control unit) 76.2 × 76.2 × 157.5 mm
(system control unit) 38.1 × 146 × 115 mm
Weight:
(FLIR turret) 24.49 kg
(electronics unit) 20.87 kg
(multifunction control unit) 0.59 kg
(system control unit) 0.45 kg

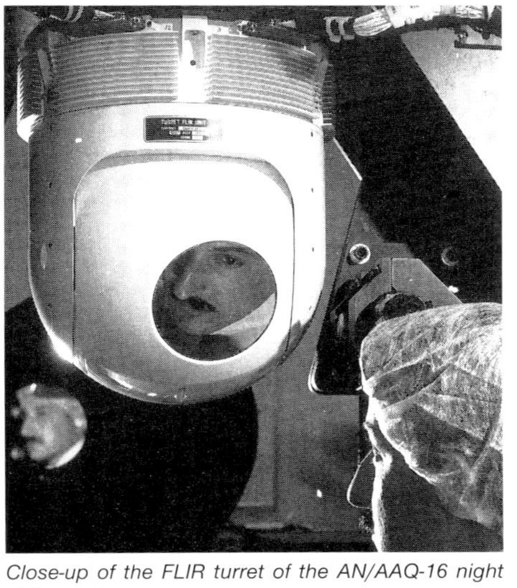

Close-up of the FLIR turret of the AN/AAQ-16 night vision system

Wavelength: 8-12 μm
Field of view:
(× 1 magnification) 30 × 40°
(× 6 magnification) 5 × 6.7°
(× 16 magnification option) 1.9 × 2.5°
Reliability: >300 h MTBF

Operational status
More than 400 systems delivered or on order for the US Army UH-60L, CH-47D, MH-47E, MH-60K; US Air Force HH/MH-60G, US Marine Corps CH-53E, Australian S-70B. Co-produced with NEC for the Japanese Defence Force.

Contractor
Raytheon Electronic Systems.

VERIFIED

AN/AAQ-17 infra-red detecting set

The AN/AAQ-17 infra-red detecting set is a multipurpose thermal imaging system. Typical missions are navigation, search and rescue, surveillance and fire control. AN/AAQ-17 systems are fitted on US Air Force AC-130A, AC-130H and AC-130U gunships, HC-130P and HC-130 tankers and C-141 special operations transports.

In the US Air Force gunship programme the AN/AAQ-17 FLIR replaces the AN/AAD-7 FLIR on the AC-130H, providing improved performance and reliability at lower cost. The AN/AAQ-17 is a derivative of the AN/AAQ-15. Both of these systems are lightweight and easily adapted to new aircraft. Originally, the AN/AAQ-15 FLIR was developed for the US Air Force HH-60 Nighthawk helicopter and has also been produced for the MC-130H Combat Talon II aircraft.

The AN/AAQ-17 consists of four LRUs: the infra-red receiver (LRU1), control-converter (LRU2), gimbal position control (LRU3) and infra-red set control (LRU4). Environmental qualification, reliability demonstrations and maintainability demonstrations are complete. The AN/AAQ-17 uses standard DoD FLIR common modules to convert long wavelength radiation into a composite TV video.

A 13.7 × 18.3° wide field of view allows navigation, area search and detection of larger targets. The 3 × 4° narrow field of view allows small target detection and target recognition. A precision gimbal provides accurate line of sight angular measurement. An adaptive gate video tracker reduces operator workload by providing hands-off automatic line of sight control. Operation of the FLIR is through manual controls or over a MIL-STD-1553B databus. This produces a flexible system adaptable to many aircraft.

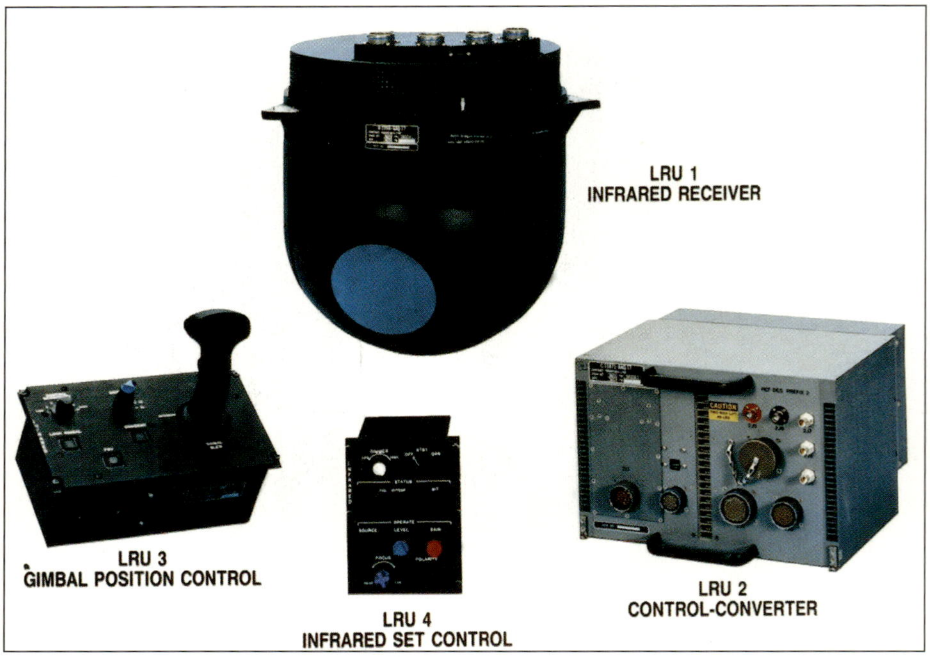

The AN/AAQ-17 infra-red detecting set 0106358

Specifications

Weight: (infra-red receiver) 43.09 kg
(control converter) 21.77 kg
(gimbal position control) 2.27 kg
(infra-red set control) 1.81 kg
Field of view: (wide) 13.7 × 18.3°
(narrow) 3 × 4°
Field of regard: (azimuth) ±200°
(elevation) +15 to −105°

Operational status

In service in AC-130A, AC-130H, AC-130U, HC-130N and HC-130P and C-141 aircraft.

Contractor

Raytheon Electronic Systems.

VERIFIED

AN/AAQ-18 forward-looking infra-red system

The AN/AAQ-18 forward-looking infra-red system is a common module update programme. The system improves reliability and readiness of the US Air Force Special Forces by replacing the AN/AAQ-10 system. System installations include the MH-53J helicopter and the MC-130E Combat Talon aircraft.

The primary mission of the AN/AAQ-18 is navigation. It is also ideal for search and rescue missions. The AN/AAQ-18 consists of 5 LRUs.

Specifications

Power supply: 115 V AC, 400 Hz, 3 phase, 3kVA
28 V DC, 2 A
Field of view:
(wide) 18.2 × 13.7°
(narrow) 4.1 × 3.1°
Field of regard:
(azimuth) ±190°
(elevation) +15 to -105°

Operational status

Operational in the MH-53J helicopter and US Special Forces MC-130E Combat Talon aircraft.

Contractor

Raytheon Electronic Systems.

VERIFIED

AN/AAQ-26 infra-red detecting set

In August 1997, US Air Force Aeronautical Systems Command awarded Raytheon Electronic Systems a contract for the fabrication of 16 AN/AAQ-26 gunship infra-red detecting sets and their installation on nine AC-130U and 7 AC-130H aircraft. The FLIR provides the aircrew with the capability to operate safely outside the threat area, at greater standoff range.

The AN/AAQ-26 FLIR is based upon the the AN/AAS-44(V) product, which is the US Navy's Lamps Mk III FLIR set.

Operational status

In production for US Air Force AC-130H and AC-130U aircraft.

Contractor

Raytheon Electronic Systems.

UPDATED

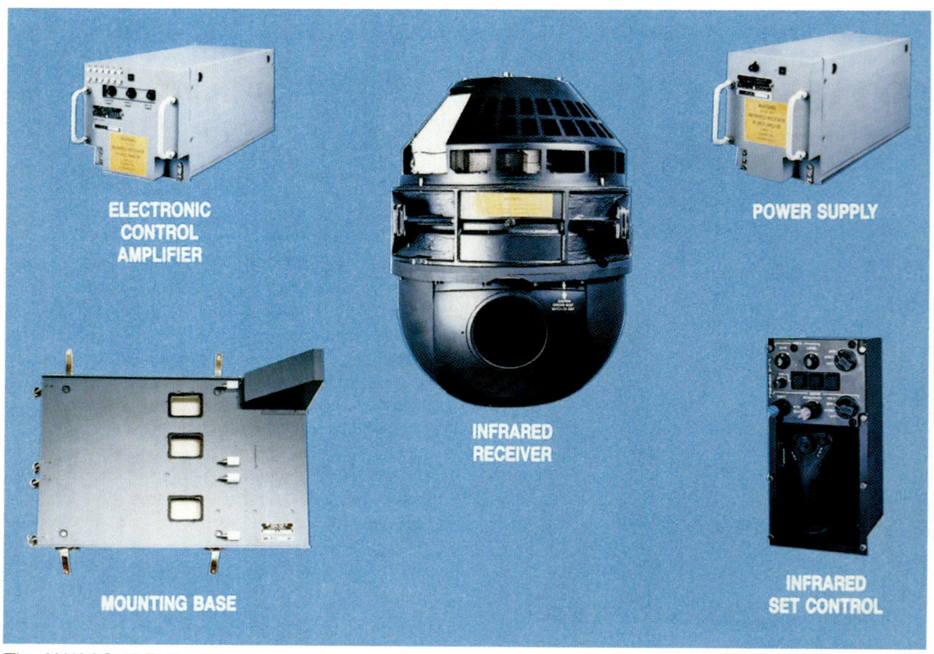

The AN/AAQ-18 FLIR system 0106359

AN/AAQ-27 (3 FOV) mid-wave infra-red imaging system

AN/AAQ-27 (3 FOV) is a third-generation, three field of view, mid-wavelength infra-red (MWIR) system for helicopter navigation, surveillance, and targeting applications. With three fields of view available, pilots can fly and navigate on low-level missions or detect and identify long-range targets from higher altitudes. Raytheon state that the system uses non-developmental production components, to provide good reliability, but that it provides higher-resolution imagery and better range performance than current long-wavelength infra-red systems.

The AN/AAQ-27 (3 FOV) is a modified version of the AN/AAQ-27 infra-red system which Raytheon (ex-Hughes) produces for the US Marine Corps MV-22 Osprey. It features substantial commonality with the combat proven AN/AAQ-16 infra-red system. The AN/AAQ-27 (3 FOV) is compatible with existing AN/AAQ-16 system mountings and is easily interchangeable for upgrade programmes.

Specifications

Detector: staring array (640 × 480 InSb); spectral band 3-5 microns
3-FOV:
30 × 40° - 1 ×
5.0 × 6.7° - 6 ×
1.3 × 1.73° - 23 ×

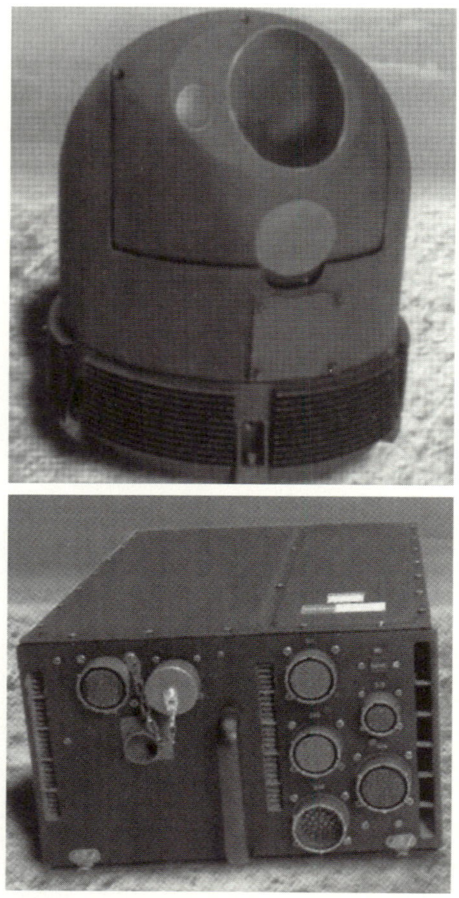

AN/AAQ-27 (3 FOV) mid-wave infra-red imaging system showing the system electronic unit (below) and turret FLIR unit (above) 0018328/0018329

Unit dimensions:
turret FLIR unit (TFU): 360.4 (height) × 303.5 mm (diameter)
systems electronics unit (SEU): 199.1 (height) × 413.5 (length) × 306.3 mm (width)
System weight: 42.27 kg
Tracker: dual-mode (centroid/correlation)
Options: eye-safe laser range-finder; laser designator/range-finder

Operational status

Selected by Kaman Aerospace International Corporation for the Royal Australian Navy SH-2G Super Seasprite helicopter programme; also reported to be part of the Royal Australian Navy's S-70B Seahawk upgrade. The 3 FOV system provides over-water navigation and long-range target detection and classification. Deliveries are scheduled for April 1999 to February 2000.

Contractor

Raytheon Electronic Systems.

VERIFIED

AN/AAQ-27 MWIR staring sensor

The AN/AAQ-27 Mid-Wave Infra-Red (MWIR) staring sensor is an advanced infra-red system that features substantial commonality with the combat-proven AN/AAQ-16B infra-red system. It uses non-developmental, in-production components to provide higher resolution imagery than current long-wave infra-red systems, and incorporates an MWIR Indium-Antimonide (InSb) staring Focal Plane Array, with 480 × 640 detector elements.

The system features a turreted FLIR weighing 22.7 kg, with a total system weight under 42.3 kg. Options include a video autotracker, third field of view and eye-safe laser range-finder.

Specifications

Field of view: 30 × 40° and 5 × 6.7°
Wavelength: 3-5 μm
Staring FPA: unprecedented imagery with less than half the aperture size
High performance staring array: 640 × 480 (InSb)
Options: third field of view; electronic zoom; dual-mode tracker; laser range-finder

Operational status

The system is in production for the US Marine Corps MV-22 Osprey, and on order for the Royal Australian Navy SH-2G helicopter.

Contractor

Raytheon Electronic Systems.

VERIFIED

AN/AAR-42 infra-red detection set

The AN/AAR-42 is a FLIR system for use on the US Navy A-7E strike aircraft. It is designed to provide a night window or bombsight which permits the pilot to perform single-seat close air support and reconnaissance missions by day or night and during poor weather conditions.

The AN/AAR-42 is installed in a pod with a gimballed FLIR unit which provides stabilised imagery on the pilot's head-up display system. It provides an azimuth coverage of ±20° and an elevation coverage from +5 to −35°. Both wide and narrow selectable fields of view are provided. The wide field of view, giving a × 1 magnification, is employed for pilot orientation, navigation update and target acquisition, while the narrow field of view, with a × 4 magnification, is used for target identification and weapon delivery. Features include automatic thermal focus compensation and sensor window de-icing.

Specifications

Weight:
(canister assembly) 95.45 kg
(servo electronics) 18.18 kg
Reliability: 390 h MTBF

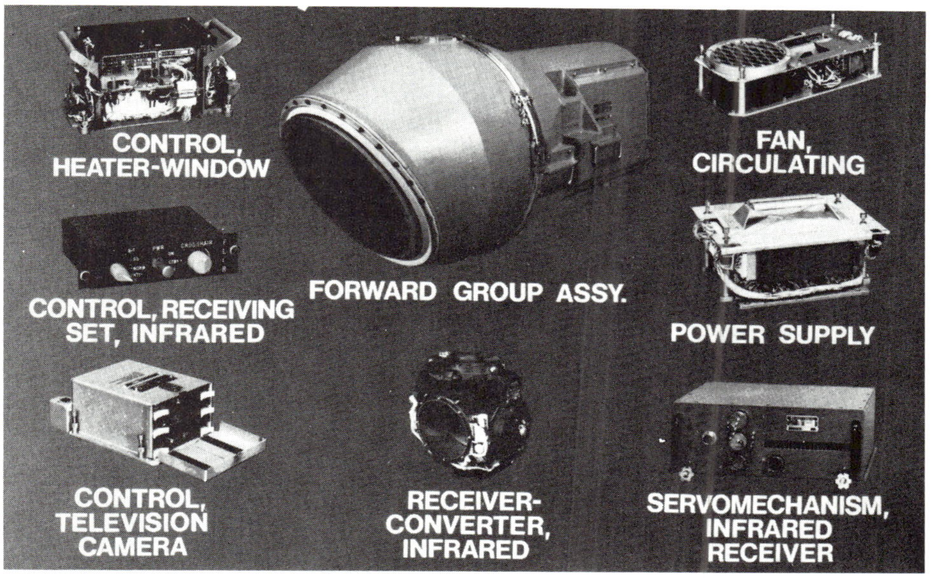

The AN/AAR-42 FLIR system

Operational status

In service. The system is currently operational on the A-7E.

Contractor

Raytheon Electronic Systems.

VERIFIED

AN/AAR-45 thermal imaging sight

The AN/AAR-45 forms a part of the Target Recognition and Attack Multisensor (TRAM) suite on the US Navy's A-6E aircraft, in which role it operates in conjunction with a Marconi Electronic Systems head-up display.

The thermal imager is mounted on a gimbal system in a pod carried beneath the aircraft and the resultant picture is displayed with flight information on the HUD. This way the pilot can fly head-up during a low-level attack by night. The system is also linked to the aircraft inertial navigation system so that the imager is always kept pointing at the target, even during manoeuvres.

Operational status

Production of the AAR-45 is complete with 91 systems delivered.

Contractor

Raytheon Electronic Systems.

VERIFIED

AN/AAR-50 Navigation FLIR (NavFLIR)

The AN/AAR-50 Navigation FLIR (NavFLIR) is a derivative of the AN/AAQ-16 night vision system installed in US Army helicopters. It uses a thermal imaging sensor to provide pilots of fixed-wing aircraft on low-level missions at night or in bad weather with a TV-like image of the terrain ahead projected on to a head-up display. The system as configured for the F/A-18 is pod-mounted in a fixed forward-staring position but could be configured in different pods for a variety of aircraft.

NavFLIR consists of four major weapon-replaceable assemblies: FLIR sensor unit, pod electronics unit, thermal control unit and pod adaptor.

Specifications

Dimensions: 1,981 × 254 mm diameter
Weight:
(pod) 73.48 kg
(adaptor) 23.13 kg
Wavelength: 8-12μm
Field of view: 19.5 × 19.5° displayed
Reliability: >410 h MTBF

Contractor

Raytheon Electronic Systems.

VERIFIED

The AN/AAR-50 NavFLIR mounted on the starboard intake of a US Navy F/A-18 0106361

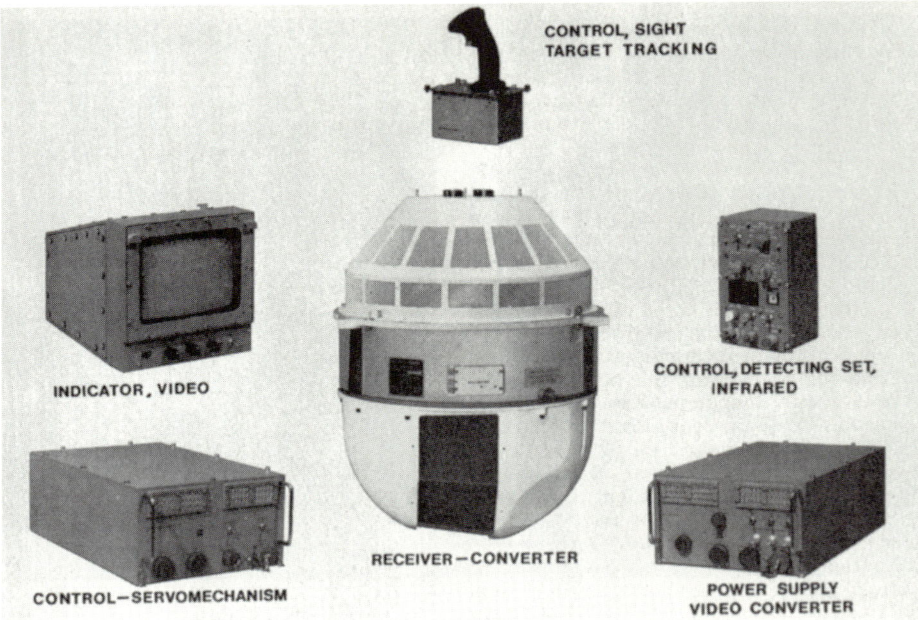

The AN/AAS-36 FLIR system is used in US Navy P-3C aircraft

AN/AAS-36 infra-red detection set

The AN/AAS-36 infra-red detection set is a FLIR system designed for US Navy P-3C maritime patrol aircraft to detect surface vessels, surfaced or snorkelling submarines and drifting survivors in darkness and limited visibility. The system was initially designed and developed to meet a P-3C update programme requirement, but the equipment has also been retrofitted to earlier P-3C and P-3B aircraft.

Production of the system commenced in 1977 following a testing, evaluation and demonstration programme which used 10 preproduction systems to assure the US Navy that design specifications were either met or exceeded. The Initial Operational Capability (IOC) was realised in 1979.

Based on US Department of Defense common modules which employ Cadmium Mercury Telluride (CMT) detectors, the AN/AAS-36 is a stand-alone system requiring only electrical power for operation. The common module infra-red receiver is mounted in an azimuth-over elevation stabilised gimbal and provides lower hemisphere coverage of ±200° in azimuth and from +16 to –82° in elevation. Additional weapon-replaceable assemblies provide system power, servo control, FLIR system control, slew commands and a real-time video display. The display presentation is on an 875-line RS-343 composite television monitor which permits the operator to identify, as well as observe, vessels.

Features include automatic optical temperature compensation, gimbal pointing outputs for servo platform slaving, self-contained stabilisation and a two field of view optical system (15 × 20° or 5 × 6.7°). A digital computer interface is available for on-line gimbal control. The system contains built-in self-test facilities which permit checkout down to weapon-replaceable assembly level and these themselves are compatible with automatic test equipment.

The system is also being supplied to many non-US operators of the P-3 for upgrading to US Navy standards. The receiver-converter weapon-replaceable assembly of the AN/AAS-36 has been fitted to Sikorsky CH-53 helicopters and Cessna Citation, Beechcraft E-90, King Air 200 and other unspecified aircraft.

Specifications

Weight: 136.36 kg
Power supply: 115 V AC, 400 Hz, 3 phase, 2.5 kVA
28 V DC, 100 W

Operational status

In service.

Contractor

Raytheon Electronic Systems.

VERIFIED

AN/AAS-37 infra-red detection set

The AN/AAS-37 equipment is a variant of the AN/AAS-36 detection set but is a more sophisticated system, being combined with a laser designation and ranging capability. Developed for the US Marine Corps OV-10D forward air control aircraft, the AN/AAS-37 provides infra-red vision for day or night operations under degraded environmental conditions, automatic target tracking and laser target-designation. The laser provides ranging and illumination of ground targets for laser-guided weapons such as the Paveway bomb or the Hellfire missile, either on an autonomous basis for aircraft equipped with the AN/AAS-37 or for a co-operating aircraft armed with these weapons.

Specifications of the FLIR sensor and associated equipment are virtually identical to those of the AN/AAS-36 system from which it was developed. It does, however, possess a number of additional features derived mainly from incorporation of the laser section. These include direct readout laser ranging and designation capability. The system can be used as a target sight aligned with the aircraft boresight by means of electronic adjustment and line of sight depression may be set by the operator for precision delivery of air-to-ground weapons. There are interfaces with aircraft systems for a radar altimeter and remote gyroscope and an accelerometer provides the system with rate-aided automatic target tracking capability using an adaptive gate centroid tracker. Offset tracking from a target or another landmark is also possible. The system's display is daylight visible.

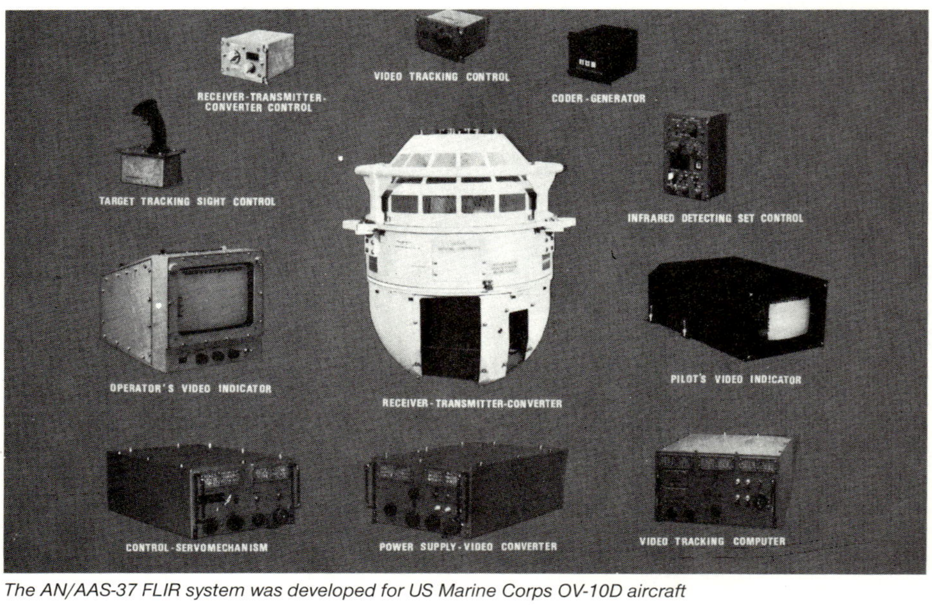

The AN/AAS-37 FLIR system was developed for US Marine Corps OV-10D aircraft

Specifications

Laser designator/ranger
Coverage:
(azimuth) ±200°
(elevation) –82 to +16°
Weight: 189 kg
Power supply: 115 V AC, 400 Hz, 3 phase, <3 kVA
28 V DC, <1,800 W

Operational status

In service.

Contractor

Raytheon Electronic Systems.

VERIFIED

AN/AAS-38A F/A-18 targeting pod

The AN/AAS-38A Forward-Looking Infra-red (FLIR) targeting pod enables pilots of US Navy and Marine Corps F/A-18 Hornet aircraft to attack ground targets day or night with a precision strike capability. The system provides TV-like infra-red imagery on a cockpit panel display and accommodates a laser range-finder/designator that can pinpoint targets for both laser-guided and conventional weapon delivery.

The pod is integrated with the aircraft's avionics system through a MIL-STD-1553 databus, allowing the pod to receive command and cue signals from the onboard mission computer and provide status and targeting information to the cockpit display and weapon delivery system.

The AN/AAS-38A consists of 12 weapon replaceable assemblies (WRAs) that can be readily accessed and replaced without the need for calibration, alignment, special tools or handling equipment.

The AN/AAS-38A configuration accepts the two laser subsystem WRAs, a laser transceiver and laser power supply to provide the aircrew with the capability for laser target designation and ranging (LTD/R).

Design qualification and flight test were successfully completed in August 1994 thus providing the US Navy a second source for AAS-38 pods and spares (see also Lockheed Martin Electronics & Missiles entry entitled NITE Hawk).

The Targeting FLIR, when integrated with the AN/AAR-50 Navigation FLIR and Night Vision Goggles, provides the pilot/aircrew with the capability to maintain situational awareness, navigate/avoid terrain, acquire/designate targets and assess battle damage for deployment of the Pave Way/GBU-24 precision-guided weapon series.

Specifications:

Dimensions: 1,830 (length) × 330 mm (diameter)
Weight: 172.7 kg
NFOV: 3 × 3°
WFOV: 12 × 12°

Field of regard:
(pitch) +30° to −150°
(roll) ±540°
Video: RS-343 875 lines

Contractor
Raytheon Electronic Systems.

VERIFIED

ATFLIR Advanced Targeting Forward-Looking Infra-Red system

The ATFLIR pod includes both infra-red targeting and navigation systems. The targeting FLIR uses the same third-generation Mid-Wave Infra-Red (MWIR) staring focal plane technology that has been used in the US Marine Corps MV-22 Osprey (the AN/AAQ-16-27 system). The navigation FLIR will be provided by Marconi Electronic Systems.

The design aim of the ATFLIR system is to achieve sufficiently accurate long-range performance for F/A-18E/F crews to be able to deliver their air-to-ground weapons from beyond the range of defensive anti-aircraft artillery and many surface-to-air missile systems.

Operational status
In development. Selected for the US Navy F/A-18E/F aircraft, and for retrofit to existing F/A-18 C/D models. The EMD contract is expected to be for 8 to 10 systems for qualification and flight testing in mid-1999, with initial low-rate production to follow, and an initial operational capability in mid-2002.

Contractor
Raytheon Electronic Systems.

VERIFIED

Cobra-Nite airborne TOW system upgrade

Cobra-Nite (C-Nite) is the designation given to the night targeting version of the airborne TOW missile system. This was developed to provide an enhanced night and adverse visibility capability for Cobra helicopters equipped with TOW missiles.

C-Nite also provides the Cobra with a target sight for use with unguided rockets and during conventional gun attacks. The system has been used effectively during exercises to monitor hostile forces and to direct ground troops from the helicopter, resulting in envelopment of the enemy by friendly forces.

Operational status
Deliveries began in 1989. Approximately 70 sets have been built for the US Army and 21 systems have been supplied to the US National Guard for installation in AH-1F helicopters. C-Nite systems are also being supplied to Bahrain and are being co-produced in Japan.

The US Army Cobra attack helicopter is equipped with the Cobra-Nite night targeting system 0106324

Contractor
Raytheon Electronic Systems.

VERIFIED

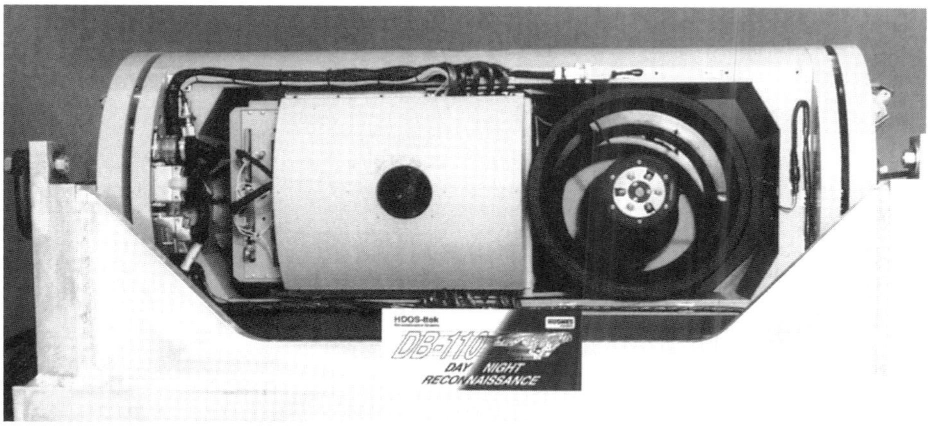

DB-110 Dual-Band reconnaissance system 0018327

DB-110 Dual-Band reconnaissance system

The DB-110 Dual-Band reconnaissance system is a powerful day/night long-range sensor, combining electro-optical and infra-red (IR) thermal imaging capabilities in a compact, lightweight, design. Modular design permits packaging as an internal fit, or within a pod (in a 275 gallon/1,500 litre fuel tank enclosure).

The high resolution achieved by the DB-110 satisfies intelligence requirements for active target detection and identification from long standoff range.

Multiple operational modes provide mission planning flexibility to interleave wide area search, spot collection, and target tracking/stereo modes. Both pre-programmed and manual control modes are provided. Control is via a MIL-STD-1553B databus. The system is compatible with digital tape recorders and digital datalinks.

Specifications
Sensor
resolution:
(EO NIIRS): 5 at 60 km
(IR NIIRS): 5 at 30 km
focal lengths:
(EO): 110 in
(IR): 55 in
dimensions: 1,270 × 470 mm (diameter)
weight: 140 kg
power: 115 V AC, 400 Hz, 669 VA
Electronics
dimensions: 4 units: 257 × 206 × 422 mm; 356 × 343 × 89 mm; 257 × 175 × 66 mm; 107 × 287 × 81 mm
weight: 28 kg
power: 115 V AC, 400 Hz, 590 VA

Optical
type: Cassegrain reflector
focal length: to fit application
(visible) 2,800 mm nominal
(infra-red) 1,400 mm nominal
aperture: 280 mm
(visible) f/10
(infra-red) f/5

Focal planes
visible (0.5 to 1.0 micron): Silicon CCD array; 5,120 × 64 line array
infra-red (2 to 5 micron): Indium antimonide (InSb) array; 512 × 484 area array; line arrays available
Data output: max data rate: 260 Mbits/s; data compression available to meet recorder or datalink budgets
Digital tape-recorder: data rate up to 240 Mbits/s continuously variable; 48 Gbytes on tape (equivalent to 20,000 nm^2)

Operation
standoff range: = 45 km
altitude: 5,000-80,000 ft
ground speed: M0.1-M1.6
field of regard:
(across line of flight) 180°
(along line of flight) ±20°
type: panoramic/sector scan (4-28°)
overlap: variable from 0-100%
performance:
(visible) up to NIIRS 6
(infra-red) up to NIIRS 5

Operational status
Selected by the UK Royal Air Force for Tornado GR. Mk 1A and GR. Mk 4/4A aircraft. The DB-110 is also marketed as part of the HISAR® Integrated Surveillance And Reconnaissance system.

Contractor
Raytheon Electronic Systems.

VERIFIED

Fast Tactical Imagery (FTI)

The FTI system is designed to transmit and receive images in near-realtime. It features an 8 mm recorder and an airborne transceiver to provide line-of-sight transmission of images.

Initial application has been on the US Navy F-14 aircraft, where it is used to transmit data from the TV camera or LANTIRN (Low Altitude Navigation Targeting InfraRed for Night) pod. The system can also share imagery with other F-14 aircraft, which can then serve as links to distant users. The FTI system supplements the F-14 Tactical Airborne Reconnaissance Pod System (TARPS).

FTI transmissions are made at a burst rate of 15 seconds per image.

Operational status
The FTI system has been developed by Raytheon's PhotoTelesis Division from a design for the US Army's AH-64 Apache helicopter. The US Navy plans to equip all of its 176 LANTIRN-equipped F-14 aircraft with the FTI system before the end of 2001.

The US Navy is also planning an enhanced Tomcat Tactical Targeting (T^3) derivative of the FTI system. This will allow GPS co-ordinates to be placed in the targeting system's cockpit image, and transmitted to other platforms for realtime targeting through a LANTIRN system software change. Lockheed Martin Electronics and Missiles is under contract for the T^3 upgrade, and expects the system to be deployed in mid-2000.

Contractor
Raytheon Electronic Systems

NEW ENTRY

Infra-Red Imaging Subsystem (IRIS)

The Infra-Red Imaging Subsystem (IRIS) is part of the AN/AAS-38 FLIR imaging system. This is a self-contained pod designed for use on the US Navy F/A-18 aircraft for target acquisition and recognition and weapon delivery. It also allows reconnaissance under day and night and adverse weather conditions.

The IRIS includes dual fields of view of 12 × 12° and 3 × 3° with automatic thermal focus. Built-in image stabilisation provides a natural horizon display to the pilot.

A key feature of IRIS is the automatic video tracker contained in the controller-processor. This tracker provides automatic target acquisition, line of sight control and offset designation for accurate weapon delivery.

The controller-processor provides video processing necessary to produce 875-line RS-343 television video. The processor adds track and field of view reticles for cockpit displays. A MIL-STD-1553 digital databus provides communications with the aircraft AN/AYK-14 mission computer. The infra-red receiver uses advanced switching regulator designs and high-density packaging to provide efficient primary to secondary power distribution, control and regulation for the IRIS and pod system.

Operational status

In service on the US Navy F/A-18 aircraft.

Contractor

Raytheon Electronic Systems.

VERIFIED

RS-700 series infra-red linescanner

The RS-700 airborne infra-red linescanner operates in the 8 to 14 μm band, where absorption by carbon dioxide and water vapour is at a minimum. The Cadmium Mercury Telluride (CMT) detectors use the common module closed-cycle cooling subsystem. The optical system focuses radiation on to the detectors to produce video electrical signals that correspond with the picture formed by the radiation pattern scanned. The system converts the video signals to visible wavelengths by light-emitting diodes for recording on film.

The RS-700 is composed of three subassemblies mechanically mounted to form a single assembly for aircraft installation. Important operational features of the equipment include: manual or automatic gain selection, manual or automatic level control, video compression of unusually hot or cold objects; continuous scanning over whole velocity/height range, adjustable hot-spot marker, event marker and built-in test equipment.

Specifications

Weight:
(without roll stabilisation) 32 kg
(with roll stabilisation) 42 kg
Scan mirror facets: 4
Optical aperture: 38.4 cm^2
Detector cooling: closed-cycle, 77°K
Detector type: CMT
Recording light source: gallium-arsenide phosphor diodes
Film width: 70 mm
Film capacity: 46 m, 70 m
Velocity/height range: 0.2-5
Thermal resolution: 0.2°C
Spatial resolution: 0.5-1.5 mrad
Field of view: 120°

Operational status

Some 180 RS-700 series linescan systems have been supplied to the air forces of Denmark, Germany, Italy, Malaysia, Saudi Arabia, Singapore, Sweden, Switzerland and the USA.

Contractor

Raytheon Electronic Systems.

VERIFIED

TIFLIR-49

TIFLIR-49 is a low weight, multiple purpose, thermal imaging sensor for navigation, surveillance, maritime search and rescue and troop-transport missions.

TIFLIR-49 includes two weapon replaceable assemblies: turret unit WRA-1 and electronics unit WRA-2. These units, together with off-the-shelf displays, controls and recorders provide a flexible, low-cost FLIR system for fixed- or rotary-wing aircraft.

Features include: a second generation focal plane array; electronic image stabilisation; dual-mode video tracker; multiple fields of view; local area processing; MIL-STD-1553B interface.

Specifications

Fields of view
(wide) 22.5 × 30°
(medium) 5 × 6.67°
(narrow) 1.3 × 1.7°
Electronic zoom: 2:1 and 4:1
Gimbal angular coverage
(azimuth) 360°
(elevation) +40 to −105°
Maximum airspeed: 300 kt

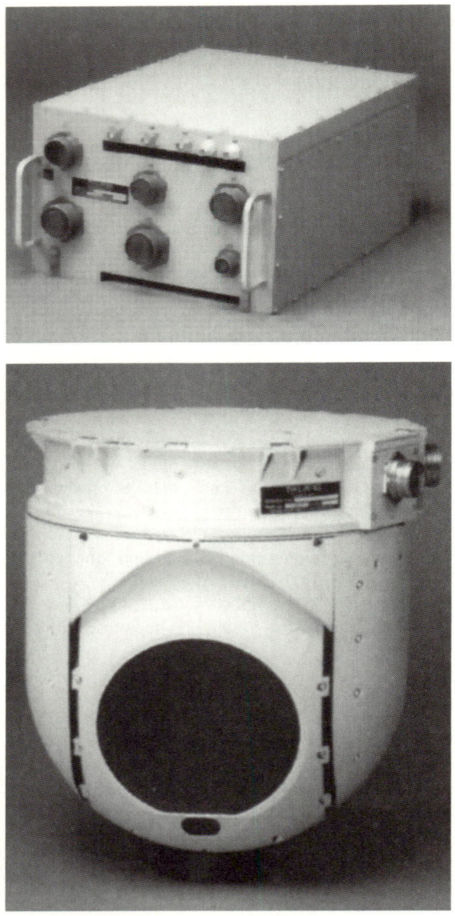

TIFLIR-49 turret unit WRA-1 (below) and electronics unit WRA-2 (above) 0018314/0018315

Operational status

Available. TIFLIR is understood to be essentially the same as the AN/AAS-44, but stripped of the laser, for international sales.

Contractor

Raytheon Electronic Systems.

VERIFIED

CA-236 36 in E-O LOROP camera

Offering high performance in a mid-range focal length electro-optical camera, the CA-236 provides long-range oblique imaging in a highly compact configuration.

Key features of the CA-236 E-O LOROP camera include: panoramic scan, spot target acquisition and multi-aspect operation; medium- to high-altitude imagery collection; digital data captured to digital tape or transmitted to a ground station via a datalink; forward motion compensation (FMC); electronic time delay integration (TDI) for haze penetration and imaging under low-contrast conditions.

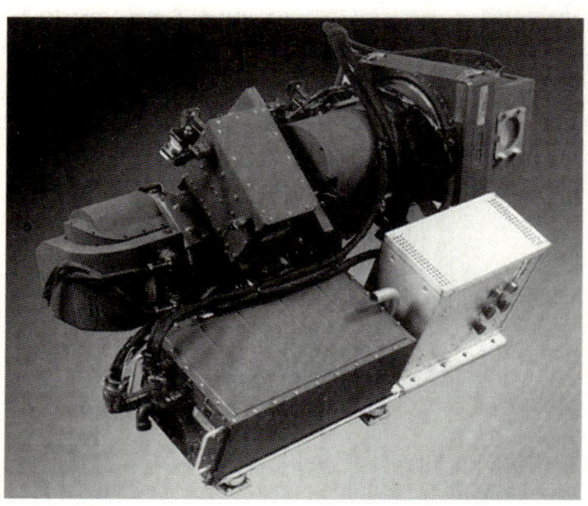

Recon/Optical Inc CA-236 E-O LOROP camera 0018324

Specifications

Lens focal length and f/no: 36 in, f/8
Operating spectrum: 510-900 nm
Detector: 32 TDI silicon; 12,064 pixels
Field of view (FoV):
4,020 active pixels, 2.52°
6,024 active pixels, 3.77°
12,048 active pixels, 7.54°
Cross-track Field of Regard (FoR): horizon-to-horizon capable, window dependant
Cross-track scan size: variable 0.5-23°
Azimuth pointing angle: ±15°

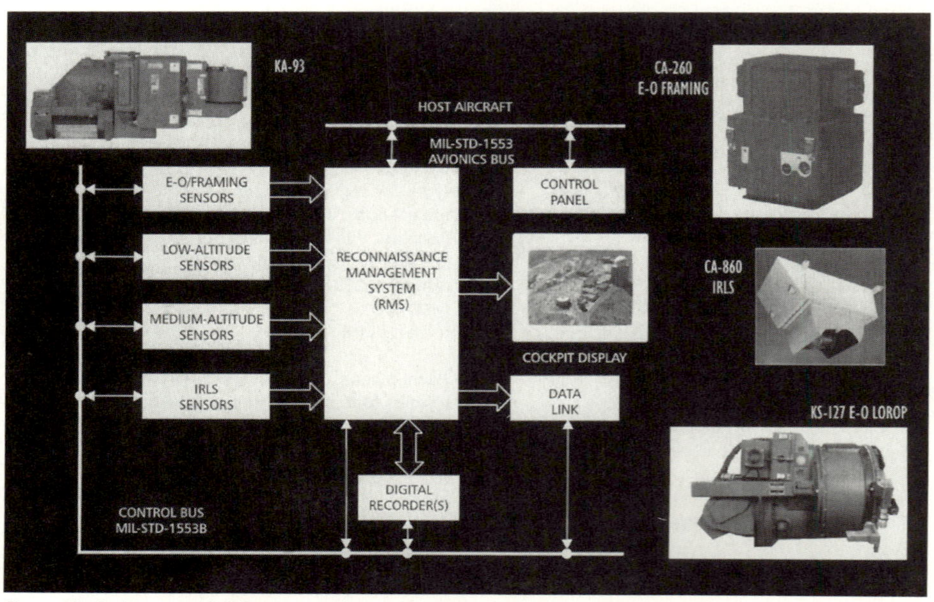

Recon/Optical Inc overall E-O reconnaissance concept 0018325

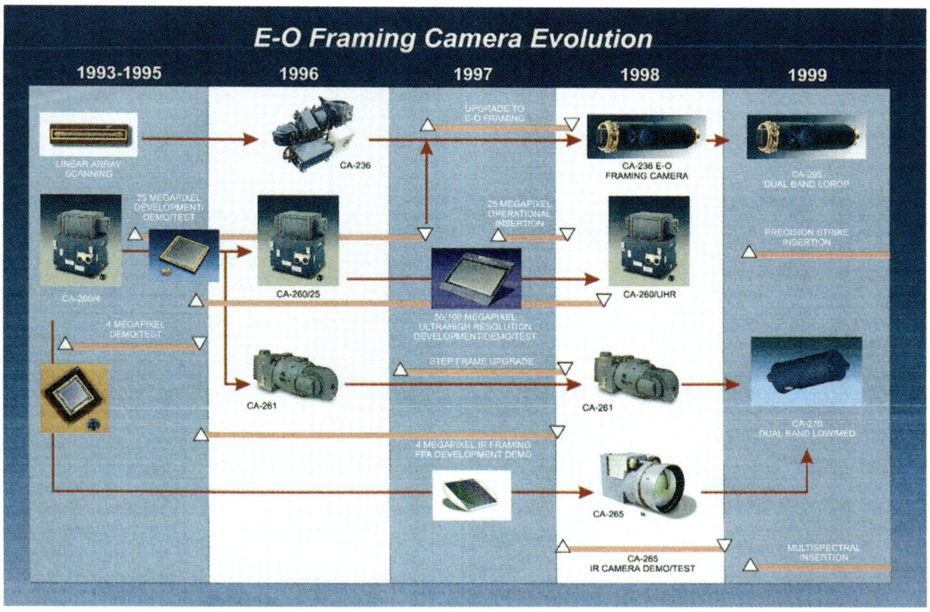

Recon/Optical, Inc E-O framing camera evolution ***1999***/0051027

An example of CA-236 E-O LOROP camera imagery at ×2.5 magnification 0018323

Pointing accuracy: <0.2° (ref to inertial space)
Camera pupil diameter: 4.5 in
Camera resolution: 11μrad/pixel
Camera output image data rate: 4,020 active pixels, 85.5 Mbit/s uncompressed
6,024 active pixels, 128 Mbit/s uncompressed
12,048 active pixels, 256 Mbit/s uncompressed
Compression: variable up to 14:1
Output image data: 8 bit ECL differential
No. of TDI stages: 1, 2, 4, 8, 16, 24, 32
Autofocus: continuous

Operational status

The CA-236 E-O LOROP camera was upgraded to E-O framing capability in 1998, with a development path to the CA-295 dual-band LOROP in 1999.

Contractor

Recon/Optical, Inc.

VERIFIED

CA-260 and CA-260/25 E-O framing reconnaissance cameras

The CA-260 E-O framing camera is designed specifically to provide near-photographic quality images while enhancing the survivability of the tactical reconnaissance platform at low to medium altitudes. It is configured for external pod or internal aircraft

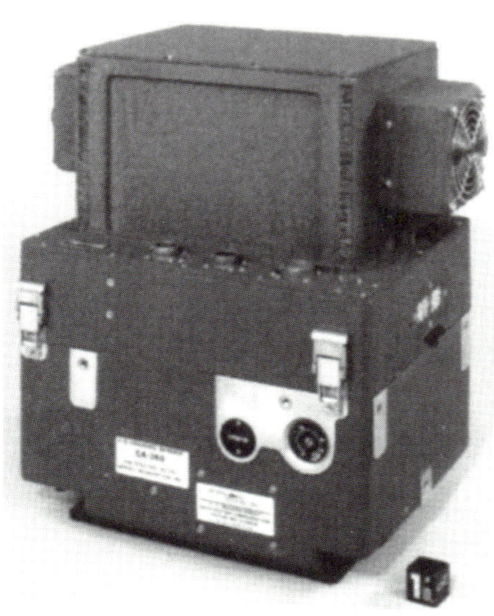

Recon/Optical Inc CA-260 E-O framing reconnaissance camera 0018322

mounting on a wide variety of reconnaissance platforms.

Advanced E-O framing technology reduces the amount of time required to cover a target area compared with conventional E-O linescan sensors. Wafer-scale processing has been used to develop the 4 M pixel or 25 M pixel array CCDs which are used in the CA-260 camera to give a wide field of view E-O image with continuous stereo coverage of targets at 56 per cent overlap. An on-chip motion compensation architecture eliminates the image blur normally associated with a wide field of view framing camera.

Because the mounting configuration and physical envelope are identical to that of the KS-87 film camera, the CA-260/4 or CA 260/25 can provide an interim E-O framing capability to users of existing KS-87 cameras.

CA-260/25 25 Mega pixel imagery using 7:1 data compression and ×16 magnification 0018321

Specifications

Lens: 1.5 in fl, f/4.5; 3.0 in fl, f/4.5; 6.0 in fl, f/4.0; 12.0 in fl, f/4.0
Field of view (FoV):
(CA-260/4) 35.8° with 1.5 in fl lens; 18.3° with 3.0 in fl lens; 9.2° with 6.0 in fl lens; 4.6° with 12.0 in fl lens
(CA-260/25) 76.9° with 1.5 in fl lens; 43.3° with 3.0 in fl lens; 22.4° with 6.0 in fl lens; 11.3° with 12.0 in fl lens
Pixel size (pitch): 0.012 × 0.012 mm
Frame rate: 2.5 frames/s max
Dimensions:
(camera) 175.3 × 261.6 × 401.3 mm
(power supply) 134.6 × 274.3 × 464.8 mm
Weight:
(camera) 27.27 kg (without lens)
(power supply) 8.18 kg
Power: 115 V AC, 400 Hz, 210 VA; 28 V DC, 140 W

Operational status

Successful engineering tests were completed in July 1993. RF-4C and F-14 TARPS demonstrations were completed in 1994. The CA-260/4 is in service on US Air Force/Air National Guard F-16s. The CA-260/25 is now in production for the US Air Force Theatre Airborne Reconnaissance System (TARS) Programme, and for Royal Danish Air Force F-16 aircraft.

Contractor

Recon/Optical, Inc.

VERIFIED

CA-261 E-O step framing reconnaissance camera

The CA-261 E-O step framing camera is designed specifically to provide photographic quality images while enhancing the survivability of the tactical reconnaissance platform at medium altitudes. It is configured for external pod or internal aircraft mounting on a wide variety of reconnaissance platforms.

The CA-261 combines the proven performance of the CA-260 25-Mpixel E-O framing camera with the stepping capability of a proven two-axis stabilised step head with de-rotation prism. This combination produces a system that captures a series of 25-Mpixel images through a 12 in focal length E-O lens in the cross track direction. These images allow for wide area coverage of up to 180° in an E-O framing format.

The captured imagery can be displayed in mosaic form to provide a 'birds eye' view of the entire area of interest. This display retains the capability to manipulate individual images in the same manner as the current CA-260. De-rotation optics are employed to eliminate image rotation. The high resolution of the CA-260 is maintained by incorporating stabilisation electronics and software developed by Recon/Optical, Inc. This combination yields residual rates <.001 rad/s even at disturbance input rates in excess of 10°/s.

Advanced E-O step framing technology reduces the amount of time required to cover a target area compared to conventional E-O linescan sensors. Recon/Optical, Inc employed wafer-scale processing to develop the 25-Mpixel array CCDs used in the CA-260 to give a wide field of view E-O image. An on-chip motion compensation architecture eliminates the

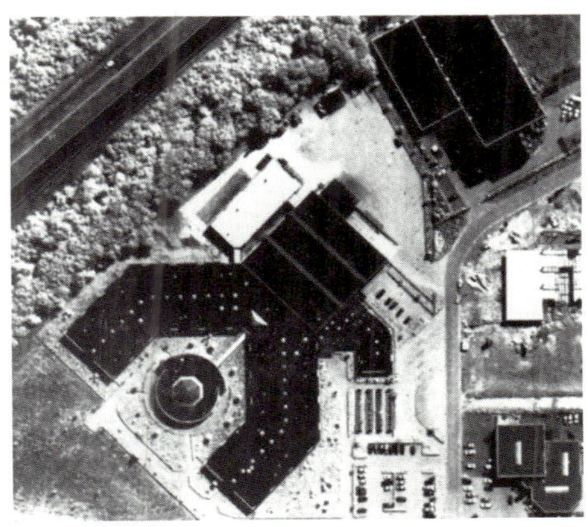

CA-261/25 E-O step framing imagery using 7:1 data compression and ×16 magnification 0018320

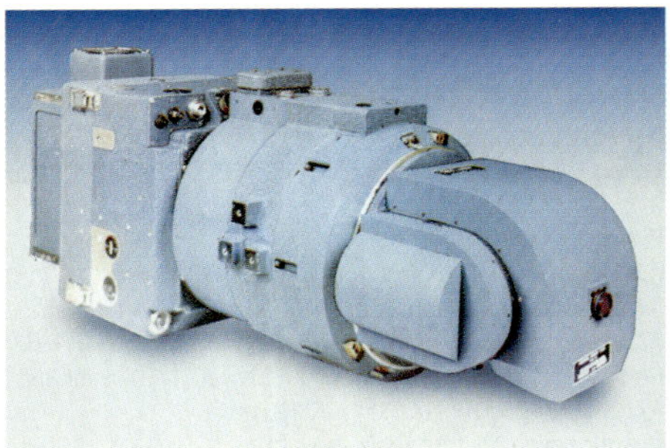

CA-261 E-O step framing camera 0106325

image blur normally associated with a wide field of view framing camera.

Four modes of operation are available:

Mode 1: Automatic. Maximum across-track coverage centred on a selected depression angle.

Mode 2: N-Select. Operator-selected number of steps and depression angle.

Mode 3: Stereo. Maximum across-track coverage with at least 56 per cent forward overlap.

Mode 4: Spot. Multi-aspect imaging of a specific target.

Specifications

Dimensions: 381 × 406.4 × 965.2 mm
Weight: 67.3 kg
Power supply: 300 W, 28 V DC; 630 VA, 115 V AC, 400 Hz, 3 Ø
Field of View (FoV): 11.3 × 11.3° each frame
Field of Regard (FoR): 180 (in cross track) × 11.3° (in line of flight)
Frame rate: 2.5 frames/s
Angular resolution: 39.4 μrad/pixel
Max V/R: 0.44 rad/s with 10% overlap

Operational status

A successful flight test was conducted in July 1997 by the Royal Netherlands Air Force on an F-16 aircraft. Imagery was captured and presented in mosaic format, demonstrating the effectiveness of the wide area coverage approach using CA-260/25 framing technology. CA-261 development is complete and the camera is in production, with selection by the Royal Danish Air Force for its F-16 modular reconnaissance pod requirement. It is also is a candidate for the Royal Netherlands Air Force for its F-16 aircraft.

Contractor

Recon/Optical, Inc.

UPDATED

CA-265 Millenium IR framing camera

The Ca-265 Millenium IR framing camera is designed for low to medium altitude operation in high-speed tactical aircraft. It incorporates ROI's highest performance 1998 technology, as an intended replacement camera for earlier conventional, down-looking Infra-Red Line Scanners (IRLS). It features vertical, forward-oblique, or side-oblique mounting. Design characteristics include: a very large frame of view; 1,968 × 1,968 Platinum Silicide (PtSi) (3-5 micron) focal plane array with patented on-chip Forward Motion Compensation (FMC); near-realtime data availablility with digital output; high resolution, long range IR

Recon/Optical CA-265 Millenium IR framing camera ***1999***/0051039

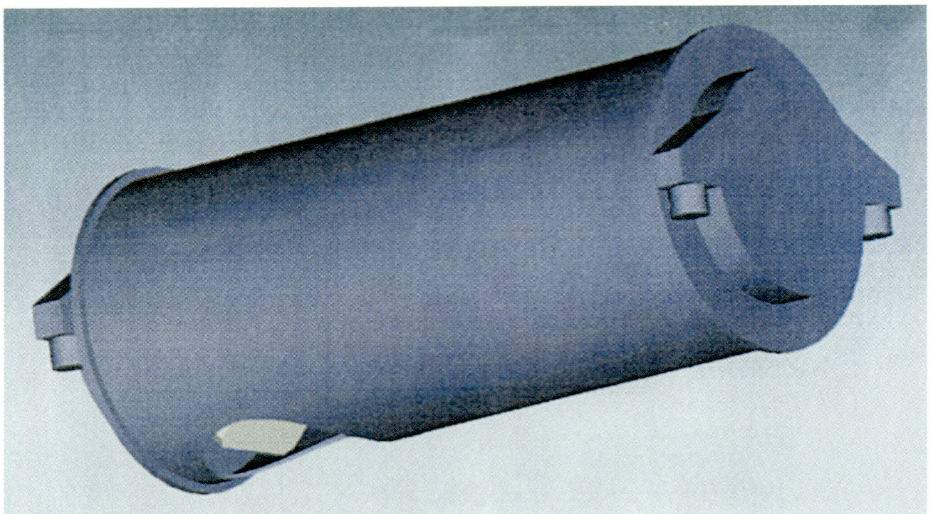

The Recon/Optical CA-270 dual-band framing camera ***2000***/0079254

imaging; electronic exposure control for wide dynamic range and imaging during manoeuvre.

Specifications

Aircraft speed: 100-580 kt
Aircraft altitude: 500-50,000 ft (200 ft with optional lens)
Angular coverage:
6 in fl lens: 21.90°
12 in fl lens: 11.96°
Operating waveband: 3-5 microns
Focal plane array: 1,968 × 1,968 PtSi
Frame rate: 2.5 frames/s
Exposure time: variable
Recorder interface: Ampex DCRsi-240 or solid-state recorder
Recorder output format: ROI standard, non-proprietary with annotation and digital sub-stamped image support
Video output (optional): RS-170
Control interface: RS-232 serial port
Dimensions:
ISU without lens: 312.4 × 325.1 × 368.3 mm (L × W × D)
VDPU: 558.8 × 259.1 × 292.1 mm (L × W × D)
power supply: 345.5 × 302.3 × 91.5 mm (L × W × D)

Operational status

Flight tested by the US Navy on a P-3 aircraft in October 1998. On order by the US Naval Research Laboratory, for delivery in January 2000.

Contractor

Recon/Optical Inc.

VERIFIED

CA-270 dual-band framing camera

The CA-270 dual-band camera captures overlapping, full-frame imagery using either visible or infra-red light, separately or simultaneously.

The CA-270 combines digital framing technology with patented electronic, on-chip graded image motion compensation to produce high-resolution reconnaissance imagery. Two imaging modules, one for the visible and one for the infra-red spectrum, provide the dual-band imaging capability. The images can be recorded on-board the aircraft on a digital data recorder, or transmitted directly to a ground station via a datalink.

Specifications

Operating spectra:
(visible/near infra-red): IR510-900 nm
(infra-red): 3.0-5.0 microns
Lens type:
(both spectra): refractive
(obscuration): none
Lens focal length and/number:
(visible spectrum): 9.0 in (228.6 mm), /5.0
(infra-red spectrum): 12.0 in (304.8 mm), /3.5
Detector arrays:
(visible spectrum): 25.4 megapixel Silicon CCD, squared format
(infra-red spectrum): megapixel array Indium Antimonide (InSb), square format
Operating modes: visible alone, infra-red alone, visible/infra-red simultaneously
Field-of-view/frame:
(visible spectrum): 12.6 × 12.6°
(infra-red spectrum): 8.5 × 8.5°
Frame rates: both spectra: up to 25 frames/s
Scan coverage rates:
(visible alone): 30°/s XLOF
(infra-red spectrum): 20°/s XLOF
(infra-red interim): 20°/s XLOF
Grid performance (all IIRS - 5):
(visible spectrum): 3.5 ft at 6.5 n mile range
(infra-red spectrum): 3.5 ft at 4.0 n mile range
(infra-red interim): 3.5 ft at 2.9 n mile range
V/H rates (10% forward overlap):
(visible alone): 0.05 rad/s max
(infra-red spectrum): 0.33 rad/s max
(simultaneous): 0.33 rad/s max
FMC: both spectra: on-chip, graded
Stabilisation: two-axis stabilised, roll and azimuth, roll axis <1.0 mr/s residual against a 25/s roll disturbance

Operational status

The CA-270 is designed to meet the requirements of the US Navy's Super Hornet Shared Reconnaissance Pod (SHARP) programme. In June 1999, Recon/Optical, Inc was awarded a contract by the US Naval Research Laboratory (NRL) to fabricate, test, deliver and flight test the CA-270 camera within 12 months.

Contractor

Recon/Optical, Inc.

VERIFIED

KS-87 reconnaissance camera

The KS-87 has interchangeable lens cones of 3, 6, 12 and 18 in focal lengths. The camera incorporates automatic exposure control and forward motion

Recon/Optical Inc KS-87 reconnaissance camera 0018316

compensation and is capable of 6 frames/s. It is the world's most widely used camera with over 4,000 cameras built to date.

Operational status

Installed in the RF-4, RF-5 and many other reconnaissance aircraft. An upgraded version has been chosen for the Open Skies treaty verification programme.

Contractor

Recon/Optical, Inc.

VERIFIED

KA-91 panoramic camera

The KA-91 is an 18 in focal length panoramic camera. It produces large-scale, wide-angle coverage photographs. The KA-91 offers a combination of features which make it ideal for medium-altitude reconnaissance. Using electronic synchronisation of all components, the camera is of modular design to simplify maintenance.

An important advantage of the KA-91 is sector scan. The camera can be set up for different across-track scan angles. The quality of the photography produced is enhanced by built-in roll stabilisation. To eliminate image blur arising from motion of the platform, the KA-91 uses translating lens forward motion compensation.

Specifications

Weight: 76 kg including magazine, film and camera body
Power: 28 V DC, 115 V AC, 400 Hz, 3Ø
Lens: 18 in (450 mm)
Film:
5 in (127 mm) aerial roll film
1,000 ft of 2.5 mm film; 700 ft of 4 mm; 500 ft of 5.2 mm

Operational status

Currently in production and in use. An upgraded version of the KA-91 is being used in the US Open Skies treaty verification programme.

Contractor

Recon/Optical, Inc.

VERIFIED

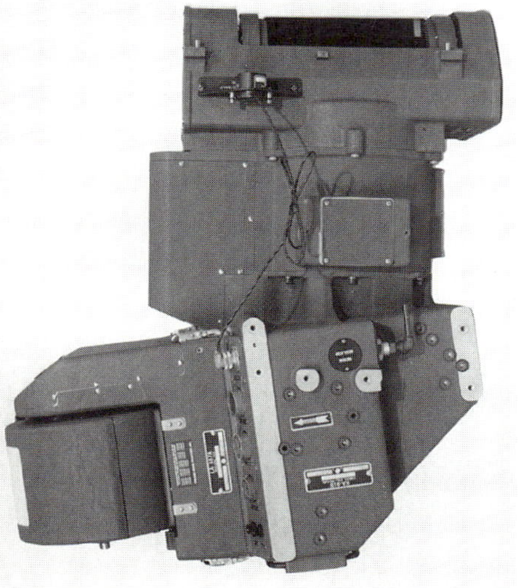

Recon/Optical Inc KA-91 panoramic camera 0018319

KA-93 panoramic camera

The KA-93 camera is a compact, prism scanning panoramic camera, which uses a 24 in f5.6 lens, for mounting internally in reconnaissance aircraft, reconnaissance pods or UAVs. It offers a combination of features which makes it ideal for medium-altitude reconnaissance.

An important advantage of the KA-93 is sector scan. The camera can be set up for six different across-track scan angles, each providing large-scale wide-angle, equivalent to the coverage provided by six framing cameras. The quality of the photography produced is

Recon/Optical Inc KA-93 panoramic camera 0018318

enhanced by built-in roll stabilisation. To eliminate image blur arising from motion of the vehicle, the KA-93 uses translating lens forward motion compensation. Using electronic synchronisation of all components, the camera is of modular design to simplify maintenance.

Specifications

Weight: 95.26 kg
Power supply: 115 V AC, 400 Hz, 3 phase, 700 VA
28 V DC, 3 A
Lens: 23 in (609.6 mm) focal length, f/5.6
Film: 5 in (127 mm) aerial roll film, 1,000 ft (305 m)

Operational status

In production and in service.

Contractor

Recon/Optical, Inc.

VERIFIED

KA-95 panoramic camera

The KA-95, Recon/Optical's most compact panoramic camera using 5 in (127 mm) film, is ideal for medium-altitude reconnaissance. Mounted internally in manned reconnaissance aircraft or externally in a pod, the camera utilises a 12 in lens which produces large-scale

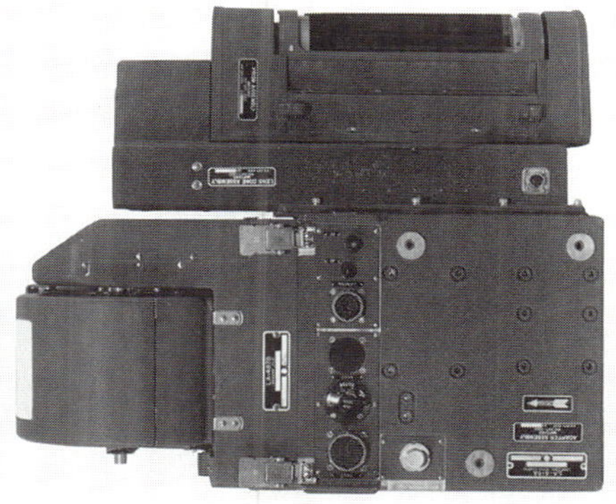

Recon/Optical Inc KA-95 panoramic camera 0018317

wide-angle photographs of up to 190° coverage, each being equivalent to the coverage provided by six framing cameras. Using electronic synchronisation of all components, the camera is of modular design to simplify maintenance.

An important advantage of the KA-95 is sector scan. The camera can be set up for six different across-track scan angles. The quality of the photography produced is enhanced by built-in roll stabilisation. To eliminate image blur arising from motion of the platform, the KA-95 uses translating lens forward motion compensation.

The KH-95B is designed for the RF-5E.

Specifications

Weight: 63.5 kg including film
Power supply: 115 V AC, 400 Hz, 3 phase;
28 V DC
Stabilisation: up to 10°/s roll
Lens: 12 in (304.8 mm) focal length, f/4
Film: 5 in (127 mm) aerial roll film, 1,000 ft 2.5 mil film

Operational status

In production and in service. The KA-95 is currently being used by a participant in the verification of the Open Skies treaty.

Contractor

Recon/Optical, Inc.

VERIFIED

KS-127 reconnaissance camera

The KS-127 is a reconnaissance camera mounted in the nose of an aircraft designed for Long-Range Oblique Photography (LOROP) missions. The film-only version is designated KS-127A, while the KS-127B is both film and E-O (electro-optic) capable. It has a 66 in (167 cm) focal length lens system, active and passive stabilisation, autofocus and thermal stabilisation.

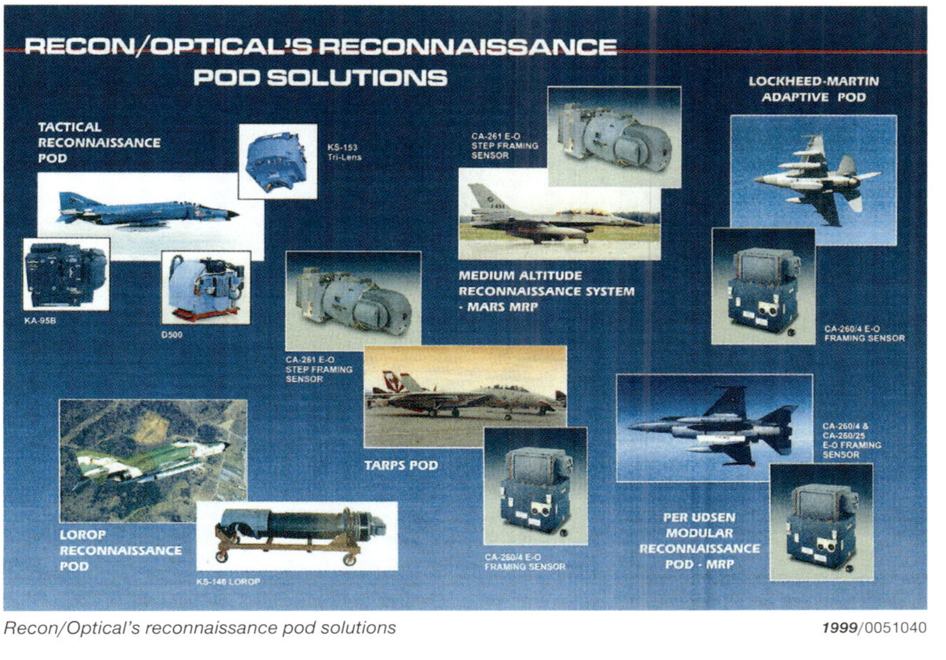

Recon/Optical's reconnaissance pod solutions ***1999***/0051040

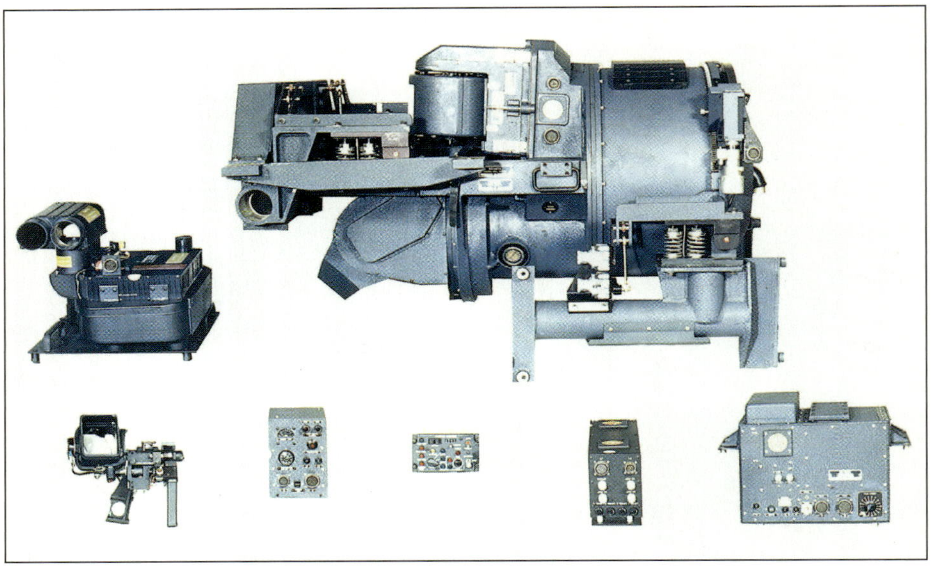

The KS-127B camera is carried by the RF-4 0106326

Typically, the KS-127B can take detailed photographs of sites over 35 km distant from an aircraft patrolling at 35,000 ft. This coverage permits missions to be conducted from safe, long-range, standoff distances. These missions eliminate the need for overflights and thereby avoid violation of foreign airspace. Aircraft survivability is also greatly improved.

Specifications

Power: 115/208 V rms, 400 Hz, 700 V A
28 V, 280 W
Lens: 1,676 mm fl, f/8.0
Film: 127 mm aerial roll film
1,000 ft of 2.5 mm film; 700 ft of 4 mm; 500 ft of 5.2 mm

Operational status

In production and service. It is currently operational in the RF-4, where a sight in the rear cockpit enables the crew to aim the camera directly at a specific target. The camera can also be upgraded to E-O operation.

Contractor

Recon/Optical, Inc.

VERIFIED

KS-146 reconnaissance camera

The KS-146 reconnaissance camera uses a 66 in focal length, f/5.6 lens to provide Long-Range Oblique Photography (LOROP). KS-146 LOROP cameras are designed for pod application. The KS-146 cameras currently in use are film cameras. Target coverage, on either side of the aircraft, can be selected from the cockpit. Typically, the KS-146 camera takes detailed pictures of sites to either the left or right side of the aircraft from 18.5 km to over 92.6 km distant from an aircraft flying at 30,000 ft or higher.

A programme is now under way to modify KS-146 film cameras to E-O cameras by using a CCD sensor array in place of the photographic film. This modification provides LOROP imagery in a digital form that can be transmitted to a ground station in real time and/or recorded on digital videotape on board the aircraft.

Specifications

Weight:
(camera) 317 kg
(total system) 385 kg
Power: 115/208 V rms, 400 Hz, 1,000 V A;
28 V, 300 W
Lens: 1,676 mm fl, f/5.6
Film: 127 mm roll film; 1,000 ft of 0.0025 in film

Operational status

In production and service. KS-146 cameras are in use by customers in Europe and Asia. These cameras can also be upgraded to E-O operation.

Contractor

Recon/Optical, Inc.

UPDATED

KS-147 reconnaissance camera

The KS-147 reconnaissance camera uses a 66 in focal length f5.6 lens to provide Long-Range Oblique Photography (LOROP). Target coverage, on either side of the aircraft, can be selected from the cockpit. The camera incorporates automatic exposure control, passive and active stabilisation, auto-focus and thermal stabilisation. Typically, the KS-147 takes detailed pictures of sites to either the left or right side of the aircraft from 18.5 km to over 92.6 km distant from an aircraft flying at 30,000 ft or higher.

A programme is now under way to modify the KS-147 to E-O camera configuration by using a CCD sensor array in place of photographic film. This modification provides LOROP imagery in a digital form that can be transmitted to a ground station in real time and/or recorded on digital tape on board the aircraft.

Specifications

Weight:
(camera) 245 kg
(total system) 270 kg
Power supply: 115/208 V AC, 400 Hz, 700 VA;
28 V DC, 280 W
Lens: 66 in (1,676.4 mm) focal length, f/5.6
Film: 5 in (127 mm) roll film, 1,000 ft of 0.0025 in film

Operational status

The KS-147 cameras presently in use are film cameras, designed for use in the RF-5E. They can also be upgraded to E-O operation

Contractor

Recon/Optical, Inc.

VERIFIED

KS-157 reconnaissance camera

The KS-157 cameras presently in use are film cameras utilising a 66 in focal length f5.6 lens designed for a business jet. The system incorporates mounted sights that are utilised to point the camera. Target coverage on either side of the aircraft can be selected from the cabin. The camera has automatic exposure control, passive and active stabilisation, auto-focus and thermal stabilisation. Missions are flown at a standoff distance of 10 to 140 km at aircraft speeds of 480 kt and altitudes of up to 50,000 ft. Detection of 3 ft-size ground objects can be achieved at 60 n miles.

A programme is now under way to modify these cameras to E-O cameras by using a CCD sensor array in place of photographic film. This modification provides LOROP imagery in a digital form that can be transmitted to a ground station in real time and/or recorded on digital videotape on board the aircraft.

Specifications

Weight:
(camera) 317 kg
(total system) 385 kg
Power supply: 115/200 V AC, 400 Hz
28 V DC
Lens: 66 in focal length, f/5.6
Film: 5 in (127 mm), 1,000 ft (305 m) capacity
Format: 114 × 114 mm

Operational status

In production and in service. They can also be upgraded to E-O operation.

Contractor

Recon/Optical, Inc.

VERIFIED

The Recon/Optical KS-157 reconnaissance camera for business jets 0106416

KS-146 (bottom) and KS-147 (top) Long-Range Oblique Photography cameras 0106327

CDU-900 Control Display Unit

The CDU-900 retains all the features of the CDU-800 series CDU/CDNU, while providing powerful new capabilities. Through utilisation of an Intel 80486 microprocessor and large-scale integration, the CDU-900 provides powerful built-in processing capability and expansion capacity for embedded functions or interfaces to external equipment.

An embedded military P(Y) code GPS receiver/processor is the first expansion module for the CDU-900. The embedded receiver consists of a GPS Embedded Module (GEM) GEM II or GEM III receiver/processor and accompanying adapters designed to fill two of the three available expansion slots in the CDU-900. Additional expansion modules planned for the CDU-900 include processor and memory modules, various digital, analogue or discrete interface modules for non-MIL-STD-1553B applications and modem/datalink modules. The standard and expansion capabilities of the CDU-900 make it suitable for the integration and control of avionics in both fixed-wing aircraft and helicopters.

Operational status

The CDU-900 with embedded GPS is integrated with the FMS-800 flight management system in the US Air Force C-5, C-9, E-4B and KC-10. The CDU-900, less the embedded GPS, is integrated in the US Air Force B-1B and KC-135. The CDU-900 is also being fitted to the US Army CH-47D, as part of the upgrade to CH-47F standard.

Latest variant of the CDU-900 family is the CDU-900G which provides a complete capability to meet US DoD GPS Integration Guidelines (GIG) for stand-alone GPS navigation, including RNAV and airways.

The CDU-900 is also manufactured in Turkey, by Aselsan Inc.

Contractor

Rockwell Collins.

UPDATED

The Rockwell Collins CDU-900 0106328

Control Display Navigation Unit (CDNU)

The Control Display Navigation Unit (CDNU) is a high-power floating point computational engine, keyboard and display system in a single product suitable for a wide variety of integrated flight management and cockpit management applications. The existing Ada software provides a full-feature lateral navigation implementation integrating GPS and air data sensors. Memory and throughput reserves are large, providing significant opportunities for growth. All application software resides in EEPROM, permitting updated software to be installed via the bus interface on the aircraft. There are seven function keys that can be user defined for specific program needs.

Current functions include flight planning with automatic sequencing, guidance computations with associated display driver, intercept of moving waypoints, integrated GPS/air data/heading navigation, internal non-volatile database containing 200 waypoints and access to online identifier database of 20,000 waypoints accessible from a data loader cartridge, integrated control and display of system tests and status, and software updates via the databus with no unit disassembly.

Specifications

Dimensions: 181 × 146 × 165.1 mm
Weight: 4.54 kg (max)
Power supply: 16-32 V DC, 30 W (max)

Operational status

In production for more than 20 US Air Force, Navy, Marine Corps, Army and Coast Guard aircraft types. Deliveries began in 1992.

Contractor

Rockwell Collins.

VERIFIED

MFD-68S Multifunction Display

The MFD-68S offers user-defined graphics for primary flight displays, engine instruments, synoptics, text, fuel management, targeting reticles and automated checklists. The 6 × 8 in (152 × 203.2 mm) active matrix liquid crystal display presents video, graphics, video with graphics overlay, split-screen video/graphics and split-screen graphics/graphics.

The MFD-268 series is offered in either a 'smart' or 'video' display to support a wide variety of architectures. The display contains internal graphics generation, analogue and digital input/output interfaces and a MIL-STD-1553B interface. The operational software can be modified over the MIL-STD-1553B bus without removal of the display from the installed position.

The MFD is programmable in the Ada language for local display format generation. A high-level graphics language simplifies tailoring of display formats.

MFD-68S multifunction display 0106329

Specifications

Dimensions:
254 (depth) × 264 (width) × 208 mm (length)
Weight: 7.71 kg nominal
Display: 152 × 203.2 mm active area

Operational status

Selected for the avionics upgrade of 58 Black Hawk helicopters of the Turkish Armed Forces, together with a dual Flight Management System (FMS).

Rockwell Collins has developed a common architecture configuration applicable to a variety of platforms including the CH-53 and the S-92.

Contractor

Rockwell Collins.

UPDATED

MFD-268S colour MultiFunction Display

The MFD-268S display presents a high-fidelity digital map and graphical data on an 8 × 6 in display surface. The 1,024 by 768 pixel resolution display provides the fidelity to crisply identify key targeting information, lethality zones and other critical data in a highly dynamic and stressed operating environment.

Operational status

The Rockwell Collins MFD-268S display started demonstrations on the Boeing B-1B in December 1998 with flight tests in 1999. The display will support the Boeing programme 'Target Planner', which enhances the ability of the crew to react to real-time events via data-linked information. This enhanced situational awareness will increase crew safety and survivability by, for example, providing the ability to adjust and re-plan the mission for new threats or target information while in flight.

In addition to this fixed-wing application, this family of displays is also being utilised on the US Army CH-47 ICH helicopter and US Marine Corps H-1 helicopter upgrades.

The MFD-268 is manufactured in Turkey, by Aselsan Inc, under the designation MFD-268E.

Contractor

Rockwell Collins.

VERIFIED

SAT-2000 Aero-I satellite communications system

Rockwell Collins' SAT-2000 Aero-I satellite communications system is designed for use with the INMARSAT Aero-I satellite service. The SAT-2000 system has a 25 W High-Power Amplifier (HPA), which

The Rockwell Collins MFD-268S colour multifunction display **1999**/0051702

will support up to six channels (one data and five voice) within the spot beams. The SAT-2000 takes advantage of the spot beam feature of the Aero-I satellites using a system which provides the benefits of telephony, fax, and real-time data communications. The 4.8 kbits/sec CODEC supports two-way voice and data traffic. Aero-I is intended for medium- and short-haul missions, but is also suitable for those long-haul missions which fly mainly within the spot beam coverage. Emergency voice service in the global beam is supported by the SAT-2000 system, (pending INMARSAT approval).

The SAT-2000 system is comprised of the SRT-2000 Satellite Data Unit (SDU) and IGA-2000 Intermediate Gain Antenna (IGA).

The SRT-2000 is a multichannel receiver-transmitter providing both voice and data channels. The SRT-2000 transmits and receives packet-mode data to/from the datalink system (CMU-900) and receives and transmits circuit mode data (voice) in analogue form to/from the flight crew headphones and microphones (via the audio management unit). Additionally, the SRT-2000 receives and transmits circuit mode data (voice, FAX, or PC MODEM) from passengers via the Cabin Telephone Unit (CTU). Thus, the SRT-2000 combines the functions of the HPA, RF Unit (RFU), Beam Steering Unit (BSU), and Satellite Data Unit (SDU) into a single LRU.

The IGA-2000 IGA allows the SAT-2000 to transmit and receive low-speed ACARS data and high-speed voice traffic efficiently. The diplexer/low-noise amplifier is integrated into the antenna to improve system performance and reduce the number of LRUs. The IGA-2000 IGA is a 6 dBi, top-mounted, electronically steered phased array antenna. The antenna complies with the INMARSAT Aeronautical System Definition Manual (SDM) requirements for intermediate gain service. The Aero-I antenna provides BITE information to the SRT-2000 and the SRT-2000 provides beam pointing information to the Aero-I antenna. The smaller size of the intermediate gain antenna in addition to the integration of the diplexer/low-noise amplifier into the antenna allows SATCOM systems to be installed in smaller aircraft.

The system is also marketed for use in long-range corporate aircraft by Rockwell Collins Business and Regional Systems, where the nomenclature is SATCOM 5000.

Specifications

Dimensions:
(SRT-2000) 8 MCU per ARINC 600
(IGA-2000) 48.2 × 110.2 × 12.6 mm (W × D × H)
Weight:
(SRT-2000) 15.9 kg
(IGA-2000) 8.2 kg
Power supply:
(SRT-2000) 115 V AC, 400 Hz or 28 V DC; 350 W (max)
(IGA-2000) 60 W
Temperature: –40 to +70°C
Altitude: up to 55,000 ft
Software: DO-178B Level D

Operational status

In production and in service on a wide variety of civilian and military aircraft.

Contractor

Rockwell Collins.

UPDATED

F-15 avionics control panels

SCI designed and manufactured many of the F-15 avionics control panels including the Integrated Communications Control Panel (ICCP), the Take Command Control Panel (TCCP), the Integrated Navigation Aids Control Panel (INACP), the Identification Friend or Foe Control Panel (IFFCP), the Main Communications Control Panel (MCCP) and the Air-to-Air Interrogator Control Panel (AAICP). In addition, SCI has produced a TeleBrief Control Panel (TBCP) for the Israeli F-15. All these avionics control panels were designed into the original F-15 cockpit in 1969 and have been in service with the F-15 for many years.

ICCP

The ICCP is the heart of the system. Using dual-microprocessor control, this unit switches audio paths, controls and displays frequencies for the aircraft radios (including 40 channels of preset non-volatile memory), generates and formats synthesised voice alerting messages, generates audio warning tones and has provision for antenna selection. Built-in test circuitry is claimed to detect 95 per cent of all electronic component failures.

MCCP

A second panel, the MCCP for main communications control, is mounted on a head-up display to provide the pilot with selection of frequency, preset channels and volume control for one of the aircraft radio sets. It also has code selection controls for IFF Mode 3/A operation, a master caution annunciator and two electronic warfare warning displays.

INACP

The ILS segment of the INACP utilises a six wire parallel output to select the ILS frequency. Controls for the ILS operating frequency are adjusted from 108.1 to 111.95 MHz. The Tacan segment utilises a serial digital data train to control the R/T channel and mode of operation. Two rotary knobs control the Tacan channel selection from 0 up to 126. X-Y selection is provided by a separate switch. A volume control is provided to control the identification tone audio. Concentric to the volume switch is a three position mode control selector which selects receive only, receive and transmit or air-to-air modes of operation.

IFFCP

The IFFCP contains three printed circuit card assemblies, controls and indicators necessary to control the functions and modes of the IFF transponder, the IFF reply evaluator and the transponder computer. The master mode control allows selection of standard or low transponder sensitivities or activates all modes in an emergency and controls and displays Mode 1 codes. It has toggle switches for enabling individual IFF modes, a Mode 4A, 4B or out switch, a Mode 4 reply select control which disables the reply signal and routes it through the audio system or display lamp, a lamp indicating a correct reply from an interrogated aircraft and a Mode 4 code hold/zero switch.

SCI avionic control panels. Top row (left to right): F-15 ICCP, F/A-18 ICS, F-15 ICCP. Middle row (left to right): F-15 TCCP, F-15 INACP, F-15 IFFCP. Bottom row (left to right): F-15 TBCP, F-15 MCCP, F-15 AAICP 0106413

AAICP

The AAICP contains the controls necessary to command the functions of the aircraft IFF evaluator and the interrogator computer. The master control switch provides automatic, normal and correct code challenges. The mode switch selects AAI Modes 1, 2, 3, 4A and 4B and an individually controlled four digit code selector.

Specifications

Dimensions:
(AAICP) 47.6 × 146 × 101.6 mm
(ICCP) 146 × 181 × 165.1 mm
(IFFCP) 85.7 × 146 × 101.6 mm
(INACP) 85.7 × 146 × 101.6 mm
Weight:
(AAICP) 0.36 kg
(ICCP) 4.4 kg
(IFFCP) 0.91 kg
(INACP) 1 kg
(MCCP) 0.95 kg

Operational status

In service in the F-15 aircraft.

Contractor

SCI Systems Inc.

VERIFIED

F-16 Voice Message Unit (VMU)

The Voice Message Unit (VMU) is a compact ruggedised voice warning system currently in service as an integral part of the F-16 avionics. The VMU is capable of monitoring twelve 28 V DC discrete and four 5 V DC differential inputs for message activation. The VMU can be arranged to react to aircraft systems sensor activity, delays, priority and a number of occurrences under software control of the VMU microprocessor. The VMU has the capacity for 12 seconds of verbal or other messages which can be broadcast over an intercom or other facilities. As the VMU is microprocessor controlled it can be adapted to monitor and provide voice/tone warnings for most aircraft and vehicle systems.

Specifications

Dimensions: 63.5 × 98.6 × 82.6 mm
Weight: 0.63 kg
Power: (max) 4 W

Operational status

The VMU is currently in service on the F-16 and interfaces with the avionic systems to provide the pilot with an audible indication of aircraft system status which has exceeded preset parameters.

Contractor

SCI Systems Inc.

VERIFIED

Tiger-Paws weapon system upgrade

Sierra Research has developed the Tiger-Paws upgrade for the F-5A/B/E/F avionics refurbishment. This integrates major avionics enhancements by replacing outdated analogue electronics with digital equipment already in use in many US and European military aircraft. Key items in the digital core include the MIL-STD-1553B databus, a smart Head-Up Display/ Weapon Aiming Computer (HUDWAC), ring laser gyro inertial navigation system and the standard central air data computer. For the radar-equipped F-5E/F, the digital core can be augmented by replacing the analogue pulse radar with a pulse Doppler coherent radar modified for compatibility with the aircraft.

Growth potential afforded by the digital system facilitates integration of a wide variety of options such as video, data transfer, stores management, radar warning receiver, multifunction displays, laser ranger and designator, FLIR, ECM, HOTAS, chaff and flares, Sidewinder control and incorporation of existing systems into a reliable modern digital system enhanced by very high MTBF.

Cockpit design can emulate the F-16 or other specified aircraft for logistics, training and operational compatibility. Installation of the Tiger-Paws digital system can be accomplished locally with kits, if required.

Similar digital enhancements are available for the T-38 Talon supersonic trainer.

Operational status

In 1991, Sierra Research received a contract, valued at US$20.7 million, to modify 15 Royal Norwegian Air Force F-5A/B aircraft with the Tiger-Paws digital core and other enhancements. This contract was completed in August 1994 and the aircraft are currently in service as F-16 lead-in trainers.

Contractor

Sierra Research, a division of SierraTech, Inc.

VERIFIED

Pylon Interface Unit (PIU)

The PIU supplies the weapons control and management functions for stores recognition, target acquisition and tracking, weapons selection, aiming, fuzing, and release for the Longbow Apache. The Load Maintenance Panel provides the ground maintenance crew with the capability to position the pylons for loading, simulation of airborne conditions for aircraft ground testing, and entry of rocket type loaded within each zone of the rocket pod.

The PIU is an intelligent MIL-STD-1553B remote terminal with a dedicated MIL-STD-1750 processor programmed with Ada software. It consists of six electronic subassemblies designed to reduce supportability costs.

Operational status

Smiths Industries Aerospace has been awarded two contracts by Boeing Helicopter Systems, Mesa, Arizona as part of the Apache modernisation programme. The contracts include supplying four Pylon Interface Units and one Load Maintenance Panel for each US Army Apache to the AH-64D configuration.

Contractor

Smiths Industries Aerospace.

VERIFIED

Smiths Industries Aerospace upgrades the US Army Apache to the AH-64D configuration providing the weapons control and management system function
0106331

AN/ASN-165 radar navigation data display set

The AN/ASN-165 is a stand-alone radar indicator system that upgrades the displays on older radars. It replaces the display subsystem on ground mapping, weather avoidance and navigation radars such as the APN-59, APS-133 and APQ-122. The aim is to improve reliability, maintainability and operational performance without replacing the entire radar. The system is designed as a drop-in replacement for the radar displays with little modification to existing structures and cabling in the aircraft. The radar's antenna and receiver/transmitter subsystems remain unchanged.

The AN/ASN-165 consists of a radar data converter, pilot indicator and navigator indicator/control panel. The radar data converter contains two independent digital scan converters that convert the radar data into a high-resolution raster video signal. Navigational data is overlayed on the radar imagery and displayed on the pilot and navigator colour indicators. Aircraft navigational systems are interfaced through a standard ARINC serial bus or dual-redundant MIL-STD-1553B databus.

The AN/ASN-165 radar navigational data display set showing (left to right) the radar data converter, navigator indicator/control panel and pilot indicator
0106332

Radar ground map, terrain-avoidance and weather imagery are typical displays. The imagery is displayed with 16 levels of shading using standard RS-170 or RS-343 video format. Multiple colours are used to indicate different weather intensities. The standard video format also permits recording of the radar imagery for mission review and training.

With the navigational interface, aircraft data, such as true heading, groundspeed and track angle error, is displayed on both the navigator's and the pilot's display. The aircraft's present position is also displayed and the data is used to calculate the position of ground targets identified by a movable cursor that can track a fixed point on the ground. The navigational interface also provides the option of stabilising the radar display to true north.

Flight plan waypoints and navigational aids can be displayed, depending on the navigation system used. Additionally, data from EW systems, such as detected threats and threat zones, can be overlayed on the radar image.

Operational status

The AN/ASN-165 is installed on the WC-130, C-130 and US Air National Guard KC-135. Variants of the system are used on the MC-130E, HC-130P/N, C-141B, RC-135 and other Special Operations and reconnaissance aircraft.

Contractor

Systems Research Laboratories.

VERIFIED

Colour AMLCD multifunction indicators

Universal Instrument Corporation manufactures a range of AMLCD, full colour, NVIS compliant, MultiFunction Display (MFD), military qualified, indicators; leading specifications are outlined in the table:

Operational status

In production and in service in military aircraft. In February 1998, Universal Instrument Corporation, as a subsidiary of Universal Avionics Systems Corporation (UASC), entered into a business alliance agreement whereby UASC will have exclusive rights to market flat-panel integrated displays manufactured by Universal Instrument Corporation to corporate and commercial aviation. The displays will be integrated with Universal flight management systems to produce a new line of flat-panel flight displays (4 × 5, 5ATI, 5 × 6 and 8 × 10 in sizes) for a new-generation of avionic suites. The design is intended for installation in new and retrofit applications, displaying primary flight and navigation data, multifunction data and engine data.

Contractor

Universal Instrument Corporation.

VERIFIED

Specifications

	Model 104	**Model 550**	**Model 570**	**Model 640**
Dimensions	8.3 × 6.2 in	4.4 × 3.3 in	4.5 × 3.5 in	5.1 × 3.8 in
Design purposes	split screen: HSI, ADI, video engine data warnings	HSI, ADI, video	EFI, HSI, ADI	HSI, ADI, video
Grey scales	64	256		64
Pixels	640 × 480 standard 800 × 600 available 800 × 600 available	640 × 480	640 × 480	640 × 480
Cooling	fan-forced cold wall construction	passive	passive	fan-forced cold wall construction
MTBF	7,500 h	3,200 h	8,000 h	8,200 h

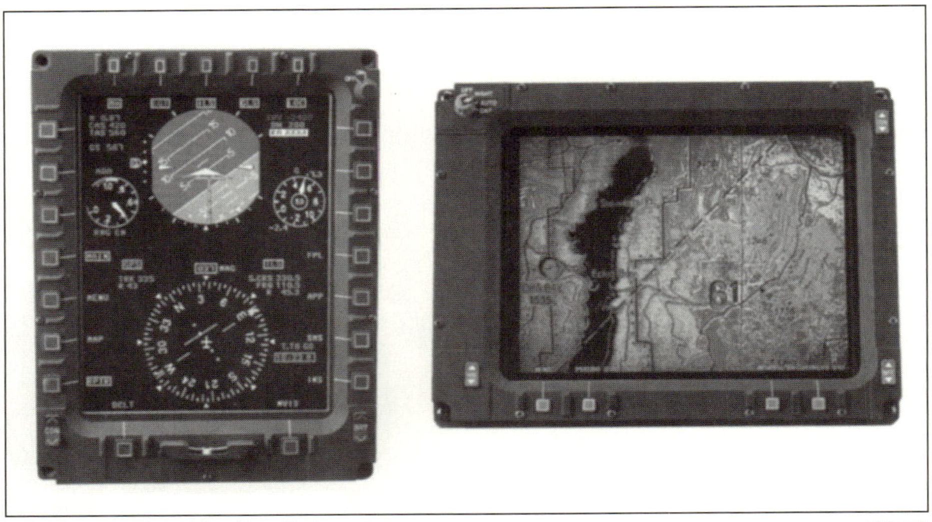

Model 104 AMLCD, MFD, video display unit 0018176

Portuguese Air Force C-130H cockpit upgraded with six Universal Instrument Corporation AMLCD indicators
0018175

Laser radar technology programmes

Laser radar ('ladar') offers benefits in ranging, imaging, and automatic target classification and recognition. The ladar functions best when used in conjunction with a wide area search sensor such as a radar or infra-red detection system. Initial efforts exploited the CO_2 technology for imaging, Doppler measurement and homing applications.

More recently, advanced solid-state laser radar systems at 1,064 nm have been investigated for their compatibility with laser-guided bomb applications, while transmitters with wavelengths greater than 1,400 nm are of interest because of the increase in eye safety associated with those wavelengths. Methods to derive the so-called eye-safe wavelengths from the designator wavelength are subjects of active research and show considerable potential to be implemented in low-cost, highly reliable multiple function systems. A multifunction system would be able to perform laser radar functions, such as imaging and ranging, and the designation function from a single laser source.

With the development of sources at the designator and eye-safe wavelengths comes the need to develop receivers with high sensitivity at these wavelengths. Recent advances in photoemissive detectors hold promise that image intensifier technology can be applied to imaging at wavelengths greater than 1,400 nm. Competing technologies include avalanche photodiode arrays based on InGaAs materials which are sensitive out to 2,000 nm and beyond. Such detector technologies have yet to reach the capabilities of detectors based on silicon, but they are maturing rapidly. Beyond just the detection process, there are investigations in readout electronics. Preliminary work to develop electronic readouts capable of supplying a three-dimensional image from a single laser pulse is under way.

Methods to shape and direct the transmitted and received laser energy are also of importance to laser radar systems. Typical methods include use of gimballed turrets and fast steering mirrors. More recently, methods of beam-steering and shaping based on optical phased-array technology have been investigated. Based on the same principles as phased array radar, optical phased-arrays allow the steering and generation of multiple beams from an aperture with no moving parts. One method of phase modulation that has been investigated is based on liquid crystal technology. These electronically adaptable optical phased arrays are still in their infancy, but hold promise for low-cost, lightweight apertures for multifunction systems.

Operational status

Solid-state laser and receiver component hardware is being fabricated and field tested, with plans for flight testing in the next few years. The US Air Force is continuing the development of systems which combine, in a single aperture, a ladar, and a passive search sensor. In addition, work is being initiated to develop modular, multifunction, open architecture electro-optical systems which will combine, through a few conformal apertures distributed about an aircraft, all of the electro-optical functions required.

Over the next three years, the US Air Force plans to develop high-power lasers suitable for long-range tactical applications. In addition to applications where these 1 and 2 μm lasers can be directly applied, they will form the pump supplies to create 2 to 5 μm laser radiation through the use of optical parametric amplifiers for multispectral target identification and chemical remote sensing applications.

Contractor

US Air Force Materiel Command, Wright Laboratory.

VERIFIED

Real-Time Information into the Cockpit (RTIC)

The Avionics Directorate, Wright Laboratory, Wright-Patterson AFB and US Air Force have demonstrated Real-Time Information into the Cockpit (RTIC) technologies through multiple programmes and flight demonstrations. A building-block approach has been used to develop and demonstrate successively more advanced RTIC technologies and capabilities. The US Air Force and US Navy have co-ordinated efforts; the US Navy is concentrating on the F/A-18, AV-8 and AH-1 platforms, while the US Air Force is concentrating on the F-15E and the F-117.

The RTIC technology has primarily been developed as a part of the Off-Board Targeting Experiments (OBTEX) programme. OBTEX has been designed to provide the warfighter with the latest information for precise targeting and situation awareness. This RTIC information has proved effective for aiding aircrew members in the acquisition and targeting of time-critical mobile targets.

OBTEX is part of a larger overall sensor-to-shooter concept which includes imaging sensors, ground processing and dissemination systems, advanced datalinks and various deep strike aircraft. The primary components of the OBTEX system are onboard image processing, the associated datalink to receive the off-board data and a ground processing element called a Joint Targeting Workstation (JTW). The JTW prepares the data by annotating the imagery with a transparent triangle over the selected target and attaching any necessary textual information. The imagery along with threat and weather information are put into a targeting package that is sent to the F-15E. The onboard processing consists of a ½ ATR box with five cards to decompress incoming data from the Joint Tactical Information Distribution System (JTIDS) and process it such that it can be displayed on the MultiPurpose Display (MPD) for execution by the aircrew. The JTIDS (LINK-16) datalink is crucial to the functionality of the OBTEX system.

Operational status

Recently, OBTEX was demonstrated as part of the Air Combat Command sponsored Project Strike II demonstration. This demonstration showed the ability to perform targeting against time-critical mobile targets by providing real-time intelligence data to an F-15E and an F-117A. The data consisted of imagery, threat updates and textual information used for targeting. The OBTEX technology included the ground node used to package and disseminate the targeting folder and the onboard processor to decompress the data and display it to the aircrew. The result of this demonstration was a much reduced targeting timeline for locating and prosecuting time-critical mobile targets. There will be a follow-on programme to OBTEX entitled Integrated RTIC/RTOC for Combat Aircraft (IRRCA) which will continue to develop advanced RTIC technologies in addition to developing Real-Time Information out of the Cockpit technology. Also, IRRCA will develop a system solution which will marry emerging technologies offering increased capabilities for offensive targeting and defensive situation awareness with applications to varying mission areas.

Based on the success of the Project Strike II programme, the IRRCA project was initiated in February 1998. A successful demonstration was conducted during June 1998. This demonstration tested the ability for a low observable aircraft to perform inflight mission replanning. The demonstration was successful in showing that mission effectiveness, for a low observable aircraft, can be enhanced using off-board information. Further tests are planned under the IRRCA programme to include the ability to send information (sensor images and mission reports out of the cockpit to command and control nodes). In addition, a major goal of the IRRCA programme is to demonstrate an interoperable communications architecture.

Contractor

US Air Force Materiel Command, Wright Laboratory.

VERIFIED

Contractors

AUSTRALIA

ASTA Components
226 Lorimer Street
Port Melbourne
Victoria 3207
Australia
Tel: (+61 3) 96 47 34 49
Fax: (+61 3) 96 46 22 53

BAE SYSTEMS Australia Ltd
Level 3, Westfield Towers
100 William Street
Sydney
New South Wales 2011
Australia
Tel: (+61 2) 93 58 29 00
Fax: (+61 2) 93 58 48 16
e-mail: sowens@baea.com.au
Web: http://www.baesystems.com

BELGIUM

BARCO n.v., Display Systems - Avionics Division
Theodoor Sevenslaan 106
B-8500 Kortrijk
Belgium
Tel: (+32 56) 23 32 11
Fax: (+32 56) 23 34 60
e-mail: avionics.bds@barco.com
Web: http://www.barco.com/display

FN Herstal S.A.
33 Voie de Liège
B-4040 Herstal
Belgium
Tel: (+32 4) 240 81 11
Fax: (+32 4) 240 86 79
e-mail: info@fnherstal.com

OIP Sensor Systems
Westerring 21
B-9700 Oudenaarde
Belgium
Tel: (+32 55) 33 38 11
Fax: (+32 55) 31 68 95
e-mail: sales@oip.be
Web: http://www.oip.be

CANADA

Array Systems Computing Inc
1120 Finch Avenue West, 8th Floor
Toronto
Ontario M3J 3H7
Canada
Tel: (+1 416) 736 09 00
Fax: (+1 416) 736 47 15
e-mail: marketing@array.ca
prodinfo@array.ca
Web: http://www.array.ca

BAE Systems, Canada Inc
600 Dr Frederik Philips Boulevard
Ville Saint-Laurent
Quebec H4M 2S9
Canada
Tel: (+1 514) 748 31 48
Fax: (+1 514) 748 31 00
e-mail: line.sabourin@baesystems-canada.com
Web: http://www.baesystems-canada.com

CAE Electronics Ltd
8585 Côte de Liesse
Saint Laurent, Quebec
Montreal H4T 1G6
Canada
Tel: (+1 514) 341 67 80
Fax: (+1 514) 341 76 99
e-mail: info@cae.com
Web: http://www.cae.com

CAL Corporation
1725 Woodward Drive
Ottawa
Ontario K2C 0P9
Canada
Tel: (+1 613) 727 17 71
Fax: (+1 613) 727 12 00
e-mail: info@calcorp.com
Web: http://www.calcorp.com

Computing Devices Canada Ltd
3785 Richmond Road
Nepean
Ontario K2H 5B7
Canada
Tel: (+1 613) 596 70 00
Fax: (+1 613) 820 50 81
e-mail: info@computingdevices.com
Web: http://www.computingdevices.com

DRS Flight Safety and Communications
A DRS Technologies Inc. Company
115 Emily Street
Carleton Place
Ontario K7C 4J5
Canada
Tel: (+1 613) 253 30 20
Fax: (+1 613) 253 30 33
Web: http://www.drs.com

Honeywell Aerospace Canada
240 Attwell Drive
Etobicoke
Rexdale
Ontario M9W 6L7
Canada
Tel: (+1 416) 675 14 11
Fax: (+1 416) 675 40 21
Web: http://www.honeywell.com/aerospace

Hughes Elcan Optical Technologies Ltd
Leitz Road
Midland
Ontario L4R 5B8
Canada
Tel: (+1 705) 526 54 01
Fax: (+1 705) 526 58 31

Litton Systems Canada
25 City View Drive
Toronto
Ontario M9W 5A7
Canada
Tel: (+1 416) 249 12 31
Fax: (+1 416) 245 03 24
Web: http://www.litton.com

Lockheed Martin Canada
6111 Royalmount Avenue
Montreal
Quebec H4P 1K6
Canada
Tel: (+1 514) 340 83 10
Fax: (+1 514) 340 84 48
e-mail: robertegery@lmco.com

MacDonald Dettweiler & Associates Ltd
Airborne Radar Division
13800 Commerce Parkway
Richmond
British Columbia V6V 2J3
Canada
Tel: (+1 604) 278 34 11
Fax: (+1 604) 278 12 85

National Research Council Canada
Flight Research Laboratory
Institute for Aerospace Research
Building U-61, Uplands
Ottawa
Ontario K1A OR6
Canada
Tel: (+1 613) 991 69 15
Fax: (+1 613) 952 72 14
e-mail: catherine.betz@nrc.ca

Northern Airborne Technology Ltd
14-1925 Kirschner Road
Kelowna
British Columbia
Canada V1Y 4N7
Tel: (+1 250) 763 22 32
Fax: (+1 250) 762 33 74
Web: http://www.northernairborne.com

Optech Inc
100 Wildcat Road
North York
Toronto
Ontario M3J 2Z9
Canada
Tel: (+1 416) 661 59 04
Fax: (+1 416) 661 41 68
e-mail: karen@optech.on.ca

Pelorus Navigation Systems Inc
5418 - 11 Street NE
Calgary
Alberta T2E 7E9
Canada
Tel: (+1 403) 730 55 55
Fax: (+1 403) 730 55 11
e-mail: spatterson@pelorus.com
Web: http://www.pelorus.com

Technisonic Industries Limited
250 Watline Avenue
Mississauga
Ontario L4Z 1P4
Canada
Tel: (+1 905) 890 21 13
Fax: (+1 905) 890 53 38
Web: http://www.til.ca

Wescam Inc
45 Innovation Drive
Flamborough
Ontario L9H 7L8
Canada
Tel: (+1 905) 689 22 31
Fax: (+1 905) 689 66 27
e-mail: snorthco@wescam.com
Web: http://www.wescam.com

CHILE

DTS Ltda
Rosas 1444
Santiago
Chile
Tel: (+56 2) 697 09 91
Fax: (+56 2) 699 33 16

CHINA, PEOPLE'S REPUBLIC

Chengdu Aero-Instrument Corporation (CAIC)
PO Box 229
Cheng Du
Sichuan 610091
China
Tel: (+86 28) 740 90 18
Fax: (+86 28) 776 94 04
e-mail: utobin@mail.sc.cninfo.net
WWW: http://www.caic-china.com

China Leihua Electronic Technology Research Institute
PO Box 3
Neijang
Sichuan 641003
China
Tel: (+86 0832) 202 38 21
Fax: (+86 0832) 202 48 22

China National Aero-Technology Import & Export Corporation
5 Liangguaching Road
PO Box 647
East City District
Beijing
China
Tel: (+86 1) 401 77 22
Fax: (+86 1) 401 53 81

China National Electronics Import & Export Corporation
Electronics Building A23
49 Fuxing Road
PO Box 140
100036 Beijing
China
Tel: (+86 10) 68 21 23 61
Fax: (+86 10) 68 22 39 16
e-mail: ceiec@ceiec.co.cn
Web: http://www. ceiec.com.cn

Southwest China Research Institute of Electronic Equipment
PO Box 429
Chadianzi Western Suburb
Chengdu 610036
Sichuan
China
Tel: (+86 28) 751 42 43
Fax: (+86 28) 751 42 43

CZECH REPUBLIC

Elektrotechnika-Tesla Kolin, a.s.
Havlickova 260
CZ-280 00 Kolin IV
Czech Republic
Tel: (+42 321) 72 46 13
Fax: (+42 321) 72 46 05

MESIT přistroje spol sro
Sokolovská 573
CZ-686 01 Uherské Hradiště
Czech Republic
Tel: (+420 632) 52 21 11
Fax: (+420 632) 55 15 46
e-mail: mo@msp.mesit.cz
Web: http://www.msp.mesit.cz

Mikrotechna Praha a.s
Barrandova 409
CZ-143 11 Prague 4
Czech Republic
Tel: (+42 2) 61 31 31 11
Fax: (+42 2) 402 56 35
e-mail: sales@mikrotechna.cz
Web: http://www.mikrotechna.com

VZLU-SPEEL Ltd
Automatic Research and Test Institute - Special Electronics
Beranovych 130
CZ-199 05 Prague 9, Letnany
Czech Republic
Tel: (+42 2) 628 16 19
Fax: (+42 2) 628 17 21
e-mail: speel@czn.cz
Web: http://www.speel.cz

DENMARK

Jorgen Andersen Ingeniorfirma A/S (JAI)
Productionswej 1
DK-2600 Glostrup
Copenhagen
Denmark
Tel: (+45) 44 91 88 88
Fax: (+45) 44 91 32 52
Web: http://www.jai.dk

TERMA Elektronik AS
Hovmarken 4
DK-8520 Lystrup
Denmark
Tel: (+45 87) 43 60 00
Fax: (+45 87) 43 60 01
e-mail: terma.hq@terma.com
Web: http://www.terma.com

TERMA Industries Grenaa AS
Fabrikvej 1
DK-8500 Grenaa
Denmark
Tel: (+45) 86 32 19 88
Fax: (+45) 86 32 14 48
e-mail: terma.ind@grenaa.terma.com
Web: http://www.terma.com

Thrane & Thrane A/S
Tobaksvejen 23A
DK-2860 Soeborg
Denmark
Tel: (+45) 39 55 88 00
Fax: (+45) 39 55 88 88
e-mail: psl@tt.dk
WWW: http://www.tt.dk

FRANCE

AATON
2 rue de la Paix
BP3002
F-38001 Grenoble Cedex
France
Tel: (+33 4) 76 42 95 50
Fax: (+33 4) 76 51 34 91
e-mail: avd@aaton.com

Aerospatiale
Division Engins Tactiques
2 rue Beranger
F-92322 Chatillan-sous-Bayneux
France
Tel: (+33 1) 47 46 21 21

Airbus Military Company
17 avenue Didier Daurat
F-31707 Blagnac Cedex
France
Tel: (+33 5) 62 11 07 82
Fax: (+33 5) 62 11 06 11
Web: http//www.airbusmilitary.com

Alkan
Rue du 8 Mai 1945
F-94460 Valenton
France
Tel: (+33 1) 45 10 86 00
Fax: (+33 1) 43 89 10 61
e-mail: alkan@compuserve.com

CEIS TM - LCD Division
4 avenue Didier Daurat
Centreda II 2e étage
BP 48
F-31702 Blagnac Cedex
France
Tel: (+33 5) 61 16 32 30
Fax: (+33 5) 61 16 32 31

Compagnie Industrielle des Lasers (CILAS)
route de Nozay
BP 27
F-91460 Marcoussis
France
Tel: (+33 1) 64 54 48 00
Fax: (+33 1) 64 54 48 19

EADS France
37 Boulevard de Montmorency
F-75781 Paris Cedex 16
France
Tel: (+33 1) 42 24 24 24
Web: http://www.eads-nv.com

EAS Europe Aero Surveillance, a subsidiary of Thomson-CSF Optrosys
Bât 45
Zone centrale
Aéroport du Bourget
BP51-93352 Le Bourget Cedex
France
Tel: (+33 1) 48 16 15 72
Fax: (+33 1) 48 16 15 73

Enertec SA
185, Avenue du Général de Gaulle
PO Box 316
F-92143 Clamart Cedex
France
Tel: (+33 1) 41 28 87 87
Fax: (+33 1) 41 28 87 00
Web: http://www.enertec.avicore.com

France Aerospace SARL
Regina-2
Place de Pyramides
F-75001 Paris
France
Tel: (+33 1) 42 60 31 10
Fax: (+33 1) 40 15 95 16

Matra BAe Dynamics SAS
20-22 rue Grange Dame Rose
BP 150
F-78141 Vélizy-Villacoublay Cedex
France
Tel: (+33 1) 34 88 30 00
Fax: (+33 1) 34 88 22 88

Monit'air
ZAC des Pres Rouz
81 rue Alain-Fournier
F-38920 Crolles
France
Tel: (+33 4) 76 08 14 39
Fax: (+33 4) 76 08 89 04

Rockwell-Collins France SA
6 avenue Didier Daurat
BP 8
F-31701 Blagnac Cedex
France
Tel: (+33 5) 61 71 77 00
Fax: (+33 5) 61 71 51 69

SAGEM SA, (Société d'Applications Générales d'Electricité et de Mécanique)
Defence and Security Division
61 rue Salvador Allende
F-92751 Nanterre Cedex
France
Tel: (+33 1) 40 70 63 63
Fax: (+33 1) 40 70 65 18
Web: http://www.sagem.com

Satori
Aéroport du Bourget
BP 151 - Batiment 66
F-93352 Le Bourget Cedex
France
Tel: (+33 1) 48 62 73 00
Fax: (+33 1) 48 64 98 56

SERPE-IESM
Société d'Etudes et de Réalisation de Protection Electronique Information Electronique Sécurité Maritime
Z.I. des Cinq Chemins
F-56520 Guidel
France
Tel: (+33 2) 97 02 49 49
Fax: (+33 1) 97 65 00 20
Web: http://www.serpe-iesm.com

Sextant
Aerodrome de Villacoublay
ZAL Bréguet
BP 200
F-78141 Vélizy Villacoublay Cedex
France
Tel: (+33 1) 46 29 70 00
Fax: (+33 1) 40 94 02 51
Web: http://www.sextant.thomson-csf.com

SFIM Industries (SAGEM Group)
13 avenue Marcel Ramolfo-Garnier
F-91344 Massy Cedex
France
Tel: (+33 1) 69 19 66 00
Fax: (+33 1) 69 19 69 19

SNECMA Control Systems
Site de Villaroche
BP 42
F-77552 Moissy Cramayel Cedex
France
Tel: (+33 1) 60 59 71 23
Fax: (+33 1) 60 59 84 44
Web: http://www.snecma.com

TEAM
10 place Vauban
Silic 127
F-94523 Rungis Cedex
France
Tel: (+33 1) 49 78 66 00
Fax: (+33 1) 49 78 66 99
e-mail: sales-marketing@team-avionics.com
Web: http://www.team-avionics.com

THALES Airborne Systems
2, Avenue Gay-Lussac
F-78851 Elancourt Cedex
France
Tel: (+33 1) 34 81 60 00
Fax: (+33 1) 34 81 52 99
Web: http://www.thalesgroup.com

Thomson-CSF Communications
See THALES

Thomson-CSF DETEXIS
See THALES

Thomson-CSF Optronique
See THALES

Thomson Marconi Sonar SAS
BP 157
F-06903 Sophia Antipolis Cedex
France
Tel: (+33 1) 92 96 30 00
Fax: (+33 1) 92 96 46 30

GERMANY

Aerodata Flugmesstechnik GmbH
Hermann-Blenk-Strasse 36
D-38108 Braunschweig
Germany
Tel: (+49 531) 235 90
Fax: (+49 531) 235 91 58
Web: http://www.aerodata.de

Bavaria Keytronic Technologie GmbH
PO Box 500272
Boschstrasse 23
D-22761 Hamburg
Germany
Tel: (+49 40) 89 69 90
Fax: (+49 40) 890 30 14

Becker Avionics Systems
Baden Air Park
Victoria Boulevard B108
D-77836 Rheinmünster
Germany
Tel: (+49) 72 29 30 50
Fax: (+49) 72 29 30 52 17
e-mail: info@becker-avionics.de
Web: http://www.becker-avionics.com

Bodenseewerk Geraetetechnik GmbH (BGT)
PO Box 101155
D-88641 Überlingen
Germany
Tel: (+49 7551) 890
Fax: (+49 7551) 89 28 22
e-mail: pr@bgt.de

DaimlerChrysler Aerospace AG
Defense and Civil Systems (Headquarters)
D-88039 Friedrichshafen
Germany
Tel: (+49 7545) 891 24
Fax: (+49 7545) 858 88
WWW: http://www.dasa.com

DaimlerChrysler Aerospace AG
Defense and Civil Systems
LFK-Lenkflugkörpersysteme GmbH
D-81663 Munich
Germany
Tel: (+49 89) 60 72 92 33
Fax: (+49 89) 60 72 55 15
WWW: http://www.dasa.com

DaimlerChrysler Aerospace AG
Military Aircraft
PO Box 80 11 60
D-81663 Munich
Germany
Tel: (+49 89) 60 70
Fax: (+49 89) 60 72 64 81
WWW: http://www.dasa.com

Diehl Luftfahrt Electronik GmbH
Fischbachstrasse 16
D-90552 Röthenbach/Pegnitz
Germany
Tel: (+49 91) 19 57 25 60
Fax: (+49 91) 19 57 24 44
e-mail: rainer.kuschel@dle.diehl.com

DLR Deutsches Zentrum für Luft- und Raumfahrt e.V.
Institut fur Hochfrequenztechnik
Oberpfaffenhofen
Postfach 1116
D-82230 Wessling
Germany
Tel: (+49 81) 53 28 23 05
Fax: (+49 81) 53 28 14 49

EADS
DaimlerChrysler Aerospace AG
Defence Electronics and Telecommunications
D-89070 Ulm
Germany
Tel: (+49 731) 392 54 16
Fax: (+49 731) 392 52 74
Web: http://www.eads-nv.com

EADS Deutschland GmbH
85521 Ottobrun
PO Box 801109
D-81663 Munich
Germany
Tel: (+49 89) 60 70
Fax: (+49 89) 60 72 64 81
Web: http://www.eads-nv.com

EADS Ewation GmbH
Wörthstrasse 85
D-89077 Ulm
Germany
Tel: (+49 731) 49 74
Fax: (+49 731) 54 34
Web: http://www.mrcm.net

ELAN Elektronische-U Anzeiger GmbH
Fritz-Ullman-Strasse 2
D-55252 Mainz-Kastel
Germany
Tel: (+49 61) 347 19 50
Fax: (+49 61) 34 17 92

ESW-Extel Systems Wedel
Gesellschaft für Ausrüstung mbH
Industriestrasse 23-33
D-22876 Wedel
Germany
Tel: (+49 41) 03 60 36 71
Fax: (+49 41) 03 60 45 03
Web: http://www.esw-wedel.de

EuroAvionics Navigationssysteme GmbH & Co
Steinegger Strasse 19
D-75233 Tiefenbronn Lehningen
Germany
Tel: (+ 49 72) 34 79 600
Fax: (+ 49 72) 34 79 112
Web: http://www.euroavionics.com

Honeywell Regelsysteme GmbH
PO Box 2010
D-63475 Maintal
Germany
Tel: (+49 61) 81 40 16 81
Fax: (+49 61) 81 40 12 89
e-mail: werner.hansli@honeywell.com
Web: http://www.honeywell.de

Liebherr-Aerospace Lindenberg GmbH
Pfaenderstrasse 50-52
D-88153 Lindenberg/Allgäu
Germany
Tel: (+49 83) 81 460
Fax: (+49 83) 81 463 77
Web: http://www.lli.liebherr.com

LITEF GmbH
Lörracher Strasse 18
Postfach 774
D-79115 Freiberg
Germany
Tel: (+49 76) 1 490 10
Fax: (+49 76) 1 490 14 80
e-mail: kempf.andrea@litef.de
Web: http://www.litef.com

MOVING TERRAIN® Air Navigation Systems AG
Moving Terrain AG
Bunzlauer Strasse 23
D-80992 Munich
Germany
Tel: (+49 89) 14 51 56
Fax: (+49 89) 14 42 96
Web: http://www.moving-terrain.de

Nord-Micro AG & Co OHG
Victor Slotosch Strasse 20
D-60388 Frankfurt am Main
Germany
Tel: (+49 69) 09 30 30
Fax: (+49 69) 09 30 32 33
e-mail: mail@nord-micro.de
Web: http//www.nord-micro.de

Peschges Variometer GmbH
Zieglerstrasse 11
D-52078 Aachen
Germany
Tel: (+49 241) 56 30 22
Fax: (+49 241) 56 39 13

Racal-Heim GmbH
Huttenstrasse 38
D-51469 Bergisch Gladbach
Germany
Tel: (+49 22) 02 93 68 30
Fax: (+49 22) 02 93 68 35

Rohde & Schwarz GmbH & Co KG
Mühldorfstrasse 15
D-81671 Munich
Germany
Tel: (+49 89) 41 29 0
Fax: (+49 89) 412 91 21 64
Web: http://www.rsd.de

SFIM Industries Deutschland GmbH (SAGEM Group)
Gottlieb-Daimler Strasse 60
D-71711 Murr
Germany
Tel: (+49 71) 448 11 40
Fax: (+49 71) 44 81 14 22

TELDIX GmbH
Postfach 10 56 08
D-69046 Heidelberg
Germany
Tel: (+49 62) 21 512 0
Fax: (+49 62) 21 512 305
e-mail: computer@teldix.de
Web: http://www.teldix.com

VDO Luftrtgeräte Werk GmbH
An der Sandelmühle 13
D-60439 Frankfurt/Main
Germany
Tel: (+49 69) 580 50
Fax: (+49 69) 580 53 99

Walter Dittel GmbH
Luftfahrtgerätebau
Erpftinger Strasse 36
D-86899 Landsberg/Lech
Germany
Tel: (+49 8191) 335 10
Fax: (+49 8191) 33 51 49
e-mail: firma@dittel.com
Web: http://www.dittel.com

Zeiss Optronik GmbH
Carl Zeiss Strasse 22
D-73447 Oberkochen
Germany
Tel: (+49 7364) 20 0
Fax: (+49 7364) 20 36 97

Z/I Imaging GmbH
Postfach 1106
D-73447 Oberkochen
Germany
Tel: (+49 7364) 20 80 02
Fax: (+49 7364) 20 29 29

INDIA

Hindustan Aeronautics Ltd
Corporate Office
PO Box 5150
15/1 Cubbon Road
Bangalore 560 001
India
Tel: (+91 80) 286 51 97
Fax: (+91 80) 286 71 40
Web: http://www.hal-india.com

ISRAEL

Astronautics C.A. Ltd
23 Hayarkon Street
PO Box 882
IL-51261 Bnei-Brak
Israel
Tel: (+972 3) 579 15 55
Fax: (+972 3) 570 44 04
e-mail: astro@astro.co.il
Web: http//www.astronautics.co.il

Controp Precision Technologies Ltd
5 Hanaggar Street
PO Box 611
Hod Hasharon
IL-45105 Israel
Tel: (+972 9) 744 06 61
Fax: (+972 9) 744 06 62
e-mail: cntrpnir@netvision.net.il
Web: http://www.controp.co.il

Elbit Systems Ltd
Advanced Technology Center
PO Box 539
IL-31503 Haifa
Israel
Tel: (+972 4) 831 53 15
Fax: (+972 4) 855 00 02
e-mail: elbit-systems@elbit.co.il
Web: http://www.elbit.co.il

Elisra Electronic Systems Ltd
48 Mivtza Kadesh Street
IL-51203 Bene Beraq
Israel
Tel: (+972 3) 617 51 11
Fax: (+972 3) 617 58 50
e-mail: misl@elisra.com
Web: http://www.elisra.com

Elop Electro Optics Industries,
an Elbit Systems Ltd company
Advanced Technology Park
KIRYAT WEIZMANN
PO Box 1165
IL-76111 Rehovot
Israel
Tel: (+972 8) 938 64 33
Fax: (+972 8) 938 62 37
e-mail: elop33@netvision.net.il
Web: http://www.elop.co.il

Israel Aircraft Industries Ltd,
Elta Electronics Industries Ltd
100, Yitzhak Hanassi Boulevard
PO Box 330
IL-77102 Ashdod
Israel
Tel: (+972 8) 857 21 55
Fax: (+972 8) 856 45 68
Web: http://www.elta-iai.com

Israel Aircraft Industries Ltd, Tamam Division
Yahud Industrial Zone
PO Box 75
IL-56100 Yahud
Israel
Tel: (+972 3) 531 53 60
Fax: (+972 3) 531 51 40
Web: http://www.iai.co.il
e-mail: infotmm@tamam.iai.co.il

Opgal Optronic Industries Ltd
PO Box 462
Industrial Centre
IL-20101 Karmiel
Israel
Tel: (+972 4) 995 39 03
Fax: (+972 4) 995 39 00
e-mail: info@opgal.com
Web: http://www.opgal.com

Rada Electronic Industries Ltd
7 Giborei Israel Street
IL-42504 Netanya
Israel
Tel: (+972 9) 892 11 11
Fax: (+972 9) 885 58 85
e-mail: nezer_rafi@rada.com
Web: http://www.rada.com

Rafael Electronic Systems Division
PO Box 2250 (80)
IL-31021 Haifa
Israel
Tel: (+972 4) 879 52 32
Fax: (+972 4) 879 40 93
e-mail: esdmarket@rafael.co.il
Web: http://www.rafael.co.il

Rafael Missile Division
PO Box 2250/30
31021 Haifa
Israel
Tel: (+972 4) 890 85 03
Fax: (+972 4) 890 62 57
Web: http://www.rafael.co.il

Rokar International Ltd
Science Based Industry Campus
Mount Hotzvim
PO Box 45049
IL-91450 Jerusalem
Israel
Tel: (+972 2) 532 98 88
Fax: (+972 2) 582 25 22
e-mail: e-mail@rokar.co.il
Web: http://www.rokar-gps.com

RSL Electronics Ltd
Ramat Gabriel Industrial Zone
PO Box 21
Migdal HaEmek
IL-10550
Israel
Tel: (+972 6) 54 75 10
Fax: (+972 6) 54 75 20
e-mail: hanan@rsl.co.il

Tadiran Electronic Systems Ltd
29 Hamerkava Street
PO Box 150
IL-58101 Holon
Israel
Tel: (+972 3) 557 72 40
Fax: (+972 3) 557 72 70
e-mail: mkt@tadsys.com.il

Tadiran Spectralink Ltd
29 Hamerkava Street
PO Box 150
IL-58101 Holon
Israel
Tel: (+972 3) 557 31 02
Fax: (+972 3) 557 31 31
e-mail: tadspl-3@netvision.net.il
WWW: http://www.yadiran-spectralink.com

ITALY

Alenia Difesa, Avionic Systems and Equipment Division
Via di S Alessandro 10
I-00131 Rome
Italy
Tel: (+39 06) 41 88 31
Fax: (+39 06) 41 88 38 00
e-mail: debenedictis@finmeccanica.it

Alenia Difesa, Avionic Systems and Equipment Division
FIAR (Fabbricaltaliana Apparecchiature
Radioelettriche)
Via GB Grassi 93
I-93-20157 Milan
Italy
Tel: (+39 02) 35 79 05 37
Fax: (+39 02) 35 79 00 74

Alenia Marconi Systems SpA
Via Tiburtina Km 12,400
I-00131 Rome
Italy
Tel: (+39 06) 41 88 31
Fax: (+39 06) 413 11 33
Web: http://www.aleniamarconisystems.com

Elettronica SpA
Via Tiburtina Km 13.700
I-00131 Rome
Italy
Tel: (+39 06) 415 41
Fax: (+39 06) 419 28 69

Elmer SpA
a Marconi Communications Company
Vialle dell' Industria 4
I-00040 Pomezia
Rome
Italy
Tel: (+39 06) 91 09 12 00
Fax: (+39 06) 91 09 14 19
e-mail: gianfranco.santoni@marconicomms.com

Italtel SpA
Via A di Tocqueville 13
I-20154 Milan
Italy
Tel: (+39 02) 438 81
Fax: (+39 02) 43 88 52 20
Web: http://www.italtel.it

Litton Italia SpA
Via Pontina Km 27.800
I-00040 Pomezia
Rome
Italy
Tel: (+39 6) 91 19 21
Fax: (+39 6) 912 25 17

Logic SpA
Via Brescia 29
I-20063 Cernusco S/N
Italy
Tel: (+39 2) 92 10 25 51
Fax: (+39 2) 92 10 25 28

Marconi SpA
via Campo nell'Elba, 3
I-00138 Rome
Italy
Tel: (+39 6) 88 69 31
Fax: (+39 6) 886 39 68

MID SpA (Marconi Italia Defence)
Via le Castello della Magliana, 75
I-00148 Rome
Italy
Tel: (+39 06) 65 96 04 51
Fax: (+39 06) 65 96 04 65

Ottico Meccanica Italiana SpA (OMI)
Via della Vasca Navale 79
I-00146 Rome
Italy
Tel: (+39 6) 54 78 81

Teleavio SrL
Via A. Negroni 1A
Genoa-Comigliano
Italy
Tel: (+39 10) 600 24 20
Fax: (+39 10) 650 84 98

JAPAN

Japan Aviation Electronics Industry Ltd
21-2 Dgenzaka 1-chome
Shibuya-Ku
Tokyo 150
Japan
Tel: (+81 3) 37 80 27 11
Fax: (+81 3) 37 80 27 33
Web: http://www.jae.co.jp

Japan Defense Agency
Boei Cho
7-45 Akasaka 9-chome
Minato-ku
Tokyo 107
Japan
Tel: (+81 3) 34 08 52 11
Fax: (+81 3) 34 08 64 80
email: info@jda.go.jp
Web: http://www.jda.go.jp

Mitsubishi Electric Corporation
Corporate Communications Dept
6-3, Marunouchi 2-chome
Chiyoda-ku
Tokyo 100-86
Japan
Tel: (+81 3) 32 10 21 21
Fax: (+81 3) 32 10 80 51

NEC Corporation
7-1 Shiba 5-chome
Minato-ku
Tokyo 108-8001
Japan
Tel: (+81 3) 34 54 11 11
Fax: (+81 3) 37 98 66 84

Shimadzu Corporation
1 Nishinokyo-Kuwabaracho
Nakagyo-ku
Koyoto 604
Japan
Tel: (+81 75) 823 11 11
Fax: (+81 75) 811 31 88
e-mail: wahasa@shimadzu.co.jp
WWW: http://www.shimadzu.co.jp/

Tokimec Inc
2-16 Minami-Kamata
Ohta-ku
Tokyo 144-8551
Japan
Tel: (+81 3) 37 32 21 11
Fax: (+81 3) 37 36 02 61
Web: http://www.tokimec.co.jp

Tokyo Aircraft Instrument Co Ltd
35-1 Izumi-Honcho 1-chome
Komae-shi
Tokyo 201
Japan
Tel: (+81 3) 34 89 11 25
Fax: (+81 3) 34 88 55 21
e-mail: kksales@mars.plala.or.jp

Toshiba Corporation
1-1 Shibaura 1-chome
Minato-ku
Tokyo 105-01
Japan
Tel: (+81 3) 34 57 31 19
Fax: (+81 3) 34 57 05 45

KOREA, SOUTH

Samsung Defense Systems
11th Floor
Samsung Main Building
250, 2Ka, Taepyung-Ro
Chung-Ku
Seoul
South Korea 100-742
Tel: (+82 2) 726 38 74
Fax: (+82 2) 726 38 28
Web: http://www.sec.co.kr

NETHERLANDS

Hollandse Signaalapparaten BV
Signaal Special Products
PO Box 241
NL-2700 AE Zoetermeer
Netherlands
Tel: (+31 79) 344 59 99
Fax: (+31 79) 344 59 38
Web: http://www.signaal.nl

NORWAY

Eidsvoll Electronics AS (EIDEL)
Gruemyra
N-2080 Eidsvoll
Norway
Tel: (+47) 63 96 42 30
Fax: (+47) 63 96 20 48
e-mail: eidel@eidel.no

Kongsberg Defence & Aerospace AS
PO Box 1003
N-3601 Kongsberg
Norway
Tel: (+47) 32 73 82 00
Fax: (+47) 32 73 85 86
Web: http://www.kongsberg.com

Navia Aviation AS
PO Box 50
Manglerud
N-0621 Oslo
Norway
Tel: (+47) 23 18 02 00
Fax: (+47) 23 18 02 10
Web: http://www.naviaav.com

POLAND

ATM Inc
Grochowska 21a
PL-04-186 Warsaw
Poland
Tel: (+48 22) 610 60 73
Fax: (+48 22) 610 41 44

PORTUGAL

CINAVE Companhia de Instrumentos de Navagação Aeronáutica, Lda
Rua do Quelhas, 27, R/c
P-1200-779 Lisbon
Portugal
Tel: (+351 1) 395 51 13
Fax: (+351 1) 390 24 74
e-mail: cinave@mail.telepac.pt

ROMANIA

Aerofina Avionics Enterprise
5 Fabrica de Glucoza Street
R-72322 Bucharest
Romania
Tel: (+40 1) 242 02 65
Fax: (+40 1) 242 09 12
e-mail: aerofina@ringier.ro

RUSSIAN FEDERATION AND ASSOCIATED STATES (CIS)

Airon Complect
Box 37
E-397
111397 Moscow
Russian Federation
Tel: (+7 095) 175 97
Fax: (+7 095) 175 97
e-mail: airon.complect@public.mtu.ru

AeroPribor-Voskhod Joint Stock Company
19 Tkatskaya Street
105318 Moscow
Russian Federation
Tel: (+7 095) 369 10 81
Fax: (+7 095) 369 76 56

All-Russia JSC Nizhegorodskaya Yarmarka
PO Box 648
420032 Kazan
Republic of Tatarstan
Tel: (+84 32) 55 71 63
Fax: (+84 32) 55 34 85

All-Russia Research Institute of Radio Equipment
Shkiperski protok 19
199106 St Petersburg
Russian Federation

AviaAvtomatika, Design Bureau of the Pribor Joint Stock Company
47 Zapolnaya st
305000 Kursk
Russian Federation
Tel: (+7 122) 264 68
Fax: (+7 122) 244 15

AviaPribor
5 Aviatzionny Per
125319 Moscow
Russia
Tel: (+7 095) 152 48 74
Fax: (+7 095) 152 26 31
e-mail: aviapribor@aviapribor.ru
Web: http://www.aviapribor.ru

Elara Cheboksary Apparatus and Instrument Plant Joint Stock Company
Moskovsky Prospect, 40
428034 Cheboksary
Russian Federation
Tel: (+7 835) 244 36 50 (Cheboksary)
Fax: (+7 835) 242 53 03 (Cheboksary)
Tel: (+7 095) 937 01 18 (Moscow)

Electroavtomatika OKB
40 Street Marshala Govorova
St Peterburg
Russian Federation
Tel: (+7 812) 252 13 98
Fax: (+7 812) 252 38 17
e-mail: alex@elavt.spb.ru

ElectroPribor Kazan Plant
20 N Ershov Street
Kazan
420045 Tatarstan
Russian Federation
Tel: (+7 8432) 76 40 01
Fax: (+7 8432) 38 89 83

ElectroPribor Voronezh Plant Joint Stock Company
20-letiya Oktyabrya 59
394006 Voronezh
Russian Federation
Tel: (+7 0732) 36 58 36
Fax: (+7 0732) 77 85 25

Elektroavtomatika
Marshala Govorova Ulitsa 40
198095 St Petersburg
Russia
Tel: (+7 812) 252 13 98
Fax: (+7 812) 252 38 17

Geophizika-NV
Matrosskaia Tishina Street
House 23, Building 2
107016 Moscow
Russian Federation
Tel: (+7 095) 269 27 42
Fax: (+7 095) 268 01 42

GosNIIAS State Research Institute for Aviation Systems
7 Victorenko Str
125319 Moscow
Russian Federation
Tel: (+7 095) 157 95 39
Fax: (+7 095) 157 01 51
e-mail: root@gosniias.msk.su
Web: http://www.gosniias.msk.ru

Joint Stock Company Pribor
47 Zapolnaya Street
305040 Kursk
Russian Federation
Tel: (+7 122) 255 72
Fax: (+7 122) 229 12

Joint Stock Company Tambovsky Zavod ElectroPribor
Morshanskoye shosse 36
392000 Tambov
Russian Federation
Tel: (+7 0752) 37 73 03

Leninetz Holding Company
212 Moskovskiy Prospect
196066 St Petersburg
Russian Federation
Tel: (+7 812) 264 32 19
Fax: (+7 812) 299 90 41

Manufacturing Stock Company AVECS
17 I-ya Ul
Yamskova Polya
Moscow
Russian Federation
Tel: (+7 812) 257 08 46
Fax: (+7 812) 257 77 32

Moscow Institute of Electromechanics and Automatics
5 Aviatsioniy Pereulok
125319 Moscow
Russian Federation
Tel: (+7 095) 152 48 74
Fax: (+7 095) 152 26 31
e-mail: comiea@adol.ru

Moscow Research-Production Complex MRPC AVIONIKA
Ul, Obraztsova, 13
103055 Moscow
Russian Federation
Tel: (+7 095) 281 33 55
Fax: (+7 095) 281 38 46

Moscow Scientific Research Institute of Instrument Engineering MNIIP
34 Kutuzov Avenue
121170 Moscow
Russian Federation
Tel: (+7 095) 249 07 04
Fax: (+7 095) 148 79 96

National Association of Avionics Producers
16 Bolshaya Monetnaya Street
197101 St Petersburg
Russian Federation
Tel: (+7 812) 233 74 59
Fax: (+7 812) 233 83 06

NAVIS
12 Tufeleva Roscha Street
109280 Moscow
Russian Federation
Tel: (+7 095) 274 63 04
Fax: (+7 095) 274 00 77

NIIAO Institute of Aircraft Equipment
Tupoleva 18
Zhukovsky-2
140160 Moscow
Russian Federation
Tel: (+7 095) 556 58 44
Fax: (+7 095) 556 23 28

NIIP State Enterprise V Tikhomirov Scientific Research Institute of Instrument Design
3 Gagarin Street
Zhukovsky
140160 Moscow Region
Russia
Tel: (+7 095) 556 23 48
Fax: (+7 095) 556 88 87

Phazotron-NIIR Scientific and Production Company
1 Electrichesky Pereulok
123557 Moscow
Russian Federation
Tel: (+7 095) 253 56 13
Fax: (+7 095) 253 04 95
e-mail: phaza@aha.ru

Pirometr St Petersburg Joint Stock Company
Bolshaya Monetnaya 16
197061 St Petersburg
Russian Federation
Tel: (+7 812) 233 74 59
Fax: (+7 812) 233 83 06
e-mail: artekpirometr@sovam.com

Polyot Research and Production Company
GSP-462
603600 Nizhny Novgorod
Russian Federation
Tel: (+7 8312) 42 21 86
Fax: (+7 8312) 42 31 57
e-mail: root@polyot.ic.sci-nnov.ru
WWW: http://www.innov.ru/polyot

Production Association RadioPribor
2 Fatkullin Street
Kazan
420022 Tatarstan
Russian Federation
Tel: (+7 8432) 37 09 01
Fax: (+7 8432) 37 40 93

Production Association Urals Optical & Mechanical Plant (PA UOMZ)
33-b Vostochnaya Street
620100 Yekaterinburg
Russia
Tel: (+7 3432) 24 18 03
Fax: (+7 3432) 24 16 80
Web: http://uomz.uralregion.ru

Ramenskoye Design Company AO RPKB
Gurjev Street 2
Ramenskoye
140103 Moscow
Russian Federation
Tel: (+7 095) 556 23 93 (Moscow)
Tel: (+7 096) 24 63 39 32 (Ramenskoye)
Fax: (+7 096) 24 63 19 72
e-mail: rpkb@space.ru

Ramensky Instrument Engineering Plant
39 Mikhalevich Street
Ramenskoye
140100 Moscow
Russian Federation
Tel: (+7 095) 501 41 11
Fax: (+7 095) 556 43 28 (Moscow)
Fax: (+7 096) 463 59 51 (Ramenskoye)

Russkaya Avionica Joint Design Bureau
M.M. Gromov LII, 140160
Zhukovsky-2
Moscow Region
Russian Federation
Tel: (+7 095) 556 52 91
Fax: (+7 095) 556 50 83

State Research Institute of Instrument Engineering
125 Prospect Mira
129226 Moscow
Russian Federation
Tel: (+7 095) 181 16 38
Fax: (+7 095) 181 33 70

Tekhpribor State Enterprise
1a Korpusnoy Proyezd
196084 St Petersburg
Russian Federation
Tel: (+7 812) 296 97 38
Fax: (+7 812) 296 95 72

Transas Marine Limited, Transas Aviation Limited
21/2 Obrukhovskoy Oborny
193019 St Petersburg
Russian Federation
Tel: (+7 812) 325 31 31
Fax: (+7 812) 567 19 01
Web: http://avia.transas.com

Ulyanovsk Instrument Design Office
10a Krylova Street
432001 Ulyanovsk
Russian Federation

Vympel State Machine Building Design Bureau
90 Volokolamskoje sh
123424 Moscow
Russian Federation
Tel: (+7 095) 491 02 39
Fax: (+7 095) 490 22 22

Zenit Foreign Trade Firm, State Enterprise P/C S.A
Zverev Krasnogorsky Zavod
8 Rechnaya Street
Krasnogorsk
143400 Moscow
Russian Federation
Tel: (+7 095) 561 33 77
Fax: (+7 095) 562 82 75

SOUTH AFRICA

ADS (Altech Defence Systems)
PO Box 432
Mount Edgecombe
4300 KwaZulu-Natal
South Africa
Tel: (+27 31) 508 11 11
Fax: (+27 31) 59 53 60

Advanced Technologies & Engineering Co (IATE)
PO Box 632
Halfway House 1685
South Africa
Tel: (+27 11) 266 76 00
Fax: (+27 11) 314 21 51
e-mail: ategen@iafrica.com

Analysis, Management & Systems (Pty) Ltd
PO Box 1980
Halfway House 1685
South Africa
Tel: (+27 11) 315 10 02
Fax: (+27 11) 315 16 45
Web: http://www.ams.co.za

Avitronics (Pty) Ltd
PO Box 8492
Centurion 0046
South Africa
Tel: (+27 12) 672 60 29
Fax: (+27 12) 672 60 20
e-mail: avitronics@grintek.com
Web: http://www.avitronics.co.za

Cumulus, Business Unit of Denel (Pty) Ltd
PO Box 8859
0046 Centurion
South Africa
Tel: (+27 12) 674 01 46
Fax: (+27 12) 674 01 99
Web: http://www.cumulus.co.za

Grintek Comms
PO Box 1463
Pretoria 0001
South Africa
Tel: (+27 12) 810 10 00
Fax: (+27 12) 803 60 39
e-mail: grintek.comms@grintek.com
Web: http://www.grintek.co.za

Grintek Electronic Systems
PO Box 10252
Centurion 0046
South Africa
Tel: (+27 12) 672 60 00
Web: http://www.grintek.co.za

Grintek System Technologies
PO Box 912-561
Silverton 0127
Republic of South Africa
Tel: (+27 12) 421 62 00
Fax: (+27 12) 349 13 08
Web: http://www.grintek.co.za

Reutech Defence Industries (RDI) (Pty) Ltd
PO Box 118
New Germany 3620
South Africa
Tel: (+27 31) 719 57 11
Fax: (+27 31) 719 57 07
e-mail: info@rdi.co.za
Web: http://www.rdi.co.za

Reutech Systems (Pty) Ltd
Reunert Defence ESD
42 James Crescent
PO Box 35
Halfway House
Midrand 1685
South Africa
Tel: (+27 11) 652 55 55
Fax: (+27 11) 652 54 71
e-mail: info@reutech.co.za
Web: http://www.reutech.co.za

Tellumat (Pty) Limited
PO Box 30451
Tokai 7966
South Africa
Tel: (+27 21) 710 29 11
Fax: (+27 21) 712 12 78
e-mail: helpdesk@tellumat.com
Web: http://www.tellumat.com

SPAIN

EADS Construcciones Aeronáuticas SA
Avda Aragón, 404
E-28022 Madrid
Spain
Tel: (+34 91) 585 70 00
Fax: (+34 91) 585 76 66
Web: http://www.eads-nv.com

ELT SA
Polig Ind La Mina
Parcela 11
E-28770 Colmenar Viejo
Madrid
Spain
Tel: (+34 91) 846 03 01
Fax: (+34 91) 846 03 02

INDRA EWS
C/Joaquin Rodrigo 11
E-28300 Aranjuez
Madrid
Spain
Tel: (+34 91) 894 88 00
Fax: (+34 91) 894 89 17
e-mail: magarcia@indra.es

INDRA EWS
Ctra de Loeches 9
E-28850 Torrejon de Ardoz
Madrid
Spain
Tel: (+34 91) 396 80 12
Fax: (+34 91) 656 58 87

Technobit SA
Polignano Industrial
E-13300 Valdepenas
Ciudad Real
Spain

SWEDEN

Ericsson Microwave Systems AB
Flöjelbergsgatan 2a
SE-431 84 Mölndal
Sweden
Tel: (+46 31) 747 00 00
Fax: (+46 31) 27 78 91
Web: http://www.ericsson.se/microwave

Ericsson Saab Avionics AB
SE-164 84 Stockholm
Sweden
Tel: (+46 8) 757 30 00
Fax: (+46 8) 752 81 72
e-mail: info@esavionics.se
Web: http://www.esavionics.se

GP&C Sweden AB
PO Box 4207
SE-171 04 Solna
Sweden
Tel: (+46 8) 627 64 34
Fax: (+46 8) 627 64 49
WWW: http://www.ssc.se

Microdata Innovation AB
Box 1178
SE-171 23 Solna
Sweden
Tel: (+46 8) 624 74 50
Fax: (+46 8) 624 74 69
e-mail: ew.equipment@microdata-innovation.se
Web: http://www.microdata-innovation.se

Polytech AB
Box 20
SE-640 32 Malmköping
Sweden
Tel: (+46 157) 217 42
Fax: (+46 157) 213 48

Saab Communication Systems AB (formerly Celsius Communications Systems AB)
Vattenkraftsvägen 8
SE-135 70 Stockholm
Sweden
Tel: (+46 8) 798 09 00
Fax: (+46 8) 798 84 33
Web: http://www.sms.se

Saab Dynamics AB
SE-581 88 Linköping
Sweden
Tel: (+46 13) 18 61 32
Fax: (+46 13) 18 60 12
e-mail: stefan.nilsson@dynamics.saab.se

SaabTech Electronics AB (formerly CelsiusTech Electronics AB)
SE-17588 Järfälla
Sweden
Tel: (+46 8) 58 08 40 00
Fax: (+46 8) 58 03 22 44
Web: http://www.celsiustech.se

SWITZERLAND

Flight Components AG
Bitzibergstrasse 5
CH-8184 Bachenbulach
Switzerland
Tel: (+41 1) 861 12 00
Fax: (+41 1) 861 17 15
e-mail: sales@flightcomponents.com
Web: http://www.flightcomponents.com

Revue Thommen AG
Hauptstrasse 85
CH-4437 Waldenburg
Switzerland
Tel: (+41 61) 965 22 22
Fax: (+41 61) 961 81 71

Vibro-Meter SA, Aerospace Division
Route de Moncor 4
PO Box 1071
CH-1701 Fribourg
Switzerland
Tel: (+41 26) 407 11 11
Fax: (+41 26) 402 36 62

TURKEY

ASELSAN Electronics Industry Inc
Microelectronics, Guidance and Electro-Optics Division
PO Box 30
Etlik
06011 Ankara
Turkey
Tel: (+90 312) 847 53 00
Fax: (+90 312) 847 53 20
Web: http://www.aselsan.com.tr

MIKES Microwave Electronic Systems Incorporated
Cankin yolu 5 km
Akyurt
TR-06750 Ankara
Turkey
Tel: (+90 312) 847 51 00
Fax: (+90 312) 847 51 14

UKRAINE

ARSENAL Central Design Office
8 Moskovskaya Str
Kiev-10 'GSP'
252601
Ukraine
Tel: (+380 44) 293 00 62
Fax: (+380 44) 293 15 09
e-mail: cdoars@gu.kiev.ua

Ukraine Research Institute of Radio Equipment
252072 Kiev
Ukraine

UNITED KINGDOM

AD Aerospace Ltd
1 Hilton Square
Pendlebury
Swinton
Manchester M27 4DB
UK
Tel: (+44 161) 727 66 00
Fax: (+44 161) 727 85 67
Web: http://www.ad-aero.co.uk

Aerosystems International, Tactical Communications Division
West Hendford
Yeovil
Somerset BA20 2AL
UK
Tel: (+44 1935) 44 30 00
Fax: (+44 1935) 44 31 11
e-mail: sales@aeroint.com
Web: http://www.aeroint.com

Alenia Marconi Systems
Eastwood House
Glebe Road
Chelmsford
Essex CM1 1QW
UK
Tel: (+44 1245) 70 27 02
Fax: (+44 1245) 70 27 00
Web: http://www.aleniamarconisystems.com

Avalon Electronics Ltd
Langhorne Park House
High Street
Shepton Mallet
Somerset BA4 5AQ
UK
Tel: (+44 1749) 34 52 66
Fax: (+44 1749) 34 52 67
Web: http://www.avalon-electronics.com

Avimo Ltd
Lisieux Way
Taunton
Somerset TA1 2JZ
UK
Tel: (+44 1823) 33 10 71
Fax: (+44 1823) 34 96 74
e-mail: aworsdell@avimo.co.uk
Web: http://www.avimo.co.uk

BAE Systems
Foxhunter Drive
Linford Wood
Milton Keynes
Bedfordshire MK14 6LA
UK
Tel: (+44 1908) 22 00 44
Fax: (+44 1908) 31 71 37

BAE SYSTEMS Avionics
Group Airport Works
Rochester
Kent ME1 2XX
UK
Tel: (+44 1634) 84 44 00
Fax: (+44 1634) 82 73 32

BAE SYSTEMS Avionics
Sensors and Communications Systems
Silverknowes
Ferry Road
Edinburgh EH4 4AD
UK
Tel: (+44 131) 332 24 11
Fax: (+44 131) 343 50 50

BAE SYSTEMS Avionics
Radar and Countermeasure Systems
The Grove
Warren Lane
Stanmore
Middlesex HA7 4LY
UK
Tel: (+44 20) 89 54 23 11
Fax: (+44 20) 84 20 39 90

BAE SYSTEMS Avionics
1 South Gyle Crescent
Edinburgh EH12 9EA
UK
Tel: (+44 131) 332 24 11
Fax: (+44 131) 314 28 39

BAE SYSTEMS Avionics
Radar and Countermeasure Systems
Building 19
Crewe Toll
Ferry Road
Edinburgh EH5 2XS
UK
Tel: (+44 131) 332 24 11
Fax: (+44 131) 343 40 11

BAE SYSTEMS Avionics
Flight and Data Systems/RCS and Identification Group
Browns Lane
The Airport
Portsmouth
Hampshire PO3 5PH
UK
Tel: (+44 1705) 22 60 00
Fax: (+44 1705) 22 71 33

BAE SYSTEMS Avionics
Sensor and Communications Systems
Christopher Martin Road
Basildon
Essex SS14 3EL
UK
Tel: (+44 1268) 52 28 22
Fax: (+44 1268) 88 31 40

BAE SYSTEMS Avionics
Clittaford Road
Southway
Plymouth
Devon PL6 6DE
UK
Tel: (+44 1752) 69 56 95
Fax: (+44 1752) 69 55 00
Web: http://www.bae.co.uk

Caledonian Airborne Systems
Caledonian House
Aberdeen Airport
Aberdeen AB21 0PD
UK
Tel: (+44 1224) 72 22 74
Fax: (+44 1224) 72 28 96

Cargo Aids Ltd
Brett Drive
De La Warr Road
Bexhill-on-Sea
East Sussex TN40 2JP
UK
Tel: (+44 1424) 21 66 11
Fax: (+44 1424) 21 66 36
e-mail: 106003.2124@compuserve.com

Chelton (Electrostatics) Ltd
Fieldhouse Lane
Marlow
Buckinghamshire SL7 1LR
UK
Tel: (+44 1628) 47 20 72
Fax: (+44 1628) 48 22 55
e-mail: mkt@chelton.co.uk
Web: http://www.chelton.com

Computing Devices
A General Dynamics Company
Castleham Road
St Leonards-on-Sea
East Sussex TN38 9NJ
UK
Tel: (+44 1424) 85 34 81
Fax: (+44 1424) 85 15 20
Web: http://www.compd.com

Defence Evaluation and Research Agency (DERA)
Ively Road
Farnborough
Hampshire GU14 OLX
UK
Tel: (+44 1252) 39 33 00
Fax: (+44 1252) 39 33 99
Web: http://www.dera.gov.uk

DRS Hadland Ltd
Harrow Yard
Akeman Street
Tring
Hertfordshire HP23 6AA
UK
Tel: (+44 1442) 82 15 00
Fax: (+44 1442) 82 15 99
e-mail: post@hadland.co.uk
Web: http://www.drs.com

DRS Rugged Systems (Europe) Ltd
Lynwood House
The Trading Estate
Farnham
Surrey GU9 9NN
UK
Tel: (+44 1252) 73 44 88
Fax: (+44 1252) 73 44 66
e-mail: juney@lynwood.com
Web: http://www.drs.com

DRS Technologies (UK) Ltd
Harrow Yard
Akeman Street
Tring
Hertfordshire HP23 6AA
UK
Tel: (+44 1442) 82 15 00
Fax: (+44 1442) 82 15 99
e-mail: Busdev.drs@mcmail.com
Web: http://www.drs.com

Enterprise Control Systems Ltd
31 High Street
Wappenham
Northamptonshire NN12 8SN
UK
Tel: (+44 1327) 86 00 50
Fax: (+44 1327) 86 00 58
e-mail: ecsltd@compuserve.com
Web: http://www.enterprisecontrol.com

Ferranti Technologies Limited
Cairo Mill
Waterhead
Oldham
Lancashire OL4 3JA
UK
Tel: (+44 161) 624 02 81
Fax: (+44 161) 624 52 44
e-mail: sales@ferranti-technologies.co.uk
WWW: http://www.ferranti-technologies.co.uk

GKN Westland Helicopters
Lysander Road
Yeovil
Somerset BA 20 2YB
UK
Tel: (+44 1935) 47 52 22
Fax: (+44 1935) 70 21 31
e-mail: info@gkn-whl.co.uk
Web http://www.gkn-whl.co.uk

Honeywell
(Normalair-Garrett) Ltd
Yeovil
Somerset BA20 2YD
UK
Tel: (+44 1935) 751 81
Fax: (+44 1935) 276 00
e-mail: angus.maclean@honeywell.com
WWW: http://www.honeywell.com

H R Smith
Street Court
Kingsland
Leominster
Herefordshire HR6 9QA
UK
Tel: (+44 1568) 70 87 44
Fax: (+44 1568) 70 87 13
Web: http://www.hr-smith.com

Litton UK Ltd, Data Systems Division
Burlington House
118 Burlington Road
New Malden
Surrey KT3 4NR
UK
Tel: (+44 208) 329 20 41
Fax: (+44 208) 329 20 42
Web: http://www.litton-dsd.co.uk

Lockheed Martin UK Government Systems
PO Box 41
North Harbour
Portsmouth
Hampshire PO6 3AU
UK
Tel: (+44 1705) 563 00 04
Fax: (+44 1705) 56 35 46
WWW: http://www.lockheedmartin.co.uk

M/A COM Ltd (UK)
Humphrys Road
Woodside Estate
Dunstable
Bedfordshire LU5 4SX
UK
Tel: (+44 1582) 47 12 00
Fax: (+44 1582) 47 22 77
Web: http://www.macom.com

Matra BAe Dynamics (UK) Ltd
Six Hills Way
Stevenage
Herts SG1 2DA
UK
Tel: (+44 1438) 31 24 22
Fax: (+44 1438) 75 33 77
Web: http://www.bae.co.uk

Meggitt Avionics
7 Whittle Avenue
Fareham
Hampshire PO15 5SH
UK
Tel: (+44 1489) 48 33 00
Fax: (+44 1489) 48 33 40
e-mail: marketing@meggitt-avionics.co.uk
Web: http://www.meggitt.com

MS Instruments plc
Electron House
Farwig Lane
Bromley
Kent BR1 3RE
UK
Tel: (+44 20) 82 90 02 00
Fax: (+44 20) 84 64 65 96
Web: http//www.msinstruments.co.uk

Octec Ltd
The Western Centre
Western Road
Bracknell
Berkshire RG12 1RW
UK
Tel: (+44 1344) 46 52 00
Fax: (+44 1344) 46 52 01
Web: http://www.octec.co.uk

Page Aerospace Ltd
Forge Lane
Sunbury-on-Thames
Middlesex TW16 6EQ
UK
Tel: (+44 1932) 78 76 61
Fax: (+44 1932) 78 03 49
Web: http://www.pageaerospace.co.uk

Penny & Giles Aerospace Ltd
6 Airfield Way
Christchurch
Dorset BH23 3TT
UK
Tel: (+44 1202) 48 17 71
Fax: (+44 1202) 48 48 46

Pilkington Optronics Ltd
See THALES

Prostab, a Smiths Industries Company
Vulcan Way
New Addington
Croydon
Surrey CR9 0BD
UK
Tel: (+44 1689) 84 34 41
Fax: (+44 1689) 84 59 70
Web: http://www.prostab.com

Racal Acoustics Ltd
See THALES

Racal Avionics Ltd
See THALES

Racal Corporate Communications Ltd
See THALES

Racal Defence Electronics Limited
See THALES

Racal Recorders Ltd
See THALES

Racal Survey Ltd
See THALES

Raytheon Systems Limited
80 Park Lane
London W1Y 3TD
UK
Tel: (+44 171) 569 55 00
Fax: (+44 171) 569 55 96
e-mail: Mike.Brown@raytheon.co.uk
Web: http://www.raytheon.co.uk

Raytheon Systems Limited
The Pinnacles
Elizabeth Way
Harlow
Essex CM19 5BB
UK
Tel: (+44 1279) 42 68 62
Fax: (+44 1279) 41 04 13
e-mail: 16613@pine.cossor.com

Raytheon Systems Limited
1 Jenner Road
Manor Royal Industrial Estate
Crawley
West Sussex RH10 2GA
UK
Tel: (+44 1293) 40 19 00
Fax: (+44 1293) 40 19 01

Ring Sights Defence Ltd
PO Box 22
Borden
Hampshire GU35 0JR
UK
Tel: (+44 1420) 47 22 60
Fax: (+44 1420) 47 83 59
e-mail: defence@ringsights.com

SIFAM Instruments Limited
Woodland Road
Torquay
Devon TQ2 7AY
UK
Tel: (+44 1803) 40 77 00
Fax: (+44 1803) 40 77 40
e-mail: info@sifam.com
Web: http://www.sifam.com

Signature Industries Ltd
Tom Cribb Road
Thamesmead
London SE28 0BH
UK
Tel: (+44 20) 83 16 44 77
Fax: (+44 20) 88 54 51 49

Simrad Optronics Limited
3 Meadowbrook Industrial Estate
Maxwell Way
Crawley
West Sussex RH10 2SA
UK
Tel: (+44 1293) 56 04 13
Fax: (+44 1293) 56 04 18
e-mail: roger.cross@simrad-optronics.co.uk

Skyforce Avionics Ltd
5 The Old Granary
Boxgrove
Chichester
West Sussex PO18 0ES
UK
Tel: (+44 1243) 78 37 63
Fax: (+44 1243) 78 39 92
e-mail: sales@skyforce.co.uk
WWW: http://www.skyforce.co.uk

SkyQuest Aviation
Unit A11
Fairoaks Airport
Chobham
Surrey GU24 8HU
UK
Tel: (+44 1276) 85 53 34
Fax: (+44 1276) 85 57 25
e-mail: info@skyquest.co.uk
Web: http://www.skyquest.co.uk

Smiths Industries Aerospace
765 Finchley Road
London NW11 8DS
UK
Tel: (+44 208) 458 32 32
Fax: (+44 208) 209 05 26
Web: http://www.smithsind-aerospace.com

Somerdata Ltd
SomerNet Business Centre
Rowdens Road
Wells
Somerset BA5 1TU
UK
Tel: (+44 1749) 67 14 81
Fax: (+44 1749) 67 14 82
Web: http://www.somerdata.com

Symmetricom Ltd
Mansard Close
Westgate
Northampton NN5 5DL
UK
Tel: (+44 1604) 58 55 88
Fax: (+44 1604) 58 55 99
e-mail: navstar@symmetricom.com
Web: http://www.symmetricom.com

Thomson Marconi Sonar Ltd
Corsair Building
Airport Works
Rochester
Kent ME1 1EJ
UK
Tel: (+44 1634) 81 61 24
Fax: (+44 1634) 81 67 91
e-mail: ann.mutter@gecm.com

TRW Aeronautical Systems, Lucas Aerospace
Stratford Road
Solihull B90 4LA
UK
Tel: (+44 121) 451 59 77
Fax: (+44 121) 451 61 11
e-mail: jackie.berger@lucasvarity.com

Ultra Electronics Controls Division
417 Bridport Road
Greenford
Middlesex UB6 8UA
UK
Tel: (+44 20) 88 13 44 44
Fax: (+44 20) 88 13 43 51
Web: http://www.uecd.co.uk

Ultra Electronics Noise & Vibration Systems
1 Cambridge Business Park
Cowley Road
Cambridge CB4 4WZ
UK
Tel: (+44 1223) 42 66 99
Fax: (+44 1223) 42 66 96
e-mail: sales@ultraquiet.com
Web: http://www.ultra-electronics.co.uk/nvs

Ultra Electronics Sonar and Communication Systems
419 Bridport Road
Greenford
Middlesex UB6 8UA
UK
Tel: (+44 20) 88 13 45 67
Fax: (+44 20) 88 13 45 68

Weston Aerospace Limited
124 Victoria Road
Farnborough
Hampshire GU14 7PW
UK
Tel: (+44 1252) 54 44 33
Fax: (+44 1252) 37 02 98
Web: http://www.westonaero.com

W Vinten Ltd
Vicon House
Western Way
Bury St Edmunds
Suffolk IP33 3SP
UK
Tel: (+44 1284) 75 05 99
Fax: (+44 1284) 75 05 98
Web: http://www.wvintenltd.com

UNITED STATES OF AMERICA

Accurate Automation Corporation
7001 Shallowford Road
Chattanooga
Tennessee 37421
USA
Tel: (+1 423) 894 46 46
Fax: (+1 423) 894 46 45
Web: http://www.accurate-automation.com

AeroSolutions
1903 Waterford Drive
Grapevine
Texas 76051
USA
Tel: (+1 817) 424 02 24
Fax: (+1 817) 251 90 25

Aerosonic Corporation
1212 North Hercules Avenue
Clearwater
Florida 34625
USA
Tel: (+1 813) 461 30 00
Fax: (+1 813) 447 59 26

Aerospace Avionics Inc
1000 MacArthur Memorial Highway
PO Box 1000
Bohemia
New York 11716
USA
Tel: (+1 516) 467 55 00
Fax: (+1 516) 467 59 39
Web: http://www.aerospace-avionics.com

Aerospace Display Systems, Inc
2321 Topaz Drive
Hatfield
Pennsylvania 19440
USA
Tel: (+1 215) 822 60 90
Fax: (+1 215) 822 79 74

AIL Systems Inc
455 Commack Road
Deer Park
New York 11729-4591
USA
Tel: (+1 631) 595 32 16
Fax: (+1 631) 595 66 39
Web: http://www.ail.com

Alliant Defense Electronics Systems Inc, Lundy Product Group
13133 34th Street North
Clearwater
Florida 34618
USA
Tel: (+1 813) 572 19 00

Alliant Defense Electronics Systems, Inc
(A wholly owned subsidiary of Alliant Techsystems, Inc)
PO Box 4648
Clearwater
Florida 34618
USA
Tel: (+1 813) 572 31 84
Fax: (+1 813) 572 24 33

Ametek Aerospace
50 Fordham Road
Wilmington
Maryland 01887
USA
Tel: (+1 508) 988 41 01
Fax: (+1 508) 988 49 44
e-mail: laura.dion@ametek.com
Web: http://www.ametek.com

Ampex Corporation
Data Systems Division
500 Broadway
Redwood City
California 94063-3199
USA
Tel: (+1 650) 367 41 11
Fax: (+1 650) 367 46 99
Web: http://www.ampex.com

Andrew SciComm
2908 National Drive
Garland
Texas 75041
USA
Tel: (+1 972) 840 49 00
Fax: (+1 972) 278 93 79
Web: http://www.andrew.com

Applied Communications Research, Inc
618 Buck Road
Holland
Pennsylvania 18966
USA
Tel: (+1 215) 364 07 57
Fax: (+1 215) 364 27 92
e-mail: sales@acrinc.com
WWW: http://www.acrinc.com

Arc Industries Inc
PO Box 033498
Indialantic
Florida 32903
USA
Tel: (+1 321) 254 19 17
Fax: (+1 321) 254 85 44

ARGOSystems (a Condor Systems business unit)
324 North Mary Avenue, PO Box 3452
Sunnyvale
California 94088
USA
Tel: (+1 408) 524 17 78
Fax: (+1 408) 524 20 23

ARINC Incorporated
2551 Riva Road
Annapolis
Maryland 21401
USA
Tel: (+1 410) 266 40 00
Fax: (+1 410) 266 23 29
e-mail: arincmkt@arinc.com
Web: http://www.arinc.com

Arnav Systems Inc
PO Box 73730
Puyallup
Washington 98373
USA
Tel: (+1 206) 848 60 60
Fax: (+1 206) 848 35 55

Artex Aircraft Supplies Inc
A Chelton Group Company
10714 South Township Road
Canby
Oregon 97013
USA
Tel: (+1 503) 266 39 59
Fax: (+1 503) 266 33 62
e-mail: info@artex.net
Web: http://www.artex.net

Astronautics Corporation of America
PO Box 523
4115 N Teutonia Avenue
Milwaukee
Wisconsin 53201
USA
Tel: (+1 414) 447 82 00
Fax: (+1 414) 447 82 31
e-mail: busdev@astronautics.com
Web: http://www.astronautics.com

Avidyne Corporation
Lincoln North Building
55 Old Bedford Road
Lincoln
Massachusetts 01773-1125
USA
Tel: (+1 781) 402 74 00
Fax: (+1 781) 402 75 99
e-mail: info@avidyne.com
Web: http://www.avidyne.com

Avionics Specialties Inc
PO Box 6400
Route 743
Charlottesville
Virginia 22906
USA
Tel: (+1 804) 973 33 11
Fax: (+1 804) 973 29 76
Web: http://www.avionics-specialties.com

AVTECH Corporation
3400 Wallingford Avenue North
Seattle
Washington 98103
USA
Tel: (+1 206) 634 25 40
Fax: (+1 206) 634 30 11

Aydin Displays
700 Dresher Road
Horsham
Pennsylvania 19044
USA
Tel: (+1 215) 657 86 00
Fax: (+1 215) 657 36 81
WWW: http://www.aydin.com

BAE Systems Aerospace Electronics Inc
305 Richardson Road
Lansdale
Pennsylvania 19446-1485
USA
Tel: (+1 215) 996 20 00
Fax: (+1 215) 996 20 88
e-mail: communications.lansdale@baesystems.com
Web: http://www.baesystems.com

BAE Systems Aerospace Inc
CNI Division
164 Totowa Road
Wayne
New Jersey 07474-0975
USA
Tel: (+1 973) 633 60 00
Fax: (+1 973) 633 64 31
Web: http://www.baesystems.com

BAE Systems, Canada Inc
Cincinnati Electronics Corporation (CE)
7500 Innovation Way
Mason
Ohio 45040-9699
USA
Tel: (+1 513) 573 61 00
Fax: (+1 513) 733 64 58
Web: http//www.cinele.com

BAE Systems Controls (formerly Lockheed Martin Control Systems)
600 Main Street
Johnson City
New York 13790
USA
Tel: (+1 607) 770 20 00
Fax: (+1 607) 770 25 67
Web: http://www.na.baesystems.com

BAE Systems, North America
700 Quince Orchard Rd
Gaithersburg
Maryland 20878-1794
USA
Tel: (+1 301) 948 75 50
Fax: (+1 301) 921 94 79
e-mail: wj.helps@wj.com
Web: http://www.wj.com

BAE Systems, North America
450 Pulaski Road
Greenlawn
New York 11740
USA
Tel: (+1 516) 216 70 00

BAE Systems, North America
Flight Systems
1434 Flightline, Building 58B
Mojave
California 93501
USA
e-mail: ldorff@tracor.com
Web: http//www.na.baesystems.com

BAE Systems, North America
Aerospace Defense Systems
6500 Tracor Lane
Austin
Texas 78725
USA
Tel: (+1 512) 929 21 26
Fax: (+1 512) 929 23 20

BAE Systems, North America
Northstar Technologies
30 Sudbury Road
Acton
Massachusetts 01720
USA
Tel: (+1 978) 897 66 00
Fax: (+1 978) 897 82 64
Web: http://www.northstarcmc.com

BAE Systems, North America
Information and Electronic Warfare Systems (formerly Sanders)
65 Spit Brook Road
Nashua
New Hampshire 03060
USA
Tel: (+1 603) 885 36 53
Fax: (+1 603) 885 36 55
Web: http://www.baesystems-iews.com

BAE Systems, North America
CNI Division
164 Totowa Road
Wayne
New Jersey 07474-0975
USA
Tel: (+1 973) 633 60 00
Web: http//www.cni.marconi-na.com

BAE Systems North America, Aircraft Controls
3400 Airport Avenue
PO Box 442
Santa Monica
California 90406
USA
Tel: (+1 310) 915 60 00
Fax: (+1 310) 915 83 84

Ball Aerospace & Technologies Corp.
1600 Commerce Street
10 Long Peaks Drive
Boulder
Colorado 80301
USA
Tel: (+1 303) 939 61 00
Fax: (+1 303) 939 61 04
Web: http://www.ballaerospace.com

BFGoodrich Aerospace Aircraft Integrated Systems
100 Panton Road
Vergennes
Vermont 05491
USA
Tel: (+1 802) 877 45 12
Fax: (+1 802) 877 41 13

BFGoodrich Aerospace Avionics Systems
5353 52nd Street SE
Grand Rapids
Michigan 49588-0873
USA
Tel: (+1 616) 949 66 00
Fax: (+1 616) 285 42 24
Web: http://www.bfgavionics.com

The Boeing Company
Military Aircraft and Missiles
Mailcode 0011312
PO Box 516
St Louis
Missouri 63166
USA
Tel: (+1 314) 234 41 87
Fax: (+1 314) 233 64 55

The Boeing Company
Military Airplanes Division - Defense & Space Group
PO Box 3707 4H-14
Seattle
Washington 98124
USA
Tel: (+1 206) 655 11 98
Fax: (+1 206) 655 93 25

The Boeing Company
Information, Space & Defense Systems Group
PO Box 3999
Seattle
Washington 98124
USA
Tel: (+1 253) 773 28 16
Fax: (+1 253) 773 39 00
e-mail: milton.j.furness@boeing.com
WWW: http://www.boeing.com

The Boeing Company
Boeing Product Support Division
PO Box 7730 MS K12-12
Wichita
Kansas 67277-7730
USA
Tel: (+1 316) 526 39 02
Fax: (+1 316) 526 76 01
e-mail: dick.ziegler@boeing.com
WWW: http://www.boeing.com

Century Flight Systems Inc
Municipal Airport
PO Box 610
Mineral Wells
Texas 76068
USA
Tel: (+1 940) 325 25 17
Fax: (+1 940) 325 25 46
e-mail: century@centuryflight.com
Web: http://www.centuryflight.com

Cincinnati Electronics Corporation
see:
BAE SYSTEMS North America

Concurrent Computer Corporation
2 Crescent Place
Oceanport
New Jersey 07757
USA
Tel: (+1 908) 870 45 00
Fax: (+1 908) 870 59 67

Condor Electronic Systems Division
996 Flower Glen Street
Simi Valley
California 93065
USA
Tel: (+1 805) 584 82 00
Fax: (+1 805) 527 83 32
Web: http://www.cwes.com

Condor Systems Inc
2133 Samaritan Drive
San Jose
California 95124
USA
Tel: (+1 408) 371 95 80
Fax: (+1 408) 371 95 89
Web: http://www.condorsys.com

Cubic Communications Inc
9535 Waples Street
San Diego
California 92121
USA
Tel: (+1 858) 643 58 00
Fax: (+1 858) 643 58 03
e-mail: denise.leger@cubic.com
Web: http://www.cubiccomm.com

Cubic Defense Systems
9323 Balboa Ave
San Diego
California 92123
USA
Tel: (+1 619) 505 28 79
Fax: (+1 619) 505 15 32
e-mail: cal.ogata@cubic.com

Cycomm International Inc
1420 Springhill Road, Suite 420
McLean
Virginia 22102
USA
Tel: (+1 703) 903 95 48
Fax: (+1 703) 903 95 28
e-mail: budh@corstone.com

Daedalus Enterprises Inc
PO Box 1869
Ann Arbor
Michigan 48106
USA
Tel: (+1 313) 769 56 49

David Clark Company Inc
360 Franklin Street
PO Box 15054
Worcester
Massachusetts 01615-0054
USA
Tel: (+1 508) 751 58 00
Fax: (+1 508) 753 58 27
Web: http://www.davidclark.com

Delco Electronics Corporation
Delco Systems Operations
6767 Hollister Avenue
Goletta
California 93117
USA
Tel: (+1 805) 961 59 03
Fax: (+1 805) 961 54 16

Diamond J, Inc
PO Box 9526
Wichita
Kansas 67277
USA
Tel: (+1 316) 945 10 10
Fax: (+1 316) 945 22 44

DNE Technologies Inc
50 Barnes Park North
PO Box 30
Wallingford
Connecticut 06492
USA
Tel: (+1 203) 265 71 51
Fax: (+1 203) 265 87 01
e-mail: dnesales@aol.com
Web: http://www.dnetech.com

dpiX, LLC
3406 Hillview Avenue
Palo Alto
California 94304-1345
USA
Tel: (+1 650) 842 96 00
Fax: (+1 650) 842 97 93
Web: http://www.dpix.com

DRS Electronic Systems Group
A DRS Technologies Inc Company
1215 South Jefferson Davis Highway
Suite 1004
Arlington
Virginia 22202
USA
Tel: (+1 703) 416 80 00
Fax: (+1 703) 416 80 10
Web: http://www.drs.com

DRS Electro-Optical Systems Group
A DRS Technologies Inc Company
2330 Commerce Park Drive NE
Palm Bay
Florida 32905
USA
Tel: (+1 321) 984 90 30
Fax: (+1 321) 984 87 46
Web: http://www.drs.com

DRS Technologies, DRS Photronics, Inc
133 Bauer Drive
Oakland
New Jersey 07436
USA
Tel: (+1 201) 337 38 00
Fax: (+1 201) 337 27 04
e-mail: photronics@drs.com
Web: http://www.drs.com

DRS Technologies, DRS Precision Echo, Inc
3105 Patrick Henry Drive
Santa Clara
California 95054
USA
Tel: (+1 408) 988 05 16
Fax: (+1 408) 727 74 91
e-mail: info-pe@drs.com
Web: http://www.drs.com

DRS Technologies, Inc
Corporate Headquarters
5 Sylvan Way
Parsippany
New Jersey 07054
USA
Tel: (+1 973) 898 15 00
Fax: (+1 973) 898 47 30
Web: http://www.drs.com

Dynamic Instruments, Inc
3860 Calle Fortunada
San Diego
California 92123
Tel: (+1 619) 278 49 00
Fax: (+1 619) 278 67 00

Electronics & Space Corporation
8100 W Florissant Avenue
St Louis
Missouri 63136
USA
Tel: (+1 314) 553 45 29
Fax: (+1 314) 553 45 55

Endevco
30700 Rancho Viejo Road
San Juan Capistrano
California 92675
USA
Tel: (+1 949) 493 81 81
Fax: (+1 949) 661 72 31
Web: http://www.endevco.com

Eventide Avionics
One Alsan Way
Little Ferry
New Jersey 07643
USA
Tel: (+1 201) 641 12 00
Fax: (+1 201) 641 16 40
Web: http://www.eventide.com

FED Corporation
Hopewell Junction
New York
USA

Flight Dynamics
16600 SW 72nd Avenue
Portland
Oregon 97224
USA
Tel: (+1 503) 443 30 00
Fax: (+1 503) 684 01 69
Web: http://www.flightdynamics.com

Flightline Electronics Inc
7525 Country Road 42
PO Box 1005
Fishers
New York 14453-0750
USA
Tel: (+1 716) 924 40 00
Fax: (+1 716) 924 57 32
Web: http://www.ultra-electronics.co.uk

Flight Visions Inc
43W752 Route 30
PO Box 250
Sugar Grove
Illinois 60554-0250
USA
Tel: (+1 630) 466 43 43
Fax: (+1 630) 466 43 58
Web: http://www.flightvisions.com

FLIR Systems Inc
16505 SW 72nd Avenue
Portland
Oregon 97224
USA
Tel: (+1 503) 684 37 31
Fax: (+1 503) 684 32 07
e-mail: sales@flir.com
Web: http://www.flir.com

Gables Engineering Inc
247 Greco Avenue
Coral Gables
Florida 33146
USA
Tel: (+1 305) 774 42 80

Garmin International
1200 East 151st Street
Olathe
Kansas 66062
USA
Tel: (+1 913) 397 82 00
Fax: (+1 913) 397 82 82
Web: http://www.garmin.com

General Atronics Corporation
1200 East Mermaid Lane
Wyndmoor
Pennsylvania 19038-7695
USA
Tel: (+1 215) 233 41 00
Fax: (+1 215) 233 99 47
Web: http://www.generalatronics.com

General Dynamics Information Systems
8800 Queen Avenue South
Bloomington
Minnesota 55431-1996
USA
Tel: (+1 612) 921 60 00
Fax: (+1 612) 921 68 69
e-mail: info@gd-is.com
Web: http://www.gd-is.com

Gentex Corporation
PO Box 315
Carbondale
Pennsylvania 18407
USA
Tel: (+1 717) 282 35 50
Fax: (+1 717) 282 85 55

GTE Government Systems Corporation
77 A Street
Needham Heights
Maryland 02194
USA
Tel: (+1 617) 449 20 00
Fax: (+1 617) 455 30 30

Hamilton Sundstrand Corporation
One Hamilton Road
Windsor Locks
Connecticut 06096
USA
Tel: (+1 203) 654 60 00
Fax: (+1 203) 654 37 73
Web: http://www.hamilton-sundstrand.com

Harris Corporation
Government Communications Systems Division
2400 Palm Bay Road NE
Palm Bay
Florida 32905
USA
Tel: (+1 321) 727 48 67
Fax: (+1 407) 727 45 00
Web: http://www.harris.com

Honeywell Aerospace, Electronic & Avionics Lighting
One Technology Center
23500 West 105th Street
Olathe
Kansas 66061
USA
Tel: (+1 913) 712 26 13
Fax: (+1 913) 712 56 97
Web: http://www.honeywell.com/aerospace

Honeywell Defense Avionics Systems
9201 San Mateo Boulevard NE
PO Box 9200
Alberquerque
New Mexico 87113
USA
Tel: (+1 505) 828 50 00
Fax: (+1 505) 828 55 00
e-mail: linn.tytler@das.honeywell.com

Honeywell Engine Systems & Accessories
717N Bendix Drive
South Bend
Indiana 46620
USA
Tel: (+1 219) 231 37 17
Fax: (+1 219) 231 33 35

Honeywell Inc, Commercial Aviation Systems
PO Box 21111
North 19th Avenue
Phoenix
Arizona 85027-2708
USA
Tel: (+1 602) 436 55 99
Fax: (+1 602) 436 53 00
Web: http://www.cas.honeywell.com

Honeywell Inc, Sensor and Guidance Products
2600 Ridgway Parkway
Minneapolis
Minnesota 55413
USA
Tel: (+1 612) 951 53 28
Fax: (+1 612) 951 53 25

Ideal Research & Development Corporation
1810 Parklawn Drive
Rockville
Maryland 20852
USA
Tel: (+1 301) 468 20 50
Fax: (+1 301) 230 08 13

ImageQuest Technologies Inc
48611 Warm Springs Boulevard
Fremont
California 94539
USA
Tel: (+1 510) 249 05 00
Fax: (+1 510) 249 05 50

International Aerospace Inc
PO Box 340
234 Garibaldi Avenue
Lodi
New Jersey 07644
USA
Tel: (+1 201) 473 00 34
Fax: (+1 212) 947 90 49

ISS Innovative Solutions & Support Inc
420 Lapp Road
Malvern
Pennsylvania 19355
USA
Tel: (+1 610) 889 98 98
Fax: (+1 610) 651 55 06
Web: http://www.innovative-ss.com

ITT Defense & Electronics Inc
ITT Aerospace/Communications Division
1919 West Cook Road
PO Box 3700
Fort Wayne
Indiana 46801
USA
Tel: (+1 219) 451 60 00
Fax: (+1 219) 451 61 26
Web: http://www.ittind.com

ITT Industries, Avionics Division
100 Kingsland Road
Clifton
New Jersey 07014-1993
USA
Tel: (+1 973) 284 01 23
Fax: (+1 973) 284 41 22
Web: http://www.ittind.com
http://www.ittavionics.com

JP Instruments
Box 7033
Huntington Beach
California 92615
USA
Tel: (+1 714) 557 54 34
Fax: (+1 714) 557 98 40

Kaiser Electronics
2701 Orchard Parkway
San Jose
California 95134
USA
Tel: (+1 408) 432 30 00
Fax: (+1 408) 432 84 40
e-mail: garciaj@kaisere.com

Kaman Aerospace International
Raymond Engineering Operations
217 Smith Street
Middletown
Connecticut 06457
USA
Tel: (+1 860) 632 45 82
Fax: (+1 860) 632 43 29
e-mail: jkrupa-ray1@kaman.com
Web: http://www.raymond-engrg.com

Kearfott Guidance & Navigation Corporation
150 Totowa Road
Wayne
New Jersey 07474-0946
USA
Tel: (+1 973) 785 60 00
Fax: (+1 973) 785 60 25
Web: http://www.kearfott.com

Kollsman Inc
220 Daniel Webster Highway
Merrimack
New Hampshire 03054
USA
Tel: (+1 603) 889 25 00
Fax: (+1 603) 889 60 80
e-mail: marcom@kollsman.com

L-3 Aviation Recorders
6000 Fruitville Road
Sarasota
Florida 34232
USA
Tel: (+1 941) 377 08 11
Fax: (+1 941) 377 55 91
Web: http://www.l-3ar.com

L-3 Communications Ocean Systems Division
15825 Roxford Street
Sylmar
California 91342-3597
USA
Tel: (+1 818) 833 27 51
Fax: (+1 818) 364 24 91

L-3 Communications Systems - East
1 Federal Street
Camden
New Jersey 08103
USA
Tel: (+1 609) 338 29 44
Fax: (+1 609) 338 27 41
e-mail: george.pavlik@l-3com.com

L-3 Communications Systems - West
PO Box 16850
Salt Lake City
Utah 84116-0850
USA
Tel: (+1 801) 594 26 36
Fax: (+1 801) 594 30 03
Web: http://www.l-3com.com

L-3 Communications Telemetry - East
PO Box 328
47 Friends Lane
Newtown
Pennsylvania 18940
USA
Tel: (+1 215) 968 42 71
Fax: (+1 215) 968 32 14
e-mail: telemetry@aydin.com

L-3 Display Systems
1355 Bluegrass Lakes Parkway
Alpharetta
Georgia 30004
USA
Tel: (+1 770) 752 70 00
Fax: (+1 770) 752 54 81
Web: http://www.l-3com.com/displays

L-3/Interstate Electronics Corporation
1001 East Ball Road
PO Box 3117
Anaheim
California 92803
USA
Tel: (+1 714) 758 05 00
Fax: (+1 714) 758 41 48
WWW: http://www.iechome.com

LH Systems, LLC
10965 Via Frontera
San Diego
California 92127
USA
Tel: (+1 858) 675 33 35
Fax: (+1 858) 675 33 45
e-mail: info@lh-systems.com
Web: http://www.lh-systems.com

Linkabit, a Titan company
3033 Science Park Road
San Diego
California 92121
USA
Tel: (+1 858) 552 95 00
Fax: (+1 858) 552 94 87
e-mail: ekb@titan.com
Web: http://www.titan.com

Litton Advanced Systems Division (Amecom)
5115 Calvert Road
College Park
Maryland 20740-3808
USA
Tel: (+1 301) 864 56 00
Fax: (+1 301) 454 98 03
e-mail: info@littonas.com

Litton Advanced Systems Division (Applied Technology)
4747 Hellyer Avenue
PO Box 7012
San Jose
California 95150-7012
USA
Tel: (+1 408) 365 47 47
Fax: (+1 408) 365 40 40

Litton Aero Products
21050 Burbank Boulevard
Woodland Hills
California 91365
USA
Tel: (+1 818) 226 20 00
Fax: (+1 818) 226 27 55
Web: http://www.@littonapd.com

Litton Electron Devices
1215 S 52nd Street
Temple
Arizona 85281-4471
USA
Tel: (+1 602) 968 44 71
Fax: (+1 602) 966 90 55

Litton Guidance & Control Systems
5500 Canoga Avenue
Woodland Hills
California 91367-6698
USA
Tel: (+1 818) 715 40 40
Web: http://littongcs.com

Litton Laser Systems Division
2787 South Orange Blossom Trail
Apopka, Florida 32703
and
PO Box 547300
Orlando, Florida 32854-7300
USA
Tel: (+1 407) 295 40 10
Fax: (+1 407) 297 48 48
Web: http://www.littonlaser.com

Litton Marine Systems
1070 Seminole Trail
Charlottesville
Virginia 22901
USA
Tel: (+1 804) 974 20 00
Fax: (+1 804) 974 22 59
e-mail: pma01@cho.litton-marine.com
Web: http://www.sperry-marine.com

Litton Special Devices
750 West Sproul Road
Springfield
Pennsylvania 19064
USA
Tel: (+1 215) 328 40 00
Fax: (+1 215) 328 40 16

Lockheed Martin Aeronautical Systems
86 South Cobb Drive
Marietta
Georgia 30063
USA
Tel: (+1 770) 494 44 11
Fax: (+1 770) 494 75 18

Lockheed Martin Aircraft & Logistic Centers
107 Frederick Street
Greenville
South Carolina 29607
USA
Tel: (+1 864) 422 62 41
Fax: (+1 864) 422 62 76

Lockheed Martin Electronics & Missiles
5600 Sand Lake Road
Orlando
Florida 32819
USA
Tel: (+1 407) 356 20 00
Fax: (+1 407) 356 29 00

Lockheed Martin Electronics Platform Integration Group
1801 State Route 17C
Owego
New York 13827
USA
Tel: (+1 607) 751 20 00
Fax: (+1 607) 751 32 59

Lockheed Martin Fairchild Systems
300 Robbins Lane
Syosset
New York 11791
USA
Tel: (+1 516) 349 22 00
e-mail: linda.mcguffie@lmco.com

Lockheed Martin Fairchild Systems
Ridge Hill
Yonkers
New York 10710
USA
Tel: (+1 516) 349 23 28
Fax: (+1 516) 349 25 89
e-mail: habel@fairchild.lmco.com
WWW: http://www.lmco.com

Lockheed Martin IR Imaging Systems
2 Forbes Road
Lexington
Massachusetts 02421
USA
Tel: (+1 781) 863 36 13
Fax: (+1 781) 863 34 96
e-mail: r.a.bell@lmco.com
WWW: http://www.lmco.com/lmiris

Lockheed Martin Naval Electronics & Surveillance Systems
Electronics Park
EP 7, MD 51
Syracuse
New York 13221-4840
USA
Tel: (+1 315) 456 32 96
Web: http://www.lmco.com/orss

Lockheed Martin Tactical Aircraft Systems
1 Lockheed Boulevard
Fort Worth
Texas 76108
USA
Tel: (+1 817) 763 40 86
Fax: (+1 817) 763 47 97
e-mail: karen.l.hagar@lmco.com
WWW: http://www.thefighterenterprise.com

Lockheed Martin Tactical Defense Systems
PO Box 85
Litchfield Park
Arizona 85340
USA
Tel: (+1 602) 925 77 04
Fax: (+1 602) 925 79 42
e-mail: jeffrey.s.lamonica@lmco.com

M/A COM Ltd
1011 Pawtucket Boulevard
PO Box 3295
Lowell
Massachusetts 01853-3295
USA
Tel: (+1 978) 442 50 00
Fax: (+1 978) 442 53 50
Web: http://www.macom.com

Magellan Systems Corporation
960 Overland Court
San Dimas
California 91773
USA
Tel: (+1 909) 394 50 00
Fax: (+1 909) 394 70 50
Web: http://www.magellangps.com

Merlin Engineering Works, Inc
(a wholly owned subsidiary of TEAC America, Inc)
1888 Embarcadero Road
Palo Alto
California 94303-3308
USA
Tel: (+1 650) 856 09 00
Fax: (+1 650) 858 23 02
Web: http://www.merlineng.com

Metrum-Datatape Inc
4800 East Dry Creek Road
Littleton
Colorado 80122
USA
Tel: (+1 303) 773 47 00
Fax: (+1 303) 773 47 62
Web: http://www.metrum-datatape.com

Microvision Inc
19910 North Creek Parkway
PO Box 3008
Bothell
Washington 98011-3008
USA
Tel: (+1 425) 415 68 47
Fax: (+1 425) 415 66 00
e-mail: info@mvis.com
Web: http://www.mvis.com

Mid-Continent Instruments
7706 E.Osie
Wichita
Kansas 67207
USA
Tel: (+1 800) 821 12 12
Fax: (+1 316) 683 18 61

Motorola Inc
Government & Systems Technology Group
PO Box 9040
8220 E Roosevelt Road
Scottsdale
Arizona 85252
USA
Tel: (+1 602) 441 40 79
Fax: (+1 602) 441 23 68

Narco Avionics Inc
270 Commerce Drive
Fort Washington
Pennsylvania 19034
USA
Tel: (+1 215) 643 29 05
Fax: (+1 215) 643 01 97

NavCom Defense Electronics Inc
4323 Arden Drive
El Monte
California 91731
USA
Tel: (+1 626) 579 86 89
Fax: (+1 626) 444 76 19
Web: http://www.navcom.com

NavSymm Positioning Systems
2300 Orchard Parkway
San Jose
California 95131
USA
Tel: (+1 888) 367 79 66
Fax: (+1 408) 428 79 98

Northrop Grumman Corporation
Electronic Sensors and Systems Sector
1580-A West Nursery Road
Linthicum
Maryland 21090
USA
Tel: (+1 410) 993 24 63
Fax: (+1 410) 993 24 81
Web: http://www.sensor.northgrum.com

Northrop Grumman Corporation
Electronic Sensors & Systems Sector - Norden Systems
10 Norden Place
PO Box 5300
Norwalk
Connecticut 06856
USA
Tel: (+1 203) 852 50 00
Fax: (+1 203) 852 78 58
e-mail: dimegfr@mail.northgrum.com
Web: http//www.northgrum.com

Northrop Grumman Corporation
Headquarters
1840 Century Park East
Los Angeles
California 90067-2199
USA
Tel: (+1 310) 553 62 62
Fax: (+1 310) 553 20 76
Web: http://www.northgrum.com

Northrop Grumman Corporation
Integrated Systems Sector
9314 West Jefferson Boulevard
Dallas
Texas 75211-9300
USA
Tel: (+1 972) 946 20 11
Web: http://www.iss.northgrum.com

Northrop Grumman Corporation
Washington Office
1000 Wilson Boulevard
Arlington
Virginia 22209-2278
USA
Tel: (+1 703) 875 84 00
Web: http://www.northgrum.com

Northrop Grumman Corporation
Logicon, Inc
2411 Dulles Corner Park
Suite 800
Herndon
Virginia 20171-3430
USA
Tel: (+1 703) 713 40 00
Web: http://www.logicon.com

Optical Imaging Systems Inc
47050 Five Mile Road
Northville
Michigan 48167
USA
Tel: (+1 734) 454 55 60
Fax: (+1 734) 207 13 50

Orbital Fairchild Defense, Orbital Sciences Corp
20301 Century Boulevard
Germantown
Maryland 20874
USA
Tel: (+1 301) 353 17 96
Fax: (+1 301) 428 64 33
e-mail: webb.cynthia@oscsystems.com

Palomar Display Products Inc
1945 Kellog Avenue
Carlsbad
California 92008
USA
Tel: (+1 760) 931 33 71
Fax: (+1 760) 931 51 98
Web: http://www.palomardisplays.com

Palomar Products Inc
Integrated Communication Systems
23042 Arroyo Vista
Ranch Santa Margarita
California 92688-2604
USA
Tel: (+1 949) 766 53 00
Fax: (+1 949) 766 53 53
Web: http://www.palpro.com

Photo-Sonics Inc
820 S Mariposa Street
Burbank
California 91506
USA
Tel: (+1 818) 842 21 41
Fax: (+1 818) 842 26 10
e-mail: pkiel@photosonics.com

Planar Advance Inc
13950 SW Karl Braun Drive
PO Box 4001
Beaverton
Oregon 97076
USA
Tel: (+1 503) 614 41 00
Fax: (+1 503) 614 41 01
e-mail: display_advance@planar.com

Planar Systems Inc
W7514 Highway V
Lake Mills
Wisconsin 53551
USA
Tel: (+1 920) 648 10 00
Fax: (+1 920) 648 10 01
e-mail: dick_barbieri@planar.com

Polhemus Inc
1 Hercules Drive
PO Box 560
Colchester
Vermont 05446
USA
Tel: (+1 802) 655 31 59
Fax: (+1 802) 655 14 39

Racal Communications Inc
See THALES

Raytheon Aircraft Integration Systems
PO Box 6056
Greenville
Texas 75402
USA
Tel: (+1 903) 457 53 99
Fax: (+1 903) 457 44 13
e-mail: suttonjp@gvl.esys.com

Raytheon Company
Aircraft Integration Systems
1301 E Collins Boulevard
Richardson
Texas 75081
USA
Tel: (+1 972) 470 26 58
Fax: (+1 972) 301 59 90
e-mail: lynnmd@gvl.esys.com

Raytheon Electronic Systems
PO Box 660023
Dallas
Texas 75266
USA
Tel: (+1 972) 272 05 15
Fax: (+1 972) 272 81 44

Raytheon Electronic Systems
Sensors and Electronic Systems
PO Box 902
El Segundo
California 90245
USA
Tel: (+1 310) 647 07 84
e-mail: jmknotts@west.raytheon.com

Raytheon Electronic Systems
One South Los Carneros
Goleta
California 93117
USA
Tel: (+1 805) 967 55 11
Fax: (+1 805) 964 04 70

Raytheon Electronic Systems
2000 E Imperial Highway
PO Box 92426
Los Angeles
California 90009
USA
Tel: (+1 310) 334 7107
Fax: (+1 310) 334 0385
e-mail: jscherer@ccgate.hac.com

Raytheon Electronic Systems
PO Box 405
2501 South Highway 121
Lewisville
Texas 75067
USA
Tel: (+1 972) 462 65 00
Fax: (+1 214) 462 65 08
e-mail: tonypr@rtis.ray.com
Web: http://www.raytheon.com

UNITED STATES OF AMERICA

Raytheon Electronic Systems
Falls Church Division
7700 Arlington Boulevard
Falls Church
Virginia 22046
USA
Tel: (+1 703) 849 15 67
Fax: (+1 703) 280 46 27

UNITED STATES OF AMERICA

Recon/Optical Inc
550 West Northwest Highway
Barrington
Illinois 60010
USA
Tel: (+1 847) 381 24 00
Fax: (+1 847) 381 24 25
e-mail: Bill_Stearns@ROI.Bourns.com

Rockwell Collins
400 Collins Road NE
Cedar Rapids
Iowa 52498
USA
Tel: (+1 319) 295 10 00
Fax: (+1 319) 295 54 29
Web: http://www.collins.rockwell.com

Rodale Electronics Inc
603 Chestnut Street
Garden City
New York 11530
USA
Tel: (+1 516) 222 00 50
Fax: (+1 516) 222 02 23
e-mail: jbclement@earthlink.net

Rogerson Kratos
a Rogerson Aircraft Corporation Subsidiary
403 S Raymond Avenue
Pasadena
California 91109
USA
Tel: (+1 714) 660 06 66
Fax: (+1 714) 660 79 55
e-mail: kirkd@rogerson.com
WWW: http://www.rogerson.com

Safe Flight Instrument Corporation
20 New King Street
White Plains
New York 10604
USA
Tel: (+1 914) 946 95 00
Fax: (+1 914) 946 78 82
e-mail: mail@safeflight.com

SCI Systems Inc
2101 West Clinton Avenue
Huntsville
Alabama 35805
USA
Tel: (+1 256) 882 48 00
Fax: (+1 256) 882 46 52
e-mail: bus.develop@scismail.sci.com
Web: http://www.sci.com

Sierra Research
485 Cayuga Road
Buffalo
New York 14225
USA
Tel: (+1 716) 631 62 00
Fax: (+1 716) 631 78 49
Web: http://www.stisierra.com

Sigtek
9075 Guildford Road
Suite C-1
Columbia
Maryland 21046
USA
Tel: (+1 410) 290 39 18
Fax: (+1 410) 290 81 46
e-mail: info@sigtek.com
Web: http://www.sigtek.com

Sigtronics Corporation
949 North Cataract Avenue, Suite D
San Dimas
California 91773
USA
Tel: (+1 909) 305 93 99
Fax: (+1 909) 305 94 99
Web: http://www.sigtronics.com

Smiths Industries Aerospace Inc
Smiths Industries Aerospace & Defense Systems Ltd
3290 Patterson Avenue
Grand Rapids
Michigan 49512-1991
USA
Tel: (+1 616) 241 70 00
Fax: (+1 616) 241 75 33
Web: http://www.smithsind-aerospace.com

SPS Signal Processing Systems
(A division of Smiths Industries Aerospace)
13112 Evening Creek Drive South
San Diego
California 92128-4199
USA
Tel: (+1 858) 679 60 00
Fax: (+1 858) 679 64 00
Web: http://www.smithsind-sps.com

S-TEC Corporation, a Meggitt Avionics Company
One S-TEC Way
Municipal Airport
Mineral Wells
Texas 76067
USA
Tel: (+1 940) 325 94 06
Fax: (+1 940) 325 39 04
Web: http://www.s-tec.com

STS Strategic Technology Systems Inc
One Electronics Drive
PO Box 3198
Trenton
New Jersey 08619
USA
Tel: (+1 609) 584 02 02
Fax: (+1 609) 584 05 05
e-mail: nowata@earthlink.net

Sunair Electronics Inc
3101 SW 3rd Avenue
Fort Lauderdale
Florida 33315
USA
Tel: (+1 954) 525 15 05
Fax: (+1 954) 765 13 22
Web: http://www.sunairhf.com

Systems & Electronics inc
201 Evans Lane
St Louis
Missouri 63121-1126
USA
Tel: (+1 314) 553 49 01
Fax: (+1 314) 553 49 49
Web: http://www.seistl.com

Systems Research Laboratories (SRL)
2800 Indian Ripple Road
Dayton
Ohio 45440
USA
Tel: (+1 513) 426 60 00
Fax: (+1 513) 426 19 84

TEAC America, Inc
7733 Telegraph Road
Montebello
California 90640
USA
Tel: (+1 213) 727 48 66
Fax: (+1 213) 727 87 08
e-mail: rburnett@teac.com
Web: http://www.teac.com

Tech Comm Inc
3650 Coral Ridge Drive
Coral Springs
Florida 33065
USA
Tel: (+1 305) 341 11 11
Fax: (+1 305) 341 67 87
e-mail: techcomm@gate.net

Tecstar Demo Systems Division
379 Science Drive
Moorpark
California 93021
USA
Tel: (+1 805) 529 18 00
Fax: (+1 805) 529 70 71
Web: http://www.tecstar.com

Teledyne Controls
8640 154th Avenue NE
Redmond
Washington 98052
USA
Tel: (+1 425) 861 69 06
Fax: (+1 425) 885 15 43

Teledyne Controls, Helicopter Systems
12333 W Olympic Boulevard
Los Angeles
California 90064
USA
Tel: (+1 310) 442 42 38
Fax: (+1 310) 442 43 24

Teledyne Electronic Technologies
12964 Panama Street
Los Angeles
California 90066
USA
Tel: (+1 310) 822 82 29
Fax: (+1 310) 574 20 15
Web: http//www.tet.com

Telephonics Corporation
815 Broad Hollow Road
Farmingdale
New York 11735
USA
Tel: (+1 631) 755 70 00
Fax: (+1 631) 755 77 27
Web: http://www.telephonics.com

Telex Communications Inc
Corporate Headquarters
12000 Portland Avenue South
Burnsville
Minnesota 55337
USA
Tel: (+1 952) 884 40 51
Fax: (+1 952) 884 00 43
Web: http://www.telex.com

Tracor Aerospace Inc
and Tracor Aerospace Electronic Systems Inc
see:
BAE SYSTEMS North America

Trimble Navigation Ltd, Aerospace Products
2105 Donley Drive
Austin
Texas 78758
USA
Tel: (+1 512) 432 04 00
Fax: (+1 512) 836 94 13
Web: http://www.trimble.com

TRW Avionics Systems Division
Space & Electronics Group
One Rancho Carmel
San Diego
California 92128
USA
Tel: (+1 619) 592 30 00
e-mail: richard.bent@trw.com
Web: http://www.trw.com

TRW Space & Electronics Groups
One Space Park
Redondo Beach
California 90278
USA
Tel: (+1 310) 812 41 05

TRW Systems & Information Technology Group
One Federal Systems Park Drive
Fairfax
Virginia 22033
USA
Tel: (+1 703) 803 54 98
Fax: (+1 703) 803 47 46
e-mail: mercedes.e.baker@trw.com

TRW Systems Integration Group
1330 Geneva Drive
PO Box 3510
Sunnyvale
California 94088
USA
Tel: (+1 408) 743 20 00
Fax: (+1 408) 752 23 00

Universal Instrument Corporation
6090-A Northbelt Parkway
Norcross
Georgia 30071
USA
Tel: (+1 770) 242 74 66
Fax: (+1 770) 242 75 33
e-mail: rmillen@avionic-displays.com
WWW:http://www.avionic-displays.com

UPS Aviation Technologies, Inc
2345 Turner Road SE
Salem
Oregon 97302
USA
Tel: (+1 503) 391 34 11
Fax: (+1 503) 364 21 38
Web: http://www.upsat.com

US Air Force Materiel Command
Department of Defense
Headquarters Aeronautical Systems Division
Wright-Patterson Air Force Base
Wright Laboratory
Ohio 45433
USA
Tel: (+1 513) 255 27 25
Fax: (+1 513) 476 40 22
e-mail: ascpa@ascpa.wpafb.af.mil

Vickers Inc, Tedeco Division
24 E Glenolden Avenue
Glenolden
Pennsylvania 19036
USA
Tel: (+1 610) 583 94 00
Fax: (+1 610) 583 39 85
e-mail: bob.fenlon@vickers-systems.com
Web: http://www.tedeco.com

Universal Avionics Systems Corporation
3260 East Universal Way
Tucson
Arizona 85706
USA
Tel: (+1 520) 295 23 00
Fax: (+1 520) 295 23 95
Web: http://www.uasc.com

VSI Vision Systems International
2711 Orchard Parkway
San Jose
California 95134
USA
Tel: (+1 408) 433 97 20
Fax: (+1 408) 432 84 49
Web: http://www.vsi-hmcs.com

Wescam
103 West North Street
Healdsburg
California 95448
USA
Tel: (+1 707) 433 30 00
Fax: (+1 707) 433 71 10
Web: http://www.wescam.com

WJ Communications, Inc
401 River Oaks Parkway
San Jose
California 95134-1916
USA
Tel: (+1 408) 577 62 00
Fax: (+1 408) 577 66 20
Web: http://www.wj.com

Wulfsberg Electronics, a Chelton Group Company
6400 Wilkinson Drive
Prescott
Arizona 86301
USA
Tel: (+1 520) 708 15 50
Fax: (+1 520) 541 76 27
e-mail: sales@wulfsberg.com
WWW: http://www.wulfsberg.com

ZETA, an Integrated Defense Technologies Company
2811 Orchard Parkway
San Jose
California 95134
USA
Tel: (+1 408) 434 36 00
Fax: (+1 408) 433 02 05
e-mail: zeta@zeta-sni.com
WWW: http://www.zeta-sni.com

INDEXES

Alphabetical Index

To help users of this title evaluate the published data, *Jane's Information Group* has divided entries into three categories.

N NEW ENTRY Information on new equipment and/or systems appearing for the first time in the title.

V VERIFIED The editor has made a detailed examination of the entry's content and checked its relevancy and accuracy for publication in the new edition to the best of his ability.

U UPDATED During the vertification process, significant changes to content have been made to reflect the latest position known to *Jane's* at the time of publication.

Items in italics refer to entries which have been deleted from this edition with the relevant page numbers from last year.

The prefixes 'Mark', 'Model', 'Series', 'System' and 'Type' have been ignored in the placing of items in this index.

B

C

F

G

H

I

Q

R

S

U

V

W

X

Y

Z

Manufacturers Index

The prefixes 'Mark', 'Model', 'Series', 'System' and 'Type' have been ignored in placing items in this index.

Honeywell Inc Commercial Aviation Systems

Honeywell Defense Avionics Systems

Honeywell Engine Systems & Accessories

Honeywell Military Avionics Systems Group

Honeywell (Normalair-Garrett) Ltd

Honeywell Regelsysteme GmbH

Honeywell Sensor & Guidance Products

Hughes Aircraft Company *see* **Raytheon Systems Company**

Hughes Elcan Optical Technologies Ltd

Ideal Research Development Corporation

ImageQuest Technologies Inc

INDRA EWS

Innovative Solutions and Support Inc

International Aerospace Inc

NOTES

NOTES

NOTES